TEXTBOOK OF
Diagnostic Microbiology

TEXTBOOK OF
Diagnostic Microbiology

Edited by

Connie R. Mahon, MS, MT(ASCP)
Associate Professor
Department of Clinical Laboratory Sciences
The University of Texas Health Science Center
 at San Antonio
San Antonio, Texas

George Manuselis, Jr., MA, MT(ASCP)
Associate Professor
Medical Technology Program
The Ohio State University
Columbus, Ohio

With 600 color illustrations

W.B. SAUNDERS COMPANY
A Division of
Harcourt Brace & Company
Philadelphia London Toronto Montreal Sydney Tokyo

W.B. SAUNDERS COMPANY
A Division of
Harcourt Brace & Company

The Curtis Center
Independence Square West
Philadelphia, Pennsylvania 19106

Library of Congress Cataloging-in-Publication Data

Textbook of diagnostic microbiology
   [edited by] Connie R. Mahon, George Manuselis, Jr.; with 49 contributors.
      p.   cm.
   Includes index.
   ISBN 0–7216–4028–1
   1. Diagnostic microbiology. I. Mahon, Connie R. II. Manuselis, George, Jr. [DNLM:
   1. Microbiological Techniques. QW 25 T355 1995]
   QR67. T49 1995   616′. 01—dc20
   DNLM/DLC   94–21550

TEXTBOOK OF DIAGNOSTIC MICROBIOLOGY   ISBN 0–7216–4028–1

Last digit is the print number: 9   8   7   6   5   4   3   2   1

To my husband, Dan, my daughter, Kathleen, and my son, Sean Patrick
CRM

To Suzanne, Kristina, Shellie, George, Katherine, Libby, and Dee
GM

For their love, support, and encouragement

Contributors

SHIRLEY ADAMS, MS, SM(ASCP), CLS
Adjunct Faculty, Greenville Technical College; Medical Technologist II, Greenville Memorial Medical Center, Greenville, South Carolina

GERALD L. ALDERSON, MD, DDS
Associate Professor of Pathology, The University of Texas Health Science Center at San Antonio; Medical Director, Clinical Virology Laboratory and Immunology Laboratory, University Hospital; Chief, Pathology and Laboratory Medicine Service, Audie L. Murphy Memorial Veterans Hospital, San Antonio, Texas

LEONA W. AYERS, MD
Associate Professor, Pathology and Allied Medical Professions, The Ohio State University College of Medicine; Director, Clinical Microbiology Laboratories, The Ohio State University Medical Center, Columbus, Ohio

JEAN BARNISHAN, BS, M(ASCP)
Clinical Instructor, School of Allied Medicine, The Ohio State University; Senior Medical Technologist, The Ohio State University Medical Center, Columbus, Ohio

AMY M. CARNAHAN, MS, SM(ASCP)
Clinical Microbiology Instructor, Department of Medical and Research Technology, University of Maryland School of Medicine, Baltimore, Maryland; Medical Technologist III, Anne Arundel Medical Center, Annapolis, Maryland

MARY CASTIGLIA, PharmD
Assistant Professor of Clinical Pharmacy, School of Pharmacy, Robert C. Byrd Health Sciences Center, West Virginia University, Morgantown, West Virginia

JAMES L. COOK, MD
Professor of Medicine, Microbiology, and Immunology, The University of Colorado Health Sciences Center; Head, Division of Infectious Diseases, National Jewish Center for Immunology and Respiratory Medicine, Denver, Colorado

VEE E. DAVISON, MS, PhD
Academy of Learning in Retirement, Branch of The University of Texas, San Antonio, Texas

JANET DUBEN-ENGELKIRK, EdD, MT(ASCP), CLS(NCA)
Director, Program in Clinical Laboratory Science, Scott and White Memorial Hospital, Temple, Texas

DENISE F. DUNBAR, BA, M(ASCP)
Section Chief, Mycobacteriology/Mycology Section, Bureau of Laboratories, Texas Department of Health, Georgetown, Texas

JANICE EISENSTADT, DO
Fellow, Department of Pathology, Section of Clinical Microbiology, Cleveland Clinic Foundation, Cleveland, Ohio; Staff Physician, Department of Internal Medicine, Infectious Disease Section, Sinai Hospital of Detroit, Detroit, Michigan

PAUL G. ENGELKIRK, PhD, MT(ASCP)
Faculty, Department of Science, Central Texas College, Killeen, Texas

ANNETTE W. FOTHERGILL, MT(ASCP), CLS(NCA)
Technical Supervisor of Fungus Testing Laboratory; Guest Lecturer for Mycology, Department of Clinical Laboratory Sciences, and for Medical Students, The University of Texas Health Science Center at San Antonio, San Antonio, Texas

SUSAN M. GIBSON, BA, M(ASCP)
Supervisor (retired), Bacteriology-Mycology Branch, Bureau of Laboratories, Texas Department of Health, Austin, Texas

LARRY J. GOODMAN, MD
Associate Professor of Medicine and Associate Dean, Medical Student Programs, Rush Medical College; Attending Physician, Section of Infectious Disease, Rush-Presbyterian-St. Luke's Medical Center, Chicago, Illinois

viii ■ CONTRIBUTORS

MARGARET GREGORY, BS, MT(ASCP)
Clinical Faculty, School of Allied Medical Professions, The Ohio State University; Medical Technologist and Laboratory Safety Officer, The University Hospitals; Columbus, Ohio

JAMES T. GRIFFITH, PhD, CLS(NCA)
Department Chairperson, Medical Laboratory Science, University of Massachusetts, Dartmouth, North Dartmouth, Massachusetts

GERRI S. HALL, PhD
Associate Professor, Departments of Medical Microbiology and Immunology and of Pathology, The Ohio State University, Columbus, Ohio; Staff Microbiologist, Section of Microbiology, Department of Clinical Pathology, Cleveland Clinic Foundation, Cleveland, Ohio

PATRICIA K. HARGRAVE, PhD
Assistant Professor, Department of Medical Technology, School of Allied Health, University of Kansas Medical Center, Kansas City, Kansas

JAMES L. HARRIS, PhD
Training Coordinator, Bureau of Laboratories, Texas Department of Health, Austin, Texas

JANET A. HINDLER, MCLS, MT(ASCP)
Senior Specialist, Department of Pathology and Laboratory Medicine, Clinical Microbiology Section, University of California at Los Angeles, Los Angeles, California

JAMES H. JORGENSEN, PhD
Professor of Pathology, Medicine, Microbiology, and Clinical Laboratory Sciences, The University of Texas Health Science Center at San Antonio; Director, Microbiology Laboratory, University Hospital, San Antonio, Texas

RAYMOND L. KAPLAN, PhD
Clinical Associate Professor, The College of Health Sciences, Georgia State University; Technical Manager, Microbiology, SmithKline Beecham Clinical Laboratories, Atlanta, Georgia

SUSAN L. KOLETAR, MD
Assistant Professor of Clinical Internal Medicine, The Ohio State University; Attending Medical Staff, The University Hospitals, Columbus, Ohio

BOO H. KWA, PhD
Associate Professor, College of Public Health, University of South Florida, Tampa, Florida

HAL S. LARSEN, MT(ASCP), CLS(NCA), PhD
Professor and Chair, Department of Clinical Laboratory Science, Texas Tech University Health Sciences Center, Lubbock, Texas

KAREN S. LONG, MS, CLS(NCA), MT(ASCP)
Assistant Professor, Medical Technology Program, West Virginia University School of Medicine, Morgantown, West Virginia

CONNIE R. MAHON, MS, MT(ASCP)
Associate Professor, Department of Clinical Laboratory Sciences, The University of Texas Health Science Center at San Antonio, San Antonio, Texas

GEORGE MANUSELIS, JR., MA, MT(ASCP)
Associate Professor, Medical Technology Program, The Ohio State University, Columbus, Ohio

MARIO J. MARCON, PhD
Clinical Associate Professor of Pediatrics and Pathology, The Ohio State University College of Medicine; Director, Clinical Microbiology/Virology Laboratories, Columbus Children's Hospital, Columbus, Ohio

DAVID P. MARMADUKE, MD
Baba Fellow of Pathology, Department of Pathology, The Ohio State University Hospitals, Columbus, Ohio; Hematopathology Fellow, University of New Mexico Cancer Center, Albuquerque, New Mexico

JEFFREY P. MASSEY, DrPH, MT(ASCP)
Chief, Molecular Epidemiology Unit, Michigan Department of Public Health, Lansing, Michigan

DEANNA A. McGOUGH, BS, MT, SM(ASCP), RM, SM(AAM)
Instructor, Department of Pathology; Administrative Director, Fungus Testing Laboratory, Department of Pathology, The University of Texas Health Science Center at San Antonio, San Antonio, Texas

DARLENE MILLER, MA, MT(ASCP)
Director of Microbiology, Bascom Palmer Eye Institute, University of Miami, Miami, Florida

DENISE A. MILLER, BS, MT(ASCP)
Staff Medical Technologist, University Hospital, San Antonio, Texas

WILLIAM F. NAUSCHUETZ, PhD
Clinical Assistant Professor, Department of Clinical Laboratory Science, The University of Texas Health Science Center at San Antonio, San Antonio, Texas; Adjunct Assistant Professor, Masters of Public Health Program at the University of Texas-El Paso, El Paso, Texas; Assistant Chief, Department of Clinical Investigation, and Chief, Infectious Disease Research Laboratory, William Beaumont Army Medical Center, El Paso, Texas

VINNIE PAWAR, PhD, MT(ASCP)
Microbiological Associates, Rockville, Maryland

MICHAEL A. PENTELLA, BS, MS
Clinical Microbiologist, Lakeland Regional Medical Center, Lakeland, Florida

RICHARD B. PRIOR, PhD
Associate Professor, The Ohio State University College of Medicine, Columbus, Ohio

THOMAS PUCHALSKI, MS, MT(ASCP)
Clinical Instructor, School of Allied Medical Professions, The Ohio State University; Senior Medical Technologist, The Ohio State University Medical Center, Columbus, Ohio

MERRILY RAUSCH, BS, MT(ASCP), CLS
Clinical Faculty, School of Allied Medical Professions, The Ohio State University; Clinical Microbiology Supervisor, The Ohio State University Medical Center, Columbus, Ohio

JOY G. REMLEY, MT(ASCP), SM
Clinical Faculty, School of Allied Medical Professions, The Ohio State University; Senior Medical Technologist, The Ohio State University Medical Center, Columbus, Ohio

ALONZO S. ROMO, BS, MBA
Evening Supervisor, Metropolitan Hospital, San Antonio, Texas

SHARON S. ROWLAND, PhD, MT(ASCP)
Assistant Professor, Department of Medical and Research Technology, University of Maryland School of Medicine, Baltimore, Maryland

LEONARD H. SCHLEICHER, Jr., MS, MT(ASCP)
Section Chief, Clinical Microbiology Laboratory, Grant Medical Center, Columbus, Ohio

PATRICIA M. SIMONE, MD
Deputy Branch Chief, Program Services Branch, Division of Tuberculosis Elimination, Centers for Disease Control and Prevention, Atlanta, Georgia

RAYMOND A. SMEGO, Jr., MD, MPH
Associate Professor of Infectious Diseases, Section of Infectious Diseases, Robert C. Byrd Health Sciences Center, West Virginia University, Morgantown, West Virginia

LINDA A. SMITH, PhD
Associate Professor, Department of Clinical Laboratory Sciences, The University of Texas Health Science Center at San Antonio, San Antonio, Texas

JOHN G. THOMAS, MS, PhD
Professor, Department of Pathology, West Virginia University School of Medicine; Director, Microbiology and Virology Clinical Laboratories, West Virginia University Hospitals, Morgantown, West Virginia

RONALD R. TRUDEL, MS, CLS(NCA)
Associate Professor of Medical Laboratory Science, University of Massachusetts, Dartmouth, North Dartmouth, Massachusetts

ROBERT G. WHIDDON, PhD
Chief of Microbiology and Immunology, Department of Pathology, Tripler Army Medical Center, Honolulu, Hawaii

Preface

Entry-level clinical laboratory scientists, clinical laboratory technicians, and clinical microbiologists need to fully understand the basic principles of clinical microbiology and how such information is used in both isolating and identifying medically important microorganisms, and then in establishing laboratory diagnoses of infectious diseases. With that building-block approach to microbiology in mind, we undertook the writing of *Textbook of Diagnostic Microbiology* to provide microbiology students with a firm theoretical foundation and a means for developing the skills that they will need. Thus, this book has been written and organized in a student-friendly way, with the text divided into three parts: basic principles and concepts of diagnostic microbiology in Part I; methods for identifying significant isolates in Part II; and the clinical and laboratory diagnosis of infectious diseases at various body sites in Part III. Since the objective of *Textbook of Diagnostic Microbiology* is to provide a foundation for students, the organisms discussed are limited to those that are medically important and commonly encountered.

This text therefore begins with a review of bacterial cell structure and physiology in Chapter 1. Chapter 2 follows with a discussion of laboratory safety measures plus methods of disinfection and sterilization. Chapter 3 describes antimicrobial mechanisms of action and discusses principles of and procedures in antimicrobial susceptibility testing, including conventional methods, such as the Kirby-Bauer method, as well as automated susceptibility procedures. Special antimicrobial susceptibility tests, such as those specific for *Haemophilus influenzae* and *Streptococcus pneumoniae,* are also described.

A discussion of quality issues in clinical microbiology, a distinctive feature of this textbook, is highlighted in Chapter 4. Also presented are general guidelines for establishing a comprehensive, continuous quality-improvement program.

Another unique section within Chapter 4 is "Putting the Laboratory Test to the Test," which discusses the principles of sensitivity and specificity and how they are used in determining the predictive values of laboratory tests.

Recovery of etiologic agents in cultures has remained the "gold standard" in microbiology, and clinicians have relied on the recovery of microorganisms to determine the probable cause of an infectious disease. *Textbook of Diagnostic Microbiology* has included a chapter on emergent technologies to show how new technologies are applied to clinical microbiology, especially as a rapid means for diagnosing infectious diseases. Chapter 5 describes conventional antibody and antigen detection methods and those that have most recently been devised, such as DNA probes, polymerase chain reaction (PCR) applications, and clinical applications of cell wall analysis.

Part I of this work also includes a chapter that introduces students to the role of normal microbial flora in health and disease, host-parasite relationships, and pathogenic mechanisms of infectious diseases.

The two most significant contributions to *Textbook of Diagnostic Microbiology* may be the chapters on direct microscopic examination of clinical specimens (Chapter 8) and utilization of colonial morphology in the initial identification of bacterial isolates (Chapter 9). We have found that both students and practitioners have difficulty in recognizing bacterial morphology on direct smear preparations and that the microscopic morphology observed on smears can aid in the identification of bacterial isolates. Similarly, recognition of colonial morphology observed on primary culture plates is helpful in the presumptive identification of isolates and then in the determination of how to proceed in the final identification. These two chapters not only illustrate the applications of these techniques but also present numerous photographs that may help the reader develop these skills.

While Part I was structured as the "backbone" of *Textbook of Diagnostic Microbiology,* Part II concentrates on the laboratory identification of significant isolates. Chapters in this section cover organisms by taxonomic groups. Diseases caused by the organisms are discussed, but emphasis is on the characteristics and isolation of, and the identification methods for, each group of organisms. Many tables summarize the major features of organisms and the schematic networks used to show relationships and differences between similar or closely related species. Photographs are used extensively to show the characteristics of a particular organism.

Special chapters have been written for anaerobes, significant fungal agents, and medically important parasites and viruses. These chapters describe special specimen collection, transport, and processing specific for these groups of organisms. The chapters are organized consistently, with each one providing learning objectives and describing general characteristics, identification methods, and the clinical infections the organisms produce.

Part III of *Textbook of Diagnostic Microbiology* uses the organ system approach to discuss the laboratory diagnosis of infectious agents. Each chapter begins with the anatomy of the organ system to be discussed and the role of the normal microbial flora at that particular site in the pathogenesis of disease. We believe that it is important for students to become comfortable with their knowledge of the normal inhabitants at a body site before they can recognize when an organism is significant in an opportunistic type of infection. It is just as important to discuss how the epidemiology of an infectious disease can serve as a diagnostic tool in the recovery of the causative agent; therefore, whenever appropriate, epidemiologic information is included in the chapters in Part III.

The discussion of infectious diseases at each body site is enhanced with case studies that describe the clinical presentation and laboratory findings related to the disease. The student is then able to correlate clinical findings and observations with the possible etiologic agent of the disease. Chapters in Part III also describe the appropriate specimen collection and rapid diagnostic methods available.

It is our intent to provide students with a textbook that is user-friendly and organized to facilitate and enhance learning. This work is not intended to replace existing reference texts written at a more comprehensive level. *Textbook of Diagnostic Microbiology* may be adopted in all types of clinical laboratory education programs and may be used by other healthcare practitioners. This book is also appropriate for those who would like a review in clinical microbiology.

We believe that the strong cast of contributors, the liberal use of full-color photographs and photomicrographs, the easy-to-use and attractive design, the special features such as case studies, and the ancillaries such as the accompanying laboratory workbook and slide set all enhance the building-block approach used in *Textbook of Diagnostic Microbiology* for clinical laboratory science, clinical laboratory technician, and microbiology students.

CONNIE R. MAHON
GEORGE MANUSELIS, JR.

During the time span in which this textbook was being prepared, there were significant changes in the taxonomy nomenclature for the Enterobacteriaceae and the nonfermentative bacilli that could not be incorporated. Some current taxonomic changes for these two large groups of organisms are included in Appendix C. However, this list of current nomenclature is directed toward more significant clinical isolates and therefore is not intended to be exhaustive.

Acknowledgments

We are grateful to all the contributing authors and to many other individuals who have helped bring this project to fruition. Many thanks to former students Wanda Jean Jenscke, Michelle Wallis, and Isabel Uriegas for their critical review of the content and organization of *Textbook of Diagnostic Microbiology* from the student's perspective; to our colleagues Denise A. Miller and Linda A. Smith for their chapter contributions and for their invaluable suggestions and comments; and to Kathy Doig and L. Michael Posey for their mentorship. Our appreciation also goes to Maureta Ott, Marilyn Soloman, Arthur Weeks, Steve Moon, John Smith, Richard Brust, Clara Murray, Peg Monigold, Eileen Buckholz, and Karen Eldredge. Finally, we wish to thank Les Hoeltzel, Manager, Developmental Editors, Gina Scala, Copy Editing Supervisor, Selma Ozmat, Senior Editor, Health-Related Professions, Cecilia Roberts, Illustration Specialist, and Karen O'Keefe, Book Designer, for their guidance and patience, and Nicole Mercier for her numerous trips to the mailroom.

CONNIE R. MAHON
GEORGE MANUSELIS, JR.

Contents

APPENDICES

PART *I*

Introduction to
Clinical Microbiology

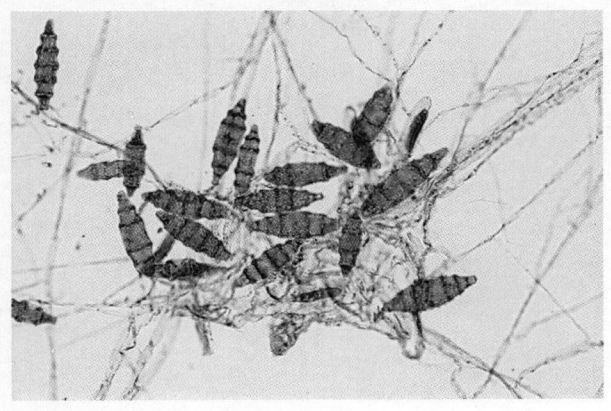

CHAPTER 1

Bacterial Cell Structure, Physiology, Metabolism, and Genetics

Sharon S. Rowland

OBJECTIVES

1. Describe microbial classification (taxonomy) and accurately apply the rules of scientific nomenclature for bacterial names.

2. List and define five methods used by epidemiologists to subdivide bacterial species.

3. Differentiate among prokaryotic, eukaryotic, and archaeobacterial cell types.

4. Compare and contrast prokaryotic and eukaryotic cytoplasmic and cell envelope structures and functions.

5. Describe the cell walls of gram-positive and gram-negative bacteria. Explain the Gram stain reaction of each cell wall type. Describe two other bacterial cell wall types, and give microbial examples of each.

6. Explain the use of the following stains in the diagnostic microbiology laboratory: Gram stain, acid-fast stains (Ziehl-Neelsen, Kinyoun, auramine-rhodamine), acridine orange, methylene blue, calcofluor white, lactophenol cotton blue, and India ink.

7. List the nutritional and environmental requirements for bacterial growth, and define the categories of media used for culturing bacteria in the laboratory.

8. Define the atmospheric requirements of obligate aerobes, microaerophiles, facultative anaerobes, obligate anaerobes, and capnophilic bacteria.

9. Describe the stages in the growth of bacterial cells.

10. Describe the importance of microbial metabolism in clinical microbiology.

11. Differentiate between fermentation and oxidation (respiration).

12. Name and compare three biochemical pathways that bacteria use to convert glucose to pyruvate.

13. Identify and compare the two types of fermentation that explain positive results with the methyl red or Voges-Proskauer test.

14. Define the following genetic terms: *genotype, phenotype, constitutive, inducible, replication, transcription, translation, genome, chromosome, plasmids, IS element, transposon, point mutations, frame-shift mutations,* and *recombination*.

15. Discuss the development and transfer of antibiotic resistance in bacteria.

16. Differentiate among the mechanisms of transformation, transduction, and conjugation in the transfer of genetic material from one bacterium to another.

17. Define the terms *bacteriophage, lytic phage, lysogeny,* and *temperate phage*.

18. Define *restriction endonuclease enzyme* and explain the use of such enzymes in the clinical microbiology laboratory.

In this chapter, the basic concepts of bacterial cell structure, physiology, metabolism, and genetics are reviewed. The reader is made aware of the practical importance of each topic to diagnostic microbiologists in their efforts to culture, identify, and characterize the microbes that cause disease in humans.

SIGNIFICANCE

Microbial inhabitants have evolved to survive in a variety of ecologic niches and growth habitats. Some grow rapidly, some slowly. Some can replicate with a minimal number of nutrients present, whereas others require enriched nutrients in order to survive. There is variation in atmospheric growth conditions, temperature requirements, and cell structure. This diversity is also found in the microorganisms that inhabit the human body, as normal flora, as opportunistic pathogens, or as true pathogens. Each microbe has its own unique physiology and metabolic pathways that allow it to survive in its particular habitat. One of the main roles of a diagnostic or clinical microbiologist is to isolate, identify, and analyze the bacteria that cause disease in humans. Knowledge of microbial structure and physiology is very important to clinical microbiologists in three areas:

- Culture of organisms from patient specimens

- Classification and identification of organisms after they have been isolated

- Prediction and interpretation of antimicrobial susceptibility patterns

Understanding the growth requirements of a particular bacterium enables the microbiologist to select the correct media for primary culture and optimize the chance of isolating the pathogen. Determination of staining characteristics, based on differences in cell wall structure, is the first step in bacterial classification. Metabolic biochemical differences between organisms form the basis for most bacterial identification systems in use today. The cell structure and biochemical pathways of an organism determine its susceptibility to various antibiotics.

The ability of microorganisms to change rapidly, to acquire new genes, and to undergo mutations presents continual challenges to diagnostic microbiologists as they isolate and characterize the microorganisms associated with humans.

CLASSIFICATION

Taxonomy

Taxonomy refers to the classification and grouping of organisms. It is based on genotypic (genetic) and phenotypic (observable) similarities and differences. The formal levels of bacterial classification, in successively smaller subsets, are kingdom, division, class, order, family, tribe, genus, and species. Bacteria have been placed in a kingdom separate from the animal and plant kingdoms, called Prokaryotae. The kingdom Prokaryotae includes unicellular organisms such as

bacteria, fungi, protozoa, and algae. Diagnostic microbiologists traditionally emphasize placement of bacterial species into three categories: the *family* (similar to a human "clan"), a *genus* (equivalent to a human last name), and a *species* epithet (equivalent to a human first name). For example, *Staphylococcus* (genus) *aureus* (species epithet) belongs to the Micrococcaceae family.

Nomenclature

Nomenclature provides naming assignments for each organism. The following standard rules for denoting bacterial names are used in this book. The family name is capitalized and has an *-aceae* ending (e.g., Micrococcaceae). The genus name is capitalized and is followed by the species name, which begins with a lowercase letter; both the genus and species should be *italicized* in print but underlined in script (e.g., *Staphylococcus aureus* or Staphylococcus aureus). Often, the genus name is abbreviated by using the first letter of the genus followed by a period and the species epithet (name) (e.g., *S. aureus*). The genus name followed by the word *species* (e.g., *Staphylococcus* species) may be used to refer to the genus as a whole. Species abbreviated *sp* (singular) or *spp* (plural) is used when the species is not specified. When bacteria are referred to as a group, their names are neither capitalized nor underlined (e.g., staphylococci). The plural of *genus* is *genera* (e.g., there are many genera within the Enterobacteriaceae family).

Classification by Phenotypic and Genotypic Characteristics

The traditional method of placing an organism into a particular genus and species is based on the similarity of all members for a number of phenotypic characteristics. In the diagnostic microbiology laboratory, this is accomplished by testing each bacterial culture for a variety of metabolic characteristics and comparing the results with those listed in established charts. In many rapid identification systems, a numerical taxonomy is used, in which phenotypic characteristics are assigned a numerical value and the derived number indicates the genus and species of the bacterium.

Epidemiologists constantly seek means of further subdividing bacterial species in order to follow the spread of bacterial infections. Species may be subdivided into *subspecies,* based on phenotypic differences (abbreviated *subsp*); *serovarieties,* based on serologic differences (abbreviated *serovar*); or *biovarieties,* based on biochemical test result differences (abbreviated *biovar*). *Phage typing* (based on susceptibility to specific bacterial phages) has also been used for this purpose. Epidemiologists also find a molecular method termed *RFLP* (restriction fragment length polymorphism) *analysis* very useful for determining differences among strains. In this method, the DNA is cut by enzymes known as *restriction endonucleases*, and the patterns of the resulting DNA fragment lengths are compared.

Current technology has allowed the analysis of genetic relatedness (DNA and RNA structure and homology) for taxonomic purposes. The analysis of ribosomal RNA (rRNA) has proved particularly useful for this purpose. The information obtained from these studies has resulted in the reclassification of some bacteria.

Classification by Cellular Type: Prokaryotes, Eukaryotes, and Archaeobacteria

Another method of classifying organisms is by cell organization. It is now recognized that organisms fall into three distinct groups based on type of cell organization and function: the prokaryotes, the eukaryotes, and the most recently identified, archaeobacteria. Bacteria are prokaryotic, whereas fungi, algae, protozoa, animal cells, and plant cells are eukaryotic in nature. The archaeobacterial cell type appears to be more closely related to eukaryotic cells than to prokaryotic cells and is found in microorganisms that grow under extreme environmental conditions. The cell envelope and enzymes of archaeobacteria have been designed for survival under stressful conditions. Because archaeobacteria are not encountered in clinical microbiology, they are not discussed further in this chapter.

In general, the interior organization of eukaryotic cells is more complex than that of prokaryotic cells (Fig. 1–1). The eukaryotic cell is usually larger and contains membrane-encased organelles or

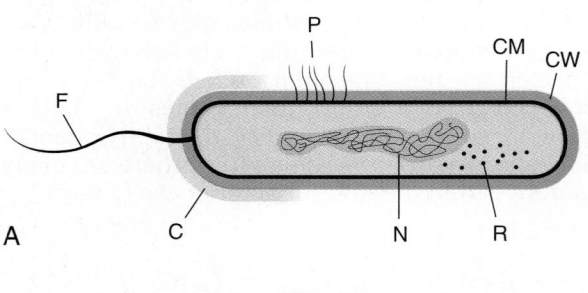

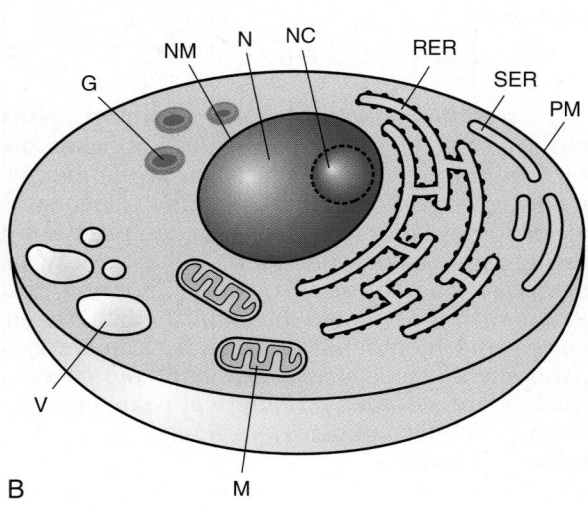

━━━━━ F i g u r e   1 – 1

Comparison of prokaryotic and eukaryotic cell organization and structures. *A,* Diagram of a prokaryotic bacterial cell. Note the location of the cell wall (CW), the cell membrane (CM), the nuclear region (nucleoid) free in the cytoplasm (N), and the ribosomes located in the cytoplasm (R). Other structures that may be present are pili or fimbriae (P), capsules or slime layers (C), and flagella (F). *B,* Diagram of a eukaryotic cell. Note the presence of the plasma membrane (PM), nuclear membrane (NM) surrounding the nucleus (N), which contains a nucleolus (NC), mitochondria (M), ribosomes attached to the rough endoplasmic reticulum (RER), the smooth endoplasmic reticulum (SER), vacuole (V), and storage granules (G).

▬▬▬ T A B L E   1 – 1

COMPARISON OF EUKARYOTIC AND PROKARYOTIC CELL ORGANIZATION

Characteristic	Eukaryote	Prokaryote
Genetic material		
Location	Contained within a membrane-bound nucleus inside the cell	Free in the cytoplasm attached to a structure called a mesosome located in the cell membrane
Form	Multiple chromosomes, which are surrounded by basic proteins called histones	A single circular piece of DNA
Replication	By mitosis and meiosis	By binary fission
Extrachromosomal DNA	In mitochondria	Plasmids, small circular pieces of DNA containing accessory information, may be present in the cytoplasm
Protein production		
Site	Rough endoplasmic reticulum, a membrane covered with ribosomes, where protein is made	No endoplasmic reticulum
	Smooth endoplasmic reticulum or Golgi complex, where secreted proteins are packaged and transported to the cell surface	Ribosomes are free in the cytoplasm or attached to the cell membrane
Ribosomes	80S in size, consisting of a 60S and a 40S subunit	70S in size consisting of a 50S and a 30S subunit
Energy production site	Within membrane-bound mitochondria	Electron transport chain is located in the cell membrane; no mitochondria present
Intracellular organelles (lysosomes)	Contain hydrolytic enzymes	Not present
Cell envelope		
Plasma membrane	Lipoprotein membrane; regulates transport	Lipoprotein membrane; regulates transport
Cell wall	Usually absent except for fungi, which contain chitin in the cell wall	Present; imparts rigidity; see text for types

compartments that serve various functions; the prokaryotic cell is noncompartmentalized. There are also differences in the processes of DNA synthesis, protein synthesis, and cell envelope synthesis and structure. Table 1–1 compares some of the differences between eukaryotic and prokaryotic cells.

Pathogenic bacteria are prokaryotic cells that infect eukaryotic hosts. Targeting antibiotic action against unique prokaryotic structures and functions inhibits bacterial growth while avoiding harm to eukaryotic host cells. This is one reason pharmaceutical companies have been so successful in developing effective antibiotics against bacterial pathogens but have been less successful in finding drugs effective against parasites, medically important fungi, and viruses, which are eukaryotic like their human hosts.

COMPARISON OF EUKARYOTIC AND PROKARYOTIC CELL STRUCTURE

Eukaryotic Cell Structure

The following structures are associated with eukaryotic cells. In the diagnostic microbiology laboratory, the eukaryotic cell type occurs in medically important fungi and in parasites.

Cytoplasmic Structures

The *nucleus* contains the DNA of the cell in the form of discrete chromosomes, which are covered with basic proteins called *histones*. The number of chromosomes in the nucleus varies according to the particular organism. A rounded refractile body called a *nucleolus* is also located within the nucleus. The nucleolus is the site of ribosomal RNA synthesis. The nucleus is bounded by a bilayered lipoprotein nuclear membrane.

The *endoplasmic reticulum* is a system of membranes that occur throughout the cytoplasm. It is found in two forms. The *rough endoplasmic reticulum* is covered with ribosomes and is the site of protein synthesis. The *smooth endoplasmic reticulum,* or *Golgi apparatus,* is the the site where secreted proteins are packaged for export to the exterior of the cell.

Eukaryotic *ribosomes,* where protein synthesis occurs, are 80S in size and dissociate into two subunits, 60S and 40S. They are attached to the rough endoplasmic reticulum.

Eukaryotic cells contain several membrane-enclosed organelles. *Mitochondria* are the main sites of energy production in the cell. Mitochondria contain their own DNA and the electron transport system that produces energy for cell functions. *Lysosomes* contain hydrolytic enzymes for degradation of macromolecules and microorganisms within the cell. *Peroxisomes* contain protective enzymes that break down hydrogen peroxide and other peroxides generated within the cell. *Chloroplasts,* found in plant cells, are the sites of photosynthesis.

Cell Envelope Structures

PLASMA MEMBRANE

The plasma membrane is a bilayered lipoprotein membrane that encircles the cell cytoplasm and regulates transport of macromolecules into and out of the cell. The presence of sterols is a trait of eukaryotic cell membranes.

CELL WALL

The function of a cell wall is to provide rigidity and strength to the exterior of the cell. *Most eukaryotic cells do not have cell walls.* Fungi, however, have cell walls principally made of polysaccharides such as chitin, mannan, and glucan. Chitin is a distinct component of fungal cell walls.

MOTILITY ORGANELLES

Cilia are short projections (3–10 μm), usually numerous, that extend from the surface and are used for locomotion. They are found in certain protozoa and in ciliated epithelial cells of the respiratory tract. *Flagella* are longer projections (>150 μm) used for locomotion by cells such as spermatozoa. The *basal body, or kinetosome,* is a small structure located at the base of cilia or flagella, where microtubule proteins involved in movement originate.

Prokaryotic Cell Structure

The bacterial cell is smaller and less compartmentalized than a eukaryotic cell. A variety of structures are, however, unique to prokaryotic cells. A general description of some of them follows.

Cytoplasmic Structures

Bacteria do not contain a membrane-bound nucleus. Their DNA consists of a single circular chromosome. This appears as a diffuse nucleoid or chromatin body *(nuclear body),* which is attached to a *mesosome,* a sac-like structure within the cell membrane.

Bacterial *ribosomes* are found free in the cytoplasm and attached to the cytoplasmic membrane. They are 70S in size and dissociate into two subunits, 50S and 30S in size.

Stained bacteria sometimes reveal the presence of granules in the cytoplasm (cytoplasmic granules). These granules are storage deposits and may consist of polysaccharides such as glycogen, lipids such as poly-β-hydroxybutyrate, or polyphosphates.

Certain genera, such as *Bacillus* and *Clostridium,* produce *endospores* in response to harsh environmental conditions. The spores occur as highly refractile bodies within the cell. Spores are visualized microscopically as unstained areas within a cell with the use of traditional bacterial stains or can be stained with specific spore stains. The size, shape, and interior location of the spore can be used as identifying characteristics.

Cell Envelope Structures

The cell envelope consists of the membrane and structures surrounding the cytoplasm. In bacteria, these are the cell membrane and the cell wall. Some species also produce capsules and slime layers.

PLASMA MEMBRANE (CELL MEMBRANE)

The cell membrane is a lipoprotein membrane that surrounds the cytoplasm. It is made of phospholipids and proteins, but does not contain sterols, unlike eukaryotic plasma membranes (except for *Mycoplasma*). The plasma membrane regulates transport across the membrane, acts as an osmotic barrier (prokaryotes have a high osmotic pressure inside the cell), and is the location of the electron transport chain, where energy is generated.

CELL WALL

The cell wall of prokaryotes is a rigid structure that maintains the shape of the cell and prevents bursting of the cell from the high osmotic pressure inside it. There are several different types of cell wall structures in bacteria, which have traditionally been categorized according to their staining characteristics. The two major types of cell walls are the *gram-positive* and the *gram-negative types* (Fig. 1–2). In addition, some mycobacteria have an *acid fast cell wall,* and mycoplasmas have *no cell wall.*

Gram-Positive Cell Wall. The gram-positive cell wall is composed of a very thick protective peptidoglycan (murein) layer. Because the peptidoglycan layer is the principal component of the gram-positive cell wall, many antibiotics effective against gram-positive organisms (such as penicillin) act by preventing synthesis of peptidoglycan. Gram-negative bacteria, which have a thinner layer of peptidoglycan and a different cell wall structure, are less affected by these antibiotics.

The *peptidoglycan* or *murein layer* consists of glycan (polysaccharide) chains of alternating *N*-acetyl-D-glucosamine (NAG) and *N*-acetyl-D-muramic acid (NAM) (Fig. 1–3). Short peptides, each consisting of four amino acid residues, are attached to a carboxyl group on each NAM residue. The chains are then cross-linked to form a thick network via a peptide bridge (varying in number of peptides) connected to the tetrapeptides on the NAM.

Other components of the gram-positive cell wall that penetrate to the exterior of the cell are *teichoic acid* (anchored to the peptidoglycan) and *lipoteichoic acid* (anchored to the plasma

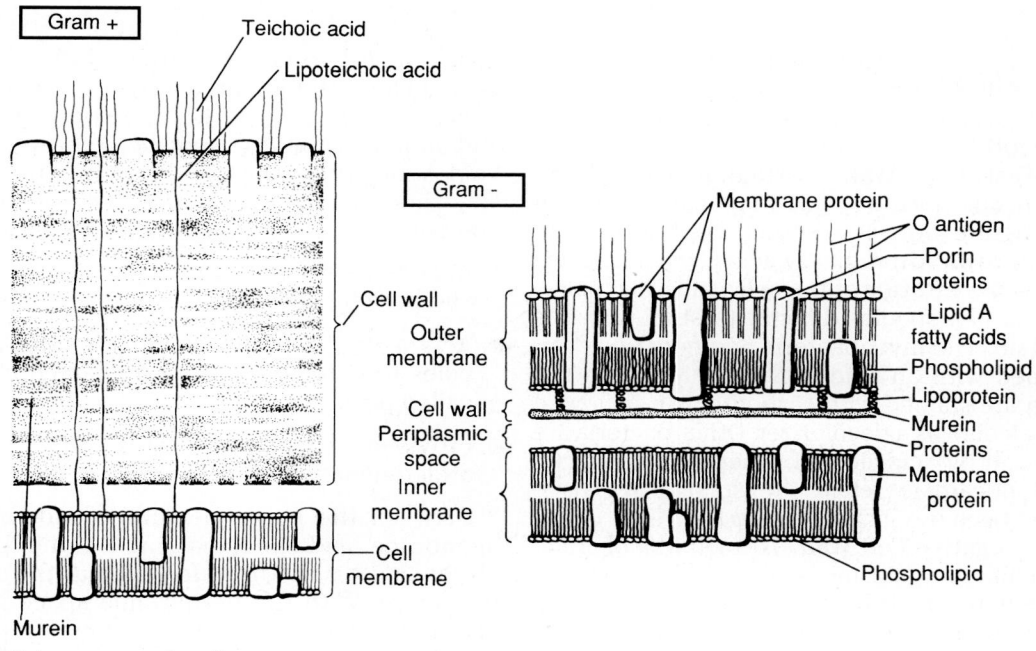

Figure 1-2

The cell envelope structure of a gram-positive *(left)* and a gram-negative *(right)* bacterium. (With permission from Schaechter M, Medoff G, Eisenstein BI: Mechanisms of Microbial Disease, 2nd ed. Baltimore: Williams & Wilkins, 1993, p 31.)

■■■■■F i g u r e   1 – 3

Diagram demonstrating the structure of the peptidoglycan layer in the cell wall of *E. coli*. NAG, *N*-acetylglucosamine; NAM, *N*-acetylmuramic acid. The amino acids in the cross-linking tetrapeptides may vary among species. (With permission from Neidhardt FC, Ingraham M, Schaechter M: Physiology of the Bacterial Cell: A Molecular Approach. Sunderland, MA: Sinauer Associates, Inc., 1990, p 17.)

membrane). These two components are unique to the gram-positive cell wall. Other antigenic polysaccharides may be present on the surface of the peptidoglycan layer.

Acid-Fast Cell Wall. Certain genera (*Mycobacterium* and *Nocardia*) have a gram-positive cell wall structure but, in addition, contain a waxy layer of glycolipids and fatty acids (mycolic acid) bound to the exterior of the cell wall. This makes *Mycobacterium* species difficult to stain with the Gram stain. The mycobacteria and nocardiae can be stained with an acid-fast stain, in which the bacteria are stained with carbolfuchsin, followed by acid-alcohol as a decolorizer. Other bacteria are decolorized by acid-alcohol, whereas mycobacteria and nocardiae retain the stain. They have therefore been designated *acid-fast bacteria.*

Gram-Negative Cell Wall. The cell wall of gram-negative microorganisms is composed of two layers. The inner peptidoglycan layer is much thinner than in gram-positive cell walls. Outside the peptidoglycan layer is an additional outer membrane unique to the gram-negative cell wall. The outer membrane contains proteins, phospho-

lipids, and lipopolysaccharide (LPS) (see Fig. 1–2). LPS contains three regions, an antigenic O–specific polysaccharide, a core polysaccharide, and an inner lipid A (also called endotoxin). The lipid A moiety is responsible for producing fever and shock conditions in patients infected with gram-negative bacteria. The outer membrane

■ Acts as a barrier to hydrophobic compounds and harmful substances

■ Acts as a sieve, allowing water-soluble molecules to enter through protein-lined channels called porins

■ Provides attachment sites that enhance attachment to host cells

Between the outer membrane and the inner membrane, and encompassing the thin peptidoglycan layer, is an area referred to as the *periplasmic space*. With the periplasmic space is a gel-like matrix containing nutrient-binding proteins and degradative and detoxifying enzymes.

Absence of Cell Wall. Prokaryotes that belong to the *Mycoplasma* and *Ureaplasma* genera are unique in that they lack a cell wall and contain sterols in their cell membranes. Because they lack the rigidity of the cell wall, they are seen in a variety of shapes microscopically. Gram-positive and gram-negative cells can lose their cell walls and grow as *L-forms* in media supplemented with serum or sugar to prevent osmotic rupture of the cell membrane.

SURFACE POLYMERS

A variety of pathogenic bacteria produce a discrete organized covering termed a *capsule*. Capsules are usually made of polysaccharide polymers, although they may also be made of polypeptides. Capsules act as virulence factors in helping the pathogen evade phagocytosis. During identification of certain bacteria by serologic typing, capsules sometimes must be removed in order to detect the somatic (cell wall) antigens present underneath them. Capsule removal is accomplished by boiling a suspension of the microorganism. *Salmonella typhi* must have its capsular (Vi) antigen removed in order for the technologist to observe agglutination with *Salmonella* somatic (O) antisera.

Slime layers are similar to capsules but are more diffuse layers surrounding the cell. They also are made of polysaccharides and serve either to inhibit phagocytosis or, in some cases, to aid in adherence to host tissue or synthetic implants.

CELL APPENDAGES

The flagellum is the organ of locomotion. *Flagella* are exterior protein filaments that rotate and cause bacteria to be motile. Bacterial species vary in their possession of flagella from none (nonmotile) to many (Fig. 1–4). Flagella that extend from one end of the bacterium are *polar.* Polar flagella may occur singly at one or both ends or multiply in tufts. Flagella that occur on all sides of the bacterium are *peritrichous.* The number and arrangement of flagella are sometimes used for identification purposes. Flagella can be visualized microscopically with special flagellum stains.

Pili (also known as *fimbriae*) are hair-like protein structures that aid in attachment to surfaces. Specialized pili known as *sex pili* are involved in bacterial conjugation and gene exchange. Bacterial pathogens often have adherence pili that allow them to attach to specific eukaryotic host cell surfaces. The proteins within the pili that aid in attachment are called *adhesins*.

BACTERIAL MORPHOLOGY

Microscopic Shapes

Bacteria vary in size from 0.4 to 2 μm. They occur in three basic shapes (Fig. 1–5):

- Cocci (spherical)
- Bacilli (rod-shaped)
- Spirochetes (helical)

Individual bacteria may form characteristic groupings. Cocci may occur singly, in pairs (*diplococci*), in chains (*streptococci*), or in clusters (*staphylococci*). Bacilli may vary greatly in size and length from very short *coccobacilli* to long *filamentous* rods. The ends may be square or rounded. Bacilli with tapered, pointed ends are termed *fusiform.* Some bacilli are curved. When a species varies in size and shape within a pure culture, the bacterium is *pleomorphic.* Bacilli may occur as single rods or in chains or may align themselves side by side (*palisading*). Spirochetes vary in length and in the number of helical turns.

Common Stains Used for Microscopic Visualization

Stains that impart color or fluorescence are needed in order to visualize bacteria under the microscope. The microscopic staining characteristics, shape, and grouping are used in the classification of microorganisms (Fig. 1–6).

Flagellar Arrangements

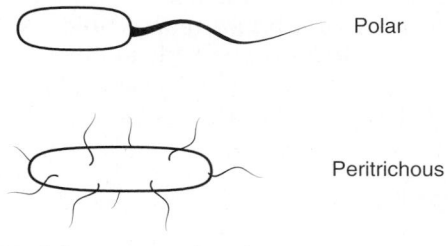

Polar

Peritrichous

■■■■ F i g u r e   1 – 4
Diagram of flagellar arrangements that occur in bacteria.

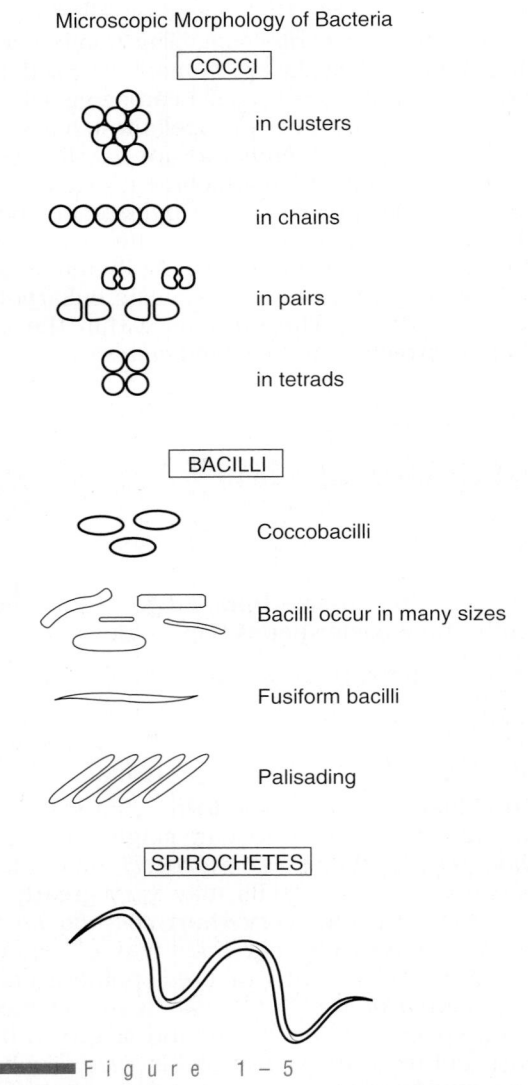

Microscopic Morphology of Bacteria

COCCI

in clusters

in chains

in pairs

in tetrads

BACILLI

Coccobacilli

Bacilli occur in many sizes

Fusiform bacilli

Palisading

SPIROCHETES

■■■■■■■ F i g u r e   1 – 5
Diagram of the microscopic shapes and arrangements of bacteria.

Gram Stain

The Gram stain is the most commonly used stain in the clinical microbiology laboratory. It places bacteria into one of the two groups: gram-positive (purple) or gram-negative (pink) (see Fig. 1–6*A* and *B*). As mentioned previously, the cell wall structure determines the Gram staining characteristics of a species. The Gram stain consists of four components: crystal violet (the primary stain), iodine (the mordant or fixative), alcohol (the decolorizer), and safranin (the counterstain).

The bacteria are initially stained purple by crystal violet, which is bound to the cell wall with the aid of iodine. When alcohol is applied to bacteria with a gram-negative type of cell wall structure, the crystal violet washes out of the cells, which then take up the pink counterstain, safranin. Gram-negative bacteria therefore appear pink under the light microscope. Bacteria with a gram-positive cell wall retain the primary crystal violet stain during alcohol treatment, and so remain purple.

Acid-Fast Stains

Acid-fast stains are used to stain bacteria that have a high lipid and wax content in their cell walls and do not stain well with traditional bacterial stains. Carbolfuchsin (a red dye) is used as the primary stain (see Fig. 1–6*C*). The cell wall is treated to allow penetration of the dye either by heat (*Ziehl-Neelsen* method) or by a detergent (*Kinyoun* method). Acidified alcohol is used as a decolorizer, and methylene blue is the counterstain. Acid-fast bacteria retain the primary stain and are red. Bacteria that are not acid-fast are blue.

Two other gram-positive genera, *Nocardia* and *Rhodococcus,* may stain acid-fast by a modified method. Acid-fast staining is used to identify a yeast, *Saccharomyces,* and coccidian parasites, such as *Isospora belli, Cryptosporidium* and other coccidia-like bodies. A fluorochrome (i.e., fluorescent) stain, *auramine-rhodamine,* has also been used to screen for acid-fast bacteria (see Fig. 1–6*D*). This stain is selective for the cell wall of acid-fast bacteria. Acid-fast bacteria appear yellow or orange under a fluorescent microscope, making them easier to find.

Acridine Orange

Acridine orange is a fluorochrome dye that stains both gram-positive and gram-negative bacteria, living or dead. It binds to the nucleic acid of the cell and fluoresces as a bright orange color. Acridine orange is used to locate bacteria in blood cultures and other specimens where it might otherwise be difficult to see any bacteria present (see Fig. 1–6*E*).

Methylene Blue

Methylene blue has been traditionally used to stain *Corynebacterium diphtheriae* for observa-

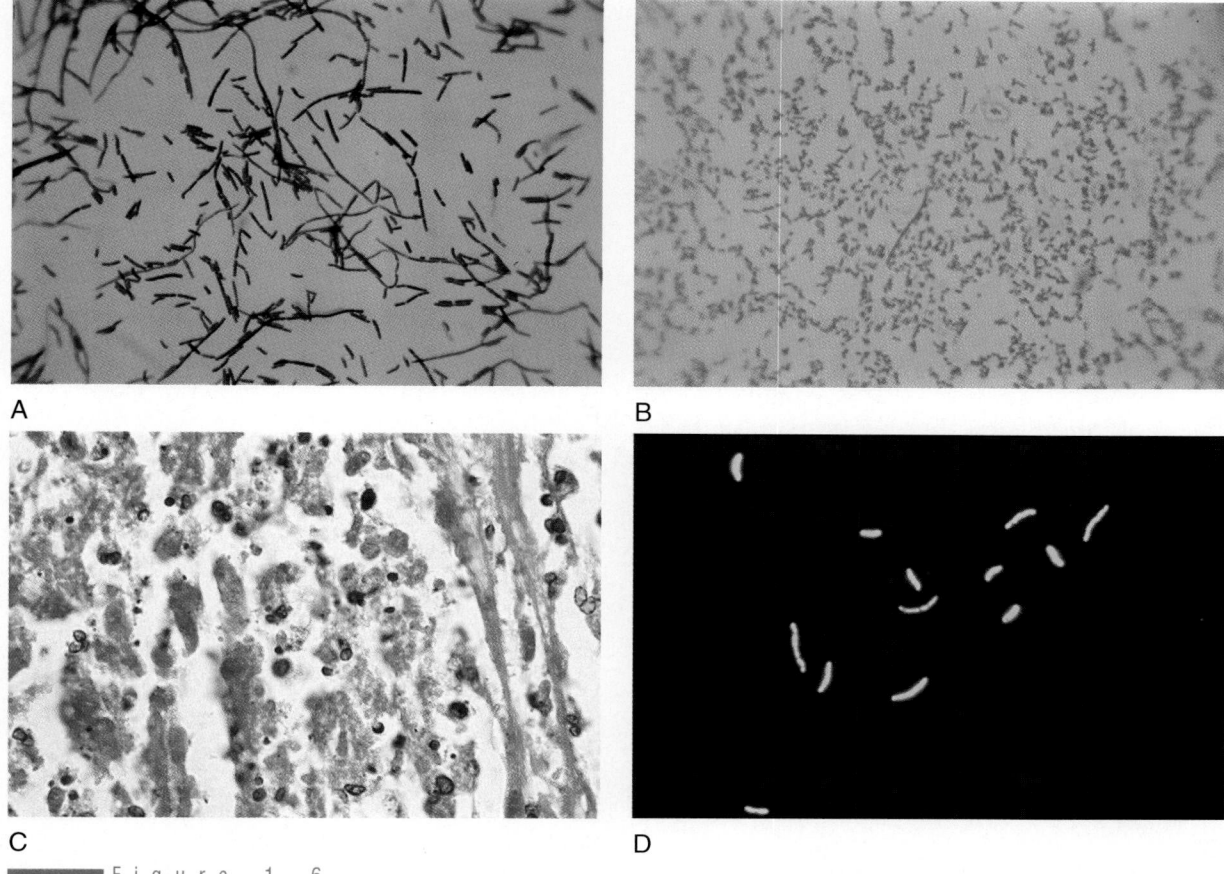

■■■■■ F i g u r e   1 – 6

A, Gram stain of *Lactobacillus* species illustrating gram-positive bacilli, singly and in chains. A few gram-negative–staining bacilli are also present. (Courtesy of Dr. Andrew G. Smith, University of Maryland School of Medicine, Baltimore.) *B,* Gram stain of *Escherichia coli* illustrating short gram-negative bacilli. (Courtesy of Dr. Andrew G. Smith, University of Maryland School of Medicine, Baltimore.) *C,* Acid-fast stain, carbolfuschin based. A sputum smear demonstrating the presence of acid-fast *Mycobacterium* species stained by the Kinyoun or Ziehl-Neelsen carbolfuchsin method. *D,* Acid-fast stain, fluorochrome based. *Mycobacterium* species stained with the acid-fast fluorescent auramine-rhodamine stain. This stain is useful for screening for the presence of acid-fast bacteria in clinical specimens. (Courtesy of Clinical Microbiology Audiovisual Study Units, Health and Education Resources, Inc., Bethesda, MD.)

tion of metachromatic granules (see Fig. 1–6*F*). It is also used as a counterstain in the acid-fast staining procedures.

Lactophenol Cotton Blue

Lactophenol cotton blue is used to stain the cell walls of medically important fungi grown in slide culture (see Fig. 1–6*G*).

Calcofluor White

Calcofluor white is a fluorochrome that binds to chitin in fungal cell walls. It fluoresces a bright apple-green or blue-white, allowing visualization of fungal structures with a fluorescent microscope.

India Ink

India ink is a negative stain used to visualize capsules surrounding certain yeasts, such as *Cryptococcus* (see Fig. 1–6*H*). The fine ink particles are excluded from the capsule, leaving a dark background and a clear capsule surrounding the yeast.

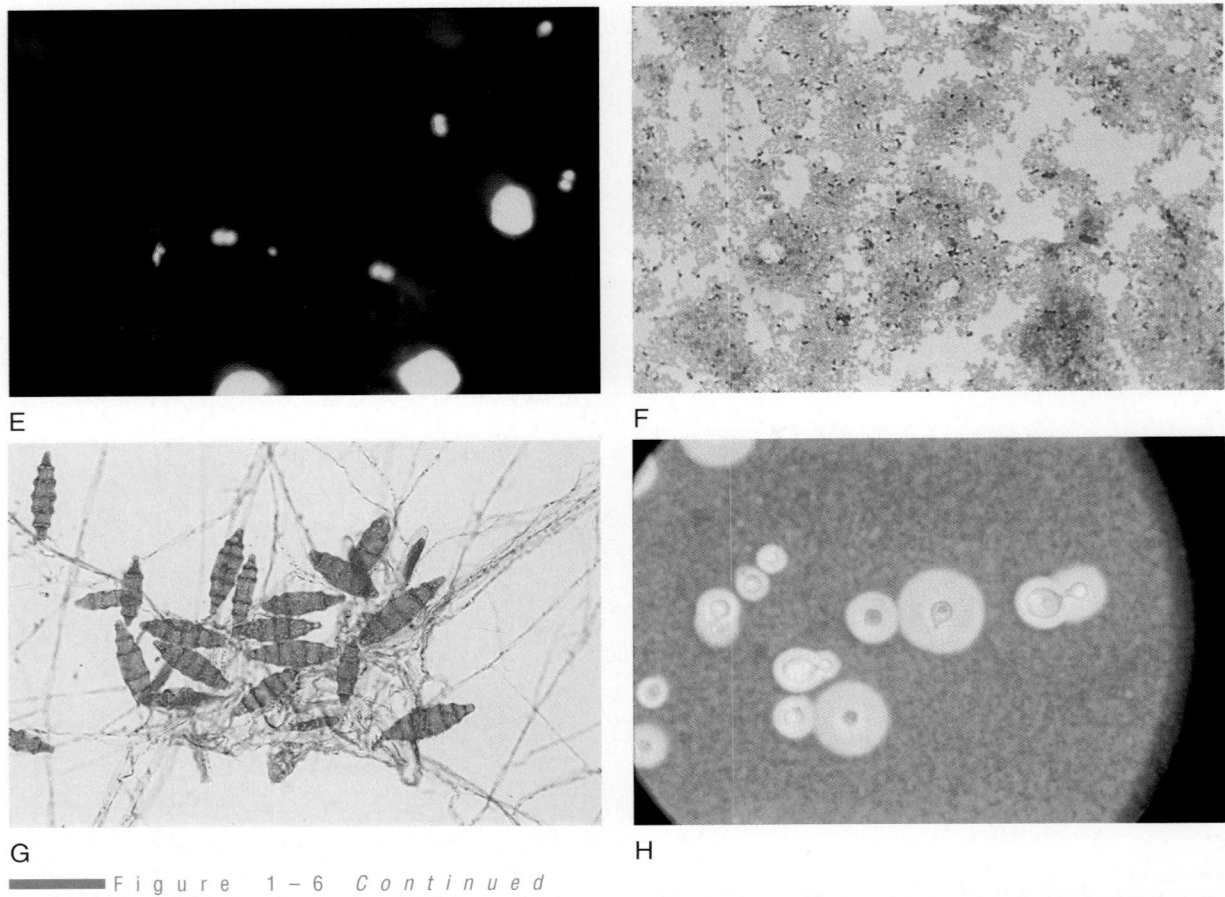

E

F

G

H

Figure 1 – 6 *Continued*

E, Acridine orange stain. A fluorescent stain demonstrating the presence of staphylococci in a blood culture broth. This stain is useful for detecting bacteria in situations where debris may mask the bacteria. (Courtesy of Dr. John E. Peters, Wilford Hall Medical Center, Lackland Air Force Base.) *F,* Methylene blue stain. A methylene blue stain demonstrating the typical morphology of *Corynebacterium diphtheriae. G,* Lactophenol cotton blue stain. Lactophenol cotton blue stained–slide of macroconidia and hyphae of the fungal dermatophyte *Microsporum gypseum. H,* India Ink. An India ink wet mount of *Cryptococcus neoformans* demonstrating the presence of a capsule. (Courtesy of Dr. Andrew G. Smith, University of Maryland School of Medicine, Baltimore.)

MICROBIAL GROWTH AND NUTRITION

All bacteria have three major nutritional needs for growth:

- A source of carbon (for making cellular constituents)

- A source of nitrogen (for making proteins)

- A source of energy (ATP) (for carrying out cellular functions)

Smaller amounts of molecules such as phosphate (for nucleic acids) and a variety of metals and ions (for enzymatic activity) must also be present. Although the basic building blocks required for growth are the same for all cells, bacteria vary widely in their ability to use different sources of these molecules.

Nutritional Requirements for Growth

Bacteria are classified into two basic groups according to how they meet their nutritional needs. Members of the first group, the *autotrophs* (lithotrophs), are able to grow simply, using CO_2 as the sole source of carbon, with only water and inorganic salts required in addition. Autotrophs obtain energy either photosynthetically *(phototrophs)* or by oxidation of inorganic compounds *(chemolithotrophs).* Autotrophs occur in environmental milieus.

The second group of bacteria, the *heterotrophs,* require more complex substances for growth. They require an organic source of carbon, such as glucose, and obtain energy by oxidizing or fermenting organic substances. Often, the same substance (for example, glucose) is used as both the carbon source and energy source.

All bacteria that inhabit the human body fall into the heterotrophic group. Within this group, however, there is great variation in nutritional needs. Bacteria such as *Escherichia coli* and *Pseudomonas aeruginosa* can use a wide variety of organic compounds as carbon sources and therefore grow on most simple laboratory media. Other pathogenic bacteria, such as *Haemophilus influenzae* and the anaerobes, are *fastidious,* requiring additional metabolites such as vitamins, purines, pyrimidines, and hemoglobin supplied in the growth medium. Some pathogenic bacteria, such as *Chlamydia,* cannot be cultured on laboratory media at all and must be grown in tissue culture or detected by other means.

Types of Growth Media

A laboratory growth medium whose contents are simple and completely defined is termed *minimal medium.* This type of medium is not usually used in the diagnostic microbiology laboratory. Media that are more complex and made of extracts of meat or soy beans are termed *nutrient media* (e.g., nutrient broth, trypticase soy broth). A growth medium that contains added growth factors, such as blood, vitamins, and yeast extract, is *enriched* (e.g., blood agar, chocolate agar). Media containing additives that inhibit the growth of some bacteria but allow others to grow are *selective* (e.g., MacConkey agar). Media that allow visualization of metabolic differences between groups or species of bacteria are *differential.* When there will be a delay between collection of the specimen and culturing the specimen, a transport medium is used. A *transport medium* is a holding medium designed to preserve the viability of microorganisms in the specimen but not allow multiplication. Stuart broth and Amies and Cary-Blair transport media are common examples.

Environmental Factors Influencing Growth

Three environmental factors influence the growth rate of bacteria and must be considered when bacteria are cultured in the laboratory:

■ pH

■ Temperature

■ Gaseous composition of the atmosphere

Most pathogenic bacteria grow best at a neutral pH, and diagnostic laboratory media for bacteria are usually adjusted to a final pH between 7.0 and 7.5.

Temperature influences the rate of growth of a bacterial culture. Microorganisms have been categorized according to their optimal temperature for growth. Bacteria that grow best at cold temperatures are called *psychrophiles* (optimal growth at 10 to 20°C). Bacteria that grow optimally at moderate temperatures are called *mesophiles* (optimal growth at 20 to 40°C). Bacteria that grow best at high temperatures are called *thermophiles* (optimal growth at 50 to 60°C). Psychrophiles and thermophiles are found environmentally in places such as the arctic seas and hot springs, respectively. Most bacteria that have adapted to humans are mesophiles that grow best near human body temperature (37°C). Diagnostic laboratories routinely incubate cultures for bacterial growth at 35°C. Some pathogenic species, however, prefer a lower temperature for growth; when these organisms are suspected, the specimen plate is incubated at a lower temperature. Fungal cultures are incubated at 30°C. The ability to grow at room temperature (25°C) or at an elevated temperature (42°C) is used as a diagnostic characteristic for some bacteria.

The bacteria that grow on humans vary in their atmospheric requirements for growth. Some require oxygen *(obligate aerobes).* Some cannot grow in the presence of oxygen *(obligate anaerobes).* Some can grow either with or without oxygen *(facultative anaerobes).* Many grow better when the atmosphere is enriched with extra carbon dioxide *(capnophilic).*

Air contains approximately 21% oxygen and 1% carbon dioxide. When the carbon dioxide content of an aerobic incubator is increased to 10%, the oxygen content of the incubator is lowered to approximately 18%. Obligate aerobes must have oxygen to grow; incubation in air or in an aerobic incubator with 10% CO_2 present satisfies their oxygen requirement. *Microaerophilic* bacteria require a reduced level of oxygen to be present in order to grow. An example of a pathogenic microaerophile is *Campylobacter,* which requires 5 to 6% oxygen. This type of atmosphere can be generated in culture jars or pouches using a commercially available microaerophilic atmosphere generating system. Obligate anaerobes must be grown in an atmosphere either devoid of oxygen or with a very reduced oxygen content. Facultative anaerobes are routinely cultured in an aerobic atmosphere, because aerobic culture is easier and less expensive than anaerobic culture; an example is *E. coli.* Capnophilic bacteria require extra carbon dioxide (5 to 10%) for growth; an example is *H. influenzae.* Because many bacteria grow better in the presence of increased carbon dioxide, diagnostic microbiology laboratories often maintain their aerobic incubators at a 5 to 10% carbon dioxide level.

Bacterial Growth

Generation Time

Bacteria replicate by binary fission, with one cell dividing into two cells. The time required for one cell to divide into two cells is called the *generation time* or *doubling time.* The generation time of a bacterium in culture can be as little as 20 minutes for a fast-growing bacterium such as *E. coli* or as long as 24 hours for a slow-growing bacterium such as *Mycobacterium tuberculosis.*

Growth Curve

If bacteria are in a balanced growth state, with enough nutrients and no toxic products present, the increase in bacterial numbers is proportional to the increase in other bacterial properties, such as mass, protein content, and nucleic acid content. Thus measurement of any of these properties can be used as an indication of bacterial growth. When the growth of a bacterial culture is plotted during balanced growth, the resulting curve shows four phases of growth: (1) a *lag phase,* during which bacteria are preparing to divide, (2) a *log phase,* during which bacteria numbers increase logarithmically, (3) a *stationary phase,* in which nutrients are becoming limited and the numbers of bacteria remain constant (although viability may decrease), and (4) a *death phase,* when the number of nonviable bacterial cells exceeds the number of viable cells. An example of such a growth curve is shown in Figure 1–7.

Determination of Cell Numbers

In the diagnostic laboratory, the number of bacterial cells present is determined in one of three ways.

■ *Direct counting under the microscope:* This method can be used to estimate the number of bacteria present in a specimen. It does not distinguish between live and dead cells.

■ *Direct plate count:* By growing dilutions of broth cultures on agar plates, one can determine the number of colony-forming units per mL (CFU/mL). This provides a count of viable cells only. This method is used in determining the bacterial cell count in urine cultures.

■ *Density measurement:* The density of a bacterial broth culture in log phase can be correlated

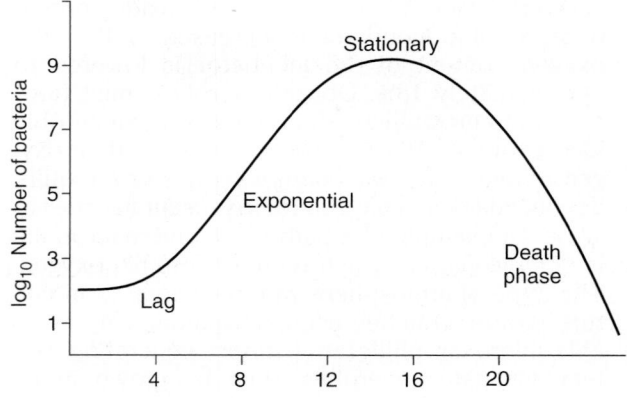

■■■■■■■ F i g u r e   1 – 7

The typical growth curve of a bacterial culture. (Modified and redrawn from Schaechter M, Medoff G, Eisenstein BI: Mechanisms of Microbial Disease, 2nd ed. Baltimore: Williams & Wilkins, 1993, p 51.)

to CFU/mL of the culture. This method is used to prepare a standard inoculum for antimicrobial susceptibility testing.

BACTERIAL BIOCHEMISTRY AND METABOLISM

Metabolism

Microbial metabolism consists of the biochemical reactions bacteria use to break down organic compounds as well as those they use to synthesize new bacterial parts from the resulting carbon skeletons. Energy for the new constructions is generated during the metabolic breakdown of the substrate.

The occurrence of all biochemical reactions in the cell depends on the presence and activity of specific enzymes. Thus, metabolism can be regulated in the cell either by regulating the production of an enzyme itself (a genetic type of regulation, in which production of the enzyme can be induced or suppressed by molecules present in the cell) or by regulating the activity of the enzyme (via feedback inhibition, in which the products of the enzymatic reaction or a succeeding enzymatic reaction inhibit the activity of the enzyme).

Bacteria vary widely in their ability to use various compounds as substrates and in the end products generated. There are a variety of biochemical pathways for substrate breakdown in the microbial world, and the particular pathway used determines the end product and final pH of the medium (Fig. 1–8). Microbiologists use these metabolic differences as phenotypic markers in the identification of bacteria. Diagnostic schemes analyze each unknown microorganism for (1) utilization of a variety of substrates as a carbon source, (2) production of specific end products from various substrates, and (3) production of an acid or alkaline pH in the test medium. Thus knowledge of the biochemistry and metabolism of bacteria is important in the clinical laboratory.

Fermentation and Respiration

Bacteria use biochemical pathways to catabolize (break down) carbohydrates and produce energy by two mechanisms—fermentation and respiration (commonly referred to as oxidation). *Fermentation* is an anaerobic process carried out by both obligate and facultative anaerobes. In fermentation, the electron acceptor is an organic compound. Fermentation is less efficient in energy generation than respiration (oxidation), because the beginning substrate is not completely reduced, and therefore, all of the energy in the substrate is not released. When fermentation occurs, a mixture of end products (such as lactate, butyrate, ethanol, and acetoin) accumulates in the medium. Analysis of these end products is particularly useful for the identification of anaerobic bacteria. End-product determination is also used in the Voges-Proskauer (VP) and methyl red tests, two important diagnostic

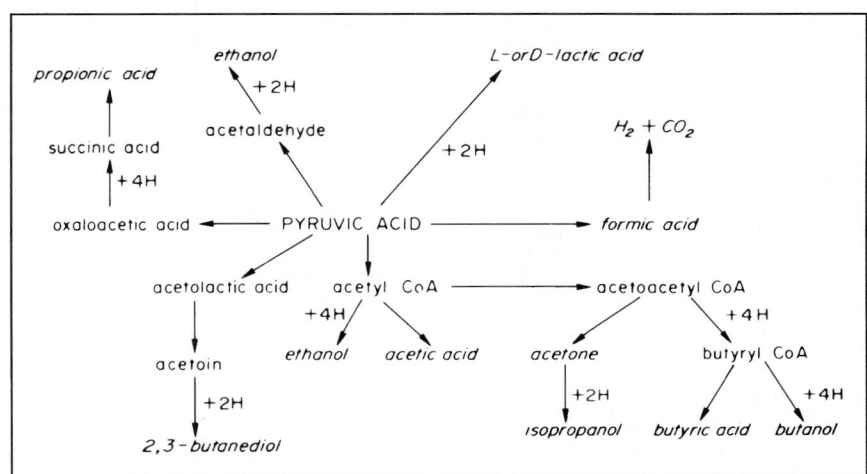

■■■■■ F i g u r e   1 – 8

The fate of pyruvate in major fermentation pathways by microorganisms. (With permission from Joklik WK, Willett HP, et al: Zinsser Microbiology, 20th ed. Norwalk, CT: Appleton & Lange, 1992, p 39.)

tests used in the identification of the Enterobacteriaceae. (The term fermentation is often used loosely in the diagnostic microbiology laboratory to indicate any type of utilization—fermentative or oxidative—of a carbohydrate—sugar—with the resulting production of an acid pH.)

Respiration is an efficient energy-generating process in which molecular oxygen is the final electron acceptor. Obligate aerobes and facultative anaerobes carry out *aerobic respiration,* in which oxygen (O_2) is the final electron acceptor. Certain anaerobes can carry out *anaerobic respiration,* in which inorganic forms of oxygen, such as nitrate and sulfate, act as the final electron acceptors.

Biochemical Pathways from Glucose to Pyruvic Acid

The starting carbohydrate for bacterial fermetations or oxidations is glucose. When bacteria use other sugars as a carbon source, they first convert the sugar to glucose, which is then processed by one of three pathways. These pathways are designed to generate pyruvic acid, a key three-carbon intermediate. The three major biochemical pathways bacteria use to break down glucose to pyruvic acid are: (1) the Embden-Meyerhof-Parnas (EMP) glycolytic pathway (Fig. 1–9), (2) the pentose phosphate pathway (Fig. 1–10), and (3) the Entner-Doudoroff pathway (see Fig. 1–10). Pyruvate can then be further processed either fermentatively or oxidatively. The three major metabolic pathways and their key characteristics are described in Table 1–2.

Anaerobic Utilization of Pyruvic Acid (Fermentation)

Pyruvic acid is a key metabolic intermediate. Bacteria process pyruvic acid further using a variety of fermentation pathways. Each pathway yields different end products, which can be analyzed and used as phenotypic markers (see Fig. 1–8). Some of the fermentation pathways used by the microbes that inhabit the human body are as follows:

- **Alcoholic fermentation:** The major end product is ethanol. This is the pathway used by yeasts when they ferment glucose to produce ethanol.

- **Homolactic fermentation:** The end product is almost exclusively lactic acid. All members of the *Streptococcus* genus and many members of the *Lactobacillus* genus ferment pyruvate using this pathway.

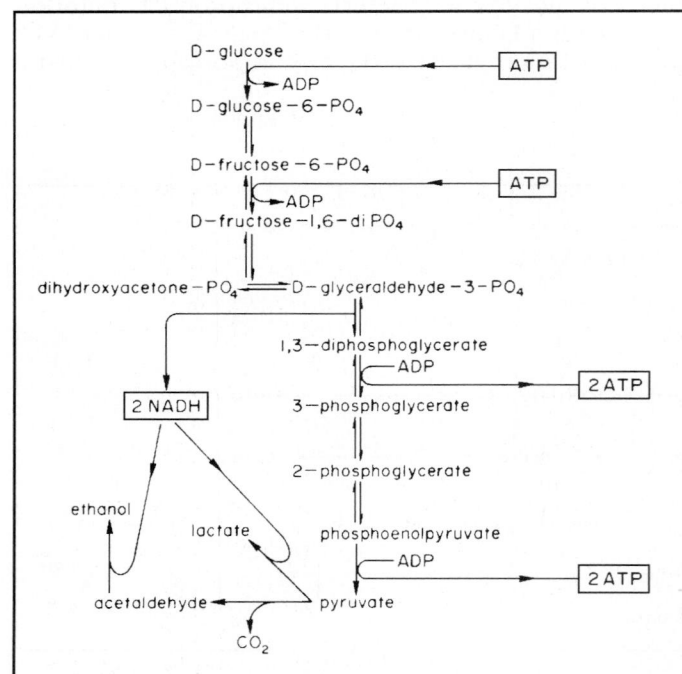

■■■■■■■ F i g u r e   1 – 9

The Embden-Meyerhof-Parnas (EMP) glycolytic pathway. (With permission from Joklik WK, Willett HP, et al: Zinsser Microbiology, 20th ed. East Norwalk, CT: Appleton & Lange, 1992, p 35.)

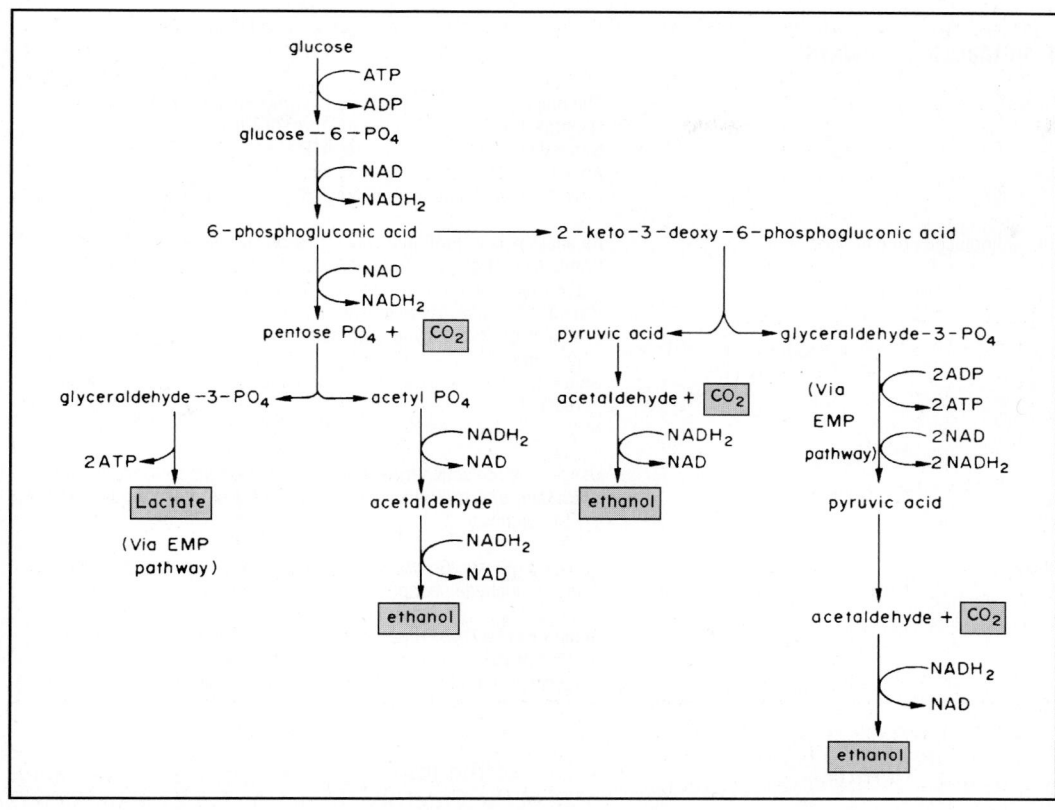

Figure 1-10

Alternative microbial pathways to the EMP pathway for glucose fermentation. The pentose phosphate pathway is on the left, and the Entner-Doudoroff pathway on the right. (With permission from Joklik WK, Willett HP, et al: Zinsser Microbiology, 20th ed. Norwalk, CT: Appleton & Lange, 1992, p 38.)

■ **Heterolactic fermentation:** Some lactobacilli use this mixed fermentation pathway, of which, in addition to lactic acid, the end products include CO_2, alcohols, formic acid, and acetic acid.

■ **Propionic acid fermentation:** Propionic acid is the major end product of fermentations carried out by *Propionibacterium acnes* and some anaerobic non–spore-forming gram-positive bacilli.

■ **Mixed acid fermentation:** Members of the genera *Escherichia*, *Salmonella*, and *Shigella* within the Enterobacteriaceae use this pathway for sugar fermentation and produce a number of acids as end products—lactic, acetic, succinic, and formic acids. The strong acid produced is the basis for the *positive*

reaction on the *methyl red test* exhibited by these organisms.

■ **Butanediol fermentation:** Members of the genera *Klebsiella*, *Enterobacter*, and *Serratia* within the Enterobacteriaceae use this pathway for sugar fermentation. The end products are acetoin (acetyl methyl carbinol) and 2,3-butanediol. Detection of acetoin is the basis for the *positive Voges-Proskauer* (VP) *reaction* characteristic of these microorganisms. Little acid is produced by this pathway. Thus, organisms that have a positive VP reaction usually have a negative reaction on the methyl red test, and vice versa.

■ **Butyric acid fermentation:** Certain obligate anaerobes, including many *Clostridium* species, *Fusobacterium,* and *Eubacterium,* produce butyric acid as their primary end product along with acetic acid, CO_2, and hydrogen.

■■■■■ T A B L E   1 – 2
THE THREE MAJOR METABOLIC PATHWAYS

Embden-Meyerhof-Parnas (EMP) glycolytic pathway	The major pathway in conversion of glucose to pyruvate Generates reducing power in the form of $NADH_2$ Generates energy in the form of ATP Anaerobic; does not require oxygen Used by many bacteria, including all members of Enterobacteriaceae
Pentose phosphate (phosphogluconate) pathway	An alternative to EMP pathway for carbohydrate metabolism Conversion of glucose to ribulose-5-phosphate, which is then rearranged into other 3-, 4-, 5-, 6-, and 7-carbon sugars Provides pentoses for nucleotide synthesis Produces glyceraldehyde-3-phosphate, which can be converted to pyruvate Generates NADPH, which provides reducing power for biosynthetic reactions May be used to generate ATP (yield is less than with the EMP pathway) Used by heterolactic fermenting bacteria, such as lactobacilli, and by *Brucella abortus,* which lacks some of the enzymes required in the EMP pathway
Entner-Doudoroff pathway	Converts glucose-6-phosphate (rather than glucose) to pyruvate and glyceraldehyde phosphate, which can then be funneled into other pathways Generates one NADPH per molecule of glucose but uses one ATP Aerobic process used by *Pseudomonas, Alcaligenes, Enterococcus faecalis,* and other bacteria lacking certain glycolytic enzymes

Aerobic Utilization of Pyruvate (Oxidation)

The most important pathway for the complete oxidation of a substrate under aerobic conditions is the Krebs or TCA (tricarboxylic acid) cycle. In this cycle, pyruvate is oxidized, carbon skeletons for biosynthetic reactions are created, and the electrons donated by pyruvate are passed through an electron transport chain and used to generate energy in the form of ATP. This cycle results in the production of acid and the evolution of CO_2 (Fig. 1–11).

Carbohydrate Utilization and Lactose Fermentation

The ability of microorganisms to use various "sugars" (carbohydrates) for growth is an integral part of most diagnostic identification schemes. The fermentation of the sugar is usually detected by acid production and a concomitant change of color due to a pH indicator present in the culture medium. In general, bacteria ferment glucose preferentially over other sugars, and therefore, glucose must not be present if the ability to ferment another sugar is being tested.

One of the important steps in classifying members of the Enterobacteriaceae family is the determination of the microorganism's ability to ferment lactose. These bacteria are classified as either lactose fermenters or lactose nonfermenters. Lactose is a disaccharide consisting of one molecule of glucose and one molecule of galactose linked together by a galactoside bond. Two steps are involved in the utilization of lactose by a bacterium. The first step requires an enzyme, β-galactoside permease, for the transport of lactose across the cell wall into the bacterial cytoplasm. The second step occurs inside the cell and requires the enzyme β-galactosidase to break the galactoside bond, releasing glucose, which can then be fermented. Thus, all organisms that can ferment lactose can also ferment glucose.

BACTERIAL GENETICS

Bacterial genetics is increasingly important in the diagnostic microbiology laboratory. New diagnostic tests have been developed that are based on identifying unique RNA or DNA sequences present in each bacterial species. The polymerase chain reaction (PCR) technique is a means of amplifying specific DNA sequences and

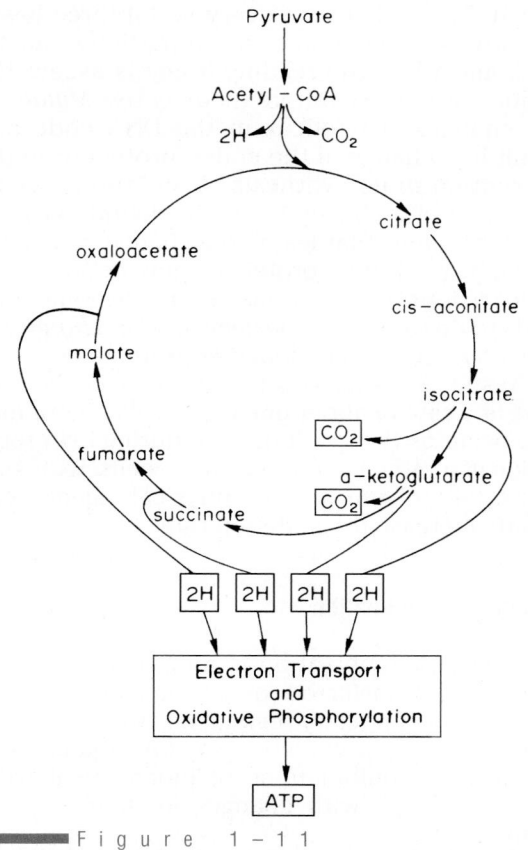

■■■■ F i g u r e 1 – 1 1

The Krebs tricarboxylic acid (TCA) cycle allowing the complete oxidation of a substrate. (With permission from Joklik WK, Willett HP, et al: Zinsser Microbiology, 20th ed. Norwalk, CT: Appleton & Lange, 1992, p 41.)

thus detecting very small numbers of bacteria present in a specimen. Genetic tests circumvent the need to culture bacteria, providing a more rapid method of identifying pathogens. Analysis of DNA structure using restriction endonuclease fragment patterns (RFLP) provides a fine tool for epidemiologists. As these techniques are perfected, they will be added to the spectrum of diagnostic tests used in the clinical microbiology laboratory.

An understanding of bacterial genetics is also necessary to understand the development and transfer of antimicrobial resistance by bacteria. The occurrence of mutations can result in a change in the expected phenotypic characteristics of an organism and provides an explanation for atypical results sometimes encountered on diagnostic biochemical tests. This section briefly reviews some of the basic terminology and concepts of bacterial genetics.

Terminology

The *genotype* of a cell is the genetic potential of the DNA of an organism. It includes all of the characteristics that are coded for in the DNA of a bacterium and that have the *potential* to be expressed. Some genes are silent genes, expressed only under certain conditions. Genes that are always expressed are *constitutive*. Genes that are expressed only under certain conditions are *inducible*. The *phenotype* of a cell consists of the genetic characteristics of a cell that actually are expressed and can be observed. The ultimate aim of a cell is to produce the proteins that are responsible for cellular structure and function and to transmit the information for accomplishing this to the next generation of cells.

Information for protein synthesis is encoded in the bacterial DNA and transmitted in the chromosome to each generation. The general flow of information in a bacterial cell is from DNA (which contains the genetic information) to messenger RNA (mRNA) (which acts as a blueprint for protein construction) to the actual protein itself (made on ribosomes containing ribosomal RNA with the aid of transfer RNA, which places specific amino acids in the growing peptide chain). *Replication* is the duplication of chromosomal DNA for insertion into a daughter cell. *Transcription* is the synthesis of single-stranded RNA (with the aid of the enzyme RNA polymerase) using one strand of the DNA as a template. *Translation* is the actual synthesis of a specific protein from the mRNA code. The term *protein expression* also refers to the synthesis (i.e., translation) of a protein.

Genetic Elements and Alterations

The Bacterial Genome

The bacterial *chromosome* (also called the *genome*) consists of a single, closed, circular piece of double-stranded DNA that is supercoiled in order to fit inside the cell. It contains all the information needed for cell growth and replication. *Genes* are specific DNA sequences that code for the amino acid sequence in one protein (i.e., one gene equals one protein). In front of each gene on the DNA strand is an untranscribed area containing a promoter region, which the RNA polymerase recognizes for transcription initiation. This area may also contain regulatory regions to which molecules may attach and cause either a decrease or an increase in transcription.

Extrachromosomal Elements

In addition to the genetic information encoded in the bacterial chromosome, many bacteria contain extra information on small circular pieces of DNA called *plasmids*. Genes that code for antibiotic resistance (and sometimes toxins or other virulence factors) are often located on plasmids. Antibiotic therapy selects for bacterial strains containing plasmids encoding antibiotic resistance genes; this is one reason antibiotics should not be over-prescribed. The number of plasmids present in a bacterial cell may vary from one (low copy number) to hundreds (high copy number). Plasmids are located in the cytoplasm of the cell and can be replicated and passed to daughter cells just like chromosomal DNA. They may also sometimes be passed from one bacterial species to another. This is one way resistance to antibiotics is acquired.

Mobile Genetic Elements

Certain pieces of DNA are mobile and may jump from one place in the chromosome to another place. These are sometimes referred to as "jumping genes." The simplest mobile piece of DNA is an *insertion sequence (IS) element*. It is about 1000 base pairs long with inverted repeats on each end. Each IS element codes for only one gene, a transposase enzyme that allows the IS element to pop into and out of DNA. Bacterial genomes contain many IS elements. The main effect of IS elements in bacteria is that when an IS element inserts itself into the middle of a gene, it disrupts and inactivates the gene. This can result in loss of an observable characteristic, such as the ability to ferment a particular sugar. *Transposons* are related mobile elements that contain additional genes. Transposons often carry antibiotic-resistance genes and are usually located in plasmids.

Mutations

A gene sequence must be read in the right "frame" in order for the correct protein to be pro-

duced. This is because every set of three bases (known as a *codon*) specifies a particular amino acid, and when the reading frame is askew, the codons are interpreted incorrectly. *Mutations* are changes that occur in the DNA code and result in a change in the coded protein or in the prevention of its synthesis. A mutation may be the result of a change in one nucleotide base (*a point mutation*) that leads to a change in a single amino acid within a protein or may be the result of insertions or deletions in the genome that leads to disruption of the gene and/or a *frameshift mutation*. Incomplete, inactive proteins are often the result. Spontaneous mutations occur in bacteria at a rate of about one in 10^9 cells. Mutations also occur as the result of error during DNA replication at a rate of about one in 10^7 cells. Exposure to certain chemical and physical agents can greatly increase the mutation rate.

Genetic Recombination

Genetic recombination is a method by which genes are transferred or exchanged between homologous (similar) regions on two DNA molecules. This has provided a way for organisms to obtain new combinations of biochemical pathways and cope with changes in their environment.

Mechanisms of Gene Transfer

There are three basic ways genetic material may be transferred from one bacterium to another. They are

- Transformation
- Transduction
- Conjugation

Transformation

Transformation is the uptake and incorporation of naked DNA into a bacterial cell (Fig. 1–12*A*). Once the DNA has been taken up, it can be incorporated into the bacterial genome by recombination. If the DNA is a circular plasmid and the recipient cell is compatible, the plasmid can replicate in the cytoplasm and be transferred to daughter cells. Cells that can take up naked DNA

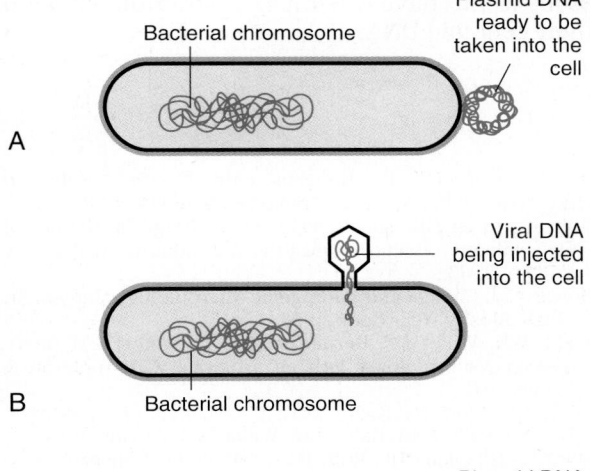

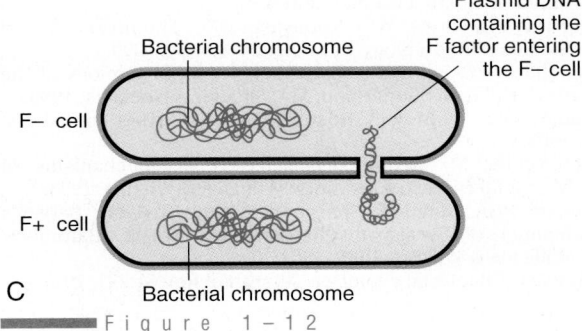

Figure 1-12

Diagram of methods of gene transfer into bacterial cells. *A,* Bacterial transformation. Free or "naked" DNA is taken up by a competent bacterial cell. After uptake, the DNA may take one of three courses: (1) it is integrated into existing bacterial genetic material, (2) it is degraded, or (3) if it is a compatible plasmid, it may replicate in the cytoplasm. *B,* Bacterial transduction. A phage injects DNA into the bacterial cell. The phage tail combines with a receptor of the bacterial cell wall and injects the DNA into the bacterium. One of two courses may then be taken. In the lytic cycle, replication of the bacterial chromosome is disrupted; phage components are formed and assembled into phage particles. The bacterial cell is lysed, releasing mature phage. In the lysogenic cycle, the phage DNA is incorporated into the bacterial genetic material, and genes encoded on the phage DNA may be expressed from this site. At a later time, the phage may be "induced," and a lytic cycle will then ensue. *C,* Bacterial conjugation. An F+ cell connects with an F– cell via sex pili. DNA is then transferred from one cell to the other.

are referred to as being *competent.* Only a few bacterial species, such as *Streptococcus pneumoniae, Neisseria gonorrhoeae,* and *H. influenzae,* do this naturally. Bacteria can be made competent in the laboratory, and transformation is the main method used to introduce genetically manipulated plasmids into bacteria, such as *E. coli,* during cloning procedures.

Transduction

Transduction is the transfer of bacterial genes by a bacteriophage (a bacterial virus) from one cell to another (see Fig. 1–12*B*). A bacteriophage consists of a chromosome (DNA or RNA) surrounded by a protein coat. When a phage infects a bacterial cell, it injects its genome into the bacterial cell, leaving the protein coat outside. The phage may then take a *lytic pathway,* in which the bacteriophage DNA directs the bacterial cell to synthesize phage DNA and phage protein and package it into new phage particles. The bacterial cell then lyses, releasing new phage, which can infect other bacterial cells. In some instances, the phage DNA instead becomes incorporated into the bacterial genome, where it is replicated along with the bacterial chromosomal DNA; this state is known as *lysogeny,* and the phage is referred to as being *temperate.* During lysogeny, genes present in the phage DNA may be expressed by the bacterial cell. An example of this in the clinical laboratory is *Corynebacterium diphtheriae.* Strains of *C. diphtheriae* that are lysogenized with a temperate phage carrying the gene for diphtheria toxin cause disease. Strains lacking the phage do not cause disease.

Under certain conditions, a temperate phage can be *induced,* the phage DNA is excised from the bacterial genome, and a lytic state occurs. During this process, adjacent bacterial genes may be excised along with the phage DNA and packaged into the new phage. The bacterial genes may then be transferred when the phage infects a new bacterium. In the field of biotechnology, phages are often used to insert cloned genes into bacteria for analysis.

Conjugation

Conjugation is the transfer of genetic material from a donor bacterial strain to a recipient strain (see Fig. 1–12*C*). It requires close contact between the two cells. In the *E. coli* system, the donor strain (F+) possesses a fertility factor (F factor) on a plasmid that carries the genes for conjugative transfer. The donor strain produces a hollow surface appendage called a sex pilus, which binds to the recipient F– cell and brings the two cells in close contact. Transfer of DNA then occurs. Both plasmids and chromosomal genes can be transferred by this method. When the F factor is integrated into the bacterial chromosome rather than a plasmid, there is a higher frequency of transfer of adjacent bacterial chromosomal genes. These strains are known as Hfr (high-frequency) strains.

Restriction Enzymes

Bacteria have evolved a system to restrict the incorporation of foreign DNA into their genomes. Specific *restriction enzymes* are produced that cut incoming foreign DNA at specific DNA sequences. The bacteria methylate their own DNA at these same sequences, so that the restriction enzymes do not cut the DNA in their own cell. Many restriction enzymes with a variety of recognition sequences have now been isolated from various microorganisms. The first three letters in the restriction endonuclease name indicate the bacterial source of the enzyme. For instance, the enzyme *Eco*RI was isolated from *E. coli,* and the enzyme *Hind*III was isolated from *H. influenzae* type d. These enzymes are used in the field of biotechnology to create sites for insertion of new genes. In clinical microbiology, epidemiologists use restriction enzyme fragment analysis to determine whether strains of bacteria have identical restriction sites in their genomic DNA.

Bibliography

De Robertis EDP, De Robertis EMF Jr: 1987. Cell and Molecular Biology. Philadelphia: Lea & Febiger, 1987.

Holt JG, Krieg NR, Sneath PHA, et al: Bergey's Manual of Determinative Bacteriology, 9th ed. Baltimore: Williams & Wilkins, 1993.

Howard BJ. Clinical and Pathogenic Microbiology, 2nd ed. St. Louis: Mosby–Year Book, 1994.

Joklik WK, Willet HP, Bernard Amos D, Wilfert CM (eds): Zinsser Microbiology, 20th ed. Norwalk, CT: Appleton & Lange, 1992.

Krieg NR, Holt JG (eds): Bergey's Manual of Systematic Bacteriology, Vol 1. Baltimore: Williams & Wilkins, 1984.

Mims CA, Playfair JHL, Roitt IM, et al: Medical Microbiology. London: Mosby Europe Limited, 1993.

Murray PR, Drew WL, Kobayashi GS, Thompson JH Jr: Medical Microbiology. St Louis: CV Mosby, 1990.

Neidhardt FC, Ingraham JL, Schaechter M: Physiology of the Bacterial Cell. Sunderland, MA: Sinauer Associates, 1990.

Sadava DE: Cell Biology. Boston: Jones & Bartlett Publishers, 1993.

Schaechter M, Medoff G, Eisenstein BI: Mechanisms of Microbial Disease. Baltimore: Williams & Wilkins, 1993.

Sneath PHA, Mair NS, Sharpe ME, Holt JG (eds): Bergey's Manual of Systematic Bacteriology, Vol 2. Baltimore: Williams & Wilkins, 1986.

Woese CR Bacterial evolution. Microbiol Rev 51:221, 1987.

CHAPTER 2

Control of Microorganisms

A. DISINFECTION AND STERILIZATION

Denise A. Miller

Safety in the laboratory cannot be overemphasized. It is also difficult to quantitate the risk of working with an infectious agent. Risk to an individual increases with the frequency and the level of contact with the agent. Therefore, it is necessary that each laboratory develop and institute an exposure control plan that will effectively minimize the risks and the overt laboratory-associated hazards to persons who may be directly or indirectly exposed to them. This chapter provides information on standard disinfection and sterilization techniques and laboratory safety guidelines for both students and clinical laboratory practitioners.

This section, on disinfection and sterilization, provides a practical overview of

■ Sterilization and disinfection

■ Chemical and physical methods commonly used

■ Principles and application of each of the methods

STERILIZATION VERSUS DISINFECTION

The use of disinfection and sterilization (as a scientific procedure) methods originated over 100 years ago, when Joseph Lister introduced the concept of aseptic surgery. Since that time, the implementation of effective sterilization and disinfection methods remains crucial in the control of nosocomial infections.

In order to fully understand the principles of disinfection and sterilization, we need to have accurate definitions of certain terms. *Sterilization* refers to the removal of all forms of life, including bacterial spores. By definition, there are no degrees of sterilization—it is an all-or-nothing process. Chemical or physical methods may be used to accomplish this form of microbial removal. *Disinfection* refers to the removal of pathogenic organisms but does not necessarily include removal of bacterial or other spores. Physical or chemical methods may be employed, but most disinfectants are chemical agents applied to inanimate objects. A disinfectant that is applied to living tissue is referred to as an *antiseptic.*

FACTORS THAT INFLUENCE THE DEGREE OF KILLING

Before discussing the particular methods, it is important to review the factors that influence the degree of killing of organisms. These factors play a significant role in the selection and implementation of the appropriate method of disinfection. They are

■ Types of organisms

■ Number of organisms present

■ Concentration of disinfecting agent

■ Amount of organic soil present

■ Nature of surface to be disinfected

Types of Organisms

Organisms vary greatly in their ability to withstand chemical and physical treatment (Fig. 2–1). This variety is due to the biochemical composition of the cells and the protective mechanism afforded by the constituents. For example, spores have coats rich in proteins, lipids, and

Schematic diagram showing the different types of organisms and their resistance to killing agents.

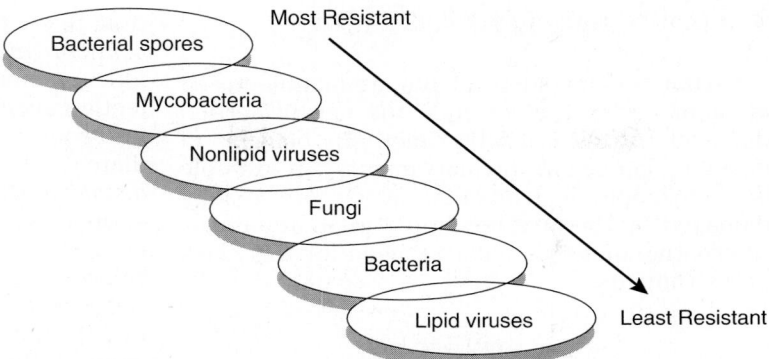

carbohydrates as well as cores rich in dipicolinic acid and calcium, all of which offer protection to spores. Cell walls of mycobacteria are rich in lipids, which may account for their resistance to chemical and environmental stresses, particularly desiccation. By contrast, viruses containing lipid-rich envelopes are more susceptible to the effects of detergents and wetting agents.

Number of Organisms

Another factor to consider is the total number of organisms present, referred to as the *microbial load.* If number of organisms is plotted against the time they are exposed to the killing agent (exposure time) logarithmically, the result is a straight line (Fig. 2–2). The death curve is logarithmic. Because the microbial load is most likely composed of organisms with varying susceptibilities to killing agents, not all the organisms die at the same time. The microbial load determines the exposure time. In general, higher numbers of organisms require longer exposure times.

Concentration of Disinfecting Agent

The concentration of a disinfecting agent is also important. Agents vary substantially among manufacturers, and it is important that manufacturers' instructions on preparation, dilution, and use be followed very carefully. Proper concentrations of disinfecting agents ensure the inactivation of target organisms and promote safe and cost-effective practices.

Organic Soil Present

Organic soil, such as blood, mucus, and pus, affects killing activity by actually inactivating the disinfecting agent. In addition, by coating the surface to be treated, organic soil prevents full contact between object and agent. For optimal killing activity, instruments and surfaces should be cleansed of excess organic material prior to disinfection.

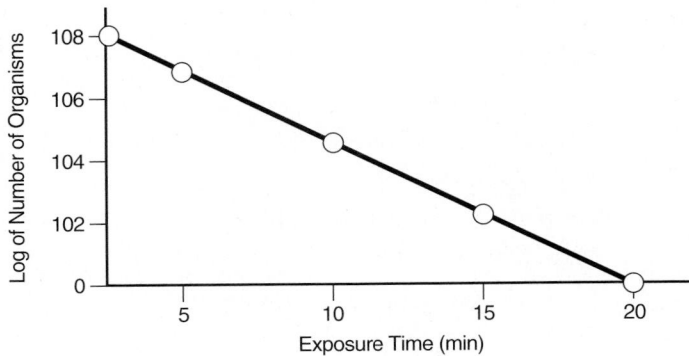

Diagram showing the effect of exposure time versus number of organisms.

Nature of Surface to Be Disinfected

Certain medical instruments are manufactured of biomaterials that exclude the use of certain disinfection or sterilization methods because of possible damage to the instruments. An example is endoscopic instruments, which are readily damaged by the heat generated in an autoclave. Alternative methods must be used for this class of instruments.

METHODS OF DISINFECTION AND STERILIZATION

We have seen how factors that affect the survival of microorganisms influence disinfection and sterilization. We will now look at how methods are selected. E.H. Spaulding categorized medical materials into three device classifications:

■ Critical materials

■ Semicritical materials

■ Noncritical materials

Critical materials are those that invade sterile tissues or enter the vascular system. These materials have the greatest chance of producing infec-tion if contaminated and require sterilization. *Semicritical materials* come into contact with mucous membranes and require high-level disin-fection agents. Finally, *noncritical materials* come into contact with intact skin and require interme-diate-level to low-level disinfection. *High-level disinfectants* have activity against bacterial endo-spores, whereas *intermediate-level disinfectants* have tuberculocidal activity but not sporicidal activity. Finally, *low-level disinfectants* have a wide range of activity against microorganisms but do not demonstrate sporicidal or tubercu-locidal activity. Table 2–1 presents a summary of these principles.

Physical Methods

As mentioned earlier, sterilization and disin-fection can be carried out by both physical and chemical methods. Although several physical methods are available, we will restrict our dis-cussion to those methods most commonly used in a laboratory or hospital setting.

Heat

Because of its reliable effects, ease of use, and economy, heat is the most common method used

■■■■■■ T A B L E   2 – 1

DEVICE CLASSIFICATION AND METHODS OF EFFECTIVE DISINFECTION

Device Classification	Disinfection Method	Killing Action Against				
		Spores	Mycobacteria	Nonlipid Viruses	Fungi	Bacteria
Critical	Sterilization: Steam Dry heat Gas Chemical Ionizing radiation	+	+	+	+	+
Semicritical	High-level disinfection: 2% glutaraldehyde Chlorine dioxide Wet pasteurization	± ± –	+ + +	+ + +	+ + +	+ + +

Noncritical	Low-level disinfection:
	Sodium hypochlorite
	Quaternary ammonium
	  compounds
	Ethyl, isopropyl
	  alcohol (70–90%)
	Phenolics
	Iodophors
		–
		–
		
		–
		–
		–
			+
			–
			
			–
			±
			–
				+
				±
				
				+
				+
				+
					+
					+
					
					+
					+
					+
						+
						+
						
						+
						+
						+

for the elimination of microorganisms. It can be used in several manners. *Moist heat* or *heat under steam pressure* is the agent used in autoclaves. Putting steam under 1 atm of pressure or 15 psi achieves a temperature of 121°C. At this temperature, all microorganisms and their endospores are destroyed in approximately 15 minutes of exposure. The time varies somewhat according to the size of the object; what is important is that the exposure time is sufficient for complete penetration of the steam at the proper temperature. An added advantage of moist heat is the shorter time required for sterilization. (Heat in water is transferred more readily to a cool body than heat in air.) Moist heat is the sterilization method of choice for heat-stable objects.

Dry heat may also be used as a sterilizing agent, although it requires much longer exposure times and higher temperatures than moist heat. This method may be used for heat-stable substances that are not penetrated by moist heat, for example, oils. Dry heat is commonly used to sterilize glassware.

Boiling and *pasteurization* are both methods that achieve disinfection but not sterilization, because neither method eliminates spores. Boiling (100°C) kills most microorganisms in approximately 10 minutes. Pasteurization, used mostly in the food industry, eliminates foodborne pathogens as well as organisms responsible for food spoilage. It is carried out at 63°C for 30 minutes. The main advantage of pasteurization is that treatment at this temperature reduces spoilage of food without affecting its taste.

Table 2–2 summarizes the applications of heat.

Filtration

Filtration methods may be used with both liquid and air. Filtration of liquids is accomplished through the use of thin membrane filters composed of plastic polymers or cellulose esters. The liquid is pulled through small, mechanically introduced pores with a vacuum. Organisms larger than the size of the pores are retained. Different pore sizes are available. Bacteria, yeasts, and molds are retained by pore sizes 0.45 µm and 0.80 µm, which may also allow passage of *Pseudomonas*-like organisms; therefore, there is also a 0.22-µm size for critical sterilizing (e.g., parenteral solutions). Membranes with pore sizes as small as 0.01 µm are capable of retaining small viruses. The most common application of filtration is in the sterilization of heat-sensitive solutions, such as vaccines and antibiotic solutions. Filtration of air is accomplished with the use of high-efficiency particulate air (HEPA) filters. These filters are able to remove microorganisms larger than 0.3 µm, and are used in laboratory hoods and in rooms of immunocompromised patients.

Radiation

Radiation may be used in two forms, ionizing and non-ionizing. Ionizing radiation, in the form of gamma rays or electron beams, is of short wavelength and high energy. This method of sterilization is used by the medical industry for the sterilization of disposable supplies, such as syringes, catheters, and gloves. Non-ionizing

■■■■■■ T A B L E   2 – 2
CONTROL OF MICROORGANISMS USING HEAT METHODS

Method	Temperature(°C)	Time Required	Applications
Boiling water (steam) Endospores survive	100	15 minutes	Kills microbial vegetative forms
Autoclave (steam under pressure)	121.6	15 minutes at 15 psi	Sterilizes; kills endospores
Pateurization Batch method	63	30 minutes	Disinfects; kills milk-borne pathogens and vegetative forms Endospores survive
Flash methods	72	15 seconds	Same; but shorter time at higher temperature
Oven (dry heat)	160–180	1.5–3 hours	Sterilizes; materials stay dry

Adapted with permission from VanDemark PJ, Batzing BL: The Microbes: An Introduction to Their Nature and Importance. Redwood City, CA: Benjamin-Cummings, 1987, p 429.

radiation in the form of ultraviolet rays is of long wavelength and low energy. Because of its poor penetrability, its usefulness is limited; it is commonly used to disinfect surfaces.

Chemical Methods

Just as physical methods are used mainly to achieve sterilization, chemical agents are used mainly as disinfectants. Some chemical agents, however, may be used to sterilize. These are known as *chemosterilizers*. All disinfectants are regulated by the U.S. Environmental Protection Agency (EPA). Chemical agents exert their killing effect by a variety of mechanisms:

■ Reaction with components of the cytoplasmic membrane

■ Denaturation of cellular proteins

■ Reaction with the thiol (−SH) groups of enzymes

■ Damage of RNA and DNA

It is important to remember that agents can exert one or a combination of actions on microorganisms. Damage to the integrity of the cytoplasmic membrane causes the cytoplasm and its contents to leak out, resulting in cell death. Denaturation of proteins effectively disrupts the metabolism of the cells. Some agents specifically react with the thiol (−SH) groups of enzymes, thereby inactivating them. Thiol groups occur usually as the amino acid cysteine. Finally, damage to RNA and DNA inhibits the replication of the organism. For ease of discussion, we group the chemical agents on the basis of chemical composition. Table 2–3 summarizes the applications of chemicals commonly used as disinfectants and antiseptics.

Alcohols

The two most effective alcohols used in hospitals for disinfection purposes are ethyl alcohol and isopropyl alcohol. Although alcohols have a wide spectrum, they are nonsporicidal. This and several other factors limit their use as disinfectants. Because the action of alcohols is greatly reduced in concentrations less than 50%, they should be used in concentrations between 60 and 90%. Contact time is also reduced owing to rapid evaporation and inability to penetrate

■■■■■ T A B L E   2 - 3

CHEMICAL AGENTS COMMONLY USED AS DISINFECTANTS AND ANTISEPTICS

Type	Agent(s)	Action(s)	Applications and Precautions
Alcohols (50–70%)	Ethanol, isopropanol, benzyl alcohol	Denature proteins; solubilize lipids	Skin antiseptics
Aldehydes (in solution)	Formaldehyde (8%), glutaraldehyde (2%)	React with NH_2, −SH, and −COOH groups	Disinfectants; kill endospores Toxic to humans
Halogens	Tincture of iodine (2% in 70% alcohol) Chlorine and chlorine compounds	Inactivates proteins Reacts with water to form hypochlorous acid (HClO); oxidizing agent	Skin disinfectants Used to disinfect drinking water Surface disinfectants
Heavy metals	Silver nitrate ($AgNO_3$) Mercuric chloride ($HgCl_2$)	Precipitates proteins Reacts with −SH groups Lyses cell membrane	Eye drop (1% solution) Disinfectant: toxic at high concentrations
Detergents	Quaternary ammonium compounds	Disrupt cell membranes	Skin antiseptics; disinfectants
Phenolics	Phenol, carbolic acid, Lysol, hexachlorophene	Denature proteins; disrupt cell membranes	Disinfectants at high concentrations; used in soaps at low concentrations
Gases	Ethylene oxide	Alkylating agent	Sterilization of heat-sensitive objects

Adapted with permission from VanDemark PJ, Batzing BL: The Microbes: An Introduction to Their Nature and Importance. Redwood City, CA: Benjamin-Cummings, 1987, p 443.

organic material. Alcohols inactivate microorganisms by denaturing proteins. They are used principally as antiseptics and to disinfect small areas, such as injection vial septa and thermometers.

Aldehydes

FORMALDEHYDE

Formaldehyde is an aldehyde generally used as formalin, a 37% aqueous solution of formaldehyde gas. Although it can be used as a chemosterilizer in high concentrations, its usefulness is limited by its adverse effects. Formaldehyde has been found to be a carcinogen, and the U.S. Occupational Safety and Health Administration (OSHA) has set worker exposure limits. In addition, it is highly irritating and is slow acting because of poor penetrability.

GLUTARALDEHYDE

Glutaraldehyde is also an aldehyde, more precisely, a saturated five-carbon dialdehyde, that has broad-spectrum activity and rapid cidal action and remains active in the presence of organic matter. Glutaraldehyde is extremely susceptible to pH changes, because it is active only in an alkaline environment. When used as a 2% solution, it is germicidal in approximately 10 minutes and sporicidal in 3 to 10 hours. Its killing activity is due to inactivation of DNA and RNA through alkylation of sulfhydryl and amino groups. Because it does not corrode lenses, metal, or rubber, it is the sterilizer of choice for medical equipment that is not heat-stable and therefore cannot be autoclaved and for material that cannot be gas-sterilized.

Halogens

IODINE

Iodine can be used as a disinfectant in one of two forms: tincture or iodophore. Tinctures are alcohol and iodine solutions, used mainly as antiseptics. An iodophore is a combination of iodine and a neutral polymer carrier that increases the solubility of the agent. This combination allows the slow release of iodine. Iodophores have the added advantage of being less irritating, nonstaining, and more stable than iodine in its pure form. Iodophores may be used as antiseptics or disin-

fectants, depending on the concentration of free iodine. The best-known iodophore is povidone-iodine, which is mainly used as an antiseptic. Iodophores are used only as disinfectants, because they are not sporicidal. Their bactericidal action is due to the oxidative effects of molecular iodine (I_2) and hypoiodic acid (HOI), both of which are found in solution.

CHLORINE AND CHLORINE COMPOUNDS

Chlorine and chlorine compounds are some of the oldest and most commonly used disinfectants. They are usually used in the form of hypochlorites, such as the liquid sodium hypochlorite (household bleach) and solid calcium hypochloride. Their killing activity is based on the oxidative effects of hypochlorous acid, formed when chloride ions are dissolved in water, on microbial enzyme systems. Hypochlorites are inexpensive and have a broad spectrum of activity; however, they are not used as sterilants owing to the long exposure time required for sporicidal action and their inactivation by organic matter. Hypochlorites are corrosive, and hypochlorite solutions are greatly influenced by the pH of the surrounding medium. They are commonly used as surface disinfectants, e.g., for tabletops and CPR mannequins. Their most important use is, of course, the chlorination of water.

Heavy Metals

Disinfectants containing heavy metals are now rarely used in clinical applications, as they have been replaced by more effective compounds. Slowly bactericidal, their action is primarily bacteriostatic. Because of the toxic effects of mercuric chloride and other mercury compounds, their use as disinfectants has declined, and they are used mainly as preservatives for paint. Silver nitrate (1% eyedrop solution) has been used as a prophylactic treatment to prevent gonococcal conjunctivitis in newborns.

Detergents: Quaternary Ammonium Compounds

Quaternary ammonium compounds are derived by substitution of the four-valence ammonium ion with alkyl halides. They are cationic, surface-active agents, or surfactants, that work by reducing the surface tension of molecules in a liquid. Their effectiveness is reduced by hard water and soap, and they are inactivated by excess organic matter. Their bactericidal action is mediated through disruption of the cellular membrane, resulting in leakage of cell contents. Because they are not sporicidal or tuberculocidal, the use of quaternary ammonium compounds is limited to disinfection of noncritical surfaces such as bench tops and floors.

Phenolics

Phenolics are molecules of phenol (carbolic acid) that have been chemically substituted, typically by halogens, alkyl, phenyl, or benzyl groups. These groups serve to increase the bactericidal effectiveness of phenol. The most common are ortho-phenylphenol and ortho-benzyl-para-chlorophenol. Phenolics have a fairly broad bactericidal spectrum but are not sporicidal. They are stable, biodegradable, and relatively active in the presence of organic soil. Their mechanism of inactivation is disruption of cell walls, resulting in precipitation of proteins. At lower concentration they are able to disrupt enzyme systems. Their main use is in the disinfection of hospital, institutional, and household environments. They are also commonly found in germicidal soaps.

Gases

Ethylene oxide is the gas most commonly used for sterilization. Because it is explosive in its pure form, it is mixed with nitrogen or carbon dioxide before use. Factors such as temperature, time, and relative humidity are extremely important in determining the effectiveness of gas sterilization. The recommended concentration is 450 to 700 mg of ethylene oxide per liter of chamber space at 55 to 60°C for 2 hours. A relative humidity of 30% has been found optimal for the destruction of spores. The killing mechanism of ethylene oxide is the alkylation of nucleic acids in the spore and vegetative cell. Gas sterilization is widely used in hospitals for materials that cannot withstand steam sterilization. This method is also used extensively by the manufacturing industry for the sterilization of low-cost thermoplastic products.

B. MICROBIOLOGY LABORATORY SAFETY

Margaret Gregory

OBJECTIVES

1. Describe the safe handling, storage, and disposal of chemicals and radioactive substances.

2. Describe the major sources of biologic hazards in the microbiology laboratory.

3. Discuss the use of the different types of biologic safety cabinets.

4. Discuss the practice of universal precautions.

5. Describe the different levels of protection and when each is applied.

6. Name the common hazardous chemicals in the laboratory.

7. Describe the methods of hazardous waste reduction.

8. Determine the different types of fire extinguishers and when each is used.

The concept of laboratory safety has changed drastically in the last decade. Prior to 1980, safety practices in most microbiology laboratories were lax. Mouth pipetting was a widely used technique, and eating, drinking, and smoking in the laboratory, although discouraged, were common. Beginning in the early 1980s, this relaxed attitude of safety among personnel changed dramatically. The impetus behind the change was the arrival in the United States of a previously unheard-of disease with an apparent 100% mortality rate. This disease became known as the acquired immune deficiency syndrome (AIDS). In addition to being a global calamity, AIDS initiated a major rethinking of employee risk for laboratory- acquired infections in hospitals around the country. Beginning with an emphasis on reducing the risks of biologic hazards (biohazards), safe-ty became a priority for laboratory personnel. The attitude "What you don't know can't hurt you," common among laboratory employees, rapidly went out of date. With the passage of time, the concept of laboratory safety expanded to include chemical, radioactive, electrical, and fire hazard protection. Currently microbiologists must pay special attention to all safety measures employed by the laboratory for their protection.

This section of Chapter 2 discusses

■ A safety program in the microbiology laboratory

■ Pertinent state and federal regulations

■ Procedures in handling biohazardous, chemical, and radioactive wastes and materials

GENERAL SAFETY PRINCIPLES

Safety in the microbiology laboratory can be best achieved by a combination of knowledge and common sense. Knowing current safety regulations, incorporating them into safety procedures manuals, and teaching the procedures to each employee through in-service education should be the duties of an assigned "safety officer." Although the provision of a "safe" work environment is ultimately the employer's responsibility, it cannot be achieved without the commitment of all persons in that environment to practice safe techniques for their own and their coworkers' protection.

Because microbiology laboratory personnel frequently deal with a variety of infectious agents—viral, bacterial, parasitic, and mycobacterial—laboratory-acquired infection is an obvious hazard. There are, however, many unseen hazards that also need to be addressed in order to provide a comprehensive and thorough safety program.

Safety Program for the Microbiology Laboratory

The comprehensive safety program for the microbiology laboratory needs to

■ Address biologic hazards

■ Describe the safe handling, storage, and disposal of chemicals and radioactive substances

■ Clearly outline the laboratory or hospital policies for correct procedures in the event of fire, natural disasters, and even bomb threats

■ Teach correct techniques for lifting and moving heavy objects and/or patients

This safety program needs to be ongoing and up to date with current federal and state regulations. Most important, it must be presented in a way that encourages each employee to incorporate these safety practices into his or her daily routines and to take responsibility for keeping the work environment safe.

HANDLING OF BIOLOGIC HAZARDS

Major Sources of Biologic Hazards

Biologic hazards in the microbiology lab come from two major sources: (1) processing of the patient specimens and (2) handling of the actively growing cultures of microorganisms. Either of these activities puts the employee at risk of potential contact with infectious agents through mucous membranes (rubbing the nose or eyes with contaminated hands), inhalation of aerosols of microorganisms, accidental ingestion (putting pens or fingers into the mouth), or needle sticks. Families of microbiology personnel and persons who work in adjacent laboratories may also be at risk. Table 2–4 lists the agents most commonly associated with laboratory-acquired infections.

Many infectious agents pose a high risk to laboratory employees. *Mycobacterium tuberculosis* has long been known to cause tuberculosis in laboratory workers exposed to aerosols created in processing sputum samples. A laboratory accident involving a spill of active *M. tuberculosis*, which could easily aerosolize through the ventilation system, is every microbiologist's nightmare. *Brucella* species and *Franciscella tularensis* are other infectious agents that can be transmitted through inhalation of aerosol created during processing of handling of specimens (e.g., blood, which may harbor this organism) or cultures of the organism. *Coccidioides immitis,* the most infectious of all the fungi, can infect several people in a room if culture plates on which the organism is growing are not sealed with tape or are opened in the absence of a biosafety hood. In the 1970's *Bacillus anthracis* was responsible for the deaths of a laboratory worker's wife and infant who were infected through handling his contaminated laboratory coat.

Hepatitis B virus (HBV) can be transmitted to laboratory workers through needle-stick injuries. The Centers for Disease Control and Prevention (CDC) estimate that approximately 12,000 health care workers become accidentally infected with this blood-borne pathogen annually. Human immunodeficiency virus (HIV) is another blood-borne pathogen that may be transmitted to laboratory personnel from contaminated specimens through needle-stick injury or other percutaneous route. Needless to say, all these infectious agents must be handled with extreme care. Laboratory workers must follow all necessary precautions to avoid exposure and minimize risks for exposure to these infectious agents.

Processing of Patient Specimens

Labeling specimens only from patients diagnosed with hepatitis or AIDS and requiring "extra precautions" for dealing with these specimens does not provide adequate protection for laboratory workers. Many patients come into emergency rooms or are admitted to the hospital with no apparent diagnosis. These patients may be in

■ T A B L E   2 – 4
MOST COMMON AGENTS ASSOCIATED WITH LABORATORY-ACQUIRED INFECTIONS

Agent	Disease	Route(s) of Infection	Source(s)
Hepatitis B virus	Hepatitis	Oral, percutaneous	Specimen
Coccidioides immitis	Coccidioidomycosis	Pulmonary	Culture
Bacillus anthracis	Human anthrax	Pulmonary	Culture
Brucella species	Brucellosis	Pulmonary, oral, percutaneous, eye	Specimen, culture
Mycobacterium tuberculosis	Tuberculosis	Pulmonary	Specimen, culture
Francisella tularensis	Tularemia	Unbroken skin, pulmonary	Culture
Shigella species	Shigellosis	Oral	Specimen, culture

■■■■■ T A B L E   2 – 5
UNIVERSAL PRECAUTIONS

1. Assume that patients are infectious for HIV and other blood-borne pathogens.

2. Place every specimen of blood or body fluid in a well-constructed container with a secure lid to prevent leaking during transport.

3. Whenever processing blood and body fluid specimens, wear gloves plus a face shield (or a mask with glasses or goggles) if there is a likelihood of spattering, and wash hands when finished processing specimens.

4. Never pipette by mouth; use pipetting aids.

5. Decontaminate all laboratory work surfaces with an appropriate chemical germicide after a spill of blood or other body fluids and when work activities are completed.

6. Limit use of needles and syringes to situations for which there are no other alternatives.

7. Decontaminate contaminated materials used in laboratory tests before reprocessing or place them into approximately labeled biohazard bags for disposal in accordance with institutional policies.

8. Always wash hands after completing laboratory activities, and remove all protective clothing before leaving the laboratory.

---

the early stages of either disease and may be asymptomatic but still contagious. Because the incidence of both hepatitis and AIDS is increasing steadily, the likelihood of exposure to one or both of these blood-borne pathogens is also increasing.

UNIVERSAL PRECAUTIONS

The CDC and the U.S. Occupational Safety and Health Administration (OSHA) jointly published recommendations entitled "Universal Precautions" which, simply stated, require that blood and body fluids from *all* patients be treated as infectious material. Table 2–5 lists the universal precautions.

Protective protocol varies from institution to institution. For example, The Ohio State University Hospitals Microbiology Laboratory has incorporated a three "Levels of Protection" scheme for laboratory workers (Table 2–6). The scheme relates the level of precaution or protection required of an employee to the potential risk of infection for the laboratory work station and task to be performed.

Figure 2–3 illustrates goggles, masks, and laboratory garments appropriate for use in laboratory work stations.

Working with Actively Growing Cultures

Many of the guidelines in place for the protection of microbiology personnel against exposure

---

■■■■■ T A B L E   2 – 6
THREE LEVELS OF PROTECTION

Level	Protection Required	Activity(ies)
0	Clothes covers (wraparound cloth garments or laboratory coats or disposable cover)	Handling materials and performing tasks that do not involve contact with or exposure to blood, body fluids, body secretions, or tissue (e.g., reading routine culture plates at the bench)
1	Fitted gloves and non–strike-through disposable clothes cover.	Presence at any laboratory work station where closed specimens are handled; specimen receiving (e.g., entering blood culture bottles into Bactec)
2	Fitted gloves, non–strike-through disposable clothes cover, and use of certified biologic safety cabinet–class II; a mask and eye cover and faceshield must be worn if safety cabinet is not available	Processing and planting specimens for culture and direct examination on slides; venting blood culture bottles; performing any activity that can generate aerosols

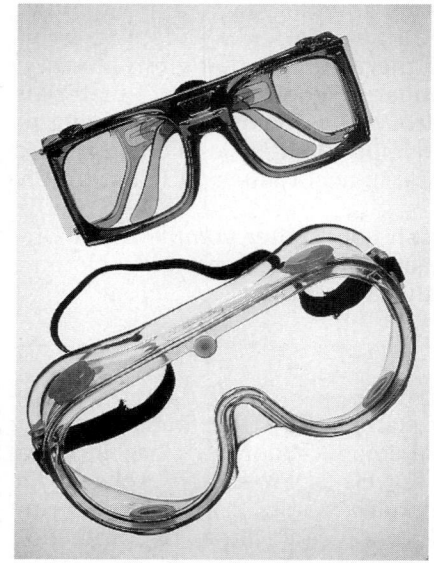

A

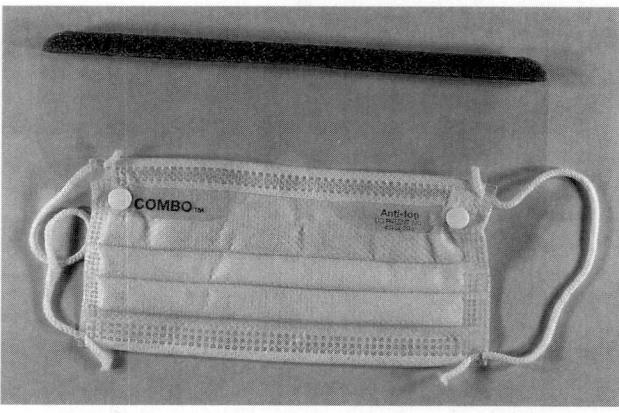

B

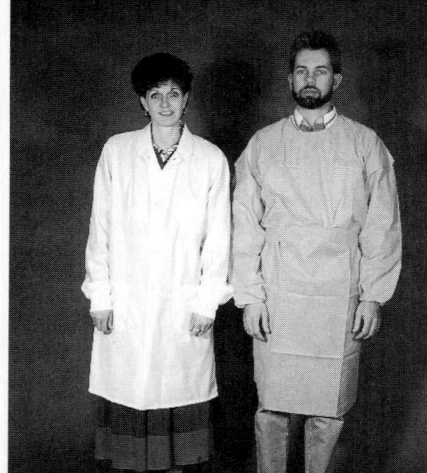

C

■■■■■■ F i g u r e 2 – 3

A, Eye protection used when chemical splashing could occur.

B, Integrated eye protection and mask.

C, Laboratory coats. The material of the white laboratory coat slows the penetration of liquids that splash or soak it. The blue coat on the right is disposable.

to blood-borne pathogens also apply to working with cultures of microorganisms at the bench. Hands must be washed frequently and kept away from the nose, mouth, and eyes. Adhesive bandages or small finger cots should be worn directly over cuts or hangnails. Any plates growing a fungus should immediately be sealed and the culture then worked on in a biosafety cabinet. Any cultures suspected of growing other potentially aerosolized infectious agents, such as *M. tuberculosis* or *Brucella* species, should also be worked on only in a biosafety cabinet. In an effort to educate those at risk for laboratory-acquired infection, CDC publishes a booklet entitled *Classification of Etiologic Agents on the Basis of Hazard,* which contains useful information on the degree of risk of biologic agents for laboratory workers.

In order to provide protection in the processing of patient specimens and handling actively growing cultures in the microbiology laboratory, a combination of environmental and engineering or equipment controls, work practices, and personal protective devices are implemented. An example of an engineering control is the use of a level II biohazard safety cabinet for plating of all specimens. This safety cabinet sterilizes the internal air containing infectious particles by passage through a high-efficiency air (HEPA) filter. Examples of work practices include frequent hand-washing (recommended after removing gloves and before leaving the laboratory), decontamination of all work stations with a 1:10 dilution of chlorine bleach (prepared fresh weekly), no eating, drinking, or smoking in the lab, and no application of cosmetics or lip balm in the laboratory. Personal protective devices are well-fitted gloves, needle resheathing devices, non–strike-through clothes covers, and face masks and goggles (or full-face shield). Figure 2–4 shows the three levels of biologic safety cabinets.

OSHA Regulations for Blood-Borne Pathogens

In 1992, the OSHA Final Rule on Blood-borne Pathogens became effective. This document contains the following requirements:

■ All personnel at risk for exposure to blood and body fluids must be identified by their employers.

■ These personnel must be properly trained in the use of all personal protective devices and/or engineering controls required for their

particular jobs and must receive counseling and follow-up treatment if an exposure occurs.

■ All employees at risk must receive annual retraining.

■ All employees who work with infectious material must be offered the hepatitis B vaccination free of charge.

DISPOSAL OF INFECTIOUS WASTE

In addition to laboratory employees who work with potentially infectious material, the general public needs to be protected from exposure to these same materials after they have been disposed of by the laboratories or hospitals. The finding of contaminated needles and other sharps along lake and ocean beaches has led to a public outcry to make hospitals accountable for their infectious waste disposal. The Medical Wastes Tracing Act was passed by the U.S. Congress in 1988 to regulate the states of New York, New Jersey, Connecticut, and Rhode Island and Puerto Rico in an effort to deal with this increasing problem.

The microbiology laboratory's safety program must follow state and local regulations for the safe disposal of its infectious wastes, usually by either autoclaving or incineration. Figure 2–5 shows the warning symbol for biohazardous materials. Warning signs containing this symbol must be placed on all biohazard wastes and material disposal containers. Figures 2–6 and 2–7 show disposal containers appropriate for contaminated and biohazardous materials.

CHEMICAL SAFETY

The scope of hazardous waste comprises more than just infectious waste. The National Committee for Clinical Laboratory Standards (NCCLS; 1986) gives the following definition of hazardous waste:

Considered in the laboratory sense, hazardous wastes are those substances which singly or in combination pose a significant present or potential threat or hazard to human health or to the environment and which singly or in combination require special handling, processing, or disposal because they are flammable, explosive, reactive, corrosive, toxic, infectious,

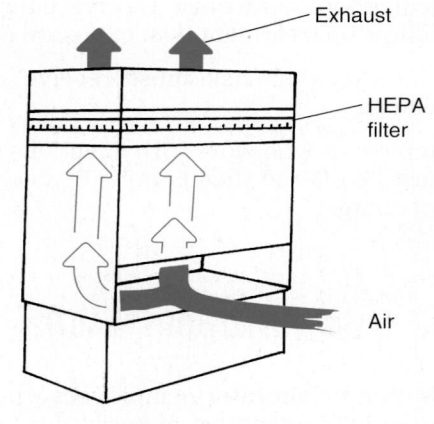

A

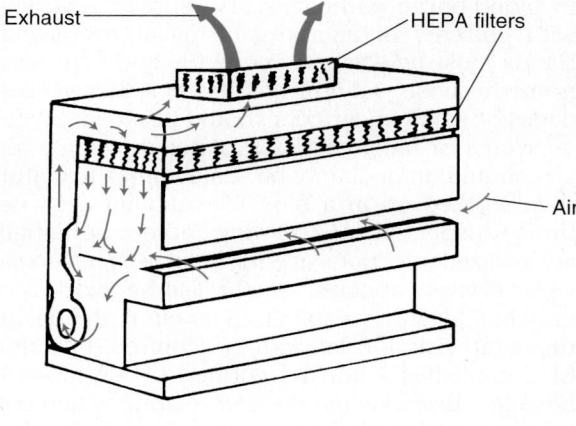

B

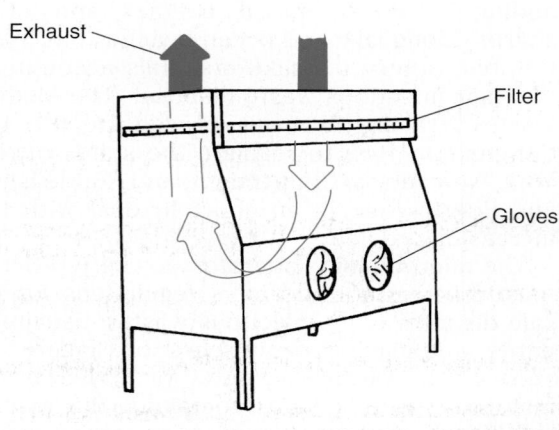

C

■■■■■ F i g u r e 2 – 4

A, The Class I biologic safety hood utilizes the same principle as the chemical fume hood. An exhaust fan moves the air inward through the open front. The air is uncirculated and is passed through a high-efficiency particulate air (HEPA) filter prior to reaching the environment. *B,* The Class II biologic safety hood is the most common in microbiology laboratories. Air is pulled inward and downward by a blower and passed up through the airflow plenum where it passes through a HEPA filter before reaching the work surface. A percentage of the remaining air is HEPA-filtered before reaching the environment. *C,* The Class III biologic safety hood is a self-contained, ventilated system for highly infectious microorganisms or materials. The closed front contains attached gloves for manipulation on the work surface.

carcinogenic, bioconcentrative-persistent in nature, potentially lethal, an irritant or a strong sensitizer.

Employee Right-to-Know

OSHA addresses employee safety with hazardous chemicals in 29 CFR 1910.1200, Hazard Communication Standard (HCS). This provides for a laboratory Chemical Hygiene Plan and a Hazard Communication, which states that all clinical laboratory personnel should have a thorough working knowledge of the hazards of the chemicals with which they come into contact, or employee right-to-know. All hazardous chemicals in the workplace need to be identified and clearly labeled with the National Fire Protection Association (NFPA) diamond stating risk for flammability, reactivity, and health (Fig. 2–8).

Material Safety Data Sheets

Material safety data sheets (MSDSs) provided by the manufacturer for every hazardous chemical indicate

- The nature of the chemical (e.g., flammable, toxic, carcinogenic)
- Precautions to take in using a chemical
- Spill clean-up procedure
- Disposal recommendations

An example of an MSDS is shown in Figure 2–9. These documents should be kept on file and available to every employee. Some of the more common hazardous chemicals found in the microbiology laboratory are listed in Table 2–7.

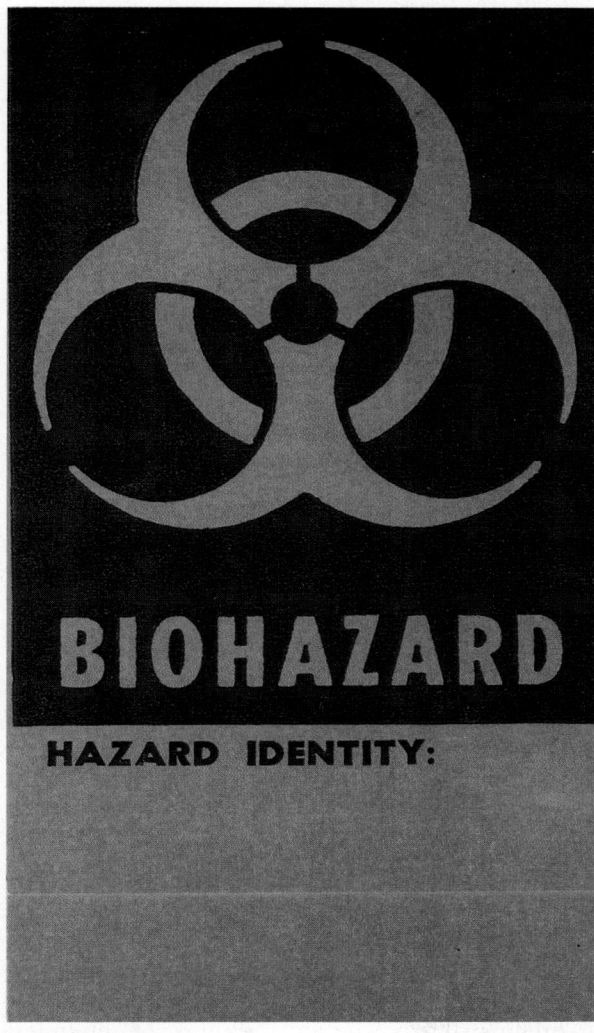

■■■■Figure 2 – 5
Biohazard symbol.

Hazardous Chemicals Inventory

The laboratory must maintain a current inventory of hazardous chemicals, which must be updated annually, and the MSDSs for those particular chemicals. The following four sources should be consulted in preparing an inventory:

■ 29 CFR Part 1910, Subpart Z, Toxic and Hazardous Substances, OSHA

■ National Toxicology Program Annual Report on Carcinogens

■ International Agency for Cancer Research Monographs

■ Manufacturers' material safety data sheets

Laboratory Safety for Hazardous Chemicals

Fume hoods must be provided to prevent inhalation of fumes. Such a hood may be vented to the outside or may be a tabletop model with filters attached that can be periodically replaced. Protective fume masks, gloves, aprons, and eyewear must be used in the case of a large spill to protect the worker from exposure. Acid spill kits and flammable spill kits should be kept in areas where such substances are used. Warning signs and symbols, as shown in Figure 2–10, must be placed in appropriate locations. Employees must be able to recognize each of these symbols and must be knowledgeable about the danger each indicates and the proper precautions that must be observed.

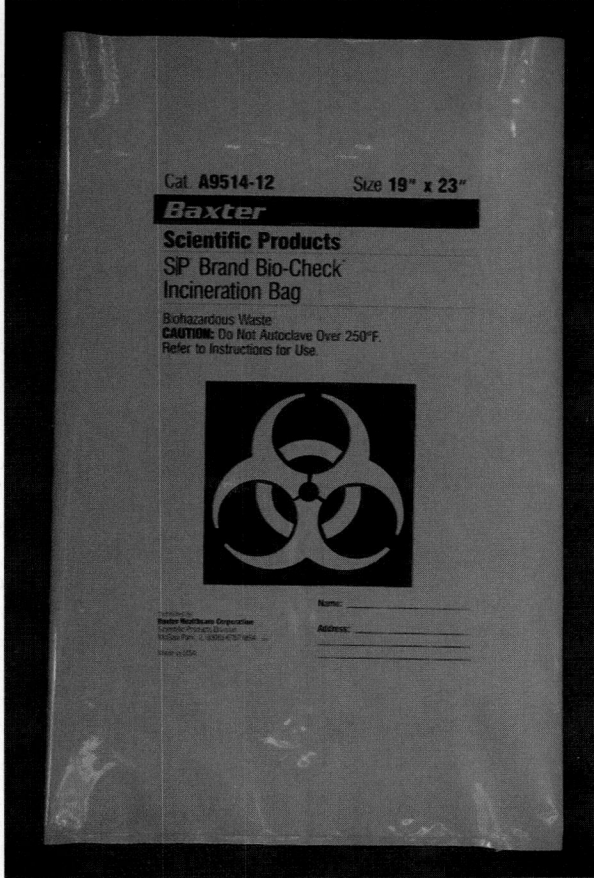

■■■■Figure 2 – 6
Biohazard bag used to dispose of culture plates and other nonsharp contaminated materials. The bag is sealed and incinerated.

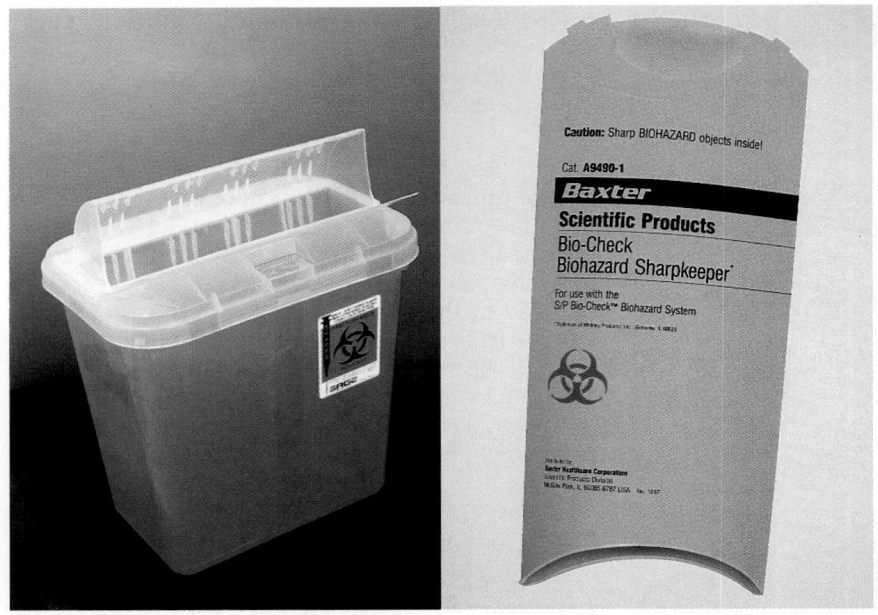

A

B

■■■■ Figure 2-7

A and *B*, Examples of Biohazard containers for disposable needles, glass slides, and other sharp materials.

HAZARDOUS WASTE REDUCTION

The EPA has also made recommendations for hazardous waste reduction through the following methods:

■■■■ Figure 2-8

Hazardous material classification symbol. (Information courtesy of Lab Safety Supply, Inc., Janesville, WI.)

- Substitute less hazardous chemicals when possible
- Develop procedures that use less of a hazardous chemical
- Recycle chemicals when possible
- Segregate infectious wastes from uncontaminated trash
- Substitute micromethodology in antibiotic susceptibility testing and identification of organisms to reduce volume of chemical reagents as well as infectious waste.

FIRE SAFETY

Bunsen burners and other open-flame burners are among the biggest sources of fire hazards in the clinical microbiology laboratory. It is important not to be careless or negligent in their use. Flammables should never be opened near a Bunsen burner in use. Another hazard associated with the Bunsen burners and other open-flame burners is the fuel. Bunsen burners use a gas. A leaking gas line may be a source of explosion or may cause illness in employees who work near the leak. If gas lines are in use, it is recommended that any lines that are not used daily be

MATERIAL SAFETY DATA SHEET

```
M a t e r i a l   S a f e t y   D a t a   S h e e t          | Page:  1
                                                              | Rev. Date
CHIRON RIBA (TM) HCV TEST SYS. 2nd GEN. ASSAY - SUBSTRATE     | 09/19/94
```

Ortho Diagnostic Systems Inc.
U.S. Route 202
Raritan, NJ 08869

Telephone Number: (800)322-6374
Emergency Phone Number: (800)424-9300

SECTION #1 - IDENTIFICATION

Product: CHIRON RIBA (TM) HCV TEST SYS. 2nd GEN. ASSAY - SUBSTRATE

Product Code: 933490

Synonyms: CHIRON RIBA HCV TEST SYS. 2nd GEN. ASSAY SUBSTRATE

This product is a component of the CHIRON RIBA (TM) HCV Test System Second Generation Assay.

SECTION #2 - HAZARDOUS CHEMICAL COMPONENTS

Component: METHANOL
CAS Number: 67-56-1   Percent of Mixture: 99.7
    ACGIH TLV-TWA: 200 ppm (skin)   OSHA PEL-TWA: 200 ppm (skin)
    ACGIH TLV-STEL: 250 ppm (skin)   OSHA PEL-STEL: 250 ppm (skin)

SECTION #3 - PHYSICAL DATA

Vapor Pressure: Not Determined
Vapor Density (Air=1): Not Determined
Specific Gravity: Not Determined
Percent Volatiles: 99.7

Appearance
_____

Colorless liquid

Odor
_____

Methanol (alcohol) odor

SECTION #4 - FIRE FIGHTING & EXPLOSION DATA

Flash Point: 52°F
Autoignition: 725°F

Flammability Class: IA

Lower Explosive Limit (%): 6%
Upper Explosive Limit (%): 36%

■■■■■■ F i g u r e   2 - 9

Material safety data sheet. (Courtesy of Ortho Diagnostic Systems, Inc., A *Johnson & Johnson* Company, Raritan, NJ.)

SECTION #4 - FIRE FIGHTING & EXPLOSION DATA Continued...

Fire and Explosion Hazards

Methanol is extremely flammable (flash point = 52°F). Its vapor may travel considerable distance to source of ignition and flash back.

Extinguishing Media

Use foam, dry chemical or carbon dioxide, as suitable for the surrounding fire.

Special Fire Fighting Instructions

Wear self-contained breathing apparatus and protective clothing that is appropriate for fighting a typical fire involving chemical materials.

SECTION #5 - EXPOSURE EFFECTS and FIRST AID

Route of Exposure - Inhalation

Overexposure may cause irritation of the nose and throat. Extensive inhalation may cause behavioral changes.

First Aid - Inhalation

Remove victim to fresh air. If necessary, get medical attention.

Route of Exposure - Skin

May cause skin irritation.

First Aid - Skin

Promptly wash exposed areas thoroughly with soap and water.

Route of Exposure - Eyes

Contact may cause eye irritation.

First Aid - Eyes

Immediately flush eyes with low pressure water. Remove contact lenses to insure a thorough flush and continue to flush with water for at least 15 minutes. If irritation persists, get medical attention.

Route of Exposure - Ingestion

May cause irritation of gastrointestinal tract. Swallowing methanol may be fatal or cause blindness, and can cause behavioral changes.

First Aid - Ingestion

Contact a physician or the Poison Control Center.

Miscellaneous Toxicological Information

Carcinogenicity: NTP: No   IARC: No   OSHA: No

The health effects noted above are based on the extrapolation of data on the pure product ingredients. To the best of our knowledge, no health effects have been identified for the product mixture under normal conditions of use, although the health effects of the product have not been thoroughly investigated.

Exposure to methanol can cause damage to the eyes, liver, heart and kidneys.

Methanol target organ data: Sense organs and special senses (optic nerve

Figure 2-9 Continued

Material safety data sheet. (Courtesy of Ortho Diagnostic Systems, Inc., A *Johnson & Johnson* Company, Raritan, NJ.)

SECTION #5 - EXPOSURE EFFECTS and FIRST AID Continued...

Miscellaneous Toxicological Information

neuropathy, visual field changes); Behavioral (headache); Lungs, thorax or respiration (dyspnea, other changes); Gastrointestinal (nausea or vomiting). TARGET ORGANS: Eyes, Kidneys.

SECTION #6 - REACTIVITY & POLYMERIZATION

Stability: Stable

Incompatible Materials

Methanol: acids, acid chlorides and anhydrides, oxidizing or reducing agents.

Hazardous Decomposition Products

Not determined.

Hazardous Polymerization: Will Not Occur

SECTION #7 - SPILL, LEAK, & DISPOSAL PROCEDURES

Steps to be Taken in The Event of Spills, Leaks, or Release

Ventilate the area. Eliminate all sources of ignition. Cover with dry-lime, sand or soda ash. Place in covered container. Wash spill site after material pickup is complete.

Waste Disposal Methods

Burn in a chemical incinerator equipped with an afterburner and scrubber, but exert extra care in igniting as methanol is highly flammable.

Dispose of in accordance with local, state and federal regulations.

SARA Title III Notifications and Information

SARA Title III - Hazard Classes: Acute Health Hazard
                                 Chronic Health Hazard
                                 Fire Hazard

SARA Title III - Section 313 Supplier Notification:

This product contains the following toxic chemicals subject to the reporting requirements of section 313 of the Emergency Planning and Community Right-To-Know Act (EPCRA) of 1986 and of 40 CFR 372:

CAS #	Chemical Name	Percent of Mixture
67-56-1 METHANOL		99.7

This information must be included on all MSDSs that are copied and distributed for this material.

Other Environmental Information

Figure 2-9 Continued

SECTION #8 - SPECIAL PROTECTIVE MEASURES

Ventilation

None normally required.

Eye Protection

Safety glasses, goggles or full-face.

Skin Protection

Latex gloves.

Respiratory Protection

None normally required. Dust respirator (disposable) recommended when ventilation is inadequate.

Other Protection

Lab coat.

Work/Hygienic Practices

CAUTION: This product is intended for use with human blood or blood elements. All blood components and products containing blood components should be handled and disposed as if capable of transmitting infectious agents.

SECTION #9 - SPECIAL PRECAUTIONS - STORAGE & HANDLING

Storage & Handling Conditions

Keep refrigerated at 2 to 8°C until use.

SECTION #10 - SHIPPING INFORMATION

Proper Shipping Name: None Required

SECTION #11 - MISC COMMENTS & REFERENCE DOCUMENTATION

Primary references used in the preparation of this document:

1. Ortho Diagnostics Product Circular
2. OSHA Z-Table, 1910.1000 (Revised Final Rule)
3. ACGIH Threshold Limit Values and Biological Exposure Indices (1991-92)
4. Methanol MSDS (Sigma Chemical Co., 4/17/90)

N/A = Not Applicable   ¯ = Approximately Equal To

DISCLAIMER OF EXPRESSED AND IMPLIED WARRANTIES

Although reasonable care has been taken in the preparation of this document, we extend no warranties and make no representations as to the accuracy or completeness of the information contained therein, and assume no responsibility regarding the suitability of this information for the user's intended purposes or for the consequences of its use. Each individual should make a determination as to the suitability of the information for their particular purpose(s).

Ortho Diagnostic Systems Inc.

Figure 2 - 9 Continued

TABLE 2–7

HAZARDOUS CHEMICALS COMMONLY USED IN THE MICROBIOLOGY LABORATORY

Flammables	Methanol
	Acetone
	Ethanol
Potential or proven carcinogens	Formaldehyde
	Aniline (crystal violet) stain
	Auramine-rhodamine (Truant) stain
Irritants and corrosives	Hydrogen peroxide
	Acids: HCl, H_2SO_4, acetic acid
	NaOH

connected to a burner and the gas burned for 5 minutes once a month to prevent "falling out" of the mercaptans (which alerts personnel to a possible leak by the smell). Other burners use a flammable liquid as the fuel, which is in itself a hazard. If at all possible, the use of burners should be replaced with other methods or techniques, such as fixing slides with methanol or using disposable loops. Other sources of ignition are heating elements, hot plates, and spark gaps in motors and light switches.

All personnel must be thoroughly trained in the procedure for responding to a fire emergency. Most institutions use the acronym *RACE,* which stands for

- **Rescue:** Remove anyone who is in danger. The safety of handicapped personnel (deaf or physically nonmobile) should be a priority.

- **Alarm:** Know where the nearest fire pull box or alarm station is located and the number to call to report the fire.

- **Contain:** Close doors to contain fire and smoke.

- **Extinguish:** Use the properly rated fire extinguisher on small fires, as described in Table 2–8. If the situation is out of control, the best course of action is to evacuate.

The fire evacuation plan must be posted, and employees should be familiar with its locations. Periodic fire drills should be conducted to ensure that all personnel will be able to react quickly and efficiently in the case of a real fire emergency. The drills should include both exit and non-exit procedures. Exit drills will help familiarize personnel with the escape routes and location of fire doors and stairwells. Non-exit procedures alert personnel to the potential for

evacuation if the fire is located elsewhere in the building. The laboratory should be kept free of clutter, and exitways should remain clear of obstructions.

Thermal Injuries

Personnel should be warned of any hot surface or situation where the potential for burns is present. Using long thermal gloves that extend

WARNING SIGNS AND SYMBOLS   POTENTIAL DANGER

Flammable

Toxic hazard
Poisonous

Carcinogenic
Cancer-causing agent

Corrosive
Harmful to mucous membranes,
  skin, eyes, or tissues

Radiation
Radioactive material present

Figure 2–10

Miscellaneous warning signs and symbols.

CLASSES OF FIRE EXTINGUISHERS

Symbol	Class	Use
A̲ (triangle)	A	Fires in ordinary combustible materials such as wood, cloth, paper, rubber, and many plastics Contains water or dry chemical to *cool*
B (square)	B	Fires in flammable liquids, gases, and greases Contains CO_2 or dry chemical to *smother*
C (circle)	C	Fires involving electrical equipment, for which the electrical nonconductivity of the extinguishing media is important Contains CO_2 or dry chemical to *smother* without damaging the equipment

to the shoulder is recommended for reaching into autoclaves or hot-air ovens. Signs should be posted warning of hot instruments or flasks that have just been sterilized. Burns may also come from extremely low temperatures found with use of liquid nitrogen or freezers that maintain temperatures of –70°C.

ELECTRICAL SAFETY

Today's microbiology laboratories contain many instruments. Each instrument must undergo regular preventive maintenance to ensure that it is functioning properly and in the best repair. Electrical cords should be checked for fraying. All cords should have grounded (three-pronged) plugs. Electrical grounding and leakage checks are required annually by the College of American Pathologists (CAP), an organization that provides accreditation to laboratories. Electrical equipment should never be placed near safety showers, owing to the risk of electrocution.

MISCELLANEOUS SAFETY CONSIDERATIONS

Table 2–9 summarizes general laboratory safety design, controls, and procedures.

Back Safety

The best way of caring for one's back is to prevent back injuries. Carrying heavy trays of culture plates, lifting heavy loads into and out of the autoclaves, and sitting or standing improperly can all contribute to back stress or injury. Some of the ways of avoiding back injuries are

- Using the legs to lift, not the back
- Asking for assistance or using a cart when a load is too heavy
- Using good posture
- Staying physically fit

Storage of Gases

Flammable and nonflammable gas cylinders in use in the laboratory must be properly secured or stored and secured in vented areas.

Because a leaking pressurized gas cylinder may become a potential "missile," care must be taken to avoid accidental breakage or removal of the pressure valve on top. The metal cap that protects this valve on top of the cylinder must be kept in place when the cylinder is not in use.

First Aid Training

All personnel should be trained in CPR and other life-saving first aid in order to be able to act quickly in an emergency involving either fellow workers or patients.

Immunizations

As required by OSHA, the hepatitis B vaccination must be offered free of charge to all personnel who are at risk for exposure to blood-borne pathogens. Microbiology personnel working with sputum specimens and/or mycobacterial cultures should be screened for exposure to *M. tuberculosis* with an annual TB skin test.

■■■■T A B L E 2 – 9
RECOMMENDED LABORATORY SAFETY DESIGN AND PRACTICES

Engineering controls and equipment	Biosafety cabinets for processing specimens
	Rigid-walled needle containers for disposal of needles
	Foot pedals at sinks for "hands-off" water flow control
	Automated blood culture systems in place of blind subcultures and Gram stains from blood bottles
	"Closed-head" centrifuges
	Fume hoods
Work practices	Frequent handwashing
	No mouth pipetting
	No eating, drinking, smoking, or applying of cosmetics in the laboratory
	Use of carts for transporting trays of cultures
	Disinfecting of work stations at end of each shift
	Storage of flammables in "Flammable" storage cupboards
Personal protective devices	Gloves
	Hospital-laundered lab coats
	Non–strike-through clothes covers
	Goggles/masks
	Full-face masks
	Needle resheathing devices
	MSDS sheets
Environmental controls	Air flow from low-risk to high-risk areas within laboratory
	Limited access to high-risk areas
	Good ventilation

SAFETY TRAINING

To be in compliance with the OSHA standard as required by Joint Commission for the Accreditation of Healthcare Organizations (JCAHO), safety training must be documented. Training must include the following safety issues:

■ **Fire:** knowledge of how and when to report fires, the location of the nearest alarm box and fire extinguishers, how to use fire extinguishers in small fires, and blocking of fire doors.

■ **Hazardous materials management:** how to use material safety data sheets.

■ **Proper storage of gases**

■ **Blood-borne pathogens program:** appropriate practice of infection control, how to handle sharps, compliance with exposure control plan, and handling of biologic spills.

Safety training programs for microbiology personnel should be conducted annually to keep safe techniques fresh in everyone's mind. New personnel should be thoroughly trained in safety procedures for the department within 2 weeks of their hiring. It is a good idea to focus on one safety topic at a time and to make the training sessions enjoyable. Safety is each employee's responsibility.

Bibliography

Baron EJ, Finegold SM (eds): Bailey & Scott's Diagnostic Microbiology, 8th ed. St. Louis: Mosby–Year Book, 1990, pp 9–11.

Block SS (ed). 1983. Disinfection, Sterilization, and Preservation, 3rd ed. Philadelphia: Lea & Febiger, 1983.

Buesching W, Neff JC, Sharma HM: Infectious hazards in the clinical laboratory: A program to protect laboratory personnel. Clin Lab Med 9 (2):351, 1989.

Favero MS, Bond WW: Sterilization, disinfection, and antisepsis in the hospital. In Balows A (ed): Manual of Clinical Microbiology, 5th ed. Washington, DC: American Society for Clinical Microbiology, 1991, pp 183–200.

Kruse RH: Microbiological Safety Cabinetry (Monograph). Lexington, KY: Medico-Biological Environmental Development Institute, Inc., Sept. 1981.

National Committee for Clinical Laboratory Standards: Proposed Guidelines GP5-P: Clinical Laboratory Hazardous Waste. Villanova, PA: NCCLS, 1986.

Richardson JH, Barkley WE: Biological safety in the clinical laboratory. In Lennette EH, Balows A, et al (eds): Manual of Clinical Microbiology, 4th ed. Washington, DC: American Society for Microbiology, 1985, pp 138–142.

Rutala WA: APIC Guidelines for Selection and Use of Disinfectants. Am J Infect Control 18:99, 1990.

The Occupational Safety & Health Administration. Washington, DC. Bloodborne Pathogens Standards, CFR 1910–1030.

Spaulding EH: Chemical disinfection and antisepsis in the hospital. J Hosp Res 9:5, 1972.

VanDemark PJ, Batzing BL: The Microbes: An Introduction to Their Nature and Importance. Redwood City, CA: Benjamin-Cummings, 1987.

CHAPTER 3

Concepts in Antimicrobial Therapy

A. ANTIMICROBIAL MECHANISMS OF ACTION

Susan L. Koletar

OBJECTIVES

1. Appreciate that understanding the basic structure of microorganisms and the specific functions of individual components is essential to understanding the actions of antimicrobial therapy.

2. Define *bacteriostatic* and *bactericidal*. Appreciate the factors that can influence ultimate outcome of specific antimicrobial activity.

3. List the major sites of action for major classes of antimicrobial agents.

4. List the classic examples of antimicrobial agents that affect structural integrity; understand how ß-lactam agents result in antibacterial activity.

5. List the agents whose primary mechanism of action is inhibition of protein synthesis. Appreciate the differences between reversible and irreversible binding at the ribosomal level.

6. List the agents whose primary mechanism of action is interference with metabolic functions, including DNA synthesis, RNA synthesis, and folic acid metabolism.

7. Describe the major mechanisms by which resistance to various antimicrobial agents can occur.

Antimicrobial therapy is a broad term for the use of chemical compounds to treat diseases caused by microorganisms. *Antibiotic* is most commonly used, often interchangeably, but the purest meaning of *antibiotic* is a chemical substance, produced by a microorganism, with the capacity to kill or inhibit other microorganisms. Antimicrobial agents may also be *synthetic compounds* (those that are completely manufactured or artificial) and *semisynthetic compounds* (naturally occurring substances that have been chemically altered). Whether naturally occurring or not, all antimicrobial agents are intended to destroy or inhibit disease-causing organisms. The specific activities of antimicrobial agents and their mechanisms of action are primarily dictated by the biologic characteristics of the microbe. For treating actual disease, the pharmacokinetic properties of the specific drugs (such as absorption, distribution, and excretion) also are important factors influencing the clinical utility of specific agents against specific disease-causing organisms. Potential problems include toxicity to the host and the development of resistance by the organisms.

Antimicrobial agents are generally categorized according to their targeted sites of action (Table 3–1). Basic activities are (1) interruption of the structural integrity by interfering with cell wall or cell membrane composition and (2) interruption of basic metabolic functions, such as protein synthesis, nucleic acid metabolism, and inhibition of essential metabolites (Fig. 3–1). Additional characterization may be made based

TABLE 3–1

COMMON MECHANISMS OF ACTION FOR DIFFERENT ANTIMICROBIAL AGENTS

Mechanism	Class of Antimicrobials
Effects on cell wall integrity	β-Lactams Vancomycin
Effects on cell membrane structure and function	Polymyxins
Inhibition of protein synthesis	Aminoglycosides Tetracyclines Macrolides Clindamycin Chloramphenicol
Inhibition of essential metabolites	Sulfonamides Trimethoprim
Interference with nucleic acid metabolism	Rifampin Quinolones Metronidazole

on whether the agent actually kills the organism (-cidal) or whether it serves to inhibit microbial growth (-static).

Properties of individual drugs and their mechanisms of action are also important to consider when antimicrobial agents are to be used in combination. There may be twice the effect of two drugs *(additive)* or perhaps no effect of combination therapy *(indifference)*. One potentially good effect of using more than one drug is *synergy*. This means

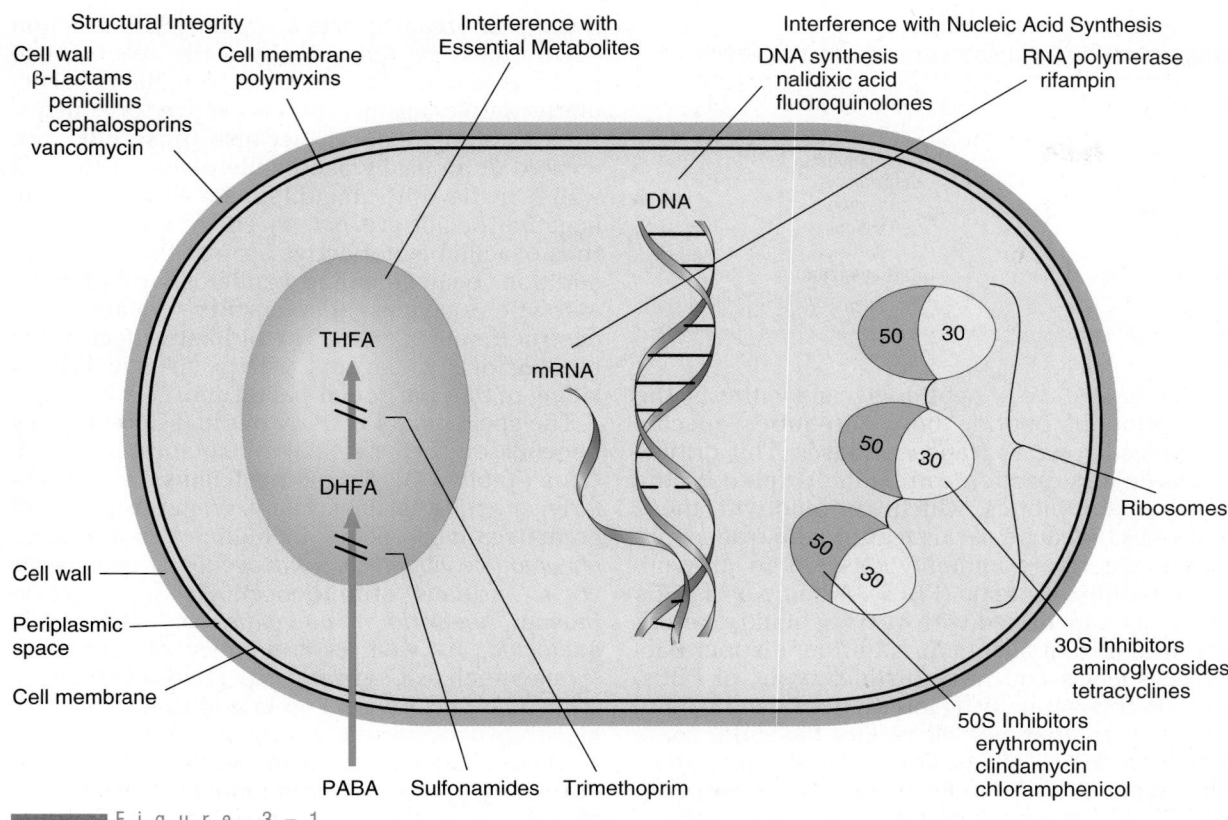

Structural Integrity
Cell wall
β-Lactams
penicillins
cephalosporins
vancomycin

Cell membrane
polymyxins

Interference with
Essential Metabolites

Interference with Nucleic Acid Synthesis
DNA synthesis
nalidixic acid
fluoroquinolones

RNA polymerase
rifampin

DNA

THFA

mRNA

DHFA

Ribosomes

Cell wall

Periplasmic
space

Cell membrane

30S Inhibitors
aminoglycosides
tetracyclines

50S Inhibitors
erythromycin
clindamycin
chloramphenicol

PABA   Sulfonamides   Trimethoprim

Figure 3–1

Primary sites of antibacterial action of major classes of antimicrobial agents. (THFA, tetrahydrofolic acid; DHFA, dihydrofolic acid.)

that the combined effect is even greater than two individual effects added together, and the overall antimicrobial activity is enhanced. A potentially bad effect is *antagonism;* one drug may counteract the other, so that the desired antimicrobial result is much less than would be expected.

The fact that the majority of our available antimicrobial armamentarium comprises antibacterial agents is good evidence of how our understanding of the basic biology of specific microorganisms affects antimicrobial development. As microbiology techniques have improved, and we have become better at determining structure and function of other microorganisms, new classes of antimicrobials, including antifungal, antiviral, and antiparasitic agents, have also emerged.

EFFECTS ON CELL WALL INTEGRITY

The bulk of the task of protecting the bacterial cytoplasmic membrane, which is of paramount importance in bacterial survival, falls to the bacterial cell wall. This structure is composed primarily of a peptidoglycan layer, the synthesis of which depends, at least in part, on specific integral enzymes. Inactivating or interfering with these enzymes structurally or functionally destroys bacterial cell walls.

β-Lactam Antibacterial Agents

The group of natural and semisynthetic agents known as β-lactam antibiotics accounts for a sizable proportion of available antibacterial agents; it comprises the penicillins, cephalosporins, carbapenems, and monobactams (Table 3–2). Their antibacterial activity is based on the basic chemical structure that they all share, the β-lactam ring. Modifications of this basic structure (Fig. 3–2) allow for varying spectra of activity ranging from very specific and narrow to very broad and inclusive.

■■■■■T A B L E   3 – 2

SOME COMMON β-LACTAM ANTIBACTERIAL AGENTS

Penicillins	Cephalosporins
Penicillin G	Cephalothin
Ampicillin	Cefamandole
Methicillin	Cefoxitin
Ticarcillin	Cefotaxime
Piperacillin	Ceftazidime
Monobactams	**Carbapenems**
Aztreonam	Imipenem

The final stage of peptidoglycan synthesis, the formation of peptide bonds, requires specific enzymes known as transpeptidases. This critical cross-linking reaction can be interrupted by the β-lactam antibiotics, which can bind with these enzymes. Because of this feature, the transpeptidases are also commonly referred to as penicillin-binding proteins (PBPs). A number of PBPs have been identified with varying affinity for different β-lactam agents. In addition, distinct bacterial species can have distinct types of PBPs, and their β-lactam drug interactions can be quite specific, so that not all agents have the same effect on all organisms. For example, a monobactam agent like aztreonam, which binds primarily to PBP 3 of gram-negative aerobes, has no activity against gram-positive or anaerobic organisms, because they do not possess that specific PBP.

Even if an organism has a specific PBP, β-lactam agents must be able to physically reach those proteins to work. For example, oxacillin is particularly active against gram-positive organisms, such as staphylococci, because those PBPs are located at an easily accessible place on the cell wall. On the other hand, similar PBPs in gram-negative bacilli are not so easily reached, and thus oxacillin is not active against that group. In addition, penicillins and cephalosporins trigger autolytic enzymes that further enhance the destruction of the cell. The ultimate effect of the binding of such agents with a specific PBP is killing of that particular bacterium.

The spectrum of activity of the β-lactam drugs depends on their particular structural modifications (Table 3–3). Simple penicillins are particularly effective against many gram-positive and gram-negative cocci, including *Streptococcus pneumoniae, Streptococcus pyogenes, Streptococcus bovis,* viridans streptococci, *Neisseria gonorrhoeae, Neisseria meningitidis,* and *Pasteurella multocida,* as well as a number of anaerobic species such as *Clostridium* sp, *Fusobacterium* sp, and some *Bacteroides* spp in addition to anaerobic streptococcus–like organisms. Ampicillin has a similar spectrum, with additional activity against enterococci, *Listeria monocytogenes,* and some gram-negative coccobacillary organisms, such as *Haemophilus influenzae* and *Haemophilus parainfluenzae, Proteus mirabilis,* and a few

■■■■■F i g u r e   3 – 2

Chemical structures of major classes of β-lactam antibiotics.

TABLE 3-3
CLINICAL UTILITY OF DIFFERENT ANTIMICROBIAL AGENTS*

Agent	Organisms Useful Against	Organisms Not Useful Against
Penicillin	Streptococci Spirochetes	Staphylococci Mycoplasmas
Oxacillin	Staphylococci[†]	Enterobacteriaceae
Piperacillin	*Pseudomonas aeruginosa*	Staphylococci
Cefazolin	Staphylococci[†]	Enterococci
Cefoxitin	Enterobacteriaceae *Bacteroides*	MRSA* Enterococci
Cefotaxime	Enterobacteriaceae	*Pseudomonas aeruginosa*
Ceftazidime	*Pseudomonas aeruginosa*	Anaerobes
Aztreonam	Enterobacteriaceae	Gram-positives
Imipenem	Many	*Xanthomonas,* MRSA*
Vancomycin	MRSA*	Gram-negatives
Aminoglycosides	Gram-negatives	Gram-positives
Tetracyclines	*Chlamydia Rickettsia*	Enterobacteriaceae
Erythromycin	*Legionella* Mycoplasmas	Enterobacteriaceae *Haemophilus influenzae*
Clindamycin	Anaerobes	Enterobacteriaceae
Sulfonamides	Enterobacteriaceae	*Pseudomonas aeruginosa*
Quinolones	Enterobacteriaceae *Pseudomonas aeruginosa*	Gram-positives Anaerobes
Metronidazole	Anaerobes	Aerobes

* MRSA, Methicillin-resistant *Staphylococcus aureus*.
† Methicillin-sensitive *Staphylococcus aureus*.

Enterobacteriaceae like *Escherichia coli,* and *Salmonella* and *Shigella* species. Carboxypenicillins and ureidopenicillins have a broader gram-negative spectrum of activity, most notably among the Enterobacteriaceae and with reasonable activity against *Pseudomonas aeruginosa.*

The generally accepted scheme of classifying cephalosporins is based on their spectrum of activity and spoken of in terms of "generations" (Table 3–4). First-generation cephalosporins are characterized by their good gram-positive but relatively modest gram-negative activity. Second-generation cephalosporins generally have better gram-negative activity owing to greater stability against certain β-lactamases (see later). Third-generation cephalosporins are thought to have less activity against gram-positive organisms but improved activity against Enterobacteriaceae; a few have particular activity against *P. aeruginosa.*

The only currently available monobactam is aztreonam. Its spectrum of activity is limited to aerobic gram-negative bacilli. It has achievable inhibitory levels against *P. aeruginosa, Serratia marcescens* and *Enterobacter* spp. Imipenem, the only available carbapenem, has the broadest spectrum of any antimicrobial agent. It has excellent activity against most major pathogens, including aerobic and anaerobic streptococcal species, methicillin-susceptible staphylococci, most Enterobacteriaceae, including *Enterobacter* sp, *Citrobacter* sp, *Serratia* sp, *Acinetobacter* sp, and *P. aeruginosa.* A new carbapenem, meropenem, is currently undergoing clinical trials.

β-Lactamase Inhibitors

In response to being exposed to β-lactam type of drugs, bacteria may produce enzymes (β-lactamases) that can hydrolyze the β-lactam ring and consequently inactivate the antimicrobial agent. One way to compensate for this potential resistance is to combine a standard β-lactam drug (e.g., ampicillin, amoxicillin, ticarcillin, piperacillin) with a *β-lactamase inhibitor,* namely clavulanic acid, sulbactam, or tazobactam. These agents alone all have weak antibacterial activity, but their combination with β-lactam drugs can be synergistic. Their main purpose in such a combination, however, is to bind with, and thus inactivate, the β-lactamases produced by a variety of organisms, including staphylococci and many aerobic and anaerobic gram-negative bacteria. The currently available and clinically used combinations include amoxicillin/clavulanate, ticarcillin/clavulanate, ampicillin/sulbactam, and, most recently, piperacillin/tazobactam.

Other Antimicrobial Agents Whose Primary Site of Action Is the Cell Wall

Another group of agents affect different aspects of peptidoglycan (cell wall) synthesis. This group includes vancomycin, bacitracin, and

■■■■■■ T A B L E   3 – 4

EXAMPLES OF CEPHALOSPORINS BY GENERATION AND SPECTRUM OF ACTIVITY

Generation	Spectrum
First	
Cephalothin	Staphylococci (methicillin sensi-
Cefazolin	tive); streptococci; few
Cephapirin	Enterobacteriaceae
Cephradine	
Second	
Cefamandole	More Enterobacteriaceae; some
Cefonicid	β-lactamase producers, includ-
Cefoxitin	ing some anaerobes
Cefotetan	
Cefuroxime	
Third	
Cefotaxime	Many Enterobacteriaceae; many
Ceftizoxime	β-lactamase producers
Ceftriaxone	
Ceftazidime	*Pseudomonas aeruginosa*
Cefoperazone	

cycloserine. Of these, vancomycin has the most established clinical value. It affects the second stage of cell wall synthesis, prior to transpeptidation, by inhibiting enzymes that are vital to the continued formation of peptidoglycan. The end result is that the bacteria cannot maintain their structural integrity. Because vancomycin is a large molecule, it can cross the cell walls only of gram-positive organisms; thus, its primary utility is against only those types of bacteria.

Bacitracin and cycloserine also affect continued formation of the basic structural units of the cell wall. Their spectra of activity are relatively narrow and their toxicities relatively broad, so their usefulness in the clinical arena is limited.

INTERRUPTION OF CELL MEMBRANE STRUCTURE AND FUNCTION

Two antibacterial agents are available whose primary site of action is the bacterial cell membrane. Their spectra of activity are relatively limited and their toxicities significant, making them moderately useful at best. Polymyxins interact with the phospholipids of the bacterial cell membrane and thereby alter the permeability and osmotic integrity. Their primary activity is against gram-negative

bacilli, and their primary clinical utility is in the form of topical preparations. Bacitracin acts by disrupting the cytoplasmic membrane, in addition to its interference with cell wall synthesis. Its primary activity is against gram-positive organisms, and it, too, is most useful as a topical agent for superficial dermatologic infections.

INHIBITION OF PROTEIN SYNTHESIS

A number of different antimicrobial agents exert their bactericidal or bacteriostatic effects by interfering with protein synthesis at the ribosomal level. This is accomplished by binding of the agent to either the 50S or 30S ribosomal subunit, and the resultant inhibition or killing of the organism depends on whether this binding is reversible or irreversible, respectively. For example, the aminoglycosides irreversibly bind to the 30S ribosomal subunit, making that integral cell structure unavailable for translation of mRNA during protein synthesis; the result is death of the cell. A secondary feature of these types of agents is that this ribosomal interaction also results in misreading of the genetic code. To be effective, aminoglycosides depend on an aerobic, energy-dependent process to reach the inside of the bacterial cell. They are especially useful in situations in which a cell wall–active agent, such as a β-lactam, is being used, because of their facilitated uptake in those instances. The spectrum of activity of the aminoglycosides is primarily aerobic gram-negative bacilli, especially Enterobacteriaceae, *P. aeruginosa*, *Acinetobacter* sp, and *Providencia* sp; they also have activity against staphylococci but should not be used as single agents to treat staphylococcal infections.

Tetracyclines are like the aminoglycosides in that they have the same site of action (30S ribosome), but their binding is reversible, and consequently, their activity is bacteriostatic. Although they can enter human cells, their antimicrobial activity is enhanced by being actively concentrated in some bacteria. Tetracyclines have a broad spectrum of activity against a number of microbial species, including gram-positive and gram-negative bacteria as well as mycoplasmas. They are particularly useful against intracellular pathogens like chlamydiae and rickettsiae.

The macrolides (erythromycin, clarithromycin, azithromycin), the lincosamide (clindamycin) and chloramphenicol also inhibit bacterial protein synthesis, their primary target being the 50S ribosomal subunit. As with the tetracyclines, the binding of these drugs is reversible. Interference or actual antagonism can occur if these agents are used in combination with each other.

Erythromycin has a relatively broad spectrum of activity, including gram-positive cocci, mycoplasmas, chlamydiae, rickettsiae, and treponemes. Its gram-negative activity is marginal at best and is clearly influenced by changes in pH, with improved potency in alkalinized environments. Erythromycin is notably the drug of choice for *Legionella pneumophila,* and probably for *Mycoplasma pneumoniae.* The more recently available agents azithromycin and clarithromycin have a similar spectrum of activity with the added feature of better activity against some gram-negative organisms such as *H. influenzae* and *Moraxella (Branhamella) catarrhalis.* Whether this expanded in vitro spectrum will be clinically relevant remains to be seen. Their primary advantages over the prototype compound are improved pharmacokinetic properties and tolerance in terms of fewer side effects.

Clindamycin has excellent activity against aerobic gram-positive organisms, including various species of clinically important streptococci as well as methicillin-susceptible staphylococci. Its claim to fame, however, is its very potent anaerobic activity. Clindamycin has also been used to treat some protozoal infections, such as babesiosis, malaria, and toxoplasmosis.

Chloramphenicol also enters bacteria by an energy-requiring process. The end result of its reversible binding to the 50S ribosome is the prevention of peptide bond formation. Although its spectrum of activity is quite broad, including both gram-positive and gram-negative aerobes and anaerobes as well as a number of rickettsial species, concerns about toxicities and significant drug interactions has limited the clinical usefulness of this once very valuable antibiotic.

INHIBITION OF ESSENTIAL METABOLITES

Folinic acid is necessary for the synthesis of bacterial DNA. Many bacteria depend on making their own folinic acid, because they do not have the ability to take up this essential metabolite. The synthesis of folinic acid occurs as the result of modification of para-aminobenzoic acid and

relies on several different enzymes (Fig. 3–3). Inhibiting one or both of these enzymes essentially prevents bacterial replication.

Sulfonamides competitively inhibit the bacterial enzyme dihydropteroate synthetase, which ultimately allows for the incorporation of para-aminobenzoic acid into dihydrofolate. For optimal efficacy, the bacteria must be exposed to a continued source of sulfonamide in order to achieve the desired result of affecting bacterial replication. The pyrimidine analog trimethoprim competitively inhibits the enzyme dihydrofolate reductase, effectively blocking the conversion to tetrahydrofolate, a precursor of folinic and other amino acids. Because these agents act at different metabolic sites, using them in combination has a synergistic effect. The spectrum of activity of these agents, when combined in a fixed preparation, is very broad and includes many gram-positive and gram-negative organisms, with the notable exception of *P. aeruginosa.*

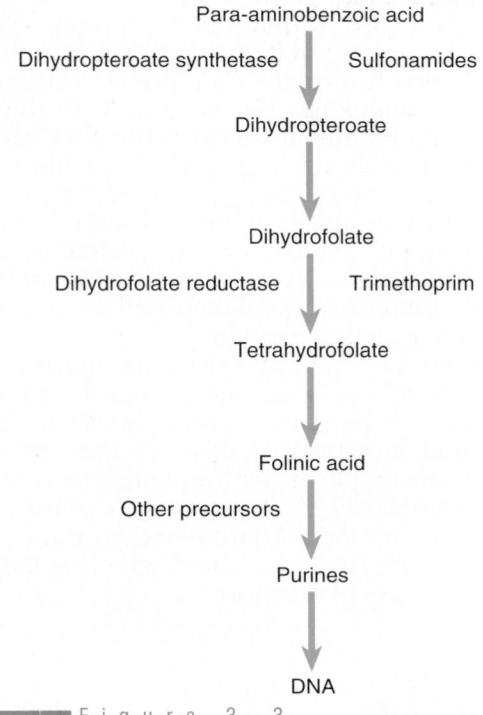

Figure 3-3

The sites of action of sulfonamides and trimethoprim and their effects on synthesis of essential amino acids and nucleic acids.

INTERFERENCE WITH NUCLEIC ACID METABOLISM

Interfering with either DNA or RNA metabolism can have devastating consequences on bacteria. Rifampin interferes with the production of mRNA by binding to one of the four subunits of DNA-directed RNA polymerase. The result is that transcription of the RNA chain is aborted at the initiation step. Rifampin also has a wide range of antibacterial activity, but its most noteworthy clinical application is still in the treatment of mycobacterial diseases, particularly *Mycobacterium tuberculosis*.

Two distinct categories of antimicrobial agents interfere with DNA synthesis: those that disrupt replication and those that disrupt the finished product. Nalidixic acid and the subsequent fluorinated quinolones (norfloxacin, ciprofloxacin, ofloxacin, lomefloxacin, enoxacin) interfere with DNA synthesis. This class of antimicrobial agents kill bacteria by rendering the essential enzyme, DNA gyrase, ineffective. This enzyme is responsible for supercoiling of DNA, allowing for the bacterial chromosome to be condensed to a size that is contained within the bacterium. DNA gyrase is composed of four subunits, two A proteins and two B proteins. The primary interaction of the quinolones seems to be with the A subunit; by inhibiting its activity, the DNA backbone cannot align itself properly, with the end result being bacterial killing. The spectrum of activity of these agents in vitro is quite broad, but gram-negative aerobes, including Enterobacteriaceae, *P. aeruginosa,* and *Haemophilus, Neisseria, Moraxella, Campylobacter, Vibrio,* and *Aeromonas* species, are most susceptible.

Metronidazole disrupts DNA by making it unstable. In its usual form, metronidazole is inactive. When it is partially reduced, however, and incoporated into DNA, it disrupts the nucleic acid. Only anaerobic bacteria and organisms that grow anaerobically, such as certain protozoa, have this ability for partial reduction of metronidazole; hence, the spectrum of activity is limited to those types of organisms.

RESISTANCE TO ANTIBACTERIAL AGENTS

In the face of continuing antibiotic challenge, bacteria have found a number of ways of adapting and thus survive. The mechanisms of resistance have developed in response to the mechanisms of action of antimicrobial drugs (Table 3–5).

For antimicrobial agents to have specific activity, they must somehow associate with the organism and reach a specified target site. One mechanism of resistance is to prevent access to that target by inhibiting uptake of the substance. For example, anaerobic bacteria are not equipped to activate the electron transport system that is necessary for aminoglycosides to reach their ribosomal site of action. Consequently, anaerobes are intrinsically resistant to aminoglycosides. Decreased accumulation can also result, owing to characteristics specific to the bacteria, such as extracellular slime or altered lipopolysaccharide, or characteristics specific to the antimicrobial agent, such as size, electronic charge, and affinity for lipids. Even in specific families of microorganisms, subtle differences may result in varied patterns of susceptibility to a single class of antibiotics. A classic example is the decreased sensitivity of *Enterobacter cloacae* to β-lactam drugs compared with *E. coli.* The blueprints for these types of mechanisms of resistance are usually encoded on the chromosomes.

The other way to prevent access to the antimicrobial's site of action is to increase excretion of the drug. This is the primary mechanism of resistance to the tetracyclines. Organisms resistant to tetracyclines do not concentrate these drugs in the cell because of an active transport exit mechanism that does not allow intracellular accumulation. The proteins responsible for this phenomenon occur as a result of plasmid encoding.

Another important mechanism of resistance is structural or functional modification of the tar-

■■■■ T A B L E 3 – 5

COMMON MECHANISMS OF RESISTANCE FOR DIFFERENT ANTIMICROBIAL AGENTS

Mechanism	Class of Antimicrobial
Prevent access to target	Aminoglycosides Tetracyclines β-Lactams
Modify target	Macrolides Quinolones Rifampin β-Lactams
Produce inactivating enzymes	β-Lactams Chloramphenicol Aminoglycosides

get site of action. Such alterations may be expressed in various ways. One such mechanism is biochemical alterations that reduce the target site's affinity for a drug. For example, in response to an enzyme, methylase, which is made from a plasmid-encoded gene in staphylococci, methylation of the 23S ribosome occurs, with the ultimate result being a modification in the 50S ribosome, the site of action of the macrolide class of antimicrobials. Along the same line of thinking, though not as clearly defined, alterations in RNA polymerase and DNA gyrase provide plausible explanations for resistance patterns seen with rifampin and the quinolones, respectively.

Other ways that the target site may be modified include the production of different proteins that can function like the targets but do not bind drug or the excessive production of targets to overwhelm the drug with competitive inhibition. Bacterial resistance to sulfonamides and trimethoprim are expressed by such mechanisms. By producing different enzymes that function in an alternative pathway or by overproducing necessary synthetase and reductase enzymes, folinic acid synthesis can be achieved.

Some organisms, on the other hand, like staphylococci, can synthesize "new" proteins that maintain their usual function as transpeptidases but do not bind even stable β-lactam drugs like methicillin. As a group, these staphylococci are known as *methicillin-resistant*. They pose difficult problems in terms of treatment, because such staphylococci do not respond well to any of the β-lactam drugs. Similar altered PBPs have been observed in a number of different species as well, and their continued presence has reinforced the ongoing search for effective antibiotics.

One of the first-recognized ways for bacteria to resist the actions of antimicrobial agents was the production of enzymes that inactivate a drug. A number of antimicrobial agents, such as aminoglycosides and chloramphenicol, have been found to be affected by inactivating enzymes, but the classic example of this phenomemon is β-lactamase production. In response to exposure to β-lactam drugs, bacteria may produce enzymes (β-lactamases) that can hydrolyze the β-lactam ring and consequently inactivate the antimicrobial agent. This was first described as a penicillinase produced by staphylococci, and since then more than a hundred such enzymes produced by a number of bacterial species have been identified. Many of these β-lactamases are the result of plasmid encoding. Pharmacologic compensations have been based on either structural manipulation of the side chain, so that the β-lactam ring is sterically protected, such as in some of the later-generation cephalosporins, or combination of the major drugs with β-lactamase inhibitor agents.

B. PROCEDURES IN ANTIMICROBIAL SUSCEPTIBILITY TESTING

Janet A. Hindler • *James H. Jorgensen*

and another battery for gram-positive bacteria. Sometimes, there is a separate battery for urine isolates, representing drugs appropriate for treating urinary tract infections. A supplemental battery that contains antimicrobial agents with enhanced spectra of activity is often included by those laboratories that encounter a significant number of highly resistant isolates. Some may elect to establish a separate battery that contains agents specific for *Pseudomonas aeruginosa,* particularly if there is insufficient space available for testing a wide variety of anti-pseudomonal agents as part of the routine gram-negative battery. Finally, unique batteries may be established for *Haemophilus* spp and *Neisseria gonorrhoeae.*

Reporting of Susceptibility Test Results

Because the identity of the isolate is often not known at the time the susceptibility test is performed, some drugs may be tested on an isolate that may be inappropriate for reporting. In such instances, a drug should never be indiscriminantly reported, because results may be misleading. Some drugs may appear active against certain species in vitro but are inappropriate for clinical use (e.g., cephalosporins against enterococci). The final reporting decision should be made once the isolate's identity is known (sometimes a preliminary identification is sufficient), and the overall antibiogram and specimen source must be taken into consideration.

As mentioned previously, reporting protocols should be developed following discussion with infectious disease clinicians, pharmacists, and others who have clinical experience with the antimicrobial therapy practices of the particular institution. A primary objective of antimicrobial therapy is to use the least toxic, most cost-effective, and most clinically appropriate agents and to refrain from more costly, broader-spectrum agents when they are unnecessary. Physician compliance with this objective is often difficult. Sometimes all drugs tested are reported, and the mechanism of control for inappropriate prescribing is outside the laboratory's purview. Alternatively, the laboratory may assist in discouraging inappropriate antimicrobial prescribing by refraining from reporting broad-spectrum agents if narrower-spectrum agents appear useful. NCCLS provides guidance for development of a such a selective reporting or cascade reporting protocol.

For several organism groups, NCCLS categorizes antimicrobial agents into four groups (see

Table 3–6). As a general guideline, it is suggested that within a particular antimicrobial class, primary (group A) agents be reported first, and that secondary (group B) agents are reported only if one of the following conditions exists:

■ The isolate is resistant to the primary agents

■ The patient cannot tolerate the primary agents

■ The infection has not responded to the primary agents *or*

■ A secondary agent would be a better clinical choice for the particular infection

For example, a primary cephalosporin, such as cephalothin or cefazolin (first-generation cephalosporin), would be a reasonable choice for a susceptible *E. coli,* and secondary cephalosporins such as cefoxitin (second-generation cephalosporin) or cefotaxime (third-generation cephalosporin) would generally not be required. An exception would occur for meningitis, because third-generation cephalosporins cross the blood-brain barrier much more effectively than their first-generation counterparts. Gentamicin is usually the aminoglycoside of choice for treating serious infections caused by gentamicin-susceptible *P. aeruginosa,* and tobramycin or amikacin may be considered for gentamicin-resistant isolates. Aminoglycosides are not effective for treating meningitis, because they do not readily cross the blood-brain barrier.

A secondary agent may also be reported if the patient has a polymicrobial infection and a secondary (but not a primary) agent would be more likely to be effective against all pathogens present. Similarly, a secondary agent may be reported if the patient has a disseminated infection and a secondary (but not a primary) agent would be more likely to be effective at all sites.

Agents with very broad-spectrum activity (group C) may be tested and reported for the reasons listed for secondary agents. In addition, group C agents would be considered for routine testing if a particular institution encounters large numbers of isolates resistant to group A and group B agents. Finally, agents listed in group D should be reported only on isolates from urine, as these drugs are clinically ineffective in treating infections other than urinary tract infections. Examples of reports generated following a selective reporting protocol are shown in Table 3–7; NCCLS Table 1 (Table 3–6) was used to determine which agents are considered primary or secondary.

SOME SUGGESTED GROUPINGS OF US FDA–APPROVED ANTIMICROBIAL AGENTS THAT SHOULD BE CONSIDERED FOR ROUTINE TESTING AND REPORTING BY CLINICAL MICROBIOLOGY LABORATORIES FOR ENTEROBACTERIACEAE

Group A:
Test and Report Routinely

ampicillin[a,b]
cefazolin[b,c]
cephalothin[b,c]
gentamicin[b]

Group B:
Test Routinely, Report Selectively

mezlocillin or piperacillin
ticarcillin

amoxicillin/clavulanic acid or ampicillin/sulbactam
ticarcillin/clavulanic acid

cefmetazole
cefoperazone[a]
cefotetan
cefoxitin

cefamandole or cefonicid or cefuroxime

Cefotaxime[a,d] or ceftizoxime[a] or ceftriaxone[a,d]

imipenem

amikacin

ciprofloxacin[a,b]

trimethoprim/sulfamethoxazole[a,b]

Group C:
Supplemental, Report Selectively

ceftazidime
azetreonam

kanamycin

netilmicin

tobramycin

tetracycline[b]

chloramphenicol[a,e]

Group D:
Supplemental for Urine Only

cinoxacin
lomefloxacin, norfloxacin[a] or ofloxacin

loracarbef

nitrofurantoin

sulfisoxazole

trimethoprim

[a] For isolates of *Salmonella* and *Shigella*, only ampicillin, a quinolone, and trimethoprim-sulfamethoxazole should be tested and reported routinely. In addition, chloramphenicol and a third-generation cephalosporin should be tested and reported for extraintestinal isolates of *Salmonella*.
[b] May also be appropriate for reporting on isolates from the urinary tract.
[c] Cephalothin can be used to represent cephalothin, cephapirin, cephradine, cephalexin, cefaclor, and cefadroxil. Cefazolin, cefuroxime, cefpodoxime, and cefprozil may be tested individually against Enterobacteriaceae because these agents may be active when cephalothin is not.
[d] Report at least one of these agents on isolates from CSF.
[e] Not routinely tested against organisms isolated from the urinary tract.
Permission to use portions of M2-A5 (Performance Standards for Antimicrobial Disk Susceptibility Tests, Fifth Edition; Approved Standard) and M7-A3 (Methods for Dilution Antimicrobial Susceptibility Tests for Bacteria that Grow Aerobically, Third Edition; Approved Standard) has been granted by the National Committee for Clinical Laboratory Standards. NCCLS is not responsible for errors or inaccuracies. The interpretive data are valid only if the methodology in M2-A5 and M7-A3 is followed. These documents and current supplements may be obtained from NCCLS, 771 E. Lancaster Avenue, Villanova, PA 19085.

EXAMPLES OF ANTIMICROBIAL AGENTS REPORTED FOLLOWING A SELECTIVE REPORTING PROTOCOL, AS SUGGESTED IN TABLE 1, NCCLS M2-A5 AND M7-A3. PRIMARY AGENTS (GROUP A) ARE IN PLAIN TYPE AND SECONDARY AGENTS (GROUP B) ARE IN BOLD TYPE.

E. coli
Source—urine

Drug	Result
ampicillin	S
cephalothin	S
gentamicin	S
nitrofurantoin	S
trimethoprim/sulfamethoxazole	S

E. coli
Source—urine

Drug	Result
ampicillin	R
ampicillin/sulbactam	R
cefoxitin	S
cephalothin	R

Drug	Result
gentamicin	S
nitrofurantoin	S
trimethoprim/sulfamethoxazole	R

Enterobacter cloacae
Source—blood

Drug	Result
amikacin	S
ampicillin	R
ampicillin/sulbactam	R
cefoxitin	R
cefotaxime	S
cephalothin	R
gentamicin	R
trimethoprim/sulfamethoxazole	R

Permission to use portions of M2-A5 (Performance Standards for Antimicrobial Disk Susceptibility Tests, Fifth Edition; Approved Standard) and M7-A3 (Methods for Dilution Antimicrobial Susceptibility Tests for Bacteria that Grow Aerobically, Third Edition: Approved Standard) has been granted by the National Committee for Clinical Laboratory Standards. NCCLS is not responsible for errors or inaccuracies. The interpretive data are valid only if the methodology in M2–A5 and M7-A3 is followed. These documents and current supplements may be obtained from NCCLS, 771 E. Lancaster Avenue, Villanova, PA 19085.

TRADITIONAL ANTIMICROBIAL SUSCEPTIBILITY TEST METHODS

Inoculum Preparation and Use of McFarland Standards

Inoculum Preparation

Inoculum preparation is the most important step in any susceptibility test. Inocula are prepared by adding cells from four to five isolated colonies of similar colony morphology to a broth medium and then allowing them to grow to the log phase. Four to five colonies, rather than a single colony, are selected in order to minimize the possibility of testing a colony that might have been derived from a susceptible mutant. Inocula may also be prepared directly by inoculating colonies grown overnight on an agar plate into broth or saline. This direct inoculum suspension preparation technique, which is preferred for bacteria that grow unpredictably in broth (e.g., fastidious bacteria), does not require incubation, but it is imperative that only fresh (16–24-hour) colonies be used.

McFarland Turbidity Standards

The numbers of bacteria tested must be standardized regardless of the method used. False-susceptible results may occur if too few bacteria are tested, and false-resistant results may follow testing of too many bacteria. The most widely used method of inoculum standardization involves McFarland turbidity standards. McFarland standards are prepared by adding specific volumes of 1% sulfuric acid and 1.175% barium chloride to obtain a barium sulfate solution with a very specific optical density. The one most commonly used is the McFarland 0.5 standard, which contains 99.5 mL of 1% sulfuric acid and 0.5 mL of 1.175% barium chloride. This solution is dispensed into tubes comparable to those used for inoculum preparation, which are sealed tightly and stored in the dark at room temperature. The McFarland 0.5 standard provides a turbidity comparable to a bacterial suspension containing 1.5×10^8 CFU/mL.

Inoculum Standardization

To standardize the inoculum, the inoculated broth and the McFarland 0.5 standard are spun thoroughly in a centrifuge and then, under adequate lighting, the two tubes are positioned side by side against a white card containing several horizontal black lines (Fig. 3–4). The turbidities are compared by looking at the black lines through the suspensions. The suspension is too dense if it is more difficult to see the lines through the inoculum suspension than through the McFarland 0.5 standard. In this case, the inoculum is diluted with additional sterile broth or saline. If the test suspension is too light, more organisms are added or the suspension is reincubated (depending on the inoculum preparation protocol) until the turbidity reaches that of the McFarland standard. The standardized inoculum suspensions should be used within 15 minutes of preparation.

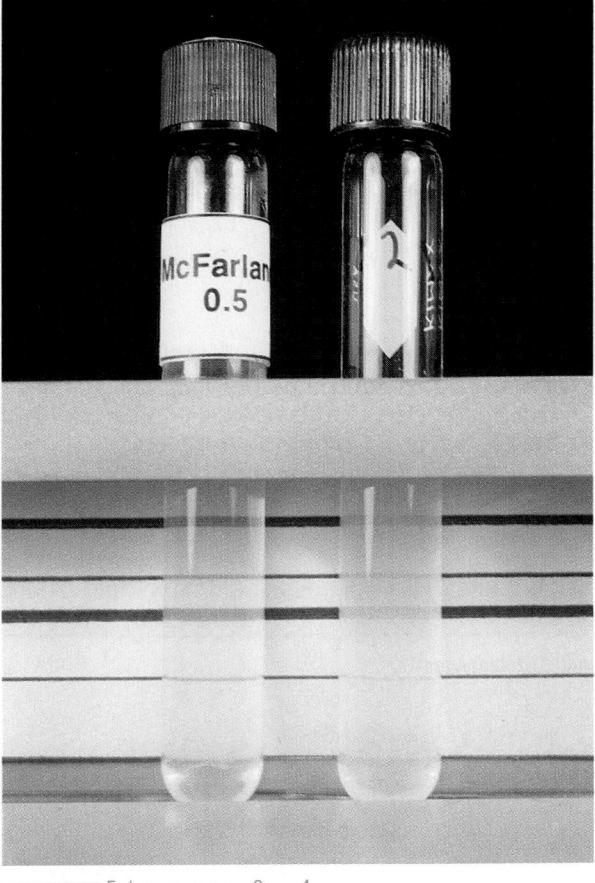

■■■■ F i g u r e 3 – 4

The tube on the left is a McFarland 0.5 turbidity standard. The tube on the right is a test bacterial suspension that has a turbidity greater than that of the McFarland standard: It is more difficult to see the black lines through the test suspension than through the McFarland standard. Sterile saline or broth must be added to the test suspension to dilute it until the turbidity matches that of the McFarland standard.

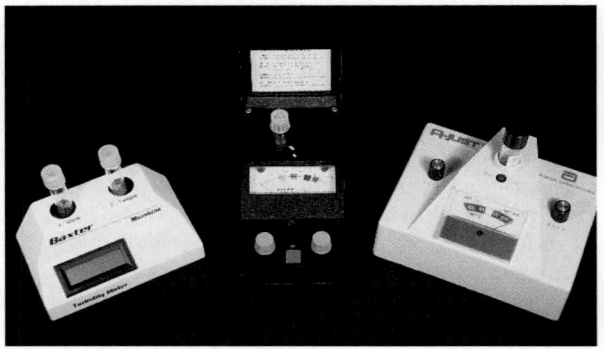

■■■■■Figure 3 – 5

Pictured are three bench-top nephelometric-type devices that can be used to standardize the turbidity of a test inoculum suspension to match that of a McFarland 0.5 (or other) turbidity standard.

An alternative to McFarland standards for visual turbidity adjustment involves use of a nephelometric or spectrophotometric device. Several commercially available bench-top instruments are available for more objective standardization of bacterial inocula in the clinical microbiology laboratory (Fig. 3–5).

Dilution Methods

Principle

Dilution antimicrobial susceptibility test methods are used to determine the *minimal inhibitory concentration (MIC)*, or the lowest concentration of antimicrobial agent required to inhibit the growth of a bacterial isolate. Varying concentrations of antimicrobial agent are added to broth or agar media. Generally, serial twofold-dilution concentrations are tested (expressed in µg/mL), and these represent concentrations that are attainable in vivo following standard dosages for each respective antimicrobial agent. Because the concentrations attainable in vivo vary with different agents, the ranges of concentrations tested vary also. For example, concentrations above 8 µg/mL for gentamicin and tobramycin cannot be safely attained in the patient, and therefore, the ranges of concentrations tested are generally 0.25 to 8.0 µg/mL. In contrast, much higher levels of the extended-spectrum penicillin ticarcillin are attainable in vivo, and a range of 4 to 128 µg/mL is commonly tested. Once the MIC is determined, the organism is interpreted as either susceptible, intermediate, or resistant to each agent with the use of a table provided in the NCCLS dilution testing document. An example is shown in Table 3–8. Note that for some antimicrobial agents, there are different MIC interpretive criteria for different organisms or organism groups. For each antimicrobial agent, the MIC breakpoint separates susceptible from resistant results. Organisms with MICs at or below the breakpoint are susceptible, and those with MICs above the breakpoint are intermediate or resistant. The NCCLS documents describe the details of performing MIC tests by broth macrodilution, broth microdilution, and agar dilution.

Antimicrobial Stock Solutions

Antimicrobial stock solutions used in MIC tests must be prepared from reference standard

■■■■■TABLE 3 – 8

MIC INTERPRETIVE STANDARDS (µG/ML) OF THREE CATEGORIES OF SUSCEPTIBILITY FOR ORGANISMS OTHER THAN *HAEMOPHILUS*, *NEISSERIA GONORRHOEAE*, AND *STREPTOCOCCUS PNEUMONIAE*

Antimicrobial Agent	Susceptible	Intermediate	Resistant
ampicillin			
when testing Enterobacteriaceae	≤ 8	16	≥ 32
when testing staphylococci and *Moraxella catarrhalis*	≤ 0.25	—	≥ 0.5
when testing *Listeria monocytogenes*	≤ 2	—	≥ 4
when testing enterococci	≤ 8	—	≥ 16
when testing streptococci and other gram positives	≤ 0.12	0.25–2	≥ 4
cefazolin	≤ 8	16	≥ 32
gentamicin	≤ 4	32	≥ 16
oxacillin			
when testing staphylococci	≤ 2	—	≥ 4

Permission to use portions of M7-A3 (Methods for Dilution Antimicrobial Susceptibility Tests for Bacteria that Grow Aerobically, Third Edition; Approved Standard) has been granted by the National Committee for Clinical Laboratory Standards. NCCLS is not responsible for errors or inaccuracies. The interpretive data are valid only if the methodology in M7-A3 is followed. These documents and current supplements may be obtained from NCCLS, 771 E. Lancaster Avenue, Villanova, PA 19085.

antimicrobial powders. Details of preparation are found in the NCCLS protocols. Stock solutions and other antimicrobial solutions must be stored frozen in non–frost-free freezers. Temperatures at or below −70°C are optimal and necessary for the more temperature labile drugs such as imipenem and clavulanic acid; however, −20°C storage is acceptable for some agents. Antimicrobial solutions must never be refrozen following thawing.

Broth Macrodilution (Tube Dilution) Tests

Broth dilution MIC tests performed in test tubes are referred to as *broth macrodilution* or *tube dilution* susceptibility tests. Generally, a twofold serial dilution, each containing 1 to 2 mL of antimicrobial agent, is prepared. Mueller-Hinton broth is the medium most commonly used for broth dilution MIC tests of nonfastidious bacteria. A standardized suspension of test bacteria is added to each dilution to obtain a final concentration of 5×10^5 CFU/mL. A growth control tube (broth plus inoculum) and an uninoculated control tube (broth only) are used in each test. After overnight incubation at 35°C, the MIC is determined visually as the lowest concentration that inhibits growth, as evidenced by the absence of turbidity.

This broth macrodilution method is impractical for use as a routine method when several antimicrobial agents must be tested on an isolate or if several isolates must be tested. Some laboratories use broth macrodilution when it is necessary to test drugs not included in their routine system or for testing fastidious bacteria that require special growth media. Additionally, this method is often used when minimum bactericidal concentration (MBC) endpoints are to be subsequently determined. The MBC test is discussed later in this chapter.

Broth Microdilution Tests

The broth macrodilution test has been adapted to multi-well microdilution trays (Fig. 3–6) for broth microdilution MIC testing. Polystyrene trays containing between 80 and 100 wells are filled with small volumes (usually 0.1 mL) of twofold dilution concentrations of anitmicrobial agent in broth. Because of the large number of wells, several dilutions of as many as 12 to 15 antimicrobial agents can be contained on a single tray that will be subsequently inoculated

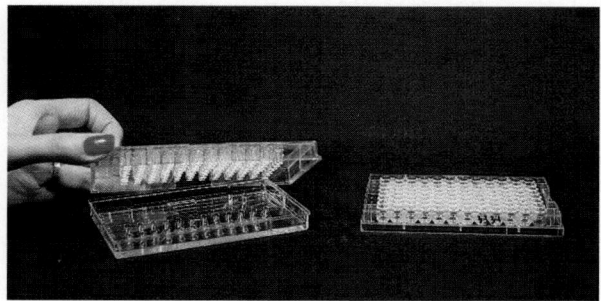

■■■ F i g u r e   3 – 6

Broth microdilution MIC tray shown with inoculum reservoir trough. Diluted inoculum suspension is placed in the reservoir trough. Then the prongs are dipped into the suspension, raised, and subsequently lowered into the wells of the broth microdilution MIC tray to inoculate all wells simultaneously.

with one bacterial isolate. The inoculum suspension is prepared and standardized as described previously. An intermediate dilution of this inoculum suspension is prepared in water, and a multi-pronged inoculator or other type of inoculating device is used to inoculate the wells to obtain a final concentration of approximately 1 to 5×10^5 CFU/mL ($1–5 \times 10^4$ CFU per 0.1-mL well). The actual dilution factor used for preparation of the intermediate dilution depends on the volume of inoculum delivered to each well by the inoculating device. An example of this calculation is illustrated in Table 3–9. A growth control well and uninoculated control well are included on each tray. Following overnight incubation at 35°C, the tray is placed on one of several types of tray-reading devices to facilitate close visual

■■■ T A B L E   3 – 9

EXAMPLE OF CALCULATIONS REPRESENTING DILUTION SCHEMA FOR PREPARATION OF INOCULA FOR BROTH MICRODILUTION MIC TESTS*

Step	Resulting Organism Concentration
1. Standardize suspension to McFarland 0.5.	1.5×10^8 CFU/mL
2. Add 0.75 mL from step 1 to 25 mL water diluent (1:33 dilution).	$4–5 \times 10^6$ CFU/mL
3. Use inoculator prong set to inoculate wells of MIC tray (each prong delivers 0.01 mL, which results in an additional 1:100 dilution).	$4–5 \times 10^4$ CFU/100-μL well $4–5 \times 10^5$ CFU/mL

* Calculations shown here are based on use of inoculator prong set (each prong delivering 0.01 mL). Dilutions in step 2 will vary, depending on the number of organisms in the initial suspension and the volume delivered to each well by the inoculating device.

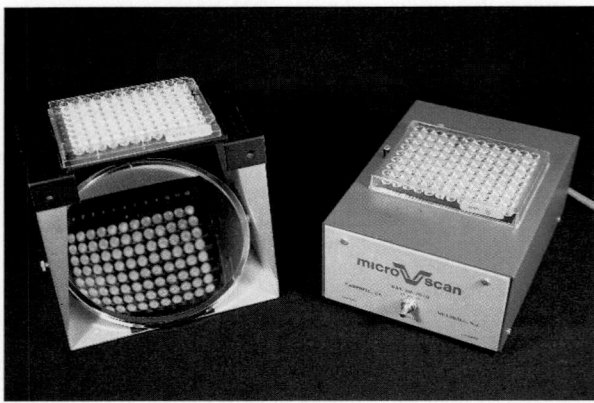

■■■■■■■ F i g u r e   3 – 7

Dynatech tray-reading stand with magnifying mirror *(left)* and a light box *(right)*. Following incubation, either of these devices is used to examine growth in the broth microdilution MIC trays. A tray is placed on the Dynatech tray reading stand, and wells are examined by looking into the magnifying mirror. A tray is placed over the hole in the light box, and light is allowed to shine through the tray to facilitate close examination of the wells.

examination of each well (Fig. 3–7). Provided that there is adequate growth in the growth control well, the MIC for a particular agent is the lowest concentration showing no obvious growth. Growth may be noted as either turbidity, a haze, or a pellet in the bottom of the well.

Some larger laboratories have dispensing devices that are used to prepare broth microdilution panels. For most laboratories, however, a practical approach to broth microdilution MIC testing involves purchasing prepared panels, either frozen or dried, from a commercial vendor. Frozen panels are thawed at room temperature prior to inoculation. For dried panels, the dried or lyophilized drugs in the wells are reconstituted at the same time the panels are inoculated. In addition to the panels, all commercial companies sell other materials needed for testing (e.g., inoculum broths, inoculum diluents, panel inoculators, and reading devices). Usually, a variety of panels containing different drugs are available for testing various organism groups (e.g., gram-positive, gram-negative), and some panels are designed to include wells containing various biochemical reagents, so that organism identification and antimicrobial susceptibility testing can be done simultaneously on a single panel. Some companies have automated or semi-automated devices to facilitate inoculation and reading. These are discussed in the following sections.

BREAKPOINT MIC PANELS

A variation of the standard broth microdilution MIC panel is the breakpoint panel, in which only one or two concentrations of each antimicrobial agent are tested on a single panel. *Breakpoint* is the term applied to the concentration of an antimicrobial that can be achieved in body fluids with optimum therapy. These concentrations typically include the concentration that defines the MIC breakpoint for susceptibility and one dilution above the breakpoint. When two concentrations are tested and there is no growth in either well, the isolate is susceptible. When there is growth in the low concentration but no growth in the high concentration, the isolate has intermediate susceptibility, and a resistant isolate grows in both wells. The qualitative interpretation—"susceptible," "intermediate," or "resistant"—is generally reported rather than an MIC value. The primary advantage of breakpoint panels is that numerous drugs can be tested on a single panel (often in combination with biochemical tests). The primary disadvantage of breakpoint panels is that a precise MIC is not obtained, because most results are either equal to or lower than the lowest concentration tested or greater than the highest concentration tested. Additionally, it is difficult to perform relevant quality control on breakpoint panels.

TRAILING AND SKIPPED WELLS

Dilution methods, particularly broth microdilution, sometimes produce an MIC endpoint that is not clear-cut, and growth in the wells may demonstrate trailing or skipped wells. *Trailing* involves heavy growth at lower concentrations followed by one or more wells that show much reduced growth, in the form of a small button or a very light haze. This commonly occurs with sulfonamides, trimethoprim, and trimethoprim/ sulfamethoxazole; the mode of action of the agents allows the bacterial cells to grow through several generations prior to inhibition. In this case, the trailing is ignored, and the endpoint is read as an 80% reduction in growth. Trailing with other drugs, however, may represent contamination and should not be ignored, unless it is known that trailing commonly occurs with the particular antimicrobial agent–organism combination.

Skipped wells involve growth at higher concentrations and no growth at one or more of the lower concentrations. This may occur as a result of contamination, improperly inoculated wells, improper concentrations of antimicrobial agent in the wells, presence of unusual resistance with the test isolate (e.g., a small resistant subpopulation), or a combination of two or more of these factors. As with trailing, each skipped well occurrence must be evaluated individually to determine whether results are reportable. If there is any doubt as to the accuracy of the results, they should not be reported, and the test should be repeated.

One of the shortcomings of the broth microdilution MIC method is its inability to produce a penicillin MIC that is consistently within the resistant range for staphylococci that are low-level β-lactamase producers. If the penicillin MIC is 0.03 µg/mL or less, the isolate may be reported as susceptible; a result 0.25 µg/mL or higher is considered resistant. An induced β-lactamase test (see below) must, however, be performed on isolates with penicillin MICs of 0.06 to 0.12 µg/mL. If the β-lactamase test result is positive, the isolate is reported as penicillin resistant; and if the result is negative, the isolate is reported as penicillin susceptible.

Agar Dilution Tests

MIC tests can also be performed using an agar dilution method. Specific volumes of antimicrobial solutions are dispensed into premeasured volumes of molten and cooled agar, which is subsequently poured into standard Petri dishes. Mueller-Hinton agar is generally used for testing aerobic isolates; however, this can be supplemented with sheep's blood (to a final concentration of 5% sheep's blood) or other nutrients for testing fastidious bacteria. A series of plates, containing varying concentrations of each antimicrobial agent, and growth control plates without antimicrobial agent are prepared. The agar is allowed to solidify, and then a standard number of test bacteria (10^4 CFU for aerobes) are "spot" inoculated onto each plate using a multipronged replicating device such as a Steer's replicator (Fig. 3–8). As many as 32 different isolates can be simultaneously inoculated onto each 100-mm round Petri dish; 100-mm square plates generally accommodate 36 isolates. Following overnight incubation, the MIC is read as the lowest concentration of antimicrobial agent that inhibits the visible growth of the test bacterium (one or two colonies are ignored).

The shelf life of agar dilution plates is only 1 week for most antimicrobial agents, because plates must be stored at 2 to 8°C, and many

■■■■■■■ F i g u r e   3 – 8

Steer's replicator. The inoculator prongs are positioned above a 36-well seed trough that contains 36 different standardized inoculum suspensions. The handle on top of the prong unit is pressed to lower the prongs into all suspensions simultaneously, and when the prongs are raised, each contains a standardized volume of inoculum. The agar plate containing a defined concentration of antimicrobial agent is positioned under the prongs (steel plate holding agar plate slides back and forth), and the prongs are carefully lowered to the agar, at which point the inocula are deposited on the agar surface. This process is repeated until all antimicrobial-containing and control agar plates (without antimicrobial agent) have been inoculated.

drugs are labile at this temperature. Because plate preparation is laborious and this procedure is practical only if large numbers of isolates are tested, agar dilution is generally performed only in research settings, although it is currently the reference method for antimicrobial susceptibility testing of anaerobes and *N. gonorrhoeae*.

Disk Diffusion Testing

Principle

The disk diffusion test, also commonly known as the Kirby Bauer test, has been widely used in clinical laboratories since 1966, when the first standardized method was described. Briefly, a McFarland 0.5 standardized suspension of bacteria is swabbed over the surface of an agar plate, and paper disks containing single concentrations of each antimicrobial agent are placed onto the inoculated surface. Following overnight incubation, the diameters of the zones produced by antimicrobial inhibition of bacterial growth are measured and the isolate is interpreted as either susceptible, intermediate, or resistant to a particular drug according to preset criteria. A photograph of an isolate of *Enterobacter aerogenes* tested by disk diffusion is shown in Figure 3–9.

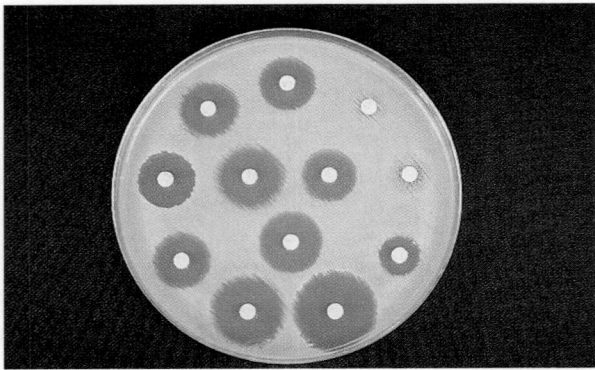

■■■■■ F i g u r e 3 – 9

Enterobacter aerogenes tested by the disk diffusion method. Zone measurements confirm that the isolate is susceptible to all agents tested except ampicillin (positioned at 1 o'clock) and cefazolin (positioned at 2 o'clock). No zones are present for either of these agents.

Establishing Zone Diameter Interpretive Breakpoints

The disk diffusion test depends on the formation of a gradient of antimicrobial concentrations as the antimicrobial agent diffuses radially into the agar. The drug concentration decreases at increasing distances from the disk. At a critical point, the amount of drug at a specific location in the medium is unable to inhibit the growth of the test organism, and a *zone of inhibition* is formed.

The zones of inhibition are related to MICs, and it is this relationship that has been used to determine the breakpoints for interpreting a particular zone measurement as indicating that the isolate is susceptible, intermediate, or resistant. To establish breakpoints for a single agent, the first step is to determine the optimum concentration of drug to incorporate into the disk. This is done by reviewing the drug's pharmacokinetic properties (e.g., antimicrobial concentration attainable in vivo) and also some of the biochemical properties (e.g., molecular size, solubility, and diffusibility in agar). Next, a sample of 150 to 200 isolates with comparable growth rates and varying susceptibility to the agent are tested by both the standard disk diffusion test and a standard dilution MIC test, and results are plotted on a graph. For each isolate, the observed MIC value is expressed in logarithmic form ($\log_2$) and plotted on the y axis, and the corresponding zone measurement is plotted on the x axis on an arithmetic scale (scattergram). A linear regression analysis is performed through the plotted points,

and a line of best fit, or regression line, is drawn. The concentrations representing therapeutically achievable blood levels following standard dosing are identified.

Results of clinical treatment situations in which the MICs of the implicated organisms and therapeutic outcomes are known are examined. The distribution of the plotted points is reviewed, and efforts are made to avoid defining breakpoints that will split a population of isolates that are probably either all susceptible or all resistant. All this information is considered, and MICs defining *susceptible, intermediate,* and *resistant* are identified. These are correlated with MIC values specified on the y axis of the graph, and the regression line is further examined to match the MIC value to its corresponding zone diameter to obtain the zone diameter breakpoints. A generic scattergram and regression analysis is shown in Figure 3–10.

The NCCLS and the U.S. Food and Drug Administration (FDA) are involved in developing zone interpretive criteria, and the procedure just described has been performed, for every antimicrobial agent for which zone interpretive criteria exist. One of the concerns in establishing zone interpretive criteria for some of the newer, very active agents, such as the fluoroquinolones, is the lack of resistant isolates to include in the analysis to provide a balanced distribution of points on the graph. This has led in some cases to define criteria for *susceptible* only; no intermediate or resistant zone interpretive criteria are specified.

Test Performance

DISK STORAGE

Most clinical laboratories follow the protocol specified by the NCCLS for disk diffusion testing, and the NCCLS document contains very explicit details for test performance. Only FDA-approved disks should be used; they must be stored properly in order to make certain that the drugs maintain their potency. For long-term storage, disks are stored at –14°C or below in a non–frost-free freezer. A working supply of disks can be stored in a refrigerator at 2 to 8°C for at least 1 week. Disks should always be stored in a tight-sealing container with desiccant. The container should be allowed to warm to room temperature prior to opening, in order to prevent condensation from occurring on the disks when warm room air contacts the cold containers.

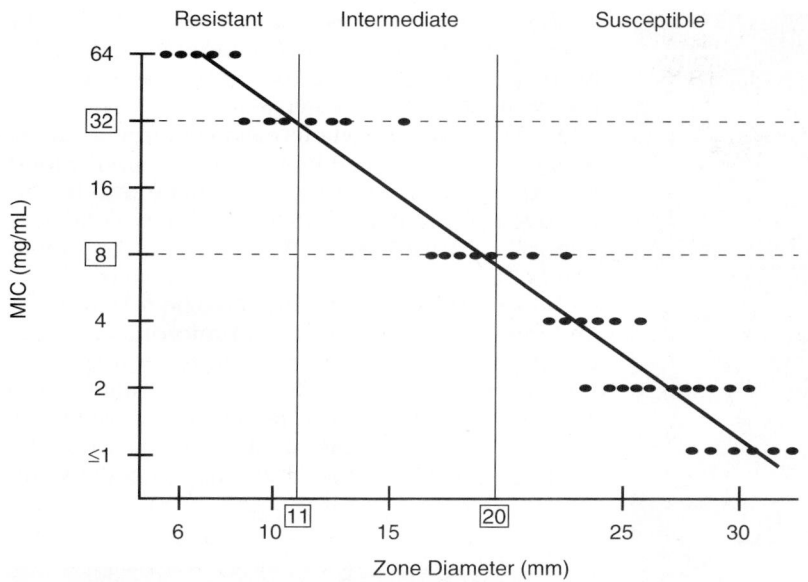

Figure 3–10

Example of a scattergram and regression analysis plot used to determine disk diffusion zone diameter interpretive breakpoints for hypothetical drug "X." Based on clinical response data, isolates with MICs ≤ 8.0 µg/mL are considered susceptible. As derived from this scattergram, corresponding zones ≥ 20 mm would be interpreted as susceptible. Isolates with MICs ≥ 32.0 µg/mL and zones ≤ 11 mm are resistant. The intermediate designation is used for isolates whose values fall between the susceptible and the resistant MICs (16 µg/mL) and zone interpretive breakpoints (12–19 mm).

INOCULATION AND INCUBATION

Inoculum suspensions are prepared using either a log-phase or direct inoculum suspension standardized to match the turbidity of a McFarland 0.5 standard, as previously described. A sterile cotton swab is dipped into the suspension, is pressed and rotated firmly against the side of the tube to express excess liquid, and then is swabbed evenly across the surface of a Mueller-Hinton agar plate. Usually, a plate 150 mm in diameter is used; it can accommodate testing of up to 12 different antimicrobial disks (placement of more than 12 disks on the plate may result in overlapping zones, which are difficult to measure and may produce erroneous results). Within 15 minutes of inoculation, the antimicrobial-containing disks are applied to the agar with a forceps for individual disks or with the aid of a multiple-disk dispenser (Fig. 3–11). The disks are pressed firmly to ensure contact with the agar. Within 15 minutes of disk placement, plates are inverted and placed in a 35°C ambient air incubator for 16 to 18 hours. Mueller-Hinton agar containing 5% sheep's blood is used for testing streptococci and sometimes other fastidious bacteria that do not grow adequately on unsupplemented Mueller-Hinton agar. Although incubation in an atmosphere of increased CO_2 is recommended for testing some fastidious bacteria, this should be avoided unless the impact on the results is known. Incubation in CO_2 results in a decreased pH, which affects the activity of some antimicrobial agents.

READING PLATES AND TEST INTERPRETATION

Following incubation, the plate is examined to make certain the test organism has grown satisfactorily. The lawn of growth must be confluent or almost confluent, and the appearance of individual colonies is unacceptable (Fig. 3–12). Provided that growth is satisfactory, the diameter

Figure 3–11

Cartridges containing antimicrobial susceptibility test disks are inserted into the dispenser *(right)*. The dispenser (which can hold up to 12 different cartridges of disks) is positioned over an inoculated plate, and light pressure is applied to the handle to simultaneously deposit one of each type of disk onto the plate. The tight-sealing container in the background contains a desiccant packet and is used for storage (at 2–8° C) of the dispenser containing a working supply of disks.

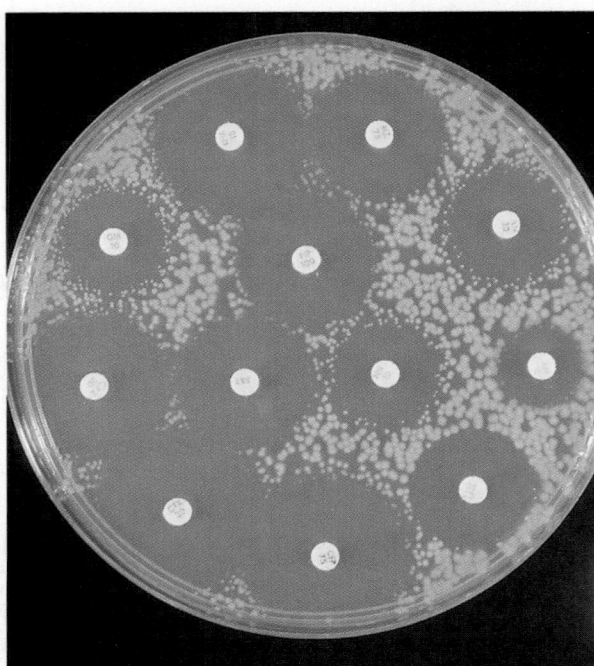

■ Figure 3–12

Escherichia coli tested by the disk diffusion method. The lawn of growth following overnight incubation shows individual colonies, representing unsatisfactory growth. The most likely explanation for the scanty growth is the use of an inoculum that either is too light or contains too many nonviable cells, resulting in larger than normal zones and potentially false susceptible results.

Transmitted light (holding plate up to light source) (Fig. 3–14) rather than reflected light must be used to examine zones for the penicillinase-resistant penicillins when testing staphylococci, and for vancomycin when testing enterococci (see later). Tests performed on media containing blood are examined from the top of the plate with the lid removed. For reading plates containing blood, it is important to read the zone of inhibition of growth and not the zone of inhibition of hemolysis.

Once zone measurements have been made, the millimeter reading for each antimicrobial agent is compared with that specified in the interpretive tables provided in the NCCLS documents, and results are interpreted as either susceptible, intermediate, or resistant. An excerpt from the chart is shown in Table 3–10. The equivalent MIC

of each inhibition zone is measured using a ruler or calipers. Plates are placed on or 2 to 3 inches above a black, nonreflecting surface, and zones are examined from the back side (agar side) of the plate illuminated with reflected light (Fig. 3–13). Tiny colonies at the zone edge and the swarm of growth into the zone that often occurs with swarming *Proteus* spp are ignored; the obvious zone is measured. As with dilution tests, the end point for the sulfonamides, trimethoprim, and trimethoprim/sulfamethoxazole is an 80% reduction of growth. Obvious colonies within a clear zone should not be ignored. These colonies may occur as a result of contamination; however, these colonies sometimes represent a small resistant subpopulation. When such colonies are noted, the original isolate should be retested, and the colonies within the zone should be subcultured to check for contamination. If repeat testing of the original isolate produces the same results, or repeat testing of the subcultured colonies from within the zone suggests resistance, the isolate should be reported as resistant.

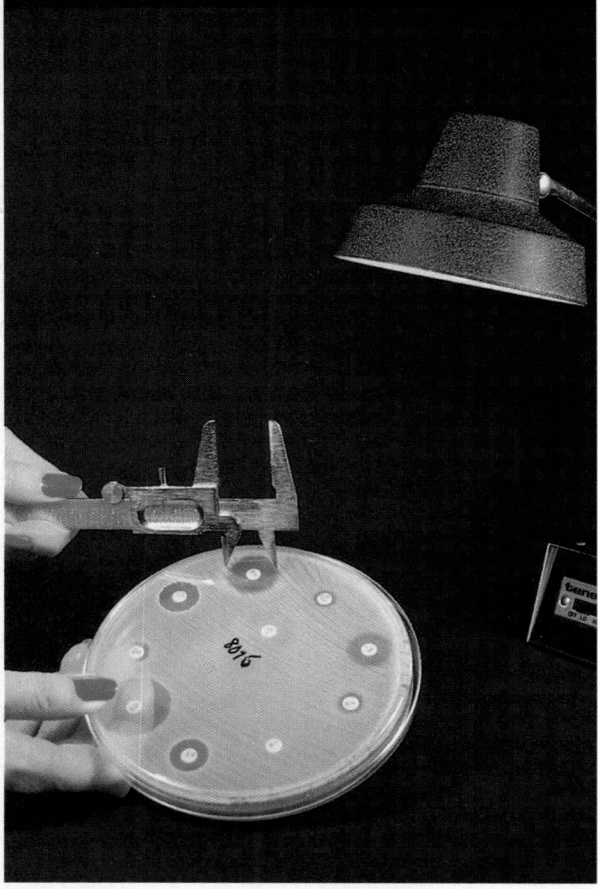

■ Figure 3–13

Routine disk diffusion tests are examined by placing the plate on or 2 to 3 inches above a black, nonreflecting surface. Reflected light is used to illuminate the plate.

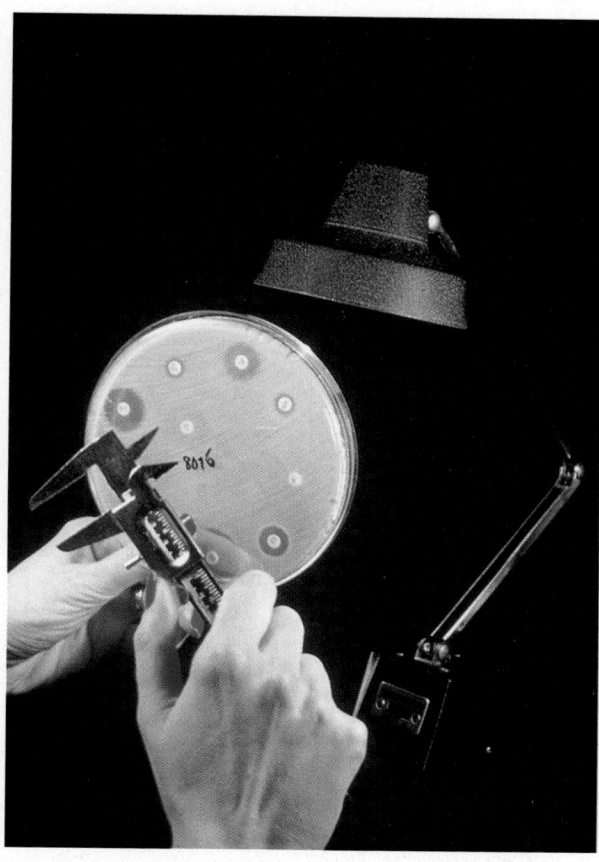

breakpoints that are used to define resistance and susceptibility are also shown. Note that as with MIC interpretive criteria, there may be several sets of interpretive criteria for some antimicrobial agents, which are specific for various organisms or organism groups.

A summary of the variables that must be carefully controlled in the performance of routine disk diffusion and broth microdilution MIC tests is listed in Table 3–11.

Modified Methods for Testing Slow-Growing or Fastidious Bacteria

Mueller-Hinton agar and broth are the standard media used for routine dilution and disk diffusion tests. These media, however, do not support the growth of all bacteria requiring antimicrobial susceptibility tests, and consequently, routine methods must be modified for testing fastidious bacteria that require supplemental nutrients, modified incubation conditions, or both.

Haemophilus influenzae and *Haemophilus* spp

Haemophilus test medium (HTM), which consists of Mueller-Hinton agar base supplemented with X (hematin) and V (nicotinamide adenine dinucleotide or NAD) factors, has been standardized for testing *Haemophilus influenzae* and other *Haemophilus* spp. HTM broth is used for

■■■■■ F i g u r e 3 – 1 4

Disk diffusion tests for staphylococci with oxacillin (or methicillin or nafcillin) and enterococci with vancomycin are examined by holding the plate up to a light source (transmitted light) for zone examination. Any growth within the zone is significant.

■■■■■ T A B L E 3 – 1 0

ZONE DIAMETER INTERPRETIVE STANDARDS AND EQUIVALENT MINIMUM INHIBITORY CONCENTRATION (MIC) BREAKPOINTS FOR ORGANISMS OTHER THAN *HAEMOPHILUS*, *NEISSERIA GONORRHOEAE*, AND *STREPTOCOCCUS PNEUMONIAE*

Antimicrobial Agent	Disk Content (µg)	Zone Diameter, nearest whole mm			Equivalent MIC Breakpoints (µg/mL)	
		Resistant	Intermediate	Susceptible	Resistant	Susceptible
Ampicillin						
when testing gram-negative enteric organisms	10	≤ 13	14–16	≥ 17	≥ 32	≤ 8
when testing staphylococci	10	≤ 28	—	≥ 29	β-Lactamase	≤ 0.25
when testing enterococci	10	≤ 16	—	≥ 17	≥ 16	—
when testing streptococci (not *S. pneumoniae*)	10	≤ 21	22–29	≥ 30	≥ 4	≤ 0.12
when testing *Listeria monocytogenes*	10	≤ 19	—	≥ 20	≥ 4	≤ 2
Cefazolin	30	≤ 14	15–17	≥ 18	≥ 32	≤ 8
Gentamicin	10	≤ 12	13–14	≥ 15	≥ 8	≤ 4
Oxacillin						
when testing staphylococci	1	≤ 10	11–12	≥ 13	≥ 4	≤ 2

Permission to use portions of M2–A5 (Performance Standards for Antimicrobial Disk Susceptibility Tests, Fifth Edition; Approved Standard) has been granted by the National Committee for Clinical Laboratory Standards. NCCLS is not responsible for errors or inaccuracies. The interpretive data are valid only if the methodology in M2-A5 is followed. These documents and current supplements may be obtained from NCCLS, 771 E. Lancaster Avenue, Villanova, PA 19085.

PRIMARY VARIABLES THAT MUST BE CONTROLLED IN PERFORMANCE OF ROUTINE DISK DIFFUSION AND BROTH MICRODILUTION MIC TESTS

Variable	Standard	Comments[†]
Inoculum	Disk diffusion: 1.5×10^8 CFU/mL Microbroth dilution: 5×10^5 CFU/mL (final concentration)	Use "adequate" McFarland turbidity standard (0.5 for disk diffusion) When preparing direct suspensions (without incubation), do not use growth from plates > 1 day old
Media Formulation	Mueller-Hinton	Prepare in-house or purchase from reliable source Perform media QC to verify acceptability prior to use for patient tests
Ca^{++}, Mg^{++} content	25 mg/L Ca^{++}, 12.5 mg/L Mg^{++}	Increased concentrations result in decreased activity of aminoglycosides against *P. aeruginosa* and decreased activity of tetracyclines against all organisms (decreased concentrations have the opposite effect)
Thymidine content	Minimal or absent	Excessive concentrations can result in false resistance to sulfonamides and trimethoprim
pH	7.2–7.4	Decreased pH can lead to decreased activity of aminoglycosides, erythromycin, and clindamycin, and increased activity of tetracyclines (increased pH has the opposite effect)
Agar depth (disk diffusion)	3–5mm	Possibility for false susceptibility if < 3mm, or false resistance if > 5 mm
Incubation Atmosphere	Humidified ambient air	CO_2 incubation decreases pH, which can lead to decreased activity of aminoglycosides, erythromycin, and clindamycin, and increased activity of tetracyclines
Temperature	34–35°C	Some MRSA may go undetected if > 35°C
Length	Disk diffusion: 16–18 hours Microbroth dilution: 16:20 hours (24 hours staphylococci with oxacillin* and enterococci with vancomycin; 24 hours sometimes needed for fastidious bacteria)	Some MRSA may go undetected if < 24 hours Some vancomycin-resistant enterococci may go undetected if < 24 hours with disk diffusion
Antimicrobial agents Disks	Use disks containing appropriate FDA/NCCLS-defined concentration of drug Proper Storage	Check NCCLS publication or FDA package insert (accompanying disks) for specifications Long-term storage in *non*–frost-free freezer at ≤–20°C in tightly sealed, desiccated container Short-term storage (at least 1 week) at 2–8°C in tightly sealed, desiccated container Allow to warm to room temperature before opening container
	Proper placement on agar	12 or fewer disks per 150-mm plate (no overlapping zones)
Solutions	Prepare from reference standard powders	Pharmacy-grade antimicrobial agents are unacceptable, as they may not show antimicrobial activity in vitro
	Proper storage	Store in *non*-frost-free freezer optimally at ≤–70°C Never refreeze
End point measurement Disk diffusion	Use reflected light, and hold plate against black background Measure zones from back of plate (except for staphylococci and oxacillin* and for vancomycin and enterococci)	Lawn must be confluent or almost confluent Ignore faint growth of tiny colonies at zone edge Trimethoprim and sulfonamides—end point at ≥ 80% inhibition Ignore swarm within obvious zone for swarming *Proteus* spp Retest colonies within zone (except staphylococci and oxacillin, enterococci and vancomycin)
	Use transmitted light for staphylococci and oxacillin* and for enterococci and vancomycin	Call "resistant" if *any* growth within zone (unless possibly artifactual or contaminated) Reproducibility is within ± 2 mm
Microbroth dilution	Use adequate lighting and reading device	MIC = lowest concentration that inhibits growth (turbidity, haze, or pellet) Sulfonamides and trimethoprim may trail (ignore trailing <2 mm buttons) Justify "skip wells" or repeat Staphylococci and penicillin—perform induced β-lactamase test if MIC is 0.06–0.12 µg/mL Reproducibility within ± 1 two-fold dilution

* Includes all penicillinase-resistant penicillins (oxacillin, methicillin, nafcillin, cloxacillin, and dicloxacillin).
[†] MRSA, methicillin-resistant *Staphylococcus aureus*.
Modified with permission from Hindler JA, Mann LM: Principles and practices for the laboratory guidance of antimicrobial therapy. In Tilton RC, Balows A, Hohnadel DC, Reiss RF (eds): Clinical Laboratory Medicine. St. Louis: Mosby, 1992, p 548.

broth dilution tests, and HTM agar is used for disk diffusion tests. The test procedures for *Haemophilus* spp are identical to those described for nonfastidious bacteria, with the exception that the disk diffusion test with HTM is incubated in an atmosphere of 5 to 7% CO_2. The NCCLS has established zone diameter and MIC interpretive criteria that are unique for this genus. For some agents, such as cefotaxime, only a susceptible range is defined. Again, this is because cefotaxime-resistant *Haemophilus* spp have not been identified, and consequently, criteria to identify such resistance could not be established. Further tests, possibly in a reference laboratory, should be performed on any *Haemophilus* spp isolate that is interpreted as other than susceptible to cefotaxime. These recommendations hold true for all drugs for which only susceptible criteria are specified by NCCLS.

Ampicillin or amoxicillin is often effective in treating localized, less serious *H. influenzae* infections; however, 20 to 50% of *H. influenzae* produce a β-lactamase that inactivates these agents. β-Lactamase–producing strains can be quickly identified by using a rapid β-lactamase test, as described later. *Haemophilus influenzae* also may be resistant to ampicillin and amoxicillin because of altered penicillin-binding proteins, but this resis- tance occurs in less than 1.0% of clinical isolates. Because most ampicillin-resistant (amoxicillin-resistant) *H. influenzae* produce β-lactamase, and because *H. influenzae* are often susceptible to alternative agents currently recommended, some laboratories do not routinely perform disk diffusion or MIC tests but perform only a β-lactamase test unless the isolate is from blood or CSF.

Streptococcus pneumoniae and Streptococcus spp

Streptococcus pneumoniae and *Streptococcus* spp require a more nutritious medium for antimicrobial susceptibility testing; they will not grow satisfactorily on unsupplemented Mueller-Hinton medium. Broth dilution tests are usually performed in Mueller-Hinton broth that has been supplemented with 2 to 5% lysed horse's blood. Agar dilution and disk diffusion tests are performed using Mueller-Hinton agar supplemented with 5% sheep's blood.

Penicillin remains the drug of choice for treating pneumococcal infections; however, resistance to penicillin as well as to other potentially useful agents is becoming widespread. Penicillin resistance is due to the presence of altered penicillin-binding proteins and is not enzymatically mediated. Consequently, β-lactamase testing is inappropriate for pneumococci.

The disk diffusion test can be used to screen for penicillin susceptibility in *S. pneumoniae*. An oxacillin disk (1 μg) rather than a penicillin disk must be used, however, for reliable detection of penicillin resistance. Following the standard disk diffusion procedure, an isolate is tested on Mueller-Hinton agar with 5% sheep's blood and incubated in CO_2. If the oxacillin zone of inhibition is 20 mm or larger, the isolate is reported as penicillin (not oxacillin) susceptible. If the oxacillin zone is 19 mm or smaller, a penicillin (not oxacillin) MIC test must be performed to clarify the level of resistance (Fig. 3–15). Penicillin MICs of 0.06 μg/mL or less are interpreted as susceptible, of 0.12 to 1.0 μg/mL as intermediate, and 2.0 μg/mL or more as resistant. It is important to determine the degree of penicillin resistance, because the recommended therapy may be different for intermediate and resistant strains. It is also important to test penicillin-resistant isolates with other clinically relevant antimicrobial agents, such as cefotaxime or ceftriaxone, and perhaps erythromycin.

Accurate penicillin susceptibility results are needed for viridans streptococci isolated from the blood of patients with bacterial endocarditis. If the isolate has a penicillin MIC of 0.12 μg/mL or less, penicillin alone is often prescribed;

■ Figure 3 - 1 5

An oxacillin (1-μg) disk is used to screen for penicillin susceptibility in *Streptococcus pneumoniae*. If the oxacillin zone of inhibition is ≥ 20 mm, the isolate is reported as susceptible. If the oxacillin zone is ≤ 19 mm, a penicillin MIC test must be performed. The isolate pictured has an oxacillin zone of approximately 15 mm, so a penicillin MIC test must be performed.

however, higher penicillin MICs (.25–2.0 µg/mL) suggest the need for concomitant therapy with an aminoglycoside. Isolates with penicillin MICs greater than 2.0 µg/mL are highly resistant, and for these vancomycin rather than penicillin is generally prescribed. Because of the critical nature of penicillin results, the disk test is not recommended in these situations, and dilution tests should be performed. As with *S. pneumoniae,* penicillin resistance in viridans streptococci is due to altered penicillin-binding proteins, and β-lactamase testing should not be performed on this group of organisms.

Neisseria gonorrhoeae and *Neisseria meningitidis*

GC agar base is supplemented with various nutrients for testing *Neisseria gonorrhoeae.* Dilution tests are performed using agar dilution, because this species has a tendency to lyse in broth media, resulting in false-susceptible results. Disk diffusion tests are performed on the same agar, and all tests are incubated in an atmosphere containing 5 to 7% CO_2. NCCLS has specified interpretive criteria unique for this species. For several agents for which resistant isolates have not yet been encountered, only susceptible criteria are available.

Although penicillin had been the drug of choice for treating uncomplicated gonorrhea for many years, the increased incidence of penicillin-resistant isolates has led to the use of ceftriaxone or a fluoroquinolone as first-line therapy. Penicillin resistance in *N. gonorrhoeae* may be due to production of a β-lactamase very similar to that produced by ampicillin-resistant *H. influenzae,* and this resistance can be readily detected with a rapid β-lactamase test. β-lactamase–producing isolates are also referred to as penicillinase-producing *N. gonorrhoeae,* or PPNG. Some *N. gonorrhoeae* are penicillin resistant due to an altered penicillin-binding protein; this resistance can be detected only with conventional dilution or disk diffusion tests. The production of altered penicillin-binding proteins is chromosomally mediated, and *N. gonorrhoeae* with altered penicillin-binding proteins are often referred to as chromosomally mediated resistant *N. gonorrhoeae,* or CMRNG. Because current therapeutic recommendations do not include penicillin and because virtually all *N. gonorrhoeae* are currently susceptible to ceftriaxone and the recommended flouroquinolones, testing is rarely indicated in the routine clinical laboratory. Some laboratories continue to per-

form β-lactamase tests, however, and public health laboratories test multiple agents against *N. gonorrhoeae* for surveillance purposes.

Except for very rare isolates that have been shown to produce β-lactamase, *Neisseria meningitidis* remains susceptible to penicillin, the drug of choice for treating infections caused by this organism. Isolates with slightly elevated penicillin MICs have, however, been observed. These have not caused major problems as yet, because the penicillin MICs are still below penicillin levels attainable in CSF, although they are higher than those for normally susceptible isolates. Meningococci are often resistant to sulfonamides and occasionally resistant to rifampin. Either of these agents is administered prophylactically to individuals who have close contact with patients with meningococcal meningitis. There are currently no standard procedures for performing either disk diffusion or MIC tests on meningococci.

Anaerobes

The reference method described by the NCCLS for testing anaerobic bacteria is an agar dilution method, and the recommended medium is Wilkens-Chalgren agar. As previously mentioned, however, agar dilution is not practical for use in the routine clinical laboratory, and a broth microdilution method is more often used. The broth microdilution procedure is similar to that used for testing aerobes except for the Wilkens-Chalgren broth, which supports the growth of anaerobes. Additionally, the number of organisms in the test inoculum is 0.5 $\log_{10}$ higher (10^6 CFU/mL) than that for testing aerobes, and trays are incubated anaerobically at 35°C for 48 hours. As with tests for aerobic bacteria, the NCCLS has defined susceptible, intermediate, and resistant criteria for interpretation of MICs for anaerobes. Additionally, several commercial companies provide broth microdilution MIC panels for testing anaerobes.

Broth disk elution was another method in widespread use in the past decade because of ease of test performance. With this method, a preset number of antimicrobial susceptibility disks (the same as those commonly used for the disk diffusion test) were dispensed into a tube of anaerobic broth. Once the drug diffused from the disks, a standardized inoculum of the test bacterium was added, and the tubes were incubated anaerobically; aerobic incubation was acceptable if the test medium was thioglycollate. It was demonstrated, however, that this method did not always produce results that compared favorably with those of reference methods. NCCLS no longer recommends the use of the broth disk elution method.

The E test, discussed in detail later, has been shown to perform satisfactorily for antimicrobial susceptibility testing of anaerobes, and this method is used in some clinical laboratories.

Additional Organism and Antimicrobial Agent Testing Concerns

Some special procedures must be employed to detect clinically significant resistance in nonfastidious bacteria.

Detection of Oxacillin (Methicillin) Resistance in Staphylococci

Oxacillin and other penicillinase-resistant penicillins, such as methicillin, nafcillin, cloxacillin, and dicloxacillin, constitute the drug class of choice for treating staphylococcal infections. Oxacillin is the class representative most commonly used to detect resistance in staphylococci and produces the most reliable results. When an isolate shows resistance to any of the penicillinase-resistant penicillins, however, it must be considered resistant to the entire group. Staphylococcal resistance to the penicillinase-resistant penicillins is due to the presence of a unique penicillin-binding protein (2a or 2′) on the surface of resistant cells. The penicillin-binding protein, which has a low affinity for penicillinase-resistant penicillins, is encoded by a gene referred to as *mec*. Detecting oxacillin resistance in isolates that possess the *mec* gene may be difficult under the standard test conditions as described previously, because staphylococci sometimes exhibit heteroresistance in their response to oxacillin. In *heteroresistant* strains, all cells in the test population have the genetic elements (the *mec* gene) for oxacillin resistance, but not all of the cells may express this resistance. Consequently, in the susceptibility test, some cells appear resistant and some appear susceptible. If too few cells appear resistant, an oxacillin-resistant strain may go undetected.

In vitro testing conditions can be modified to enhance the expression of oxacillin resistance; they are as follows:

- Preparation of inocula using the direct inoculum suspension procedure

- Incubation of tests at temperatures no greater than 35°C

- Obtaining final test readings after a full 24 hours of incubation

- Supplementation of Mueller-Hinton broth or agar with 2% NaCl for dilution tests

The extended incubation allows the slower-growing resistant subpopulation sufficient time to grow to detectable numbers. In addition, test plates should always be examined very closely. For disk tests, zones of inhibition must be examined by using transmitted light (holding plate up to the light source; see Fig. 3–14), and *any* growth is considered significant. It is not uncommon to observe a "haze" of growth within the inhibition zone for oxacillin-resistant isolates (Fig. 3–16).

The clinical significance of oxacillin-resistant or methicillin-resistant *S. aureus* (MRSA) is heightened by the fact that these isolates are usually resistant to other antistaphylococcal agents (clindamycin, erythromycin, tetracycline, and sometimes gentamicin and trimethoprim/sulfamethoxazole), with the exception of vancomycin. Sometimes, oxacillin-resistant staphylococci appear susceptible in vitro to other β-lactam agents, such as the cephalosporins; however, these are clinically ineffective. Consequently, all oxacillin-resistant staphylococci must be reported as resistant to all β-lactam agents (including cephalosporins, β-lactam/β-lactamase inhibitor combinations, and imipenem) regardless of the in vitro test results.

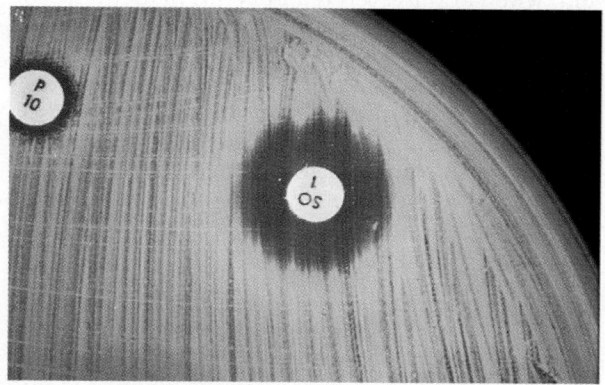

■ F i g u r e 3 – 1 6

The oxacillin zone for heteroresistant oxacillin-resistant *Staphylococcus aureus* often shows a haze of growth within the zone of inhibition. This haze is significant, and the isolate here is oxacillin resistant.

An oxacillin screen plate that contains Mueller-Hinton agar supplemented with 4% NaCl and 6 µm/mL oxacillin has been used to detect oxacillin-resistant staphylococci. To perform the oxacillin screen test, a McFarland 0.5 suspension is prepared as for the disk diffusion test. A swab is dipped into this suspension and streaked over an area of approximately 2 × 5 cm or deposited as a "spot" on the agar surface. Following overnight incubation at 35°C *any* growth is an indication that the isolate is oxacillin resistant.

Some staphylococci have a more subtle and less common type of oxacillin resistance that is unrelated to the presence of the *mec* gene. The resistance mechanism in these isolates is due to either hyperproduction of β-lactamase or the presence of altered penicillin-binding proteins (not related to 2a or 2′). These strains generally have MICs right above (or zones of inhibition right below) the breakpoint for susceptibility, and they are sometimes referred to as *borderline-resistant* isolates. Additionally, these isolates are usually susceptible to other antistaphylococcal agents. The clinical response of isolates with borderline oxacillin resistance to penicillinase-resistant penicillins as well as to other β-lactam agents is not as well defined as for the *mec* gene–positive isolates. Isolates with borderline resistance generally do not grow on oxacillin screen plates.

Enterococci

Ampicillin or penicillin is effective for treating uncomplicated enterococcal infections (e.g., urinary tract infections). These cell wall–active agents are only bacteriostatic against enterococci and, when administered alone, are inadequate for treating serious infections, such as endocarditis, that require bactericidal therapy. To obtain a bactericidal effect, ampicillin or penicillin (or vancomycin in the penicillin-allergic patient) must be given in combination with an aminoglycoside, usually gentamicin and sometimes streptomycin.

DETECTION OF HIGH-LEVEL AMINOGLYCOSIDE RESISTANCE IN ENTEROCOCCI

Enterococci are inherently resistant to low concentrations of aminoglycosides, precluding their use as single agents for treatment of enterococcal infections. This low-level resistance is due to poor drug uptake by the enterococcal cells. For isolates with low-level aminoglycoside resistance, a synergistic interaction occurs when an aminoglycoside is administered together with a cell wall–active agent such as ampicillin, penicillin, or vancomycin. Sometimes, however, enterococci develop high-level aminoglycoside resistance, in which the particular aminoglycoside does not demonstrate synergism with the cell wall–active agent (ampicillin, penicillin, or vancomycin). High-level aminoglycoside resistance in enterococci is usually due to enzymatic inactivation of the drugs, and the enzymes that destroy gentamicin also destroy tobramycin, amikacin, kanamycin, and netilmicin, as shown in Table 3–12. Consequently, none of these agents would be used in treating infections caused by enterococci with high-level gentamicin resistance. If the isolate does not have concomitant high-level streptomycin resistance, however, streptomycin could be used, although some isolates have high-level resistance to both gentamicin and streptomycin. These latter isolates represent a significant therapeutic dilemma when encountered in serious infections, and the most appropriate strategy for treating infections caused by them has yet to be defined.

In vitro tests are used to detect high-level aminoglycoside resistance. Disk diffusion tests have been described that utilize special disks containing very high concentrations of gentamicin (120 µg/mL) or streptomycin (300 µg/mL). More commonly, routine agar or broth dilution types of tests are used. Gentamicin is tested at concentrations of 500 µg/mL and streptomycin at 2000 µg/mL (agar) or 1000 µg/mL (broth). The tests are performed as described for routine dilution tests, and growth at

■■■ T A B L E   3 – 1 2

AMINOGLYCOSIDES REPRESENTED WHEN HIGH CONCENTRATIONS OF GENTAMICIN AND STREPTOMYCIN ARE TESTED TO DETERMINE HIGH-LEVEL AMINOGLYCOSIDE RESISTANCE IN ENTEROCOCCI

Agent Tested	Aminoglycoside(s) Represented
Gentamicin*	Gentamicin Amikacin Kanamycin Netilmicin Tobramycin
Streptomycin[†]	Streptomycin

* An isolate that has high-level resistance to gentamicin also has high-level resistance to the other aminoglycosides listed; enzymes that inactivate gentamicin also inactivate the other agents.
† An isolate that has high-level resistance to streptomycin has high-level resistance to this agent only (unless it also has high-level resistance to gentamicin as indicated by testing high concentrations of gentamicin).

the high concentration indicates that the isolate has high-level resistance to the agent tested.

DETECTION OF AMPICILLIN AND PENICILLIN RESISTANCE IN ENTEROCOCCI

Ampicillin and penicillin resistance in enterococci may be due to altered penicillin-binding proteins and, rarely, to β-lactamase production. Resistance due to altered penicillin-binding proteins, which is readily detected by routine dilution and disk diffusion tests, occurs infrequently in *Enterococcus faecalis* but is common in *Enterococcus faecium*. Because of difficulties in obtaining a "resistant" result with the standard inocula used in routine dilution and disk diffusion tests, a rapid β-lactamase test is needed to identify β-lactamase–producing enterococci.

DETECTION OF VANCOMYCIN RESISTANCE IN ENTEROCOCCI

The incidence of vancomycin resistance in enterococci is increasing. Isolates that are highly resistant to vancomycin can be readily detected as vancomycin resistant when tested by conventional antimicrobial susceptibility test methods. Some isolates, however, may have a more subtle type of vancomycin resistance whereby MICs or inhibition zone measurements are just slightly above or below, respectively, the susceptible breakpoints. Hence, dilution tests must be viewed closely, and inhibition zones in disk diffusion testing must be examined using transmitted (rather than reflected) light; any growth within the zone should be considered significant.

The lower-level vancomycin resistance appears to be inherent in some infrequently encountered enterococcal species, including *Enterococcus gallinarum* and *Enterococcus casseliflavus*. Additionally, *Leuconostoc* spp, *Pediococcus* spp and *Lactobacillus* spp inherently demonstrate high-level vancomycin resistance, and these bacteria should not be confused with the morphologically similar enterococci.

AUTOMATED ANTIMICROBIAL SUSCEPTIBILITY TEST METHODS

Principles of Technologies Used

The automated susceptibility testing instruments that are currently available represent a choice of several different levels of automation. Some instruments interpret growth end points of broth microdilution panels only when they are placed into an automated reader device, whereas certain other instruments provide hands-off incubation and reading functions for microdilution trays or special cards in an incubator/reader device. The instruments that offer the highest level of automation generally accomplish these tasks through the use of robotics to move the trays or cuvettes during the incubation/reading sequences in the instrument or to add reagents to certain test wells for biochemical tests.

Instruments also vary in the optical methods used for examining the test wells of the antimicrobial-containing trays or cards. Most current instruments utilize the principle of turbidimetric detection of bacterial growth in a broth medium by use of a photometer to examine the test wells. The determination of antimicrobial susceptibility based upon lack of development of tubidity (suppression of growth) or, conversely, an indication of resistance based on an increase in turbidity in the presence of an antimicrobial agent is the same principle as that used when manual interpretation of growth endpoints is performed. The second means of growth detection, used by two instruments, is the detection of hydrolysis of a fluorogenic growth substrate incorporated in a special test medium. With this technology, growth is detected by a fluorometer as emission of a fluorescent signal when a microorganism consumes fluorophore-labeled substrate in the test medium.

Instruments for antimicrobial susceptibility testing may function to provide assistance in the interpretation of test results following a conventional overnight incubation period, or they may allow results to be determined in a shortened analysis period of 4 to 8 hours. Instrumentation may allow the interpretation of antimicrobial susceptibility test endpoints sooner than manual readings, because of the greater sensitivity of the instruments' optical systems to detect subtle changes in microbial growth. All of the instruments rely heavily on microprocessor-controlled functions and utilize personal computer hardware to provide final printed reports and to store and retrieve data on antimicrobial susceptibility. Most of the instruments may also be used to perform identifications of gram-negative or gram-positive bacteria and may be able to merge and print identification and antimicrobial susceptibility results into a single report.

Currently Available Automated Systems

Automated Reader Devices for Broth Microdilution Susceptibility Tests

Virtually every manufacturer of broth microdilution antimicrobial susceptibility testing panels offers a view box or other device to facilitate manual interpretation of results following incubation. Those products that feature freeze-dried antimicrobial panels also offer a mechanized device to simplify hydration and inoculation of trays. In addition, most manufacturers offer an instrument-assisted reader device that allows the technologist to record the results of manual readings of the panels by use of a video display screen resembling the configuration of the tray, or, alternatively by use of a touch-sensitive template that overlies the microdilution tray (e.g., Baxter MicroScan TouchSCAN, Pasco Data Management System). These reader devices facilitate data recording and provide a computer-printed report as well as long-term data storage. The personal computers included with these systems provide the capability of data storage and retrieval for periodic generation of cumulative susceptibility profiles for various organisms that have been tested.

Automated photometers, which interpret growth patterns in panels by turbidimetric analysis, represent the next level of instrumentation available from several manufacturers; the API UniScept Autoreader (BioMerieux Vitek, Hazelwood, MO), the Baxter MicroScan AutoSCAN-4 (Baxter Healthcare Corp., West Sacramento, CA), and the autoSceptor (Becton Dickinson Microbiology Instrument Systems, Towson, MD). These instruments provide automated interpretation of the results of broth microdilution susceptibility tests following overnight incubation in a standard laboratory incubator. They are also configured with personal computers for report printing and data storage. Therefore, this level of automation involves only the final, reading step of the microdilution susceptibility panels. A published evaluation of one of these instruments has shown that the interpretation of MIC end points can be accomplished by the instrument with reasonable accuracy.

Automated Instruments That Provide More Rapid Test Results

Only two currently available instruments are capable of generating rapid (4 to 8 hours) sus-

ceptibility test results. Three previously marketed instruments (the Autobac, the Abbott MS-2 or Avantage, and the Sensititre 5 h fluorogenic system) are no longer available. The Sensititre fluorogenic substrate test with AutoReader (Radiometer America Inc., Westlake, OH) was the first system to be marketed in the U.S. that incorporated fluorogenic substrate hydrolysis as its method of detecting bacterial growth during an antimicrobial susceptibility test. The use of a fluorometric detection system was intended for testing of Enterobacteriaceae and some *S. aureus* using either MIC or breakpoint formats following a 5-hour or 18-hour incubation period. Problems were reported, however, with the 5-hour version during testing of *Proteus mirabilis, P. aeruginosa,* MRSA, coagulase-negative staphylococci, and enterococci. Marketing of the 5-hour version of the Sensititre has been discontinued, although the 18-hour fluorogenic substrate version of the instrument is still available.

AutoSCAN Walkaway

The Baxter MicroScan AutoSCAN Walkaway (Baxter Healthcare Corp., West Sacramento, CA) consists of a large, self-contained incubator/reader unit with capacities for either 40 or 96 test panels and a personal computer with video display terminal and printer (Fig. 3–17). The Walkaway utilizes standard-size microdilution trays, which are hydrated and inoculated with a hand-operated inoculator device, then placed in one of the positions in the incubator module. The type of test to be performed is indicated on

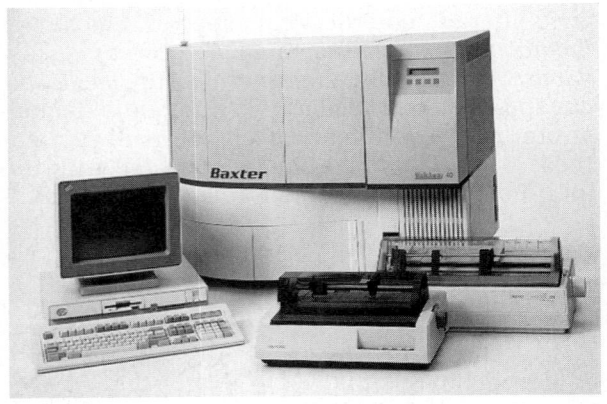

■■■■■ F i g u r e   3 – 1 7
AutoSCAN[R]–W/A or for 96 (pictured) Walkaway™ instrument. (Courtesy of Baxter Diagnostics Inc., MicroScan, West Sacramento, CA.)

an instrument-readable bar-code label placed on the end of each tray. The instrument then incubates the trays for the appropriate period (depending on the type of panel and organism), robotically positions the trays to add reagents if needed, and moves them under the central photometer/fluorometer station to perform the final readings of growth endpoints at the conclusion of the tests.

The Walkaway offers a choice of either overnight incubation with conventional photometric detection of turbidity for MIC or breakpoint testing of gram-positive and gram-negative bacteria or rapid (3.5–7 hours for gram-negative; 3.5–15 hours for gram-positive) readings of special MIC or breakpoint panels incorporating fluorogenic substrates. Special "combo" trays are available that allow susceptibility and organism identification in the same tray using either conventional substrates (overnight incubation) or special fluorogenic substrates that allow identification of some gram-negative bacilli in only 2 hours.

VITEK SYSTEM

The Vitek System (BioMerieux Vitek, Hazelwood, MO) was originally designed for use in the U.S. space exploration efforts of the 1970s as an onboard test system for spacecraft exploring other planets for life. Because of its original design intention, it was highly automated and relatively compact (Fig. 3–18). Small plastic reagent cards (similar in size to a credit card) contain microliter quantities of various biochemical test media in 30 test wells for organism identification (Fig. 3–19). Other cards contain various

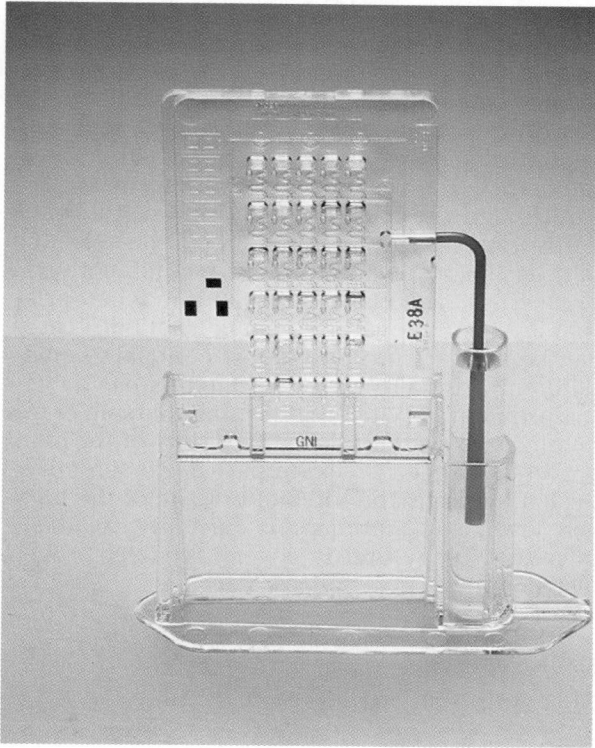

■ F i g u r e   3 – 1 9

Vitek test card containing dried antimicrobials connected by a transfer straw to a tube containing a standardized inoculum suspension of bacteria to be tested. These are placed into an evacuating chamber, which allows the inoculum to be transferred into the card. All wells containing varying concentrations of various antimicrobial agents are reconstituted during this inoculation process. Finally, the inoculated cards are placed into the Vitek instrument for processing.

concentrations of a variety of antimicrobial agents for susceptibility testing. The Vitek can be configured to accommodate 30, 60, 120, or 240 cards. The susceptibility cards allow either quantitative MIC or qualitative susceptible/intermediate/resistant results for most rapidly growing gram-positive and gram-negative aerobic bacteria in a period of 4 to 10 hours.

Vitek hardware consists of a filling module for inoculation of the cards, an incubator/reader module that incorporates a carousel to hold the test cards, a robotic system to manipulate the cards, a photometer for measurement of optical density and biochemical reaction color changes in the cards, and a computer module with video display terminal and printer for viewing and printing results. Vitek also offers an information management system for storing and retrieving test data for a variety of statistical reports. The

■ F i g u r e   3 – 1 8

Vitek instrument. (Courtesy of BioMerieux Vitek, Hazelwood, MO.)

Vitek uses kinetic (every 60 minutes) measurements of growth in the presence of antimicrobial agents to provide analysis of growth curves, leading to computer algorithm–derived MIC values. The technologist can choose to have susceptibility test results printed using either qualitative (S, I, R) or quantitative (MIC) formats.

Although the fixed configuration of the Vitek cards imposes some limitations on the selection of drugs for testing, Vitek offers a combination of two cards (called Flex cards) at a reduced price that are tested together; results from both cards can be merged into one report containing many different drugs. To some degree, this has solved the problem for Vitek that is inherent in all of the commercial systems, i.e., inflexibility of the standard test panels. Owing in part to its aerospace design heritage, the Vitek offers one of the highest levels of automation currently available in microbiology and is a very reliable, proven instrument.

Newer Nonautomated Antimicrobial Susceptibility Test Methods

Three relatively new products have been made available that differ slightly or significantly in principle from the test methods described thus far in this chapter. Whereas one of them utilizes an instrument to initialize a susceptibility test, the other two tests are interpreted manually.

Alamar System

The Alamar System (Alamar, Sacramento, CA) utilizes special oversize but lightweight microdilution trays configured with 168 wells, each containing a filter paper disk attached to the bottom of the tray well. The disks contain different twofold concentrations of up to 20 different antimicrobial agents in the same tray for testing gram-positive or gram-negative bacteria. The disks are also impregnated with a redox indicator (Alamar Blue), which is blue when the bacterial inoculum is first added but turns bright pink when bacterial growth occurs in the well (Fig. 3–20). MICs are interpreted manually by noting the last blue well in a two-fold antimicrobial dilution series as evidence that growth has been inhibited. In addition to the simplicity of interpreting the growth end points by use of the indicator, the use of the paper disks allows the manufacturer to offer custom-configured test panels very easily and at a competitive price. Data given in the manufacturer's package insert indicate excellent correlation between MICs determined using the Alamar panels and those generated by a conventional agar dilution method.

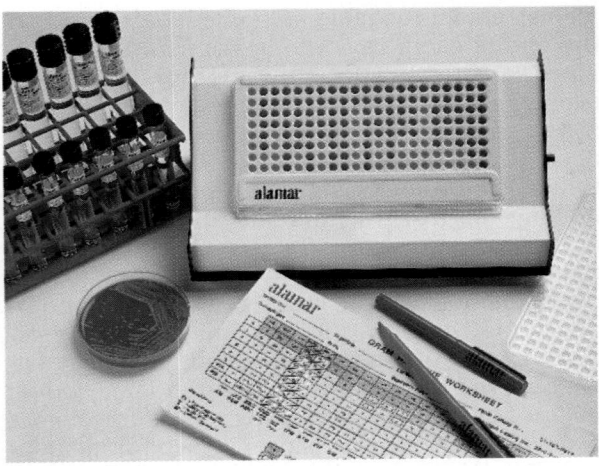

■ F i g u r e   3 – 2 0

Alamar MIC test system. The MIC panels are inoculated in a manner similar to that described for the standard broth microdilution test procedure. Following overnight incubation, the pink wells indicate growth and the blue no growth. The MIC is read as the lowest concentration that shows a blue color.

E Test

Both the E test (AB Biodisk, Solna, Sweden) and the Spiral Gradient Endpoint System (discussed later) utilize the principle of establishing an antimicrobial density gradient in an agar plate as a means of determining antimicrobial susceptibility. The E test utilizes very thin plastic test strips that are impregnated on the under surface with an antimicrobial concentration gradient and are marked on the upper surface with a concentration index or scale. The strips may be placed in a radial fashion on the surface of an agar plate that has been inoculated in a manner similar to that for a disk diffusion test. After overnight incubation, the test results are read by viewing the plates from the top side with the lids removed. The antimicrobial gradient that forms in the agar around the E test strips gives rise to elliptical inhibitory areas with each strip. The MIC is determined where the growth ellipse intersects the E test strip (Fig. 3–21). The E test shares with the disk diffusion test the intrinsic flexibility of

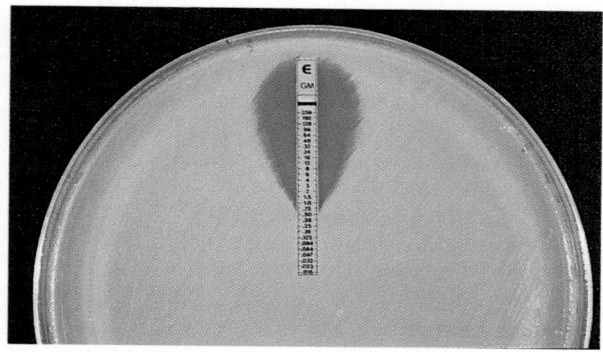

■■■■■ F i g u r e   3 – 2 1

Escherichia coli tested with an E test gentamicin strip. The gentamicin MIC (where the ellipse crosses the gradient) is 0.75 µg/mL.

drug selection and testing because selected strips are applied to the surface of test plates. The cost of E test strips is much greater than that of disks, however, and represents one of the main limitations of this product. Initial published studies have indicated that MICs determined using the E test compare favorably (within one twofold dilution interval) with those determined by conventional methods.

The E test may be especially useful for testing fastidious organisms such as *H. influenzae, S. pneumoniae,* and anaerobic bacteria. This is in part due to the fact that the strips can be placed on special enriched media or in special incubation atmosphere (e.g., anaerobic or increased CO_2), and the fact that fewer antimicrobial agents may need to be tested against fastidious organisms. Consequently, the relatively high cost of the E test strips would be minimized.

Spiral Gradient Endpoint System

The Spiral Gradient Endpoint System (Spiral Biotech Co., Bethesda, MD) involves the generation of an antimicrobial test gradient in an agar plate in a spiral fashion, beginning in the center of the plate and moving concentrically outward by use of the Spiral Gradient instrument. Test isolates are then applied to the surface of the plate in a radial orientation perpendicular to the antimicrobial gradient. Following 24 hours or more of incubation, quantitative growth endpoints are determined by measuring the distance of growth of the test organisms from the edge of the plate (Fig. 3–22). This method may be especially useful for testing anaerobic bacteria, because it negates

the problems of trailing or ill-defined growth end points that occur with some drugs using conventional broth or agar test methods. The considerable cost and complexity of this instrument, however, will likely restrict its use to only the largest and most specialized laboratories.

INTERPRETATION OF IN VITRO ANTIMICROBIAL SUSCEPTIBILITY TEST RESULTS

A number of elements are important in performing any in vitro antimicrobial susceptibility test. This chapter has described the methodology for performing several types of tests. The procedural steps of each method must be followed explicitly in order to obtain reproducible results. A susceptibility test should never be performed using an unstandardized inoculum or a mixed culture. For that reason, direct susceptibility tests that incorporate the use of a patient's infected body fluid cannot be recommended, even to obtain a presumptive result.

The results of a susceptibility test must be interpreted in the laboratory prior to communicating a report to a patient's physician. In the

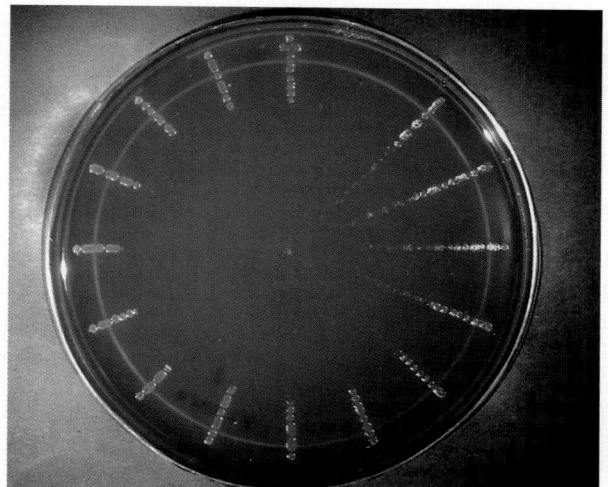

■■■■■ F i g u r e   3 – 2 2

Fifteen anaerobe isolates tested on a plate containing clindamycin. Clindamycin is deposited onto the plate in concentric circles using the Spiral Gradient instrument. The highest concentration of antimicrobial agent is in the center of the plate, and the lower concentrations are at the periphery. After the drug dries, each isolate is inoculated in a manner that looks like spokes on a wheel. The value representing the distance from the edge of the plate to the place where growth begins is inserted into an equation to calculate the MIC of each isolate.

case of a disk diffusion test, the inhibition zone size must be interpreted using a table of values that relates the diameter of the zone to a category of susceptibility, which is sometimes related to the identity of the isolate for correct interpretation. The table used for such interpretations must represent the most up-to-date criteria that have been reviewed and accepted by the NCCLS. It is important to bear in mind that the NCCLS documents are updated frequently, usually once per year. Use of old or outdated NCCLS tables could represent a serious shortcoming in the reporting of patients' results.

The inhibition zone size and MIC interpretive criteria published by the NCCLS are established by careful analysis of three kinds of data—microbiologic data (e.g., a comparison of MICs versus zone sizes on a large number of bacterial strains), pharmacokinetic data (e.g., serum, CSF, urine, and other secretion and tissue levels of an antimicrobial agent), and results of clinical studies obtained during the phase prior to FDA approval and marketing of an antimicrobial agent. Thus, MIC interpretive criteria are not based simply on a comparison of serum levels of an antimicrobial agent and MIC values. Zone size interpretive criteria are, however, based in large part on direct correlations of MICs and zone sizes.

Whether based on determination of an MIC or on interpretation of a disk diffusion zone size, the three categories of susceptibility currently recommended by the NCCLS should be interpreted in the same manner. If the MIC or zone size is interpreted as "susceptible" using the latest NCCLS tables, the clinical interpretation of the result is that the patient's infecting organism should respond to therapy with that antimicrobial agent using the dosage recommended normally for that type of infection and that species. Conversely, a MIC or zone size interpreted as "resistant" should not be inhibited by the normally achievable concentrations of the antimicrobial agent based on the dosages normally used with that drug. An "intermediate" result indicates that a microorganism falls into a range of susceptibility in which the MIC approaches or exceeds the level of antimicrobial agent that can ordinarily be achieved and for which clinical response is likely to be less than with a susceptible strain. Exceptions can occur if the antimicrobial agent is highly concentrated in a body fluid, such as urine, or if higher than normal doses of the antimicrobial agent can be safely administered (e.g., some penicillins and cephalospor-

ins). At times, the "intermediate" result means that certain of the variables in the susceptibility test may not have been well controlled and that the values have fallen into a "buffer zone" separating susceptible from resistant strains. Certain other specific aspects of susceptibility test reporting are detailed in the NCCLS tables, such as refraining from reporting results for antimicrobial agents that do not penetrate into the cerebrospinal fluid on isolates from patients who have meningitis. Additionally, results for antimicrobial agents that are only useful for treating urinary tract infections must not be reported on isolates from specimens other than urine.

Lastly, there is no objective evidence that the reporting of an MIC result is any more relevant clinically than reporting of a category (S, I, R) result in the majority of infections. Perhaps reporting of MIC results could aid a physician in selecting from among a group of similar drugs for therapy of infective endocarditis or osteomyelitis, in which therapy is likely to be protracted. For virtually all other infections, however, category results provide the clinician with the information necessary to select appropriate therapy. Only those physicians trained in infectious diseases are likely to be familiar with expected MICs for the multitude of antimicrobial agents currently available. Thus, if MIC results are to be reported, it is essential to also include appropriate interpretive criteria with the results.

METHODS OF DETECTING ANTIMICROBIAL-INACTIVATING ENZYMES

β-Lactamase Tests

β-lactamases are enzymes that selectively destroy β-lactam molecules by attacking the β-lactam ring component of the molecule (Fig. 3–23). Production of β-lactamase is a significant mechanism contributing to β-lactam resistance in certain organisms, such as *H. influenzae, N. gonorrhoeae, Moraxella catarrhalis, Staphylococcus* spp, *Enterococcus* spp, and some *Bacteroides* spp. Simple β-lactamase tests are performed in the clinical laboratory to identify β-lactamase production in these organisms, and a positive reaction means that the β-lactam agent(s) commonly used to treat infections caused by them (primarily ampicillin, amoxicillin and penicillin)

β-Lactam ring

R-CONH

CH_3
CH_3
S

N

O

CO_2^-

Penicillin

β-Lactamase →

R-CONH

CH_3
CH_3
S

N
H

O
O^-

CO_2^-

Penicilloic acid

F i g u r e 3 – 2 3
β-Lactamase hydrolyzes the β-lactam ring portion of the penicillin molecule. The hydrolysis results in the formation of penicilloic acid, which does not have antibacterial activity.

would be ineffective. Although many other organisms, such as Enterobacteriaceae and *Pseudomonas* spp, produce a variety of different types of β-lactamases, the β-lactamase tests as currently performed in clinical laboratories cannot predict resistance to all β-lactam agents that might be considered for therapy here, and the β-lactamase test should not be used.

Several methods may be used to detect β-lactamase production. The most commonly employed β-lactamase test utilizes the chromogenic cephalosporin nitrocefin. Cefinase disks commercially available from Becton Dickinson Microbiology Systems (Cockeysville, MD) are filter paper disks impregnated with nitrocefin. A disk is moistened, and a loopful of organisms is rubbed onto it. Within 10 minutes (or within 60 minutes for staphylococci), the area where the organisms are deposited will turn brownish red in the presence of β-lactamase–producing organisms. No color change occurs with β-lactamase–negative organisms (Fig. 3–24).

The other types of β-lactamase test methods, acidimetric and iodometric, are based on the detection of penicilloic acid, which results from the action of β-lactamase on penicillin. The acidimetric method uses citrate-buffered penicillin and phenol red as a pH indicator. When colonies of a β-lactamase–positive organism are added to the solution, the penicilloic acid present results in a drop in pH, causing a color change from red to yellow. In the iodometric method, a solution of phosphate-buffered penicillin and starch-iodine complex is used. With β-lactamase–positive organisms, penicilloic acid reduces iodine and prevents it from combining with starch. A positive reaction is colorless, and a negative reaction is purple.

All of the species mentioned previously in this section, except staphylococci, produce β-

lactamase constitutively, meaning that the same amount of enzyme is produced regardless of exposure to an inducing agent. An inducing agent is simply a β-lactam agent that stimulates production of the enzyme. Production of β-lactamase in staphylococci is inducible, and exposure to an inducing agent (β-lactam agent) is often required in the laboratory to obtain enough of the enzyme to be detected with conventional β-lactamase tests.

Testing organisms that have been exposed to an inducing agent can be accomplished by using growth from the periphery of a zone surrounding a β-lactam disk (e.g., oxacillin disk). Here, the bacteria at the zone edge have been exposed to β-lactam molecules. Alternatively, the test can be

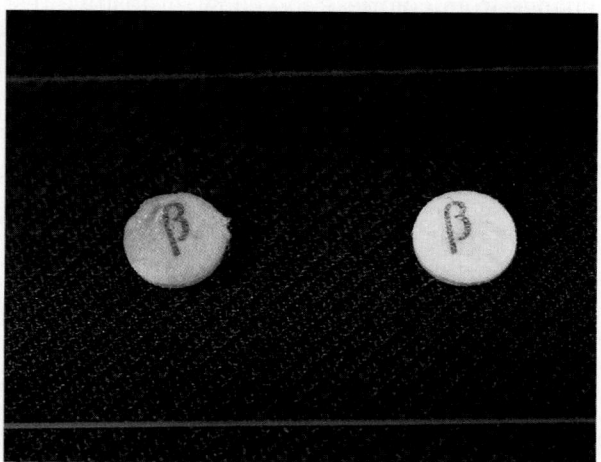

F i g u r e 3 – 2 4
Cefinase β-lactamase disk test. Cells from several colonies of *Haemophilus influenzae* were applied to a moistened disk. This photo shows results following testing of two different isolates. Within 10 minutes, the disk on the left turned brown-red (positive), and that on the right maintained a light yellow color (negative).

performed on bacteria growing in a well of a broth microdilution tray that contains a sub-inhibitory concentration (low concentration that does not inhibit visual growth) of a β-lactam agent.

The rapidity of β-lactamase tests makes them attractive. A positive reaction confirms resistance to ampicillin, amoxicillin, and penicillin, and a negative reaction indicates that the test organisms do not produce β-lactamase, although they may be resistant to these agents through an alternative mechanism. Resistance due to other mechanisms may be detected only with conventional types of dilution or disk diffusion tests. With organisms in which the rate of resistance due to alternative mechanisms is very low, such as *M. catarrhalis,* supplemental tests may not be routinely warranted in the clinical laboratory.

Chloramphenicol Acetyltransferase Tests

Chloramphenicol-resistant bacteria usually produce the enzyme chloramphenicol acetyltransferase (CAT). A rapid CAT test has been useful when chloramphenicol was a drug of choice in treating bacterial meningitis caused by *H. influenzae* (particularly in strains that were confirmed to be β-lactamase producers). CAT tests can be performed using either a tube test or commercially available disks (Remel, Lenexa, KS) impregnated with CAT test reagents. For either method, the presence of CAT results in a change from colorless to yellow. A positive test result means that the organism is chloramphenicol resistant; however, a negative test result is inconclusive, because the organism may be chloramphenicol resistant by an alternative mechanism. Conventional tests are necessary to confirm the latter. The availability of third-generation cephalosporins for treating *H. influenzae* meningitis has reduced the need for this rapid assay.

In addition to *H. influenzae,* the CAT test has been used for *S. pneumoniae* and *Salmonella* spp. At this time, however, the CAT test is used primarily in research settings.

QUALITY CONTROL OF ANTIMICROBIAL SUSCEPTIBILITY TESTS

Quality control of antimicrobial susceptibility tests involves testing standard reference strains that have defined antimicrobial susceptibility (or resistance) to the drugs tested. It is important to use strains that are representative of the types of patient isolates that are tested in the respective laboratory. Additionally, the quality control strains should represent varying degrees of susceptibility (or resistance). Ideal quality control strains for MIC tests have what is known as *on-scale MIC endpoints* for the drugs tested. An on-scale end point falls within the range of concentrations tested.

The NCCLS has identified ATCC (American Type Culture Collection) strains that are useful for quality control testing (Table 3–13). The NCCLS documents also include tables that define acceptable results (zone measurements for disk diffusion tests or MICs for dilution tests) for these strains; examples are shown in Tables 3–14 and 3–15, respectively.

The procedure followed in testing quality control reference strains must be identical to that used for testing patient isolates. If results do not fall within the defined acceptable limits, corrective action must be taken to determine the reason for the out-of-control observation prior to reporting of any patient results. Quality control testing is recommended each day that patient tests are performed; however, the frequency of quality control testing can be reduced to weekly if a laboratory can demonstrate acceptable proficiency in performing the test. Proficiency consists of obtaining acceptable results with each antimicrobial agent/quality control strain combination following test performance for 30 consecutive test days. Quality control must always be performed when new lots of materials are put into use, and test materials must never be used beyond their stated expiration dates.

There are other, less obvious components of a quality control program for antimicrobial susceptibility tests. Supplemental quality control strains may be periodically tested to validate acceptable performance of specific antimicrobial agent/organism combinations that may be only modestly controlled with the routine reference strains. An MRSA strain could be included to make certain the test system can detect heteroresistant strains. Similarly, an ampicillin-resistant *Enterobacter cloacae* might be included to make certain the system can detect ampicillin resistance. Supplemental quality control strains are sometimes used for troubleshooting specific problems or training new employees. Another component of a quality control program involves including mechanisms to ensure that personnel

■■■■ T A B L E 3 – 1 3

STRAINS COMMONLY USED FOR QUALITY CONTROL OF ROUTINE ANTIMICROBIAL SUSCEPTIBILITY TESTS

Test	QC Strain(s) Used	Comments
Antimicrobial susceptibility of gram-positive organisms	*Staphylococcus aureus* ATCC 25923 *S. aureus* ATCC 29213 *S. aureus* ATCC 38591	β-Lactamase negative for disk diffusion tests β-Lactamase positive for MIC tests Oxacillin resistant; currently not an NCCLS-recommended QC strain
Antimicrobial susceptibility of gram-negative organisms	*Escherichia coli* ATCC 25922 *Pseudomonas aeruginosa* ATCC 27853 *E. coli* ATCC 35218	β-Lactamase positive for testing β-lactam/β-lactamase inhibitor combination agents only
Antimicrobial susceptibility of *Haemophilus* spp	*Haemophilus influenzae* ATCC 49247 *H. influenzae* ATCC 49766 *H. influenzae* ATCC 10211	Ampicillin resistant, non–β-lactamase producing Ampicillin susceptible Used by media manufacturers to assess growth-supporting capabilities of the medium
Antimicrobial susceptibility of *Neisseria gonorrhoeae*	*N. gonorrhoeae* ATCC 49226	β-Lactamase negative
Antimicrobial susceptibility of anaerobes	*Bacteroides fragilis* ATCC 25285 *Bacteroides thetaiotamicron* ATCC 29741 *Eubacterium lentum* ATCC 43055	
Assessment of acceptability of medium (low thymine and thymidine content) for testing sulfonamides, trimethoprim, and trimethoprim/sulfamethoxazole	*Enterococcus faecalis* ATCC 29212	

■■■■ T A B L E 3 – 1 4

CONTROL LIMITS FOR MONITORING ANTIMICROBIAL DISK SUSCEPTIBILITY TESTS—ZONE DIAMETER (MM) LIMITS FOR INDIVIDUAL TESTS ON MUELLER-HINTON MEDIUM WITHOUT BLOOD OR OTHER SUPPLEMENTS

Antimicrobial Agent	Disk Content (μg)	*E. coli* ATCC 25922	*S. aureus* ATCC 25923	*P. aeruginosa* ATCC 27853	*E. coli* ATCC 35218
ampicillin	10	16–22	27–35	—	—
amoxicillin/ clavulanic acid	20/10	19–25	28–36	—	18–22
cefazolin	30	23–29	29–35	—	—
gentamicin	10	19–26	19–27	16–21	—

Permission to use portions of M2-A5 (Performance Standards for Antimicrobial Disk Susceptibility Tests, Fifth Edition; Approved Standard) has been granted by the National Committee for Clinical Laboratory Standards. NCCLS is not responsible for errors or inaccuracies. The interpretive data are valid only if the methodology in M2-A5 is followed. These documents and current supplements may be obtained from NCCLS, 771 E. Lancaster Avenue, Villanova, PA 19085.

■■■■ T A B L E 3 – 1 5

ACCEPTABLE QUALITY CONTROL RANGES OF MICS (μG/ML) FOR REFERENCE STRAINS

Antimicrobial Agent	*E. coli* ATCC 25922	*S. aureus* ATCC 29213	*P. aeruginosa* ATCC 27853	*E. coli* ATCC 35218
ampicillin	2–8	0.25–1	—	—
amoxicillin/clavulanic acid	2/1–8/4	—	—	4/2–16/8
cefazolin	1–4	0.25–1	—	—
gentamicin	0.25–1	0.12–1	.5–2	—

Permission to use portions of M7-A3 (Methods for Dilution Antimicrobial Susceptibility Tests for Bacteria that Grow Aerobically, Third Edition; Approved Standard) has been granted by the National Committee for Clinical Laboratory Standards. NCCLS is not responsible for errors or inaccuracies. The interpretive data are valid only if the methodology in M7-A3 is followed. These documents and current supplements may be obtained from NCCLS, 771 E. Lancaster Avenue, Villanova, PA 19085.

performing testing are proficient in their tasks. Self-assessment checklists and supervisory review of reported results are examples of such mechanisms. Satisfactory performance on proficiency survey specimens and utilization of relevant testing strategies are also quality control parameters.

The most widely used supplemental quality control measure is the use of antibiograms to verify results generated on patient isolates. An *antibiogram* is the overall antimicrobial susceptibility profile of a bacterial isolate to a battery of antimicrobial agents. Certain species have "typical" antibiograms, which can be used to verify the identification as well as the susceptibility results generated on the isolate (Table 3–16). For example, *P. aeruginosa* is typically resistant to ampicillin, cefazolin (and other first- and second-generation cephalosporins), and trimethoprim/sulfamethoxazole; however, it is often susceptible to gentamicin (and other aminoglycosides), extended-spectrum penicillins (e.g., mezlocillin) and ciprofloxacin. In contrast, *E. coli* is generally susceptible to all of the aforementioned antimicrobial agents. Table 3–17 shows several atypical antibiograms suggesting the result highlighted is erroneous. When atypical antibiograms are encountered, the results must be verified. Verification procedures include the following:

■■■ T A B L E   3 – 1 6
EXAMPLES OF TYPICAL ANTIBIOGRAMS FOR SEVERAL GRAM-NEGATIVE SPECIES*

Antimicrobial Agent	*Escherichia coli*	*Enterobacter cloacae*	*Proteus mirabilis*	*Pseudomonas aeruginosa*	*Xanthomonas maltophilia*
Amikacin	S	S	S	S	R
Ampicillin	S	R	S	R	R
Ampicillin/sulbactam	S	R	S	R	R
Cephalothin	S	R	S	R	R
Cefoxitin	S	S-R	S	R	R
Cefotaxime	S	S-R	S	S-R	R
Ceftazidime	S	S	S	S	S-R
Ciprofloxacin	S	S	S	S	R
Gentamicin	S	S	S	S	R
Imipenem	S	S	S	S	R
Mezlocillin	S	S	S	S	R
Nitrofurantoin	S	S	R	R	R
Tobramycin	S	S	S	S	R
Trimethoprim/ sulfamethoxazole	S	S	S	R	S

* Results indicated represent the typical response found in the majority of clinical isolates; however, these can vary significantly. R, resistant; S, susceptible; S-R, variable result.

■■■ T A B L E   3 – 1 7
EXAMPLES OF "PROBLEM" ANTIBIOGRAMS HIGHLY SUSPICIOUS FOR TECHNICAL ERRORS*

Antimicrobial Agent	*Escherichia coli*[†]	*Enterobacter cloacae*[‡]	*Pseudomonas aeruginosa*[§]	*Xanthomonas maltophilia*[¶]
Amikacin	R	S	S	R
Ampicillin	S	R	S	R
Cephalothin	S	S	R	R
Cefoxitin	S	R	R	R
Cefotaxime	S	R	S-R	R
Gentamicin	S	S	S	R
Tobramycin	S	S	S	R
Trimethoprim/ sulfamethoxazole	S	S	R	R

* R, resistant; S, susceptible; S-R, variable result.
[†] It is very unusual for an isolate to be resistant to amikacin and susceptible to gentamicin and tobramycin, because amikacin is the most active of these three aminoglycosides.
[‡] Third-generation cephalosporins (e.g., cefotaxime) are usually more active than second-generation cephalosporins (e.g., cefoxitin), which in turn are more active than first-generation cephalosporins (e.g., cephalothin) against the Enterobacteriaceae. Consequently, this antibiogram is unusual.
[§] Ampicillin does not have activity against *P. aeruginosa,* and this antibiogram is unusual.
[¶] *X. maltophilia* shows nearly universal susceptibility to trimethoprim/sulfamethoxazole, which is the drug of choice for infections caused by this very resistant species. This antibiogram is unusual.

- Re-examination of the disk diffusion plate, MIC tray, etc., to make certain results were properly interpreted and that the materials were not overtly defective (e.g., empty well in tray)
- Checking previous reports to see whether the particular patient had an isolate with an atypical antibiogram (that was verified) previously
- Repeating the test, if necessary (sometimes it is necessary to repeat the identification tests as well as the antimicrobial susceptibility tests to verify the atypical results; sometimes testing with an alternative method is useful)

Accrediting agencies require hospitals to compile cumulative antibiogram statistics to help monitor antimicrobial resistance within the institution. Cumulative antibiograms are generated by examining the overall incidence of susceptibility (or resistance) for a given antimicrobial agent/organism combination (e.g., the percentage of *E. coli* isolates that are susceptible to ampicillin). Cumulative antibiograms are compiled regularly (e.g., monthly, quarterly, or yearly depending on the institution's policies), and newly tabulated data are compared with data from previous intervals and sometimes with that of other institutions. Upon review of such data, an increase in the incidence of MRSA, for example, from 15% in January to 25% in February, might suggest a problem with nosocomial transmission of MRSA among patients. An investigation by infection control personnel into the reason behind such observations would be warranted.

SELECTING AN ANTIMICROBIAL SUSCEPTIBILITY TEST METHOD

Clinical microbiology laboratories can choose from among several manual or instrument-assisted methods to perform their routine antimicrobial susceptibility testing: the disk diffusion (or Kirby-Bauer) test, broth microdilution (with or without use of an instrument or growth indicator), and rapid automated instrument methods. The E test and Spiral Gradient end-point methods may also be useful for certain fastidious or anaerobic bacteria.

There has been a trend away from use of the disk diffusion test in favor of either broth microdilution or automated instrument methods. This trend has been noted in the results of the proficiency testing surveys conducted by the College of American Pathologists (CAP) (Table 3–18). The Kirby-Bauer test may be regaining some of its popularity, however, because of its inherent flexibility and cost-effectiveness. One of the problems most often mentioned with the commercial microdilution or automated systems is the inflexibility of the standard antimicrobial batteries or test panels. With the current availability of more than 50 antimicrobial agents in the U.S. and the diversity that exists among antimicrobial agent formularies in different hospitals, it is virtually impossible for manufacturers to provide standard test panels that fit every hospital's needs. Thus, the inherent flexibility of the disk diffusion test allows a laboratory to test any 12 antimicrobial agents deemed appropriate on a 150-mm Mueller-Hinton agar plate. Other assets of the disk diffusion procedure are that it is one of the best standardized of all tests and that its performance is continually updated by the NCCLS consensus efforts. The interpretive category results (susceptible, intermediate, resistant) of the disk diffusion test should be readily understood by all physicians who use them. The latter has not always been the case with MIC results. In fact, a survey of U.S. infectious disease physicians indicated that MIC results may be misinterpreted by some physicians, and that S, I, R category results may be preferable.

As stated previously, commercial microdilution susceptibility test products have become the most popular among U.S. clinical laboratories. Advantages of this method include its quantitative nature, i.e., an MIC rather than a strict category result, the fact that MICs may be determined with some organisms for which the disk test may not be standardized, and the attraction of the computerized hardware systems available from several of the manufacturers. Indeed, the computerized data management systems that

TABLE 3–18

MOST COMMONLY USED ANTIMICROBIAL SUSCEPTIBILITY TEST METHODS (based on 1991 CAP Proficiency Survey D-A)

Method Used	Frequency of Use (%)
Broth microdilution*	46.6
Disk diffusion	31.8
Rapid automated	19.7
Other methods	2.0

*More than 99% were commercially prepared panels.

accompany some of these instruments are extremely useful to some laboratories, which may not possess a laboratory information system. An MIC method should not be chosen, however, on the grounds that MICs are more valuable to physicians. As stated previously, many physicians who are not infectious diseases specialists are not familiar with MIC values for the myriad of contemporary antimicrobial agents.

Another problem with automated susceptibility test instruments has been limitations of instrument quality control procedures, largely because control strains used with the systems often result in many offscale values, i.e., MIC values less than or equal to the lowest concentration or greater than the highest concentration tested by the instrument. The instruments are mechanically and optically complex devices that must function properly for reproducible test results. Each manufacturer describes routine maintenance and function checks to prevent or detect overt malfunctions. The lack of on-scale control values means, however, that the potency of antimicrobials and functioning of the instrument may not be determined with the level of precision that clinical microbiologists have come to expect with the disk diffusion susceptibility test.

When a serious mechanical failure occurs, only tests from instruments that utilize conventional microdilution trays and photometric analysis of growth patterns after overnight incubation can be completed by manual incubation and interpretation. The instrument methods that incorporate rapid test interpretation or use fluorogenic substrate analysis cannot be manually interpreted. Therefore, laboratories must maintain a back-up testing procedure, such as disk diffusion or manual overnight broth microdilution testing in order to avoid delays in generating patients' results.

A laboratory may choose to perform rapid, automated antimicrobial susceptibility testing in order to generate test results more rapidly than can be accomplished by manual methods or to reduce the amount of labor required to perform susceptibility tests. Provision of important laboratory results one day sooner than by conventional methods is a logical advancement in patient care. Despite this self-evident statement, however, there is little objective evidence that rapid susceptibility test results reduce mortality or morbidity. This situation may be due in part to the fact that physicians have come to expect antimicrobial susceptibility test results on the

second day after a specimen is submitted, rather than late in the afternoon of the day after submission. Thus, there may be a period of transition (and education) required for full advantage to be taken of this approach.

One of the previous shortcomings of rapid susceptibility testing methods has been some sacrifice in the ability to detect certain inducible or otherwise subtle antimicrobial resistance mechanisms. The instruments most notorious for such problems are no longer marketed, however, and manufacturers of the remaining instruments have made significant strides to correct earlier problems. Nevertheless, it is important to emphasize that accuracy should not be sacrificed in an effort to generate a rapid susceptibility result.

The concept that automated susceptibility instruments reduce labor requirements through greater efficiency has been overly optimistic. Obviously, the greatest savings would involve use of panels that allow antimicrobial susceptibility testing and organism identification on the same panel, but a review of current CAP workload values for both manual and automated susceptibility test methods reveals that only minimal labor savings are provided by current instrumentation. These modest labor savings may be most meaningful to large clinical microbiology laboratories that perform large numbers of tests daily. The magnitude of the labor savings realized in clinical chemistry or hematology through automation has yet to be manifest in clinical microbiology.

In an era of cost containment, it is important to recognize that the most economical susceptibility test method currently available is the disk diffusion test. The microdilution and rapid automated instrument methods offer several direct or ancillary benefits, as described previously, but, with rare exception, they are more costly than the disk diffusion test.

Bibliography

Baker CN, Stocker SA, et al: Comparison of the E test to agar dilution, broth microdilution and agar diffusion susceptibility testing techniques by using a special challenge set of bacteria. J Clin Microbiol 29:533, 1991.
Bauer AW, Kirby WMM, et al: Antibiotic susceptibility testing by a standardized single disk method. Am J Clin Pathol 45:493, 1966.
Citron DM, Ostovari MI, et al: Evaluation of E test for susceptibility testing of anaerobic bacteria. J Clin Microbiol 29:2197, 1991.
College of American Pathologists: Workload Recording Method and Personnel Management Manual, 1992 ed., Northfield, IL: College of American Pathologists, 1992.

DeLencaster H, Figueiredo SA, et al: Multiple mechanisms of methicillin resistance and improved methods for detection in clinical isolates of *Staphylococcus aureus.* Antimicrob Agents Chemother 35:632, 1991.

Doern GV, Scott DR, Rashad AL: Clinical impact of rapid antimicrobial susceptibility testing of blood culture isolates. Antimicrob Agents Chemother 21:1023, 1982.

HIll GB, Schalkowsky S: Development and evaluation of the spiral gradient endpoint method for susceptibility testing of anaerobic gram-negative bacilli. Rev Infect Dis 12(suppl 2): S200, 1990.

Hindler JA, Mann LM: Principles and practices for the laboratory guidance of antimicrobial therapy. In Tilton RC, Balows A, Hohnadel DC, Reiss RF (eds): Clinical Laboratory Medicine. St. Louis, MO: CV Mosby, 1992.

Jones RN, Edson DC, et al: Antimicrobial susceptibility testing trends and accuracy in the United States. Arch Pathol Lab Med 115:429, 1991.

Jorgensen JH: Update on mechanisms and prevalence of antimicrobial resistance in *Haemophilus influenzae*. Clin Infect Dis 14:1119, 1992.

Klugman KP: Pneumococcal resistance to antibiotics, Clin Microbiol Rev 3:171, 1990.

Lorian V (ed.): Antibiotics in Laboratory Medicine, 3rd ed. Baltimore, Williams & Wilkins, 1991.

Murray BE: The life and times of the enterococcus. Clin Microbiol Rev 3:46, 1990.

National Committee for Clinical Laboratory Standards: Methods for Antimicrobial Susceptibility Testing of Anaerobic Bacteria, 3rd ed. Approved Standard M11-A3. Villanova PA: NCCLS, 1993.

National Committee for Clinical Laboratory Standards: Methods for Dilution Antimicrobial Susceptibility Testing for Bacteria That Grow Aerobically. 3rd ed. Approved Standard M7-A3. Villanova, PA: National Committee for Clinical Laboratory Standards, 1993.

National Committee for Clinical Laboratory Standards: Performance Standards for Antimicrobial Disk Susceptibility Tests, 5th ed. Approved Standard M2-A5. Villanova PA: National Committee for Clinical Laboratory Standards, 1993.

National Committee for Clinical Laboratory Standards: Development of In Vitro Susceptibility Testing Criteria and Quality Control Parameters. Tentative Standard M23-T2. Villanova, PA: National Committee for Clinical Laboratory Standards, 1992.

Neumann MA, Sahm DF, et al: Cumitech 6A: New Developments in Antimicrobial Agent Susceptibility Testing: a Practical Guide. Coord. ed, JE McGowan Jr. Washington, DC: American Society for Microbiology, 1991.

Thornsberry C (section ed). Antimicrobial agents and susceptibility tests. In Balows, A, Hausler WJ Jr, Herrmann, KL, et al (eds): Manual of Clinical Microbiology, 5th ed. Washington, DC: American Society for Microbiology, 1991.

C. SPECIAL ANTIMICROBIAL SUSCEPTIBILITY TESTS

MINIMUM BACTERICIDAL CONCENTRATION (MBC) TEST
Controlling Test Variables
Interpretation Concerns

TIME-KILL ASSAYS

SYNERGY TESTS

SERUM BACTERICIDAL TEST

MOLECULAR PROBES FOR IDENTIFYING DETERMINANTS OF ANTIMICROBIAL RESISTANCE

MEASUREMENT OF ANTIMICROBIAL AGENTS IN SERUM AND BODY FLUIDS
Biologic Assays
Immunoassays
Chromatographic Assays

Janet A. Hindler

Several special antimicrobial susceptibility tests are generally performed only in specialized laboratories and used in only a few defined clinical settings. They are listed in Table 3–19.

MINIMUM BACTERICIDAL CONCENTRATION (MBC) TEST

Minimum inhibitory concentration (MIC) tests identify the amount of antimicrobial agent required to inhibit the growth or multiplication of a bacterial isolate. If a concentration of antimicrobial agent that exceeds the MIC is attained at the infection site, the drug generally inhibits multiplication of the bacteria so that the patient's immune defense mechanisms are no longer overwhelmed. The immune defense mechanisms (e.g., phagocytic cells, antibody) work in concert with antimicrobial agents to eradicate infecting bacteria; this explains why inhibitory concentrations of drug at the infection site are generally sufficient for treating most infections.

In immunosuppressed patients, however, and in patients with serious infections such as endocarditis and osteomyelitis, the immune defense mechanisms are suboptimal. Inhibitory concentrations of drug may not be sufficient, and it is important to obtain bactericidal concentrations of antimicrobial agents at the infection site to effect cure. For many types of infections, the bactericidal capacity of a specific antimicrobial regimen can be predicted on the basis of previous experience. For example, most β-lactam antimicrobial agents are bactericidal for *Escherichia coli,* provided that their MIC is in the susceptible range. On the other hand, the bactericidal activity of β-lactams and other cell wall–active agents (e.g., vancomycin) against *Staphylococcus aureus* is less predictable. When a serious *S. aureus* infection occurs in a patient with poor immune defense mechanisms, an in vitro determination of the amount of antimicrobial agent required to kill as well as inhibit the isolate may be helpful; the *minimum bactericidal concentration (MBC)* test can be used for this purpose. The National Committee for Clinical Laboratory Standards (NCCLS) has described several procedures for

■■■■■■■ T A B L E   3 – 1 9

SPECIAL ANTIMICROBIAL SUSCEPTIBILITY TESTS

Test	Purpose
Antimicrobial level test (assay)	Measure of the amount of antimicrobial agent in serum or body fluid
Minimum bactericidal concentration (MBC) test	Measure of the lowest concentration of antimicrobial agent that kills a bacterial isolate
Serum bactericidal test (SBT)	Measure of the highest dilution or titer of a patient's serum that is inhibitory to and the highest dilution or titer that is bactericidal to the patient's own infecting bacterium
Synergy test	Measure of the susceptibility of a bacterial isolate to a combination of two antimicrobial agents
	MICS and often MBCs for each antimicrobial agent alone and in combination are determined
Time-kill assay	Measure of the rate of killing of bacteria by an antimicrobial agent as determined by examining the number of viable bacteria remaining at various intervals following exposure to the agent

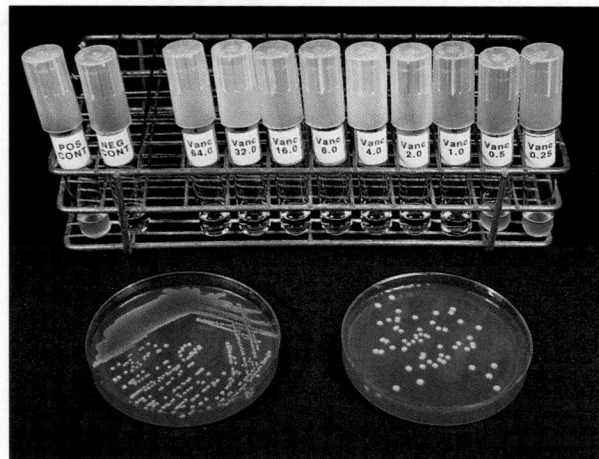

■■■■■■F i g u r e   3 – 2 5

Broth macrodilution test showing vancomycin and *Staphylococcus aureus*. The MIC is 1.0 µg/mL. The purity plate shows a pure culture. The colony count plate shows 53 colonies, which means that 5.3×10^5 CFU/mL bacteria were in the test tubes immediately after inoculation of the MIC test. For the colony count plate, immediately after inoculation the growth control tube was diluted 1:1000, and 0.1 mL was plated. Now, 0.01 mL wil be plated from each clear tube (tubes containing 1.0 through 128 µg/mL) for the MBC determination. Because it was shown that the actual colony count in the MIC test was 5.3×10^5 CFU/mL, growth of five or fewer colonies would indicate a 3 $\log_{10}$ decrease or 99.9% killing. By definition, the concentration of drug in the respective tube would be considered bactericidal.

assessing bactericidal activity; however, unlike disk diffusion and MIC tests, a standardized MBC test has not been in use for a sustained period. In the past, numerous methodologic variations have existed that compromised the use of the test results.

The MBC test is performed in conjunction with a broth macrodilution or broth microdilution MIC test. The antimicrobial agent concentrations that show inhibition (at and above the MIC) may or may not have killed the bacteria in the test inoculum (Fig. 3–25). Following the MIC determination, a 0.01-mL aliquot of each clear tube or well is subcultured to an agar medium to determine the MBC or the lowest concentration of antimicrobial agent that is needed to kill the test bacterium. The numbers of colonies that grow on subculture are compared with the actual number of organisms inoculated into the MIC test to determine the extent of bactericidal activity at each antimicrobial concentration. If the numbers of colonies on a subculture plate total less than 0.1% of the initial inoculum (indicating ≥ 99.9% killing), a bactericidal effect has by defintion been achieved.

As described earlier, the final number of bacteria in each tube (or well) immediately after inocula-

tion of the MIC test is approximately 5×10^5 CFU/mL (colony-forming units per mL). For the MBC test, however, an actual colony count must be performed on the test inoculum at the time the MIC test is inoculated. A small aliquot from the growth control tube or well is diluted in saline or broth to obtain a countable number of colonies; the final dilution is plated to an agar medium. Generally, a 1:1000 dilution is performed (0.01mL from the growth control tube is diluted in 10 mL), and 0.1 mL is spread over the surface of an agar plate. Following overnight incubation, the number of colonies that have grown on the colony count plate are noted. Because this count represents a 1:10,000 dilution, the count is multiplied by 10^4 to determine the number of bacteria in the original growth control (and antimicrobial agent) solutions. A calculation is performed to determine the number of colonies representing 99.9% of the test inoculum in the MIC test, because the *MBC end point* is defined as the lowest concentration of antimicrobial agent that kills 99.9% of the test bacteria.

In the example shown in Figure 3–25, the actual count on the colony count plate is 53; therefore, the number of bacteria in each tube of the MIC test immediately after inoculation was 5.3×10^5 CFU/mL (calculated by multiplying 53 times the dilution factor, which is 10^4). A 99.9% killing (or 0.1% survival) would be accomplished if 5 or fewer colonies grow upon subculture of each clear well or tube following reading of the MIC test (Fig. 3–26). In this example, subcultures from all tubes containing 2.0 µg/mL vancomycin or

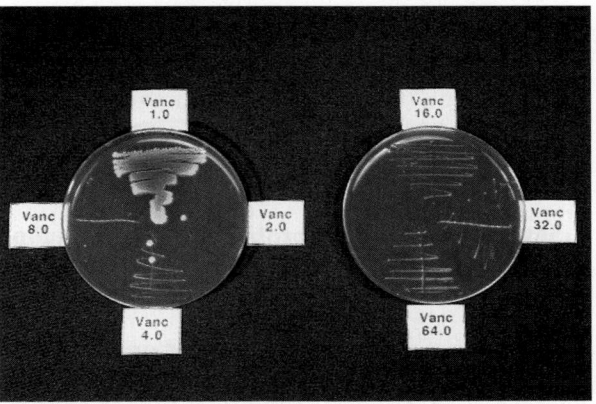

■■■■■■F i g u r e   3 – 2 6

Subculture plates from the MIC test of vancomycin and *Staphylococcus aureus* shown in Figure 3–25. Subcultures from tubes containing 8.0 to 128 µg/mL show no growth. Subcultures from the tubes containing 2.0 µg/mL and 4.0 µg/mL show one and two colonies, respectively, indicating > 99.9% killing. More than five colonies have grown from the 1.0 µg/mL tube, so the MBC is 2.0 µg/mL.

more grew fewer than 5 colonies. Consequently, the MBC is 2.0 μg/mL. The 99.9% end point (or three $\log_{10}$ reduction in growth of the original inoculum) is an arbitrary value with 95% confidence limits, although its clinical relevance has not been rigorously confirmed.

Controlling Test Variables

MBC tests are subject to more technical pitfalls than MIC tests, and several variables must be rigidly controlled during MBC testing. The first involves inoculum. Because many antimicrobial agents exert a bactericidal effect only on growing cells, bacteria in the mid-logarithmic phase of growth must be used as the inoculum to prevent falsely elevated MBCs. The inoculum preparation methods described for MIC tests that use stationary phase growth are unacceptable for MBC tests.

Secondly, during inoculation for MIC tests, care must be taken to ensure that all bacteria in the test inoculum are deposited directly into the antimicrobial solution. It has been shown that if this is not done, bacteria may stick to the wall of the tube or well above the meniscus of the antimicrobial solution and may remain viable during incubation of the MIC portion of the test. These cells (which have not been exposed to antimicrobial agent) may then be inadvertently transferred during the subculture step, ultimately resulting in falsely elevated MBCs.

Third, the volume subcultured following reading of the MIC test must be large enough to contain sufficient inoculum but small enough to prevent carryover of large amounts of antimicrobial agent to continue to exert an antibacterial effect. Usually, 10 μL (0.01 mL) is recommended.

Interpretation Concerns

Several interpretive problems, most common with β-lactam agents, have been associated with MBC tests, and they relate to technical or biologic issues. Sometimes there are more colonies growing on subcultures at higher drug concentrations than at lower concentrations. This decreased bactericidal activity at higher concentrations is referred to as a *paradoxic* or *Eagle effect*. Sometimes, small numbers (but slightly greater than 0.1% of the test inoculum) of bacteria grow on several subculture plates (persis-

ters). This may occur if some bacteria are metabolically inactive at the time of testing; however, when the persisting colonies are retested, their MICs are comparable to those orginally obtained. Finally, tolerance to the intrinsic bactericidal effect of an antimicrobial agent is demonstrated when the numbers of colonies growing on subculture plates exceed the 0.1 cutoff for several successive drug concentrations above the MIC. *Tolerance* is generally defined as an MBC:MIC ratio of 32 or greater. Tolerance has been associated with a defect in bacterial cellular autolytic enzymes.

TIME-KILL ASSAYS

Bactericidal activity of antimicrobial agents can also be assessed by performance of in vitro time-kill assays. Briefly, test bacteria in the mid-logarithmic growth phase are inoculated into several tubes of broth containing varying concentrations of antimicrobial agent and a growth control tube without drug. These tubes are incubated at 35°C. Then, small aliquots are removed at specific time intervals (e.g. at 0, 4, 8, and 24 hours), diluted to obtain countable numbers of colonies, and plated to agar for colony count determinations. The number of bacteria remaining in each sample is plotted over time to determine the rate of antimicrobial agent killing. Generally, a three or more $\log_{10}$ reduction in bacterial counts in the antimicrobial suspensions as compared with the growth control indicates an adequate bactericidal response. Because this test is very labor intensive, it is usually performed only in research settings.

SYNERGY TESTS

Some types of infections require therapy with a combination of two or more antimicrobial agents. Enterococcal endocarditis, for example, requires use of a penicillin (or vancomycin) and an aminoglycoside for reliable killing of the organism. A broad-spectrum cephalosporin and an aminoglycoside are often prescribed for gram-negative sepsis in neutropenic patients. Goals of combination therapy are to obtain broad-spectrum coverage, enhance antibacterial activity through synergistic interactions, and minimize resistance development. For most infections requiring combination therapy, single-agent MIC results and previous experience in treating similar types of infections are sufficient to guide selection of antimicrobial agent. In unusual situations, however, in which the patient may not be responding to what would appear to be an adequate regimen, unusual organisms or resistance properties are encountered, or host factors preclude use of certain agents, in vitro synergy tests may be warranted.

In vitro synergy tests may be performed using a broth dilution checkerboard method or time-kill assays. The checkerboard assay is a type of two-dimensional test; all steps are performed as for single agents, but two agents are tested in each well or tube.

Checkerboard synergy tests are usually performed in broth microdilution MIC trays. A wide variety of combinations of concentrations are tested by dispensing drugs in a two-dimensional "checkerboard" format, and each drug tested in the combination is also tested by itself. A combination is *synergistic* if its antibacterial activity is significantly greater than that of the single agents; i.e., when the MIC for each drug in the combination is less than or equal to one-fourth of the single-agent MICs. Conversely, *antagonism* is defined as the activity of the combination less than (and MICs are greater than) that of the single agents. In *indifference,* the activity of the combination is equal to that of the single agents (Fig. 3–27).

Time-kill assays can also be used to study synergistic interactions by testing a combination of drugs in a single tube and each drug individually in additional tubes. Several drug concentrations alone and in combination are usually examined. If subsequent colony counts reveal a two or more $\log_{10}$ reduction in the combination tube counts at 24 hours compared with the most active single-agent tube count, synergy has been demonstrated (Fig. 3–28). A change of less than ten-fold (increase or decrease) in colony counts from the combination tube compared with the most active single-agent tube represents indifference.

The NCCLS has not addressed synergy testing, and a number of methodologic variations exist.

SERUM BACTERICIDAL TEST

In the late 1940s, Schlichter and MacLean described a test that measured the effectiveness

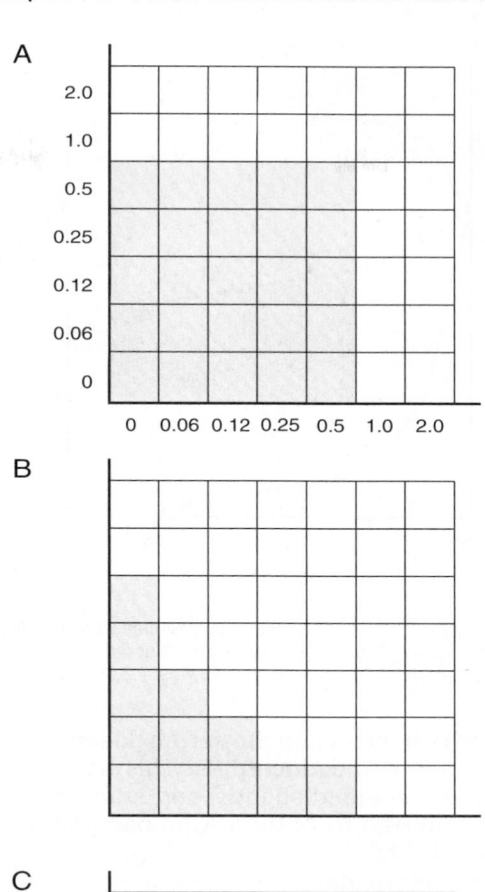

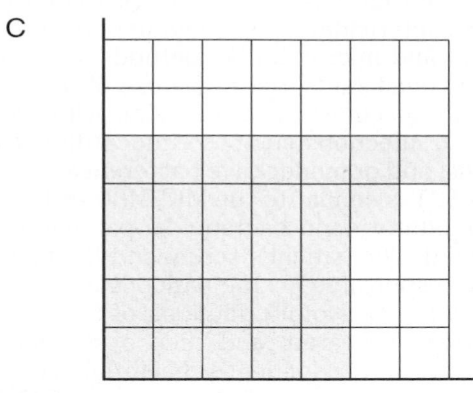

■■■■ F i g u r e   3 – 2 7

Assessment of antimicrobial combinations with the checkerboard method. Panels *A, B,* and *C* depict the results of testing combinations of two drugs (diluted in geometric twofold increments along the x and y axes; drug A along x axis and drug B along y axis). Shading indicates visible growth, and concentrations are expressed as multiples of the minimum inhibitory concentration. *A,* indifference; *B,* synergism; *C,* antagonism. (Modified and redrawn with permission from Eliopoulos GM, Moellering RC, Jr: Antimicrobial combinations. In Lorian V [ed]: Antibiotics in Laboratory Medicine, 3rd ed. Baltimore: Williams & Wilkins, 1991, p 437.)

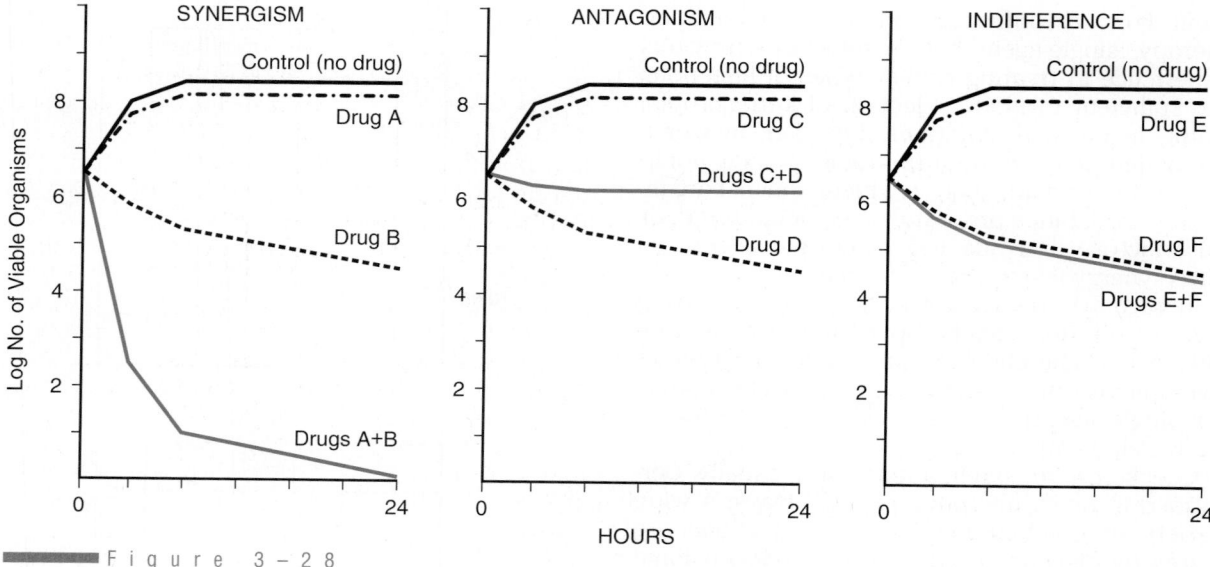

Effects of antimicrobial combinations as measured with the killing-curve method. A + B = synergism; C + D = antagonism; E + F = indifference. (Modified and redrawn from Eliopoulos GM, Moellering RC, Jr: Antimicrobial combinations. In Lorian V [ed]: Antibiotics in Laboratory Medicine, 3rd ed. Baltimore: Williams & Wilkins, 1991, p 443.)

with which penicillin in serum killed bacteria associated with endocarditis. This test was subsequently modified slightly, and standardized; it is now referred to as the serum bactericidal test (SBT). NCCLS has published procedures for serum bactericidal tests, including broth macrodilution and microdilution methods. Some clinical data available support the use of the SBT to evaluate specimens from patients with serious bacterial infections such as endocarditis, osteomyelitis, and gram-negative bacteremia.

The SBT is similar to the MIC/MBC test, in that both inhibitory and bactericidal parameters are evaluated. The patient's serum and the bacterial isolate responsible for the patient's infection are needed. Serial twofold dilutions of the patient's serum are prepared, and then a standardized inoculum of the patient's bacterial isolate is added to each dilution. Following overnight incubation, the tubes or wells are examined to determine the highest dilution of patient's serum that is inhibitory to the bacteria. Subsequently, as with the MBC tests, all tubes or wells showing inhibition are subcultured to an agar medium to determine the highest dilution that kills the bacteria. All of the potential technical pitfalls mentioned for the MBC test also apply to the SBT test.

SBT results relate to the amount of antimicrobial agent and any other antibacterial factors

(e.g., antibody, opsonins, complement) present in the patient's serum. Timing the collection of serum is critical, and generally both trough and peak titers are tested (Table 3–20). Older literature holds that a peak bactericidal titer of 1:8 or greater indicates that therapy was adequate. NCCLS states that a trough bactericidal titer of 1:32 or greater and a peak bactericidal titer of 1:64 or greater correlated with bacteriologic cure in patients with endocarditis. The NCCLS interpretive guidelines are applicable only if NCCLS methods are followed in performing the test. As with the MBC tests, technical complexity limits the widespread use of the SBT.

■■■■■■ T A B L E   3 – 2 0

GUIDELINES FOR OBTAINING SERUM SPECIMENS FOR THE SERUM BACTERICIDAL TEST AND ANTIMICROBIAL ASSAYS

Trough*	Obtain 0–30 minutes prior to next dose.
Peak*	Obtain either • 30–60 minutes after completion of a 30-minute intravenous (IV) infusion. • 60 minutes after an intramuscular (IM) injection. • 90 minutes after an oral dose (varies by specific drug).

* Ideally, collect trough and peak specimens around the same dose.

MOLECULAR PROBES FOR IDENTIFYING DETERMINANTS OF ANTIMICROBIAL RESISTANCE

Molecular methods are being used throughout the clinical microbiology laboratory to identify certain microorganisms. Several probes directed toward antimicrobial resistance genes have been developed; at this time, however, their use is confined to research settings. A probe directed toward the *mec* gene, which codes for oxacillin (methicillin) resistance in staphylococci, has been studied. Probes directed toward genes responsible for other resistance mechanisms (e.g., β-lactamases, aminoglycoside-modifying enzymes, tetracycline resistance factors) have also been described. A concern about using probes to confirm resistance involves the ability of some bacteria to contain specific resistance genes that may not be expressed. In these cases, the clinical significance of the presence of the resistance genes is questionable. Use of molecular methods to study resistance is in its infancy, and further studies will undoubtedly elucidate the clinical correlation of various findings.

MEASUREMENT OF ANTIMICROBIAL AGENTS IN SERUM AND BODY FLUIDS

The amount of antimicrobial agent in serum or other body fluid can be measured by a variety of antimicrobial assay procedures. Antimicrobial assays are performed for those antimicrobial agents for which the therapeutic concentration is very close to the toxic concentration. Assay results often lead to modification of subsequent doses to prevent accumulation of excessive drug concentrations that may be harmful to the patient. A patient's renal and hepatic status greatly influences in vivo levels of some antimicrobial agents. The antimicrobial agents with the greatest toxic risks and those most commonly monitored are the aminoglycosides, vancomycin, and chloramphenicol. For evaluation of antimicrobial levels, it is recommended that both trough and peak samples be assayed as for the serum bactericidal test (Table 3–20).

Biologic Assays

Antimicrobial assays were initially performed by a biologic assay method, and bioassays are still sometimes used today when it is important to focus on the amount of biologically active drug present rather than the amount of "chemical" present. The bioassays utilize a specific strain of bacteria (indicator organism) that is susceptible to the drug to be assayed, and the test is performed in either broth or agar. The antibacterial activity of the patient's specimen against this bacterium is compared with that of solutions containing defined concentrations of the antimicrobial being assayed, to determine the concentration in the patient's specimen through use of standard dose-response curves.

Immunoassays

Radioimmunoassay (RIA), fluorescent immunoassay, fluorescent polarization immunoassay, and enzyme immunoassay (EIA) procedures have all been used to measure antimicrobial agents in serum and other body fluids. The basic principles of these assays are similar, in that they all utilize antibodies directed against the specific antimicrobial agents to be assayed. Because of the nature of these tests, they are often performed in the chemistry or therapeutic drug monitoring sections of the laboratory. Several commercial manufacturers offer various types of immunoassay kits for performing gentamicin, tobramycin, amikacin, vancomycin, and chloramphenicol assays.

Chromatographic Assays

Various chromatographic methods, including gas-liquid, thin-layer, and paper chromatography, have been used on occasion for antimicrobial assays. The most widely employed chromatographic method, however, has been high-performance liquid chromatography (HPLC). Chromatographic methods are used primarily to measure levels of antimicrobial agents for which commercial immunoassay kits are not available. These tests are usually performed in research settings.

Bibliography

Berenbaum MC: A method for testing synergy with any number of agents. J Infect Dis 137:122, 1978.

1991 CAP Surveys, Bacteriology Survey Set D-A. Northfield, IL: College of American Pathologists, 1991.

Chapin-Robertson K, Edberg SC: Measurement of antibiotics in human body fluids: Techniques and significance. In Lorian V (ed): Antibiotics in Laboratory Medicine, 3rd ed. Baltimore: Williams & Wilkins, 1991, pp 295–366.

Eliopoulos GM, Moellering RC: Antimicrobial combinations. In Lorian V (ed): Antibiotics in Laboratory Medicine, 3rd ed. Baltimore: Williams & Wilkins, 1991, pp 432–492.

Halbert DN: DNA probes for the detection of antibiotic resistance genes. Clin Microbiol Newsletter 10:33, 1988

Moody J: Synergism tests. In Isenberg HI (ed): Clinical Microbiology Procedures Handbook. Washington, DC: American Society for Microbiology, 1992, pp. 5.18.1–5.18.28.

National Committee for Clinical Laboratory Standards: Methodology for the Serum Bactericidal Test: Tentative Guideline M21-T. Villanova, PA: NCCLS, 1992.

National Committee for Clinical Laboratory Standards: Methods for Determining Bactericidal Activity of Antimicrobial Agents: Tentative Guideline M26-T. Villanova, PA: NCCLS, 1992.

Schlicter JG, MacLean H: A method for determining the effective therapeutic level in the treatment of subacute bacterial endocarditis with penicillin: A preliminary report. Am Heart J 34:209, 1947.

Stratton CW, Cooksey RC: Susceptibility tests; special tests. In Balows A, et al (eds): Manual of Clinical Microbiology, 5th ed. Washington, DC: American Society for Microbiology, 1990, pp 1153–1165.

CHAPTER 4

Quality Improvement in the Microbiology Laboratory

A. QUALITY ISSUES IN CLINICAL MICROBIOLOGY

GENERAL GUIDELINES FOR ESTABLISHING QUALITY CONTROL
   Temperatures
   Thermometer Calibration
   Equipment QC
   Culture Media Performance
   Reagent QC
   Antimicrobial Susceptibility QC
   Personnel Competency
   Use of Stock Cultures
   QC Manual

CONTINUOUS QUALITY IMPROVEMENT (CQI)
   Mission Statement
   Indicators of QI: Process versus Outcome
   The Ten-Step Plan
   Problem/Action Form
   The Customer Concept
   Fixing the Process
   Benchmarking
   Commercially Purchased Monitors

Merrily Rausch

1. Define *quality control* (QC) as it applies in the clinical microbiology laboratory.

2. Discuss the general guidelines for establishing a QC program, describing how to monitor equipment maintenance and performance, culture media and reagent performance, personnel competency, use of stock cultures, and the development and updating of procedure manuals.

3. Describe proper documentation and institution of appropriate corrective action.

4. Define *continuous quality improvement* (CQI), and describe how it differs from QC.

5. Discuss the ten-step plan for establishing quality monitors.

6. Describe the customer concept.

7. Define *benchmarking*.

The issue of quality in the laboratory is complex. The emphasis and the terminology have changed tremendously in recent years. Laboratories have always taken measures to control the testing performed on patient specimens. This effort has been termed *quality control*; it is defined as those measures designed to ensure the medical reliability of laboratory data. Examples are checking media and reagents with specific organisms to see whether expected results are obtained and documenting that instrumentation meets all operating parameters previous to its use on patient samples.

Laboratorians now realize that quality control is only a small part of the issue of quality. Even when the laboratory has effectively controlled media, reagents, and instruments, the quality of the test result is poor if the specimen degraded before arriving in the laboratory but was still tested. Suppose a specimen contained the wrong patient name; again the media can be top quality, the incubator temperature accurate, and the technologist very competent, but if the results are recorded on the wrong patient chart, quality patient care management does not exist for the intended patient.

The actual laboratory testing is called an *analytical activity*. It is now important to realize that preanalytical, analytical, and postanalytical activities all affect quality. A quality outcome can be interrupted or destroyed at any point in the process. Table 4–1 attempts to clarify these three stages by giving examples of each kind of activity.

What has always been known as *quality control* relates only to analytical activities. *Quality assurance* was used to define those measures taken to ensure high-quality patient care. Quality assurance involves monitoring and regulating preanalytical, analytical, and postanalytical activities, and it is measured completely by patient outcome. It did not take long for managers to realize that "assurance" was probably the wrong word because we cannot actually assure the quality. The Joint Commission for the Accreditation of Healthcare Organizations (JCAHO) now uses the terms quality assessment and improvement, often shortened to *quality improvement* (QI). The focus changed when the terminology changed. QI is a more proactive approach that becomes the responsibility of all employees. It involves correcting whole processes instead of just pieces of processes. Any complicated, multifaceted process should always be able to be improved.

Prevention also became a bigger emphasis than detection. Terminology continued to change. For a while, it was *total quality management* (TQM),

■ TABLE 4–1

THREE STAGES OF ACTIVITIES THAT AFFECT OUTCOME OF LABORATORY TESTING

Stage	Activities
Preanalytical	Test ordering
	Order transcription
	Patient preparation
	Specimen collection
	Specimen identification
	Specimen transport
Analytical	Sample testing
Postanalytical	Result transcription
	Result delivery
	Result review
	Action taken on basis of result

and then *continuous quality improvement* (CQI). *Continuous improvement* means trying to improve every single day even if a problem does not exist. The emphasis is still on processes, with organization-wide commitment, so quality control is actually a piece of continuous quality improvement.

This section presents two major quality issues:

■ Applications of quality control

■ Continuous quality improvement

GENERAL GUIDELINES FOR ESTABLISHING QUALITY CONTROL

All quality control activities that take place must be recorded or there is no proof of their existence. All record sheets must list tolerance limits, when applicable, so that the person recording the results will always know whether the value being recorded is acceptable. Corrective action must also be recorded when any measurement falls outside a tolerance limit. The responsibility for quality control may rest mainly with one person, but in reality, everyone must participate for a program to be successful. A quality control program for a laboratory must include procedures for control of the following items: temperatures, equipment, media, reagents, susceptibility testing, and personnel.

Temperatures

Daily temperature checks are required on all temperature-dependent equipment:

■ Incubators

■ Heating blocks

■ Water baths

■ Refrigerators

■ Freezers

Incubator and refrigerator thermometers are easier to read if they are permanently immersed in glycerol. This helps to avoid temperature fluctuations when the door is opened to read the thermometer. Prior to being placed in use, each thermometer must be checked against a reference thermometer from the National Bureau of Standards (NBS). It is most efficient to check a large batch of thermometers at the same time at all temperature ranges that are likely to be used. It is common in clinical microbiology to test all thermometers at $0°C$, $37°C$, and $56°C$. The NBS thermometer comes with certification papers that list correction factors to be used at various temperature ranges. These correction factors are applied to all values obtained on individual laboratory thermometers. Laboratories arbitrarily determine the acceptable temperature variance. For most routine work, thermometers that vary by $1°C$ or more from the reference thermometer are discarded.

Thermometer Calibration

Thermometers are calibrated by batch upon arrival in the laboratory. Procedure 4–1 outlines the calibration procedure. Once the thermometer has passed calibration and is placed in use, repeat calibration of the thermometer should not be necessary. Alternatively, thermometers already checked against an NBS thermometer can be purchased.

Equipment QC

Equipment used in the clinical microbiology laboratory must be tested for proper performance at intervals appropriate for each piece. This may involve checking the percentage of CO_2 in an incubator daily or measuring the rpm of a centrifuge twice a year. Sometimes, frequency of testing is dictated by a regulatory agency, and other times, it is arbitrary. Table 4–2 gives examples of some laboratory equipment, the type of testing done, and the frequency of testing.

A preventive maintenance program must be established as an additional control measure. Preventive maintenance performed on equipment generally involves tasks such as oiling and cleaning, replacing filters, and recalibrating instruments. Keeping an instrument in top shape and functioning at the proper level will increase its lifetime and help control the quality of the results. Figure 4–1 shows an example of a preventive maintenance log sheet.

Culture Media Performance

All prepared media must be quality controlled to document their performance and sterility. Records must be maintained for 2 years. The

PROCEDURE 4-1. THERMOMETER CALIBRATION PROCEDURE

Labeling

1. Create a calibration log sheet.

2. Etch a number on the back of each thermometer using a diamond-tipped marker.

Calibration Method

1. Place the reference thermometer and the thermometer(s) to be calibrated in an ice bath.

2. When the reference thermometer reads 0°C, take the reading of all thermometer(s) being calibrated, and record the readings on the log sheet.

3. Repeat steps 1 and 2 at all other applicable temperatures, such as 37°C and 56°C.

Calculation of Correction Factor

1. Correct for the difference in readings between the reference thermometer (reference temperature) and that of each thermometer being calibrated, by subtracting the higher reading from the lower reading. If the thermometer being calibrated reads less than the reference thermometer, assign a plus (+) sign to the result of the subtraction step. This value must be added to the thermometer reading to equal the reading that would be taken from the reference thermometer. Conversely, if the reading of the thermometer being calibrated is greater than the reference thermometer, assign a minus (–) sign to the result. This value must then be subtracted from that thermometer's reading to equal the reference value.

2. Correct for the difference between the reading of the reference thermometer and the "real" temperature, using the process just described. Add the correction factor for the reference thermometer to the correction factor calculated in step 1 for each thermometer being calibrated.

3. Determine the correction factor at each temperature for each thermometer, and record it on the log sheet for the thermometer. The log sheet should be kept for the life of the thermometer.

4. Tolerance limits are set by each laboratory, but are usually 1°C. Discard all thermometers that exceed the established tolerance limit.

Examples of correction factors are shown in the table below:

"Real" Temperature	Reference Temperature	Correction Factor
0°	+0.1°	–0.1
37°	36.9°	–0.1
56°	55.9°	–0.1

■ TABLE 4-2
FREQUENCY OF EQUIPMENT TESTING

Equipment	Test Type	Frequency
Incubator	Temperature, CO_2	Daily
GasPak jar	Anaerobiasis, catalyst heated	Each use
Anaerobe chamber	Anaerobiasis, humidity, temperature	Daily
Biohazard hood	Air flow (done by specialist)	Annually or any time hoods are moved
Centrifuge	RPM check	Every 6 months
Microscope	Cleaned and adjusted	Four times per year or as needed
Autoclave	Temperature	Each load
	Spore testing	Weekly
Balance	Accuracy of weights	Annually

criteria are established by the National Committee for Clinical Laboratory Standards (NCCLS) and listed in document M22-A. Commercial media are always tested by the manufacturer. The laboratory must obtain a statement of quality control from the manufacturer for all media the laboratory will not retest. Only certain kinds of media must be retested by the user, usually because of complexity or history of failure rate. Media that require retesting are chocolate agar, selective media for pathogenic *Neisseria,* and *Campylobacter* media. Figure 4–2 lists specialty commercial media that have been retested by the laboratory.

Media that are not quality controlled by the laboratory should still undergo daily observation for moisture, sterility, and color:

Preventive Maintenance

MEDIA-REAGENTS-SMALL EQUIPMENT	JANUARY	FEBRUARY	MARCH	APRIL	MAY	JUNE
MEDIA, REAGENTS:						
1. Check refrigerators/freezers for outdated material	1/21/94 af	2/24/94 LH	3/28/94 DS	4/28/94 SP	5/27/94 MH	6/30/94 JJ
2. Check for "received" and "opened" dating	1/21/94 af	2/24/94 LH	3/28/94 DS	4/28/94 SP	5/27/94 MH	6/30/94 JJ
THERMOMETERS: Calibrate by batch upon arrival.						
PIPETTORS: Calibrate			3/10/94 MS			
PH METER: (Rooms 337 and 354)						
1. Clean the exterior	1/21/94 af	2/24/94 LH	3/28/94 DS	4/28/94 SP	5/27/94 MH	6/30/94 JJ
2. Replace water in the electrode holder	1/21/94 af	2/24/94 LH	3/28/94 DS	4/28/94 SP	5/27/94 MH	6/30/94 JJ
3. Check the AgCl level of the electrode	1/21/94 af	2/24/94 LH	3/28/94 DS	4/28/94 SP	5/27/94 MH	6/30/94 JJ
EYEWASHES: Flush						
(rooms 332, 337, 339, 342, 351(2), 354)	1/21/94 af	2/24/94 LH	3/28/94 DS	4/28/94 SP	5/27/94 MH	6/30/94 JJ
STAINING SINK: Clean	1/20/94 af	2/24/94 LH	3/28/94 DS	4/28/94 SP	5/27/94 MH	6/30/94 JJ
GROUNDING: Check annually						6/15/94 RB
REVIEWED BY: M Rausch Page 1 of 5						
YEAR: 1994						
Preventive maintenance is to be performed in the months highlighted for each item listed.						

Figure 4 – 1

Preventive maintenance log sheet.

Commercial Media Performance Retesting

Medium Tested	Mfg	Lot number	Exp. date	Sterility	Test Organism(s)	Result Pass/Fail	Action taken	Date	Tech
Chocolate II	BD	L3RTPO	5/23/94	OK	H. influenzae N. meningitidis	Pass	–	3/8/94	MR
GC-Lect	BD	–							
Jembec-Neiss.	BD	A 3NENC	4/3/94	OK	N. gonorrhoeae N. meningitidis S. epidermidis C. albicans E. coli	Pass	–	3/8/94	MR
Campy bld	BD	LINDHK	4/28/94	OK	C. jejuni E. coli	Pass	–	3/8/94	LM
Campy thio	BD	H4EOAJ	2/1/95	OK	C. jejuni E. coli	Pass	–	3/8/94	LM

Reviewed by: _____ Date: _____

Figure 4 – 2

Commercial media that have been retested by the laboratory.

■ **Moisture:** Plates should be free of moisture before use but should never show signs of drying around the edges.

■ **Sterility:** Plates should be free of contaminants.

■ **Color:** Blood-based plates should not show signs of hemolysis, and any other plate that deviates from the normal color should not be used.

Results of daily media observation must be recorded and must include lot numbers. Figure 4–3 shows an example of a daily media observation log, which helps ensure that good-quality media are used on all patient samples. Corrective action must be taken when a medium does not meet standards. This can be documented on a separate record known as a media failures log (Table 4–3).

When a medium does need to be quality controlled because it was prepared "in-house" (in the laboratory) or because it is complex, several basic rules must be followed:

■ All media must be tested previous to use.

■ Each medium must be tested with organisms expected to grow or give a positive reaction as well as with organisms expected either not to grow or to give a negative reaction.

■ The medium should be tested for sterility and pH.

■ The organisms selected for QC should represent the most fastidious organisms for which the medium was designed.

■ Testing techniques should be different for primary plating media than for biochemical or subculture media. Primary plating media should be tested with dilute suspensions of organisms, whereas biochemical media can be tested with undiluted organisms.

■ QC testing should be performed according to NCCLS recommendations.

■ Expiration dates must be established.

Figure 4–4 shows a log sheet used for testing media prepared in-house.

Reagent QC

With few exceptions, reagents should be tested each day of use with both positive and negative controls. Reagents that are documented to have consistent and dependable results may be tested less frequently. Gram stain, for instance, is commonly tested weekly instead of daily. Some reagents may be tested more than once a day. When more than one vial of disks such as bacitracin disks is in use at the same time in large laboratories, each vial should be tested daily. This is because each vial is refrigerated at night but usually left at room temperature during the day and therefore has the opportunity to degrade while in use. Examples of reagents that should undergo QC in microbiology are as follows:

■ All stains

■ Bacitracin

■ β-Lactamase

■ CAMP

■ Catalase

■ Coagulase

■ $FeCl_3$

■ Gelatin

■ Germ tube

■ Hippurate

■ Kovács

■ Nitrate

■ Optochin

■ Oxidase

■ L-Pyroglutamyl-β-naphthylamine (PYR)

■ Typing sera

■ Voges-Proskauer (VP)

■ X and V strips

Figure 4–5 shows a variety of testing that might be done at an individual workstation daily.

Antimicrobial Susceptibility QC

The National Committee for Clinical Laboratory Standards (NCCLS) provides guidelines for control of susceptibility testing. The recommended control organisms are specific strains from the American Type Culture Collection (ATCC) (Table 4–4). In addition to the organisms listed in Table 4–4, specific strains of *Haemophilus*

DATE	MEDIUM	LOT NO.	MOISTURE	STERILITY	COLOR	COMMENT	INITIALS
3/4/94	BAP	A2RUW0	✔	✔	✔		MR
	MAC	A4RUUH	✔	✔	✔		MR
	CHOC	A4RUUX	✔	✔	✔		MR
	CNA	A1RUWB	✔	✔	✔		MR
	HE	AZRCUF	✔	✔	✔		MR
	CIN	AZNEJE	✔	✔	✔		MR
	PD	OSU PREP: 1/6/94	✔	✔	✔		MR
	GC-LECT	K3RTNZ	✔	✔	✔		MR
	SCH	A4NETG	✔	✔	✔		MR
	SCH-GV	A3NENN	✔	✔	✔		MR
	CDC ANA	AZRWAW	✔	✔	✔		MR
	SMAC	OSU PREP: 2/15/94	✔	✔	✔		MR

A ✔ in each column denotes medium is OK.

■■■■■ F i g u r e   4 – 3
Daily media observation log.

■■■■■ T A B L E   4 – 3
MEDIA FAILURES LOG

Date:	2/14/95
Media:	TMS slants
Lot #:	In-house preparation 2/13/95
Expiration Date:	6 months from preparation
Quantity:	2 racks
Failure:	Failure to give proper reaction with *S. epidermidis*. *S. aureus* and other coagulase-neg. staphs are OK.
Action Taken:	QC repeated. *S. epidermidis* failed. Memo sent to all techs and all tubes discarded. New TMS prepared.
Technologist:	MAR

■■■■■ T A B L E   4 – 4
RECOMMENDED CONTROL ORGANISMS FOR SUSCEPTIBILITY TESTING

Organism	Susceptibility Test(s)
Escherichia coli, ATCC 25922	Gram-negative drugs
Escherichia coli, ATCC 35218	β-Lactamase-inhibitor drugs
Staphylococcus aureus, ATCC 25923	Gram-positive drugs— Kirby-Bauer test
Staphylococcus aureus, ATCC 29213	Gram-positive drugs— minimal inhibitory concentration
Pseudomonas aeruginosa, ATCC 27853	Monitors Ca^{++}, Mg^{++} content*
Enterococcus faecalis, ATCC 29212	Monitors thymidine [†]

*As Ca^{++} and Mg^{++} concentrations increase, *P. aeruginosa* becomes more resistant to the aminoglycosides.
[†]Increases in thymidine cause false resistance to certain drugs, such as sulfonamides, trimethoprim, and trimethoprim/sulfamethoxazole.

IN-HOUSE MEDIA QUALITY CONTROL

DATE	MEDIA AND LOT NO.	ORGANISM PLATED	QC PLATED DATE	QC READ DATE	QC PASSED/FAILED	INIT
3/8/94	TMA	S. aureus S. epidermidis	3/9/94	3/10/94	Pass	MR
3/8/94	Beta toxin	S. agalactiae ⊕ S. pyogenes ⊖	3/9/94	3/9/94	Pass	MR
3/8/94	PSE	E. faecalis	3/9/94	3/10/94	Pass	MR
3/8/94	TSI	P. aeruginosa C. freundii	3/9/94	3/10/94	Pass	MR

■■■■■■ F i g u r e   4 – 4
Log sheet for testing media prepared in-house.

influenzae and *Neisseria gonorrhoeae* are used to test patient isolates of these organisms.

In any susceptibility system, there are many variables to control that can affect the accuracy of results; some of these variables are:

■ Antibiotic potency

■ Agar depth (Kirby-Bauer test)

■ Evaporation (microtiter dilution)

■ Cation content

■ pH

■ Thymidine content

■ Instrument failure

■ Inoculum concentration

■ Temperature

■ Moisture (Kirby-Bauer test)

■ Difficulty in determining end points

Careful storage of degradable supplies and precision in implementation of recommended procedures are mandatory to obtain accurate and reproducible susceptibility results.

Susceptibility testing is usually conducted daily. Control organism results must be evaluated previous to determining endpoints on patient isolates. Minimal inhibitory concentration (MIC) values must be within one log dilution of the expected MIC based on NCCLS guidelines. If precision can be demonstrated with 30 consecutive days of susceptibility testing using NCCLS guidelines, QC organisms may be tested weekly instead of daily.

BLOOD CULTURE WORKSTATION QUALITY CONTROL

Month _____

Year _____

DATE	INITIALS	GAS PAK JAR anaerobic	HEAT CATALYST	CHANGE DESICCANT date (weekly)	ACRIDINE ORANGE pos	neg	NaDesoxy/ WELLCOGEN pos	neg	HEATING BLOCKS (35–37°C) #20	#17
1	LM	OK	✔		N	D	+	−	35°	36°
2	Ad	OK	✔		N	D	+	−	35°	36°
3	Ad	OK x 2	✔		N	D	N	D	35°	36°
4	MCl	OK	✔		+	−	N	D	36°	36°
5	MCl	OK	✔		N	D	N	D	35°	36°
6	LH	OK x 2	✔		N	D	N	D	35°	36°
7	Ad	OK x 2	✔	✔	N	D	+	−	36°	36°
8	Ad	OK x 2	✔		N	D	+	−	36°	37°
9	LH	OK	✔		N	D	+	−	36°	36°
10	Ad	OK	✔		N	D	N	D	36°	36°
11	Cl	OK	✔		+	−	N	D	36°	36°
12	MG	OK	✔		+	−	+	−	36°	36°
13	DS	OK x 2	✔		N	D	N	D	36°	36°
14	DS	OK	✔	✔	N	D	N	D	36°	35°
15	DD	OK	✔		N	D	N	D	35°	36°
16	af	OK	✔		N	D	N	D	36°	36°
17	af	OK	✔		N	D	+	−	36°	36°
18	af	OK	✔		N	D	+	−	36°	36°
19	MG	OK	✔		N	D	N	D	36°	35°
20	MG	OK	✔		N	D	+	−	38°/36°	36°
21	MG	OK x 2	✔	✔	+	−	N	D	36°	36°
22	MG	OK x 2	✔		N	D	N	D	36°	36°
23	MG	OK x 2	✔		N	D	N	D	36°	36°
24	JR	OK	✔		+	−	N	D	36°	36°
25	JR	OK	✔		+	−	N	D	36°	36°
26	JR	OK x 2	✔		+	−	N	D	35°	36°
27	JR	OK	✔		N	D	+	−	35°	36°
28	JR	OK	✔	✔	N	D	N	D	36°	36°
29	LH	OK	✔		N	D	+	−	36°	36°
30	LH	OK x 2	✔		N	D	N	D	36°	36°
31	LH	OK x 2	✔		+	−	+	−	36°	36°

REVIEWED BY: Mbtt

DATE: 4/4//94

■■■■■■ F i g u r e   4 – 5

Variety of testing that can be done at an individual workstation daily.

Personnel Competency

Even the personnel must pass quality control repeatedly, using a variety of techniques. One popular technique is proficiency testing, whereby carefully designed samples are given to technologists as "unknowns" for the purpose of determining competency in identifying them. Proficiency samples may be purchased commercially or prepared internally. Technologists may perform testing without knowing they are working on a proficiency sample or may clearly understand that the sample is a proficiency test. In either case, samples should not receive any "special" treatment.

Another form of personnel quality control is to prohibit all technologists from "signing out" or finalizing their own work. If each technologist's results are reviewed by another technologist, mistakes are likely to be caught previous to release.

Employee competency has always played a large role in quality. The Clinical Laboratory Improvement Act of 1988 (CLIA) has mandated that the competency of each employee be determined and verified upon employment. Reverification must take place annually. Proof of competency must be maintained in each employee's personnel file. A person may be qualified to prepare a slide for staining but may not be able to stain it; or a person may be able to prepare and stain a slide but not competent at reading or interpreting the smear. As part of CLIA requirements, all tests or analytes have been assigned a complexity rating. In microbiology, all tests are either moderately complex or highly complex. Personnel must meet certain educational requirements before being permitted to perform at each level of complexity.

In addition to meeting educational requirements, a person's competency must be observed and documented for each test performed. The many agencies involved in accreditation and inspection have different requirements and interpretations of competency verification, making this a complicated task for all laboratories.

The requirement for ongoing continuing education (CE) programs for all employees is yet another form of quality control. These programs may teach theory or new techniques or may be case studies or simply training on new instrumentation. It is important to document completion of all CE programs. Training for new instrumentation should be documented in the individual employee's personnel competency file also.

Use of Stock Cultures

In order to operate a quality control program, stock cultures must be maintained by all laboratories. They are available from many sources:

- Commercial sources
- Proficiency testing isolates
- Patient isolates
- American Type Culture Collection (ATCC) organisms

When quality control testing appears to have failed, it is more often the stock culture that has failed than the test itself. Organisms may mutate with repeated subculturing. It is advantageous to grow a stock culture in a large volume of broth, then divide it among enough small freezer vials to last a year. With this technique, a new vial can be removed from the freezer weekly, so that organisms do not have to be continually subcultured. It may be necessary to subculture an organism twice after thawing to return it to a healthy state. Media selection for freezing is up to individual laboratories but should not contain sugars. If organisms utilize sugars while being maintained, the acid products that result may kill the organisms with time. Popular media choices for stock cultures are:

- Schaedler broth with glycerol
- Skim milk
- Chopped meat (anaerobes)
- Tryptic soy agar deeps (at room temperature)
- Cysteine-tryptic agar (CTA) without carbohydrates)

If organisms are stored in a freezer, they should be kept at −70°C. Alternative storage methods include freezing in liquid nitrogen and lyophilization.

QC Manual

All rules and procedures for quality control should be available to employees at the workstation in written form in a QC manual. The manual must be reviewed and signed at least annually and revised as needed by a supervisor.

CONTINUOUS QUALITY IMPROVEMENT (CQI)

Accrediting agencies such as the Joint Commission on Accreditation of Healthcare Organizations (JCAHO) and the College of American Pathologists (CAP) put strong emphasis on quality improvement in their accreditation checklists. The Clinical Laboratory Improvement Act (CLIA) also mandates use of written quality improvement policies in laboratories that include all three phases of testing. Every laboratory must have a plan for improvement. In order to improve quality effectively, it is necessary for *all* employees to understand the plan and take active roles.

Mission Statement

Creating a short mission statement for all employees to learn can be an effective tool for uniting everyone behind the same cause. It can be as simple as "Working Together For A Healthy America," used at The Ohio State University Medical Center, or "The Right Thing, The Right Way, The First Time, Every Time," used by Intermountain Healthcare. Putting the statement on the backs of employee identification badges is one way to focus on and emphasize the mission so that all become involved and understand that they have roles in the mission and must make CQI a part of everyday life. Problems are not to be viewed as true problems, but as opportunities for improvement.

Indicators of QI: Process Versus Outcome

Many types of monitors or indicators can be incorporated into a quality improvement program. Accrediting agencies generally check that a variety of types are used. Some monitors are ongoing data collections with no suspected problems. Results are compiled and evaluated routinely. This procedure establishes a trend and makes problems easy to spot as disruptions in the trend. These are often thought of as *process monitors.*

Outcome monitors are measurements of the result of a process. An example is the complications a patient experiences as the result of a process.

Other monitors may be created in response to a suspected problem. Data may be collected for a short period to resolve a specific issue. These monitors are called *focused monitors.*

The Ten-Step Plan

The JCAHO has developed a method for establishing quality monitors that is known as the Ten-Step Plan. When a laboratory uses this system, a plan for each monitor should be written in the Ten-Step Plan format previous to implementation. The format is shown in Table 4–5.

Table 4–6 gives an example of a Quality Improvement Monitor for Corrected Reports written in the JCAHO Ten-Step Plan format. The JCAHO does not require that monitors be designed using the Ten-Step Plan, but it is an effective tool. No specific tools are necessary. In fact, anyone who knows the problem and knows the solution should not monitor it, but just fix it!

Problem/Action Form

A more simplistic approach to monitoring or documenting quality issues is a problem/action form. This is most commonly used to document issues that are quickly resolved, but it could also be used for long-term monitor summation. The form is a brief statement consisting of the following information:

■ Date

■ Problem

■■■■■ T A B L E   4 – 5

THE JCAHO TEN-STEP PLAN FOR ESTABLISHING QUALITY MONITORS

1. Assign responsibility.
2. Delineate the scope of care.
3. Identify important aspects of care.
4. Identify indicators and criteria.
5. Set thresholds (acceptable expectations).
6. Determine how to collect and organize data.
7. Evaluate data.
8. Take action (correction for improvement).
9. Assess the effectiveness.
10. Communicate results.

Adapted from Accreditation Manual for Hospitals. Vol 1: Standards, Oakbrook Terrace, IL: Joint Commission on Accreditation of Healthcare Organizations, 1993, pp 12–13.

■ T A B L E   4 - 6
MONITOR FOR CORRECTED REPORTS

Assign Responsibility	A medical technologist in the Clinical Microbiology Laboratory will monitor all corrected reports with "major" errors (see definition).
Scope of Care	Clinical Microbiology Laboratory
Important Aspects of Care	Reliability of reported results
Indicators and Criteria	The number and kinds of corrected laboratory reports need to be evaluated to determine (1) the effect they may have on patient management and (2) the feasibility of reducing the number of corrected reports.
Thresholds	99% of all specimens received should have no corrected reports.
Data Collection	Count the corrected reports. Categorize the types of errors. Evaluate the timing of corrections.
Evaluation	With each corrected report, a brief "Problem" will be stated. A "Solution" will follow each problem listed. These solutions are agreed upon by those medical technologists present at monthly meetings specifically called for this purpose.
Action	Minutes from the monthly meetings will be distributed to all medical technologists. Inservice, education, and retraining will be done if deemed necessary according to a solution.
Assessment	This will be an ongoing monitor that will be reviewed monthly until otherwise stated.
Report	Data will be accumulated and readily available in the "Corrected Report" notebook and will be presented biannually to the Clinical Laboratories Quality Assurance Committee.

■ Evaluation

■ Corrective action

■ Outcome

The form may be signed by the person submitting it, and additional documentation may be attached as necessary. Table 4–7 illustrates the use of the problem/action form.

The Customer Concept

Laboratories must focus on the *customer concept*. Who are the customers and what is a customer's perception of quality? Patients are not the only customers. Anyone who looks to the laboratory for a service is a customer. Doctors, nurses, insurers, and patients are all customers. Each customer may view quality differently and may have different expectations. Laboratorians may view quality with an emphasis on accuracy, whereas a physician views it as turnaround time, the patient as compassion and relief from pain, and the insurance company as cost-effectiveness. Customer satisfaction needs to be surveyed to determine perceptions.

Fixing the Process

When patient outcome is less than desirable, the process must be evaluated and fixed. The focus is on the process, not on an individual. The primary rule to follow is that there be no finger-pointing, no fault-finding. It is important to note that preanalytical and postanalytical activities usually take place outside the laboratory and require cross-functional teams to evaluate and correct the process. Department representatives can easily become defensive and territorial. Barriers to

■ T A B L E   4 - 7
CLINICAL MICROBIOLOGY PROBLEM/ACTION REPORT

Date	1/15/95
Problem	A blood culture needed to be drawn on a patient at the urgent care facility, but nobody was trained to draw blood cultures. The patient had to drive from the urgent care facility to the hospital to have the blood drawn by a phlebotomist at the hospital who was trained in the appropriate techniques.
Evaluation	No employees at the urgent care facility have been trained to collect blood cultures. The same is true of the Clinic Outpatient Lab. This is the second occurrence in about 2 months.
Corrective action	Seriously ill patients should not have to travel from one facility to another to have their blood drawn. Laboratory Administration was informed of this situation.
Outcome	All outpatient sites will receive training and written instructions for the proper collection of blood for culture. Patients will no longer have to drive to the hospital for this service if they are already at an outpatient facility.

Submitted by M Rausch
Attached documentation? No

cooperation must be removed. Cross-functional teams need to include a trained facilitator. A facilitator is most effective when he or she has no vested interest in the process being evaluated. The facilitator's role on the team is to use problem-solving training and experience to help the team brainstorm and keep on track. In the formation of a cross-functional team, the people "in the trenches" must be represented, rather than all team members being supervisory personnel. If the issue is transportation of specimens, for example, the team should include, at minimum, a transporter, a specimen processor, a staff nurse, a medical student or physician representative, appropriate supervisory personnel, and a facilitator. The most accurate and meaningful brainstorming ideas usually come from the people who perform the tasks, not those who designed the process.

When is a process fixed? It is not fixed just because an explanation of why something went wrong is available. It is fixed only when the problem is prevented from happening again. For example, a laboratory result was never recorded on a patient's chart. Investigation determined that the patient was transferred to another unit while the report was being generated. Having this explanation is not good enough. To fix the problem, a process must be established to ensure that charting follows every patient in a timely manner every time.

Benchmarking

A benchmark is a reference point. *Benchmarking* is seeking an industry or profession's best practices to imitate and improve. It was initially practiced in business and industry but has now become an important part of hospital quality management programs. It is done willingly and above board. A hospital may join a large group of other hospitals that all share operating statistics. Productivity and cost-effectiveness are two large categories commonly used in a benchmarking comparison. The best performers in a group of hospitals are highlighted. Hospitals can individually and anonymously see where they are by comparison with other hospitals' statistics, then contact the best performers to evaluate their differences. There is a code of conduct to be followed when benchmarking that includes both ethics and etiquette. Hospitals generally benchmark other hospitals but with time will also realize that there is a lot to be learned from other successful industries.

Commercially Purchased Monitors

The College of American Pathologists (CAP) has put together a national quality assurance assessment program called Q-Probes. Laboratories throughout the country subscribe to an annual series of monitors. At least one of these monitors each year has a microbiology focus. Q-Probes have covered topics such as adequacy of sputum cultures, turnaround time for spinal fluid Gram stain results, blood culture contamination rates, and appropriateness of ordering patterns for stool specimens. The method of data collection is precisely outlined, and all worksheets and data forms are provided. Information is returned to subscribers in a manner that enables institutions to benchmark.

It has been said that quality is a journey, not a destination. There are always more things to monitor than there are people or resources to monitor them. The focus is usually on high-volume, high-risk, and problem-prone issues. One of the most important factors in a successful QI program is that it originates at the top of the organization, with the chief executive officer, and involves all employees working together. It is important to point out team successes and give group rewards. A QI program does not materialize overnight. It takes a tremendous amount of planning, education, and resources. The process is forever changing. A nursery rhyme from *The Quality Quest* describes the process well:

> *Good, Better, Best*
> *Never let it rest,*
> *Until the good becomes better*
> *And the better becomes best.*

Bibliography

August MJ, Hindler JA, Huber TW, Sewell DL: Quality Control and Quality Assurance Practices in Clinical Microbiology. Cumitech 3A. Washington, DC: American Society for Microbiology, 1990.
Bartlett RC: Quality control in clinical microbiology. In Balows A, et al (eds): Manual of Clinical Microbiology, 5th ed. Washington, DC: American Society for Microbiology, 1991.
Clark GB: Quality assurance: An administrative means to a managerial end. Clin Lab Manage Rev Part 1, 4:7–15, 1990; Part II, 4:224–252, 1990; Part III, 5:463–475, 1991.
Clinical Laboratory Improvement Act of 1988: Rules and Regulations. Fed Reg, February 28, 1992; 57.
College of American Pathologists: Standards for Laboratory Accreditation. Northfield, IL: College of American Pathologists, 1993.

Continuous Quality Improvement: Can it work in hospitals? AHA Teleconference, March 12, 1992.

Deming WE: Quality, Productivity, and Competitive Position. Cambridge, MA: MIT Center for Advanced Engineering Study, 1982.

Eisenberg HD (ed): Clinical Microbiology Procedures Handbook, Secs 12 and 13. Washington, DC: American Society for Microbiology, 1992.

Joint Commission on Accreditation of Healthcare Organizations: Accreditation Manual for Hospitals. Vol 1: Standards. Oakbrook Terrace, IL: JCAHO, 1993.

Leibor W: The Quality Quest. Chicago, IL: American Hospital Publishing, 1991.

National Committee for Clinical Laboratory Standards: 1991 Performance Standards for Antimicrobial Susceptibility Testing, M100-S3. Villanova, PA: NCCLS, 1991.

National Committee for Clinical Laboratory Standards. 1990 Performance Standards for Antimicrobial Disk Susceptibility Tests, 4th ed. Approved Standard; M2-A4. Villanova, PA: NCCLS, 1990.

National Committee for Clinical Laboratory Standards: 1990 Methods for Dilution Antimicrobial Susceptibility Tests for Bacteria That Grow Aerobically, 2nd ed. Approved Standard; M7-A2. Villanova, PA: NCCLS, 1990.

B. PUTTING THE LABORATORY TEST TO THE TEST

ANALYTICAL ANALYSIS OF TESTS
  Analytical (Technical) Sensitivity
  Analytical (Technical) Specificity
  Accuracy

CLINICAL ANALYSIS OF TESTS
  Clinical (Diagnostic) Sensitivity
  Clinical (Diagnostic) Specificity

OPERATIONAL ANALYSIS OF TESTS
  Prevalence of Disease
  Incidence of Disease
  Predictive Values of Tests
    Positive Predictive Value
    Negative Predictive Value
    Example
    Clinical Applications of Positive and Negative
      Predictive Values
  Efficiency of Tests
  Other Concepts

CHOOSING A LABORATORY TEST

Richard B. Prior

The development of new diagnostic tests in clinical microbiology is being directed toward the detection of antigens more so than the detection of antibodies. Although culture still remains the gold standard for diagnostic purposes, the newer, noncultural antigen detection tests offer simplicity and speed for detection of the etiologic agent. In addition, the immune status of the patient does not affect test results. These noncultural antigen detection tests are being designed for physician's office use and, as such, are simple for nonlaboratory personnel to perform. Because the test results are available quickly, the physician can make an informed decision regarding patient management.

These tests can be misused, however, or their test results misinterpreted. Tests with a high sensitivity and specificity may be promoted as extremely reliable diagnostic tests. In certain clinical situations, however, such tests may not add meaningful information to the diagnosis. This section reviews the pertinent definitions related to evaluating tests and presents several clinical examples to illustrate the basic concepts.

ANALYTICAL ANALYSIS OF TESTS

The following terms relate to the actual test and are not used in the critical evaluation of a test in the clinical setting. *Analytical sensitivity and specificity* should not be confused with *clinical sensitivity and specificity.*

Analytical (Technical) Sensitivity

The *analytical sensitivity* of a test refers to its ability to detect a particular analyte or a small change in its concentration. For example, the enzyme immunoassay (EIA) is more sensitive than the precipitation test in detecting antibody. EIA can detect lower concentrations of antibody.

Analytical (Technical) Specificity

A test's analytical specificity refers to its ability *not* to react with substances other than the analyte of interest.

Accuracy

The degree of conformity of a measurement to a standard or a true value is *accuracy.* It is a measure of analytical capability. For example, with the Performance Standards for Antimicrobial Disk Susceptibility Tests, a mean value of several observations is compared to a predetermined standard value or range. If the mean (or range) of control limits found in standard tables is exceeded, a technical systematic error (variation) exists that may lead to misinterpretation of test results.

Accuracy is a function of two characteristics: precision and bias. *Precision* is the measure of exactness or the degree of refinement with which a test is performed. It is the dispersion of repeated observations caused by random errors. The precision (reproducibility) of a test is usually monitored by means of the range (maximum minus minimum) within sets of observations. *Bias* is the mean difference of test results from an accepted reference method caused by systematic errors.

Typically, precision and bias are determined by studies whereby several investigators or laboratories perform repeated testing of several specimens of known values. Then, test results are analyzed by statistical methods to determine the test's precision and bias within and among laboratories.

CLINICAL ANALYSIS OF TESTS

Clinical (Diagnostic) Sensitivity

Clinical (diagnostic) sensitivity refers to the proportion of positive test results obtained when a test is applied to patients known to have the disease. Thus, it is the frequency of positive test results in patients *with* the disease (*true-positive* results). For example, if 100 patients who have gonorrhea are tested for that disease and the test yields positive results in 95 and is negative in the other 5, then the sensitivity of the test is 95%. Five of these 100 patients had *false-negative* test results.

The sensitivity of a particular test does not change, provided that the test is performed correctly. The sensitivity is determined by multicenter clinical trials whereby the test results, obtained by performing the test according to a specific protocol, are compared with a gold standard. In microbiology, the standard is usually culture, which in many cases is not absolute. Thus, the reported sensitivity for a certain test may vary among laboratories. Evaluating new diagnostic tests is made difficult because of these imperfect standards.

Clinical (Diagnostic) Specificity

Clinical (diagnostic) specificity refers to the proportion of negative results obtained when a test is applied to patients known to be free of the disease. Thus, it is the frequency of negative test results in patients *without* the disease (*true-negative* results). For example, if 100 patients without gonorrhea are tested for that disease and the test yields negative results in 90 and is positive in the other 10, then the specificity of the test is 90%. Ten of these 100 patients had *false-positive* test results.

As with sensitivity, the specificity does not change, provided that the test is performed correctly. The specificity is also determined by clinical trials whereby the test results are confirmed by a more definitive test or procedure.

OPERATIONAL ANALYSIS OF TESTS

Although the sensitivity and specificity do not change for a given test, the prevalence of the disease and the positive and negative predictive values of a test do change. In clinical medicine, these parameters are extremely important to know when evaluating a particular test result.

Prevalence of Disease

Prevalence is the frequency of the disease at a designated point in time in the population being tested. For example, during influenza season, the prevalence of "strep" throat in children attending daycare centers will be higher than the prevalence in children who do not attend such centers. The role prevalence plays in determining the positive and negative predictive values of a test can be seen in the formulas below. To properly interpret test results, the clinician must have an understanding or estimate of the prevalence of the disease in the population being tested. Prevalence can be estimated on the basis of clinical experience and on information provided by the local and state health departments as well as information periodically provided by the Centers for Disease Control and Prevention (CDC) in Atlanta, GA.

Incidence of Disease

Incidence refers to the number of new cases per year and is a measure of events. The relation between prevalence (P) and incidence (I) can be seen in the equation $P = I \times D$, where D is duration of disease from onset (diagnosis) to termination.

For chronic diseases such as cancer, the prevalence is greater than the incidence; for acute diseases, such as gonorrhea, the prevalence is less than the incidence. Prevalence is used to compute the predictive values of tests.

Predictive Values of Tests

A test can have both a positive and a negative predictive value. The three elements needed to compute the positive and negative predictive values of a test are:

■ Sensitivity of the test

■ Specificity of the test

■ Prevalence of the disease being tested

Although predictive value theory has only recently been used in clinical medicine, the formulas and concepts are not at all new. The formulas for calculating positive and negative predictive values are commonly referred to as *Bayes theorem,* which was published posthumously in 1763.

Positive Predictive Value

The *positive predictive value* (PPV) of a test is the probability that a patient with a positive test result *does* have the disease. It is the likelihood that a patient in whom a test is positive actually has the disease, sometimes is referred to as the *diagnosability* of a test. The positive predictive value can be computed as follows:

$$PPV = \frac{(P)\ (Se)}{(P)\ (Se) + (1-P)\ (1-Sp)} \times 100$$

where:
   PPV = positive predictive value of test
   P = prevalence of the disease being tested
   Se = sensitivity of test
   Sp = Specificity of test.

Negative Predictive Value

The *negative predictive value* (NPV) of a test is the probability that a patient with a negative test result *does not have* the disease. It is the likelihood that a patient in whom a test is negative is free from the disease, sometimes is referred to as the *excludability* of a test. The negative predictive value can be computed as follows:

$$NPV = \frac{(1-P)\ (Sp)}{(1-P)\ (Sp) + (P)\ (1-Se)} \times 100$$

where:
   NPV = negative predictive value of test
   P = prevalence of the disease being tested
   Se = sensitivity of test
   Sp = Specificity of test.

Example

To illustrate the concepts of predictive values, consider a certain diagnostic test that has a sensitivity (Se) of 95% and a specificity (Sp) of 95%. In a primary care hospital where the prevalence of the disease being tested is 1%, the positive predictive value of the test is only 16%. This is calculated using the PPV equation, as follows:

$$PPV = \frac{(0.01)\ (0.95)}{(0.01)\ (0.95) + (1-0.01)\ (1-0.95)} \times 100$$
$$= 0.16\ \text{or}\ 16\%$$

The clinician has only a 16% certainty that a patient with a positive test result actually has the disease. If, however, the same test is used in a tertiary care hospital where the prevalence of the disease may be 50%, the positive predictive value increases to 95% (calculated using the same equation). The clinician in this case has a 95% certainty that a patient with a positive test result actually has the disease. Therefore, prevalence of the disease has a great influence on the predictive value of a test.

Clinical Applications of Positive and Negative Predictive Values

GROUP A STREPTOCOCCUS TESTING OF THROAT SAMPLES

Acute pharyngitis is one of the most common conditions seen by primary care physicians. Although most of the infections are caused by viruses, 5 to 15% of cases have a bacterial etiology, usually group A β-hemolytic streptococci (*Streptococcus pyogenes*). To test for these organisms, 28 to 36 million throat cultures are performed annually in the United States.

Noncultural tests have been developed and approved for use in the private office setting to evaluate patients with acute pharyngitis for the presence of group A streptococci. The reported sensitivities and specificities of these tests vary considerably, however, ranging from the low sixties to the high nineties. Although speculative, explanations for such variations may be the difficulty in obtaining a good throat sample, especially in children, and the use of nonperfect culture methods as standards.

Assuming that a certain "strep" test has reported sensitivity and specificity of 90 and 98%, respectively, and that the estimated prevalence for "strep" throat in acute pharyngitis cases is 5%, the positive and negative predictive values would be 70.3 and 99.5%, respectively. With a positive test result, there is approximately a 30% chance that the patient does not have a streptococcal infection. If the test result is negative, there is a greater than 99% chance that the patient is not infected. If the prevalence in the population being tested increases to 15%, the positive predictive value increases to 88.8%, but the negative predictive value decreases slightly, to 98.2%.

Usually, negative test results are more reliable in predicting the absence of disease than positive test results are in predicting the presence of disease. In this example, a negative result would indicate the absence of group A β-hemolytic streptococci with a high degree of certainty, and the patient could be spared the administration and cost of penicillin. Although clinical judgment is very important, these tests can be helpful to the clinician in the proper evaluation and management of the patient with acute pharyngitis.

DIRECT DETECTION OF *CHLAMYDIA TRACHOMATIS* IN URETHRAL AND CERVICAL SPECIMENS

Chlamydia is the most prevalent STD in the United States, with estimates of from 3 to 10 million new cases occurring per year. Many of the individuals infected are asymptomatic, and the proper diagnosis and treatment are essential to prevent the spread of this disease and its associated complications.

The growth cycle of the bacterium *C. trachomatis* is unique, because it is an obligate intracellular parasite that requires living cells for cultivation. It takes several days to obtain results, however, and the sensitivity of the culture is only about 85%. As a result, newer noncultural methods have been developed and tested. Several tests are now available for use, including tests utilizing the EIA, the immunofluorescence assay (IF), and DNA probes. Each has a considerable variance in reported sensitivities and specificities, most likely because of the nature of the organism and the imperfect culture standard to which the results are compared. In addition, each test requires specialized equipment that must be maintained and calibrated.

Table 4–8 shows the positive and negative predictive values for the IF and DNA probe tests applied directly to cervical samples in a population of obstetrics and gynecology clinic patients in whom the prevalence of chlamydial cervicitis is estimated to be 5%. The sensitivities and specificities for the two tests are similar, but the positive predictive value is better for the DNA probe test. Thus, patients with positive results of the DNA probe test are more likely to have chlamydial cervicitis than with positive results of the IF test. On the other hand, both tests identify patients with negative test results as not having chlamydial cervicitis with a high degree of certainty (99.5%).

Table 4–8 also shows the predictive values if the same tests are used in a sexually transmitted dis-

ease (STD) clinic where the prevalence of chlamydial cervicitis is estimated to be 30%. The positive predictive values increase significantly, especially for the IF test, and both tests would function well as diagnostic tools. The negative predictive values drop for both tests, however, to 95.8%. Thus, approximately 4% of patients with negative test results could be infected with *C. trachomatis;* this rate may not be acceptable owing to the nature of the disease.

Efficiency of Tests

The *efficiency* of a test indicates the percentage of patients who are correctly classified as having disease or not having disease. The efficiency is calculated using the following equation:

$$\text{Efficiency} = \frac{(TP + TN)}{(TP + FP + TN + FN)} \times 100$$

where:
TP = number of patients with true-positive results
TN = number of patients with true-negative results
FP = number of patients with false-positive results
FN = number of patients with false-negative results.

Other Concepts

Additional concepts, such as the medical decision-making analysis of tests, "benefit-cost analysis," combination testing, and nondisease factors affecting laboratory test results, are beyond the

■■■■T A B L E   4 – 8

COMPARISON OF THE IF AND DNA PROBE TESTS TO DETECT CHLAMYDIAL CERVICITIS IN PATIENT POPULATIONS WITH 5% AND 30% PREVALENCES OF THE DISEASE*

Test	Sensitivity	Specificity	5% Prevalence		30% Prevalence	
			PPV	*NPV*	*PPV*	*NPV*
IF	90.0	98.0	70.3	99.5	95.1	95.8
DNA probe	89.8	99.5	90.4	99.5	98.7	95.8

*The positive (PPV) and negative (NPV) predictive values were computed using equations (see text). All values are given in percentages.

scope of this discussion. For those wishing to examine these in detail, however, an excellent discussion with appropriate tables may be found in the "Statistics" chapter by Robert Galen in *Gradwohl's Clinical Laboratory Methods and Diagnosis* (see Bibliography). The tables for the positive and negative predictive values in this reference are not needed, because predictive values are easy to compute using a simple hand calculator. Anyone with access to a computer can set up a spreadsheet incorporating the predictive value formulas. When the disease prevalence and test sensitivity and specificity are entered, the computer program calculates the predictive values for the specified prevalence.

CHOOSING A LABORATORY TEST

Increasing the sensitivity and specificity of a test, by improving its chemistry or methodology, for example, will improve its predictive values. In fact, companies are constantly trying to improve their tests' performance in order to increase the predictive values. The ideal test would have a sensitivity and a specificity of 100%, in which case, according to the PPV and NPV equations, the positive and negative predictive values would be 100% and would not change regardless of the prevalence of the disease. In practice, however, tests do not have 100% sensitivity and 100% specificity at the same time. Therefore, the clinician must decide what factors are important in the management of the patient. Criteria that the clinician should use to decide whether to use a test include the prevalence of the disease, the severity of the disease, the cost of the test, the amenability of the disease to treatment, and the consequences of treatment.

Tests with high sensitivities, preferably 100%, should be used when the disease is serious and should not be missed, or when the disease is treatable. Such diseases are the treatable STDs (gonococcal and chlamydial infections). Furthermore, tests with sensitivities of 100% will also yield negative predictive values of 100% regardless of the disease prevalence. Excluding diseases is just as important as diagnosing them,

especially with STDs, and such tests would be valuable in this clinical setting.

Tests with high specificities, preferably 100%, should be used when the disease is serious but not treatable or when the knowledge that the disease is absent would have psychological or public health value. Such disease conditions include human immunodeficiency virus and occult cancers. Furthermore, tests with specificities of 100% yield positive predictive values of 100% regardless of the disease prevalence.

Performance of a test is also affected by the clinical judgment of the clinician. Ordering a test places the patient clinically suspected of having a particular disease in a new population in which the prevalence or probability of the disease is high. Thus, the test will perform much better, because the predictive values will be higher in the population with the higher prevalence.

New laboratory tests as well as old ones should always be "put to the test." Understanding and using the concepts of predictive values and the effect of prevalence on those values is important for the proper use of a test and interpretation of the results. The development of rapid noncultural tests for the detection of infectious diseases makes the use of these concepts even more important. A test result can no longer be considered as simply "positive" or "negative" but must be interpreted in view of the concepts presented in this discussion. When test results are interpreted properly, better patient care is achieved.

Bibliography

Galen R: Statistics. In Sonnenwirth AC, Jarett L (eds): Gradwohl's Clinical Laboratory Methods and Diagnosis, 8th ed. St. Louis: CV Mosby, 1980, pp 41–68.

Gullen WH, Bearman JE: Put laboratory tests to the test! Patient Care, February 15, 1980, pp 74–93.

Vecchio TJ: Predictive value of a single diagnostic test in unselected populations. N Eng J Med 274:1171–1173, 1966.

Washington JA II, Doern GV: Assessment of new technology. Balows A, et al (eds): In Manual of Clinical Microbiology, 5th ed. Washington, DC: American Society for Microbiology, 1991, pp 44–48.

Weigert HT, Weigert O: The impact of disease prevalence on the predictive value of laboratory tests in primary care. J Fam Pract 8:1199–1203, 1979.

CHAPTER 5

Emergent Technologies

Mario J. Marcon

INTRODUCTION

Since the pioneering work of Robert Koch in the 1870s and 1880s, microbiologists have relied on culture isolation of microorganisms to establish the etiology of infectious diseases. Koch's techniques for isolating organisms in pure culture, along with the earlier invention and improvement of the microscope, allowed scientists to discover the causal organisms for most of the major bacterial human diseases before the turn of the century. Even today, we recognize that isolation in pure culture and biochemical and/or serologic identification of viable organisms is still the gold standard for diagnosis of infectious diseases. Moreover, isolation is necessary if standard antimicrobial susceptibility testing is to be performed on the organism.

There are a number of limitations to the classic approach of culture isolation and identification of an organism to establish diagnosis of an infectious disease. In some cases, the agent cannot be cultivated on artificial media, such as is the case with *Treponema pallidum,* the causative agent of syphilis. In the case of viral diseases and bacterial diseases caused by obligate intracellular organisms (e.g., chlamydial and rickettsial agents), isolation of the organism requires animal inoculation or the use of cell culture, a labor-intensive and expensive method that may not be available to all laboratories. Still other cultivable organisms may be relatively labile in transport conditions, may require a long incubation period, or may be so fastidious as to render isolation unreliable. Collectively, these limitations make it difficult for clinicians to make early patient care decisions on the basis of culture results.

Serologic diagnosis, the demonstration of a significant rise in serum antibody titer (*seroconversion*) to a component(s) of the infecting organism, has been used with great success as an adjunct or alternative procedure to culture isolation for diagnosis of infectious diseases. Because it may take 2 weeks or longer for specific antibody to appear, serology is most often useful for diagnosing chronic infections (e.g., hepatitis, acquired immune deficiency syndrome [AIDS]), infections by noncultivable agents (syphilis), or for retrospective diagnosis or epidemiologic studies rather than acute care medical decision-making. Moreover, individual patients respond differently to antigenic stimuli. Patients with congenital or acquired immunodeficiencies may not respond to antigenic stimuli, or they may respond with limited antibody production (no seroconversion). Additionally, a serologic response may not be specific for the test agent, thus contributing to difficulties in interpreting serologic tests.

The ability to directly detect an intact microorganism or one of its chemical components in a patient's tissue, body fluid, or other specimen offers the potential for rapid diagnosis of an infectious disease. This is because the need for culture isolation or the development of an antibody response is bypassed. Of course, many of the classic, nonspecific staining and wet preparation methods coupled with microscopy (light, darkfield, phase-contrast) can be classified as direct, noncultural microbial detection methods; they include Gram stain for bacteria, KOH preps for fungi, iodine mounts or trichome stains for protozoan parasites, and Giemsa stain for viral or chlamydial inclusions. These methods generally lack specificity (except perhaps for protozoan identification), because they do not identify the agent to the species level. Additionally, they lack sensitivity compared with culture, particularly if a specimen concentration method is not used. This is because a relatively large number of organisms (for bacteria, $\geq 10^4$ per mL or per gram of specimen) must be present in unconcentrated clinical material before nonspecific staining methods yield positive results.

Over the past two decades, a number of direct microbial antigen detection methods that identify a specific component of an organism (generally a protein or carbohydrate) in a patient's specimen have found their way into the clinical laboratory. The most popular of these methods, the immunoassays, depend on the specific interactions of labeled antibodies with microbial antigens. Other direct detection methods rely on chromatographic detection of a specific microbial component or metabolic product or on the demonstration of the presence in the specimen of a unique microbial toxin either by assaying for specific cytotoxicity in vitro or by performing animal inoculation. The detection of microbial nucleic acid sequences in clinical specimens, either directly or following in vitro amplification, has also received quite a bit of attention because of the potential for increased sensitivity and specificity in direct detection assays. This chapter discusses these noncultural approaches for diagnosis of infectious diseases, along with rapid manual and instrument-assisted methods.

A. DIRECT MICROBIAL ANTIGEN DETECTION

Mario J. Marcon

OBJECTIVES

1. Describe the principles of the following antigen detection methods:

- Particle agglutination.
- Precipitin test.
- Fluorescence-labeled antibody test.
- Enzyme-labeled antibody test.

2. Differentiate each of the following solid phase immunoassay methods:

- Enzyme immunoassay (EIA).
- Fluorescent immunoassay (FIA).
- Radioimmunoassay (RIA).
- Optical immunoassay (OIA).

3. Describe the current clinical applications of direct antigen detection methods in each of the following infectious diseases:

- Respiratory tract infections.
- Meningitis and sepsis.
- Gastrointestinal infections.
- Sexually transmitted diseases.
- Blood-borne and body fluid–borne diseases.

4. List possible future applications of direct antigen detection in the early diagnosis of infectious diseases.

DEFINITIONS

Direct microbial antigen detection is the process by which microbial-specific structural components, known as *antigens,* are identified in specimens obtained from an infected host. The antigens are usually high-molecular-weight (HMW) proteins or polysaccharides that are considered foreign to the host. Antigen detection does not involve culture of the clinical specimen, although antigen detection methods can be used to identify microorganisms once they have been recovered in culture. The method depends on the fact that some microbial components are chemically unique and form areas on the molecule known as antigenic *determinants.* There may be many antigenic determinants, either identical or different in chemical composition, on a large antigen molecule. Antigens can be recognized by and can combine specifically with antibody molecules to form stable products; the interaction

occurs between the antigenic determinant and the antigen binding site. Thus, *antibodies* can be defined simplistically as proteins that react specifically with antigens.

The basic process of antigen detection involves reacting the clinical specimen, either directly or following some initial preparation, with an antibody preparation specific for an antigen of interest. If the microbial antigen is present in the specimen, an antigen-antibody interaction occurs. This interaction can then be detected by a number of techniques, often taking advantage of the fact that antibody molecules are polyvalent—have two or more antigen-binding sites per molecule—and can combine with two or more antigen molecules (Fig. 5–1). Although our discussion of antigen detection methods is limited to microbial antigens, the same techniques are broadly applicable to a number of analytes, such as serum proteins, hormones, and drugs, as well as for the detection and quantitation of serum antibodies.

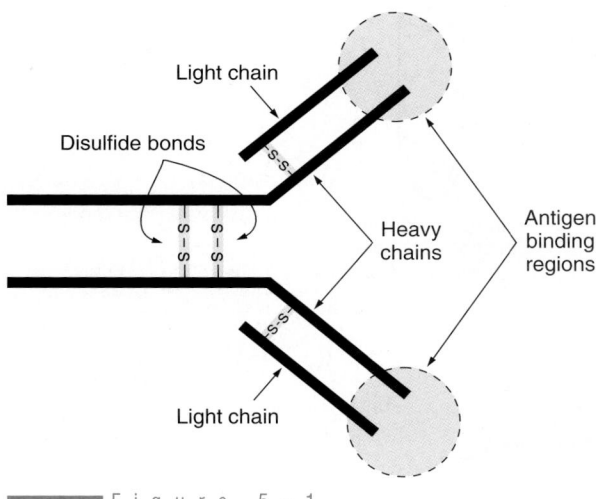

Light chain

Disulfide bonds

Heavy chains

Antigen binding regions

Light chain

■■■■■ F i g u r e   5 – 1

Structure of an antibody molecule of the immunoglobulin G (IgG) class.

HISTORICAL PERSPECTIVE

The origins of direct microbial antigen detection lie in the early 1900s, when investigators discovered that the serum and urine of patients with typhoid fever contained a soluble substance (*precipitin*) that would precipitate when mixed with rabbit anti-*Salmonella* antiserum. At about the same time, the cerebrospinal fluid (CSF) of patients with meningococcal meningitis was found to contain a similar precipitin when reacted with antiserum against *Neisseria meningitidis*. Shortly thereafter, the urine and serum of patients with pneumococcal pneumonia were found to contain similar precipitins.

The next significant development in antigen detection occurred around 1950, when fluorescent-labeled antibody tests for visualizing whole organisms in specimens became practical. The 1950s and 1960s also saw considerable progress in detecting soluble antigens, particularly as it relates to two organisms, *Cryptococcus neoformans* and hepatitis B virus.

We can consider the modern era of direct antigen detection to have begun in the mid-1970s. In the time since then, we have seen a growing interest by physicians in the rapid diagnosis of infectious diseases. The number of methods and applications of direct microbial antigen detection also increased dramatically. Perhaps just as important as the newer methods for antigen detection, the ability to produce large quantities of antibody with high purity, *avidity* (the strength of

binding of antibody to antigen), and *specificity* (the ability of antigen-binding sites to discriminate among closely related antigenic determinants) was developed. This development was largely the result of *cell hybridoma* technology, the ability to fuse a single plasma B cell producing an antibody of interest with a different cell type (commonly, mouse myeloma cells) that has been transformed and thus able to grow and divide in culture forever (*immortalized*). The progeny of these cells, if appropriately selected, can be grown in cultures or animals and can secrete large amounts of the antibody of interest. The antibody produced by such cells, *monoclonal antibody,* is so named because it originates from a single B cell or clone and is thus chemically homogeneous (Fig. 5–2). Monoclonal antibodies are highly specific for a single antigenic determinant and less likely to cross-react with chemically related antigens than *polyclonal antibodies,* which recognize multiple antigenic determinants. The latter are commonly produced by immunizing animals with the antigen of interest and then isolating and purifying the antibody from the animal's serum (see Fig. 5–2). These antibodies are generally heterogeneous in nature because of the variability associated with the immune response; they may lack avidity or specificity compared with monoclonal antibodies. Many of the methods and commercial products discussed here utilize monoclonal, as well as traditional polyclonal, antibodies.

ANTIGEN DETECTION METHODS

Particle Agglutination

Latex Agglutination

Latex agglutination takes advantage of the fact that latex (polystyrene) beads, which can be manufactured to a number of diameters, will bind antibody molecules—(immunoglobulin (Ig)G or IgM, monoclonal or polyclonal)—either nonspecifically, through noncovalent interactions, or specifically, by covalent bonding to chemical groups placed on the surface of the beads during their manufacture. Latex beads commonly used in diagnostic tests are about one micron in diameter (Fig. 5–3), and each bead can be charged with thousands of antibody molecules. The antibody-charged latex beads form a homogeneous, milky suspension when thoroughly mixed. Interaction with specific antigen results in antigen-

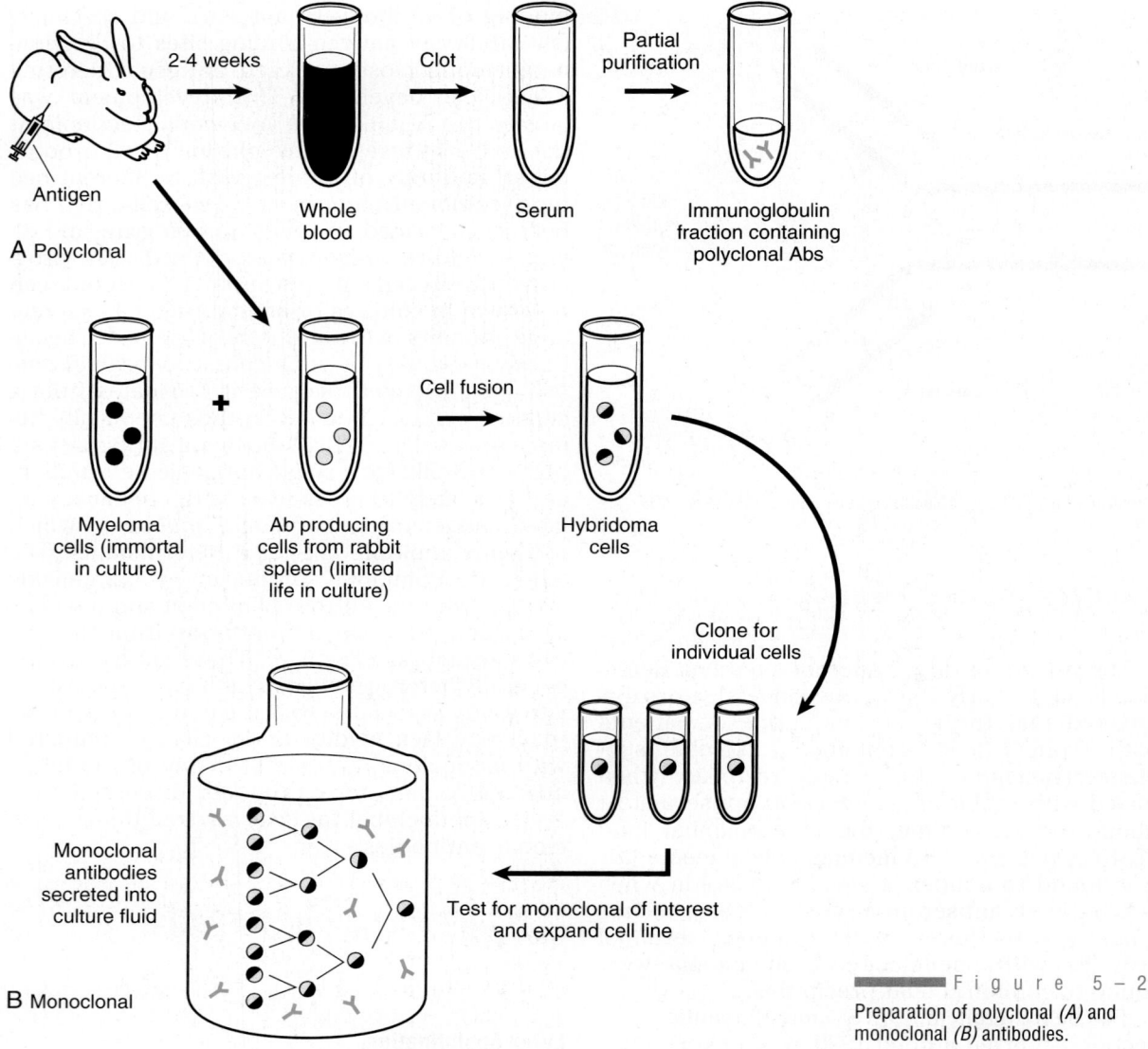

2-4 weeks

Clot

Partial
purification

Antigen

Whole
blood

Serum

Immunoglobulin
fraction containing
polyclonal Abs

A Polyclonal

+

Cell fusion

Myeloma
cells (immortal
in culture)

Ab producing
cells from rabbit
spleen (limited
life in culture)

Hybridoma
cells

Clone for
individual cells

Monoclonal
antibodies
secreted into
culture fluid

Test for monoclonal of interest
and expand cell line

B Monoclonal

F i g u r e   5 – 2
Preparation of polyclonal (A) and
monoclonal (B) antibodies.

antibody binding; however, this primary binding between antigen and antibody molecules does *not* produce a visible agglutination (clumping) reaction. If the antigen molecules contain multiple antigenic determinants (polyvalent antigen) for binding to antibody (as is commonly the case for HMW polysaccharide and protein antigens), the antigen molecules can be cross-linked by the antibody-charged latex beads. It is this secondary cross-linking of antigen and antibody, resulting in a macromolecular lattice, that precipitates out of solution and results in a visible agglutination reaction (Fig. 5–4).

The reaction is generally conducted on a treated cardboard or glass slide using liquid specimen and latex volumes of about 50μL each (Fig. 5–5). The reagents are mixed thoroughly, and then the slide is rocked or rotated by hand or with a mechanical device for a specified time (often 2 or 3 or up to 10 minutes) before being read using appropriate lighting and the naked eye. The reaction depends on many variables, including latex particle size, avidity of antibody, type of antibody (monoclonal or polyclonal), pH and ionic strength of the test specimen, reaction temperature, and concentration of antigen in the

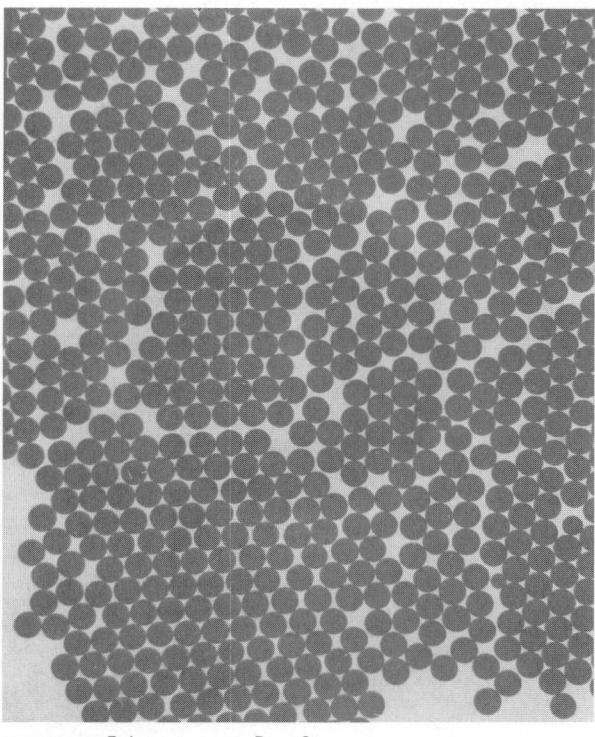

■■■■■ F i g u r e   5 – 3

Transmission electron micrograph of a latex suspension. Each latex bead is about 1 micron in diameter. (Courtesy of Interfacial Dynamics Corp., Portland, OR.)

specimen. Levels of detection of microbial polysaccharide or protein antigens have been as low as 0.1ng/mL. The strength and rapidity of the reaction varies depending on all these reaction conditions and, most importantly, on the concentration of antigen in the specimen. With high antigen concentration, a maximum agglutination reaction (4+ on a 1+ to 4+ scale) may occur in only a few seconds. In the absence of specific antigen, the latex suspension remains homogeneous and milky. Although most applications of latex agglutination use white latex beads, colored latexes are also available.

Some important controls must be run along with the test latex and patient's specimen (Table 5–1). These controls include (1) a positive antigen control (a solution containing the antigen of interest), (2) a negative antigen control (a solution *not* containing the antigen), and (3) a control to detect the presence of factors in patients' specimens responsible for nonspecific agglutination reactions. This last control usually involves testing the patient's specimen with a latex suspension charged with an immunoglobin (generally obtained from the same animal species in which the specific antibody was made) whose specificity is not directed to the test antigen. A nonspecific agglutination reaction occurs when the patient's specimen reacts with both the test and the control latex; when such reactions occur, the test is uninterpretable. A positive test result requires that the test latex but not the control latex agglutinate with the patient's specimen.

A number of specimen pretreatment procedures can be used to eliminate or minimize nonspecific agglutinations, presumably by removing or inactivating factors in the specimen responsible for these reactions. These procedures include specimen centrifugation to remove particulate material, heat treatment to inactivate protein constituents (acceptable when test antigen is a heat-stable polysaccharide), and passing the specimen through a membrane filter. When the test specimen contains very high protein content, such as serum, and the antigen of interest is a heat-stable polysaccharide, it may be necessary to first treat the specimen with ethylenediamine tetraacetic acid (EDTA) before heating, or polysaccharide antigen would be trapped in the coagulated protein and removed from solution when the specimen is centrifuged.

Latex agglutination tests for microbial antigen, like other laboratory tests, are subject to false-negative and false-positive reactions when compared with culture. False-negative reactions (negative antigen test result, positive culture) may be due to the presence of antigen in the specimen at concentrations below the test detection limit. False-positive reactions (positive antigen test result, negative culture) are more difficult to explain; they may be due to the presence of cross-reacting antigens shared with other microorganisms or host cellular components. It is important to remember that latex agglutination tests do not require that viable microbes be present in the specimens. Either whole living or dead organisms, microbial components such as cell wall or membrane fragments, or soluble antigen such as bacterial capsular polysacchride may be detected by latex agglutination. Thus, an apparent false-positive antigen test result may really represent a true-positive result for disease in a patient with a negative culture.

Some specimens, such as urine, may be concentrated before testing. Concentration procedures usually involve membrane filtration of the specimen using a filter that will hold back or exclude HMW antigenic materials while allowing

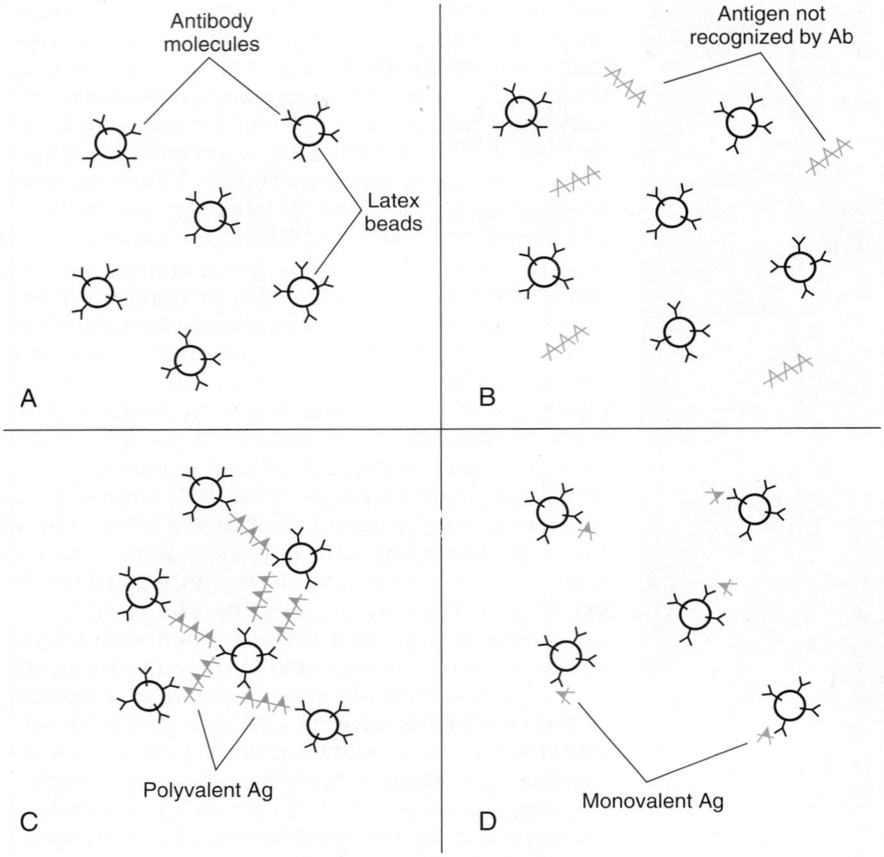

Antibody
molecules

Latex
beads

A

Antigen not
recognized by Ab

B

Polyvalent Ag

C

Monovalent Ag

D

■ Figure 5 – 4

Illustration of latex agglutination reactions with soluble antigen. Antibody-coated latex particles form a homogeneous suspension *(A)*. When mixed with a solution containing antigen not recognized by the antibody *(B)*, the antibody-coated latex particles remain in suspension (no agglutination). Agglutination occurs when the antibody-coated latex combines with specific polyvalent antigen, allowing cross-linking of molecules *(C)*, but not when the specific antigen is monovalent *(D)*.

■ Figure 5 – 5

Commercial latex agglutination test slide showing reaction of controls and a patient's cerebrospinal fluid with latex reagents specific for *Haemophilus influenzae* b (Hib), *Streptococcus pneumoniae* (Sp), group B *Streptococcus* (GBS), and five different serogroups (groups C and W135, groups A and Y, and group B) of *Neisseria meningitidis* (Nm). Note the positive reactions with each test latex reagent and positive control (P), as well as the positive reaction with the patient's specimen and Hib test latex. (Courtesy of Becton Dickinson and Company, Cockeysville, MD.)

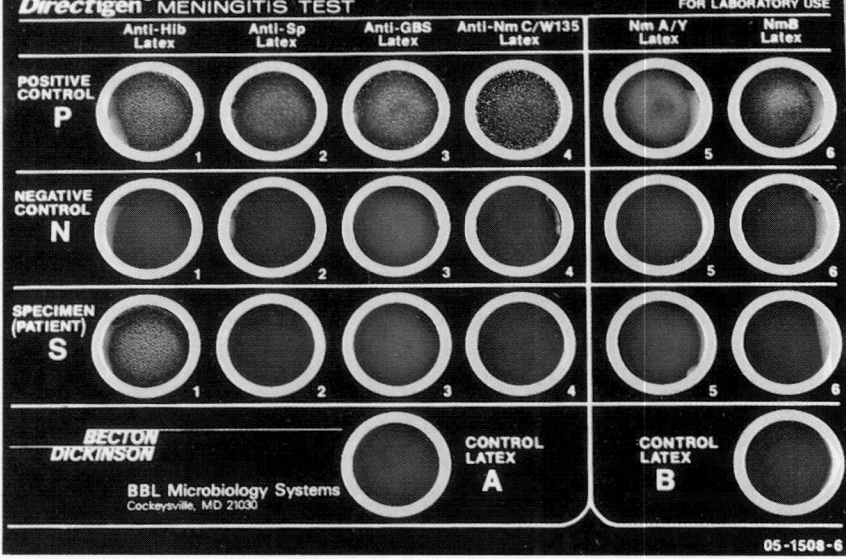

■■■■ T A B L E   5 – 1

INTERPRETATION OF LATEX AGGLUTINATION TEST CONTROLS AND PATIENT SPECIMENS

Reaction Number	Test Specimen	Agglutination Reaction		Interpretation
		Test Latex	*Control Latex*	
1	Positive antigen control	+	0	Control OK
2	Negative antigen control	0	0	Control OK
3	Patient A	+	0	Positive test
4	Patient B	0	0	Negative test
5	Patient C	+	+	Nonspecific agglutination

+, agglutination; 0, no agglutination.

water and small compounds to pass through. These concentration methods increase the sensitivity of the test.

When a commercial system for antigen detection is used, it is important that the manufacturer's recommendations be followed exactly and that reagents from different manufacturers not be used interchangeable. It is also important that a system not be used for an application for which it has not been evaluated and approved by the manufacturer and regulatory agencies.

The advantages of the latex agglutination test are availability of good-quality reagents in complete kit form, good sensitivity, relative rapidity, and ease of performance. Disadvantages include subjectivity in reading end points and the fact that some tests are not as "rapid" as an ordering physician might expect. Still, it is one of the most widely used antigen detection methods. Also, improvements in latex bead technology are continually being made. For example, latex beads with magnetized cores have been developed. Once charged with antibody molecules, they can be used for detection and physical separation of antigens.

Staphylococcal Coagglutination

Similar to latex agglutination, *staphylococcal coagglutination* or, simply, *coagglutination* utilizes particle-bound antibody to enhance the visibility of antigen-antibody reactions. Instead of using a latex bead to bind antibody, formalin-killed but intact *Staphylococcus aureus* cells, which contain a large amount of protein A in the cell wall, are used (typically Cowan I strain of *S. aureus*). This protein has the capacity to bind antibody molecules by means of a receptor located in the Fc portion

(base of the heavy immunoglobulin chain) of the IgG antibody, leaving the antigen-binding sites of the molecule available to react with specific antigen (Figs. 5–6 and 5–7).

It has been estimated that each staphylococcal cell has about 80,000 antibody binding sites. The actual number of antibody molecules bound to the staphylococcal cell is limited by stearic hindrance. Coating is sufficient, however, to render the product of clinical utility in direct antigen tests.

Most of the observations made regarding latex agglutination are also true for coagglutination. Reactions are prepared by mixing antibody-sensitized staphylococcal cells with a solution containing the antigen of interest on a slide or card (Fig. 5–8). Coagglutination procedures appear more susceptible to nonspecific agglutination reactions, and thus, specimen preparation is important. This is particularly true for testing serum specimens, probably because staphylococcal cells may bind human IgG in test serum specimens and subsequently be agglutinated by the presence of rheumatoid factor (IgM anti-IgG) in the serum. Coagglutination may also be less sensitive in detecting microbial antigens. It is now less commonly used than latex agglutination for direct testing but is still a popular method for identifying culture isolates.

Liposome-Mediated Agglutination

One of the newest developments in agglutination technology involves the use of *liposomes,* single-lipid bilayer membranes that form closed vesicles under appropriate conditions. In their manufacture, antigen or antibody molecules may be incorporated into the surface of the membrane

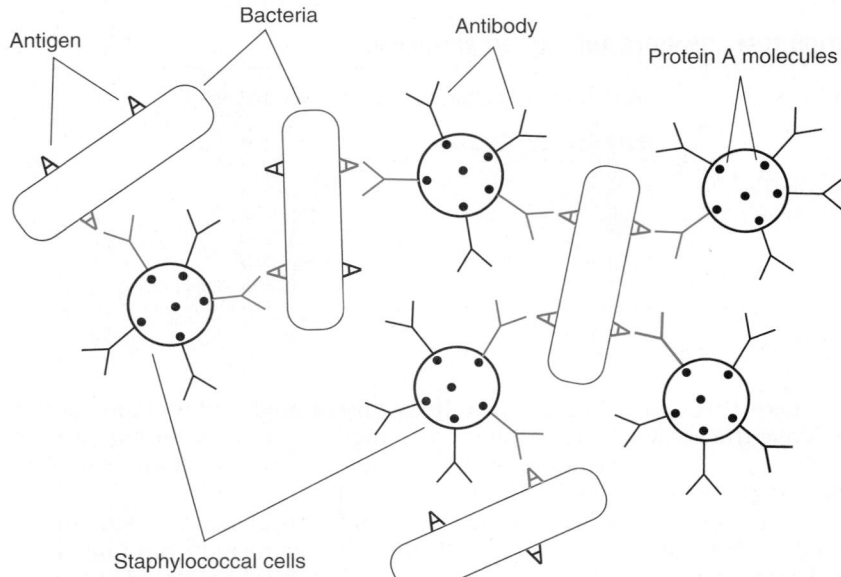

Staphylococcal cells

■ F i g u r e   5 – 6
Diagram of coagglutination reaction with whole bacterial cell antigen.

and thus be available for interaction with the corresponding molecule (Fig. 5–9). In addition, the liposome vesicle may be constructed with a chemical dye or bioactive molecule trapped in the interior. The colored dye allows easy visual detection of lattice formation and agglutination between liposome-bound antibody and antigen. Alternatively, combining dye-containing liposomes and latex beads, both of which contain the reactive antibody on the surface, may increase the sensitivity of latex agglutination. Perhaps the greatest potential advantage of liposome technology is in its application to immunoassays (other than particle agglutination) that make use of the ability of the liposome to carry reactive chemicals. The use of liposomes in antigen detection assays will grow as their cost to manufacturers decreases and standardization improves. A comparison of particle agglutination methods is shown in Table 5–2.

Precipitin Tests

Tube and Agar Precipitin

The basic chemical principles that govern antigen-antibody interactions are the same for both agglutination and precipitin reactions, with the one exception that in the latter, neither antigen

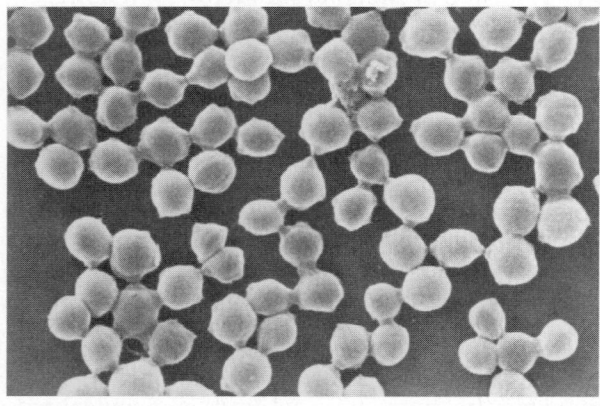

■ F i g u r e   5 – 7
Scanning electron micrograph of a coagglutination reaction showing cross-linking of the staphylococcal cells by soluble antigen. (Courtesy of Boule Diagnostics AB, Huddinge, Sweden.)

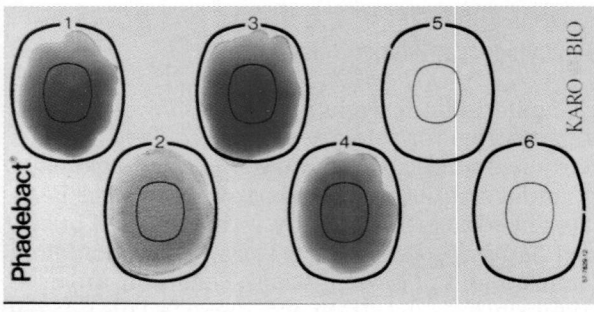

■ F i g u r e   5 – 8
Commercial coagglutination test card showing negative (1, 3, 4) and positive (2) reactions. The blue color is an indicator dye added to make the agglutination reaction easier to read against a white background. (Courtesy of Boule Diagnostics AB, Huddinge, Sweden.)

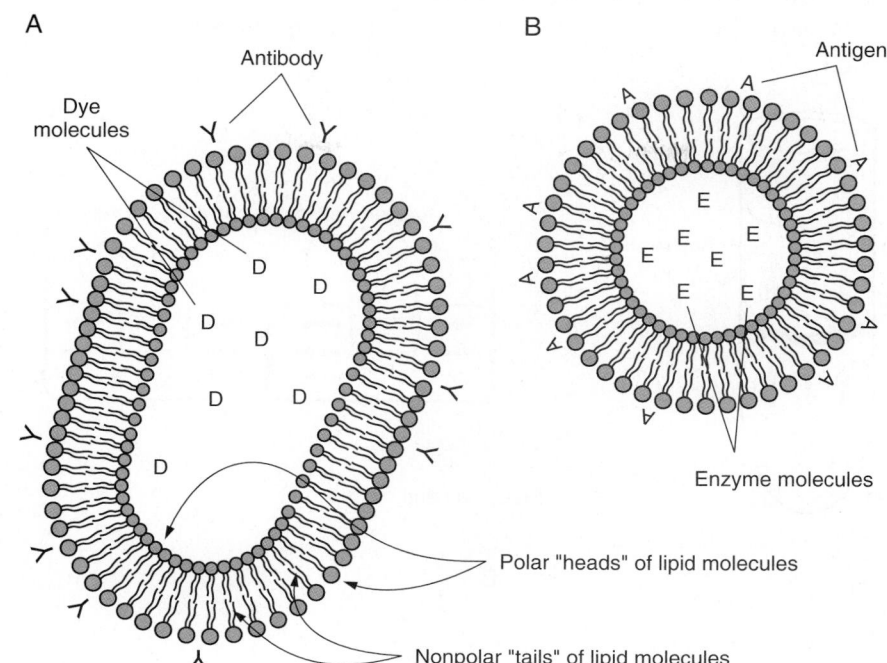

Figure 5-9

Diagram of liposome particles showing bilipid layer structure. Either antibody *(A)* or antigen *(B)* can be attached to the surface of the liposome. The interior of the liposome can carry indicator or reporter molecules (e.g., dyes, enzymes).

nor antibody is bound to a particulate carrier; thus, the reaction occurs between two soluble components. Lattice formation and the development of an observable secondary reaction (precipitin) take place more slowly than with an agglutination reaction. Soluble antigens and antibodies may combine to form visible precipitins in solution in a test tube (Fig. 5–10*A*); however, reactions that are developed in agar or agarose are more stable and easier to achieve (see Fig. 5–10*B*). Antibody-antigen reactions in agar are commonly performed by a technique called *double immunodiffusion* or Ouchterlony gel diffusion. In this technique, cylindrical holes or wells a few millimeters in diameter and appropriately distanced are cut out of the agar in Petri dishes. The fluid specimen containing the soluble antigen to be detected is placed in one well, and a known antibody-containing solution is placed in an adjacent well. The antigen and antibody molecules in solution diffuse out of the wells and through the porous agar. If antibody specific for antigen is present, the two components combine in the agar, where they meet at a point of optimal concentrations of each component and precipitate out of solution, producing a visible precipitin band. Diffusion in gel is a slow process, and not generally amenable to rapid diagnosis. It is more commonly used to detect serum antibodies.

Counterimmunoelectrophoresis

Counterimmunoelectrophoresis (CIE) is a modification of the principle of immunodiffusion in

TABLE 5-2

COMPARISON OF PARTICLE AGGLUTINATION METHODS

Agglutination Method	Particle Type	Nature of Antibody-Binding to Particle	Diagnostic Utility
Latex agglutination	Latex Beads	Nonspecific adsorption or chemical coupling	Most widely used agglutination method for direct detection
Staphylococcal coagglutination	Formalin-fixed staphylococci	Fc portion of antibody molecule combines with protein A on staphylococcal cell wall	Commonly used for antigenic identifiation of bacteria in culture
Liposome-mediated agglutination	Liposomes	Chemical coupling	Newer method; limited applications of commercial tests

Tube precipitin   Agar gel diffusion

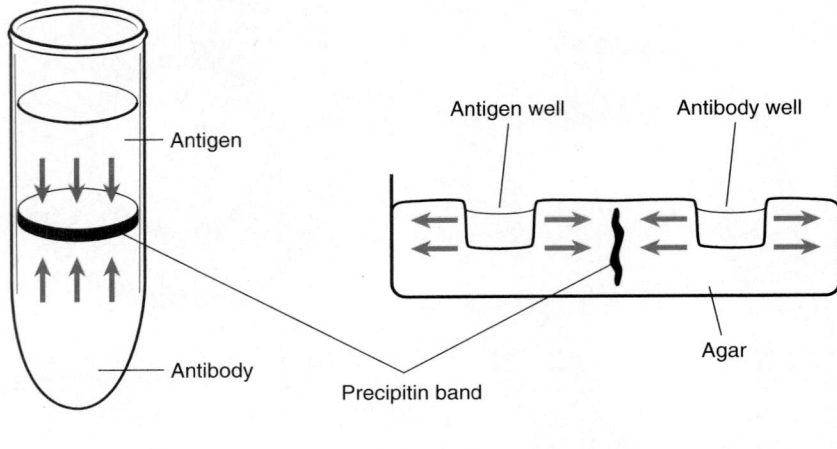

A   B

■ Figure 5 - 1 0

Comparison of tube *(A)* and agar gel diffusion *(B)* precipitin tests.

gels that dramatically speeds up migration of soluble antigens and antibodies. Antigen and antibody solutions are placed in adjacent wells cut out of agar supported on a glass or plastic surface. The gel is placed in an alkaline buffer–containing electrophoresis chamber, and an electric field is applied (Fig. 5–11). Under the buffer conditions chosen, most microbial antigens have a net negative charge (anions) and thus migrate in the gel toward the positively charged electrode (anode). Antibody molecules, on the other hand, are very weakly negatively charged or neutral under alkaline buffer conditions. They do not migrate significantly in the charged field of the electrophoresis chamber, but rather are carried toward the cathode (negative electrode) by the effect of buffer ions, a phenomenon known as *endosmosis*. If specific antigen and antibody are present, they will meet in the gel at some point of optimum proportions or equilibrium and form a visible precipitin band (Fig. 5–12). The process may occur in 1 hour or less, compared with days required for precipitin bands to form in gels by passive diffusion. CIE plates are generally read immediately after electrophoresis and then again after washing to remove nonspecific precipitins and overnight refrigeration to allow the specific bands to intensify.

Although very popular in the 1970s for a number of applications, CIE has been replaced by particle agglutination tests and immunoassays for rapid microbial antigen detection. It is not really a rapid method when one considers that overnight refrigeration is required, it is cumbersome to set up, and some antigens are not detected by standard buffer systems.

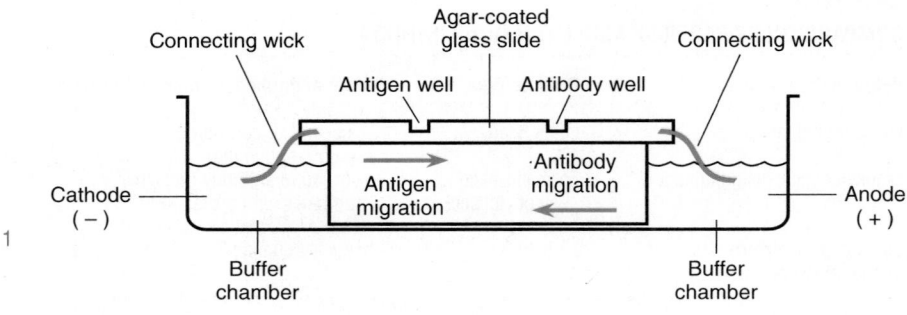

■ Figure 5 - 1 1

Diagram of counterimmunoelectrophoresis apparatus.

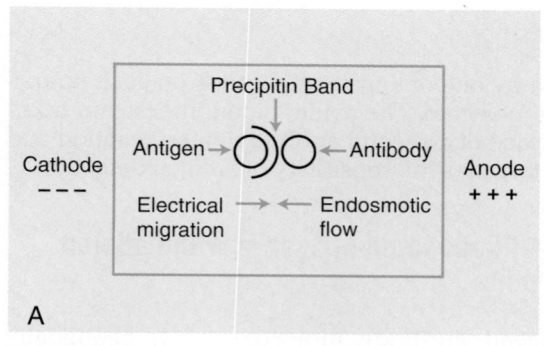
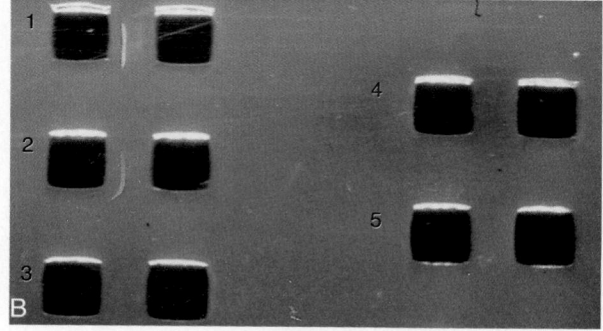

■■■■■ F i g u r e 5 – 1 2

Diagram (A) and actual photograph (B) showing precipitin band formation between antigen and antibody wells of a counterimmunoelectrophoresis test. Note strong (1), weak (2), and no (3–5) bands in B.

Microscope-Assisted Labeled-Antibody Staining

Fluorescence-Labeled Antibody

The combining of antibodies with specific fluorescent dyes enables the detection and localization of microbial antigens in tissues and infected cells applied to glass slides. Detection techniques may be direct or indirect. In the direct fluorescent antibody (FA) test, the antigen-specific antibody (monoclonal or polyclonal) is directly labeled with a fluorescent tag (fluorescein isothiocyanate is commonly used) and reacted with the test specimen (Fig. 5–13A). With the indirect or sandwich FA method, the antigen-specific antibody is unlabeled, and its binding to specific antigen in the test specimen is detected with a second labeled antibody that is specific for the first or primary antibody (see Fig. 5–13B). In either case, the slide is viewed with a microscope properly equipped with a light source and filters to excite and view the fluorescent tag used. Indirect FA may offer an advantage in sensitivity over direct FA, because multiple labeled antibody molecules may bind to each primary antibody bound to specific antigen.

Specimens that may be evaluated using fluorescence-labeled antibody techniques include tissue thin sections, body fluids, and swab specimens. Either whole microbial cellular antigens or antigens associated with infected mammalian cells may be detected. The specimen must be transferred and fixed to a glass slide, with a fixative such as formalin, acetone, ethanol, or methanol. Fixation also renders mammalian cell membranes permeable to chemical reagents and thus allows intracellular antigens to be stained. Fluorescent antibody–stained specimens are often counterstained with a nonspecific fluorescent dye that stains all background material (Fig. 5–14). Counterstaining is often helpful in assessing the amount of cellular material and the overall quality and adequacy of the specimen. Disadvantages of FA

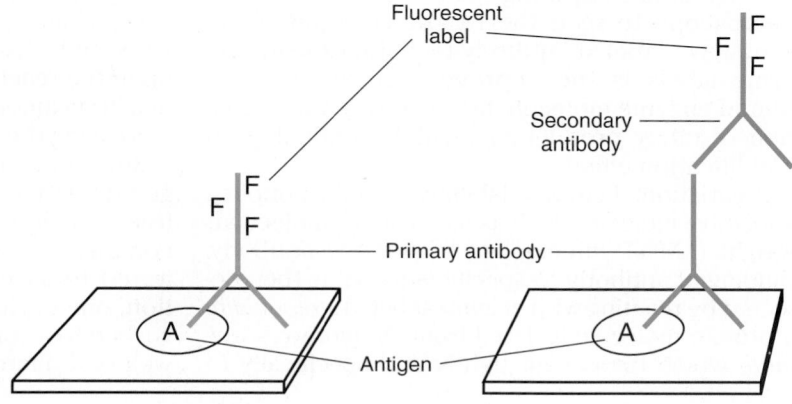

■■■■■ F i g u r e 5 – 1 3

Diagram of direct (A) and indirect (B) fluorescent antibody tests for antigen detection.

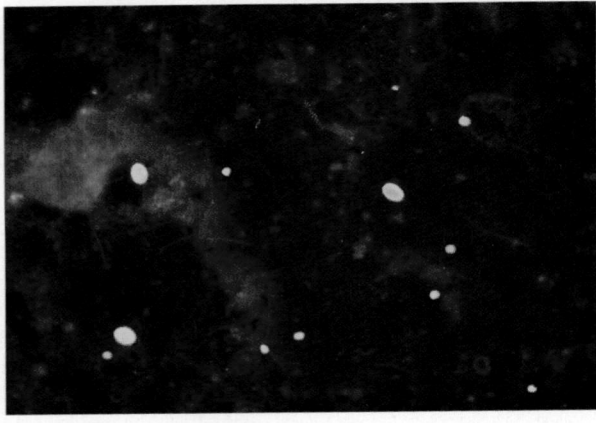

Figure 5–14

Direct fluorescent antibody–stained cells of *Giardia lamblia* (three larger apple-green, oval cells) and *Cryptosporidium* sp (smaller cells) in stool. (Courtesy of Meridian Diagnostics, Cincinnati, OH.)

include the requirement for a fluorescent microscope and the significant subjective interpretation involved in reading the slides. In addition, the FA procedure is time consuming and not easily amenable to automation.

Enzyme-Labeled Antibody

The FA technique may be modified by labeling the primary or secondary antibody with an enzyme instead of a fluorescent tag. The enzyme selected is capable of converting a colorless substrate to a colored end product. Both horseradish peroxidase and calf intestine alkaline phosphatase are commonly used for this purpose. The specific antigen-antibody interaction is detected by the addition of the enzyme substrate and development of the colored product of the enzyme reaction. This technique offers the advantage of not requiring the use of a fluorescent microscope to read the reactions. Another advantage of labeled antibody techniques using enzyme labels is the improved sensitivity; each bound enzyme molecule may catalyze the formation of many product molecules, producing an amplification effect.

A variation of enzyme labeling involves coupling a reactive molecule such as *biotin,* a low-molecular-weight (LMW) vitamin, to the primary antibody. Binding of antibody to specific antigen is then detected by reacting with enzyme-labeled *strepavidin,* a protein molecule isolated from *Streptomyces avidinii,* which binds very tightly and specifically to

biotin by one of four reactive sites on each strepavidin molecule. The avidin-biotin interaction takes the place of a second antibody in the reaction and may increase the sensitivity of antigen detection.

Solid-Phase Immunoassays with Labeled Reagents

Tagged antibody molecules with chemically active groups as specific reagents for direct microbial antigen detection have been available for some time. Fluorescent dyes and enzymes have traditionally been used for the detection of intact microbial cells or host cell–associated microbial antigens in tissues or fluid specimens. Newer methods of tagged antibody immunoassays have been developed so that soluble, non–cell-associated microbial antigens can be identified without the use of a microscope either with the naked eye or with optical instruments (spectrophotometers). These newer methods are much more amenable to automation and to testing of large numbers of samples. Labeling reagents used for this purpose are enzymes, radioisotopes, and light-sensitive (fluorescent, chemiluminescent, or bioluminescent) probes. Enzyme immunoassays are the most widely used of the systems and are discussed at length here.

Enzyme Immunoassay

SOLID-PHASE ENZYME IMMUNOASSAY

Enzyme immunoassay (EIA) depends on the fact that enzyme molecules can be covalently coupled to antibody molecules in such a way that both enzymatic and antigen-binding activities of the respective molecules are preserved. If an enzyme that catalyzes the production of visible (colored) end product from substrate is chosen, then specific antibody-antigen interactions can be measured. Because the enzyme molecule is *not* used up in the reaction, it may catalyze the production of a large amount of colored product, thus greatly increasing the sensitivity of the reaction.

Most EIA procedures used for microbial antigen detection require physical separation of the free, labeled antibody from antibody-antigen complexes. For this reason, the assays are referred to as *heterogeneous.* To facilitate separation, one of the reactants is usually bound or adsorbed to a solid support such as a tube, bead, or well of a microdilution tray. Thus the term *solid-*

phase enzyme immunoassay, or *enzyme-linked immunosorbent assay* (ELISA), was developed.

The basic ELISA procedure for antigen detection, known as the *double-antibody sandwich ELISA* method, involves four steps (Fig. 5–15*A*):

1. The solid support (bead, tube well, etc.) must be coated with an appropriate antibody in solution. This is often done by incubation of the solution overnight at 4°C, followed by thorough washing. At this point, the antibody-sensitized solid support may be dried and stored at 4°C for several months.

2. When one is ready to perform the assay, the test sample is added to the solid phase, and any antigen present is allowed to bind for a time (about 1 hour) at room temperature.

3. After washing-out of unbound sample, a second enzyme-labeled antibody with the same antigenic specificity as the bound (capture)

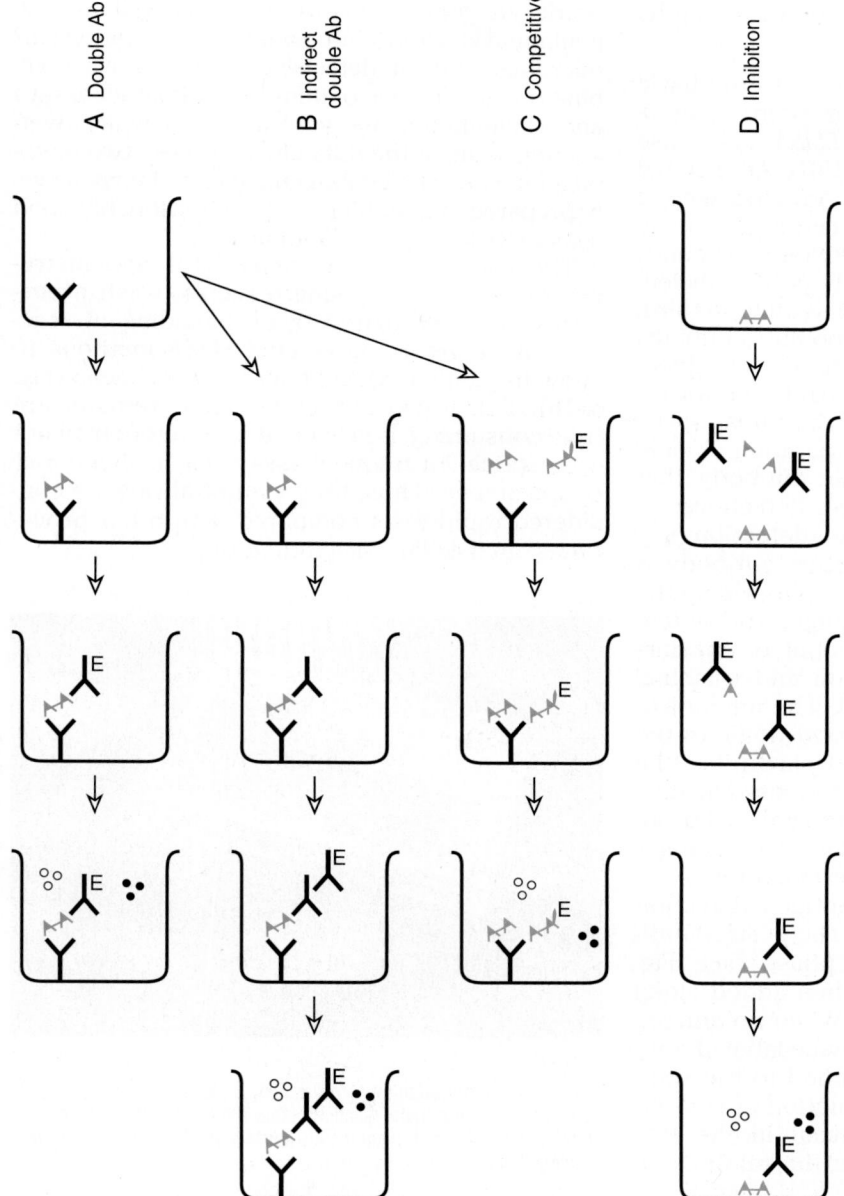

■■■■■ F i g u r e   5 – 1 5
Principle of various ELISA assays for antigen. The first three methods (double Ab, indirect double Ab, and competitive ELISA) start with specific antibody bound to the solid phase, whereas inhibition ELISA begins with bound antigen. Arrows (→) separate steps in the procedures where washing of the solid phase takes place. Y, antibody; ▲, antigen; E, enzyme; ○, enzyme substrate; ●, enzyme product.

antibody is added and allowed to react with the captured antigen. Alkaline phosphatase and horseradish peroxidase are the two most commonly used enzyme labels. After another incubation period, usually at room temperature, the solid support is washed again to remove unreacted enzyme-labeled antibody.

4. An appropriate enzyme substrate is added, and the enzymatic reaction is allowed to proceed for a fixed period, terminated, and the colored product is read visually or optically with a spectrophotometer. Specific ELISA readers capable of measuring end products at various wavelengths are available.

There are a number of variations to the basic ELISA method just described. For example, an *indirect double-antibody sandwich ELISA* makes use of a third antibody (see Fig. 5–15B). After initial antigen capture by bound antibody, the second antigen-specific antibody, produced in an animal species different from that of the capture (bound) antibody, is added; this antibody is not labeled. After appropriate incubation and washing, a third, enzyme-labeled, anti-immunoglobulin antibody, directed against the second antibody, is added. Finally, addition of enzyme substrate allows for detection of antigen present in the initial specimen. This method allows for preparation of only one labeled anti-immunoglobulin antibody that can be used for a variety of antigen detections.

In *competitive ELISA* for antigen detection (see Fig. 5–15C), the primary or capture antibody is first adsorbed to the solid phase, followed by the addition of test specimen (thought to contain antigen) along with a fixed amount of enzyme-labeled antigen. After incubation and washing, the enzyme substrate is added. The amount of colored product is inversely proportional to the amount of antigen in the test specimen, i.e., the more antigen in the specimen, the less enzyme-labeled antigen bound and the less colored product. Competitive ELISA assays are readily adaptable to quantitative antigen determinations.

Inhibition ELISA assays for antigen detection are performed by first coupling the purified antigen of interest to the solid phase (see Fig. 5–15D). The test specimen is then added along with enzyme-labeled antibody. When no antigen is present in the specimen, enzyme-labeled antibody binds to the antigen attached to the solid phase and is available for production of colored product. When antigen is present in the test specimen, however, the antigen in solution in-

hibits enzyme-labeled antibody from binding to the solid-phase antigen, Thus, the amounts of both bound antibody and colored product are reduced when enzyme substrate is added.

Although this technique is theoretically straightforward, many practical aspects of ELISA assays potentially may limit their effectiveness. These include the quality and purity of antigen and antibody reagents available, the thoroughness of wash procedures, the choice of solid phase, the method of reading endpoints, and the appropriate choice and use of controls. Commercial companies have produced good-quality reagents as well as self-contained kits for a wide variety of applications for microbial antigen detection. ELISA assays combine the advantages of fluorescent immunoassays and radioimmunoassays (see later) while overcoming some of the difficulties of these two methods. Enzyme-labeled reagents can be inexpensively prepared, are stable, and have reasonably good sensitivity in most applications.

The use of microtitration plates and instrumentation for dispensing reagents, wash procedures, and automated optical reading of reactions have greatly facilitated ELISA methods to allow handling large numbers of specimens (Fig. 5–16). ELISA procedures in some formats are time consuming, however, and therefore may not be practical for frequent assay of a small number of specimens. Thus, they cannot always be considered rapid tests compared with other procedures such as latex agglutination.

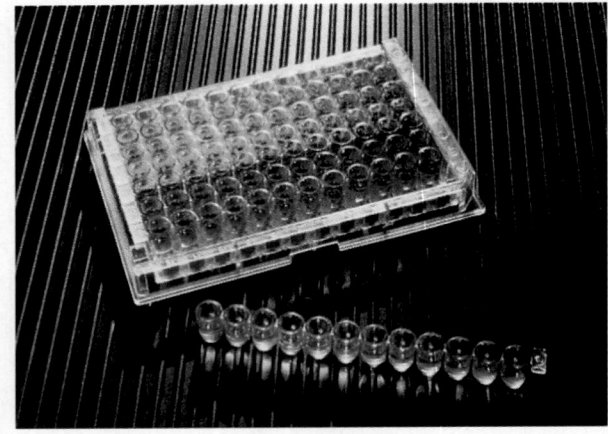

■■■■ F i g u r e 5 – 1 6
EIA test in microtitration tray *(top)* and strip *(bottom)* showing wells with product of enzymatic reaction before (blue) and after (yellow) addition of an acidic stop solution. (Courtesy of Meridian Diagnostics, Cincinnati, OH.)

AVIDIN-BIOTIN–ENHANCED EIA

Just as biotin-tagged antibodies can be used in direct detection of intact microbes or infected cells in tissue, soluble antigens may be detected in ELISA assays by reacting avidin-coupled enzyme molecules with biotin-labeled antibody bound to antigen captured to a solid phase. The avidin-biotin interaction takes the place of primary antibody–secondary antibody interaction. The increase in sensitivity over standard EIA methods results from the fact that multiple biotin molecules may be bound to antibody.

MEMBRANE-BOUND SOLID-PHASE EIA

A newer format for ELISA, known as *membrane-bound EIA,* has resulted in greater speed and simplicity of enzyme immunoassays. These improvements are derived by immobilizing antibody onto the surface of porous membranes that permit a higher ratio of surface area to sample volume, so that more sample is exposed to immobilized antibody. Such a test typically uses a disposable plastic device (cassette) consisting of a small chamber to which fluid sample can be added (Fig. 5–17). The bottom of the chamber contains an antibody-coated membrane (typically, nitrocellulose or nylon) and an underlying absorbent bed

of material such as cellulose acetate. When sample is added to the chamber, the absorbent material below serves to pull fluid through the antibody-coated membrane by a wicking action and thus presents a relatively large volume of sample (which may contain antigen) to the antibody. Because the rate of binding of the antigen is proportional to its concentration in solution near the membrane surface, this action increases the rate and extent of binding. Bound antigen may then be detected by a variety of methods using a second enzyme-labeled antibody. One variation of this method utilizes antibody-coated, dye-containing liposomes instead of enzyme to detect bound antigen.

These "flow-through" or "immunoconcentration" ELISA methods have become popular because they can easily be performed as rapid single tests and because positive and negative controls can be incorporated at different sites on the same membrane used for patient sample testing (Fig. 5–18).

EMIT

A modification of the ELISA assay, known as EMIT, or *enzyme-multiplied immunoassay technique,* is commonly used in the clinical laboratory

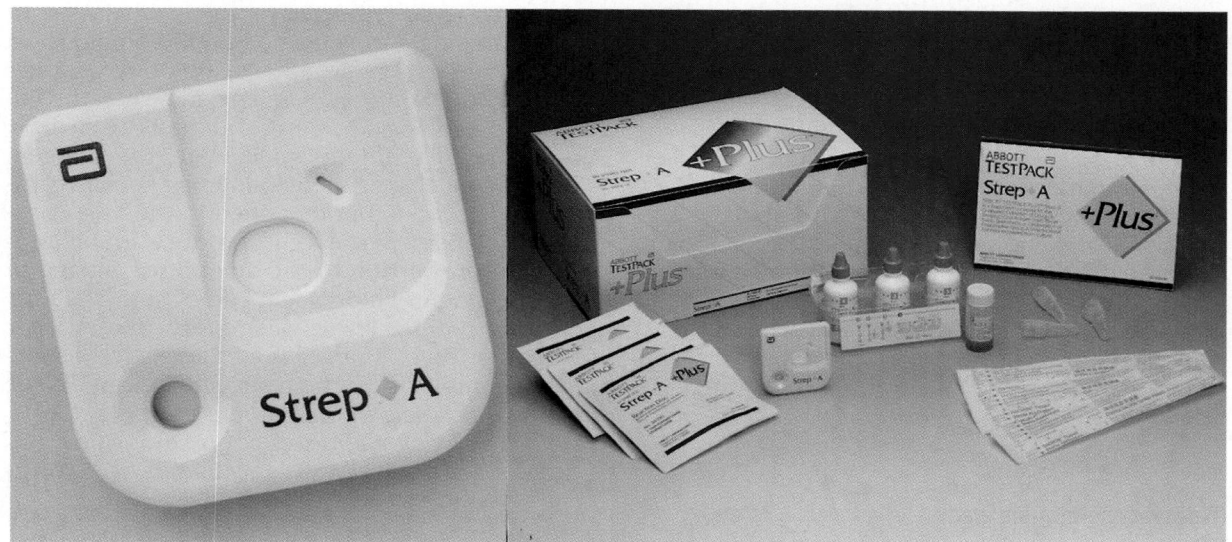

A   B

Figure 5–17

TestPack Strep A kit components *(A)* and individual cassette *(B)* of membrane-bound EIA test for group A streptococcal polysaccharide antigen. (Reproduction has been granted with approval of Abbott Laboratories, all rights reserved by Abbott Laboratories.)

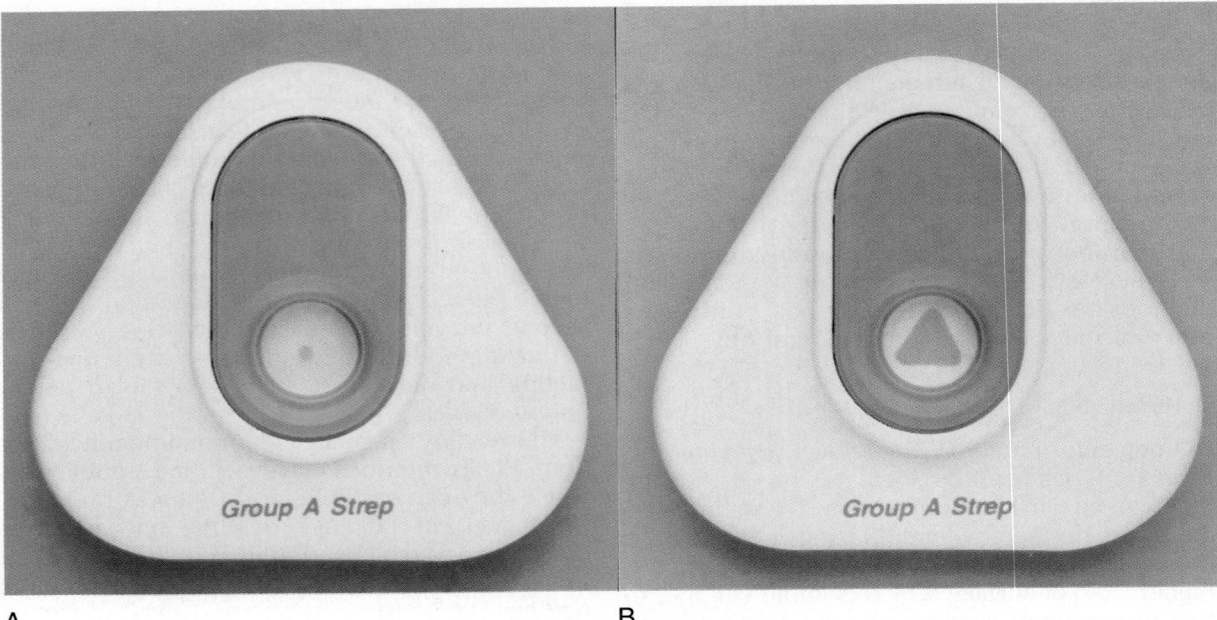

A   B

■■■■■ F i g u r e   5 – 1 8

Color PAC™ devices showing negative *(A)* and positive *(B)* reactions of liposome enhanced EIA for group A streptococcal polysaccharide antigen. (Courtesy of Becton Dickinson and Company, Cockeysville, MD.)

for assay of low LMW substances, such as drugs of abuse, hormones, and chemotherapeutic agents, including antibiotics; however, EMIT is not commonly used for microbial antigen detection. It is a homogenous assay that does not use a solid support phase and does not require washing and separation steps. Instead, it is based on the observation that an antigenic molecule may be linked to an enzyme in such a way that the enzyme activity is altered (generally decreased) when the molecule combines with specific antibody. The reaction mix contains the test specimen, enzyme-labeled antigen, antibody, and enzyme substrate. If the test specimen contains the specific antigen, it combines with the antibody and leaves the enzyme-antigen complex free and active to react with substrate and produce end product.

Fluorescent Immunoassay

Fluorescent molecules, or *fluorochromes,* can be used in a manner analogous to enzymes in direct detection assays for soluble antigens. One commercially available heterogenous fluorescent immunoassay (FIA) utilizes competitive binding

of a labeled antigen and unlabeled antigen in a patient specimen with antibody bound to a solid phase. Another system, analogous to the homogenous EMIT-type assays, takes advantage of the fact that a fluorogenic compound bound to an antigen can be converted into a fluorescing compound by the addition of an enzyme when the fluorogenic compound is free but not when it is bound to an antibody specific to the antigen. Thus, a clinical specimen containing specific antigen would bind antibody added to the assay system and allow fluorescent product to be produced in proportion to the amount of antigen in the specimen. Fluorescent detection systems generally require fewer wash steps and shorter incubation times than ELISA-type assays, reducing the complexity of the tests. Still newer assays, such as *fluorescent polarization assays* and *time-resolved immunofluorescence assays,* have increased sensitivity and specificity of fluorescence-based assays even further.

Radioimmunoassay

Radioimmunoassay, or RIA, is less commonly used today for microbial antigen detection because of the refinements in EIA and similar

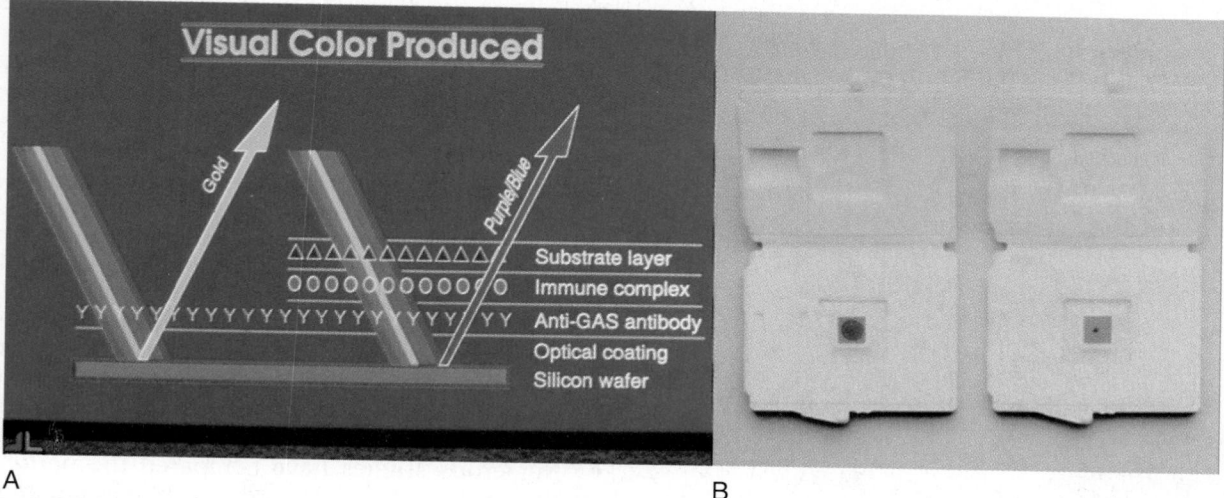

Figure 5–19

Optical ImmunoAssay (OIA™) for group A streptococcus. *A*, Principle of Biostar Strep A OIA™ test showing color change from gold to purple following attachment of immune complex containing group A streptococcal antigen. *B*, Test cassettes showing a positive reaction on the left and negative on the right. (Courtesy of BioStar, Inc., Boulder, CO.)

assays. RIA was at one time, however, the method of choice for hepatitis A and B virus antigen and antibody assays, and it continues to be used in some clinical laboratories for this purpose and more commonly for drug and hormone detection. RIA for antigen detection, as generally performed, is analogous to competitive ELISA assays for antigen; that is, labeled antigen competes with unlabeled antigen, which may be present in the patient specimen, for binding to specific antibody. The main difference is that a radioisotope such as radioactive iodine (^{125}I) or carbon (^{14}C) is used to label the antigen instead of an enzyme. Instruments that measure radioactivity are used to detect specifically bound, labeled antigens. The hazards of working with radioisotopes, along with the problems associated with their disposal, have encouraged the use of alternative methods such as ELISA. Nonetheless, RIA is generally considered to be more sensitive than ELISA for many applications in the clinical laboratory.

Optical Immunoassay

It has long been known that antigen-antibody interactions occurring on inert surfaces can be detected by changes in light reflectance due to alteration in thickness of the reactants on the surface. Thus, it is possible to detect an antigen or an antigen-antibody complex by allowing the sample containing antigen to react with an ap-

propriate reflecting surface to which specific antibody is attached. Light reflecting off the surface film containing bound antibody only is viewed as one color; however, antigen binding increases the thickness of the film, causing the surface to appear a different color.

A commercial company has developed a practical application for optical immunoassay (OIA)—the detection of group A streptococcal antigen in throat swab specimens (Fig. 5–19). The OIA method promises to be useful for a number of other applications for antigen detection.

CURRENT CLINICAL APPLICATIONS

A wide variety of infectious diseases can be diagnosed using commercially available antigen detection test reagents. Some of these are listed in Table 5–3 and described here.

Respiratory Tract Infections

Streptococcal Pharyngitis

One of the most widely used applications of direct antigen tests, popularized in the 1980s, is for the detection of group A β-hemolytic streptococci (*Streptococcus pyogenes*) (GAS) in throat swab specimens for the diagnosis of

■■■■■■■T A B L E 5 – 3

SOME REPRESENTATIVE INFECTIOUS DISEASES AND THE DIRECT ANTIGEN DETECTION METHODS COMMERCIALLY AVAILABLE TO DETECT THEM

Type of Infection	Test Methods*
Bacterial	
Group A streptococcal pharyngitis	LA, EIA, liposome-enhanced immunoassay, OIA
Bacterial meningitis[†]	LA, CoA
Clostridium difficile colitis	LA, EIA
Chalmydial urethritis	EIA, FA
Pertussis	FA
Legionnaires' disease	FA, RIA
Fungal	
Cryptococcal meningitis	LA, EIA
Parasitic	
Giardiasis	FA, EIA
Cryptosporidiosis	FA, EIA
Pneumocystis pneumonia	FA
Trichomonas vaginitis	FA
Viral	
Rotavirus gastroenteritis	LA, EIA
Hepatitis B infection	EIA
Respiratory virus infection[††]	FA, EIA
Herpes simplex virus infection	FA, EIA
Cytomegalovirus infection	FA
AIDS	EIA

*EIA, enzyme immunoassay; FA, fluorescent antibody test; LA, latex agglutination test; RIA; radioimmunoassay; OIA, optical immunoassay.

[†]Includes *Haemophilus influenzae* b, *Streptococcus pneumoniae*, *Neisseria meningitidis*, and *Streptococcus agalactiae*.

[††]Includes influenza, parainfluenza, and respiratory syncytial viruses.

streptococcal pharyngitis. The main advantage of rapid direct antigen testing for GAS over standard throat culture procedures is that results are available while the patient is in the physician's office and antibiotic therapy can be given immediately rather than waiting 24 to 48 hours for culture results. This is more convenient for the physician and patient alike. Furthermore, early antibiotic administration shortens the patient's illness, reduces the likelihood of secondary infections or spread to close contacts, and allows the patient to return to work or school sooner.

Currently, over 20 manufacturers produce diagnostic test kits for direct GAS detection. The majority of these kits use either latex agglutination or some version of solid-phase immunoassay. Both types of tests contain reagents to perform initial extraction of the group A carbohydrate antigen detected in the assay from the cell wall of the organism. This is most commonly achieved by exposing the organism to nitrous acid, but it can also be accomplished by enzymatic digestion. Once the solubilized antigen preparation has been pH adjusted in a buffer solution, the specimen is tested according to the specific procedure of the manufacturer. A number of EIA tests, particularly those that use membrane-bound antibody to speed up the reaction with antigen and facilitate the washing procedure, have become popular. Because of the colored product, they are generally easier to read than latex agglutination tests. A variation of the membrane-bound immunoassay for GAS uses colored, dye-filled liposomes bound to antibody as a detection reagent; this reagent binds to antigen captured on the surface of the membrane. An optical immunoassay, which uses the change in reflection of light from a surface to detect GAS antigen bound to the surface, is also available.

Numerous studies have compared the performance of direct GAS tests against one another and against the gold standard—throat culture. A conservative estimate of their accuracy suggests sensitivity and specificity in the range of 60 to 80% and 90 to 95%, respectively. These values vary by manufacturer, individual performing the test, and accuracy of the throat culture procedure used for comparison. Overall, the sensitivities of the rapid GAS antigen tests are not sufficiently high to permit them to be used without an accompanying culture. Initial studies of the optical immunoassay (discussed previously), however, suggest that its sensitivity and specificity may exceed the ranges listed for other direct GAS tests and may be even more sensitive than routine culture methods for detecting GAS from throat swabs. These data need to be confirmed by other studies. At present, the best overall approach for diagnosis of streptococcal pharyngitis would couple a rapid direct test with a throat culture performed only when the direct test result is negative. Positive direct test results would lead to immediate initiation of antibiotic therapy. It must be remembered that the most important reason for diagnosing and treating GAS pharyngitis is to prevent nonsuppurative sequelae such as rheumatic heart disease. Delay in initiation of therapy for a few days (while awaiting culture results) will not increase the patient's risk of developing such complications.

Whooping Cough (Pertussis)

Bordetella pertussis, the agent of whooping cough, is a slow-growing, fastidious, gramnegative coccobacilli. Recovery of the organism in culture from nasopharyngeal swabs or aspirates may take up to 7 days, even when optimal media and culture conditions are used. A direct FA test may be performed on smears prepared from these specimen; it has been reported to be positive in about 70% of culture-positive specimens when performed by well-trained technologists. Thus, culture should always be performed along with direct FA tests for *B. pertussis.* Results of direct FA tests are generally available on the same day as specimen collection, thus allowing early initiation of antibiotic therapy and isolation of the patient to prevent secondary spread of this highly contagious organism.

Legionnaires' Disease

Legionella pneumophila and other *Legionella* spp cause acute lobar pneumonia with multisystem involvement (legionnaire's disease) and a less serious disease known as Pontiac fever. Early diagnosis and appropriate antibiotic therapy are important, particularly in immunosuppressed patients. Disease may occur as community-acquired individual cases or outbreaks and as nosocomial infections. A variety of antigen detection methods have been evaluated and compared with isolation in culture. A number of direct polyclonal and monoclonal FA tests, performed on lower respiratory tract secretions or lung tissue, are commercially available for *L. pneumophila* serogroups 1 through 6 and other less common species. Sensitivity of these tests has generally ranged from 50 to 75%. Specificity has been a problem with some reagents, owing to antigens shared with *Pseudomonas* spp and other organisms. A number of immunoassays, including ELISA and RIA, have been developed and evaluated in research laboratory settings. Some of these are designed to detect soluble antigen in serum or urine; at least one RIA assay to detect urine antigen is commercially available. Reports suggest that the sensitivity of these assays is in the 80 to 90% range, with even higher specificity. At least one commercially latex agglutination (LA) test is available for *Legionella* antigen detection in urine; however, this test is not as sensitive as the immunoassays.

Respiratory Virus Infections

A number of respiratory viruses can be directly detected in nasopharyngeal or throat washings or swabs. They include respiratory syncytial virus (RSV), influenza A and B, parainfluenza 1, 2, and 3, and adenovirus. Rapid direct detection of RSV infection is particularly important because (1) it causes serious lower respiratory tract disease (bronchiolitis and pneumonia) in young children, often requiring hospitalization, (2) the organism may be difficult to culture, and (3) specific antiviral therapy (ribavirin) is available. Direct and indirect FA and EIA tests are available for RSV and the other agents listed. They are highly sensitive and specific for RSV but are not as sensitive for the other agents.

Pneumonia in Immunocompromised Patients

Cytomegalovirus (CMV) and *Pneumocystis carinii* (PC), a protozoan parasite, are important pulmonary pathogens in a number of patient populations, including those who have undergone solid organ or bone marrow transplantation and those with cancer or AIDS. Monoclonal antibodies to CMV and PC are available for direct immunofluorescence testing of respiratory tract secretions or lung tissue. This technique is not sensitive compared with culture for CMV but is very helpful when positive. Because there are no widely available culture methods for PC, direct detection by fluorescent antibody or nonspecific staining methods is the primary laboratory diagnostic tool used for this organism.

Meningitis and Sepsis

Bacterial Meningitis and Sepsis

Clinical laboratories have regularly used antigen detection testing of CSF and other body fluids to detect organisms causing bacterial meningitis for about 20 years. Techniques have included CIE, popular in the 1970s, EIA, coagglutination, and, most recently, latex agglutination; the last method is most commonly used today, and several manufacturers produce latex-based kits for this purpose. The bacterial agents detected in the kits include *Haemophilus influenzae* b, *N. meningitidis*, *Streptococcus pneumoniae,* and *Streptococcus agalactiae* (group B streptococcus). Although antibodies in each kit vary somewhat, they are all designed to detect capsular polysaccharide antigens of the organisms. For example, the capsule of *H. influenzae* b consists of repeating subunits of ribose and ribitol, and is referred to as polyribosephosphate (PRP). When this organism causes infection, the PRP antigen is shed by growing bacterial cells into body tissues and fluids. *Haemophilus influenzae* b (as well as the other organisms listed) most often causes bacteremia before the organism invades the central nervous system. Thus, the organism and antigens may be detected in the serum before or at the same time they are found in the CSF in cases of meningitis. Furthermore, as circulating organisms and soluble antigens are trapped, degraded, and released by phagocytic cells of the liver and spleen, these products are cleared from the body by filtration in the kidney and excreted in the urine; thus, urine can be examined for the presence of antigen in addition to CSF in suspected meningitis. Urine is also useful as a test specimen either alone or in addition to serum in cases of bacteremia and focal infections other than meningitis. It may be concentrated by ultrafiltration (removing water but not microbial antigens) to increase test sensitivity.

The clinical utility of antigen tests for diagnosing meningitis varies with a number of factors, including the manufacturer of the kit, the specific organism, and the specimen type tested. For example, overall accuracy of these tests is generally greatest for *H. influenzae* b antigen detection (sensitivity and specificity about 90%), with group B streptococcus a close second. Accuracy of pneumococcal and meningococcal antigen testing is variable and generally low, however. There may be little difference in the relative sensitivity of CSF antigen testing and CSF gram staining for detecting bacterial agents of meningitis. In patients with suspected meningitis, these tests should always be performed along with cultures, and appropriate antibiotic therapy should *never* be withheld pending culture results in patients with negative antigen tests results (unlike the situation with antigen testing for group A streptococcus). The greatest clinical utility of these antigen tests probably occurs when testing specimens from patients who have received antibiotic therapy before cultures were collected, thus reducing or eliminating the likelihood of recovering the organism.

Cryptococcal Meningitis

The clinical utility of antigen detection for *C. neoformans,* an important fungal pathogen causing meningitis, pneumonia, and disseminated disease, is considerably different. Antigen testing of CSF, commonly done by LA, is considerably more sensitive than India ink direct examination of CSF. In addition, quantitative antigen detection, which consists of titrating CSF antigen by performing serial dilutions, is an important prognostic indicator of clinical response to antifungal therapy. An EIA test also has been marketed for the detection of this organism.

Gastrointestinal Tract Infections

Clostridium difficile–Associated Enteric Disease

Direct microbial antigen detection methods have been applied to the diagnosis of bacterial, viral and parasitic agents of gastrointestinal tract infections. *Clostridium difficile,* an anaerobic gram-positive rod, causes pseudomembranous colitis and antibiotic-associated diarrhea in individuals receiving antibiotics and other chemotherapeutic agents that alter bowel flora. The organism produces at least two exotoxins, toxin A and B, which are involved in the disease process and can be detected in stool specimens from patients. The standard laboratory test for this organism involves detection of toxin B using an in vitro cell culture assay. A number of EIA tests for both toxins A and B have been marketed. They have the advantage of not requiring cell culture techniques and appear to be almost as sensitive and specific as the cell culture assay. A latex agglutination test is also available for detection of a *C. difficile*–associated antigen.

Viral Gastroenteritis

The most important identifiable agents of viral gastroenteritis are human rotaviruses, enteric adenoviruses, and the Norwalk and Norwalk-like viruses. These agents are either very difficult or impossible to culture, and thus, antigen detection methods are very important for diagnosis. Rotaviruses are the most important agents in this group because they commonly infect very young infants, causing vomiting and diarrhea and resulting in dehydration that may be life-threatening in patients at this age. Both EIA and LA tests are available for rotavirus detection in stool specimens; EIA appears to be more sensitive than LA. In either case, soluble antigen must be extracted from the particulate matter in stool before testing; this is generally accomplished by mixing the stool with extract buffer solution, clarifying by centrifugation, and testing the supernatant fluid. At least one EIA test is available for detection of enteric adenoviruses, but there are currently no commercially available reagents for detecting Norwalk virus.

Giardiasis and Cryptosporidiosis

Fluorescent antibody direct detection reagent kits are available for identification of *Giardia lamblia* and *Cryptosporidium* spp in stool specimens. The reagents can be used on fresh or formalin-fixed stool specimens. These parasitic agents of gastrointestinal tract disease (giardiasis and cryptosporidiosis) are most commonly detected by microscopic examination of concentrated and nonimmunologically stained stool specimens; FA staining appears to be as good as or better than these methods. One company has marketed a *Giardia/Cryptosporidium* direct FA test kit that contains pooled monoclonal antibodies for both organisms in one reagent. Several commercial EIA test kits are also available for detection of soluble *Giardia* or *Cryptosporidium* antigen in stool.

Sexually Transmitted Diseases

Chlamydial Infections

Chlamydia trachomatis, an obligate intracellular bacterium, infects the genital tract, causing urethritis, cervicitis, and other infections. In addition, neonates born to infected mothers are at risk of developing conjunctivitis and pneumonia. Although isolation in cell culture is still considered the gold standard for diagnosis, a large number of direct antigen detection tests are commercially available. Most of these assays are EIA or FA based. The numerous published evaluations of these products suggest moderately good sensitivity and very good specificity when they are used on symptomatic patients at high risk for infection (e.g., patients visiting a sexually transmitted disease clinic). The tests do not, however, perform as well when used in low-prevalence populations or asymptomatic patients. In addition, they should never be used in children in whom evidence of chlamydial infection subsequent to sexual abuse is being sought.

Herpes Simplex Virus Infection

Genital herpes simplex virus infection, characterized by painful vesicular lesions, can be diagnosed by FA staining of scrapings of the lesions or by EIA of soluble antigens obtained by swabbing of the vesicle fluid. Neither one of these methods is as sensitive as culture.

Trichomonas Vaginitis

Trichomonas vaginalis causes a common parasitic infection presenting as vaginitis. Diagnosis is usually accomplished by observing the organism in wet-mount smears prepared from vaginal secretions. A relatively new commercial FA test is designed to detect this organism, but few data are available comparing FA with wet-mount or culture techniques.

Blood-Borne and Body Fluid–Borne Diseases

Hepatitis B Virus Infection

Both hepatitis B virus (HBV) and human immunodeficiency virus type 1 (HIV-1) are agents for which culture is generally not available; thus, noncultural methods are the mainstays of laboratory diagnosis. Both measurement of serologic response to viral antigens and direct viral antigen detection are important.

For HBV, EIA methods have largely replaced RIA methods. Viral antigens can be found in the blood of infected patients. Both hepatitis B surface antigen and hepatitis B e antigen can generally be detected early in acute infection and may also be detected in patients who are chronically infected. These antigen tests have important diagnostic and prognostic value.

AIDS

Although diagnosis of HIV-1 infection is generally accomplished by serology, EIA detection of the viral p24 antigen has been used to detect active viral replication and may have some value in detecting early infection.

FUTURE APPLICATIONS

Significant progress has been made in development and improvement of direct antigen detection tests; however, many of the methods have not replaced, but instead are used in addition to, cultural techniques. Such applications are justifiable when rapid patient care decisions are based on antigen detection methods or when culture methods fail to isolate the organism. More important, however, are applications for antigen detection for organisms that are difficult or impossible to culture. There is little doubt that the variety and quality of antigen detection reagents will increase in the future. With the parallel improvement and availability of microbial nucleic acid amplification and detection techniques for diagnosis of infectious diseases, laboratory scientists will have to make wise choices regarding the most cost-effective and clinically relevant use of these reagents.

Bibliography

Baron EJ, Finegold SM (eds): Nontraditional methods for identification and detection of pathogens or their products. In Bailey and Scott's Diagnostic Microbiology, 8th ed. St. Louis: CV Mosby, pp 127–141.

Buck GE: Noncultural methods for detection and identification of microorganisms in clinical specimens. Pediatr Clin. North Am 36:95, 1989.

de Macario EC, Macario AJL: Monoclonal antibodies for bacterial identification and taxonomy: 1985 and beyond. Clin Lab Med 5:531, 1985.

Fung JC, Tilton RC: Rapid methods in microbiology for detecting infectious diseases. Lab Med 1874, 1987.

Kohler RB (ed): Antigen Detection to Diagnose Bacterial Infections. Boca Raton, FL: CRC Press, 1986.

Rosebrock J: Labeled-antibody techniques: Fluorescent, radioisotopic, immunochemical. In Balows A, Hausler WJ, et al (eds): Manual of Clinical Microbiology, 5th ed. Washington, DC: American Society for Microbiology, 1991, pp 79–86.

Rytel MW (ed): Rapid Diagnosis in Infectious Disease. Boca Raton, FL: CRC Press, 1979.

Thompson KD, Van Enk RA: Changing technologies in immunodiagnosis: Enzyme immunoassays versus immunofluorescence assays. Clin Microbiol Newsl 10:73, 1991.

Tinghitella TJ, Edberg SC: Agglutination tests and Limulus assay for diagnosis of infectious diseases. In Balows A, Hausler WJ, Herrmann KL, et al (eds): Manual of Clinical Microbiology, 5th ed. Washington, DC: American Society for Microbiology, 1991, pp 61–72.

Tom BH, Six HR: Liposomes and Immunology. New York: Elsevier Science, 1980.

Valkiers GE, Barton R: Immunoconcentration—a new format for solid phase immunoassays. Clin Chem 31:1427, 1985.

Voller A, Bidwell DW, Bartlett A: The Enzyme Linked Immunosorbent Assay (ELISA): A Guide with Abstracts of Microplate Applications. Alexandria, VA: Dynatech Laboratories, 1979.

Wicher K (ed): Microbial Antigenodiagnosis. Boca Raton, FL: CRC Press, 1987.

Yolken RH: Enzyme immunoassays for the detection of infectious agents in body fluids: Current limitations and future prospects. Rev Infect Dis 4:35, 1982.

B. SEROLOGIC DIAGNOSIS OF INFECTIOUS DISEASES

IMMUNE RESPONSE TO INFECTIOUS AGENTS
  Host Resistance to Infection
    Physical and Chemical Barriers
    Natural Immunity
    Acquired Immunity
  Nature of the Immune Response to Infectious Agents
  Antigens and Antibodies
    Characteristics of Antigens
    Classification and Characteristics of Antibodies
  Primary and Secondary Antibody Responses

INTERPRETING SEROLOGIC TEST DATA
  Acute and Convalescent Antibody Titers
  Antibody Specificity and Cross-Reactivity
  False-Negative and False-Positive Serologic Test Results
  Value of Serologic Tests
    Population Studies
    Immune Status Testing
    Congenital Infections
    Infections After the Newborn Period

ANTIBODY DETECTION METHODS AND APPLICATIONS
  Particle Agglutination Assays
    Direct, Natural Particle Agglutination
    Indirect, Carrier Particle Agglutination
  Precipitation Assays
    Double Immunodiffusion
    Counterimmunoelectrophoresis
    Flocculation
  Complement Fixation Test
  Neutralization tests
    Virus Neutralization
    Antistreptolysin-O Test
    Treponema pallidum Immobilization
  Microscope-Assisted Labeled-Reagent Techniques
    Indirect Fluorescent Antibody Test
    Fluorescent Antibody tests with Enhanced Sensitivity
  Immunoassays with Labeled Reagents
    Enzyme Immunoassays
    Other Immunoassays
  Western Blotting

FUTURE APPLICATIONS

Mario J. Marcon

OBJECTIVES

1. Describe how physical and chemical barriers protect the host from infectious agents.

2. Differentiate the dynamics of the humoral from the cell-mediated immune response.

3. Explain how both types of immune responses are used as indictors of infectious process.

4. Describe antigens. Classify and characterize the different types of antibodies.

5. Differentiate the primary from the secondary antibody response.

6. Discuss the importance of acute and convalescent antibody titers. Interpret significant results.

7. Recognize the significance of serologic tests in the following situations:

- Population studies.

- Immune status testing.

- Congenital infections.

- Infections beyond the newborn period.

8. Describe the principles and applications of each of the following serologic methods:

■ Particle agglutination assays.

■ Precipitation assay.

■ Complement fixation.

■ Neutralization.

■ Indirect FA test.

■ Indirect ELISA.

■ IgM antibody capture ELISA.

■ Western blotting.

Serologic testing plays an important role in the diagnosis of a wide variety of infectious diseases. Often, the first practical laboratory tests for newly discovered infectious agents are based on serology. For example, three recently discovered diseases—acquired immune deficiency syndrome (AIDS), hepatitis C virus infection, and Lyme disease—are diagnosed in the laboratory most frequently with serologic tests. These three diseases are all caused by infectious agents (human immunodeficiency virus type 1, hepatitis C virus, and *Borrelia burgdorferi,* respectively) that are difficult to cultivate in routine clinical laboratories. Although currently available serologic tests for these and other agents are not perfect, significant improvements in quality and methodology of tests have been made over the past several years. Many of these improvements have resulted from advancements in preparation of antibodies and antigens of high purity. In addition, assay techniques such as immunoassays with labeled antibodies or antigens have notably improved. A wide variety of commercial kits of high quality are available. In this section, we discuss the aspects of the immune response that are important in understanding serologic testing. We also review principles of antibody detection methods and current clinical applications. Many of these applications are based on the principles already presented in the preceding section on antigen detection.

IMMUNE RESPONSE TO INFECTIOUS AGENTS

Host Resistance to Infection

Physical and Chemical Barriers

The human has evolved a complex system of defense mechanisms to prevent infectious agents from gaining access to and replicating in the body. The skin and mucous membranes of the respiratory, digestive, and urogenital systems provide a *physical barrier* to penetration by many microorganisms. This is particularly evident in the keratinized outer layer of the skin. Furthermore, the secretions of the mucous membranes and the ciliated epithelial cells of the respiratory tract promote trapping and removal of microorganisms. In addition, many body secretions provide a *chemical barrier* to infection. For example, the acidic pH of the stomach and vagina, the presence of enzymes such as *lysozyme* in saliva and tears, and oils produced by the sebaceous glands of the skin have chemical properties useful in inhibiting invasion by pathogenic microorganisms. Interestingly, the normal flora of these sites survive and contribute to the host resistance to pathogens.

Natural Immunity

Once the physical and chemical barriers to infection have been penetrated, a number of *non-*

specific mechanisms of immunity or natural immunity become operational. These mechanisms are so named because they do not depend on recognition of a specific infectious agent but operate to some degree against all foreign agents. These consist of cellular mechanisms, including phagocytic cells such as neutrophils, and humoral (fluid) mechanisms, such as activation of serum complement proteins. Phagocytic cells can ingest and kill microorganisms, whereas activated complement components contribute to a wide variety of immunologic events, including promotion of attachment and engulfment of bacteria by neutrophils (opsonization) and attraction of neutrophils to sites of infection (chemotaxis). Collectively, these immunologic defense mechanisms, along with host tissue damage caused by the invading organisms, combine to produce an acute inflammatory response in the host. The importance of natural immune mechanisms rests in their rapid response to invading organisms. However, these mechanisms are primarily effective against extracellular bacterial pathogens, playing only a minor role by themselves in immunity to intracellular bacterial pathogens, viruses, and fungi.

Acquired Immunity

Unlike the nonspecific mechanisms characteristic of natural immunity, the specific or acquired immune response to infectious agents depends on a complex system of organs, tissues, and effector cells by which the body recognizes and responds to specific foreign substances or antigens of the agents, and then records the event in immunologic memory for future encounters with the antigen. Like natural immunity just discussed, acquired immune mechanisms consist of both a cellular arm and a humoral arm, which are closely intertwined. The key effector cell for both of these arms is the lymphocyte.

Lymphocytes originate in the bone marrow from stem and progenitor cells, and mature and take up residence in various body tissues and organs, including the thymus, lymph nodes, and spleen. They are a diverse group of cells that can be classified into two major types—T (thymus-derived) cells and B (bone marrow–derived) cells—on the basis of cell surface markers. The uniqueness of these cells lies in the presence of specific cell surface receptor molecules that recognize and bind a unique antigen, activating the cell to divide, differentiate, and secrete a number of effector substances. The

millions of lymphocytes found in the body have been pre-engineered during embryogenesis to recognize a vast array of substances as foreign while learning which substances constitute self. Thus, the result of encounter with antigen is an expanded clone or clones of activated lymphocytes.

HUMORAL IMMUNITY

B lymphocytes play the predominant role in acquired humoral immunity. Each B cell has surface receptors that recognize only one type of antigen. After antigen binding, the B cell undergoes multiple divisions, and the resulting cells, known as plasma cells, actively secrete proteins known as immunoglobulins or antibodies. All of the antibody molecules derived from a single clone of B cells are of a single specificity (recognize a unique antigen), identical to the receptor molecule on the original activated B cell. These antibody molecules circulate in the blood stream and lymphatics, bathe body tissues, and are capable of binding to infectious agents or substances to aid the host in eliminating them from the body. The detection and quantification of these antibody molecules, obtained from a patient's serum, constitute the primary goal of diagnosing infectious diseases through serologic methods.

CELL-MEDIATED IMMUNITY

In contrast to humoral immunity, the primary effector cell in cell-mediated immunity is the T lymphocyte. The T cell does not secrete antibody molecules; however, the result of antigen binding, activation, cell division, and differentiation is the production of a number of low-molecular-weight (LMW) proteins known as lymphokines. Lymphocytes effect their immunologic function through direct cell-to-cell contact or through the activity of the lymphokines on other cells, such as macrophages. The measurement and diagnostic significance of cell-mediated immunity are beyond the scope of this book, and cellular immune function tests generally are not performed in microbiology or microbial serology laboratories.

Nature of the Immune Response to Infectious Agents

Although both humoral immunity and cell-mediated immunity are important in protecting the human from a wide variety of infectious agents, each contributes differently, in terms of overall importance, according to the type of pathogen and virulence mechanisms. Immunity to *extracellular* bacterial pathogens, such as *Staphylococcus aureus* and *Streptococcus pyogenes,* is mediated primarily by antibody functioning either alone (neutralization of toxins and extracellular bacterial enzymes) or with complement and neutrophils (chemotaxis and phagocytosis of bacterial cells). Immunity to *intracellular* bacterial pathogens, such as *Mycobacterium tuberculosis,* is primarily cell mediated, through the activities of T lymphocytes, lymphokines, and macrophages. If antibody is produced, it plays little role in eliminating this pathogen, because the pathogen is sequestered (hidden) intracellularly, where antibody cannot reach.

Viral infections often elicit both humoral and cell-mediated immune responses. Antibody may bind directly to and neutralize viral particles (render the virus noninfectious—unable to infect other cells) when they are found free in the blood stream or other body fluids. For example, central nervous system infection by *arboviruses* (which cause encephalitis) or by certain *enteroviruses* (which cause meningitis) may be prevented if neutralizing antibody against these organisms is present when the virus reaches the blood stream but before it enters the central nervous system. Some viruses, however, cause infections that spread by cell-to-cell transmission (herpes simplex virus, which causes cold sores); they would not be subject to the neutralizing effect of antibodies. In these cases, cell-mediated immunity plays a predominant role in eliminating the agent. Immunity to both fungal and parasitic infections is also primarily cell-mediated; antibody plays little or no role in prevention of or recovery from infection due to these agents.

It should be remembered that although antibodies may *not* have protective value for certain infectious agents, they may nonetheless have diagnostic and prognostic value in serologic tests. The remainder of this chapter focuses on humoral immunity and the detection and significance of antibodies.

Antigens and Antibodies

Characteristics of Antigens

Antigens are relatively high-molecular-weight (HMW) substances, usually proteins or polysac-charides (less commonly, lipids or nucleic acids either alone or complexed to proteins or polysaccharides), that can combine specifically with antibody molecules. Antigens that are recognized as foreign substances by a host are said to be immunogenic (antigenic), capable of eliciting an immune response. Often, the terms *antigen* and *immunogen* are used interchangeably, but the former term really refers to the antibody-binding properties of the molecule, whereas the latter term refers to the antibody-eliciting property of the molecule.

A number of factors determine whether a chemical compound will be an effective immunogen in a given host. First and foremost, the compound must be recognized as foreign or *nonself* by the host's immune system. This is a complex event largely related to the size and structural complexity of the compound. The larger and more chemically diverse a substance, the greater the likelihood it will be recognized as foreign. In addition, the compound must be present in sufficient quantity and have ample stability to remain in the host for enough time to "expose" the immune system. It should be remembered that some large molecules, such as complex proteins, may be of sufficient diversity to contain two or more different antigenic determinants. Complex polysaccharides may also contain multiple antigenic determinants (multivalent antigen), but these are generally identical because of the repeating subunit nature of polysaccharides.

Microorganisms contain a wide array of molecules capable of eliciting an immune response in the host. Immunogenic substances may be structural or nonstructural components of the microorganism. For example, many bacteria have polysaccharide or protein *capsules* that coat the organism. Because these capsules are located on the cell surface, they are often the first antigenic determinants recognized by the host. Additional structural antigenic components are the bacterial cell wall and membrane proteins and polysaccharides. Nonstructural antigens include bacterial enzymes and toxins that may be found within the cell or released into the host tissues. Because many bacterial components are "hidden" deep within the cell, antibody responses to some of them may not develop until the cell has been partially degraded by the host's natural immunity (phagocytic cells, complement, etc.). Thus, a single infecting bacterial species might stimulate the production of a wide array of antibodies.

Classification and Characteristics of Antibodies

Antibody molecules, found in serum and other body fluids and secretions, may be classified into one of five distinct immunoglobulin groups or classes. The classes differ from one another in several ways, including chemical structure, serum concentration, half-life, and functional activity.

Immunoglobulin G (IgG) class antibodies constitute about 70 to 75% of the total serum immunoglobulin pool. Their half-life in serum is about 3 to 4 weeks, and IgG can cross the maternal placenta to the fetus, possibly conferring some protection in both the prenatal and postnatal periods. Structurally, IgG is about a 150,000-MW protein molecule consisting of four polypeptides (two identical light chains and two identical heavy chains) bridged by several disulfide bonds (see Fig. 5–1). Although the amino acid sequence of some regions of the polypeptides is nearly identical among all IgG molecules (conserved regions), one of the ends of each polypeptide is highly variable. These variable regions create two active sites for antigen binding on each IgG molecule (see Fig. 5–1). Thus, IgG antibodies are said to be bivalent—capable of binding two antigen molecules.

Antibodies of the IgM class account for 10 to 15% of serum immunoglobulins. Their half-life in serum is about 5 days, and IgM cannot cross the placenta. A developing fetus in the second or third trimester, as well as a newborn, may respond to an infectious agent with an IgM antibody response. IgM antibody molecules are very large; the whole molecule has a molecular weight

T A B L E   5 – 4
COMPARISON OF IgG AND IgM

Property	Immunoglobulin Class	
	IgG	IgM
Molecular weight (daltons)	150,000	900,000
Number of 4-polypeptide subunits	1	5
Number of antigen-binding sites	2	10
Serum concentration (mg/dL)	800–1600	50–200
Percentage of total immunoglobulin	75	10
Ability to cross placenta	+	–
Half-life (days)	23–25	5–8

of about 900,000 daltons and consists of five basic subunits—each composed of two heavy chains and two light chains (similar to an IgG molecule) and linked to another polypeptide chain (J chain) by disulfide bonds (Fig. 5–20). Thus, the IgM molecule has ten antigen-binding sites available. Both IgG and IgM antibodies are commonly assayed in a variety of serologic tests. The differences in size and configuration of IgG and IgM molecules result in differences in functional activity of the molecules in serologic tests (Table 5–4).

Immunoglobulin A (IgA) antibodies represent 15 to 20% of the total serum immunoglobulin pool. IgA constitutes the predominant immunoglobulin class in certain body secretions, such as saliva, tears, and intestinal secretions. Because of this association of IgA with mucosal surfaces, it provides protection against microorganisms invading at those sites. Serum IgA occurs primarily as two subunits (each similar to an IgG molecule) linked together by a J chain. When found in secretions, however, the molecule also contains a secretory component that stabilizes the molecule. Although significant increases in serum IgA may occur in association with certain infections, the function of serum IgA is unclear, and few serologic tests for the diagnosis of infectious disease are designed to specifically detect IgA antibody.

The remaining two immunoglobulin classes, IgD and IgE, are found in very low concentrations in serum (<1%). IgE antibody levels rise during infection by a number of parasites and may play a role in eliminating these infectious agents from the host. Although total serum IgE levels may rise during parasitic infection, IgE-specific serologic tests for the diagnosis of parasitic agents have not been developed. The role of serum IgD antibodies during infection is not known.

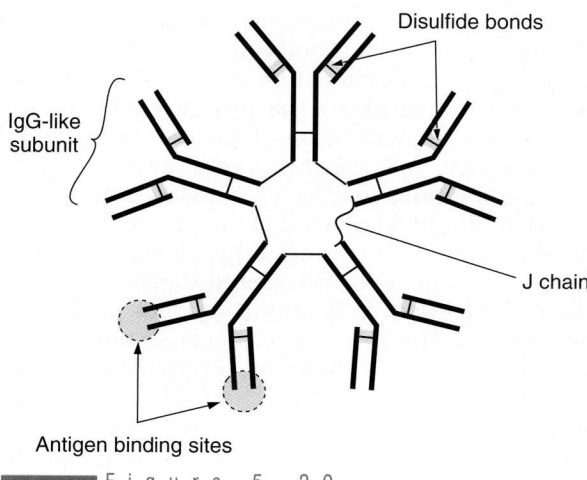

Disulfide bonds

IgG-like subunit

J chain

Antigen binding sites

F i g u r e   5 – 2 0
Structure of an IgM molecule.

Primary and Secondary Antibody Responses

After exposure to an infectious agent, the host's acquired humoral immunity may respond through the production of various classes of antibody directed to one or more antigens associated with the agent. If the host has *not* been previously exposed to the antigen(s), a *primary immune response,* characterized by the relatively rapid appearance of IgM antibodies, occurs. IgM antibody levels usually peak in 1 to 2 weeks, followed by a gradual decline to undetectable levels over the next few months. At the time when IgM levels have nearly peaked, IgG (and in some cases IgA) antibodies become detectable and continue to increase for about 1 month, surpassing peak IgM levels. IgG levels remain elevated for months and then decline slowly, often persisting at low but detectable levels for years (Fig. 5–21).

A subsequent exposure to the same antigen elicits a *secondary* or *anamnestic immune response,* characterized by a rapid increase in IgG antibody associated with higher levels, a prolonged elevation, and a more gradual decline (see Fig. 5–21). IgM antibody synthesis plays a minor role in a secondary immune response. Serologic tests that are designed to separately detect IgG and IgM antibodies take advantage of the differences in IgM production between a primary and a secondary immune response. Thus, a positive test result for IgM antibody is considered indicative of a primary current or very recent infection, whereas the presence of IgG antibody alone suggests a previous infec-tion or exposure. Similarly, the presence of significant levels of IgM antibody (with or without IgG) in a newborn suggests in utero infection (IgM can be synthesized by the fetus and cannot cross the placenta), whereas IgG antibody only in the newborn is indicative of passive maternal transfer of IgG across the placenta, not in utero infection.

INTERPRETING SEROLOGIC TEST DATA

Acute and Convalescent Antibody Titers

Unless a serologic test is designed to measure IgM-specific antibody for diagnosing a current infection or is being used to determine previous infections or immunization (immune status) by testing the IgG level in a single serum specimen, serodiagnosis of an infectious disease requires measurement of IgG antibody concentration on both *acute-phase* and *convalescent-phase* serum specimens. Specific IgG antibody is usually not detected in serum collected during the acute phase of the illness (within 1 week of manifestation of symptoms). A significant rise in IgG detected during the convalescent (recovery) phase (usually 2 weeks later) is diagnostic for infection and is referred to as *seroconversion.* Although seroconversion usually occurs within 2 to 3 weeks after onset of illness, it may be delayed in certain patients or types of infection.

Because some IgG antibody may already be present in a patient's acute-phase serum specimen, it is generally necessary to quantitate the concentration of antibody in both the acute-phase and convalescent-phase specimens. The concentration or *titer* of the antibody is the reciprocal of the highest serum dilution that reacts in the serologic test. Sera are generally tested as two-fold dilutions in a series of tubes. A four-fold rise (two doubling dilutions) in antibody titer between acute-phase and convalescent-phase serum specimens is considered diagnostic for a current infection. It is important that both sera be tested at the same time, because most serologic tests have an inherent variability that can alter the titer by at least two-fold. Testing the sera at the same time reduces this variability.

In practice, it is not uncommon to test a single serum specimen for IgG antibody to attempt to diagnose current or recent infection. In many cases,

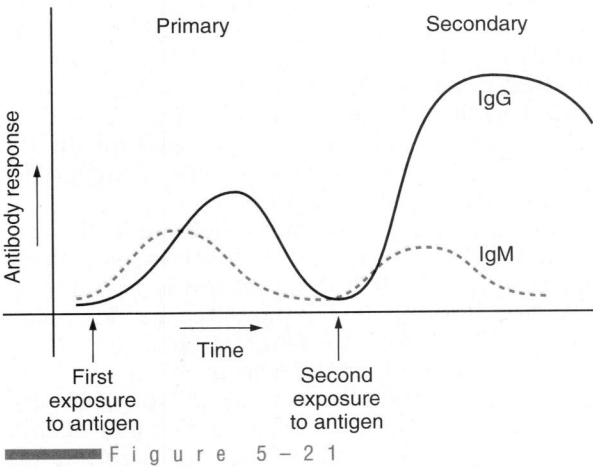

Figure 5 – 2 1
Primary and secondary antibody responses.

the presence of IgG antibody is difficult or impossible to evaluate; it may represent past infection, either clinically apparent or subclinical (without overt symptoms). In certain other cases, however, such as infections that are rare (rabies) or will have been present a relatively long time when symptoms first appear (AIDS), the presence of IgG antibody in a single serum may be diagnostic.

Antibody Specificity and Cross-Reactivity

Most antigen-antibody reactions show high specificity; that is, the antigen-binding sites on the antibody molecule react with specific antigenic determinants and not with other antigens containing different determinants. Some antigens associated with different microorganisms are closely related, however, and a host may respond by producing antibody not only to the invading organism but also to antigenically closely related organisms. These antibodies are said to *cross-react,* which may lead to misinterpretation of serologic tests. Because of this phenomenon, it is often best to perform a battery of serologic tests using the organisms known to show cross-reactivity and to do so on paired sera. An example of this type of testing would be a fungal battery consisting of tests for histoplasmosis, blastomycosis, and coccidioidomycosis.

False-Negative and False-Positive Serologic Test Results

An ideal serologic test should have 100% diagnostic accuracy; that is, it should be positive on all specimens from patients with infection and negative on all specimens from patients without infection by the specific agent. In practice, serologic tests, like other laboratory tests, fall short of this ideal.

A false-negative result is defined as a negative result for a patient who really is infected. It may occur for a number of reasons. For example, a patient may *not* have an intact immune system, and therefore may not be able to respond to an antigenic stimulus. This might be the case in an individual with a congenital or acquired immunodeficiency disease or in a patient receiving either immunosuppressive therapy after organ transplantation or cancer chemotherapy. In addition, neonates may not always respond to an infec-

tious agent because their immune systems are not fully mature. For some infections, such as Lyme disease and legionnaires' disease, antibody titers may not rise until months after acute infection, thus leading to apparent false-negative serologic test results. It is important to remember that not all individuals react the same way to antigenic stimuli, owing to the inherent genetic differences in the immune system among individuals.

A different type of false-negative result may occur in assays designed specifically to detect IgM antibody. It might be a result of competition for antigen-binding sites between IgG and IgM antibody in a serum specimen from a patient with high levels of IgG and relatively low levels of IgM (Fig. 5–22). For example, IgM antibody tests are sometimes performed on the serum of a newborn to diagnose a suspected in utero infection. A newborn's serum specimen might contain maternal IgG to a specific infectious agent as well as low levels of fetal IgM. The IgG antibody molecules bind to antigen in the assay and block IgM binding, thus producing a false-negative IgM result. Most assays designed to detect IgM antibody include some initial procedure to physically separate IgG from IgM or to "capture" IgM in the assay and then remove IgG (see later).

A *false-positive* serologic test result is a positive result for a patient who is not infected by the specific agent for which the test is designed. It might occur from the production of cross-reacting antibody, as discussed previously, or from the reactivation of a latent organism due to infection by a different organism. For example, influenza virus A infection may cause reactivation of latent cytomegalovirus (CMV) with a concomitant rise in CMV antibody. False-positive IgM antibody assays may also occur (see Fig. 5–22). These are due to the presence of rheumatoid factor activity in the serum. *Rheumatoid factor* is IgM antibody (produced in some individuals) that reacts with the individual's own IgG when the IgG is bound to antigen (complexed). IgM rheumatoid factor cannot be readily differentiated from organism-specific IgM in some serologic tests. Thus, if both organism-specific IgG and rheumatoid factor (but not organism-specific IgM) are present in a patient's serum specimen, the serologic test result may be falsely positive for organism-specific IgM antibody. Finally, individuals receiving intravenous immunoglobulin, a product prepared by pooling large quantities of plasma from multiple volunteer donors, may show specific antibody to a variety of infectious agents because of passive

A True positive

B False positive

C False negative

■■■■■■F i g u r e 5 – 2 2

Causes of false-positive and false-negative IgM assays. *A*, True-positive IgM assay. *B*, False-positive IgM assay. Rheumatoid factor (RF) binds to antigen-specific IgG and is detected with labeled anti-IgM antibody. *C*, False-negative IgM assay. Excess antigen-specific IgG inhibits antigen-specific IgM from binding.

transfer, not active infection. Laboratory personnel must be aware of this and any therapy that may be of significance in interpreting serologic test results for a specific patient specimen.

Value of Serologic Tests

Population Studies

Serologic tests for a specific infectious agent or a battery of agents may be performed to determine the percentage of individuals previously exposed or infected with the agent(s) in a geographic area. This information provides investigators and public health officials with information about how widespread an infectious agent is in a given area. For example, such studies have shown that the fungus *Histoplasma capsulatum* (which causes histoplasmosis), found in soil, is widely distributed in the Ohio River Valley and that the bacterium *B. correlia burgdorferi* (which causes Lyme disease), transmitted by a tick vector, is common in the upper Midwest, New England, and Middle Atlantic states. Similarly, serologic studies performed on animals that are reservoirs for human disease may alert public health officials that the disease (for example, viral encephalitis) may be active in the area and may pose a threat to humans.

Immune Status Testing

There are several situations in which it may be important to determine whether an individual is immune (either through previous infection or immunization) to a specific infectious disease. For example, many state health agencies require that individuals must undergo serologic testing for syphilis before being issued a marriage license. Also, health care facilities may require prospective employees to show evidence of immunity to varicella-zoster virus (which causes chickenpox). This is because if infected, they may transmit the virus during the incubation period (before symptoms develop) to susceptible patients. Infection with varicella-zoster in a newborn or immunosuppressed patient may be life-threatening. Similarly, female employees at risk for becoming pregnant may be screened for rubella or CMV infection, both agents that may cause significant morbidity or mortality to the developing fetus if contracted during pregnancy. It is generally recommended that all women of childbearing age be tested for their rubella immune status and, if it is negative, that they be vaccinated before considering pregnancy.

An additional situation in which immune status testing might be considered is whole organ or bone marrow transplantation. Cytomegalovirus may reside latently in leukocytes of donor tissue and may cause significant disease in a nonimmune recipient. Thus, it is recommended that a CMV-negative transplant recipient receive tissue and blood products from a CMV-negative donor.

Congenital Infections

Serology is often used to attempt to establish the diagnosis of congenital infection (acquired in utero) in a newborn, because some agents have the ability to cross the placenta and cause infection of the fetus. Such infections may cause only minimal symptoms in the mother during pregnancy, and thus may go undiagnosed at the time. If the mother is nonimmune, however, the infection may be significant to the fetus. The agents most commonly tested for are the TORCH agents—*Toxoplasma gondii* (which causes toxoplasmosis), rubella virus, CMV, and herpes simplex virus—and *Treponema pallidum* (which causes syphilis). Testing the mother's serum for IgG antibody is useless unless she was screened prenatally or early in pregnancy and found to be negative. Because the TORCH agents are common infectious agents, many individuals have antibody remaining from previous infections. Similarly, testing for IgG antibody in the newborn is of no value, because maternal IgG crosses the placenta and is present in neonatal serum. Testing for maternal IgM antibody is likewise of little value (unless infection occurred near the time of delivery), because it may be undetectable by 1 or 2 months following infection. Thus, as previously discussed, IgM antibody detection on neonatal serum is the method of choice for serologic diagnosis of congenital infection by one of the TORCH agents.

Infections After the Newborn Period

Beyond the newborn period, serologic diagnosis generally requires testing paired sera to detect a four-fold rise in antibody. More and more commercial diagnostic tests are designed, however, to be performed without serial dilution (on undiluted serum or after a single serum dilution) and establishment of titers; in these cases, the determination of a single positive result suggests, but does not always prove, current infection. Generally, if serologic diagnosis of current infection on a single serum specimen is to be attempted, an IgM assay as well as an IgG-specific assay should be done. If both assays are *negative,* the patient has probably not been infected. If both assays are positive, or just IgM antibody is detected, the patient has probably been recently infected. If just IgG antibody is present, the patient has probably been infected in the past and does not have a current infection. It is very important that laboratorians completely understand all the performance characteristics of a commercial serologic test system.

ANTIBODY DETECTION METHODS AND APPLICATIONS

Many of the test methods described in the discussion of antigen detection are useful for antibody detection. These include particle agglutination tests such as latex agglutination, precipitation tests (Ouchterlony diffusion and counterimmunoelectrophoresis), enzyme immunoassays, and microscope-assisted, labeled-reagent methods. In addition, a number of methods have been developed uniquely for antibody detection. Antibody tests for a wide variety of infectious agents are available commercially (Table 5–5). All of these methods, along with specific diagnostic applications, are discussed in the remainder of this section.

Particle Agglutination Assays

Agglutinating antibodies react with antigens on the surface of microscopic particles to form visible clumps of particles. Such agglutination tests can be performed on the surface of glass slides or in test tubes. Agglutination tests can be classified as either *direct* (natural particle) or *indirect* (carrier particle) agglutination.

Direct, Natural Particle Agglutination

Serologic tests based on direct or natural particle agglutination take advantage of antigens that are naturally occurring on biologic cells. Most commonly, these cells are whole bacteria or erythrocytes.

■■■■■■ T A B L E 5 – 5

EXAMPLES OF SOME COMMERCIALLY AVAILABLE SEROLOGIC TESTS FOR THE DIAGNOSIS OF INFECTIOUS DISEASE

Organisms	Serologic Test Method(s) Available*
Blastomyces dermatitidis	CF, ID, EIA
Bordetella pertussis	EIA
Borrelia burgdorferi	IFA, EIA, WB, IgM cap
Cytomegalovirus	EIA, LA, IFA
Helicobacter pylori	EIA
Hepatitis virus A, B, C	EIA
Herpes simplex virus	EIA, LA, IFA
Human immunodeficiency virus types 1, 2	EIA, IFA, WB
Histoplasma capsulatum	CF, ID, EIA
Influenza virus A, B	EIA
Legionella pneumophila	IFA, EIA
Mycoplasma pneumoniae	CF, EIA, IFA
Respiratory syncytial virus	EIA
Rubella virus	EIA, IFA, HAI
Streptococcus pyogenes	NT
Toxoplasma gondii	EIA, IFA, LA
Treponema pallidum	IHA, IFA

*CF, complement fixation; EIA, enzyme immunoassay; HAI, hemagglutination inhibition; ID, immunodiffusion; IFA, indirect fluorescent antibody; IgM cap, IgM capture; IHA, indirect hemagglutination; LA, latex agglutination; NT, neutralization; WB, Western blot.

WHOLE BACTERIAL CELL AGGLUTINATION

In whole bacterial cell agglutination, surface antigens that make up the bacterial cell wall or capsule function to allow cross-linking and visible agglutination of the cells in the presence of specific antibody. The so-called febrile agglutinin tests, for the detection of antibodies to *Brucella* sp, *Salmonella* sp, and *Francisella tularensis*, are based on bacterial agglutination. In some cases, an organism different from the suspected infectious agent may carry a chemically related antigen and may be more easily used in agglutination tests than the actual infectious agent. This approach is used in the Weil-Felix test to determine infection due to certain rickettsial agents by testing for agglutinating antibodies to various bacteria in the genus *Proteus*. Both febrile agglutinin and Weil-Felix agglutinin tests have been generally replaced in the United States by more sensitive and specific methods of diagnosis.

DIRECT HEMAGGLUTINATION

A special kind of natural particle agglutination test, known as hemagglutination, detects antibodies to naturally occurring antigens present on the surface of erythrocytes. During infection with *Mycoplasma pneumoniae*, for example, antibodies that can agglutinate human erythrocytes at 4°C but not at 37°C develop in some patients. These *cold-agglutinating antibodies*, which rise very quickly, may suggest *M. pneumoniae* infection, but the test lacks sensitivity and specificity.

Another example of a direct hemagglutination test is the detection of heterophile antibodies in the acute stage of infectious mononucleosis (due to Epstein-Barr virus infection). Heterophile antibodies are antibodies developed in one mammalian species that react with surface antigens on cells of an unrelated mammalian species. In the case of infectious mononucleosis, human heterophile antibody reacts with horse, ox, and sheep erythrocytes. In the appropriate clinical setting, and when performed by the appropriate method, heterophile antibody testing is diagnostically very useful and is commonly used in clinical laboratories today. It may be performed as a tube test or, more commonly, as a slide or card agglutination test (Monospot, Ortho Diagnostics, Raritan, NJ; Mono-Test, Wampole Laboratories, Cranbury, NJ: Fig. 5–23).

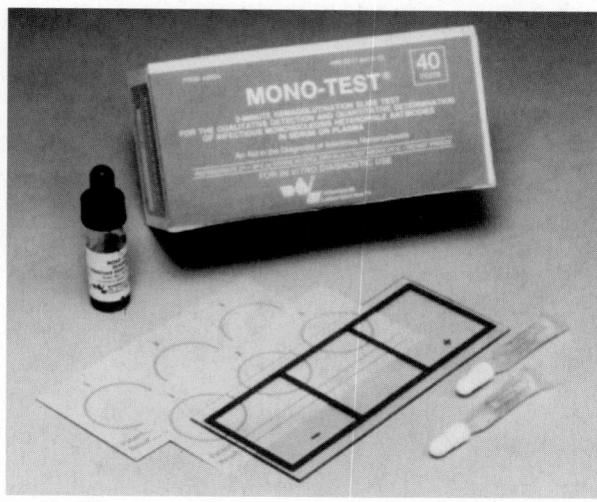

Figure 5 – 2 3

Mono-Test for the detection of heterophile antibodies produced during infectious mononucleosis. (Courtesy of Wampole Laboratories, Division of Carter-Wallace, Inc., Cranbury, NJ.)

HEMAGGLUTINATION INHIBITION

A serologic test that has great historic significance to diagnostic serology, particularly to viral serology, is the hemagglutination inhibition (HI) test. This test takes advantage of the fact that a number of viral agents, including rubella and influenza viruses, have surface antigens that can agglutinate erythrocytes from certain mammalian species. Antibodies present in sera can bind to the virus, thus inhibiting this agglutination reaction. The hemagglutination inhibition antibody titer is the highest dilution of the patient's serum that completely inhibits agglutination of the erythro-

cytes by the virus. Most HI tests have been replaced by other methods for determining viral antibodies, particularly enzyme immunoassays.

Indirect, Carrier Particle Agglutination

A variety of antigens can be passively or chemically coupled to naturally occurring particles such as erythrocytes or to synthetic particles such as latex beads. In this arrangement, the particle can then be agglutinated by specific antibody found in a patient's serum.

INDIRECT HEMAGGLUTINATION

In indirect or passive hemagglutination (IHA), microbial antigens are attached to erythrocytes after chemical treatment of the cells with tannic acid, chromic chloride, glutaraldehyde, or another substance that promotes cross-linking of the antigens. The sensitized cells can then be reacted with patient's serum to determine agglutinating antibody. One example of an IHA test is for the detection of antibodies to streptococcal extracellular antigens (Streptozyme, Wampole Laboratories, Cranbury, NJ) In this procedure, aldehyde-fixed sheep erythrocytes are sensitized with group A streptococcal exoantigens. When patient serum containing specific antibody is reacted with the cells on a slide, a positive agglutination reaction occurs (Fig. 5–24). Another important diagnostic application for the IHA test is the microhemagglutination assay for *T. pallidum* (MHA-TP), the causative agent of syphilis. In this test, specific treponemal antibodies are detected using treponemal antigen–coated erythrocytes. The MHA-TP

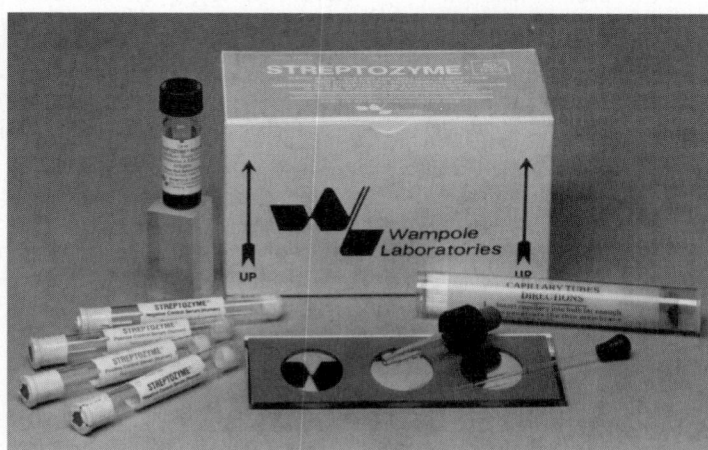

Figure 5 – 2 4

Streptozyme test for the detection of antibodies to streptococcal products. (Courtesy of Wampole Laboratories, Division of Carter-Wallace, Inc., Cranbury, NJ.)

test is generally used as a second or confirmatory test in patients who have tested positive on a syphilis screening test to detect nontreponemal antibodies (VDRL, RPR; see later).

LATEX AGGLUTINATION

Latex beads, which can bind antibodies and can be used to detect microbial antigens in clinical specimens, can also be sensitized with antigen and used in agglutination tests to detect serum antibodies. Commercial serologic tests based on latex agglutination are available for a number of viral infections, including CMV, rubella, and infectious mononucleosis (heterophile antibody), as well as a few bacterial (streptococcal, mycoplasmal antibodies), fungal (candidal antibody), and parasitic (toxoplasmal antibody) infections.

Precipitation Assays

Unlike agglutination assays, which rely on visible clumping of sensitized, insoluble particles when antigen and antibody are cross-linked, precipitation assays rely on transforming soluble antigens and antibody into insoluble, visible products that precipitate in a limited space. Those methods with diagnostic significance for serologic tests are double immunodiffusion, counterimmunoelectrophoresis, and flocculation tests.

Double Immunodiffusion

As described previously in this chapter, double or Ouchterlony immunodiffusion tests are performed in agar or agarose gels and rely on passive diffusion of antigen and antibody. For antibody detection, a known crude or purified antigen extract of a microorganism is placed in one well of the agar plate, and patient's serum in an adjacent well. If the patient has precipitating antibodies to the antigen(s), one or more precipitin lines will develop between the wells. Because the test relies on passive diffusion of molecules, reactions may take 48 to 72 hours to develop. Double immunodiffusion is commonly used today to detect antibody to a battery of fungal pathogens, including *H. capsulatum, Blastomyces dermatitidis,* and *Coccidioides immitis.*

Counterimmunoelectrophoresis

Counterimmunoelectrophoresis (CIE) is a variation of double immunodiffusion; it adds an electric current to the gel to allow known antigens and a patient's serum containing antibodies to migrate and form precipitin bands more quickly. CIE was fully discussed earlier as it related to antigen detection. It is not commonly used for serologic diagnosis today, because more sensitive and less cumbersome methods are available.

Flocculation

Flocculation tests are a variation of precipitation tests that also have some properties of agglutination assays. In these tests, because of the chemical nature of the antigen (not a truly soluble antigen), the antigen-antibody reaction forms a macroscopically or microscopically visible clump or precipitate of fine particles that remain in suspension. The most important applications for flocculation tests in antibody detection relate to syphilis serology, and these are discussed in detail.

Following infection with *T. pallidum,* the body reacts by developing two different types of antibodies—*specific* or *treponemal* antibodies, directed against *T. pallidum* antigens, and *nonspecific* or *nontreponemal* antibodies, directed against normal host tissue. These nontreponemal antibodies are also referred to as *reagin* antibodies.

The most commonly used treponemal tests for the diagnosis of syphilis are the MHA-TP, already discussed, and the fluorescent treponemal antibody absorption test (FTA-ABS; see later). These tests are very sensitive and specific, but they are not suited for testing (screening) large numbers of individuals, because (1) they are technically demanding and time consuming and (2) like other diagnostic tests, they would result in decreased diagnostic accuracy if applied to a large population with a disease of low prevalence. They are best suited for confirmatory testing.

Nontreponemal tests are technically easier and more rapid to perform; therefore, they are the tests of choice for syphilis screening. The most commonly used nontreponemal tests today are the VDRL (Venereal Disease Research Laboratory) and the RPR (rapid plasma reagin) tests.

The VDRL test is a microscopic flocculation test run on heat-inactivated serum and performed on glass slides. The test uses cardiolipin-lecithin-cholesterol particles as antigen and may be performed as both a qualitative and quantitative test. It may also be used to test cerebrospinal fluid to diagnose neurosyphilis and to follow titers in a newborn to diagnose congenital syphilis. Because of the extreme attention to reagent preparation and quality control required, the VDRL test is being replaced by tests like the RPR.

The RPR test is a commercially available nontreponemal test that uses cardiolipin-lecithin-cholesterol antigen with carbon particles to allow macroscopic determination of flocculation (Fig 5–25). Sera are tested (without heating) on cardboard cards. Newborn serum and cerebrospinal fluid should not be tested by this method.

Because both the VDRL and RPR tests are not specific for treponemal antigens, they are subject to false-positive results, the rate of which may be as high as 10 to 30% for all sera tested. False-positive results occur most commonly in conjunction with autoimmune disorders (lupus erythematosus, rheumatic fever), infectious diseases (infectious mononucleosis, hepatitis), in pregnancy, and in old age. It is imperative that positive VDRL or RPR test results be followed by specific confirmatory treponemal antibody tests.

Complement Fixation Test

Complement fixation (CF) tests are versatile assays that can be used for both antibody and antigen detection. They have been used for many years, particularly for antibody detection, but have gradually been replaced by more rapid and sensitive assays. CF tests are broadly applicable for a variety of infectious agents, however, and are still used in reference and public health laboratories for certain agents.

Although a discussion of the serum complement protein system is beyond the scope of this chapter, it is important to note that the system plays a vital role in a number of immunologic functions. The proteins involved in the system are usually found in an inactive form; however, once activated, the proteins become involved in a chemical cascade that involves production of enzymes with a variety of biologic activities. Activation of complement occurs by the classical pathway when specific IgG or IgM antibody combines with antigen and exposes a complement-binding site on the antibody molecule. It is the cell-lysing ability of activated complement components that is important in CF tests.

The CF test requires both an indicator system and a test system (Fig. 5–26). The *indicator system* typically consists of a combination of sheep erythrocytes, rabbit antibody to sheep erythrocytes, and guinea pig complement (a menagerie of biologics). When these three components are present and active, the rabbit antibody first combines with the sheep erythrocytes, and complement is subsequently bound (fixed) to the antibody-antigen complex, resulting in activation of complement by the classical pathway; this

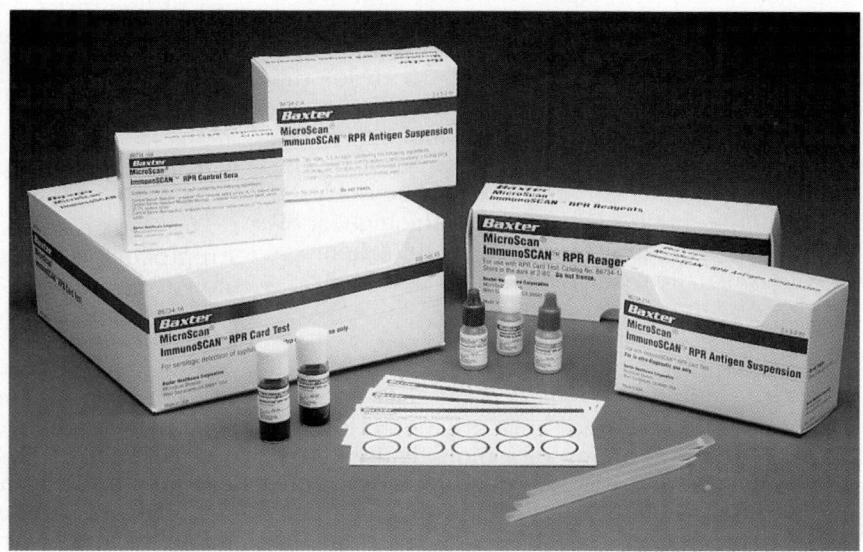

■■■■■Figure 5–25
RPR card test for detection of nontreponemal antibodies. (Courtesy of Baxter U. S. Distribution, Division of Baxter Healthcare, Inc., Deerfield, IL.)

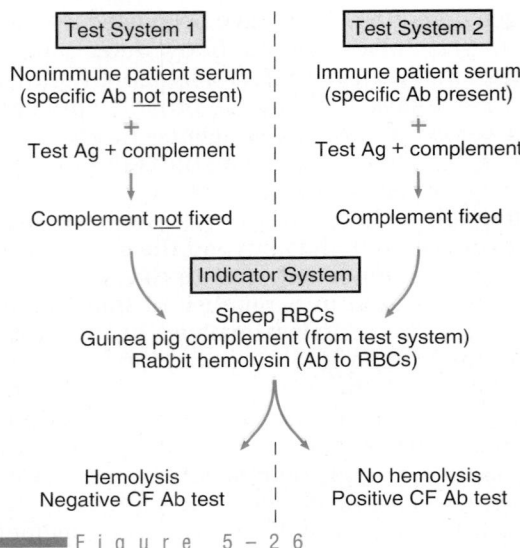

■■■■■F i g u r e   5 – 2 6
Principles of complement fixation test.

results in cell lysis. The *test system* consists of a mixture of known antigen and the patient's serum that may or may not contain specific antibody.

The CF test is performed as a two-step procedure (see Fig. 5–26). The patient's serum is serially diluted in test tubes, and each dilution is mixed with a known amount of antigen (test system). A fixed amount of complement is then added to each tube, and the mixture is incubated for a short time. Next, the sheep erythrocytes and rabbit antibody are added to each tube, and the tubes are incubated again (indicator system). If the patient's serum had specific antibodies to the test antigen, complement would be fixed by the test system and *not* available to lyse the erythrocytes of the indicator system. If the patient's serum did not have specific antibody to the test antigen, complement would be fixed by the indicator system, resulting in erythrocyte lysis. The CF antibody titer is considered the highest dilution of patient serum resulting in 100% hemolysis, and is usually read spectrophotometrically.

Neutralization Tests

Neutralization tests for antibody detection are based on the interaction of a biologically active antigen with antibodies that can block or render inactive the biologic activity of the antigen. These neutralizing antibodies may play an important role not only in serologic tests but also in functioning as protective antibodies in vivo. For example, immunity to a number of viruses depends on circulating antibodies that may neutralize, or destroy the infectivity of, viruses that reach the blood stream. Neutralization probably occurs because the antibody binds to the viral particle and blocks subsequent attachment of the virus to receptor sites on target cells. Neutralizing antibodies to viruses can be measured in vitro by the use of cell cultures. Other examples of neutralization assays are the antistreptolysin-O (ASO) neutralization test to detect antistreptococcal antibodies and the *Treponema pallidum* immobilization (TPI) test to measure antitreponemal antibodies.

Viral Neutralization

In viral neutralization, a patient's serum is serially diluted, and each dilution is mixed with a standard amount of a known virus suspected of causing disease in the patient. The virus-serum mixtures are then inoculated to a series of cell culture tubes or flasks. If the patient's serum contains neutralizing antibody to the virus, the antibody will block viral infection and cytopathic effect (CPE) on the cells. The neutralizing antibody titer is the highest dilution of the patient's serum to completely block CPE. For diagnosis of current disease, acute-phase and convalescent-phase sera should be tested. Viral neutralization assays are highly sensitive and specific, but they are also technically demanding and require the use of live virus and cell culture. They are not commonly used serologic tests today, but neutralization tests are valuable in identifying an unknown virus with known antibody.

Antistreptolysin-O Test

Streptococcus pyogenes, or group A β-hemolytic streptococcus, causes a number of acute, common pyogenic (associated with pus, or neutrophil response) infections, including pharyngitis and skin infections. In addition, the organism is responsible for certain nonsuppurative (nonpyogenic) diseases, such as acute rheumatic fever and poststreptococcal glomerulonephritis, which occur weeks after the acute infectious process and are thought to be autoimmune or immune complex–mediated diseases; that is, the host develops antibodies to streptococcal components that cross-react with normal host tissue. Although the pyogenic infections are best diagnosed by isolation of the organism in culture, the nonsuppurative diseases occur at a time when the organism may no longer be present; thus, serologic diagnosis is usually performed.

The antistreptolysin-O or ASO antibody test is commonly used to demonstrate serologic response to *S. pyogenes*. This test measures the ability of a patient's serum to neutralize the erythrocyte-lysing ability of a specific streptococcal enzyme, streptolysin O. Very high titers of anti-streptolysin-O antibody usually develop in most cases of rheumatic fever, but the titers and percentages of patients with poststreptococcal glomerulonephritis who develop ASO antibodies is significantly lower. Thus, when these diseases are suspected in patients who do *not* show an elevation in ASO titers, additional serologic tests directed at other streptococcal enzymes should be performed; the most valuable of these are antibody to DNAase B and antibody to hyaluronidase. These latter tests are also performed as enzyme neutralization assays.

Because serodiagnosis is usually attempted late after acute infection, it may be difficult to show seroconversion in streptococcal neutralization tests. Thus, "upper limit of normal" titers have been established; these vary by patient age and geographic location. These upper limits should not discourage an attempt to show seroconversion, however. A number of screening tests, based on particle agglutination, are also available to detect antibodies to streptococcal enzymes.

Treponema pallidum immobilization

The TPI test, like the MHA-TP and FTA-ABS tests mentioned previously, is a specific treponemal antibody test. Because it requires live organisms, however, it is generally performed only in research laboratories evaluating newer treponemal tests. The TPI test measures the ability of a patient's serum to neutralize the motility (immobilize) of *T. pallidum* organisms and is read microscopically.

Microscope-Assisted Labeled-Reagent Techniques

The basic principles that were described previously for the immunomicroscopic detection of antigens with labeled antibodies (fluorescent or

enzymatic labels) also apply to the detection of antibodies. The most commonly used test method for this purpose is the indirect fluorescent antibody (IFA) test. This method has been used for many years and still remains the method of choice for serologic detection of a number of infectious diseases. In addition, the method can be used to differentially measure IgG and IgM antibodies. The IFA test and some variations that increase the sensitivity of the basic technique are discussed.

Indirect Fluorescent Antibody Test

In the basic IFA test for antibody, whole microbial cells (bacteria, fungi, protozoan parasites) or virus-infected mammalian cells are washed and then fixed at a given density to glass slides (Fig 5–27A). Fixation is usually accomplished with methanol, ethanol, or acetone at room temperature or lower. The patient's serum is then serially diluted, applied to different fixed-cell preparations, and incubated for a short time (10–30 minutes) at 22 to 37°C in a humid environment to prevent drying; this procedure allows antigen-antibody

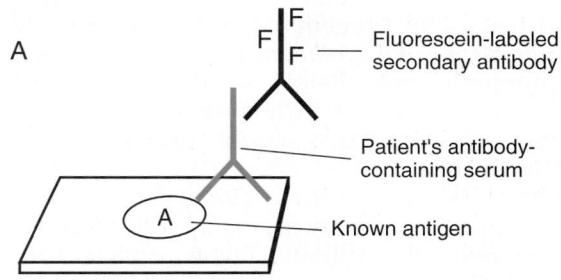
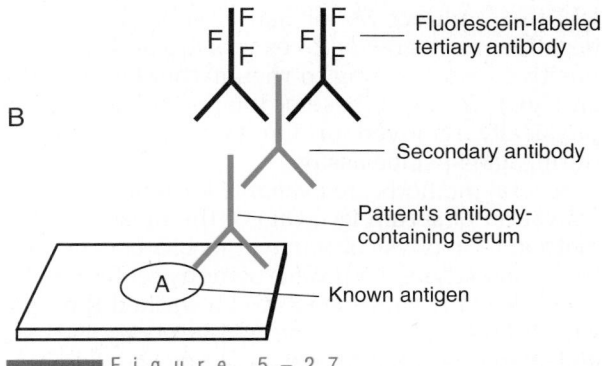

■■■■ Figure 5 – 27
Indirect fluorescence antibody *(A)* and double indirect fluorescence antibody *(B)* tests for antibody detection.

binding to occur. The slides are then washed repeatedly with a buffered solution to remove unbound antibody, and a second, nonhuman antibody, which is fluorescein-labeled and reactive with human immunoglobulin (IgG, IgM, or total), is applied. The slides are incubated again as described, and then washed, air-dried, and viewed using a microscope fitted with a light source and filters to excite the fluorescein (FA microscope). If specific antibody to the microbial antigen is present in the patient's serum, the antibody will bind and permit the fluorescein-labeled antibody to form secondary binding. The cells will fluoresce and be scored on a semiquantitative scale as positive (1+ to 4+) or negative. The FA titer is given as the reciprocal of the highest dilution of the patient's serum giving a minimum level of fluorescence (often at 2+ or greater).

Indirect FA tests are commonly used to detect antibody to a number of infectious agents, including the TORCH agents, Epstein-Barr virus, *Legionella pneumophila*, *B. burgdorferi*, *Rickettsia rickettsii*, *M. pneumoniae*, and *T. pallidum*. As mentioned previously, the detection of *T. pallidum* using the classic fluorescent treponemal antibody absorption (FTA-ABS) test is perhaps the most widely used and important application of IFA tests. This procedure uses the pathogenic Nichols strain of *T. pallidum* as test antigen and the nonpathogenic Reiter strain in a pretest absorption procedure to remove cross-reacting serum antibodies that could produce false-positive results.

Indirect FA tests are also useful in detecting IgM or IgG specific antibody, by using a fluorescein-labeled second antibody highly specific for human IgM or IgG, respectively. As discussed previously, IgM-specific assays are particularly important in several clinical settings. As with any IgM test, indirect FA for IgM is subject to false-negative results owing to excess IgG and to false-positive results owing to rheumatoid factor (IgM anti-IgG). To avoid these problems, IgG should be physically removed or functionally inactivated during IgM-specific assays.

Several methods are available for removal or inactivation of serum IgG. One of the most popular methods for physical removal of IgG uses miniature ion-exchange chromatography columns to trap IgM while allowing IgG to be washed through with a buffer solution. The IgM antibody is then collected by elution from the column with a lower-pH buffer; these columns are available commercially. Another method of removing IgG from whole serum utilizes adsorption to protein A found in the cell wall of staphylococci. Protein A binds most subclasses of human IgG, thus facilitating their removal. Yet another useful method utilizes anti–human IgG to inactivate or remove (requires precipitation and centrifugation) IgG for IgM specific assays. This method has the additional advantage of removing rheumatoid factor, because the latter will bind to the IgG–anti-IgG complexes.

The utility of indirect FA tests is limited by a number of factors, including the labor-intensive nature of the procedures and the subjectivity of reading endpoints. The ability to visualize and evaluate reactions, however, gives a high level of certainty to the procedure. Thus, newer serologic assays are often evaluated with indirect FA procedures. Indirect FA tests are also quickly adaptable to studying serologic response to newly discovered infectious agents.

Fluorescent Antibody Tests with Enhanced Sensitivity

A number of variations of the indirect FA test have been developed. The basic objective in most of these procedures has been to increase sensitivity while maintaining specificity of the assay. One example of such a procedure is the *double indirect* FA test (see Fig. 5–27B). This method uses a second, unlabeled antibody to bind to microbial-specific antibody in the patient's serum. The fluorescein-labeled antibody is directed at the second antibody. The additional step amplifies the reaction, because multiple molecules of the second antibody may bind to the patient's antibody, thus providing additional binding sites for the fluorescein-labeled antibody. Another example of enhanced FA test sensitivity uses a biotin-labeled second antibody. The binding of this reagent to the patient's antibody is detected using fluorescein-labeled avidin. Each biotin molecule binds four avidin molecules, thus increasing the total amount of bound fluorescein and, ultimately, the sensitivity of the assay. These and other microscopic techniques using fluorescent as well as enzymatic labels will continue to play an important role in serologic diagnosis of infectious diseases.

Immunoassays with Labeled Reagents

Immunoassays for the detection of serum antibody are performed in a manner similar to

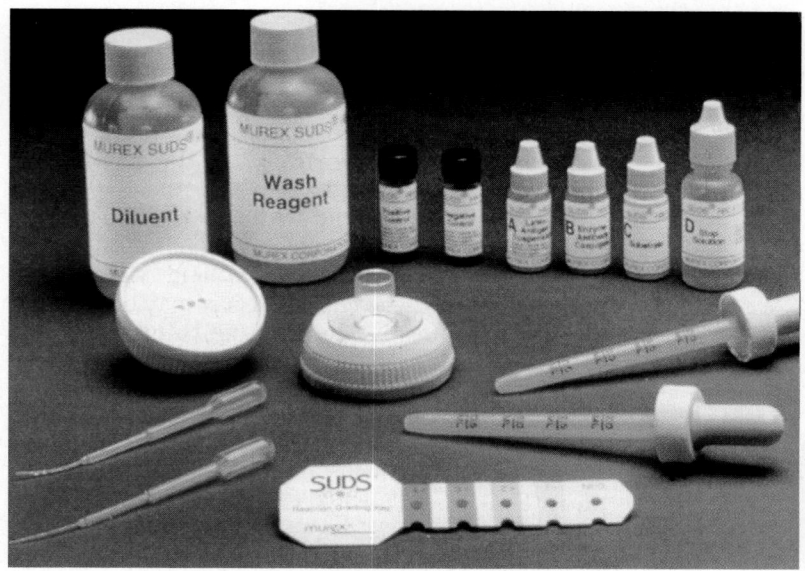
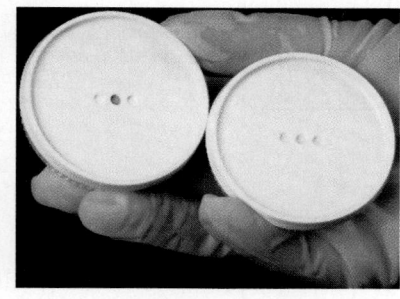

A   B

Figure 5–28

Single-use diagnostic system (SUDS®) for the detection of antibody to HIV-1 by EIA. (Courtesy of Murex Corp., Norcross, GA.) *A,* Components of the test kit. *B,* Cartridges showing positive *(left)* and negative *(right)* reactions.

immunoassays for microbial antigen detection, except that the roles of antigen and antibody are reversed. Most serum antibody assays are performed as solid-phase assays using antigen-coated tubes or wells. Ninety-six-well microtitration plates have become popular for this purpose because of the large number of tests that can be run and the small serum volume required. Single-test, single-serum dilution cassettes are also available from commercial sources for the low-volume or infrequent test setting. In either case, patient antibody is bound to the microbial antigen–coated surface, and the antibody is detected using a labeled (radioactive, fluorescent, enzymatic) anti–human immunoglobulin. Enzyme immunoassays (EIAs), or more specifically, solid-phase EIAs (enzyme-linked immunosorbent assay, ELISA), are the most popular types of immunoassay in use today.

Enzyme Immunoassays

Enzyme immunoassays are widely used for the serologic diagnosis of infectious diseases. They have become popular for a number of reasons. These include (1) the availability of commercial EIA kits for a large number of infectious agents; (2) the adaptability of EIA tests to automation, thus allowing more tests to be performed in shorter times; and (3) the objective interpretation of test results with colored end products that can be read spectrophotometrically. Many commercial products are tailored to suit high-volume laboratories; others are designed for single-use applications on individual patient specimens (Fig. 5–28).

INDIRECT ELISA

Most EIAs used in the clinical laboratory for antibody detection are performed as indirect ELISA tests (Fig. 5–29). In this procedure, the antigen of interest is first attached to the solid phase. The patient's serum is then added and allowed to incubate. After a wash procedure, antibody bound to the solid phase is detected with an enzyme-labeled, anti-immunoglobulin, secondary antibody. This secondary antibody may be anti–human IgG, IgM, or a combination of both. It should be remembered, however, that ELISA tests for IgM antibody are susceptible to the same problems of false-positive and false-negative results as IFA assays for IgM. One of the methods to physically separate patient's IgG and IgM antibodies should also be employed for ELISA tests for IgM.

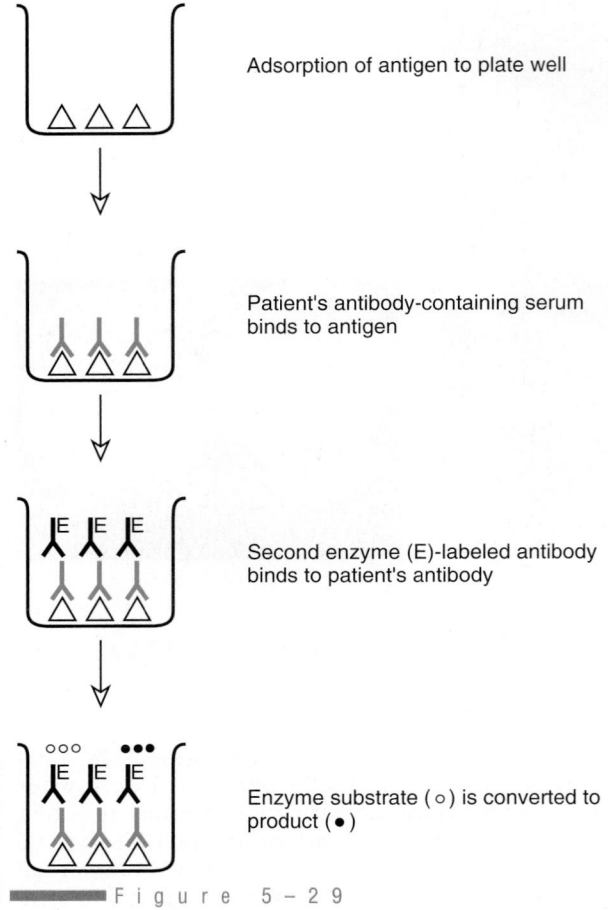

Adsorption of antigen to plate well

Patient's antibody-containing serum binds to antigen

Second enzyme (E)-labeled antibody binds to patient's antibody

Enzyme substrate (○) is converted to product (●)

■ F i g u r e   5 – 2 9

Principle of indirect solid-phase EIA (ELISA) for antibody detection.

IgM ANTIBODY CAPTURE ELISA

One of the alternative approaches that can be used to detect IgM antibody is referred to as *IgM antibody capture ELISA* (Fig. 5–30). In this procedure, the solid phase is first coated with animal antibody specific for human IgM. The patient's serum is added and incubated, and then the tube or plate is washed. At this point, IgM antibody molecules, regardless of specificity, should be bound to the solid phase, and antibodies of other immunoglobulin classes should be washed away. The next step in the IgM capture assay involves adding the antigen of interest and can be performed in one of two ways. The antigen can be directly labeled with an enzyme or can be added unlabeled. In either case, the antigen is bound to the solid phase if specific IgM antibody is present. With labeled antigen, the assay is completed by added enzyme substrate and mea-

suring the colored product. Alternatively, unlabeled antigen bound to IgM is detected by adding a secondary enzyme-labeled antibody directed against the specific antigen, followed by addition of enzyme substrate. The amount of colored product is proportional to the amount of specific IgM captured on the solid phase.

The IgM antibody capture assay has the advantage of eliminating the need for separation of IgG that might compete with IgM and yield false-negative results. The capture assay may be susceptible to false-positive results due to IgM rheumatoid factor bound to the solid phase. Modifications of the assay can be employed to reduce the possibility of this problem. Although IgM capture assays offer several advantages, they have not become readily available from commercial sources. Indirect ELISA tests for IgM employing IgG-IgM separation steps are much more widely available.

Other Immunoassays

The basic principles of immunoassays utilizing fluorochromes or radiolabels have been described earlier in the discussion of antigen detection. Fluorescent immunoassays (FIAs) for antibody detection use fluorochrome-labeled anti–human immunoglobulin to detect antigen-specific antibody bound to the solid phase. The bound fluorochrome gives a fluorescent signal that may be detected by fluorometry. The assay has the advantage of not requiring the addition of enzyme substrate and the washing steps needed in ELISA. Some additional problems, however, are associated with FIA, and only a few commercial companies offer antibody detection tests using this assay. Similarly, radioimmunoassays (RIAs) for antibody detection, although very sensitive and specific, are not commonly available for serodiagnosis. Because of the hazards and disposal problems associated with radioisotopes, most commercial companies have replaced RIAs with EIAs.

Western Blotting

Although serologic test methods such as IIF assays and EIAs generally provide excellent sensitivity and specificity in most clinical applications, they are limited in their ability to resolve the complex antibody response occurring during infection by most microbes. Because most antigens used in these assays are crude microbial extracts, a posi-

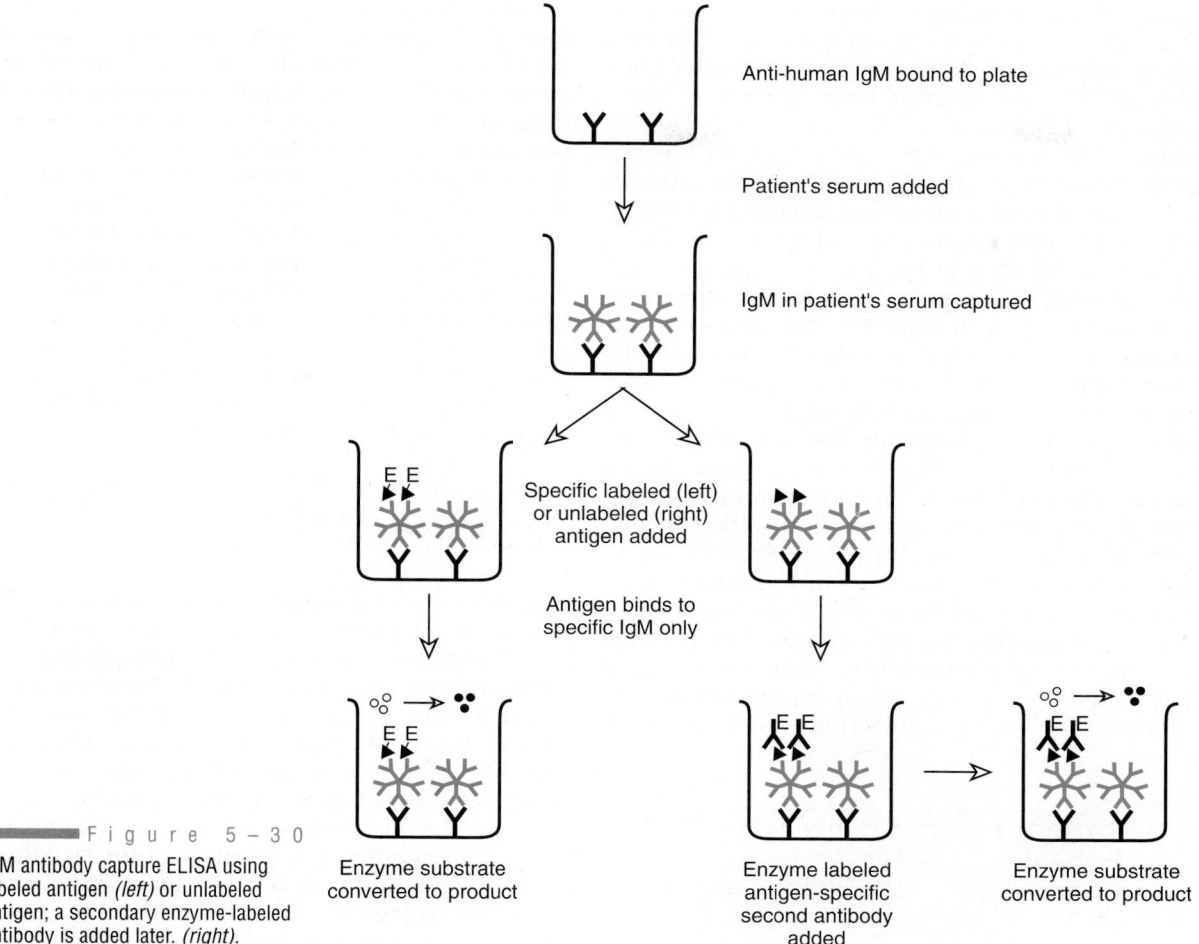

Anti-human IgM bound to plate

Patient's serum added

IgM in patient's serum captured

Specific labeled (left) or unlabeled (right) antigen added

Antigen binds to specific IgM only

Enzyme substrate converted to product

Enzyme labeled antigen-specific second antibody added

Enzyme substrate converted to product

Figure 5–30

IgM antibody capture ELISA using labeled antigen *(left)* or unlabeled antigen; a secondary enzyme-labeled antibody is added later. *(right).*

tive result may represent an antibody response to one or to many antigens. EIA tests may be designed to detect antibody to a number of individual microbial antigens, but this requires the expensive and labor-intensive process of purifying antigens and running multiple EIAs. As an alternative, the technique of immunoblotting or Western blotting was developed. This technique allows for characterization of multiple antibodies to a microorganism by first utilizing an electrophoretic separation of the microbial antigens and transfer to a nitrocellulose membrane. Western blotting has gained importance and is extensively used as a confirmatory test for the presence of antibodies to human immunodeficiency virus type 1 (HIV-1) in patients whose sera have been repeatedly reactive in EIA tests. It is also being applied to serologic diagnosis of Lyme disease and other infections in which the development of antibody to a specific microbial antigen or antigens has diagnostic significance.

In the Western blotting procedure, a crude microbial antigen preparation (e.g., a partially purified virus preparation) is first heated with a detergent such as sodium dodecyl sulfate (SDS). The SDS releases individual polypeptides from complex proteins. The polypeptides are then separated according to their molecular weight by polyacrylamide gel electrophoresis with SDS (SDS-PAGE). The separated polypeptides form invisible protein "bands" in the gel. At this point, the protein bands are electrophoretically transferred to an inert filter membrane support, usually made of nitrocellulose. The nitrocellulose membrane thus becomes a solid matrix to which the separated protein antigens tightly adhere and upon which an antibody-antigen reaction can occur and be visualized—much like a microtitration plate well with an ELISA essay. A number of commercial kits for Western blotting contain nitrocellulose membrane strips on which the protein antigens of interest have already been electrophoretically separated. For analysis, the nitrocellulose membrane is immersed in a patient's serum sample (usually diluted) and allowed to incubate. Specific antibodies in the sample, if present, will bind to unique antigenic determinants in the protein bands. After incubation, the nitrocellulose membrane is thoroughly washed to remove unbound antibody, and a labeled anti–human immunoglobulin (secondary antibody) is allowed to react with the membrane. The secondary antibody is commonly labeled with an enzyme or biotin. The detection of an antigen-antibody reaction at any particular protein band is then accomplished by the addition of the enzyme substrate (or enzyme-labeled strepavidin followed by enzyme substrate) and final wash steps (Fig. 5–31).

Western blotting has added a new dimension of versatility and specificity to immunoassays. It has helped to resolve false-positive EIA results (particularly for HIV-1 infection) due to antibodies cross-reacting to other viruses, autoimmune disorders, technical error, and other, undetermined reasons. It has also helped to dissect and analyze the antibody response to individual proteins in complex mixtures. Western blotting is a procedure, however, that must be well controlled and interpreted using strict criteria. It should also be recognized that antigens detected by other methods may not be detected by Western blotting techniques, because the polypeptide antigens are not native, intact proteins (SDS treatment denatures the proteins), and some antigens may not be transferred well to the membranes. Nevertheless, Western blotting will continue to play an important role in immunoserology of selected infectious agents.

FUTURE APPLICATIONS

In this age of emphasis on nucleic acid amplification and detection techniques for the diagnosis of infectious diseases, one must question the future role of immunoserology in the clinical laboratory. Clearly, serology will remain important as a method of determining immune status. For the diagnosis of active infection, however, it is likely that nucleic acid detection techniques will become the methods of choice for diagnosing certain infections for which culture or direct antigen detection is unreliable, insensitive, slow, or cumbersome, and for which serology is now the method of choice. These may include infections such as AIDS and Lyme disease. IFA and EIA techniques are likely to play an important role in the clinical laboratory at least through the 1990s. Automation of EIA techniques will be particularly important during this era of cost containment in medicine.

Bibliography

Aloisi RM: Principles of Immunology and Immunodiagnostics. Philadelphia: Lea & Febiger, 1988.

Baron EJ, Finegold SM (eds): Diagnostic immunological principles and methods. In Bailey and Scott's Diagnostic Microbiology, 8th ed. St Louis: CV Mosby, 1990, pp 157–170.

Barrett J: Textbook of Immunology. St Louis: CV Mosby, 1988.

Chernesky MA, Ray CG, Smith TF: In Drew WL (ed): Laboratory Diagnosis of Viral Infections: Cumitech 15. Washington DC: American Society for Microbiology, 1982, pp 9–12.

Conroy JM, Stevens RW, Hechemy KE: Enzyme immunoassay. In Balows A, Hausler WJ, et al (eds): Manual of Clinical Microbiology, 5th ed. Washington DC: American Society for Microbiology, 1991, pp 87–92.

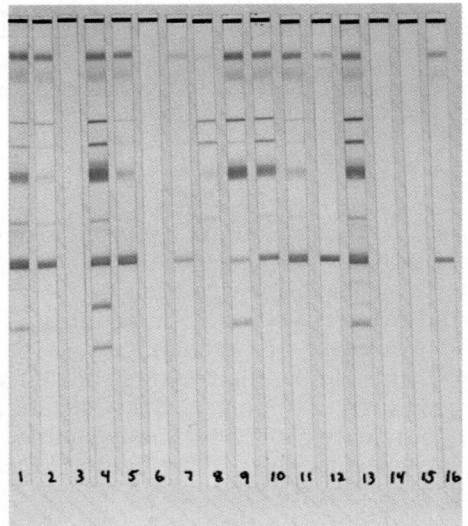

■■■■■ F i g u r e   5 – 3 1

Western blot for the detection of antibody to HIV-1. (Courtesy of Organon Teknika Corp., Durham, NC.) Strips 1 and 4 are high-positive; strips 2 and 5 are low-positive; and strips 3 and 6 are negative controls. Strips 7 through 13 and strip 16 are positive reactions of patient sera. Strips 14 and 15 are negative reactions of patient sera. Gp160 represents a viral glycoprotein with a molecular weight of 160,000 daltons; gp41 and p24 represent other viral proteins.

Coonrod JD: Immunologic diagnosis. In Hoeprich PD, Jordan MC (eds): Infectious Diseases, 4th ed. Philadelphia: JB Lippincott, 1989, pp 166–174.

Eisen HN: Antibody-antigen reactions. In Davis BD, Dulbecco R, et al (eds): Microbiology, 4th ed. Philadelphia: JB Lippincott, 1990, pp 249–276.

Eisen HN: Immunoglobulins and immunoglobulin genes. In Davis BD, Dulbecco R, et al (eds): Microbiology, 4th ed. Philadelphia: JB Lippincott, 1990, pp 277–318.

James K: Immunoserology of infectious diseases. Clinical Microbiol Rev 3:132, 1990.

Landay A, Folds JD (section eds): Section 9: Immunology. In Isenberg HD (ed): Clinical Microbiology Procedures Handbook. Washington DC: American Society for Microbiology, 1992, pp 9.1.9–9.15.7.

Lee FK: Detection and qualitation of specific immunoglobulin responses. In Balows A, Hausler WJ, et al (eds): Manual of Clinical Microbiology, 5th ed. Washington DC: American Society for Microbiology, 1991, pp 105–110.

Peter JB: The Use and Interpretation of Tests in Medical Laboratory Immunology, 5th ed. Los Angeles: Specialty Laboratories, Inc, 1989.

Proffitt MR: Blotting techniques for the diagnosis of infectious diseases. Clin Microbiol Newsl 16:121, 1990.

Rose N, deMacario EC, et al (eds): Manual of Clinical Immunology, 4th ed. Washington DC: American Society for Microbiology, 1990.

Thompson KD, VanEnk RA: Changing technologies in immunodiagnosis: Enzyme immunoassays versus immunofluorescence assays. Clin Microbiol Newsl 10:73, 1991.

Turgeon ML: Immunology and Serology in Laboratory Medicine. St Louis: CV Mosby, 1990.

Yolken RH: Immunoassays for the diagnosis of infectious diseases. In Wilcheck M, Bauer EA (eds): Methods in Enzymology, Vol 184. San Diego: Academic Press, 1990, pp 529–537.

C. RAPID METHODS AND AUTOMATION IN THE MICROBIOLOGY LABORATORY

WHAT IS *RAPID?*

MICROSCOPIC METHODS FOR RAPID DETECTION

RAPID BIOCHEMICAL TESTS PERFORMED ON ISOLATED COLONIES FROM SOLID MEDIA

RAPID ENZYMATIC TESTS USING CHROMOGENIC SUBSTRATES
Minitek
RapID NH

Rapid E
Quadferm
Staph-Ident
RapID ANA II

AUTOMATED IDENTIFICATION SYSTEMS
Microscan Systems
Vitek

EVALUATION OF RAPID METHODS

Vinnie Pawar • Michael A. Pentella

Until the late 1970s, microbiologists relied on the growth and isolation of bacteria in broth culture and agar media. Once bacteria are cultured in vitro, their properties can be used for identification. Isolation and identification of the infectious agent from clinical samples not only require incubation time (24–48 hours) to produce growth but also involve many variables in the isolation of the pathogen.

Rapid diagnosis of infectious diseases and early institution of therapy in hospitalized patients therefore remained major challenges for the clinical microbiology laboratory. All too often, microbiology results merely confirmed the physician's diagnosis. With the use of rapid methods, i.e., reporting of test results in a time frame to have optimum impact on patient care, the role of microbiology has become proactive rather then retrospective. Rapid and accurate reporting of results means early diagnosis and immediate institution of appropriate therapy.

Continued prominence of gram-negative bacilli as major pathogens and the emergence of antibiotic-resistant bacteria require the laboratory to provide fast, accurate, and cost-effective identification of clinical isolates.

This section describes the major concepts and applications of rapid methods and automation in the microbiology laboratory currently available. Other sections in this chapter describe emerging technologies that also provide rapid results.

WHAT IS *RAPID*?

For decades, microbiologists relied on the ability of an organism to ferment sugars, degrade amino acids, and produce unique end products for identification purposes. Diagnosis of an infectious disease has been a complex, laborious, and frequently slow process.

In the late 1950s and 1960s, traditional biochemical tests became miniaturized. Smaller test tubes

and molded plastic vessels were introduced. These changes made testing more convenient but did not improve turnaround times in reporting results. In the 1970s, microbiologists began to rely on computerized databases, so that numerous results could be considered simultaneously and the most statistically probable result could be regarded as the identification of the unknown organism. This development improved reliability of the results but still did not improve reporting turnaround times. In the mid to late 1970s, automated instruments for identification and susceptibility testing gradually appeared. They shortened turnaround time, yielded greater precision, improved productivity, and provided accurate results.

The term *rapid method* encompasses a wide variety of procedures and techniques and has been loosely applied to any procedure affording results faster than the conventional method. There are rapid methods for microscopy, biochemical identification, antigen detection, and antibody detection. *Rapid* is definitely a relative term used to describe time and depends on the procedures being compared. For example, an enzyme immunoassay (EIA) method to detect *Clostridium difficile* toxin is rapid compared with the 48-hour tissue culture assay. The radiometric detection of *Mycobacterium tuberculosis* within 2 weeks is rapid compared with 6 weeks required to grow the organism on agar slants.

Microscopic procedures using common stains and fluorescent antibody to detect specific organisms are rapid methods. So are conventional procedures used for initial differentiation or presumptive identification of certain groups of organisms. Some of these procedures have been modified to provide immediate results that may lead to a presumptive identification.

Rapid identification of clinical isolates also involves commercially packaged identification kits and fully automated instruments. These manufactured kits are usually miniaturized test systems that employ chromogenic or fluorogenic substrates to assess preformed enzymes. Reaction end points may be reached after 2 to 6 hours of incubation, although some may require an overnight incubation. The capabilities of these kits and systems vary widely. Certain systems may still require manual reading by the technologist, whereas others may utilize mechanized reading through spectrophotometer, numerical coding, and computerized databases. Some of these instruments incubate, read, and interpret the enzymatic test results.

MICROSCOPIC METHODS FOR RAPID DETECTION

Direct microscopic examination of body fluids provides immediate results and is valuable to the clinician. For example, India ink and Gram stain results from a spinal fluid specimen along with other laboratory results, such as those from blood chemistry and hematology, may therapeutically be the most helpful information a physician may obtain from the laboratory. Table 5–6 shows common stains that are used for direct smear examination of clinical samples.

RAPID BIOCHEMICAL TESTS PERFORMED ON ISOLATED COLONIES FROM SOLID MEDIA

Table 5–7 summarizes the principles, modes of action, and applications of certain old, established procedures for quick presumptive differentiation between groups of organisms or presumptive identification of bacterial species. These tests may also provide direction about additional tests needed for definitive identification.

RAPID ENZYMATIC TESTS USING CHROMOGENIC SUBSTRATES

Rapid enzymatic tests using chromogenic substrates produce reaction end points in 2 to 4 hours. Plastic cupules, reaction chambers, or filter paper strips contain desiccated or dehydrated chemical enzyme substrates. A suspension of bacterial cells or a loopful of isolated colony is added to the

■■■■■■ T A B L E   5 – 6
MICROSCOPIC METHODS FOR RAPID DETECTION OF PATHOGENS

Stain	Application	Expected Results	Comments
Gram stain	Provides presumptive diagnosis of bacterial pneumonia, meningitis	White blood cells (WBC) are detected Gram stain morphology of common bacterial agents is recognized	Requires skill and experience for proper evaluation Does not include other infectious agents
Acridine orange	Useful in differentiating true organisms from artifacts, especially in blood cultures	Fluorochrome stain binds the nucleic acid of organisms Organisms show orange fluorescence against a dark background Both viable and nonviable organisms will fluoresce	Requires a fluorescent microscope
Darkfield microscopy	Direct examination of suspected primary syphilitic lesions	Motile spirochetes are visualized	Requires experience
Calcofluor white	Direct examination of spinal fluids and other body fluids for fungal elements	Yeast and fungi will show white fluorescence	Fluorescent stain that binds cell wall of fungi and yeast but not bacteria or inflammatory cells Preferred to India ink and KOH preparations
India ink	Direct smear examination of spinal fluid, blood, and urine for *Cryptococcus*	*Cryptococcus* possess a polysaccharide capsule that appears as halo against the black background	Reverse stain Capsule does not stain Requires experience to differentiate WBC from yeast
KOH 10%	Direct smear examination of skin, hair, and nails for dermatophytes	Visualizes fungal elements	KOH with gentle heating digests protein in samples to allow visualization

■■■■ T A B L E   5 – 7
RAPID BIOCHEMICAL TESTS PERFORMED ON ISOLATED COLONIES

Test	Bacterial Enzyme	Mode of Action	Applications
Indole	Trytophanase	Hydrolysis of substrate tryptophane to indole develops a red color upon addition of paradimethyl amino-benzaldehyde	Positive reaction (red color) identifies *Escherichia coli, Proteus vulgaris* Aids in the identification of anaerobes
Spot indole (DMAC)	Tryptophanase	Organism from blood agar or any tryptophane-containing medium is placed on a swab, and reagent is added Hydrolysis of tryptophane to indole is indicated by the production of blue color upon addition of paradimethyl amino-cinnamaldehyde	Aids in the identification of anaerobes
ONPG	β-Galactosidase	An ester linkage of ortho-nitrophenyl moieties to various carbohydrates Hydrolysis results in release of yellow orthonitrophenol	Determines lactose fermentation (yellow color) in slow lactose fermenters Differentiates *Neisseria lactamica* from pathogenic *Neisseria*
Oxidase	Cytochrome *c* oxidase	A blue compound is produced when tetramethyl-para-phenylenediamine reacts with cytochrome *c*	Differentiation of nonfermenters Aids in the identification of *Neisseria* sp., *Aeromonas* sp, *Vibrio* sp, and *Campylobacter* sp
Catalase	Catalase	Breakdown of hydrogen peroxide into oxygen and water, resulting in rapid production of bubbles	Differentiation of staphylococci from streptococci, and of *Listeria* from streptococci
Bile solubility		Autocatalyzes colony in the presence of the surfactant sodium deoxy-cholate (bile salts)	Presumptive identification of *Streptococcus pneumonia* in sputum, blood, and CSF cultures
PYR (*L*-pyrrolidonyl-*B*-naphtylamide)	*L*-pyrroglutamyl-amino-peptidase	Hydrolysis of amide substrate with formation of the free *B*-naphthy-lamine, which combines with cinnamaldehyde, a reagent, to form a bright red color	Identification of group A streptococci Differentiates *Enterococcus* from group D streptococci
Rapid urease	Urease	Rapid hydrolysis of urea by enzyme urease releases the end product ammonia The alkalinity causes the phenol red indicator to change from yellow to red	Screening test for *Cryptococcus, Proteus, Klebsiella,* and *Yersinia enterocolitica*
Rapid hippurate		Enzymatic hydrolysis of hippurate visualized by addition of triketohydrindene hydrate	Speciation of *Streptococcus agalactiae, Campylobacter jejuni, Listeria*

reaction substrate or rubbed off to the reaction area. The substrate, when hydrolyzed, produces either a color reaction or a byproduct that can be detected by the addition of color-developing reagent.

Methods based on enzyme substrates have certain advantages over conventional methods. Because enzyme methods base their reactions on preformed enzymes and are growth independent, endpoints are reached in 2 to 4 hours. Also, the tests are sensitive, so the endpoints are easy to interpret.

Substrate tests have certain limitations, however. They include factors that play an important role in producing false-positive and false-negative results, such as:

■ Concentration and stability of substrate

■ Accessibility of enzyme

■ Age of the inoculum

Several rapid manual methods for bacterial and fungal identification are commercially available. The methods described here are a few examples of commercially available kits.

Minitek

The Minitek Enterobacteriaceae III and Neisseria (BBL Microbiology Systems, Cockeysville, MD) are currently available for identification of organisms after 4 hours of incubation. The systems utilize 20-well plastic plates to which paper disks impregnated with substrates are added. The kits contain either basic 14 or 4 tests for Enterobacteriaceae III or Neisseria, respectively. The disks are inoculated with appropriate bacterial suspension.

For the Enterobacteriaceae III system, mineral oil is overlaid on urea and H_2S/indole wells. After incubation, reagents are added to the assays to detect production of indole, acetoin, phenylalanine deaminase activity, and nitrate reduction. The Neisseria kit uses glucose, maltose, sucrose, and o-nitrophenyl-β-galactoside (ONPG) tests. Both kits are incubated for 4 hours at 35 to 37°C. Reactions are read and interpreted by the technologist. Identification is made by using differential charts or by means of a computer. The user is also able to select the number and type of tests desired, making these two systems flexible.

RapID NH

The RapID NH (Innovative Diagnostic Systems, Inc, Decatur, GA), used to identify Neisseria and Haemophilus species, consists of a plastic panel containing 10 test cavities with 12 dehydrated conventional and chromogenic substrates for the performance of phosphatase, nitrate reduction, prolyl-p-nitroanilid, ∂-glutamyl-p-nitroanilid, ONPG, reaszurin reduction, glucose, sucrose, indole, urease, ornithine decarboxylase, and β-lactamase reactions. A suspension of test organism equivalent to a McFarland No. 3 turbidity standard is inoculated, and the panel is incubated for 4 hours at 35 to 37°C. Reactions in some test cavities are read before and after appropriate reagents are added. Species identification from test scores is made with differential chart or ID Code compendium.

RapID E

The RapID E test (bioMerieux Vitek, Hazelwood, MO) for the identification of Enterobacteriaceae species consists of a plastic strip of 20 microtubes containing dehydrated substrates. Substrates are reconstituted by the addition of a suspension of test organism in medium provided with the system. The reactions are read after 4 hours of incubation at 35 to 37°C. Profile numbers are constructed and interpreted with the Identification Code Book.

Quadferm

Quadferm (bioMerieux Vitek, Hazelwood, MO), for the identification of Neisseria species and β-lactamase production, consists of a plastic strip of five microtubes containing dehydrated substrates. Substrates are reconstituted by the addition of a suspension of test organism equivalent to a McFarland No. 3 turbidity standard. The panel is incubated for 4 hours at 35 to 37°C. Identification is made from a differential chart.

Staph-Ident

Staph-Ident (bioMerieux Vitek, Hazelwood, MO) is used for identification of Staphylococcus species. It consists of a plastic strip with microcupules containing dehydrated conventional and chromogenic substrates. Tests included are β-glucosidase, β-glucuronidase, β-galactosidase, phosphatase, urease, salicin fermentation, mannose, mannitol, and trehalose. The inoculum is prepared in a sterile physiologic saline. The suspension is inoculated into the microcupules, and the strips is incubated for 5 hours at 35 to 37°C. The reactions are recorded after color-developing reagent is added to the β-galactosidase test cupule. Identification is made using the Staph-Ident Profile Register.

RapID ANA II

The RapID ANA II system (Innovative Diagnostic Systems, Decatur, GA), used to identify anaerobic bacteria, consists of 10 reaction wells. Eight of the wells are bifunctional—two separate tests contained in each. Well 1 of the panel detects the increase in pH produced by the hydrolysis of urea. Wells 2 through 9 detect enzymatic hydrolysis of nitrophenyl carbohydrate or phosphoester derivatives. Bacterial inoculum is prepared in physiologic saline. Inoculation of the test panel rehydrates the substrates. Panels are incubated for 4 hours at 35 to 37°C. After incubation, identification is made from reactions of tests, and scores are determined using the IDS Identification Compendium.

AUTOMATED IDENTIFICATION SYSTEMS

Automated instruments offer laboratories the ability to perform same-day identification and susceptibility testing. Valuable information is available 1 day sooner than from conventional methods. This information has a positive impact for the patient only if the physician is apprised of the results the same day they are reported, so he or she can implement changes in antibiotic therapy if necessary. Systems are available for a wide range of needs, from those of small community hospital laboratories to those of large reference laboratories. In today's cost-conscious health care environment, clinical laboratories are continually pressured to reduce costs and increase productivity. With automation, workflow can be streamlined, reporting standardized, and data management capabilities enhanced. Microscan Walk/Away and Vitek are two lines of automated instruments that yield rapid identification and susceptibility test results.

Microscan Systems

The Microscan Touchscan, Autoscan, and Autoscan Walk/Away models (Baxter Travenol, West Sacramento, CA) are designed to identify gram-negative bacteria, nonfermenters, gram-positive bacteria, *Haemophilus, Neisseria,* anaerobic bacteria, and yeasts and to perform antimicrobial susceptibility testing. These automated instruments use turbidity, colorimetry, and fluorescence principles. Test trays are available with dried test substrates or antimicrobial minimum inhibitory concentration (MIC) panels or both. The chromogenic substrates are incorporated in panels designed to identify *Haemophilus,* yeast, and anaerobes; the fluorogenic substrates are incorporated in panels designed to identify gram-positive and gram-negative bacteria. The systems provide a 2-hour identification of gram-negative and gram-positive bacteria, and a 4-hour susceptibility test of organisms on chromogenic panels. Inoculation is performed manually with a 95-prong transfer lid, and incubation takes 18 to 24 hours at 35°C. The Walk/Away model automatically reads plates. The chromogenic panels are read on the Autoscan reader, and the data are processed by the computer to yield the most probable species identities and their likelihood percentages.

Vitek

The Vitek system (bioMerieux Vitek, Hazelwood, MO) is designed to identify gram-positive bacteria, gram-negative bacteria, *Haemophilus, Neisseria,* yeasts, and anaerobic bacteria using specific cards. Each test card contains 30 wells. The cards are automatically filled from a standardized suspension of bacteria prepared in a test tube, and are then placed in an appropriately controlled incubator. The principle of operation is based on turbidity. Depending on the organism, species identification and antimicrobial susceptibility data can be available in 4 to 6 hours or 18 hours. The system consists of a filler-sealer module for automatic inoculation of cards, a reader incubator for automatic reading of cards, a data terminal, and a printer.

EVALUATION OF RAPID METHODS

Most new rapid and automated procedures are designed to provide results with greater speed and precision than traditional methods. Whether greater efficiency and productivity are achieved is debatable; savings are minimal. Automated systems help reduce analytic errors. Theoretically, early reporting of results may shorten length of hospital stay for a patient and therefore cut hospitalization costs. For this to occur, the microbiology department must work closely with the pharmacy and provide the information to the physician as efficiently as possible. The efficacy of rapid or automated procedures depends on the following concepts established by Sherris and Ryan (1985):

■ **Reference point:** Results of automated and rapid susceptibility test procedures must be related to those of nationally accepted reference or standard procedures to ensure reproducible results with other methods. New procedures for identification must be referable to genus and species determinations by generally accepted taxonomic criteria and methods.

■ **Accuracy:** Whenever new automated methods are evaluated, results are compared with those of reference procedures or taxa. Methods are compared first, using stock cultures and a large number of sequential or selected clinical isolates from a group of laboratories. When bacterial identifications or susceptibility test categories are compared, the results are expressed in degrees of error and reproducibility according to the individual species and antibiotics. Another important consideration in evaluating automated procedures is recognizing tests or organisms that should be excluded in routine use or for which the procedures must be supplemented with other tests.

■ **Quality control:** Quality control procedures must be performed to ensure performance of the instrument. A carefully selected panel of stock strains representative of those that will be tested in practice must be used to control bacterial identification. Tests should be performed during introduction of every new batch of growth modules or reagents and also at intervals between such occurrences. Quantitative susceptibility test results can be controlled by interval testing of strains of known performance and established control limits.

Bibliography

Aldridge KE, Kogos C, et al: Comparison of rapid identification assays for *Staphylococcus aureus*. J Clin Microbiol 19:703, 1984.

Aldridge KE, Stratton CW, et al: Comparison of the Staph-Ident system with a conventional method for species identification of urine and blood isolates of coagulase-negative staphylococci. J Clin Microbiol 17:516, 1983.

Altwegg M: Performance of two four-hour identification systems with atypical strains of Enterobacteriaceae. Eur J Clin Microbiol, 2:529–533, 1983.

Appelbaum PC, Lawrence RB: Comparison of three methods for identification of pathogenic *Neisseria* species. J Clin Microbiol 9:598, 1979.

Brown WJ: Modification of the Rapid Fermentation test for *Neisseria gonorrhoeae*. Appl Microbiol 27:1027, 1974.

Doern GV, Earls JE, et al: Species identification and biotyping of staphylococci by the API Staph-Ident system, J Clin Microbiol 17:260, 1983.

Ferraro MJB, Edelblut MA, Kunz LJ: Accurate automation identification of selected Enterobacteriaceae at four hours. J Clin Microbiol 13:151, 1981.

Givini F, Husson MO, et al: Evaluation of AutoSCAN-4 for identification of members of the family Enterobacteriaceae. J Clin Microbiol 26:1586, 1988.

Jorgensen JH: Selection criteria for an antimicrobial susceptibility testing system. J Clin Microbiol 31:2841, 1993.

Kloos WE, Wolfshohl JF: Identification of *Staphylococcus* species with API Staph-Ident system. J Clin Microbiol 16:509, 1982.

Matsen JM: 1991. Concept of Rapid Reporting in Microbiology: Rapid Methods and Automation, 2nd ed. New York: Springer-Verlag, 1991.

Morse SA, Bartenstein L: Adaptation of Minitek system for the rapid identification of *Neisseria gonorrhoeae*. J Clin Microbiol 3:8, 1976.

O'Hara CM, Tenover FC, Miller JM: Parralel comparison of accuracy of API 20E, Vitek GNI, Microscan Walk/Away Rapid ID, and Becton-Dickinson Cobas Micro ID-E/NF for identification of member of the family Enterobacteriaceae and common gram-negative, non-glucose fermenting bacilli. J Clin Microbiol 31:3165, 1993.

Peiffer S, Cox M: Enzymatic reactions of *Clostridium difficile* in aerobic and anaerobic environments with the RapID-ANA-II identification system. J Clin Microbiol 31:279, 1993.

Reddick A: A simple carbohydrate fermentation test for identification of the pathogenic *Neisseria*. J Clin Microbiol 2:72, 1975.

Robinson MJ, Oberhofer TR: Identification of pathogenic *Neisseria* species with the RapID NHJ system. J Clin Microbiol 17:400, 1983.

Ruoff KL, Ferraro MJ, et al: Automated identification of gram-positive bacteria. J Clin Microbiol 17:400, 1982.

Sherris JC, Ryan KJ. Rapid Methods and Automation in Microbiology—1981. New York: Springer-Verlag, 1985.

Stoakes L, Schieven BC, et al: Evaluation of MicroScan Rapid Pos Combo Panels for identification of staphylococci. J Clin Microbiol 30:93, 1992.

Woolfrey BF, Lally RT, Quall CO: Evaluation of the AutoSCAN-3 and Sceptor systems for Enterobacteriaceae identification. J Clin Microbiol 17:807. 1983.

D. IDENTIFICATION OF MICROORGANISMS USING CHROMATOGRAPHIC TECHNIQUES

Leona W. Ayers

OBJECTIVES

1. Describe the chemical technique of chromatography.

2. Compare and contrast the various types of chromatography.

3. List the instrument parts necessary for a functional gas chromatograph.

4. Compare and contrast the possible analytes for chemotaxonomy.

5. Given a list of microorganisms, select the types of fatty acids that characterize or are most closely associated with a particular genus or species.

6. Describe the application of statistical cluster analysis to chemotaxonomy, and identify two functions used to demonstrate relatedness of microorganism CFAs.

7. List the functional instrument components of an automated chromatographic system for chemotaxonomy using microbial constituent cellular fatty acids.

8. Given a list of computer software functions that could support chemotaxonomy using long-chain CFAs, be able to select three of the critical functions.

Technical advances in chromatography methods, equipment, and columns and in associated computer software make routine chromatography of microbial cellular fatty acids (CFAs) for chemotaxonomy and epidemiology clinically feasible. Today, in medical diagnostic laboratories serving large or tertiary-care hospitals and in microbiology reference services such as those found in commercial, state, and federal laboratories, this single chemical method is used to identify and relate a wide variety of species. Chromatographic methods used to assist in the identification of organisms by isolating and measuring selected metabolic byproducts or chemical components also continue to be useful. Conventional packed-column gas-liquid chromatography (GLC) to identify the metabolic short-chain fatty acids from spent glucose is the most common clinical application. Other chromatographic techniques for selected bacterial cellular components, such as identification of teichoic acids by GLC, of isoprenoid quinones by thin-layer chromatography (TLC), of cellular decomposition products by pyrolysis-GLC (PGLC), and of mycolic acids by high-pressure liquid chromatography (HPLC), are well described and routinely used to characterize taxonomic relationships among organisms. No matter the application, chromatography is an important method for diagnostic, chemotaxonomic, and research microbiology.

DEFINITION

Chromatography (from *chroma,* meaning "color," and *graphein,* meaning "to write") is an analytical method to separate individual components within a mixture from one another and to quantify the separated components. Separations are accomplished using two immiscible phases, one mobile and the second stationary. The components distribute between the two phases to produce a physical partitioning based on molecular size, shape, or affinity for the stationary phase. The mobile phase can be a gas or a liquid, and the stationary phase can be a solid or a liquid. Chromatographic methods are named according to these mobile and stationary phases (Table 5–8).

Initially, other techniques such as infrared spectroscopy and mass spectrometry must be used to identify the components that are separated by the chromatography. Once the components have been identified, however, known standards and computer software can be used to

━━━ T A B L E 5 - 8
CHROMATOGRAPHIC METHODS

Phases		Notation
Mobile	**Stationary**	
Gas	Liquid	GLC
Gas	Solid	GSC
Liquid	Solid	LSC
Liquid	Liquid	LLC
Liquid	Solid (thin-layer)	TLC

efficiently identify the "chromatographed" components.

VOCABULARY OF CHROMATOGRAPHY

An extensive vocabulary is associated with the science of chromatography. Following is a list of commonly encountered terms that, if properly understood, facilitate the study of this analytical technique and its application to microorganism identification. The list should be studied and then used as a reference for reading the subsequent discussion.

Baseline: Chromatograph detector record tracing of the chromatogram resulting from only eluent or carrier gas emerging from the column.

Capillary column: Open tubular column (OTC) with a small internal diameter (0.2–1.0 mm) in which the inner walls are used to support the stationary phase.

Carrier gas: Gas chromatographic mobile or moving phase that transports the sample through the column.

Chromatogram: Visible plot of the chromatograph detector response to effluent concentration seen as baseline and peaks marked for retention times (Fig. 5–32).

Chromatograph: *(Verb):* To separate sample components by partitioning. *Noun:* instrument that carries out a chromatographic separation.

Chromatography: Technique for the separation and quantitation of chemical components from a mixture in which two phases, one stationary and the other mobile, are used.

Column: Metal, plastic, or glass tube packed or coated on its inner surface with the material through which the sample components and mobile phase (carrier gas) flow and where the chemical separation occurs. The column is housed within the chromatograph.

Column material: Stationary phase within the column used to effect the separation. The configuration of the column and the characteristics of the material determine the type of component separation possible.

Component: Compound or molecule in the sample mixture that is to be separated.

Dendogram: A statistical tool that shows the linear relationship of unweighted pair matchings based on fatty acid compositions.

Derivatization: Process by which compounds are chemically altered to make them more volatile or less polar (e.g., addition of methyl side chain = methylation).

Detector: Chromatograph device that registers the signals from the presence of a separated component eluted from a chromatographic column.

Detector response: The chromatograph detector signal produced by the sample, which varies with the nature and quantity of the sample components.

Eluent: Gas (mobile phase) used for separation by elution.

Elution: Process of transporting a sample component through and out of the analytical column by use of the carrier gas or mobile phase.

FAME: Fatty acid methyl esters prepared for fatty acid identification by gas-liquid chromatography.

Injection point (t_0): Starting point on the chromatogram, corresponding to introduction of the sample into the chromatographic system.

Injection port: Chromatograph part where the sample is introduced through a syringe into the column.

Injection temperature: Temperature of the chromatographic system at the injection point.

Integrator: Chromatograph electrical or mechanical device employed for a continuous mathematical summation of the detector output with respect to time.

Internal standard: Pure compound added to a sample in known concentration to eliminate the need to measure the sample size by quantitative analysis.

Liquid phase: Nonvolatile liquid with special solubility characteristics adsorbed onto the solid support or coated onto the walls of columns.

Mobile phase: Carrier gas or gas phase.

Partition chromatography: Another term for gas-liquid chromatography.

Peak area: "Band area"; area enclosed between the chromatogram peak and the peak base, representing quantity.

Peak height: The distance between the peak (band) maximum and the peak base.

Peak resolution (R_s): Degree of separation of two peaks in terms of their average peak widths.

Quantitative analysis: Measurement of the concentration or the absolute weight of one or more components of the sample.

Retention time (absolute) (t_R): Amount of time elapsed from injection of the sample into the column to the recording of the peak maximum of the component band (peak); "time on the column."

Sample: Gas or liquid mixture of components injected into the chromatographic system for separation and quantitation.

Sample injector: Chromatograph device used to introduce liquid or gas samples into the column.

Separation: Time elapsed between elution of two successive components, measured on the chromatogram as the distance between the recorded bands or peaks.

Separation temperature: Temperature of the chromatographic column necessary for component separation.

Solute: Synonym for components in a sample.

Solvent: Synonymous with liquid phase.

Stationary phase: Liquid phase in gas-liquid chromatography, and the granular solid adsorbent in gas-solid chromatography.

Temperature programming: Function by which the temperature of the column is changed systematically during part or all of the chromatographic process to enhance separation of components.

Two-Dimensional Plot (2-D) of Principle Components: A two-dimensional depiction of the relatedness of CFA profiles using principal component analysis.

INSTRUMENTATION

Gas Chromatographs

Gas chromatographs are relatively simple instruments that house or are attached to the following devices: sampler, injection port, column, and detector. Temperatures of the injection port, oven (column housing), and detector are independently controlled, usually by programmable devices. Carrier gases (mobile phases) are hydrogen, helium, and nitrogen and require that the sample be volatilized into the gas stream. The differences in gas chromatographs relate to the type of sampler, manual versus automated, type of column, type of detector, and sophistication of the control software, attached integrator, and plotter/printer.

Sample can be introduced by hand through the injection port, or the chromatograph can be automatically supplied with samples by attaching an automated injector with attached sample tray. With the latter, the sample information can be entered in numbered sequence into a computer, and the samples loaded into the sample tray in the same sequence for automatic processing. After loading and starting the processing, the technologist can leave the instrument unattended.

GC Columns

Columns are packed open tubular (OTC) or capillary configurations. Packed columns can be constructed of glass or metals and range in diameter from 1.6 to 10 mm and in length up to 3 meters. These columns are packed with the stationary phase according to their intended use. Capillary columns, constructed of heat-resistant glass or fused silica, have an internal diameter of less than 1 mm and range in length up to 150 meters. Component separations provided by these small-bore, long columns are greater with less retention times than those provided by the larger-bore, short columns. Historically, multiple extraction procedures and packed columns were required to accomplish the results of present-day single capillary column analysis.

GC Detectors

Sensitivity to eluted components and cost of GC detectors vary significantly. The flame ionization detector is the most widely used detector because of its high sensitivity to organic compounds and insensitivity to inorganic gases as well as its concentration response range. The thermal conductivity detector is chosen by many diagnostic laboratories for its good response but relatively low cost. The choice of a flame ionization detector, electrolytic conductivity detector, electron capture detector, flame photometric detector, photoionization detector, thermal conductivity detector, or thermionic emission detector depends on the specific analytic application. Other detectors are chosen for narrow applications requiring specific component detection or increased sensitivity.

High-Pressure Liquid Chromatographs

High-pressure liquid chromatographs (HPLC) have essentially the same devices as GCs, with the exception of a liquid reservoir and powerful pump to produce the liquid mobile phase at very high pressure to carry the sample through a column. This requires leakproof connections for the column and pump function. HPLC separation does not require that the components be volatile, and therefore is appropriate for separation of components with high molecular weights, high boiling points, and thermal stability of the analytes.

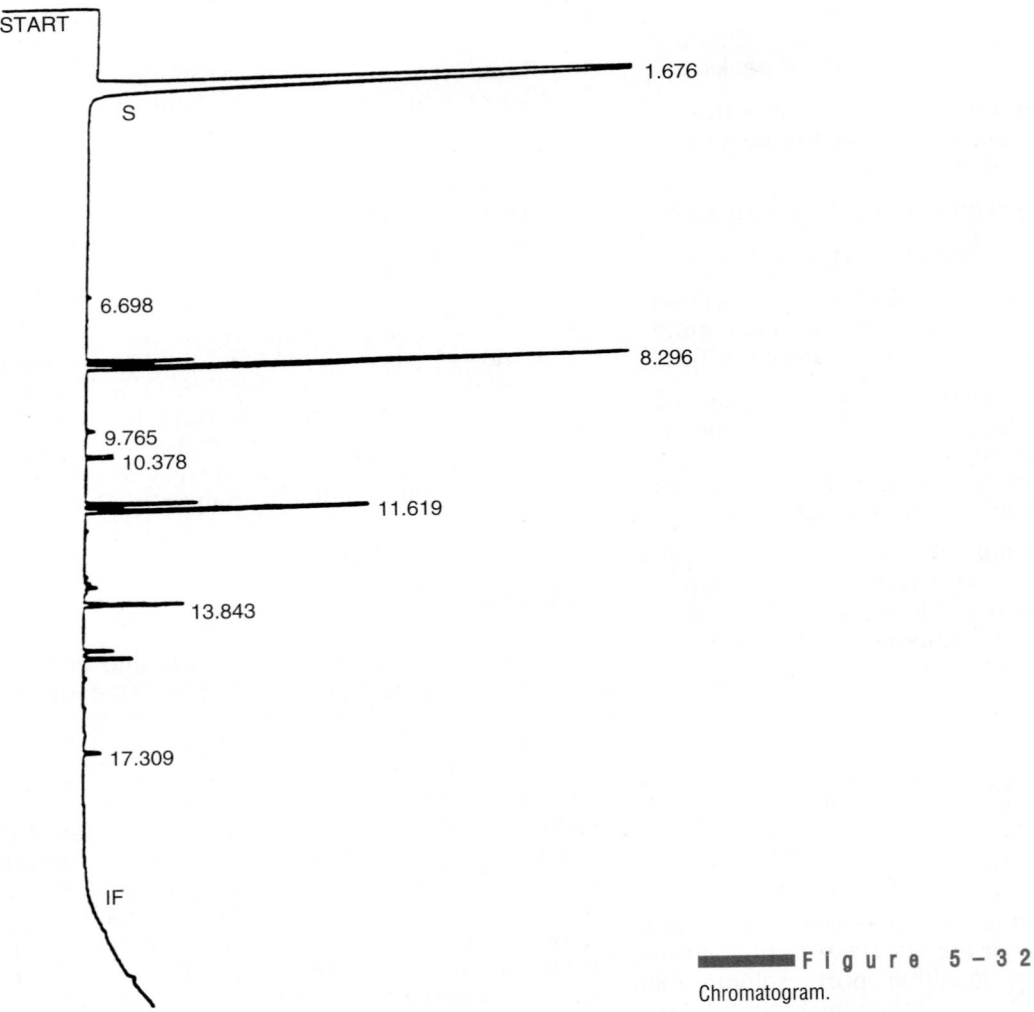

Figure 5-32

Chromatogram.

HPLC Columns

HPLC columns must be constructed of metal, usually stainless steel, to tolerate the high pressures required to force the mobile phase liquid through the column. Diameters range from 4 to 6 mm; lengths range from 5 to 25 cm. Stationary phases are supported by silicas and aluminas and are either coated or chemically bonded to the solid support.

HPLC Detectors

Broad or narrow applications of HPLC detectors are available. Commonly chosen are the flu-orescence detector, refractive index detector, and ultraviolet (UV) absorption detector. UV detectors provide variable-wavelength detection but are limited to analysis of components that absorb at given wavelengths.

Thin-Layer Chromatography

Thin-layer chromatography (TLC) is not instrument dependent. TLC plates with coated or bonded phases are commercially available for a variety of separations. Qualitative analyses can be easily performed, although quantitation may be done with scanning densitometers.

ANALYTES FOR CHROMATOGRAPHY

Carbohydrates

Sample preparation and the lack of a good derivatization technique have prevented sugars and amino sugars from becoming important chromatographic analytes. Carbohydrates are important structural components of microorganisms and are defined as part of organism characterization.

Isoprenoid Quinones

Common compounds in microbes, isoprenoid quinones, occur in different structural forms, which assist in taxonomy. Isolated from microbial membranes, they are soluble in lipid solvents and can be extracted into mixtures and separated by TLC or HPLC. The molecules are too large to be separated by GLC.

Whole Microbial Cells

Microorganisms can be broken down into a crude component mixture by rapid heating of a cell mass with an electric current *(pyrolysis)*. The resulting component mixture is injected onto a GLC column swept by a carrier gas. The cleavage fragments in the component mixture are poorly reproducible, and peaks are difficult to identify. Internal standards are not directly usable; identi-

fication of peaks may require mass spectrometry. Despite the difficulties inherent in this technique, pyrolysis has appeal because of the simple sample preparation.

Fatty Acids

Hundreds of different fatty acids are found within the spent growth medium of microorganisms and within their cellular membranes. The great variety of acids makes this analyte ideal for chemotaxonomy. For over 50 years, scientists have studied fatty acids and associated the various fatty acids with specific organisms. The most extensive data have been accumulated for bacteria, but the technique also applies to other organisms with cell membranes.

Nomenclature of Fatty Acids

Common fatty acids (FAs) have conventional names that reflect their original source. These conventional names are not used in this discussion of FAs but are of interest in understanding original terminology so that it may be recognized when seen in scientific reports (Table 5–9).

The different chemical configurations of FAs and their relationships to the hydrocarbons from which they are derived must be understood. The types of configurations that characterize FAs and their related alcohols, along with commonly used chemical notations, are summarized in Table 5–10.

T A B L E   5 – 9

NAMING CONVENTIONS FOR COMMON FATTY ACIDS

Type	Systematic Name	Chemical Notation	Historical Name
Metabolic			
Volatile	Propanoic	$C_3H_6O_2$	Propionic
	Hexanoic	$C_6H_{12}O_2$	*n*-Caproic
Nonvolatile	Butanedioic	$C_4H_6O_4$	Succinic
Long-chain			
Saturated	Dodecanoic	C12:0	Lauric
	Tetradecanoic	C14:0	Myristic
	Hexadecanoic	C16:0	Palmitic
	Octadecanoic	C18:0	Stearic
	Eicosanoic	C20:0	Arachidic
Unsaturated	*cis*-9-Hexadecanoic	C16:1ω7c	Palmitoleic
	cis-9 Octadecanoic	C18:1ω9c	Oleic
	cis-11-Octadecanoic	C18:1ω7c	Vaccenic
Branched-chain	10-Methyl octadecanoic	10 ME-C18:0	Tuberculostearic
Hydroxy	3-Hydroxytetradecanoic	3OH-C14:0	β-hydroxy-myristic
Cyclopropane	*cis*-11, 12-Methylene octadecanoic	C19:0cyclo 11, 12	Lactobacillic

■■■■ T A B L E   5 – 1 0

CHEMICAL CONFIGURATION OF FATTY ACIDS, ESTERS, AND ALCOHOLS

		Chemical Notation
Hydrocarbon		
Straight-chain		C12:0
Saturated Acid		
Normal straight-chain		n-C12:0
Methylated Acid		
Methyl ester		C12:0 FAME
Dimethyl acetal		C12:0 DMA
Branched Acids		
Iso-chain		C15:0 iso
Anteiso-chain		C15:0 anteiso
Hydroxy acids		
Alpha position		C15:0 2OH
Beta position		C15:0 3OH
Alcohol		
Straight chain		C15:0 N alcohol

Table continued on following page

TABLE 5–10 Continued

CHEMICAL CONFIGURATION OF FATTY ACIDS, ESTERS, AND ALCOHOLS

Monounsaturated Acids | **Chemical Notation**

cis position — C16:1ω7c

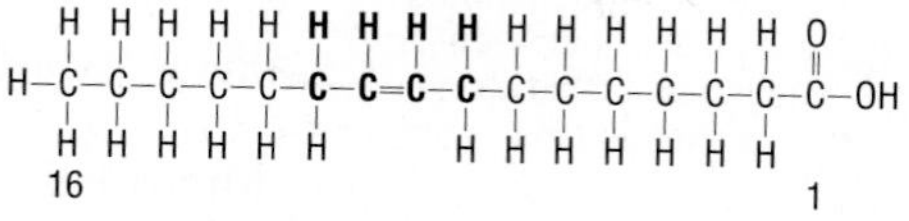

trans position — C16:ω7t

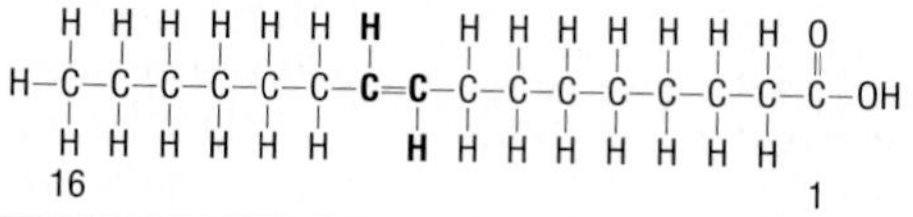

Cyclo Acids

Cyclopropane — C20:0ω cyclo 11–12

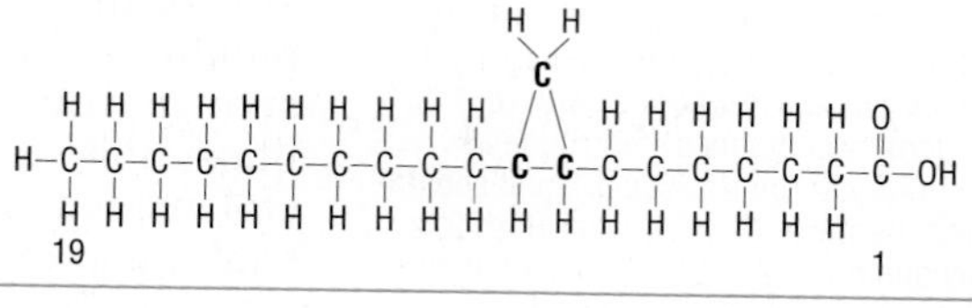

Complex Fatty Acids

Mixed groups — C18:0 iso 2-OH

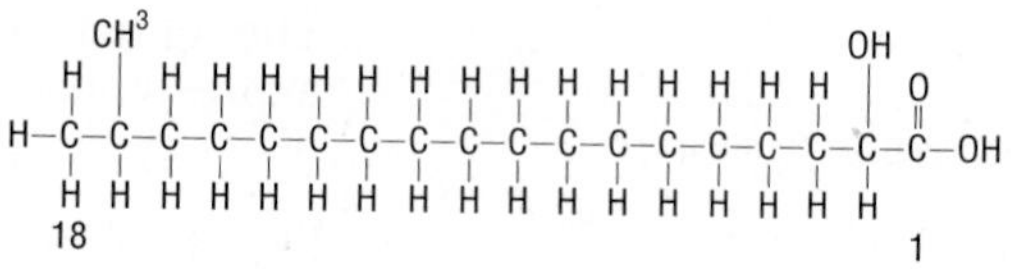

alpha (α)—acid end.
omega (ω)—opposite the α end. The symbol ω indicates the location from the omega end of an unsaturated bond or of a carbon added at the unsaturated double bond position.
iso—methyl groups substitute for hydrogen atoms as branches from the second-to-last carbon in the chain.
anteiso—methyl groups substitute for hydrogen atoms as branches from the third-to-last carbon in the chain.
alpha—hydroxyl group added in the two-carbon position.
beta—hydroxyl group added in the three-carbon position.
cis (c)—remaining hydrogens at the double bond are on the same side of the compound.
trans (t)—remaining hydrogens at the double bond are on the opposite side of the compound.

Metabolic Short-Chain Fatty Acids

Metabolic pathways of organisms are genetically stable or conserved traits of microbes that are not subject to the strain variations found with many phenotypic traits. Short-chain fatty acids, both volatile and nonvolatile, accumulate as metabolic waste from spent carbohydrate substrates in broths. Although metabolic FAs are stable for the species, the amounts produced from the substrate can vary, depending on the carbohydrate-to-peptone concentrations in the growth medium. Fermentative organisms prefer the carbohydrate substrate, whereas nonfermenters produce iso-acids from the peptones early in incubation. Fermenters grown in a high-peptone medium, however, will have an undesirable proportion of iso-acids compared with being grown in low-peptone medium. Growth in a glucose–low-peptone medium for a limited time, usually 24 hours, yields the most reproducible production of metabolic acids from glucose and other carbohydrate substrates.

VOLATILE METABOLIC ACIDS

The chain length of a volatile metabolic fatty acid is usually seven carbons or less. Included in this group of acids are formic, acetic, propionic, *n*-butyric, isobutyric, *n*-valeric, isovaleric, *n*-caproic, isocaproic, and heptanoic acids. These acids can be readily chromatographed.

NONVOLATILE METABOLIC ACIDS

Lactic, pyruvic, succinic, dicarboxylic, and phenylacetic acids are nonvolatile. Methylation is required to chromatograph mixtures of these acids (Table 5–11).

■■■■■■■■ T A B L E   5 – 1 1
CHROMATOGRAPHY OF METABOLIC SHORT-CHAIN FATTY ACIDS

Sample Types	Anaerobic bacteria incubated to turbidity in glucose-peptone or lactate or threonine broths (48 hours) or infected body fluids or bacterial vaginosis fluids.

Volatile Acids

Component and Standard Mixture:	Formic, acetic, propionic, *n*-butyric, isobutyric, isovaleric, *n*-valeric, isocaproic, and heptanoic acids.
Acidify Broth pH 2:	Add 2.0 drops of 50% H_2SO_4 to 5 mL of broth/fluid.
Extract:	Transfer 1 mL of broth/fluid culture (PYG) to stoppered tube, add 0.1 mL 50% H_2SO_4 and 1.0 mL methy-t-butyl-ether, mix well (tight cap) with rocking motion twenty times, centrifuge at 1500–2000 rpm for liquid-liquid separation. Remove the ether layer (top) to a second tube, and add a pinch of anhydrous $CaCl_2$ to remove trace water. Allow to stand a few minutes before injection.

Nonvolatile Acids

Component and Standard mixture:	Lactic, succinic, and phenylacetic acids.
Methylation:	Transfer 1 mL of spent broth/fluid to stoppered tube and add 1 mL methanol and 0.1 mL 50% H_2SO_4. Stopper tube and mix vigorously. Heat to 60°C for 30 minutes or let stand overnight at room temperature.
Extract:	Add 0.5 mL chloroform to cool broth/fluid (fume hood), cap, mix well, and centrifuge to separate chloroform/-water layers.
Gas Chromatograph:	Column SP–1220 with Flame Ionization Detector: Inject 1.0 μ of acidified broth supernatant for volatile acids; inject 1 to 14 μL for nonvolatile acids. Column SP–1000 with Thermal Conductivity Detector: Inject 5–15 μ of Extract 1 for volatile acids; inject 1 to 14 μL for nonvolatile acids.
Chromatograms:	Analyze the standard mixture peak separations to verify sample peaks.
Comments/Notes:	Sample injection volume varies with GC column and operating conditions. Determine performance using standards.

CHROMATOGRAPHY OF METABOLIC ACIDS

Both volatile and nonvolatile metabolic acids can be identified in a single chromatographic analysis by converting all short-chain fatty acids and some alcohols from the component mixture to volatile derivatives using butyl-trifluoroacetyl and the appropriate gas chromatograph equipped with a flame ionization detector.

Metabolic acids are analyzed routinely in most large diagnostic laboratories to identify obligate anaerobes (see Chapter 19).

Cellular Long-Chain Fatty Acids

Microbial cellular fatty acids (CFAs) come from the lipoteichoic acids of gram-positive bacteria, lipopolysaccharides in gram-negative bacteria cell walls, the complex lipid lipid A, and the phospholipids in cell outer membranes (Fig. 5–33). Carbon chain lengths included are between nine (C9:0) and twenty carbons (C20:0). Chain lengths greater than 20 carbons cannot be volatilized and require a different chromatographic method. Anaerobes contain unique CFAs called *plasmalogens,* phospholipid analogs that result in a dimethylacetal instead of the usual methyl ester upon derivatization. Microorganisms contain from less than 5 to more than 20 different fatty acids within their cell walls, depending on the size of the genome. The constitutive cellular fatty acids are stable expressions of the cellular genetics.

NATURAL GENETIC DIVERGENCE OF SPECIES

The natural divergence of microbial species over time has provided the variety in microbial metabolism and cell-wall construction that is the basis for the efficiency of chemotaxonomy using fatty acids. Large numbers of strains must be examined to determine the boundaries between FA profiles. When quantitative methods are applied to proportions of cellular fatty acids from closely related species, small numbers of strains may identify the extremes of the same statistical cluster, thus falsely suggesting two clusters rather than one. Strains from different geographic locations must be included to appreciate the extent of divergence and the number of individual CFA clusters for each species. Divergence for those species that are widely spread across the animal kingdom and within the general environment appears to be greater than that appreciated with primary pathogens, which have had fewer opportunities to mutate over time.

Closely related but different species from the same environmental niche can have more similar cellular fatty acids than the same species from different ecosystems. This characteristic can cause confusion in the construction of statistical clusters (Fig. 5–34). The same species from different hosts or environments have variations in fatty acid proportions ranging from moderate to pronounced.

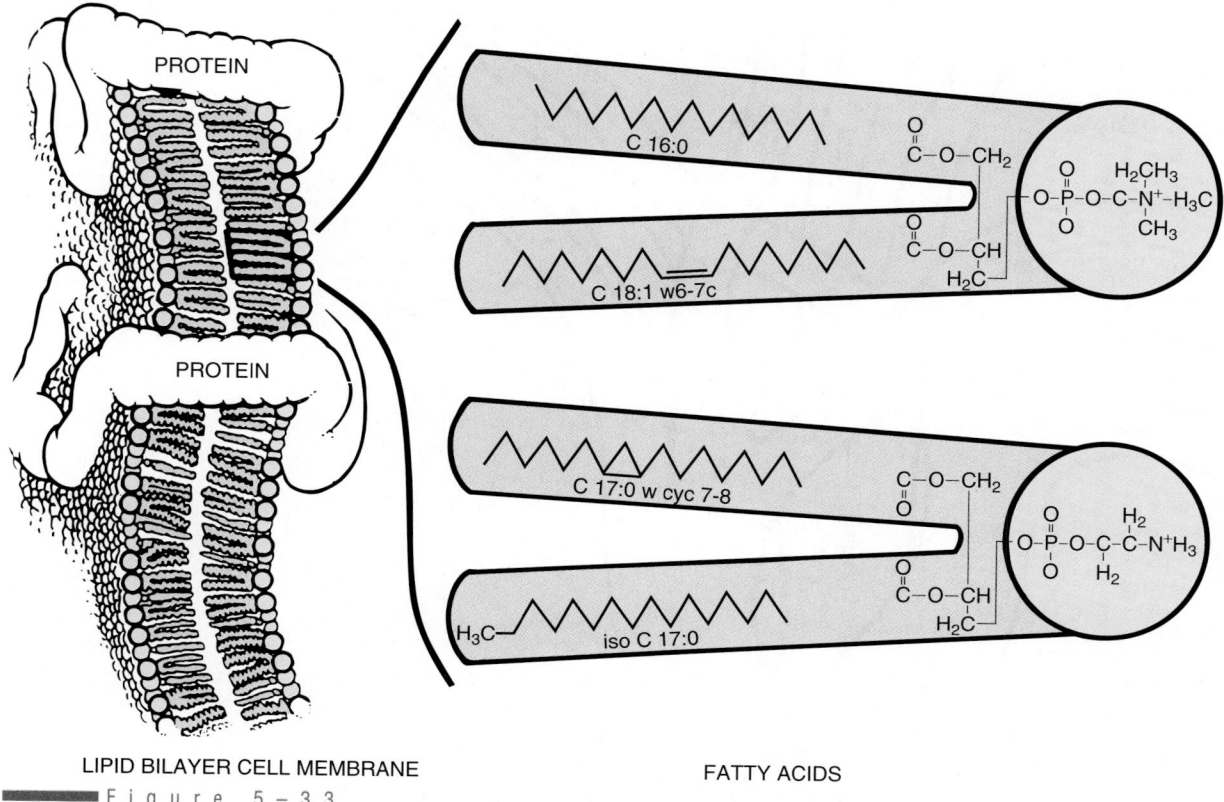

LIPID BILAYER CELL MEMBRANE   FATTY ACIDS

Figure 5 – 3 3

Long-chain fatty acid components of lipid membrane. (Adapted with permission from Welch DF: Applications of cellular fatty acid analysis. Clin Microbiol Rev 4:422, 1991; and adapted from Rose AH: Chemical Microbiology. New York: Plenum Publishing Corp, 1968, pp 4–56.)

PREPARATION OF COMPONENT MIXTURE FOR LONG-CHAIN CELLULAR FATTY ACIDS

The component mixture preparation for longer-chain FAs is relatively simple but requires the attention to detail associated with all chemical analytical techniques. This type of organism identification is a departure from the qualitative techniques used for biochemical phenotyping of isolates in most microbiology laboratories (Fig. 5–35).

Cultivation. Organisms are harvested when the growth is complete but before the initiation of lipid storage. Harvest is usually optimal by 24 hours with rapidly growing species. Inoculation must be done on or in a medium that supports rapid growth to maturity, preferably within 24 hours, and provides consistent CFA production. Some species produce proportionally more iso-acids when fatty acid precursors are available in the growth medium. The growth medium must have standard ingredients so that there is little variation from lot to lot.

Uniform typical colonies and small colony variants should be harvested. Small colony types have minor changes in their cell walls that can affect the percentage of individual acids recovered. Resting cells have thick multi-layered walls and must be cultured to produce the uniform cell walls found during synchronous growth. Cultures stored on agar slants, lyophilized, or frozen cannot be harvested until colonies are uniform, usually requiring at least three subcultures. Organisms recently isolated from infected patients do not require multiple subcultures. Slow growth rates related to suboptimal medium or temperature produce variations in cell-wall construction that cause maldistribution of CFAs.

Media should be evaluated for uniform growth before analysis. Unfortunately, a growth medium adequate for one species may not be adequate for another (see section on staphylococci). Growth-associated variations in CFAs emphasize the importance of standard cultivation when the culture

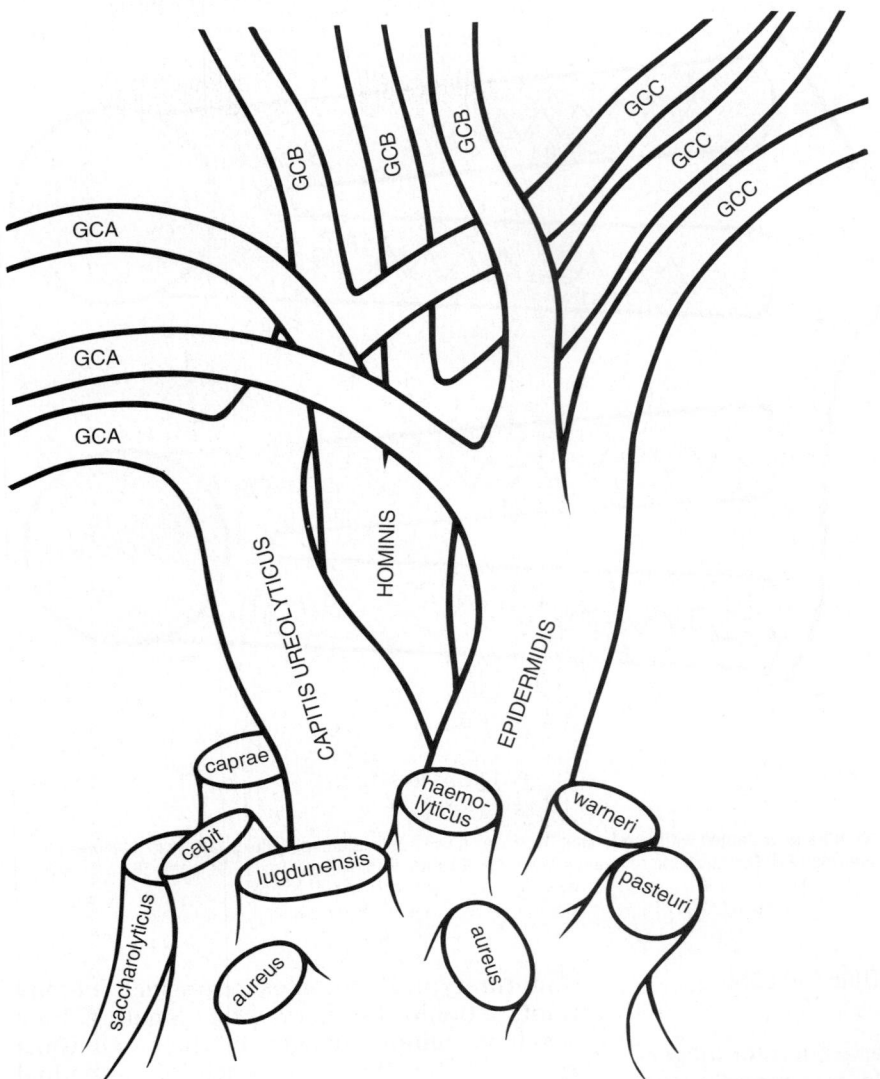

GCB
GCB
GCB
GCC
GCC
GCC
GCA
GCA
GCA

CAPITIS UREOLYTICUS
HOMINIS
EPIDERMIDIS

caprae
capit
lugdunensis
haemo-lyticus
warneri
pasteuri
saccharolyticus
aureus
aureus

STAPHYLOCOCCI

■■■■■ Figure 5–34

Genetic divergence within closely related species produces a variety of fatty acid subgroups. Similarities in fatty acid composition may be greater between species than within the same species for some subgroups.

growth is used as the analyte in quantitative chromatography. The type of growth conditions used must always be noted in the report of results.

Harvesting. One heaping loopful (40–50 mg) of live wet cells or one flat loopful of mycobacteria is ample for processing. Cells are collected from the area of growth with separated colonies. The loaded loop is inserted into a labeled test tube, and the cells are wiped off the loop onto the bottom of the tube by rapid spinning of the loop between the thumb and forefinger. Harvesting should be accomplished quickly to prevent continued growth at room temperature.

Saponification. The cells are killed to release the fatty acids by treatment with methanolic NaOH and heat. CFAs are hydrolyzed from the cell lipids and form sodium salts.

Methylation. Volatile fatty acid methyl esters (FAMEs) are formed from the CFA sodium salts by treatment of the sample with methanol and strong acid under moderate heat. Too much time under these conditions can alter FAME profiles by degrading cyclo acids.

Extraction. The FAME component is moved from the acidic aqueous phase into an organic ether solvent phase using a liquid-liquid extraction.

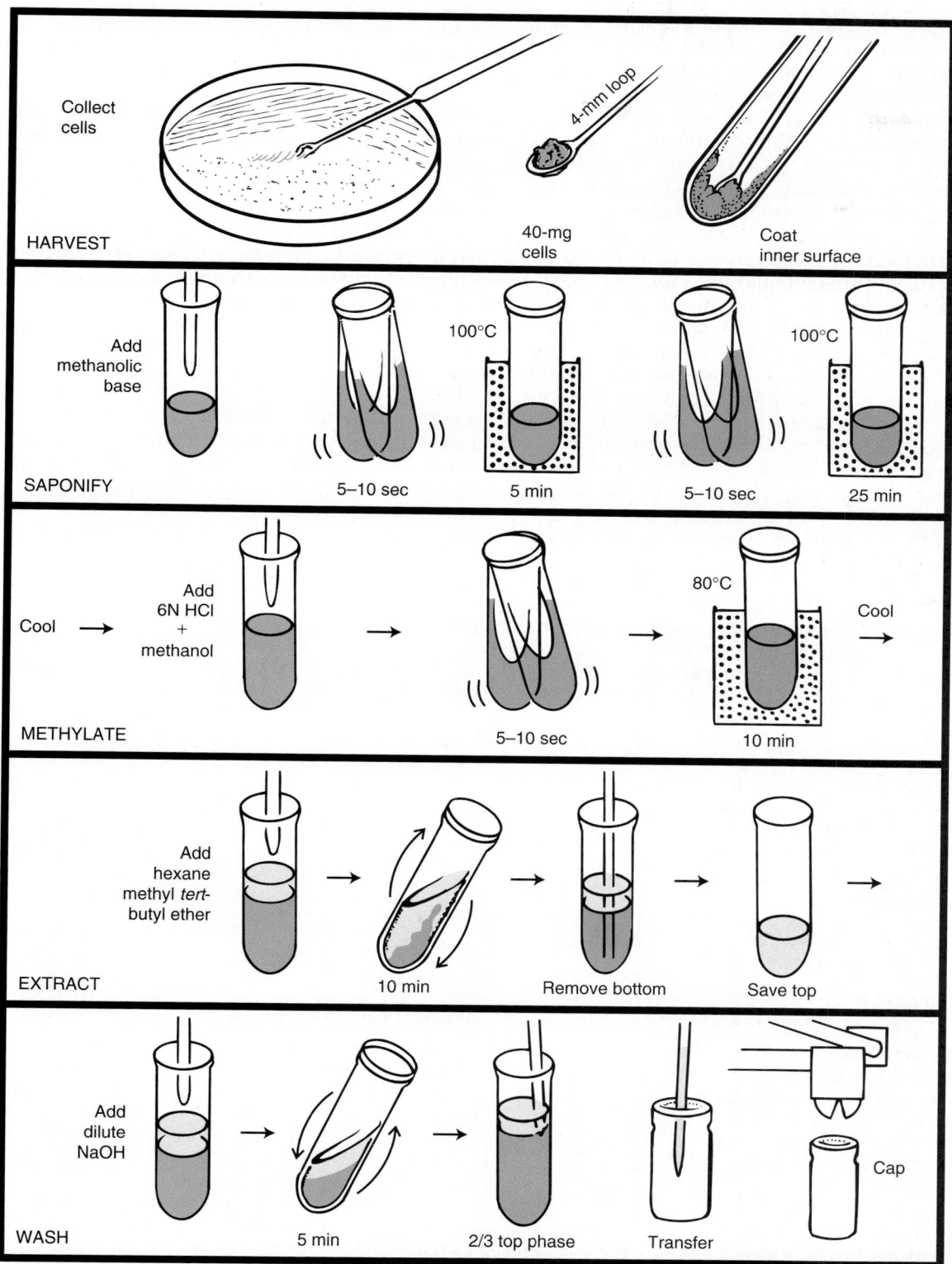

Figure 5-35

Component mixture preparation for long-chain CFAs.

Washing. Free fatty acids and residual reagent are removed from the organic extract using dilute NaOH. Centrifugation clears the solvent phase. Failure to remove these chemical remnants can result in loss of hydroxy acids.

Chromatography. FAME components are separated and quantitated by GLC using a single, fused, silica capillary column and a flame ionization detector. The mixture (solute) of FAMEs is vaporized at a heated injection port and swept by carrier gas (the mobile phase) over a large absorbent or adsorbent surface (solid or stationary phase) inside the column. The components of the solute move at different rates and become separated along the great length of the column. The components are measured by the detector (flame ionization) and traced as a chromatogram. A standard mixture of straight-chain saturated fatty acids from 9 to 20 (C9:0–C20:0) carbon length and representative hydroxy acids are chromatographed with each sample run. The peaks of the sample can be inferred from comparing the retention times of the standard with the new chromatogram.

CELLULAR FATTY ACID CHEMOTAXONOMY

Organisms with small genomes have small families of acids, whereas those with large genomes can have large families, with over 20 different acids represented in a species. CFAs that are highly conserved in prokaryotes, such as straight-chain C16:0, vary in proportion rather than in presence. In some species, CFAs change in their proportions with differences in strain and culture conditions (Table 5–12).

Anaerobes

Large numbers of FAMEs are derived from anaerobic species. Clinically important fusobacteria can be distinguished from *Bacteroides* sp by the absence of branched FAMEs and the presence of C:14:0 3OH and C16:0 3OH acids. The gram-positive anaerobe *Propionibacterium* is

■■■■ T A B L E 5 – 1 2
DISTRIBUTION OF FATTY ACID TYPES AMONG SELECTED BACTERIA*

Organisms	Culture Medium	No. Acids	Straight chain	Branch	OH Hyroxy	Δcyclo	Unsat. chain	TBSA	Prominent or Unusual
Gram-Positive Cocci									
Aerococcus viridans	B	10	L	–	–	–	L	–	20:1ω9c
Enterococcus faecalis	B	10	L	–	–	+	L	–	19:0 cycloω8c
Micrococcus luteus	B	13	+	L	–	–	+	–	16:1ω7c
Stomatococcus mucilaginosus	B	12	+	L	–	–	+	–	14:0 iso
Staphylococcus aureus	B	20	+	L	–	–	+	–	17:0 anteiso
Gram-Positive Bacilli									
Corynebacerium diphtheriae	B	8	L	–	–	–	L	–	16:1ω7c
Listeria monocytogenes	B	11	L	L	–	–	+	–	15:0 anteiso
Mycobacterium tuberculosis	M	12	L	–	–	–	L	L	18:0 10 Me
Nocardiopsis	T	17	+	L	–	–	L	+	18:0 10 Me
Rhodococcus equi	T	17	L	–	+	–	L	L	18:0 10 Me
Gram-Negative Bacilli									
Acinetobacter baumannii	T	12	L	–	+	–	–	L	16:1ω7c
Bacteroides fragilis	S	22	+	L	L	–	–	–	17:0 iso 30H
Campylobacter jejuni	C		L	–	–	L	+	–	19:0 cycloω8c
Franciscella tularensis	C	13	+	–	L	–	+	–	18:0 30H
Fluoribacter bozemanae	Y	15	+	L	–	+	+	–	15:0 anteiso
Pseudomonas aeruginosa	T	10	L	–	+	+	L	–	10:0 30H
Bartonella henselae	C	4	L	–	–	–	–	–	18:1ωc
Salmonella typhi	T	11	+	–	–	–	L	L	14:0 2OH

*T, tryptocase soy broth agar; B, 5% sheep's blood agar; M, Middlebrook 7H10; S, special anaerobe broth; C, CHOC agar; Y, BCYE (buffered charcoal yeast extract agar); L, large amount.

+, present

–, absent

characterized by large amounts of branched C15:0 and C17:0 acids. *Propionibacterium acnes* is separated from others in the genera by relatively small proportions of branched acid C15:0 anteiso. Anaerobes as a group are more difficult to use as analytes because of the problem of cultivating an adequate cell mass for harvesting.

Coryneforms

Coryneform bacteria generally contain large quantities of C16:0, C18:0, and C18:1. Selected genera or species contain tuberculostearic acid (TBSA). This group is not taxonomically well characterized and therefore is difficult to identify by matching CFAs to named species.

Enterics

Enteric species primarily contain saturated cyclopropane and C14:0 3OH acids. Strain differences as well as species differences occur, but the close genetic relatedness of many strains limits the value of chemotaxonomy.

Micrococci

Like the staphylococci, micrococci feature the branched C15:0 and C17:0 acids, but they differ in having no straight-chain C20:0 acid.

Mycobacteria

Constitutive CFAs of mycobacteria (not mycolic acids) include large quantities of C16:0, C18:1ω9c, and methyl branched C18:0 tuberculosteric acid (TBSA). *Mycobacterium gordonae* does not contain TBSA, but instead contains a methyl branched C:14 acid. Compare, in the following box, the similarities as well as the differences in the family of CFAs characteristic for the two species grown on Middlebrook 7H10.

Mycobacterium tuberculosis	*Mycobacterium gordonae*
—	12:0
14:0	14:0
—	**2-ME-14:0**
15:0	—
16:1 ω6c	16:1 ω6c
16:0	16:0
17:0	17:0
18:1 ω9c	18:1 ω9c
18:0	18:0
10Me-18:0 TBSA	—
20:0	20:0

Pseudomonads

Hydroxy acids are prominent features of pseudomonads. The absence of branched acids in pseudomonads has influenced the move of species such as *Xanthomonas maltophilia* to other genera.

Staphylococci

Most staphylococci have the same family of fatty acids, which includes straight-chain C18:0 and C20:0. If staphylococci are grown on blood agar, a large quantity of branched-chain acids (C15:0 iso and C15:0 anteiso, C17:0 iso and C17:0 anteiso, C19:0 iso and C19:0 anteiso) replace the straight chains, particularly C20:0. Differences among species and GC subgroups are quantitative when they are cultivated under the same conditions. When the cultivation conditions are different, the quantitative distribution of acids can be markedly different. Note, in the following box, the difference in *S. epidermidis* CFAs with variation in media and temperature of cultivation.

S. epidermidis, 5% sheep's blood agar at 35°C		*S. epidermidis*, tryptocase soy broth agar (TSBA) at 28°C	
CFA	%	CFA	%
—		13:0 ISO	1.5
14:0 ISO	1.3	14:0 ISO	3.5
—		14:0	1.9
15:0 ISO	7.3	15:0 ISO	11.2
15:0 ANTEISO	35.2	15:0 ANTEISO	27.9
16:0 ISO	1.1	16:0 ISO	0.8
16:0	3.1	16:0	4.0
17:0 ISO	5.9	17:0 ISO	2.9
17:0 ANTEISO	10.7	17:0 ANTEISO	2.6
18:0 ISO	0.9	—	
18:1 ω9c	3.4	—	
18:0	13.1	18:0	12.4
19:0 ISO	3.6	19:0 ISO	2.4
19:0 ANTEISO	4.4	19:0 ANTEISO	0.8
20:0	**6.6**	20:0	**27.7**

Streptococci

All streptococci contain C16:1ω5, which is absent in enterococci, aerococci, and other look-alikes.

Primary Pathogens and Low-Incidence Isolates

The ability to kill the pathogen prior to taxonomic manipulation and to match the CFA profile are significant advantages when one is handling organisms that can cause serious or contagious infections. The families of CFAs and individual CFAs may be distinct for this type of isolate. The patterns of CFAs for a variety of clinical isolates are compared in Table 5–12.

AUTOMATED CHEMOTAXONOMY

Although many important observations, such as number and type of CFA, can be made from viewing the chromatogram directly, the best use of this technique relies on computer management of peak naming, peak pattern matching, and cluster analysis with matching for relatedness.

Modern chromatography, when paired with computer management, offers a single method to provide accurate identification for a wide variety of species. A single commercial system with this configuration is currently available (Midi, Inc; see Bibliography). In this system, the data from the calibration FA mixture is converted to equivalent chain length (ECL) data for automatic naming of microbial fatty acids by special software. The external standard is used to compute an ECL value for each compound in a run. Peak naming is computer software–generated using a comput-

er dictionary corrected according to the ECL generated by each standard. Computer software for this system manages the identification of strains in a run by using a covariance matrix, principal component analysis, and pattern recognition. Identification library software uses calculation of ratios between fatty acid peak amounts and the principal component base to match the new isolate with previously stored strains of the same species, ecovar, or biovar.

The commercial equipment and accompanying software configuration (Fig. 5–36) provide for a single chromatographic technique to automate identification of aerobic and anaerobic bacteria, mycobacteria, some rickettsiae, yeasts, and fungi. The reports generated include a listing of the FAME components, quality control calculations, and a similarity index for the identification provided (Table 5–13).

The information provided allows the technologist to check for technical accuracy of the run, as well as for consistency between the family of CFAs identified and the identification based on a match with library entries. The isolate CFA peak data are stored so that relatedness of strains can be studied later by the use of statistical analysis.

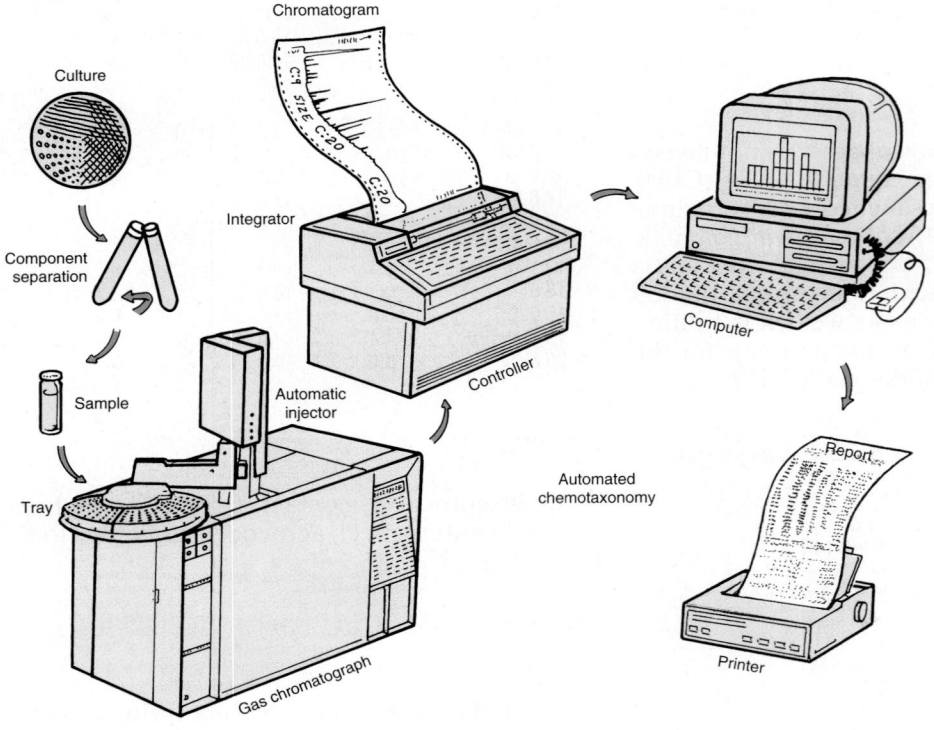

■■■■■ F i g u r e   5 – 3 6

Configuration of instruments for automated microbial identification by cellular fatty acid.

■■■■ T A B L E 5 – 1 3
AUTOMATED CHEMOTAXONOMY REPORT COMPONENTS

ID: 21270 STAP-UN (M11551) Date of run: 21-DEC-93 14:01:26
Bottle: 7 Sample [AEROBE] Date edited: 05-JAN-94 12:08:00

Retention Time	ECL	Name	%
0.000	3.607	Solvent Park.	. . .
5.455	12.612	13:0 ISO.	0.40
8.191	14.621	15:0 ISO.	43.93
8.329	14.711	15:0 ANTEISO	14.67
10.416	16.000	16:0	2.96
11.499	16.630	17:0 ISO.	13.81
11.658	16.722	17:0 ANTEISO	6.29
12.137	17.001	17:0	0.52
13.398	17.721	Sum in Feature 6	1.50
13.483	17.770	18:1ω9C	2.35
13.883	17.998	18:0	5.82
14.990	18.633	19:0 ISO.	2.42
15.161	18.731	19:0 ANTEISO	0.55
15.631	19.001	19:0	0.57
16.319	19.400	20:4ω6,9,12,15c. . . .	0.60
16.893	19.733	20:2ω6,9c.	0.98
17.354	20.000	20:0	2.65
*****	. . .	Feature 6 . . . . 18:2ω6,9c/18:0 ANTEISO	1.50

Total Area	Named Area	% Named	Total Amt
160392	160392	100.00	154200

Similarity Index

BA3524 [Rev 1.50]
Staphylococcus. 0.926
S. schleiferi 0.926
S. s. schleiferi 09.26
S. s. s. GC B subgroup 0.926
S.s. s. TYPE STRAIN. 0.506
S. s. s. GC D subgroup. .0.472
S. intermedius 0.498
S. i. GC G subgroup 0.498
S. i. GC A1 subgroup. 0.493
S. i. A2 subgroup. 0.473

Modified from Microbial Identification System; Operating Manual, Version 4.
Newark, DE: MIDI, Inc., 1993.

ACCURACY AND PRECISION OF MATCHING LIBRARIES

To match new isolates with *accuracy* to known strains through a matching library requires that the statistical clusters used be properly constructed from well-characterized strains and that the cluster to be matched not overlap other clusters. Overlapping of clusters prevents clear separation of strains on the basis of generation of similarity indexes. All clusters should be examined by dendrogram and 2-D plot methods to ensure proper separation between clusters.

Statistical clustering of isolates for the purpose of determining relatedness is enhanced if isolates are batch-processed in a consistent manner, harvested carefully, and processed quantitatively. To accomplish matching of single strains, the CFA analysis must be performed from rigidly controlled organism cultivation and must be the same as that used for comparison library entries. The chemical analysis must be performed as an analytical technique to achieve *precision* in results.

The gas chromatograph is rigidly controlled with external standards to show a high level of peak naming. Control strains run with each chromatographed "sample batch" should "hit" the quality control statistical cluster target within a specified range. The variation of the analysis in any one laboratory can be defined by the scatter within the statistical cluster created by repeated runs of the same quality control strain.

FUTURE OF CHROMATOGRAPHIC METHODS

The automated analysis of cellular fatty acids with direct naming of the organism from prepared libraries or automated chemotaxonomy is the most promising clinical area of development. Given adequate growth conditions, optimal temperature, and strictly controlled analysis, organisms with cell walls yield reproducible FA profiles, both qualitatively and quantitatively. Much correlation remains to be done between this method and other methods used to relate organisms to one another. Of particular interest will be the correlation with DNA hybridization and chromosomal DNA analysis by ribotyping and field-inversion gel electrophoresis or pulsed-field gel electrophoresis. With improved taxonomy and the availability of better-defined "standard strains," this chromatographic method for chemotaxonomy should become a standard feature of reference clinical microbiology laboratories. A variety of other chromatographic methods and microbial analytes will continue to play an important role in industry and microbiologic research.

Bibliography

GENERAL

Lambert M, Moss C: Comparison of the effects of acid and base hydrolysis on hydrozy and cyclopropane fatty acids in bacteria. J Clin Microbiol 18:1370, 1983.
MIDI, Inc: Microbial Identification System; Operating Manual, Version 4. Newark, DE: MIDI, Inc, 1993.

Miller L: Single derivation method from routine analysis of bacterial esters, including hydroxy acids. J Clin Microbiol 16:584, 1982.

Moss C: Gas-liquid chromatography as an analytical tool in microbiology. J Chromatogr 203:337, 1981.

Moss C, Dees S, Guerrant G: Gas-liquid chromatography of bacterial fatty acids with fused-silica capillary column. J Clin Microbiol 12:127, 1980.

O'Leary W: The chemistry of microbial lipids. Crit Rev Microbiol 4:41, 1975.

Shaw N: Lipid composition as a guide to the classification of bacteria. Adv Appl Microbiol 17:63, 1974.

Tornabene T: Lipid analysis and the relationship to chemotaxonomy. Methods Microbiol 18:209, 1985.

Welch DF: Applications of cellular fatty acid analysis. Clin Microbiol Rev 4:422, 1991.

DATA PROCESSING

Eurola E, Lehtonen O: Optimal data processing procedure for automatic bacterial identification of cellular fatty acids. J Clin Microbiol 26:1745, 1988.

Romesburg H: Cluster Analysis for Researchers. Malbar, FL: Robert E Krieger Publishing, 1990.

SELECTED ORGANISM REFERENCES

Anaerobes

Cundy K, Willard K, et al: Comparison of traditional gas chromatography (GC), headspace GC, and the Microbial Identification Library GC System for the identification of *Clostridium difficile.* J Clin Microbiol 29:260, 1991.

Tuner K, Baron, E, et al: Cellular fatty acids in *Fusobacterium* species as a tool for identification. J Clin Microbiol 30:3225, 1992.

Brucella, Francisella

Coloe P, Sinclair A, et al: Differentiation of *Brucella ovis* from *Brucella abortus* by gas-liquid chromatographic analysis of cellular fatty acids. J Clin Microbiol 19:896, 1984.

Jantzen E, Berdal B, Omland T: Cellular fatty acid composition of *Francisella tularensis.* J Clin Microbiol 10:928, 1979.

Campylobacter

Lambert M, Patton C, et al: Differentiation of *Campylobacter* and *Campylobacter*-like organisms by cellular fatty acid composition. J Clin Microbiol 25:706, 1987.

Corynebacterium

Bernard K, Bellefeuille M, Ewan E: Cellular fatty acid composition as an adjunct to the identification of asporogenous, aerobic gram-positive rods. J Clin Microbiol 29:83, 1991.

Leptospira

Cacciapuoti B, Ciceroni L, Barbini D: Fatty acid profiles, a chemotaxonomic key for the classification of strains of the family Leptospiraceae. Int J Syst Bacteriol 41:295, 1991.

Mycobacteria

Butler WR, Thibert L, Kilburn JO: Identification of *Mycobacterium avium* complex strains and some similar species by high-performance liquid chromatography. J Clin Microbiol 30:2698, 1992.

Luquin M, Ausina V, et al: Evaluation of practical chromatographic procedures for identification of clinical isolates of mycobacteria. J Clin Microbiol 29:120, 1991.

Nonfermenters

Veys A, Callewaert W, et al: Application of gas-liquid chromatography to the routine identification of nonfermenting gram-negative bacteria in clinical specimens. J Clin Microbiol 27:1538, 1989.

Bartonella, Rickettsiae

Moss C, Holzer G, et al: Cellular fatty acid compositions of an unidentified organism and a bacterium associated with cat scratch disease. J Clin Microbiol 28:1071, 1990.

Tzianabos T, Moss C, McDade J: Fatty acid composition of *Rickettsieae.* J Clin Microbiol 13:603, 1981.

Staphylococci

Kotilainen P, Huovinen P, Eerola E: Application of gas-liquid chromatographic analysis of cellular fatty acids for species identification and typing of coagulase-negative staphylococci. J Clin Microbiol 29:315, 1991.

Streptococci

Bosley G, Wallace P, et al: Phenotypic characterization, cellular fatty acid composition, and DNA relatedness of aerococci and comparison to related genera. J Clin Microbiol 28:416, 1990.

Vibrio

Urdaci M, Marchand M, Grimont P: Characterization of 22 *Vibrio* species by gas chromatography analysis of their cellular fatty acids. Res Microbiol 141:437, 1990.

E. DIAGNOSTIC APPLICATIONS OF DNA PROBES

Gerri S. Hall

OBJECTIVES

1. Discuss the principle of nucleic acid hybridization.

2. Describe the different hybridization formats.

3. Name the most common DNA/RNA probe labels.

4. Discuss how probe technology provides rapid diagnosis of infectious diseases.

5. Compare DNA probe technology with the conventional methods for detecting infectious agents.

6. List the advantages and disadvantages of DNA probes in the clinical microbiology laboratory.

PRINCIPLE OF NUCLEIC ACID HYBRIDIZATION

To understand probe technology and its applications in the clinical microbiology laboratory, one needs to understand the principle of hybridization. *Nucleic acid hybridization* is formation of stable double-stranded nucleic acid molecules from complementary single-stranded molecules. The single-stranded molecules can be RNA or DNA, and the resultant hybrids formed can be DNA-DNA, RNA-RNA, or DNA-RNA.

A *probe* is a labeled or marked single-strand sequence of nucleic acid that is complementary to the nucleic acid sequence to be detected and can be either RNA or DNA. The detected nucleic acid sequence is referred to as the *target,* and it

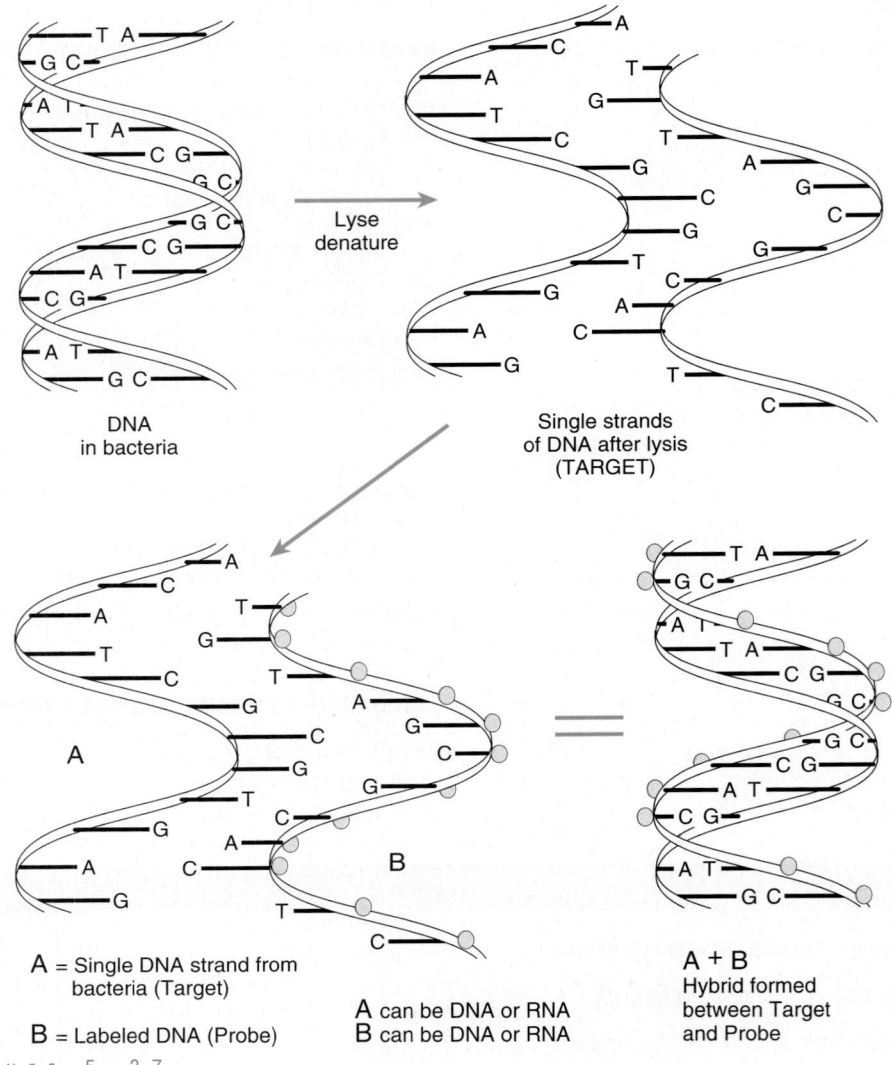

DNA
in bacteria

Lyse
denature

Single strands
of DNA after lysis
(TARGET)

A

B

A + B

A = Single DNA strand from
bacteria (Target)

B = Labeled DNA (Probe)

A can be DNA or RNA
B can be DNA or RNA

A + B
Hybrid formed
between Target
and Probe

Figure 5–37

Hybridization format.

too can be DNA or RNA. The target is nucleic acid located within a specimen, in colonies on a plate or in a broth culture, for example.

The nucleic acid probe is traditionally constructed from specific nucleic acid fragments (sequence) of the organism (target), or the probe may be a synthetically produced oligonucleotide of specific sequences. In a probe test, the target and probe must be allowed to come together and hybridize, and the conditions whereby this may occur are referred to as the *hybridization format or method* (Fig. 5–37).

Hybridization Formats

Solid-Support Hybridization

Table 5–14 lists various hybridization formats available for probe testing. The conventional approach to hybridization assays has been solid-support hybridization—that is, the target nucleic acid is secured to a membrane (Fig. 5–38*A*). The most common example is a filter paper made of nitrocellulose or nylon fibers. The probe is overlaid over the solid support for a time to allow

━━━━T A B L E   5 – 1 4
HYBRIDIZATION FORMATS FOR DNA PROBES

Solid-support hybridization
- Filter
- Modified filter: microtiter tray or test tube
- Colony hybridization
- Sandwich hybridization

In-Solution hybridization
In Situ hybridization
Southern hybridization

hybridization to occur—that is, to allow the target and probe nucleic acids to come together. Hybridization will occur if nucleic acid sequences in the target are complementary to the probe. Washing of the filter or solid support allows for removal of the nonbound probe (Fig. 5–38). Depending upon the manner in which the nucleic acid is added to the filter paper, the resultant reaction can be referred to as a dot, a spot, or a slot (Fig. 5–39).

A modification of solid-support hybridization method can be made by applying the probe to a microtiter tray or to the inside coating of a test tube. Solution containing the target nucleic acids is added. After a specific time, the tube is washed, and detection is visualized on the walls of the tray or tube. Elution of the potential hybrids may be performed and measured in the eluate either inside or outside the solid support.

Nucleic acid extracted from colonies on a plate can be transferred from the plate to a nitrocellulose filter or other solid support. Hybridization can be performed directly on the filter paper, resulting in detection of target material in specific patterns.

Another type of solid-support hybridization is *sandwich hybridization*. This format utilizes two probes, one unlabeled and attached to the solid support and the other labeled probe and added in solution after the target nucleic acid has been allowed to hybridize with the first probe. The target nucleic acid is thus "sandwiched" between the two probes (see Fig. 5–38*B*). This method may help to reduce background nonspecific binding but tends to be a less sensitive format for hybridization. It is more suitable for processing of specimens or other crude samples of bacteria, cell lysates, or secretions.

In-Solution Hybridization

With an in-solution hybridization format, both target and probe nucleic acids move about freely in solution (Fig. 5–40). The reaction kinetics occur five- to ten-fold faster than solid support hybridization. There is, however, a need for differential separation and removal of the nonbound probe from the hybridized probe after the hybridization step. Various methods have been utilized for this purpose, including the use of hydroxyapatite or micromagnetic beads, each of which attracts the double-stranded, bound nucleic acids, allowing for the selective removal of unbound probe. A company producing commercial probes (Gen-Probe, San Diego, CA) has introduced use of chemical hydrolysis to remove the unbound probe, providing differential measurement of bound probe only. These probes are discussed later in this section.

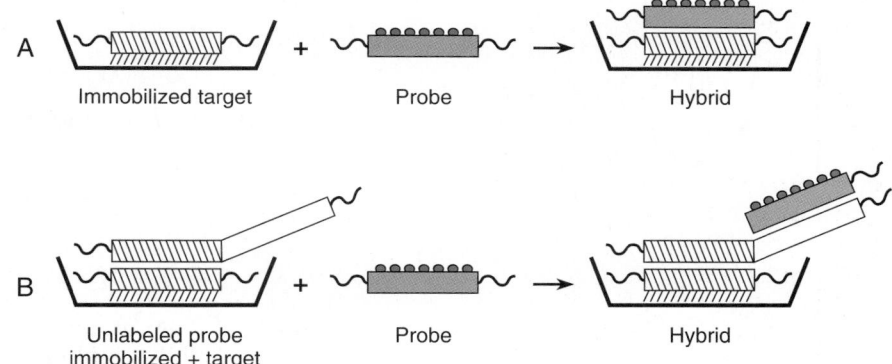

━━━━F i g u r e   5 – 3 8
A, Solid-state hybridization. *B,* Sandwich hybridization.

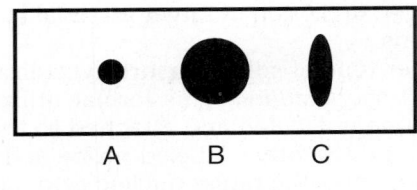

Solid Support Hybridization

A = Dot
B = Spot
C = Slot

■■■■■ F i g u r e   5 – 3 9

Resultant reactions of solid support hybridization. *A,* Dot; *B,* spot; *C,* slot.

IN SITU HYBRIDIZATION

In situ hybridization is the format used to mediate detection of specific nucleic acid sequences within structurally intact cells or tissues. The probe is applied to formalin-fixed or paraffin-embedded tissues that may contain the target DNA. A positive reaction provides localization of the target nucleic acid within the cellular and subcellular detail of the tissue. Its main use presently is in virology, for the detection of viruses such as herpes simplex virus (HSV), human papilloma virus (HPV), cytomegalovirus (CMV), and Epstein-Barr virus (EBV); however, it may have other applications in nonvirologic diagnosis.

SOUTHERN HYBRIDIZATION

Southern hybridization is the format used primarily in research for detection of very specific nucleic acid fragments. DNA must be released and purified, and then cleaved with appropriate enzymes called *restriction endonucleases.* The resultant DNA fragments are then separated by electrophoresis and transferred to nitrocellulose or nylon filters for probing. The specific labeled probe is added. When the reaction is positive, bands can be detected that relate to the specific DNA fragments with sequences identical to the probe used. This method is often used in research or industry for the manufacture of specific probes.

Probe Labels

In order to detect the hybridization that may have occurred between probe and target DNA or RNA, a label or marker is needed. In most cases, this involves prior labeling of the probe with any of the markers listed in Table 5–15. The con-

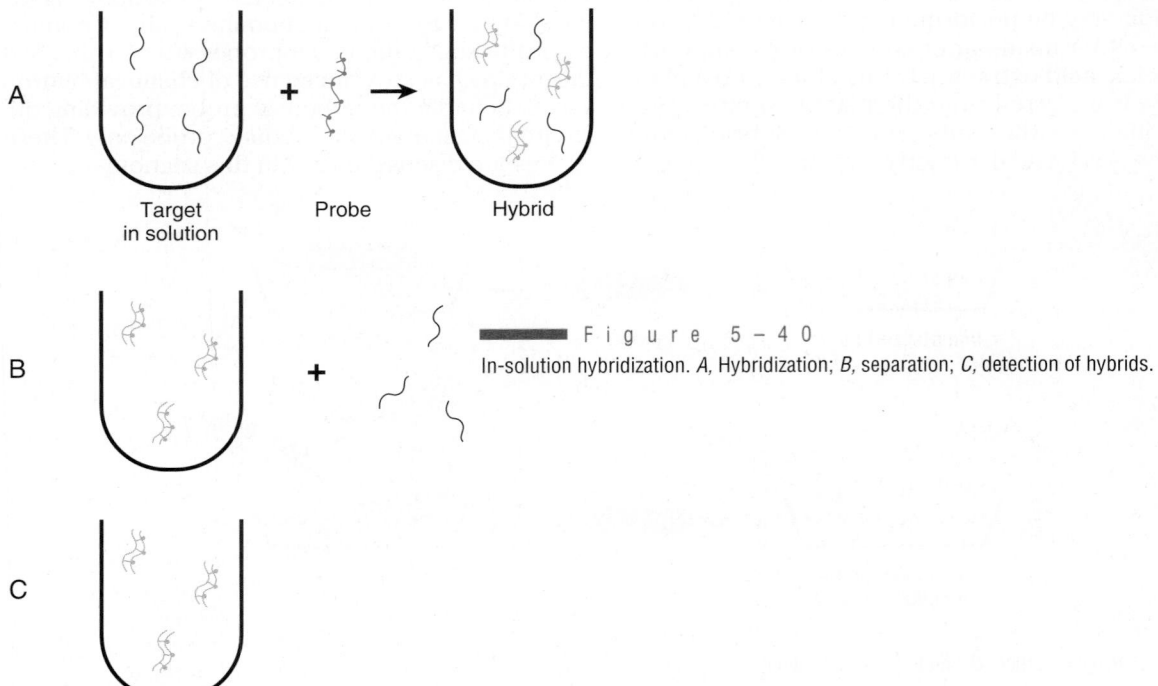

■■■■■ F i g u r e   5 – 4 0

In-solution hybridization. *A,* Hybridization; *B,* separation; *C,* detection of hybrids.

■■■■■ T A B L E   5 – 1 5
LABELS FOR DNA/RNA PROBES

Radioactive labels
^{32}P
^{125}I
^{35}S
Biotinylated nucleotides (biotin-avidin probes)
Enzyme-conjugated probes
  Alkaline phosphatase
  Horseradish peroxidase
Probes directly labeled with antibody
Chemiluminescence
Chemical (e. g., digoxigenin)
Fluorescein labels

---

ventional hybridization technique in research always utilized radioactive probes to maximize sensitivity of the detection. Radioisotopes such as radioactive phosphorus (^{32}P), iodine (^{125}I), and sulfur (^{35}S) are most commonly employed. Use of these detectors is not always convenient for the clinical microbiology laboratory, because they have short half-lives, require elaborate equipment or lengthy incubations, and need to be handled appropriately for proper disposal of the radioactive materials.

Biotin-Avidin

Biotinylated nucleotide probes were first utilized as replacements of radioisotopic labels. Biotin can be incorporated into DNA in the form of biotinylated nucleotide derivatives, e.g., deoxyuridine triphosphate, using DNA polymerase. A biotin-binding protein, most commonly avidin, is added after hybridization of the target with biotinylated probe. An enzyme is linked to the avidin, e.g., alkaline phosphatase or horseradish peroxidase. With addition of appropriate chromogenic substrates for these enzymes, a colored product results if hybridization occurred initially. The product can be read manually or in a colorimeter. Biotin-labeled probes have a very long shelf-life, in contrast to radioactively labeled probes, but are often less sensitive by about a tenfold weaker signal. They are, however, more convenient for the clinical laboratory and offer an opportunity for in situ probing.

Other Labels

Haptens may be added to the DNA probe. One such method for this is via modification of the guanine residues with *N*-acetoxy-*N*-2-acetylaminofluorene (AAA or AAF). Guanosineacetylaminofluorene can then be added after hybridization for detection. Chemi-probe (Orgenics Ltd, Columbia, MD) utilizes a *sulfone* label in the cytidine residues of the nucleic acid. Specific monoclonal antibodies can then be added to bind with high affinity to the sulfonated cytosine residues.

Another hapten, *digoxigenin* (steroid hapten isolated from the foxglove plant), is employed by (Boehringer Mannheim Corporation, Indianapolis, IN) and sold in the Genuis nonradioactive labeling kit. The probes can be used for filter hybridization or for in situ experiments. Detection of these probes can be performed by purchasing, from the same company, a detection kit consisting of anti-digoxigenin antibody conjugated with, for example, alkaline phosphatase, peroxidase, fluorescein, or rhodamine. Depending upon the substrate added, visualization is done by colorimetric, chemiluminescent, or microscopic means.

Use of these labels allows for nonradioactive detection of hybrids with probes that are stable for up to 1 year and with sensitivities that can detect as little as 1 pg of DNA in 1 hour. Subpicogram level detection requires longer incubations or detection.

Chemiluminescence

Chemiluminescent labeling may be the most sensitive method for probe detection, perhaps even exceed the sensitivity of radioactive labeling. Gen-Probe (San Diego, CA) has utilized acridinium ester–labeled probes for this purpose. The labeled probes are commercially available for a variety of bacteria and fungi (Table 5–16). After hybridization, hydrogen peroxide is added to the probe-target complex. This causes release of light as acridinium ester molecules are hydrolyzed. The amount of light is measured in the Gen-Probe Leader I Luminometer as relative light units (RLUs). Other commercially available products are described later in this discussion.

Tropix (Bedford, MA) utilizes chemiluminescent detection along with antibody-labeled probes. For example, AMPPD is a chemiluminescent substrate for alkaline phosphatase, which dephosphorylates AMPPD to form dioxetane anions, which in turn fragment into adamantanone and the excited state of meta-oxybenzoate-anion, a light emitter. AMPPD has an indefinite shelf-life at 4°C and can provide a sensitive means of detection. Other chemiluminescent

■■■■■■ T A B L E 5 – 1 6

ORGANISMS FOR WHICH Accuprobe DNA PROBE PRODUCTS ARE AVAILABLE TO CONFIRM CULTURE RESULTS

Bacteria	
Gram-negative	*Neisseria gonorhoeae*
	Campylobacter jejuni
	Campylobacter coli
	Campylobacter lari
	Haemophilus influenzae
Gram-positive	Group A *Streptococcus*
	Group B *Streptococcus*
	Streptococcus pneumoniae
	Enterococcus spp
	Staphylococcus aureus
	Listeria monocytogenes
Mycobacteria	*Mycobacterium tuberculosis* complex
	Mycobacterium avium complex
	Mycobacterium avium
	Mycobacterium intracellulare
	Mycobacterium gordonae
Fungi	*Histoplasma capsulatum*
	Blastomyces dermatitidis
	Cryptococcus neoformans
	Coccidioides immitis

labels are oxalate esters (American Cyanamid, NJ) and Luminal and its derivatives (Amersham, MA).

Other probes may also be labeled directly with fluorescent groups, such as fluorescein and rhodamine, or directly with antibodies, such as peroxidase and phosphatase. One company, Syngene (a division of Molecular Biosystems, San Diego, CA), manufactures alkaline phosphatase–labeled mycobacterial probes that can be detected visually in a dot-blot format (SNAP). Use of this nonradioactive probe for culture confirmation allows for a rapid (about 4.5 hours) detection of mycobacterial DNA, specifically that of *Mycobacterium tuberculosis* complex and *Mycobacterium avium* complex. These products are discussed later. SNAP probes are also distributed by Digene (Silver Springs, MD).

Probe Target

Target material for probe technology can be DNA for RNA. Most commercial probes utilize DNA as target, with a DNA or RNA probe. Gen-Probe (San Diego, CA) utilizes ribosomal RNA (rRNA) as the target for chemiluminescent probes. The advantage of RNA is the increased sensitivity, due to the higher number of copies of rRNA as compared with DNA per cell.

APPLICATIONS OF PROBE TECHNOLOGY

Probes for Culture Confirmation

Bacteria

Several acridinium ester–labeled probes (e.g., Accuprobe, Gen-Probe, San Diego, CA) can be purchased for culture confirmation of a variety of microorganisms. A list of gram-positive and gram-negative bacteria that can be identified by Accuprobe products is given in Table 5–16. Daly and colleagues (1991) reviewed the performance of these probes. The procedure used to perform the probe tests for bacteria is shown in Figure 5–41*A*, with modification as noted for use on *Mycobacterium* sp. (see Fig. 5–41*B*), Culture confirmation can be obtained in 1.5 to 2 hours or less. Application of the culture confirmation probes to positive blood culture bottles isolates has been reported by Davis and Fuller (1991). The Leader I Luminometer (Gen-Probe, San Diego, CA) is required to obtain results, which are recorded as relative light units (RLUs). The RLU values needed for a positive result may vary according to the test kit but are included in the package inserts sold with the tests.

PROBES FOR GASTROINTESTINAL PATHOGENS

Another manufacturer, Diagnostic Hybrids (Athens, OH), produces a probe under the name Hybriwix. Scholl and colleagues (1990) reported results for Hybriwix culture confirmation of presumptive *Salmonella* isolates from a variety of selective and nonselective culture media. The procedure requires 30 to 120 minutes. It is unclear whether any of the Hybriwix products for bacteria are commercially available or will be in the future. Hybriwix probes for viruses are also under development; they are discussed later in the section on viruses.

Noncommercially available probes for bacterial culture confirmation include probes for enteric pathogens, e.g., *Escherichia coli*. Biotinylated DNA probes have been used to hybridize with bacterial colony blots made on Whatman No. 541 filter paper to detect enterotoxigenic (ETEC), enteroin-

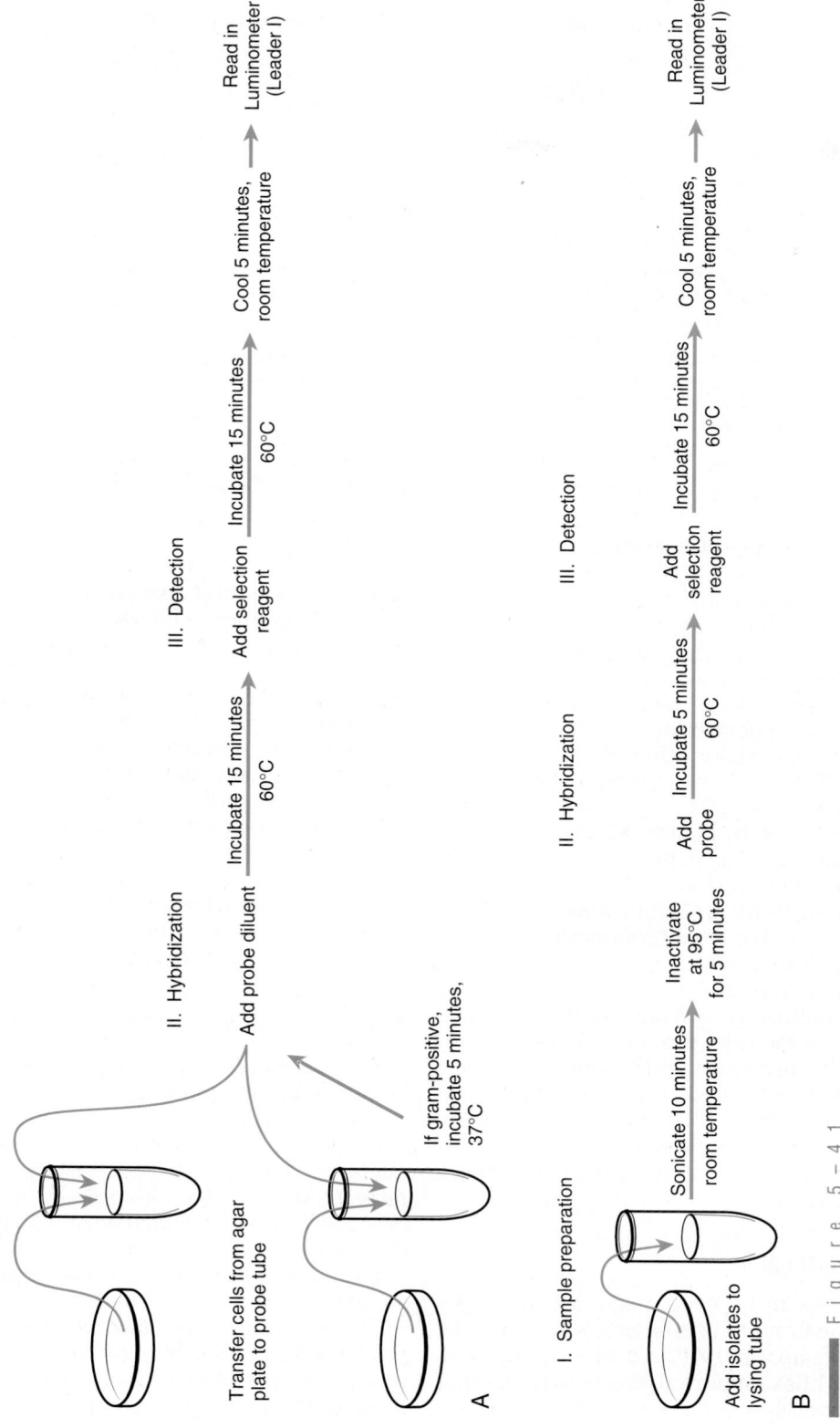

I. Sample preparation

II. Hybridization

III. Detection

Transfer cells from agar plate to probe tube

Add probe diluent

Incubate 15 minutes 60°C

Add selection reagent

Incubate 15 minutes 60°C

Cool 5 minutes, room temperature

Read in Luminometer (Leader I)

If gram-positive, incubate 5 minutes, 37°C

A

I. Sample preparation

II. Hybridization

III. Detection

Add isolates to lysing tube

Sonicate 10 minutes room temperature

Inactivate at 95°C for 5 minutes

Add probe

Incubate 5 minutes 60°C

Add selection reagent

Incubate 15 minutes 60°C

Cool 5 minutes, room temperature

Read in Luminometer (Leader I)

B

▶ F i g u r e 5 – 4 1

Culture identification by Accuprobe. *A,* Gram-positive and gram- negative organisms. *B,* Mycobacteria and fungi.

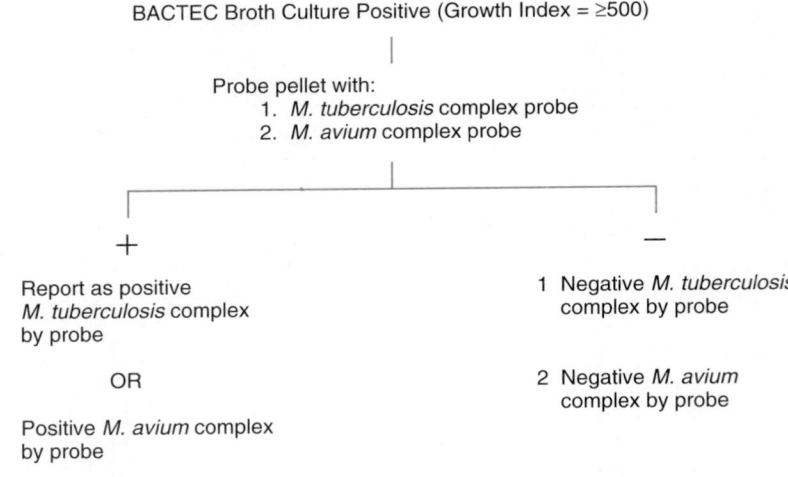

■■■■■ F i g u r e   5 – 4 2

Gen-probe analysis of BACTEC positive mycobacterial cultures.

vasive (EIEC), enteropathogenic (EPEC), entero-haemorrhagic (EHEC), and diffuse adherence (DAEC) *E. coli* strains causing diarrhea. Results appear to be reproducible, economical, and easy to interpret. Use of such nonradioactive probes will be beneficial to developed as well as underdeveloped countries. Many other research probes for *E. coli,* such as those described by Gomes and associates (1987) for EIEC, by Sommerfelt et al (1988) for ETEC, and by Echeverria et al (1987) for both *E. coli* and *Shigella* sp, have been reported in the literature. Escheverria et al (1987) reported using probes directly on stool specimen blots with a ^{32}P radioactive label (Sommerfelt and colleagues [1988] for ETEC).

Another enteric pathogen, *Campylobacter,* can be identified culturally by means of the previously mentioned Accuprobe product. *Campylobacter jejuni, Campylobacter coli,* and *Campylobacter lari* can be probed within 30 minutes with good results. Because these three are the most common *Campylobacter* species in stool specimens, their quick identification may be of benefit to the clinical laboratory.

PROBES FOR ANAEROBES

Anaerobic bacteria have been identified by probes, although none is commercially available. Dix and associates in 1990 and Moncla and colleagues in 1991 have reported on a DNA probe that can detect anaerobes and other bacteria associat-

ed with periodontal disease. Microprobe (Bothell, WA) manufactures culture-confirmation, alkaline phosphatase–labeled probes for the identification of *Bacteroides gingivalis, Bacteroides intermedius* types I and II, *Bacteroides forsythus, Fusobacterium nucleatum, Wolinella recta, Eikenella corrodens, Haemophilus aprophilus, Actinobacillus actinomycetemcomitans,* and *Streptococcus intermedius.* Direct detection of *B. forsythus* in plaque samples using this same probe was found to be 100% specific and sensitive compared with culture. DNA probes labeled with ^{32}P have also been described for culture confirmation of *Bacteroides* sp, from nonperiodontal specimens, but these are not known to be commercially available.

PROBES FOR *MYCOBACTERIUM* SP

Two commercial probes are available for the culture confirmation of *Mycobacterium* sp. Accuprobe (Gen-Probe, San Diego, CA) utilizes the same format as previously described for other bacteria to detect *M. tuberculosis* complex *(M. tuberculosis, Mycobacterium bovis, mycobacterium africanum, Mycobacterium microti)* and *M. avium* complex *(M. avium* and *Mycobacterium intracellulare).* A *Mycobacterium gordonae* probe is also available.

For all, isolates from solid medium or broth (BACTEC) can be probed in 1.5 to 2 hours. Many reports in the literature have verified the specificity of the probe. Sensitivity is not in question with solid medium isolates, and testing from a broth (i.e., BACTEC) has a good sensitivity provided that the growth index (GI), an indirect measurement of amount of the organisms, is adequate (GI ≥ 100). Accuprobe also produces probes individually for *M. avium* and *M. intracellulare.* Figures 5–42 and 5–43 suggest how such probes can be used for identification in the routine laboratory.

SNAP for *Mycobacterium* sp (Syngene, San Diego, CA) is an alkaline phosphatase–labeled probe for *M. tuberculosis* complex and *M. avium* (MAI) complex. The format used in SNAP is a dot blot. Results are read visually after color development. Cultures from solid or liquid media can be utilized. A report by Lim and colleagues (1991) has shown SNAP to be a very sensitive and specific probe. Compared with the Accuprobe, more MAI isolates were detected by Syngene's SNAP probe. Work at the Cleveland Clinic has also demonstrated a difference in detection of MAC isolates. The differences between Accuprobe and SNAP were speculated to be due to the presence of an "X" sequence in the latter product; this sequence has been added to the Accuprobe *M. avium* complex probe and will be available shortly to users of the probe. The two probes will then probably be equivalent in identification capabilities.

The Accuprobe requires 1.5 to 2 hours, and the SNAP requires 2.5 to 3 hours. The Accuprobe also requires an instrument for detection; SNAP utilizes no instruments for detection, although the method of lysis for SNAP is via mechanical breakage, which requires a "mini Bead-beater." The advantage of using either of the mycobacterial probes is speed of identification. When coupled with a method of isolation, (e.g., BACTEC) that decreases the time for detection, the identification of mycobacterial isolates can be greatly reduced in time. Table 5–16 lists the *Mycobacterium* spp that can be confirmed by Accuprobe. As of the final writing of this paper, SNAP products are no longer available from Syngene.

Fungi

Accuprobes (Gen-Probe, San Diego, CA) for fungal identification are available for *Histoplasma capsulatum, Blastomyces dermatitidis, Cryptococcus neoformans,* and *Coccidioides immitis* (Table 5–16). The same format for hybridization, i.e., in-solution hybridization using an acridinium ester label and chemiluminescent detection, is employed. A DNA probe for *Candida albicans* was reported by Cheung and Hudson (1988), but it was found to be genus specific but not species specific.

Reagan and associates (1990) used a rRNA probe from *Saccharomyces cerevisiae* to epidemiologically type strains of *Candida* sp isolated from blood and other cultures. Although this

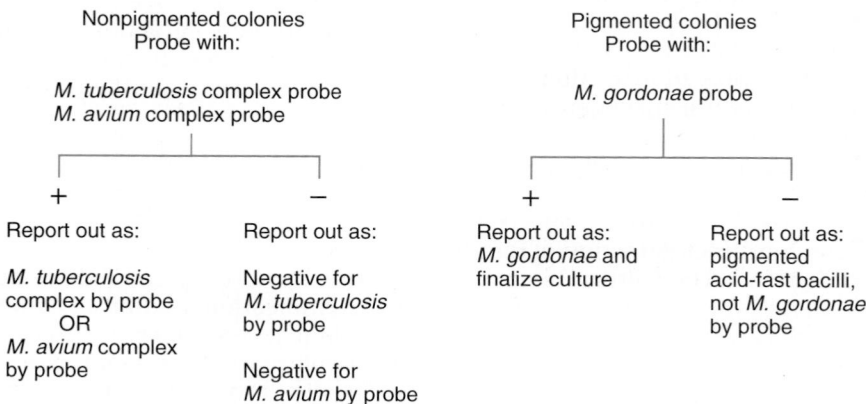

Complete identification (if negative alone) and perform susceptibility testing as required

Figure 5-43

Identification of mycobacterial culture positive on solid media by DNA probe.

accomplishment is of interest epidemiologically, the probe is not commercially available.

Probes for Rapid Diagnosis of Infectious Diseases

Respiratory Infections

A ^{125}I-labeled probe for *Legionella* sp is available commercially from Gen-Probe (San Diego, CA). It can be applied directly to clinical specimens from the respiratory tract, such as sputum, transtracheal aspirates, and lung tissue. The sensitivity of the probe is similar to that of a direct fluorescent antibody stain for *Legionella*, but offers the advantage of detection of all species of *Legionella*.

For *Mycoplasma pneumoniae* detection in clinical specimens, a direct specimen probe was available from Gen-Probe, but it has been removed from the market.

Direct specimen probes for viral respiratory pathogens are available from a variety of manufacturers, including ENZO Diagnostics, Inc. (Syosset, NY) and Digene (Silver Springs, MD). In situ probes for Epstein-Barr virus, cytomegalovirus, and adenovirus are available, but some are for research use only.

Sexually Transmitted Diseases

Commercial probes are available for the diagnosis of *Neisseria gonorrhoeae* and *Chlamydia trachomatis*. The PACE system (Gen-Probe, San Diego, CA) utilizes a chemiluminescent detection method, described earlier for bacteria. The same specimen collection tube can be used to collect one specimen for both procedures. After collection of the cervical or urethral specimen, cells are lysed, releasing the target nucleic acid, if present. At this point, all organism is killed, and thus, if culture is desired in conjunction with probe assay, separate specimens would have to be collected.

The PACE system for direct detection of microbial organisms differs from the cultural confirmation probe from Gen-Probe (Accuprobe). In the PACE system procedure, after hybridization with specific probe, the unbound probe is removed via a magnetic bead separation step. The magnetic beads that are added will attract the bound probe (probe-target complex, if present), and when tubes are placed on a magnetic separation rack and then decanted, the unbound probe is

eliminated in the supernatant. The remaining material is resuspended and read in the luminometer as with the Accuprobe. Results for the PACE probes look promising.

For the identification of *Trichomonas vaginalis* and *Gardnerella vaginalis* in clinical specimens, a Microprobe (Bothell, WA) has been developed and may be commercially available in the future. Probes for *Mobiluncus* sp, an organism believed to be associated with bacterial vaginosis, have also been described.

For the diagnosis of herpes simplex virus and human immunodeficiency virus, probes are available commercially from a variety of manufacturers, including PathoGene (ENZO, Syosset, NY) and HSV Disk (Diagnostic Hybrids, Inc., Athens, OH). Human papilloma viruses (HPVs) can be detected by means of dot blot hybridization or in situ methods manufactured commercially as Bio Pap (ENZO Diagnostics, Inc., Syosset, NY), Vira Pap (Digene Diagnostics, Inc., Silver Springs, MD), and DAKO Biotinylated HPV Probes (DAKO Corporation, Carpentera, CA).

Miscellaneous Infectious Diseases

A probe was available for use in screening urine for the presence of urinary tract pathogens (Gen-Probe); however, it was removed from the market. As mentioned previously, probes for the diagnosis of enteric pathogens have been used directly on clinical specimens. In addition, noncommercially available probes for the detection of enteroviruses have been reported in the literature.

Probes have also been described in the literature for direct detection of *Plasmodium* sp, *Borrelia burgdorferi,* hepatitis B virus, hepatitis C virus, and a variety of other infectious diseases.

IMPLEMENTATION OF PROBES IN THE CLINICAL MICROBIOLOGY LABORATORY

Probe technology can modify our approach to the identification and/or isolation of microbes from clinical specimens. They can be used for culture confirmation, as mentioned earlier. Probes are particularly applicable to those organisms that require expensive materials for identification by conventional methods and to organisms for which conventional methods are lacking. The more obvious advantage of probes in clinical laboratories is for direct detection of pathogens in clinical specimens, obviating the need for culture. Applications of probe technology are listed in Table 5–17.

The advantages of using a probe in the laboratory are:

■ High specificity

■ Ability to detect nonviable or antigen-negative infections

■ Speed

■ Use of standardized, commercially available products

Some problems and disadvantages of probes need to be addressed as well. First, specimens often still need to be cultured (1) for susceptibility testing, (2) to detect the presence of polymicrobial infections, and (3) to increase sensitivity, because most probes, without amplification, will not be as sensitive as culture. Second, probes are usually more costly than biochemical methods. Third, practically speaking, probes would have to be run only on "highly suspicious" specimens and for organisms that are "always" pathogenic. For example, a probe for *E. coli* in stool is not practical, cost efficient, or usable, but a probe for the verotoxin of enterohemorrhagic *E. coli* might be extremely beneficial and cost efficient.

Table 5–18 lists some issues to consider before adopting a probe for use in the laboratory. First on any list should be the reliability and performance of the test. If the test does not work, there is no question that it should not be

TABLE 5–17
APPLICATIONS OF PROBE TECHNOLOGY

Outside clinical microbiology
  Genetic defect detection
  Food microbiology
  Plant pathology
Within clinical microbiology
  • Direct detection of microbes in clinical specimens
  • Organisms not able to be cultivated
  • Organisms with long incubation times
  • Diseases with unknown etiologies
  Identification of cultural isolates
  Stain identification
  Identification of toxins, virulence factors
  Identification of resistance markers
  Identification of microbes in environmental specimens
  Identification of resistance factors

TABLE 5–18
ISSUES TO CONSIDER BEFORE ADOPTING A PROBE

Reliability (PVP/PVN comparison)
Speed
Cost
Need for culture
Difficulty of procedure
Availability of equipment/personnel
Practicality

used. If there is a good gold standard and the test compares well with it, then consideration can be given to using the probe if costs are justified. If the prevalence of the disease entity in the population is very low, the probe must demonstrate good specificity and a very high positive predictive value, or many false-positive diagnoses will be made. Speed of the procedure is important as a justification for cost, perhaps especially for identifications for which earlier diagnosis may save nosocomial infection transmission or enable an earlier discharge of the patient. If specimens must be batched because of low volume of the test material or the cost of performing many quality control procedures, one must first consider how much time is really being saved with the probe and whether or not the demand for the test warrants the use of any procedure at all. Equipment costs are not extensive for probe technology, but for some tests, instruments must be purchased, so the cost and space requirements must be considered.

In general, probes should not be employed in any laboratory simply because they are available, but rather because they may meet a need better than other available methods. A reasonable amount of time should be spent "up front" to investigate thoroughly the effectiveness of the probe, its cost demands, and the practicality of implementing the probe procedure in the laboratory. This time will save much frustration in the future. Amplification methods in conjunction with probe technology are about to be made readily available commercially; soon as many products will be developed as the clinical laboratories perhaps need or want. Care must be taken, however, to use all of these molecular techniques with good microbiologic knowledge and forethought.

Bibliography

Cheung LL, Hudson JB: Development of DNA probes for *Candida albicans.* Diagn Microbiol Infect Dis 10:171, 1988.

Daly JA, Clifton NL, et al: Use of nonradioactive DNA probes in culture confirmation tests to detect *Streptococcus agalactiae, Haemophilus influenzae* and *Enterococcus* spp. from pediatric patients with significant infections. J Clin Microbiol 29:80, 1991.

Davis TE, Fuller DD: Direct identification of bacterial isolates in blood cultures by using a DNA probe. J Clin Microbiol 29:2193, 1991.

Dix K, Watanabe SM, et al: Species-specific oligodeoxy-nucleotide probes for the identification of peridontal bacteria. J Clin Microbiol 28:319, 1990.

Echeverria P, Taylor DN, et al: Comparative study of synthetic oligonucleotide and cloned polynucleotide enterotoxin gene probe to identify enterotoxingenic *Escherichia coli.* J Clin Microbiol 25:106, 1987.

Ellner PD, Kiehn TE, et al: Rapid detection and identification of pathogenic mycobacteria by combining radiometric and nucleic acid probe methods. J Clin Microbiol 26:1349, 1988.

Gicquelais KG, Baldini MM, et al: Practical and economical method for using biotinylated DNA probes with bacterial colony blots to identify diarrhea-causing *Escherichia coli.* J Clin Microbiol 28:2485, 1990.

Gomes TAT, Toledo MRF, et al: DNA probes for identification of enteroinvasive *Escherichia coli.* J Clin Microbiol 25:2025, 1987.

Hillier SL, Briselden AM: Evaluation of a rapid oligonucleotide test for direct detection of *Gardnerella vaginalis* and *Trichomonas vaginalis* from vaginal specimens. Abstract presented at American Society for Microbiology meeting, New Orleans, May 1992.

Iwen PC, Blair TMH, Woods GL: Comparison of the Gen-Probe PACE 2 system, direct fluorescent-antibody and cell culture for detecting *Chlamydia trachomatis* in cervical specimens. Am J Clin Pathol 95:578, 1991.

Kejian G, Bowden DS: Digoxigenin-labeled probes for the detection of hepatitis B virus DNA in serum. J Clin Microbiol 29:506, 1991.

Kiehn TE, Edwards FF: Rapid identification using a specific DNA probe of *Mycobacterium avium* complex from patients with acquired immunodeficiency syndrome. J Clin Microbiol 25:1551, 1987.

Kuritza AP, Getty CE, et al: DNA Probes for identification of clinically important *Bacteroides* species. J Clin Microbiol 23:343, 1986.

Lees MI, Newnan DM, Garland SM: Comparison of a DNA probe assay with culture for the detection of *Chlamydia trachomatis.* J Med Microbiol 35:159, 1991.

Lim SD, Todd J, et al: Genotypic identification of pathogenic *Mycobacterium* species by using a nonradioactive oligonucleotide probe. J Clin Microbiol 29:1276, 1991.

McLaughlin GL, Decrind C, et al: Optimization of a rapid non-isotopic DNA probe assay for *Plasmodium falciparum* in the Gambia. J Clin Microbiol 29:1517, 1991.

Moncla BJ, Motley ST, et al: Use of synthetic oligonucleotide DNA probes for identification and direct detection of *Bacteroides forsythus* in plaque samples. J Clin Microbiol 29:2158, 1991.

Panke ES, Yank LI, et al: Comparison of Gen-Probe DNA probe test and culture for the detection of *Neisseria gonorrhoeae* in endocervical specimens. J Clin Microbiol 29:883, 1991.

Pascule AW, Veto GE, et al: Laboratory and clinical evaluation of a commercial DNA probe for the detection of *Legionella* sp. J Clin Microbiol 27:2350, 1989.

Popovic-Uroic T, Patton CM, et al: Evaluation of an oligonucleotide probe for identification of *Campylobacter* species. Lab Med 22:533, 1991.

Reagan DR, Pfaller MA, et al: Characterization of the sequence of colonization and nosocomial candidemia using DNA fingerprinting and a DNA probe. J Clin Microbiol 28:2733, 1990.

Roberts MC, Hillier SL, et al: Nitrocellulose filter blots for species identification of *Mobiluncus curtisii* and *Mobiluncus mulieris.* J Clin Microbiol 20:826, 1984.

Rotbart HA: Nucleic acid detection systems for enteroviruses. Clin Microbiol Rev 4:156, 1991.

Scholl DR, Kaufmann C, et al: Clinical application of novel sample processing technology for the identification of salmonellae by using DNA probes. J Clin Microbiol 28:237, 1990.

Sommerfelt H, Kalland KH, et al: Cloned polynucleotide and synthetic oligonucleotide probes used in colony hybridization are equally efficient in the identification of enterotoxigenic *Escherichia coli.* J Clin Microbiol 26:2275, 1988.

Tenover FC, Carlson L, et al: DNA probe culture confirmation assay for identification of thermophilic *Campylobacter* species. J Clin Microbiol 28:1284, 1990.

Wise DJ, Weaver TL: Detection of Lyme disease bacterium, *Borrelia burgdorferi,* by using the polymerase chain reaction and a nonradioisotopic gene probe. J Clin Microbiol 29:1523, 1991.

F. DIAGNOSTIC APPLICATIONS OF POLYMERASE CHAIN REACTION

Jeffrey P. Massey

OBJECTIVES

1. Describe the basic principles of the PCR amplification process.

2. Apply PCR techniques to the detection of infectious agents.

3. Discuss the strengths and weaknesses of PCR.

4. Be aware of *specimen carryover* and how to minimize the problem.

The goal of the clinical microbiology laboratory is to provide the clinician with evidence of the presence or absence of an infectious agent that may be causing a particular illness. Traditionally, this diagnosis relies upon laboratory techniques that may or may not directly detect the pathogen. Both in vitro culture and serologic techniques are commonly used in the diagnostic laboratory. In vitro culture techniques permit a direct identification of the causative agent. Serology, on the other hand, provides an indirect measure of infection. Techniques such as enzyme immunoassay (EIA) and immunofluorescence assay (IFA) detect antibody that results from exposure to an infectious agent.

There are several infectious agents for which reliable in vitro culture systems or serologic techniques are unavailable. Examples are *Mycobacterium leprae* and human papilloma virus (HPV). Because the laboratory is unable to provide definitive evidence of such infections, diagnosis depends primarily upon the clinical picture.

Limitations of DNA Probe Technology

The previous section discussed nucleic acid (DNA) hybridization. Ideally, this technology identifies fastidious organisms directly from the patient without their prior isolation in culture. These techniques need to be highly sensitive and specific in a clinical setting to avoid cross-reactivity with closely related, but distinct organisms. Unfortunately, most applications of DNA probe technology lack the sensitivity required for routine use in the diagnostic laboratory. Although commercial DNA hybridization kits are available, they usually require that the agent first be amplified in culture. These kits work best for infectious agents that grow well in culture.

History and Development of PCR

Fastidious organisms would be proper targets for DNA probe technology if their initial amplification by culture was not essential. The development of the *polymerase chain reaction (PCR)* in 1985 allows the researcher to amplify the target to detectable limits without the need of in vitro culture procedures. Because PCR results in the exponential generation of short, discrete DNA molecules, large amounts of starting material are not essential to generate a positive result. Instead, one molecule of target DNA may be sufficient to prove the infectious nature of the disease.

PCR has become an essential methodology in many basic and applied research laboratories. Modification of the basic procedure allows one to clone and sequence DNA, introduce specific mutations into a known DNA sequence, and even to characterize an unknown sequence of DNA that lies close to a region with a known sequence. Through the use of PCR, a variety of infectious diseases, genetic diseases (for example, sickle cell anemia), and chronic illnesses (for example, chronic myeloid leukemia) have been characterized.

It is possible to adapt this methodology for use in the diagnostic microbiology laboratory. PCR may be useful in the identification of organisms currently undetectable by standard techniques. In addition, PCR more rapidly identifies certain organisms. Some pathogens require weeks of culture prior to identification, but PCR gives a result in 1 to 2 days. This technology, therefore, has the capability of greatly enhancing the usefulness of the clinical microbiology laboratory.

Definitions

Amplicon: The final product of PCR containing the target sequence of interest. This short, discrete DNA molecule is defined at each of its four ends by the primer sequences used in the amplification.

Anneal: The process by which oligonucleotides "attach" to targeted DNA sequence.

Autoradiograph: The image formed on x-ray film resulting from the radioactive emissions of ^{32}P.

Complementary: Certain nucleotidases that align opposite each other in the two strands of DNA: G pairs with C; A pairs with T.

Cycle: A set of three different temperature settings, performed sequentially, for DNA denaturation, primer annealing, and primer extension.

Denaturation: The use of high temperature (94°C) to separate the double-stranded DNA into single-stranded DNA.

DNA: Deoxyribose nucleic acid, composed of a double helix consisting of pair nucleotides and a ribose backbone.

DNA amplification: The process by which a single molecule of DNA is manipulated to increase its concentration exponentially.

DNA polymerase: A specific enzyme that catalyzes the formation of additional strands of double-stranded DNA for the DNA template.

DNA template: The starting ds DNA.

End labeling: The addition of a radioactively labeled phosphate group to the 5′ end of a oligonucleotide probe.

Extension: The process by which primed DNA is synthesized by the action of *Taq* polymerase.

Genomic DNA: DNA contained in a cell's chromosome.

Nucleotides: Building blocks of the nucleic acids of which DNA is composed. Four bases make up DNA: A, adenine; T, thymine; C, cytosine, and G, guanine.

Oligonucleotides: Short strands of single-stranded DNA of defined sequence, less than 50 bases in length. Examples are the primers and probes used in PCR.

PCR: Polymerase chain reaction; a primer-mediated, temperature-dependent technique for the enzymatic amplification of a specific DNA sequence.

Primer: A short DNA sequence that anneals to a specific area of the target DNA. DNA polymerase initiates synthesis from this point.

Primer-dimer: PCR product that is composed mainly of the primers used in the amplification. These products are formed independently of the DNA template and accumulate exponentially.

Probe: In PCR, a radioactively labeled, single-stranded DNA oligonucleotide, usually 40 to 45 bases long, which is used to specifically identify complementary regions of the amplified PCR product.

***Taq* Polymerase:** A heat-stable DNA polymerase isolated from the bacterium *Thermus aquaticus.*

Target DNA: A sequence of known nucleotides specifically chosen for amplification.

BASIC PCR METHODOLOGY

Theory of Amplification

PCR is a primer-mediated, temperature-dependent technique for the enzymatic amplification

TABLE 5–19

REACTION STEPS REQUIRED FOR AMPLIFICATION OF DNA BY PCR

Step	Temperature (°C)	Action
Denaturation	94	Double-stranded DNA broken into single strands (dsDNA → ssDNA)
Primer annealing	55	Attachment of oligonucleotide primers to complimentary regions on ssDNA.
Extension	72	Synthesis of dsDNA (5′ → 3′); catalyzed by *Taq* DNA polymerase

of a specific DNA sequence. A typical reaction involves a total of 25 to 40 cycles. Each cycle is a sequential series of three different temperature settings. Table 5–19 identifies the action of each step in the amplification process. The methodology is similar to the in vivo mode of DNA replication. The following components are required:

■ A short single-stranded DNA (ssDNA) molecule known as an oligonucleotide *primer* that initiates the DNA for replication

■ A DNA polymerase enzyme that catalyzes the formation of addition molecules of double-stranded DNA (dsDNA)

■ The individual building blocks of DNA, the deoxynucleotide bases (dATP, dCTP, dGTP, and dTTP or dUTP)

Table 5–20 lists the role of each component of the reaction.

Nucleic acid sequence information is available for most organisms of clinical importance. Careful evaluation of this information determines the tar-

TABLE 5–20

COMPONENT REQUIRED FOR AMPLIFICATION OF DNA BY PCR

Reagent	Usual Concentration	Action
KCl	50 mmol	Provides proper salt concentration
Tris-HCl, pH 8.3	10 mmol	Maintains proper pH for efficient action of enzyme
Gelatin	0.01%	Stabilizes enzyme
$MgCl_2 \cdot 6H_2O$	1.5–2.5 mmol	Provides cation required for proper functioning of enzyme
Deoxynucleotides (dATP, dCTP, dTTP or dUTP, dGTP)	200 µmol	Building blocks of DNA required for synthesis of new DNA strands
Sense strand primer	50–100 pmol	Anneals to (+) ssDNA; provides starting point for DNA replication
Anti-sense strand primer	50–100 pmol	Anneals to (−) ssDNA; provides starting point for DNA replication
Template DNA	1 µg	Contains target DNA to be amplified
Enzyme (*Taq* or other thermostable DNA polymerase)	2.5 U	Catalyzes formation of dsDNA
UNG (uracil-*n*-glycosylase)	5 U	Cleaves PCR products that contain UTP

get for amplification. Microorganisms typically contain unique regions in their genetic material that is characteristic of a particular strain. Primers designed to recognize these unique regions permit highly specific identification of a single strain or species. Conversely, the amplification may be intentionally nonspecific. If primers are chosen that flank a region shared by a diverse group of organisms, the resulting amplification would yield the detection of more than one type of organism.

Proper selection of a pair of oligonucleotide primers is essential to the ultimate success of the amplification. Primers are short pieces (18 to 28 nucleotides in length) of single-stranded DNA. They flank the two sides of the target sequence and serve to define the ends of the DNA that has been targeted for amplification.

Primer Characteristics

The general characteristics of primer design are as follows:

- Primers are 18 to 28 nucleotides long with a 50 to 60% guanine plus cytosine composition.

- The temperature at which the primer dissociates from the DNA template, the melting temperature (T_m), must be between 55 and 80°C.

- T_m should be the same for both members of the primer pair.

- Runs of three or more cystosines or guanines at the 3′ end of the primer should be avoided.

- Regions that are complementary at the 3′ end of each primer should be avoided.

Adherence to these requirements will eliminate many of the artifacts that lower amplification efficiency. The most common artifact is the *primer-dimer*. As the name implies, a primer-dimer is essentially a dimerization of the two PCR primers. Dimerization forms a short double-stranded product that contains complementary regions for both primers. A primer-dimer will form when the 3′ ends of the primers are complementary to each other. Because of these complementary regions, the primers will more readily anneal to each other than to the DNA template. Once primer-dimers develop, they accumulate very efficiently in the reaction mixture. Other causes of the primer-dimer effect are a high number of cycles (30 to 40), a low annealing temperature, and a high concentration of enzyme or primers.

DNA Polymerases Used in PCR

Amplification of the target DNA depends on the action of a *DNA polymerase*. The function of this enzyme is to catalyze the formation of a new strand of double-stranded DNA identical in content to the original DNA template. The mechanism of this replication process is similar to manner in which DNA replicates in vivo:

- A dsDNA molecule template is heated and separated into single-stranded DNA.

- The primer anneals to its complementary sequence on the template DNA.

- The DNA polymerase starts at the 3′ end of the primer and reads the original DNA sequence base-by-base in a 5′ to 3′ direction.

- A complementary strand of dsDNA is formed as the DNA polymerase adds the appropriate deoxynucleotide (dNTP).

Klenow Fragment of *Escherichia coli*

The Klenow fragment of *Escherichia coli* DNA polymerase I was the enzyme used in the development of PCR. This enzyme functions best at 37°C but is heat labile. Inactivation of the Klenow fragment occurs at 94°C, the temperature required for denaturation of DNA. As a result, all enzymatic activity was lost at the end of each cycle, so each successive cycle required the addition of fresh enzyme. This extra step was time consuming, was costly in terms of enzyme, and led to a low efficiency of amplification.

Taq DNA Polymerase

Taq DNA polymerase is the enzyme most commonly used for PCR. Heat stable, it was isolated from *Thermus aquaticus,* a bacterium that resides in hot springs. It retains full activity after repeated denaturing steps. Because the enzymatic activity of *Taq* DNA polymerase is highest at 72°C, both the primer annealing and primer extension steps can occur at higher temperatures. This higher temperature permits a more efficient amplification, because ideally only perfect primer annealings have the proper stability for extension. It is necessary to add enzyme only at the start of each reaction. Because the enzyme continues to be active even after multiple cycles, automation of the process is possible. Programmable heating and cooling blocks permit the sequential, rapid temperature changes essential to PCR.

Other DNA Polymerases

Additional heat-stable DNA polymerases are commercially available and may prove to be more useful than *Taq* DNA polymerase. They include Vent polymerase, isolated from *Thermococcus litoralis* (available from New England Biolabs, Beverly, MA), and *Pfu* DNA polymerase, isolated from *Pyrococcus furiosus* (available from Stratagene, LaJolla, CA). Both enzymes have higher thermostability than *Taq*. *Taq* retains 50% activity after 40 minutes at 95°C, whereas both Vent and *Pfu* retain full activity after 2 hours at 98°C. In addition, these two enzymes possess a 3′-5′ exonuclease component, which confers proofreading ability. This proofreading ability allows the enzyme to remove a misincorporated base and replace it with the proper one. As a result, both enzymes demonstrate greater fidelity of DNA synthesis than *Taq* DNA polymerase.

Cycling Conditions

One complete cycle consists of three different settings: a denaturation step, an annealing step, and an extension step.

Heating the DNA template to 94°C initiates the procedure. This step, termed *template denaturation,* separates the DNA into two single strands through disruption of the hydrogen bonds holding the dsDNA together.

Primer annealing then occurs. A primer anneals, or attaches, to complementary regions on the target DNA. The complementary nature of DNA is an essential component of this process. DNA is composed of four bases: G (guanine), C (cytosine), A (adenine), and T (thymidine). A double-stranded DNA molecule results from the specific pairing of these bases. Adenine pairs with thymidine through the formation of two hydrogen bonds, whereas cytosine pairs with guanine through the formation of three hydrogen bonds.

Following denaturation, the ssDNAs remain separate until the strands reanneal at a lower temperature. Before this takes place, however, the oligonucleotide primers specifically anneal to complementary regions on the ssDNA. Depending on the melting temperature (T_m) of the primer, this annealing occurs between 40 and 60°C. The primers are added in such a high concentration that the ssDNAs more readily anneal to the primers than to their own complementary strand. The annealing of the oligonucleotide primer and the target DNA results in a stable double-stranded hybrid.

Once the primers have annealed, increasing the temperature to 72°C allows for extension of the primed ssDNA strands. The site where the primer has annealed serves as a starting point for replication. *Taq* DNA polymerase starts at the 3′ end of the primer and uses the ssDNA as a template to incorporate complementary bases. The DNA polymerase reads the template DNA base by base in a 5′ to 3′ direction and incorporates a complementary base into the growing PCR product.

The first cycle results in the formation of two long products. These are molecules defined at one end by a sequence that corresponds to one of the primers. The other three ends of the long products extend indefinitely and represent genomic DNA of varying length (Fig. 5–44). The end of the molecule that contains the primer sequence represents a strong stop to DNA replication. The polymerase dissociates from the PCR product when it reaches this point. In subsequent cycles, each of these newly synthesized DNA fragments serves as a template for additional extension products.

These long products contribute to the formation of two intermediate products at the end of the second cycle. The intermediate products that form during this cycle contain specific primer sequences at three of the four ends of the molecule.

When an intermediate product is denatured, one of the single-stranded products is defined at both ends by primer sequences. This is the template from which the short, discrete product of interest (the *amplicon*) results. The third cycle therefore results in the formation of two molecules that contain the target sequence and are defined at all four ends by primer sequences.

Beginning with the fourth cycle, the amount of the amplicon doubles with each subsequent cycle. At the same time, the long and intermediate products accumulate linearly. After 25 cycles of this exponential expansion, the amount of amplicon is significantly larger than the amounts of other products of amplification.

Under optimal conditions, the amount of

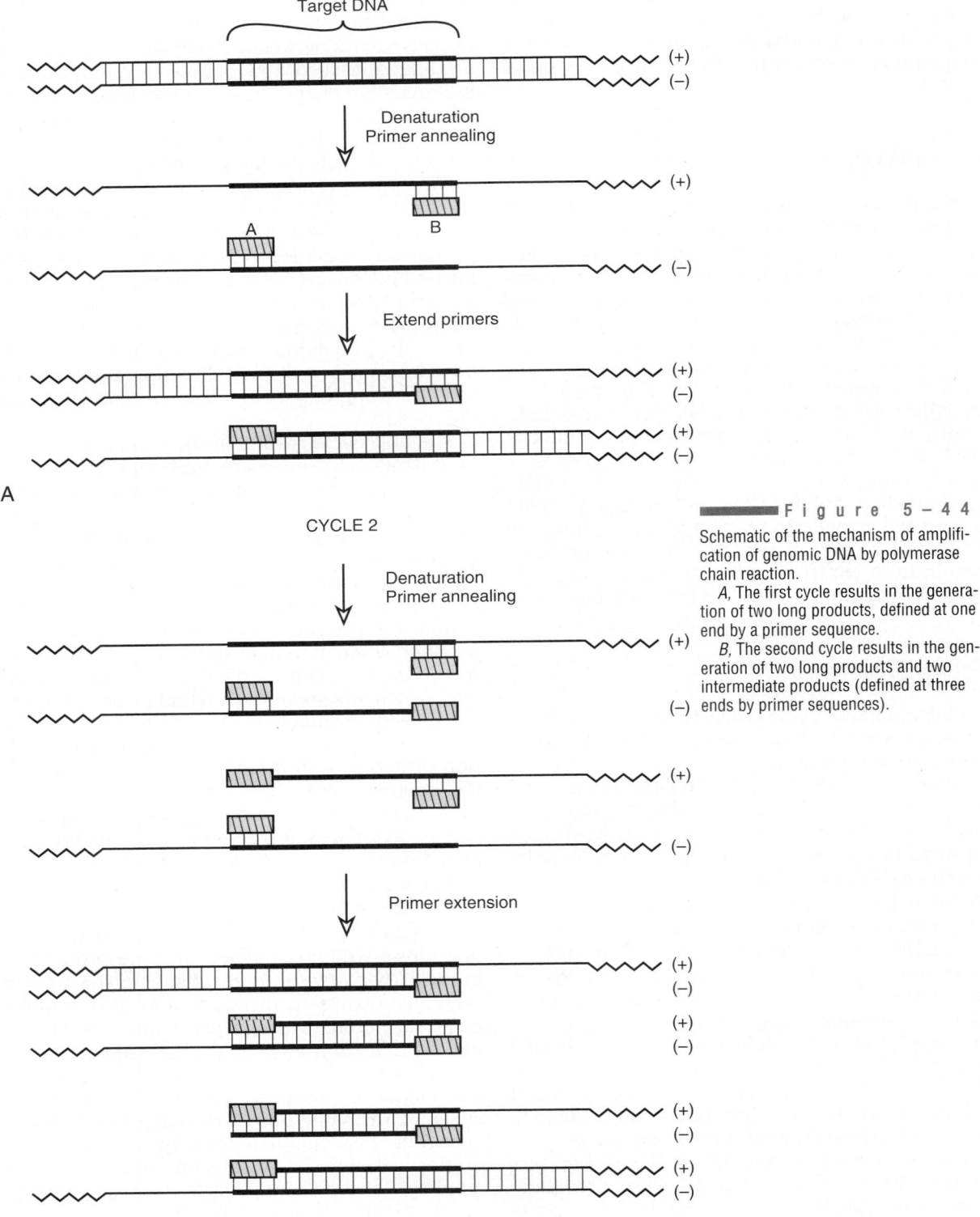

CYCLE 1

Target DNA

Denaturation
Primer annealing

Extend primers

A

CYCLE 2

Denaturation
Primer annealing

Primer extension

B

Figure 5−44
Schematic of the mechanism of amplification of genomic DNA by polymerase
chain reaction.
  A, The first cycle results in the generation of two long products, defined at one
end by a primer sequence.
  B, The second cycle results in the generation of two long products and two
intermediate products (defined at three
ends by primer sequences).

CYCLE 3

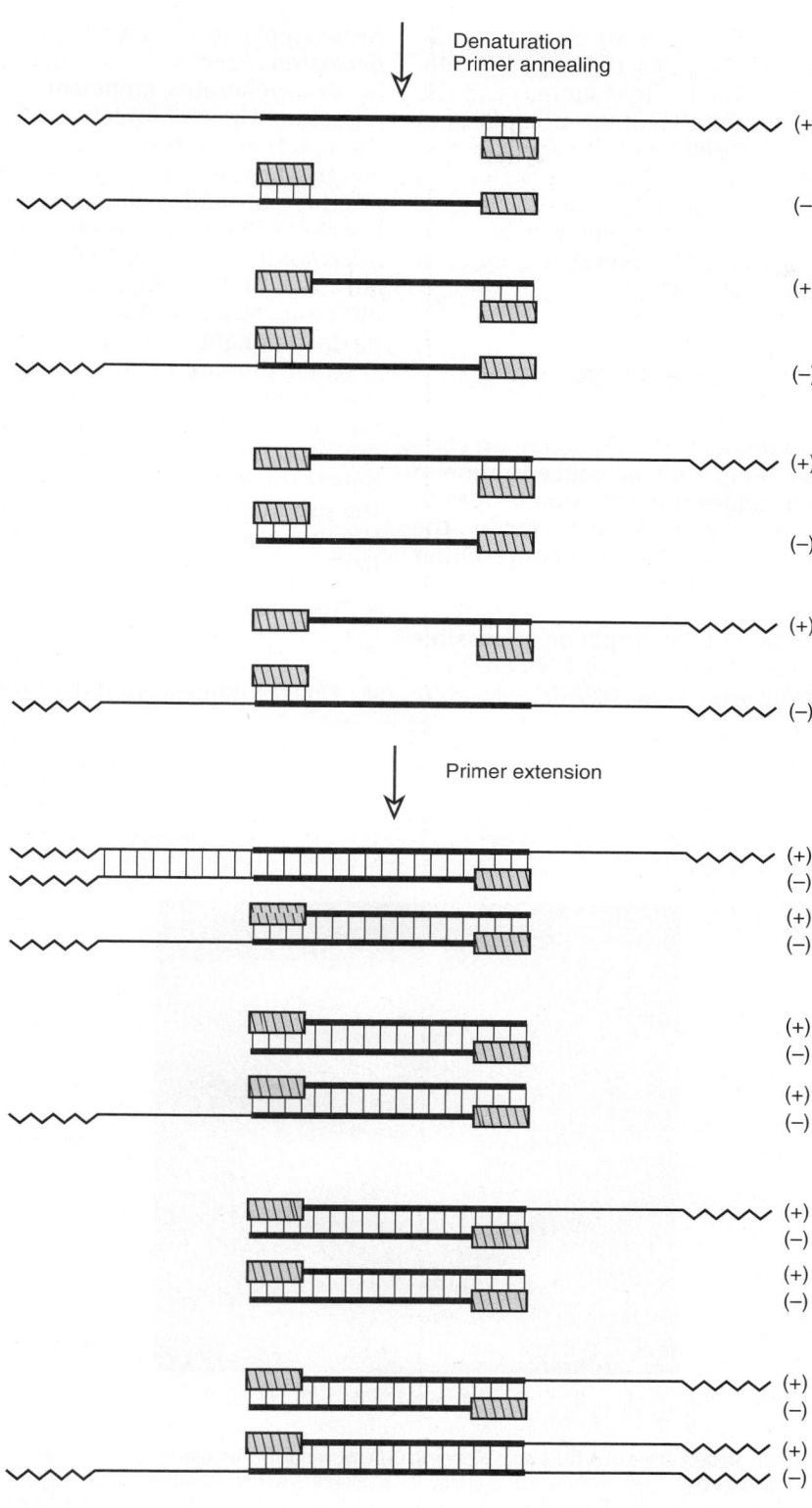

C

Figure 5-44 Continued

C, The third cycle results in the generation of two short products defined at all four ends by primer sequences. (Redrawn by permission of the New England Journal of Medicine, 322:179, 1990.)

amplicon doubles after each successive cycle. An amplification of 25 cycles that is 100% efficient results in a 3.35×10^7 fold increase of the target sequence. No amplification is capable of achieving this level of efficiency, however, owing to primer and enzyme exhaustion, a plateau of efficiency is reached after 20 to 25 cycles. Thereafter, each successive cycle will be even less efficient. As a result, the overall efficiency of the procedure is usually 60 to 85%.

DETECTION OF THE AMPLICON

It is essential to detect the amplicon upon completion of the reaction. The final concentration of the amplicon determines the detection format. In some applications it is possible to detect the product directly from the reaction mixture. Other applications require the use of DNA probe technology.

Direct visualization of the amplicon is possible when multiple copies of the target DNA are present in the cell from which the DNA is extracted.

An example is the HLA-DQα locus. Because all nucleated blood cells contain this gene, the amplification generates sufficient material for direct detection. Upon completion of PCR, an aliquot of the reaction mixture is resolved by agarose gel electrophoresis. The gel is then stained with ethidium bromide, which becomes trapped within a dsDNA molecule through a process known as *intercholation*. As seen in Figure 5–45, direct visualization of the amplicon is possible because ethidium bromide fluoresces upon exposure to ultraviolet light.

When the cell contains only a single copy of the target—as is the case with the human immunodeficiency virus (HIV) provirus—ethidium bromide staining is not sensitive enough to detect the PCR product. Specific identification of the product in this case is possible only through the use of standard probing techniques that utilize one of the following formats:

- Radioisotopic detection
- Colorimetric detection
- Chemiluminescent detection.

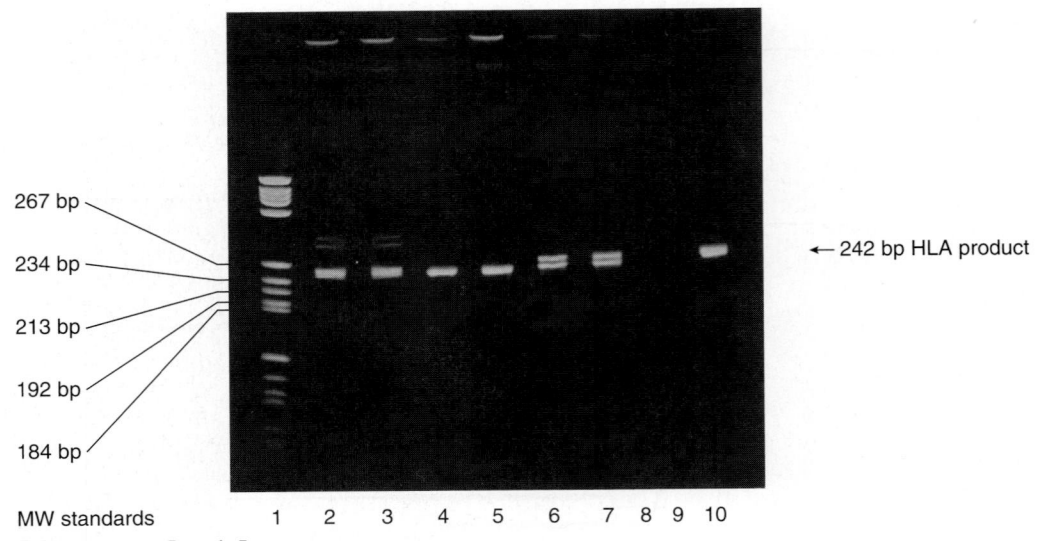

267 bp
234 bp
213 bp
192 bp
184 bp

MW standards   1 2 3 4 5 6 7 8 9 10

← 242 bp HLA product

Figure 5 – 4 5

Detection of the 242 bp PCR product of the HLA-DQα locus. Genomic DNA was amplified through the use of primers specific for the HLA-DQα locus (GH26 and GH27; Jackson et al, 1990). Following amplification, the amplicons were resolved on 5% polyacrylamide gel electrophoresis. After staining of DNA with ethidium bromide, the product was visualized by ultraviolet light on a transilluminator. Lane 1: Molecular weight standards. Lanes 2–10: HLA products of genomic DNA preparations. Lanes 2–8, 10: Positive for HLA-DQα. Lane 9: Negative for HLA-DQα.

Radioisotopic Detection

Liquid Hybridization

The most commonly used radioisotopic detection format is liquid hybridization. This technique uses an oligonucleotide *probe* that is complementary to a 35- to 45-base region within the target DNA. T4 polynucleotide kinase is used to label the 5′ end of this probe with radioactive phosphorus 32 (^{32}P). Hybridization of the probe with the amplicon occurs in the liquid phase. Separation of the hybridization product from unhybridized probe occurs on a polyacrylamide gel, as shown in Figure 5–46. The gel is dried and exposed to a piece of x-ray film. The high-energy emissions of the radioisotope that expose the x-ray film result in a picture of the hybridization product. This procedure is known as *autoradiography*. Only hybridization products that contain the radiolabeled probe will generate the characteristic pattern seen in Figure 5–47.

Solid-Phase Hybridization

An alternative to liquid hybridization is solid-phase hybridization. This technique employs a solid support such as nylon membrane or nitrocellulose. As with liquid hybridization, this technique uses a probe labeled at its 5′ end with ^{32}P. Examples of this format are the Southern blot, dot blot, and slot blot techniques.

In the Southern blot technique, the amplicon is first resolved by agarose gel electrophoresis. It is then denatured and transferred by vacuum or capillary action to a piece of nylon membrane. The membrane serves to immobilize the denatured amplicon and provides a surface for hybridization with the ^{32}P-labeled probe. Again, the hybridization product is visualized through autoradiography.

Dot blot and slot blot techniques are similar to the Southern blot. In these techniques, however, the amplicon is denatured and then applied directly to the nylon membrane. Hybridization occurs as before on the surface of the membrane. Figure 5–48 shows an example of a dot blot procedure.

Colorimetric Detection

Some laboratories use alternative detection formats to avoid the use of radioisotopes. They use traditional Southern blot or dot blot methodology with colorimetric detection formats. These formats use probes labeled with alkaline phosphatase. Hybridization results in the immobilization of this enzyme on the surface of the nylon membrane. An essential component of this technique is nitroblue tetrazolium (NBT), a chromogenic substrate for alkaline phosphatase. Following immobilization of the enzyme on the nylon membrane, addition of NBT causes an insoluble precipitate to form. An observable color reaction results.

Enzyme-Linked Oligonucleotide Sorbent Assay

An alternative colorimetric method is referred to as *enzyme-linked oligonucleotide sorbent assay* (ELOSA). This methodology is similar to the ELISA technique described earlier in this chapter. The difference is that the target for detection is the amplicon, not an antigen or antibody. This procedure uses biotin-labeled primers in the amplification of target DNA. The amplicons resulting from this reaction will therefore contain biotin.

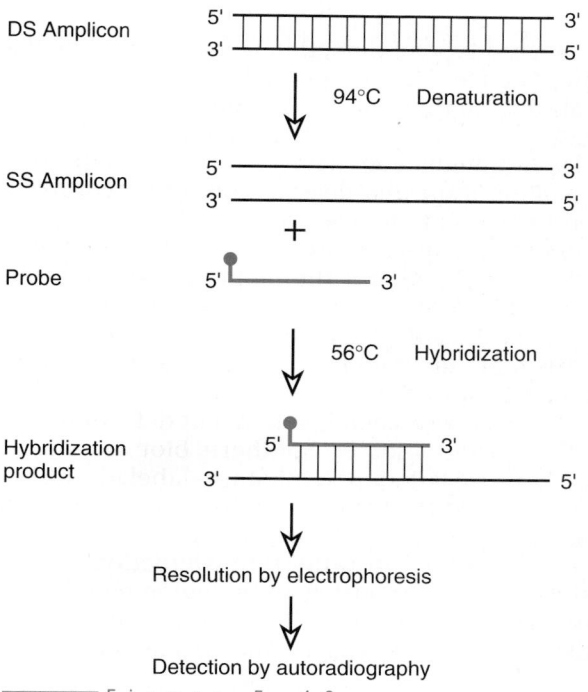

DS Amplicon

5′
3′

94°C   Denaturation

SS Amplicon

5′   3′
3′   5′

+

Probe

5′ ⌐_____ 3′

56°C   Hybridization

Hybridization product

5′   3′
3′   5′

Resolution by electrophoresis

Detection by autoradiography

Figure 5–46

Schematic diagram of the reaction steps required for detection of the PCR product by liquid hybridization.

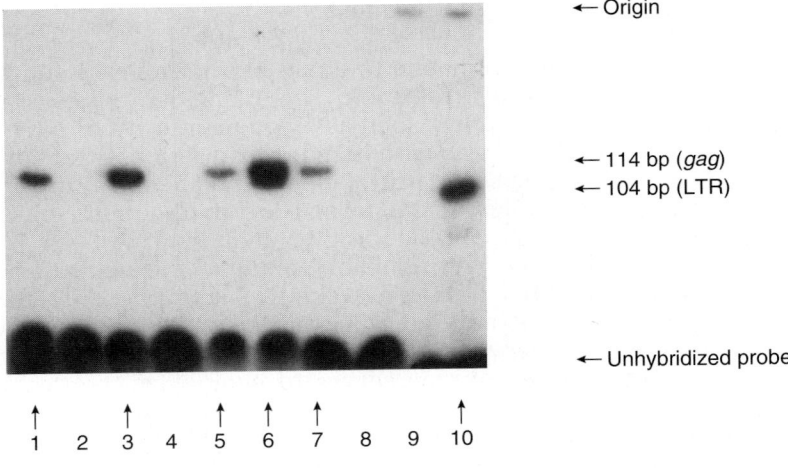

■■■■■ F i g u r e   5 – 4 7

Detection of HIV-1 amplified proviral DNA by liquid hybridization. This autoradiograph shows the hybridization products characteristic of a positive PCR reaction. Lanes 1–8: 114 bp product generated with *gag* (SK38/SK39) primers. Lanes 9 & 10: 104 bp product generated with LTR (SK29/SK30) primers. lanes 1, 3, 5–7, 10: Specimens from HIV-seropositive individuals. Lanes 2, 4, 8, 9: Specimens from HIV-seronegative individuals. Unhybridized probe is visualized at the bottom of the gel in each lane.

Following denaturation, the biotinylated amplicon is added to a microtiter plate coated with a detector probe. The ensuing hybridization results in the immobilization of the hybridization product in the walls of the microtiter wells. After washing to remove free biotin-labeled primers, the next step is the addition of alkaline phosphatase conjugated–avidin molecules. Since avidin has a high affinity for biotin, a complex forms that immobilizes alkaline phosphatase in the microtiter well. As before,

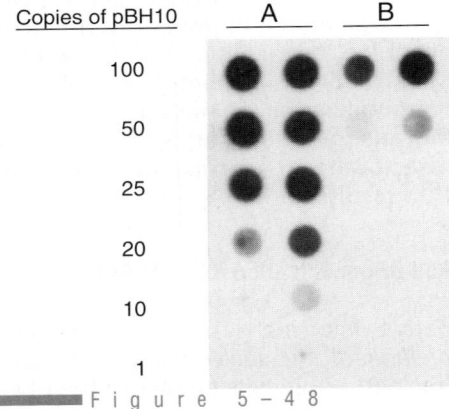

■■■■■ F i g u r e   5 – 4 8

Dot blot analysis of serial dilutions of cloned HIV-1 template. Amplicons generated through the use of *gag* or LTR (Ou et al, 1988)-specific primers were detected with ^{32}P-labeled probes following hybridization on nylon membrane and autoradiography. *A, gag;* 10 copies of pBH10 (Hahn, 1984) detected. *B*, LTR: 20 copies of pBH10 detected.

color development results after addition of NBT. This is indicative of a positive result.

Chemiluminescence

A third type of detection format uses chemiluminescence. This is a procedure in which chemical degradation of an acridinium ester (AE) results in the emission of a detectable light particle. A luminometer detects the light emissions resulting from this degradation. The hydrolysis protection assay uses an AE-labeled oligonucleotide probe. Hybridization of the single-stranded probe and the amplicon protects the AE group from subsequent alkaline hydrolysis. The protected AE group then generates a detectable chemiluminescent signal upon addition of an appropriate substrate.

Alternatively, chemiluminescent detection may utilize a dot blot or Southern blot technique with an alkaline phosphatase–labeled probe. Following solid-phase hybridization, a chemiluminescent signal occurs upon addition of a chemiluminescent substrate. Detection of the signal follows exposure of the nylon membrane to x-ray film and detection by autoradiography. Exposure of the film is due to chemiluminescent emission of light particles, whereas this is not true with radioisotopes.

CLINICAL APPLICATIONS OF PCR

The detection of many pathogens is possible through the use of PCR (Table 5–21). Because sequence information is readily available, the design of specific primers and probes is possible. Table 5–21 presents a brief summary of documented PCR applications to organisms of clinical importance. The bibliography at the end of this chapter contains the references for these applications.

The choice of which application a clinical laboratory adapts for use must be carefully considered. One of the following criteria must be met:

■ **PCR would supplement existing technology:** In many cases, current culture techniques are

sufficient to identify a bacterial pathogen. PCR, however, can be used to enhance the diagnosis. An example is the use of PCR in the detection of enterotoxigenic strains of *E. coli,* which contain a characteristic heat-labile toxin. PCR primers that flank this gene permit a specific detection of this organism. Related organisms that lack this gene are not detected.

■ **PCR would replace existing technology:** An example would be in the detection of cytomegalovirus (CMV). CMV requires a lengthy culture period, up to 4 weeks, before a negative result may be reported. PCR generates a result in 1 to 2 days.

■ **PCR represents new diagnostic technology:** PCR is of most benefit in the diagnosis of fastidious organisms that cannot be detected by conventional culture techniques, such as human papillomavirus.

Technical Considerations for Implementation of PCR

The following features are of importance in the implementation of a PCR application in clinical microbiology.

Determination of the Appropriate Sample Source

Careful selection of the type of specimen to be collected is essential. Typically, the specimen requirement for PCR is the same as that for culture.

Construction of Primers and Probes

Thorough DNA sequencing will show the most conserved regions of the genome. Primers designed to these areas are necessary to achieve a high degree of sensitivity and specificity.

Optimization of Amplification Conditions

There is a complex interaction of temperature settings, primer and enzyme concentrations, and ionic strength in this procedure. The most important component is magnesium concentration. The optimal activity of *Taq* DNA polymerases depends on magnesium. If the magnesium concentration is too low, the enzyme does not function. If the magnesium concentration is too high,

TABLE 5–21

CLINICALLY IMPORTANT PATHOGENS THAT HAVE BEEN IDENTIFIED THROUGH THE USE OF PCR

Viruses	Parvovirus
	Herpes simplex virus
	Rotavirus
	Papillomavirus
	Human herpesvirus 6
	Dengue virus
	Varicella-zoster virus
	Rubella virus
	Adenovirus
	Rhinovirus
	Hepatitis B virus
Bacteria	*Borrelia burgdorferi*
	Vibrio cholerae
	Helicobacter pylori
	Rickettsia tsutsugamushi
	Staphylococcus aureus
	Chlamydia trachomatis
	Shigella dysenteriae
	Chlamydia psittaci
	Treponema pallidum
	Enterotoxigenic *Escherichia coli*
	Mycobacterium tuberculosis
	Legionella pneumophila
Parasites	*Plasmodium falciparum*
	Toxoplasma gondii
	Entamoeba histolytica
	Pneumocystis carinii
	Trypanosoma cruzi

the enzyme is overly active, leading to the generation of nonspecific amplification products. An imbalance in any of the other parameters also leads to the generation of excessive amounts of primer-dimers and nonspecific amplification products, ultimately resulting in a decrease in efficiency.

Choice of Detection Format

The use of colorimetric or chemiluminescent detection formats is preferable for routine use in the clinical laboratory. Although radioisotopic detection formats are the most sensitive, they also present a health risk and a disposal problem. The nonisotopic formats previously described show promise for use in the clinical laboratory. These techniques compare favorably in sensitivity with radioisotopes and are safer to work with.

Interpretation of Test Results

Before adoption into the clinical laboratory, the clinical usefulness of PCR must be well established. Establishment of positive and negative predictive values is required. Because the technique does not depend on viable organisms to achieve a positive result, the detection of specific DNA sequences in a patient may or may not be clinically significant.

Quality Assurance and Quality Control

PCR must be monitored by an appropriate quality assurance program. Each test should include positive, negative, and reagent controls. In addition, each laboratory should initiate a procedure to validate the performance of new lots of reagents, including the primers and probes.

It is also essential to demonstrate the integrity of a DNA sample prior to amplification. Specimen integrity is confirmed through the amplification of a gene sequence normally found in all individuals regardless of their infection status. If the DNA sample is excessively damaged during processing or if it contains an inhibitor to *Taq* DNA polymerase (such as hemoglobin), a false-negative reaction could occur.

The detection of human immunodeficiency virus (HIV) proviral DNA by PCR involves the amplification of genomic DNA. Amplification of the HLA-DQα locus assesses the integrity of the DNA sample. All samples, regardless of their HIV infection status, have a positive reaction for the HLA primers.

To illustrate, assume that a sample contains an inhibitor to *Taq* DNA polymerase. This specimen will test negative for the HLA primers. The inhibitor will also interfere with the amplification of HIV-specific primers and lead to a false-negative result. In this scenario, the negative HIV result is meaningless, and the test must be repeated on a new sample. An example of negative HLA test is shown in Figure 5–45. Lane 9 contains DNA contaminated with excessive amounts of red blood cells during the cell lysis procedure. The remaining lanes show DNA preparations of high integrity. The negative HLA result was a result of inhibition of *Taq* DNA polymerase activity by the hemoglobin released during the cell lysis procedure.

Setting Up a PCR Laboratory

The PCR laboratory should consist of three distinct work areas (Fig. 5–49). In order to avoid the contamination problems mentioned earlier, each area should be dedicated to a single procedure. Specimen preparation occurs in the first area, reagent preparation and PCR set-up in the second area, and amplification and detection in the third area.

The entire procedure can be performed in a single room if proper precautions are taken. The following practices will diminish the potential for contamination.

■ Each area should have dedicated supplies and reagents (see Fig. 5–49)

■ Color coding of reagents and supplies identifies those that belong to a particular area

■ Reagents, supplies, and equipment should never be taken from one area to another; three sets of pipettors are therefore essential

■ The workflow must be unidirectional from "clean" (pre-PCR) to "dirty" (post-PCR), as shown in Figure 5–49.

■ Dedicated labcoats and gloves should be worn at each worksite; when moving to a new area, workers should put on new gloves and labcoats.

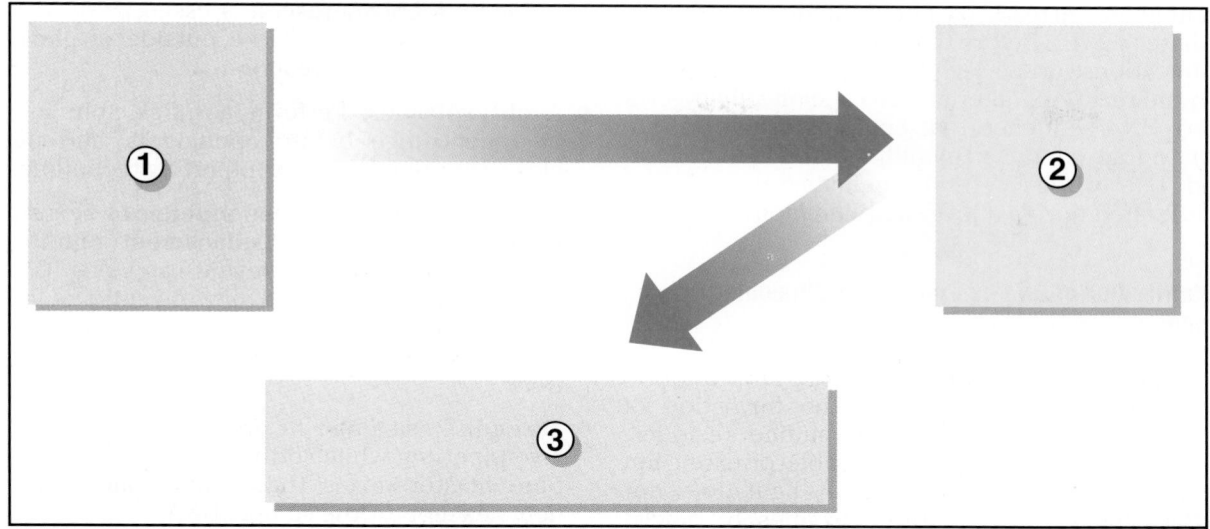

■ F i g u r e   5 – 4 9

Unidirectional workflow in a PCR laboratory. Dedicated equipment and reagent requirements in a PCR laboratory are as follows:

Area 1: Sample preparation:

■ Positive-displacement pipettes or pipettors with aerosol-resistant tips

■ Gloves and laboratory coat

■ Refrigerator, freezer, water bath or dry-heat block, laminar flow biosafety cabinet

■ Cell lysis reagents

Area 2: Reagent preparation and PCR set-up:

■ Amplification reagents and supplies

■ Positive-displacement pipettes or pipettors with aerosol-resistant tips

■ Laminar-flow biosafety cabinet or dead air box

■ Gloves and laboratory coat

■ Refrigerator and freezer

■ Water bath or dry-heat block

Area 3: Amplification and detection:

■ Thermal cycler

■ Pipettors with aerosol-resistant tips

■ Detection equipment (electrophoresis unit, incubator, plate washer, plate reader, water bath)

■ Refrigerator and freezer

■ Reagents and supplies for detection

■ Pre-PCR reagents should be stored separately from post-PCR reagents.

Specimen Carryover: The Major Pitfall of PCR

Careful attention to technique is essential to performing PCR. Because of the high sensitivity of the procedure, the introduction of even a minute amount of a positive template into a negative specimen can lead to the generation of a false-positive result. An amplification consisting of 25 cycles yields up to 10^{13} copies/mL of the amplicon. A 0.1-μL aliquot of this mixture contains 10^9 amplicons. By comparison, there are only 1.4×10^5 copies of a single-copy gene in a microgram of human DNA. As a result, such contamination results in a false-positive reaction. It is not surprising that the biggest problem associated with the successful use of PCR is this potential for specimen carryover.

This carryover can be controlled through a number of different techniques, as follows.

Biochemical Sterilization of Previously Amplified Material

Commercial kits are now available (Carryover Prevention Kit, Cetus Corp., Norwalk, CT) that cause the enzymatic degradation of previously amplified material. The methodology incorporates deoxyuridine triphosphate (dUTP) into the amplicon instead of deoxythymidine triphosphate (dTTP). Before the start of subsequent amplifications, uracil-*N*-glycosylase (UNG) is added to the reaction material. UNG specifically digests any DNA sequence that contains uracil. A template that contains thymidine will not be affected. Use of this method eliminates any product generated during a prior amplification.

Sterilization of DNA by Exposure to Ultraviolet Light

The exposure of double-stranded DNA to ultraviolet (UV) light results in the formation of dimers between adjacent thymidine residues. These altered bases are incapable of extension by the enzyme. Exposure to UV light does not affect the action of either enzymes or single-stranded primers. It is therefore possible to treat all reagent mixtures with UV light prior to the addition of template DNA in order to eliminate any contaminating DNA. It is also preferable to treat the work area with UV light.

Strict Adherence to Proper Laboratory Technique

It is essential to prevent the introduction of previously amplified material into future reaction mixtures. To this end, the following suggestions have been made.

■ **Physically isolate PCR preparations and products:** A separate clean room and biological safety cabinet are required for set-up of the reaction. Processing of amplified material must be performed in a totally different work area.

■ **Autoclave solutions:** Autoclaving degrades DNA into fragments of low molecular weight.

■ **Aliquot reagents:** Do not repeatedly open containers of buffers, enzymes, nucleotides, or primer stocks. Instead, prepare single-use aliquots and discard any remaining material after set-up of the reaction.

■ **Use disposable gloves and change gloves often during set-up:** DNA may splash onto gloves when the tops of microcentrifuge tubes are opened. Frequent changing of gloves reduces the possibility of transferring DNA between specimens. The powder that is used in latex gloves also interferes with the action

of *Taq* DNA polymerase. It is essential to wash any excess powder off the outside of gloves prior to setting up a reaction.

■ **Avoid splashes:** Perform a quick spin in a microcentrifuge before opening it, and pay close attention to careful pipetting techniques.

■ **Use positive-displacement pipettes or aerosol-resistant tips on air-displacement pipettes:** These are essential to prevent carryover. DNA contaminates the barrel of conventional air-displacement pipettes. The contaminating DNA is easily transferred to subsequent reactions.

■ **"Premix" reagents:** In order to avoid excessive pipetting while setting up a reaction, prepare master mixes that contain all components except for the sample DNA.

■ **Add DNA last:** Adding the DNA as the last step decreases the chance of carryover.

■ **Choose positive and negative controls carefully:** Positive controls that contain large amounts of template DNA should be avoided, because their use would increase the likelihood of cross-contamination. Instead, use the lowest dilutions that generate a positive result.

SUMMARY

Polymerase chain reaction is primarily a tool of the research laboratory. With the development of UNG to eliminate carryover, the integration of the technology into the clinical laboratory is now feasible. The UNG technique is an effective means of controlling most instances of contamination. The availability of nonisotopic detection formats also enhances the potential for introduction of PCR into the clinical laboratory. Finally, the ease with which the technique is automated permits the development of commercial kits.

It is important to realize that the use of UNG does not totally prevent the threat of carryover. Sporadic cases of carryover still occur, in which positive DNA is accidentally introduced into adjacent tubes. Even though UNG treatment cannot prevent these instances of sporadic contamination, it is possible to control for such occurrences. Performing each amplification in duplicate with careful attention to technique reveals a false-positive reaction due to carryover. A true-positive result occurs in both amplifications, and a true-negative result consists of no reaction in either amplification. A sample that yields a positive result in one reaction but a negative result in another indicates sporadic carryover. The test needs to be repeated in duplicate before a result is reported.

The chance of sporadic carryover can be controlled but not totally eliminated. It will be important, therefore, for the clinical laboratory to pay careful attention to this crucial problem before adopting the polymerase chain reaction for clinical use.

Bibliography

Allard U, Girones R, et al: Polymerase chain reaction for detection of adenoviruses in stool samples. J Clin Microbiol 28:2659, 1990.

Clewley JP: Polymerase chain reaction assay of Parvovirus B19 DNA in clinical specimens. J Clin Microbiol 27:2647, 1989.

Conway B, Adler KE, et al: Detection of HIV-1 DNA in crude cell lysates of peripheral blood mononuclear cells by the polymerase chain reaction and nonradioactive oligonucleotide probes. J Acquir Immune Defic Syndr 3:1059, 1990.

de Lomas JG, Sunzeri FJ, Busch MP: False-negative results by polymerase chain reaction due to contamination by glove powder. Transfusion 32:83, 1992.

Eckert KA, Kunkel TA: DNA polymerase fidelity and the polymerase chain reaction. PCR Methods Appl 1(1):17, 1991.

Eggerding FA, Peters J, Lee RK, Inderlied CB: Detection of rubella virus gene sequences by enzymatic amplification and direct sequencing of amplified DNA. J Clin Microbiol 29:945, 1991.

Furuya Y, Yoshida Y, et al: Specific amplification of Rickettsia tsutsugamushi DNA from clinical specimens by polymerase chain reaction. J Clin Microbiol 29:2628, 1991.

Gama RE, Horsnell PR, Hughes PJ, et al: Amplification of rhinovirus specific nucleic acids from clinical samples using the polymerase chain reaction. J Med Virol 28:73, 1989.

Gelfand DH, White TJ: Thermostable DNA polymerases. In Innis MA, Gelfand DH, Sninsky JJ, White TJ (eds): PCR Methods Appl. San Diego, CA: Academic Press, 1990.

Hahn BH, Shaw GM, et al: Molecular cloning and characterization of the HTLV-III virus associated with AIDS. Nature 312:166, 1984.

Ho S, Hoyle JA, et al: Direct polymerase chain reaction test for detection of Helicobacter pylori in humans and animals. J Clin Microbiol 29:2543, 1991.

Innis MA, Gelfand DH: Optimization of PCRs. In Innis MA, Gelfand DH, Sninsky JJ, White TJ (eds): PCR Protocols: A Guide to Methods and Applications. San Diego, CA: Academic Press, 1990.

Jackson MP: Detection of Shiga toxin-producing Shigella dysenteriae type 1 and Escherichia coli by using polymerase chain reaction with incorporation of digoxigenin-11-dUTP. J Clin Microbiol 29:1910, 1991.

Jackson JB, MacDonald KL, et al: Absence of HIV infection in blood donors with indeterminate Western blot tests for antibody to HIV-1. N Engl J Med 322:217, 1990.

Johnson WM, Tyler SD, et al: Detection of genes for enterotoxins, exfoliative toxins, and toxic shock syndrome toxin 1 in Staphylococcus aureus by the polymerase chain reaction. J Clin Microbiol 29:426, 1991.

Kain K, Lanar DE: Determination of genetic variation with Plasmodium falciparum by using enzymatically amplified DNA from filter paper disks impregnated with whole blood. J Clin Microbiol 29:1171, 1991.

Kaltenboeck B, Kousoulas KG, Storz J: Detection and strain differentiation of Chlamydia psittaci mediated by a two-step polymerase chain reaction. J Clin Microbiol 29:1969, 1991.

Kaneko S, Feinstone S, Miller RH: Rapid and sensitive method for the detection of serum hepatitis B virus DNA using the polymerase chain reaction technique. J Clin Microbiol 27:1930, 1989.

Kitada K, Oka S, et al: Detection of Pneumocystis carinii sequences by polymerase chain reaction: Animal models and clinical application to noninvasive specimens. J Clin Microbiol 29:1985, 1991.

Kondo K, Hayakawa Y, et al: Detection by polymerase chain reaction amplification of human herpesvirus 6 DNA in peripheral blood of patients with exanthem subitum. J Clin Microbiol 28:970, 1990.

Kwok S, Higuchi R: Avoiding false positives with PCR. Nature 339:237, 1989.

Ling LL, Keohavong P, et al: Optimization of the polymerase chain reaction with regard to fidelity: Modified T7, Taq, and Vent DNA polymerases. PCR Methods Appl 1:63, 1991.

Malloy DC, Nauman R, Paxton H: Detection of Borrelia burgdorferi using the polymerase chain reaction. J Clin Microbiol 28:1089, 1990.

Morita K, Tanaka M, Igarashi A: Rapid identification of dengue virus serotypes by using polymerase chain reaction. J Clin Microbiol 29:2107, 1991.

Moser DR, Kirchhoff LV, Donelson JE: Detection of Trypanosoma cruzi by DNA amplification using the polymerase chain reaction. J Clin Microbiol 27:1477, 1989.

Mull LM: Current status of polymerase chain reaction assays in clinical research of human immunodeficiency virus infections. AIDS Updates 3:1, 1990.

Mullis K, Faloona F, et al: Specific enzymatic amplification of DNA in vitro: The polymerase chain reaction. Cold Spring Harbor Symp Quantit Biol 51:263, 1986.

Mullis KB: The polymerase chain reaction in an anemic mode: How to avoid cold oligodeoxyribonuclear fusion. PCR Methods Appl 1:1, 1991.

Noordhoek G, Wolters E, et al: Detection by polymerase chain reaction of treponema pallidum DNA in cerebrospinal fluid from neurosyphilis patients before and after antibiotic treatment. J Clin Microbiol 29:1976, 1991.

Ochman H, Ajioka JW, et al: Inverse polymerase chain reaction. In Ehrlich HA (ed): PCR Technology: Principles and Applications for DNA Amplification. New York: Stockton Press, 1989.

Olive DM: Detection of enterotoxigenic Escherichia coli after polymerase chain reaction amplification with a thermostable DNA polymerase. J Clin Microbiol 27:261, 1989.

Olive DM, Simsek M, Al-Mufti S: Polymerase chain reaction for detection of human cytomegalovirus. J Clin Microbiol 27:1238, 1989.

Ostergaard L, Birkelund S, Christiansen G: Use of polymerase chain reaction for detection of Chlamydia trachomatis. J Clin Microbiol 28:1254, 1990.

Ou CY, McDonough SH, et al: Rapid and quantitative detection of enzymatically amplified HIV-1 using chemiluminescent oligonucleotide probes. AIDS Res Hum Retroviruses 6:1323, 1990.

Patel R, Fries J, et al: Sequence analysis and amplification by polymerase chain reaction of a cloned DNA fragment for identification of *Mycobacterium tuberculosis*. J Clin Microbiol 28:513, 1990.

Peter JB: The polymerase chain reaction: Amplifying our options. Rev Infect Dis 13:166, 1991.

Puchhammer-Stockl E, Popow-Kraupp T, et al: Detection of varicella-zoster DNA by polymerase chain reaction in the cerebrospinal fluid of patients suffering from neurological complications associated with chicken pox or herpes zoster. J Clin Microbiol 29:1513, 1991.

Rozenberg F, Lebon P: Amplification and characterization of herpesvirus DNA in cerebrospinal fluid from patients with acute encephalitis. J Clin Microbiol 29:2412, 1991.

Saiki RK, Gelfand DH , et al: Primer-directed enzymatic amplification of DNA with a thermostable DNA polymerase. Science 239:487, 1988.

Shibata DK, Arnheim N, Martin WJ: Detection of human papilloma virus in paraffin-embedded tissue using the polymerase chain reaction. J Exp Med 167:225, 1988.

Shirai H, Nishibuchi M, et al: Polymerase chain reaction for detection of the cholera enterotoxin operon of *Vibrio cholerae*. J Clin Microbiol 29:2517, 1991.

Starnbach MN, Falkow S, Tompkins LS: Species-specific detection of *Legionella pneumophila* in water by DNA amplification and hybridization. J Clin Microbiol 27:1257, 1989.

Tachibana H, Ihara S, et al: Differences in genomic DNA sequences between pathogenic and nonpathogenic isolates of *Entamoeba histolytica* identified by polymerase chain reaction. J Clin Microbiol 29:2234, 1991.

Van De Ven E, Melchers W, et al: Identification of *Toxoplasma gondii* infections by BI gene amplification. J Clin Microbiol 29:2120, 1991.

Wilde J, Eiden J, Yolken R: Removal of inhibitory substances from human fecal specimens for detection of group A rotaviruses by reverse transcriptase and polymerase chain reaction. J Clin Microbiol 28:1300, 1990.

CHAPTER 6

Host-Parasite Interaction

A. NORMAL MICROBIAL FLORA

Hal S. Larsen

1. Define the following terms: *parasitism, normal* or *indigenous flora, commensal, symbiont, opportunist, resident flora, transient flora,* and *carrier.*

2. Describe some factors that determine the nature of the normal flora at various body sites.

3. List the predominant flora of various body sites in a healthy individual.

4. Describe the role of the normal microbial flora in the pathogenesis of infectious disease.

5. Describe the role of the normal flora in host defense against infectious diseases.

6. Describe how the human host protects itself from microbial invasion by the following:

   ■ Epithelial linings in the different organ systems.

   ■ Secretion of fluids that contain antibacterial substances.

   ■ Phagocytes

INTRODUCTION

Origin of Normal Flora

The fetus is in a sterile environment until birth. During the first few days of life, the newborn is introduced to the many and varied microorganisms present in the environment. Each organism has the opportunity to find an area on or in the infant that it is adapted to. Those that find their niche colonize various anatomic sites and become the predominant organisms. Others are transient or fail to establish themselves at all. Within a short time following birth, the infant's microbial flora is similar to that of other individuals.

Characteristics of Normal Microbial Flora

Microorganisms that are commonly found on or in body sites of healthy persons are termed *normal flora.* The different body sites may have the same or different normal flora, depending upon local conditions. It is important to remember that *normal flora* may be a very nebulous term, because what may be normal in one individual is pathogenic in another, especially with the use of immunosuppressants or in debilitating disease. Local conditions select for those organisms that are suited for growth in a particular area. For example, the environment found on the skin surface is different from that found in the mouth.

The microorganisms that colonize an area for months or years at a time are termed *resident flora.* This is in contrast to *transient flora,* which are present at a site temporarily. Transient flora come to visit but not usually to live or stay. Organisms that constitute the normal flora can be classified as *parasites* (live at the expense of the host), *symbionts* (benefit the host), or *commensals* (have neutral effect on the host).

Some pathogenic organisms may establish themselves in a host without manifesting symptoms. These hosts are capable of transmitting the infection. They are termed *carriers,* and the condition is called the *carrier state.* The carrier state may be acute (short in duration) or chronic (lasting for months, years, or permanently). An example of a chronic carrier state is found in *Salmonella typhi* infection. This organism may establish itself in the bile duct and be excreted in the stool over a period of years. In contrast, *Neisseria meningitidis* may be found in the throat of asymptomatic individuals during an outbreak of meningitis. After a few days or weeks at most, these individuals no longer harbor the organism, so the carrier state was acute. The most transient of carrier states is inoculation of a person's hands or fingers with an organism (e.g., *Staphylococcus aureus* from a wound) that is carried only until the hands are washed.

A number of factors determine which microorganisms are able to colonize the various body sites: the nutritional status of the site, oxidation-reduction potentials, antibody and other antibacterial substances, pH, and interference by already established organisms. These conditions may change with age, nutritional status, disease states, and drug or antibiotic effects. The changes may predispose an individual to infection by normal flora (i.e., opportunistic infection). For example, two groups at increased risk for gram-negative rod pneumonia are diabetics and alcoholics. Antibiotics may reduce a particular population of bacteria (e.g., gram-positive), allowing the proliferation of other organisms such as *Candida albicans.* An increase in age brings with it a decrease in the effectiveness of the immune response. As a result, the incidence of infection by opportunistic organisms increases.

NORMAL FLORA AT DIFFERENT BODY SITES

The human host is colonized by approximately 100 different species of microorganisms. The effectiveness of the various host defenses is evidenced by the relatively low incidence of infection in uncompromised individuals by members of the normal flora. The fact remains, however, that infections caused by normal flora are commonly seen in the clinical microbiology laboratory. It is helpful, if not essential, that the medical technologist have an understanding of the types of microorganisms found at the various body sites.

Normal Flora of the Skin

The skin contains a wide variety of microorganisms. Most of these are found on the most superficial layers of cells and the upper parts of hair follicles. Some, such as *Propionibacterium acnes,* colonize the deep sebaceous glands. Superficial antisepsis of the skin does not eliminate this

TABLE 6-1
MICROORGANISMS FOUND ON THE SKIN

Common	Less Common
Candida spp	*Streptococcus* spp
Micrococcus spp	*Acinetobacter calcoaceticus*
Staphylococcus spp	*Bacteroides* spp
Clostridium spp	Gram-negative rods (fermenters and
Lactobacillus	nonfermenters)
Diphtheroids	*Moraxella* spp

TABLE 6-2
MICROORGANISMS FOUND IN THE MOUTH

Common	Less Common
Staphylococcus epidermidis	*Staphylococcus aureus*
Streptococcus mitis	*Enterococcus*
Streptococcus sanguis	*Eikenella corrodens*
Streptococcus salivarius	*Fusobacterium nucleatum*
Streptococcus mutans	*Candida albicans*
Peptostreptococcus	
Veillonella	
Lactobacillus	
Actinomyces israelii	
Bacteroides spp	
Prevotella/Porphyromonas	
Bacteroides oralis	
Treponema denticola	
Treponema refringens	

organism, which may be found as a contaminant in those culture specimens that require invasive procedures (e.g., blood, cerebrospinal fluid), as a result of contamination of the needle. More organisms are found in moist areas than in dry areas.

Normal skin has a number of mechanisms to prevent infection. These include the mechanical separation of microorganisms from the tissues, fatty acids that inhibit many microorganisms, excretion of lysozyme by sweat glands, and the desquamation of the epithelium. Washing may decrease the number of skin bacteria by 90%, but the numbers return to normal within a few hours. What is important, however, is that washing removes transient pathogens that may cause infection through a variety of mechanisms. Table 6–1 lists the microorganisms most commonly found on the skin. Other organisms have been isolated from the skin but are found only occasionally or rarely and are therefore not listed.

Normal Flora of the Mouth

The mouth contains large numbers of bacteria, with *Streptococcus* being the predominant species. Many organisms bind to the buccal mucosa and tooth surfaces. Bacterial plaques that develop on teeth may contain as many as 10^{11} streptococci per gram. Plaque also results in a low oxidation-reduction potential at the tooth surface; this supports the growth of strict anaerobes, particularly in crevices and in the areas between the teeth. A partial list of microorganisms found in the mouth is found in Table 6–2.

■■■■■ T A B L E   6 – 3
MICROORGANISMS FOUND IN THE NOSE AND NASOPHARYNX

Common	Less Common
Staphylococcus aureus	*Streptococcus pneumoniae*
Staphylococcus epidermidis	*Moraxella catarrhalis*
Diphtheroids	*Haemophilus influenzae*
Haemophilus parainfluenzae	*Neisseria meningitidis*
Streptococcus spp	*Moraxella* spp

Normal Flora of the Respiratory Tract

The respiratory tract consists of the mouth (previously discussed), nasopharynx, oropharynx, nose, trachea, bronchi, and lungs. The trachea, bronchi, and lungs are protected by the action of ciliary epithelial cells and by the movement of mucus. The tissues of these structures are normally sterile as a result of this protective action. The organisms found in the mouth, nasopharynx, oropharynx, and nose are similar but show some differences. The most common microorganisms found in the *nose* are *S. aureus* and *Staphylococcus epidermidis*. The population of the *nasopharynx* mirrors that of the nose, although the environment is different enough from that of the nose to select for several additional organisms. Those organisms found as normal flora of the nose and nasopharynx are outlined in Table 6–3.

The *oropharynx* contains a mixture of *Streptococcus* species. A number of species of the viridans group can be isolated; they are: *S. mitis, S. mutans, S. milleri, S. sanguis,* and *S. salivarius.* In addition, diphtheroids and *Moraxella catarrhalis* can be readily isolated. Hospitalized patients often show colonization with gram-negative rods. The normal flora of the oropharynx is listed in Table 6–4.

Normal Flora of the Gastrointestinal Tract

The esophagus, stomach, small intestine, and colon compose the gastrointestinal tract. Microorganisms usually do not multiply in the esophagus and stomach but are present in ingested food and as transient flora. Most are destroyed in the stomach. Those that survive generally are protected by being enmeshed in food and move to the small intestine. The small intestine contains few microorganisms. They are, however, prevalent in the colon, where a count between 10^8 and 10^{11} bac-

teria per gram of solid material is normal. Although we usually think of the facultative anaerobes as being the predominant organisms, anaerobes far outnumber the facultative gram-negative rods, making up over 90% of the microbial flora of the large intestine. The population may be altered by antibiotics. In some cases, certain populations or organisms are eradicated or suppressed and other members of the normal flora are able to proliferate. This can be the cause of a severe necrotizing enterocolitis *(Clostridium difficile),* diarrhea *(C. albicans, S. aureus),* or other superinfection. The bacteria constituting the normal intestinal flora also carry out a variety of metabolic degradations and syntheses that appear to play a role in the health of the host. A summary of the organisms found in the gastrointestinal tract is shown in Table 6–5.

Normal Flora of the Genitourinary Tract

The kidneys, bladder, and fallopian tubes are normally free of microorganisms. The urethra is colonized in its outermost segment by those organisms found on the skin. During childbearing years, the vagina is colonized with *Lactobacillus* as well as anaerobic gram-negative rods and gram-positive cocci. Many organisms are inhibited by the low pH (4–5) of the vagina. Microorganisms expected to be isolated from the genitourinary tract are listed in Table 6–6.

The normal flora have beneficial effects. The development of immunologic competence de-

■■■■■ T A B L E   6 – 4
MICROORGANISMS FOUND IN THE OROPHARYNX

Common	Less Common
α-Hemolytic and nonhemolytic streptococci	*Streptococcus pyogenes*
Diphtheroids	*Neisseria meningitidis*
Staphylococcus aureus	*Haemophilus influenzae*
Staphylococcus epidermidis	Gram-negative rods
Streptococcus pneumoniae	
Streptococcus mutans	
Streptococcus milleri	
Streptococcus mitis	
Streptococcus sanguis	
Streptococcus salivarius	
Moraxella catarrhalis	
Haemophilus parainfluenzae	
Anaerobic streptococci	
Bacteroides spp	
Prevotella/Porphyromonas	
Bacteroides oralis	
Fusobacterium necrophorum	

■ T A B L E   6 – 5
COMMON MICROORGANISMS FOUND IN THE GASTROINTESTINAL TRACT

Bacteroides spp
Clostridium spp
Enterobacteriaceae
Eubacterium spp
Fusobacterium spp
Peptostreptococcus
Peptococcus
Staphylococcus aureus
Enterococcus

■ T A B L E   6 – 6
MICROORGANISMS FOUND IN THE GENITOURINARY TRACT

Common	Less Common
Lactobacillus	Group B streptococci
Bacteroides	Enterobacteriaceae
Clostridium	*Acinetobacter*
Peptostreptococcus	*Candida albicans*
Staphylococcus aureus	
Staphylococcus epidermidis	
Enterococcus	
Diphtheroids	

pends upon the normal flora. The immune system is constantly "primed" by contact with the normal flora. Animals born and raised in a germ-free environment have a poorly functioning immune system. Exposure to otherwise innocuous organisms can be fatal to such animals. The normal flora produces conditions at the microenvironmental level that block colonization by extraneous pathogens. When the composition of the normal flora is altered (e.g., by antibiotic therapy with broad-spectrum antibiotics), other organisms capable of causing disease may fill the void. *Candida albicans* may greatly multiply and cause diarrhea or infections in the mouth or vagina. *Clostridium difficile* produces a colitis as a result of its proliferation following antibiotic therapy.

The normal flora play an important role in both health and disease. Eradication of the normal flora may have profound negative effects, and yet, many common infections are caused by members of the normal flora. Knowledge of these organisms is important to the understanding of the collection and processing of specimens as well as the identification of isolates.

B. PATHOGENESIS OF INFECTION

INTRODUCTION
  Pathogenicity
    Pathogens
    Opportunistic Pathogens
  Virulence

HOST RESISTANCE FACTORS
  Physical Barriers
  Cleansing Mechanisms
  Antimicrobial Substances
  Normal Flora
  Phagocytosis
    Chemotaxis
    Attachment
    Ingestion
    Killing
  Inflammation
  Immune Responses

INFECTIOUS AGENT FACTORS

  Adherence
  Proliferation
  Tissue Damage
    Exotoxins
    Endotoxins
  Invasion
  Dissemination

ROUTES OF TRANSMISSION
  Airborne Transmission
  Transmission by Food and Water
  Close Contact
  Cuts and Bites
  Arthropods
  Zoonoses

EPIDEMIOLOGY
  Definitions
  Surveillance and Reporting

Hal S. Larsen

INTRODUCTION

Pathogenicity

The relationship that exists between the human host and the microbial world is exceedingly complex. Each is in constant interaction with the other as well as with numerous additional influences, all of which affect the host-microbe relationship. Factors such as nutrition, stress, genetic background, and other diseases present all have an impact on the outcome of the meeting of human and microbe. The relationship is an equilibrium that is in constant motion and subject to change. This portion of Chapter 6 discusses the role of each of these factors in host-parasite relationships and how each interacts with the others to prevent or initiate a disease process.

Pathogens

A *pathogen* is a microbe that can cause disease in a susceptible host. Years ago, this definition applied to relatively few organisms. It was obvious that certain bacteria, e.g., *Yersinia pestis* and *Bacillus anthracis,* were pathogenic in nearly all situations. Others, such as *Serratia marcescens* and *Leuconostoc,* were considered never to be pathogenic. Our understanding of host-parasite interactions and our ability to isolate, grow, and identify organisms have improved steadily over the years. In addition, patient populations have changed owing to a longer life span, highly invasive medical procedures, transplants, and so on. As a result, organisms that are found as normal flora and in the environment are being seen with increasing frequency in clinical settings. Our definition of *pathogen,* therefore, must be expanded to apply to virtually any microorganism when conditions for infection are met. We must also broaden our view of the human host. The potential pathogen list for a 20-year-old healthy college student is much smaller than for a 90-year-old person, a transplant recipient, or a 20-year-old college student with acquired immune deficiency syndrome (AIDS).

Opportunistic Pathogens

The majority of microorganisms that humans encounter cause disease only if there is a significant change in host resistance or within the organism itself. These normally innocuous organisms are called *opportunistic pathogens.* They are usually part of the normal flora, but the classic pathogens may also be found among the opportunistic organisms when host defenses weaken. The infections caused by these organisms are *opportunistic infections.* Infections due to these organisms seldom occur in healthy individuals. In fact, there is the assumption that the infected individual is not "normal" (in terms of body defenses) when infection by an opportunistic pathogen occurs. Table 6–7 lists some of the common opportunistic microorganisms.

AN ABBREVIATED LIST OF OPPORTUNISTIC MICROORGANISMS

Conditioning Compromising Host Defenses	Organism(s)
Foreign bodies (catheters, shunts, prosthetic heart valves)	*Staphylococcus epidermidis* *Propionibacterium acnes* *Aspergillus* spp *Candida albicans* Viridans streptococci *Serratia marcescens* *Pseudomonas aeruginosa*
Alcoholism	*Streptococcus pneumoniae* *Klebsiella pneumoniae*
Burns	*Pseudomonas aeruginosa*
Hematoproliferative disorders	*Cryptococcus neoformans* Varicella-zoster virus
Cystic fibrosis	*Pseudomonas*
Immunosuppression (drugs, congenital, disease)	*Candida albicans* Pneumocystis carinii Herpes simplex virus *Aspergillus* spp Diphtheroids Cytomegalovirus *Staphylococcus* *Pseudomonas*

When an infection is the result of medical treatment or procedures, it is termed an *iatrogenic infection*. For example, many patients who have indwelling urinary catheters develop a urinary tract infection. The catheter was a necessary procedure in the medical treatment of the individual, but its use has resulted in an infection. Patients who are given immunosuppressive drugs because they have received a transplant are more susceptible to infection. Because any infection in such a patient would probably be due to the physician-ordered drug therapy, it would be an iatrogenic infection.

Virulence

Virulence is the relative ability of a microorganism to cause disease, or the degree of pathogenicity. It is usually measured by the numbers of microorganisms necessary to cause infection in the host. Those organisms that can establish infection with a relatively low infective dose are considered more virulent than those that require high numbers for infection. This generalization is somewhat misleading, in that there is often a difference in the severity of disease between different organisms. If a microorganism requires a relatively high infective dose but the disease it causes is often fatal, we tend to think of the microorganism as highly virulent. On the other hand, a different organism may require a low infective dose but produces a relatively mild disease. A number of organism characteristics or factors contribute to virulence: capsules, toxins, enzymes, cell wall receptors, pili, and others. These *virulence factors* allow the pathogen to evade or overcome host defenses and cause disease. Many virulence factors are well defined, such as diphtheria and cholera toxins, the capsule of *Streptococcus pneumoniae,* and the pili of *Neisseria gonorrhoeae.* The exact role, if any, of other factors in disease production (e.g., coagulase, streptokinase, lipase, and IgA protease) is unclear.

HOST RESISTANCE FACTORS

Physical Barriers

Healthy skin is a very effective barrier against infection. The stratified and cornified epithelium presents a mechanical barrier to penetration by most microorganisms. Organisms that can cause infection by penetrating the mucous membrane epithelium usually cannot penetrate unbroken skin. Only a few microorganisms are capable of entering the body by way of intact skin. Some of these, as well as others that normally enter when the skin barrier is compromised, are listed in Table 6–8.

Most of the organisms listed in Table 6–8 require help in breaking the skin barrier (e.g., animal or arthropod bite). Those capable of penetrating normal, healthy skin are few and include *Leptospira, Francisella tularensis, Treponema,* and some fungi. Even these organisms probably require microscopic breaks in the skin surface. It is clear that healthy, intact skin is the primary mechanical barrier to infection. The skin also has substantial numbers of normal flora. These organisms are usually not pathogens. Some, however, such as *Staphylococcus aureus,* commonly cause infections. The normal flora contributes to a low pH, competition for nutrients, and production of bactericidal substances by the resident organisms. These conditions serve to prevent colonization by transient organisms. Additionally, the low pH resulting from long-chain fatty acids secreted

■■■■■T A B L E   6 – 8
MICROORGANISMS THAT INFECT SKIN OR ENTER THE BODY VIA SKIN

Microorganisms	Disease	Comments
Arthropod-borne viruses	Various fevers, encephalitides	150 distinct viruses, transmitted by infected arthropod bites
Rabies virus	Rabies	Bites from infected animals
Vaccinia virus	Skin lesion	Vaccination against smallpox
Rickettsieae	Typhus, spotted fevers	Infestation with infected arthropods
Leptospira	Leptospirosis	Contact with water containing infected animal urine
Staphylococci	Boils, impetigo	Most common skin invaders
Streptococci	Impetigo, erysipelas	
Bacillus anthracis	Cutaneous anthrax	Systemic disease following local lesion at inoculation site
Treponema pallidum and *pertenue*	Syphilis, yaws	Warm, moist skin is more susceptible
Yersinia pestis	Plague	Bite from infected rat flea
Plasmodium	Malaria	Bite from infected mosquito
Dermatophytes	Ringworm, athlete's foot	Infection restricted to skin, nails, and hair

With permission from Mims CA: The Pathogenesis of Infectious Disease. New York: Academic Press, 1977.

by sebaceous glands ensures that relatively few organisms can survive and prosper in the acid environment of the skin.

Cleansing Mechanisms

Normally the term *cleansing* brings to mind a liquid. One of the most effective cleansing mechanisms humans have, however, is the desquamation of the skin surface. The keratinized squamous epithelium or outer layer of skin is being continuously shed. Many of the microorganisms colonizing the skin are disposed of with the sloughing of the epithelium.

More obvious is the cleansing action of the fluids of the eye and the respiratory, digestive, urinary, and genital tracts. The eye is continually exposed to microorganisms, and it is not surprising that this organ has some highly developed antimicrobial mechanisms. Tears bathe the cornea and sclera. This not only lubricates the eye but washes away foreign matter from the surface, including infectious agents. Additionally, tears contain IgA and lysozyme.

The respiratory tract is also continuously exposed to microorganisms. It is protected by nasal hairs, ciliary epithelium, and mucous membranes. There is a continuous flow of mucus from the membranes lining the nasopharynx that traps particles and microbes and sweeps them along to the oropharynx, where they are either expectorated or swallowed. The trachea is lined with ciliary epithelium. These cells have hair-like extensions (cilia) that sweep particles and organisms upward toward the oropharynx. This mate-

rial is then expectorated or swallowed. The purpose of these mechanisms is to prevent infectious agents and other particles from reaching the bronchioles and lungs. Under normal conditions, they are very effective, and the air moving into and out of the lungs is sterile.

Bacteria are swallowed, either as part of the normal flora of the mouth and upper respiratory tract or in liquids and food. Most bacteria are easily destroyed by the low pH found in the stomach. Some bacteria, however, are able to survive and pass into the small intestine. The number of bacteria in the intestine increases as the distance from the stomach increases. The number of bacteria in the distal portion of the colon is extremely high. The organisms do not normally gain entrance to the body from the intestine. The mucous secretions and peristalsis serve to prevent the organisms from attaching to the intestinal epithelium. Additionally, secretory antibody and phagocytic cells lining the mucosa defend against infection.

The genitourinary tract is cleansed by the voiding of urine. Consequently, only the outermost portions of the urethra have a microbial population. The vagina contains a large population of organisms as part of the normal flora. The acidity of the vagina, resulting from the breakdown of glycogen by the resident flora, tends to inhibit transient organisms from colonizing.

Antimicrobial Substances

A variety of substances produced in the human host have antimicrobial activity. Some are produced as part of phagocytic defense and are discussed later. Others, such as fatty acids, HCl in the stomach, and secretory IgA have already been mentioned. A substance that plays a major role in resistance to infection is *lysozyme*. This is a low-molecular-weight (approximately 20,000 daltons) enzyme that hydrolyzes the peptidoglycan layer of bacterial cell walls. In some bacteria, the peptidoglycan layer is directly accessible to lysozyme. These bacteria are killed by the enzyme alone. In other bacteria, the peptidoglycan layer is exposed after other agents have damaged the cell wall (e.g., antibody and complement, hydrogen peroxide). In these cases, lysozyme acts with the other agents to cause death of the infecting bacteria. Lysozyme is found in serum and tissue fluids as well as tears, breast milk, saliva, and sweat.

Antibodies, especially secretory IgA, are found in mucous secretions of the respiratory, genital, and digestive tracts. They may serve as opsonins, thereby enhancing phagocytosis, or they may fix complement and neutralize the infecting organism.

Serum also contains low-molecular-weight cationic proteins termed β-*lysins*. These proteins are lethal against gram-positive bacteria and are released from platelets during coagulation. The site of action is the cytoplasmic membrane.

These antimicrobial substances and systems work best together. A combination of antibody, complement, lysozyme, and β-lysin is significantly more effective in killing bacteria than each alone or than any combination in which one or more are missing.

Proliferation of viruses is inhibited by *interferon*. The interferons are a group of cellular proteins induced in eukaryotic cells in response to virus infection or other inducers. Uninfected cells that have been exposed to interferon are refractory to virus infection. A number of bacteria, viruses, and their products induce interferon production. The interferon produced binds to the surface receptors on noninfected cells. This binding stimulates the cell to synthesize enzymes that inhibit viral replication over a period of several days. The antiviral effect of interferon is only one action it exhibits. One type of interferon plays an important role in the immune response. It inhibits cell proliferation and tumor growth while enhancing phagocytosis by macrophages, activity of natural killer cells, and generation of cytotoxic T cells.

Normal Flora

Nonpathogenic microorganisms compete with pathogens for nutrients and space. This competition lessens the chance that the pathogen will colonize the host. Some normal flora species produce *bacteriocins,* substances that inhibit the growth of closely related bacteria. These proteins are produced by a variety of both gram-positive and gram-negative bacteria and appear to give the secreting bacterium an advantage, because they can eliminate other bacteria that would compete for nutrients and space. Some species of bacteria produce metabolic byproducts that result in a microenvironment hostile to potential pathogens. Vitamins and other essential nutrients are synthesized by certain bacteria in the intestine and appear to contribute to the overall health of the host.

Phagocytosis

Phagocytosis is an essential component in the resistance of the host to infectious agents. It is the primary mechanism in the host defense against extracellular bacteria as well as a number of viruses and fungi. The polymorphonuclear neutrophils (PMNs) and macrophages (monocytes in the peripheral blood) are the body's first line of defense.

The stem cells for neutrophils arise in the bone marrow, where they differentiate to form mature neutrophils. During this maturation, the cells synthesize myeloperoxidase, proteases, cathepsin, lactoferrin, lysozyme, and elastase. These products are incorporated into membrane-bound vesicles called *lysosomes.* The lysosomes contain the enzymes and other substances necessary for the killing and digestion of the engulfed particles. They show up as azurophilic granules on a Wright stain. The PMN also has receptors on the cell membrane for some complement components that stimulate cell motion, the metabolic burst, and secretion of the lysosome contents into a phagosome. The PMN is an end-stage cell and has a circulating half-life of 6 to 7 hours. It may migrate to the tissues, where its half-life is less than a week.

Macrophages also originate in the bone marrow. They circulate as monocytes for 1 to 2 days and then migrate through the blood vessel walls into the tissues and take up residence in specific tissues as part of the reticuloendothelial system. These cells are widely distributed in the body (see Table 6–9) and play a central role in specific immunity as well as in nonspecific phagocytosis.

Four activities must occur for phagocytosis to take place and be effective in host defense: (1) migration to the area of infection (chemotaxis), (2) attachment of the particle to the phagocyte, (3) ingestion, and (4) killing.

Chemotaxis

The PMNs circulate through the body, followed by movement into the tissues by an action called *diapedesis*—movement of the neutrophils between the endothelial cells of the blood vessels into the tissues. The body is under constant surveillance by these and other phagocytic cells. When an infection occurs, massive numbers of PMNs accumulate at the site. This accumulation is not a random event but rather is a directed migration of PMNs into the area needing their services. This migration is called *chemotaxis* (a chemical "taxi" or chemically caused movement). Several substances serve as *chemotactic agents*. These are certain components of complement, a number of bacterial products, and products from damaged tissue cells as well as products from responding immune cells. It appears that the initial contact of the PMN with an invading organism may be random. As the organism causes the body's defense mechanisms to respond via inflammation, however, directed migration of phagocytes occurs (chemotaxis). The speed and magnitude of this response are easily visualized by recalling how quickly a splinter or similar injury becomes infected and how much pus is produced. Figure 6–1 illustrates chemotaxis.

Attachment

One of the most effective defenses bacteria have against phagocytosis is the capsule. This structure prevents attachment of the organism to the neutrophil's membrane, which must occur before ingestion can take place. Attachment is facilitated by specific antibodies to the microorganism. The neutrophil membrane has a variety of receptors; these include the Fc portion of IgG1, IgG3, and the C3b component of complement. In addition, these three factors can, and do, bind to

TABLE 6–9
TISSUE DISTRIBUTION OF MONOCYTES/MACROPHAGES

Cell Name	Tissue Distribution
Monocyte	Blood
Kupffer cell	Liver
Alveolar macrophage	Lung
Histiocyte	Connective tissue
Peritoneal macrophage	Peritoneum
Microglial cell	Central nervous system
Mesangial cell	Kidney
Macrophage	Spleen, lymph nodes

the invading microorganism. The result is that the invading microorganisms are coated with one or more of these factors. The receptor on the PMN for the particular factor coating the bacterium binds to the factor and forms a bridge that brings the particle into close physical contact with the leukocyte membrane. The coating of the bacterium with antibody or complement components results in enhanced phagocytosis by the PMN. This process or phenomenon, called *op-sonization,* is illustrated in Figure 6–2.

The antibody and complement components are termed *opsonins.* Opsonization can be accomplished by three different types of responses: (1) IgG1 or IgG3 binds to the organism, (2) the antibody response is insufficient for opsonization but complement is fixed on the surface of the organism, or (3) the alternative complement pathway is activated by the endotoxin or polysaccharides of the organism.

Ingestion

The next step of phagocytosis is ingestion. This process occurs rapidly following attachment and is outlined in Figure 6–3. The cell membrane of the phagocytic cell invaginates and surrounds the attached particle. The particle is taken into the cytoplasm and enclosed within a vacuole called a phagosome. The phagosome fuses with lysosomes, which are vacuoles containing hydrolytic enzymes. The lysosomes release their contents into the phagosome, a process called *degranulation.* The list of enzymes found within the lysosomes is long—more than 60 enzymes, including proteases, lipases, RNase, DNase, peroxidase, and acid phosphatase. Several of these are important in the killing and digestion of the engulfed bacterial cell.

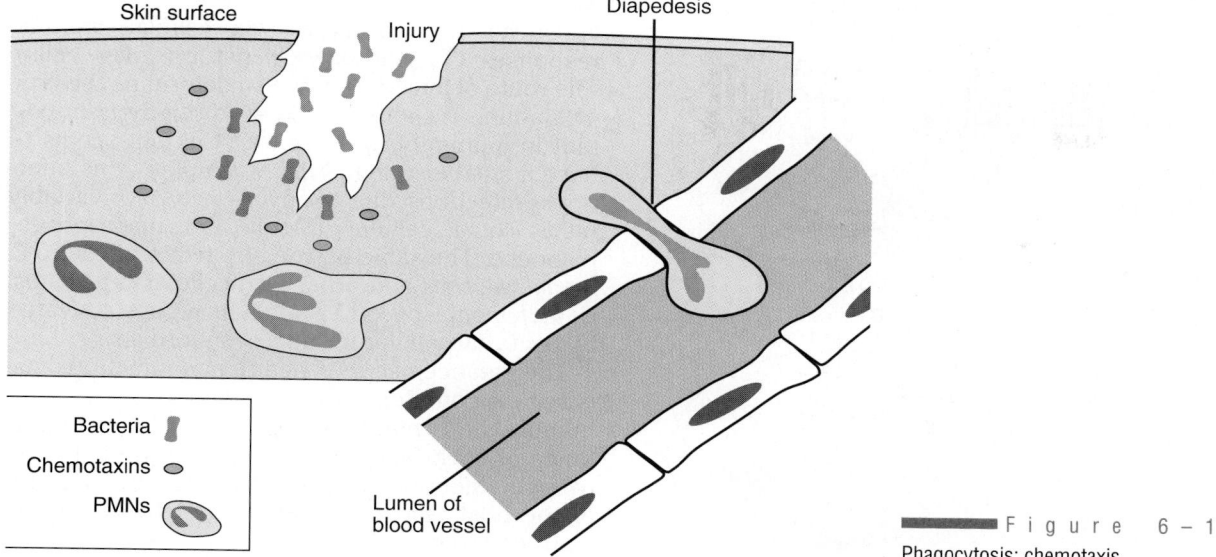

Bacteria

Chemotaxins

PMNs

Lumen of
blood vessel

━━━ F i g u r e   6 – 1
Phagocytosis: chemotaxis.

Killing

The phagocytosis of a particle triggers a significant increase in the metabolic activity of the neutrophil or macrophage. This increase is termed a *metabolic* or *respiratory burst*. The cell demonstrates increases in glycolysis, the hexose monophosphate shunt pathway, oxygen utilization, and production of lactic acid and hydrogen peroxide. The hydrogen peroxide produced at this time diffuses from the cytoplasm into the phagosome. It acts in conjunction with other compounds to exert a bactericidal effect. In addition, other enzymes from the lysosome have antimicrobial action. They include lactoferrin, which chelates iron and prevents bacterial growth, lysozyme, and several basic proteins. The usual result, depicted in Figure 6–4, is that a phagocytosed organism is quickly engulfed, killed, and digested. Organisms that are "intracellular" (e.g., *Mycobacterium tuberculosis, Listeria monocytogenes, Brucella*) are able to survive phagocytosis and, in fact, may actually multiply within the phagocyte. Obviously, other defense mechanisms must play a major role in immunity to these intracellular organisms.

The importance of phagocytosis is seen in patients with defects in the numbers of function of phagocytic cells. Such patients have frequent infections in spite of possessing high levels of serum antibody.

Microorganisms have developed a number of ways of countering phagocytosis. A list of these organisms, along with the types of interference demonstrated and the responsible factors, is given in Table 6–10. Many of the organisms listed in Table 6–10 are common isolates. This fact is not surprising, because they have developed a means to interfere with phagocytosis, thereby increasing their pathogenicity.

Inflammation

Inflammation is the body's response to injury or foreign body. A hallmark of inflammation is the accumulation of large numbers of phagocytic

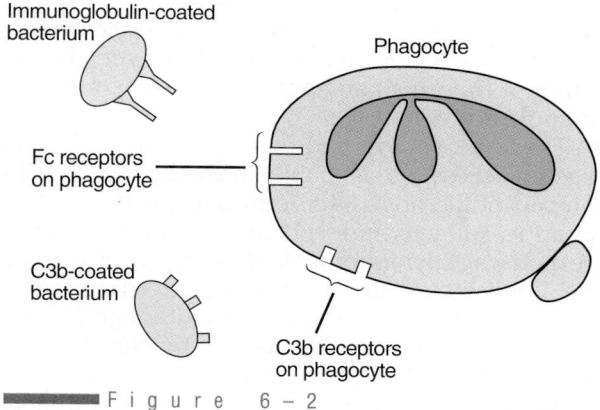

Immunoglobulin-coated
bacterium

Phagocyte

Fc receptors
on phagocyte

C3b-coated
bacterium

C3b receptors
on phagocyte

━━━ F i g u r e   6 – 2
Phagocytosis: attachment.

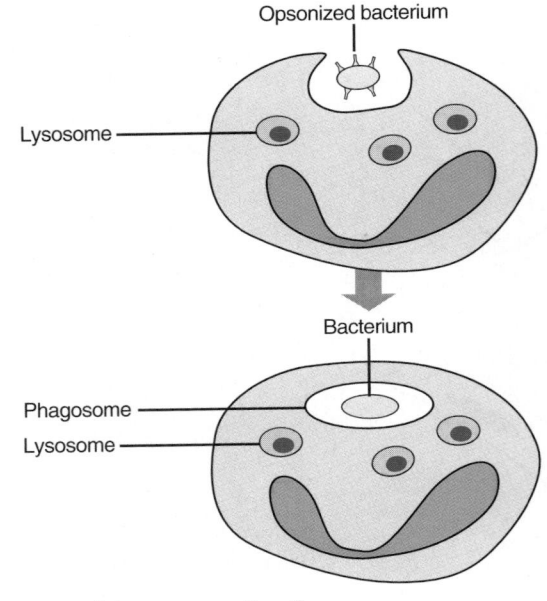

Figure 6 – 3

Phagocytosis: ingestion.

cells. These leukocytes release mediators or cause other cell types to release mediators. The mediators cause erythema as a result of greater blood flow, edema from an increase in vascular permeability, and continued phagocyte accumulation, resulting in pus. The enzymes released by the phagocytes digest the foreign particles, injured cells, and cell debris. After the removal of the invading object, the injured tissue is repaired.

Immune Responses

A discussion of the responses of the immune system to infection is beyond the scope of this chapter. A brief discussion, however, in an attempt to furnish an appreciation of its role and complexity, is in order.

The wide variety and complexity of infectious agents necessitate flexibility in the immune mechanisms of the host. Microorganisms with polysaccharide antigens induce predominantly humoral responses (B lymphocyte). These antigens tend to be T-cell independent. Protein antigens cause a response by both T and B cells. Although a number of the parasite's antigens are virulence factors, seldom does an antibody response to a particular factor result in immunity. An exception to this is antibody to M protein (anti–M protein) of *Streptococcus pyogenes*. Anti– M protein confers

immunity to the host. Viral and fungal infections are dealt with primarily via cellular immune responses (T lymphocytes, natural killer cells). The route of infection may also determine the type of immune response elicited. Antibody is important in immunity to influenza virus but seems to have a smaller role in herpes simplex virus infections. Infections due to myco-bacteria invariably result in a cellular (T-cell and macrophage) response. The same is true of infections due to *L. monocytogenes* and other intracellular organisms. Complement and other substances are activated by bacterial cellular products (endotoxin).

The balance between health and infectious disease is complex and is mediated by both humoral and cellular factors. The relative importance of each factor depends upon the parasite, route of infection, condition and genetic make-up of the host, and other factors yet to be clearly characterized.

Table 6–11 summarizes the defenses used by the human host against infection.

INFECTIOUS AGENT FACTORS

Adherence

Most infectious agents must attach to host cells before infection occurs. In some diseases due to

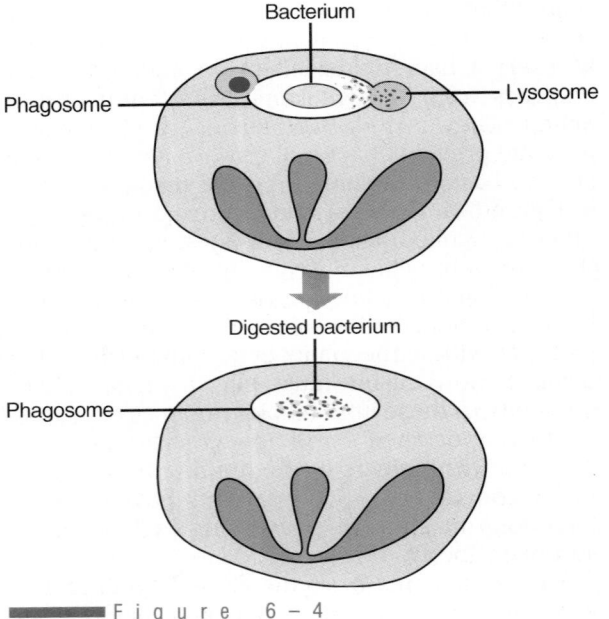

Figure 6 – 4

Phagocytosis: killing.

■■■■■■■ T A B L E 6 – 1 0
TYPES OF INTERFERENCE WITH PHAGOCYTIC ACTIVITIES

Microorganisms*	Type of Interference†	Mechanism (or Responsible Factor)
Streptococci	Kill phagocyte	Streptolysin induces lysosomal discharge into cell cytoplasm
	Inhibit chemotaxis	Streptolysin
	Resist phagocytosis	M substance
	Resist digestion	
Staphylococci	Kill phagocyte	Leucocidin induces lysosomal discharge into cell cytoplasm
	Inhibit opsonized phagocytosis	Protein A blocks Fc portion of Ab
	Resist killing	Cell wall mucopeptide
Bacillus anthracis	Kill phagocyte	Toxic complex
	Resist killing	Capsular polyglutamic acid
Haemophilus influenzae	Resist phagocytosis (unless Ab present)	Polysaccharide capsule
Streptococcus pneumoniae		
Klebsiella pneumoniae	Resist digestion	
Pseudomonas aeruginosa	Resist phagocytosis (unless Ab present)	"Surface slime" (polysaccharide)
	Resist digestion	
Escherichia coli	Resist phagocytosis (unless Ab present)	{ O antigen (smooth strains)
		{ K antigen (acid polysaccharide)
	Resist killing	K antigen
Salmonella typhi	Resist phagocytosis (unless Ab present)	Vi antigen
	Resist killing	
Cryptococcus neoformans	Resist phagocytosis	Polysaccharide capsule
Treponema pallidum	Resist phagocytosis	Cell wall
Yersinia pestis	Resist killing	Protein-carbohydrate cell wall
Mycobacteria	Resist killing and digestion	Cell wall structure
	Inhibit lysosomal fusion	?
Brucella abortus	Resist killing	Cell wall substance
Toxoplasma gondii	Inhibit attachment to PMN	?
	Inhibit lysosomal fusion	?

*Often it is only the virulent strains which show the type of interference listed.
†Sometimes the type of interference listed has been described only in a particular type of phagocyte (polymorph or macrophage) from a particular host, but it generally bears a relationship to pathogenicity in that host.
With permission from Mims CA: The Pathogenesis of Infectious Disease. New York: Academic Press, 1977.

exotoxins (e.g., botulism, staphylococcal food poisoning), adherence is not important. In virtually all other cases, however, the bacterium, virus, or fungus requires adherence to the host cell before infection and disease progress. The cell surface structures that mediate attachment are called *adhesins*. The host cells must possess the necessary receptors for the adhesins. If the host or the infectious agent undergoes a mutation that changes the structure of the adhesin or the receptor, adherence likely will not take place and the virulence of the infectious agent is affected.

Virus infections depend upon the cell's maintaining an appropriate receptor for the virus particle. Infection of the cell occurs only if attachment is the initial event. The main adhesins in bacteria are the pili (fimbriae) and surface polysaccharides. Pili enable bacteria to adhere to host cell surfaces. For example, the strains of *Escherichia coli* that cause traveler's diarrhea use their pili to adhere to cells of the small intestine, where they secrete a toxin that causes the disease symptoms. Similarly, pili are essential for

gonococci to infect the epithelial cells of the genitourinary tract. Antibodies to the pili of *N. gonorrhoeae* are protective by preventing the organism from attaching to the epithelial cells.

Proliferation

In order to establish itself and cause disease, a pathogen must be able to replicate following attachment to host cells. Numerous host factors work to prevent proliferation. Secretory antibody, lactoferrin, and lysozyme have been mentioned previously. To be successful in establishing infection, infectious agents must be able to avoid or overcome these local factors. For example, lactoferrin competes with bacteria for free iron; meningococci can utilize lactoferrin as a source of iron. They are not inhibited by the presence of lactoferrin and, in fact, are able to utilize it for growth. On the other hand, the nonpathogenic *Neisseria* are usually unable to utilize the iron in lactoferrin and are inhibited by its presence.

━━━━ T A B L E   6 – 1 1

SUMMARY OF DEFENSES OF THE HUMAN OR ANIMAL HOST TO INFECTION AND EVASIONARY MECHANISMS ATTRIBUTED TO VARIOUS MICROORGANISMS

Host Defense	Mechanism of Evasion by Microorganism	Example
Hydrodynamic flow	Attachment	Pili, surface proteins, lipoteichoic acid Pseudomembrane of diphtheria
Mucus barrier	Attachment Penetration	Mannose-sensitive pili
Deprivation of essential nutrients	Systems of high-affinity uptake	Iron metabolism
Lysozyme in secretions	Resistance to lysis	Substitution of peptidoglycan
Surface immunoglobulins	Absent or low immunogenicity	Hyaluronic acid, capsules
	Antigenic heterogenicity	Pili, capsules, lipopolysaccharide (LPS), M protein, etc.
	Masking of antigens	Capsules, IgA-binding proteins
	Destruction	IgA protease
Unbroken surface (epithelial cell surface)	Penetration	*Neisseria gonorrhoeae, Shigella*
Unknown defenses in lymphatics (intercellular space)		*N. gonorrhoeae, Shigella*
Serum defenses		
Recognition by antibody	Antigenic heterogenicity	Pili, capsules, LPS, M protein, etc.
	Masking of antigen	Capsules, Ig-binding proteins
	Destruction of antibody	
	Antigenic variation	Borreliae
Complement system	Failure to activate alternative pathway	Sialic acid capsules
	Inactivation of complement components	Cleavage of C3b in empyema fluids
	Resistance to bacteriolysis	ColV plasmid, etc.
	Formation of abscess	*Bacteroides fragilis* capsule
Localization		
Fibrin trapping	Fibrinolysis	*Streptococcus*
Abscess formation	Collagenase, elastase	*Pseudomonas, Clostridium*
Secondary immune response	Nonspecific B-cell activation	LPS, lipoprotein, etc.
	Inhibition of delayed hypersensitivity	Anergy of miliary tuberculosis
	Rapidly fatal (toxin)	Anthrax, plague, *Clostridium*
Phagocytosis	Inhibition of chemotaxis	*Brucella, Salmonella, Neisseria, Staphylococcus, Pseudomonas*
	Inhibition of attachment and ingestion	Capsules, M protein, Ig-binding proteins, gonococcal pili
	Inhibition of metabolic burst	*Salmonella typhi*
	Inhibition of degranulation	Mycobacteriaceae
	Resistance to permeability inducing cationic protein	Gram-positive cell wall, smooth LPS, polyanionic capsules
	Resistance to oxidative attack	Catalase, superoxide dismutase, carotenoid pigments
	Escape from phagosome	*Mycobacterium bovis, Legionella pneumophila*
	Destruction of phagocyte	*Streptococcus pneumoniae, Streptococcus pyogenes, Staphylococcus aureus, Pseudomonas aeruginosa*

Adapted with permission from Gotschlich EC: Thoughts on the evolution of strategies used by bacteria for evasion of host defenses. Rev Infect Dis 5:S779, 1983.

Several pathogens (*Haemophilus influenzae, N. gonorrhoeae, Neisseria meningitidis*) produce an IgA protease that degrades the IgA found at mucosal surfaces. Other pathogens (influenza virus, *Borrelia*) circumvent host antibodies by shifting key cell-surface antigens. The host produces antibodies against the "old" antigens, which are no longer effective.

An extremely important event in the life of an invading pathogen is phagocytosis. Evasion of phagocytosis is essential for most pathogens to be able to survive and multiply. The most common characteristic of bacteria that allows for such evasion is a polysaccharide capsule. Many of those possessing a capsule are highly virulent until its removal, at which point their virulence becomes extremely low. Some pathogens are able to survive phagocytosis. Those that do have developed several methods to prevent being killed. Some prevent fusion of phagosomes and lysosomes. Others have a resistance to the effects of the lysosomal contents, and still others escape into the cytoplasm.

Tissue Damage

Generally speaking, disease from infection is noticeable only if tissue damage occurs. This damage may be from toxins, either exotoxins or endotoxins, or from inflammatory substances that cause host-driven, immunologically mediated damage.

Exotoxins

Many of the bacterial exotoxins are highly characterized. Most are composed of two subunits: the first is nontoxic and serves to bind the toxin to the host cells; the second is toxic. It is common for the toxin gene to be encoded by phage, plasmids, or transposons. Therefore, only those carry the extrachromosomal DNA coding for the toxin gene produce toxin. This is why isolates of *Corynebacterium diphtheriae* have to be tested for toxin production as well as identified as to genus and species. Other pathogenic bacteria show similarities. Table 6–12 lists many of the bacterial exotoxins that are important in disease production.

Endotoxins

Gram-negative bacteria have endotoxins. Endotoxins are composed of the lipopolysaccharide portion of the cell wall. The toxicity is due to the lipid A portion of the lipopolysaccharide. The effects of endotoxin consist of dramatic changes in blood pressure, clotting, body temperature, circulating blood cells, metabolism, humoral immunity, cellular immunity, and resistance to infection.

Endotoxin stimulates the fever centers in the hypothalamus. An increase in body temperature

occurs within an hour after exposure. Exposure to endotoxin also causes hypotension. Severe hypotension occurs within 30 minutes following exposure. Septic or endotoxic shock is a serious and potentially life-threatening problem. Unlike shock caused by fluid loss, such as that seen in severe bleeding, septic shock is unaffected by fluid administration. The endotoxin also initiates coagulation, which can result in intravascular coagulation. This process depletes clotting factors and activates fibrinolysis so that fibrin split products accumulate in the blood. These fragments are anticoagulants and can cause serious bleeding. Another feature of patients with endotoxic shock is severe neutropenia, which can occur within minutes following exposure. It results from sequestration of neutrophils in capillaries of the lung and other organs. Leukocytosis follows neutropenia because neutrophils are released from the bone marrow.

Endotoxin has a wide variety of effects on the immune system. It stimulates proliferation of B lymphocytes in some animal species, activates macrophages, activates complement, and has an adjuvant effect with protein antigens. It also stimulates interferon production and causes changes in carbohydrates, lipids, iron, and sensitivity to epinephrine. It is not difficult to see that a severe infection with gram-negative bacteria can lead to serious and often life-threatening problems.

A comparison of bacterial exotoxins and endotoxins is given in Table 6–13.

Invasion

Pathogens exhibit invasion to some degree or another. *Invasion* is the process of penetrating and growing in tissues. With some organisms, the invasion is localized and involves only a few layers of cells. With others, it involves deep tissues; the gonococcus, for example, is an invasive organism that may infect the fallopian tubes.

Dissemination

Dissemination is the spread of organisms to distant sites, i.e., organs and tissues. With some organisms, such as *Salmonella,* dissemination is an important aspect of the disease. Other organisms, such as *C. diphtheriae,* do not spread beyond their initial site of infection, yet the dis-

■■■■■ T A B L E   6 – 1 2
EXOTOXINS OF PATHOGENIC BACTERIA

Bacterium	Disease Caused in Man	Toxins
Bacillus anthracis	Anthrax	Complex lethal edema-producing toxin
Bordetella pertussis	Whooping cough	lethal, dermonecrotizing toxin
Clostridium botulinum	Botulism	**6 type-specific lethal neurotoxins**
C. oedematiens	Gas gangrene	1. alpha, lethal, dermonecrotizing
		2. beta, lethal, dermonecrotizing, hemolytic
		3. gamma, lethal, dermonecrotizing, hemolytic
		4. delta, hemolytic
		5. epsilon, lethal, hemolytic
		6. zeta, hemolytic
C. perfringens	Gas gangrene and enteritis necroticans	1. **alpha, lethal, dermonecrotizing, hemolytic**
		2. beta, lethal
		3. gamma, lethal
		4. delta, lethal
		5. epsilon, lethal, dermonecrotizing
		6. eta, lethal (?)
		7. iota, lethal, dermonecrotizing
		8. theta, lethal, cardiotoxic, hemolytic
		9. kappa, lethal, proteolytic
		10. **enterotoxin**
C. septicum	Gas gangrene	Alpha, lethal, hemolytic
C. sordellii	Gas gangrene	1. edema-producing toxin
		2. hemorrhagic toxin
C. tetani	Tetanus	1. **tetanospasmin, lethal, neurotoxic**
		2. neurotoxin, nonspasmogenic
		3. tetanolysin, lethal, cardiotoxic, hemolytic
Corynebacterium diphtheriae	Diphtheria	**Diphtheria toxin, lethal, dermonecrotizing**
Escherichia coli	Diarrhea	1. **heat-labile enterotoxin**
		(2. heat-stable enterotoxin)
Pseudomonas aeruginosa	Pyogenic infections	**Exotoxin A**
Staphylococcus aureus	Pyogenic infections, enterotoxemia	1. alpha, lethal, dermonecrotizing, hemolytic
		2. beta, lethal, hemolytic
		3. gamma, lethal, hemolytic
		4. delta, hemolytic
		5. **exfoliating toxin**
		6. **enterotoxin**
Streptococcus pyogenes	Pyogenic infections, scarlet fever, rheumatic fever	1. Dick toxin, erythrogenic, nonlethal
		2. streptolysin O, lethal, hemolytic, cardiotoxic
		3. streptolysin S, lethal, hemolytic
Vibrio cholerae	Cholera	**Cholera toxin, lethal, enterotoxic**
Salmonella typhimurium	Enteritis	**Enterotoxin?**
Shigella sp	Dysentery	**Enterotoxin**
Yersinia pestis	Plague	Murine toxin

Boldface indicates toxins that produce harmful effects of infectious disease.

Adapted with permission from Braude AI, Davis CE, Fierer J (eds): Infectious Diseases and Medical Microbiology. Philadelphia: WB Saunders, 1986, p 43.

ease they produce is serious and often fatal. Certain organisms that survive phagocytosis may be disseminated rapidly to many body sites, but the organisms themselves are not invasive. The phagocyte simply carries the organism, but the bacterium itself is incapable of penetrating tissues. *Clostridium perfringens* is an example of a highly invasive organism that may not necessarily disseminate.

It should be apparent that infectious agents have a wide variety of mechanisms that allow them to cause disease. Some, such as capsules and toxins, are common to more than one organism. Others tend to be specialized, such as the tissue tropism of the gonococcus. It is tempting to propose some simple explanations of how infectious diseases occur. Microorganisms produce many extracellular factors that appear to aid in infection; however, the exact role of most of these is unknown. Also, although our knowledge of pathogenesis has grown dramatically over the past few years, there are only a half-

■■■■■ T A B L E   6 – 1 3
DIFFERENCES BETWEEN BACTERIAL EXOTOXINS AND ENDOTOXINS

	Exotoxins	Endotoxins
Parent organisms	Gram-positive and gram-negative	Gram-negative
Within or without parent organism	Within and without	Within
Chemical nature	Simple protein	Protein-lipid-polysaccharide
Stability to heating (100°C)	Labile	Stable
Detoxification by formaldehyde	Detoxified	Not detoxified
Neutralization by homologous antibody	Complete	Partial
Biologic activity	Individual to toxin	Same for all toxins
Toxicity compared with strychnine as 1	100 to 1,000,000	0.1

Modified with permission from Braude AI, Davis CE, Fierer J (eds): Infectious Diseases and Medical Microbiology. Philadelphia: WB Saunders, 1986, p 42.

dozen or so organisms whose precise mode and method of infection are known. Our knowledge and understanding of pathogenesis still lack a great deal.

ROUTES OF TRANSMISSION

The route by which a pathogen may be transmitted to a susceptible host is often explained by the characteristics of that pathogen. Although some organisms may be naturally transmitted by more than one route, most have a limited number of routes. These routes can be characterized as airborne, via food and water (ingestion), through close contact (include sexual transmission), through cuts and bites, and via arthropods; animal diseases that can infect humans are transmitted through animal contact (zoonoses). A summary of the routes of transmission is given in Table 6–14.

Airborne Transmission

Respiratory spread of infectious disease is common. Often, the respiratory secretions are aerosolized by coughing, sneezing, and talking. Very small particles, referred to as *droplet nuclei,* are the residue from the evaporation of fluid from larger droplets and are light enough to

remain airborne for long periods. Pathogens that are spread through the air generally must be resistant to drying and inactivation by ultraviolet light. Some infectious agents may be transmitted by dust particles that have become airborne. As discussed earlier in this chapter, the body has a number of defenses against airborne infectious agents. The nasal turbinates, oropharynx, and larynx provide a twisting, mucous-lined passageway that makes direct access to the lower respiratory tract mechanically difficult. In addition, the lower portions of the respiratory tract contain ciliary epithelium that sweeps organisms upward. In order for a microorganism to cause disease, it must circumvent these defenses, penetrate the mucous layer, and attach to the epithelium. The host also produces secretory IgA, lysozyme, alveolar macrophages, and other humoral factors that act on a pathogen that manages to get beyond the physical defenses.

Respiratory tract infections are the most common reason that patients of all ages seek medical attention. Although most upper respiratory tract infections are self-limiting and can be treated by the patient with over-the-counter medications, some are more serious. Streptococcal sore throat, sinusitis, otitis media, acute epiglottitis, and diphtheria can be serious, even life-threatening. Viral diseases causing the common cold and infectious mononucleosis are usually not life-threatening but can result in much discomfort and time lost from work or school.

Although all the diseases just mentioned can be spread via aerosols, some may also be transmitted via the fingers and hands, especially true for the common cold–causing rhinovirus. The fingers and hands are contaminated with infectious nasal secretions because of hand-to-nose contact. The infectious virus particles are passed from the infected individual to a susceptible recipient via hand-to-hand or hand-to-face contact. The recipient transmits the virus picked up from the hands of the infected individual by touching the face and nose. In this case, the disease is transmitted via the respiratory route, but it is not in the normal, classic manner of respiratory transmission.

Transmission may also result from contact with inanimate objects contaminated with the infectious agent. For example, a door knob is contaminated by the hand and fingers of an infected individual, and the virus is transmitted to a susceptible person's hand and fingers when that person opens the door. Control of such transmission is often as simple as frequent handwashing.

■■■■ T A B L E   6 - 1 4
COMMON ROUTES OF TRANSMISSION*

Route of Exit	Route of Transmission	Example
Respiratory	Aerosol droplet inhalation	Influenza virus; tuberculosis
	Nose or mouth → hand or object → nose	Common cold (rhinovirus)
Salivary	Direct salivary transfer (e.g., kissing)	Oral-labial herpes; infectious mononucleosis
	Animal bite	Rabies
Gastrointestinal	Stool → hand → mouth and/or	Enterovirus infection; hepatitis A
	stool → object → mouth	
	Stool → water or food → mouth	Salmonellosis; shigellosis
Skin	Skin discharge → air → respiratory tract	Varicella; poxvirus infection
	Skin to skin	Human papillomavirus (warts); syphilis
Blood	Transfusion or needle prick	Hepatitis B; cytomegalovirus infection; malaria; human immunodeficiency virus (AIDS)
	Insect bite	Malaria; relapsing fever
Genital secretions	Urethral or cervical secretions	Gonorrhea; herpes simplex; *Chlamydia* infection
	Semen	Cytomegalovirus infection
Urine	Urine → hand → catheter	Hospital-acquired urinary tract infection
	Urine → aerosol (rare)	Tuberculosis
Eye	Conjunctival	Adenovirus
Zoonotic	Animal bite	Rabies
	Contact with carcasses	Tularemia
	Arthropod	Plague; Rocky Mountain spotted fever; Lyme disease

*The examples cited are incomplete, and in some cases more than one route of transmission exists.
Reprinted by permission of the publisher from Sherris JC (ed): Medical Microbiology: An Introduction to Infectious Diseases, 2nd ed. New York: Elsevier, 1990, p 184. Copyright 1990 by Elsevier Science Publishing Co., Inc.

Infections of the lower respiratory tract are less common but more serious than those of the upper respiratory tract. The organisms causing these infections have managed to bypass host defenses, or the host defenses have been compromised (e.g., by alcoholism, heavy smoking), allowing the pathogen access to the deeper portions of the respiratory tract.

The most common microorganism causing lower respiratory tract infection of individuals over age 30 is *S. pneumoniae*. The pneumococcus is often seen in aspiration pneumonia, a common type of hospital-acquired pneumonia. Pneumococcal pneumonia begins very suddenly and is a very serious, life-threatening disease, particularly in elderly patients. Among younger people, the common causes of pneumonia are *Mycoplasma* and viruses. The onset of these pneumonias is more gradual than with pneumococcal pneumonia, and the outcome is far more favorable.

In chronic lower respiratory tract infections, the survival of the infecting agent within phagocytes plays a role in the pathogenic mechanism. *Mycobacterium tuberculosis,* the agent of tuberculosis, a chronic debilitating infection, is the classic example of an intracellular pathogen. This organism is highly virulent, is invasive, and survives well and multiplies within phagocytes.

Transmission by Food and Water

Transmission of gastrointestinal infections is usually a result of ingestion of contaminated food or water. In some situations, infection occurs via the fecal-oral route.

The digestive tract is colonized with vast numbers of different microorganisms. Under normal conditions, the normal flora maintains a harmless relationship with the host. Gastric enzymes and juices in the stomach act to prevent survival of most organisms, but many survive and colonize the small intestine and colon.

Gastrointestinal infections result from organisms that are able to survive the harsh conditions of the stomach and competition with the normal flora and to produce damage to the tissues of the gastrointestinal tract. This damage is a result of either a preformed toxin or disruption of the normal functioning of the intestinal cells by invasion of the pathogen or production of a toxin within the intestine.

Organisms that can cause disease by means of a preformed toxin include *Clostridium botulinum, Bacillus cereus,* and *S. aureus.* The severity of disease ranges from a mild diarrhea to a rapidly fatal intoxication. Food poisoning by *B. cereus* and *S. aureus* is relatively common and is self-limiting. Botulism, due to *C. botulinum,* although rare, can be life-threatening.

Other bacteria produce a toxin after infection of the intestinal tract. Generally, in order to be effective as a disease producer, an organism must survive, adhere to, and colonize the intestinal mucosa and either produce a toxin or invade deeper tissues. A commonly seen cause of diarrhea and intestinal infection is *E. coli.* This organism is a member of the normal intestinal flora; however, some strains of *E. coli* produce cytotoxins that cause alterations in the biochemical activity of the intestinal epithelial cells, resulting in problems with fluid and electrolyte control by the intestinal cells. These strains of *E. coli,* referred to as *enterotoxigenic,* are a common cause of "traveler's diarrhea" as well as other intestinal problems.

Probably the classic intestinal pathogen is *Vibrio cholerae,* the cause of cholera. This organism produces an enterotoxin that causes the outpouring of fluid from the cells into the lumen of the intestine. Massive amounts (up to 20 liters per day) can be lost. Other intestinal pathogens are *Clostridium difficile, Shigella, Aeromonas hydrophila, Campylobacter jejuni,* and *Salmonella.* The infective dose, severity, and incidence of disease vary with the agent.

A number of viruses also cause diarrheal disease. They multiply within the cells of the intestinal mucosa and affect the normal functioning of the cells. Viral agents in this category include hepatitis, rotavirus, adenovirus, coxsackievirus, and Norwalk-like agents. The incidence of diarrhea due to these agents is very high, especially in situations of very close contact (daycare centers, nurseries, military camps). Numerous parasites, such as *Fasciolopsis buski, Giardia lamblia, Entamoeba histolytica,* and *Balantidium coli,* also infect the gastrointestinal tract.

Close Contact

All of the routes of transmission require what could be called close contact. Obviously, for a respiratory pathogen to be transmitted via aerosols, the susceptible host must be relatively close. For this discussion, however, *close contact* refers to passage of organisms by salivary, skin, and genital contact. Two prominent infections passed by direct transfer of saliva (e.g., kissing) are herpes simplex and infectious mononucleosis. Skin-to-skin transfer of infectious disease is not as common as some of the other routes, but

diseases such as warts (human papillomavirus), syphilis, and impetigo result when material from infectious lesions inoculates a susceptible host's skin. The list of sexually transmitted diseases is a long one. In North America, the most commonly transmitted venereal diseases or agents are gonorrhea, herpes simplex, hepatitis, *Chlamydia,* syphilis, *Trichomonas,* and AIDS.

Cuts and Bites

The classic example of a bite wound infection is rabies. In fact, human rabies is relatively rare. Of more concern with animal, and especially human, bites is infection by the normal flora of the mouth. Dog and cat bite infections often yield *Pasteurella multocida,* but the possibilities are extensive. Human bites are extremely dangerous because they are difficult to treat and because the normal human oral flora consists of many different organisms in extremely high numbers.

Arthropods

Infection as a result of a tick, flea, or mite bite is a common occurrence in many parts of the world. The diseases spread by arthropods include malaria, relapsing fever, plague, Rocky Mountain spotted fever, Lyme disease, typhus, and untold numbers of regional hemorrhagic fevers. In most cases, the infectious agent multiplies in the arthropod, which then feeds off a human host and transmits the microorganism.

Zoonoses

The route of transmission known as *zoonosis* depends upon contact with animals or animal pro-ducts. Certain diseases of animals may also infect humans in contact with them. These diseases may be passed by animal bites (rabies), arthropod vectors (plague), contact with secretions (brucellosis), and contact with animal carcasses and products (tularemia, listeriosis). The diseases are transmitted by routes already discussed. The common factor is that, regardless of the route, the disease is a disease of animals that is transmitted to humans. A partial list of zoonotic diseases and infecting organisms is given in Table 6–15.

■■■■T A B L E   6 – 1 5
ZOONOSES

Disease	Organism
Anthrax	*Bacillus anthracis*
Brucellosis	*Brucella* spp
Erysipeloid	*Erysipelothrix rhusiopathiae*
Leptospirosis	*Leptospira interrogans*
Tularemia	*Francisella tularensis*
Ringworm	*Trichophyton, Microsporum*
Lyme disease	*Borrelia burgdorferi*
Plague	*Yersinia pestis*
Rocky Mountain spotted fever	*Rickettsia rickettsii*
Yellow fever	Flavivirus
Encephalitis	Alphavirus
Colorado tick fever	Orbivirus
Leishmaniasis	*Leishmania* spp
Rabies	Rhabdovirus
Blastomycosis	*Blastomyces dermatitidis*
Tuberculosis	*Mycobacterium bovis*
Q fever	*Coxiella burnetii*
Ornithosis	*Chlamydia psittaci*
Gastroenteritis	*Campylobacter, Salmonella*
Listeriosis	*Listeria monocytogenes*
Giardiasis	*Giardia lamblia*
Toxoplasmosis	*Toxoplasma gondii*
Tapeworms	*Taenia saginata*
	Taenia solium
	Diphyllobothrium latum
	Echinococcus
Trichinosis	*Trichinella spiralis*

EPIDEMIOLOGY

Epidemiology is the study of the occurrence, distribution, and causes of disease and injury. This definition includes not only infectious diseases but also problems as diverse as gunshot injuries and heart attacks. The emphasis in this section is on infectious diseases, but the principles of epidemiologic study are the same regardless of the disease or problem studied.

Definitions

Several terms are helpful in describing the incidence and effect of disease in a population.

Carrier. A carrier is a person or animal who harbors and spreads a microorganism that causes disease but who does not become ill. There are two types of carriers. A *casual carrier* harbors the microorganism temporarily for a few days or weeks; examples of microorganisms maintained in this fashion are *C. diphtheriae, N. meningitidis,* and *S. pyogenes.* A *chronic carrier*

remains infected for a relatively long time, sometimes throughout life; the typhoid bacillus may be carried chronically.

Endemic. When an organism or disease is constantly present in a population, that disease or organism is termed endemic. It is indigenous to a geographic area or population. Some microorganisms or diseases are endemic for one geographical area but not for another. Schistosomiasis is endemic to parts of the Middle East, but not for many other parts of the world. Cholera is endemic for large portions of the globe. In the United States, we have a more-or-less constant incidence of common colds, streptococcal pharyngitis, and gonorrhea.

Epidemic. When a disease affects a significantly large number of people at the same time in a geographic area, it is said to be an epidemic. The number of cases per given time is not the only measure. Ten cases of diphtheria in a short time could be considered an epidemic because diphtheria is not often seen. Ten cases of streptococcal pharyngitis during that same period are probably normal, however, because it is an endemic disease. A classic example of an epidemic is influenza. There are epidemics of varying degrees every year. Periodically (in intervals of 10 to 20 years), there are worldwide epidemics of influenza affecting tens of millions of people; they are termed *pandemics.* Currently, cholera is in its seventh pandemic; it has spread throughout world regions and even continents.

Incidence Rate. The number of times a new event occurs in a given period is called the incidence rate. It is usually given as cases or infections per 1000 or 100,000 population. It allows a prevalence comparison between diseases or infections.

Incubation Period. The time between exposure to a pathogen and the onset of symptoms is the incubation period. This is often difficult to determine because individuals often have difficulty pinpointing the date or time of exposure. An individual may be infectious during the incubation period; this situation presents a difficult public health problem, because no symptoms are present to identify the infectious person but transmission of the organism is taking place.

Index Case. The first case of a disease, which serves as a source of infection, is the index case.

Morbidity Rate. The rate at which an illness occurs, the morbidity rate, is the number of cases of a disease in a specified population during a defined time interval. The morbidity rate can be a measure of the infectiousness of an organism.

Mortality Rate. The mortality rate is the number of deaths due to a disease in a population.

Organisms that are highly virulent often have a high mortality rate.

Nosocomial Infections. An infection acquired during hospitalization is termed a nosocomial infection.

Reservoir. The reservoir is the source of an infection. It may be a person, an animal, or something in the environment.

Surveillance. Surveillance is the collection of data pertaining to disease occurrence. It is carried out at various levels (physician, city, county, state, federal, international) with a system for reporting to public health agencies.

Surveillance and Reporting

Certain infectious diseases are required by law to be reported to public health authorities. Generally, these are diseases that have significant impact on the population (venereal diseases) or have the potential for grave consequences (anthrax, plague). There is a reporting system in place in the United States, by which local public health authorities report to a state health agency, which in turn reports to the Centers for Disease Control and Prevention (CDC) at the national level. The CDC works closely with worldwide health agencies such as the World Health Organization (WHO) as well as the national health agencies of other countries. These organizations monitor the incidence of disease to determine the current health status of the world's population.

At the local level, surveillance and reporting may consist of investigating an outbreak of food poisoning at a church picnic. The state might become involved if it was determined that the origin of the food poisoning was a vendor who distributes to other areas of the state. If the vendor is a multi-state distributor, the national authorities might be included. Legionellosis, AIDS, Lyme disease and many others provide examples of full case studies in the methods and procedures used to investigate and monitor an epidemic as well as the organization of the public health system. These methods and procedures can be complex and cannot be discussed here, but they involve statistical analysis of minute details of a disease outbreak. The clinical laboratory plays a significant role in providing support for these investigations.

Bibliography

Braude AI, Davis CE, Fierer J (eds): Infectious Diseases and Medical Microbiology. Philadelphia: WB Saunders, 1986.

Gotschlich EC: Thoughts on the evolution of strategies used by bacteria for evasion of host defenses. Rev Infect Dis 5:S778, 1983.

Mims CA: The Pathogenesis of Infectious Disease. New York: Academic Press, 1977.

Sherris JC (ed): Medical Microbiology: An Introduction to Infectious Diseases. New York: Elsevier, 1990.

Spitznagel JK: Microbial interactions with neutrophils. Rev Infect Dis 5:S806, 1983.

CHAPTER 7

General Concepts in Specimen Collection and Handling

Merrily Rausch • *Joy G. Remley*

Adequate collection, transport, and processing of clinical specimens for microbiologic evaluation are important requirements in the diagnosis of infectious disease. The medical support team meets these requirements through cooperative efforts. This team may consist of individuals with varied knowledge and backgrounds. Nevertheless, all team members—clinical practitioners, nurses, and laboratory practitioners—must be instructed in the basic principles of specimen collection and the requirements of rapid transport, proper storage, and preservation of different specimen types. It is the responsibility of the laboratory practitioner to recognize and reject suboptimal specimens and to inform and educate the other team members. There must be ongoing communication and education among all members of the team for the team to function properly and to allow for rapid diagnosis and treatment of infectious disease.

This chapter discusses the following:

■ The basic principles of specimen collection

■ Methods for processing clinical specimens for optimal recovery of common bacterial pathogens as well as unusual organisms

■ Methods of preservation and proper use of transport media

■ Selection of culture media

■ Methods for processing nonroutine samples.

BASIC PRINCIPLES OF SPECIMEN COLLECTION

The laboratory can make accurate and useful determinations only if a specimen has been collected appropriately. The following basic principles of specimen collection may serve as guidelines to ensure appropriate collection processes:

■ If at all possible, a culture specimen should be taken in the acute phase of the infection and before antibiotics are administered.

■ The collection process is initiated by a written order followed by selection of a site to culture. Failure to select an appropriate site to culture leads to misleading culture results and may adversely affect patient management.

■ It is important to culture the infecting agents while avoiding the usual flora and colonizing organisms.

■ Test results must always be compared with suspected diagnosis carefully. For example, if gas is observed in the tissues of an amputation site, and the suspected diagnosis is gas gangrene caused by *Clostridium perfringens,* the laboratory practitioner should be suspicious of inadequate site selection when the organism is not seen on direct examination or recovered in culture. Failure to recover the organism could be due to a true absence of the organism or to inadequate sampling of the site.

Appropriate Collection Techniques

Aspirates and Tissues

Aspirates of sterile body fluids are collected using sterile technique and generally present the laboratory with few problems relating to quality and quantity of specimen. A piece of tissue is another excellent specimen, provided that the sample is taken from along the active line of infection. Tissue samples must be protected from drying during transport. Several drops of sterile saline may be used to prevent the tissue from drying.

Lesions, wounds, and abscesses seem to present the most problems related to quantity and quality. Because infection typically elicits a white cell response in the host, an aspirate of pus is the ideal specimen. In situations in which multiple sites appear infected, an aspirate of a fluctuant, unopen wound is far superior to a swab from an open draining wound.

Swabs

Swabs are used only as a last resort and are considered "second rate." The volume of specimen collected by swabbing a wound tends to be inadequate, and the attempt to extract the material from the swab is often ineffective. If a swab must be used instead of an aspirate or piece of tissue, four basic rules must be followed: (1) clean the wound, (2) explore the wound, (3) obtain fresh culture material, and (4) obtain adequate quantity of material.

CLEAN THE WOUND

All organisms like to feed on pus. These organisms may consist of endogenous flora from the skin and mucous membranes, contaminants from the air, and organisms from the patient's or caregivers' hands. The goal should be to grow only those organisms responsible for the production of the pus. Cleaning the wound kills organisms and also clears any nonviable anaerobic organisms from the wound surface. A 3% solution of hydrogen peroxide can be used to clean the wound. One advantage of hydrogen peroxide is that it kills organisms only for as long as it bubbles. In contrast, povidone-iodine is advantageous as a wound packing because it releases iodine molecules hour after hour, decreasing the numbers of organisms in the wound. A swab con-taminated with povidone-iodine may adversely affect the laboratory's ability to grow organisms associated with infection.

EXPLORE THE WOUND

Exploration of the wound, with emphasis on finding any existing sinus tracts, will better define its extent. Organisms found deep in sinus tracts may be very different from those found on the surface and may require special antibiotic therapy for their eradication. Nasopharyngeal swabs, which contain a very small head on a flexible wire, are ideal for probing sinus tracts.

OBTAIN FRESH CULTURE MATERIAL

Material submitted for culture must always be fresh. Pus should never be obtained from slow-draining lines, bags, or bottles, where material has been sitting for a time. Many specimens are polymicrobic. If pus sits in a container, detection of slow-growing organisms in a specimen taken from it may be difficult or impossible owing to overgrowth of more rapidly growing organisms. This result may affect subsequent decisions about therapy. For the same reason, swab collection systems should contain holding media to protect organisms without permitting rapid multiplication during transport.

OBTAIN ADEQUATE QUANTITY OF MATERIAL

Quantity of material is equally as important as freshness. Wounds with a small opening may need to be sampled with a smaller swab, such as the nasopharyngeal swab, in order to obtain material from inside them. When smaller swabs are substituted, more of them will be necessary in order to gather adequate amounts of material. It may also be necessary to allow the swab to remain in a wound for a time while one "milks" the wound to obtain adequate specimen. The common practice of dabbing at a wound does not permit the necessary saturation of the swab.

Thorough evaluation of a wound *always* includes both smear and culture. A minimum of two swabs per culture is required. The value of the smear is to determine the adequacy of the specimen, to identify classic pathogens, and to determine the need for the use of special media, special incubation atmospheres, and length of incubation. It is likely that organisms such as *Actinomyces* species and *Nocardia* species would not be grown unless they were first seen on a smear, because cultures would not routinely be held long enough. When a single swab is received, the smear must be eliminated. Culture is still a more sensitive and definitive test. Additionally, quantity of culture material or numbers of swabs must be determined on the basis of the number of tests requested. A request for acid-fast, fungal, viral, and routine bacterial cultures cannot be fulfilled using just two swabs.

Need for Repeat Cultures

It is important to remember that even with therapy, wounds do not generally change overnight. Repetitious culturing of wounds is a waste of time and money. Most wounds can be easily accessed and sampled thoroughly, and should not require repeat culture for 4 to 5 days. Monitoring the situation with a Gram stain is far more economically efficient than repeat culturing.

PATIENT EDUCATION AND PREPARATION

Patient Education: Patient-Collected Samples

Not all specimens are collected by hospital personnel. Specimens such as urine, sputum, and stool are commonly obtained by the patient and most often require extensive patient education. Patients should never be asked whether they know how to collect a particular type of specimen. The answer "Yes" does not always mean that a patient has learned correct techniques. Attaching printed instructions in multiple languages to a collection device does not ensure that people will read them. The most effective educational tool appears to be giving the patient a preprinted sheet of instructions using simple language and pictures to help the patient follow along and understand as the procedure is verbally described to them. Table 7–1 has more extensive patient teaching tips.

Urine

Explanation of urine collection must include instructions for skin preparation and an explanation of "midstream" collection. Patients must be

■ T A B L E   7 – 1

PATIENT TEACHING TIPS

Deal with patient's immediate concerns first.
Assess current understanding and build on it.
Create learning environment—patient needs to know he is being taught.
Establish mutual learning outcomes with patient. Keep sessions short; separate long and complex material into short segments.
Avoid too much detail; use examples patient can relate to; teach skills in logical progression.
Have patient use as many senses as possible. This includes hearing, seeing, touching, writing, speaking and doing. Patients remember 90% of what they say and do, but only 10% of what they hear. Learning requires active participation.
Have patient practice new skills; expect a demonstration by the patient of skill taught.
Document what patient said or did to demonstrate learning.
Think teaching; incorporate it into your cares.

Reprinted with the permission of Abbott Northwestern Hospital. ©1986 Patient Education Department, Abbott Northwestern Hospital, Minneapolis, MN.

asked to begin to void without collecting the first portion of the specimen and to collect the middle portion (midstream). This technique helps to eliminate contaminating organisms. The technique should also be used by personnel collecting catheterized specimens, in order to eliminate organisms carried up the urethra during catheterization. Having separate urine collection instructions for males and females helps clarify the task for patients.

Sputum

Lower respiratory tract specimens are among the most difficult specimens to collect adequately and may require the help of a respiratory therapist. The patient should be instructed to remove dentures, rinse the mouth, gargle with water, and then expectorate with the aid of a deep cough. First morning specimens may be more successful owing to the volume of secretions that collect in the lungs overnight.

If the specimen is being collected for the purpose of diagnosing disease but appears normal, it should not be sent to the laboratory; another specimen should be collected. Infectious secretions exhibit such abnormalities as pus, blood, and dark flecks.

If patients are unable to expectorate a respiratory specimen, the specimen may be aspirated with the aid of a Leuken trap (Fig. 7–1). If secretions are too thick to aspirate, saline nebulization may be used to loosen them, as long as care is taken to avoid excessive dilution of the specimen.

■Figure 7–1

Leuken's trap.

Stools

Kits should be provided for the collection of stool specimens. Three-vial kits are recommended. Each such kit contains a vial with a preservative for parasites, a vial with a preservative for enteric pathogens, and a clean vial for examination of the raw specimen and any additional testing. It is important not to exceed the recommended amount of stool specimen per vial in order to maintain a 1:3 ratio of stool to preservative. It is also important to thoroughly mix formed specimens with the preservative to protect the organisms or parasites. Specimens should not be collected within 4 days of the patient's receiving barium. Barium will appear as a white chalky substance in the specimen. Patients should also be instructed to avoid contamination of the stool with urine. Multiple specimens may be necessary, particularly for diagnosing parasitic infections. It is important that each specimen be collected on a different day to take into account the life cycles of parasites.

Patient or Site Preparation

Preparation of the patient or the specimen site previous to specimen collection contributes to optimal recovery of infecting agents. Removal of an eschar before culture or biopsy of certain wounds or burns allows for more accurate wound assessment. Removing the mucous plug before obtaining a specimen for cervical culture increases the chances for recovery of cervical pathogens. Table 7–2 lists examples of how

■■■■■T A B L E 7 – 2
PATIENT OR SITE PREPARATION

Site	Preparation
Blood cultures	Skin preparation: alcohol, 2% tincture of iodine, alcohol
Body fluids (sterile)	Skin preparation as for blood cultures
Catheter tips	Iodine preparation previous to insertion
	Remove ointment or blood with alcohol before removal
Drainage sites	Aspirate or tissue is first choice
	Must get to source of drainage
Eye	Cleanse skin around eye
	Collect purulent material on swab or do corneal scrapings and place on media at bedside
	Use topical anesthetic only for corneal scraping
Fungal scrapings	Clean well with 70% alcohol or gauze soaked in sterile water or saline, then scrape periphery with scalpel
Genitalia	
Cervix/vagina	Use speculum
	Remove mucous plug before culturing
	Unroof vesicles before scraping the bases
Urethra	Instruct patient not to urinate for 1 hr previous to collection
	Use Dacron swab
	Insert swab 2–4 cm for 2–3 seconds and rotate gently while withdrawing swab
Lesion/wound/abscess	
Lesion/abscess	Unopen abscess: Skin preparation as for blood cultures
	Open wounds: Remove packing, clean with 3% H_2O_2
Burn wounds	Remove eschar before culturing or taking biopsy specimen
Biopsy	Skin preparation as for blood cultures
Sputum	Have patient remove dentures, rinse mouth and gargle with water
	May enlist aid of respiratory therapy
Stool	Avoid contamination with urine and barium
Urine	Cleanse perineal area with antiseptic towelette before collection, as per instructions on collection kit

patients or sites may be prepared prior to speci-men collection for each body site.

PRESERVATION, STORAGE, AND TRANSPORT OF SPECIMENS

Rarely is there an ideal situation in which patients come to the laboratory for a culture, allowing specimens to be processed immediately without transport delay. Transportation and preservation of the specimen is a problem whether the specimen comes from a nursing station three floors above the laboratory or from an outpatient facility, nursing home, or doctor's office across town. Major concerns are:

■ Overgrowth of rapidly growing organisms

■ Death of organisms due to changes in tempera-ture and pH

■ Inaccurate quantitation of organisms in urine and quantitative tissue

■ Loss of organisms due to drying

■ Protection from oxygen

■ Protection from clotting

■ Safety of the transporter

Maintenance or protection of the primary speci-men can be aided by the use of preservatives, anti-coagulants, holding media, and even culture media.

Use of Preservatives

Two major specimen types protected by preser-vatives are urine and stool. Boric acid is used in commercial products to maintain accurate urine colony counts. Phosphate-buffered saline (PBS) helps to preserve feces so that fragile organisms such as *Shigella* species will not die because of temperature and pH changes. Formalin, polyvinyl alcohol (PVA), and Schaudinn solution all help to preserve parasite trophozoites and cysts so they will remain in recognizable form.

Use of Anticoagulants

The use of anticoagulants is important for any specimen that might clot, because organisms bound up in clotted material are difficult to grow. Sodium polyanethol sulfonate (SPS) is the most common anticoagulant used for microbiology specimens. It is important that the concentration of SPS does not exceed 0.025% (w/v); some *Neisseria* species and anaerobic streptococci are inhibited by concentrations of 0.025 to 0.05%. Vacutainer tubes (Becton Dickinson) containing SPS are convenient for bone marrow aspirates or synovial fluids that tend to clot.

Heparin is another anticoagulant that can be used selectively in microbiology, but it is routine-ly used for viral culture as opposed to bacterial cultures. Heparin may inhibit the growth of gram-positive organisms and yeast. Citrate and ethylenediaminetetraacidic acid (EDTA) should not be used for microbiology specimens.

Use of Holding and Transport Media

Swab collection systems contain a holding medium that maintains viability of the organisms but does not permit an increase or decrease in numbers of organisms. Modified Stuart transport medium and Cary-Blair transport medium are commonly used. Studies have shown that many transport media perform equally well at refriger-ator temperature, but they should be tested against one another at room temperature when one is attempting to select a superior product.

Unlike other specimens, blood is usually placed in a broth culture medium immediately upon col-lection instead of in a preservative or a holding medium. The major reason is the need to quickly grow and identify the organism present in a poten-tially life-threatening situation such as bacteremia. The numbers of organisms present in bacteremia are usually low, and efforts must be made to enhance these numbers quickly. Organisms gener-ally multiply quickly in broth culture.

Unprotected Specimens

Many specimen types are transported without special protection. They include sputums, body fluids, tissues, catheters, medical devices, and specimens for sterility culture. The need to process such specimens without delay upon their arrival in the laboratory should be reinforced.

Storage of Specimens

Is it possible to process all specimens without delay upon arrival? Probably not! The laboratory

TABLE 7–3
STABILITY OF ORGANISMS

Fragile Organisms	Hardy Organisms
Streptococcus pneumoniae	Enterics
Neisseria gonorrhoeae	Pseudomonads
Neisseria meningitidis	Enterococcus
Salmonella	Other Streptococcus
Shigella	Staphylococcus
Haemophilus influenzae	Yeast
Anaerobes	Fungi
Mycoplasma	Mycobacteria
Viruses	Legionella
Chlamydia	Clostridium difficile toxin
Unpreserved parasites	

must make decisions regarding the storage of specimens that cannot be processed immediately. Decisions should be based on knowledge of organisms found in each type of specimen and the stability of those organisms. Table 7–3 separates organisms that are *fragile* (susceptible to temperature, environmental, and pH changes) from those that are *hardy* (not readily affected by these changes).

Should specimens be stored at refrigerator temperature, room temperature, or incubator temperature? Urine, viral blood specimens, catheters, and swabs should be refrigerated, but bacterial blood specimens, cerebrospinal fluid (CSF), and cultures processed onto agar plates at the bedside should be placed in an incubator. Room temperature is most appropriate for specimens such as hair and nails for isolation of fungus.

Special problems arise when one is deciding how to store respiratory and stool cultures. Respiratory specimens are likely to contain *Streptococcus pneumoniae*. Although this organism is fragile and may not tolerate the cold well, a respiratory specimen is also likely to be polymicrobic. Overgrowth becomes a problem without refrigeration. Stool specimens may contain *Shigella*, a very fragile, temperature-sensitive organism, but also contain many other species of gram-negative rods, making overgrowth an even bigger problem than with respiratory specimens. Most laboratories choose to refrigerate both specimen types if processing will be delayed, even though organism viability is not guaranteed. Although the Gram stain is often helpful in diagnosing *S. pneumoniae* even when it is unconfirmed by culture, the Gram stain does not differentiate *Salmonella* and *Shigella* from other gram-negative rods in a stool specimen. The best solution is to process stool specimens without delay.

Mailing Etiologic Agents

Mailing or shipping of etiologic agents or biohazardous materials is governed by federal regulations devised by the U.S. Department of Health and Human Services (Fig. 7–2). An *etiologic agent* is a viable microorganism or its toxin that causes, or may cause, human disease. Such material must be placed in a securely closed, watertight container such as a vial or test tube, known as the primary container. The primary container is placed in a secondary container with sufficient absorbent material (e.g., paper towels) between them to absorb the contents in case of breakage. The secondary container is then placed in an approved mailing container. An etiologic agent label must be placed on the mailing container. This label contains a biohazard symbol and a phone number for The Centers for Disease Control, Atlanta, Georgia (CDC), allowing for notification in the event of leakage.

The regulations also address such things as mailing of serum samples, maximum volumes of samples to be mailed, and lists of dangerous pathogens that must be sent by registered mail or an equivalent system. The regulations are subject to change. At present, additional or enhanced regulations have been proposed and await approval. Regulations are published in the *Federal Register*.

SAFETY

Protection of the Specimen Transporter

Protection of the transporter and of the laboratory personnel from infectious agents in the specimen is as important as protection of the specimen. Leaking specimens and specimens with needles attached present the greatest hazards. In compliance with the rules of Universal Precautions instituted by CDC, all specimens must be transported in leakproof secondary containers. Plastic bags with permanent seals and separate pouches on the outside for requisitions are recommended. After arrival in the laboratory, the bag is opened by tearing along a semiperforated line. Zip-lock bags are not preferred, because they are difficult to close and may leak during transport.

Transporting personnel should refuse to transport routine specimens without the protection of a secondary container. Refusing to accept

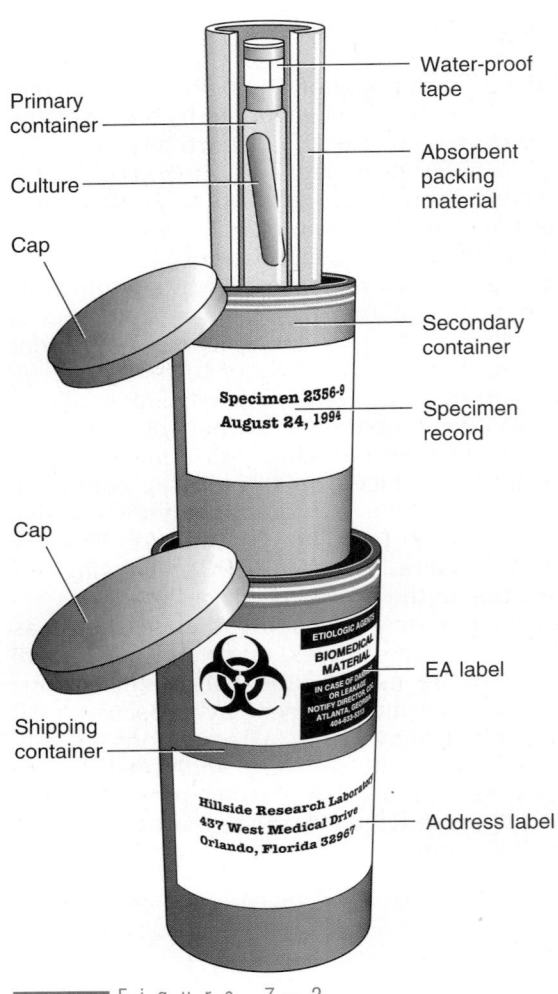

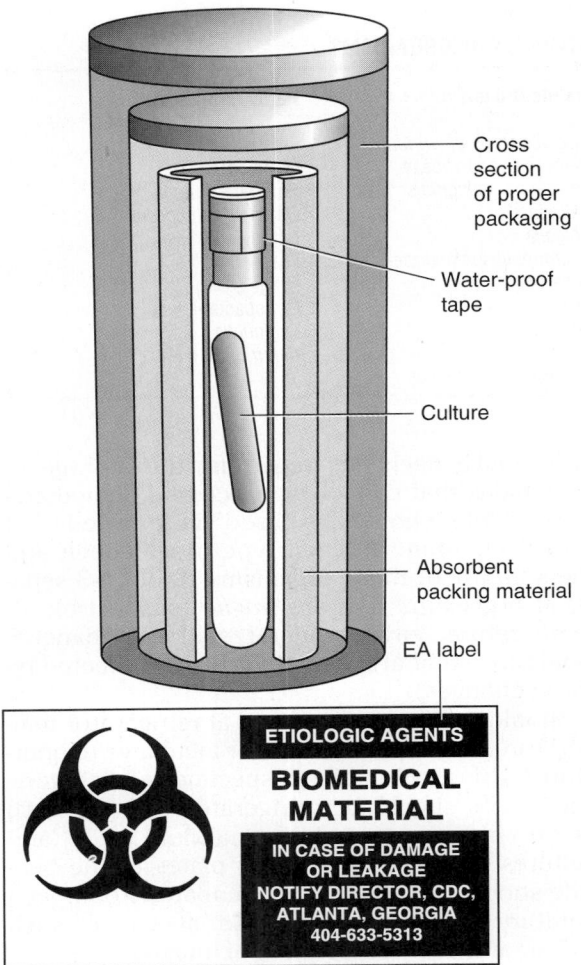

ETIOLOGIC AGENTS

BIOMEDICAL
MATERIAL

IN CASE OF DAMAGE
OR LEAKAGE
NOTIFY DIRECTOR, CDC,
ATLANTA, GEORGIA
404-633-5313

■■■■■■ F i g u r e   7 – 2
Packaging and labeling of etiologic agents.

syringes with needles attached is also appropriate. Needle sticks are among the most common and most hazardous laboratory accidents. They are probably also the most avoidable. A needle must be replaced with a tight-fitting rubber stopper or a stopcock to put resistance on the plunger. Another option would be to transfer the aspirated material to another type of container or sterile screw-capped tube, provided there is enough material to fill the tube (to eliminate air that could mix with the specimen and harm any anaerobes present).

As with everything else, there is an exception to the rule. There are times when a valuable aspirate is so small that most of it is in the needle instead of the syringe. When this happens, the specimen can be transported "individually"

(with needle attached) to the microbiology laboratory and placed in the hands of a laboratory employee to process immediately. Under no circumstances should such a specimen be sent via routine transportation mechanisms.

Protection of the Specimen Processor

All health care workers must adhere to strict safety guidelines in daily practice. Laboratory personnel are at risk for accidental infection if they do not use common sense and take a defensive attitude. All individuals handling patient specimens must wear protective clothing and must open specimens only in a biohazard hood. It is the responsibility of all medical staff to alert the laboratory and other health care workers when extremely dangerous pathogens are to be considered. Pathogens such as *Brucella, Francisella tularensis,* hepatitis B virus, and *Mycobacterium tuberculosis* are extremely hazardous and must be handled only in laboratories with sufficient safety procedures. Microbiology personnel must routinely participate in safety education.

LABELING AND REJECTION OF SPECIMENS

Requisitions

It is important that the laboratory requisition includes enough information to enable the laboratory to do the best job.

All that the laboratory knows about the patient is learned from the requisition. All microbiology requisitions should include information about the source, the diagnosis and/or history, and the test requested.

Source

The source should be specific. It is not enough to list wound or drainage. The source should include the anatomic location and the type of wound. Is it from the head or the leg? Is it a bite, a puncture wound, a surgical incision, or a skin abrasion?

Diagnosis/History

The diagnosis can sometimes help the technologist suspect growth of a specific organism that may require special media, a different incubation environment, or a longer incubation time. Additionally, knowing that the patient has osteomyelitis or pyelonephritis can affect the extent to which an organism is characterized and can determine the need for susceptibility testing. Knowledge of a patient's antibiotic therapy is useful for correlation of test results.

Test Requested

Every test requested should have a written order on the patient's chart. Most laboratories require a separate requisition for each test requested. Test requisitioning may also be electronically ordered from terminals at remote sites such as nursing stations. Computer screens are usually formated to require all information necessary to complete the ordering of a specific test or tests. A properly completed requisition is an important part of specimen collection.

Unacceptable Specimens

The information on the requisition must match the information on the specimen label. If names or sources do not match, the specimen should be re-collected. The laboratory should not assume the responsibility or liability for questionable information.

Perhaps the best approach to this issue is to divide specimens into two groups, invasive and noninvasive. A *noninvasive* specimen, such as urine, sputum, stool, and wounds, must always be re-collected if it is received unlabeled or there is a misidentification. An *invasive* specimen, such as blood culture, sterile body fluid, amniotic fluid, or operating room specimen, may be processed if the person responsible for the error comes to the laboratory and signs a laboratory release form (Fig. 7–3). This places all responsibility and liability on the person signing the form. The person's responsibility may be emphasized by giving him or her a copy of the signed release form.

All rejected specimens require a documented phone call to the collection site to expedite recollection of a specimen. This should be followed by written documentation to verify receipt of the specimen in the laboratory and the reason for its rejection.

Other suboptimal specimens that must be rejected are as follows:

- Leaking specimens
- Syringes with needles attached
- Stools contaminated with urine or barium
- Anaerobes on inappropriate sources
- Unpreserved specimens older than 2 hours
- Refrigerated blood cultures
- Dried-up specimens
- Specimens in formalin

Processing a suboptimal specimen will always yield suboptimal results.

T · H · E
OHIO
STATE
UNIVERSITY
HOSPITALS

The Ohio State University
Form 11123

LABORATORY RELEASE

UNACCEPTABLE SPECIMEN: TRANSFER OF RESPONSIBILITY

PATIENT IDENTIFICATION

SPECIMEN

NAME _____

MEDICAL RECORD NO. ☐☐☐☐☐☐☐☐☐

HOSPITAL NO. (4 DIGIT) ☐☐☐☐

NURSING STATION _____

ATTENDING DR. _____

SPECIMEN TYPE _____

REQUISITION

NAME _____

MEDICAL RECORD NO. ☐☐☐☐☐☐☐☐☐

HOSPITAL NO. (4 DIGIT) ☐☐☐☐

NURSING STATION _____

ATTENDING DR. _____

SPECIMEN TYPE _____

TEST(S) REQUESTED: _____

PROBLEM

DATE _____ TIME _____ REPORTED BY _____ SUPERVISOR REVIEW _____

☐ INVASIVE   ☐ NONINVASIVE (STATE EXCEPTION IN COMMENT AREA.)

☐ NO LABEL   ☐ LABEL AND REQUISITION NOT MATCHED.

☐ INCOMPLETE INFORMATION   ☐ (SUSPECTED) MISDRAW OR MISIDENTIFICATION.

COMMENTS: _____
_____
_____
_____
_____
_____

(continue on separate sheet.)

UNRESOLVED (NOT PROCESSED)

UNIT NOTIFIED: _____ NAME: _____

STAFF DECLINED RESPONSIBILITY (NAME) _____

RELEASE

THIS SPECIMEN WAS OBTAINED FROM:

NAME: _____

MEDICAL RECORD NO. ☐☐☐☐☐☐☐☐☐

LAB USE ONLY

I AM ASSUMING FULL RESPONSIBILITY FOR PROPER
IDENTIFICATION OF THIS SPECIMEN

DATE _____ TIME _____

_____
(PRINT FULL NAME/TITLE)

_____
(SIGNATURE)

WHITE—QA COMMITTEE   YELLOW—LABORATORY   PINK—STAFF

Figure 7 – 3

Laboratory release form. (Courtesy of Ohio State University Hospitals.)

PROCESSING OF CLINICAL SAMPLES FOR OPTIMAL ORGANISM RECOVERY

Prioritization During Processing

All specimens require prompt processing after arrival in the laboratory. It is often impossible to process every specimen as it is received. Laboratory staffing and specimen load may have a significant impact on timely processing. The laboratory must set guidelines for prioritizing specimens on the basis of a number of variables. A four-level scheme of prioritization may be used. The first level requires immediate attention, whereas processing of fourth-level specimens may be delayed. Table 7–4 lists clinical samples and how each can be prioritized in a four-level system.

Level 1 specimens need to be classified as "critical" owing to their invasive nature or severity of disease. They require immediate processing.

Level 2 specimens are unprotected and may quickly degrade or have rapid overgrowth of contaminating flora, changing the nature of such a specimen. There is a need to quickly provide an optimal growth environment for fastidious organisms that may be found in these specimens.

Level 3 specimens require quantitation. Urine, catheter tips, and quantitative tissue biopsy

specimens are level 3. Delay in processing level 3 specimens may adversely affect the accuracy of quantitation.

If processing of level 2 and level 3 specimens is postponed, some type of protection or preservation must be initiated. This may mean refrigerating urine specimens or placing blood culture bottles in the incubator until a spinal fluid specimen is processed. Placing a small undercounter refrigerator at the site of specimen accessioning is convenient for the processor and makes it more likely that urine specimens will be refrigerated during peak workload times.

Level 4 specimens are all other protected specimens arriving in the laboratory in holding media. Processing of level 4 specimens may be delayed in order to process higher-level specimens.

It must be emphasized that all specimens need to be processed in a timely manner. Batch processing should be avoided when possible.

Gross Examination of Specimens

Observing and documenting the gross appearance of a specimen may provide useful information to both the microbiologist and the physician. It is important to document for the written report the physical characteristics of the specimen:

■ Was it a swab or an aspirate?
■ Was the stool formed or liquid?
■ Was the specimen bloody?
■ What was the volume?
■ Was the fluid clear or cloudy?

Observations should also be made to determine the adequacy of the specimen:

■ Is there evidence of improper collection or transport?
■ Is the container leakproof and sterile?
■ Does the transport medium cover the swab or is the swab dry?
■ Is there enough specimen to perform all tests requested?
■ Is the fluid clotted?
■ Is there evidence of barium in the stool?

Gross appearance may determine the need for special processing.

TABLE 7–4
LEVELS OF SPECIMEN PRIORITIZATION

Level	Description	Specimens
1	Critical/invasive	Cerebrospinal fluid Brain Blood Heart valves Pericardial fluid Amniotic fluid Bronchoalveolar lavage Vitreous/aqueous fluids
2	Unpreserved May degrade or overgrow	Sputum Tissue Stool Body fluids not listed for level 1 Drainage from wounds Pus Bone
3	Accuracy of quantitation affected	Urine Quantitative tissue Catheter tip
4	Protected/preserved	Swabs in holding medium (aerobic and anaerobic)

■ Should anaerobe cultures be performed owing to purulence, foul smell, gas, or sulfur granules in the specimen?

■ Are worms or proglottids present in the specimen?

Direct Examination Techniques

Direct examination, regardless of the stain used, may be utilized to (1) determine the quality of the specimen, (2) aid in diagnosis of infectious disease, (3) guide routine culture interpretation, and (4) dictate the need for nonroutine processing.

Lack of material for direct examination makes all of these goals impossible or extremely difficult to achieve. Because the direct smear should be used as a guide to routine culture evaluation and interpretation whenever possible, the culture plates should not be viewed until the direct smear has been evaluated. Direct microscopic examination is a requirement in critical situations, such as meningitis, to guide therapy choices when therapy must be initiated before culture results are available.

Smear Preparation

Specimens may be received in many forms. Preparation of the direct smear depends on the type of material received. Techniques will vary according to whether the specimen is a tissue, swab, or fluid.

TISSUES

When tissue is received, a touch preparation may be made by cutting a fresh piece of tissue and touching it several times to different areas of the slide. Caution must be used with large pieces of tissue to select a representative area.

SWABS

If swabs are submitted, the swab must be carefully rolled back and forth across a dry, clean slide.

ASPIRATES AND BODY FLUIDS

When the specimen is an aspirate or a body fluid, at least four techniques are available for smear preparation: single drop, centrifuged sediment, layered, and cytocentrifuged (alone or with additives).

Single-Drop Smear. A sterile pipette may be used to place a drop of fluid on a slide. This technique is useful if high numbers of organisms are likely or the specimen is thick. Care must be taken to keep the smear thin enough to read.

Centrifuged Sediment Smear. Fluids in which small numbers of organisms may be present may be centrifuged at 1,500g for 15 minutes, and the sediment used to prepare the slide.

Layered Smear. Low-volume fluids such as CSF may not provide a sediment. An alternative technique would be to use the majority of the fluid by layering it onto a slide. A drop of fluid may be placed on a slide and allowed to dry, and another drop added on top of it, producing a layered smear. Care should be taken not to spread the drops out. This technique is advantageous for CSF smears for mycobacteria.

Cytocentrifuged Smear. An alternative to spinning the entire specimen is cytocentrifugation. Cytocentrifugation—with devices such as the Cytospin 2 (Shandon Inc, Pittsburgh, PA, Fig. 7–4) or the Aerospray Gram Slide Stainer/Cytocentrifuge (Wescor, Inc, Logan, UT)—is a technique used to concentrate a small amount of body fluid (0.1–0.5 mL) directly onto a circular area of a microscopic slide. This technique concentrates both cellular material and organisms in a monolayer and provides a small area to scan. The blotter used with the device absorbs the excess fluid not centrifuged onto the slide. This technique may be used for many types of body fluids, including bronchial alveolar lavage. Up to 12 slides may be spun at once, and different stains may be applied and examined. It has been noted by some researchers that the cytocentrifuge slide prepara-

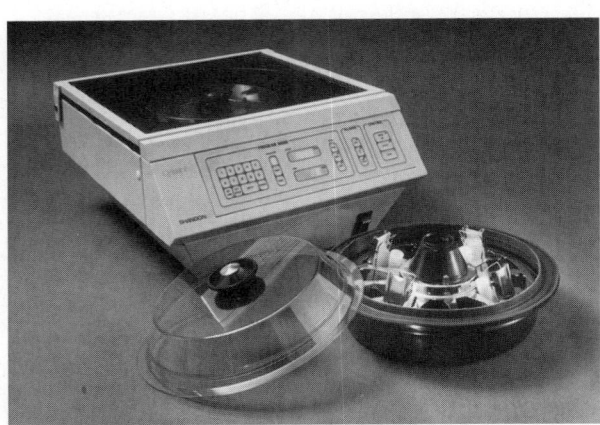

■■■ F i g u r e   7 – 4
Cytospin 2 (Shandon, Inc., Pittsburgh, PA).

tion may increase sensitivity of the CSF Gram stain up to 2 logs over conventional concentrated smears. This is a significant increase in sensitivity that enhances visualization of the small numbers of organisms sometimes found in CSF.

Bacterial and cellular morphology appears to be better in cytocentrifuge preparations. In our experience, the cytocentrifuge preparation has been useful for all body fluids and has proved to be more sensitive than smears made from concentrated fluids. Owing to the superiority of cytocentrifuge smears, a cytocentrifuge is a justified piece of equipment for all microbiology laboratories. Figure 7–5 illustrates a direct smear of a CSF prepared by cytocentrifugation and by the single drop method. Observe the concentration of both cells and organisms with clearing of the background material in the cytocentrifuge smear.

Additives. Other substances may be added to the body fluid. Sterile albumin may be added to clear CSF to help minimal cellular material adhere to the slide. When the cytocentrifuge procedure is utilized for extremely mucoid bronchial alveolar lavage specimens, dithiothreitol solution (Sputo-lysin Test, Behring Diagnostics Inc, Somerville, NJ) may be added as a mucolytic agent, before the specimen is placed in the cytocentrifuge funnel.

When Direct Smear Is Not Useful

Direct examination is not appropriate or useful for some types of specimens. A Gram-stained smear from a throat swab or nasopharyngeal swab

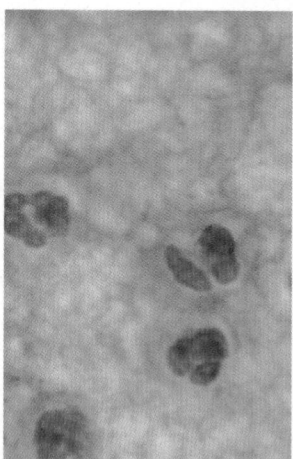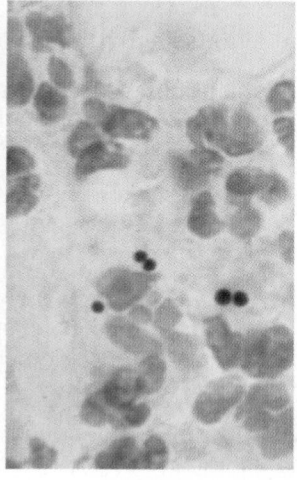

■■■■■ F i g u r e   7 – 5

Comparison of cytocentrifuged smear *(right)* and single-drop smear *(left)*.

cannot differentiate pathogenic from nonpathogenic streptococci. Many laboratories do not perform direct smears from urine specimens because culture results are available within 24 hours. Unless the urine smear is specified as a "stat" procedure, its results may not be available before the culture results. A direct urine smear may provide useful information only if it is made and read immediately after collection. Its ability to detect organisms is in the range of 10^5 organisms per mL. If one organism is seen on smear, it represents 100,000 organisms per mL. Urine smears should be made using a loopful of unspun urine. Direct examination techniques are included in Table 7–5.

Primary Inoculation of Routine Specimens

Selection of Primary Culture Media for Routine Specimens

Selection of aerobic primary media is somewhat standardized for the routine bacterial culture. Individual laboratories may favor one medium over another, however, on the basis of past experience, patient population considerations, or other special circumstances. Several basic goals must be met in selecting routine media regardless of the specific choice.

TYPES OF CULTURE MEDIA

In order to meet these goals, an understanding of medium types is needed. Culture media may be divided into categories defined by ability to support bacterial growth. These categories are as follows:

Nonselective Media. A nonselective medium will support the growth of most nonfastidious bacteria, usually in agar plate form. Sheep blood agar is the standard nonselective medium used in the United States.

Selective Media. Selective media support the growth of one type of bacteria over another. A selective medium may contain inhibitory substances to prevent the growth of some bacteria. Such media may be created by adding antibiotics or other inhibitory chemicals to nonselective media. MacConkey agar is selective for enteric gram-negative bacilli, and CNA (Columbia agar with colistin and nalidixic acid) is selective for gram-positive organisms.

Differential Media. A differential medium allows grouping of bacteria on the basis of

■■■■■ T A B L E   7 – 5
INITIAL PROCESSING TECHNIQUES FOR BACTERIAL SPECIMENS

Specimen Type	Cytocentrifuge Smear	Routine Smear	Liquefaction	Filtration	Inoculum Counter-Streak
Amniotic fluid	✓				
CSF	✓				✓
Synovial fluid	✓				
Other body fluids	✓			If >10 ml	If <10 ml
Bronchial lavage	✓				
Bile		✓			
Bone marrow		✓			
CAPD	✓			If >10 ml	
Peritoneal/ascitic	✓				
Pericardial fluid	✓				If <10 ml
Lesion/wound/abscess on swabs		✓	Centrifuge		
Organ/tissue/biopsy		Touch prep	"Stomach"		
Aspirate/drainage		✓			
Bone		✓	Grind		
Urine		✓			
Organ perfusate	✓			If >10 ml	If <10 ml
Medical device soak solution				If >10 ml	If <10 ml

different characteristics displayed on the medium. Media may be differential and nonselective; for example, sheep blood agar can differentiate organisms on the basis of hemolysis. Media may be differential and selective; for example, Mac-Conkey agar differentiates gram-negative bacilli on the basis of lactose fermentation.

Enriched Media. In an enriched medium, growth enhancers have been added to nonselective agar to allow fastidious organisms to flourish. Chocolate agar is an enriched medium.

Enrichment Broth. Enrichment broth is designed to encourage the growth of small numbers of a particular organism while suppressing other flora present. Gram-negative (GN) broth and selenite broth, the two most common enrichment broths, are used in stool cultures to enhance recovery of *Salmonella* and *Shigella*.

Broth Media. A broth medium may enhance small numbers of most aerobes, anaerobes, and microaerophils and some fastidious organisms. Thioglycolate (THIO) broth is an example of a primary broth medium.

REQUIREMENTS FOR ROUTINE PRIMARY PLATING MEDIA

The routine primary plating media chosen for use in a laboratory should include the following items:

■ A nonselective agar plate.

■ An enriched medium for fastidious organisms, if the source is a normally sterile body fluid or a site in which fastidious organisms are expected.

■ A selective and differential medium for enteric gram-negative bacilli for most routine bacterial cultures.

■ A selective plate for gram-positive organisms for specimens from anatomic sites in which mixed gram-positive and gram-negative organisms are suspected.

■ Additional selective media for specific pathogens as needed. Body site specimens from which pathogenic *Neisseria* may be isolated require a selective *Neisseria* plate.

■■■■■ T A B L E   7 – 6

SELECTION OF PRIMARY CULTURE MEDIA FOR ROUTINE SPECIMENS*

Specimen Type	Gram Stain	BAP	Choc CO_2	MAC	PD	CNA	THIO Broth	TM (CO_2)	Special Medium; Comments
Body fluids	✓	✓	✓	✓	✓		✓		
Amniotic fluid	✓	✓	✓	✓	✓		✓	✓	
CSF	✓	✓	✓		✓		✓		
Synovial fluid	✓	✓	✓	✓	✓		✓	✓	
Bile	✓	✓	✓	✓	✓		✓		
Bone marrow	✓	✓	✓	✓	✓		✓		
CAPD <10 ml	✓	✓	✓	✓	✓		✓		
>10 ml	✓		Filter				✓		
Peritoneal/ascitic	✓	✓	✓	✓	✓		✓		
Pericardial	✓	✓	✓	✓	✓		✓		
General sources Bone	✓	✓	✓	✓	✓		✓		
Lesion/wound/abscess	✓	✓	✓	✓	✓	✓	✓		
Wound aspirate	✓	✓	✓	✓	✓	✓	✓		
Tissue/biopsy	✓	✓	✓	✓	✓		✓		
Eye/ocular	✓		✓		✓		✓		Bedside inoculation Infants: add TM (CO_2)
Aqueous/vitreous	✓	✓	✓		✓		✓		
Urogenital Urine Catheter/void				✓		✓			
Upper tract	✓	✓	✓	✓	✓	✓	✓		
Genital: male or female	✓	✓	✓	✓	✓	✓	✓	✓	
Placenta	✓	✓	✓	✓	✓		✓		
IUD	✓								
Urogenital screens *Neisseria*			✓					✓	
Group B beta streptococcus		✓							SBM
Yeast					✓				
Gardnerella						✓			
Gastrointestinal Feces	✓			✓			CT 4°		CIN (25°), CHE, SMAC CBA or 0.45-μm filter on BAP (42°C, 10% CO_2)
Duodenal aspirate	✓	✓	✓	✓	✓	✓			
Gastric aspirate (infant)	✓	✓	✓	✓	✓		✓		
Gastric biopsy	Touch prep								Skirrow
Respiratory sources Throat		✓	✓						

Table continued on following page

■ T A B L E 7 – 6
SELECTION OF PRIMARY CULTURE MEDIA FOR ROUTINE SPECIMENS* *(Continued)*

Specimen Type	Gram Stain	BAP	Choc CO_2	MAC	PD	CNA	THIO Broth	TM (CO_2)	Special Medium; Comments
Nasopharyngeal/nasal		✓	✓						
Sputum	✓	✓	✓	✓					
Bronchial lavage	✓	✓	✓	✓	✓				
Bronchial brush	✓	✓	✓	✓	✓		✓		
Sinus aspirate	✓	✓	✓	✓	✓	✓	✓		
Pleural fluid	✓	✓	✓	✓	✓		✓		
Lung tissue	✓	✓	✓	✓	✓		✓		
Respiratory screens Group A beta streptococcus		✓							
Staphylococcus aureus		✓							MS or MS + oxac
Neisseria			✓					✓	
Yeast				✓					

*BAP, sheep blood agar; CBA, Campylobacter blood agar; Choc, chocolate agar; CIN, cefsulodin-irgasor-novobiocin agar; CNA, Columbia colistin–nalidixic acid agar; CT, Campylobacter THIO broth; HBT, human blood bilayer medium; HE, Hektoen-enteric medium; IUD, intrauterine device; MS, mannitol salt agar; MS + oxac, mannitol salt agar with oxacillin; PD, potato dextrose agar with antibiotics; SBM, special broth medium; SMAC, sorbitol-MacConkey agar; THIO, thioglycolate; TM, Thayer-Martin or other Neisseria-selective agar.

■ A broth medium. Some laboratories have chosen to eliminate broth media in routine processing. Other laboratories include a thioglycolate broth when plating specimens of body fluids, tissues, lesions, wounds, and abscesses. A broth should be considered at least for plating normally sterile body fluids or specimens from blood- or brain-related sites.

■ A potato dextrose agar plate for yeast isolation.

Table 7–6 lists selection of primary media for specific anatomic sites. Equivalent media may be substituted.

Selection of Temperature and Environmental Conditions

Once media selection is made, temperature and environmental conditions must also be considered.

■ Environmental conditions are chosen according to the growth requirements of the indigenous flora or pathogen(s) suspected for the body site from which the specimen is taken.

■ Fastidious organisms may require increased CO_2 or an anaerobic environment for growth.

■ Most routine bacterial culture plates are incubated at 35 to 37°C for 48 hours.

■ Broth cultures, when included, are routinely held 5 to 7 days.

Primary Isolation Media for Unusual and Fastidious Bacteria

Temperature requirements and length of incubation may vary for individual organisms. Unusual organisms may require special processing and selection of a medium beyond the routine. It is helpful if the clinician indicates to the laboratory that an unusual organism is suspected. Table 7–7 lists a variety of unusual or fastidious organisms, recommended media, and incubation requirements.

Initial Processing Techniques for Primary Inoculation

LIQUEFACTION OF SPECIMENS

As previously stated, most specimens arrive in the laboratory in one of three forms; swab, tissue, or liquid (fluid). The objective of the labora-

■■■■■■■ T A B L E 7 – 7

UNUSUAL AND FASTIDIOUS BACTERIA: PRIMARY ISOLATION MEDIA

Organism	Recommended Isolation Media*	Special Considerations	Specimen Selection
Anaerobes	Anaerobic blood agar supplemented with vitamin K and hemin (Schaedler, CDC anaerobic blood agar)	Prereduce all anaerobic media and hold 6–7 days Nonselective	Aspirate Tissues Sterile body fluids
	Kanamycin/vancomycin laked blood agar (KV laked blood)	Selective for *Bacteroides*	
	Phenylethyl alcohol sheep blood agar (PEA), anaerobic	Inhibits facultative gram-negative rods and inhibits *Proteus* swarming	
	Bacteroides bile esculin agar (BBE)	Optional; presumptive identification of *Bacteroides fragilis* group	
	Thioglycolate broth with vitamin K and hemin	Recommended for body fluids and tissues only	
Actinomyces	See Anaerobic media selection (see above)	Hold broth 3 weeks	
Bordetella pertussis	Bordet-Gengou Charcoal–horse blood agar (Regan-Lowe)	Bedside inoculation; incubate at high humidity and 35°C in air for 6–7 days; CO_2 not recommended Direct fluorescent antibody test—supplement to culture	Use calcium alginate swabs; collect 2 nasopharyngeal swabs Bronchial washing Throat
Brucella	Vented biphasic or broth blood culture bottles for blood and body fluids (Castenada)	Potentially dangerous pathogen; process all cultures in biohazard hood with gloves and protective clothing Hold blood bottles 30 days at 35–37°C in CO_2 and subculture every 4–5 days Hold 10 days at 35–37°C in 10% CO_2	Bone marrow Blood
	Modified Thayer-Martin or 5% sheep blood agar, *Brucella* agar with 5% serum	Use in potentially contaminated specimens	Abscesses/tissues
Clostridium botulinum	Anaerobic media (see above)	Reference lab test Toxin assay for diagnosis	Feces Serum, feces
Corynebacterium-diphtheriae	Nonselective 5% sheep blood agar	Not fastidious 35°C in air	Throat Nasopharyngeal Skin
	Loeffler serum slant	Rabbit serum enhances growth	
	Potassium tellurite medium	Need toxigenicity testing to confirm a diagnosis	
	Tinsdale or cysteine-tellurite blood agar	Supports growth of only some strains	
Francisella	Commercial chocolate agar Modified yeast extract agar	Potentially dangerous pathogen, highly infectious by aerosol or penetration of unbroken skin; process all cultures in biohazard hood with gloves and protective clothing Increased CO_2 required; 3–5 days required for visible growth	Lymph node aspirate Sputum Throat Bronchial washing Ulcer biopsy (advancing edge of lesion)
Haemophilus-ducreyi (chancroid)	Fresh commercial chocolate agar Addition of vancomycin in contaminated cultures	35°C, 3–5% CO_2 with high humidity; hold for 5 days	Genital lesion
Leptospira	Fletcher's medium Tween 80 bovine serum Albumin medium	28–30°C for 6 weeks; requires dark-field microscopy Remains viable in urine for 1 hour only Important to collect multiple specimens Reference lab test Serology	Blood & CSF–1st week Urine–after 2nd week

Table continued on following page

■■■ T A B L E 7 – 7

UNUSUAL AND FASTIDIOUS BACTERIA: PRIMARY ISOLATION MEDIA (Continued)

Organism	Recommended Isolation Media*	Special Considerations	Specimen Selection
Nocardia	Grows well on routine media 5% sheep blood agar, BHI Grows well on BCYE	Hold for 5–7 days in 3–5% CO_2	
Streptobacillus moniliformis (rat bite fever)	Serum-supplemented media (rabbit, calf or horse serum)	Fastidious microaerophilic Practical laboratory tests not available Reference lab test inhibited by SPS High humidity at 35–37°C in 10% CO_2 Use citrate as anticoagulant	Blood—specimen of choice Joint fluid Abscess fluid Lymph node
Vibrio	TCBS (thiosulfate-citrate-bile-sucrose agar) Grows well on routine stool media	Selective and differential; not cost-effective to use as routine medium in many parts of the country	Stool Lesion

*Representative media listed—equivalent media may be substituted.
BCYE, buffered charcoal yeast extract agar; BHI, brain-heart infusion agar; SPS, sodium polyanethol sulfonate.

tory should be to convert all specimens for culture to a liquid form (liquefy) without significant dilution of the organisms. Liquefaction permits equal distribution of organisms onto each medium inoculated. To accomplish this task, swabs are placed in 0.5 to 1.0 mL of broth, then centrifuged to produce an even suspension of organisms. A sterile pipette is used to dispense inoculum onto plates and into broth. A separate swab is used to make the smear. Tissues are ground in tissue grinders or "stomached" in a Stomacher

■■■ F i g u r e 7 – 6

Stomacher Lab Blender. (Courtesy of Seward Medical, Ltd., Tekmar, Cincinnati, OH).

Lab Blender (Seward Medical, Ltd, Tekmar, Cincinnati, OH) in small amounts of broth. The broth-tissue suspension is then processed (Fig. 7–6). It is important to remember, however, that the direct smear must be made from the original material prior to liquefaction.

Other specimens, such as normally sterile body fluids, pus, urine, and even sputum, are inoculated directly onto selected media. Extremely tenacious or mucoid specimens may be digested with substances such as Sputolysin previous to media inoculation. Methods used to increase the recovery of organisms from body fluids include inoculum counter-streak technique and filtration.

INOCULUM COUNTER-STREAK TECHNIQUE

The inoculum counter-streak technique is used when the numbers of organisms are expected to be low but the volume of specimen received is not adequate for filtration (Fig. 7–7). The technique uses a larger volume of inoculum than conventional processing, thereby increasing the sensitivity of isolation. If the entire specimen is used, any existing organisms should be recovered. A sterile pipette is used to inoculate the plate from edge to edge across its center. A loop is then used to counter-streak the plate perpendicular to the inoculum line.

The number and type of plate media inoculated may be modified, depending on the volume of fluid received. The objective is to use a large volume of specimen. This technique is a good

PROCEDURE 7–1. NALGENE FILTERING PROCEDURE

Processing should take place in a biohazard hood.

1. Attach a Nalgene filter unit to the Nalgene hand vacuum pump. Attach one end of the tubing to the vacuum pump and the other end to the container to be evacuated.

2. Remove the lid and fill the container with the fluid to be filtered.

3. Manually pump the hand vacuum until all fluid has passed through the filter. Air is exhausted at the rate of 15 mL per stroke for the smaller pump and 36 mL per stroke for the larger pump.

4. Remove the filter following the manufacturer's instructions, and place it on the surface of an agar plate. Chocolate agar is recommended.

If a fluid is extremely cloudy, filter clogging may be a problem. If the filter does clog, a set of agar plates should be inoculated using the inoculum counter-streak technique of processing described previously.

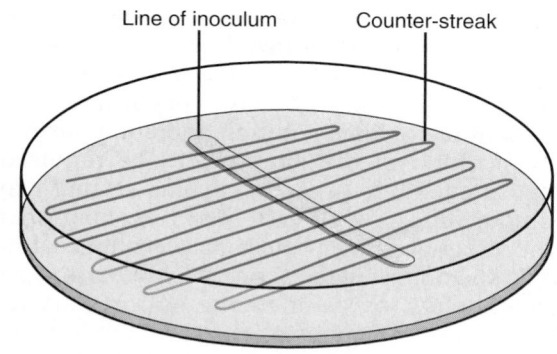

Line of inoculum   Counter-streak

■■■■ F i g u r e 7 – 7
Inoculum counter-streak technique.

choice for processing spinal fluids, because a CSF specimen could be plated to one plate or to three plates, depending on the volume received. If there is only enough specimen for one plate, chocolate agar should be inoculated. Media priority should be established ahead of time.

FILTRATION

Some body fluids and sterility cultures require concentration by filtration when large quantities are received. When the specimen consistency is thin enough to avoid filter clogging, filtration with a Nalgene filter unit is recommended.

Filtration is an excellent technique for processing continuous ambulatory peritoneal dialysis (CAPD) fluids, because of the ability to easily obtain a large volume of specimen and the likelihood that the specimen will have low numbers of organisms.

For attempting to recover *Campylobacter* from stool specimens, a filtration technique is both cost-effective and efficient. Liquefied stool is passed through a 0.45-μm membrane filter (HT-450 filter; Gelman Sciences, Ann Arbor, MI). The filter is placed on a blood agar plate at 42°C in 10% CO_2. After 30 minutes, the filter is removed, and the blood agar plate is returned to the microaerophilic environment, permitting *Campylobacter* to grow luxuriantly.

Processing Nonroutine Specimens

Routine specimens are defined as those specimens that have standardized processing procedures already established in the microbiology laboratory. Examples are urine, stool, sputum, and wound specimens.

In contrast, *nonroutine specimens* may be processed often by the laboratory but have no standardized, established processing procedure. They include vein grafts (both real and artificial), multiple-lumen catheters, heart valves, implant soak solutions, perfusates, water samples, and equipment suspected of playing a role in nosocomial infection. As technology advances, laboratories are receiving many more requests to culture new specimen types that have no established culture procedure. Hospitals with transplant programs may find it necessary to culture food previous to consumption by immunosuppressed patients. Even when standardized procedures do not exist, a standardized thought process can be developed to ensure that even the strangest specimen is appropriately cultured. This process begins by asking the series of questions listed below.

■ **Is the specimen likely to contain low numbers or high numbers of organisms?** If there are low numbers of organisms, concentration of the specimen will be advantageous. When few organisms are anticipated, large amounts of specimen will yield better results.

■ **If the number of organisms is extremely low, is it important to enhance them?** Some specimens, such as perfusates, are being checked for sterility. In this case, the presence of even one organism is significant. Other specimens such as foods may have a tolerance limit of a given number or kinds of bacteria; they will need to be processed using a quantitative or semiquantitative procedure so that colony counts may be determined. The use of a broth is important when there is a need to enhance growth, but it can confuse results for specimens requiring colony counts.

■ **Does the specimen contain any preservatives or growth inhibitors that must be counteracted?** Sometimes, the effects of a preservative can be eliminated or reduced by dilution or by the use of a specific medium such as standard method agar or D/E neutralizing broth.

■ **What is a reasonable amount to culture? Are all areas of the specimen homogeneous, or will the portion chosen for culture affect the results?** If a 12-cm piece of vein is received for culture, it may contain plaque on only a relatively small portion. Sampling one spot may not be representative of the whole. It would be necessary to sample and grind small pieces from multiple sites along the 12-cm piece to be efficient.

■ **Are the organisms to be found in a specific specimen likely to be fastidious or nonfastidious?** This affects the choice of medium, temperature(s), environment, and length of incubation.

■ **Is there any normal flora associated with the specimen?** The presence of normal flora might make special collection techniques important. It also dictates rapid or protected transportation and timely processing.

■ **Is the objective to select a single agent from a mixed culture?** An enriched or selective medium is helpful for this type of isolation. If there is an outbreak of methicillin-resistant staphylococci, for example, wherein equipment or personnel specimens are being cultured, it is helpful to put an appropriate amount of methicillin in the screening medium to eliminate methicillin-sensitive staphylococci.

■ **Is there a need to culture both external and internal surfaces?** This issue becomes important for devising beneficial methods of culturing catheters and other inanimate objects.

Types of Nonroutine Specimens

IMPLANT SOAK SOLUTION AND PERFUSATES

Previous to implantation, valves, corneas, organs, and other medical devices are maintained in soak solutions. At present, there seems to be no standardized procedure for culturing a soak solution. If contamination exists, it would be expected to be at a very low level. A large volume of soak solution and some form of concentration of the soak would be required, because even one organism in this setting may be important. A broth with heavy inoculation, a cytocentrifuge smear, and a large volume of filtered specimen placed on a chocolate agar plate should provide adequate opportunity for low numbers of organisms to grow. The Nalgene filtering procedure previously described works well for this type of specimen. Perfusates are processed in the same manner as implant soak solutions.

WATER STERILITY SPECIMENS

A specimen of water from a source such as a whirlpool, a still, or reagent water also requires concentration in order to provide adequate opportunity to detect microbes. The Millipore Sampler is one product designed for this purpose. The sampler is designed to use 18 mL of water to fill the sample case, but only 1 mL of water is actually absorbed.

The National Committee for Clinical Laboratory Standards (NCCLS) Document C3-A2, Vol. 11, No. 13, entitled "Preparation and Testing of Reagent Water in the Clinical Laboratory" and approved in August 1991, outlines other options for water testing.

PROCEDURE 7–2. PROCEDURE FOR USING A MILLIPORE SAMPLER

1. Pour water into the sample case to the upper line (18 mL).

2. Insert paddle, close case tightly, and lay case on a flat surface with grid side facing down.

3. Do not agitate, and do not exceed 30 sec-onds or medium might be lost through the membrane.

4. Remove paddle from case. Empty water.

5. Insert paddle back into case and incubate with grid side down.

EQUIPMENT

Contaminated equipment may be extremely difficult to culture owing to lack of access to the contaminated site. An attempt should be made to culture any reservoir where moisture might collect or where proper cleaning might be difficult. If culturing dry surfaces is necessary, the use of a moistened swab may be beneficial. Cleansing and assembling techniques should be reviewed at the time of culturing. Single-use, disposable equipment is preferred when possible.

INTRAUTERINE DEVICES (IUDS)

Intrauterine devices are usually cultured for the purpose of identifying *Actinomyces*. A Gram stain of the material should adequately identify the presence of this organism. Cultures require lengthy incubation and manipulation when the information obtained from them is no more helpful in guiding patient management than the smear results.

VASCULAR CATHETER TIPS

Vascular catheter tips are submitted to the laboratory for culture to aid in diagnosis of catheter-related infections. Standardization of the procedure is important.

In 1977, Maki and colleagues[1] determined that diagnosis of catheter site infection and catheter-related septicemia could be made on the basis of organism colony counts after rolling a 5- to 7-cm segment of the catheter across a blood agar plate. More than 15 colonies indicated site infection, whereas catheter-related septicemia correlated with more than 1,000 colonies. Many laboratories elect to use this method because of its ease of use. It is far more valuable in predicting catheter-related infection than use of a broth culture. If a catheter containing even one colony is placed in a broth, it will yield a "positive" culture. Even the Maki method demonstrated that only 1.6% of positive catheter cultures accurately predicted septicemia, giving this method a very low positive predictive value.

In the 1980's and 1990's, newer methods were developed that increase the positive predictive value. They involve techniques that culture both the internal and external surfaces of the catheter. Sheretz and colleagues[2] used a 3-minute sonication procedure to remove organisms from both the internal and external walls of the catheter. The catheter is placed in 10 mL of broth and sonicated in a water bath for 1 minute, centrifuged 15 seconds, and plated in two dilutions. For central venous and arterial catheters, sonication has been shown to be superior to the roll plate method of Maki and associates. Many laboratories may hesitate to change to this technique owing to the time, equipment, and manipulation involved.

Results of catheter tip culture can be adversely affected by delay in transport and processing, because the tip is an unprotected specimen. As the surface area dries, organisms may die, yielding inaccurate results. If the catheter tip is extremely long or is transported with an excessive amount of blood on it, the colony count and its interpretation may be significantly affected. If care is not taken during their removal, catheters may become contaminated with multiple skin organisms, again adding to the interpretive confusion and the cost of processing a specimen. With these issues in mind, it may not be appropriate to identify and test susceptibility of all organisms found in quantities greater than 15 colonies on a culture. Laboratories should set rules based on clinical relevance and cost-effectiveness for highly mixed cultures.

Bibliography

Bobadilla M, Sifuentes J, Garcia-Tsao G: Improved method for bacteriological diagnosis of spontaneous bacterial peritonitis. J Clin Microbiol 27:2145, 1989.

Chapin-Robertson K, Dahlberg SE, Edberg SC: Clinical and laboratory analysis of cytospin-prepared Gram stains for recovery and diagnosis of bacteria from sterile body fluids. J Clin Microbiol 30:377, 1992.

Costello MJ, Morrow SL, et al: Guidelines for specimen collection, transportation, and test selection. Lab Med 24:19, 1993.

Department of Labor, Occupational Safety and Health Administration: Occupational exposure to bloodborne pathogens; proposed rule and notice of hearing. Federal Register, pp 23041–23139, May 30, 1989.

Doyle PW, Crichton EP, et al: Clinical and microbiological evaluation of four culture methods for the diagnosis of peritonitis in patients on continuous ambulatory peritoneal dialysis. J Clin Microbiol 27:1206, 1989.

Fody ED: Clinical Laboratory Handbook for Patient Preparation and Specimen Handling, Fascicle III. Northfield, IL: College of American Pathologists, 1985.

Fuller D, Davis T, et al: Comparison of BACTEC Plus 26 and 27 Media with and without Fastidious Organism Supplement with conventional methods for culture of sterile body fluids. J Clin Microbiol 32:1488, 1994.

Gill VJ, Nelson NA, et al: Optimal use of the cytocentrifuge for recovery and diagnosis of *Pneumocystis carinii* in bronchoalveolar lavage and sputum specimen. J Clin Microbiol 26:1641, 1988.

Isenberg HD, Schoenknecht FD, et al: Cumitech 9, Collection and Processing of Bacteriological Specimens, Washington, DC: American Society for Microbiology, 1979.

Isenberg HD, Washington JA, II, et al: Manual of Clinical Microbiology, 5th ed. Washington, DC: American Society for Microbiology, 1991, pp 15–28.

Maki DG, Weise CE, Sarafin HW: A semiquantitative culture method for identifying intravenous catheter–related infection. N Engl J Med 296:1305, 1977.

Males BM, Walshe JJ, et al: Addi-Chek filtration, Bactec, and 10-ml culture methods for recovery of microorganisms from dialysis effluent during episodes of peritonitis. J Clin Microbiol 23:350, 1986.

National Committee for Clinical Laboratory Standards: Protection of Laboratory Workers from Infectious Disease Transmitted by Blood, Body Fluids, and Tissue; Tentative Guideline. NCCLS Document M29-T. Villanova, PA: National Committee for Clinical Laboratory Standards, 1989.

Pezzlo M: Clinical Microbiology Procedures Handbook. Washington, DC: American Society for Microbiology, 1992, pp 1.1.1–1.18.37.

Pezzlo M: Specimen collection: The role of the clinical microbiologist. Clinical Microbiology Updates. Hoechst-Roussel Pharmaceuticals Inc. 3(4), 1991.

Rossett W, Hodges GR: Antimicrobial activity of heparin. J Clin Microbiol 11:30, 1980.

Shanholtzer CJ, Schaper PJ, Peterson LR: Concentrated Gram stain smears prepared with a cytospin centrifuge. J Clin Microbiol 16:1052, 1982.

Sheretz RJ, Raad II, et al: Three-year experience with sonicated vascular catheter cultures in a clinical microbiology laboratory. J Clin Microbiol 28:76, 1990.

Weinstein MP: Clinical evaluation of a urine transport kit with lyophilized preservative for culture, urinalysis, and sediment microscopy. Diagn Microbiol Infect Dis 3:501, 1985.

Yeung C, May J, Hughes R: Infection rate for single-lumen versus triple-leumen subclavian catheters. Infect Control Hosp Epidemiol 9:154, 1988.

CHAPTER **8**

Microscopic Examination of Infected Materials

Leona W. Ayers

Direct microscopy for visualization of microorganisms has been possible for just over 200 years but was not a practical reality until Koch established the germ theory of disease in the 1880s. By 1880, a Scottish surgeon had published his direct observations of cluster-forming cocci in purulence from human disease. He named these cocci *Staphylococcus.* In 1884, Christian Gram developed the stain that today allows us to directly examine a pus specimen for the gram-positive cocci of *Staphylococcus.* Differential staining and microscopy underpin the laboratory diagnosis of infectious diseases.

This underpinning in the diagnostic microbiology laboratory is the ability to combine the rapid response of direct specimen examination with culture isolation and antibiotic susceptibility testing in order to:

- Confirm that the material submitted for study is representative

- Identify the cellular components and debris of inflammation and thereby establish the probability of infection

- Identify specific infectious agents using direct visual detection supported by appropriate culture isolations and immune antibody techniques

- Provide antibiotic susceptibilities of isolated pathogens to guide treatment

- Develop epidemiologic data

In upward of 88% of instances, the physician has a correct idea about the diagnosis after taking a patient history and performing a physical examination. In the remaining instances, in which the diagnosis is not evident, help comes from laboratory or radiologic studies. With infectious diseases, the physician has an idea of the likely etiology from the rate of symptom progression and is able to evaluate the extent of the infectious process. Pressure is great on the physician to begin immediate treatment of patients who are symptomatic. Specimens are collected and sent to the laboratory to confirm the idea that the physician has about the patient's illness. The ability of the laboratory to respond to the physician in a timely manner with useful results is key to keeping the treatment moving in the correct direction or to changing treatment direction if the physician's presumptive diagnosis proves to have been incorrect. The diagnostic microbiology laboratory has the opportunity to respond to the physician during the treatment decision-making or early in presumptive therapy. Direct visualization of pathogens becomes primary or direct evidence to confirm or refute the physician's initial clinical impression. If this impression is incorrect, reconsideration is facilitated, and additional studies can be undertaken as needed. Culture results are usually too late to alter presumptive therapy. At best, they confirm the correctness of the therapeutic choices already made and implemented.

PREPARATION OF SAMPLES

The preparation of samples for routine, bright-field microscopy has the objective to prepare material in a manner that facilitates adequate examination within a reasonable time. For smears, specimens should be examined grossly to determine the best approach (Table 8–1). Both thick, but not opaque, and thin (monolayer) smear areas should be produced by the smear process chosen.

Smears from Swabs

Smears should not be prepared from a spent swab after it has been used to inoculate culture media. Ideally, if the sample can be collected only on swabs, two swabs are submitted. Smears from swabs are prepared by rolling the swab back and forth over contiguous areas of the glass slide to deposit a thin layer of sample material (Fig. 8–1). This preserves the morphology and relationships of the microorganisms and cellular elements. The

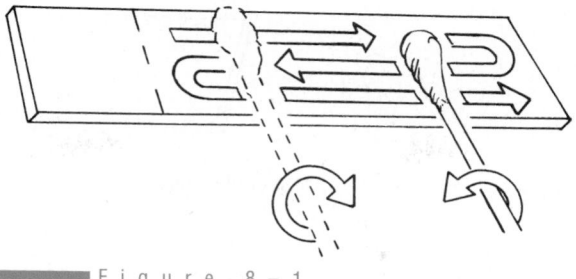

Figure 8 – 1

Smear preparation of a sample collected on a swab. The swab should be rolled back and forth across the slide to completely deposit the sample.

swab should *never* be rubbed back and forth across the slide, because important material on the opposite side of the swab might not be deposited and because smear elements could be broken up.

Smears from Thick Liquids or Semisolids

Swabs can also be used as the tool for preparation of smears from thick liquid or semisolid specimens such as feces (Fig. 8–2). The swab is immersed in the specimen for several seconds and then used to prepare a thin spread of material on the glass slide for staining and viewing. This swab method of preparation is adequate but may produce less desirable results than other methods.

■■■ T A B L E   8 – 1

PREPARING INFECTED MATERIALS FOR VISUAL EXAMINATION

Preparation	Specimen/Organism Type
For gross examination	
Wet preparation	Parasites
	Materials >1 mm in size
For microscopic examination	
Wet preparation (direct or sedimented)	Fluids or semisolids
Cytocentrifuged (direct or presedimented)	Clear or slightly turbid fluids
Smear	Clear or slightly turbid fluids
Drop	Pus/fluid
	Tissue homogenate
	Swab rinse
Pellet	Blood culture
	Dilute specimen
Rolled	Swabbed material
Imprint (touch preparation)	Tissue

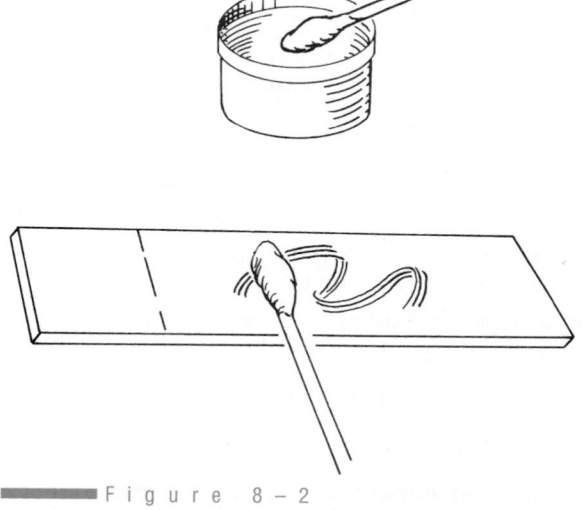

Figure 8 – 2

Smears from opaque thick liquids or semisolids, such as stool, can be made using a swab to sample and smear the material.

Drop or place the sample onto the surface of a labeled glass slide

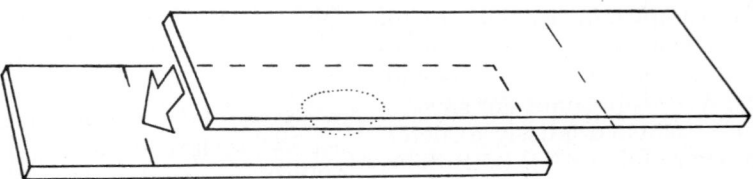

Place the second slide face down over the material

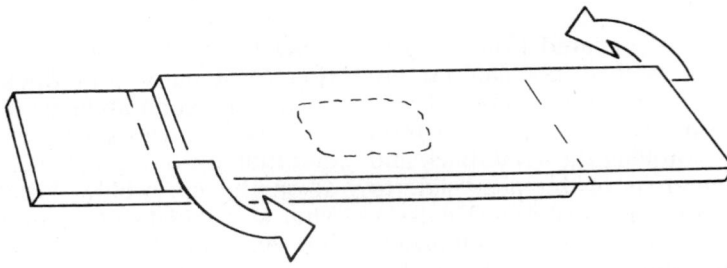

Press to flatten or crush the material, and rotate the two glass surfaces against each other

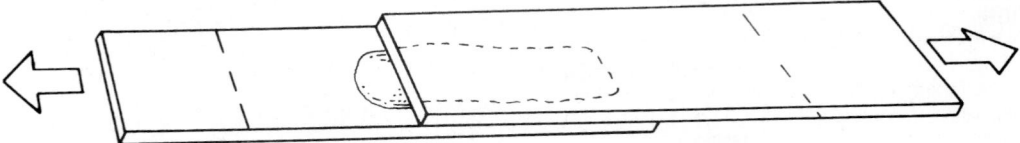

Pull to spread

■■■■■■ Figure 8 – 3

Preparation of smears from thick, granular or mucoid samples.

Smears from Thick, Granular, or Mucoid Materials

Opaque material must be thinly spread so that a monolayer of material is deposited in some areas. Having both thick and thin areas is most desirable. Granules within the material must be crushed so that their make-up can be assessed. A better presentation of granules is possible if granules or grains are "fished" from the surrounding materials and crushed on a separate slide using the technique shown in Figure 8–3. Granules that are to hard to crush between two glass slides probably do not represent infectious materials. More likely, they are small stones or foreign bodies. Examination using a dissecting microscope may help to characterize the nature of hard granules.

The following steps should be used to prepare a smear from thick, granular, or mucoid materials:

1. Place a portion of the sample on the labeled slide, and press a second slide, label down, onto the sample to flatten or crush the components.

2. Rotate the two glass surfaces against each other so that the shear forces break up the material.

3. Once the material is flattened and sufficiently thinned, pull the glass slides smoothly away from each other to produce two smears.

4. If the material is still too thick, repeat the first three steps with another (third) glass slide. The best smear or both smears can be retained for staining.

Slides of material from sites that are difficult to sample, from scant samples, or from patients with critical illnesses should not be discarded until the culture evaluation is complete.

Smears from Thin Fluids

"Thin" specimens of fluids such as urine, cerebrospinal fluid (CSF), and transudates should be dropped but not spread on the slide. The area of sample drop should be marked on the reverse side of the slide using a wax pencil or placed within the circle or well of a premarked slide.

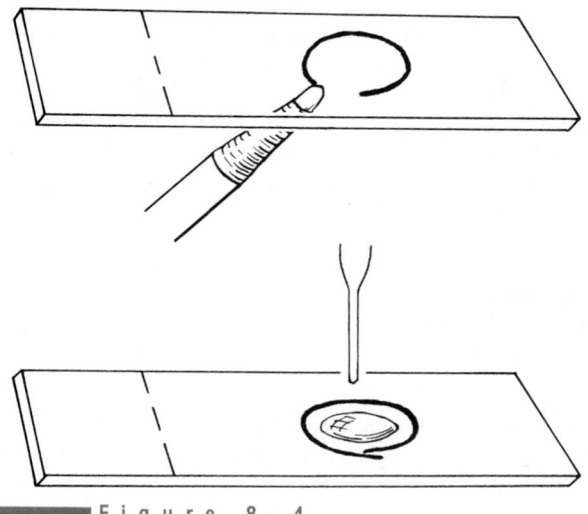

Figure 8–4
Smears from thin fluids can be prepared by placing a single drop of fluid or resuspended fluid sediment onto a well-marked area of the slide.

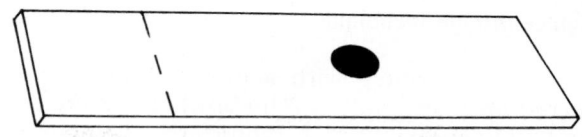

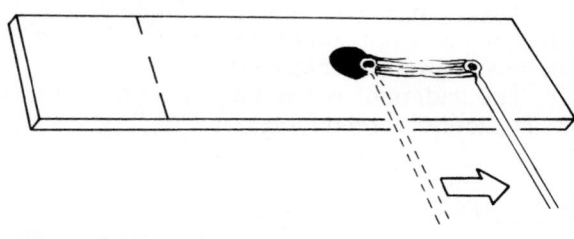

Figure 8–5
Cytocentrifuge preparations deposit the concentrated sample within a limited area for viewing. If the deposit is too heavy, a portion of the material may be smeared to produce a thin area.

Cytocentrifugation is preferred for this type of specimen, if available.

Thin fluid can be prepared by drawing up a small quantity of the fluid or a resuspended sediment of the fluid into a pipette and depositing it as a drop of fluid onto a clearly marked area on the slide (Fig. 8–4). The material may not be grossly visible after staining because of a low protein or cell count. The fluid should not be spread over a larger area of the slide unless it is turbid. Turbid or thick fluids can be more efficiently prepared by the previously described method.

Cytocentrifuge Preparations

Cytocentrifugation is an excellent method for preparing nonviscous fluids such as CSF and bronchoalveolar lavage fluids. The cytocentrifugation process deposits cellular elements and microorganisms from the specimen onto the surface of a glass slide as a monolayer. The cellular elements are deposited within a discrete area for easy viewing (Fig. 8–5). The protein is dissipated into a filter pad, leaving the background clearer for viewing gram-negative morphotypes. Cell morphology is good, and the concentrating effect shortens viewing time and increases the volume of cellular material reviewed.

Cytocentrifuge Technique

A cytocentrifuge with a closed bowl is preferred for microbiology. The bowl can be loaded and unloaded within a biohazard chamber to avoid possible infectious aerosols. The steps of the technique are as follows:

1. Small aliquots of fluid (0.1–0.2 mL) are placed in the cytocentrifuge holders.
2. The material is spun for 10 minutes.
3. The slide is removed. If the deposit of cells is too heavy, a portion of the cellular deposition can be smeared (see Fig. 8–5).
4. The sediment is fixed and decontaminated in 70% alcohol for 5 minutes.

STAINS

Staining imparts an artificial coloration to the smear materials that allows them to be visualized using the magnification provided by a microscope. There are many types of stains, each with specific applications. Stains can be categorized as *simple stains* (directed toward coloring the forms and shapes present), *differential stains* (coloring specific components of those elements present) and *diagnostic antibody or DNA probe–mediated stains* (directed specifically at an organism identification). Stains most commonly used in the diagnostic laboratory are listed in Table 8–2. Four of the stains—Gram, acid-fast, calcofluor white, and rapid modified Wright-Giemsa—should be available in all diagnostic microbiology laboratories (Procedures 8–1 to 8–4). Most of the other stains are directed toward specific organism groups and should be available where needed.

MICROSCOPES

Examination of specimens should begin with gross visual inspection and proceed to the level of magnification needed to visualize the pathogen or to determine that no pathogen is present. The tests ordered provide a guide to the examination, but a routine approach to specimen management should include an examination procedure that will discover unexpected pathogens. In most diagnostic microbiology laboratories, this consists of visual inspection at the time of smear and culture preparation and microscopic examination

■■■■ T A B L E   8 – 2

STAINS FOR INFECTED MATERIALS

Stains	Applications
General Morphology	
Wright-Giemsa	Bronchoalveolar lavages
	Tzanck preparations
	Samples with complex cellular backgrounds (visualizes bacteria, yeast, parasites, and viral inclusions)
Selected Morphology	
Leifson	Flagella
Methylene blue	Metachromatic granules of *Corynebacterium diphtheriae*
Acid-fast stains	
Ziehl-Neelsen	Sediments for mycobacteria (concentrated smears)
	Partial acid-fastness of *Nocardia* spp
Fluorochrome	Sediments for mycobacteria (concentrated smears) (auramine and rhodamine) Preferred acid-fast stain
Kinyoun	Acid-fast stain modification of Ziehl-Neelsen method for cryptosporidia and cyclospora parasites in stool specimens
Calcofluor white stain	Bronchoalveolar fungi and some parasitic cysts
	Differentiates them from background materials of similar morphology
Gram stain	
Traditional	Routine stain for diagnostic area
	Yeast differentiated from all other organisms
Enhanced	Provides the same differential staining but enhances red-negative organisms by staining the background material a green to gray-green
Genus (Species)-Specific Stains	
Antibody or DNA probe stains	Used for the specific identification of selected pathogens, such as *Chlamydia trachomatis, Bordetella pertussis, Legionella pneumophila;* herpes simplex virus, varicella-zoster virus, cytomegalovirus, adenovirus, and respiratory viruses

of a Gram-stained preparation for structures too small to be seen with the unaided eye.

Microscopes vary both in their ability to resolve small structures and in their modifications. Microscopes are divided into two basic types, compound light microscopes, with common resolving limits of 1 to 10 μm and enlargements up to 2000×, and electron microscopes, with enlargements upward of 1,000,000× (Table 8–3). The microbiology laboratory uses several modifications of the compound light microscope, but the workhorse of the laboratory is the brightfield microscope.

TERMINOLOGY FOR DIRECT EXAMINATIONS

The microscopist must have a consistent vocabulary for the description of materials seen
Text continued on page 268

PROCEDURE 8–1. GRAM STAIN

Principles: This staining method was developed empirically by the Danish bacteriologist Christian Gram in 1884. The sequential steps provide for crystal violet (hexamethyl-*p*-rosanaline chloride) to color all the cells and background material a deep blue, and for Gram's iodine to provide the larger iodine element to replace the smaller chloride in the stain molecule. Bacteria with thick cell walls containing teichoic acid retain the crystal violet–iodine complex dye after decolorization and appear deep blue; they are *gram-positive*. Other bacteria with thinner walls containing lipopolysaccharides do not retain the dye complex; they are *gram-negative*. The alcohol/acetone decolorizer damages these thin lipid walls and allows the stain complex to wash out. All unstained elements are subsequently counterstained red by safranin dye. The differential ability of the Gram stain makes it useful in microbial taxonomy. The quickness and ease with which the method can be performed makes it an ideal choice for the clinical laboratory setting.

Application: The Gram stain is used routinely and by request in the clinical microbiology laboratory for the primary microscopic examination of specimens submitted for smear and culture. It is ideally suited for those specimen types in which bacterial infections are strongly suspected but may be used to characterize any specimen. Cerebrospinal fluid, sterile fluids, expectorated sputum or bronchoalveolar lavages, and wounds and exudates are routinely stained directly. Urine and stool may not be routinely stained directly. Samples sent for focused screening cultures are not usually stained. The Gram stain is regularly used to characterize bacteria growing on culture media.

Procedure:

1. Dry the material on the slide so that it does not wash off during the staining procedure. Adherence can be improved by fixation in 70 to 95% alcohol or by gently warming the slide to remove all water from the material.

2. Place the smear on a staining rack, and overlay the surface of the material to be stained with the stains in sequence as shown in Figure 8–6.

3. Place the smear in an upright position in a staining rack, allowing the excess water to drain off and the smear to dry. Never blot a critical smear. Never put immersion oil on a smear until it is completely dry.

4. Examine the stained smear using the low-power objective; then select an area to examine more closely using a 40 to 60¥ oil objective; suspicious areas are evaluated using the 100¥ oil objective of the microscope.

Results: Gram-positive bacteria stain dark blue to blue-black. All other elements stain a safranin red. Individual structures absorb a different amount of safranin, so some will have prominent staining (strong avidity) and others will be weakly stained (low avidity). Among the gram-negative bacteria, the enterics have strong avidity and will stain a bright red; pseudomonads are less avid and will stain moderately well. Anaerobic bacilli and other thin-walled gram-negative organisms, such as *Borrelia, Legionella,* and *Spirillum,* stain weakly. Always check the quality of the stain before moving to interpretation.

Precautions: The Gram stain reaction may vary from the expected in a number of well-recognized circumstances. If the crystal violet is rinsed too vigorously before it is complexed with the iodine, it will wash away and leave poor or no staining of gram-negative organisms. If the decolorization is too vigorous or prolonged, the gram-positive complex will be removed, and the normally gram-positive organisms will not stain. If the decolorization is insufficient, organisms may be falsely gram-positive, and organisms in the thicker areas of the sample may be obscured. If the safranin is left on the slide for a prolonged period (minutes), the gram-positive complex will be leached from the positive cells; however,

Procedure continued on next page

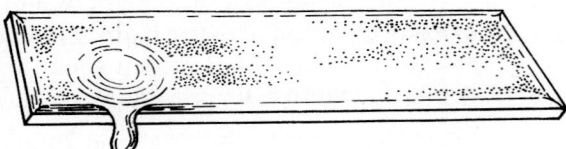

Flood the slide with crystal violet and
allow to stand for 30 seconds

The crystal violet stain is not tied into
the organism until the iodine is added.
Any rinsing between the crystal violet
and the iodine steps must be very brief.

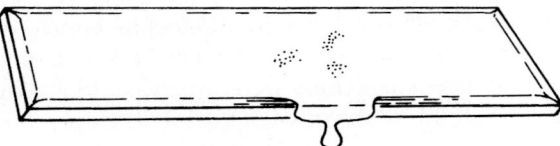

Flood with Gram's iodine and allow
to stand for 30–60 seconds

The dilute iodine solution can be used to
wash away the crystal violet, and no water
rinse employed.

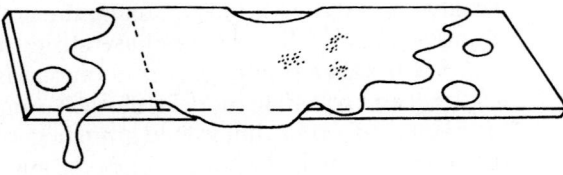

Decolorize the slide with acetone or absolute
alcohol or a mixture of the two decolorizers
and wash immediately with water

Acetone is a more rapid decolorizer and
may give better results, but the reaction
must be stopped with water as soon as the
purple color disappears.

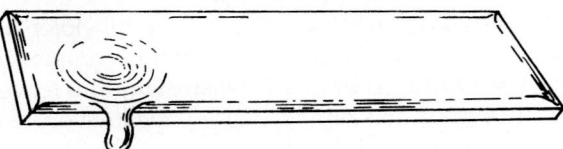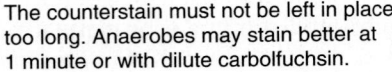

Flood with safranin or dilute carbolfuchsin
or neutral red for 30 seconds to 1 minute (anaerobes).
Rinse very lightly with water

The counterstain must not be left in place
too long. Anaerobes may stain better at
1 minute or with dilute carbolfuchsin.

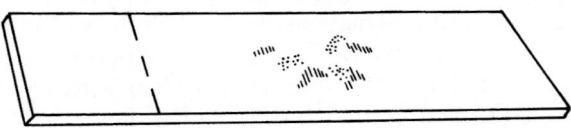

Check the staining reactions before
proceeding with smear interpretation.

■ F i g u r e 8 – 6
Gram stain technique.

PROCEDURE 8–1. (CONTINUED)

failure to leave the safranin in place for sufficient time will result in failure to stain gram-negative bacteria and background materials. Gram stain characteristics may be atypical in antibiotic-treated and dead or degenerating organisms. Typical morphotypes should be sought. Any sample that raises questions about quality of stain or method should be restained.

PROCEDURE 8–2. ACID-FAST STAINING OF MYCOBACTERIA

Principles: The primary stain binds to mycolic acid in the cell walls of the mycobacteria and is retained after the decolorizing step with acid alcohol. The counterstain does not penetrate the mycobacteria to affect the color of the primary stain.

Application: The direct smear examination is a valuable diagnostic procedure for the detection of mycobacteria in clinical specimens.

Materials:

Difco TB auramine-rhodamine T Stain

Carbolfuchsin stain (prepared by laboratory)

0.5% acid alcohol (0.5% HCl in 70% ethanol)

2% acid alcohol (prepared by laboratory)

1% sulfuric acid (partial acid-fast)

0.5% aqueous potassium permanganate solution. 0.3% aqueous methylene blue solution (prepared by laboratory)

Microscope slides 1 × 3 inches

Sterile water

Procedure:

Fluorescent Stain

1. Cover smears with Difco TB auramine-rhodamine T stain, and stain for 25 minutes.

2. Wash in running tap water.

3. Move slides to slide rack on acid alcohol collection container.

4. Flood smears with 0.5% acid alcohol and decolorize for 2 minutes.

5. Wash smears in running tap water.

6. Move slides to original staining rack.

7. Flood smears with potassium permanganate counterstain for 4 minutes.

8. Wash smears in running tap water.

9. Air dry.

10. Examine smears with the 16× and 40× objectives of the fluorescent microscope equipped with a filter system comparable to a BG-12 exciter filter and an OG-1 barrier filter. Examine each smear for 3 to 5 minutes.

Kinyoun Stain

1. Cover smears with carbolfuchsin and stain for 5 minutes.

2. Wash slides with running tap water.

3. Move slides to staining rack on acid alcohol collection container.

4. Decolorize with acid alcohol until no more color appears in the washings.

5. Wash slides with running tap water, and move them to original staining rack.

6. Flood slides with methylene blue counterstain for 1 minute.

7. Wash with running tap water, drain, and air dry.

Procedure continued on next page

PROCEDURE 8–2. (CONTINUED)

8. Examine smears with 100¥ oil immersion lens.

Modified Kinyoun Stain (Partial Acid-Fast)

1. Flood the slides with carbolfuchsin stain for 5 minutes.

2. Rinse with running tap water.

3. Flood slides with 70% ethanol, and rinse with tap water. Repeat until excess red dye is removed.

4. Move slides to rack on acid collection container.

5. Continuously drop 1% sulfuric acid on the smear until the washing becomes colorless.

6. Rinse with running tap water.

7. Move slides to original staining rack.

8. Counterstain with methylene blue for 30 seconds.

9. Rinse with running tap water and air dry.

10. Examine smears with 100× oil immersion objective.

Ziehl-Neelsen Stain

1. Cover smear with a piece of filter paper cut slightly smaller than the slide.

2. Layer filter paper with carbolfuchsin stain. With Bunsen burner, heat the smears gently until steaming occurs. Stain for 5 minutes without additional heating.

3. Proceed as for Kinyoun method, beginning with step 2.

Results:

Fluorescent Stain

Mycobacteria stain bright orange. Count the number of acid-fast bacilli seen on the smear and report as follows.

Number of Acid-Fast Bacilli	Report
1–20	Number seen
21–80	Few
81–300	Moderate
300+	Numerous

Kinyoun and Ziehl-Neelsen Stains

Mycobacteria stain red, whereas the background material and non–acid-fast bacteria stain blue.

PROCEDURE 8–3. CALCOFLUOR WHITE STAIN/FUNGI-FLUOR KIT (POLYSCIENCES, ARRINGTON, PA)

Principle: Calcofluor white is a colorless dye that binds to cellulose and chitin. It fluoresces when exposed to long-wavelength ultraviolet and short-wavelength visible light. Special filters are required for optimal use.

Application: Calcofluor white may be used as a specific stain for rapid screening of clinical specimens for fungal elements. This stain may be useful when morphology is ambiguous and the nonspecific staining of other techniques such as Grocott's methenamine silver (GMS) gives confusing results.

Materials:

Stock Solution

A 1% (w/v) aqueous solution of calcofluor white is prepared by dissolving the powder in distilled water with gentle heating. The stock solution is stable for 1 year at room temperature.

Working Solution

0.1% calcofluor white containing 0.01% to 0.08% Evans blue as a counterstain.

PROCEDURE 8 – 3 . (CONTINUED)

Procedure:

1. Add 1 to 2 drops of working calcofluor white solution or solution A (Fungi-Fluor) to fixed smear or imprint for 1 to 2 minutes.

2. Coverslip or rinse and dry.

3. Examine specimen on fluorescent microscope using the following set of filters: G 365, LP 450, FT 395.

4. Add Fungi-Fluor solution B if quenching of nonspecific staining is desired. (The quenching with solution B may be excessive [1:4 dilution preferred].)

Results: Yeast cells, pseudohyphae, and hyphae display a bright apple-green or blue-white fluorescence. The central body of the *Pneumocystis* cyst also fluoresces; with quenching (Fungi-Fluor solution B), the cysts of *Pneumocystis* are visible.

PROCEDURE 8 – 4 . RAPID MODIFIED WRIGHT-GIEMSA STAIN

Principle: The Wright-Giemsa stain is available in a modification that requires only 1 to 3 minutes. This neutral dye is a combination of basic thiazine dyes and acid eosin that attach to oppositely charged sites on proteins. The results are metachromatic.

Application: Wright-Giemsa (modified) is a rapid stain for smears and imprints to fully stain background materials and cells and a wide variety of microorganisms.

Precautions: Avoid getting reagents in eyes or on skin or clothing; if this does occur, flush with copious quantities of water. Use with adequate ventilation. If stain is discarded into sink, flush with large volumes of water to prevent azide build-up, which may react with lead and copper plumbing to form highly explosive metal azides.

Procedure:

1. Prepare smear or imprint. Alcohol fix.

2. Dip slide in Fixative solution five times for 1 second each time. Allow excess to drain.

3. Dip slide in solution I five times for 1 second each time. Allow excess to drain.

4. Dip slide in solution II five times for 1 second each time. Allow excess to drain.

5. Rinse slide with tap water.

6. Allow to dry. Examine.

Note: The intensity of each stain may be altered by increasing or decreasing dips in Solutions I and II. Never use fewer than three dips of 1 full second each.

Results: Blood cells stain as with Wright stain. The cytoplasm is basophilic. The chromatin of white cells is purple. Bacteria are blue. Parasitic protozoan nuclei are red.

■■■■■■ T A B L E 8 – 3
OBSERVING MICROBIAL PATHOGENS

Tools	Magnification (×)	Application
Eyes	0	Gross examination
Magnifying glass	5	Gross examination
Dissecting microscope	2.5–30	Gross detailed examination and manipulation
Compound light microscope		
Brightfield	10–2,000	Cells stained
Darkfield	10–400	Cells not readily stained for brightfield microscopy
Phase-contrast	10–400	Living or unstained cells
Fluorescence	10–400	Preparations using fluorochrome stains, which can directly stain cells or can be connected to antibodies that attach to cells
Electron microscopes		
Transmission electron	150 to 10 million	Determine ultrastructure of cell organelles
Scanning electron	20–10,000	Determine surface shapes and structures

in viewed samples. This vocabulary must be shared by the microbiology and medical communities, so that when observations are reported, everyone will be able to understand the implications of the descriptions. Common observations can be coded so that they are consistent among observers. The use of computers for recording coded observations and generating reports of the findings further extends the need for uniform terminology. Only unusual findings should be individually described in a report.

The background of the sample being evaluated should be described in sufficient detail to convey the composition of the material. The presence of cells representing a response to injury supports the probability of infection and directs attention toward specific types of pathogens. Common morphotype descriptions and the most prevalent associated species are shown in Table 8–4. Examples of useful descriptive phrases with quantitation are listed in Table 8–5.

Microorganisms can be described in such a way that, based on prevalence, the description implies the identification of the organism. For example, the observation of a gram-negative bacillus, small and pleomorphic, from the spinal fluid of an infant implies that *Haemophilus influenzae* is the infecting agent.

EXAMINATION OF PREPARED MATERIAL

A limited number of microbial pathogens are regularly encountered by the technologist or microbiologist from commonly sampled infected sites. The simple Gram stain or acid-fast stain is the fastest and least expensive method for presumptive diagnosis in these common clinical settings (see Fig. 8–6). Organisms are readily seen, because upward of 10^5 CFU of infecting organisms per mL are commonly present in clinically evident infections.

Two types of infection are important to distinguish, those caused by a *single* species and those caused by *multiple* species, or *polymicrobial*. The single agents of infection or those causing classic infections are easily recognized by microscopy and require limited interpretations. Infections caused by common single species, or by classic infectious agents, include *Streptococcus pneumoniae* pneumonia, *Staphylococcus aureus* abscesses or pyodermas, *H. influenzae* tracheobronchitis or meningitis in infants, *Clostridium perfringens* gas gangrene, *Nocardia* lung abscesses, and gonococcal urethritis.

Polymicrobial presentations in smears require more interpretation and must take into account smear background, the morphology of the organisms, and the anatomic location of the suspected infection as well as accompanying clinical symptoms. Polymicrobial infections usually arise from displaced normal or altered flora, and culture will yield the same species that can be isolated in culture from uninfected but contaminated specimens. These infections usually represent displacement of environmental, skin, oropharyngeal, gastrointestinal, or vaginal flora into tissues, with subsequent infection. Commonly encountered infections of this type are surgical wound (skin flora) infection, aspiration (oropharyngeal flora) pneumonia, perirectal (fecal flora) abscesses, and tubo-ovarian (vaginal flora) abscess.

Characterization of Background Materials

One should always look at the slide material with the unaided eye before beginning the microscopic examination. The distribution and consistency of the material should be noted. The microscope's low-power objective (2.5–10×)

■■■■■T A B L E   8 – 4
GRAM STAIN MORPHOLOGY AND ASSOCIATED ORGANISMS

Morphotype Description	Most Common Organisms
Bacteria	
COCCI	
Gram-positive cocci	Aerococcus, Enterococcus, Leuconostoc, Pediococcus, Planococcus, Staphylococcus, Stomatococcus, Streptococcus
Gram-positive cocci	
Pairs	Staphylococcus, Streptococcus, Enterococcus
Tetrads	Micrococcus, Staphylococcus, Peptostreptococcus
Groups	Staphylococcus, Peptostreptococcus, Stomatococcus
Chains	Streptococcus, Peptostreptococcus
Clusters, intracellular	Microaerophilic Streptococcus, viridans streptococci, Staphylococcus
Encapsulated	Streptococcus pneumoniae, Streptococcus pyogenes (rarely), Stomatococcus mucilaginosus
Gram-positive diplococci (lancet-shaped)	Streptococcus pneumoniae
Gram-negative diplococci	Pathogenic Neisseria, Moraxella (Branhamella) catarrhalis
BACILLI	
Gram-positive bacilli	
Small	Listeria monocytogenes, Corynebacterium
Medium	Lactobacillus, anaerobic bacilli
Large	Clostridium, Bacillus
Diphtheroid	Corynebacterium, Propionibacterium, Rothia
Pleomorphic, gram-variable	Gardnerella vaginalis
Beaded	Mycobacteria, antibiotic-affected lactobacilli, and corynebacteria
Filamentous	Anaerobic morphotypes, antibiotic-affected cells
Filamentous, beaded, branched	Actinomycetes, Nocardia, Nocardiopsis, Streptomyces, Rothia
Bifid or V forms	Bifidobacterium, brevibacteria
Gram-negative coccobacilli	Bordetella, Haemophilus (pleomorphic)
Masses	Veillonella
Chains	Prevotella, Veillonella
Gram-negative bacilli	
Small	Haemophilus, Legionella (thin with filaments), Actinobacillus, Bordetella, Brucella, Francisella, Pasteurella, Capnocytophaga, Prevotella, Eikenella
Bipolar	Klebsiella pneumoniae, Pasteurella, Bacteroides
Medium	Enterics, pseudomonads
Large	Devitalized clostridia/bacilli
Curved	Vibrio, Campylobacter
Spiral	Campylobacter, Helicobacter, Gastrobacillum, Borrelia, Leptospira, Treponema
Fusiform	Fusobacterium nucleatum
Filaments	Fusobacterium necrophorum (pleomorphic)
Yeast/Fungi/Algae	
Yeast	
Small	Histoplasma, Torulopsis
Medium	Candida
With capsules	Cryptococcus neoformans
Thick-walled, broad-based bud	Blastomyces
Hyphae	
Septate	Fungi
Aseptate	Zygomycetes
With arthroconidia	Coccicioides
With branches 45-degree angle	Aspergillus
Pseudohyphae	Candida
Spherule (endospores)	Coccidioides
Sporangia with endospores	Protothecae
Viruses	
Single or multinuclear cells with intranuclear inclusions	Herpes virus, measles virus
Enlarged cells with intranuclear inclusions or cytoplasmic inclusions	Cytomegalovirus
Cells with dark "smudged" nuclei	Adenovirus

DESCRIPTIONS OF BACKGROUND MATERIAL

Cells/Structures	Associations
Amorphous debris (light, moderate, or heavy)	Necrosis, heavy protein fluid
Black particulate debris	Smoke inhalation, crack cocaine
Charcot-Leyden crystals	Eosinophils
Epithelials with contaminating bacteria	Passage of specimen through contaminated area during collection
Curschmann spirals present (sputum)	Bronchospasm, obstruction, asthma
Epithelial cells (light, moderate, or heavy)	Epithelial surface involved or adjacent to collection site
Intracellular organisms	Inferred association in infection
Local material (light, moderate, heavy)	Reflection of collection site
Mononuclear cells present	Chronic inflammation
Mucus (light, moderate, heavy)	Irritation of glandular surface
Purulence (none, light, moderate, heavy)	Acute inflammation, exudation
Red blood cells	Trauma, hemorrhage

should be used first to evaluate the general content of the material on the slide. Specimens can be homogeneous or heterogeneous and may contain pathogens evenly distributed throughout the specimen or limited to one visual field. A mental inquiry checklist should be followed until the habit of searching a slide systematically is developed. Items that should be included in such a checklist appear here and on following pages in **bold type** preceded by a bullet (■).

■ *Is there evidence of contamination by normal (resident) microbial flora?* One should look for squamous epithelial cells, bacteria without the cells of inflammation, food, or other debris. Does this material constitute the entire sample, or is a representative sample also available in a manner that it can be recognized? Contamination of specimens not collected from sterile sites diminishes the value of culture studies.

■ *Is necrotic (amorphous) debris in the background?* Infection with organisms such as *C. perfringens* and *Nocardia* spp may elicit few polymorphonuclear neutrophils (PMNs). The inflammatory cells that do migrate into the area of infection can be lysed. Leukopenic patients may also have few inflammatory cells within their inflammatory debris. Amorphous debris is usually the remains of tissue mixed with the breakdown products or fluids of inflammation and should always be searched for organisms. *Mycobacterium tuber-*

culosis organisms stain poorly as beaded gram-positive bacilli or not at all with Gram stain; they can appear within necrotic debris as negative images.

■ *Are unexpected structures present?* The characteristic coiled structure of a Curschmann's spiral is more easily recognized on low power. This structure must not be confused with parasitic larvae, which can also be found in sputum using low-power magnification. Large granules, grains, or fungal forms such as spherules or fungal mats can be best recognized at low power.

Search for Microorganisms

After the full extent of the material has been examined on low power, a representative area should be selected for viewing with the oil-immersion lens. A 40× or 60× lens is preferred for scanning, and a 100× lens is used for final evaluation. In infection, the organism will be intimate with the purulence or necrotic debris. All grains or granules and the background should be examined carefully. The delicate gram-positive filaments of *Nocardia* spp may blend into the background. *Haemophilus influenzae* may be present in large numbers, hidden within the mucus, in acute exacerbations of chronic bronchitis. Intracellular and extracellular forms should be noted. Strict criteria for microbial morphotypes should be maintained. The examiner must not be distracted by precipitated gram-positive stain, keratohyaline granules, or other artifacts. Organisms should be evaluated for shape, size, and Gram reaction. Because cell wall–damaged bacteria, antibiotic-treated bacteria, or dead bacteria may appear falsely gram-negative, their shapes and sizes are critical "co-characteristics." The classic misleading smear example is the observation of gram-negative, lancet-shaped diplococci mixed with the predominant gram-positive forms. The inexperienced observer may misinterpret this as mixed infection rather than as the simple presence of dead pneumococci in a classic infection.

■ *Examine more than one area of the smear.* More than one organism should be found if possible. It is rare not to be able to find more than one, because in infection, organisms are usually distributed throughout the specimen. Care should be taken in the interpretation of very low numbers of bacteria, especially in the absence of inflammation or necrosis and in specimens from nonsterile sites. Small numbers of organisms in samples from sterile sites must be seriously considered. Additional smears can be made and examined, however, if the likelihood of contamination is high.

■ ***Do not overinterpret the findings.*** Specific diagnosis should be limited to a small number of instances in which the smear is classic in its presentation and extent of infection is not an issue. If acid-fast bacteria are suspected, the acid-fast stain should be performed before an opinion is rendered. If a fungal element is not clearly gram-positive, a calcofluor stain should be performed. Both of these follow-up stains can be performed on the decolorized, Gram-stained preparation.

Evaluation of Choice of Antibiotic

The symptomatic patient with suspected infection is most likely going to be treated with antibiotics. The physician will expect the laboratory to affirm or reject the antibiotic choice made after the presumptive diagnosis. Antibiotic choice should be kept in mind as the smear is viewed. There are a number of important observations.

■ ***Is there evidence of purulence?*** Remember that purulence with red blood cells, neutrophils, protein background, and necrosis reflects acute inflammation. Mononuclear cells, including lymphocytes, monocytes, and macrophages, reflect chronic inflammation. Patients who are cytopenic will not have the cellular response seen in normal individuals. Purulence, blood, and necrosis can be present if traumatic tissue damage occurs in the absence of infection.

■ ***Is there a single most probable etiologic microorganism?*** If so, is the morphology sufficiently characteristic to presume identification? Is there a specified antibiotic for treatment of infections with this agent? Will antibiotic susceptibility testing be necessary or will identification of the suspected pathogen be sufficient?

■ ***Is the infection monomicrobial or polymicrobial?*** Are morphotypes present that characterize the source of the organisms? Can the mixture of organisms be characterized?

■ ***If antibiotics are to be used, will the suspected pathogens be susceptible?*** Specific considerations should be made about the likelihood of:

1. Penicillinase or β-lactamase producers such as *Staphylococcus aureus, Haemophilus,* and the gonococci

2. Enterococci, which have limited susceptibility to antibiotics and may therefore require synergistic antibiotic action for killing

3. Species resistant to aminoglycosides, penems, and cephems, such as the *Bacteroides fragilis* group and *Pseudomonas aeruginosa*

4. Fungal organisms, which will be unresponsive to antibacterial antibiotics

Direct Examination Summary

Once this stepwise evaluation of the smear has been completed, the information yielded, along with the clinical setting, allows for reasonable management of infected patients until the subsequent steps of culture isolation, susceptibility testing, or antibody or antigen detection can take place.

Initiation of Special Handling for Unsuspected or Special Pathogens

The direct microscopic examination of infected materials, along with specimen site and historical information, may suggest modifications in routine culture techniques to allow the isolation of a suspected pathogen. Modifications might involve ordering other culture tests, adding special media, increasing incubation time, or changing incubation temperature or atmosphere. If recognition or suspicion of such pathogens does not occur from smear or history, the isolation of certain pathogens will not be made, and diagnosis will be delayed or missed.

Some pathogens will not grow in culture and are usually unsuspected. A common one is the nematode parasite *Strongyloides stercoralis,* which can be seen in the sputum and bronchoalveolar fluid of patients with a hyperinfestation syndrome. Visual recognition can lead to a request for stool examination for parasite load and prompt treatment. Untreated hyperinfestations are associated with death of the patient.

Other pathogens that will grow in culture but not in routine bacterial culture can be recognized on smear and redirected for appropriate culture.

Organisms such as *Legionella, Mycobacterium,* viruses, and *Bartonella* must be placed on appropriate media for culture confirmation of infection.

GRADING OR CLASSIFYING MATERIALS

Microscopic examination also immediately discloses that a specimen is unlikely to be helpful in diagnosis or culture management. The specimen may be just blood rather than infected material, or may be oropharyngeal surface debris or some other normal surface material. Processing and culture interpretation of such nonrepresentative specimens may lead to delayed treatment because of a falsely negative culture or to inappropriate antibiotic treatment owing to a falsely positive culture from growth of normal flora or antibiotic-altered flora.

Several grading or classification systems have been derived to aid the technologist in arriving at decisions relating to culturing the specimen or to interpretation of growth from culture of a specimen. Most evaluations are aimed at specimens such as sputum, for which collection is complicated by contamination with throat and mouth flora because culture alone may be misleading. The objective is to separate the representative sample from the contaminated sample prior to culture or prior to culture evaluation. Barlett's method for scoring sputum and the Murray-Washington method for contamination assessment document the association of 10 to 20 squamous epithelial cells (SECs) per 10× microscopic field with unacceptable specimens and 10 to 25 PMNs per 10× field with significant specimens. Heineman's method emphasizes the ratio between the SECs and PMNs.

The Ohio State University Medical Center diagnostic microbiology laboratory uses the documented observations related to SECs and PMNs but attempts to coordinate observations related to background materials, which consist of local materials, contaminating materials, and purulence, and to describe the relationship of microorganisms to these background materials. The body site of the sample and the classification of the smear together determine the extent of culture evaluation.

Contaminating Materials

Contaminating materials (Fig. 8–7) are recognized as those not coming from the collection site, not contributed by the inflammatory response from the tissues, or not likely to contain the infecting organism. They have usually been added to the specimen in the course of collection from or through a nonsterile area. The most bothersome contaminating materials are those that contain microorganisms that will grow in culture and may confuse the culture evaluation. Expectorated sputum, which collects in the lower lung airways and is expelled through the mouth, is the most common contaminated specimen managed by the clinical laboratory.

Criteria. <25 polymorphonuclear cells per low-power field (LPF) and >10 epithelial cells and/or mixed bacteria per LPF.

Gram Smear Report. Quantitate the contaminating materials as 1+ (light), 2+ (moderate), 3+ (moderately heavy), or 4+ (heavy).

Culture Identification Guidelines. "New culture" using careful collection technique should be requested. If culture is requested, organism identification should be limited to brief evaluation for *S. aureus,* streptococci or viridans streptococci, *Lactobacillus,* diphtheroids, and β-hemolytic streptococci. Gram-negative rods are reported as enterics (coliforms, non–lactose fermenters, or *Proteus* spp [spreading colonies]), pseudomonads (oxidase-positive). Pathogenic *Neisseria* (identified), *Haemophilus* spp (smear only), yeast (note only), and any primary pathogen.

Antibiotic Susceptibility Testing. None appropriate except on primary pathogens.

Local Materials

Criteria. <25 polymorphonuclear leukocytes (WBCs) per LPF and <10 contaminating epithelial cells per LPF along with cellular and fluid elements *local* to the area sampled.

The local constituents may vary as follows:

- Respiratory secretions (Fig. 8–8)—mucus, alveolar pneumocytes (macrophages), ciliated columnar cells, goblet cells, and occasionally metaplastic epithelial cells (smaller than normal squamous epithelial cells)

- Cerebrospinal fluid—few cellular elements

- Cavity fluid—macrophages, few mixed WBCs, mesothelial cells, and proteinaceous fluid

- Wounds—blood and proteinaceous fluids

- Amniotic fluid—anucleate squamous cells and heavy proteinaceous fluid

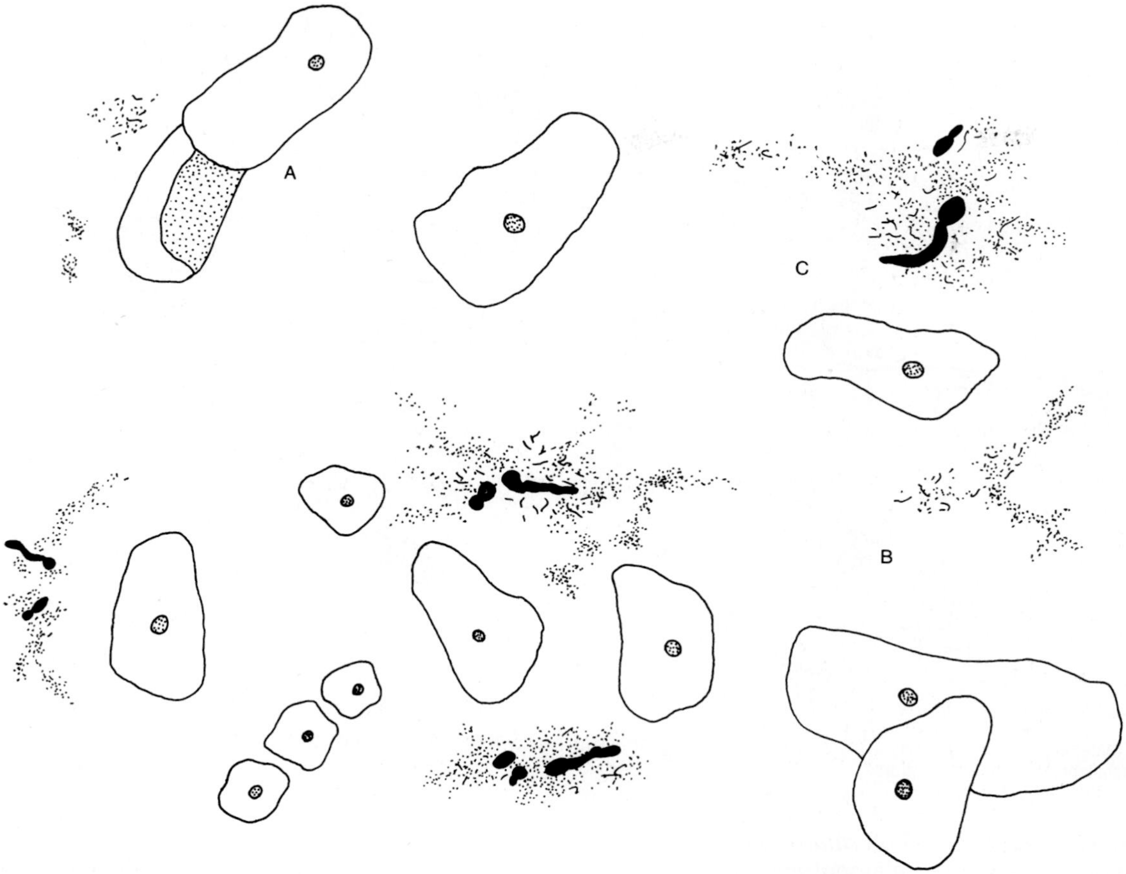

■ F i g u r e   8 – 7

Contaminating materials. Squamous epithelial cells *(A)*, bacteria *(B)*, and yeast *(C)* from a sample labeled "expectorated sputum."

■ Cervix—mucus, columnar epithelial cells, goblet cells, metaplastic squamous epithelial cells, and leukocytes (vary with menstrual cycle)

■ Prostatic secretions or semen—spermatozoa and mucus

Gram Smear Report. Quantitate local microbial flora as 1+ to 4+ (see "Contaminating Materials").

Culture Identification Guidelines. The designation "normal flora" or a brief presumptive description of colony-type growth may be used (see "Contaminating Materials").

Antibiotic Susceptibility Testing. None appropriate.

Purulence

Criteria. >25 polymorphonuclear leukocytes (WBCs) per LPF and no or few (<10) epithelial cells with mixed bacteria per LPF. Mucus (Fig. 8–9*A*) and/or heavy proteinaceous material may be present.

Gram Smear Report. Quantitate only organisms intimately associated with the WBCs (Fig. 8–9*B*), mucus, or proteinaceous exudate. Use the following system: 1+ (≤1 organism per oil immersion field [OIF]), 2+ (few organisms per OIF), 3+ (moderate number per OIF), and 4+ (many per OIF). Quantitate contaminating materials separately—should be ≤1 + (none or few).

Culture Identification Guidelines. Correlate colony growth with Gram-stained smear. Identify: *S. pneumoniae* from viridans streptococci, β-hemolytic streptococci (Lancefield groups A, B, and D; C, F, and G if indicated clinically), *S. aureus* (inventory for epidemiology), *H. influenzae, Haemophilus parainfluenzae* (test for β-lactamase production), *Haemophilus aphrophilus,* pathogenic *Neisseria,* gram-negative bacillic

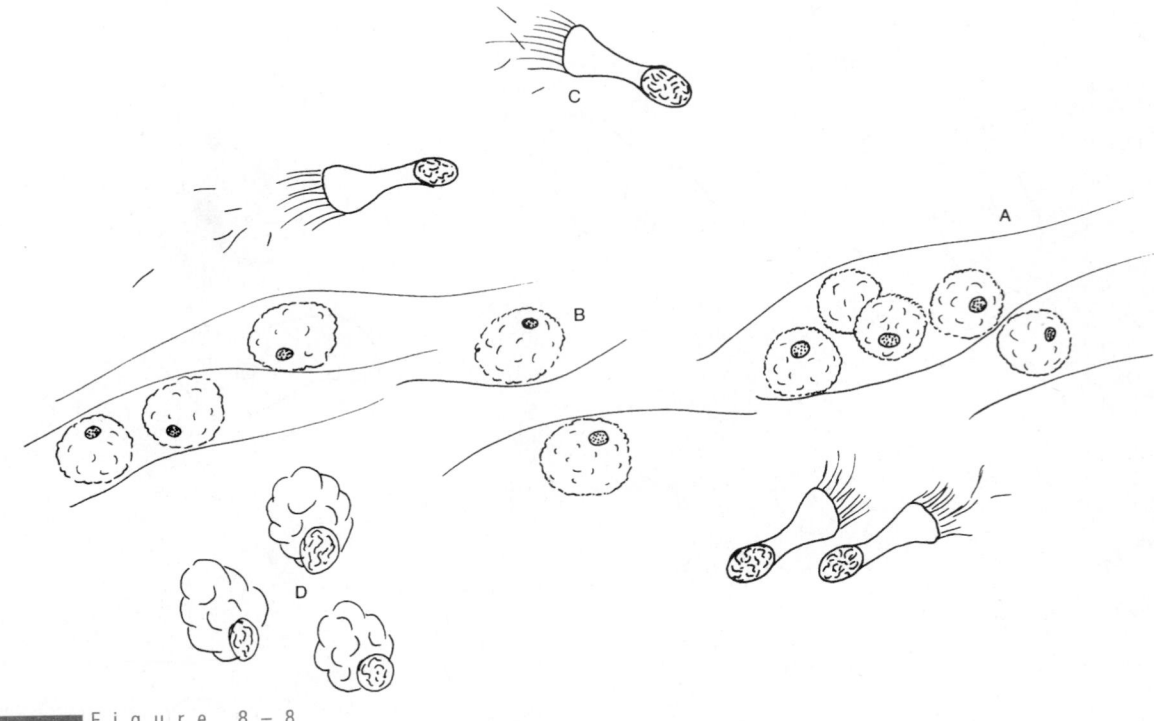

■■■■■ F i g u r e   8 – 8

Local materials from a respiratory secretion. Mucus *(A)*, alveolar macrophages *(B)*, ciliated columnar cells *(C)*, and goblet cells *(D)* from a sample labeled "aspirated sputum."

yeast (*Cryptococcus neoformans* only; note presence of other genera), filamentous fungi (transparent tape preparation in biosafety hood), and other organisms as indicated by smear findings.

Antibiotic Susceptibility Testing. *Staphylococcus aureus,* gram-negative nonfastidious bacilli, and other organisms as appropriate or specifically requested.

Mixed Materials

Mixed materials consist of purulent exudate, contaminating materials, and local materials in a single smear. A smear may be of *type I,* in which the various types of materials appear in layers (Fig. 8–10), or *type II,* in which the various material types are intermingled (Fig. 8–11).

Criteria. >25 polymorphonuclear leukocytes (WBCs) per LPF and >10 epithelial cells and/or contaminating bacteria per LPF. Local secretions may also be present. Two patterns of mixed secretions are identified. Type I consists of a layering of the specimen so that the purulent exudate (Fig. 8–10*A*) is physically separated from the contaminating secretions (Fig. 8–10*B*) or the local materials (Fig. 8–10*C*). Type II is an intimate mixture of all elements so that microbial associations cannot be reasonably determined (Fig. 8–11).

Gram Smear Report. For type I mixed materials, quantitate only those organisms intimately associated with purulent exudate. Also quantitate the amounts of other elements, contaminating materials, and local materials. For type II mixed materials, the smear should be quantitated as 1+ to 4+ (see "Contaminating Materials").

Culture Identification Guidelines. A "new culture" *may* be requested for both types of mixed specimens, but a new specimen *must* be requested for the type II specimen, because of the presence of purulence and uninterpretable culture results. With the type I smear, the culture may be interpreted as a purulent exudate for only those organism morphotypes intimately associated with the WBCs. Culture evaluation of the specimen with a type II smear should await a new specimen. If a new culture specimen cannot be obtained, evaluation should proceed as for contaminating materials.

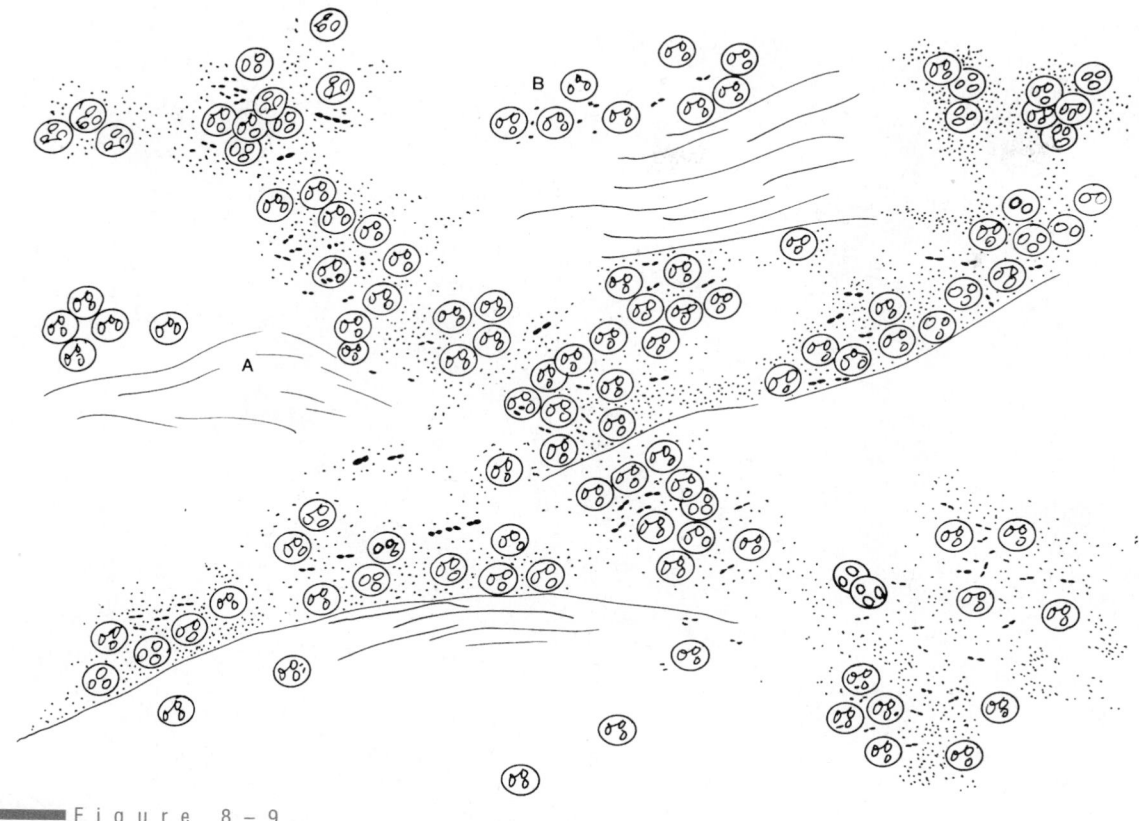

■■■■■ F i g u r e   8 – 9

Purulent material. Mucus *(A)* and neutrophils *(B)* from a sample labeled "expectorated sputum."

Antibiotic Susceptibility Testing. Use purulence guidelines for testing organisms that appear significant.

REPORTS OF DIRECT EXAMINATIONS

Reports of the results of direct specimen examination should be made available as soon as they are completed. The availability of computer-managed reporting utilizing direct physician access through computer terminals in patient care areas as well as paper reporting facilitates immediate reporting of results. Reports of direct examinations should be simple and should include all information elements needed by the physician to understand the report. The report lists the type of material that was submitted and clearly states whether the microbes seen are of significance. The format for computer-managed microscopic reports from direct specimen examinations at The Ohio State University is shown in the samples given in Table 8–6.

Only those elements that are useful in characterizing the specimen should be included in the report. Interpretative comments may also be included when, on the basis of specimen type, background materials, and organism morphology, there is little doubt about the nature of the process or the offending infectious agent. All telephone reports of direct examinations should be recorded in the report.

EXAMPLES OF SAMPLE OBSERVATIONS AND REPORTS

Study the examples of direct observations and the associated reports and comments given in Plates I to XIV. Practice to determine whether you can make a similar report using only the observation provided. Then read each report to see whether you obtain similar impressions of the specimen and the pathogen from the observation and the written report.

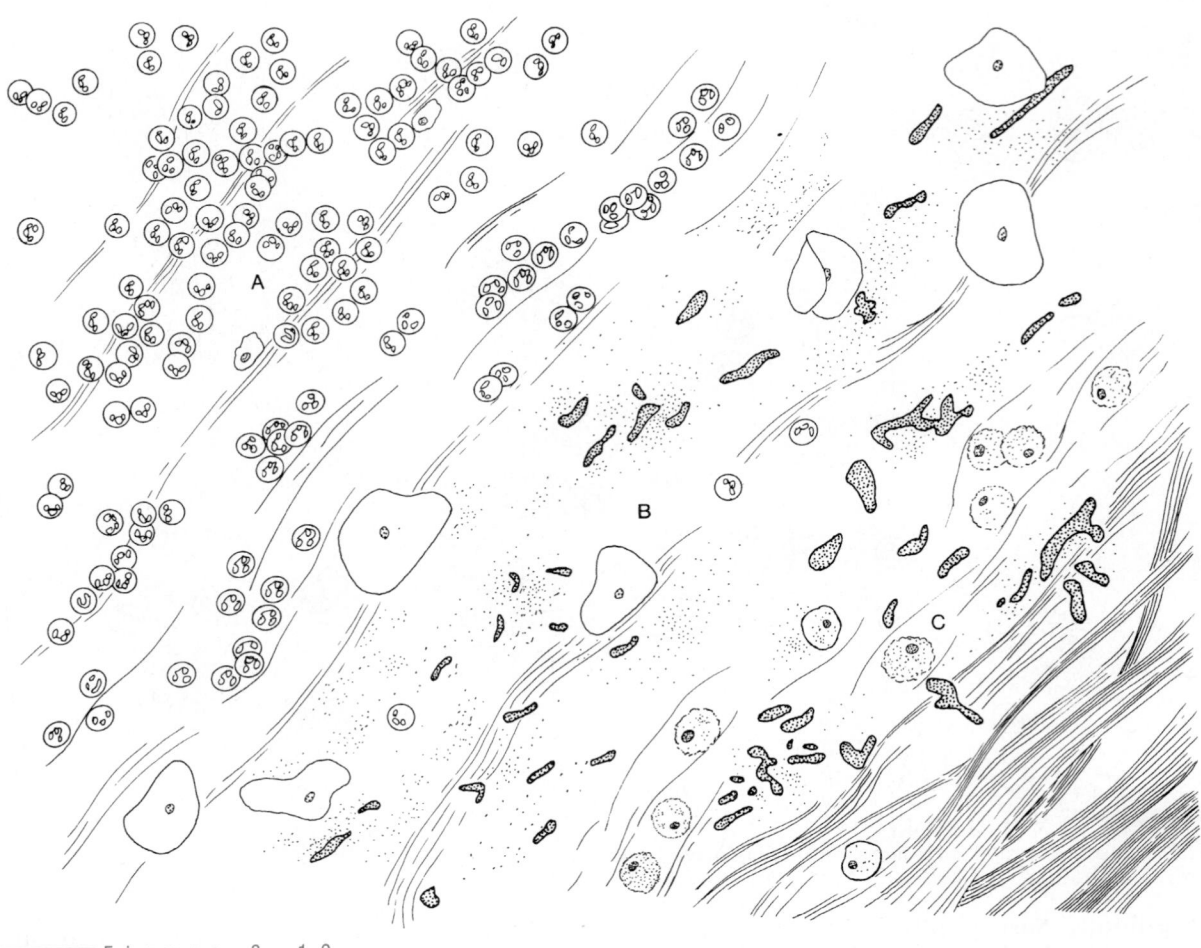

■■■■■■F i g u r e   8 – 1 0

Mixed materials, type I, layered. Purulence *(A)*, layered against contaminated epithelial cells *(B)* and against local materials *(C)*, from a sample labeled "expectorated sputum."

QUALITY CONTROL IN DIRECT MICROSCOPIC INTERPRETATIONS

Quality control issues, such as quality of specimens submitted and adequacy of culture interpretation, can be monitored using the results of direct examination. Quality control practice that monitors both the smear and culture interpretation should be an ongoing work activity. Correlation between the two results should be made for each patient. Explanations for discrepant results should be sought within the work material. This repeated inspection of results enables each observer to practice self-education and to improve his or her skills of observation. Review of these quality control activities allows corrections to be made in specimen collection, specimen management, and culture management.

Quality improvement activities are often suggested by results of quality control monitoring of direct specimen examination. For example, it may be documented that sputum specimens of poor quality are consistently submitted from certain doctors' offices, clinics, or nursing stations in the hospital. This observation becomes the basis for planning corrections that move outside the laboratory to involve the community of patients served.

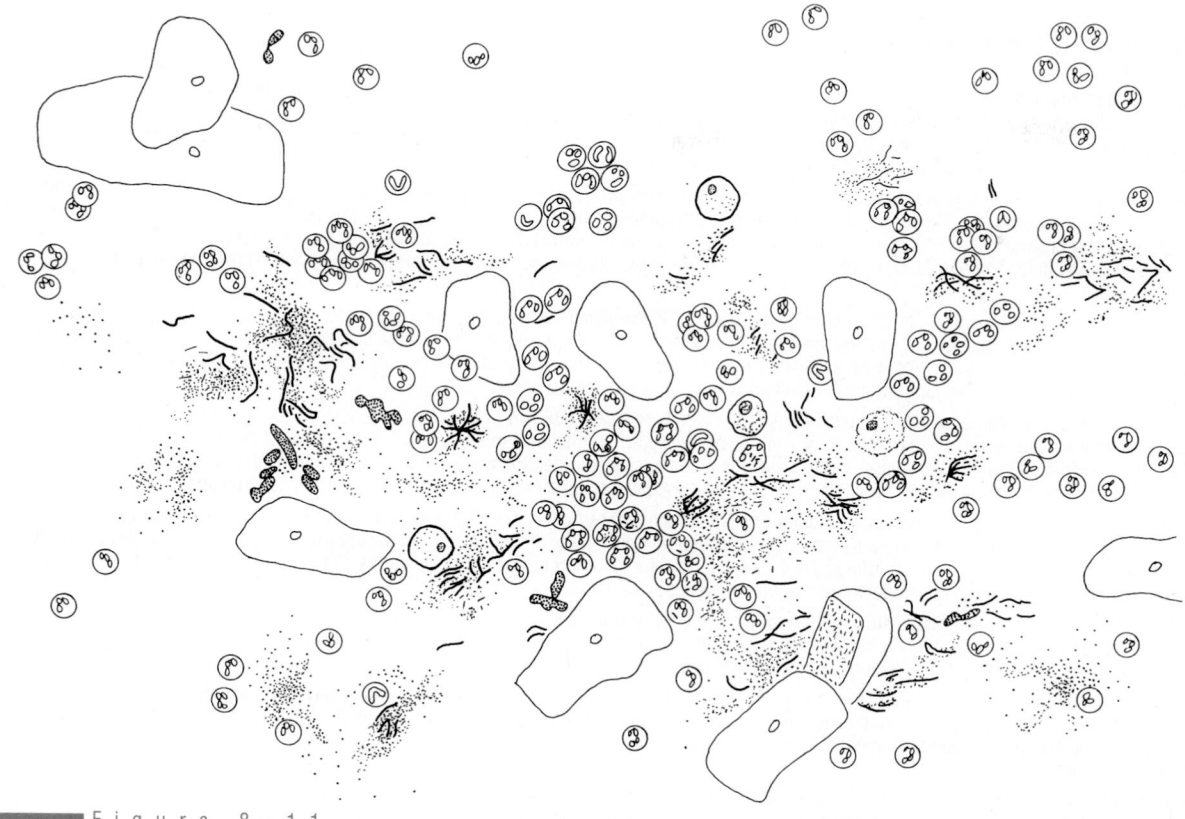

■■■ F i g u r e 8 – 1 1
Mixed materials, type II, intermingled. Mixed purulence, contaminated epithelial cells, and local materials from a sample labeled "expectorated sputum.

■■■ T A B L E 8 – 6

SAMPLE REPORTS OF DIRECT MICROSCOPIC EXAMINATIONS

RESPIRATORY CULTURE	ACC. NO. XXX0

SOURCE:
MICROSCOPIC:
	SPUTUM: EXPECTORATED
	PURULENCE HEAVY
	CONTAMINATING BACTERIA, YEAST
	    AND EPITHELIALS HEAVY
	GRAM-POSITIVE DIPLOCOCCI: CON-
	    SISTENT WITH PNEUMOCOCCI

CALLED TO DR. DOE AT 1800 PM 4/13/99

RESPIRATORY CULTURE	ACC. NO. XXXXX
SOURCE: MICROSCOPIC:	SINUS: ETHMOID CONTENTS PURULENCE MODERATE LOCAL MATERIALS LIGHT RED BLOOD CELLS PRESENT GRAM-NEGATIVE BACILLI: CONSIS- TENT WITH PSEUDOMONAS

CALL TO DR. DOE AT 1135 AM 4/5/99

SUMMARY

Historically, the process of infectious disease diagnosis began with observations from direct specimen examination. We should continue the process of direct specimen examination for all of those instances in which this activity provides the opportunity to support or refute clinical diagnoses or to discover unsuspected diagnoses.

For Bibliography see page 306

■ P l a t e I

Direct Examination Showing Local and Contaminating Materials

A, Expectorated sputum, smear, Gram stain, light microscopy, low-power view (LPV). Purulence none. Contaminating bacteria and epithelial cells heavy. No pathogens seen. Please submit carefully collected sample of lower respiratory tree material. The sample is saliva, not sputum. There could be several reasons for submission of this sample to the laboratory. The patient could have been poorly directed and simply "spit" into the collection container, or the patient's cough may not be productive of sputum.

B, Amniotic fluid, cytocentrifuge preparation, Gram stain, light microscopy, medium-power view (MPV). Purulence moderate. Local materials moderate. No organisms seen. The presence of purulence (neutrophils or "polys") indicates a process suspicious for infection. The absence of organisms in this normally sterile fluid is a critical observation. Squamous epithelial cells are local to this specimen type and confirm that the sample is amniotic fluid. The blue keratohyalin granules must never be mistaken for bacteria. Compare and contrast the clinical importance of *A* and *B* smears.

C, Expectorated sputum, smear, Gram stain, light microscopy, LPV. Purulence none. Local materials moderate. Contaminating bacteria and epithelials heavy. No pathogens seen. The bolus of sputum, consisting of mucus with entrapped alveolar macrophages, confirms that lower respiratory tree material is present. There is no evidence of an infectious process. The sputum is heavily coated by contaminating materials from the oropharynx or mouth. Contaminating organisms will grow in a routine sputum culture.

D, Aspirated sputum, smear, Gram stain, light microscopy, high-power view (HPV). Purulence none. Local materials moderate. No organisms seen. The alveolar macrophages and mucus (pink-stained background) are the local materials from the surface of the tracheobronchial tree. This smear confirms that sputum was sampled and that there is no suspicion for infection and no evidence of significant contamination. Routine bacterial culture of this specimen will still grow insignificant oral flora because culture is more sensitive than direct examination.

E, Cerebrospinal fluid (CSF), cytocentrifuge preparation, Gram stain, light microscopy, HPV. Purulence moderate. No organisms seen. CSF is a sterile fluid and normally does not have purulence. The presence of neutrophils is critical. Careful observation for bacteria is mandatory. Acridine orange stain may be helpful in clinical settings in which bacteria are low in number and gram-negative. Cytocentrifuged sediments commonly have a concentration of organisms sufficient for routine microscopy ($\geq 10^5$/mL).

F, Expectorated sputum, smear, Gram stain, light microscopy, MPV. Purulence heavy. Local materials—Curschmann spiral light. No organisms seen. The Curschmann spiral is material local to the tracheobronchial tree but is not normal, so it is specifically reported. This spiral may present in a variety of sizes depending on the size of bronchus involved.

G, Trauma eye, vitreous aspirate, smear, Gram stain, light microscopy, HPV. Purulence none. No organisms seen. Local materials light. The light protein background and the pigment-containing cell are normal material local to the vitreous of the eye. This local material confirms that the sample is representative. The ability to see this brown pigment cell in smear material is related to the eye trauma. The important emphasis is the absence of purulence.

H, Eye, vitreous aspirate, smear, Gram stain, light microscopy, HPV. Purulence light. Amorphous debris moderate. Local materials light. No organisms seen. This smear suggests that there has been ongoing injury. The pigment has been phagocytized and is seen within a large macrophage. Pigment must never be mistaken for bacteria, and bacteria must never be overlooked if mixed with local materials.

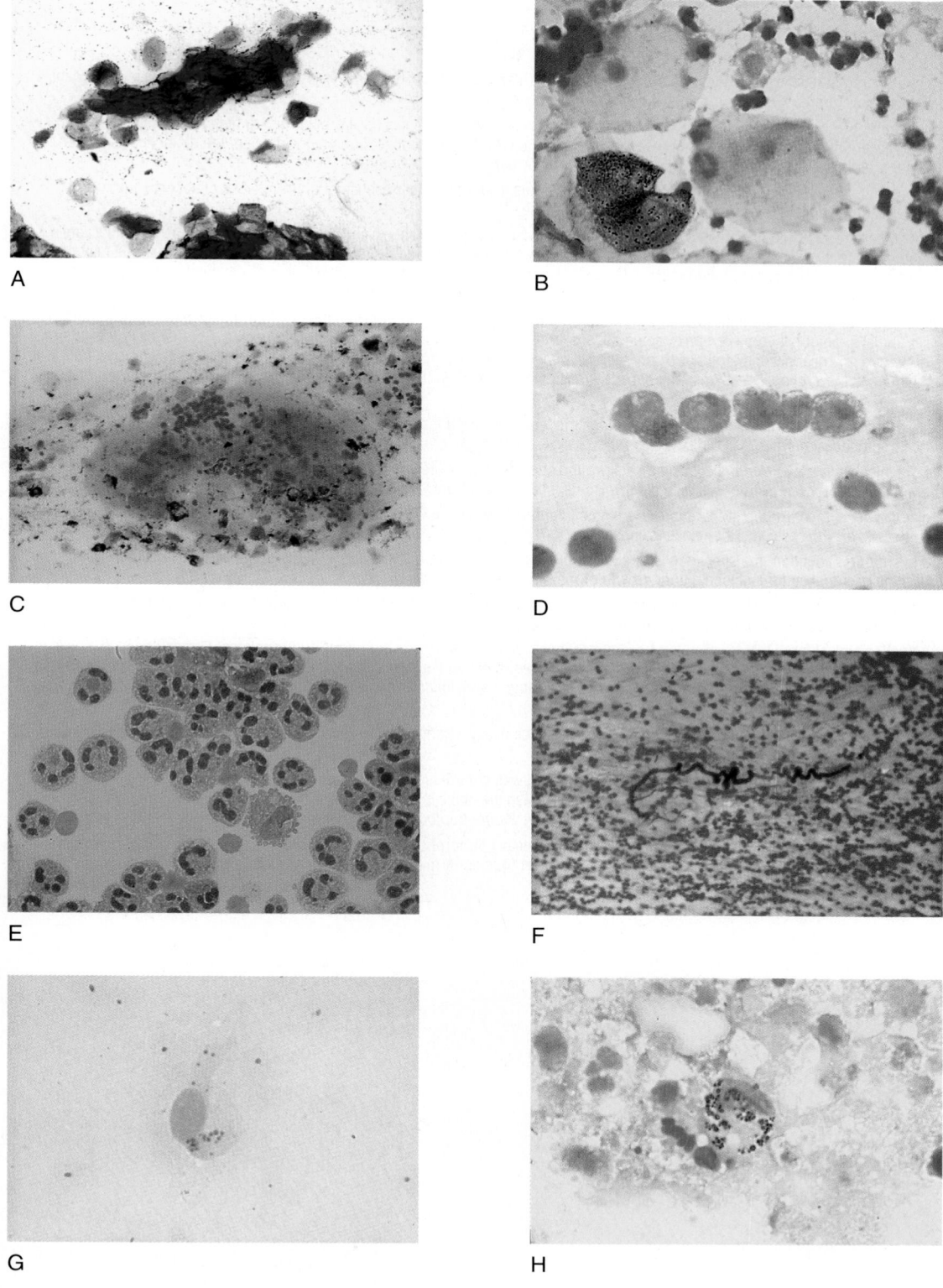

━━━━━■ P l a t e I I

Direct Examination in Common Bacterial Infections

A, Expectorated sputum, smear, Gram, stain, light microscopy, MPV. Purulence light. Amorphous debris moderate. Gram-positive diplococci, encapsulated, extracellular. *Impression:* Pneumococcal disease.

This is a typical smear presentation for early pneumococcal pneumonia. The pneumococci have proliferated to high numbers, and the lung is responding with increased mucus and fluid release. There is early migration of the neutrophils, but phagocytosis of the diplococci is limited. Routine bacterial culture of this sample should yield a heavy growth of *Streptococcus pneumoniae*.

B, Expectorated sputum, smear, Gram stain, light microscopy, HPV. Purulence heavy. The presence of gram-positive diplococci, intracellular morphology suggest antibiotic effect. *Impression:* Pneumococcal disease.

This is a typical smear presentation of a treated but unresolved pneumococcal pneumonia. Neutrophils cover the field, the diplococci are largely intracellular and partially digested, and the background amorphous material is gone. Routine bacterial culture of this sample may be negative for typical colonies of *Streptococcus pneumoniae*. A few colonies may be found by a careful search among the contaminating normal flora colonies.

C, Aspirated sputum, smear, Gram stain, light microscopy, MPV. Purulence light. Amorphous debris heavy. Gram-positive cocci, pairs, encapsulated, extracellular. Initial antibiotic therapy can be directed toward streptococci and staphylococci (*Stomatococcus*). Routine bacterial culture isolated a pure growth of an encapsulated strain of *Streptococcus pyogenes*. The heavy amorphous background is protein-rich edema fluid from the capillary bed damaged by *S. pyogenes* toxins. The patient subsequently died from the infection despite correct antibiotic therapy and a correct diagnosis.

D, Wound, smear, Gram stain, light microscopy, MPV. Purulence moderate. Amorphous debris moderate. Gram-positive cocci, chains, extracellular. *Impression:* Streptococcal disease.

The presence of typical chains of *Streptococcus* on a background showing purulence with poorly preserved "polys" and amorphous debris is suggestive of hemolytic streptococci with tissue cytotoxicity. Routine bacterial culture yielded a pure growth of *Streptococcus pyogenes*.

E, Bronchoalveolar lavage, cytocentrifuge preparation, Gram stain, light microscopy, HPV. Purulence moderate. Gram-positive cocci, pairs, short chains, groups, intracellular. *Impression:* Staphylococcal disease.

The background supports infection, the specimen is directly from the alveolar spaces of the lung, and the organism morphology is typical for staphylococci. Routine bacterial culture yielded a pure growth of methicillin-resistant *Staphylococcus aureus* (MRSA).

F, Expectorated sputum, smear, Gram stain, light microscopy, MPV. Purulence light. Local materials moderate. Gram-positive cocci, pairs, groups, intracellular and extracellular. *Impression:* Staphylococcal disease.

Smear morphology is typical for staphylococci, but no staphylococcal colonies were present on the culture plates. *Stomatococcus mucilaginosus* colonies were present in high numbers. Careful correlation between direct and culture examinations demonstrated this organism to be the probable cause of infection. The presumptive report implying or suggesting staphylococci followed by a *negative* culture report without explanation raises doubts about competence of the laboratory.

G, Abscess aspirate, smear, Gram stain, light microscopy, MPV. Purulence heavy. Gram-positive cocci, groups, extracellular. *Impression:* Staphylococcal disease.

Aerobic and anaerobic culture plates were negative at 24 hours. There was culture isolation of *Staphylococcus aureus* from this same abscess 5 days previously. The patient was being treated with clindamycin. Because the cocci in the smear did not appear antibiotic damaged, another species of gram-positive coccus was sought. The anaerobic culture grew *Peptostreptococcus* organisms, which are clindamycin resistant.

H, Bone aspirate, smear, Gram stain, light microscopy, MPV. Purulence heavy. Amorphous debris moderate. Gram-positive cocci, remnants, intracellular. Morphology suggests prominent antibiotic effect. Culture records confirmed a *Staphylococcus aureus* osteomyelitis currently under antibiotic treatment. Bacterial culture was negative.

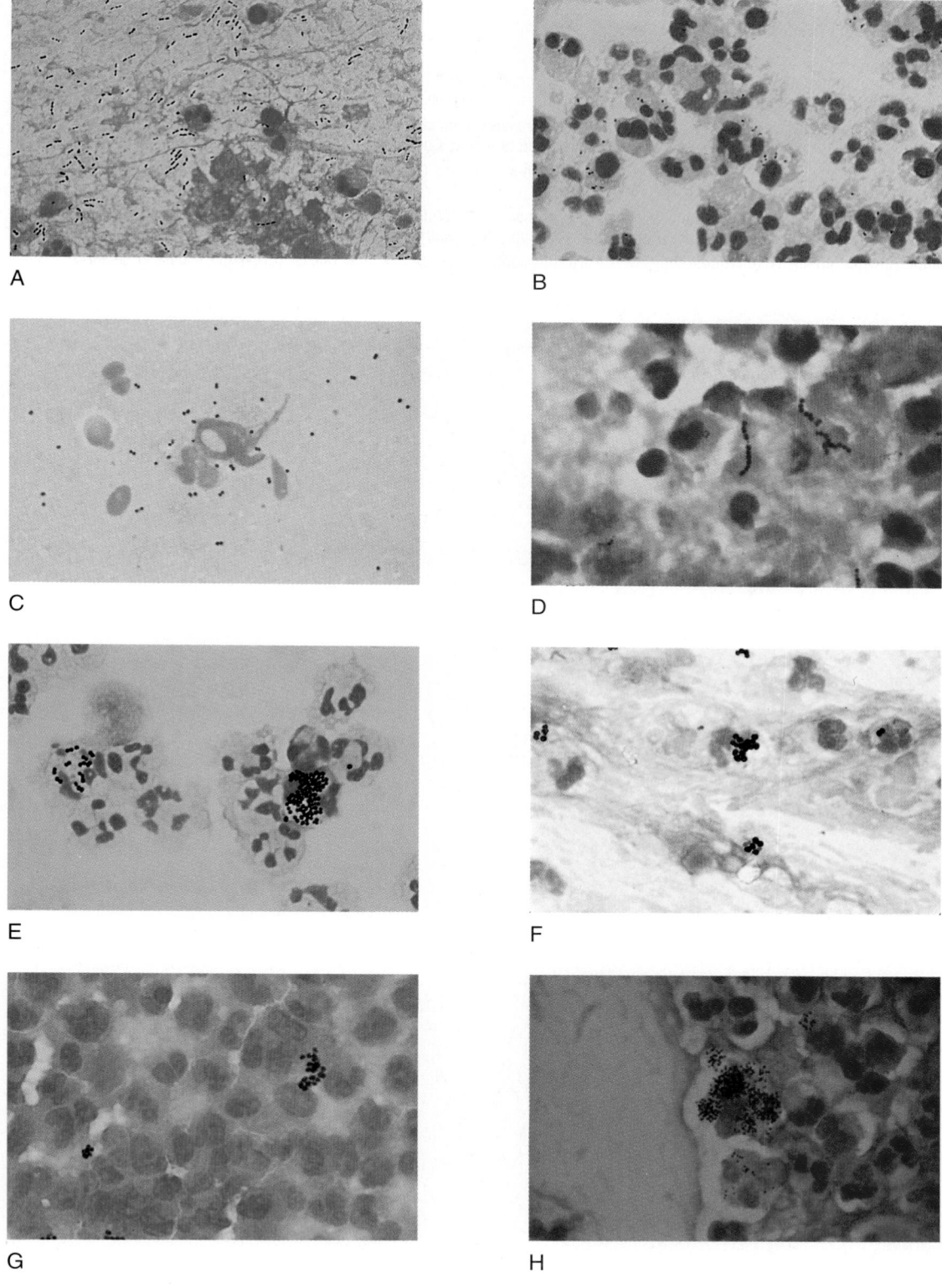

■■■■■■■Plate III

Direct Examination in Gram-Positive Bacillary Infections

A, Amniotic fluid, cytocentrifuge, Gram stain, light microscopy, MPV. Purulence light. Local materials moderate. Gram-positive bacilli, small. Morphology consistent with *Listeria monocytogenes. Impression*: Congenital listeriosis.

B, Expectorated sputum, smear, Gram stain, light microscopy, MPV. Purulence moderate. Local materials moderate. Gram-positive bacilli, diphtheroid. Morphology suggests coryneform infection. Routine bacterial culture grew *Corynebacterium pseudodiphtheriticum.*

C, Urine, cytocentrifugation, Gram stain, light microscopy, MPV. Purulence moderate. Gram-positive bacilli, medium, long, chaining. Morphology consistent with *Lactobacillus* spp. *Impression*: Cystitis.

D, Amniotic fluid, cytocentrifuge preparation. Gram stain, light microscopy, HPV. Purulence light. Local materials moderate. Gram-positive bacilli, medium, long, intracellular. Morphotype consistent with *Lactobacillus* spp. *Impression*: Amnionitis.

E, Amniotic fluid, smear, Gram stain, light microscopy, MPV. Purulence light. Local materials moderate. Gram-positive bacilli, large. Morphology consistent with *Clostridium perfringens. Impression*: Amnionitis.

F, Wound cellulitis, smear, Gram stain, light microscopy, MPV. Purulence none. Amorphous debris moderate. Gram-positive bacilli, large. Gram-negative bacilli, large. Morphology consistent with *Clostridium perfringens. Impression*: Gas gangrene. Note that the growth rate of this organism is rapid and that both viable (gram-positive) and nonviable (gram-negative) bacilli can be present in the smear material.

G, Colony from blood agar, smear, Gram stain, light microscopy, HPV. Gram-positive bacilli, diphtheroid, variably Gram-staining. Morphotype consistent with *Gardnerella vaginalis* (see III*H*).

H, Amniotic fluid, smear, Gram stain, light microscopy, HPV. Local material heavy. Gram-positive bacilli, diphtheroid, variably Gram-staining. Morphotype consistent with *Gardnerella vaginalis. Impression:* Amnionitis.

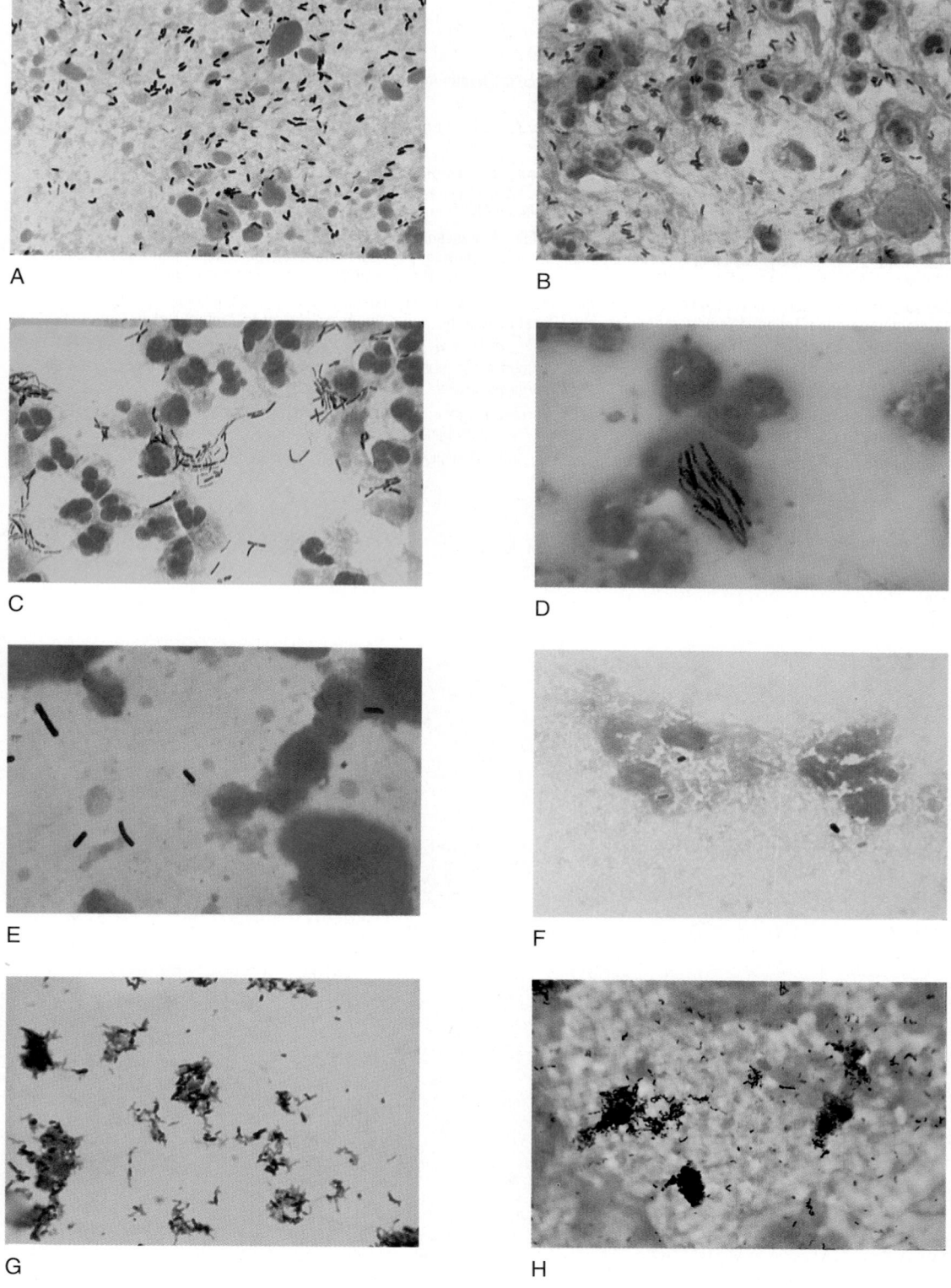

A

B

C

D

E

F

G

H

■■■■■■■ P l a t e I V
Direct Examination in Uncommon Gram-Positive Bacilli

A, Abscess aspirate, smear, Gram stain, light microscopy, HPV. Purulence moderate. Amorphous debris light. Gram-positive bacilli, beaded. Suspect mycobacteria: initiate additional testing (see IV*B*).

B, Abscess aspirate, smear, acid-fast stain (Ziehl-Neelsen), light microscopy, MPV. Purulence moderate. Acid-fast bacilli, numerous. Mycobacterial cultures grew *Mycobacterium kansasii*.

C, Expectorated sputum, smear, Gram stain, light microscopy, MPV. Amorphous debris heavy. Bacillary shapes, negative image. Oil removed from smear, decolorized with acid alcohol and immediately restained with Ziehl-Neelsen acid-fast stain. Acid-fast bacilli numerous. Specimen recovered; acid-fast culture requested by laboratory. Suspicious for tuberculosis.

D, Expectorated sputum, concentrated smear, fluorochrome acid-fast stain, fluorescent microscopy, MPV. Typical acid-fast bacteria, numerous. *Impression:* Tuberculosis. The patient had been placed in respiratory isolation following physical examination and history and was immediately begun on antituberculosis therapy following receipt of the direct examination report. *Mycobacterium tuberculosis* was identified from culture.

E, Expectorated sputum, smear, Gram stain, light microscopy, MPV. Purulence light. Local materials light. Amorphous debris moderate. Gram-positive bacilli, medium, beaded. Follow-up acid-fast stain positive. Chest radiograph with right upper lobe mass. Culture isolation of *Rhodococcus equi*.

F, Blood culture, sediment smear, Gram stain, light microscopy, MPV. Blood. Gram-positive bacilli, branched, beaded. Morphology consistent with *Actinomyces* or *Propionibacterium* spp. Culture isolation of *Actinomyces israelii*.

G, Expectorated sputum, smear, Gram stain, light microscopy, MPV. Purulence light. Local materials light. Amorphous debris moderate. Gram-positive bacilli, branched, beaded. Morphotype consistent with *Nocardia* or *Actinomyces*.

H, Expectorated sputum, smear, partial acid-fast stain, light microscopy, MPV. Partially acid-fast bacilli, branched, beaded. Morphology consistent with *Norcardia* spp. *Impression:* Nocardiosis.

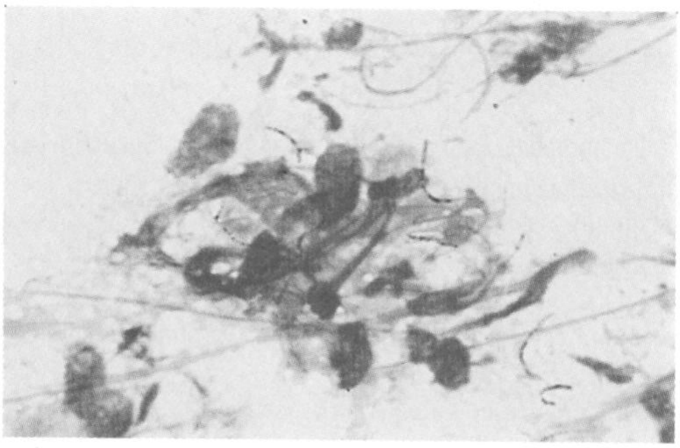

A

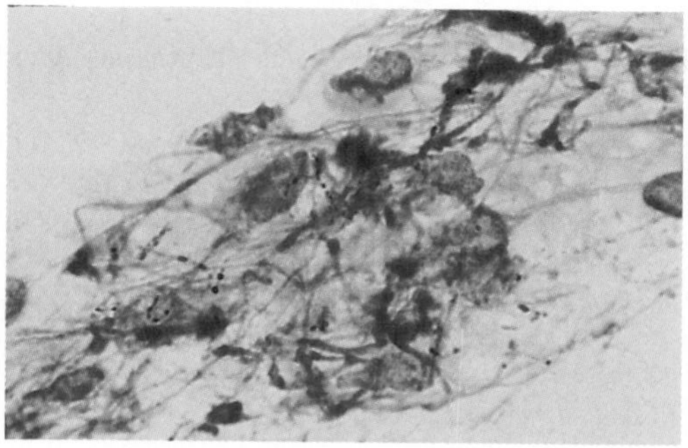

B

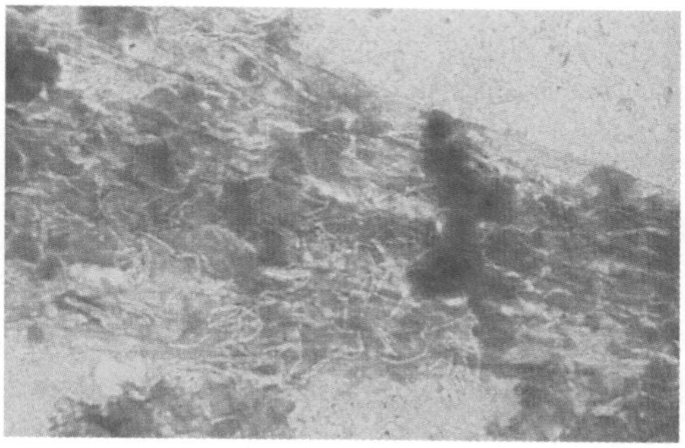

C

D

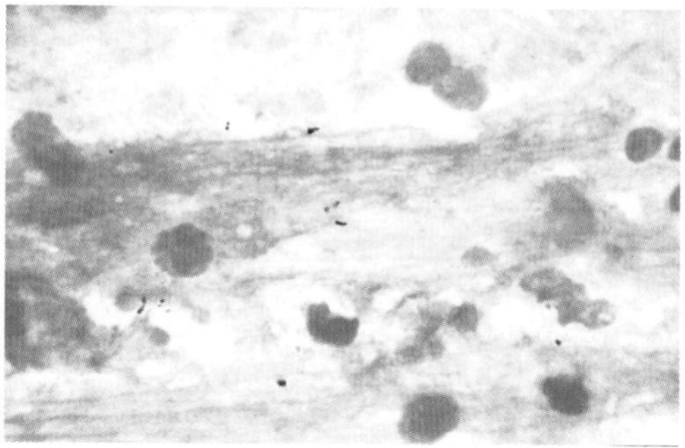

E

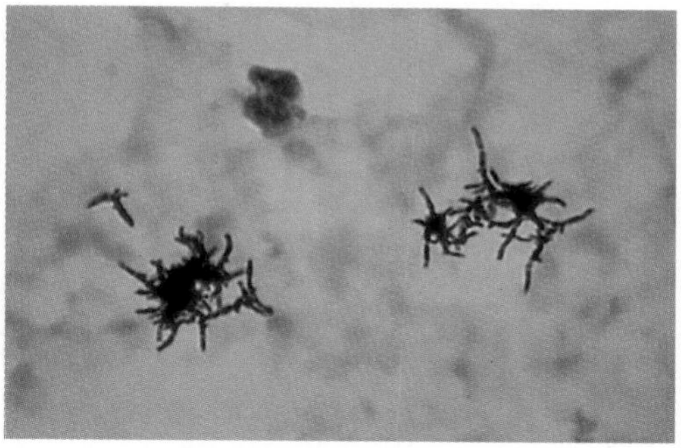

F

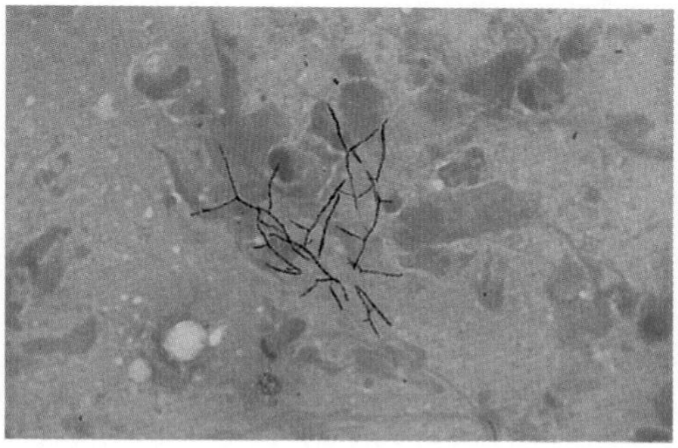

G

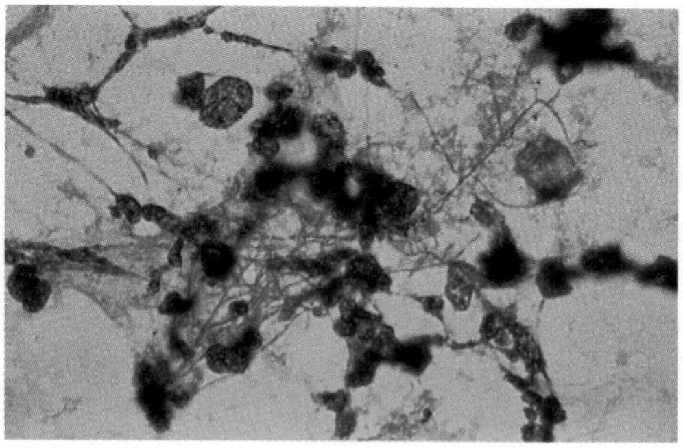

H

■■■■■■ P l a t e   V

Direct Examination in Gram-Positive Bacilli with Filaments and Branches

A, Jaw abscess aspirate, smear, Gram stain, light microscopy, LPV. Purulence heavy. Granules present. Suspicious for *Actinomyces* (see V*B*).

B, Jaw abscess aspirate, smear, Gram stain, light microscopy, HPV. Purulence heavy. Gram-positive bacilli, filamentous, beaded, branched, partial acid-fast stain negative. Morphology consistent with *Actinomyces* spp. *Impression:* Actinomycosis (lumpy jaw).

C, Cutaneous sinus tract aspirate, smear, Gram stain, light microscopy, HPV. Purulence heavy. Gram-positive bacilli, filamentous, beaded, branched, partial acid-fast stain negative. Morphology consistent with *Actinomyces* spp (see V*D*). *Impression:* Actinomycosis.

D, Cutaneous sinus tract aspirate, colonies on anaerobic blood agar plate. Mixed colony morphotypes. Molar tooth colonies. Morphology consistent with *Actinomyces israelii*. *Impression*: Mixed anaerobic infection–actinomycosis.

E, Expectorated sputum, smear, Gram stain, light microscopy, HPV. Purulence heavy. Local materials moderate. Grain present. Gram-positive bacilli, filamentous, beaded, branched. Suspicious for *Nocardia* or *Actinomyces* (see V*F*).

F, Expectorated sputum, chalky white colonies on plate with 5% sheep's blood agar at 5 days' incubation. Routine sputum culture. Morphology consistent with *Nocardia*, *Nocardiopsis*, or *Streptomyces* spp.

G, Sinus tract granule, crush-pull smear, Gram stain, light microscopy, HPV. Amorphous debris moderate. Gram-positive bacilli, filamentous, beaded, branched, partial acid-fast negative. Gram-positive bacilli, regular. Gram-negative bacilli, small. Gram-positive cocci. Morphology suggests mixed anaerobic infection with actinomycetes. *Impression:* Actinomycosis. Culture isolation of *Actinomyces naeslundii*.

H, Surgical biopsy of abnormal area in jawbone, smear, Gram stain, light microscopy, HPV. Amorphous debris heavy. Gram-positive bacilli, filamentous, beaded, branched, partial acid-fast stain negative. Morphology suggests actinomycete. Aerobic and anaerobic cultures negative.

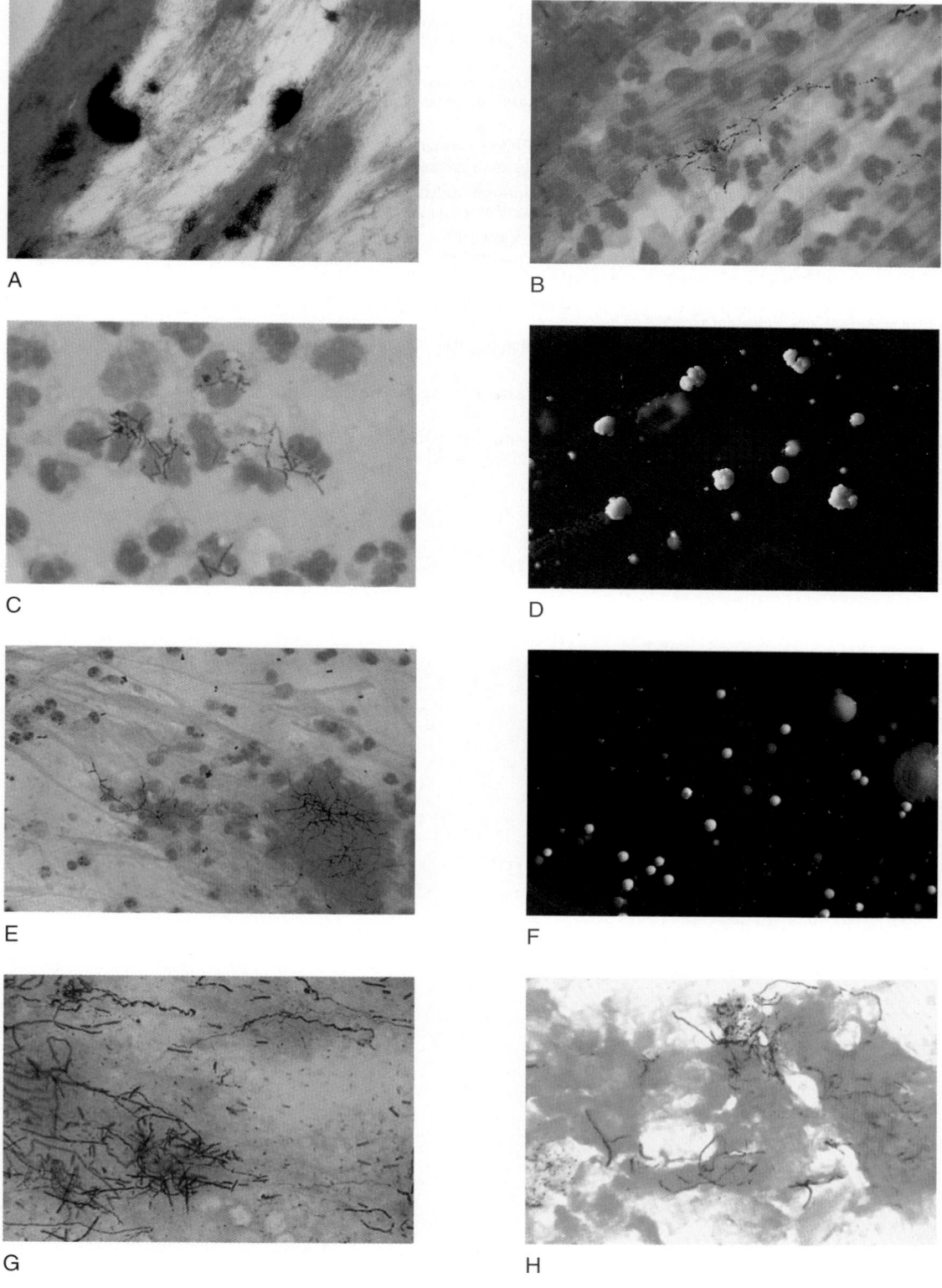

■■■■■■■ P l a t e   V I

Direct Examination in Selected Gram-Negative Bacterial Infections

A, Expectorated sputum, smear, Gram stain, light microscopy, MPV. Mixed materials, type I, layered. Contaminating bacteria and epithelial cells moderate. Purulence light. Gram-negative diplococci. Morphology suggests pathogenic *Neisseria* or *Moraxella* spp. Routine bacterial culture isolated *Moraxella* (*Branhamella*) *catarrhalis*.

B, Expectorated sputum, smear, Gram stain, light microscopy, MPV. Purulence moderate. Local materials moderate. Gram-negative diplococci, intracellular, extracellular. Morphology suggests pathogenic *Neisseria* or *Moraxella* spp. Routine bacterial culture isolated *Neisseria meningitidis*.

C, Expectorated sputum, smear, Gram stain, light microscopy, HPV. Purulence moderate. Local materials heavy. Mucus present. Gram-negative coccobacilli, intracellular, extracellular. Morphology consistent with *Haemophilus influenzae*.

D, Expectorated sputum, smear, Gram stain, light microscopy, HPV. Purulence heavy. Local materials moderate. Gram-negative bacilli, small, pleomorphic, intracellular, extracellular. Gram-positive diplococci, encapsulated, intracellular, extracellular. Morphology consistent with *Haemophilus influenzae* and *Streptococcus pneumoniae*. *Impression:* Polymicrobial infection.

E, Bronchoalveolar lavage, cytocentrifuge preparation. Wright-Giemsa stain, light microscopy, HPV. Purulence none. Lymphocytes present. Local materials moderate. Small bacilli, numerous (see VI*F*).

F, Bronchoalveolar lavage, cytocentrifuge preparation, Gram stain, light microscopy, HPV. Purulence none. Lymphocytes present. Local materials moderate. Gram-negative bacilli, small (see VI*G*).

G, Bronchoalveolar lavage, cytocentrifuge preparation, Wright-Giemsa stain, light microscopy, HPV. Purulence none. Lymphoctyes present. Local materials moderate. Gram-negative bacilli, small (see VI*H*).

H, Bronchoalveolar lavage, cytocentrifuge preparation. Wright-Giemsa stain, light microscopy, HPV. Purulence none. Lymphocytes present. Local materials moderate. Ciliated columnar epithelial cells with numerous small bacilli adherent to cilia. Morphology consistent with *Bordetella pertussis*. *Impression:* Whooping cough.

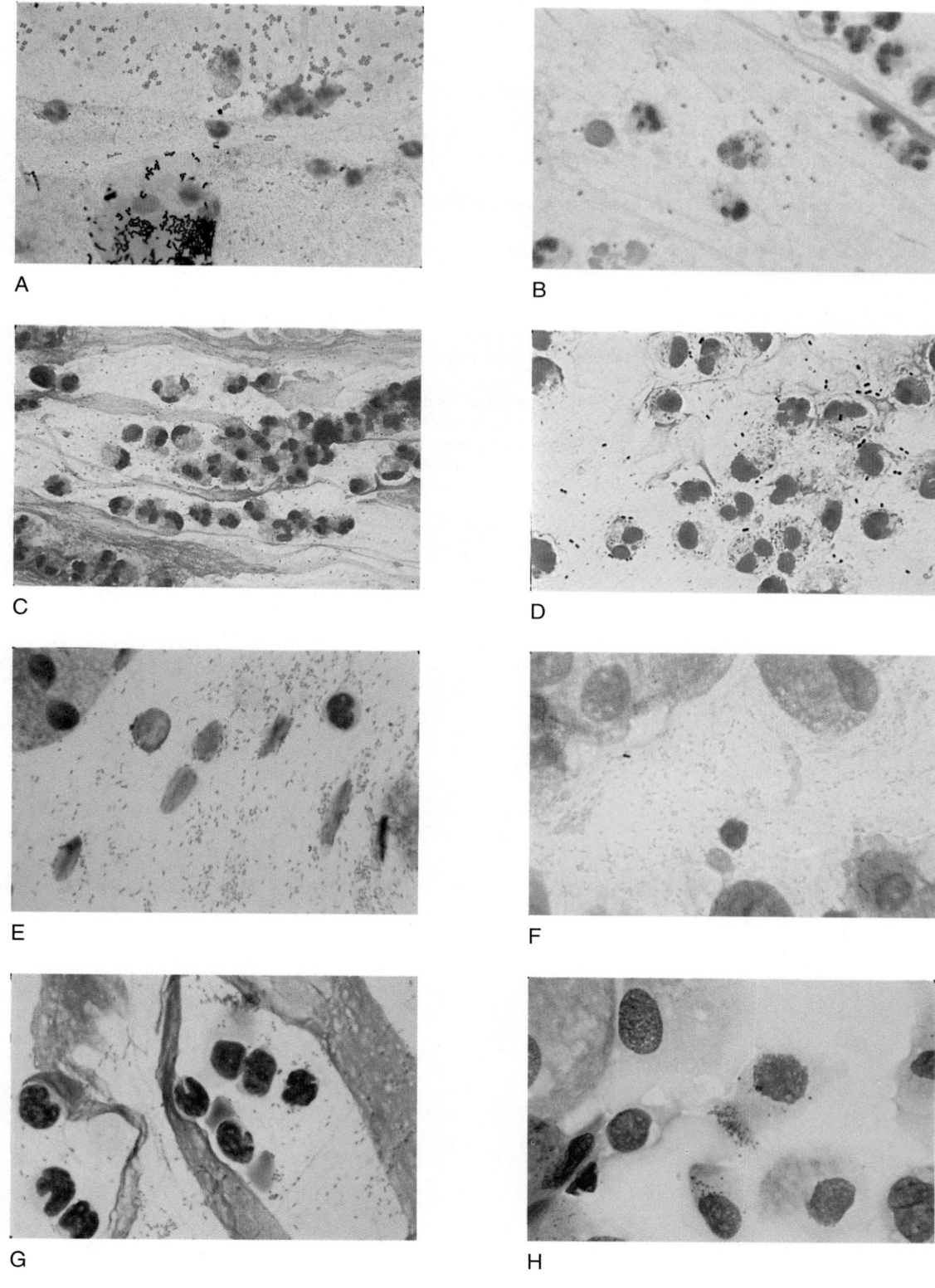

■ P l a t e   V I I

Direct Examination in Selected Gram-Negative Bacillary Infections

A, Cerebrospinal fluid (CSF), drop smear, Gram stain, light microscopy, HPV. Purulence moderate. Gram-negative coccobacilli, chains. Morphotype suggests *Bacteroides* spp. *Impression*: Gram-negative bacillary meningitis.

B, Amniotic fluid, cytocentrifuge preparation, Gram stain, light microscopy, HPV. Purulence moderate. Local materials moderate. Gram-negative coccobacilli. Gram-negative bacilli, filamentous, medium, fusiform. Morphology suggests gram-negative bacillary anaerobic infection. *Impression*: Amnionitis, mixed anaerobic bacteria.

C, Bronchoalveolar lavage, cytocentrifuge preparation, Gram stain, light microscopy, HPV. Purulence moderate. Local materials light. Gram-negative bacilli, small, intracellular within phagocytic vacuoles (see VII*D*).

D, Bronchoalveolar lavage, cytocentrifuge preparation, direct fluorescent antibody (DFA). *Legionella pneumophila*, polyvalent antisera, fluorescent microscopy, HPV. Immunofluorescence positive. *Impression*: Legionnaires' disease.

E, Expectorated sputum, smear, Gram stain, light microscopy, MPV. Purulence light. Local materials light. Mucus moderate. Gram-negative bacilli, medium. *Impression*: Enteric bacillary infection.

F, Urine, direct drop smear, Gram stain, light microscopy, MPV. Purulence heavy. Gram-negative bacilli, medium. Gram-positive cocci. Urine culture grew *Escherichia coli* and *Enterococcus faecalis*. The smear is consistent with a bacterial density of 10^5 colony forming units per mL of urine.

G, Decubitus skin ulcer, smear, Gram stain, light microscopy, HPV. Purulence moderate. Amorphous debris heavy. Gram-negative bacilli, medium, encapsulated. Yeast. Morphology suggests *Klebsiella pneumoniae*. *Impression*: Enteric bacillary disease.

H, Expectorated sputum, smear, Gram stain, light microscopy, HPV. Purulence moderate. Mucus moderate. Gram-negative bacilli, aberrant, encapsulated. Morphology suggests antibiotic-affected *Klebsiella pneumoniae*. *Impression*: Enteric bacillary disease, partially treated.

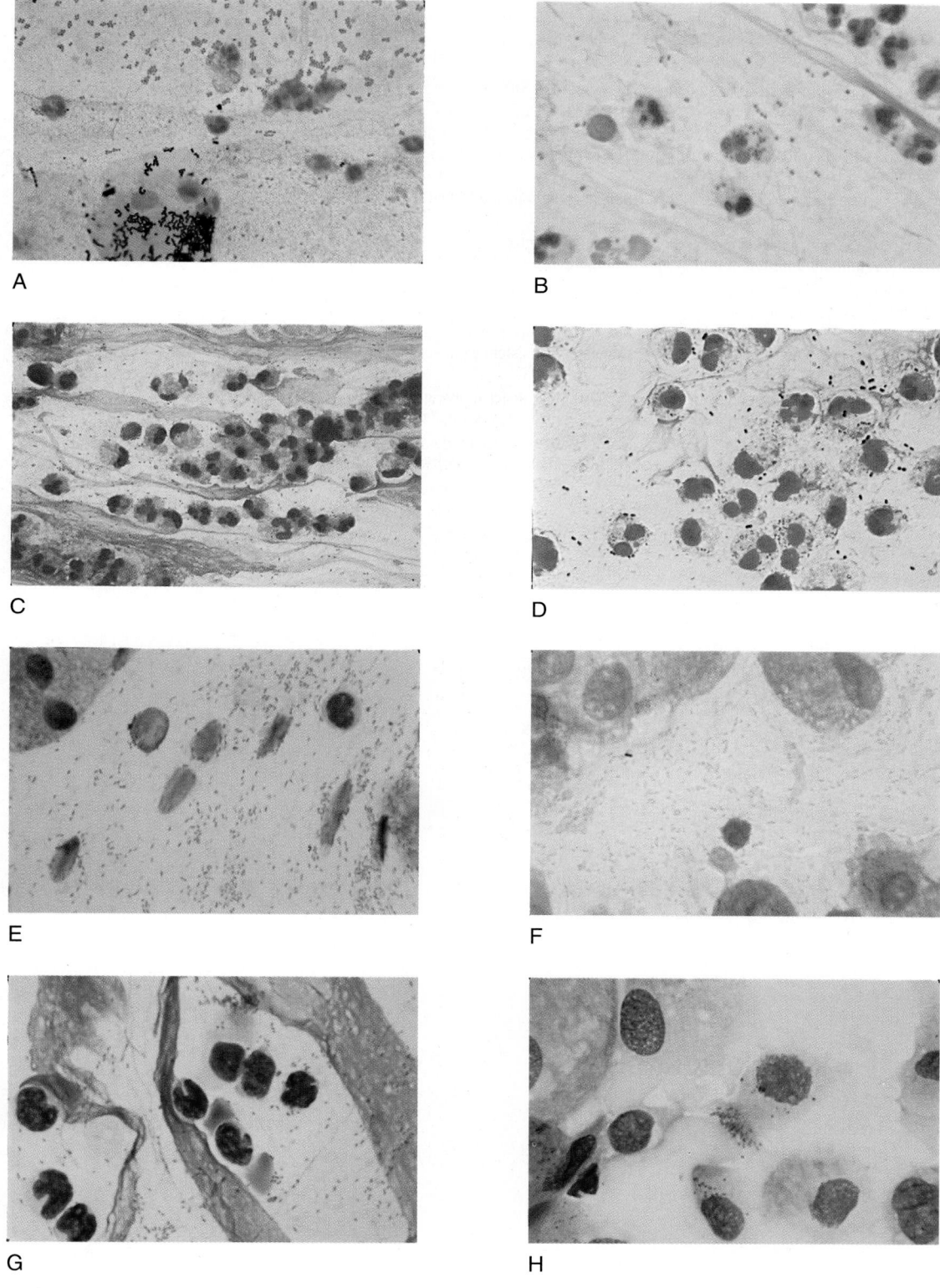

A

B

C

D

E

F

G

H

Direct Examination in Selected Gram-Negative Bacillary Infections

A, Cerebrospinal fluid (CSF), drop smear, Gram stain, light microscopy, HPV. Purulence moderate. Gram-negative coccobacilli, chains. Morphotype suggests *Bacteroides* spp. *Impression*: Gram-negative bacillary meningitis.

B, Amniotic fluid, cytocentrifuge preparation, Gram stain, light microscopy, HPV. Purulence moderate. Local materials moderate. Gram-negative coccobacilli. Gram-negative bacilli, filamentous, medium, fusiform. Morphology suggests gram-negative bacillary anaerobic infection. *Impression*: Amnionitis, mixed anaerobic bacteria.

C, Bronchoalveolar lavage, cytocentrifuge preparation, Gram stain, light microscopy, HPV. Purulence moderate. Local materials light. Gram-negative bacilli, small, intracellular within phagocytic vacuoles (see VII*D*).

D, Bronchoalveolar lavage, cytocentrifuge preparation, direct fluorescent antibody (DFA). *Legionella pneumophila*, polyvalent antisera, fluorescent microscopy, HPV. Immunofluorescence positive. *Impression*: Legionnaires' disease.

E, Expectorated sputum, smear, Gram stain, light microscopy, MPV. Purulence light. Local materials light. Mucus moderate. Gram-negative bacilli, medium. *Impression*: Enteric bacillary infection.

F, Urine, direct drop smear, Gram stain, light microscopy, MPV. Purulence heavy. Gram-negative bacilli, medium. Gram-positive cocci. Urine culture grew *Escherichia coli* and *Enterococcus faecalis*. The smear is consistent with a bacterial density of 10^5 colony forming units per mL of urine.

G, Decubitus skin ulcer, smear, Gram stain, light microscopy, HPV. Purulence moderate. Amorphous debris heavy. Gram-negative bacilli, medium, encapsulated. Yeast. Morphology suggests *Klebsiella pneumoniae. Impression:* Enteric bacillary disease.

H, Expectorated sputum, smear, Gram stain, light microscopy, HPV. Purulence moderate. Mucus moderate. Gram-negative bacilli, aberrant, encapsulated. Morphology suggests antibiotic-affected *Klebsiella pneumoniae. Impression:* Enteric bacillary disease, partially treated.

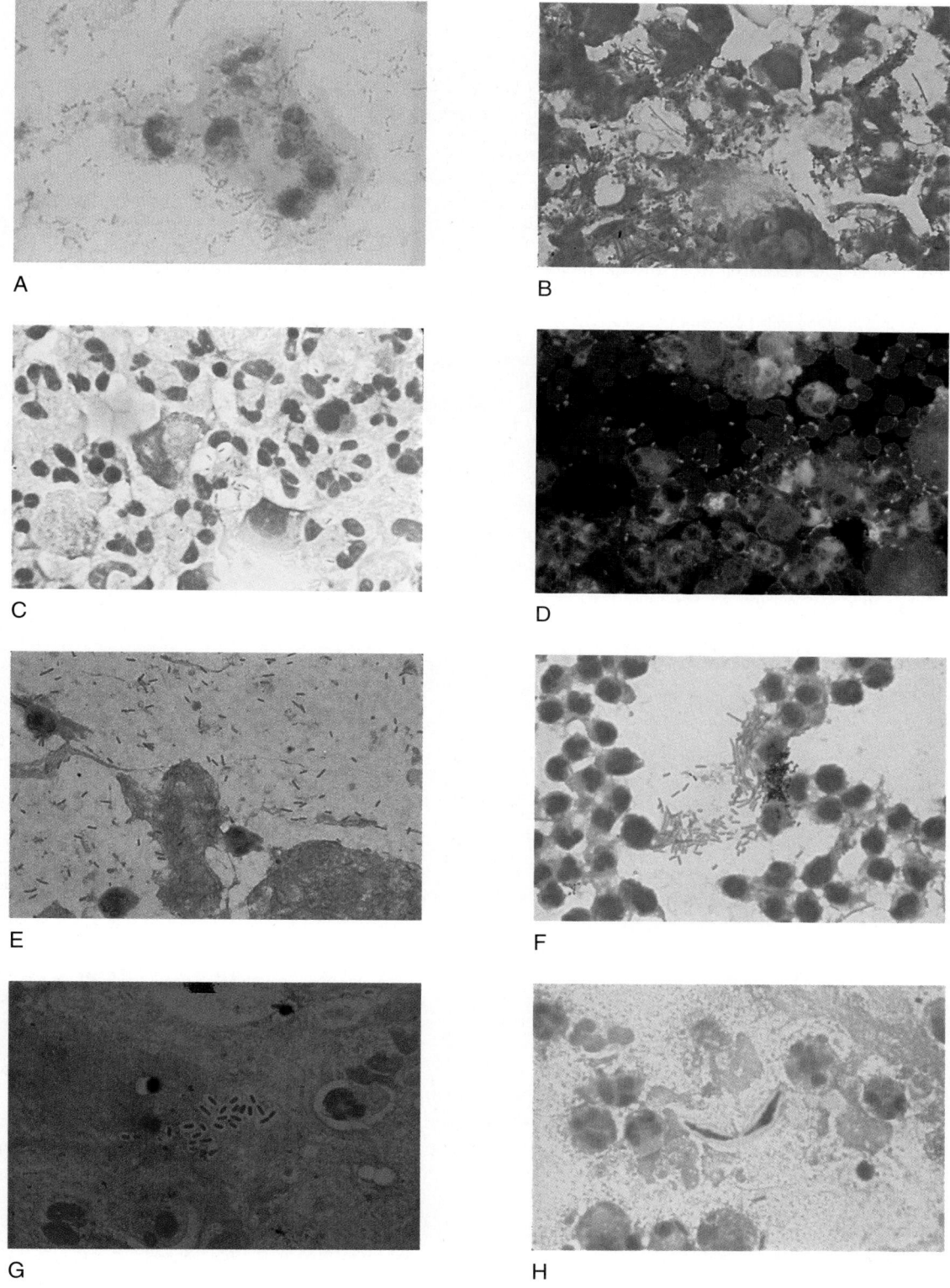

A

B

C

D

E

F

G

H

Direct Examination in Selected Gram-Negative Bacillary Infections

A, Peripheral blood, smear, Wright/Giemsa stain, light microscopy, MPV. Bacilli, medium, bipolar staining (see VIII*B*).

B, Peripheral blood, smear, Gram stain, light microscopy, HPV. Local materials moderate. Gram-negative bacillus, medium, with prominent bipolar staining. Suspicious for *Pasteurella pestis*. *Impression:* Bubonic plague.

C, Expectorated sputum, smear, Gram stain, light microscopy, HPV. Purulence none. Local materials none. Mucus present. Gram-negative bacilli, regular, chains. Morphotype suggests *Pseudomonas aeruginosa*. *Impression: Pseudomonas* infectious disease.

D, Expectorated sputum, smear, Gram stain, light microscopy, HPV. Purulence light. Local materials none. Mucus present. Gram-negative bacilli, regular, enveloped in prominent slime layer. Morphotype suggests mucoid *Pseudomonas aeruginosa*. *Impression:* Pseudomonas infectious disease.

E, Amniotic fluid, cytocentrifuge preparation, Gram stain, light microscopy, HPV. Purulence light. Local material moderate. Gram-negative bacilli, fusiform. Morphology suggests *Fusobacterium nucleatum*. *Impression:* Amnionitis, anaerobic bacteria.

F, Amniotic fluid, cytocentrifuge preparation, Gram stain, light microscopy, HPV. Purulence moderate. Local material moderate. Gram-negative bacilli, medium and long forms. Morphology suggests *Fusobacterium* spp. *Impression*: Amnionitis, anaerobic bacteria.

G, Colonies on blood agar medium, subculture from blood culture, smear, Gram stain, light microscopy, HPV. Local material light. Gram-negative bacilli with spirals, gull-wings. Morphology suggests *Campylobacter* spp.

H, Amniotic fluid, drop smear, Gram stain, light microscopy, HPV. Purulence moderate. Local materials moderate. Gram-negative bacilli, spiral. *Impression*: Amnionitis.

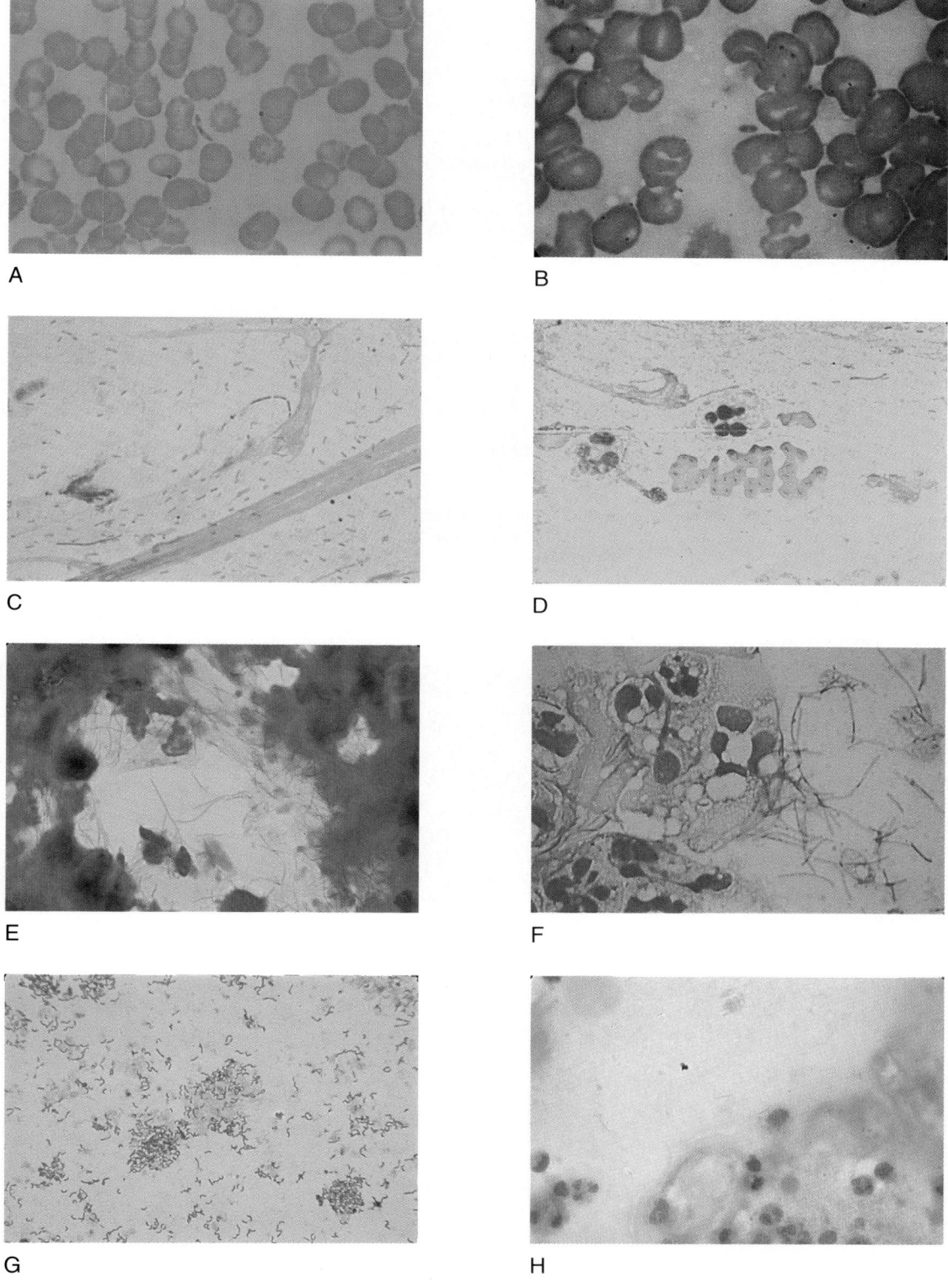

━━━━━ P l a t e   I X

Direct Examination in Polymicrobial Infections

A, Buccal space abscess, smear, Gram stain, light microscopy, HPV. Purulence heavy. Gram-positive cocci, pairs, chains. Gram-positive bacilli, small, diphtheroid, medium, branched. Gram-negative coccobacilli. Morphology suggests polymicrobial infection from mouth flora. *Impression:* Polymicrobial infection, oropharyngeal flora.

B, Maxillary sinus aspirate, smear, Gram stain, light microscopy, HPV. Purulence heavy. Gram-positive cocci, chains. Gram-negative coccobacilli, large masses. Morphotype suggests mixed infection with streptococci and anaerobic Gram-negative coccobacilli. *Impression:* Polymicrobial infection, aerobic and anaerobic species.

C, Cervix, smear, conventional Gram stain, light microscopy, MPV. Purulence heavy. Gram-positive cocci, pairs. Gram-negative coccobacilli. Gram-negative filaments. Trichomonads (*Trichomonas vaginalis*) (compare with IX*D*). *Impression:* Trichomoniasis with mixed aerobic and anaerobic bacterial flora.

D, Cervix, smear, *enhanced* Gram stain, light microscopy, MPV. Purulence heavy. Gram-positive cocci, pairs. Gram-negative coccobacilli. Gram-negative filaments. Trichomonads (*Trichomonas vaginalis*) (compare with IX*C*). *Impression:* Trichomoniasis with mixed aerobic and anaerobic bacterial flora.

E, Eye, vitreous aspirate, smear, Gram stain, light microscopy, MPV. Purulence moderate. Gram-positive diplococci, encapsulated, lancet-shaped. Gram-negative bacilli, small, pleomorphic. Morphotype suggests mixed infection with *Streptococcus pneumoniae* and *Haemophilus influenzae.* *Impression:* Vitritis, mixed infection.

F, Wound, smear, Gram stain, light microscopy, MPV. Purulence heavy. Gram-negative bacilli, medium. Gram-positive cocci, pairs. *Impression:* Enteric bacillary infectious disease.

G, Drainage, ruptured appendix, smear, light microscopy, HPV. Purulence heavy. Gram-positive bacilli, large and medium forms. Gram-negative bacilli, small, bipolar. Gram-positive cocci. Morphology suggest polymicrobial infection with fecal flora. *Impression:* Polymicrobial infection, fecal flora.

H, Aspirated sputum, smear, Gram stain, light microscopy, MPV. Purulence light. Local material light. Gram-positive bacilli, large. Gram-negative bacilli, medium, intracellular. Gram-positive cocci, pairs, chains. Morphology suggests polymicrobial infection with fecal flora. *Impression:* Polymicrobial infection, fecal flora.

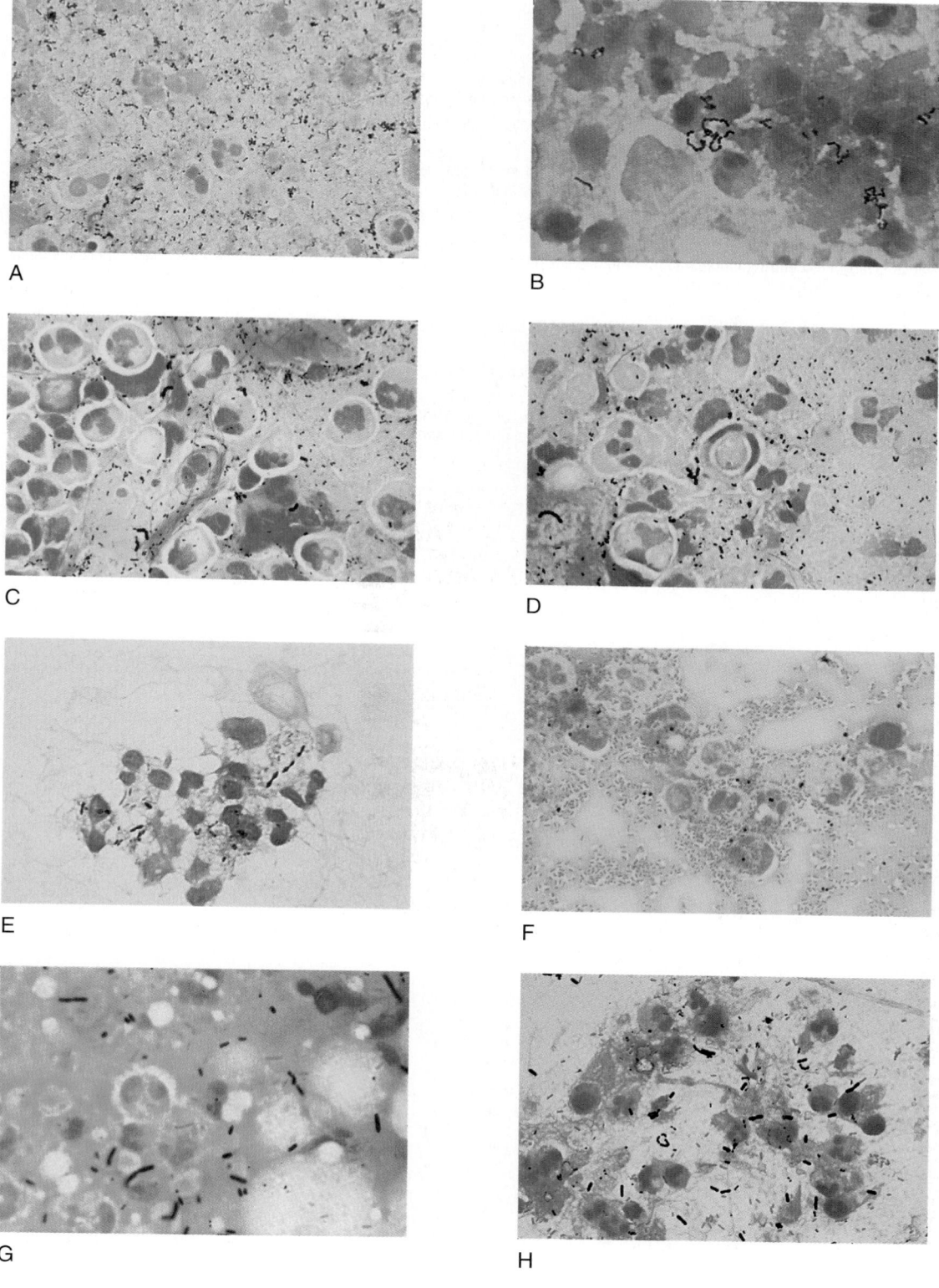

■ P l a t e X

Direct Examination in Fungal Infections

A, Bronchoalveolar lavage, cytocentrifuge preparation, Gram stain, light microscopy, HPV. Purulence light. Local materials moderate. Gram-positive yeast with buds. Morphotype consistent with *Candida* spp. *Impression:* Candidiasis.

B, Expectorated sputum smear, Gram stain, light microscopy, HPV. Purulence light. Local materials moderate. Gram-positive pseudohyphae. Morphotype consistent with *Candida* spp. *Impression:* Candidiasis.

C, Bronchoalveolar lavage, cytocentrifuge preparation, Gram stain, light microscopy, HPV. Local materials moderate. Red blood cells present. Gram-variable yeast with capsules. Morphology suggests *Cryptococcus* (see X*D*).

D, Bronchoalveolar lavage, cytocentrifuge preparation, calcofluor white stain, fluorescence microscopy, HPV. Fluorescent yeast, small, with capsules. Morphology suggests *Cryptococcus neoformans. Impression:* Cryptococcosis.

E, Bronchoalveolar lavage, cytocentrifuge preparation, Wright-Giemsa stain, light microscopy, HPV. Local materials heavy. Yeast, small, budding, intracellular. Morphotype suggests *Histoplasma.* Note the large number of yeast and the absence of reactive cells in this specimen from an AIDS patient (see X*F*).

F, Bronchoalveolar lavage, cytocentrifuge preparation, calcofluor white stain, light microscopy, HPV. Fluorescent yeast (2–4 µm), small, budding. Morphology suggests *Histoplasma capsulatum. Impression:* Histoplasmosis.

G, Expectorated sputum, smear, calcofluor white stain, fluorescence microscopy, MPV. Yeast (8–20 µm), round, thick-walled, broad-based bud. Morphology suggests *Blastomyces dermatitidis. Impression:* Blastomycosis.

H, Expectorated sputum, smear, calcofluor white stain, fluorescence microscopy, MPV. Eosinophils. These cells are another component that stains with calcofluor white. The granules from ruptured eosinophils stain brightly and should not be interpreted as remnants of fungi or parasites.

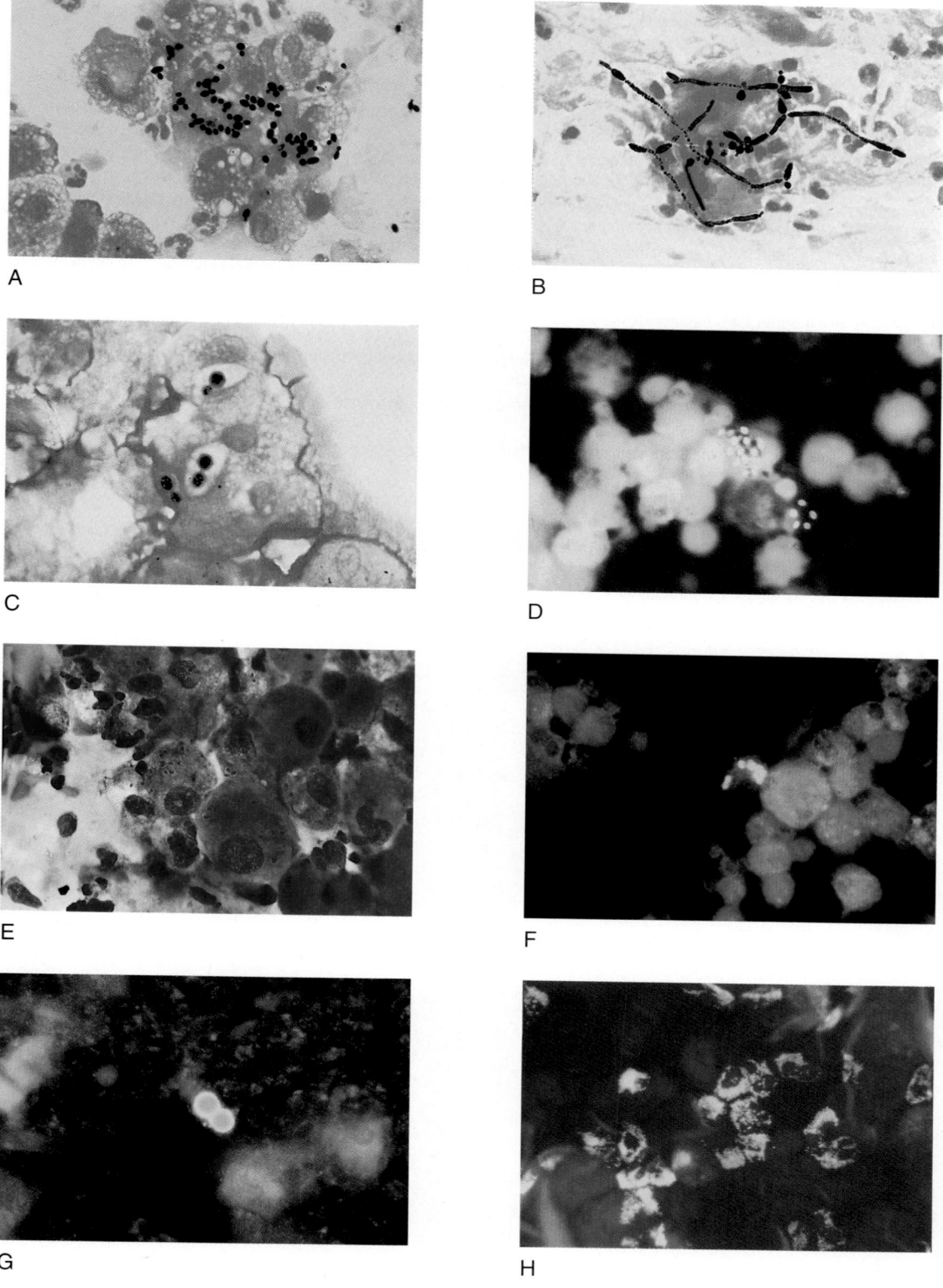

■ P l a t e   X I

Direct Examination in Fungal Infections

A, Skin scales, scrapings, KOH wet preparation, light microscopy, HPV. Hyphae present, septate, thin. Morphology suggests dermatophyte.

B, Skin scales from XI*A*, scrapings, calcofluor white stain, fluorescence microscopy, HPV. Hyphae, thin. Morphology suggests dermatophyte. *Impression:* Dermatophytosis.

C, Expectorated sputum, smear, Gram stain, light microscopy, HPV. Purulence light, Local materials heavy. Gram-variable hyphae present (3–10 μm), septate, branched 45-degree angle. Morphology suggests *Aspergillus* spp (see XI*D*).

D, Expectorated sputum, smear, calcofluor white stain, fluorescence microscopy, MPV. Fungal hyphae present (3–10 μm), septate, branched 45-degree angle. Morphology suggests *Aspergillus* spp. *Impression:* Aspergillosis.

E, Expectorated sputum, smear, Gram stain, light microscopy, HPV. Purulence none. Local materials moderate. Gram-positive conidia (2–4 μm), in chain. Morphotype suggests *Aspergillus* spp in cavity with air interface. *Impression:* Cavitary aspergillosis. Care must be taken not to mistake these conidia for streptococci or for yeast (see XI*F*).

F, Expectorated sputum, smear, Gram stain, light microscopy, HPV. Purulence none. Local materials moderate. Gram-negative conidia (2–4 μm), sporulating. Care must be taken not to confuse these conidia with yeast germ tubes.

G, Brain abscess, smear, toluidine blue stain, light microscopy, MPV. Fungal hyphae present (3–10 μm), septate, branched 45-degree angle. Morphology suggests *Aspergillus* spp. *Impression:* Cerebral aspergillosis.

H, Soft tissue abscess, smear, calcofluor white stain, fluorescence microscopy, HPV. Fungal hyphae, septate, branched chlamydospores. *Impression:* Mycosis.

These hyphae were not clearly visible on the Gram stain smear but stain brightly here. A dermatophyte was isolated in culture.

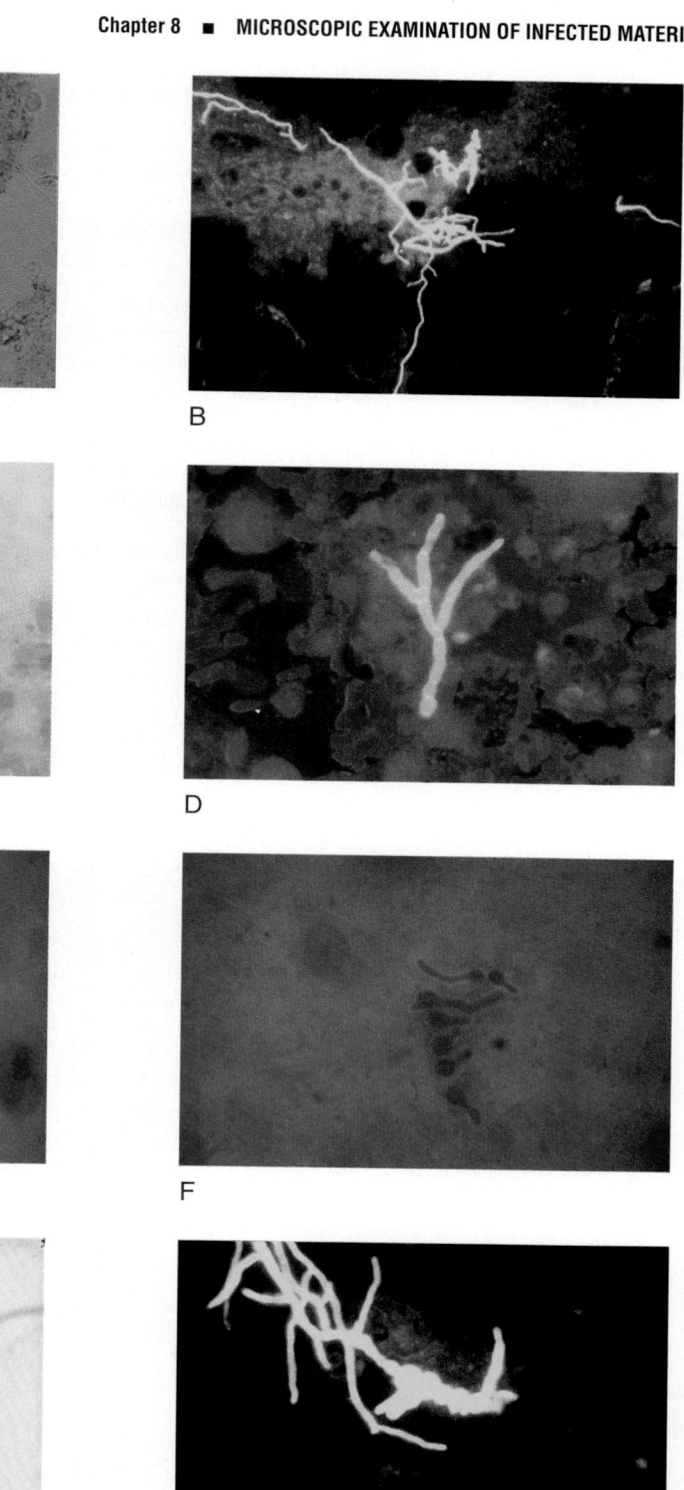

■ P l a t e X I I

Direct Examination in Parasitic Infections

A, Cornea, scraping, Wright-Giemsa stain, light microscopy, HPV. Purulence none. Local materials moderate. Parasitic pre-cyst (13 μm). Morphology consistent with *Acanthamoeba* spp.

B, Cornea, scraping from XII*A*, calcofluor white stain, fluorescence microscopy, MPV. Polyhedral parasitic cyst (13 μm). Morphology consistent with *Acanthamoeba* cyst. *Impression: Acanthamoeba* keratitis.

C, Bronchoalveolar lavage, cytocentrifuge preparation, Wright-Giemsa stain, light microscopy, HPV. Purulence light. Local materials light. Amorphous debris light. Crescent-shaped cells with central nucleus (see XII*D*).

D, Bronchoalveolar lavage, cytocentrifuge preparation, acridine orange stain, fluorescence microscopy, HPV. Crescent-shaped cells composed of RNA. Morphology consistent with trophozoites (tachyzoites) of *Toxoplasma gondii. Impression*: Toxoplasmosis.

E, Bronchoalveolar lavage, cytocentrifuge preparation, Gram stain, light microscopy, HPV. Purulence none. Local materials moderate. Alveolar cast composed of gram-negative matrix and intracystic bodies. Morphology consistent with *Pneumocystis carinii* (see XII*E*).

F, Bronchoalveolar lavage, cytocentrifuge preparation, calcofluor white stain, fluorescence microscopy, HPV. Fluorescent cysts with coccoid bodies. Morphology consistent with *Pneumocystis carinii. Impression:* Pneumocystosis.

G, Corneal ulcer, smear, Gram stain, light microscopy, HPV. Purulence light. Local materials light. Gram-positive spores, medium (4–5 μm). Morphology suggests spores of microsporidia (see XII*H*).

H, Eye, cornea, biopsy, transmission electron microscopy, 29,000X. Parasitic spores medium (3 x 5 μm). Thick endospore layer. Nine polar tubes present. Paired, abutted nuclei. Morphology consistent with *Nosema ocularum. Impression:* Microsporidial keratitis.

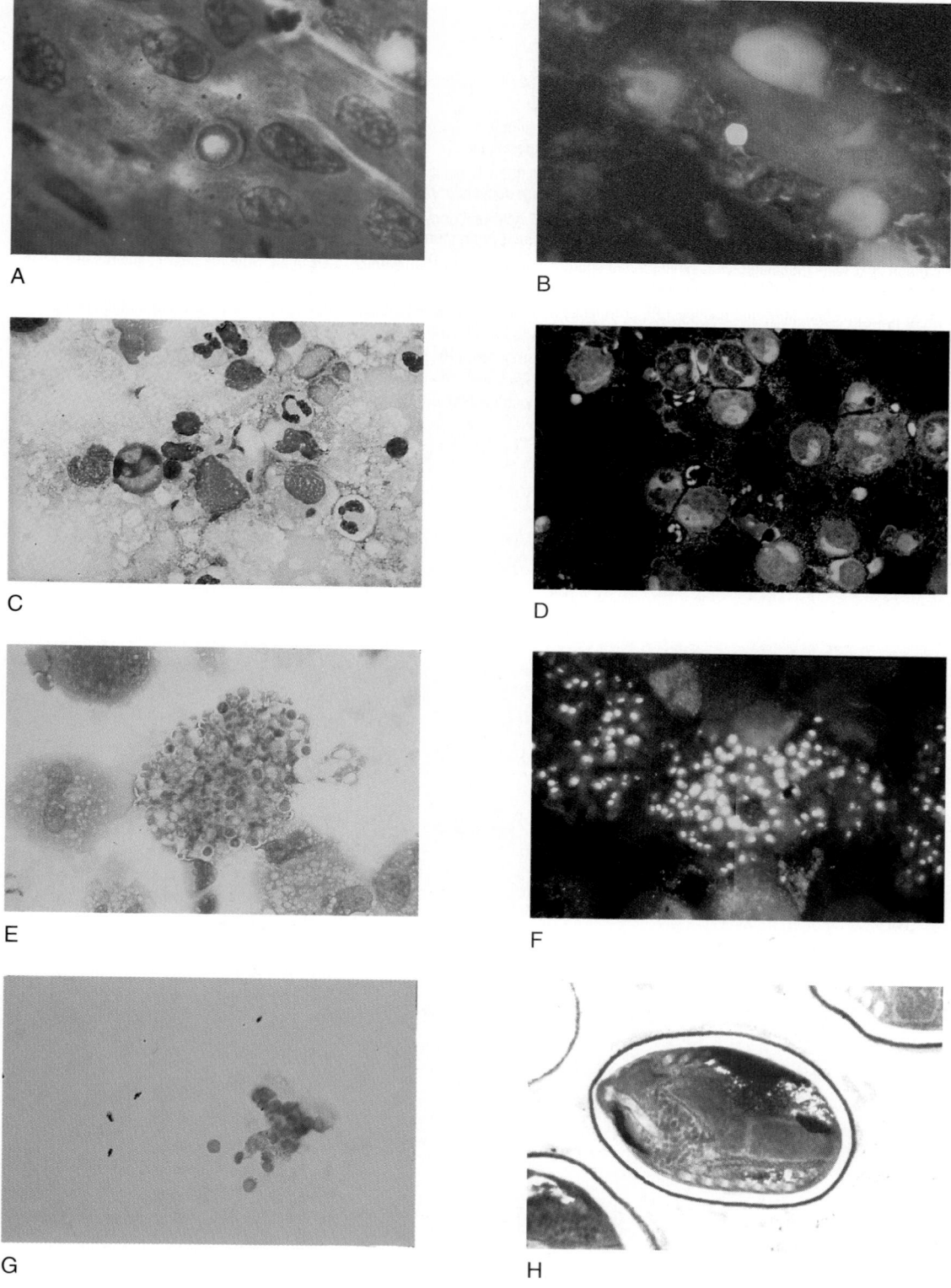

■ P l a t e   X I I I

Direct Examination in Parasitic Infections

A, Diarrheic stool, smear, modified Weber stain, light microscopy, 2000X oil. Purulence none. Parasite spores small (1.5 x 0.9 μm). Morphology consistent with enterocytozoon (see XIII*B*).

B, Diarrheic stool from XIB, concentration, transmission electron microscopy, 57,000X. Endospore layer and polar tubes present. Morphology consistent with *Enterocytozoon bieneusi. Impression:* Intestinal microsporidiosis.

C, Watery, frothy diarrheic stool, smear, acid-fast stain, MPV. Purulence none. Local materials heavy. Acid-fast oocysts (4–6 μm) (see XIII*D*). The measurement is taken to clearly separate this oocyst from the 8 to 10 μm oocysts of *Cyclospora* spp.

D, Watery, frothy diarrheic stool from XIII*C*, smear, acid-fast stain, HPV. Acid-fast oocysts (4–6 μm). Sporulated oocysts containing four sporozoites. Morphology consistent with *Cryptosporidium parvum. Impression:* Cryptosporidiosis.

E, 5% Sheep's blood agar plate inoculated with expectorated sputum, 24-hour incubation with 5% CO_2 in air. Note the heavy bacterial growth in the area of primary inoculation with thin trails of colonies lacing the surface of the agar (see XIII*F*).

F, Aspirated sputum, Gram stain, light microscopy, LPV. Purulence none. Local materials moderate. Mucus present. Coiled nematode larvae. Morphology consistent with *Strongyloides stercoralis. Impression: Strongyloides* hyperinfestation syndrome.

G, Parasite, pubic hair from surgical patient, unstained mount, light microscopy, LPV. Three pair of legs are identified with characteristic claws at tips. Morphology consistent with *Phthirus pubis* (crab louse). *Impression:* Louse infestation.

H, Muscle tissue, directly viewed. Encysted calcified larvae. Morphology consistent with *Trichinella spiralis. Impression:* Trichinosis.

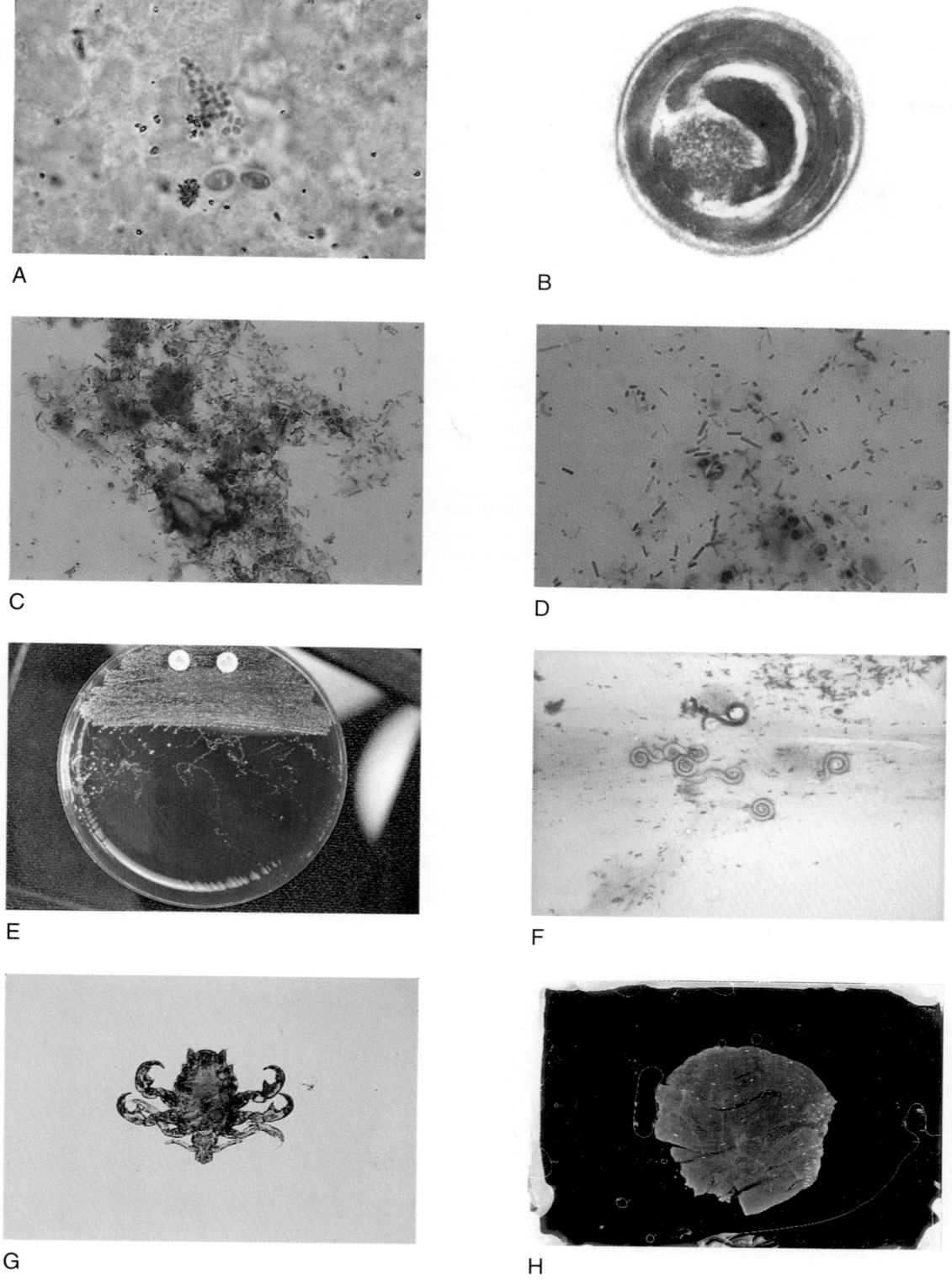

A

B

C

D

E

F

G

H

■■■■■P l a t e   X I V

Direct Examination in Viral Infections

A, Skin, vesicle fluid, Tzanck preparation, hematoxylin and eosin (H&E) stain, light microscopy, MPV. Purulence moderate. Local materials light. Multinucleated epithelial cells present. Intranuclear inclusions present. Morphology consistent with herpes viral inclusions. *Impression:* Herpes simplex infection.

B, Bronchoalveolar lavage, cytocentrifuge preparation, rapid Wright-Giemsa stain, light microscopy, MPV. Purulence light. Local materials light. Red blood cells present. Multinucleated epithelial cells present. Intranuclear inclusions present. Morphology consistent with herpes viral inclusions. *Impression:* Herpes simplex infection. Compare with XIV *A.* Note the change in appearance of the herpes-infected cells with the change in the type of fixation and stain. The H&E stain more clearly shows the "ground-glass" appearance of the nuclear inclusion rimmed by the cell nuclear chromatin. The rapid Wright's stain provides an adequate visual presentation and is more time efficient.

C, Skin, vesicle fluid, Tzanck preparation, Wright/Giemsa stain, light microscopy, HPV. Purulence light. Multinucleated epithelial cells present. Intranuclear inclusions present. Morphology consistent with herpes viral inclusions. *Impression:* Varicella-zoster infection.

D, Skin, vesicle fluid, Tzanck preparation, antibody stain for herpes simplex virus, fluorescent microscopy, MPV. Immunostaining positive. Herpes simplex infection, confirmed.

E, Bronchoalveolar lavage, cytocentrifuge preparation, Wright/Giemsa stain, light microscopy, MPV. Purulence light. Blood moderate. Local materials light. Enlarged pneumocyte with intranuclear inclusion. Morphology consistent with cytomegalovirus. Observe the characteristic nuclear changes for cytomegalovirus. The cell and the nucleus are enlarged, the nucleus is granular, and the nuclear membrane is indistinct. Blood is an indication of capillary damage. *Impression:* Cytomegalovirus disease.

F, Bronchoalveolar lavage, cytocentrifuge preparation, Wright/Giemsa stain, light microscopy, MPV. Purulence light. Blood moderate. Local materials light. Enlarged pneumocyte with intracytoplasmic inclusions. Morphology consistent with cytomegalovirus. The large, regular-sized, magenta cytoplasmic viral inclusions, when present, are characteristic of this virus. *Impression:* Cytomegalovirus disease.

G, Bronchoalveolar lavage, cytocentrifuge preparation, Wright/Giemsa stain, light microscopy, HPV. Purulence light. Local materials moderate. Amorphous debris moderate. No organisms seen. Suspicious for viral infection (see XIV*H*).

H, Bronchoalveolar lavage, cytocentrifuge preparation, IFA adenovirus stain, fluorescence microscopy, HPV. Prominent specific fluorescent staining of infected cells. *Impression:* Adenovirus infection. The necrosis and cellular debris associated with adenovirus within the bronchi can be easily overlooked, because the necrotic, virus-infected "smudge" cells may not be recognized. Specific immunostaining should be performed on the basis of clinical suspicion and compatible background material.

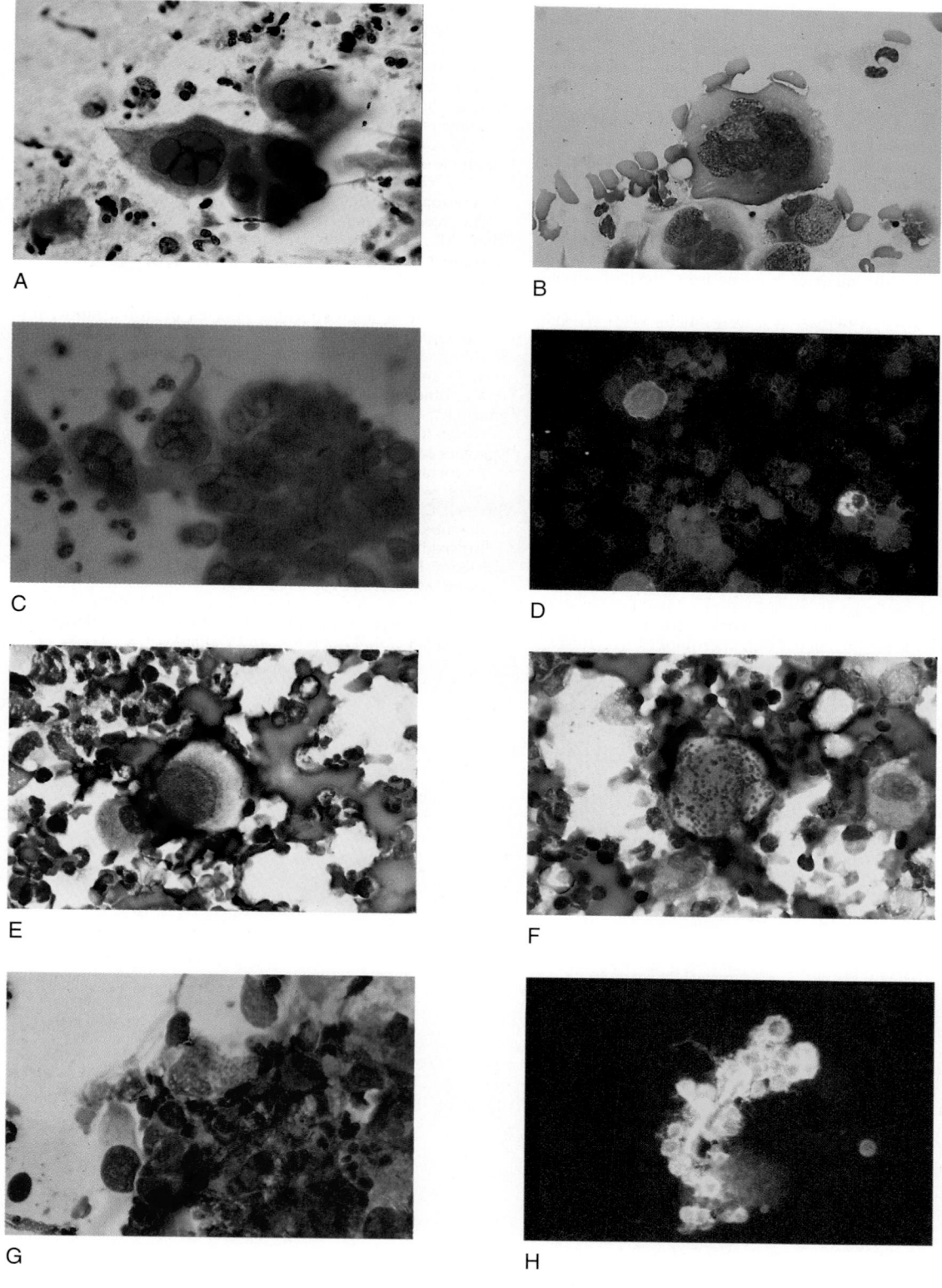

Bibliography

Balows A, Hausler WJ, et al (eds): Manual of Clinical Microbiology, 5th ed. Washington, DC: American Society for Microbiology, 1991.

Bartlett RE: Medical Microbiology: Quality, Cost, and Clinical Relevance. New York: John Wiley & Sons, 1974, pp 27–28.

Broaddus C, Dake MD, et al: Bronchoalveolar lavage and transbronchial biopsy for the diagnosis of pulmonary infections in the acquired immunodeficiency syndrome. Ann Intern Med 102:747, 1985.

Chapin-Robertson K, Dahlver SE, Edberg SC: Clinical and laboratory analysis of Cytospin-prepared Gram stains for recovery and diagnosis of bacteria from sterile body fluids. J Clin Microbiol 30:377, 1992.

Cordonnier C, Bernaudin JF, et al: Diagnostic yield of bronchoalveolar lavage in pneumonitis occurring after allogeneic bone marrow transplantation. Am Rev Respir Dis 132:1118, 1985.

Golden JA, Hollander H, et al: Bronchoalveolar lavage as the exclusive diagnostic modality for *Pneumocystis carinii* pneumonia. Chest 90:18, 1986.

Goswitz JJ, Willard KE, et al: Utility of slide centrifuge Gram's stain versus quantitative culture for diagnosis of urinary tract infection. Am J Clin Pathol 99:132, 1993.

Heineman HS, Chawla JK, Lofton WM: Misinformation from sputum cultures without microscopic examination. J Clin Microbiol 6:518, 1977.

Kokoskin E, Gyorkos TW, et al: Modified technique for efficient detection of microsporidia. J Clin Microbiol 32:1074, 1994.

La Scolea LJ Jr, Dryja D: Quantitation of bacteria in cerebrospinal fluid and blood of children with meningitis and its diagnostic significance. J Clin Microbiol 19:187, 1984.

Lewis JF, Alexander J: Microscopy of stained urine smears to determine the need for quantitative culture. J Clin Microbiol 4:372, 1976.

Magee CM, Rodeheaver GT, et al: A more reliable Gram staining technic for diagnosis of surgical infections. Am J Surg 130:341, 1975.

Mengel M: The use of the cytocentrifuge in the diagnosis of meningitis. Am J Clin Pathol 84:212, 1985.

Murray PR, Washington JA: Microscopic and bacteriologic analysis of expectorated sputum. Mayo Clin Proc 50:339, 1975.

Ognibene FP, Shelhamer J, et al: The diagnosis of *Pneumocystis carinii* pneumonia in patients with the acquired immunodeficiency syndrome using subsegmental bronchoalveolar lavage. Am Rev Respir Dis 129:929, 1984.

Olson ML, Shanholtzer CJ, et al: The slide centrifuge Gram stain as a urine screening method. Am J Clin Pathol 96:454, 1991.

Ryan NJ, Sutherland G, et al: A new trichrome-blue stain for detection of microsporidial species in urine, stool, and nasopharyngeal specimens. J Clin Microbiol 31:3264, 1993.

Shanholtzer CJ, Schaper PJ, Peterson LR: Concentrated Gram stain smears prepared with a Cytospin centrifuge. J Clin Microbiol 16:1052, 1982.

Smith JW, Barlett MS: Laboratory diagnosis of *Pneumocystis carinii* infection. Clin Lab Med 2:383, 1982.

Spengler M, Rodeheaver GT, et al: The Gram stain—the most important diagnostic test of infection. JACEP 7.12:434, 1978.

Stover DE, Zaman MB, et al: Bronchoalveolar lavage in the diagnosis of diffuse pulmonary infiltrates in immunosuppressed host. Ann Intern Med 101:1, 1984.

Van Scoy RE: Bacterial sputum cultures, a clinician's viewpoint. Mayo Clin Proc 52:39, 1977.

Vestal AL: Procedures for the Isolation and Identification of Mycobacteria. Publication No. CDC75-8230. Atlanta: US Dept of Health, Education, and Welfare, 1975.

CHAPTER 9

Utilization of Colonial Morphology for the Presumptive Identification of Microorganisms

IMPORTANCE OF COLONIAL MORPHOLOGY AS A DIAGNOSTIC TOOL

INITIAL OBSERVATION AND INTERPRETATION OF CULTURES

GROSS COLONY CHARACTERISTICS USED TO DIFFERENTIATE AND PRESUMPTIVELY IDENTIFY MICROORGANISMS
Hemolysis
α Hemolysis
β Hemolysis
Size
Form or Margin
Elevation
Density
Color
Consistency
Pigment
Odor

COLONIES WITH MULTIPLE CHARACTERISTICS

GROWTH OF ORGANISMS IN LIQUID MEDIA

George Manuselis, Jr.

OBJECTIVES

1. Describe how growth on blood, chocolate, and MacConkey agars is used in the preliminary identification of isolates.

2. Differentiate α hemolysis from β hemolysis.

3. Describe how gross colony characteristics are used in the presumptive identification of microorganisms.

4. Utilizing colonial morphology, differentiate among the following microorganisms:

   - *Staphylococcus* and *Streptococcus.*
   - *Streptococcus agalactiae* and *Streptococcus pyogenes.*
   - α-Hemolytic *Streptococcus* and *Streptococcus pneumoniae.*
   - *Neisseria* species and *Staphylococcus.*
   - Yeast and *Staphylococcus.*
   - "Diphtheroids" and *Staphylococcus.*
   - Lactose fermenters from lactose nonfermenters.
   - *Proteus* species from other Enterobacteriaceae.

The mastery of *colonial morphology* (colony characteristics and form) and interpretation of gram-stained smears from clinical specimens and microbial colonies cannot be overemphasized. Although gram-stained smears provide initial identification of microorganisms by microscopic characterization, growth characteristics of microorganisms on certain types of laboratory media facilitate description of colonial morphology for their identification processes.

An analogy serves to illustrate this concept. Close your eyes and imagine the physical characteristics of a parent, relative, or friend. It may be the height, weight, shape, color or style of hair, eyes, freckles, color of skin, even the voice or laugh that makes the person distinctive even in a crowd or when his or her back is facing you. In the same manner, many specific microorganisms have characteristics that distinguish them in a crowd of other genera or species.

This chapter explains how characterization of colonies on culture media facilitates presumptive identification of commonly isolated organisms. It discusses the characteristics that are used to describe morphology of certain groups of organisms and how these characteristics are used to differentiate one species from a closely related species and one genera from another.

IMPORTANCE OF COLONIAL MORPHOLOGY AS A DIAGNOSTIC TOOL

In many ways, the usefulness of colonial morphology extends the capabilities of the microbiologist and, ultimately, the clinical laboratory. The ability to provide a presumptive identification by colonial morphology may:

■ **Provide a presumptive diagnosis to the physician in time of critical need.** Even in this age of rapid identification systems, incubation times and procedures can be protracted. In a critical situation, the microbiologist makes an educated judgment about the presumptive identity prior to performing diagnostic procedures.

■ **Enhance the quality of patient care through rapid reporting of results and by increasing this cost-effectiveness of laboratory testing.** This may best be illustrated by using sputum cultures as an example. Because the upper respiratory tract contains many indigenous organisms, to identify every organism present in culture would be a time-consuming, cost-prohibitive, and insurmountable task. It is, therefore, important to be able to differentiate potential pathogens from the "normal" inhabitants of the upper respiratory tract and to direct the diagnostic work-up only to potential pathogens. Moreover, potential pathogens are presumptively identified by colonial characteristics, and preliminary reporting will initiate immediate therapy.

■ **Play a signficant role in quality control, especially of automated procedures and other commercially available identification systems.** When commercial and automated systems are used, a mixed inoculum will produce biochemical test results or erroneous interpretation of reactions that significantly alters the identification. The ability of the microbiologist to determine whether the inoculum is mixed and to ascertain whether the results generated by a commercial or automated system correlate with the suspected identification of the organism is an important component of quality control, accomplished by being able to recognize organisms by their colonial characteristics.

INITIAL OBSERVATION AND INTERPRETATION OF CULTURES

Generally, microbiologists observe the colonial morphology of organisms isolated on primary culture after 18 to 24 hours of incubation. Incubation time may certainly vary according to when the specimen is received and processed in the laboratory, which may affect the "typical" morphology of a certain isolate. For example, young cultures of *Staphylococcus aureus* may appear smaller and may not show the distinct β hemolysis older cultures produce. In addition, the microbiologist must be aware of factors that may significantly alter the colonial morphology of growing microorganisms. These factors include the medium's ingredients, its inhibitory nature, and antibiotics present in the medium.

The interpretation of primary cultures, commonly referred to as *plate reading*, is actually a comparative examination of microorganisms growing on a variety of culture media. Many specimens, such as sputum and wounds, that arrive in the clinical laboratory are plated on blood agar (BAP), chocolate agar (CHOC), and MacConkey (MAC). Therefore, as a culture set from a specimen, growth on these three culture media illustrates the concept of comparative colonial examination or plate reading.

First, the ability to determine which organisms grow on selective and nonselective media aids the microbiologist in making an initial distinction between gram-positive and gram-negative isolates. BAP and CHOC support the growth of a variety of fastidious and nonfastidious organisms, gram-positive as well as gram-negative bacteria. Although BAP supports fastidious organisms, highly fastidious species such as *Haemophilus* and *Neisseria gonorrhoeae* do not grow on it. Chocolate agar provides nutritional growth requirements to support highly fastidious organisms such as *Haemophilus* and *N. gonorrhoeae*. Therefore, a gram-negative bacillus that grows on CHOC but not on BAP or MAC will be suspected to be a *Haemophilus* species, whereas gram-negative diplococci with the same growth pattern will be suspected to be *N. gonorrhoeae* (Fig. 9–1). With this initial interpretation, the microbiologist is able to provide a presumptive identification and determine how to proceed in identifying the isolated organisms.

Second, MacConkey agar, which inhibits gram-positive organisms and some fastidious gram-negative organisms, such as *Haemophilus* and *Neisseria* species, supports most gram-negative rods, especially the Enterobacteriaeceae. Growth on BAP and CHOC but not on MAC, therefore, is indicative of a gram-positive isolate or of a fastidious gram-negative bacillus or coccus.

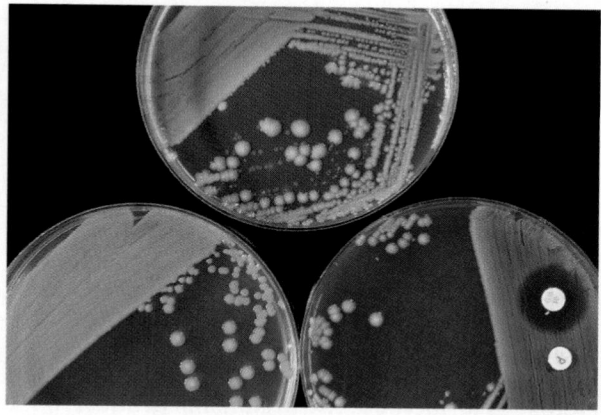

■■■■ F i g u r e   9 – 1

Clockwise from the top: chocolate agar (CHOC), blood agar (BAP), MacConkey agar (MAC). The large colonies growing on all three plates are gram-negative rods (enterics). These gram-negative rods grow larger, gray, and mucoid on BAP and CHOC. Notice the smaller grayish brown fastidious colonies of *Haemophilus* growing on CHOC, which are not growing on BAP or MAC.

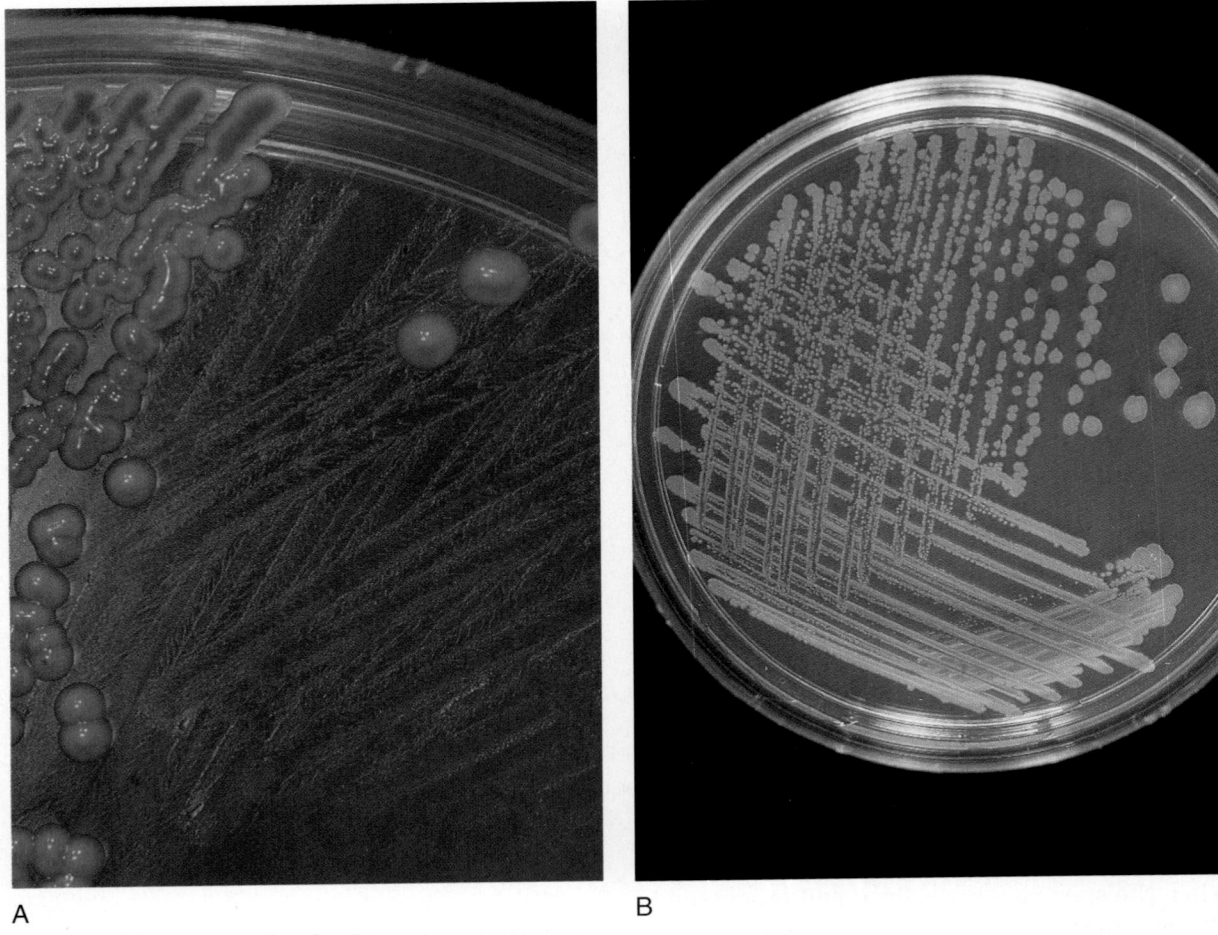

A   B

■■■■■ F i g u r e   9 – 2

A, Example of lactose-fermenting gram-negative rods producing pink colonies on MAC. *B,* Example of non–lactose-fermenting gram-negative rods producing colorless colonies on MAC.

Gram-negative rods are better described on MacConkey agar, because these organisms produce similar-looking colonies on BAP and CHOC media: large, gray, and mucoid. MAC is best used, however, to differentiate lactose fermenters from lactose nonfermenters. Lactose fermenters are easily detected because of the color change they produce on the media; as the pH changes when lactose is fermented, the organisms produce dark pink to red colonies (Fig. 9–2*A*). Colonies of nonfermenters remain clear and colorless (see Fig. 9–2*B*). This differentiation is particularly important in screening for enteric pathogens from stool cultures. Most enteric pathogens do not ferment lactose.

Certain enteric pathogens produce a characteristic colony on MAC that is helpful in presumptive identification. *Escherichia/Citrobacter*–like organisms produce a dry, pink colony with a surrounding "halo" of pink, precipitated bile salts (Fig. 9–3). *Klebsiella/Enterobacter*–like organisms produce large, mucoid pink colonies that occasionally have cream-colored centers (Fig. 9–4).

Microorganisms grow on culture media in the same proportion or concentration in which they are present in the clinical specimen. Because many specimens are polymicrobic, this trait can be beneficial in identifying different colony types. The reader should remember that this is a comparative analysis of the growth on the three types of culture media.

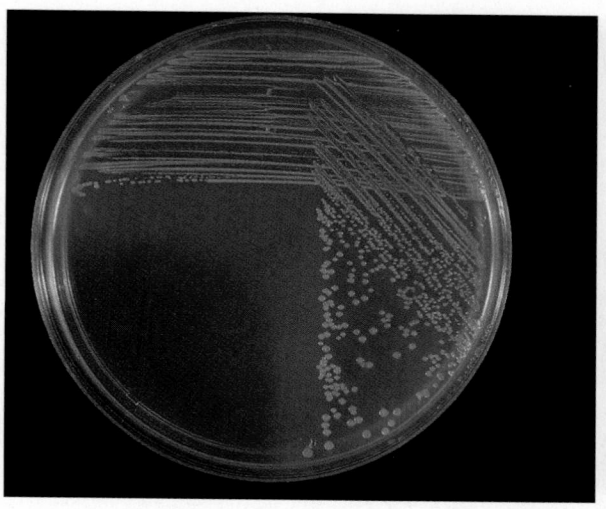
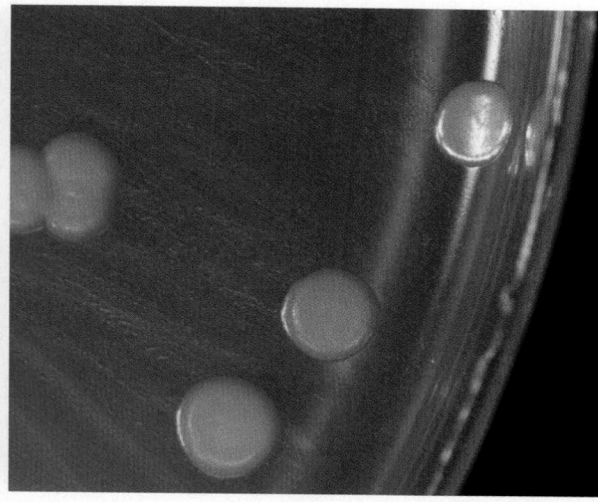

A

B

■■■■■■■ F i g u r e   9 – 3

A, Lactose-fermenting *Escherichia/Citrobacter*–like organisms growing on MAC. Notice the dry appearance of the colony and the pink precipitate of bile salts extending beyond the periphery of the colonies. *B,* Close-up of dry, flat *Escherichia/Citrobacter*–like lactose fermenters growing on MAC. Compare with Figure 9–4*B.*

GROSS COLONY CHARACTERISTICS USED TO DIFFERENTIATE AND PRESUMPTIVELY IDENTIFY MICROORGANISMS

By observing the colonial characteristics of the organisms that have been isolated, the microbiologist is able to make an educated guess about the identification of the isolate. The following descriptions are routinely used to examine colony characteristics.

Hemolysis

On blood agar, hemolysis is a reaction observed in the media immediately surrounding or underneath the colony. Often, the colony has to be removed with a loop in order to visualize the hemolytic pattern. Hemolysis on blood agar is helpful in the presumptive identification, particularly of streptococci. It is important to determine whether true hemolysis is present or whether discoloration of the media is due to growth of the organism on the plate. Proper technique requires the passing of bright light through the bottom of the plate (*transillumination*) in order to determine whether the organism is hemolytic (Fig. 9–5).

Although there are many types of hemolysis, only α hemolysis and β hemolysis are illustrated in this chapter.

α Hemolysis

α Hemolysis is partial clearing of blood around the colony that results in a green discoloration of the medium. Examples of organisms that produce α hemolysis include *Streptococcus pneumoniae* and certain viridans streptococci. (For a comparison of the colonial morphology of these two organisms, see Figure 9–23).

β Hemolysis

β Hemolysis is complete clearing of blood around the colonies due to the complete lysis of red blood cells. Certain organisms such as group A β-hemolytic streptococci (*Streptococcus pyogenes*), produce a wide clear zone of hemolysis, whereas others, such as group B β-hemolytic streptococci (*Streptococcus agalactiae*) and *Listeria monocytogenes,* produce a narrow, diffuse zone of hemolysis. These features are helpful

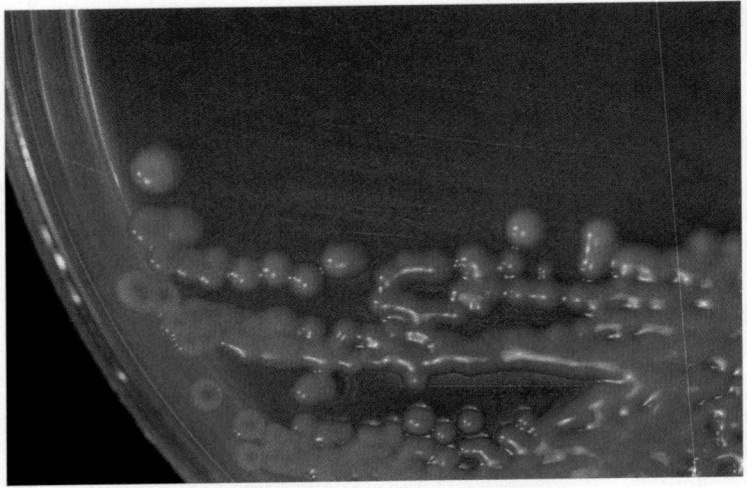

A

B

Figure 9 – 4

A, Klebsiella/Enterobacter–like lactose fermenters growing on MAC. Notice the pink, heaped, mucoid appearance. *B,* Close-up of *Klebsiella/Enterobacter*–like colonies on MAC. Notice the mucoid, heaped appearance and the slightly cream-colored center after 48 hours' growth.

hints in the identification of certain species of bacteria. (For a comparison of the colonial characteristics of group A and group B *Streptococcus,* see Figure 9–24). Chocolate agar does not display true hemolysis, because the red cells in the medium have already been lysed. Organisms that are α-hemolytic or β-hemolytic on blood agar usually show a green coloration around the colony on chocolate agar (Fig. 9–6). This coloration, therefore, should not be mistaken for a hemolytic characteristic.

Size

Colonies may be described as large, medium, small, or pinpoint. Rarely, however, does a microbi-

ologist take a ruler and actually measure a colony. Size is generally a visual comparison between genera or species. For example, gram-positive bacteria, in general, produce smaller colonies than gram-negative bacteria. *Staphylococcus* species are usually larger than *Streptococcus* species. Figure 9–7 shows colonies of gram-negative rods in comparison with gram-positive cocci.

Form or Margin

The edge of the colonies should be observed and the form described as *smooth, filamentous, rough* or *rhizoid,* or *irregular* (Fig. 9–8). Colonies of *Bacillus anthracis* on visual examination are

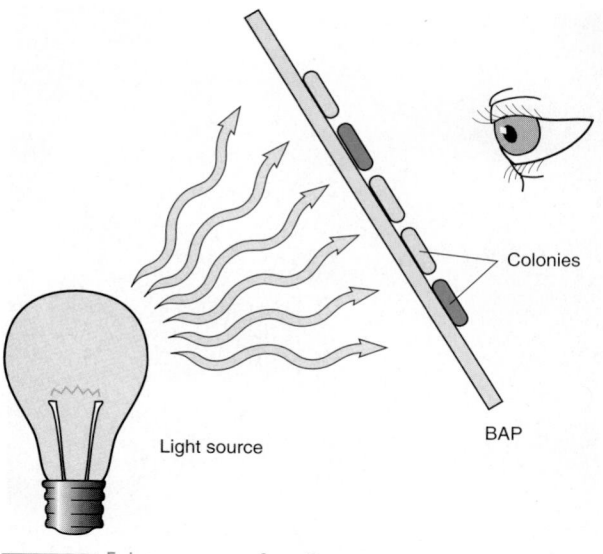

Colonies

Light source

BAP

■ Figure 9 – 5

The use of transillumination to determine whether the colonies are hemolytic.

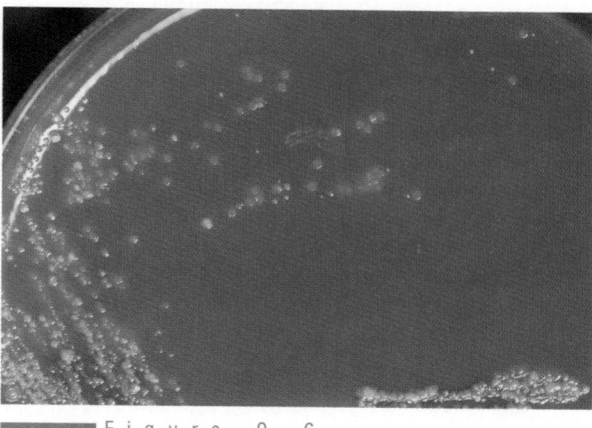

■ Figure 9 – 6

Chocolate agar does not display true hemolysis, because the red cells in the medium have already been lysed. Bacteria that are hemolytic on BAP usually have a green coloration around the colony on CHOC.

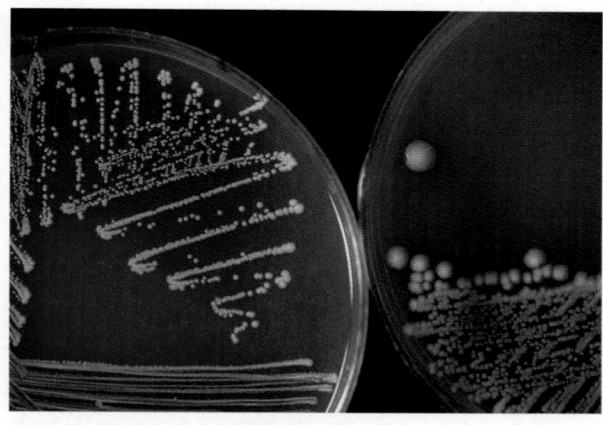

■ Figure 9 – 7

Large, gray, mucoid colonies on the *right* BAP are gram-negative rods. On the *left* BAP, the smaller, white colonies are gram-positive cocci.

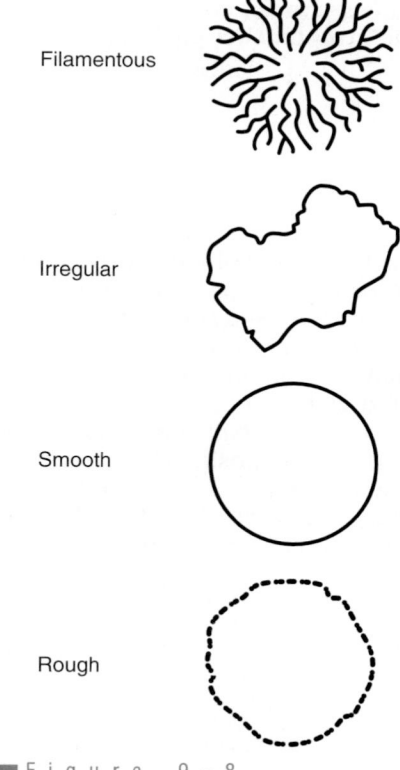

Filamentous

Irregular

Smooth

Rough

■ Figure 9 – 8

Illustration of form or margin to describe colonial morphology.

described as "Medusa heads" because of the filamentous appearance. Certain species, such as *Proteus* sp, may swarm on nonselective agar such as blood or chocolate. *Swarming* is a hazy blanket of growth on the surface. Figure 9–9 shows swarming colonies of *Proteus* sp. Diphtheroids produce colonies that have rough edges (Fig. 9–10), whereas yeasts produce colonies that are described as *stars* or colonies *with feet or pedicles.* (For a comparison of the colonial morphology of yeast and *Staphylococcus,* see Figure 9–25).

Elevation

The elevation should be determined by tilting the culture plate and looking at the side of the colony (Fig. 9–11). Elevation may be *raised, convex, flat, umbilicate* (depressed center, an "inny"), or *umbonate* (raised or bulging center, an "outy"). *Streptococcus pneumoniae* typically produces umbilicate colonies unless the colonies are mucoid because of the presence of polysaccharide capsule. *Staphylococcus aureus* typically produces convex colonies. In comparison, β-hemolytic streptococci generally produce flat colonies.

Density

The density of the colony can be *transparent, translucent,* or *opaque.* β-Hemolytic streptococci, except for group B (*S. agalactiae*), are described as translucent. *Streptococcus agalactiae* produces colonies that are semiopaque, with the organisms concentrated at the center of the colony, sometimes described as a *bull's-eye* colony. Staphylococci and other gram-positive bacteria are usually opaque. Most gram-negative rods are also opaque. *Bordetella pertussis* is described as shiny, like a *half-pearl,* on blood-containing media.

Color

Color, in contrast with pigmentation, is a term used to describe in general a particular genus. Colonies may be *white, gray, yellow,* or *buff.* Coagulase-negative staphylococci are white (Fig.

■ F i g u r e   9 – 9

Swarming colonies of *Proteus* sp.

9–12), whereas *Enterococcus* species may be gray. Certain *Micrococcus* species and *Neisseria* (nonpathogenic) species are yellow or off-white (Fig. 9–13). Diphtheroids are buff. Most gram-negative rods are gray.

Consistency

Consistency is determined by touching the colony with a sterile loop. Colony consistency may be *brittle* (splinters), *creamy* (butyrous), *dry,* or

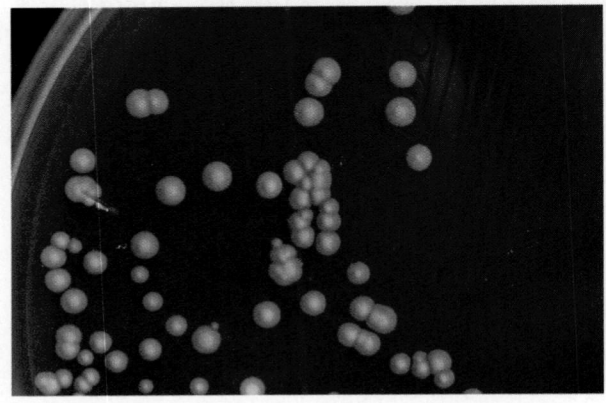

■ F i g u r e   9 – 1 0

"Diphtheroid" colonies with rough edges, dry appearance, and umbonate center growing on BAP.

Flat

Raised

Convex or "dome"

Umbilicate

Umbonate

■ F i g u r e   9 – 1 1

Illustration of elevations to describe colonial morphology.

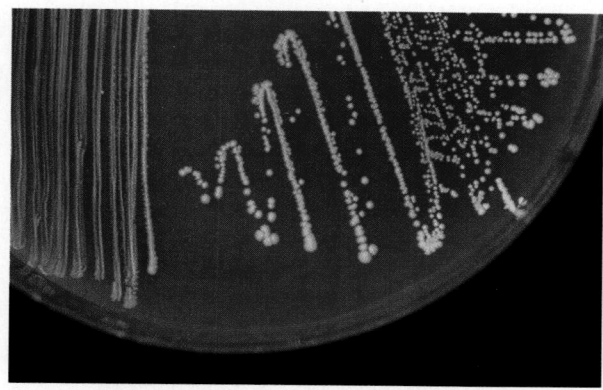

■ F i g u r e   9 – 1 2

Example of white colonies of coagulase-negative *Staphylococcus*.

waxy; occasionally, the entire colony adheres (*sticky*) to the loop. *Staphylococcus aureus* is creamy, whereas certain *Neisseria* species are sticky. *Nocardia* sp produces colonies that are brittle, crumbly, and wrinkled, resembling bread crumbs on a plate. Diphtheroid colonies are usually dry and waxy. Most β-hemolytic streptococci are dry (except for mucoid types), and when pushed by a loop, the whole colony remains intact.

Pigment

Pigment production is an inherent characteristic of a specific organism confined generally to the colony. Examples of organisms that produce pigment are:

■ *Pseudomonas aeruginosa*—green, sometimes a metallic sheen (Fig. 9–14)

■ *Serratia marcescens*—brick-red (Fig. 9–15)

■ *Kluyvera*—blue

■ *Chromobacterium violaceum*—purple

■ *Prevotella melaninogenica*—brown-black (anaerobic)

Pigment production for these organisms is variable.

Odor

For safety reasons, it is not a good idea to inhale at the surface of the plate to check for odor. Odor should be determined when the lid of the culture plate is removed and its odor dissipates into the surrounding environment. Examples of microorganisms that produce distinctive odors are:

■ *Staphylococcus aureus*—old sock (stocking that has been worn continuously for a few days without washing)

■ *Pseudomonas aeruginosa*—fruity or grape-like

■ *Proteus mirabilis*—putrid

■ *Haemophilus*—musty basement

■ *Nocardia*—freshly plowed field

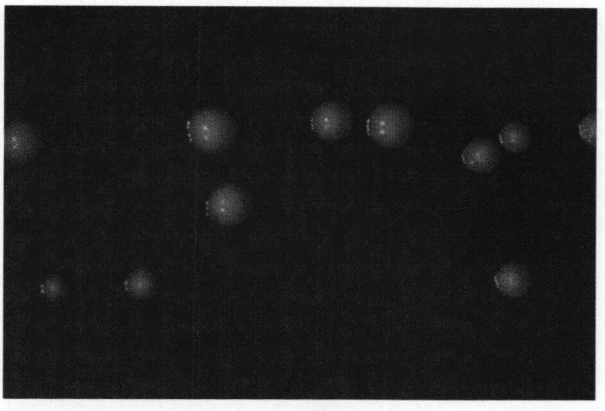

■ F i g u r e   9 – 1 3

Example of the yellow colonies characteristic of certain nonpathogenic species of *Neisseria*.

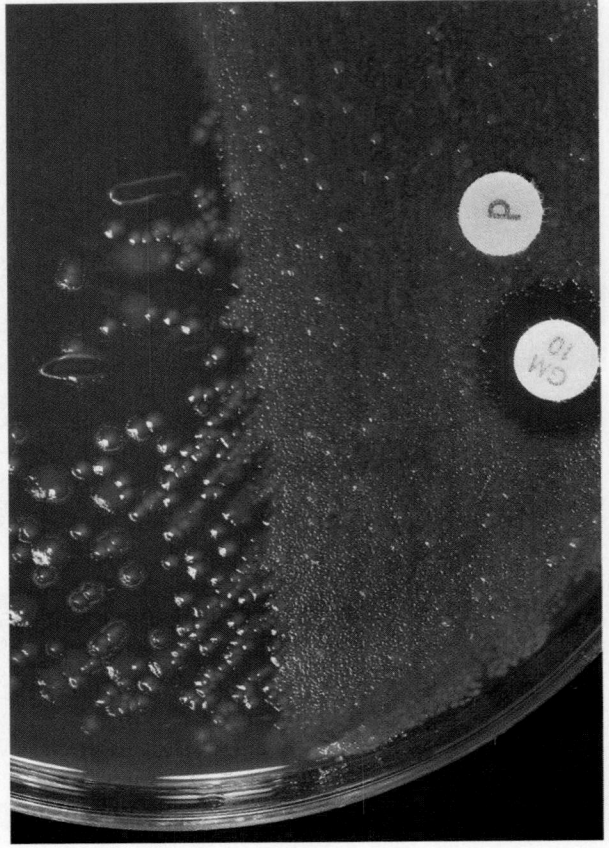

A

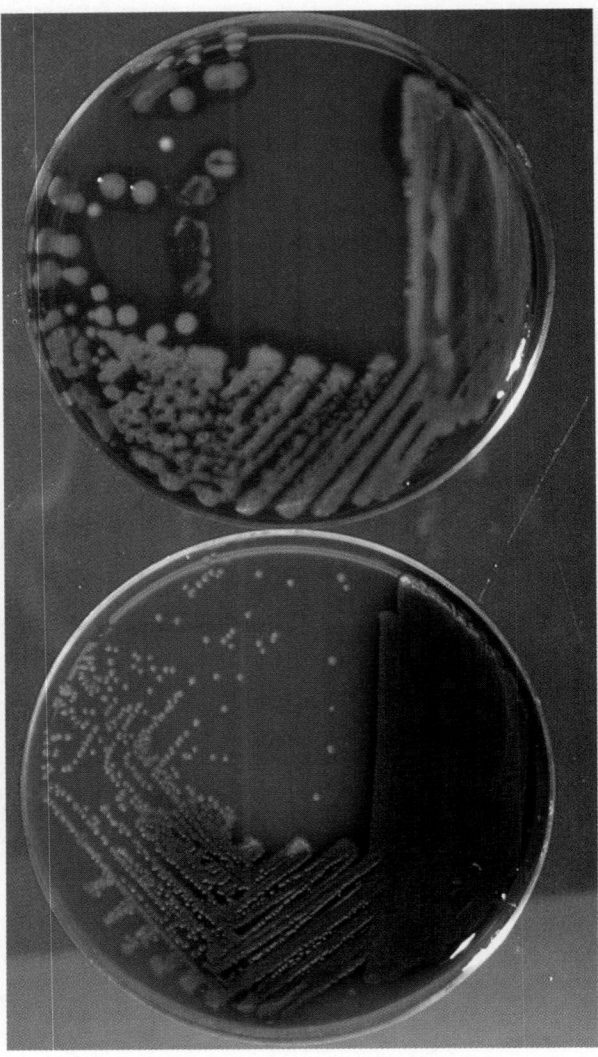

B

■■■■■■■■ F i g u r e   9 – 1 4

A, Pseudomonas aeruginosa illustrating the metallic sheen of the colonies. *B,* Green pigmentation of *Pseudomonas aeruginosa* on BAP.

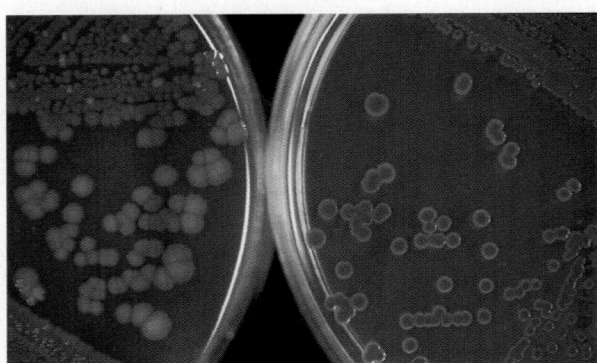

■■■■■■■■ F i g u r e   9 – 1 5

Brick-red pigment of *Serratia marcescens,* which is evident on the MAC on the *right.* This brick-red pigment should *not* be confused with lactose fermentation. The pigment is slightly visible on CHOC (left).

COLONIES WITH MULTIPLE CHARACTERISTICS

In addition to the organisms already mentioned, other bacteria fit in multiple descriptive categories of colonial morphology. *Bacillus cereus* forms large, rough, greenish, hemolytic colonies on blood agar (Fig. 9–16). *Eikenella corrodens* forms a small, "fuzzy-edged" colony with an umbonate center on blood or chocolate agar (Fig. 9–17).

GROWTH OF ORGANISMS IN LIQUID MEDIA

Important clues to an organism's identification can also be detected by observing the growth of the organism in liquid media such as thioglycolate. *Streamers* or vines and *puff balls* are associated with certain species of streptococci (Fig. 9–18). *Turbidity* (and usually gas if the media contains glucose) is produced by enterics (Fig. 9–19). Yeast and *Pseudomonas* produce scum at the sides of the tube (Figs. 9–20 and 9–21). In addition, yeast occasionally grows below the surface, in the microaerophilic area of the media (Fig. 9–22).

Figures 9–23 (*S. pneumoniae* and α-hemolytic *Streptococcus*), 9–24 (*S. pyogenes* and *S. agalactiae*), and 9–25 (*Staphylococcus* and yeast) show the differentiation between various organisms by colonial morphology.

Text continued on page 321

Figure 9–17

Small, "fuzzy-edged," umbonate center–appearing colony of *Eikenella corrodens* on CHOC. This organism has the tendency to "pit" the agar.

A B

Figure 9–18

A, "Vine" or "streamer" effect exhibited by certain species of *Streptococcus* when growing in thioglycollate. Notice the effect is more prevalent toward the bottom of the tube. *B,* Example of the "puffed balls" effect exhibited by certain streptococcal species when growing in thioglycolate.

Figure 9–16

Large, rough, greenish-appearing, hemolytic colonies of *Bacillus cereus* on BAP.

■ F i g u r e   9 – 1 9

Turbidity produced by enterics when growing in thioglycolate. Notice the gas bubbles at the surface of and in the middle of the medium.

■ F i g u r e   9 – 2 1

Illustration of *Pseudomonas* producing surface "scum" at the sides of the thioglycolate. Occasionally, *Pseudomonas aeruginosa* produces a diffusible green pigment and a metallic sheen at the surface.

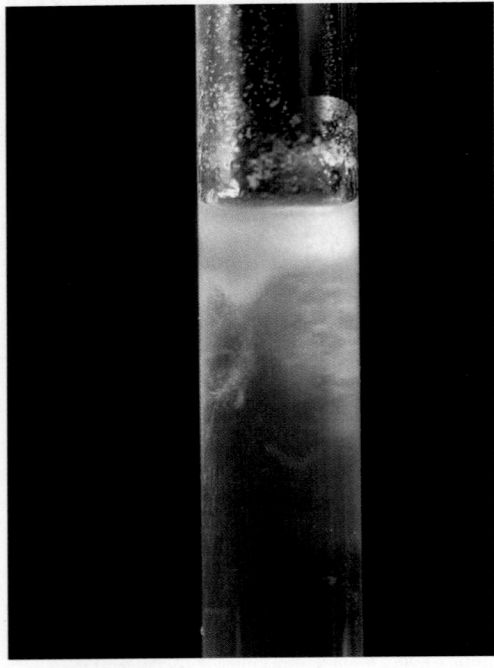

■ F i g u r e   9 – 2 0

Production of "scum" by yeast at the surface of the thioglycolate.

■ F i g u r e   9 – 2 2

Yeast growing in the microaerophilic area of thioglycolate.

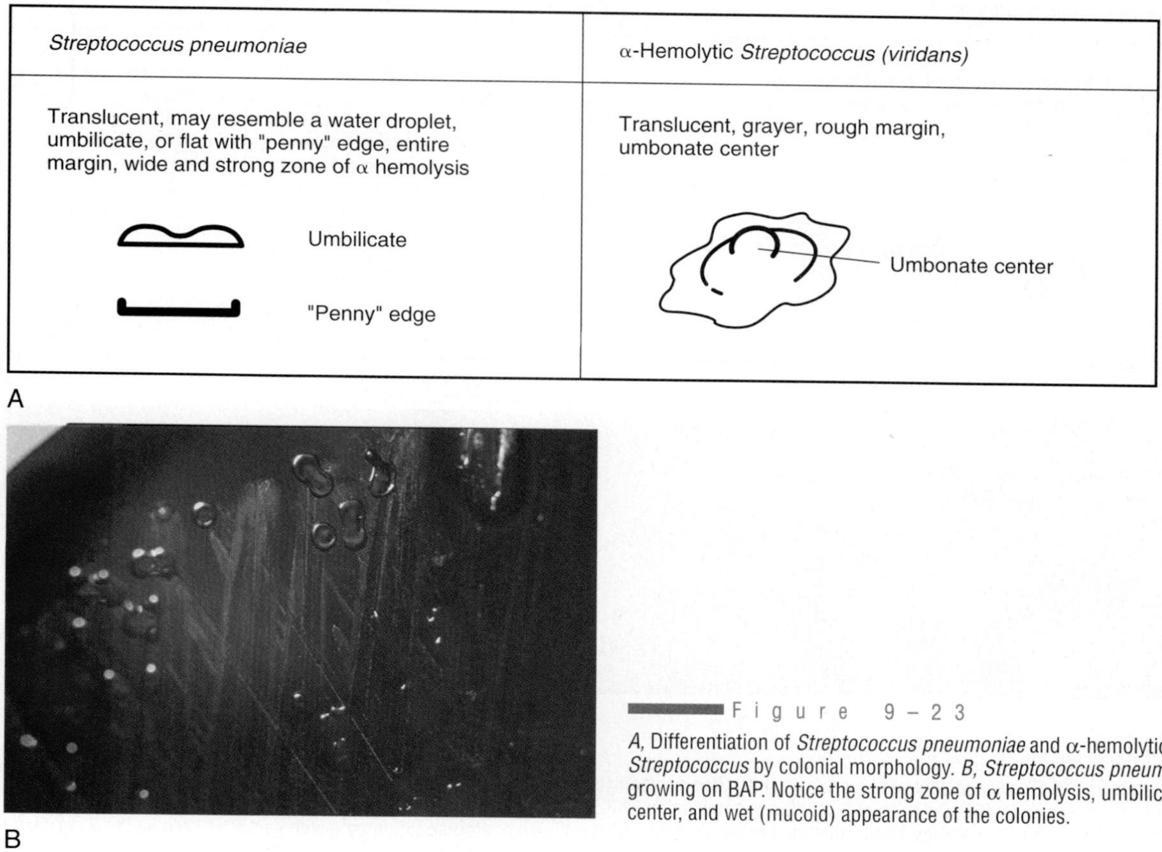

Streptococcus pneumoniae	α-Hemolytic *Streptococcus (viridans)*
Translucent, may resemble a water droplet, umbilicate, or flat with "penny" edge, entire margin, wide and strong zone of α hemolysis	Translucent, grayer, rough margin, umbonate center

Umbilicate

"Penny" edge

Umbonate center

A

B

Figure 9–23

A, Differentiation of *Streptococcus pneumoniae* and α-hemolytic *Streptococcus* by colonial morphology. *B, Streptococcus pneumoniae* growing on BAP. Notice the strong zone of α hemolysis, umbilicate center, and wet (mucoid) appearance of the colonies.

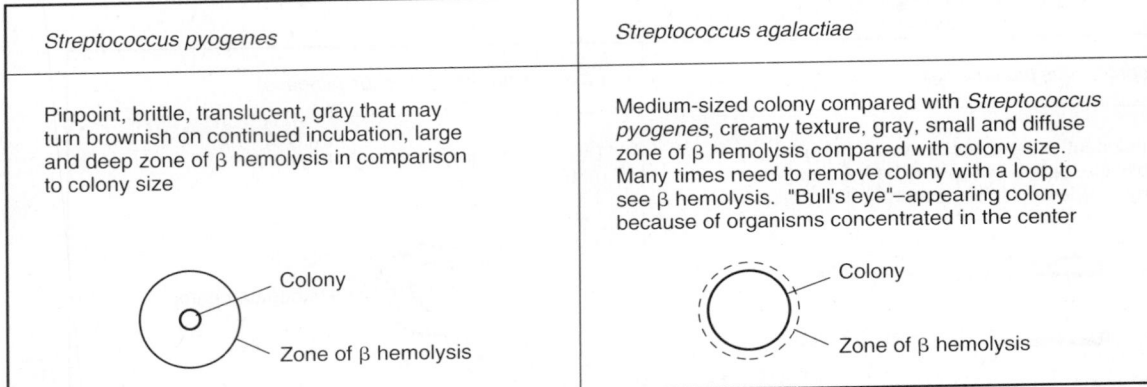

Streptococcus pyogenes	*Streptococcus agalactiae*
Pinpoint, brittle, translucent, gray that may turn brownish on continued incubation, large and deep zone of β hemolysis in comparison to colony size	Medium-sized colony compared with *Streptococcus pyogenes*, creamy texture, gray, small and diffuse zone of β hemolysis compared with colony size. Many times need to remove colony with a loop to see β hemolysis. "Bull's eye"–appearing colony because of organisms concentrated in the center

A

B   C   D

■■■■■■■■ F i g u r e   9 – 2 4

A, Differentiation of *Streptococcus pyogenes* and *S. agalactiae* by colonial morphology. *B,* Pinpoint colony of *S. pyogenes* exhibiting large, deep zone of β hemolysis on BAP. *C,* Colonies of *S. agalactiae* growing on BAP. This organism produces a larger colony and a smaller, more diffuse zone of hemolysis than *S. pyogenes*. Notice that the hemolysis is not evident in this photograph. Compare with part *B. D,* Colonies of *S. agalactiae* growing on BAP. Through the use of transillumination, the hemolytic pattern is now evident; hemolysis is diffuse and remains close to the periphery of the colony. The same colonial morphology is produced by *Listeria monocytogenes,* a gram-positive rod. Compare with part *B.*

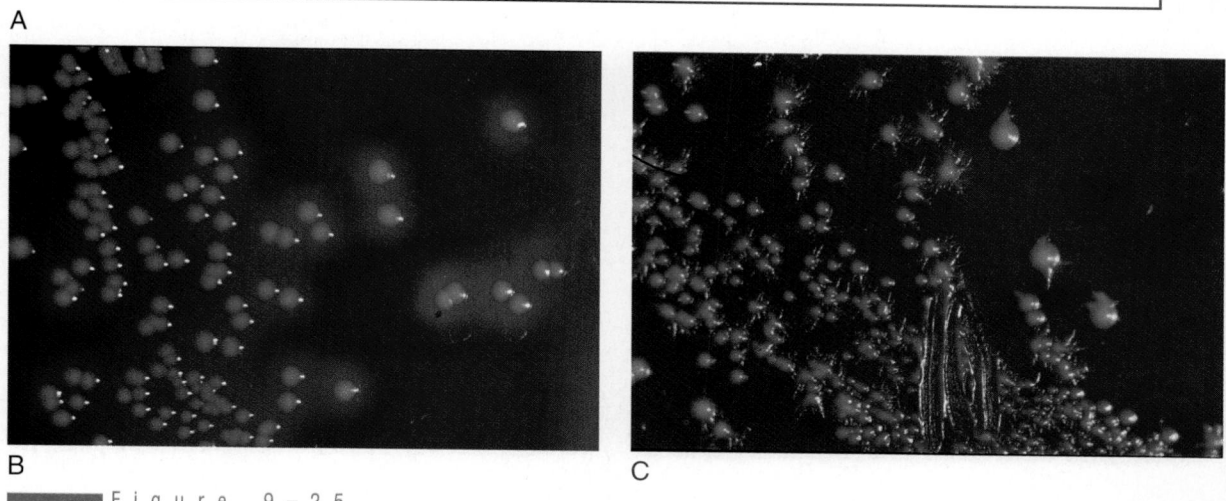

Staphylococcus	Yeast
Large, flat, or convex or possesses an umbonate center after 24 hours of incubation, shiny, moist, creamy, white to yellowish. *S. aureus* is usually β-hemolytic	Smaller than staphylococci, convex, grows upward more than outward, creamy, white, dull surface. *Candida albicans* usually displays tiny projections at the base of the colony after 24 hours of incubation

A

B   C

■■■■ F i g u r e   9 – 2 5

A, Differentiation between *Staphylococcus* and yeast by colonial morphology. *B,* Large, white, convex, shiny, moist, β hemolytic colonies of *S. aureus* growing on BAP. *C,* "Heaped" or convex, white, dull appearance and butyrous texture of *Candida albicans* on BAP. Notice the tiny projections or "feet" at the edge of the colonies.

CONCLUSION

It is important to remember that the colonial morphology described in this chapter is not infallible. Variations do occur quite frequently. Therefore, the morphologies described are general characteristics for any given organism. The identification process must include Gram stain and biochemical reactions in addition to colonial morphology.

Microbiologists become frustrated when changes in colony morphology, gram staining, and biochemical reactions occur in microorganisms that produce characteristic features. Many times, organisms exhibit characteristics far different from those previously described for them. The ability to recognize these differences and changes in characteristics make this discipline a challenge.

Bibliography

Koneman EW, et al: Color Atlas and Textbook of Diagnostic Microbiology, 4th ed. Philadelphia: JB Lippincott, 1992, pp 33–36.

Le Beau LJ: Effective lighting systems for photography of microbial colonies. J Biol Photographic Assoc 44:4, 1976.

PART II

Laboratory Identification of Significant Isolates

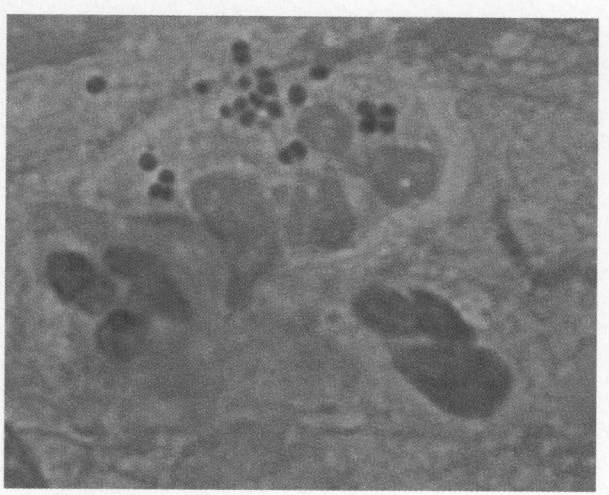

CHAPTER *10*

Staphylococcus

Hal S. Larsen • *Connie R. Mahon*

1. Give the general characteristics of the genus *Staphylococcus*.

2. Differentiate between *Staphylococcus* and other gram-positive cocci.

3. Describe the virulence factors associated with the staphylococci.

4. Describe the clinical infections associated with *Staphylococcus*.

5. Name the differential tests that may be used to identify the clinically relevant *Staphylococcus* species.

6. State which characteristics should be utilized to identify a *Staphylococcus*-like organism isolated from a clinical sample.

7. Explain why methicillin resistance is a serious clinical problem.

Gram-positive cocci are not uncommon isolates in the clinical microbiology laboratory. Although most are members of the indigenous microbial flora, some species are definite agents of serious infectious disease. This chapter discusses the most commonly encountered staphylococci: their characteristics, the infections they produce, and their laboratory identification. Infections caused by *S. aureus*, *S. epidermidis*, and *S. saprophyticus* are emphasized.

GENERAL CHARACTERISTICS

The staphylococci are catalase-producing, gram-positive cocci that belong to the family Micrococcaceae. When seen on stained smears, they are spherical cells (0.5–1.5 μm) that appear singly, in pairs, and in clusters that have been described as "bunches of grapes." These organisms are nonmotile and non–spore forming, and most are facultatively anaerobic, except for *S. saccharolyticus* which is an obligate anaerobe. Colonies produced after 18 to 24 hours of in-cubation appear creamy, white or light gold, and "buttery looking." Some species produce β-hemolytic zones around the colonies. *Staphylococcus* is a common isolate in the clinical laboratory and is responsible for several suppurative types of infections. This organism is found as a normal inhabitant of the skin and mucous membranes of humans and other animals.

Species of staphylococci are initially differentiated by the coagulase test. Coagulase-producing (coagulase-positive) staphylococci are *S. aureus*,

S. intermedius, *S. delphini*, and some strains of *S. hyicus* and *S. schleiferi*. With the exception of *S. aureus*, these are animal species and thus are rarely isolated from human samples. Consequently, for the vast majority of clinical laboratory situations, coagulase-positive isolates from human sources are considered to be *S. aureus*. *Staphylococcus aureus* is responsible for cutaneous infections such as boils, carbuncles, and purulent abscesses. Toxin-induced diseases, such as food poisoning, scalded skin syndrome, and toxic shock syndrome, are also associated with this organism.

Staphylococci that do not produce coagulase are referred to as *coagulase-negative staphylococci*. The most clinically significant species in this group are *S. epidermidis* and *S. saprophyticus*. *Staphylococcus epidermidis* has been known to be responsible for a variety of hospital-acquired infections, whereas *S. saprophyticus* is mainly associated with urinary tract infections in young females who are sexually active.

Currently, there are 33 recognized species of coagulase-negative staphylococci. The groups listed in Table 10–1, consisting of species of coagulase-negative staphylococci, are established. Most of these species are human isolates usually found as inhabitants of the skin and mucous membranes. Certain species are found in very specific sites. Others have been isolated from animals and animal products.

In addition to the genus *Staphylococcus*, the family Micrococcaceae includes the nonpathogenic *Planococcus* and *Micrococcus*. Some investigators, however, have written that the three genera are not as closely related as previously thought. Micrococci are catalase-producing,

■ T A B L E 1 0 – 1

GROUPINGS OF COAGULASE-NEGATIVE STAPHYLOCOCCI AND THEIR CLINICAL SOURCE AND SIGNIFICANCE

Species	Source*
S. epidermidis group	
S. epidermidis	**Human**, animal
S. haemolyticus	**Human**
S. hominis	Human
S. capitis subsp *capitis*	Human
S. capitis subsp *ureolyticus*	Human
S. caprae	Human, animal
S. auricularis	Human
S. saccharolyticus	Human
S. warneri	Human, animal
S. pasteruri	Animal, human
S. saprophyticus group	
S. saprophyticus	**Human**
S. cohnii subsp *cohnii*	Animal, human
S. cohnii subsp *urealyticum*	Animal, human
S. xylosus	Animal, human
S. arlettae	Animal
S. equorum	Animal, human
S. gallinarum	Animal
S. kloosii	Animal
S. lentus	Animal
S. simulans species group	
S. simulans	Animal, human
S. carnosus	Animal
S. intermedius group	
S. schleiferi subsp *schleiferi*	**Animal**, human
S. sciuri group	
S. sciuri	Animal, human
S. lentus	Animal, human
S. vitulus	Animal
S. hyicus group	
S. chromogenes	**Animal**
Unspecified species group	
S. caseolyticus	Animal
S. felis	Animal
S. hyicus	Animal
S. lugdunensis	**Human**, animal
S. muscae	Animal
S. piscifermentans	Animal

*__Boldface__ indicates common in human or veterinary disease.
Courtesy of Leona W. Ayers, M.D.

coagulase-negative, gram-positive cocci found in the environment and as residents of the normal skin. Micrococci, especially *M. luteus,* have a tendency to produce a yellow pigmented colony. It is not unusual to isolate them from clinical samples. Because they are considered non-pathogens, they need to be differentiated from the potentially pathogenic coagulase-negative staphylococci. Differentiating characteristics of *Micrococcus* and *Staphylococcus* are shown in Table 10–2.

CLINICALLY SIGNIFICANT SPECIES

Staphylococcus aureus

Virulence Factors

Enterotoxins, cytolytic toxins, and cellular components such as protein A have been described as being responsible for the pathogenicity and virulence of *S. aureus.*

ENTEROTOXINS

Staphylococcal enterotoxins are heat-stable exotoxins that cause diarrhea and vomiting in humans. Enterotoxins A and D are resistant to gastric and digestive juices and have been associated with food poisoning. Enterotoxin F, previously described as pyrogenic exotoxin C is produced by *S. aureus* phage group I and causes toxic shock syndrome. This enterotoxin is now referred to as toxic-shock syndrome toxin–1 (TSST-1).

EXFOLIATIVE TOXIN

Produced by phage group II, exfoliative toxin is also known as epidermolytic toxin. It causes the epidermal layer of the skin to slough off and is known to cause scalded skin syndrome, or Ritter disease.

■ T A B L E 1 0 – 2

DIFFERENTIATION OF STAPHYLOCOCCI AND MICROCOCCI IN THE ROUTINE LABORATORY

Test	*Staphylococcus*	*Micrococcus*
Anaerobic acid production from glucose	+	–[a]
Growth on furoxone– Tween 80–oil red 0 agar	–	+
Aerobic acid production from glycerol in the presence of erythromycin	+	–
Resistance to bacitracin (0.04 UI)	R(+)[b]	S
Lysozyme (50 μg disk)	R	S
Lysostaphin test	S(–)[b]	R

[a] *M. kristinae, M. varians* are positive.
[b] Some strains show opposite reaction.

Reprinted by permission of the publisher from Schumacher-Perdreau F: Clinical significance and laboratory diagnosis of coagulase-negative staphylococci. Clin Microbiol Newsl 13:97–101, 1991. Copyright 1991 by Elsevier Science Inc.

CYTOLYTIC TOXINS

Staphylococcus aureus produces other extracellular proteins that affect red blood cells and leukocytes. These hemolysins and leukocidins are cytolytic toxins with properties different from those of previously described toxins. *Staphylococcus aureus* produces α-, β-, and δ-hemolysins. In addition to hemolyzing red blood cells, α-toxin (α-hemolysin) is capable of destroying platelets and causing severe tissue damage. β-Hemolysin, also referred to as "hot-cold lysin," acts on the sphingomyelin of red blood cell membranes. The "hot-cold" feature identified with this toxin is seen as an enhanced hemolytic activity upon incubation at 37°C and subsequent exposure to cold (4°C). δ-Hemolysin, although it has been described to cause injury in most cells in culture and to leukocytes, is considerably less lethal than α- or β-lysin.

Staphylococcal leukocidin (Panton-Valentine leukocidin), is an exotoxin lethal to polymorphonuclear leukocytes. Although its exact role in staphylococcal infections is unclear, it has been implicated as contributing to the invasiveness of the organism by suppressing phagocytosis.

ENZYMES

Several enzymes are produced by both coagulase-positive and coagulase-negative staphylococci. Examples are coagulase, hyaluronidase, and lipase. Coagulase is mainly produced by *S. aureus*. Whereas the exact role of coagulase in pathogenicity remains uncertain, it is considered a virulence marker. Many strains of *S. aureus* produce hyaluronidase. This enzyme hydrolyzes hyaluronic acid present in the intracellular ground substance that makes up connective tissues, permitting the easy spread of infection. Lipases are produced by both coagulase-positive and coagulase-negative staphylococci. Lipases act on substances present on the surface of the skin, particularly fats and oil secreted by the sebaceous glands. This activity allows colonization of the organisms in the area.

PROTEIN A

Protein A is one of several cellular components that have been identified in the cell wall of *S. aureus*. Probably the most significant role of protein A in infections caused by *S. aureus* is its ability to bind the Fc portion of the immunoglobulin, thereby avoiding phagocytosis.

Epidemiology

Staphylococcus aureus inhabits the anterior nares of human carriers. Nasal carriage in patients admitted to the hospital is common. Because close contact among patients and hospital personnel is not unusual, transfer of organisms is likely to take place. Consequently, increased colonization in patients and hospital workers frequently occurs. In addition to nasal carriage, colonization in the perineum is also common among carriers.

Hospital outbreaks may occur in infant nurseries, in burn units, and among patients who have undergone surgery or other invasive procedures. Transmission of *S. aureus* may occur by direct contact with unwashed contaminated hands and by fomites. Community-acquired infections caused by *S. aureus* are usually associated with poor hygiene and fomite transmission and develop from individuals who are carriers themselves.

Infections Caused by *S. aureus*

As with most infections, the development of staphylococcal infection is determined by the virulence of the strain, the size of the inoculum, and the status of the host's immune system. Any event that compromises the host's ability to resist infection encourages colonization and infection. Individuals with normal defense mechanisms are able to combat the infection more easily than those with impaired immune systems. Once the organism has transgressed the initial barriers, it activates the host's immune system for an acute inflammatory response, which leads to the proliferation of polymorphonuclear and phagocytic cells. The organisms are able to resist the action of inflammatory cells, however, by eliciting toxins and enzymes, thereby establishing a focal lesion.

SKIN AND WOUND INFECTIONS

Infections caused by *S. aureus* are suppurative and pyogenic. Typically, the abscess is filled with pus and surrounded by necrotic tissues and damaged leukocytes. Some of the common skin infections caused by *S. aureus* are boils, carbuncles, furuncles, folicullitis, and bullous impetigo. These opportunistic infections occur usually as a result of previous skin injuries such as cuts, burns and surgical wounds. Bullous impetigo caused by *S. aureus* is different from streptococcal impetigo, in that staphylococcal pustules are larger and surrounded by a small zone of erythema. Bullous impetigo is a highly contagious infection easily spread by direct contact, by fomites, or by autoinoculation. Boils and furuncles are superficial abscesses, although furuncles may progress into deeper tissues. Multiple lesions may develop from a furuncle, resulting in colonies of lesions described as carbuncles.

Staphylococcal infections may also occur secondary to skin diseases of different etiologies. Dry, irritated skin in combination with poor personal hygiene encourages the development of infection. Some of these infections manifest because of increased colonization of the organisms in hair follicles, sebaceous glands, and sweat glands when these are blocked. Immunocompromised individuals, particularly those who are receiving chemotherapy, who are debilitated by chronic illnesses, or who have undergone instrumentation, are easily predisposed to staphylococcal infections.

FOOD POISONING

Staphylococcus aureus produces enterotoxins that have been identified and associated with gastrointestinal upset, most commonly enterotoxins A and D. The source of contamination is usually an infected food handler. Staphylococcal food poisoning occurs when the individual ingests food contaminated with enterotoxin-producing strains. The heat-stable toxins are preformed in inadequately refrigerated rich foods such as creamy sauces and mayonnaise. The enterotoxins do not cause any detectable odor or change in the appearance or taste of the food. Symptoms appear rapidly, approximately 2 to 8 hours after ingestion of the food, and resolve within 6 to 8 hours. There is no fever associated with this condition, but nausea, vomiting, abdominal pain, and severe cramping are common. Headaches may also be present.

SCALDED SKIN SYNDROME

Scalded skin syndrome, or Ritter disease, is an extensive exfoliative dermatitis that occurs primarily in newborns and previously healthy young children. This syndrome is caused by staphylococcal exfoliative or epidermolytic toxin produced by phage group II staphylococci, prob-

ably present at a lesion distant from the site of exfoliation. The disease has also been recognized in adults. Cases of staphylococcal scalded skin syndrome (SSSS) in adults are reported to occur most commonly among patients with chronic renal failure and those whose immune systems are compromised. Whereas the mortality rate is low in cases seen among children (0–7%), the rate in adults is as high as 50%.

The severity of the disease varies from a localized skin lesion, in the form of bullous impetigo, to a more extensive generalized condition. Bullous impetigo manifests as a localized lesion that contains seropurulent material. This lesion may progress to the generalized form: cutaneous erythema followed by profuse peeling of the epidermal layer of the skin. The typical pattern in which the erythema occurs is origination from the face, neck, axillae, and groin and then extension to the trunk and extremities. The duration of the disease is brief, about 2 to 4 days. Incidence of spontaneous recovery among children is high.

The toxin is metabolized and excreted by the kidneys. Investigators believe that this may be the reason why the incidence of SSSS is higher among children less than 5 years and among adults with chronic renal failure and impaired immune systems.

Toxic epidermal necrolysis (TEN), a clinical manifestation with multiple etiologies, has a very similar initial presentation to that of SSSS. Differential diagnosis between SSSS and TEN must be made, because therapies of these two forms of exfoliative disease differ. Whereas TEN is resolved by the administration of steroid therapy, steroids aggravate SSSS. Prompt antimicrobial therapy should be initiated, particularly in adult cases of SSSS, in which the mortality rate is high.

TOXIC SHOCK SYNDROME

Toxic shock syndrome (TSS) is a multisystem disease characterized by high fever, hypotension, and shock. It was first described by Todd in 1978. The syndrome has since been attributed to a toxin released by certain strains of *S. aureus,* referred to as toxic shock syndrome toxin–1 (TSST-1). Although the first cases of TSS occurred in seven children of both sexes, an increased incidence of the infection affecting predominantly young menstruating women was observed 2 years later. The association of TSS with tampon use was then established. The syndrome was also seen in males as well as in nonmenstruating females. It was later found that any wound due to a strain of *S. aureus* that produces TSST-1 may cause TSS.

The initial clinical presentation of TSS consists of high fever, rash, and signs of dehydration, particularly if the patient has had watery diarrhea and vomiting for several days. In extreme cases, patients may be severely hypotensive and in shock. A rash is found predominantly on the trunk but is also spread all over the body.

Laboratory findings include an elevated leukocyte count with the differential blood count showing an increase in band forms and metamyelocytes. The number of platelets is decreased and although there is no evidence of bleeding, disseminated intravascular coagulation is likely. The effects of dehydration on the kidneys are manifested by elevations in creatinine and blood urea nitrogen. Cultures of focal lesions may yield *S. aureus,* but blood cultures are usually negative.

Supportive therapy to replace vascular volume loss is given, along with an appropriate β-lactamase–resistant antimicrobial. *S. aureus* does not need to be isolated to confirm the diagnosis of TSS.

Most patients with TSS recover, although 2 to 5% of cases may be fatal. Recurrence of TSS in menstruating women is as high as 60%. Multiple recurrences that become less severe and shorter in duration have led investigators to hypothesize that immunity to the organism or to the effect of the toxin is eventually developed. Preventive measures such as avoidance of tampon use, use of low-absorbency tampons, and more frequent tampon change have greatly decreased the risk of TSS.

OTHER INFECTIONS

Staphylococcal pneumonia has been known to occur as a bacterial infection secondary to influenza A. The infection, although rare, has a high mortality rate. Staphylococcal pneumonia, which develops as either a contiguous lower respiratory tract infection or a complication of bacteremia, is characterized by multiple abscesses and focal lesions in the pulmonary parenchyma. Infants and immunocompromised patients, such as the elderly and patients receiving chemotherapy or immunosuppressants, are most affected.

Staphylococcal bacteremia leading to secondary pneumonia and endocarditis has been observed among intravenous drug addicts. Fever is the most striking symptom of endocarditis, which must be suspected in an IV drug abuser presenting with fever. The organisms gain entrance into the blood, originating from a focal lesion that could be present on the skin or in the respiratory or genitourinary tract.

Staphylococcal osteomyelitis occurs as a manifestation secondary to bacteremia. The infection develops when the organism is present in a wound or other focus of infections and gains entrance into the blood. It may lodge in the diaphysis of the long bones and establish an infection. Symptoms, include fever, chills, swelling, and pain around the affected area.

Septic arthritis, seen in children and in patients with a history of rheumatoid arthritis and IV drug abuse, has also been attributed to *S. aureus.* The organisms may be recovered from aspirated joint fluid.

Staphylococcus epidermidis

The role of *S. epidermidis* as an etiologic agent of disease has become increasingly evident. Infections caused by *S. epidermidis* are predominantly hospital-acquired. Some of the predisposing factors are instrumentation procedures such as catheterization, prosthetic heart valve implantation, and immunosuppressive therapy. *S. epidermidis* is probably the most common cause of hospital-acquired urinary tract infections. Prosthetic valve endocarditis due to *S. epidermidis* has also increased, from a reported 25% of cases in the 1970's to approximately 50% in the 1980's, with about 70% mortality. Other infections due to *S. epidermidis* occur in intravascular catheters, CSF shunts, and other prosthetic devices. Septicemia has been reported in patients who are immunocompromised.

Infections associated with instrumentation, such as the use of indwelling catheters and prosthetic devices, are often caused by isolates shown to produce an adherence factor described as *slime.* Slime-producing strains of *S. epidermidis* are able to adhere and form colonies on the surface of these devices. Slime-producing strains also utilize this adherence factor to inhibit the action of lymphocytes and neutrophils. A laboratory test to detect slime production has been described, but its value in the treatment of the clinical infection is questionable.

Staphylococcus saprophyticus

Staphylococcus saprophyticus has been associated with urinary tract infections in young, sexually active females. This species is found to adhere more effectively to the epithelial cells lining the urogenital tract than other coagulase-negative staphylococci. It is rarely found in other skin areas or mucous membranes. When present in urine cultures, *S. saprophyticus* may be found in low numbers and yet considered significant.

Other Coagulase-Negative Staphylococci (CNS)

Other species of CNS are found as normal flora in humans and animals. They are not commonly seen as pathogens, although their role in some infections is well established. Therefore, they cannot be automatically discarded as contaminants in all cases.

Staphylococcus haemolyticus is the second most commonly isolated of the CNS following *S. epidermidis*. It has been reported in wounds, bacteremia, endocarditis, and urinary tract infections. Of notable interest with this organism is the emergence of vancomycin resistance in some isolates.

Two species that are uncommonly seen but have established themselves as opportunistic pathogens are *S. lugdunensis* and *S. schleiferi*. A wide range of infections have been associated with these organisms (i.e., endocarditis, septicemia, and prosthesis infections).

LABORATORY DIAGNOSIS

Specimen Collection and Handling

Proper specimen collection, transport, and processing are essential elements in the correct diagnosis and interpretation of any bacterial culture result. Clinical materials collected from infected sites should be transported to the laboratory without delay to prevent drying, to maintain the proper environment, and to minimize the growth of contaminating organisms. Although the recovery of staphylococci requires no special procedures, specimens should be taken from the actual site of infection after appropriate cleansing of the surrounding area to avoid contamination by the normal skin flora.

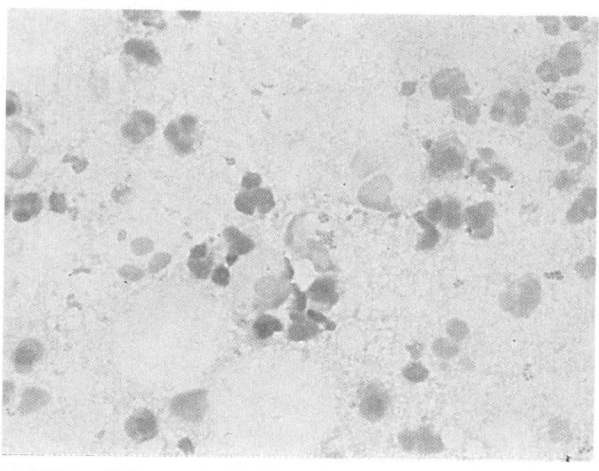

■ F i g u r e   1 0 – 1
Gram-stained direct smear from an exudate showing white blood cells and gram-positive cocci in clusters.

Microscopic Examination

Microscopic examination of stained smears prepared directly from clinical samples (Fig. 10–1) provides information that is helpful in the early diagnosis and treatment of the infection and must always be performed. Numerous gram-positive cocci, along with polymorphonuclear cells in purulent exudates, joint fluids, aspirated secretions, and other body fluids, would be easily seen. Culture should be done regardless of the results of the microscopic examination, because the genus or species cannot be appropriately identified by microscopic morphology alone. Microscopic morphology is shown in Figure 10–2.

Isolation and Identification

Staphylococci grow easily on routine laboratory culture media, particularly sheep's blood agar. A fluid medium such as thioglycolate is usually included in the primary isolation scheme. A selective medium such as mannitol salt agar, Columbia colistin–nalidixic acid agar (CNA), or phenylethyl alcohol (PEA) agar, can be used on heavily contaminated specimens.

Cultural Characteristics

Staphylococci produce round, smooth, white, creamy colonies on blood agar after 18 to 24

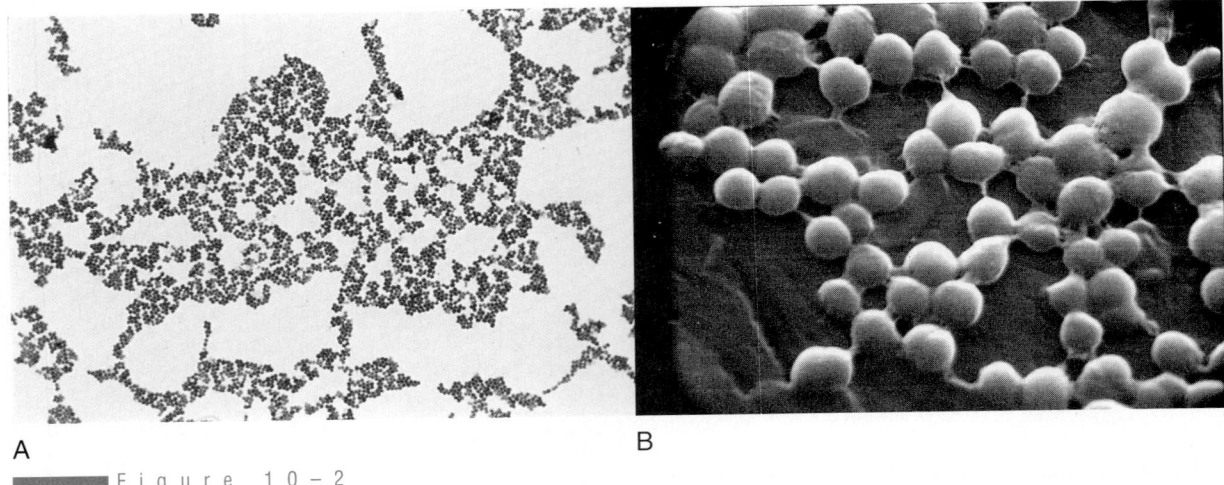

A   B

■■■■Figure 10-2

A, Microscopic morphology of *Staphylococcus* sp on Gram stain. *B,* Scanning electron micrograph showing the typical "clusters" of staphylococci.

hours of incubation at 35 to 37°C. *Staphylococcus aureus* may produce hemolytic zones around the colonies (Fig. 10–3), which may produce pigment (yellow, tan, orange). *S. epidermidis* colonies are usually small to medium-sized, nonhemolytic, white colonies. *S. saprophyticus* forms slightly larger colonies, with approximately 50% of the strains producing a yellow pigment. Identification of staphylococci on the basis of colony morphology alone is not recommended.

Identification Methods

Staphylococci have been traditionally differentiated from micrococci on the basis of oxidation-

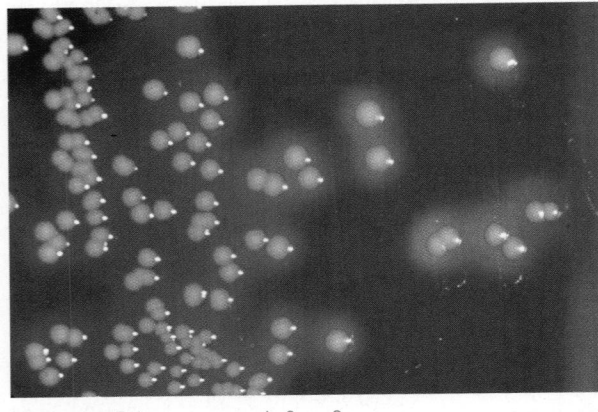

■■■■Figure 10-3

Staphylococcus aureus growing on sheep's blood agar showing hemolytic, creamy, buttery-looking colonies.

fermentation (O/F) reactions produced on O/F glucose medium. Staphylococci ferment glucose, whereas micrococci fail to produce acid under anaerobic conditions. The O/F tests, however, do not sufficiently discern certain weak acid producers, such as *Micrococcus kristinae,* and those staphylococci that fail to grow or produce acid anaerobically *(S. saprophyticus, S. auricularis, S. hominis, S. xylosus,* and *S. cohnii).* Tests to differentiate micrococci from staphylococci are shown in Table 10–2. Table 10–3 outlines the key characteristics for differentiating *Staphylococcus* from other gram-positive cocci.

Staphylococcus aureus is identified by the coagulase test (Fig. 10–4). The cell-bound coagulase, also referred to as *clumping factor,* clots human, rabbit, or pig plasma and is considered a major marker for *S. aureus.* Pig plasma forms a stronger clot and is less susceptible to autolysis than rabbit plasma. The enzyme is easily detected by the slide method and is used to screen colonies that morphologically resemble *S. aureus.* A suspension of the suspected organism is prepared on a glass slide and mixed with a drop of rabbit plasma. If clumping occurs, the isolate is identified as *S. aureus.* Isolates that do not produce cell-bound coagulase should be tested for extracellular free coagulase by the tube method. *Free coagulase* is an extracellular enzyme that causes a clot to form when bacterial cells are incubated with plasma (Fig. 10–5). The clot formed in the tube may have a tendency to autolysis, giving the appearance of a negative result. The microbiologist should look for

■■■■■ T A B L E   1 0 – 3

DIFFERENTIATION OF THE GENUS *STAPHYLOCOCCUS* FROM SOME OTHER GRAM-POSITIVE COCCI*

Characteristic	*Staphylococcus*	*Enterococcus*	*Streptococcus*	*Aerococcus*	*Planococcus*	*Stomatococcus*	*Micrococcus*
Strict aerobe	–			–	+	–	+
Facultative anaerobe	d	+	+	+		+	
Motility	–	d	–		–	+	–
Growth on NaCl agar				–	+		
5% NaCl	+	+	d				
6.5% NaCl	+	+	d	+	+	–	+
12% NaCl	d	(±)	–	+	+	–	+
Catalase	+	–	–	–	+	–	d
Benzidine test	+	–	–	–	+	±	+
Anaerobic acid from glucose	d	+	+	(+)	+	+	+
Lysostaphin (200 µg/mL)	–	+	+	+	–	+	–
Erythromycin (9.04 µg/mL)	+	+	–	ND	+	+	+[†]
Bacitracin (0.04-U disk)	+	+	d	–	ND	ND	–[‡]
Furazolidone (100-µg disk)	–	–	–	–	–	–	+

*+, 90% or more species or strains positive; ±, 90% or more species or strains weakly positive; –, 90% or more species or strains negative; d, 11 to 89% of species or strains positive; (), delayed reaction; ND, not determined.

[†]Some strains of *M. luteus*, *M. roseus*, and *M. sedentarius* demonstrate susceptibility to lysostaphin, presumably because of contaminating levels of endo-β-*N*-acetylglucosaminidase activity.

[‡]A few *Micrococcus* strains demonstrate high-level (MIC ≥ µg/mL) erythromycin resistance.

Modified with permission from Kloos WE, Lambe DE: Staphylococcus. In Balows A, et al (eds): Manual of Clinical Microbiology, 5th ed. Washington, DC: American Society for Microbiology, 1991, p 223.

clot formation after 4 hours of incubation at 37°C. If no clot formation appears, the tube should be left at room temperature to incubate overnight, and checked the following day. Because 5% of *S. aureus* do not produce cell-bound coagulase, any negative slide coagulase test result must be confirmed with the tube method. Table 10–4 groups the coagulase-positive staphylococci,

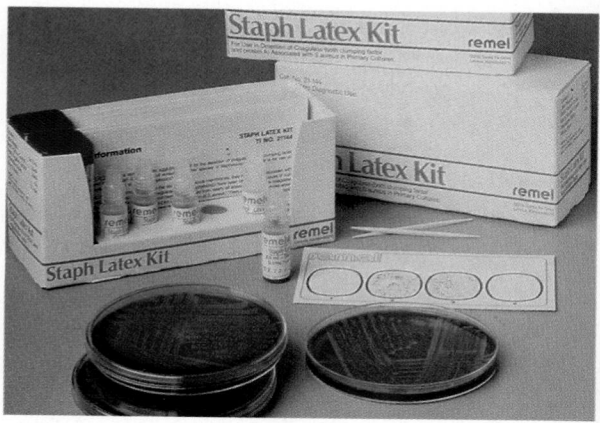

■■■■ F i g u r e   1 0 – 4

Slide coagulase test (Staph Latex) detects cell wall–bound "clumping factor." Latex agglutination method is available commercially. (Courtesy of Remel.)

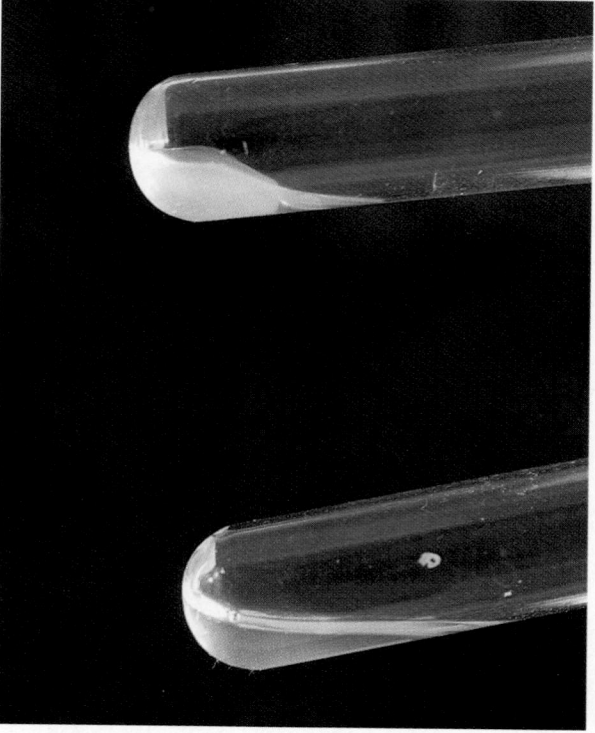

■■■■ F i g u r e   1 0 – 5

Tube coagulase test detects extracellular enzyme "free coagulase."

■ T A B L E 1 0 – 4

GROUPINGS OF COAGULASE-POSITIVE STAPHYLOCOCCI AND THEIR CLINICAL SOURCE AND SIGNIFICANCE

Species	Source*
S. aureus group	
S. aureus	**Human**, animal
S. aureus subsp anaerobius	Animal
S. hyicus group	
S. hyicus	**Animal**
S. intermedius group	
S. intermedius	**Animal**
S. schleiferi subsp coagulans	**Animal**
S. delphini	Animal

*__Boldface__ indicates common in human or veterinary disease.
Courtesy of Leona W. Ayers, M.D.

their clinical source, and significance. The microbiologist must be aware that staphylococci other than *S. aureus* produce bound or free coagulase (Table 10–5).

Isolates that do not produce either bound or free coagulase are reported as coagulase-negative staphylococci (Fig. 10–6). Urine isolates that are coagulase negative are further tested to presumptively identify *S. saprophyticus*. Presumptive identification of *S. saprophyticus* is accomplished by testing for novobiocin susceptibility using a 5-μg novobiocin disk (Fig 10–7). *Staphylococcus saprophyticus* is resistant to novobiocin, but most other coagulase-negative staphylococci are susceptible. Figure 10–8 shows the schema for the identification of clinically significant staphylococci.

Although *S. epidermidis* and *S. saprophyticus* are the most clinically signficant species of coagulase-negative staphylococci, other species are taking on greater clinical importance. Table 10–5 outlines some key tests for identification of the clinically significant species of *Staphylococcus*, including coagulase-negative isolates.

Rapid Methods of Identification

There are a number of rapid test kits on the market for differentiating *S. aureus* from the

■ T A B L E 1 0 – 5

KEY TESTS FOR IDENTIFICATION OF THE MOST CLINICALLY SIGNIFICANT *STAPHYLOCOCCUS* SPECIES*

Test	S. aureus	S. epidermidis	S. haemolyticus	S. lugdunensis	S. schleiferi	S. saprophyticus	S. intermedius	S. hyicus
Colony pigment[†]	+	–	d	d	–	d	–	–
Staphylocoag-ulase	+	–	–	–	–	–	+	d
Clumping factor[†]	+	–	–	(+)	+	–	d	–
Heat-stable nuclease	+	–	–	–	+	–	+	+
Alkaline phosphatase	+	+	–	–	+	–	+	+
Pyrrolidonyl arylamidase[†]	–	–	+	+	+	–	+	–
Ornithine decarboxylase	–	(d)	–	+	–	–	–	d
Urease[†]	d	+	–	d	–	+	+	d
β-Galactosidase[†]	–	–	–	–	(+)	+	+	–
Acetoin production	+	+	+	+	+	+	–	–
Novobiocin resistance[†]	–	–	–	–	–	+	–	–
Polymyxin B resistance[†]								
Acid (aerobically from)				d	–	–	–	+
D-Trehalose	+	–	+	+	d	+	+	+
D-Mannitol	+	–	d	–	–	d	(d)	–
D-Mannose	+	(+)	–	+	+	–	+	+
D-Turanose	+	(d)	(d)	(d)	–	+	d	–
D-Xylose	–	–	–	–	–	–	–	–
D-Cellubiose	–	–	–	–	–	–	–	–
Maltose	+	+	+	+	–	+	(±)	–
Sucrose	+	+	+	+	–	+	+	+

*+, 90% or more strains positive; ±, 90% or more strains weakly positive; –, 90% or more strains negative; d, 11 to 89% of strains positive; (), delayed reaction.
[†]Descriptions are the same as for Table 10–3.
Modified with permission from Kloos WE, Lambe DE: Staphylococcus. In Balows A, et al (eds): Manual of Clinical Microbiology, 5th ed. Washington, DC: American Society for Microbiology, 1991, p 230.

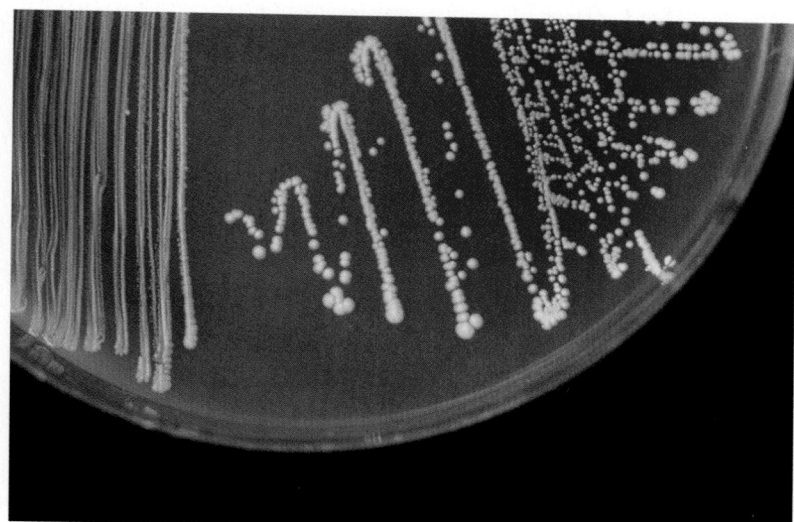

Coagulase-negative staphylococci growing on sheep's blood agar showing nonhemolytic, white, creamy colonies.

coagulase-negative staphylococci. Among these are SERO STAT Staph latex slide test (Scott Lab), Staphylatex (American Microscan), Accu-Staph (Carr-Scarborough), and Staph Latex (Remel) (see Fig. 10–4). These kits utilize plasma-coated latex particles. The plasma detects both clumping factor (with fibrinogen) and protein A in the cell wall of *S. aureus* (IgG). Another test kit available is the Staphyloslide Test (BBL), which detects clumping factor using sensitized sheep's red blood cells with fibrinogen, Staphase (API), a miniaturized coagulase test.

Rapid identification of staphylococci may be performed using the API-Staph Ident. Identification is made on the basis of positive reactions observed in microcupules that contain substrates for carbohydrate utilization. A four-digit profile code is produced from the reactions. Species identification is derived from a computer-based profile book.

ANTIMICROBIAL SUSCEPTIBILITY

Infections by staphylococcal strains that do not produce β-lactamase can be treated with penicillin. The incidence of penicillin resistance, however, especially in *S. aureus,* is so high (85–90%) that other antibiotics must often be used. There is considerable variability in the susceptibility patterns of staphylococcal isolates. Therefore, it is extremely important to perform antimicrobial susceptibility tests on all isolates, especially those from serious infec-

tions. Various β-lactamase–resistant penicillins have been developed to treat infections caused by penicillin-resistant staphylococcal species, especially *S. aureus*. The most notable of these is methicillin. As with the initial success of penicillin in the treatment of staphylococcal infections, methicillin treatment appeared to solve the problem of β-lactamase production by clinical isolates. Unfortunately, the solution was temporary.

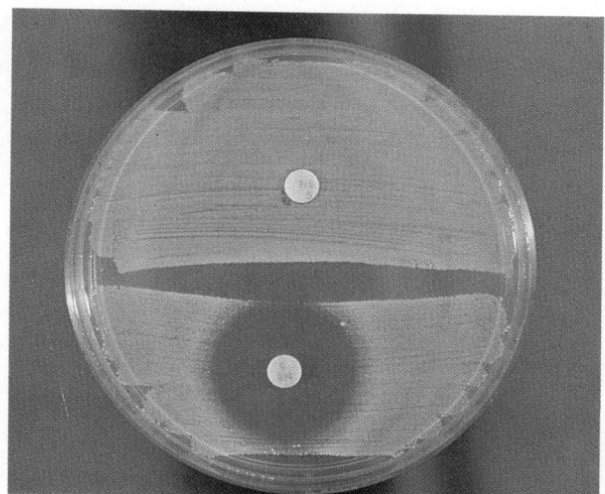

Novobiocin susceptibility test to differentiate coagulase-negative staphylococci isolate from urine samples. *Staphylococcus saprophyticus* is resistant to novobiocin, depicted by *no* zone of inhibition around the disk.

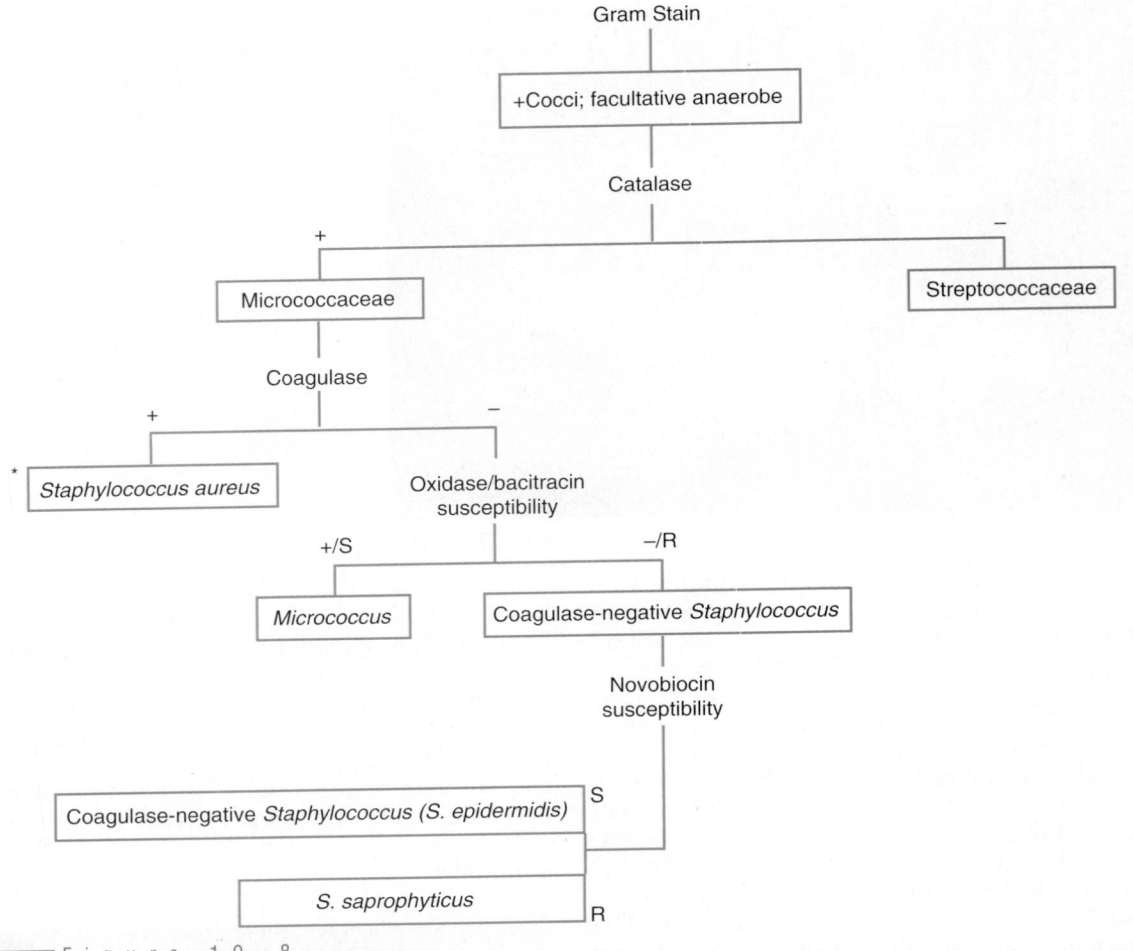

■ Figure 10-8

Schema for the identification of staphylococcal species. *Note:* Other *Staphylococcus* spp that are coagulase positive besides *S. aureus* include *S. lugdunensis* (many times latex test positive), *S. intermedius* (tube positive and latex positive), and *S. delphini* (tube positive and latex positive). (Modified and redrawn with permission from Kloos WE, Jorgensen JH: Staphylococci. In Lennette EH, et al [eds]: Manual of Clinical Microbiology, 4th ed. Washington, DC: American Society for Microbiology, 1985, p 147.)

METHICILLIN-RESISTANT STAPHYLOCOCCI

One of the major problems that has concerned clinicians since the mid-1980's is the evolution of methicillin-resistant strains of *S. aureus* (MRSA) and *S. epidermidis* (MRSE). The increased number of resistant strains being isolated has been seen worldwide. Some isolates of MRSA are also resistant to multiple agents, including aminoglycosides. Previously limited to large institutions, outbreaks of MRSA are now quite common in all hospital settings in the United States. Transmission of these strains has been attributed to cross-contamination among infected patients and carriers. Hospital personnel have been implicated as possible human vectors. Similarly, MRSE has also emerged in hospitalized patients and patients who have undergone prosthetic heart valve surgery.

Vancomycin remains the antimicrobial of choice for endocarditis due to MRSA and MRSE. Because of adverse effects related to this drug, its use is reserved for patients with systemic infections such as bacteremia, endocarditis, and pneumonia. It has been reported that administration of vancomycin in combination with rifampin or gentamicin enhances the therapeutic effects of vancomycin against MRSE.

Prevention and control of nosocomial infections due to MRSA and MRSE depend on proper hand-washing techniques and adherence to established infection control measures by health care providers.

Bibliography

Baddour LM, Christensen GD: Prosthetic valve endocarditis due to small-colony staphylococcal variants. Rev Infect Dis 9:1168, 1987.

Bergdoll MS: Enterotoxins. In Easmon CSF, Adlam C (eds): Staphylococci and Staphylococcal Infections. New York: Academic Press, 1983.

Berkley SF, Hightower AW, et al: The relationship of tampon characteristics to menstrual toxic shock syndrome. JAMA 258:917, 1987.

Bonventr, PF, Weckbach L, et al: Production of staphylococcal enterotoxin F and pyrogenic exotoxin C by Staphylococcus aureus isolates from toxic shock syndrome–associated sources. Infect Immun 40:1023, 1983.

Christensen GD, Simpson WA, et al: Adherence of slime-producing strains of Staphylococcus epidermidis to smooth surfaces. Infect Immun 37:318, 1982.

Davis P, Chesney PJ, et al: Toxic-shock syndrome: Epidemiologic features, recurrence, risk factors, and prevention. N Engl J Med 303:1429, 1980.

Dunne WM, Franson TR: Coagulase-negative staphylococci: The Rodney Dangerfield of pathogens. Clin Microbiol Newsl 8:37, 1986.

Elias PM, Fritsch P, Epstein E: Staphylococcal scalded skin syndrome. Arch Dermatol 113:207, 1977.

Eng RHK, Wang C, Person A, et al: Species identification of coagulase-negative staphylococcal isolates from blood cultures. J Clin Microbiol 15:439, 1982.

Fleurette J, Bès M, et al: Clinical isolates of Staphylococcus lugdunensis and S. schleiferi: Bacteriological characteristics and susceptibility to antimicrobial agents. Res Microbiol 140:107, 1989.

Friedman L, Brown AE, et al: Staphylococcus epidermidis septicemia in children with leukemia and lymphoma. Am J Dis Child 138:715, 1984.

Fritsch P, Elias P, Varga J: The fate of staphylococcal exfoliation in newborn and adult mice. Br J Dermatol 95:275, 1976.

Garbe P, Arko RJ, et al: Staphylococcus aureus isolates from patients with nonmenstrual toxic shock syndrome. JAMA 253:2538, 1985.

Gill VJ, Selepak AT, Williams ED: Species identification and antibiotic susceptibilities of coagulase-negative staphylococci isolated from clinical specimens. J Clin Microbiol 18:1314, 1984.

Goldberg, N, Ahmed T, et al: Staphylococcal scalded skin syndrome mimicking acute graft-vs-host disease in a bone marrow transplant recipient. Arch Dermatol 125:85, 1989.

Haller PR: Infections in intravenous drug abusers. Postgrad Med 83:95, 1988.

Hebert GA, Crowder CG, et al: Characteristics of coagulase-negative staphylococci that help differentiate these species and other members of the family Micrococcaceae. J Clin Microbiol 26:1939, 1988.

Helgerson S, Mallery B, Foster L: Toxic shock syndrome in Oregon. JAMA 252:3402, 1984.

Holt JG, et al (eds): Gram-positive cocci. In Bergey's Manual of Determinative Bacteriology, 9th ed. Baltimore: Williams & Wilkins, 1994.

Jeljaszewicz J, Ciborowski P (eds): The Staphylococci. Proceedings of the VIth International Symposium on Staphylococci and Staphylococcal Infections. New York: Gustav Fischer Verlag, 1991.

Jorgensen JH: Laboratory and epidemiologic experience with methicillin-resistant Staphylococcus aureus in the USA. Eur J Clin Microbiol 5:693, 1986.

Karchmer AW, Archer GL, Dismukes WE: Staphylococcus epidermidis causing prosthetic valve endocarditis: Microbiologic and clinical observations as guides to therapy. Ann Intern Med 98:447, 1983.

Kloos WE, Bannerman TL: Update on clinical significance of coagulase-negative staphylococci. Clin Microbiol Rev 7:117, 1994.

Kloos WE, Musselwhite MS: Distribution and persistence of Staphylococcus and Micrococcus species and other aerobic bacteria on human skin. Appl Microbiol 30:381, 1975.

Kloos WE, Wolfshoh JF: Identification of Staphylococcus species with the API STAPH-IDENT system. J Clin Microbiol 16:509, 1982.

Koontz F, Pfaller M: Coagulase-negative staphylococci: Why and when to do what. Clin Microbiol Newsl 11:125, 1989.

Levine G, Norden CW: Staphylococcal scalded-skin syndrome in an adult. N Engl J Med 287:1339, 1972.

Macdonald A, Smith G (eds): The Staphylococci. Proceedings of the Alexander Ogston Centennial Conference. Aberdeen, Scotland: Aberdeen University Press, 1981.

Maple PA, Hamilton-Miller JM, Brumfitt W: World-wide antibiotic resistance in methicillin-resistant Staphylococcus aureus. Lancet 1(8637):537, 1989.

Marrie TJ, Kwan C, et al: Staphylococcus as a cause of urinary tract infections. J Clin Microbiol 16:427, 1982.

Marsik FJ, Brake S: Species identification and susceptibility to 17 antibiotics of coagulase-negative staphylococci isolated from clinical specimens. J Clin Microbiol 15:640, 1982.

Melish M, Glasgow L: Staphylococcal scalded skin syndrome: The expanded clinical syndrome. J Pediatr 78:958, 1971.

Morrison V, Oldfield E: Postoperative toxic shock syndrome. Arch Surg 118:791, 1983.

Petitti D, Reingold A, Chin J: The incidence of toxic shock syndrome in northern California. JAMA 255:368, 1986.

Pfaller M, Herwaldt LA: Laboratory, clinical, and epidemiological aspects of coagulase-negative staphylococci. Clin Microbiol Rev 1:281, 1988.

Reingold AL, Hargrett NT, et al: Non-menstrual toxic shock syndrome. Ann Intern Med 96(part 2):871, 1982.

Roberson JR, Fox LK, et al: Evaluation of methods for differentiation of coagulase-positive staphylococci. J Clin Microbiol 30:3217, 1992.

Schleivert PM, Shands KN, et al: Identification and characterization of an exotoxin from Staphylococcus aureus associated with toxic shock syndrome. J Infect Dis 143:509, 1981.

Schumacher-Perdreau F: Clinical significance and laboratory diagnosis of coagulase-negative staphylococci. Clin Microbiol Newsl 13:97, 1991.

Schwalbe RS, Stapleton JT, Gilligan PH: Emergence of vancomycin resistance in coagulase-negative staphylococci. N Engl J Med 316:927, 1987.

Sheagren JN: *Staphylococcus aureus:* The persistent pathogen, part 1. N Engl J Med 310:1368, 1984.

Sheagren JN: *Staphylococcus aureus:* The persistent pathogen, part 2. N Engl J Med 310:1437, 1984.

Sorrell TC, Packham DR, et al: Vancomycin therapy for methicillin-resistant *Staphylococcus aureus.* Ann Intern Med 97:344, 1982.

Todd J: Toxins, test tubes, and tampons. Am J Med 84:579, 1988.

Wiseman G: The hemolysins of *Staphylococcus aureus.* Bacteriol Rev 39:317, 1987.

Streptococcaceae

Hal S. Larsen

The organisms included in the Streptococcaceae are gram-positive cocci that are usually arranged in pairs or chains. Most are facultative anaerobes. The growth requirements can be complex, with the use of blood or enriched medium necessary for their isolation. Their role in human disease ranges from well-established and common to rare but increasing. Clinical microbiologists have ample opportunity to become well acquainted with the members of the Streptococcaceae family.

STREPTOCOCCUS AND *ENTEROCOCCUS:* GENERAL CHARACTERISTICS

Members of *Streptococcus* and *Enterococcus* are gram-positive spherical cells arranged in chains or pairs (Fig. 11–1). Compared with cells of other gram-positive cocci, however, the cells of *Streptococcus* and *Enterococcus* appear somewhat more elongated than spherical. They are more likely to appear in chains when grown in broth culture.

The metabolism is fermentative, with lactic acid the primary product. A key characteristic of both genera is that they are catalase negative. Growth is poor on nutrient media such as trypticase soy agar. On media enriched with blood, serum, or glucose, however, growth is more pronounced. The colonies are usually small and somewhat transparent. In addition, some species require increased carbon dioxide for good growth.

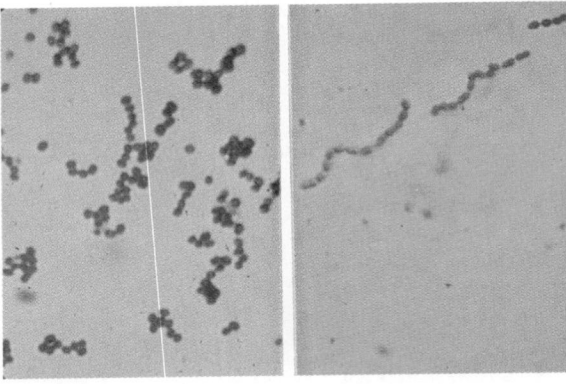

■ F i g u r e   1 1 – 1
Gram stain of *Streptococcus. Left*, Culture. *Right*, Liquid medium.

Certain other gram-positive cocci resemble the streptococci. These genera, which include *Aerococcus, Lactobacillus, Leuconostoc,* and *Pediococcus,* are also considered later in this chapter. This chapter discusses the characteristics, clinical infections associated with each species, and their laboratory diagnosis.

The role of the Streptococcaceae in disease has been known for over 100 years. The range of infections caused by these organisms is wide and well-studied. As we have also seen with other organisms, however, the roles in disease of the normal flora, the previously unknown or poorly characterized species, and the saprophytes are becoming more prominent.

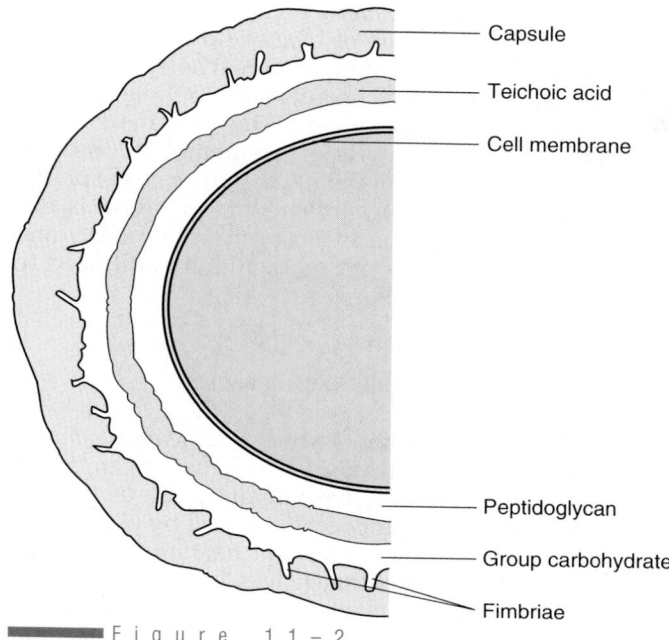

Capsule

Teichoic acid

Cell membrane

Peptidoglycan

Group carbohydrate

Fimbriae

■ Figure 11–2
Schematic representation of streptococcal cell wall.

Cell Wall Structure

As a group, the Streptococcaceae possess a typical gram-positive cell wall consisting of mucopeptide (peptidoglycan) and teichoic acid as well as a capsule in young cultures. All streptococci except the viridans group have a layer of C carbohydrate, which can be utilized to serologically classify an isolate. A schematic diagram of the streptococcal cell wall is shown in Figure 11–2. Other cell-wall antigens are present in specific C carbohydrate groups and are explained in the discussions of the individual groups.

Classification

Several different approaches to classification have been used. Of these, four remain useful to the practicing technologist. These four classification schemes are (1) hemolytic pattern on sheep's blood agar, (2) physiologic characteristics (type of infection, site of origin, etc.), (3) serologic group or type of C carbohydrate present (Lancefield classification), and (4) biochemical characteristics. Each scheme is used to some degree, and the identification process for a streptococcal isolate in the clinical laboratory may utilize features from each scheme.

Hemolytic Patterns

The medical technologist often makes an initial classification on the basis of the hemolytic pattern of the isolate on sheep's blood agar. Although hemolysis patterns can be helpful during the initial work-up of an isolate, it must be kept in mind that many species of streptococci may show more than one type of hemolytic pattern. The types of hemolysis possible are outlined in Table 11–1.

When the red blood cells in the agar surrounding the colony are completely lysed, the resulting area is clear. It is referred to as *beta* (β) *hemolysis* (Fig. 11–3). A partial lysis of the red cells results

■ TABLE 11–1
TYPES OF HEMOLYSIS

Hemolysis	Description
Alpha (α)	Partial lysis of red blood cells around colony
	Greenish discoloration of area around colony
Beta (β)	Complete lysis of red blood cells around colony
	Clear area around colony
Nonhemolytic	No lysis of red blood cells around colony
	No change in agar
Alpha-prime (α′) or wide zone	Small area of intact red blood cells around colony surrounded by a wider zone of complete hemolysis

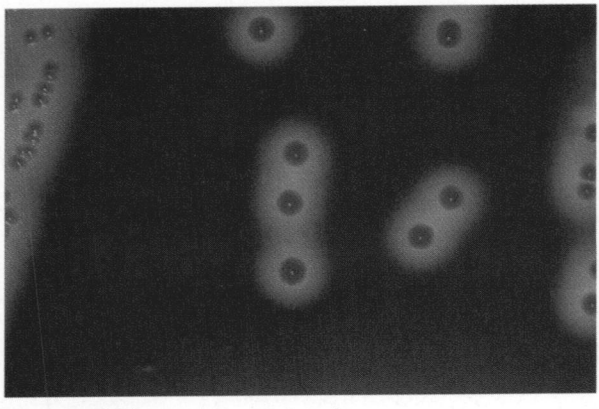

A β-hemolytic streptococcal colony on blood agar.

in a greenish discoloration of the area surrounding the colony. It is termed *alpha (α) hemolysis* (Fig. 11–4).

When the red cells immediately surrounding the colony are unaffected, there is no change in the color of the agar; such a colony is termed *nonhemolytic*. Some references term this result as gamma (γ) hemolysis. Because there is no lysis of the red cells, however, γ *hemolysis* is confusing and is not used in this chapter. Some isolates belonging to the viridans group produce what is called *wide-zone* or α-*prime hemolysis*. The colonies are surrounded by a very small zone of no hemolysis and then a wider zone of β hemolysis. This reaction may be mistaken for β hemolysis at first glance. The use of a dissecting microscope or the scanning objective shows the narrow zone of intact red cells and the wider zone of complete hemolysis.

Physiologic Characteristics

Streptococci have also been classified according to physiologic characteristics. This classification divides the species into four groups: pyogenic streptococci, lactic acid streptococci, en-terococci, and viridans streptococci. The *pyogenic streptococci* are those that produce pus; these organisms are mostly β-hemolytic and constitute the majority of the Lancefield groups. The *lactic acid streptococci* are nonhemolytic organisms often found in dairy products; they are part of Lancefield group N. The *enterococci* comprise those species found as part of the flora of the human intestine; this group of organisms is now

part of the genus *Enterococcus*. The *viridans streptococci* are not part of Lancefield's classification because they do not have a C carbohydrate; they are widely found as normal flora in the upper respiratory tract of humans. The viridans streptococci are α-hemolytic or nonhemolytic and are often seen as opportunistic pathogens. For the most part, this physiologic classification is historical. Nevertheless, the terms *enterococci* and *viridans streptococci* remain and are still used to describe clinical isolates.

The Lancefield Classification Scheme

The most commonly used classification scheme was developed in the 1930's by Rebecca Lancefield. She found that the C carbohydrate can be extracted from the streptococcal cell wall by placing the organisms in dilute acid and heating for 10 minutes. The soluble antigen was used to immunize rabbits to obtain antisera to the various C carbohydrate groups. After first recognizing the antigen in β-hemolytic streptococci, Lancefield was able to divide the streptococci into serologic groups (designated by letters). Organisms in group A possess the same antigenic C carbohydrate, those in group B have the same C carbohydrate, and so on.

Some of the groups contain several different streptococcal species, whereas others have only one species. For example, *S. pyogenes* is the only member of group A. Therefore, the terms *S. pyogenes* and *group A streptococci* describe the same

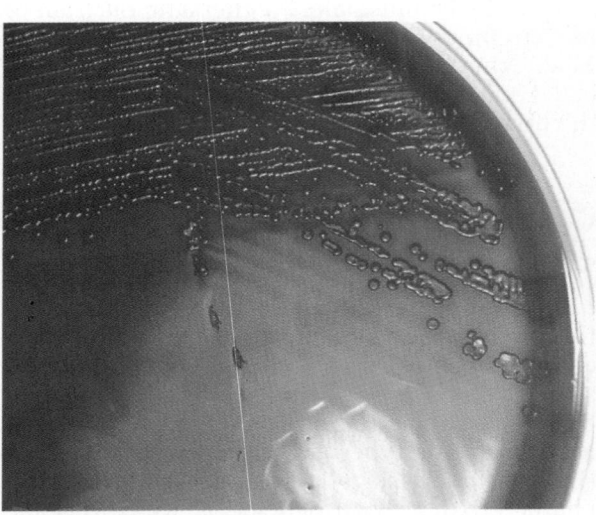

An α-hemolytic streptococcal colony on blood agar.

■■■■■ T A B L E   1 1 – 2

CLASSIFICATION OF *STREPTOCOCCUS* AND *ENTEROCOCCUS*

Species	Lancefield Group Antigen	Hemolysis Type(s)	Common Terms	Disease Association(s)
S. pyogenes	A	β*	Group A strep	Rheumatic fever; scarlet fever; pharyngitis; glomerulonephritis; pyogenic infections
S. agalactiae	B	β*	Group B strep	Neonatal sepsis; meningitis; puerperal fever; pyogenic infections
S. equisimilis	C	β	Group C strep	Pharyngitis (?); impetigo; pyogenic infections
S. equi subsp *zooepidemicus*				
S. equi				
S. bovis	D	α, none	Nonenterococcus	Endocarditis; urinary tract infections; pyogenic infections
S. equinus				
E. faecalis	D	α, β, none	Enterococcus	Urinary tract infections; pyogenic infections
E. faecium				
E. durans				
Other species	F, G	β, (α, none)		Pyogenic infections
S. pneumoniae	—	α	Pneumococcus	Pneumonia; meningitis; pyogenic infections
S. anginosus	—	α, none	Viridans strep	Endocarditis; dental caries; abscesses in various tissues
S. sanguis				
S. mitis				
S. mutans				
etc.				

*Occasionally, isolates are found that are nonhemolytic.

organism, because practically, there are no other organisms in this Lancefield group. The same is true of group B, the only member of which is *S. agalactiae*. The terms *group B streptococci* and *S. agalactiae* are therefore interchangeable. The other Lancefield groups contain varying numbers of species. Group D streptococci contain several species, including *S. bovis* and *S. equinus*.

In addition, streptococcal species other than those that produce β hemolysis are found to possess C carbohydrate. Some are found as normal flora in animals or as animal pathogens, and others may be found in both humans and animals. The Lancefield groups most commonly seen in human infections are A, B, C, D, F, and G, although not all Lancefield groups commonly cause human infection.

Classification of *Streptococcus* and *Enterococcus* is shown in Table 11–2.

Biochemical Identification

Biochemical identification can be performed even by small laboratories. Although definitive identification requires a large number of characteristics or perhaps serologic methods, presump- tive identification can be accomplished relatively easily with a few key tests and characteristics. Presumptive identification, in the great majority of cases, possesses a high enough rate of accuracy to be useful to the clinician and does not require the exhaustive additional tests that are needed to meet the criteria for definitive identification, especially for those species in groups A, B, and D as well as *S. pneumoniae* and *Enterococcus*. Speciation of the viridans streptococci, however, does require a considerable increase in the number of tests. The medical technologist must evaluate the needs of the clinicians and patient population served, the cost of an expanded identification scheme, the resources and abilities of the laboratory, and the usefulness of the data produced.

Table 11–3 outlines the biochemical characteristics used to presumptively identify selected members of the Streptococcaceae.

SUSCEPTIBILITY TESTING

Susceptibility to various antimicrobials can be useful in the early differentiation of gram-positive cocci. *Bacitracin* has been used for many years to differentiate group A streptococci from other groups of β-hemolytic streptococci. A low

BIOCHEMICAL IDENTIFICATION OF *STREPTOCOCCUS* AND RELATED ORGANISMS*

Characteristic	*S. pyogenes*	*S. agalactiae*	Other β-hemolytics†	*Enterococcus*	Group D streptococci	*S. pneumoniae*	Viridans streptococci	*Aerococcus*	*Pediococcus*	*Leuconostoc*
Hemolysis type	β	β	β	α, β, none	α, none	α	α, none	α, none	α, none	α, none
Susceptibility to										
Vancomycin	S	S	S	S(R)	S	S	S	S	R	R
Bacitracin	S	R‡	R‡	R	R	S	R‡	S		
SXT	R	R	S	R	V	S	S			
Optochin	R	R	R	R	R	S	R			
Hydrolysis of										
Hippurate	–	+	–	–‡	–	–	–‡	+	+	
PYR	+	–	–	+	–	–	–	+	–	–
CAMP	–	+	–	–	–	–	–	–		
Leucine aminopeptidase	+	+	+	+	+	+	+‡	–	+	–
Bile esculin	–	–	–	+	+	–	–‡	V	+	V
Growth in 6.5% NaCl	–	–	–	+	–	–	–	+	V	V

*R, resistant; S, susceptible; S(R), greater percentage susceptible; V, variable; +, present; –, absent.
†β-hemolytic groups other than A, B, and D.
‡Exceptions may occur.

PROCEDURE 11–1. BACITRACIN SUSCEPTIBILITY

Purpose:	To differentiate *S. pyogenes* from other β-hemolytic groups.
Principle:	Group A streptococci are susceptible to low levels (0.04 U) of bacitracin, whereas other groups are resistant. Rare strains of group A streptococci are resistant (approximately 1%), whereas some strains of groups, B, C, and G streptococci are sensitive (5–10%). Sensitivity to bacitracin *presumptively* identifies an isolate as *S. pyogenes*. This procedure was designed for use only with pure cultures.
Specimen:	Isolated colonies on sheep's blood agar.
Media:	5% sheep's blood agar plate.
Reagent:	Bacitracin disk (0.04 U).
Procedure:	1. Streak surface of agar plate to obtain confluent growth.
	2. Aseptically place bacitracin disk onto inoculated surface. Press down gently on the disk to ensure complete contact with the agar surface.
	3. Invert and incubate plates at 35°C for 18 to 24 hours.
Interpretation:	A positive result = Any zone of inhibition around the bacitracin.
	A negative result = Uniform lawn of growth up to the edge of the disk (Fig. 11–5).
Controls:	Positive: Group A streptococcus (bacitracin susceptible).
	Negative: Group B streptococcus (bacitracin resistant).

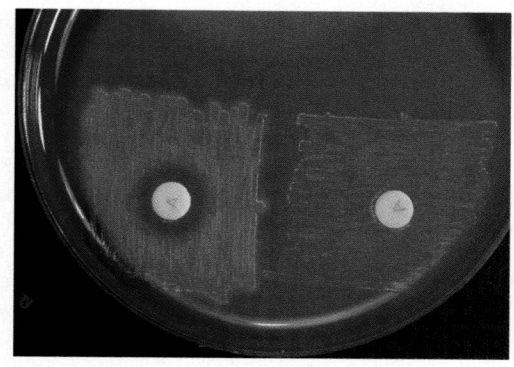

Figure 11–5

Group A streptococcus on blood agar showing susceptibility to bacitracin. *Left*, Positive; *right*, Negative.

concentration of bacitracin (0.04 U) selectively inhibits the growth of group A streptococci. Procedure 11–1 shows the presumptive identification of *S. pyogenes* using bacitracin susceptibility.

Susceptibility to *SXT* (sulfamethoxazole and trimethoprim) can be used in conjunction with that to bacitracin to improve the accuracy of group A identification (Fig. 11–5). Groups A and B are resistant to SXT, whereas groups other than A and B are sensitive. Figure 11–6 shows resistance of group A β-hemolytic streptococci to SXT. The pattern shown by *S. pyogenes* when both bacitracin and SXT disks are used is susceptibility to bacitracin and resistance to SXT. The pattern shown by *S. agalactiae* (group B) is resistance to both bacitracin and SXT. If the blood agar contains SXT, a bacitracin disk may be placed directly on the primary inoculum, because growth of the majority of interfering normal flora will be inhibited but *S. pyogenes* and *S. agalactiae* will grow. This method is helpful to screen for group A streptococci from throat cultures. The sheep's blood agar containing SXT is inoculated with the throat swab, and a bacitracin disk is placed onto the agar. Those β-hemolytic colonies that grow and are susceptible to bacitracin are presumptively identified as *S. pyogenes*.

Vancomycin is an effective antimicrobial for treating infections caused by gram-positive organisms. Until the mid-1980's, resistance to vancomycin was rarely seen in clinical isolates. Vancomycin resistance is now being seen more commonly, although it is still not widespread. Gram-positive isolates should be tested routinely for vancomycin susceptibility. Table 11–3 lists the usual patterns of vancomycin susceptibility for a number of gram-positive cocci. In particu-

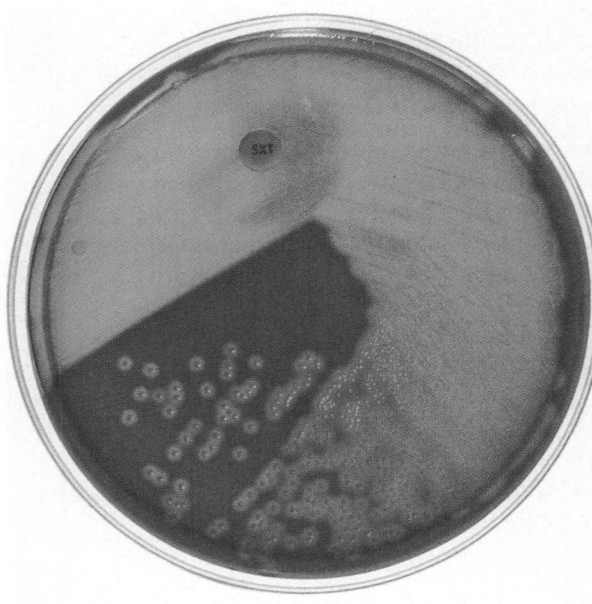

■■■ F i g u r e   1 1 – 6

Group A streptococcus on blood agar showing resistance to SXT disk.

lar, *Pediococcus* and *Leuconostoc* are resistant to vancomycin. Some *Streptococcus* isolates and a significant number of *Enterococcus* isolates demonstrate resistance to this antimicrobial.

The differentiation of *S. pneumoniae* from other α-hemolytic streptococcal isolates can be accomplished by the use of *optochin* susceptibility. This is done in the same fashion as the bacitracin disk method. The suspected isolate is inoculated onto a blood agar plate. A disk saturated with a solution of optochin (ethylhydrocuprein hydrochloride) is placed onto the inoculated area. A zone of inhibition around the disk is presumptive evidence of *S. pneumoniae*. Other α-hemolytic species of streptococci are resistant to optochin.

BILE SOLUBILITY

Another characteristic that correlates well with optochin susceptibility is bile solubility. The test for bile solubility takes advantage of the very active autocatalytic enzymes that *S. pneumoniae* possesses. Under the influence of a bile salt, the organism's cell wall lyses during cell division. A suspension of *S. pneumoniae* in a solution of sodium deoxycholate lyses and the solution becomes clear. Other α-hemolytic organisms do not lyse, and the solution remains cloudy.

HYDROLYSIS

A useful test to differentiate group B streptococci from other β-hemolytic streptococci is *hippurate hydrolysis*. *Streptococcus agalactiae* possess the enzyme hippuricase, which hydrolyzes sodium hippurate to form sodium benzoate and glycine. This hydrolysis can be detected by adding Ninhydrin, which reacts with the α-amino groups to form a purple color. Procedure 11–2 describes the test.

A test that provides a high probability for presumptive separation of group A enterococci and enterococci from the other streptococcal species is the *PYR hydrolysis test* (Fig. 11–7). It is more specific than bacitracin susceptibility. There are a number of commercial systems on the market, and the reader is referred to the specific package insert. The PYR test takes advantage of the fact that *S. pyogenes* and *Enterococcus* spp are able to hydrolyze the substrate PYR. (Substrates utilized in the PYR test are: L-pyrrolidonyl-β naphthylamide, and L-pyroglutamic acid-β naphthylamide). It is as specific as bile esculin and NaCl broth (see later) for *Enterococcus* and more specific than bacitracin for *S. pyogenes*.

Following hydrolysis of the substrate by the peptidase, the resulting β-naphthylamide produces a red color upon the addition of 0.01% cinnamaldehyde reagent (0.01% *p*-dimethylaminocinnamaldehyde). The substrate is usually impregnated into paper disks, which are moistened and inoculated with the suspected isolate. After 2 minutes to allow for hydrolysis, the cinnamaldehyde reagent is dropped onto the disk. A pink or cherry red color appears within 1 minute if the reaction is positive. A negative reaction is indicated by no color change. The genera that are PYR positive include *Enterococcus*, *Aerococcus*, and *Gemella*. The only member of *Streptococcus*

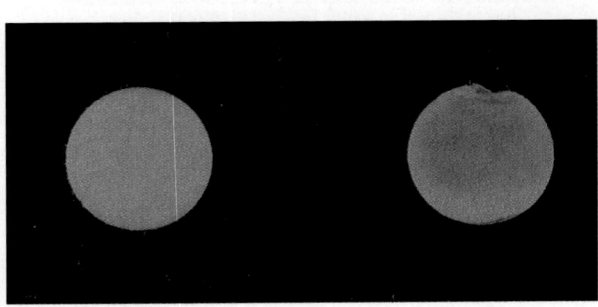

■■■ F i g u r e   1 1 – 7

The PYR test for *Streptococcus pyogenes* and *Enterococcus*. A positive test gives a bright red color change.

PROCEDURE 11–2. HIPPURATE HYDROLYSIS

Purpose:	To differentiate *S. agalactiae* from other β-hemolytic streptococci.
Principle:	The enzyme, hippuricase, hydrolyzes hippuric acid to form sodium benzoate and glycine. Subsequent addition of Ninhydrin results in the release of ammonia from the oxidative deamination of the α-amino group in glycine as well as the reduced form of Ninhydrin, hydrindantin. The ammonia reacts with residual Ninhydrin and hydrindantin to give a purple-colored complex. Some isolates of group D streptococci also hydrolyze hippurate; however, these isolates are less likely to be β-hemolytic, and their colony morphology is different from that of group B streptococci. An isolate that is hippurate positive and bile esculin negative has a very high probability of being *S. agalactiae*.
Specimen:	Isolated colonies on sheep's blood agar.
Reagents:	*Sodium hippurate (1%):*

Sodium hippurate	1 g
Distilled water	100 mL

Dispense 0.5-mL aliquots in small capped vials. Store at −10°C. Storage life is 6 months.

Ninhydrin reagent:

Ninhydrin	3.5 g
Acetone-butanol mixture (1:1)	100 mL

Store at room temperature. Storage life is 12 months.

Procedure:	1. Inoculate the solution of sodium hippurate heavily with colonies 18-24 hours old until a milky suspension is obtained. 2. Incubate tube for 2 hours at 35°C. 3. Add 0.2 mL of Ninhydrin reagent. 4. Mix and incubate for 10 to 15 minutes.
Interpretation:	Positive result = Deep purple color (indicates hippurate hydrolysis). Negative result = No color change. A very slight purple color should be interpreted as negative.
Controls:	Positive: *S. agalactiae.* Negative: *S. pyogenes.*

that is PYR positive is *S. pyogenes.* The PYR test is an excellent addition to the test menu for the presumptive identification of the gram-positive cocci.

CAMP TEST

Another test that is used to presumptively iden-tify group B streptococci is the CAMP test. *CAMP* is an acronym based on the first letters of the surnames of the individuals who first described the reaction: Christie, Atkins, and Munch-Petersen. The CAMP test can be performed three ways. The first is with the use of a β-lysin–producing strain of *S. aureus,* and the second is with the use of a disk

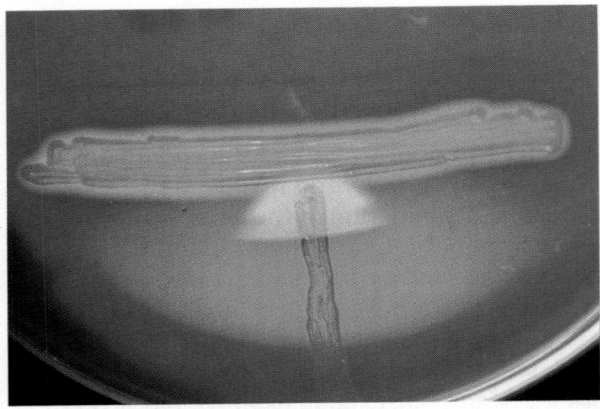

■■■■■Figure 11–8

The CAMP test for presumptive identification of group B streptococci. *Streptococcus agalactiae* shows the classic arrow shape near the staphylococcus streak.

In addition, a rapid CAMP test utilizing the extracted β-lysin from the strain of *S. aureus* that produces it is currently in use. It involves placing a drop of the extracted β-lysin on the area of confluent growth of the suspected group B streptococci. After incubation at 37°C for at least 20 minutes, enhanced hemolysis is observed (Fig. 11–9).

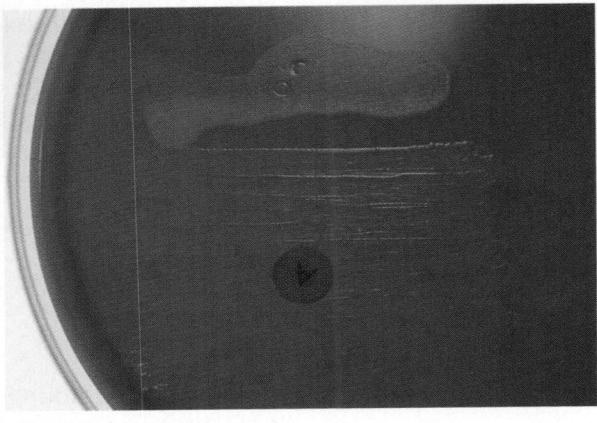

■■■■■Figure 11–9

A modification of the CAMP test showing the enhanced hemolysis produced by *Streptococcus agalactiae* when a drop of extracted hemolysin is placed on the colony.

impregnated with the β-lysin. Both methods take advantage of the enhanced hemolysis that takes place when the β-lysin and the hemolysins produced by group B streptococci are in contact. The result is a characteristic arrow-shaped hemolysis pattern (Fig. 11–8). Procedure 11–3 describes how to perform the CAMP test.

PROCEDURE 11–3. CAMP TEST

Purpose:	To differentiate *S. agalactiae* from other β-hemolytic streptococci.
Principle:	Group B streptococci produce a CAMP factor that enhances the lysis of sheep red cells by staphylococcal β-lysin. A positive reaction can be observed in 5 to 6 hours with incubation in carbon dioxide (18 hours with incubation in ambient air).
Specimen:	1. Isolated colonies on sheep's blood agar. 2. β-lysin–producing *S. aureus* on blood agar.
Medium:	Sheep's blood agar plate.
Procedure:	1. Inoculate *S. aureus* along a line down the center of the agar plate. 2. Inoculate the streptococcal isolate along a thin line 2 cm long and perpendicular to, but not touching, the *S. aureus* streak. 3. Incubate plate at 35°C for 18 hours.
Interpretation:	Positive result = Arrowhead-shaped area of enhanced hemolysis where the two streaks (staphylococcal and streptococcal) approach each other. Negative result = No enhanced hemolysis.
Controls:	Positive: *S. agalactiae*. Negative: *S. pyogenes*.

BILE ESCULIN AND NaCl TESTS

Two tests that have been mainstays in identification schemes for group D streptococci and *Enterococcus* are the bile esculin test (Procedure 11–4 and Fig. 11–10) and growth in 6.5% NaCl (Procedure 11–5).

Group D streptococci and *Enterococcus* can grow in the presence of 40% bile and hydrolyze esculin. The resulting product is a black complex in the agar, indicating a positive reaction.

Growth in 6.5% sodium chloride broth is used to identify the non–β-hemolytic, catalase-negative, gram-positive cocci (Fig. 11–11). *Enterococcus, Aerococcus,* and some species of *Pediococcus* and *Leuconostoc* grow in a 6.5% NaCl broth when incubated for 24 hours. Group D streptococci, however, do not grow in a 6.5% NaCl broth.

LEUCINE AMINOPEPTIDASE TEST

Leucine aminopeptidase (LAP) is helpful in differentiating *Aerococcus* and *Leuconostoc* from other gram-positive cocci. As Table 11–3 outlines, *Streptococcus, Enterococcus,* and *Pediococcus* are LAP positive. *Aerococcus* and *Leuconostoc* are LAP negative.

Noncultural Identification

Identification of streptococci, particularly group A, can be made by the direct detection of the group-specific antigen, from either isolated colonies or, in some cases, a direct clinical specimen such as a throat swab.

Identification from isolated colonies can be accomplished by extracting the C carbohydrate by means of acid or heat. The extract containing the specific group carbohydrate is then utilized in a capillary precipitin test or a slide agglutination test. In the capillary precipitin test, antiserum to the specific group carbohydrate is overlayed by the solution containing the streptococcal extract. After 5 to 10 minutes, the interface between the extraction solution and the antiserum is examined carefully for a white precipitate, which indicates a positive reaction. Each extract can be tested with a number of antisera specific to the various group antigens.

The slide agglutination test utilizes a carrier particle for the group-specific antiserum. In the coagglutination test, the specific antiserum is conjugated to killed *S. aureus* cells containing high amounts of protein A. Another carrier parti-

---

PROCEDURE 11–4. BILE ESCULIN HYDROLYSIS

Purpose:	To differentiate group D streptococci and *Enterococcus* from other gram-positive cocci.
Principle:	Group D streptococci and *Enterococcus* grow in the presence of bile and also hydrolyze esculin to esculetin and glucose. Esculetin diffuses into the agar and combines with ferric citrate in the medium to give a black complex.
Specimen:	Isolated colonies on blood agar.
Media:	Bile esculin agar.
Procedure:	1. Pick one or two isolated colonies from the blood agar plate and inoculate to a bile esculin agar slant.
	2. Incubate at 35°C for 18 to 24 hours. (Note: A positive result is often seen within 4 hours. A negative result should be incubated for an additional 24-hour period.)
Interpretation:	Positive result = Blackening of the agar slant.
	Negative result = No blackening of the agar slant. Growth alone does not constitute a positive result.
Controls:	Positive: Group D *Streptococcus.*
	Negative: Viridans *Streptococcus.*

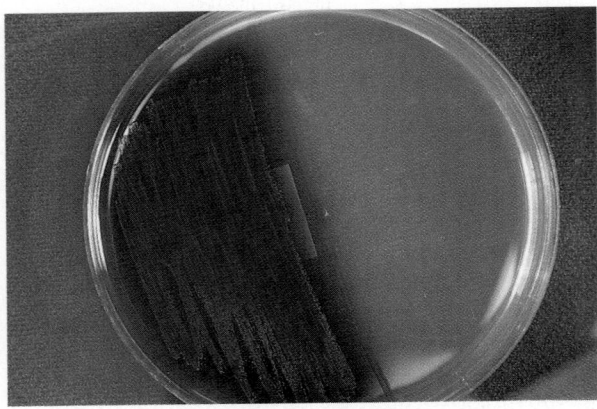

■ F i g u r e   1 1 – 1 0

The bile esculin test. A positive test *(left)* shows blackening of the agar. Right side is negative.

cle often used is latex. Both types of particles serve as holders for the antiserum. When an antibody (carrier particle + specific group carbohydrate antiserum)–antigen (group carbohydrate extraction solution) reaction occurs, it is visualized macroscopically as agglutination of the particles (Fig. 11–12).

These methods give a definitive identification as to the Lancefield group the isolate belongs to, but, when the group comprises several species, the methods do not give a species iden-

tification. Of course, if the results indicate that the isolate belongs to group A or B, then it can be identified as *S. pyogenes* or *S. agalactiae,* respectively. The cost of these methods is much higher than that of the standard cultural approaches. Consequently, few clinical laboratories routinely utilize them to identify all streptococcal isolates.

There are also over two dozen different products available, many of which utilize an ELISA system, to detect group A streptococci from throat swabs. These are designed primarily for use in the clinic and the physician's office. Although a positive finding allows the physician to treat for *S. pyogenes* without waiting for culture results, a negative result necessitates a follow-up throat culture. This is because the sensitivity of direct detection methods for group A streptococci has not been high enough to ensure confidence that the negative result is not a false-negative. Additionally, the cost of using direct detection methods can be significantly higher than that of culture only. If the direct test is negative and a subsequent culture is performed to confirm the result, the total cost is almost twice as much as a culture only. The direct detection method for group A streptococci can be a valuable diagnostic tool in the right context; however, the cost and sensitivity must be weighed carefully against desired and actual results.

PROCEDURE 11–5. NaCl TEST

Purpose:	To differentiate those gram-positive cocci that will grow in 6.5% NaCl from those that are inhibited by this salt concentration.
Principle:	*Enterococcus, Aerococcus,* and some species of *Pediococcus* and *Leuconostoc* can withstand a higher salt concentration than other gram-positive cocci.
Specimen:	Isolated colonies on sheep's blood agar.
Medium:	6.5% NaCl broth (nutrient broth base).
Procedure:	1. Pick one or two isolated colonies from the blood agar plate and inoculate into 5 mL of NaCl broth.
	2. Incubate tube at 35°C for 3 days. Check for growth daily.
Interpretation:	Positive result = Turbidity.
	Negative result = No turbidity.
Controls:	Positive: *Enterococcus faecalis.*
	Negative: Viridans *Streptococcus.*

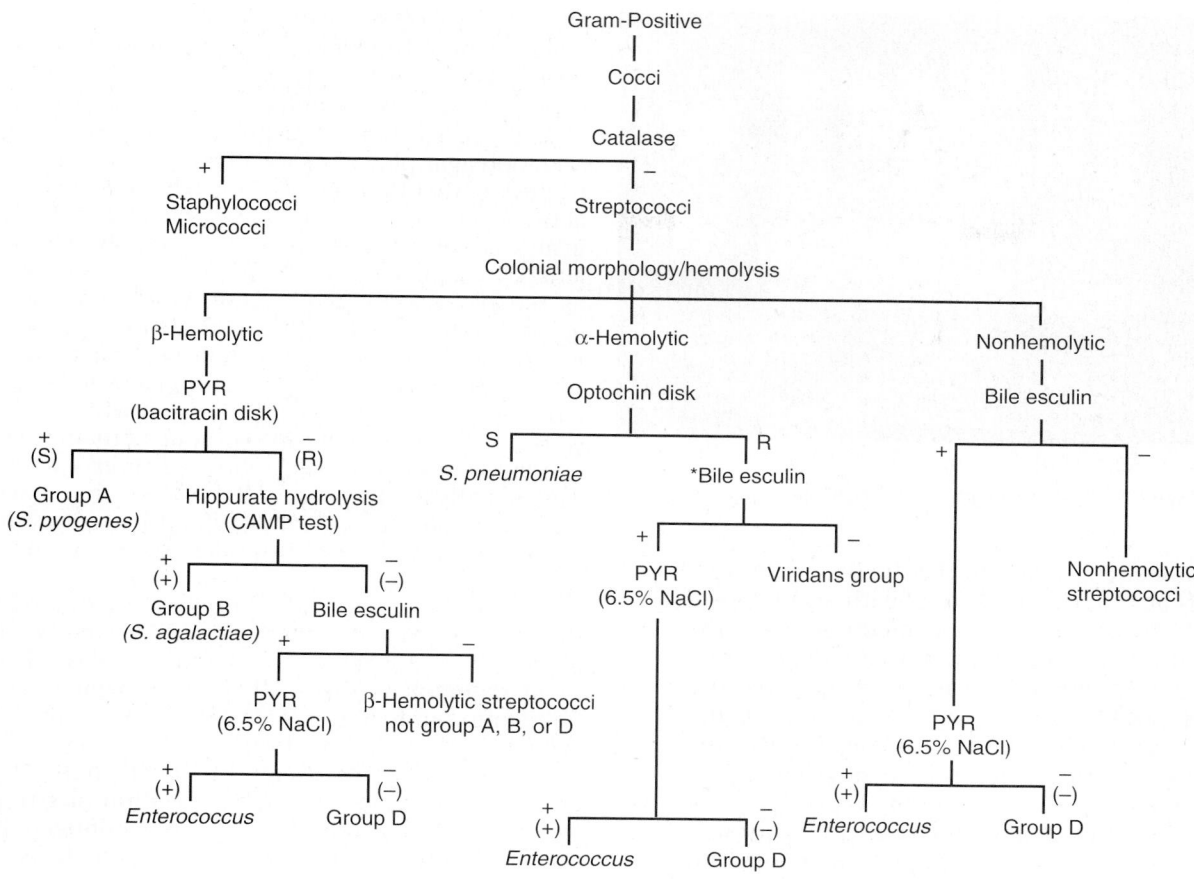

*Perform additional tests if isolate is from nonrespiratory source.

Figure 11-11

Schematic diagram for the presumptive identification of gram-positive cocci.

CLINICALLY SIGNIFICANT STREPTOCOCCI AND THEIR ASSOCIATED DISEASES

Streptococcus pyogenes (Group A Streptococci)

Antigenic Structure

The structure of group A streptococci is illustrated in Figure 11–13. *Streptococcus pyogenes* shows a cell wall structure similar to that of other streptococci and gram-positive bacteria. The group antigen is unique, placing the organism in Lancefield group A. An antigen that is not found in the other Lancefield groups is referred to as the *M protein*. This is a molecule that is attached to the peptidoglycan of the cell wall and extends to the surface of the cell wall. The M protein is essential to virulence.

Virulence Factors

In addition to M protein, there are a number of extracellular products produced by *S. pyogenes*, including hemolysins, toxins, and enzymes.

The best-defined virulence factor is M protein. There are more than 80 different serotypes of M protein, which are labeled M5, M10, M23, and so on. Resistance to infection with *S. pyogenes* appears to be related to the presence of

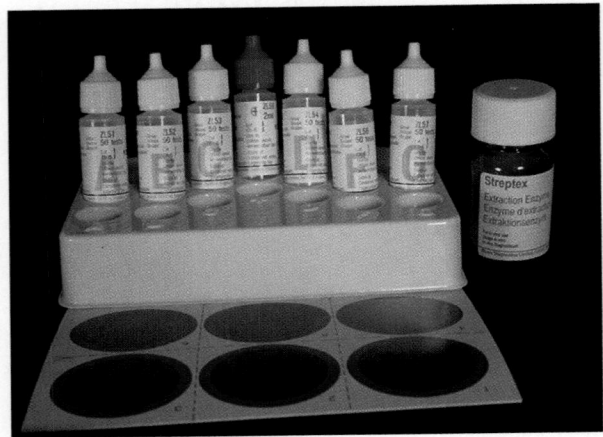

■ F i g u r e   1 1 – 1 2
Slide agglutination test for grouping streptococci.

type-specific antibodies to the M protein. This means that an individual with antibodies against M5 is protected from infection by *S. pyogenes* with the M5 protein but remains unprotected against infection with the roughly 80 remaining M protein serotypes. The M protein molecule causes the streptococcal cell to resist phagocytosis. It also plays a role in adherence of the bacterial cell to mucosal cells.

Other products produced by *S. pyogenes* are streptolysin O, streptolysin S, deoxyribonucle-

ase, streptokinase, hyaluronidase, and erythrogenic toxin. Although all of these products have been postulated to play a role in virulence, the exact role each has in infection is not clear.

A hemolysin that is responsible for hemolysis on blood agar plates is *streptolysin O*. The *O* refers to the fact that this hemolysin is *oxygen labile*. It is active only in the reduced form, which is achieved in an anaerobic environment. This hemolysin lyses leukocytes, platelets, and other tissue cells as well as red blood cells. Streptolysin O is highly antigenic, and the infected host readily forms antibodies to the hemolysin. These can be measured in the antistreptolysin-O test to determine whether an individual has had a recent infection with *S. pyogenes*. A second hemolysin is produced that is oxygen stable and so is referred to as *streptolysin S*. The hemolysis seen around colonies that have been incubated aerobically is due to streptolysin S. It is nonantigenic. Streptolysin S also lyses leukocytes.

Streptococcus pyogenes secretes four different deoxyribonucleases (DNases), A, B, C, and D. All strains form at least one deoxyribonuclease. The most common is DNase B. These enzymes are antigenic, and antibodies to DNase can be detected following infection.

Filtrates of group A streptococci cause the lysis of fibrin clots, through the action of streptokinase on plasminogen. The plasminogen is converted into a protease (plasmin), which lyses

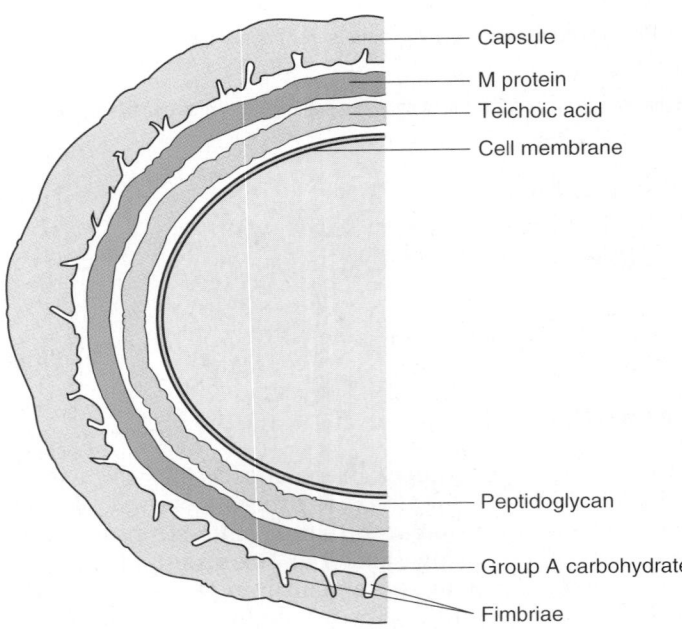

■ F i g u r e   1 1 – 1 3
A schematic representation of the group A streptococcal cell wall.

the fibrin. Antibodies to streptokinase can be detected following infection but are not specific indicators of group A infection, because groups C and G also form streptokinase.

Another name for hyaluronidase is *spreading factor.* This enzyme was so named because it acts to solubilize the ground substance of mammalian connective tissues (hyaluronic acid). It was postulated that the bacteria utilize this enzyme to separate the tissue and then spread the infection. There is no real evidence, however, that hyaluronidase favors the spread of *S. pyogenes* through the tissues.

Some strains of *S. pyogenes* cause a red spreading rash referred to as *scarlet fever* (see section on pyodermal infections). This condition is caused by erythrogenic toxin. It is a protein exotoxin, and three types (A, B, and C) are known. The role, if any, in pathogenesis is unknown.

Clinical Infections

Infections due to *S. pyogenes* are common and include pharyngitis, scarlet fever, skin infections, and other septic infections. In addition, rheumatic fever and glomerulonephritis may occur as a result of infection with group A streptococci.

BACTERIAL PHARYNGITIS

The most common clinical expressions of group A streptococcal infection are pharyngitis and tonsillitis. The majority of cases of bacterial pharyngitis are due to infection with *S. pyogenes.* Other groups, particularly C and G, have the capability to produce significant acute pharyngitis but are much less commonly seen.

"Strep throat" is most commonly seen in children between 5 and 15 years of age. After an incubation period of 1 to 4 days, there is an abrupt onset of illness, with sore throat, malaise, fever, and headache. It is not unusual to see nausea, vomiting, and abdominal pain as well. The tonsils and pharynx are inflamed. The cervical lymph nodes are swollen and tender. The disease ranges in intensity, however, and all of these symptoms may not be seen. In fact, it is not unusual to isolate a nearly pure culture of *S. pyogenes* from the throat of a child who complains only of a mild sore throat and fever. The symptoms subside within 3 to 5 days unless complications, such as peritonsillar abscesses, occur. The disease is spread by droplets and close contact. Although clinical criteria have been proposed, the diagnosis of streptococcal sore throat relies on a throat culture. Approximately one

third of those complaining of sore throat have a throat culture positive for *S. pyogenes.*

PYODERMAL INFECTIONS

Skin infections with group A streptococci result in the syndrome of impetigo, cellulitis, erysipelas, wound infection, or gangrene. *Impetigo,* a localized skin disease, begins as small vesicles that progress to weeping lesions. The lesions crust over after several days. Impetigo is usually seen on very young children (2 to 5 years) and affects exposed areas of the skin. Inoculation of the organism occurs through minor abrasions or insect bites. *Erysipelas* is an infection of the skin and subcutaneous tissues. It is an acute spreading skin lesion that is uncommonly seen. The lesion is intensely erythematous with a plainly demarcated, but irregular edge. It is most often seen in elderly patients. *Cellulitis* may develop following deeper invasion by streptococci. The infection can be serious, even life-threatening, with bacteremia or sepsis present. In patients with peripheral vascular disease or diabetes, cellulitis may lead to gangrene.

As mentioned earlier, infection with strains of *S. pyogenes* that produce the erythrogenic toxin may result in *scarlet fever.* This condition, which appears within 1 to 2 days following infection, is characterized by a diffuse red rash that appears on the upper chest and spreads to the trunk and extremities. The rash disappears over the next 5 to 7 days and is followed by desquamation.

POSTSTREPTOCOCCAL INFECTION

There are two serious complications of group A streptococcal disease, rheumatic fever and acute glomerulonephritis.

Rheumatic fever is a complication of *S. pyogenes* pharyngitis. It is characterized by fever and inflammation of the heart, joints, blood vessels, and subcutaneous tissues. Attacks usually begin within a month after infection. The most serious result is a chronic, progressive damage to the heart valves. Repeated infections may produce further valve damage. By 1980, many clinicians considered rheumatic fever eradicated. There was a resurgence of rheumatic fever during the late 1980's, however, and it is once again a significant problem. The cause of this resurgence is not completely clear. It appears to be a result of a number of factors, including clinical management of bacterial pharyngitis, epidemiologic changes, and bacteriologic factors. The pathogenesis of rheumatic fever is poorly understood. Several theories have been proposed, including antigenic cross-reactivity between streptococcal antigens and heart tissue, direct toxicity due to bacterial exotoxins, and actual invasion of the heart tissues by the organism. Most evidence favors cross-reactivity as being responsible for the effects.

Another potential result of infection with group A streptococci is *acute glomerulonephritis* (AGN). In contrast to rheumatic fever, AGN may occur after a cutaneous or pharyngeal infection. It is more common in children than adults. The pathogenesis appears to be a result of immunologic mechanisms. Circulating immune complexes are found in the serum of patients with AGN, and it is postulated that these antigen-antibody complexes deposit in the glomerulus. Complement is subsequently fixed, and an inflammatory response causes damage to the glomerulus, resulting in impairment of kidney function.

INVASIVE STREPTOCOCCAL INFECTIONS

Streptococcal Toxic Shock Syndrome. There have been a number of reports in recent years of group A streptococcal infections associated with toxic shock. The initial streptococcal infection was severe (pharyngitis, peritonitis, cellulitis, wound infections), and symptoms developed that were very similar to those of staphylococcal toxic shock syndrome. All isolates were group A streptococcus, and all produced the toxin associated with scarlet fever. It has been proposed that this toxin plays a major role in the pathogenesis of this disease.

ANTIMICROBIAL THERAPY

The group A streptococci are susceptible to penicillin, which remains the drug of choice. For patients allergic to penicillin, erythromycin can be used. For patients who have a history of rheumatic fever, prophylactic doses of penicillin are given to prevent any recurrent infections that may cause additional damage to the heart valves.

Laboratory Diagnosis

Colonies of *S. pyogenes* on blood agar are small, transparent, and smooth with a well-defined area of β hemolysis. A Gram stain will reveal gram-positive cocci with some short chains. Examination of Gram stains of upper respiratory speci-

mens or skin swabs is of little value, because these areas have considerable amounts of normal gram-positive cocci.

An essential step in the diagnosis of streptococcal pharyngitis is proper sampling. The tongue should be depressed, and the swab rubbed over the posterior pharynx and each tonsillar area. If exudate is present, it should also be touched with the swab. Care should be taken to avoid the tongue and uvula.

Transport media are not required for normal conditions. The organism is resistant to drying and can be recovered from swabs after several hours of holding. A blood agar plate is inoculated and streaked for isolation. Incubation should be at 35 to 37°C either in ambient air or under anaerobic conditions. Studies have shown that the normal respiratory flora tend to overgrow the β-hemolytic streptococci when incubated in increased CO_2. Several selective media, such as sheep's blood agar containing trimethoprim-sulfamethoxazole, have been recommended for better recovery of β-hemolytic streptococci from throat cultures. The plate is observed after 24 hours for the presence of β-hemolytic colonies. If none are found, incubation should continue for 24 hours longer before the culture is read as negative. Suspect colonies can be Lancefield-typed using serologic methods, which will give a definitive identification, or biochemical tests can be performed. The correlation between the presumptive identification using biochemical methods and the definitive serologic method is high. Therefore, many laboratories utilize the less expensive biochemical methods. A key test that should be done is bacitracin susceptibility or PYR hydrolysis. *S. pyogenes* is susceptible to bacitracin and hydrolyzes PYR, whereas the other β-hemolytic groups are resistant to bacitracin and are PYR negative.

When the isolate's origin is not from the throat (i.e., blood, sputum), additional tests should be part of the early identification scheme. In this case, hippurate hydrolysis or the CAMP test, bile esculin test, and NaCl broth growth should also be included. The reactions shown by some of the catalase-negative, gram-positive cocci to various biochemical tests are outlined in Table 11–3.

Some immunologic tests used to detect past infection with *S. pyogenes* include antistreptolysin-O, anti-DNase, antistreptokinase, and anti-hyaluronidase titers.

Streptococcus agalactiae (Group B Streptococci)

Antigenic Structure

All strains of *S. agalactiae* have the group B–specific antigen, an acid-stable polysaccharide located in the cell wall. Additionally, there are three major serotypes, labeled I, II, and III. These type-specific antigens are capsular polysaccharides and can be detected by precipitin tests. The terminal position in the repeating unit is composed of sialic acid, which is an important virulence factor.

Virulence Factors

The capsule is another important virulence factor in group B streptococcal infections. Antibodies against the type-specific antigens protect mice against strains of *S. agalactiae* with the homologous polysaccharide in the capsule. The capsule prevents phagocytosis but is ineffective after opsonization. Sialic acid appears to be the most significant component of the capsule. Studies with mutant strains of *S. agalactiae* that lacked the sialic acid component of the capsule showed that loss of capsular sialic acid was associated with loss of virulence. The capsular sialic acid appears to be a critical virulence determinant. It is postulated that the sialic acid in the surface of the bacterial cell inhibits activation of the alternative pathway of complement.

Other products produced by *S. agalactiae* include a hemolysin, the CAMP factor, neuraminidase, deoxyribonuclease, hyaluronidase, and protease. There is no evidence that any of these products plays a role in the virulence of this organism.

Clinical Infections

Group B streptococci have been known for many years as the cause of mastitis in cattle. It was not until Lancefield defined streptococcal classification that their role in human disease was recognized.

Streptococcus agalactiae is a significant cause of infections in the newborn. Two clinical syndromes are used to describe neonatal group B streptococcal disease: early-onset infection and late-onset infection. Most infections of infants occur in the first 3 days after birth, usually within 24 hours. This infection is commonly associated with obstetric complications and premature birth. It is often a pneumonia or meningitis with bacteremia. The mortality rate is high, and death usually occurs if treatment is not started quickly. The most important determining factor seems to be the presence of group B streptococci in the vagina of the mother.

Late-onset infection occurs between 1 week and 3 months after birth and usually is seen as a meningitis. This infection is uncommonly associated with obstetric complications. Also, the organism is rarely found in the mother's vagina prior to birth. The mortality rate is considerably less than that of early-onset disease, but it is high enough to be of serious concern.

The incidence of group B streptococcal infections drops dramatically after the neonatal period. In adults, infection affects two types of patients. The first is the young, previously healthy woman who becomes ill after childbirth or abortion; endometritis and wound infections are most common. The second type of patient is the elderly person with a serious underlying disease or immunodeficiency.

The drug of choice for treating group B infections is penicillin, although this group is less susceptible to penicillin than group A streptococci. The clinical response to antibiotic therapy is often poor in spite of the heavy doses given. Some clinicians recommend a combination of ampicillin and an aminoglycoside for treating group B streptococcal infections.

Laboratory Diagnosis

Group B streptococci grow on blood agar as grayish white mucoid colonies surrounded by a small zone of β hemolysis (Fig. 11–14). Group B streptococci are gram-positive cocci that form short chains in clinical specimens and longer chains in culture. Presumptive identification is based on biochemical reactions. The most useful test is hippurate hydrolysis or the CAMP test. Table 11–3 lists the reactions of *S. agalactiae* in various tests. These tests enable the organism to be readily differentiated from other β-hemolytic streptococcal isolates. Figure 11–15 demonstrates how bacitracin and the CAMP test can be used to differentiate *Streptococcus*. The definitive identification can be made by extracting the group antigen and reacting it with specific

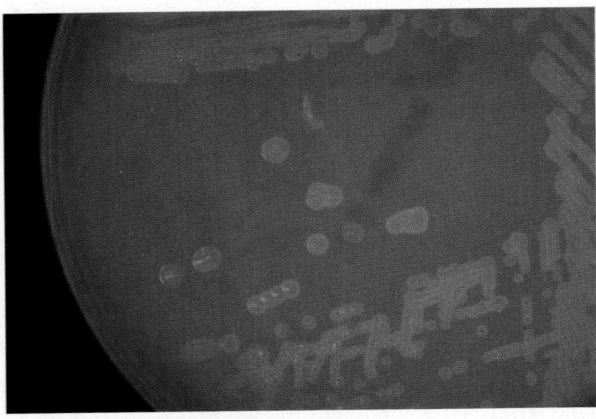

■■■■■ F i g u r e   1 1 – 1 4

Streptococcus agalactiae colony growing on blood agar.

anti–group B antisera in a precipitin or agglutination procedure.

Other Groups

Groups C and G

In spite of their different Lancefield antigens, groups C and G are now believed to belong to the same species. The name *Streptococcus dysgalactiae* has been proposed for strains isolated from human infections. Isolates from animal sources include *S. equi* and *S equi* subsp *zooepidemicus* (group C), and *S. canis* (group G).

Group D

The group D streptococci include *S. bovis* and *S. equinus*. Until the mid-1980's, the group D streptococci were subdivided into the enterococcal and nonenterococcal groups, with the understanding that those found in the intestinal tract were part of the enterococcal group. Both groups were bile esculin positive, but the nonenterococcal organisms would not grow in a nutrient broth with 6.5% NaCl. As more became known about the molecular characteristics of each of these subgroups, however, the enterococcal group was placed in a new genus, *Enterococcus*, but the nonenterococcal group remained part of the group D streptococci. Although *S. bovis* is considered a nonenterococcal isolate, it can be found in the intestinal tract.

The group D streptococci may produce bacterial endocarditis, urinary tract infections, and other diseases, such as abscesses and wound infections. An association has been made between bacteremia due to *S. bovis* and the presence of gastrointestinal tumors. Isolation of *S. bovis* from a blood culture may be the first indication that the patient has an occult tumor. It is important to distinguish the group D streptococci from *Enterococcus,* because group D is susceptible to penicillin, whereas *Enterococcus* is usually resistant.

The group D streptococci can be presumptively identified as indicated in Table 11–3. The differentiation of nonhemolytic streptococci is outlined in Figure 11–16. Normally, hemolysis is absent (Fig. 11–17) or α hemolysis is seen, although isolates can be β-hemolytic on occasion. A key reaction of this group is that it is positive for bile esculin but fails to grow in 6.5% NaCl broth. In addition, it can be separated from *Enterococcus* with the PYR test, because group D is negative but *Enterococcus* is positive. Serotyping should be done to identify an isolate as *S. bovis* because it cannot be positively distinguished from some of the viridans group on the basis of biochemical tests alone (see Table 11–3).

Enterococcus

As already mentioned, *Enterococcus* was previously referred to as group D *Streptococcus* enterococcus. This genus is found in the intestinal tract. The species found in this genus include *E. faecalis,* which is the most common isolate, *E. faecium, E. avium,* and *E. durans*. They share a

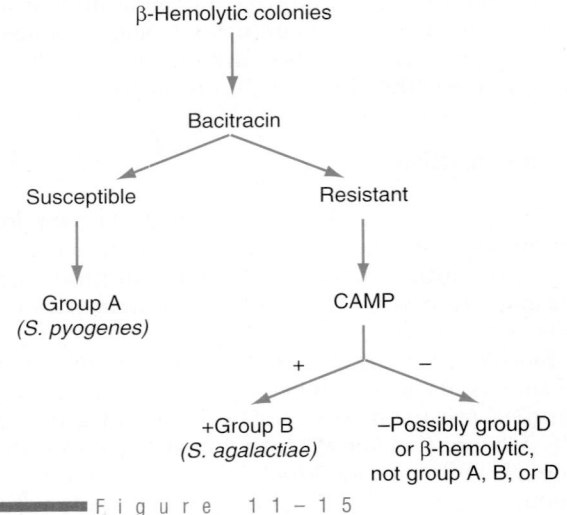

■■■■■ F i g u r e   1 1 – 1 5

Schematic diagram for differentiation of group A from group B streptococci.

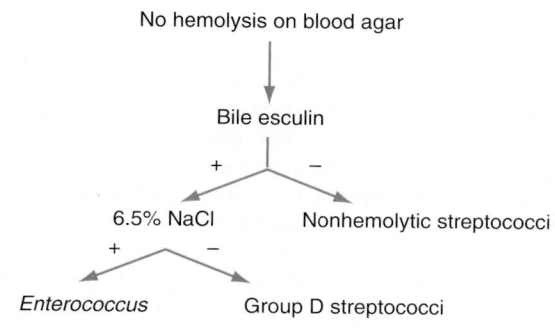

No hemolysis on blood agar

↓

Bile esculin

+ / −

6.5% NaCl   Nonhemolytic streptococci

+ / −

Enterococcus   Group D streptococci

■ F i g u r e   1 1 – 1 6

Schematic diagram for identification of nonhemolytic streptococci.

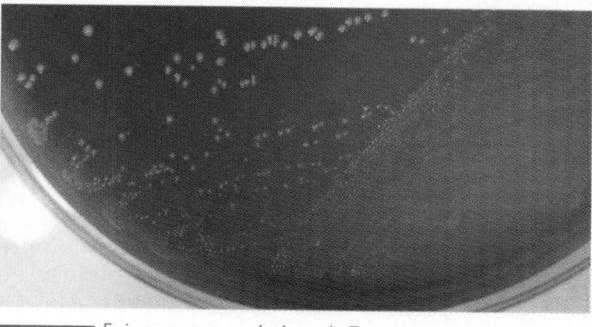

■ F i g u r e   1 1 – 1 7

Enterococcus sp growing on blood agar.

number of characteristics with the group D streptococci, including the group D antigen. They show resistance to several of the commonly used antibiotics, so differentiation with *Streptococcus* and susceptibility testing is important. The diseases caused by *Enterococcus* are similar to those seen with group D streptococcal infection.

It is not difficult to differentiate between *Enterococcus* and group D isolates. In addition to being positive for bile esculin, *Enterococcus* grows in 6.5% NaCl broth and is PYR positive. The use of bile esculin, PYR, and 6.5% NaCl to differentiate *Enterococcus* from group D streptococcus is shown in Figure 11–18. It may be worth mentioning that the catalase test result may be confusing when one is trying to differentiate *Enterococcus* species from catalase-producing *Staphylococcus* species. *Enterococcus* species can give a weakly positive (slight bubbling) catalase test reaction on a culture 24 to 48 hours old.

Streptococcus pneumoniae

Also known as the pneumococcus, *S. pneumoniae* is often isolated from a variety of infections.

Antigenic Structure

The cell wall of *S. pneumoniae* contains an antigen, referred to as *C substance*, that is similar to the C carbohydrate of the various Lancefield groups. A β-globulin in human serum, called the *C-reactive protein* reacts with this C substance to form a precipitate. This is a chemical reaction and not an antigen-antibody combination. The amount of C-reactive protein increases during inflammation and infection. The key antigen of the pneumococcus is the capsular antigen. There are some 82 different capsular types based on chemical variations of the capsular polysaccharide. Isolates from certain sources, for example, cases of lobar pneumonia, show a predominance of a particular capsular type or types. The capsule is antigenic and can be identified with appropriate antiserum. In the presence of specific anticapsular serum, the capsule swells (quellung reaction). This reaction not only allows for identification of *S. pneumoniae* but serves to specifically serotype the isolate as well.

Virulence Factors

The characteristic of *S. pneumoniae* that is clearly associated with virulence is the capsular polysaccharide. Laboratory strains that have lost the ability to produce a capsule are nonpathogenic. In addition, opsonization of the capsule renders the organism nonvirulent. Several toxins are produced, including a hemolysin, an immunoglobulin A protease, neuraminidase, and hyaluronidase. None of these has been shown to have a role in disease production.

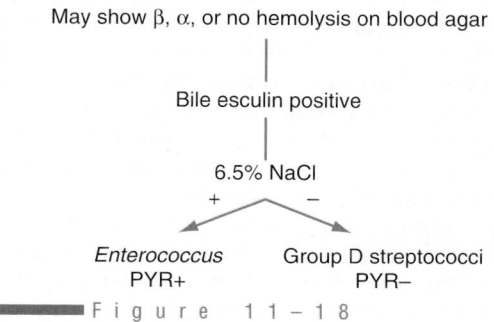

May show β, α, or no hemolysis on blood agar

|

Bile esculin positive

|

6.5% NaCl

+ / −

Enterococcus   Group D streptococci
PYR+   PYR−

■ F i g u r e   1 1 – 1 8

Schematic diagram for differentiation of group D streptococci from *Enterococcus*.

Clinical Infections

Streptococcus pneumoniae is an important human pathogen, causing pneumonia, sinusitis, otitis media, bacteremia, and meningitis. It is a common isolate in the clinical microbiology laboratory, both as a pathogen and as a member of the normal respiratory flora.

The most common cause of bacterial pneumonia, it is especially prevalent in the elderly as well as in patients with underlying disease. Of the more than 80 capsular serotypes, about a dozen account for the majority of pneumococcal pneumonia cases. Pneumonia due to *S. pneumoniae* is not usually a primary infection but rather is a result of disturbance of the normal defense barriers. Such predisposing conditions as alcoholism, anesthesia, malnutrition, and viral infections of the upper respiratory tract may lead to pneumococcal disease. The most common type of pneumococcal pneumonia is lobar pneumonia. This form is characterized by a very sudden onset with chills, dyspnea, and cough. The infection begins with aspiration of respiratory secretions, which often contain pneumococci. The infecting organisms in the alveoli stimulate an outpouring of edema fluid, which serves to facilitate the spread of the organism to adjacent alveoli. The process stops when the fluid reaches fibrous septa that separate the major lung lobes. This accounts for the "lobar" distribution of the infection, and hence the name. The majority of isolates from pneumococcal lobar pneumonia are types 1, 2, and 3.

In order for an individual to contract pneumococcal pneumonia, the organism must be present in the nasopharynx, the individual must be deficient in specific circulating antibody against the capsular type of the colonizing strain of *S. pneumoniae*, and there must be some predisposing factor such as a viral infection. The disease may be complicated by a pleural effusion that is usually sterile (empyema). The laboratory may receive fluid from a pleural aspirate for culture. An infected effusion contains many white cells and pneumococci, which are visible on Gram stain. Even with antibiotics, the mortality is relatively high (5–10%); without therapy, however, the mortality rate approaches 50%.

Other upper respiratory tract infections caused by *S. pneumoniae* include sinusitis and otitis media. *Streptococcus pneumoniae* is the most common isolate in children under the age of 3 years with recurrent otitis media.

The pneumococcus is the most common cause of bacterial meningitis in adults but also affects all age groups. Meningitis usually follows other infections with *S. pneumoniae*, such as otitis media and pneumonia. The course of the disease is rapid, and the mortality rate is near 40%. Direct smears of the spinal fluid often reveal leukocytes and numerous gram-positive cocci in pairs.

Pneumococci may also be involved in other infections, such as endocarditis, peritonitis, and bacteremia. Bacteremia often occurs during the course of a serious infection. Consequently, samples for blood culture are often taken simultaneously with sputum or a fluid aspirate.

Pneumococcal infections are usually treated with penicillin, because the majority of isolates are susceptible. Some strains have shown resistance, and these are treated with erythromycin or chloramphenicol.

A vaccine containing the polysaccharide capsular material of the most commonly encountered types is available. It is recommended for those at risk of developing pneumococcal disease (e.g., asplenic individuals, elderly, patients with cardiac or pulmonary disease). The vaccine has been successful in reducing the incidence and severity of pneumococcal disease.

Laboratory Diagnosis

MICROSCOPY

The cells are characteristically seen on Gram stain as gram-positive cocci in pairs (diplococci). The ends of the cells are slightly pointed, giving them an oval or lancet shape (Fig. 11–19). The cocci may occur singly, in pairs, or in short chains but are most often seen as pairs. As the culture ages, the Gram reaction becomes variable, with gram-negative cells seen. The capsule can be demonstrated by using a capsule stain.

CULTURAL CHARACTERISTICS

The nutritional requirements of *S. pneumoniae* are complex. Media such as brain-heart infusion agar, trypticase soy agar with 5% sheep's blood, or chocolate agar are necessary for good growth. Some isolates require increased CO_2 for growth during primary isolation. The organism can utilize a wide range of carbohydrates and is a facultative anaerobe. Isolates produce a significant zone of α hemolysis surrounding the colonies. Young cultures have a round, glistening, wet, mucoid, dome-shaped appearance (Fig. 11–20).

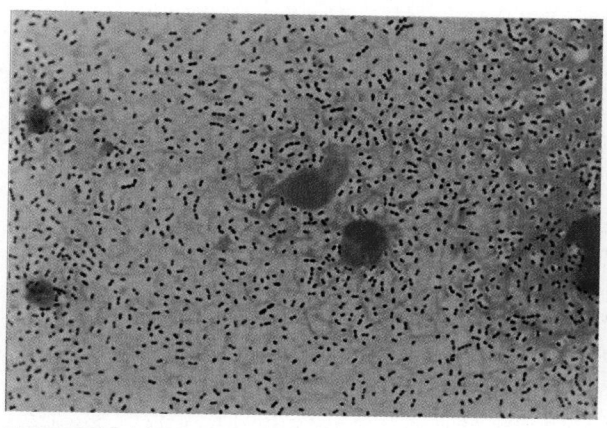

■ Figure 11-19

Gram stain of *Streptococcus pneumoniae*. Direct smear of sputum from a patient with pneumonia caused by *S. pneumoniae*.

■ Figure 11-20

Streptococcus pneumoniae colonies on blood agar. The colonies demonstrate a characteristic mucoid appearance.

As the colonies become older, autolytic changes result in a collapse of each colony's center, giving it the appearance of a coin with a raised rim. The tendency of *S. pneumoniae* to undergo autolysis can make it difficult to keep isolates alive. Clinical isolates and stock cultures require frequent subculturing (every 2 to 3 days) in order to ensure viability. The colonies may closely resemble those of the viridans streptococci, but a presumptive differentiation is not difficult to make.

IDENTIFICATION

The greatest problem in laboratory diagnosis is distinguishing *S. pneumoniae* from the viridans streptococci. The procedures used most commonly to accomplish this are optochin susceptibility, bile solubility, and the quellung reaction.

Optochin susceptibility is the most commonly used procedure of the three just listed. It takes advantage of the fact that *S. pneumoniae* is susceptible to optochin (ethylhydrocuprein hydrochloride) whereas other α-hemolytic species are resistant. A filter paper disk impregnated with optochin is placed on a blood agar plate that has been inoculated with the suspected *S. pneumoniae*. The plate is incubated at 35°C for 18 to 24 hours. If growth is inhibited around the disk, the isolate is susceptible to optochin and is presumed to be *S. pneumoniae*. The viridans streptococci are resistant to optochin and will grow up to the edge of the optochin disk (Fig. 11–21). The differentiation of α-hemolytic streptococci is outlined in Figure 11–22.

The bile solubility test evaluates the ability of *S. pneumoniae* to lyse in the presence of bile salts.

It correlates with optochin susceptibility; that is, an isolate that is *S. pneumoniae* will be optochin susceptible and bile soluble. This test also differentiates pneumococcus from the viridans streptococci. *S. pneumoniae* has an autolytic amidase that hydrolyzes the peptidoglycan cell wall layer. This enzyme is activated by surface-active agents, such as bile or bile salts (and detergents such as Dreft), resulting in lysis of the organisms. When a heavy suspension of *S. pneumoniae* is added to a solution of sodium deoxycholate, the cloudiness of the broth clears after incubation at 35°C for 3 hours. A suspension of viridans streptococci remains cloudy.

Although currently seldom used, the quellung reaction identifies an isolate as *S. pneumoniae* as well as determines its capsular type. This reac-

■ Figure 11-21

S. pneumoniae on blood agar showing susceptibility to optochin.

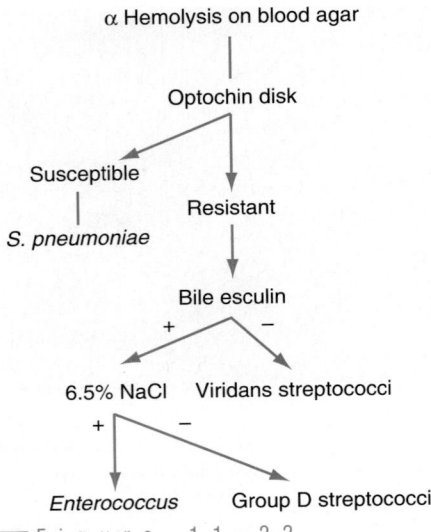

■■■■■ F i g u r e 1 1 - 2 2

Schematic diagram for the differentiation of α-hemolytic streptococci from *Streptococcus pneumoniae*.

tion can be used to identify the organism directly from sputum, spinal fluid, and other sources. The antipneumococcal serum is mixed with the material (clinical specimen or isolated colony) along with methylene blue and then examined using the oil-immersion objective. A positive reaction occurs when the pneumococci are mixed with homologous capsular antiserum: the capsule becomes more refractile and swollen.

Viridans Streptococci

Traditionally, the viridans group includes those α-hemolytic streptococci that lack Lancefield group antigens and do not meet the criteria for *S. pneumoniae*. The term *viridans* (green) is not entirely correct, because the group also includes

nonhemolytic species. Identification of the viridans streptococci to the species level is a difficult task, partly because there is not widespread agreement on a classification scheme. This lack of agreement among taxonomists results in confusion for the student and the medical technologist at the bench. Those species commonly isolated from human infections are listed in Table 11–4.

Clinical Infections

The viridans streptococci are oropharyngeal commensals. The facts that they are normal upper respiratory flora and that they are relatively susceptible to most antibiotics have kept them from being identified down to the species level. Although their virulence is low, they cause disease if host defenses are compromised. Viridans streptococci are the most common cause of subacute bacterial endocarditis. They have also been implicated in meningitis, dental caries, abscesses, osteomyelitis, and empyema. Those that cause endocarditis produce dextran, which may allow the organism to adhere to the damaged vascular endothelium. Generally, the course of endocarditis is very slow; symptoms may be present for weeks or months. Individuals whose heart valves have been damaged by rheumatic fever are especially susceptible to endocarditis due to viridans streptococci.

Treatment of viridans streptococci infections is with penicillin. Although some resistant strains have been reported, most remain susceptible.

Laboratory Diagnosis

Viridans streptococci show typical *Streptococcus* characteristics on Gram stain. The colonies produced are small and are surrounded by a

■■■■■ T A B L E 1 1 - 4
COMMONLY ISOLATED SPECIES OF VIRIDANS STREPTOCOCCI

	Mannitol	Sorbitol	Voges-Proskauer	Arginine	Esculin	Urease
S. mutans	+	+	+	−	+	−
S. salivarius	−	−	+	−	+	+/−
S. sanguis	−	−	−	+	+	−
S. mitis	−	−	−	−	−	−
S. anginosus	−	−	+	+	+	−

Modified from Facklam RR, Washington JA: *Streptococcus* and related catalase-negative gram-positive cocci. In Balows A, et al (eds): Manual of Clinical Microbiology, 5th ed. Washington, DC: American Society for Microbiology, 1991, p 250.

zone of α hemolysis. Some isolates are non-hemolytic. They are bile insoluble and optochin resistant, characteristics that distinguish them from *S. pneumoniae*. The lack of β hemolysis separates the viridans streptococci from groups A, B, C, and G. Perhaps the two groups that might be confused with the viridans streptococci are *Enterococcus* and group D *Streptococcus*. Growth in 6.5% NaCl broth differentiates the viridans group from *Enterococcus,* because the viridans streptococci fail to grow (see Fig. 11–22). Some strains of *S. salivarius* can be misidentified as *S. bovis,* however, because a significant number of *S. salivarius* isolates are bile esculin positive. The *S. milleri* group is composed of strains that may have A, C, F, G, or no Lancefield antigen. These are minute colony types showing α, β, or no hemolysis. A positive Voges-Proskauer test result and a negative PYR test result identify a β-hemolytic isolate as *S. milleri. Streptococcus milleri* has been isolated from abscesses and other pyogenic infections but is rarely involved in endocarditis. A single name, *Streptococcus anginosus,* has been proposed for those organisms belonging to the *S. milleri* group.

The speciation of members of the viridans streptococci group involves performing a considerable number of biochemical tests. This makes it difficult for identifying isolates to the species level without the use of a commercial system developed specifically for that purpose. Several commercial systems identify the viridans streptococci, but no one system can identify all possible species. Also, such systems are relatively expensive.

Nutritionally Variant Streptococci

A subgroup of the viridans streptococci that are nutritionally deficient have been isolated from patients with endocarditis and otitis media. These bacteria are also known as *pyridoxal-dependent* or *vitamin B₆–dependent, thiol-dependent,* and *symbiotic* streptococci. Pyridoxal is not present in most liquid and solid bacteriologic media. This group of organisms were found as satellites around an organism that produces pyridoxal. Nutritionally variant streptococci (NVS) may be seen as satellites around colonies of staphylococci, *Escherichia coli, Kelbsiella* spp, *Enterobacter* spp, and yeasts.

The NVS colonies are small, measuring 0.2 to 0.5 mm in diameter. On Gram-staining, the morphology can vary from that of classic gram-positive streptococci to gram-negative or gram-variable pleomorphic forms. The nutritional state of the organisms determines the state of morphology, with those in close proximity to pyridoxal-producing helper bacteria showing typical *Streptococcus* morphology. As the optimal concentrations or required nutrients decrease, the cells become pleomorphic, even showing globular and filamentous forms.

Many manufacturers of bacteriologic media now include pyridoxal (vitamin B₆) in their media used to grow *Streptococcus* in order to detect NVS. Isolates of NVS have been reported to resemble existing species of viridans streptococci rather than to form a separate species. Most isolates are susceptible to penicillin, but resistant strains have appeared. Susceptibility to other antibiotics (i.e., erythromycin, penicillin in combination with an aminoglycoside) is the norm.

Nutritionally variant streptococci are part of the normal oral flora and may cause endocarditis. Cases of otitis media and wound infections have also been found that were caused by NVS.

When a laboratory finds positive Gram reactions but negative cultures, NVS should be considered, especially if the Gram stain showed variable morphology and staining characteristics. The specimen should be cultured on a pyridoxal-supplemented medium or plated with a staphylococcal streak, and great care should be taken with inspection of the plates. The NVS are identified by their requirement for pyridoxal or by demonstrating satellitism. In addition, the NVS are bile esculin negative, do not grow in 6.5% NaCl, and are PYR positive. The viridans streptococci are PYR negative.

STREPTOCOCCUS-LIKE ORGANISMS

Streptococcus-like organisms are genera that resemble *Streptococcus* and *Enterococcus* in microscopic and colonial morphology. There is a particular resemblance to the viridans streptococci on blood agar. They may be recognized when antibiotic susceptibility testing of a "streptococcal" isolate shows it to be vancomycin resistant. Because the streptococci show universal sensitivity to vancomycin, further investigation is done, which shows the isolate to be a nonstreptococcal organism. The vancomycin-resistant, gram-positive coccus isolate is likely a member of *Leuconostoc* or *Pediococcus. Aerococcus* is normally susceptible to vancomycin.

Aerococcus

The only recognized species is *Aerococcus viridans*. It is a gram-positive, nonmotile, spherical cell that tends to form tetrads when grown in broth media. On blood agar the colonies are α-hemolytic and resemble viridans streptococci. *Aerococcus* has been isolated from dust, meat, raw vegetables, and various environmental sources, including hospital rooms. It has also been found in the upper respiratory tract and on the skin. It had been considered an airborne contaminant but in later years has been shown to cause endocarditis, urinary tract infections, and other infections. Aerococcal infection is rare and is normally found in immunocompromised patients. As with so many former "nonpathogens," the incidence can be expected to increase as more is known of the organism.

Aerococcus differs from streptococci in microscopic appearance. Unlike *Streptococcus,* it does not form chains but rather tends to form tetrads. Detection usually depends on whether microscopy is performed and the tetrad arrangement of the cells is observed. Table 11–3 lists some of the biochemical characteristics of *Aerococcus* in comparison with the other gram-positive cocci. It is PYR positive and bile esculin variable, and it grows in 6.5% NaCl broth. Because of these characteristics, it may easily be confused with *Enterococcus;* however, *Enterococcus* is LAP positive.

Leuconostoc

Leuconostoc resembles *Streptococcus* and appears as coccobacilli in pairs and occasional

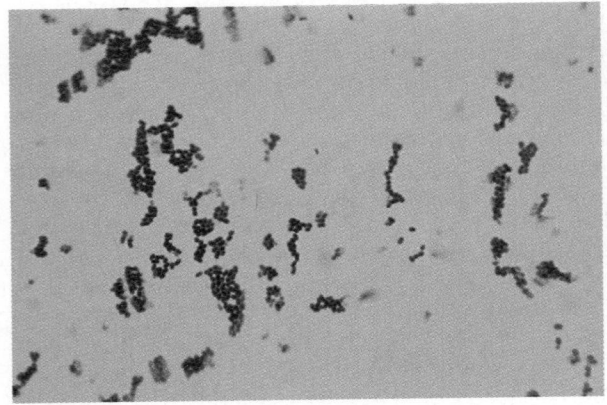

Gram stain of *Leuconostoc* spp.

■ F i g u r e   1 1 – 2 4
Leuconostoc spp colonies growing on blood agar. *Leuconostoc* spp may produce α hemolysis and may resemble viridans streptococci.

short chains (Fig. 11–23). It also resembles the viridans streptococci (Fig. 11–24) or *Enterococcus* on blood agar. Isolates are usually resistant to vancomycin. Three species of *Leuconostoc* have been isolated from human infections; they are *L. lactis, L. mesenteroides,* and *L. paramesenteroides*. The organisms are found on plants, vegetables, dairy products, wine, and sugar solutions. They have been isolated from cases of meningitis, endocarditis, and septicemia. Like *Enterococcus,* they may grow in 6.5% NaCl and they hydrolyze esculin, but they are PYR and LAP negative. The formation of gas from glucose metabolism distinguishes *Leuconostoc* from *Streptococcus.*

Pediococcus

Members of *Pediococcus* form round cells arranged in pairs, tetrads, and clusters. There are eight species, of which *P. acidilactici* and *P. pentosaceus* have been isolated from human infections. Pediococci have been isolated from saliva, stool, urine, and wounds. They are found on plants and in alcoholic beverages. Pediococci are also used in the bioprocessing and biopreservation of cheese, meats, vegetables, and soy products. In addition, they have importance as silage additives. They are very rarely seen as causing human disease, but a number of cases of septicemia and bacteremia have been documented. *Pediococcus* appears to be universally resistant to vancomycin. To confirm an organism as *Pediococcus,* a Gram stain is very helpful. Cocci found in clusters and tetrads are consistently seen. The bile esculin and LAP reactions are positive, the PYR reaction is negative, and growth is variable in 6.5% NaCl broth. In addition, the organism grows at 45°C.

Gemella

Gemella isolates are similar in colonial morphology and habitat to the viridans streptococci. Strains have been isolated from cases of endocarditis and from wounds and abscesses.

Bibliography

Barreau C, Wagener G: Characterization of *Leuconostoc lactis* strains from human sources, J Clin Microbiol 28:1728, 1990.

Bartter T, Dascal A, et al: Toxic strep syndrome: A manifestation of group A streptococcal infection. Arch Intern Med 148:1421, 1988.

Bosley GS, Wallace PL, et al: Phenotypic characterization, cellular fatty acid composition, and DNA relatedness of aerococci and comparison to related genera. J Clin Microbiol 28:416, 1990.

Campos JM: Noncultural diagnosis of group A streptococcal pharyngitis. Clin Microbiol Newsl 9:152, 1987.

Chenoweth C, Schaberg D: The epidemiology of enterococci. Eur J Clin Microbiol Infect Dis 9:80, 1990.

Christensen JJ, Vibits H, et al: *Aerococcus*-like organism, a newly recognized potential urinary tract pathogen. J Clin Microbiol 29:1049, 1991.

Christensen JJ, Korner B, Kjaergaard H: *Aerococcus*-like organism—an unnoticed urinary tract pathogen. Acta Pathol Microbiol Immunol Scand 97:539, 1989.

Colman G: *Streptococcus* and *Lactobacillus*. In Parker MT, Duerden BI (eds): Topley & Wilson's Principles of Bacteriology, Virology and Immunity. Volume 2: Systematic Bacteriology, 8th ed. Philadelphia: BC Decker, 1990.

Dascal AS, Gioseffini S, et al: PYR-positive *Aerococcus* spp. Clin Microbiol Newsl 12:38, 1990.

Facklam RR, Washington JA II: *Streptococcus* and related catalase-negative gram-positive cocci. In Balows A, et al (eds): Manual of Clinical Microbiology, 5th ed. Washington, DC: American Society for Microbiology, 1991.

Facklam RR, Collins MD: Identification of *Enterococcus* species isolated from human infections by a conventional test scheme. J Clin Microbiol 27:731, 1989.

Fischetti VA: Streptococcal M protein: Molecular design and biological behavior. Clin Microbiol Rev 2:285, 1989.

Friedland IR, Snipelisky M, Khoosal M: Meningitis in a neonate caused by *Leuconostoc* sp. J Clin Microbiol 28:2125, 1990.

Gerber MA: Comparison of throat cultures and rapid strep tests for diagnosis of streptococcal pharyngitis. Pediatr Infect Dis J 8:820, 1989.

Ginsburg I: *Streptococcus*. In Braude AI, Davis CE, Fierer J (eds): Infectious Diseases and Medical Microbiology, 2nd ed. Philadelphia: WB Saunders, 1986.

Golledge CL: Bacteremia due to *Leuconostoc* species. Clin Microbiol Newsl 11:29, 1989.

Golledge CL, Stingemore N, et al: Septicemia caused by vancomycin-resistant *Pediococcus acidilactici*. J Clin Microbiol 28:1678, 1990.

Gray BM: Streptococcal Infections. In Evans AS, Brachman PS (eds): Bacterial Infections of Humans, Epidemiology and Control, 2nd ed. New York: Plenum Medical Book Co, 1991.

Hamoudi AC, Dribar MM, et al: Clinical relevance of viridans and nonhemolytic streptococci isolated from blood and cerebrospinal fluid in a pediatric population. Am J Clin Pathol 93:270, 1989.

Hayden GF, Murphy TF, Hendley JO: Non-group A streptococci in the pharynx. Am J Dis Child 143:794, 1989.

Kellogg JA: Suitability of throat culture procedures for detection of group A streptococci and as reference standards for evaluation of streptococcal antigen detection kits. J Clin Microbiol 28:165, 1990.

Klein JO: Diagnosis of streptococcal pharyngitis: An introduction. Pediatr Infect Dis J 8:813, 1989.

Mastro TD, Spika JS, et al: Vancomycin-resistant *Pediococcus acidilactici:* Nine cases of bacteremia. J Infect Dis 161:956, 1990.

Meier FA, Centor RM, et al: Clinical and microbiological evidence for endemic pharyngitis among adults due to group C streptococci. Arch Intern Med 150:825, 1990.

Murray BE: The life and times of the enterococcus. Clin Microbiol Rev 3:46, 1990.

Park JW, Grossman O: *Aerococcus viridans* infection. Clin Pediatr 29:525, 1990.

Roddey OF, Mauney CU, et al: Comparison of immediate and delayed culture for isolation of group A streptococci. Pediatr Infect Dis J 8:710, 1989.

Ross PW: Streptococcal diseases. In Parker MT, Duerden BI (eds): Topley & Wilson's Principles of Bacteriology, Virology and Immunity. Volume 3: Bacterial Diseases, 8th ed. Philadelphia: BC Decker, 1990.

Ruoff KL: Nutritionally variant streptococci. Clin Microbiol Rev 4:184, 1991.

Ruoff KL: Recent taxonomic changes in the genus *Enterococcus*. Eur J Clin Microbiol Infect Dis 9:75, 1990.

Ruoff KL, de la Maza L, et al: Species identities of enterococci isolated from clinical specimens. J Clin Microbiol 28:435, 1990.

Ruoff KL: Gram-positive vancomycin-resistant clinical isolates. Clin Microbiol Newsl 11:1, 1989.

Ruoff KL: An update on streptococcal taxonomy. Clin Microbiol Newsl 10:1, 1988.

Shulman ST: Streptococcal pharyngitis: Clinical and epidemiologic factors. Pediatr Infect Dis J 8:816, 1989.

Swenson JM, Facklam RR, Thornsberry C: Antimicrobial susceptibility of vancomycin-resistant *Leuconostoc, Pediococcus,* and *Lactobacillus* species. Antimicrob Agents Chemother 34:543, 1990.

Turner JC, Hayden GF, et al: Association of group C β-hemolytic streptococci with endemic pharyngitis among college students. JAMA 264:2644, 1990.

Wasilauskas BL: Viridans streptococci: Methods and rationale for species identification. Clin Microbiol Newsl 9:125, 1987.

Wessels MR, Rubens CE, et al: Definition of a bacterial virulence factor: Sialylation of the group B streptococcal capsule. Proc Natl Acad Sci USA, 86:8983, 1989.

CHAPTER 12

Corynebacterium and Other Non–Spore-Forming Gram-Positive Rods

Hal S. Larsen

1. List the general characteristics of *Corynebacterium* spp, *Listeria monocytogenes,* and *Erysipelothrix rhusiopathiae.*

2. Explain the clinical significance of these organisms. Describe how the infections they cause are acquired.

3. Describe the distinctive features of the microscopic morphology in direct smears from clinical specimens and/or primary media of these organisms.

4. Name the media required for the isolation of the major pathogens discussed.

5. Describe the colonial morphology of each of the pathogens discussed.

6. List the differential procedures used to diagnose infections caused by *Corynebacterium* spp.

7. Differentiate *Listeria monocytogenes* from other non–spore-forming gram-positive rods and from streptococci.

8. Describe the motility patterns of *Listeria monocytogenes* in hanging drop and in semisolid medium.

9. Differentiate *Erysipelothrix rhusiopathiae* from other non–spore-forming gram-positive rods.

The gram-positive non–spore-forming rods comprise a heterogeneous group of organisms that include *Corynebacterium, Arcanobacterium, Rhodococcus, Listeria,* and *Erysipelothrix.* Many of these are poorly characterized, hence classification often changes. A wide range of clinical conditions results from infection with these organisms. Although their frequency of isolation in the clinical laboratory is not as great as that of many other organisms (i.e., *Escherichia coli, Staphylococcus aureus, Streptococcus*), members of this group are often seen. The frequency of isolation is increasing, and some of the lesser-known members (i.e., *Rhodococcus, Arcanobacterium*) are becoming more prominent.

CORYNEBACTERIUM

General Characteristics

The genus *Corynebacterium* consists of a large group of bacteria, animal and human pathogens as well as free-living saprophytes and plant pathogens. They are found worldwide in fresh and salt water, soil, and air. The corynebacteria are closely related to mycobacteria and nocardiae; these three groups collectively may be referred to as the CMN group. They have common cell wall structures. The classification of the diphtheroids (or coryneforms) is not well characterized. Consequently, there is a low rate of identification for clinical isolates. Even when sent to a reference laboratory, 30 to 50% of the

coryneform-like isolates are unable to be identified to the species level.

The premier pathogen of the group is *C. diphtheriae,* which has been extensively studied and is well characterized. Other species that produce disease in humans are *C. bovis* (not recognized as a species in the genus *Corynebacterium* by *Bergey's Manual of Systematic Bacteriology*), *C. ulcerans* (not recognized as a species in the genus *Corynebacterium* by *Bergey's Manual of Systematic Bacteriology*), *C. equi* (*Rhodococcus equi*), *C. haemolyticum* (*Arcanobacterium haemolyticum*), *C. xerosis, C. jeikeium* (formerly CDC Group JK), *C. pseudodiphtheriticum,* and *C. pseudotuberculosis.*

The nondiphtheria corynebacteria are commonly isolated from clinical specimens. They are usually dismissed as contaminants. As with many previously "nonpathogenic" organisms, however, the coryneforms are being isolated from a variety of body sites, especially in immunocompromised patients.

Corynebacterium diphtheriae

Physiology

The diphtheria bacillus, like other corynebacteria, is a facultative anaerobe. It grows best under aerobic conditions; with the optimal growth temperature of 37°C, although multiplication occurs within the range of 15 to 40°C. Growth requirements are complex, with eight amino acids being essential. The organism ferments glucose and maltose; producing acid but not gas. It does not produce urease but reduces nitrate to nitrite.

The growth medium markedly affects morphology and toxin production. In stained smears, diphtheria bacilli characteristically appear in palisades or as individual cells lying at sharp angles to one another in V and L formations. When grown on suboptimal media such as Loeffler coagulated serum, the cells are pleomorphic and stain irregularly with methylene blue. Club-shaped swellings and beaded forms are common. The amount of toxin produced is influenced by the growth conditions. The iron content of the medium must be growth rate limiting for full toxin production. The addition of iron to iron-starved cultures inhibits toxin production very quickly.

Corynebacterium diphtheriae is readily killed by heat and most of the usual disinfectants. It is, however, resistant to drying and remains viable in the environment for weeks.

Virulence Factors

The major virulence factor associated with *C. diphtheriae* is the diphtheria toxin. This toxin is produced by those strains of *C. diphtheriae* that are infected with a temperate bacteriophage, which carries the structural gene *(tox)* for diphtheria toxin. Nontoxigenic strains can be converted to *tox*⁺ by infection with the appropriate bacteriophage. Only toxin-producing *C. diphtheriae* causes the infection diphtheria; however, *C. ulcerans* and *C. pseudotuberculosis,* which belong to what is called the "*C. diphtheriae* group," may also produce the toxin when they become infected with the *tox*-carrying bacteriophage.

Diphtheria toxin is a simple protein of 62,000 daltons. It is exceedingly potent, being lethal for humans in amounts of 130 ng per kg of body weight. The toxicity of the toxin is due to its ability to block protein synthesis in eukaryotic cells. The toxin is secreted by the bacterial cell and is nontoxic until exposure to trypsin. The trypsinization results in two polypeptide fragments, A and B, which are linked together by a disulfide bridge. Both fragments are necessary for cytotoxicity. Fragment A is responsible for the toxicity; fragment B binds to receptors on the eukaryotic cells and mediates the entry of fragment A into the cytoplasm following cleavage by cellular enzymes. On reaching the cytoplasm, fragment A disrupts protein synthesis. Fragment A splits

nicotinamide adenosine dinucleotide (NAD) to form nicotinamide and adenosine diphosphoribose (ADPR). ADPR binds to and inactivates elongation factor 2 (EF-2), an enzyme required for elongation of polypeptide chains on ribosomes. The reaction can be summarized as follows:

$$NAD^+ + EF\text{-}2 \rightleftharpoons ADPR\text{-}EF\text{-}2 + nicotinamide + H^+$$
$$\text{active} \qquad\qquad \text{inactive}$$

Toxin production depends on a lysogenic state in which the bacterium is infected with a bacteriophage that carries the *tox*⁺ gene, which codes for the production of diphtheria toxin.

Production of the toxin in vitro depends on a number of environmental conditions: an alkaline pH (7.8–8.0), oxygen, and, most importantly, iron concentration in the medium. The amount of iron needed for optimal toxin production is less than that needed for optimal growth. Strangely, the toxin is released in significant amounts only when the available iron in the culture medium is exhausted.

Clinical Infections

Diphtheria, which occurs in two forms, respiratory and cutaneous, is found worldwide but is uncommon in North America and Western Europe. Those cases that occur are invariably in unimmunized populations. Humans are the only natural hosts for *C. diphtheriae*. The organism is carried in the upper respiratory tract and is spread by droplet infection or hand-to-mouth contact. The incubation period averages 2 to 5 days. The illness begins gradually and is characterized by low-grade fever, malaise, and a mild sore throat. The most common site of infection is the tonsils or pharynx. The organisms rapidly multiply on the epithelial cells and trigger an inflammatory reaction. The infecting toxigenic strain of *C. diphtheriae* produces toxin locally, causing tissue necrosis and exudate formation. This combination of cell necrosis and exudate forms a very tough gray to white pseudomembrane that attaches to the tissues. It may appear on the tonsils and then spread downward into the larynx and trachea. There is the potential for suffocation if the membrane spreads and blocks the air passage or if it is dislodged, perhaps as the result of sampling for a throat culture.

The toxin is also absorbed and produces systemic effects. The systemic manifestations of the

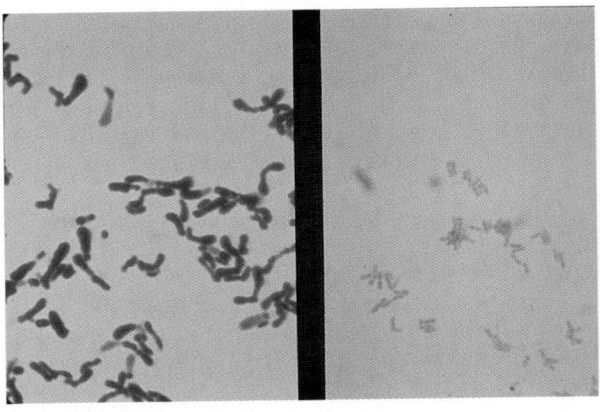

■ F i g u r e   1 2 – 1

Left, Microscopic Gram stain of diphtheroid. *Right*, Microscopic Loeffler methylene blue stain of *Corynebacterium* spp. (Courtesy of Cathy Bissonette.)

disease involve the kidneys, heart, and nervous system, although all tissues possess the receptor for the toxin and may be affected. Death is often a result of cardiac failure. Another effect of the toxin is a demyelinating peripheral neuritis, which may result in paralysis following the acute illness.

Other nonrespiratory sites may be infected, although much less often than the upper respiratory tract. In the cutaneous form of diphtheria, which is prevalent in the tropics, the toxin is also absorbed systemically, but systemic complications are less common than from upper respiratory infections with *C. diphtheriae*.

Diphtheria is treated by prompt administration of antitoxin. Commercial diphtheria antitoxin is produced in horses. Consequently, hypersensitivity to horse serum precludes its administration. Approximately 10% of patients who receive the antitoxin develop an allergic reaction to the horse serum. Antibiotics have no effect on toxin that is already circulating, but they do serve to eliminate the focus of infection as well as to prevent the spread of the organism. The drug of choice is penicillin. Erythromycin is used for penicillin-sensitive individuals.

Laboratory Diagnosis

MICROSCOPY

Corynebacterium diphtheriae is a gram-positive, nonsporulating, nonmotile bacillus. The organism is highly pleomorphic and appears in palisades or as individual cells lying at sharp angles to another in V and L formations. This particular arrangement associated with *C. diphtheriae* has been described by westerners as "Chinese characters" (Fig. 12–1), although it may be demonstrated by other *Corynebacterium* species. The organisms often stain irregularly, especially when stained with methylene blue, giving them a beaded appearance. The metachromatic areas of the cell, which stain more intensely than other parts, are called *Babès-Ernst granules*. They represent accumulation of polymerized polyphosphates. Their presence indicates accumulation of food reserves and varies with the type of medium and the metabolic state of the cells.

CULTURAL CHARACTERISTICS

Although *C. diphtheriae* will grow on nutrient agar, better growth is obtained on a medium containing blood or serum, such as Loeffler serum agar or Pai slant. Characteristic microscopic morphology is demonstrated well when organisms are grown on Loeffler medium. On blood agar (Fig. 12–2), the organism may have a very small zone of hemolysis.

Tinsdale agar, which contains sheep's blood, bovine serum, cystine, and potassium tellurite is used as both a selective and differential medium. When grown on Tinsdale agar, corynebacteria form black or brownish colonies. This appearance

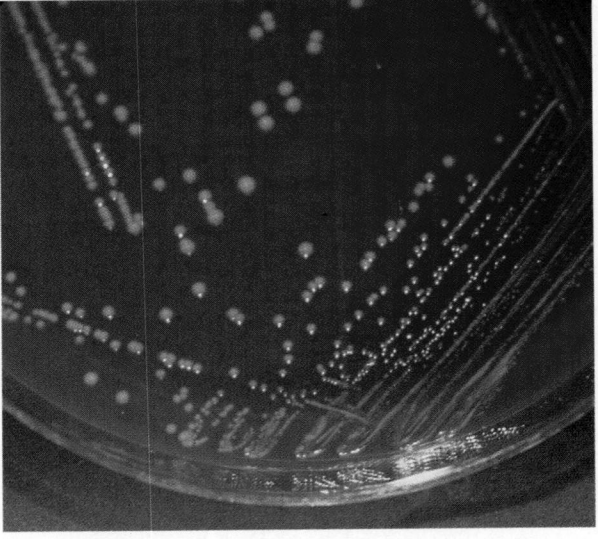

■ F i g u r e   1 2 – 2

Corynebacterium diphtheriae growing on BAP. (Courtesy of Cathy Bissonette.)

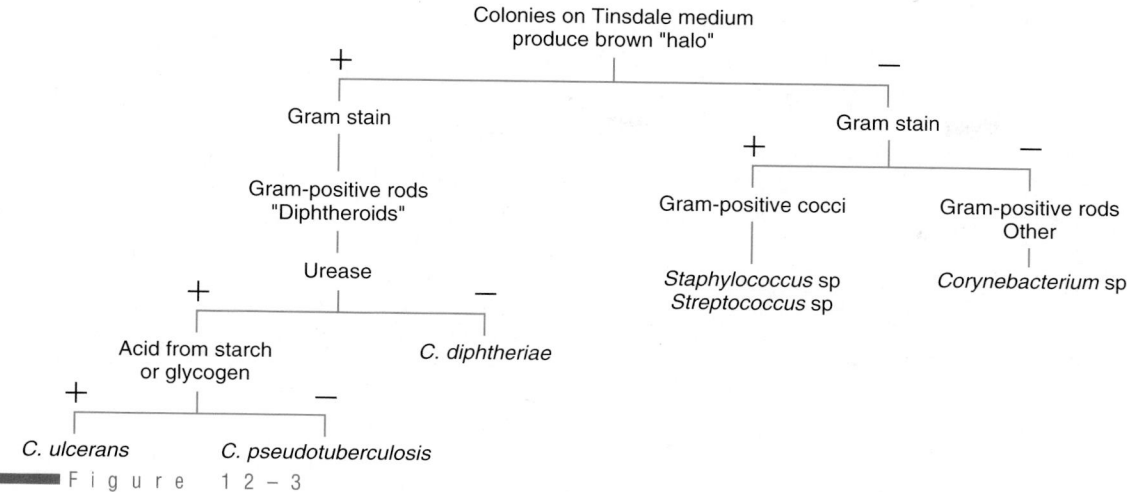

■■■ F i g u r e   1 2 – 3

A schematic diagram of the presumptive identification of *Corynebacterium diphtheriae.*

is not unique for *C. diphtheriae,* however, so care must be taken not to presumptively identify other genera that produce black colonies (*Staphylococcus* and *Streptococcus*) as *Corynebacterium.* A brown halo surrounding the colony is a useful differentiating feature, because only *C. diphtheriae, C. ulcerans,* and *C. pseudotuberculosis* produce a brown halo on Tinsdale agar. Media containing tellurite (e.g., cystine-tellurite agar) can be used to isolate *C. diphtheriae.* Tellurite salts present in the medium inhibit the growth of most normal respiratory flora but *C. diphtheriae* is able to grow. Colonies of *C. diphtheriae* and *C. ulcerans* are gray or gray-black. *Listeria* may develop gray colonies.

Three biotypes of *C. diphtheriae* can be distinguished by their growth characteristics on Tinsdale agar: *mitis, intermedius,* and *gravis.* The colonies of each differ in size and color. Practically, however, there is little reason to classify an isolate by biotype, because there appears to be little, if any, correlation between disease severity and biotype.

IDENTIFICATION

The biochemical identification of medically important corynebacteria is outlined in Table 12–1. Tinsdale medium is useful for differentiations because only *C. diphtheriae, C. ulcerans,* and *C. pseudotuberculosis* form a brown halo. *Corynebacterium diphtheriae* is distinguished from the other two species by its lack of urease production. A schematic diagram to presumptively identify *C. diphtheriae* is shown in Figure 12–3.

TEST FOR TOXIGENICITY

The identification of an isolate as *C. diphtheriae* does not mean that the patient has diphtheria. Diagnosis of diphtheria depends on showing that the isolate produces diphtheria toxin. This can be done by either in vivo or in vitro testing. In vivo testing is rarely done because the in vitro methods are reliable, less expensive, and free from the need to use animals. The procedure used for in vivo testing is discussed only to give historical perspective.

The in vivo test to determine toxin production by an isolate is carried out in guinea pigs. Approximately 24 hours prior to the test, one guinea pig is injected with diphtheria antitoxin. The next day, the protected guinea pig and an untreated guinea pig are injected with a suspension of the suspected organism prepared from Loeffler slants. If the isolate produces diphtheria toxin, the untreated animal will die within 3 to 5 days, and the antitoxin-treated animal will survive.

The in vitro diphtheria toxin detection procedure is an immunodiffusion test first described by Elek. In the Elek test, organisms (controls and unknowns) are streaked on media of low iron content to optimize toxin production. The organisms are each streaked in a single straight line parallel to each other and 10 mm apart on an Elek plate. A filter paper strip impregnated with diphtheria antitoxin is laid along the center of the plate on a line at right angles to the lines of control and unknown organisms (Fig. 12–4). The plate is then incubated at 35°C and examined after 18, 24, and 48 hours. Lines of precipitation

T A B L E 1 2 – 1
IDENTIFICATION OF CORYNEBACTERIA*

Characteristic	C. diphtheriae	C. ulcerans	C. pseudotuberculosis	C. xerosis	C. jeikeium (Gp JK)	C. urealyticum (Gp D-2)	C. pseudodiphtheriticum	C. striatum	C. kutscheri
Tinsdale halo	+	+	+	–	–	–	–	–	–
Catalase	+	+	+	+	+	+	+	+	+
β Hemolysis	V	V	+	–	–	–	–	–	–
Nitrate reduction	+	–	V	+	–	–	+	+	+
Urease	–	+	+	–	–	+	+	–	+
Hydrolysis									
Gelatin	–	+†	–	–	–	–	–	–	–
Esculin	–	–	–	–	–	–	–	–	+
Carbohydrate Fermentation									
Glucose	+	+	+	+	+	–	–	+	+
Maltose	+	+	+	+	V	–	–	–	+
Sucrose	–	–	–	+	–		–	+	+

*+, present or positive; –, absent or negative; V, variable.
†25°C.

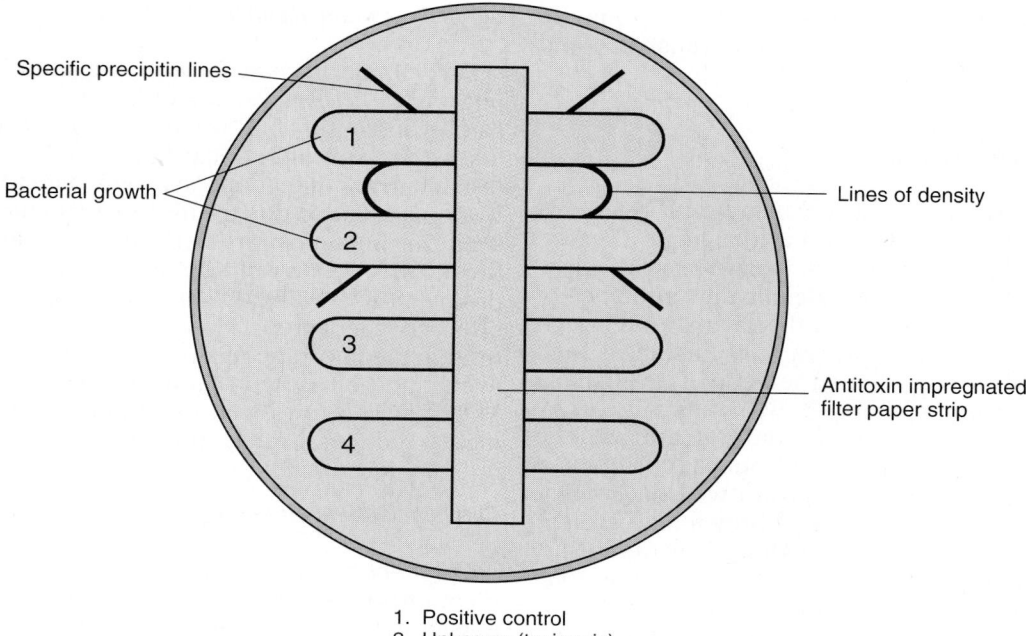

1. Positive control
2. Unknown (toxigenic)
3. Negative control
4. Unknown (nontoxigenic)

■ F i g u r e   1 2 – 4

Colonial Elek test for toxin-producing strains of *Corynebacterium diphtheriae.*

are best seen by transmitted light against a dark background. The white precipitin lines start about 4 to 5 mm from the filter paper strip and are at an angle of about 45 degrees to the line of growth. If an isolate is positive for toxin production, and it is placed next to the positive control, the toxin line of the positive control should join the toxin line of the positive unknown to form an *arch of identity* (see Fig. 12–4). The Elek test requires that reagents and antisera be carefully controlled and titrated. For this reason, and because of the difficulty of the test, it is performed only in certain reference laboratories.

Other Corynebacteria

Corynebacterium jeikeium (CDC Group JK)

Named after Johnson and Kaye, who first reported human infections caused by it, *C. jeikeium* shows a high degree of antimicrobial resistance. These organisms appear to be part of the normal skin flora. Infections have been limited to those patients who are immunocompromised or have undergone invasive procedures. The presence of catheters or prosthetic devices also contributes to infection with *C. jeikeium,* and in fact, it is the most common cause of diphtheroid prosthetic valve endocarditis in adults. *Corynebacterium jeikeium* has been reported to be typically resistant to a wide range of antimicrobials. Most strains are susceptible to vancomycin but show variable susceptibility to tetracycline and erythromycin.

Corynebacterium urealyticum (CDC Group D-2)

Corynebacterium urealyticum (CDC Group D-2) has now been described as a urinary pathogen. The species is a slow-growing organism, so cultures must incubate at least 48 hours before growth is detected. Urine isolates with very small, nonhemolytic, white colonies that have characteristic diphtheroid morphology are likely *C. urealyticum* suspects. The isolate should be catalase positive and should rapidly produce urease (within minutes following inoculation on a Christensen urea slant) in order to be presumptively called *C. urealyticum.* In addition, *C. urealyticum* does not ferment glucose. It has also shown resistance to a wide variety of antimicrobials, such β-lactam, aminoglycoside, and macrolide.

Corynebacterium ulcerans

A veterinary pathogen causing mastitis in cattle and other domestic and wild animals, *C. ulcerans* has been isolated from patients with diphtheria-like illness. A significant number of isolates produce the diphtheria toxin, although the amount of toxin elaborated is much less than by *C. diphtheriae*. Human infections are usually acquired through contact with animals or by ingestion of unpasteurized dairy products. The organism has been isolated from skin ulcers and exudative pharyngitis. It grows well on Loeffler and Tinsdale media, giving a brown halo around the colonies on Tinsdale medium. The organisms grow well on blood agar and show a narrow zone of hemolysis. Unlike *C. diphtheriae*, *C. ulcerans* does not reduce nitrate. It gives a positive gelatin reaction at room temperature. It may be that this organism, which has not been fully accepted as a species, has been classified as a subgroup of *C. diphtheriae*.

Corynebacterium pseudotuberculosis

Closely resembling *C. diphtheriae* on blood agar and Gram morphology, *C. pseudotuberculosis* also gives a brown halo on Tinsdale medium. It is urease positive and gelatin negative. It produces a dermonecrotic toxin that causes death of a variety of cell types. Like *C. ulcerans*, *C. pseudotuberculosis* is a veterinary pathogen. Human infections have usually been associated with contact with sheep. The organism is susceptible to penicillin and erythromycin.

Corynebacterium xerosis

Corynebacterium xerosis is commonly found on skin and mucocutaneous sites. Opportunistic infections associated with this organism include prosthetic valve endocarditis, bacteremia associated with intravenous catheters, postsurgical wound infections, and pneumonia. Human infection with *C. xerosis* is rare, and invariably, affected patients are immunosuppressed. This organism grows well on blood agar and forms pigmented (yellow to tan) colonies.

Corynebacterium pseudodiphtheriticum

Part of the normal flora of the human naso-pharynx, *C. pseudodiphtheriticum* very rarely causes infection, but when infection occurs, it takes the form of endocarditis. Respiratory and urinary tract infections as well as cutaneous wound infections due to this organism have been seen in immunosupressed patients, including those with acquired immune deficiency syndrome (AIDS). *C. pseudodiphtheriticum*, unlike other corynebacteria, does not show the characteristic pleomorphic morphology. The cells stain evenly and often lie in parallel rows (palisades). The species grows well on standard laboratory media, reduces nitrate, and hydrolyzes urea.

Corynebacterium striatum

Corynebacterium stratum is a slow-growing, pleomorphic species that often produces a yellowish green soluble pigment. It is found in the naso-pharynx and is a rare cause of infection.

Corynebacterium kutscheri

Primarily an animal pathogen, *C. kutscheri* microscopically resembles *C. diphtheriae*, but the colonies are yellow to gray on blood agar.

Arcanobacterium

Arcanobacterium haemolyticum was formerly known as *Corynebacterium haemolyticum*. This organism produces small colonies on sheep's blood agar that demonstrate a narrow zone of hemolysis after 24 hours of incubation that is similar in appearance to the β-hemolytic streptococci. It has been recovered from patients with pharyngitis and so must be distinguished from *C. diphtheriae* and *C. ulcerans* as well as group A streptococcus. *Arcanobacterium haemolyticum* is catalase negative, whereas *C. diphtheriae* and *C. ulcerans* are catalase positive. A Gram stain of the isolated colony in question quickly eliminates the possibility of group A streptococcus, because *A. haemolyticum* is a gram-positive rod.

Rhodococcus

Rhodococcus equi (formerly known as *Corynebacterium equi*) may demonstrate filaments, some with branching. The colonies resemble *Klebsiella* on blood agar and form a salmon-pink pigment upon prolonged incubation, especially at room temperature. The species may be partially acid fast. Biochemical identification is difficult because it does not ferment carbohydrates and shows a variable reaction to a number of characteristics (i.e., nitrate reduction, urease). A major clue to identification of *Rhodococcus* is the salmon-pink pigment, followed by a Gram stain showing characteristic diphtheroid gram-positive rods with traces of branching.

Rhodococcus equi is found in soil. It is the cause of respiratory infection in animals. Human infection is rare, although an increased incidence in immunosuppressed patients, particularly AIDS patients, is being reported.

Undesignated CDC Coryneform Groups

Several coryneform bacilli remain characterized by CDC group numbers and letters while awaiting proper species designation. All of these organisms have been isolated from a wide range of clinical samples, and they should be regarded as potential nosocomial pathogens or as opportunistic pathogens in the immunocompromised patient.

Rothia dentocariosa

A member of the normal human oropharyngeal flora, *R. dentocariosa* may be found in saliva and in supragingival plaque. It has been isolated from patients with endocarditis. Microscopically, this organism resembles coryneform bacilli, producing gram-positive short rods but also producing branching filaments that resemble those of facultative actinomycetes. When placed in broth, however, the species produces coccoid cells, a characteristic differentiating it from the *Actinomyces*. *Rothia* is catalase and nitrate positive, nonmotile, and urease negative.

LISTERIA MONOCYTOGENES

General Characteristics

The genus *Listeria* comprises seven species, of which *L. monocytogenes* is the only human and animal pathogen. It is a coccobacillus that often appears coccoid in culture. It is gram-positive, aerobic, and nonsporulating. *Listeria monocytogenes* grows well on routine laboratory media, producing β-hemolysis on sheep's blood agar. Colonies are small, smooth, and translucent, and they show a very narrow ring of hemolysis. They may be confused with group B streptococci because the resemblance is striking. *Listeria monocytogenes* is widespread in the environment. It has been recovered from soil, water, vegetation, and animal products. It has also been isolated from crustaceans, flies, and ticks. It has long been known to cause illness in many species of wild and domestic animals, including sheep, cattle, swine, horses, dogs, cats, rodents, birds, and fish.

The first human infection was described relatively recently (1926). Since then, the incidence has increased somewhat, and listeriosis is now recognized as an uncommon but serious infection primarily of neonates, pregnant women, and immunocompromised hosts. Infection may also occur in healthy individuals.

Physiology

The optimal growth temperature for *L. monocytogenes* is 30 to 35°C, but growth occurs over a wide range (0.5–45°C). It can grow in a high salt concentration (up to 10% NaCl). A source of carbohydrate is essential for growth. The organism prefers a slightly increased CO_2 tension for isolation. It is catalase positive, which differentiates it from *Streptococcus,* and motile at room temperature, which, along with β hemolysis, excludes the corynebacteria. In wet mount preparations, *L. monocytogenes* exhibits "tumbling motility" when viewed microscopically. The use of motility medium demonstrates the characteristic "umbrella" pattern when the organism is incubated at room temperature (25°C) but not at 35°C (Fig. 12–5).

Listeria monocytogenes gives a positive CAMP reaction, similar to that of group B streptococci when *Staphylococcus aureus* is utilized to augment hemolysis. A more pronounced CAMP reaction is seen with *L. monocytogenes* when *Rhodococcus equi* is used in place of *S. aureus.*

Virulence Factors

Listeria monocytogenes produces a number of products that have been proposed as virulence factors. These include hemolysin (listeriolysin O), catalase, superoxide dismutase, phospholipase C,

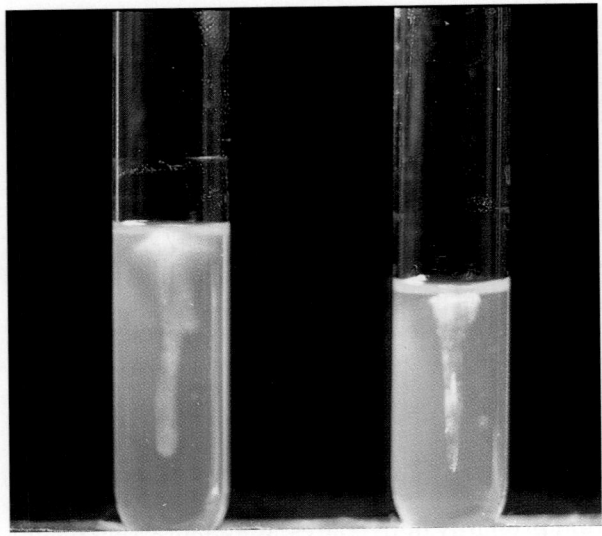

■■■■■ Figure 12-5

Umbrella motility: Listeria. Motility test for *Listeria monocytogenes* showing the typical "umbrella" pattern when this organism is incubated at room temperature. Tube on the left is positive. Tube on right is negative control.

and a surface protein, p60. Protein p60 induces phagocytosis via increased adhesion and penetration into mammalian cells. Listeriolysin O damages the phagosome membrane, effectively preventing killing of the organism by the phagocyte. The correlation between listeriolysin O production and virulence is strong. Nonhemolytic isolates are found to be avirulent and demonstrate no intracellular spread of the organism.

Clinical Infections

The infectious dose and the portal of entry of listeriosis have not been determined, but animal studies as well as analysis of human outbreaks seems to indicate that the ingestion of contaminated food with subsequent systemic spread via the intestine is likely. The clinical manifestations of listeriosis differ among patient groups. Infections of newborns and immunocompromised adults are the most common, but disease in healthy individuals, particularly in pregnant women, also occurs.

DISEASE IN PREGNANT WOMEN

During pregnancy, listeriosis is most commonly seen during the third trimester. It has been postulated that *L. monocytogenes* is responsible for spontaneous abortion and stillborn neonates. A pregnant woman with listeriosis may experience a flu-like illness, with fever, headache, and myalgia. At this point, the organism is in the blood stream and has seeded the uterus and fetus. It may progress and result in premature labor or septic abortion within 3 to 7 days. It appears that the infection is often self-limited, because the source of the infection is eliminated when birth occurs.

DISEASE IN THE NEWBORN

Infection of the neonate with *L. monocytogenes* is extremely serious. Fatality rates are high, approaching 50% if the fetus is born alive. There are two forms of neonatal listeriosis: early-onset and late-onset. Early-onset listeriosis results from an intrauterine infection that can cause illness at or shortly after birth. The result is most often sepsis. Early-onset disease may be associated with aspiration of infected amniotic fluid.

Late-onset disease occurs several days to weeks after birth. Affected infants are generally full term and healthy at birth. The disease is most likely to present as meningitis. The fatality rate is lower than in early-onset infection, although it also is a very serious, often fatal infection.

DISEASE IN THE IMMUNOSUPPRESSED HOST

Invasive listeriosis occurs most commonly in persons who are immunosuppressed or elderly, and particularly in patients receiving chemotherapy. The most common manifestations are central nervous system infection and endocarditis. Affected patients usually present with meningitis, meningoencephalitis, or sepsis. Diagnosis is made by culturing *L. monocytogenes* from the blood or spinal fluid.

Infection of apparently healthy individuals may occur via the intestinal tract when they eat food contaminated with *L. monocytogenes*. Outbreaks have occurred as a result of eating contaminated cheese, coleslaw, and chicken. The manifestation in these cases is nearly always meningitis, and the fatality rate is high. The antibiotics that have been effectively used to treat listeriosis are the penicillins, aminoglycosides, and macrolides. Resistance is not common, although some strains are resistant to one or more antibiotics.

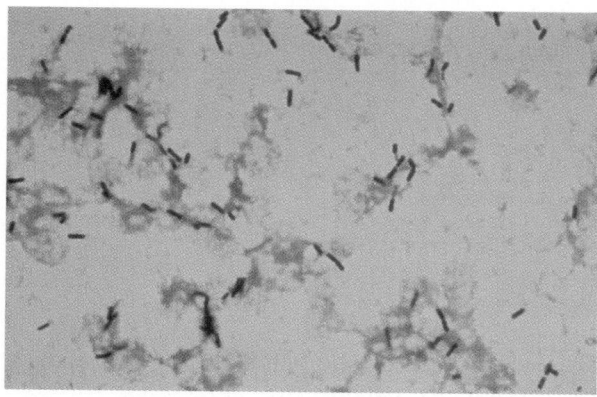

■ F i g u r e   1 2 – 6

Microscopic Gram stain of *Listeria monocytogenes* in the blood. (Courtesy of Cathy Bissonette.)

Laboratory Diagnosis

MICROSCOPY

In direct smears (Fig. 12–6), *L. monocytogenes* appears as a gram-positive coccobacillus. With subculturing, it tends to form coccoid forms. Older cultures often appear gram variable. They may be found singly, in short chains, or in palisades. Depending upon the culture conditions, *L. monocytogenes* may resemble *Streptococcus* when found in the coccoid form, and *Corynebacterium* when the bacillus forms prevail. Organisms are not usually seen on the spinal fluid smear.

CULTURAL CHARACTERISTICS

Listeria monocytogenes grows well on blood agar and chocolate agar as well as on nutrient agars and broths such as brain-heart infusion and thioglycolate. The colonies are small, round, smooth, and translucent. They are surrounded by a narrow zone of β hemolysis, which may be visualized only if the colony is removed. The colonies and hemolysis are similar to those seen with *Streptococcus agalactiae* or group B streptococci (Fig. 12–7A). Growth is normally complete in 1 or 2 days.

Because *L. monocytogenes* grows at 4°C, an unusual characteristic, a technique called *cold enrichment* may be used to isolate the organism from clinical specimens. This technique calls for inoculation of the specimen into broth and incubation at 4°C for several weeks. Subcultures are made at weekly intervals and examined for *L. monocytogenes*. The length of time required for isolation using this technique lessens its importance in the clinical setting, because treatment must begin early in the infectious process.

IDENTIFICATION

The diagnosis of listeriosis depends on isolating *L. monocytogenes* from blood, cerebrospinal fluid, or swabs of lesions. Table 12–2 lists the characteristics of *L. monocytogenes* and other gram-positive bacteria.

Listeria monocytogenes is differentiated from group B streptococci by a positive catalase test,

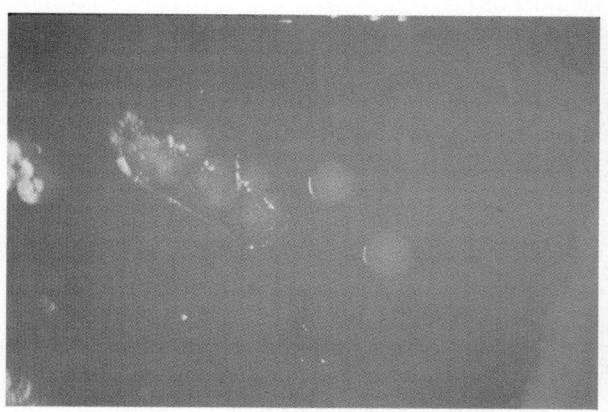

A

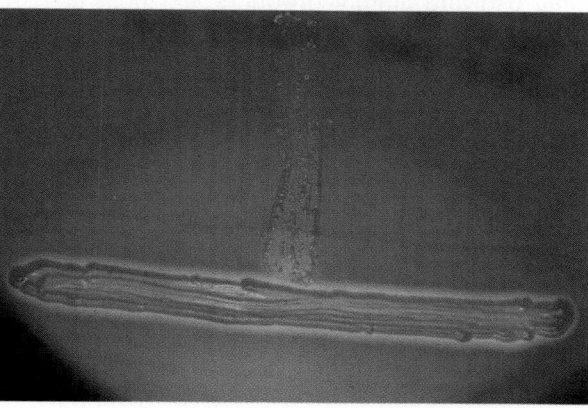

B

■ F i g u r e   1 2 – 7

A, β hemolysis: Listeria on BAP. Colonial *Listeria monocytogenes* growing on blood agar with colonial morphology similar to that of group B β-hemolytic streptococci. (Courtesy of Cathy Bissonette.) *B*, Conventional CAMP test with *Listeria monocytogenes* showing "block" hemolysis at the junction with the *Staphylococcus*.

■■■■■ T A B L E   1 2 – 2
DIFFERENTIATION OF *LISTERIA MONOCYTOGENES* AND OTHER GRAM-POSITIVE BACTERIA*

Organism	Catalase	Esculin Hydrolysis	Motility	β Hemolysis	Growth in 6.5% NaCl
Listeria monocytogenes	+	+	+	+	+
Corynebacterium spp	+	–	V	V	V
Streptococcus agalactiae	–	+	–	+	–
Enterococcus spp	–†	+	–	V	+

*+, present or positive; –, absent or negative; V, variable.
†May be weak.

motility, and a negative hippurate hydrolysis test. It gives a positive CAMP reaction that is similar to that of *S. agalactiae* (see Fig. 12–7*B*). The positive CAMP reaction distinguishes *L. monocytogenes* from the other *Listeria* species, which are CAMP negative. Motility at room temperature is a characteristic umbrella-shaped pattern (see Fig. 12–5).

ERYSIPELOTHRIX RHUSIOPATHIAE

General Characteristics

Erysipelothrix rhusiopathiae is the only species in the genus. It is a gram-positive, nonsporulating, pleomorphic rod that has a tendency to form long filaments. It is found worldwide and is a commensal or a pathogen in a very wide variety of vertebrates and invertebrates. Domestic swine are the major reservoir. Human cases are relatively rare, with infections resulting from occupational exposure. Those individuals whose work involves handling fish and animal products are most at risk. The usual route of infection is through cuts or scratches on the skin. The organism is resistant to salting, pickling, and smoking and survives well in environmental sources such as water, soil, and plant material.

Clinical Infections

Erysipelothrix rhusiopathiae produces three types of disease in humans: septicemia, endocarditis, and erysipeloid. Systemic infection is very uncommon and rarely develops from localized infection. Endocarditis has been seen in patients who have had valve replacements, but also in individuals with apparently normal heart valves.

Erysipeloid is a localized skin infection that resembles streptococcal erysipelas. The lesions are usually seen on the hands or fingers because the organisms usually are inoculated through work activities. The incubation period is 1 to 4 days. The infected area is painful and swollen and gives rise to a characteristic lesion: a sharply defined, slightly elevated, purplish red zone that spreads peripherally as discoloration of the central area fades. Low-grade fever, arthralgia, lymphangiitis, and lymphadenopathy may occur. A mild form of the disease lasts 2 to 4 weeks, but it may continue for months. Erysipeloid is a self-limiting infection that normally heals within 3 to 4 weeks. A problem is that second attacks can occur and relapses are common.

Antibiotic therapy is effective, with penicillins, cephalosporins, erythromycin, and clindamycin being useful. The incidence of *Erysipelothrix rhusiopathiae* infections is low.

Laboratory Diagnosis

MICROSCOPY

Erysipelothrix rhusiopathiae is a thin, rod-shaped, gram-positive organism that may form long filaments (Fig. 12–8). It is arranged singly, in short chains, or in a V shape similar to that seen with the corynebacteria.

CULTURAL CHARACTERISTICS

The specimen received by the clinical laboratory is from a tissue biopsy or aspirates from skin lesions. These should be inoculated to a nutrient broth with 1% glucose and incubated in 5% CO_2 at 35°C. Subcultures should be made daily to blood agar plates. On blood agar, the colonies are nonhemolytic or α hemolytic and are pinpoint after 24 hours of incubation. After 48 hours of incubation, two distinct colony types

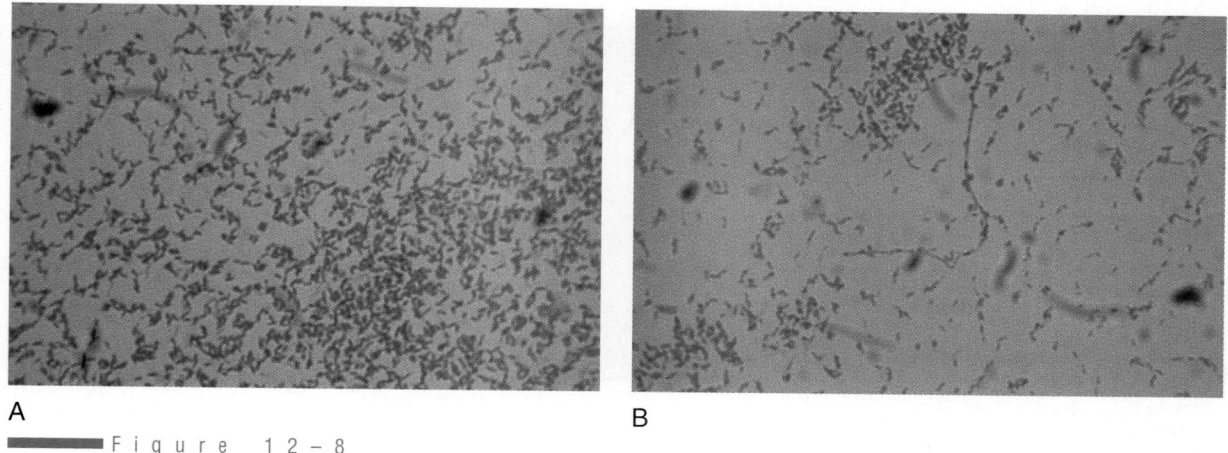

A

B

■■■■■■ F i g u r e   1 2 – 8

A, Gram stain of *Erysipelothrix rhusiopathiae* at 24 hours. *B*, Gram stain of *E. rhusiopathiae* at 72 hours showing the tendency to form long filaments, which are easily decolorized. (Courtesy of Cathy Bissonette.)

are seen. A smaller, smooth form is transparent, glistening, and convex with entire edges. The larger, rough colonies are flatter with a matte surface, curled structure, and irregular edges.

IDENTIFICATION

A comparison of *Erysipelothrix* with *Listeria* is shown in Figure 12–9. Table 12–3 lists the characteristics of *Erysipelothrix* and other related gram-positive bacilli. Identification is based on the Gram stain, hydrogen sulfide production, lack of motility, indole and catalase activities, and a negative Voges-Proskauer reaction. An additional test that can be used to differentiate *Erysipelothrix* from *L. monocytogenes* is susceptibility to neomycin; *Erysipelothrix* is resistant and *L. monocytogenes* is susceptible.

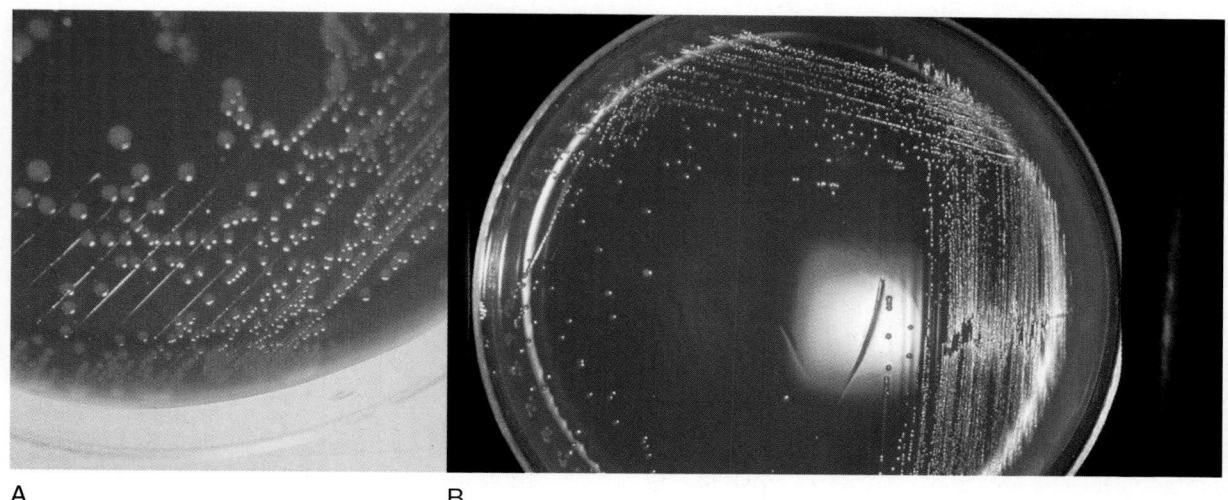

A

B

■■■■■■ F i g u r e   1 2 – 9

Comparison of colony morphology of *Listeria* (*A*) and *Erysipelothrix* (*B*) growing on blood agar. (Courtesy of Cathy Bissonette.)

▬▬▬ T A B L E 1 2 - 3

CHARACTERISTICS OF *LISTERIA, CORYNEBACTERIUM, ERYSIPELOTHRIX,* AND OTHER RELATED GRAM-POSITIVE BACILLI*

Organism	Catalase	Motility	Esculin Hydrolysis	Acid from Glucose	H₂S from TSI Agar	Hemolysis Type	Nitrate Reduction	Urease Production
Corynebacterium spp	+	−	V	V	−	V	V	V
Listeria monocytogenes	+	+[†]	+	+	−	β	−	−
Erysipelothrix rhusiopathiae	−	−	−	+	+	None, α	−	−
Kurthia spp	+	+	−	−	−	None	−	−
Lactobacillus spp	−	−	−	−	−	None	−	−
Arcanobacterium haemolyticum	−	−	−	+	−	β	−	−
Rhodococcus spp	+	−	−	−	−	None	V	+
Rothia dentocariosa	+	−	+	+	−	None	+	−
Oerskovia	+	+	+	+	−	None	V	V

*+, positive; −, negative; V, variable.
[†]Motile at 25°C.

Bibliography

Brooks R, Joynson DHM: Bacteriological diagnosis of diphtheria. Association of Clinical Pathologists Broadsheet No. 125:576, 1990.

Camilli A, Goldfine H, Portnoy DA: *Listeria monocytogenes* mutants lacking phosphatidylinositol-specific phospholipase C are avirulent. J Exp Med 173:751, 1991.

Clarridge JE: When, why, and how far should coryneforms be identified? Clin Microbiol Newsl 8:32, 1986.

Cossart P, Vicente MF, et al: Listeriolysin O is essential for virulence of *Listeria monocytogenes:* Direct evidence obtained by gene complementation. Infect Immun 57:3629, 1989.

Courtieu AL: Latest news on listeriosis. Comp Immun Microbiol Infect Dis 14:1, 1991.

Dietrich MC, Watson DC, Kumar ML: *Corynebacterium* group JK infections in children. Pediatr Infect Dis J 8:233, 1989.

Lachica RV: Same-day identification scheme for colonies of *Listeria monocytogenes.* Appl Environ Microbiol 56:1166, 1990.

Lipsky BA, Goldberger AC, et al: Infections caused by nondiphtheria corynebacteria. Rev Infect Dis 4:1220, 1982.

Mounier J, Ryter A, et al: Intracellular and cell to cell spread of *Listeria monocytogenes* involves interaction with F-actin in the enterocyte cell line Caco-2. Infect Immun 58:1048, 1990.

Reboli A, Farrar WE: *Erysipelothrix rhusiopathiae:* An occupational pathogen. Clin Microbiol Rev 2:354, 1989.

Ronci-Koenig TJ, Tan JS, et al: Infections due to *Corynebacterium* Group D2. Arch Intern Med 150:1965, 1990.

Schuchat A, Swaminathan B, Broome CV: Epidemiology of human listeriosis. Clin Microbiol Rev 4:169, 1991.

Weinheimer LA, Sass E: Clinically significant *Rhodococcus equi* isolates from AIDS patients at Parkland Memorial Hospital. Clin Lab Sci 5:140, 1992.

CHAPTER 13

Aerobic Gram-Positive Bacilli

Hal S. Larsen

1. Give the general characteristics of the genus *Bacillus*.

2. Describe the clinical infections associated with *Bacillus anthracis*.

3. Name the differential tests that may be employed to identify *Bacillus anthracis*.

4. Discuss the clinical significance of other spore-forming gram-positive bacilli.

5. Differentiate *Bacillus anthracis* from other saprophytic gram-positive spore-forming rods.

6. Describe the clinical infections associated with *Nocardia, Actinomadura,* and *Streptomyces*.

7. Differentiate infections caused by *Nocardia, Actinomadura,* and *Streptomyces* from those caused by fungal agents.

8. Describe the types of clinical specimens that are likely to contain *Bacillus, Nocardia, Actinomadura,* and *Streptomyces*.

9. Describe the microscopic morphology and colonial appearance of *Bacillus, Nocardia, Actinomadura,* and *Streptomyces*.

10. Name the differential tests that may be used to identify the organisms listed, and describe how each isolate is differentiated from closely related species.

This chapter discusses a large group of bacteria that are commonly encountered in the microbiology laboratory. Most are found in the environment and can be easily isolated from water and soil. The majority are not highly pathogenic but are being isolated from clinical infections with increasing frequency. Bacteria that belong to the gram-positive aerobic bacilli include the spore-forming *Bacillus* as well as the filamentous *Nocardia, Actinomadura,* and *Streptomyces*. Figure 13–1 shows a schematic diagram for the presumptive identification of aerobic gram-positive bacteria, including spore-formers, non–spore-formers, and gram-positive cocci.

BACILLUS

General Characteristics

Members of the genus *Bacillus* are aerobic, gram-positive, rod-shaped organisms that form endospores (Fig. 13–2). On Gram stain, however, spores do not stain, and appear only as "empty spaces." Spore stain, such as shown in Figure 13–3, is used to demonstrate the presence of spores. More than 50 species of *Bacillus* are widely found in the environment. Members of the genus can be isolated from soil from all climates—the subarctic and desert regions, thermal springs, fresh and salt water, and plant materials. The temperature range for growth depends on the species but includes temperatures as low as −5°C and as high as 75°C. The survival of *Bacillus* in nature is due to the formation of the endospores, which are resistant to conditions to which vegetative cells are intolerant.

Most species grow well on blood agar as well as other common enriched media. As a result, *Bacillus* species are found as contaminants in specimens from a number of sources. The different species show a wide variety of metabolic characteristics. They are catalase positive and form endospores under aerobic conditions. They can be divided into three morphologic groups on the basis of the location and size of the endospore (Table 13–1). In group 1, the spores are oval or cylindrical and are located centrally or terminally. The key characteristic placing a species in group I is the spore. Regardless of location, it does not distend the vegetative cell. Another way to describe this is to say that the sporangium is not swollen. Organisms belonging to group II have oval spores that are central or terminal and cause the vegetative cell to swell (i.e., swollen sporangia). Morphologic group III consists of species showing round, terminal, swollen sporangia.

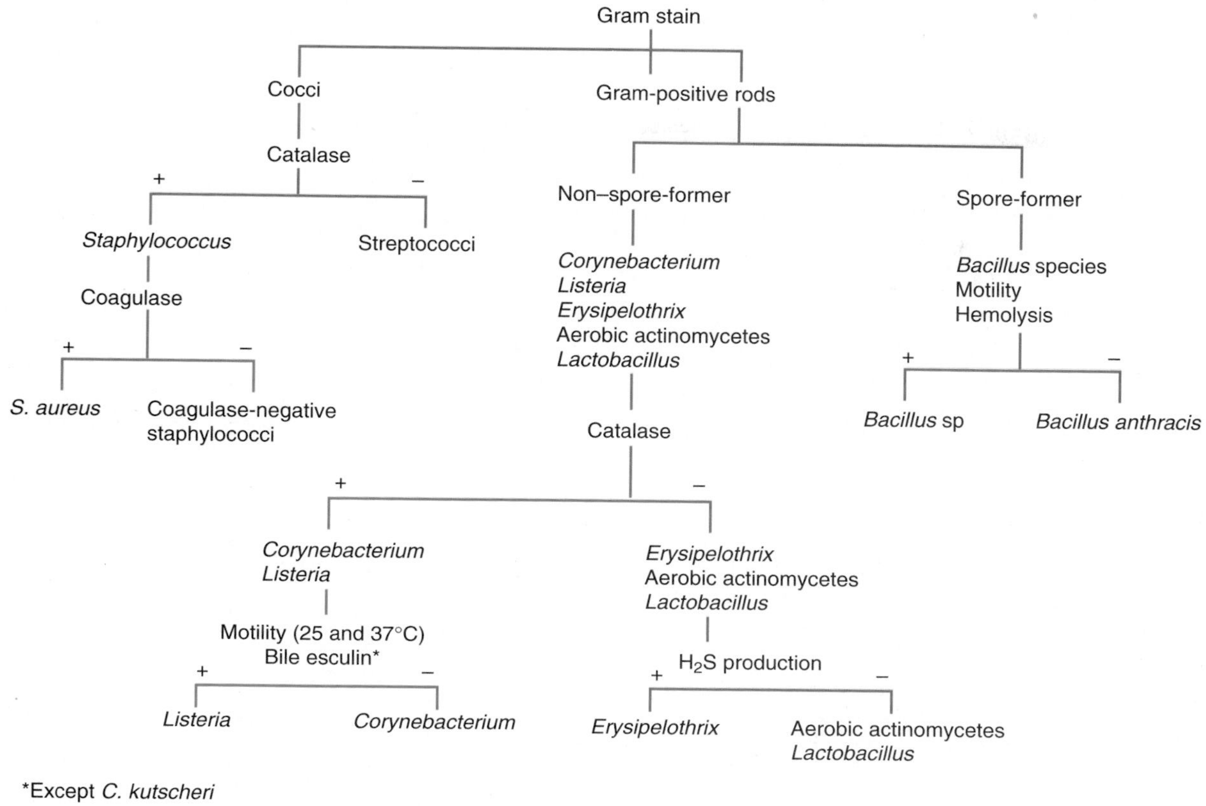

Gram stain

Cocci

Gram-positive rods

Catalase

+

–

Staphylococcus

Streptococci

Coagulase

+

–

S. aureus

Coagulase-negative
staphylococci

Non–spore-former

Spore-former

Corynebacterium
Listeria
Erysipelothrix
Aerobic actinomycetes
Lactobacillus

Bacillus species
Motility
Hemolysis

+

–

Bacillus sp

Bacillus anthracis

Catalase

+

–

Corynebacterium
Listeria

Erysipelothrix
Aerobic actinomycetes
Lactobacillus

Motility (25 and 37°C)
Bile esculin*

H_2S production

+

–

+

–

Listeria

Corynebacterium

Erysipelothrix

Aerobic actinomycetes
Lactobacillus

*Except *C. kutscheri*

■ F i g u r e   1 3 – 1

Schematic diagram for the identification of gram-positive bacteria.

Colony characteristics vary considerably among the species and are often influenced by the type of medium used. Pigment formation is seen in a number of species. Pigment colors range from pink to blue-black and may vary according to growth conditions and substrates. Most species, however, are unpigmented. Historically, *Bacillus anthracis,* the causative agent of anthrax, has been the most

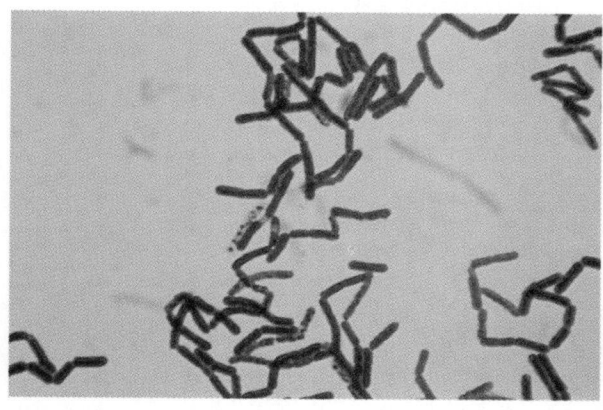

■ F i g u r e   1 3 – 2

Gram stain of *Bacillus.* (Courtesy of Cathy Bissonette.)

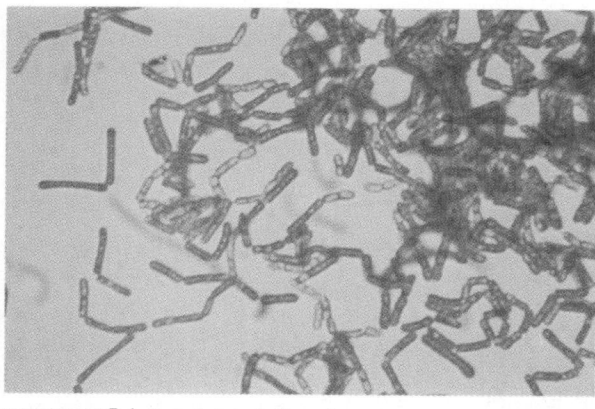

■ F i g u r e   1 3 – 3

Spore stain of *Bacillus.* (Courtesy of Cathy Bissonette.)

TABLE 13-1

CHARACTERISTICS OF SOME SPECIES OF *BACILLUS*

Characteristics	Group I						Group II					Group III
	ANTHRACIS	CEREUS†	COAGULANS	LICHENIFORMIS	MEGATERIUM	SUBTILIS	ALVEI	BREVIS	CIRCULANS	MACERANS	POLYMYXA	SPHAERICUS
Morphologic												
Gram stain	+	+	+	+	+	+	v	v	v	v	v	v
Size‡	−	−	s	s	l	s	s	s	s	s	s	s
Motility	−	−	s	+	v	+	+	+	v	+	+	+
Spore shape§	o	o	o	o	o	o	o	o	o	o	o	r
Spore location¶	c	c	cts	c	c	c	cts	cts	cts	t	cts	t
Swelling of vegetative cell	−	−	v	−	−	−	+	+	+	+	+	+
Capsule	+‖	−	−	+	v	−	−	−	−	−	+	−
Biochemical												
Catalase	+	+	+	+	+	+	+	+	+	+	+	+
Acid** from:												
Arabinose	−	−	v	+	v	+	−	−	+	+	+	−
Glucose	+	+	+	+	+	+	+	+	+	+	+	+
Mannitol	−	−	v	+	v	+	−	v	+	+	+	−
Xylose	−	−	v	+	v	+	−	−	+	+	+	−
Salicin	−	+	n	n	v	n	n	n	+	n	n	n
Gelatin liquefaction	+	+	−	+	+	+	+	+	+	+	+	v
Hemolysis on sheep blood agar††	−	+	n	n	n	+	n	n	n	n	n	n
Lecithinase activity	+	+	−	−	−	−	−	−	−	−	−	v
Anaerobic growth‡‡	+	+	+	+	−	−	+	−	v	+	+	−
Growth at pH <6	+	v	+	+	+	+	−	v	v	+	+	v
Growth in 7% NaCl	+	v	−	+	+	+	−	−	v	−	−	v
Growth at 50°C	−	−	+	+	−	+	−	+	+	v	−	−

*Variable reactions indicated by v; n indicates reaction not reported in the manuals.
†*B. cereus* var. *mycoides* is not listed as a legitimate species in Bergey's Manual but is often mentioned in the literature and seen in the laboratory. This variant grows as large, rhizoid colonies whose outgrowths may cover the agar surface. The variant usually is not motile, in contrast to *B. cereus*.
‡l represents large (width 1.0 to 1.5 μm and length 2 to 5 μm); s represents small (width 0.6 to 0.8 μm and length 1.7 to 3 μm).
§o represents oval; r, round.
¶c represents central; t, terminal; and s, subterminal.
‖Demonstration of capsule requires special growth conditions.
**In media containing $(NH_4)_2HPO_4$, and the appropriate sugar.
††Presence of hemolysis eliminates possibility of *B. anthracis*, but lack of hemolysis may be characteristic of other species also.
‡‡In broth containing glucose.
Modified with permission from Braude AI, Davis CE, Fierer J (eds): Infectious Diseases and Medical Microbiology. Philadelphia: WB Saunders, 1986, p 271.

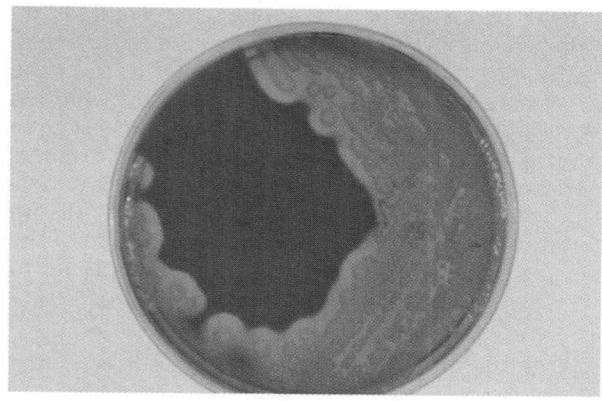

■ F i g u r e   1 3 – 4

Bacillus cereus colony on blood agar. (Courtesy of Cathy Bissonette.)

important member of this genus; however, anthrax is now rarely seen in the United States. Other species, most notably *Bacillus cereus,* are seen in the clinical laboratory as pathogens and contaminants. Several species are pathogenic for insects (Fig. 13–4).

Bacillus anthracis

Physiology

Bacillus anthracis is a gram-positive or gram-variable rod. It is catalase positive and aerobic or facultative. Spore formation takes place aerobically; this feature, along with a positive catalase test, differentiates *Bacillus* from *Clostridium.* It is non-motile, distinguishing it from other members of *Bacillus.* Although *B. anthracis* ferments glucose, it fails to ferment mannitol, arabinose, and xylose. It produces lecithinase, and so an opaque zone can be seen around colonies growing on egg-yolk agar. This species grows in high salt (7% NaCl) and low pH (<6). Unlike *B. cereus, B. anthracis* is susceptible to penicillin (10 U/mL). It is nonhemolytic on sheep's blood agar. A summary of characteristics used to differentiate the various species of *Bacillus* is found in Table 13–1. Characteristics especially important to differentiate *B. anthracis* from the closely related *B. cereus* are found in Table 13–2.

Virulence Factors

The virulence of *B. anthracis* depends upon a glutamic acid capsule and a complex toxin. The capsule, which protects the organism from phagocytosis, is a polypeptide of D-glutamic acid. This particular isomer of glutamic acid is resistant to hydrolysis by host proteolytic enzymes because it is the "unnatural" form. Although the capsule is necessary for virulence, antibodies against the capsule do not confer immunity.

The anthrax toxin consists of three factors:

■ Edema factor (EF)

■ Protective antigen (PA)

■ Lethal factor (LF)

The effect of edema factor and lethal factor is seen when either is combined with protective antigen. Edema results from the combination of protective antigen with edema factor, whereas death occurs when protective antigen and lethal factor combine. The toxin increases vascular permeability as well as interferes with phagocytosis. The genes for the synthesis of the toxin are located on a plasmid. If a virulent isolate is repeatedly cultured in vitro, the plasmid is eventually lost, and the organism is no longer virulent.

Clinical Infections

Anthrax is a common disease in livestock worldwide. The disease is not spread from animal to animal, but rather from animals feeding on plants contaminated with the spores. Humans are infected primarily as a result of contact with the animals or animal products. The incidence of human anthrax in the United States is very low;

■ T A B L E   1 3 – 2

DIFFERENTIATION OF *BACILLUS ANTHRACIS* AND *BACILLUS CEREUS*

Characteristic	B. anthracis	B. cereus
Hemolysis on sheep blood agar	−	+
Motility	−	+
Lecithinase production	+	+
Fermentation of salicin	−	+/−
Growth in penicillin (10 U/mL) agar	−	+
String of pearls reaction	+	−
Gelatin hydrolysis	−	+
Growth on phenylethyl alcohol agar (PEA)	−	+
Pathogenicity for mice by injection	+	−

All cultures incubated at 36 to 37°C.

Modified with permission from Braude AI, Davis CE, Fierer J (eds): *Infectious Diseases and Medical Microbiology.* Philadelphia: WB Saunders, 1986, p 276.

less than 5 cases per year are reported. Worldwide, however, cases number several thousand. The disease is enzootic in many parts of the world, including Central and South America. A number of names have been given to infections with *B. anthracis*. The majority refer to occupational associations. Terms such as "woolsorter's disease" and "ragpickers' disease" were used to describe infection with the spores of *B. anthracis* as a result of handling contaminated animal fibers, hides, and other animal products.

In humans, there are three forms of anthrax:

- Cutaneous
- Inhalation or pulmonary
- Gastrointestinal

All three forms of infection result from wound contamination, inhalation, or ingestion of spores, which germinate within the host tissue.

CUTANEOUS ANTHRAX

Wounds contaminated with anthrax spores acquired through skin cuts, abrasions, or insect bites may manifest the cutaneous form. The overwhelming majority of anthrax cases in the world are cutaneous. In this form of anthrax, a small pimple or papule appears at the site of inoculation 2 to 3 days following exposure. A ring of vesicles develops; the vesicles coalesce to form an erythematous ring. A small dark area appears in the center of the ring and eventually ulcerates and dries, forming a depressed black necrotic central area known as an eschar ("black eschar," "malignant pustule"). The lesion is painless and does not produce pus unless it becomes secondarily infected with a pyogenic organism. The eschar is normally 1 to 3 cm in diameter, although it may be more extensive. The eschar begins to heal after 1 to 2 weeks. The lesion dries, separates from the underlying base, and falls off, leaving a scar. Usually the infection remains localized, but there may be regional lymphangitis and lymphadenopathy. If septicemia occurs, symptoms of fever, malaise, and headache are seen. Normally, in uncomplicated cases, no systemic symptoms are present.

INHALATION OR PULMONARY ANTHRAX

Pulmonary anthrax, or "woolsorter's disease" as it is sometimes known, is acquired when spores are inhaled into the pulmonary parenchyma. The infection begins as a nonspecific illness consisting of mild fever, fatigue, and malaise 2 to 5 days following exposure to the spores. It resembles an upper respiratory tract infection such as that seen with colds and "flu." This initial, mild form of the disease lasts 2 to 3 days. It is followed by a sudden severe phase in which respiratory distress is common. The severe phase of the disease is extremely serious. The respiratory problems (dyspnea, cyanosis, pleural effusion) are followed by disorientation, then coma, and death. The course of the severe phase (onset of respiratory symptoms to death) may last for only 24 hours.

GASTROINTESTINAL ANTHRAX

The gastrointestinal form of anthrax occurs when the spores are inoculated into a lesion on the intestinal mucosa following ingestion of the spores. The symptoms of gastrointestinal anthrax include abdominal pain, nausea, anorexia, and vomiting. Bloody diarrhea may also occur. Because this form of the disease is difficult to diagnose, the fatality rate is higher than in the cutaneous form. Fortunately, gastrointestinal anthrax accounts for less than 1% of the total cases worldwide; it has never been reported in the United States.

Complications and Treatment

Approximately 5% of patients with anthrax (cutaneous, pulmonary, gastrointestinal) develop meningitis. The symptoms are typical of any bacterial meningitis and occur very rapidly. Unconsciousness and death, if they occur, follow 1 to 6 days after initial exposure.

Recovery from infection appears to confer immunity. An effective vaccine is available for those who are at risk for occupational exposure. In addition, vaccines are available for veterinary use.

Most isolates of *B. anthracis* are sensitive to penicillin, which is the drug of choice for treatment of anthrax. The organism is also sensitive to many of the broad-spectrum antibiotics, including gentamicin, erythromycin, tetracycline, and chloramphenicol.

Laboratory Diagnosis

MICROSCOPY

Bacillus anthracis is a large, square-ended, gram-positive rod found singly or in chains. The spores are seen as unstained spots within the cells. When they are in chains, the ends of the single cells fit snugly together. This, together with the unstained central spore, gives the appearance of bamboo rods. Only young cultures are gram positive. As the cells become older or if they are under nutritional stress, they become gram variable.

CULTURAL CHARACTERISTICS

On blood agar, colonies of *B. anthracis* are nonhemolytic, large (3–5 mm), gray, and flat with an irregular margin due to outgrowths of long filamentous projections of bacteria that can be seen with a dissecting microscope. The term *Medusa head* has been used to describe the colony of *B. anthracis*. Colonies hold tightly to the agar surface and when the edges are lifted with a loop, they stand upright without support. This has been described as having the appearance or characteristics of beaten egg whites.

IDENTIFICATION

Caution should always be used in working with an isolate suspected of being *B. anthracis*. Work should be done in a bacteriologic biologic safety hood, and the area should be disinfected when the work is completed.

An isolate with the appropriate colonial and microscopic morphology may be suspected of being *B. anthracis* if it (1) is nonhemolytic on blood agar, (2) is nonmotile, and (3) produces lecithinase. A presumptive identification can be made by inoculating the suspected isolate onto agar containing penicillin (0.05 to 0.5 U/mL). After incubation for 3 to 6 hours at 37°C, the areas of inoculation are examined microscopically for the presence of large spherical bacilli in chains; this phenomenon is referred to as a *string of pearls*. The isolate should be forwarded to the state health laboratory for confirmatory identification. A summary of characteristics useful in the differentiation of *B. anthracis* from other members of the genus is given in Tables 13–1 and 13–2. Species identification of *Bacillus* can be accomplished by the use of the API 20E and 50CH systems. (Analytab Products, Plainview, NY).

Other *Bacillus* Species

Bacillus cereus

Bacillus cereus is a relatively common cause of food poisoning as well as opportunistic infections in the susceptible host. Food poisoning due to *B. cereus* takes two forms, diarrheal and emetic.

The diarrheal syndrome, usually associated with ingestion of meat or poultry, is characterized by an incubation period of 8 to 16 hours. Afflicted individuals suffer abdominal pain and diarrhea. About 25% of individuals experience vomiting. Fever is uncommon. The average duration of the illness is 24 hours. The diarrheal form is clinically indistinguishable from the diarrhea caused by *Clostridium perfringens*.

The emetic form has the predominant symptoms of abdominal cramps and vomiting. Diarrhea is present in about one third of those affected. This form has been associated with ingestion of fried rice, particularly when prepared in Oriental restaurants. The average duration of the illness is 9 hours. For both the diarrheal and emetic forms of *B. cereus* food poisoning, the illness is usually mild and self-limiting. The two forms of illness are caused by two distinct enterotoxins produced by *B. cereus*. A comparison of the enterotoxins is shown in Table 13–3.

B. cereus is similar to *B. anthracis* in many ways, both morphologically and metabolically.

■■■■■ T A B L E 1 3 – 3

COMPARISON OF ENTEROTOXINS PRODUCED BY *BACILLUS CEREUS*

Characteristic	Type of Enterotoxin	
	Diarrheal	Emetic
Clinical syndrome		
Incubation period	8 to 16 hr	1 to 5 hr
Diarrhea	Very common	Fairly common
Vomiting	Occasional	Very common
Duration of illness	12 to 24 hr	6 to 24 hr
Foods implicated	Meat products, soups, vegetables, puddings, sauces	Fried or boiled rice
Enterotoxin		
Molecular weight	ca. 50,000	<5,000
Stability to heat	−	+
Fluid accumulation in ligated rabbit ileal segment	+	−
Increased vascular permeability in guinea pig or rabbit skin	+	−
Lethal for mice after intravenous injection	+	−
Stimulation of adenylate cyclase–cAMP system in intestinal epithelial cells	+	−
Response when fed to rhesus monkeys	Diarrhea	Vomiting

Modified with permission from Braude AI, Davis CE, Fierer J (eds): Infectious Diseases and Medical Microbiology. Philadelphia: WB Saunders, 1986, p 274.

Differentiation between *B. cereus* and *B. anthracis* is outlined in Table 13–2. The organism can be grown aerobically at 37°C on blood agar. A β-hemolytic frosted-glass–appearing colony that is aerobic, spore forming, gram positive, and motile, ferments salicin, and is lecithinase positive is likely *B. cereus*. It also differs from *B. anthracis* in being resistant to penicillin and nonpathogenic for mice.

Culture of the suspected food may be done to quantitate and isolate *B. cereus*. If more than 10^5 *B. cereus* per gram of food are present and other pathogens are absent, then food poisoning by this organism is confirmed. The stool of patients with food poisoning may also be examined for *B. cereus*, although it must be remembered that the organism is part of the normal fecal flora. To confirm the organism as the cause of the disease, the viable counts from the stool should also be at least 10^5 per gram. Quantitative cultures must be done because *B. cereus* may be found in small numbers in a significant proportion of healthy people.

Opportunistic infections caused by *B. cereus*, particularly of the eye, are becoming more common. *Bacillus* species are contaminants of illicit drug paraphernalia and have been documented as causes of meningitis, septicemia, osteomyelitis, and a number of other types of infections.

Unlike *B. anthracis*, *B. cereus* is resistant to penicillin and many of the other cell-wall antibiotics (e.g., ampicillin, cephalothin, methicillin). Treatment with a clindamycin and gentamicin combination has been successful.

Bacillus subtilis

Infections by other members of the genus are rare. They occasionally cause gastrointestinal illness. They are seen, however, as contaminants. One of the more commonly seen species is *Bacillus subtilis*. The colonies of *B. subtilis* are large and may be pigmented. Pigment colors range from pink and yellow to orange or brown. Identification can be made using the API 20E and 50CH systems.

AEROBIC ACTINOMYCETES

The genera *Nocardia* and *Actinomyces* are similar morphologically to fungi. These organisms demonstrate filamentous hyphae in culture that are similar microscopically to those seen with fungi. Upon close inspection, however, differences are observed. These genera are true bacteria and can be differentiated from fungi as well as from each other. Although they are not commonly seen in most laboratories, they are responsible for significant human diseases.

Nocardia

General Characteristics

The organisms in this genus are aerobic, gram-positive bacilli that often form branched hyphae. The hyphae are easily disrupted into rods and cocci. Even though *Nocardia* is, strictly speaking, gram positive, organisms will often stain gram variable and may be weakly acid fast. The colonial and microscopic morphology, as well as the types of infections caused, resemble those of the fungi, but these organisms are true bacteria. *Nocardia* species grow well on standard nonselective media. Growth may take a week or more. The organisms in this genus are commonly found in soil. Generally, infections caused by *Nocardia* are seen in immunocompromised patients. Reports of infection in patients with no apparent illness or immunosuppressive therapy are increasing, however. The species of medical importance are *N. asteroides*, *N. brasiliensis*, and *N. caviae*. Of the total number of aerobic actinomycetes received by the Centers For Disease Control (CDC) from October 1985 through February 1988, 27% were *N. asteroides* and 6% were *N. brasiliensis* (McNeil et al, 1990). The majority of isolates received for identification were from sputum and wounds.

Physiology

The growth requirements of *Nocardia* are not as well defined as those of many other medically important bacteria. These organisms show an oxidative-type metabolism and as a genus, they utilize a wide variety of sugars. They do not require specific growth factors as do *Haemophilus* and *Francisella*. *Nocardia* species grow well on most common nonselective laboratory media at temperatures between 25 and 37°C, although it may take 3 to 6 days or longer for growth to occur. Some characteristics of the aerobic actinomycetes are listed in Table 13–4.

■■■■■ T A B L E 1 3 – 4

SOME CHARACTERISTICS OF SELECTED AEROBIC ACTINOMYCETES OF MEDICAL IMPORTANCE

Characteristic	Nocardia asteroides	Nocardia caviae	Nocardia brasiliensis	Actinomadura madurae	Streptomyces griseus
Aerial hyphae	+	+	+	v	+
Acid fastness	v	v	v	–	–
Rifampin resistance	+	+	+	v	v
Hydrolysis of					
Esculin	+	+	+	+	v
Casein	–	–	+	+	+
Hippurate	v	v	–	–	+
Xanthine	–	+	–	–	v
Starch	v	v	v	+	+
Tyrosine	–	v	+	+	+
Acid from					
Arabinose	–	v	–	+	v
Cellobiose	–	–	–	+	+
Glucose	+	+	+	+	+
Inositol	–	+	+	+	+
Mannitol	–	v	+	v	v
Xylose	–	–	–	+	+
Growth at 10°C	–	v	v	–	+

+, 90% or more of the strains tested were positive.

–, 90% or more of the strains tested were negative.

v, variable.

Modified with permission from Mishra SK, Gordon RE, Barnett DA: Identification of nocardiae and streptomycetes of medical importance. J Clin Microbiol 11:730, 1980.

Virulence Factors

The role of such factors as toxins and extracellular proteins in nocardiosis is unclear. No virulence factors have been identified, although virulence has been correlated with alterations in the components in the cell envelope. The precise role of the various cell-wall molecules in virulence is unknown. *Nocardia* produces a superoxide dismutase and catalase that may give the organism resistance to oxidative killing by phagocytes. It also produces an iron-chelating compound called nocobactin. A correlation has been reported between the amount of nocobactin produced by the organism and its virulence.

Clinical Infections

Nocardia species are found worldwide in soil and on plant material. Infection occurs by two routes: pulmonary and cutaneous. Pulmonary infection by *Nocardia* is due to the inhalation of the organism, which is present in dust or soil. The disease appears to be associated with impaired host defenses, because most persons seen with nocardiosis have an underlying disease or compromised immune defenses. Even so, a significant number of seemingly normal patients show an infection with *Nocardia* and no obvious immune impairment.

Infection with *Nocardia* is a very serious situation. Approximately 40% of the diagnoses are made at autopsy. The mortality rate is high, and those who survive often suffer significant tissue damage.

PULMONARY FORM

The majority of pulmonary infections are due to *N. asteroides.* The most common manifestation of infection is a confluent bronchopneumonia that is usually chronic but may be acute or relapsing. The disease generally progresses more rapidly than tuberculosis but is measured in months rather than years. In the acute form, which is often seen in patients with underlying immune defects, the time course is a matter of weeks.

The initial lesion in the lung is a focus of pneumonitis that advances to necrosis. The abscesses that form may extend into the tissue and coalesce with each other. Extensive tissue involvement and damage result. Unlike with some pneumonias, there is little inflammatory response or scarring and no encapsulation of the abscesses. There is no granuloma formation. Dissemination to other organs, especially the brain, may occur, with reports of involvement of virtually every organ in the literature. The sputum is thick, sticky, and purulent. Unlike with infection by the anaerobic actinomycetes, there are no sulfur granules or sinus tract formation.

CUTANEOUS FORM

The second route of infection is cutaneous, or inoculation of the organism into the skin or subcutaneous tissues. *Nocardia brasiliensis* is the most frequent cause of this form of nocardiosis, which is usually seen in the hands and feet as a result of outdoor activity. The trauma most likely is minor, such as from a thorn or wood sliver.

The infection begins as a localized subcutaneous abscess that is invasive and quite destructive of the tissues and underlying bone. These lesions are termed *mycetomas* (certain species of fungi also form mycetomas). Mycetomas are characterized by swelling, draining sinuses and granules. About half of the mycetomas seen clinically are caused by the actinomycetes, and the remaining half are caused by fungi. As the infection progresses, burrowing sinuses open to the skin surface and drain pus. The pus may be pigmented and contain *sulfur granules* (Fig. 13–5), which are masses of filamentous organisms bound together by calcium phosphate. They often appear yellow or orange and have a distinct granular appearance.

Treatment of *Nocardia* infection often involves drainage and surgery as well as antimicrobials. The organisms are resistant to penicillin but susceptible to sulfonamides. Antifungal agents, of course, have no activity against *Nocardia;* this fact underscores the importance of laboratory diagnosis, because many of the clinical manifestations of both pulmonary and cutaneous infection are shared with other organisms, including fungi. This organism represents a classic example of a situation in which laboratory results are absolutely essential to proper antimicrobial treatment.

Laboratory Diagnosis

MICROSCOPY

The gram-positive branching filaments characteristic of *Nocardia* are often seen in sputum and exudates or aspirates from skin or abscesses. The specimen often contains coccobacillary bodies as well. The Gram reaction may be weak or irregular, causing a "beading" appearance similar to the appearance of chains of gram-positive cocci. This morphology may easily confuse the

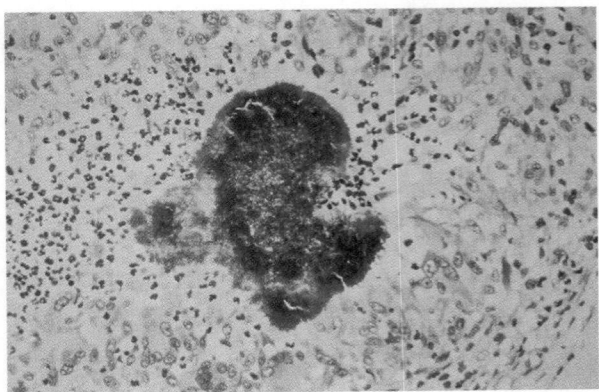

■ F i g u r e 1 3 – 5
Gross appearance of sulfur granules collected from draining sinus tracts. These granules contain masses of filamentous organisms with pus materials.

microscopist. If the isolate is also acid fast, the possibility of its being *Norcardia* is high.

Wet mounts should also be performed. Granules may be seen in specimens from cutaneous infection. Sampled tissue and pus from the draining sinuses are the specimens of choice for direct examination. The granules may be visualized by separating them from the pus with an inoculating needle and then washing in sterile saline. The granules of *N. asteroides, N. brasiliensis,* and *N. caviae* are soft, white to cream colored, and 0.5 to 1 mm in size. They may be crushed between two glass slides to visualize the branching and cellular morphology, comprised of gram-positve, interwoven, thin (0.5 to 1.0 mm in diameter) filaments. The granules may also be used to inoculate the appropriate growth media. The granules of a fungal mycetoma (eumycotic mycetoma) are composed of broad, interwoven, septate hyphae that are wider (2 to 5 mm) than those of the actinomycetes.

CULTURAL CHARACTERISTICS

Colonies of *Nocardia* may have a chalky, matte, or velvety appearance and may be pigmented. They have a dry, crumbly appearance that is likened to that of bread crumbs. Table 13–5 outlines the colonial appearance of the aerobic actinomycetes.

IDENTIFICATION

Nocardia grows well on standard nonselective laboratory media, including those used for fun-

■■■■■■ T A B L E   1 3 – 5
COLONIAL AND MICROSCOPIC APPEARANCE OF AEROBIC ACTINOMYCETES[a]

Genus and Species	Macroscopic Appearance[b]	Microscopic Appearance[c]
Actinomadura		Fine, intertwining, branched filaments with delicate aerial hyphae; nonfragmenting and may form short chains of spores
A. madurae	Waxy, heaped, folded, membranous, and tough; white, tan, pale orange, pink, or red	
A. pelletieri	Heaped, irregular, waxy, and granular; areas of bright and dark red; sparse aerial hyphae	
Mycobacterium		Do not show aerial filaments
Nocardia (N. asteroides, N. brasiliensis, N. caviae)	Orange colonies, glabrous, heaped, and folded; may also be white to pink with aerial hyphae; dry, crumbly, and adherent	Fine, intertwining, branched filaments with delicate aerial hyphae; may have hyphal fragmentation to produce spores from aerial hyphae
Nocardiopsis dassonvillei	Orange colonies, glabrous, heaped, and folded; may also be white to pink with aerial hyphae; dry, crumbly, and adherent	
Rhodococcus		Do not show aerial filaments
Streptomyces somaliensis	Leathery, heaped, and folded; wide range of pigmentation from cream to brown-black; white aerial hyphae	Fine, intertwining, branched filaments with delicate aerial hyphae; nonfragmenting and often form chains of spores

[a]On Sabouraud dextrose agar at 25° to 37°C.
[b]A dissecting microscope is best for macroscopic observation.
[c]Slide cultures must be set up if the original culture shows no aerial hyphae.
With permission from Howard BJ, et al (eds): Clinical and Pathogenic Microbiology. St. Louis: CV Mosby, 1987, p 556.

gal cultures. Media containing antibiotics used for isolating fungi should not be used, however, because *Nocardia* is sensitive to many of the antimicrobial agents used in these media. The possibility of isolating *Nocardia* is increased by the *paraffin bait technique,* which takes advantage of the fact that *Nocardia* uses paraffin as an energy source and other aerobic bacteria do not.

An isolate showing branching filaments that are gram postive and partially acid fast should be suspected of belonging to the genus *Nocardia* (Fig. 13–6). Tentative identification of the aerobic actinomycetes is outlined in Table 13–4. Confirmation is normally performed by a reference laboratory experienced in identification of the actinomycetes.

Other Actinomycetes

Actinomadura

The aerobic actinomycetes belonging to *Actinomadura* include the species *Actinomadura madurae, Actinomadura pelletieri,* and *Actinomadura dassonvillei.* They were formerly classified as members of *Nocardia.* They are etiologic agents for mycetoma, which is identical to that caused by *Nocardia.* The colonial appearance is outlined in Table 13–5. The microscopic and colonial morphology of *Actinomadura* is very similar to that of *Nocardia,* and it would be difficult, if not impossible, for an inexperienced technologist to detect a difference. Differentiation can be made using metabolic variations, as shown in Table 13–4. *Actinomadura madurae* is cellobiose and xylose positive, whereas the *Nocardia* do not produce acid from these two carbohydrates. Treatment parallels that for infections with *Nocardia.*

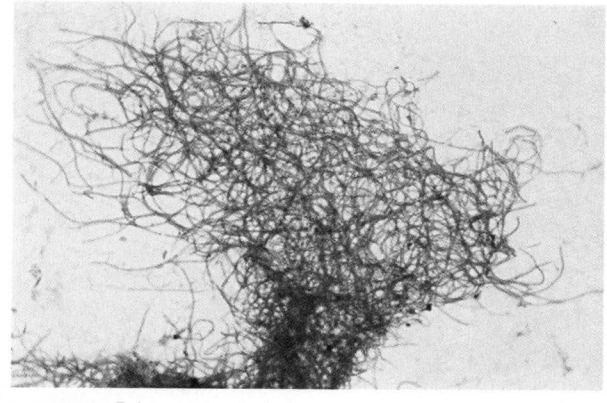

■■■■■■ F i g u r e   1 3 – 6
Acid-fast stain of *Nocardia* showing partially acid-fast appearance.

Streptomyces

The genus *Streptomyces* comprises approximately 23 species. According to Mishra and colleagues (1980), *Streptomyces griseus* is the third most common aerobic actinomycete, after *N. asteroides* and *N. brasiliensis*. *Streptomyces* species are also soil microorganisms and resemble the other aerobic actinomycetes, previously discussed, with regard to morphology and the diseases they cause. Tables 13–4 and 13–5 compare the colonial morphology and metabolic characteristics of *Streptomyces* with those of other members of the aerobic actinomycetes. Isolates from this genus may also need to be identified by reference laboratories.

Bibliography

Beaman BL: *Nocardia:* Pathogenesis and host resistance. In Ortiz-Ortiz L, Bojalil LF, Yakoleff V (eds): Biological, Biochemical, and Biomedical Aspects of Actinomycetes. London: Academic Press, 1984.

Bowden GH, Goodfellow M: The Actinomycetes: *Actinomyces, Nocardia* and Related Genera. In Parker MT, Duerden BI (eds): Topley & Wilson's Principles of Bacteriology, Virology and Immunity, Volume 2: Systematic Bacteriology, 8th ed. Philadelphia: BC Decker, 1990.

Braude AI, Davis CE, Fierer J (eds): Infectious Diseases and Medical Microbiology. Philadelphia: WB Saunders, 1986.

Knudson GB: Treatment of anthrax in man: History and current concepts. Mil Med 151:71, 1986.

Land G, McGinnis MR, et al: Aerobic pathogenic *Actinomycetales.* In Balows A, Hausler WJ, Jr et al (eds): Manual of Clinical Microbiology, 5th ed. Washington, DC: American Society for Microbiology, 1991.

McNeil MM, Brown JM, et al: Comparison of species distribution and antimicrobial susceptibility of aerobic actinomycetes from clinical specimens. Rev Infect Dis 12:778, 1990.

Mishra SK, Gordon RE, Barnett, DA: Identification of nocardiae and streptomycetes of medical importance. J Clin Microbiol 11:728, 1980.

Shawar RM, More DG, LaRocco MT: Cultivation of *Nocardia* spp. on chemically defined media for selective recovery of isolates from clinical specimens. J Clin Microbiol 28:508, 1990.

Turnbull P: Anthrax. In Parker MT, Duerden BI (eds): Topley & Wilson's Principles of Bacteriology, Virology and Immunity, Volume 3: Bacterial Diseases, 8th ed. Philadelphia: BC Decker, 1990.

Turnbull P, Kramer J, Melling J: *Bacillus.* In Parker MT, Duerden BI (eds): Topley & Wilson's Principles of Bacteriology, Virology and Immunity, Volume 2: Systematic Bacteriology, 8th ed. Philadelphia: BC Decker, 1990.

Turnbull PCB, Kramer JM: *Bacillus.* In Balows A, et al (eds): Manual of Clinical Microbiology, 5th ed. Washington, DC: American Society for Microbiology, 1991.

Williams RP: *Bacillus anthracis* and other aerobic spore-forming bacilli. In Braude AI, Davis CE, Fierer J (eds): Infectious Diseases and Medical Microbiology, 2nd ed. Philadelphia: WB Saunders, 1986.

CHAPTER 14

Neisseria

Karen S. Long • *John G. Thomas* •
Jean Barnishan

OBJECTIVES

1. List some general characteristics of the *Neisseria* genus.

2. Discuss the function of pili as virulence factors.

3. Define *AHU strains* of *Neisseria gonorrhoeae,* and discuss their role in gonococcal infections.

4. Discuss potential complications of asymptomatic gonococcal infections in women.

5. Define *ophthalmia neonatorum.*

6. List some risk groups for epidemic meningococcal meningitis.

7. Discuss clinical findings in meningococcemia.

8. Discuss the importance and need for correct collection and transportation of specimens for *Neisseria gonorrhoeae.*

9. Compare and contrast the usefulness of the direct Gram stain in the diagnosis of gonorrhea in men and women.

10. List two selective media for *Neisseria gonorrhoeae* and *Neisseria meningitidis.*

11. State the reference method for carbohydrate utilization in the *Neisseria* species.

12. Discuss some drawbacks of nonculture detection methods for *Neisseria gonorrhoeae.*

13. Indicate the reason for ß-lactamase testing of all *Neisseria gonorrhoeae* isolates.

14. Discuss the pathogenic significance of *Moraxella (Branhamella) catarrhalis.*

15. Differentiate the "pathogenic" and "nonpathogenic" species of *Neisseria.*

The family Neisseriaceae currently contains four genera: *Neisseria, Moraxella, Acinetobacter,* and *Kingella.* Characteristics of the family and differential characteristics of these four genera are shown in Figure 14–1. At present, the genus *Neisseria* contains 12 species and biovars that can be isolated from humans; the genus *Moraxella* subgenus *Branhamella* contains one such species.

DNA studies have demonstrated that *Neisseria* are more closely related to *Eikenella, Simonsiella, Alysiella,* and other Centers for Disease Control (CDC) groups; so the family designation may be changed in the future. This chapter discusses only the morphologically and biochemically similar species of *Neisseria* and *Moraxella (Branhamella) catarrhalis.*

have optimal growth in a moist atmosphere. The natural habitat of *Neisseria* species are the mucous membranes of the respiratory and urogenital tracts. Table 14–1 shows the pathogenicity and host range of *Neisseria* and *Moraxella* species.

Neisseria gonorrhoeae and *Neisseria meningitidis* are the primary human pathogens of the genus. *Neisseria gonorrhoeae* is always pathogenic, but *Neisseria meningitidis* may be found as a commensal inhabitant of the upper respiratory tract of carriers. All other *Neisseria* species are considered opportunistic pathogens. It is important, however, to recognize and differentiate these species from *N. gonorrhoeae* and *N. meningitidis* in isolates from clinical specimens. Pathogenic *Neisseria* are fastidious organisms, requiring enriched media for optimal recovery.

GENERAL CHARACTERISTICS

Essentially all species of *Neisseria* are aerobic gram-negative diplococci, cytochrome oxidase and catalase positive; *Neisseria elongata,* which is catalase negative and rod shaped, is the only known exception. *Neisseria* are capnophilic and

PATHOGENIC *NEISSERIA*

Neisseria gonorrhoeae

Humans are the only natural host for *N. gonorrhoeae,* the agent of gonorrhea. Gonorrhea is an

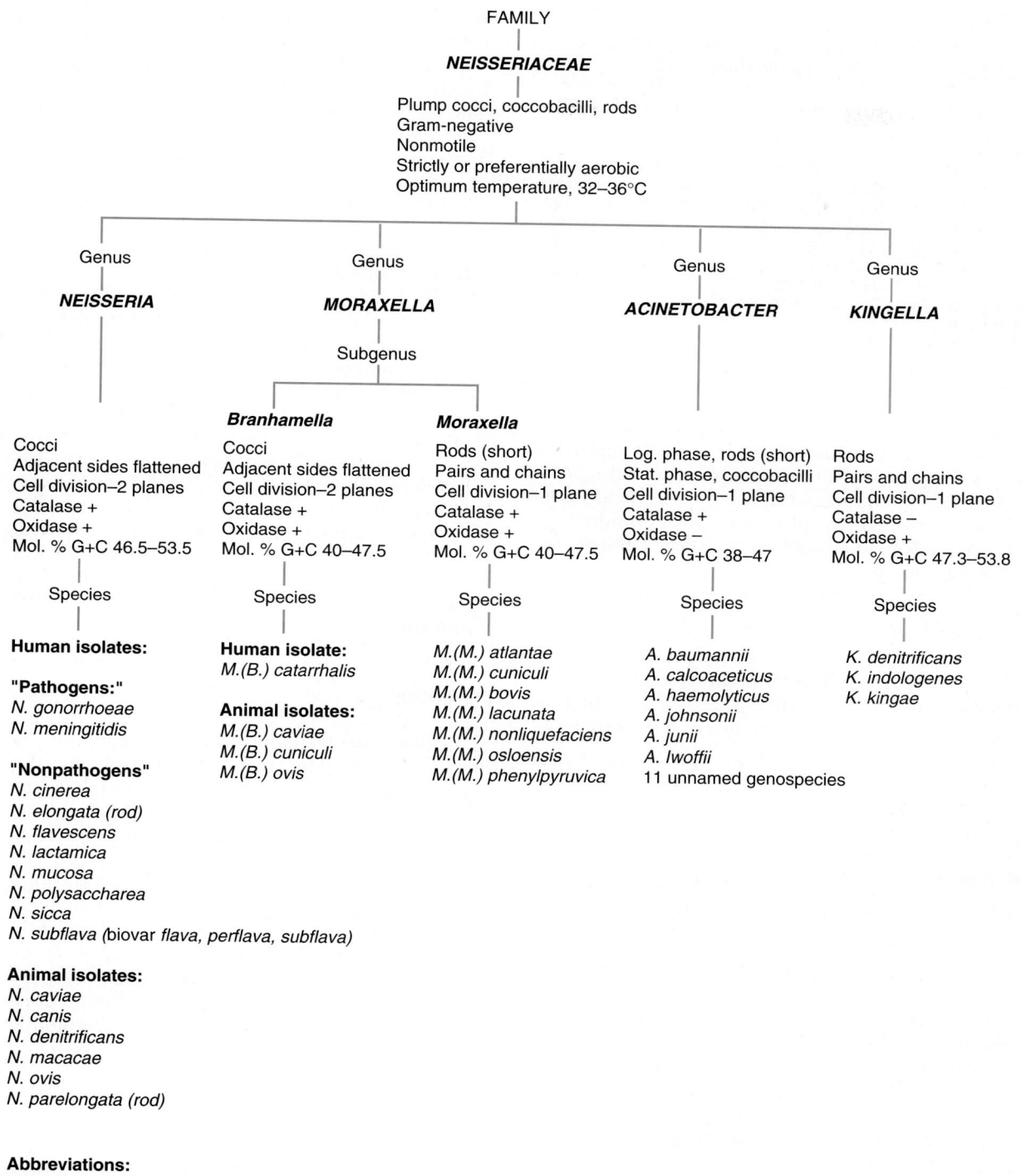

FAMILY

NEISSERIACEAE

Plump cocci, coccobacilli, rods
Gram-negative
Nonmotile
Strictly or preferentially aerobic
Optimum temperature, 32–36°C

Genus **NEISSERIA**

Genus **MORAXELLA**

Genus **ACINETOBACTER**

Genus **KINGELLA**

Subgenus

Branhamella

Moraxella

NEISSERIA

Cocci
Adjacent sides flattened
Cell division–2 planes
Catalase +
Oxidase +
Mol. % G+C 46.5–53.5

Species

Human isolates:

"Pathogens:"
N. gonorrhoeae
N. meningitidis

"Nonpathogens"
N. cinerea
N. elongata (rod)
N. flavescens
N. lactamica
N. mucosa
N. polysaccharea
N. sicca
N. subflava (biovar *flava, perflava, subflava*)

Animal isolates:
N. caviae
N. canis
N. denitrificans
N. macacae
N. ovis
N. parelongata (rod)

Branhamella

Cocci
Adjacent sides flattened
Cell division–2 planes
Catalase +
Oxidase +
Mol. % G+C 40–47.5

Species

Human isolate:
M.(B.) catarrhalis

Animal isolates:
M.(B.) caviae
M.(B.) cuniculi
M.(B.) ovis

Moraxella

Rods (short)
Pairs and chains
Cell division–1 plane
Catalase +
Oxidase +
Mol. % G+C 40–47.5

Species

M.(M.) atlantae
M.(M.) cuniculi
M.(M.) bovis
M.(M.) lacunata
M.(M.) nonliquefaciens
M.(M.) osloensis
M.(M.) phenylpyruvica

ACINETOBACTER

Log. phase, rods (short)
Stat. phase, coccobacilli
Cell division–1 plane
Catalase +
Oxidase –
Mol. % G+C 38–47

Species

A. baumannii
A. calcoaceticus
A. haemolyticus
A. johnsonii
A. junii
A. lwoffii
11 unnamed genospecies

KINGELLA

Rods
Pairs and chains
Cell division–1 plane
Catalase –
Oxidase +
Mol. % G+C 47.3–53.8

Species

K. denitrificans
K. indologenes
K. kingae

Abbreviations:
Log. phase = logarithmic growth phase.
Stat. phase = stationary growth phase.
Mol. % G+C = DNA guanine + cytosine ratio.

■■■■ F i g u r e 1 4 – 1

Characteristics of the family and differential characteristics of the four genera. The genus *Neisseria* currently contains 12 species and biovars isolated from humans. The genus *Moraxella,* subgenus *Branhamella* contains one species isolated from humans.

■■■■■T A B L E 1 4 – 1

PATHOGENICITY AND HOST RANGE FOR SPECIES OF *NEISSERIA* AND *MORAXELLA (BRANHAMELLA)*

Species	Pathogenicity	Infected Host
N. gonorrhoeae	Primary pathogen	Humans only
N. meningitidis	Primary pathogen	Humans only
N. lactamica	Opportunistic pathogen	Warm-blooded animals
N. sicca	Opportunistic pathogen	Warm-blooded animals
N. subflava	Opportunistic pathogen	Warm-blooded animals
N. mucosa	Opportunistic pathogen	Warm-blooded animals
N. flavescens	Opportunistic pathogen	Warm-blooded animals
N. cinerea	Opportunistic pathogen	Warm-blooded animals
N. polysaccharea	Opportunistic pathogen	Warm-blooded animals
N. elongata	Opportunistic pathogen	Warm-blooded animals
M. catarrhalis	Opportunistic pathogen	Humans only

acute pyogenic infection of columnar and transitional epithelium; infection may be established at any site where these cells are found. Gonococcal infections occur primarily in the urethra, endocervix, anal canal, pharynx, and conjunctiva. Disseminated infections from the primary site may also occur.

Virulence Factors

The pathogenic *Neisseria* species have several characteristics that contribute to their virulence. These virulence factors include the presence of:

■ Capsule

■ Pili

■ Cell-wall proteins

■ Lipopolysaccharide (endotoxin)

■ IgA protease that cleaves IgA on mucosal surfaces

A schematic diagram of the cellular structure of *N. gonorrhoeae* is shown in Figure 14–2.

Neisseria gonorrhoeae has been divided into five morphologically distinct colony types, T1 through T5, based on the presence or absence of pili, the fine hair-like projections that are important in the initial attachment of the organism to host tissues. Pili also inhibit phagocytosis of the organism by neutrophils and aid in the exchange of genetic material from cell to cell. Types T1 and T2, which possess pili, are virulent forms, whereas T3 through T5, devoid of pili, are avirulent strains. Piliated organisms usually predominate

when first isolated from uncomplicated genitourinary infections, but upon subculture, pili are lost and colony types T3 through T5 appear. Antigenic variation allows the gonococcus to regain its pili, contributing to the organism's ability to evade the defenses of the immune system.

The capsule, cell wall proteins (proteins I to III), and the lipopolysaccharides act to prevent phagocytosis of the organism. Moreover, the lipopolysaccharide endotoxin is a major in vivo virulence factor of gram-negative bacteria that mediates damage to body tissues.

Protein I (PI), representing about 60% of the total weight of the outer membrane, demonstrates antigenic variability and has been used for enzyme-linked immunosorbent assay (ELISA) and coagglutination tests for gonococcal serotyping. Several distinct PII proteins are detectable; the gonococcus can rapidly change the expression of this protein, which may account for recurrent infections. Strains with high-molecular-weight (HMW) PI and expressing PII are usually found in symptomatic genital infections; whereas low-molecular weight PI strains that lack PIIs are found in disseminated infections. Protein III is believed to be the major binding site on the outer membrane for IgG-blocking antibody.

Epidemiology

Infections are transmitted most commonly by sexual contact. The primary reservoir is the asymptomatic carrier. Since 1965, gonorrhea has been the most commonly reported sexually transmitted bacterial infection in the United States. Reported cases increased steadily through the 1960s and 1970s but declined somewhat through the 1980s. The fear of acquired immune deficiency syndrome (AIDS) and a subsequent reduction in high-risk sexual behavior are thought to contribute to this decline. Actual numbers of infected individuals are probably much higher than reported, owing to a large reservoir of asymptomatic carriers and other unreported cases.

Clinical Infections

DISEASE IN THE MALE

Gonorrhea has a short incubation period, approximately 2 to 7 days after acquiring the organism. In men, acute urethritis, usually result-

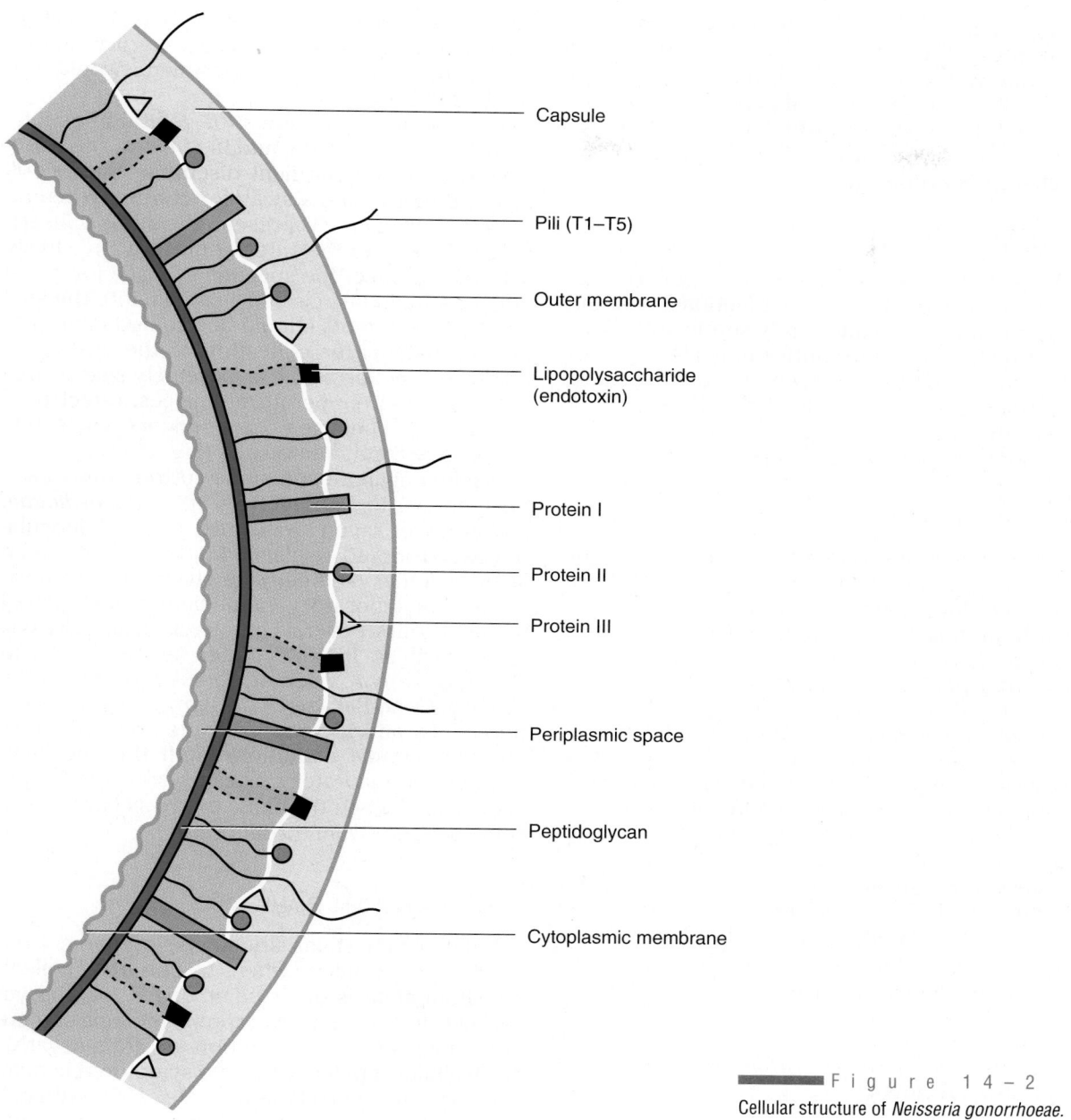

Capsule

Pili (T1–T5)

Outer membrane

Lipopolysaccharide
(endotoxin)

Protein I

Protein II

Protein III

Periplasmic space

Peptidoglycan

Cytoplasmic membrane

Figure 14–2
Cellular structure of *Neisseria gonorrhoeae*.

ing in purulent discharge and dysuria, is the common manifestation. Asymptomatic infection in men is uncommon; only 3 to 5% of cases may be asymptomatic, whereas 95% show acute infections. *Neisseria gonorrhoeae* strains with a nutritional requirement for arginine, hypoxanthine, and uracil (AHU strains) are often isolated from asymptomatic men. Complications in males

include ascending infections such as prostatitis and epididymitis.

DISEASE IN THE FEMALE

The endocervix is the most common site of infection in women, resulting in vaginal discharge and dysuria. Up to 50% of cases in women may be asymptomatic, however. Symptoms of infection, when present, include dysuria, lower abdominal pain, and vaginal bleeding. Untreated gonococcal cervicitis is a major cause of pelvic inflammatory disease (PID) in women, which may cause sterility, ectopic pregnancy, or perihepatitis (Fitz-Hugh–Curtis syndrome).

DISSEMINATED INFECTIONS

Blood-borne dissemination of *N. gonorrhoeae* occurs in less than 1% of all infections, resulting in purulent arthritis and rarely septicemia. Fever and a rash on the extremities may also be present. The majority of disseminated gonococcal infections are attributed to the AHU strains and occur in women, having been acquired from infected but asymptomatic males.

INFECTIONS IN OTHER SITES

Other conditions associated with *Neisseria gonorrhoeae* include anorectal and oropharyngeal infections. Infections in these sites are more common in homosexual males but can also occur in women. Most infections are asymptomatic or have nonspecific symptoms. Pharyngitis is the chief complaint in symptomatic oropharyngeal infections, whereas discharge, rectal pain, or bloody stools may be seen in rectal gonorrhea. Approximately 30 to 60% of females with genital gonorrhea have concurrent rectal infection.

Newborns can acquire ophthalmia neonatorum, a gonococcal eye infection, during vaginal delivery through an infected birth canal. This condition, which can result in blindness if not immediately treated, is rare in the United States because application of antimicrobial eye drops or silver nitrate is legally required at the birth of every infant. Ocular infections can occur in adults owing to inoculation of the eye with infected genital secretions or, rarely, as a result of a laboratory accident.

Laboratory Diagnosis

SPECIMEN COLLECTION, TRANSPORT, AND PROCESSING

Specimens collected for the recovery of *N. gonorrhoeae,* as noted previously, may come from typical sources or from other sites, such as the rectum, blood, pharynx, and joint fluid. It is important, however, that the laboratory be noti-fied when cultures from sites such as the rectum or throat are requested, because normal laboratory protocols for such specimens would not recover the organism.

The specimen of choice for genital infections in males is the urethra, and in females, the endocervix. In males, purulent discharge can be collected directly onto a swab for culture. When no apparent discharge is present, the swab is inserted 2 to 3 cm into the anterior urethra and slowly rotated to collect the specimen. Swabs for rectal culture should be inserted 4 to 5 cm into the anal canal. Disinfectants should be avoided in preparing the patient for collection of the specimen. Because *N. gonorrhoeae* is extremely susceptible to drying and temperature changes, direct plating of the specimen to gonococcal-selective media gives optimum results (Fig. 14–3).

Calcium alginate and some cotton swabs have been shown to be inhibitory to *N. gonorrhoeae,* so Dacron or rayon swabs are preferred. Inoculated swabs should be placed in a transport system such as Amies medium with charcoal, transported to the laboratory immediately, and plated within 6 hours. Several commercial transport systems, such as JEMBEC plates (*J*ames *E*. *M*artin *B*iological *E*nvironmental *C*hamber) (Fig. 14–4), Bio-Bag, Gono-Pak, and Transgrow, contain selective media and a carbon dioxide atmosphere to provide optimal conditions until the specimen reaches the laboratory. These systems are especially useful when the clinic or physician office is some distance from the laboratory.

DIRECT MICROSCOPIC EXAMINATION

Direct Gram stains should be prepared from urogenital specimens when the culture is collected. Gram stain is not recommended for pharyngeal specimens, in which saprophytic *Neisseria* are often present. Demonstration of gram-negative intracellular diplococci from a symptomatic male with discharge correlates at a rate of 95% with culture and is presumptive evidence of gonococcal infection (Fig. 14–5*A*). Because women have vaginal and cervical saprophytic flora that resemble gonococci, direct Gram stain correlates in only 50 to 70% of cases with culture. The direct Gram stain may be helpful in a symptomatic woman with discharge, but culture is necessary for confirmation.

A Gram stain with more than five polymorphonuclear neutrophils (PMNs) per field but no bacteria may suggest nongonococcal urethritis

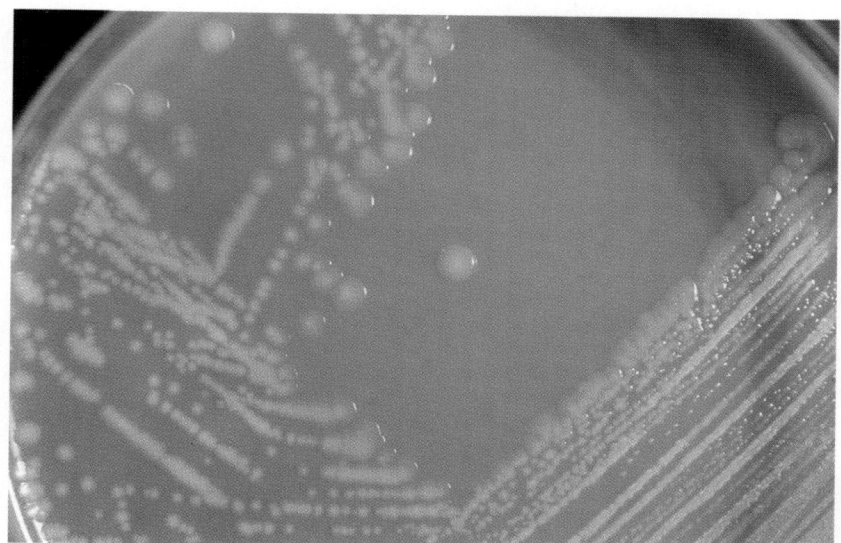

■ F i g u r e   1 4 – 3
Virulent *Neisseria gonorrhoeae* after 24
hours of growth on modified Thayer-
Martin (MTM) agar.

with organisms such as *Chlamydia trachomatis* or *Ureaplasma urealyticum* (Fig. 14–5*B*).

CULTURE

Cultivation of *N. gonorrhoeae* requires the use of chocolate agar, but this enriched medium also supports the growth of many other organisms found as saprophytes in specimens collected for recovery of gonococci. To prevent overgrowth of the normal flora and to enhance the recovery of the pathogenic species, a selective medium containing inhibitors for gram-negative and gram-positive organisms and yeast is used. Commonly used selective media are described in Table 14–2.

New York City, a transparent agar, has the added advantage of supporting the growth of the possible urogenital pathogens *Mycoplasma hominis* and *U. urealyticum*. Over the past several years, reports have noted that up to 10% of *N. gonorrhoeae* strains are sensitive to vancomycin. In order to recover these sensitive strains, it is good practice to include a nonselective chocolate agar as a primary plating medium.

It must be noted that several other genera will grow on selective gonococcal media. Some of these are *Acinetobacter* sp, *Capnocytophaga* sp, and *Kingella denitrificans*. These are differentiated from *N. gonorrhoeae* by the oxidase and catalase tests.

All specimens received in the laboratory for recovery of *Neisseria* should be held at room temperature and plated as soon as possible. Media should be warmed to room temperature before inoculation, because *Neisseria* are suscep-

tible to cold temperatures. Specimens received on swabs should be rolled in a Z pattern on the media and cross-streaked with a loop to facilitate growth of isolated colonies.

INCUBATION

Inoculated plates should be incubated at 35°C in a 3 to 5% carbon dioxide atmosphere. This is easily accomplished by use of a candle extinction jar (Fig. 14–6) or a commercially available CO_2 incubator. Sufficient humidity is provided by the moisture evaporating from the media in a closed jar; most commercially available CO_2 incubators are automatically humidified, or a

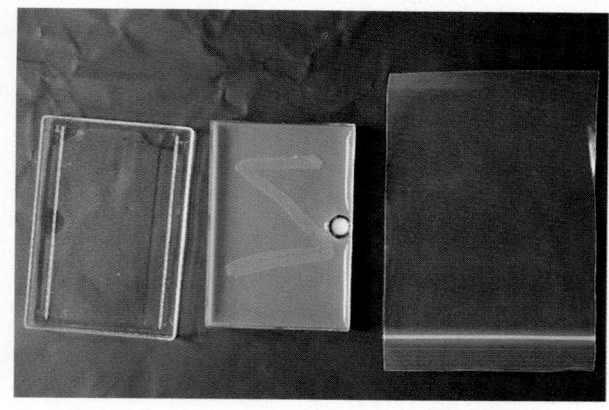

■ F i g u r e   1 4 – 4
Twenty-four hour growth of *Neisseria gonorrhoeae* on a JEMBEC plate streaked in a characteristic "Z" pattern.

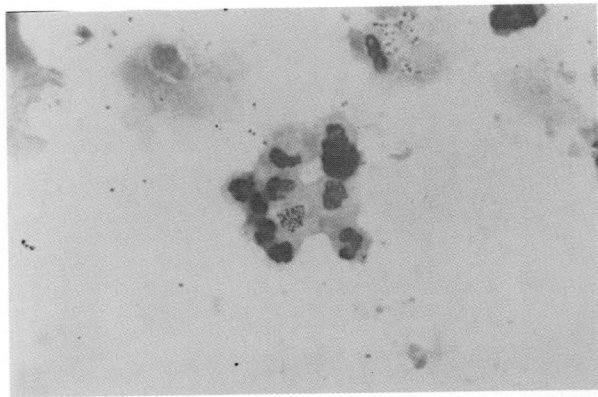

A

Figure 14 – 5

A, Direct Gram-stained smear of male urethral discharge showing intracellular and extracellular gram-negative diplococci, which is diagnostic of *Neisseria gonorrhoeae. B,* A direct smear with more than five PMNs per fluid but no bacteria may suggest nongonococcal urethritis (NGU).

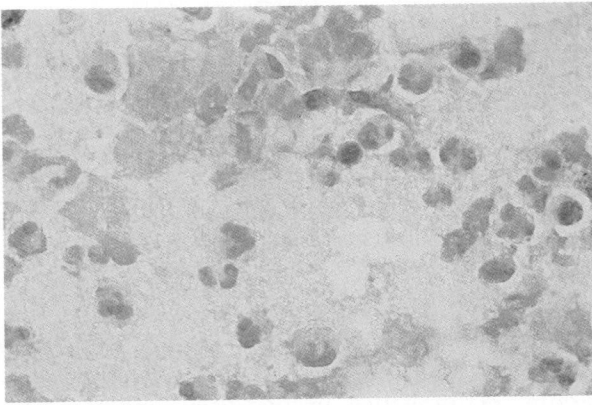

B

pan of water can be placed in the bottom. Scented or colored candles may be inhibitory to the gonococci, so only white wax candles are used in the candle extinction jar.

IDENTIFICATION
Presumptive Identification
Microscopic Morphology. The Gram stain must be performed on all suspected *Neisseria* isolates to verify the appearance of gram-negative kidney bean–shaped diplococci. Some gram-negative rods, such as *Kingella* and *Acinetobacter,* are occasionally able to grow on gonococcal-selective media. In order to differentiate these from the gram-negative diplococci, the organism can be streaked to a plate with a 10-U penicillin disk added (Fig. 14–7). After growth, the edge of the zone of inhibition is stained to visualize the morphology.

Colonial Morphology. Cultures are examined daily for growth and held for 72 hours. Colonies of *N. gonorrhoeae* on chocolate or selective agar

are small, gray, translucent, and raised after 24 to 48 hours of incubation. As noted previously, five colony types of *N. gonorrhoeae* have been described: T1 and T2 have pili and are considered virulent types; these colonies are smaller and raised, and they appear bright in reflected light. Types T3 through T5 do not have pili and usually grow as larger, flatter colonies. The AHU (atypical) strains produce smaller colonies that grow more slowly; they are often more difficult to identify with biochemical methods.

The gonococci are able to produce autolytic enzymes that may make the isolate nonviable, so primary plates should not be incubated in the candle jar once sufficient growth has been obtained. A fresh subculture of the organism should be used for identification tests.

Oxidase Test. The filter paper or direct plate oxidase test must be done on all isolates. In the filter paper method, oxidase reagent (1% dimethyl- or tetramethyl-*p*-phenylene-diamine-

■■■■■ T A B L E   1 4 – 2

SELECTIVE MEDIA FOR THE ISOLATION OF *NEISSERIA GONORRHOEAE* AND *NEISSERIA MENINGITIDIS*

Selective Medium	Inhibitory Agents	Suppressed Organisms
Thayer-Martin (TM)	Vancomycin	Gram-positive
	Colistin	Gram-negative
	Nystatin	Yeast
Modified Thayer-Martin (MTM)	Vancomycin	Gram-positive
	Colistin	Gram-negative
	Nystatin	Yeast
	Trimethoprim	Swarming *Proteus*
Martin-Lewis (ML)	Vancomycin	Gram-positive
	Colistin	Gram-negative
	Anisomycin	Yeast
	Trimethoprim lactate	Swarming *Proteus*
New York City (NYC)	Vancomycin	Gram-positive
	Colistin	Gram-negative
	Amphotericin B	Yeast
	Treimethoprim lactate	Swarming *Proteus*
GC-LECT	Vancomycin	Gram-positive
	Lincomycin	Gram-positive
	Colistin	Gram-negative
	Amphotericin B	Yeast
	Trimethoprim lactate	Swarming *Proteus*
		Capnocytophaga spp

have advantages and disadvantages. The selection of a particular method depends on factors such as the demographic profile of the patients, sensitivity and specifity of the method with low- or high-prevalence groups, cost of materials and technical time, and number of tests performed.

Carbohydrate Utilization Methods. The traditional method for the identification of *Neisseria* species has been by carbohydrate utilization in cystine trypticase agar (CTA), containing 1% of the individual carbohydrate with phenol red as an indicator. A yellow color is produced in 24 to 72 hours if the organism utilizes the particular carbohydrate, as shown in Figure 14–9 and described in Procedure 14–1. Many problems are associated with this method, however, and it has largely been replaced by newer, faster, and more accurate methods.

dihydrochloride) is placed on filter paper, and a colony from the plate is rubbed onto the reagent with an applicator stick or a non-nichrome needle. A purple color should develop in 10 seconds in a positive reaction in a fresh isolate (Fig. 14–8A). Alternatively, the oxidase reagent may be dropped directly on a colony. The colony turns pink then black in a positive reaction (Fig. 14–8B). If subculture of the positive colony is needed, it must be done while the colony is still pink; when black, the organism is no longer viable.

Definitive Identification. It was once acceptable to make a presumptive diagnosis of gonorrhea if an oxidase-positive, gram-negative diplococcus was recovered from gonococcal-selective media, but this procedure is no longer recommended. Oxidase-positive, gram-negative diplococci, such as *Neisseria cinerea* and *N. meningitidis,* as well as *M. catarrhalis* can grow on selective media from sites where gonorrhea is expected. These organisms would be incorrectly reported as *N. gonorrhoeae* if no further identification were done.

Many different methods are currently used for the speciation of *Neisseria* and *Moraxella (Branhamella)* or for confirmation only of *N. gonorrhoeae* isolates. Both culture and nonculture tests are available for the detection of *N. gonorrhoeae* and are listed by type in Table 14–3. All

■■■■■ F i g u r e   1 4 – 6

Candle extinction jar with inoculated modified Thayer-Martin agar plates.

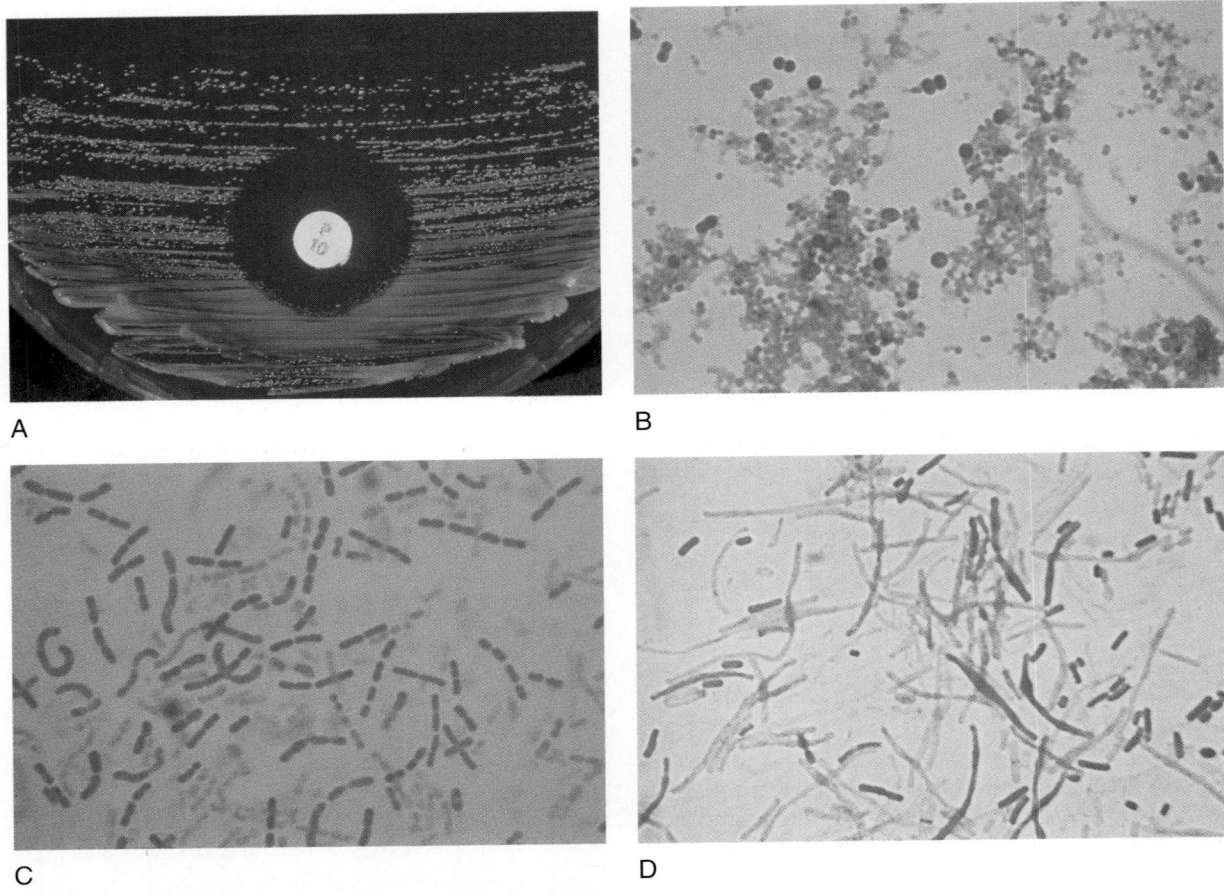

■ F i g u r e 1 4 - 7

A, In order to differentiate some gram negative rods from the gram-negative diplococci, the organism can be streaked to a plate and a 10-unit penicillin disk added. After growth, the edge of the zone of inhibition is stained to visualize microscopic morphology. *B,* The microscopic morphology of *Neisseria gonorrhoeae* remains a gram-negative diplococci using the 10-U penicillin disk. *C,* The gram-negative rod microscopic morphology of *Kingella* after the penicillin disk test. *D,* The elongated gram-negative rod microscopic morphology of *Acinetobacter* after the penicillin disk test.

The rapid carbohydrate degradation tests require pure cultures, but can be read in 2 to 4 hours rather than the 24 to 72 hours needed for the CTA carbohydrate test. These rapid tests also detect acid production from various carbohydrates, but they are based on the presence of preformed enzymes for carbohydrate utilization rather than on bacterial growth. Problems noted with these methods, however, include:

■ Weak acid production from glucose by certain strains of *N. gonorrhoeae*

■ Misidentification of sucrose-negative strains of *Neisseria subflava* as *N. meningitidis*

■ Species of *N. cinerea* that give positive glucose reactions

Additional tests such as superoxol or colistin susceptibility are needed when acid production is equivocal. The superoxyl test uses 30% H_2O_2 and is performed in the same way as the catalase test. Colonies of *N. gonorrhoeae* produce immediate, vigorous bubbling. *N. meningitidis* and *N. lactamica* produce weak delayed bubbling in this reagent. Other *Neisseria* spp produce weak delayed bubbling or none at all.

Chromogenic substrate tests detect enzymes that hydrolyze the substrates and produce colored end products. Only strains that are isolated on selective media should be tested, but the advantage of these tests is the identification of *Neisseria* strains with aberrant carbohydrate utilizations. Problems noted with these tests include misidentification of nonpathogenic species, such

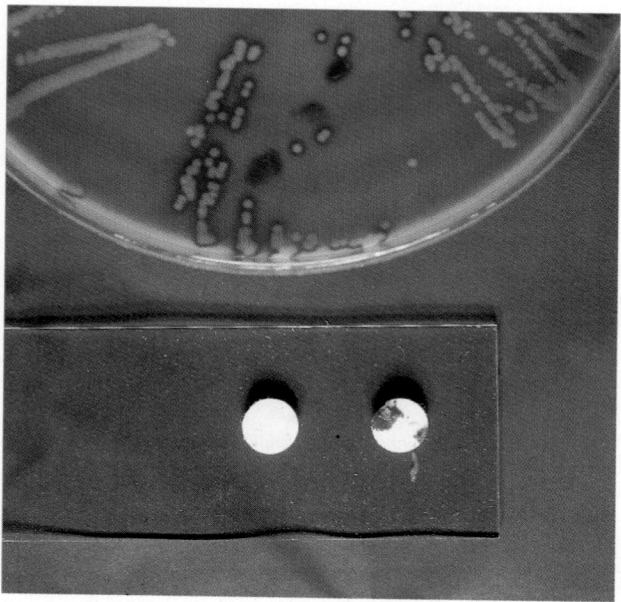

A

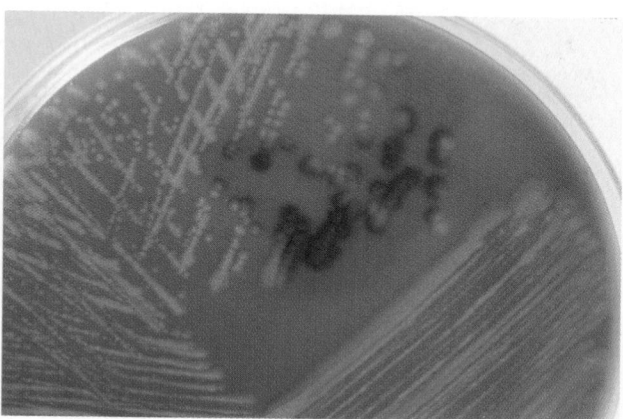

B

■■■■■■■Figure 14 – 8

A, The oxidase disk test. Negative control is on the left and the positive reaction (purple color) is on the right. *B,* Example of a positive oxidase reaction when the reagent is dropped directly on the colony.

as *N. cinerea, N. sicca, N. subflava,* and *N. mucosa,* as *N. gonorrhoeae* or *N. meningitidis.*

The multitest conventional-chromogenic enzyme methods combine enzyme substrate tests with other biochemical tests and allow identification of strains isolated on selective or nonselective media. These tests can also speciate other genera, such as *Haemophilus.*

Characteristics of significant species of *Neisseria, Moraxella (Branhamella),* and *Kingella* are shown in Table 14–4. Key differentiating reactions of the major pathogens include utilization of glucose only by *N. gonorrhoeae,* whereas *N. meningitidis* utilizes both glucose and maltose. *Moraxella (Branhamella) catarrhalis* is asaccha-rolytic but, unlike the *Neisseria* species, is DNase and butyrate esterase positive.

Immunologic Methods. Immunologic methods employ monoclonal antibodies for the identification of *N. gonorrhoeae.* These methods do not require pure or viable organisms and can be done from the primary plates. Immunologic methods include coagglutination and fluorescent antibody testing.

The *coagglutination tests* use monoclonal antibodies attached to a carrier particle; agglutination occurs with *N. gonorrhoeae.* There is no reported cross-reaction with *N. cinerea,* but rare isolates of *Neisseria lactamica* have been reported as *N. gonorrhoeae.*

■■■■ T A B L E   1 4 - 3

SELECTED TEST METHODS FOR IDENTIFICATION OF *NEISSERIA* AND RELATED SPECIES

Method Type and Name	Manufacturer(s)	Principle	Notes
Conventional Cystine trypticase agar (CTA) with carbohydrates (1%)	Various	Acid production from carbohydrate utilization	Requires pure culture Must be incubated 24–48 hr
Rapid carbohydrate degradation		Acid production from carbohydrate utilization; detected by preformed enzymes	Requires pure culture Read in 2–4 hr
MINITEK	BBL Microbiology Systems (Cockeysville, MD)		
QuadFERM +	API Analytab Products (Plainview, NY)		Includes acidometric β-lactamase
Neisseria-Kwik Gonobio-Test	Micro-biologics (St. Cloud, MN) I.A.F. Production, Inc. (Laval, Quebec, Canada)		
Chromogenic substrate Gonochek II	E.Y. Laboratories (San Mateo, CA)	Detects enzyme production	Confirms isolates only from selective media
Modified conventional/ chromogenic enzyme		Combines enzyme substrate with other biochemical tests Speciates *Neisseria* and *Haemophilus*	Isolates may be from selective or nonselective media
Neisseria-Haemophilus Identification (NHI)	Vitek Systems, Inc. (Hazelwood, MO)		
RapID NH System	Innovative Diagnostic Systems, Inc. (Atlanta, GA)		
Microscan HNID Panel	Baxter Healthcare Corp., Microscan Division, (W Sacramento, CA)		
Immunologic coagglutination		Monoclonal antibodies used to detect *N. gonorrhoeae* Slide agglutination	Does not require pure or viable culture
Phadebact Monoclonal GC Omni Test	Pharmacia Diagnostics (Piscataway, NJ)		
Gonogen II	New Horizons, Diag Corp. (Columbia, MD)		
FA monoclonal antibody Syva Microtrak *Neisseria gonorrhoeae* Culture confirmation test	Syva Co. (Palo Alto, CA)	Fluorescent-labeled mono- clonal antibodies used to detect *N. gonorrhoeae*	Does not require pure culture Requires UV microscope
Nonculture tests			Not recommended for rectal or pharyngeal sites Cannot idenfity β-lactamase– producing strains Provides no isolate for susceptibility testing
Gonozyme	Abbott Laboratories (N. Chicago, IL)	Enzyme method (EIA) for direct detection of gonococcal antigens from specimen	
Neisseria gonorrhoeae PACE 2	Gen-Probe (San Diego, CA)	DNA probe for direct detection gonococcal rNA from specimen	Uses a chemiluminescent labeled, single-stranded DNA probe
AccuProbe Neisseria Culture Confirmation Test	Gen-Probe (San Diego, CA)	Nucleic acid hybridization for confirmation of *N. gonorrhoeae* culture	Uses a chemiluminescent labeled, single-stranded DNA probe

The *fluorescent antibody (FA) method* uses monoclonal antibodies that recognize epitopes on the principal outer membrane protein (PI) of *N. gonorrhoeae*. The method is extremely sensitive and specific; there has been no demonstrated cross-reactivity with *N. cinerea* or other *Neisseria* species. The FA method has the added advantage of microscopically confirming the morphologic appearance of the diplococci.

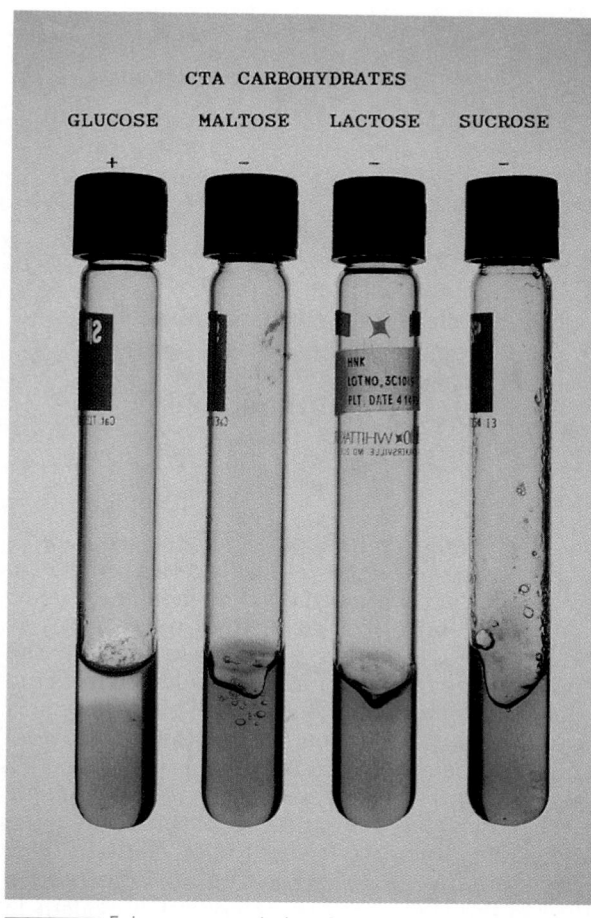

CTA CARBOHYDRATES

GLUCOSE	MALTOSE	LACTOSE	SUCROSE
+	–	–	–

■ F i g u r e   1 4 – 9

Conventional CTA sugars. A yellow color at the top of the media is considered positive. Acid is produced in the glucose tube only, identifying this organism as *Neisseria gonorrhoeae.*

Nonculture Methods. Alternatives to culture for *N. gonorrhoeae* are available. These methods detect gonococcal antigen or nucleic acid directly in cervical or urethral exudates. Currently available are enzyme-linked immunosorbent assays (ELISAs) and nucleic acid probe technology.

The *nucleic acid probe test* is a nonisotopic chemiluminescent DNA probe that hybridizes specifically with ribosomal RNA (rRNA) of *N. gonorrhoeae.* These methods are rapid and sensitive, but they do have some drawbacks. Some disadvantages of the nucleic acid probes include the following:

■ Lesser sensitivity than cervical culture in females, so should be used in high-risk populations only

■ Not approved for pharyngeal or rectal specimens, and should not be used in children or sexual abuse cases

■ Cannot identify a *N. gonorrhoeae* infection produced by a β-lactamase–producing strain

■ Do not allow for recovery of an organism to be used for susceptibility testing

■ False-positive results reported with some strains of *N. lactamica* and *N. cinerea*

Probe technology can be very expensive. This method is probably not cost-effective if it is the only type of probe test used in the laboratory or if low volumes of *N. gonorrhoeae* tests are requested. Added to the expense is the need to repeat the probe test when equivocal results are obtained.

As noted previously, it is highly recommended that in low-prevalence populations, such as children, or in cases of suspected sexual abuse direct specimen detection methods not be used. The specimens should be cultured onto selective gonococcal media, and isolates must be identified to species level by the use of two procedures with different principles. The organism should also be frozen in case further testing is needed. With various methods, isolates such as *N. cinerea, M. catarrhalis,* and *K. denitrificans,* all of which closely resemble *N. gonorrhoeae* biochemically, have been incorrectly reported as *N. gonorrhoeae.* Potentially serious medical, social, and legal ramifications could obviously result.

Additional and Future Methods. As mentioned previously, some strains of *N. gonorrhoeae* have a specific need for one or more nutritional factors to be included in artificial media for growth to occur. These strains are called auxotypes; of the approximately 30 known auxotypes, the most common is AHU, which requires arginine, hypoxanthine, and uracil. The AHU strains are usually highly susceptible to penicillin, are commonly recovered from patients with disseminated gonococcal infection, and can cause asymptomatic urethritis in males.

The auxotyping procedure is labor intensive and is not recommended for the routine clinical microbiology laboratory.

Detection of *N. gonorrhoeae* antigens in clinical specimens has also been accomplished utilizing a *polymerase chain reaction (PCR) amplification and hybridization technique.* The method is highly sensitive and specific, and is now available

---

PROCEDURE 14–1. CTA CARBOHYDRATE METHOD

1. Prepare a heavy inoculum in saline of a pure isolate from a fresh subculture of the organism. The subculture must *not* be from a selective agar.

2. Inoculate the top portions only of the following CTA media: glucose, maltose, lactose, and sucrose, plus a carbohydrate-free CTA control.

3. Incubate tightly capped tubes at 35 to 37°C in a non–carbon dioxide atmosphere.

Incubation in an acid CO_2 atmosphere can produce the appearance of carbohydrate degradation in all tubes and, hence, false-positive reactions.

4. A yellow color at the top of the media in 24 to 48 hours is considered positive. A bright yellow color throughout the media usually indicates contamination with other organisms.

---

commercially. A PCR method may be available shortly to allow detection of both *N. gonorrhoeae* and *C. trachomatis* from the same specimen.

Antimicrobial Resistance

Until 1976, almost all strains of *N. gonorrhoeae* in the United States were susceptible to penicillin. The first plasmid-mediated penicillinase-producing gonococcal (PPNG) strains were isolated in that year, largely imported from Southeast Asia or Africa. By 1980, more than half of the reported PPNG cases were of domestic origin. Penicillinase-producing gonococcal strains are now endemic in New York City, Los Angeles, and Florida, with as many as 30% of isolates in some places exhibiting this type of resistance. It is, therefore, important that all *N. gonorrhoeae* isolates recovered be tested for β-lactamase (penicillinase) production. The β-lactamase test should be performed from isolates on the primary culture plate, because the plasmid may be lost on subculture.

The CDC has developed a scheme to define PPNG prevalence in various locations. If less than 1% of gonorrhea cases reported are PPNG, the area is defined as *nonendemic*. An *endemic* area is characterized as 1 to 3% of cases being PPNG, whereas *hyperendemic* areas are defined as having a greater than 3% incidence of PPNG. This method of surveillance reporting is still evolving.

MECHANISMS OF RESISTANCE

There are several means by which antimicrobial resistance may develop. In addition to plasmid-mediated penicillin resistance, in which the organism acquires a new plasmid with genes for β-lactamase production, chromosome-mediated resistance was initially noted in the United States in 1983. These isolates are β-lactamase negative; the resistance mutants are randomly selected from the gonococcal population. Resistance of these strains is due to a combination of mutations at several chromosomal loci. Chromosomal resistance to penicillin or tetracycline is usually low level. High-level tetracycline resistance is plasmid mediated and was first observed in the United States in 1985.

Spectinomycin resistance, first reported in 1981 in the United States, is not yet prevalent in this country. Resistance to spectinomycin is due to a chromosomal mutation resulting in high-level resistance to the antibiotic. The mechanisms of resistance of *N. gonorrhoeae* are summarized in Table 14–5.

Treatment

Ceftriaxone is currently recommended for gonococcal therapy by the U. S. Public Health Service. This agent is active against organisms with plasmid or chromosomal resistance to other antimicrobials and cures uncomplicated gonorrhea of the pharynx and rectum as well as urogenital sites. As chromosomal resistance to penicillin increases, however, resistance to ceftriaxone may occur.

In 1990, the National Committee for Clinical Laboratory Standards (NCCLS) issued revised guidelines for dilution antimicrobial susceptibility tests (Document M7-A2). A disk susceptibility revi-

CHARACTERISTICS OF SIGNIFICANT SPECIES OF *NEISSERIA*, *BRANHAMELLA*, AND *KINGELLA**

Characteristic	N. gonorrhoeae	N. meningitidis	N. lactamica	N. cinerea	N. sicca	N. flavescens	Moraxella (Branhamella)	Kingella
Pigment on nutrient agar	–	–	–	–	+	+	–	–
Catalase on 3% H_2O_2	+	+	+	+	+	+	+	–
Superoxol (30% H_2O_2)	+	–	–	–	–	+	–	–
Growth on								
MTM, ML, NYC	+	+	+	d	–	–	d	+
Nutrient medium @ 35°C	–	–	+	+	+	+	+	–
Acid production from								
Glucose	+	+	+	+†	+	–	–	+
Maltose	–	+	+	–	+	–	–	–
Sucrose	–	–	–	–	+	–	–	–
Lactose	–	–	+	–	–	–	–	–
Fructose	–	–	–	–	+	–	–	–
DNase	–	–	–	–	–	–	+	–
Reduction of								
NO₃	–	–	–	–	–	–	+	+
NO₂‡	–	d	d	+	+	+	+	–
Tributyrin hydrolysis	–	–	–	–	–	–	+	NT
Enzymes produced								
β-*D*-Galactosidase	–	–	+	–	–	NT	–	–
γ-Glutamylamino-peptidase	–	+	–	–	–	NT	–	–
Hydroxyprolylamino-peptidase	+	–	–	+	NT	+	–	+

*+, most strains (>90%) positive; –, most strains (>90%) negative; d, some strains positive, some negative; MTM, modified Thayer-Martin agar; ML, Martin-Lewis agar; NT, not tested; NYC, New York City medium.
†Occasional strains may give weak glucose reactions in some rapid carbohydrate tests.
‡0.1% (W/V) nitrate.

■ T A B L E   1 4 – 5
RESISTANCE OF *NEISSERIA GONORRHOEAE* IN THE UNITED STATES

Type	Acronym*	First Observed	Mechanism
Plasmid-mediated penicillin resistance	PPNG	1976	Plasmid codes for β-lactamase production
Chromosome-mediated resistance	CMRNG	1983	Selection of mutants for low-level penicillin and tetracycline resistance
Plasmid-mediated, high-leval tetracycline resistance	TRNG	1985	Plasmid codes for tetracycline resistance
Chromosome-mediated spectinomycin resistance	—	1981	Mutation causes high-level resistance

*PPNG, penicillinase-producing *N. gonorrhoeae;* CMRNG, chromosomally mediated resistant *N. gonorrhoeae;* TRNG, tetracycline-resistant *N. gonorrhoeae.*

sion (Document M100-S4) was released in 1992. These documents have expanded considerably the number of reportable antimicrobials. The NCCLS guidelines should be followed in the performance of susceptibility testing, because criteria have been developed specifically for testing of *N. gonorrhoeae* isolates, including appropriate control strains.

Neisseria meningitidis

Although the meningococcus *N. meningitidis* can be found in the nasopharynx and oropharynx of 3 to 30% of asymptomatic individuals, it is an etiologic agent of endemic and epidemic meningitis, meningococcemia, and pneumonia. *Neisseria meningitidis* has also been recovered from urogenital and rectal sites as a result of oral-genital contact.

Virulence Factors

The virulence factors associated with *N. meningitidis* include pili, the polysaccharide capsule, and endotoxin production. Many virulent strains produce IgA$_1$ protease, an enzyme that aids invasiveness. Of the 13 meningococcal serogroups, the encapsulated strains A, B, C, Y, and W-135 are most often associated with epidemics. Group A strains are often incriminated in worldwide epidemics. Serogroups B and C are most common in the United States, with Group B frequently involved in community-acquired disease. Serogroup Y primarily causes meningococcal pneumonia, whereas W-135 is often responsible for invasive disease.

Clinical Infections

The primary source of epidemic meningitis are respiratory droplets from asymptomatic carriers, especially among close contacts in closed populations such as college dormitories and military barracks. Epidemic meningitis most often occurs in young adults. It is characterized by abrupt onset of frontal headache, stiff neck, and, sometimes, fever. Meningococcemia, or sepsis, may occur with or without meningitis and carries a 25% mortality rate, even if treated. Petechial skin lesions may develop during bacteremic spread, and thrombosis is common. In some cases, the disease spreads rapidly, causing disseminated intravascular coagulation (DIC), septic shock, hemorrhage in the adrenal glands (Waterhouse-Friderichsen syndrome; Fig. 14–10), and death. Individuals with a deficiency in complement components C5 to C8 are at increased risk of meningococcemia. Meningococcal pneumonia usually affects older individuals with underlying pulmonary problems.

Treatment

The drug of choice for treatment is penicillin, but rifampin or a sulfonamide is recommended as prophylaxis for close contacts.

Laboratory Diagnosis

SPECIMEN COLLECTION AND TRANSPORT

Culture specimens for *N. meningitidis* may come from a wide variety of sterile and nonsterile sites. These include cerebrospinal fluid (CSF), blood, nasopharyngeal swabs and aspirates, and, less commonly, sputum and urogenital sites. Collection and transport should be performed as specified by the laboratory for the various specimen types.

When commercial blood culture systems are used, data from the manufacturer should be con-

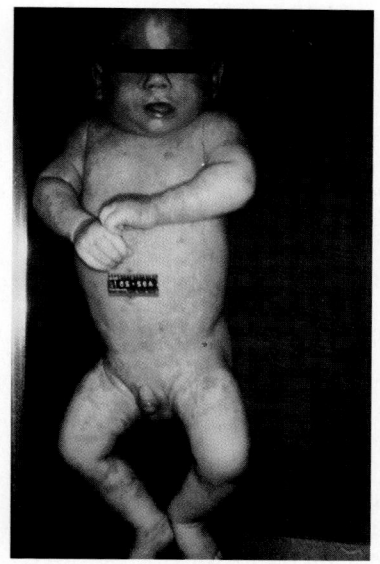

A

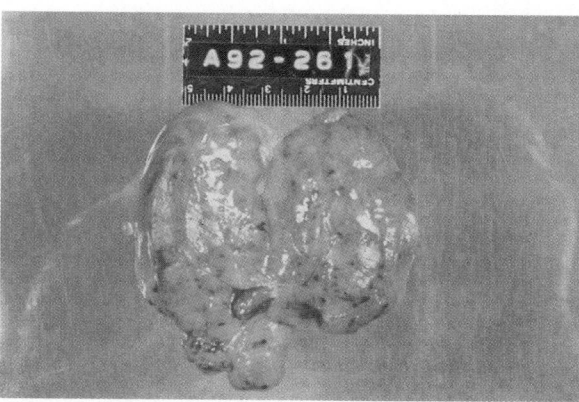

B

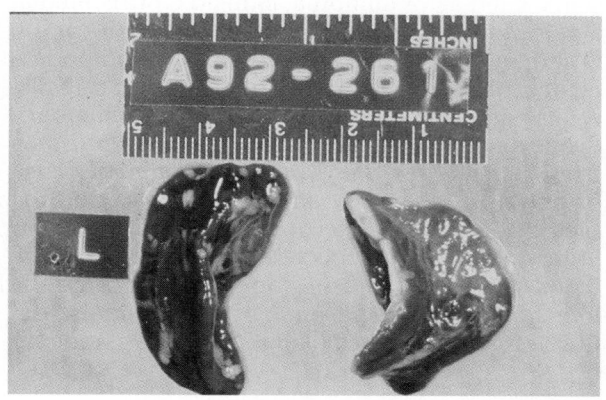

C

Figure 14–10

A, Petechial skin rash associated with meningococcemia in a baby. *B and C,* Waterhouse-Friderichsen syndrome illustrating hemorrhage in the adrenal glands.

sulted to see whether *N. meningitidis* can be routinely recovered or if techniques such as blind subculture (subculture to chocolate agar of bottle with no apparent visual growth) are required.

DIRECT MICROSCOPIC EXAMINATION

On Gram-stained smears from specimens such as CSF, the meningococci appear as intracellular and extracellular gram-negative diplococci (Fig. 14–11). Encapsulated strains may have a halo around the organisms. The highest yield of positive CSF Gram stains is obtained when specimens are concentrated. A number of investigators have reported that concentration by cytocentrifugation is superior to that by traditional centrifuga-

tion and has the potential to increase Gram stain detection by 10 to 100 fold.

In a patient with disseminated meningococcemia who has petechiae from hemorrhage of surface blood vessels, an impression Gram stain smear is often positive for gram-negative diplococci.

CULTURE

Culture using selective and nonselective media for the isolation of *N. meningitidis* and *M. catarrhalis* should, like cultures for *N. gonorrhoeae,* be incubated under carbon dioxide. Both these species, as well as the saprophytic *Neisseria* species, will grow on sheep's blood agar as well as chocolate agar. *Neisseria meningitidis* will grow on

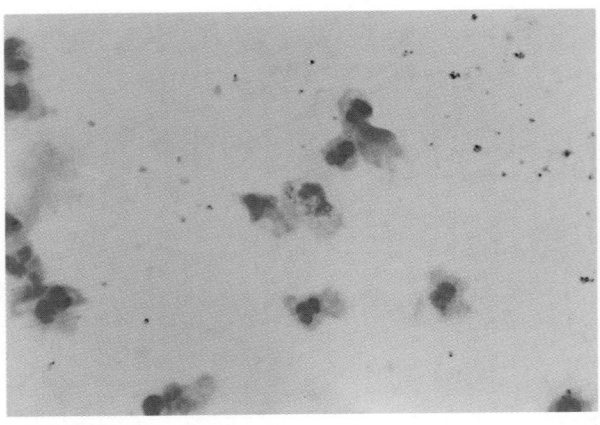

■ F i g u r e   1 4 – 1 1

Direct Gram-stained smear of CSF illustrating intracellular and extra-cellular gram-negative diplococci of *Neisseria meningitidis.*

gonococcal-selective agars and produce small gray, sometimes mucoid, convex colonies on sheep's blood or chocolate agar (Fig. 14–12). *Moraxella (Branhamella) catarrhalis* is usually inhibited by the colistin in the medium, but some species resistant to colistin may grow. *Neisseria lactamica,* a generally nonpathogenic species that may mimic *N. meningitidis,* can also grow on selective media. The lactose-positive characteristic may be delayed or nonexistent. A rapid *o*-nitro-phenyl-β-D-galactopyranoside (ONPG) test (which detects lactose utilization) is usually positive in 30 minutes for *N. lactamica.*

Some species of the saprophytic *Neisseria* may be yellow and often produce dry colonies on sheep's blood and chocolate agar.

IDENTIFICATION

The oxidase and catalase tests should be performed on all isolates. Any of the carbohydrate utilization tests described earlier can be used to speciate *N. meningitidis.* Optimal results in these tests are obtained when a fresh subculture of the organism is used. Serogrouping the meningococci is most commonly done by slide agglutination.

Immunologic Methods. Immunologic methods such as latex agglutination and counterimmuno-electrophoresis are commercially available in kit form and are used to detect the group-specific surface antigens of *N. meningitidis.* These bacterial antigen tests, when performed on CSF, blood, or urine, often allow more rapid detection of a

causative organism, but they should not replace culture and Gram stain.

Moraxella (Branhamella) catarrhalis

Moraxella catarrhalis, formerly known as *Neisseria catarrhalis,* was transferred to the genus *Branhamella* in 1974. The species was reassigned as a subgenus of *Moraxella* in 1984 and is currently called *Moraxella (Branhamella) catarrhalis.*

Growth on chocolate agar is seen in Figure 14–13*A.*

Clinical Infections

Isolated only from humans, *M. catarrhalis* is a normal commensal of the respiratory tract. This organism has become an opportunistic pathogen and has been associated with a number of infections, such as pneumonia, sinusitis, otitis media, and systemic disease. Predisposing factors in the pathogenesis include advanced age, immunodeficiency, neutropenia, and chronic debilitating diseases such as chronic obstructive pulmonary disease. *Moraxella (Branhamella) catarrhalis* has been reported as the third most common cause of acute otitis media and sinusitis in children (see Fig. 14–13*B*). The presence of intracellular gram-negative diplococci in these types of specimens may alert the microbiologist to possible infection with *M. catarrhalis.* Most isolates produce ß-lactamase, making them resistant to

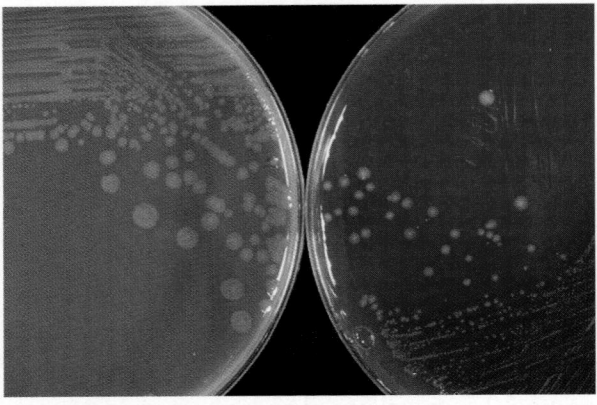

■ F i g u r e   1 4 – 1 2

Growth of *Neisseria meningitidis* after 48 hours on blood agar (BAP) and chocolate agar (CHOC). Of the classic pathogens, only the meningococcus grows on BAP and CHOC.

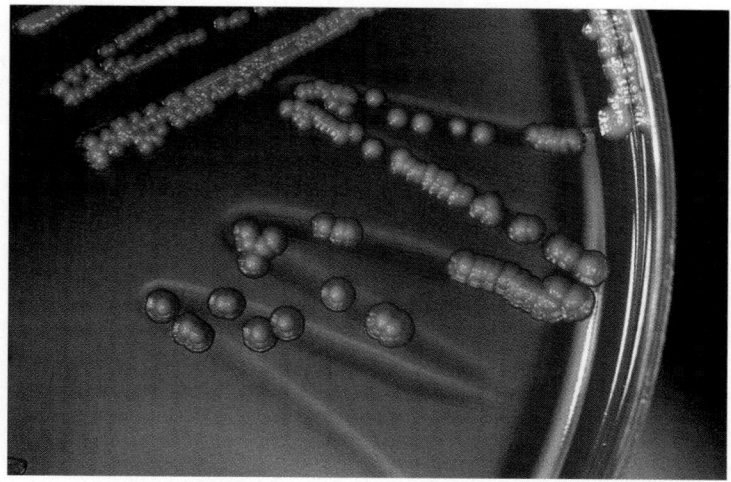

A

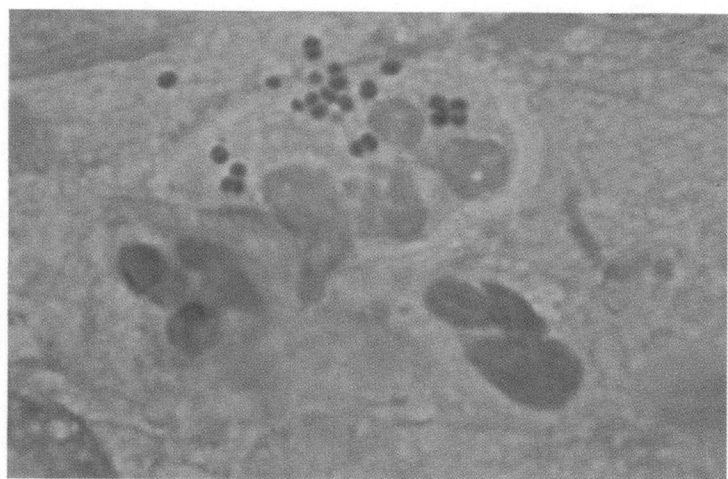

B

■■■■■■ F i g u r e 1 4 - 1 3

A, Growth of *Moraxella catarrhalis* after 48 hours, illustrating the wagon-wheel appearance on chocolate agar. *B*, Direct smear of an otitis media specimen illustrating intracellular gram negative diplococci. The organism was identified biochemically as *M. catarrhalis* from cultures.

ampicillin and amoxacillin, which are commonly used for otitis media.

Laboratory Diagnosis

SPECIMEN COLLECTION AND IDENTIFICATION

Specimen collection for *M. catarrhalis* should be determined by the laboratory according to the various specimen types. The organism will grow on both sheep's blood and chocolate agars, producing smooth, opaque, gray to white colonies. *Moraxella (Branhamella) catarrhalis* is usually inhibited on gonococcal selective agars by the colistin in the media, but some species resistant to this antimicrobial may grow. Like *Neisseria* species, *M. catarrhalis* is both oxidase and catalase positive. The organism is asaccharolytic in carbohydrate degradation tests, and it may be differentiated from *Neisseria* species by positive DNase and butyrate esterase reactions (see Table 14–4).

NONPATHOGENIC *NEISSERIA*

Other *Neisseria* species exist as normal inhabitants of the upper respiratory tract. Referred to as *commensals, saprophytes,* or *nonpathogens,* these species are occasionally isolated from the genital tract. The commensal *Neisseria* species rarely cause disease but have sporadically been implicated in meningitis, endocarditis, prosthetic valve infections, bacteremia, pneumonia, empyema, bacteriuria, osteomyelitis, and ocular infections (Table 14–6). One must keep in mind that the incidence of infections caused by the commensal *Neisseria* spp is extremely low and that the main reason for the microbiologist to be familiar with these organisms is to accurately separate them from the pathogenic species.

Identification

In the clinical laboratory, when isolated from respiratory specimens, the commensal *Neisseria* are usually identified only by Gram stain and gross colony morphology and called "*Neisseria* species" or "normal oral flora." Further identification by biochemical tests is not done. When they are isolated from selective agar medium or sterile body sites, differentiation from the pathogenic *Neisseria* may be required. Identification of the *Neisseria* species is traditionally based on:

■ Gram stain

■ Colony morphology

■■■■ T A B L E   1 4 – 6

INFECTIONS REPORTED TO BE CAUSED BY *NEISSERIA* SPECIES OTHER THAN *N. GONORRHOEAE* AND *N. MENINGITIDIS*

Infection	*Neisseria* species
Meningitis	*N. lactamica*
	N. sicca
	N. subflava
	N. mucosa
	N. flavescens
Endocarditis	*N. sicca*
	N. subflava
	N. mucosa
Prosthetic valve infection	*N. sicca*
Bacteremia	*N. lactamica*
	N. flavescens
	N. cinerea
Pneumonia	*N. sicca*
Empyema	*N. mucosa*
Bacteriuria	*N. subflava*
Osteomyelitis	*N. sicca*
Ocular infection	*N. mucosa*

■ Catalase

■ Oxidase

■ Growth on selective agar media for pathogenic *Neisseria*

■ Acid production from carbohydrates

These tests and observations do not always adequately differentiate all the commensal species from one another or from the pathogens. In addition, insufficient test parameters and equivocal carbohydrate reactions have led to confusion between the pathogenic and commensal *Neisseria* spp. Additional tests used to help further differentiate these organisms are:

■ Growth on nutrient agar at 35°C

■ Growth on blood or chocolate agar at 22°C

■ Reduction of nitrate and nitrite

■ DNase activity

Table 14–7 lists the colonial morphology and primary isolation sites of the *Neisseria* species and related organisms. Table 14–8 lists the traditional and additional tests used to identify the *Neisseria* species and some related genera. The organisms are divided into 3 groups:

■ Group 1: Traditional pathogens.

■ Group 2: Commensal *Neisseria* that may grow on selective medium.

■ Group 3: Commensal *Neisseria* species that do not usually grow on selective medium.

Groups 2 and 3 are further divided on the basis of their activities in carbohydrates. Saccharolytic organisms are able to split carbohydrates, whereas asaccharolytic organisms are unable to do so.

Neisseria polysaccharea

Neisseria polysaccharea was first described by a group of French investigators in 1974, who isolated the organism from throats of healthy children while conducting meningococcal carriage rate surveys. The organism produces large amounts of extracellular polysaccharide when grown in media containing 1 or 5% sucrose; thus the species name, polysaccharea. *Neisseria* polysaccharea's colonial morphology and carbohydrate utilization (glucose and maltose; but rarely sucrose positive) have led

---

====== TABLE 14−7

COLONIAL MORPHOLOGY AND PRIMARY ISOLATION SITES OF *NEISSERIA* AND RELATED ORGANISMS

Organism	Colonial Morphology*	Primary Isolation Sites
N. gonorrhoeae	Small (0.5–1.0 mm), grayish white, translucent, raised with entire edge Usually easily emulsified Smaller than *N. meningitidis* Up to 5 different colonial morphologies from primary culture	Male: urethra Female: endocervix Laboratory should be notified about looking for this organism from other sites so appropriate media can be used
N. meningitidis	1–2 mm, bluish gray (serogroup B may be yellowish) Serogroup A & C may be mucoid, translucent, convex with smooth glistening surface, and may be greenish cast in agar around colonies Usually easily emulsified	Nasopharynx and oropharynx (carriers) Spinal fluid: meningitis Blood: meningococcemia Lower respiratory tract: meningococcal pneumonia
N. polysaccharea	Small (sometimes yellowish) translucent, raised Resembles *N. gonorrhoeae* colony	Nasopharynx of infants and children
N. lactamica	Small, grayish white (often with a yellow ring), translucent, slightly butyrous Resembles *N. meningitidis* but smaller	Nasopharynx of infants and children Rarely found in adults
N. cinerea	Small (1.0–1.5 mm), grayish white, translucent, raised with entire edge and slightly granular Resembles *N. gonorrhoeae*	Nasopharynx
K. denitrificans	Small (≤ 0.5 mm), gray, semitransparent, convex, may pit the agar Colonial morphology of those that do not pit the agar similar to *N. gonorrhoeae*	Upper respiratory tract
M. catarrhalis	3–5 mm, grayish white, opaque 48-hr colony may have elevated center, thinner wave-like periphery (wagon wheel) Often granular, difficult to emulsify Colony can be swept across plate intact (hockey puck)	Upper respiratory tract
N. mucosa	Large (up to 4 mm), grayish to buff yellow, translucent and mucoid, smooth surface, entire edge Viscous (sticky) consistency	Nasopharynx
N. sicca	Large (up to 3 mm), grayish white, opaque, deeply wrinkled, dry, irregular (bread crumb) colony Firmly adherent, difficult to impossible to emulsify	Nasopharynx, saliva, sputum
N. subflava biovars	0.5–2 mm, pale greenish yellow to yellow, smooth surface with entire edge Often adherent	Naso-oropharynx
N. flavescens	Colonies similar to *N. meningitidis* but with golden yellow pigment	Pharynx (not often isolated from clinical specimens)
N. elongata	Large (up to 3 mm), grayish white with yellowish tinge, low convex to almost flat Corroding of agar may occur Clay-like colony, difficult to emulsify	Nasopharynx

*On chocolate agar at 24 to 48 hours.

---

to its misidentification as the pathogenic *N. meningitidis*. Differential tests to separate *N. polysaccharea* from *N. meningitidis* are ability to grow on nutrient agar at 35°C and production of polysaccharide from 1 or 5% sucrose (see Table 14–8). Additional differential tests to separate *N. polysaccharea* from *N. subflava* biovar *subflava* (refer to Table 14–8) are growth on blood or chocolate agar at 22°C and lack of yellow pigment production (see Table 14–7).

Neisseria cinerea

Neisseria cinerea was first described in 1906 but subsequently misclassified as a subtype of *Moraxella (Branhamella) catarrhalis* (*Neisseria pseudocatarrhalis*). It was called *Neisseria cinerea* in 1939. The organism has received considerable attention in the past 10 years because of its misidentification as *N. gonorrhoeae* in some com-

■■■■■ T A B L E   1 4 – 8

DIFFERENTIAL TESTS FOR COMMENSAL *NEISSERIA* AND RELATED GENERA*

Organism	Growth on ML, MTM, or NYC	Catalase	Oxidase	GLUCOSE	MALTOSE	LACTOSE (ONPG)	SUCROSE	FRUCTOSE
				Traditional Tests — Acid produced from				
Group 1: Traditional pathogens								
N. gonorrhoeae	+	+	+	+	−	−	−	−
N. meningitidis	+	+	+	+	+	−	−	−
Group 2: Commensal species— possible growth on selective agar media								
Saccharolytic								
K. denitrificans	v	−	+	(+)	−	−	−	−
N. lactamica	+	+	+	+	+	+	−	−
N. polysaccharea	+	+	+	+	+	−	v	−
Asaccharolytic								
N. cinerea	v	+	+	−	−	−	−	−
M. catarrhalis	v	+	+	−	−	−	−	−
Group 3: Commensal species—no growth on selective agar media								
Saccharolytic								
N. mucosa	−	+	+	+	+	−	+	+
N. sicca	−	+	+	+	+	−	+	+
N. subflava biovars:								
subflava	−	+	+	+	+	−	−	−
flava	−	+	+	+	+	−	−	+
perflava	−	+	+	+	+	−	+	+
Asaccharolytic								
N. flavescens	−	+	+	−	−	−	−	−
N. elongata	−	−	+	−	−	−	−	−

*+, positive; −, negative; (+), positive (delayed); v, variable; ML, Martin-Lewis agar; MTM, modified Thayer-Martin agar; NYC, New York City medium.

mercial identification systems. Although *N. cinerea* is glucose negative in CTA sugars, in some commercial kits the glucose was read as positive, making it biochemically identical to *N. gonorrhoeae*. The colonial morphology of this organism is also similar to the T3 colonies of *N. gonorrhoeae* (Fig. 14–14). Useful tests for differentiation of *N. cinerea* from *M. catarrhalis* are reduction of nitrate and negative DNAse reaction (see Table 14–8). Useful observation for differentiation from *N. flavescens* is lack of yellow pigment production (see Table 14–7).

Kingella denitrificans

CDC first described *Kingella denitrificans* in 1972 and gave it the designation TM-1 because of its isolation from throat cultures plated on Thayer-Martin medium in carrier surveys of *N. meningitidis* and *N. lactamica*. In 1976, it was placed in the genus *Kingella* and given the species name *denitrificans* because of its ability to reduce nitrate. This organism is normal flora in the upper respiratory tract and rarely causes disease, although there are several reports of endocarditis due to it. *Kingella's* colonial morphology (if it does not pit the agar) and carbohydrate utilization (only glucose positive) have led to its misidentification as a pathogenic *Neisseria* (*N. gonorrhoeae*). The Gram stain is the most definitive differential test to separate these organisms. Although *Kingella* is a gram-negative rod, at times coccoidal forms may predominate, and the penicillin disk test discussed earlier will reveal its true rod form. Additional differential tests include negative catalase reaction, ability to grow on agar medium at various temperature, and reduction of nitrate (see Table 14–8).

Growth on Blood or Chocolate Agar at 22°C	Additional Tests			
		Reduction of		
	Growth on Nutrient Agar at 35°C	NITRATE	NITRITE	DNase
−	−	−	−	−
−	−	−	v	−
+	+	+	v	−
v	+	−	v	−
−	+	−	v	−
−	+	−	+	−
v	+	+	+	+
+	+	+	+	
+	+	−	+	−
+	+	−	v	−
+	+	−	v	−
+	+	−	v	−
+	+	−	+	−
+	+	−	+	−

Neisseria lactamica

Neisseria lactamica was reported as early as 1934 but did not become widely recognized as being separate from *N. meningiditis* until 1968. It is commonly found in the nasopharynx of infants and children and, like *N. polysaccharea,* is commonly encountered in meningococcal carrier surveys. The carriage rate of this species in children appears to peak at about 2 years of age and to steadily decline from there. It is rarely isolated from adults. It is the only *Neisseria species* that utilizes lactose; thus its species designation, *lactamica.*

Neisseria lactamica can be misidentified as *N. meningitidis* because of its similar colony morphology (a little smaller), its carbohydrate reactions (glucose, maltose, but rarely sucrose posi-

Figure 14–14

Colonial morphology of *Neisseria cinerea* on blood agar (48-hour culture).

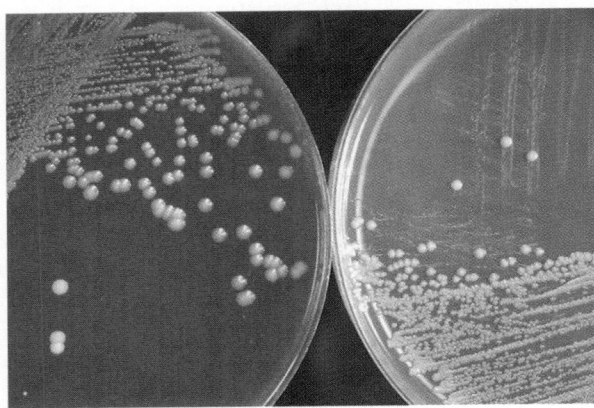

■■■■■ F i g u r e   1 4 – 1 5

Culture of *Neisseria lactamica* after 48 hours on blood agar and chocolate agar. This organism resembles *N. meningitidis*.

tive), and some cross-reaction with meningococcal typing sera (Fig. 14–15). The definitive test for differentiation from *N. meningitidis* and all other *Neisseria species* is utilization of lactose, or positive ONPG reaction.

Neisseria mucosa

The colonies of *Neisseria mucosa* are large, often adherent to the agar, and very mucoid, giving the species its name. It is usually isolated from the nasopharynx of children or young adults. It has also been isolated from the airways of dolphins. This organism has been documented to cause pneumonia in children. It has the same carbohydrate pattern as *N. sicca* and *N. subflava* biovar *perflava,* but differs from these species in its ability to reduce nitrite to nitrogen gas, its colonial morphology, and its lack of pigment production (see Tables 14–7 and 14–8).

Neisseria sicca

The colonies of *Neisseria sicca* are usually dry, wrinkled, adherent, and "breadcrumb like" (see Table 14–7 and Fig. 14–16). The word *sicca* in Latin means "dry." *Neisseria sicca* and *N. subflava* biovar *perflava* are usually the two most common *Neisseria* species found in the respiratory tract of adults. Differentiation of this organism from *N. mucosa* and *N. subflava* biovar *perflava* has been discussed (see *"Neisseria mucosa"*).

Neisseria subflava

Neisseria subflava's species name means "less yellow" (Fig. 14–17). It consists of three biovars that differ from one another by their carbohydrate utilization patterns. Differentiation of this species from *N. polysaccharea* is through its ability to grow on blood or chocolate agar at 22°C.

Neisseria flavescens

Neisseria flavescens (*flavescens* means "yellow") is a yellow-pigmented *Neisseria* that is asaccharolytic (does not utilize carbohydrates). It can be differentiated from *N. cinerea* by its ability to grow on blood or chocolate agar at 22°C.

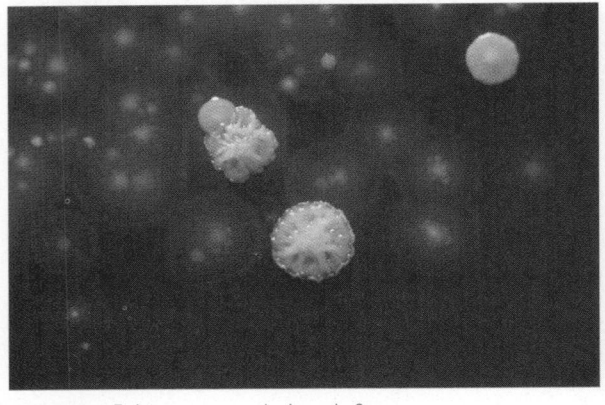

■■■■■ F i g u r e   1 4 – 1 6

Dry, wrinkled "breadcrumb-like" colonial morphology of *Neisseria sicca* on blood agar (48-hour culture).

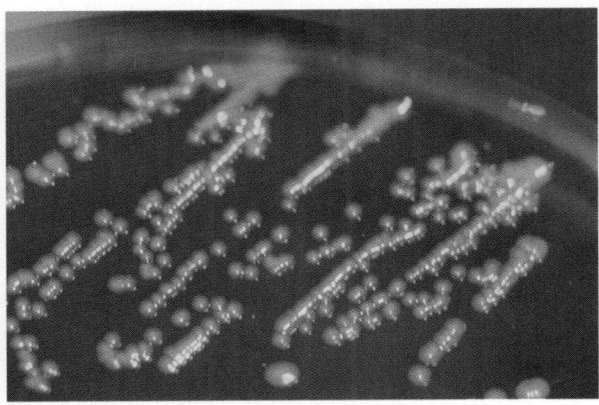

■■■■■ F i g u r e   1 4 – 1 7

Yellow pigmentation of *Neisseria subflava* on blood agar (48-hour culture).

Neisseria elongata

Neisseria elongata is unique among the members of the *Neisseria* species, in that it is a rod and is catalase negative (sometimes weakly positive). It can be differentiated from *K. denitrificans* by its inability to reduce nitrate.

Bibliography

Arditi M, Yogev R: Common etiological agents of bacterial meningitis. In Shuman ST, Phair JP, Sommers HM (eds): The Biologic and Clinical Basis of Infectious Diseases, 4th ed. Philadelphia: WB Saunders, 1992, pp 350–356.

Catlin BW: *Branhamella catarrhalis:* An organism gaining respect as a pathogen. Clin Microbiol Rev 3:293, 1990.

Chapin-Roberston K, Reece EA, Edberg SC: Evaluation of the Gen-Probe Pace II assay for the direct detection of *Neisseria gonorrhoeae* in endocervical specimens. Diagn Microbiol Infect Dis 15:645, 1992.

Dillon JR, Carballo M, Pauze M: Evaluation of eight methods for identification of pathogenic *Neisseria* species: Neisseria-Kwik, RIM-N, Gonobio-Test, Minitek, Gonochek II, Gonogen, Phadebact Monoclonal GC OMNI Test, and Syva MicroTrak Test. J Clin Microbiol 26:493, 1988.

Dolter J, Bryant L, Janda JM: Evaluation of five rapid systems for the identification of *Neisseria gonorrhoeae.* Diagn Microbiol Infect Dis 13:267, 1990.

Granato PA, Franz MR: Use of the Gen-Probe PACE system for the detection of *Neisseria gonorrhoeae* in urogenital samples. Diagn Microbiol Infect Dis 13:217, 1990.

Joklik WK, Willet HP, et al: Zinsser Microbiology, 20th ed. Norwalk, CT: Appleton & Lange, 1992, pp 451–453.

Judson FN: Gonorrhea. Med Clin North Am 74:1353, 1990.

Judson FN: Management of antibiotic resistant *Neisseria gonorrhoeae.* Ann Intern Med 110:5, 1989.

Knapp JS: Historical perspectives and identification of *Neisseria* and related species. Clin Microbiol Rev 1:415, 1988.

Lauer BA, Masters HB: Toxic effect of calcium alginate swabs on *Neisseria gonorrhoeae.* J Clin Microbiol 26:54, 1988.

Lussier M, Brodeur BR, Winston S: Detection of *Neisseria gonorrhoeae* by dot-enzyme immunoassay using monoclonal antibodies. J Immunoassay 10:373, 1989.

Morello JA, Janda WM, Doern GV: *Neisseria* and *Branhamella.* In Balows A, et al (ed): Manual of Clinical Microbiology, 5th ed. Washington, DC: American Society for Microbiology, 1991, pp 258–276.

Panke ES, Yang LI, et al: Comparison of Gen-Probe DNA probe test and culture for the detection of *Neisseria gonorrhoeae* in endocervical specimens. J Clin Microbiol 29:883, 1991.

Waitz JA: Methods for Dilution Antimicrobial Susceptibility Tests for Bacteria That Grow Aerobically, 2nd ed. Approved Standard, Document M7-A2. Villanova, PA: National Committee for Clinical Laboratory Standards, 1990.

Waitz JA: Performance Standards for Antimicrobial Disk Susceptibility Tests, 5th ed. Approved Standard, Document M100-S4. Villanova, PA: National Committee for Clinical Laboratory Standards, 1992.

Welch WD, Cartwright G: Fluorescent monoclonal antibody compared with carbohydrate utilization for rapid identification of *Neisseria gonorrhoeae.* J Clin Microbiol 26:293, 1988.

Whittington WL, Knapp JS: Trends in resistance of *Neisseria gonorrhoeae* to antimicrobial agents in the United States. Sex Trans Dis 15:202, 1988.

Whittington WL, Rice RJ, et al: Incorrect identification of *Neisseria gonorrhoeae* from infants and children. Pediatr Infect Dis 7:3, 1988.

CHAPTER **15**

Haemophilus and Other Fastidious Gram-Negative Rods

A. *HAEMOPHILUS, PASTEURELLA, BRUCELLA,* AND *FRANCISELLA*

George Manuselis, Jr. •
Jean Barnishan •
Leonard H. Schleicher, Jr.

██████   OBJECTIVES

1. Characterize each of the following bacterial species by morphology, habitat, and nutritional requirements.

2. Describe the modes of transmission of each of the organisms characterized.

3. Explain the clinical significance of these organisms when isolated in the clinical laboratory.

4. Name the appropriate specimens for the recovery of these organisms.

5. Determine the appropriate culture media required to isolate each of the organisms.

6. Describe the microscopic and colonial morphology.

7. Describe the methods of identification currently used to diagnose infections caused by these organisms.

This chapter describes miscellaneous gram-negative bacilli that are fastidious. Most of these organisms require special nutrients and environmental growth factors for isolation and identification.

The organisms to be covered in this chapter include the following genera:

■ *Haemophilus*

■ *Legionella*

■ *Bordetella*

■ *Pasteurella*

■ *Brucella*

■ *Francisella*

Two of the genera, *Haemophilus* and *Pasteurella*, belong to the family Pasteurellaceae. Characteristically, members of Pasteurellaceae are gram-negative, pleomorphic, coccoid to rod-shaped cells that are nonmotile and aerobic or facultatively anaerobic, form nitrites from nitrates, and are oxidase and catalase positive.

The genera *Pasteurella*, *Brucella*, and *Francisella* are etiologic agents of true zoonotic infections. (For a more detailed description of these three genera see Chapter 34.)

Legionella species are facultative intracellular parasites that have a wide range of pathogenic potential in humans. *Legionella pneumophila*, the most common isolated species, gained immediate notoriety as the causative agent of a relatively fatal respiratory infection at the 1976 American Legion Convention in Philadelphia. The new family, Legionellaceae, and the new genus, *Legionella*, were named after this famous outbreak.

Three of the four recognized species of the genus *Bordetella* are associated with human disease. *Bordetella pertussis* and *B. parapertussis* are primary human respiratory pathogens that cause whooping cough or pertussis, and *B. bronchiseptica,* an opportunistic human pathogen, causes pneumonia and wound infections. These three species are implicated mostly in diseases of childhood.

Figure 15–1 depicts the prevalence of these fastidious organisms in relation to other gram-negative bacilli found in clinical specimens.

HAEMOPHILUS

General Characteristics

The genus *Haemophilus* consists of gram-negative, pleomorphic coccobacilli or rods that may vary microscopically from small coccobacilli in direct smears of clinical material to long filaments occasionally seen in smears of colony growth. They are nonmotile and aerobic or facultatively anaerobic, ferment carbohydrates, are generally oxidase and catalase positive, reduce nitrates to nitrites, and are obligate parasites on the mucous membranes of humans and animals.

In accordance with *Bergey's Manual of Systematic Bacteriology*, there are 10 species of *Haemophilus* associated with humans: *H. influenzae*, *H. parainfluenzae*, *H. haemolyticus*, *H. parahaemolyticus*, *H. aphrophilus*, *H. paraphrophilus*, *H. paraphrohaemolyticus*, *H. aegyptius*, *H. segnis*, and *H. ducreyi*. In addition, there are six species associated with animals and three species of

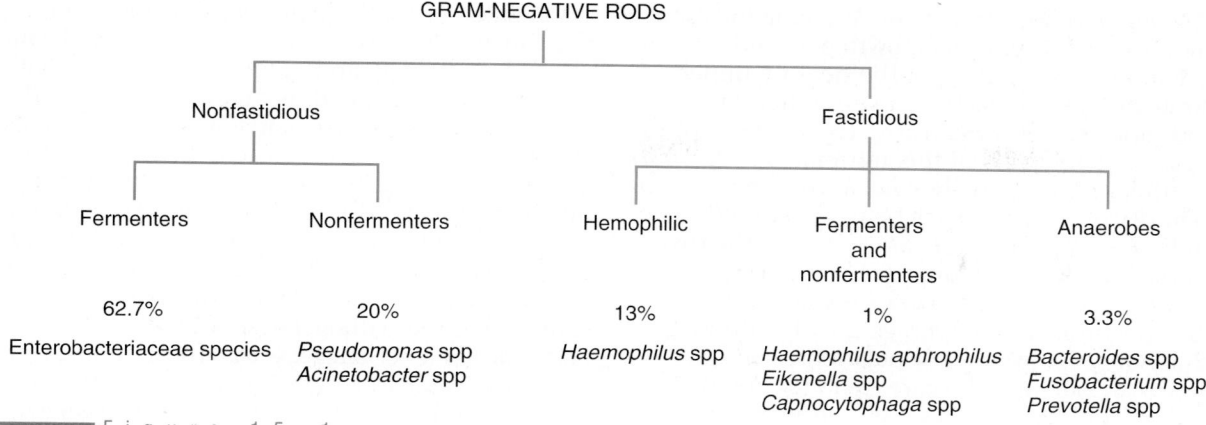

GRAM-NEGATIVE RODS

Nonfastidious

Fastidious

Fermenters   Nonfermenters

Hemophilic   Fermenters   Anaerobes
                  and
                  nonfermenters

62.7%   20%   13%   1%   3.3%

Enterobacteriaceae species   *Pseudomonas* spp   *Haemophilus* spp   *Haemophilus aphrophilus*   *Bacteroides* spp
                             Acinetobacter spp   *Eikenella* spp   *Fusobacterium* spp
                                                                           Capnocytophaga spp   *Prevotella* spp

■ F i g u r e 1 5 – 1

Prevalence of gram-negative rods isolated from cultures in a large tertiary hospital. (Data from Clinical Microbiology Laboratory, OSU Medical Center, 1992–1994.)

uncertain status. Most members of the genus are nonpathogenic or produce opportunistic infections. Therefore, the emphasis of this section is on the major pathogenic species, *H. influenzae, H. aegyptius,* and *H. ducreyi.*

The genus name *Haemophilus* is derived from the Greek words meaning "blood-lover." As the name implies, *Haemophilus* requires preformed growth factors present in blood: X factor (hemin, hematin) and/or V factor (nicotinamide-adenine dinucleotide [NAD]). Traditionally, a small, gram-negative bacillus (coccobacillus) is assigned to this genus, based on its requirements for the X and/or V factor. Species with the prefix "para" require V factor only for growth. In addition, the production of hemolysis on 5% horse's or rabbit's blood agar is an important differential characteristic. Although certain species are also hemolytic on a sheep's blood agar plate, the organisms will not grow in pure culture on this medium.

Both X and V factors are found within red blood cells and are important for in vitro growth. Most laboratories use blood agar containing sheep erythrocytes prepared by commercial sources. Only X factor is directly available in this medium. *Haemophilus* species that are V factor dependent do not grow because the red cells are still intact and the sheep blood contains enzymes (NADases) that hydrolyze V factor. To alleviate this problem most clinical laboratories use chocolate agar to facilitate the recovery of *Haemophilus* species from clinical specimens. The lysing of the red cells by heat in the preparation of chocolate agar releases both X and the V growth factors and inactivates the enzymes that hydrolyze V factor.

A phenomenon that helps in the recognition of *Haemophilus* species that require V factor is satellitism. Satellitism occurs when an organism such as *Staphylococcus aureus, Streptococcus pneumoniae,* or *Neisseria* species produces V factor (NAD) as a byproduct of its metabolism. The *Haemophilus* isolate obtains X factor from the sheep's blood agar and V factor from one of these organisms. On sheep's blood agar plate, tiny colonies of *Haemophilus* may be seen growing or engaging in "satellitism" around the V factor–producing organism. Figure 15–2 illustrates *H. influenzae* satellitism around colonies of *S. aureus.* Except for *H. aphrophilus* and *H. ducreyi,* all clinically significant

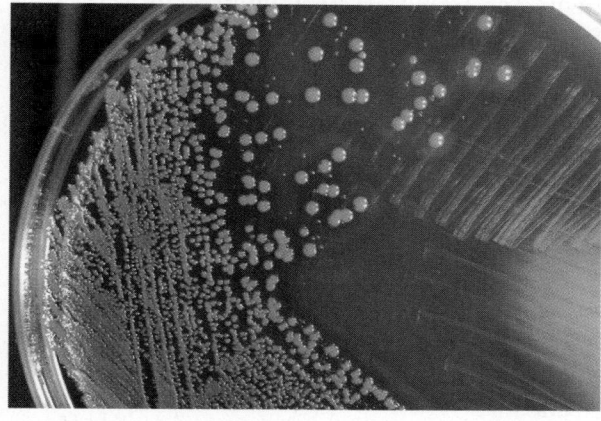

■ F i g u r e 1 5 – 2

Haemophilus influenzae satellitism around and between the large, white, hemolytic staphylococci. The small, gray, glistening colony is *H. influenzae.*

Haemophilus species require V factor for growth and display this unusual growth pattern.

The indigenous flora of the healthy upper respiratory tract consists of many different genera and species of organisms (see Chapter 26). Approximately 10% of this normal bacterial flora in adults consists of *Haemophilus* spp, with the majority of the organisms being *H. parainfluenzae* and nonencapsulated *H. influenzae*. Of the two, *H. parainfluenzae* is the predominant species.

Colonization trends begin in infancy with encapsulated strains and average 2 to 6% throughout childhood. In selected populations, such as children who attend day care centers, colonization may reach as high as 60%. Nonencapsulated strains of *H. influenzae* in healthy children average 2% of the normal bacterial flora.

Haemophilus influenzae

Historical Perspective

Influenza, commonly referred to as the "flu," is a viral infectious disease characterized by acute inflammation of the upper airways. Symptoms often progress to intense inflammation of the mucous membranes lining the nose (coryza), headache, bronchitis, and severe generalized muscle pain (myalgias).

Haemophilus influenzae was named erroneously from the influenza pandemic that ravaged the world between 1889 and 1890. The basis for this assumption was the frequent isolation of this bacillus from the nasopharynx of influenza patients and from postmortem lung cultures. After viral culture techniques became available, it became apparent that the disease influenza was caused by a virus and that the actual role of *H. influenzae* was that of a secondary invader.

Virulence Factors

Haemophilus influenzae, the major clinical pathogen of the species, has a wide range of pathogenic potential. The following virulence factors may play a role in the invasiveness of this organism and in the initiation of infection:

■ Capsule

■ IgA proteases

■ Outer membrane proteins and lipooligosaccharide (LOS)

■ Adherence

Capsule. Of all the potential virulence factors, the capsule, if present, plays the most significant role. The serologic grouping of *H. influenzae* into six serotypes, a, b, c, d, e, and f, is based on the capsular polysaccharide. Most invasive infections caused by encapsulated strains of *H. influenzae* belong to capsular serotype b (Hib) and occur primarily in young children. In contrast with the other serotypes, a, c, d, e, and f, the serotype b capsule is a unique polymer composed of ribose, ribitol, and phosphate. The capsule plays a significant role in the pathogenesis of invasive disease. Scientific evidence suggests that the antiphagocytic property and anticomplementary activity of the type b capsule are important factors in virulence.

It is important to emphasize in this discussion that not all strains of *H. influenzae* are encapsulated. Therefore, two patterns of disease attributed to *H. influenzae* emerge. The first is invasive disease caused by encapsulated (typable) strains predominantly of serotype b in which bacteremia (hematogenous spread) plays a significant role. The second pattern of disease is a more localized infection caused by the contiguous spread of unencapsulated (nontypable) strains and occurs within or in close proximity to the respiratory tract. Examples of invasive disease include meningitis, cellulitis, arthritis, and epiglotittis, which are found in the pediatric population. Examples of localized infection include pneumonia and sinusitis, which occur primarily in the adult population.

Exceptions to the encapsulated and unencapsulated categorizations and to whether these organisms appear in the adult or pediatric population do occur. Unencapsulated strains can cause meningitis in the adult population, especially in the immunocompromised or debilitated person, and can cause neonatal sepsis and invasive lower respiratory tract infections in children.

The other serotypes, a, c, d, e, and f, rarely cause disease in humans. Reports of infections with these serotypes have included pneumonia and bacteremia caused by serotypes a, d, and f in immunocompromised adults, and neonatal sepsis caused by serotype c.

IgA Proteases. Secretory IgA is present on human mucosal surfaces in organ system areas for which *H. influenzae* has a predilection. Of all the species of *Haemophilus, H. influenzae* is the only one that produces this enzyme. Since this enzyme has the ability to cleave secretory IgA, its production may contribute to the organism's virulence potential.

Adherence. The role of adherence as a virulence factor is not well defined. Studies indicate that most unencapsulated strains are adherent to human epithelial cells, whereas most serotype b strains are not. The lack of this adherent capability in type b organisms may explain the tendency for type b strains to cause systemic infections. The presence of this adherent capability by unencapsulated strains may explain the tendency for these strains to cause more localized infections.

Outer Membrane Components and LOS. Although the role of these antigens is not well defined, antibody directed against these antigens may play a significant role in human immunity. Each one of these components may be responsible for a specific capability, such as invasiveness, attachment, and antiphagocytic function. LOS has been shown to have a paralyzing effect on the sweeping motion of ciliated respiratory epithelium.

Clinical Manifestations of *Haemophilus influenzae* Infections

INFECTIONS CAUSED BY ENCAPSULATED (TYPABLE) STRAINS

Meningitis. Serotype b is a common cause of pediatric meningitis in children between the ages of 3 months and 6 years. Blood stream invasion and bacteremic spread follow colonization, invasion, and replication of this organism in the respiratory mucous membranes. Headache, stiff neck, and other meningeal signs are usually preceded by mild respiratory disease.

Epiglottitis. Serotype b is the most common cause of this syndrome. The manifestations include rapid onset, acute inflammation, and intense edema that may cause complete airway obstruction, requiring an emergency tracheostomy. The peak incidence in children occurs between the ages of 2 and 4 years.

Arthritis. The bacteremic spread of serotype b is the most common cause of arthritis seen in children younger than 2 years of age.

Cellulitis. Serotype b is the usual culprit in cases of cellulitis in children younger than 2 years of age. The most common infected site is the cheek, although infection may occur in other areas of the upper extremities. Cellulitis of the cheek is characterized by rapid onset, pain, edema, and a reddish-blue color on the inflamed area. Pediatricians usually diagnose the disease based on the age of the child, symptoms, appearance, and proximity of the cellulitis to the oral mucosa.

Other infections caused by typable strains include acute pharyngitis and pneumonia. Pneumonia in children is usually caused by Hib. The mean age at infection is approximately 14 months.

With the advent of the Hib vaccine, pediatric hospital laboratories are reporting a significant decrease in invasive disease.

INFECTIONS CAUSED BY NONENCAPSULATED (NONTYPABLE) STRAINS

Infections usually caused by nontypable strains of *H. influenzae* include otitis media, bronchitis, sinusitis, pneumonia in elderly patients, and genital tract infections. Table 15–1 summarizes infections caused by *H. influenzae* in different population groups.

■■■■■ T A B L E   1 5 – 1

INFECTIONS CAUSED BY *HAEMOPHILUS INFLUENZAE* IN DIFFERENT POPULATION GROUPS

Invasive Type b[1] Pediatric Population	Unencapsulated Pediatric Population	Unencapsulated Adult Population

Meningitis
  CSF 50–95% culture positive
  Blood 50–95% culture positive
Cellulitis
  Skin 75–90% culture positive
  Blood 50–75% culture positive
Epiglottitis
  Blood 90–95% culture positive
Conjunctivitis
  Eye 50–75% culture positive
  Blood < 10% culture positive
	Otitis media
	  Tympanocentesis 50–70% culture positive
		Pneumonia, bronchitis
		  Sputum 25–75% culture positive
		  Blood 10–30% culture positive
		Sinusitis
		  Sinus aspirate 50–75% culture positive

[1]May cause invasive disease in immunocompromised and debilitated adults. Modified with permission from Murray PR: Pasteurellaceae. In Murray PR, Drew LW, Kobayashi GS, Thompson JH (eds): Medical Microbiology. St. Louis: Mosby–Year Book, 1990, p 143.

Infections Associated With Other *Haemophilus* Species

Haemophilus aegyptius is a species that is genetically similar to *H. influenzae*. Because of their similar identifying characteristics, it is difficult to differentiate *H. influenzae* from *H. aegyptius* in the clinical laboratory. Formerly a separate species of the genus, *H. aegyptius* is now known as *H. influenzae* biogroup *aegyptius*. Both organisms can cause conjunctivitis in primarily the pediatric population. *Haemophilus influenzae* biogroup *aegyptius* causes both a more acute, contagious, purulent conjunctivitis, commonly referred to as "pinkeye," and a severe systemic disease known as Brazilian purpuric fever (BPF) in hot climates. BPF is characterized by recurrent or concurrent conjunctivitis, high fever, vomiting, petechiae, purpura, septicemia, and shock. The mortality rate for BPF may reach as high as 70%.

Haemophilus ducreyi is the etiologic agent of chancroid, a highly communicable sexually transmitted disease. After an incubation period of approximately 4 to 14 days, a nonindurated, painful lesion with an irregular edge develops, generally on the genitalia or perianal areas. The most common sites of infection are on the penis of males or on the labia or within the vagina of females. Suppurative (forming or discharging pus), enlarged, draining, inguinal lymph nodes (buboes) are common in the majority of infected patients.

Haemophilus parainfluenzae, *H. aphrophilus*, and *H. paraphrophilus* have a very low incidence of pathogenicity and have been implicated most often as causative agents of endocarditis.

Laboratory Diagnosis

Specimen Processing and Isolation

Haemophilus species have been associated with many diseases in humans. Almost any specimen submitted for routine bacteriology examination may harbor this organism. Common sources include blood, cerebrospinal fluid, middle-ear exudate, joint fluids, upper and lower respiratory tract specimens, swabs from conjunctivae, vaginal swabs, and abscess drainage. *Haemophilus* species die rapidly in clinical specimens: therefore, prompt transportation and processing is vital for the isolation of this organism. This is especially true of genital specimens submitted for *H. ducreyi*. *Haemophilus ducreyi* is extremely fastidious. Specimens from genital sites submitted for the isolation of this organism should first be cleaned with sterile gauze moistened with sterile saline. A cotton swab, premoistened with sterile phosphate-buffered saline (PBS), should then be used to collect material from the base of the ulcer. Processing in the laboratory must take place within 10 minutes of collection for maximum recovery.

It is important to remember that most conventional media do not support the growth of *Haemophilus* species because of the lack of one or more growth factors. When attempting to isolate *H. influenzae*, a common medium to use is chocolate agar incubated between 33 and 37°C with 5 to 10% carbon dioxide added. Chocolate agar contains both X factor (hemin) and V factor (NAD), which are required growth factors for *H. influenzae*. It has been shown that chocolate agar supplemented with bacitracin (300 mg/L) is an excellent medium for the isolation of *Haemophilus* species from respiratory specimens. The bacitracin is added to reduce overgrowth of normal respiratory flora. Growth on chocolate agar is usually seen after 18 to 24 hours of incubation. However, specimens submitted for *H. ducreyi* should be held for at least 9 days, and specimens for *H. aegyptius* should be held for at least 4 days before reporting a negative result.

Because of their fastidious nature, specimens submitted for *H. ducreyi* and *H. aegyptius* must be plated using special media. For *H. aegyptius*, chocolate agar supplemented with 1% IsoVitaleX (BBL Microbiology Systems, Cockeysville, MD) is required. For *H. ducreyi*, the use of GC agar (GIBCO Laboratories, Grand Island, NY) containing 1 to 2% hemoglobin, 5% fetal calf serum, 10% CVA enrichment, and 3 μg of vancomycin has been proved to be reliable. The use of vancomycin in this media helps reduce commensal flora from genital specimens and allows the visualization of *H. ducreyi*. The culture plates for the recovery of this organism should be incubated in a carbon dioxide atmosphere containing high humidity. High humidity is provided by placing the plates, along with sterile gauze moistened with sterile water, in an open plastic bag ("baggy") in the incubator. Sterile gauze is recommended in lieu of regular nonsterile paper towels to decrease fungal contamination of the plates.

Microscopic Morphology

As is common with most of the species of *Haemophilus*, the microscopic morphology varies

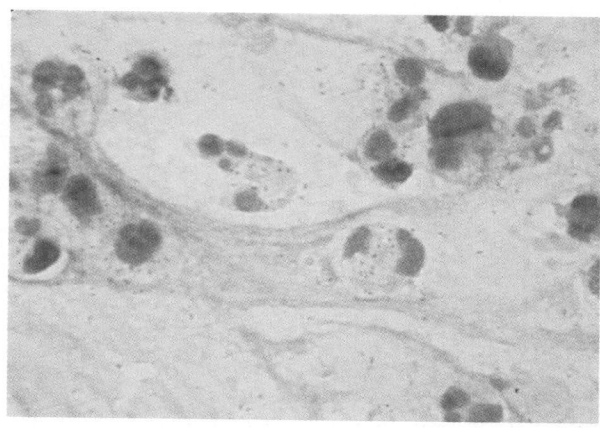

■■■■■■ F i g u r e   1 5 – 3

Direct smear of *H. influenzae* in CSF in a case of meningitis. Note the intracellular and extracellular gram-negative coccobacilli.

from a small gram-negative coccobacilli to long filaments. The coccobacillary or small, regular rod microscopic morphology is the more predominant form found in clinical specimens. Encapsulated forms of *H. influenzae* may be observed in Gram-stained direct smears as clear, nonstaining areas ("halos") surrounding the organisms in purulent secretions. Figure 15–3 illustrates the microscopic morphology of *H. influenzae* in a direct smear of cerebrospinal fluid from a patient with meningitis. Figure 15–4 is an example of a Gram stain of an isolated colony of *H. influenzae*.

Gram stains of genital lesions caused by *H. ducreyi* may show gram-negative coccobacilli arranged in groups commonly referred to as the "school of fish" formation (Fig. 15–5).

Colony Morphology

As mentioned previously (see Chapter 9), most clinical specimens are plated on a variety of culture media and read as a culture set after 24 hours of incubation. Usually blood agar plate, chocolate agar, and MacConkey agar are inoculated simultaneously from clinical specimens from areas of the human body where *Haemophilus* may be isolated. Colonies of *Haemophilus* on chocolate agar appear translucent, moist, smooth, and convex, with a distinct "mousy" or "bleach-like" odor. *Haemophilus influenzae* produces a more grayish-appearing colony. Figure 15–6 shows the typical colony morphology of *H. influenzae*. The encapsulated strains grow larger and more mucoid than the nonencapsulated strains.

The first clue that the isolate may belong to the *Haemophilus* group is the growth of gram-negative pleomorphic coccobacilli on chocolate agar, with no growth on sheep's blood agar in pure culture. Because of their fastidious nature, *Haemophilus* species do not grow on MacConkey agar or other enteric isolation media.

Laboratory Identification

There are several tests that can be used in the clinical laboratory for the identification of *Haemophilus* isolates. These include testing for growth factors (X and V), traditional biochemicals, hemolysis on rabbit's or horse's blood agar, oxidase, and catalase. The porphyrin test is an alternative method for the determination of X factor requirements. In place of traditional biochemicals, there are several commercial systems that can be used to identify and biotype *Haemophilus* species. They include the Minitek System (BBL Microbiology Systems), the RapID-NH (Innovative Diagnostics, Norcross, GA), HNID (MicroScan, West Sacramento, CA), and NHI (BioMerieux Vitek, Hazelwood, MO). Latex agglutination (Murex Diagnostics, Inc., Norcross, GA; Wampole Laboratories, Cranbury, NJ; BBL Microbiology Systems) and coagglutination tests (Boule Diagnostics, Huddings, Sweden) have proved to be specific and sensitive in the detection of Hib in cerebrospinal fluid and, to a slightly lesser extent, in other body fluids. Of all the species that require V factor, *H. segnis* is the only organism that is oxidase negative.

Testing for X and V factor requirements using impregnated strips is the traditional approach.

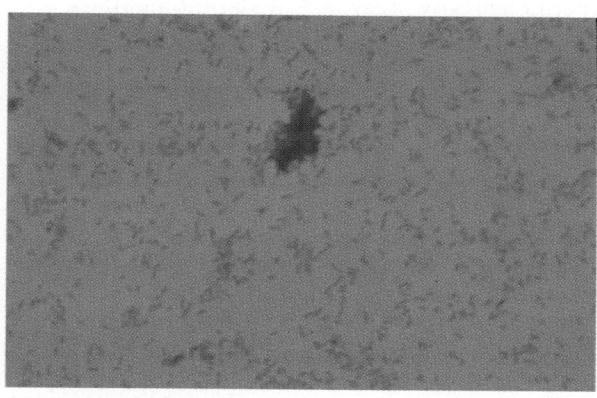

■■■■■■ F i g u r e   1 5 – 4

Gram stain of a *Haemophilus influenzae* colony. Note the slightly more elongated rods.

■ Figure 15–5

"School of Fish" formation of *H. ducreyi.*

This method entails making a suspension of the isolate in trypticase soy broth. Care must be taken not to transfer any X factor–containing media to the broth (carryover may produce erroneous or less than definitive results). The organism suspension is then inoculated onto media devoid of X and V factors. Mueller-Hinton agar or trypticase soy agar are acceptable. Allow the plate to dry. Using sterile forceps, place a strip containing X factor on the plate. Press the strip onto the agar using the forceps. Place the V factor strip parallel to the X factor strip approximately 15 mm away. Again, using the forceps, press the V factor strip onto the agar surface. Incubate at 35 to 37°C in 5 to 10% carbon dioxide for 18 to 24 hours. Observe plates for growth around the strips. Figure 15–7 illustrates the reactions obtained when testing for X and V factor requirements using impregnated strips. Figures 15–8 through 15–10 illustrate the actual identification of unknowns using the impregnated strips. When using this traditional approach it is important to know under what culture conditions the organisms were isolated. *Haemophilus*

organisms are facultative anaerobes. When they are grown anaerobically they do not require heme but still require NAD. If a *Haemophilus influenzae* organism was incubated anaerobically, it could be misidentified as *H. parainfluenzae.*

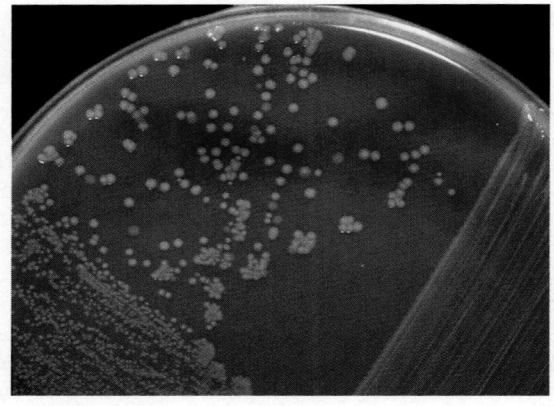

■ Figure 15–6

Example of *H. influenzae* growing on CHOC agar. Notice the gray mucoid colonies characteristic of encapsulated strains.

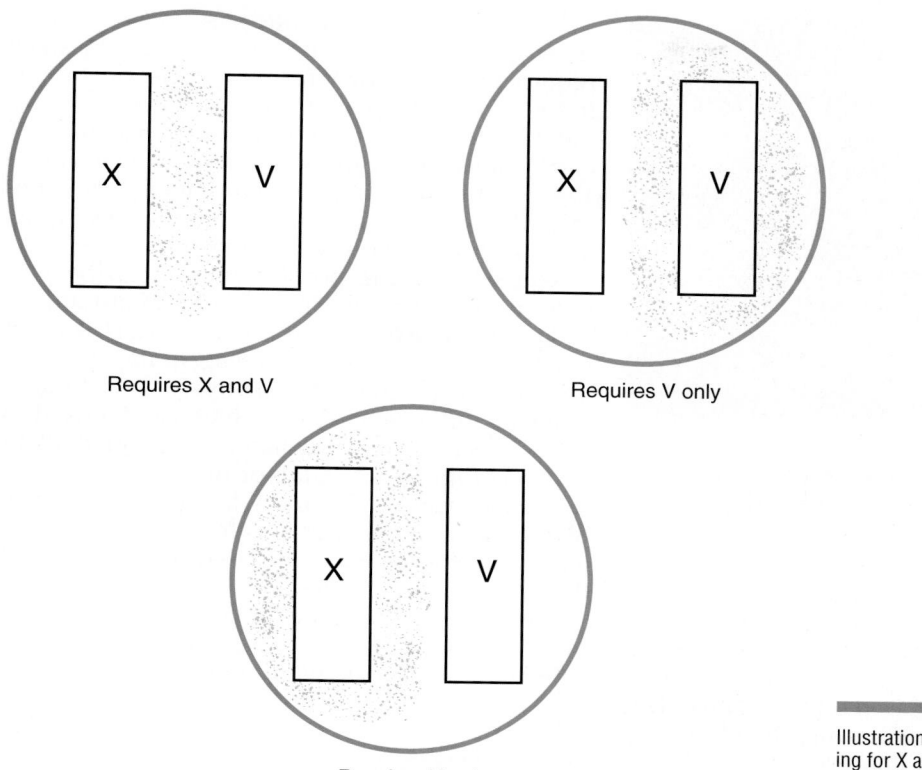

Requires X and V

Requires V only

Requires X only

■■■■■ F i g u r e   1 5 – 7

Illustrations of reactions obtained when testing for X and V factor requirements using impregnated strips.

The porphyrin test is an alternative method for differentiating the heme-producing strains of *Haemophilus* species. The principle of the test is based on the ability of the organism to produce enzymes that convert δ-aminolevulinic acid (ALA) into porphyrins or protoporphyrins. Porphobilinogen can be detected by the addition of *p*-dimethylaminobenzaldehyde (Kovacs reagent). After the addition of Kovacs reagent, a red color forms in the lower aqueous phase if porphobilinogen is present. Porphyrins can be detected using an ultraviolet light (Wood lamp). Porphyrins

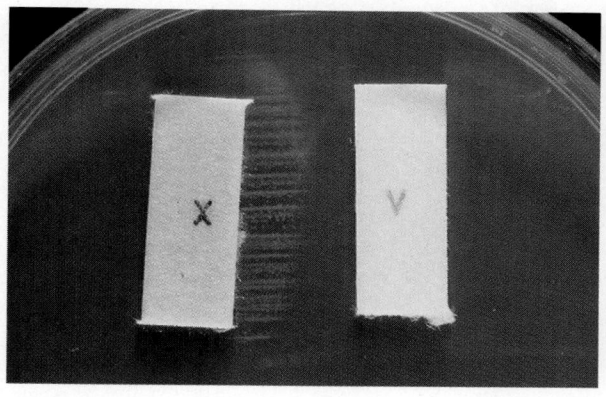

■■■■■ F i g u r e   1 5 – 8

This organism would be identified as *H. influenzae* because it is utilizing both X and V factors.

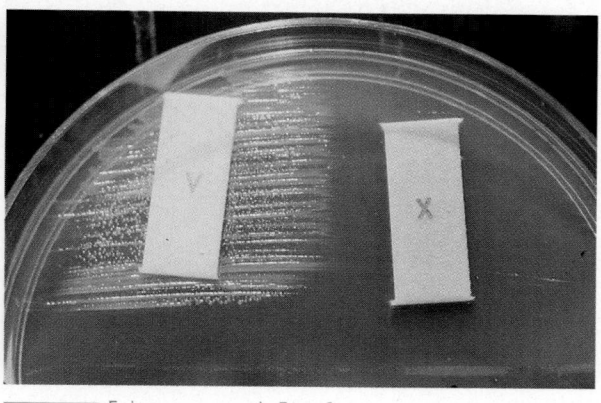

■■■■■ F i g u r e   1 5 – 9

This organism is utilizing V factor only. It would be identified as *H. parainfluenzae*.

■■■■■ F i g u r e   1 5 – 1 0

This organism is positive for X factor only. The probable species is *H. aphrophilus* because this species may appear to be hemin dependent on initial isolation.

fluoresce reddish orange under ultraviolet light. The main advantage of this test is that X factor (heme) is not required; therefore, the problem of carryover is eliminated. The disadvantage is that primary identification is based on a "negative" test result. In the case of *H. influenzae,* the ultraviolet light test is negative (no fluorescence), and after the addition of Kovacs reagent there is no color change. *Haemophilus* species that can biosynthesize their own heme are porphyrin positive (X strip negative); species that are porphyrin negative cannot synthesize their own heme and therefore are found to be X factor positive when the impregnated strip is used. Figure 15–11 illustrates a positive porphyrin test.

Haemophilus aphrophilus is not a "true" *Haemophilus* because it does not have a strict requirement for X or V factor. It appears at times to be X factor dependent by impregnated strips; but the porphyrin test is positive, meaning *H. aphrophilus* can synthesize its own heme. Figure 15–12 is an example of *H. aphrophilus* that is not X and/or V factor dependent and is growing all over the trypticase soy agar plate.

Reagents for the porphyrin test can be purchased commercially from Remel Microbiology Products (Lenexa, Kansas). The A.L.A. Disk is a 1 to 6 hour test that uses ALA reagent impregnated on a filter paper disk. Alternatively, Remel offers a *Haemophilus* Quad Plate, which can be used to identify *Haemophilus* isolates. The Quad Plate contains four zones: X factor only, V factor only, X and V factors, and X and V factors with horse's blood. The *Haemophilus* isolate may be identified based on the factor(s) required for growth and the presence of hemolysis.

Table 15–2 lists the differential tests for *Haemophilus* species. Table 15–3 lists the differential tests for *Haemophilus* biogroups.

Treatment

The current recommended treatment of life-threatening illness caused by *H. influenzae* is cefotaxime, ceftriaxone, or cefuroxime. Alternative drugs include trimethoprim-sulfamethoxazole,

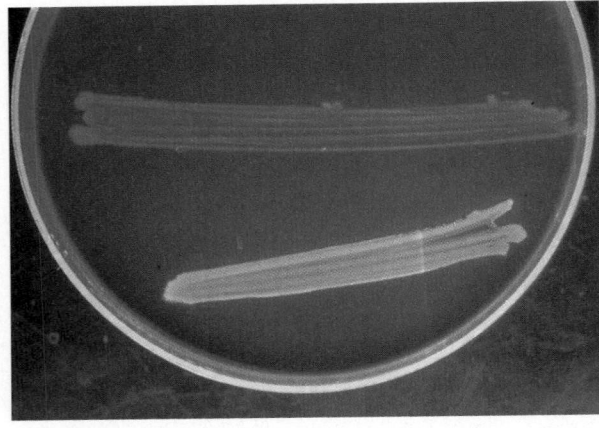

■■■■■ F i g u r e   1 5 – 1 1

Under ultraviolet light, the organism on the bottom is exhibiting a positive porphyrin reaction. The organism on the top is porphyrin negative.

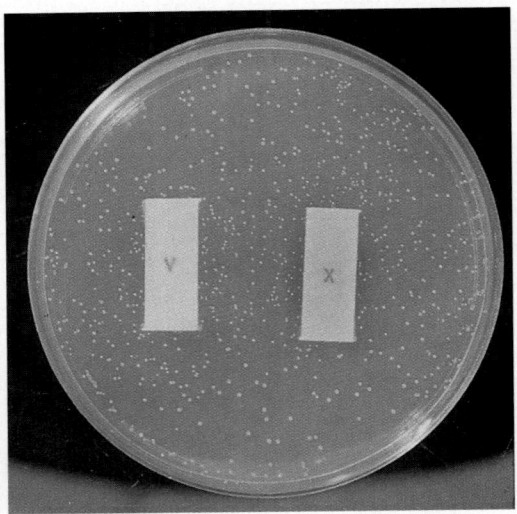

■■■■■ F i g u r e   1 5 – 1 2

A *Haemophilus aphrophilus* organism that is not X factor dependent and is growing all over the trypticase soy agar plate.

■■■■■■■ T A B L E   1 5 – 2

DIFFERENTIAL TESTS FOR *HAEMOPHILUS* SPECIES

	Oxidase	Catalase	Hemolysis (Horse, Rabbit Blood)	Carbon Dioxide Enhances Growth	TSI	Glucose	Pyruvate	Sucrose	Mannose	Fructose	Mannitol	Maltose	Xylose	Lactose	Nitrate	Esculin	Ornithine	Indole	Urea
Factor X+, V+, Porphyrin−																			
Haemophilus influenzae	+	+	−	−	NG	+	+	−	−	V	−	−	+	−	+	−	see biotype chart		
Haemophilus influenzae subsp *aegypticus*	+	+	−	−	NG	V	+	−	−	V	−	−	+	−	+	−	−	−	+
Haemophilus haemolyticus	+	+	+	−	NG	+	−	−	−	V	−	−	V	−	+	−	−	V	+
Factor X−, V+, Porphyrin+																			
Haemophilus parainfluenzae	+	V	−	V	NG	+	−	+	V	+	−	+	−	−	+	−	see biotype chart		
Haemophilus segnis	−	V	−	−	NG	W	−	W	−	W	−	W	−	−	+	−	−	−	−
Haemophilus paraphrophilus	+	−	−	+	A/A	+	−	+	+	+	−	+	V	+	+	−	−	−	−
Haemophilus parahaemolyticus	+	V	+	−	NG	+	−	+	−	+	−	+	−	−	+	−	V	−	+
Haemophilus paraphrohaemolyticus	+	+	+	+	A/A	+	−	+	−	+	−	+	−	−	+	−	−	−	+
Factor X+, V−, Porphyrin−																			
Haemophilus ducreyi	−	−	+/−	+/−	NG	−	−	−	−	−	−	−	−	−	+	−	−	−	−
Factor X−, V−, Porphyrin+																			
Haemophilus aphrophilus *	+/−	−	−	+	A/A	+	V	+	V	+	−	+	−	+	+	−	−	−	−

*On initial isolation, may appear to be hemin dependent.
V, variable; W, weak reaction; A, acid; -, + = 90% or greater; NG, no growth.

imipenem, and ciprofloxacin. Also effective is chloramphenicol. Because of the increased resistance to ampicillin, and the possibility of resistance to chloramphenicol, combination therapy with ampicillin and chloramphenicol may also be used for initial therapy. Non–life-threatening *H. influenzae* infection may be treated with amoxicillin-clavulanate, an oral second- or third-generation cephalosporin, trimethoprim-sulfamethoxazole, or ampicillin-sulbactam. For the treatment of *H. ducreyi,* erythromycin, ceftriaxone, or a fluoroquinolone is recommended. Infections with *H. aphrophilus* can be treated with penicillin and gentamicin, or cephalothin and gentamicin.

Resistance to ampicillin by *Haemophilus* species is due to enzyme production (β-lactamase) or, to a lesser extent, altered penicillin binding proteins. Several rapid tests to detect β-lactamase production are available, including the chromogenic cephalosporin and acidimetric tests.

Either the chromogenic cephalosporin test (Cefinase, Becton-Dickinson, Cockeysville, MD) or the acidimetric (Beta-Test, Medical Wire and Equipment Co., Wiltshire, England) can be used. A positive β-lactamase test means that the *Haemophilus* species is resistant to ampicillin and amoxicillin.

TABLE 15-3

DIFFERENTIAL TESTS FOR *HAEMOPHILUS* BIOGROUPS

				Distribution of Biotypes			
	Ornithine	Indole	Urea	Meningitis & Epiglottitis	Ear Infection	Conjunctivitis	Upper Respiratory Tract
Haemophilus influenzae							
Biotype II	−	+	+	●	●●	●●	●
Biotype III	−	−	+	●	●	●●	●
Biotype IV	+	−	+	●	●	●	●
Biotype V	+	+	−	●	●	●	●
Biotype VI	+	−	−	●	●	●	●
Biotype VII	−	+	−	●	●	●	●
Biotype VIII	−	−	−	●	●	●	●
Haemophilus parainfluenzae							
Biotype I	+	−	−				
Biotype II	+	−	+				
Biotype III	−	−	+				

−, + = 90% or greater; ● = 0–25%; ●● = 26–50%; ●●● = 51–75%; ●●●● = 76–100%.

In the chromogenic cephalosporin test, a disk impregnated with Cefinase is moistened with one drop of water. Using a sterile loop, several colonies are smeared onto the disk surface, or forceps can be used to wipe the moistened disk across the colonies. If the β-lactam ring of the Cefinase is broken by the enzyme, a color reaction occurs. A positive result is a red color on the area where the culture was applied. The reaction occurs usually within 5 minutes.

In the acidimetric test, a strip impregnated with benzylpenicillin, and a pH indicator, bromcresol purple, is moistened with one or two drops of sterile distilled water. Using a sterile loop, several colonies are smeared onto the test strip. If the β-lactamase ring of the benzylpenicillin is broken by the enzyme, penicilloic acid is formed, causing a drop in pH. This drop in pH is demonstrated by a color change from purple (negative) to yellow (positive) on the strip within 5 to 10 minutes.

PASTEURELLA

The genus *Pasteurella* is briefly summarized in this chapter. For a more detailed description of the microscopic and colony morphology, clinical infections, specimen processing, and laboratory diagnosis, see Chapter 34.

General Characteristics

The genus *Pasteurella* consists of gram-negative, pleomorphic coccobacilli that vary microscopically from ovoid to short rods to filamentous forms; bipolar staining is common. They are nonmotile and catalase positive; most isolates are oxidase positive and facultatively anaerobic; and they ferment glucose with weak to moderate acid production. *Pasteurella* may exhibit weak acid production from glucose without gas. In triple sugar iron agar (TSI) it gives the appearance of what is referred to as a "sick" TSI reac-

Figure 15-13

The TSI reactions of other gram-negative rods are compared with the weak ("sick") acid production of *Pasteurella multocida*. *A*, Uninoculated. *B, Pasteurella multocida. C, Citrobacter* (a member of Enterobacteriaceae). *D*, Nonfermenter.

tion. Figure 15–13 compares the TSI reactions of other gram-negative rods to the "sick" acid production of *Pasteurella multocida*.

The various species form grayish colonies that are similar in appearance. All species grow on blood agar plate and chocolate agar; species vary in their ability to grow on MacConkey agar, although most cannot. The ability to grow on blood agar plate without "feeder" organisms or in pure culture and the characteristic bipolar staining of the organism may help differentiate these species from *Haemophilus* species on first observation. Figure 15–14 presents an example of the colony morphology of *Pasteurella*.

Pasteurella species parasitize primarily the mucous membranes of the upper respiratory and gastrointestinal tracts of mammals and birds. They are rare inhabitants of the upper respiratory tract of humans. These organisms can be primary pathogens or secondary invaders and are pathogenic for a wide range of hosts. Humans acquire infections primarily through animal exposure.

Pasteurella multocida

There are currently more than 17 species of *Pasteurella*. The most commonly isolated species is *Pasteurella multocida*. This species now includes three subspecies: subspp *multocida, septica,* and *gallicida*. *Pasteurella multocida* consists of five serogroups, A, B, D, E, and F, which are based on capsular antigens.

Pasteurella multocida produces nonhemolytic colonies that may be mucoid in appearance on

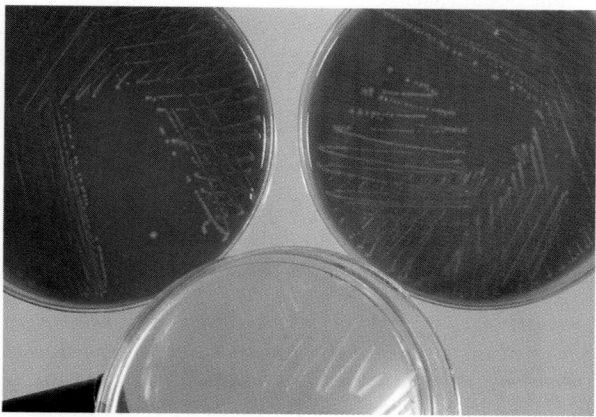

━━━━━F i g u r e   1 5 – 1 4

Pasteurella multocida growing on blood agar plate and CHOC. The MacConkey agar plate is negative for growth.

blood agar plate after 24 hours of incubation. After 48 hours, a narrow green to brown halo can surround the colony.

The most common form of infection caused by *P. multocida* seen in humans is cellulitis. This infection occurs primarily from bites or scratches inflicted by dogs and cats. The organisms are usually transmitted from the animal to the bite site or by the animals' licking preexisting abrasions.

BRUCELLA

The genus *Brucella* is briefly discussed here. For a more detailed description of its characteristics and clinical infections see Chapter 34.

General Characteristics

The genus *Brucella* consists of six species: *B. melitensis, B. abortus, B. canis, B. suis, B. neotomae,* and *B. ovis.* Four species that originate in animal reservoirs are pathogenic to humans (although the remaining two have caused rare infections): *B. melitensis, B. abortus, B. canis,* and *B. suis. Brucella melitensis* is the most common cause of brucellosis.

In the animal hosts, *Brucella* species can induce spontaneous abortion secondary to bacteremia in pregnant females. Urine and milk contain the infective organisms. The brucellae primarily infect humans through contact with infected animals and animal products; occupational exposure and ingestion of contaminated milk are the major means of infection.

The brucellae are strict aerobes, have a gram-negative coccobacillary microscopic morphology, and are non–spore-forming, nonmotile, non-encapsulated, intracellular parasites. Using traditional biochemical tests, they are oxidase positive and catalase positive, require carbon dioxide for growth (some species), are hydrogen sulfide positive using the more sensitive lead acetate method (most species), and are urease positive within 2 hours.

Blood and bone marrow cultures and acute and convalescent sera for serologic testing are usually submitted for examination. Serologic tests are the primary means of diagnosis of brucellosis. Since *Brucella* organisms are considered to be a type 3 biohazard, isolates should be sent to appropriately equipped reference laboratories.

FRANCISELLA

Francisella tularensis

Francisella tularensis is a highly infectious, strictly aerobic, nonmotile, small, gram-negative coccobacillus. In addition, it is a facultatively intracellular parasite. *Francisella tularensis* causes tularemia, an acute, febrile, granulomatous disease characterized by rapid onset and flu-like symptoms. Like *Pasteurella* and *Brucella, F. tularensis* is considered a zoonotic infection (see Chapter 34).

Two biovars cause disease in North America, *F. tularensis* biovar *tularensis* (type A) and *F. tularensis* biovar *palearctica* (type B). Type A is the more virulent strain, and humans are usually infected by rabbits, sheep, and ticks. Type B is more often transmitted by rodents and mosquitoes.

Since *Francisella* is a type 3 biohazard, isolates should be sent to reference laboratories for processing and identification. Serologic testing is most frequently used for confirmation.

Bibliography

Baron EJ, Finegold S: Diagnostic Microbiology, 8th ed. St. Louis: Mosby–YearBook, 1990.

Balows A, Hauser WJ, et al (eds): Manual of Clinical Microbiology, 5th ed. Washington, DC: American Society for Microbiology, 1991.

Isenberg HD: Clinical Microbiology Procedures Handbook. Washington, DC: American Society for Microbiology, 1992.

Joklik WK, Willett HP, et al (eds): Zinsser Microbiology, 20th ed. East Norwalk, CT: Appleton & Lange, 1992.

Koneman EW, Allen SD, et al: Color Atlas and Textbook of Diagnostic Microbiology, 4th ed. Philadelphia: JB Lippincott, 1992.

Mandell GL, Douglas RG, Bennett JE (eds): Principles and Practices of Infectious Diseases, 3rd ed. New York: Churchill Livingstone, 1990.

Murray PR, Drew LW, et al: Medical Microbiology. St. Louis: Mosby–YearBook, 1990.

Sanford JP: Guide to Antimicrobial Therapy. Dallas: Antimicrobial Therapy, Inc., 1993.

B. *LEGIONELLA*

EPIDEMIOLOGY

CLINICAL INFECTIONS
    Legionnaires' Disease
    Pontiac Fever

LABORATORY DIAGNOSIS
    Specimen Collection and Handling

    Direct Microscopic Examination
        Nonspecific Stains
        Direct Fluorescent Antibody
    DNA Probe
    Culture and Identification
    Urine Antigen Test
    Antibody Detection

Thomas Puchalski

In 1976, during an American Legion convention in Philadelphia, 221 persons became ill with pneumonia, and 34 of them died of a mysterious disease. *Legionella pneumophila,* the agent of this outbreak, became the first named member of the family Legionellaceae. Currently, the genus includes more than 30 species and 50 sero-groups. *Legionella pneumophila* contains 14 sero-groups.

The isolation of *Legionella* species from patient specimens and environmental sources has been recently made possible for most microbiology laboratories with the advent of commercial media and other diagnostic methods currently available. An understanding of the microscopic and colonial morphology of the *Legionella* species, their nutritional requirements, the pathogenesis of infection, and the advantages and disadvantages of the various laboratory tests available to diagnose infections enables the technologist and clinical microbiologist to arrive at a reliable laboratory diagnosis.

Legionella organisms are ubiquitous gram-negative rods acquired by humans primarily through the inhalation of aerosols. Clinically, infected patients may present with a wide variety of conditions, ranging from asymptomatic infection to life-threatening disease. The laboratory diagnosis of *Legionella* infection depends on the results of procedures such as the following:

- Direct fluorescent antibody technique
- Nucleic acid probe
- Urine antigen detection
- Serologic detection of antibodies
- Isolation using special media

The isolation and presumptive identification of the most common *Legionella* species can be accomplished by the microbiology laboratory using commercially available media, reagents, and kits.

EPIDEMIOLOGY

Most members of Legionellaceae are found worldwide, occurring naturally in aquatic sources, such as lakes, rivers, hot springs, and mud. Since *Legionella* species can tolerate chlorine concentrations below 2 to 3 mg per L, they resist water treatment and subsequently gain entry into and colonize human-made water supplies. Hot water systems, cooling towers, and evaporative condensers are major artificial reservoirs. Other sources include cold water systems, ornamental fountains, whirlpools, humidifiers, and industrial process waters. The factors that contribute to the ability of *Legionella* to colonize these sources include the following:

- The ability to multiply over the temperature range of 20 to 43°C and survive for varying periods at 40 to 60°C
- The capacity to adhere to pipes, rubber, plastics, and sediment and persist in piped water systems even when flushed
- The ability to survive and multiply within free-living protozoa and in the presence of commensal bacteria and algae

Legionella are transmitted to human hosts from these environmental sources primarily via aerosolized particles, like those produced by normal tap water pressure. Most outbreaks of disease originate from potable water distribution contamination. Other means of transmission include aspiration of contaminated water or secretions, direct inoculation by respiratory therapy equipment, and immersion of wounds in contaminated water. Transmission between humans has not been demonstrated.

CLINICAL INFECTIONS

Legionella species are facultative intracellular organisms that produce a wide range of infections in humans. Some species have been recovered from environmental sources only. The clinical manifestations of *Legionella* infections include pneumonia with or without extrapulmonary involvement (legionnaires' disease), a nonpneumonic flu-like febrile illness (Pontiac fever), and asymptomatic infection. These differing presentations may be influenced by several factors, including the organism's ability to enter and survive and multiply within the host cells, especially bronchoalveolar macrophages, and the ability to produce proteolytic enzymes. In addition, host factors, such as a suppressed immune system, chronic lung disease, alcoholism, and heavy smoking, predispose individuals to legionnaires' disease. The mode of transmission and the number of infecting organisms in the inoculum also probably play a role in the clinical features of the infection. When *Legionella* infections are diagnosed early, they can usually be treated successfully with erythromycin or a combination of erythromycin and rifampin. Alternative drugs include doxycycline, minocycline, trimethoprim-sulfa-methoxazole, azithromycin, and clarithromycin. Ciprofloxacin is the drug of choice in transplant recipients.

Legionnaires' Disease

Legionnaires' disease can present in three major patterns:

- sporadic cases, usually in the community—the most common presentation

■ epidemic outbreaks characterized by short duration and low attack rates

■ endemic nosocomial disease characterized by continual cases of disease over long duration

Pneumonia is the predominant manifestation of the disease, representing 5% of community-acquired pneumonia and up to 30% of cases of nosocomial pneumonia. The mortality rate is reported at 15 to 30% and may approach 50% in patients with nosocomial pneumonia if the correct diagnosis is not made early. *Legionella pneumophila* serogroup 1 accounts for most cases. Other *L. pneumophila* serogroups, commonly 4 and 6, and *L. micdadei* are more frequently implicated in clinical infections than the other *Legionella* species.

The incubation period for legionnaires' disease is 2 to 10 days. Patients typically present with a nonproductive cough, fever, headache, and myalgia. Later, as pulmonary infiltrates develop, sputum may be bloody or purulent. Rales, dyspnea, and shaking chills are clinical manifestations of progressing disease.

Dissemination via the circulatory system may lead to extrapulmonary infections with or without pneumonia. Infections of the kidneys, liver, heart, central nervous system, lymph nodes, spleen, and bone marrow as well as cutaneous abscesses have been described. Bacteremia, renal failure, liver function abnormalities, watery diarrhea, nausea, vomiting, headache, confusion, lethargy, and other central nervous system abnormalities have been associated with these infections.

Pontiac Fever

The nonpneumonic form of *Legionella* infection, Pontiac fever, usually has an incubation period of about 2 days. Patients are previously healthy individuals who complain of flu-like symptoms of fever, headache, and myalgia that last 2 to 5 days and then spontaneously subside. The incidence of Pontiac fever in the general population is unknown. *Legionella pneumophila* is responsible for most cases of this illness.

LABORATORY DIAGNOSIS

Several methods such as direct examination, culture, and antigen and antibody detection are available for the laboratory diagnosis of infec-

tions caused by *Legionella* species. Most laboratories utilize more than one method to maximize their diagnostic capabilities.

Specimen Collection and Handling

Specimens for culture and direct examination commonly include sputum, bronchoalveolar lavage (BAL), and bronchial washings. Ingram and Plouffe recommend that sputum purulence screens not be used and that all sputum specimens be accepted for culture. Transtracheal aspiration, lung tissue, blood, wound and abscess material, and pleural, peritoneal, and pericardial fluids may also be submitted. Water from environmental sources may be cultured for epidemiologic investigation. Urine is collected for antigen detection.

Respiratory secretions and body fluids (except blood) are submitted in sterile, leak-proof containers. Small pieces of tissue may be overlaid with sterile water. Saline or buffer should not be used in processing or transporting specimens because of the inhibitory effects of sodium on *Legionella*. Refrigerate these specimens if there is a delay of more than 2 hours between collection and processing, as overgrowth of contaminating flora may inhibit growth of *Legionella* when transport of the specimens is prolonged. When samples are to be transported to a reference laboratory, place the samples on wet ice. Freeze the specimens at −70°C if processing will be delayed for several days.

For blood cultures, the Isolator (Wampole Laboratories, Cranbury, NJ) system, which utilizes the lysis centrifugation method, is preferred. Approximately 10 mL of blood is collected in a special lysis centrifugation tube and transported to the laboratory for processing. *Legionella* can also be isolated from routine or Bactec blood culture bottles.

Collect at least 50 to 100 mL of water for culture in a sterile, leak-proof container.

Urine specimens for antigen testing are collected in sterile, leak-proof containers and assayed within 24 hours of collection. If testing is delayed, specimens should be stored at 2 to 8°C or frozen at −20°C.

Direct Microscopic Examination

The rapid and accurate diagnosis of *Legionella* infections is challenging to the clinician because clinical symptoms, radiologic presentations, and standard laboratory tests infrequently implicate the causative agent. Direct examination of stained specimens offers a rapid method for detecting *Legionella*.

Nonspecific Stains

When visualized microscopically in clinical specimens, *Legionella* organisms are pleomorphic, weakly staining gram-negative rods approximately 1 to 2 μm by 0.5 μm in size. The organisms may be found within macrophages and segmented neutrophils and extracellularly (Fig. 15–15). The staining intensity of the organisms can be enhanced by extending the safranin counterstaining time to at least 10 minutes. Other stains including Diff-Quik (Baxter Scientific, Kansas City, MO) and Giemsa may be used to facilitate detection of the organisms. However, all of these nonspecific staining methods are most useful for examination of specimens from normally sterile sites.

Direct Fluorescent Antibody (DFA) Test

The direct fluorescent antibody test (Procedure 15–1) is one of the most rapid laboratory procedures available to specifically detect the more common species of *Legionella* in clinical samples from the lower respiratory tract or from extrapulmonary sites. The test is especially useful if the result is positive; however, a negative result does not rule out *Legionella* infection. Available are fluorescein isothiocyanate (FITC)– labeled

conjugates that detect all known sero-groups of *L. pneumophila* (Genetic Systems, Seattle, WA) and *L. pneumophila* groups 1 to 7, *L. bozemanii, L. dumoffii, L. gormanii, L. longbeachae* groups 1 and 2, *L. micdadei,* and *L. jordanis* (SciMedx, Denville, NJ). The conjugate binds to the antigen in the cell membrane of the organisms, and these antigen-antibody complexes are detected using a fluorescence microscope when the ultraviolet blue light excites the FITC. The organisms thus appear as bright yellow to green, short or coccobacillary rods with intense peripheral staining (Fig. 15–16).

The specificity of the DFA test has been reported as 94 to 99%. Some species of *Pseudomonas, Bacteroides, Corynebacterium,* and other bacteria may, however, cross-react with the polyvalent conjugates. The microscopist must have experience with the morphologic and staining features of *Legionella,* so atypical organisms will not be misidentified. Some laboratories prefer not to use polyvalent reagents for direct specimen examination because of these cross-reactions.

The sensitivity of DFA testing for direct specimen examination is approximately 25 to 80% when compared with that of culture. To visualize organisms, approximately 10,000 to 100,000 organisms per mL of specimen must be present. Other factors such as excessively thick smears, which tend to obscure organisms, and the technical skill of the observer may influence the sensitivity of the test.

In the clinical microbiology laboratory at The Ohio State University Hospitals, the DFA is performed when requested by the clinician or when (1) suspicious organisms are seen, or (2) many neutrophils and no organisms are seen on Gram stained or specially stained smears. Because the sensitivity of the DFA test is less than that of culture, specimens should not be submitted for DFA testing only.

The DFA test also provides a useful method of confirming that an isolate is *Legionella* and of identifying the more common species and serogroups of the genus.

DNA Probe

A DNA probe is commercially available in kit form to detect the presence of all known species and serogroups of *Legionella* in clinical specimens (GenProbe, San Diego, CA). The I^{125}-labeled, single-stranded DNA probe combines with the target organism's ribosomal RNA, and

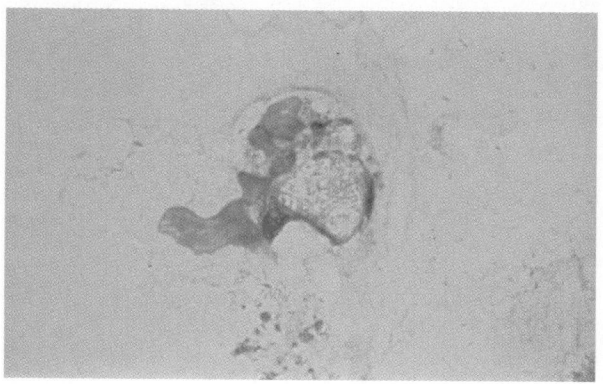

■■■■■ F i g u r e 1 5 – 1 5
Gram stain of specimen demonstrating intracellular and extracellular *L. pneumophila.* ×1,000.

PROCEDURE 15–1. DIRECT FLUORESCENT ANTIBODY STAIN

Materials:

1. FITC-labeled polyvalent and monovalent conjugates

2. Buffered glycerol mounting medium, pH 9

3. Well slides—two or three wells per slide

Method:

1. Prepare thin smears using a two- or three-well slide. Two smears per patient specimen can increase the sensitivity of direct testing.

    With a flamed loop, select purulent or bloody areas of sputum, and spread this material evenly over the wells.

    Centrifuge liquid specimens in excess of 2 mL at 3,000 rpm for 10 minutes, decant all but 0.5 mL of the supernatant, and resuspend the sediment using a sterile Pasteur pipette to make the smears.

    Using sterile instruments, cut a fresh face of tissue, and press and squeeze the tissue against the slide. Alternatively, grind the tissue with sterile distilled water to make a 10% homogenate, and spread this over the wells with a sterile wooden applicator stick.

    A suspicious colony from the culture plate is emulsified in a small portion of sterile distilled water to make a barely turbid suspension, and 1 drop of the suspension is placed in each well with a sterile Pasteur pipette.

2. Air-dry the smears, and heat-fix briefly.

3. Flood the smears with 10% neutral formalin for 10 minutes. Rinse with distilled water, and air-dry.

4. Using a moisture chamber, proceed with the staining procedure with a polyvalent or monovalent conjugate per instructions included with the reagents.

5. Air-dry slides.

6. Add a small drop of buffered glycerol, pH 9, to the smears, and cover with a coverslip.

7. Examine with a fluorescence microscope fitted with a mercury lamp and an FITC blue excitation filter combination, such as an excitation filter of 440 to 490 nm and a barrier filter at 520 nm.

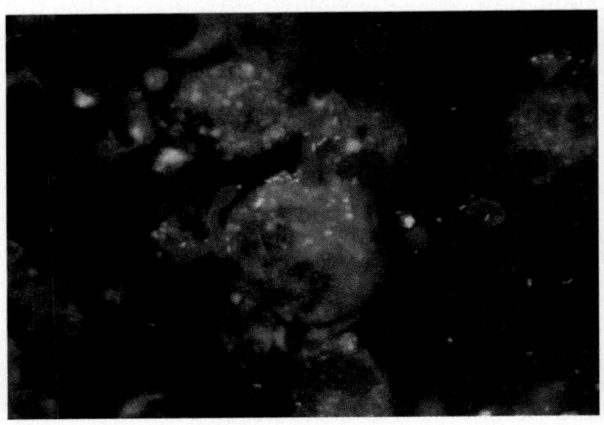

A   B

■■■■■ Figure 15–16

A, *Legionella pneumophila* in specimen smear stained by DFA technique. ×450. B, *Legionella pneumophila* in specimen smear stained by DFA technique. ×1,000. Note intense peripheral staining of the organisms.

the labeled DNA-RNA hybrids are detected in a gamma scintillation counter.

With a specificity of 99 to 100% and a sensitivity of 70 to 75% compared with culture, the test can be used as a substitute for DFA. Finkelstein and colleagues reported that the diagnostic performance of the DNA probe assay was superior to that of DFA testing, even though the probe test sensitivity was 31 to 67%, depending on the diagnostic criteria. The polymerase chain reaction (PCR) technique increases the sensitivity of the probe; however, the technique is not currently available for clinical use. Advantages of the probe are that the procedure is less tedious and time-consuming than that of the DFA and is free from subjective interpretation. Also, the probe can reliably identify all *Legionella* colonies to the genus level. However, inherent disadvantages of the radioactive probe (short shelf life, expensive equipment, disposal restrictions) limit its use to high-volume laboratories.

Culture and Identification

The single most important test for legionnaires' disease is culture of the organism. *Legionella* species are fastidious, aerobic bacteria that are unable to grow on sheep's blood agar and require the amino acid L-cysteine for growth. The organisms may appear as tiny colonies on chocolate agar that contains L-cysteine. However, buffered charcoal yeast extract (BCYE) is the recommended medium for *Legionella* isolation and is available commercially as nonselective and semiselective media. Semiselective BCYE contains polymyxin B, anisomysin, and either vancomycin (PAV) or cefamandole (PAC) and improves recovery of *Legionella* from highly contaminated specimens. However, growth of some *Legionella* is inhibited by semiselective media; therefore, BCYE with and without antibiotics should be used for culture. Tzielan and coworkers recommend the use of three media, including BCYE without antimicrobial agents as well as PAC and PAV, to provide maximal sensitivity.

Acid treatment of specimens contaminated with bacteria before inoculation also enhances isolation of *Legionella* (Fig. 15–17). In this procedure, an aliquot of the specimen is first diluted 1:10 with a 0.2 N KCl-HCl solution and allowed to stand for 5 minutes. Then the medium is inoculated with a portion of the acid-treated specimen (Procedure 15–2).

Inoculated medium is incubated at 35 to 37°C in air for at least 7 days. Usually within 3 to 5 days *Legionella* are visible as grayish-white or blue-green, convex, glistening colonies measuring approximately 2 to 4 mm in diameter. When these colonies are viewed with a dissecting microscope illuminated from above, they show a characteristic appearance (Fig. 15–18). The central portion of young colonies appears light gray and granular, like ground glass, while the periphery of the colony has pink and/or light blue or bottle-green bands with a furrowed appearance. Plates should be examined daily, as older colonies lose these characteristic features and may be mistaken for other bacteria. Plates with suspicious colonies

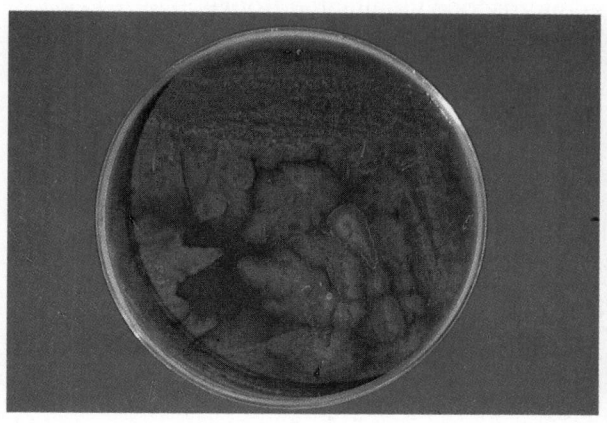

A   **B**

■■■■■■■ F i g u r e   1 5 – 1 7

A, Nonselective BCYE agar plate directly inoculated with sputum specimen. Note overgrowth of respiratory flora. (Courtesy of Richard Brust.)
B, Selective BCYE agar plate inoculated with same sputum specimen, which has been acid-washed prior to inoculation. Much of the respiratory flora has been eliminated. *Legionella* colonies are the smallest ones in the first quadrant. (Courtesy of Richard Brust.)

PROCEDURE 15–2. *LEGIONELLA* ISOLATION

Materials:

1. Buffered charcoal yeast extract (BCYE) nonselective and selective plates

2. KCl/HCl buffer, pH 2.2

   Add 5.3 mL of 0.2 N HCl and 25 mL of 0.2 N KCl to 100 mL of distilled water. Adjust pH to 2.2 with HCl or KCl.

   Dispense 4.5 mL into small screw-capped tubes, and autoclave for 10 minutes at 121°C.

   Store under refrigeration. Expiration 6 months.

Method:

1. Prior to medium inoculation, perform the following techniques to obtain the proper inoculum:

   Centrifuge liquid specimens in excess of 2 mL at 3,000 rpm for 10 minutes and 50-mL water samples at 3,000 rpm for 30 minutes. Use the sediment as inoculum.

   Homogenize pieces of tissue in 1 mL of sterile distilled water using a sterile tissue grinder.

   Prepare concentrates from Isolator or other blood culture bottles.

2. Inoculate one BCYE plate and one selective BCYE plate with an aliquot of the sample, and streak for isolation. Note that for water samples this step should be omitted. Proceed to step 3.

3. Additionally, specimens likely to be contaminated with bacterial flora are processed as follows:

   Add 0.5 mL of specimen to 4.5 mL of sterile HCl-KCl buffer pH 2.2 in a sterile screw-capped tube. Place three to four sterile glass beads in the tube if the specimen is excessively mucoid. Recap the tube, mix the specimen with the buffer, and break up the mucus using a vortex mixer.

   Let the suspension stand for 5 minutes.

   Pipet 0.1 mL of the suspension to a nonselective BCYE plate and 0.1 mL to a selective BCYE plate.

   Spread the inoculum over the surface of each plate.

4. Incubate the plates aerobically at 35 to 37°C for 7 days.

5. Examine cultures daily using a dissecting microscope illuminated from above.

---

can also be illuminated with a long-wave ultraviolet light (366 nm) and examined for differences in colonial autofluorescence to detect possible mixed *Legionella* infections (Table 15–4).

The physical and biochemical properties of *Legionella* are listed in Table 15–5. However, biochemical testing has limited value in the presumptive identification of isolates to the species level. The confirmation of the suspected colonies and their presumptive identification as *Legionella* species can be achieved in the microbiology laboratory using growth requirement for L-cysteine testing, Gram stain, direct fluorescent antibody tests, and nucleic acid probe (Fig. 15–19). The fol-

lowing methods are used to presumptively identify the more common *Legionella* species isolated in the clinical microbiology laboratory at The Ohio State University:

1. Prepare a smear from the suspicious colony growing on BCYE medium, and perform the Gram stain. *Legionella* organisms are thin, gram-negative rods that may show size variation from 2 to 20 μm in length.

2. Subculture the isolate to BCYE with L-cysteine and to either sheep's blood agar or BCYE without L-cysteine. *Legionella* grows only on BCYE medium supplemented with L-cysteine.

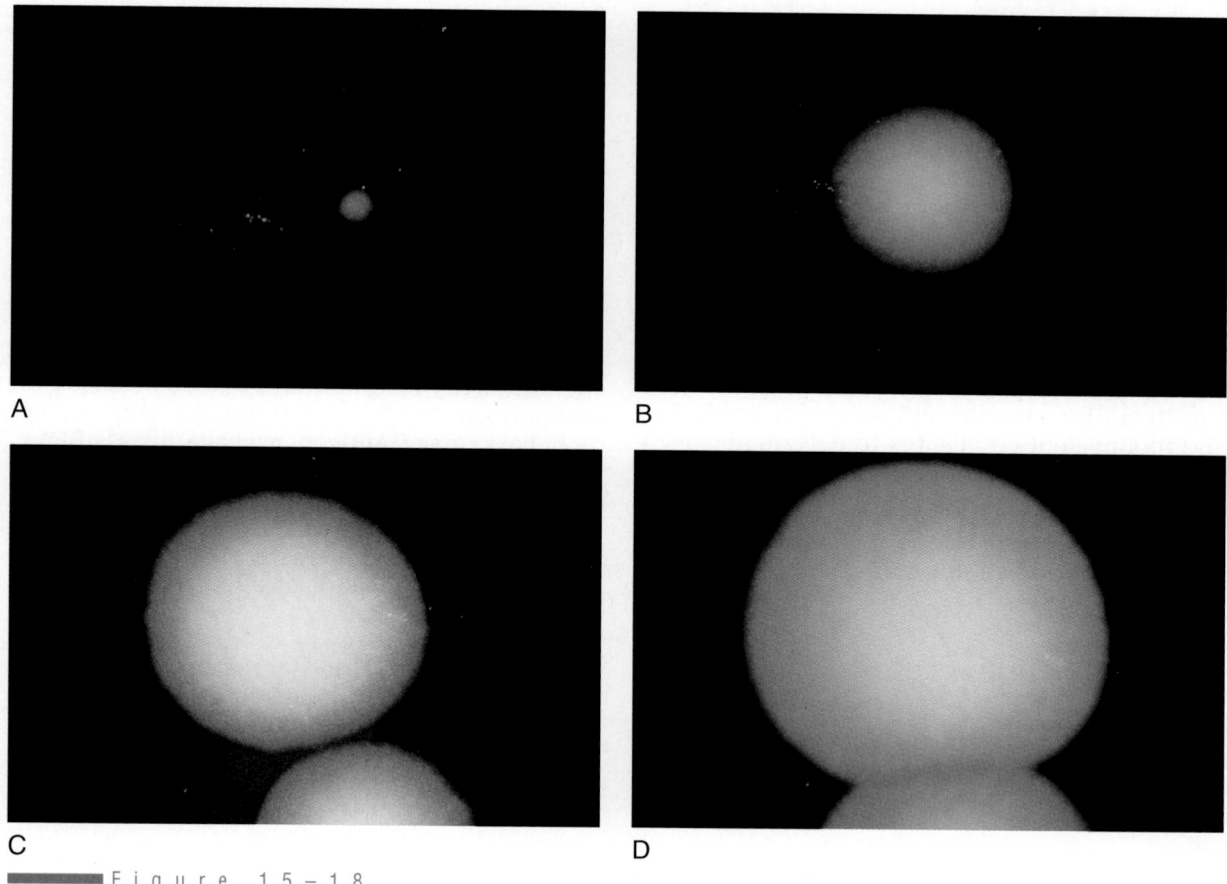

■■■■■ Figure 15–18

A, Legionella pneumophila colony on BCYE agar after 3 days of incubation, viewed with a dissecting microscope. ×20. *B,* Same colony after 4 days of incubation. *C,* Same colony after 5 days of incubation. *D,* Same colony after 7 days of incubation.

■■■■■ TABLE 15–4

GROUPING OF *LEGIONELLA* SPECIES BASED ON COLONIAL AUTOFLUORESCENCE[a]

Negative		Blue-White	Yellow-Green	Red
L. pneumophila	L. santicrusis	L. dumoffii	L. birminghamensis	L. rubrilucens
L. micdadei	L. israelensis	L. bozemanii		L. erythra
L. longbeachae	L. moravica	L. gormanii		
L. jordanis	L. brunensis	L. anisa (variable)		
L. oakridgensis	L. quinlivanii	L. parisiensis		
L. wadsworthii	L. lansingensis	L. cherrii		
L. fairfieldensis	L. adelaidensis	L. steigerwaltii		
L. feeleii	L. sainthelensi	L. tucsonensis		
L. maceachernii	L. geestiana			
L. hackeliae	L. quarteirensis			
L. cincinnatiensis	L. nautarum[b]			
L. jamestowniensis	L. worsleiensis[b]			
L. spiritensis	L. londiniensis[b]			

[a]Colonies are exposed to long-wave ultraviolet light (366 nm) (Mineralight Lamp, UVP Inc., San Gabriel, CA 91778).
[b]Proposed new species.
Modified from Rodgers FG, Pasculle W: *Legionella.* In Balows A, Hausler WJ, Jr, et al (eds): Manual of Clinical Microbiology, 5th ed. Washington, DC: American Society for Microbiology, 1991, p 443.

■ T A B L E 1 5 – 5
COMMON PHENOTYPIC CHARACTERISTICS OF *LEGIONELLA*

Gram-negative rod
Require L-cysteine for primary isolation
No growth on unsupplemented blood agar
Reduction of nitrate to nitrite: negative
Urease: negative
Acid from D-glucose: negative
Catalase or peroxidase: weakly positive

Data from Koneman EW, Allen SD, et al: Color Atlas and Textbook of Diagnostic Microbiology, 4th ed. Philadelphia: JB Lippincott, 1992, p 362.

3. Prepare smears from colonies that require L-cysteine for growth, and test with polyvalent and monovalent conjugates to determine specific species and serogroup.

Definitive identification using latex agglutination, gas chromatography, ubiquinone analysis, and DNA hybridization can be achieved by reference laboratories.

Urine Antigen Test

A radioimmunoassay method is commercially available for the detection of *L. pneumophila* serogroup 1 soluble antigen in urine (Binax, South Portland, ME). The tubes contained in the kit are coated with polyclonal rabbit antibody specific for the antigen, which is captured when the urine sample is incubated in the coated tube. After aspirating the sample and washing the tube to remove any remaining unbound sample, radio-labeled polyclonal rabbit antibody to the *L. pneumophila* serogroup 1 antigen is added to the tube and incubated. After another wash step, the presence of bound radioactivity is measured with a gamma scintillation counter.

This assay has several favorable characteristics. The sensitivity of the test is 90 to 94%, and the specificity ranges from 97 to 100% based on comparisons with culture results (Equate Legionella Urinary Antigen package insert, Binax, Inc., South Portland, ME). A urine sample is often easier to obtain than sputum and is an ideal specimen early in the disease, since urine antigen may be detectable as early as 3 days after symptoms of clinical disease begin. The cost of the test is relatively low.

However, there are some drawbacks associated with the test. Urine antigen excretion may continue for up to a year after infection; therefore, a positive test result may not indicate current infection. Prolonged antigenuria has been associated with immunosuppression, renal fail-

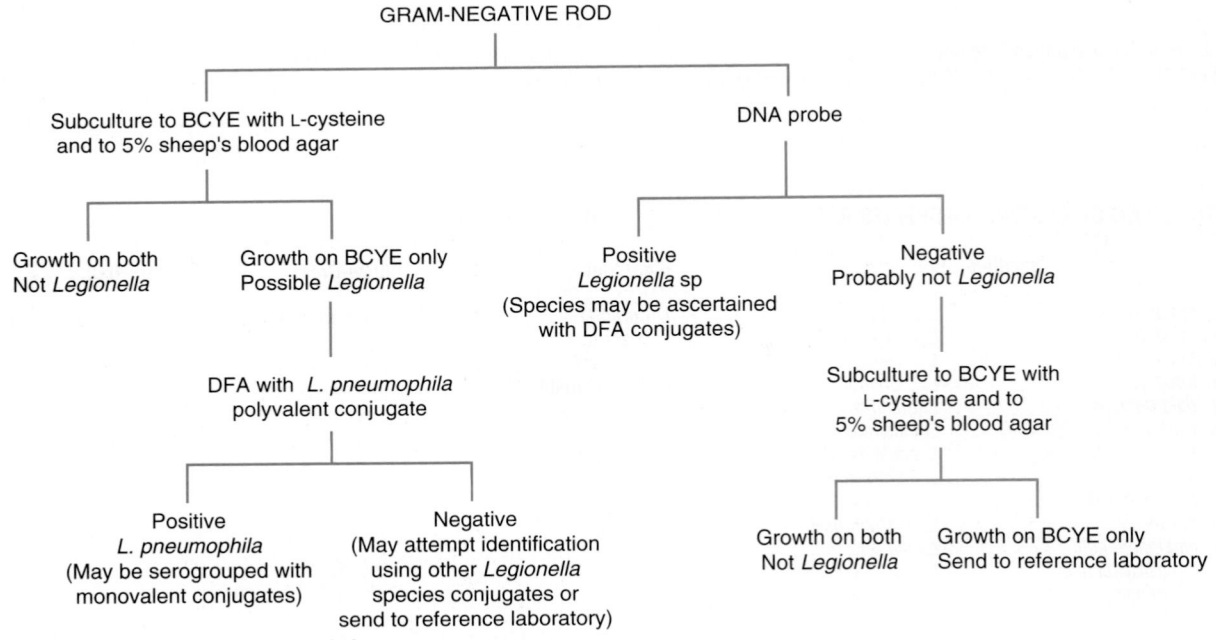

■ F i g u r e 1 5 – 1 9
Presumptive identification of *Legionella* using commercial reagents.

ure, and chronic alcoholism. Also, early treatment with erythromycin may decrease antigen excretion in some patients. Finally, the serospecificity of the test limits its breadth in the diagnosis of all *Legionella* infections.

Antibody Detection

The indirect fluorescent antibody (IFA) technique is the method of choice for the serologic diagnosis of legionnaires' disease. The method involves primarily two steps. First, the antibodies in the patient's serum attach to *Legionella* bacterial cell antigen on the test slide. Then, anti-human FITC-labeled globulin reacts with the antibody attached to the antigen. When this mixture is examined with a fluorescence microscope illuminated with ultraviolet blue light, the FITC emits fluorescence and the bacterial cell walls appear yellow-green.

The sensitivity of IFA is reported as 75 to 80%, with a specificity of 90 to 100%. The specificity of the test is enhanced when paired sera from patients with symptoms of legionellosis are tested. A fourfold rise in titer to at least 1:128 from the acute serum phase, obtained within 1 week of onset of symptoms, to the convalescent serum phase, 3 to 6 weeks later, is evidence of recent infection. However, patients with disease may not demonstrate a rise in titer for 8 weeks or longer after symptoms commence.

A single antibody titer of at least 1:256 in conjunction with compatible illness indicates a probable case of *Legionella* infection according to the definition of the Centers for Disease Control and Prevention (CDC). However, antibody titers may remain elevated months to years after acute infection; thus a single titer of 1:256 or greater is more suggestive of infection during an outbreak than in sporadic cases.

Cross-reacting antibodies in infections with gram-negative rods, *Mycoplasma,* and *Chlamydia* are known. The cross-reactivity is less a factor with formalin-killed antigen than with heat-killed antigen.

SUMMARY

Legionella species are ubiquitous, aquatic, gram-negative rods acquired by humans primarily through inhalation of aerosols. Clinically, infected patients may present with a variety of conditions ranging from asymptomatic infection to life-threatening systemic disease. The definitive diagnosis of *Legionella* infection depends on the results of laboratory procedures such as direct examination of specimens by the fluorescent antibody technique and urinary antigen detection, serologic detection of antibodies, isolation using special media, and identification of isolates by colonial characteristics, growth requirements, and differential staining with FITC-labeled conjugates. The isolation and presumptive identification of the most common *Legionella* species can be accomplished by the microbiology laboratory using commercially available media, reagents, and kits.

Bibliography

Bangsborg JM, Jensen BN, et al: Legionellosis in patients with HIV infection. Infection, 18:342, 1990.

Benson RF, Thacker WL, et al: *Legionella adelaidensis*, a new species isolated from cooling tower water. J Clin Microbiol 29:1004, 1991.

Berlin OGW, Drew WL, et al: Unclassified or unusual but easily cultivated etiologic agents of infectious disease. In Baron EJ, Finegold SM (eds): Bailey and Scott's Diagnostic Microbiology, 8th ed. St. Louis: Mosby–Year Book, 1990.

Birtles RJ, Harrison TG, et al: Evaluation of urinary antigen ELISA for diagnosing *Legionella pneumophila* serogroup 1. Infection 43:685, 1990.

Buesching WJ, Brust RA, Ayers LW: Enhanced primary isolation of *Legionella pneumophila* from clinical specimens by low-pH treatment. J Clin Microbiol 17:1153, 1983.

Dennis PJ, Brenner DJ, et al: Five new *Legionella* species isolated from water. Syst Bacteriol 43:329, 1993.

Edelstein PH, Edelstein MAC: Evaluation of the Merifluor-*Legionella* immunofluorescent reagent for identifying and detecting 21 *Legionella* species. Clin Microbiol 27:2455, 1989.

Fang G, Yu VL: Other *Legionella* species. In Mandell GL, Douglas RG, and Bennett JE (eds): Principles and Practice of Infectious Diseases, 3rd ed. New York: Churchill Livingstone, 1990.

Finkelstein R, Brown P, et al: Diagnostic efficacy of a DNA probe in pneumonia caused by *Legionella* species. Med Microbiol 38:183, 1993.

Goetz A, Yu VL: Screening for nosocomial legionellosis by culture of the water supply and targeting of high-risk patients for specialized laboratory testing. Am J Infect Control 19:63, 1991.

Gray JJ, Ward KN, et al: Serological cross-reaction between *Legionella pneumophila* and *Citrobacter freundii* in indirect immunofluorescence and rapid microagglutination tests. J Clin Microbiol 29: 200, 1991.

Hart CA, Makin T: *Legionella* in hospitals: A review. J Hosp Infect 18 (Suppl A): 481, 1991.

Holliday MG: Use of latex agglutination technique for detecting *Legionella pneumophila* (serogroup 1) antibodies. J Clin Pathol 43:860, 1990.

Horbach I, Fehrenbach FJ: Legionellosis in heart transplant recipients. Infection 18:361, 1990.

Ingram JG, Plouffe JF: Danger of sputum purulence screens in culture of *Legionella* species. J Clin Microbiol 32:209, 1994.

Koneman EW, Allen SD, et al (eds): Color Atlas and Textbook of Diagnostic Microbiology, 4th ed. Philadelphia: JB Lippincott, 1992.

Nichol KL, Parenti CM, Johnson JE: High prevalence of positive antibodies to *Legionella pneumophila* among outpatients. Chest 100:663, 1991.

Ramsey MK, Roberts GH: *Legionella pneumophila:* The organism and its implications. Lab Med 23:244, 1992.

Rodgers FG, Pasculle W: *Legionella.* In Balows A, Hausler WJ, Jr, et al (eds): Manual of Clinical Microbiology, 5th ed. Washington, DC: American Society for Microbiology, 1991.

Ruf B, Schurmann D, et al: Prevalence and diagnosis of *Legionella pneumophila:* A 3-year prospective study with emphasis on application of urinary antigen detection. J Infect Dis 162:1341, 1990.

Thacker WL, Benson RF, et al: *Legionella fairfieldensis* sp. nov. isolated from cooling tower waters in Australia. J Clin Microbiol 29:475, 1991.

Thacker WL, Dyke JW, et al: *Legionella lansingensis* sp. nov. isolated from a patient with pneumonia and underlying chronic lymphocytic leukemia. J Clin Microbiol 30:2398, 1992.

Tzielan CL, Vickers RM, et al: Growth of 28 *Legionella* species on selective culture media: A comparative study. J Clin Microbiol 31:2764, 1993.

Vickers RM, Stout JE, et al: Cefamandole-susceptible strains of *Legionella pneumophila* serogroup 1: Implications for diagnosis and utility as an epidemiological marker. J Clin Microbiol 30:537, 1992.

Wilkinson IJ, Sangster N, et al: Problems associated with identification of *Legionella* species from the environment and isolation of six possible new species. Appl Environ Microbiol 56:796, 1990.

World Health Organization: Epidemiology, prevention, and control of legionellosis: Memorandum from a WHO meeting. Bull World Health Organ 68(2):155, 1990.

Yu VL: *Legionella pneumophila* (legionnaires' disease). In Mandell GL, Douglas RG, Bennett JE (eds): Principles and Practices of Infectious Diseases, 3rd ed. New York: Churchill Livingstone, 1990.

Yu VL: Personal Communication, *Legionella* Update, 1993.

C. *BORDETELLA*

GENERAL CHARACTERISTICS

EPIDEMIOLOGY

VIRULENCE FACTORS

CLINICAL INFECTIONS

LABORATORY DIAGNOSIS
   Specimen Collection and Transport
   Direct Fluorescent Antibody Test
   Nucleic Acid Detection
   Culture and Identification
   Serology

ANTIMICROBIAL SUSCEPTIBILITY

Mario J. Marcon

GENERAL CHARACTERISTICS

Members of the genus *Bordetella* are small, gram-negative rods or coccobacilli. At least four species are recognized: *B. pertussis, B. parapertussis, B. bronchiseptica,* and *B. avium.* All are obligately aerobic bacteria, grow best at 35 to 37°C, and are relatively inactive in biochemical test systems. *Bordetella pertussis* and *B. parapertussis* are primary human pathogens of the respiratory tract, causing whooping cough or pertussis; the latter organism is usually associated with a milder form of the disease. Both of these organisms are fastidious to primary isolation, requiring special collection and transport systems as well as culture media. *Bordetella bronchiseptica* and *B. avium* are respiratory tract pathogens of wild and domestic birds and mammals and are generally nonfastidious and recoverable on routine microbiologic culture media. *Bordetella bronchiseptica* is also an opportunistic human pathogen, causing pneumonia and wound infections.

EPIDEMIOLOGY

Infections due to *Bordetella* species are acquired through the respiratory tract via the aerosol route. The bordetellae are uniquely adapted to adhere to and replicate on ciliated respiratory epithelial cells. The organisms remain localized to the respiratory tract, but toxins and other virulence factors are produced and have systemic effects.

Pertussis is one of the most highly communicable diseases of childhood, particularly problematic in developing countries that do not have access to vaccination. Even in the United States and other developed nations, where vaccine is readily available, periodic pertussis outbreaks occur every few years and isolated cases occur at all times. Major epidemics have occurred in countries following well-documented decreases in vaccination rates. It is largely unclear how *B. pertussis* is maintained in the human population. It is thought that adults who have been vaccinated or previously infected become transiently colonized with the organism when exposed to it. They may or may not experience mild respiratory tract symptoms, but they do provide a reservoir for transmitting the organism to susceptible individuals.

VIRULENCE FACTORS

Bordetella pertussis produces a variety of virulence factors that play a role in the pathogenesis of disease. Pertussis toxin (PT), also known as lymphocytosis-promoting factor, is a protein exotoxin. It produces a wide variety of biologic responses in vivo, and antibody to PT is thought to be very important in immunity to clinical pertussis. *Bordetella parapertussis* and *B. bronchiseptica* contain the structural gene for PT, but it is generally not transcribed and translated to PT. Filamentous hemagglutinin (FHA) is a protein component of *B. pertussis* that does not have toxigenic activity but is thought to be involved in the adherence of the organism to the ciliated epithelial cells of the upper respiratory tract. Antibody to this protein is also thought to be important in immunity, probably by interfering with initial attachment of the organism. Adenyl cyclase (AC) is an enzymatic protein secreted by *B. pertussis* cells. It has been suggested that this enzyme enters phagocytic cells, such as macrophages and polymorphonuclear neutrophils, and induces high levels of cyclic AMP in these cells. Cyclic AMP inhibits bactericidal functions, and thus *B. pertussis* organisms may be more likely to survive and cause disease in the respiratory tract. Tracheal cytotoxin (TC) is a small glycopeptide that appears to be toxic to the ciliated epithelial cells of the respiratory tract. Destruction of these cells impairs normal airway clearance and is probably important in allowing the organism to remain in the respiratory tract for a relatively long period. The role of other potential virulence factors, such as lipopolysaccharide (endotoxin), hemolysins, and surface fimbriae or pili, in the pathogenesis of pertussis is less well understood. All the virulence factors described earlier (except PT) are also produced by *B. parapertussis* and *B. bronchiseptica.*

CLINICAL INFECTIONS

Classic pertussis or whooping cough due to *B. pertussis* occurs following exposure to the organism through the respiratory tract and a 1- to 2-week incubation period. The initial symptoms are generally nonspecific and resemble the "common cold" or "flu." These include sneezing, mild cough, runny nose, and perhaps conjunctivitis. At this stage, known as the *catarrhal phase* of the disease, the infection is highly communicable because of the large number of organisms in the respiratory tract. However, cultures are not often performed at this stage because the symptoms are nonspecific. The catarrhal phase may last for 1 to 2 weeks and is followed by the *paroxysmal phase* of the disease. The hallmark of this phase is the sudden attack of severe, repetitive coughing followed by the characteristic "whoop" at the end of the coughing spell. The whooping sound is caused by the rapid gasp for air following the prolonged bout of coughing. Coughing spells may occur many times a day and are sometimes followed by vomiting. Young children may experience apnea, become cyanotic, and require aid in maintaining a patent airway. Many of these symptoms may be either absent or altered in very young infants, partially immunized children, or adolescents and adults; in addition *B. parapertussis* generally causes disease with milder symptoms. The *convalescent phase* of disease generally begins within 4 weeks of onset with a decrease in frequency and severity of the coughing spells. Complete recovery may require weeks or months.

LABORATORY DIAGNOSIS

Current laboratory diagnosis of pertussis generally employs culture isolation with or without DFA testing. For a variety of reasons, culture lacks sensitivity and DFA testing is probably no better than 60 to 70% sensitive when compared with culture; DFA tests may also lack specificity. Although serologic diagnosis identifies more cases, it is not generally utilized because it has not been standardized and is not widely available. Some newer tests, such as PCR amplification and detection of *B. pertussis* DNA, show significant promise for accurate and rapid diagnosis.

Specimen Collection and Transport

The specimen of choice for culture and DFA testing of *Bordetella* species is secretion collected from the posterior nasopharynx; throat culture specimens yield reduced sensitivity. *Bordetella pertussis* has been recovered in a few cases from bronchial lavage disease and transbronchial secretions from patients with acquired immunodeficiency syndrome (AIDS). Nasopharyngeal specimens may be collected either by aspiration

of the posterior nasopharynx with suction through a small catheter or as pernasal swab specimens. The swabs should be of either calcium alginate or Dacron (polyester) with a flexible wire shaft. Generally, two swabs are collected, one through each external naris; the swabs should be inserted as far back as possible into the nasopharynx, rotated, held a few seconds, and then gently withdrawn. In practice, swabs are used more commonly than aspiration methods.

Nasopharyngeal swab specimens may be plated directly onto culture media at the bedside or transferred to an appropriate transport system for delivery to the laboratory. Bedside plating is practical only when the patient is located close to the processing laboratory, so fresh culture media can be provided at short notice to medical personnel. Otherwise, a transport system based on the expected delay to culture should be selected. For a transport time of only several hours, a swab collection and transport system consisting of Amies transport medium with charcoal may be used. Alternatively, the swab may be expressed into a solution of 1% casein hydrolysate medium. In either case, the specimen should be transported to the laboratory at room temperature and transferred to culture medium on the day of collection. In any situation requiring overnight or several day transport, half-strength charcoal–horse's blood agar transport medium containing 40 mg/L of cephalexin (Regan-Lowe transport medium) should be used. This medium is prepared in screw-capped containers and is inoculated by streaking the surface and then submerging the swab in the agar and leaving it in place. The inoculated Regan-Lowe medium may be sent to the processing laboratory immediately or following incubation for 1 or 2 days at 35°C (preferred method). The processing laboratory may test any growth on the transport medium directly for *Bordetella* spp and use the swab to inoculate an isolation medium.

Direct Fluorescent Antibody Test

Although the DFA test is commonly used along with culture, the lack of sensitivity diminishes the clinical utility of a negative result. The test is generally helpful if the result is positive, but falsely positive test results may occasionally occur, even in experienced hands. Thus, the DFA test should always be used in conjunction with, not as a replacement for, culture. Slides for DFA testing may be prepared directly from swab specimens or following expression of the material from the swab to a solution of 1% casein hydrolysate. At least two slides should be prepared; these are dried, heat-fixed, and stained on the same day of collection or stored at −70°C and heat-fixed immediately before staining. Specific fluorescent-labeled conjugates for both *B. pertussis* and *B. parapertussis* can be used; however, *B. pertussis* occurs much more frequently, and it may be advantageous to stain both slides with this reagent. On microscopic examination, the organisms appear as small, fat rods or coccobacilli with intense peripheral yellow-green fluorescence and darker centers. Other organisms found in the nasopharynx, such as *Staphylococcus, Streptococcus,* and *Neisseria* spp, may stain evenly but weakly. Readers must learn to ignore these organisms. A smear may be deemed positive on the basis of as few as 10 typical organisms; however, it is not unusual to find hundreds of organisms in a well-collected and well-prepared specimen. Good specimens also contain leukocytes, ciliated epithelial cells, and strands of mucus.

Nucleic Acid Detection

Detection of *B. pertussis* and *B. parapertussis* DNA in nasopharyngeal specimens holds great promise for laboratory diagnosis of pertussis. Diagnosis by DNA detection would circumvent many of the problems associated with specimen transport and culture isolation. Moreover, isolation of the organism for antimicrobial susceptibility testing is not required because the organism is predictably susceptible to erythromycin. Several studies have already shown the clinical utility of PCR for amplification of *B. pertussis* DNA and subsequent detection. Although various *B. pertussis* genes have been targeted for amplification, a conserved, repeated chromosomal DNA sequence has been successfully used by several investigators. The technique of PCR needs to be developed by commercial companies before widespread application of this technology to pertussis diagnosis can be realized.

Culture and Identification

Since the original development of Bordet-Gengou potato infusion agar with glycerol and sheep's blood, a number of alternative media for recovery of *B. pertussis* have been tested. Although a few media and modifications can be used successfully, charcoal agar supplemented with 10% horse's blood and 40 mg/L of cephalexin has become most popular. This medium is identical in composition to the transport medium of Regan and Lowe described previously except that it contains agar at full strength. The medium has a shelf life of up to 8 weeks and is commercially available or can be easily prepared in house. Care should be taken to ensure the appropriate concentration of cephalexin in the medium. Some strains of *B. pertussis* have been reported to be inhibited at 40 mg/L or above. For this reason, it may be advisable to also plate a medium without cephalexin. However, it is probably more important to use media with cephalexin to prevent overgrowth of *Bordetella* spp by more rapidly growing organisms.

Plates for the recovery of *Bordetella* spp should be incubated at 35°C in air for a minimum of 7 days. It is important to ensure adequate moisture during this period to prevent plates from drying out. Most isolates of *B. pertussis* are detected in 3 to 5 days, while *B. parapertussis* is detected a day or so sooner. A stereomicroscope should be used to detect the colonies before they become visible to the unaided eye. On charcoal–horse's blood agar, the colonies are smooth, glistening, and silver, becoming whitish-gray with age (Fig. 15–20).

■■■■ F i g u r e   1 5 – 2 0

Five-day-old colonies of *B. pertussis* on charcoal–horse's blood agar (incident light from lower right corner).

Suspicious colonies should be Gram-stained; small gram-negative rods or coccobacilli should be further screened for *B. pertussis* and *B. parapertussis* using agglutinating antisera or fluorescein-labeled antisera. For the fluorescent antibody test from a plate isolate, slides should be carefully prepared so that organisms are well dispersed on the slide. This ensures that individual cells are well-stained and fluorescence will be easier to interpret. The agglutination test requires a larger amount of growth, and thus subculture from the primary isolation plate is sometimes required. For this reason, the fluorescent antibody test is often preferred and is just as convenient if available for DFA testing.

When the results of fluorescent staining or agglutination are clear-cut, confirmatory testing is not required. *Bordetella pertussis* and *B. parapertussis* can be adequately identified and separated with these serologic reagents alone. If initial serologic tests yield equivocal results, suspicious organisms should be subcultured to charcoal–horse's blood, routine blood agar, and chocolate agar plates (the last for recovery of *Haemophilus* spp that may break through on *Bordetella* media). Growth patterns should be observed, serologic tests repeated, and additional biochemical tests performed (Table 15–6) to confirm the identification. Table 15–7 summarizes the laboratory diagnosis of pertussis by culture.

Serology

Serologic diagnosis of pertussis has been used extensively to study outbreaks of disease and to

■ T A B L E 1 5 – 6
DIFFERENTIAL CHARACTERISTICS OF *BORDETELLA* SPECIES INFECTING HUMANS

Characteristics	Species		
	B. pertussis	*B. parapertussis*	*B. bronchiseptica*
Growth on			
Charcoal–horse's blood	+(3–5 d)	+(2–3 d)	+(1–2 d)
Blood agar	–	+	+
MacConkey agar	–	–	+
Catalase	+	+	+
Oxidase	+	–	+
Urease production	–	+(24 hr)	+(4 hr)
Nitrate reduction	–	–	+
Motility	–	–	+

d, days.

■ T A B L E 1 5 – 7
SUMMARY OF LABORATORY DIAGNOSIS OF PERTUSSIS BY CULTURE

1. Using calcium alginate or Dacron, flexible wire swabs, collect 2 pernasal specimens from the posterior pharynx
2. Transfer to transport system a, b, or c:
   a. 1% casein hydrolysate (1–2 hr holding time)
   b. Amies transport medium with charcoal (2–8 hr holding time)
   c. Half-strength charcoal–horse's blood with cephalexin (incubate 1–2 days at 35°C before transport to laboratory)
3. Prepare and read 2 smears for DFA test for *B. pertussis*
4. Inoculate charcoal–horse's blood agar with cephalexin. Incubate in moist chamber (without carbon dioxide) at 35°C for 7 days
5. Examine plates at 2 days, and daily thereafter, for typical colonies; use stereomicroscope
6. Screen *B. pertussis/parapertussis*–like colonies with Gram stain, fluorescent antibody test, or agglutinating antisera
7. Confirm identification of equivocal results by subculture, repeat antisera tests, and biochemical tests

document seroconversion following immunization. Unfortunately, the tests have not been well standardized and applied to routine diagnosis of pertussis. At present, there is no licensed commercial test available in the United States. Most recent interest in pertussis serology has involved enzyme-linked immunosorbent assays (ELISAs). Optimal diagnostic sensitivity requires paired sera, and testing for multiple immunoglobulin class–antigen combinations (e.g., IgG to FHA and PT, along with IgA to FHA). Because paired sera are required, diagnosis cannot be obtained quickly. ELISAs for IgA antibody to FHA from a single serum specimen or from nasopharyngeal aspirates show some promise for a more rapid serologic test.

ANTIMICROBIAL SUSCEPTIBILITY

Erythromycin is the drug of choice for treatment of pertussis as well as for prophylaxis of individuals exposed to patients with disease. Minimum inhibitory concentrations to erythromycin have remained constant (≤ 0.25 µg/mL) over the last 30 years. Although erythromycin is important for eradication of the organism and prevention of secondary cases, it has clinical efficacy only if treatment is started during the catarrhal phase of disease. A number of newer antimicrobials including fluoroquinolones (ciprofloxacin) and macrolides (azithromycin, clarithromycin) are active in vitro, but clinical trials have not been done. There is no need for laboratories to perform routine susceptibility tests to *B. pertussis* or *B. parapertussis;* patients are treated for 14 days with erythromycin. The susceptibility of *B. bronchiseptica* is not predictable, although the organism is usually susceptible to the aminoglycosides. Laboratory susceptibility tests should be done to support the clinical management of patients with opportunistic *B. bronchiseptica* infection.

SUMMARY

Bordetella pertussis and *B. parapertussis* are small, fastidious, gram-negative rods that invade and cause respiratory tract disease in humans. The organisms are maintained in the human population by transient colonization and asymptomatic infection of adults, whereas most disease occurs in young, unimmunized or partially immunized children. Clinical disease often presents with classic signs and symptoms, many of which can be attributed to a variety of virulence factors. Definitive diagnosis is dependent on recovery and identification of the organism from nasopharyngeal specimens; however, the fastidious nature of the organism and other factors often preclude successful isolation. Other laboratory tests with present or potential utility include DFA, nucleic acid detection by PCR, and serology. Erythromycin remains the drug of choice for treatment and prophylaxis. *Bordetella bronchiseptica* is an opportunistic human pathogen causing respiratory tract and wound infections.

Bibliography

Campbell PB, Masters PL, Rohwedder E: Whooping cough diagnosis: A clinical evaluation of complementing culture and immunofluorescence with enzyme-linked immunosorbent assay of pertussis immunoglobulin A in nasopharyngeal secretions. J Med Microbiol 27:247, 1988.

Daughterty MP, Dolter J, et al: Processing of specimens for isolation of unusual organisms. In Isenberg H (ed): Clinical Microbiology Procedures Handbook. Washington, DC: American Society for Microbiology, 1992, pp. 1.18.13–1.18.17.

Ewanowich CA, Chui LWL, et al: Major outbreak of pertussis in northern Alberta, Canada: Analysis of discrepant direct fluorescent-antibody and culture results by using polymerase chain reaction methodology. J Clin Microbiol 31:1715, 1993.

Friedman RL: Pertussis: The disease and new diagnostic methods. Clin Microbiol Rev 1:365, 1988.

Gilchrist MJR: Pertussis: Pathophysiology and prevention. Clin Microbiol Newsl 12:17, 1990.

Gilchrist MJR: Laboratory diagnosis of pertussis. Clin Microbiol Newsl 12:49, 1990.

Glare EM, Paton JC, et al: Analysis of repetitive DNA sequence from *Bordetella pertussis* and its application to the diagnosis of pertussis using the polymerase chain reaction. J Clin Microbiol 28:1982, 1990.

Halperin SA: Interpretation of pertussis serologic tests. Pediatr Infect Dis J 10:791, 1991.

Hoppe JE: Methods for isolation of *Bordetella pertussis* from patients with whooping cough. Eur J Clin Microbiol 7:616, 1988.

Kurzynski TA, Boehm DM, et al: Comparison of modified Bordet-Gengou and modified Regan-Lowe media for the isolation of *Bordetella pertussis* and *Bordetella parapertussis*. J Clin Microbiol 26:2661, 1988.

Kurzynski TA, Boehm DM, et al: Antimicrobial susceptibilities of *Bordetella* spp. isolated in a multicenter pertussis surveillance project. Antimicrob Agents Chemother 32:137, 1988.

Onorato IM, Wassilak SGF: Laboratory diagnosis of pertussis: The state of the art. Pediatr Infect Dis J 6:145, 1987.

Regan J, Lowe F: Enrichment medium for the isolation of *Bordetella*. J Clin Microbiol 6:303, 1977.

Sneed JO: Laboratory diagnosis of pertussis. In Isenberg H (ed): Clinical Microbiology Procedures Handbook. Washington, DC: American Society for Microbiology, 1992, pp 1.14.9–1.14.1.13.

Sutcliffe EM, Abbott JD: Selective medium for the isolation of *Bordetella pertussis* and *parapertussis*. BMJ 6:732, 1972.

Thomas MG: Epidemiology of pertussis. Rev Infect Dis 11:255, 1989.

Weiss AA: The genus *Bordetella*. In Balows A, Truper HG, et al (eds): The Prokaryotes: A Handbook on the Biology of Bacteria, Vol 3, Ecophysiology, Isolation, Identification, Applications, 2nd ed. New York: Springer-Verlag, 1992, pp 2530–2543.

Woolfrey BF, Moody JA: Human infections associated with *Bordetella bronchiseptica*. Clin Microbiol Rev 4:243, 1991.

Enterobacteriaceae

Connie R. Mahon •
George Manuselis Jr.

OBJECTIVES

1. Give the general characteristics of organisms that belong to the family Enterobacteriaceae.

2. Describe the antigenic structures of this group of organisms, and explain how these structures are used for identification.

3. Given the organism's characteristic growth on nonselective and selective differential media; presumptively identify the isolate.

4. Explain the different reactions that may be observed in the triple sugar iron (TSI) agar.

5. Describe the reactions involved and the products of metabolism tested in the following miscellaneous reactions:

   ■ Indole test
   ■ Methyl red and Voges-Proskauer tests
   ■ Nitrate reduction test
   ■ Citrate utilization
   ■ Urease production
   ■ Phenylalanine deaminase
   ■ Decarboxylase test
   ■ Lysine iron agar

6. Given the key reactions for identification, place an unknown organism in its proper tribe or genus.

The family Enterobacteriaceae includes several genera and species; clinical isolates in general acute care facilities consist primarily of *Escherichia coli, Klebsiella pneumoniae,* or *Proteus mirabilis.* It is important, however, to be aware of the other species, as they also cause infectious diseases.

This chapter is divided into three major areas. The first discusses the clinically significant enteric species that cause opportunistic infections, the second covers the primary intestinal pathogens and their related human infections, and the third describes the methods of identification of these organisms. New genera and biotypes are briefly discussed (see Appendix C for recent changes in nomenclature for the Enterobacteriaceae).

GENERAL CHARACTERISTICS

The family Enterobacteriaceae, often referred to as "enterics," consists of a large number of diverse organisms. Described in this section are the general characteristics of the members in this family: their morphology, classification, virulence and antigenic factors, and clinical significance. Members of this family have four major features:

■ All ferment glucose

■ All reduce nitrates to nitrites

■ None produce cytochrome oxidase

■ All except *Klebsiella, Shigella,* and *Yersinia* are motile; with a rare exception, the flagellar arrangement is peritrichous if the organism is motile

Microscopic and Colonial Morphology

Members of the family Enterobacteriaceae are gram-negative, non–spore-forming, facultatively anaerobic bacilli. On gram-stained smears, they may appear as coccobacilli or straight rods. Colonial morphology on nonselective media, such as sheep's blood agar or chocolate agar, is of little value in their initial identification. With the exception of certain members (e.g., *Klebsiella*) that produce characteristically large and very mucoid colonies, all members of this family produce large, moist, gray colonies on nonselective media and are therefore indistinguishable.

A wide variety of differential and selective media, such as MacConkey agar, and highly selective media, such as Hektoen (HE) agar and

XLD (xylose-lysine deoxycholate) agar, are available for the presumptive identification of enteric pathogens. These media contain one or more carbohydrates, such as lactose and sucrose, which show the ability of the species to ferment any carbohydrate. Fermentation is indicated by a color change on the medium, which results from a drop in pH detected by a pH indicator incorporated into the medium. Nonfermenting species are differentiated by lack of color change, and colonies retain the original color of the medium. Species that produce hydrogen sulfide (H_2S) may be readily distinguished when placed on HE or XLD agar. HE and XLD agars contain sodium thiosulfate and ferric ammonium citrate, which produce blackening of H_2S-producing colonies. These features have been used to initially differentiate and characterize certain genera. Definitive characterization and identification depend on the biochemical reactions and serologic antigenic structures demonstrated by the particular species.

Classification

The use of tribes in classifying the members in this family was proposed by Ewing in 1963 and has since been continued and extended in subsequent editions of *Edwards and Ewing's Identification of Enterobacteriaceae.* In classifying species into tribes, Ewing grouped bacterial species with similar biochemical characteristics. Within the tribes, species are further classified into their respective genera. Differentiation of each genus and definitive identification of species are based on biochemical characteristics. Table 16–1 lists the bacterial species in the family Enterobacteriaceae and their respective tribes. Table 16–2 shows the biochemical features that differentiate the tribes. Although the concept of using tribes in the classification of bacteria has not been used in *Bergey's Manual of Systematic Bacteriology,* this classification has been an effective way of placing species in groups based on similar biochemical features.

Virulence and Antigenic Factors

The virulence of the members of Enterobacteriaceae is controlled by a number of factors, such as the ability to colonize, adhere, produce various toxins, and invade tissues. Some species also possess plasmids that may mediate resistance to antimicrobials.

Many members of this family possess antigens that can be used for identifying different serologic groups. These antigens include the following:

- O, or somatic, antigen, a heat-stable antigen located in the cell wall
- H, or flagellar, antigen, a heat-labile antigen found in the flagellum
- K, or capsular, antigen, a heat-labile polysaccharide found in certain species, such as the K1 antigen of *E. coli* and Vi antigen of *Salmonella typhi*

Clinical Significance

Members of the family Enterobacteriaceae are ubiquitous in nature. With the exception of a few species, most are present in the intestinal tract of animals and humans as commensal flora. Some species exist as free-living organisms in soil, water, and sewage, while others are known to be plant pathogens.

Based on the clinical infections they produce, members of the family Enterobacteriaceae may be divided into two categories: (1) opportunistic pathogens and (2) primary intestinal pathogens.

The opportunistic pathogens are often a part of the normal intestinal flora of both humans and animals. Outside their habitat, these organisms may produce serious extraintestinal opportunistic infections. For example, *E. coli,* a member of the normal bowel flora, may cause fatal meningitis in the newborn, and septicemia or wound and urinary tract infections in persons of any age group. Other organisms found in the environment can be equally devastating if infections develop in a compromised host or in wounds contaminated with soil or water.

The primary intestinal pathogens, *Salmonella, Shigella,* and *Yersinia enterocolitica,* are considered true pathogens, as they are not commensal flora in the gastrointestinal tract of humans. These organisms produce infections that result from the ingestion of contaminated food and water. Table 16–3 describes the diseases commonly associated with members of Enterobacteriaceae.

■ T A B L E   1 6 – 1

CLASSIFICATION OF THE FAMILY ENTEROBACTERIACEAE

Tribe	Genus	Species	Tribe	Genus	Species
I. Escherichieae	I. *Escherichia*	coli blattae vulneris fergusonii hermannii		II. *Enterobacter*	aerogenes cloacae agglomerans amnigenus sakazakii gergoviae dissolvens nimipressuralis asburiae taylorae hormaechei
	II. *Shigella*	dysenteriae flexneri boydii sonnei		III. *Hafnia* IV. *Serratia*	alvei marcescens liquefaciens rubidaea fonticola odorifera plymuthica ficaria
II. Edwardsielleae	I. *Edwardsiella*	tarda hoshinae ictaluri			
III. Salmonelleae	I. *Salmonella* Subgroup I	(Most serotypes) typhi choleraesuis paratyphi A gallinarum pullorum	VI. Proteeae	I. *Proteus*	mirabilis vulgaris penneri myxofaciens
	Subgroup II Subgroup III (*Arizona*) Subgroup IV Subgroup V			II. *Morganella* III. *Providencia*	morganii alcalifaciens stuartii rettgeri rustigianii
IV. Citrobacteriaceae	I. *Citrobacter*	freundii diversus amalonaticus	VII. Yersinieae	I. *Yersinia*	pseudotuberculosis pestis enterocolitica frederiksenii kristensenii intermedia ruckeri aldovae
V. Klebsielleae	I. *Klebsiella*	pneumoniae ozaenae oxytoca rhinoscleromatis planticola terrigena ornithinolytica			
			VIII. Erwinieae	I. *Erwinia*	amylovora carotovora

Modified from Ewing WH: Edwards and Ewing's Identification of Enterobacteriaceae, 4th ed. East Norwalk, CT: Appleton & Lange, 1986, pp 2–3.

OPPORTUNISTIC MEMBERS OF THE FAMILY ENTEROBACTERIACEAE AND ASSOCIATED INFECTIONS

Escherichia coli

E. coli, the most significant species in the genus *Escherichia,* is recognized as an important potential pathogen in humans. A gram-negative bacillus, it is a common isolate from the colon flora. *E. coli* has a distinctive colony morphology on certain laboratory media, such as MacConkey agar. Although it may appear as a non–lactose-fermenter or as a mucoid colony, *E. coli* usually produces a dry, pink (lactose positive) colony with a surrounding pink area of precipitated bile salts on MacConkey agar. Figure 16–1*A* to *C* illustrates the different colonial morphologies of *E. coli* growing on MacConkey agar. In addition, *E. coli* may appear as a β-hemolytic colony on blood agar plate (BAP).

Most strains of *E. coli* are motile and generally possess both sex pili and adhesive fimbriae. The

TABLE 16-2

BIOCHEMICAL CHARACTERISTICS OF TRIBES OF ENTEROBACTERIACEAE

Test or Substrate	Escherichieae	Edwardsielleae	Citrobacteriaceae	Salmonelleae	Klebsielleae	Proteeae	Yersinieae
Hydrogen sulfide (TSI agar)	−	+	+ or −	+	−	+ or −	−
Urease	−	−	(+w) or −	−	− or (+)	+ or −	+
Indole	+ or −	+	− or +	−	−	+ or −	+ or −
Methyl red	+	+	+	+	−	+	+
Voges-Proskauer	−	−	−	−	+	+	+
Citrate (Simmons)	−	−	+	+	+	d	−
KCN	−	−	+ or −	−	+	+	−
Phenylalanine deaminase	−	−	−	−	−	+	−
Mucate	d	−	+ or −	d	− or +	+	−
Mannitol	+ or −	+	+	+	+	− or +	+

Key
+   90% or more positive within 1 or 2 days.
(+)   positive reaction after 3 or more days (decarboxylase tests: 3 or 4 days).
−   no reaction (90% or more) in 30 days.
+ or −   most cultures positive; some strains negative.
− or +   most strains negative; some cultures positive.
+ or (+)   most reactions occur within 1 or 2 days; some are delayed.
d   different reactions, +, (+), −.
w   weakly positive reaction.

Note: *Salmonella* biosers *typhi* and *paratyphi* A and some rare bioserotypes fail to utilize citrate in Simmons medium. Cultures of bioser *paratyphi* A and some rare bioserotypes may fail to produce hydrogen sulfide; an occasional strain of almost any serotype of *Salmonella* may be hydrogen sulfide negative. Some cultures of *P. mirabilis* may yield positive Voges-Proskauer tests.
Modified from Ewing WH: Edwards and Ewing's Identification of Enterobacteriaceae, 4th ed. East Norwalk, CT: Appleton & Lange, 1986, p 43.

organism also possesses O, H, and K antigens. *E. coli* O groups have shown cross-reactivity with similar antigens in other members of Enterobacteriaceae, notably with *Shigella*. Typing for H antigens is useful in completing the serogrouping of a particular strain. The capsular K antigen, located on the bacterial surface, often masks the O antigen during bacterial agglutination by specific antiserum. Some *E. coli* K antigens are identical to capsular antigens of other species. The K1 antigen has been found to be identical to the capsular antigen in group B *Neisseria meningitidis*, suggesting a virulence property of K antigens.

Characteristically, *E. coli* does the following:

■ Ferments glucose, lactose, trehalose, and xylose

■ Has positive indole and methyl red tests

■ Does not produce H_2S, DNase, urease, or phenylalanine deaminase

■ Does not grow in the presence of potassium cyanide

■ Cannot utilize citrate as a sole source of carbon

■ May be motile or nonmotile

■ Produces a negative result with Voges-Proskauer test

TABLE 16-3

BACTERIAL SPECIES AND THE INFECTIONS THEY COMMONLY PRODUCE

Bacterial Species	Diseases
Escherichia coli	Bacteriuria, septicemia, neonatal sepsis, meningitis, and diarrheal syndrome
Shigella	Diarrhea, dysentery
Edwardsiella	Diarrhea, wound infection, septicemia, meningitis, enteric fever
Salmonella	Septicemia, enteric fever, diarrhea
Citrobacter	Opportunistic and nosocomial infections (wound, urinary)
Klebsiella	Bacteriuria, pneumonia, septicemia
Enterobacter	Opportunistic and nosocomial infections, wound infection, septicemia, bacteriuria
Serratia	Opportunistic and nosocomial infection, wound infection, septicemia, bacteriuria
Proteus	Wound infection, septicemia, bacteriuria
Providencia	Opportunistic and nosocomial infections, wound infection, septicemia, bacteriuria
Morganella	Opportunistic and nosocomial infections
Yersinia	
pestis	Plague
pseudotuberculosis	Mesenteric adenitis, diarrhea
enterocolitica	Mesenteric adenitis, diarrhea
Erwinia	Wounds contaminated with soil or vegetation
Pectobacterium	Wounds contaminated with soil or vegetation

Modified from Washington J: Laboratory Procedures in Clinical Microbiology, 2nd ed. New York: Springer-Verlag, 1981, p 181.

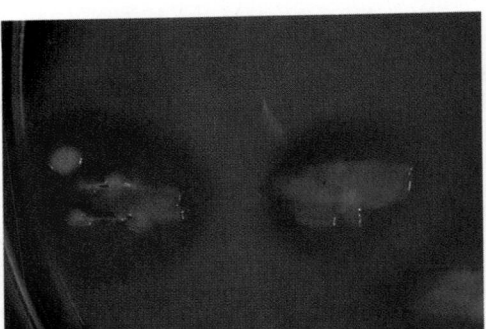
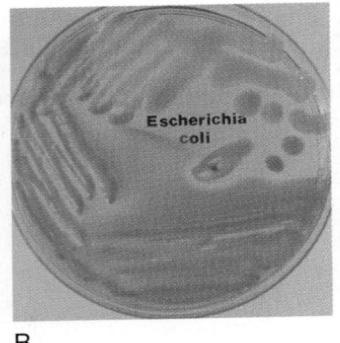
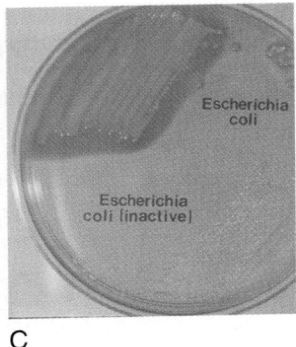

A

B

C

■■■■■■ F i g u r e   1 6 – 1

A, The typical dry, lactose-positive *Escherichia coli* growing on MacConkey agar. Note the pink precipitate surrounding the individual colonies.
B, Mucoid colonies of *E. coli* growing on MacConkey agar. (Courtesy of Jean Barnishan.) *C,* Non–lactose-fermenting (inactive) *E. coli* compared to typical *E. coli* on MacConkey agar. (Courtesy of Jean Barnishan.)

Clinical Infections

First described by Theodore Escherich in 1885, *E. coli* was considered a nonharmful member of the colon flora. Since then, *E. coli* has been associated with a wide range of diseases and infections, including meningeal (particularly in the newborn), gastrointestinal, urinary tract, wound, and bacteremic infections in all age groups.

GASTROINTESTINAL INFECTIONS

E. coli may cause several types of diarrheal illnesses. There are five major categories of diarrheogenic *E. coli,* based on definite virulence factors, clinical manifestations produced, epidemiology, and different O:H serotypes. These include the following:

■ Enteropathogenic (EPEC)

■ Enterotoxigenic (ETEC)

■ Enteroinvasive (EIEC)

■ Enterohemorrhagic (EHEC) serotype O157:H7

■ Enteroadherent (EAEC)

The serotypes associated with these categories and the features associated with the intestinal infections produced by these strains are summarized in Chapter 28.

Enteropathogenic *E. coli*. Whereas the enteropathogenic *E. coli* (EPEC) strain has been known to cause infantile diarrhea since the 1940's, its pathogenic role has remained controversial over the last two decades. Certain O serogroups of EPEC were identified in the late 60's and 70's as a cause of diarrhea, but only certain H types within each O serogroup were connected to the intestinal infections. O serogrouping could not, however, differentiate these *E. coli* strains from strains of normal flora. In 1978, Levine and colleagues attempted to settle the dispute concerning the pathogenic role of EPEC by challenging volunteers with EPEC strains that lacked the toxins of ETEC and the invasiveness of EIEC. This study showed that these EPEC strains caused distinct diarrhea. Subsequent studies further showed the adhesive property of EPEC strains, a characteristic not seen in ETEC or EIEC strains. The pathogenesis of enteroadherent EPEC is fully described in Farthing and Keusch's textbook *Enteric Infections*.

Diarrheal outbreaks due to EPEC have occurred in hospital nurseries and day care centers. Cases in adults are rarely seen. The illness is characterized by low-grade fever, malaise, vomiting, and diarrhea. The stool contains large amounts of mucus, but gross blood is not usually present.

Detection of diarrheal illness due to EPEC depends primarily on the index of suspicion of the clinician. In cases of severe diarrhea in children younger than 1 year, infection with EPEC should be suspected. Serologic typing with pooled antisera may be performed to identify EPEC serotypes. However, serotyping for EPEC has been used chiefly for epidemiologic studies rather than for diagnostic purposes.

Enterotoxigenic *E. coli*. Enterotoxigenic (ETEC) strains are associated with diarrhea of infants and adults in tropical and subtropical climates, especially in developing countries, where it is one of the major causes of infant bacterial diarrhea. In the United States and other developed countries, ETEC diarrhea is the most common cause of a diarrheal disease sometimes referred to as "traveler's diarrhea." This diarrheal illness is often acquired by travelers from industrialized countries when they visit developing, or Third World, countries. ETEC infection is commonly acquired by consuming contaminated food or water. Poor hygiene, inadequate sources of drinking water, and lack of proper sanitation are major contributing factors in the spread and transmission of the disease. A high infective dose of the organisms (10^6 to 10^{10} organisms) is necessary to initiate disease in an immunocompetent host. Protective mechanisms such as stomach acidity have been described as inhibiting colonization and initiation of disease; those suffering from achlorhydria seem to be at greater risk than are normal individuals.

Colonization of ETEC on the proximal small intestine has been recognized to be mediated by adhesion fimbriae that permit ETEC to bind to specific receptors on the microvilli. Once enterotoxigenic strains of E. coli are established, they may release into the small intestine one or both of the toxins they produce: a heat-labile toxin (LT) and a heat-stable toxin (ST). The action of the LT is similar to that of Vibrio cholerae toxin. Two fragments (A and B) make up the LT; the B fragment binds on the receptor site present on the GM_1 ganglioside of the intestinal mucosa. This binding facilitates the entry of the A fragment, which then acts on adenyl cyclase, activating the conversion of adenosine triphosphate (ATP) to cyclic adenosine, monophosphate (cAMP). The accumulation of cAMP in the intestinal mucosa initiates the hypersecretion of electrolytes and fluids into the lumen, resulting in a watery diarrhea. The ST, on the other hand, stimulates guanylate cyclase, causing the increased production of cyclic guanosine monophosphate (cGMP) and subsequent hypersecretion.

The usually self-limiting disease caused by ETEC is characterized by nonbloody, watery diarrhea, nausea, abdominal cramps, and low-grade fever. There is no evidence of mucosal penetration or invasion. The illness may last from 1 to 5 days. Diagnosis of ETEC infections is made primarily by the characteristic presenting symptoms and the isolation of solely lactose-fermenting organisms on differential media. Testing for toxins or colonizing factors remains in the research and reference laboratories, and there is no justification for its use in

the clinical laboratory for diagnostic purposes. Enzyme-labeled oligonucleotide probes have been reported to detect ETEC in fecal specimens. This method is undergoing further testing to determine its efficacy. ETEC infections must be differentiated, however, from other diarrheal illnesses which may appear similar (see Chapter 28).

Enteroinvasive E. coli. Enteroinvasive strains of E. coli (EIEC) are strains very different from the strains of EPEC and ETEC. Enteroinvasive strains produce dysentery, with direct penetration, invasion, and destruction of the intestinal mucosa. This diarrheal illness is very similar to that produced by Shigella. The EIEC infections seem to occur in adults and children alike. Direct transmission of EIEC from person to person via the fecal-oral route with subsequent occurrence of an outbreak was reported by Harris and colleagues in 1985.

The clinical infection is characterized by fever, severe abdominal cramps, malaise, and watery diarrhea accompanied by toxemia. Scanty stool containing pus, mucus, and blood follows the watery diarrhea. The organisms may be easily misidentified because of its similarity to Shigella.

EIEC strains may be nonmotile and do not ferment lactose; cross-reactions between Shigella and EIEC O antigens have been seen. Isolates may be mistaken for nonpathogenic E. coli; although EIEC do not decarboxylate lysine, more than 80% of E. coli decarboxylate lysine. For these reasons, cases of diarrheal illness due to EIEC may be underreported. Although EIEC and Shigella have been found to be similar in morphology and in clinical presentation, the infective dose of EIEC necessary to produce disease is much higher than that of Shigella.

The enteroinvasiveness of EIEC has to be demonstrated for definitive identification. The tests currently available to determine the invasive property of EIEC are not widely used in most clinical microbiology laboratories. The Sereny test, which determines the organisms' ability to produce keratoconjunctivitis in the guinea pig, is one of the assays previously used to determine the virulence of both Shigella and EIEC. DNA probes to identify EIEC strains have been studied and compared with the Sereny test, with comparable results. Other more recent developments in detecting invasiveness include monolayer cell cultures with Hep-2 cells.

Recently, DNA probes to screen stool samples for EIEC have been developed, which eliminate the need for various other tests to identify EIEC.

Enterohemorrhagic *E. coli*. In 1982, the O157:H7 strain of *E. coli* was first recognized during an outbreak of hemorrhagic diarrhea and colitis. The enterohemorrhagic *E. coli* (EHEC) strain serotype O157:H7 has since been associated with hemorrhagic diarrhea, colitis, and hemolytic-uremic syndrome (HUS). HUS is characterized by low platelet count, hemolytic anemia, and kidney failure.

The classic illness caused by EHEC is characterized by a watery diarrhea that progresses to a bloody diarrhea and crampy abdominal pain, with low-grade fever or no fever at all. The diarrheic stool contains no leukocytes, which differentiates it from *Shigella* dysentery or EIEC strain infection. The infection is potentially fatal, especially in young children in day care centers and schools and among the elderly in nursing homes. Processed meats, such as undercooked hamburger served at fast food restaurants, unpasteurized milk, and apple cider, have been implicated in the spread of infection.

E. coli O157:H7 produces two cytotoxins: verotoxins I and II. Verotoxin I is a phage-encoded cytotoxin identical to the Shiga toxin produced by *Shigella dysenteriae* type 1. This verotoxin shows damage on vero cells (African green monkey kidney cells), hence the term *vero*toxin. It also reacts with and is neutralized by the antibody against Shiga toxin. In contrast with verotoxin I, the second cytotoxin (verotoxin II) produced by some strains of O157:H7 is not neutralized by the antibody to Shiga toxin.

In the laboratory, verotoxin-producing *E. coli* may be identified by one of three methods:

■ Stool culture on highly differential medium, with subsequent serotyping

■ Finding the verotoxin in stool filtrates

■ Demonstration of a fourfold or greater increase in verotoxin-neutralizing antibody titer

Stool culture for *E. coli* O157:H7 may be performed using MacConkey agar containing sorbitol instead of lactose. *E. coli* O157:H7 does not ferment sorbitol in 48 hours, a characteristic that differentiates it from most *E. coli*. The use of this differential medium facilitates the primary screening of *E. coli* O157:H7, which ordinarily would not be distinguished from other *E. coli* on lactose-containing MacConkey or other routine enteric agar. *E. coli* O157:H7 appears colorless on sorbitol–MacConkey agar. While isolation of other non–sorbitol-fermenting organisms may

occur in up to 15% of cultures, *E. coli* O157:H7, when present, produces a heavy growth. A latex agglutination test for rapid presumptive detection of *E. coli* O157:H7 has also been reported useful; isolates must be tested with the negative control to detect nonspecific agglutination.

The commercially available MUG assay (4-methylumbelliferyl β-D-glucuronide) is a biochemical test that may be used to screen for O157:H7, in addition to testing for sorbitol fermentation. *Escherichia coli* O157:H7 rarely produces the enzyme β-glucuronidase, whereas 92% of the other strains do. If the enzyme is present, MUG is cleaved, and a fluorescent product is formed that is detectable with ultraviolet light.

Sorbitol-negative colonies are subsequently subcultured for serotyping using *E. coli* O157:H7 antiserum. Following the serotyping with O157:H7 antiserum, the isolates are tested for verotoxin production. Investigative reports have shown that all *E. coli* O157:H7 strains have so far been found to produce high levels of verotoxin. Verotoxin-producing strains may be detected using Vero cell tissue culture assays, which are more sensitive than HeLa cells. Whereas other groups of bacteria that produce verotoxin have been detected, they produce low to moderate levels of verotoxin.

Free verotoxin present in stool samples has been detected even in samples that yielded negative culture results. It had been previously reported that patients with hemorrhagic colitis shed the organism for only a brief period of time; nevertheless, verotoxin may still be detected when stool culture no longer yields the organism. A fourfold increase in verotoxin-neutralizing antibody titer has been demonstrated in patients with hemolytic-uremic syndrome and in whom verotoxin or verotoxin-producing *E. coli* had been detected.

Enteroadherent *E. coli*. Enteroadherent *E. coli*, most recently termed enteroaggregative *E. coli* (EAggEC), cause diarrhea by adhering to the mucosal surface of the intestine. These organisms produce symptoms such as watery diarrhea, vomiting, dehydration, and, occasionally, abdominal pain.

OTHER *E. COLI* INFECTIONS

Urinary Tract Infections. *E. coli* is known to be the most common cause of urinary tract and kidney infections in humans. The *E. coli* that cause urinary tract infections usually originate in the large intestine as resident or transient members of the colon flora and may exist as a dominating or minority strain. Strains that cause urinary tract infections are believed to be selected from the fecal flora because of their special adaptation to the urinary tract epithelial mucosa. Moreover, strains isolated from urinary tract infections and acute pyelonephritis in immunocompetent hosts differ from those isolated in hosts compromised by instrumentation (e.g., catheterization) or by other defects of the urinary tract.

E. coli strains that cause acute pyelonephritis in immunocompetent hosts have been shown to be dominating and resident members of the colon flora. These isolates from pyelonephritis infections, which belong to certain serotypes, were also shown to be resistant to the bactericidal activity of serum. Isolates from immunocompromised hosts, on the other hand, consist of a wide variety of strains. Similarly, strains isolated from asymptomatic bacteriuria (ABU) cannot be attributed to a particular population of the colon flora isolates. In addition, isolates from ABU are susceptible to the bactericidal activity of the serum. No specific serotype has been identified.

Among the factors that contribute to the virulence of urethrogenic E. coli is the capability of the organism to adhere to the epithelial cells lining the urinary tract. This capability is mediated by adhesins. Another factor is the ability to produce hemolysins and aerobactin. Hemolysins kill leukocytes and inhibit phagocytosis and chemotaxis. Aerobactin is an extracellular iron chelator.

Septicemia and Meningitis. E. coli remains one of the most common causes of septicemia and meningitis among neonates, accounting for approximately 40% of the cases of gram-negative meningitis. Similar infections due to this organism are uncommon among older children.

The newborn usually acquires the infection in the birth canal just before or during delivery, when the mother's vagina is heavily colonized. Infection may also result if contamination of the amniotic fluid takes place. Although several strains of E. coli have been identified, based on serotypes and enterotoxin production, and associated with diarrheic illnesses, these strains have not been associated with neonatal sepsis or meningitis.

The capsular antigen K1 present in certain strains of E. coli has been the most documented virulence-associated factor in neonatal meningeal infections. E. coli K1 antigen is also immunochemically identical with the capsular antigen of group B Neisseria meningitidis. The association

of K1 antigen was established when E. coli strains possessing capsular K1 antigen were isolated from neonates with septicemia or meningitis. Fatality rates for infants with meningitis caused by E. coli K1 strains also were higher than those for infants infected with non–K1 strains.

In addition to the neonatal population, E. coli also remains as a clinically significant isolate in blood cultures from adults. E. coli bacteremia in adults may result primarily from a genitourinary tract infection or from a gastrointestinal source.

Other *Escherichia* Species

E. hermannii, formerly called E. coli atypical or enteric group II, is a yellow-pigmented organism that has been isolated from spinal fluid, wounds, and blood. Reports of isolating E. hermannii from foodstuffs such as raw milk and beef, the same sources of E. coli O157:H7, have been published. However, its clinical significance is not fully established.

The newest species added to this genus, *Escherichia vulneris,* has been isolated from humans with infected wounds. More than half of the strains of E. vulneris may also produce yellow-pigmented colonies. Figure 16–2 compares the colonial morphology of E. hermannii with that of E. vulneris. Escherichia blattae is an indole-negative species currently found only in the feces of cockroaches.

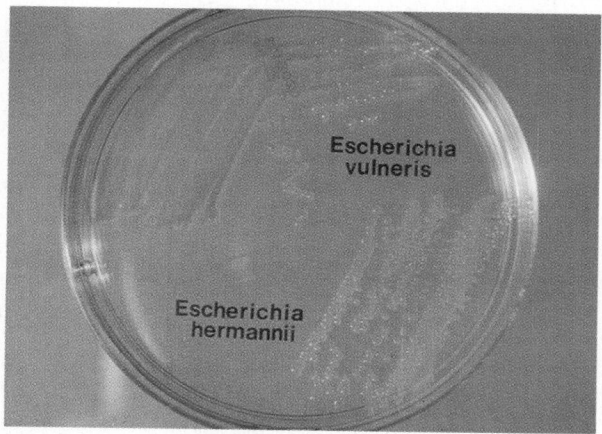

■■■■ F i g u r e 1 6 – 2

Comparison of the colonial morphology of E. vulneris and a yellow pigmented E. hermannii on MacConkey agar. Escherichia vulneris may also produce a yellow-pigmented colony, but the yellow is more prevalent in E. hermannii. (Courtesy of Jean Barnishan.)

Klebsiella, Enterobacter, Serratia, and Hafnia

Members of these genera are usually found in the intestinal tract of humans and animals or free-living in soil, water, and plants. These bacterial species have been associated with a wide variety of opportunistic and nosocomial infections, particularly pneumonia and wound and urinary tract infections.

Klebsiella, Enterobacter, Serratia, and *Hafnia* demonstrate variable biochemical reactions. Characteristic of these genera are the following:

■ Some species are motile, and some are non-motile

■ Most grow on Simmons sodium citrate and in potassium cyanide broth

■ None produce hydrogen sulfide

■ None (rare) deaminate phenylalanine

■ Few hydrolyze urea slowly

■ All give a negative reaction with the methyl red test and a positive reaction with the Voges-

Proskauer test. With a few exceptions, indole is not produced

Klebsiella. Usually found in the gastrointestinal tracts of humans and animals, the genus *Klebsiella* consists of several species, namely *K. pneumoniae, K. oxytoca, K. ozaenae,* and *K. rhinoscleromatis.* The absence of motility distinguishes *Klebsiella* species from other members of the family Enterobacteriaceae. Differential features of *Klebsiella* species are shown in Table 16–4.

Klebsiella pneumoniae is the most commonly isolated species. *K. pneumoniae* has the distinct feature of possessing a polysaccharide capsule. The capsule offers the organism protection against phagocytosis and antimicrobial absorption, thus contributing to its virulence. This capsule is also responsible for the moist, mucoid colonies characteristic of *K. pneumoniae.* Occasionally evident in direct smears from clinical materials, this capsule is sometimes helpful in providing a presumptive identification. Figure 16–3 illustrates the mucoid appearance of *K. pneumoniae* on MacConkey agar.

Colonization of gram-negative rods in the respiratory tract of hospitalized patients, particu-

■■■■ T A B L E   1 6 – 4
DIFFERENTIATION OF COMMON SPECIES WITHIN THE GENUS *KLEBSIELLA*

Test or Substrate	*K. pneumoniae* Sign	% +	(% +)	*K. oxytoca* Sign	% +	(% +)	*K. ozaenae* Sign	% +	(% +)	*K. rhinoscleromatis* Sign	% +	(% +)
Urease	+	95.4	(0.1)	+	90		d	14.5ʷ	(14.8)	−	0	
Indole	−	0		+	100		−	0		−	0	
Methyl red	− or +	11.3		+	100		+	97.7		+	100	
Voges-Proskauer	+	93.7		+	96		−	0		−	0	
Citrate (Simmons)	+	96.8	(0.6)	+	95		d	28.1	(32.4)	−	0	
Gelatin (22°C)	−	0	(0.2)	(+) or −	64		−	0		−	0	
Lysine decarboxylase	+	97.2	(0.1)	+	99		− or +	35.8	(6.3)	−	0	
Malonate	+	92.5		+	100		−	6		+ or −	50	
Mucate	+	92.8		+	95		− or +	25		−	0	
Sodium alginate (utilization)	+ or (+)	88.5	(9.2)	nd			− or (+)	0	(11)	−	0	
Gas from glucose	+	96		+	100		d	55.5	(9.4)	−	0	
Lactose	+	98.7	(1)	+	100		d	26.2	(61.3)	d	6	(70)
Dulcitol	− or +	33		+ or −	53		−	0		−	0	
Organic acid media												
Citrate	+ or −	64.4		nd			− or +	18		−	0	
D-Tartrate	+ or −	67.1		nd			− or +	39		−	0	

Key
+   90% or more positive within 1 or 2 days.
(+)   positive reaction after 3 or more days (decarboxylase tests: 3 or 4 days).
−   no reaction (90% or more).
+ or −   most cultures positive; some strains negative.
− or +   most strains negative; some cultures positive.
+ or (+) most reactions occur within 1 or 2 days; some are delayed.
d   different reactions, +, (+), −.
nd   no data.
w   weakly positive reaction.
Modified from Ewing WH: Edwards and Ewing's Identification of Enterobacteriaceae, 4th ed. East Norwalk, CT: Appleton & Lange, 1986, p 370.

■ Figure 16 – 3

Mucoid appearance of *Klebsiella pneumoniae* on MacConkey agar.

larly of *Klebsiella pneumoniae*, increases with the length of hospital stay. *K. pneumoniae* is a frequent cause of lower respiratory tract infections among hospitalized patients and in other immunocompromised hosts, such as newborns, the aged, and seriously ill patients on respirators. Other infections commonly associated with *K. pneumoniae* involving immunocompromised hosts are wound infections, urinary tract infections, and bacteremia. There have also been reports of nosocomial outbreaks of *Klebsiella* infections resistant to multiple antibiotics in newborn nurseries. These outbreaks have been attributed to the plasmid transfer of antimicrobial resistance. Other *Klebsiella* species have been associated with a number of infections. *K. oxytoca* is identical to *K. pneumoniae* except for its production of indole. It produces infections similar to those caused by *K. pneumoniae*. *K. ozaenae* has been isolated from nasal secretions and cerebral abscesses.

K. rhinoscleromatis has been isolated from patients with rhinoscleroma, an infection of the nasal cavity that manifests as an intense swelling and malformation of the entire face and neck. Cases of rhinoscleroma have been reported in Africa and South America. Both *K. ozaenae* and *K. rhinoscleromatis* are not considered true species but are viewed as biochemically inactive strains of *K. pneumoniae*. Organisms of *Klebsiella* group 47, proposed name *K. ornithinolytica* (indole- and ornithine decarboxylase–positive) and *K. planticola,* have been isolated from the urine, respiratory tracts, and blood of humans. *Klebsiella terrigena* has been found in soil and water isolates but has not been implicated in human diseases.

***Enterobacter* Species.** The genus *Enterobacter* is composed of 12 species, one of which consists of two biotypes. Clinically significant *Enterobacter* species that have been isolated from clinical samples include *E. cloacae, E. aerogenes, E. agglomerans, E. gergoviae, E. sakazakii,* and *E. hormaechei* (newly proposed name). Members of this genus are characterized as motile. The colony morphology of many of the species resembles that of *Klebsiella* when growing on MacConkey agar. *Enterobacter* species grow on Simmons citrate and in potassium cyanide broth; the methyl red is negative and the Voges-Proskauer test is positive. Unlike *Klebsiella*, however, *Enterobacter* species usually produce ornithine decarboxylase; lysine decarboxylase is produced by most species but not by *E. agglomerans* or *E. cloacae.*

Enterobacter cloacae and *E. aerogenes* are the two most common isolates from this genus. Distinguishing characteristics between *E. cloacae, E. aerogenes,* and *K. pneumoniae* are shown in Table 16–5. These two species have been isolated from wounds, urine, blood, and spinal fluid. *Enterobacter agglomerans,* which may produce a yellow-pigmented colony, gained notoriety with a nationwide outbreak of septicemia due to contaminated intravenous fluids. It is closely related to *Plantoea agglomerans,* which is primarily a plant pathogen. Figure 16–4 depicts a yellow-pigmented *E. agglomerans. Enterobacter gergoviae* is found in respiratory samples but is rarely isolated from blood cultures. *Enterobacter sakazakii,* a yellow-pigmented *Enterobacter* species, has been documented as a pathogen in neonates causing meningeal and bacteremic infections. It has also been isolated from cultures taken from brain abscesses and respiratory and wound infections. Figure 16–5 illustrates the colonial morphology of *E. sakazakii. Enterobacter hormaechei* has been isolated from human sources such as blood, wounds, and sputum.

Enterobacter amnigenus biotypes 1 and 2 and *E. intermedius* are found naturally in soil and water. *Enterobacter dissolvens* and *E. nimipressuralis* are newly recognized species with unknown clinical significance. *Enterobacter asburiae* is similar biochemically to *E. cloacae* and has been isolated from blood, urine, feces, sputum, and wounds.

***Serratia* Species.** The genus *Serratia* is composed of *S. marcescens, S. liquefaciens, S. rubidaea, S. odorifera, S. plymuthica, S. ficaria,* and *S. fonticola. Serratia* species are opportunistic pathogens associated with nosocomial outbreaks. With the

TABLE 16-5

DIAGNOSTIC FEATURES OF *ENTEROBACTER CLOACAE*, *ENTEROBACTER AEROGENES*, AND *KLEBSIELLA PNEUMONIAE*

Test or Substrate	E. cloacae Sign	%+	(%+)	E. aerogenes Sign	%+	(%+)	K. pneumoniae Sign	%+	(%+)
Urease	+^{w} or −	74.6		−	5^w		+	95.4	(0.1)
Motility	+	92.4		+	91.7		−	0	
Lysine decarboxylase	−	0		+	97.5		+	97.3	(6.3)
Arginine dihydrolase	+	92.4	(2)	−	0		−	0	
Ornithine decarboxylase	+	93.7	(1.3)	+	95.9	(0.8)	−	0	(0.2)
Gelatin (22°C)	(+)	0.6	(94.2)	(+) or −	0	(61.2)	−	0	
Adonitol, gas	− or +	21.7	(1.3)	+	94.2		d	84.4	(0.3)
Inositol									
Acid	d	13	(8)	+	96.7		+	97.2	(0.9)
Gas	−	4.1	(1.5)	+	93.4		+	92.5	(1.5)
Jordan D-tartrate	− or +	27.4		+ or −	78.3		+	94.4	
Sodium alginate (utilization)	−	0		−	0		+ or (+)	88.9	(8.9)

Key: +, 90% or more positive within 1 or 2 days. (+), positive reaction after 3 or more days (decarboxylase tests: 3 or 4 days). − , no reaction (90% or more) in 30 days. + or −, most cultures positive; some strains negative. − or +, most strains negative; some cultures positive. + or (+), most reactions occur within 1 or 2 days; some are delayed. d, different reactions, +, (+), −. w, weakly positive reaction.
Modified from Ewing's Identification of Enterobacteriaceae, 4th ed. East Norwalk, CT: Appleton & Lange, 1986, p 384.

exception of *S. fonticola, Serratia* species ferment lactose slowly (positive for orthonitrophenyl galactoside [ONPG]) and are differentiated from the tribe by their ability to produce extracellular DNase. *Serratia* species are also known for their resistance to a wide range of antimicrobials. Susceptibility tests must be performed on each isolate to determine appropriate antimicrobial therapy.

Serratia marcescens and *S. rubidaea* produce a characteristic pink to red pigment especially

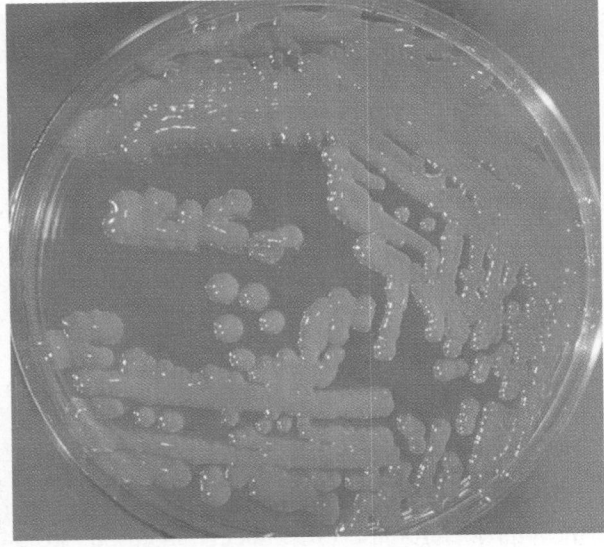

Figure 16-5
Mucoid, yellow-pigmented colonies of *E. sakazakii* growing on brain-heart infusion agar. (Courtesy of Jean Barnishan.)

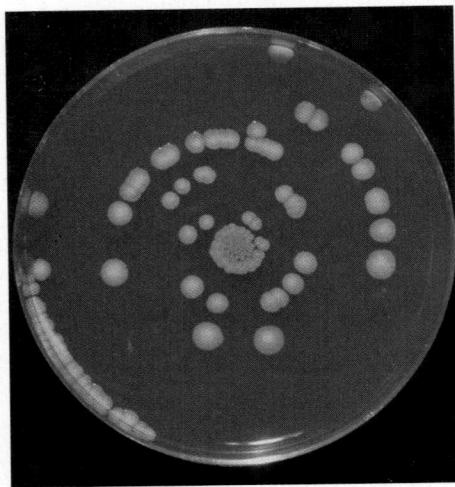

Figure 16-4
Illustration of a yellow-pigmented *E. agglomerans* blood agar plate. (Courtesy of Jean Barnishan.)

when the cultures are left at room temperature. Figure 16-6*A* and *B* illustrates the pigmentation of *S. marcescens* and *S. rubidaea.*

Serratia marcescens is the species that is usually considered clinically important. It has frequently been found in hospital-acquired infections of the urinary or respiratory tract and in bacteremic outbreaks in nurseries and cardiac surgery and burn units. Contamination of anti-

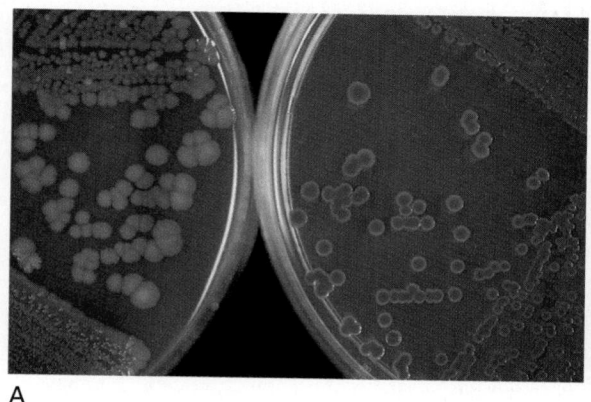

Figure 16-6

A, Example of brick-red pigment of *S. marcescens* when growing on MacConkey agar. *B,* Pinkish red pigmentation of *S. rubidaea* growing on MacConkey agar.

septic solution used for joint injections has resulted in an epidemic of septic arthritis. *S. plymuthica* osteomyelitis was found following a motorcycle accident. *Serratia odorifera,* as the species name implies, gives off a dirty, musty odor like potatoes and contains two biogroups. *Serratia odorifera* biogroup 1 is isolated predominately from the respiratory tract and is positive for sucrose, raffinose, and ornithine. In addition, biogroup 1 may be indole positive (60%). *Serratia odorifera* biogroup 2 is negative for sucrose, raffinose, and ornithine and has been isolated from blood and cerebrospinal fluid. Biogroup 2 may also be indole positive (50%). *Serratia liquefaciens* and *S. rubidaea* have also been isolated from human sources.

Hafnia. The genus *Hafnia* is composed of one species, *Hafnia alvei.* However, there are two distinct biotypes recognized: *H. alvei* and *H. alvei* biotype 1. Biotype 1 grows in the beer wort of breweries and has not been isolated clinically. *Hafnia* has been isolated from a number of anatomical sites in humans and in the environment. *Hafnia* species are not known to cause gastroenteritis but are occasionally isolated from stool cultures. A delayed positive citrate reaction is a major characteristic of *Hafnia* species.

Proteus, Morganella, and *Providencia*

Members of these genera are normal intestinal flora and are recognized as opportunistic pathogens. Members of the tribe are differentiated from other members of the family Enterobacteriaceae by their ability to deaminate phenylalanine and oxidatively deaminate lysine. All fail to ferment lactose.

Proteus. The genus *Proteus* consists of four species: *P. mirabilis, P. vulgaris, P. penneri,* and *P. myxofaciens. Proteus mirabilis* and *P. vulgaris* are widely recognized human pathogens. Both species have been isolated from urine, wounds, and ear and bacteremic infections. *Proteus mirabilis* and *P. vulgaris* are easily identified in the clinical laboratory because of their characteristic colony morphology. Both species produce swarming colonies on nonselective media, such as sheep's blood agar. The colonies also produce a distinct odor, sometimes described as "burned chocolate." Both species also produce hydrogen sulfide and hydrolyze urea. *Proteus mirabilis* is differentiated from *P. vulgaris* by the indole and ornithine decarboxylase tests; *P. mirabilis* does not produce indole from tryptophan and is ornithine positive, whereas *P. vulgaris* produces indole and is ornithine negative. *Proteus vulgaris* is sucrose positive and gives an acid/acid reaction in triple sugar iron agar.

Formerly a *P. vulgaris* strain, *P. penneri* is a newly recognized species that also swarms on nonselective media. *Proteus myxofaciens,* a species that has been isolated only from gypsy moths, is characterized by the large amount of slime it produces.

Morganella. The genus *Morganella* has only one species, *M. morganii,* formerly known as *Proteus morganii. Morganella morganii* has been implicated in diarrheal illness, but its role as an etiologic agent of diarrheal disease remains to be further examined. It is, however, a documented cause of urinary tract infections and has been isolated from other human body sites.

Providencia. The genus *Providencia* consists of four species: *P. alcalifaciens, P. stuartii, P. rettgeri* (formerly *Proteus rettgeri*), and *P. rustigianii. Providencia rettgeri* is a documented pathogen of the urinary tract and has caused occasional nosocomial outbreaks. Similarly, *P. stuartii* has been incriminated in nosocomial outbreaks in burn units and has been isolated from urine cultures. Infections caused by *P. stuartii* and *P. rettgeri,* especially in immunocompromised patients, are particularly difficult to treat because of their resistance to antimicrobials.

Providencia alcalifaciens is usually found in the feces of children with diarrhea; however, its role as a cause of diarrhea has not been proved. The new species *P. rustigianii* is rarely isolated, and its pathogenicity also remains unproven. Table 16–6 shows the differentiating characteristics of *Proteus, Providencia,* and *Morganella.*

Edwardsiella

The genus *Edwardsiella* is composed of three species: *E. tarda, E. hoshinae,* and *E. ictaluri. Edwardsiella tarda* is the only recognized human pathogen. Members of this genus are negative for urea and positive for lysine decarboxylase, hydrogen sulfide, and indole and do not grow on Simmon citrate.

Edwardsiella tarda is an opportunist, causing bacteremic and wound infections. Its pathogenic role in cases of diarrhea remains a controversy. *Edwardsiella hoshinae* has been isolated from snakes, birds, and water. *Edwardsiella ictaluri* causes enteric septicema in fish.

Erwinia and Pectobacterium

Both species are plant pathogens and are not significant in human infections. *Erwinia* grows poorly at 37°C and fails to grow on selective media, such as eosin–methylene blue agar, MacConkey agar, and other differential media typically used for the isolation of enterics. Identification of these organisms will be more for academic interest than for the evaluation of their significance as causative agents of infection.

Citrobacter

Earlier classifications of the family Enterobacteriaceae included the genus *Citrobacter* under

■ T A B L E 1 6 – 6
DIFFERENTIATING CHARACTERISTICS OF SPECIES OF *PROTEUS, PROVIDENCIA,* AND *MORGANELLA*

Test	Proteus* penneri	Proteus mirabilis	Proteus vulgaris	Providencia alcalifaciens	Providencia stuartii	Providencia rettgeri	Morganella morganii
Indole	–	–	+	+	+	+	+
Methyl red	+	+	+	+	+	+	+
Voges-Proskauer	–	– or +	–	–	–	–	–
Simmons citrate	–	+ or (+)	d	+	+	+	–
Christensen urea	+	+ or (+)	+	–	– or +	+	+
H₂S (TSIA)	– (70%)	+	+	–	–	–	+
Ornithine decarboxylase	–	+	–	–	–	–	+
Phenylalanine deaminase	+	+	+	+	+	+	+
Acid produced from							
Sucrose	+	d	+	d	d	d	–
Mannitol	–	–	–	–	d	+	–
Salicin	–	–	d	–	–	+	–
Adonitol	–	–	–	+	–	+	–
Rhamnose	–	–	–	–	–	+ or –	–
Maltose	+	–	+	–	–	–	–
Xylose	+	+	+ or (+)	–	–	– or +	–
Arabitol	–	–	–	–	–	+	–
Swarms	+	+	+	–	–	–	–

a Adapted from Brenner et al.[15]
*Esculin-negative.
Modified from Washington J: Laboratory Procedures in Clinical Microbiology, 2nd ed. New York: Springer-Verlag, 1981, p 189.

the tribe Salmonelleae, which formerly had consisted of *Salmonella, Citrobacter,* and *Arizona.* However, recent changes in the classification and nomenclature of bacterial species belonging to the tribe Salmonelleae (*Edwards & Ewing's Identification of Enterobacteriaceae,* 4th edition) have caused the reclassification of *Citrobacter* into its own tribe and of *Arizona* as a subspecies of *Salmonella.* The genus *Citrobacter* consists of *C. freundii, C. diversus,* and *C. amalonaticus.* Most *Citrobacter* species hydrolyze urea slowly and ferment lactose, producing colonies on MacConkey agar that resemble those of *E. coli* (see Fig. 16–1*A*). All grow on Simmons citrate and give positive reactions in the methyl red test.

Citrobacter freundii can be isolated in diarrheal stool cultures, and although it is a known extraintestinal pathogen, its pathogenic role in intestinal disease is not established. *Citrobacter freundii* has been associated with infectious diseases acquired in hospital settings; urinary tract infections, pneumonias, and intraabdominal abscesses have been reported. A report involving *C. freundii* in a case of endocarditis in an intravenous drug abuser was published by Plantholf and Trofa in 1987, the third case reported in medical literature and the first to be cured by antimicrobial therapy alone. One of the previously reported cases of *C. freundii* endocarditis required aortic valve replacement when antimicrobial therapy failed; the other case was fatal.

As most *C. freundii* (80%) produce hydrogen sulfide and some strains (50%) fail to ferment lactose, the colonial morphology of *C. freundii* on primary selective media may be easily mistaken for that of *Salmonella* species when isolated from stool cultures. It is therefore important to be able to differentiate *C. freundii* from *Salmonella.* Differentiation can be done by using a minimal number of biochemical tests, such as urea hydrolysis and lysine decarboxylase. Most *C. freundii* (70%) hydrolyze urea, but all (100%) fail to decarboxylate lysine; while *Salmonella* species fail to hydrolyze urea and most decarboxylate lysine.

Citrobacter diversus is a pathogen documented as the cause of nursery outbreaks of neonatal meningitis and brain abscesses.

Citrobacter amalonaticus is frequently found in feces, but there is no evidence that it is a causative agent of diarrhea. It has been isolated from sites of extraintestinal infections, such as blood and wounds.

PRIMARY INTESTINAL PATHOGENS AND RELATED HUMAN INFECTIONS

Salmonella and *Shigella* produce gastrointestinal illnesses in humans. *Salmonella* species inhabit the gastrointestinal tracts of animals. Humans acquire the infection by ingesting the organisms in contaminated animal food products or insufficiently cooked poultry, milk, eggs, and dairy products. Other *Salmonella* species are found only in humans, and infections are transmitted by human carriers.

Infections caused by *Shigella* species are associated with human carriers responsible for spreading the disease; no animal reservoir has been identified. *Shigella* dysentery usually indicates improper sanitary conditions and poor personal hygiene.

Yersinia infections, on the other hand, are transmitted by a wide variety of wild and domestic animals. *Yersinia* infections range from gastrointestinal disease, to mediastinal lymphadenitis, to fulminant septicemia and penumonia.

Salmonella

Members of the genus *Salmonella* produce significant infections in humans and in certain animals. *Salmonella* organisms are gram-negative, facultatively anaerobic rods that morphologically resemble other enteric bacteria. On selective and differential media used primarily to isolate enteric pathogens, Salmonellae produce clear, colorless, non–lactose-fermenting colonies; colonies with black centers are seen if the media contain indicators for hydrogen sulfide production. The biochemical features for the genus include the following:

- They do not ferment lactose

- They are negative for indole, the Voges-Proskauer test, phenylalanine, and urease

- Most produce hydrogen sulfide on triple sugar iron agar

- They do not grow in potassium cyanide

Classification. Until recently, the genus *Salmonella*, a large and complex group of organisms, comprised three biochemical discrete species: *S. enteritidis, S. choleraesuis,* and *S. typhi.* Genetic studies have shown, however, that bacterial species in the genus *Salmonella* are very

■ T A B L E 1 6 – 7

BIOCHEMICAL DIFFERENTIATION OF SELECTED MEMBERS OF THE *SALMONELLA* GROUP

Test	Reaction of[a]:			
	S. cholerae-suis	*S. para-typhi A*	*S. typhi*	Other[b]
Arabinose fermentation	–	+	–	+
Citrate utilization	V	–	–	+
Glucose gas production	+	+	–	+
Lysine decarboxylase	+	–	+	+
Ornithine decarboxylase	+	+	–	+
Rhamnose fermentation	+	+	–	+
Trehalose fermentation	–	+	+	+

[a] Symbols: –, ≤9% of strains positive; V, 10 to 89% of strains positive; +, ≥90% of strains positive.
[b] Typical strains in serogroups A through E.
Data from Farmer JJ, Davis BR, et al: J Clin Microbiol 21:46, 1985.

closely related and that only one species, *Salmonella enterica,* must be designated. Additional changes in the *Salmonella* classification include the classification of the genus into seven subgroups with designated subspecies. Subgroup I includes species that cause infections in humans. Members of subgroup I have very similar biochemical characteristics with the exception of *S. typhi, S. choleraesuis,* and *S. paratyphi.* These bacterial species are less active biochemically and are the most serious pathogens for humans, causing enteric fevers. Table 16–7 shows the characteristic features of *S. typhi, S. choleraesuis,* and *S. paratyphi.* Table 16–8 shows the seven *Salmonella* subgroups.

Species in subgroups II, III, and IV are usually found in cold-blooded animals, as well as in rodents and birds, which serve as their natural hosts. In addition to these changes, *Arizona,* which used to belong to its own genus, has become a member of the genus *Salmonella* and has been reclassified into subgroup III. *Arizona* infection may cause symptoms identical to those of *Salmonella* infections and may be transmitted to humans from pet turtles, snakes, and fish.

Virulence Factors. Factors responsible for the virulence of *Salmonella* have been the subject of speculation and still remain questionable. The role of fimbriae in adherence in initiating intestinal infection has been raised. It is apparent that fimbriated strains appear more virulent than nonfimbriated strains.

Another factor that contributes to the invasiveness of *Salmonella* is its ability to traverse intestinal mucosa. Specific factors that mediate this mechanism have not been established. Last, enterotoxin produced by certain *Salmonella* strains that cause gastroenteritis has been implicated as a significant virulence factor.

Antigenic Structures. Salmonellae possess antigens similar to those of other enterobacteria. The somatic O antigens and flagellar H antigens are the primary antigenic structures used in serologic grouping of *Salmonella.* A few strains may possess capsular K surface antigens, designated as Vi antigen. The serologic identification of the Vi antigen is important in identifying *S. typhi.*

Figure 16–7 shows the antigenic structures used in serologic grouping and their locations.

The heat-stable O antigen of *Salmonella* as is the case with other enteric bacteria, is the lipopolysaccharide (LPS) located in the outer membrane of the cell wall. There are many different O antigens present among the subspecies of *Salmonella*; more than one O antigen may also be found in a particular strain. The O antigens are designated by Arabic numbers.

Unlike the O antigens, flagellar antigens are proteins that are heat-labile and are treatable with ethanol or acid. The H antigens of *Salmonella* may occur in two phases; phase 1, the specific phase, and phase 2, the nonspecific phase. Phase 1 flagellar antigens occur only in a small number of serotypes and determine the immunologic identity of the particular serotype. Phase 1 antigens agglutinate only with homologous antisera. Phase 2 flagellar antigens, on the other hand, occur among several strains. Shared by numerous serotypes, phase 2 antigens react with heterologous antisera.

The heat-labile Vi (coined from the term virulence) antigen is a surface polysaccharide capsular antigen found in *S. typhi* and a few other strains of *Salmonella* subgroup I. The capsular antigen plays a significant role in preventing phagocytosis of the organism. The Vi antigen most often blocks the O antigen during serologic typing but may be removed by heating.

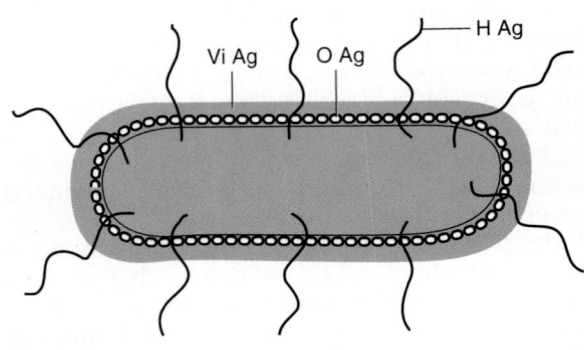

■ F i g u r e 1 6 – 7
The antigenic structures of *Salmonella* used in serologic typing.

Clinical Infections

In humans, salmonellosis may occur in several forms:

■ An acute gastroenteritis or food poisoning characterized by vomiting and diarrhea

■ Typhoid fever, the most severe form of enteric fever, caused by *S. typhi;* other enteric fevers are caused by other *Salmonella* species (i.e. *S. paratyphi, S. choleraesuis*)

■ Nontyphoidal bacteremia

■ Carrier state that follows *Salmonella* infection

Humans acquire the infection by ingesting the organisms in food, water, and milk contaminated with human or animal excreta. With the exception of *S. typhi* and *S. paratyphi*, salmonellae infect various animals that serve as reservoirs, and sources of human infections. *Salmonella typhi* and *S. paratyphi* have no known animal reservoirs, and infections seem to occur only in humans. Carriers are often the source of infection.

***Salmonella* Gastroenteritis.** One of the most common forms of "food poisoning," gastrointestinal infection caused by *Salmonella* results from the ingestion of the organisms through contaminated food. The *Salmonella* strains associated with this infections are usually those found in animals; most in the United States belong to the serotypes of *Salmonella enteritidis*. Consequently, the source of the infection has been attributed primarily to poultry, milk, eggs, and egg products as well as to handling pets. Insufficiently cooked eggs and domestic fowl, such as chicken, turkey, and duck, are common sources of infection.

▄▄▄▄▄▄▄ T A B L E   1 6 – 8

PROPERTIES OF THE SEVEN *SALMONELLA* SUBGROUPS[a]

Property or Test	Salmonella Subgroup						
	1	2	3a	3b	4	5	6
DNA Hybridization group of Crosa et al.[b]	1	2	3	4	5	Not studied	Not studied
Genus according to Ewing (1986)	*Salmonella*	*Salmonella*	*Arizona*	*Arizona*	*Salmonella*	*Salmonella*	*Salmonella*
Salmonella subgenus name formerly used	I	II	III	III	IV		
Subspecies according to Le Minor et al [c]	*choleraesuis*	*salamae*	*arizonae*	*diarizonae*	*houtenae*	*bongori*	*indica*
Flagella are usually monophasic (Mono) or diphasic (Di)	Di	Di	Mono	Di	Mono	Mono	Di
Usually isolated from humans and warm-blooded animals	+	–	–	–	–	–	–
Usually isolated from cold-blooded animals and the environment	–	+	+	+	+	+	+
Pathogenic for humans	++++	+	+	+	+	+?	+?
Differential tests[d]							
Dulcitol fermentation	96	90	0	1	0	92	62
Lactose fermentation	1	1	15	85	0	0	25
ONPG[e] test	2	15	100	100	0	92	50
Malonate utilization	1	95	95	95	0	0	0
Growth in KCN medium	1	1	1	1	95	100	0
Mucate fermentation	90	96	90	30	0	85	100
Gelatin hydrolysis[f]	–	+	+	+	+	–	+
D-Gelacturonic acid fermentation	–	+	–	+	+	100	100
Lysis by bacteriophage O1	+	+	–	+	–	46	88
D-Sorbitol fermentation	+	+	+	+	96	100	0

[a]Data from Le Minor et al. (references 93 and 94 in Farmer, Davis, et al. [1985]).

[b]Crosa JH, Brenner DJ, Ewing WH, Falko S: Molecular relationships among the salmonelleae. J Bacteriol 115:307–315, 1973.

[c]Le Minor L, Popoff MY, Laurent B, Hermant D: Individualisation d'une septieme sous-espèce de *Salmonella: S. choleraesuis* subsp. *indica* subsp. nov. Ann Microbiol (Paris) 137B:211–217, 1986.

[d]Numbers indicate percent positive after 2 days of incubation and are based on actual Centers for Disease Control and Prevention data; symbols are based on the data of Le Minor et al.: +, 90% or more positive; –, 10% or fewer positive.

[e]ONPG, *o*-Nitrophenyl-*p*-D-galactopyranoside.

[f]Rapid film method at 37°C (almost all strains are negative by the tube method at 22°C within 2 days).

Modified with permission from Balows A, Hausler WJ, Jr, Herrmann KL, et al: Manual of Clinical Microbiology, 5th ed. Washington, DC: American Society for Microbiology, 1991, p. 372.

Cooking utensils such as knives, pans, and cutting boards used in preparing the contaminated meat can have spread the contamination to other food. Direct transmission from person to person has been reported in institutions. *Salmonella* gastroenteritis, although referred to as food poisoning, occurs when a sufficient number of organisms contaminate food that is maintained under inadequate refrigeration, thus allowing growth and multiplication of the organisms. The infective dose necessary to initiate the disease is higher than that required for shigellosis. Approximately 10^6 bacteria may initiate infection, but infections resulting from lower infective doses have been reported.

The symptoms of intestinal salmonellosis, which may appear 8 to 36 hours after ingestion of contaminated food, include nausea, vomiting, fever, and chills, accompanied by watery diarrhea and abdominal pain. The role of enterotoxins in the pathogenesis of *Salmonella* infection remains unclear.

Most cases of *Salmonella* gastroenteritis are self-limiting. Symptoms disappear usually within a few days, with little or no complications. Those who suffer from sickle cell disease and other hemolytic disorders, ulcerative colitis, and malignancy seem to be more susceptible to *Salmonella* infection. The infection may be more severe in the very young, the elderly, and those suffering from other underlying disease. Antimicrobial therapy is usually not indicated in uncomplicated cases. Antimicrobial therapy is believed to prolong the carrier state. Antidiarrheal agents are also restricted in cases of salmonellosis, as these agents may encourage adherence and further invasion. In cases of dehydration, fluid replacement therapy may be indicated.

Dissemination may occasionally occur; in such cases, antimicrobial therapy is required. The antimicrobials of choice include chloramphenicol, ampicillin, and trimethoprim-sulfamethoxazole. Nevertheless, susceptibility testing must be performed.

Typhoid Fever and Other Enteric Fevers. The clinical features of enteric fevers include the following:

■ Prolonged fever

■ Bacteremia

■ Involvement of the reticuloendothelial system, particularly the liver, spleen, intestines, and mesentery

■ Dissemination to multiple organs

Enteric fever caused by *Salmonella typhi* has been known as typhoid fever, a febrile disease that results from the ingestion of food contaminated with the organisms originating from infected individuals or carriers. *Salmonella typhi* does not have a known animal reservoir; therefore, humans are the only source of infection. Other enteric fevers include paratyphoid fevers, which may be due to *S. paratyphi* serovar A, another strict human pathogen, *S. paratyphi* serovar B and *S. paratyphi* serovar C. Other serovars that have been implicated in cases of septicemia in humans are those subspecies of *S. choleraesuis*. The clinical manifestations of paratyphoid fevers are similar to those of typhoid fever but are less severe, and the fatality rate is lower. Therefore, the clinical features of typhoid fever are discussed here in greater detail.

Typhoid fever occurs more often in tropical and subtropical countries, where foreign travelers easily acquire the infection. Improper disposal of sewage, poor sanitation, and lack of a modern water system have caused outbreaks of typhoid fever when the organisms reach a water source. This is uncommon in the United States and other developed countries, where water is purified and treated and handling of wastes is greatly improved. Pasteurization of milk has also diminished the incidence of waterborne typhoid in industrialized countries. Carriers, particularly food handlers, are important sources of infection anywhere in the world. Direct transmission through fomites is also possible. Laboratory workers in the microbiology laboratory have contacted typhoid fever while working with the organisms. Typhoid fever develops approximately 9 to 14 days following ingestion of the organisms. The onset of symptoms depends on the number of organisms ingested; the larger the inoculum, the shorter the incubation period. Characteristically, during the first week of the disease, the patient develops fever, accompanied by malaise, anorexia, lethargy, myalgia, and a continuous dull frontal headache.

When the organisms are ingested, they seem to be resistant to gastric acids and, on reaching the proximal end of the small intestine, subsequently invade and penetrate the intestinal mucosa. At this time, the patient experiences constipation rather than diarrhea. The organisms gain entrance into the lymphatic system and are sustained in the mesenteric lymph nodes. They eventually reach the blood stream and are further spread to the liver, spleen, and bone marrow, where they are immediately engulfed by mononuclear phagocytes. The organ-

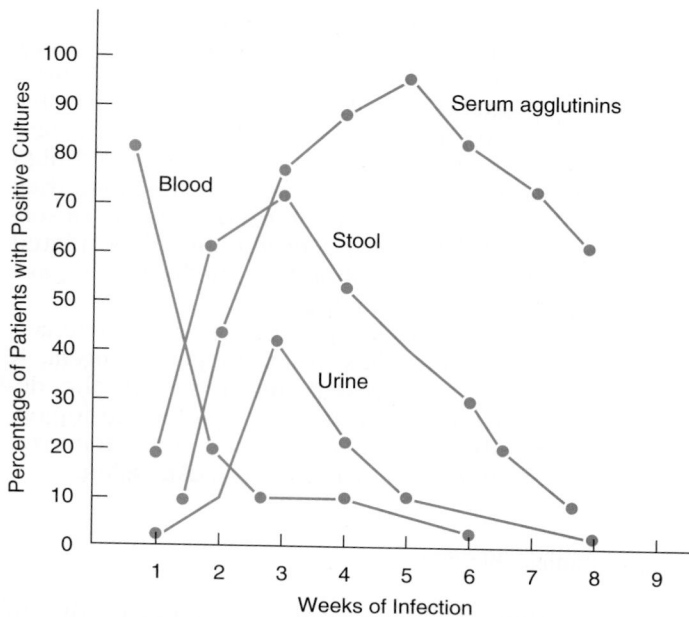

Figure 16-8
Culture and serologic diagnosis of typhoid fever.
(Modified and redrawn from Konemann E, Allen S, et al:
Color Atlas in Diagnostic Microbiology, 3rd ed.
Philadelphia: JB Lippincott, 1988.)

isms multiply intracellularly; later, they are released into the blood stream for the second time. The febrile episode becomes more evident during this release of the organisms into the circulatory system. At this time, the organisms may easily be isolated from the blood. Figure 16–8 shows the course of typhoid fever.

During the second and third weeks of the disease, the patient experiences sustained fever with the prolonged bacteremia. The organisms invade the gallbladder and Peyer's patches of the bowel. They also reach the intestinal tract via the biliary tract. "Rose spots" (blanching, rose-colored papules around the periumbilical region) appear during the second week of fever.

Involvement of biliary system sites initiates gastrointestinal symptoms as the organisms reinfect the intestinal tract. The organism now exists in large numbers and may be isolated from the stool. The gallbladder becomes the foci of long-term carriage of the organism, occasionally reseeding the intestinal tract and shedding the organisms in the feces. Necrosis in the gallbladder leading to necrotizing cholecystitis, and necrosis of the Peyer's patches leading to hemorrhage and perforation of the bowel, may occur as serious complications. Pneumonia and thrombophlebitis are other complications that occur in typhoid fever, as well as meningitis, osteomyelitis, endocarditis, and abscesses.

***Salmonella* Bacteremia.** *Salmonella* bacteremia, with and without extraintestinal foci of infection caused by nontyphoidal *Salmonella,* is characterized primarily by prolonged fever and intermittent bacteremia. The most commonly associated strains of *Salmonella* are *S. typhimurium, S. paratyphi A* and *B,* and *S. choleraesuis.*

Salmonella infection has been observed among two different groups of the population: (1) young children, who experience fever and gastroenteritis with brief episodes of bacteremia; and (2) adults, who experience transient bactermia during episodes of gastroenteritis or develop symptoms of septicemia without gastroenteritis. The latter manifestations were observed among patients who had underlying illnesses, such as malignancies and liver disease. There is a risk of metastatic complications that could be more severe than the bacteremia itself, even in individuals who do not have underlying diseases. Cases of septic arthritis may also occur in patients who had asymptomatic salmonellosis.

Carrier State. Individuals who recover from the infection may harbor the organisms in the gallbladder, which becomes the site of chronic carriage. Such individuals excrete the organisms in their feces either continuously or intermittently; nevertheless, they become an important source of infection for susceptible persons. The carrier state may be terminated by antimicrobial therapy if gallbladder infection is not evident. Otherwise, cholecystectomy has been the only solution to the chronic state of enteric carriers.

Shigella

The genus *Shigella* is very closely related to the genus *Escherichia* and belongs to the tribe Escherichieae. *Shigella* species, however, are not members of the normal gastrointestinal flora, and all *Shigella* species can cause bacillary dysentery. The genus *Shigella* is named after the Japanese microbiologist Kiyoshi Shiga, who first isolated the organism in 1896. The organism, descriptively name *Shigella dysenteriae,* caused the enteric disease bacillary dysentery. Dysentery was characterized by the presence of blood, mucus, and pus in the stool. The disease occurred in an epidemic dimension.

The genus consists of four species that are biochemically similar. *Shigella* species are also divided into four major O antigen groups and must be identified by serologic grouping. The four species and their respective serologic groups are as follows:

■ *S. dysenteriae* (group A)

■ *S. flexneri* (group B)

■ *S. boydii* (group C)

■ *S. sonnei* (group D)

There are several serotypes within each species with the exception of *S. sonnei,* which has only one serotype. *Shigella sonnei* is the most common isolate in the United States.

Characteristics of *Shigella* species are the following:

■ They are nonmotile

■ Except for certain types of *S. flexneri,* they do not produce gas from glucose

■ They do not hydrolyze urea

■ When cultured on triple sugar iron agar, they do not produce hydrogen sulfide

■ They do not decarboxylate lysine

Unlike *Escherichia, Shigella* do not utilize acetate or mucate as a source of carbon. Table 16–9 shows the biochemical characteristics of *Shigella* species. *Shigella sonnei* is unique in its ability to decarboxylate ornithine and slowly ferments lactose, forming pink colonies on MacConkey agar after 48 hours of incubation. *Shigella sonnei* also yields a positive reaction in ONPG. Figure 16–9 illustrates the growth of *S. sonnei* on MacConkey agar after 24 and 48 hours of incubation. On dif-

ferential and selective media used primarily to isolate intestinal pathogens, shigellae generally appear as clear, non–lactose-fermenting colonies.

Shigellae are fragile organisms. They are susceptible to the various effects of physical and chemical agents, such as disinfectants and high concentrations of acids and bile. Because they are susceptible to the acid pH of stool, feces suspected of containing *Shigella* should be plated immediately onto laboratory media to increase recovery of the organism.

All *Shigella* species possess O antigens, and certain strains may possess K antigens. *Shigella* K antigens, when present, interfere with the detection of the O antigen during serologic grouping. The K antigen is heat-labile and may be removed by boiling the organism in a cell suspension.

Clinical Infections

Although all *Shigella* species can cause dysentery, species vary in epidemiology, mortality rate, and severeness of disease produced. In the United States, *S. sonnei* is the predominant isolate, followed by *S. flexneri.*

In the United States and other industrialized countries, shigellosis is probably underreported, as most patients are not hospitalized and usually recover from the infection without culture to recover the etiologic agent. *Shigella sonnei* infection is usually a short, self-limiting disease characterized by fever and watery diarrhea.

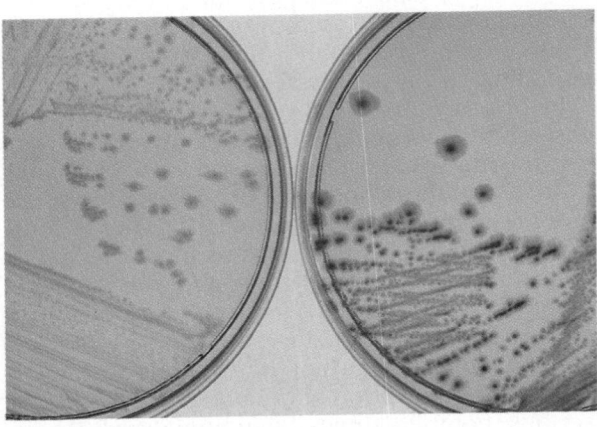

■■■■ F i g u r e   1 6 – 9

Left, Lactose-negative appearance of *S. sonnei* growing on MacConkey agar at 18 to 24 hours of incubation. *Right,* Lactose-positive appearance of *S. sonnei* growing on MacConkey agar after 48 hours of incubation.

T A B L E   1 6 – 9

BIOCHEMICAL AND SEROLOGICAL DIFFERENTIATION OF *SHIGELLA* SPECIES

Test	Reaction of[a]:			
	S. dysenteriae	*S. flexneri*	*S. boydii*	*S. sonnei*
Mannitol fermentation	–	+	+	+
ONPG	V	–	V	+
Ornithine decarboxylase	–	–	–	+
Serogroup	A	B	C	D

[a] Symbols: –, ≤ 9% of strains positive; V, 10 to 89% of strains positive; +, ≥ 90% of strains positive.

From Balows A, Hausler WJ, et al (eds): Manual of Clinical Microbiology, 4th ed. Washington, DC: American Society for Microbiology, 1985, p 273.

The demographics of *S. flexneri* infection have changed during recent years, from the disease's affecting mostly young children to its producing infections in young adults (approximately 25 years old). This observation was made simultaneously with the recognition of the "gay bowel syndrome" in homosexual men, in which *S. flexneri* has been the leading isolate. Conversely, in developing countries, *S. dysenteriae* type 1 and *S. boydii* are the most common isolates. *Shigella dysenteriae* type 1 remains the most virulent species, with significant morbidity and high mortality. There have been reports of mortality rates of 5 to 10%, and perhaps even higher, due to *S. dysenteriae* type 1, particularly among undernourished children during epidemic outbreaks.

Humans are the only known reservoir of *Shigella*. Transmission may occur by direct person-to-person contact, and spread may take place via the fecal-oral route, with carriers as the source. Shigellae may also be transmitted by flies, fingers, and food or water contaminated by infected persons.

Personal hygiene plays a major role in the transmission of *Shigella*. Young children in day care centers, people living in crowded and less than adequate housing, and people who participate in anal-oral sex are most likely affected. Children younger than 10 years of age seem to be most affected; those 1 year of age and younger are the most susceptible.

The infection is highly communicable because of the low infective dose required to produce the disease. It has been reported that fewer than 200 bacilli are needed to initiate the disease in some healthy individuals.

Bacillary dysentery caused by *Shigella* is marked by penetration of intestinal epithelial cells following attachment of the organisms to mucosal surfaces. Local inflammation, shedding of the intestinal lining, and formation of ulcers follow the epithelial penetration. The clinical manifestations of shigellosis vary from asymptomatic to severe forms of the disease. The initial symptoms, marked by high fever, chills, abdominal cramps, and pain accompanied by tenesmus, appear approximately 24 to 48 hours after ingestion of the organisms. The organisms, which originally multiplied in the small intestine, move toward the colon, where they may be isolated 1 to 3 days after the infection develops. Bloody stools containing mucus and numerous leukocytes follow the watery diarrhea, as the organisms invade the colonic tissues and cause an inflammatory reaction.

In dysentery caused by *S. dysenteriae* type 1, patients experience more severe symptoms. Bloody diarrhea that progresses to dysentery may appear within a few hours to a few days. Patients suffer from extremely painful bowel movements, which contain predominantly mucus and blood. In young children, abdominal pain is quite intense, and rectal prolapse may result from excessive straining.

Although the effects of *Shigella* toxin have been implicated as the mechanism responsible for the signs of the disease, the connection between the toxin hypothesis and the symptoms remains unclear. However, it has been reported that the detectable toxin levels produced by *S. dysenteriae* type 1 are higher than those produced by other *Shigella* species.

Severe cases of shigellosis may become life-threatening as extraintestinal complications develop. One of the most serious complications is ileus, an obstruction of the intestines, with marked abdominal dilatation, possibly leading to toxic megacolon. While *Shigella* organisms infrequently penetrate the intestinal mucosa and disseminate to other body sites, in 1985 Struelens and colleagues reported that as many as 4% of severely ill hospitalized patients in Bangladesh suffered from bacteremia caused by *dysenteriae* type 1. *Shigella flexneri* bacteremia and bacteremia due to other enteric organisms occur, presumably predisposed by ulcers initiated by the shigellae.

Other complications include seizures, which may occur during any *Shigella* strain infection, and HUS, a complication exclusively associated with *S. dysenteriae* type 1 shigellosis.

Yersinia

The genus *Yersinia* currently consists of 11 named species. *Yersinia* is a relatively new genus added to Enterobacteriaceae. The species *Y. pestis* and *Y. pseudotuberculosis* were previously classified in the genus *Pasteurella. Yersinia pestis* is the causative agent of plague, a life-threatening disease of rodents transmitted to humans by fleas. *Yersinia pseudotuberculosis* and *Y. enterocolitica* have caused sporadic cases of mesenteric lymphadenitis in humans, especially in children, and generalized septicemic infections in immunocompromised hosts. *Yersinia enterocolitica* produces an infection that mimicks appendicitis. It has also been found to be the cause of diarrhea in a number of community outbreaks. The other members of the genus *Yersinia* are found in water, soil, and lower animals; occasionally isolates have been found in wounds and the urine of humans. Evidence that other species, in addition to *Y. enterocolitica,* have caused intestinal disease has not been found. *Yersinia ruckeri* is well documented as the causative agent of red mouth disease in fish.

Yersinia pestis

The causative agent of the ancient disease plague still exists in areas where reservoir hosts are found. Plague, caused by *Y. pestis,* is a disease primarily of rodents. It is transmitted to humans by bites of fleas, its most common and effective vector. In humans, plague can occur in two forms: the bubonic, or glandular, from and the pneumonic form. The bubonic form usually results from the bite of an infected insect vector. Characteristic symptoms appear 2 to 5 days after infection. The symptoms include high fever with painful regional lymph nodes as buboes begin to appear.

Pneumonic plague occurs secondary to the bubonic plague when organisms proliferate in the blood stream and respiratory tract. Subsequent epidemic outbreaks may arise from the respiratory transmission of the organisms. The fatality rate in pneumonic plague is high if patients remain untreated.

Yersinia pestis is a gram-negative, short, plump rod. When stained with methylene blue or Wayson stain, it shows intense staining at each end of the bacillus, referred to as "bipolar staining," which gives it a "safety-pin" appearance.

Yersinia pestis may be isolated on routine culture medium. Although it grows at 37°C, it has a preferential growth temperature of 25 to 30°C. A *Yersinia pestis*–specific DNA probe for plague surveillance has been studied. If this DNA probe is proved successful, it may be applicable for laboratory diagnostic testing.

Yersinia enterocolitica

Human infections due to *Y. enterocolitica* have occurred worldwide, predominantly in Europe, although cases in the northeastern United States and Canada have been reported. It is the most commonly isolated species of *Yersinia.* The organisms have been found in a wide variety of animals, including domestic swine, cats, and dogs. The infection may therefore be acquired from contact with household pets. The role of the pig as a natural reservoir has been greatly emphasized in Europe. Other animal reservoirs however, have also been identified, and cultures from environmental reservoirs, such as water from streams, have yielded the organism. Human infections have also been reported following the ingestion of contaminated food and possibly water. Other sources of infection include contaminated food, such as market meat and vacuum-packed beef. A major concern regarding the potential risk of transmitting infection with this organism is its ability to survive in cold temperatures; food refrigeration becomes an ineffective preventive measure. In addition, *Y. enterocolitica* sepsis associated with the transfusion of contaminated packed red blood cells has been reported. *Yersinia enterocolitica* infections manifest in several forms: an acute enteritis, an appendicitis-like syndrome, arthritis, and erythema nodosum. Acute enteritis, the most common form of the infection, is characterized by acute gastroenteritis with fever accompanied by headaches, abdominal pain, nausea, and diarrhea. Stools may contain blood. This form of infection, which often afflicts infants and young children between the ages of 1 and 5 years, is usually mild and self-limiting.

The clinical form that mimics acute appendicitis occurs primarily in older children and adults. It presents with severe abdominal pain and fever; the abdominal pain is concentrated in the right lower quadrant. Enlarged mesenteric lymph nodes and inflamed ileum and appendix are common findings in cases of *Y. enterocolitica* infections.

Arthritis is a common extraintestinal form of *Y. enterocolitica* infection, usually following a gastrointestinal episode or an appendicitis-like syndrome. This form of yersiniosis has been reported more often in adults than in children. The arthritis form exhibits characteristics similar to those of the arthritis seen in other bacterial infections, that is, in infections with *Shigella, Salmonella,* and *N. gonorrhoeae* and in acute rheumatic fever.

Erythema nodosum is an inflammatory reaction characterized by tender, red nodules that may be accompanied by itching and burning. The areas involved include the anterior portion of the legs, while some patients have reported nodules on their arms. The reported cases have shown the syndrome to be more common in female patients than in male ones.

The incidence of generalized infection among adults involved those with underlying diseases, such as liver cirrhosis, diabetes, acquired immunodeficiency syndrome (AIDS), leukemia, aplastic anemia, and other hematologic conditions. Cases of liver abscess and acute infective endocarditis caused by *Y. enterocolitica* have also been reported. *Yersinia enterocolitica* morphologically resembles other *Yersinia* species, which appear as gram-negative coccobacilli with bipolar staining. The organism also grows on routine isolation media, such as blood agar and MacConkey agar. It has an optimal growth temperature of 25 to 30°C. Cold enrichment has provided better recovery of *Y. enterocolitica*. Appropriate cultures on a specific *Yersinia* media at 25°C should be performed in diarrheal outbreaks of unknown etiology. These organism grow better with cold enrichment, and motility is clearly noted at 25°C but not at 35°C. Fecal samples suspected of containing this organism are inoculated on isotonic saline and kept at 4°C for 1 to 3 weeks.

During recent years, a selective medium to detect the presence of *Y. enterocolitica* was introduced. CIN agar employs cefsulodin, irgasan, novobiocin, bile salts, and crystal violet as the inhibitory agents. This medium, which inhibits normal colon organisms better than MacConkey agar, provides more opportunities to recover *Y. enterocolitica* from feces. This selective medium has since been modified, and media manufacturers such as Difco Bacto (Detroit, MI) Yersinia selective agar base have added a differential property to the medium by adding mannitol. Fermentation of mannitol results in a localized drop in pH around the colony. The drop in pH causes the pH indicator neutral red to turn red at the center of the colony, and the bile starts to precipitate. Nonfermentation of mannitol produces a colorless, translucent colony.

Yersinia pseudotuberculosis

Yersinia pseudotuberculosis, like *Y. pestis,* is a pathogen primarily of rodents, particularly of guinea pigs. In addition to farm and domestic animals, birds are also natural reservoirs; turkey, geese, pigeons, doves, and canaries have yielded positive cultures for this organism.

Yersinia pseudotuberculosis causes a disease characterized by caseous swellings called pseudotubercles. The disease is often fatal in animals. Human infections, which are rare, are associated with close contact with infected animals or their fecal material or ingestion of contaminated drink and foodstuff. When the organisms are ingested, they spread to the mesenteric lymph nodes, producing a generalized infection. The clinical manifestations include septicemia accompanied by mesenteric lymphadenitis, a presentation similar to appendicitis.

Yersinia pseudotuberculosis appears as a typical-looking plague bacillus. It may be differentiated from *Y. pestis* by its motility at 18 to 22°C, production of urease, and ability to ferment rhamnose. Table 16–10 shows differentiating characteristics between *Yersinia* species.

New Genera and Biotypes

Budivicia

Budivicia aquatica is a group of organisms found by DNA hybridization to be closely related. They are not as closely related to the other members of Enterobacteriaceae but do qualify to belong to the family. These organisms are usually found in water; however, they occasionally occur in clinical specimens.

Buttiauxella

Buttiauxella agrestis, the only species of *Buttiauxella,* has been isolated from water but not from human specimens. Biochemically, these organisms are similar to both *Citrobacter* species

■■■■ T A B L E 1 6 – 1 0
DIFFERENTIATION WITHIN THE GENUS *YERSINIA*

Test	Y. pestis	Y. enterocolitica	Y. pseudo-tuberculosis
Indole	–	d	–
Methyl red	+	+	+
Voges-Proskauer			
25°C	–	d	–
37°C	–	–	–
Motility			
25°C	–	+	+
37°C	–	–	–
β-Galactosidase	+	+	+
Christensen urea	–	+	+
Phenylalanine deaminase	–	–	–
Ornithine decarboxylase	–	+	–
Acid produced from			
Sucrose	–	+	–
Lactose	–	–	–
Rhamnose	–	– or +[a]	+
Melibiose	–	– or +[a]	+
Trehalose	–	+ or –	+
Cellobiose	–	+	–

[b]Test results at 25°C.
Modified from Washington J: Laboratory Procedures in Clinical Microbiology, 2nd ed. New York: Springer-Verlag, 1985, p 190.

and *Kluyvera* species, but DNA hybridization distinctly differentiates *Buttiauxella* from both genera.

Cedecea

The genus *Cedecea* is composed of five species: *C. davisae, C. lapagei, C. neteri,* and *Cedecea* species types 3 and 5. Most have been recovered from sputum, blood, and wounds. Of the five, *C. davisae* is the most commonly isolated species.

Ewingella

Ewingella americana is the only species of this genus. *Ewingella* was formerly called enteric group 40. Most isolates have come from human blood cultures or respiratory specimens. *Ewingella* was first thought to be related to *Cedecea* species; however, DNA hybridization confirmed the finding of a new genus.

Kluyvera

The genus *Kluyvera* is made up of two closely related species, *K. ascorbata* and *K. cryocrescens.*

Both have been found in respiratory, urine, and blood cultures and may produce a blue-violet pigment usually but not exclusively on non–blood-containing media. Both species resemble *E. coli* colonies growing on MacConkey agar. Figure 16–10*A* to *C* illustrates the colony characteristics of *Kluyvera* spp. Cephalothin and carbenicillin disk susceptibility tests separate the two species: *K. cryocrescens* shows large zones of inhibition; *K. ascorbata* has small zones. In addition, *K. ascorbata* does not ferment D-glucose at 5°C, whereas *K. cryocrescens* ferments D-glucose at this temperature.

Koserella

The name *Koserella trabulsii* was proposed in 1985 to replace the designation enteric group 45. These organisms were first thought to be another species of *Hafnia,* but DNA hybridization showed a 15% relatedness, which was not sufficient to include these organisms in that genus. They are biochemically similar to *Hafnia* but differ primarily by yielding negative Voges-Proskauer test results. Koserellae have been isolated from human specimens, but further study will be required to determine their significance in human disease.

Leminorella

Leminorella is proposed as a genus for the enteric group 57, with two species, *L. grimontii* and *L. richardii.* These organisms produce hydrogen sulfide and have shown weak reactions with *Salmonella* antisera. However, complete biochemical tests will differentiate *Leminorella* species from *Salmonella; Leminorella* species are relatively inactive. The clinical significance of these organisms is unknown; however, they have been isolated from urine, feces, and water.

Moellerella

The new genus *Moellerella,* with one species, *M. wisconsensis,* was formerly called enteric group 46. *Moellerella* is positive for citrate, methyl red, lactose, and sucrose; is negative for lysine, ornithine, arginine decarboxylase, and indole; and resembles *E. coli* growing on enteric media. The clinical significance of this organism has not been established, although it has been isolated from feces in two cases of diarrhea.

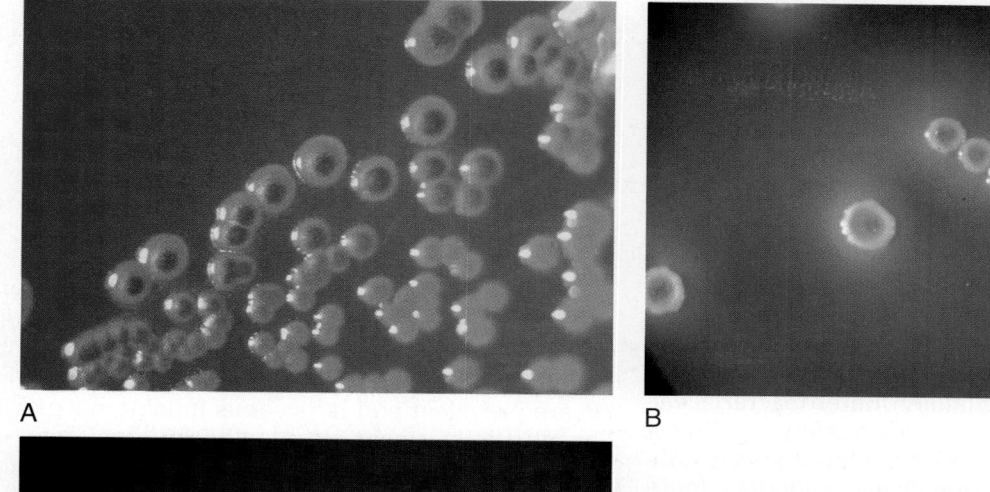

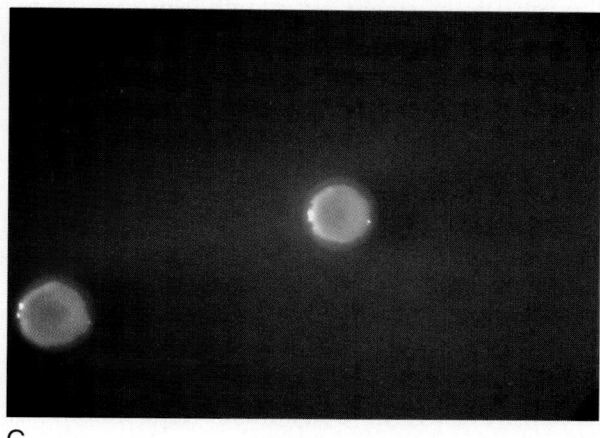

■■■■■■■ F i g u r e   1 6 – 1 0

A, Blue-violet pigment of *Kluyvera* growing on blood agar plate. The species of this genus resemble the colony morphology of *E. coli* growing on MacConkey agar. *B,* Appearance of *K. cryocrescens* growing on MacConkey agar. *C,* Appearance of *K. ascorbata* growing on MacConkey agar.

Obesumbacterium

The genus *Obesumbacterium proteus* biogroup 2 are more closely related to *Escherichia blattae* than to other members of Enterobacteriaceae. These are slow-growing organisms, fastidious at 36°C, and have not been found in human specimens.

Rahnella

Rahnella aquatilis is the name given to a group of water bacteria. These organisms have no single characteristic that distinguishes them from the other members of Enterobacteriaceae. They resemble *Enterobacter agglomerans;* however, they can be distinguished by a weak phenylalanine reaction; the fact that they are negative for KCN, gelatin, lysine, ornithine; motility; and their lack of yellow pigmentation.

Tatumella

Tatumella ptyseos is the only species of the genus *Tatumella,* the name given to group EF-9. This organism is unusual to Enterobacteriaceae in several ways: stock cultures may be kept frozen in sheep's blood or may be freeze-dried but will die in a few weeks on agar slants; show more biochemical reactions at 25°C than at 35°C; are motile at 25°C but not at 35°C; and demonstrate large 15- to 36-mm zones of inhibition around penicillin disks. In addition, *Tatumella* organisms are slow-growing, produce tiny colonies, and are relatively nonreactive in laboratory media. These organisms have been isolated from human sources, especially sputum, and may be a rare cause of infection.

Xenorhabdus

The genus *Xenorhabdus* is composed of *X. nematophilus* and *X. luminescens,* organisms that grow best at 25°C. *Xenorhabdus luminescens* DNA group 5 contains all the human clinical isolates. Colonies are yellow-pigmented, and the organism has been isolated from wounds and blood. *Xenorhabdus nematophilus,* which has been isolated from nematodes, has not been found in human specimens.

LABORATORY DIAGNOSIS

Specimen Collection and Transport

Members of the family Enterobacteriaceae may be isolated from a wide variety of clinical samples. Most often these bacterial species are isolated with other organisms, including more fastidious pathogens. Therefore, to ensure isolation of both opportunistic and fastidious pathogens, laboratories must provide appropriate collection and transport media, such as Cary-Blair, Amies, or Stuart media. Microbiology personnel must encourage immediate transport of clinical samples to the laboratory for processing, regardless of the source of the clinical sample.

Isolation and Identification

To determine the clinical significance of the isolate, the microbiologist must consider the site or origin. Generally, enteric opportunistic organisms isolated from sites that are normally sterile are highly significant. However, one should carefully examine organisms recovered from, for example, the respiratory tract, urogenital tract, stool, and wounds in open sites that are inhabited by other indigenous microflora.

Members of the family Enterobacteriaceae are routinely isolated from stool cultures; therefore, complete identification should be directed only toward true intestinal pathogens. On the other hand, sputum cultures from hospitalized patients may contain enteric organisms that may require complete identification.

Direct Microscopic Examination

Unlike the case with gram-positive bacteria, in which microscopic morphology may essentially provide a presumptive identification, the microscopic characteristics of enterics are indistinguishable from other gram-negative bacteria.

However, smears prepared directly from cerebrospinal fluid, blood, other body fluids, or exudates from an uncontaminated site may be examined microscopically for the presence of gram-negative bacteria. Although this examination is nonspecific for enteric organisms, this presumptive result may aid the clinician in the preliminary diagnosis of the infection, and appropriate therapy can be instituted immediately.

On the other hand, direct smears prepared from samples, such as sputum, that contain indigenous microbial flora do not provide valuable information, since their significance cannot be fully assessed unless the gram-negative bacteria are prevalent and indigenous inhabitants are absent. Direct smear examination of stool samples is not particularly helpful in identifying enteric pathogens but may reveal the inflammatory cells. This information is helpful in determining whether the gastrointestinal disease is toxin-mediated or an invasive process.

Culture

Media. Most laboratories utilize a wide variety of nonselective media such as blood agar plate and chocolate agar as well as selective media, such as MacConkey agar, to recover enteric organisms from wound, respiratory tract secretions, urine, sterile body fluids, and blood. On chocolate agar or on blood agar plate, enteric bacteria produce large, grayish, smooth colonies. On blood agar plate colonies may be hemolytic or nonhemolytic; if hemolytic, β hemolysis is usually produced. On a selective medium, such as MacConkey agar, lactose-fermenting species produce pink to red colonies as acid is produced from lactose fermentation; crystal violet is precipitated, and neutral red turns red in an acid pH. Non–lactose-fermenting (NLF) species produce clear, colorless colonies on MacConkey agar. Bile and crystal violet are added to the medium to inhibit the growth of gram-positive bacteria.

Stool specimens contain enteric organisms as normal colon flora; therefore, in processing stool samples, laboratories may develop their own protocol for the maximum recovery of enteric pathogens. Most microbiologists inoculate stool samples on highly selective media, such as Hektoen enteric (HE) or XLD agar, in addition to MacConkey agar. An enrichment broth has been traditionally inoculated to enhance recovery; this practive is slowly being phased out of the protocol.

On HE agar, lactose-fermenting species produce yellow colonies, while NLF species produce green colonies. Certain species of *Proteus* that are NLF and produce hydrogen sulfide appear green with black centers of HE agar. *Citrobacter freundii* usually produces yellow colonies with black centers. On XLD agar, NLF species like *Salmonella* (which produces lysine decarboxylase) produce red colonies with black centers (Fig. 16–11).

Environmental Requirements. Members of the family Enterobacteriaceae are facultatively anaerobic and grow at an optimal temperature of 35 to 37°C, preferably without carbon dioxide. Certain species may grow at low temperatures (1–5°C for *Serratia* and *Yersinia*) or tolerate high temperatures (45–50°C for *E. coli*). Colonies become visible on nonselective and differential media after 18 to 24 hours of incubation.

Identification

Currently, there are several ways of identifying members of the family Enterobacteriaceae. Certain laboratories may still prefer to use conventional biochemical tests in tubes, while other may prefer miniaturized or automated commercial identification systems. Clinical laboratories that utilize conventional biochemical tests in tubes may find it cumbersome to test an isolate with all of the biochemical tests available. Therefore, most of these laboratories develop identification tables and protocol that suit their needs and capabilities. These tables are based on the key features necessary to identify each particular genus and certain species. Figure 16–12 shows an example of a schematic diagram for the identification of commonly isolated enterics using conventional biochemical tests.

Table 16–11 shows the differentiating characteristics of the species, biogroups, and enteric groups of the Enterobacteriaceae.

To identify an isolate, the microbiologist must first determine whether the isolate belongs to the family Enterobacteriaceae. All members of the family (1) are oxidase-negative, (2) utilize glucose fermentatively, and (3) reduce nitrates to nitrites.

Gram-negative isolates, especially non–lactose-fermenters, should be tested for cytochrome oxidase production. This may be accomplished by wetting a tweezer-held disk with oxidase reagent and touching the disk to the colony. A black color change on the disk indicates a positive oxidase test (*non*-Enterobacteriaceae). The oxidase test is best performed off of blood agar plate. Touching an oxidase-impregnated disk to a colony on highly selective media such as CIN may give a false-negative reaction, while touching MacConkey agar (differential medium) may give the appearance of a false-positive reaction.

Regardless of the identification system utilized, the microbiologist may presumptively determine utilization of carbohydrates by observing the colonial morphology of the isolate on a differential or selective medium, such as MacConkey, HE, or XLD.

The traditional biochemical tests to perform include the following:

■ TSI or Kligler iron agar (KIA) to determine glucose and lactose, or sucrose, utilization (sucrose in TSI only) and hydrogen sulfide production

■ Lysine-iron agar (LIA) to determine lysine decarboxylase activity

■ Urea test to determine hydrolysis of urea

■ Simmons citrate as the sole source of carbon

■ SIM (sulfide, indole, motility) or MIO (motility, indole, and ornithine) test

■ Carbohydrate fermentation

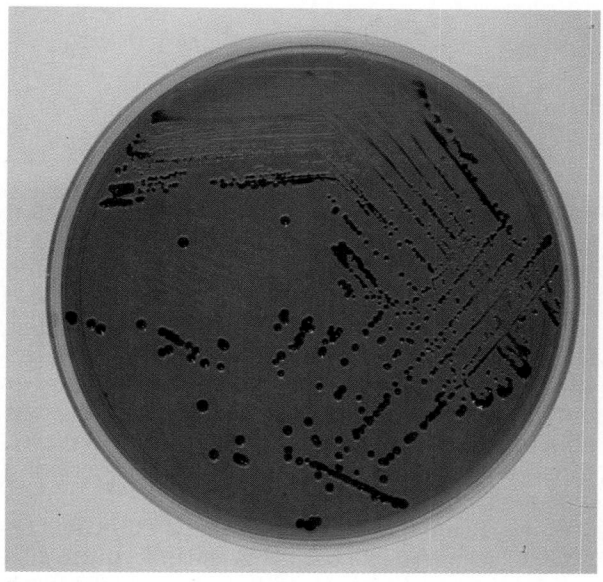

■ F i g u r e   1 6 – 1 1
Hydrogen sulfide–producing colonies of *Salmonella* growing on XLD. (Courtesy of American Society for Clinical Laboratory Science, Education and Research Fund, Inc. © 1982.)

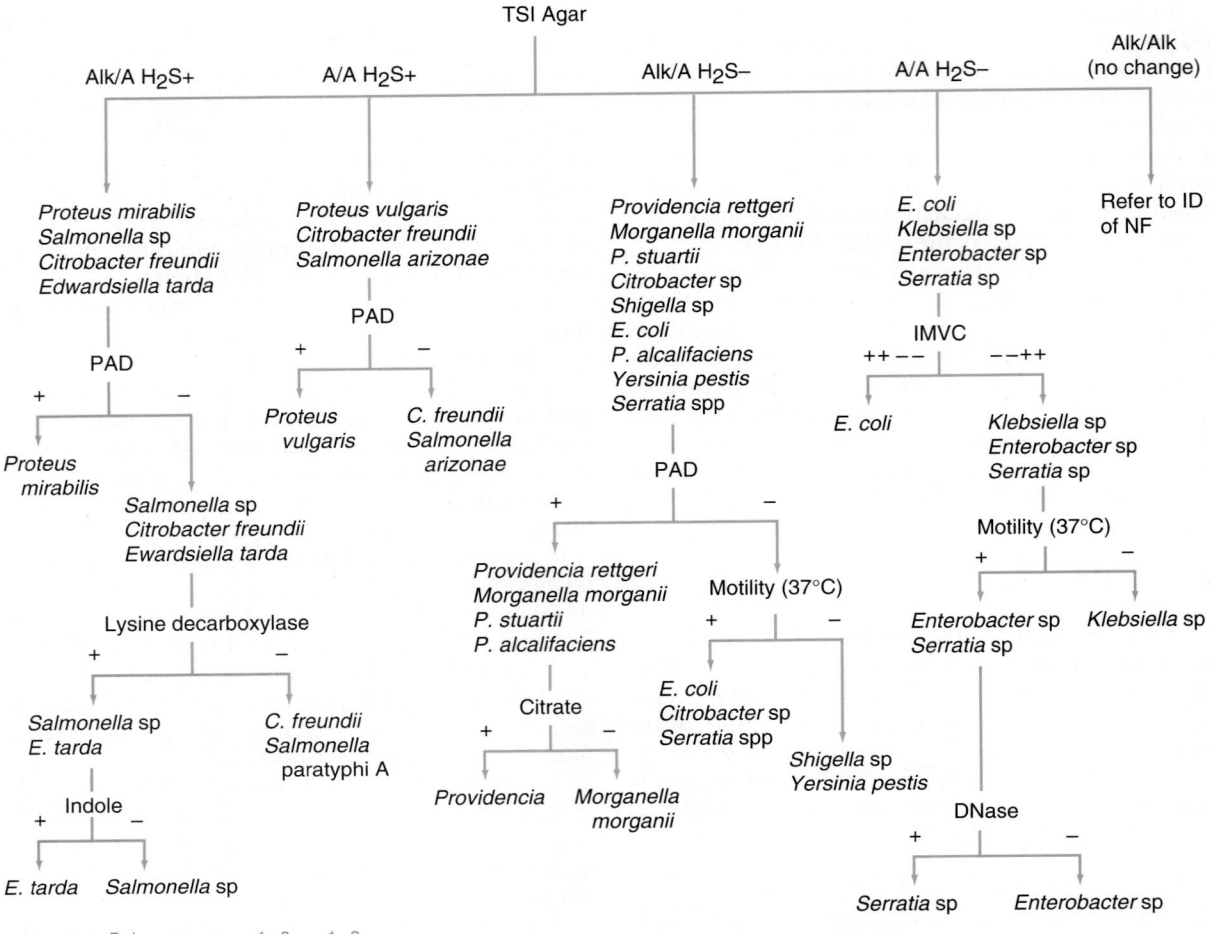

Figure 16–12

Flow chart for the presumptive identification of commonly encountered Enterobacteriaceae on TSI agar. (Data from Koneman E, Allen S, et al: Color Atlas in Diagnostic Microbiology, 3rd ed. Philadelphia: JB Lippincott, 1988.)

BIOCHEMICAL PRINCIPLES AND REACTIONS ON CONVENTIONAL MEDIA

Lactose Fermentation and Utilization of Carbohydrates

Lactose is a disaccharide carbohydrate that consists of glucose and galactose connected by a galactoside bond. Glucose and galactose are released when this bond is cleaved by the enzyme β-galactosidase. Lactose degradation has been used to initially differentiate those bacterial species that are capable of fermenting lactose (LF) and those that are NLF species.

Two enzymes are necessary for a bacterium to take up lactose. These are β-galactoside per-

mease, which serves as a transport enzyme that facilitates entry of the lactose molecule through the bacterial cell wall, and β-galactosidase, the enzyme that hydrolyzes lactose into glucose and galactose. Glucose becomes available for bacterial metabolism. Needless to say, certain bacterial species may be able to utilize carbohydrates only in their simplest form, glucose, and are not able to attack the disaccharide lactose. Similarly, those bacterial species incapable of fermenting glucose cannot utilize lactose.

Since all members of the family Enterobacteriaceae ferment glucose, this carbohydrate has not been the choice for differential media such as MacConkey or eosin–methylene blue agar. Non–lactose-fermenting species (typical of true enteric pathogens such as *Salmonella* and *Shigel-*

la) are not easily distinguished from lactose-fermenting enterics (common enteric flora). This is the reason that lactose—and not glucose—is incorporated into most gram-negative differential media and highly selective enteric media.

By definition, lactose-fermenters possess both β-galactoside permease and β-galactosidase, and non–lactose-fermenters do not possess β-galactosidase. Some bacterial species lack β-galactoside permease but possess the β-galactosidase. These bacterial species, termed late lactose-fermenters or slow lactose-fermenters, eventually cleave the lactose molecule but may take longer because of the deficiency in β-galactosidase.

Triple Sugar Iron (TSI) Agar

KIA or TSI agar is very useful in the presumptive identification of enterics, particularly in screening for intestinal pathogens. The formulas for KIA and TSI are identical except that TSI contains sucrose in addition to glucose and lactose. Lactose is present in a concentration 10 times that of glucose (10:1 ratio). In TSI, sucrose is present in concentration 10 times that of glucose. Ferrous sulfate and sodium thiosulfate are added to detect the production of hydrogen sulfide. Phenol red is used as the pH indicator, which is yellow below the pH of 6.8. Uninoculated medium is red, since the pH is buffered at 7.4. Both KIA and TSI are useful in detecting the ability of the microorganism to produce gas from the fermentation of sugars; to detect the fermentation of glucose and lactose in KIA and glucose, lactose, and/or sucrose in TSI; and to detect the production of hydrogen sulfide gas.

Both KIA and TSI agars are poured on a slant (see Fig. 16–13). This sets up reaction chambers within the tube; the slant portion is the aerobic chamber being exposed to oxygen; the butt, or deep portion, is the anaerobic chamber, being protected from the air. To inoculate TSI agar, pick a well-isolated colony with an inoculating needle and stab the butt (all the way to the bottom) at the base of the slant. Streak the slant by "fish tailing" (moving the needle back and forth) as the needle is being pulled out of the butt.

Based on the reactions shown on the TSI agar, the microbiologist determines whether the isolate ferments glucose only (typical of true enteric pathogens) or whether it ferments glucose, lactose, and/or sucrose (typical of most opportunistic enterics).

The reaction patterns are written with the slant results first, followed by the butt reaction, separated by a backslash (slant reaction/butt reaction), and it is important that the reactions be read within an 18- to 24-hour incubation period. The reaction patterns shown below are an integral part of the identification schema for the family Enterobacteriaceae.

Reactions on KIA or TSI Agar
A. No Fermentation
Alkaline slant/alkaline butt (ALK/ALK or K/K) or alkaline slant/no change (ALK/no change or K/NC).

These reactions are typical of organisms that are not members of the Enterobacteriaceae. Although nonenteric bacilli are unable to ferment either lactose or glucose, these organisms can degrade the peptones present in the medium, resulting in the production of alkaline byproducts and changing the indicator to a deep red color.

B. No Lactose (Sucrose) Fermentation: Glucose Only
Acid slant/acid butt (A/A) (8–12 hours).

KIA and TSI agar contain glucose in a 0.1% concentration. The acid produced from this concentration of glucose is enough to change in indicator to yellow throughout the medium in this time frame. Reading the results after less than 12 hours of incubation gives the false appearance of an organism capable of fermenting glucose and lactose (sucrose). For this reason, KIA or TSI agar must be incubated for 18–24 hours.

Alkaline slant/acid butt (Alk/acid or K/A) (18 to 24 hours).

This reaction is typical of an organism that ferments only glucose. After 18 to 24 hours the glucose concentration is depleted in the slant and the butt. The organism begins oxidative degradation of the peptones in the slant, resulting in alkaline byproducts that change the indicator to a deep red color. Fermentation (anaerobic) of glucose in the butt produces larger amounts of acid, overcoming the alkaline effects of peptone degradation; therefore the butt remains acid (yellow).

C. Lactose (Sucrose) Fermentation (18–24 Hours)
Acid/acid (A/A).

The lactose present in TSI and KIA and the sucrose present in TSI is 10 times the concentration of glucose. The Enterobacteriaceae attack the simple sugar (glucose) first, and then the lactose (sucrose). The acid production from the fermentation of the additional sugar(s) is sufficient to keep both the slant and the butt acid (yellow) when examined at the end of 18 to 24 hours.

Text continued on page 480

■■■■ TABLE 16-11

BIOCHEMICAL REACTIONS OF THE NAMED SPECIES, BIOGROUPS, AND ENTERIC GROUPS OF THE FAMILY ENTEROBACTERIACEAE*

SPECIES	Indole production	Methyl red	Voges-Proskauer	Citrate (Simmons)	Hydrogen sulfide (TSI)	Urea hydrolysis	Phenylalanine deaminase	Lysine decarboxylase	Arginine dihydrolase	Ornithine decarboxylase	Motility (36°C)	Gelatin hydrolysis (22°C)	Growth in KCN	Malonate utilization	D-Glucose, acid	D-Glucose, gas	Lactose fermentation	Sucrose fermentation	D-Mannitol fermentation	Dulcitol fermentation
Buttiauxella																				
B. agrestis	0	100	0	100	0	0	0	0	0	100	100	0	80	60	100	100	100	0	100	0
Cedecea																				
C. davisae†	0	100	50	95	0	0	0	0	50	95	95	0	86	91	100	70	19	100	100	0
C. lapagei†	0	40	80	99	0	0	0	0	80	0	80	0	100	99	100	100	60	0	100	0
C. neteri†	0	100	50	100	0	0	0	0	100	0	100	0	65	100	100	100	35	100	100	0
Cedecea sp. 3†	0	100	50	100	0	0	0	0	100	0	100	0	100	0	100	100	0	50	100	0
Cedecea sp. 5†	0	100	50	100	0	0	0	0	50	50	100	0	100	0	100	100	0	100	100	0
Citrobacter																				
C. freundii†	5	100	0	95	80	70	0	0	65	20	95	0	96	15	100	95	50	30	99	55
C. diversus†	99	100	0	99	0	75	0	0	65	99	95	0	0	90	100	98	35	45	100	50
C. amalonaticus†	100	100	0	85	0	80	0	0	85	95	98	0	95	0	100	97	50	15	100	0
C. amalonaticus biogroup 1†	100	100	0	1	0	45	0	0	85	100	99	0	96	0	100	93	19	100	100	4
Edwardsiella																				
E. tarda†	99	100	0	1	100	0	0	100	0	100	98	0	0	0	100	100	0	0	0	0
E. tarda biogroup 1†	100	100	0	0	0	0	0	100	0	100	100	0	0	0	100	50	0	100	100	0
E. hoshinae	13	100	0	0	0	0	0	100	0	95	100	0	0	100	100	35	0	100	100	0
E. ictaluri	0	0	0	0	0	0	0	100	0	65	0	0	0	0	100	50	0	0	0	0
Enterobacter																				
E. aerogenes†	0	5	98	95	0	2	0	98	0	98	97	0	98	95	100	100	95	100	100	5
E. cloacae†	0	5	100	100	0	65	0	0	97	96	95	0	98	75	100	100	93	97	100	15
E. agglomerans†	20	50	70	50	0	20	20	0	0	0	85	2	35	65	100	20	40	75	100	15
E. gergoviae†	0	5	100	99	0	93	0	90	0	100	90	0	0	96	100	98	55	98	99	0
E. sakazakii†	11	5	100	99	0	1	50	0	99	91	96	0	99	18	100	98	99	100	100	5
E. taylorae†	0	5	100	100	0	1	0	0	94	99	99	0	98	100	100	100	10	0	100	0
E. amnigenus biogroup 1†	0	7	100	70	0	0	0	0	9	55	92	0	100	91	100	100	70	100	100	0
E. amnigenus biogroup 2	0	65	100	100	0	0	0	0	35	100	100	0	100	100	100	100	35	0	100	0
E. intermedius	0	100	100	65	0	0	0	0	0	89	89	0	65	100	100	100	100	65	100	100
Escherichia-Shigella																				
E. coli†	98	99	0	1	1	1	0	90	17	65	95	0	3	0	100	95	95	50	98	60
E. coli, inactive†	80	95	0	1	1	1	0	40	3	20	5	0	1	0	100	5	25	15	93	40
Shigella, serogroups A, B, and C†	50	100	0	0	0	0	0	0	5	1	0	0	0	0	100	2	0	0	93	2
S. sonnei†	0	100	0	0	0	0	0	0	2	98	0	0	0	0	100	0	2	1	99	0
E. fergusonii†	98	100	0	17	0	0	0	95	5	100	93	0	0	35	100	95	0	0	98	60
E. hermannii†	99	100	0	1	0	0	0	6	0	100	99	0	94	0	100	97	45	45	100	19
E. vulneris†	0	100	0	0	0	0	0	85	30	0	100	0	15	85	100	97	15	8	100	0
E. blattae	0	100	0	50	0	0	0	100	0	100	0	0	0	100	100	100	0	0	0	0
Ewingella																				
E. americana†	0	84	95	95	0	0	0	0	0	0	60	0	5	0	100	0	70	0	100	0
Hafnia																				
H. alvei†	0	40	85	10	0	4	0	100	6	98	85	0	95	50	100	98	5	10	99	0
H. alvei biogroup 1	0	85	70	0	0	0	0	100	0	45	0	0	0	45	100	0	0	0	55	0
Klebsiella																				
K. pneumoniae†	0	10	98	98	0	95	0	98	0	0	0	0	98	93	100	97	98	99	99	30
K. oxytoca†	99	20	95	95	0	90	1	99	0	0	0	0	97	98	100	97	100	100	99	55
Klebsiella group 47 indole positive, ornithine positive†	100	96	70	100	0	100	0	100	0	100	0	0	100	100	100	100	100	100	100	10
K. planticola†	20	100	98	100	0	98	0	100	0	0	0	0	100	100	100	100	100	100	100	15
K. ozaenae†	0	98	0	30	0	10	0	40	6	3	0	0	88	3	100	50	30	20	100	2
K. rhinoscleromatis†	0	100	0	0	0	0	0	0	0	0	0	0	80	95	100	0	0	75	100	0
K. terrigena	0	60	100	40	0	0	0	100	0	20	0	0	100	100	100	80	100	100	100	20
Kluyvera																				
K. ascorbata†	92	100	0	96	0	0	0	97	0	100	98	0	92	96	100	93	98	98	100	25
K. cryocrescens†	90	100	0	80	0	0	0	23	0	100	90	0	86	86	100	95	95	81	95	0
Moellerella																				
M. wisconsensis	0	100	0	80	0	0	0	0	0	0	0	0	70	0	100	0	100	100	60	0
Morganella																				
M. morganii†	98	97	0	0	5	98	95	0	0	98	95	0	98	1	100	90	1	0	0	0
M. morganii biogroup 1†	100	95	0	0	41	100	100	100	0	95	0	0	91	5	100	91	0	0	0	0
Obesumbacterium																				
O. proteus biogroup 2	0	15	0	0	0	0	0	100	0	100	0	0	0	100	100	0	0	0	0	0
Proteus																				
P. mirabilis†	2	97	50	65	98	98	98	0	0	99	95	90	98	2	100	96	2	15	0	0
P. vulgaris†	98	95	0	15	95	95	99	0	0	0	95	91	99	0	100	85	2	97	0	0
P. penneri†	0	100	0	0	30	100	99	0	0	0	85	50	99	0	100	45	1	100	0	0
P. myxofaciens	0	100	100	50	0	100	100	0	0	0	100	100	100	0	100	100	0	100	0	0

Salicin fermentation	Adonitol fermentation	myo-Inositol fermentation	D-Sorbitol fermentation	L-Arabinose fermentation	Raffinose fermentation	L-Rhamnose fermentation	Maltose fermentation	D-Xylose fermentation	Trehalose fermentation	Cellobiose fermentation	α-Methyl-D-glucoside fermentation	Erythritol fermentation	Esculin hydrolysis	Melibiose fermentation	D-Arabitol fermentation	Glycerol fermentation	Mucate fermentation	Tartrate, Jordan	Acetate utilization	Lipase (corn oil)	DNase at 25°C	Nitrate → nitrite	Oxidase, Kovacs	ONPG†	Yellow pigment	D-Mannose fermentation
100	0	0	0	100	100	100	100	100	100	100	0	0	100	100	0	60	100	60	0	0	0	100	0	100	0	100
99	0	0	0	0	10	0	100	100	100	100	5	0	45	0	100	0	0	0	0	91	0	100	0	90	0	100
100	0	0	0	0	0	0	100	0	100	100	0	0	100	0	100	0	0	0	60	100	0	100	0	99	0	100
100	0	0	100	0	0	0	100	100	100	100	0	0	100	0	100	0	0	0	0	100	0	100	0	100	0	100
100	0	0	0	0	100	0	100	100	100	100	50	0	100	100	100	0	0	0	50	100	0	100	0	100	0	100
100	0	0	100	0	100	0	100	100	100	100	0	0	100	100	100	0	0	0	50	50	0	100	0	100	0	100
5	0	3	98	100	30	99	99	99	99	55	5	0	0	50	0	98	95	90	80	0	0	99	0	95	0	100
20	98	0	99	100	0	100	100	100	100	99	40	0	2	0	100	98	93	75	75	0	0	100	0	96	0	100
40	0	0	100	100	5	99	99	99	100	100	5	0	10	5	0	70	98	85	75	0	0	99	0	100	0	100
0	0	0	100	100	100	100	100	100	100	100	70	0	0	100	0	55	100	93	82	0	0	100	0	100	0	100
0	0	0	0	9	0	0	100	0	0	0	0	0	0	0	0	30	0	25	0	0	0	100	0	0	0	100
0	0	0	0	100	0	0	100	0	0	0	0	0	0	0	0	0	0	0	0	0	0	100	0	0	0	100
50	0	0	0	13	0	0	100	0	100	0	0	0	0	0	0	65	0	0	0	0	0	100	0	0	0	100
0	0	0	0	0	0	0	100	0	0	0	0	0	0	0	0	0	0	0	0	0	0	100	0	0	0	100
100	98	95	100	100	96	99	99	100	100	100	95	0	98	99	100	98	90	95	50	0	0	100	0	100	0	95
75	25	15	95	100	97	92	100	99	100	99	85	0	30	90	15	40	75	30	75	0	0	99	0	99	0	100
65	7	15	30	95	30	85	89	93	97	55	7	0	60	50	50	30	40	25	30	0	0	85	0	90	75	98
99	0	0	0	99	97	99	100	99	100	99	2	0	97	97	97	100	2	97	93	0	0	99	0	97	0	100
99	0	75	0	100	99	100	100	100	100	100	96	0	100	100	0	15	1	1	96	0	0	99	0	100	98	100
92	0	0	1	100	0	100	99	100	100	100	1	0	90	0	0	1	75	0	35	0	0	100	0	100	0	100
91	0	0	9	100	100	100	100	100	100	100	55	0	91	100	0	0	35	9	0	0	0	100	0	91	0	100
100	0	0	100	100	0	100	100	100	100	100	100	0	100	100	0	0	100	0	0	0	0	100	0	100	0	100
100	0	0	100	100	100	100	100	100	100	100	100	0	100	100	0	100	100	100	0	0	0	100	0	100	0	100
40	5	1	94	99	50	80	95	95	98	2	0	0	35	75	5	75	95	95	90	0	0	100	0	95	0	98
10	3	1	75	85	15	65	80	70	90	2	0	0	5	40	5	65	30	85	40	0	0	98	0	45	0	97
0	0	0	30	60	50	5	30	2	80	0	0	0	0	50	0	10	0	30	2	0	0	100	0	2	0	100
0	0	0	2	95	3	75	90	2	100	5	0	0	0	25	0	15	10	90	0	0	0	100	0	90	0	100
65	98	0	0	98	0	92	96	96	96	96	0	0	46	0	100	20	0	96	96	0	0	100	0	83	0	100
40	0	0	0	100	40	97	100	100	100	97	0	0	40	0	8	3	97	35	78	0	0	100	0	98	98	100
30	0	0	1	100	99	93	100	100	100	100	25	0	20	100	0	25	78	2	30	0	0	100	0	100	50	100
0	0	0	0	100	0	100	100	100	75	0	0	0	0	0	0	100	50	50	0	0	0	100	0	0	0	100
80	0	0	0	0	0	23	16	13	99	10	0	0	50	0	99	24	0	35	10	0	0	97	0	85	0	99
13	0	0	0	95	2	97	100	98	99	15	0	0	7	0	0	95	0	70	15	0	0	100	0	90	0	100
55	0	0	0	0	0	0	0	0	70	0	0	0	0	0	0	0	0	30	0	0	0	100	0	30	0	100
99	90	95	99	99	99	99	98	99	99	98	90	0	99	99	98	97	90	95	75	0	0	99	0	99	0	99
100	99	98	99	98	100	100	100	100	100	100	100	2	100	99	98	99	93	98	90	0	0	100	0	100	1	100
100	100	95	100	100	100	100	100	100	100	100	100	0	100	100	100	100	100	96	100	95	0	100	0	100	0	100
100	100	100	92	100	100	100	100	100	100	100	100	0	100	100	100	100	100	100	62	0	0	100	0	100	1	100
97	97	55	65	98	90	55	95	95	98	92	70	0	80	97	95	65	25	50	2	0	0	80	0	80	0	100
98	100	95	100	100	90	96	100	100	100	100	0	0	30	100	100	50	0	50	0	0	0	100	0	0	0	100
100	100	80	100	100	100	100	100	100	100	100	100	0	100	100	100	100	100	100	20	0	0	100	0	100	0	100
100	0	0	40	100	98	100	100	99	100	100	98	0	99	99	0	40	90	35	50	0	0	100	0	100	0	100
100	0	0	45	100	100	100	100	91	100	100	95	0	100	100	0	5	81	19	86	0	0	100	0	100	0	100
0	100	0	0	0	100	0	30	0	0	0	0	0	0	0	100	75	10	0	30	10	0	90	0	90	0	100
0	0	0	0	0	0	0	0	10	0	0	0	0	0	0	0	5	0	95	0	0	0	90	0	5	0	98
0	0	0	0	0	0	0	0	0	0	0	0	0	0	0	0	100	0	100	0	0	0	91	0	0	0	95
0	0	0	0	0	0	15	50	15	85	0	0	0	0	0	0	0	0	15	0	0	0	100	0	0	0	85
0	0	0	0	0	1	1	0	98	98	1	0	0	0	0	0	70	0	87	20	92	50	95	0	0	0	0
50	0	0	0	0	1	5	97	95	30	0	60	1	50	0	0	60	0	80	25	80	80	98	0	0	1	0
0	0	0	0	0	0	1	0	100	100	55	0	80	0	0	0	55	0	85	5	45	40	90	0	0	1	0
0	0	0	0	0	0	0	0	100	0	0	0	100	0	0	0	100	0	100	0	100	50	100	0	0	0	0

Table continued on following page

■■■■■■■■ T A B L E   1 6 – 1 1

BIOCHEMICAL REACTIONS OF THE NAMED SPECIES, BIOGROUPS, AND ENTERIC GROUPS OF THE FAMILY ENTEROBACTERIACEAE* Continued

SPECIES	Indole production	Methyl red	Voges-Proskauer	Citrate (Simmons)	Hydrogen sulfide (TSI)	Urea hydrolysis	Phenylalanine deaminase	Lysine decarboxylase	Arginine dihydrolase	Ornithine decarboxylase	Motility (36°C)	Gelatin hydrolysis (22°C)	Growth in KCN	Malonate utilization	D-Glucose, acid	D-Glucose, gas	Lactose fermentation	Sucrose fermentation	D-Mannitol fermentation	Dulcitol fermentation
Providencia																				
P. rettgeri†	99	93	0	95	0	98	98	0	0	0	94	0	97	0	100	10	5	15	100	0
P. stuartii†	98	100	0	93	0	30	95	0	0	0	85	0	100	0	100	0	2	50	10	0
P. alcalifaciens†	99	99	0	98	0	0	98	0	0	1	96	0	100	0	100	85	0	15	2	0
P. rustigianii†	98	65	0	15	0	0	100	0	0	0	30	0	100	0	100	35	0	35	0	0
Rahnella																				
R. aquatilis†	0	88	100	94	0	0	95	0	0	0	6	0	0	100	100	98	100	100	100	88
Salmonella																				
Subgroup 1 serotypes†—most	1	100	0	95	95	1	0	98	70	97	95	0	0	0	100	96	1	1	100	96
S. typhi†	0	100	0	0	97	0	0	98	3	0	97	0	0	0	100	0	1	0	100	0
S. choleraesuis†	0	100	0	25	50	0	0	95	55	100	95	0	0	0	100	95	0	0	98	5
S. paratyphi A†	0	100	0	0	10	0	0	0	15	95	95	0	0	0	100	99	0	0	100	90
S. gallinarum†	0	100	0	0	100	0	0	90	10	1	0	0	0	0	100	0	0	0	100	90
S. pullorum†	0	90	0	0	90	0	0	100	10	95	0	0	0	0	100	90	0	0	100	0
Subgroup 2 strains†	2	100	0	100	100	0	0	100	90	100	98	2	0	95	100	100	1	1	100	90
Subgroup 3a strains† (Arizona)	1	100	0	99	99	0	0	99	70	99	99	0	1	95	100	99	15	1	100	1
Subgroup 3b strains† (Arizona)	2	100	0	98	99	0	0	99	70	99	99	0	1	95	100	99	85	5	100	1
Subgroup 4 strains†	0	100	0	98	100	2	0	100	70	100	98	0	95	0	100	100	0	0	98	0
Subgroup 5 strains†	0	100	0	100	100	0	0	100	100	100	100	0	100	0	100	80	0	0	100	100
Serratia																				
S. marcescens†	1	20	98	98	0	15	0	99	0	99	97	90	95	3	100	55	2	99	99	0
S. marcescens biogroup 1†	0	100	60	30	0	0	0	55	4	65	17	30	70	0	100	0	4	100	96	0
S. liquefaciens group†	1	93	93	90	0	3	0	95	0	95	95	90	90	2	100	75	10	98	100	0
S. rubidaea†	0	20	100	95	0	2	0	55	0	0	85	90	25	94	100	30	100	99	100	0
S. odorifera biogroup 1†	60	100	50	100	0	5	0	100	0	100	100	95	60	0	100	0	70	100	100	0
S. odorifera biogroup 2†	50	60	100	97	0	0	0	94	0	0	100	94	19	0	100	13	97	0	97	0
S. plymuthica†	0	94	80	75	0	0	0	0	0	0	50	60	30	0	100	40	80	100	100	0
S. ficaria†	0	75	75	100	0	0	0	0	0	0	100	100	55	0	100	0	15	100	100	0
"Serratia" fonticola†	0	100	9	91	0	13	0	100	0	97	911	0	70	88	100	79	97	21	100	91
Tatumella																				
T. ptyseos†	0	0	5	2	0	0	90	0	0	0	0	0	0	0	100	0	0	98	0	0
Xenorhabdus																				
X. luminescens (25°C)	50	0	0	50	0	25	0	0	0	0	100	50	0	0	75	0	0	0	0	0
X. nematophilus (25°C)	40	0	0	0	0	0	0	0	0	0	100	80	0	0	80	0	0	0	0	0
Yersinia																				
Y. enterocolitica†	50	97	2	0	0	75	0	0	0	95	2	0	2	0	100	5	5	95	98	0
Y. frederiksenii†	100	100	0	15	0	70	0	0	0	95	5	0	0	0	100	40	40	100	100	0
Y. intermedia†	100	100	5	5	0	80	0	0	0	100	5	0	10	5	100	18	35	100	100	0
Y. kristensenii†	30	92	0	0	0	77	0	0	0	92	5	0	0	0	100	23	8	0	100	0
Y. pestis†	0	80	0	0	0	0	0	0	0	0	0	0	0	0	100	0	0	0	97	0
Y. pseudotuberculosis†	0	100	0	0	0	95	0	0	0	0	0	0	0	0	100	0	0	0	100	0
"Yersinia" ruckeri	0	97	10	0	0	0	0	50	5	100	0	30	15	0	100	5	0	0	100	0
Enteric group 17†	0	100	2	100	0	60	0	0	21	95	0	0	97	3	100	95	75	100	100	0
Enteric group 41†	100	100	0	0	0	50	0	0	0	0	100	0	100	50	100	100	100	100	100	100
Enteric group 45†	0	100	0	100	0	0	0	100	22	100	100	0	78	0	100	89	0	0	100	0
Enteric group 57†	0	70	0	40	100	0	0	0	0	0	0	0	30	0	100	60	0	0	0	50
Enteric group 58†	0	100	0	85	0	70	0	100	0	85	100	0	100	85	100	85	30	0	100	85
Enteric group 59†	10	100	0	100	0	0	30	0	60	0	100	0	80	90	100	100	80	0	100	0
Enteric group 60†	0	100	0	0	0	50	0	0	0	100	75	0	0	100	100	100	0	0	50	0
Enteric group 63	0	100	0	0	0	0	0	100	0	100	65	0	0	0	100	100	0	0	100	0
Enteric group 64	0	100	0	50	0	0	0	0	50	0	100	0	100	100	100	50	100	0	100	0
Enteric group 68†	0	100	50	0	0	0	0	0	0	0	100	0	100	0	100	0	0	100	100	0
Enteric group 69	0	0	100	100	0	0	0	0	100	100	100	0	100	100	100	100	25	100	100	100

*Each number gives the percentage of positive reactions after 2 days' incubation at 36°C (except *Xenorhabdus*, which was incubated at 25°C). The vast majority of these positive reactions occur within 24 hrs. Reactions that become positive after 2 days are not considered.

†Known to occur in clinical specimens.

‡ ONPG, o-nitrophenyl-β-galactopyranoside.

From Farmer JJ II, Davis BR, Hickman-Brenner FW, et al: Biochemical identification of new species and biogroups of Enterobacteriaceae isolated from clinical specimens. J Clin Microbiol 21:46, 1985. Reprinted by permission.

Salicin fermentation	Adonitol fermentation	myo-Inositol fermentation	D-Sorbitol fermentation	L-Arabinose fermentation	Raffinose fermentation	L-Rhamnose fermentation	Maltose fermentation	D-Xylose fermentation	Trehalose fermentation	Cellobiose fermentation	α-Methyl-D-glucoside fermentation	Erythritol fermentation	Esculin hydrolysis	Melibiose fermentation	D-Arabitol fermentation	Glycerol fermentation	Mucate fermentation	Tartrate, Jordan	Acetate utilization	Lipase (corn oil)	DNase at 25°C	Nitrate 1 nitrite	Oxidase, Kovacs	ONPG	Yellow pigment	D-Mannose fermentation
50	100	90	1	0	5	70	2	10	0	3	2	75	35	5	100	60	0	95	60	0	0	100	0	5	0	100
2	5	95	1	1	7	0	1	7	98	5	0	0	0	0	0	50	0	90	75	0	10	100	0	10	0	100
1	98	1	1	1	1	0	1	1	2	1	0	0	0	0	0	15	0	90	40	0	0	100	0	1	0	100
0	0	0	0	0	0	0	0	0	0	0	0	0	0	0	0	5	0	50	25	0	0	100	0	0	0	100
100	0	0	94	100	94	94	94	94	100	100	0	0	100	100	0	13	30	6	6	0	0	100	0	100	0	100
0	0	35	95	99	2	95	97	97	99	5	2	0	5	95	0	5	90	90	90	0	2	100	0	2	0	100
0	0	0	99	2	0	0	97	82	100	0	0	0	0	100	0	20	0	100	0	0	0	100	0	0	0	100
0	0	0	90	0	1	100	95	98	0	0	0	1	0	45	1	0	0	85	1	0	0	98	0	0	0	95
0	0	0	95	100	0	100	95	0	100	5	0	0	0	95	0	10	0	0	0	0	0	100	0	0	0	100
0	0	0	1	80	10	10	90	70	50	10	0	1	0	0	0	0	50	100	0	0	10	100	0	0	0	100
0	0	0	10	100	1	100	5	90	90	5	0	0	0	0	0	0	0	0	0	0	0	100	0	0	0	100
5	0	5	100	100	0	100	100	100	100	0	8	0	15	8	0	25	96	50	95	0	0	100	0	15	0	95
0	0	0	99	99	1	99	98	100	99	1	1	0	1	95	1	10	90	5	90	0	2	100	0	100	0	100
0	0	0	99	99	1	99	98	100	99	1	1	0	1	95	1	10	30	20	75	0	2	100	0	100	0	100
60	5	0	100	100	0	98	100	100	100	50	0	0	0	100	5	0	0	65	70	0	0	100	0	0	0	100
0	0	0	100	100	0	100	100	100	100	0	0	0	0	75	0	0	100	0	100	0	0	100	0	0	0	100
95	40	75	99	0	2	0	96	7	99	5	0	1	95	0	0	95	0	75	50	98	98	98	0	95	0	99
92	30	30	92	0	0	0	70	0	100	4	0	0	96	0	0	92	0	50	4	75	82	83	0	75	0	100
97	5	60	95	98	85	15	98	100	100	5	5	0	97	75	0	95	0	75	40	85	85	100	0	93	0	100
99	99	20	1	100	99	1	99	99	100	94	1	0	94	99	85	20	0	70	80	99	99	100	0	100	0	100
98	50	100	100	100	100	95	100	100	100	100	0	0	95	100	0	40	5	100	60	35	100	100	0	100	0	100
45	55	100	100	100	7	94	100	100	100	100	0	7	40	96	0	50	0	100	65	65	100	100	0	100	0	100
94	0	50	65	100	94	0	94	94	100	88	70	0	81	93	0	50	0	100	55	70	100	100	0	70	0	100
100	0	55	100	100	70	35	100	100	100	0	8	0	100	40	100	0	0	17	40	77	100	92	8	100	0	100
100	100	30	100	100	100	76	97	85	100	6	91	0	100	98	100	88	0	58	15	0	0	100	0	100	0	100
55	0	0	0	0	11	0	0	9	93	0	0	0	0	25	0	7	0	0	0	0	0	98	0	0	0	0
20	0	30	99	98	5	1	75	70	98	75	0	0	25	1	40	90	0	85	15	55	5	98	0	95	0	100
92	0	20	100	100	30	99	100	100	100	100	0	0	85	0	100	85	5	55	15	55	0	100	0	100	0	100
100	0	15	100	100	45	100	100	100	100	96	77	0	100	80	45	60	6	88	18	12	0	94	0	90	0	100
15	0	15	10	77	0	0	100	85	100	100	0	0	0	0	45	70	0	40	8	0	0	100	0	70	0	100
70	0	0	50	100	0	1	80	90	100	0	0	0	50	20	0	50	0	0	0	0	0	85	0	50	0	100
25	0	0	0	50	15	70	95	100	100	0	0	0	95	70	0	50	0	50	0	0	0	95	0	70	0	100
0	0	0	50	5	5	0	95	0	95	5	0	0	0	0	0	50	0	30	0	30	0	75	0	50	0	100
0	0	0	0	0	0	25	0	0	0	0	0	0	0	0	0	0	0	50	0	0	0	0	0	0	50	100
0	0	0	0	0	0	0	0	0	0	0	0	0	0	0	0	0	0	60	0	0	20	20	0	0	60	80
100	0	0	100	100	70	5	100	97	100	100	95	0	95	0	0	11	21	30	87	0	0	100	0	100	0	100
100	100	0	0	100	100	100	100	100	100	100	0	0	100	100	100	0	100	100	0	0	0	100	0	100	100	100
11	0	0	0	100	22	100	100	100	100	100	0	0	55	80	0	0	0	13	55	0	0	89	0	80	0	100
0	0	0	0	90	0	0	0	90	0	0	0	0	0	0	0	0	60	100	0	0	0	100	0	0	0	0
100	0	0	100	100	0	100	100	100	100	55	0	0	0	0	0	30	0	60	45	0	0	100	0	100	0	100
100	0	0	0	100	0	100	100	100	100	10	0	100	0	10	0	10	60	50	50	0	0	100	0	100	0	100
0	0	0	25	0	75	0	0	100	0	0	0	0	0	0	0	75	0	75	0	0	0	100	0	100	0	100
100	0	0	100	0	100	100	100	100	100	100	65	0	100	0	0	0	65	0	0	0	0	100	0	100	0	100
100	100	0	100	0	100	100	100	100	100	0	0	0	100	0	100	0	50	0	0	0	0	100	0	100	0	100
50	0	0	0	0	0	50	0	100	0	0	0	0	0	0	0	50	0	0	0	0	100	100	0	0	0	100
100	0	0	100	100	100	100	100	100	100	100	100	0	100	100	0	0	100	0	25	0	0	100	100	100	100	100

Hydrogen Sulfide Production

1. Bacterium (acid environment) + sodium thiosulfate = H_2S gas

2. H_2S + ferric ions = ferrous sulfide (black precipitate)

D. Hydrogen Sulfide Production

Alkaline slant/acid butt, H_2S in butt (Alk/Ac H_2S or K/A H_2S) or acid slant/acid butt, H_2S in butt (A/A H_2S).

Hydrogen sulfide production is another system that is helpful in differentiating the Enterobacteriaceae. There are two indicators present in the medium: sodium thiosulfate and ferrous sulfate. Because the visualization of H_2S production is a two-step process (see above), both indicators must be present. In addition, H_2S is a colorless gas, therefore the second indicator is necessary to visually detect its production.

E. Gas Production (Aerogenic) or No Gas Production (Nonaerogenic)

The production of gas results in the formation of bubbles or splitting of the media in the butt or complete displacement of the media from the bottom of the tube. Figure 16–13 illustrates the reactions on TSI agar.

Orthonitrophenyl Galactopyranoside (ONPG)

Organisms that are slow or late lactose-fermenters appear as nonfermenting colonies on primary isolation medium. When placed on TSI or KIA slants, these species produce similar results after 18 to 22 hours of incubation, raising suspicions that the species is an intestinal enteric pathogen. The ONPG test determines whether the organism is a slow or late lactose-fermenter (one that lacks the enzyme permease but possesses β-galactosidase) or a true non–lactose-fermenter. The ONPG is structurally similar to lactose, except that glucose is replaced with ONPG, and ONPG is more easily transported through the bacterial cell wall. β-Galactosidase acts on the ONPG (a colorless compound), cleaves it into galactose and orthonitrophenol, and the compound turns yellow. The compound remains colorless if the organism is a non–lactose-fermenter (Fig. 16–14).

Glucose Metabolism and Its Metabolic Products

Lactose degradation results in glucose and galactose available for bacterial consumption. Glucose is metabolized via the Embden-Meyerhof pathway, producing several intermediate byproducts, including pyruvic acid. Further degradation of pyruvic acid produces mixed acids as final end products. There are two separate pathways, however, that enterics take: the mixed acid fermentation pathway or the butylene glycol pathway.

The methyl red (MR) and Voges-Proskauer (VP) tests detect the end products of glucose fermentation. Each test detects products from a different pathway.

Methyl Red Test

If glucose is metabolized by the mixed acid fermentation pathway, acidic end products are pro-

■ Figure 16 - 13

Triple sugar iron agar reactions of Enterobacteriaceae. *Left to right,* Tube 1, uninoculated; tube 2, A/A gas; tube 3, K/A; tube 4, K/A H_2S; tube 5, A/A H_2S gas.

Methyl Red Test Reaction

Glucose → pyruvic acid → mixed acid fermentation (pH 4.4)

↓

Red color with methyl red indicator

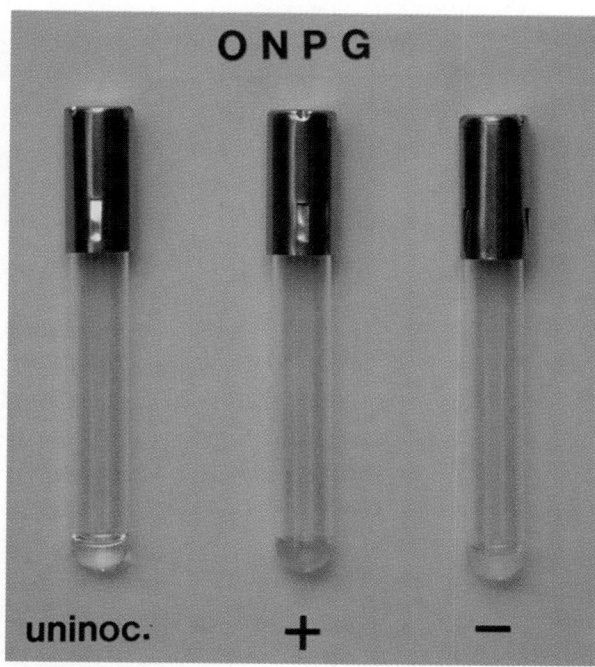

ONPG

uninoc.   +   —

ONPG test. (Courtesy of American Society for Clinical Laboratory Science, Education and Research Fund, Inc. © 1982.)

duced, which results in a low pH. Red color (indicating a low pH; see above reaction) develops after addition of the indicator methyl red.

Voges-Proskauer Test

Carbon compounds are degraded to butylene glycol and acetylmethylcarbinol (acetoin), which is further converted to diacetyl. Diacetyl in the presence of potassium hydroxide and α-naphthol forms a red complex (see reaction on p. 482). The pH remains at a relatively neutral level. Figure 16–15A to C illustrates the MRVP test.

Miscellaneous Reactions

INDOLE PRODUCTION

Indole is one of the degradation products of the amino acid tryptophan. Organisms that possess the enzyme tryptophanase are capable of deaminating tryptophan, with the formation of the intermediate degradation products of indole, pyruvic acid and ammonia. A red color develops after the addition of para-dimethylaminobenzaldehyde (PDAB) (Fig. 16–16).

CITRATE UTILIZATION TEST

The citrate utilization test determines whether the organism can utilize sodium citrate as the sole source of carbon for metabolism. The alkaline pH that results from use of citrate turns the indicator in the medium from green to blue. It is

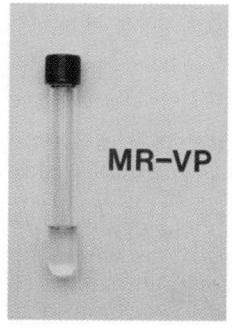

MR–VP

A

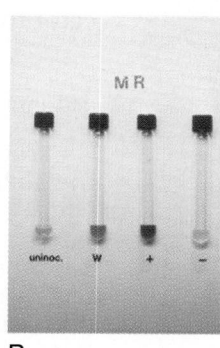

M R

uninoc.   w   +   —

B

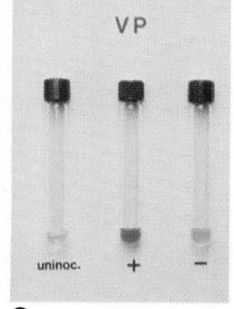

VP

uninoc.   +   —

C

A, The methyl red–Voges-Proskauer test is inoculated and incubated overnight. Then it is split equally into two parts, one part for the methyl red test, the other for the Voges-Proskauer test. *B,* Methyl red test. *C,* Voges-Proskauer test. (Courtesy of American Society for Clinical Laboratory Science, Education and Research Fund, Inc. © 1982.)

Voges=Proskaver Test Reaction

Glucose $\rightarrow$ pyruvic acid $\rightarrow$ acetoin $\rightarrow$ diacetyl + KOH + α-naphthol $\rightarrow$ red complex
$$\rightarrow \text{butylene glycol}$$

important to keep the inoculum light, since dead organisms can be a source of carbon, producing a false-positive reaction (Fig. 16–17).

UREASE PRODUCTION TEST

The urease test determines whether the organism hydrolyzes urea, releasing a sufficient amount of ammonia to produce a color change. A positive test result is indicated by a bright pink color (Fig. 16–18).

MOTILITY TEST

The motility test medium has agar concentrations of 0.4% or less, to allow free spread of organism. A single stab into the medium is made. After overnight incubation, movement away from the stab line or a hazy appearance throughout the medium indicates a motile organism.

MALONATE TEST

The malonate test determines whether the organism is capable of utilizing sodium malonate

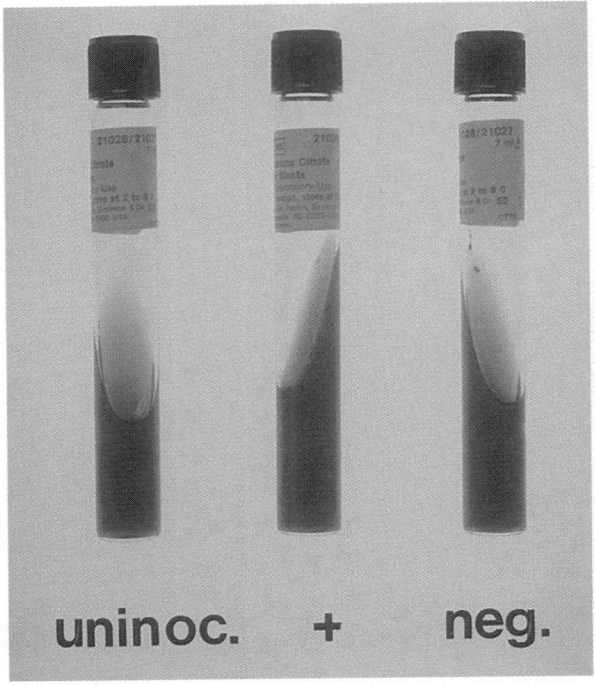

■■■ F i g u r e   1 6 – 1 7

Citrate utilization test. (Courtesy of American Society for Clinical Laboratory Science, Education and Research Fund, Inc. © 1982.)

as its sole source of carbon. A positive test results in increased alkalinity, changing the indicator to blue (Fig. 16–19).

CARBOHYDRATE FERMENTATION TESTS

Carbohydrate fermentation tests determine the ability of a microorganism to ferment a specific 1% concentration of carbohydrate incorporated into a basal medium such as purple broth base, producing acid or acid with gas when the test result is positive. Examples of sugars used to differentiate bacteria include maltose, rhamnose, lactose, sucrose, raffinose, and arabinose. The polyhydric alcohols (which end in "ol") that are collectively called "sugars" include adonitol, dulcitol, mannitol, and sorbitol. A yellow color is considered a positive reaction.

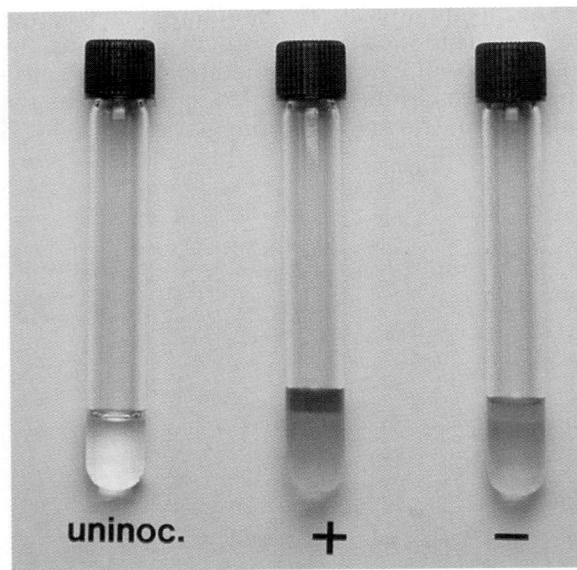

■■■ F i g u r e   1 6 – 1 6

Indole broth. (Courtesy of American Society for Clinical Laboratory Science, Education and Research Fund, Inc. © 1982.)

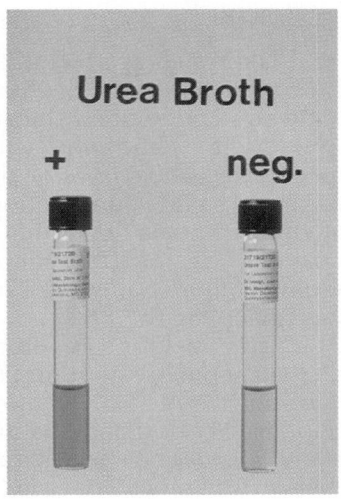

Figure 16-18
Urease test. (Courtesy of American Society for Clinical Laboratory Science, Education and Research Fund, Inc. © 1982.)

PHENYLALANINE DEAMINASE

This test determines whether the organism possesses the enzyme that deaminates phenylalanine to phenylpyruvic acid (see p. 484 for biochemical reaction). Addition of a ferric chloride reagent results in a green color if phenylpyruvic acid is present. This is helpful in initial differentiation of *Proteus, Morganella,* and *Providencia* from the rest of the Enterobacteriaceae (Fig. 16–20).

DECARBOXYLASE TESTS

Decarboxylase tests determine whether the bacterial species possess enzymes capable of decarboxylating (attacking the carboyl group of an amino acid) specific amino acids in the test medium. The three amino acids commonly used to test for Enterobacteriaceae are lysine, ornithine, and arginine. Specific amine products and carbon dioxide are products of decarboxylation. Degradation of the amino acids and their specific end products are shown on page 1984.

Arginine conversion involves a two-step process. First, arginine is converted to citrulline by arginine dihydrolase. Next, citrulline is further converted to ornithine, which is decarboxylated to putrescine.

The test to detect decarboxylation of Enterobacteriaceae contains the Moeller decarboxylase base medium. This base contains glucose, peptone, pH indicator, bromcresol purple and cresol red, and the specific amino acid. The role of glucose in the medium is important because decarboxylases are inducible enzymes produced in an acid pH. In addition, by definition, all members of Enterobacteriaceae are glucose-positive, providing a growth stimulus. The uninoculated medium is purple. A control tube containing only the base medium without the amino acid is tested along with the test organism to determine the viability of the organism. For decarboxylation to take place, two conditions must be met: an acid pH and an anaerobic environment. The control tube determines whether sufficient acid is produced. Both tubes are inoculated with the test organism, are overlayed with a layer of sterile mineral oil, which creates anaerobic conditions, and then are incubated.

During the first few hours of incubation, Enterobacteriaceae organisms attack the glucose first, changing the pH to acid. The change in pH in the medium changes the purple color to yellow. If the organism produces the specific decarboxylase and the amino acid in the medium is attacked, release of the amine products causes an alkaline pH shift. This results in a purple (positive result) color in the broth medium. If the organism does not possess the specific decarboxylase, the medium remains yellow (negative result). The control tube remains the original color (Fig. 16–21).

Modifications of the decarboxylase test to detect other biochemical reactions are also routinely

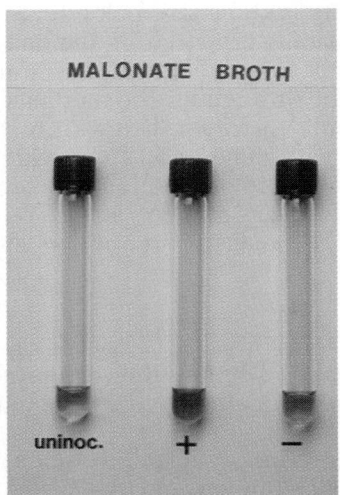

Figure 16-19
Malonate test. (Courtesy of American Society for Clinical Laboratory Science, Education and Research Fund, Inc. © 1982.)

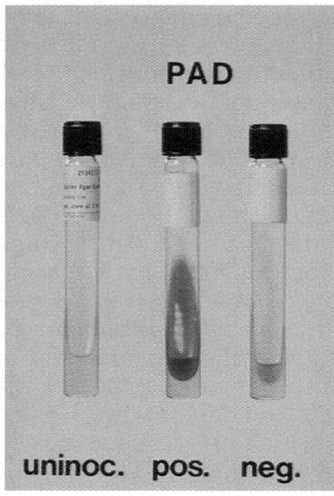

PAD

uninoc. pos. neg.

■■■■■F i g u r e   1 6 – 2 0

Phenylalanine deaminase test. (Courtesy of American Society for Clinical Laboratory Science, Education and Research Fund, Inc. © 1982.)

used. Examples of these include MIO (motility, indole, ornithine) and lysine iron agar (LIA) tests.

MIO (MOTILITY, INDOLE, ORNITHINE) TEST

MIO is a semisolid agar medium used to detect motility and production of indole and ornithine decarboxylase. MIO is useful in differentiating *Klebsiella* species from *Enterobacter/Serratia* species. Motility is shown by a clouding of the medium or spreading growth from the line of inoculation. Ornithine decarboxylation is indicated by a purple color throughout the medium. Since MIO is a semisolid medium, it does not have to be overlayed with mineral oil to provide anaerobic conditions. Indole production is detected by the addition of Kovacs reagent; a pink to red color is formed in the reagent area if the test result is positive.

LYSINE IRON AGAR SLANT

The LIA test is a tubed agar slant. It contains lysine as the specific amino acid, glucose, ferric ammonium citrate, and sodium thiosulfate. LIA is used primarily to determine whether the bacterial species decarboxylates or deaminates lysine. Hydrogen sulfide production is also detected in this medium. LIA is inoculated in the same manner as the TSI agar slant. LIA is most useful in conjunction with TSI in screening stool specimens for the presence of enteric pathogens. LIA is helpful in differentiating *Salmonella* (lysine-positive) from *Citrobacter* species (lysine-negative).

LIA is also useful in differentiating *Proteus, Morganella,* and *Providencia* species from the rest of the members of Enterobacteriaceae. This group of enterics deaminate (attack the NH_2 group instead of the carboxyl group) amino acids. In the LIA slant, deamination of lysine turns the original purple color slant to a plum or reddish-purple color; the butt remains yellow. The various reactions in LIA are shown in Figure 16–22.

NITRATE REDUCTION TEST

The nitrate reduction test determines whether the organism has the ability to reduce nitrate to nitrite and further reduce nitrite to nitrogen.

The organism is inoculated into a nutrient broth containing a nitrogen source. After 24 hours of incubation, sulfanilic acid and *N,N*-dimethyl-1 naphthylamine (NNDN) is added. A red color indicates the presence of nitrite (see the reaction on p. 486). If no color develops, this may indicate that nitrate has not been reduced or *that nitrate has been further reduced to nitrogen gas (N_2), nitric oxide (NO), or nitrous oxide (N_2O), which the reagents will not be able to detect.* To determine whether the test has produced a true-negative result or whether the lack of color production was due to reduction beyond nitrate, add a small amount of zinc dust. Zinc dust reduces nitrate.

Deamination of Phenylalanine
Phenylalanine → phenylalanine deaminase → phenylpyruvic acid + ($FeCl_3$) Green

Degradation of Amino Acids and Their Specific End Products
Lysine (amino acid) → lysine decarboxylase → cadaverine (amine) + CO_2

Ornithine → ornithine decarboxylase → putrescine

Arginine → arginine dihydrolase → citrulline → ornithine → putrescine

cal laboratories because of its potential lethal effects on personnel. The principle of the test is to determine the ability of the organism to live and grow in KCN. The enzymes of certain bacteria, such as *Citrobacter freundii*, are resistant to KCN and therefore grown in the medium; others such as *Salmonella* are sensitive to KCN and do not grow in the medium.

SCREENING STOOL CULTURES FOR PATHOGENS

Because of the mixed microbial flora of fecal specimens, efficient screening methods should be utilized for the recovery and identification of stool pathogens. Enteric pathogens include the following: *Salmonella, Shigella, Aeromonas, Campylobacter, Yersinia, Vibrio,* and *E. coli* O157:H7. All fecal specimens should be screened for *Salmonella, Shigella,* and *Campylobacter* (see Chapter 17). In addition, many laboratories are routinely screening for *E. coli* O157:H7. Screening routinely for the remaining organisms may not be cost-effective; therefore, these organisms should be addressed on the basis of patient history (e.g., travel near coastal areas, where certain organisms are endemic) and gross description of the specimen (bloody or watery). It should be noted that any method screening for *Salmonella* and *Shigella* also screens for *Aeromonas* and *Plesiomonas* (see Chapter 17), and therefore additional media are not needed.

Fecal pathogens are generally non–lactose-fermenters. These organisms appear as clear or colorless and translucent colonies on MacConkey agar. In addition, many laboratories use selective media, such as HE agar, which detects lactose and sucrose fermentation and H_2S production in

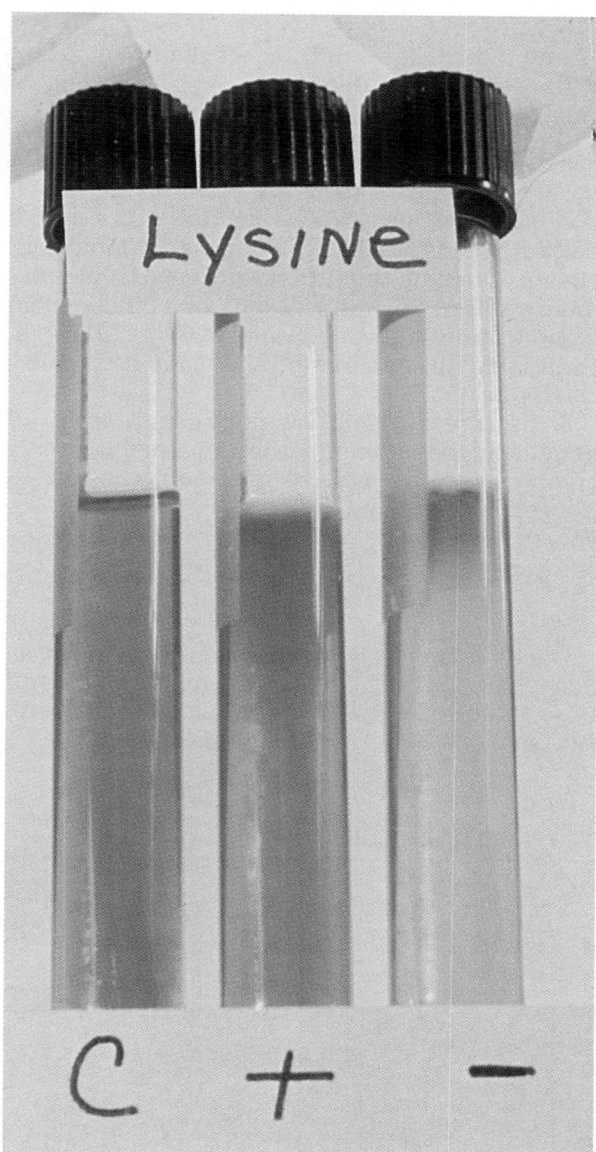

■ Figure 16-21

Lysine decarboxylase test using semisolid media (C, control). In the negative reaction, the organism lacks the decarboxylase enzyme; therefore, the medium remains yellow after 24 hours of incubation. The slight purple at the top of the negative tube is due to diffused oxygen, since decarboxylation is an anaerobic process.

Therefore, development of a red color after the addition of zinc confirms a true-negative test result.

POTASSIUM CYANIDE TEST

The KCN test is mentioned briefly here in a historical context. The test is not performed in clini-

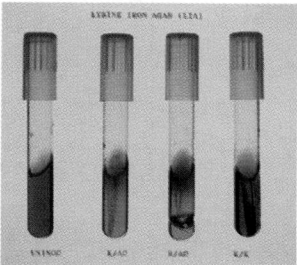

■ Figure 16-22

Lysine iron agar reactions (R, deamination of lysine indicated by a red color). The tube on the right is lysine decarboxylase–positive and shows a small amount of H_2S.

Nitrate Reduction Test Reaction

Nutrient broth with 0.1% potassium nitrate → nitrate reductase → nitrite + sulfanilic acid + N,N-dimethyl-1-naphthylamine → diazo red dye

conjunction with MacConkey agar. Since media selection varies in many laboratories, only MacConkey agar will be used here as an example.

However, many of the bacteria that compose common fecal flora also appear as non–lactose-fermenters, for example, *Proteus*, *Providencia*, *Serratia*, *Citrobacter*, and *Pseudomonas* as well as the late lactose-fermenting organisms. For this reason, it is necessary to set up screening tests to differentiate these organisms from stool pathogens. An easy approach is to take a well-isolated, NLF colony and perform a screening battery consisting first of an oxidase test and then of inoculation of LIA and TSI agar. The following will serve as an example. After 18 to 24 hours of incubation, the results of the reactions are for the TSI, K/AG, and for the LIA, K/A. Using Table 16–12 and tracing the TSI result to where it intersects the LIA result, you see that this organism could be *Salmonella*, *Shigella*, *Aeromonas*, *E. coli*, *Morganella*, *Enterobacter*, or *Citrobacter*. If the oxi-

dase test performed the previous day yielded a negative result, *Aeromonas* is eliminated as a potential pathogen. Since there are other potential pathogens in this category, additional biochemical testing must be performed to identify the organism. When the results of a biochemical identification indicate *Salmonella* or *Shigella*, serotyping must be performed using a commercial typing kit.

If the screening battery pinpoints a group of organisms that are nonpathogens, the process is complete and the culture is discarded (after 48 hours of incubation).

SEROLOGIC GROUPING

Once an isolate is biochemically identified as *Salmonella* or *Shigella* sp, serologic grouping of the isolate for the O serogroups (somatic antigen) must be performed for confirmation.

■■■ T A B L E   1 6 – 1 2
STOOL CULTURE SCREENING FOR ENTERIC PATHOGENS UTILIZING TSI AND LIA IN COMBINATION

LIA Reactions	TSI Reactions							
	K/A H$_2$S	K/AG H$_2$S	K/AG	K/A	A/AG H$_2$S	A/AG	A/A	K/K
R/A		*P. vulgaris* *P. mirabilis*	*M. morganii* Providencia	*M. morganii* Providencia	*P. vulgaris* *P. mirabilis*		*Providencia*	
K/K H$_2$S	* Salmonella Edwardsiella	* Salmonella Edwardsiella	* Salmonella	* Salmonella	* Salmonella			
K/K	* Salmonella		Hafnia Klebsiella Serratia	* Salmonella Plesiomonas[†] Hafnia Serratia		Klebsiella Enterobacter E. coli	Serratia	Pseudomonas[†]
K/A H$_2$S		Citrobacter			Citrobacter			
K/A			* Salmonella Shigella Aeromonas[†] E. coli Morganella Enterobacter Citrobacter	* Shigella Yersinia Aeromonas[†] E. coli Morganella Enterobacter		* Aeromonas[†] E. coli Citrobacter Enterobacter	* Aeromonas[†] Yersinia E. Coli Enterobacter	

[*]Results of TSI and LIA reactions in this category indicate a potential pathogen; additional tests must be performed.
[†]oxidase positive.
K, alkaline; A, acid; G, gas; R, deamination (red slant).
Data from the Microbiology Laboratory, OSU Hospitals, and Maureta Ott.

Salmonella Species

Based on the common O antigens, *Salmonella* may be placed into major groups designated by capital letters. There are approximately 60 O antigenic groups; however, 98% of *Salmonella* isolates from humans belong to serogroups A through G.

Laboratories may report isolates as *Salmonella* groups (A–G) when specific serotyping is not available; isolates may be sent to a reference laboratory for serotyping. If an isolate is suspected of being one of these three species, *S. typhi, S. choleraesuis,* or *S. paratyphi A,* because of its medical implications, it must be biochemically identified and serologically confirmed. Their identification is very important in providing proper therapy for the patient and in limiting any possible complications that may develop. Measures may also be taken to prevent an epidemic outbreak.

It is imperative for any laboratory performing bacteriology to be able to serologically identify *S. typhi,* particularly, and other members of *Salmonella* in O groups A through G. Other isolates can be identified as "biochemically compatible with *Salmonella*" and submitted to a reference laboratory for further testing.

To perform serologic grouping by slide technique, first prepare a saline emulsion from a pure culture of the organism. Serologic typing is best performed on a colony taken from a pure culture growing on nonselective media, such as blood agar plate, although TSI or MacConkey agar can be used for presumptive serologic identification. A slide with wells is easy to use for the agglutination test. A regular microscope slide may be used by marking separate squares with a wax pencil. Place one drop of antisera on the appropriately labeled slide. Add one drop of bacterial emulsion to each drop of antisera. Antisera kits usually consist of a polyvalent A through G, Vi, and serogroups A, B, C, C_2, D, E, and G. In the event of a positive agglutination in the Vi antisera with no agglutination in the other groups, the emulsion should be heated to 100°C for 10 minutes to inactivate the capsular Vi antigen. The emulsion is then cooled and retested with antisera A to G. If the organism agglutinates with group D antisera, it can be reported as *S. typhi.* Larger laboratories usually maintain antisera to serotype *Salmonella* for all the somatic types. H antigen or flagella typing is usually performed in a reference laboratory that provides epidemiologic information of the common source in outbreaks.

Shigella Species

Similar serogrouping procedures may be used in the serologic testing for *Shigella* species. Serologic grouping of *Shigella* species is based on the O antigen present. *Shigella* species may belong to one of the four serogroups: A, B, C, and D. *Salmonella dysenteriae, S. flexneri, S. boydii,* and *S. sonnei* correspond to these serogroups, respectively. The O antisera used for serogrouping are polyvalent, containing several serotypes within each group (with the exception of group D, which contains only a single serotype).

If agglutination fails, the suspension must be heated to remove the capsular antigen that may be present, and subsequently the agglutination test procedure is repeated.

The members of the family Enterobacteriaceae present a challenge to the microbiologist when isolated from infected sites. The microbiologist must determine the clinical significance of the isolate, particularly those that produce opportunistic infections. Intestinal pathogens, when isolated, must be reported to the clinician as soon as possible. With the development of rapid diagnostic tests to identify these organisms, treatment can be provided when appropriate, and prevention of the spread of disease may be easily accomplished.

Bibliography

Arduino MJ, Bland LA et al: Growth and endotoxin production of *Yersinia enterocolitica* and *Enterobacter agglomerans* in packed erythrocyes. J Clin Microbiol 27:1483, 1989.

Azad MA: Colonic perforation in *Shigella dystenteriae* 1 infection. Pediatric Infect Dis 5:103, 1986.

Brenner DJ, Davis BR, et al: Atypical biotypes of *E. coli* found in clinical specimens and description of *E. hermanii* sp. nov. J Clin Microbiol 15:703, 1982.

Brenner DJ, McWhorter AC, et al: *Escherichia vulneris:* A new species of Enterobacteriaceae associated with human wounds. J Clin Microbiol 15:1133, 1982.

Boedeker EC: Enteroadherent (enteropathogenic) *E. coli.* In Farthing MJ, Keusch GT (eds): Enteric Infection: Mechanisms, Manifestation, and Management. New York: Raven Press 1988.

Boileau C, D'Hauteville H, Sansonetti P: DNA hybridization technique to detect *Shigella* sp. and enteroinvasive *E. coli.* J Clin Microbiol 20:959, 1984.

Brown JE, Griffin DE, et al: Purification and biological characterization of Shiga toxin from *Shigella dysenteriae* 1. 36:996, 1982.

Butler T: Plague and other *Yersinia* infections. In Greenough WB, Merigan TC (eds): Current Topics in Infectious Disease. New York: Plenum Med. Book Co., 1983.

Bitar R, Tarpley J: Intestinal perforation in typhoid fever: A historical and state-of-the-art review. Rev Infect Dis 7:257, 1985.

Bader M, Pedersen AHB, et al: Veneral transmission of shigellosis in Seattle–King County. Sex Trans Dis 4:89, 1977.

Baron EJ, Finegold S: Bailey & Scott's Diagnostic Microbiology, 8th ed. St. Louis: CV Mosby, 1990.

Candy DCA, Stephen J: Salmonella. In Farthing MJ, Keusch GT (eds): Enteric Infection: Mechanisms, Manifestations, and Management. New York: Raven Press, 1988.

Cherubin CE, Neu HC, et al: Septicemia with non-typhoid Salmonella. Medicine 53:365, 1974.

Carbonetti NH, Boonchai S, et al: Aerobactin-mediated iron uptake by E. coli isolates from human extraintestinal infections. Infect Immun 51:966, 1986.

Chmel H: Serratia odorifera biogroup 1 causing an invasive human infection. J Clin Microbiol 26:1244, 1988.

Cohen JI, Rodday P: Yersinia enterocolitica bacteremia in a patient with the acquired immunodeficiency syndrome. Am J Med 86:254, 1989.

Cravioto A, Gross RJ, et al: An adhesive factor found in strains of Escherichia coli belonging to the traditional infantile enteropathogenic serotypes. Curr Microbiol 3:95, 1979.

Donahue AR, Kelley MA, et al: Enzyme-linked immunosorbent assay for Shigella toxin. J Clin Microbiol 24:65, 1986.

Echeverria P, Seriwatana J, et al: Plasmids coding for colonization factor antigens I and II, heat-labile enterotoxin, and heat-stable enterotoxin A2 in Esherichia coli. Infect Immun 51:626, 1986.

Eden CS, Hausson S, et al: Host-parasite interaction in the urinary tract. J Infect Dis 157:421, 1988.

Eden CS, Hanson LA, et al: Variable adherence to normal human urinary-tract epithelial cells of Escherichia coli strains associated with various forms of urinary tract infection. Lancet 2:490, 1976.

Ewing WH: Edwards & Ewing's Identification of Enterobacteriaceae, 4th ed. New York: Elsevier, 1986.

Farmer JJ, Howard BJ, Weissfeld AS: Enterobacteriaceae. In Howard BJ, Klaas J, et al (eds): Clinical and Pathogenic Microbiology. St. Louis: CV Mosby, 1987.

Farmer JJ, McWhorter AC, et al: The Salmonella-Arizona group of Enterobacteriaceae: Nomenclature, classification, and reporting. Clin Microbiol Newsl 6:63, 1984.

Farmer JJ, Davis BR, et al: Biochemical identification of new species and biogroups of Enterobacteriaceae isolated from clinical specimens. J Clin Microbiol 21:46, 1985.

Farthing MJ, Keusch GT (eds): Enteric Infection: Mechanisms, Manifestations, and Management. New York, Raven Press, 1988.

Freeman BA: Burrows Textbook of Microbiology, 22nd ed. Philadelphia: WB Saunders, 1985.

Fukushima H, Gomyoda M, et al: Yersinia pseudotuberculosis infection contracted through water contaminated by a wild animal. J Clin Microbiol 26:584, 1988.

Gomes TA, Toledo MR, et al: DNA probes for identification of enteroinvasive Escherichia coli. J Clin Microbiol 25:2025, 1987.

Griffin PM, Ostroff SM, et al: Illnesses associated with Escherichia coli O157:H7 infections: A broad clinical spectrum. Ann Intern Med 109:705, 1988.

Harris JR, Mariano J, et al: Person-to-person transmission in an outbreak of enteroinvasive Escherichia coli. Am J Epidemiol 122:245, 1985.

Hayes P, Wells JG, Griffin PM: Isolation and identification of Escherichia coli O157:H7. BACTinews, published by REMEL, Vol 2, No 2. Lenexa, KS, April 1994.

Horowitz H, Nadelman R, et al: Serratia plymuthica sepsis associated with infection of central venous catheter. J Clin Microbiol 25:1562, 1987.

Jacobs J, Jamaer JJ, et al: Yersinia enterocolitica in donor blood: A case report and review. J Clin Microbiol 27:1119, 1989.

Karmali M, Petric M, et al: The association between idiopathic hemolytic-uremic syndrome and infection by verotoxin-producing E. coli. J Infect Dis 151:775, 1985.

Keusch GT, Bennish M: Shigella. In Farthing MJ, Keusch GT (eds): Enteric Infection: Mechanisms, Manifestations, and Management. New York: Raven Press, 1988.

Khanna R, Levendoglu H: Liver abscess due to Yersinia enterocolitica. Dig Dis Sci 34:636, 1989.

Knutton S, Lloyd D, McNeish A: Identification of a new fimbrial structure in enterotoxigenic E. coli (ETEC) serotype O148:H28 which adheres to human intestinal mucosa: A potentially new human ETEC colonization factor. Infect Immun 55:86, 1987.

Koneman E, Allen S, et al: Color Atlas in Diagnostic Microbiology, 3rd ed. Philadelphia: JB Lippincott, 1988.

Korhonen, TK, Valtonen MV, et al: Serotypes, hemolysin production, and receptor recognition of Escherichia coli strains associated with neonatal sepsis and meningitis. Infect Immun 48:486, 1985.

Kelly M, Brenner D, Farmer JJ: Enterobacteriaceae. In Lennette EH (ed): Manual of Clinical Microbiology, 4th ed. Washington, DC: American Society for Microbiology, 1985.

Kelly M, Brenner D, Farmer JJ: Enterobacteriaceae. In Balows A, Hausler W, et al (eds): Manual of Clinical Microbiology, 5th ed. Washington, DC: American Society for Microbiology, 1991.

Levine M, Nalin DR, et al: Escherichia coli strains that cause diarrheae but do not produce heat-labile or heat-stable enterotoxins and are noninvasive. Lancet 1:1119, 1978.

Levine M, Nalin DR, et al: Immunity to enterotoxigenic Escherichia coli. Infect Immun 17:78, 1979.

Levine M, Kaper BJ, et al: New knowledge on pathogenesis of bacterial enteric infections as applied to vaccine development. Microbiol Rev 47:510, 1983.

Levine MM: Escherichia coli that cause diarrhea: Enterotoxigenic, enteropathogenic, enteroinvasive, enterohemorrhagic, and enteroadherent. J Infect Dis 155:377, 1987.

Lew PD, Baker AS, et al: Intra-abdominal Citrobacter infections: Association with biliary or upper gastrointestinal source. Surgery 95:398, 1984.

Lipsky AB, Hook EW, et al: Citrobactor infections in humans: Experience at the Seattle VA Medical Center and review of the literature. Rev Infect Dis 2:746, 1980.

McDonough KA, Shwan TG, et al: Identification of a Yersinia pestis–specific DNA probe with potential for use in plague surveillance. J Clin Microbiol 26:2515, 1989.

Mathewson JJ, Cravioto A: HEp-2 cell adherence as an assay for virulence among diarrheagenic E. coli. J Infect Dis 159:1057, 1989.

March B, Ratnam S: Latex agglutination test for detection of Escherichia coli serotype O157. J Clin Microbiol 27:1675, 1989.

March SB, Ratnam S: Sorbitol-MacConkey medium for detection of Escherichia coli O157:H7 associated with hemorrhagic colitis. J Clin Microbiol 23:869, 1986.

Medon PP, Lanser JA, et al: Identification of enterotoxigenic Escherichia coli isolates with enzyme-labeled synthetic oligonucleotide probes. J Clin Microbiol 26:2173, 1988.

Nakashima A, MaCarthy M, et al: Epidemic septic arthritis caused by Serratia marcesens and associated with benzalkonium chloride antiseptic. J Clin Microbiol 25:1014, 1987.

O'Brien AD, Lively TA, et al: Purification of Shigella dysenteriae 1 (Shiga)-like toxin from Escherichia coli O157:H7 strain associated with hemorrhagic colitis. Lancet 2:573, 1983.

Pai CH, Kelly JK: Shiga-like toxin producing *E. coli*. In Farthing MJ, Keusch GT (eds): Enteric Infection: Mechanisms, Manifestations, and Management. New York: Raven Press 1988.

Pai C, Ahmed N, et al: Epidemiology of sporadic diarrhea due to verocyto-toxin-producing *E. coli*: A 2-yr prospective study. J Infect Dis 157:1054, 1988.

Pai C, Gordon R, et al: Sporadic cases of hemorrhagic colitis associated with *E. coli* O157:H7. Ann Intern Med 101:738, 1984.

Palmer SR, Rowe B: Investigation of outbreaks of *Salmonella* in hospitals. BMJ 287:891, 1983.

Pickering, LK, Bartlett AV, Woodward WE: Acute infectious diarrhea among children in day care. Rev Infect Dis 8:539, 1986.

Plantholt SJ, Trofa AF: *Citrobacter freundii* endocarditis in an intravenous drug abuser. S Med J 80:1439, 1987.

Raj P: Pathogenesis and laboratory diagnosis of *Escherichia coli*–associated enteritis. Clin Micro Newsletter 15:89, 1993.

Ratnam S, Mercer E, et al: A nosocomial outbreak of diarrheal disease due to *Yersinia enterocolitica* serotype 0:5 biotype 1. J Infect Dis 145:242, 1982.

Riley LW, et al: Hemorrhagic colitis associated with a rare *E. coli* serotype. N Engl J Med 308:681, 1982.

Robins-Browne RM: *Yersinia enterocolitica*. In Farthing MJ, Keusch GT (eds): Enteric Infection: Mechanisms, Manifestations, and Management. New York: Raven Press, 1988.

Rothbaum R, McAdams AJ, et al: A clinicopathologic study of enteroadherent *E. coli*: A cause of protracted diarrhea in infants. Gastroenterology 83:441, 1982.

Rubin R, Weinstein L: Salmonellosis: Microbiologic, Pathologic, and Clinical Features. New York: Stratton International Medical Book Corp, 1977.

Sack RB: Enterotoxigenic *E. coli*: Identification and characterization. J Infect Dis 142:279, 1980.

Sack B, Sack D, et al: Enterotoxigenic *E. coli* isolated from food. J Infect Dis 135:313, 1977.

Sansonetti PJ: Enteroinvasive E. coli. In Farthing MJ, Keusch GT (eds): Enteric Infection: Mechanisms, Manifestations, and Management. New York: Raven Press, 1988.

Sarf LD, McCracken GH Jr, et al: Epidemiology of *E. coli* K1 in healthy and diseased newborns. Lancet 1:1099, 1975.

Strampfer M, Schoch P, Cunha B: Cerebral abscess caused by *Klebsiella ozaenae*. J Clin Microbiol 25:1553, 1987.

Struelens MJ, Patte D, et al: *Shigella* septicemia: Prevalence, presentation, risk factors, and outcome. J Infect Dis 152:784, 1985.

Symonds J: Haemorrhagic colitis and *Escherichia coli* O157: A pathogen unmasked. BMJ 296:875, 1988.

Taylor D, Echeverria P, et al: Clinical and microbiologic features of *Shigella* and enteroinvasive *Escherichia coli* infections detected by DNA hybridization. J Clin Microbiol 26:1362, 1988.

Wanke CA, Guerrant RL: Enterotoxigenic *E. coli*. In Farthing MJ, Keusch GT (eds): Enteric Infection: Mechanisms, Manifestations, and Management. New York: Raven Press, 1988.

Washington J: Laboratory Procedures in Clinical Microbiology, 2nd ed. New York: Springer-Verlag, 1985.

Watanakunakorn C: Acute infective endocarditis due to *Yersinia enterocolitica*. Am J Med 86:723, 1989.

Weisfeld A, McNamara A: Enterobacteriaceae. In Howard B (ed): Clinical and Pathogenic Microbiology, 2nd ed. St. Louis: CV Mosby, 1994.

Wilhelm I, Quiros BD, et al: Epidemic outbreak of *Serratia marcescens* infection in a cardiac surgery unit. J Clin Microbiol 25:1298, 1987.

Wood K, Morris JG Jr, et al: Comparison of DNA probes and the Sereny test for identification of invasive *Shigella* and *Escherichia coli* strains. J Clin Microbiol 24:498, 1986.

Zbinden R, Blass R: *Serratia plymuthica* osteomyelitis following a motorcycle accident. J Clin Microbiol 26:1409, 1988.

CHAPTER 17

Vibrio, Aeromonas, Plesiomonas, and Campylobacter

Amy M. Carnahan • *Raymond L. Kaplan*

1. Describe the general characteristics of each of the organisms reviewed in this chapter. Discuss the similarities and differences between each group of organisms.

2. Discuss the various infections associated with each organism and how the infections are acquired.

3. Describe the microscopic and colonial morphology characteristic of these species.

4. Describe how to differentiate the bacterial species that cause gastrointestinal illnesses from other diarrheal agents.

5. Describe the appropriate specimen collection, transport, and processing for maximum recovery of each of these organisms.

6. Give the selective media that are appropriate for their recovery.

7. Describe the biochemical tests that will presumptively identify these groups of organisms.

8. Name the confirmatory tests commonly used to identify these isolates.

This chapter covers agents of diarrheal diseases and other infections caused by species of *Vibrio*, *Aeromonas*, *Plesiomonas*, and *Campylobacter*. *Helicobacter pylori*, a newly recognized agent, is also discussed.

VIBRIO

The genus *Vibrio* resides in the family Vibrionaceae and encompasses more than 30 species, although to date only 12 of these species have been implicated in human infections. These microorganisms are commonly found in a wide variety of aquatic environments, including fresh water, brackish or estuarine water, and marine or salt water. Pandemics (worldwide epidemics) of cholera, a devastating diarrheal disease caused by *Vibrio cholerae*, have been documented since 1817, and we are currently in the midst of the seventh such pandemic. An eighth cholera epidemic due to a new serogroup, O/139, has recently emerged in India and is rapidly spreading. We have also witnessed a general increase in the number of reported cases of *Vibrio* infections caused by other species and originating from both gastrointestinal and extraintestinal sources. The various reasons for this significant rise in the isolation and identification of *Vibrio* clinical isolates include the following:

■ Increased travel to either coastal or cholera-endemic areas

■ Increased consumption of seafood (particularly uncooked)

■ Increased use of recreational water facilities, which encourages aquatic exposure

■ Larger populations of immunocompromised individuals

■ Increased awareness of the existence and significance of these organisms in the clinical microbiology laboratory

General Characteristics

Microscopic Morphology

Vibrio species are facultatively anaerobic, asporogenous, gram-negative rods that measure approximately 0.5 μm by 1.5 to 3.0 μm. These organisms possess polar, sheathed flagella when grown in broth, but they can produce peritrichous, unsheathed flagella when grown on solid media. They have been described classically as "curved" gram-negative rods, but this morphology is often seen only in the initial Gram stain of the clinical specimen (Fig. 17–1A and B). Vibrios usually appear as small, straight gram-negative rods, but they can be highly pleomorphic, especially under suboptimal growth conditions.

Physiology

All 12 clinically significant species are oxidase-positive and able to reduce nitrate to nitrite, except for *V. metschnikovii*. All species are generally susceptible to the vibriostatic compound

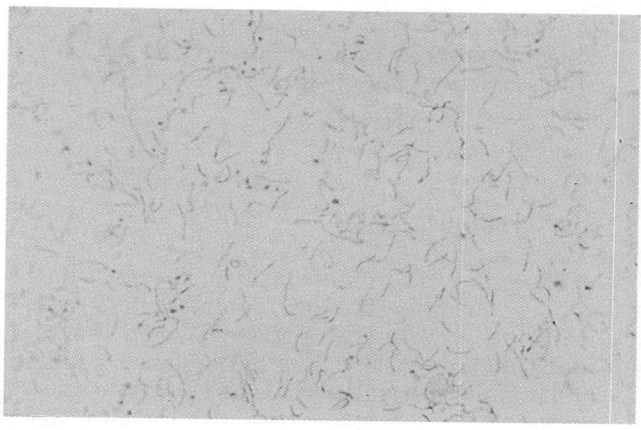

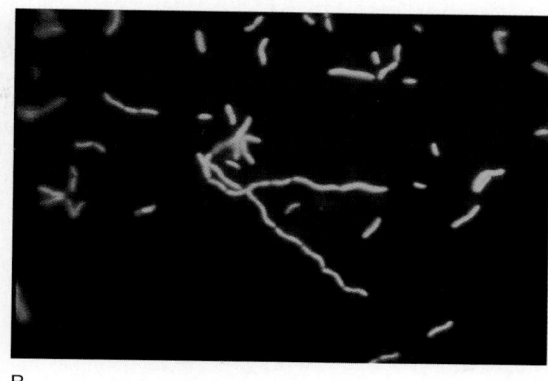

B

A

■■■■■■Figure 17-1

A, Microscopic morphology of *Vibrio* sp on Gram-stained smear. *B*, Acridine orange stain of *V. cholerae*.

O/129 (2, 4-diamino-6, 7-diisopropyl pteridine), exhibiting a zone of inhibition to a 150 µg Vibriostat disk (Oxoid, USA) on either a Mueller-Hinton or a trypticase soy agar (Fig. 17–2). Most vibrios also exhibit a positive string test that is observed as a mucoid "stringing" reaction after emulsification of colonies in 0.5% sodium desoxycholate.

All species, except for *V. cholerae* and *V. mimicus*, are halophilic, or salt-loving, and require the addition of Na⁺ for their growth and accurate identification. Vibrios can be differentiated from the closely related genera *Aeromonas* and *Plesiomonas* by means of these key biochemical and growth requirement characteristics, as shown in Table 17–1.

Antigenic Structure

Very little is known about the antigenic structure of *Vibrio* species, except for *V. cholerae*, *V. parahaemolyticus*, and some limited work on *V. fluvialis* and *V. vulnificus*. All strains of *V. cholerae* share a common flagellar (H) antigen. The cholera vibrios are further divided into six serogroups on the basis of their somatic (O) antigen. Strains of *V. cholerae* that agglutinate in O1 antisera are associated with epidemic cholera. Within the *V. cholerae* serogroup are three subtypes: Ogawa (A, B), Inaba (A, C), and Hikojima (A, B, C). Those strains that phenotypically resemble *V. cholerae* but fail to agglutinate in O1 antisera are referred to as non-O1 *Vibrio cholerae* strains.

The species of *V. parahaemolyticus* can also be serotyped by means of its O (somatic) and K (capsule) antigens, but this is generally used only in outbreaks or in large epidemiologic studies conducted by reference laboratories.

Clinical Disease Spectrum

Vibrio species can originate from a number of clinical sources, and most species have been implicated in more than one disease process, ranging from mild gastroenteritis to cholera and from simple wound infections to fatal septicemia. Table 17–2 shows the various clinical infections associated with *Vibrio* species. The four major *Vibrio* species likely to be encountered in the clinical laboratory are *V. cholerae* (O1 and non-O1), *V. parahaemolyticus*, *V. vulnificus*, and *V. alginolyticus*.

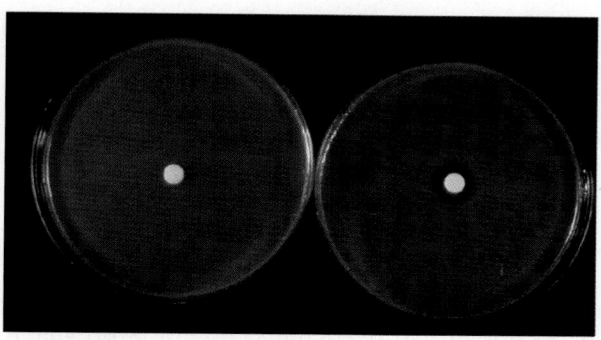

■■■■■■Figure 17-2

O/129 susceptibility test for *Vibrio* sp.

■■■■ T A B L E 1 7 – 1

SALIENT FEATURES FOR THE IDENTIFICATION OF *VIBRIO*, *AEROMONAS*, AND *PLESIOMONAS*

	Vibrio	*Aeromonas*	*Plesiomonas*
Gram reaction	–	–	–
Oxidase activity	+	+	+
Resistance to 0/129*			
10 µg	+/–	+	+/–
150 µg	–	+	–
Growth in nutrient broth with:			
0% NaCl	–/+	+	+
6.5% NaCl	+	–	–
Acid from:			
Glucose	+	+	+
Inositol	–	–	+
Mannitol	+	+/–	–
Sucrose	+/–	+/–	–
Gelatin liquefaction	+	+	–

+, Most strains positive; –, most strains negative; +/– or –/+, predominant reaction first.

*Vibriostatic agent (2, 4-diamino-6, 7-diisopropylpteridine)

Modified and reprinted by permission of the publisher from Carnahan AM: Update on *Aeromonas* identification. Clin Microbiol Newsl 13(22):169, 1991. Copyright 1991 by Elsevier Science Publishing Co., Inc.

Because most laboratories, except those in coastal areas, have a fairly low frequency of isolation of *Vibrio* species, a good medical history is extremely important. Often the best indication of a possible *Vibrio* infection is the presence of certain recognized factors, such as the following:

■ Recent consumption of raw seafood (especially oysters)

■ Recent immigration or foreign travel

■ Gastroenteritis with cholera-like or "rice-water" stools

■ Accidental trauma incurred during contact with fresh or marine water or associated products (e.g., shellfish, hooks)

Vibrio cholerae

Epidemiology

Vibrio cholerae O1 is the causative agent of cholera, also known as Asiatic cholera or epidemic cholera. It has been a disease of major public health significance for centuries. Most epidemics occur in developing countries, where it is endemic; in particular, cholera is prevalent in the Bengal region of India and in Bangladesh.

However, there has been a fairly recent increase in the number of cases of cholera in the United States, particularly off the Gulf Coast, as well as in South America.

Clinical Infection

Cholera is an acute diarrheal disease that is spread mainly through contaminated water. However, improperly preserved and handled foods, including fish and seafood, milk, ice cream, and unpreserved meat, have been responsible for outbreaks. The disease manifests in acute cases as a severe gastroenteritis, accompanied by vomiting and followed by diarrhea. The stools produced by cholera patients are described as "rice-water" stools because they are liquid and watery and contain numerous flecks of mucus. The number of stools may be as many as 10 to 30 per day. If left untreated, cholera can result in such a rapid fluid and electrolyte loss that it leads to dehydration, hypovolemic shock, metabolic acidosis, and death in a matter of hours.

■■■■ T A B L E 1 7 – 2

CLINICAL INFECTIONS ASSOCIATED WITH VIBRIOS

Species	Clinical Infection	Frequency
V. cholerae O1	Cholera, gastroenteritis, wound infections, bacteremia	Common
V. cholerae O139	Cholera	Relatively common
V. cholerae non-O1	Gastroenteritis, septicemia, ear infections	Relatively common
V. parahaemolyticus	Gastroenteritis, wound infections	Common
V. vulnificus	Septicemia, wound infections	Common
V. alginolyticus	Wound infections, ear infections, conjunctivitis, respiratory infections, bacteremia	Common
V. mimicus	Gastroenteritis, ear infections	Uncommon
V. damsela	Wound infections	Uncommon
V. fluvialis	Gastroenteritis	Uncommon
V. furnissii	Gastroenteritis	Uncommon
V. hollisae	Gastroenteritis	Uncommon
V. cincinnatiensis	Meningitis	Rare
V. metschnikovii	Septicemia, peritonitis	Rare
V. carchariae	Wound infections	Rare

Data modified from Bottone EJ, Janda JM: *Vibrio*. In Howard B, et al (eds): *Clinical and Pathogenic Microbiology*, 1st ed. St. Louis: CV Mosby, 1987.

This devastating clinical scenario is the result of a powerful enterotoxin known as cholera toxin or choleragen. Once ingested, the cholera organisms colonize the small intestine, where they multiply and produce choleragen. The toxin consists of two toxic A subunits and five multiple binding B subunits. There is an initial binding to the GM_1-ganglioside receptor on the cell membrane via the B subunits. Then the A2 subunit facilitates the entrance of the A1 subunit. Once inside the cell, the active A1 subunit stimulates the production of adenylate cyclase through the inactivation of the G protein. This leads to an accumulation of cAMP along the cell membrane, which stimulates a hypersecretion of electrolytes (NA^+, K^+, HCO_3^-) and water out of the cell and into the lumen of the intestine, as shown in Figure 17–3. The net effect is that the gastrointestinal tract's absorptive ability is overwhelmed, resulting in the massive outpouring of rice-water stools.

Treatment and management of cholera are best accomplished by the administration of copious amounts of intravenous or oral fluids to replace those lost. The administration of antibiotics can shorten the duration of diarrhea and thereby reduce fluid losses.

Epidemic *V. cholerae* O1 strains occur in two biogroups: classic and El Tor. El Tor has been the predominant biogroup in the last two pandemics. Recent studies in Bangladesh, however, indicate a rapidly occurring reemergence of the classic biogroup. The El Tor biogroup differs from the classic in that El Tor is Voges-Proskauer–positive, hemolyzes erythrocytes, is inhibited by polymyxin B (50 μg), and is able to agglutinate chicken red blood cells. The two biogroups also have different phage susceptibility patterns.

There have been a number of cases of *Vibrio* infections reported involving other serogroups, such as *V. cholerae* non-O1. These strains are phenotypically similar to toxigenic *V. cholerae* O1, but most lack the cholerae toxin gene and appear to cause a milder form of gastroenteritis or cholera-like disease. These non-O1 strains have also been implicated in a variety of extraintestinal infections, including cholecystitis, ear infections, cellulitis, and septicemia.

Vibrio parahaemolyticus

Epidemiology

Vibrio parahaemolyticus is the second most common *Vibrio* species implicated in gastroenteritis. It was first recognized as a pathogen in Japan in 1950, when it was the culprit in a large food-poisoning outbreak; even today, it is the number

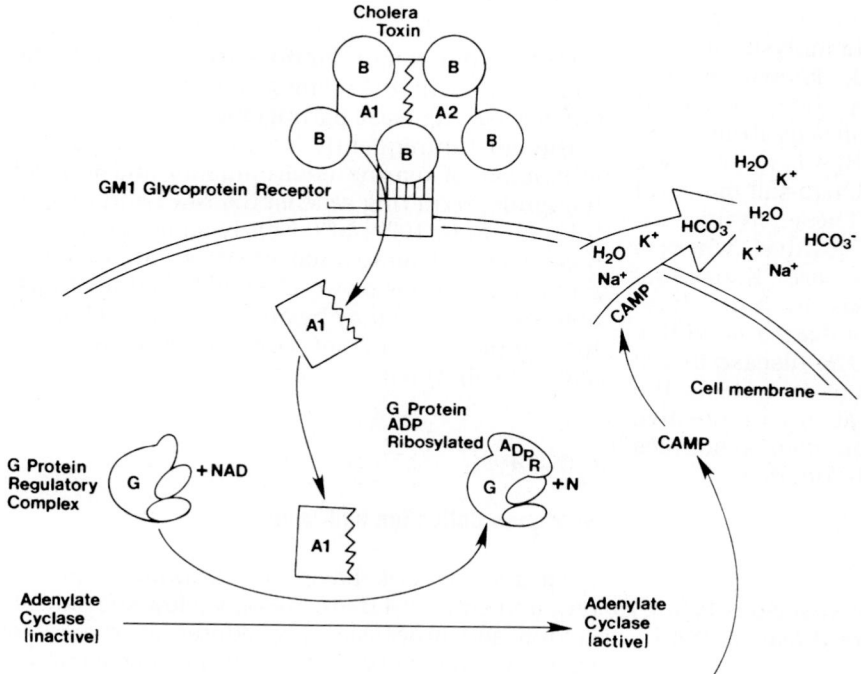

■■■■■Figure 17–3

The action of cholera toxin. The complete toxin is shown binding to the GM_1-ganglioside receptor on the cell membrane via the binding (B) subunits. The active portion (A_1) of the A subunit enters the cell and inactivates the G protein by ADP-ribosylation. Because the G protein acts to return adenylate cyclase from its active to inactive form, the net effect is persistent activation of adenylate cyclase. The increased adenylate cyclase activity results in accumulation of cyclic adenosine 3', 5'-monophosphate (cAMP) along the cell membrane. The cAMP causes the active secretion of sodium (Na^+), chloride (Cl^-), potassium (K^+), bicarbonate (HCO_3^-), and water out of the cell into the intestinal lumen. (Used with permission from Ryan KJ: *Vibrio* and *Campylobacter*. In Sherris J [ed]: Medical Microbiology: An Introduction to Infectious Diseases, 2nd ed. New York: Elsevier Science Publishing Co., 1990, p 386.)

one cause of "summer diarrhea" in Japan. It has also been isolated in Europe, the Baltic area, Australia, Africa, Canada, and nearly every coastal state in the United States. Like other vibrios, *V. parahaemolyticus* is found in aquatic environments, but it appears to be limited to coastal or estuarine areas, despite a halophilic requirement for 1 to 8% NaCl. *Vibrio parahaemolyticus* has a definite association with at least 30 different marine species, including oysters, clams, crabs, lobsters, scallops, sardines, and shrimp. Hence, most cases of gastroenteritis can be traced to recent consumption of raw, improperly cooked, or recontaminated seafood, particularly oysters.

Clinical Infection

The gastrointestinal disease caused by *V. parahaemolyticus* is generally self-limited. Patients have watery diarrhea, moderate cramps or vomiting, and little if any fever. Symptoms begin approximately 24 to 48 hours after ingestion of contaminated seafood.

Vibrio parahaemolyticus has occasionally been isolated from extraintestinal sources, such as wounds, and ear and eye infections, and even in a case of pneumonia. Invariably, the patient has a history of recent aquatic exposure or a water-associated traumatic injury to the infected site.

The pathogenesis of *V. parahaemolyticus* is not as clear-cut as in the case of *V. cholerae* and the production of choleragen. However, there is a possible association between hemolysin production and virulence potential, known as the Kanagawa phenomenon. It has been observed that most clinical *V. parahaemolyticus* strains produce a heat-stable hemolysin that is able to lyse human erthrocytes in a special high-salt mannitol medium (Wagatsuma agar). These strains are considered to be Kanagawa-positive, whereas most environmental isolates are Kanagawa-negative. There are exceptions to both these observations, however, and the exact role of this hemolysin in pathogenesis of the disease is still not understood. Also important to note is the recent emergence of atypical urease-positive *V. parahaemolyticus* strains from clinical sources along the Pacific coast of North America.

Vibrio vulnificus

After cholera, the second most serious type of Vibrio-associated infections are those caused by *V. vulnificus*.

Epidemiology

This vibrio species can be found in marine environments on the Atlantic, Gulf, and Pacific coasts of North America. Until 1976, *V. vulnificus* was commonly referred to as the lactose-positive *Vibrio*.

Clinical Infection

Infections caused by *V. vulnificus* generally fall into two categories: primary septicemia and wound infections. The former is surmised to occur via a gastrointestinal route following the consumption of shellfish, especially raw oysters. Those patients with liver dysfunction and syndromes that result in increased serum levels of iron (e.g., hemochromatosis, cirrhosis, thalassemia major, hepatitis) are particularly predisposed to this scenario. Within hours, septicemia can develop, with a mortality rate of 40 to 60%. Those patients presenting with wound infections with *V. vulnificus* invariably have a history of some type of traumatic aquatic wound that often presents as a cellulitis. The majority of such patients are usually not immunocompromised, have experienced a mild to severe injury to the infected site, and do not live in a coastal area.

Vibrio alginolyticus

Of the four major *Vibrio* species likely to be encountered in the clinical laboratory, *V. alginolyticus* is the least pathogenic for humans and is the most infrequently isolated. It is a common inhabitant of marine environments and a strict halophile, requiring at least 3% NaCl and able to tolerate up to 10% NaCl. Nearly all isolates originate from extraintestinal sources, such as eye and ear infections or wound and burn infections, and the organism may be an occupational hazard for people in constant contact with seawater, such as fishermen or sailors.

Laboratory Diagnosis

Specimen Collection and Transport

Vibrios do not have any fastidious growth requirements, and there are only a few special collection and processing procedures necessary to ensure the recovery of vibrios from clinical material. Whenever possible, body fluids, pus, or tissues should be submitted, but swabs are acceptable if they are transported in an appropriate holding medium, such as Cary-Blair, to prevent desiccation. Buffered glycerol saline is not recommended as a transport or holding medium, because the glycerol is toxic for vibrios. Even strips of blotting paper soaked in liquid stool and placed in airtight plastic bags are considered viable specimens for up to 5 weeks. Stool specimens should be collected as early as possible in the course of the illness, before the administration of any antimicrobial agents, and plated as quickly as possible.

Culture Media

The salt concentration (0.5%) in most commonly used laboratory media, such as nutrient agar or blood agar, is sufficient to support the growth of any vibrios present. On blood-containing agars, such as sheep's blood or chocolate agar, vibrios produce medium to large colonies that appear smooth, opaque, and iridescent with a greenish hue. The sheep's blood agar plate should also be examined for the presence of β, α, or γ hemolysis. On MacConkey agar, the pathogenic vibrios usually grow as non–lactose-fermenters. However, lactose-fermenting species such as *V. vulnificus* may be overlooked and incorrectly considered to be members of the Enterobacteriaceae, such as *Escherichia coli*. It is therefore imperative to determine the oxidase activity of any suspicious *Vibrio*-like colonies. This can be accomplished by either directly testing colonies from sheep's blood or chocolate agar plates with fresh oxidase reagent or subculturing any suspicious lactose-fermenting colonies on MacConkey to a fresh sheep's blood agar plate for next-day testing. This is necessary because lactose-positive colonies from selective-differential media such as MacConkey or CIN (cefsulodin-irgasin-novobiocin) agar may give false-negative oxidase reactions.

If a selective medium is warranted, either because of the clinical history (exposure to seafood or seawater) or for geographic reasons (coastal area resident or recent foreign travel), TCBS (thiosulfate citrate bile salts sucrose) agar is the most available and widely used selective medium. It differentiates sucrose-fermenting (yellow) species (Fig. 17–4) such as *V. cholerae, V. alginolyticus, V. fluvialis, V. furnissii, V. cincinnatiensis, V. metschnikovii, V. carchariae,* and some *V. vulnificus* from the non–sucrose-fermenting (green) vibrios, that is *V. mimicus, V. parahaemolyticus,*

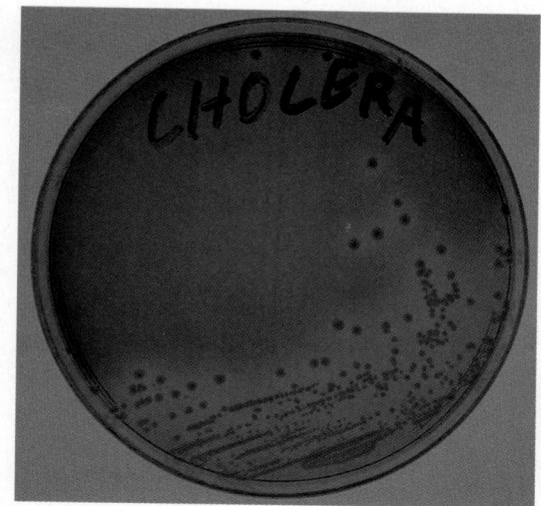

Figure 17–4

Vibrio cholerae on TCBS agar.

V. damsela, and most *V. vulnificus.* Although TCBS generally inhibits all other organisms, it should be monitored with stringent quality-control measures, as there is great variation in performance from lot to lot, and not all *Vibrio* species grow on TCBS. If an enrichment procedure is desired to enhance isolation of vibrios, alkaline peptone water with 1% NaCl (pH 8.5) can be inoculated (at least 20 mL volume) and incubated for 5 to 8 hours at 35°C before subculturing to TCBS.

Presumptive Identification

There are several key tests that can aid in the initial identification of a *Vibrio* isolate. Vibrios can be easily confused with other genera, including *Aeromonas, Plesiomonas,* Enterobacteriaceae, and even *Pseudomonas.* Their general resistance to the vibriostatic agent O/129 (150 μg) distinguishes them from *Aeromonas.* Inability to ferment inositol (except for *V. cincinnatiensis*) separates them from *Plesiomonas;* positive oxidase activity separates them from the Enterobacteriaceae; and a fermentative metabolism separates them from the oxidative *Pseudomonas* (see Table 17–1).

Definitive Identification

After the presumptive identification of *Vibrio,* there are a number of useful biochemical tests that aid in identifying the majority of clinical isolates to the species level. Kelly and colleagues recently

T A B L E 1 7 – 3

EIGHT KEY DIFFERENTIAL TESTS TO DIVIDE THE 12 CLINICALLY SIGNIFICANT *VIBRIO* SPECIES INTO SIX GROUPS

Reactions of the Species in:

	Group 1		Group 2	Group 3	Group 4	Group 5			Group 6			
Test	V. cholerae	V. mimicus	V. metschnikovii	V. cincinnatiensis	V. hollisae	V. damsela	V. fluvialis	V. furnissii	V. alginolyticus	V. parahaemolyticus	V. vulnificus	V. carchariae
Growth in nutrient broth:												
With no NaCl added	+	+	–	–	–	–	–	–	–	–	–	–
With 1% NaCl added	+	+	+	+	+	+	+	+	+	+	+	+
Oxidase	+	+	–	+	+	+	+	+	+	+	+	+
Nitrate → nitrite	+	+	–	+	+	+	+	+	+	+	+	–
myo-Inositol fermentation	–	–	–	+	–	–	–	–	–	–	–	–
Arginine dihydrolase	–	–			–	+	+	+	–	–	–	–
Lysine decarboxylase	+	+			–	+			+	+	+	+
Ornithine decarboxylase	+	+			–							

Key test results are boxed.
Used with permission from Kelly MT, Hickman-Brenner FW, Farmer JJ III: *Vibrio.* In Balows A, et al (eds): Manual of Clinical Microbiology, 5th ed. Washington, DC: American Society for Clinical Microbiology, 1991, p 389.

■■■■■■ T A B L E   1 7 – 4
DIFFERENTIATION OF THE FOUR MAJOR CLINICAL *VIBRIO* SPECIES

Test or Property	Reaction				
	V. cholerae (Classical)	*V. cholerae* (El Tor)	*V. parahaemolyticus*	*V. vulnificus*	*V. alginolyticus*
Growth in nutrient broth with:					
0% added NaCl	+	+	–	–	–
3% NaCl	+	+	+	+	+
6% NaCl	–	–	+	+	+
8% NaCl	–	–	+	–	+
10% NaCl	–	–	–	–	+
Sucrose fermentation	+	+	–	–	+
Lactose fermentation	–	–	–	+	–
Voges-Proskauer	–	+	–	–	+

+, Most strains positive; –, most strains negative.

outlined eight key differential tests to divide the 12 clinically significant *Vibrio* species into six groups as an initial identification step (Table 17–3). Some of the additional tests necessary to identify the four major clinical *Vibrio* species are summarized in Table 17–4. However, it is important to note that with the halophilic or salt-loving vibrios, it is often necessary to add at least 1% NaCl to most biochemical media to obtain reliable reaction results.

Rapid or Semiautomated Identification Systems. Although these systems contain databases of the more commonly encountered clinical vibrios, they are generally inadequate for the accurate identification of these species, particularly the less common species. In particular, their inoculating suspensions should contain at least 0.85% NaCl, such as that found with the API-20E identification strips (BioMerieux Vitek, Inc., Hazelwood, MO). Even then, the halophilic vibrios may grow poorly if at all, and they are often confused with other genera, such as *Aeromonas*. Most authors advocate identification with conventional biochemicals, supplemented with additional NaCl where warranted; alternatively, the rapid-system identification should be confirmed at the reference laboratory level.

Serology. It is generally considered sufficient for most clinical laboratories below the level of reference laboratory to simply screen their presumptive *V. cholerae* isolates with commercially available polyvalent O1 antiserum. However, there have been reports of non-O1 *V. cholerae* misidentified as *V. cholerae* O1 and vice versa. In any case, all isolates with a presumptive identification of *V. cholerae* should be promptly reported to the appropriate public health authorities, and the isolate should be forwarded to the health department and/or a reference laboratory for serogrouping and cholera toxin testing, if warranted.

Antimicrobial Susceptibility

Fortunately, both Mueller-Hinton agar and broth contain sufficient salt to support the growth of the *Vibrio* species most often isolated from clinical specimens. The recommended methods to be employed are standardized disk diffusion (Bauer-Kirby) or dilution susceptibility testing methods.

In general, most strains of *V. cholerae* are susceptible to tetracycline, gentamicin, chloramphenicol, and nalidixic acid, with tetracycline being the drug of choice. Increased resistance from the acquisition of plasmids is relatively uncommon in the United States, but there are reports of multiply resistant strains from both Asia and Africa.

AEROMONAS

The genus *Aeromonas* consists of ubiquitous oxidase-positive, glucose-fermenting, gram-negative rods that are widely distributed in freshwater, estuarine, and marine environments worldwide. They are frequently isolated from retail produce sources and animal meat products. Aeromonads are responsible for a diverse spectrum of disease syndromes among a variety of warm- and cold-blooded animals, including fish, reptiles, amphibia, mammals, and humans.

Until recently, they resided in the family Vibrionaceae. However, phylogenetic evidence from molecular studies resulted in the proposal of a separate family Aeromonadaceae. There are currently 10 proposed species of *Aeromonas*, but only seven species have been isolated with any frequency from clinical specimens (see Table 17–5).

General Characteristics

Aeromonads are straight, not curved, rods (1.1–4.4 μm by 0.4–1.0 μm), and most are motile by means of a single polar flagellum. To date, two species, *A. salmonicida* and *A. media*, are generally considered nonmotile and not pathogenic for humans. The range of temperature for growth of all aeromonads is from 0°C to 45°C, with a seasonal pattern of increased isolation from May through October. Most clinical mesophilic strains can be cultured from 4°C to 42°C, and the psychrophilic fish pathogen *A. salmonicida* usually grows only below 37°C.

Antigenic Structure. There are currently two serotyping systems developed for mesophilic aeromonads based on the somatic O antigens. The first serotyping schema was developed by Sakasaki and Shimada and established 45 groups among the mesophilic species. A later study by Cheasty et al. established 10 additional serogroups, but at present, serotyping is recommended only at the reference laboratory level for epidemiologic purposes.

Clinical Infection

Gastroenteritis. Although aeromonads have been epidemiologically implicated as causative agents of enteric disease since the 1960's, they are still not firmly established as enteric pathogens. Despite numerous case-control studies and case reports, we still lack an animal model to fulfill Koch's postulates and are awaiting a successful human volunteer trial. Nevertheless, there are sufficient clinical associations to warrant the screening of stool specimens for the presence of aeromonads, followed by further identification to the level of species. The medical history of patients displaying diarrhea and harboring aeromonads often, but not always, involves aquatic exposure, such as an association with untreated well water or consumption of seafood, particularly raw oysters or clams.

There are currently five diarrheal presentations observed in patients in whom *Aeromonas* has been isolated from their stools:

1. An acute, secretory diarrhea often accompanied by vomiting
2. An acute, dysenteric form of diarrhea with blood and mucus
3. A chronic diarrhea usually lasting more than 10 days
4. A choleraic type that includes "rice-water" stools
5. The nebulous syndrome commonly referred to as traveler's diarrhea

Most cases are self-limiting, but in the pediatric and geriatric populations, supportive therapy and antimicrobials are often indicated.

The species *A. hydrophila*, *A. veronii* (biovars sobria and veronii), and *A. trota* are most strongly associated with cases of gastroenteritis to date, with a possible pathogenic role for *A. caviae* among pediatric patients.

Wound Infection. This is the second most common type of *Aeromonas* infection. It invariably involves a recent traumatic aquatic exposure and generally occurs on the extremities. The most common presentation is cellulitis, although there are a few instances of myonecrosis with or without gas gangrene, and even a rare case of ecthyma gangrenosum associated with sepsis. Most aeromonad wound isolates are *A. hydrophila*, *A. veronii* biovar veronii, or *A schubertii*. The latter species has been isolated almost exclusively from aquatic wound infections and blood.

The most interesting association between aeromonads and wound infections involves the species *A. hydrophila* and the use of leeches for medicinal therapy following plastic surgery. These patients end up with serious aeromonad wound infections following the use of leech therapy to relieve venous congestion. It appears that the leech, *Hirudo medicinalis*, has a symbiotic relationship with the aeromonads within its gut, where the organisms aid in the enzymatic digestion of the blood ingested by the leech.

Septicemia. Aeromonad sepsis appears to be the most invasive type of *Aeromonas* infection and has a strong association with the species *A. veronii* biovar sobria (formerly identified as *A. sobria* in clinical isolates). Such patients are most likely to be immunocompromised, with a history of liver disease or dysfunction, hematologic malignancies, hepatobiliary disorders, or traumatic injuries. Also at high risk are individuals diagnosed with leukemia, lymphoma, or myeloma. Although the original source of infection has not been elucidated, it is surmised that it is the gastrointestinal tract, the biliary tract, or, more rarely, the respiratory tract.

Miscellaneous Extraintestinal Infections. Aeromonads have also been implicated in cases of osteomyelitis, meningitis, pelvic abscesses, otitis, cystitis, endocarditis, peritonitis, cholecystitis, and endophthalmitis in both healthy and immunocompromised individuals.

Laboratory Diagnosis

Culture Media

Aeromonads grow quite readily on most media used for both routine cultures and stool cultures. After 24-hour incubation at 35–37°C, aeromonads appear as large round, raised, opaque colonies with an entire edge and a smooth, often mucoid surface. Often, an extremely strong odor is present, and pigmentation ranges from translucent and white to buff in color. Hemolysis is variable on blood agar media, with most species displaying β hemolysis. Although aeromonads grow on nearly all enteric media, they are often overlooked on MacConkey agar because a large number of aeromonads ferment lactose. A study by Kelly and colleagues of several media suggests that the combined use of ampicillin blood agar and a modified CIN agar may yield the highest recovery of aeromonads. However, the incorporation of ampicillin in the blood agar may inhibit some *A. caviae* strains as well as the newly proposed species, *A. trota*, whose hallmark is susceptibility to ampicillin. The use of an enrichment broth is not generally considered necessary. However, if such a medium is warranted for detecting chronic cases or asymptomatic carriers, alkaline peptone water is recommended. This can be inoculated, incubated overnight, and subsequently subcultured to plate media.

Presumptive Identification

Because the genera *Aeromonas*, *Plesiomonas*, and *Vibrio* are so closely related, it is necessary to first determine that an isolate is indeed an aeromonad (see Table 17–1). The simplest way of screening for aeromonads is to perform a test for oxidase activity on growth from a blood agar plate. This distinguishes aeromonads from Enterobacteriaceae. Other advantages to the use of a blood agar medium are the simultaneously ability to test for indole production and observe for any hemolysis. The presence and type of hemolysis are often the only clues to an infection involving more than one species of *Aeromonas*. To distinguish the aeromonads from *Vibrio* and *Plesiomonas*, testing for sensitivity to the vibriostatic agent O/129 is recommended in lieu of a salt tolerance, since *V. cholerae* and *V. mimicus* grow quite well without additional salt. Finally, the ability to ferment glucose distinguishes *Aeromonas* from *Pseudomonas*.

Definitive Identification

Definitive identification is accomplished with a small number of biochemical tests and antimicrobial markers using a dichotomous key, Aerokey II (Fig. 17–5). When used in conjunction with the additional tests from Table 17–5, the clinical microbiologist should be able to identify nearly all *Aeromonas* isolates to the level of species. However, it should be noted that the esculin hydrolysis test requires the use of an agar-based medium, not a broth. Further, any antimicrobial resistance markers and susceptibility studies should be determined by the standard Bauer-Kirby method, because of possible discrepancies in β-lactamase detection by rapid minimal inhibitory concentration (MIC) methods.

Rapid or Semiautomated Identification Systems. Although these systems can identify an isolate as belonging to the *Aeromonas hydrophila* or *A. hydrophila* complex, the majority (with the exception of API-20E) are currently inadequate for identification to the species level. This is due to a lack of sufficient discriminatory markers to detect interspecies differences as well as poor correlation between conventional test results and rapid or miniaturized versions, specifically esculin hydrolysis, decarboxylase reactions, and sugar fermentation.

The presence of species-related disease syndromes (e.g., *A. hydrophila* and *A. schubertii* from aquatic wounds, *A. veronii* biovar sobria from septicemia, *A. caviae* from pediatric diarrhea, and *A. hydrophila* from medicinal leech therapy cases), coupled with differences in antimicrobial susceptibilities among the species, strongly

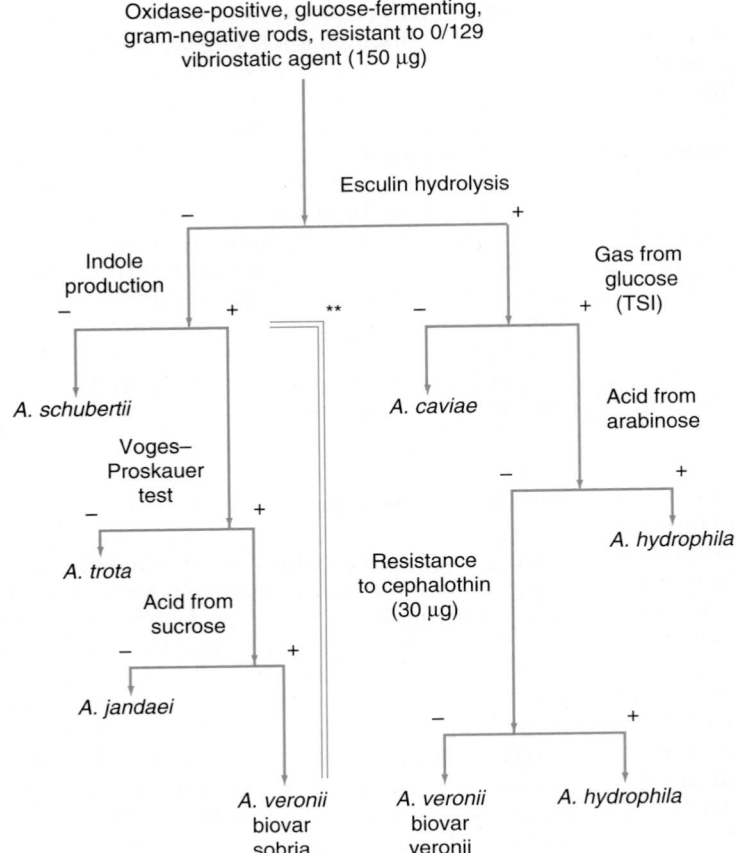

** Aerokey II can be modified to end here with an identification of *A. veronii* biovar sobria

Figure 17-5

Aerokey II—identification key for clinical *Aeromonas* species. (Modified and redrawn with permission from Carnahan AM, Behram S, Joseph SW: Aerokey II—a flexible key for identifying clinical *Aeromonas* species. J Clin Microbiol 29:2843, 1991.)

suggests that conventional identification and antimicrobial susceptibilities should be performed on clinical aeromonad isolates.

Antimicrobial Susceptibility

Although most cases of *Aeromonas*-associated gastroenteritis are self-limited, antimicrobial therapy is often indicated. It is always warranted in wound infection and septicemia. As a genus, aeromonads are nearly uniformly resistant to penicillin, ampicillin, carbenicillin, and cephalothin. Exceptions are the susceptibility of *A. veronii* biovar sobria to cephalothin and the susceptibility of *A. trota* to ampicillin. Otherwise, aeromonads are generally susceptible to third-generation cephalosporins, trimethoprim-

sufamethoxazole, aminoglycosides, chloramphenicol, and quinolones.

PLESIOMONAS

The genus *Plesiomonas* currently resides in the family Vibrionaceae and includes a single species, *Plesiomonas shigelloides*. Like the vibrios and aeromonads, these organisms are oxidase-positive, glucose-fermenting, facultatively anaerobic gram-negative rods that are motile by polar flagella. However, recent phylogenetic studies have presented evidence that the ancestry of *Plesiomonas* is actually closer to the family Enterobacteriaceae, particularly the genus *Proteus*, and it has been proposed that *Plesiomonas* be moved to this genus.

■■■■■ T A B L E 1 7 – 5

DIFFERENTIAL CHARACTERISTICS FOR MESOPHILIC CLINICAL *AEROMONAS* SPECIES

Characteristic	A. hydrophila	A. veronii biogroup sobria	A. veronii biogroup veronii	A. caviae	A. schubertii	A. jandaei	A. trota
Esculin hydrolysis	+	–	+	+	–	–	–
Voges-Proskauer	+	+	+	–	V	+	–
Pyrazinamidase activity	+	–	–	+	–	–	–
Arginine dihydrolase	+	+	–	V	+	+	+
Fermentation:							
Arabinose	V	–	–	+	–	–	–
Cellobiose	–	–	+	V	–	–	+
Mannitol	+	+	+	+	–	+	+
Sucrose	+	+	+	+	–	–	–
Susceptibility:							
Ampicillin	R	R	R	R	R	R	S
Carbenicillin	R	R	R	R	R	R	S
Cephalothin	R	S	S	R	S	R	R
Colistin*	V	S	S	S	S	R	S
Decarboxylase:							
Lysine	+	+	+	–	+	+	+
Ornithine	–	–	+	–	–	–	–
Indole	+	+	+	+	–	+	+
H₂S†	+	+	+	–	–	+	+
Glucose (gas)	+	+	+	–	–	+	+
Hemolysis (5% sheep erythrocytes)	+	+	+	V	+	+	V

+ = positive, – = negative, V = variable, R = resistant, S = susceptible.
*MIC single dilution 4 µg/mL.
†H₂S from gelatin-cysteine-thiosulphate media.
Modified with permission from Carnahan AM, et al: *Aeromonas trota* sp. nov., an ampicillin-susceptible species isolated from clinical specimens. J Clin Microbiol 29(6):1209, 1991.

Epidemiology

These microorganisms are found in both soil and aquatic environments, but because of an intolerance to increased NaCl and a minimum growth temperature of 8°C, they are generally found only in the fresh and estuarine waters of tropical and subtropical climates. Like the genus *Aeromonas,* they are widely distributed among both warm- and cold-blooded animals, including dogs, cats, pigs, vultures, snakes, lizards, fish, newts, and shellfish. In recent years, they have emerged as a potential cause of enteric disease in humans and have also been isolated from a number of extraintestinal infections. Cases are probably underreported because of the organism's similarity to *E. coli* on most ordinary enteric media. But increased laboratory awareness of the existence of *Plesiomonas,* coupled with the knowledge of previously outlined recreational, immunologic, and gastronomic risk factors, has resulted in a gradual but steady increase in the number of reported cases, mostly among the adult population.

General Characteristics

Microscopic Morphology

Plesiomonads are straight (0.8–1 µm by 3 µm) gram-negative rods that can occur singly, in pairs, or even in short chains or filamentous forms. They do not form spores or capsules and are motile by monotrichous or two to five lophotrichous polar flagella.

Antigenic Structure

The genera *Plesiomonas* and *Shigella* share both biochemical and antigenic features, and plesiomonads often cross-agglutinate with *Shigella sonnei, S. dysenteriae,* and even *S. boydii,* hence the species name *shigelloides.* However, unlike *Shigella, Plesiomonas* appears to possess a much lower virulence potential, with a low symptomatic carriage rate among humans. Plesiomonads can be serotyped by their somatic O antigens (50 groups) and their flagellar H antigens

(17 groups), based on a schema by Aldova. Some of the 107 serovars are ubiquitous, and others are confined only to certain regions.

Clinical Infections

Gastroenteritis. Unlike the genus *Aeromonas,* there have been at least two well-documented outbreaks of diarrheal disease in Japan caused by *Plesiomonas.* The most common vehicle of transmission is the ingestion of contaminated water or food, particularly uncooked seafood such as oysters or shrimp. However, Koch's postulates have not yet been fulfilled with a suitable animal model, and only a few possible virulence features have been defined to explain this organism's association with enteric disease. There are at least three major clinical types of gastroenteritis caused by *Plesiomonas:*

1. The more common watery or secretory diarrhea
2. A second subacute or chronic disease that lasts between 14 days and 2 to 3 months
3. A more invasive, dysenteric form that resembles colitis

On the average, 25 to 40% of all patients present with fever and/or vomiting, and the single most common clinical symptom for all such patients is abdominal pain. Most cases are self-limited, but antimicrobial therapy is indicated in severe and prolonged cases.

Extraintestinal Infection. Due to the organism's wide dissemination among the animal population, it is readily apparent that occupational exposure can be a source of such infections for veterinarians, zookeepers, aquaculturists, fish handlers, and athletes participating in water-related sports. Finally, there is the factor of the immune status of the patient, with the more serious infections such as bacteremia and meningitis usually occurring only in severely immunocompromised patients or neonates.

Laboratory Diagnosis

Culture Media

Plesiomonas grows quite readily on most media routinely used in the clinical laboratory. After 18 to 24 hours incubation at 35°C, shiny, opaque, nonhemolytic colonies appear, with a slightly raised center and a smooth and entire edge. However, since the majority of strains are lactose-fermenters, the easiest screening procedure is an oxidase test performed on colonies from nonselective media such as sheep's blood agar or chocolate agar. Although a specialized medium is not recommended for the detection of plesiomonads from stool specimens, because certain strains are inhibited on EMB (eosin–methylene blue) or MacConkey agar, IBB (inositol brilliant green bile salts) agar can be employed to enhance their isolation. Plesiomonad colonies are white to pink on this medium, and most coliform colonies are either green or pink. In addition, the agar used should not contain ampicillin, since plesiomonads are often susceptible to ampicillin and will therefore be inhibited.

Identification

Plesiomonas can be presumptively differentiated from nearby genera with several key tests (see Table 17–1). Their positive oxidase activity separates them from the Enterobacteriaceae, their sensitivity to the vibriostatic agent O/129 (150 μg) separates them from *Aeromonas,* and their ability to ferment glucose separates them from all *Aeromonas* and nearly all *Vibrio* species.

Most rapid identification systems include *Plesiomonas shigelloides* in their databases and appear to be able to identify it with a fairly high degree of accuracy. This is in large part due to its unique profile of positive ornithine and lysine decarboxylases and arginine dihydrolase reactions, combined with the fermentation of inositol.

As previously mentioned, there is a serotyping schema for *Plesiomonas,* although this is employed when an isolate is referred to the reference laboratory. What is important to reinforce is the possibility of cross-reactivity with *Shigella* antigens, particularly in some rapid serotyping kits, and the absolute necessity of an initial determination of an isolate's oxidase reaction from noninhibitory media.

Antimicrobial Susceptibility

Although most cases of plesiomonad gastroenteritis are self-limited, antimicrobial therapy is indicated in patients with severe or chronic gastroenteritis. Likewise, extraintestinal infections, particularly among neonates, often require antimicrobial therapy. Studies have shown a general resistance to the penicillin class of antibiotics, but penicillins combined with a β-lactamase inhibitor as well as trimethoprim-sulfamethoxazole are active. There are reports of resistance to more than one aminoglycoside (e.g., gentamicin, tobramycin, and amikacin), but the new quinolones appear to be an effective therapy.

CAMPYLOBACTER AND *CAMPYLOBACTER*-LIKE ORGANISMS

Campylobacter species and *Campylobacter*-like organisms, which include *Helicobacter* and *Wolinella,* have recently undergone certain changes in taxonomy. Based on RNA (rRNA) sequence studies *W. recta* and *W. curva* have been found to be similar to the campylobacters; although they may appear to be strict anaerobes, they have grown under a microaerophilic environment. Hence, *W. recta* and *W. curva* have been transferred to the genus *Campylobacter* as *C. rectus* and *C. curvus.* As a result of these hydridization studies, a new genus, *Arcobacter,* and a new family, Campylobacteraceae, have been proposed. A few species have been designated in the new genera *Helicobacter* and *Arcobacter.* These recent changes in taxonomic status are shown in Table 17–6. This table also includes other *Campylobacter* species that have been identified, most of which are rarely isolated from human specimens.

Epidemiology

Campylobacter species have been known to cause abortion in domestic animals, such as cattle, sheep, and swine. Although these organisms were suspected of causing human infections earlier, campylobacters were not established as human pathogens until sensitive isolation procedures and isolation media were developed.

Today, the most common cause of bacterial gastroenteritis worldwide is *Campylobacter jejuni.* The transmission of campylobacterioses has been attributed to direct contact by exposure to animals and handling infected pets, such as dogs, cats, and birds, and indirectly by the consumption of contaminated water and dairy products and improperly cooked poultry. Person-to-person transmission has also been reported. *Campylobacter* species are also sexually transmitted.

■■■■ T A B L E 1 7 – 6

MEMBERS OF THE FAMILY CAMPYLOBACTERACEAE AND OTHER *CAMPYLOBACTER*-LIKE ORGANISMS

Current Designation	Previous Designation and Synonyms
Campylobacter jejuni ssp *jejuni*	*Campylobacter fetus* ssp *jejuni*
C. jejuni ssp *doylei*	*C. fetus* ssp *jejuni*
C. coli	*C. fetus* ssp *jejuni*
C. lari	*C. laridis*
C. fetus ssp *fetus*	*C. fetus* ssp *intestinalis*
C. fetus ssp *venerealis*	*C. fetus* ssp *fetus*
C. hyointestinalis	None
C. sputorum biovar *sputorum*	*C. sputorum* ssp *sputorum*
C. sputorum biovar *bubulus*	*C. sputorum* ssp *bubulus*
C. sputorum biovar *fecalis*	*C. fecalis*
C. upsaliensis	None
C. mucosalis	*C. sputorum* ssp *mucosalis*
C. concisus	None
C. curvus	*Wolinella curva*
C. rectus	*W. recta*
Arcobacter cryaerophilus	*C. cryaerophila*
A. nitrofigilis	*C. nitrofigilis*
A. butzleri	*C. butzleri*
Helicobacter pylori	*C. pylori*
H. mustelae	*C. pylori* ssp *mustelae*
H. felis	None
H. muridarum	None
H. fennelliae	*C. fennelliae*
H. cinaedi	*C. cinaedi*

Modified from Kaplan RL, Weissfeld AS: *Campylobacter, Helicobacter,* and related organisms. In Howard B, et al (eds): Clinical and Pathogenic Microbiology. St.Louis: CV Mosby, 1994, pp 453–460. Reprinted with permission.

Although the first reported cases of *Campylobacter* gastroenteritis in humans involved primarily children, later investigations showed that the diarrheal disease also occurs in adults. The population that most often manifests the disease includes children younger than 1 year of age and adults between 20 and 29 years of age. Among the other *Campylobacter* species that cause gastrointestinal disease (enteric campylobacters) are *C. coli* and *C. lari.*

Campylobacter fetus ssp *fetus* has been isolated most frequently from blood cultures and is rarely associated with gastrointestinal illness. The majority of infections occur among immunocompromised and elderly patients.

Other *Campylobacter* species have not been associated with human disease or recovered from human samples. Table 17–7 summarizes the clinical significance of *Campylobacter, Arcobacter,* and *Helicobacter* species.

Helicobacter pylori. Until recently, *H. pylori* was an unnamed organism. Within the past few years, it has been strongly associated with

■ T A B L E 1 7 - 7

CAMPYLOBACTER SPECIES AND THEIR CLINICAL SIGNIFICANCE

Campylobacter Species	Clinical Significance
C. fetus ssp *fetus*	Bacteremia in immunocompromised patients
C. fetus ssp *venerealis* *C. sputorum* biovar *bubulus* *C. sputorum* biovar *fecalis* *C. mucosalis* *A. nitrofigilis* *H. mustelae* *H. muridarum* *H. felis*	Rarely involved in human infections
H. pylori	Common cause of duodenal ulcers and type B gastritis; possibly a risk factor in gastric carcinoma
C. hyointestinalis	Enteric disease in swine; occasionally associated in human enteric illness
C. lari	Enteritis very similar to that caused by *C. jejuni*
C. upsaliensis	Potential pathogen in humans, causing gastrointestinal illness and bacteremia in both immunocompetent and immunocompromised patients
C. concisus	Involved in periodontal disease; has also been recovered from individuals with gastrointestinal illness
C. rectus	Associated with periodontal disease; has been recovered from patients with root canal infections and Crohn's disease
C. jejuni	Most common cause of bacterial diarrhea worldwide
H. fennelliae	Recovered from homosexual men presenting with proctitis, proctocolitis, and enteritis
H. cinaedi	Recovered from blood of homosexual patients with AIDS and from blood and feces of children and adult females
A. butzleri	Associated with diarrheal disease in humans and in children with recurring gastrointestinal illness (abdominal cramps)

gastric and duodenal ulcers. The significance of these organisms had been questioned as early as the beginning of the century, and although the organisms were previously found in human gastric tissue, it was difficult to assess their significance because the samples were taken at autopsy. Therefore, it was not apparent whether the organisms were present before the patients died.

Currently, *H. pylori* is recognized as the major cause of chronic superficial gastritis (type B gastritis). *H. pylori* has been identified in more than 75% of gastric ulcer patients.

H. pylori has been reported to colonize 20 to 40% of the adult population in the United States. In developing countries in Africa, Asia, and South America, the incidence is reported to be as high as 80 to 100%. This greater incidence is attributed to poor sanitary conditions. The exact means of transmission have not been identified and there is no human reservoir for this organism.

General Characteristics

Campylobacter species are non–spore-forming, curved, gram-negative rods, showing an S-shaped, "seagull-wing" appearance. These organisms are oxidase-positive and motile, using a single polar flagellum. They exhibit a characteristic "darting" motility on hanging drop preparations or when visualized under phase-contrast microscopy. Campylobacters require selective media and a microaerophilic environment for growth and isolation.

Campylobacters were formerly classified with the vibrios because of their characteristic microscopic morphology, but DNA homology studies have shown that *Campylobacter* species do not belong with the vibrios. In addition, unlike the vibrios, which are fermentative, campylobacters are nonfermentative. *Bergey's Manual* lists five different species for the genus *Campylobacter*: *C. fetus*, *C. jejuni*, *C. coli*, *C. sputorum*, and *C. concisus*. *Campylobacter fetus* contains two subspecies: *C. fetus* ssp *fetus* and *C. fetus* ssp *venerealis*. Formerly designated as subspecies of *C. fetus*, *C. jejuni* and *C. coli* are now recognized as separate species. In recent years, several other *Campylobacter* species have been described.

C. pyloridis or *C. pylori* was first referred to as a *Campylobacter*-like organism, but *Helicobacter pylori* is now the designated nomenclature for this newly recognized organism. The genus *Helicobacter* includes other renamed *Campylobacter* species as well as newly described organisms (see Table 17–6).

Clinical Infection

Patients infected with *Campylobacter jejuni* present with a diarrheal disease that begins with mild abdominal pain within 2 to 10 days after ingestion of the organisms. Cramps and bloody diarrhea may follow the initial presentation. Patients may experience fever and chills and, rarely, nausea and vomiting. In most patients, the illness is self-limited and usually resolves in 2 to 6 days. Untreated patients may remain carriers for several months. Other enteric *Campylobacter* infections (i.e., those caused by *C. coli* and *C. lari*) present with similar clinical manifestations.

Helicobacter pylori. Once acquired, *H. pylori* colonizes the stomach for a long time. Unlike other gastrointestinal pathogens, *H. pylori* causes a low-grade inflammatory process, producing a chronic superficial gastritis. Although it does not invade the gastric epithelium, the infection is recognized by the host immune system, which initiates an antibody response. The antibodies produced are not protective, however.

H. pylori is also now recognized as a major cause of type B gastritis, a condition formerly associated primarily with stress and chemical irritants. In addition, based on recent data, the strong association between long-term *H. pylori* infection and gastric cancer has raised more questions regarding the clinical significance of this organism. There is speculation that long-term *H. pylori* infection that leads to chronic gastritis is an important risk factor for gastric carcinoma.

Laboratory Diagnosis

Specimen Collection and Transport

C. fetus ssp *fetus* may be recovered from several routine blood culture media. *Campylobacter* species that cause enteric illness are isolated from stool samples and rectal swabs. If delay in processing the stool is anticipated, it can be placed in a transport medium such as Cary-Blair to maintain the viability of the organisms. A common stool transport medium, buffered glycerol-saline, is toxic to enteric campylobacters and should therefore be avoided.

Helicobacter pylori may be recovered from gastric biopsy materials. Samples must be transported quickly to the laboratory. Stuart medium can be used to maintain the viability of the organisms if delay in processing is anticipated. Tissue samples may also be placed in cysteine Brucella broth with 20% glycerol and frozen at −70°C.

Culture Media

An enriched selective agar, Campy BAP (blood agar plate), is the most commonly used medium in the United States to isolate *C. jejuni* and other enteric campylobacters. This commercially available medium contains Brucella agar base, 10% sheep's blood, and a combination of antimicrobials: vancomycin, trimethoprim, polymyxin B, amphotericin B, and cephalothin. Other selective media that have been successful in recovering *Campylobacter* species are Butzler medium and Skirrow medium. Table 17–8 shows the composition of each of these selective media.

Medium V, a modification of the original Butzler medium, contains cefoperazone, rifampin, colistin, and amphotericin B; it seems to inhibit normal colon flora better than the original formulation. CVA medium has been reported to provide better suppression of fecal flora, even when this medium is incubated at 37°C. Incubation at 37°C allows the recovery of *Campylobacter* species that are inhibited at 42°C.

C. fetus ssp *fetus* and *C. rectus* and *C. curvus* can be isolated using routine culture media.

To recover *H. pylori,* a combination of a nonselective medium, such as chocolate agar, and a selective medium, such as Skirrow, may be used. It is important that the inoculated medium be fresh and moist and that the culture be incubated in an environment with increased humidity.

Incubation. There is a double purpose for incubating stool cultures at 42°C to recover *C.*

■ TABLE 17–8

SELECTIVE MEDIA FOR THE CULTIVATION OF *CAMPYLOBACTER* SPECIES

Medium	Base	Antimicrobial Agent
Campy BAP	Brucella agar 10% sheep red blood cells	Vancomycin Trimethoprim Polymyxin B Amphotericin B Cephalothin
Skirrow	Oxoid blood agar base Lysed, defibrinated horse red blood cells	Vancomycin Trimethoprim Polymyxin B
Butzler	Thioglycolate fluid with agar added 10% sheep red blood cells	Bacitracin Novobiocin Actidione Colistin Cefazolin

jejuni. First, *C. jejuni* and other enteric campylobacters grow optimally at 42°C. Second, growth of colon organisms is inhibited at this higher temperature. *C. fetus* ssp *fetus,* on the other hand, is a rare stool isolate, and growth is suppressed at 42°C; therefore, to isolate this organism, media should be incubated at 37°C.

Enteric *Campylobacter* species require a microaerophilic and capnophilic environment. The ideal atmospheric environment for these organisms contains a gas mixture of 5% to 10% O_2 and 10% CO_2. Except for *C. rectus* and *C. curvus,* a strict anaerobic environment does not support the growth of most *Campylobacter* species.

Any of the methods described below can be used to obtain the required environment for campylobacters.

With the CAMPY PAK II system (Becton Dickinson Microbiology Systems, Cockeysville, MD), plates are placed in a jar (usually one similar to that used for anaerobic cultures), and a gas-generating envelope is activated when water is added. Once the envelope is activated, it takes approximately an hour to achieve the ideal atmospheric environment. Gas Generating Kit System BR56 (Oxoid, USA) is similar to CAMPY PAK II.

An evacuation replacement system similar to that used to obtain a strict anaerobic condition may also be used. The anaerobic jar is evacuated to a pressure of 15 inches Hg at least twice and refilled each time with one of the following gas mixtures: 10% CO_2, 90% N_2; 5% CO_2, 10% H_2, 85% N_2; or 10% CO_2, 10% H_2, 80% N_2.

The BioBag Type Cfj (Becton Dickinson) consists of a single plate bag and a generator ampule. The inoculated plate is placed in the bag and heat-sealed. Once the generator ampule is crushed, it provides an atmosphere of 5% to 10% O_2 and 8% to 10% CO_2. The plate can be examined at any time without being removed from the bag.

A candle jar may be used if none of the above-mentioned systems can be obtained. However, this method provides the least ideal environmental condition. The incubation time should be extended to 72 hours to more efficiently isolate enteric *Campylobacter* species. This procedure allows facultative organisms present on the medium to reduce the O_2 tension created by the candle to a more suitable concentration for campylobacters.

Another method that relies on the presence of facultative organisms such as *Proteus* and *E. coli* to reduce the oxygen content in the environment is the Fortner principle. In this method, a Campy BAP is inoculated and placed in a bag with a plate inoculated with the facultative organism. It is necessary to incubate the plates for at least 72 hours to allow growth of *Campylobacter* species, which does not take place until the proper environment is achieved.

Other systems such as Kaplan's Poly Bag system and Bacti-Gas Station (Scott Laboratories, Inc., Kiskeville, RI) may also be considered as alternatives. The inexpensive Poly Bag system holds up to eight inoculated plates in one bag. The bag is charged with the gas mixture (5% O_2, 10% CO_2, 85% N_2) several times and then tied off with a rubber band. The Bacti-Gas Station comes with reusable environmental bags and a gas cylinder to provide the gas mixture.

Presumptive Identification

Microscopic Morphology. *Campylobacter* species are curved, gram-negative rods that measure approximately 0.2 to 0.5 µm by 0.5 to 5.0 µm. On Gram-stained smears, these organisms stain faintly. For better visualization, carbolfuchsin is recommended as a counterstain; if safranin is used, counterstaining should be extended to 2 to 3 minutes. Enteric campylobacters may appear as long spirals, S shapes, or seagull-wing shapes. These organisms may appear as coccobacilli in smears prepared from older cultures. Phase-contrast or darkfield microscopy of fresh stool samples may show the characteristic "darting" motility typical of enteric campylobacters.

Arcobacter species have a microscopic morphology similar to that of *Campylobacter* species. *Helicobacter pylori* also appears similar to campylobacters, but ultrastructural study has shown that *Helicobacter* has multiple flagella at one pole, unlike the single polar flagellum of campylobacters.

Colonial Morphology. The typical colonial morphology of *C. jejuni* and other enteric campylobacters is moist, "runny looking," and spreading. Colonies are usually nonhemolytic; some are round and raised, and others may be flat. *Campylobacter fetus* ssp *fetus* produces smooth, convex, translucent colonies. A tan or slightly pink coloration is observed in some enteric campylobacter colonies. Other *Campylobacter* species produce colonies similar to those of *C. jejuni.* Although most do not produce pigment, *C. mucosalis* and *C. hyointestinalis* can produce a dirty yellow pigment.

BIOCHEMICAL TESTS TO DIFFERENTIATE *CAMPYLOBACTER*, *ARCOBACTER*, AND *HELICOBACTER* SPECIES

Species	Catalase	Nitrate Reduction	Urease	H$_2$S Production (TSI)	Hippurate Hydrolysis	Indoxyl Acetate Hydrolysis	Growth at			Susceptibility to	
							15°C	25°C	42°C	Nalidixic Acid (30 μg)	Cephalothin (30 μg)
C. jejuni ssp jejuni	+	+	–	–	+	+	–	–	+	S	R
C. jejuni ssp doylei	V	–	–	–	V	+	–	–	–	S	S
C. coli	+	+	–	–	–	+	–	–	+	S	R
C. lari	+	+	V	–	–	–	–	–	+	R	R
C. fetus ssp fetus	+	+	–	–	–	+	–	+	–	R	S
C. hyointestinalis	+	+	–	+	–	–	–	+	+	R	S
C. upsaliensis	W	+	–	–	–	+	–	–	+	S	S
C. concisus	–	+	–	+	–	–	–	–	+	R	R
C. curvus	–	+	–	+	–	+	–	–	+	S	ND
C. rectus	–^{w}	+	–	+	–	+	–	–	W	S	ND
Arcobacter butzleri	+	V	–	–	–	+	+	+	W	V	R
Helicobacter pylori	+	–	+	–	–	–	–	–	V	R	S
H. fennelliae	+	–	–	–	–	+	–	–	V	S	S
H. cinaedi	+	+	–	–	–	V	–	–	–	S	S

TSI, Triple sugar iron agar slant; +, positive result; –, negative result; W, weak; V, variable result; ND, not determined; S, susceptible; R, resistant; W, mostly negative, some weak.

Definitive Identification

Isolates from stool and rectal swabs can be presumptively identified as *Campylobacter* species by observing the characteristic Gram-stained microscopic morphology, the characteristic motility in a hanging-drop preparation using phase-contrast microscopy or darkfield microscopy, and positive oxidase and catalase reactions. The microscopic morphology is very important, as it differentiates *Campylobacter* from other bacterial species (i.e., *Aeromonas* and *Pseudomonas*) that are oxidase-positive and can grow at 42°C in a microaerophilic environment. To observe the typical motility, organisms should be suspended in *Brucella* or tryptic soy broth. Distilled water and saline seem to inhibit motility.

Table 17–9 lists the biochemical tests most useful in definitively identifying the most commonly encountered *Campylobacter, Helicobacter,* and *Arcobacter* species.

Helicobacter pylori may be presumptively identified in a gastric biopsy specimen by testing for the presence of a rapidly acting urease reaction. The collected tissue sample may be placed onto a Christensen's urea medium and incubated at 37°C for 2 hours. A rapid color change suggests the presence of *H. pylori*. Urease activity may also be detected by the "urea breath test." The breath test is reportedly both sensitive and specific and is recommended for monitoring therapy. In this test, the patient is given ^{14}C-labeled urea to drink. Urea degraded by the urease activity of *H. pylori* releases $^{14}CO_2$, which is detected in the exhaled breath by a scintillation counter.

Another detection method, although not specific for *H. pylori,* is to stain the gastric biopsy tissue using Giemsa, Gram, or silver stain to visualize the bacterium.

Serology. Latex agglutination tests are now available for rapid identification of colonies of enteric campylobacters on primary isolation media. The Meritec-Campy (jcl) test (Meridian Diagnostics, Inc., Cincinnati) and Campyslide (Becton Dickinson) can detect the presence of *C. jejuni, C. coli,* and *C. lari*. Campyslide also identifies *C. fetus* ssp *fetus*. However, neither system differentiates between the isolated *Campylobacter* species.

Serologic assays for the detection of antibodies to *H. pylori* are also currently available. Specific antibodies in serum may be detected by enzyme-linked immunosorbent assay (ELISA) or by indirect immunofluorescent assay (IFA) methods. These methods have been reported to be reasonably sensitive and specific indicators of *H. pylori* infections.

Antimicrobial Susceptibility

Antimicrobial susceptibility testing for *Campylobacter* species is not routinely performed in the clinical microbiology laboratory and is not standardized. The drug of choice for treating intestinal campylobacteriosis is erythromycin, although most patients recover without antimicrobial intervention. Gentamicin is used to treat systemic infections. Tetracycline, erythromycin, and chloramphenicol may be used as substitutes for gentamicin.

Several drug modalities for treating *H. pylori* infections are currently being studied. The most common thought is that triple therapy, which may include amoxicillin, metronidazole or tetracycline, and bismuth, is most effective in eradicating the organism.

Bibliography

Aldova E: Experience with serology of *Plesiomonas Shigelloides* O-antigenic stucture. J Hyg Epidemiol Microbiol Immunol 29:201; 1985.

Altwegg M: *Aeromonas caviae.* An enteric pathogen? Infection 13:228, 1985.

Blaser MJ, Berkowitz ID, et al: *Campylobacter* enteritis: Clinical and epidemiologic features. Ann Intern Med 91:179, 1979.

Bottone EJ, Janda, JM: *Vibrio.* In Howard B, et al (eds): Clinical and Pathogenic Microbiology, 1st ed. St. Louis: CV Mosby, 1987, pp. 359–365.

Brenden RA, Miller MA, Janda JM: Clinical disease spectrum and pathogenic factors associated with *Plesiomonas shigelloides* infections in humans. Rev. Infect Dis 10:303, 1988.

Butzler JP, De Boeck M, Goosens H: New selective medium for isolation of *Campylobacter jejuni* from fecal specimens. Lancet 2:818, 1983.

Butzler JP, Skirrow MB: *Campylobacter* enteritis. Clin Gastroenterol 8:737, 1979.

Carnahan A: Update on *Aeromonas* identification. Clin Microbiol Newsl 13:169, 1991.

Carnahan A, Behram S, Joseph SW: Aerokey II—a flexible key for identifying clinical *Aeromonas* species. J Clin Microbiol 29:2843, 1991.

Chan FTH, Mackenzie AMR: Enrichment medium and control system for isolation of *C. fetus* subsp. *jejuni* from stools. J Clin Microbiol 15:12, 1982.

Clark RB, Janda JM: *Plesiomonas* and human disease. Clin Microbiol Newsl 13:49, 1991.

Cover TL, Blaser MJ: *Helicobacter pylori* and gastroduodenal disease. Annu Rev. Med 43:135, 1992.

Dekeyser P, Gossvin-Detrain M, et al: Acute enteritis due to related vibrio: First positive stool cultures. J. Infect Dis 125:390, 1972.

Edmonds P, Patton CM, et al: *Campylobacter hyointestinalis* associated with human gastrointestinal disease in the United States. J Clin Microbiol 25:685, 1987.

Fliegelman RM, Petrak RM, et al: Comparative in vitro activities of twelve antimicrobial agents against *Campylobacter* species. Antimicrob Agents Chemother 27:429, 1985.

Fox JG, Childers T, et al: *Campylobacter mustelae,* a new species resulting from the elevation of *Campylobacter pylori* subsp *mustelae* to species status. Int J Syst Bacteriol 39:301, 1989.

Haapasalo M: *Bacteroides buccae* and related taxa in necrotic root canal infections. J Clin Microbiol 24:940, 1986.

Han YH, et al: *Wolinella recta, Wolinella curva, Bacteroides ureolyticus* and *Bacteroides gracilis* are microaerophiles, not anaerobes. Int J Syst Bacteriol 41:218, 1991.

Hoek FJ, Noach LA, et al: Evaluation of the performance of commercial test kits for detection of *Helicobacter pylori* antibodies in serum. J Clin Microbiol 30:1525, 1992.

Janda JM: Pathogenic *Vibrio* spp: An organism group of increasing medical significance. Clin Microbiol Newsl 9:49, 1987.

Janda JM: Recent advances in the study of the taxonomy, pathogenicity, and infectious syndromes associated with the genus *Aeromonas.* Clin Microbiol Rev 4:397, 1991.

Janda JM, Duffey PS: Mesophilic aeromonads in human disease: Current taxonomy, laboratory identification, and infectious disease spectrum. Rev Infect Dis 10:980, 1988.

Janda JM, Powers C, et al: Current perspectives on the epidemiology and pathogenesis of clinically significant *Vibrio* spp. Clin Microbiol Rev 1:245, 1988.

Joseph SW, Colwell, RR, Kaper JB: *Vibrio parahaemolyticus* and related halophilic vibrios. Crit Rev Microbiol 10:77, 1982.

Kaplan RL: *Campylobacter.* In Lennette EH, et al (eds): Manual of Clinical Microbiology, 3d ed. Washington, DC: American Society for Microbiology, 1980, pp 235–241.

Kaplan RL, Barrett JE: Monograph: *Campylobacter* 1981. Kansas City, MO: Marion Scientific, 1981.

Kaplan RL, Weissfeld AS: *Campylobacter, Helicobacter,* and related organisms. In Howard B, et al (eds): Clinical and Pathogenic Microbiology. St. Louis: Mosby-Year Book, 1994, pp 453–460.

Karmali MA, Fleming PC: Application of the Fortner principle to isolation of *Campylobacter* from stools. J Clin Microbiol 10:245, 1979.

Kelly MT, Hickman-Brenner FW, Farmer JJ III: Vibrio. In Balows A, et al (eds): Manual of Clinical Microbiology, 5th ed. Washington, DC: American Society for Microbiology, 1991, pp 384–395.

Kiehlbauch JA, Brenner DJ, et al: *Campylobacter butzleri* sp. nov. isolated from humans and animals with diarrheal illness. J Clin Microbiol 29:376, 1991.

Morris JG, and Cholera Laboratory Task Force: *Vibrio cholerae* O139 Bengal. In Wachsmuth I, et al (eds): *Vibrio cholerae* and Cholera. Molecular and Global Perspectives. Washington DC: American Society for Microbiology, 1994, pp 95–102.

Pai CH, Sorger S, et al: *Campylobacter* gastroenteritis in children. J Pediatr 94:589, 1979.

Paisley JW, Mirrett S, et al: Darkfield microscopy of human feces for presumptive diagnosis of *Campylobacter fetus* subsp *jejuni* enteritis. J Clin Microbiol 15:61, 1982.

Parsonnet J, Friedman GD, et al: *Helicobacter pylori* infection and risk of gastric carcinoma. N Engl J Med 325:1127, 1991.

Patton CM, Shaffer N, et al: Human disease associated with "*Campylobacter upsaliensis*" (catalase-negative or weakly positive *Campylobacter* species) in the United States. J Clin Microbiol 27:66, 1989.

Penner JL: The genus *Campylobacter.* A decade of progress. Clin Microbiol Rev 1:157, 1988.

Pugina P, Benzi G, et al: An outbreak of *Arcobacter (Campylobacter) butzleri* in Italy. Microbiol Ecol Health Dis 4(S):S94, 1991.

Quinn TC, Goodell SE, et al: Infections with *Campylobacter jejuni* and *Campylobacter*-like organisms in homosexual men. Ann Intern Med 101:187–192, 1984.

Sack, RB, Tilton RC, Weissfield AS: *Cumitech* 12. In Rubin SJ (ed): Laboratory Diagnosis of Bacterial Diarrhea. Washington, DC: American Society for Microbiology, 1980.

Segreti J, Gootz TD, et al: High-level quinolone resistance in clinical isolates of *Campylobacter jejuni.* J Infect Dis 165:667, 1992.

Shimada T, Kosako Y: Comparison of two O-serotyping systems for mesophilic *Aeromonas* spp. J Clin Microbiol 29:197, 1991.

Skirrow MB: *Campylobacter* enteritis. A "new" disease. Br Med J 2:9, 1977.

Smibert RM: Genus *Campylobacter* Sebald and Bernon 1963, 907. In Krieg NR, Holt JG (eds): Bergey's Manual of Systemic Bacteriology, Vol. 1. Baltimore: Williams and Wilkins, 1984, pp 111–118.

Soltesz V, Zeeberg B, et al: Optimal survival of *Helicobacter pylori* under various transport conditions. J Clin Microbiol 30:1453, 1992.

Steele TW, Owen RJ: *Campylobacter jejuni* subsp. *doylei* subsp. nov., a subspecies of nitrate-negative campylobacters isolated in central and south Australia. J Clin Microbiol 24:562, 1988.

Talley NJ, Newell DG, et al: Serodiagnosis of *Helicobacter pylori.* Comparison of enzyme-linked immunosorbent assays. J Clin Microbiol 29:1635, 1991.

Tanner ACR, Badger S, et al: *Wolinella* gen. nov., *Wolinella succinogenes (Vibrio succinogenes* Wolin et al) comb. nov., and description of *Bacteroides gracilis* sp. nov., *Wolinella recta* sp. nov., *Campylobacter concisus* sp. nov., and *Eikenella corrodens* from humans with periodontal disease. Int J Syst Bacteriol 31:432, 1981.

Tauxe RV, Patton CM, et al: Illness associated with *Campylobacter laridis,* a newly recognized *Campylobacter* species. J. Clin Microbiol 21:222, 1985.

Totten PA, Fennell CL, et al: *Campylobacter cinaedi* (sp. nov.) and *Campylobacter fennelliae* (sp. nov.): Two *Campylobacter* species associated with enteric disease in homosexual men. J Infect Dis 151:131 1985

Vandamme P, De Ley J: Proposal for a new family Campylobacteraceae. Int J Syst Bacteriol 41:451, 1991.

Vandamme P, Falsen E, et al: Identification of *Campylobacter cinaedi* isolated from blood and feces of children and adult females. J Clin Microbiol 28:1016:1020, 1990.

Vandamme P, Falsen E, et al: Revision of *Campylobacter, Helicobacter* and *Wolinella* taxonomy: Emendation of generic descriptions and proposal of Arcobacter gen. nov. Int J Syst Bacteriol 41:88, 1991.

von Graevenitz A, Altwegg M: *Aeromonas* and *Plesiomonas.* In Balows A (ed): Manual of Clinical Microbiology, 5th ed. Washington, DC: American Society for Microbiology, 1991, pp 396–401.

Warren JR, Marshall B: Unidentified curved bacilli on gastric epithelium in active chronic gastritis (letter). Lancet 1:1273, 1983.

CHAPTER 18

Nonfermenting Gram-Negative Bacilli and Miscellaneous Gram-Negative Rods

Gerri S. Hall

1. Describe the general characteristics of nonfermentative gram-negative rods.

2. Differentiate the metabolic pathways utilized by nonfermentative and fermentative organisms.

3. Describe how nonfermentative organisms cause infections.

4. Recognize the initial clues to nonfermentative organisms.

5. Describe the typical reactions and characteristic features of most of the commonly encountered nonfermentative organisms.

This chapter discusses the group of miscellaneous organisms that are becoming more clinically significant because of the increasing numbers of patients who are immunocompromised. The groups of organisms discussed in this chapter are the pseudomonads, *Acinetobacter* species, *Xanthomonas maltophilia* and other oxidase-negative oxidizers, Hacek members, and *Capnocytophaga* species (see Appendix C for recent changes in nomenclature for the Enterobacteriaceae and gram-negative nonfermentative bacilli).

Aerobic gram-negative bacilli can be divided into at least two large groups, those that ferment carbohydrates (fermentative or fermenters) and those that do not ferment (nonfermentative or nonfermenters). The pathways that these organisms utilize to break down carbohydrates are described in Figure 18–1.

GENERAL CHARACTERISTICS OF NONFERMENTERS

Nonfermentative organisms that can break down carbohydrates oxidatively are also referred to as *oxidizers* or *saccharolytic,* whereas those organisms that are not able to break down carbohydrates either fermentatively or oxidatively are referred to as *biochemically inert* or *nonoxidizers.* Nonfermenting gram-negative bacilli and coccobacilli are ubiquitous in the environment. They are found in soil, water, plants, decaying vegetation, and foodstuffs; in hospital, they are isolated from nebulizers, dialysate fluids, saline, and catheter devices.

They are variably resistant to agents such as chlorhexidine and quaternary ammonium compounds.

Clinical Infections

Nonfermenters account for approximately 15% of all isolates of gram-negative bacilli found in the clinical microbiology laboratory. Clinically, there are differences among infections caused by each species; however, there are some common disease manifestations and risk factors. Some of the disease manifestations associated with nonfermenting organisms are septicemia, meningitis, osteomyelitis, and wound infections usually following surgery or trauma. Risk factors for development of infection by these organisms are listed in Table 18–1. Immunosuppression, foreign body implantation, and traumatic breaks in a host barrier are the main events preceding infection with one of these organisms.

Biochemical Characteristics

In general, nonfermenters vary in their biochemical and morphologic characteristics. One common feature of this group of organism, how-

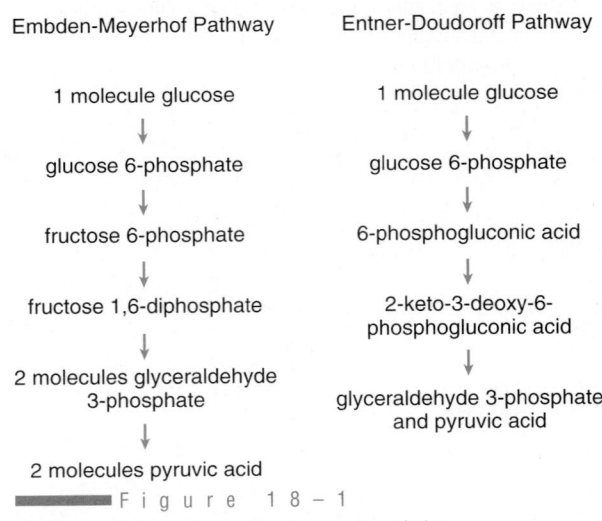

Embden-Meyerhof Pathway

1 molecule glucose
↓
glucose 6-phosphate
↓
fructose 6-phosphate
↓
fructose 1,6-diphosphate
↓
2 molecules glyceraldehyde 3-phosphate
↓
2 molecules pyruvic acid

Entner-Doudoroff Pathway

1 molecule glucose
↓
glucose 6-phosphate
↓
6-phosphogluconic acid
↓
2-keto-3-deoxy-6-phosphogluconic acid
↓
glyceraldehyde 3-phosphate and pyruvic acid

■ Figure 18–1

Two fermentation pathways for glucose degradation.

RISK FACTORS FOR DISEASES CAUSED BY NONFERMENTA-TIVE GRAM-NEGATIVE BACILLI

Immunosuppression

Diabetes mellitus
Cancer
Steroids
Transplantation

Trauma

Gun shot, knife wounds, punctures
Surgery
Burns

Foreign Body Implantation

Catheters: urinary or blood stream
Prosthetic devices: joints, valves
Corneal implants or contact lenses

Infused Fluids

Dialysate
Saline irrigations

ever, is their nonreactivity in triple sugar iron (TSI) agar or Kligler iron agar (KIA) (Table 18–2). A fermenter typically produces an acid (yellow) butt with an acid or alkaline (red) slant on TSI or KIA in 24 hours. A nonfermenter (either an oxidizer or nonoxidizer) produces no change (red) in the butt and slant or may produce an alkaline (red) slant (Fig. 18–2). In addition to these types of organisms, some "true" fermenters are fastidious and do not easily acidify the butt or slant of a TSI like other fermenters but do show reactions if other, more sensitive media are employed.

Characteristically, nonfermenters and fastidious fermenters often produce either weak or small amounts of acids from carbohydrates. In media, such as TSI agar, that contain large amounts of peptones (2.0%), whatever acids produced are neutralized or "masked" by the alkaline reaction from peptone utilization. To detect small amounts of acids produced, whether fermentatively or oxidatively, Hugh and Leifson developed a medium that contains the same amount of carbohydrates (1%) found in the TSI and KIA media but a lower amount of peptone (0.2%) in the oxidation-fermentation (OF) medium (1%) (Fig. 18–3).

Table 18–2 shows the differences in reactions among these groups of organisms. When isolates are tested in Hugh-Leifson OF medium, two tubes are inoculated: one is overlayed with sterile mineral oil to create an anaerobic environment (closed); the other is left aerobic, without mineral oil overlay (open). When acid is produced in both open and closed tubes, the isolate is determined to be a fermenter; whereas the absence of acid production in the closed tube indicates that the organism is a nonfermenter. The open tube may or may not show acidity. No acid production in the open tube may indicate that the organism is a nonoxidizer or is asaccharolytic.

Initial Clues to Nonfermenters

Initial clues to the presence of a nonfermentative organism in the clinical laboratory are as follows:

■ Long, thin gram-negative bacilli or coccobacilli

■ Oxidase-positive reaction (although reaction is variable in some and absent in others)

■ Nonreactive in 24 hours in commercial kit systems for the identification of Enterobacteriaceae

BIOCHEMICAL REACTIONS CHARACTERISTIC OF NONFERMENTERS ON TSI, KIA, AND OF MEDIA

	Triple Sugar Iron (TSI) Agar	Kligler Iron Agar (KIA)	Hugh-Leifson Oxidation-Fermentation (OF) Medium
Carbohydrates (concentration)	Glucose (0.1%) Lactose (1%) Sucrose (1%)	Glucose (0.1%) Lactose (1%)	Glucose or other carbohydrate being tested (1%)
Peptone	2%	2%	0.2%
Fermenter	Acid butt Acid or alkaline slant	Acid butt Acid or alkaline slant	Open tube: Acid Sealed tube: Acid
Nonfermenter			
Oxidizer	Alkaline butt Alkaline slant	Alkaline butt Alkaline slant	Open tube: Acid Sealed tube: No acid
Nonoxidizer (asaccharolytic)	Alkaline butt Alkaline slant	Alkaline butt Alkaline slant	Open tube: No acid Sealed tube: No acid

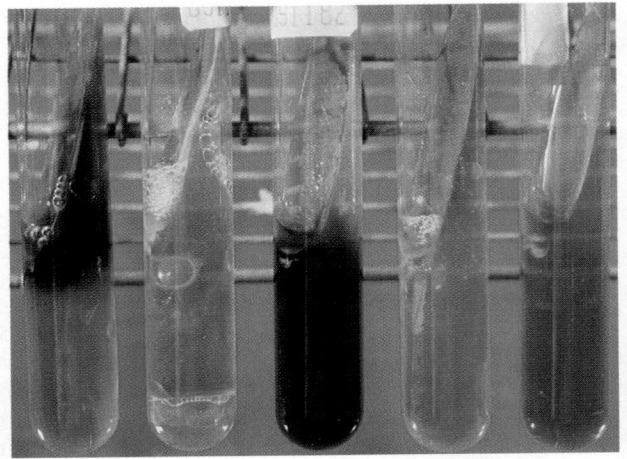

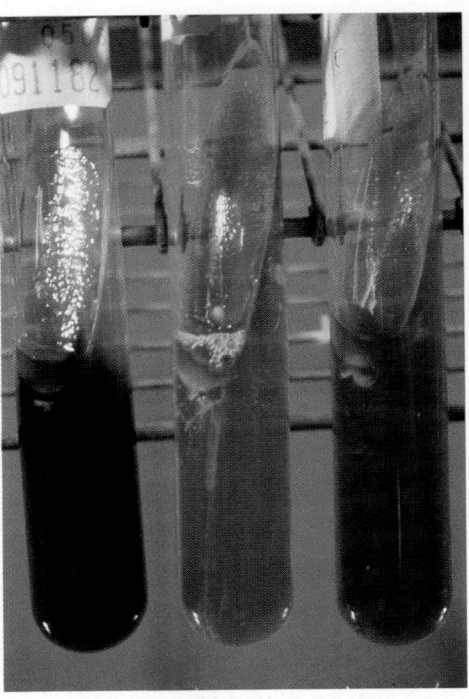

■ Figure 18–2

Reactions in triple sugar iron (TSI) agar. *A, Left to right*: A/A, H$_2$S, gas: glucose and sucrose fermenter, e.g., *Proteus* sp.; A/A, no H$_2$S, gas: glucose, sucrose, and lactose fermenter, e.g., *E. coli*; A/A, H$_2$S, no gas: glucose, sucrose, and lactose fermenter, e.g., *Citrobacter freundii*; ALK/A, no H$_2$S, no gas: glucose fermenter only, e.g., *Shigella* sp.; ALK/ALK, no gas, no H$_2$S: nonfermenter, e.g., *Pseudomonas* sp. *B, Left to right*: A/A, H$_2$S, no gas: lactose-fermenter with H$_2$S; ALK/A, no gas, no H$_2$S: glucose-fermenter, non–lactose-fermenter; ALK/ALK, no gas, no H$_2$S: nonfermenter.

■ TSI: nonreactive; i.e., no acid production in slant or butt

■ Resistance to antibiotics; i.e., aminoglycosides, cephalosporins, imipenem, penicillins

The nonfermenters may be organized into smaller groups based on reactions to three tests that are commonly performed to facilitate easy identification: (1) growth on MacConkey agar, (2) oxidase reaction, and (3) glucose oxidation (OF test). The eight possible combinations of results are then used to group the nonfermenters as shown in Figure 18–4. Included in this figure are some isolates of nonfermentative gram-negative bacilli not always considered to be with the nonfermenters. They are, however, biochemically very similar, i.e., *Brucella* sp, *Bordetella* sp, and are discussed in Chapter 15. Figure 18–4 will be cited in the following detailed discussions of the nonfermenter groups most commonly encountered in the clinical microbiology laboratory, to allow the reader to become familiar with the "initial clues" for each organism.

Identification Methods

For the identification of nonfermentative gram-negative bacilli, conventional tube biochemical testing, kit systems, or a combination of the two approaches can be used. The majority of clinical isolates are *Pseudomonas aeruginosa*, *Acinetobacter* sp, or *X. maltophilia*. For these most common isolates, most of the kit systems and automated identification systems perform adequately, or a limited number of conventional biochemicals can be used to identify these organisms. For the remainder of the nonfermenters, decisions must first be made as to which need to be identified.

The decision to identify organisms depends on the site from which they are isolated; that is, were they isolated from a sterile site in which the nonfermenter is the only isolate or from a nonsterile site in which three or four other bacteria are also present? In the former case, it may be decided that definitive identification and susceptibility testing are required. Use of a kit system with or without a conventional biochemical scheme should be utilized. In the latter case, a genus identification may be appropriate and may be achieved through use of a few biochemical tests, for example, oxidase, growth on MacConkey agar, glucose utilization, indole, and motility. Definitive identification of every nonfermenter can be very time consuming and costly and may not contribute much to the diagnosis of the disease. Figure 18–5 gives an example of results

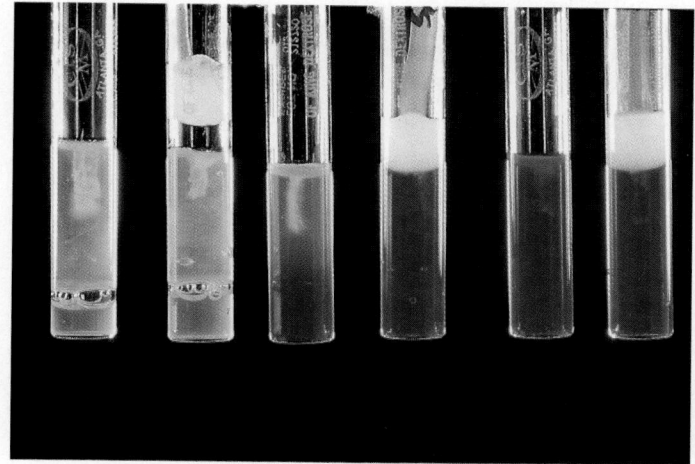

A

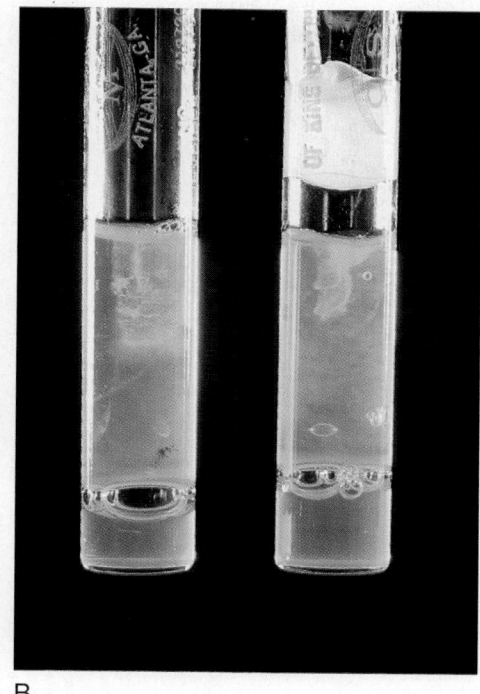

B

C

D

Figure 18-3

Reactions in oxidative fermentation (OF) media. *A, Left to right*: Fermenter: open and sealed tubes positive for acid production; nonfermenter: open tube positive for acid production, sealed tube negative for acid production; nonfermenter/nonoxidizer: open and sealed tubes negative for acid production. In *B*, fermenter is in OF media. In *C*, oxidizer/nonfermenter is in OF media. In *D*, nonfermenter/nonoxidizer is in OF media.

obtained for a large number of nonfermentative gram-negative bacilli using 20 biochemical or morphologic characteristics. If needed, thought should be given to sending out the isolate to a reference laboratory that is set up more readily and cost-effectively for these identifications.

Commercial kits and automated systems available for the identification of nonfermenters are listed in Table 18–3. The systems listed are those most commonly used in clinical laboratories.

Scott Laboratories, Inc., offers packaged tube media for the identification of nonfermenters along with a program to enable the technologist to choose tests to run and to interpret the results as compared with the database. Chemical analysis of nonfermenters by means of gas-liquid chromatography (GLC) has also been employed by some laboratories.

A number of comparison studies of various "kit" and automated systems for identification of

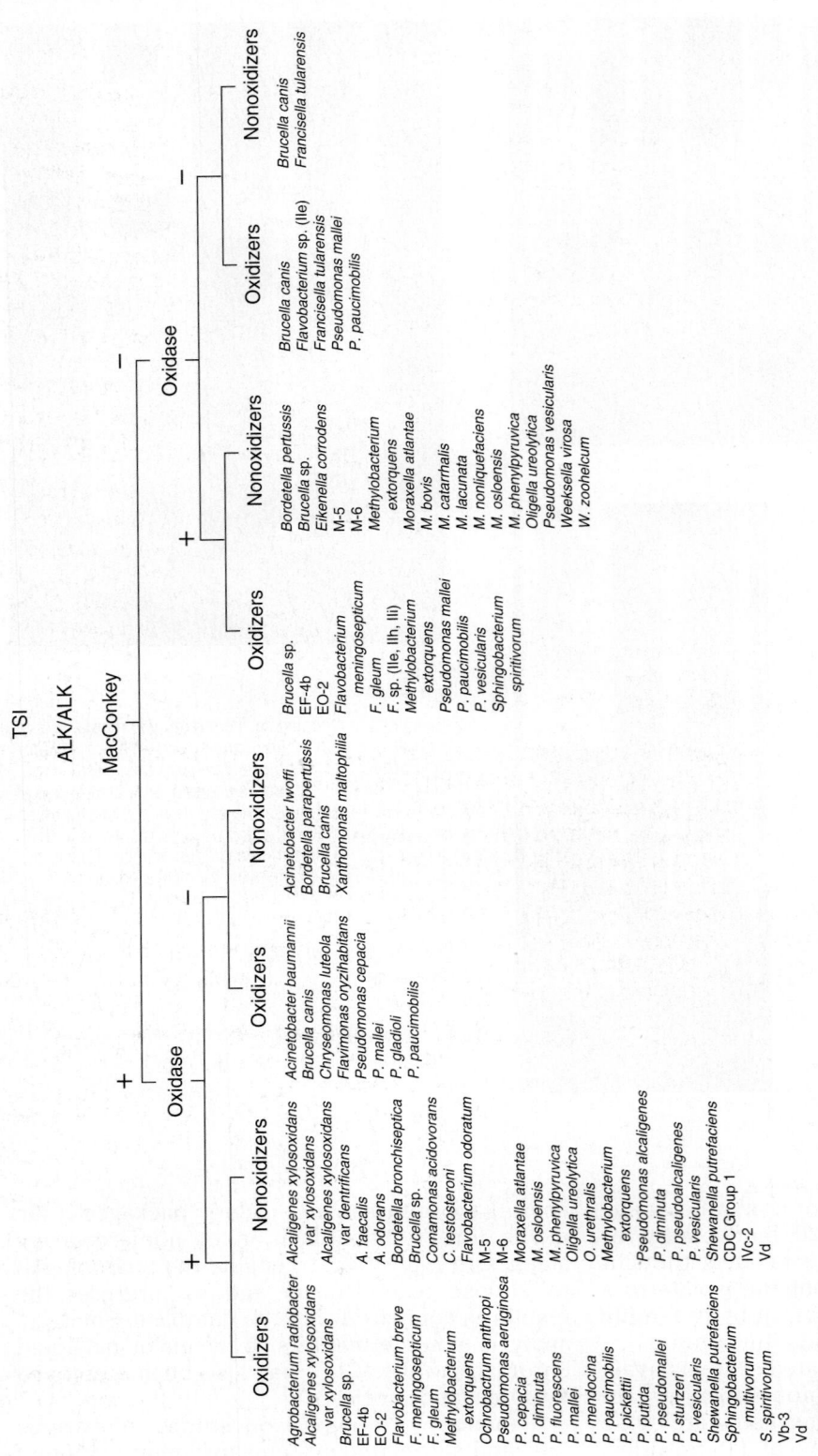

Figure 18-4
Grouping of nonfermenters based on eight possible results.

the nonfermenters have been conducted; they are listed in the Bibliography.

MOST COMMONLY ENCOUNTERED NONFERMENTATIVE ORGANISMS

The Pseudomonads

The genus *Pseudomonas* accounts for a large percentage of all nonfermenters isolated in the clinical microbiology laboratory. The characteristics common to the genus are as follows:

- Gram-negative bacillus or coccobacillus
- Motile with polar or polar tufts of flagella (except *Pseudomonas mallei*)
- Oxidase and catalase positive
- Usually grows on MacConkey agar
- Usually oxidizes carbohydrates

Table 18–4 lists the species currently included in the genus *Pseudomonas*.

Pseudomonas aeruginosa

Pseudomonas aeruginosa is the single most common and reported isolate of the genus. It is an uncommon part of normal flora; for example, only 4 to 12% of humans carry it as part of normal fecal flora. It may, however, account for 5 to 15% of all nosocomial infections.

CLINICAL INFECTIONS

The clinical diseases documented to be caused by *P. aeruginosa* include:

- Bacteremia with ecthyma gangrenosum of the skin as a hallmark
- Wound infections
- Pulmonary disease, especially among individuals with cystic fibrosis
- Nosocomial urinary tract infections
- Endocarditis
- Infections after burns
- Rarely, central nervous system diseases, including meningitis

Other conditions associated with *P. aeruginosa* infection are otitis externa, in particular, in swimmers or divers, and a necrotizing skin rash, referred to as *jacuzzi* or *hot tub syndrome*, that develops in users of these recreational facilities. The organisms have a propensity to invade vascular walls of vessels, which further their spread in the body. *Pseudomonas aeruginosa* accounts for 6.2% of all bacteremias and for up to 74% of nosocomial bacteremias. One study by Bisbe and colleagues (1988) looked at prognostic factors associated with *P. aeruginosa* bacteremia and found that independent poor prognostic factors included septic shock, granulocytopenia, inappropriate antibiotic therapy, and the presence of septic metastases. Figure 18–6 shows the Gram stain of a bronchial specimen containing *P. aeruginosa*.

VIRULENCE FACTORS

Pseudomonas aeruginosa may produce a variety of factors that lend to its pathogenicity, including endotoxin, proteases, hemolysin, "slime," lecithinase, elastase, coagulase, DNase, and exotoxin.

IDENTIFYING CHARACTERISTICS

Most isolates are β-hemolytic on sheep's blood agar and produce a characteristic "green metallic sheen," due to the presence of the pigment pyocyanin. On MacConkey agar (Fig. 18–7), pyocyanin causes colonies of *P. aeruginosa* to appear blue-green. Some strains appear mucoid on blood agar, others may not (Fig. 18–8). Pyocyanin is easy to detect on most media, with most strains, but can be specifically determined in media such as Sellers and fluorescent-nitrate (FN) medium. No other strain of a gram-negative nonfermenter produces pyocyanin, so its production is a characteristic used to differentially identify *P. aeruginosa*. Pyocyanin typing can be used to differentiate between strains of *P. aeruginosa* that are of epidemiologic interest. About 4% of clinical strains, however, are apyocyanogenic.

Most *P. aeruginosa* strains also produce pyoverdin, a fluorescent pigment that can be detected on Sellers or FN medium. Pyoverdin is produced by other species of *Pseudomonas*, including *Pseudomonas fluorescens* and *Pseudomonas putida*.

Many strains of *P. aeruginosa* produce a fruity, grape-like odor owing to the presence of 2-aminoacetophenone. Other key characteristics of *P. aeruginosa* are as follows:

	MOTILE, STRONGLY SACCHAROLYTIC NONFERMENTERS										MOTILE, WEAK OR NONSACCHAROLYTIC NONFERMENTERS										
	Pseudomonas aeruginosa	*P. fluorescens/putida*	*P. cepacia*	*Alcaligenes xylosoxidans subsp xylosoxidans*	*Xanthomonas maltophilia*	*Pseudomonas pseudomallei*	*P. stutzeri*	*P. paucimobilis*	*P. mendocina*	*P. pickettii*	*P. acidovorans*	*P. pseudoalcaligenes*	*Alcaligenes faecalis*	*A. xylosoxidans subsp denitrificans*	*Oligella ureolytica*	*Bordetella bronchiseptica*	*P. alcaligenes*	*P. diminuta*	*P. vesicularis*	*P. testosteroni*	*Shewanella putrefaciens*
Oxidase	+	+	+	+	−	+	+	+	+	+	+	+	+	+	+	+	+	+	+	+	+
Pyocyanin	+/−	−	−	−	−	−	−	−	−	−	−	−	−	−	−	−	−	−	−	−	−
Fluorescein	+	−/+	−	−	−	−	−	−	−	−	−	−	−	−	−	−	−	−	−	−	−
Glucose	+	+	+	+/−	+/−	+	+	+/−	+	+	−	−/+	−	−	−	−	−	−	+	−	−
Xylose	+	+	+/−	+	−	+	−/+	+/−	+	+	−	−	−	−	−	−	−	−	−	−	−
Mannitol	+/−	+/−	+/−	−	−	+	−/+	−	−	−	+/−	−	−	−	−	−	−	−	−	−	−
Lactose	−	−	+/−	−	−	+	+/−	+/−	−	−	−	−	−	−	−	−	−	−	−	−	−
Maltose	−	−	+/−	−	+	+	+/−	+/−	−	−	−	−/+	−	−	−	−	−	−	+	−	−
42°C	+	−	+/−	+	+/−	+	+	−	+	+/−	−	+	+/−	−/+	−	+	+/−	−/+	+/−	+/−	+/−
Esculin	−	−	−/+	−	+	+/−	−	+	−	−	−	−	−	−	−	−	−	−	+	−	−
Urea	+	−/+	+/−	−	+/−	−/+	−/+	−/+	+/−	+	−	−	−/+	−	+	+	−/+	+/−	−	−	+/−
DNase	−	−	−	−	+	−	−	+	−	−	−	−	−	−	−	−	−	+	+	−	+
ONPG	−	−	+/−	−	+	−	−	+/−	−	−	−	−	−	−	−	−	−	−	−	−	−
Indole	−	−	−	−	−	−	−	−	−	−	−	−	−	−	−	−	−	−	−	−	−
Motility	+	+	+	+	+	+	+	−	+	+	+	+	+	+	+	+	+	+	+	+	+
Flagella	1	>1	>1	P	>1	>1	1	1	1	1	>1	1	P	P	P	P	1	1	1	>1	1
H₂S	−	−	−	−	−	−	−	−	−	−	−	−	−	−	−	−	−	−	−	−	+
N₂ Gas	+/−	−	−	−	−	+	+	−	+	−/+	−	−	+	+/−	+	−	−	−	−	−	−
Pigment	B,F,G	F	Y	−	Y	−	B,Y	Y	−	−	−	−	−	−	−	−	−	−	B,Y	−	B
Growth on MAC	+	+	+	+	+	+	+	−	+	+	+	+	+	+	+	+	+	+	+	−/+	+/−	+

Figure 18–5

Biochemical and morphologic characteristics of selected nonfermentative gram-negative bacilli. (Modified from data supplied by Ohio State University Hospital.)

| | NONMOTILE, PIGMENTED, INDOLE-POSITIVE NONFERMENTERS | | | | | NONMOTILE COCCOBACILLI | | |
| | | | | | | | Oxidase-Negative | |
	Flavobacterium meningosepticum	*F. indologenes*	*F. odoratum*	*Weeksella zoohelcum*	*W. virosa*	*Moraxella species*	*Acinetobacter lwoffii*	*A. baumannii*
Oxidase	+	+	+	+	+	+	−	−
Pyocyanin	−	−	−	−	−	−	−	−
Fluorescein	−	−	−	−	−	−	−	−
Glucose	+/−	+/−	−	−	−	−	−	+
Xylose	−	−/+	−	−	−	−	−	+
Mannitol	−/+	−	−	−	−	−	−	−
Lactose	−	−	−	−	−	−	−	+
Maltose	−/+	+/−	−	−	−	−	+/−	−
42°C	+/−	−/+	−	−	−/+	−/+	−	+
Esculin	+	+	−	−	−	−	−	−
Urea	−/+	−/+	+	+	−/+	−	−/+	−/+
DNase	+	+	+	+	+	−	−	−/+
ONPG	+/−	−/+	−	−	−	−	−	−
Indole	+	+/−	−	−/+	+	−	−	−
Motility	−	−	−	−	−	−	−	−
Flagella	−	−	−	−	−	−	−	−
H₂S	−	−	−	−	−	−	−	−
N₂ Gas	−	−	−/+	−	−	−	−	−
Pigment	Y	Y	Y	B	B	−	−	−
Growth on MAC	+	+/−	+	−	−	+/−	+	+

+, 85% or more of the isolates are positive.

−, 85% or more of the isolates are negative.

+/−, 50% or more of the isolates are positive.

−/+, 50% or more of the isolates are negative.

Flagella

1, Polar monotrichous.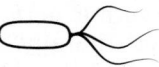

>1, Polar tuft (>1 flagellum).

P, Peritrichous.

Pigment
B, brown; G, green; Y, yellow; F, fluorescent;
−, none present

■ T A B L E   1 8 - 3

COMMERCIAL KITS AND AUTOMATED INSTRUMENTS FOR THE IDENTIFICATION OF NONFERMENTERS

Manual Kits

Rapid NFT (Analytab Products, Plainview, NY)
API-20E (Analytab Products, Plainview, NY)
API-20NE (API Systems, SA, Mantalieu, France)
Oxi-Ferm (Roche Diagnostics, Nutley, NJ)
Minitek (BBL Microbiology Systems, Cockeysville, MD)
Titertek-NF (Flow Laboratories, Meckenheim, Germany)
RapID NF Plus System (Innovative Diagnostics Systems, Inc., Atlanta, GA)
Sceptor (BBL, Microbiology Systems, Cockeysville, MD)
Fox (Micro Media Systems, Cleveland, OH)

Automated Systems

Autobac II (Organon Teknika, Durham, NC)
Vitek AMS (Vitek Bio-Merieux Systems, Hazelwood, MO)
MicroScan (American MicroScan Division, Baxter Healthcare, Sacramento, CA)
Quantum II (Abbott Diagnostics, Irving, TX)
Unisept 20E (Analytab Products, Plainview, NY)
Sensititre (Sensititre, Salem, NH)

■ Gluconate production from glucose via oxidation

■ Alcohol dehydrogenase (ADH) positivity

■ Growth at 42°C

■ Cetrimide positivity

■ Citrate positivity

■ Acetamide utilization

TREATMENT

Pseudomonas aeruginosa is usually resistant to many antibiotics, including penicillin, ampicillin, many cephalosporins, and chloramphenicol. It is also usually susceptible to the aminoglycosides, semisynethic penicillins such as piperacillin, and azlocillin, the third-generation cephalosporins, ceftazidime, and imipenem. Treatment of severe infections with *P. aeruginosa* most often requires double-drug therapy, for example, ceftazidime with tobramycin or piperacillin with tobramycin.

Pseudomonas fluorescens and **Pseudomonas putida**

These two organisms are members of the fluorescent group of pseudomonads, along with *P. aeruginosa,* that produce pyoverdin. Neither *P. putida* or *P. fluorescens* produces pyocyanin or grows at 42°C, differentiating them from *P. aeruginosa.* As to differentiating between the two, *P.*

putida is gelatin hydrolysis negative and *P. fluorescens* is positive. They are both of low virulence, rarely causing clinical diseases. Both have been isolated from respiratory cultures, contaminated blood bank products, urine, cosmetics, and hospital equipment, and fluids. Both have been documented, although rarely, as causes of urinary tract infections, abscesses (postsurgical), empyema, septic arthritis, and other wound infections. They are, in general, susceptible to aminoglycosides, polymyxin, cefoperazone, and piperacillin but resistant to carbenicillin.

Xanthomonas maltophilia

Xanthomonas maltophilia is the third most common nonfermentative gram-negative bacillus isolated in the clinical laboratory. Prior to 1983, it was believed to be a member of the pseudomonads, but has been reclassified as a member of the plant pathogen genus *Xanthomonas.* Isolates are fairly ubiquitous in the environment, being common to water, sewage, and plant materials; likewise, the organism is common to the hospital environment, where it is found as a contaminant in blood-drawing equipment, disinfectants, transducers, and the like. *Xanthomonas maltophilia* is not considered part of the normal human flora, but it may colonize hospitalized patients, especially those who undergo proce-

■ T A B L E   1 8 - 4

MEMBERS OF *PSEUDOMONAS* GENUS

Species	Characteristic(s) of Interest or Distinction
P. aeruginosa	Blue-green pigment
	Grape-like odor
	Most common isolate
P. alcaligenes	
P. cepacia	Often in patients with cystic fibrosis
P. diminuta	
P. fluorescens	Fluorescent pigments
P. gladioli	
P. mallei	Animal pathogen (glanders)
P. mendocina	
P. paucimobilis	Yellow pigment
P. petucinogena	
P. pickettii	
P. pseudoalcaligenes	
P. pseudomallei	Cause of melioidosis
P. putida	Fluorescent pigment
P. stutzeri	Wrinkled colonies
P. vesicularis	
Pseudomonas, CDC Group 1	

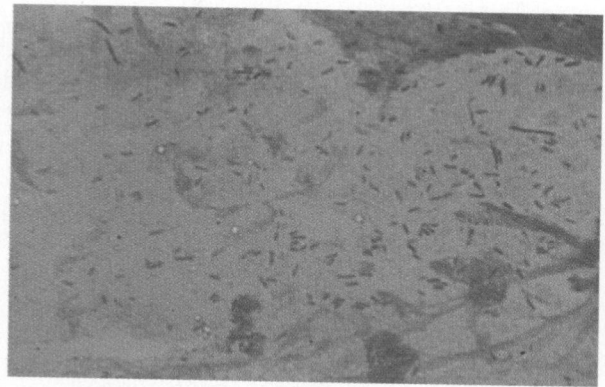

■■■■Figure 18 – 6
Gram stain of bronchial specimen positive for a nonfermentative gram-negative bacillus, e.g., *Pseudomonas aeruginosa*.

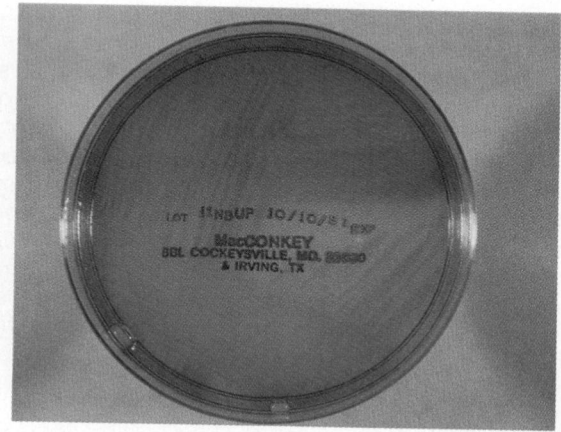

■■■■Figure 18 – 7
P. aeruginosa on MacConkey agar. Note blue-green pigment.

dures that expose them to contaminated equipment or to antibiotics to which *X. maltophilia* may be resistant. These antimicrobials include cephalosporins, penicillins, imipenem, and aminoglycosides. The isolates often show multiple resistance to these agents.

Clinically, *X. maltophilia* isolated from clinical specimens is initially regarded as a saprophyte or colonizer. With the increased use of agents to which it alone is innately resistant, however, more and more reports of disease production can be attributed to this organism. There are literature reports of pneumonia, endocarditis (especially in a setting of prior IV drug abuse and/or heart surgery) wound infections, including cellulitis and ecthyma gangrenosum, bacteremia, and, rarely, meningitis and urinary tract infections. Almost all of these infections have occurred in a nosocomial setting, i.e., are hospital acquired. Pseudoinfections have also occurred as a result of contaminated collection tubes or cups, for example, blood collection tubes.

In a 1990 review, Elting and Bodey documented 149 cases of bacteremia that were ascribed to *X. maltophilia*. The single most important risk factor in the affected individuals was the presence of a venous catheter. A good response to therapy was demonstrated in most of the patients presenting with the bacteremia, unless they had concomitant pneumonia or shock.

IDENTIFYING CHARACTERISTICS

Isolates of *X. maltophilia* can be recognized as nonfermentative gram-negative bacilli with a negative oxidase reaction. In addition, they are positive for catalase, DNase, esculin, and gelatin hydrolysis and for the presence of lysine decarboxylase.

Acinetobacter Species

The genus *Acinetobacter,* in the family Neisseriaceae, consists of 12 DNA hybridization groups (genospecies). Two genospecies are the most common clinical isolates: *Acinetobacter baumannii* (previously referred to as *A. calcoaceticus* var *anitratus*) and *Acinetobacter Iwoffii* (previously referred to as *A. calcoaceticus* var. *Iwoffii*). The genus has had many synonyms over the years, including *Mima, Herellea,* and even *Diplococcus.*

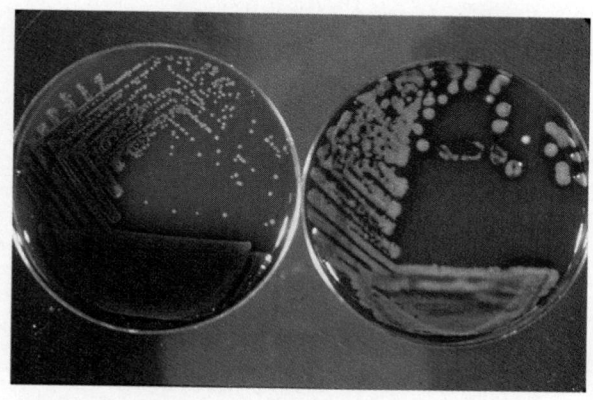

■■■■Figure 18 – 8
P. aeruginosa on sheep blood agar. *Left,* Mucoid colonies. *Right,* Nonmucoid colonies. Note discoloration of media, especially on the right.

The *Acinetobacter* species are ubiquitous in the environment (soil, water, milk, frozen soups) and in the hospital (ventilators, humidifiers, catheters). About 25% of adults have skin colonization with an *Acinetobacter* spp, and 7% carry the organism in their pharynx. Hospitalized patients may become easily colonized if they were not already harboring the organisms, and thus, *Acinetobacter* may be isolated nonsignificantly from urine, feces, vaginal secretions, and many different types of respiratory specimens. As many as 45% of tracheostomy sites may be colonized.

CLINICAL INFECTIONS

These organisms are opportunistic and account for 1 to 3% of all nosocomial infections. *Acinetobacter* species are second only to *P. aeruginosa* in frequency of isolation in the clinical microbiology laboratory. Diseases for which they have been reported responsible include:

- Urinary tract infections
- Pneumonia and/or tracheobronchitis
- Endocarditis (with up to a 22% associated mortality)
- Septicemia
- Meningitis (often as a complication of intrathecal chemotherapy for cancer)
- Cellulitis, most often as a result of contaminated indwelling catheters, trauma, burns, or introduction of a foreign body

Eye infections due to *Acinetobacter*, including endophthalmitis, conjunctivitis, and corneal ulcerations, have been reported in the literature. Most of the disease produced by members of the genus *Acinetobacter* are caused by *A. baumannii* (Fig. 18–9). *Acinetobacter lwoffii* is less virulent, and its isolation most often indicates contamination or colonization rather than infection.

Isolates of *Acinetobacter* spp, particularly *A. baumannii,* are often resistant to many antimicrobials, including penicillins and first- and second-generation cephalosporins. They demonstrate variable susceptibility to aminoglycosides, imipenem, and aztreonam in vitro.

IDENTIFYING CHARACTERISTICS

Members of the genus *Acinetobacter* have the following cultural and biochemical traits in common:

- Coccobacilli
- Oxidase negative

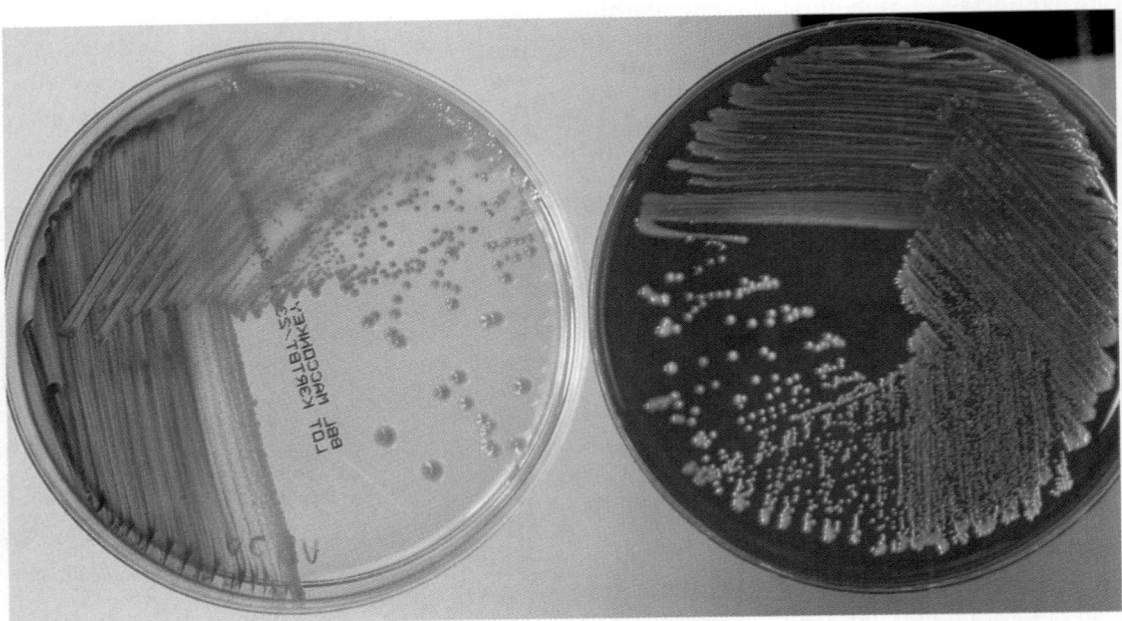

■■■■■ F i g u r e   1 8 – 9

Acinetobacter baumannii is saccharolytic, which may cause it to resemble a lactose-fermenting organism on MAC.

- Catalase positive

- Nonmotile

They possess few growth requirements and thus are capable of growing on most laboratory media, including MacConkey agar. One characteristic feature of this organism is the purplish hue it produces on MacConkey agar. Isolates usually grow better at 30°C than 37°C and at a pH of 5.5 to 6.0. *A. baumannii* are saccharolytic and *A. lwoffii* are nonsaccharolytic species. The purplish hue produced on this medium may cause the species to resemble a lactose-fermenting organism.

Other Common Pseudomonads

Pseudomonas cepacia

Pseudomonas cepacia has been known as *P. multivorans, P. kingii,* and EO-1. It is a plant pathogen that commonly attacks onion bulbs.

Clinically, *P. cepacia* is a low-grade, nosocomial pathogen that has most often been associated with pneumonia in cystic fibrosis (CF) patients. It has also been reported to cause endocarditis, specifically in drug addicts, pneumonitis, urinary tract infections, osteomyelitis, dermatitis, and other wound infections from use of contaminated water. It has been isolated from irrigation fluids, anesthetics, nebulizers, detergents, and disinfectants. In patients with cystic fibrosis, much work has been done to document the association of *P. cepacia* and increased severity of disease and/or death.

Pseudomonas cepacia is usually susceptible to chloramphenicol, ceftazidime, and trimethoprim-sulfamethoxazole but resistant to aminoglycosides, polymyxins, many cephalosporins, and penicillins. Development of resistance may be quick.

The organism grows well on most laboratory media but may lose viability on blood agar in 3 to 4 days without appropriate transfers. Selective media are available to increase recovery of *P. cepacia* from respiratory specimens in patients with CF. Other isolates of *Pseudomonas* sp may also be recovered, but a selective medium does help to reduce normal flora overgrowth. The preferred temperature is 30°C. Unlike other members of the genus *Pseudomonas, P. cepacia* may be oxidase negative. It utilizes glucose, maltose, lactose, and mannitol and is lysine decarboxy-

lase (LDC) positive, ornithine decarboxylase (ODC) positive, and ADH negative. Isolates are motile by means of polar tufts of flagella. It does not fluoresce like *P. aeruginosa,* but does produce a nonfluorescing yellow or green pigment that may diffuse into the media. Colonies of *P. cepacia* are nonwrinkled, and this trait may be used to differentiate isolates from *P. stutzeri,* which also produces a yellow pigment.

Pseudomonas stutzeri

An initial clue to the identity of *Pseudomonas stutzeri* is its macroscopic appearance—a wrinkled, leathery, adherent colony that may produce a light yellow or brown pigment. Like most members of the family Pseudomonadaceae, it is a saprophyte but has been found to produce a variety of diseases in the immunosuppressed patient (septicemia, pneumonia, and endocarditis) and in the surgical patient (wound infections, septic arthritis, conjunctivitis, or urinary tract infections). Isolates in vitro are usually susceptible to aminoglycosides, trimethoprim-sulfamethoxazole, ampicillin and polymyxin, tetracyclines, quinolones, and third-generation cephalosporins but resistant to chloramphenicol and the first- and second-generation cephalosporins.

Pseudomonas paucimobilis

A yellow-pigmented pseudomonad, *Pseudomonas paucimobilis* does not grow on MacConkey agar and requires more than 48 hours for culture on blood agar. Although isolates have been found to produce esterases, endotoxin, lipases, and phosphatases, inherent virulence is limited, and most isolates have to be regarded as colonizers or contaminants. This organism can be isolated environmentally from water, including swimming pools, as well as from hospital equipment and laboratory supplies.

Reports in the literature have documented *P. paucimobilis* infections: seven cases of peritonitis associated with chronic ambulatory peritoneal dialysis (CAPD), eight cases of septicemia, meningitis, leg ulcer, empyema, and splenic and brain abscesses. Isolates demonstrate variable resistance to antibiotics, although most are susceptible to aminoglycosides, quinolones, third-generation cephalosporins such as ceftazidime, ceftriaxone, and ceftizoxime, sulfamethoxazole-trimethoprim (SXT), and ampicillin.

Less Common Pseudomonads

Pseudomonas pseudomallei

Clinically, *Pseudomonas pseudomallei* is important because it can cause a condition referred to as *melioidosis,* an aggressive granulomatous pulmonary disease, following ingestion, inhalation, or inoculation of the organisms with further metastatic abscess formation in lungs and other viscera. Overwhelming septicemia may occur. The incubation period in vivo may be very prolonged. The organisms are found in water and muddy soils in Southeast Asia (including Vietnam and Thailand), Northern Australia, and Mexico. Local infections, including orbital cellulitis, dacrocystitis, and draining abscesses, may occur. Although isolates may be susceptible in vitro to a variety of antibiotics, including SXT, chloramphenicol, tetracycline, semisynthetic penicillins, and ceftazidime, the clinical response to therapy is usually slow, and relapses are common. Individuals who have traveled to endemic areas are at risk for development of diseases with *P. pseudomallei,* and this organism should be considered especially where there is isolation of a nonfermentative, wrinkled colony (Fig. 18–10) that demonstrates bipolar staining on Gram-stained smears. *P. stutzeri,* which may also appear as wrinkled colonies, does not utilize lactose, in contrast to *P. pseudomallei,* which oxidizes lactose.

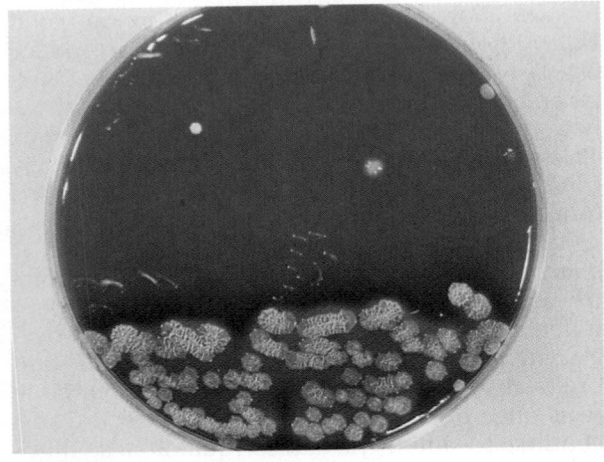

■ F i g u r e 1 8 – 1 0
P. pseudomallei on sheep blood agar.

Pseudomonas diminuta

Pseudomonas diminuta has been found in blood, cerebrospinal fluid (CSF), urine and wounds, usually as contaminants. The isolates are motile and possess a very short wavelength on their single polar flagellum. *Pseudomonas diminuta* oxidizes glucose; is indole negative and oxidase positive and may produce a brown, water-soluble pigment on heart infusion agar with tyrosine added. In vitro, the organism demonstrates resistance to ampicillin, cefoxitin, and nalidixic acid.

Pseudomonas gladioli

A plant pathogen, *Pseudomonas gladioli* resembles *P. cepacia* and may be mistaken for it. Isolates have been found in cystic fibrosis patients, but these organisms are more susceptible to antimicrobials than *P. cepacia.* A yellow pigment may be produced. These organisms are motile by means of one or two polar flagella and are catalase and urease positive. Biochemical characteristics include:

- Oxidize glucose
- Mannitol positive
- Decarboxylase negative
- Oxidase and nitrate reduction to gas variable but usually negative
- Positive for growth on MacConkey agar
- 100% resistant to polymyxin.

Pseudomonas mendocina

Pseudomonas mendocina resembles a nonpigmented *P. aeruginosa* but is acetamide and 2-ketogluconate negative. The organisms are motile by means of a single polar flagellum, oxidize glucose and xylose, grow on MacConkey agar, and are positive for oxidase and ADH, and are nonproteolytic. They are found in soil and water and are rarely isolated from humans; if so they are usually contaminants.

Pseudomonas pickettii (CDC Va-1)

Isolates of *Pseudomonos pickettii* can be found contaminating sterile hospital fluids and, as such, may be isolated from human specimens such as those from urine, nasopharynx, abscesses, wounds, and blood, usually as colonizers or contaminants. These isolates are: oxidase, catalase, and urease positive, grow on MacConkey agar, reduce nitrate, oxidize glucose and xylose and are motile by means of a single polar flagellum. They are susceptible to most agents, except for the aminoglycosides and polymyxin.

Pseudomonas pseudoalcaligenes, Pseudomonas alcaligenes, and CDC Group 1

One group of pseudomonads, consisting of *Pseudomonas pseudoalcaligenes, Pseudomonas alcaligenes,* and CDC Group 1, are asaccharolytic and oxidase positive. They all grow on MacConkey agar and may or may not reduce nitrates. They differ in their ability to grow at 42°C; *P. pseudoalcaligenes* does, *P. alcaligenes* does not, and CDC Group 1 shows variable growth at this temperature.

Pseudomonas vesicularis

Like *P. diminuta, Pseudomonas vesicularis* is a slender rod, with short-wavelength polar flagella. About 15% of *P. vesicularis* strains may produce a yellow pigment, and some may produce a tan pigment when grown on media with tyrosine. *Pseudomonas vesicularis* is oxidase positive and oxidizes glucose and maltose. Isolates have been found clinically in CSF, blood, urine, and eye specimens, but like *P. diminuta* and other members of miscellaneous pseudomonads, they usually are colonizers or contaminants.

Pseudomonads and Other Nonfermenters Whose Names Have Been Changed

Table 18–5 lists the present taxonomy of many nonfermenters to help clarify old vs. new terminology. Table 18–6 summarizes observed characteristics that are common to most nonfermenters.

Shewanella putrefaciens

Pseudomonas putrefaciens is now called *Shewanella putrefaciens* and is a member of "Unknown

■■■■■ T A B L E   1 8 – 5
TAXONOMIC CHANGES FOR SOME NONFERMENTERS

New Taxonomy	Old Taxonomy
Alcaligenes faecalis	*Alcaligenes odorans/faecalis*
Alcaligenes xylosoxidans subsp denitrificans	*Alcaligenes denitrificans,* CDC Group Vc
Alcaligenes xylosoxidans subsp xylosoxidans	*Achromobacter xylosoxidans,* CDC Group III a,b
Chryseomonas luteola	*Pseudomonas luteola,* CDC Group Ve-1
Comamonas acidovorans	*Pseudomonas acidovorans*
Comamonas testosteroni	*Pseudomonas testosteroni*
Flavimonas oryzihabitans	*Pseudomonas oryzihabitans*
Flavo sp IIe	CDC Group IIe
Flavo sp IIh	CDC Group IIh
Flavo sp IIi	CDC Group IIi
Flavobacterium gleum	CDC Group IIb
Flavobacterium indologenes	CDC Group IIa
Methylobacterium spp	*Pseudomonas mesophilica*
Ochrobactrum anthropi	*Achromobacter* spp biovar 1,2; Vd-1,2
Oligella ureolytica	CDC Group IV-e
Oligella urethralis	*Moraxella urethralis*
Pseudomonas gladioli	*Pseudomonas marginata*
Pseudomonas mendocina	CDC Vb-2
Pseudomonas pickettii	*P. thomasii,* Va-1,2
	P. pseudoalcaligenes, biovar 2
Psychrobacter immobilis	*Micrococcus cryophilus*
Shewanella putrefaciens	*Pseudomonas putrefaciens*
Sphingobacterium multivorum	*Flavobacterium multivorum,* IIk-2
Sphingobacterium spiritivorum	*Flavobacterium spiritivorum,* IIk-3
Weeksella virosa	CDC Group IIf (Flavo IIf)
Weeksella zoohelcum	CDC Group IIj (Flavo IIj)
Xanthomonas maltophilia	*Pseudomonas maltophilia*

Modified with permission from Gilardi GI: Update on taxonomy of nonfastidious glucose-nonfermenting gram-negative bacilli. Clin Microbiol Newsl 12:73, 1990.

RNA Homology Group Affiliation." *Shewanella putrefaciens* produces profuse H_2S on TSI agar, possibly resembling H_2S producers of the family Enterobacteriaceae. The oxidase test should differentiate *Shewanella* from the other group of organisms. This isolate also produces reddish brown or pink colonies that are often mucoid. Rarely pathogenic, the isolates can be obtained from abscesses and traumatic ulcers, but usually as colonizers. Any environmental source, such as stagnant water, natural gas, or petroleum brine, and spoiled dairy products may contain *S. putrefaciens*. These organisms are usually susceptible to ampicillin, tetracycline, chloramphenicol, erythromycin, and the aminoglycosides.

Comamonas species

Comamonas species used to be referred to as *Pseudomonas acidovorans* or *Pseudomonas testosteroni*. They resemble vibrios or spirillum-like bacteria, produce alkalinity in a blue color OF media, most commonly are motile by multitrichous polar

■■■■■T A B L E   1 8 – 6
CHARACTERISTICS COMMON TO GROUPS OF NONFERMENTERS

Pigmentation	
Yellow	*Flavobacterium* sp (fermenter)
	Pseudomonas paucimobilis
	Chryseomonas luteola
	Flavimonas oryzihabitans
	Sphingobacterium sp
	Pseudomonas stutzeri (light yellow)
Pink	*Methylobacterium* sp
Purple (MacConkey agar)	*Acinetobacter* sp
Blue-green	*Pseudomonas aeruginosa*
Violet	*Chromobacterium violaceum* (fermenter)
Lavender to lavender-green	*Flavobacterium* sp
(blood agar)	*Xanthomonas maltophilia*
Tan (occasionally)	*Pseudomonas stutzeri*
	Shewanella putrefaciens
Wrinkled colonies	*Pseudomonas stutzeri*
	Pseudomonas pseudomallei
Odors	
Sweet	*Alcaligenes faecalis*
	Flavobacterium odoratum
	Pseudomonas aeruginosa (grapes)
Popcorn	EO-4
	EF-4ab
Nonmotile	*Acinetobacter* sp
	Moraxella sp
	Flavobacterium sp (fermenter)
	Sphingobacterium sp (may "glide")
	Oligella sp (non-*ureolytica*)
Oxidase negative	*Acinetobacter* sp
	Xanthomonas maltophilia
	Chryseomonas sp
	Flavimonas sp
	Pseudomonas cepacia (+/−)
H₂S positive	*Shewanella putrefaciens*

Thanks to Anne Morrissey (Cleveland, OH), George Manuselis, and Connie Mahon.

flagella (see Fig. 18–11), and accumulate hydroxybutyrate. Ubiquitous in soil and water, *Comamonas* species are rarely isolated from clinical specimens but have been found in hospital equipment and fluids. Rarely, isolates of this genus have been reported to cause nosocomial bacteremia, corneal ulcerations, endocarditis in intravenous drug abusers, sepsis, and pyoarthrosis. *Comamonas acidovorans* may be more resistant than other members of the genus, having demonstrated resistance to aminoglycosides.

Methylobacterium extorquens

Methylobacterium extorquens used to be called *Pseudomonas mesophilica*. Isolates produce a characteristic pink to coral pigment, prefer a lower temperature for growth (25–35°C), produce distinctive vacuoles, and utilize methanol (hence the newer nomenclature). These isolates are often first seen on fungal media, such as

Sabouraud agar, and do not grow as well on blood or chocolate agar. Buffered charcoal yeast extract is also a good medium for their recovery.

Epidemiologically, *M. extorquens* are isolated from soil, sewage, water and hospital nebulizers and can be isolated from clinical specimens such as throat swabs and bronchial and even blood specimens. Clinically, they have been reported to cause bacteremia (one case) and skin ulcers. Contaminated tap water has been implicated as a cause of positive blood cultures in a bone marrow transplant patient receiving irrigations. The only other pink-pigmented nonfermentative bacilli are a group of coccoid bacteria not otherwise named.

MISCELLANEOUS NONFERMENTING GRAM-NEGATIVE BACILLI

Nonfermenters with Peritrichous Flagellation

Alcaligenes Species

The family Alcaliginaceae includes species of *Alcaligenes* and *Bordetella*. The *Alcaligenes* genus includes the following species: *A. faecalis* (*odorans*), *A. piechaudii*, *A. xylosoxidans* subsp *xylosoxidans* and subsp *denitrificans*.

Isolates of *Alcaligenes* are found in water (swimming pools, tap water, dialysis fluids) and are resistant to disinfectants such as chlorhexidine and quaternary ammonium compounds. They are isolated in specimens from hospitalized patients such as urine, feces, sputum, and wounds. *Alcaligenes faecalis* has been isolated from blood of patients with and without septicemia. *Alcaligenes xylosoxidans* subsp *xylosoxidans* has been associated in cases of otitis media, meningitis, pneumonia, surgical wound infections, urinary tract infections, and peritonitis; and 21 cases of bacteremia due to this organism have been reported. There is also a report of in utero transmission of the organism in a case of neonatal meningitis.

Alcaligenes piechaudii was reported to be isolated from an ear discharge of a diabetic man. Gram-stained smears of the exudate were positive for gram-negative bacilli and gram-positive cocci. Both coagulase-negative staphylococci and *A. piechaudii* were repeatedly isolated. The patient recovered once the diabetes was stabilized.

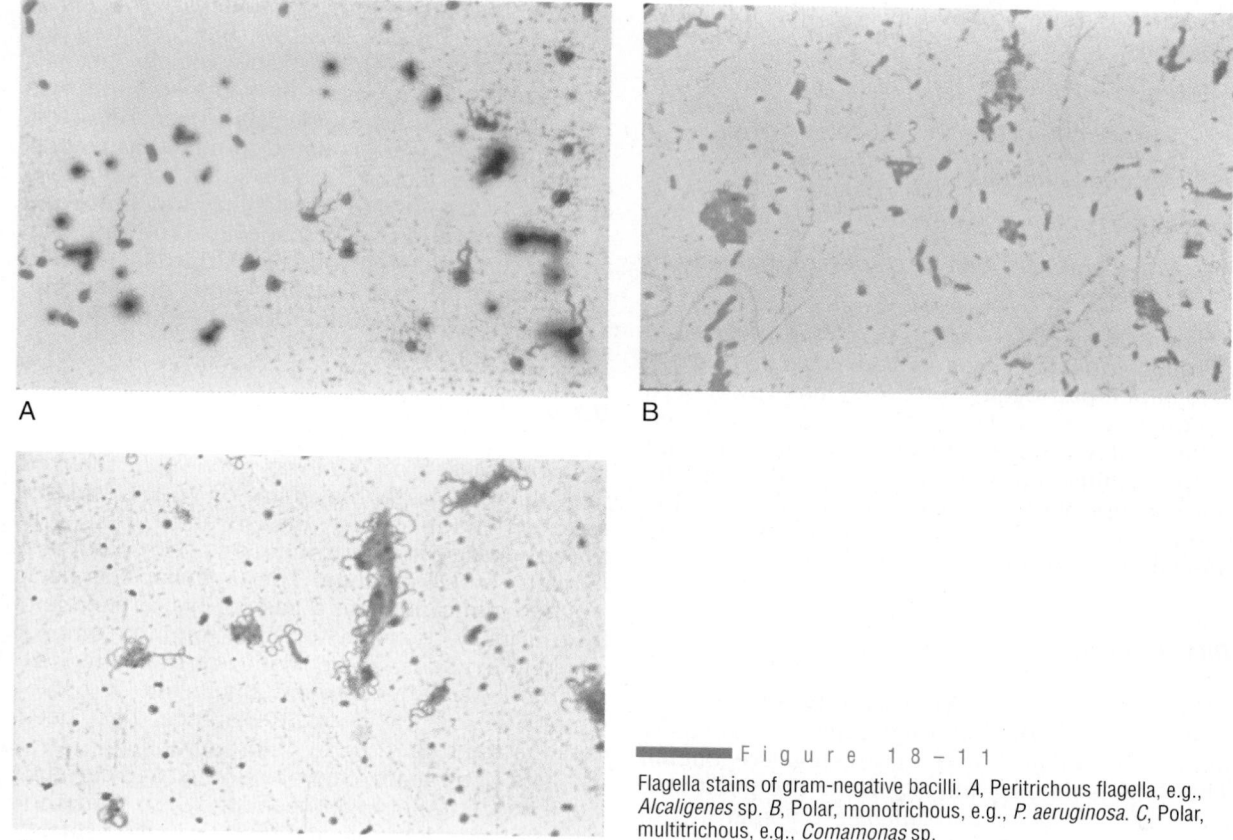

A

B

C

■■■■■■■■Figure 18-11

Flagella stains of gram-negative bacilli. *A*, Peritrichous flagella, e.g., *Alcaligenes* sp. *B*, Polar, monotrichous, e.g., *P. aeruginosa*. *C*, Polar, multitrichous, e.g., *Comamonas* sp.

All *Alcaligenes* species possess peritrichous flagella (Fig. 18–11), are oxidase positive, usually are nonsaccharolytic, and are obligately aerobic and gram-negative. These organisms usually grow well on most laboratory media, including MacConkey agar. In OF media, most are nonoxidative and produce a deep blue color at the top, except for *A. xylosoxidans* subsp *xylosoxidans*, which produces an acid reaction in both glucose and xylose; hence, the name. All species reduce nitrates to nitrites; both the subspecies of *A. xylosoxidans* and *A. faecalis* can further reduce nitrites to nitrogen gas. Isolates of *Alcaligenes* species are negative for indole production and for esculin and gelatin hydrolysis.

Isolates of members of the genus *Alcaligenes* are usually susceptible to SXT, piperacillin, ticarcillin, carbenicillin, ceftazidime, cefoperazone, and quinolones (although variably). Resistance to the aminoglycosides is common. In addition, *A. piechaudii* may be susceptible to amoxicillin.

Agrobacterium Species

The genus *Agrobacterium* is a group of nonfermentative, peritrichously motile plant pathogens that are ubiquitous in soil and water. Three species are known to infect plants: *A. tumefaciens,* the cause of crown gall, *A. rhizogenes,* the cause of hairy root disease, and *A. rubi,* the cause of cane gall. A variant of *A. tumefaciens,* the pathovar *Agrobacterium radiobacter* is the only member of the genus isolated from human clinical specimens. It has been found in sputum, pleural fluid, synovial fluid, and urine. At one time, this isolate was referred to as Vd-3. Infections by the organism have included prosthetic valve endocarditis, septicemia (with pneumonia), and three cases of catheter-related infection.

Isolates of *A. radiobacter* can be recognized by the characteristic production of 3-ketolactone, which is due to the oxidation of lactose at the C-3 glycerol moiety. They are detected in the clinical laboratory more commonly by the following reactions:

- Oxidase positive
- H_2S production (using lead acetate strips)
- Peritrichous flagella
- Positive esculin hydrolysis
- Positive urease test (to differentiate from *Alcaligenes* sp)

There is a related group of yellow-pigmented *Agrobacterium*-like, gram-negative bacilli that are also plant pathogens, but they have not been adequately characterized or named as yet. Isolates of *Agrobacterium* sp are usually resistant to penicillins, chloramphenicol, and cefazolin but susceptible to second- and third-generation cephalosporins, polymyxin, SXT, imipenem, and the aminoglycosides.

Oligella ureolytica

Oligella ureolytica (CDC Group IVe) belongs to a genus of bacteria that are nonoxidative, usually nonmotile, small, paired, gram-negative bacilli. They are isolated clinically from the genitourinary tract. *Oligella ureolytica* is unusual in the genus, because most isolates are motile by means of peritrichous flagella; hence the relationship to *Alcaligenes* sp as well as to another genus, *Taylorella*. Isolates of *O. ureolytica* are urea positive and reduce nitrates to nitrites. They are also phenylalanine deaminase (PDA) positive, which helps differentiate them from *Alcaligenes* sp. Clinically, there have been urine isolates of this organism as well as one report of a blood culture isolate from a patient with obstructive uropathy in which the isolate was believed to be significant. Another species of *Oligella, O. urethralis,* is described later in the section "*Moraxella, Moraxella*-Like Genera, and *Oligella* Species."

CDC Group IVc-2

CDC Group IVc-2 organisms are morphologically similar to *O. ureolytica* but differ biochemically in that they are not able to reduce nitrates to nitrites and are PDA negative. These organisms are most like the *Alcaligenes eutrophus* species. CDC IVc-2 is a motile, oxidase-positive, catalase-positive, asaccharolytic, gram-negative bacillus. Isolates are urease positive but negative for gelatin, esculin, and indole. They usually grow on MacConkey agar. Cases of septicemia in two patients undergoing CAPD have been reported. The organisms were isolated in mixed as well as pure cultures. Isolates are resistant to aminoglycosides, vancomycin, ampicillin, and first- and second-generation cephalosporins. They are usually susceptible to ciprofloxacin, third-generation cephalosporins, piperacillin, and doxycycline, although not many isolates have been tested.

Ochrobactrum anthropi

Ochrobactrum anthropi was at one time referred to as *Achromobacter* biovar 1, 2 or Vd-1, 2. Isolates are positive for oxidase, urease, and H_2S in addition to possessing peritrichous flagella. They are oxidative in OF medium. The organism has been isolated clinically from a case of bacteremia and from urinary tract infections in immunocompromised individuals. There is one report in which *O. anthropi* was judged to be significant in osteochondritis of the foot. Resistance has been demonstrated in vitro to chloramphenicol, tetracyclines, aztreonam, cephalosporins, and carbenicillin. Susceptibility has been shown to trimethoprim-sulfamethoxazole and the aminoglycosides.

Nonmotile, Gram-Negative Bacilli

Flavobacterium Species

The genus *Flavobacterium* consists of isolates of gram-negative weakly fermentative bacilli that are ubiquitous in soil and water. They are not considered part of the normal human flora. Because isolates are often found contaminating hospital equipment, patients may become colonized readily while in the hospital. *Flavobacterium* sp are thus responsible for a share of nosocomial nonfermentative, gram-negative infections. Even though they are weak fermenters, the reactions are usually delayed, and the isolates initially appear to be nonfermenters; hence their discussion in this chapter.

Flavobacterium sp are long, thin bacilli, often with bulbous ends. They are nonmotile and often possess a yellow intracellular pigment (especially prominent in IIb) (Fig. 18–12). On media with blood, a lavender-green discoloration of the agar

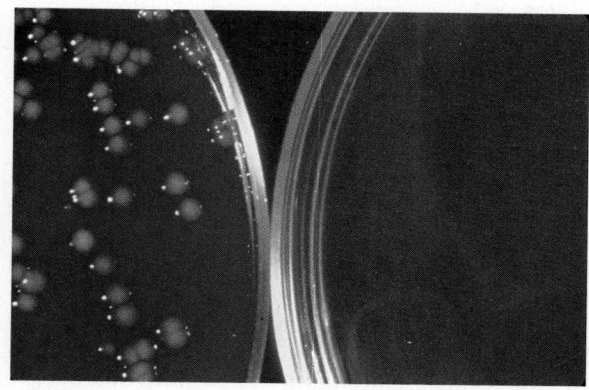

■■■■■ F i g u r e   1 8 – 1 2

Flavobacterium meningosepticum. Note absence of growth on MacConkey plate and growth with yellow pigment on blood agar.

may occur owing to the proteolytic activity of the organisms. Some species release a characteristic fruity odor. Most are DNase positive, oxidase positive, and gelatin hydrolysis positive. All except *F. odoratum* are indole positive, a very distinctive characteristic that separates *Flavobacterium* from other species of nonfermenters. Clinically important species of the genus include *F. indologenes, F. meningosepticum, F. odoratum, F. breve* and unnamed species: IIb, IIe, IIh, and IIi. The species IIb is more common in clinical specimens than all other flavobacteria.

Most diseases produced by members of the genus *Flavobacterium* that have been reported are due to *F. meningosepticum.* The disease is typically a meningitis or septicemia in a newborn, especially in conjunction with prematurity. In adults, *F. meningosepticum* may cause pneumonia, endocarditis, bacteremia, and meningitis. Some of the species in vitro are susceptible to penicillin, which is also an unusual characteristic of gram-negative bacilli, including the nonfermenters. The response to other antibiotics is variable, and an in vitro susceptibility test is needed.

Two of the prior unnamed *Flavobacterium* species, IIf and IIj, have been renamed as members of the genus *Weeksella.* These isolates are nonsaccharolytic and indole positive. *Weeksella virosa* colonies may be mucoid or "slimy" and possess a yellow-green pigment. Isolates have been found from genitourinary specimens and will grow on Thayer-Martin or other *Neisseria gonorrhoeae*–selective media. *Weeksella zoohelcum* is nonmucoid, although its colonies may be sticky; it is urease positive and otherwise similar

to *W. virosa.* It has been isolated from many sources, in particular from dog bite wounds.

Sphingobacterium Species

Isolates of another genera, *Sphingobacterium,* were at one time regarded as members of the genus *Flavobacterium.* These isolates are aflagellate but do produce a gliding motility characteristic of the genus. They are oxidase, catalase, and esculin positive. Unlike the *Flavobacterium* sp, they are indole negative. What is unique to the group is the presence of sphingophospholipids in the cell wall. They are truly fermenters and oxidize some carbohydrates. Growth on MacConkey agar is sparse to none, but these organisms will grow in the presence of 40% bile. Two species, *S. spiritovorum* and *S. multivorum,* are very similar biochemically, but *S. spiritovorum* produces acid from mannitol, ethanol, and rhamnose, and *S. multivorum* does not. Both produce a yellow pigment similar to that of isolates of *Flavobacterium* sp. Clinically, *S. multivorum* has been isolated from blood of patients with septicemia and from cases of peritonitis. *Sphingobacterium spiritovorum* has also been isolated from clinical specimens and from hospital environments.

There is another species in the genus *Sphingobacterium, S. mizutae,* formerly called *Flavobacterium mizutaii.* They are oxidase positive, indole negative, and esculin positive. *Sphingobacterium mitzutae* has been isolated from a case of meningitis in premature birth. Most isolates of *Sphingobacterium* sp are sensitive to trimethoprim-sulfamethoxazole and resistant to penicillins, cephalosporins, aminoglycosides, clindamycin, and polymyxin.

Chryseomonas and *Flavimonas*

Chryseomonas luteola (formerly CDC Ve-1) and *Flavimonas oryzihabitans* (formerly CDC Ve-2) are gram-negative, nonfermentative, oxidase-negative bacilli. They both are catalase positive and nonmotile, grow on MacConkey agar, and often produce a yellow pigment. *Flavimonas* species specifically may produce wrinkled or rough colonies at 48 hours. *Chryseomonas luteola,* possessing multitrichous flagella, is positive for esculin, arginine, urease, and nitrate. *Flavimonas* sp is negative for these biochemicals and possesses a single polar flagellum. Both oxidize glucose and other sugars.

The natural habitat of species of *Flavimonas* and *Chryseomonas* is unknown, although *Flavimonas* has been found in Japanese rice paddies and has been isolated from hospital drains and respiratory therapy equipment. These two organisms are rarely isolated from humans, but the literature reports of their isolation from wounds, abscesses, and blood cultures, as well as peritoneal fluid, both CAPD and other. Both have been implicated in cases of bacteremia and peritonitis, and possibly meningitis, although many times in association with each other or with other bacteria. *Chryseomonas luteola* has been recovered as the only isolate from a case of prosthetic valve endocarditis and subdiaphragmatic abscess. *Flavimonas* sp has been isolated from eye cultures and was described as the cause of "sticky eye" in one patient who responded to therapy appropriate for *Flavimonas*. There appears to be higher risk for infection by these organisms in the presence of foreign materials, (i.e., catheters), corticosteroid use, and/or immunocompromised states.

Both *C. luteola* and *F. oryzihabitans* are susceptible to aminoglycosides, third-generation cephalosporins, ureidopenicillins, and quinolones. *Flavimonas* sp is resistant to first- and second-generation cephalosporins but usually susceptible to penicillin. *Chryseomonas luteola* is usually sensitive to cephalosporins and penicillin. Both demonstrate variable susceptibilities to tetracycline, chloramphenicol, and SXT.

Moraxella, *Moraxella*-Like Genera, and *Oligella* Species

Members of the genus *Moraxella* are strongly oxidase positive, nonmotile, coccobacillary to bacillary gram-negative bacilli. They are, in general, biochemically inert with regard to carbohydrate oxidation and alkalinization. They are strictly aerobic and most often susceptible to penicillin, an unusual characteristic for the nonfermenters. These isolates are opportunists that reside on mucous membrane of humans and lower animals and can be isolated as such from the respiratory tract, urinary tract, and eyes; however, they rarely cause disease in humans. Members of the genus *Moraxella* commonly encountered include:

■ *M. nonliquefaciens*

■ *M. lacunata*

■ *M. osloensis*

■ *M. phenylpyruvica*

■ *M. atlantae*

Moraxella nonliquefaciens is the most commonly isolated member of the genus. It often resides as normal flora in the respiratory tract and rarely causes disease in humans. Rare cases of bacteremia, keratitis, and endophthalmitis caused by *M. nonliquefaciens* have been reported. The organism is gelatin hydrolysis negative and urease negative, and does not usually grow on MacConkey agar. Its phenylalanine deaminase (PD) reaction is negative as well. *Moraxella osloensis* is very similar morphologically and biochemically to *M. nonliquefaciens;* unlike the latter, however, it grows and produces an alkaline reaction in acetate medium and acidifies ethanol. *Moraxella osloensis* is found as normal flora in the genitourinary tract.

Moraxella phenylpyruvica has been isolated from urine, blood, CSF, and the genitourinary tract. Unlike the others, it is urease positive and it is also PD positive. Very similar to *M. phenylpyruvica* is *M. atlantae*, a more fastidious member of the genus that will, however, grow on MacConkey agar. It may also spread and pit the agar surfaces of the medium on which it grows. It is gelatin negative, PD negative, and nitrate negative. *Moraxella lacunata,* a common conjunctival isolate, is a small coccobacillus that is usually gelatin positive, urease negative, and unable to grow on MacConkey agar. It may be PD positive.

M-5 and M-6 are two closely related strains of *Moraxella* that are oxidase positive, nonmotile, asaccharolytic coccobacilli. M-5 has been isolated from wounds of animal bites. It is usually PD positive and gelatin positive and often produces a water-soluble yellow-tan pigment. It is nitrate negative but may reduce nitrites. The tentative name of these isolates is *Neisseria parelongata*. M-6 has been isolated from numerous body sites and implicated in cases of endocarditis and osteomyelitis. Unlike the other members of the *Moraxella* and *Moraxella*-like genera, M-6 is catalase negative. It usually does not grow on MacConkey agar. Its tentative name is *Neisseria elongata*. M-6 may be resistant to penicillin but, like most other isolates, is sensitive to the aminoglycosides, ampicillin, chloramphenicol, and many other antimicrobial agents.

Related to the genus *Moraxella* is the genus *Oligella*. Members of this genus include *O. urethralis,* formerly called *M. urethralis*. These are

PD, nitrate, and nitrite positive with gas formation, and gelatin negative. They are small coccoid organisms, most often isolated from the gastrointestinal tract. They usually do not grow on MacConkey agar, and unlike members of the genus *Moraxella,* they alkalinize citrate when the test is performed in an aerobic low-glucose peptone (ALP) slant. Unlike other members of the genus *Oligella (O. ureolytica),* O. urethralis does not oxidize sugars and is nonmotile (see earlier discussion under *Oligella ureolytica).*

Other Gram-Negative Nonfermenters

This section discusses a miscellaneous group of nonfermentative bacilli and fastidious gram-negative bacilli that are difficult to place in specific groups.

EO-2 and *Psychrobacter*

EO-2

Isolates of EO-2 are coccoid, vacuolated, or peripherally staining O-shaped gram-negative bacteria in pairs, short chains, or packets. They are nonmotile, are mucoid on agar plates, and oxidize glucose, xylose, lactose, and mannitol. They do not always grow on MacConkey agar, are oxidase and catalase positive, and reduce nitrates to nitrites. EO-2 isolates are negative biochemically for indole, decarboxylases, esculin, and gelatin but positive for urease. Isolates have been obtained from urine, eye, blood, pleural fluid, CSF, lung, throat, and genitourinary tract specimens. Their clinical significance is unclear.

PSYCHROBACTER SPECIES

Psychrobacter immobilis are cold-loving (psychrotrophic) nonmotile, oxidase-positive, oxidative diplococci. They resemble *Moraxella* species and, like them, are penicillin susceptible. They have been associated with fish, processed meat, and poultry. Clinically, they have been isolated from the eye of a newborn, who had acquired the infection nosocomially via a water source. There was also a report of *P. immobilis* isolation from the blood and CSF of a 2-day-old infant. Isolates grow well at 5 to 25°C but not at 35 to 37°C. They are nitrate positive and can grow on Thayer-Martin medium.

Dysgonic Fermenters (DF)

Strains of gram-negative bacilli referred to by the CDC as dysgonic fermenters (DF) are a group of slowly growing, fastidious bacilli that do ferment sugars but require the addition of increased CO_2 to the medium and/or an environment for isolation. DF-2 and DF-2–like are two such strains. DF-1 are now called *Capnocytophaga* sp and are discussed in this chapter along with the HACEK organisms. DF-2 and DF-2–like have been named by CDC as members of the genus *Capnocytophaga* as well. DF-2 is now named *C. canimorsus,* and DF-2–like is *C. cynodegmi.* Like other *Capnocytophaga* sp, these two species are oxidase and catalase positive with gliding motility. Both are negative for indole and gelatin and are usually nitrite positive. Both ferment glucose and lactose. *Capnocytophaga canimorsus* ferments galactose and glycogen most of the time, whereas *C. cynodegmi* does so less often; *C. cynodegmi* ferments melibiose, raffinose, and sucrose, and *C. canimorsus* does not.

Capnocytophaga canimorsus can cause sepsis, with or without endocarditis, cellulitis, and meningitis. Affected patients are most likely to have asplenia or alcoholism, or to be receiving cortiocosteroids. There is often an associated prior dog or cat bite in the majority of clinical cases. Endophthalmitis following corneal transplantation has also been reported.

Capnocytophaga spp are susceptible to β-lactams, except for the first- and second-generation cephalosporins, which have intermediate to poor activity (except for cefoxitin). There have been reports of β-lactamase–producing strains, but the addition of a β-lactamase inhibitor can restore good activity. DF isolates are also susceptible to quinolones, metronidazole, clindamycin, tetracycline, and imipenem but demonstrate resistance to aminoglycosides, vancomycin, and semisynthetic penicillins.

DF-3

DF-3 is a short rod to coccobacillary, nonmotile, fastidious, gram-negative fermentative bacillus. It requires increased CO_2 like members of *Capnocytophaga* and addition of serum to enhance the fermentations. It can have a sweet to bitter odor on solid medium and will not grow on MacConkey agar. Indole reaction may be weakly positive, esculin reaction is positive, and the isolates are negative for nitrate and the decarboxylases. Fermentation occurs with glucose, lactose, maltose, and raffinose. Isolates have been reported from blood, urine, wounds, peritoneal abscesses, and stool specimens. In the last, isolation is often associated with patients who have hypogammaglobulinemia with chronic diarrhea, but the role of DF-3 in disease is unclear. Isolates are susceptible to carbenicillin, chloramphenicol, tetracycline, clindamycin, and SXT, but resistant to ß-lactams, quinolones, metronidazole, vancomycin, and erythromycin.

Eugonic Fermenters (EF)

Eugonic fermenters are a group of nonfastidious fermenters, as distinguished from unnamed dysgonic fermenters, which are "fastidious" fermenters. The abbreviation for this group of gram-negative bacilli is EF, and EF-4a and EF-4b are discussed here. Both have been isolated from infected areas following dog and cat bites, respectively. Isolates are nonmotile, short rods to coccobacilli, oxidase and catalase positive. EF-4a is a glucose fermenter, closely resembling *Pasteurella* sp, but it ferments only glucose. It may produce a yellow to tan pigment, reduces nitrate to gas, may grow on MacConkey agar, and may liquefy gelatin or grow at 42°C. It is negative for indole and urea. EF-4a colonies give off a characteristic "popcorn-like" odor, a trait shared with EF-4b.

EF-4b is an oxidizer of glucose, not a fermenter, usually associated with cat bites or scratches. In addition, EF-4b is gelatin negative and does not reduce nitrates to gas. Both are susceptible to β-lactams, chloramphenicol, tetracycline, and aminoglycosides.

HB-5

HB-5, still unnamed, are capnophilic (CO_2-requiring) coccobacilli and rods that are facultatively anaerobic, fastidious fermenters. They ferment glucose and fructose and are catalase positive, oxidase positive or negative, and indole positive. They may grow on MacConkey agar and are nonmotile. Isolates have been obtained from placenta, amniotic fluid, blood, rectal sites, abscesses, and urogenital specimens.

Chromobacterium Species

Chromobacterium violaceum is a fermentative gram-negative bacillus that may be oxidase posi-tive. It is motile with polar flagella and, as its name implies, produces a violet pigment about 91% of the time. The pigment is violacein, an ethanol-soluble, water-insoluble pigment. The presence of the pigment may hamper proper oxidase reactions. Isolates are usually indole negative, but nonpigmented strains may be indole positive. Isolates ferment glucose and, variably, sucrose; grow on MacConkey agar and most enteric media; reduce nitrate; and grow at 42°C. On Gram stain, the organisms may appear as curved bacilli that may resemble vibrios if the oxidase reaction is positive. When the oxidase reaction is negative, it may resemble enterics.

Reservoirs of *C. violaceum* are soil and water. They are found more commonly in tropical and subtropical climates. They are opportunists, attacking the immunocompromised patient with neutrophil deficits, usually as a result of contamination of wounds with water or soil. They have been isolated from cases of osteomyelitis and abscesses as well as from blood, urine, and gastrointestinal infections. *Chromobacterium violaceum* are sensitive to chloramphenicol, tetracycline, and SXT, but resistant to cephalosporins and variably resistant to penicillin and the aminoglycosides. Many patients do not do well, however, even with adequate therapy, because of their underlying disease.

HACEK Group of Gram-Negative Bacilli

HACEK is an acronym consisting of the first initial of each genus represented in the group:

■ *Haemophilus* sp (*H. aprophilus, H. influenzae,* and *H. parainfluenzae*)

■ *Actinobacillus actinomycetemcomitans*

■ *Cardiobacterium hominis*

■ *Eikenella corrodens*

■ *Kingella* sp

This group of gram-negative bacilli have in common the need for an environment with enhanced CO_2 (a candle extinction jar or incubator with 5 to 10% CO_2) and the predilection for attachment to heart valves (usually damaged or prosthetic) that makes them major causes of endocarditis. Members of this group include both fermenters and nonfermentative gram-negative bacilli. All of the members are usual normal flora of the oral cavity, allowing for their introduction in the blood stream and resultant infections. All are opportunists as well and require a immunocompromised host or host tissue. The genus *Haemophilus* is described in Chapter 15 of this textbook and is not further reviewed here. Although not a strict member of the HACEK group, another gram-negative bacillus, *Capnocytophaga* sp, is included in this series of discussions because it is similar in its requirement for CO_2 to enhance growth and its isolation from blood cultures. Unlike the other members of the group, it is not as commonly involved in endocarditis as is septicemia in the granulocytopenic patient.

Actinobacillus actinomycetemcomitans

Actinobacillus actinomycetemcomitans is a member of a genus that includes animal pathogens or animal endogeneous flora that in general do not routinely cause infections in humans. Other members of the genus include *A. suis* (pigs); *A. lignieresii* (sheep and cattle); *A. equuli* (horses); *A. capsulatus,* and *A. hominis.* All are small rods to coccoid gram-negative bacilli that are nonmotile. *Actinobacillus actinomycetemcomitans* is found as mouth flora, but clinically they have been isolated from blood, lung tissue, abscesses of the mouth, and sinuses. This organism rarely causes osteomyelitis, meningitis, pericarditis, and urinary tract infections. Although rare, *A. actinomycetemcomitans* is commonly isolated from the blood as the causative agent of bacterial endocarditis with an insidious and protracted presentation.

Actinobacillus actinomycetemcomitans is fastidious as all members of the HACEK group are, requiring increased CO_2 at least for initial isolation from clinical specimens. It is a fermenter, although the addition of serum to the carbohydrate-containing tubes is often necessary to demonstrate this. The isolates may require greater than 24 hours for isolation; distinctive star formation in the center of the colonies is often seen at 48 hours (Fig. 18–13). In broth, the organism is granular and may adhere to the sides of the tube. Isolates are catalase positive, variable in the oxidase reaction, do not grow on MacConkey agar, and are negative for urease, indole, esculin, and citrate. Fermentations are positive for glucose (with or without positive gas), and somewhat variable for xylose, mannitol, and maltose. The isolates do not ferment lactose or sucrose. They demonstrate sensitivity to penicillin in vitro, although this agent is not always successful clinically. In addition, isolates are susceptible to

aminoglycosides, third-generation cephalosporins, quinolones, SXT, rifampin, chloramphenicol, and tetracycline. Resistance is common to vancomycin and erythromycin. Usual treatment for endocarditis is with a penicillin and an aminoglycoside.

Cardiobacterium hominis

Cardiobacterium hominis (IID, *Pasteurella*-like), a pleomorphic, nonmotile, fastidious, gram-negative bacillus, is another member of the HACEK group. Gram stains often show false-positive reaction. The organisms tend to form rosettes, swellings, and, in yeast extract, stick-like structures. They grow on blood showing α-hemolysis and chocolate agar but do not grow on MacConkey agar. On agar, pitting may be produced. *Cardiobacterium hominis* is a fermenter, but as with *A. actinomycetemcomitans,* reactions may be weak and serum may be needed to enhance them. This organism ferments glucose, mannitol, sucrose (unlike *A. actinomycetemcomitans*), and maltose. Isolates are oxidase positive, catalase negative, and indole positive, the latter two traits helping to further differentiate them from *Actinobacillus* sp. They are negative for urease, nitrate, gelatin, and esculin.

Cardiobacterium hominis is normal flora of the nose, mouth, and throat and may be present in the gastrointestinal tract. The usual clinical manifestation is that of endocarditis, often presenting

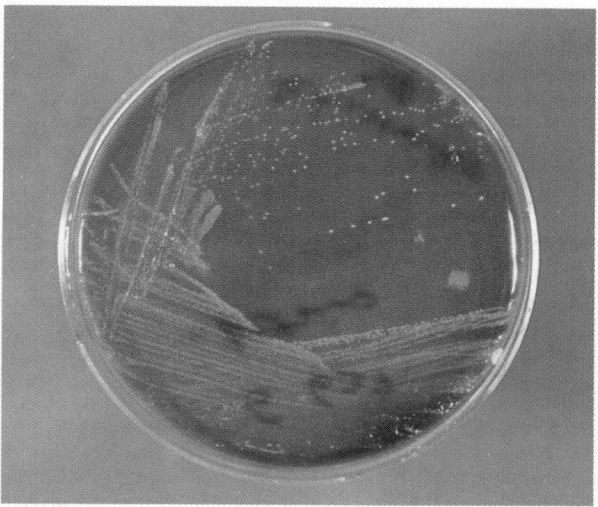

■■■ F i g u r e   1 8 – 1 3
Actinobacillus actinomycetemcomitans on blood agar.

with very large vegetations and no demonstratable fever. Rarely, isolates cause meningitis. Sensitivity can be seen to β-lactams, chloramphenicol, and tetracycline with variable response to aminoglycosides, erythromycin, clindamycin, and vancomycin. Usual modes of therapy include penicillin and an aminoglycoside.

Eikenella corrodens

Eikenella corrodens (HB-1) are fastidious gramnegative coccobacilli that grow best under conditions of increased CO_2 and hemin with or without added cholesterol. They are nonmotile, oxidase positive, and nonsaccharolytic and hence are very similar to the *Moraxella* sp. Unlike the latter, however, they are catalase negative and often produce a yellow pigment. About 45% of the isolates of *E. corrodens* pit or corrode the surface of the agar. In broth medium, they may adhere to the sides of the tube and produce granules. A bleach-like odor from the agar surface may be obvious . They do not usually grow on MacConkey or eosin–methylene blue agar, and they are LDC and ODC positive and ADH negative. Characteristically, they are resistant to clindamycin.

Eikenella corrodens are part of the normal flora of the oral and bowel cavities. Most infections associated with these organisms have been mixed and occur often as a result of trauma, especially after human fights or bites (i.e., after the skin has been broken by human teeth). They have been reported as the cause of meningitis, empyema, pneumonia, osteomyelitis, arthritis, and postoperative tissue infections. In drug addicts, they have been implicated in cellulitis as a result of direct inoculation of the organisms into the skin after oral contamination of needle paraphernalia. *Eikenella corrodens,* another member of the HACEK group, shows a predilection for attachment to heart valves and thus causes endocarditis.

In vitro, isolates demonstrate sensitivity to penicillin, ampicillin, cefoxitin, chloramphenicol, carbenicillin, and imipenem. Resistance is seen with clindamycin and the aminoglycosides.

Kingella Species

Members of the genus *Kingella* are coccobacillary to short rods that exist in pairs or short chains. They are nonmotile but may demonstrate a "twitching" motility. They are nutritionally fastidious, oxidase positive, catalase negative fermenters of glucose and other sugars.

Kingella denitrificans may grow on Thayer-Martin medium and, if it does not pit the agar as many strains do, may resemble *Neisseria gonorrhoeae.* Gram stain morphology as a rod when isolated on blood agar should aid in distinguishing *Kingella* sp. It is positive for glucose fermentation and nitrate reduction and may grow at 42°C. It is negative for urease, indole, esculin, gelatin, and citrate and does not grow on MacConkey agar. This species is rarely isolated as a pathogen but has been associated with bacteremia and abscesses.

Kingella indologenes differs from *K. denitrificans* in being positive for indole and fermentation of maltose and sucrose. The rods are more plump than those of *K. denitrificans.* Eye infections with this species have been reported.

Kingella kingae is a weakly fermentating member of the genus. It is negative for sucrose but positive for glucose and maltose; it may produce a yellow-brown pigment, unlike other *Kingella* spp, and may pit the agar. It is biochemically inactive otherwise. Isolates have been obtained clinically from blood, bone, joint fluid, urine, and wounds. Most isolates are from children less than 5 years old. Isolates of *Kingella* are usually susceptible to most agents, including penicillin.

Capnocytophaga Species

Capnocytophaga species, including *C. ochracea (Bacteroides ochraceus)*, *C. gingivalis,* and *C. sputigena,* are fastidious, pitting fermentative gram-negative bacillus that are normal flora of the mouth and oral cavities. They are thin and often fusiform on Gram stain, resembling *Fusobacterium* sp, and may produce a gliding motility on agar surfaces, although flagella are usually absent. Like HACEK organisms, they may require an enhanced CO_2 atmosphere for isolation and an extra day of incubation for culture. On agar surfaces, they often produce a yellow-orange pigment and are nonhemolytic. They ferment (although TSI tubes may be negative without enrichment) sucrose, glucose, maltose, and lactose. They are negative for most biochemical reactions, although they may reduce nitrates and may hydrolyze esculin.

Common sites of clinical isolation include blood cultures from the granulocytopenic patient who has oral ulcers (source of the *Capnocytophaga*), juvenile periodontal disease, and endocarditis. *Capnocytophaga ochracea,* the most common isolate clinically, is susceptible to ampicillin, penicillin, clindamycin, ureidopenicillins, piperacillin, imipenem, erythromycin, tetracycline, chloramphenicol, third-generation cephalosporins, quinolones, and metronidazole. It is resistant to trimethoprim and aminoglycosides and shows variable response to aztreonam and first- and second-generation cephalosporins.

Bibliography

Baron EJ, Finegold SM: Bailey & Scott's Diagnostic Microbiology, 8th ed. St. Louis: CV Mosby, 1990.

Bergogne-Berezin E, Joyl-Guillou ML: An underestimated nosocomial pathogen *Acinetobacter calcoaceticus.* J Antimicrob Chemother 16:535, 1985.

Bisbe J, Gatell JM, et al: *Pseudomonas aeruginosa bacteremia:* Univariate and multivariate analyses of factors influencing the prognosis in 133 episodes. Rev Infect Dis 10:629, 1988.

Bouvet PJM, Grimont PAD: Taxonomy of the genus Acinetobacter with the recognition of *Acinetobacter baumannii* sp. nov., *Acinetobacter haemolyticus* sp. nov., *Acinetobacter johnsonii* sp. nov. and *Acinetobacter junii* sp. nov. and emended descriptions of *Acinetobacter calcoaceticus* and *Acinetobacter Iwoffii.* Int J System Bacteriol 36:228, 1986.

Carson LA, Tablan OC, et al: Comparative evaluation of selective media for isolation of *Pseudomonas cepacia* from cystic fibrosis patients and environmental sources. J Clin Microbiol 26:2096, 1988.

Clark WA, Hollis DG, et al: Identification of Unusual Pathogenic Gram-Negative Aerobic and Facultatively Anaerobic Bacteria. Centers for Disease Control, 1984.

Elting LS, Bodey GP: Septicemia due to *Xanthomonas* species and non-*aeruginosa Pseudomonas* species: Incidence of catheter related infections. Medicine 69:296, 1990.

Forlenza SW: *Capnocytophaga:* An update. Clin Microbiol Newsl 13:89, 1991.

Freney J, Hansem W, et al: Septicemia caused by *Sphingobacterium multivorum.* J Clin Microbiol 25:1126, 1987.

Fyfe JAM, Harris G, Govan GRW: Revised pyocin typing method for *Pseudomonas aeruginosa.* J Clin Microbiol 29:47, 1984.

Gilardi GL: *Pseudomonas* and related genera. In Balows A (ed) Manual of Clinical Microbiology, 5th ed. Washington DC: American Society for Microbiology, 1991, pp 429–441.

Gilardi GL: Update on taxonomy of nonfastidious glucose-nonfermenting gram-negative bacilli. Clin Microbiol Newsl 12:73, 1990.

Glupczynski Y, Hansen W, et al: *Pseudomonas paucimobilis* peritonitis in patients treated with peritoneal dialysis. J Clin Microbiol 20:1255, 1984.

Goldman DA, Klinger JD: *Pseudomonas cepacia:* Biology, mechanisms of virulence, epidemiology. J Pediatr 108:806, 1986.

Hammerberg O, Bialkowska-Hobrzanska O, Gopaul D: Isolation of *Agrobacterium radiobacter* from a central venous catheter. Eur J Clin Microbiol Infect Dis 10:450, 1992.

Hearn YR, Gardner RM: *Achromobacter xylosoxidans.* An unusual neonatal pathogen. Am J Clin Pathol 96:211, 1991.

Kerr KG, Corps CM, Hawkey PM: Infections due to *Xanthomonas maltophilia* in patients with hematologic malignancy. Rev Infect Dis 13:762, 1991.

Khardori N, Elting L, et al: Nosocomial infections due to *Xanthomonas maltophilia (Pseudomonas maltophilia)* in patients with cancer. Rev Infect Dis 12:997, 1990.

Kitch T, Jacobs MR, Applebaum PC: Evaluation of the 4H NF plus method for identification of 86 gram-negative nonfermentative rods [abstract #149]. American Society for Microbiology General Meeting, Dallas, TX, 1991.

Kostman JR, Solomon F, Fekete T: Infections with *Chryseomonas luteola* (CDC Group Ve-1) and *Flavimonas oryzihabitans* (CDC Group Ve-2) in neurosurgical patients. Rev Infect Dis 13:223, 1991.

Lampe AS, van der Reijden TKJ: Evaluation of commercial test systems for the identification of nonfermenters. Eur J Clin Microbiol 3:301, 1984.

Levett PN, Garrett DA, Wickramasuriya T: *Flavimonas oryzihabitans* as a cause of ocular infection. Eur J Clin Microbiol Infect Dis 10:594, 1991.

Lloyd-Puryear M, Wallace D, et al: Meningitis caused by *Psychrobacter immobilis* in an infant. J Clin Microbiol 29:2041, 1991.

Lorian V (ed): Antibiotics in Laboratory Medicine. Baltimore: Williams & Wilkins, 1986.

Marshall WF, Keating MR, et al: *Xanthomonas maltophilia:* An emerging nosocomial pathogen. Mayo Clin Proc 64:1097, 1989.

McGowan JE, Del Rio C: Other gram negative bacilli. In Mandell GL, Douglas RG, Bennett JE (eds): Principles and Practice of Infectious Diseases, 3rd ed. New York: Churchill Livingstone, 1990, pp 1782–1793.

Moss CW, Wallace PL, et al: Cultural and chemical characterization of CDC group EO-2, M-5, and M-6 *(Moraxella)* Species, *Oligella urethralis, Acinetobacter* Species and *Psychrobacter immobilis.* J Clin Microbiol 26:484, 1988.

Oberhofer TR: Use of the API 20E, Oxi/Ferm, and Minitek systems to identify nonfermentative and oxidase positive fermentative bacteria: Seven years of experience. Diagn Microbiol Infect Dis 1:241, 1983.

Peel MM, Hibberd AJ, et al: *Alcaligenes piechaudi* from chronic ear discharge. J Clin Microbiol 26:1580, 1988.

Pickett MJ, Hollis DG, Bottone EJ: Miscellaneous gram-negative bacteria. In Balows A (ed): Manual of Clinical Microbiology, 5th ed. Washington, DC: American Society for Microbiology, 1991, pp 410–428.

Rhoden DL, O'Hara CM: Evaluation of the Update Quantum II System for the identification of gram-negative bacilli. J Clin Microbiol 27:2420, 1989.

Rossau R, Kersters K, et al: *Oligella,* a new genus including *Oligella urethralis* comb. nov. (formerly *Moraxella urethralis*) and *Oligella ureolytica* sp. nov. (formerly CDC Group IVe): Relationship to *Taylorella equigenitalis* and related taxa. Int J System Bacteriol 37:198, 1987.

Stoloff AL, Gillies ML: Infections with *Eikenella corrodens* in a general hospital: A report of 33 cases. Rev Infect Dis 8:50, 1986.

Tablan OC, Chorba TL, et al: *Pseudomonas cepacia* colonization in patients with cystic fibrosis: Risk factors and clinical outcome. J Pediatr 107:382, 1985.

Tenover FC, Mizuki TS, Carlson LG: Evaluation of AutoSCAN-W/A automated microbiology system for the identification of non-glucose-fermenting gram-negative bacilli. J Clin Microbiol 28:1628, 1990.

Truant AL, Starr E, et al: Comparison of AMS Vitek, MicroScan and Autobac Series II for the identification of gram-negative bacilli. Diagn Microbiol Infect Dis 12:211, 1989.

Veys A, Callewaert W, et al: Application of gas-liquid chromatography to the routine identification of nonfermenting gram-negative bacteria in clinical specimens. J Clin Microbiol 27:1538, 1989.

Welch DF, Muszynski MJ, et al: Selective and differential medium for recovery of *Pseudomonas cepacia* from the respiratory tracts of patient with cystic fibrosis. J Clin Microbiol 25:730, 1987.

Yabuuchi E, Kaneko T, et al: *Sphingobacterium* gen nov., *Sphingobacterium apiritivorum* comb nov., *Sphingobacterium multivorum* comb nov., *Sphingobacterium mizutae* sp. nov., and *Flavobacterium indologenes* sp. nov.: Glucose nonfermenting gram-negative rods in CDC groups IIk-2 and IIb. Int J System Bacteriol 33:580, 1983.

Yu PKW, Rolfzen MA, et al: Application of QuadFERM+ for the identification of fastidious gram-positive and gram-negative bacilli. Diagn Microbiol Infect Dis 14:185, 1991.

Zapardiel J, Blu G, et al: Peritonitis with CDC Group IVc-2 bacteria in a patient on continuous ambulatory peritoneal dialysis. Eur J Clin Microbiol Infect Dis 10:509, 1991.

CHAPTER *19*

Anaerobes of Clinical Importance

Paul G. Engelkirk •
Janet Duben-Engelkirk

1. Differentiate obligate anaerobes from facultative organisms.

2. Describe how anaerobes, as part of normal flora, initiate and establish infection.

3. Given the clues to an anaerobic infection (signs and manifestations), give the most probable etiologic agent of the following:

   ■ Wound botulism
   ■ Tetanus
   ■ Gas gangrene
   ■ Actinomycosis
   ■ Lung abscess
   ■ Peritonitis

4. Give the bacteriologic indicators used to recognize anaerobes as the possible causative agent.

5. Describe the clinical infections associated with the following organisms and how they are acquired and manifested:

   ■ *Clostridium* species
   ■ Anaerobic, non–spore-forming, gram-positive bacilli
   ■ *Actinomyces*
   ■ *Bacteroides* spp
   ■ *Fusobacterium* spp
   ■ Gram-positive cocci
   ■ *Veillonella* spp

6. Describe the laboratory methods of performing cultures and identifying anaerobes:

   ■ Acceptable/unacceptable specimens
   ■ Culture environments
   ■ Isolation media
   ■ Identification systems

7. Describe the acceptable methods for performing anaerobic antimicrobic susceptibility tests.

8. Given the microscopic, colony morphology, and key reactions in the differentiating tests used to identify anaerobic isolates, identify the most notable species of the following:

   ■ *Clostridium*
   ■ Anaerobic, gram-positive, non–spore-forming bacilli
   ■ Gram-negative, non–spore-forming bacilli
   ■ Anaerobic cocci

Anaerobic bacteria are significant for a variety of reasons. They are important in human and veterinary medicine because they play a role in serious, often fatal, infections and intoxications. They can be involved in infectious processes in virtually any organ or tissue of the body and, consequently, can be recovered from most clinical specimens.

This chapter describes the following:

■ Anaerobes of clinical importance and their role in disease

■ Proper techniques for selecting, collecting, transporting, and processing clinical specimens for anaerobic bacteriology

■ Procedures for identifying and testing the susceptibility of anaerobic isolates

IMPORTANT CONCEPTS IN ANAEROBIC BACTERIOLOGY

Anaerobes Defined

In addition to Gram reaction and cellular morphology, bacteria are commonly classified in the clinical microbiology laboratory on the basis of their relationship to oxygen (O_2) and carbon dioxide (CO_2). A bacterial isolate can be classified into one of six groups:

1. Obligate aerobe
2. Microaerophilic aerobe
3. Facultative anaerobe
4. Aerotolerant anaerobe
5. Microaerotolerant anaerobe
6. Obligate anaerobe

Table 19–1 shows the classification of bacteria based on their relationship to oxygen and carbon dioxide. This type of classification is easily accomplished by comparing an organism's ability to multiply on a solid medium, usually blood agar, in four different environments:

1. Air (about 21% O_2 and 0.03% CO_2)
2. CO_2 incubator (about 15% O_2 and 5–10% CO_2)
3. Microaerophilic system (5% O_2)
4. Anaerobic system (0% O_2)

In general, anaerobes are organisms that do not require oxygen for life and reproduction. In addition, oxygen's direct toxic effect may prohibit the growth of these organisms in environments in which oxygen is present. Although the term "anaerobe" is used throughout this chapter as a synonym for anaerobic bacteria, one must understand that other types of anaerobic microorganisms, such as certain fungi and protozoa, also exist.

Obligate anaerobes, which grow only in the absence of molecular oxygen, vary in their sensitivity to oxygen and can be classified as *moderate anaerobes* and *strict anaerobes.* A moderate anaerobe, such as *Bacteroides fragilis, Fusobacterium nucleatum,* or *Prevotella melaninogenica,* is unable to multiply in an atmosphere containing more than 2 to 8% O_2. Moderate anaerobes can tolerate exposure to air for several hours on the surface of blood agar but require an anaerobic environment for multiplication. Strict anaerobes, such as *Clostridium haemolyticum, C. novyi* type B, and certain treponemes, cannot multiply

in the presence of more than 0.5% O_2 and are killed by only a few minutes' exposure to air. Fortunately, strict anaerobes are seldom associated with human infections.

Why Are They Anaerobes?

Several theories have been advanced to explain the anaerobic requirement of certain microorganisms. Common concepts that have been well-studied include oxygen toxicity, absence of protective enzymes, and bacteriostatic and bactericidal effects of oxygen.

Oxygen Toxicity. As stated by Margulis and Sagan, "Oxygen is toxic because it reacts with organic matter. It grabs electrons and produces so-called free radicals: highly reactive, short-lived chemicals that wreak havoc with the carbon, hydrogen, sulfur, and nitrogen compounds at the basis of life. Oxygen breaks down or renders useless the small metabolites—food—that otherwise become components in cellular systems. Oxygen combines with the enzymes, proteins, nucleic acids, vitamins, and lipids that are vital to cell reproduction."

Molecular oxygen itself can be toxic to some anaerobes, but substances produced when oxygen becomes reduced are even more toxic. During oxidation-reduction reactions, molecular oxygen is reduced in a stepwise manner by the addition of electrons, as shown in the following equations:

$$O_2 + e^- \longrightarrow O_2^- \text{ (superoxide anion)}$$

$$O_2^- + e^- + 2H^+ \longrightarrow H_2O_2 \text{ (hydrogen peroxide)}$$

$$H_2O_2 + e^- + H^+ \longrightarrow H_2O + OH^\bullet \text{(hydroxyl radical)}$$

$$OH^\bullet + e^- + H^+ \longrightarrow H_2O$$

Absence of Protective Enzymes. Organisms that use oxygen have one or more enzymes that protect them from superoxide anions and their toxic derivatives. The most important of these is a family of enzymes known as superoxide dismutases (SODs). SODs are found in every type of cell that uses oxygen as a final electron acceptor as well as in many that do not. As shown in the following equation, SODs catalyze the conversion of two superoxide anions into a molecule of O_2 and a molecule of hydrogen peroxide (H_2O_2):

CLASSIFICATION OF BACTERIA ON THE BASIS OF THEIR RELATIONSHIP TO OXYGEN AND CARBON DIOXIDE

Category	Requirement	Examples
Obligate aerobe	Atmosphere of 15–21% O_2 (as found in a CO_2 incubator or air)	Mycobacteria, fungi
Microaerophile	O_2 in concentrations lower than that found in ambient air	*Neisseria, Campylobacter* spp
Facultative anaerobe	Multiplies equally well in the presence or absence of O_2	Enterobacteriaceae, most staphylococci, streptococci
Aerotolerant anaerobe	Anaerobic system and a microaerophilic environment	Most strains of *Propionibacterium, Clostridium* spp
Obligate anaerobe	Strict anaerobic environment	Most *Bacteroides* spp, many *Clostridium, Eubacterium, Fusobacterium* spp, *Peptostreptococcus* spp, *Porphyromonas* spp, most strains of *Veillonella parvula*
Capnophile	Increased concentrations of CO_2	Some anaerobes, *Neisseria, Haemophilus* spp

Modified from Engelkirk PG, Duben-Engelkirk J, Dowell VR Jr: Principles and Practice of Anaerobic Bacteriology, 1992; used with permission of Star Publishing Company, Belmont, CA.

$$O_2^- + O_2^- \xrightarrow{\text{SOD}} O_2 + H_2O_2$$

(two superoxide anions)   (oxygen) (hydrogen peroxide)

In the presence of the enzyme catalase, the resulting H_2O_2 can be converted to water and oxygen, as shown in the equation below.

Thus, the presence of SODs not only reduces the concentration of superoxide anions but also decreases the production of other toxic derivatives of oxygen. If anaerobes lacked SOD, this would explain their unique susceptibility to oxygen and its reduction products. Unfortunately, this is not the complete explanation. Although some anaerobes lack SODs, many obligate anaerobes produce them in varying quantities.

Likewise, absence of catalase is not the answer. Although most obligate anaerobes do not produce catalase, thus are highly susceptible to the toxic effects of hydrogen peroxide, some (e.g., *Bacteroides fragilis* and *Propionibacterium acnes*) produce this enzyme. In fact, a catalase test is useful in identifying certain anaerobes.

Bacteriostatic and Bactericidal Effects of O_2. A current theory to explain the anaerobic requirement of obligate anaerobes suggests that oxygen damage occurs in a two-phase process. The first phase (phase 1) is *bacteriostatic,* inhibiting the growth or multiplication of bacteria; the second (phase 2) is *bactericidal,* killing the bacteria.

In phase 1, when anaerobes are exposed to oxygen, electrons that would usually be available for metabolic functions are diverted to the reduction of molecular oxygen, with a consequent decrease in the amount of energy available for growth and synthesis of new cell material. This results in a slowing down or complete cessation of growth—a bacteriostatic effect. If the period of exposure to oxygen is brief, the effect may be reversible. If the anaerobes are placed back into an anaerobic environment at this point, electrons would again be available for normal metabolic processes, energy production, and cell growth. However, should the anaerobes remain in the presence of oxygen, phase 2 would occur. Phase 2 is the lethal, irreversible effect of oxygen toxicity due to the previously mentioned superoxide anions, hydroxyl radicals, and hydrogen peroxide. Phase 2, therefore, has a bactericidal effect.

Strict anaerobes may also require an environment that has a low oxidation-reduction (redox) potential. This may be due in part to the fact that certain enzymes that are essential for bacterial growth require fully reduced sulfhydryl (–SH) groups to be active. Reducing agents such as thioglycollate, cysteine, and dithiothreitol are often added to microbiologic media to obtain a low redox potential.

In vivo, bacteria have a tendency to lower the redox potential at their site of growth. Conse-

$$2\,H_2O_2 \xrightarrow{\text{Catalase}} 2\,H_2O + O_2$$

(two hydrogen peroxide molecules)   (two water molecules)   (oxygen)

quently, anatomical sites colonized with mixtures of organisms frequently provide conditions favorable to the growth of obligate anaerobes.

Where Anaerobes Are Found

Based on scientific evidence, many scientists believe that anaerobes originated about 3 to 4 billion years ago in warm, shallow waters, where they were protected from the sun's deadly ultraviolet rays. It is thought that life on this planet remained anaerobic for hundreds of millions of years.

Today, anaerobes are found only in specific ecologic niches. They can be found in soil, in freshwater and saltwater sediments, and as components of the microbial flora of humans and other animals. Anaerobes that exist outside of the bodies of animals are referred to as *exogenous anaerobes* and cause exogenous types of infections. Anaerobes that exist inside the bodies

of animals *(indigenous microflora)* are referred to as *endogenous anaerobes* and usually are the source of endogenous infections.

Anaerobic infections of exogenous origin are usually caused by gram-positive spore-forming bacilli such as *Clostridium.* Clostridia or their toxins initiate infection when spores are ingested through contaminated food or gain access through open wounds contaminated with soil. Although less frequently encountered in human diseases, *Sarcina ventriculi* (an endospore-forming, gram-positive coccus), *Fusobacterium ulcerans, Desulfovibrio desulfuricans,* and *Desulfomonas* spp gain access to the body via soil-contaminated wounds and may cause exogenous infections.

By far, the anaerobes most frequently isolated from infectious processes in humans are those of endogenous origin. Table 19–2 shows endogenous anaerobes that are commonly encountered in human infections. Although many different species of anaerobes can potentially be isolated

TABLE 19–2
ENDOGENOUS ANAEROBES COMMONLY INVOLVED IN HUMAN INFECTIONS

Infection	Anaerobe	Comments
Actinomycosis	*Actinomyces israelii,* other *Actinomyces* spp, *Propionibacterium acnes*	—
Antibiotic-associated diarrhea and pseudomembranous colitis	*Clostridium difficile, C. perfringens* less often	—
Bacteremia	*Bacteroides* spp, *Fusobacterium* spp, *Peptostreptococcus* spp	75% of clinically significant anaerobic bacteremias involve *Bacteroides* spp and *Fusobacterium* spp; of these 80–90% are due to the *Bacteroides fragilis* group
Brain abscess	*Bacteroides* spp, *Fusobacterium* spp, *Clostridium* spp (infrequently)	These infections are often polymicrobial
Complication of Vincent's angina (necrotizing ulcerative gingivitis)	*Fusobacterium necrophorum*	—
Endocarditis	*Bacteroides* spp, gram-positive cocci, non–spore-forming, gram-positive bacilli	Anaerobes are uncommon isolates
Eye infections	*Peptostreptococcus* spp, *Clostridium* spp, *Bacteroides* spp, *Actinomyces* spp	—
Infections of the female urogenital tract	Gram-positive cocci, *Bacteroides* spp, *Clostridium* spp	—
Intraabdominal infections, liver abscess, peritonitis	*Bacteroides fragilis* group, other *Bacteroides* spp, *Fusobacterium* spp, *C. perfringens,* other *Clostridium* spp, *Peptostreptococcus* spp, *Actinomyces* spp	Frequently polymicrobial
Myonecrosis (gas gangrene)	*C. perfringens, C. novyi, C. septicum*	80–95% of the cases
Oral, sinus, and dental infections	*Peptostreptococcus* spp, *Porphyromonas* spp, *Wolinella* spp, *Fusobacterium* spp	Often polymicrobial
Perineal and perirectal infections	*B. fragilis* group, other *Bacteroides* spp, *Fusobacterium* spp, *Clostridium* spp, *Peptostreptococcus* spp, *Eubacterium* spp, *Actinomyces* spp	—
Pleuropulmonary infections, aspiration pneumonia	*Porphyromonas* spp, *F. nucleatum, Peptostreptococcus* spp, *B. fragilis* group, *Actinomyces* spp, *Eubacterium* spp	—

from human clinical specimens, the number of species routinely isolated is relatively small. Approximately two-thirds of clinically significant anaerobe-associated infectious processes involve the following anaerobes or groups of anaerobes as shown in Figure 19–1:

1. Members of the *Bacteroides fragilis* group
2. Pigmented species of *Porphyromonas* and *Prevotella*
3. *Fusobacterium nucleatum*
4. *Clostridium perfringens*
5. Anaerobic cocci

Anaerobes of endogenous origin can contribute to an infectious disease in any anatomic site of the body if suitable conditions exist for colonization and penetration of the bacteria. Consequently, these are the most critical anaerobes that microbiologists must isolate from clinical materials, identify, and test for susceptibility or resistance to appropriate antimicrobial agents.

Anaerobes at Specific Anatomical Sites

Anaerobes outnumber aerobes at mucosal surfaces, such as the linings of the oral cavity, gastrointestinal (GI) tract, and genitourinary (GU) tract. These heavily colonized surfaces are the usual portals of entry into the tissues and blood stream for endogenous anaerobes. Under ordinary circumstances, microorganisms that are members of the microbial flora do not cause disease. Indeed, many can actually be beneficial.

However, when some of these organisms gain access to usually sterile body sites (e.g., blood stream, brain, lungs), they can cause serious or even fatal infections. For these reasons, clinicians and microbiologists should suspect anaerobe involvement in infectious processes that occur at or near mucosal surfaces. Knowledge of the composition of the microflora at specific anatomical sites is useful for predicting the particular organisms most likely to be involved in infectious processes that arise at or adjacent to those sites. Because some anaerobes have fairly predictable susceptibility patterns, such knowledge may also be of value to physicians considering empiric antimicrobial therapy.

Furthermore, the finding of site-specific organisms at a distant and/or unusual site can serve as a clue to the underlying origin of an infectious process. For example, the isolation of oral anaerobes from a brain abscess may suggest the presence of an oral lesion. Table 19–3 summarizes the variety of endogenous anaerobes that may be found at specific body sites.

Respiratory Tract. Ninety percent of the bacteria present in saliva, nasal washings, and gingival and tooth scrapings are anaerobes; the number equals or exceeds that of aerobic organisms. Gram-negative bacilli and anaerobic cocci are the anaerobes occurring in the highest numbers. These particular anaerobes should be suspected as participants in any infectious process occurring in the oral cavity and in cases of aspiration pneumonia. An oral nidus (origin) should be suspected whenever certain oral anaerobes such

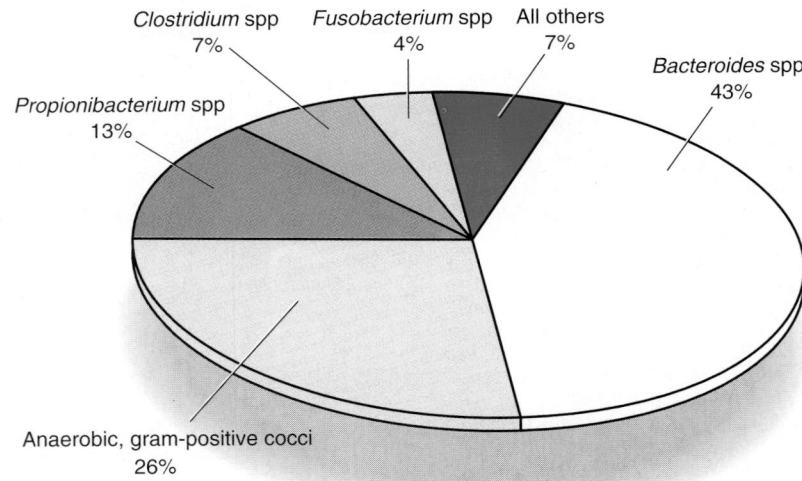

■Figure 19–1

Frequency of isolation of anaerobes from clinical specimens. (Data from Brook [1988]. Redrawn from Engelkirk PG, Duben-Engelkirk J, Dowell VR Jr: Principles and Practice of Anaerobic Bacteriology, 1992; used with permission of Star Publishing Company, Belmont, CA.)

■ T A B L E 1 9 - 3
ENDOGENOUS ANAEROBES AT VARIOUS ANATOMICAL SITES

Site	Anaerobe
Oral cavity	*Bacteroides* spp
	Fusobacterium spp
	Peptostreptococcus spp
	Veillonella spp
	Gram-positive non–spore-forming bacilli
Upper respiratory tract	*Actinomyces* spp
	Bacteroides spp
	Fusobacterium spp (esp. *F. necrophorum*)
	Peptostreptococcus spp
	Veillonella spp
	Propionibacterium spp
	Eubacterium spp
Skin	*Propionibacterium* spp
	Peptostreptococcus spp
	Eubacterium spp
Urethra	*Bacteroides* spp
	Fusobacterium spp
	Peptostreptococcus spp
Vagina	*Lactobacillus* spp
	Peptostreptococcus spp
	Bacteroides spp
	Propionibacterium spp
Colon	*Bacteroides* spp
	Bifidobacterium spp
	Clostridium spp
	Eubacterium spp
	Lactobacillus spp
	Peptostreptococcus spp

as *Fusobacterium nucleatum* and *Porphyromonas* spp are recovered from the blood stream or from abscesses located far from the oral cavity. Certain oral anaerobes produce volatile and foul-smelling metabolic byproducts that undoubtedly contribute significantly to the odor of "bad breath" and exudates from oral lesions and aspiration pneumonia.

Expectorated sputum and oral specimens collected by swab are unacceptable for anaerobic bacteriology. Because they would contain indigenous anaerobes of the oral cavity, it would be difficult to differentiate organisms causing the disease from contaminants present in the specimen.

Skin. The indigenous members of the skin flora include anaerobes. *Propionibacterium acnes,* a common skin colonizer, is frequently isolated from blood cultures. Its presence in blood culture bottles often represents "contamination" from the patient's skin, although its presence can also signify a true bacteremia. *P. acnes,* like coagulase-negative staphylococci, may cause endocarditis. Other anaerobes found on the skin include *Peptostreptococcus* and *Eubacterium* spp.

Superficial wound or abscess specimens aspirated by needle and syringe are much better specimens for anaerobic bacteriology than material collected by swabs; the latter are often contaminated with anaerobes of the skin microflora.

Genitourinary Tract. Anaerobic bacteria that colonize the distal urethra would be recovered if specimens of voided or catheterized urine were cultured for anaerobes. Although anaerobes can be a rare cause of infectious diseases of the urinary tract, their presence in such specimens generally indicates contamination of the urine with organisms flushed from the distal urethra. Similarly, 50% of the bacteria in cervical and vaginal secretions are anaerobes. Whenever anaerobes are recovered from vaginal and cervical swabs, neither the microbiologist nor the physician can distinguish the indigenous microflora contaminants from the organisms actually contributing to the patient's infectious process. For this reason, GU tract swabs and voided or catheterized urine specimens are unacceptable for anaerobic bacteriology.

Gastrointestinal Tract. Of the estimated 500 species of bacteria that inhabit the human body, approximately 300 to 400 live in the colon. Microflora studies have found that anaerobes outnumber aerobes by a factor of 1,000:1. Although *B. fragilis* is the most common species of anaerobic bacteria isolated from human soft-tissue infection and anaerobic bacteremia, this species accounts for less than 1% of the human intestinal microflora. *Bacteroides vulgatus, B. thetaiotaomicron,* and *B. distasonis* are among the most common species of bacteria isolated from human feces. Other species commonly inhabiting the GI tract include *Bifidobacterium, Clostridium, Eubacterium,* and *Peptostreptococcus* spp.

Stool specimens are not routinely cultured for anaerobes. However, if a patient is suspected of having pseudomembranous colitis and/or antibiotic-associated diarrhea caused by *Clostridium difficile,* diarrheal specimens from that patient should be examined.

Factors That Predispose Patients to Anaerobic Infections

Factors that commonly predispose the human body to anaerobic infections include trauma of mucous membranes or skin, vascular stasis, tissue necrosis, and decrease in the redox potential of tissue.

The precise mechanisms by which anaerobic bacteria cause disease are not known. However, anaerobes produce or possess a variety of enzymes, capsules, and adherence factors thought to play a role in pathogenicity. These are collectively referred to as virulence factors (Table 19–4). In addition to those listed in Table 19–4, β-lactamases, chondroitin sulfatase, fibrinolysin, heparinase, leukocidins, and endotoxin have all been suggested as possible virulence factors of anaerobes.

Generally, infectious diseases involving anaerobic bacteria follow some type of trauma to protective barriers such as the skin and mucous membranes. Trauma at these sites allows anaerobes of the indigenous microflora (or soil anaerobes, in some cases) to gain access to deeper tissues. Vascular stasis (blockage of blood flow) prevents oxygen from entering a particular site. This results in an environment conducive to growth and multiplication of any anaerobes that might be present at that site. Similar results may occur in the presence of tissue necrosis and when redox potential in tissue is decreased. Table 19–5 lists examples of conditions that predispose a patient to anaerobic infections.

Indications of Anaerobe Involvement in Human Disease

Infectious processes involving anaerobes are usually purulent, but the absence of leukocytes does not rule out the possibility that anaerobes are contributing to the process. Some of the more serious infectious processes in humans are caused by anaerobes that produce cytotoxins and other histotoxic virulence factors that contribute to the necrotizing process by destroying neutrophils, macrophages, and other cells.

Table 19–6 contains a list of indications of anaerobe involvement in infectious processes. Although any of these indicators should alert the physician to the possible involvement of anaerobes, most are not specific for anaerobes. For example, large quantities of gas in the specimen might be due to organisms such as *Escherichia coli* or a mixture of enteric bacteria other than anaerobes. Similarly, the foul odor usually associated with specimens containing anaerobes could be absent.

Many of the infectious processes involving anaerobes are polymicrobial, consisting of mixtures of obligate anaerobes or mixtures of obligate or

■■■■■■ T A B L E   1 9 – 4
POTENTIAL VIRULENCE FACTORS OF ANAEROBIC BACTERIA

Potential Virulence Factor	Possible Role of the Virulence Factor	Anaerobes Known or Thought to Possess the Virulence Factor
Polysaccharide capsules	Promote abscess formation; serve an antiphagocytic function	*Bacteroides fragilis, Porphyromonas gingivalis,* and some other anaerobic gram-negative bacilli
Adherence factors	Certain pili, fimbriae, fibrils enable organisms to adhere to cell surfaces	*B. fragilis* and *P. gingivalis* in humans; *B. nodosus* in sheep
Clostridial toxins/exoenzymes		
Collagenases	Catalyze the degradation of collagen	
Cytotoxins	Toxic to specific types of host cells	
DNAses	Destroy DNA	
Enterotoxins	Toxic to cells of the intestinal mucosa	
Hemolysins	Liberate hemoglobin from red blood cells by lysing the cells	
Hyaluronidase	Catalyze the hydrolysis of hyaluronic acid, the cement substance of tissues	
Lipases	Catalyze the hydrolysis of ester linkages between the fatty acids and glycerol of triglycerides and phospholipids	Certain *Clostridium* spp
Necrotizing toxins	Cause necrosis (death) of cells	
Neuraminidases	Destroy neuraminic acid, a sialic acid found on cell surfaces	
Neurotoxins	Destroy or disrupt nerve tissue	
Permeases (e.g., ADP-ribosyl-transferases)	Alter the physiology of cells in such a way as to cause severe fluid loss and ionic imbalance	
Phospholipases	Catalyze the splitting of phospholipids, lecithinases	
Proteases	Split proteins by hydrolysis of peptide bonds	
Proteinases	Split the interior peptide bonds of proteins (endopeptidases)	

Modified from Engelkirk PG, Duben-Engelkirk J, Dowell VR Jr: Principles and Practice of Anaerobic Bacteriology, 1992; used with permission of Star Publishing Company, Belmont, CA.

TABLE 19-5

EXAMPLES OF CONDITIONS THAT PREDISPOSE A PATIENT TO ANAEROBE-ASSOCIATED INFECTIONS OR DISEASES

Predisposing Condition	Explanation
Human or animal bite wounds	Endogenous anaerobes of the oral cavity enter the wound
Aspiration of oral contents into the lungs after vomiting	Endogenous anaerobes of the oral cavity and other materials (mucus, stomach contents, etc.) enter the lungs
Tooth extraction, oral surgery, or traumatic puncture of the oral cavity	Endogenous anaerobes of the oral cavity gain entrance to traumatized tissue and the blood stream
Gastrointestinal tract surgery or traumatic puncture of the bowel	Endogenous anaerobes of the gastrointestinal tract gain access to traumatized tissue and the blood stream
Genital tract surgery or traumatic puncture of the genital tract	Endogenous anaerobes of the genital tract gain access to traumatized tissue and the blood stream
Introduction of soil into a wound	Clostridial spores in the soil enter the tissues via a traumatic wound, resulting in colonization and multiplication of the bacteria

Modified from Engelkirk PG, Duben-Engelkirk J, Dowell VR Jr: Principles and Practice of Anaerobic Bacteriology, 1992; used with permission of Star Publishing Company, Belmont, CA.

microaerotolerant anaerobes and facultative organisms. Symbiotic relationships frequently exist between some of the bacteria involved in polymicrobial infections, which can act synergistically in the production of disease.

SPECIMEN SELECTION, COLLECTION, TRANSPORT, AND PROCESSING

Specimen Quality

When a physician suspects an infection involving anaerobes, he or she must properly select a clinical specimen and arrange for its proper collection and rapid transport to the laboratory. Although the selection, collection, and transport of clinical specimens are procedures routinely performed outside the laboratory by nonlaboratory professionals, laboratory practitioners can play a major role in accomplishing these steps, which are vital to the successful outcome of an anaerobic culture. The laboratory must provide physicians and other health care workers with proper education, written guidelines, and appropriate devices for collection and transport.

When clinical laboratory professionals provide in-service education to those who select, collect, and transport specimens for anaerobic bacteriology, a rationale must be given for proper specimen selection. The consequences of working up improper or incorrectly collected or transported specimens should be outlined. The laboratory must also develop criteria for the rejection of inappropriate specimens, with the cooperation of hospital medical staff. For example, when a specimen is not acceptable, particularly when collected at surgery, the physician must be notified before the sample is discarded. The reason for rejection should always be stated.

Many types of specimens are acceptable for anaerobic culture; these are listed in Table 19–7. Table 19–8 contains a list of specimens not recommended for anaerobic culture, most of which are collected by swab.

TABLE 19-6

INDICATIONS OF INVOLVEMENT OF ANAEROBES IN INFECTIOUS PROCESSES

Indication	Rationale
Infection is in close proximity to a mucosal surface	Anaerobes are the predominant microflora at mucosal surfaces
Infection persists despite aminoglycoside therapy	Aminoglycosides are ineffective against most anaerobes
Presence of foul odor	*Porphyromonas* and *Fusobacterium* spp produce foul-smelling metabolic end products
Presence of large quantity of gas	*Clostridium* spp produce large quantities of gas during metabolism
Presence of black color or brick-red fluorescence	Pigmented species of *Porphyromonas* and *Prevotella* produce a pigment that fluoresces brick-red under long-wave ultraviolet light; turns dark brown to black
Presence of sulfur granules	Sulfur granules are often present in patients with actinomycosis
Distinct morphologic characteristics in Gram-stained preparations	The morphology of certain anaerobes is somewhat (but not completely) distinctive: Certain anaerobes such as *Bacteroides* and *Fusobacterium* are quite pleomorphic; *F. nucleatum* is fusiform; *Clostridium* spp are often large, gram-positive rods that may or may not contain spores

Data from Finegold et al (1986).
Modified from Engelkirk PG, Duben-Engelkirk J, Dowell VR Jr: Principles and Practice of Anaerobic Bacteriology, 1992; used with permission of Star Publishing Company, Belmont, CA.

▪▪▪▪ T A B L E   1 9 – 7

ACCEPTABLE SPECIMENS FOR ANAEROBIC BACTERIOLOGY

Anatomic Source	Specimens and Recommended Methods of Collection
Central nervous system	Cerebrospinal fluid, carefully aspirated abscess material, tissue from biopsy or autopsy
Dental/ENT specimens	Carefully aspirated material from abscesses, biopsied tissue
Localized abscesses	Needle and syringe aspiration of closed abscesses
Decubitus ulcers	Aspiration of pus from beneath skin flaps or from deep pockets, following thorough cleansing of the area with antiseptic
Sinus tracts or draining wounds	Aspiration by syringe through a small plastic catheter introduced as deeply as possible through a decontaminated skin orifice
Deep tissue or bone	Specimens obtained during surgery from depths of wound or underlying bone lesion
Pulmonary	Percutaneous transtracheal aspiration: aspirate obtained by direct lung puncture; pleural fluid obtained by thoracentesis Biopsied tissue: "sulfur granules" from draining fistula; if properly used, a double catheter with bronchial brush may be suitable for bronchoscopic specimens
Intraabdominal	Aspirate from abscess, ascitic fluid, biopsied tissue
Urinary tract	Suprapubic bladder aspiration
Female genital tract	Aspirate from loculated abscess; culdocentesis specimen, preferably following decontamination of the vagina using povidone-iodine; a double catheter with bronchial brush or a sterile swab may be used for uterine cavity specimens
Other	Blood, bone marrow, aspirated synovial fluid, biopsied tissue from any normally sterile site

Data from Finegold et al (1986) and Sutter et al (1985).
Modified from Engelkirk PG, Duben-Engelkirk J, Dowell VR Jr: Principles and Practice of Anaerobic Bacteriology, 1992; used with permission of Star Publishing Company, Belmont, CA.

Specimen Transport and Processing

Regardless of the type of specimen being submitted for anaerobic bacteriology, it must be transported and processed as rapidly as possible and with minimum exposure to oxygen. Specimens are usually collected from a warm, moist environment that is low in oxygen. Thus, it is important to avoid "shocking" the anaerobes by exposing them to oxygen or permitting them to dry out. In addition, the amount of time they remain at room temperature should be minimized.

Aspirates. Specimens collected by needle and syringe are better for anaerobic bacteriology than those collected by swab. They are less likely to be contaminated by indigenous microflora, including endogenous anaerobes. Following aspiration of the specimen, any air present in the syringe and needle should rapidly be expelled. To prevent production of a potentially infectious aerosol, place an alcohol-soaked gauze pad over the needle while cautiously expelling the air.

Transporting specimens within a needle and syringe assembly is no longer acceptable because of the hazard of accidental skin puncture. The aspirate should be injected into some type of oxygen-free transport tube or vial, preferably one containing a prereduced, anaerobically sterilized (PRAS) transport medium. PRAS media are prepared by boiling (to remove dissolved oxygen),

▪▪▪▪ T A B L E   1 9 – 8

UNACCEPTABLE SPECIMENS FOR ANAEROBIC BACTERIOLOGY

Clinical Specimen	Brief Explanation
Exudates and other material collected by swabs from superficial wounds, abscesses, burns, cysts, or ulcers	Such specimens would be contaminated with skin flora anaerobes (e.g., *Propionibacterium acnes* and anaerobic gram-positive cocci)
Vaginal, cervical, or urethral swabs	Such specimens would be contaminated with a variety of vaginal flora anaerobes or those that colonize the distal urethra (e.g., *Bacteroides* spp and *Fusobacterium* spp)
Respiratory tract specimens collected by swab, nasotracheal or orotracheal suction, or bronchoscope; expectorated sputum	Such specimens would be contaminated with oral flora anaerobes (e.g., *Porphyromonas* spp, *Fusobacterium* spp, *Veillonella* spp)
Stool specimens, rectal swabs	Such specimens would be contaminated with a variety of gastrointestinal tract flora anaerobes (e.g., *Bacteroides* spp, *Clostridium* spp, *Eubacterium* spp); fecal specimens may be cultured when specific pathogens (e.g., *C. difficile* or *C. botulinum*) are being sought.
Voided or catheterized urine	Such specimens would be contaminated with endogenous anaerobes that colonize the distal urethra (e.g., *Bacteroides* and *Fusobacterium* spp)

Data from Sutter et al (1985).
Modified from Engelkirk PG, Duben-Engelkirk J, Dowell VR Jr: Principles and Practice of Anaerobic Bacteriology, 1992; used with permission of Star Publishing Company, Belmont, CA.

autoclaving (to sterilize the mixture), and replacing any air with an oxygen-free gas mixture.

Once in the laboratory, aspirates in transport containers should be vortexed to ensure even distribution of the material, especially when the sample is grossly purulent. Using a sterile Pasteur pipette, add one drop of purulent material or two to three drops of nonpurulent material to each plate and streak it in such a manner as to obtain well-isolated colonies. Also inoculate 0.5 to 1 mL of the specimen into the bottom of a tube of enriched thioglycollate broth. Spread one drop evenly over an alcohol-cleaned glass slide for Gram staining.

Swabs. In those rare instances in which swabs are deemed necessary, use commercially available, oxygen-free swabs. Such containers are available from a number of commercial sources. Although not widely used, the Accu-CulShure collection instrument (Technology for Medicine, Inc., Pleasantville, NY) is a uniquely designed swab that exposes the swab only at the collection site (e.g., the cervix) and protects it from indigenous microflora contamination both before and after collecting the specimen. Swabs submitted for anaerobic bacteriology should always be transported in oxygen-free transport containers.

On its arrival in the laboratory, place the swab in a tube containing about 1 mL of sterile thioglycollate broth. Vortex the swab vigorously to remove the clinical material and then press it firmly against the inner wall of the tube to remove as much liquid as possible. Inoculate the remaining liquid suspension as previously described for an aspirate.

Tissue. Tissue specimens collected by biopsy or at autopsy from usually sterile sites are acceptable specimens for anaerobic culture. Small pieces of tissue can be placed in oxygen-free transport tubes or vials containing PRAS medium to keep the tissue moist. When inserting either swabs or small pieces of tissue into an anaerobic transport container, care must be taken not to tip the container. This would cause the heavier-than-air, oxygen-free gas mixture to be displaced by room air, thus defeating the primary purpose of using such a transport medium.

Larger tissues present greater transport problems. To resolve this, large tissues may be transported in pouches containing an oxygen-free atmosphere to prevent exposure to oxygen while en route to the laboratory. Such bags or pouches are available commercially from Becton Dickinson

(Bio-Bags and GasPak Pouches; Cockeysville, MD) and Remel (Anaerobe Pouches; Lenexa, KS) and described in more detail in a later section of this chapter.

To process the tissue or bone fragments in the laboratory, add 1 mL of sterile thioglycollate broth to a sterile tissue grinder. Homogenize the piece of tissue or bone fragment until a thick suspension is obtained. Ideally, this procedure can be performed within an anaerobic chamber. If a chamber is not available, the grinding must be accomplished as quickly as possible at the workbench. Inoculate the suspension as described previously for an aspirate.

Blood. A discussion of the variety of commercially available blood culture systems is beyond the scope of this chapter. Blood must be cultured in such a manner as to recover any and all bacteria or yeasts that may be present. This usually requires aseptic inoculation of both an anaerobic (unvented) and aerobic (vented) bottle. Investigations have shown that single-bottle systems are less efficient in isolating certain organisms (including anaerobes) than two-bottle systems. Once inoculated, blood culture bottles should be rapidly transported to the laboratory, where they are incubated at 35 to 37°C. It is important to note that blood for culture must be carefully collected so as to minimize contamination with skin flora. This is usually accomplished by meticulous preparation of the venipuncture site with a bactericidal agent, such as tincture of iodine or an iodophor.

Quality-assurance considerations relating to specimens are summarized in Table 19–9.

Processing Clinical Samples for Maximum Recovery of Anaerobic Pathogens

To ensure that results are clinically significant, only properly selected, collected, and transported specimens should be processed in the anaerobic bacteriology laboratory. In processing a specimen for anaerobic bacteriology, primary emphasis should be placed on safety, adherence to prescribed procedures, and speed.

Ideally, once a specimen arrives in the laboratory, it is immediately placed in an anaerobic chamber to prevent further exposure of clinical materials to oxygen. Anaerobic chambers allow all steps in the processing of a specimen to be performed in an oxygen-free environment. In

■ T A B L E   1 9 – 9
SPECIMEN QUALITY ASSURANCE

Component	Comments
Specimen selection	Persons responsible for selecting specimens must be familiar with the types of specimens that are suitable for anaerobic bacteriology.
Request slips	Request slips must include patient demographics, specimen type and source, date and time of collection, the requesting physician, and antimicrobial agents the patient may have already taken.
	The time of collection is especially important, because it will determine how quickly the specimen is transported to the laboratory.
	The type of specimen alerts the microbiologists as to the anaerobes to expect, which serves as a guide to media selection.
Specimen collection	Collect with an attempt to minimize contamination of the specimen with organisms of the indigenous microflora.
	Minimize exposure of the specimens to oxygen.
Specimen handling	Label specimen containers with the patient's name, identification number, source of specimen, and date and time of collection.
	In the laboratory, check the name on the specimen container against the name on the request slip to ensure that specimens have not been inadvertently switched.
Specimen transport	Transport specimens rapidly to minimize exposure to oxygen and time spent at room temperature.
	Use appropriate transport devices whenever a delay is anticipated between collection and processing.
	Clinical laboratory personnel must provide the appropriate containers and must instruct health care practitioners in their proper use.
Written laboratory directives/guidelines	Written guidelines to clinicians should include the following: appropriate and inappropriate specimens, specimen collection and transport procedures, and policy regarding specimen duplication.
	Criteria for rejection should include unlabeled specimen container, inappropriate specimen for patient's condition, incorrectly collected and/or transported specimen, too much delay between collection and arrival in the lab, and dried out samples.
Education of clinicians	Clinical laboratory professionals should participate in educating health care practitioners who are involved in patient assessment, test ordering, and specimen collection regarding anaerobic bacteriology.

Modified from Engelkirk PG, Duben-Engelkirk J, Dowell VR Jr: Principles and Practice of Anaerobic Bacteriology, 1992; used with permission of Star Publishing Company, Belmont, CA.

those laboratories not equipped with anaerobic chambers, holding systems may be used (described later in this chapter). To comply with mandatory infectious disease safety policies, fol-

low the appropriate safety precautions. Wear disposable latex gloves when handling clinical specimens containing potentially infectious agents and use a laminar flow safety cabinet whenever it is appropriate to do so.

In addition to the initial processing steps previously described, the following procedures should also be performed:

■ Macroscopic examination of the specimen

■ Preparation of Gram-stained smears for microscopic examination

■ Inoculation of appropriate plated and tubed media, including media specifically designed for culturing anaerobes

■ Anaerobic incubation of inoculated media

Macroscopic Examination of Specimen

Examine each specimen received in the anaerobic bacteriology section macroscopically, and record pertinent observations on some type of a worksheet. Some of the characteristics to note during the macroscopic examination are listed in Table 19–10.

Direct Microscopic Examination of Specimen

Direct smears for Gram stain must be prepared on all specimens received in the laboratory. Examination of a thin Gram-stained smear is one of the most important diagnostic procedures performed in anaerobic bacteriology laboratories. In some laboratories, smears are stained before inoculation of media; microscopic observations can then serve as a guide to selecting the media to be inoculated. In other laboratories, a routine battery of media is inoculated before staining and examining the smear, thus eliminating any further delay in media inoculation. The latter approach is recommended.

Gram-stained smears must be examined for several reasons:

■ The Gram stain reveals the various types and relative number of microorganisms present. The presence of multiple distinct morphologic forms suggests that a polymicrobic infectious process is present. This should be reported to the requesting physician.

■ Certain morphotypes may provide a presumptive identification of organisms and may serve

■ **T A B L E   1 9 – 1 0**

CHARACTERISTICS TO NOTE DURING THE MACROSCOPIC EXAMINATION OF A SPECIMEN

Questions to Ask	Comments
Is it an appropriate specimen?	Inappropriate specimens should be rejected.
Was it submitted in an appropriate transport container?	Improperly transported specimens should be rejected.
How old is the specimen? Are the date and time of collection recorded on the accompanying request slip?	Specimens that are too old may be cause for rejection.
Is there evidence that the specimen has dried out during transit?	Specimens that have dried out during transport should be rejected.
Does the specimen have a foul odor?	Many anaerobes, especially *Fusobacterium* and *Porphyromonas* spp, have foul-smelling metabolic end products; however, the lack of a foul odor does not exclude anaerobes.
Does the specimen fluoresce brick-red when exposed to a Wood's lamp? (A Wood's lamp emits long-wave [366 nm] ultraviolet light.)	Pigmented species of *Porphyromonas* and *Prevotella* produce substances that fluoresce under long-wave UV light prior to becoming darkly pigmented. Although a brick-red fluorescence is presumptive evidence of these organisms, some members of this group fluoresce colors other than brick-red.
Is the necrotic tissue or exudate black?	Such discoloration may be due to the pigment produced by pigmented species of *Porphyromonas* and *Prevotella*.
Does the specimen contain sulfur granules?	Such granules are associated with actinomycosis, a condition caused by *Actinomyces* spp, *Propionibacterium propionicus,* and closely related organisms, such as *Propionibacterium acnes*.
Is the specimen bloody?	Such information should be included in a preliminary report to the requesting physician.
Is the specimen purulent?	Such information should be included in a preliminary report to the attending physician.

Modified from Engelkirk PG, Duben-Engelkirk J, Dowell VR Jr: Principles and Practice of Anaerobic Bacteriology, 1992; used with permission of Star Publishing Company, Belmont, CA.

as a guide to media selection. For example, if large, gram-positive bacilli are seen, clostridia may be suspected. An egg-yolk agar to detect lecithinase or lipase may be added in the primary set-up. Thin, gram-negative bacilli with tapered ends may be fusobacteria. Extremely pleomorphic, gram-negative bacilli with bizarre shapes are suggestive of *Fusobacterium mortiferum* or *F. necrophorum*. Tiny, round to oval, gram-negative cocci with a tendency to stain gram-variable are suggestive of *Veillonella* species, whereas gram-negative coccobacilli may be *Bacteroides, Porphyromonas,* or *Prevotella* spp.

■ The Gram stain often reveals the presence of leukocytes. However, certain anaerobes produce necrotizing toxins that destroy leukocytes. Thus, the absence of leukocytes in a Gram-stained smear does not rule out the involvement of anaerobes and should never be used as a criterion for rejecting a wound specimen.

■ Finally, the Gram stain can serve as a quality-control technique. Failure to isolate certain organisms observed in the Gram-stained smear might indicate that problems exist with the anaerobic technique being used. However, failure to recover certain morphotypes could also indicate that those particular organisms were dead or that the patient was receiving antimicrobial agents that inhibited growth of the organisms on the plated media.

Direct smears for Gram stain should be methanol-fixed rather than heat-fixed. Methanol fixation preserves the morphology of leukocytes and bacteria better than heat fixation. Gram-negative anaerobes frequently stain a very pale pink when safranin is used as the counterstain and are thus easily overlooked in Gram-stained smears of clinical specimens and blood cultures. To enhance the red color of gram-negative anaerobes, use of basic fuchsin as the counterstain, or counterstaining with safranin for 3 to 5 minutes, is recommended.

In addition to the Gram stain, some laboratories routinely examine wet mounts of clinical materials using regular transmitted light, phase-contrast microscopy, or darkfield illumination. These procedures aid in the detection of motile organisms and refractile spores.

Inoculation of Appropriate Plated and Tubed Media

The choice of media for use in the anaerobic bacteriology laboratory is an extremely important aspect of successful anaerobic bacteriology. Anaerobes have special nutritional requirements for vitamin K, hemin, and yeast extract, and all primary isolation media for anaerobes should contain these three ingredients. Recommendations of different authorities in the area of anaerobic

■■■■ T A B L E 1 9 – 1 1

PRIMARY SET-UP MEDIA RECOMMENDED FOR RECOVERY OF ANAEROBES

Medium	Supports Growth of	Comments
Anaerobic blood agar plate (Brucella blood agar [BRU/BA] is the most popular choice)	Virtually all obligate and facultative anaerobes, when incubated anaerobically	An enriched medium containing sheep's blood for enrichment and detection of hemolysis, vitamin K_1 (required by some *Porphyromonas* spp), and hemin (which enhances growth of some *Bacteroides* spp, including members of the *B. fragilis* group); similar excellent media (e.g., CDC anaerobe agar, enriched brain heart infusion blood agar, and Schaedler blood agar) are also available for use; variations have been reported in the ability of these different blood agar media to support growth of certain anaerobes (Sheppard et al 1990)
Bacteroides bile esculin (BBE) agar plate	Bile-tolerant *Bacteroides* spp, when incubated anaerobically; some strains of *Fusobacterium mortiferum, Klebsiella pneumoniae*, enterococci, and yeast may grow to a limited extent	A selective medium containing gentamicin (which inhibits most aerobic organisms), 20% bile (which inhibits most anaerobes), and esculin; used primarily for rapid isolation and presumptive identification of members of the *B. fragilis* group, which grow well on BBE (due to their bile-tolerance) and turn the originally light-yellow medium to brown (due to esculin hydrolysis)
Kanamycin-vancomycin-laked blood (KVLB) agar plate	*Bacteroides* and *Prevotella* spp, when incubated anaerobically; yeasts and kanamycin-resistant, facultative, gram-negative bacilli will also grow	A selective medium containing kanamycin (which inhibits most facultative gram-negative bacilli), vancomycin (which inhibits most gram-positive organisms and vancomycin-sensitive strains of *Porphyromonas* spp), and laked blood (which accelerates production of brown-black pigmented colonies by certain *Prevotella* spp; used primarily for rapid isolation and presumptive identification of pigmented species of *Prevotella;* a similar medium that substitutes paromomycin for kanamycin inhibits growth of kanamycin-resistant, facultative, gram-negative bacilli
Phenylethyl alcohol (PEA) agar plate (also called phenylethanol agar or phenethyl alcohol agar)	Virtually all obligate anaerobes (both gram-positive and gram-negative) and gram-positive, facultative anaerobes, when incubated anaerobically	A selective medium containing phenylethyl alcohol; used primarily to suppress the growth of any facultative, gram-negative bacilli (e.g., *Enterobacteriaceae*) that might be present in the clinical specimen, especially swarming *Proteus* spp
Anaerobic broth (e.g., thioglycollate [THIO], chopped meat)	Virtually all types of bacteria grow in THIO: obligate aerobes and microaerophiles near the top, obligate anaerobes at the bottom, and facultative anaerobes throughout	Because obligate anaerobes can be overgrown by more rapidly growing facultative organisms present in the specimen and killed by their toxic metabolic by-products, THIO serves only as a backup source of culture material (e.g., in the event that there is no growth on plated media due to a jar failure or the presence of antimicrobial agents in the specimen); chopped meat carbohydrate broth can be used in place of THIO; broth cultures should never be relied on exclusively for isolating anaerobes from clinical material

Data from Sutter et al (1985).
Modified from Engelkirk PG, Duben-Engelkirk J, Dowell VR Jr: Principles and Practice of Anaerobic Bacteriology, 1992; used with permission of Star Publishing Company, Belmont, CA.

bacteriology with regard to specific media to be included in the primary isolation set-up of anaerobe cultures vary slightly.

Media for Use in the Primary Isolation Set-up. Table 19–11 lists the primary media recommended for recovery of anaerobes. Although these media are designed for anaerobes, they also support the growth of most aerobes. No one medium exists that supports all common species of anaerobes while inhibiting all aerobes. In addition to those listed in Table 19–11, other nonselective and selective media may be helpful. Table 19–12 lists media designed for specific anaerobes to be used in conjunction with the primary media.

The ideal media for use in the culture of anaerobes are those that have never been exposed to oxygen or those that have been exposed only briefly. Such media include freshly prepared media and those stored under anaerobic conditions from the time they were made. Media exposed to air for extended periods may contain toxic substances, produced as a result of the reduction of molecular oxygen. Such media may also have redox potentials above that required for anaerobes to initiate growth. Although it is not practical to prepare fresh media each time there is a need for it, freshly prepared media can be stored within an anaerobic chamber or holding system until used.

An alternative is to use commercial media that have been prepared, packaged, shipped, and stored under anaerobic conditions (i.e., not exposed to oxygen until they are inoculated, or not exposed at all when plates are inoculated within an anaerobic chamber). Such PRAS media (shown in Figure 19–2) are available from

■■■■■ F i g u r e   1 9 – 2

Prereduced, anaerobically sterilized (PRAS) plated media. PRAS plated media are manufactured, packaged, shipped, and stored under anaerobic conditions. They are packaged as single plates, as several plates of one particular type of medium (e.g., BRU/BA plates), or as sets of media (e.g., the variety of media used in the primary set-up of one specimen). Initial exposure of these media to oxygen occurs when the gas-impermeable foil pouches are opened at the bench at the time of specimen inoculation. Exposure to oxygen is totally avoided when the pouches are opened within an anaerobic chamber. (Photograph courtesy of Anaerobe Systems, San Jose, CA.)

Anaerobe Systems (San Jose, CA). Growth is initiated quickly on PRAS media, and many anaerobes produce sufficient growth after only 24 hours of incubation. Studies have demonstrated that these media perform better than fresh media or other commercially available media.

Reducing agents, such as palladium chloride, are sometimes added to media prior to autoclaving in an attempt to "prereduce" them. However, care must be taken in selecting the type of reducing agents, as some are toxic to certain anaerobes. Media containing reducing agents are available from a variety of commercial sources. Storing them in gas-permeable cellophane sleeves and/or at refrigerator temperatures may decrease their shelf life. In all likelihood, there is a limit to the length of time that reducing agents can maintain sufficiently low oxidation-reduction potentials when they are in constant contact with oxygen. Additional studies are needed to determine the extent to which the toxic substances mentioned earlier (e.g., superoxide anions and hydrogen peroxide) can actually be removed from media by placing such plates in an anaerobic atmosphere prior to inoculation. Readers wanting additional information about reducing agents and culture media should refer to the introductory chapters of the book by Smith and Williams.

■■■■■ T A B L E   1 9 – 1 2

ADDITIONAL SELECTIVE AND NONSELECTIVE MEDIA

Medium	Special Use
Cycloserine cefoxitin fructose agar (CCFA)	*Clostridium difficile*
Bacteroides gingivalis blood agar (BGBA)	Dental cultures for *Prevotella gingivalis*
Egg-yolk agar (EYA)	*Clostridium*
Cadmium sulfate–fluoride–acridine trypticase agar (CFAT)	Dental cultures for *Actinomyces*
Josamycin, vancomycin, norfloxacin (JVN)	*Fusobacterium* and *Leptotrichia*
Lactobacillus selective media (LBS)	*Lactobacillus*
Peptostreptococcaceae selective agar (PS)	*Peptostreptococcus*

Additional Media for Aerobic Incubation. In addition to the battery of selective anaerobic media to be incubated anaerobically, a variety of plated media to be incubated aerobically in a CO_2 incubator is also inoculated. The specific media vary somewhat from one laboratory to another and depend on the specimen type; nevertheless, blood, MacConkey, and chocolate agar plates are usually included. Some laboratories also routinely include a phenylethylalcohol (PEA) plate or colistin nalidixic acid (CNA) plate. Figure 19–3 shows a typical plating protocol for a wound sample and the kinds of results that might be expected.

Inoculation Procedures. In laboratories not equipped with an anaerobic chamber, inoculation of appropriate plated and tubed media could be performed in an area with a suitable nitrogen gas holding system. Such a holding system, which may employ a jar, a box, or other small chamber, allows uninoculatd plates to be held under anaerobic conditions until needed. Inoculated plates should be held under near-anaerobic conditions until placed into an anaerobic jar or bag. Care should be taken to ensure that inoculated plates do not remain in the holding jar at room temperature for extended periods (i.e., not longer than 1 hour). Also, because the holding jar should remain as anaerobic as possible, care should be taken to minimize convection currents whenever freshly inoculated plates are added to the jar.

Some microbiologists believe that it is better to batch-process specimens than to process each specimen as it arrives in the laboratory; this relieves their concern that freshly inoculated plates will remain at room temperature in a holding system that may contain some oxygen. Batch processing is certainly an acceptable alternative for aspirates and specimens received in proper transport containers (i.e., those kept moist under anaerobic conditions). It would be unacceptable, however, to batch-process improperly submitted specimens (e.g., dry swabs) or clinical materials apt to contain rapidly growing bacteria. The delay would further expose anaerobes to molecular oxygen, increase the likelihood of specimens drying out, and decrease the probability of recovering anaerobes.

Anaerobic Incubation of Inoculated Media

After specimens are rapidly processed and inoculated onto the appropriate media, the inoc-ulated plates must be incubated anaerobically at 35 to 37°C. The most common and practical choices for anaerobic incubation systems for clinical laboratories are anaerobic chambers, anaerobic jars, and anaerobic bags or pouches. The choice of system will be influenced by a number of factors, including financial considerations, the number of anaerobic cultures performed, and space limitations.

Anaerobic Chambers. The ideal anaerobic incubation system is an anaerobic chamber, which provides an oxygen-free environment for inoculating media and incubating cultures. Identification and susceptibility tests can also be performed within the chamber, if necessary.

A number of anaerobic chambers are available commercially. Some models, called "glove boxes," are fitted with airtight rubber gloves (Fig. 19–4). The microbiologist inserts his or her hands into the gloves and manipulates specimens, plates, and tubes inside the chamber. Glove boxes have traditionally been constructed of flexible vinyl. They are available from companies such as Coy Corporation (Grass Lake, MI), Forma Scientific (Marietta, OH), and Lab-Line Instruments (Melrose Park, IL).

Gloveless anaerobic chambers are also available commercially (Fig. 19–5) (Anaerobe Systems, San Jose, CA). Airtight rubber sleeves that fit snugly against the user's bare forearms are used in place of gloves, enabling the microbiologist to work within an anaerobic environment with his or her bare hands. However, to comply with mandatory infectious disease safety precautions, it is recommended that disposable latex gloves be worn if clinical specimens are being processed within gloveless chambers. Subsequent operations may be performed with bare hands. The newer gloveless models have a dissecting microscope mounted on the front of the rigid Plexiglas chamber. This enables the user to observe colony morphology within the chamber, eliminating the need to remove the plates from the chamber and eliminating exposure of the colonies to oxygen.

All anaerobic chambers contain the following:

■ Catalyst

■ Desiccant

■ Hydrogen gas (5–10%)

■ Carbon dioxide gas (5–10%)

■ Nitrogen gas (80–90%)

■ Indicator

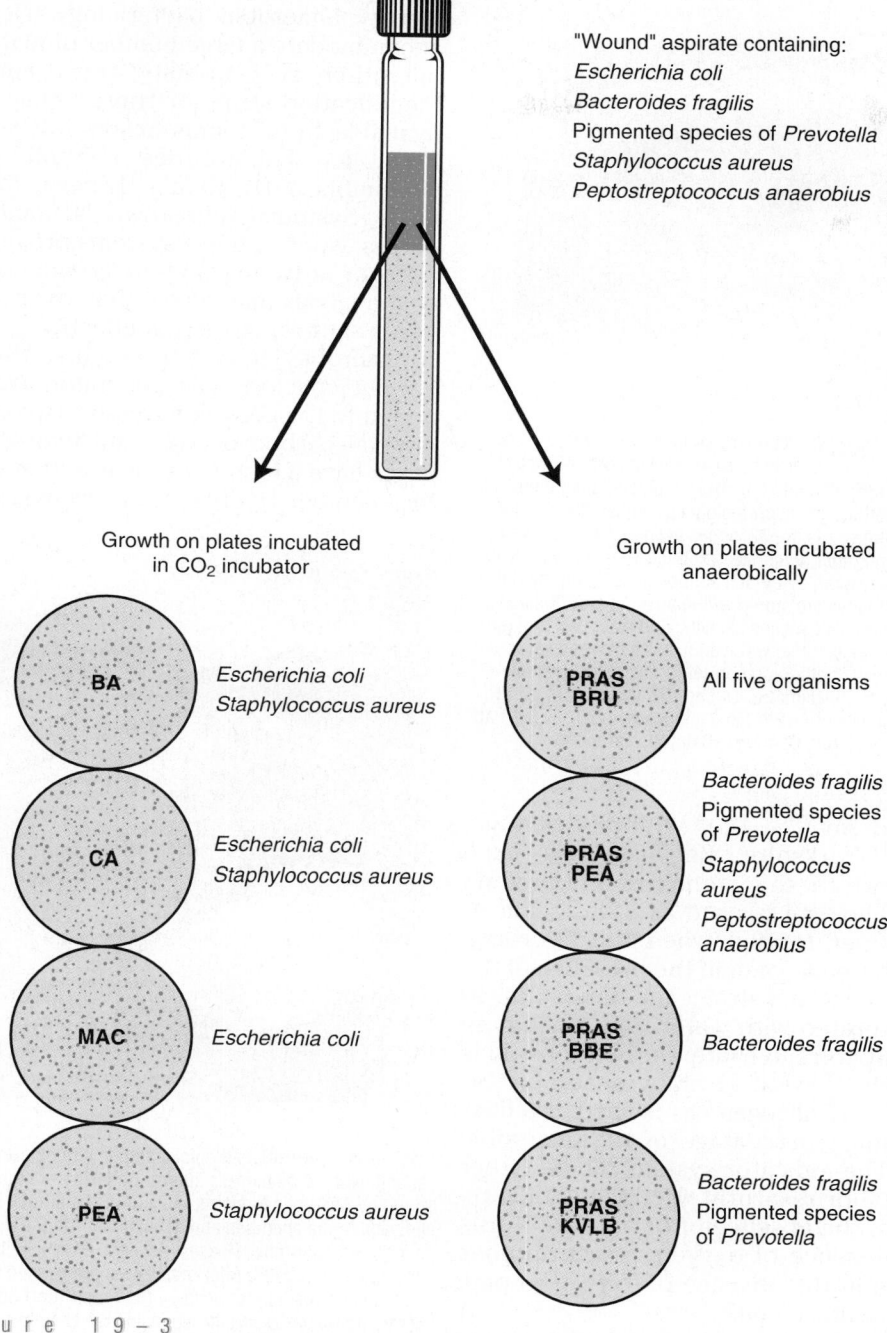

"Wound" aspirate containing:
Escherichia coli
Bacteroides fragilis
Pigmented species of *Prevotella*
Staphylococcus aureus
Peptostreptococcus anaerobius

Growth on plates incubated
in CO₂ incubator

BA — *Escherichia coli* / *Staphylococcus aureus*

CA — *Escherichia coli* / *Staphylococcus aureus*

MAC — *Escherichia coli*

PEA — *Staphylococcus aureus*

Growth on plates incubated
anaerobically

PRAS BRU — All five organisms

PRAS PEA — *Bacteroides fragilis* / Pigmented species of *Prevotella* / *Staphylococcus aureus* / *Peptostreptococcus anaerobius*

PRAS BBE — *Bacteroides fragilis*

PRAS KVLB — *Bacteroides fragilis* / Pigmented species of *Prevotella*

Figure 19-3

Culture results that might be obtained from the primary isolation set-up of a hypothetical "wound" specimen. This diagram illustrates the media and atmospheric conditions that would support growth of various organisms contained in the specimen. (BA, blood agar; BBE, Bacteroides bile-esculin agar; BRU, Brucella agar; CA, chocolate agar; KVLB, kanamycin-vancomycin laked blood agar; MAC, MacConkey agar; PEA, phenylethyl alcohol blood agar; PRAS medium.) (Redrawn from Engelkirk PG, Duben-Engelkirk J, Dowell VR Jr: Principles and Practice of Anaerobic Bacteriology, 1992; used with permission of Star Publishing Company, Belmont, CA.)

Anaerobic chambers contain a catalyst (usually palladium-coated alumina pellets) that removes residual oxygen from the atmosphere within the chamber. With time, the catalyst pellets become inactivated by gaseous metabolic end products produced by the anaerobes. A

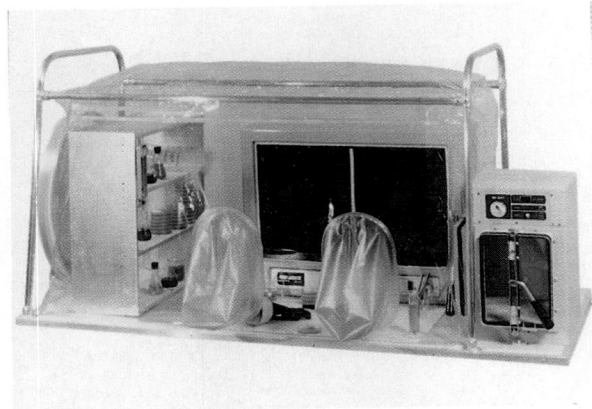

■ F i g u r e   1 9 – 4

"Glove Box" type of anaerobic chamber manufactured by Coy Laboratory Products, Inc. The flexible, clear vinyl chambers are available in three lengths (36, 59, and 78 inches), including one fitted with two pairs of gloves so that two microbiologists can use the chamber simultaneously. All three sizes are 32 inches deep and 40 inches high. Specimens, plates, and other items are introduced into and removed from a chamber by way of an automatic air lock (located on the right). Inoculated plates and tubes are stored within an incubator (located in the rear). The chamber also contains shelving and a catalyst box (out of view). Glove ports can be fitted with various size gloves to accommodate different users. (Photograph courtesy of Coy Corporation, Grass Lake, MI). (From Engelkirk PG, Duben-Engelkirk J, Dowell VR Jr: Principles and Practice of Anaerobic Bacteriology, 1992; used with permission of Star Publishing Company, Belmont, CA.)

product called Anatox (Don Whitley Scientific, Shipley, West Yorkshire, England) has been shown to absorb these metabolites and prolong catalyst life. Silica gel is used as a desiccant to absorb the water formed when the hydrogen combines with free oxygen in the presence of the catalyst. The silica gel desiccant turns blue to pink when saturated with water, and it needs to be heated daily to rejuvenate. Carbon dioxide is required for the growth of many anaerobic organisms, and inert nitrogen gas is used as a filler for the remaining percentage of the anaerobic atmosphere. The indicator system can be either methylene blue or resazurin. Methylene blue remains white in the absence of oxygen and turns blue in the presence of oxygen; resazurin goes from colorless in the absence of oxygen to pink in the presence of oxygen.

Anaerobic Jars. For small laboratories, where the volume of anaerobic cultures may not justify the purchase of anaerobic chambers, alternative systems are available. One such alternative is the GasPak jar, manufactured by Becton Dickinson (Cockeysville, MD) (Fig. 19–6). The jars have been used in clinical laboratories for many years,

enabling even small laboratories to perform satisfactory anaerobic bacteriology. Newer models accommodate a large number of plates as well as microtiter susceptibility trays and anaerobic identification strips or trays. Other systems are available from Adams Scientific (West Warwick, RI), Difco Laboratories (Detroit, MI), Oxoid, (Columbia, MD), Remel (Lenexa, KS), EM Diagnostic Systems (Gibbstown, NJ), and other companies. None of these systems provide all the features or advantages of anaerobic chambers, and cost analysis may reveal that, over time, a chamber is actually more cost-effective.

Anaerobic jar systems such as the GasPak utilize an envelope gas generator. When water is added to the GasPak envelope, two gases are generated—carbon dioxide and hydrogen. The two gases have a function similar to that in the anaerobic chamber. Hydrogen is explosive, and if the cat-

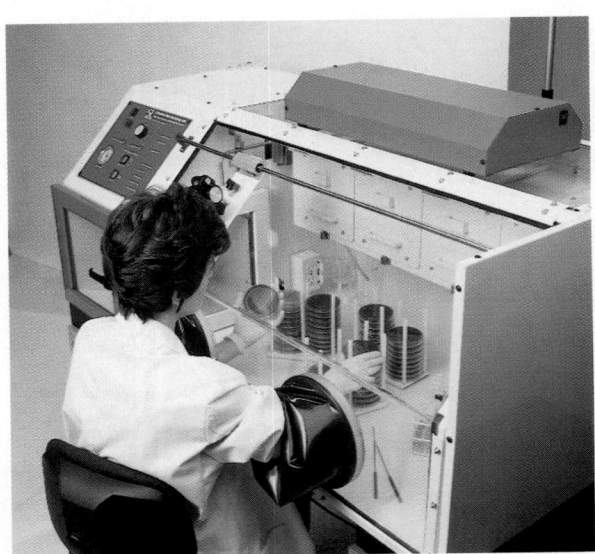

■ F i g u r e   1 9 – 5

"Gloveless" type of anaerobic chamber with dissecting microscope attachment. This stainless steel and Plexiglas chamber was manufactured by Anaerobe Systems. An automatic air lock is located on the left side of the chamber, and an incubator on the right. In place of rubber gloves are rubber sleeves, the ends of which fit snugly around the user's bare forearms. After the sleeves are flushed with an oxygen-free gas mixture (using foot pedals), the round "port hole–type" entry doors are removed and placed inside the chamber. Instruments constantly monitor and adjust temperature and anaerobic conditions within the chamber. A high level of relative humidity is constantly maintained within the chamber. The BACTRON I, BACTRON II, and BACTRON IV (shown here) are 48, 62, and 88.5 inches wide, respectively; all three models are 31 inches deep and 27 inches high. The incubator in the BACTRON IV has a 600-plate capacity. (Photograph courtesy of Anaerobe Systems, San Jose, CA, and Sheldon Manufacturing, Inc., Cornelius, OR.)

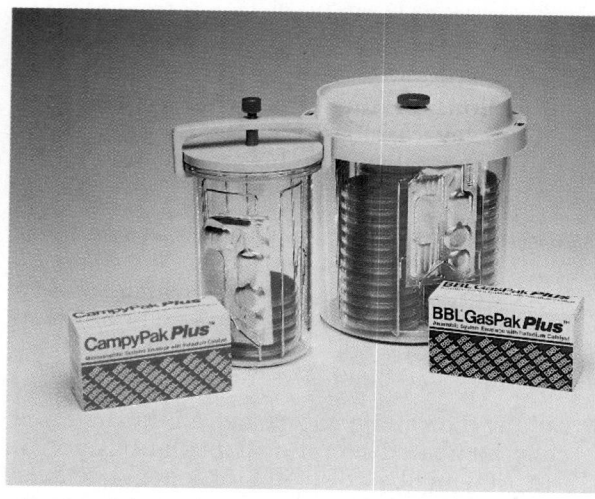

■Figure 19-6

Anaerobic jars. This photograph depicts two of the many different types of anaerobic jars that are available commercially. Inoculated plates, gas-generating envelopes, and an indicator strip are placed within a jar, which contains fresh or rejuvenated catalyst pellets. Many of the commercially available, gas-generating packets (e.g., GasPak Plus) contain "built-in" catalyst and indicator. Water is added to the packet, the jar is sealed, and carbon dioxide and hydrogen gases are released from the packet. If the catalyst is performing properly, the hydrogen combines with the oxygen in the jar to form water vapor. The indicator system lets the user know that anaerobic conditions were achieved. As can be seen in this photograph, the jars can also be used to culture for microaerophilic organisms. (Photograph courtesy of Becton Dickinson Microbiology Systems, Cockeysville, MD.)

alyst is not functioning properly, hydrogen gas will be present in the jar; therefore, following incubation, one should never open the jar in the vicinity of an open flame. A methylene blue indicator strip is always added to the jar to verify that an anaerobic atmosphere has been achieved. If the catalyst performs properly, water vapor will be present on the inside of the jar and the indicator strip will be colorless. Some of the newer gas-generating packets contain a built-in indicator. Failure to achieve anaerobic conditions could be due to a "poisoned" catalyst or a crack in the jar, lid, or "O" ring.

A "poisoned" catalyst results from the gases (particularly H_2S) produced by anaerobes. The reusable catalyst pellets can be rejuvenated after every use by heating them in a 160°C oven for a minimum of 2 hours. Some of the newer gas-generating packets contain a built-in supply of fresh catalyst, thus eliminating the need to maintain a container of rejuvenated catalyst within the jar.

The major disadvantage of any anaerobic jar system is that the plates have to be removed from the jar to be examined. This, of course, exposes the colonies to oxygen, which is especially hazardous to the anaerobes during their first 48 hours of growth. For this reason, a suitable holding system should always be used in conjunction with anaerobic jars. Plates can be removed from the anaerobic jar, placed in an oxygen-free holding system, removed one by one for rapid microscopic examination of colonies, and then quickly returned to the holding system. Plates should never remain in room air on the open bench.

Anaerobic Bags or Pouches. Other alternatives to an anaerobic chamber are anaerobic bags or pouches (Fig. 19–7). Such products are available from Becton Dickinson (Type A BioBag and GasPak Pouch, Cockeysville, MD) and Remel (Anaerobic Pouch, Lenexa, KS). These products will be collectively referred to as bags. One or two inoculated plates are placed into a bag, an oxygen-removal system is activated, and the bag is sealed and incubated. Theoretically, the plates can be examined for growth without removing the plates from the bags—thus without exposing

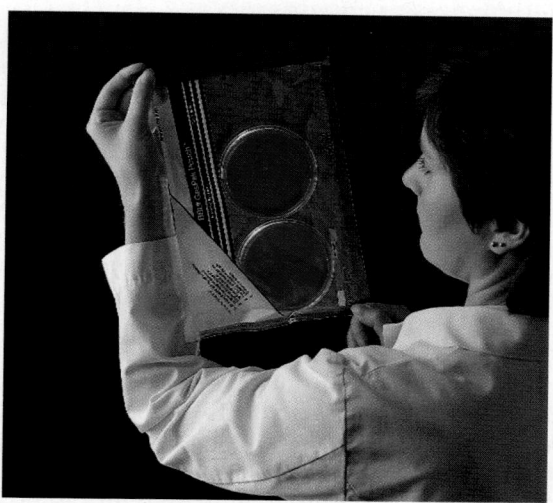

■Figure 19-7

Anaerobic pouch. This photograph depicts the GasPak Pouch, one of the commercially available anaerobic bag or pouch systems. After inoculated plates are added to the pouch, an oxygen-removal system is initiated. The pouch is then tightly sealed and placed in a non-anaerobic incubator. In some cases, the bags or pouches permit observation of growth directly through the plastic, thus eliminating the need to remove plates from the bag or pouch to make such observations. Identification and susceptibility testing systems requiring anaerobic incubation can be incubated within some of the bags or pouches. Bags or pouches are also useful for transporting specimens anaerobically to the laboratory. (Photograph courtesy of Becton Dickinson Microbiology Systems, Cockeysville, MD.)

the colonies to oxygen. However, with some of the products, a water-vapor film on the inner surface of the bag or the lid of the plate can sometimes obscure vision. In such cases, the plates must be removed from the bag to observe for growth, and a new bag and oxygen-removal system must be used whenever additional incubation is required. As with the anaerobic jar, plates must be removed from the bags in order to work with the colonies at the bench. Thus, an anaerobic holding system should be used in conjunction with any of the anaerobic bags.

Any of these bags are also useful transport devices. For example, a biopsy specimen can first be placed in a sterile screw-capped tube containing sterile saline, which is then placed in one of these bags. The oxygen-removal system is activated, and the specimen is transported to the laboratory within the bag.

PROCEDURES FOR IDENTIFYING ANAEROBIC ISOLATES

Preliminary Procedures

When to Examine Primary Plates. Plates incubated in an anaerobic chamber can be examined at any time without exposing the colonies to oxygen; however, those in anaerobic jars, bags, or pouches are exposed to the potentially damaging effects of oxygen whenever the containers are opened. Thus, when should anaerobic jars, bags, or pouches be opened? Should they be opened after 24 or 48 hours of incubation? If they are opened at 24 hours, slow-growing anaerobes will be too small to work with. More rapidly growing anaerobes, however, such as some *Clostridium* and *Bacteroides* spp, could be worked up at 24 hours, especially when appropriate PRAS media are used for the primary isolation set-up. Thus, waiting until 48 hours to open jars or bags could delay identification of clinically important anaerobes by a day.

If a holding system is available at the anaerobe workstation and is used in conjunction with jars, bags, or pouches, exposure to oxygen will be minimized, and plates can be removed and examined at 24 hours and then placed immediately into the holding system. As soon as all plates are examined, those requiring additional incubation are returned to an anaerobic system and incubated for an appropriate period before reexamination. Cultures are routinely held 5 to 7 days to

allow growth of particularly slow-growing anaerobes and up to 3 weeks whenever *Actinomyces* spp are suspected.

Indications of the Presence of Anaerobes. There are several clues that alert the microbiologist that anaerobes may be present on the primary plates. These include:

■ A foul odor upon opening an anaerobic jar or bag; some anaerobes, especially *Clostridium difficile, Fusobacterium,* and *Porphyromonas* spp, produce foul-smelling metabolic end products that are readily apparent when the jars, bags, or pouches are opened.

■ Colony morphotypes present on the anaerobically incubated blood agar plates but not on the CO_2-incubated blood or chocolate agar plates.

■ Good growth (more than 1 mm in diameter) of gray colonies on the BBE (Bacteroides bile-esculin) plate, characteristic of members of the *B. fragilis* group.

■ Colonies on either the KVLB (kanamycin-vancomycin laked blood) agar or anaerobic blood agar plate (e.g., BRU/BA [Brucella/blood agar] plate) that fluoresce brick-red under ultraviolet light or are brown to black in ordinary light, characteristic of pigmented *Porphyromonas* and *Prevotella* spp, although most strains of *Porphyromonas* will not grow on KVLB due to their sensitivity to vancomycin.

■ Double zone of hemolysis on blood agar, suggestive of *Clostridium perfringens.*

Processing Colonies Suspected of Being Anaerobes. When anaerobes are suspected, the following steps must be performed and recorded for each colony morphotype present on the anaerobic blood agar plate to initiate presumptive identification of the isolates:

■ Describe colony morphology and whether growth occurred on each of the selective (BBE, KVLB) and nonselective anaerobic blood agar plate (BRU/BA) media used.

■ Describe Gram-stain reaction and cell morphology.

■ Inoculate pure culture/subculture plate; add appropriate disks.

■ Set up aerotolerance test.

Colony Morphology. Enumerate, describe, and record each colony morphotype and growth characteristics on a worksheet. The use of a dissecting microscope to observe the fine details of the colonies and to pick colonies for isolation is recommended. Figure 19–8 depicts the appearance of colonies as seen through a dissecting microscope. Growth of a particular morphotype can be semiquantitated using terms such as "light," "moderate," and "heavy" or by using some type of coding system (e.g., 1+, 2+). Use Table 19–13 as a guide for characteristic features to note when examining colonies of anaerobic isolates.

Gram Reaction and Cell Morphology. Prepare a smear for Gram stain. Record the gram reaction and cell morphology on a worksheet. The Gram-stain reaction and morphologic appearance of the organism aid in the presumptive identification of isolates, as shown in a schematic diagram in Figure 19–9.

Pure Culture/Subculture Plate and Special-Potency Disks. Although the Gram stain is helpful in the initial identification of an anaerobic isolate, certain species of *Clostridium* may stain pink and thus appear to be gram-negative bacilli. To determine the true Gram-stain reaction of the isolate, special-potency disks are used. These disks, in addition to determining the true Gram-stain reaction, may provide a presumptive identification of the suspected anaerobic organism.

Based on the Gram-stain reaction and morphology of the organism, special-potency disks are added to the subculture plate. The disks are placed on the heavily inoculated area of the plate (usually the first quarter or first third of the

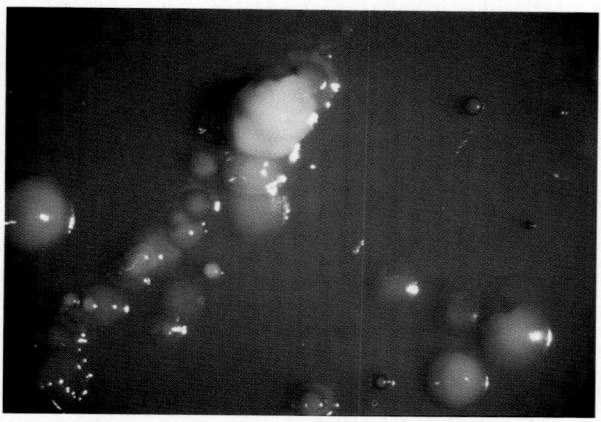

■■■■■■ F i g u r e   1 9 – 8

Anaerobic blood agar plate from an intrauterine device culture as seen through a dissection microscope. The heavy mixture of sizes and types of colonies makes the task of isolating and identifying colonies very difficult without the enhancement obtained with a dissecting microscope.

■■■■■ T A B L E   1 9 – 1 3

CHARACTERISTICS TO NOTE WHEN EXAMINING COLONIES OF ANAEROBIC BACTERIA

Characteristic	Suspected Anaerobe
Color/pigment	
Brown-black colonies	If a gram-negative rod, may be a pigmented species of *Porphyromonas* or *Prevotella;* if a gram-positive coccus, may be *Peptococcus niger*
Red, pink, tan, yellow	If a gram-positive rod, may be an *Actinomyces* sp
Surface (e.g., glistening or dull)	
Density (e.g., opaque, translucent, transparent)	Members of the *Bacteroides ureolyticus* group (gram-negative bacilli) typically have translucent to transparent colonies
Consistency (e.g., butyrous [butter-like], viscous, membranous, brittle)	—
Form, elevation, margin	—
Fluorescence under ultraviolet light	
Brick-red	If a gram-negative rod, may be a pigmented species of *Porphyromonas* or *Prevotella*
Red	If a gram-negative coccus, may be *Veillonella* spp
Chartreuse	If a gram-negative rod, may be a subspecies of *Fusobacterium nucleatum;* if a gram-positive rod, may be *Clostridium difficile*
Pitting of the agar	May be *B. ureolyticus, B. gracilis,* or *Wolinella* spp (all gram-negative bacilli)
Double zone of hemolysis	*Clostridium perfringens* (a gram-positive rod)
Extensive swarming	May be *Clostridium tetani* (terminal spores) or *C. septicum* (subterminal spores); both are gram-positive bacilli
Odor	
Like a horse stable	*C. difficile* (a gram-positive rod)
Sweet, unpleasant	If a gram-positive coccus, may be *Peptostreptococcus anaerobius*
Spider-like appearance (having thin, wooly filaments, originating at a single point)	If a gram-positive rod, may be a young colony of *Actinomyces israelii* or *Propionibacterium propionicus*
Molar tooth appearance	If a gram-positive rod, may be an older colony of *Actinomyces israelii* or *Propionibacterium propionicus*
White, bread-crumb appearance	If a gram-negative rod, may be a subspecies of *F. nucleatum*
Ground-glass appearance	If a gram-negative rod, may be a subspecies of *F. nucleatum*
Fried-egg appearance	If a gram-negative rod, may be *F. mortiferum* or *F. varium,* although other anaerobes also produce such colonies

Modified from Engelkirk PG, Duben-Engelkirk J, Dowell VR Jr: Principles and Practice of Anaerobic Bacteriology, 1992; used with permission of Star Publishing Company, Belmont, CA.

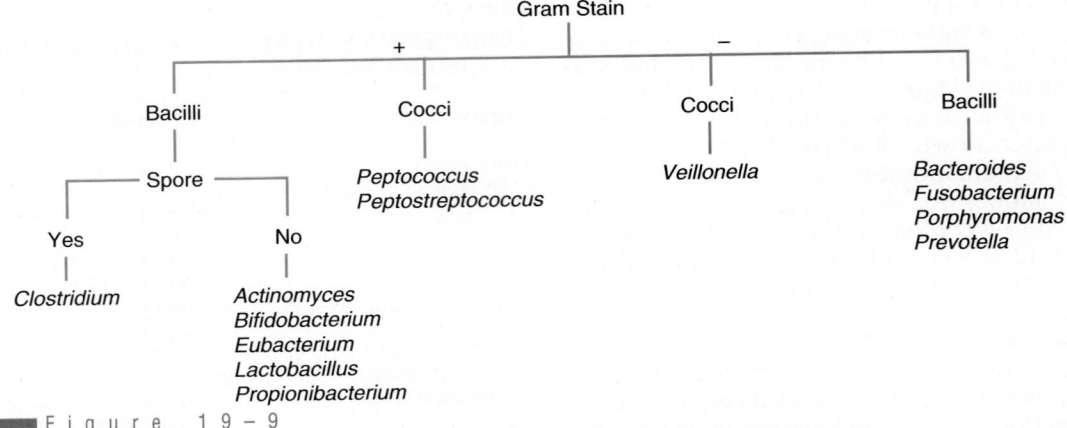

Figure 19-9

Schematic diagram for the Initial Identification of anaerobic isolates based on Gram stain morphology.

plate, depending on the particular manner of streaking). The specific disks to be added are shown in Figure 19–10. Interpretation of special-potency antimicrobial disk results is shown in Table 19–14.

Aerotolerance Testing. The aerotolerance test determines whether an isolate is a strict anaerobe or a facultative anaerobe. Incubating the suspected isolate in both aerobic and anaerobic environments determines the true atmospheric requirements of the organism. The subculture plate incubated anaerobically may serve as one of the plates used in the aerotolerance test. Suspected colonies should also be inoculated onto the following additional plates:

■ A blood agar plate for incubation in a non-CO_2 incubator. This plate enables differentiation between aerobic and capnophilic organisms.

■ A chocolate agar plate for incubation in a CO_2 incubator.

With the exception of the subculture plate, several different isolates can be tested on each plate. Incubate all aerotolerance test plates appropriately for 24 to 48 hours. Following incubation, examine the aerotolerance test plates for growth. Theoretically, an anaerobe should grow only on the anaerobically incubated plates. However, some aerotolerant anaerobes (e.g., certain *Clostridium, Actinomyces, Propionibacterium, Bifidobacterium,* and *Lactobacillus*) may grow on the CO_2-incubated chocolate agar plate, but they usually grow better on the anaerobically incubated plate. A facultative, noncapnophilic organism will grow on all plates, but a capnophilic aerobe should grow only on the CO_2-incubated chocolate agar plate. *Haemophilus influenzae,* however, will grow on the anaerobic

plate and the CO_2-incubated chocolate agar plate but not on the aerobically incubated blood agar plate. Table 19–15 shows how the aerotolerance test is interpreted.

Identification of Anaerobic Isolates

There are various methods available to microbiologists for identifying anaerobic isolates. Table 19–16 lists several options from which the microbiologist may choose. The method and level of identification usually depend on the size and capabilities of the laboratory. There are several reasons why an anaerobic isolate must be identified beyond the microscopic and colonial morphology.

■ Identification of an anaerobic isolate can often indicate to the physician the probable source of the infectious process. For example, *Clostridium septicum* bacteremia suggests colon cancer or other diseases of the colon. Similarly, regardless of the site of isolation, recovery of certain pigmented *Porphyromonas* species suggests an oral origin, whereas isolation of members of the *Bacteroides fragilis* group suggests an intestinal origin.

■ Empirical therapy is likely to be more effective if the physician's choice of antimicrobial agent(s) is based on knowledge of the infecting organism(s).

■ Over time, the identification of anaerobes provides a database of information regarding the role of certain anaerobic bacteria in infectious processes, in much the same way that susceptibility testing results provide a database of

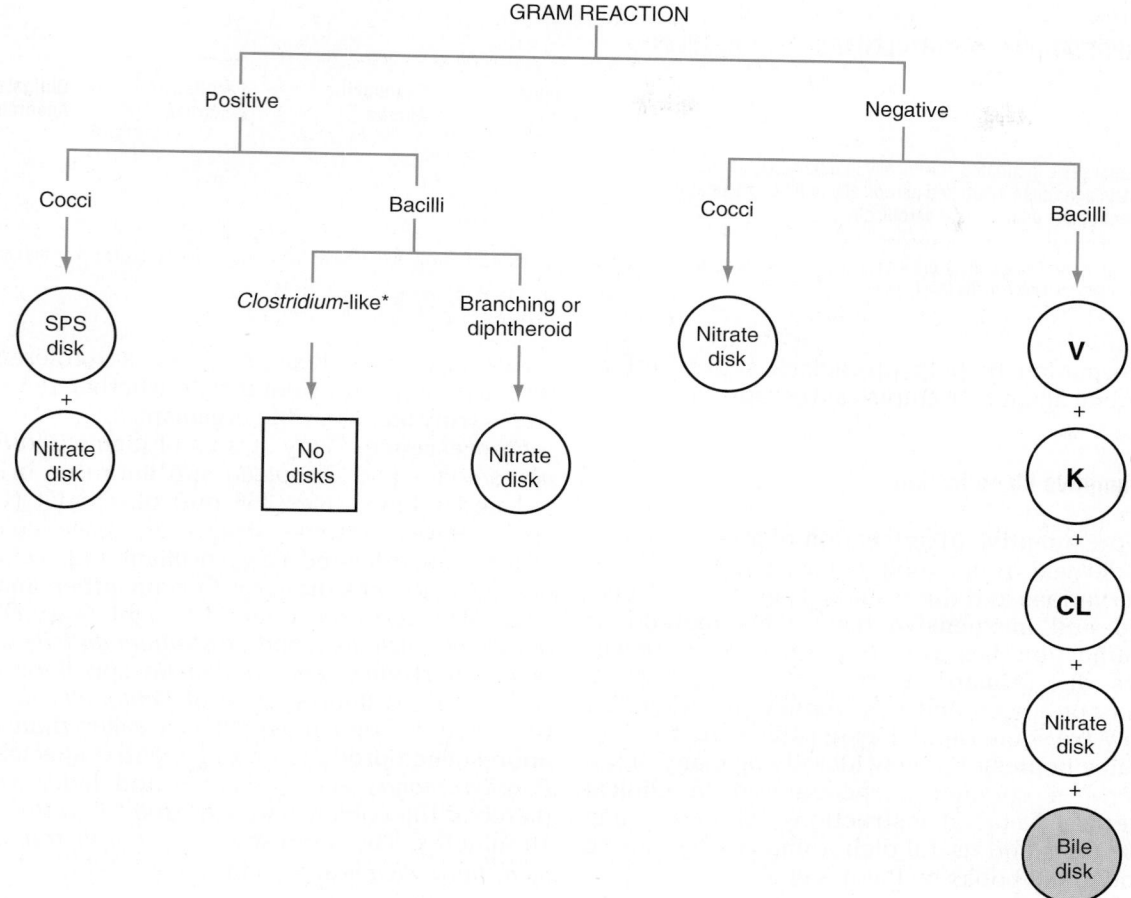

GRAM REACTION

*Large, unbranched, gram-positive rods, with or without spores.
(Not all clostridia are large, and not all clostridia will stain as
gram-positive rods.)

Figure 19–10

Disks to add to the pure culture/subculture plate. (SPS, sodium polyanethol sulfonate; V, vancomycin [5 µg]; K, kanamycin [1 mg]; CL, colistin [10 µg]). Based on Sutter et al, 1985. (Modified and redrawn from Engelkirk PG, Duben-Engelkirk J, Dowell VR Jr: Principles and Practice of Anaerobic Bacteriology, 1992; used and redrawn with permission of Star Publishing Company, Belmont, CA.)

TABLE 19–14

INTERPRETATION OF SPECIAL-POTENCY ANTIMICROBIAL DISK RESULTS

	Vancomycin (5 µg)	Colistin (10 µg)	Kanamycin (1 mg)
Gram-positive organisms (and vancomycin-sensitive *Porphyromonas* spp)	S	—	—
Most gram-negative organisms	R	—	—
Fusobacterium spp (and *Bacteroides ureolyticus* group)	R	S	S
Most *Bacteroides* and *Prevotella* spp	R	V	R
Bacteroides fragilis group	R	R	R

S = sensitive.
R = resistant.
V = variable.
Data from Sutter et al (1985).
Modified from Engelkirk PG, Duben-Engelkirk J, Dowell VR Jr: Principles and Practice of Anaerobic Bacteriology, 1992; used with permission of Star Publishing Company, Belmont, CA.

■■■■■ T A B L E   1 9 – 1 5
INTERPRETATION OF AEROTOLERANCE TEST RESULTS

	Obligate Aerobe	Capnophilic Aerobe	Facultative Anaerobe	Obligate Anaerobe
Blood agar plate incubated aerobically in a non-CO_2 incubator	+	–	+	–
Chocolate agar plate incubated aerobically in a CO_2 incubator	+	+	+	–
Blood agar plate incubated anaerobically	–	–*	+	+

* Haemophilus influenzae will grow on an anaerobically incubated Brucella blood agar plate but can be differentiated from an anaerobe by its growth on the chocolate agar plate incubated in the CO_2 incubator.

information to help physicians select antimicrobial agents for empirical therapy.

Presumptive Identification

A presumptive identification of a bacterium is one derived from simple colony and Gram-stain observations and the results of several relatively rapid and inexpensive tests. This method of identification has gained popularity in recent years, due primarily to the ever-increasing emphasis on speed and cost reduction. Described below are some rapid, inexpensive tests that are of value in presumptively identifying many of the anaerobes commonly encountered in clinical materials. Detailed instructions for performing these tests and useful dichotomous keys can be found in the books by Balows et al.

Aerotolerance. Results of the aerotolerance test may be used to determine whether the isolate is truly an anaerobic organism.

Fluorescence. Many strains of pigmented *Porphyromonas* and *Prevotella* spp fluoresce brick-red under long-wave (366 nm) ultraviolet (UV) light. However, some strains fluoresce colors other than brick-red (e.g., brilliant red, yellow, orange, and pink-orange). Certain other anaerobes also fluoresce under UV light (e.g., *Fusobacterium nucleatum* and *Clostridium difficile* fluoresce chartreuse, and *Veillonella* spp fluoresce red). The red fluorescence of *Veillonella* is culture medium–dependent; it is weaker than the fluorescence produced by pigmented species of *Porphyromonas* and *Prevotella* and fades completely if the colonies are exposed to air for 5 to 10 minutes. The fluorescence of *Porphyromonas asaccharolytica* is shown in Figure 19–11.

■■■■■ T A B L E   1 9 – 1 6
OPTIONS AVAILABLE FOR IDENTIFYING ANAEROBIC BACTERIA

Identification Technique	Time to Obtain Results	Extent of Identification Capability
Presumptive identification based on colony morphology, Gram-stain observations, and results of simple tests (e.g., disks, catalase, spot indole)	Same day that the pure culture/subculture (PC/SC) plate is available	Limited capability; many clinically encountered anaerobes cannot be identified
Definitive identification		
Commercially available, preexisting enzyme-based identification systems (e.g., ANIDENT, MicroScan, RapId ANA II, Vitek ANI card)	Same day that the PC/SC plate is available (most systems require 4 hr incubation)	Most of the commonly encountered, clinically significant anaerobes can be identified
Commercially available biochemical-based identification systems (e.g., API 20A, Minitek, Sceptor)	24–48 hr after the PC/SC plate is available	Most commonly encountered saccharolytic anaerobes can be identified, but many asaccharolytic anaerobes cannot be
Cellular fatty acid analysis by high-resolution gas-liquid chromatography (GLC) (e.g., the MIDI system)	24–48 hr after the PC/SC plate is available	Most clinically encountered anaerobes can be identified
Conventional tubed biochemical tests and fatty acid analysis by GLC	24–72 hr after the PC/SC plate is available	Most clinically encountered anaerobes can be identified

Modified from Engelkirk PG, Duben-Engelkirk J, Dowell VR Jr: Principles and Practice of Anaerobic Bacteriology, 1992; used with permission of Star Publishing Company, Belmont, CA.

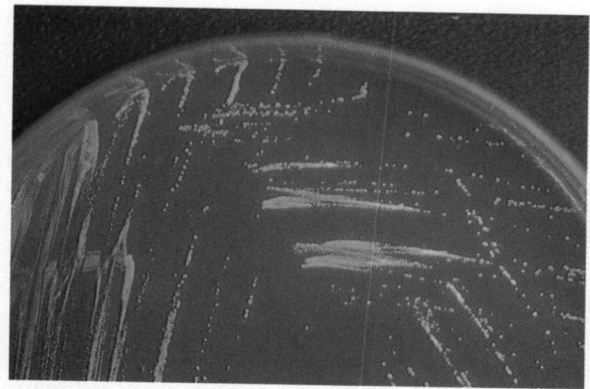

■ F i g u r e   1 9 – 1 1

Example of the brick-red fluorescence seen when colonies of certain *Porphyromonas* and *Prevotella* species are subjected to long-wave ultraviolet light; *Porphyromonas asaccharolytica,* 72-hour-old culture shown here. (Photograph courtesy of Star Publishing Co., Belmont, CA.)

Special-Potency Antimicrobial Disks. To verify that the organism is truly a gram-negative bacillus as opposed to a *Mobiluncus* species or a species of *Clostridium* (e.g., *C. ramosum* or *C. clostridioforme*) that may stain gram-negative, add special-potency disks to anaerobic blood agar subcultures. Disks of the proper potency must be pressed firmly to the surface of the plate to ensure uniform diffusion of the agent into the medium. Disk results are analyzed in a stepwise manner (see Table 19–14). The disk results for *C. ramosum* are shown in Figure 19–12.

Sodium Polyanethol Sulfonate (SPS) Disk. Add an SPS disk to the anaerobic blood agar subculture plate whenever the Gram stain reveals the isolate to be a gram-positive coccus. The SPS disk is used to presumptively identify *Peptostreptococcus anaerobius,* which is susceptible to SPS.

Nitrate Disk. Perform a nitrate reduction test using a disk that is a miniaturized version of the conventional nitrate reduction test. This determines an organism's ability to reduce nitrate.

Bile Disk. This disk is used to determine an organism's ability to grow in the presence of relatively high concentrations (20%) of bile. Add a bile disk to the anaerobic blood agar subculture plate whenever the Gram stain reveals the isolate to be a gram-negative bacillus. Good growth on the BBE plate and growth in 20% bile broth also indicate bile resistance. An anaerobic, gram-negative bacillus that is a member of the *Bacteroides fragilis* group fits this description.

Catalase Test. Using a plastic, disposable inoculating loop or a wooden applicator stick, remove some of the growth from the subculture plate and rub it onto a small area of a glass microscope slide. Perform a catalase test by adding a drop of 15% hydrogen peroxide. Watch for the production of bubbles of oxygen gas. Among other uses, the catalase test is of value in differentiating aerotolerant strains of *Clostridium* (catalase-negative) from *Bacillus* species (catalase-positive).

Spot Indole Test. Saturate a small piece of filter paper with the spot indole reagent p-dimethylaminocinnamaldehyde (DMCA). Using an inoculating loop, remove some of the growth from the subculture plate and rub it onto the saturated area. Rapid development of a blue or green color indicates a positive test (i.e., production of indole from the amino acid tryptophan), whereas a pink or orange color indicates a negative test. The spot indole test can also be performed directly on a pure culture plate. DMCA was compared to Kovac and Ehrlich reagents for the detection of indole production by anaerobes. It proved to be the most sensitive; the Kovac reagent was the least sensitive of the three.

Motility test. Motility may be determined using either very young (4 to 6 hours old) broth cultures or 24- to 48-hour colonies on agar. Motile, gram-negative anaerobes include some *Campylobacter* (e.g., *C. concisus, C. curvus, C. rectus*) and *Mobiluncus* species, among many others.

Lecithinase and Lipase Reactions. An egg-yolk agar plate is used to determine the activities of these enzymes. These reactions are of value in identifying many species of clostridia.

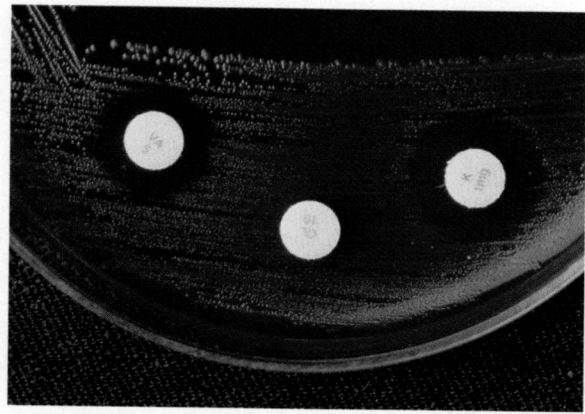

■ F i g u r e   1 9 – 1 2

Typical special-potency antimicrobial disk results for *Clostridium ramosum;* susceptible to vancomycin (disk at left) and kanamycin (disk at right), and resistant to colistin (center disk). (Photograph courtesy of Anaerobe Systems, San Jose, CA.)

Presumpto Plates. Many of the previously mentioned observations and tests are incorporated into specialized types of plated media, collectively referred to as Presumpto plates or CDC (Centers for Disease Control and Prevention) differential agar media. Three types of quad plates called Presumpto 1, Presumpto 2, and Presumpto 3 are available. These plates contain tests for the determination of growth on 20% bile; lipase and lecithinase production; casein, esculin, gelatin, and starch hydrolysis; hydrogen sulfide and indole production; DNase activity; and glucose, lactose, mannitol, and rhamnose fermentation. Formulations for Presumpto plate media can be found in a 1977 CDC publication by Dowell and colleagues. Presumpto plates are available from several commercial sources, including Remel (Lenexa, KS) and Carr-Scarborough Microbiologicals Inc. (Stone Mountain, GA).

Definitive Identifications

Although many laboratories base their identification of anaerobic organisms on presumptive identification criteria, others perform additional procedures to obtain definitive identification. Many anaerobic isolates cannot be identified using presumptive tests. Whenever such anaerobes are encountered, definitive identification is required. A wide variety of techniques exist for making definitive identification:

■ PRAS and non-PRAS tubed biochemical test media

■ Biochemical-based and preexisting enzyme-based minisystems

■ Gas-liquid chromatographic analysis of metabolic end products

■ Cellular fatty acid analysis by gas-liquid chromatography (GLC)

The majority of today's clinical microbiology laboratories use one of the commercially available biochemical-based or preexisting enzyme-based minisystems for making definitive identifications, but it is important to remember that none of them will identify all the anaerobes that could potentially be isolated from clinical specimens. Most are designed to identify those anaerobes discussed earlier that are most frequently encountered in clinical specimens. It is far more important for a system to identify such organisms than to identify obscure anaerobes that are only rarely isolated from clinical specimens or involved in infectious processes.

Conventional Tubed Biochemical Identification Systems. The use of tubed biochemical systems originated from studies of rumen contents and sewage sludge by R. E. Hungate in the 1950's and 1960's. Major components of the Hungate technique were oxygen-free, flame-sterilized gassing jets or cannulae, agar-containing media on the inner surfaces of tightly stoppered glass tubes or "roll tubes," and a variety of tightly stoppered test tubes containing PRAS biochemicals. Roll tubes were inoculated under flowing oxygen-free gas; the tightly stoppered, inoculated tubes served as their own anaerobic culture chambers.

In general, conventional or traditional systems for the identification of anaerobic isolates employ large test tubes containing a variety of PRAS or non-PRAS biochemical test media. Such media can be either prepared in the laboratory or purchased from commercial sources (e.g., Adams Scientific, West Warwick, RI; Carr-Scarborough, and Remel).

Citron and associates described a modified "gold standard" PRAS biochemical method for identifying bile-resistant *Bacteroides* spp. Their abbreviated protocol involves preparing inoculum directly from pure culture plates. PRAS biochemical tubes containing arabinose and trehalose are inoculated using a tuberculin syringe. A catalase test is included for identifying indole-negative species. Rhamnose, salicin, sucrose, trehalose, and xylose are added for identifying indole-positive species. Tests are incubated overnight, and a pH indicator (bromothymol blue) is added after growth occurs. The investigators reported that 98% of 189 clinical isolates were correctly identified using this protocol.

Biochemical-Based Minisystems. The first commercially available alternatives to conventional tubed media were the biochemical-based identification systems manufactured by Analytab Products (API20A; Plainview, NY) and BBL (Minitek; Cockeysville, MD), which are still in use in some laboratories. These systems provide many of the same tests as the conventional system but in the form of a small plastic strip or tray; hence the term "minisystem."

The biochemical-based minisystems are easier and faster to inoculate than a conventional system. Although they can be inoculated aerobically, they require anaerobic incubation. The larger model BBL anaerobic jars and some of the commercially available bags and pouches can be used to incubate biochemical-based minisystem trays and strips if an anaerobic chamber is not available. After 24 to 48 hours of incubation, test results are read, a code number is generated for each isolate, and the numbers are looked up in a compendium codebook. It is important to note that the databases from which the codebooks are developed do not contain all the anaerobes that can potentially be isolated from clinical specimens.

Preformed Enzyme-Based Minisystems. Many of the newer commercial systems are based on the presence of preformed enzymes. Because these minisystems do not depend on enzyme induction, there is virtually no lag time, and results are available in 4 hours. The small plastic panels or cards are easy to inoculate, can be inoculated at the bench, and do not require anaerobic incubation. Most of the systems generate code numbers, which are looked up in a manufacturer-supplied codebook. Like the biochemical-based minisystems, these systems are primarily of value for identifying commonly isolated anaerobes. The databases from which the codebooks are developed do not contain all the anaerobes that can potentially be isolated from clinical specimens. Examples of preexisting enzyme-based minisystems include ANI Card (Vitek Systems, Hazelwood, MO), AN-IDENT (Analytab Products, Plainview, NY), Rapid Ana II (Innovative Diagnostic Systems, Atlanta, GA), and Rapid Anaerobe Identification Panel (Baxter Healthcare Corp., MicroScan Division, West Sacramento, CA). In general, the systems use the same or similar substrates. They contain a number of nitrophenyl and naphthylamide compounds, which are colorless substances that produce yellow or red products, respectively, in the presence of appropriate enzymes.

Gas-Liquid Chromatography (GLC). Laboratories that must routinely definitively identify anaerobic isolates that cannot be identified using one of the previously mentioned commercially available minisystems may wish to incorporate GLC into the identification protocol. Like other types of chromatographic techniques (e.g., gas-solid chromatography, thin-layer chromatography, ion-exchange chromatography), GLC uses the chemical and physical properties of a particular component to separate it from other components in a mixture. All chromatographic techniques have a mobile phase and a stationary

phase. In GLC, an inert gas serves as the mobile phase and a liquid serves as the stationary or separating phase. The unknown mixture is usually prepared for separation, volatilized, and carried as a gas through a packed column. Acids are recovered in a specific order, based on their chemical and physical properties. As the column separates the acids, they are identified based on their position on the chromatograph by their relative retention time as compared to a standard. Quantitation of an unknown is based on the area under the peak with respect to a known standard. Figure 19–13 depicts the Capco gas-liquid chromatograph (Dodeca, Fremont, CA), specifically manufactured for and commonly used in anaerobic bacteriology laboratories.

Identifying Anaerobes via Metabolic End Product Analysis by GLC. Identification of anaerobes by GLC analysis of metabolic end products was pioneered by W. E. C. Moore, E. P. Cato, L. V. Holdeman, and others at the Virginia Polytechnic Institute (VPI) in Blacksburg. Their GLC procedure, details of which can be found in the *VPI Anaerobe Laboratory Manual,* is still widely used in laboratories throughout the world. The VPI guidelines for GLC were used for many years at the CDC but were modified by G. L. Lombard, V. R. Dowell, Jr., and their colleagues.

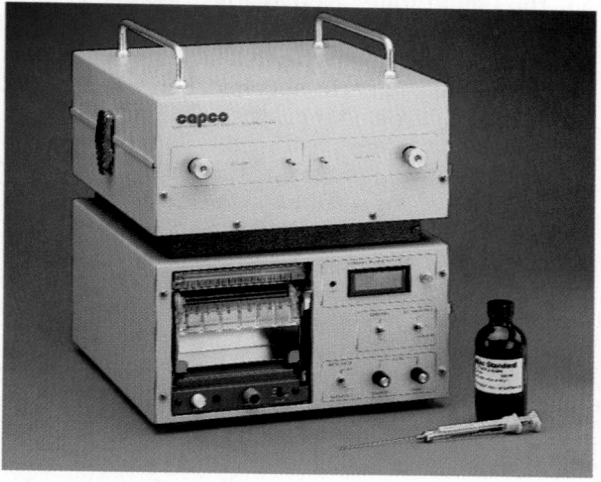

■■■■■ F i g u r e   1 9 – 1 3

The Capco model 700A anaerobe identification system, manufactured by Dodeca. The "turnkey system," which costs approximately $7,000, consists of a compact gas-liquid chromatograph with integral recorder and all accessories and supplies (except a tank of helium carrier gas) needed to analyze anaerobic bacteria on the basis of their metabolic products. The dual-column chromatograph is 12.5 inches wide, 13.75 inches deep, and 14.0 inches high and weighs only 26 lb. (Photograph courtesy of Dodeca, Fremont, CA.)

The CDC guidelines for GLC can be found in *Gas-Liquid Chromatography Analysis of the Acid Products of Bacteria.* A brief summary of the CDC procedure follows.

The anaerobe to be tested is grown in a tube of peptone–yeast extract–glucose medium (PYG), which is available commercially from Adams Scientific (West Warwick, RI), Carr-Scarborough, Remel, and other media manufacturers. This medium also contains cysteine hydrochloride, resazurin, and a salt solution. Aliquots of the PYG culture are used to analyze volatile and nonvolatile acids. Short-chain volatile acids produced by anaerobes include formic, acetic, propionic, isobutyric, isovaleric, valeric, isocaproic, butyric, caproic, and heptanoic.

Volatile acids are identified by comparing the elution times of products in the ether extract with those of known acids in a standardized volatile acid mixture that has been chromatographed in the same manner on the same day. Figure 19–14*A* and *B* shows chromatographs of a volatile acid standard and *Fusobacterium,* respectively.

Nonvolatile, low-molecular-weight aliphatic and aromatic acids produced by anaerobes are identified by comparing the elution times of products in the chloroform extract with those of known acids in a standardized, nonvolatile acid mixture that has been chromatographed in the same manner on the same day. Nonvolatile, low-molecular-weight aliphatic and aromatic acids produced by anaerobes include pyruvic, lactic, oxalacetic, oxalic, malonic, fumaric, succinic, benzoic, phenylacetic, and hydrocinnamic acids. In Figure 19–14*C* and *D* chromatographs show the nonvolatile acid standard and *Actinomyces,* respectively.

Identifying Anaerobes via Cellular Fatty Acid Analysis by High-Resolution GLC. Cellular fatty acid analysis is another method of identifying anaerobes. The term "cellular fatty acids" refers to fatty acids and related compounds (aldehydes, hydrocarbons, dimethyl acetals) that are present within organisms as cellular components. Cellular fatty acids are coded for on bacterial chromosomes, as opposed to plasmids, and are not affected by simple mutations or plasmid loss. Thus, the fatty acid composition of a particular organism is relatively stable when grown under certain growth conditions (medium, incubation temperature, and time). Although fatty acid profiles can be identified manually, computerized, high-resolution gas chromatography and specialized software programs are now available to analyze cellular fatty acids of unknown bacteria and compare the results to patterns of known species.

First grown in pure culture in PYG broth or another standardized medium, the bacterial cells are removed by centrifugation and then saponified to release the fatty acids from the bacterial lipids. After extraction, the methyl esters are analyzed by GLC. A chromatogram depicting the unknown organism's fatty acid composition and a comparison to a database or "library" of fatty acids of known anaerobes are generated as a computer printout or report. The report includes computer-generated statistical values or "similarity indices," which are based on deviations in the unknown organism's fatty acid composition from the known profile in the library. Because no subjective interpretations are required, the identifications are objective and highly reproducible.

Using GLC and related techniques, investigators have been attempting for many years to directly identify anaerobes and other organisms in clinical specimens. Johnson and colleagues, for example, described a method for identifying *Clostridium difficile* in stool specimens using culture-enhanced GLC. In this technique, GLC is used to detect four distinctive peaks produced by *C. difficile.* In the future, assays of this type (i.e., those performed directly on clinical specimens) will probably be modified to enable use of instrumentation.

FREQUENTLY ENCOUNTERED ANAEROBES AND THEIR ASSOCIATED DISEASES

Taxonomically, anaerobic bacteria encountered in human clinical specimens may be divided into two major groups: those capable of forming endospores (spore-formers) and those incapable of forming endospores (non–spore-formers). The presence or absence of spores, coupled with the Gram-staining reaction and cellular morphology, is often helpful in making the initial or presumptive identification of anaerobic bacteria and in determining the appropriate identification tests to be performed.

Gram-Positive Spore-Forming Anaerobic Bacilli

All spore-forming anaerobic bacilli are classified in the genus *Clostridium* and are collectively referred to as clostridia. Although all clostridia

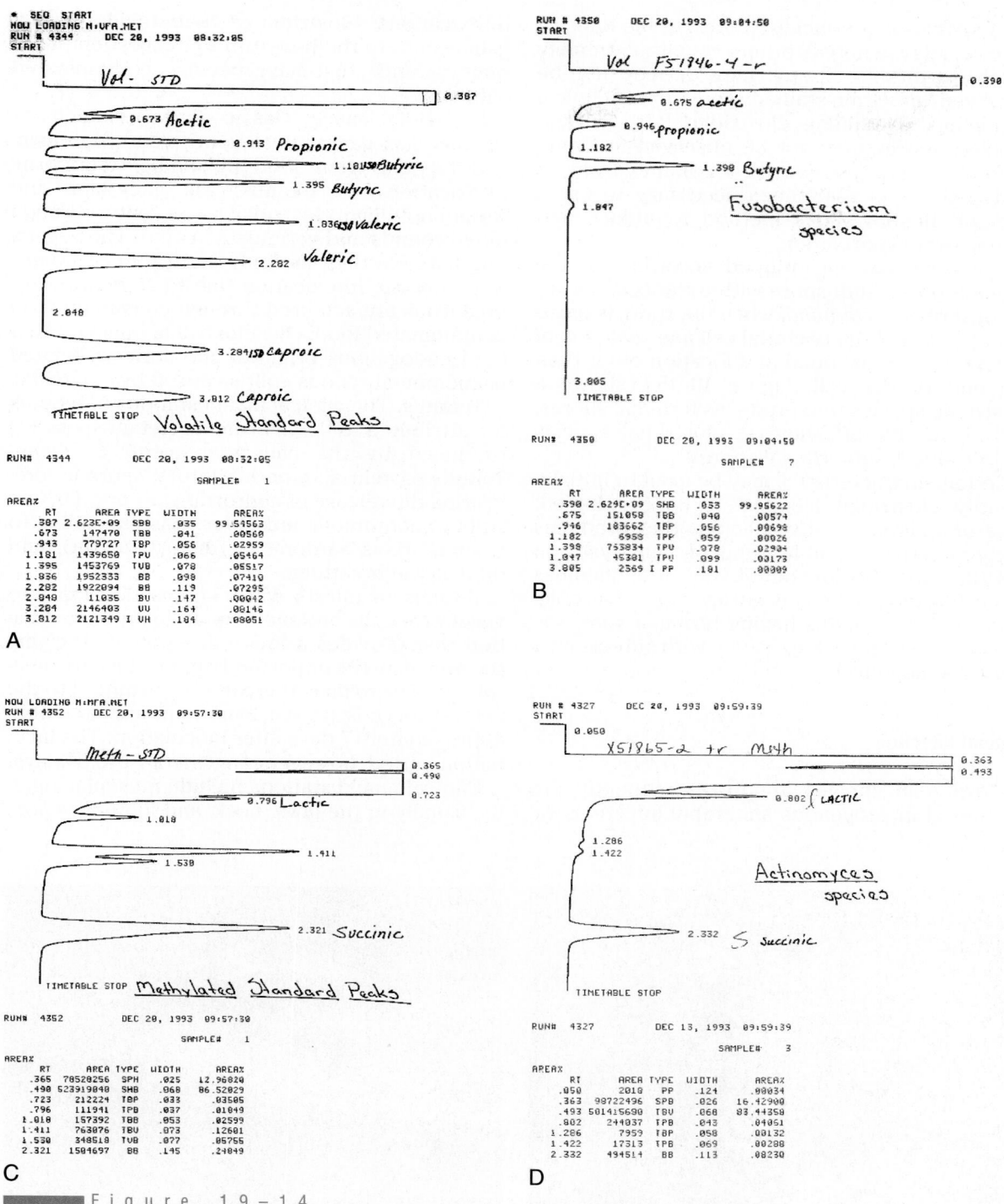

Figure 19-14

A, Volatile acid standard chromatograph. Elution times of known acids and their peaks are depicted in the extract of the standard solution. *B,* Volatile acid chromatograph of unknown identified as *Fusobacterium* species after comparison of the retention times of the acids with those for the standard solution. Numbers located near the acid peaks are retention times. *C,* Methylated acid standard chromatograph. Elution times of known acids and their peaks are depicted in the extract of the standard solution. *D,* Methylated acid chromatograph of an unknown identified as *Actinomyces* species after comparison of the retention times of the acids with those of the standard solution. Numbers located near the acid peaks are retention times in minutes. RT, retention time.

are capable of producing spores, some species do so readily, whereas others require extremely harsh conditions. Spores may or may not be observed in Gram-stained smears of clinical specimens containing clostridia. With certain species, spores may not be observed in Gram-stained smears of *Clostridium* colonies from an agar plate. It is sometimes necessary to use a heat or alcohol shock method to induce and demonstrate sporulation.

Clostridia may be grouped according to the location of the endospore within the cell. Spores are described as *terminal* when the spore is located at the end of the bacterial cell and *subterminal* when the spore is found at a location other than the end of the cell. Figure 19–15*A* shows a clostridial species that produces terminal spores, and Figure 19–15*B* shows a clostridial species that produces subterminal spores.

Certain characteristics may be used to initially identify clostridial isolates. A boxcar-shaped, anaerobic, gram-positive bacillus that produces a characteristic double zone of hemolysis on BRU/BA can be presumptively identified as *C. perfringens*. A heavily swarming, anaerobic, gram-positive bacillus having terminal spores is probably *C. tetani,* whereas one with subterminal spores is most likely *C. septicum.*

Clinical Infection

Clostridium species are most frequently encountered in exogenous anaerobic infections or intoxications. Clostridia or their toxins usually gain access to the body through ingestion or via open wounds that have become contaminated with soil.

Clostridia cause classic diseases such as tetanus, gas gangrene (myonecrosis), botulism, and *C. perfringens* food poisoning (foodborne intoxication). In tetanus, gas gangrene, and wound botulism, clostridial spores enter through open wounds and germinate in vivo. The vegetative bacteria then multiply and produce toxins. In foodborne intoxication due to *C. perfringens,* organisms are acquired through consumption of contaminated food. One clostridial infection that is of endogenous origin is antibiotic-associated pseudomembranous colitis caused by *C. difficile.*

Tetanus. The clinical manifestations of tetanus are attributed to the neurotoxin (tetanospasmin) produced by the causative agent, *C. tetani.* Tetanospasmin acts on inhibitory neurons, preventing the release of neurotransmitters. This results in continuous muscular spasms leading to trismus, risus sardonicus (distorted grin), and difficulty in breathing.

Tetanus manifests when spores develop into vegetative cells and multiply at a site of inoculation that provides a low redox potential, giving the organism the opportunity to produce the neurotoxin. The toxin is thereafter transmitted to the central nervous system. Symptoms appear within approximately 7 days after inoculation. The incubation period, however, ranges from 1 to 54 days.

Clinical manifestations include muscular rigidity, usually in the jaws, neck, and lumbar region.

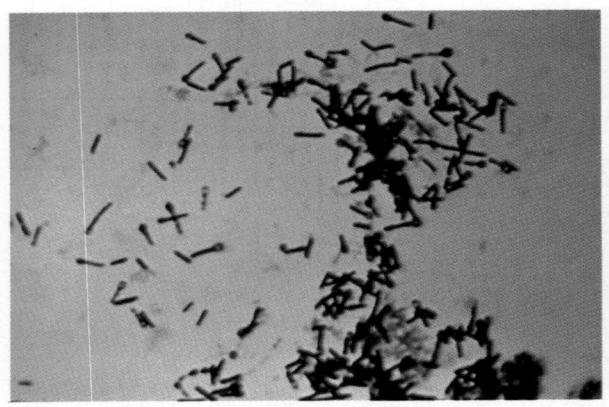

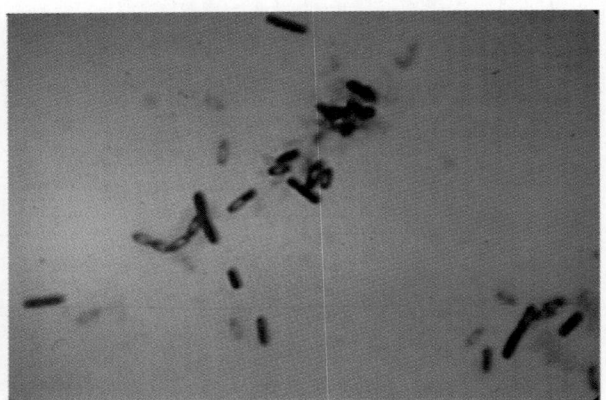

A
B

F i g u r e   1 9 – 1 5

A, Classification of some clinically encountered clostridia by endospore location: Gram-stained appearance of terminal spores of *C. tetani. B,* Gram-stained appearance of subterminal spores of *C. sordellii.* (Photomicrographs courtesy of Suzette L. Bartley, James D. Howard, and Ray Simon, Centers for Disease Control and Prevention, Atlanta, GA.)

Difficulty in swallowing results from muscular spasms in the pharyngeal area. Rigidity of the abdomen, chest, back, and limbs may also occur.

As a result of the widespread use of DPT (diphtheria-pertussis-tetanus) vaccine, tetanus is no longer a common disease in the United States. Tetanus in the newborn, or tetanus neonatorum, often results from the use of contaminated instruments, from umbilical infections, or in cases of septic abortions. Therapy for tetanus requires antitoxin injection, muscle relaxants, and constant nursing care, in addition to supportive therapy.

Myonecrosis or Gas Gangrene. Myonecrosis usually occurs when organisms contaminate wounds, through either trauma or surgery. *C. perfringens, C. histolyticum, C. septicum, C. novyi,* and *C. bifermentans* can cause myonecrosis. Under favorable conditions, the organisms are able to grow, multiply, and release potent exotoxins. In gas gangrene, clostridial exotoxins, such as α toxin produced by *C. perfringens,* cause necrosis of the tissue and allow deeper penetration by the organisms. The onset and spread of myonecrosis are rapid, and the affected extremities often require amputation.

Clinical manifestations of myonecrosis include pain and swelling in the affected area. Bullae, serous discharge, discoloration, and definite tissue necrosis are observed. Treatment of gas gangrene involves extensive surgical débridement of involved tissue, antimicrobial therapy, and sometimes the use of hyperbaric oxygen.

Botulism. Occasionally, intoxication (poisoning) may follow ingestion of preformed toxins produced by exogenous anaerobes. Foodborne botulism results from the ingestion of preformed botulin, a toxin produced in the food by *C. botulinum.* Botulin is an extremely potent toxin; only a small amount can produce paralysis and death. Botulin attaches to the neuromuscular junction of affected nerves, preventing the release of acetylcholine, which results in a flaccid type of paralysis and death.

Common food sources include home-canned vegetables, home-cured meat such as ham, fermented fish, and other preserved foods. Clinical manifestations develop as early as 2 hours or as late as 3 to 8 days following ingestion of food containing the botulin toxin, which is absorbed in the small intestine. Weakness and paralysis are the main features of botulism. Double vision, impaired speech, and difficulty in swallowing are common features. Respiratory paralysis may occur in severe cases. Treatment of foodborne botulism involves the use of antitoxin and supportive care.

Infant botulism, unlike that in adults, follows ingestion of *C. botulinum* spores. The most common associated food source in infant botulism is contaminated honey. The spores germinate, colonize in the colon, and subsequently produce toxins in vivo.

Wound botulism is the result of contamination of wounds with spores of *C. botulinum,* with subsequent germination, multiplication, and production of toxins in vivo. Clinical manifestations that develop in wound botulism are similar to those of foodborne intoxication. Wound botulism is usually treated with penicillin.

***Clostridium perfringens* Food Poisoning.** *C. perfringens* food poisoning, a relatively mild and self-limited GI tract illness, usually follows ingestion of enterotoxin-producing *C. perfringens* in contaminated food. Following an 8- to 12-hour incubation period, the patient experiences diarrhea and crampy abdominal pain for about 24 hours. Other than fluid replacement, therapy is usually unnecessary.

Antibiotic-Associated Diarrhea and Pseudomembranous Colitis. *C. difficile* is the most common but not the sole cause of antibiotic-associated diarrhea and pseudomembranous colitis. This organism is found as part of the GI flora of many individuals. Following antimicrobial therapy, many organisms of the GI flora other than *C. difficile* are killed, thus allowing *C. difficile* to multiply and produce abundant quantities of two types of toxins: toxin A, an enterotoxin; and toxin B, a cytotoxin. Bloody diarrhea with associated necrosis of colon mucosa is seen in patients with pseudomembranous colitis.

In addition, *C. difficile* is a common cause of nosocomial (hospital-acquired) infection. The organism is frequently transmitted among hospitalized patients and is often present on the hands of hospital personnel who are caring for such patients.

Table 19–17 shows other clinically encountered *Clostridium* species and associated infections.

Laboratory Diagnosis

Table 19–18 summarizes the laboratory confirmation of major clostridial diseases.

Microscopic Morphology. Clostridial cell wall structure is similar to that of other gram-positive bacteria. However, some species appear gram-variable, and some routinely stain gram-negative.

■■■■■T A B L E 1 9 – 1 7
CLINICALLY ENCOUNTERED *CLOSTRIDIUM* SPECIES

Clostridium Species	Associated Infections
C. baratii	War wounds; peritonitis; eye, ear, and prostate infections
C. bifermentans	Wounds, abscesses, bacteremia
C. botulinum	Urinary tract, lower respiratory tract, pleural cavity, and abdominal infections; wounds; abscesses; bacteremia
C. cadaveris	Abscesses, wounds
C. clostridioforme	Abdominal, cervical, scrotal, and pleural infections; septicemia; peritonitis; appendicitis
C. difficile	Antibiotic-associated diarrhea, pseudomembranous colitis, bacteremia (rarely), pyogenic infections; also isolated from the hospital environment
C. hastiforme	War wounds, bacteremia, abdominal abscesses
C. histolyticum	War wounds, gas gangrene; has been isolated from the gingival plaque of institutionalized and primitive populations
C. innocuum	Commonly isolated from infections of the gastrointestinal tract; empyema
C. limosum	Bacteremia, peritonitis, pulmonary infections
C. novyi	Wounds, gas gangrene
C. paraputrificum	Bacteremia, peritonitis, wounds, appendicitis
C. perfringens	Infections derived from colonic contents (e.g., peritonitis, intraabdominal abscesses, and infections of soft tissues below the waist); involved in about 80% of cases of gas gangrene; bacteremia
C. putrefaciens	Bacteriuria (pregnant women with bacteremia)
C. putrificum	Abscesses, wounds, bacteremia
C. ramosum	Infections of the abdominal cavity, genital tract, lung, biliary tract; bacteremia
C. septicum	Bacteremia, suppurative infections, necrotizing enterocolitis, gas gangrene
C. sordellii	Wounds, penile lesions, bacteremia, abscesses, infections of the abdomen and vagina
C. sporogenes	Bacteremia, endocarditis, central nervous system and pleuropulmonary infections, penile lesions, infected war wounds, other pyogenic infections
C. subterminale	Bacteremia (rarely); empyema; infections of the biliary tract, soft tissue, and bone
C. tertium	Appendicitis, brain abscess, infections related to the intestinal tract and soft tissue, infected war wounds, bacteremia; has been isolated from gingival sulcus of patients with periodontitis
C. tetani	Infected gums and teeth, corneal ulcerations, infections of mastoid and middle ear, intraperitoneal infections, tetanus neonatorum, post partum uterine infection, various soft-tissue infections related to trauma (including abrasions and lacerations) and use of contaminated needles; has been isolated from the hospital environment

Note: Virtually all *Clostridium* spp listed here have been isolated from fecal specimens of apparently healthy persons. Such isolates could represent transient rather than permanent residents of the normal colonic flora. Eighteen additional *Clostridium* species isolated from feces but not from clinical specimens or other anatomic sites have been excluded from this list.
Data from Sneath et al (1986).
Modified from Engelkirk PG, Duben-Engelkirk J, Dowell VR Jr: Principles and Practice of Anaerobic Bacteriology, 1992; used with permission of Star Publishing Company, Belmont, CA.

■■■■■T A B L E 1 9 – 1 8
LABORATORY CONFIRMATION OF SELECTED CLOSTRIDIAL DISEASES

Disease	Laboratory Confirmation Procedures
Botulism	
Foodborne	Isolation of *C. botulinum* from the stool specimen or detection of botulin toxin in the serum, stool, or epidemiologically implicated food
	Botulin toxin is initially identified by toxicity to mice; the specific toxin type is determined by neutralization tests using type-specific antitoxins
Infant	Detection of *C. botulinum* and/or botulin toxin in the stool of symptomatic infants
Wound	Isolation of *C. botulinum* from a wound specimen or detection of botulin toxin in the patient's serum or stool
C. perfringens food poisoning	Requires two or more findings ■ >10^5 colony CFU of *C. perfringens*/gram of implicated food ■ Median *C. perfringens* spore count >10^6/gram of stool from affected patients (such counts have been found in healthy individuals) ■ Isolation of the same type of *C. perfringens* from stools of affected patients and from the suspected food (isolates are not always serotypable) ■ Presence of *C. perfringens* enterotoxin in stool specimens from ill persons and its absence in stool specimens from well persons
C. difficile–induced diarrhea	Tissue culture cytotoxin assay (CTA); other methods include CIE, ELISA, latex agglutination, and a dot immunobinding assay; positive latex results should be confirmed by CTA, and stools yielding negative dot immunobinding assay results should be further tested using CTA
Tetanus	Laboratory confirmation rarely required; when *C. tetani* is isolated from wounds of patients with tetanus, toxicity and neutralization tests can be performed by intramuscular injection of culture supernatant or whole culture into untreated mice and mice protected with tetanus antitoxin; failure to isolate *C. tetani* from wound cultures does not eliminate the possibility of tetanus

Modified from Engelkirk PG, Duben-Engelkirk J, Dowell VR Jr: Principles and Practice of Anaerobic Bacteriology, 1992; used with permission of Star Publishing Company, Belmont, CA.

C. *ramosum* and C. *clostridioforme* routinely appear as pink-staining bacilli in Gram-stained preparations. Special-potency antimicrobial disks should always be used to determine the true gram reaction of a pink-staining anaerobic bacillus. Figure 19–12 shows how the special-potency antimicrobial results can be used to determine the true gram reaction of a pink-staining bacillus.

Colonial Morphology. Whenever clostridia are suspected, either clinically or as a result of Gram-stain observations (i.e., the presence of large gram-positive bacilli, with or without spores), certain media and procedures are added in addition to those routinely used for anaerobic cultures.

Egg-Yolk Agar (EYA). EYA is useful for detecting enzymes (i.e., lecithinase, lipase, and proteolytic enzymes) produced by some clostridia. Lecithinase-positive organisms produce a colony that is surrounded by a wide zone of opacity. This opacity is actually in the medium and not a surface phenomenon. The appearance of the lecithinase reaction is shown in Figure 19–16. Lecithinase-positive clostridia include C. *bifermentans,* C. *sordellii,* C. *perfringens,* C. *novyi* type A, and C. *baratii.* It must be noted, however, that a few organisms other than *Clostridium* species also cause this reaction.

Lipase-positive organisms produce a colony that is covered with an iridescent, multicolored sheen, sometimes described as resembling the appearance of gasoline on water or mother-of-pearl. This multicolored sheen may also appear on the surface of the agar in a narrow zone

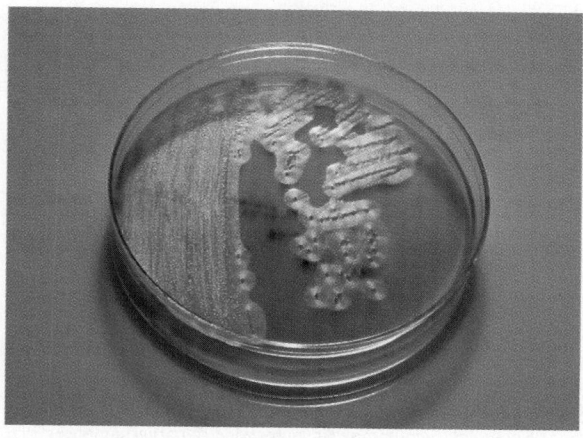

Positive lipase reaction on egg yolk agar (EYA). The reaction occurs on the *surface* of colonies and the surrounding medium. *Fusobacterium necrophorum* is shown here. (Photograph courtesy of Anaerobe Systems, San Jose, CA.)

around the colony. In contrast to the lecithinase reaction, the lipase reaction is essentially a surface phenomenon. The appearance of the lipase reaction is depicted in Figure 19–17. Examples of lipase-positive clostridia are C. *botulinum,* C. *novyi* type A, and C. *sporogenes.* Other anaerobes that are lipase-positive include *Fusobacterium necrophorum, Prevotella intermedia,* and some isolates of P. *loescheii.*

Organisms that produce proteolytic enzymes have a completely clear zone (often quite narrow) around their colonies. Proteolysis is best observed by holding the plate up to a strong light source. It is reminiscent of the complete clearing seen with β-hemolytic organisms on BRU/BA plates.

Direct Nagler Test. If C. *perfringens* is suspected (gram-positive boxcar-shaped bacilli seen on Gram stain), a direct Nagler test can be performed as part of the primary isolation set-up. This test employs an EYA plate and a reagent known as C. *perfringens* type A antitoxin. This reagent inhibits the lecithinase reaction produced by C. *perfringens* and three other species of *Clostridium.* Although the test is used to presumptively identify C. *perfringens,* a positive test result is not specific for this organism; C. *baratii,* C. *bifermentans,* and C. *sordellii* are also Nagler test–positive.

Brucella/Blood Agar (BRU/BA). Extensive swarming on the BRU/BA plate is characteristic of C. *septicum* and C. *tetani.* A double zone of hemolysis (an inner zone of complete β hemolysis and an

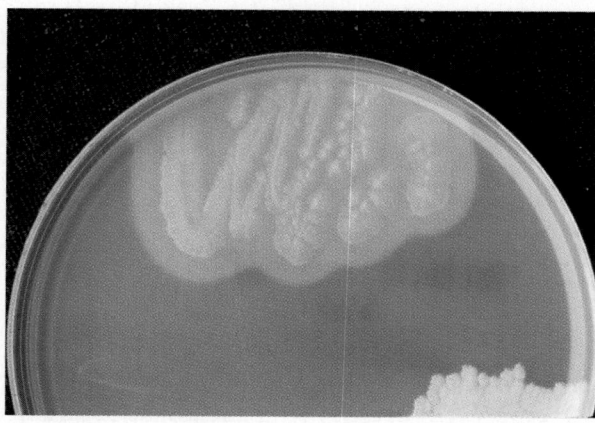

Positive lecithinase reaction on egg yolk agar (EYA). The reaction occurs *within* the agar. *Clostridium perfringens* is shown here. (Photograph courtesy of Anaerobe Systems, San Jose, CA.)

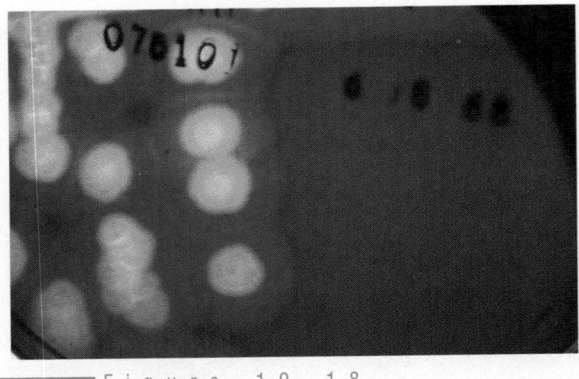

■■■■■ F i g u r e   1 9 – 1 8

Double zone of hemolysis produced by *Clostridium perfringens;* inner zone of *complete* beta-hemolysis; outer zone of *partial* hemolysis. (Photograph courtesy of Anaerobe Systems, San Jose, CA.)

outer zone of partial hemolysis) that appears within 24 hours of incubation is characteristic of *C. perfringens,* as shown in Figure 19–18.

    Cycloserine-Cefoxitin-Fructose Agar (CCFA). Whenever a patient is suspected of having antibiotic-associated diarrhea or pseudomembranous colitis, a CCFA plate may be inoculated with the patient's fecal specimen. CCFA is a selective and differential medium for the recovery and presumptive identification of *C. difficile,* the most common cause of those two entities. On CCFA, *C. difficile* produces yellow ground-glass colonies, and the originally pink agar turns yellow in the vicinity of the colonies. The appearance of *C. difficile* on CCFA is shown in Figure 19–19. Although other organisms may grow on CCFA, their colonies are smaller and do not resemble the characteristic colonies of *C. difficile. C difficile* has a characteristic horse-stable odor, and colonies on BRU/BA fluoresce chartreuse under UV light. Organisms that are isolated should then be tested for toxin production on appropriate cell lines.

    Bartley and Dowell reported that modified cycloserine mannitol agar (mCMA) and modified cycloserine mannitol blood agar (mCMBA) are superior to modified CCFA (mCCFA) for recovery of *C. difficile* from fecal specimens. In addition, mCCFA fails to inhibit growth of a variety of commonly encountered gram-positive and gram-negative intestinal organisms, whereas mCMBA and mCMA inhibit all such organisms except *Serratia liquefaciens.*

Identification

    Identification of clinically encountered clostridia using biochemical characteristics is summarized in Table 19–19.

Gram-Positive Non–Spore-Forming Anaerobic Bacilli

    Non–spore-forming, anaerobic, gram-positive bacilli include *Actinomyces, Bifidobacterium, Eubacterium, Mobiluncus, Lactobacillus,* and *Propionibacterium* spp (Table 19–20). Many species are found as part of the indigenous microflora (normal flora) of humans and other animals; therefore, infections that result are opportunistic. The microscopic morphology of these organisms varies, ranging from very short rods to long, branching filaments. Although certain species of *Lactobacillus* are anaerobic, their role in infectious processes is uncertain, and they will not be described here.

Clinical Infection

    Actinomycosis. *Actinomyces* spp and related anaerobic bacteria, such as *Bifidobacterium, Eubacterium,* and *Propionibacterium,* are the usual cause of actinomycosis. This group of organisms has been associated with actinomycosis of the brain, orofacial region, pleuropulmonary region, and genital organs.

    Actinomycosis is a chronic, granulomatous, infectious disease characterized by the development of sinus tracts and fistulae, which erupt to the surface and drain pus containing sulfur granules (small colonies of bacteria). Examinations of wet mounts and Gram-stained preparations of pus from draining sinuses are useful diagnostic

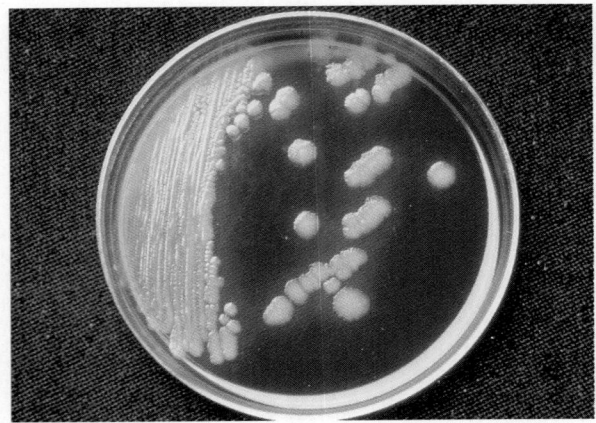

■■■■■ F i g u r e   1 9 – 1 9

The appearance of *Clostridium difficile* on cycloserine-cefoxitin-fructose agar (CCFA); yellow, ground-glass colonies and a yellowing of the medium around the colonies. (Photograph courtesy of Anaerobe Systems, San Jose, CA.)

CHARACTERISTICS OF SOME CLINICALLY ENCOUNTERED CLOSTRIDIA

Species	Relationship to Oxygen	Swarming on BRU/BA	Double Zone of Hemolysis	Chartreuse Fluorescence	Gram Reaction	Position of Spores	Motility	Flagella	Indole	Indole Derivatives	Esculin Hydrolysis	Lecithinase	Lipase	Proteolysis in Milk	Gelatin Hydrolysis	DNase	Glucose	Lactose	Mannitol	Rhamnose	Urease	Major Acids Produced in PYG
C. bifermentans	MA, AN	–	–	–	+	ST	+	PF	+	–	–	+	–	+	+	–	+	–	–	–	–	A
C. clostridioforme	AN	–	–	–	$-^{+}$	ST	$-^{+}$	PF	–	–	+	–	–	–	–	–	+	$+^{-}$	–	$+^{-}$	–	A, IB, B, IV, V, IC
C. difficile	AN	–	–	+	+	ST	+	SSF^{PF}	–	–	$+^{-}$	–	–	–	–	$-^{+}$	+	–	+	–	–	A, P, B
C. novyi type A	AN	–	+	–	+	$-^{ST}$	$+^{-}$	PF^{NF}	–	–	$+^{-}$	+	+	–	+	NG	$+^{-}$	–	–	–	–	A, B
C. perfringens	MA, AN	–	+	–	+	$-^{ST}$	–	–	–	–	V	+	–	+	+	+	+	+	–	–	–	A, B
C. ramosum	AN	–	–	–	–	(T)	–	–	–	–	+	–	–	–	–	+	+	+	$+^{-}$	V	–	A
C. septicum	MA, AN	+	–	–	+	ST	+	PF	–	–	+	–	–	–	+	+	+	+	–	+	–	A, B
C. sordellii	MA, AN	–	–	–	+	T^{ST}	+	PF	+	+	–	+	–	+	V	V	+	–	–	–	+	A, B
C. sphenoides	AN	–	–	–	–	ST	+	PF	+	+	+	–	–	–	–	–	+	+	$+^{-}$	+	–	A
C. sporogenes	AN	–	–	–	$+^{-}$	ST	+	PF	–	–	+	$-^{+}$	+	+	$+^{-}$	$-^{+}$	+	+	+	–	–	A, IB, B, IV
C. tertium	FA, AT	–	–	–	$+^{-}$	T	+	PF	–	–	+	–	–	–	–	–	+	+	+	–	–	A, B
C. tetani	AN	+	–	–	+	T	+	PF	$+^{-}$	–	–	–	–	+	+	+	–	–	–	–	–	A, P, B

AN = obligate anaerobe
MA = microaerotolerant anaerobe
AT = aerotolerant anaerobe
FA = facultative anaerobe
ST = subterminal
T = terminal

NF = no flagella
SSF = single subpolar flagellum
PF = peritrichous flagella
V or ( ) = variable
NG = no growth
A = acetic acid

P = propionic acid
IB = isobutyric acid
B = butyric acid
IV = isovaleric acid
V = valeric acid
IC = isocaproic acid

(T) = variable, but usually terminal
SSF^{PF} = most strains SSF, but some PF
PF^{NF} = most strains PF, but some NF
$+^{-}$ = most strains positive, but some negative
$-^{+}$ = most strains negative, but some positive
T^{ST} = mostly terminal, occasionally subterminal
$-^{ST}$ = usually not observed, but subterminal when seen

Data from Bartley (1990) and Dowell (1989), using CDC media.
Modified from Engelkirk PG, Duben-Engelkirk J, Dowell VR Jr: Principles and Practice of Anaerobic Bacteriology, 1992; used with permission of Star Publishing Company, Belmont, CA.

■■■■■ T A B L E   1 9 – 2 0

CLINICALLY ENCOUNTERED GRAM-POSITIVE NON–SPORE-FORMING ANAEROBIC BACILLI

Species	Body Site	Clinical Significance
Actinomyces		
A. israelii	Oral cavity, tonsillar crypts, dental plaque, intestinal and female genital tracts	Principal agent of human cervicofacial, thoracic, and abdominal actinomycosis; lacrimal canaliculitis and conjunctivitis; dacrocystitis; cervicitis and endometritis in women using intrauterine or vaginal contraceptive devices; "sulfur granules" are produced
A. meyeri	Periodontal sulcus	Infrequently isolated from brain abscesses and pleural fluid; less often from abscesses of cervicofacial area, hip, hand, foot, spleen, and bite wounds
A. odontolyticus	Oral cavity	Actinomycosis (rare cause); lacrimal canaliculitis; periodontitis (possibly) and dental caries (possibly)
A. viscosus	Oral cavity, caries, dental plaque, calculus	Periodontal disease; cervicovaginal secretions of women with and without IUDs; uninfected conjunctiva and cornea; cervicofacial and abdominal cases of actinomycosis (occasionally); lacrimal canaliculitis and other infections of the eye; cervicitis and endometritis of women using IUDs (possibly)
Bifidobacterium		
B. bifidum	Colon, vagina	—
B. breve	Colon, vagina	Has been isolated from clinical specimens
B. dentium	Colon, vagina, oral cavity, dental caries, dental plaque	Has been isolated from clinical specimens
B. infantis	Colon (infants), vagina	—
B. longum	Colon	Has been isolated from clinical specimens
B. pseudocatenulatum	Colon (infants)	—
Eubacterium		
E. aerofaciens	Colon	Occasionally isolated from blood cultures and various infections, including subacute bacterial endocarditis, renal abscess fluid, and appendiceal abscess
E. alactolyticum	Oral cavity	Frequently isolated from infected sites; dental calculus and gingival crevice in periodontal disease, root canals, purulent pleurisy, jugal cellulitits, postoperative wounds; abscesses of the brain, lung, intestinal tract, and mouth
E. brachy	Oral cavity	Subgingival samples and supragingival tooth scrapings from persons with periodontal disease, lung abscesses
E. combesii	Unknown	Has been isolated from various infections
E. contortum	Colon, vagina	Blood, abdominal aortic aneurysm, wounds
E. lentum	Colon	Blood, postoperative wounds, and various kinds of abscesses (brain, rectal, scrotal, pelvic)
E. limosum	Colon	Rectal and vaginal abscesses, blood, wounds
E. moniliforme	Colon	Blood and various other infections
E. nitritogenes	Colon	Has been isolated from various infections
E. nodatum	Oral cavity	Subgingival samples and supragingival tooth scrapings from persons with periodontal disease
E. rectale	Colon	—
E. saburreum	Dental plaque and gingival crevice	—
E. tenue	Unknown	Abscess following abortion, knee synovial fluid, blood

■■■■■T A B L E 1 9 – 2 0

CLINICALLY ENCOUNTERED GRAM-POSITIVE NON–SPORE-FORMING ANAEROBIC BACILLI *Continued*

Species	Body Site	Clinical Significance
E. ventriosum	Colon	Mouth abscess, infections of the neck, purulent pleurisy, pulmonary abscesses, bronchiectasis
E. yurii	Unknown	Periodontal pockets, subgingival dental plaque
Mobiluncus*		
M. curtisii subsp. *curtisii* subsp. *holmesii*	Vagina	Thought to play a role in bacterial vaginosis
M. mulieris	Vagina	Thought to play a role in bacterial vaginosis
Propionibacterium		
P. acnes	Skin, colon	Acne vulgaris, wounds, blood, pus, and soft-tissue abscesses
P. avidum	Moist areas of the skin (e.g., vestibule of the nose, axilla, perineum, sinuses)	Has been isolated from infected sinuses, chronically infected wounds, and submaxillary abscesses but is probably not the primary cause of these infections
P. granulosum	Oily areas of skin (e.g., forehead or between shoulder blades)	May play some part in the pathogenesis of acne but probably not otherwise pathogenic
P. propionicus	Oral cavity, cervicovaginal secretions of healthy women, secretions from uninfected conjunctiva and cornea	A cause of actinomycosis and lacrimal canaliculitis

Note: Ten *Eubacterium* spp and three *Bifidobacterium* spp isolated from feces but not from clinical specimens or other anatomic sites have been excluded from this list.
*Although *Mobiluncus* spp have a typical gram-positive cell wall, they stain pink to gram-variable with Gram-staining procedures.
Data primarily from Sneath et al (1986).
Modified from Engelkirk PG, Duben-Engelkirk J, Dowell VR Jr: Principles and Practice of Anaerobic Bacteriology, 1992; used with permission of Star Publishing Company, Belmont, CA.

procedures for demonstrating the non–spore-forming gram-positive bacilli that frequently exhibit branching in clinical materials (Fig. 19–20).

Bacterial Vaginosis. Some investigators believe that bacterial vaginosis (BV) involves endogenous anaerobes of the vagina such as *Mobiluncus* spp (especially *M. curtisii* spp *curtisii*), which are curved, motile, gram-positive bacteria. In all likelihood, BV is a synergistic infectious process involving a variety of organisms, including *Bacteroides* and *Peptostreptococcus* spp in addition to *Mobiluncus* spp and *Gardnerella vaginalis*.

Clinical features of BV include a gray-white, homogenous, malodorous vaginal discharge with small bubbles, little or no discomfort, and no inflammation. Although *Mobiluncus* spp have a typical gram-positive cell wall structure, they frequently stain pink or gram-variable with Gram stain (Fig. 19–21).

Laboratory Diagnosis

The clinically encountered anaerobic, non–spore-forming, gram-positive bacilli and their

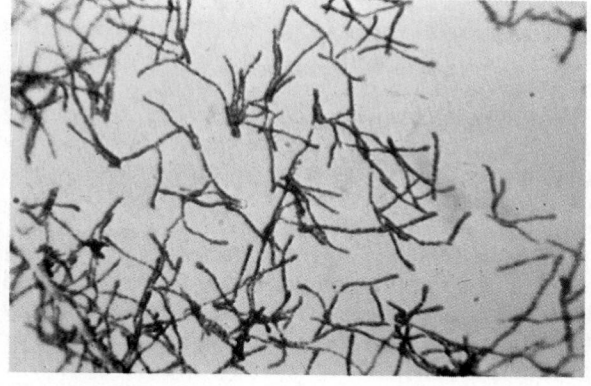

■■■■■F i g u r e 1 9 – 2 0
Gram-stained appearance of *Actinomyces israelii*, illustrating the term *Actinomyces*-like.

clinical significance are shown in Table 19–20.

***Actinomyces* spp.** *Actinomyces* spp are straight to slightly curved rods with varying lengths, from short rods to long filaments. Short rods may have clubbed ends and may be seen in diphtheroid arrangements, short chains, or small clusters.

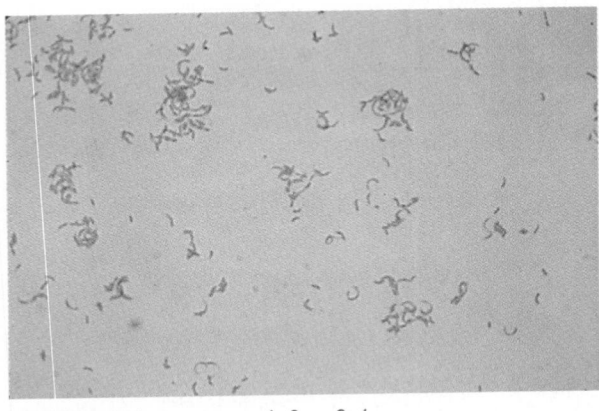

■■■F i g u r e   1 9 – 2 1

Gram-stained appearance of *Mobiluncus curtisii* subsp *curtisii,* illustrating the curved morphology of this organism. (Photomicrograph courtesy of Suzette L. Bartley, James D. Howard, and Ray Simon, Centers for Disease Control and Prevention, Atlanta, GA.)

Longer rods and filaments may be straight or wavy and branched. Although the *Actinomyces* are gram-positive, irregular staining may cause a beaded or banded appearance. The typical branching, filamentous, Gram-stained appearance of an *Actinomyces* species depicted in Figure 19–20 is referred to as "*Actinomyces*-like."

Investigators at the Centers for Disease Control and Prevention (CDC) Anaerobic Bacteria Branch recently found that members of the genus *Actinomyces* are seldom obligate anaerobes. However, some are quite fastidious, requiring special vitamins, amino acids, and hemin for adequate growth.

Young *Actinomyces* colonies are frequently spider-like or wooly, whereas older colonies may resemble a raspberry or a molar tooth. Depending on the species, colonies may be red, pink, tan, yellow, white, or grayish.

***Bifidobacterium* spp.** *Bifidobacterium* spp are variable in shape, ranging from coccobacilli to long, branching rods. The ends of the cells may be pointed, bent, club-shaped, spatulated, or bifurcated (forked). Cells may appear singly or in chains, as star-like aggregates, "V" arrangements, or "palisade" clusters. Colonies of *Bifidobacterium* spp are convex, entire, cream to white, smooth, glistening, and soft.

***Eubacterium* spp.** *Eubacterium* spp may be observed as either uniform or pleomorphic (variable in shape) gram-positive rods. They may be coccoid, diphtheroidal, or filamentous and range in width from thin to plump. *Eubacterium*

colonies are usually small, circular, and convex, with an entire margin.

***Mobiluncus* spp.** *Mobiluncus* spp are found in the healthy vagina but increase dramatically in number in bacterial vaginosis. They are curved, motile bacilli and, although they stain pink or gram-variable with the Gram stain, they are *not* gram-negative organisms. Their cell walls are structurally similar to gram-positive organisms and lack lipopolysaccharide (LPS). Like gram-positive organisms, *Mobiluncus* spp are susceptible to vancomycin and resistant to colistin. The Gram-stained appearance of *Mobiluncus* is depicted in Figure 19–21.

***Propionibacterium* spp.** *Propionibacterium* spp are pleomorphic rods that frequently appear as diphtheroids. The Gram-stained appearance of *P. acnes* is shown in Figure 19–22. Because *P. acnes* is a common member of the skin flora, it is frequently isolated from blood culture bottles and often dismissed as a skin contaminant. However, *P. acnes,* like coagulase-negative staphylococci, can cause subacute bacterial endocarditis and bacteremia and is thus not always a contaminant. A catalase-positive, spot indole–positive, anaerobic gram-positive diphtheroid can be presumptively identified as *P. acnes.*

Like *Actinomyces* spp, *P. propionicus* (formerly *Arachnia propionica*) can cause actinomycosis. This organism varies considerably in size and shape, ranging from coccoid and short diphtheroidal rods to long, branched filaments. Individual cells may be of uneven diameter and may have distended or clubbed ends.

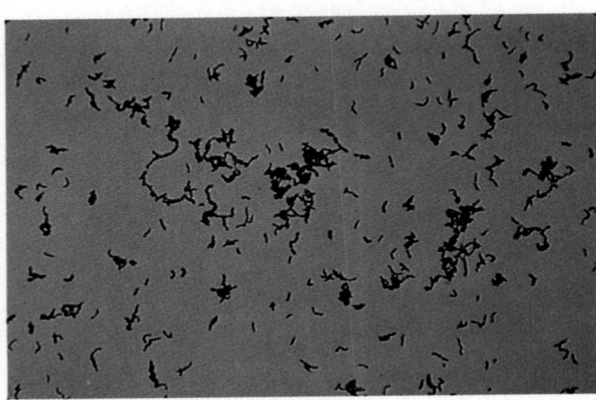

■■■F i g u r e   1 9 – 2 2

Gram-stained appearance of *Propionibacterium acnes,* illustrating the term diphtheroid. (Photomicrograph courtesy of Suzette L. Bartley, James D. Howard, and Ray Simon, Centers for Disease Control and Prevention, Atlanta, GA.)

Identification

Identifying characteristics of clinically encountered non–spore-forming gram-positive bacilli are shown in Table 19–21.

Anaerobic Gram-Negative Bacilli

Many of the non–spore-forming, anaerobic, gram-negative bacilli involved in human infectious processes are found as members of the indigenous microflora. Tables 19–22 and 19–23 list the clinically encountered anaerobic gram-negative non–spore-forming bacilli, their habitat, and clinical significance. At first glance, the variety of such organisms appears overwhelming. Fortunately, only a few of the genera are commonly encountered in clinical specimens, which include the four major groups: *Bacteroides fragilis* group, *Porphyromonas* spp, *Prevotella* spp, and *Fusobacterium* spp.

Clinical Infection

Anaerobic gram-negative bacilli are frequently found in mixed infections, causing abscesses and

TABLE 19–21
CHARACTERISTICS OF SOME CLINICALLY ENCOUNTERED GRAM-POSITIVE NON–SPORE-FORMING ANAEROBIC BACILLI

Species	Relationship to Oxygen	48-hr Colony <1 mm Diameter	Red Pigment	β-Hemolysis	Rough Colonies	Branched Rods	Catalase	Indole	Gelatin Hydrolysis	Proteolysis in Milk	Glucose	Lactose	Mannitol	Rhamnose	Major Acid Products in PYG
Actinomyces spp															
A. israelii	FA, AT, MA, AN	+	–	–	+	+	–	–	–	–	+	$+^-$	$+^-$	$+^-$	A, L, S
A. meyeri	FA, AT, MA, AN	$-^+$	–	–	–	+	–	–	–	–	+	$-^+$	–	–	A, L, S
A. naeslundii	FA, AT, MA, AN	$-^+$	–	–	$-^+$	+	–	–	–	–	+	$-^+$	–	–	A, L, S
A. odontolyticus	FA, AT, MA, AN	–	+	–	–	+	–	–	–	–	+	$-^+$	–	–	A, L, S
A. pyogenes	FA, AT	–	–	+	–	–	–	–	+	+	+	–	–	–	A, L, S
A. viscosus	FA, AT, MA, AN	–	–	–	$-^+$	+	+	–	–	–	+	$+^-$	–	–	A, L, S
Bifidobacterium spp															
B. dentium	FA, AT, MA, AN	–	–	–	–	–	–	–	–	–	+	+	–	–	A, L
Eubacterium spp															
E. alactolyticum	AN	+	–	–	–	–	–	–	–	–	V	–	–	–	A, B, C
E. lentum	AN	–	–	–	–	–	V	–	–	–	–	–	–	–	(A)
E. limosum	AN	–	–	–	–	–	–	–	–	–	+	–	+	–	A, B
Propionibacterium spp															
P. acnes	FA, AT, MA, AN	–	–	–	–	–	+	+	+	+	+	–	$-^+$	–	A, P, (L), (S)
P. propionicus	FA, AT, MA, AN	+	–	–	+	+	–	–	–	–	+	$+^-$	–	–	A, P, L

AN = obligate anaerobe   FA = facultative anaerobe   P = propionic acid   L = lactic acid
MA = microaerotolerant anaerobe   V or ( ) = variable   B = butyric acid   S = succinic acid
AT = aerotolerant anaerobe   A = acetic acid   C = caproic acid
$+^-$ = most strains are positive, but some are negative
$-^+$ = most strains are negative, but some are positive
Data from Bartley (1990) and Dowell (1989) using CDC media.
Modified from Engelkirk PG, Duben-Engelkirk J, Dowell VR Jr: Principles and Practice of Anaerobic Bacteriology, 1992; used with permission of Star Publishing Company, Belmont, CA.

■■■■■ T A B L E   1 9 – 2 2

CLINICALLY ENCOUNTERED *BACTEROIDES, PORPHYROMONAS,* AND *PREVOTELLA* SPECIES

Species	Normal Flora Site	Clinical Significance
Bile Tolerant		
***Bacteroides fragilis* group**		
B. caccae	Colon	Rarely isolated from clinical specimens
B. distasonis	Colon (common)	Occasionally isolated from clinical specimen
B. fragilis	Colon	The most common anaerobic species isolated from soft-tissue infections
B. merdae	Colon	—
B. ovatus	Colon	Occasionally isolated from clinical specimens
B. stercoris	Colon	
B. thetaiotaomicron	Colon (common)	Frequently found in clinical specimens
B. uniformis	Colon	Various clinical specimens
B. vulgatus	Colon (common)	Occasionally isolated from infections
Other *Bacteroides* species		
B. eggerthii	Colon	Occasionally isolated from infections
B. splanchnicus	Colon, vagina	Occasionally isolated from infections
Bile Sensitive		
Pigmented		
Prevotella		
P. corporis	Unknown	Various clinical specimens
P. denticola	Gingival crevice	Various clinical specimens
P. intermedia	Gingival crevice	Specimens from head, neck, and pleural infections; occasionally isolated from blood, abdominal, and pelvic sites
P. loescheii	Gingival crevice	
P. melaninogenica	Gingival crevice	Various clinical specimens
Porphyromonas		
P. asaccharolytica	Unknown	Various clinical specimens
P. endodontalis	Unknown	Dental root canals
P. gingivalis	Mouth	—
Nonpigmented *Prevotella*		
P. bivia	Vagina	Infections of the urogenital or abdominal region; occasionally isolated from mouth, blood, chest fluid, breast abscess
P. buccae	Gingival crevice	Chest drainage, blood, sinus aspirate (sinusitis), peritoneal fluid, mandibular cyst
P. buccalis	Oral cavity	—
P. disiens	Vagina, mouth	Abdominal and urogenital infections
P. heparinolytica	Unknown	Periodontitis lesions
P. oralis	Gingival crevice	Infections of the oral cavity and the upper respiratory and genital tracts
P. oris	Gingival crevice	Systemic infections; nonoral isolates from face, neck, and chest abscess and drainage; abdominal wound drainage; peritoneal fluid, blood, spinal fluid
P. oulora	Gingival crevice	—
P. veroralis	Oral cavity	—
P. zoogleoformans	Gingival sulcus	—
Nonpigmented Pitting *Bacteroides* Species		
B. gracilis	Gingival crevice	—
B. ureolyticus	Buccal cavity, intestinal and urogenital tracts	Infections of the respiratory and intestinal tracts; has been isolated from blood following tooth extractions

■■■■■■■T A B L E   1 9 – 2 2
CLINICALLY ENCOUNTERED *BACTEROIDES*, *PORPHYROMONAS*, AND *PREVOTELLA* SPECIES *Continued*

Species	Normal Flora	Clinical Significance
Nonpitting *Bacteroides* Species		
B. capillosus	Colon, mouth	Cysts and wounds
B. coagulans	Colon, urogenital tract	Occasionally isolated from clinical specimens
B. forsythus	Oral cavity	—
B. galacturonicus	Colon	—
B. pectinophilus	Colon	—

Data primarily from Krieg and Holt (1984).
Modified from Engelkirk PG, Duben-Engelkirk J, Dowell VR Jr: Principles and Practice of Anaerobic Bacteriology, 1992; used with permission of Star Publishing Company, Belmont, CA.

sepsis. *B. fragilis* is often isolated from soft-tissue infections. *Porphyromonas asaccharolytica* is widely distributed in human tissues and fluids, whereas *P. gingivalis* and *P. endodontalis* are associated primarily with oral infections. *Fusobacterium* spp are most often found in mixed infections and are associated with gum and other oral lesions.

Laboratory Diagnosis

***Bacteroides* spp.** One of the most clinically important genera of anaerobic gram-negative bacilli is the genus *Bacteroides*. Organisms in this genus have generated considerable interest due to their frequent involvement in infectious processes and their resistance to antimicrobial agents used to treat anaerobe-associated infections. *Bacteroides* spp are found as part of indigenous microflora of the oral cavity and the gastrointestinal and genitourinary tracts.

The genus *Bacteroides* can be divided into species that are bile tolerant and those that are bile sensitive. In the past, bile-sensitive species were subdivided into pigmented and nonpigmented species. Recently, however, most pigmented species of bile-sensitive *Bacteroides* have been reclassified into the genera *Porphyromonas* and *Prevotella*. Many nonpigmented species of *Bacteroides* have also been transferred to the genus *Prevotella*.

***Bile-Tolerant Bacteroides* spp.** As shown in Table 19–22, bile-tolerant *Bacteroides* spp include members of the *B. fragilis* group and two other species, *B. eggerthii* and *B. splanchnicus*. Members of the *B. fragilis* group (which includes *B. fragilis* and eight closely related species) are especially pathogenic. Figure 19–23 illustrates the frequency of isolation of the various members of the *B. fragilis* group. *B. fragilis* is the most

common species of anaerobic bacteria isolated from infectious processes of soft tissue and anaerobic bacteremia.

It is important to identify members of the *B. fragilis* group to the species level because of species-to-species variability in both virulence and drug resistance. Penicillin-resistant strains of *B. fragilis* and other members of the *B. fragilis* group have been encountered for a number of years, but there are also reports of resistance to tetracycline, cefotaxime, cefoperazone, moxalactam, clindamycin, and other antimicrobials among isolates of this group. Although members of the *B. fragilis* group were the first anaerobes reported to produce β lactamases, it is now known that many other anaerobes are also capable of producing them.

Gram-stained smears of *Bacteroides* spp colonies reveal gram-negative coccobacilli or bacilli, but cells in broth cultures are frequently pleomorphic. The Gram-stained appearance of a typical *Bacteroides* species is shown in Figure 19–24.

Colonies of the *B. fragilis* group on the BBE agar plate are gray, and a minimum of 1 mm in diameter. The originally light-yellow medium turns brown in the area around the colonies. Good growth is the result of bile tolerance, and browning of the medium is due to esculin hydrolysis. A dark precipitate (stippling) in the medium around the areas of heavy growth is suggestive of the species *B. fragilis,* although some strains of *B. ovatus* also cause stippling. The appearance of *B. fragilis* group organisms on BBE agar is shown in Figure 19–25.

Caution must be taken, however, when interpreting results on BBE agar. *B. vulgatus,* a member of the *B. fragilis* group, does not hydrolyze esculin and therefore does not produce a brown discoloration of the medium. *B. splanchnicus* and *B. eggerthii,* which are currently not members of

■■■■■ T A B L E   1 9 – 2 3

CLINICALLY ENCOUNTERED GRAM-NEGATIVE ANAEROBIC BACILLI (OTHER THAN *BACTEROIDES*, *PORPHYROMONAS*, AND *PREVOTELLA* SPECIES)

Species	Normal Flora Site	Clinical Significance
Anaerobiospirillum succiniciproducens	Unknown	Blood, feces from diarrhetic patients
Anaerorhabdus furcosus	Colon (infrequent)	Infected appendix, lung, and abdominal abscess
Bilophila wadsworthia	Colon	Intraabdominal specimens from patients with gangrenous and perforated appendicitis
Butyrivibrio		
B. crossotus	Colon	—
B. fibrisolvens	Colon	Eye infection
Campylobacter concisus	Oral cavity	Periodontal disease, wounds
Centipeda periodontii	Oral cavity	Periodontal disease
Desulfomonas pigra	Colon	Peritoneal fluid, pylonidal cyst abscess, ruptured sigmoid colon
Desulfovibrio vulgaris	Colon	Pleural fluid
Fusobacterium		
F. alocis	Gingival sulcus	Submandibular abscess, mouth ulcer
F. gonidiaformans	Intestinal, urogenital tracts	Various types of infections
F. mortiferum	Colon	Blood and various clinical specimens
F. necrogenes	Colon	—
F. nechrophorum	Body cavities	Necrotic lesions, abscess, blood
subsp *funduliforme*		—
subsp *nechrophorum*		—
F. nucleatum	Gingival margin and sulcus	Infections of the upper respiratory tract and pleural cavity; occasionally from wounds and other types of infections
F. periodonticum	Oral cavity	—
F. prausnitzii	Colon (very common)	—
F. pseudonecrophorum	Unknown	—
F. russii	Colon	—
F. sulci	Gingival sulcus	—
F. ulcerans	None	Cutaneous tropical ulcers
F. varium	Colon	Purulent infections of the upper respiratory tract, surgical wounds, and peritonitis
Leptotrichia buccalis	Oral cavity, femal periurethral region	—
Megamonas hypermegas	Colon	—
Mitsuokella		
M. dentalis	Oral cavity	Dental root canals
M. multiacidus	Colon	Occasionally isolated from clinical specimens
Selenomonas		
S. artemidis	Gingival crevice	—
S. dianae	Gingival crevice	—
S. flueggei	Gingival crevice	—
S. noxia	Gingival crevice	—
S. sputigena	Gingival crevice	Transtracheal aspirate and pleural fluid; relationship to periodontal disease unknown
Succinimonas amylolytica	Unknown	Wound drainage
Succinovibrio dextrinosolvens	Oral cavity (rare), colon (rare)	Septicemia
Tissierella praeacuta	Colon	Infrequently from lung abscess, gangrenous lesions, blood
Wolinella		
W. curva (*Campylobacter curvus*)	Oral cavity	Lesions in oral cavity, blood; pathogenicity unknown
W. recta (*Campylobacter rectus*)	Gingival crevice	Periodontal pockets, necrotic dental root canals; pathogenicity unknown

Data primarily from Krieg and Holt (1984).
Modified from Engelkirk PG, Duben-Engelkirk J, Dowell VR Jr: Principles and Practice of Anaerobic Bacteriology, 1992; used with permission of Star Publishing Company, Belmont, CA.

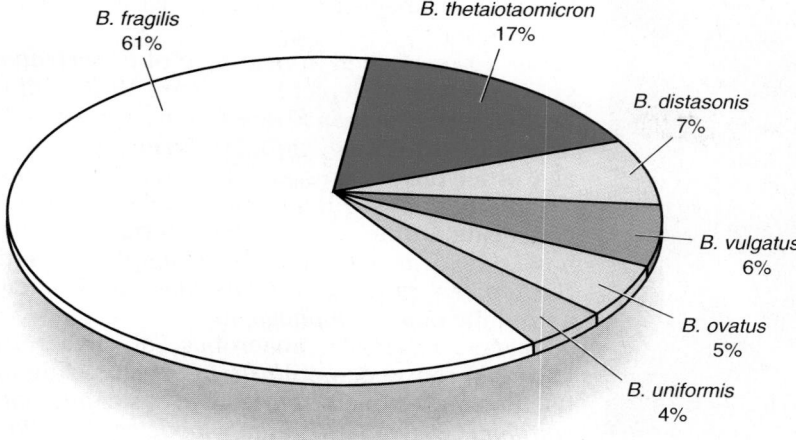

B. fragilis
61%

B. thetaiotaomicron
17%

B. distasonis
7%

B. vulgatus
6%

B. ovatus
5%

B. uniformis
4%

■■■■■■■ Figure 19–23

Frequency of isolation of members of the *Bacteroides fragilis* group from clinical specimens. From: Goldstein and Citron (1988). (Redrawn from Engelkirk PG, Duben-Engelkirk J, Dowell VR Jr: Principles and Practice of Anaerobic Bacteriology, 1992; used with permission of Star Publishing Company, Belmont, CA.)

the *B. fragilis* group, are bile-resistant and esculin hydrolysis–positive; thus, colonies of these organisms have the same appearance on BBE agar as members of the *B. fragilis* group.

Depending on the commercial source, age, and storage conditions of the medium, other organisms such as *Fusobacterium mortiferum, Klebsiella pneumoniae, Enterococcus* spp, and yeasts may also grow on BBE agar. Their colony size (which tends to be smaller), Gram-stain morphology, and aerotolerance will aid in the recognition of these other organisms.

Bile-Sensitive Pigmented Species. Certain species of *Prevotella,* namely *P. corporis, P. intermedia, P. loescheii, P. melaninogenica,* and some strains of *P. denticola,* produce protoporphyrin, a dark pigment that causes their colonies to become brown to black with age. *Prevotella* spp appear as gram-negative coccobacilli or bacilli, very similar to *Bacteroides* spp.

Colony pigmentation (brown to black) may take 2 to 3 weeks of incubation before it becomes evident on routine blood agar plates, but it appears sooner on media containing laked blood (blood that has been frozen and then thawed). For this reason, always subject nonpigmented colonies on KLVB agar (or BRU/BA) to long-wave UV light, such as a Wood's lamp, to detect the typical brick-red fluorescence of pigment-producing *Prevotella* spp. The brick-red fluorescence of such pigmented species under UV light is similar to that shown in Figure 19–11.

Some species of pigmented *Prevotella* fluoresce colors other than brick-red, such as brilliant red, yellow, orange, and pink-orange, and some do not fluoresce at all. Only brick-red fluorescence allows presumptive identification of the pigmented *Prevotella* group.

Porphyromonas asaccharolytica, P. gingivalis, and *P. endodontalis* also produce a dark brown to black pigment, protoheme, when grown for 6 to 10 days on blood agar plates. Most strains of these fastidious, obligate anaerobes require hemin and menadione, a vitamin K derivative, for

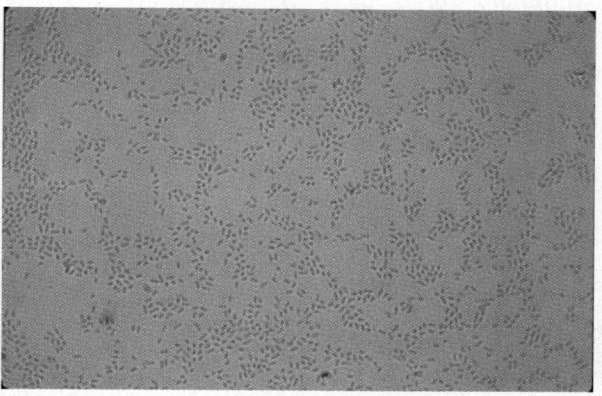

■■■■■■■ Figure 19–24

Gram-stained appearance of *Bacteroides thetaiotaomicron,* illustrating the typical appearance of *Bacteroides* spp. (Photomicrograph courtesy of Suzette L. Bartley, James D. Howard, and Ray Simon, Centers for Disease Control and Prevention, Atlanta, GA.)

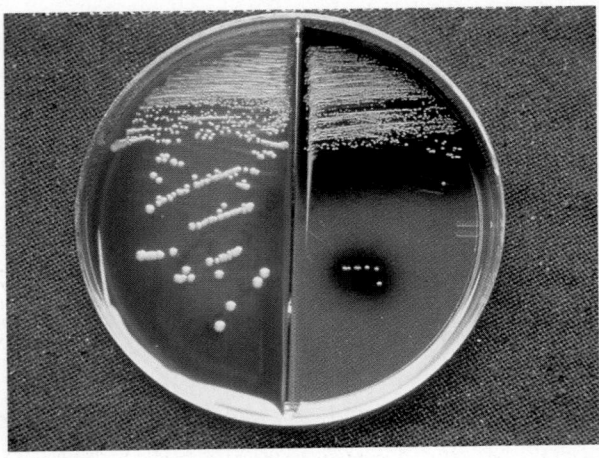

■■■■ F i g u r e   1 9 – 2 5

Appearance of *Bacteroides fragilis* on a KVLB agar/BBE agar biplate; browning of the BBE medium is the result of esculin hydrolysis. (Photograph courtesy of Anaerobe Systems, San Jose, CA.)

growth. *Porphyromonas* also produces brick-red fluorescence and other colors under UV light, similar to *Prevotella,* but certain species do not fluoresce at all. Because most *Porphyromonas* strains are susceptible to vancomycin, they will not grow on media containing 5 mg or more vancomycin per milliliter, such as KVLB agar.

Bile-Sensitive Nonpigmented Species. Bile-sensitive nonpigmented *Bacteroides* spp are listed in Table 19–22. This table shows those species that pit the agar and those that are nonpitting. Recent reports have shown that the so-called pitting anaerobes of the *B. ureolyticus* group (which includes *B. ureolyticus, B. gracilis, Wolinella curva,* and *W. recta*) are actually microaerophiles rather than obligate anaerobes. Moreover, *W. curva* and *W. recta* have recently been reclassified as *Campylobacter curvus* and *C. rectus,* respectively.

Not all strains of these pitting organisms actually pit agar, and among those strains that do pit, not all colonies will appear to be pitting. Thus, they may resemble a mixed culture. Nonpigmented *Prevotella* spp include *P. bivia, P. buccae, P. buccalis, P. disiens, P. oralis, P. oris, P. oulora,* and *P. veroralis.*

Fusobacterium. *Fusobacterium* is often described microscopically as long, thin, and tapered rods, a morphology characteristically referred to as "fusiform." It is important to note, however, that only *F. nucleatum* ssp *nucleatum* has cells that are consistently fusiform in shape, and clinically encountered bacteria that are

fusiform in shape are not necessarily *Fusobacterium* spp.

The Gram-stained appearance of *F. nucleatum* ssp *nucleatum* is depicted in Figure 19–26. Other fusobacteria, such as *F. mortiferum,* appear pleomorphic, exhibiting globular forms, swellings, and other bizarre shapes. The pleomorphism of *F. mortiferum* is depicted in Figure 19–27. Organisms other than fusobacteria may also have fusiform-shaped cells; examples include *Bacteroides gracilis, B. forsythus,* and microaerophilic *Capnocytophaga* spp.

Certain *Vibrio*-like anaerobes, or "anaerobic vibrions" (curved, motile, gram-negative staining bacilli), have been reclassified as *Anaerobiospirillum, Butyrivibrio, Campylobacter, Desulfovibrio, Selenomonas, Succinovibrio,* and *Mobiluncus* spp. Of these, *Campylobacter curvus, C. rectus,* and *Mobiluncus* spp are encountered most frequently in clinical materials—the *Campylobacter* from oral, GI, and vaginal specimens and the *Mobiluncus* spp from vaginal specimens. Although cells of *C. curvus* are curved, *C. rectus* cells are straight rods.

Identification

A schematic diagram that leads to the presumptive identification of gram-negative non–spore-forming bacilli is shown in Figure 19–28. Additional identification information can be found in Tables 19–24 and 19–25.

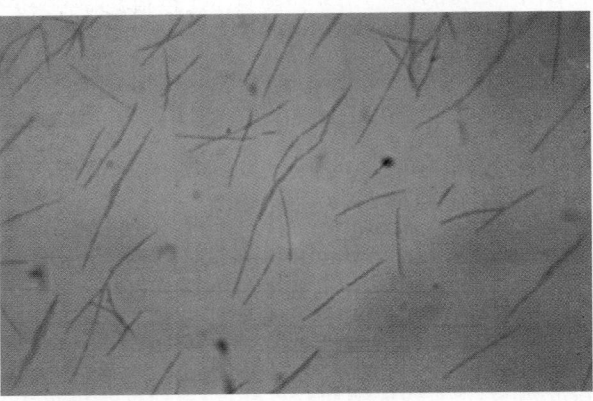

■■■■ F i g u r e   1 9 – 2 6

Gram-stained appearance of *Fusobacterium nucleatum* subsp *nucleatum,* illustrating the fusiform morphology of this organism. (Photomicrograph courtesy of Suzette L. Bartley, James D. Howard, and Ray Simon, Centers for Disease Control and Prevention, Atlanta, GA.)

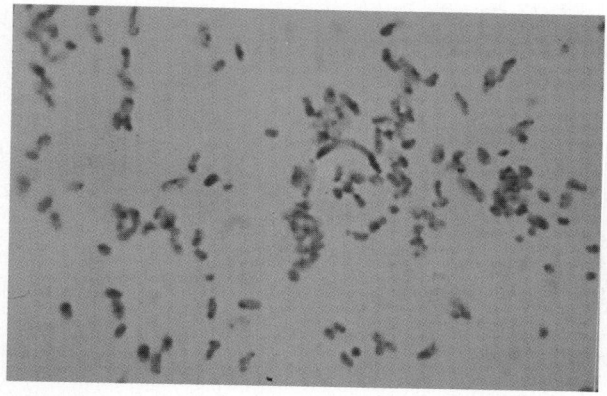

■ F i g u r e 1 9 – 2 7

Gram-stained appearance of *Fusobacterium mortiferum*, illustrating pleomorphism. (Photomicrograph courtesy of Suzette L. Bartley, James D. Howard, and Ray Simon, Centers for Disease Control and Prevention, Atlanta, GA.)

Anaerobic Cocci

Clinical Infection

Anaerobic cocci are isolated from a wide variety of infections, including brain abscess, aspiration pneumonia, lung abscess, gingivitis, and other periodontal diseases. Table 19–26 describes the various species of anaerobic cocci, where they are found, and infections they cause.

Gram-Positive Cocci

Most of the gram-positive anaerobic cocci previously classified as *Peptococcus* spp have been renamed *Peptostreptococcus* spp, with the exception of *Peptococcus niger*. Most of the anaerobic gram-positive cocci isolated from clinical specimens are in the genus *Peptostreptococcus* (Fig.

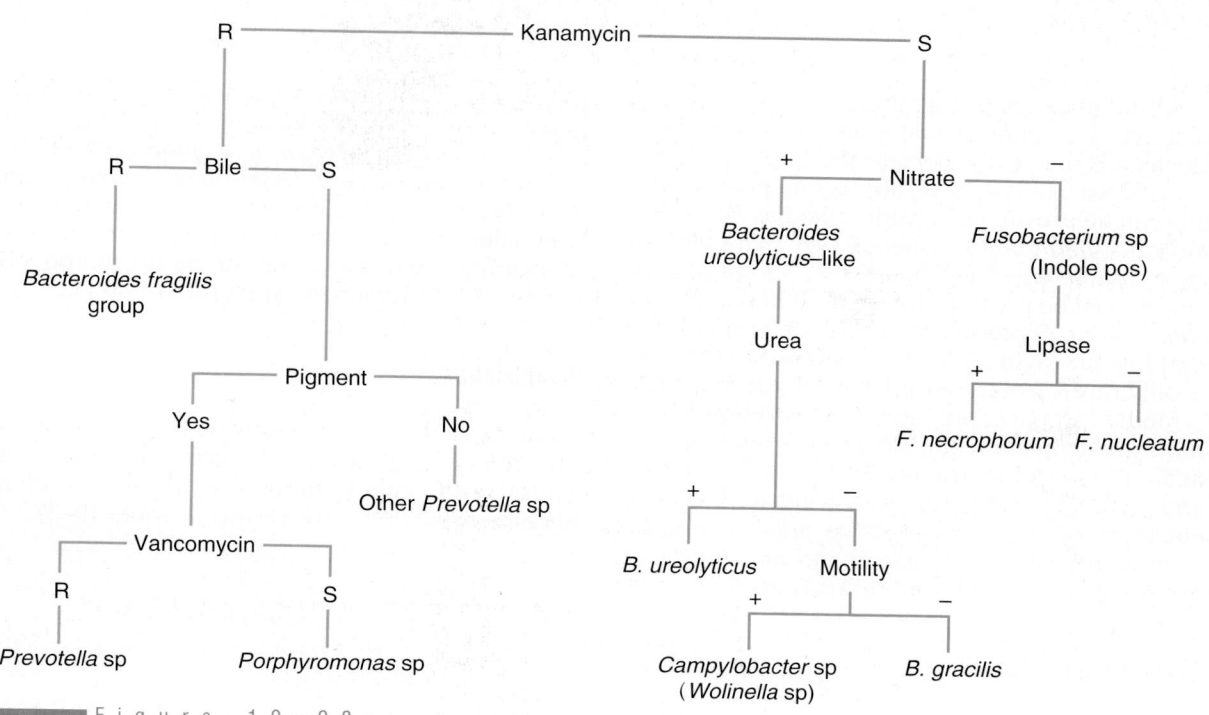

■ F i g u r e 1 9 – 2 8

Schematic diagram for the presumptive identification of anaerobic gram-negative non–spore-forming rods.

TABLE 19–24

CHARACTERISTICS OF SOME CLINICALLY ENCOUNTERED NONMOTILE GRAM-NEGATIVE ANAEROBIC BACILLI

Species	Brown-Black Pigmentation	Red Fluorescence	Pitted Colonies	Colonies ≤ 1mm Diameter	Catalase	Oxidase	Indole	Inhibited by Bile	Esculin Hydrolysis	Lipase	DNase	Glucose	Lactose	Gelatin Hydrolysis	Proteolysis in Milk	Urease	Rifampin (15 µg)	Acids Produced in PYG — Butyric	Succinic	Phenylacetic
Bacteroides spp																				
B. gracilis	−	−	−⁺	+	−	−	−	+	−	−	−	−	−	−	−	−	S	−	+	−
B. ureolyticus	−	−	+⁻	+	−	+	−	+	−	−	−	−	−	−	−	+	S	−	+	−
Fusobacterium spp																	R	+	−⁺	−
F. mortiferum	−	−	−	−	−	−	−	−	+	−	−	+	+⁻	−⁺	V	−	S	+	−	−
F. necrophorum	−	−	−	−	−	−	+	+	−	+	−	+	−	−	−	−	S	+	−	−
F. nucleatum	−	−	−	−	−	−	+	+	−	−	−	−	−	−	−	−	R	+	−	−
F. varium	−	−	−	−	−	−	+⁻	−	−	−	−	+	−	−	−	−		+	−	−
Porphyromonas spp																				
P. asaccharolytica	+	+	−	−	−	−	−	+	+	−	−	−	−	+	+	−	S	+	+	−
P. endodontalis	+	+⁻	−	−	−	−	−	+	+	−	−	−	−	+	+	−	S	+	+	−
P. gingivalis	+	−	−	−	−	−	−	+	+	−	−	−	−	+	+	−	S	+	+	+
Prevotella spp																				
P. intermedia	+	+	−	−	−	−	+	+	−	+	+	+	−	+	+	−	S	−	+	−

V = variable
R = resistant
S = sensitive
+⁻ = most strains are positive, but some are negative
−⁺ = most strains are negative, but some are positive

Data from Bartley (1990) and Dowell (1989), using CDC media.
Modified from Engelkirk PG, Duben-Engelkirk J, Dowell VR Jr: Principles and Practice of Anaerobic Bacteriology, 1992; used with permission of Star Publishing Company, Belmont, CA.

19–29). Other anaerobic gram-positive cocci encountered in clinical materials are *Coprococcus, Gemella, Sarcina,* and certain *Streptococcus* spp.

An SPS-sensitive, anaerobic, gram-positive coccus can be presumptively identified as *Peptostreptococcus anaerobius,* whereas an SPS-resistant, spot indole–positive, anaerobic, gram-positive coccus can be presumptively identified as *P. asaccharolyticus. Peptococcus niger* produces colonies that become light gray when exposed to air, but it is only rarely isolated from clinical specimens.

Media preparation, age, and storage conditions are critical for the isolation of clinically significant anaerobic gram-positive cocci. The size and arrangement (e.g., pairs, tetrads, chains, clusters) of the gram-positive anaerobic cocci vary with growth conditions and are, therefore, not reliable criteria for identification.

Gram-Negative Cocci

Although several genera of anaerobic gram-negative cocci (see Table 19–26) are found in the indigenous microflora, only *Veillonella* spp are implicated as pathogens. *Veillonella* cocci are very small (0.3–0.5 mm in diameter) and inhabit the oral cavity. Other genera of gram-negative cocci (*Acidaminococcus* and *Megasphaera*) live in the gastrointestinal tract but are only rarely isolated from clinical specimens. A nitrate-positive, anaerobic, gram-negative coccus can be presumptively identified as a *Veillonella* spp (Table 19–27).

Identification

A schematic diagram showing the presumptive identification of anaerobic cocci is shown in Figure 19–30. Other characteristics used to identify anaerobic cocci are shown in Table 19–27.

SUSCEPTIBILITY AND β-LACTAMASE TESTING

Antimicrobial susceptibility testing of anaerobes is currently controversial. The information contained in this section should not to be con-

Chapter 19 ■ ANAEROBES OF CLINICAL IMPORTANCE 585

TABLE 19-25

CHARACTERISTICS OF SOME CLINICALLY ENCOUNTERED BILE-TOLERANT, SACCHAROLYTIC *BACTEROIDES* SPECIES (the *B. FRAGILIS* GROUP)

Species	Catalase	Indole	DNase	Penicillin (2 U Disk)	Rifampin (15 µg Disk)	Kanamycin (1 mg Disk)	Arabinose	Cellobiose	Mannitol	Rhamnose	Salicin	Trehalose	Acids Produced in PYG Butyric	Succinic
B. caccae	−	−	+	R	S	R	+	+−	−	V	+	+	−	+
B. distasonis	+−	−	−	R	S	R	−	−	−	+	+	+	−	+
B. fragilis	+−	−	−	R	S	R	−	−	−	+	+	+	−	+
B. merdae	−	−	−	R	S	R	−	+	−	−	−	−	−	+
B. ovatus	+	+	+	R	S	R	−	+	−	+	+	+	−	+
B. stercoris	−	+	+	R	S	R	+	+	+	+	+	+	−	+
B. thetaiotaomicron	+	+	+	R	S	R	−+	−	−+	+	−+	+	−	+
B. uniformis	−	+	+	R	S	R	+	+	−	−+	+−	+	−	+
B. vulgatus	−+	−	−	R	S	R	+	−	−	+	+−	−	−	+

V = variable
R = resistant
S = sensitive
+− = most strains positive, but some are negative
−+ = most strains negative, but some are positive
Data from Bartley (1990) and Dowell (1989), using CDC media.
Modified from Engelkirk PG, Duben-Engelkirk J, Dowell VR Jr: Principles and Practice of Anaerobic Bacteriology, 1992; used with permission of Star Publishing Company, Belmont, CA.

sidered definitive. Issues that clinicians and microbiologists have not agreed on include the following:

■ Clinical significance of the anaerobic isolate

■ Predictability of anaerobe susceptibilities

■ Cost of susceptibility testing

■ Difficulty of performing susceptibility testing on anaerobes

Questions frequently asked are as follows:

■ Should susceptibility testing of anaerobes be performed at all in clinical microbiology laboratories?

■ Which isolates should be tested?

■ Which antimicrobial agents should be tested?

■ Which method should be used?

These are all valid questions, and each is addressed in this section.

Traditionally, the isolation, identification, and susceptibility testing of obligate anaerobes have been quite slow when compared to other groups of bacteria. As a result, antimicrobial therapy of anaerobe-associated infectious pro-cesses has frequently been initiated on an empirical basis: When physicians suspected the presence of anaerobes, they selected an antimicrobial agent that they thought would be effective. Although not a very scientific ap-proach, empiric therapy has undoubtedly saved many lives.

Anaerobe Resistance to Antimicrobial Agents

Although anaerobe susceptibility patterns were once thought to be quite predictable, the results of numerous studies have shown otherwise. Resistance of anaerobes to antimicrobial agents is no longer limited to the *Bacteroides fragilis* group; it occurs among many different anaerobes and involves a variety of antimicrobial agents. In addition to members of the *B. fragilis* group, especially resistant anaerobes include *B. gracilis*, *Clostridium ramosum*, and *Fusobacterium varium*.

Table 19–28 reflects susceptibility patterns obtained at the Wadsworth Veterans Administration Medical Center in Los Angeles. Locally determined patterns should be used to guide empirical therapy at other institutions.

■■■■■ T A B L E   1 9 – 2 6

CLINICALLY ENCOUNTERED ANAEROBIC COCCI

Genus	Normal Flora Site	Clinical Significance
Gram-Positive Cocci		
Streptococcus		—
S. hansenii	Colon	—
S. pleomorphus	Colon	—
Coprococcus	Three species have been isolated from feces but not from clinical specimens	
Gemella	Unknown	—
Peptococcus niger	Umbilicus, vaginal area	Only occasionally isolated from human samples; more common among veterinary specimens
Peptostreptococcus		
P. anaerobius	Vagina, colon	Isolated from a wide variety of clinical specimens, including abscesses of the brain, jaw, pleural cavity, ear, pelvic region, urogenital area, and abdominal region; also from blood, spinal fluid, joint cultures, and specimens from cases of osteomyelitis; gingival crevice of persons with gingivitis or periodontal diseases
P. asaccharolyticus	Vagina	Vaginal discharge, skin abscess, peritoneal abscess
P. hydrogenalis	Unknown	Has been isolated from feces and vaginal discharge
P. magnus	Genital tract; rare in colon or gingival crevice	Wounds; abdominal abcess; peritoneal, appendiceal, and urogenital sites; more frequent in anatomic sites below diaphragm
P. micros	Genital tract; infrequently in healthy gingival crevice and intestinal tract	Frequently isolated from clinical specimens, including brain, lung, jaw, head, and neck; bite abscesses, spinal fluid, blood, and abscesses at other body sites; often a major component of the gingival sulcus in periodontal disease
P. prevotii	Skin, vagina, tonsils	—
P. productus	Colon (one of the more predominant members of the fecal flora)	—
P. tetradius	Vagina	Vaginal discharge and various purulent infections
Gram-Negative Cocci		
Acidaminococcus fermentans	Colon	Rarely isolated from clinical specimens
Megasphaera elsdenii	Colon	Rarely isolated from clinical specimens
Veillonella		
V. atypica		
V. dispar	Oral cavity	Head, neck, dental, and pulmonary infections; bite wounds
V. parvula		

Data primarily from Krieg and Holt (1984).
Modified from Engelkirk PG, Duben-Engelkirk J, Dowell VR Jr: Principles and Practice of Anaerobic Bacteriology, 1992; used with permission of Star Publishing Company, Belmont, CA.

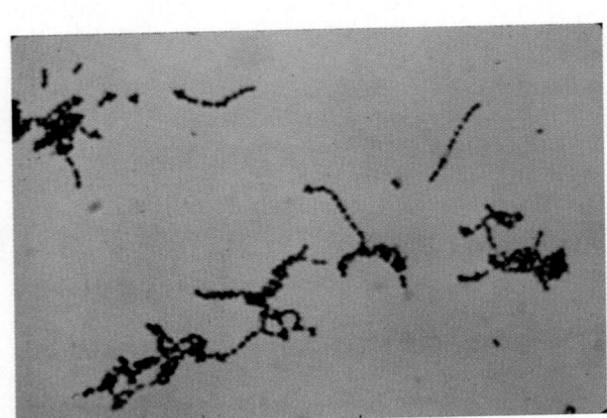

■■■■■ F i g u r e   1 9 – 2 9

Gram-stained appearance of a *Peptostreptococcus* sp illustrating the chain formation that occurs with some species. (Photomicrograph courtesy of Suzette L. Bartley, James D. Howard, and Ray Simon, Centers for Disease Control and Prevention, Atlanta, GA.)

TABLE 19–27

CHARACTERISTICS OF SOME CLINICALLY ENCOUNTERED ANAEROBIC COCCI

Species	Relationship to Oxygen	Gram-Stain Reaction	Catalase	Indole	Esculin Hydrolysis	Urease	Nitrate Reduction	Glucose	Lactose	Acids Produced in PYG
Peptostreptococcus spp										
P. anaerobius*	AN, MA	+	−	−	−	−	−$^+$	+	−	A, IB, IV, IC
P. asaccharolyticus	AN	+	−$^+$	+	−	−	−	−	−	A, B
P. indolicus†	AN	+	−	+	−	−	+	−	−	A, P, B
P. magnus	AN	+	−$^+$	−	−	−	−	−	−	A, P, B
P. micros	AN	+	−	−	−	−	−	−	−	A
P. prevotii	AN	+	−$^+$	−	−	−	−	−	−	A
P. tetradius	AN	+	−	−	−	+	−	+	−	A, P, B
Sarcina spp										
S. ventriculi‡	AN, MA	+	V	−	+	+	+	+	+	A, B, L
Staphylococcus spp										
S. saccharolyticus	AN, MA, AT	+	+	−	−	−	+	+	−	A
Veillonella spp										
V. parvula	AN, MA	−	+	−	−	−	+$^-$	−	−	A, P

*Inhibited by SPS disk.
†Coagulase-positive.
‡Forms endospores.
AN = obligate anaerobe; MA = microaerotolerant anaerobe; AT = aerotolerant anaerobe; V = variable; A = acetic acid; P = propionic acid; IB = isobutyric acid; B = butyric acid; IV = isovaleric acid; IC = isocaproic acid; L = lactic acid; +$^-$ = most strains positive, but some are negative; −$^+$ = most strains negative, but some are positive.
Modified from Engelkirk PG, Duben-Engelkirk J, Dowell VR Jr: Principles and Practice of Anaerobic Bacteriology, 1992; used with permission of Star Publishing Company, Belmont, CA.

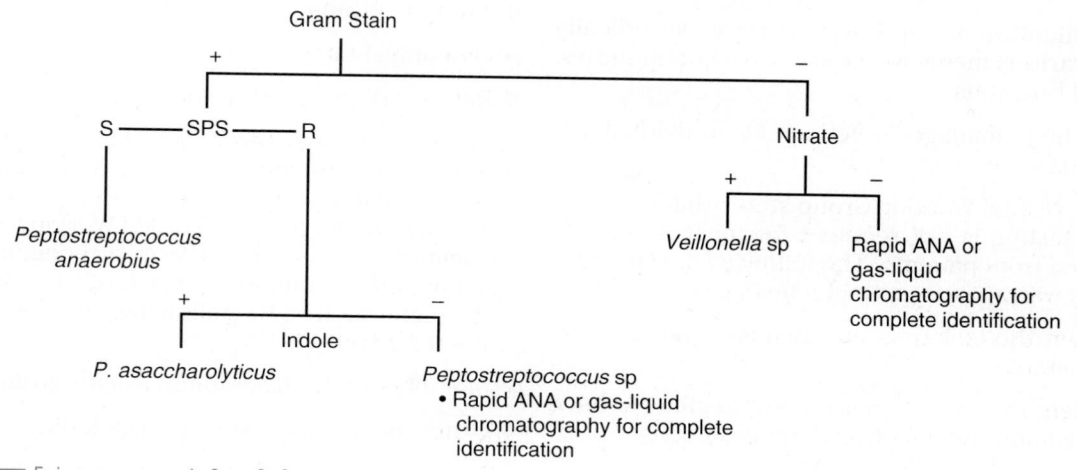

Figure 19–30
Schematic diagram for the presumptive identification of anaerobic cocci.

■■■■T A B L E 1 9 – 2 8

PERCENTAGE OF ANAEROBES SUSCEPTIBLE AT BREAKPOINT CONCENTRATIONS USING A BRUCELLA BLOOD AGAR DILUTION TECHNIQUE

Antimicrobial Agent	Breakpoint (µg/ml)	Bacteroides fragilis	Bacteroides fragilis group	Bacteroides gracilis	Fusobacterium spp	Anaerobic cocci	Clostridium spp	Non–Spore-Forming Gram-Positive Bacilli
Ampicillin/sulbactam*	16	100	100	Not tested	97	100	100	100
Cefoperazone	32	57	51	67	88	100	100	84
Cefotaxime	32	50	50	78	100	100	100	90
Cefotetan	32	85	56	Not tested	81	100	100	100
Cefoxitin	32	92	75	78	99	100	65	95
Ceftizoxime	32	43†	45†	89	100	100	100	87
Chloramphenicol	16	100	100	100	100	100	100	97
Clindamycin	4	93	81	61	92	97	90	86
Imipenem	8	100	100	100	95	100	100	100
Metronidazole	16	100	100	93	100	98.5	99	63
Moxalactam	16	78	55	78	77	100	100	73
Pencillin G	8	5	6	67	100	100	100	97
Piperacillin‡	128	84	85	78	99	100	100	100

* Amoxicillin/clavulanic acid and ticarcillin/clavulanic acid have comparable activities against anaerobes.
† When a broth microdilution technique is used, 90–95% of strains are susceptible at breakpoint.
‡ Carbenicillin, ticarcillin, and mezlocillin have similar activities against anaerobes.
Data from Finegold (1990).
Modified from Engelkirk PG, Duben-Engelkirk J, Dowell VR Jr: Principles and Practice of Anaerobic Bacteriology, 1992; used with permission of Star Publishing Company, Belmont, CA.

Susceptibility Testing of Anaerobes

When to Test

The National Committee for Clinical Laboratory Standards (NCCLS) Working Group on Anaerobic Susceptibility Testing cites the following major reasons for susceptibility testing of anaerobic isolates:

■ To determine patterns of susceptibility of anaerobes to new antimicrobial agents

■ To monitor susceptibility patterns periodically in various medical centers, local communities, and hospitals

■ To help manage infections in individual patients

The NCCLS Working Group states that susceptibility testing is not required for most anaerobes isolated from patients. The following specific situations warrant susceptibility testing of anaerobes:

■ When the isolate is an organism known to be resistant

■ When the usual therapeutic regimens have failed and the infectious process persists

■ When the role of antimicrobial agents is pivotal in determining the outcome

■ When there exists no precedent for empiric therapy

The NCCLS Working Group also suggests that the following specific infectious processes warrant susceptibility testing of anaerobic isolates:

■ Brain abscess

■ Endocarditis

■ Infection of a prosthetic device or vascular graft

■ Joint infections

■ Osteomyelitis

■ Refractory or recurrent bacteremia

Because many anaerobe-associated infectious processes are polymicrobial, deciding which anaerobic isolates warrant susceptibility testing is frequently difficult. The NCCLS Working Group suggests that anaerobes that are recognized as virulent and/or commonly resistant to antimicrobial agents, such as the following, should be considered for testing:

■ Members of the *Bacteroides fragilis* group

■ Other *Bacteroides* species (including *B. gracilis*)

■ *Porphyromonas* and *Prevotella* spp

- *Clostridium perfringens*
- *Clostridium ramosum*
- *Clostridium septicum*
- Certain *Fusobacterium* spp (including *F. varium*)
- *Bilophila wadsworthia*

Antimicrobial Agents to Be Tested

Initiation of antimicrobial therapy is usually empirical and is based on the following knowledge:

- The nature and location of the infectious process. Special characteristics of specific diseases significantly affect the approach to therapy. In bacterial endocarditis, for example, a bactericidal drug must be used. The blood-brain barrier makes certain drugs unsuitable and others less effective for treating brain abscess or meningitis (clindamycin, for example, does not cross the blood-brain barrier).

- The pathogens anticipated in infectious processes of the type being treated. Aspiration pneumonia, for example, would be expected to involve mixed flora of the oral cavity. Intra-abdominal infectious processes would be expected to involve mixed gastrointestinal flora. Female genitourinary infectious processes would be expected to involve mixed vaginal flora.

- Factors that might have modified the anticipated flora.

- Typical susceptibility patterns of the expected flora.

- Gram-stain results.

- The severity of the infection.

The clinical outcome in any given case will be influenced by a multitude of factors, including the speed with which antimicrobial therapy is initiated; whether the agent crosses the blood-brain barrier; the concentration of agent achieved in serum, tissues, and body cavities or spaces; the degree of protein binding that occurs; and the effect of microbial and other enzymes on the agent.

Recommendations regarding antimicrobial therapy of anaerobe-associated diseases change frequently due to the introduction of new drugs, shifting resistance patterns among anaerobes and other pathogens, and the results of local as well as large-scale, multicenter investigations. Although Finegold and coworkers provide examples of currently recommended therapeutic regimens, they advise clinicians that the choice of regimen depends on a number of factors, including the type of infection, whether it is community- or hospital-acquired, the seriousness of the infection, whether complications exist, the general health of the patient, and last—but certainly not least—the specific organisms that are isolated and their resistance patterns.

The selection of agents to be used in susceptibility testing of anaerobes will certainly be influenced by their availability in the hospital pharmacy, and should include the following:

Penicillin G. Penicillin G has a narrow anaerobic spectrum of activity, low toxicity (some hypersensitivity reactions and neurotoxicity), good cerebrospinal fluid (CSF) penetration, and very good bactericidal activity. Although penicillin can be administered both orally and parenterally, the parenteral route is usually preferred for anaerobic infections because higher dosages and blood levels are facilitated.

Broad-Spectrum Penicillins. The carboxypenicillins (e.g., ticarcillin) have a broad spectrum of activity, are bactericidal, have relatively low toxicity (some hypersensitivity reactions, neurotoxicity, bleeding, and fluid or electrolyte problems), and good CSF penetration. Piperacillin and the ureidopenicillins (e.g., mezlocillin and azlocillin) have a broad spectrum of activity, relatively low toxicity (some hypersensitivity reactions, neurotoxicity, and bleeding), good CSF penetration, and very good bactericidal activity. All these drugs are administered parenterally because they are poorly absorbed by the oral route. They are very good for systemic infections when given parenterally.

Combination Agents. Combination agents consist of β-lactam antibiotics (some penicillin-like molecule) plus a β-lactamase inhibitor that renders an antibiotic resistant to degrading enzymes (lactamases) produced by the microorganism. The combination drug amoxicillin/clavulanate (drug/inhibitor) has low toxicity (some hypersensitivity reactions and pseudomembranous colitis) and very good bactericidal activity. It may be administered orally but not parenterally. Ticarcillin/clavulanate and piperacillin/tazobactam have a broad anaerobic spectrum of activity, relatively low toxicity (some hypersensitivity reactions, neurotoxicity, and bleeding), unpredictable CSF penetration, and very good bactericidal activity. They may be administered parenterally but not orally.

Other β-Lactam Antimicrobials. Cefoxitin is bactericidal, has a broad spectrum of anaerobic activity, and low toxicity (some hypersensitivity reactions and pseudomembranous colitis). Cefoxitin may be administered parenterally but not orally.

Imipenem has a broad spectrum of activity, relatively low toxicity (some hypersensitivity reactions and neurotoxicity), seemingly good CSF penetration, and very good bactericidal activity. Imipenem may be administered parenterally but not orally.

Other Antimicrobial Agents. Chloramphenicol is a bacteriostatic agent that usually has a broad spectrum of activity, high toxicity (aplastic anemia), and excellent CSF penetration. Chloramphenicol can be administered both orally and parenterally.

Clindamycin has a broad anaerobic spectrum of activity, low to moderate toxicity (some pseudomembranous colitis), poor CSF penetration, and moderate bactericidal activity. Clindamycin can be administered both orally and parenterally.

Metronidazole has low toxicity (some neurotoxicity), excellent CSF penetration, and excellent bactericidal activity. Metronidazole can be administered both orally and parenterally. The drug has no aerobic spectrum of activity.

Problems in Susceptibility Testing of Anaerobic Isolates

There is, unfortunately, no general agreement as to a "gold standard" technique, because problems are encountered with all available methods. The problems include a lack of reproducibility, failure of some anaerobes to grow on or in particular media, difficulty in reading endpoints with certain methods, and a lack of comparability between methods. Cost and procedure complexity are other factors that inhibit laboratories from performing susceptibility testing of anaerobes. Of far greater concern are the accuracy of the methods and their correlation with the in vivo or clinical situation.

Susceptibility Testing Options

A variety of methods exist for susceptibility testing of anaerobes. These methods have been described in Chapter 3 of this textbook. The agar dilution method using Wilkens-Chalgren agar is the reference method recommended by the NCCLS. However, the procedure is not practical, and its use is limited to large institutions.

The broth dilution method, whether the macro or micro method is used, utilizes W-C broth or Brucella broth enriched with hemin, $NaHCO_3$, and vitamin K_1. Macrodilution is reported to be more dependable than microdilution, but it is less practical due to the commercial availability of frozen or lyophilized microtiter MIC (minimal inhibitory concentration) trays for susceptibility testing of anaerobes.

Two new methods, Spiral Gradient Endpoint (SGE) and PDM Epsilometer (E test), described in Chapter 3, attempt to offer alternative methods for anaerobic susceptibility testing (Figs. 19–31 and 19–32). The SGE method has been found useful for susceptibility testing of anaerobes. Although SGE concentrations correlate well with conventional agar dilution MIC values and may offer a reliable alternative to the conventional agar dilution method, SGE has not been endorsed by the NCCLS. The E test reportedly offers a reliable method for susceptibility testing of anaerobic bacteria.

To date, none of these susceptibility testing methods has been established as fully reliable for predicting the clinical or bacteriologic outcome of a given anaerobe-associated infectious process.

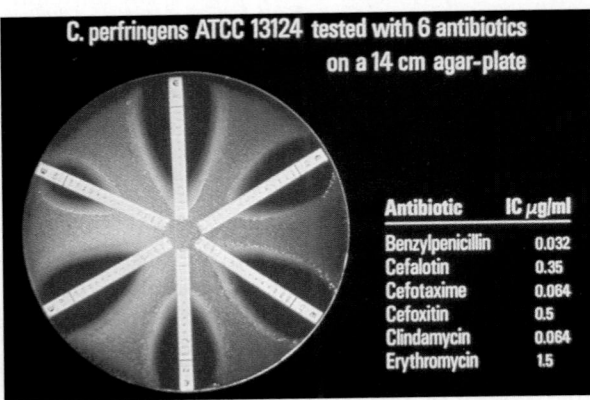

Antibiotic	IC μg/ml
Benzylpenicillin	0.032
Cefalotin	0.35
Cefotaxime	0.064
Cefoxitin	0.5
Clindamycin	0.064
Erythromycin	1.5

■■■■■ F i g u r e   1 9 – 3 1

E test inhibition ellipse *(Clostridium perfringens);* This photograph depicts results obtained when a strain of *C. perfringens* (ATCC 13124) was tested against six different antimicrobial agents using the Epsilometer or E test method. (Photograph courtesy of AB Biodisk North America, Culver City, CA.)

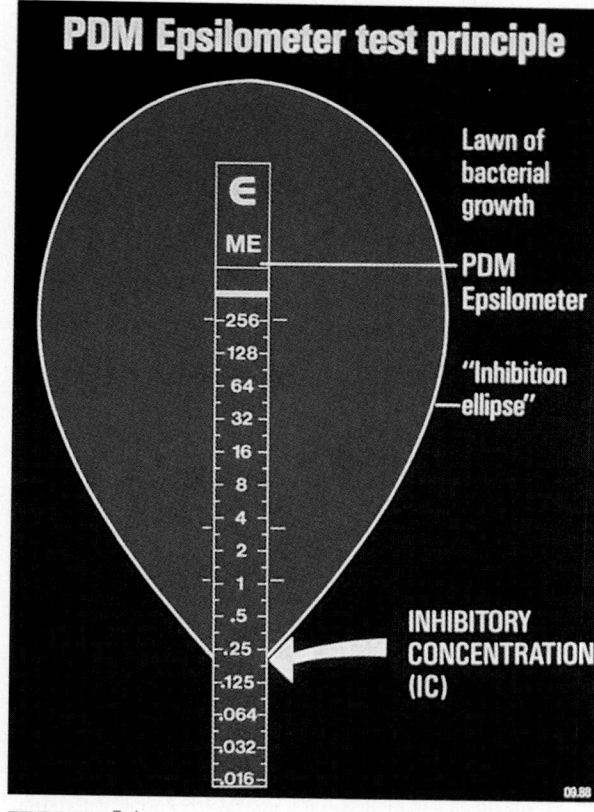

PDM Epsilometer test principle

Lawn of bacterial growth

PDM Epsilometer

"Inhibition ellipse"

INHIBITORY CONCENTRATION (IC)

■■■ F i g u r e   1 9 – 3 2
E test principle. If organism growth was inhibited to any extent by the antimicrobial agent, an inhibition ellipse (such as shown here) is seen. With the use of the interpretive scale on the carrier strip, an inhibitory concentration (IC) is determined and recorded. (Photograph courtesy of AB Biodisk North America, Culver City, CA.)

Quality-Assurance Considerations Pertaining to Susceptibility Testing

An NCCLS-approved method must be used for susceptibility testing of anaerobes. Whichever procedure is selected, it must be performed exactly in accordance with NCCLS guidelines and/or the manufacturer's instructions. Appropriate quality-control (QC) organisms must be used to monitor the accuracy of the method each time it is performed. The QC organisms currently recommended by the NCCLS are:

■ *Bacteroides fragilis* (ATCC 25285).

■ *Bacteroides thetaiotaomicron* (ATCC 29741).

■ *Eubacterium lentum* (ATCC 43055).

This list of organisms is provided solely to illustrate QC standards in effect at press time. Readers should use only the most recent edition of any NCCLS standards. These organism may be obtained directly from the American Type Culture Collection or companies such as Adams Scientific (West Warwick, RI), Anaerobe Systems (San Jose, CA), and Remel (Lenexa, KS).

β-Lactamase Testing

β-lactamases are enzymes that destroy the β-lactam ring of penicillins (penicillinases) and cephalosporins (cephalosporinases), thus rendering these antibiotics ineffective. The first anaerobes shown to produce β-lactamases were members of the *B. fragilis* group. Those produced by *B. fragilis* are primarily cephalosporinases. It is now known that some strains of *Bacteroides* spp, *Prevotella* spp, *Fusobacterium*, and *Clostridium* produce β-lactamases as well. The ability of an anaerobe to produce β-lactamase enzymes can be determined using simple, commercially available methods (see Chapter 3).

The absence of β-lactamase does not necessarily mean that an organism will be susceptible to β-lactam antibiotics. Anaerobes can be resistant to such agents by mechanisms other than β-lactase (e.g., by altering the number or type of penicillin-binding proteins or blocking penetration of the drug into the active site via alteration of the bacterial outer membrane pores). Anaerobes known to be resistant to β-lactam drugs by mechanisms other than production of β-lactamase include *Bacteroides gracilis, Bilophila wadsworthia,* and some strains of *B. distasonis* and *B. fragilis.* Thus, β-lactamase testing may be used as an adjunct to susceptibility testing but never as a replacement for it. It is not necessary to test all clinically significant anaerobes for β-lactamase production.

TREATING ANAEROBE-ASSOCIATED DISEASES

Finegold states that treatment of anaerobe-associated infectious processes involves a three-pronged attack:

1. Create an environment in which anaerobes cannot proliferate. Useful measures include removing dead tissue (débridement), draining pus, eliminating obstructions, decompressing tissues,

releasing trapped gas, and improving circulation in and oxygenation of tissues. In lesser infections, surgical therapy may be all that is required.

2. Arrest the spread of anaerobes into healthy tissues. Antimicrobial agents play an important role here.

3. Neutralize toxins produced by the anaerobes when such toxins are present. Specific antitoxins can be used.

Bibliography

Appelbaum PC, Spangler SK, Jacobs MR: Beta-lactamase production and susceptibilities to amoxicillin, amoxicillin-clavulanate, ticarcillin, ticarcillin-clavulanate, cefoxitin, imipenem, and metronidazole of 320 non-*Bacteroides fragilis* *Bacteroides* isolates and 129 fusobacteria from 28 U.S. centers. Antimicrob Agents Chemother 34:1546, 1990.

Appelbaum PC, Spangler SK, Jacobs MR: Evaluation of two methods for rapid testing for beta-lactamase production in *Bacteroides* and *Fusobacterium*. Eur J Clin Microbiol Infect Dis 9:47, 1990.

Appleman MD, Heseltine PNR, Cherubin CE: Epidemiology, antimicrobial susceptibility, pathogenicity, and significance of *Bacteroides fragilis* group organisms isolated at Los Angeles County–University of Southern California Medical Center. Rev Infect Dis 13:12, 1991.

Balows A, Hausler WJ Jr, et al (eds): Manual of Clinical Microbiology, 5th ed. Washington, DC: American Society for Microbiology, 1991.

Bartley SL: Personal communication, 1990.

Bartley SL, Dowell VR Jr: Comparison of media for the isolation of *Clostridium difficile* from fecal specimens. Lab Med 22:335, 1991.

Brazier JS: Appraisal of Anotox, a new anaerobic atmospheric detoxifying agent for use in anaerobic cabinets. J Clin Pathol 35:233, 1982.

Brazier JS, Riley TV: UV red fluorescence of *Veillonella* spp. J Clin Microbiol 26:383, 1988.

Brook I: Recovery of anaerobic bacteria from clinical specimens in 12 years at two military hospitals. J Clin Microbiol 26:1181, 1988.

Citron DM, Baron EJ, et al: Short prereduced anaerobically sterilized (PRAS) biochemical scheme for identification of clinical isolates of bile-resistant *Bacteroides* species. J Clin Microbiol 28:2220, 1990.

Citron DM, Ostovari MI, et al: Evaluation of the E test for susceptibility testing of anaerobic bacteria. J Clin Microbiol 29:2197, 1991.

Cuchural GJ Jr, Tally FP, et al: Comparative activities of newer beta-lactam agents against members of the *Bacteroides fragilis* group. Antimicrob Agents Chemother 34:479, 1990.

Dowell VR Jr: Media for selective isolation of *Clostridium difficile*. Les anaerobies microbiologie-pathologie, symposium international, 1981, pp 30–32.

Dowell VR Jr: Personal communication, 1989.

Dowell VR Jr, Hawkins TM: Laboratory Methods in Anaerobic Bacteriology: CDC Laboratory Manual. Atlanta: Centers for Disease Control, 1974.

Dowell VR Jr, Lombard GL: Procedures for Preliminary Identification of Bacteria. Atlanta: Centers for Disease Control, 1984.

Dowell VR Jr, Lombard GL, et al: Media for Isolation, Characterization, and Identification of Obligately Anaerobic Bacteria. Atlanta: Centers for Disease Control, 1977.

Downes J, Mangels JI, et al: Evaluation of two single-plate incubation systems and the anaerobic chamber for the cultivation of anaerobic bacteria. J Clin Microbiol 28:246, 1990.

Engelkirk PG, Duben-Engelkirk J, Dowell VR Jr: Principles and Practice of Clinical Anaerobic Bacteriology. Belmont, CA: Star Publishing, 1992.

Finegold SM: Anaerobes: Problems and controversies in bacteriology, infections, and susceptibility testing. Rev. Infect Dis 12 (Suppl 2):S223, 1990.

Finegold SM: Therapy of anaerobic infections. In Finegold SM, George WL (eds): Anaerobic Infections in Humans. San Diego: Academic Press, 1989, pp 793–818.

Finegold SM, Baron EJ, et al: A Clinical Guide to Anaerobic Infections. Belmont, CA: Star Publishing, 1992.

Finegold SM, George WL (eds): Anaerobic Infections in Humans. San Diego: Academic Press, 1989.

Finegold SM, George WL, Mulligan ME: Anaerobic Infections. Chicago: Year Book Medical Publishers, 1986.

Goldstein EJC, Citron DM: Annual incidence, epidemiology, and comparative in vitro susceptibilities to cefoxitin, cefotetan, cefmetazole, and ceftizoxime of recent community-acquired isolates of the *Bacteroides fragilis* group. J Clin Microbiol 26:2361, 1988.

Han Y-H, Smibert RM, Krieg NR: *Wolinella recta, Wolinella curva, Bacteroides ureolyticus,* and *Bacteroides gracilis* are microaerophiles, not anaerobes. Int J Syst Bacteriol 41:218, 1991.

Hentges DJ: Anaerobes as normal flora. In Finegold SM, George WL (eds): Anaerobic Infections in Humans. San Diego: Academic Press, 1989, pp 37–53.

Hill G: Spiral gradient endpoint method compared to standard agar dilution for susceptibility testing of anaerobic gram-negative bacilli. J Clin Microbiol 29:975, 1991.

Hill GB, Schalkowsky S: Development and evaluation of the spiral gradient endpoint method for susceptibility testing of anaerobic gram-negative bacilli. Rev Infect Dis 12 (Suppl 2):S200, 1990.

Holdeman LV, Cato EP, Moore WEC (eds): VPI Anaerobe Laboratory Manual, 4th ed. Blacksburg: Virginia Polytechnic Institute and State University, 1977 (with 1991 update).

Johnson LL, McFarland LV, et al: Identification of *Clostridium difficile* in stool specimens by culture-enhanced gas-liquid chromatography. J Clin Microbiol 27:2218, 1989.

Krieg NR, Holt JG (eds): Bergey's Manual of Systematic Bacteriology, Vol 1. Baltimore: Williams & Wilkins, 1984.

LaRocco M, Robinson A: Evaluation of three commercial tests for rapid detection of beta-lactamase in anaerobic bacteria. Eur J Clin Microbiol 4:593, 1986.

Loesche WJ: Oxygen senstivity of various anaerobic bacteria. Appl Microbiol 18:723, 1969.

Lombard GL, Dowell VR Jr: Comparison of three reagents for detecting indole production by anaerobic bacteria in microtest systems. J Clin Microbiol 18:609, 1983.

Lombard GL, Dowell VR Jr: Gas-Liquid Chromatography Analysis of the Acid Products of Bacteria. Atlanta: Centers for Disease Control, 1982.

Mangels JI, Cox ME, Lindberg LH: Methanol fixation: An alternative to heat fixation of smears before staining. Diagn Microbiol Infect Dis 2:129, 1984.

Mangels JI, Douglas BP: Comparison of four commercial Brucella agar media for growth of anaerobic organisms. J Clin Microbiol 27:2268, 1989.

Margulis L, Sagan D: Microcosmos: Four Billion Years of Microbial Evolution. New York: Summit Books, 1986.

McFarland LV, Mulligan ME, et al: Nosocomial acquisition of *Clostridium difficile* infection. N Engl J Med 320:204, 1989.

Moore HB: Rapid methods in microbiology: IV. Presumptive and rapid methods in anaerobic bacteriology. Am J Med Technol 47:705, 1981.

Murray PR, Citron DM: General processing of specimens for anaerobic bacteria. In Balows A, Hausler WJ Jr, et al (eds): Manual of Clinical Microbiology, 5th ed. Washington, DC: American Society for Microbiology, 1991, pp 488–504.

NCCLS:M11-A-3. Methods for Antimicrobial Susceptibility Testing of Anaerobic Bacteria: Approved Standard, 3rd ed. Villanova, PA: National Committee for Clinical Laboratory Standards, 1993.

Nord CE, Hedberg M: Resistance to beta-lactam antibiotics in anaerobic bacteria. Rev Infect Dis 12 (Suppl 2):S231, 1990.

Rolfe RD, Finegold SM: *Clostridium difficile:* Its Role in Intestinal Disease. San Diego: Academic Press, 1988.

Sheppard A, Cammarata C, Martin DH: Comparison of different medium bases for the semiquantitative isolation of anaerobes from vaginal secretions. J Clin Microbiol 28:455, 1990.

Slots J, Reynolds HS: Long-wave UV light fluorescence for identification of black-pigmented *Bacteroides* spp. J Clin Microbiol 16:1148, 1982.

Smith LDS, Williams BL: The Pathogenic Anaerobic Bacteria, 3d ed. Springfield, IL: Charles C. Thomas, 1984.

Sneath PHA, Mair NS, et al (eds): Bergey's Manual of Systematic Bacteriology, Vol 2. Baltimore: Williams & Wilkins, 1986.

Sutter VL, Citron DM, et al: Wadsworth Anaerobic Bacteriology Manual, 4th ed. Belmont, CA: Star Publishing, 1985.

Sutter VL, Finegold SM: Antibiotic disk susceptibility tests for rapid presumptive identification of gram-negative bacilli. Appl Microbiol 21:13, 1971.

Wexler HM: Susceptibility testing of anaerobic bacteria: Myth, magic, or method? Clin Microbiol Rev 4:470, 1991.

The Spirochetes

Richard B. Prior

1. Describe the general characteristics of the genus *Leptospira*.

2. List the clinical infections caused by *Leptospira*.

3. Describe the diagnostic tests used to identify *Leptospira* in the clinical laboratory.

4. State the general characteristics of the genus *Borrelia*.

5. Describe the etiologic agent and the arthropod vector of relapsing fever.

6. Describe the etiologic agent and the arthropod vector of Lyme disease.

7. Describe the classic skin lesion, erythema chronicum migrans (ECM), of Lyme disease.

8. Name the four species of *Treponema* that are pathogenic for humans.

9. Describe the primary, secondary, and tertiary clinical manifestations of syphilis.

10. Describe the diagnostic tests used to identify *Treponema pallidum* in the clinical laboratory.

11. Characterize the three nonvenereal *Treponema* diseases.

The order Spirochaetales contains two families, the Leptospiraceae and the Spirochaetaceae. The Leptospiraceae family contains one genus, *Leptospira*, and the Spirochaetaceae family contains two, *Borrelia* and *Treponema*. These three genera include the causative agents of important human and zoonotic diseases.

The spirochetes are slender, flexuous, helically shaped, unicellular bacteria ranging from 0.1 to 3.0 μm wide and from 5 to 250 μm in length with one or more complete turns in the helix. They differ from other bacteria in that they have a flexible cell wall around which several fibrils are wound. These fibrils, termed the *periplasmic flagella* (also known as the axial fibrils, axial filaments, endoflagella, and periplasmic fibrils), are responsible for the motility. A multilayered outer sheath similar to the outer membrane of other gram-negative bacteria completely surrounds the protoplasmic cylinder (the cytoplasmic and nuclear regions enclosed by the cytoplasmic membrane–cell wall complex, and periplasmic flagella).

The spirochetes are motile organisms exhibiting three types of motion in liquid media: locomotion, rotation about their longitudinal axis, and flexion. They are free living or survive in association with animal and human hosts, and some species are pathogenic. In addition, they are chemoheterotrophic and can utilize carbohydrates, amino acids, long-chain fatty acids, or long-chain fatty alcohols as carbon and energy sources. Metabolism can be anaerobic, facultatively anaerobic, or aerobic, depending on the species. They are gram negative and reproduce like other bacteria, i.e., via transverse fission.

LEPTOSPIRA

General Characteristics

The typical organism of the genus *Leptospira* is a tightly coiled, thin, flexible spirochete, 0.1 μm wide and 5 to 15 μm long (Fig. 20–1). In contrast to both the *Treponema* and the *Borrelia* (see below), the spirals are very close together, so that under the microscope, the organism is often seen as a chain of cocci. One or both ends of the organism have hooks rather than tapering off as do the other two genera in the group. Motion is rotational. Two species, *Leptospira interrogans* and *Leptospira biflexa,* are recognized for the so-called pathogenic and saprophytic leptospiras, respectively.

Like the *Treponema* and *Borrelia*, under electron microscopy *Leptospira* show a long axial filament covered by a very fine sheath. All species have two periplasmic flagella. The organisms cannot be stained readily, but they can be impregnated with silver and visualized in that manner. Unstained cells are not visible by brightfield

■■■■■■■F i g u r e   2 0 – 1
Scanning electron micrograph of *Leptospira interrogans* isolated from blood of a patient. The tight coils and bent ends are characteristic of this organism. × 2,500.

microscopy but are visible by darkfield, phase-contrast, and immunofluorescent microscopy.

Leptospira are obligately aerobic and can be grown in artificial media (e.g. Fletcher semisolid and Stuart liquid media) at 30°C. Generation time is from 6 to 16 hours. Animal inoculation methods are particularly useful for the isolation of strains from tissues or body fluids containing contaminating microorganisms. The laboratory animals of choice for leptospirosis are weaning hamsters and young guinea pigs.

The antigenic composition of leptospiras is used for taxonomic purposes, with the serovar as the basic taxon. *Leptospira interrogans* can be divided into multiple serogroups and serovars (serotypes). The etiologic agents of leptospirosis include approximately 200 different serovars, and these are determined on the basis of their agglutinogenic properties. Disease in the United States has been caused by more than 10 different serovars, the most common of which are *L. icterohemorrhagiae* and *L. canicola*.

Virulence Factors and Pathogenicity

The various serogroups of *L. interrogans* are parasitic or pathogenic for a wide range of wild and domestic animals and humans, but the mechanism of pathogenicity is not known. Several biologic properties of the pathogenic strains of *Leptospira* may play a role, including the failure of human macrophages and polymorphonuclear neutrophils (PMNs) to ingest virulent *Leptospira* in the presence of normal serum. In addition, the production of a soluble hemolysin

by some virulent strains, production of cell-mediated sensitivity to leptospiral antigen, and the finding of small amounts of endotoxin in some strains contribute to its pathogenicity. With regard to the endotoxin, the clinical findings in animals with leptospirosis suggest the presence of endotoxemia.

Clinical Infections

Leptospirosis is a zoonotic disease in humans and is primarily associated with occupational exposure. Work with animals or in rat-infested surroundings poses infectious hazards for veterinarians, dairy workers, swineherds, slaughterhouse workers, miners, sewer workers, fish and poultry processors, and so on. In the United States, the majority of cases result from recreational exposures and occur most commonly in the summer. In 1992, 54 cases of leptospirosis were reported in the United States; however, many cases may go unrecognized and unreported.

In the natural host, *Leptospira* live in the lumen of renal tubules, from which the organisms are shed into the urine. Dogs, rats, and other rodents are the principal animal reservoirs of *Leptospira*. Hosts acquire infections directly by contact with the urine of carriers or indirectly by contact with bodies of water contaminated with the urine of carriers. *Leptospira* can survive in neutral or slightly alkaline waters for months.

Human infections are usually associated with occupational exposure of individuals who have worked with animals. Transmission is zoonotic. The organisms most likely enter the human host through small breaks in the skin or intact mucosa. The initial sites of early multiplication are unknown. Nonspecific host defenses do not contain the *Leptospira*, and a leptospiremia occurs during the acute illness. *Leptospira* usually infect the kidneys, and the major renal lesion is an interstitial nephritis with associated glomerular swelling and hyperplasia without involving the glomerulus. Late manifestations of the disease may be caused by the host immunologic response to the infection.

The incubation period of leptospirosis is usually 10 to 12 days but ranges from 3 to 30 days after inoculation. The severity of the disease depends on many factors, such as the infecting strain and the general health of the host. Clinical illness is usually abrupt, with nonspecific, influenza-like constitutional symptoms such as fever, chills, headache, severe myalgia, and malaise. The subsequent course is protean, frequently biphasic, and often results in hepatic, renal, and central nervous system involvement. The most characteristic physical finding is conjunctival suffusion, but this is seen in less than half of the patients. Duration of the illness varies from less than 1 week to 3 weeks.

Homologous agglutinating antibodies develop in all patients with a leptospiral bacteremia. Immunoglobulin M (IgM) antibodies are detected within a week after onset of disease and may persist in high titers for many months. A month or more after the onset of illness, IgG antibodies can be detected in some patients. Convalescent serum contains protective antibodies, and the ability of this serum to protect against disease correlates with the titer of agglutinating antibody.

Laboratory Diagnosis

Cultural Characteristics

Leptospira can be easily cultured from the blood or cerebrospinal fluid (CSF) during the acute phase of the disease. After the first week of the disease, there may be intermittent shedding of the spirochetes in the urine, which can also be cultured. Isolation of *Leptospira* can be accomplished by direct inoculation of laboratory media or by animal inoculation.

Microscopy

Although direct demonstration of *Leptospira* in clinical specimens during the first week of the disease by darkfield or phase-contrast microscopy, fluorescent antibody testing, or silver impregnation staining is possible, it is successful only in a small percentage of cases.

Serologic Tests

A macroscopic slide agglutination test is available for rapid screening for the presence of leptospiral antibody. For the determination of serovar-specific antibody, a more sensitive microscopic agglutination test is used. These tests require the maintenance of appropriate living organisms and as such are found in larger referral laboratories only.

An unequivocal laboratory diagnosis of a current case of leptospirosis can be established by the isolation of the organism from blood or CSF or by the demonstration of significant rises in antibody titer in two or more properly timed serum samples. From the viewpoint of the clinician, however, the management and treatment of leptospirosis do not depend on the infecting serovar. Thus, a laboratory diagnosis of leptospirosis per se serves as confirmation of the clinical diagnosis.

Treatment and Prevention

The *Leptospira* are sensitive to penicillin, streptomycin, tetracycline, doxycycline, and the macrolide antibiotics in vitro. Although its effectiveness is unproven, penicillin is considered beneficial and alters the course of the disease if treatment is initiated before the fourth day of illness. When started after the fourth day of illness, penicillin usually does not alter the course of the disease. Doxycycline in adults appears to shorten the course of the illness and also to reduce the incidence of convalescent leptospiruria.

If the patients are hospitalized, blood and body fluid (including urine) precautions are warranted for the duration of hospitalization.

Protective clothing such as boots and gloves should be worn for occupational exposure. Also, rodent control is indicated. Drainage of waters known to be contaminated also helps reduce the risk of exposure. Short-term prophylaxis may be accomplished by administration of doxycycline weekly for high-risk occupational groups with short-term exposure.

Vaccination of dogs and livestock prevents the disease but not the infection and leptospiruria in animals. For humans in endemic areas, vaccines have been effectively used in veterinary medicine.

BORRELIA

General Characteristics

The genus *Borrelia* comprises several species of spirochetes that are morphologically similar but that have different pathogenic properties and host ranges. The most important species from the human disease standpoint is *Borrelia recurrentis*, which causes relapsing fever. A recently recognized species, *Borrelia burgdorferi*, has been shown to be the etiologic agent of Lyme disease and related disorders. All borreliae are arthropod borne.

The borreliae are gram-negative, highly flexible organisms varying in thickness from 0.2 to 0.5 μm and in length from 3 to 20 μm. The spirals vary in number from three to ten per organism and are much less tightly coiled than those of *Leptospira* (Fig. 20–2). Unlike the *Leptospira* and *Treponema*, the borreliae stain easily and can be examined under the regular light microscope. Electron-microscopic pictures show the same general patterns as are seen with the *Treponema*; long periplasmic flagella coated with a sheath of protoplasm and a periplasmic sheath. From 15 to 20 periplasmic flagella per cell are seen. The borreliae multiply by binary fission.

The borreliae can be cultivated easily in their arthropod vectors or in a variety of vertebrate hosts, such as small rodents. Several species, including *B. recurrentis*, have been cultivated in vitro using Kelly medium. Borreliae are micro-aerophilic and require long-chain fatty acids for growth.

Borrelia recurrentis

Borrelia recurrentis causes relapsing fever, an acute infection characterized by febrile episodes that subside spontaneously but tend to recur over a period of weeks. The relapses are due to the mutability of the antigens associated with the *Borrelia*. These organisms undergo several antigenically distinct variations within a host during the course of a single infection. Each succeeding episode is usually shorter than and not as severe as the preceding one.

Relapsing fever can be either tick or louse borne. The tick-borne disease is transmitted by a large variety of soft-shelled ticks of the genus *Ornithodoros*. Borreliae that cause tick-borne relapsing fevers are widely distributed throughout the eastern and western hemispheres. Transmission of the tick-borne spirochetes to a vertebrate host takes place via infected saliva or infectious body fluids of the tick.

The louse-borne disease is transmitted by human-to-human spread via the body louse, *Pediculus humanus humanus*. Humans are the only reservoir of this agent. The borreliae infect the hemolymph of the louse. Unlike with the tick-borne disease, transmission of the louse-borne disease to humans does not occur by the bite via saliva but by the contamination of the bite wound with the infectious hemolymph. *Borrelia* escape the louse to infect the host when the louse is traumatized or smashed by scratching.

Clinical Infections

A constant spirochetemia occurs during the entire course of borreliosis. The spirochetemia worsens during the febrile periods and wanes between recurrences. Specific pathogenic factors are ill defined, but there appears to be a heat-stable pyrogen present in the organisms that does not act like classic endotoxin.

Following an incubation period of 2 to 15 days, massive spirochetemia develops, with sudden high fever, rigors, severe headache, muscle pains, and weakness. The febrile period lasts about 3 to 7 days and terminates abruptly with the development of an adequate immune response. Then, the disease relapses several days

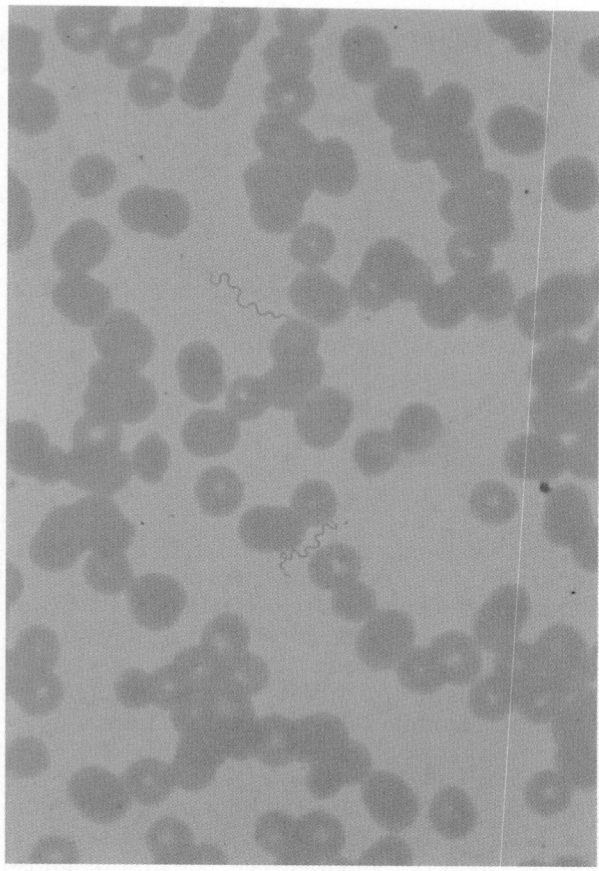

■ Figure 20–2
Appearance of *Borrelia recurrentis* in blood. Giemsa stain. × 850.

to several weeks later, following the same general course but with less severity. The symptoms and severity depend on the immune status of the patient, geographic location, strain of *Borrelia*, and phase of the epidemic. There are also differences in the clinical manifestations between the tick-borne and louse-borne forms. In the tick-borne disease both the primary febrile attack and the afebrile interval are shorter, and the relapses are longer and more numerous.

Laboratory Diagnosis

MICROSCOPY

During the febrile period, diagnosis of borreliosis is readily made by Giemsa or Wright staining of blood smears. This is the only spirochetal disease in which the organisms are visible in blood with brightfield microscopy. The appearance of the spirochete among the red cells is characteristic (see Fig. 20–2).

CULTURAL CHARACTERISTICS

The infecting borreliae can also be recovered using cultural and animal inoculation procedures. For culture, Kelly medium is used, and *B. recurrentis*, *B. hermsii*, *B. parkeri*, *B. turicatae*, and *B. hispanica* have been cultivated. For animal inoculation, suckling Swiss mice or suckling rats are the animals of choice for both the tick-borne and louse-borne borreliae.

The ability of the relapsing fever spirochetes to change their antigenic structure makes the serodiagnosis of these diseases difficult.

Treatment and Prevention

Borreliae are susceptible to many antibiotics. Tetracyclines, however, are the drugs of choice. Tetracyclines reduce the relapse rate and rid the central nervous system of spirochetes. Treatment also includes general supportive measures, such as maintenance of fluid and electrolyte balance.

Relapsing fever can be prevented by control of exposure to the arthropod vectors. For tick-borne relapsing fever, exposure control includes wearing protective clothing, cleaning of rodent-infested cabins, use of insecticides, and the use of repellents. For louse-borne relapsing fever, control is best achieved by the application of good personal and public standards of hygiene.

Borrelia burgdorferi

Epidemiology

Borrelia burgdorferi has been recognized as the agent of Lyme disease, first described after an outbreak among children in Lyme, Connecticut, in 1975. *Borrelia burgdorferi* is transmitted via the *Ixodes* ticks. In the United States from 1975 to 1979, about 500 cases of Lyme disease, the majority from the Northeast, were reported. In 1982, as a result of increased surveillance, 487 cases, mostly from the Northeast and Midwest, were reported. In 1992, 9,895 cases of Lyme disease were reported in the United States.

Onset of illness is generally between May and November, with the majority of onsets occurring during June through September, when people are involved in outdoor activities. Patients of all ages and both sexes may be affected.

Clinical Infections

Lyme disease is characterized in its early stages by symptoms of fever, headache, generalized muscle pain, and perhaps fatigue and weight loss. About 60% of patients exhibit erythema chronicum migrans (ECM), the classic skin lesion that is normally found at the site of the tick bite. It begins as a red macule that expands to form a large annular erythema with partial central clearing. Late manifestations of the disease involve the joints and the cardiac and neurologic systems. Arthritis is the most common manifestation, occurring weeks to years later.

Laboratory Diagnosis

Early diagnosis and antibiotic treatment are important for preventing neurologic, cardiac, and joint abnormalities that can occur late in the disease. Culture of *B. burgdorferi* from patients permits definitive diagnosis, but it lacks sensitivity. In addition, the spirochetes are rarely seen on direct examination of blood or skin transudate specimens.

Because culture and direct visualization of the borreliae are often negative, antibody detection has been the only laboratory method useful for the diagnosis of Lyme disease. The first diagnostic tests are the determination of the immunofluorescent antibody titer. The antibody titer usually reaches a peak between the third and sixth week after onset of the disease. IgM antibody response to *B. burgdorferi* begins during the first 10 days after infection but does not peak until 3 to 6 weeks after the onset of the disease. Peak levels of IgG antibodies are not observed until after months or years of illness. (For the limited number of commercially available antibody assays for the diagnosis of Lyme disease, sensitivity ranges from 73 to 96% and specificity ranges from 75 to 100%. Current research involves the direct detection of spirochetal antigens in patients.)

Treatment and Prevention

Tetracycline is the drug of choice for patients 9 years and older. For patients under 9 years old, penicillin is recommended.

Awareness of the tick and the diseases it can transmit are the most important control measures. Protective clothing should be worn in areas where tick exposure is intense, and attached ticks should always be removed.

TREPONEMA

Treponema comprises four species that are pathogenic for humans: *Treponema pallidum* subsp *pallidum,* the causative agent for syphilis; *T. pallidum* subsp *pertenue,* the causative agent of yaws; *T. carateum,* the causative agent of pinta; and *T. pallidum* subsp *endemicum,* the causative agent of endemic syphilis. At least six non-pathogens have been identified as part of the normal flora, and they are particularly prominent in the oral cavity.

Treponema pallidum subsp *pallidum* is the most studied of the treponemes. The *T. pallidum* species is a fine spiral organism, measuring about 0.1 to 0.2 μm in thickness and 6 to 20 μm in length. It is difficult to visualize under the normal lighting system of an ordinary microscope but can be seen very easily using darkfield microscopy. One of the fine characteristics of the organism is that the spirals are regular and angular with 4 to 14 spirals per organism, with an amplitude of 0.2 to 0.3 μm and a wavelength of about 1.1 μm (Fig. 20–3). Three periplasmic flagella are inserted into each end of the cell. The ends are pointed and covered with a sheath. The cells are motile with graceful flexuous movements in liquid media.

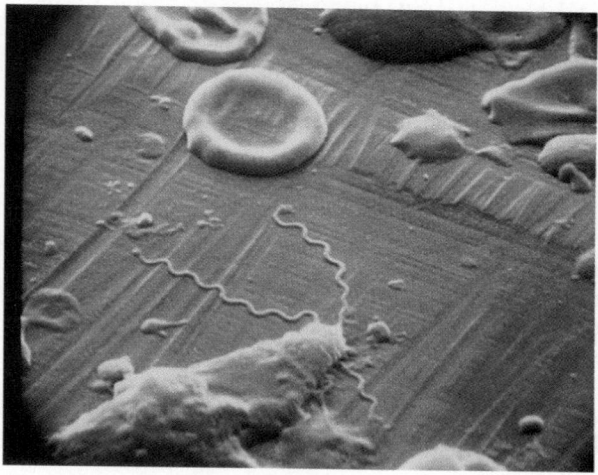

Figure 20–3

Scanning electron micrograph of *Treponema pallidum* (Nichols stain). Two treponemes are shown adjacent to an erythrocyte. × 2,500.

The four pathogenic treponemes cannot be grown in vitro, although some limited growth has now been achieved in primary cell culture. Animal inoculation is not routinely performed for patient diagnosis. When necessary, the rabbit (testes) is the animal of choice for growing *T. pallidum.* Generation time is long, approximately 30 hours. *Treponema pallidum* is microaerophilic and survives for a longer period in an atmosphere of 3 to 5% oxygen. Metabolic information is limited because of the extreme difficulty of obtaining sufficient organisms for study.

Treponema pallidum dies rapidly on drying and is susceptible to a wide range of disinfecting agents.

Syphilis

Syphilis is a disease of blood vessels and of the perivascular areas. The causative agent, *T. pallidum,* is exclusively a human pathogen under natural conditions. Syphilis was first recognized in Europe at the end of the fifteenth century, when it reached epidemic proportions. Two theories are proposed concerning its origin. The first suggests that Columbus's crew acquired syphilis in the West Indies and brought it from the New World back to Europe. The second suggests that the disease was endemic in Africa for centuries and may have been transported to Europe via the migration of armies and civilians. The venereal transmission of syphilis was not recognized until the eighteenth century. The causative agent for syphilis was not discovered until 1905.

Syphilis (or lues) was also known by a variety of names, including "French disease," "Italian disease," and "The Great Pox." *Syphilis* comes from a poem written in 1530, which describes the plight of a mythical shepherd afflicted with the "French disease" as punishment for cursing the gods. The poem is a compendium of knowledge of the time regarding the disease, and it also recognized the venereal nature of the infection. The shepherd's name was Syphilus.

Epidemiology

Syphilis ranks as the third most commonly reportable communicable disease in the United States, exceeded only by gonorrhea and chickenpox. There was a rapid decrease in cases of syphilis following World War II, with the availability of penicillin. Since the mid 1980s, however, there has been an increase in the total number of cases, including primary and secondary cases. In 1992, the total rate of primary and secondary syphilis for the United States was 13.7 per 100,000 population. There were 33,973 cases of primary and secondary syphilis and 3,850 cases of congenital syphilis (patients <1 year old) reported. For all stages of syphilis, there were 112,581 cases reported in 1992. The eastern and southern states have much higher rates than other regions of the country. By sex, there are more cases reported for males than for females. By age, the highest primary and secondary syphilis case rates occur in the group 20 to 29 years old, for which 14,128 cases were reported in 1992. The next highest rate was for the group 30 to 39 years old, with 10,623 cases reported. The group 15 to 19 years old had 3,828 cases reported. By race, the vast majority of cases of primary and secondary syphilis occurred in the black non-Hispanic population, accounting for 28,606 of the total 34,102 cases reported in 1992.

Transmission

Syphilis is acquired from direct sexual contact with an individual who has an active primary or secondary syphilitic lesion. The genital organs—the vagina and cervix in females and the penis in males—are the usual sites of inoculation. It can also be acquired by nongenital contact with a lesion (on the lip, for example) or by transplacental transmission to the fetus (congenital syphilis).

Pathogenesis

Following invasion through a break in the epidermis or possible penetration through intact mucous membranes, the organisms undergo rapid multiplication and then disseminate throughout the body. The natural course of syphilis can be broken down into primary, secondary, and tertiary stages on the basis of clinical manifestations.

PRIMARY DISEASE

After inoculation, the spirochetes multiply rapidly and disseminate to local lymph nodes and to other organs via the blood stream. The primary lesion develops 10 to 90 days after infection, as a result of an inflammatory response to the infection at the site of the inoculation. This lesion, known as the *chancre*, is typically a single erythematous lesion that is nontender and firm with a clean surface and raised border. The lesion is teeming with treponemes and is extremely infectious. It may be inapparent in females, in whom it commonly is situated on the cervix or vaginal wall. It can also be in the anal canal of either sex and remain undetected. There are no systemic signs or symptoms at this time.

SECONDARY DISEASE

Two to 10 weeks following the primary lesion, the patient may experience secondary disease, with clinical findings of fever, sore throat, generalized lymphadenopathy, headache, and rash. (The rash is unusual, in that it can also occur on the palms and soles.) All secondary lesions of the skin and mucous membranes are highly infectious. The stage may last for several weeks and may relapse. It may also be mild and go unnoticed by the patient.

TERTIARY DISEASE

After the secondary stage, the disease is cured spontaneously in about 25% of cases. In another

25%, the disease becomes latent and produces no further clinical manifestations, whereas the remaining 50% of untreated cases develop tertiary manifestions some 2 to 20 years later. These manifestations include: the development of granulomatous lesions (gummas) in skin, bones, and liver (benign tertiary syphilis), degenerative changes in the central nervous system (neurosyphilis), and syphilitic cardiovascular lesions, particularly aortitis and aortic valve insufficiency. Patients with tertiary disease are usually not infectious. In the United States, tertiary disease is not often seen because most patients are adequately treated with antibiotics before the tertiary stage is reached.

Congenital Syphilis

Treponemes can cross the placenta and be transmitted from an infected mother to the fetus. A multisystem disease, congenital syphilis is very severe and mutilating. All pregnant women should have serologic examinations for syphilis early in pregnancy. Cases of congenital syphilis occur, however, despite efforts to screen, identify, and treat infected women during pregnancy. In 1992, 3,850 civilian cases of congenital syphilis were reported in the United States in patients less than 1 year old.

Immunity

During the early stages of syphilis, the infected person rapidly becomes immune to re-infection, although the immunity is short lived if the patient is treated successfully. Both IgM and IgG antibodies are produced and are detectable by the time the chancre appears. With successful treatment, the IgM antibody titers drop within 1 or 2 years, but the IgG titers persist for the lifetime of the patient. Ironically, if the patient is not treated, the stages of syphilis evolve in spite of the humoral antibody response.

An individual with active syphilis appears to be resistant to superinfection with *T. pallidum*; if early syphilis is treated adequately and the infection is eradicated, however, the individual becomes again fully susceptible.

An antibody-like substance called *reagin* is also produced in patients infected with syphilis. Although it is not specific to syphilis, detection of reagin is the basis of the nontreponemal tests.

Laboratory Diagnosis

DARKFIELD MICROSCOPIC EXAMINATION

Darkfield microscopy is a key aid to the detection of *T. pallidum* in primary and secondary lesions. The treponemes appear white against a dark background. In primary syphilis, demonstration of treponemes in material from the chancre is a very effective diagnostic test; with the serologic tests providing about 76% and 86% sensitivity for the nontreponemal and treponemal tests, respectively. Motile organisms rotate around their longitudinal axis, and they also bend, snap, and flex along their length.

Nevertheless, the darkfield microscopic examination requires considerable skill and experience. The test is very prone to misinterpretation, especially in the examination of specimens from oral lesions, where nonpathogenic spirochetes are numerous. In addition, the lesion material must be examined at the time of collection. If darkfield microscopy is not available, collecting a sample and sending it to a referral laboratory is useless. The patient has to be sent to the referral laboratory where the darkfield microscope is located to obtain reliable results.

SEROLOGIC TESTS FOR SYPHILIS

Most cases of syphilis are diagnosed serologically. There are two major types of serologic tests, those that detect reagin and are termed *nontreponemal tests*, and those that detect specific treponemal antibody and are termed *treponemal tests*. Both lack high sensitivity in the primary stage. Both, however, have sensitivities of nearly 100% in secondary syphilis, and the treponemal tests retain a very high sensitivity into the tertiary stage.

Nontreponemal Tests. The nontreponemal tests detect reaginic antibodies and are useful as screening tests, in monitoring therapy (titers drop following successful therapy), and in detecting reinfection. They are biologically nonspecific and are known to react in a variety of diseases and conditions other than syphilis, thus giving false-positive reactions. The antigen employed is a cardiolipin-lecithin complex made from beef hearts.

Two nontreponemal tests are widely used today. They are the VDRL (Venereal Disease Research Laboratory) and RPR (Rapid Plasma Reagin) tests. These tests provide similar clinical information and have similar advantages, i.e., they are inexpensive to perform, they demonstrate rising and falling reagin antibody titers, and they correlate somewhat with the clinical status of the patient.

The VDRL test uses a cardiolipin antigen, which is mixed with the patient serum. Flocculation occurs in a positive reaction and is observed microscopically. The RPR test uses a suspension of carbon particles to which the cardiolipin antigen is affixed, and coagglutination of the carbon particles, when mixed with the patient serum, is observed visually without the need of a microscope.

Treponemal Tests. The treponemal tests detect antibodies specific for treponemal antigens. They are helpful in the detection of late infections and in the confirmation of positive nontreponemal test results. Titers detected with the treponemal tests remain high and usually do not drop in response to therapy as the nontreponemal test results do. Thus, they are not useful in following therapy or in detecting re-infection. The antigens used are spirochetes derived from rabbit testicular lesions. The two most commonly used treponemal tests are the fluorescent treponemal antibody absorption (FTA-ABS) test and the microhemagglutination test for *T. pallidum* antibody (MHA-TP).

In the FTA-ABS test, the patient's serum is absorbed with extracts of a cultivated treponema that is not *T. pallidum* to remove any nonspecific treponemal antibody. Then, the absorbed serum is placed on a slide which has *T. pallidum* organisms affixed to it. After any specific antibody is allowed to react with the organisms, nonbound constituents in the serum are removed by washing. The presence of anti–*T. pallidum* antibody is then detected by application of a fluourescein-labeled anti–human globulin serum and examination of the slide with an ultraviolet (UV) microscope. Positive results are indicated by bright fluorescence of the *T. pallidum* organisms.

The MHA-TP test is an indirect hemagglutination test using *T. pallidum* antigens absorbed to tanned erythrocytes. It is simpler to perform than the FTA-ABS test, does not require an expensive UV microscope, and is only slightly less sensitive.

Treatment and Prevention

Penicillin is the drug of choice for treating patients with syphilis. It is the only proven therapy that has been widely used for patients with neu-

rosyphilis, congenital syphilis, and syphilis during pregnancy. *Treponema pallidum* is exquisitely sensitive to penicillin, with a minimal inhibitory concentration of approximately 0.004 U penicillin per mL. Resistant strains have not occurred. Owing to the long generation time of *T. pallidum* and its ability to locate in interstitial spaces, where peak antibiotic concentrations may be difficult to achieve, successful therapy requires that maintenance levels of penicillin (0.03 u/mL) be maintained for 7 to 10 days. Long-acting penicillin such as benzathine penicillin is preferred. Dosages vary according to the stage of disease. Alternative regimens for penicillin-allergic patients who are not pregnant include doxycycline and tetracycline. A typical Jarisch-Herxheimer reaction may occur within hours following treatment and is probably due to the sudden release of endotoxin from the spirochetes.

For a discussion on interpretation of serologic tests for syphilis and the recommendations for appropriate therapy, please refer to the *Sexually Transmitted Diseases Treatment Guidelines* published by the Centers for Disease Control (see references).

Educating people about sexually transmitted diseases, identifying and treating all sexual contacts of persons infected with syphilis, and reporting each case to the health authorities for contact investigation help control the spread of syphilis. In addition, pregnant females should have serologic examinations early and late in the pregnancy. Serologic screening of high-risk populations should also be performed.

Other forms of prevention include the proper use of barrier contraceptives, such as condoms.

Other Treponemal Diseases

Three nonvenereal treponemal diseases occur in different geographic locations. These treponematoses are found in developing countries where hygiene is poor, little clothing is worn, and direct skin contact is common owing to overcrowding. All three diseases have primary and secondary stages, and tertiary manifestations may develop. These infections are rarely transmitted by sexual contact, and congenital infections do not occur. The three nonvenereal treponemal diseases are:

■ Yaws

■ Pinta

■ Endemic syphilis or bejel.

Yaws

Yaws is a spirochetal disease caused by *T. pallidum* subsp *pertenue*. It is endemic in the humid, tropical belt: the tropical regions of Africa, parts of South America, India, and Indonesia, and many of the Pacific Islands. It is not seen in the United States. The course of yaws resembles that of syphilis, but tertiary stages are rare. It is readily treated with penicillin.

Pinta

Pinta, caused by *T. carateum,* is found in the tropical regions of Central and South America. It is acquired by person-to-person contact and is rarely transmitted by sexual intercourse. Tertiary disease is uncommon. Treatment with penicillin is highly efficacious.

Endemic Syphilis (Bejel)

Bejel is caused by *T. pallidum* subsp *endemicum* and closely resembles yaws in clinical manifestations and epidemiology. It is found in the Middle East and in the arid, hot areas of the world. The primary and secondary lesions occur in the oral cavity and often go unnoticed. Tertiary lesions are rare. Poor hygienic conditions are important in perpetuating these infections. Bejel is transmitted by direct contact.

Bibliography

Balows A, Hausler WJ, et al, (eds): *Manual of Clinical Microbiology*, 5th ed. Washington, DC: American Society for Microbiology, 1991, pp 554–571.

Centers for Disease Control and Prevention; *1993 Sexually Transmitted Diseases Treatment Guidelines.* MMWR 42:01, 1993.

Centers for Disease Control and Prevention; *Summary of notifiable diseases, United States, 1992.* MMWR 41:55, 1992.

Fister RD, Weymouth LA, et al: Comparative evaluation of three products for the detection of *Borrelia burgdorferi* antibody in human serum. *J Clin Microbiol* 27:2843, 1989.

Holmes KK, Mårdh P-A (eds.): Sexually Transmitted Disease, 2nd ed. New York: McGraw-Hill, pp 205–262.

Jaffe HW: The laboratory diagnosis of syphilis, new concepts. *Ann Intern Med* 83:846, 1975.

Joklik WK, Willett HP: et al (eds.): *The spirochetes. In Zinsser Microbiology*, 19th ed. Norwalk, CT: Appleton & Lange, 1988, pp 555–571.

Krieg NR, Holt JG (eds.): *The spirochetes. In Bergey's Manual of Systematic Bacteriology*, Vol 1. Baltimore; Williams & Wilkins, 1984, pp 38–70.

Larsen SA, Hunter EF, et al (eds): A Manual of Tests for Syphilis. APHA, 1990.

Lukehart SA, Hook EW, et al: Invasion of the central nervous system by *Treponema pallidum*: Implications for diagnosis and treatment. *Ann Intern Med* 109:855, 1988.

Musher DN: How much penicillin cures syphilis? [editorial] *Ann Intern Med* 109:849, 1988.

Sherris JC, Ryan KJ. et al (eds.): *Spirochetes. In Medical Microbiology, An Introduction to Infectious Diseases.* New York: Elsevier, 1984, pp 280–290.

Steere AC, Bartenhagen NH, et al: The early clinical manifestations of Lyme disease. *Ann Intern Med* 99:76, 1983.

Steere AC, Hutchinson GJ, et al: Treatment of the early manifestions of Lyme disease. *Ann Intern Med* 99:22, 1983.

Syphilis, a synopsis. Public Health Service Publication No. 1660. Washington, DC: U. S. Government Printing Office, 1968.

CHAPTER 21

Chlamydia, Mycoplasma, and Ureaplasma

A. CHLAMYDIA

John G. Thomas • *Karen S. Long*

OBJECTIVES

1. List the members of the family Chlamydiaceae.

2. Discuss the unique growth cycle of *Chlamydia,* describing elementary and reticulate bodies.

3. Describe the characteristics of *Chlamydia* and distinguish the genus from rickettsiae, viruses, and bacteria.

4. List the major diseases associated with each of the three species of *Chlamydia.*

5. Describe the modes of transmission for each species of the genus.

6. Define the appropriate cultures for detection of the genus *Chlamydia.*

7. List, in descending order (from most important to least important), the appropriate assays for antigen detection, including direct fluorescence, EIA, nucleic acid probe, and PCR.

8. Discuss the general types of therapy used in the management of chlamydial infections, emphasizing the obligate intracellular parasitic relationship of Chlamydiaceae.

9. Discuss the importance of serologic investigations as well as the limitations of complement fixation and microimmunofluorescence and the corresponding rise and fall of these markers in disease.

10. Interpret the serologic results and significance of antibody titers.

11. Discuss the problems with serologic cross-reactivity among the three species.

For some time, the organisms associated with selected ocular diseases were so unique that they were unclassified; in older textbooks they were classified as the TRIC agents, an acronym for trachoma and inclusions conjunctivitis. These unclassified organisms shared selected characteristics with rickettsiae, bacteria, viruses, and mycoplasma (Table 21–1), although initially they were thought to be more closely associated with viral pathogens than with bacteria. Today, members of the genus *Chlamydia* are recognized as being unique and, therefore, are classified into their own order (Chlamydiales) and family (Chlamydiaceae). This discussion describes the unique growth cycle of *Chlamydia* and the human diseases caused by *Chlamydia.*

There are three species in the genus *Chlamydia: C. trachomatis, C. psittaci,* and the newest member, *C. pneumoniae. Chlamydia pneumoniae* was previously referred to as *C. trachomatis,* strain TWAR. As shown in Table 21–2, initial differentiation of the *Chlamydia* species was based on selected characteristics of the growth cycle, susceptibility to sulfa drugs, and DNA relatedness.

■■■■■■T A B L E   2 1 – 1
COMPARATIVE PROPERTIES OF MICROORGANISMS

	Organisms				
Characteristic	*Chlamydia*	*Viruses*	*Bacteria*	*Rickettsiae*	*Mycoplasma*
DNA and RNA	+	−	+	+	+
Obligate intracellular parasites	+	+	−	−	−
Peptidoglycan in cell wall	+	−	+	−	−
Growth on nonliving medium	−	−	+	−	+
Contain ribosomes	+	−	+	+	+
Sensitivity to					
Antibiotics	+	−	+	+	+
Interferon	+	+	−	−	−
Binary fission (replication)	+	−	+	+	+

■ T A B L E   2 1 - 2
INITIAL DIFFERENTIATION OF *CHLAMYDIA* SPECIES (1988)

Properties	C. trachomatis	C. psittaci	C. pneumoniae
Inclusion morphology	Round, vacuolar	Variable shape, dense	Round, dense
Glycogen in inclusions	+	−	−
Elementary body morphology	Round	Round	Pear-shaped
Sulfa drug sensitivity	+	−	−
DNA relatedness (against *C. pneumoniae*)	10%	10%	100%

Additional properties of the *Chlamydia* species have helped further differentiate the three species on the basis of natural host, major diseases, and number of antigenic variants, i.e., serovars (Table 21–3).

GENERAL CHARACTERISTICS

Chlamydiae have a unique growth cycle, because they are deficient in independent energy metabolism and are, therefore, obligate intracellular parasites. Replication involves two distinct forms of the organism: an *elementary body* (EB), which is infectious, and a *reticulate body* (RB), also called "initial body," that is noninfectious. The life cycle begins when the small elementary body infects the host cell by inducing energy-requiring active phagocytosis. In vivo, host cells are primarily the nonciliated, columnar or transitional epithelial cells that line the conjunctiva, respiratory tract, urogenital tract, and rectum. During the next 8 hours, they organize into larger, reticulating initial bodies, which then divert the cells' synthesizing functions to their own metabolic needs and begin to multiply by binary fission (Fig. 21–1). About 24 hours after infection, the dividing organisms begin reorganizing into infective elementary bodies. At about 30 hours, multiplication ceases,

and by 35 to 40 hours, the disrupted host cell dies, releasing new elementary bodies that can infect other host cells, and the cycle continues (Fig. 21–2). The life cycle of chlamydiae is diagrammatically shown in Figure 21–3.

The elementary body has an outer membrane that is similar to that of many gram-negative organisms. The most prominent component of this membrane is the major outer membrane protein, often called MOMP. The MOMP is a transmembrane protein that contains both species-specific and subspecies-specific epitopes that can be defined by monoclonal antibodies.

The chlamydial outer membrane also contains a large lipopolysaccharide (LPS). This extractable LPS (with ketodeoxyoctonate [KDO]) is shared by all members of the genus and is the primary antigen detectable in genus-specific tests and serologic assays for *Chlamydia*.

Further antigenic characterization of similar LPSs have, additionally, separated *C. trachomatis* into 15 serovars and *C. psittaci* into 10 serovars. Most recently, the 15 *C. trachomatis* serovars have been reorganized into 10 immunotypes and grouped into 4 antigenic pools called C complex, D complex, G and F complex, and K complex.

Chlamydia trachomatis is unique in that it carries 10 stable cryptic plasmids whose function is currently unknown. This unique characteristic is a major reason for the applications of nucleic

■ T A B L E   2 1 - 3
PROPERTIES OF THE THREE *CHLAMYDIA* SPECIES

Property	C. trachomatis	C. psittaci	C. pneumoniae
Natural hosts	Humans	Birds Lower animals	Humans
Major human diseases	Sexually transmitted diseases, trachoma Lymphogranuloma venereum	Pneumonia FUO	Pneumonia Pharyngitis Bronchitis
Number of serovars	15	10	1

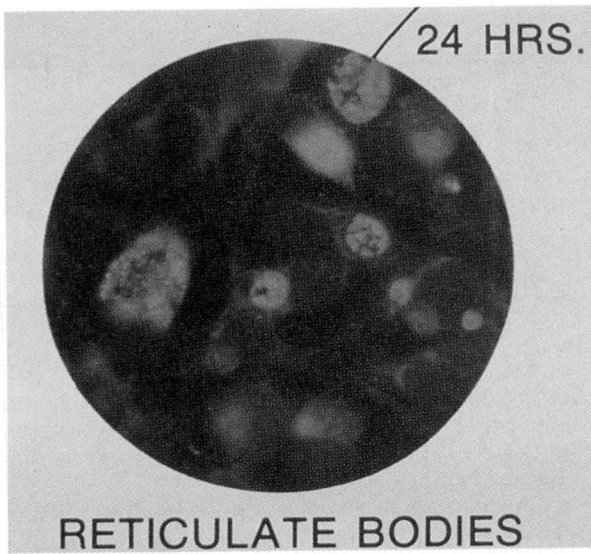

RETICULATE BODIES

■■■■■ F i g u r e   2 1 – 1

Chlamydia growth cycle highlighting reticulate bodies (RBs), sometimes referred to as initial bodies. (Courtesy of Syva-Microtrak, Palo Alto, CA.)

acid amplification by polymerase chain reaction (PCR) and identification by hybridization, discussed later.

CHLAMYDIA PNEUMONIAE

Chlamydia pneumoniae is the newly recognized third species of *Chlamydia*. It was formerly known as *Chlamydia* species, strain TWAR. As shown in Table 21–4, TWAR was originally identified in 1965 from a conjunctival culture of a child (TW) enrolled in a Taiwan trichoma vaccine study. In 1983 at the University of Washington, a similar organism was isolated in HeLa cells from a pharyngeal specimen of a college student (AR). Today, *C. pneumoniae* is recognized as an important respiratory pathogen. It is known to be the cause of acute respiratory disease, pneumonia, and pharyngitis (Table 21–5). It has also been isolated from patients with otitis media, effusion, pneumonia with pleural effusion, and aseptic pharyngitis. Most recently, it has been identified as an important factor in asthma. Infection with *C. pneumoniae* has been established as a risk factor for coronary heart disease and Guillain-Barré syndrome. There also appears to be a relationship between sarcoidosis and *C. pneumoniae*,

but considerable work needs to be done to establish the existence and degree of this relationship. A single *C. pneumoniae* serovar has been found to date.

Clinical Infections

Infection with *C. pneumoniae* is thought to be fairly common, although probably 90% of the infections are asymptomatic or mildly symptomatic. In adults, antibodies have been demonstrated in more than 50% of infections, but in sharp counterdistinction, there is virtually no antibody detectable in children before the age of 5 years. It is thought that the attack rate is highest between the ages of 6 and 20 years, with a particular emphasis in college-age students. Unlike the case with a number of viral respiratory diseases, there seems to be no seasonal incidence, although some Scandinavian data have indicated the possibility of epidemics every 4 to 6 years. Reinfection with *C. pneumoniae* appears to be very common and can be either milder or more severe than the initial infection. The epidemiologic features of *C. pneumoniae* infection is as follows:

■ Common infection; antibodies present in 50% of adults

■ Virtually no antibody present before 5 years of age

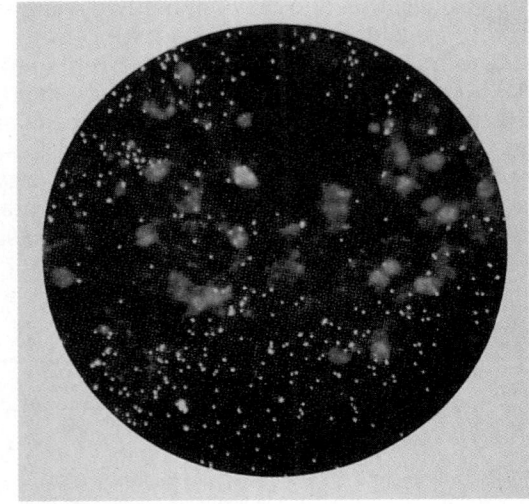

■■■■■ F i g u r e   2 1 – 2

Elementary bodies (EBs) and cells in *Chlamydia trachomatis*–positive direct specimen. (Courtesy of Syva Microtrak, Palo Alto, CA.)

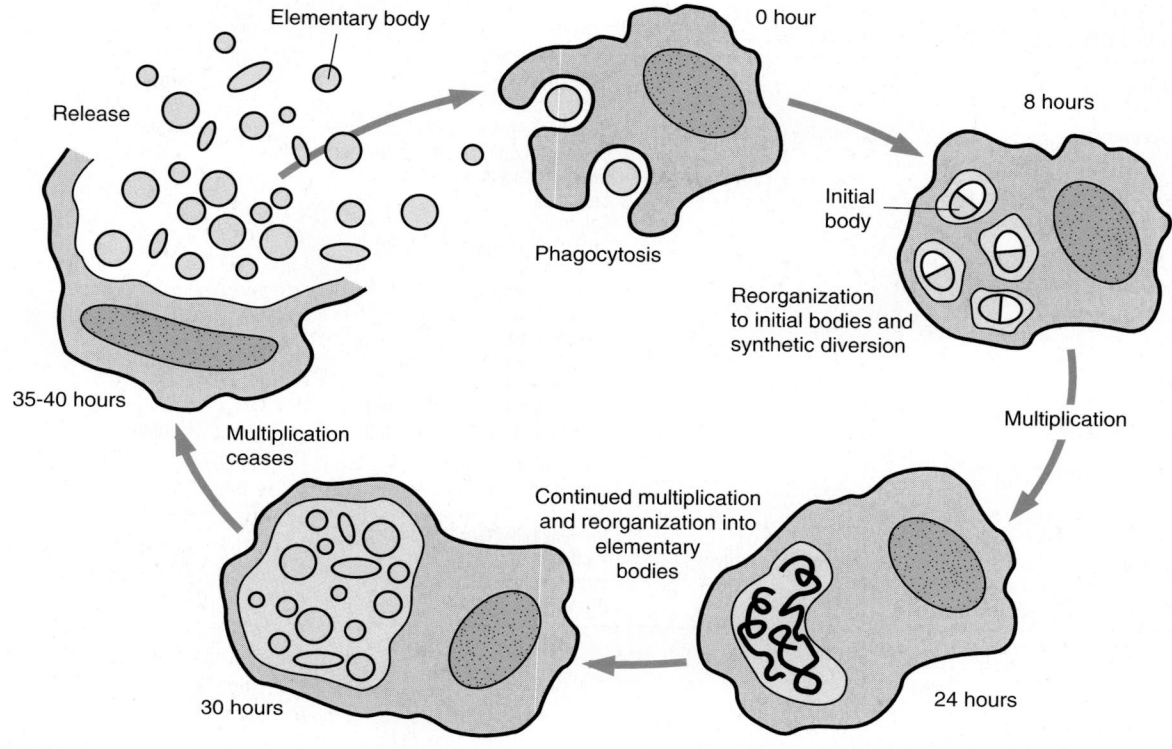

Figure 21-3
Life cycle of *Chlamydia*.

- Attack rate highest between 6 and 20 years of age

- No seasonal incidence; epidemics have been reported every 4 to 6 years

- Frequent reinfection

The clinical picture is varied, but a key characteristic in college-age students is a biphasic clinical course. This infection produces a prolonged sore throat (5 to 7 days) and hoarseness, followed by flu-like lower respiratory symptoms (8 to 15 days). Because of its striking clinical similarity to bacterial pharyngitis, the result of a streptococcal antigen test ordered for clinical confirmation is often thought to be falsely negative. The second phase of the biphasic illness often results in pneumonia and bronchitis but is rarely accompanied by a sinusitis. Fever is relatively uncommon, and radiographs show isolated pneumonitis. Approximately one of nine infections results in this pneumonia. *Chlamydia pneumoniae* is now recognized as the third most common cause of infectious respiratory disease.

It accounts for approximately 6 to 10% of outpatient and hospitalized cases of pneumonia. The clinical features may be summarized as follows:

- Estimated to account for 6 to 10% of outpatient and hospitalized cases of pneumonia (third most common cause)

- 90% of infections are asymptomatic or mildly symptomatic

TABLE 21-4
HISTORY OF *C. PNEUMONIAE* (STRAIN TWAR)

Strain	Year	Comments
TW-183	1965	Isolated from a conjunctiva culture of a child in trachoma vaccine study in Taiwan
10L-207	1967	British isolated *Chlamydia* strain from conjunctiva of an Iranian child with trachoma Similar to TW-183
	1978	Retrospective serologic studies of pneumonia epidemic in teenagers and young adults in Finland supports *Chlamydia* agent as etiology
AR-39	1983	University of Washington, first pharyngeal isolate isolated in HeLa cells from college student

■■■■T A B L E   2 1 – 5
HUMAN DISEASES CAUSED BY *CHLAMYDIA*

Species	Serovars*	Disease	Target
C. trachomatis	A, B, Ba, C	Hyperendemic blinding trachoma	Human
	D, E, F, G, H, I, J, K	Inclusion conjunctivitis (adult and newborn)	Human
		Nongonococcal urethritis	
		Cervicitis	
		Salpingitis	
		Pelvic inflammatory disease	
		Endometritis	
		Acute urethral syndrome	
		Proctitis	
		Epididymitis	
		Pneumonia of newborns	
		Perihepatitis	
		(Fitz-Hugh–Curtis syndrome)	
	L_1, L_2, L_3	Lymphogranuloma venereum	Human
C. pneumoniae (strain TWAR)	1	Pneumonia, bronchitis	
		Pharyngitis	
		Sinusitis	
		Influenza-like febrile illness	
C. psittaci	10 unidentified serotypes (?)	Psittacosis	Bird
		Endocarditis	
		Abortion	

*Predominant serovars associated with disease.

■ Pneumonia and bronchitis, rarely accompanied by sinusitis

■ Biphasic illness; prolonged sore throat and hoarseness, followed by lower respiratory (flu-like) symptoms

■ Fever relatively uncommon

■ Chest radiograph shows isolated pneumonitis

■ One in nine infections results in pneumonia

■ Sarcoidosis relationship ? ? ? ?

The mode of transmission, the incubation period, and the infectiousness of *C. pneumoniae* infections are still largely unknown. It is known, however, that there is no animal reservoir or animal vector. It is not associated with sexual transmission, because there is no association between the prevalence of the antibody and an increase in the number of sexual partners, early age of first intercourse, or a history of sexually transmitted diseases (STDs), particularly gonococcal infection. Table 21–6 organizes known data about *C. pneumoniae* and highlights those situations and/or populations at risk that would benefit from detection of *C. pneumoniae*, usually by serologic methods.

Laboratory Diagnosis

Chlamydia pneumoniae may be cultured on selected cell lines and visualized with fluorescein-conjugated monoclonal antibodies. Established cell lines H 292 and Hep-2 from the human respiratory tract are more sensitive than human lines (HL), but the traditional HeLa 229 cells are more sensitive than McCoy or L cells. Monoclonal antibodies specific for the *C. pneumoniae* strain are used to identify inclusions in cell culture. It should be noted that a genus-reactive monoclonal antibody will identify *C. pneumoniae* inclusions but cannot differentiate this organism from the other chlamydial species.

Attempts to culture *C. pneumoniae,* if undertaken, should take into account the organism's lability. *Chlamydia pneumoniae* seems to be considerably more labile than *C. trachomatis,* although its viability is relatively stable at 4°C. Specimens used for detection of chlamydial infections are listed in Table 21–7.

An indirect fluorescent antibody method has been reported for detecting *C. pneumoniae* in respiratory secretions; the antibody reacts with the MOMP (Fig. 21–4). This same antibody can be used to identify infected culture monolayers.

■ TABLE 21-6
WHO TO EVALUATE FOR *C. PNEUMONIAE*

Population/Situation	Evaluation Methods*	Comments
Surveillance to establish baseline prevalence, both qualitative and quantitative		
Pneumonias requiring hospitalization (age 6–20)	*C. pneumoniae*–specific IgM and IgG: acute and convalescent, use micro-IF	12% antibody prevalence
Pharyngitis in college students	IgM, single visit	9% antibody prevalence
Retrospective, undiagnosed outbreaks in young adults, college or military	CF or micro-IF, IgG-specific	
Selected patients		
Serious pneumonia, undiagnosed; clinically presents like *Mycoplasma pneumoniae*	*C. pneumoniae*–specific IgM and IgG by micro-IF	Rather than repeat cultures for similar respiratory pathogens, i.e., *M. pneumoniae,* establish etiology and impact on diagnosis-related group reimbursement

*Micro-IF, microimmunofluorescence.

Given the difficulty of, and lack of standardization for, isolation of *C. pneumoniae,* serologic tests have been the method of choice for detection. A complement fixation (CF) test that uses a genus-specific antigen has been the traditional assay most often employed for *C. pneumoniae* detection. In initial investigations to determine prevalence and reinfection rate, a four-fold rise in CF titer suggested recent chlamydial infection but did not distinguish among the three known species. The present method of choice is microimmunofluoroescence (micro-IF), which is more sensitive and specific than CF. Further, it does not cross-react with *C. trachomatis* and *C. psittaci.* Micro-IF can also distinguish an IgM from an IgG

response. Single titer evaluations, although not diagnostic, may be suggestive. An IgM titer of greater than 1:32 or an IgG single titer greater than 1:512 may suggest *C. pneumoniae* as a recent etiologic agent, warranting further evaluation. An IgG titer greater than or equal to 1:16 but less than 1:512 is evidence of past infection or exposure.

Two antibody response patterns have been identified for *C. pneumoniae* infections. In the primary response, most often seen in adolescents, university students, and military trainees, CF antibodies usually appear first. By micro-IF, *C. pneumoniae*–specific IgM does not appear until 3 weeks after onset of symptoms, and often *C. pneumoniae*–specific IgG does not reach

■ TABLE 21-7
APPROPRIATE SPECIMENS FOR DETECTION OF CHLAMYDIAL INFECTIONS

Clinical Manifestation/Site of Infection	Specimen Site/Type	Comments
Inclusion conjunctivitis and trachoma	Conjunctival swab, scraping with spatula or tears	Specimen collection in neonates is difficult
Urethritis	Urethral swab	In males, >4 cm and do not use discharge
Epididymitis	Epididymis aspirate	
Cervicitis	Endocervical swab	Remove exudate first
Salpingitis	Follopian tube (lumen) or biopsy	
Lymphogranuloma venereum	Bubo or cervical lymph node aspirate	
Infant pneumonia	Throat swab, nasopharyngeal aspirate, or lung tissue	
STD, male sex partner	Urine	Noninvasive diagnostic procedure EIA antigen detection is 80% accurate; PCR, 98%
Psittacosis	Sputum, lung tissue	
Chlamydia pneumoniae pneumoniae or pharyngitis	Sputum, throat or lung tissue	Tissue culture isolation and direct immunofluorescence are relatively new and need further evaluation
STD, result clarification	Rectal, vaginal swabs	May be used for supplemental information and in clarifying previous isolates or diagnostic dilemmas

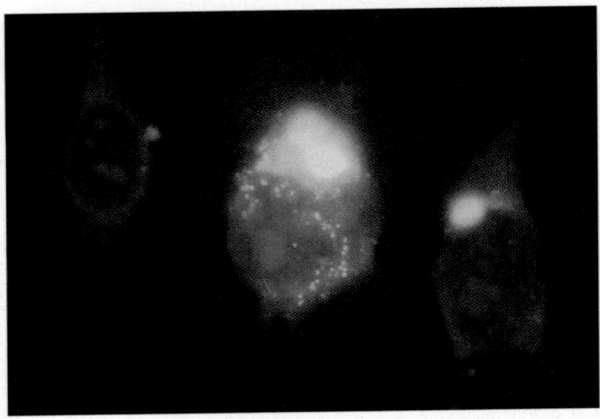

Chlamydia pneumoniae detection from direct sputum smear using fluorescent labeled monoclonal antibody, highlighting cytoplasmic inclusions (magnification × 700). (Courtesy of DAKO Reagents, Carpinteria, CA.)

tive value for these serologic assays as well as isolation, detection, and identification.

CHLAMYDIA TRACHOMATIS

Chlamydia trachomatis is the most common sexually transmitted bacterial pathogen in the United States. There are between 4 and 10 million new cases each year. Only genital warts caused by human papillomavirus is a more common STD in the United States. *Neisseria gonorrhoeae* is a distant third, with approximately 3 to 4 million new cases per year.

Clinical Infections

Trachoma

diagnostic levels for 6 to 8 weeks. Therefore, the traditional convalescent serum obtained approximately 14 to 21 days after onset does not contain micro-IF–detectable *C. pneumoniae* antibody. In contrast, during reinfection, a CF antibody change is not detected, but by micro-IF, an IgG titer of 1:512 or more can appear within 2 weeks. IgM levels may be detectable but are low.

Table 21–8 summarizes these serologic findings. Table 21–9 addresses the situations in which these tests may be most applicable, identifying the population group(s) at greatest risk. This information provides the greatest predic-

The human diseases caused by *C. trachomatis* and the corresponding associated serotypes are listed in Table 21–5. Initially, *C. trachomatis* was associated with hyperendemic blindness and trachoma. This worldwide epidemic was associated with serotypes A, B, Ba, and C. These organisms are shed in feces and spread from person to person, people to fomites, and flies to people. These serotypes are most frequently found near the equator and are seen in climates with high temperature and high humidity but not in the United States, which has a temperate climate. In India and Egypt, the usual history of the infection ends in blindness

■ T A B L E 2 1 – 8

DETECTION OF *CHLAMYDIA* SPECIES BY VARIOUS SEROLOGIC METHODS

Species of Chlamydia	Serologic Findings*			
	CF	IgM	Micro-IF†	IgG
C. trachomatis serovars (15 serovars)				
A–C	+	+		+
D–K	(>1:16)	Newborn		
L1–L3 (LGV)	>1:64			
C. pneumoniae (strain TWAR, only known strain)	4-fold rise			4-fold rise
	+	>1:32		or >1:512
				(single specimen)
C. psittaci (10 serovars)	≥32			
	+	4-fold rise		>1:512
	4-fold rise			single specimen
	(A/C)	(A/C)		

*CF, complement fixation (using LPS), where >1:16 is indicative of previous exposure; A/C, acute/convalescent sera.
†Detects antibodies (IgM/ and total) against 10 major immunotypes using 4 antigenic pools: C complex: C-V-H-L, B complex: B-E-D, G-F complex: K.

■■■■■ T A B L E   2 1 – 9

APPROPRIATE *CHLAMYDIA TRACHOMATIS* ASSAYS FOR SELECTED PATIENT POPULATIONS*

Assay	Prenatal		Newborn		Clinics†		LEGAL APPLICABILITY (rape or child abuse)?	Test of Cure?
	LOW-RISK	HIGH-RISK	EYE	THROAT	LOW-RISK	HIGH-RISK		
Culture	A/B	A	B	A	A/B	B	Yes	Yes
Non-culture methods								
DFA	B	A	A	A	B	A	No	No
EIA	A/B	A	A	A	A/B	A	No	No
Probe	B	A	IUO	IUO	B	A	No	No
PCR	A	A	IUO	IUO	A	A	IUO	IUO
Serology								
CF	B, LGV	B, LGV	NA	NA	B, LGV	B, LGV	No	No
EIA	B	B	A	A	B	B	NA	NA
Micro-IF	NA	B	A	A (IgM)	B	B	No	No

*A, most useful, stands alone; B, probable, but needs verification or complementary assay employing conjugate recognizing different *C. trachomatis* macromolecules, i.e., LPS (EIA) vs MOMP (DFA) or competition assay for DNA probes; C, least useful; IUO, investigational use only; LGV, lymphogranuloma venereum; NA, not available.

†A *low-risk* population is defined as one with a <5% incidence, such as in an obstetrics-gynecology or family practice patient group (birth control, annual gynecologic examination, etc.). A *high-risk* population is defined as one with a >10% incidence, such as those in STD clinics, university/college student health centers, and emergency department patients.

in adults. Prevention includes either or both antibiotic treatment and a simple surgical procedure on the eyelid to stop the continual abrasion of the cornea. Figure 21–5 shows a patient with trachoma.

Lymphogranuloma Venereum

Chlamydia trachomatis serotypes L_1, L_2, and L_3 cause lymphogranuloma venereum (LGV), a sexually transmitted disease that is often seen in immigrants from and travelers to endemic parts of the world. In LGV, the patients have inguinal and anorectal symptoms (Fig. 21–6).

In the United States, *C. trachomatis* serovars D to K are associated with chlamydial STD infections. These isolates are associated with temperate climates (seroprevalence studies, however, have shown how commonly the infection may manifest in different clinical forms.) The *C. trachomatis* serovars A to C apparently are not primary STDs but cause infections of the eye, pharynx, and, less frequently, the male and female genital tracts in temperate climates.

Chlamydia in the Newborn

Infants can be infected with *Chlamydia* as they travel through an infected birth canal. Chlamydial infection in an infant delivered by cesarean section is a rarity, and infection from seronegative mothers has not been reported. Infants suffering from chlamydial infection experience conjunctivitis, nasopharyngeal infections, and pneumonia (Fig. 21–7). The portal of entry is ocular, with colonization of the oropharynx a necessary event prior to infection. Twenty to 25% of babies born to *Chlamydia*-culture–positive mothers develop conjunctivitis, 15 to 20% develop nasopharyngeal infection, and 3 to 18% develop pneumonia. Otitis media is a less frequent infection. Babies may also be colonized in the vagina and the rectum. Clinically, it is believed that pneumonia in babies younger than 6 months of age is *C. trachomatis*

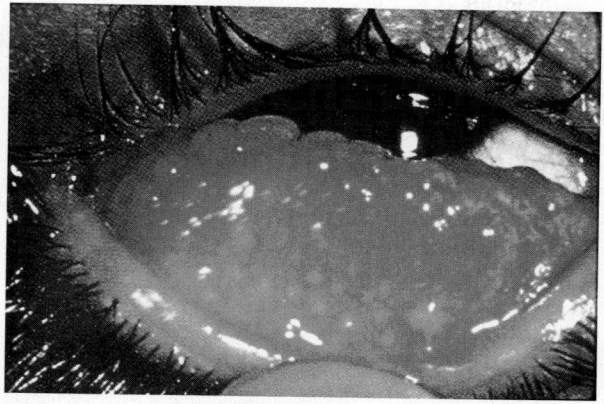

■■■■ F i g u r e   2 1 – 5

Conjunctival scarring and hyperendemic blindness caused by *Chlamydia trachomatis* in ocular infections.

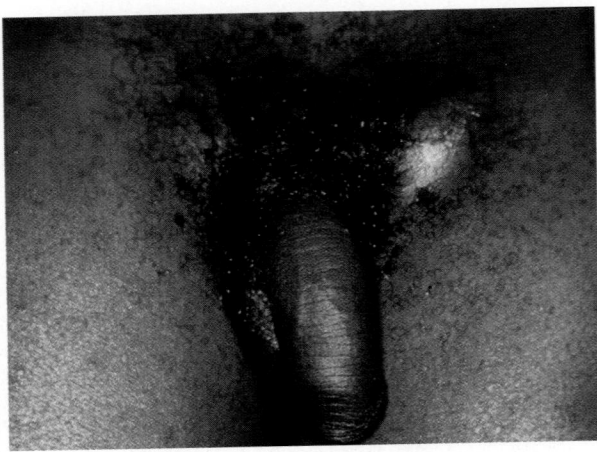

■ F i g u r e 2 1 – 6

Inguinal swelling and lymphatic drainage due to *C. trachomatis* serovars L₁, L₂, or L₃, that is, LGV (lymphogranuloma venereum).

associated, unless proven otherwise. This pneumonia can also occur as a mixed infection, often concomitantly with gonococcus, cytomegalovis, *Pneumocystis,* and other viruses. Table 21–10 shows selected features associated with neonatal inclusion conjunctivitis.

Sexually Transmitted Diseases

Infections in the adult male include nongonococcal urethritis (NGU), epididymitis, and prostatitis. Between 45 and 68% of female partners of men with *Chlamydia*-positive NGU yield chlamydial isolates from the cervix. Approximately 50% of current male partners of women with a cervical chlamydial infection are also infected.

Infections in the adult female include urethritis, follicular cervicitis (leukorrhea, hypertrophic cervical erosion), endometritis, proctitis, salpingitis, pelvic inflammatory disease (PID), and perihepatitis, also called the Fitz-Hugh–Curtis syndrome. Infections can be persistent and subclinical as well as acute and demonstrable. Salpingitis can lead to scarring and dysfunction of the ova ductile transport system, resulting in infertility or ectopic pregnancy. In the United States, this is a major cause of sterility.

Laboratory Diagnosis

There are numerous methods for laboratory diagnosis of *C. trachomatis*. These methods vary in their sensitivity, specificity, and positive predictive value. Methods to detect *C. trachomatis* infections may be classified as follows:

- Antigen detection
- Antibody detection
- Cell culture
- Immunofluorescence (IF)
- Enzyme immunoassay (EIA)
- Nucleic acid probes
- Nucleic acid amplification (PCR)

The most appropriate tests or combination of assays utilized depend on the following factors:

- Knowledge of the population at risk
- Capability and facilities available for testing
- Cost of materials
- Ability to batch specimen types
- Experience of technologist

Prevalence in the population to be tested is a very important criterion in determining which method or combination of methods should be used. For any assay, the positive predictive value increases (assuming optimum technical condi-

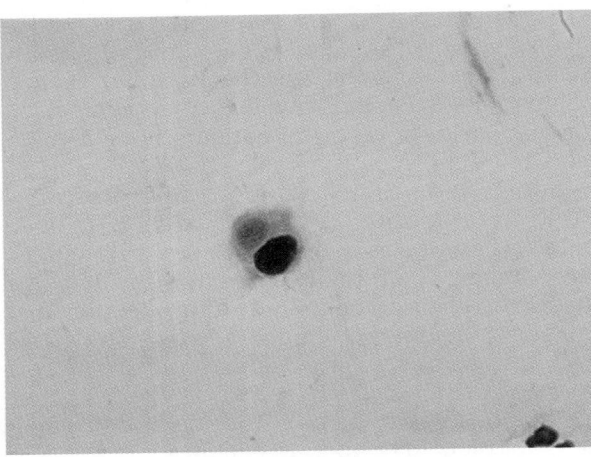

■ F i g u r e 2 1 – 7

Giemsa stain showing inclusion body from ocular swab of 7-day-old neonate who was discharged, but then was readmitted with fever, weight loss, lack of eating, and "fussiness." At 3 days after delivery, *Neisseria gonorrhoeae* was isolated from ocular discharge, although the patient had been given silver nitrate. Eye cultures confirmed the presence of *Chlamydia trachomatis* and diagnosis of neonatorum ophthalmia.

T A B L E   2 1 – 1 0

INCLUSION CONJUNCTIVITIS IN THE NEONATE CAUSED BY *CHLAMYDIA TRACHOMATIS*

Incubation period	4–5 days
Signs	Edematous eyelids
Discharge	Copious, yellow
Course	Untreated, weeks to months
Complications	Corneal panus formation, conjunctival scarring
Giemsa stain of conjunctival scraping	Juxtanuclear inclusion

tions) when the prevalence of the disease in the population is high.

The type of specimen selected for laboratory processing depends on the symptoms of the patient and the clinical presentation. Whatever the source, however, the specimen should consist of infected epithelial cells and not exudate. Further, it is important to remember that plastic or metal swabs are superior to wooden swabs, which are toxic to cell culture, if that method is employed for isolation. Table 21–7 lists the optimum specimens for detection of *Chlamydia* in patients with a variety of clinical manifestations.

Direct Microscopic Examination

Direct specimen examination by cytologic methods primarily emphasize trachoma and inclusion conjunctivitis. Various investigators have estimated this method as nearly 95% sensitive, but it is technically demanding and influ-

enced by the quality of the specimen from the neonate. Although this method is difficult to use with large batches of endocervical specimens, it does offer rapidity in selected cases, particularly in detecting ocular infection in newborns. When direct fluorescent antibody (DFA) testing is used for endocervical or urethral specimens, the characteristic "green apple" elementary body is suggestive but needs verification by alternative methods employing a different epitope. Direct specimen examination offers one additional important advantage: It allows for immediate quality control of the specimen, revealing whether the cells present in the field are columnar epithelial. Figure 21–8 shows inclusion bodies demonstrated by direct examination of cytologic stains of endocervical smears.

Cell Culture

Chlamydial cell culture (until the development of PCR) was considered the gold standard for detecting infection with *C. trachomatis,* but its utility has been limited because of the inherent technical complexity, time and specimen handling requirements, expense, and labile nature of the organism. Even under the most stringent and optimal conditions, isolation of chlamydiae is only approximately 80% sensitive. Cell lines used for the detection of chlamydiae include McCoy, HeLa 229, BHK-21, and Buffalo Green Monkey kidney. The cell lines are grown on coverslips in 1-dram shell vials or on the surface of multiwell cell culture dishes containing cyclohexamine.

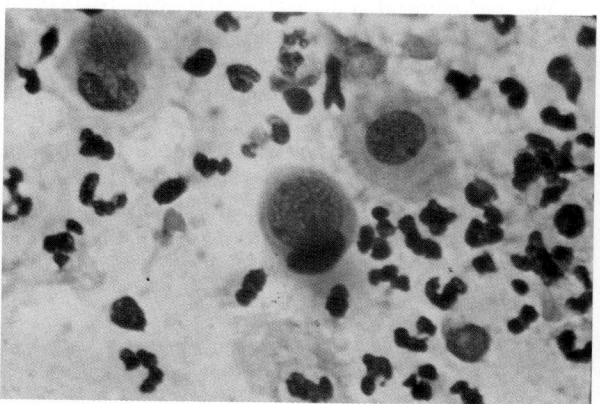
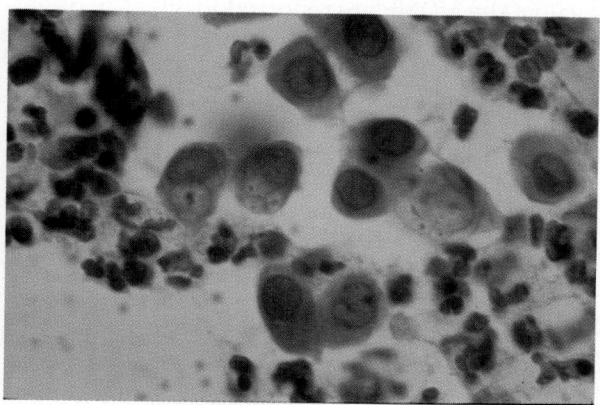

A

B

■ F i g u r e   2 1 – 8

A and *B*, Cytologic examination of endocervical specimens demonstrating inclusion bodies consistent with *Chlamydia trachomatis.* Papanicolaou stain; *B* × 600.

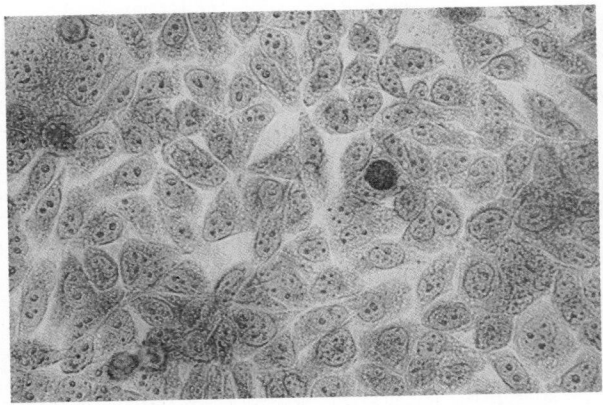
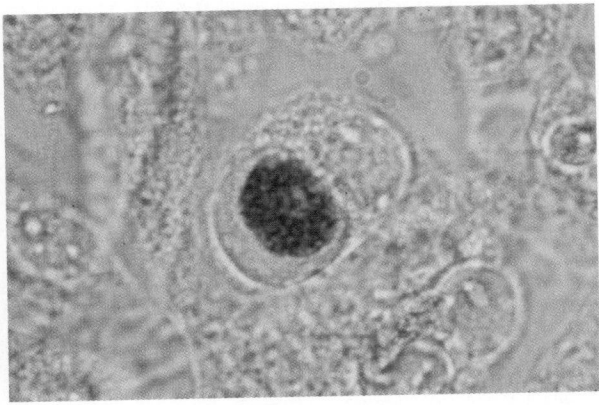

Figure 2 1 – 9

A and *B*, Iodine-stained inclusion bodies from *Chlamydia trachomatis*–infected McCoy cells. *A*, X40; *B*, X100. Note the size and half-moon shape of the inclusion.

The "shell vial technique" has also been utilized and found to be more sensitive than the micro-well method, because multiple blind passes are not necessary to maximize the isolation rate in a 1-dram vial. The specimen is centrifuged onto the cell monolayer and incubated for 72 hours. Fluorescein conjugate and monoclonal antibodies or an iodine or Giemsa stain can be used to detect the chlamydial inclusions (Fig. 21–9).

The fluorescent monoclonal antibody stain is considered the most sensitive. There are a number of commercially available fluorescent antibodies. Some researchers use species-specific monoclonal antibodies that bind to the MOMP, whereas others use the genus-specific antibody, which binds to a LPS component. Monoclonal antibodies against the MOMP are reported to offer the brightest fluorescence, with consistent elementary body morphology and less nonspecific staining than monoclonal antibodies against the LPS. There have been a great many publications determining the specificity, sensitivity, and positive and negative predictive values of the fluorescent monoclonal antibody method. Published reports show a sensitivity range of 70 to 95% and a specificity range of 92 to 99%.

Enzyme Immunoassay

The most commonly employed rapid antigen assay for the detection of *C. trachomatis* is the EIA. Depending on the manufacturer, the EIA detects either the outer membrane LPS chlamydial antigen or the MOMP. A plethora of commercial kits are available, all having similar advantages. These include the ability to screen large volumes of specimens, objective results, test performance in 3 to 5 hours, and the ability to use various specimen types, including urine for males, which may be very important when evaluating partner contact. A summary of the published sensitivity, specificity, and negative and predictive values as well as test specimens is shown in Table 21–11. Basically, however, none of them equals the sensitivity of culture, and most are significantly less. Discrepancies in sensitivity could be based on differences in sample size, disease prevalence, population characteristics, collection sampling techniques, and laboratory standards.

One additional caution must be observed when EIA is employed for chlamydial antigen detection. A positive result must be considered preliminary and should be verified, because antigen detection methods may give a false-positive result when used in low-prevalence (<5%) populations. Verification of a positive specimen can be made on the same specimen using a monoclonal blocking antibody and a repeated EIA test using DFA or, alternatively, competitive binding for probe assays.

Nucleic Acid Probes/Amplification

The newest advances in *Chlamydia* identification have dealt with detection of nucleic acids. Initially, there was only one probe commercially available, a nonisotopically labeled DNA probe

■■■■■T A B L E   2 1 - 1 1
DETECTION CAPABILITIES OF VARIOUS METHODS FOR *C. TRACHOMATIS*

	SENS* %	SPEC* %	PPV* %	NPV* %	Specimen Site Cervical-Urethral	Rectal	Urine	Eye	False +/−	Reported Cross-Reactivity	Comments
1. Culture	70–80	100	73–98	90–100	+	+	+	+	False −	None	Labor-intensive gold standard for specificity
2. Nonculture DFA	70–95	92–98	73–98	95–99	+	+	−	+	False −/+	Staphylococcus	Screen only; experience in FA needed
EIA	72–95	90–99	45–92	95–99	+	−	LA†	LA†	False +/−	Streptococcus, GC, *Acinetobacter*	Verify with complementary assay
Probe	60–90	92–98	85–99	98–99	+	−	−	−	False +/−	Selected bacteria	Screen and verify by repeat culture
Nucleic acid amplification PCR	93–99	99–100	85–95	100	+	−	+	RUO‡	False −	None reported	No verification necessary
3. Serology					See Table 21–8					Depends on antigen used	Several methods: CF, MIF

SENS = sensitivity; SPEC = specificity; PPV = positive predictive value; NPV = negative predictive value.
*RANGE: Low to high prevalence as described in the text.
†LA = Limited availability.
‡RUO = Research use only.

that detected urogenital *C. trachomatis* rRNA (Gen-Probe, Pace 2, San Diego, CA). Although the sensitivity, specificity, and positive and negative predictive values have been in the same range as or slightly less than those reported for EIA (see Table 21–9), the DNA probe has the advantage of detecting two sexually transmitted diseases (*C. trachomatis* and *N. gonorrhoeae*) from one sample.

Most recently, the PCR has been reported as both an amplification and a detection method for *C. trachomatis*. Initial reports are very encouraging and suggest an increase in detection from urogenital specimens of greater than 15 to 30% over established EIA and tissue culture methods. Greater recovery is also seen in male urine specimens. Sensitivity, specificity, and predictive values all approach 100%. On June 6, 1993, the U.S. Food and Drug Administration (FDA) approved the first commercial PCR assay for *C. trachomatis* detection, marketed by Roche Diagnostic, Roche Diagnostic Systems, Branchburg, NJ, under the tradename Amplicor.

Additional observations must be noted. The concomitant use of a rapid screen for white cells (leukocyte esterase test) in urogenital specimens may heighten suspicion, demonstrate leukor-

rhea, and target specimens that would have the highest probability of detecting a true STD infectious agent. Conversely, there are significant, selected clinical limitations for the use of rapid screening methods employing EIA, DFA, or nucleic acid hybridization. As established by the Centers for Disease Control and Prevention (CDC), these assays cannot be used in rape or child abuse cases, nor should they be used in the measurement of treatment success. Because established appropriate antibiotics have a predictable 95% cure rate, the false-positive rate (due to less than 5% potentially uncured) would obscure follow-up results and suggest treatment failure. Polymerase chain reaction (PCR) may also be most useful in low-risk patient populations because it has essentially no false-positive result.

Antibody Detection

Serologic assays can be used in the detection of *C. trachomatis* infections. Historically, these were thought to be limited and problematic. Many individuals have chlamydial antibodies from previous infections, and because chlamydial infections tend to be localized, they do not cause the traditional fourfold rise in antibody titer between acute and convalescent specimens. Today, the interpretation and significance of serologic assays are being reevaluated, and serologic testing is growing as a complementary diagnostic tool in certain selected scenarios such as the following.

■ After reorganization of the serologic classification of *C. trachomatis* from 15 serovars to 10 immunotypes, four antigenic pools or complexes of *C. trachomatis* are now recognized. These are C complex, D complex, G and F complex, and K complex. With micro-IF, when a specific IgM response to a different pool of *C. trachomatis* immunotypes is observed, new infections can be detected in patients who have had previous infections with other immunotypes.

■ Ascending infections by *C. trachomatis* involving fallopian tubes and additional organs of the upper female genital tract are almost never detected by endocervical cultures. Hence, patients at risk for chronic infections would be missed with the standard screening methods employing a cervical swab. The best serologic screening of patients at risk, particularly as part of a prenatal work-up, may be for the immunotypes now defined for *C. trachomatis* utilizing the four pools listed previously. Prenatal patient sera are often kept for approximately 1 year following initial serologic evaluations—for example, for hepatitis B virus, human immunodeficiency virus, and the TORCH agents. This is a serum reservoir that needs to be recognized and better utilized.

■ Complement fixation detects genus-reactive antibody, including elevated levels of antibody in systemic infections, such as LGV. Diagnosis of LGV is supported by CF titers of 1:64 or more, as demonstrated in Table 21–8. It must be noted, however, that CF is generally not useful in nonsystemic chlamydial conjunctivitis or routine urogenital tract infections.

Micro-IF detects antibodies to chlamydial EBs; they are serotype-specific antibodies. Hence, high levels of chlamydial IgM by micro-IF are diagnostic of systemic *C. trachomatis* infection in infants. In fact, because same-day diagnosis is possible, IgM micro-IF is the method of choice for diagnosis of *C. trachomatis* pneumonia in infants, preferable even to culture. Further, infants with inclusion conjunctivitis normally do not have detectable IgM antibodies unless they have a systemic infection. Chlamydial IgG is generally not useful in infants, because rising titers are seldom observed, and when they are, they probably reflect maternal antibody.

Results Reporting

With such great latitude in current testing choices, it is very important for each laboratory to clearly report and define results. Some key points in the development of an approach to ordering and reporting results of tests for *C. trachomatis* and related organisms in a patient specimen are as follows:

■ Agreeing in advance with the Obstetrics/ Gynecology and Emergency departments on which organisms are associated with which-clinical syndrome, then testing accordingly, using profiles.

■ Reporting which tests were performed and which were not for each patient profile.

■ Reporting unusual observations. Pure isolates of *Pseudomonas, Haemophilus, Neisseria meningitidis,* and yeast are not normal, and the physician needs to be aware of their presence.

CHLAMYDIA PSITTACI

Chlamydia psittaci is the cause of psittacosis, also known as ornithosis and parrot fever. Diagnosis of psittacosis is usually based on a history of exposure to psittacines and a fourfold rise in CF antibody to the chlamydial group LPS antigen. In U.S. adults, several hundred cases of pneumonia each year are due to *C. pneumoniae,* compared with only a few hundred due to *C. psittaci.* Retrospective serologic testing of sera taken from patients with acute respiratory disease have shown that many people previously thought to have *C. psittaci* infections due to "transient bird exposure" were, in fact, actually infected with the *C. pneumoniae* organism. Hence, misdiagnosis of *C. psittaci* is a problem, and one needs to know the tests that are most appropriate for differentiating these chlamydial isolates (see Table 21–8).

Isolation of *C. psittaci* in culture, although diagnostic, is difficult and is not routinely employed nor recommended in the United States. Therefore, almost all diagnoses of *C. psittaci* are based on serologic evaluation. As shown in Table 21–8, a single CF titer of greater than 1:32 is suggestive of acute illness in a symptomatic patient during an outbreak of psittacosis. The rise in CF antibodies is usually not demonstrable until the acute illness is over, however, and is often weak or absent if appropriate antibiotic therapy is given. This is most often a "rule-out" disease. If *C. pneumoniae*– and *C. trachomatis*–specific IgG and IgM are not detected by micro-IF and a fourfold rise in chlamydial antibodies is detected by CF, then *C. psittaci* should be strongly suspected. A good history is paramount in evaluating bird exposure, incubation time, and disease process.

Bibliography

Altaie S, Merer F, et al: Evaluation of two ELISA's for detecting *Chlamydia trachomatis* from endocervical swabs. Diagn Microbiol Infect Dis 15:579, 1992.

Bass CA, Jungkind DL, et al: Clinical evaluation of a new polymerase chain reaction assay for detection of *Chlamydia trachomatis* in endocervical specimens. J Clin Microbiol 31:2648, 1993.

Campbel LA, Kuo C-C, et al: Serological responses to *Chlamydia pneumoniae* infection. J Clin Microbiol 28:1261, 1990.

Centers for Disease Control: False-positive results with use of *Chlamydia* test in the evaluation of suspected sexual abuse. MMWR 39:932, 1991.

Chapin-Robertson K: Use of molecular diagnostics in sexually transmitted disease: Critical assessment. Diagn Microbiol Dis 16:173, 1993.

Fonte CE, Carlisle J, Lazerson J: Vaginal discharge in a two month old infant. Hosp Pract Oct:39, 1987.

Glaser JB, Hammerschlag MR, MacCormick WM: Sexually transmitted diseases in the victims of sexual assault. N Engl J Med 315:625, 1986.

Grayston JT, Krio C, et al: A new *Chlamydia psittaci* strain, TWAR, isolated in acute respiratory tract infections. N Engl J Med 315(3)161, 1986.

Grayston JT, Wang S-P, Kuo, et al: Current knowledge of *Chlamydia pneumoniae*, strain TWAR, an important cause of pneumonia and other acute respiratory diseases. Eur J Clin Microbiol 8:191, 1989.

Huovinen P, Lahtonen R, et al: Pharyngitis in adults: The presence and coexistence of viruses and bacterial organisms. Ann Intern Med 110:612, 1989.

LaScolea LJ Jr: The value of noncultured chlamydial diagnostic tests [editorial]. Clin Micro Newsl 13:21, 1991.

Loeffelholz M, Lewinski C, et al: Detection of *Chlamydia trachomatis* in endocervical specimens by polymerase chain reaction. J Clin Microbiol 30:2847, 1992.

Moncada J, Schacter J, et al: Evaluation of Syva's enzyme immunoassay for the detection of *Chlamydia trachomatis* in urogenital specimens. Diagn Microbiol Infect Dis 15:663, 1992.

Nettleman MD, Jones RB, et al: Cost effectiveness of culturing for *Chlamydia trachomatis:* A study in a clinic for sexually transmitted diseases. Ann Intern Med 105:189, 1986.

Phillips RS, et al: Criteria for selective testing of chlamydia. Am J Med 186:515, 1989.

Schacter J: Why we need a program for the control of *Chlamydia trachomatis* [editorial]. NEJM. 320(12):802, 1989.

Sillus M, White P: Rapid identification of *Chlamydia psittaci* and TWAR *(C. pneumoniae)* in sputum samples using an amplified enzyme immunoassay [letters to the editor]. J Clin Pathol 43:260, 1990.

Stagano S, Raysfield DM, et al: Infant pneumonitis associated with cytomegalovirus, *Chlamydia, Pneumocystis,* and *Ureaplasma:* A prospective study. Pediatrics 68:322, 1981.

B. *MYCOPLASMA* AND *UREAPLASMA*

John G. Thomas • *Karen S. Long*

OBJECTIVES

1. Describe the general characteristics of *Mycoplasma* and how they differ from other bacterial species.

2. Describe the clinical diseases caused by *Mycoplasma pneumoniae*, *Mycoplasma hominis*, and *Ureaplasma urealyticum*.

3. Identify the preferred stains for demonstration of the mycoplasmas.

4. Discuss the possible roles of *Mycoplasma hominis* and *Ureaplasma urealyticum* in infections of low-birth-weight and high-risk neonates.

5. Discuss the potential effect of *Mycoplasma fermentans* [*incognitus* strain] on the immune system and its possible role as a cofactor in HIV-1 disease.

6. List and discuss the various diagnostic methods for *Mycoplasma pneumoniae* infection.

7. Name two selective media for detection of the mycoplasmas.

This discussion highlights a group of organisms once thought to be viruses because of their size. Mycoplasmas are the smallest free-living organisms in nature; *Mycoplasma* and *Ureaplasma* are two of the six genera in the family Mycoplasmataceae (Table 21–12). Of the approximately 70 species of *Mycoplasma* and *Ureaplasma* identified in plants and animals, three species are known to be human pathogens:

- *Mycoplasma pneumoniae,* which causes respiratory disease

- *Mycoplasma hominis,* associated with urogenital tract disease

- *Ureaplasma urealyticum,* associated with urogenital tract disease

In the laboratory, mycoplasmas are common and hard-to-detect contaminants of cell cultures; they may also be emerging or opportunistic pathogens for selected patient populations.

GENERAL CHARACTERISTICS

Mycoplasmas are pleomorphic organisms that do not possess a cell wall (Fig. 21–10), a characteristic that makes them resistant to cell-wall–active antibiotics such as penicillins and cephalosporins. Because of the absence of cell wall, they were originally grouped under the general term *cell wall deficients* (CWDs). Table 21–13 compares features of three genera known to be pathogenic for humans. Generally, mycoplasmas are slowly growing, highly fastidious, facultative

■ T A B L E   2 1 - 1 2
DIVERGENT ECOSYSTEMS INHABITED BY GENERA OF THE FAMILY MYCOPLASMATACEAE

Ecosystem	Mycoplasma	Ureaplasma	Acholeplasma	Spiroplasma	Thermoplasma	Anaeroplasma
Soil and grasses	–	–	–	+	–	–
Crops and plants	–	–	–	+	–	–
Mown hay	–	–	–	+	–	–
Water	–	–	+	–	+	–
Deciduous trees	–	–	–	+	–	–
Humans	+	+	–	–	–	–
Cattle	+	+	–	–	–	+

anaerobes requiring complex media containing cholesterol and fatty acids for growth; important exceptions include aerobic *M. pneumoniae* and the more rapidly growing *M. hominis*. On solid media, mycoplasma form colonies with the center of the colony embedded beneath the surface, giving the classic "fried egg" appearance (Fig. 21–11).

The first Mycoplasma was isolated in the late 1800s from a cow with pleuropneumonia. *Mycoplasma pneumoniae* has been called the Eaton agent, named after the researcher who first isolated it. Until recently, *M. pneumoniae* has been known as a pleuropneumonia-like organism (PPLO). The mycoplasmas adhere to the epithelium of mucosal surfaces in the respiratory and urogenital tracts and are not eliminated by mucous secretions or the passage of urine. Figure 21–12 shows electron micrographs depicting ciliated tracheal epithelial cells prior to and following *M. pneumoniae* infection. Figure 21–13 is an electron micrograph demonstrating the shape of *M. pneumoniae* and its orientation of attachment by a "two-pronged end" structure. Mycoplasma species indigenous to humans are shown in Table 21–14.

CLINICAL INFECTIONS

Mycoplasma pneumoniae

Mycoplasma pneumoniae may cause bronchitis, pharyngitis, or a relatively common respiratory infection known as primary atypical pneumonia or walking pneumonia. The organism does not occur as a normal commensal; therefore, its isolation (when successful) is always significant and pathognomonic. *Mycoplasma pneumoniae*

causes approximately 20% of reported pneumonias in the general population and up to 50% in military settings. It is unusual to find just one case; there is often a cluster of cases owing to transmission by droplet. School-aged children and young adults are especially susceptible to infection. Clinical disease is uncommon in very young children and older adults. Other groups at risk include closed-in populations such as prisoners, college students, and military personnel. Epidemics are known to occur in these populations. Infection is not considered seasonal, but many cases occur in autumn and early winter. Outbreaks have also been noted when adolescents return to school in the fall. Transmission is probably via aerosol droplet spray produced in coughing.

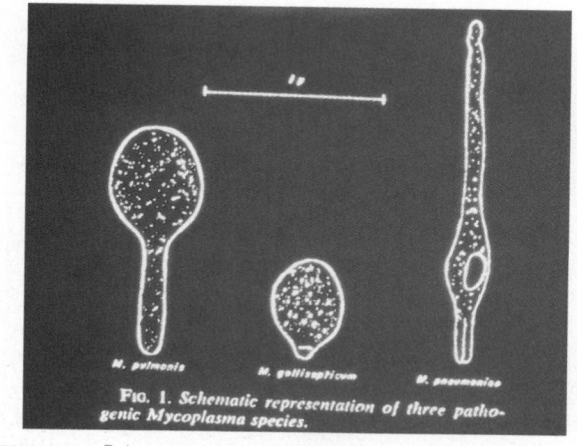

FIG. 1. *Schematic representation of three pathogenic Mycoplasma species.*

■ F i g u r e   2 1 - 1 0

Schematic representation of three *Mycoplasma* species demonstrating varied shape and pleomorphism within genera. *Left, M. pulmonis; center, M. gallisepticum; right, M. pneumoniae.*

■■■■ T A B L E   2 1 – 1 3

CWDS AND PPLOS: HUMAN PATHOGENS IN THE FAMILY MYCOPLASMATACEAE

Feature	*Mycoplasma*	*Ureaplasma*	*Acholeplasma*
Cell-wall deficient	+	+	+
Gram stain	–	–	–
Penicillin susceptible	–	–	–
Urease activity	–	+	–
Induced in hypertonic solution and penicillin, lysozyme, or salts	–	–	–
Derivation: Exists in nature as free-living organism	+	+	+
Contains cell wall components but lacks true cell wall	+	+	+
Reverts to parental form (or re-establishes cell wall) when induction is eliminated	–	–	–
Independent replication in vitro 6a sterol (medium) i.e., cell-free medium	+	+	–
Pleomorphic shape	+	+	+
Other shared characteristics	Smaller than bacteria—close in size to myxoviruses Limited metabolic activity (i.e., fastidious) Lower GC (guanidine/cytosine) ratio than most bacteria Many mycoplasma contain DNAase Smaller genome than bacteria		

CWD = cell wall deficient; PPLO = Pleuropneumonia-like organism.

Many infections are completely asymptomatic or very mild; only approximately 3 to 10% of patients demonstrate clinically apparent pneumonia (Fig. 21–14). The incubation period is usually 2 to 3 weeks, and early symptoms are nonspecific, consisting of headache, low-grade fever, malaise, and anorexia. Sore throat, dry cough, and earache are accompanying symptoms. Extrapulmonary complications, including cardiovascular, central nervous system, dermatologic, and gastrointestinal problems, are rare but do occur. *Mycoplasma pneumoniae* is not associated with infections of the urogenital tract. It has, however, been associated with extrapulmonary disease and implicated as a coinfection or cofactor in epidemic group A meningococcal meningitis and infant pneumonitis.

Mycoplasma hominis and *Ureaplasma urealyticum*

Mycoplasma hominis and *Ureaplasma urealyticum* are both associated with infections in the

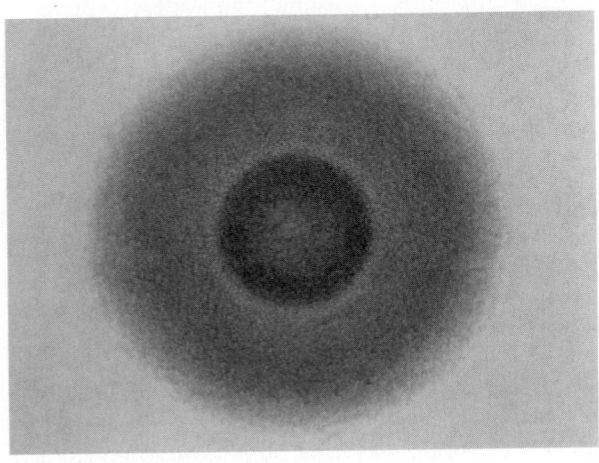

■■■ F i g u r e   2 1 – 1 1

Typical large *Mycoplasma* colony showing "fried egg" appearance. (Courtesy of Bionique Testing Laboratories, Inc., Saranac Lake, NY.)

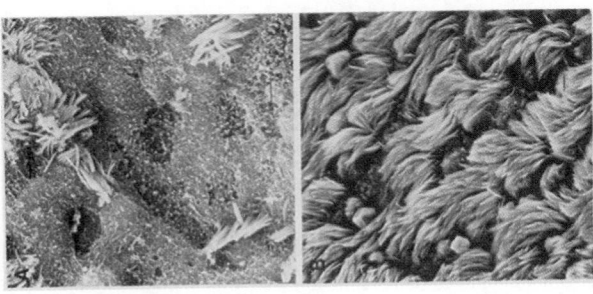

■■■ F i g u r e   2 1 – 1 2

Electron micrograph of effect of *M. pneumoniae* on ciliated tracheal cells. *Left*, infected animal model; *right*, uninfected.

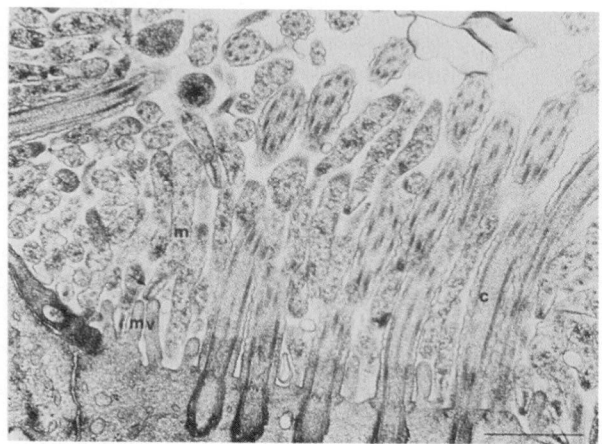

■■■■■F i g u r e   2 1 – 1 3

Electron micrograph of *M. pneumoniae* attaching by specific attachment features to ciliated trachea; mv, microvilli.

urogenital tract. They are, however, frequently isolated from asymptomatic individuals, making interpretation of a positive culture difficult. Because they are opportunistic pathogens, the immune status of the host is an important factor in the occurrence and severity of disease. In addition, it has been reported that among sexually active individuals, the rate of colonization is directly related to the number of sexual partners. Higher rates of colonization have been noted in African-American men and women and in adults of lower socioeconomic status. The organisms do not persist in infants colonized at birth. Table 21–15 summarizes the known association of genital mycoplasmas with urogenital and newborn diseases.

Mycoplasma hominis is found in the lower genitourinary tracts of approximately 50% of healthy adults and has not been reported as a cause of nongonococcal urethritis (NGU). The organism may, however, invade the upper genitourinary tract and cause salpingitis, pyelonephritis, pelvic inflammatory disease, or postpartum fevers.

Ureaplasma urealyticum does not cause disease in the female lower genital tract but has been associated with approximately 10% of cases of nongonococcal urethritis in men and with upper female genitourinary tract disorders. *Ureaplasma urealyticum* has been recovered from greater than 60% of normal, sexually active females. *Ureaplasma* has been associated with reproduction disorders, chorioamnionitis, congenital pneumonia, and the development of chronic lung disease in premature infants. Although it is not a primary cause of chronic lung disease, Cassell and associates (1988) reported *U. urealyticum* as the most common organism isolated from tracheal aspirates of low-birth-weight infants with respiratory disease. Fourteen percent of infections were in neonates delivered by caesarean section, thus indicating that infection occurred in utero and not during passage through the birth canal.

Both *M. hominis* and *U. urealyticum* can be transmitted to the fetus at delivery and have been reported as the most common organisms recovered from the cerebrospinal fluid of certain high-risk newborns, including preterm and low-birth-weight babies. It has been recommended that culture for these organisms be attempted when the CSF specimen from a newborn with

■■■■■T A B L E   2 1 – 1 4

MYCOPLASMA SPECIES INDIGENOUS TO HUMANS

Species	Usual Habitat	Reported Frequency	Colony Morphology*
M. salivarium	Oropharynx	Very common	Large, fried egg
M. orale	Oropharynx	Common	Small, spherical
M. buccale	Oropharynx	Uncommon	Large, fried egg
M. faucium	Oropharynx	Uncommon	Large, fried egg
M. lipophilum	Oropharynx	Rare	Large, fried egg
M. pneumoniae	Oropharynx	Common (with disease)	Small, spherical, granular
M. fermentans	Oropharynx	Rare	Small, fried egg, spherical
	Urogenital tract	Uncommon	
M. hominis	Oropharynx	Uncommon	Large, fried egg, vesiculated peripheral zone
	Urogenital tract	Common (9–50% women)	
M. genitalium†	Urogenital tract	Common?	
U. urealyticum	Urogenital tract	Very common (81% women; 30–50% men)	Tiny, spherical, fried egg, granular

*Relative sizes: large = >100 nm; small = 50–100 nm; tiny = < 50 nm.
†Cross-reacted with *M. pneumoniae* DNA probe (when previously available).

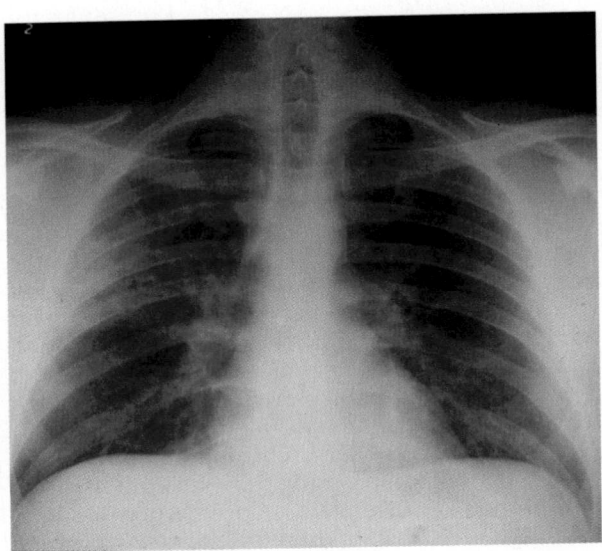

Typical chest radiograph of a patient with a 3-week course of "atypical" pneumonia. Note nonspecific interstitial pneumonia, patchy infiltrate delineated by "feathery outline."

evidence of meningitis is negative for both Gram stain and traditional bacteriology culture (Table 21–16).

Mycoplasma hominis may be isolated from culture in 24 to 48 hours. This does not include blood cultures, where its growth is inhibited by sodiumpolyanethole sulfonate (SPS), an additive often found in blood culture media. Identification is often made by typing methods employing immunofluorescence and observation of plate media using a stereoscope. *Ureaplasma urealyticum* has also been reported to cause chronic inflammatory diseases such as arthritis and cystitis in hypogammaglobulinemic patients.

Lastly, it has been reported that these isolates have been intermittently associated with patients who are immunosuppressed or have endocarditis, sternal wound infections or arthritis.

Emerging *Mycoplasma* Pathogens

The pathogenicity of *Mycoplasma genitalium*, first isolated in 1980, has not been established, but the organism has been associated with some cases of nongonococcal urethritis and pelvic inflammatory disease. Its prevalence is not known, but it may be primarily a resident of the gastrointestinal tract that occurs secondarily in the genitourinary or respiratory tract. *Mycoplasma genitalium* is very difficult to recover from culture and may require 2 to 3

■■■■■ T A B L E   2 1 – 1 5

SUMMARY OF THE ASSOCIATION OF GENITAL MYCOPLASMAS WITH UROGENITAL AND NEWBORN DISEASES

Disease	M. hominis	U. urealyticum	Comments
Nongonococcal urethritis	None	Strong	Ureaplasmas cause some cases, but the proportion is unknown
Prostatitis	Weak	None	An association with a few cases of chronic disease has been reported; a causal relation is unproven.
Epididymitis	None	None	Mycoplasmas are not an important cause
Reiter disease	None	None	The role of ureaplasmas should be studied
Bartholin gland abscess	Weak	None	M. hominis may cause some disease but is not an important cause
Vaginitis and cervicitis	None	None	M. hominis often associated with disease, but a causal relation is unproven
Pelvic inflammatory disease	Strong	Weak	M. hominis causes some cases, but the proportion is unknown
Postabortal fever	Strong	None	M. hominis is responsible for some cases, but the proportion is unknown
Postpartum fever	Strong	None	Recent work indicates M. hominis may be a major cause
Urinary calculi	None	Weak	Ureaplasmas cause calculi in male rats, but no convincing evidence exists that they cause natural human disease
Pyelonephritis	Strong	None	M. hominis causes some cases
Involuntary infertility	None	Weak	Ureaplasmas are associated with altered motility of sperm
Repeated spontaneous abortion and stillbirth	None	Weak	Maternal and fetal infections have been associated with spontaneous abortion, but a causal relation is unproven
Chorioamnionitis	None	Strong	An association exists, but a causal relation is unproven
Low birth weight	None	Strong	An association exists, but a causal relation is unproven
Neonatal infections, including sepsis, pneumonia, meningitis	Strong	Strong	Further clarification is needed, but importance is growing in a selected prenatal population

■■■■T A B L E 2 1 - 1 6

SUMMARY OF THE RECENT ASSOCIATION OF GENITAL MYCOPLASMAS WITH NEONATAL DISEASE

Condition/Target Population	Isolates		Comments
	M. hominis	*U. urealyticum*	
Neonatal period, including preterm delivery, very low birth weight. Clinical signs compatible with			These findings need further clarification because most neonatal infections resolve without therapy, but in low socioeconomic groups, "diagnostic work-up" of neonates should now include CSF and blood cultures for detection of *Mycoplasma*. This includes low birth weight and preterm neonates, in whom traditional CSF cell counts and cultures would be negative.
Meningitis (CSF)	+	+	
Pneumonia (trachea)	+	+	
Sepsis (blood)	+	+	

months of incubation. Understanding this organism is important, because it cross-reacts with *M. pneumoniae* in some direct fluorescent antibody tests, causing a false-positive result.

Mycoplasma fermentans [*incognitus* strain] is a newly described Mycoplasma that closely resembles *Mycoplasma fermentans* and is most likely a new strain rather than a new species. It is more fastidious than *M. fermentans,* growing only on modified SP-4 culture medium, with colonies visible in 10 to 14 days. Recent interest in *M. fermentans* [*incognitus* strain], sometimes called Lo's *Mycoplasma* in recognition of its discoverer, has centered on its relationship to patients infected with the human immunodeficiency virus type 1 (HIV-1). The association of this *Mycoplasma* with the development of acquired immune deficiency syndrome (AIDS) is not clear at present, but *M. fermentans* [*incognitus* strain] may be able to suppress the immune system by reducing the number of T lymphocytes or "turning on" T lymphocytes to produce cytokines. Using polymerase chain reaction (PCR) technology in a serologic study of *Mycoplasma* in AIDS patients, Lo et al (1989) reported coinfection of HIV and *Mycoplasma fermentans* [*incognitus* strain] in the majority of patients studied. The organism has been identified in thymus, liver, spleen, lymph nodes, brain, and kidney of HIV-1–positive patients. It has, not surprisingly, been called "the AIDS-related *mycoplasma.*"

LABORATORY DIAGNOSIS

Specimen Collection and Transport

Owing to the lack of a cell wall, all mycoplasmas are extremely sensitive to drying. Therefore, specimens for culture should be delivered immediately to the laboratory. Swabs should be placed in a transport medium such as trypticase soy broth with 0.5% albumin and 400 U/mL of penicillin or SP4 medium (sucrose phosphate buffer, *Mycoplasma* base, horse serum [20%], and neutral red) (see Fig. 21–15). Upon arrival in the laboratory, the specimens should be frozen at −70°C if immediate plating is not possible.

Isolation of *Mycoplasma pneumoniae* from respiratory sites is attempted infrequently because recovery from culture is difficult (sensitivity is approximately 40%). Growth may take several weeks, and technical expertise is necessary. Diagnosis is usually established serologically, traditionally with acute and convalescent sera collected 2 to 3 weeks apart to demonstrate a four-fold rise in titer. *Mycoplasma hominis* and *Ureaplasma* are less stringent in their growth requirements but require cholesterol for synthesis of plasma membranes.

Optimally, serum samples for serologic testing should be collected at the onset of symptoms and 2 to 3 weeks later for acute and convalescent measurements; often, however, this is not practical. With newer methods, single serum samples collected during the disease may rule out the infection or suggest additional evaluations. A schematic representation of classic clinical and corresponding diagnostic manifestations of *M. pneumoniae* is shown in Table 21–17. As noted earlier, many of the early symptoms are nonspecific, and a thorough understanding of the disease process is necessary for interpretation of both serum and culture results.

Culture

Media

Several media have been developed for the recovery of mycoplasmas. They include supplemented PPLO agar, Shephard A-7B agar (for *U.*

■ TABLE 21-17

MAJOR CLINICAL AND CORRESPONDING DIAGNOSTIC MANIFESTATION OF *M. PNEUMONIAE**

Manifestation	Days After Onset						
	5	**10**	**15**	**20**	**25**	**35**	**40**
Headache and malaise	+1	+3	+3	+2	+1		
Dry cough	+2	+4	+4	+1			
Chest soreness	+3	+3	+1				
Fever†							
104°F							
102°F							
100°F							
Chest radiograph	+2	+3	+2	+2	+1		
Mycoplasma culture with or without antibiotic treatment	+	+	+	+	+	+	+
Cold agglutinin (titer)	≤1:8	1:32	1:64	1:320	1:320	1:64	1:16
Complement Fixation (titer)	≤1:8	≤1:8	1:32	1:64	1:256	1:256	1:128
Mycoplasma-specific Ig							
IgM	–	+	+	+	+	+	+
IgG	–	–	+	+	+	+/–	–

*+4, most severe; +1, least severe; +, present or positive; –, absent or negative.
† ■, patient treated without appropriate antibiotic; ▯, patient treated with optimum antibiotic therapy, dose, and duration.

urealyticum), SP4, Hayflick's biaphasic medium, E agar, and New York City agar, a selective agar for *Neisseria gonorrhoeae* that also supports the growth of *M. hominis* and *U. urealyticum*. *Mycoplasma pneumoniae* and *M. genitalium* require glucose (their major source of energy), *M. hominis,* arginine, and *U. urealyticum,* urea. *Ureaplasma* requires liquid medium to have a pH of 6.0 and is difficult to maintain in culture because death occurs rapidly when the urea is depleted. Figure 21–15 presents a schematic representation of media and methods used in the traditional isolation and identification of *Mycoplasma*.

Identification

Mycoplasma-like colonies are stained with Dienes stain, because the organisms do not stain well with Gram or acridine orange (AO) stain. It is performed by placing a small block of the agar plate on a glass slide, covering the colony with the stain, adding a coverslip, and examining microscopically under low power. *Mycoplasma* have a typical "fried egg" appearance, the periphery staining a light blue and the center dark blue (Fig. 21–16). *Mycoplasma* almost universally show a mixed colony presentation on primary isolation when examined with a stereoscope (Fig. 21–17).

A direct plate immunofluorescent (IF) method can also be used. Fluorescent-labeled anti–*M. pneumoniae* antibody is flooded on colonies on the plate; the plate is then washed and examined for immunofluorescence. Another fluorochrome method used to identify *Mycoplasma*-infected eukaryotic tissue culture is the Chen method. It uses a DNA fluorochrome stain (Hoechst 33258), which highlights *Mycoplasma* as small ovoid bodies distributed throughout the glacial acetic acid–fixed cell culture. Figure 21–18 shows Vero cells (a monkey cell line) artificially infected with *Mycoplasma orale* (Fig. 21–18*A*) and *M. salivarium* (Fig. 21–18*C*), respectively. Note the differences in morphotypes and distribution. Vero cell nuclei, rich in DNA, fluoresce with Hoechst 33258 stain in the negative control (Fig. 21–18*B*) as well as in the infected cell cultures. This method offers a unique way for diagnostic and clinical virology laboratories to perform quality control on their continuous cell cultures, as required by the College of American Pathologists for laboratory accreditation (personal communication, Bionique Testing Laboratories, Saranac Lake, NY, 1993).

Mycoplasma hominis requires as few as 2 days to form colonies on solid media, whereas *M. pneumoniae* may take approximately 20 days to show typical growth. The characteristic of guinea pig red blood cells (0.4%) adhering to colonies of *M. pneumoniae* and not *M. hominis,* is another standard assay that helps distinguish the two species (see Table 21–18). Further, guinea pig cells do not adhere to large-colony mycoplasma, which are common inhabitants of the upper respiratory tract.

Ureaplasma colonies, once called "T-strain mycoplasma" (T for tiny), are extremely small and difficult to see with the naked eye; hence,

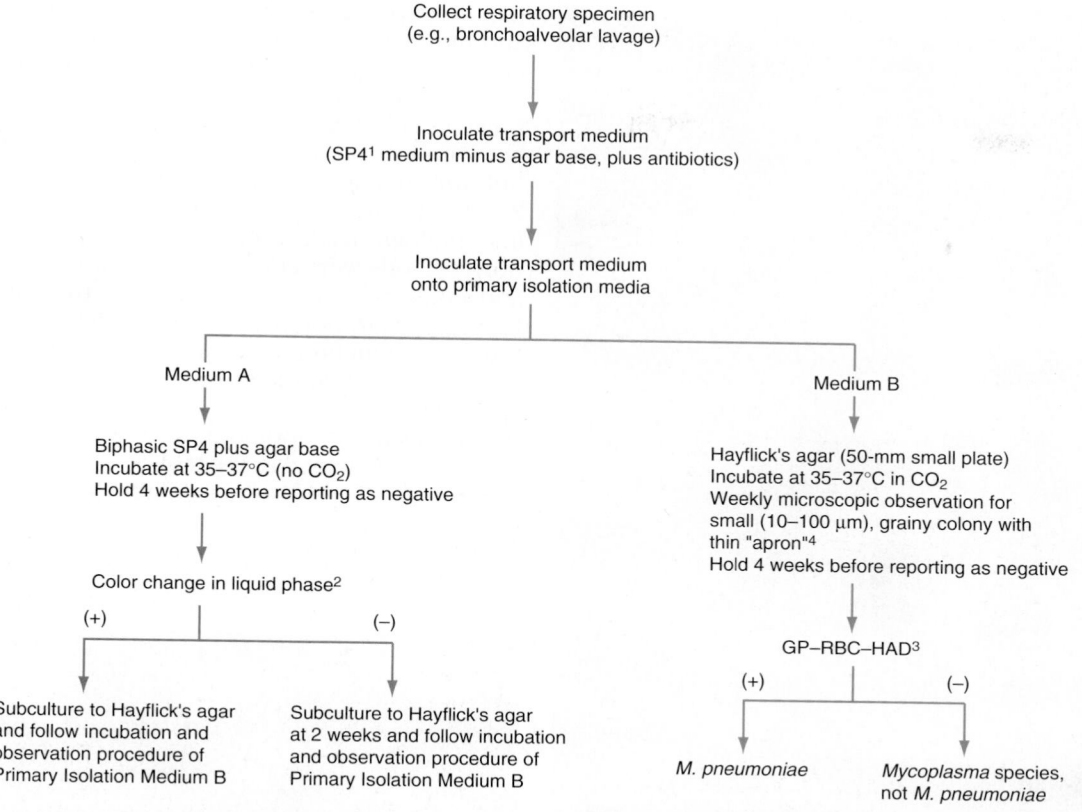

Collect respiratory specimen
(e.g., bronchoalveolar lavage)

↓

Inoculate transport medium
(SP4[1] medium minus agar base, plus antibiotics)

↓

Inoculate transport medium
onto primary isolation media

Medium A

↓

Biphasic SP4 plus agar base
Incubate at 35–37°C (no CO_2)
Hold 4 weeks before reporting as negative

↓

Color change in liquid phase[2]

(+) | (−)

Subculture to Hayflick's agar and follow incubation and observation procedure of Primary Isolation Medium B

Subculture to Hayflick's agar at 2 weeks and follow incubation and observation procedure of Primary Isolation Medium B

Medium B

↓

Hayflick's agar (50-mm small plate)
Incubate at 35–37°C in CO_2
Weekly microscopic observation for small (10–100 μm), grainy colony with thin "apron"[4]
Hold 4 weeks before reporting as negative

↓

GP–RBC–HAD[3]

(+) | (−)

M. pneumoniae

Mycoplasma species, not *M. pneumoniae*

---

[1]SP4 = Sucrose phosphate buffer, *Mycoplasma* base, horse serum (20%), neutral red. Medium stabilizes and decontaminates specimen. Storage at −70°C for repeat testing is recommended.

[2]Color change: positive = yellow color with no gross turbidity; negative = red color.

[3]GP–RBC–HAD = Guinea pig red blood cell hemadsorption. β-Hemolysis test for presumptive identification of *M. pneumoniae* may be used in lieu of GP–RBC–HAD.

[4]Thin colony periphery.

Note: Diene's stain can be used for detection of *Mycoplasma* species on Hayflick's agar; plate immunofluorescence using labeled antibody can be used for identification.

■■■ Figure 21–15

Flow diagram for *Mycoplasma* isolation utilizing classic methods.

Mycoplasma cultures on solid media should always be examined with a stereoscopic microscope. Figure 21–19 shows both *M. hominis* and *U. urealyticum* grown on New York City agar; the arrow points to the "tiny" ureaplasma. Urease activity of ureaplasma may also be detected on solid agar containing urea and manganese chloride (U9B urease color test medium). Urease-positive colonies are a dark golden-brown color owing to the deposition of manganese dioxide. Both *M. hominis* and *U. urealyticum* require cholesterol for synthesis of plasma membranes and other undetermined growth factors; horse serum (20% v/v) is the traditional source.

Although uncommon, extragenital *M. hominis* infections are emerging, and this organism should be considered whenever many PMNs are seen on Gram stain but the routine bacterial culture is sterile. The isolate grows well anaerobically and will appear as pinpoint (0.05-mm), clear, glistening, raised colonies on Columbia colistin–nalidixic acid (CNA) agar or anaerobic blood agar (CDC formula) in 48 hours. Under these anaerobic conditions, the colonies do not display the typical "fried egg" morphology characteristic of mycoplasma. The anaerobic plate should be examined utilizing oblique light. These colonies that do not Gram stain should be sub-

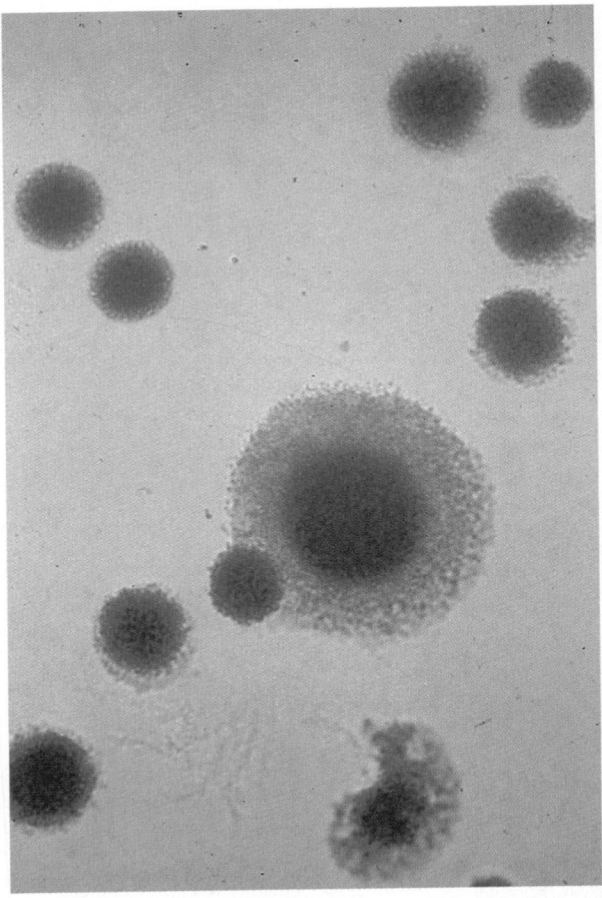

Figure 21–16
Diene's stain of *Mycoplasma* colonies demonstrating typical "fried egg" appearance.

cultured to A7 medium, on which they demonstrate typical "fried egg" growth, stain positive with Dienes stain, and hydrolize arginine if they are *Mycoplasmas* (personal communication, A. William Pasculle, 1992).

Serologic Diagnosis

The cold agglutinin antibody titer has been used for many years as an indicator of primary atypical pneumonia (see Table 21–17), but it is both insensitive and nonspecific for *Mycoplasma pneumoniae*. Approximately 50% of patients with primary atypical pneumonia produce a detectable cold agglutinin antibody titer.

Until recently, the most commonly used technique for demonstration of *M. pneumoniae* antibodies was micromethod complement fixation (CF), which was time consuming and had inherent

technical problems. Enzyme immunoassay (EIA) and indirect hemagglutination methods (IHA) are now available for detection of serum antibodies, in some cases detecting either IgM or IgG. Polymerase chain reaction (PCR) methods have been investigated and found to be very specific but are not yet commercially available. Table 21–18 highlights selected features of these immunologic assays and other methods. Detection methods were added for comparative analysis and completeness. It is important to remember that demonstration of a significant rise in antibody titer in conjunction with culture isolation is preferable for definitive diagnosis. The table also highlights another point: Given the number of assays, no one assay is optimum, although methods of choice are indicated in the table.

Serologic methods are available for *M. hominis* or *U. urealyticum*, but they are generally performed by reference laboratories only.

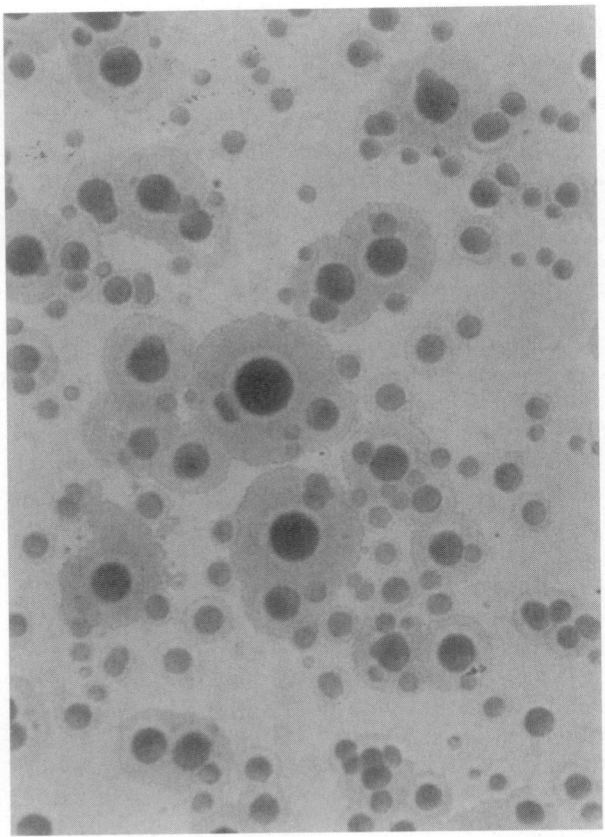

Figure 21–17
Typical mixed sizes of *Mycoplasma* on primary isolation media: *M. salivarium*. (Courtesy of Bionique Testing Laboratories, Inc., Saranac Lake, NY.)

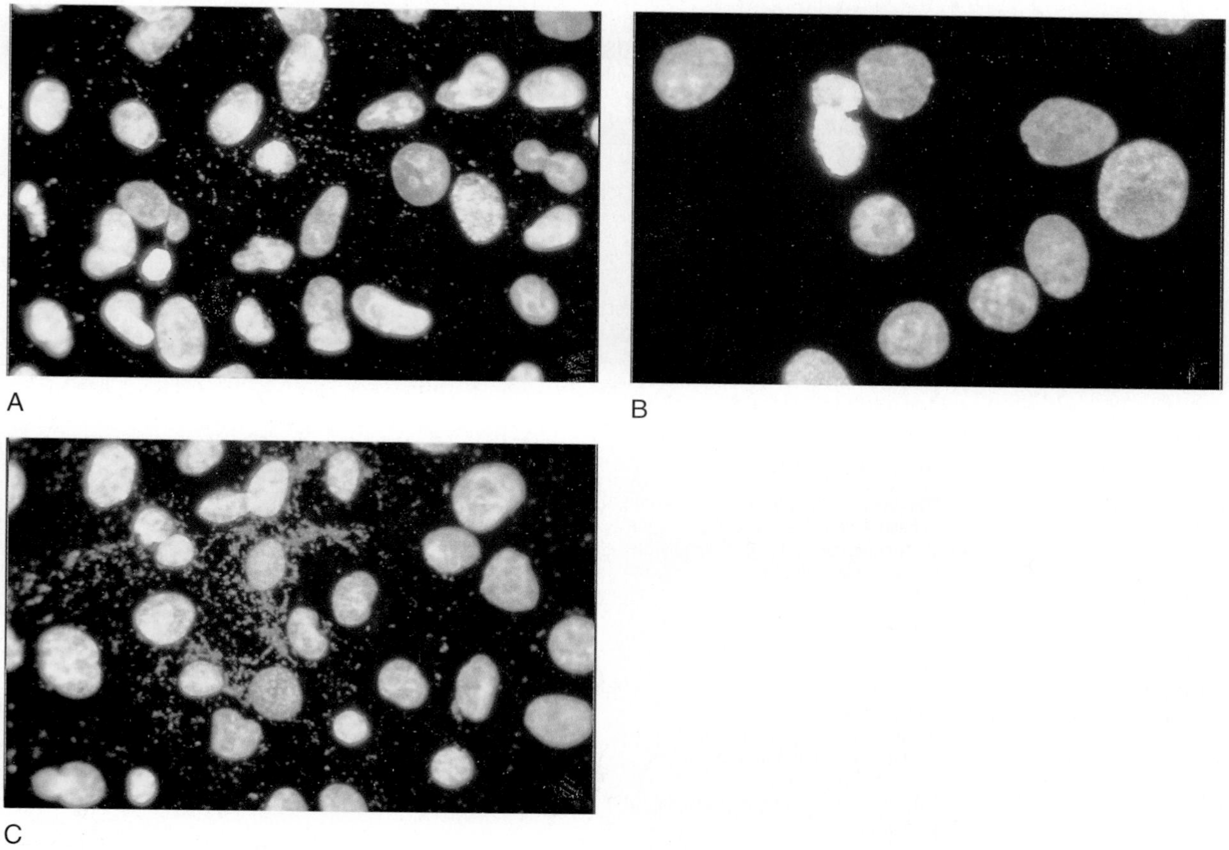

A

B

C

Figure 2 1 – 1 8

Identification of *Mycoplasma*-infected eukaryotic cell culture employing DNA-fluorochrome stain (Hoechst no. 33258 stain). *A, M. orale; B,* uninfected Vero cell culture highlighting the DNA-rich nucleus; *C, M. salivarium.* The mycoplasma appear as small pinpoint fluorescent bodies throughout the background. (Courtesy of Bionique Testing Laboratories, Inc., Saranac Lake, NY.)

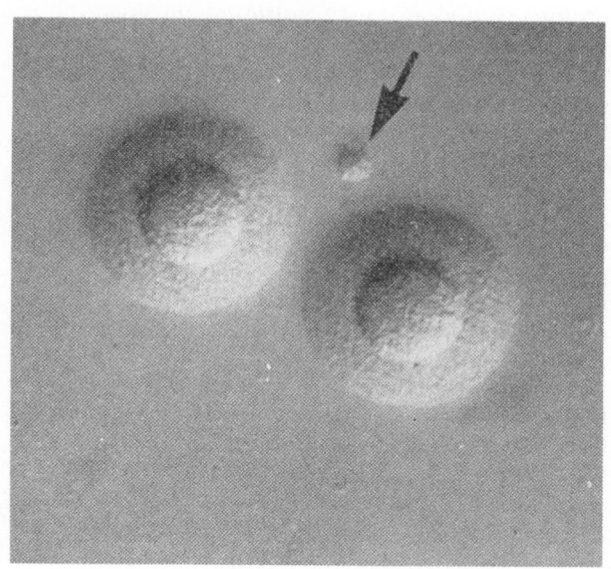

Figure 2 1 – 1 9

Mixed isolation of *Mycoplasma hominis* and *Ureaplasma urealyticum* showing why *U. urealyticum* was originally called "T" for "tiny-strain" *(arrow).*

■ T A B L E 2 1 – 1 8

COMPARATIVE FEATURES OF VARIOUS LABORATORY METHODS USED IN THE DETECTION OF *M. PNEUMONIAE*, *M. HOMINIS*, AND *U. UREALYTICUM*

Detection Method	M. pneumoniae	M. hominis	U. urealyticum
Nonserologic			
Culture	Traditional methods difficult but new "kit" formats simplify protocols	Method of choice, CNA plate, but must differentiate "infection" from "colonization"	Method of choice using urease detection, simultaneously
Indirect immunofluorescence	Respiratory antigen for early-stage infection Research use only, but is promising		
Latex agglutination	Method of choice using polyclonal antibody specific for *M. pneumoniae* in respiratory exudates		
Serologic			
Cold Agglutinins	Historical only >1:16 single titer		
Complement fixation (CF)	Traditional assay *but* <50% seroconvert; need 4-fold rise between acute and convalescent sera >1:32 single titer may be suggestive		
Indirect hemagglutination		Method of choice but need composite antigen recognizing 7 serovars of *M. hominis*	
Latex agglutination	IgM/IgG		
Enzyme immunoassay (EIA)	Method of choice *M. pneumoniae* reactive IgM, IgG, and IgA, *but* IgM may remain elevated for 1 yr	IgG only	
Immunoblot	IgM, IgG, *M. pneumoniae*-specific		
Polymerase chain reaction	Research only		

DNA Probe Technology

Because attempted culture of *M. pneumoniae* is tedious and slow, and traditional serologic techniques required the comparison of acute and convalescent sera collected over 2 to 3 weeks, a nucleic probe was developed for rapid diagnosis to expedite prompt initiation of therapy (Gen-Probe, Inc., San Diego, CA). The system was designed to detect ribosomal RNA from *M. pneumoniae* in throat swabs and also to be used to confirm an isolate recovered from culture. The probe assay had a reported sensitivity and specificity of more than 95% compared with other methods but was removed from the market in 1992 at the vendors' request.

Although rapid and sensitive, probe technology had some other disadvantages. It was generally not cost effective for small laboratories with infrequent requests, because the isotope had a short shelf-life and a gamma counter was required.

ANTIMICROBIAL SUSCEPTIBILITY

No agent is mycoplasmacidal for *M. pneumoniae*, but erythromycins (particularly the latest generation) and tetracycline can shorten the duration of symptoms in patients with respiratory infections. Because of side effects, tetracycline is used only for treatment of adults. *M. hominis*, which is much more resistant than *M. pneumoniae*, is usually resistant to erythromycin and susceptible to lincomycin, whereas *U. ureaplasma* is generally resistant to lincomycin and sensitive to erythromycin. Both organisms are usually sensitive to tetracycline, but some high-level resistance is emerging. Standard methods have not been developed for susceptibility testing for mycoplasma, and protocols have varied considerably from laboratory to laboratory. With reported antimicrobial resistance, newer broad-spectrum antimicrobials and new strains, i.e., *M.*

fermentans [incognitus strain], however, methods are being reevaluated. Susceptibility testing is usually recommended for clinically significant *M. hominis,* employing broth dilution; although several methods have been reported in the literature, these isolates should be forwarded to a reference laboratory.

INTERPRETATION OF LABORATORY RESULTS

Finally, a word about interpreting and reporting results for *Mycoplasma* generally and *Ureaplasma* specifically. *Mycoplasma pneumoniae* detected by any method from pulmonary or nonpulmonary specimens should be considered significant and a pathogen. *Mycoplasma hominis* is not as clearcut; differentiation from colonization and infection requires detailed clinical analysis and potentially repeat cultures, given its propensity for rapid detection utilizing anaerobic media (CNA), i.e., within 48 hours. *Ureaplasma urealyticum* is the most difficult to assess. In urogenital specimens, it has been reported to colonize up to 70% of males and 45% of females with no apparent

infection. Its isolation is absolutely not indicative of pathogenicity, and it is incumbent upon the laboratory to educate the physician, usually including a statement with culture results suggesting its potential for colonization versus pathogenicity. In nonurogenital specimens, particularly CSF isolates, it is reasonable to evaluate cautiously the clinical significance of ureaplasma.

Table 21–19 is a "reality check." Respiratory specimens received in the laboratory often provide limited clinical information. Specimens are processed and inoculated onto the most appropriate media given the most likely candidate for the disease process, clinical presentation, age of patient, and seasonability, recognizing that there is a certain predictability with selected pathogens. Table 21–19 highlights a scheme in which selected stains, cultures, and nonculture techniques may be used for the detection of several pathogens: *Mycloplasma, Chlamydia,* viruses, mycobacteria, fungi, *Legionella,* and *Pneumocystis pneumoniae.* All respiratory specimens should be stored at −70°C, including an acute serum. The patient would not be charged until the specimen was processed for the most likely pathogen following initial evaluation. This table complements Table 21–18.

■■■■ T A B L E   2 1 – 1 9
LABORATORY DETECTION OF FREQUENT RESPIRATORY NONBACTERIAL PATHOGENS

	Epidemiologic Factors					Laboratory Methods		
Age	Age/Organism Frequently Involved*	Disease	Season	Specimen Source	Stains	Culture	Nonculture	
Newborn	1, 3	Pneumonia, aseptic work-up		Tracheal suction	Gram	Traditional, plus mycoplasmal		
Grade school	2, 3	Atypical pneumonia	Fall	Sputum	Gram, acid-fast bacillus	Traditional, plus mycoplasmal	EIA, mycoplasmal, IgM	
College student	1, 2	Biphasic disease with pharyngitis and, later, bronchitis	Spring?	Sputum	Direct FA	Cell culture, mycoplasmal	Cold agglutinin	
Adult	2, 3, 4	Pneumonia or immunocompromised		Sputum Bronchoalveolar lavage specimen	Direct FA, acid-fast bacillus Gomori methenamine silver, toluidine blue, and/or calcofluor white	Cell culture Traditional plus fungal, acid-fast bacillus	IgG, mycoplasmal	

*1 = *Chlamydia pneumoniae* (strain TWAR); 2 = *Mycoplasma pneumoniae* (outbreak) and/or *M. hominis*; 3 = viral (outbreak) adenovirus/RSV/influenza (seasonal); 4 = other: AFB, fungus, legionnaire bacterium (LDB), *Pneumocystis carinii* pneumonia (PCP).

Bibliography

Barker CE, Sillis M, Wreghitt TG: Evaluation of Serodia Myco II particle agglutination test for detecting *Mycoplasma pneumoniae* antibody: Comparison with u-capture ELISA and indirect immunofluorescence. J Clin Pathol 43:163, 1990.

Baseman JB, Dallo SF, et al: Isolation and characterization of *Mycoplasma genitalium* strains from the human respiratory tract. J Clin Microbiol 26:2266, 1988.

Bernet C, Garret M, et al: Detection of *Mycoplasma pneumoniae* by using the polymerase chain reaction. J Clin Microbiol 27:2492, 1989.

Cassell GH, Crouse DT, et al: Association of *Ureaplasma urealyticum* infection of the lower respiratory tract with chronic lung disease and death in very-low-birth-weight infants. Lancet 2:240, 1988.

Cassell GH, Crouse DT, et al: Does *Ureaplasma urealyticum* cause respiratory disease in newborns? Pediatr Infect Dis J 7:535, 1988.

Cassell GH, Waites KB, et al: *Ureaplasma urealyticum* intrauterine infection: Role in prematurity and disease in newborns. Clin Microbiol Rev 6:69, 1993.

Cassell GH, Watson HL, et al: Protein antigens of genital mycoplasmas. Rev Infect Dis 10:S391, 1988.

Chen TR: In situ detection of *Mycoplasma* contamination in cell culture by fluorescent Hoechst 33258 stain. Exp Cell Res 104:255, 1977.

Cunningham K: Role of genital mycoplasma in neonatal disease. Clin Microbiol Newsl 12:147, 1990.

Dular R, Kajioka R, Kasatiya S: Comparison of Gen-Probe commercial kit and culture technique for the diagnosis of *Mycoplasma pneumoniae* infection. J Clin Microbiol 26:1068, 1988.

Huovinen P, Lahtonen R, et al: Pharyngitis in adults: The presence and coexistence of viruses and bacterial organisms. Ann Intern Med 110:612, 1989.

Kenny GE: Mycoplasmas. In Balows A (ed): Manual of Clinical Microbiology, 5th ed. Washington, DC: American Society for Microbiology, 1991, pp 478–482.

Limb DI: Mycoplasmas of the human genital tract. Med Lab Sci 46:146, 1989.

Lo SC, Shih JW, et al: Virus-like infectious agent (VLIA) is a novel pathogenic mycoplasma: *Mycoplasma incognitus.* Am J Trop Med Hyg 41:586, 1989.

Mansel JK, Rosenow EC, et al: *Mycoplasma pneumoniae* pneumonia. Chest 95:639, 1989.

Marjaana Kleemola SR, Karjalainen JE, Raty RKH: Rapid diagnosis of *Mycoplasma pneumoniae* infection: Clinical evaluation of a commercial probe test. J Infect Dis 162:70, 1990.

Martinez OV, Chan J, et al: *Mycoplasma hominis* septic thrombophlebitis in a patient with multiple trauma: A case report and literature review. Diagn Microbiol Infect Dis 12:193, 1989.

McMahon DK, Dummer JS, et al: Extragenital *Mycoplasma hominis* infections in adults. Am J Med 89:275, 1990.

Moore PS, Hierholzer J, et al: Respiratory viruses in *Mycoplasma* as cofactors for epidemic group A meningococcal meningitis. JAMA 264:1271, 1990.

Mycoplasma still putative cofactor in AIDS. ASM News 57:402, 1991.

Risi GF, Sanders CV: The genital mycoplasmas. Obstet Gynecol Clin North Am 16:611, 1989.

Sasaki T, Sasaki Y, et al: Evidence that Lo's Mycoplasma (*Mycoplasma fermentans* incognitus) is not a unique strain among *Mycoplasma fermentans* strains. J Clin Microbiol 30:2435, 1992.

Sillis M: The limitations of IgM assays in the serological diagnosis of *Mycoplasma pneumoniae* infections. J Med Microbiol 33:253, 1990.

Stagno S, Brasfield D, et al: Infant pneumonitis associated with cytomegalovirus, *Chlamydia, Pneumocystis,* and *Ureaplasma:* A prospective study. Pediatrics 68:322, 1981.

Tilton RC, Dias F, et al: DNA probe versus culture for detection of *Mycoplasma pneumoniae* in clinical specimens. Diagn Microbiol Infect Dis 10:109, 1988.

Waites KB, Crouse DT, et al: Chronic *Ureaplasma urealyticum* and *Mycoplasma hominis* infections of central nervous system in preterm infants. Lancet 1:17, 1988.

Waites KB, Duffy LB, et al: Mycoplasmal infections of cerebrospinal fluid in newborn infants from a community hospital population. Pediatr Infect Dis 9:241, 1990.

Yajko DM, Balston E, et al: Evaluation of PPLO, A7B, E, and NYC agar media for the isolation of *Ureaplasma urealyticum* and *Mycoplasma* species from the genital tract. J Clin Microbiol 19:73, 1984.

Zahawi MF, Kerns AM, et al: A study of three blood culture media for isolating genital mycoplasmas from obstetrical and gynecologic patients. J Infect 21:143, 1990.

CHAPTER **22**

Mycobacterium tuberculosis and Other Nontuberculous Mycobacteria

Continued

Janice Eisenstadt • Gerri S. Hall •
Susan M. Gibson • Denise F. Dunbar

OBJECTIVES

1. Describe the general characteristics of mycobacteria and differentiate them from other groups of organisms.
2. Discuss the safety precautions to be followed while working in a mycobacteriology laboratory.
3. Describe the appropriate specimen collection and processing procedures to recover mycobacteria from clinical samples.
4. Explain why clinical samples for mycobacterial isolation require digestion and decontamination procedures.
5. Name the different digestion reagents and their proper uses.
6. Describe the principle and procedures for the stains used to demonstrate mycobacteria in clinical samples and isolates.
7. List the different culture media used for the isolation of mycobacteria.
8. Discuss the different tests used to identify mycobacteria.
9. Describe new methods of detecting mycobacterial species in samples.
10. Discuss the clinical disease caused by *Mycobacterium tuberculosis*.
11. Discuss the clinical significance of nontuberculous mycobacteria.

The genus *Mycobacterium* consists of 54 species. The most familiar of the species are *M. tuberculosis* and *M. leprae,* the causative agents of tuberculosis (TB) and leprosy, respectively. Both diseases have long been associated with chronic illness and social stigma. Until 1985, there had been a steady decline in the incidence of tuberculosis in the United States, but the growing number of immunocompromised patients worldwide has lead to the resurgence of tuberculosis as well as to an explosion of disease caused by mycobacteria other than the tubercle bacillus (MOTT). Table 22–1 shows the usual clinical significance of *Mycobacterium* species isolates.

These epidemiologic changes have led to certain challenges within the mycobacteriology laboratory, including rapid identification of all clinically significant mycobacteria and antimicrobial susceptibility testing of *Mycobacterium* species. Fortunately, new developments in the field of clinical mycobacteriology are helping to meet these challenges.

■■■■■ T A B L E 2 2 – 1
USUAL CLINICAL SIGNIFICANCE OF *MYCOBACTERIUM* SPECIES ISOLATES

Pathogen	*M. tuberculosis* *M. bovis* *M. ulcerans*
Often pathogen, potential pathogen	*M. kansasii* *M. marinum*
Potential pathogen	*M. avium* complex *M. fortuitum* *M. chelonae* *M. abscessus* *M. simiae* *M. scrofulaceum* *M. szulgai* *M. xenopi* *M. malmoense* *M. haemophilum* *M. asiaticum* *M. genavense*
Usual saprophyte, rare pathogen	*M. gordonae* *M. flavescens* *M. gastri* *M. nonchromogenicum* *M. terrae* *M. triviale* *M. phlei* *M. smegmatis* *M. vaccae* *M. thermoresistibile*

These methods may eliminate the need for lengthy culturing for isolation and protracted biochemical methods of identification. Further developments in the application of molecular biology to mycobacteriology may, even more, diminish the time required to increase accuracy, reproducibility, rapidity, and ease of performance and allow cost containment.

This chapter discusses:

■ The general characteristics of mycobacteria

■ Safety precautions that must be observed in the mycobacteriology laboratory

■ Appropriate specimen collection and handling procedures

■ Laboratory procedures for isolation and identification of mycobacterial isolates

■ Clinically significant *Mycobacterium* species and the infections they produce.

GENERAL CHARACTERISTICS

Mycobacteria are slender, slightly curved or straight rod-shaped organisms 0.2 to 0.4 × 2 to 10 µm in size. They are nonmotile and do not form spores. The cell wall structure has a very high lipid content; thus, mycobacterial cells resist staining with commonly used basic aniline dyes at room temperature. Mycobacteria do take up dye with increased staining time or with application of heat; however, they resist decolorization with up to 3% hydrochloric acid, and some also resist decolorization with 95% ethanol. These characteristics, referred to as *acid fastness* and *acid-alcohol fastness,* respectively, are basic to distinguishing mycobacteria from other genera and species.

Mycobacteria are strictly aerobic and grow more slowly than most bacteria pathogenic for humans. The most rapidly growing species generally grow on simple media in 2 to 3 days at temperatures of 20 to 40°C. Most mycobacteria associated with disease require 2 to 6 weeks of incubation on complex media at specific optimum temperatures. The growth of *M. tuberculosis* is enhanced by an atmosphere of 5 to 10% carbon dioxide and a growth medium with a pH of 6.5 to 6.8. One of the mycobacteria pathogenic for humans, *M. leprae,* fails to grow in vitro.

Rate of growth, colony morphology, pigmentation, nutritional requirements, optimum incubation temperature, and biochemical test results are traditional characteristics for differentiating species within the genus *Mycobacterium.* More rapid techniques currently available include a radiometric culture system (BACTEC), which can be used for isolation, susceptibility testing, and distinguishing *M. tuberculosis* from the nontuberculous mycobacteria using a selective growth inhibitor. Additionally, a limited number of species-specific nucleic acid probes offer rapid identification of culture isolates. Researchers have developed polymerase chain reaction (PCR) assays, which increase the sensitivity of nucleic acid probes. Other techniques, such as thin-layer chromatography, gas-liquid chromatography, and, more recently, high-performance liquid chromatography, have been used to distinguish mycobacterial species. The application of PCR and chromatography methods has slowly moved out of the research and larger reference mycobacteriology laboratories to routine laboratories and can be expected to be increasingly adopted in the near future.

SAFETY CONSIDERATIONS

The serious nature of tuberculosis disease and the usual airborne route of infection require that special safety precautions be used by anyone handling mycobacterial specimens. The incidence of tuberculosis infection, i.e., skin test positivity, in those who work in the mycobacteriology laboratory is at least three times higher than among other laboratory personnel in a given institution. The hazard of working in a mycobacteriology laboratory is, however, small when the laboratory is well designed, appropriate equipment is available, and precautions are followed closely.

Personnel Safety

The administration of the microbiology laboratory must ensure that each employee is (1) provided with adequate safety equipment, (2) trained in safe laboratory procedures, (3) informed of the hazards associated with the procedures, (4) prepared for action following an unexpected accident, and (5) monitored regularly by medical personnel. Laboratory personnel must be responsible for using appropriate safety equipment and following established procedures. The laboratory worker should maintain optimal health. A skin test (Mantoux test) with purified protein derivative (PPD) of tuberculin should be administered upon employment and thereafter regularly to persons previously skin test negative (nonreactors). Individuals known to be previously skin test positive (reactors) should be counseled regularly and should be referred for medical evaluation if their health status changes.

Proper Ventilation

Laboratory design and ventilation play an important role in mycobacteriology laboratory safety. Ideally, the mycobacteriology laboratory should be separate from the remainder of the laboratory and should have a nonrecirculating ventilation system. The area in which specimens and cultures are processed should have negative air pressure in relation to other areas; that is, the airflow should be from clean areas, such as corridors, into less clean areas. Six to 12 room air changes per hour effectively remove 99% or more of airborne particles within 30 to 45 minutes. A much higher number of room air changes per hour may cause problems of air turbulence within the biologic safety cabinets.

Proper Use of Biologic Safety Cabinet

Because the route of infection by mycobacteria is primarily through inhalation, it is essential that the dispersal of organisms into the air be minimized and that inhalation of airborne bacilli be avoided. The biologic safety cabinet is the single most important equipment in a mycobacteriology laboratory. There are various types of biologic safety cabinets, including Class I negative-pressure cabinets and Class II vertical-laminar-flow cabinets. Proper installation, maintenance, and testing are essential to their performance. Each cabinet should be tested and recertified at least yearly by trained personnel with special monitoring equipment. Processing raw specimens or transferring viable cultures outside a safety cabinet should not be permitted.

Ultraviolet (UV) light has a killing effect on microorganisms. After the safety cabinet work area has been cleaned with disinfectant, the UV light within the cabinet should be used to further eliminate contamination of surfaces and airborne bacteria. Because of the hazards to skin and eyes associated with excess UV light, the UV light should be turned on only when the cabinet is not in use.

Use of Proper Disinfectant

To prevent the dispersal of infectious aerosols into the laboratory area, all potentially infectious materials should be tightly covered when outside the biologic safety cabinet. Specimens should be centrifuged in aerosol-free safety carriers, and the tubes should be removed from the safety carriers only inside the biologic safety cabinet. Covering the work surface with a towel or absorbent pad soaked with a phenolic disinfectant reduces the accidental creation of infectious aerosols. Specimens taken out of the safety cabinet for transport to a decontamination area must be covered.

In the selection of a disinfectant for the mycobacteriology laboratory, the product brochure should be consulted to make certain that the disinfectant is bactericidal for mycobacteria (tuberculocidal). The following types of disinfectants are suitable for cleaning work areas, but some have limitations on their use.

- **Phenol-soap mixtures** containing orthophenol or other phenolic derivatives are effective with contact periods of 10 to 30 minutes.

- **Sodium hypochlorite** at a concentration of 0.1 to 0.5% (e.g., a 1:50 to 1:10 dilution of most household bleaches). The solution should be made fresh daily, and the contact time should be 10 to 30 minutes. Sodium hypochlorite loses effectiveness in the presence of a large amount of protein material.

- **Formaldehyde,** 3 to 8%; alkaline glutaraldehyde, 2%. Contact time should be at least 30 minutes or longer.

- **Phenol,** 5%, with a contact time of 10 to 30 minutes is adequate. Phenol is irritating to the skin and hazardous to the eyes.

Other Precautions

For sterilizing a wire inoculating loop, an electric incinerator should be used within the biologic safety cabinet. An alcohol-sand flask can also be used to clean the waxy culture material from the wire before flaming in a Bunsen burner. Disposable sterile applicator sticks or plastic transfer loops are very convenient and efficient for making smears and transfers. Splashproof discard containers must be used to prevent aerosol formation and possible cross-contamination of samples.

Protective clothing provides an extra measure of safety to the individual working in the mycobacteriology laboratory. Gloves and laboratory coats or gowns are essential. Face masks or respirators are recommended. Some laboratory directors believe that caps and shoe covers should also be worn.

SPECIMEN COLLECTION AND PROCESSING

Successful isolation of mycobacteria from clinical specimens begins with properly collected and handled specimens. Whenever possible, diagnostic specimens should be collected before the initiation of therapy. All specimens should be transported to the laboratory and ideally should be processed as soon as possible after collection. If immediate transport is not possible, the specimen should be refrigerated, but no longer than overnight. Delays in processing lead to false-negative cultures and increased bacterial contamination.

Each specimen should be confined to a single collection in an individual collection container recommended by the laboratory that is to provide the requested diagnostic service. The most commonly recommended containers are a sterile, wide-mouth jar with tightly fitted screw-cap lid and a sterile, disposable (nonpolystyrene), 50-mL centrifuge tube with a leakproof screw-cap lid. Special receptacles containing a 50-mL centrifuge tube for sputum collection are available commercially.

The spectrum of illness caused by *Mycobacterium* spp is so broad that almost any site may yield an acceptable specimen. Each specimen type, even when properly collected, transported, and processed, may have an intrinsic maximal yield. This can be the result of tubercle burden at the collection site or of environmental effects like pH that may affect recovery. Emphasis should be placed on collecting the number and types of specimens that, when transported and processed correctly, maximize the diagnostic yield. Table 22–2 lists the types of clinical samples acceptable for mycobacteriologic culture.

Sputum and Other Respiratory Secretions

Although a variety of clinical specimens may be submitted to the laboratory to recover *M. tuberculosis* (MTB) and MOTT, respiratory secretions, such as sputum and bronchial aspirates, are the most commonly collected. The number of specimens necessary to obtain culture confirmation and perform susceptibility testing is related to the frequency of smear positivity. According to Bates (1979), "When at least two of the first three sputum smears are positive, then three specimens usually are enough to confirm the diagnosis. If none, or only one of the first three sputum smears is positive, a larger number of specimens is needed for culture confirmation." Smear positivity and culture yield vary with the extent of the disease, i.e., whether there

■■■■■ T A B L E 2 2 – 2
ACCEPTABLE SPECIMENS FOR PROCESSING

Respiratory Specimens

Spontaneously expectorated sputum
Normal saline–nebulized, induced sputum
Transtracheal aspirate
Bronchioalveolar lavage
Bronchioalveolar brushing
Laryngeal swab
Nasopharyngeal swab

Body Fluids

Pleural fluid
Pericardial fluid
Joint aspirate
Gastric aspirate
Peritoneal fluid
Cerebrospinal fluid
Stool
Urine
Pus

Body Tissues

Blood
Bone marrow biopsy/aspirate
Solid organ
Lymph node
Bone
Skin

is cavitary or noncavitary pulmonary disease or endobronchial or laryngeal disease.

The sputum samples should be 5- to 10-mL samples produced by a deep cough or expectorated sputum induced by inhalation of an aerosol of hypertonic saline. When sputum is not obtainable, bronchoscopy may be performed, at which time samples such as bronchial washing, bronchoalveolar lavage (BAL), or transbronchial biopsy specimens are obtained. Brushings appear to be more commonly diagnostic than washing or biopsy. This may be the result of an inhibitory effect on the mycobacteria of the volumes of lidocaine used in adults during bronchoscopy or of dilution of the specimen with saline. Often, patients are able to produce sputum for several days after bronchoscopy; these samples should be collected and examined.

Gastric Aspirates and Washings

Gastric aspirates should be obtained in the morning after an overnight fast. Sterile water, 30 to 60 mL, is instilled either orally or via nasogas-

tric tube aspiration. Prolonged exposure to gastric acid kills mycobacteria and diminishes culture yield. Specimen processing should be done expeditiously, or the specimen should be neutralized with sodium carbonate or another buffer salt to pH 7.0.

Children with primary pulmonary tuberculosis typically have closed caseous lesions with relatively small numbers of organisms, resulting in a low yield from analysis of sputum. Abadco and colleagues (1992) have suggested that gastric lavage is better than bronchoalveolar lavage for the detection of mycobacteria in children. They reported that a bacteriologic diagnosis of childhood pulmonary TB could be made in 10% versus 50%, respectively, when BAL was compared with three morning gastric lavage specimens. This has not appeared to be the case in adults. Gastric lavage may offer a diagnostic alternative only in those unable to expectorate sputum and in whom BAL might be contraindicated.

Urine

For examination of urine, a first morning midstream specimen is preferred. The entire volume of voided urine, or a minimum of 15 mL, is collected in a sterile container. A specimen may be collected through an indwelling catheter with a sterile needle and syringe. Urine specimens should be refrigerated during the interval between collection and processing; specimens should be processed promptly.

As a general rule, pooled specimens collected over 12 to 24 hours are not recommended. Such specimens are more subject to contamination and may contain fewer viable tubercle bacilli.

Stool

Examination of stool specimens for the presence of acid-fast organisms can be useful in the diagnosis of disseminated mycobacterial disease in patients with acquired immune deficiency syndrome (AIDS) and in individuals at risk for AIDS. Frequently, the number of organisms found in the bowel in these patients is very high, although it has been reported that 68% of *M. avium* complex (MAC) culture-positive stool specimens are acid-fast smear negative. Stool specimens should be collected in clean containers without any preservative and sent directly to the laboratory for processing. If processing within a few hours is not possible, the specimen should be kept frozen at –20°C until processing time. Culture of feces for mycobacteria from patients other than those with or at risk for AIDS is usually not warranted.

Blood

Mycobacteremia, once considered rare, is now often seen in patients with AIDS but less frequently in other immunocompromised hosts. The Isolator lysis-centrifugation system (Wampole Laboratories, Cranbury, NJ) with subsequent transfer into the radiometric culture isolation system, and BACTEC 12B or direct inoculation of blood into BACTEC 13A Medium (Becton Dickinson Diagnostic Instrument Systems, Towson, Maryland) have been shown to be effective for the collection and culture of blood samples. The two collection systems have been considered equivalent, although one report suggests that the Isolator blood culture system inhibits growth in the BACTEC 12B, with diminished recovery of *M. avium* complex, particularly from blood specimens.

Tissue and Other Body Fluids

At times, tissue and other body fluids may be needed for microscopic examination and culture. Cerebrospinal fluid (CSF) specimens, whenever possible, should be from large-volume spinal taps to increase diagnostic yield. Diagnosis of tuberculous meningitis is very difficult. Peritoneal (ascitic fluid) smears are also rarely positive for acid-fast bacilli. Culture of large volumes and collection of the specimen into BACTEC bottles or other mycobacterial liquid media may help maximize yield in this and other dilute specimens.

When noninvasive techniques have failed to provide a diagnosis, surgical procedures may need to be considered. Specimens may need to be obtained from the lung, pericardium, lymph nodes, bones, joints, bowel, or liver. The tissue or fluid should be collected aseptically and placed in a sterile container. If the tissue is not processed immediately, a small amount (10 to 15 mL) of sterile saline should be added to prevent dehydration. It may be necessary to collect fluid containing fibrinogen (pleural, pericardial, peritoneal) into a container with an anticoagulant. The amount of fluids recommended for culture vary—2 mL for cerebrospinal fluid, 3 to 5 mL for exudates and for pericardial and synovial fluids, and 10 to 15 mL for abdominal and chest fluids. Immediate processing of these samples is important. When tissue is collected, histologic evaluation may reveal caseating or noncaseating granuloma formation with the presence of multinucleated giant cells. These histologic changes would be consistent with but not specific for mycobacterial disease.

DIGESTION AND DECONTAMINATION OF SPECIMENS

Most clinical specimens contain mucin or organic debris that surround the bacteria within the sample. An abundance of nonmycobacterial organisms, as well as possible mycobacteria, make up the microflora of these specimens. When placed into culture medium, the abundant nonmycobacterial organisms can quickly overgrow the more slowly growing mycobacteria. The purposes of the digestion-decontamination process are (1) to liquefy the sample through digestion of the proteinaceous material and (2) to allow the chemical decontaminating agent to contact and kill the nonmycobacterial organisms. The high lipid content in the cell walls of mycobacteria makes them somewhat less susceptible to the killing action of various chemicals. With liquefaction of the specimen, the surviving mycobacteria can be concentrated with centrifugation. Additionally, liquefying the mucin enables the mycobacteria to contact and utilize the nutrients of the medium to which they are subsequently inoculated.

Specimens that contain mucus and need both digestion and decontamination are sputum, gastric washing, bronchoalveolar lavage, bronchial washing, and transtracheal aspirate. Voided urine, autopsy tissue, abdominal fluid, and any fluid known to be contaminated need decontamination only. Specimens from normally sterile sites, such as blood, CSF, synovial fluid, and biopsy tissue from deep organs, do not need decontamination. Sterility needs to be strictly maintained in collection and transport. Stool decontamination is especially difficult and may require repeated attempts at decontamination.

Decontamination and Digestion Agents

Each laboratory should determine that the proper balance between rate of recovery of mycobacteria and the suppression of contaminating growth is being maintained. Failure to isolate mycobacteria from patients with signs and symptoms of classic mycobacterial disease may indicate that the decontamination is too harsh. On the other hand, if more than 5% of all specimens cultured are contaminated, the decontamination procedure may be inadequate. The bactericidal action of a decontaminating agent is influenced by the concentration of the chemical agent, exposure time, and temperature; therefore, alterations in any of these factors may increase or decrease the bactericidal effect. In general, what is considered acceptable in this delicate balance is that between 2 and 5% of mycobacterial cultures are bacterially contaminated.

The optimum decontamination procedure requires an agent that is mild and yields growth of mycobacteria while controlling contaminants. The use of selective or antibiotic-treated media may diminish the need for harsh decontamination procedures.

Sodium Hydroxide

Sodium hydroxide (NaOH), usual concentration 2, 3 or 4%, serves as both a digestant and a decontaminating agent. It must be used cautiously, because it is only slightly less harmful to the mycobacteria than to the contaminating organisms. The selection of the concentration of sodium hydroxide to be used in the laboratory depends on the type, age, and transport of the specimens received by that laboratory. Lower concentrations of NaOH result in higher recovery rates for some mycobacterial species that are more susceptible to the alkali.

N-Acetyl-ʟ-Cysteine

A combination of a liquefying agent, such as *N*-acetyl-ʟ-cysteine (NALC) or dithiothreitol, and sodium hydroxide is commonly used. The liquefying agent has no inhibitory effect on bacterial cells; however, liquefaction of the sample allows the decontaminating chemical to come into uniform contact with the contaminating bacteria more readily. When contamination can be controlled with a lower concentration of sodium hydroxide, the recovery of mycobacteria is indirectly improved using the milder procedure, because fewer mycobacteria are lost in the process. A modified procedure decreases the NaOH from 2 to 1.5%.

Procedure 22–1 describes NALC-NaOH digestion and decontamination.

Cetylpyridium Chloride and Sodium Chloride

Sputum samples that may be in transport for more than 24 hours, as during transport to a reference laboratory, may be mixed at the time of collection with equal volumes of 1% cetylpyridium chloride (CPC) and 2% sodium chloride to effect decontamination, liquefaction, and then concentration of the specimen. The advantages are that the procedure increases yield of MOTT with less contamination; however, it has been reported that the decontaminating agent cetylpyridium chloride remains active on agar-based media and that this residual drug may inhibit the growth of *Mycobacterium* spp. As a consequence, specimens processed in this manner should be planted only on egg-based media.

Benzalkonium Chloride

Another digestant-decontamination procedure utilizes benzalkonium chloride (Zephiran) combined with trisodium phosphate (Z-TSP). Trisodium phosphate liquefies sputum rapidly but requires long exposure to decontaminate the specimen. The addition of benzalkonium chloride shortens the exposure time and effectively destroys many contaminants, with little bactericidal effect on the tubercle bacilli.

The addition of phosphate buffer to digested specimens results in greater isolation of mycobacteria and, in general, produces larger colonies. Like CPC, Zephiran is bacteriostatic for tubercle bacilli, necessitating either neutralization before planting or use of egg-based media to make use of its inherent neutralizing capacity.

Oxalic Acid

Oxalic acid, 5%, is used to decontaminate specimens contaminated with *Pseudomonas aeruginosa,* such as sputum specimens from patients with cystic fibrosis.

PROCEDURE 22–1. NALC–SODIUM HYDROXIDE (NALC-NaOH) DIGESTION-DECONTAMINATION

Reagents

NALC-NaOH Digestant: Combine equal volumes of 2.94% sodium citrate dihydrate (= 0.1M) and 4% sodium hydroxide (NaOH). Just before use, add 0.5 g of powdered NALC per 100 mL of mixture. Store refrigerated when not in use; discard after 24 hours.

Phosphate buffer, 0.067 M, pH 6.8, or water, sterile, distilled, 30–40 mL per specimen.

Bovine albumin fraction V, 0.2% in saline, sterile.

Procedure

1. Working within the biologic safety cabinet, transfer 10 mL, or total specimen if volume is less than 10 mL, to 50-mL screw-cap centrifuge tube. If specimen volume is greater than 10 mL, select most purulent-appearing material. Add an equal volume of NALC-NaOH digestant to each sample.

2. Tighten caps and mix on a test tube mixer until liquefied (5 to 20 seconds per tube), inverting each tube to ensure that NALC-NaOH solution contacts any untreated particles on the upper part of the tube.

3. Let tubes stand at room temperature for 15 minutes. If more decontamination is desired, increase the concentration of NaOH rather than the time the specimen is exposed to the digestion-decontamination mixture.

4. Dilute the digested-decontaminated specimens to the 50-mL mark with sterile distilled water or sterile phosphate buffer to minimize the continuing action of the NaOH and lower the specific gravity of the specimen. Tighten caps and invert or swirl to mix.

5. Centrifuge at 3000g for 15 minutes (or the appropriate combination of relative centrifugal force and time to give 95% sedimentation) using aerosol-free safety cups or in an aerosol-controlled vented centrifuge.

6. Holding the tube so that the sediment is on the upper side of the tube, pour off supernatant into a splashproof discard container of disinfectant.

7. Holding the tube in a horizontal position to keep the sediment as dry as possible, use a sterile applicator stick to remove a small part of the sediment and place it on a marked microscope slide. The smear should be about 1 × 2 cm.

8. Resuspend sediment in 1–2 mL of sterile 0.2% bovine albumin solution. If media will be inoculated immediately, the sediment may be resuspended in sterile water or sterile saline.

9. An optional step is to prepare a 1:10 dilution using 0.5 mL of the resuspended sediment in 4.5 mL of sterile water. Dilution decreases the concentration of toxic substances that may inhibit growth of mycobacteria. Inoculate diluted and undiluted specimens to solid media.

Sulfuric Acid

Rigorous decontamination with 4% sulfuric acid is used for the specimens that are persistently contaminated after alkaline decontamination. Such specimens include urine and other body fluids. Less careful timing is required when more gentle processing methods are employed. All specimens must be either neutralized after digestion and decontamination or inoculated on media with an inherent neutralizing capacity.

A digestant's toxicity is not insignificant. Investigations have shown that the NALC-NaOH with 1% NaOH and Z-TSP digestion-decontamination procedures result in the death of 28 to 33% of the mycobacteria in clinical specimens. When a specimen is treated for 15 minutes with 1 to 2% or occasionally 3% NaOH with NALC or dithiothreitol, loss of mycobacteria could be as much as 10^4. This is regarded as a relatively mild decontamination procedure when compared with the 60 to 70% killing of mycobacteria when 4% NaOH is used.

CONCENTRATION

The specific gravity of the tubercle bacilli ranges from 1.07 to 0.79. Because of the low specific gravity of the acid-fast bacilli, a low centrifugal force has a buoyant rather than a sedimenting effect. Excess mucus will compound this effect. Treatment with mucolytic agents such as NALC splits mucoprotein, allowing greater sedimentation. Kent and Kubica (1985) suggest that a 95% sedimenting efficiency should be the goal for recovery. Therefore, concentration centifugation speeds must be at least 3,000g to maximize recovery.

Lower g force necessitates longer centrifuge time. The consequences of longer centrifuge time are prolonged exposure to the toxic effects of both the digestion-decontamination agents used and the higher temperatures generated by unrefrigerated centrifuges. Alternatives to high-speed centrifugation concentration have been explored. In summary, the digestion-decontamination agent used, its concentration, the length of exposure of the agent to the specimen, and the centrifugation speed and temperature all affect the recovery of *Mycobacterium*.

STAINING FOR ACID-FAST BACILLI

Irregular uptake of stain may give the mycobacteria a beaded appearance, and when Gram stained, *Mycobacterium* spp stain faintly or not at all.

The acid-fast smear examination of the patient's specimen is a simple and rapid procedure that can be performed routinely in the clinical microbiology laboratory. Stains in the clinical laboratory are prepared from digested, decontaminated, and concentrated specimens. Staining is based on the increased lipid content of the cell wall of *Mycobacterium* and their resistance to acids, alkali, and drying.

The conventional acid-fast staining methods, Ziehl-Neelsen and Kinyoun, utilize carbolfuchsin solutions for the primary stain, acid-alcohol as a decolorizing agent, and a methylene blue coun-

terstain. The Ziehl-Neelsen staining procedure involves the application of heat with the carbol-fuchsin stain, whereas the Kinyoun acid-fast stain is a cold stain. Slides are examined using a 100× oil immersion objective on a light microscope for 15 minutes, viewing a minimum of 300 fields before a slide is called negative.

The auramine or auramine-rhodamine fluorochrome stains, considered more sensitive than the carbolfuchsin stains, have proved to be very reliable and specific (Sommers and McClatchy, 1983). Fifteen to 18% of all culture-positive specimens had smears that were positive on auramine-rhodamine stain but negative on Kinyoun or Ziehl-Neelsen stain. A fluorescence microscope equipped with an appropriate filter system is needed for the examination of a fluorochrome-stained smear. The smear is examined with a 25× or 40× objective under a mercury vapor lamp with a strong blue filtered light. Positive stains reveal bright yellow-orange bacilli against a dark background. Because the stain is examined under lower power, less time is involved in scanning the slide.

In the interpretation of a smear as positive for acid-fast organisms, it must be realized that organisms other than *Mycobacterium* may stain at least partially acid fast. *Nocardia* sp, *Legionella micdadei,* and *Rhodococcus* sp may all appear acid fast.

The sensitivity of the direct acid-fast smear examination for the diagnosis of mycobacterial infection is lower than that of culture; however, it is the easiest and quickest test that can be performed. Results of the examination can provide preliminary information for the clinician. The smear is most sensitive when the patient is able to produce sputum with larger numbers of bacilli when the patient is the most contagious. The 10^5 bacilli per mL found in sputum from patients with cavitary tuberculosis are easily detected microscopically with standard concentration techniques. A minimum of 10^3 organisms per mL is thought to be necessary for a positive smear. Patients with noncavitary disease excrete low numbers of organisms, and excretion may be intermittent.

Overall sensitivity of acid-fast stain examination is low. In a review of clinical specimens submitted to the VA Medical Center in Seattle, 33% of culture-positive specimens were smear positive overall. Respiratory specimens that were culture positive for MTB were positive on acid-fast staining 65% of the time. In this study, it appeared that acid-fast smear microscopic examination was more sensitive at detection of specimens with MTB than with MOTT. This is believed to be related to the greater number of organisms seen in MTB disease. False-negative acid-fast smear results may be due to overzealous decontamination, loss from concentration techniques, organisms obscured by a too thick smear, over-decolorizing of the smear, poor counterstaining, and lack of observer proficiency in reading stains.

The smear does not reflect the viability of the acid-fast bacilli stained, although fluorochrome dye may stain nonviable organisms better than carbolfuchsin. Attempts have been made to produce a fluorescent viability stain. Smears may remain positive for a short period after adequate therapy is initiated, but cultures will become negative within a few weeks. With treatment, the cell wall changes so as not to stain as well with carbolfuscin but still to stain with fluorochromes. This is one explanation for a "false-positive" smear, i.e., one for which the companion culture is negative. Other absolute causes of false-positive smears are lack of proper decolorization and laboratory contamination as a result of MOTT contaminating water supplies, harsh decontamination, delayed processing, not maximizing yield by using multiple media, specimen not held long enough, and overgrowth with bacteria.

Parrot and associates (1971) found that "the number of organisms seen on smear had a bearing on the culture results. There was a 50% chance of a negative culture when only 1 organism per 100 fields was seen." Multiple specimens increase the yield of acid-fast smears. When patients with pulmonary TB submitted more than one specimen, 96% had at least one positive sample. Of these patients with at least one positive smear, 89% had a positive culture. The relative frequency of false-positive smears depends on the prevalence of TB in the laboratory's population and the absolute frequency of false-positive smears. A relative increase in the frequency of TB results in an increase in the reliability of acid-fast smears. Because of a somewhat higher incidence of false-positive reactions with fluorochrome staining, some laboratories confirm stains that are in dispute with Ziehl-Neelsen or Kinyoun staining.

Most laboratories report prevalence rates for acid-fast smears as 1 to 8% for all clinical specimens. It should be emphasized that a positive acid-fast stain reaction is only a presumptive diagnosis of mycobacterial disease that, if at all possible, must be substantiated by culture.

CULTURE MEDIA AND ISOLATION METHODS

Mycobacteria are strictly aerobic and grow more slowly than most bacteria pathogenic for humans. The generation time of the mycobacteria is more than 12 hours, that of *M. tuberculosis* having the longest replication time at 20 to 22 hours. The most rapidly growing species generally grows on simple media in 2 to 3 days at temperatures of 20 to 40°C. Most mycobacteria associated with disease require 2 to 6 weeks of incubation on complex media at specific optimum temperatures. The growth of *M. tuberculosis* is enhanced by an atmosphere of CO_2 between 5 and 10% for the few first weeks of incubation. This organism has a narrow range of tolerance to CO_2. Mycobacteria require a pH between 6.5 and 6.8 for the growth medium, and they grow better at higher humidity. One of the mycobacteria pathogenic for humans, *M. genavense,* grows very limitedly whereas another, *M. leprae,* fails to grow in vitro.

The many different media available for the recovery of mycobacteria from a clinical specimen are variations of three general types (Table 22–3): inspissated egg medium, serum albumin agar medium, and liquid medium. Within each general type, there are nonselective formulations and formulations that have been made selective by the addition of antimicrobial agents. Because some isolates do not grow on a particular agar and each type of culture medium offers certain advantages, a combination of culture media is generally recommended for primary isolation.

Egg-Based Media

The basic ingredients in an inspissated egg medium, such as Löwenstein-Jensen (LJ), Petragnani, and American Thoracic Society (ATS) media, are fresh whole eggs, potato flour, and glycerol, with slight variations in defined salts, milk, and potato (see Table 22–3). Each contains malachite green to suppress the growth of gram-positive bacteria. Löwenstein-Jensen medium is most commonly used in clinical laboratories and is intermediate in its concentration of malachite green. Petragnani medium contains twice as much malachite green as LJ and is more commonly used by laboratories in which the specimens are expected to have higher contamination. ATS medium, which has a lower concentration of

◾◾◾TABLE 22-3
MYCOBACTERIAL CULTURE MEDIA

Medium	Composition	Inhibitory Agents
American Thoracic Society (ATS)	Fresh whole eggs, potato flour, glycerol	Malachite green (0.02%)
Löwenstein-Jensen (LJ)	Fresh whole eggs, defined salts, glycerol, potato flour	Malachite green (0.025%)
Petragnani	Fresh whole eggs, egg yolks, whole milk, potato, potato flour, glycerol	Malachite green (0.052%)
Middlebrook 7H10	Defined salts, vitamins, cofactors, oleic acid, albumin, catalase, glycerol, glucose	Malachite green (0.00025%)
Middlebrook 7H11	Defined salts, vitamins, cofactors, oleic acid, albumin, catalase, glycerol, 0.1% casein hydrolysate	Malachite green (0.0001%)
Middlebrook 7H9, 7H12 (BACTEC 12B)	Broth base, casein hydrolysate, bovine serum albumin, catalase, C-14–labeled palmitic acid, deionized water	Polymyxin B Amphotericin B Nalidixic acid Trimethoprim Azlocillin
Middlebrook 7H9, 7H13 (BACTEC 13A)	Broth base, casein hydrolysate, bovine serum albumin, catalase C-14–labeled palmitic acid, SPS (sodium polyanethol sulfonate), polysorbate 80	Polymyxin B Amphotericin B Nalidixic acid Deionized water Trimethoprim Azlocillin
Gruft (modification of LJ)	Fresh whole eggs, defined salts, glycerol, potato flour, RNA	Malachite green Penicillin Nalidixic acid
Mycobactosel (BBL) LJ	Fresh whole eggs, defined salts, glycerol, potato flour	Malachite green Cycloheximide Lincomycin Nalidixic acid
Middlebrook 7H10 (selective)	Defined salts, vitamins, cofactors, oleic acid, albumin, catalase, glycerol, glucose	Malachite green Cycloheximide Lincomycin Nalidixic acid
Mitchison's selective 7H11	Defined salts, vitamins, cofactors, oleic acid, albumin, catalase, glycerol, glucose, casein hydrolysate	Carbenicillin Amphotericin B Polymyxin B Trimethoprim lactate

malachite green than LJ, is best used with specimens that are only lightly contaminated or are uncontaminated. Selective media that contain antibiotics, such as Gruft modification of LJ and Mycobactosel (Becton Dickinson Microbiology Systems, Cockeysville, MD), are sometimes used in combination with nonselective media to increase isolation of mycobacteria from contaminated specimens. The nonselective egg-based media offer the advantage of a long shelf-life of 1 year and the disadvantage that distinguishing early growth from debris is sometimes difficult.

Serum- or Agar-Based Media

Serum albumin agar media, such as Middlebrook 7H10 and 7H11 agars, are prepared from a basal medium of defined salts, vitamins, cofac-

tors, glycerol, malachite green, and agar combined with an enrichment consisting of oleic acid, bovine albumin, glucose, and beef catalase (Middlebrook OADC enrichment). Middlebrook 7H11 also contains 0.1% casein hydrolysate, which improves recovery of isoniazid-resistant strains of *M. tuberculosis*. The addition of antimicrobial agents to either 7H10 or 7H11 makes the medium more selective by suppressing the growth of contaminating bacteria. Mitchison's selective 7H11 contains polymyxin B, amphotericin B, carbenicillin, and trimethoprim lactate (see Table 22–3).

In contrast to the opaque egg-based media, clear agar-based media can be examined using a dissecting microscope for early detection of growth and colony morphology. Drug susceptibility tests may be performed on agar-based media without the alteration of drug concentrations that occurs with egg-based media. When specimens are inoculated to Middlebrook 7H10 and 7H11 media and incubated in an atmosphere of 10% carbon dioxide and 90% air, 99% of the positive cultures are detected in 3 to 4 weeks, earlier than for those detected on egg-based media.

Certain precautions should be followed in the preparation, storage, and incubation of Middlebrook media. Excess heat and exposure of the prepared media to light can both result in the release of formaldehyde, which is toxic to mycobacterial growth.

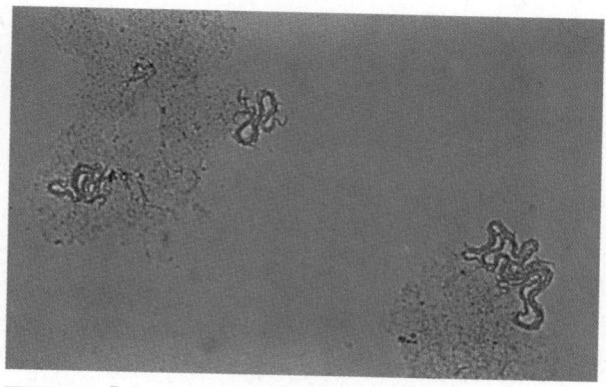

Figure 22–1

Early microcolonies.

Microcolony Plates

Microscopic examination of thinly poured 7H11 agar allows earlier detection of mycobacterial growth with appraisal of colonial morphology. These plates (Fig. 22–1) inoculated with 0.1 mL of concentrated specimen are, on the average, positive within 11 days. When this medium is combined with conventional media, rapid recovery time is combined with high sensitivity.

Liquid Media

Mycobacterium spp grow more rapidly in liquid media. Middlebrook 7H9 broth is a nonselective liquid medium used for subculturing stock strains, picking single colonies, and preparing inoculum for in vitro testing.

The most sensitive and rapid primary isolation liquid media are Middlebrook 7H12 and 7H13 (BACTEC 12B and 13A, Becton Dickinson Diagnostic Instruments Systems, Towson, MD). The BACTEC system is an automated radiometric culture system for detecting the growth of *Mycobacterium*.

The BACTEC broth contains a ^{14}C-labeled substrate (palmitic acid) that is metabolized by mycobacteria, liberating radioactive carbon dioxide ($^{14}CO_2$) into the headspace of the vial. The amount of $^{14}CO_2$ liberated is detected by the BACTEC 460 instrument and interpreted as a "growth index." It is assumed that the release of CO_2 denotes growth of the organism.

The recommended inoculum for the BACTEC 12B 4-mL vial is 0.5 mL of decontaminated concentrated specimen. Antibiotics supplied by the manufacturer—polymyxin B, amphotericin B, nalidixic acid, trimethoprim, and azlocillin (PANTA) recon-

stituted into polyoxyethylene stearate, a growth-enhancing agent—are added to each vial at the time of inoculation. The Middlebrook 7H13 medium (BACTEC 13A) was introduced for the culturing of larger volumes of blood or bone marrow. Its components are similar to those of the BACTEC 12B vial, except that an anticoagulant, sodium polyanethol sulfonate (SPS), and polysorbate 80 have been added. Five milliliters of blood may be added directly to this 30-mL vial.

A number of studies have shown that the radiometric BACTEC isolation method significantly improves the isolation rate of mycobacteria and reduces the recovery time compared with conventional isolation media. The BACTEC vials should be read within 4 days of inoculation. Negative vials should be retested every 3 to 4 days for the first 2 weeks and then once weekly for the remaining 6 weeks. With the BACTEC method, mycobacteria may be detected in clinical specimens in less than 2 weeks.

In general, the time required for isolation and identification of *M. tuberculosis* is reduced from 6 weeks to 3 weeks. The time in which the radiometric system will detect growth reflects the quantity of viable *Mycobacterium* species in the submitted sample, which is indirectly reflected by smear positivity.

In smear-positive samples, *M. tuberculosis* may be detected as early as 7 to 8 days and *M. avium* as early as 5 to 8 days. In smear-negative samples, MTB and MOTT are usually detected in 14 to 28 days and 8 to 12 days, respectively, by the BACTEC system. These detection times are an obvious improvement in the recovery time from conventional media. In conventional agar cultures, recovery from smear-positive TB specimens

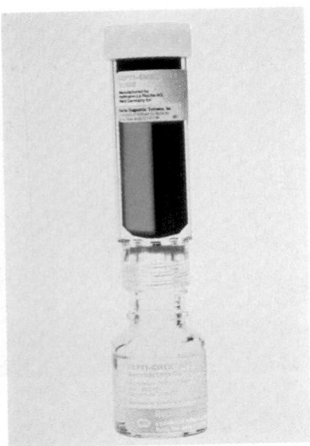

■ Figure 22–2
Septi-Chek AFB.

may require 16 days, and from smear-negative specimens, 26 days. By the end of the fourth week, 96.8% of all effectual positives will have been detected, and by the end of the fifth week, 98.8%. The greater sensitivity of the BACTEC system has resulted in a higher yield from smear-negative specimens.

The disadvantage of the BACTEC system is that when the BACTEC vial is positive before a companion agar or egg medium culture, there is no colony morphology or pigmentation to suggest that the growth is of a mycobacterial species other than tuberculosis. The BACTEC system also may not be as good in recovery of all species of *Mycobacterium*, specifically *M. fortuitum* and *M. avium* complex. False-positive results from cross-contamination have been reported.

The BACTEC system can also be used for the antimicrobial susceptibility testing of *M. tuberculosis* (see later). Currently, it is not recommended for sensitivity testing of MOTT. Under development are automated, nonradiometric microbial detection systems designed for the recovery of mycobacteria from clinical specimens.

Other Culture Media for Recovery of Mycobacteria

A chocolate agar plate should be included in the primary isolation media for skin and other body surface specimens specifically for the recovery of *M. haemophilum*, which requires hemoglo-bin or hemin for growth. The plate should be incubated at 30°C, the optimum temperature for recovery of this organism and for *M. marinum*.

A biphasic media system for the detection and isolation of mycobacteria, Septi-Chek AFB, is available commercially (BBL Septi-Chek AFB, Becton Dickinson Microbiology Systems, Cockeysville, MD) (Fig. 22–2). The system consists of a bottle containing Middlebrook 7H9 broth with an atmosphere of 5 to 8% CO_2, an enrichment consisting of growth-enhancing factors and antimicrobial agents, and a paddle with agar media. One side of the paddle is covered with nonselective Middlebrook 7H11 agar. One of the two sections on the reverse side of the paddle contains a modified egg-based medium for differentiating *M. tuberculosis* from other mycobacteria, and the other contains chocolate agar for the detection of contaminating bacteria.

The biphasic media system provides for growth for rapid identification and drug susceptibility testing without the need for routine subculturing. The biphasic media is a sensitive detection system that avoids the use of radioactivity and a CO_2 incubator. Detection times are shorter than with conventional agar but significantly longer than with the BACTEC system.

Isolator Lysis-Centrifugation System

Isolator (Wampole Laboratories, Cranbury, NJ) is a collection system that contains saponin to liberate intracellular organisms. After treatment with the saponin, the sample is inoculated into mycobacteria media plates or tubes. The system allows for higher yields and shorter recovery times for mycobacteria. It offers the advantage of yielding isolated colonies and the ability to quantitate mycobacteremia, which may be useful in monitoring the effectiveness of therapy in disseminated *M. avium* complex infection. Isolator concentrate can also be inoculated into BACTEC recovery systems.

For maximum recovery of mycobacteria, many laboratories use a battery of one or more units of an egg-based medium, one agar medium, and the radiometric broth method for primary isolation. A selective medium is often reserved for specimens in which heavy contamination is anticipated.

Procedure 22–2 outlines the inoculation and incubation of isolation media.

PROCEDURE 22-2. INOCULATION AND INCUBATION OF ISOLATION MEDIA FOR *MYCOBACTERIUM*

1. After resuspending the decontaminated specimen in 1–2 mL of 0.2% bovine albumin (or sterile distilled water or phosphate buffer), mix on a vortex mixer. If the BACTEC system is used for isolation, phosphate buffer is the reagent of choice for resuspension of patient's specimen.

2. Using a sterile plugged capillary pipette or a disposable plastic transfer pipette, transfer approximately 0.25–0.5 mL of specimen to each unit of solid media.

3. Using an allergist syringe, inoculate 0.5 mL of specimen to one BACTEC 12B vial that has been previously tested to establish CO_2 atmosphere and supplemented with 0.1 mL of PANTA Plus (a supplement of antimicrobial agents).

4. Leave caps loose on tubes and bottles of egg-based medium to allow adequate carbon dioxide gas exchange when placed in a CO_2 incubator. An atmosphere of 5–10% CO_2 stimulates growth of all mycobacteria, but it is necessary only during the log phase of growth, the first 7 to 10 days after inoculation.

5. Place tubes and bottles in the incubator on a slant with inoculated surface up to allow distribution and absorption of the inoculum over the entire slant. After overnight absorption of inoculum, tubes may be placed upright. Middlebrook agar plates should be incubated upright and can be placed in CO_2-permeable bags to retard dehydration of the plates. Incubation temperature is 37°C. Incubate BACTEC 12B vials at 37°C in air.

6. Specimens from skin lesions should be inoculated to a duplicate set of media and incubated at 30°C for the recovery of *M. marinum,* and inoculated to a hemin-containing medium for the recovery of *M. haemophilum.*

7. Solid culture media should be examined twice weekly for the first 4 weeks and at weekly intervals thereafter.

8. BACTEC 12B vials should be read on the BACTEC instrument within 4 days of inoculation. The remaining negative vials should be tested every 3 to 4 days for the first 2 weeks and then once a week for the remaining recommended 6 weeks.

9. All cultures should be maintained until there is no evidence of growth at 8 weeks. Prolonged incubation, i.e., 10–12 weeks or longer, may sometimes be necessary. (Kent and Kubica, 1985.)

IDENTIFICATION

Laboratory Levels or Extents of Service

A change in the distribution of mycobacterial laboratory testing led to development of the concepts of *levels of service* by the American Thoracic Society (ATS) and *extents of service* by the College of American Pathologists (CAP) to maintain quality of service.

Extents of service, as defined by ATS, are as follows:

1. Specimen collection only. No mycobacteriologic procedures performed, all specimens sent to another laboratory.

2. Acid-fast stain and/or inoculation only. Identification by reference laboratory.

3. Isolation and definitive identification of *M. tuberculosis,* preliminary grouping of nontuberculous *Mycobacterium,* with definitive identification at a reference laboratory.

4. Definitive identification of all mycobacterial isolates with assistance in the selection of therapy, with or without drug susceptibility testing.

The CAP's *levels of service* are as follows:

Level I. Specimen collection only. No mycobacteriologic procedures performed. All specimens sent to another laboratory.

Level II. Perform microscopy. Isolate and identify and sometimes perform susceptibility tests for *M. tuberculosis.*

Level III. Perform microscopy. Isolate, identify, and perform susceptibility testing for all species of *Mycobacterium.*

A facility's selection of a level of service depends on the volume of specimens submitted, the ability to perform the requested tests according to comfort and training in performance of each requested test, as well as the time, effort, and funds allocated for the service.

The Centers for Disease Control and Prevention (CDC) has developed separate training courses designed for each level of service as well as manuals for reference.

Procedures available within a particular mycobacteriology laboratory then vary with the level of service of that laboratory. In addition, the methodology within laboratories varies. The CDC, in collaboration with the Association of State and Territorial Public Health Laboratory Directors, surveyed mycobacterial laboratories for their practices in isolation, identification, and susceptibility testing of MTB. Those laboratories that used conventional mycobacteriologic methods for culturing and identification were not able to report results as quickly as those that have already incorporated newer methodologies, amount of time needed being 43 versus 22 days, respectively.

Phenotypic Characteristics

Colony morphology, growth rate, optimum growth temperature, and photoreactivity are phenotypic characteristics that may help speciate mycobacteria. These characteristics do not allow for definitive identification but are presumptive and help in the selection of other, more definitive tests. Table 22–4 summarizes the identification characteristics of clinically important mycobacteria.

Colony Morphology

Colonies of mycobacteria are generally distinguished as having either a smooth and soft or a rough and friable appearance. Colonies of *M. tuberculosis* that are rough often also exhibit a prominent patterned texture referred to as *cording* (curved strands of bacilli); this texture is the result of tight cohesion of the bacilli. Colonies of *M. intracellulare* may appear to have a dense center, looking like a "fried egg."

Growth Rate and Recovery Time

Growth rate and recovery time depend on the species of mycobacteria but are also influenced by the media used, the temperature of incubation, and the initial innoculum size. The range in recovery time is wide, from 3 to 60 days. Mycobacteria are generally categorized as having visible growth in less than or more than 7 days. *Rapid growers* are able to produce colonies in less than 7 days upon subculture to a nonselective media. The inoculum should be sufficiently small to produce isolated colonies. Microscopic agar examination for microcolonies allows earlier detection of growth.

Temperature

The optimum temperature and range at which a mycobacterial species may grow may be extremely narrow, especially at the time of initial incubation. *Mycobacterium marinum, M. ulcerans,* and *M. haemophilum* grow best at 30 to 32°C and poorly, if at all, at 35 to 37°C. At the other extreme, *M. xenopi* grows best at 42°C.

Photoreactivity

Mycobacterium have classically been categorized according to their photoreactive characteristics. Those species that produce carotene pigment upon exposure to light are referred to as *photochromogens.* Those species that produce pigment in the light or the dark are referred to as *scotochromogens.* Color may range from pale yellow to orange. Growth temperature may influence the photoreactive characteristics of a species. Other species are *nonchromogenic, nonphotoreactive,* like *M. tuberculosis.* These colonies are a buff color.

Biochemical Identification

A panel of biochemical tests can identify most mycobacteria isolates, but because growth of *Mycobacterium* is so very slow, this may take several weeks to accomplish. Progress in molecular technology has diminished the frequency with which biochemical tests are routinely done in the identification of mycobacteria. Because mycobacterial species may show only quantitative differences in enzymes used in biochemical identification, no single biochemical test should be relied upon for the identification of a species, and for expediency, all necessary biochemical tests should be set up at one time. The biochemical tests are based on the enzymes the organisms possess, the substances that their metabolism produce, and the inhibition of their growth on exposure to selected biochemicals. Appropriate positive and negative controls should be included for each biochemical test.

Niacin Accumulation

Most mycobacteria possess the enzyme that will convert free niacin to niacin ribonucleotide. Accumulation of niacin is detected as nicotinic acid. Nicotinic acid reacts with cyanogen bromide in the presence of an aniline to form a yellow pigmented compound (Fig. 22–3). Reagent-impregnated strips have eliminated the need to handle and dispose of cyanogen bromide, which is both caustic and toxic. Cyanogen bromide must be alkalinized with sodium hydroxide before disposal.

Ninety-five percent of *M. tuberculosis* isolates produce free niacin (nicotinic acid) because the species lacks the niacin-converting enzyme that most other mycobacteria possess. This test is the most commonly used biochemical test for the identification of MTB. This test may be negative when performed on young cultures with few colonies. It is recommended that the test be done on egg agar cultures 3 to 4 weeks old and with at least 50 colonies. Tests that yield negative results may need to be repeated in several weeks. The test should not be performed on scotochromogenic or rapidly growing species because rarely, *M. simiae,* BCG (bacillus of Calmette-Guérin) strain of *M. bovis, M. africanum, M. marinum, M. chelonae,* and *M. bovis* may be positive. Results are most consistent when the test is performed on egg media.

Nitrate Reduction

Nitrate reduction is characteristic of 97% of *M. tuberculosis.* The production of nitroreductase, which catalyzes the reduction of nitrate to nitrite, is relatively uncommon among other *Mycobacterium,* but a positive result may be seen in *M. kansasii, M. szulgai,* and *M. fortuitum,* as well as *M. tuberculosis.* Sulfanilamide and *N*-naphthylenediamine added to a solution of bacteria in sodium nitrate turn red in the presence of nitrite (Fig. 22–4). When no color change occurs, however, either no reaction has occurred or the reaction has gone beyond nitrite. The addition of zinc detects free nitrate resulting in a pink color change in a true-negative reaction. The nitrate reduction test differentiates *M. tuberculosis* from the scotochromogens and MAC.

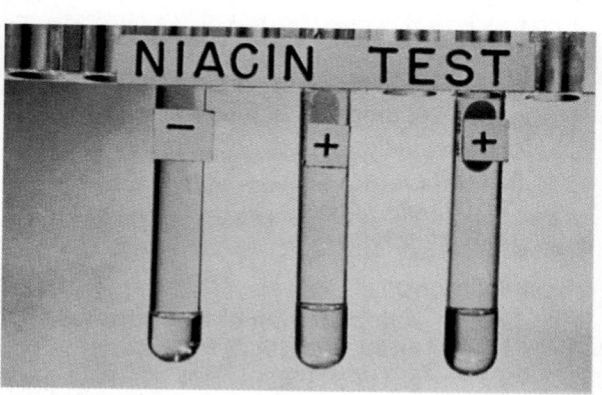

■ F i g u r e   2 2 – 3

Niacin test.

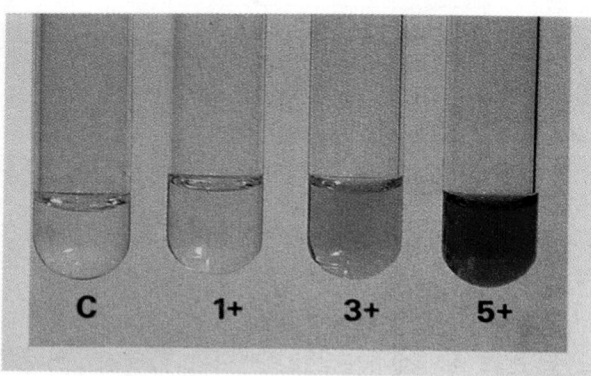

■ F i g u r e   2 2 – 4

Nitrate reduction test.

TABLE 22-4

IDENTIFICATION OF CLINICALLY IMPORTANT MYCOBACTERIA[1]

Descriptive Term	SPECIES (subspecies)	Colony Morphology	Temp. Growth Range (°C)	Growth Rate (R=Rapid S=Slow)	Pigment (Photo, Scoto, Nonphoto)	Arylsulfatase 3 Day	Arylsulfatase 2 Wk	Sodium Citrate	Inositol	Mannitol	Catalase Semi-Quant >45	Catalase Heat Stable (68°C)	Iron Uptake	Growth on MacConkey	Niacin	Nitrate Reduction	Pyrazin amidase[2]	Growth on 5% NaCl	Tellurite Reduction	Growth on TCH	Tween Hydrolysis + 10 day	Tween Opacity (1 week)	Urease[3]
Rapid Growers	M. fortuitum biovar fortuitum	(87) Smooth Rough (13)	22-40	R	N	+ (97)	+ (99)	- (99)	- (99)	- (99)	>45 (98)	+ (94)	+ (99)	+ (96)	- (99)	+ (99)		+ (75)	+ (97)	+ (99)	+/- (50)		+ (93)
	M. fortuitum biovar peregrinum	Smooth	22-37	R	N	+ (97)	+ (99)	- (99)	- (99)	+ (99)	>45 (98)	+ (94)	+ (99)	+ (96)	- (99)	+ (99)		+ (75)	+ (97)	+ (99)	+/- (50)		+ (93)
	M. fortuitum[5] 3rd. biovariant complex	Smooth	22-37	R (97)	N	+ (97)	+ (99)	+ (99)	+ (99)	+ (99)	>45 (98)	+ (94)	+ (99)	+ (96)	- (99)	+ (99)		+ (75)	+ (97)	+ (99)	+/- (50)		+ (93)
	M. chelonae subsp chelonae	Smooth Rough 60/40	22-35	R (97)	N (99)	+ (96)	+ (98)	- (99)	- (99)	- (99)	>45 (97)	-/+ (53/47)	- (98)	+ (96)	- (92)	- (99)		(99)	+ (85)	+ (98)	-/+ (46/54)	+ (91)	+ (99)
	M. chelonae subsp abscessus		22-40	R (97)	N (99)	+ (96)	+ (98)	- (99)	- (99)	- (99)	>45 (97)	-/+ (53/47)	- (98)	+ (96)	- (92)	- (99)		+ (90)	+ (85)	+ (98)	-/+ (46/54)	+ (91)	+ (99)
	M. chelonae (turtle-like)		22-30	R (97)	N (99)	+ (96)	+ (98)	+ (99)	- (99)	- (99)	>45 (97)	-/+ (53/47)	+/- (Tan)(99)	+ (96)	- (92)	- (99)		- (99)	+ (85)	+ (98)	-/+ (46/54)	+ (91)	+ (99)
	M. fallax[4] (looks like M. tb.)	Rough	30-37	R (30°) S (37°)	N=99%	- (99)		- (99)	- (99)	- (99)	- (66)	- (74)	- (99)	- (99)	- (99)	+ (99)	+ (98)	- (99)		+ (99)	+ (99)		- (96)
	Other Rapid Growers	Smooth or Rough	17-52		P=4% S=38% N=58%	- (72)	+ (75)	- (70)	+ (59)	+ (60)	+ (66)	- (63)	- (56)	- (52)	- (95)	- (51)	-	+ (52)	- (87)	+ (99)	+ (75)	V	+ (66)
TB Complex	M. ulcerans	Smooth, Rough	25-33	S	N	-	-					+		-	-	-	-	-		+	-		-
	M. tuberculosis	Rough, Cords	33-39	S	N=99%	- (99)	- (93)				- (99)	- (99)	+ (99)	- (99)	+ (98)	+ (99)	+ (98)	- (99)	- (70)	- (92)	+ (68)	- (83)	+ (98)
	M. africanum	Rough	35-38	S	N	- (99)	-				-	- (99)	- (99)	-	- (95)	-		- (99)		Var.	- (84)		+ (99)
	M. bovis	7H10 = R LJ = S	35-38	S	N=99%	- (99)	- (87)				- (97)	- (92)	+ (99)	- (99)	- (99)	- (94)	- (92)	- (99)	- (99)	- (94)	+ (98)	- (75)	+ (99)
	M. avium complex	Smooth Rough Trans.	22-45	S=99%	N=85% S=12% P=3%	- (76)	+ (51)				- (99)	+ (76)	- (92)	- (68)	- (99)	- (92)	+ (51)	- (99)	+ (55)	+ (99)	- (99)	- (99)	- (99)
	M. xenopi	(99) Smooth X	35-45	S	N=80% S=20%	+ (76)	+ (99)				+/- (99)	+ (83)	- (94)	- (99)	- (94)	- (94)	+ (86)4d (99)7d	- (99)	- (82)	+ (99)	+ (99)		- (99)
Nonphotochromogens	M. shimoidei	Rough	30-45	S	N	-	-				+	+		-	-	-	- (99)4d (63)7d	- (99)	- (59)	+	+		-
	M. gastri	Smooth	25-40	S=80% R=20%	N=99%	- (96)	+ (99)				- (99)	- (99)	-	- (84)	- (99)	- (99)	+ (63)4d (88)7d	- (99)		+ (99)	+ (99)	- (99)	+ (79)
	M. terrae complex	Smooth few Rough	22-37	S=77% R=23%	N=95% S=4% P=1%	- (99)	- (51)				+ (99)	+ (96)		- (77)	- (99)	+ (72)	+ (60)4d (90)7d	- (94)	+ (74)	+ (99)	+ (99)	- (99)	- (91)
	M. triviale	Rough	22-37	S	N=99%	- (66)	- (99)				+ (93)	+ (99)		- (99)	- (97)	+ (99)		+ (99)	- (75)	+ (99)	+ (99)	- (93)	- (86)
	M. malmoense	Smooth	22-37	S	N=90% S=10%	- (99)	- (89)				- (99)	- (69)			- (99)	- (99)		- (99)	- (99)	+ (99)	+ (97)		(67)
	M. haemophilum (Needs hemin)	Rough	25-35	S	N	-						-							+ (71)		-		

Notes: No "X" Colony (99) (M. avium complex); No "X" Colony (99) (M. gastri / M. terrae complex).

	Colony	Growth Temp (°C)	Pigment	Pigment																	
Photochromes																					
M. simiae	Smooth	22-37		P=91% N=9%	−(99)	−(99)	−(84)		+(99)	+(95)	−(99)	+(85)	+(85)		−(99)	+(85)	+(99)	+(99)	−(85)		+(64)
M. kansasii	Smooth Rough	25-40	S=98% R=2%	P=99% S=<1% N=<1%	−(99)	+(55)	+(99)	+(95)		+(99)	+(95)		+(99)	−(99)	−(99)	(99)	−(99)	+(84)	+(99)	+(99)	−(64)	+(97)
M. marinum	Smooth	25-35	R=61% S=39%	P=99%	−(66)	+(99)	−(79)	+(58)		−(99)	−(80)	−(96)	+	−(97)		+(79)	+(99)	+(99)	+(99)		+(99)
M. asiaticum	Smooth	33-37	S=99%	P=99%	−(99)	−(67)	+(91)	+(91)		+(91)	−(99)	−(99)		−(99)		−(70)	+(99)	+(99)	+(91)		−(99)
Scotochromogens																					
M. scrofulaceum	Smooth Rough	22-37	S=95% R=5%	S=97% P=3%	−(99)	−(64)	+(93)	+(96)		+(93)	−(99)	−(85)	+(60) 4d +(99) 7d	−(99)		+(57)	+(99)	+(98)	−(99)		+(99)
M. szulgai	Smooth Rough	22-37	S (99) 37° 25°	S P (99) 37° 25°	−(99)	+(62)	+(99)	+(81)		+(99)	−(99)	+(99)	+	−(99)		±(56)	+(99)	+(99)	±(50)	−(99)	+(94)
No "X" Colony −(99)																					
M. gordonae	Smooth	22-37	S=99% R=1%	S=99%	−(99)	+(57)	+(92)	+(97)		+(99)	−(99)	+(95)	+(67) 4d +(78) 7d	−(99)		−(77)	+(98)	+(99)	+(99)	−(99)	−(85)
M. flavescens	Smooth	25-42	R=59% S=41%	S=99%	+(84)	+(78)	+(85)	+(99)		+(99)	−(99)	+(96)	+(99)	−(71)		−(67)	+(99)	+(97)	+(99)	+(99)	+(71)

[1] Test reactions listed as "+" or "−", followed by pecentage of strains reacting as indicated. If no percentage is given, or space is blank, insufficient data were available or test is of no apparent value.

[2] Pyrazinamidase data (Wayne Method) from both Wayne and Hawkins. Unless indicated, results are those at 4 days. Strains of *M. tuberculosis* resistant to PZA are often pyrazinamidase negative. Within *M. terrae* complex, *M. nonchromogenicum* usually "+" and *M. terrae* usually "−".

[3] Urease data (Murphy-Hawkins Disk Method) primarily from Dr. Jean E. Hawkins.

[4] Data on *M. fallax* from Levy-Frebault et al (I.J.S.B. 33:336, 1983).

[5] Several strains in this complex have varied in carbon sources as follows: "+" sodium citrate, "+" mannitol, "−" inositol.

From Kent PT, Kubica G.P: Public Health Mycobacteriology: A Guide for the Level III Laboratory. Atlanta: Centers for Disease Control, 1985, p 125.

Figure 22 – 5

Catalase test.

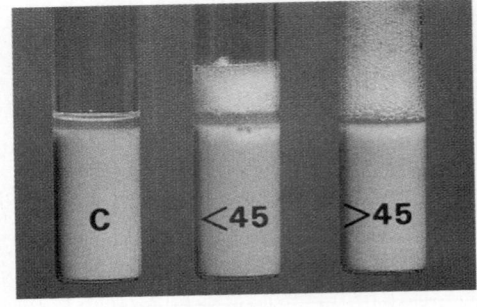

Figure 22 – 6

Semiquantitative catalase test.

Catalase

Catalase is an enzyme that can split hydrogen peroxide into water and oxygen. Mycobacteria are catalase positive with the exception of some isoniazid-resistant *M. tuberculosis*, *M. bovis*, *M. gastri*, and *M. kansasii*. Catalase quantity and heat stability are, however, species dependent. Semi-quantitation of catalase production uses the addition of Tween 80 (a detergent) and hydrogen peroxide to a 2-week-old culture. The reaction is monitored after 5 minutes, and the resulting column of bubbles is measured (Figs. 22–5 and 22–6). The column size is recorded as greater than or less than 45 mm. Catalase heat stability may be evaluated after the specimen is heated to 68°C for 20 minutes.

Hydrolysis of Tween 80

Some mycobacteria possess a lipase that can split the detergent Tween 80 into oleic acid and polyoxyethylated sorbitol. The time required for the hydrolysis is variable. The products of hydrolysis alter optical rotation of transmitted light, causing a pink color change. This test is helpful in distinguishing scotochromogenic and nonphotochromogenic mycobacteria.

Iron Uptake

Some mycobacteria are able to convert ferric ammonium citrate to an iron oxide. After growth of the isolate appears on an egg-based medium slant, rusty-brown colonies will appear in a positive reaction upon the addition of 20% aqueous solution of ferric ammonium citrate, as a result of the iron uptake (Fig. 22–7). The test is most useful in distinguishing *M. chelonae*, which is generally negative, from other rapid growers, which are positive.

Arylsulfatase

Arylsulfatase is an enzyme possessed by most members of the genus *Mycobacterium*. This enzyme hydrolyzes the bond between the sulfate group and the aromatic ring structure in compounds with the formula $R-OSO_3H$. Tripotassium phenolphthalein sulfate is such a molecule, from which phenolphthalein is liberated upon exposure to arylsulfatase. The liberation of phenolphthalein causes a pH change in the presence of sodium bicarbonate, indicated by a pink color change. The *M. fortuitum-chelonae* complex (see later), *M. xenopi*, and *M. triviale* have rapid aryl-sulfatase activity that can be detected in 3 days. *Mycobacterium marinum* and *M. szulgai* exhibit activity with 14-day incubation.

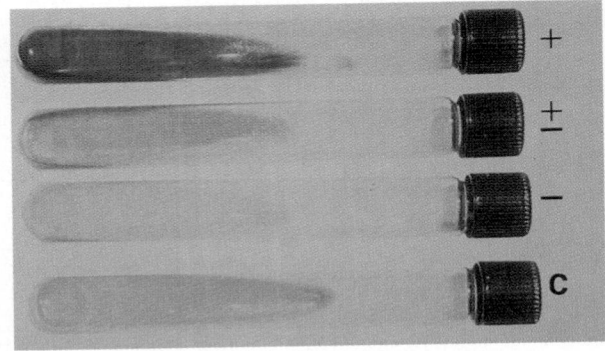

Figure 22 – 7

Iron uptake.

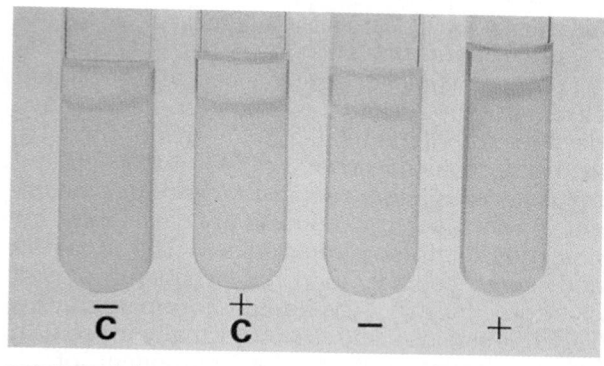

■ Figure 22-8

Pyrazinamidase test.

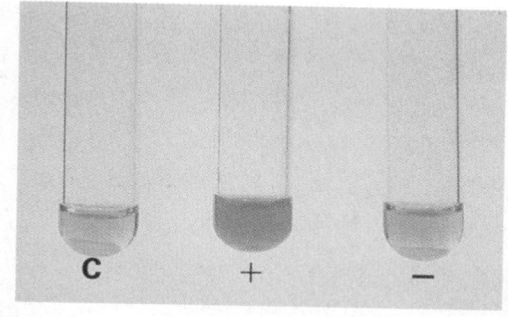

■ Figure 22-9

Urease test.

Pyrazinamidase

The presence of pyrazinamidase allows deamination of pyrazinamide to pyrazinoic acid and ammonia in 4 days, producing a red pigment (Fig. 22-8). This test may be useful in distinguishing *M. marinum* from *M. kansasii,* and *M. bovis* from *M. tuberculosis.*

Urease

Detection of urease activity may be used to distinguish *M. scrofulaceum* from *M. gordonae* (Fig. 22-9).

Inhibitory Tests

NAP

NAP (p-nitro-aceytlamino-β-hydroxypropiophenone) is a precursor in the synthesis of chloramphenicol. Since 1960, NAP has been known to selectively inhibit *M. tuberculosis* complex. Laszlo and Eidus (1978) later confirmed that 99.5% of more than 5,000 strains they tested were inhibited by NAP. The radiometric application of this selective inhibition evolved soon after.

It is recommended that the test be performed either with isolated MTB colonies or with broth from a BACTEC medium with a growth index of 50 to 100 or more. An aliquot of the original BACTEC medium is transferred to the BACTEC-NAP vial which contains 5 mg of NAP and no antibiotics. Growth is then closely monitored and compared with the original BACTEC vial, which serves as a growth control. A 20% increase or decrease in growth index is considered significant. Four to 6

days more are required for identification if such indirect NAP testing is performed. Testing should be done with positive and negative controls.

Direct specimen BACTEC-NAP testing, eliminating a prior isolation, has been done and has been reported to identify *M. tuberculosis* in an average of 7.8 days, but with diminished sensitivity.

TCH

TCH (thiophene-2-carboxylic acid hydrazide) distinguishes *M. bovis* from *M. tuberculosis. Mycobacterium bovis* is susceptible to lower levels of TCH than MTB. Variability in inhibition will exist, depending on the concentration of the inhibitory agent and the temperature of incubation. This selective inhibitory agent has also been applied to radiometric systems.

GROWTH IN 5% NaCl

High salt concentration in egg-based media inhibits the growth of most mycobacterium. *Mycobacterium flavescens, M. triviale,* and most rapidly growing *Mycobacterium* are exceptions that do grow in 5% NaCl.

TELLURITE REDUCTION

Reduction of colorless potassium tellurite to black metallic tellurium within 3 to 4 days is a characteristic of *M. avium* complex (Fig. 22-10).

GROWTH ON MacCONKEY AGAR

Mycobacterium fortuitum-chelonae complex can grow on MacConkey agar without crystal violet, whereas most other mycobacteria cannot.

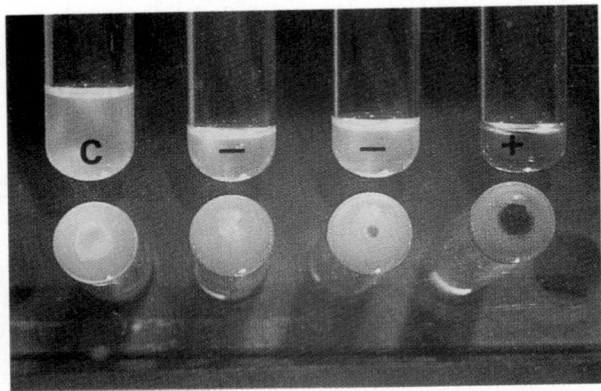

■Figure 22–10
Test for tellurite reduction.

Slow vs. Rapid Growth

Figures 22–11 and 22–12 show schematic diagrams for the identification of slowly growing and rapidly growing *Mycobacterium* species.

Chromatography

The cell walls of *Mycobacterium* spp contain long-chain fatty acids called *mycolic acids* that may be detected *chromographically*. The type and quality of mycolic acids are specific to species. Chromatographic species identification of *Mycobacterium* has been of longstanding interest, and methods have changed as technology has evolved. Earlier methods, such as column chromatography and thin-layer chromatography, have been replaced by gas-liquid chromatography and, most recently, high-performance liquid chromatography. Current methods allow sufficient amount of mycolic acid to be easily extracted from small quantities of bacterial cultures. Tisdale and associates (1979) developed a basic saponification–lipid extraction method that allows correct identification of most mycobacterial species using chromatograms and colony characteristics. Acid methanolysis and higher temperatures have improved mycolic acid separation.

Species identifications made with high-performance liquid chromatography (HPLC) have been shown to agree well with biochemical and probe identifications. Chromatography is rapid and highly reproducible, but the initial equipment cost is high.

Detection of Other Mycobacterial Products

Frequency-pulsed electron capture gas-liquid chromatography and mass spectrometry have been used to detect small quantities of the fatty acid tuberculostearic acid (TSA) in the CSF and serum as a rapid diagnosis of tuberculous meningitis. Selective procedures are available for detecting femtomole quantities of TSA in serum and cerebrospinal fluid by frequency-pulsed electron capture gas-liquid chromatography. Tuberculostearic acid is not normally present in human tissues, but it is a component of all *Mycobacterium* and Actinomycetales.

Amplification for *M. tuberculosis*

DNA Hybridization

The use of nucleic acid hybridization techniques allows rapid identification of certain common mycobacterial species. The technology involves the binding of a ^{125}I- or acridine ester–labeled piece of nucleic acid, the *probe*, to complementary nucleic acid when present. The associated pair form stable double-stranded complexes that may then be detected. The probe consists of single-stranded DNA complement to ribosomal RNA. Because each cell contains more RNA than DNA, the sensitivity of the test is increased by using a probe to detect RNA rather than DNA. Acridine ester–bound probe is detected via differential hydrolysis with chemiluminescence and reported in relative light units (RLU). The sensitivity of recognition is about 10^4 organisms per mL. It is estimated that 10^5 to 10^6 (some say as many as 10^7) organisms per mL are required for detection by the Gen-Probe (San Diego, CA) procedure. According to the manufacturer, 3 to 6×10^8 organisms per mL are required. In the experience of Gonzalez and colleagues (1987), the amount of organism required for testing "corresponded in the case of MTB to a single colony of at least 1 mm in diameter or in the case of MAIC to a barely visible film of growth on the surface of the slant." Most positive results are well above the cut-off value of 10% hybridization. When a probe is used on a contaminated specimen, the obtained percentage hybridization may incorrectly fall below accepted cut-off hybridization levels, leading to falsely negative results.

DNA hybridization identification can be applied to growth on conventional agar as well

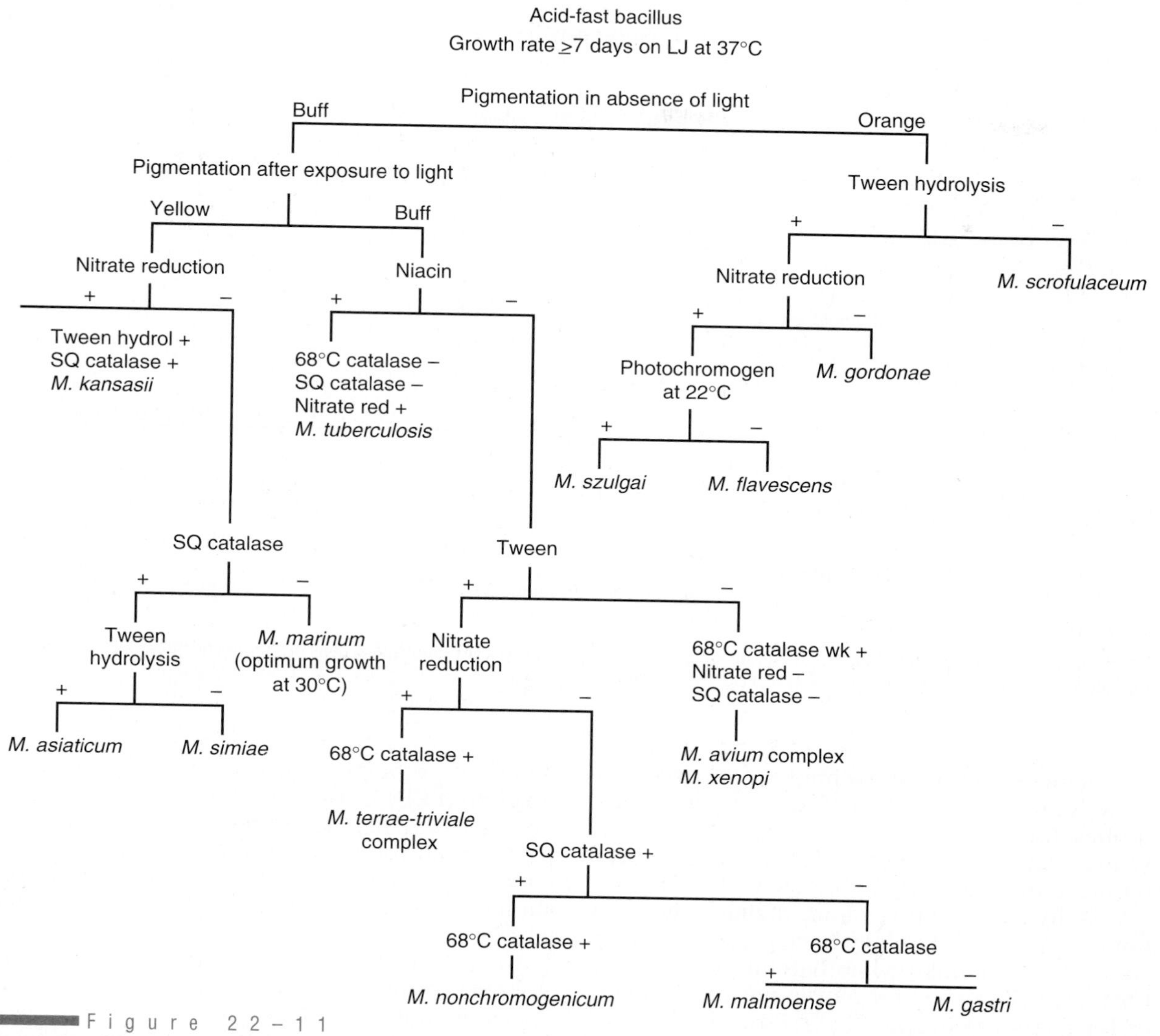

Acid-fast bacillus
Growth rate ≥7 days on LJ at 37°C

Pigmentation in absence of light

Buff — Orange

Pigmentation after exposure to light — Tween hydrolysis

Yellow — Buff

Nitrate reduction — Niacin

+ −

Tween hydrol +
SQ catalase +
M. kansasii

+ −

68°C catalase −
SQ catalase −
Nitrate red +
M. tuberculosis

Tween hydrolysis:
+ → Nitrate reduction
+ → Photochromogen at 22°C
+ → *M. szulgai*   − → *M. flavescens*
− → *M. gordonae*
− → *M. scrofulaceum*

SQ catalase
+ Tween hydrolysis
   + *M. asiaticum*   − *M. simiae*
− *M. marinum* (optimum growth at 30°C)

Tween
+ Nitrate reduction
   + 68°C catalase +
      M. terrae-triviale complex
   − SQ catalase +
      + 68°C catalase +
         M. nonchromogenicum
      − 68°C catalase
         + *M. malmoense*   − *M. gastri*
− 68°C catalase wk +
   Nitrate red −
   SQ catalase −
   M. avium complex
   M. xenopi

Figure 22-11

Schematic diagram for the identification of slowly growing *Mycobacterium* species. Exceptional reactions occur. Organisms should be subjected to a battery of morphologic and physiologic tests before a final identification is made. SQ, semiquantitative.

as to growth in radiometric liquid media such as BACTEC. The combination of radiometric detection and DNA hybridization identification using the probe technology allows rapid recovery and identification. Waiting for the early log phase of growth or an index reading of 500 to 999 increases the sensitivity of probe identification. Ellner and associates (1988), using either BACTEC 12B medium or Middlebrook 7H11 medium and the Gen-Probe, found that 95% of TB cultures could be identified within 4 weeks. Seven weeks were required for identification with growth on LJ medium.

Commercially available mycobacterial probes include those for *M. tuberculosis* complex, *M. avium* complex (as well as for each species independently), *M. kansasii*, and *M. gordonae*. The probe for *M. tuberculosis* complex detects *M. tuberculosis, M. bovis, M. africanum,* and *M. microti.* The first three of these species are human pathogens. Because *M. bovis* and *M. africanum* are rare in the United States, distinguishing between them and *M. tuberculosis* is not generally necessary. When clinically appropriate, the three can be distinguished biochemically (see later) as well as by antimicrobial sensitivities.

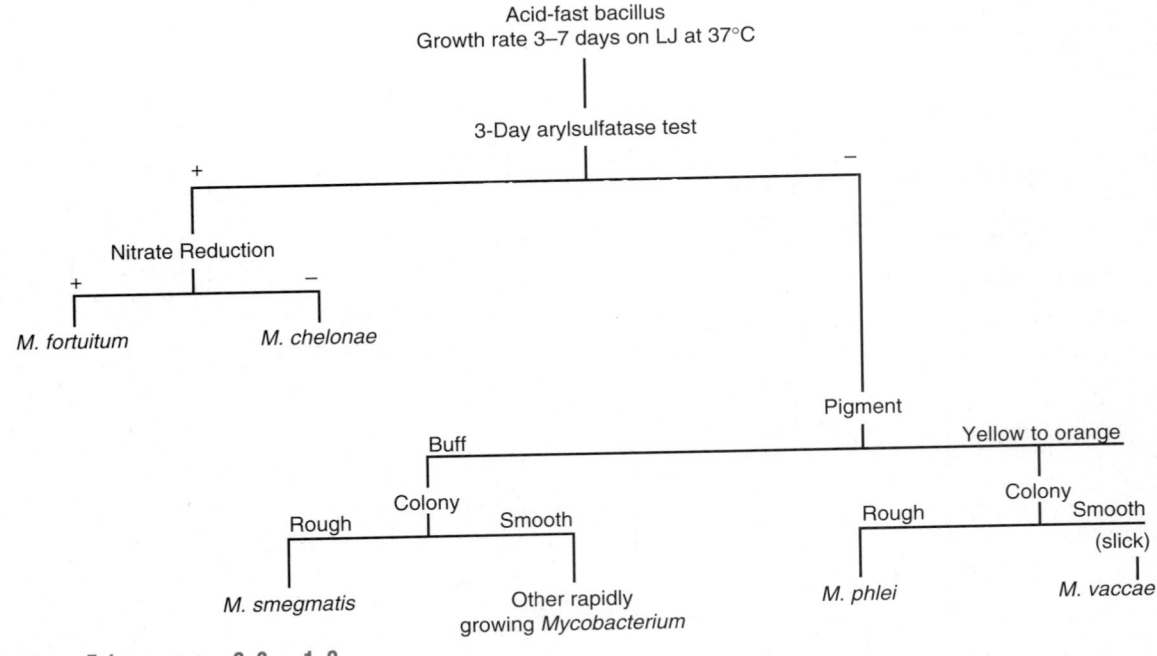

Figure 22–12

Schematic diagram for the identification of rapidly growing *Mycobacterium* species. Exceptional reactions occur. Organisms should be subjected to a battery of morphologic and physiologic tests before final identification is made.

Specificity of probe technology is excellent. However, Chopin-Robertson has reported probes for *M. tuberculosis* hybridizing with *M. xenopi* and *M. terrae*. MAC probes have also been reported to incorrectly hybridize with MTB.

DNA hybridization is rapid, taking only a few hours, but the technology is costly. Cost can be somewhat contained when batching is possible. Currently, the probe reagent is not recommended for direct use in specimens. Probe technology is rapidly developing, however, and biotin-labeled probes and probes for direct use on submitted specimens are in development. In addition, DNA hybridization technology can be applied after DNA amplification for a direct test (see later).

Use of genetic probes for culture confirmation of *M. tuberculosis* has greatly reduced the time needed for its identification. Direct specimen testing on the same day as collection would be ideal, however, to reduce the chance of transmission and to determine appropriate therapy for the patient. Culture confirmation probes are not sensitive enough to use directly on specimens without prior cultural amplification. There are numerous reports of use of polymerase chain reaction (PCR) testing for direct specimen detection of *M. tuberculosis,* especially in sputum samples. Within the next few years, at least two different commercial amplification procedures should be available for clinical and/or reference laboratories.

A system developed by Roche Diagnostic Systems (Montclair, NJ) utilizes PCR of the digested sputum sample, followed by hybridization with probes in microtiter trays. One study in the literature has shown the method's sensitivity to be 83.5%, specificity 99%, and PVP (Predictive Value Positive) 94.2%. For smear-positive *M. tuberculosis* specimens, positivity was 94%, but for smear-negative specimens, PCR positivity was only 62%. There were 71 false-positives out of 948 culture-negative specimens.

Another amplification assay, Gen-Probe Amplified Mycobacteria Tuberculosis Direct Test (Gen-Probe, San Diego, CA), requires 5 hours for completion and utilizes ribosomal RNA amplification and acridinum ester probe in-solution hybridization. In one study, the sensitivity was 82%, specificity 99%, PVP 97%, and PVN (Predictive

Value Negative) 96%. In a study that compared these two amplification assays, the two procedures compared favorably with each other.

It is not known how either of these assays will fit into the normal workflow of a clinical laboratory. In areas where the prevalence of *M. tuberculosis* is low, it would be very costly to process all submitted specimens. In areas with higher prevalence, the assay on smear-positive specimens would quickly confirm the identification of *M. tuberculosis,* and on-smear negative specimens, it would aid in obtaining a quicker (1-day) result for at least a majority of specimens. There will probably be other amplification methods as well in the future.

To date, the application of PCR and chromatography methods has been limited to research and larger reference mycobacteriology laboratories. Their utilization in routine laboratory studies is under investigation and can be expected to be adopted in the near future.

Serology

The history of serologic testing in mycobacterial disease parallels developments in immunologic testing. Despite these continued efforts, "serological testing still has not become applicable to the clinical laboratory" (David and Selin, 1980). A lack of antigen specificity and the weakness of antibody response to illness have been insurmountable obstacles.

Serology has been suggested to be useful in certain situations, and a specific serologic test for the detection of lepromatous leprosy using a phenolic glycolipid has been developed. The development of monoclonal antibodies and increasing sophistication in serologic testing may alter the historic uselessness of mycobacterial serology.

Susceptibility of *M. tuberculosis*

The CDC currently recommend that all isolates of *M. tuberculosis* be tested for in vitro susceptibility to isoniazid, rifampin, ethambutol, and streptomycin. (Pyrazinamide is to be considered as well.) Likewise, the testing should be repeated if a patient's cultures for *M. tuberculosis* remain positive after 3 months of therapy. The methods

for performing susceptibility tests may vary, but all are based on the premise that in any population of *M. tuberculosis,* there is a potential development of resistance to a single agent at a fairly well-defined rate. For example, with isoniazid, the chance that a resistant isolate will develop is 1 in 10^5; for streptomycin, 1 in 10^6, and for both, 1 in 10^{11}. In a patient with pulmonary tuberculosis, the pulmonary cavity may contain 10^7 to 10^9 total bacteria. Random drug resistance has a good chance of developing, especially with the use of only one drug. Hence, treatment of tuberculosis has always required the use of at least two drugs, and often more.

In clinical correlations with in vitro data, if 1% of a patient's bacilli are resistant to a particular drug, treatment fails. Laboratory tests then must demonstrate the rate of resistant organisms. To do so, the inoculum of bacilli used is adjusted to enable this 1%—usually 100 to 300 colony-forming units (CFU) per mL—to be determined. The test is often referred to as the *proportion method,* because it allows one to predict the possibility that the 1% is resistant or not. The methods for determining such resistance include agar dilution, disk elution, and the BACTEC system.

The *agar dilution* method involves dispensing of media (usually Middlebrook 7H10) containing appropriate drugs into quadrants of a Petri dish. Acid-fast bacilli are then prepared in an inoculum to yield about 100 to 300 CFU/mL, which entails preparation of a barely turbid broth culture, which is then diluted 10^{-2} and 10^{-4}. The two dilutions provide for a set of plates that should be countable (i.e., 100 to 300 CFU on the control plate). Alternatively, for the *disk elution* method, filter paper disks with appropriate concentrations of antibiotics are added to quadrants of Petri dishes, and agar is poured over them. In either case, a control plate is set up in each run of drugs so that the numbers of colonies on the test quadrants can be counted and compared with the number on the control quadrant. If the test growth is less than 1% of the control growth, the organism is said to be susceptible; if greater, resistant. By this method, results can be obtained in 2 to 3 weeks, depending on growth of the organism. Plates are kept at 37°C for incubation.

A laboratory utilizing the BACTEC 460 system for isolation of *Mycobacterium* spp, can also use it for susceptibility tests. The same proportion method can be employed in the BACTEC. The major difference is that the control BACTEC vial contains a 1:100 dilution of the inoculum used for the drug-containing vials, so that a 1% population is achieved in the control. Any growth above that in the drug vials is considered a resistance marker.

The BACTEC bottles are read for 4 to 5 days after inoculation. After a growth index (GI) of 30 is achieved in the growth control vial, the readings over the next 2 days are recorded. If the change in GI reading in a drug vial exceeds that in the growth vial, the organism is said to be resistant; if it does not, susceptible.

Table 22–5 lists drugs and their concentrations to be used in susceptibility tests. The susceptibility test is usually performed on pure culture isolates, identified as *M. tuberculosis*. A direct susceptibility may be performed, however, using clinical specimens that are positive on smear, provided that appropriate dilutions are made on the basis of numbers of acid-fast bacilli seen. The advantage of this direct assay is a quicker susceptibility report—within 2 to 3 weeks of culture, rather than 5 to 7 weeks or more. If, however, cultures are not thoroughly decontaminated, overgrowth is a problem. Likewise, the mycobacteria isolated may be other than *M. tuberculosis*.

Because multidrug-resistant *M. tuberculosis* cases are increasing in the United States, there is a need to rapidly isolate *M. tuberculosis* and determine its susceptibility in vitro, so that appropriate therapy may be employed. To this end, the laboratory is being required to perform the most efficient and rapid tests possible. Researchers are looking for more rapid methods, involving amplification, use of luciferase enzymes, and detection of resistance genes. It is hoped that these methods will be available for laboratory use soon.

MYCOBACTERIUM TUBERCULOSIS COMPLEX

Mycobacterium tuberculosis complex consists of *M. tuberculosis*, *M. bovis*, *M. africanum*, and *M. microti*. *M. africanum* has been associated with human cases of tuberculosis in tropical Africa. *M. microti* is not associated with human disease. These two latter species are rarely encountered in the laboratory in the United States and therefore are not included in the following discussion.

Mycobacterium tuberculosis

Mycobacterium tuberculosis was first described by Robert Koch in 1882; however, the disease tuberculosis is one of the oldest documented communicable diseases. Today in the world, there are over 1 billion cases of tuberculosis infection, with 8 to 10 million new cases of disease and 3 million deaths attributed to tuberculosis per year. A person is "infected" with *M. tuberculosis* when he or she is exposed to the organism. Whether or not a person has the disease tuberculosis is determined by his or her cellular immunity, the amount of exposure and the virulence of the strain.

In the United States, prior to 1985, there was a continual decline in the number of tuberculosis cases per year at a rate of about 5% decrease per year. In 1985, this decline ended, and currently, the number of documented tuberculosis cases per year in the United States is increasing (over 18% cumulative increase since 1985). This increase has been attributed to several factors, including the epidemic of AIDS, increased use of intravenous drugs, and greater spread among inhabitants of closed environments, such as nursing homes, correctional facilities, and shelters for the homeless. In addition, more cases are associated with immigration from endemic areas or higher numbers of migrant workers from such areas in certain states in the United States.

■■■■ T A B L E   2 2 – 5
ANTITUBERCULOSIS DRUGS AND THEIR RECOMMENDED CONCENTRATIONS

Drug	Concentration (μg/mL)*
Primary	
Isoniazid	0.2
Isoniazid	1.0
Streptomycin	2.0
Streptomycin	10.0
Rifampin	1.0
Ethambutol	5.0
Secondary†	
Ethionamide	5.0
Capreomycin	10.0
Cycloserine	20.0
Kanamycin	5.0
Pyrazinamide (at pH 5.5)	25.0

*7H10 medium.
†Adapted with permission from M24-P, "Antimycobacterial Susceptibility Testing; Proposed Standard," NCCLS, 771 E. Lancaster Avenue, Villanova, PA 19085.

Tuberculosis: Clinical Disease

Clinically, tuberculosis is usually a disease of the respiratory tract. Tubercle bacilli are acquired from persons who have active disease and are excreting viable bacilli by means of coughing, sneezing, or talking. Airborne droplet nuclei, 1 to 10 μm, enter the respiratory tract of an exposed individual and are deposited in the lung alveoli. The exposed individual is said to be "infected" with *M. tuberculosis*. After infection, in most individuals, *M. tuberculosis* organisms are phagocytized in alveolar macrophages. They are still at this time capable of intracellular multiplication. Within 4 to 6 weeks, in a person with adequate cellular immunity, T cells arrive with macrophage-activating polypeptides called *lymphokines*. This enables the white blood cells in the area of infection to destroy the intracellular mycobacteria.

There is then a regression and healing of the primary lesion and any disseminated foci of infection by the *M. tuberculosis* organisms. The pathologic features of this infection are the result of the hypersensitivity and the local concentration of the antigen (bacterium). If there is little antigen and a great deal of hypersenstivity reaction, a hard tubercle or granuloma may be formed. The *granuloma* is an organization of lymphocytes, macrophages, giant cells, fibroblasts, and capillaries. With this granuloma formation, healing occurs, with fibrosis, encapsulation, and scar formation as a reminder of the past infection.

If the antigen load and hypersensitivity reaction are both high, tissue necrosis from enzymes of degenerating macrophages may occur, the tissue response is less organized, and no granuloma is formed. Without granuloma or necrosis, lesions may heal without obvious pathology. With necrosis, a caseous material may be present at the site of the primary lesion, as a result of solid or semisolid amorphous material being laid down at the site of necrosis.

After this healing of first-degree infection, the bacilli are not totally eradicated, and there is, in the infected individual, a potential for reactivation of tuberculosis disease. The risk of reactivation of tuberculosis is about 3.3% during the first year after a positive tuberculin skin test and a total of 5 to 15% thereafter in the person's lifetime. Progression from infection to active disease varies with age and with the intensity and duration of exposure. Malnutrition, with or without other factors such as alcoholism, homelessness, incarceration, immunosuppression, and AIDS, can contribute greatly to the progression to active tuberculosis.

Clinical diagnosis of primary tuberculosis is usually limited to detection of a positive tuberculin skin test using purified protein derivative (PPD). Children may demonstrate a nonproductive cough and fever with or without shortness of breath; these symptoms are unusual in adults. Chest radiographs are usually normal, although rarely there may be infiltrates without cavitation in the anterior segment of the upper, middle, or lower lobe with hilar or paratracheal adenopathy. Along with these limited clinical findings there is a paucity of bacteriologic findings. If sputum or bronchial washings are cultured during the primary infection, the yield is only 25 to 30% positive. A small percentage of individuals who are infected with tuberculosis develop progressive pulmonary disease. This is most often due to a failed cellular immune response and hence a failure to stop multiplication of the bacilli. In the very young or the elderly who are primarily infected and in people with an underlying immunodeficiency, massive lymphohematogeneous dissemination may occur and lead to meningeal or miliary (disseminated) tuberculosis. In addition, 10% of young adults may progress to active disease from their primary infection. This will look like reactivation tuberculosis in the elderly, and the only way to differentiate it is by finding a new PPD positivity.

Reactivation Tuberculosis

Reactivation tuberculosis occurs when there is an alteration or a diminution of the cellular immune system in the infected host that favors replication of the bacilli and progression to disease. The symptoms of disease are slow in developing and consist of fever, shortness of breath, night sweats and chills, fatigue, anorexia, and weight loss. Twenty percent of individuals may have no symptoms, but the majority of patients eventually present with cough, chest pain, and productive sputum. Hemoptysis occurs in 25% of cases. The radiographs of patients with reactivation tuberculosis reveal a patchy or confluent consolidation with increased linear densities extending to the hilum; thick-walled cavities without air-fluid levels usually are found in apical or posterior segments of the upper lobe or superior segment of the middle lobe of the lung. If there is bronchogenic spread of the bacilli, multiple alveolar densities will be seen; rarely is there enlargement of the lymph nodes. If the disease has been chronic, fibrosis, loss of lung volume, and calcifications will be demonstrated.

The PPD test may be negative in up to 25% of these cases; diagnosis is confirmed by smear and culture of sputum, gastric aspirates, or bronchoscopy specimens. Fiberoptic bronchoscopy has been found to yield a 95% recovery; post-bronchoscopic sputa are usually positive as well.

In any case of pulmonary tuberculosis disease, there may be complication if diagnosis and treatment are delayed. These include empyema, pleural fibrosis, massive hemoptysis, adrenal insufficiency (rare), and hypercalcemia (up to 25% of cases). In patients who have AIDS and tuberculosis with drug-susceptible bacilli, the risk of progression to disease from infection is quite high, although the clinical findings may vary from those in the non-AIDS patient with reactivation tuberculosis. The diagnosis is usually made by culture and smears, with a rate of sensitivity similar to that in the non-AIDS patient.

Extrapulmonary Tuberculosis

Extrapulmonary tuberculosis occurred much less commonly than pulmonary tuberculosis (<15%) prior to the AIDS epidemic; however, as cases of pulmonary TB in the United States declined, the number of cases of extrapulmonary tuberculosis remained constant. Cases of extrapulmonary disease have increased since 1988 because it is a common presentation in the HIV-infected individual, although most often found in association with pulmonary disease.

Miliary tuberculosis refers to the seeding of many organs outside the pulmonary tree with acid-fast bacilli through hematogenous spread. This usually occurs shortly after primary pulmonary disease but can take place anywhere in the course of acute or chronic tuberculosis. The most common sites of spread of *M. tuberculosis* in its miliary form are spleen, liver, lungs, bone marrow, kidney, adrenal gland, and eyes, usually in that order of preference. Overall, children account for the majority of cases of miliary TB, but it is also a most common form of tuberculosis disease in the HIV-infected individual. The mortality is 20% or higher in most literature series; the finding of meningitis is an extremely bad prognostic indicator.

Other forms of extrapulmonary tuberculosis are: pleural, lymphadenitis, gastrointestinal, skeletal, meningeal, peritoneal, genitourinary, and miscellaneous infections. Pleurisy, an unexplained pleural effusion with mononuclear pleurocytosis, manifests as cough, fever and chest pain, resembling the presentation of bacterial pneumonia; it occurs in about 5% of all cases of tuberculosis. In endemic areas, pleurisy is a presentation in the young individual; in the United States, middle-aged to older persons are most affected. Resolution is common. Acid-fast bacilli are rarely seen in pleural fluid, but cultures may be positive in 20 to 50% of cases; pleural biopsies offer a higher yield of microbiologic diagnosis.

Lymphadenitis is usually a disease of children, presenting as painless head or neck swellings. Lymph node involvement, particularly mediastinal, has been a common extrapulmonary manifestation in the AIDS patients.

Genitourinary tuberculosis can involve the kidneys and genital organs. Renal tuberculosis accounts for 2% of all cases of tuberculosis and manifests as typical urinary tract symptoms and sterile pyuria. Cultures may be positive in up to 80% of cases. Male genital tuberculosis usually presents as a scrotal mass and occurs most often along with renal tuberculosis; in females, hematogenous spread is usually the source.

Skeletal tuberculosis of the spine is referred to as Pott disease. Back pain is the most common presenting characteristic. Cultures of bone and tissue are needed to confirm the diagnosis. Peripheral skeletal bones and joints may also be involved, with the hip and knee being the most common sites.

Meningitis due to *M. tuberculosis* is usually due to a rupture of a tubercle into the subarachnoid space, not usually to hematogenous spread. In childhood, it occurs rarely after primary pulmonary infection. Most of the infection occurs at the base of the brain, and patients may develop very thick, gelatinous, mass-like lesions there. With more chronic disease, a fibrous mass may surround cranial nerves. Involvement of arteries may cause infarctions. CSF examination usually reveals elevated protein, decreased glucose, and predominance of lymphocytes. Prognosis with this form of extrapulmonary tuberculosis is not good.

Virtually any organ of the body can be infected by *M. tuberculosis,* and other, uncommon manifestations are: gastrointestinal infection, peritonitis, cutaneous tuberculosis, laryngitis, otitis, and involvement of urea, adrenal glands, eyes, and breast. Up to 70% of HIV-infected patients may present with extrapulmonary tuberculosis alone or, most often, in combination with pulmonary disease. The most common extrapulmonary sites in this population are disseminated (miliary), lymph node involvement (especially mediastinal), genitourinary, and intraabdominal. Bacteremia is not uncommon.

Identification of *M. tuberculosis*

Colonies of this slowly growing species are thin, flat, spreading, and friable with a rough appearance. The colonies are classically described as being buff in color (Fig. 22–13). Elaboration of cord factor may result in characteristic cord formation. Isoniazid-resistant strains may have somewhat longer recovery times. Optimum growth occurs at 35 to 37°C. Colonies are not photoreactive.

Biochemically, *M. tuberculosis* is characteristically positive for niacin accumulation, reduction of nitrate to nitrite, and production of catalase, which is destroyed after heating. Isoniazid-resistant strains may not produce catalase at all. *Mycobacterium tuberculosis* is inhibited by NAP. This species can be distinguished from *M. bovis* by the inhibition of *M. bovis* by TCH and by pyrazinamidase activity.

Treatment of Tuberculosis

For pulmonary tuberculosis, treatment usually involves a 9-month course of therapy with isoniazid and rifampin, usually once per day the first month and two times a week thereafter. Many regimens also include a 2- to 8-week initial course of streptomycin or ethambutol. Most individuals clear their sputum of acid-fast bacilli within the first 2 months. Pyrazinamide (PZA) may be added to the regimen if there is a suspicion of lowered cellular immunity and a need to obtain bactericidal levels of antituberculous activity intracellularly in macrophages. PZA is usually recommended for a shorter course—initially along with isoniazid and rifampin.

For cases of resistance to isoniazid or rifampin, second-line antituberculosis drugs are capreomycin, cycloserine, kanamycin, amikacin, ciprofloxacin, and ethambutol. With the numbers of cases of multidrug-resistant *M. tuberculosis*

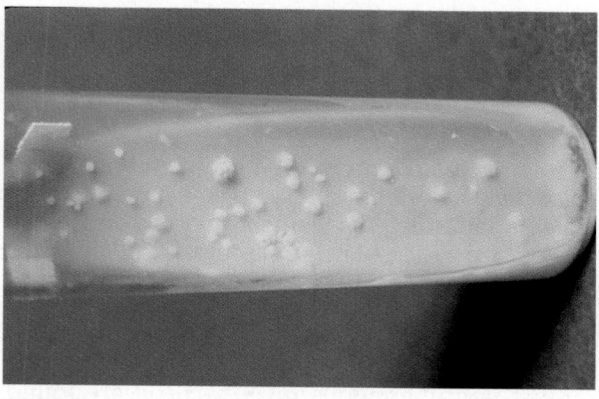

■■■■■Figure 22–13
M. tuberculosis growing on Löwenstein-Jensen (LJ) medium.

increasing, newer agents are being tested in vitro to determine their efficacy.

The two reasons for drug failure are:

■ Lack of patient compliance with drug therapy

■ Resistance of the isolates

If compliance is an issue, directly observed therapy (DOT) is recommended to ensure proper medication. Otherwise, resistance may be assumed and tested for in vitro.

Mycobacterium bovis

Mycobacterium bovis produces tuberculosis primarily in cattle but also in other ruminants as well as dogs, cats, swine, parrots, and humans. The disease in humans closely resembles that caused by *M. tuberculosis* and is treated similarly. In some areas of the world, a significant percentage of the cases of tuberculosis are due to *M. bovis,* but in the United States, the number of isolates of this organism is very low.

Mycobacterium bovis is closely related taxonomically to *M. tuberculosis* and belongs to the *M. tuberculosis* complex. It grows very slowly on egg-based medium, producing small, granular, rounded, white colonies with irregular margins after 21 days of incubation at 37°C. Growth is nonpigmented. On Middlebrook 7H10 medium, colonies are similar to those of *M. tuberculosis* but slower to mature. Most strains of *M. bovis* are niacin negative, do not reduce nitrate, and do not grow in the presence of TCH, characteristics that distinguish the species from most strains of *M. tuberculosis.*

NONTUBERCULOUS MYCOBACTERIA: CLINICAL SIGNIFICANCE AND DIFFERENTIATION

There is a large group of mycobacteria, excluding *M. tuberculosis* complex and *M. leprae,* that normally inhabit the environment and can cause disease, often resembling tuberculosis, in humans. These organisms are sometimes referred to as *atypical mycobacteria* or *mycobacteria other than the tubercle bacillus* (MOTT). The term nontuberculous mycobacteria is used here.

Most nontuberculous mycobacteria are found in soil and water. They have been commonly implicated as opportunistic pathogens in patients with underlying lung disease, immunosuppression, or percutaneous trauma. AIDS has contributed greatly to the incidence and awareness of nontuberculous mycobacterial disease. Chronic pulmonary disease resembling tuberculosis is the usual clinical presentation associated with these organisms, although a few species are more often associated with cutaneous infections. Infections caused by the nontuberculous mycobacteria are not considered transmissible from person to person. Regional differences in the incidence of nontuberculous mycobacterial disease are quite striking.

Mycobacterium avium Complex

Epidemiology

Organisms within the *M. avium* complex (MAC) are commonly found in the environment and have been recovered from soil, water, house dust, and other environmental sources. *Mycobacterium avium* is a cause of disease in poultry and swine, but animal-to-human transmission has not been shown to be an important factor in human disease. Environmental sources, especially natural waters, seem to be the reservoir for human infections. A large increase in *M. avium* complex infections has occurred in the past decade, primarily owing to the number of infections in patients with AIDS.

Clinical Infections

Pulmonary disease resulting from MAC infection presents a clinical picture similar to that of tuberculosis: cough, fatigue, weight loss, low-grade fever, and night sweats. Radiologic examination demonstrates cavitary disease in most patients, whereas solitary nodules or diffuse infiltrates may be observed in others. Disseminated disease has become more common, usually occurring in immunocompromised patients or in patients with hematologic abnormalities. MAC infections are now the most common systemic bacterial infection in patients with AIDS.

The clinical outcome of *M. avium* complex lung disease is unpredictable; thus, the management of affected patients can be difficult. Observation, therapy for underlying pulmonary disease (e.g., bronchodilators, broad-spectrum antibiotics, smoking cessation), and periodic sputum cultures may be all that is required for most patients. For patients with significant symptoms and advanced or progressive radiographic disease, multidrug therapy is indicated. For children with cervical lymphadenitis due to MAC, excisional surgery without chemotherapy is usually successful. A combination of surgical excision and chemotherapy is the usual treatment for adults with localized, nonpulmonary disease.

Most cases of disseminated disease in immunosuppressed patients without AIDS responds to multidrug regimens. Multidrug therapy consisting of ethambutol, rifampin (or rifabutin), clofazimine, and an injectable aminoglycoside has resulted in symptomatic and clinical improvement in most but not all patients with AIDS. The treatment of MAC disease may best be directed by physicians experienced in pulmonary or mycobacterial disease.

Laboratory Diagnosis

Because the two species within the MAC complex, *M. avium* and *M. intracellulare,* are so similar, most laboratories do not distinguish between them, but report isolates of both species as *M. avium* complex. On primary isolation media, these organisms grow slowly, producing thin, transparent or opaque, homogeneous, smooth colonies. A small proportion of strains may exhibit rough colonies. Usually the colonies are nonpigmented, but they may become yellow with age. Rarely are the colonies pigmented from the onset of detectable growth. Optimum growth temperature is 37°C. On microscopic examination, the cells are short and coccobacillary, uniformly stained without beading or banding.

Long, thin, beaded bacilli resembling *Nocardia* species may be seen in stains of very young cultures or under certain other conditions. MAC species are inactive in the majority of physiologic tests used to identify the mycobacteria. Exceptions are the production of a heat-stable catalase and the ability to grow on media containing TCH at 2 µg/mL (see Table 22–4). Nonisotopic nucleic acid probes are available for the identification of MAC as well as the two individual species.

Susceptibility Testing

In laboratory tests, members of the *M. avium* complex are generally resistant to the relatively low concentrations of antituberculosis drugs currently used for testing *M. tuberculosis.* Treatment recommendations have been based largely on empiric data rather than on in vitro susceptibility testing. For this reason, routine agar dilution susceptibility testing with the antituberculosis agents, as currently performed for testing *M. tuberculosis,* is not recommended for *M. avium* complex isolates. Currently, the usefulness of testing at higher drug concentrations than used for *M. tuberculosis* or determination of minimum inhibitory concentrations is being evaluated. In vitro susceptibility studies using combinations of drugs have shown significant synergism between drugs. The significance of in vitro tests in predicting clinical response and recommendations for testing individual isolates have yet to be determined.

Mycobacterium kansasii

Epidemiology

Mycobacterium kansasii, along with *M. avium* complex, is one of the two most common causes of nontuberculous mycobacterial pulmonary disease in humans. In the United States, most cases of *M. kansasii* infections have been reported from the southern states of Texas, Louisiana, and Florida, from Illinois and Missouri in the Midwest, and from California. *Mycobacterium kansasii* strains have been isolated from water, yet the natural source of human infection is not clear. As with other nontuberculous mycobacteria, infections are not normally considered contagious from person to person.

Clinical Infections

The most common manifestation is chronic pulmonary disease involving the upper lobes, usually with evidence of cavitation and scarring. Extrapulmonary infections, including lymphadenitis, skin and soft-tissue infection, and joint infection, have been reported occasionally. Disseminated *M. kansasii* infection rarely occurs in the immunocompetent but has been reported in severely immunocompromised patients, particularly those with AIDS.

For treatment of pulmonary disease caused by *M. kansasii,* a multidrug regimen of isoniazid, rifampin, and ethambutol is currently recommended. Isolates of *M. kansasii* are resistant to pyrazinamide; therefore, this drug is not an alternative choice for treatment.

Laboratory Diagnosis

Cells of *M. kansasii* are long rods with distinct cross-banding. This slowly growing organism has an optimal growth temperature of 37°C. Colonies are smooth to rough with wavy edges and dark centers when grown on Middlebrook 7H10 agar. Some cording can usually be seen with low-power magnification. Colonies grown in the dark are nonpigmented or buff; when grown in light or exposed to light, colonies become yellow (photochromogenic) (Fig. 22–14). With prolonged exposure to light, most strains form dark red crystals of β-carotene on the surface of and inside the colony. Rarely, scotochromogenic and nonchromogenic strains are isolated. Most strains are strongly catalase positive (>45 mm); less commonly isolated are strains that are low catalase producers (<45 mm).

Characteristics that distinguish this species are a growth rate similar to that of *M. tuberculosis* at 37°C, strong photochromogenic properties, ability to hydrolyze Tween 80 in 3 days, strong nitrate reduction and catalase production, and pyrazinamidase production (see Table 22–4). A nonisotopic nucleic acid probe for the identification of *M. kansasii* isolates is available commercially.

When tested in vitro using the current drug concentrations recommended for *M. tuberculosis,* most strains of *M. kansasii* are susceptible to rifampin and ethambutol, partially resistant to isoniazid and streptomycin, and resistant to pyrazinamide.

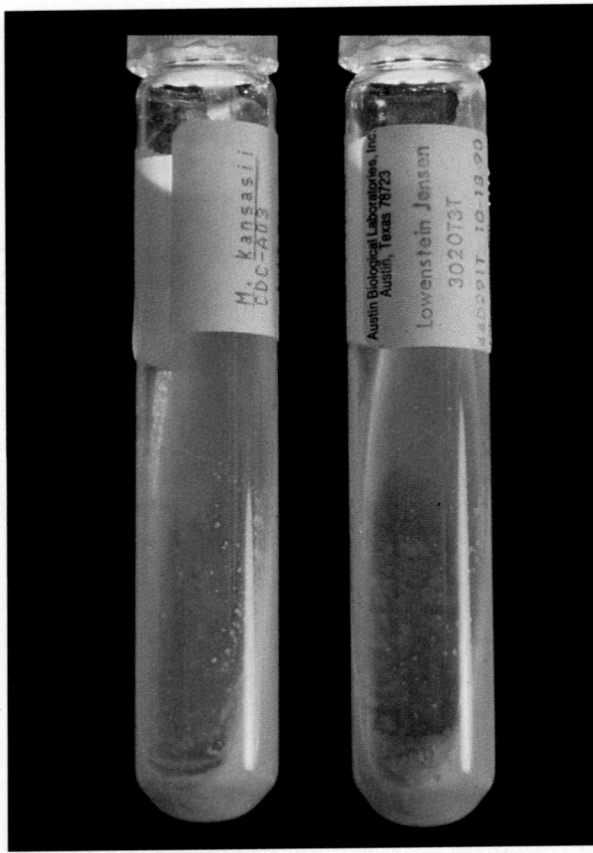

■■■■■ F i g u r e   2 2 – 1 4

M. kansasii on LJ medium showing photoreactivity. *Left,* Before exposure to light; *right,* after exposure to light.

Mycobacterium fortuitum-chelonae Complex

The *Mycobacterium fortuitum-chelonae* complex is made up of the species *M. fortuitum* and its biovariants, *M. chelonae* and its subspecies, and *M. abscessus* (formerly *M. chelonae* subsp *abscessus*). An organism that is acid fast and nonpigmented, grows in less than 7 days at its optimum temperature, is arylsulfatase positive at 3 days, and grows at 38°C on MacConkey agar without crystal violet can be placed in this complex. Species within the complex can be differentiated by additional biochemical tests, however.

Mycobacterium fortuitum and *M. chelonae* are commonly found in the same types of infections, yet the two species vary in their susceptibility to antimicrobial agents, *M. fortuitum* generally being more susceptible. For this reason, determination

of species and subspecies of clinically significant isolates may be warranted.

Mycobacterium fortuitum

Common in the environment, *M. fortuitum* has been isolated from water, soil, and dust. The organism has been implicated frequently in infections of the skin and soft tissues, including localized infections and abscesses at the site of puncture wounds. Infections associated with long-term use of intravenous and peritoneal catheters, injection sites, and surgical wounds following mammoplasty and cardiac bypass procedures have been reported. A variety of other infections have been associated with *M. fortuitum*. Differences in susceptibility to antimicrobial agents occur within the three subspecies; thus, in vitro susceptibility testing is often recommended for clinically significant isolates. A test method utilizing a broth microdilution minimum inhibitory concentration determination has been well studied for the rapidly growing mycobacteria.

After 3 to 5 days of incubation at 37°C, colonies of the rapidly growing *M. fortuitum* appear rough or smooth and either nonpigmented, creamy white, or buff. Microscopic examination of growth on cornmeal-glycerol and Middlebrook 7H11 agars after 1 to 2 days of incubation reveals colonies with branching, filamentous extensions, and rough colonies with short aerial hyphae. On microscopic examination, cells are pleomorphic, ranging from long, tapered to short, thick rods. Most cultures, especially older cultures, tend to decolorize and appear partially acid fast on any of the acid-fast staining techniques. Additional characteristics that distinguish *M. fortuitum* from other rapidly growing mycobacteria are the positive 3-day arylsulfatase test and the reduction of nitrate (see Table 22–4). Three biovariants of *M. fortuitum* exist: biovariant *fortuitum,* biovariant *peregrinum* (proposed *M. peregrinum* sp. Nov., nom.rev.), and a biovariant referred to as "third group" (*M. fortuitum* subsp. *acetamidolyticum*). Separation of these three biovariants can be made on the ability of the isolate to utilize mannitol, inositol, or sodium citrate as a sole source of carbon.

Mycobacterium chelonae

Mycobacterium chelonae is found in the environment and is associated with many of the same opportunistic infections as *M. fortuitum.* Both species have been associated with a variety of infections of the skin, lungs, bone, central nervous system, and prosthetic heart valves as well as with disseminated disease. *Mycobacterium chelonae* exhibits more resistance to antimicrobial agents than *M. fortuitum* but sometimes is susceptible to amikacin and a sulfonamide.

Microscopically, young cultures of *M. chelonae* are strongly acid fast, with pleomorphism ranging from long, tapered to short, thick rods. This rapidly growing mycobacterium produces rough or smooth, nonpigmented to buff colonies within 3 to 5 days of incubation at 37°C. Unlike *M. fortuitum, M. chelonae* does not produce extensive filamentous branching colonies on cornmeal-glycerol agar. A positive 3-day arylsulfatase test, no reduction of nitrate, and growth on MacConkey agar without crystal violet are additional characteristics that differentiate *M. chelonae* from other rapidly growing mycobacteria (see Table 22–4). *Mycobacterium chelonae* subsp *chelonae* can be distinguished from *M. abscessus* (formerly designated as *M. chelonae* subsp *abscessus*) because the former does not grow in the presence of 5% sodium chloride and is able to utilize citrate as a sole source of carbon. An unnamed mycobacterial species, referred to as *M. chelonae*–like organism (MCLO), does not grow in the presence of 5% sodium chloride, usually utilizes citrate and mannitol as a sole carbon source, and gives an unusual tan (±) reaction on iron uptake medium.

Mycobacterium marinum

Mycobacterium marinum has been implicated in diseases of fish and has been isolated from aquariums. Cutaneous infections in humans have occurred when traumatized skin came in contact with inadequately chlorinated fresh water or with salt water. Outbreaks of cutaneous lesions in lifeguards have been reported. The typical presentation of a tender red or blue-red subcutaneous nodule, or "swimming pool granuloma," usually occurs on the elbow, knee, toe, or finger. In some cases, an abscess develops at the primary site of inoculation, with secondary ascending spread along the lymphatics. Treatment modalities that have been used include simple observation of minor lesions, surgical excision, antituberculosis drug therapy, and the use of single antibiotic agents. In the standard in vitro susceptibility testing, as currently used to test *M. tuberculosis, M. marinum* is susceptible to rifampin and ethambutol, resistant to isoniazid and pyrazinamide, and partially resistant, or intermediate, to streptomycin.

Cells of *M. marinum* are moderately long to long rods with cross-barring. Colonies of this slowly growing organism are smooth to rough (Fig. 22–15) and wrinkled on inspissated egg medium, but may be smooth when grown on Middlebrook 7H10 or 7H11 agar. *Mycobacterium marinum* is photochromogenic, i.e., young colonies grown in the dark may be nonpigmented or buff, whereas colonies grown in or exposed to light develop a deep yellow color. Growth is optimum at incubation temperatures of 30 to 32°C. This preference for growth at the lower incubation temperatures, along with photochromogenicity, are clues to the identification of *M. marinum.* Some strains of *M. marinum* produce niacin; however, none reduces or produces nitrate heat-stable catalase. The organisms hydrolyze Tween 80, and produce urease and pyrazinamidase (see Table 22–4).

Mycobacterium scrofulaceum

The most common form of disease associated with *M. scrofulaceum* is cervical lymphadenitis in children. The infection manifests in one or more enlarged nodes, often adjacent to the mandible and high in the neck, with little or no pain. Patients are usually treated by surgical incision and drainage; antituberculosis drugs usually are not necessary. Pulmonary infections caused by *M. scrofulaceum* have been reported. *Mycobacterium scrofulaceum* is resistant to isoniazid,

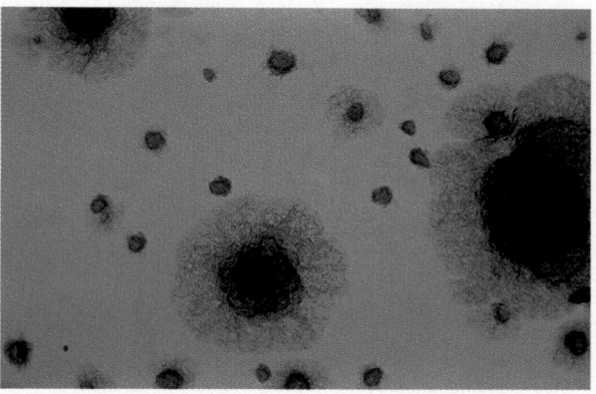

■ F i g u r e 2 2 – 1 5
M. marinum on Middlebrook 7H10 growing rough colonies.

streptomycin, ethambutol, and *p*-aminosalicylic acid when tested in vitro using the current procedure for testing *M. tuberculosis.*

On microscopic examination, *M. scrofulaceum* is a uniformly stained, acid-fast, medium to long rod. The organism grows slowly (4 to 6 weeks) at incubation temperatures ranging from 25 to 37°C. Colonies are smooth with dense centers and pigmentation from light yellow to deep orange. The organism is scotochromogenic (Fig. 22–16), that is, pigment is produced when cultures are incubated in the absence of light and may darken when exposed to light. Members of this species do not hydrolyze Tween 80 nor reduce nitrate, but they do produce urease and are high (>45 mm) catalase producers (see Table 22–4). These characteristics aid in differentiating this organism from other slowly growing scot-

ochromogens, including certain strains of *M. avium* complex, *M. gordonae,* and *M. szulgai.*

Organisms with characteristics of both *M. scrofulaceum* and *M. avium* complex have been isolated from clinical specimens. These organisms are referred to as *M. avium-intracellulare-scrofulaceum,* or MAIS group. The potential for pathogenicity of these organisms has yet to be determined.

Mycobacterium xenopi

Mycobacterium xenopi has been recovered from hot and cold water taps, including water storage tanks of hospitals, and from birds. The organism was first isolated from an African toad and was considered nonpathogenic for humans until recently. Isolation of *M. xenopi* is relatively uncommon in the United States, yet it has been reported as one of the most commonly found nontuberculous mycobacteria in southeast England. The reported human cases of *M. xenopi* infection are mostly slowly progressive pulmonary infections in individuals with predisposing conditions. Preexistent lung disease, alcoholism, malignancy, and diabetes mellitus are some of the conditions associated with reported *M. xenopi* infections. The pulmonary infections presented clinical pictures similar to those seen in patients with *M. tuberculosis, M. kansasii,* or *M. avium* complex infection. Disseminated and extrapulmonary infections have been reported. Strains of *M. xenopi* are susceptible to the quinolones (ciprofloxacin and ofloxacin); some isolates are susceptible to vancomycin, erythromycin, or cefuroxime. In vitro susceptibility to antituberculosis drugs is variable, with resistance to ethambutol only being the most common pattern.

On acid-fast–stained smears, *M. xenopi* are long, filamentous rods. Colonies of this slowly growing mycobacterium on Middlebrook 7H10 agar are small with dense centers and filamentous edges. Microscopic observation (low-power magnification) of colonies growing on cornmeal-glycerol agar reveals distinctive round colonies with branching and filamentous extensions. Aerial hyphae are usually seen in rough colonies. Young colonies on cornmeal agar show a "bird's nest" appearance, with stick-like projections. Optimal growth temperature is 42°C; the organism grows more rapidly at this temperature than at 37°C and fails to grow at 25°C. This organism has been classified with the nonphotochro-

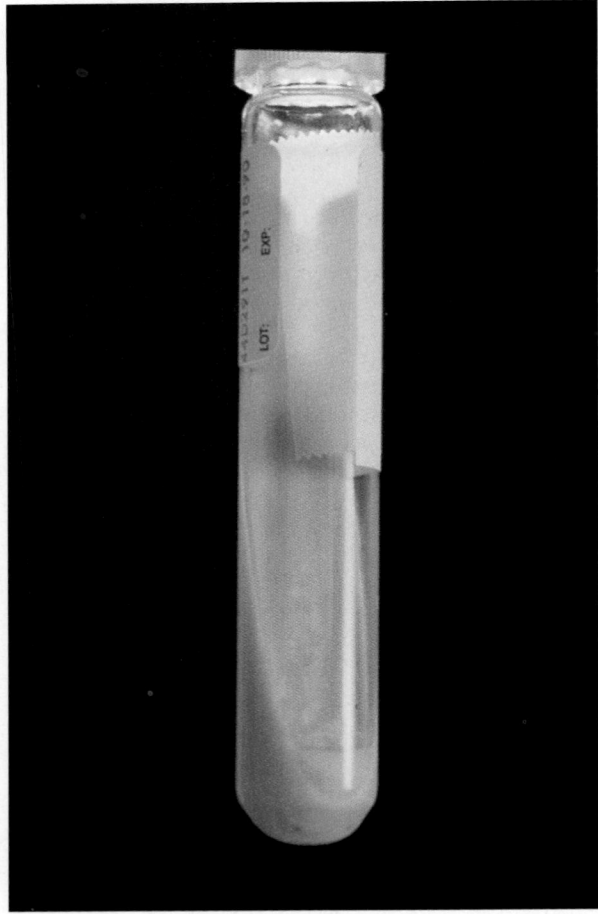

■■■■■ F i g u r e   2 2 – 1 6
M. scrofulaceum on LJ medium.

mogenic group; however, colonies frequently are bright yellow on primary isolation when incubated in the absence of light and when exposed to light. Distinctive characteristics, in addition to optimum growth at 42°C and yellow scotochromogenic pigment, are negative reactions for niacin accumulation and nitrate reduction and positive reactions for the production of heat-stable catalase, arylsulfatase, and pyrazinamidase (see Table 22–4).

Mycobacterium celatum

A newly described species, *Mycobacterium celatum* sp nov, is a slowly growing, nonphotochromogenic organism very similar in biochemical characteristics to *M. xenopi*. Mycolic acid patterns of this new species, as determined by HPLC, are most similar to those of *M. xenopi* but are distinct from other described species. In contrast to *M. xenopi*, *M. celatum* grows best at 37°C and poorly at 45°C, produces large colonies on Middlebrook 7H10 agar, and is usually resistant to rifabutin.

Mycobacterium szulgai

Of the reported infections with *M. szulgai*, the most common manifestation is pulmonary disease similar to tuberculosis. Extrapulmonary infections, including lymphadenitis and bursitis, have also been reported. This organism is much more susceptible than *M. avium* complex to the conventional antituberculosis drugs.

On microscopic examination of an acid-fast–stained smear, cells of *M. szulgai* are medium to long rods with some cross-barring. When the organism is cultured on egg-based medium at 37°C, smooth and rough colonies are observed. At 37°C, yellow to orange pigment develops in the absence of light and intensifies with exposure to light (scotochromogenic). Colonies grown at 22°C are nonpigmented or buff in the absence of light and develop yellow to orange pigment with light exposure (photochromogenic). Characteristics that differentiate *M. szulgai* from other slowly growing mycobacteria are slow hydrolysis of Tween 80, positive nitrate reduction, and inability to grow in the presence of 5% sodium chloride. The last characteristic, along with photochromogenicity at 22°C, distinguishes *M. szulgai* from *M. flavescens* (see Table 22–4).

Mycobacterium malmoense

Pulmonary disease associated with *M. malmoense* has been reported more commonly outside the United States, i.e., in Sweden, England, Wales, and Scotland. Reports of cases of chronic pulmonary disease and cervical adenitis in the United States have appeared, however. In laboratory studies using conventional antituberculosis drug resistance testing, *M. malmoense* is resistant to isoniazid, streptomycin, *p*-aminosalicylic acid, and rifampin and is susceptible to ethambutol and cycloserine.

Mycobacterium malmoense appears as short coccobacilli without cross-bands on acid-fast–stained smears. Colonies are smooth (Fig. 22–17), glistening, and opaque with dense centers. Color is nonpigmented to buff; exposure to light does not produce pigment (nonphotochromogenic). Growth rate is slow; optimum growth temperature is 37°C; growth at 22°C may require as much as 7 weeks of incubation. Some strains may require longer incubation (up to 12 weeks) before colonies become visible. For this reason, some investigators suspect that *M. malmoense* may be under-reported in the United States, because most laboratories incubate cultures for 6 weeks, 2 to 6 weeks less than the incubation period needed for some strains of this organism. The increase in the isolation of *M. malmoense* may be attributed to the implementation of radiometric culture techniques in larger laboratories.

Differential characteristics of *M. malmoense* are no accumulation of niacin, absence of nitrate reduction, ability to hydrolyze Tween 80, and the usual production of a heat-stable (68°C) catalase. *Mycobacterium malmoense* can be differentiated

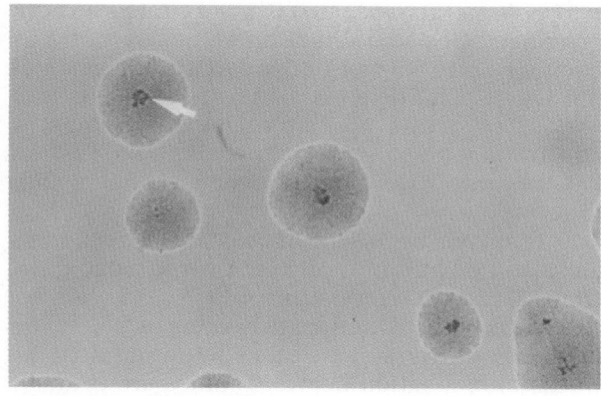

■■■■■■■F i g u r e   2 2 – 1 7
M. malmoense on Middlebrook 7H10 medium.

from biochemically similar *M. gastri* on the basis of the urease-negative and pyrazinamidase-positive reactions of *M. malmoense* (see Table 22–4).

Mycobacterium simiae

The original strains of *M. simiae* were isolated from the lymph nodes of monkeys. Although the organism has been recovered from tap water, there seems to be significant geographical variation in the incidence. For example, *M. simiae* is rarely isolated in most parts of the United States, but in parts of Texas, it is a relatively common isolate. Infrequent cases of human infection from *M. simiae* have been reported and are most often pulmonary disease in patients with preexisting lung damage. Most isolates are resistant to most antituberculosis drugs in vitro.

Cells of *M. simiae* appear as short coccobacilli. When they are grown on inspissated egg medium at 37°C, smooth colonies appear in 10 to 21 days. Colonies on Middlebrook 7H10 agar are thin, transparent or tiny, and filamentous. The species is usually photochromogenic, being nonpigmented or buff when incubated in the absence of light and yellow when exposed to light. Development of the yellow pigment may require prolonged incubation, whereas some strains may fail to produce pigment on exposure to light. Differential biochemical characteristics are the accumulation of niacin, negative nitrate, reduction, and high level (>45 mm) of heat-stable catalase (see Table 22–4).

Mycobacterium ulcerans

Mycobacterium ulcerans is a very rare cause of mycobacteriosis in the United States but has been reported more frequently in other parts of the world. The disease manifests as a painless nodule under the skin after previous trauma. A shallow ulcer develops that may be quite severe. Patients rarely develop fever or systemic symptoms.

The acid-fast cells of *M. ulcerans* are moderately long without beading or cross-banding. Optimal growth temperature is 30 to 33°C with little growth at 25°C and usually none at 37°C. The organism grows slowly, often requiring 6 to 12 weeks of incubation before colonies are evident. Colonies are smooth and rough, and nonpigmented or lightly buff, and they do not develop pig-

ment with exposure to light (nonphotochromogenic). *Mycobacterium ulcerans* produces a heat-stable catalase but is inert to most other conventional biochemical tests.

Mycobacterium haemophilum

The rare infections associated with *M. haemophilum* occur primarily in patients who are immunocompromised. Cases have been reported in patients with Hodgkin's disease and AIDS. Submandibular lymphadenitis, subcutaneous nodules, painful swellings, ulcers progressing to abscesses, and draining fistulas are often the clinical manifestations.

A unique characteristic of this organism is its requirement for hemoglobin or hemin for growth. Isolation of this species is accomplished on media supplying the needed growth supplement, such as chocolate agar, Mueller-Hinton agar with 5% Fildes enrichment, and Löwenstein-Jensen medium containing 2% ferric ammonium citrate. Successful isolation on Middlebrook 7H10 agar with an X-factor disk planted in the inoculated area has been reported (Vadney and Hawkins). Optimum growth temperature is 28 to 32°C; little or no growth occurs at 37°C. Colonies are rough to smooth and nonpigmented. Microscopically, the cells are strongly acid-fast, short, occasionally curved bacilli without banding or beading and arranged in tight clusters or cords.

Mycobacterium gordonae

Mycobacterium gordonae is commonly found in water taps and soil and is commonly referred to as the *tap-water bacillus*. This organism is frequently found in clinical specimens as a causal resident but is rarely implicated in disease. Infections reported in the literature have been isolated cases of meningitis secondary to ventriculoatrial shunts, hepatoperitoneal disease, endocarditis in a prosthetic aortic valve, synovitis, cutaneous lesions of the hand, and possible cases of pulmonary disease. In laboratory in vitro antituberculosis drug resistance tests, *M. gordonae* is resistant to isoniazid, streptomycin, and *p*-aminosalicylic acid but susceptible to rifampin and ethambutol.

Growth of *M. gordonae* appears on egg-based medium after 10 to 14 days as smooth, yellow-orange colonies pigmented with both absence of and exposure to light (scotochromogenic). Optimum growth temperature range is 22 to 37°C. Differential characteristics are negative nitrate reduction, ability to hydrolyze Tween 80, and the production of heat-stable catalase (see Table 22–4). Isolates of *M. gordonae* can be rapidly identified with the use of a commercially available nucleic acid probe specific to the species.

Mycobacterium asiaticum

Mycobacterium asiaticum is a normally saprophytic *Mycobacterium* species that very rarely causes human infection. The cells are acid-fast coccoid rods. Growth on inspissated egg medium after 15 to 21 days at 37°C is dysgonic and smooth. Members of this species are usually photochromogenic, showing no pigment or buff when cultured in the absence of light and yellow when exposed to light. Occasional strains fail to develop pigment after exposure to light. *Mycobacterium asiaticum* fails to reduce nitrate and produces a high level (>45 mm) of heat-stable catalase. No accumulation of niacin and the hydrolysis of Tween 80 are useful characteristics in differentiating this organism from *M. simiae* (see Table 22–4).

Mycobacterium thermoresistibile

Mycobacterium thermoresistibile is uncommonly isolated from clinical specimens. The species is classified as a rapidly growing mycobacterium, yet on primary isolation, it may grow slowly because of its preference for higher incubation temperatures. Thus, it may be mistaken for a slowly growing scotochromogen. On subculture on egg-based medium, the organism produces smooth to rough yellow colonies within 3 to 5 days. Distinctive characteristics of this species are the ability to grow at 52°C and a negative iron uptake test (see Table 22–4).

Mycobacterium terrae-triviale Complex

There have been a few reported cases of human infection associated with the *M. terrae-triviale* complex. Rare cases of septic arthritis, synovitis, osteomyelitis, respiratory infection, and tenosynovitis of the fingers, hands, and wrists have been reported. Isolates from the sputum and gastric lavage of humans are not uncommon and are often considered casual residents.

Microscopically, members of the *M. terrae-triviale* complex appear as acid-fast, short to medium coccobacilli. Both species, *M. terrae* and *M. triviale,* grow slowly at an optimum temperature of 37°C and are nonphotochromogenic. The colonies of *M. triviale* on egg-based medium are rough, dry, and heaped, whereas the colonies of *M. terrae* tend to be smoother. Characteristics that differentiate the complex from other mycobacteria are hydrolysis of Tween 80, reduction of nitrate, and the presence of a heat-stable catalase (see Table 22–4).

Mycobacterium nonchromogenicum

Mycobacterium nonchromogenicum is another species that is most commonly saprophytic but on rare occasions pathogenic. Primary lung disease due to *M. nonchromogenicum* has been reported. The cells are acid-fast, moderately long rods. After incubation at 37°C, colonies of this slowly growing, nonphotochromogenic species are smooth to rough, and white to buff. Positive pyrazinamidase and negative nitrate reduction reactions differentiate *M. nonchromogenicum* from the *M. terrae-triviale* complex (see Table 22–4). Because of the similarities among *M. nonchromogenicum, M. terrae,* and *M. triviale,* some observers classify the three species in an *M. terrae* complex.

Mycobacterium flavescens

Isolation of *M. flavescens* from human clinical specimens usually represents contamination or colonization. The cells are coccoid rods that show irregular acid fastness with acid-fast stains. Growth rate is intermediate with soft, yellow-orange, butyrous colonies appearing on egg-based medium after 7 to 10 days at 37°C. Some strains grow well at temperatures up to 45°C. The species is scotochromogenic, pigmentation appearing when isolates are cultivated in the absence and in the presence of light. This organism can grow on routine bacteriologic media, as can other of the more rapidly growing mycobacterial species. Other differential characteristics are the ability to hydrolyze Tween 80, reduction of nitrate, and tolerance to 5% sodium chloride (see Table 22–4).

Mycobacterium smegmatis

Commonly considered saprophytic, *M. smegmatis* has been implicated in rare cases of pulmonary, skin, soft-tissue, and bone infections. Microscopically, on acid-fast stain, cells are long and tapered or short rods with irregular acid fastness. Occasionally, rods are curved with branching or Y-shaped forms; swollen with deeper staining, beaded or ovoid forms are sometimes seen. Colonies appearing on egg medium after 2 to 4 days are usually rough, wrinkled, or coarsely folded; smooth, glistening, butyrous colonies may also be seen. Colonies on Middlebrook 7H10 agar are heaped and smooth or rough with dense centers. Pigmentation is rare or late; colonies appear nonpigmented, creamy white, or buff to pink in older cultures. In addition to the rapid growth rate and the nonpigmented, rough colony form, characteristics valuable in the identification of this organism are its negative arylsulfatase reaction, positive iron uptake, ability to reduce nitrate, and growth in the presence of 5% sodium chloride and on MacConkey agar without crystal violet (see Table 22–4).

Mycobacterium phlei

At one time, *M. phlei* was common to the environment, particularly in hay and grass, but it is now only occasionally encountered in human clinical specimens. This organism is classified as a rapidly growing mycobacterium, with growth appearing on egg-based medium after 3 to 5 days at 37°C. Growth also takes place at incubation temperatures of 45°C and 52°C. Colonies are usually rough or coarsely wrinkled with a deep yellow to orange pigment. Occasional strains are smooth and butyrous. Typical strains incubated on Middlebrook 7H10 agar show heaped, rough colonies with dense centers. Microscopically, *M. phlei* is short, coccoid rods that stain irregularly with the acid-fast stain, i.e., some cells within the smear do not appear acid fast. Other characteristics, in addition to the rough, orange-pigmented colonies, that differentiate *M. phlei* from other rapidly growing mycobacteria are a negative arylsulfatase reaction, positive iron uptake, and inability to grow on MacConkey agar without crystal violet (see Table 22–4).

Mycobacterium vaccae

Mycobacterium vaccae, a saprophytic organism that is occasionally isolated from clinical specimens, is a rapidly growing (3 to 5 days) mycobacterium. Colonies on egg-based medium are smooth, moist, and shiny with buff to orange pigment. The organism may be buff at 35°C incubation but turns pigmented at cooler temperatures. Colony morphology on Middlebrook 7H10 agar is varied, often "slick" in appearance; there may be amorphous or poor growth on this medium. Cells are short, coccoid rods staining irregularly acid fast in older cultures. *Mycobacterium vaccae* is differentiated from other arylsulfatase-negative, rapidly growing mycobacteria primarily on the basis of colony morphology and pigmentation (see Table 22–4).

Mycobacterium gastri

Mycobacterium gastri has been found in soil and very rarely in human gastric lavage and sputum specimens. When present in clinical specimens, it has generally been regarded as a casual resident not associated with disease. Cells are moderately long to long rods with cross-barring common. Colonies on inspissated egg medium at 37°C are slowly growing, rough, and nonpigmented or buff in both absence and presence of light (nonphotochromogenic). On Middlebrook 7H10 agar, colonies are dense and rough and may have wavy edges. Differentiation of *M. gastri* from other slowly growing, nonphotochromogenic mycobacteria is based on the following characteristics: no niacin accumulation, ability to hydrolyze Tween 80, negative nitrate reduction, and low level of heat-labile catalase (see Table 22–4).

Mycobacterium paratuberculosis

Mycobacterium paratuberculosis is the causative agent of Johne disease, an intestinal infection occurring as a chronic diarrhea in cattle, sheep, goats, and other ruminants. A *Mycobacterium* species that most closely resembles *M. paratuberculosis* has been reported as being isolated from samples taken from resected terminal ileum of three patients with Crohn's disease. *Mycobacterium paratuberculosis* is difficult to cultivate because of its very slow growth rate (3 to 4 months) and its need for a mycobactin-supplemented medium for primary isolation. (Mycobactin is an iron-binding hydroxymate compound produced by other mycobacterial species.)

Mycobacterium genavense

A proposed new species, *Mycobacterium genavense*, has been reported as a cause of disseminated infections in patients with AIDS. This slowly growing, fastidious mycobacterium has been recovered in BACTEC cultures but failed to grow on subculture to routine solid media. Dysgonic growth was obtained when subculture medium Middlebrook 7H11 agar was supplemented with mycobactin J. In a reported study (Coyle) of isolates from seven AIDS patients, the subculture growth from the supplemented medium consistently yielded positive tests for semiquantitative and heat-stable-catalase, pyrazinamidase, and urease.

MYCOBACTERIUM LEPRAE

Mycobacterium leprae is the causative agent of Hansen disease (leprosy), an infection of the skin, mucous membranes, and peripheral nerves. The disease is rare in the United States and other Western countries, yet it remains a major problem in other parts of the world. In the United States, the reported cases are generally from areas with a warm climate, including California, Texas, Louisiana, Florida, Hawaii, and Puerto Rico.

There are two major forms of the disease, tuberculoid leprosy and lepromatous leprosy, but there is no distinct separation between these two manifestations. With tuberculoid leprosy, skin lesions and nerve involvement producing areas of anesthesia may occur. Spontaneous recovery often occurs with tuberculoid leprosy. On the other hand, lepromatous leprosy is slowly progressive, malignant, and, if untreated, life threatening. It is characterized by skin lesions and progressive, symmetric nerve damage. Lesions of the mucous membranes of the nose may lead to destruction of the cartilaginous septum, resulting in nasal and facial deformities. Current therapy usually consists of a combination of diaminodiphenylsulfone (dapsone), clofazimine, and rifampin.

Laboratory diagnosis of leprosy depends on the microscopic demonstration of acid-fast bacilli that cannot be cultured from skin biopsy

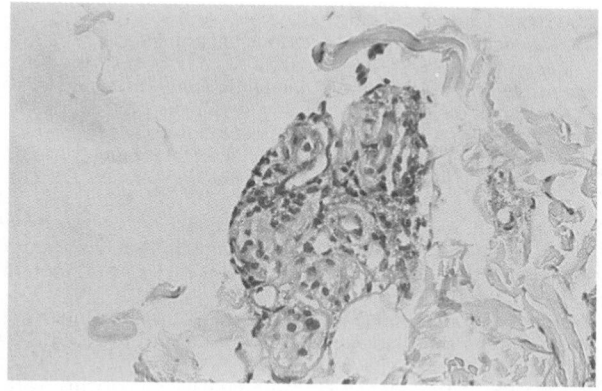

Figure 22-18

M. leprae from a skin biopsy from a patient with lepromatous leprosy (acid-fast smear stained with Ziehl-Neelsen stain).

specimens. Organisms are extremely rare and may not be detected in skin scrapings or biopsy specimens from patients with tuberculoid leprosy. Acid-fast bacilli are, however, usually abundant in samples from patients with lepromatous leprosy (Fig. 22–18). *Mycobacterium leprae* is a rod-shaped bacterium usually 1 to 7 μm in length and 0.3 to 0.5 μm wide. After examination of the entire smear, the number of organisms present per oil immersion field (1000×) is reported as the *bacteriologic index* (BI). The number of "solid-staining" cells per 100 total bacilli examined is reported as the *morphologic index* (MI). "Solid-staining" cells are those with dense, uniform staining of the entire bacillus with even sides and rounded ends in which the length of the bacillus is at least five times the width of the bacillus. The BI and the MI aid the clinician in determining the progress of the disease. The definitive laboratory diagnosis is the development of disease in laboratory mice following inoculation of patient biopsy material to the mouse footpad.

Bibliography

Abadco DL, Steiner P: Gastric lavage is better than bronchoalveolar lavage for isolation of *Mycobacterium tuberculosis* in childhood pulmonary tuberculosis. Pediatr Infect Dis J 11:735, 1992.

Abe C, Hirano K, et al: Detection of *Mycobacterium tuberculosis* in clinical specimens by polymerase chain reaction and Gen-Probe amplified *Mycobacterium tuberculosis* direct test. J Clin Microbiol 31:3270, 1993.

Abe C, Hosojima S, et al: Comparison of MB-Check, BACTEC, and egg-based media for recovery of Mycobacteria. J Clin Microbiol 30:878, 1992.

American Thoracic Society: Diagnosis and treatment of disease caused by nontuberculous mycobacteria. Am Rev Respir Dis 142:940, 1990.

American Thoracic Society: Diagnostic standards and classification of tuberculosis. Am Rev Respir Dis 142:725, 1990.

American Thoracic Society: Levels of laboratory services for mycobacterial diseases: Official Statement of the American Thoracic Society. Am Rev Respir Dis 128:213, 1983.

Anargyros P, Astill DSJ, Lim FSL: Comparison of improved BACTEC and Lowenstein-Jensen media for culture of mycobacteria from clinical specimens. J Clin Microbiol 28:1288, 1990.

Arnold LJ, Hammond PW, et al: Assay formats involving acridinium-ester-labeled DNA probes. Clin Chem 35:1588, 1989.

Barnes PF, Bloch AB, et al: Tuberculosis in patients with human immunodeficiency virus infection. N Engl J Med 324:1644, 1991.

Bates JH: Diagnosis of tuberculosis. Chest 76:757, 1979.

Beam RE, Kubica GP: Stimulatory effects of carbon dioxide on the primary isolation of tubercle bacilli on agar-containing medium. Am J Clin Pathol 50:395, 1968.

Biswas SK, Singh PP, et al: Detection of tubercle bacilli in sputum smear by flotation method. Indian J Chest Dis Allied Sci 29:8, 1987.

Bloch AB, Cauther GM, et al: Nationwide survey of drug resistant tuberculosis in the United States. JAMA 271:665, 1994.

Böttger EC, Hirschel B, Coyle MB: Mycobacterium genavense sp. nov. Int J Syst Bacteriol 43:841, 1993.

Butler WR, O'Connor SP, et al: Mycobacterium celatum sp. nov. Int J Syst Bacteriol 43:539, 1993.

Chopin-Robertson K, Dahlberg S, et al: Detection and identification of Mycobacterium directly from BACTEC bottles by using a DNA-rRNA probe. Diagn Microbiol Infect Dis 17:203, 1993.

Clarridge JE, Shawar RM, et al: Large scale use of polymerase chain reaction for detection of Mycobacterium tuberculosis in a routine mycobacteriology laboratory. J Clin Microbiol 31:2059, 1993.

Coyle MB, Carlson LDC: Laboratory aspects of "Mycobacterium genavense," a proposed species isolated from AIDS patients. J Clin Microbiol 30:3206, 1992.

Daniel TM: New approaches to the rapid diagnosis of tuberculous meningitis. J Infect Dis 155:602, 1986.

David HL, Selin MJ: Immune response to mycobacteria. In Rose NR, (ed): Manual of Clinical Immunology, 2nd ed. Washington, DC: American Society for Microbiology, 1980, pp 520–525.

Davis TE: Mycobacterium tuberculosis: A renewed challenge for the clinical microbiology laboratory. Clin Microbiol Newsl 14:97, 1992.

Desmond EP: Clinical features and laboratory identification of Mycobacterium genavense. Clin Lab Newsletter 16:49, 1994.

DesPrez RM, Heim CR: Mycobacterium tuberculosis. In Mandell GL, Douglas RG, Bennett J (eds): Principles and Practice of Infectious Disease, 3rd ed. New York: Churchill Livingstone, 1985, pp 1877–1906.

Eisenach KD, Sifford MD, et al: Detection of Mycobacterium tuberculosis in sputum samples using a polymerase chain reaction. Am Rev Respir Dis 144:1160, 1991.

Ellner PD, Kiehn TE, et al: Rapid detection and identification of pathogenic mycobacteria by combining radiometric and nuclei acid probe methods. J Clin Microbiol 26:1349, 1988.

Fisher JF, Ganapathy M, et al: Utility of Gram's and Giemsa stains in the diagnosis of pulmonary tuberculosis. Am Rev Respir Dis 141:511, 1990.

Forbes BA, Hicks KES: Direct detection of Mycobacterium tuberculosis in respiratory specimens in a clinical laboratory by polymerase chain reaction. J Clin Microbiol 31:1688, 1993.

Gill VJ, Park CH, et al: Use of lysis-centrifugation (Isolator) and radiometric (BACTEC) blood culture systems for the detection of mycobacteremia. J Clin Microbiol 22:543, 1985.

Gonzalez R, Hanna BA: Evaluation of Gen-Probe DNA hybridization systems for the identification of Mycobacterium tuberculosis and Mycobacterium avium-intracellulare. Diagn Microbiol Infect Dis 8:69, 1987.

Good RC, Mastro TD: The modern mycobacteriology laboratory: How it can help the clinician. Clin Chest Med 10:315, 1989.

Gordin F, Slutkin G: The validity of acid-fast smears in the diagnosis of pulmonary tuberculosis. Arch Pathol Lab Med 114:1025, 1990.

Gross WM, Hawkins JE: Radiometric selective inhibition tests for differentiation of Mycobacterium tuberculosis, Mycobacterium bovis, and other Mycobacteria. J Clin Microbiol 21:565, 1985.

Guerrant GO, Lambert MA, Moss CW: Gas-chromotographic analysis of mycolic acid cleavage products in mycobacteria. J Clin Microbiol 13:899, 1981.

Guthertz LS, Lim SD, et al: Curvilinear-gradient high performance liquid chromatography for identification of mycobacteria. J Clin Microbiol 31:1876, 1993.

Hobby GL, Holman AP, et al: Enumeration of tubercle bacilli in sputum of patients with pulmonary tuberculosi. Antimicrob Agents Chemother 4:94, 1973.

Huebner RE, Good RC, Tokars JI: Current practices in mycobacteriology: Results of a survey of state public health laboratories. J Clin Microbiol 31:771, 1993.

Inderlied CB, Kemper CA, Bermudez LEM: The Mycobacterium avium complex. Clin Microbiol Rev 6:266, 1993.

Isenberg HD, D'Amato RF, et al: Collaborative feasibility study of a biphasic system (Roche Septi-Chek AFB) for rapid detection and isolation of Mycobacteria. J Clin Microbiol 29:1719, 1991.

Jonas V, Alden MJ, et al: Detection and identification of Mycobacterium tuberculosis directly from sputum sediments by amplification of rRNA. J Clin Microbiol 31:2410, 1993.

Kent PT, Kubica GP: Public Health Mycobacteriology: A Guide for the Level III Laboratory. Atlanta: US Dept Health and Human Services, Public Health Service, Centers for Disease Control, 1985.

Kiehn TE, Gold JWM, et al: Mycobacterium tuberculosis bacteremia detected by the Isolator lysis-centrifugation blood culture system. J Clin Microbiol 21:647, 1985.

Kirihara JM, Hillier SL, Coyle MB: Improved detection times for Mycobacterium avium complex and Mycobacterium tuberculosis with the BACTEC radiometric system. J Clin Microbiol 22:841, 1985.

Koneman EW, Allen SD, et al: Color Atlas and Textbook of Diagnostic Microbiology, 4th ed. Philadelphia: JB Lippincott, 1992, pp 703–756.

Kubica GP, Wayne LG: The Mycobacteria: A Sourcebook, Part A and Part B. New York: Marcel Dekker, 1984.

Kusunoki S, Takayuki E: Proposal of Mycobacterium peregrinum sp. nov., nom.rev., and elevation of Mycobacterium chelonae sub sp. abscessus (Kubica et al.) to special status:

Mycobacterium abscessus comb. nov. Int J Syst Bacteriol 42:240, 1992.

Kvach JT, Veras JR: A fluorescent staining procedure for determining the viability of mycobacterial cells. Int J Lepr 50:183, 1982.

Laszlo A, Eidus L: Test for differentiation of *M. tuberculosis* and *M. bovis* from other mycobacteria. Can J Microbiol 24:754, 1978.

Leelaraasamee A, Bovornkittii S: Immunodiagnosis of tuberculosis: A review. Asian Pac J Allergy Immunol 7:57, 1989.

Lefford MJ: Immune response to Mycobacteria: Diagnostic methods. In Rose NR, de Macario EC, Fahey JL, Friedman H, Penn GM (eds): Manual of Clinical Laboratory Immunology, 4th ed. Washington, DC: American Society for Microbiology, 1992.

Lipsky BA, Gates JA, et al: Factors affecting the clinical value of microscopy for acid-fast bacilli. Rev Infect Dis 6:214, 1984.

Martin C, Levy-Frebalult VV, et al: False positive result of *Mycobacterium tuberculosis* complex DNA probe hybridization with a *Mycobacterium terrae* isolate. Eur J Clin Microbiol Infect Dis 12:309, 1993.

Morgan MA, Horstmeier CD, et al: Comparison of a radiometric method (BACTEC) and conventional culture media for recovery of mycobacteria from smear-negative specimens. J Clin Microbiol 18:384, 1983.

Morgan MA, Doerr KA, et al: Evaluation of the p-nitro-alpha-acetylamino-beta-hydroxypropriophenone differential test for identification of *Mycobacterium tuberculosis* complex. J Clin Microbiol 21:634, 1985.

Musical CE, Tice LS, et al: Identification of mycobacteria from culture by using the Gen-Probe rapid diagnostic system for *Mycobacterium avium* complex and *Mycobacterium tuberculosis* complex. J Clin Microbiol 26:2120, 1988.

Narain R, Rao MSS, et al: Microscopy positive and microscopy negative cases of pulmonary tuberculosis. Am Rev Respir Dis 103:761, 1971.

National Committee for Clinical Laboratory Standards: Antimycobacterial Susceptibility Testing. Proposed Standard Document M24-P, Vol. 10. Villanova, PA: NCCLS, 1990.

Nolte FS, Metchock B, et al: Direct detection of *Mycobacterium tuberculosis* in sputum by polymerase chain reaction and DNA hybridization. J Clin Microbiol 31:1777, 1993.

Park CH, Hixon DL, et al: Rapid recovery of mycobacteria from clinical specimens using automated radiometric technic. Am J Clin Pathol 81:341, 1984.

Parrot RJ, Augier B, et al: La microscopie en fluorescence du bacille tubérculeux. I: Sensibilité de la method. Rev Tuberc Pneumol 34:197, 1971.

Rickman TW, Moyer NP: Increased sensitivity of acid-fast smears. J Clin Microbiol 11:618, 1980.

Roberts GD, Koneman EW, Kim YK: *Mycobacterium.* In Balows A, et al (eds): Manual of Clinical Microbiology, 5th ed. Washington, DC: American Society for Microbiology, 1991, pp 304–339.

Roberts GD, Goodman NL, et al: Evaluation of the BACTEC radiometric method for recovery of mycobacteria and drug susceptibility testing of *Mycobacterium tuberculosis* from smear-positive specimens. J Clin Microbiol 18:689, 1983.

Roberts GD, Thompson GP: Bacteriology and bacteriologic diagnosis of tuberculosis. In Schlossberg D (ed): Tuberculosis 3rd ed. New York: Springer-Verlag, 1994, pp 23–33.

Roberts MC, McMillan C, et al: The use of p-nitro-alpha-acetylamino-beta-hydroxypro-priophenone in the differentiation of mycobacteria. Am Rev Respir Dis 38:759, 1960.

Roberts MC, McMillan C, Coyle, MB: Whole chromosomal DNA probes for rapid identification of *Mycobacterium tuberculosis* and *Mycobacterium avium* complex. J Clin Microbiol 25:1239, 1987.

Roland RP: False-positive acid-fast smear [letter]. Lancet 2:1063, 1975.

Runyon EH: Identification of mycobacterial pathogens utilizing colony characteristics. Am J Clin Pathol 54:578, 1970.

Shafer RW, Kim DS, et al: Extrapulmonary tuberculosis in patients with human immunodeficiency virus infection. Medicine 70:384, 1991.

Smithwick RW: Laboratory Manual for Acid-Fast Microscopy, 2nd ed. Atlanta: US Dept Health, Education, and Welfare, Public Health Service, Center for Disease Control, 1976.

Smithwick RW, Stratigos CB: Acid-fast microscopy on polycarbonate membrane filter sputum sediments. J Clin Microbiol 13:1108, 1981.

Sommers HM, McClatchy JK: CUMITECH 16: Laboratory Diagnosis of the Mycobacterioses. Washington, DC: American Society for Microbiology, 1983.

Spark RP, Fried ML: Negative BACTEC 460-TB cultures: How long to incubate? AM J Clin Pathol 89:213, 1988.

Stager CE, Libonati JP, et al: Role of solid media when using in conjunction with the BACTEC system for mycobaterial isolation and identification. J Clin Microbiol 29:154, 1991.

Stager CE, Davis JR, Siddiqi SH: Identification of *Mycobacterium tuberculosis* in sputum by direct inoculation in the BACTEC-NAP vial. Diagn Microbiol Infect Dis 10:67, 1988.

Strong BE, Kubica GP: Isolation and Identification of *Mycobacterium tuberculosis:* A Guide for the Level II Laboratory. Atlanta: US Dept Health, Education, and Welfare, Public Health Service, Centers for Disease Control, 1981.

Telenti A, Imboden P, et al: Detection of rifampin-resistant mutations in *Mycobacterium tuberculosis.* Lancet 341:647, 1993.

Telenti A, Imboden P, et al: Direct, automated detection of rifampin-resistant *Mycobacterium tuberculosis* by polymerase chain reaction and single-stranded conformation polymorphism analysis. Antimicrob Agents Chemother 37:2054, 1993.

Tenover FC, Crawford JT, et al: The resurgence of tuberculosis: Is your laboratory ready? J Clin Microbiol 31:767, 1993.

Tisdale PA, Roberts GD, Anhalt JP: Identification of clinical isolates of *Mycobacterium* with gas-liquid chromotography alone. J Clin Microbiol 10:506, 1979.

Thibert L, LaPierre S: Routine application of high-performance liquid chromatography for identification of mycobacteria. J Clin Microbiol 31:1759, 1993.

Toossi Z, Ellner JJ: Tuberculosis. In Gorbach SL, Bartlett JG, Blacklow NR (eds): Infectious Diseases. Philadelphia. WB Saunders, 1992, pp 1238–1246.

Tsukamura MI, Yano I, Imaeda T: *Mycobacterium fortuitum* subspecies *acetamidolyticum,* a new subspecies of *Mycobacterium fortuitum.* Microbiol Immunol 30:97, 1986.

Vadney FS, Hawkins JE: Evaluation of a simple method for growing *Mycobacterium haemophilum.* J Clin Microbiol 22:884, 1985.

Vannier AM, Tarrand JJ, Murray PR: Mycobacterial cross contamination during radiometric culturing. J Clin Microbiol 26:1867–1868, 1988.

Wayne LG, Kubica GP: Family Mycobacteriaceae. Chester 1897, 63, genus *Mycobacterium* (Lehmann and Neuman 1896, 363). In Sneath PHA, Mair NS, Sharpe ME, Holt JG (eds): Bergey's Manual of Systematic Bacteriology, Vol 2. Baltimore, Williams & Wilkins, 1986, pp 1436–1457.

Welch DF, Guruswamy AP, et al: Timely culture for mycobacteria which utilizes a microcolony method. J Clin Microbiol 31:2178–2184, 1993.

Whittier PS, Westfall K, et al: Evaluation of the Septi-Chek AFB system in the recovery of mycobacteria. Eur J Clin Microbiol 11:915–918, 1992.

Wilson SM, McNerney R et al: Progress toward a simplified polymerase chain reaction and its application to diagnosis of tuberculosis. J Clin Microbiol 31:766, 1993.

Woods GL, Washington JA II: Mycobacteria other than *Mycobacterium tuberculosis:* Review of microbiologic and clinical aspects. Rev Infect Dis 9:275, 1987.

Zhang Y, Heym B, et al: The catalase-peroxidase gene and isoniazid resistance of *Mycobacterium tuberculosis.* Nature 358:591, 1992.

Medically Significant Fungi

Continued

Deanna A. McGough •
Linda A. Smith • *James L. Harris* •
Annette W. Fothergill •
Alonzo S. Romo

OBJECTIVES

1. Describe the general characteristics of fungi.

2. List and describe the growth requirements of fungi.

3. Define the terms associated with fungal structures.

4. Classify fungi into their respective classes.

5. Describe asexual reproduction and sexual reproduction of fungi.

6. Describe the appropriate specimen collection procedures, staining methods, and culture techniques used in the mycology laboratory.

7. Characterize the following different types of mycoses, defining the tissues they affect:

   ■ Superficial.
   ■ Cutaneous.
   ■ Subcutaneous.
   ■ Systemic.
   ■ Opportunistic saprobic.

8. Differentiate the etiologic agents of these mycoses.

9. List the common opportunistic saprobes associated with infections in immunocompromised hosts.

Fungi constitute an extremely diverse group of organisms. Fungi are generally classified as either *molds* or *yeasts*. Some have been recognized as classic pathogens, and many others have been recognized as saprobes that exist in the environment. With the advent, however, of chemotherapy, radiation therapy, and newly recognized diseases such as acquired immunodeficiency syndrome (AIDS) that affect the immune system, the line between pathogen and saprobe has been blurred. Therefore an organism can no longer routinely be considered insignificant if it is not a classic pathogen. The isolation of all organisms, especially in the immunocompromised patient, must initially be considered a significant finding and evaluated in light of the patient history and physical examination results.

This chapter discusses the following:

■ General characteristics of fungi, including the basic terminology relating to fungal structures

■ Taxonomy and classification of fungi

■ Specimen collection and processing appropriate for fungal recovery

■ Methods of isolation and identification

■ Various types of clinical infections associated with the most commonly encountered fungi, including agents of superficial infections, dermatophytoses, subcutaneous and systemic infections, opportunistic saprobes, and yeasts

GENERAL CHARACTERISTICS

The characteristics of the fungi are different from those of plants or bacteria. Fungi are eukaryotic, that is, they possess a true nucleus with nuclear membrane and mitochondria. Bacteria are prokaryotic (lack these structures). Unlike plants, fungi possess no chlorophyll and must absorb nutrients from the environment. In addition, the cell walls of fungi are made of chitin, whereas those of the plant are made of cellulose.

Most fungi are aerobes that grow best at a neutral pH, although they tolerate a wide range of pH values. A moist environment is necessary for growth of the organism, but spores and conidia survive in dry conditions.

Yeasts versus Molds

Fungal infections (mycoses) are caused by either yeasts or molds. Yeasts are single vegetative cells that typically form a smooth, creamy bacterial-like colony without aerial mycelia. Because the microscopic morphology of all yeasts is similar, identification of isolates is typically based on results of biochemical testing. Yeasts reproduce by budding, with subsequent production of a blastoconidium (daughter cell), as shown in Figure 23–1. This process involves lysis of the yeast cell wall so that a blastoconidium can form. As this structure enlarges, the nucleus of the parent cell undergoes mitosis. Once the new nucleus is passed into the daughter cell, a septum forms and the daughter cell breaks free (Fig. 23–1).

Most molds, on the other hand, have a "fuzzy" or woolly appearance that is due to the mycelium (Fig. 23–2). The mycelium is made up of many long strands of tube-like structures called hyphae, which may be aerial or vegetative. The aerial mycelium extends above the surface of the colony and is responsible for the macroscopic appearance of the colony. In addition, it may support the reproductive structures of the organism. These reproductive structures can be used to identify the different fungal genera. The vegetative mycelium extends downward into the medium to absorb nutrients.

In some fungal species, the vegetative mycelium may form a number of specialized structures, which provide additional aids to identification. As shown in Figure 23–3*A,* antler, racquet, and spiral hyphae may be formed in the vegetative mycelium. Antler hyphae have swollen, branching tips that resemble moose antlers. Racquet hyphae contain enlarged, club-shaped areas. Spiral hyphae have tight coils throughout. Rhizoids (see Fig. 23–3*B*), root-like structures arising from the

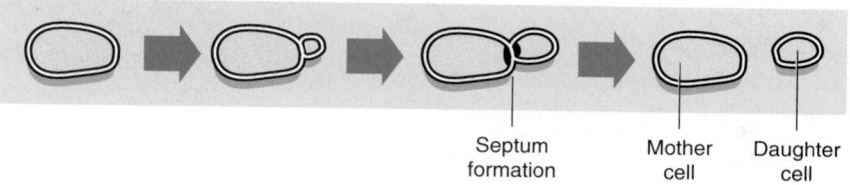

■■■ F i g u r e   2 3 – 1

Formation of blastoconidia in yeast.

Septum formation   Mother cell   Daughter cell

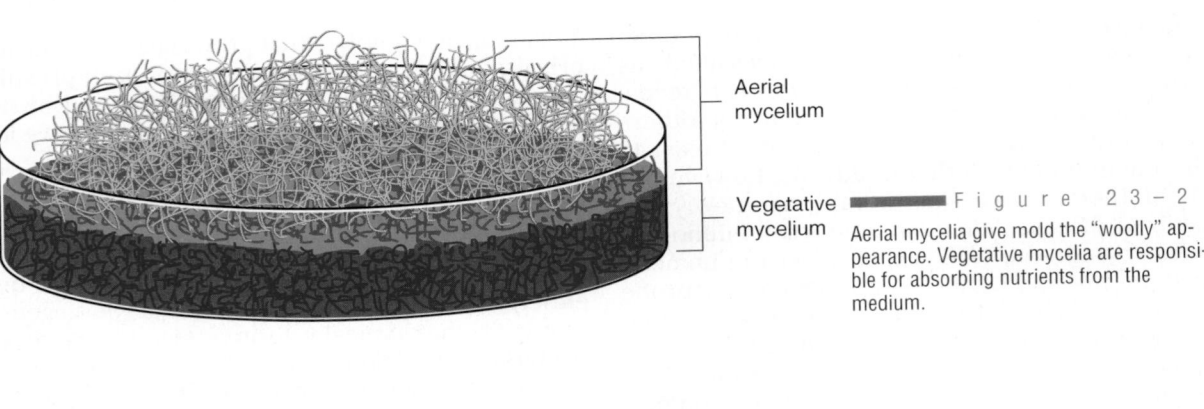

Aerial
mycelium

Vegetative
mycelium

Figure 23–2

Aerial mycelia give mold the "woolly" appearance. Vegetative mycelia are responsible for absorbing nutrients from the medium.

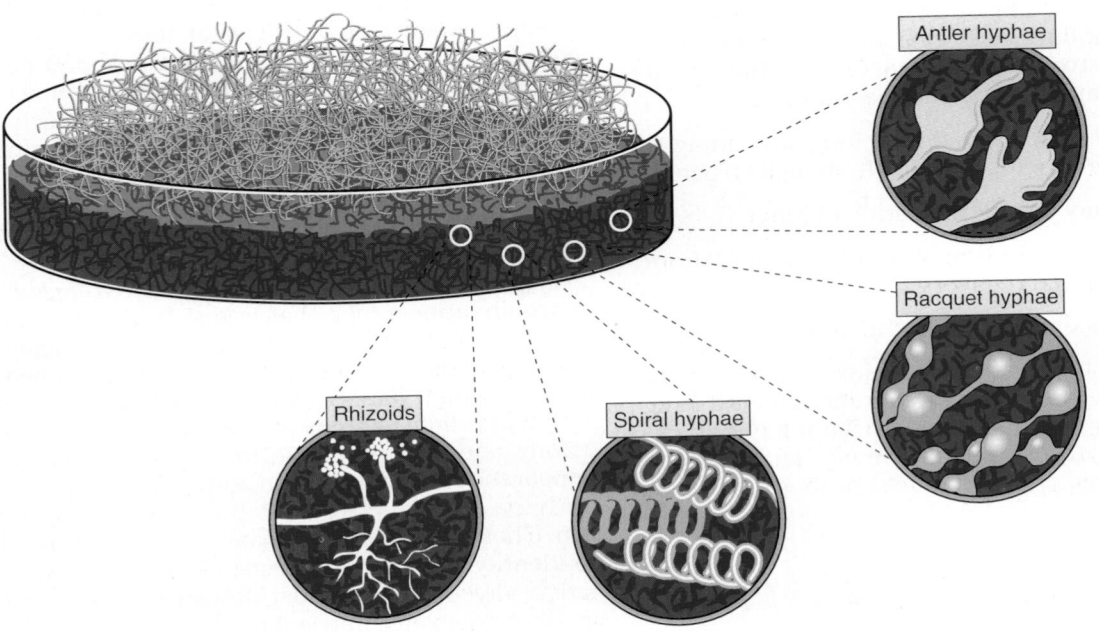

Antler hyphae

Racquet hyphae

Rhizoids

Spiral hyphae

A

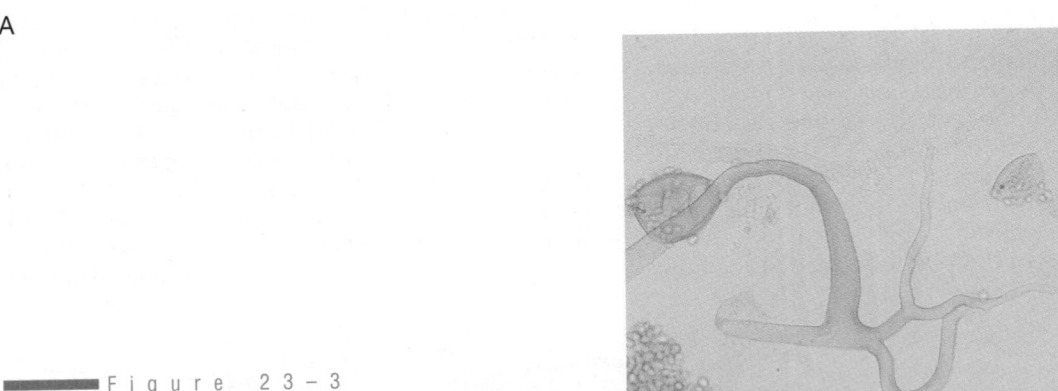

Figure 23–3

A, Specialized structures that are formed in vegetative mycelia by certain fungal species. B, *Rhizopus* sp, showing rhizoids.

B

vegetative hyphae, may be seen in some of the Zygomycetes.

Septate versus Nonseptate

Hyphae are also differentiated into *septate* or *nonseptate* (sparsely septate). Septate hyphae (Fig. 23–4A) show frequent cross-walls between the cells. Aseptate hyphae, which are actually sparsely septate (see Fig. 23–4B), have few cross-walls that appear at irregular intervals.

Hyaline versus Dematiaceous

Another characteristic of fungal species that may help in the initial identification is pigmentation. *Hyaline* hyphae are lightly pigmented (see Fig. 23–4A), whereas *dematiaceous* hyphae are darkly pigmented (Fig. 23–5). This dark pigmentation is due to the presence of melanin in the cell wall.

Dimorphism

Dimorphism is a characteristic of some of the most pathogenic fungi. *Dimorphic* means that an isolate exists in two morphologic states, a yeast or tissue phase and a mold phase. The yeast or tissue state is seen in vivo or when the organism is grown at 37°C with increased CO_2. The mold phase is seen when the organism is grown at room temperature (25°C).

Blastomyces dermatitidis is a good example of this type of organism. In patient tissue, the organism is seen as a broad-based budding yeast. In culture, the organism grows as a white, fluffy mold. To confirm diagnosis of a dimorphic fungus, the mold must be converted to the tissue form. The mold is transferred to a brain-heart infusion (BHI) agar and incubated at 35°C in 10% CO_2. Conversion to the tissue or yeast form confirms the identification of the dimorphic fungus. Classic dimorphic fungal species include *B. dermatitidis, Coccidioides immitis, Histoplasma capsulatum* var *capsulatum,* and *Sporothrix schenckii.*

Reproduction

Fungi may reproduce asexually or sexually. Asexual reproduction results in the formation of conidia following mitosis. Asexual reproduction is carried out by specialized fruiting structures

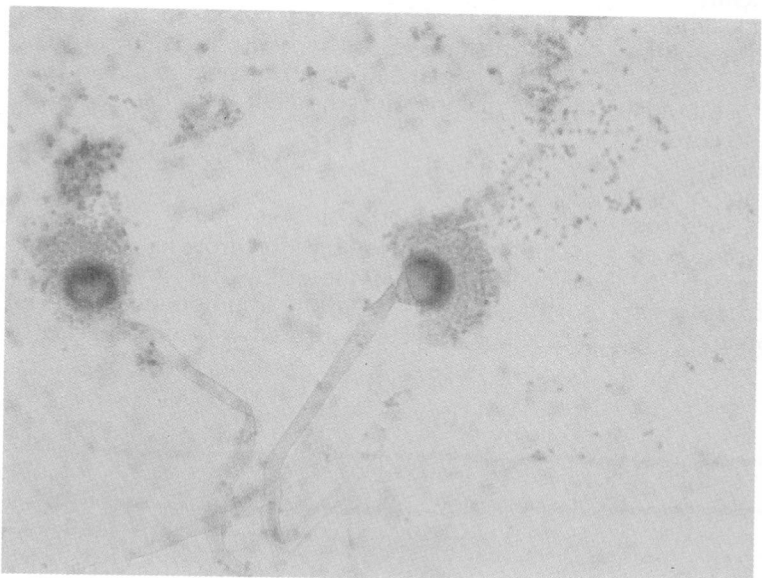
A

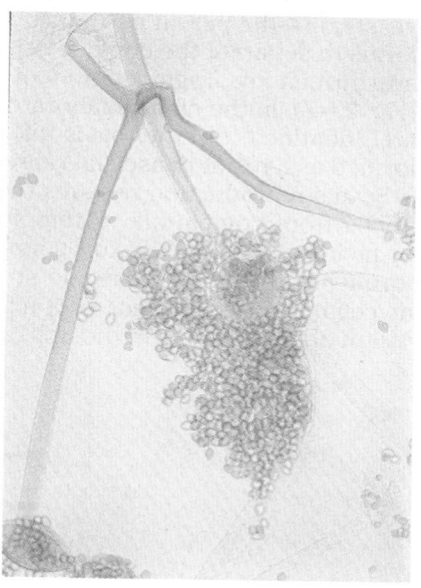
B

Figure 23–4

A, Aspergillus fumigatus as an example of septate hyaline fungi. Note the crosswalls along the hyphae in *A*, whereas *B* shows hyphae that appear sparsely or irregularly septate.

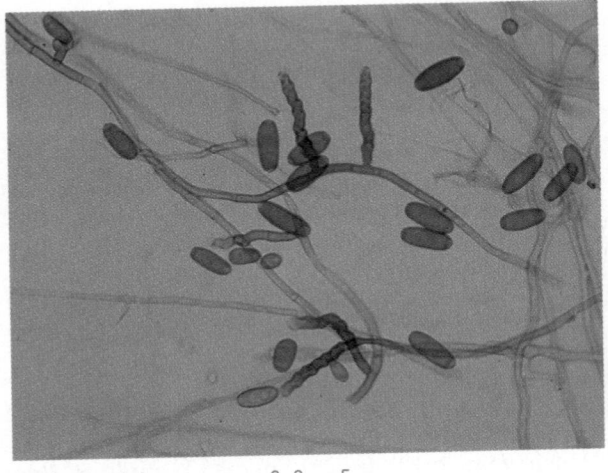

■ F i g u r e   2 3 – 5

Bipolaris sp as an example of a dematiaceous fungus. Note the dark pigmentation, which is due to the presence of melanin in the cell wall.

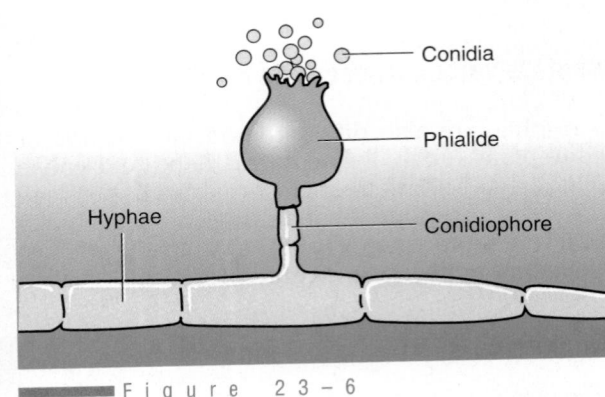

■ F i g u r e   2 3 – 6

An example of asexual reproduction is the production of phialoconidia. Conidia are formed from conidiogenous cells like phialide (a vase-like structure). Phialoconidia are "blown out" of the phialide.

known as conidiogenous cells. These structures form conidia, which contain all the genetic material necessary to create a new fungal colony. Two common conidiogenous cells are the *phialide* (vase-like structures that produce phialoconidia) (Fig. 23–6) and the *annellide* (ringed structures that produce annelloconidia). Both "blow out" their conidia and form them blastically like yeasts do—the parent cell enlarges and a septum forms to separate the conidial cell. Arthroconidia are formed by fragmentation of fertile hyphae (Fig. 23–7). In the clinical laboratory, most molds are identified on the basis of the structures formed as a result of asexual reproduction.

Sexual reproduction requires the joining of two nuclei from compatible mating strains, followed by meiosis (Fig. 23–8). Sexual spores are produced through fertilization of female structures on the mycelium by the nucleus of a male structure or fusion of the contents of the tips of hyphae.

TAXONOMY

There are four groups of fungi in which most of the etiologic agents of clinical infections are found. They consist of the divisions Zygomycota, Ascomycota, Basidiomycota, and the Form-Division Fungi Imperfecti. The assignment of an organism to one of these groups is based on the type of colony produced, the type of mycelium present, the characteristics of spores or conidia, and the method of spore development.

Zygomycetes

Zygomycetes are rapidly growing organisms normally found in the soil. They are often opportunistic pathogens in immunocompromised hosts. Zygomycetes produce profuse, gray to

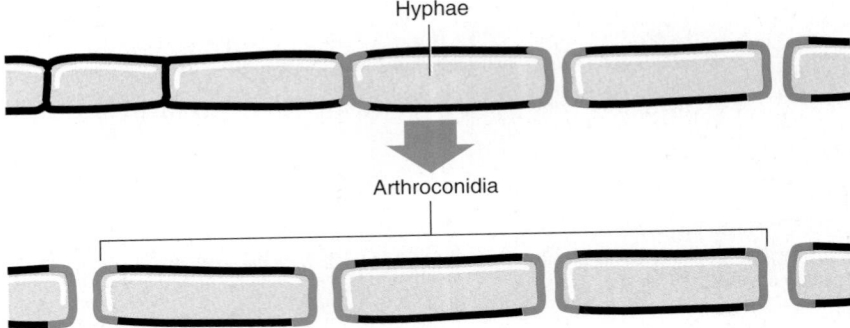

■ F i g u r e   2 3 – 7

Arthroconidia, another form of asexual reproduction, are formed by fragmentation of fertile hyphae.

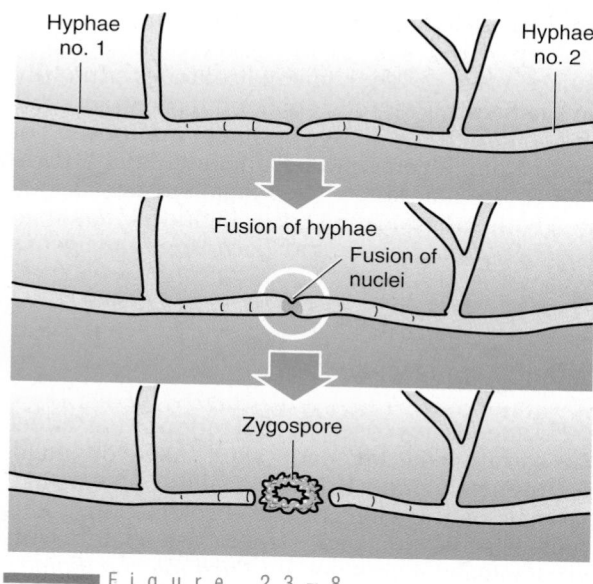

<figure>Figure 23 – 8

Sexual reproduction is fertilization of female structures by compatible male mating strain that may occur in the same medium to produce sexual spores.</figure>

between compatible mating strains with the production of zygospores. Common Zygomycetes are *Mucor, Rhizopus,* and *Absidia.*

Ascomycetes

Ascomycetes have septate mycelium and produce asexual spores (conidia) from structures known as conidiophores/conidiogenous cells, which originate in the mycelium. Sexual spores known as ascospores are produced in a structure referred to as an ascus. Representative organisms in this group include *Geotrichum* spp, *Microsporum* spp, *Trichophyton* spp, and *Pseudallescheria boydii.*

Basidiomycetes

Basidiomycetes are rarely isolated in the clinical laboratory. The only major pathogen is *Filobasidiella neoformans,* the perfect (sexual) form of *Cryptococcus neoformans,* var. *neoformans.*

Fungi Imperfecti

The Form-Division Fungi Imperfecti contains the largest number of organisms that are etiologic agents of cutaneous, subcutaneous, and systemic mycoses. The name is derived from the fact that the method of sexual spore production

white aerial mycelium characterized by the presence of aseptate or sparsely septate hyphae. Zygomycetes reproduce both asexually and sexually. Asexual reproduction of Zygomycetes is characterized by the presence of sporangiophores and sporangiospores. The asexual spores (sporangiospores) are produced in a structure known as a sporangium that develops from a hypha (Fig. 23–9). Sexual reproduction occurs

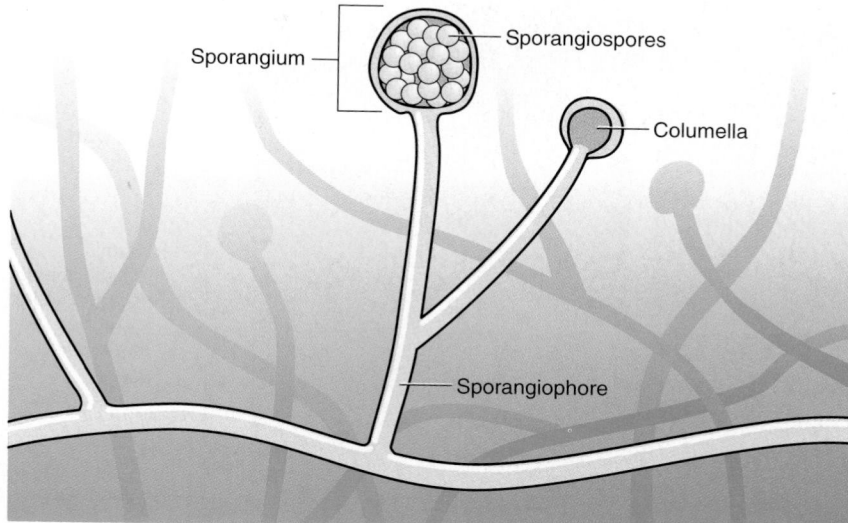

<figure>Figure 23 – 9

Asexual reproduction by Zygomycetes is characterized by the production of spores (sporangiospores) from within a sporangium.</figure>

has not been identified for these organisms at this time. Therefore, only asexual reproduction is known. In general, these organisms have septate hyphae and produce conidia from condiophores/conidiogenous cells.

CLINICAL SITES OF INFECTION

One may also classify mycoses as superficial, cutaneous, subcutaneous, or systemic, according to the tissues affected. Figure 23–10 shows the different layers of tissues where fungal infections may occur.

Superficial Mycoses

Superficial mycoses are infections confined to the outermost layer of skin and/or hair. The clinical infection pityriasis versicolor is character- ized by discoloration or depigmentation and scaling of the skin. The agents of tinea nigra cause brown or black macular patches primarily on the palms. Piedra is confined to the hair shaft and characterized by nodules composed of hyphae and a cement-like substance that attaches it to the hair shaft. The fungal species associated with superficial mycoses include *Malassezia furfur, Piedraia hortae,* and *Trichosporon beigelii.*

Cutaneous Mycoses

Cutaneous mycoses are infections that affect the keratinized layer of skin, hair, or nails. Symptoms include itching, scaling, or ring-like patches of the skin; brittle, broken hairs; and thick, discolored nails. Genera associated with cutaneous infection include *Trichophyton, Epidermophyton,* and *Microsporum.*

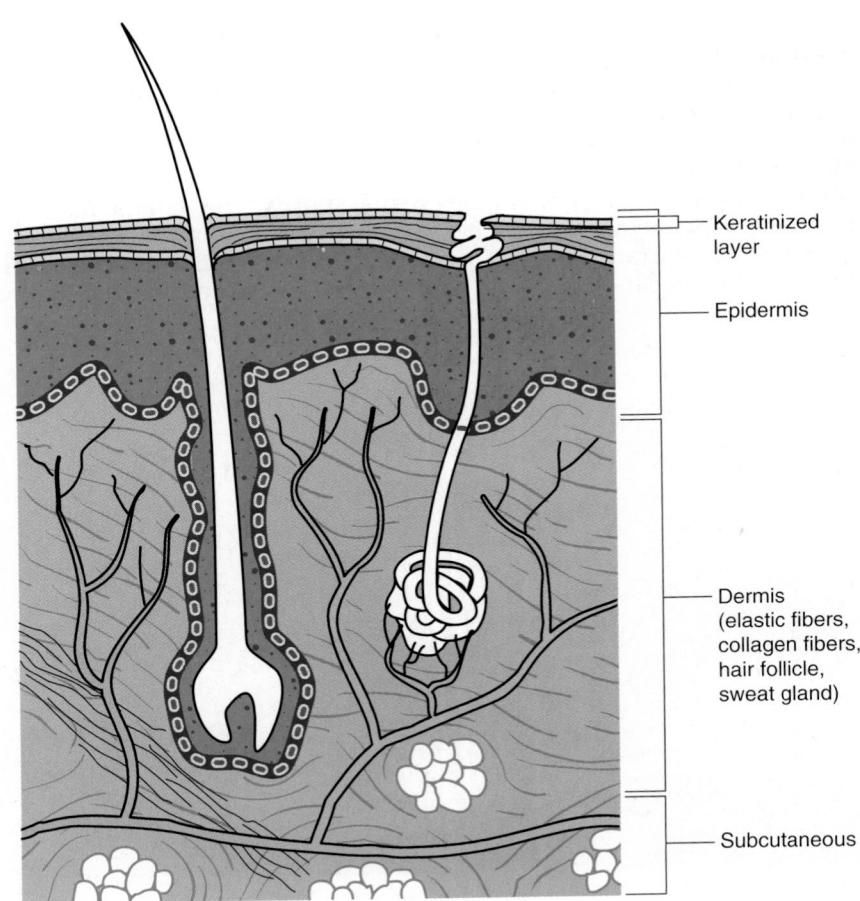

Keratinized layer

Epidermis

Dermis (elastic fibers, collagen fibers, hair follicle, sweat gland)

Subcutaneous

■■■■■ F i g u r e   2 3 – 1 0

Diagram of the layers of skin and tissues where fungal infections may occur.

Subcutaneous Mycoses

Subcutaneous mycoses involve the deeper skin layers, including muscle and connective tissue. Except in certain patient populations, there is usually no dissemination through the blood to major organs. Characteristic clinical features include progressive, nonhealing ulcers and the presence of draining sinus tracts. In tropical areas, agents such as *Phialophora* spp and *Cladosporium* sp cause chromoblastomycosis, which is characterized by draining sinus tracts and tissue destruction. *Sporothrix schenckii* commonly presents a lymphocutaneous form but dissemination into systemic sporotrichosis may occur.

Systemic Mycoses

Systemic mycoses are those infections that affect internal organs or deep tissues of the body. Often the initial site of infection is the lung, from which the organism disseminates hematogenously to other organs. Pulmonary infiltrates may be seen on radiography. Generalized symptoms include fever and fatigue. Chronic cough and chest pain may also accompany these infections. Etiologic agents of systemic mycoses include the genera *Histoplasma, Coccidioides,* and *Blastomyces.* Other fungal agents not previously associated with dissemination have been implicated in systemic fungal infections.

SPECIMEN COLLECTION, HANDLING, AND TRANSPORT

Collection of appropriate specimens is the primary criterion for accurate diagnosis of mycotic infections. All specimens for mycology should be transported and processed as soon as possible. Many of the pathogenic fungi grow slowly. Therefore, any delay in processing compromises specimen quality and decreases the probability of isolating the causative agent by allowing overgrowth by contaminating bacteria or rapidly growing molds. All work with clinical materials must be carried out under a biologic safety hood. In addition, all laboratories should maintain a protocol for rejection of unsatisfactory or improperly labeled specimens.

Although almost any tissue or body fluid can be submitted for fungal culture, the most common specimens are respiratory secretions, hair, skin, nails, tissue, blood or bone marrow, and cerebrospinal fluid (CSF). Table 23–1 represents the predominant culture sites for recovery of etiologic agents.

Hair, skin, or nails submitted for dermatophyte culture are generally contaminated with bacteria and/or rapidly growing fungi. With these types of specimens, media containing antibiotics should be inoculated. The following procedures are recommended for collecting and processing clinical samples usually submitted for fungal studies.

Hair

Sterile forceps should be used to pull affected hair. A Wood lamp may be useful in identifying infected hairs. Hairs infected with fungi such as *Microsporum audouinii* fluoresce when a Wood lamp is shone on the scalp. A less useful method involves cutting the hairs close to the scalp with sterile scissors. Hairs are placed directly into a sterile Petri dish. A few pieces of hair are inoculated onto fungal medium and placed in a 25°C incubator.

■■■■■ T A B L E 2 3 – 1

PREDOMINANT CULTURE SITES FOR RECOVERY OF ETIOLOGIC AGENTS*

Infection	Respiratory	Blood	Bone Marrow	Tissue	Skin	Mucus	Bone
Blastomycosis	+				+	+	
Histoplasmosis	+	+	+		+		+
Coccidioidomycosis	+				+	+	
Paracoccidioidomycosis	+				+	+	
Sporotrichosis	+			+	+		
Chromoblastomycosis				+	+		
Eumycotic mycetoma				+	+		
Phaeohyphomycosis				+	+		

*Organisms may be recovered from multiple sites in disseminated infections.

Skin

Skin samples are scraped from the outer edge of a surface lesion. Skin must be cleaned with 70% isopropanol before sampling. A potassium hydroxide (KOH) wet mount is prepared with some of the scrapings, and the remainder are inoculated directly on the agar. KOH will break down tissue and allow fungal hyphae to be seen.

Nails

Nail specimens may be submitted as either scrapings or cuttings and occasionally as a complete nail. Nails are cleaned with 70% isopropyl alcohol before scraping the surface. Deeper scrapings should be obtained to prepare a KOH preparation and inoculate media. Sterile scissors are used to cut complete nails into small thin strips, which are used to inoculate media.

Blood and Bone Marrow

Blood from septicemic patients can harbor a wide variety of fungal pathogens as well as opportunistic saprobes. When fungal agents are suspected, blood should be drawn into a lysis centrifugation system such as the Isolator tube (Wampole, Cranbury, NJ). The lysis of white blood cells and red blood cells releases organisms, which are then concentrated into a sediment during centrifugation. The concentrate obtained is then inoculated onto routine culture media.

Bone marrow may be plated directly onto media.

Cerebrospinal Fluid

Cerebrospinal fluid and other sterile body fluids should be concentrated by centrifugation prior to inoculation. One drop of the concentrate is used for India ink preparations, and the remainder is inoculated to media. If more than 5 mL is submitted, the CSF may be filtered through a membrane filter, and portions of the filter placed on media. Use of media with antibiotics may not be needed, because the body fluids are sterile.

Abscess Fluid and Wound Exudates

Abscess fluid and material from wounds may be plated directly on the media. Tissue should be gently minced before inoculation onto media.

While grinding of tissue has been recommended, this process may destroy fragile fungal elements and prevent recovery of etiologic agents, particularly if a zygomycete is present. Grinding of tissue may be necessary for KOH and calcofluor white preparations. When large sections of tissue are submitted, suspicious areas, such as purulent or discolored sections, are selected for grinding and subsequent culture.

Respiratory Specimens

Respiratory tract secretions (sputum, transtracheal aspirates) and pleural lavage fluids are commonly submitted because many fungal infections have a primary focus in the lungs. Sputa should be obtained from a deep cough shortly after the patient arises in the mornings. If the patient cannot produce sputum, a nebulizer may be used to induce sputum. All sputum specimens should be collected into a sterile, screw-top container.

The specimens may be inoculated onto media with a sterile pipette if the material is not too viscous. With viscous materials such as a thick tracheal aspirate, either a cotton swab may be used to inoculate the material onto the media, or the specimen may be digested with trypsin and concentrated prior to inoculation. In addition to nonselective media, a medium with antibiotics should be used to prevent bacterial overgrowth. A KOH preparation should also be made. On occasion, mucolytic agents such as N-acetyl-L-cysteine may be used.

Urogenital and Fecal Specimens

Specimens such as urine, feces, and vaginal secretions are often received for bacteriologic culture and on occasion grow a yeast that requires identification. Urine submitted specifically for fungal culture should be centrifuged and the sediment used to inoculate media. A first-voided morning urine specimen is preferred.

METHODS OF IDENTIFYING FUNGAL AGENTS

Direct Microscopic Examination of Specimens

The direct examination of clinical material for fungal elements serves several purposes. First, it provides a rapid report to the physician, which may in turn result in the early initiation of treatment. Second, in some cases, specific morphologic characteristics may provide a clue to the genus of the organism. In turn, any special media indicated for species identification can be inoculated immediately. Third, direct examination may provide evidence of infection in spite of negative cultures. Such a situation may occur with specimens from patients who are on antifungal therapy, which may inhibit growth in vitro even though the infection may still be present in the patient.

Although the Gram stain performed in the routine microbiology laboratory often gives the first evidence of infections with yeast, other direct stains give more specific information concerning a mold infection. The types of direct examination used in identification of fungal infections include wet preparations such as the potassium hydroxide (KOH) preparation, India Ink, and calcofluor white. Histologic stains, such as the periodic acid–Schiff (PAS) stain, Grocott-Gomori methenamine–silver nitrate (GMS) stain, and hematoxylin and eosin (H&E) stain, may be useful.

KOH Preparation

A 10 to 20% solution of potassium hydroxide is useful for detecting fungal elements in skin, hair, nails, and tissue. In this procedure, KOH is mixed in equal proportions with the specimen on a slide. The slide is then coverslipped and heated gently. Any fungi present will be visible, because the KOH dissolves keratin and other cellular material in the specimen.

KOH with Calcofluor White

If a fluorescent microscope is available, a drop of calcofluor white (fluorescent dye) can be added to the KOH preparation prior to coverslipping. Calcofluor white binds to polysaccharides present in the chitin of the fungus or to cellulose. Fungal elements fluoresce apple green or blue-white, depending on the combination of filters used. Therefore, any element with a polysaccharide skeleton fluoresces. The actual fungal structure must be seen before a positive preparation is reported.

India Ink

India ink preparations may be used to examine cerebrospinal fluid for the presence of the encapsulated yeast *Cryptococcus neoformans*. A drop of India ink is mixed with a drop of sediment from a centrifuged spinal fluid specimen, and the preparation is examined on high power. With this negative stain, budding yeast surrounded by a large clear area against a black background (Fig. 23–11) is presumptive evidence of *C. neoformans*. White blood cells and other artifacts may resemble encapsulated organisms; therefore, careful examination is necessary. Many laboratories, however, now use the latex agglutination test for cryptococcal antigen in place of the India ink examination.

Tissue Stains

Common tissue stains used in the histology section for detection of fungal elements are the periodic acid–Schiff (PAS) stain, the Grocott-Gomori methenamine–silver nitrate (GMS) stain, the Giemsa stain, and the Masson-Fontana stain. Giemsa stain is used primarily to detect *H. capsulatum* in blood or bone marrow (Fig. 23–12). PAS attaches to polysaccharides in the fungal wall and stains pink. The Masson-Fontana method stains melanin in the cell wall and identifies the presence of a dematiaceous fungus. Acid-fast stains are used primarily to differentiate *Nocardia* species from other actinomycetes. Table 23–2 lists the characteristic fungal reactions seen with selected stains.

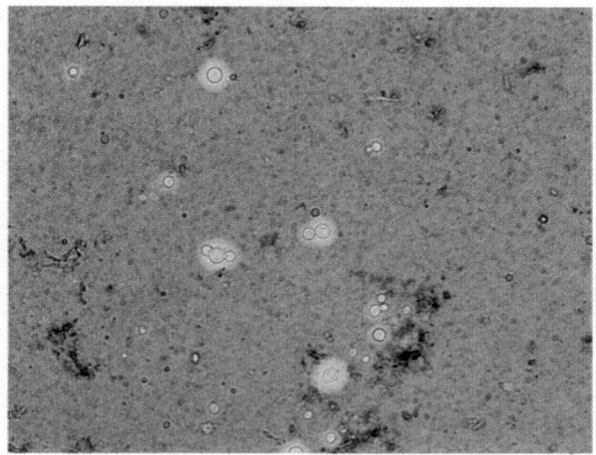

■■■■■■■ F i g u r e   2 3 – 1 1

India ink preparation is used primarily to examine cerebrospinal fluid for the presence of the encapsulated yeast *Cryptococcus neoformans*. This is an India ink preparation from an exudate containing encapsulated budding yeasts.

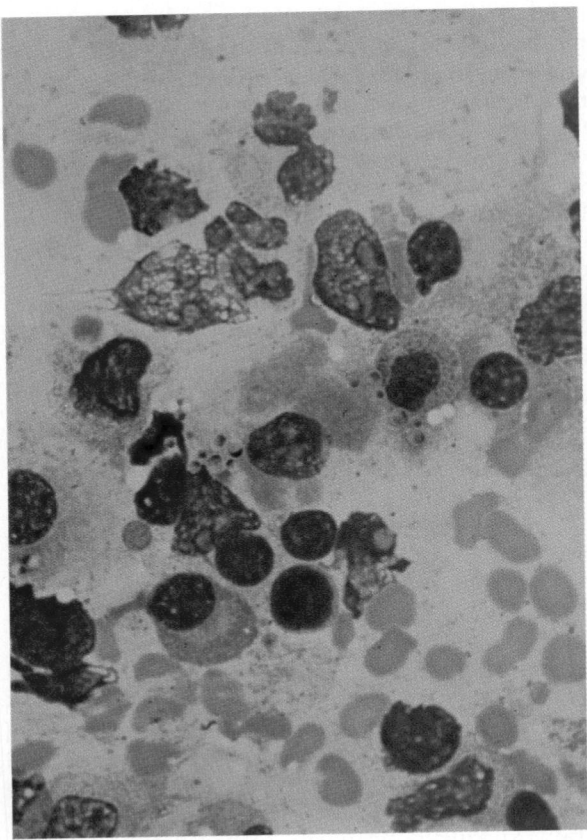

■■■■■■ F i g u r e   2 3 – 1 2

Giemsa stain is used primarily to detect organisms in tissue samples. This is a bone marrow stained with Giemsa, showing the yeast phase of *Histoplasma capsulatum* var *capsulatum*.

Culture

Culture Media

In general, fungi do not share the broad range of nutritional and environmental needs that characterize bacteria; and therefore, relatively few types of standard media are needed for primary isolation. These include Sabouraud dextrose agar (SDA), SDA with antibiotics, and brain-heart infusion (BHI) agar enriched with blood and antibiotics. The pH of the Emmons modification of SDA is close to neutral and is a more efficient medium for primary isolation than the original formulation.

The antimicrobials usually included in SDA with antibiotics are chloramphenicol and cycloheximide. Chloramphenicol inhibits bacterial growth; cyclohexamide inhibits the saprobic fungi. Table 23–3 shows the expected growth results with some of the standard fungal media.

Incubation

Most laboratories routinely incubate fungal cultures at room temperature (25–30°C). Fungi

■■■■■■ T A B L E   2 3 – 2

STAINING CHARACTERISTICS OF FUNGI

Stain	Color of Fungal element	Background color
Periodic acid–Schiff	Magenta	Pink or green
Grocott-Gomori methenamine–silver nitrate	Black	Green
Giemsa	Purple-blue yeast with clear halo	Pink-purple
India ink	Yeast with clear halo	Black
KOH	Refractile	Clear
KOH-calcofluor	Fluorescent	Dark
Masson-Fontana	Brown	Pink-purple

■■■■■■ T A B L E   2 3 – 3

SUMMARY OF PRIMARY FUNGAL CULTURE MEDIA

Medium*	Expected Growth Results
At 25°C	
SDA	Initial isolation of pathogens and saprobes
	Dimorphic fungi may exhibit their mycelial phase
SDA with antibiotics	Saprobes are generally inhibited on this medium
	Dermatophytes and most of the fungi considered primary pathogens grow
BHI	Initial isolation of pathogens and saprobes
BHI with antibiotics	Recovery of pathogenic fungi
	Dermatophytes not usually recovered
Inhibitory mold agar	Initial isolation of pathogens except dermatophytes
Cyclohexamide	Primary recovery of dermatophytes
At 37°C	
SDA	The yeast form of dimorphic fungi and other organisms grow
	Dermatophytes grow poorly
BHI with blood	Yeasts such as *Cryptococcus* grow well
	The yeast form of *Histoplasma capsulatum* takes up some of the heme pigment in the medium and becomes light tan with a grainy wrinkled texture

*SDA, Sabouraud dextrose agar; BHI, brain-heart infusion agar.

grow optimally at this temperature, but bacteria have a slower growth rate. If the etiologic agent suspected is a dimorphic fungus, cultures should also be incubated at 37°C. Cultures are generally maintained for 4 to 6 weeks and should be examined weekly or twice a week for growth. Zygomycetes such as *Mucor* and *Rhizopus* are rapidly growing molds that may fill the tube with aerial mycelium within several days, whereas with slow-growing organisms such as *Fonsecaea* and *Phialophora,* it may be 2 weeks or more before growth is seen.

Data that should be recorded about an isolate include the number of days until first visible growth, whether mold or yeast forms, media on which isolated, temperature at which growth occurred, and morphology of the colonies.

BEGINNING THE IDENTIFICATION

Although the number of fungal species known exceeds 100,000, the number typically implicated in human disease is around 100. The number routinely seen is much lower. Therefore, identification at least to the genus level is usually possible. The traditional starting place is to decide whether the isolate is a yeast or a mold.

Gross Examination of the Culture

Once an organism has grown, it must be examined for characteristic gross and microscopic morphologic structures so that identification can be made. Figure 23–13 presents a schematic guideline to how identification may be made. Gross morphologic traits, such as color, texture, and growth rate, are initial observations that should be made. Rapidly growing organisms such as the Zygomycetes usually appear within 1 to 3 days, whereas intermediate growers may take 5 to 9 days, and slow growers up to 2 weeks. Pigment on the reverse side of the colony or in the aerial mycelium can be noted but is not always helpful, especially in the dematiaceous fungi.

Microscopic Examination for Fungal Structures

The most common procedure for microscopic examination is a direct mount of the fungus isolate. A tease mount or cellophane tape preparation may be prepared. On occasion, a slide culture may be prepared, when the initial isolate fails to show conidial production. Characteristics that should be observed are:

■ Septate versus nonseptate hyphae

■ Hyaline or dematiaceous hyphae

■ The types, size, shape, and arrangement of conidia

Tease Mount

For the tease mount, two teasing needles are used to remove a portion of the mycelium from the middle third of the colony. This is placed into a drop of lactophenol cotton blue (LPCB) on a slide and gently teased apart. The LCPB is used to stain tease mounts or cellophane tape mounts from cultures. The combination of lactic acid, phenol, and the blue dye kills, preserves, and stains the organism. The preparation is coverslipped and examined under low (10×) and high (40×) power for characteristic conidial structures. Hyaline hyphae take up the LPCB but dematiaceous fungi retain their dark color. The major disadvantage of the procedure is the disruption of conidia during the teasing process.

Cellophane Tape Preparation

Cellophane tape preparation involves gently placing a section of transparent tape, sticky side down, on top of the colony and then removing it. The tape is placed onto a drop of LPCB on a slide and examined. An advantage of this procedure is that the conidial arrangement is retained. A major disadvantage is the potential contamination of the colony. A coverslip is not needed if the cellophane tape technique is used, because the tape serves as coverslip.

Slide Culture

Slide cultures are useful for demonstrating natural morphology of fungal structures and for encouraging conidiation in some poorly fruiting fungi. Several methods have been devised for constructing these culture chambers, and Procedure 23–1 incorporates the best features of several of these.

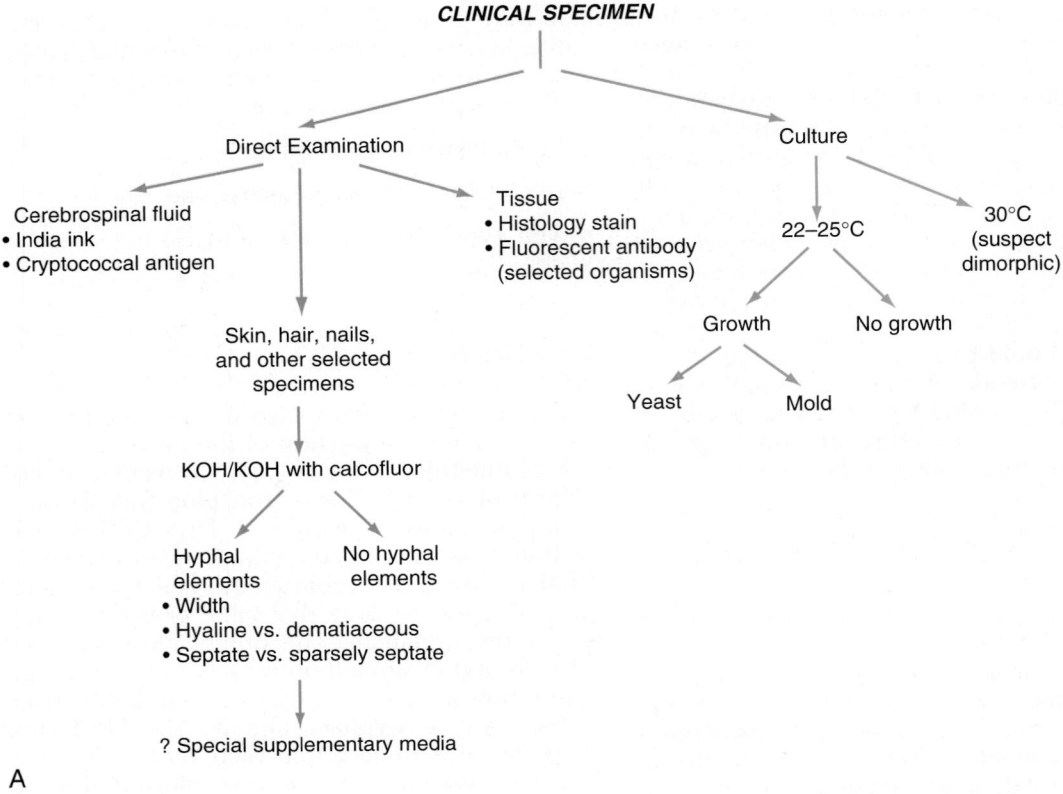

A

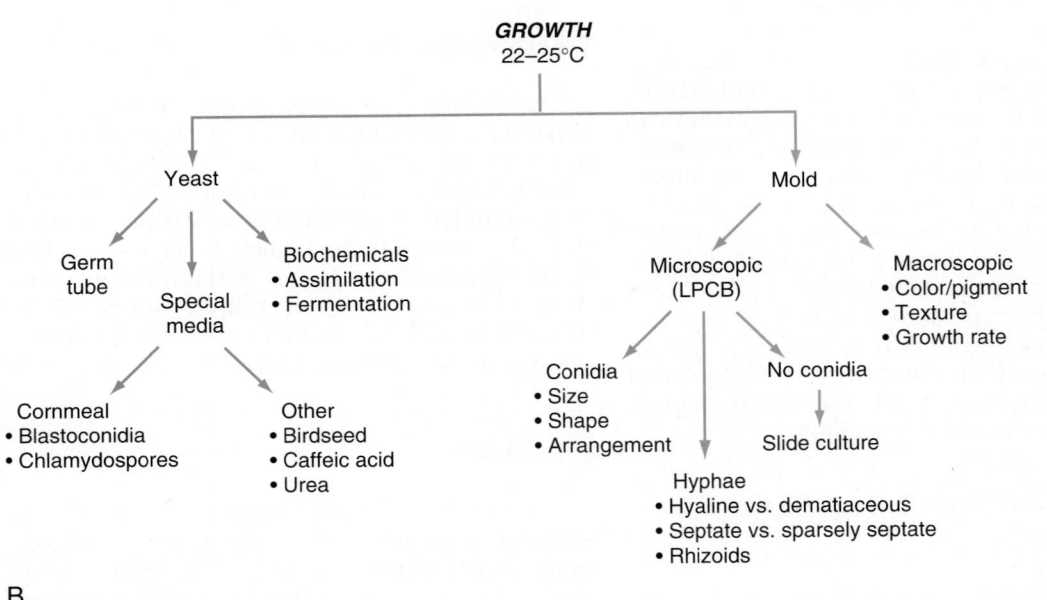

B

■■■■ F i g u r e   2 3 – 1 3

A and *B,* Schematic guideline for the identification of fungal isolates.

PROCEDURE 23–1. CONSTRUCTING SLIDE CULTURE CHAMBERS FOR FUNGI

Prepare and take apart slide cultures within a properly operating biologic safety hood. Fungi suspected of being pathogens are not recommended for observation in slide cultures.

1. Into a 60-mm sterile Petri disk, pour 7 to 8 mL of sterile, 1.5% water agar, and allow to solidify.

2. Lay a sterile coverglass onto the agar. Place on the glass a 5- to 8-mm-square block of nutrient medium (e.g., potato dextrose agar, potato agar, V8 Juice agar, or other sporulation-inducing medium).

3. Inoculate the sides of the block with the desired fungus, and cover the block with a second sterile coverglass and replace the dish lid.

4. Incubate this completed slide culture chamber at room temperature (25°C) for 5 to 10 days. The progress of growth may be monitored by inverting the chamber and examining the fungus on the microscope stage using the low-power objective.

5. When conidia or spores are evident, carefully lift the coverglasses away from the nutrient medium. Mount each coverglass separately in a drop of LPCB on a microscope slide for examination.

SAFETY ISSUES

Universal precautions apply to the mycology laboratory. Because of the additional hazard of airborne conidia, a class II biologic safety cabinet should be used to reduce personnel exposure to fungal elements. Specimen processing and plating must be done under a properly maintained and operating hood. The use of a hooded electric incinerator is recommended to eliminate the hazards of open gas flames and to contain particles emitted when loops or needles are incinerated. The cabinet should be checked daily to see that none of the airflow inlets and outlets is blocked by supplies, incinerators, or waste disposal jars.

Use of Petri dishes in the mycology laboratory is hazardous, and screw-top tubes are recommended. Screw-capped tubes tend to show less media dehydration than Petri dishes and are more easily handled and stored. There is less chance for release of airborne conidia. On the other hand, Petri dishes have a greater surface area for colony isolation and are easier to manipulate when making preparations for microscopic examination.

AGENTS OF SUPERFICIAL MYCOSES

Four fungal agents are commonly associated with superficial mycoses:

■ *Malassezia furfur*

■ *Piedraia hortae*

■ *Trichosporon beigelii*

■ *Phaeoannellomyces werneckii*

Superficial mycoses are fungal diseases that affect only the cornified layers (stratum corneum) of the epidermis. Patients who suffer from superficial fungal infections do not show any overt symptomatology, because the fungal agents do not activate any tissue response or inflammatory reaction. Patients usually seek medical attention for the cosmetic effects the fungal agents bring about.

Malassezia furfur

Malassezia furfur causes tinea versicolor or pityriasis versicolor, a disease chracterized by patchy lesions or scaling of varying pigmentation. Also described as "fawn-colored liver spots," pityriasis may involve the chest, trunk, or abdomen. *Malassezia furfur* has also been implicated in disseminated infections in patients receiving high-dose lipid replacement therapy, particularly in infants.

Malassezia furfur is a common endogenous skin colonizer. Although reasons for overgrowth by *Malassezia furfur* resulting in the clinical manifestations are still unknown, it appears to be related to squamous cell turnover rates. This is evidenced by the higher incidence of tinea versicolor among persons receiving corticosteroid therapy, which decreases the rate of squamous epithelial cell turnover. Some investigators have identified genetic influence, poor nourishment, and excessive sweating as other factors that contribute to the overgrowth of the organism on the skin.

Malassezia furfur is identified in potassium hydroxide (KOH) preparations of skin scrapings or by observing yellow fluorescence on a Wood's lamp examination of the infected body site. Microscopic examination of the direct smear in KOH preparation shows budding yeasts, approximately 4 to 8 μm, along with septate, sometimes branched hyphal elements. This microscopic appearance has gained *M. furfur* the term "spaghetti and meatballs" fungi (Fig. 23–14). Skin scrapings are seldom cultured. If desired, *M. furfur* can be cultured using a suitable agar medium with fatty acid (oleic acid) overlay.

Recommended treatment using 1% selenium sulfide, usually found in shampoo (e.g., Selsun Blue) provides a temporary remedy. The infection usually recurs when treatment is stopped.

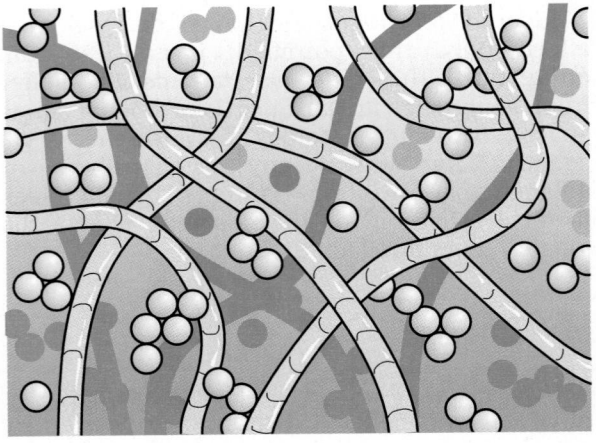

■■■ F i g u r e 2 3 – 1 4

Diagram of the typical "spaghetti and meatballs" appearance of *Malassezia furfur.*

Piedraia hortae

Piedraia hortae is the causative agent of black piedra, an infection that occurs on scalp hair. The disease is endemic in tropical areas of Africa, Asia, and Latin America. The organism *P. hortae* produces hard, dark brown to black gritty nodules (Fig. 23–15) made of asci containing eight ascospores. When infected hair shafts are removed and placed on 10 to 20% KOH, the nodules may be crushed open to reveal the asci. Thick-walled rhomboid cells containing ascospores are seen.

Piedraia hortae grows slowly on Sabouraud dextrose agar at room temperature. It forms dark brown colonies that microscopically show dematiaceous septate hyphae. Treatment usually consists of removal of infected hair shafts and application of topical fungicides.

Trichosporon beigelii

Trichosporon beigelii causes white piedra, which also occurs on the hair shaft. White piedra is characterized by a soft mycelial mat around facial and genital hair, less commonly scalp hair (Fig. 23–16). White piedra is endemic in tropic areas of South America, the Far East, and the Pacific.

Widely distributed in nature, *T. beigelii* is also occasionally found as part of the normal skin flora. It has also been recognized as an opportunistic systemic pathogen. Although rare, systemic disease caused by this fungus is fatal and occurs most often in the immunocompromised host, commonly people who have hematologic disorders or malignancies or are undergoing chemotherapy.

Diagnosis is made by finding hyphal elements within the shaft nodule and budding blastoconidia and athroconidia in culture. *Trichosporon beigelii* grows rapidly on primary fungus media. The colonies are cream-colored and yeast-looking. Colonies eventually becomes wrinkled as they mature. Microscopically, *T. beigelii* produces both arthroconidia and blastoconidia (Fig. 23–17). Identification is confirmed by biochemical reactions; *T. beigelii* does not ferment carbohydrates or utilize potassium nitrate but does assimilate glucose, galactose, sucrose, maltose, and lactose, features differentiating it from other *Trichosporon* species.

McManus et al (1985) have reported the role of latex agglutination test for cryptococcal antigen in diagnosing disseminated *Trichosporon* infec-

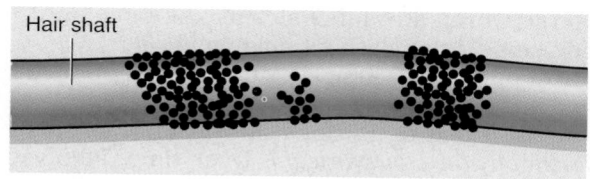

■ F i g u r e   2 3 – 1 5

Diagram of black piedra on a hair shaft. Black piedra is caused by *Piedraia hortae.*

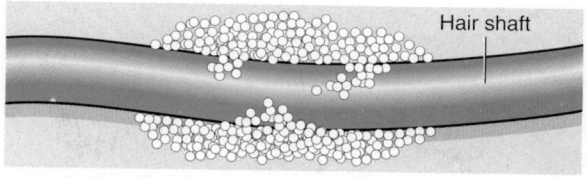

■ F i g u r e   2 3 – 1 6

Diagram of white piedra caused by *Trichosporon beigelii.* Note presence of intertwined hyphae and arthroconidia.

tions. They described the presence of a heat-stable antigen produced by *T. beigelii* that shares antigenic sites with the polysaccharide capsule of *C. neoformans.* This false-positive result, although misleading the mycologic diagnosis, has nevertheless allowed the early institution of appropriate therapy.

Phaeoannellomyces werneckii

Tinea nigra, characterized by brown to black nonscaly macules that occur most often on the palms of the hands and soles of the feet, is caused by *Phaeoannellomyces werneckii.* The disease is endemic in the tropic areas of Central and South America, Africa, and Asia. It involves no inflammatory or other tissue reaction to the infecting fungus. The clinical presentation is so similar to other conditions, however, especially malignant melanoma, that misdiagnosis could essentially occur. The diagnosis of malignant melanoma has resulted in unnecessary surgical procedures.

Proper diagnosis of tinea nigra can be made by direct examination of skin scrapings placed in 10 to 20% KOH. Microscopic examination shows septate hyphal elements and budding cells (Fig. 23–18). When grown in culture medium, *P. werneckii* produces shiny, moist, yeast-looking colonies that start with a brownish coloration that eventually turns olive to greenish black. Microscopic examination of the yeast-looking colony shows budding blastoconidia, whereas the older mycelial portion of the colony shows hyphae with blastoconidia in clusters. Annelloconidia are seen in older mold colonies. Treatment consists of the application of keratolytic agents.

AGENTS OF DERMATOPHYTOSES

Three genera of fungi, *Microsporum, Trichophyton,* and *Epidermophyton,* are etiologic agents of dermatophytoses. Species within these genera

are keratinophilic, that is, they are adapted to grow on hair, nails, and cutaneous layers of skin that contain the scleroprotein keratin. Infection of deep tissue by these fungi is rare, but occasionally, extensive inflammation and nail bed involvement may result.

Epidemiology

Most of the agents of dermatophytoses live freely in the environment, but a few have adapted almost exclusively to living on human host tissues, and these are very rarely recovered from any other source. Distribution of many dermatophyte species is worldwide, whereas others are found only in restricted geographic regions. Approximately 43 species of dermatophytes and dermatophyte-like fungi have been described, and just over two dozen of these have been documented to cause human infection.

Those dermatophytes that primarily inhabit the soil are termed *geophilic.* Most geophilic fungi produce large numbers of conidia and

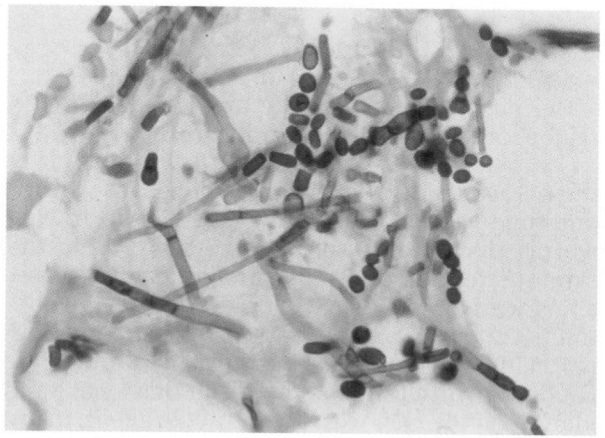

■ F i g u r e   2 3 – 1 7

Microscopic appearance of *T. beigelii* on LPCB preparation, showing the presence of both blastoconidia and arthroconidia.

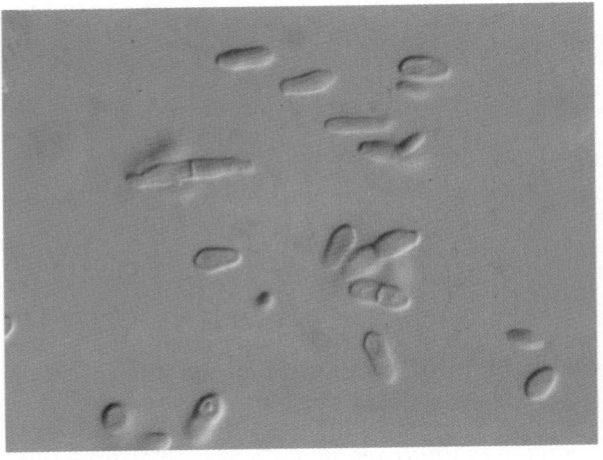

■ F i g u r e   2 3 – 1 8

Microscopic structures of *Phaeoannellomyces werneckii,* showing characteristic budding blastoconidia. *P. werneckii* causes tinea nigra.

therefore are among the most readily identified species. *Zoophilic* dermatophytes are typically adapted to life on animals and are not commonly found living freely in soil or on dead organic substrates. They often cause infections in their animal hosts and may be spread as disease agents to humans. Fewer conidia are produced by zoophilic fungi than by geophilic species. A few dermatophytes have become adapted exclusively to human hosts and are termed *anthropophilic.* Although they are encountered almost always as agents of human disease, the infections are seldom inflammatory. Species identification may be quite difficult because most anthropophilic species produce few conidia.

Clinical Infections

Dermatophytoses usually involve a restricted region of the host, and traditionally, these diseases are named with respect to the portion of the body affected. Because the infections were at one time believed to be the result of burrowing worms that formed ring-shaped patterns in the skin, the term *tinea* was applied to each disease, along with a Latin term for the body site. We continue to describe the various forms of "ringworm" in these terms, as shown in Table 23–4. Each ringworm lesion is the result of a local inoculation on the skin with the etiologic agent; many lesions represent many sites of infection. Lesions enlarge with time, usually with most inflammation occurring at the growing edge. Some cases

of ringworm are subclinical, exhibiting only a dry, scaly lesion without inflammation.

Not only are diverse sites on the host involved, but certain speices cause distinctive lesions. The notable example is tinea imbricata, caused by *Trichophyton concentricum.* Over time, involved portions of the trunk develop diagnostically distinctive concentric rings of scaling tissue. In some forms of ringworm, there is a persistent allergic reaction, the dermatophytid, which is manifested in the formation of sterile, itching lesions on body sites distant from the point of infection. Symptoms of dermatophyte infections vary from slight to moderate and occasionally severe.

Infections Involving Hair and Hair Follicles

Different body sites manifest different symptoms. Infections in the scalp, where hair follicles are the initiation sites, may be among the most severe and disfiguring forms of the disease. Tinea favosa, or favus, begins as an infection of the hair follicle by *Trichophyton schoenleinii* and progresses to a crusty lesion made up of dead epithelial cells and fungal mycelia. Crusty, cup-shaped flakes called scutula are formed. Hair loss and scar tissue formation commonly follow.

Two distinct forms of tinea capitis, gray-patch ringworm and black dot ringworm, are caused by different species of dermatophyte. Gray-patch ringworm is a common childhood disease easily spread among children. The fungus primarily colonizes the outer portion of hair shafts, the so-called "ectothrix" hair involvement. The lesions are seldom inflamed, but luster and color of the hair shaft may be lost. *Microsporum audouinii* and *Microsporum ferrugineum* are agents of this form

■ T A B L E   2 3 – 4

VARIOUS FORMS OF DERMATOPHYTOSES AND THE RESPECTIVE AFFECTED SITES

Type of Ringworm	Site Affected
Tinea capitis	Head
Tinea favosa	Head (distinctive pathology)
Tinea barbae	Beard
Tinea corporis	Body—glabrous skin
Tinea manuum	Hand
Tinea unguium	Nails
Tinea cruris	Groin
Tinea pedis	Feet
Tinea imbricata	Body (distinctive lesion)

of disease. Black-dot ringworm is an endothrix hair involvement. The hair follicle is the initial site of infection, and fungal growth continues within the hair shaft, causing it to weaken. The brittle, infected hair shafts break off at the scalp, leaving the "black dot" stubs. *Trichophyton tonsurans* and *Trichophyton violaceum* are the most common fungi implicated in this form of ringworm.

Infections Involving the Nail and Nail Bed

Onychomycosis is most often caused by dermatophytes but also may be the result of infection by other fungi. These nail and nail bed infections may be among the most difficult dermatophytoses to treat. Long-term, costly therapy with griseofulvin is considered the best but may still result in unsatisfactory resolution of the disease. Some common agents that infect the nail are the *Trichophyton* species *T. rubrum, T. mentagrophytes,* and *T. tonsurans* as well as *Epidermophyton floccosum.*

Athlete's Foot

Among the shoe-wearing human population, tinea pedis or ringworm of the foot is a common disease, particularly of men. The gender disparity may be related more to shoe styles than to any physiologic differences between the sexes. Various sites on the foot may be involved, but most often, tinea pedis affects the toe webs or toenails. In more severe cases, the sole of the foot may develop extensive scaling with fissuring and erythema. The disease may progress around the sides of the foot from the sole, giving rise to use of the term "moccasin foot," descriptive of the shape of the lesion. Infections of the glabrous skin range from mild with only minimal scaling and erythema to severely inflamed lesions.

Systemic Infections

Immunocompromised persons may suffer systemic dermatophyte infections.

Treatment

Successful treatment of dermatophytic skin infections is usually accomplished with keratinolytic agents, which remove outer layers of skin along with the fungal elements. Systemic treatment with griseofulvin is appropriate for more extensive tinea corporis lesions. Recurring infections are common with most types of ringworm, even with the best therapies. Nail bed infections are often resistant to therapy and may at best remain chronic problems.

Commonly Encountered Dermatophytes

Trichophyton mentagrophytes

In many parts of the world, *Trichophyton mentagrophytes* is the most commonly isolated dermatophyte, and both anthropophilic and zoophilic strains are known. *Trichophyton mentagrophytes* infects skin, hair, and nails. Colonial morphology varies with the extent of conidia production. The granular colony form has a powdery appearance owing to abundant microconidia formed, usually in clusters. In the downy form, conidia are less abundant, and the microscopic morphology may resemble that of *Trichophyton rubrum.* Compared with other dermatophytes, *T. mentagrophytes* is a relatively rapid-growing fungus. Macroconidia are thin walled, smooth, and cigar shaped with four to five cells separated by parallel cross-walls. These conidia measure about 7 by 20 to 50 µm and are produced singly on undifferentiated hyphae. Microconidia (Fig. 23–19) are globose to tear shaped and measure 2.5 to 4 µm.

Trichophyton rubrum

Trichophyton rubrum is recognized as the most common species isolated from many forms of

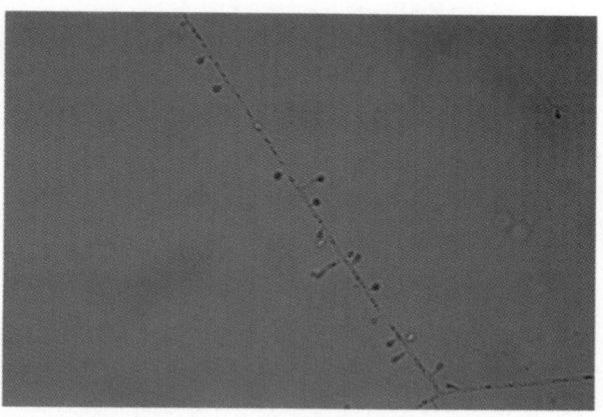

■■■■■ F i g u r e   2 3 – 1 9

Trichophyton mentagrophytes, showing globose, teardrop-shaped microconidia.

ringworm. It is a relatively slow-growing anthropophilic dermatophyte found wherever human populations live. Skin and nails are common sites infected by *T. rubrum*, but hair is seldom involved. The colonies usually remain hyaline, but with age, they may develop pink to deep burgundy wine–colored pigment on the reverse side. Although *T. rubrum* is known to produce three- to eight-celled cylindrical macroconidia measuring somewhat smaller than those of *T. mentagrophytes*, these are seldom seen in clinical isolates. A typical microscopic picture of *T. rubrum* contains clavate or peg-shaped microconidia (Fig. 23–20) formed along undifferentiated hyphae, and even these may be sparse. Pigmentation studies, 5-day urease production, and hair perforation tests are important tools for distinguishing between *T. mentagrophytes* and *T. rubrum*.

Trichophyton tonsurans

Trichophyton tonsurans, which infects skin, hair, and nails, has become the leading cause of tinea capitis in children in many parts of the world, including the United States. When grown on Sabouraud dextrose agar, *T. tonsurans* colonies usually form a rust-colored pigment on the colony's reverse. This dermatophyte is anthropophilic and typically produces only small numbers of peg-shaped microconidia. *Trichophyton tonsurans* has an absolute requirement for thiamine, a feature helping to distinguish it from *T. mentagrophytes* and *T. rubrum*.

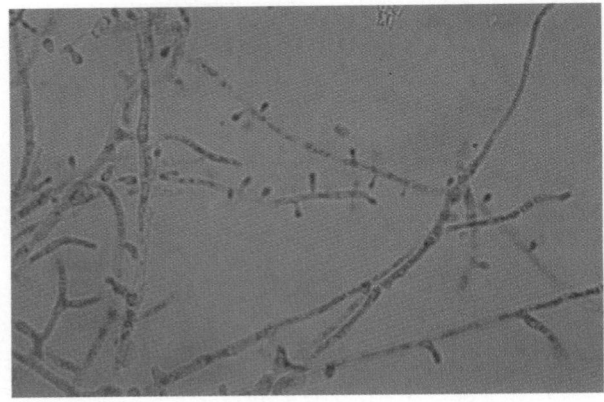

■■■■■■■F i g u r e   2 3 – 2 0
Trichophyton rubrum, showing clavate or peg-shaped microconidia.

Microsporum canis

Microsporum canis, a relatively rapid-growing zoophilic dermatophyte found worldwide, is the species of *Microsporum* most commonly isolated from humans. It primarily infects skin and hair. Freshly isolated cultures typically produce many macroconidia and microconidia, and the reverse side of the colony usually develops a lemon-yellow pigment, especially on potato dextrose agar. Macroconidia are spindle shaped with echinulate, thick walls; they measure 12 to 25 by 35 to 110 μm and have 3 to 15 cells (Fig. 23–21). The tapering, sometimes elongated, spiny distal ends of macroconidia are key features that distinguish this species. Microconidia are abundantly formed by most isolates, and these may be the only conidia present in cultures that have been serially transferred.

Microsporum gypseum

Microsporum gypseum is a rapidly growing geophilic species found in soils all over the world. Infections in humans are not common, but sites involved are primarily skin and hair. Abundant macroconidia and microconidia produced by most isolates of this species result in a powdery, granular appearance on colony surfaces. Colonies that form tan to buff conidial masses are typical of fresh isolates, but this species tends to develop pleomorphic tufts of white, sterile hyphae in aging cultures and after serial transfers. Abundant brown to red pigment may form beneath some strains, but others remain colorless. The fusiform, moderately thick-walled conidia (Fig. 23–22) measure 8 to 15 by 25 to 60 μm and may have as many as six cells. In some isolates, the distal end of the macroconidium may bear a thin, filamentous tail that is longer than the rest of the conidium. The cell walls are moderately thick.

Microsporum audouinii

A slow-growing anthropomorphic dermatophyte, *M. audouinii* was responsible for most of the gray-patch ringworm of children until a few decades ago, when *T. tonsurans* replaced it. *Microsporum audouinii* commonly infects hair and skin but seldom the nails. Colonies of

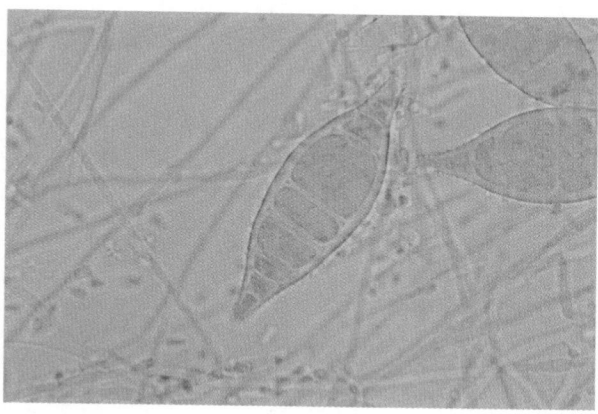

Figure 23-21

Microsporum canis, showing spindle-shaped, echinulate macroconidia with thick walls and tapered ends, which are key features in the identification of this species.

M. audouinii appear cottony-white and generally form little or no pigment on the reverse. Conidia are only rarely produced. Some isolates form chlamydoconidium-like swellings terminally on hyphae.

Epidermophyton floccosum

Epidermophyton floccosum is an anthropomorphic dermatophyte species infecting skin and nails. Colonies of *E. floccosum* are yellow to yellow-green. *Macroconidia* (describing the larger of two sizes of conidia) is used in reference to the conidia of *E. floccosum* even though this fungus is unique among the dermatophytes in its lack of microconidia. The smooth, thin-walled macroconidia are produced in clusters or singly. The distal end of the conidium is broad, or spatulate, and reminiscent of a beaver's tail. Occasionally, conidia may be single celled, but usually they are separated into two to five cells by perpendicular cross-walls. *Epidermophyton* isolates are notorious for developing pleomorphic tufts of sterile hyphae in older cultures.

Laboratory Diagnosis

Specimen Collection and Processing

The affected site should be washed or swabbed briefly with alcohol before collection of the specimen. Properly collected hair, skin scrapings, nail scrapings, or nail clippings are possible sources of the infecting fungi.

Direct Microscopic Examination

Initial diagnosis of a dermatophyte infection of the skin may be made in a skin scraping mounted in potassium hydroxide. These preparations may be modified by incorporating Parker Quink blue-black ink (Parker Pen USA, LTD, Janesville, WI) or calcofluor white fluorescent brightener into the KOH to enhance detection and observation of fungal elements in the epithelial cells. The KOH acts to dissolve the skin cells, leaving intact the hyaline, septate hyphae, which appear similar among all the various etiologic agents of ringworm.

Culture

Isolation of dermatophytes in pure culture is fundamental in correctly identifying them to genus and species. Short clippings of suspected infected hair, skin scrapings, or nail scrapings are placed on the surface of suitable inhibitory and noninhibitory agar media. Nail clippings should be stuck into the medium or distributed on the agar. It is generally inappropriate to plant whole nails onto the medium surface. Rather, scrapings or clippings of the most suspicious-looking, often deeper, thicker portions of the nail should be removed and placed in culture. Skin

Figure 23-22

Microsporum gypseum, showing fusiform, moderately thick-walled macroconidia containing several cells.

scrapings taken from the border of the lesion increases the chance of recovering viable fungal elements. Scalp lesions may be inspected under a Wood lamp, a source of near ultraviolet wavelength light. Infection of hair by a select few dermatophytes *(Microsporum canis, M. audouinii, M. ferrugineum,* and *T. tonsurans)* cause hair shafts to fluoresce, and the ultraviolet light aids in detection of these infections.

Identification

Differential tests for dermatophytes include nutritional studies, pigment formation on special media, hair perforation tests, and urease production.

HAIR PERFORATION TEST

In the hair perforation test, sterile 5- to 10-mm hair fragments are floated on sterile water supplemented with a few drops of sterile, 10% yeast extract. Conidia or hyphae from the dermatophyte in question are inoculated onto the water surface. Hair shafts are removed and microscopically examined in LPCB at weekly intervals for up to 1 month. *Trichophyton rubrum,* which may be morphologically similar to *T. mentagrophytes,* usually causes only surface erosion of hair shafts in this test, whereas *T. mentagrophytes* usually forms perpendicular penetration pegs in the hair shafts (Fig. 23–23). Some laboratories also use this test to distinguish penetration-capable *M. canis* from *Microsporum equinum,* which does not penetrate hair.

UREASE TEST

Another test used to help differentiate *T. mentagrophytes* from *T. rubrum* is the 5-day urease test. Tubes of Christensen urea agar are very lightly inoculated with the dermatophyte and held for 5 days at room temperature. Most isolates of *T. mentagrophytes* demonstrate urease production within that time, whereas most *T. rubrum* isolates require more than 5 days to give a positive reaction.

THIAMINE REQUIREMENT

Some dermatophytes cannot synthesize certain vitamins and therefore do not grow on vitamin-free media. Although several vitamin deficiencies are recognized in fungi, the test for

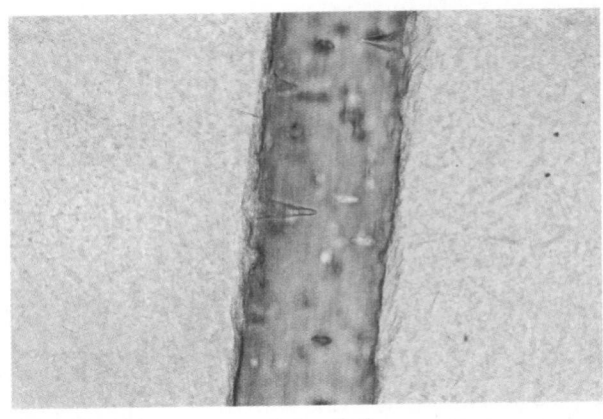

■■■■■Figure 2 3 – 2 3
A positive hair perforation test shows penetration of the fungal agent in the hair shaft. This is the typical reaction by *T. mentagrophytes,* whereas *T. rubrum* causes only surface erosion of hair shaft.

thiamine requirement is perhaps the single most useful nutritional test for dermatophytes. Tubes of media with and without the vitamin are inoculated with a tiny, medium-free portion of hyphae or conidia and observed for growth after 10 to 14 days. Great care must be exercised to avoid transfer of culture media with the inoculum, because even minuscule amounts of vitamin carried over with it can adequately supply the requirement and thereby mask an otherwise negative reaction.

GROWTH ON RICE GRAINS

Poorly sporulating isolates of *M. canis* may be difficult to differentiate from *M. audouinii,* a species that typically forms few spores. Sterile, nonfortified rice is inoculated lightly with hyphae of the isolate under study. After 10 days of incubation at room temperature, the medium is observed for growth. *Microsporum canis* and virtually all other dermatophytes grow well and usually form many conidia, whereas *M. audouinii* does not grow.

OTHER TESTS

By way of a single developmental process, dermatophytes typically form two sizes of reproductive cells, macroconidia and microconidia. Both of these are anamorphic or asexual conidia and their distinctive size, shape, and surface features make them valuable structures for identification of species. Some dermatophytes are known to also have teleomorphic (sexual) stages, in which ascospores are the reproductive cells. Teleomorphs in this group of organisms are not observed in routine laboratory studies of patient specimens, because dermatophytes are heterothallic, requiring combination of two distinct mating types. Although a few reference laboratories perform mating tests with known "tester strains," this is not a procedure regularly used in clinical laboratories.

AGENTS OF SUBCUTANEOUS MYCOSES

Subcutaneous mycoses are fungal diseases affecting subcutaneous tissues. These mycoses are usually the result of the traumatic implantation of foreign objects into the deep layers of the skin, permitting the fungus to gain entry into the host. The etiologic agents responsible are organisms commonly found in soil or on decaying vegetation. Organisms inciting subcutaneous mycoses belong to a variety of genera in the Form-Class Hyphomycetes. Although some are moniliaceous (hyaline or light colored), many are dematiaceous, producing darkly pigmented colonies and containing melanin in their cell walls. The infections are commonly chronic in nature and usually incite the development of lesions at the site of trauma.

Subcutaneous fungal infections may be grouped together by the disease processes they cause or the etiologic agents involved. This section discusses the following diseases and their respective etiologic agents:

■ **Sporotrichosis:** *Sporothrix schenckii*

■ **Chromoblastomycosis:** *Phialophora compacta, Phialophora verrucosa, Cladosporium carrioni, Fonsecaea pedrosoi,* and *Rhinocladiella aquaspersa*

■ **Eumycotic mycetoma:** *Pseudallescheria boydii, Acremonium falciforme, Madurella mycetomatis, Madurella grisea,* and *Exophiala jeanselmei*

■ **Phaeohyphomycosis:** *Exophiala* species and *Wangiella dermatitidis*

Other subcutaneous mycoses rarely seen in the United States are lobomycosis, subcutaneous zygomycosis, and rhinosporidiosis.

Sporotrichosis

Epidemiology

Sporothrix schenckii is commonly recovered from the soil and is associated with decaying vegetation. It is endemic in warm, arid areas such as Mexico and also in moist, humid regions such as Brazil, Uruguay, and South Africa. In temperate countries such as France, Canada, and the United States, most cases of sporotrichosis are associated with gardening, particularly with exposure to rose thorns (rose handler's disease) and sphagnum moss.

Clinical Infections

The most commonly seen presentation of *Sprorthrix schenckii* infection is lymphocutaneous sporotrichosis. This chronic infection is characterized by nodular and ulcerative lesions along the lymph channels that drain the primary site of inoculation. Less commonly seen disease entities include fixed cutaneous sporotrichosis, in which the infection is confined to the site of inoculation and mucocutaneous sporotrichosis, a relatively rare condition. Primary and secondary pulmonary sporotrichosis as well as extracutaneous and disseminated forms of the disease may also occur. Several serologic procedures are available for the diagnosis of sporotrichosis, the most useful being immunodiffusion and latex slide agglutination.

Laboratory Diagnosis

SPECIMEN COLLECTION

Clinical samples submitted for examination include aspirates from cutaneous nodules, pus, exudate, and material from curettage or swabbing of open lesions.

DIRECT MICROSCOPIC EXAMINATION

Direct examination of tissue may reveal *S. schenckii* as small, cigar-shaped yeast. Although the organism may occasionally be seen in a gram-stained smear, wet mount of materials are often unrewarding because of the small numbers of organisms present.

CULTURE

Because *S. schenckii* is dimorphic, cultures are examined at 25°C and 37°C. This fungus grows well on most culture media, including those containing cycloheximide (Mycosel [BBL Diagnostics, Cockeysville, MD], Mycobiotic [Difco Laboratories, Detroit, MI]).

Characteristics at 25°C. The colonial morphology of *S. schenckii* can be quite variable. At room temperature, colonies are often initially white, glabrous, and yeast-like and turn darker and becoming more mycelial as they mature. Microscopic examination reveals thin, delicate hyphae bearing conidia developing in a "rosette" pattern at the ends of delicate conidiophores (Fig. 23–24A). Dark-walled conidia are also produced along the sides of the hyphae.

Characteristics at 37°C. Demonstration of dimorphism is important for specific identification of *S. schenckii*. To induce mycelial to yeast conversion, the fungus is inoculated on blood agar tubes and incubated at 37°C. The formation of yeast colonies may require several subcultures (see Fig. 23–24B). Complete conversion seldom occurs, but a portion of the colony will develop cigar-shaped yeastlike cells.

Chromoblastomycosis

Epidemiology

Also known as verrucous dermatitidis and chromycosis, chromoblastomycosis occurs worldwide but is most common in tropical and subtropical regions of the Americas and Africa. In the United States, most cases have occurred in Texas and Louisiana. Several organisms are responsible for the disease, and particular organisms appear to reside in specific areasof endemicity throughout the world. Chromoblastomycosis is caused by several etiologic agents, namely *P. compacta, P. verrucosa* (Fig. 23–25A), *C. carrioni* (see Fig. 23–25B), *F. pedrosoi,* and *Rhinocladiella aquaspersa.*

Clinical Infections

Chromoblastomycosis is a chronic mycosis of the skin and subcutaneous tissue that develops over a period of months or, more commonly, years. It is mostly asymptomatic in the absence of secondary complications, such as bacterial infections, carcinomatous degeneration, and elephantiasis. Lesions are usually confined to the extremities, often the feet and lower legs, and are a result of trauma to these areas. Lesions of chromoblastomycosis most often appear as verrucous nodules that may become ulcerated and crusted. Longstanding lesions have a cauliflower-like surface. Brown, round sclerotic bodies, which are nonbudding structures occurring singly or in clusters, are seen in tissues. These sclerotic bodies reproduce by dividing in various planes, resulting in multicellular forms. Occasionally, short hyphal elements may also be seen.

Laboratory Diagnosis

At present, serologic evaluation of patients for chromoblastomycosis does not provide a practical adjunct to diagnosis.

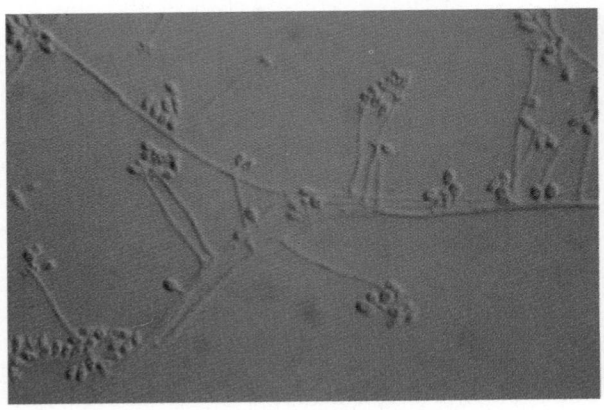

A

B

■■■■■■Figure 23–24

A, Mold phase of *Sporothrix schenckii,* indicating hyaline conidia borne at the ends of conidiophore in "rosettes" as well as dematiaceous conidia along the sides of the hyphae (Nomarski optics, ×625). *B,* Yeast phase of *Sporothrix schenckii,* showing cigar-shaped yeast cells typical of the species.

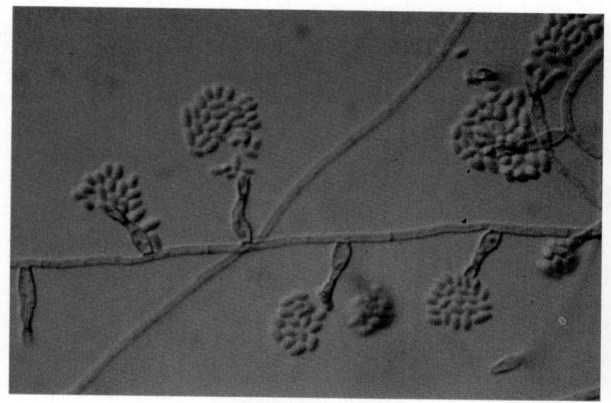

A

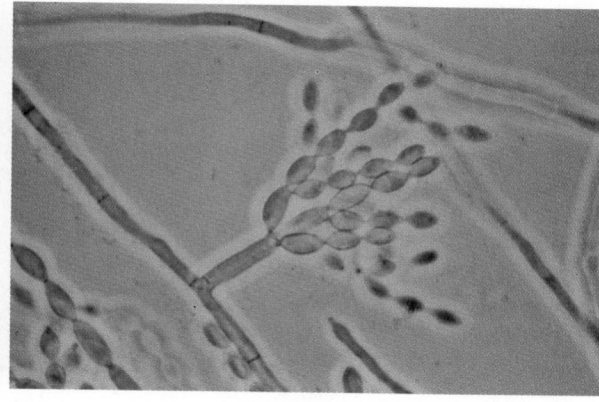

B

■■■■■ F i g u r e   2 3 – 2 5

A, Conidia of *Phialophora verrucosa* at the tips of phialides with collarettes (Nomarski optics, ×1,250). *B,* Conidial arrangement of *Cladosporium carrionii.* (Courtesy of Dr. Michael McGinnis.)

SPECIMEN COLLECTION

Small punch biopsies of affected skin areas should be submitted for histopathology and culture. The presence of sclerotic bodies in tissue, combined with a compatible clinical picture, is diagnostic. Samples are inoculated onto the usual fungal media.

MICROSCOPIC EXAMINATION

The microscopic morphology of each of the agents is described in Table 23–5 and shown in Figure 23–25. Etiologic agents are identified on the basis of characteristic structures, such as arrangement of conidia and how conidia are borne.

CULTURE

Organisms inciting chromoblastomycosis are darkly pigmented or dematiaceous molds. Growth is moderate to slow, and colonies are velvety to woolly, gray-brown to olivaceous black. Species are not differentiated by colonial morphologies, because they all produce similar characteristics.

Eumycotic Mycetoma

Epidemiology

Mycetomas are indolent infections of the subcutaneous tissues that arise at the site of inocu-

lation. Mycetomas may be caused by either fungi or bacteria. Those caused by bacteria are referred to as *actinomycotic mycetomas,* and those caused by fungal agents are referred to as

■■■■■ T A B L E   2 3 – 5

MICROSCOPIC MORPHOLOGY OF FUNGI CAUSING CHROMOBLASTOMYCOSIS

Organism	Microscopic Morphology
Phialophora verrucosa	Conidiogenous cells dematiaceous, flask-shaped phialides with collarettes Conidia oval, one-celled, occur in balls at tips of phialides
Fonsecaea pedrosoi	Primary one-celled conidia formed on sympodial conidiophores Primary conidia function as conidiogenous cells to form secondary one-celled conidia Some conidia similar to those seen in *Cladosporium,* some like those in *Rhinocladiella,* and others like those in *Phialophora* sp
Fonsecaea compactum	Similar to *F. pedrosoi* but with more compact conidial heads Conidia are subglobose rather than ovoid
Cladosporium carrioni	Erect conidiophores bearing branched chains of one-celled, brown blastoconidia Conidium close to tip of conidiophore termed "shield cell" Fragile chains
Rhinocladiella aquaspersa	Conidiophores erect, dark, bearing conidia only on upper portion near the tip Conidia elliptical, one-celled, produced sympodially

eumycotic mycetoma. Although clinical manifestations are similar for the two types of mycetomas, etiology must be determined for appropriate therapy. Mycetomas occur in tropical and subtropical areas but may also be seen in temperate zones. The disease is endemic in India, Africa, and South America.

Although mycetoma is an uncommon mycosis in the United States, the following species are the most commonly incriminated agents: *P. boydii, A. falciforme, M. mycetomatis, M. grisea,* and *Exophiala jeanselmei.*

Clinical Infections

Mycetomas are generally confined to the extremities, although they may be observed at other anatomic sites. The lesions are made up of granulomas and abscesses that drain to the outside through sinus tracts. The pus from these lesions contains granules (grains) that are composed of compact mycelial masses. The characteristic granules formed by the different genera of fungi that incite mycetomas are listed in Table 23–6. These lesions, which are initially confined to the subcutaneous tissue, frequently proliferate to involve the musculature and, in advanced cases, to cause severe destruction of the bone.

Laboratory Diagnosis

SPECIMEN COLLECTION

Grains or granules from draining sinus tracts should be examined for color (light or dark), size, and texture (soft or hard) and should be cultured to recover the etiologic agent. Table 23–7 lists the fungal agents known to cause eumycotic mycetoma according to the color of the grains or granules they form.

■ T A B L E   2 3 – 6

DESCRIPTION OF GRANULES SEEN IN EUMYCOTIC MYCETOMAS

Fungus	Color	Size (mm)	Texture
Pseudallescheria boydii	White	0.5–1.0	Soft
Acremonium falciforme	White	0.2–0.5	Soft
Madurella mycetomatis	Black	0.5–5.0	Hard
Madurella grisea	Black	0.3–0.6	Soft
Exophiala jeanselmei	Black	0.2–0.3	Soft

■ T A B L E   2 3 – 7

FUNGI KNOWN TO INCITE EUMYCOTIC MYCETOMAS*

Black grains	*Exophiala jeanselmei*
	Curvularia lunata
	Curvularia geniculata
	Madurella grisea
	Madurella mycetomatis
	Leptosphaeria senegalensis
	Leptosphaeria tompkinsii
	Pyrenochaeta mackinnonii
	Pyrenochaeta romeroi
	Corynespora cassicola
	Pseudochaetosphaeronema larense
	Plenodomus avramii
White grains	*Pseudallescheria boydii*
	Fusarium solani
	Fusarium moniloforme
	Aspergillus nidulans
	Acremonium falciforme
	Acremonium kiliense
	Acremonium recifei
	Neotestudina rosatii
	Phialophora cyanescens
	Polycytella hominis

*This list is not all inclusive.

DIRECT MICROSCOPIC EXAMINATION

Direct microscopic examination of the granules immediately differentiates eumycotic from actinomycotic mycetomas. Figure 23–26*A* shows the branching filamentous rods of Actinomycetes in contrast to the hyphal elements (see Fig. 23–26*B*) seen in eumycotic infections. To recover the eumycotic agent, samples are inoculated onto routine fungal media.

CULTURE

Pseudallescheria boydii colonies grow rapidly and produce white to dark gray colonies on potato dextrose agar at 25 and 35°C. The anamorph (asexual form of the fungus), *Scedosporium apiospermum,* produces oval conidia singly at the tips of conidiogenous cells (cells that make conidia) known as annellides (Fig. 23–27). Some strains may produce the teleomorph (sexual form of the fungus) in the laboratory and may form cleistothecia containing ascospores (Fig. 23–28). This phenomenon occurs in fungi that are homothallic, i.e., require only one mating strain to produce the sexual form of the fungus.

Acremonium falciforme is a moniliaceous (light-colored) mold. Colonies grow slowly, and are grayish brown, becoming grayish violet. This organism produces mucoid clusters of single or

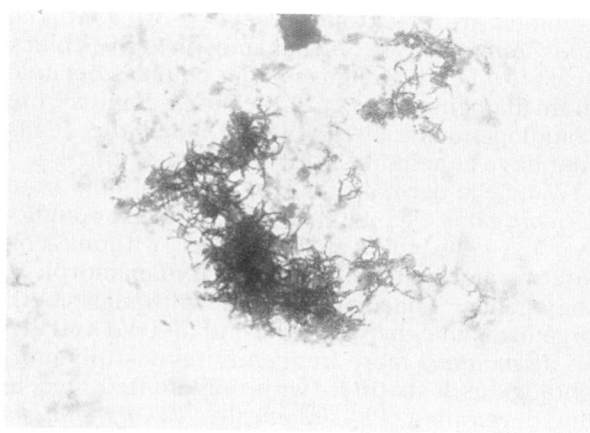

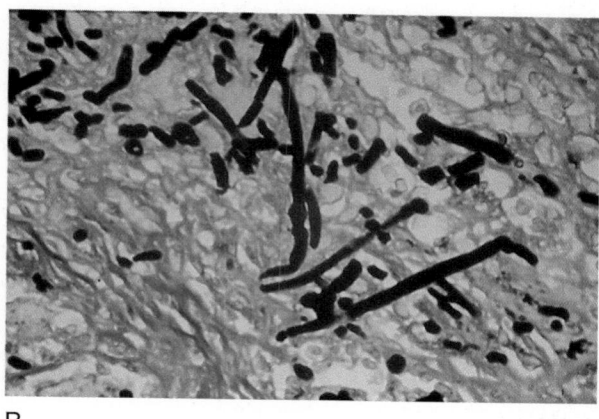

B

A

■■■■■■■ F i g u r e   2 3 – 2 6

Actinomycotic mycetoma, showing fine-branching, filamentous rods in tissue sample (*A*) compared with the hyphal elements (*B*) seen in eumycotic infections.

two-celled, slightly curved conidia borne from phialides at the tips of long, unbranched, multiseptate conidiophores, that are held together in mucoid clusters at the apices.

Madurella mycetomatis grows slowly. It is white initially, then becomes yellow, olivaceous, or brown with a characteristic diffusable brown pigment. Although about half of the isolates produce conidia from the tips of phialides, many remain sterile.

Madurella grisea grows slowly, produces olive brown to black colonies, and may produce a reddish brown pigment. *Madurella mycetomatis* grows best at 37°C and may grow at 40°C, but 30°C is the optimum temperature for *M. grisea.*

Exophiala jeanselmei produces olivaceous to black colonies that are initially yeast-like but become velvety at maturity. Their conidia are also borne from annellides, and they aggregate in masses at the tips of the conidiophore, as seen in Figure 23–29.

Subcutaneous Phaeohyphomycosis

Phaeohyphomycosis is a mycotic disease caused by darkly pigmented fungi or fungi that have melanin in their cell walls. This term was coined by Ajello and colleagues in 1974 to separate several clinical infections caused by dematiaceous

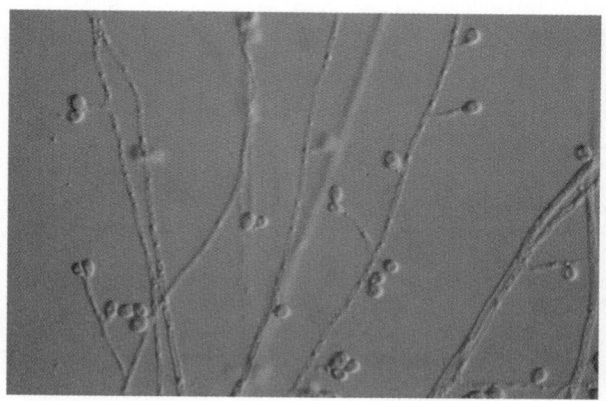

■■■■■■■ F i g u r e   2 3 – 2 7

The *Scedosporium apiospermum* synanamorph of *Pseudallescheria boydii* (Nomarski optics, ×625).

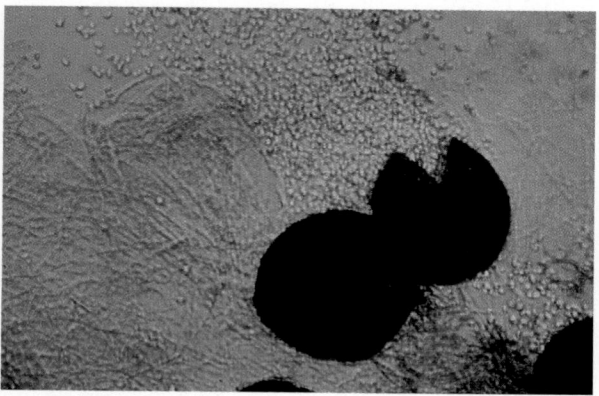

■■■■■■■ F i g u r e   2 3 – 2 8

Sexual structures (cleistothecia containing ascospores) of *Pseudallescheria boydii* (Nomarski optics, ×325).

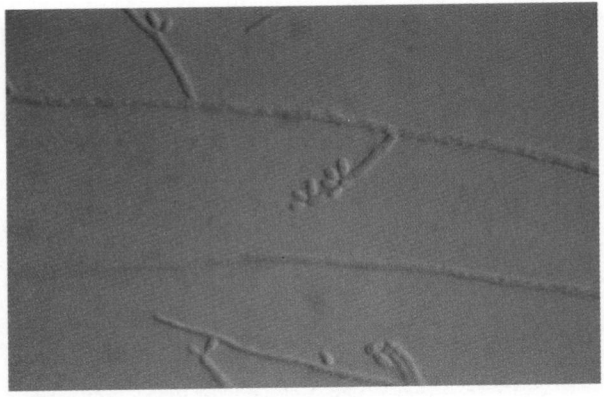

■■■■Figure 23–29

Conidia of *Exophiala jeanselmei* borne at the tips of annellides (Nomarski optics, ×1,250).

colonies are black and yeast-like owing to the *Phaeoannellomyces* synanamorph (the black yeast form of the fungus). Older colonies become more filamentous and produce their conidia from conidiogenous cells known as annellides (cells that have annellations or rings at their apices).

Wangiella dermatitidis (Fig. 23–30) differs from *E. jeanselmei* (see Fig. 23–29) by having conidiogenous cells that are phialides without collarettes and by having its yeast synanamorph in the genus *Phaeococcomyces*. Although both organisms may be yeast-like and mucoid initially, *W. dermatitidis* more frequently retains this morphology as it matures, with only limited mycelium developing. Physiologically, *W. dermatitidis* grows at 40°C and is nitrate negative; *E. jeanselmei* fails to grow at 40°C and is nitrate positive.

fungi from those distinct clinical entities known as chromoblastomycosis. In tissue, these fungi may form yeast-like cells that are solitary or in short chains, or hyphae that are septate, branched, or unbranched, and often swollen to toruloid. Agents responsible for these mycoses are organisms commonly found in nature, encompassing many genera of Hyphomycetes, Coelomycetes, and Ascomycetes.

Fungi that appear to be regularly associated with this condition include *Exophiala* species and *Wangiella dermatitidis*. A more complete list of genera associated with subcutaneous phaeohyphomycosis is found in Table 23–8.

Laboratory Diagnosis

SPECIMEN COLLECTION

Biopsy specimens of phaeohyphomycosis lesions should be submitted to the histology laboratory for fungal stains and to the mycology laboratory for culture.

MICROSCOPIC EXAMINATION

In addition to the usual stains, such as GMS, H&E and PAS, one may wish to include the Masson-Fontana stain to detect melanin in the cell walls, thus documenting the dematiaceous nature of the etiologic agent.

CULTURE

Exophiala jeanselmei has been previously described as an agent of mycetoma. Young

AGENTS OF SYSTEMIC MYCOSES

Organisms that cause classic, systemic fungal diseases have historically been categorized together because they share several characteristics, such as mode of transmission, dimorphism, and systemic dissemination. Although the term *systemic* generally refers to the organisms described here, it must be understood that any fungus, given an immunocompromised host, has the potential to become invasive and to disseminate to sites far removed from the portal of entry of usual site of infection.

The four classic agents of systemic mycoses are

■ *Blastomyces dermatitidis*

■ *Histoplasma capsulatum* var *capsulatum*

■ *Coccidioides immitis*

■ *Paracoccidioides brasiliensis*

■■■■TABLE 23–8

DEMATIACEOUS GENERA INCITING SUBCUTANEOUS PHAEOHYPHOMYCOSIS*

Alternaria	Fonsecaea
Anthopsis	Mycocentrospora
Bipolaris	Oidiodendron
Chaetomium	Phaeosclera
Cladosporium	Phialophora
Curvularia	Phoma
Dactylaria (Ochroconis)	Ulocladium
Exophiala	Xylohypha

*This list is not all inclusive.

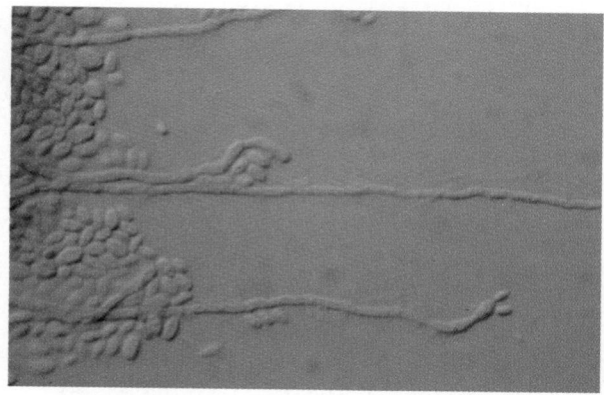

Figure 23-30

Conidia of *Wangiella dermatitidis* borne at the tips of phialides as well as the black yeast synanamorph (Nomarski optics, ×1,250).

These organisms are dimorphic, which refers to their ability to grow in the mold form in their natural environment or in the laboratory at 25 to 30°C as shown in Table 23–9 or in the yeast or spherule form when incubated at 35 to 37°C on enriched media. Each agent has a fairly well-defined area of endemicity, and the diseases they all incite are contracted by the inhalation of infectious conidia. Table 23–10 provides a summary of systemic mycoses, their agents, and characteristics. All laboratory procedures to recover and identify these agents must be performed under a certified biologic safety cabinet.

Blastomyces dermatitidis

Epidemiology

Blastomyces dermatitidis causes blastomycosis, also known as Gilchrist disease, North American blastomycosis, and Chicago disease. It occurs primarily in North America and parts of Africa. In the United States, it is endemic in the Mississippi and Ohio river basins. Sporadic point-source outbreaks have also occurred in the St. Lawrence River basin. The natural reservoir has not been unequivocally established, although the organism has been recovered from the soil and from some natural environments. Apparently, only a very narrow range of conditions supports its growth. In areas where the organism appears endemic, natural disease occurs in dogs and horses, and the disease process mimics that seen in human infections.

Clinical Infections

Blastomycosis occurs when the conidia are inhaled. It is most prevalent in middle-aged men, as are other systemic mycoses, presumably owing to men's greater occupational exposure to the soil. Although patients with primary infection may exhibit flu-like symptoms, most are asymptomatic and cannot accurately define the

■ TABLE 23-9
MORPHOLOGY OF SYSTEMIC FUNGI AT 25°C

Fungus	Macroscopic Morphology	Microscopic Morphology

B. dermatitidis	Slow to moderate growth
	  White to dark tan
	Young colonies tenacious,
	  older colonies glabrous to woolly
	Spicules in center of colony
		Oval, pyriform, to
		  globose smooth conidia borne
		  on short, lateral hypha-like conidiophores

H. capsulatum *	Slow growth
	  White to dark tan with age
	  Woolly, cottony, or granular
		Microconidia small,
		  one-celled, round, smooth (2–5 μm)
		Tuberculated macroconidia large, round (7–12 μm)
		Hypha-like conidiophores

C. immitis	Rapid growth
	  White to tan to dark gray
	  Young colonies tenacious, older colonies cottony
	  Tend to grow in concentric rings
		Alternating one-celled,
		  "barrel-shaped" arthroconidia with disjunctor cells

P. brasiliensis	Slow growth
	  White to beige
	  Colony glabrous, leathery, flat to wrinkled,
	  folded or velvety
		Colonies frequently
		  only produce sterile hyphae
		Fresh isolates may produce conidia
		  similar to those of *B. dermatitidis*

* *Histoplasma capsulatum* var *capsulatum* (teleomorph *Ajellomyces capsulatus*). *Histoplasma capsulatum* var *duboisii* is endemic in central Africa and is not discussed in this chapter.

■■■■■T A B L E   2 3 - 1 0
SUMMARY OF SYSTEMIC MYCOSES

Fungus	Ecology	Clinical Disease	Tissue Form
B. dermatitidis	Mississippi and Ohio River valleys	Primary lung Chronic skin/bone Systemic, multiorgan	Large yeast (8–12 μm) Broad-based bud
H. capsulatum*	Ohio, Missouri, and Mississippi River valleys Bird and bat guano Alkaline soil	Primary lung Asymptomatic Immunodeficient hosts prone to disseminated disease	Small, oval yeast (2–5 μm) in histiocytes, phagocytes
C. immitis	Semiarid regions: southwest U.S., Mexico, Central & South America In soil	Primary lung Asymptomatic Secondary cavitary Progressive pulmonary Multisystem	Spherules (30–60 μm) containing endospores
P. brasiliensis	Central and South America In soil	Primary lung Granulomatous Ulcerative nasal and buccal lesions Lymph node involvement Adrenals	Thick-walled yeasts (15–30 μm) Multiple buds "Mariner's wheel"

* Histoplasma capsulatum var *capsulatum* (teleomorph *Ajellomyces capsulatus*). *Histoplasma capsulatum* var *duboisii* is endemic in central Africa and is not discussed in this chapter.

time of onset. When the primary disease fails to resolve, pulmonary disease may ensue, with cough, weight loss, chest pain, and fever. Progressive pulmonary or invasive disease may follow, resulting in ulcerative lesions of the skin and bone. In the immunodeficient patient, multiple organ systems may be involved, and the course may be rapidly fatal.

Laboratory Diagnosis

At present, there are still no practical, reliable serologic procedures for the diagnosis or prognosis of blastomycosis. Diagnosis requires identification of the organism in tissue or isolation and identification in culture.

SPECIMEN COLLECTION

In pulmonary disease, the first morning sputum may be submitted for direct microscopic examination and culture. Specimens must be processed immediately. Bronchial washings and other pulmonary secretions may also be examined for cytology.

Exudative material from cutaneous lesions and tissue must also be examined directly for yeasts.

DIRECT MICROSCOPIC EXAMINATION

Microscopic examination of tissue or purulent material in cutaneous skin lesions may reveal large, spherical, refractile yeast cells, 8 to 15 μm in diameter, with a double-contoured wall and buds connected by a broad base (Fig. 23–31). Potassium hydroxide (10%) or calcofluor white may be used to aid examination for the presence of the yeast cells.

CULTURE

Although demonstration of characteristic yeast cells in direct smear examination is usually diagnostic, cultures should also be performed, particularly when smears are negative. Because

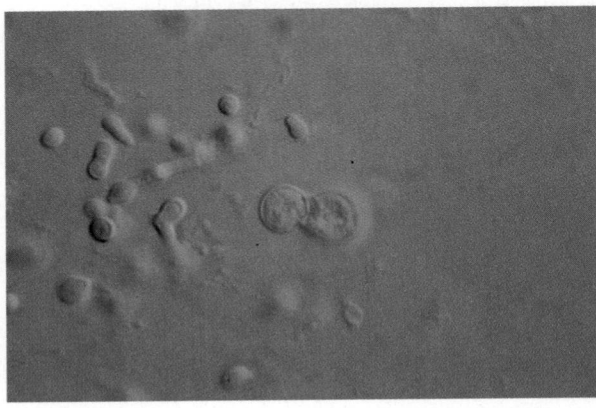

■■■■■F i g u r e   2 3 - 3 1

Conversion of the mold phase of *Blastomyces dermatitidis* to the "broad-based bud" yeast form (Nomarski optics, ×1,250).

the yeast phase is susceptible to cycloheximide, nonselective media such as Sabouraud dextrose agar or brain-heart infusion should be employed. Heavily contaminated material should be plated on similar media containing antibacterial agents such as chloramphenicol, gentamicin, streptomycin, or penicillin. Two sets of cultures are set up, one to be incubated at 25°C and the other at 37°C.

Characteristics at Room Temperature. In culture at 25°C, the organism may produce a variety of colonial morphologies. Colonies may be white, tan, or brown and may be fluffy to glabrous, growing in concentric rings. Frequently, raised areas termed *spicules* or *prickles* are seen in the centers of the colonies.

Microscopically, the anamorphic or asexual form of the fungus produces conidia borne on short lateral branches that are ovoid to dumbbell-shaped and vary in diameter from 2 to 10 μm. They often resemble the microconidia of *H. capsulatum* var *capsulatum* (Fig. 23–32).

The teleomorph or sexual form of *B. dermatitidis* was described in 1967 and named *Ajellomyces dermatitidis*. It occurs only in rigidly controlled environments by mating of isolates with tester strains to produce gymnothecia containing ascospores. This teleomorph does not occur in the routine laboratory because the species is heterothallic (requires two mating strains to produce the sexual form).

Characteristics at 37°C. When grown at 37°C on suitable media (Table 23–11), *B. dermatitidis* produces characteristic broad-based yeast cells (see Fig. 23–31). This conversion process is necessary to identify this dimorphic fungus. The organism may also be identified by utilizing the

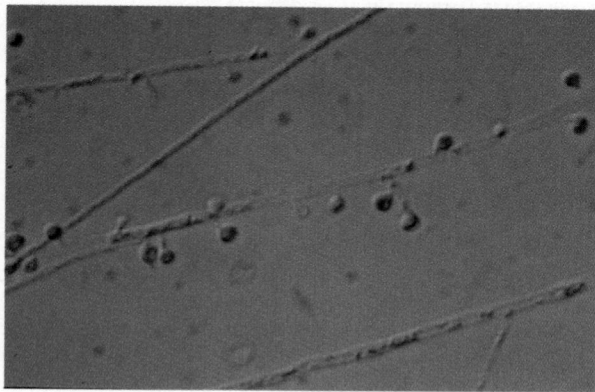

■■■■F i g u r e   2 3 - 3 2

Mold phase of *Blastomyces dermatitidis* grown on potato flakes agar (Nomarski optics, ×1,250).

■■■■T A B L E   2 3 - 1 1

MOLD TO YEAST CONVERSION IN DIMORPHIC FUNGI*

Fungus	Culture Media & Temperature	Yeast Form
B. dermatitidis	Blood agar, 37°C	Large yeast (8–12 μm) Blastoconidia attached by broad base
H. capsulatum	Pines medium, glucose-cysteine-blood, or BHI-blood, 37°C	Small, oval yeast (2–5 μm)
P. brasiliensis	BHI-blood, 37°C	Multiple blastoconidia budding from single, large yeast (15–30 μm)

* *Coccidioides immitis* may be converted to the spherule phase in a modified Converse medium at 40°C in 5–10% CO_2. Exoantigen testing is preferred over this procedure in the routine clinical laboratory.

exoantigen technique of Kaufman and Standard (1987). Exoantigens, cell-free antigens produced by the mycelial forms of dimorphic fungi (Table 23–12), are detected by precipitin lines of identity in immunodiffusion tests of concentrated, thimevosal-killed culture supernatant fluid (Fig. 23–33).

Histoplasma capsulatum var *capsulatum*

Epidemiology

Histoplasma capsulatum var *capsulatum* causes histoplasmosis, also known as reticuloendothelial cytomycosis, cave disease, spelunker's disease, and Darling disease. Histoplasmosis occurs worldwide. The highest endemicity in the United States occurs in the Ohio, Missouri, and Mississippi river deltas. This organism resides in soil containing a high nitrogen content, particularly in areas heavily contaminated with bat and bird guano. Skin testing of long-term residents in endemic areas

■■■■T A B L E   2 3 - 1 2

EXOANTIGENS IDENTIFIED IN DIMORPHIC FUNGAL PATHOGENS*

Fungus	Exoantigen(s)
Blastomyces dermatitidis	A
Histoplasma capsulatum	h, m
Coccidioides immitis	F, HL, HS
Paracoccidioides brasiliensis	1, 2, 3

*Exoantigens are identified by forming lines of identity with reference antigen-antibody complexes (see Fig. 23–33).

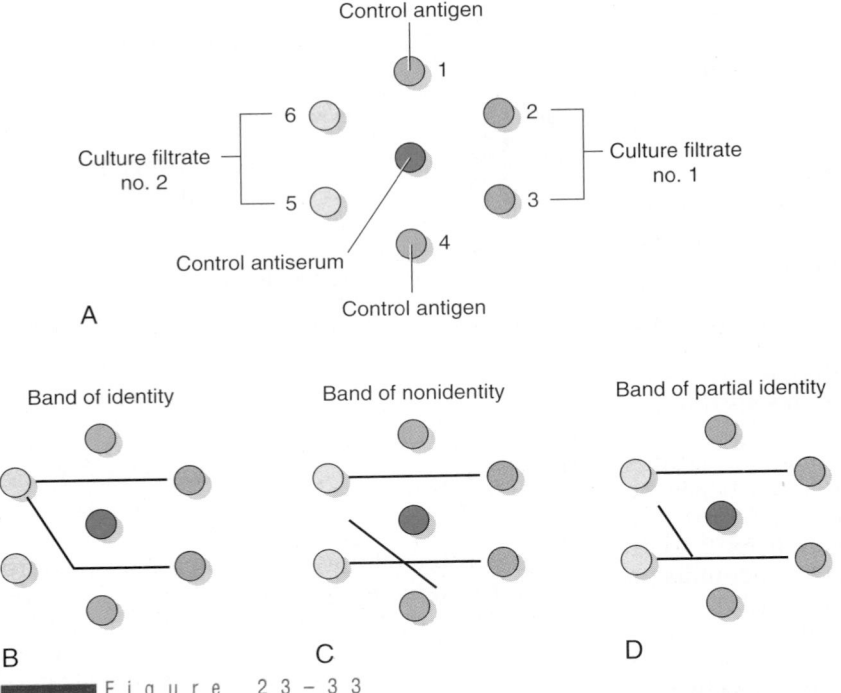

■■■■■ F i g u r e 2 3 – 3 3
How the exoantigen immunodiffusion test is set up (*A*); a band of identity (*B*); a band of nonidentity (*C*); and a band of partial identity (*D*).

indicates that approximately 80% of the population have been infected. *Histoplasma capsulatum* var *duboisii,* which is endemic in Central Africa, causes a clinically distinct form of disease. Another variant, *H. capsulatum* var *farciminosum,* causes epizootic lymphangitis in horses and mules. These two forms of the organism are not discussed in this chapter. *Histoplasma capsulatum,* like *B. dermatitidis,* is also a heterothallic ascomycete, which produces the teleomorph state, *Ajellomyces capsulatus,* when mated with appropriate tester strains.

Clinical Infections

Histoplasmosis is acquired by the inhalation of the microconidia of *H. capsulatum* var *capsulatum.* The microconidia are phagocytized by macrophages present in the pulmonary parenchyma. In the host with intact immune defenses, the infection is limited and is usually asymptomatic, the only sequelae being areas of calcification in the lungs, liver, and spleen. With heavy exposure, however, acute pulmonary disease may occur. In the mild form of the disease, viable organisms remain in the host, quiescent for years; this is the presumed source of reactivation

in individuals with abrogated immune systems. In immunocompromised individuals, *H. capsulatum* may cause a progressive and potentially fatal disseminated disease. Chronic pulmonary histoplasmosis in patients with chronic obstructive pulmonary disease may also occur. Other various manifestations of the disease are mediastinitis, pericarditis, and mucocutaneous lesions.

Laboratory Diagnosis

Unlike blastomycosis, serologic procedures for the diagnosis of histoplasmosis may be an adjunct to culture methods. Tests that may be employed include skin tests (delayed or immediate sensitivity reactions) complement fixation, immunodiffusion, latex agglutination, counterimmunoelectrophoresis, radioimmunoassay (detection of circulating antibody), and fluorescent antibody (FA) microscopy (to detect either viable or nonviable fungal elements in tissue sections). Currently, the most useful serologic applications for the diagnosis of histoplasmosis appears to be the combination of complement fixation and immunodiffusion tests, in which rising titers in serial dilutions are considered significant.

SPECIMEN COLLECTION

A variety of specimens, such as centrifuged sputum and bronchoscopic fluids from patients with cavitary disease, may yield *H. capsulatum.* In AIDS patients, the organisms are usually numerous in bone marrow aspirates and are frequently even present in peripheral blood smears. Exudates from mucocutaneous lesions, liver, and spleen may also be submitted for recovery of the organism. Clinical specimens are inoculated on two sets of Sabouraud dextrose agar, brain-heart infusion agar, or inhibitory mold agar media and incubated 25°C and 37°C.

DIRECT MICROSCOPIC EXAMINATION

Careful examination of direct smear preparations of specimens for histoplasmosis frequently reveals the small yeast cells of *H. capsulatum,* particularly in the opportunistic infections seen in the immunodeficient host. The yeast cells measure 2 to 3 by 4 to 5 μm. When smears are stained with Giemsa or Wright stain, the yeast cells are commonly seen within monocytes and macrophages occurring in significant numbers, as shown in Figure 23–34*A*, bone marrow smear stained with Giemsa. These small cells, when found in tissue (see Fig. 23–34*B*), resemble the blastoconidia of *Torulopsis glabrata* but may be differentiated by FA techniques or culture. Yeast cells of *H. capsulatum* should also be distinguished from the small yeast form of *B. dermatitidis,* yeast cells of *Cryptococcus neoformans,* and the released endospores from spherules of *Coccidioides immitis.*

CULTURE

Histoplasma capsulatum grows as a white to brownish mold. Early growth of the mycelial culture produces round to pyriform microconidia measuring 2 to 5 μm. As the colony matures, large echinulate or tuberculate macroconidia, a characteristic of the species, are formed (Fig. 23–35). Although microconidia may resemble *Chrysosporium* spp and the macroconidia resembles *Sepedonium* spp, neither saprobe produces two types of conidia, nor are they dimorphic. Conversion of the mold form to the yeast form, utilizing brain-heart infusion agar incubated at 37°C, is confirmation for *H. capsulatum.* Although complete conversion is seldom noted, a combination of forms is sufficient for identification. Identification of the isolates is confirmed by exoantigen procedures detecting two exoantigens (see Table 23–11) produced by this species, H and M antigens. Demonstration of these exoantigens is necessary for identification of isolates that fail to produce characteristic conidia.

Coccidioides immitis

Epidemiology

Coccidioides immitis causes coccidioidomycosis, also called Posada disease, coccidioidal granuloma, valley fever, desert rheumatism, valley bumps, and California disease. It resides in a narrow ecologic niche known as the Lower Sonoran Life Zone, characterized by low rainfall

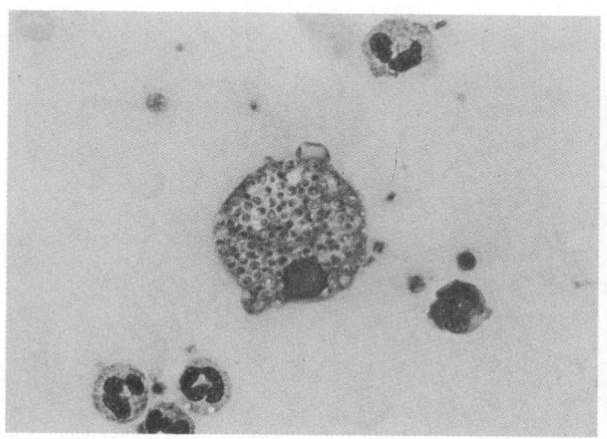
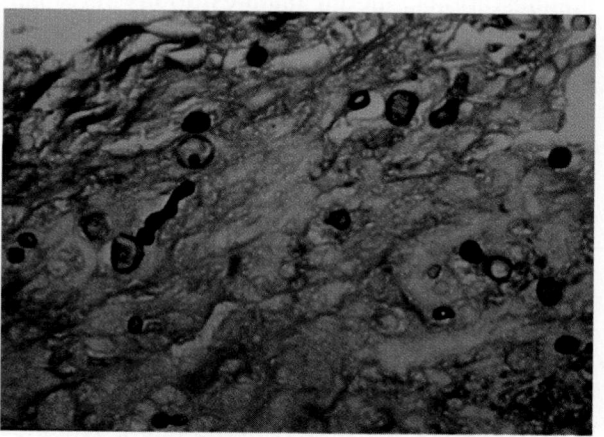

A   B

■■■■Figure 23 – 34

A, Bone marrow aspirate stained with Giemsa, showing the yeast cells of *Histoplasma capsulatum* var *capsulatum* inside the monocytes (×1,200). *B,* Tissue phase of *Histoplasma capsulatum* var *capsulatum.* (GMS stain, ×1,200).

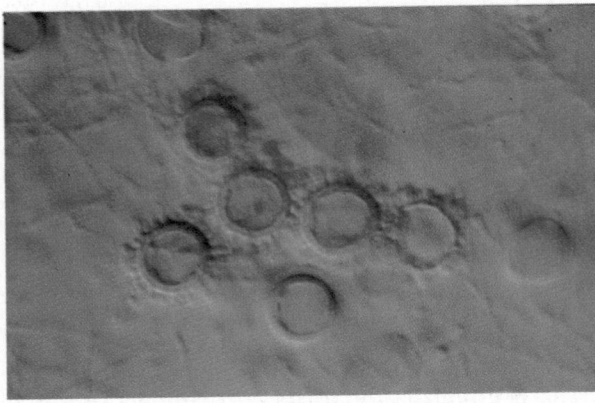

Figure 23–35

Large tuberculate macroconidia of *Histoplasma capsulatum* var *capsulatum* (Nomarski optics, ×1,250).

and semiarid conditions. Highly endemic areas include the San Joaquin Valley of California, the Maricopa and Pima counties of Arizona, and southwestern Texas. Outside the United States, high areas of endemicity are found in northern Mexico, Guatemala, Honduras, Venezuela, Paraguay, Argentina, and Columbia.

Clinical Infections

Coccidioides immitis is probably the most virulent of all human mycotic agents. The inhalation of only a few arthroconidia produces primary coccidioidomycosis. Clinical infections include asymptomatic pulmonary disease and allergic manifestations. Allergic manifestations may manifest as toxic erythema, erythema nodosum (desert bumps), erythema multiforme (valley fever), and arthritis (desert rheumatism). Primary disease usually resolves without therapy and confers a strong, specific immunity to reinfection, which is detected by the coccidioidin skin test.

In symptomatic patients, fever, respiratory distress, cough, anorexia, headache, malaise, and myalgias may manifest for 6 weeks or longer. The disease may then progress to secondary coccidioidomycosis, which can include nodules, cavitary disease, or progressive pulmonary disease. Single or multisystem dissemination follows in approximately 1% of this population. Filipinos and African Americans run the highest risk of dissemination, with meningeal involvement as a common sequela. The sex distribution ratio for clinically apparent disease is approximately 9:1

(male/female), which also holds true for other classic systemic diseases. The exception is in pregnant women, in whom the dissemination rate equals or exceeds that for men.

Laboratory Diagnosis

Several serologic procedures exist for initial screening, confirmation, and prognostic evaluation of coccidioidomycosis. The combined use of the immunodiffusion test and the latex particle agglutination test detects approximately 93% of the cases. The complement fixation (CF) and tube precipitin tests may also be employed for diagnosis as well as for confirmation. Prognostic studies commonly employ serial CF titers.

SPECIMEN COLLECTION

Respiratory specimens such as sputum, tracheal aspirates, and lung biopsy tissue may be submitted for suspected cases. CSF and blood cultures are performed for patients suspected of having disseminated forms.

DIRECT MICROSCOPIC EXAMINATION

After inhalation, the barrel-shaped arthroconidia, which measure 2.5 to 4 by 3 to 6 μm, round up as they convert to spherules. At maturity, the spherules (30 to 60 μm) produce endospores by a process known as *progressive cleavage*: Rupture of the spherule wall releases the endospores, which in turn form new spherules (Fig. 23–36). Direct smear examination of secretions may reveal the spherules containing the

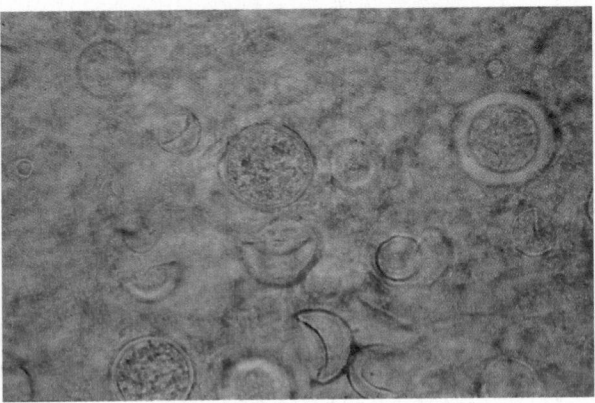

Figure 23–36

Spherules of *Coccidioides immitis* in tissue (×300).

endospores. Caution must be exercised, however, when diagnosis is made by histopathologic means only. Small, empty spherules may resemble the yeast cells of *B. dermatitidis,* and the endospores can be confused with the cells of *C. neoformans, H. capsulatum,* and *P. brasiliensis.*

CULTURE

Coccidioides immitis produces a variety of mold morphologies at 25°C. Initial growth, which occurs within 3 to 4 days, is white to gray, moist, and glabrous. It rapidly develops abundant aerial mycelium, and the colony appears to enlarge in a circular "bloom." Mature colonies usually become tan to brown to lavender. Microscopically, fertile hyphae arise at right angles to the vegetative hyphae produce alternating (separated by a disjunctor cell), hyaline arthroconidia. When released, conidia have an annular frill at both ends. As the culture ages, the vegetative hyphae also fragment into arthroconidia (Fig. 23–37). *Malbranchea* spp resemble *C. immitis* but fail to convert to spherules in infected animals or special culture media. *Malbranchea* spp also lack lines of identity in the exoantigen tests, which are specific for *C. immitis.*

Paracoccidioides brasiliensis

Epidemiology

Paracoccidioides brasiliensis is the causative agent of paracoccidioidomycosis (South American blastomycosis, Brazilian blastomycosis, Lutz-Splendore-Almeida disease, paracoccidioidal granuloma), a chronic, progressive fungal disease that is endemic in Central and South America. Geographic areas of highest incidence are typically humid, high-rainfall areas with acidic soil conditions. As with other systemic mycoses, the sex distribution for clinically significant disease is approximately 9:1 (male/female).

Clinical Infections

Although the primary route of infection is pulmonary, and is usually inapparent and asymptomatic, subsequent dissemination leads to the formation of ulcerative granulomatous lesions of the buccal, nasal, and occasionally, gastrointestinal mucosa. A concomitant striking lymph node involvement is also evident. Although *P. brasiliensis* has a rather narrow range of temperature tolerance, as evidenced by its predilection for growth in cooler areas of the body (nasal and oropharyngeal), dissemination to other organs, particularly the adrenals, occurs with diminished host defenses.

Laboratory Diagnosis

A wide variety of samples may be submitted to the laboratory for recovery of *P. brasiliensis.* They include sputum, bronchoalveolar lavage, pus from draining lymph nodes, scrapings from ulcers and biopsy tissue. Direct microscopic examination of cutaneous and mucosal lesions demonstrates the characteristic yeast cells. The typical budding yeast measures 15 to 30 μm in diameter with multipolar budding at the periphery, resembling a mariner's wheel (Fig. 23–38).

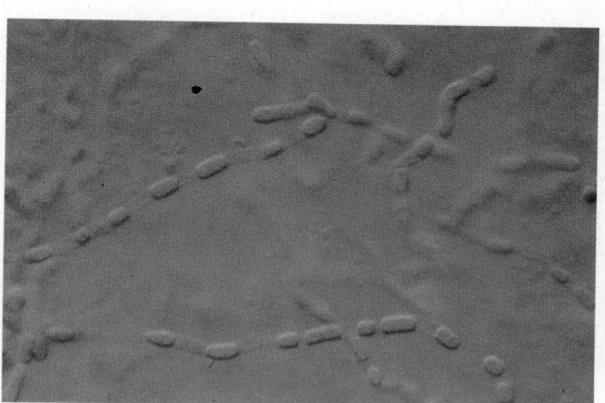

■ F i g u r e   2 3 – 3 7

Mold phase of *Coccidioides immitis,* 25°C (Nomarski optics, ×1,250).

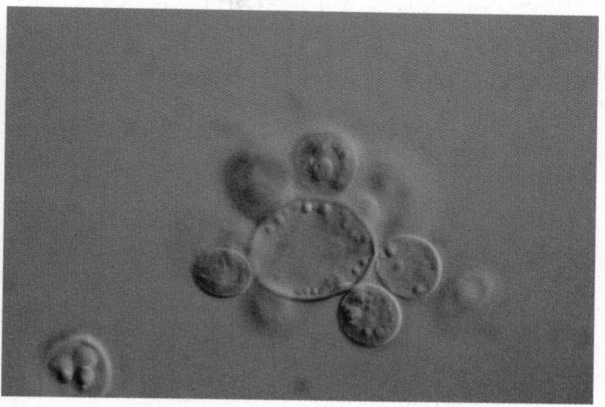

■ F i g u r e   2 3 – 3 8

Yeast phase ("mariner's wheel") of *Paracoccidioides brasiliensis* with multipolar budding (Nomarski optics, ×1,250).

These "daugther" cells (2 to 5 μm) are connected by a narrow base, unlike the broad-based attachment in *B. dermatitidis*. Many buds of various sizes may occur, or there may be only a few buds, giving the appearance of a "Mickey Mouse cap" to the yeast cell.

Paracoccidioides brasiliensis produces a variety of mold morphologies when grown at 25°C. Flat colonies are glabrous to leathery, wrinkled to folded, floccose to velvety, pink to beige to brown with a yellowish-brown reverse, resembling those of *B. dermatitidis*. Microscopically, the mold form produces small (2 to 10 μm in diameter), one-celled conidia, generally indistinguishable from those observed with the mold phase of *B. dermatitidis* or the microconidia of *H. capsulatum*. On BHI-blood agar at 37°C, the mycelial phase rapidly converts to yeast phase. Both complement fixation and agar gel immunodiffusion procedures are available for serodiagnosis.

AGENTS OF OPPORTUNISTIC FUNGAL INFECTIONS: THE SAPROBES

Saprobe and *saprophyte* have been used to describe free-living microorganisms in the environment that are not of concern as agents of human disease. Toward the end of this century, the line between saprobic and parasitic or pathogenic organisms is increasingly blurred. The major reason for this development is the growing number of persons with minor or major defects in their immune systems. For several decades, medical science has made advances in life-sustaining and life-lengthening treatments. A serious side effect of procedures such as organ transplants and cancer chemotherapy is the short- or long-term insult to the host defenses. Magnifying the problem greatly during the last two decades has been the spread of acquired immune deficiency syndrome (AIDS). All these persons constitute the prime targets for infection by a wide variety of microorganisms, including the recognized pathogenic fungi and a growing list of fungi heretofore regarded as harmless.

The types of disease caused by these fungi are as varied as the species, and sometimes more so, because a given fungus may cause multiple disease forms. Various wounds from surgical procedures are ideal points of inoculation for these saprobes to become opportunistic agents of disease, especially in the compromised host. Skin and nail bed infections as well as severe respiratory infections may be caused by a variety of fungi in the AIDS patient. What follows is a discussion of the most common saprobes that have been associated with opportunistic infections at different body sites. Characteristic morphologic features of each fungal species are described.

- ■ *Absidia*: Rapid-growing zygomycete with erect sproangiophores that terminate in a columella surrounded by a sporangium. Sporangiophores are formed in clusters on the intermediate portion of the stolons, and rhizoids are formed at the ends of the stolons.

- ■ *Acremonium* (Figure 23–39): Hyaline, rapid-growing, flat colonies that develop irregular fluffiness with age. Slight pigmentation in shade of pink or yellow. Straight phialides yield ovoid or cylindric conidia that in some species have a slight curve. Many species accumulate conidia in a ball, and a few form delicate chains of elongated conidia.

- ■ *Alternaria* (Figure 23–40): Dematiaceous, rapid-growing fungus. Short conidiophores bear conidia in chains which lengthen in acropetal fashion. Multicelled conidia have angular cross-walls and taper toward the distal end.

- ■ *Aureobasidium*: Moderately rapid-growing, yeast-like fungus. Young cultures are off-white to pink, but with age, many cells become enlarged, thick-walled and very darkly pigmented. Short hyphae may be formed and may give rise to

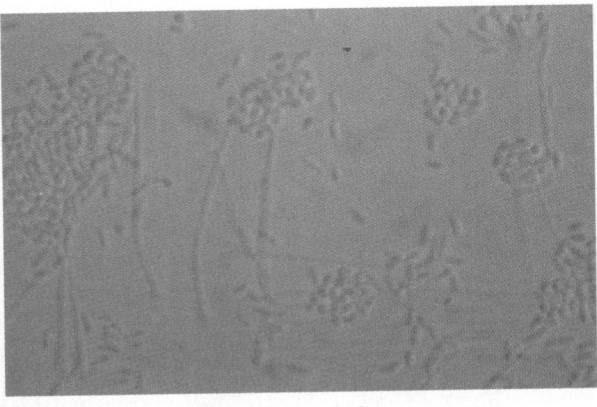

■ F i g u r e   2 3 – 3 9

Acremonium.

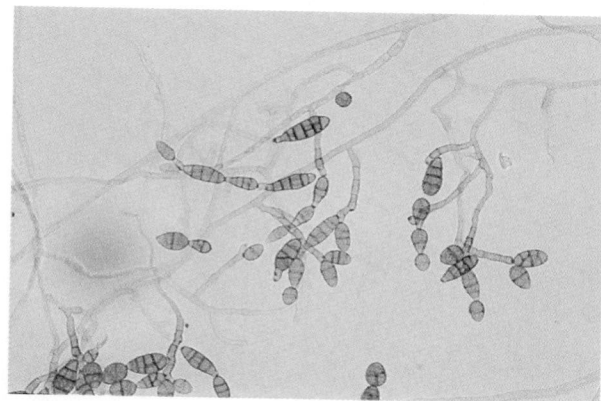

■ Figure 23–40

Alternaria.

buds, both synchronously and asynchronously. Many of these cells become pigmented and usually result in black colonies.

■ ***Aspergillus* (Figure 23–41):** Perhaps the most commonly encountered genus of fungi in the clinical laboratory. Many species exist, and they are differentiated by their conidial and conidiophore morphology and color. The erect conidiophore arises from a "foot cell" within a vegetative hypha and terminates in a swelling or vesicle on which are borne phialides in a distinctive pattern. These produce phialoconidia in long chains. Conidia of some species are separated easily, whereas others remain in chains, these chains may align in very straight, parallel columns. Most of the colony color in *Aspergillus* species lies in the conidia. Color ranges from

black to white, and includes yellow, brown, green, gray, pink, beige, and tan. Some species also form diffusible subsurface pigments on a variety of media.

■ ***Beauveria* (Figure 23–42):** Hyaline, moderately rapid-growing, fluffy colonies, sometimes developing a powdery surface reminiscent of *Trichophyton mentagrophytes.* Abundant, single-celled, tear-shaped sympoduloconidia are formed on sympodulae, which taper extremely from a rather swollen base. Conidiophores may cluster in some isolates to form radial tufts.

■ ***Chaetomium* (Figure 23–43):** Moderately rapid growing, dematiaceous ascomycete. Young colonies may be dirty gray; with age, they develop numerous perithecia, which are ornamented with straight or curled elaters. The asci are evanescent, so at maturity, the pigmented, lemon-shaped ascospores are released within the perithecium.

■ ***Chrysosporium* (Figure 23–44):** Hyaline fungus with moderate growth rate that, with age, may develop light shades of pink, gray, or tan pigment. Simple, wide-based, single-celled conidia are produced on nonspecialized cells. The conidiogenous cell disintegrates or breaks to release the conidia.

■ ***Cladosporium* (Figure 23–45):** Slow-growing to moderately rapid-growing dematiaceous fungus. Brown to olive to black hyphae and conidia. Conidiophores are erect and may branch into

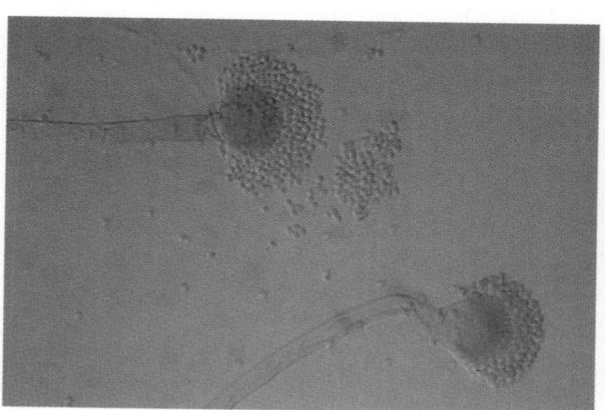

■ Figure 23–41

Aspergillus.

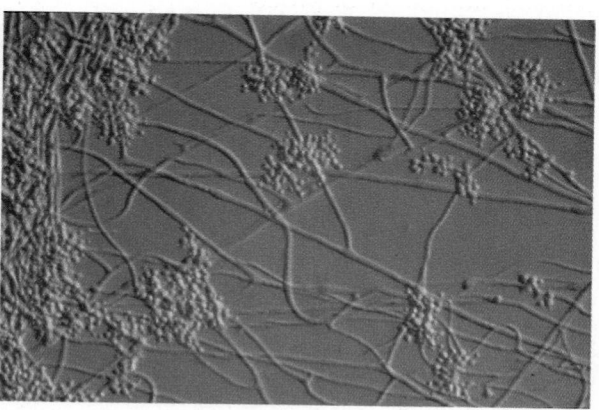

■ Figure 23–42

Beauveria.

■ F i g u r e   2 3 – 4 3

Chaetomium.

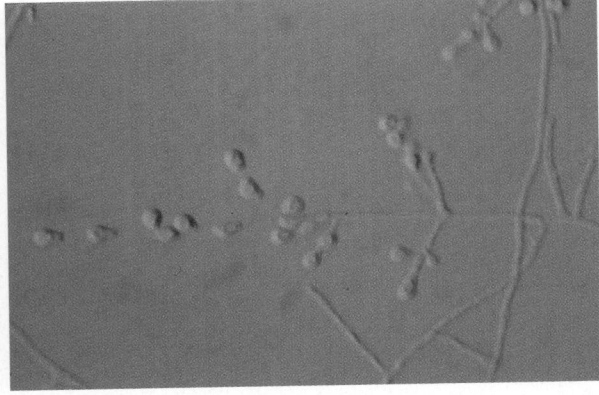

■ F i g u r e   2 3 – 4 4

Chrysosporium.

several conidiogenous cells. Spherical to ovoid conidia form blastically on the end of each previously formed conidium. Branched conidium-bearing cells may dislodge, and the three scars on each of these cells give them somewhat the appearance of a shield. Generally, conidial chains of the saprophytic species break up easily, whereas those of pathogenic species remain connected.

■ ***Cunninghamella* (Figure 23–46):** Rapid-growing zygomycete that forms a cottony colony. Sporangiophores are erect, branching into several vesicles that bear sporangioles. These may be covered with long, fine spines.

■ ***Curvularia* (Figure 23–47):** Rapid-growing dematiaceous fungus that forms a cottony, dirty

gray to black colony. The multicelled conidia are produced on sympodulae. This genus is among the easier to identify because of the frequently crescent-shaped conidia with three to five cells of unequal sizes and usually slight pigmentation differences.

■ ***Epicoccum*:** Moderately rapid-growing dematiaceous fungus with yellow to orange hyphae that give rise to brown to black multicelled conidia in sporodochial clusters. The conidial cross-walls lie in diverse planes.

■ ***Fusarium* (Figure 23–48):** Rapid-growing hyaline fungus that may develop various colors with age, ranging from rose to mauve to purple to yellow. Normally abundant macroconidia and microconidia are produced on vegetative hyphae.

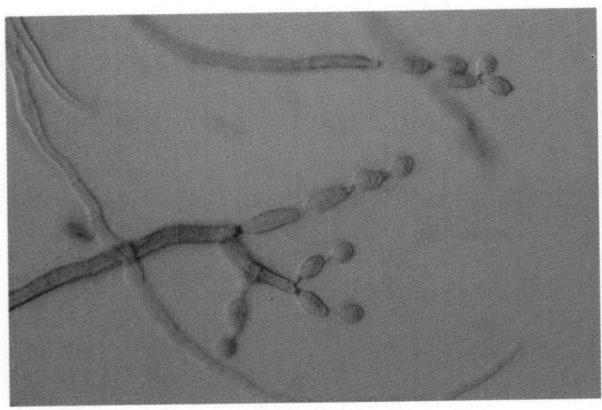

■ F i g u r e   2 3 – 4 5

Cladosporium.

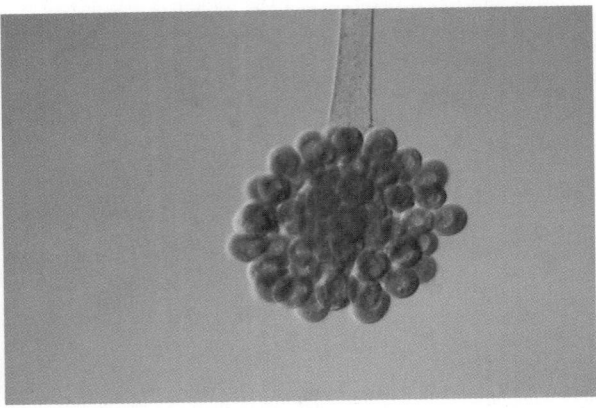

■ F i g u r e   2 3 – 4 6

Cunninghamella.

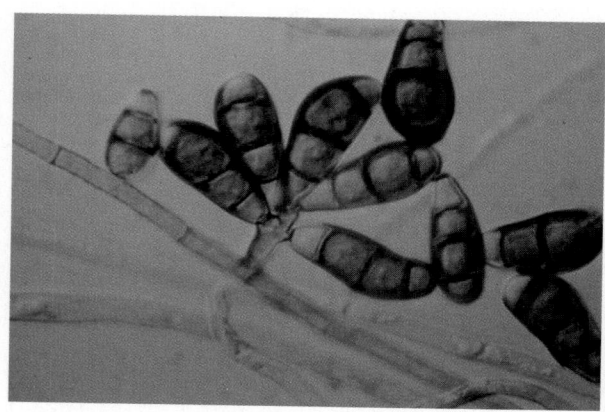

■ F i g u r e   2 3 – 4 7

Curvularia.

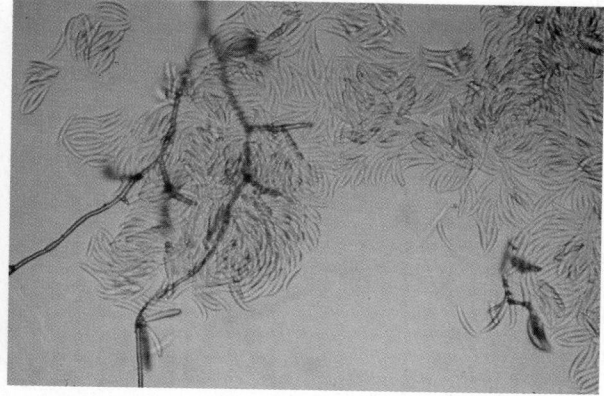

■ F i g u r e   2 3 – 4 8

Fusarium.

These may cluster in sporodochia or may be formed singly. Macroconidia typically are multi-celled and crescent-shaped.

■ *Geotrichum* (Figure 23–49): Rapid-growing, yeast-like, hyaline fungus that forms arthroconidia from vegetative hyphae.

■ *Mucor* (Figure 23–50): Rapid-growing zygomycete that forms cottony, dirty white colonies. Sporangiospores are formed in sporangia on erect sporangiophores. Rhizoids, typical of some zygomycetes, are not found in *Mucor*.

■ *Nigrospora*: Rapid-growing fungus with hyaline hyphae that turn gray to black with age. Conidia are dense, black ovoid cells formed on slightly swollen conidiophores.

■ *Paecilomyces* (Figure 23–51): Rapid-growing, usually very flat colony covered with conidia that are pastel tan, brownish-gold, or lavender. Green or blue-green colors are not seen. Care must be taken to avoid confusion between *Paecilomyces* and *Penicillium* species. Phialides of *Paecilomyces* are generally longer and more obviously tapered, and they may be singly formed or arranged in a verticillate pattern, on which long chains of spindle-shaped or somewhat cylindrical conidia are formed.

■ *Penicillium* (Figure 23–52): Rapid-growing, commonly seen fungus with colonies most often in shades of green or blue-green. Conidiophores are erect, sometimes branched, with metulae bearing one or several phialides on which oval to ovoid conidia are produced in long, loose chains.

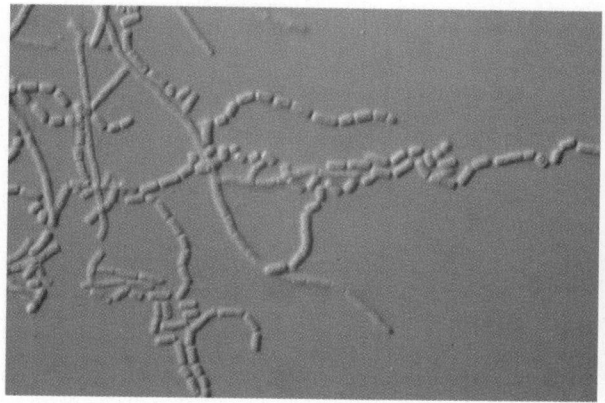

■ F i g u r e   2 3 – 4 9

Geotricum.

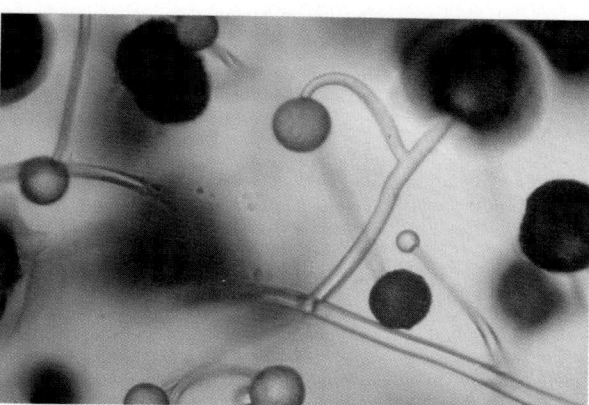

■ F i g u r e   2 3 – 5 0

Mucor.

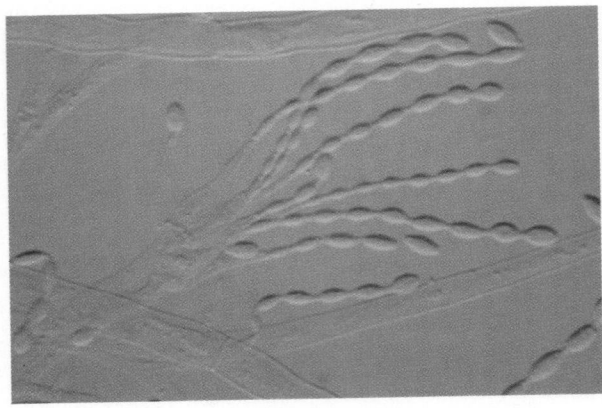

■ Figure 23-51

Paecilomyces.

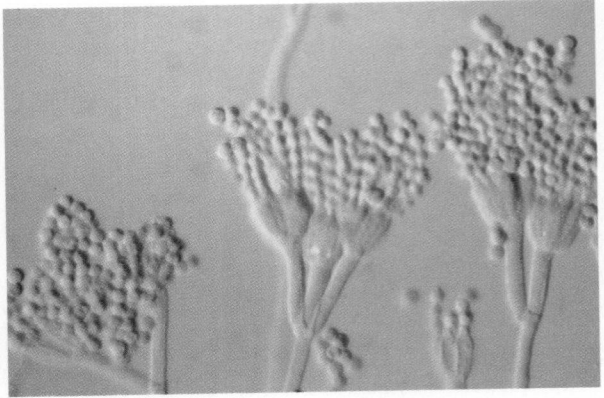

■ Figure 23-52

Penicillium.

■ ***Phoma* (Figure 23–53):** Moderately rapid-growing gray to brown colony that produces pycnidia-organized black fruiting bodies that are globose and lined inside with short conidiophores. Large numbers of hyaline conidia are generated in the pycnidium and flow out a small apical papilla.

■ ***Pithomyces* (Figure 23–54):** Rapid-growing dematiaceous colonies that produce dark, somewhat barrel-shaped conidia singly on simple short conidiophores. Conidia have both transverse and longitudinal cross-walls and often echinulate surfaces.

■ ***Rhizopus* (Figure 23–55):** Rapidly-growing zygomycetous fungus with erect sporangiophores terminated by dark sporangia and spo-

rangiospores. At the base of the sporangiophore are brown rhizoids. Separate clusters of sporangiophores are joined by stolons, arching filaments that terminate at the rhizoids.

■ ***Scopulariopsis* (Figure 23–56):** Moderately rapid-growing colonies covered by tan to buff conidia. The clusters of conidiophores are annellides that increase in length as conidia are formed. The truncate-based conidia tend to remain in chains on the annellides.

■ ***Syncephalastrum* (Figure 23–57):** Rapid-growing zygomycete with erect sporangiophores. Each sporangiophore has a large columella on which merosporangia, containing stacks of sporangiospores, are formed.

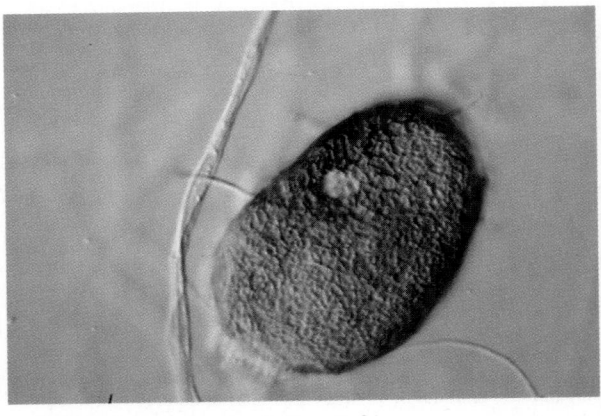

■ Figure 23-53

Phoma.

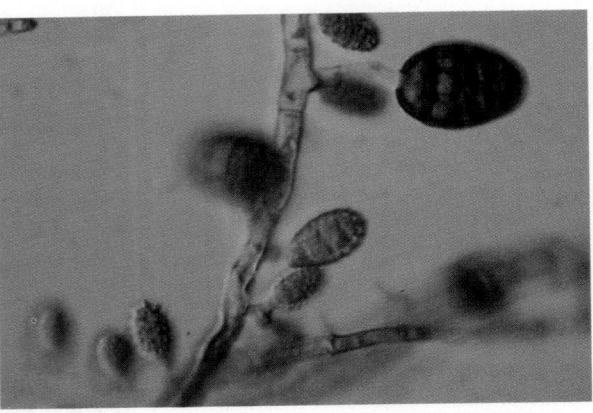

■ Figure 23-54

Pithomyces.

■ *Trichoderma* (Figure 23–58): Rapid-growing hyaline hyphae that give rise to yellow-green to green patches of conidia formed on clusters of tapering phialides. Conidia may remain clustered in balls at the phialide tips.

■ *Ulocladium* (Figure 23–59): Rapid-growing dematiaceous fungus bearing dark, multicelled conidia on sympodulous conidiophores. Conidia have angular cross-walls and, in some species, echinulate surfaces.

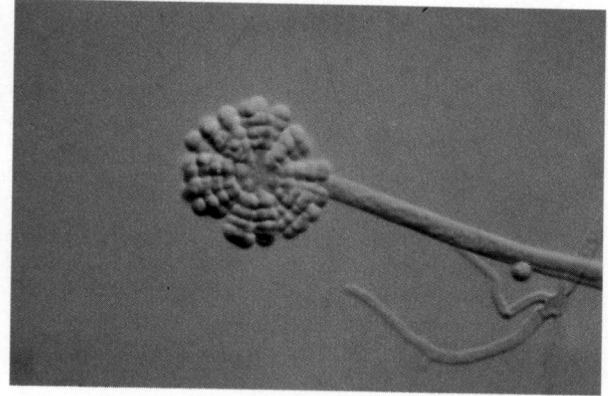

■ Figure 23 – 57

Syncephalastrum.

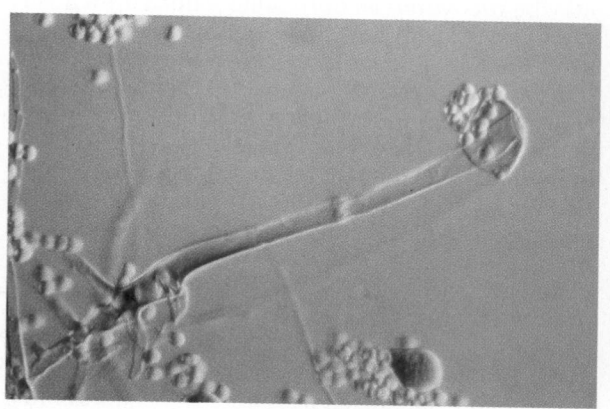

■ Figure 23 – 55

Rhizopus.

■ Figure 23 – 58

Trichoderma.

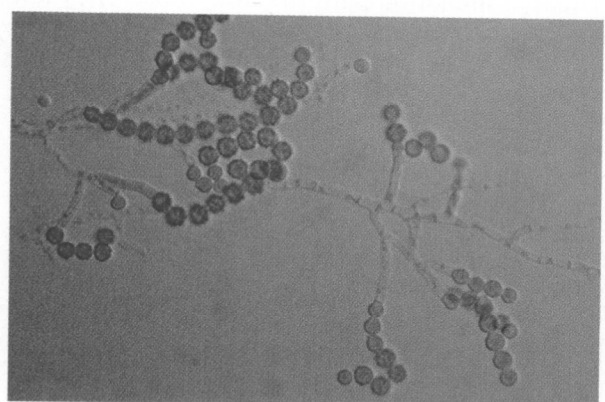

■ Figure 23 – 56

Scopulariopsis.

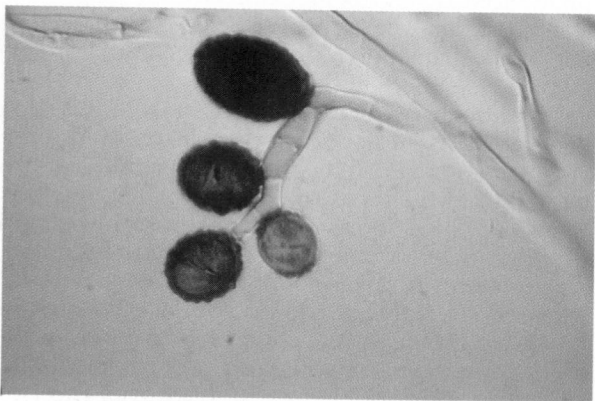

■ Figure 23 – 59

Ulocladium.

AGENTS OF YEAST INFECTIONS

The escalating incidence of yeast and yeast-like fungi isolated from patient specimens has increased the importance of identifying yeast isolates to the species level. With greater immunosuppression, the variety of organisms implicated in disease also expands. *Candida albicans* has become the fourth most common cause of blood-borne infection in the United States today. Isolation of other yeasts from clinical samples, including *Candida tropicalis, Candida parapsilosis,* and *Candida krusei,* is also increasing. Infections caused by these yeasts are extremely aggressive and difficult to treat.

Yeast fungi can be classified in one of two groups—yeasts and yeast-like fungi. Isolates that reproduce sexually, either by forming ascospores or basidiospores, are termed *yeasts.* The majority of isolates that are not capable of sexual reproduction or whose sexual state has not yet been discovered are correctly called *yeastlike fungi.* For ease of discussion, all isolates are referred to here as *yeasts.*

General Characteristics

Molds and yeasts are very different morphologically, but some of the macroscopic characteristics used as aids in identifying molds can also be used to identify yeasts. The most common characteristics noted are color and colony texture. The color of a yeast colony ranges from white to cream or tan, with a few species in the pink to salmon spectrum. Some yeasts isolates referred to as *dematiaceous* yeasts, are darkly pigmented owing to melanin in their cell walls. Dematiaceous yeasts are associated with several species of the polymorphic fungi and are discussed elsewhere in this chapter.

The actual texture of the yeast colonies also varies. For example, *Cryptococcus* spp tend to be very mucoid and may flow across the plate, a trait shared by some bacterial isolates, such as *Klebsiella* spp. Some yeasts are butter-like, and others range in texture from velvety to wrinkled. Strain-to-strain variation in texture may be noted within a species, but the microbiologists should be aware of *phenotypic switching;* this phenomenon is noted when two colony types occur upon subculture. Further DNA testing proves the two types to be the same organism. Switching occurs most often with *T. beigelii* but may also be seen in other isolates.

Clinically Significant Yeast Species

Candida albicans and *Cryptococcus neoformans* are two of the more widely recognized clinically significant yeasts. A number of other organisms, however, including other *Candida* spp, *Rhodotorula,* and *Torulopsis* sp, have also been implicated in clinical infections.

Candida Species

Candida spp not only are commonly present as normal flora of the mucosa, skin, and digestive tract but also are the most notorious agents of yeast infection. Clinical disease ranges from superficial skin infections to disseminated disease.

Candida albicans currently reigns as the premier cause of yeast infection in the world. This isolate may be recovered as normal host flora from a variety of sites, including skin, oral mucosa, and vagina. When host conditions are altered, however, this isolate is capable of causing disease in virtually any site. One of the most widely recognized manifestations of *C. albicans* infection is thrush. In individuals with an intact immune system, infections are localized and limited. Thrush is also recognized as an indicator of immunosuppression. Among individuals infected with human immunodeficiency virus (HIV) as well as those receiving prolonged antimicrobial therapy or other chemotherapeutic agents, thrush manifests as a serious and, in some cases, disseminated infection.

Candida tropicalis is probably the second most common *Candida* species to incite disease. Infections associated with *C. tropicalis* tend to be aggressive and difficult to treat with traditional antifungal therapy. This organism has different sugar assimilation patterns from those of *C. albicans* and, therefore, can easily be differentiated.

Other notable species of *Candida* are *C. parapsilosis, C. guilliermondi* and *C. lusitaniae. Candida parapsilosis* has become a major cause of hospital outbreaks of nosocomial infections. This organism, like *C. tropicalis,* is refractory to traditional antifungal therapy. These two isolates are identified by the differences in their carbohydrate assimilation patterns and other secondary testing procedures. Table 23–13 shows important differentiating characteristics among *Candida* species and other yeasts.

Cryptococcus Species

Cryptococcus spp are important causes of meningitis and pulmonary disease. *Cryptococcus*

■■■■■ T A B L E   2 3 – 1 3

DIFFERENTIATING CHARACTERISTICS OF YEAST ISOLATES*

	Temperature Growth at			Cornmeal Agar			Cyclohexamide	Urea	Nitrate
	37°C	42°C	45°C	Pseudohyphae	True Hyphae	Arthroconidia			
Candida									
C. albicans	+	+	+	+	+	–	+	–	–
C. guilliermondi	+	+	–	+	–	–	+	–	–
C. krusei	+	+	–	+	–	–	+	–	–
C. lusitaniae	+	+	+	+	–	–	–	v	–
C. parapsilosis	+	–	–	+	–	–	v	–	–
C. stellatoidea	+	+	+	+	+	–	–	–	–
C. tropicalis	+	+	+	+	–	–	v	–	–
Cryptococcus									
C. albidus	–	–	–	–	–	–	–	+	+
C. neoformans	+	–	–	–	–	–	–	+	+
Trichosporon beigelii	+	v	–	+	+	+	–	+	–
Torulopsis glabrata	+	+	+	–	–	–	–	–	–

*+, positive; –, negative; v, variable.

neoformans, the most noted pathogen in this group, has become one of the major causes of opportunistic infection in AIDS patients. The organism is commonly found in soil contaminated with pigeon droppings and is most likely inhaled prior to clinical infections. Common manifestations include meningitis, pneumonia, and bacteremia.

Cryptococcus spp are surrounded by a capsule that produces the characteristic mucoid colonial appearance. The capsule can be detected surrounding the budding yeast in spinal fluid with the aid of India ink (see Fig. 23–11). The ink stains the CSF while leaving clear halos around individual yeast cells. The use of India ink preparation is being replaced by the latex agglutination test for cryptococcal antigen, because of the former's low sensitivity and rate of detection. The latex agglutination test is being recommended for routine use in most clinical microbiology laboratories.

Cryptococcus spp are noted for not producing true hyphae or pseudohyphae on cornmeal agar. Although all species of the genus are urease positive, nitrate reaction varies. Production of phenol oxidase is a feature differentiating *C. neoformans* from other *Cryptococcus* species. Sugar assimilations also vary for each species. *Cryptococcus neoformans* may be differentiated by using the characteristics described in Table 23–13.

Rhodotorula Species

Rhodotorula spp are noted for their bright, salmon-pink color. They are closely related to the cryptococci, in that they bear a capsule and are urease positive. Some species are also nitrate positive. They are not common agents of disease but have been known to cause opportunistic infection.

Torulopsis glabrata

Torulopsis glabrata is another aggressive, somewhat refractory agent in yeast infection. This organism was at one time placed in the genus *Candida,* but it was removed because it lacked the ability to form either true hyphae or pseudohyphae on cornmeal agar. Many taxonomists believe that this criterion was not sufficient for creating a new taxon and therefore still refer to this organism as *Candida glabrata.* In this chapter, *Torulopsis* stands as correct taxonomy, because hyphal production is an important feature in yeast identification and should be considered critical. This organism is capable of assimilating only glucose and trehalose. It is negative for all secondary testing.

Trichosporon beigelii

Trichosporon beigelii is most commonly regarded as a cause of white piedra—a yeast overgrowth on the hair shafts of individuals lacking proper personal hygiene. It is also an emerging agent of disseminated infection that has occurred in major outbreaks among cancer patients. Treatment is complicated by the fact that *T. beigelii* tends to be resistant to amphotericin

B (AMB), the choice for treatment of life-threatening fungal infection. *Trichosporon* sp are noted for their production of arthroconidia as well as blastoconidia on cornmeal agar.

Malassezia furfur

Malassezia furfur, a lipophilic organism, requires long-chain fatty acids to live. It is a common skin colonizer and, when present, may cause a condition known as tinea versicolor—as hyperpigmented or hypopigmented areas on the skin. This condition usually causes no discomfort but is cosmetically undesirable. *Malassezia furfur* entered the infectious diseases realm when major outbreaks of infection were noted in neonatal intensive care units where premature infants were receiving high-lipid formula intravenously. The organism, presumably originating from the skin, gains access via catheters, thrives in the lipid rich formula, and ultimately causes fungemia. Infection subsides when the feeding lines are removed, and treatment is not usually necessary. Additional reports have linked *Malassezia furfur* to similar syndromes in anorexic patients receiving high-lipid diets through intravenous lines. This organism should be considered when yeasts are seen in blood culture bottles or culture material but no organisms are recovered. The previously mentioned technique of overlaying streaked culture plates with olive oil provides the lipids required, and growth should be detected in 3 to 5 days.

Methods of Yeast Identification

Tests used in the identification of yeast range from simple tests, such as production of germ tubes, urease, and characteristic structures on cornmeal agar, to carbohydrate assimilation.

Germ Tube Production

The germ tube test is probably the most basic and easiest test to perform for identification of yeasts. Figure 23–60 shows a schematic diagram of how the germ tube test may be used to presumptively identify yeasts species. *Candida albicans* is identified by its germ tube production (Fig. 23–61). The standard procedure requires the use of serum or plasma. Expired fresh-frozen plasma, negative for both hepatitis B and HIV, from the blood bank is useful in this test and can be stored at 4°C indefinitely. Many other liquid media (e.g., brain heart infusion, trypticase soy broth, or nutrient broth) have been used successfully as alternative media. One colony of the yeast culture is added to 0.5 mL of serum or plasma, incubated for 2.5 to 3 hours at 35°C, and viewed microscopically at the end of the incubation period for germ tube production. Care must

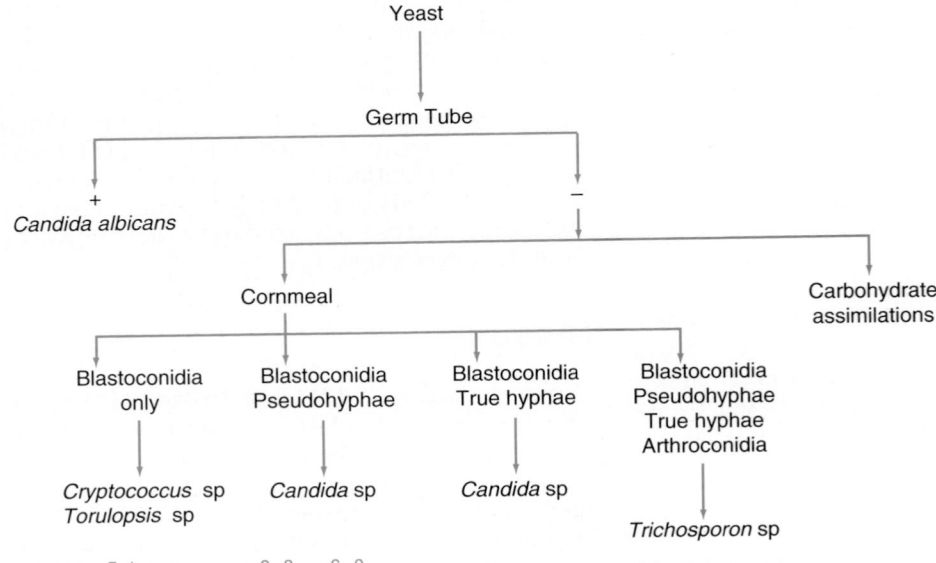

■■■■ F i g u r e　2 3 – 6 0

Schematic diagram showing how germ tube test may be used to presumptively identify yeasts.

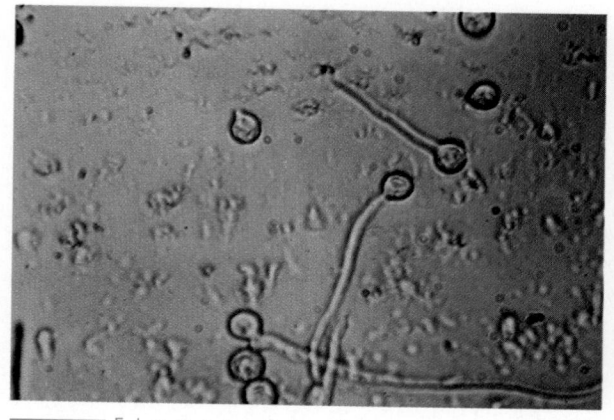

■ F i g u r e   2 3 – 6 1

Germ tube production by *Candida albicans*. A positive germ tube has no constriction at its base.

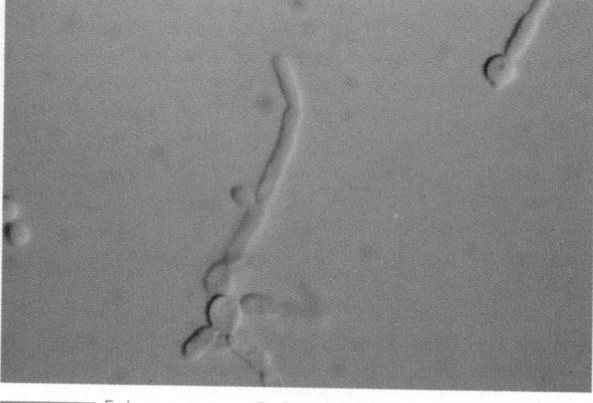

■ F i g u r e   2 3 – 6 2

Candida tropicalis shows constriction at the base of the germ tube.

be taken not to over-incubate the tube, because other agents are capable of forming germ tubes with extended incubation.

After the incubation period, an accurate identification of *C. albicans* can be made when true germ tubes are present. True germ tubes lack constriction at their bases, where they attach to the mother cell. If a constriction is present at the base of a germ tube, the yeast is not *C. albicans*. Such constricted germ tubes are more in keeping with *C. tropicalis* (Fig. 23–62). *Candida stellatoidea* is also capable of germ tube production, but many taxonomists believe that this organism is more correctly classified as a variety of *C. albicans*, which is not capable of assimilating sucrose.

Negative germ tube results should lead the technologist to other procedures for identification. It should be noted that only rarely do *C. albicans* isolates yield a negative germ tube test result.

Positive and negative controls should be set up in conjunction with all testing. A known germ tube–positive isolate of *C. albicans* can serve as the positive control; *Cryptococcus* sp works well as a negative control.

Carbohydrate Assimilation

Sugar fermentation tests, although valuable, are time and labor intensive, thus making them impractical for the routine microbiology laboratory. Carbohydrate assimilation tests, however, can be readily performed as part of the routine bench procedures. Assimilation tests identify which carbohydrates a yeast can utilize as a sole source of carbon. Assimilation patterns may be determined from methods as sophisticated as the automated identification system or as simple as the various manual procedures and commercial kits. The individual laboratory should adopt the method that can be practically implemented into its particular working environment.

API 20C

Although many such kits are available for yeast identification, the API 20C yeast identification system remains the gold standard for assimilation testing. In this method, a series of freeze-dried sugars are placed into wells on a plastic strip. Unknown yeast isolates are suspended in an agar basal medium, pipetted into the wells, and incubated at 30°C for 72 hours. As sugars are assimilated, the wells become turbid with growth. Wells remain clear when the sugar is not assimilated. A code is derived from the assimilation patterns and matched against a computerized database. Identifications are accompanied by a percentage, which indicates the probability that the identification is correct. Although this test is very reliable, other auxiliary testing should accompany assimilation results before a final identification is made.

Automated systems are also available for yeast identification. Many of these systems use enzyme reactions as well as assimilation reactions to aid in yeast identification.

Cornmeal Agar Morphology

Yeast morphology on cornmeal agar is second in importance to sugar assimilations in determining proper yeast identification. A cornmeal agar plate is inoculated with two parallel 1-cm streaks. The streaks are then covered with a coverslip and allowed to incubate at room temperature for 48 hours. The Petri dish can be placed on the microscope stage and viewed with a 20× objective for production of hyphae, pseudohyphae, arthroconidia, or blastoconidia.

Recognition of one of these four different types of morphology is a very important clue to yeast identification. Blastoconidia are the characteristic budding yeast forms most often seen on direct mounts. *Candida albicans* produces chlamydoconidia along with hyphae, as shown in Figure 23–63. Pseudohyphae (Fig. 23–64) occur when the blastoconidia germinate to form a filamentous mat. The cross-walls help determine whether the structures are true hyphae or pseudohyphae. Cross-walls of pseudohyphae are constrictions, not true septations, whereas true hyphae remain parallel at cross-walls, with no indentation. The fourth morphology type is arthroconidia. These begin as true hyphae but break apart at the cross-walls with maturity. Rectangular fragments of hyphae should be accompanied by blastoconidia in order for an isolate to be considered a yeast.

Potassium Nitrate Assimilation

Potassium nitrate assimilation patterns provide additional valuable information for separating the clinically significant yeasts. Use of the modified KNO₃ agar described by Pincus and colleagues (1988) is a fairly rapid, easy, and accu-

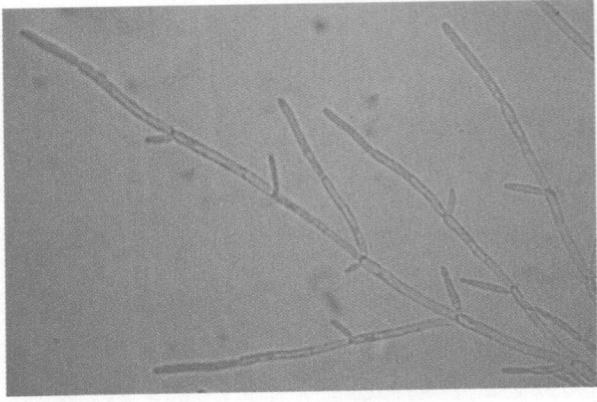

■ Figure 23–64

Pseudohyphae occur when the blastoconidia germinate and form a filamentous mat.

rate method to determine nitrate assimilations. A positive KNO_3 assimilation result turns the agar medium blue, and a negative result turns the medium yellow. Control organisms that may be used are *Cryptococcus albidus* (positive) and *C. albicans* (negative).

Urease Test

Yeast isolates producing the enzyme urease can easily be detected with a simple urea agar. This fairly rapid, easily read test aids in differentiation between *Cryptococcus* and *Rhodotorula* species. Media for detecting urea hydrolysis may be obtained commercially. Positive test results turn the media bright fuchsia pink, whereas negative results cause little if any change. *Cryptococcus albidus* can be used as a positive control, and *C. albicans* as a negative control.

Temperature

Temperature studies also offer additional information for yeast identification. *Cryptococcus* spp have weak growth at 35°C and no growth at 42°C. Several *Candida* spp have the ability to grow well at temperatures as high as 45°C.

Any one of these tests alone, with the exception of the germ tube test, is not sufficient for proper identification, but when they are used in concert, proper identification is often easily accomplished. Armed with the foregoing procedures, the technologist should be able to identify the most commonly encountered yeast isolates.

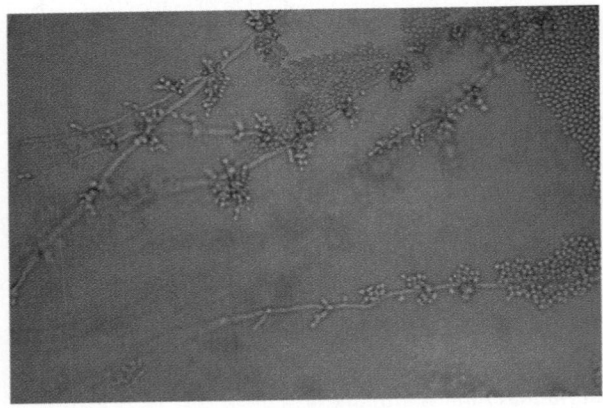

■ Figure 23–63

Candida albicans on cornmeal agar showing typical chlamydospores.

Bibliography

Ajello L (ed): *Coccidioidomycosis:* Current Clinical and Diagnostic Status. New York: Stratton Intercontinental Medical Book Corp, 1977.

Ajello L, Georg LK, et al: A case of phaeohyphomycosis caused by a new species of *Phialophora*. Mycologia 66:490, 1974.

Bonner JR, Alexander WJ, et al: Disseminated histoplasmosis in patients with acquired immune deficiency syndrome. Arch Intern Med 144:2178, 1984.

Cooper BH, Silva-Hutner M: Yeast of medical importance. In Lennette EH, Balows A, et al (eds): Manual of Clinical Microbiology, 4th ed. Washington, DC: American Society for Microbiology, 1985.

Denton JF, DiSalvo AF: Isolation of *Blastomyces dermatitidis* in soil associated with a large outbreak of blastomycosis in Wisconsin. N Engl J Med 31:529, 1964.

Denton FF, DiSalvo AF: Additional isolations of *Blastomycosis dermatitidis* from natural sites. Am J Trop Med Hyg 28:697, 1979.

Einstein HE, Catanzaro A: Coccidioidomycosis. Proc. 4th Int. Conf., Washington, DC, National Foundation of Infectious Diseases, 1985.

Ellis MB: Dematiaceous Hyphomycetes. Kew, Surrey, England: Commonwealth Mycological Institute, 1971.

Goodman NL, Larsh HW: Environmental factors and growth of *Histoplasma capsulatum* in soil. Mycopathol Mycol Appl 33:145, 1967.

Greer DL, Restrepo, A: The epidemiology of paracoccidioidomycosis. In Al-Doory Y (ed): The Epidemiology of Human Mycotic Diseases. Springfield, IL: Charles C Thomas, 1975.

Holdane DJ, Robart E: A comparison of calcofluor white, potassium hydroxide and culture for the laboratory diagnosis of superficial fungal infections. Diagn Microbiol Infect Dis 13:337, 1990.

Huppert M: Serology of Coccidioidomycosis. Mycopathologia 41:107, 1970.

Kaufman L, Standard PG: Improved version of the exoantigen test for identification of *Coccidioides immitis* and *Histoplasma capsulatum* cultures. J Clin Microbiol 8:42, 1978.

Kaufman L, Standard PG: Specific and rapid identification of medically important fungi by exoantigen detection. Annu Rev Microbiol 41:209, 1987.

Kedes LH, Siemski J, Braude AI: The syndrome of the alcoholic rose gardener: Sporotrichosis of radial tendon sheath-report of a case with amphotericin B. Ann Intern Med 61:1139, 1964.

Klein BS, Vergeront JM, et al: Isolation of *Blastomyces dermatitidis* in soil associated with a large outbreak of blastomycosis in Wisconsin. N Engl J Med 31:529, 1986.

Kwon-Chung KJ, Bennett JE: Medical Mycology. Malvern, PA: Lea & Febiger, 1992.

Manos NE, Ferebee SH, Kerschbaum WF: Geographic variation in the prevalence of histoplasmin sensitivity. Dis Chest 29:649, 1956.

Matsumoto T, Padhye AA, et al: Critical review of human isolates of *Wangiella dermatitidis*. Mycologia 76:232, 1984.

McDonough ES, Lewis AL: *Blastomyces dermatitidis:* Production of the sexual stage. Science 156:528, 1969.

McGinnis MR: Chromoblastomycosis and phaeohyphomycosis: New concepts, diagnosis and mycology. *J AM Acad Dermatol* 8:1, 1983.

McGinnis MR: Laboratory Handbook of Medical Mycology. New York: Academic Press, 1980.

McManus EJ, Jones JM: Detection of a *Trichosporon beigelii* antigen cross-reactive with *Cryptococcus neoformans* capsular polysaccharide in serum from a patient with disseminated *Trichosporon* infection. J Clin Microbiol 21:681, 1985.

Pincus DH, Salkin IF, et al: Modification of potassium nitrate assimilation test for identification of clinically important yeasts. J Clin Microbiol 26:366, 1988.

Rebell G, Taplin D: Dermatophytes, Their Recognition and Identification. Coral Gables, FL: University of Miami Press, 1970.

Restrepo MA, Robledo M, et al: The gamut of paracoccidioidomycosis. Am J Med 61:33, 1976.

Rippon JW: Medical Mycology: The Pathogenic Fungi And The Actinomycetes, 3rd ed. Philadelphia: WB Saunders, 1988.

Sekhon AS, Bogorous MS, et al: Blastomycosis: Report of three cases from Alberta and a review of Canadian cases. *Mycopathologia* 65:53, 1979.

Stevens D: Coccidioidomycosis: A Test. New York: Plenum Press, 1981.

Walsh TJ, Melcher GP, et al: Disseminated trichosporonosis resistant to amphotericin B. J Clin Microbiol 28:1616, 1990.

Wheat LJ, French MLV, et al: The diagnostic laboratory tests for histoplasmosis: Analysis of experience in a large urban outbreak. Ann Intern Med 97:680, 1982.

Diagnostic Parasitology

Linda A. Smith

1. Cite the major considerations in collection and handling of specimens for identification of intestinal and blood and tissue parasites.

2. Describe the general procedure for performing the direct wet mount, concentration procedures, and permanent stained smears.

3. List the stages of parasites found with each of the following: direct wet mount, concentration, permanent stained smears.

4. Identify the general characteristics of major phyla of parasites.

5. For the major human pathogens, describe mechanism of pathogenesis, method of infection, clinical symptoms, prevention, and treatment.

6. For each organism, describe morphology, life cycle including infective stage and diagnostic stage, and usual procedure for identification.

Parasites have always contributed to human morbidity and mortality. In the United States and other developed countries, however, they are not often regarded as major causes of disease. In the last few years, a number of factors have contributed to greater awareness of the importance of considering a parasite as the underlying etiologic agent for a patient's clinical condition in this country. These factors include the increasing number of immunocompromised patients who are susceptible to infections with known pathogens as well as with opportunistic organisms, the higher numbers of persons who travel to countries with less than ideal sanitation and a large number of endemic parasites, and the larger population of immigrants from areas with endemic parasites.

When a clinician is confronted with an infection that may be due to an intestinal or blood and tissue parasite, the patient's symptoms and clinical history, including travel, are significant data to be gathered and shared with the clinical laboratory scientist. The laboratorian and clinician should collaborate to make sure that the appropriate specimen is properly collected and handled prior to and during the clinical work-up. Knowledge of common pathogens as well as nonpathogens that exist in specific geographical regions and for a given body site is necessary to ensure identification and, if necessary, therapy.

Parasitic infections can be difficult to diagnose because patients present with nonspecific clinical symptoms that can be attributed to a number of disease agents. Detection and identification of a parasite depend not only on the adequacy of the submitted specimen but also on procedures established by the clinical laboratory, including criteria for specimen collection, handling and transport, and laboratory methods utilized. This chapter:

■ Presents a general overview of sample collection, handling and transport, and quality assurance

■ Describes procedures such as the preparation of blood films, wet mounts, concentration methods, and staining methods

■ Discusses the major medically important parasites, their epidemiology and life cycle, and the clinical infections they produce

■ Presents the diagnostic features that characterize these agents and provide differentiation from nonpathogens

Readers are referred to standard parasitology references for detailed procedures, reagent preparation, and a comprehensive description of parasites that have been implicated in human disease.

GENERAL CONCEPTS IN PARASITOLOGY LABORATORY METHODS

Fecal Specimens

Collection, Handling, and Transport

A single stool specimen may not be sufficient to isolate an intestinal parasite, because many intestinal organisms shed eggs or cysts on an irregular schedule. It has traditionally been recommended that for optimal detection of intestinal parasites, a series of three stool specimens should be collected within 10 days but spaced a day or two apart. This procedure of examining each specimen submitted is very time consuming and labor-intensive. Recent articles suggest that in some circumstances, pooling of the three formalin-preserved specimens gives a recovery rate comparable to that of individual examination of formalin-preserved stools. In addition, whether the cases are inpatient or outpatient and the presence of symptoms should dictate whether three specimens are needed.

The appropriate collection container is clean, dry, and waterproof, such as a half-pint waxed cardboard or plastic container with lid. Commercial systems that incorporate collection container and preservatives are also available. Stool specimens should never be collected from bedpans or the toilet bowl; such a practice might contaminate the specimen with urine or water, resulting in the destruction of trophozoites or introduction of free-living protozoa. As an alternative, the specimen should be collected on a clean piece of waxed paper or newspaper and transferred to the container. Another alternative is to use one of the disposable collection containers that can be fitted under the toilet bowl rim. The specimen should be submitted as soon as possible after passage. Identification noted on the container should include the patient name and the date and time the stool was collected. The laboratory requisition should include the same information and any additional pertinent clinical data, such as the suspected diagnosis.

Stool specimens for parasites should be collected before a barium series or before the start of antimicrobial therapy. Antimicrobials can decrease the number of organisms present. If the patient has undergone a barium series, stool examination should be delayed for 7 to 10 days, because barium obscures organisms. If a purged specimen is to be collected, it is recommended that a saline or phosphosoda purgative be used, because mineral oil droplets will interfere with identification of parasites, especially protozoan cysts. The second or third specimen after the purge is more likely to contain trophozoites that inhabit the cecum.

Preservation

Several methods are available for stool preservation if the specimen will not be delivered immediately to the laboratory. Choice of a preservative is usually governed by the procedure to be performed. Regardless of preservative used, the ratio of three parts fixative to one part feces should be maintained for optimal fixation. The time that the stool was passed and the time that it was placed in the fixative should be written on the laboratory requisition and the container. A commercially available two-vial system using polyvinyl alcohol (PVA) fixative in one vial and 10% formalin in the other vial is the most commonly used system. The system comes with patient instructions and a self-sealing plastic bag for transport. PVA fixative, which consists of mercuric chloride (for fixation) and polyvinyl alcohol (a resin to increase adhesion of the stool to the slide), is used when a permanently stained smear will be made. Concern about disposal of hazardous mercury compounds has led to development and evaluation of a zinc sulfate–based compound for PVA. Evaluation of stains made from stools preserved with the zinc-based compound showed a less sharp morphology but overall good agreement with identifications made from trichrome-stained smears using traditional PVA preservative.

Formalin (10%) can be used when either a wet amount or concentration procedure (either sedimentation or flotation) will be performed. The fixative sodium acetate–acetic acid formalin (SAF) can be used for preserving fecal specimens when both concentration procedures and permanent stains will be used. Vials of Merthiolate-iodine-formalin (MIF) can also be used to preserve trophozoites, cysts, larvae, and helminth eggs for wet mount or concentration procedures. This preservative is not routinely used for permanently stained smears.

Examination of the Fecal Specimen

Any stool specimen submitted to the microbiology laboratory should be carefully handled because it is a potential source of infection. Specimens should be opened and handled in designated areas that provide protection for the technologist.

MACROSCOPIC EXAMINATION

The examination of an unpreserved stool specimen should include macroscopic (gross) as well as microscopic procedures. The initial laboratory procedure is the macroscopic examination. During gross examination, intact worms or proglottids may be identified on the surface of the stool.

Gross examination of the specimen reveals the consistency (liquid, soft, formed) of the stool sample. Consistency may help determine the type of preservation to be used, may indicate the forms of parasites that may be expected to be present, or may dictate the immediacy of examination. Figure 24–1 shows the relationship between stool consistency and protozoan stage. For example, a soft or liquid stool specimen or a purged specimen primarily contains motile protozoan trophozoites; hence, purged specimens should be examined immediately after passage. Soft or liquid specimens should be examined within one-half hour of passage to ensure motility of the organisms. If examination will be delayed, a portion should also be placed into a fixative such as PVA, so that permanent stained smears for definitive identification can be made.

Gross examination also reveals the color of the stool specimen. A normal stool sample usually appears brown. Stool that appears black may indicate bleeding in the upper gastrointestinal tract, whereas the presence of fresh blood may indicate bleeding in the lower portion of the intestinal tract. Any portion of the stool that contains blood or blood-tinged mucus should be selected for wet mount preparations and be placed in preservative.

A formed stool specimen should be examined within 2 to 3 hours of passage if held at room temperature; however, examination may be delayed up of 24 hours after passage if the specimen is placed in the refrigerator. A portion of the formed stool should be placed into formalin for concentration procedures as well as into PVA or permanently stained smears. The specimen should not be placed into a 37°C incubator, which would increase the rate of disintegration of organisms present and enhance overgrowth by bacteria.

MICROSCOPE EXAMINATION

Several diagnostic methods can be used in the microscopic examination of a fecal specimen:

■ Direct wet-mount examination (stained and unstained) of fresh stool specimens

■ Concentration procedures with wet-mount examination of the concentrate

■ Preparation of permanently stained smears

In general, the concentration and permanent staining procedures should be performed on all specimens.

Direct Wet Mount. The direct wet mount of unpreserved fecal material is primarily used to detect the presence of motile protozoan trophozoites in a fresh liquid or soft stool or from sigmoidoscopy material. A direct wet mount of a formalin-preserved stool specimen or a formed stool specimen may demonstrate helminth eggs or larvae and protozoan cysts. Because of the low diagnostic yield and labor-intensiveness of a wet mount from stool specimens, however, it has been suggested that routine use of this practice be discontinued on formed specimens.

The direct wet-mount procedure uses a 3 × 2-inch glass slide on which a drop of physiologic saline (0.85%) has been placed at one end and a drop of iodine (Dobell and O'Connor solution, D'Antoni solution, or a 1:5 dilution of Lugol solution) at the other end. A small amount (2 mg) of feces is added to each drop and mixed well. Each preparation should be covered with a No. 1, 22-mm square coverslip. The preparation should be thin enough so that newsprint can be read through it and should not overflow beyond the edges of the coverslip. If the specimen has been preserved in 10% formalin, the drop of saline may be omitted from the unstained preparation.

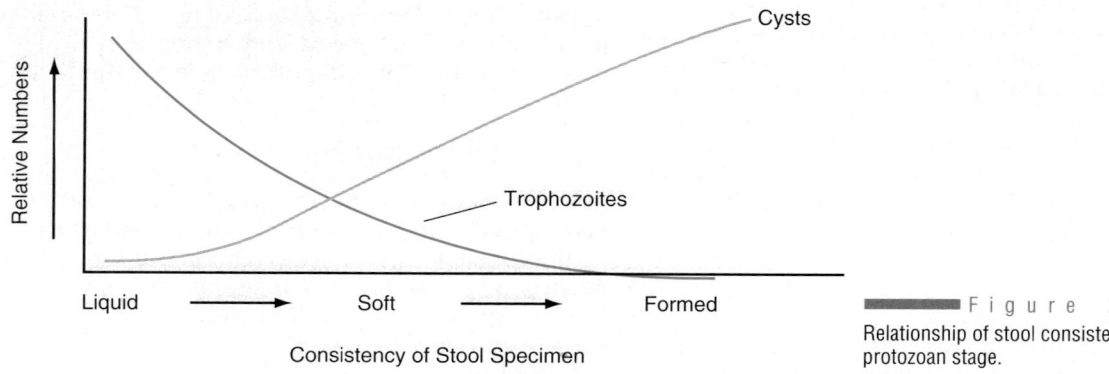

■■■■■ F i g u r e   2 4 – 1
Relationship of stool consistency to protozoan stage.

PVA-preserved specimens are not acceptable for wet mounts, because the PVA becomes cloudy when exposed to air.

The saline preparation is useful for detection of helminth eggs or larvae and refractile protozoan cysts. Iodine emphasizes nuclear detail and glycogen masses. Stains such as buffered methylene blue have been used to enhance nuclear morphology in trophozoites but may inhibit motility and cause the organism to round up.

Reading the wet mount involves thorough examination of each coverslipped preparation at low power, starting at one corner and following a systematic vertical or horizontal pattern until the entire preparation has been examined. A high-power objective is used to identify any suspicious structures. Oil immersion should not be used on a wet preparation unless the preparation has been sealed with either clear nail polish or Vaspar (50-50 mixture of petroleum jelly and paraffin). Sealing the preparation also prevents rapid drying out and allows further examination.

Concentration Techniques. Concentration techniques are designed to concentrate the parasites present into a small volume of fluid and remove as much debris as possible. Either fresh or formalin-preserved stool specimens may be used in the procedure. The concentrate may then be examined either unstained or stained with iodine. Protozoan trophozoites do not survive the procedure; protozoan cysts, helminth larvae, and helminth eggs are usually detected using this method.

Sedimentation and flotation methods, both of which are based on the difference in specific gravity between the parasites and the concentrating solution, are used to concentrate parasites into a small volume for easier detection. In sedimentation methods, the organisms are concentrated in sediment at the bottom of the centrifuge tube. In flotation methods, the organisms are suspended at the top of a high-density fluid. Overall, sedimentation methods concentrate a greater diversity of organisms, including cysts, larvae, and eggs.

The formalin-ether method, once the classic sedimentation procedure, has been replaced by the formalin–ethyl acetate (FEA) method to avoid the safety hazards associated with use of ether. A number of manufacturers now market self-contained fecal concentration kits. Although these kits offer the advantage of disposability and a cleaner preparation owing to the filtration used, initial studies indicate that their use is more expensive and time consuming than the traditional formalin-ether technique.

The zinc sulfate method is the usual flotation procedure. Although the zinc sulfate method yields less fecal debris in the finished preparation than the FEA method, the zinc sulfate causes operculated eggs to open or collapse. It also tends to distort protozoan cysts. Infertile *Ascaris lumbricoides* eggs and *Schistosoma* sp eggs may be missed if this procedure is used. Owing to their high density, these eggs sink to the bottom of the tube. Most organisms also tend to settle after about 30 minutes. Therefore, the examination should be made as soon as possible after the procedure has been completed to ensure optimum recovery of organisms.

Special flotation procedures, such as the Sheather sugar flotation, have been used in detection of specific organisms such as *Cryptosporidium* sp. Although oocysts of *Cryptosporidium* sp can be detected with either the formalin–ethyl acetate or the zinc sulfate method, Sheather flotation allows for better visibility of the oocysts because of the greater refractility of the oocyst against the background solution. Fresh or formalin-fixed feces can be used in this procedure. A new concentration method that involves laying sediment from a FEA concentration procedure over saturated sodium chloride has been described. This method increases separation of oocysts from fecal debris and enhances detection. It has particular application for oocysts of organisms such as *Cryptosporidium* spp.

Permanently Stained Smears. Permanently stained smear preparations should be made of all stool specimens to detect and identify protozoan trophozoites and cysts. Characteristics needed for identification of the protozoa, including nuclear detail, size, and internal structures, are visible in a well-made and properly stained smear. Permanent stains commonly used include iron hematoxylin and trichrome (Wheatley modification of the Gomori stain). The stain of choice in most laboratories is the trichrome stain, because results are somewhat less dependent on technique and the procedure is less time consuming. A trichrome stain can be performed on a smear made from a fresh stool specimen fixed in Schaudinn fixative or from one that has been preserved in PVA. Although laboratories have traditionally prepared the stain in house, some manufacturers now provide prepared, pre-packaged stains and reagents for this procedure. Specimens preserved in SAF do not stain well with trichrome and should be stained with iron hematoxylin.

To prepare a trichrome stained smear on a fresh specimen, applicator sticks are used to smear a thin film of stool across a 1 × 3-inch slide, with care taken to ensure that the stool extends to the sides of the slide. The smear is placed immediately in Schaudinn fixative; it must not be allowed to dry before fixation. For PVA-fixed specimens, several drops of specimen are placed on a paper towel to drain excess fluid; the material on the paper towel is collected to prepare the smear in the same way as for a fresh specimen. This specimen is allowed to air dry thoroughly before staining.

In a well-stained trichrome smear, the cytoplasm of protozoan cysts and trophozoites stains a blue-green, although *Entamoeba coli* often takes on a purple color. Nuclear chromatin, karyosomes, chromatoidal bars, and red blood cells stain a dark red-purple. Eggs and larvae stain red; background debris and yeasts stain green. In an iron hematoxylin stain, the organisms stain gray-black, nuclear material stains black, and background material is light blue-gray. With either stain, poor fixation of fecal material results in poorly staining or nonstaining organisms.

Smears should be examined by first scanning for thick and thin areas using lower-power of magnification (10× or 40× objective). Thin areas should be selected and observed under oil immersion (100× objective) for examination and identification of organisms. It should take approximately 10 to 15 minutes to adequately examine selected areas. Organisms that stain lightly and may be difficult to identify are *Entamoeba hartmanni, Dientamoeba fragilis, Endolimax nana, Chilomastix mesnili,* and *Giardia lamblia.*

Procedures for Detection of Specific Parasites

CELLOPHANE TAPE PREPARATION FOR PINWORM

The life cycle of the pinworm *(Enterobius vermicularis)* includes migration of the female out the anus at night to lay eggs in the perianal area. Therefore, a fecal specimen is not the optimal specimen for detection of infection with this organism. Instead, the cellophane tape preparation is routinely used for detection of suspected pinworm infections. This procedure involves swabbing the child's perianal area with a tongue blade covered with cellophane tape (sticky side out). The sampling should take place first thing

in the morning, before the child uses the bathroom or is bathed. After the sample has been taken, the sticky side of the tape is placed onto a microscope slide and scanned at low and high power fields of magnification for the characteristically shaped eggs. Commercial kits are now available to reduce manipulation of the specimen by caregivers.

MODIFIED ACID-FAST PROCEDURE FOR *CRYPTOSPORIDIUM* SP

Use of a Kinyoun modified acid-fast stain enhances detection of oocysts of *Cryptosporidium parvum* and *Isospora belli*. With this procedure, the oocysts appear as magenta-stained organisms against a blue background. The use of a stain combining iron hematoxylin with carbol fuchsin for simultaneous staining of *Isospora* sp, *Cryptosporidium* sp, and protozoa has been reported.

STAINS FOR *PNEUMOCYSTIS CARINII*

Stains that are used in identification of *P. carinii* include Gomori methenamine silver and Giemsa stains. The silver stain is used to identify cysts of the organism, which will appear as black, rounded or punched-in balls against a greenish background. The Giemsa stain demonstrates the trophozoites and intracystic bodies within the cyst, but the outside wall of the cyst will not stain.

Other Specimens Examined for Intestinal Parasites

Duodenal Aspirates

Material obtained from the Enterotest (HDC Corp., San Jose, CA) or from duodenal aspirates may be submitted in cases of suspected giardiasis or strongyloidiasis when clinical symptoms are suggestive but routine stool examinations are negative. This material may be examined by direct wet mount for trophozoites or may be placed in PVA for preparation of permanently stained smears. Eggs of *Fasciola hepatica* or *Opisthorchis sinensis* as well as oocysts of *Cryptosporidium* spp or *I. belli* may also occasionally be recovered.

Sigmoidoscopy Specimens

Scrapings or aspirates obtained by sigmoidoscopy may be used to diagnose amebiasis or cryptosporidiosis. These specimens are examined immediately for motile trophozoites, and a portion of the sample placed in PVA fixative so that permanently stained smears can be prepared for examination.

Urine, Vaginal, or Urethral Specimens

Eggs of *Schistosoma haematobium*, eggs of *E. vermicularis,* and trophozoites of *Trichomonas vaginalis* can be detected in the sediment of a urine specimen. *Trichomonas vaginalis* can also be detected in a wet mount of vaginal or urethral discharge. A plastic envelop method for culturing *T. vaginalis* has been developed. Dry ingredients in the culture bag are rehydrated, the patient specimen is added, and growth of the organism can be observed within 3 days.

Sputum

In cases of *Strongyloides stercoralis* hyperinfection, the filariform larvae may be seen in a direct wet mount of sputum. Eggs of the lung fluke *Paragonimus westermani* can also be identified in a sputum wet mount. For the patient with suspected pulmonary abscess due to *Entamoeba histolytica,* the sputum specimen should be examined as a permanently stained smear.

Examination of Specimens for Blood and Tissue Parasites

Blood Smears

Examination of a blood smear stained with Giemsa or Wright stain is the most common method to detect malaria, *Babesia* sp, *Trypanosoma* spp, and some species of microfilaria. Although motile organisms such as *Trypanosoma* and microfilariae can be detected on a wet preparation of a fresh blood specimen under low- and high-power magnification, identification is made on the basis of characteristics seen on a permanently stained smear. Concentration methods using membrane filters can be used to detect *Trypanosoma* sp or microfilariae but are rarely performed in the clinical laboratory. Tissue parasites such as *Trichinella spiralis*, *Leishmania* spp, *Pneumocystis carinii,* and *Toxoplasma gondii* can be identified by examination of tissue biopsy or by serologic methods.

COLLECTION AND PREPARATION OF THE BLOOD SPECIMEN

Blood taken directly from a finger stick should be used for a malarial smear because it tends to give the best staining characteristics. Blood collected in ethylenediaminetetra-acetic acid (EDTA) gives adequate staining if processed within 1 hour. With Giemsa stain, the cytoplasm of the parasite stains bluish and the chromatin red to purple-red. If malarial stippling is present, it will appear as discrete pink-red dots. Giemsa staining gives the best morphologic detail but is a time-consuming procedure. Wright stain has a shorter staining period, but color intensity for differentiation of parasites is not as good as that with Giemsa stain.

PROCEDURE FOR IDENTIFICATION OF THE ORGANISM

For suspected cases of blood parasites, both a thick film and a thin film should be made. Both preparations can be made on the same slide or on separate slides. Because the two preparations are treated differently prior to staining, however, use of two slides is more efficient. Giemsa stain provides the best staining of the organisms and should be used on both thick and thin films. Wright stain cannot be used for a thick film because the methanol content will fix red cells.

A thick film is best for detection of parasites, because organisms are concentrated in a relatively small area. The thick film is made by pooling several drops of blood on the slide and then spreading it into a 1.5-cm area. A too thick film peels from the slide; thickness is optimal when newsprint is barely visible through the drop of blood before it dries. The blood should be allowed to dry for at least 6 hours before staining. It should not be fixed with methanol prior to staining; fixing prevents hemoglobin from being released from the red cell. Use of Giemsa stain automatically lakes hemoglobin from unfixed red cells. Initial scanning of the stained smear at 10× detects microfilariae. At least 100 oil immersion fields should be examined before a negative result is reported.

Species identification should be made from a thin film, because the characteristics of the parasite and the red blood cells can be seen. The thin film is made in the same way as that for a differential count. It should be fixed in methanol for 1 minute and air-dried prior to staining with Giemsa stain. The entire smear should be scanned at 10× for detection of large organisms such as microfilariae; then at least 100 oil immersion fields must be examined for the presence of organisms such as *Trypanosoma* sp or for intracellular organisms such as *Plasmodium* sp or *Babesia* sp. In a symptomatic patient, several blood smears from samples collected at approximately 6-hour intervals over 36 to 48 hours should be examined before a final negative diagnosis is made.

Biopsy Specimens

Biopsy specimens are usually needed to diagnose infections with *Leishmania* spp because the organisms are intracellular. Depending on the species present, the amastigote stage can be detected in tissues such as skin, liver, spleen, and bone marrow. Cutaneous lesions should be sampled below the edges of the ulcer; surface samples do not yield infected cells.

Cerebrospinal Fluid

Viable organisms in suspected cases of amebic meningitis or sleeping sickness can occasionally be seen in a spinal fluid (CSF) specimen. The trypomastigote is visible because of the motion of the flagellum and undulating membrane. It requires a skillful microscopist, however, to discern amebic motility in a field of neutrophils. If amebic meningoencephalitis due to *Naegleria fowleri* is suspected, the CSF can be cultured. Nonnutrient agar is seeded with an *Escherichia coli* overlay, and the spinal fluid sediment is inoculated onto the media. The specimen is sealed and incubated at 35°C. The medium is examined daily for thin tracks in the bacterial growth, which indicate that amebae have been feeding on the bacteria.

Immunologic Diagnosis

Parasites that invade tissue are the primary organisms to stimulate antibody production. Many serologic tests are useful if invasive methods cannot be used for identification. In most cases, however, tests for antibody serve only as epidemiologic markers. Current tests detect antibody that may persist after acute infection and are not useful in endemic areas. They may, however, be useful for diagnosis in a person who has traveled to an endemic area and is now symptomatic. Another disadvantage of antibody tests is that they may have a large number of cross-reactions, which limit their diagnostic usefulness. In addition, serologic tests used by reference laboratories such as the Centers for Disease Control

(CDC) are not commercially available. Immunoassay or fluorescent antibody tests for antibodies to *T. gondii* or *E. histolytica* (extraintestinal infections) are available for use in clinical laboratories. In contrast, tests for parasitic antigens provide information about current infection.

The parameters that should be considered by a laboratory in selecting methods to be used include not only cost but also diagnostic yield, patient population, relative incidence of the parasite in the area, and number of specimens to be processed.

Enzyme Immunoassay

Enzyme immunoassay (EIA) methods for the detection of antibody to intestinal parasites are rarely used in the clinical laboratory because of the difficulty in obtaining antigen, cross-reactivity of antibodies, and poor sensitivity and specificity. The major use of the enzyme immunoassay has been to detect antigens from specific parasites. Giardiasis is often difficult to detect because the cysts are shed irregularly. Enzyme immunoassays to detect the presence of *Giardia*-specific antigen in stool are available. Tests using monoclonal antibody to detect both *Giardia* and *Cryptosporidium* are also available.

Fluorescent Antibody Techniques

Fluorescent antibody (FA) techniques using monoclonal antibodies have been developed to detect *Cryptosporidium* oocysts in fecal specimens. These methods are more expensive than the modified acid-fast procedure but demonstrate greater sensitivity, especially when only rare oocysts are present. An FA combination reagent for *G. lamblia* and *Cryptosporidium* antigens has been developed. The monoclonal antibodies eliminate false-positive and false-negative results. Such procedures are useful in screening large numbers of specimens during epidemiologic studies. Fluorescent antibody techniques for *P. carinii* are also available.

Quality Assurance in the Parasitology Laboratory

Quality assurance procedures in the parasitology laboratory are similar to those in other sections of the laboratory. An updated procedure manual, controls for staining procedures, records of centrifuge calibration, ocular micrometer calibration, and refrigerator and incubator temperatures should be available. Reagents and solutions should be properly labeled.

In addition, the parasitology laboratory should have:

- A textbook collection, including reference texts and atlases

- A set of 2 × 2 Kodachrome slides of common parasites

- A set of clinical reference specimens, including permanently stained smears and formalin-preserved feces

The department should also be enrolled in an external proficiency testing program. There should be an ongoing internal proficiency testing program to enhance identification skills of the technologists, especially if a full-time parasitologist is not employed. It has been shown that approximately twice as many parasites are detected when a single technologist staffed the parasitology department as when technologists rotate through the department.

Ocular Micrometer

Size is an important diagnostic criterion for parasites, and use of a properly calibrated ocular micrometer ensures accurate measurement of organisms. The micrometer should be calibrated for each objective on the microscope.

The micrometer consists of two separate parts: the stage micrometer, a 0.1-mm line, which is ruled in 0.01-mm units, and the ocular micrometer, which is ruled into 100 units but has no value assigned to the units. Values for each ocular unit can be calculated by using the stage micrometer according to Procedure 24–1.

MEDICALLY IMPORTANT PARASITIC AGENTS

Medically important parasites can be found in phyla representing single-celled organisms such as the protozoa and complex, multicelled organisms such as tapeworms and roundworms. Table 24–1 lists the characteristics of the classes in which most medically important human parasites are found; they are described in the remainder of this chapter.

Protozoa

Intestinal Amebae

In general amebae present the most difficult challenge with regard to identification. Their

---

PROCEDURE 24–1. CALIBRATION OF THE OCULAR MICROMETER

1. Insert the ocular micrometer in the eyepiece of the microscope so that the zero of the scale is on the left side. The etched side of the micrometer should be facing you.

2. Place the calibrated stage micrometer on the stage and focus on low power (10×).

3. While using low power, align the lefthand zero of the stage micrometer with the lefthand zero of the ocular micrometer. Do not move the stage micrometer after this point.

4. Scan the two scales until a division line on the ocular micrometer directly aligns with a division line on the stage micrometer.

5. Count the number of stage units and ocular units at this point. Divide the number of stage units by the number of ocular units, and multiply the result by 1,000. This gives the value (in micrometers) for one ocular unit on low power.

6. Repeat the procedure at high power and with oil immersion to get the value of one ocular unit at each of those magnifications.

To calculate the size of an organism, count the number of ocular units, multiply by the value for an ocular unit at that magnification, and report the value in micrometers.

━━━━━━■ T A B L E 2 4 – 1

CHARACTERISTICS OF PHYLA OF MEDICALLY IMPORTANT PARASITES

Organisms	Characteristics
Phylum: Sarcomastigophora	
Subphylum: Sarcodina (ameba)	Single celled
	Move by pseudopodia
	Trophozoite and cyst stages
	Asexual reproduction
Subphylum: Mastigophora (flagellates)	Single celled
	Most move by action of flagella
	Trophozoite and cyst stages for intestinal organisms
	Asexual reproduction
	Some blood flagellates
Phylum: Ciliaphora (ciliates)	Single celled
	Move by action of cilia
	Trophozoite and cyst stages
	Asexual reproduction
Phylum: Apicomplexa (sporozoa)	Single celled
	Usually inhabit tissue/blood cells
	Insects and other mammals are involved as part of life cycle
	May have both sexual and asexual life cycles
Phylum: Platyhelminthes (flatworms)	
Class: Trematoda (flukes)	Multicelled and bilaterally symmetric
	Most hermaphroditic
	Egg, miracidium, cercaria, and adult are stages in life cycle
	Fish, snails, crabs are involved as intermediate hosts in life cycle
Class: Cestoda (tapeworms)	Multicellular, ribbon-like body
	Hermaphroditic
	Eggs, larval stages, and adult worms are stages in life cycle
	Mammals and insects are involved as intermediate hosts in life cycles
Phylum: Aschelminthes	
Class: Nematoda (roundworms)	Adults of both sexes
	Eggs, larval forms, and adult worms are life cycle stages
	Some may have free-living forms or require intermediate hosts

average size range is smaller than most other parasitic organisms, and they must be distinguished from artifacts and cells that appear in the clinical specimen. The intestinal amebae discussed in this section are:

- *Entamoeba histolytica*

- *E. hartmanni*

- *E. coli*

- *Endolimax nana*

- *Iodamoeba bütschlii*

Entamoeba histolytica is recognized as a true pathogen; the remainder of the organisms listed are considered nonpathogens. *Entamoeba polecki,* which is rarely isolated in the United States, is not discussed. The organism *Blastocystis hominis* is also included in this section, although it is not recognized as a true ameba. There is also a question about its pathogenicity.

GENERAL CHARACTERISTICS OF AMEBAE

Species identification, whether in the cyst or trophozoite stage, often rests on the following characteristics: size, number of nuclei, nuclear structure, and presence of specific internal structures. In a wet preparation, the motility of the trophozoite may aid in identification. Overall, however, the permanently stained smear is the best preparation for identification of the amebae.

All these organisms live in the large intestine. With the possible exception of *B. hominis,* all possess both a trophozoite stage and a cyst stage. The trophozoite is the motile, feeding stage that reproduces by binary fission. The cyst is a resistant stage that is infective for humans. Multiplication of nuclei in the cyst stage also serves a reproductive function.

TREATMENT

Treatment is given only for *E. histolytica* infections; treatment for nonpathogens is not usually indicated. Luminal amebicides such as metronidazole are given to carriers in nonendemic areas to prevent the invasive phase and to decrease risk of transmission. In endemic areas with a high risk of re-infection, treatment may not be indicated. Patients with invasive amebiases are treated with systemic drugs as well as luminal amebicides.

LIFE CYCLE

The life cycle of amebae is relatively simple, with no intermediate hosts and direct fecal-oral transmission in food or water via the cyst stage. Humans ingest the infective cyst, and organisms excyst in the intestinal tract and multiply by binary fission. Trophozoites colonize the cecal area. Figure 24–2 illustrates a generalized life cycle for amebae as well as the extraintestinal phase of *E. histolytica.* Table 24–2 shows the comparison of the trophozoites and cysts of the common intestinal amebae.

Entamoeba histolytica

Entamoeba histolytica is found worldwide, especially in the tropics and subtropics. It is the major amebic pathogen for humans and ranks third, behind malaria and schistosomiasis, as a cause of death, accounting for an estimated 40,000 to 100,000 deaths per year. Prevalence varies according to socioeconomic level and sanitary practices; infection is more common in poorly developed areas or crowded institutions. The organism has also been identified as a sexually transmitted agent in the homosexual population.

Pathogenesis. Pathogenicity of the organism is reflected in its ability to cause invasive intestinal amebiasis and extraintestinal amebic infections. The mechanism of invasion consists of the following steps:

1. Adherence to the mucous layer of the intestine, which is mediated by an adherence lectin (*N*-acetyl *D*-galactosamine).
2. Disruption of the intestinal barrier by secretion of proteolytic enzymes.
3. Invasion of the epithelial cells.
4. Lysis of intestinal epithelial cells.

The organism also demonstrates resistance to host immune defense mechanisms, including phagocytosis and complement-mediated cell lysis.

The characteristic lesion in the intestinal mucosa, referred to as the flask-shaped ulcer of *E. histolytica,* is the result of lysis of the intestinal mucosa. The lesion demonstrates a pinpoint ulceration on the mucosal surface and a gradual

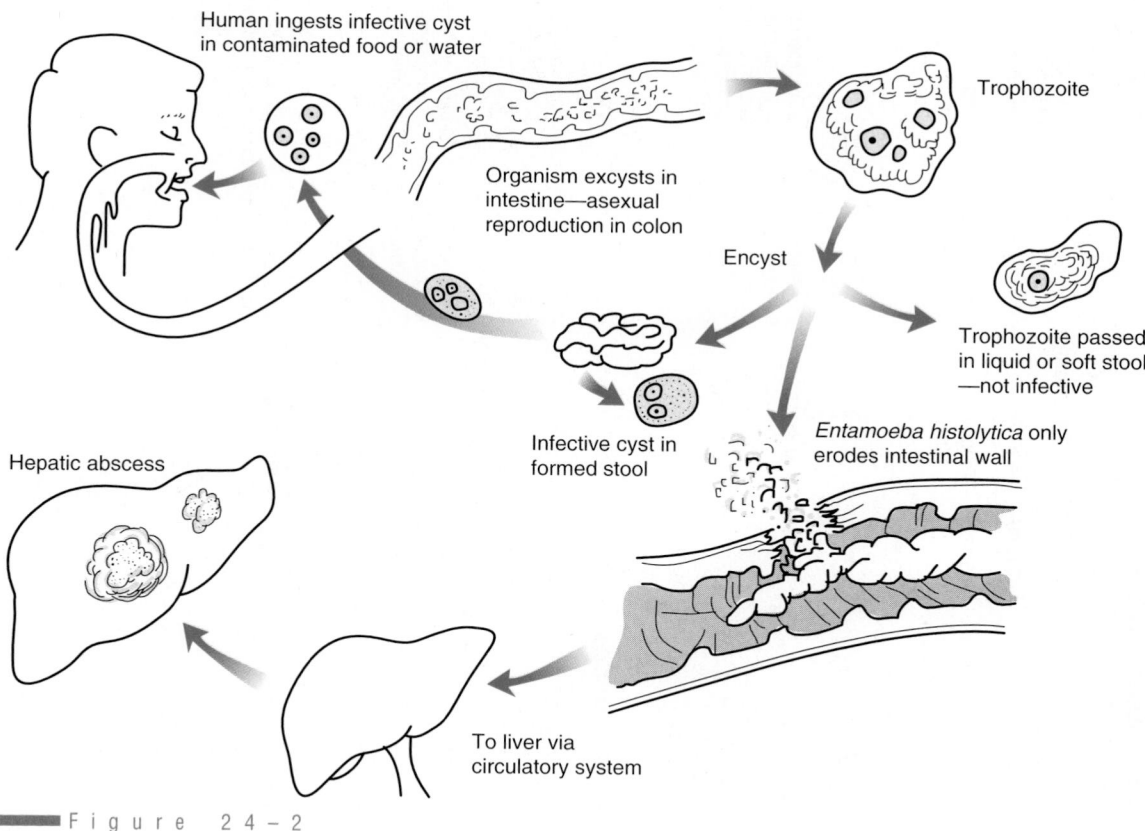

■■■■■■■ F i g u r e   2 4 – 2
Generalized life cycle of intestinal amebae.

T A B L E 2 4 – 2
COMPARISON OF AMEBAE

Organism	Trophozoite			Trophozoite and Cyst Nuclear Structure	Cyst			
	Size (μm)	Motility	Cytoplasm		Size (μm) and Shape	No. of Nuclei in Mature Cyst	Chromatoidal Bars	Glycogen Vacuole
Entamoeba histolytica	15–25	Progressive, directional	Finely granular May contain ingested RBCs	Small, central karyosome Fine, evenly distributed peripheral chromatin	10–20, round	4	Rounded Elongated Usually seen	Not usually seen Diffuse in young cyst
Entamoeba coli	15–50	Nondirectional	Vacuolated Ingested bacteria	Large, eccentric karyosomes Coarse, uneven peripheral chromatin	15–25, round	8	Elongated Splintered ends Not always seen	Not seen
Entamoeba hartmanni	4–12	Nondirectional	Finely granular	Small, central karyosomes Fine, evenly distributed peripheral chromatin	5–10, round	4	Rounded ends Elongated Not always present	Not seen
Endolimax nana	5–12	Nondirectional	Vacuolated May contain ingested bacteria	Large, irregularly shaped karyosome No peripheral chromatin	5–12, oval	4	Not present	Not seen
Iodamoeba bütschlii	6–20	Nondirectional	Vacuolated May contain ingested bacteria	Large karyosome surrounded by achromatic granules No peripheral chromatin	6–15, oval or irregular	1	Not present	Single, defined

widening of the lesion in the submucosal areas as the parasite invades the tissue. The organism may completely erode the intestinal mucosa and enter the circulation. When this occurs, the organ most commonly colonized is the right lobe of the liver, because organisms are trapped in the venules of liver. Patients with hepatic abscesses may exhibit symptoms such as fever and pain in the upper right quadrant. Lung abscesses may be seen as the result of penetration of the diaphragm by amebae from hepatic abscesses or from hematogenous spread. Other sites, such as the perianal area and bladder, may be colonized, with resulting tissue destruction.

Some patients with *E. histolytica* develop an ameboma (amebic granuloma), a tumor-like lesion that forms in the submucosa of the intestine. This represents an area of chronic lysis and infiltration with neutrophils, lymphocytes, and eosinophils.

In recent years, the pathogenicity of *E. histolytica* and its ability to cause extraintestinal infections have been investigated. Studies using electrophoretic isoenzyme patterns (zymodemes) show that pathogenic strains differ from noninvasive strains. Zymodemes of pathogenic and nonpathogenic *E. histolytica* have been identified. Initial studies show that strains in cyst passers (see next section) and in most homosexual patients are nonpathogenic, whereas several strains from areas with high rates of endemic disease are pathogenic.

Clinical Infections and Symptoms. Persons infected with *E. histolytica* may be asymptomatic, generally when only the lumen of the intestine is colonized. Symptomatic clinical infections may be differentiated into acute and chronic forms. In acute infections, the trophozoite form is passed in the stool. Patients with amebic dysentery have vague abdominal symptoms, such as tenderness, cramping, fever, and up to 20 diarrheic stools per day containing blood and mucus. In severe cases, the patient may shed pieces of intestinal mucosa.

In chronic infections, on the other hand, the cyst is the usual form passed, although during episodes of diarrhea, the trophozoite may be passed in the stool. Patients who have chronic infections are usually asymptomatic and are referred to as *cyst passers*.

Patients with hepatic abscesses may have hepatomegaly or dull pain or tenderness in the upper right abdominal quadrant or may be asymptomatic. Fever, weight loss, increased white blood cell count, or elevated liver enzymes may be present. Invasion of the lung may cause the patient to have chest pain, dyspnea, and a productive cough.

Laboratory Diagnosis. Patients with diarrhea are most likely to have trophozoites in the stool specimen, which may be seen in wet mounts or trichrome-stained smears. Sigmoid biopsies may be used to demonstrate the characteristic morphology of the intestinal ulcers or to identify trophozoites in tissue when none can be isolated from the stool specimen.

Amebic ulcers of the liver are often detected by ultrasound or radiographic tests. Subsequent aspiration of the abscess may yield motile trophozoites as well as necrotic material composed of lysed cells. Serologic methods to detect antibody to *E. histolytica* are available and are positive in more than 90% of patients with extraintestinal disease. These antibodies rise after tissue invasion but are not protective. DNA probes for the diagnosis of *E. histolytica* are being developed. In addition, monoclonal antibody tests to detect pathogenic from nonpathogenic zymodemes are under development.

Characteristics of the Trophozoite. In a direct saline wet mount of a *diarrheic stool,* the trophozoite of *E. histolytica* may exhibit a progressive, directional motility by extending long thin pseudopods. The size of the organism ranges from 10 to 50 μm, averaging 15 to 25 μm. The organism is refractile, and the characteristic bull's-eye nucleus, consisting of a small central karyosome and even, fine peripheral chromatin, may be only slightly visible. In a trichrome-stained smear, the cytoplasm of the organism appears clean and free of ingested bacteria and vacuoles. Finely granular nuclear chromatin, which is evenly distributed on the nuclear membrane, and the small central karyosome stain a dark purple-red. Ingested red cells are diagnostic for *E. histolytica* trophozoites but may not be seen in all organisms. Figure 24–3 shows trichrome-stained trophozoite of *E. histolytica,* with the trophozoite in Fig. 24–3*B* demonstrating an ingested red blood cell.

Characteristics of the Cyst. The average size of the cyst is 10 to 20 μm (Fig. 24–4). Cysts of the organism may have one to four nuclei each with a small central karyosome and fine, evenly distributed peripheral chromatin. The cytoplasm may contain cigar-shaped chromatoidal bars with rounded ends. These bars are composed of ribonucleic acid. In an iodine wet mount, the nuclei appear as yellowish refractile bodies within

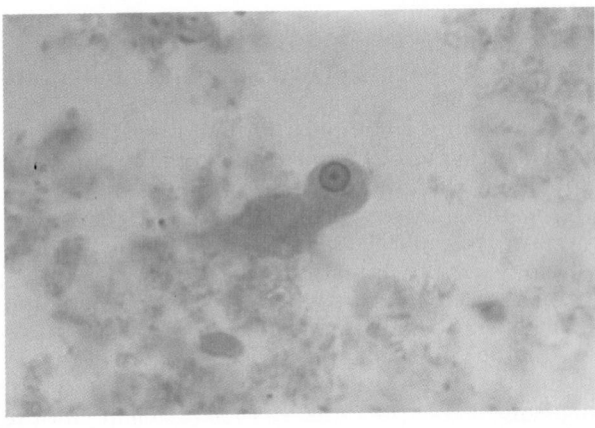

A

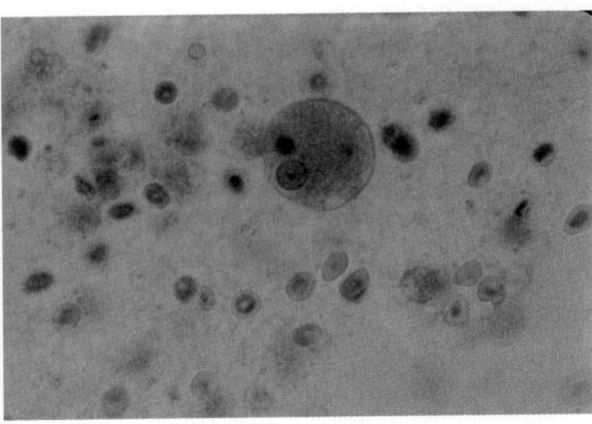

B

■■■■ F i g u r e   2 4 – 3

A, Entamoeba histolytica trophozoite (trichrome stain). *B, Entamoeba histolytica* trophozoite. Notice darkly staining ingested red blood cell near nucleus (trichrome stain).

the cyst; chromatoidal bars do not take up stain and appear as colorless areas. With trichrome stain, the cyst is light green-gray; nuclear material and chromatoidal bars stain a dark purple-red. Young cysts may show discrete glycogen masses that stain light brown in an iodine wet mount, but in the more mature cyst, the glycogen is diffuse. Cysts may be killed by drying, temperatures over 55°C, superchlorination, or addition of iodine to drinking water.

Entamoeba hartmanni

Entamoeba hartmanni, once known as "small race" *Entamoeba histolytica,* is a non-pathogen. It generally resembles *E. histolytica* in a trichrome-stained smear but is more likely to have an eccentric karyosome or uneven peripheral chromatin resembling that of *Entamoeba coli.* Size is a major determinant to differentiate *E. histolytica* and *E. hartmanni.* Average size of the trophozoites of *E. hartmanni* is 4 to 12 μm; trophozoites with an average size measuring greater than 12 μm are identified as *E. histolytica.* Cysts measure 5 to 10 μm; those of 10 μm or more are identified as *E. histolytica.* Figure 24–5 shows the trichrome-stained trophozoite, and Figure 24–6 shows the cyst of *E. hartmanni* with three nuclei visible.

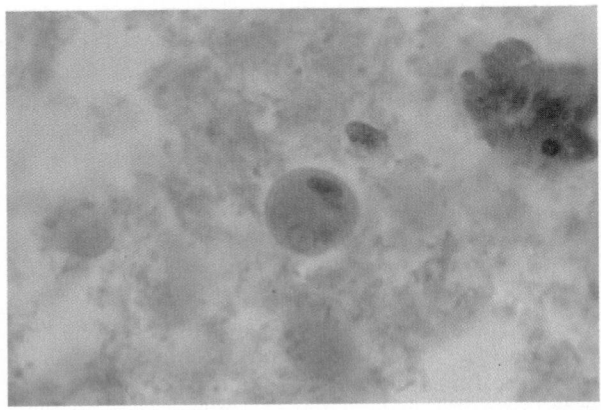

■■■■ F i g u r e   2 4 – 4

E. histolytica cyst with round-ended chromatoidal bars. Two nuclei visible (trichrome stain).

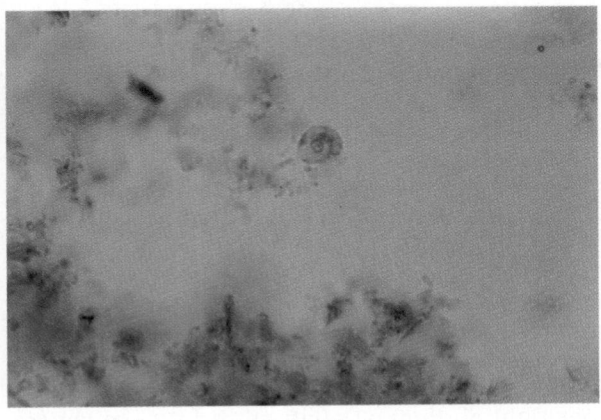

■■■■ F i g u r e   2 4 – 5

Entamoeba hartmanni trophozoite (trichrome stain).

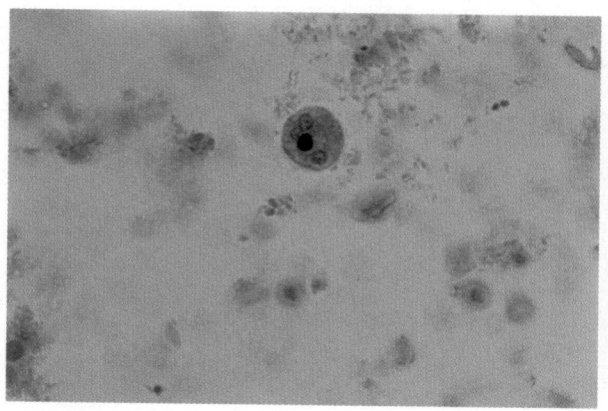

■Figure 24-6
Entamoeba hartmanni cyst (trichrome stain).

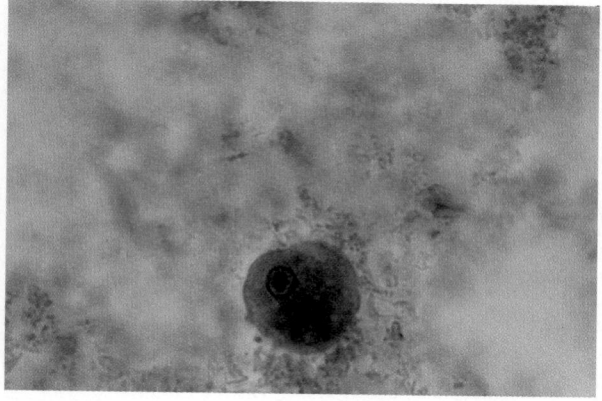

■Figure 24-7
Entamoeba coli trophozoite. Notice darkly staining, highly vacuolated cytoplasm (trichrome stain).

Entamoeba coli

Entamoeba coli is a commonly found intestinal commensal transmitted by ingestion of cysts in fecally contaminated food or water. The average size of the trophozoite is 15 to 50 μm with most measuring 25 μm (Fig. 24–7). The nuclear structure is characterized by the presence of a large eccentically placed karyosome and coarse, uneven peripheral chromatin on the nuclear membrane. Motility of the trophozoite in a wet preparation is sluggish and nondirectional. In a permanently stained preparation, the cytoplasm of the trophozoite may stain a purplish-gray and contains vacuoles and ingested materials.

The mature cyst contains eight nuclei; the immature cyst may have one or two large nuclei with a large glycogen vacuole. Chromatoidal bars, when present, have a pointed, splintered appearance. The average size of the cyst is 15 to 25 μm. Figure 24–8A shows an MIF wet mount of a cyst of *E. coli,* and *B* shows a trichrome-stained cyst.

Endolimax nana

The trophozoite of *Endolimax nana* has a large karyosome with no peripheral chromatin on the nuclear membrane. Size of the trophozoite ranges from 5 to 12 μm with the average being less than 10 μm (Fig. 24–9). The cytoplasm is granular and vacuolated. In a wet preparation, the motility is sluggish. In a wet mount, it may be difficult to distinguish the large karyosome of

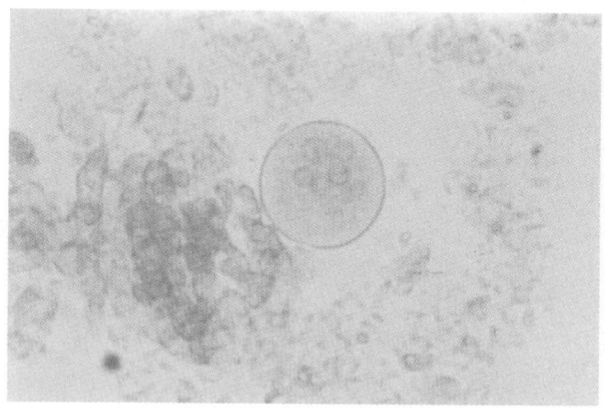

A

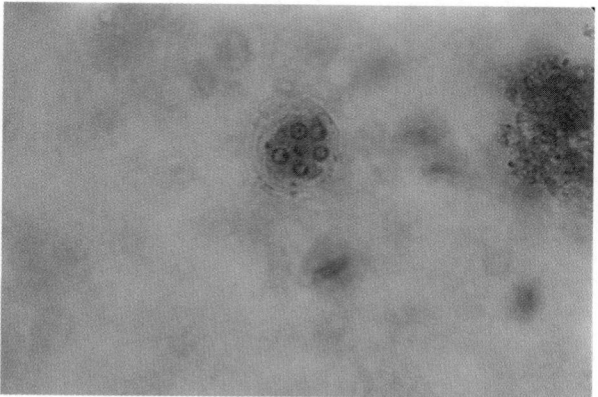

B

■Figure 24-8
A, Entamoeba coli cyst (MIF wet mount). *B, Entamoeba coli* cyst with five nuclei visible (trichrome stain).

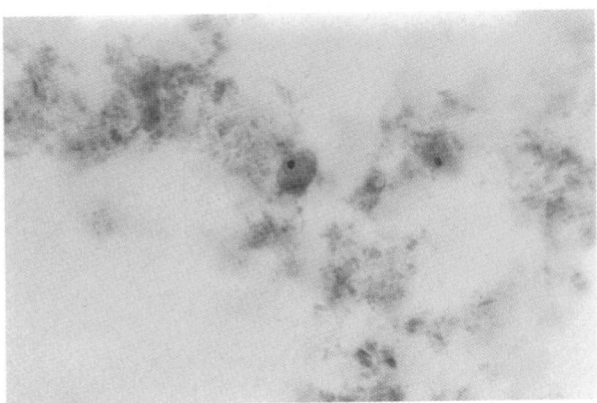

■■■■ F i g u r e   2 4 – 9
Endolimax nana trophozoite (trichrome stain).

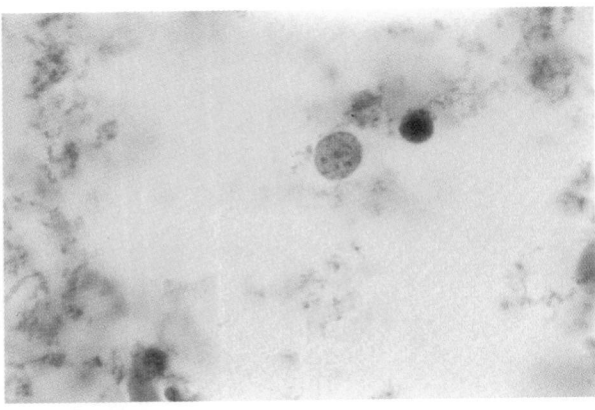

■■■■ F i g u r e   2 4 – 1 0
Endolimax nana cyst (trichrome stain).

E. nana from the karyosome of *E. hartmanni,* and the organisms may be misidentified. The cyst of *E. nana* is oval or spherical, is 5 to 12 μm, and has up to four large karyosomes (Fig. 24–10).

Iodamoeba bütschlii

Iodamoeba bütschlii is less commonly encountered than *E. coli* or *E. nana*. The nucleus is composed of a single, irregularly shaped karyosome surrounded by achromatic granules and a thin nuclear membrane with no peripheral chromatin. The trophozoites of *I. bütschlii,* which are 6 to 20 μm, show a vacuolated cytoplasm in a permanently stained smear (Fig. 24–11). The oval cyst is 6 to 15 μm (average 9 to 10 μm) and contains a single large karyosome and a large, well-defined glycogen vacuole. The vacuole stains dark brown in an iodine wet mount and appears empty in a permanently stained smear. Figure 24–12 demonstrates a trichrome-stained cyst of *I. bütschlii*.

Blastocystis hominis

Blastocystis hominis, a protozoan once thought to be a yeast, has also come to prominence as a possible cause of diarrhea in humans, although controversy concerning its pathogenicity still exists. Not considered a common cause of diarrheal disease, this organism nevertheless has been found in patients with diarrhea who have no other intestinal pathogens. These patients often present with a history of travel abroad. Some authors suggest that *B. hominis* has no role as a pathogen, but others suggest that the organism be considered a pathogen in symptomatic patients if it is present in a count of more than 5 per high-power field and no other known enteric pathogens can be found.

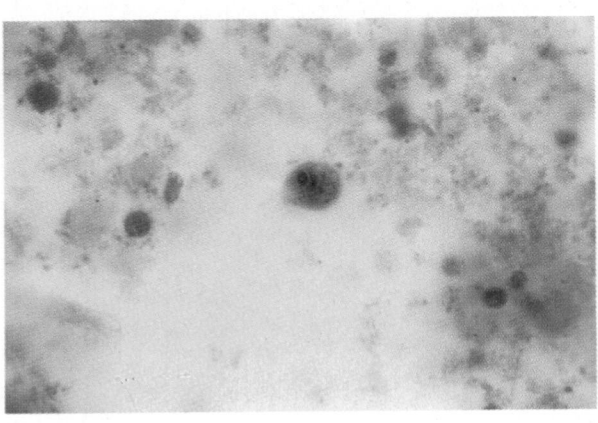

■■■■ F i g u r e   2 4 – 1 1
Iodamoeba bütschlii trophozoite (trichrome stain).

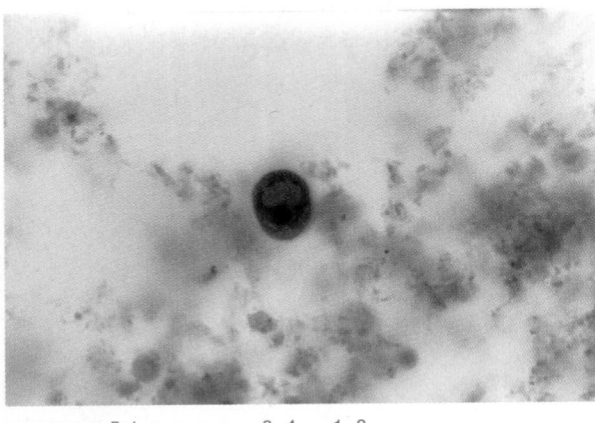

■■■■ F i g u r e   2 4 – 1 2
Iodamoeba bütschlii cyst with prominent glycogen vacuole (trichrome stain).

The organism exists in ameboid, granular, and spherical forms, with the spherical form being most commonly identified. There is no identifiable cyst stage. Average size for the spherical form is 5 to 15 μm, but up to 20% of organisms are smaller than 5 μm. The organism has a layer of cytoplasm lining the inner wall with as many as four nuclei present, usually pushed to the side, and a large central body. In an iodine mount, the cytoplasm stains brown and the central area does not stain. On trichrome stain, the cytoplasm stains dark green and the central area may stain pale to intensely green with the nuclei a dark purple-black (Fig. 24–13).

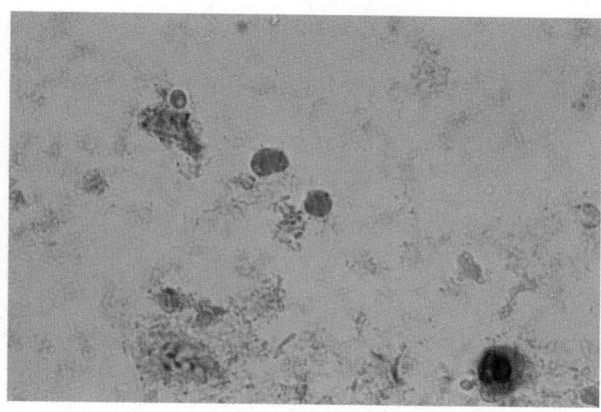

■Figure 24–13
Blastocystis hominis spherical form (trichrome stain).

Tissue Amebae

Two organisms, *Naegleria fowleri* and *Acanthamoeba* spp, have been identified as the organisms most commonly associated with tissue invasion in humans. The ameboflagellate *N. fowleri* is the etiologic agent of primary amebic meningoencephalitis (PAM), a rapidly fatal condition involving the central nervous system. *Acanthamoeba* spp have been associated with a more chronic condition, granulomatous amebic encephalitis (GAE), and with amebic keratitis. A comparison of the central nervous system (CNS) infections caused by tissue amebae is given in Table 24–3.

Naegleria fowleri

Primary amebic meningoencephalitis (PAM) has been reported from countries all over the world and occurs in children and young adults with no predisposing condition. A common factor is the report of recent swimming or other water-related activities in warm, artificial lakes or brackish or muddy water. The life cycle of *N. fowleri* is relatively simple, consisting of three stages: a free-living amebic trophozoite, a transient flagellate form, and an environmentally resistant cyst.

The organism enters the nasal cavity, colonizes in the amebic form within the nasal cavity, penetrates the cribriform plate, moves along the olfactory nerve, and invades the central nervous system.

Clinical Infection. The incubation period is usually 2 to 3 days but may range up to 2 weeks. Clinically, the disease cannot be distinguished from bacterial meningoencephalitis. Initial symptoms include severe bifrontal headache, fever, stiff neck, and nausea and vomiting. The organism multiplies within brain tissue, and within 2 to 4 days, the patient may suffer drowsiness, confusion, and seizures, and progress into a coma. The

■TABLE 24–3
COMPARISON OF CNS INFECTIONS CAUSED BY AMEBAE

	Primary Amebic Meningoencephalitis	Granulomatous Amebic Encephalitis
Etiologic agent	*Naegleria fowleri*	*Acanthamoeba* spp
Stages in		
CSF	Trophozoite	Trophozoite
Brain biopsy	Trophozoite	Trophozoite and cyst
Characteristics	Trophozoite 10–12 μm	Trophozoite 10–45 μm
	Large karyosome	Spine-like pseudopod
	Broad pseudopods	Cyst
		15–20 μm
		Wrinkled double wall
Entry	Nasal passage olfactory nerve to CNS	Lungs and skin with hematogenous spread to CNS
Clinical course	Fulminant (death within 1 week of onset)	Slow and chronic
Population at risk	Children to young adults, healthy (history of water activities in stagnant, warm water)	Immunocompromised

disease is usually fatal within 1 week of the appearance of clinical symptoms. The possibility of cure depends on early diagnosis. Aggressive therapy with intravenous and intrathecal amphotericin B has been used. Rifampin, miconazole, and tetracycline have been used in addition to amphotericin B.

Laboratory Diagnosis. Diagnosis can be made by finding motile trophozoites in the spinal fluid. The trophozoite is 10 to 12 μm and moves by extending large, broad pseudopods. The nucleus contains a large central karyosome that may be surrounded by a halo. The spinal fluid contains many segmented neutrophils and red blood cells; it also has elevated protein and decreased glucose values. Cysts, which are 10 μm with a round, smooth double wall, are not seen in clinical or biopsy specimens. Histologic preparations of the brain at autopsy show inflammatory lesions containing many segmented neutrophils, eosinophils, and the trophozoites.

A method that isolates this organism consists of overlaying nonnutrient agar with *Escherichia coli* and inoculating it with a drop of the spinal fluid sediment. Plates are examined daily for clearing of the agar in thin tracks, which indicate that trophozoites have fed on the bacteria. The trophozoite stage can be converted to the flagellate stage by adding one drop of spinal fluid sediment to 1 mL of distilled water and incubating at 37°C. The flagellate form occurs in 2 to 20 hours.

Acanthamoeba spp

Acanthamoeba spp, also soil and water organisms, cause granulomatous amebic encephalitis (GAE), which occurs primarily in immunosuppressed or debilitated patients. Unlike PAM, this condition is characterized by a hematogenous spread to the central nervous system from a primary inoculation site in either the lungs or skin. The incubation time is unknown but may extend from months to years. Symptoms include drowsiness, seizures, hemiparesis, headache, stiff neck, and personality disorders. The trophozoite is rarely seen in spinal fluid; brain biopsy demonstrates both cyst and trophozoite. Histologic preparations of the brain at autopsy show inflammatory lesions containing many segmented neutrophils, eosinophils, and trophozoites. Therapeutic agents for GAE are generally not established because most infections have been diagnosed at autopsy.

Amebic keratitis is another condition associated with *Acanthamoeba* spp that has been identified since the 1980s. The primary group at risk for developing this condition are persons wearing contact lenses, especially the soft and extended-wear types. A common factor in these infections is the preparation of homemade saline solution, although some patients also report a history of corneal trauma or wearing the contact lenses during swimming. Patients experience photophobia, blurred vision, inflammation, ring infiltrates, and pain. Because of the similarity in tissue damage, the infection may initially be confused with herpes simplex viral infection. Phase microscopy of direct wet-mount preparations of corneal scrapings may show the cyst. Permanent stains such as trichrome and Giemsa may also demonstrate presence of the trophozoite in clinical specimens. Isolation of *Acanthamoeba* may be performed in a similar manner to that for *N. fowleri,* except that corneal scrapings serve as the inoculum. Amebic keratitis has been treated with propamidine isethionate, neomycin–polymixin B, gramicidin, and clotrimazole. Corticosteroids are used to decrease in-flammation. Many patients still lose the sight in the affected eye despite treatment.

Acanthamoeba spp possess only two stages, the resistant cyst and the motile trophozoite. The cyst is approximately 15 to 20 μm, spherical, and double walled, with the walls having a wrinkled appearance. The trophozoite ranges from 10 to 45 μm (average size 20 μm) and has a single nucleus with a central prominent endosome. Blunt pseudopods and characteristic spine-like projections of the cytoplasm (acanthopodia) may also be seen on a wet mount. The method described to recover *N. fowleri* in culture may also be used to recover *Acanthamoeba* from corneal scrapings.

The Ciliates

Balantidium coli

Only one ciliate, *Balantidium coli,* is considered a pathogen for humans. The true host for this organism is the hog, and humans serve as accidental hosts. The organism lives in the large intestine, where it may cause mucosal lesions but not extraintestinal infections. Most persons with this infection are asymptomatic, but the organism may cause a self-limiting diarrhea with nausea, vomiting, and abdominal tenderness.

Life Cycle and Morphology. The life cycle is similar to that of the amebae, with the cyst being the infective stage for humans. The organism is quite large and covered with short cilia (Fig. 24–14). The oval trophozoite demonstrates two

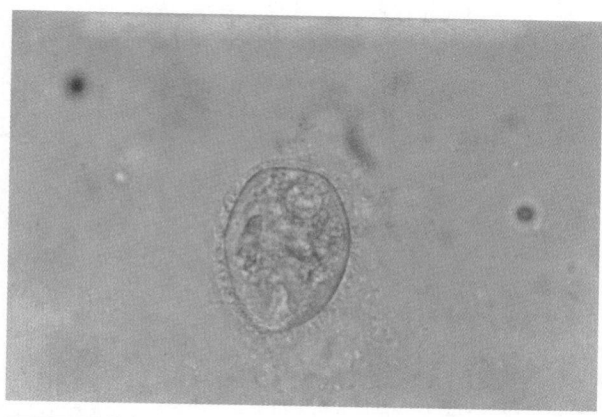

Figure 24-14
Balantidium coli trophozoite wet mount.

size ranges: 45 to 60 by 30 to 40 µm and 90 to 120 by 60 to 80 µm. The cilia-lined cytostome is located at the slightly pointed anterior end. The cytoplasm contains food vacuoles. A small opening at the posterior, the cytopyge, is used to expel the contents of food vacuoles. In a wet mount, the cilia are seen propelling the organism with a rotary motion.

The rounded, thick-walled cyst averages 45 to 75 µm. Cilia may be seen retracted within the cyst wall. Both stages are characterized by the presence of two nuclei: a kidney bean–shaped macronucleus and a small round micronucleus that is usually situated in the small curvature of the macronucleus.

Pathogenic Intestinal and Urogenital Flagellates

The flagellates constitute another major group of parasites that may inhabit the intestinal tract. The life cycle is relatively simple, and resembling that of the amebae (Fig. 24–15). Most flagellates possess both cyst and trophozoite stages. *Dientamoeba fragilis, T. vaginalis,* and *Trichomonas hominis* lack a cyst stage, however, and the trophozoite of these organisms serves as the infective stage.

The intestinal organisms covered in the section are:

- *Giardia lamblia*
- *Dientamoeba fragilis*
- *Chilomastix mesnili*
- *Trichomonas hominis*

Giardia lamblia was originally considered the only pathogenic intestinal flagellate. In recent years, however, *D. fragilis* has been identified as a potential pathogen. *Trichomonas vaginalis,* an inhabitant of the genitourinary tract in both men and women, is also discussed in this section. Table 24–4 shows the characteristics of the trophozoite and cyst stages of the intestinal and genitourinary flagellates.

Nonpathogenic organisms, such as *Retortomonas intestinalis* and *Enteromonas hominis,* that have low infection or detection rates are not discussed. Readers are referred to a standard parasitology text for information on such organisms.

Giardia lamblia

Giardia lamblia has a worldwide distribution and is often the etiologic agent of outbreaks of gastroenteritis and traveler's diarrhea. In the United States, it is the most commonly reported intestinal parasite. Animals such as the beaver may serve as

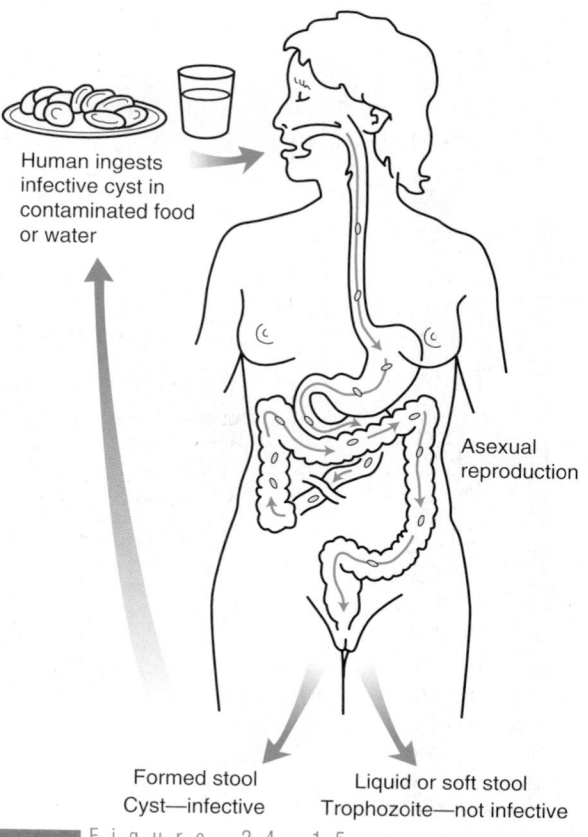

Human ingests infective cyst in contaminated food or water

Asexual reproduction

Formed stool
Cyst—infective

Liquid or soft stool
Trophozoite—not infective

Figure 24-15
Generalized life cycle of intestinal flagellates.

T A B L E 2 4 – 4

COMPARISON OF INTESTINAL AND UROGENITAL FLAGELLATES

Organism	Trophozoite				Cyst		
	Size (μm)	Motility	No. of Nuclei	Other Features	Size (μm) and Shape	No. of Nuclei	Other Features
Giardia lambia	9–21 × 5–15	"Falling leaf"	2	Sucking disk, ventral surface Parabasal bodies and axonemes	8–12, oval	4	Cytoplasm retracted from cyst wall Fibrils and flagella inside cyst
Chilomastix mesnili	10–20 × 3–10	Rotary	1	Spiral groove Cytostome	6–10, lemon-shaped	1	Anterior of cyst has "nipple-like" protrusion
Trichomonas hominis	6–14	Jerky	1	Undulating membrane the entire length of organism Axostyle through body			No cyst stage for this organism
Dientamoeba fragilis	5–12	Nondirectional	2 (20% have 1)	Nucleus made of 4–8 clustered granules Resembles ameba			No cyst stage for this organism
Trichomonas vaginalis	7–23 × 5–12; average, 15–18	Jerky, nondirectional	1	Undulating membrane half length of body Found in urine			No cyst stage for this organism

reservoirs and may be a source of infection for backpackers who drink from streams or rivers. *Giardia lamblia* has also been isolated in outbreaks of diarrhea in nurseries and day care centers as a result of person-to-person contact as well as in food-borne and water-borne infections. Along with *E. histolytica, G. lamblia* has been identified as a sexually transmitted pathogenic protozoan in the male homosexual population.

Pathogenesis. *Giardia lamblia* lives in the duodenal area of the small intestine. Although it does not invade the mucosal surface, it does attach to the surface of columnar epithelial cells. Possible pathologic mechanisms associated with the organism are adherence and damage to the intestinal mucosa and interference with absorption of nutrients and irritation. There is no enterotoxin activity, but atrophy of the intestinal villi can be seen in heavy infections. Patients with deficiencies in secretory immunoglobulin A (IgA) often present with a more severe infection.

Clinical Symptoms. In most patients, the acute infection manifests as a self-limiting diarrhea with malaise, cramps, nausea, and abdominal tenderness after an incubation period of 12 to 14 days. Explosive, foul-smelling diarrhea is present. Patients with secretory IgA deficiency or achlorhydria seem not only to be more prone to the infection but also to develop chronic infection. In these cases, there may be a malabsorption-like syndrome with weight loss, fatigue, anorexia, and steatorrhea with large amounts of gas. Metronidazole is the drug given for infection.

Laboratory Diagnosis. Feces serve as the usual diagnostic specimen, but shedding of the cysts is irregular, and multiple specimens often fail to detect the organism. In cases in which clinical symptoms persist and the organism cannot be demonstrated, a duodenal aspirate or the Enterotest (HDC Corp., San Jose, CA) may be used to isolate the organism. In the Enterotest, the patient swallows a gelatin capsule containing a weighted string. One end of the string has been taped to the side of the patient's mouth; the weighted end is carried into the upper small intestine. After about 4 hours, the string is brought up, and part of the mucus adhering to the surface is stripped off and examined on a wet mount for motile trophozoites, and the remainder is placed in a fixative for a permanently stained smear. Serologic tests using monoclonal antibodies to detect *G. lamblia* antigens in stool have also been developed.

Characteristics of the Trophozoite. The trophozoites of *G. lamblia* are pear-shaped and bilaterally symmetrical, and they measure approximately 9 to 21 by 5 to 15 µm. They show a characteristic "falling leaf" motility in a wet mount. In a permanently stained smear, the binucleate organism has been described as having an "old man" appearance (Fig. 24–16). Two oval nuclei, each with a large central karyosome, are on each side of the midline. Four pair of flagella, midline axonemes, and two median bodies posterior to the nuclei are also present. There is a large ventral sucking disk, which the organism uses to attach to the intestinal wall. The organism often stains faintly with trichrome stain.

Cysts. Cysts of *G. lamblia* are oval and approximately 8 to 12 by 7 to 10 µm. There are up to four nuclei, and the cytoplasm is often pulled away from the cyst wall. On a permanently stained smear, the retracted flagella and other internal structures give a cluttered appearance to the cyst (Fig. 24–17).

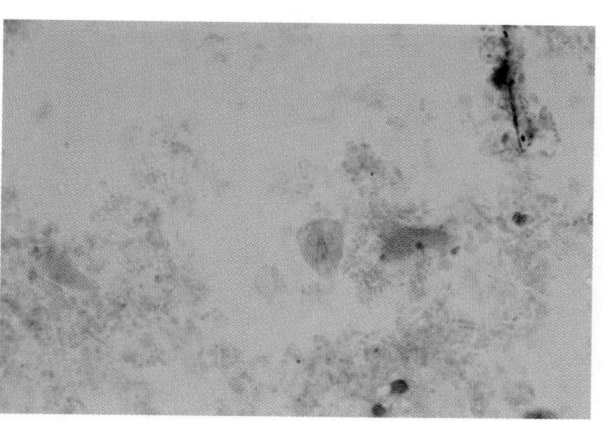

■ F i g u r e 2 4 – 1 6

Giardia lamblia trophozoite (trichrome stain).

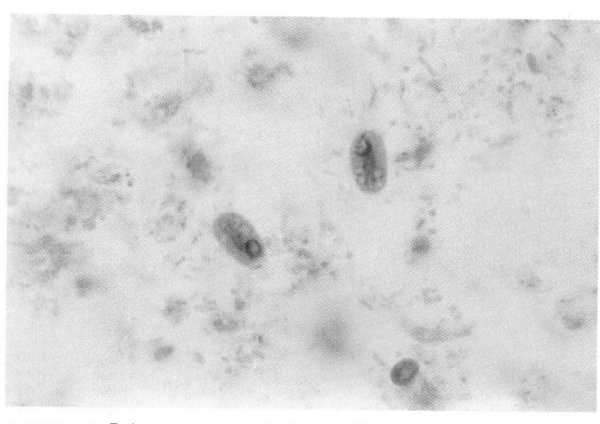

■ F i g u r e 2 4 – 1 7

Giardia lamblia cyst (trichrome stain).

Dientamoeba fragilis

Dientamoeba fragilis is another flagellate organism of the large intestine that is now recognized as a cause of gastrointestinal illness. Patients infected with the organism are usually asymptomatic but may have abdominal pain or tenderness and diarrhea. The organism lacks a cyst stage; it appears that the trophozoite may be transmitted to humans by being ingested on a helminth egg, especially that of *E. vermicularis.*

The morphology of *D. fragilis* closely resembles that of the amebae, but studies of ultrastructure indicate that it belongs to the subphylum Mastigophora. The organism is characteristically described as being binucleate, with 50 to 80% of the organisms demonstrating this characteristic. The nuclear membrane has no peripheral chromatin, and the karyosome consists of four to eight discrete granules. Size of the trophozoite ranges from 5 to 12 μm, and the cytoplasm contains many food vacuoles and bacteria (Fig. 24–18). This organism can be difficult to see on a trichrome-stained smear, because its outline often is indistinct and blends into the background.

Trichomonas vaginalis

Trichomonas vaginalis, a pathogen of the urogenital tract in men and women, causes trichomoniasis, one of the most common sexually transmitted diseases. The organism lacks a cyst stage, and the trophozoite stage is infective via sexual contact. In women, the infection is primarily localized in the vagina, resulting in itching and the production of a frothy, creamy vaginal discharge as well as dysuria. Men infected with *T.*

vaginalis are usually asymptomatic and serve as carriers, although they may develop nonspecific urethritis with a milky discharge that lasts up to 4 weeks. Infections in either sex are usually treated with metronidazole. Treatment of both sexual partners is suggested to obtain optimum cure. Long-term immunity is not developed after an acute infection, and re-infection can occur.

Laboratory Diagnosis. Diagnosis in women may be made by finding the trophozoite in urine or vaginal discharge; in men, the trophozoite is seen in urine or prostatic secretions. The organism has four free anterior flagella and an undulating membrane that extends half the length of the body. In a wet mount, the trophozoite has a characteristic jerky motility, and the motion of the flagella and undulating membrane may be seen. In a stained preparation, the pear-shaped organism shows the presence of an axostyle, single nucleus, and chromatic granules extending the length of the axostyle. The average size of *T. vaginalis* is 5 to 18 μm. Figure 24–19 shows the trophozoite in a Papanicolaou (Pap) smear. Although most clinicians rely on wet-mount preparations, the method has relatively low sensitivity. Culture is the most sensitive and definitive method of detection, although specimens must be held for up to 7 days for results. There are no commercially available kits for serologic diagnosis of trichomoniasis.

Nonpathogenic Intestinal Flagellates

Chilomastix mesnili and *Trichomonas hominis* are intestinal nonpathogens that must be differentiated from pathogenic flagellates. *Chilomastix mesnili* trophozoites are pear-shaped and approximately 10 to 20 μm long by 3 to 10 μm wide. The

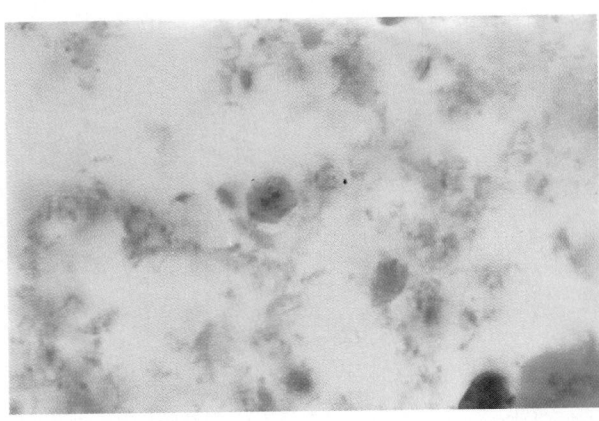

■ F i g u r e 2 4 – 1 8

Dientamoeba fragilis binucleate trophozoite (trichrome stain).

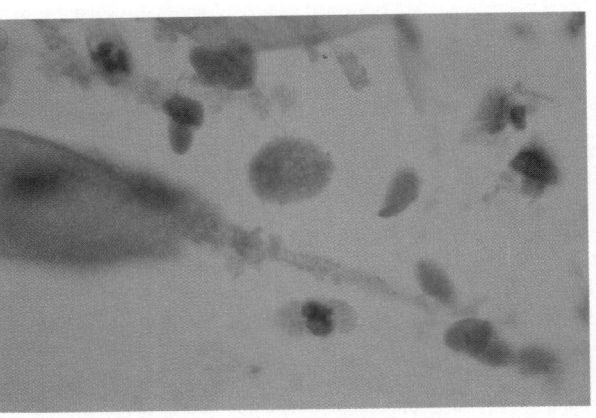

■ F i g u r e 2 4 – 1 9

Trichomonas vaginalis trophozoite.

cytostome and nucleus are prominent in the anterior of the organism, and a spiral groove encircles the body of the organism. The nucleus has a small central karyosome and is surrounded by fibrils that curl around the cytostome to give a "shepherd's crook" appearance. The cytostome is elongate and rounded at the anterior and posterior. The cyst of *C. mesnili,* which measures 6 to 10 μm, is lemon shaped with an anterior nipple. The nucleus, cytostome, and curved fibrils are visible in a stained smear (Fig. 24–20).

Trichomonas hominis is a small organism not usually identified in the stool. The trophozoite is 6 to 14 μm long with a prominent axostyle extending through the posterior of the organism, four anterior flagella, and an oval nucleus with a small karyosome. The undulating membrane extends the length of the organism and is joined to the body along the costa.

Blood and Tissue Flagellates

The hemoflagellates in the genera *Leishmania* and *Trypanosoma* differ in several ways from the

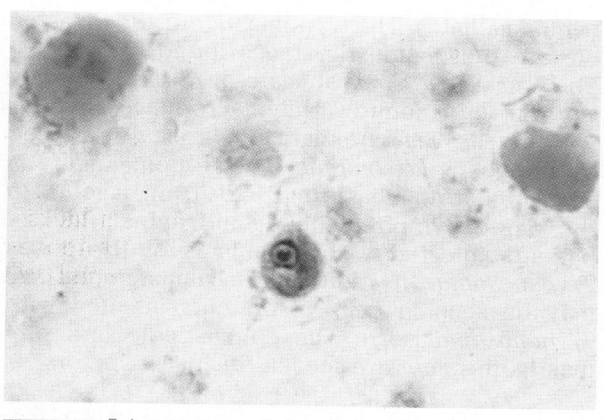

■ Figure 24–20
Chilomastix mesnili cyst (trichrome stain).

intestinal flagellates. First, they are transmitted by insect vectors, which are necessary for completion of the life cycle. Second, these organisms have different life cycle stages for diagnosis. Figure 24-21 shows the four life cycle stages of the

HUMAN

Tissue
(*Leishmania* sp and *Trypanosoma cruzi*)

Kinetoplast   Nucleus

Amastigote (flagellum retracted)

Blood/Cerebrospinal Fluid
(*Trypanosoma* sp)

• Free flagellum
• Undulating membrane attached posterior to nucleus

INSECT

Sand fly
(vector for *Leishmania* sp)

Promastigote

• Free flagellum

Tsetse fly
Reduviid bugs

• Free flagellum
• Undulating membrane attached anterior to nucleus

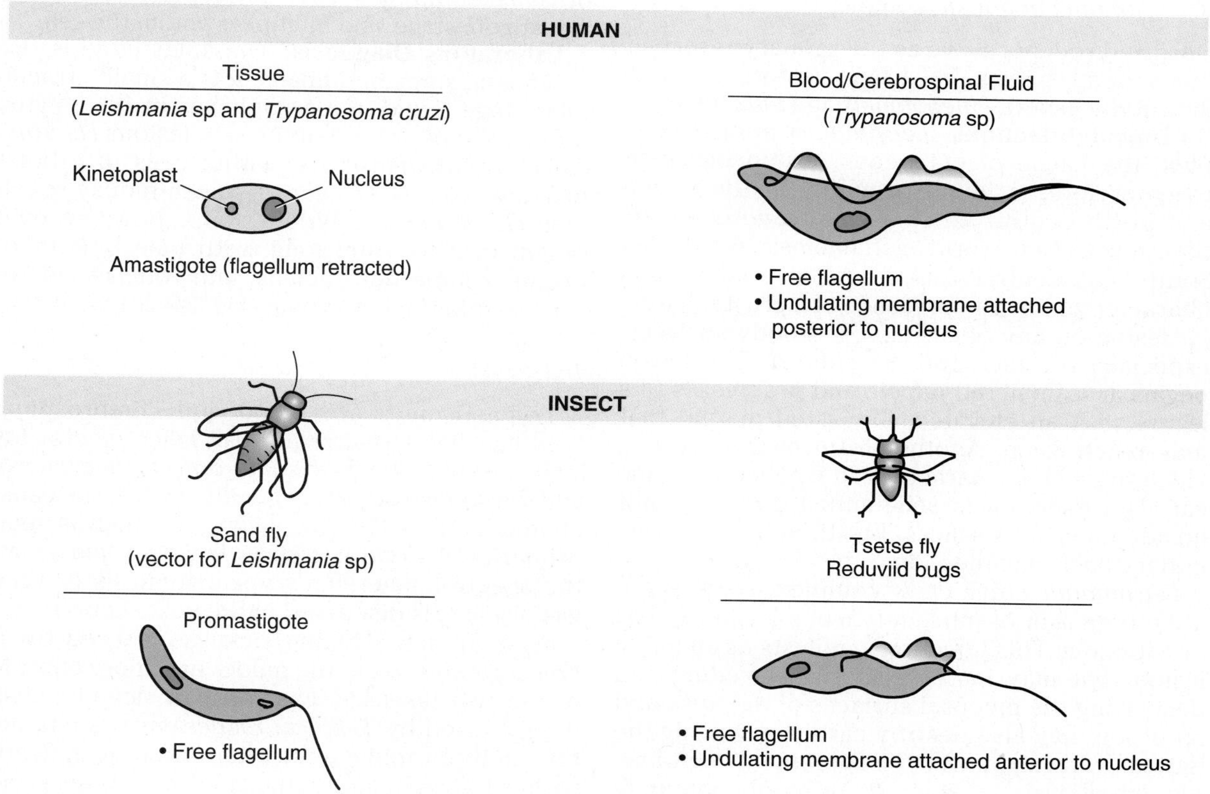

■ Figure 24–21
Life cycle stages of the blood and tissue flagellates.

hemoflagellates. The trypomastigote and amastigote are the diagnostic stages found in humans. The amastigote is an obligate intracellular organism, 2 to 3 μm, found within macrophages, liver or spleen cells, or bone marrow in diseases caused by the genus *Leishmania.* The trypomastigote, a flagellated form measuring 15 to 20 μm, is found in the blood, lymphatic fluid, and spinal fluid of patients infected with organisms of the genus *Trypanosoma.* In addition, the amastigote stage may be seen in cells of patients infected with *Trypanosoma cruzi.* The epimastigote and promastigote stages are seen in the insect vectors.

Leishmania

The genus *Leishmania* contains several complexes of species that cause disease in humans. These complexes are:

- *Leishmania tropica* complex
- *Leishmania braziliensis* complex
- *Leishmania mexicana* complex
- *Leishmania donovani* complex

Dogs and rodents serve as the primary reservoir hosts for all species. The insect vectors are sandflies of the genera *Phlebotomus* and *Lutzomyia.*

Clinical Infections. *Leishmania tropica* complex, the cause of cutaneous leishmaniasis or Oriental sore, is found primarily in the Orient and North Central Africa. *L. mexicana* complex, the cause of New World leishmaniasis, is found in South and Central America. The condition is characterized by the presence of a crusted circular lesion on any of the exposed body surfaces, especially the face and extremities. The lesion begins as a small red papule and progresses to a lesion with an elevated, indurated margin that may reach 8 cm. Another form of the disease, chiclero ulcer, is characterized by lesions on the ear. The infection is self-limiting and does not invade mucosal surfaces, but there may be secondary bacterial infection.

Leishmania braziliensis complex is the causative organism of mucocutaneous leishmaniasis or espundia. This infection manifests as an initial lesion that may increase in size, invading and destroying the mucosal surfaces of the nose and mouth. It may also destroy cartilage, leaving the patient with significant disfigurement. *Leishmania braziliensis* complex is primarily found in Mexico, South America, and Central America.

The most severe infection, visceral leishmaniasis or kala-azar, is endemic in parts of South America, Africa, Southern Europe, and Asia. The etiologic agents are organisms of the *Leishmania donovani* complex. In this disease, organisms spread through the lymphatics and invade organs of the reticuloendothelial system, including the liver, spleen, lymph nodes, and bone marrow. Patients with kala-azar exhibit malaise, anorexia, headache, and fever. In addition, they may show hepatomegaly and splenomegaly with invasion of *Leishmania*-containing macrophages into the bone marrow. Kidneys and heart may also be affected. If untreated, the disease is often fatal within 2 years. Standard therapy for all leishmanial infections is use of pentavalent antimony compounds.

Life Cycle. Figure 24–22 shows the generalized life cycle for *Leishmania* sp. The organism is ingested as an amastigote when the insect takes a blood meal. It develops as a promastigote in the gut of the insect and migrates to the salivary glands when mature. The promastigote is transmitted to the human through the salivary glands of the insect when it takes a blood meal. The promastigote is taken up by a macrophage, converts to the amastigote stage, and multiplies within the cell.

Laboratory Diagnosis. The amastigote is the diagnostic stage in humans. It is a small intracellular stage found in macrophages or histiocytes around the periphery of the skin lesions (*L. tropica* or *L. braziliensis*) or within cells of a bone marrow aspirate or liver or spleen biopsy specimen (*L. donovani*). Wright stain shows an oval organism 2 to 5 μm long, with pale blue cytoplasm, a large red nucleus, and rod-like kinetoplast within the cytoplasm (Fig. 24–23).

Trypanosoma

Trypanosomes are blood and cerebrospinal fluid flagellates that require an insect vector for transmission. *Trypanosoma brucei rhodesiense* and *Trypanosoma brucei gambiense* are the causative agents of sleeping sickness, which is seen primarily in Central Africa. *Trypanosoma cruzi,* the agent of American trypanosomiasis or Chagas disease, is discussed in the next section.

West African sleeping sickness, caused by *T. brucei gambiense,* is the milder and more chronic of the two diseases. East African sleeping sickness, caused by *T. brucei rhodesiense,* is characterized by a rapid course often resulting in death within 1 year. In addition, game animals serve as important reservoir hosts of *T. rhodesiense.*

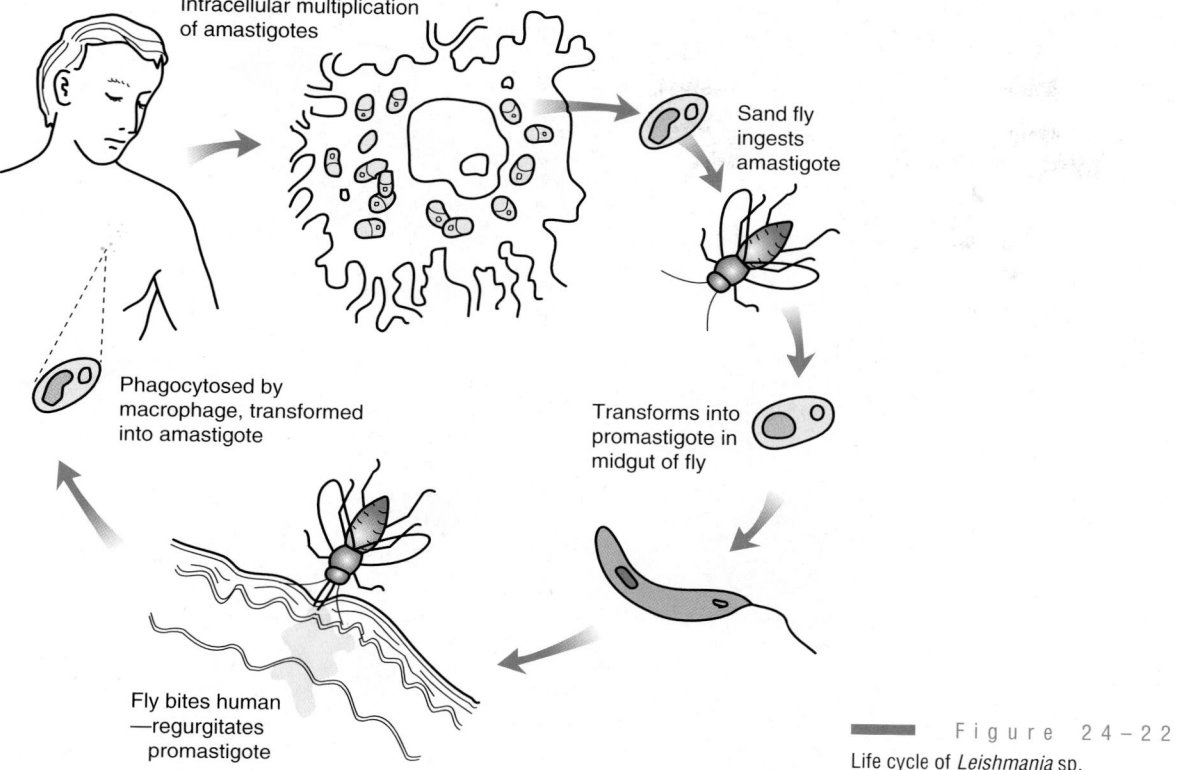

Intracellular multiplication of amastigotes

Sand fly ingests amastigote

Phagocytosed by macrophage, transformed into amastigote

Transforms into promastigote in midgut of fly

Fly bites human —regurgitates promastigote

Figure 24–22
Life cycle of *Leishmania* sp.

Life Cycle. The tsetse fly (*Glossina* sp) is the biologic vector for agents of sleeping sickness. Figure 24–24 shows a generalized life cycle for these agents. The fly ingests the trypomastigote stage when it takes a blood meal from a human. The organisms migrate to the insect gut and develop into an epimastigote, which, when mature, migrates to the salivary gland. There it develops into an infective metacyclic trypomastigote, which is transmitted to humans in saliva when the fly bites.

Clinical Infections. Initial symptoms include a local reaction at the site of the insect bite within 2 to 3 days. As the trypomastigotes enter the blood and lymphatics, the patient experiences fever, headache, joint and muscle pain, enlarged lymph nodes, especially in the posterolateral triangle of the neck (Winterbottom sign). Edema in the legs and arms and around the eyes is possible. As the trypomastigotes invade the central nervous system, the patient develops severe headaches, mental dullness, and apathy and may experience coordination problems, altered reflexes, and paralysis. Eventually, the patient has convulsions, lapses into a coma, and dies.

Laboratory Diagnosis. The diagnostic stage in humans is the trypomastigote, which is usually seen in a Wright-stained blood smear. The organism, however, can also be isolated from lymphatic fluid and cerebrospinal fluid. The trypomastigote is 15 to 20 μm with a single large nucleus and a posterior kinetoplast to which is attached the flagellum of the undulating membrane (Fig. 24–25). Species of *Trypanosoma* cannot be differentiated on a blood smear; diagnosis is based on clinical symptoms as well as geographic area.

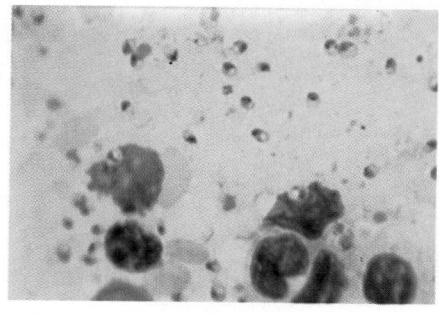

Figure 24–23
Amastigotes of *Leishmania* sp.

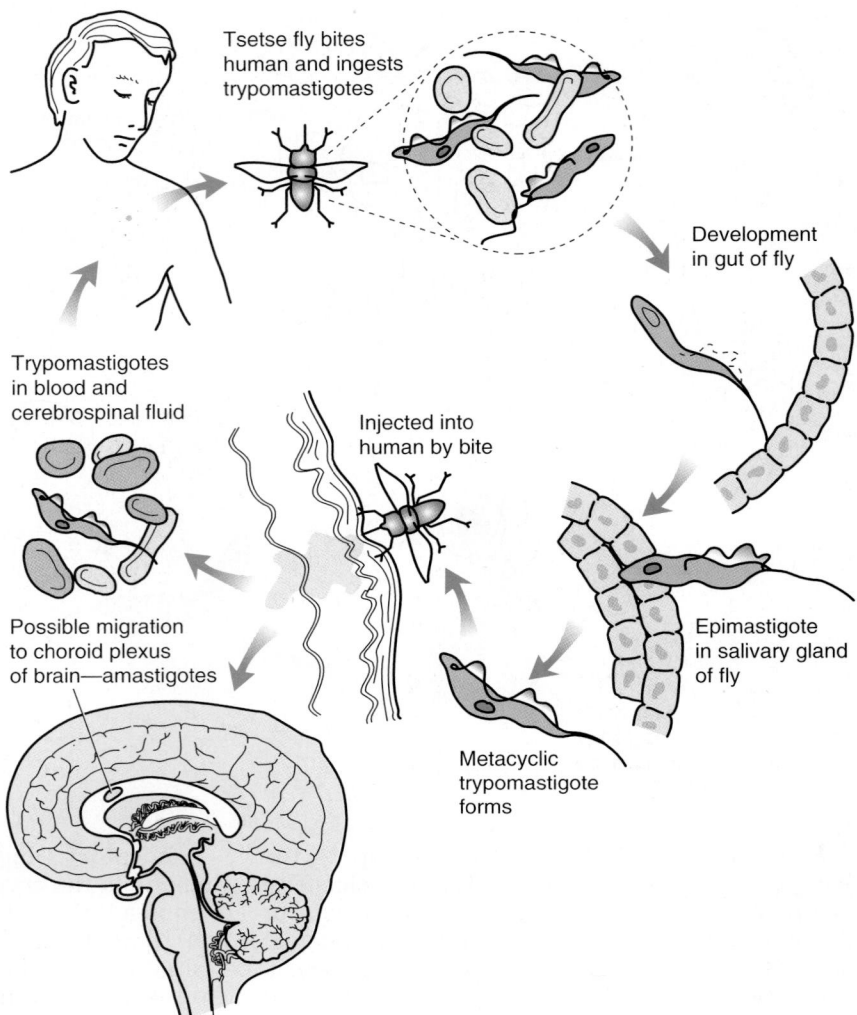

Figure 24–24

Life cycle of the etiologic agents of sleeping sickness (*Trypanosoma gambiense* and *T. rhodesiense*).

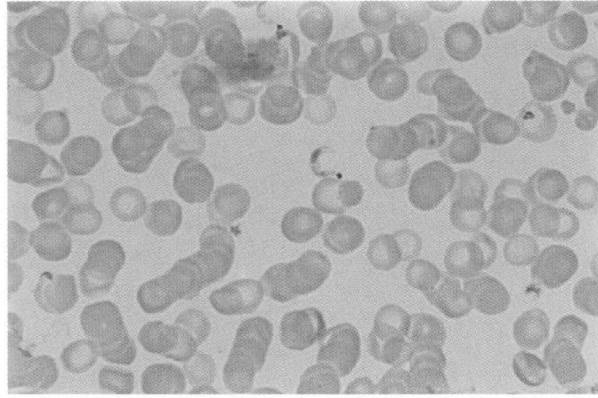

Figure 24–25

Trypanosoma sp. Trypomastigote in a blood smear.

Trypanosoma cruzi

Chagas disease is a zoonotic infection found primarily in rural areas of Mexico, Central America, and South America. It is caused by *T. cruzi,* which is transmitted by the insect known as the triatomid bug, reduviid bug, or kissing bug *(Triatoma* sp or *Panstrongylus* sp*)*. These insects live within mud or thatch walls of a dwelling during the day and come out at night to take a blood meal from the human inhabitants. Although insect transmission is most common, the organism has been transmitted by blood transfusion and congenital infection.

Life Cycle. Transmission of the organism to humans occurs when the insect defecates in the area surrounding its bite. Organisms in the feces are scratched into the bite and invade the blood stream as trypomastigotes. The trypomastigote enters a cell, transforms into an amastigote, multiplies, breaks out of the cell, and invades other cells. Cells of the cardiac muscle and skeletal muscle are most commonly infected. Figure 24–26 shows the life cycle of *T. cruzi*.

Clinical Infection. Acute infection with this organism usually occurs in children and may involve multiple organ systems. The incubation period ranges from 2 to 4 weeks. Symptoms during the acute stage vary and may include: the presence of a chagoma (ulcerative skin lesion), fever, a unilateral edema around the eye (Romaña's sign), lymphadenitis, hepatospleno-

megaly, malaise, muscular pains, and diarrhea and vomiting. Acute myocarditis, which progresses into a chronic form, also develops in many cases and may progress to congestive heart failure. Chronic disease in some patients may manifest as megacolon, megastomach, or megaesophagus.

Laboratory Diagnosis. The primary diagnostic stage in the blood is the trypomastigote. It is an elongate structure 15 to 20 μm long that often appears in a C or U shape. Like the other trypomastigotes, it shows a single large nucleus midbody and a posterior kinetoplast to which is attached the undulating membrane. Occasionally, the suggestion of a flagellum can be seen at

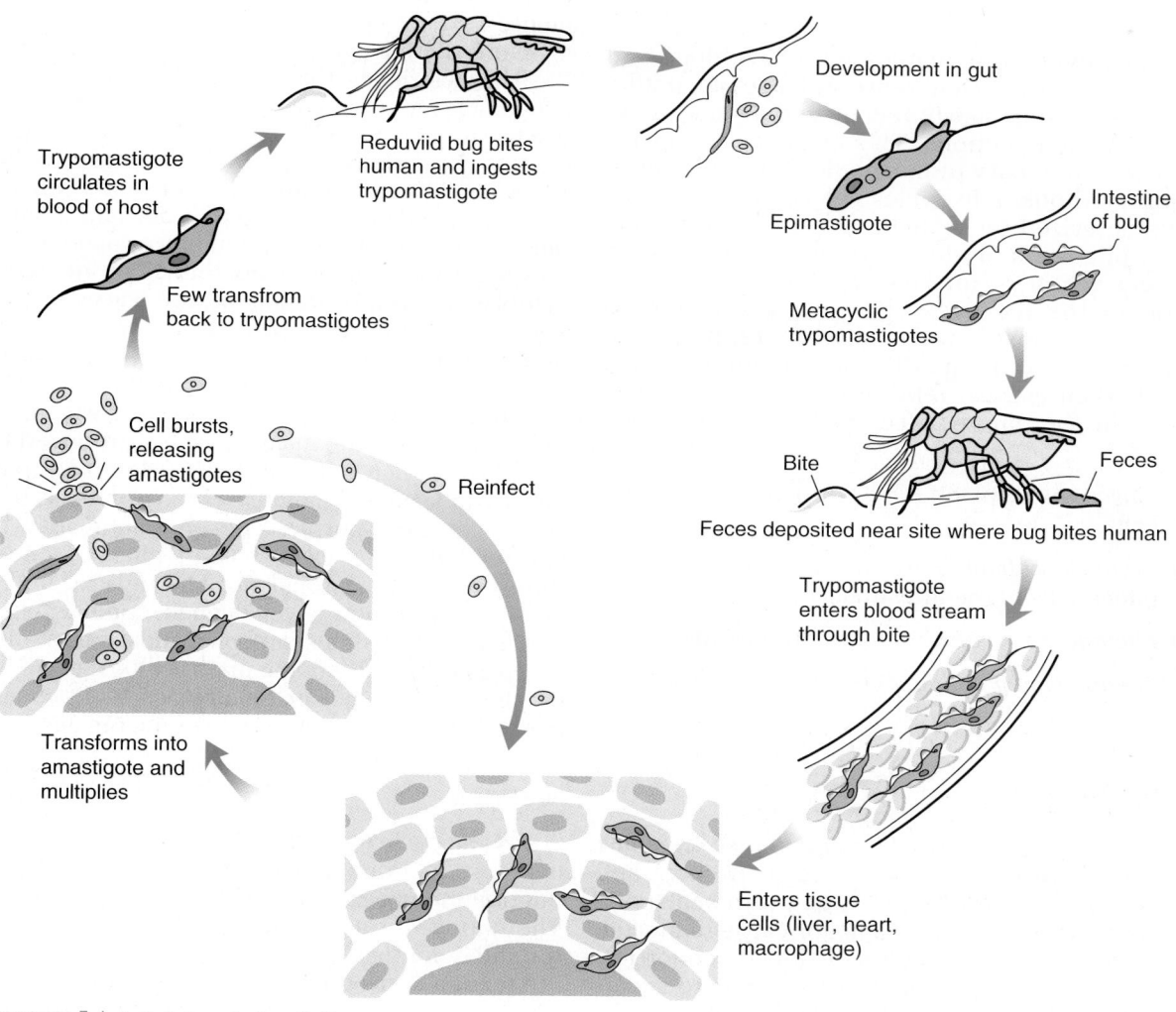

Trypomastigote circulates in blood of host

Reduviid bug bites human and ingests trypomastigote

Development in gut

Epimastigote

Intestine of bug

Metacyclic trypomastigotes

Few transfrom back to trypomastigotes

Cell bursts, releasing amastigotes

Reinfect

Bite

Feces

Feces deposited near site where bug bites human

Trypomastigote enters blood stream through bite

Transforms into amastigote and multiplies

Enters tissue cells (liver, heart, macrophage)

Figure 24 – 26

Life cycle of *Trypanosoma cruzi.*

the anterior end. In a cardiac or other tissue biopsy specimen, the organism can be seen in the amastigote stage. The morphology of all trypomastigotes is similar; therefore, a complete patient history is necessary to determine the species present.

Xenodiagnosis is used in Central and South America as a diagnostic method. A laboratory-raised triatomid bug is allowed to feed on patients suspected of harboring *T. cruzi*. When the insect is returned to the laboratory, its feces is examined on a regular basis for the presence of trypomastigotes. Presence of this stage in the insect's feces indicates that the patient was infected.

Apicomplexa

The phylum Apicomplexa contains blood and tissue parasites that represent age-old pathogens as well as newly recognized agents of opportunistic infections. This group of organisms shows a diversity of morphology and transmission methods. Life cycles are complex and are characterized by sexual and asexual reproduction phases. In addition, some may require an insect vector or intermediate host for completion of the life cycle. Humans may serve as definitive hosts when sexual reproduction takes place in human tissues; as intermediate host when asexual reproduction occurs. Organisms in this group infect many different body sites:

■ *Plasmodium* sp and *Babesia* sp infect red blood cells.

■ *Cryptosporidium parvum* and *Isospora belli* infect cells of the intestinal tract.

■ *Pneumocystis carinii* is found in the lung.

■ *Toxoplasma gondii* infects a variety of different organs.

Plasmodium spp

Malaria caused by *Plasmodium* sp remains endemic throughout the world in tropical and subtropical countries and, along with schistosomiasis and amebiasis, is a major cause of mortality in people in underdeveloped countries. Between 1 and 2 million deaths are due to malaria yearly. *Plasmodium vivax, Plasmodium ovale, Plasmodium malariae,* and *Plasmodium falciparum* are the etiologic agents of human malaria.

Plasmodium vivax has the widest geographic distribution and is the one most likely to occur in temperate climates. *Plasmodium ovale* is confined to Africa; *P. falciparum* and *P. malariae* have similar distributions throughout Africa and tropical countries. Infection with *P. malariae* is less common and less severe than that with *P. falciparum.* The CDC's Malaria Surveillance document (1990) noted that infections with malaria increased 8% between 1989 and 1990 in the United States. Of the 1,102 reported cases, *P. vivax* was the etiologic agent in 48%, *P. falciparum* in 41%. Most cases, however, were imported.

Although insect transmission of malaria is most common, there are reports of transmission through blood transfusion and infected needles. Transplacental infection has also been documented.

Before the life cycle of *Plasmodium* is detailed, the characteristics that distinguish the erythrocytic stages in a Giemsa-stained smear should be discussed. Figure 24–27 shows a sketch of these stages. The earliest stage is the ring-form trophozoite, in which the organism presents with a prominent red-purple chromatin dot and a small blue ring of cytoplasm surrounding a vacuole. The growing trophozoite is characterized by increased cytoplasm, disappearance of the vacuole, and appearance of malarial pigment in the cytoplasm of the organism. The immature schizont is characterized by a splitting of the chromatin mass. The mature schizont contains merozoites, which are individual chromatin masses, each surrounded by cytoplasm. Microgametocytes (male) have pale blue cytoplasm and a diffuse chromatin mass that stains pale pink-purple. The chromatin may be surrounded by a clear halo. Pigment is seen throughout the cytoplasm. The macrogametocytes (female) show a well-defined, compact chromatin mass that stains dark pink and a darker blue cytoplasm. The chromatin mass is often set eccentrically in the organism. Small masses of pigment may be clumped near the edge of the organism.

LIFE CYCLE

The life cycle of *Plasmodium* involves both sexual reproduction (sporogony) and asexual reproduction (schizogony), as shown in Figure 24–28. The *Anopheles* mosquito serves as biologic vector and definitive host.

Asexual Reproduction. Asexual reproduction, which occurs in the human, has an exoerythrocytic phase that takes place in the liver and an erythrocytic phase that takes place in red blood cells.

Exoerythrocytic Phase. The human serves as intermediate host and acquires the infection when the female mosquito takes a blood meal and injects the infective sporozoites with salivary secretions. At this time, sporozoites enter the human circulation and initiate schizogony. Sporozoites take approximately 30 to 60 minutes to reach the liver, where they begin the exoerythrocytic cycle by penetrating parenchymal cells. Maturation through the trophozoite and schizont phases results in production of merozoites. Each schizont contains 10,000 to 30,000 merozoites. Release of mature merozoites from liver cells and invasion of red blood cells signal the beginning of the erythrocytic phase. Generally, there is only one cycle of merozoite production in the liver before red cells are invaded. Organisms such as *P. vivax* and *P. ovale,* however, may persist in the liver in a dormant stage known as hypnozoite, which accounts for the relapse of the disease after many years. Neither *P. malariae* nor *P. falciparum* has a persistent liver phase, although recrudescence in untreated persons with either of these organisms may be the result of a continued subclinical erythrocytic infection.

Erythrocytic Phase. Merozoites have structures that are selectively adhesive for the red blood cell membrane and that attach to receptors on the membrane. Once merozoites have attached, endocytic invagination of the red cell membrane allows the organism to enter the red cell within a vacuole. *Plasmodium vivax* may use antigens of specific blood groups such as those in the Duffy system as receptors, whereas *P. falciparum* may simply attach to a portion of the membrane itself.

Once inside the cell, the organism feeds on hemoglobin and goes through a maturation sequence—ring-form trophozoite to mature trophozoite. Malarial pigment is formed in the growing trophozoite as a result of incomplete metabolism of the hemoglobin. Unutilized hematin is combined with protein and deposited in the cytoplasm of the organism. Once the organism has reached the mature trophozoite stage, the chromatin begins to divide (schizont). When the chromatin split has been completed, each chromatin mass is surrounded by its own bit of malarial cytoplasm (merozoite). For each malarial species, there is a typical number of merozoites, which may be used as an identifying characteristic. When the red cell ruptures, merozoites are released to invade other red cells. There are two possible outcomes at this point. One is that the merozoite enters a cell and

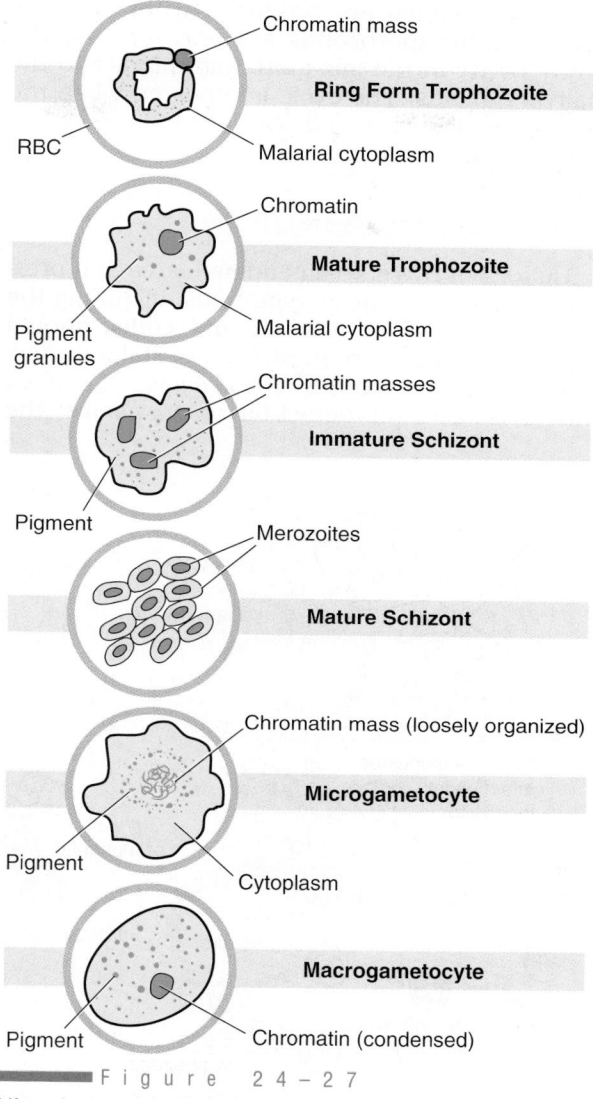

Figure 24–27
Life cycle stages of malaria.

repeats development into a schizont; the other is that it enters a cell and develops into one of the sexual stages—either the microgametocyte or the macrogametocyte.

Sexual Reproduction. Sporogony, which takes place in the mosquito, results in production of sporozoites infective for humans. Both microgametocyte and macrogametocyte are infective for the female mosquito when she takes a blood meal. In the insect, stomach exflagellation by the male and subsequent fertilization results in formation of an ookinete that migrates through the gut wall and forms an oocyst on the exterior gut wall. Sporozoites are produced within the oocyst. Mature sporozoites are released into the body cavity of the mosquito and migrate to the salivary glands. The cycle is repeated when the female injects sporozoites into a human as she takes her blood meal.

DIAGNOSIS AND IDENTIFICATION

History of travel to an endemic area and presence of classic clinical symptoms, including the malarial paroxysm of fever and chills, should alert a clinician to request Giemsa-stained thick and thin blood smears for malaria. The thick smear is used for detection of organisms; the thin smear for their identification. Trophozoites, schizonts, and gametocytes may be seen in the blood smear. Table 24–5 gives the general characteristics of trophozoites and schizonts of the malarial species. The laboratory identification of malarial species involves examination of both the red blood cell (RBC) morphology and the parasite's characteristics.

Plasmodium vivax. *Plasmodium vivax* has a tertian life cycle pattern—that is, it takes approximately 48 hours for the life cycle to complete itself. The invasion of a new group of red cells begins on the third day. *Plasmodium vivax* usually invades young RBCs (reticulocytes) and is therefore characterized by enlarged red cells,

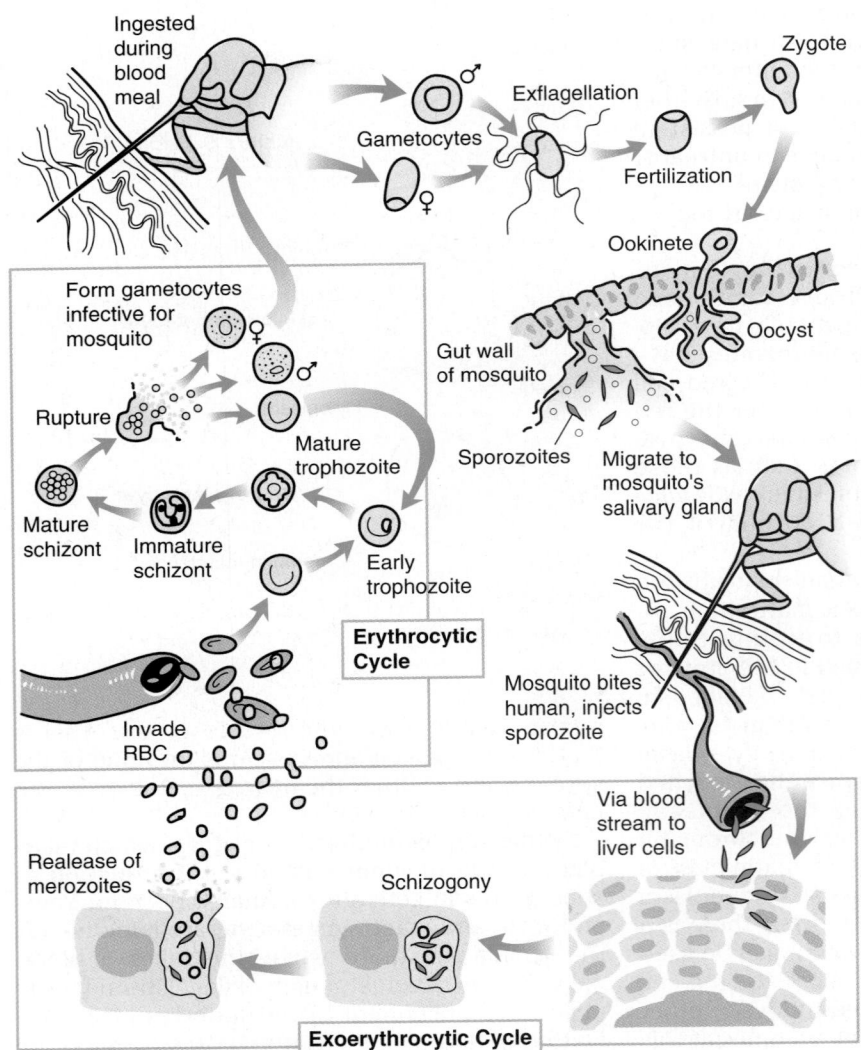

Figure 24–28
Life cycle of *Plasmodium* sp.

■■■■■■ T A B L E   2 4 – 5
COMPARISON OF MALARIAL SPECIES

Plasmodium Species	RBC Morphology	Trophozoite	Number of Merozoites in Schizont	Reproductive Cycle (hours)	Other Characteristics
P. vivax	Enlarged (1.5–2×) Schüffner stippling	Ameboid Large vacuoles Golden-brown pigment	12–24 Average: 16	48	Wide range of stages in peripheral blood
P. malariae	Normal size May have dark hue Ziemann dots (rare)	Compact May assume "band form" across cell Coarse, dark brown pigment	6–12 Average: 8 with "daisy petal"–like arrangement around clumped pigment	72	—
P. ovale	Enlarged, oval Fringed edge Schüffner stippling	Compact Coarse, dark pigment (resembles P. malariae)	6–16 Average: 8	48	—
P. falciparum	Normal size Multiple infections Maurer dots (rare)	Small, delicate Double chromatin dots Appliqué forms Dark pigment	12–36 Average: 20–24	Irregular, 36–48	Crescent-shaped gametocyte Ring and gametocyte stages only in peripheral blood

often 1.5 to 2 times normal. A fine pink stippling known as *Schüffner stippling* may be present in the cell. The young trophozoite is characterized by its ameboid appearance; by maturity, it usually fills the RBC, and golden brown malarial pigment is present. The mature schizont contains 12 to 24 merozoites, with an average of 16. Gametocytes are rounded and fill the cell. Macrogametocytes are often difficult to differentiate from mature trophozoites. Figure 24–29*A* shows a trophozoite; *B*, a schizont; *C*, a macrogametocyte; and *D*, a microgametocyte of *P. vivax.*

Plasmodium malariae. *Plasmodium malariae,* on the other hand, usually invades older red cells, perhaps accounting for the occasional darker appearance of the invaded red cell. The life cycle is characterized as quartan, with reproduction every 72 hours and invasion of new red cells every fourth day. The trophozoite is compact and may assume a characteristic "band" appearance, in which it stretches across the diameter of the red cell (Fig. 24–30*A*). Dark, coarse brown-black pigment is present. Occasionally, a few pink cytoplasmic dots called Ziemann dots may be seen. The mature schizont contains 6 to 12 merozoites, with an average of 8. Merozoites may be arranged in a characteristic loose "daisy petal" arrangement around the clumped pigment; however, they may also be randomly arranged. Figure 24–30*B* shows a schizont of *P. malariae.*

Plasmodium ovale. *Plasmodium ovale,* the least commonly seen of the species, exhibits char-

acteristics of a *P. vivax*-infected cell but looks like a *P. malariae* organism. In infections with *P. ovale,* the RBC is enlarged and may assume an oval shape with fimbriated or fringe-like edges. Schüffner stippling is seen less commonly than with *P. vivax.* The parasite remains compact with dark brown pigment and has a range of 6 to 12 merozoites in the mature schizont. It also exhibits a tertain life cycle. Figure 24–31 shows a trophozoite of *P. ovale.*

Plasmodium falciparum. Although identified as having a tertian life cycle, *P. falciparum* usually demonstrates an asynchronous life cycle, with rupture of the red cells taking place at irregular intervals ranging from 36 to 48 hours. The life cycle stages seen in peripheral blood are usually limited to the ring-form trophozoite and the gametocyte. Other stages mature in the venules and capillaries of the major organs. *Plasmodium falciparum* invades red cells of any age and, for this reason, often exhibits the highest parasitemia—reaching 50% in some cases. Wedge-shaped dots known as Maurer clefts may be present, but they occur uncommonly and require excellent staining to be visualized. The ring forms of *P. falciparum* (Fig. 24–32*A*) are more delicate than those of other species and often have two chromatin dots. Appliqué forms of the parasite appear to be external to the RBC membrane, and multiple ring forms in a single cell are common. The mature trophozoite is small and compact and may have dark brown pigment. The schizont has from 8 to 36

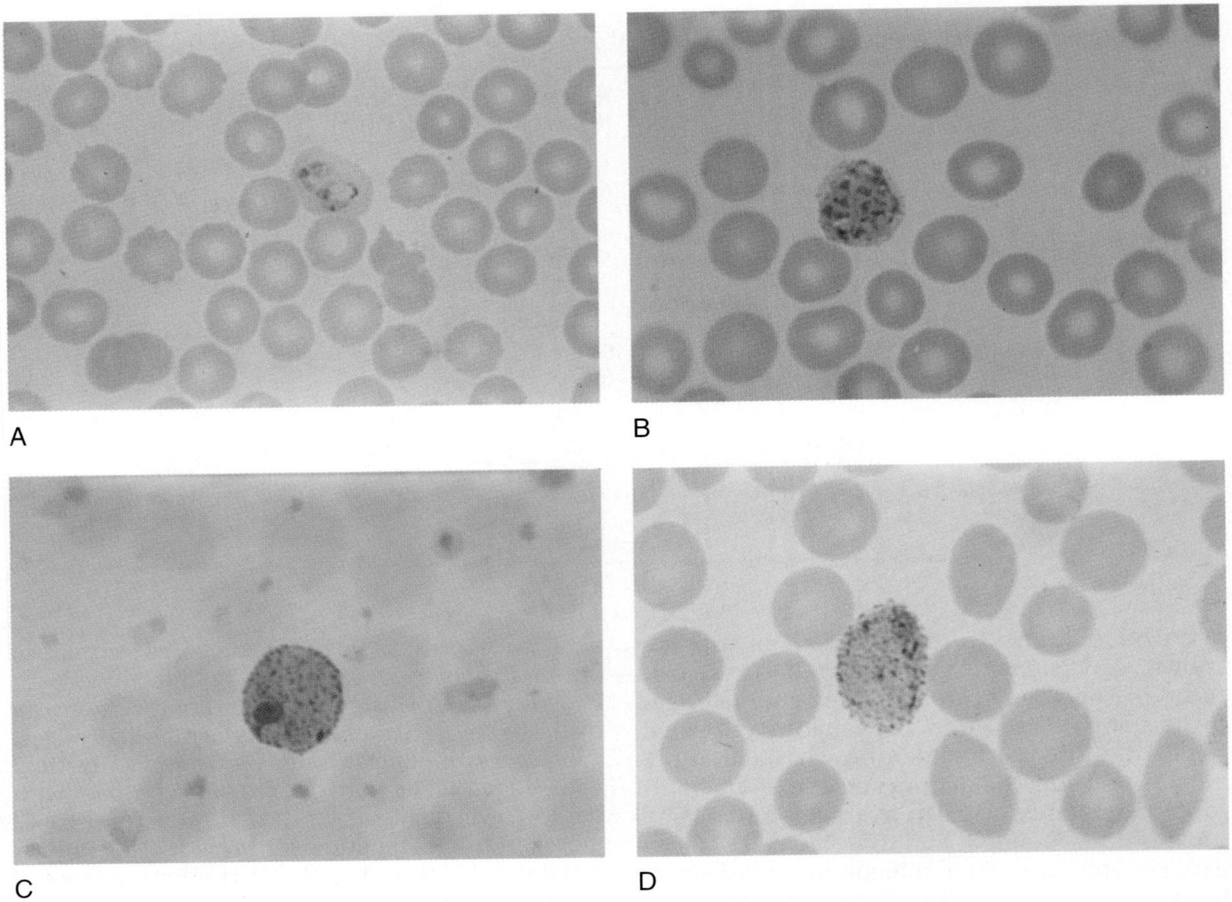

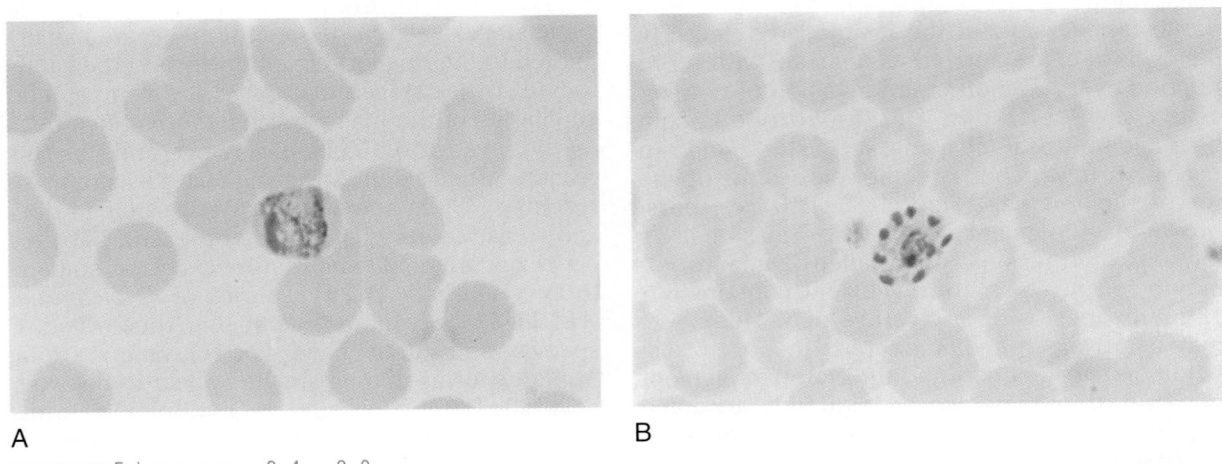

A

B

C

D

■■■■ F i g u r e 2 4 – 2 9

A, Plasmodium vivax trophozoite. *B, P. vivax* immature schizont. *C, P. vivax* macrogametocyte. *D, P. vivax* microgametocyte.

A

B

■■■■ F i g u r e 2 4 – 3 0

A, Plasmodium malariae trophozoite "band form." *B, P. malariae* schizont.

merozoites, with an average of 20 to 24. Gametocytes have a characteristic banana or crescent shape (see Fig. 24–32*B*).

SYMPTOMS AND PATHOLOGY

The malarial paroxysm is the primary symptom associated with the erythrocytic cycle. It is linked to rupture of the red cells and release of merozoites, malarial metabolites, and endotoxin-like substances into the blood stream. The prodromal phase of the paroxysm involves headache, bone pain, nausea, or flu-like symptoms. A shaking chill (10 to 15 minutes) initiates the paroxysm and is followed by a fever of up to 40°C. The paroxysm lasts from as few as 2 hours up to 20 hours. When the fever finally breaks, the patient begins to sweat profusely. This cycle repeats itself at regular intervals, depending on the species of malaria present. Long-term infections with any malarial organism may result in damage to liver and spleen by deposits of malarial pigment (hemozoin).

Infections with *P. vivax* and *P. ovale* generally do not have the range of complications seen with other species. Infections with *P. malariae* may lead to nephrotic syndrome owing to deposition of circulating immune complexes of malarial antigen and antibody on the basement membrane of the glomerulus, which causes an autoimmune reaction against the basement membrane.

The most severe form of malaria, however, is caused by *P. falciparum,* with the primary complication being the development of cerebral malaria. Between 20 and 50% of deaths due to *P. falciparum* are caused by this CNS complica-

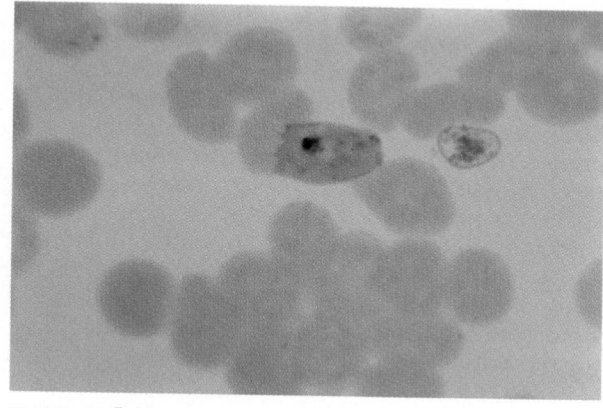

■Figure 24–31
Plasmodium ovale trophozoite.

tion. The parasite-infected red blood cell develops sticky knobs that mediate adhesion to the endothelial cells of the capillary walls. Blood flow is slowed, causing a decrease in oxygen delivery to the tissues with resultant tissue anoxia. The patient has severe headaches, may be confused, and ultimately lapses into a coma. Renal failure is due to tubular necrosis resulting from the decreased blood flow.

A second but less common complication of infection with *P. falciparum* is blackwater fever. It usually develops in patients with repeated infections and those on quinine therapy. It may be mediated by an antigen-antibody reaction caused by development of an autoantibody against the red cell. The black appearance of the urine that gives the syndrome its name is due to

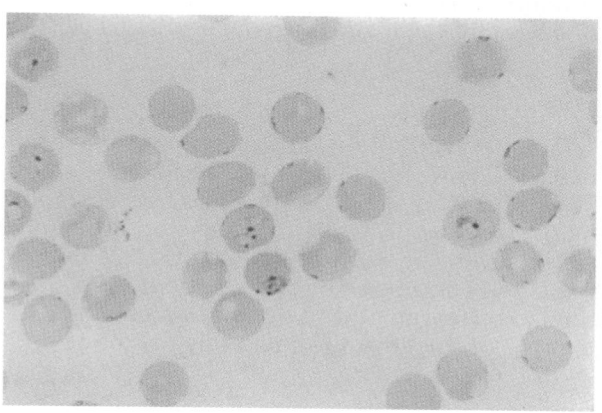
A

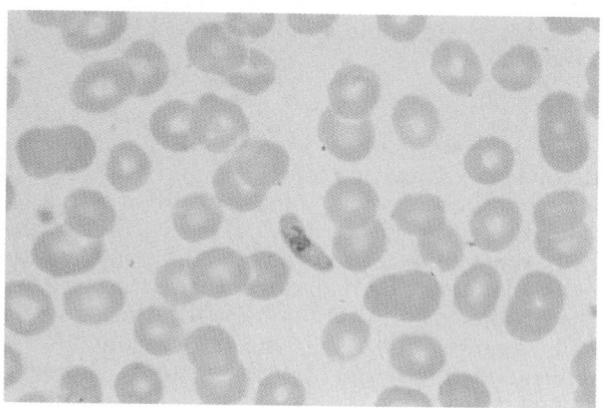
B

■Figure 24–32
A, Plasmodium falciparum ring form trophozoites. *B, P. falciparum* gametocyte.

massive intravascular hemolysis and resulting hemoglobinuria.

Although there is no complete immunity to malaria, patients in endemic areas develop antibody against asexual stages, which helps decrease the parasite load and the severity of illness. In addition, reports of antibodies directed against sporozoites and gametes indicate that they may also help decrease the rate of infection and severity of illness. Patients with hemoglobinopathies such as sickle cell disease are somewhat protected against severe malaria, because the parasite cannot exist in these cells.

TREATMENT

Chloroquine remains the primary drug for prophylaxis and treatment of malaria and is effective against all asexual stages of malaria and all gametocytes except mature *P. falciparum*. A number of strains of *P. falciparum,* however, have become resistant to the drug. Pyrimethamine-sulfadoxine can be given in addition to chloroquine to persons traveling to areas with endemic *P. falciparum*–resistant strains. Some strains of *P. falciparum* and *P. vivax,* however, have become resistant to this drug also. Mefloquine may be used as prophylaxis and in treatment of multidrug-resistant strains of *P. falciparum.* There is evidence, however, that resistance to this drug is developing in the malarious areas of Southeast Asia and Africa. Quinine use has been reestablished in some cases of multidrug-resistant *P. falciparum.*

Primaquine, effective against hyponozoites that persist in the liver, is used to treat persons infected with *P. vivax* and *P. ovale* to prevent relapses of these infections.

LABORATORY DIAGNOSIS

The classic method for diagnosis of malaria involves examination of thick and thin blood smears. In the thick smear which is used to detect the presence of malarial parasites, the red cells have been destroyed, so that only white cells, platelets, and malarial parasites are visible. Distortions of morphology and the lack of red cells require that a thin smear be examined to determine the actual species present. The thin smear is used for species identification, in that morphology of the RBC as well as that of the organism can be seen.

Several articles have described a procedure using a modified microhematocrit tube coated with acridine orange stain for detection and identification of malaria. This method, the quantitative buffy coat (QBC) system, demonstrates the presence of parasites, but a thin smear must still be made for definitive identification. DNA probes have been proposed as another method for diagnosing malaria. The extended time and costly equipment involved do not make the procedure cost effective or efficient at this time, especially for field work. Serologic tests for antibody to malaria are not useful in an endemic area but may be useful for persons who have traveled to an endemic area and are presenting with clinical symptoms of malaria.

Babesia microti

Babesia microti is another intraerythrocytic human parasite. *Babesia* spp have been known to infect cattle and other animals, but the first cases to be reported in humans occurred in the 1950s. These initial cases were limited to splenectomized patients, but since then, cases of babesiosis have been reported in patients who are not asplenic. The first documented cases in the United States were in the Martha's Vineyard and Nantucket Island area, where a hard tick (*Ixodes dammini*) served as vector. As with malaria, cases of perinatal and transfusion-transmitted babesiosis have been reported.

There have also been reports indicating simultaneous transmission of babesiosis and lyme borreliosis, because the tick vectors are the same for the two causative organisms. Patients manifest clinical symptoms of one disease but show a concurrent rise and fall in antibody titers to both *Babesia* and the Lyme borreliosis organism.

SYMPTOMS

Patients with babesiosis may present with malaria-like symptoms such as fever, chills, sweats, and myalgia. Many cases are asymptomatic, however. Anemia may develop if hemolysis is severe or prolonged. The clinical course tends to be more severe if the patient is splenectomized. Treatment with quinine sulfate and clindamycin is effective in eliminating the infection.

LIFE CYCLE

As with *Plasmodium,* there are alternating sexual and asexual reproduction stages of the life cycle of *Babesia.* Production of the infective stage (sexual reproduction) takes place in the tick, and schizogony (asexual reproduction) in the red blood cells of the human. There is no exoerythrocytic cycle in humans. There also is no identifiable gametocyte stage in the red cell. Transovarian transmission can occur in the tick, allowing the life cycle to persist without an intermediate host.

DIAGNOSIS

The diagnosis of babesiosis is made from a stained thin blood smear. The organisms appear as small, delicate, ring-form trophozoites, about 1 to 2 μm with a prominent chromatin dot and faintly staining cytoplasm. More mature trophozoites may appear as pyriform organisms in single, double, or the classic tetrad (Maltese cross) formation within the red cell. The morphology may initially be confused with that of *P. falciparum*. Unlike in malaria however, extracellular organisms are seen. Lack of pigment and absence of other life cycle stages serve as additional keys to differentiate *Babesia* from *P. falciparum*. Serologic studies can be used to quantitate antibody titers. Although transfusion-transmitted babesiosis has been reported, there is no test for screening blood donors. Figure 24–33*A* shows the small, compact, ring-like trophozoites of *B. microti;* 24–33*B* demonstrates the tetrad formation characteristic of the organism.

Toxoplasma gondii

Toxoplasmosis is a worldwide disease that can manifest as a wide range of clinical symptoms and complications. Serologic studies indicate that more than 50% of people in the United States have antibodies to this organism.

CLINICAL INFECTION

Patients with acute infections may be asymptomatic or may present with mild flu-like or mononucleosis-like symptoms, including low-grade fever, lymphadenopathy, malaise, and muscle pain. Once the acute infection is resolved, the organism enters a relatively inactive stage in the tissues.

The organism may also be transmitted congenitally. Children who acquire *T. gondii* this way may present with a range of serious complications, including mental retardation, microcephaly, seizures, hydrocephalus, retinochoroiditis, and blindness. The earlier in the pregnancy that the fetus is exposed to the infection, the more likely that severe complications will result. Infections acquired later in pregnancy may result in the child's being asymptomatic at birth but developing complications later in childhood.

Immunosuppressed patients, particularly those with leukemia or lymphoma and those undergoing chemotherapy, may suffer from a primary infection or reactivation of a latent infection. Either manifests as a fulminating encephalitis and results in rapid death of the patient. There are a number of reports of the death of patients with AIDS due to disseminated toxoplasmosis, which manifests as central nervous system symptoms. Computed tomography scans may demonstrate lesions in the brain that represent *Toxoplasma* cysts. Pulmonary toxoplasmosis may also be present in conjunction with the CNS infections.

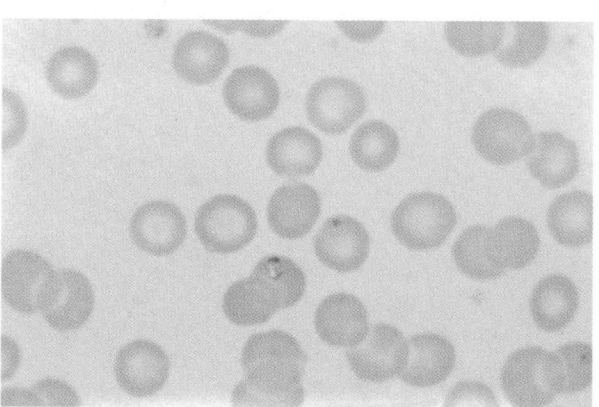

A

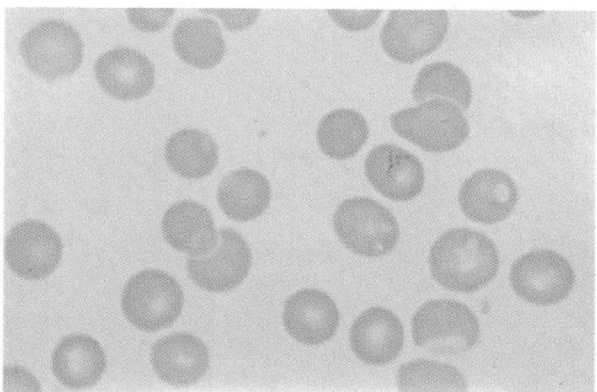

B

Figure 24–33

A, Babesia microti trophozoite. *B, Babesia microti* tetrad form.

TRANSMISSION AND LIFE CYCLE

The household cat and other members of the family Felidae serve as definitive hosts for the organism. Sexual and asexual reproduction occur in the intestine of the cat; only asexual reproduction occurs in humans. The result of sexual reproduction is the oocyst, which is passed in cat feces. The oocyst requires 2 to 5 days in the environment to mature and become infective.

There are two life cycle stages in the human—the tachyzoite and the bradyzoite. Actively motile and reproducing forms are referred to as *tachyzoites*. They are crescent shaped, are 4 to 6 μm long, and have a prominent nucleus (Fig. 24–34A). Figure 24–34B shows lung tissue containing free tachyzoites as well as a number of intracellular forms. Bradyzoites are the slowly growing and reproducing forms found within a cyst-like structure during the dormant phase of the infection.

There are basically three ways in which humans can acquire infection with *T. gondii,* as shown in the life cycle illustrated in Figure 24–35:

■ Ingestion or inhalation of the oocyst, which is shed in cat feces

■ Ingestion of undercooked meat containing the cyst with bradyzoites

■ Congenital transmission

In the first method, when a human ingests the infective oocyst, sporozoites present in the oocyst are liberated in the intestine, penetrate the intestinal wall, gain access to the circulation, and migrate to the organs. These tachyzoites invade cells, multiply, and eventually cause cells to rupture and release tachyzoites to invade other cells. The immune system, in particular the T cells, respond when tachyzoites invade the tissues. This immune response results in formation of a large cyst-like structure that contains the slowly growing and reproducing bradyzoites. At this point, the infection remains in a dormant state, unless the immune system is compromised. In the immunocompromised patient, the bradyzoites are released from the cyst and become active tachyzoites that invade multiple organs, especially the central nervous system.

In the second method, a human ingests raw or undercooked meat containing the cyst with bradyzoites. The cyst wall is dissolved and bradyzoites are liberated. Studies show that these organisms are resistant to digestive tract enzymes for about 6 hours, during which time they convert to tachyzoites and invade the intestinal wall. They then gain access to the circulation and subsequently invade various organs.

Congenital transmission occurs when tachyzoites in the maternal circulation cross the placenta and enter the fetal circulation and tissues.

LABORATORY DIAGNOSIS

Identification of the tachyzoite or pseudocysts with bradyzoites in tissue is very difficult, because no single organ is invaded. Antibodies to the organism show rapid rise in infection, and test for antibodies are most commonly used for

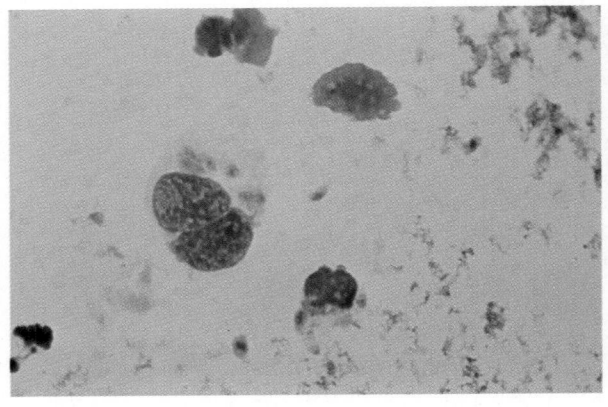

A

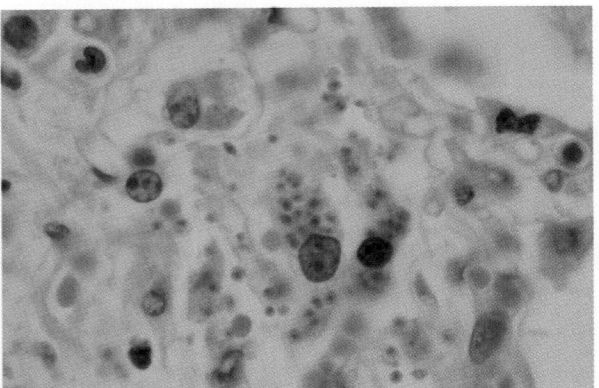

B

■■■■F i g u r e   2 4 – 3 4
A, Toxoplasma gondii tachyzoites. *B, T. gondii* tachyzoites in lung tissue.

diagnosis. The Sabin-Feldman dye test was the first antibody test developed. It is not used in clinical laboratories because it requires live organisms. Indirect fluorescent antibody tests and enzyme immunoassay tests are routinely used for

diagnosis. Because most persons have an antibody titer to the organism, interpretation of the titer must be linked to clinical symptoms. A rise in titer between acute and convalescent specimens may indicate acute infection. An IgM-specific test

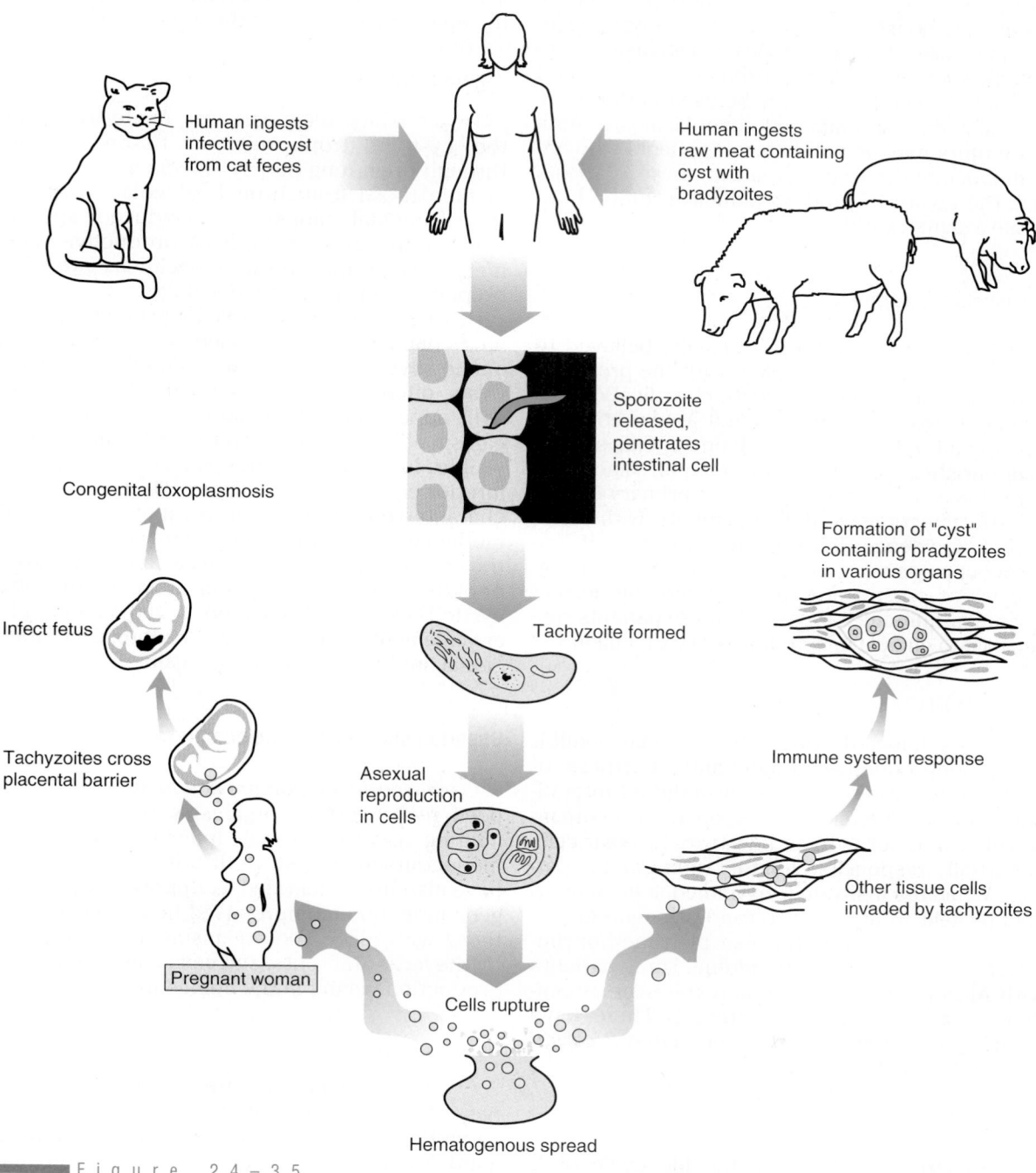

Human ingests infective oocyst from cat feces

Human ingests raw meat containing cyst with bradyzoites

Sporozoite released, penetrates intestinal cell

Congenital toxoplasmosis

Infect fetus

Tachyzoite formed

Formation of "cyst" containing bradyzoites in various organs

Tachyzoites cross placental barrier

Immune system response

Asexual reproduction in cells

Pregnant woman

Other tissue cells invaded by tachyzoites

Cells rupture

Hematogenous spread

Figure 24–35

Life cycle of *Toxoplasma gondii.*

may also be used to diagnose acute infections and may be useful in diagnosis of toxoplasmosis in pregnant women with suspected exposure to the organism or in the neonate with suspected congenital infection. IgM titers usually peak within the first month of infection. In disseminated toxoplasmosis, histologic stains of biopsy materials may demonstrate the cyst with bradyzoites or, in some cases, the tachyzoites. Noninvasive technology, such as magnetic resonance imaging and computed tomography, may be used in diagnosis of suspected disseminated toxoplasmosis. Antibody titers may be unreliable in patients who are immunocompromised, because these patients lack the ability to produce sufficient antibody to cause a significant rise in titer.

Pneumocystis carinii

Pneumocystis carinii was originally believed to be a yeast, was then classified with the protozoa, and now is being reevaluated for classification as a fungus. It was initially identified as the etiologic agent in interstitial plasma cell pneumonia seen in malnourished or premature infants. Since the early 1980s, it has been one of the primary opportunistic infections found in patients with AIDS. Higher incidence is also due to use of immunosuppressive drugs in patients with malignancies or organ transplants. Underlying defects in cellular immunity apparently make patients susceptible to clinical infection with the organism.

CLINICAL INFECTION

Patients infected with *P. carinii* may exhibit fever, nonproductive cough, and shortness of breath. Chest radiographs show a diffuse interstitial infiltrate. The immune response to the organism after it attaches to and destroys alveolar cells is partially responsible for this radiographic pattern. When the infiltrate is examined, it is found to contain cells from the alveoli and plasma cells.

Treatment includes the use of trimethoprim-sulfamethoxazole or pentamidine. Some patients with AIDS have been given aerosolized pentamidine as a prophylactic treatment. There have been reports, however, of disseminated *P. carinii* infection in such patients.

LIFE CYCLE

There are two stages in the life cycle of *P. carinii*—the trophozoite, which is 1 to 4 μm in

size and is irregularly shaped, and the cyst, which is a thick-walled sphere of 6 to 8 μm containing up to eight intracystic bodies. Transmission of the organism is known to occur through the respiratory route, with the cyst being the infective stage. The intracystic bodies are released from the cyst in the lung and multiply on the epithelial cells lining the lung.

LABORATORY DIAGNOSIS

Traditionally, diagnosis was made by finding the cyst or trophozoite in tissue obtained through open lung biopsy. Specimens now used include those from bronchoalveolar lavage, or transbronchial biopsy, and induced sputum. Sputum, however, is the least productive specimen. Lavage and sputum specimens are often prepared using the cytocentrifuge.

Histologic stains such as Giemsa and Gomori methenamine silver are used. With the methenamine silver stain, the cyst wall will stain black. Cysts often having a punched-out, "Ping-Pong ball" appearance. With the Giemsa stain, the organism appears round and the cyst wall is barely visible. Intracystic bodies are seen around the interior of the organism. Figure 24–36A shows the characteristic black-staining cyst of *P. carinii* with methenamine silver stain. Figure 24–36B shows the cyst as stained with Giemsa stain. The exterior of the cyst does not pick up stain, but the intracystic bodies can be demonstrated as a circular arrangement within the cyst. A fluorescent monoclonal antibody stain has also gained wide use.

Opportunistic Intestinal Apicomplexa

Cryptosporidium parvum and *Isospora belli* have been recognized as organisms that cause self-limiting gastrointestinal infection in the immunocompetent host. Both organisms, however, have also been identified as opportunistic pathogens in immunocompromised hosts, particularly those with AIDS. The organisms are transmitted by the fecal-oral route, and sexual and asexual reproduction occurs in the human intestinal tract.

Cryptosporidium parvum

Cryptosporidium spp are recognized animal parasites that were initially seen in humans as zoonotic infections in veterinarians and other animal handlers. They are obligate intracellular parasites transmitted by ingestion of the infec-

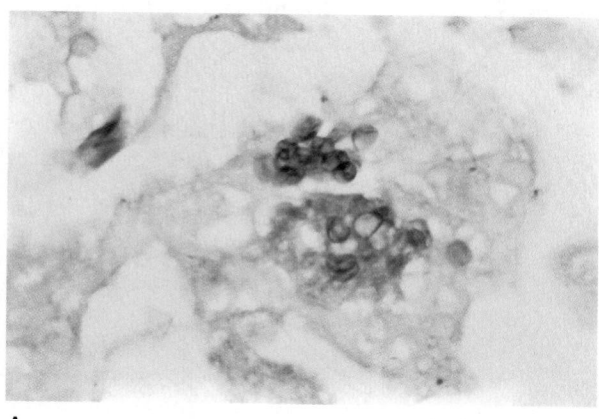
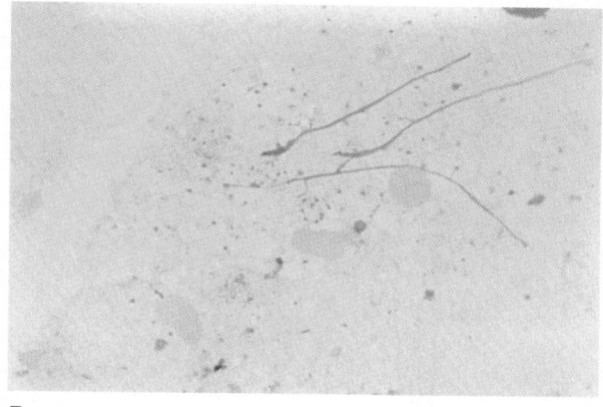

A   B

Figure 24-36

A, *Pneumocystis carinii* cysts (silver stain). B, *P. carinii* (Giemsa stain). Notice circular arrangement of intracystic bodies within the faint outline of cyst wall in center of field.

tive oocyst. There have also been reports of outbreaks in the general population and in day care centers. In the 1980s, cryptosporidiosis was identified as a major opportunistic infection in patients with AIDS. The primary organism associated with human outbreaks is *C. parvum*.

Clinical Infection. In immunocompetent patients, *C. parvum* causes a profuse, watery diarrhea along with mild to severe nausea and vomiting, headache, and cramps. The onset is rapid, but the infection is self-limiting, and symptoms resolve in several weeks owing to a cell-mediated immune response. In patients with AIDS, the infection may be life-threatening. The diarrhea is cholera-like, with bits of mucus and little fecal material. The organism alters osmotic pressure in the gut, with a resulting influx of fluid. Fluid loss has been reported to range from 3 to 6 L/day to as much as 17 L/day. In addition to weight loss, patients show signs of dehydration and electrolyte imbalance. In chronic cases, the intestinal villi may be injured or destroyed, and the patient shows signs of malabsorption syndrome. There is no totally effective antimicrobial for this infection although spiramycin has been used successfully in the early stages of diarrhea in AIDS patients.

Life Cycle. The sexual and asexual life cycles of *C. parvum* occur in the same host. Life cycle stages, as shown in Figure 24–37, develop under the brush border of the intestinal mucosal epithelial cells. Infective oocysts may be ingested in contaminated water or food or passed by person-to-person contact. Ingestion of the infective oocyst initiates the asexual cycle (merogony) with release of sporozoites. These attach to and penetrate the

intestinal mucosal border and mature into trophozoites. Once trophozoites have matured, the development of meronts with merozoites begins. The nucleus and cytoplasm divide to form individual organisms known as merozoites. When the meront ruptures, merozoites are released and penetrate other cells, either to continue asexual reproduction or to transform into microgametes or macrogametes of the sexual reproductive cycle (gametogony). Fertilization of the macrogamete results in formation of the oocyst, which contains four sporozoites. Two types of oocysts may be formed—the thin-walled oocyst, which ruptures within the intestine and results in autoinfection, and the thick-walled oocyst, which is infective when passed in the stool.

Key parts in the life cycle of this organism that contribute to its pathogenicity areas follows:

1. The oocysts are infective when passed.
2. There is potential for continual autoinfection when thin-walled oocysts rupture in the intestine.
3. Patients may continue to shed oocysts for a time after the diarrhea ceases.

Laboratory Diagnosis. The small size (4 to 6 μm) and round shape of the oocyst make detection difficult with routine concentration procedures. The oocyst is refractile and may resemble a yeast or red cell. The use of Sheather sugar flotation for concentration improves detection, but other intestinal parasites are not easily identified using this method. In addition, one study showed that concentration of a stool by formalin–ethyl acetate method may lead to a significant decrease

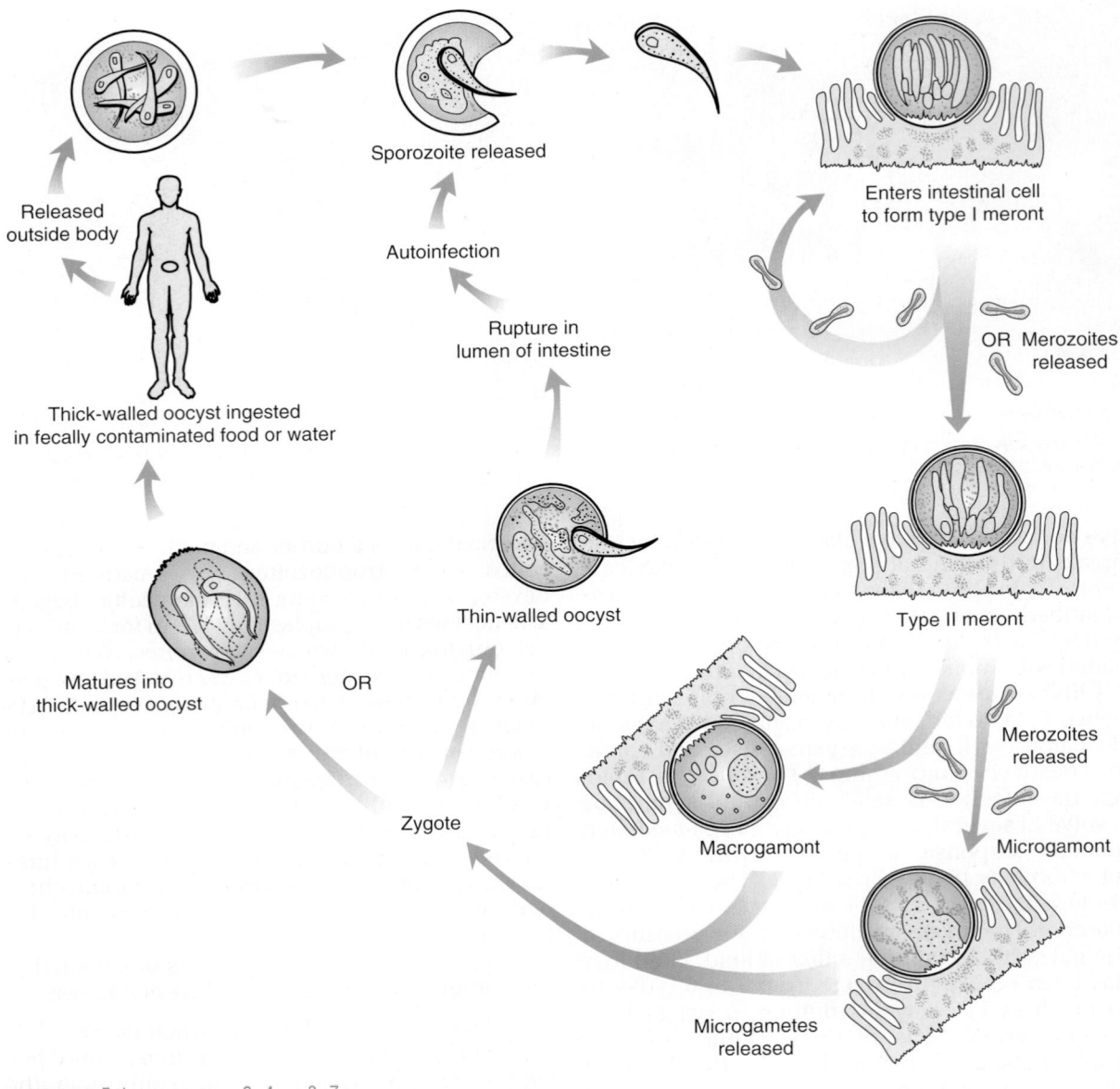

Released
outside body

Sporozoite released

Enters intestinal cell
to form type I meront

Autoinfection

OR Merozoites
released

Rupture in
lumen of intestine

Thick-walled oocyst ingested
in fecally contaminated food or water

Thin-walled oocyst

Type II meront

Matures into
thick-walled oocyst

OR

Zygote

Macrogamont

Merozoites
released

Microgamont

Microgametes
released

■■■■ F i g u r e   2 4 – 3 7
Life cycle of *Cryptosporidium* sp.

in the number of oocysts seen. A newer concentration method involving use of traditional formalin–ethyl acetate and subsequent overlay of sediment with saturated sodium chloride has been described. This method improves recovery rate of oocysts.

Trichrome and iron hematoxylin are not useful permanent stains for identification of *Cryptosporidium* sp. The recommended detection methods when *Cryptosporidium* infection is suspected

are the modified Ziehl-Nielsen acid-fast stain and an indirect fluorescent antibody test using a monoclonal antibody directed against *Cryptosporidium*. With the acid-fast stain, the organism stains as a bright red sphere, which distinguishes it from yeasts, which stain green. Figure 24–38 shows an acid-fast stain of *Cryptosporidium* sp. Studies indicate that the monoclonal antibody test demonstrates greater sensitivity and specificity than the modified acid-fast stain. With any of the permanent stains, more oocysts are seen if the stool is liquid, fewer if the stool is formed. Biopsy of the intestine is not routine, but various life cycle stages can be seen within the intestinal cells.

Isospora belli

Isospora belli is seen with less frequency than *C. parvum,* and acute infections with *I. belli* are usually clinically indistinguishable from infections with *Cryptosporidium* sp.

Clinical Infection. Most patients infected with *I. belli* are asymptomatic, but symptoms such as low-grade fever, headache, diarrhea, and abdominal pain may be present. The infection is self-limiting and usually resolves in several weeks. The immunocompromised patient infected with *I. belli* may present with watery diarrhea and concurrent weight loss. Treatment with trimethoprim-sulfamethoxazole has been effective in eliminating diarrhea, but patients often

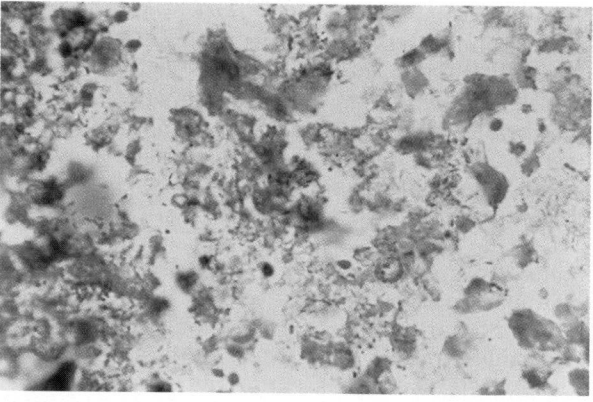

Figure 24–38

Cryptosporidium sp oocysts (modified acid-fast stain).

show recurrence of infection when therapy is discontinued.

Life Cycle. The life cycle of *Isospora belli* is similar to that of *Cryptosporidium* sp but occurs within the cytoplasm of epithelial cells of the small intestine. The oocyst of this organism is not infective when passed in the feces and requires 24 to 48 hours of existence outside the body before it is infective.

Laboratory Diagnosis. The mature oocyst of *I. belli* is oval, 20 to 33 by 10 to 19 μm with a hyaline cell wall. The immature oocyst usually shows a single sporoblast; the mature oocyst has two sporocysts, with four elongated sporozoites within each. Both may be seen in wet mounts. The modified acid-fast stain has been used to increase identification capabilities. The oocyst wall does not stain but often shows a faint outline owing to precipitated stain, and the sporoblasts or sporocysts stain dark red. Figure 24–39 shows the characteristic appearance of an acid-fast stain of the oocyst of *I. belli.*

Microsporidia

A newly described group of organisms in the phylum Microspora have been linked to infections in immunocompromised hosts. These organisms, referred to as microsporidia, are obligate intracellular parasites common to invertebrates and other animals. Five genera—*Nosema, Encephalitozoon, Pleistophora, Microsporidium,* and *Enterocytozoon* —have been implicated in human infection. Organs infected included eyes, the CNS, and the gastrointestinal tract. Intestinal infections cause symptoms similar to those of cryptosporidiosis.

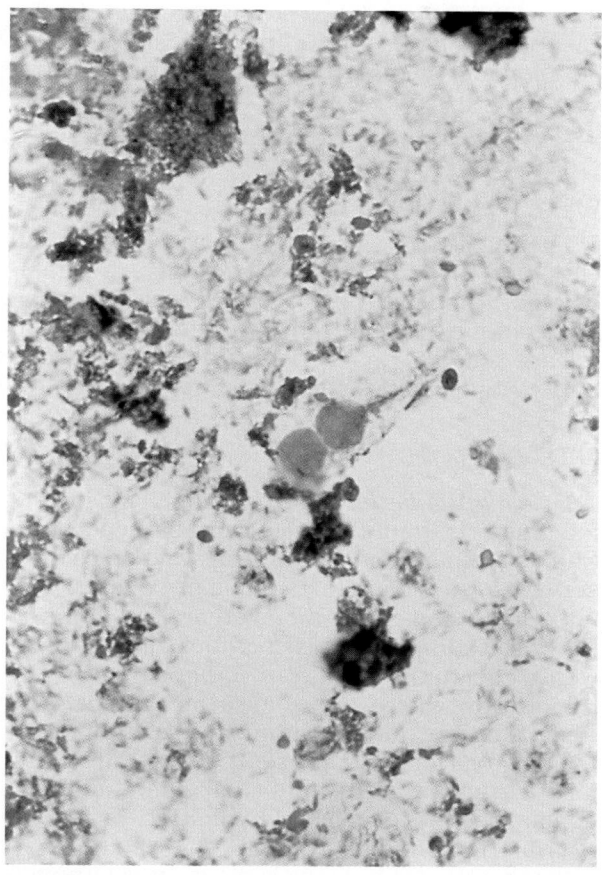

Figure 24–39

Isospora belli oocysts (modified acid-fast stain).

The organisms reproduce by a combination of asexual and sexual methods. Once the spore is ingested, it gains access to host cells by inserting a coiled tube (polar filament) through the cell membrane. Within the cell, there may be binary fission or schizogony and a sexual cycle, which results in production of spores. These spores are then released into the stool or other host tissue.

Identification has been limited to Giemsa-stained tissue sections or electron-microscopic examination of biopsy specimens. Because of the staining characteristics and small size of the spore, light-microscopic examination of stool has not proved useful. Detection methods such as modifications of the trichrome method to enhance staining and indirect fluorescent antibody to detect spores in the stool are under development.

Helminths

Helminth infections in the human are caused by flukes, tapeworms, or roundworms. In general, the adults of these organisms are macroscopic and produce eggs. Humans may become infected by directly ingesting the egg, by ingesting larvae in an intermediate host, or through direct larval penetration of skin. Adults do not multiply in the human body; therefore, the number of adults present is related to the number of infective stages ingested. Pathology and severity of infection are related to the number of adults present, commonly referred to as the *worm burden.* Generally, the patient with only a few adults is asymptomatic, whereas a patient with a large number of adults exhibits clinical symptoms. The majority of the organisms are in the intestinal tract, but species also inhabit the liver, lungs, lymphatics, and blood vessels.

Flukes

Flukes (trematodes) are members of the phylum Platyhelminthes, or flatworms. Most infections are seen in persons from the Orient, Africa, South America, and some areas of the Caribbean. Sizes of the adults range from several millimeters to almost 8 cm. With the exception of the blood flukes, adult flukes are dorsoventrally flattened and have an oral sucker at the anterior end and a ventral sucker located midline, posterior to the anterior sucker. Flukes are hermaphroditic. In all species, eggs must reach water to mature and have a snail species as the first intermediate host.

LIFE CYCLE

Figure 24–40 shows a generalized life cycle of the liver, lung, and intestinal flukes. The miracidium (first-stage larva) is ingested while within the egg or is released from eggs and penetrates the snail, which serves as the first intermediate host. Within the snail tissue, there is a complex development of germinal tissue resulting first in sporocysts, which contain undifferentiated germinal structures, and then in rediae, which contain partially differentiated germinal material. The cercaria (second-stage larva) develops within the redia and is released into the water. The cercaria then attaches to aquatic vegetation or invades the flesh of aquatic organisms. At this stage, the organism is referred to as a metacercaria and is infective for humans. With the exception of the schistosomes, which infect humans by direct cercarial penetration, infection occurs when a human ingests the metacercaria in raw or undercooked aquatic animals or on water vegetation. Prevention includes adequate cooking of water vegetation, fish, and crustaceans. In the case of the blood flukes, persons should wear clothing and shoes to prevent cercarial penetration. The drug of choice for treating fluke infections, regardless of body site, is praziquantel.

LABORATORY DIAGNOSIS

The egg is the primary diagnostic stage. All eggs but those of the schistosomes are operculated; therefore, zinc sulfate concentration should not be used because eggs might open and release contents or sink. Table 24–6 shows a comparison of the characteristics of the fluke eggs.

INTESTINAL FLUKES

Fasciolopsis buski, known as the giant intestinal fluke, is found in the Far East, including China, Vietnam, and India. Dogs and pigs may serve as reservoir hosts. Humans acquire the infection by ingesting metacercaria on freshwater vegetation such as bamboo shoots and water chestnuts. Adults of *F. buski* live in the duodenum, where they cause mechanical and toxic damage. Inflammation and ulceration of the mucosa may be present. Heavy infections may result in persistent diarrhea, anorexia, edema, ascites, nausea and vomiting, or intestinal obstruction.

Finding the adult or egg is diagnostic, although the egg is more commonly seen. The adult is flat-

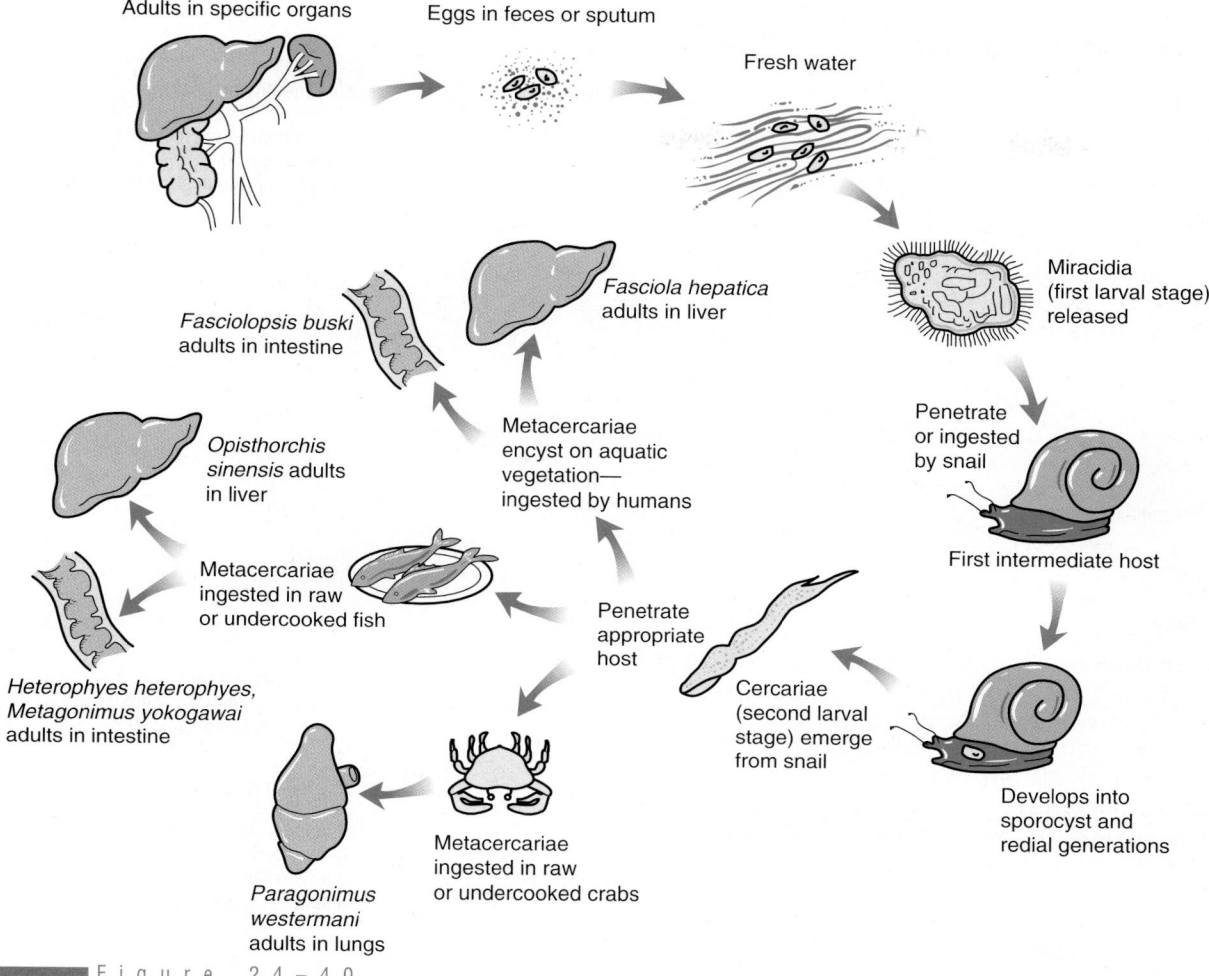

Adults in specific organs Eggs in feces or sputum

Fresh water

Miracidia (first larval stage) released

Fasciola hepatica adults in liver

Fasciolopsis buski adults in intestine

Opisthorchis sinensis adults in liver

Metacercariae encyst on aquatic vegetation— ingested by humans

Penetrate or ingested by snail

First intermediate host

Metacercariae ingested in raw or undercooked fish

Penetrate appropriate host

Cercariae (second larval stage) emerge from snail

Heterophyes heterophyes, Metagonimus yokogawai adults in intestine

Develops into sporocyst and redial generations

Metacercariae ingested in raw or undercooked crabs

Paragonimus westermani adults in lungs

Figure 24–40

Life cycle of liver, lung, and intestinal flukes.

tened, is 2 to 7 cm long, and lacks the cephalic cone seen in *Fasciola hepatica*. Adults are usually not seen in a stool specimen unless it is a purged specimen. Eggs are yellow-brown, average 130 to 140 by 80 to 85 μm, and have a small, relatively inconspicuous operculum. They are unembryonated when passed (Fig. 24–41).

Metagonimus yokogawai and *Heterophyes heterophyes* are two small flukes found in the Far East and Mideast. Humans acquire infection with these organisms by ingesting the metacercaria in undercooked or raw fish. Adults live in the small intestine and produce few symptoms. A patient with a heavy worm burden may have diarrhea, colic, and stools with a large amount of mucus. Adults of both species are small (1 to 2 mm) and delicate. Eggs serve as the primary diagnostic

stage. They are 28 to 30 μm long with a vase or flask shape. They are embryonated and operculated with inconspicuous shoulders at the operculum. Eggs of these species resemble each other as well as those of *O. sinensis*.

LIVER FLUKES

Fasciola hepatica, the sheep liver fluke, is seen in the major sheep-raising areas of the world, including parts of the southwestern United States. In sheep, the organism causes a disease known as liver rot, which is characterized by destruction of the liver. Humans acquire the infection by ingesting metacercaria on raw water vegetation, especially watercress. The larvae reach the liver by migrating through the intestinal wall and

━━━ T A B L E   2 4 – 6
COMPARISON OF FLUKE EGGS

Organism	Average Size (μm) and Shape	Other Identifying Features
Fasciola hepatica	130 × 60–90 Ellipsoidal	Small, indistinct operculum Yellow-brown color Unembryonated when passed
Fasciolopsis buski	130 × 80–85 Ellipsoidal	Cannot be distinguished from *F. hepatica* Unembryonated when passed
Paragonimus westermani	80–118 × 48–60 Oval	Brown, thick shell Slightly flattened operculum Shoulders at operculum Unembryonated when passed
Opisthorchis sinensis	29–35 × 12–19 Vase-shaped	Domed operculum Prominent shoulders Knob at end opposite operculum Embryonated when passed
Heterophyes heterophyes and *Metagonimus yokogawai* ***Note:*** **These 2 species are virtually indistinguishable**	28–30 × 15–17 Vase-shaped	Operculated Shoulders not distinct Small knob Embryonated Similar to *O. sinensis*
Schistosoma mansoni	115–175 × 45–75 Oval	Lateral spine No operculum Embryonated when passed
Schistosoma haematobium	110–170 × 40–70 Oval	Rounded anterior Terminal spine Embryonated when passed
Schistosoma japonicum	60–95 × 40–60 Round to slightly oval	Small, inconspicuous, hooked lateral spine Embryonated when passed

peritoneal cavity. Adults live in the biliary passages and gallbladder and rarely cause overt symptoms because infections are light. Tissue damage during migration through the liver may result in an inflammatory reaction, secondary infections, and fibrosis in the biliary ducts. Heavy infections may induce diarrhea, abdominal pain, hepatomegaly, cirrhosis, and liver obstruction with resulting jaundice.

Adults are approximately 3 cm long and have a prominent cephalic cone. The unembryonated and operculated eggs are carried in the bile to the intestinal tract and are passed in the feces. The size range is 130 to 150 by 60 to 90 μm. They are virtually indistinguishable from eggs of *F. buski* (see Fig. 24–41). Eggs meeting these characteristics should be reported as "*F. buski/F. hepatica* eggs seen."

The Chinese liver fluke, *Opisthorchis sinensis,* is geographically limited to the Far East, where dogs and cats serve as reservoir hosts. The adults live in the distal bile ducts. As with *F. hepatica,* light infections produce few or no symptoms. Repeated or heavy infections may cause inflammation due to mechanical irritation, fever, diarrhea, pain, fibrotic changes, or obstruction of the bile duct. Humans acquire the infection by ingesting the metacercaria in raw, undercooked, or pickled fish. Diagnosis is made by finding the egg in a stool specimen or, occasionally, in duodenal aspirates. Adults are thin, tapered at both ends, and 1.0 to 2.5 cm long. The egg is 29 to 35 μm long, embryonated when passed, flask shaped, and operculated with prominent shoulders at the operculum and a knob at the opposite end (Fig. 24–42).

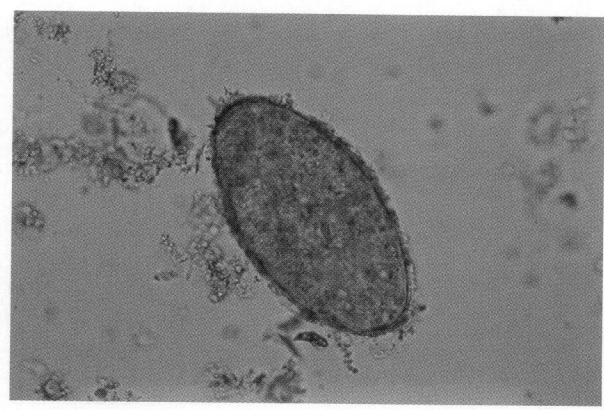

Figure 24–41

Fasciola hepatica/Fasciolopsis buski egg.

LUNG FLUKE

Organisms of the genus *Paragonimus* usually infect tigers, leopards, dogs, and foxes. *Paragonimus westermani,* the lung fluke, is primarily found in humans in the Orient. Humans acquire the infection by ingesting metacercaria in raw, pickled, or undercooked freshwater crabs or crayfish. The metacercaria excysts in the intestine and burrows through the intestinal wall and diaphragm, eventually entering the lung. The host shows few symptoms during this migration. Major symptoms associated with lung habitation are nonspecific and may include an inflammatory response, persistent cough, chest pain, and hemopytsis. Adults are reddish brown and approximately 1 cm long, and they live within capsules in the bronchioles.

Sputum is the primary diagnostic specimen. Eggs are expelled from the capsule into the bronchioles and carried upward in the sputum. Eggs may be found in the feces if they have been coughed up and subsequently swallowed. Eggs are broadly oval, 80 to 118 by 48 to 60 μm, with a flattened operculum and slight shoulders. They are unembryonated when passed. There is also a thickening of the shell at the end opposite the operculum (Fig. 24–43). These eggs can appear similar to those of *Diphyllobothrium latum* and must be carefully examined when seen in the feces. A wet mount of sputum demonstrates the egg in some patients.

BLOOD FLUKES

The blood flukes, *Schistosoma* sp, differ from other flukes in that:

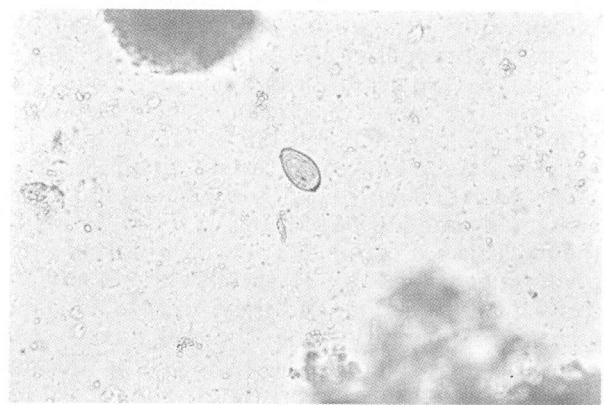

Figure 24–42

Opisthorchis sinensis egg.

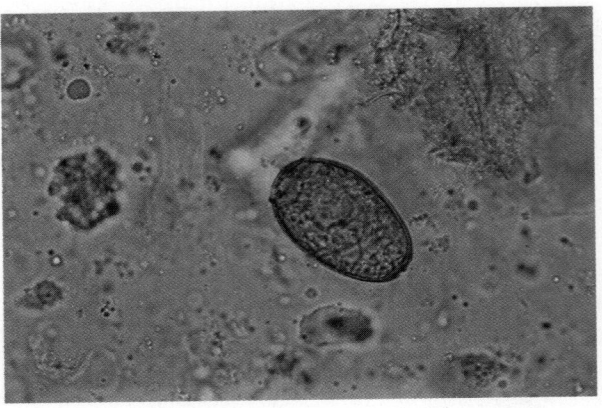

■■■■■■ F i g u r e   2 4 – 4 3
Paragonimus westermani egg.

- There is both a male and female
- The eggs are unoperculated
- Humans are infected by direct cercarial penetration of the skin
- Flukes are cylindrical rather than dorsoventrally flattened

The female lives in an involuted chamber (the gynecophoral canal) that extends the length of the male.

The three primary species of schistosomes pathogenic to man are *S. mansoni, S. haematobium,* and *S. japonicum. Schistosoma mansoni,* which is most commonly found in Africa, parts of South America, the West Indies, and Puerto Rico, lives in venules of the large intestine. *Schistosoma japonicum,* which is commonly found in the Far East, including Japan, China, and the Philippines, lives in venules of the small intestine. This species, unlike the other two, has many mammalian reservoir hosts. *Schistosoma haematobium,* which is primarily found in the Nile Valley, the Mideast, and East Africa, lives in the veins surrounding the bladder. Adults measure 7 to 20 mm for males and 7 to 26 mm for females.

Clinical Manifestations. Schistosomiasis (bilharziasis), has a prevalence of about 200 million cases worldwide. The term is used to describe symptoms caused by any of the schistosomes. Patient symptoms are related to phases of the fluke's life cycle. Cercarial penetration may cause local dermatitis, including irritation, redness, and rash, that persists for about 3 days. Larval migration through the body causes generalized symptoms such as urticaria, fever, and malaise, which

may last up to 4 weeks. Adults in and of themselves cause little damage, because they acquire host antigens on their surface that decrease the host's immune response. Egg production and migration through the tissues, however, are responsible for the most acute damage. After the adult female dilates the vein to lay eggs, the vein contracts, and aided by secretion of enzymes, the eggs begin to penetrate vessel walls and tissue. Eggs subsequently find their way into the lumen of the intestine or bladder. The egg spines cause trauma to the tissues and walls of the vessels during the early stage of acute infections and may result in hematuria *(S. haematobium)* or diarrhea *(S. mansoni* and *S. japonicum).*

In chronic infections, the eggs remaining in the tissue induce granuloma formation, which leads to thickening and fibrotic changes. Scarring of the veins, development of ascites, pain, anemia, hypertension, hepatomegaly, and splenomegaly are also seen. In urinary schistosomiasis, microscopic bleeding into the urine is present during the acute phase. In chronic stages, dysuria, urine retention, and urinary tract infections are present.

Life Cycle. The life cycles of all three schistosomes are identical (Fig. 24–44). The eggs are embryonated when passed, and the miracidium is released when the egg reaches water. After penetrating the snail (first intermediate host), sporocysts and then cercaria are produced during a 6-week period. Cercaria migrate from the snail into water. When the cercaria encounter the skin of a human, they shed their forked tails, secrete enzymes, and begin penetration. Once in the veins, they are referred to as schistosomula and circulate until they reach the lungs or enter the liver, where maturation is completed. The paired adult flukes use the portal system to reach veins of the intestine or bladder.

Laboratory Diagnosis. Diagnosis is made by finding embryonated eggs in the feces *(S. mansoni* and *S. japonicum)* or in the urine *(S. haematobium).* The egg of *S. mansoni* (Fig. 24–45*A*) is yellowish, elongated, and 115 to 175 by 45 to 75 μm, and has a prominent lateral spine. *Schistosoma haematobium* eggs (see Fig. 24–45*B*) are elongated and 110 to 170 by 40 to 70 μm, and they possess a terminal spine. *Schistosoma japonicum* eggs (see Fig. 24–45*C*) are round and 60 to 95 by 40 to 60 μm, and they possess a small, curved, rudimentary spine. The best time to collect eggs in urinary schistosomiasis is during peak excretion time in early afternoon (12 PM to 2 PM). Hatching tests and biopsies may also be used in diagnosis of schistosomiasis. Serodiag-

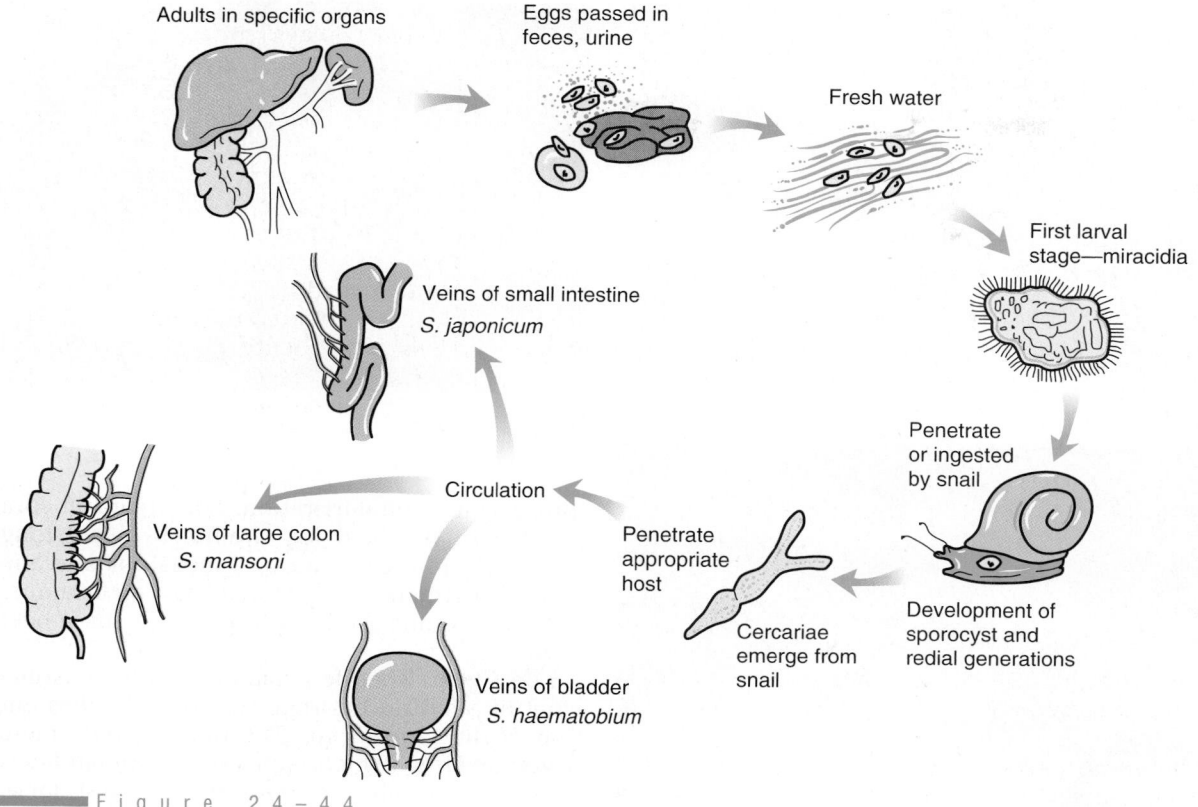

■ F i g u r e 2 4 – 4 4
Life cycle of blood flukes (Schistosoma sp).

nosis may be useful to diagnose infection in patients from nonendemic countries who develop symptoms after visiting endemic areas.

SCHISTOSOMAL DERMATITIS

Penetration of humans by cercaria of the flukes of birds and other mammals causes a dermatitis commonly referred to as *swimmer's itch*. Foreign proteins elicit a tissue reaction characterized by small papules 3 to 5 mm in diameter, edema, erythema, and intense itching. Symptoms last about a week and disappear as cercaria die and degenerate.

Tapeworms

Tapeworms (cestodes) are the second group of human parasites in the phylum Platyhelminthes. They show a wide range of sizes, from 3 mm to 10 m, generally require intermediate hosts in their life cycle, and are hermaphroditic. They are ribbon-like organisms whose method of growth involves addition of segments called proglottids, each of which, when mature, produces eggs that are infective for the intermediate host. The scolex, or head, has suckers and, in some species, hooklets as a means of attachment to the intestinal mucosa. The neck is directly behind the scolex and serves as the origin for the proglottids. Unless the scolex is detached by treatment, the infection will continue. Gravid proglottids at the distal end of the organism contain eggs to be discharged into the feces.

Eggs of most of the tapeworms contain a hexacanth embryo (oncosphere) that is infective for the intermediate host. Transmission to humans involves ingestion either of a larval stage called the cysticercus, cysticercoid, or plerocercoid larva in raw or undercooked meat or fish or of insects harboring the larval stage. This larval stage contains an invaginated scolex of the tapeworm inside a protective membrane. Figure 24–46 shows a general diagram of the tapeworm. Diagnosis of tapeworm infection is usually made

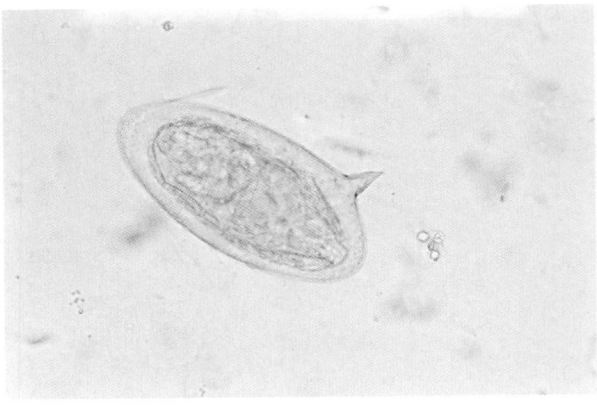

A

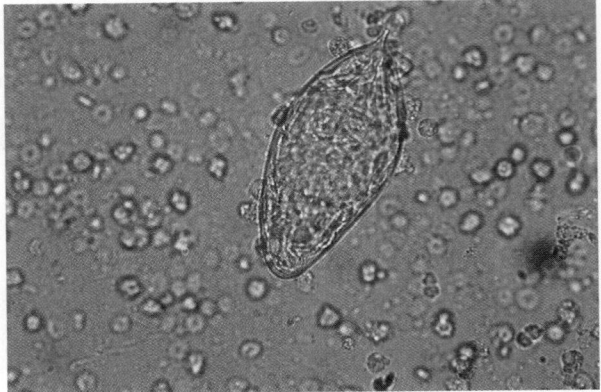

B

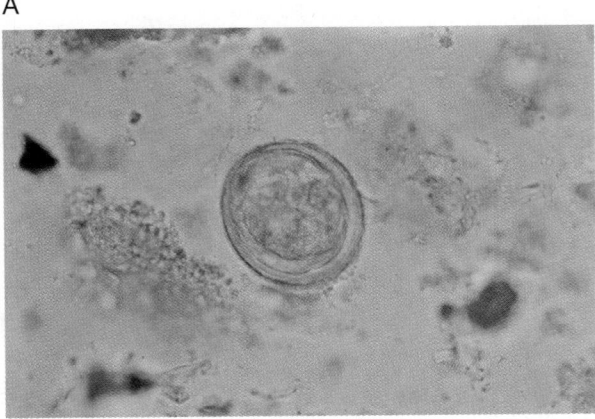

C

■■■■■F i g u r e   2 4 – 4 5

A, Schistosoma mansoni egg. *B, S. haematobium* egg. Notice RBCs in background. *C, S. japonicum* egg.

by finding the eggs in feces, although proglottids can be used if passed intact. Table 24–7 gives a comparison of the characteristics of the tapeworm eggs.

Diphyllobothrium latum

Diphyllobothrium latum, the fish tapeworm, is found worldwide in areas where the population eats pickled or raw freshwater fish. In the United States, it is primarily seen in the areas around the Great Lakes. Fish-eating mammals in endemic areas may also be infected. Humans usually harbor only a single worm, which attaches in the jejunum and may reach a length of up to 10 m. Most infected persons demonstrate no clinical symptoms; others present with vague gastrointestinal symptoms, including nausea and vom-

iting and intestinal irritation. The organism competes with the host for vitamin B_{12}, and long-term infection may lead to a megaloblastic anemia. Treatment of choice is niclosamine, although quinacrine hydrochloride has been used.

Life Cycle. The life cycle of *D. latum* is somewhat of a hybrid between that of the flukes and that of the tapeworms. The operculated, unembryonated egg, which is passed in human feces, must reach water to mature. The first larval stage (coracidium) is ingested by a copepod and develops into a procercoid larva within the copepod. When the infected copepod is ingested by a fish, the larva leaves the fish's intestine and invades the flesh, where it develops into a plerocercoid larva, which consists of a scolex with a thin, ribbon-like portion of tissue.

Humans ingest the plerocercoid larvae by eating raw or undercooked fish. The scolex is released in the intestine, where it develops into an adult worm. Figure 24–47 shows the life cycle of *D. latum.*

Laboratory Diagnosis. The scolex, proglottid, or egg may be used as a diagnostic finding in a fecal specimen. The egg is unembryonated when passed, operculated, and yellow-brown (Fig. 24–48). It is about 58 to 76 by 40 to 50 μm and has a small, knob-like protuberance at the end opposite the operculum. The knob may not be seen on all eggs, so size and lack of shoulders must be used to distinguish the egg from that of *P. westermani.* The proglottid is wider than it is long, with a characteristic rosette-shaped or coiled uterus. The scolex, which is 2 to 3 mm long, is elongated and has two sucking grooves, one located on the dorsal surface and the other on the ventral surface.

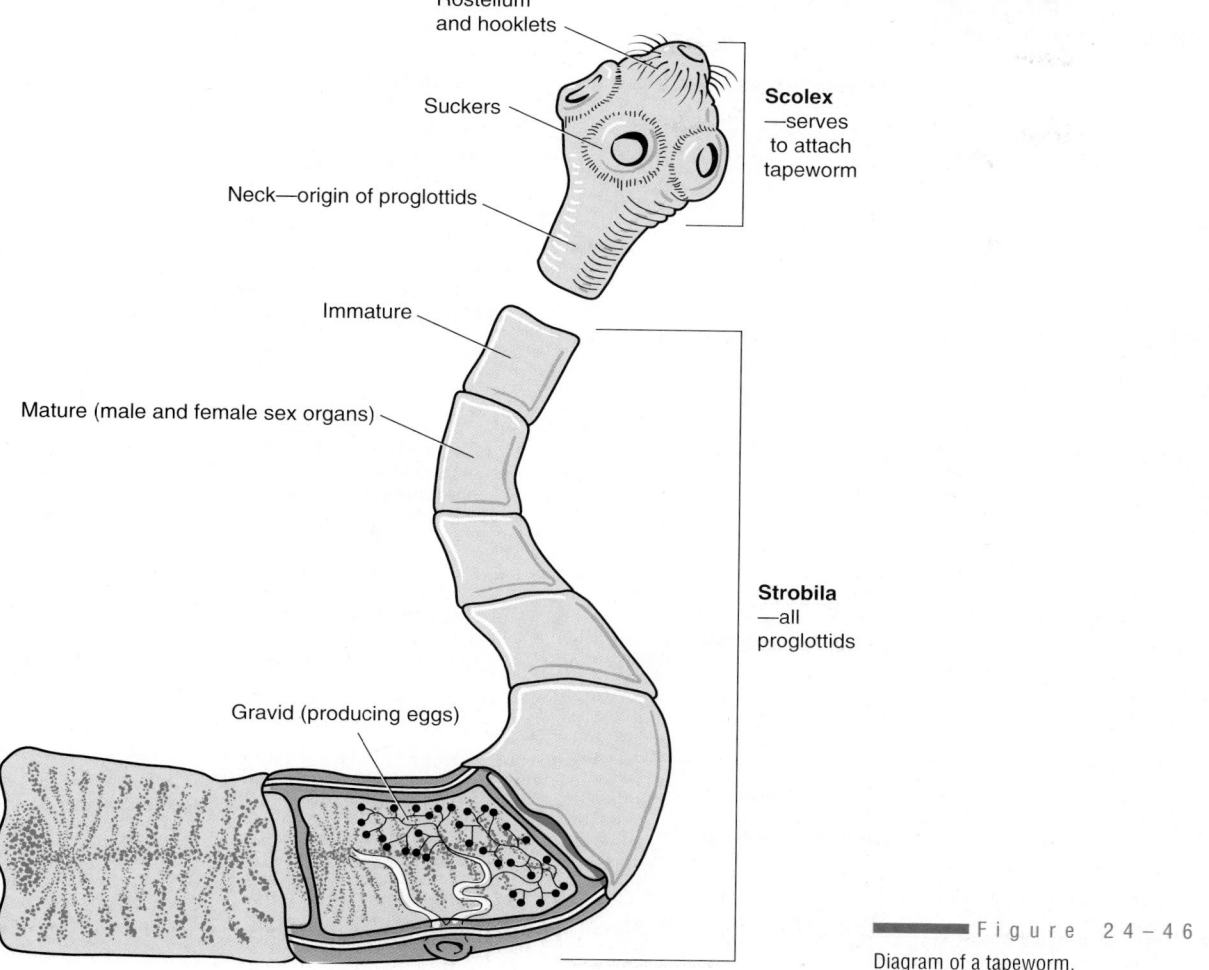

■Figure 24–46
Diagram of a tapeworm.

Taenia species

Two *Taenia* species infect humans—*Taenia saginata,* the beef tapeworm, which is found primarily in beef-eating countries of the world, and *Taenia solium,* the pork tapeworm, which is found in those areas of the world, such as Latin America, with a high consumption of pork. Both organisms attach to the intestinal mucosa of the small intestine. Adults of *T. saginata* may reach a length of 10 m, whereas those of *T. solium* may reach only 7 m.

Clinical Manifestations. Infection with the adult tapeworm of either species usually causes few clinical symptoms, although vague abdominal pains, indigestion, and loss of appetite may be present. The proglottids are motile and, if broken off in the intestinal tract, may actively migrate out the anus.

The major complication of infection with *T. solium* is cysticercosis, in which the human becomes the intermediate host and harbors the larvae in tissues. This infection is discussed in the section on tissue infections.

Life Cycle. As previously mentioned, the life cycles of the two *Taenia* organisms are identical except for the fact that humans may also serve as intermediate hosts for *T. solium.* Embryonated eggs are passed in human feces and ingested by the intermediate host. The oncosphere is freed in the intestinal tract, migrates through the intestinal wall, and gains access via the circulatory system to muscles of the host, where it transforms into a cysticercus. When humans ingest raw or undercooked meat, the scolex within the cysticercus is freed, attaches in the human small intestine, and

■■■■ T A B L E   2 4 – 7
COMPARISON OF TAPEWORM EGGS

Organism	Average Size (μm) and Shape	Other Identifying Features
Diphyllobothrium latum	58–76 × 40–50 Oval	Inconspicuous operculum Small knob at end opposite operculum Unembryonated when passed
Taenia sp	30–45 Round	Thick, brown, radially striated shell Embryonated with 6-hooked oncosphere when passed
Hymenolepis nana	30–47 Oval	2 membranes—inner has 2 polar knobs, from which extend 4 polar filaments into space between inner and outer membranes Embryonated with 6-hooked oncosphere when passed
Hymenolepis diminuta	50–75 Round to slightly oval	2 membranes—inner has very slight polar knobs No polar filaments Embryonated with 6-hooked oncosphere when passed
Dipylidium caninum	20–40 (each egg) Round Resembles Taenia sp	Eggs passed in a packet of 15–25 Eggs embryonated with 6-hooked oncosphere when passed

matures into the adult tapeworm within 10 weeks. Figure 24–49 shows the life cycle of *Taenia* spp.

Laboratory Diagnosis. Diagnosis of *Taenia* sp infection can be made by finding the egg, scolex, or proglottid in the feces. The egg is yellow-brown, round, and surrounded by a thick wall with radial striations, and it measures 30 to 43 μm. The egg is embryonated with a six-hooked oncosphere (hexacanth embryo) when passed in the feces (Fig. 24–50). Eggs of these species are indistinguishable and must be reported as "*Taenia* sp eggs." Gravid proglottids may be seen in the stool specimen and can be used to differentiate the two organisms. Proglottids of *T. solium* show 7 to 13 primary uterine branches on each side of the main uterine trunk, whereas proglottids of *T. saginata* show 15 to 20 per side. The scolex, if found, can also be used to distinguish the organisms. The scolex of *T. saginata* is less than 5 mm long and has four suckers, whereas that of *T. solium* has a rostellum with a double row of 25 to 30 hooklets in addition to the four suckers.

HYMENOLEPIS SPP

The dwarf tapeworm, *Hymenolepis nana,* is found worldwide and is a common tapeworm in children. Light infections are usually asymptomatic; large numbers of worms may cause abdominal pain, diarrhea, irritability, and headache. Infections with this organism are easily transmitted among children, because an intermediate host is not required. Direct fecal-oral transmission of the egg, development of the cysticercoid in the intestinal tissue of the host, and reentry into the lumen for development into an adult characterize the life cycle (Fig. 24–51). An insect vector may serve as intermediate host.

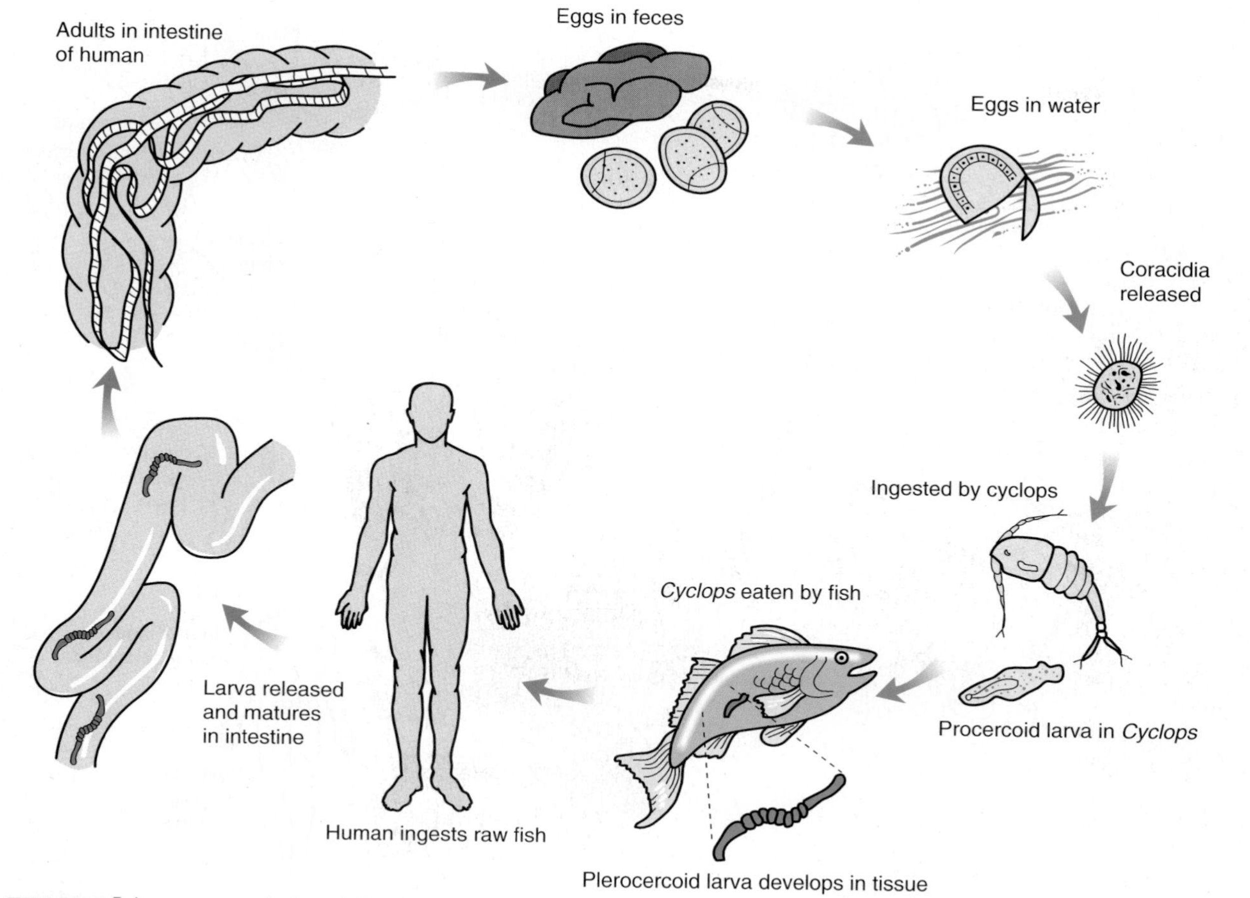

Adults in intestine
of human

Eggs in feces

Eggs in water

Coracidia
released

Ingested by cyclops

Cyclops eaten by fish

Procercoid larva in Cyclops

Larva released
and matures
in intestine

Human ingests raw fish

Plerocercoid larva develops in tissue

■ Figure 24–47

Life cycle of *Diphyllobothrium latum.*

Laboratory Diagnosis. The adult is 40 mm long and has a small scolex with four suckers and a rostellum with spines. The primary method of diagnosis is finding the egg in a stool specimen. The egg of *H. nana* is spherical to oval, measures to 30 to 47 μm, and has a grayish color. The hexacanth embryo is contained within an inner membrane, and the area between the inner membrane and egg wall contains two polar thickenings from which four to eight polar filaments extend (Fig. 24–52).

Hymenolepis diminuta, the rat tapeworm, is less commonly seen, and the eggs must be distinguished from those of *H. nana.* Infection with *H. diminuta* is acquired by ingesting fleas that contain the infective cysticercoid. The adult is 20 to 60 cm long. The egg is 50 to 75 μm, gray or straw colored, and oval. An inner membrane with inconspicuous

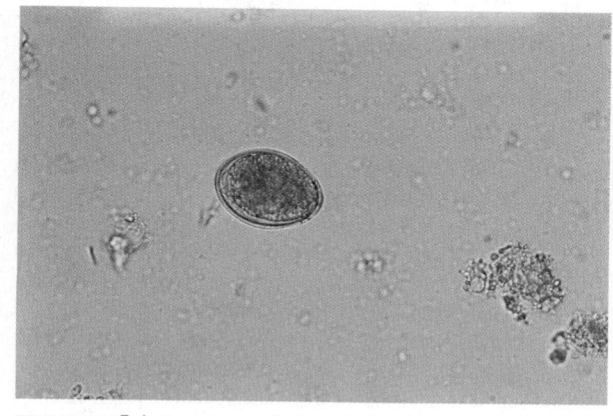

■ Figure 24–48

Diphyllobothrium latum egg.

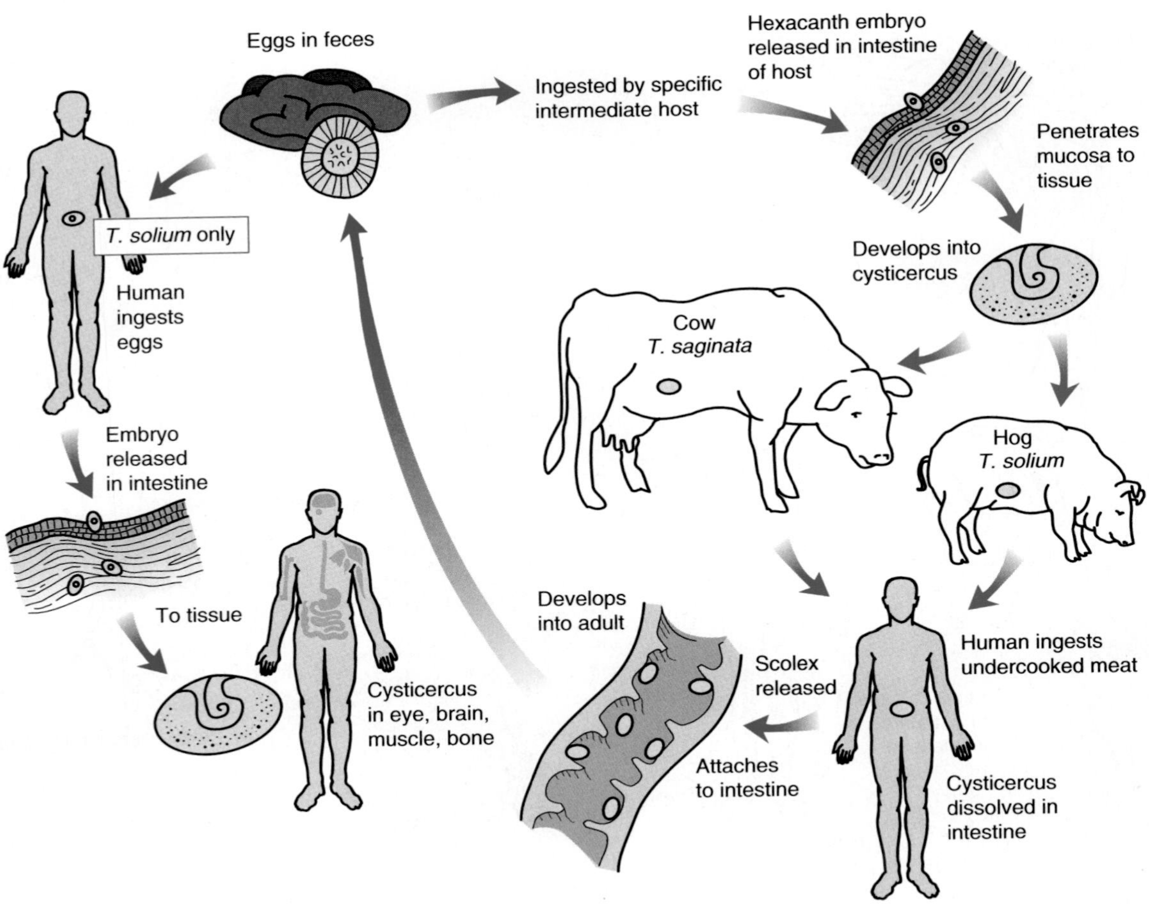

Eggs in feces

Hexacanth embryo released in intestine of host

Ingested by specific intermediate host

Penetrates mucosa to tissue

T. solium only

Human ingests eggs

Develops into cysticercus

Cow *T. saginata*

Embryo released in intestine

To tissue

Hog *T. solium*

Cysticercus in eye, brain, muscle, bone

Develops into adult

Scolex released

Human ingests undercooked meat

Attaches to intestine

Cysticercus dissolved in intestine

■ **Figure 24–49**

Life cycle of *Taenia* sp.

polar thickenings but no polar filaments surrounds the oncosphere (Fig. 24–53).

Dipylidium caninum

The human serves as an accidental host for *Dipylidium caninum,* the dog tapeworm. Children are most often infected by ingesting fleas containing the larval stage. The infections are usually asymptomatic. The proglottid may be seen in human feces and is characterized by its pumpkin seed shape, twin genitalia, and the presence of two genital pores, one on each side of the proglottid. The eggs are characteristically seen in packets of 15 to 25 eggs. Individual eggs are 20 to 40 μm and may resemble those of *Taenia* sp (Fig. 24–54).

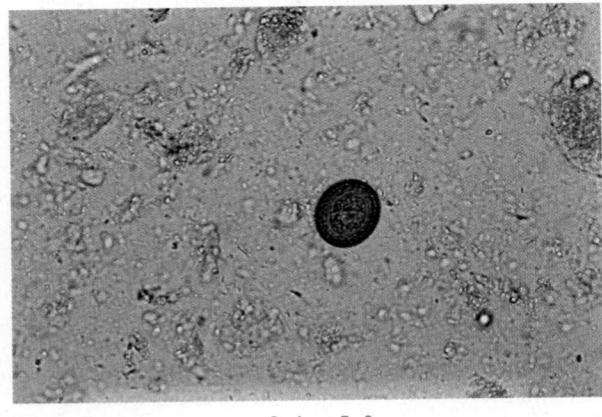

■ **Figure 24–50**

Taenia sp egg.

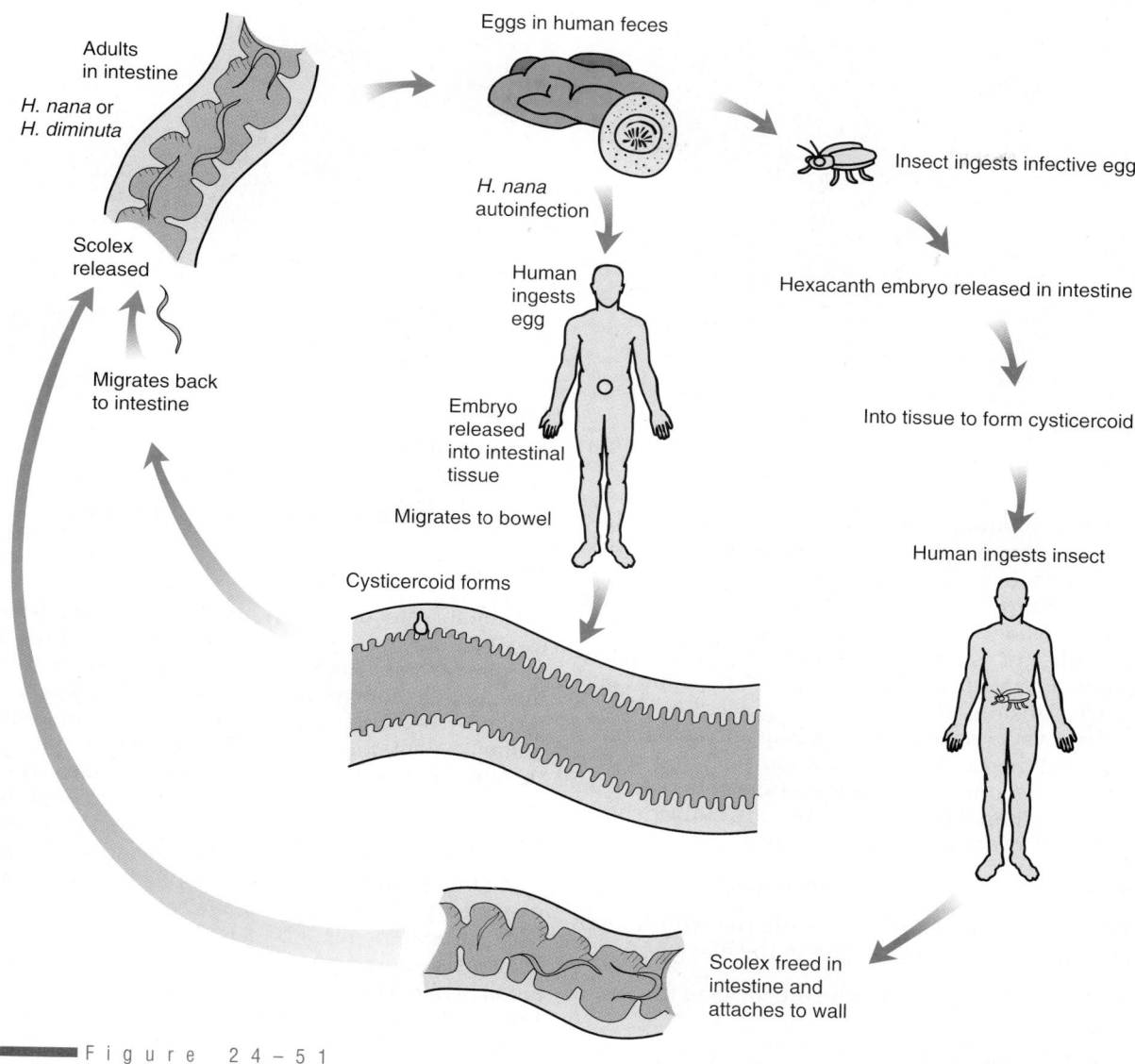

Adults
in intestine

H. nana or
H. diminuta

Scolex
released

Migrates back
to intestine

Eggs in human feces

H. nana
autoinfection

Human
ingests
egg

Embryo
released
into intestinal
tissue

Migrates to bowel

Cysticercoid forms

Insect ingests infective egg

Hexacanth embryo released in intestine

Into tissue to form cysticercoid

Human ingests insect

Scolex freed in
intestine and
attaches to wall

Figure 24 – 51

Life cycle of *Hymenolepis nana.*

Tissue Infections with Cestodes

Cysticercosis, sparganosis, and hydatid cyst disease are the major diseases caused by the tissue stage of the tapeworm. They originate when a human accidentally becomes the intermediate host for the parasite.

CYSTICERCOSIS

Cysticercosis results when a human ingests the eggs of *Taenia solium,* the pork tapeworm.

The hexacanth embryo is released in the intestine, penetrates the intestinal wall, and enters the circulation to develop as a cysticercus in any tissue or organ. The living organism may elicit a host-tissue reaction, resulting in production of fibrous capsule. Once the larva has died, there is increased inflammatory reaction and eventual calcification. The most commonly infected sites are striated muscle, eye, and brain.

Light infections usually cause no clinical symptoms. Symptoms, when present, depend on the organ affected. Muscular pain, weakness, and

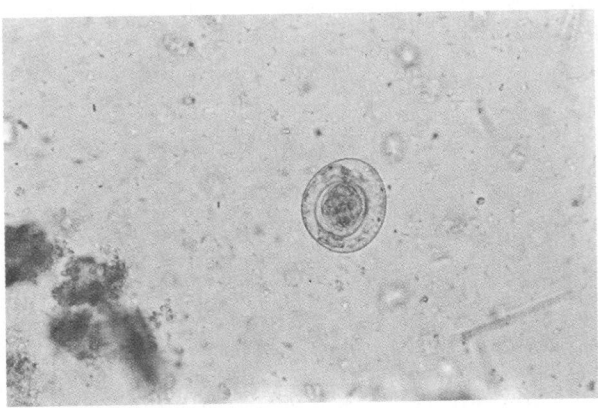

■ Figure 24 – 52

Hymenolepis nana egg.

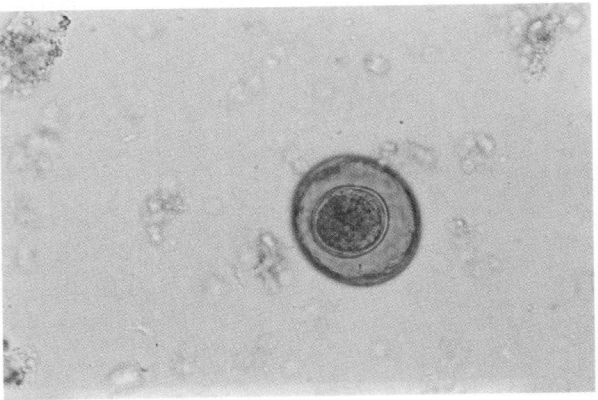

■ Figure 24 – 53

Hymenolepis diminuta egg.

cramps characterize infections of the striated muscle. Neurocysticercosis may be manifested by headaches, signs of meningitis or brain tumor, convulsions, or a variety of motor and sensory problems. In the eye, a cysticercus forms in the vitreous or supretinal space. Retinal detachment, intraorbital pain, flashes of light, and blurred vision may occur.

The cysticercus is oval, translucent, and about 2 to 5 mm in size. It contains an invaginated scolex containing four suckers and a circle of hooklets on the rostellum. Diagnosis of the infection may be made by a variety of methods, including:

■ Radiography to detect calcified cysts

■ Examination of the eye with the ophthalmoscope to detect cysticerci in the eye

■ Imaging techniques to locate larva in the brain

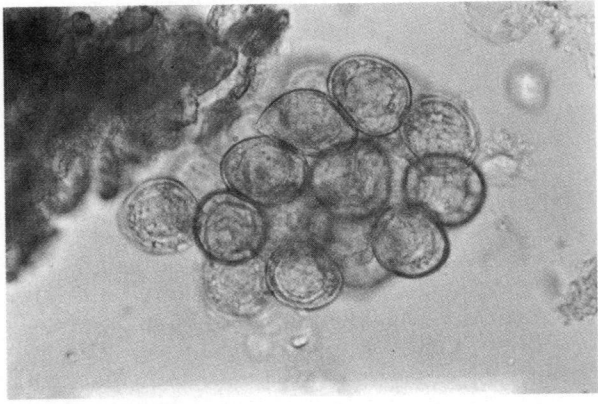

■ Figure 24 – 54

Dipylidium caninum egg packet.

■ Biopsy and histologic staining of tissue.

SPARGANOSIS

Human infection with the plerocercoid larva (sparganum) of a dog or cat tapeworm can result in sparganosis. Humans acquire the infection by ingesting a copepod containing the procercoid larva, by ingesting reptiles, amphibians, or other animals containing the plerocercoid larva, or through invasion by the plerocercoid larva when the raw tissue from the second intermediate host is used as a poultice. The disease is most common in Southeast Asia. A commonly infected site is the eye, after a poultice has been applied to relieve an eye infection. The organism may also cause migratory subcutaneous nodules, itching, and pain. Diagnosis of sparganosis is made by finding a small, white, ribbon-like organism with a rudimentary scolex. Size varies from a few millimeters to 40 cm. The organism may be removed surgically.

ECHINOCOCCOSIS

Echinococcosis is an infection by *Echinococcus granulosus* that normally involves the dog or other member of the family Candidae as the host of the adult tapeworm. Sheep and other herbivores are the usual host at the larval stage, the hydatid cyst. The disease is primarily seen in sheep-raising areas of the world, including Australia, southern South America, and parts of the southwestern United States.

The adult worm is about 5 mm long and contains only three proglottids. The eggs are found in the feces of the dog or other definitive host and resemble those of *Taenia* sp.

A human becomes an intermediate host by accidentally ingesting the eggs of *E. granulosus* containing the hexacanth embryo. The oncosphere penetrates the intestinal mucosa, enters the circulation, and usually lodges in the liver. The embryo develops a central cavity-like structure lined with a germinal membrane, from which brood capsules and protoscolices (hydatid sand) develop. The hydatid cyst's size is limited by the organ in which it develops. In bone, a limiting membrane never develops, so the cyst fills the marrow and eventually erodes the bone.

Symptoms vary according to the organ infected. Pressure from increasing size of a cyst may cause necrosis of surrounding tissue. Rupture of the cyst liberates large amounts of foreign protein that may elicit an anaphylactic response. In addition, freed germinal epithelium may serve as source of new infection.

Diagnosis may be made by radiologic examination, ultrasound, or other imaging techniques. Aspiration of the cyst contents usually reveals the presence of protoscolices.

Roundworms

Human roundworms include those that occupy the intestinal tract and those that occupy blood and tissue. These organisms, found worldwide, may be transmitted by soil or may require an insect vector. Intestinal roundworms are the most common of all the helminths that infect humans in the United States. Infected persons are found in highest numbers in the warm, moist areas of the Southeast and in areas with poor sanitation. These organisms are characterized by the presence of two sexes and a life cycle that may involve larval migration throughout the body. The organism may be transmitted to humans by ingestion of an infective egg or by larval penetration through the skin. Patients may be asymptomatic or symptomatic, and the severity of symptoms is related to worm burden, nutritional status and age of the host, and duration of infection in the host. The roundworms discussed in this section are *Enterobius vermicularis*, *Trichuris trichiura*, *Ascaris lumbricoides*, hookworm, and *Strongyloides stercoralis*. Blood and tissue roundworms are discussed in a separate section. Table 24–8 gives a comparison of diagnostic characteristics of the eggs and larvae of intestinal roundworms.

Enterobius vermicularis

Enterobius vermicularis, commonly called the pinworm, is a worldwide parasite commonly detected in children, especially those 5 to 10 years old. Enterobiasis is often found in families or in crowded conditions where the eggs may be easily transmitted. The eggs are resistant to drying and are easily spread in the environment. Adult worms live in the large intestine, although they have occasionally been found in the appendix or vagina.

Clinical Manifestations. Although infection with *E. vermicularis* is often asymptomatic, the patient may experience loss of appetite, abdominal pain, loss of sleep, and nausea and vomiting. Anal pruritus is due to migration of the female to the perianal area. Treatment, if given, is usually pyrantel pamoate, mebendazole, or piperazine citrate. The treatment is repeated in 2 weeks to eliminate organisms that matured as a result of ingestion of eggs remaining in the environment.

Life Cycle. The life cycle of this organism (Fig. 24–55) is characterized by migration of the female out the anus during the night to lay eggs in the perianal area. The eggs are infective with a third-stage larva within several hours after being laid, and transmission involves inhalation or ingestion of the infective egg. Direct anal-mouth transmission is common in children. Retroinfection, in which the hatched larva reenters the intestine to mature into an adult, may also occur.

Laboratory Diagnosis. A fecal specimen is unsatisfactory because eggs are laid outside the body in the perianal area and are rarely present in the stool. The cellophane tape preparation is considered the diagnostic method of choice. The procedure must be done as soon as the child arises in the morning—before going to the bathroom or bathing—in order to recover eggs. A piece of cellophane tape is placed sticky side out over the edge of a tongue blade. The perianal area is touched with the sticky side of the tape, and then the tape is placed onto a glass slide. Adaptations of this procedure using paddles with a sticky surface are available commercially. The adult female may occasionally be seen in this preparation. Eggs may also be seen in a urine specimen.

The adult female measures 8 to 13 mm long and has a long pointed tail and three cuticle lips with alae at the anterior end. The less commonly seen male is 2 to 5 mm long with a curved posterior. The egg is oval, colorless, and slightly flattened on one side. It measures approximately 50 to 60 by 20 to 30 μm. The egg is usually seen embryonated with a C-shaped larva (Fig. 24–56).

■■■■■■ T A B L E   2 4 – 8
COMPARISON OF ROUNDWORMS EGGS AND LARVAE

Organism	Average Size (μm) and Shape	Other Identifying Features
Ascaris lumbricoides Fertile	45–75 × 35–50 Oval	Bile-stained shell Bumpy, mammillated In 1-cell stage when passed Some eggs may be decorticated (lack mammillated coat)
Infertile	85–95 × 43–47 Oval (some bizarrely shaped)	Mammillated Thin shell Undifferentiated internal granules
Enterobius vermicularis	50–60 × 20–30 Oval, flattened on one side	Colorless shell Usually embryonated with C-shaped larva
Trichuris trichiura	50–55 × 22–23 Barrel shaped	Bile-stained, thick shell Hyaline polar plugs Unembryonated when passed
Hookworm Egg	50–60 × 35–40 Broadly oval	Thin shell, colorless In 4–8-cell stage when passed
Rhabditiform larva	250–300	Long buccal capsule Inconspicuous genital primordium
Filariform larva	500	Pointed tail Esophageal-intestinal ratio 1:4
Strongyloides stercoralis	Egg rarely seen—resembles that of hookworm	
Rhabditiform larva	200–250	Short buccal capsule Prominent genital primordium
Filariform larva	500	Notched tail Esophageal-intestinal ratio 1:1

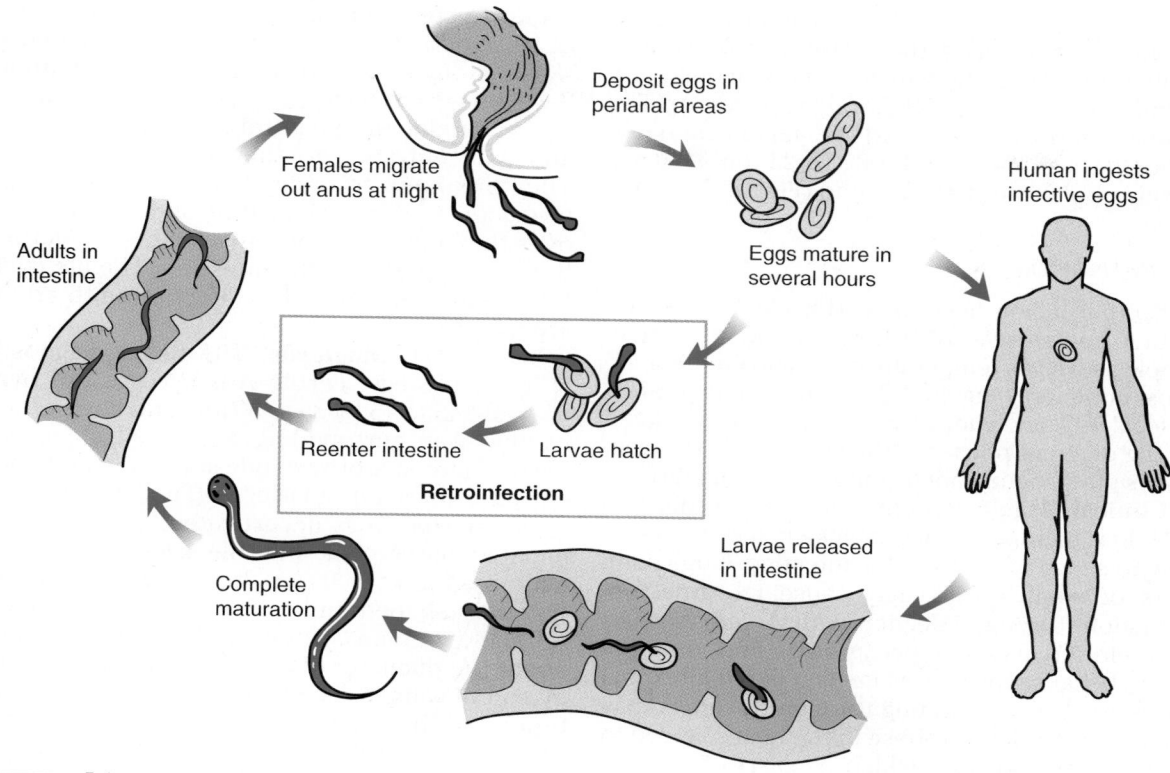

Females migrate
out anus at night

Deposit eggs in
perianal areas

Human ingests
infective eggs

Adults in
intestine

Eggs mature in
several hours

Reenter intestine   Larvae hatch

Retroinfection

Larvae released
in intestine

Complete
maturation

■ F i g u r e   2 4 – 5 5

Life cycle of *Enterobius vermicularis*.

Trichuris trichiura

Trichuris trichiura, commonly referred to as the whipworm, is found worldwide, especially in areas with a moist, warm climate. It occurs in the southeastern United States, often as a co-infection with *A. lumbricoides*.

Clinical Manifestations. Light infections with *T. trichiura* rarely cause symptoms; heavy infections result in bleeding, weight loss, abdominal pain, nausea and vomiting, and chronic diarrhea. Inflammation of the mucosa occurs as the adults thread themselves through it. Prolonged, heavy infection may result in diarrhea with blood-tinged stools. Rectal prolapse is the result of heavy infections in undernourished children. Hypochromic anemia may occur with inadequate iron and protein intake in the presence of chronic infection. Treatment is usually with mebendazole.

Life Cycle. Eggs are passed in the feces and require at least 14 days in warm, moist soil for embryonation to occur. Humans acquire infection by ingesting the infective egg. The larva is

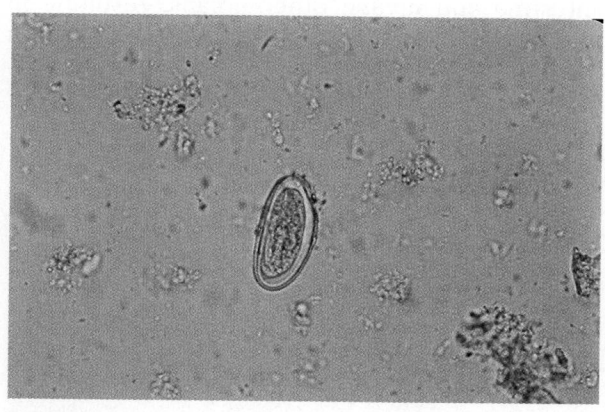

■ F i g u r e   2 4 – 5 6

Enterobius vermicularis egg.

released in the small intestine and undergoes several molts before maturing into an adult worm in the cecum.

Laboratory Diagnosis. The egg and occasionally the adult of *T. trichiura* may be seen in the fecal specimen. The adult male measures 30 to 45 mm and has a thin anterior and a thick, coiled posterior. The female is 30 to 50 mm with a thin anterior and thick, straight posterior. The brown barrel-shaped egg is unembryonated when passed, 50 to 55 by 22 to 23 μm, and has hyaline polar plugs at each end (Fig. 24–57).

Ascaris lumbricoides

Up to 1 billion persons worldwide are infected with *A. lumbricoides*. The organism can be found in tropic as well as temperate areas, and children are most commonly infected. Transmission is primarily fecal-oral, and clinical symptoms may be related to the different phases of the life cycle. The organism is often found concurrently with whipworm.

Clinical Manifestations. Abdominal discomfort, loss of appetite, and colicky pains are due to the presence of adults in the intestine. Large numbers of adult worms may cause intestinal obstruction. Chronic infection with *A. lumbricoides* in children may hamper growth and development, because the worms feed on liquid intestinal contents. Larva migrating through the lungs may cause an immune response in the host referred to as Loeffler syndrome, which consists of asthma, edema, pneumonitis, and eosinophilic infiltration. On some occasions, fever or other disease conditions may cause the adults to migrate from the intestine and invade other organs, resulting in peritonitis, liver abscess, or secondary infection in the lungs. Treatment is piperazine citrate, which relaxes the worm and allows peristalsis to carry it out of the intestinal tract.

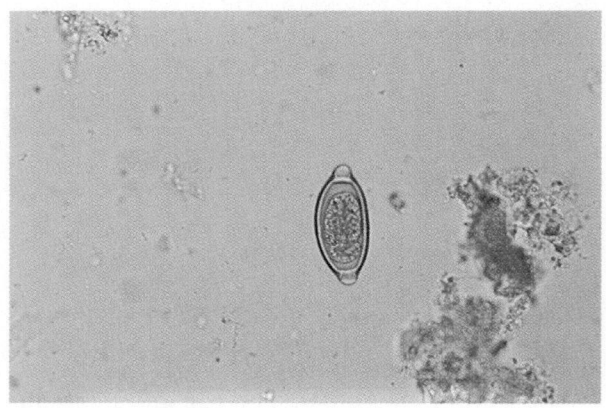

■■■■ F i g u r e   2 4 – 5 7

Trichuris trichiura egg.

Life Cycle. Figure 24–58 shows the life cycle of *A. lumbricoides*. Eggs that are deposited in warm, moist soil become infective within about 2 weeks. After the egg is ingested, larvae hatch in the duodenum, penetrate the intestinal wall, and gain entrance to the circulatory system. They break out from capillaries into the lungs, travel up the bronchial tree and trachea and over the epiglottis, and are swallowed. Maturation is completed in the intestine. The life cycle takes about 30 days from infection until adults are mature.

Laboratory Diagnosis. The usual diagnostic stage is the egg. Fertile *Ascaris* eggs are oval, measure 45 to 75 by 35 to 50 μm, and have a thick hyaline wall surrounding a one-cell stage embryo. There is a brown, bile-stained, mammillated outer layer on most eggs (Fig. 24–59). Some eggs, referred to as decorticated, lack the mammillated outer coat. Infertile eggs, whose size may range up to 90 μm, are often elongated or bizzarely shaped and contain a mass of highly refractile granules. Adults 15 to 35 cm long and about the diameter of a lead pencil may be seen in a stool sample. The female has a straight posterior and the male has a curved posterior. Both have three anterior lips with small tooth-like projections.

HOOKWORM

Two species of hookworm, *Necator americanus* (New World) and *Ancylostoma duodenale* (Old World), infect humans. *Ancylostoma duodenale* is seen in southern Europe and northern Africa along the Mediterranean as well as in parts of southeast Asia and South America, and *Necator americanus* has a geographic distribution in Africa, southeast Asia, and South and Central America and is endemic in rural areas of southeastern United States. Adults of the two species can be differentiated, but the eggs are identical. These worms live in the small intestine and attach to the mucosa by means of teeth *(A. duodenale)* or cutting plates *(N. americanus)*. Once attached, they secrete anticoagulants and ingest blood as a source of nourishment. There seems to be a racial distribution, with infections more prevalent in whites than in African Americans.

CLINICAL MANIFESTATIONS

Clinical symptoms vary according to the phase of the life cycle and worm burden. A small red

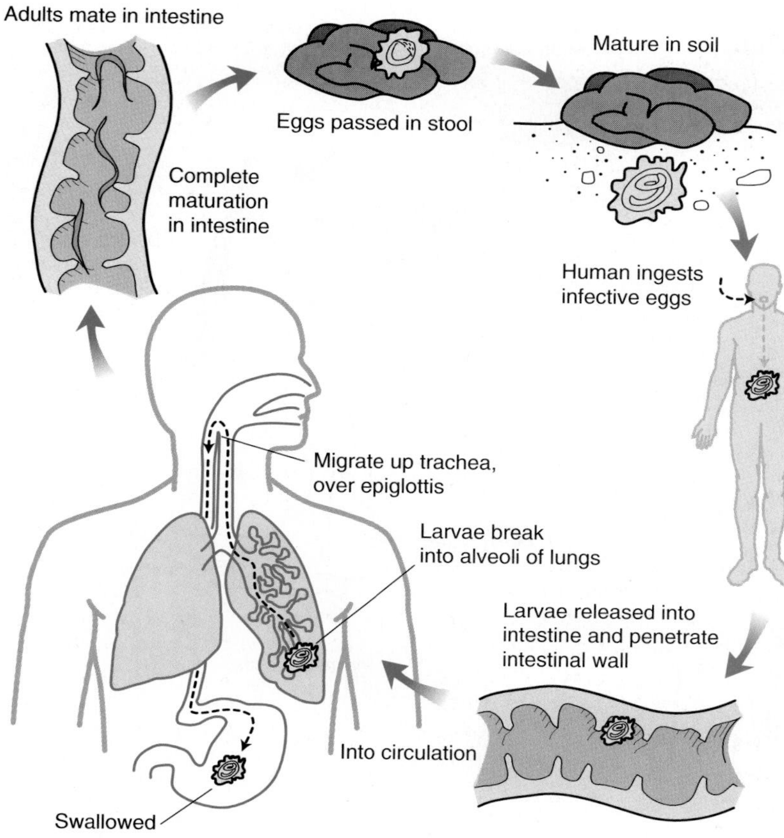

Adults mate in intestine

Eggs passed in stool

Mature in soil

Complete maturation in intestine

Human ingests infective eggs

Migrate up trachea, over epiglottis

Larvae break into alveoli of lungs

Larvae released into intestine and penetrate intestinal wall

Into circulation

Swallowed

Figure 24–58

Life cycle of *Ascaris lumbricoides.*

itchy papule referred to as ground itch develops at the site of larval penetration. If large numbers of larvae are present during the lung phase of migration, the patient may experience bronchitis, but there is no host sensitization as seen with *Ascaris* larva. The most severe symptoms are associated with the adult, including nonspecific symptoms such as diarrhea, fever, and nausea and vomiting. Blood loss, ranging from 0.03 to 0.2 mL of blood per worm per day, is primarily due to the ingestion of blood as a source of nourishment. Hemorrhages at the site of attachment, however, also contribute to total blood loss. Chronic heavy infection with hookworm may lead to microcytic hypochromic anemia. The mental and physical development of a child may be affected by chronic heavy infections. Treatment is usually with mebendazole, which blocks glucose uptake by the organism, or pyrantel pamoate, which paralyzes the worm so that it can be expelled. Supportive therapy including iron and protein supplements may be needed in severe cases.

LIFE CYCLE

Figure 24–60 shows the life cycle of the hookworm. Once eggs are deposited in warm moist soil, the first-stage rhabditiform larva develops

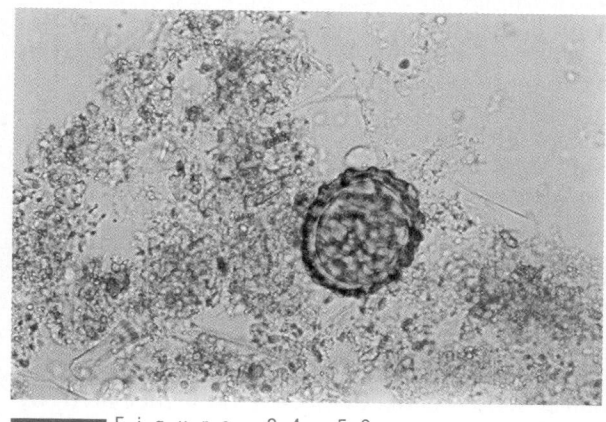

Figure 24–59

Ascaris lumbricoides egg, fertile.

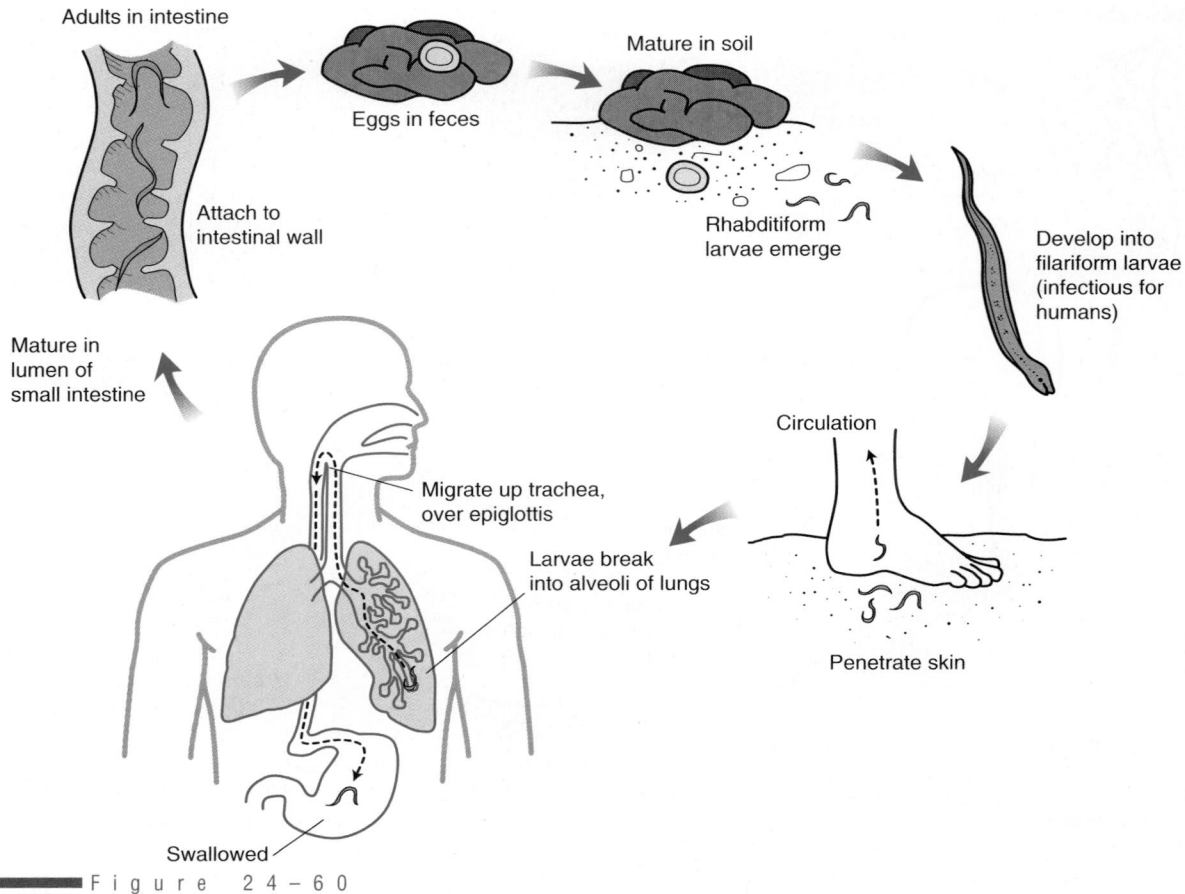

F i g u r e   2 4 – 6 0

Life cycle of hookworm.

within 1 to 2 days and feeds on bacteria in the soil. A nonfeeding infective filariform larva forms within a week. Humans are infected when the filariform larvae penetrate their skin. The organisms enter the circulation and break out of the capillaries into the lung, then migrate up the bronchial tree, over the epiglottis, and into the digestive tract. After additional larval molts, the worms attach to the mucosa in the small intestine. Eggs are produced about 5 weeks after skin penetration by the filariform larva.

LABORATORY DIAGNOSIS

Adult hookworms are rarely seen in the stool specimen; the egg and the rhabditiform larva are considered the usual diagnostic stages. Adult males may be distinguished and speciated by examination of the copulatory bursa in the posterior. The eggs and rhabditiform larvae of the two species are indistinguishable; therefore, the laboratory can report only "hookworm" when a characteristic egg or larva is found in a stool specimen. The egg is oval, colorless, thin shelled, and 50 to 60 μm and usually contains an embryo in the four to eight cell stage of cleavage (Fig. 24–61). The rhabditiform larva must be differentiated from that of *S. stercoralis.* Hookworm is 250 to 300 μm long and has a long buccal capsule and small, inconspicuous genital primordium (Fig. 24–62*A*). Figure 24–62*B* shows a close-up of the buccal capsule. The filariform larva must also be distinguished from that of *S. stercoralis.* Hookworm filariform larvae are about 500 μm, with a pointed tail and a esophageal-intestinal ratio of 1:4.

Strongyloides stercoralis

Strongyloides stercoralis, known as the threadworm, inhabits the small intestine but is also

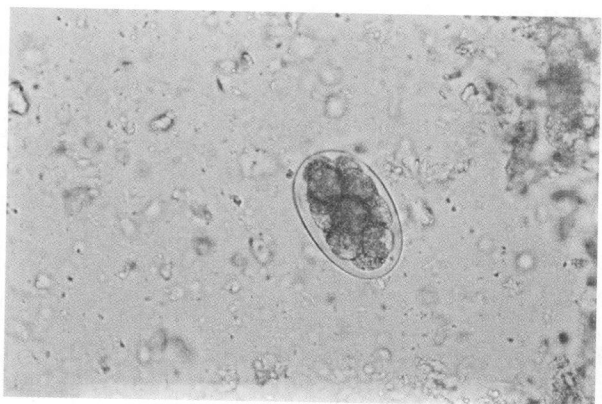

Figure 24–61
Hookworm egg.

capable of existing as a free-living worm. It is endemic in the tropics and subtropics and has been identified in a number of U.S. military veterans who served in Vietnam and other Southeast Asian countries. In the United States, most cases are found in persons living in Appalachia or in rural areas of the Southeast.

CLINICAL MANIFESTATIONS

Although many patients with *S. stercoralis* are asymptomatic, patients may experience nausea and vomiting and sharp, stabbing pains that resemble those of an ulcer. Chronic mild diarrhea may be present. In comparison with hookworm, *S. stercoralis* larval penetration of the skin does not cause a prominent papule, and migration through the lungs rarely elicits pneumonitis.

In contrast to the mild symptoms in the immunocompetent host, immunocompromised patients may develop severe infections referred to as disseminated strongyloidiasis or hyperinfection. In this population, large numbers of the filariform larvae develop in the intestine and migrate from the intestine into other organs such as the liver, heart, and central nervous system, causing a fulminating, often fatal infection. Death is due to complications resulting in respiratory failure. Respiratory failure has been reported in patients who are on chemotherapy or who have malignancies, chronic debilitating disease, or lymphoma, but is not yet commonly encountered in AIDS patients.

LIFE CYCLE

The life cycle of *S. stercoralis* can take one of three forms—direct, which is similar to that of hookworm, indirect, which involves a free-living phase, or autoinfection (Fig. 24–63). In the direct life cycle, the fertile egg hatches in the intestine and develops into the rhabditiform larva, which is passed in the stool. This larval form develops into a filariform larva, which is infective for humans by direct penetration. Once the larva has penetrated the skin, it enters the circulation, breaks out from capillaries in the lung, migrates up the bronchial tree and over the epiglottis, and enters the digestive tract, where it matures into the adult worm. In some patients, development into the filariform larva occurs in the intestine. These filariform (infective) larvae then penetrate the mucosa, enter the circulation, and return to

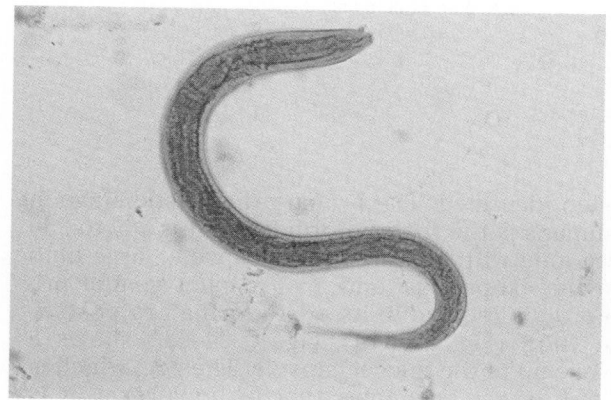

Figure 24–62
A, Hookworm rhabditiform larva. Notice long buccal capsule and lack of genital primordium. *B,* Hookworm rhabditiform larva, buccal capsule.

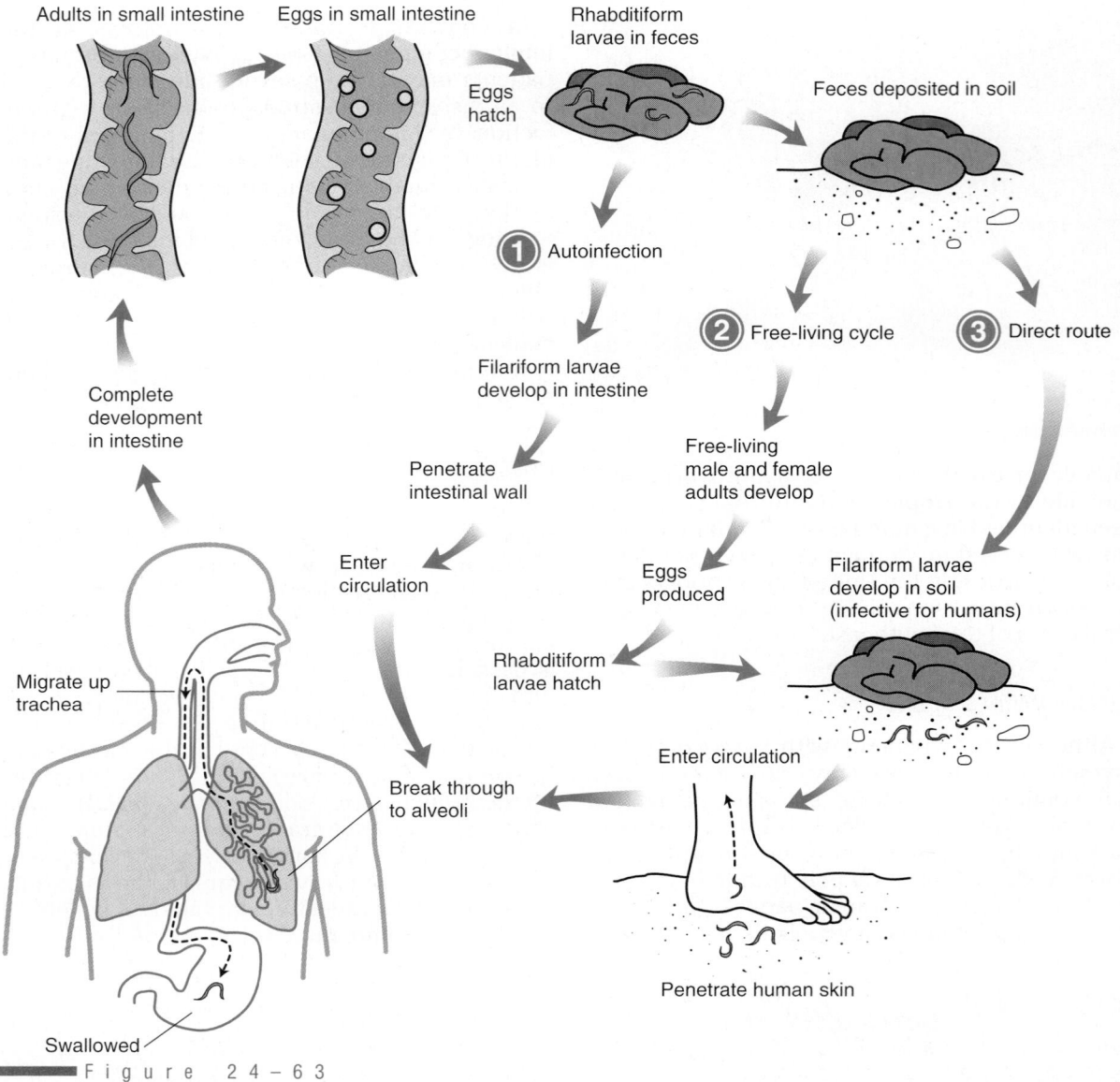

Adults in small intestine   Eggs in small intestine   Rhabditiform larvae in feces

Feces deposited in soil

Eggs hatch

1 Autoinfection

2 Free-living cycle   **3** Direct route

Complete development in intestine

Filariform larvae develop in intestine

Penetrate intestinal wall

Free-living male and female adults develop

Filariform larvae develop in soil (infective for humans)

Enter circulation

Eggs produced

Migrate up trachea

Rhabditiform larvae hatch

Enter circulation

Break through to alveoli

Swallowed

Penetrate human skin

■ F i g u r e   2 4 – 6 3
Life cycle of *Strongyloides stercoralis*.

the intestine to develop into adults. This part of the life cycle, termed autoinfection, may allow an initial infection to persist for years. In the indirect life cycle, the rhabditiform larvae develop into free-living male and females that produce eggs. At any point, the free-living cycle may revert and result in production of infective filariform larvae.

LABORATORY DIAGNOSIS

The female threadworm is small (2.5 mm) and rarely seen in a stool specimen. No male has been identified. The primary diagnostic stage in humans is the rhabditiform larva. It is 200 to 250 µm long with a short buccal capsule, large bulb in the esophagus, and a prominent genital primordium located in its posterior half to posterior third (Fig. 24–64A). Figure 24–64B shows a close-up of the buccal capsule. The egg, which is rarely seen, resembles that of hookworm. It is thin shelled, measures 54 by 32 µm, and often is segmented. The filariform larva has a notched tail, is 500 µm long, and has an esophageal-intestinal ratio of 1:1. Filariform larvae may be

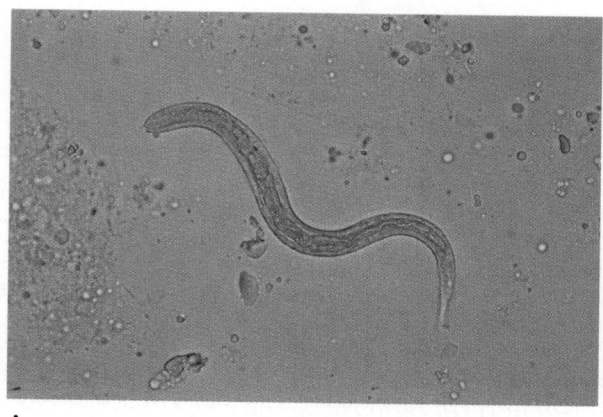
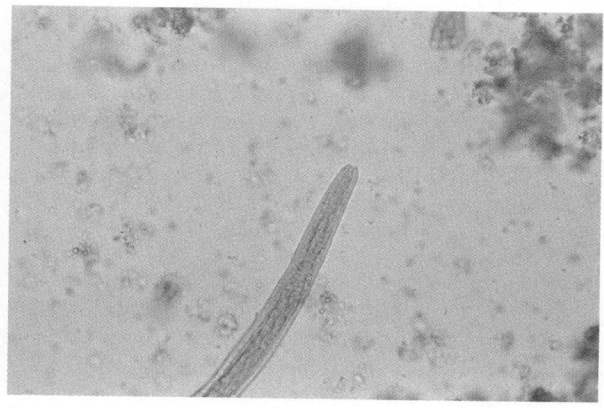

A

B

■ Figure 24–64

A, Strongyloides stercoralis rhabditiform larva. Notice short buccal capsule and prominent genital primordium. *B, S. stercoralis* rhabditiform larva, buccal capsule.

identified in the sputum of patients with hyperinfection.

If clinical symptoms suggest *Strongyloides* infection but multiple stool specimens are negative for the presence of the larvae, a duodenal aspirate or the Enterotest (HDC Corp., San Jose, CA) may be used for diagnosis. Treatment to eliminate the worm usually requires albendazole. Mebendazole can be used but requires a longer course of treatment, because it is not as effective against the larval stage.

Blood and Tissue Infections with Roundworms

Trichinella spiralis

Trichinosis is the infection of muscle tissue with the larval form of *Trichnella spiralis,* a helminth whose adult stages live in the human intestine. Humans acquire the infection by eating undercooked meat, particularly pork, that contains the larval forms. The larvae are released from the tissue capsule in the intestine and mature into adults. The female produces liveborn larvae that penetrate the intestinal wall, enter the circulation, and are carried to all areas of the body.

Clinical Manifestations. During the intestinal phase there are few symptoms, but the patient may experience diarrhea and abdominal discomfort. The majority of symptoms occur during the migration and encapsulation period, and their severity depends on the number of parasites, the tissues invaded, and general health of the patient. Common symptoms are: periorbital edema, fever, muscular pain or tenderness, headache, and general weakness. Eosinophilia of 40 to 80%

is common in most patients. The larvae encapsulate only in striated muscle; in other organs, they instigate an inflammatory reaction and are eventually absorbed. Encapsulation of the larva is usually completed within 3 weeks of migration, and calcification occurs within 18 months. Patients with symptoms should be treated with analgesics and other general supportive treatment. Steroids are rarely given.

Laboratory Diagnosis. Diagnosis is often based on clinical symptoms and patient history because of the difficulty of recovering adults or larvae in a stool specimen. Biopsy of muscle tissue and identification of the encapsulated coiled larva is the definitive diagnostic method. Figure 24–65 shows a biopsy of a muscle containing the larva of *T. spiralis.* Specimens from large muscles such as the deltoid and gastrocnemius should be

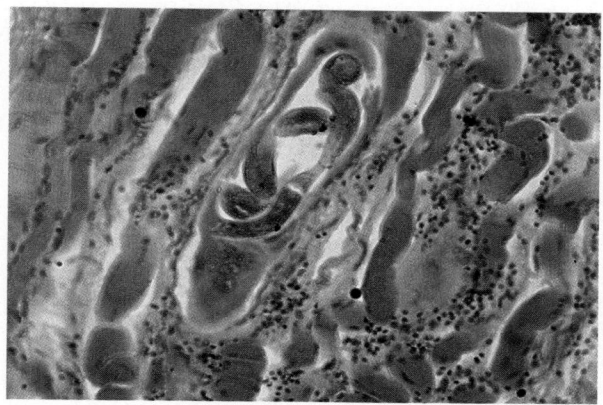

■ Figure 24–65

Trichinella spiralis larva, biopsy.

examined for presence of viable larvae and stained for histologic examination. Presence of calcified larva on radiography indicates previous infection. Serologic tests are available. An intra-dermal skin test yields a positive reaction within 3 weeks of infection.

LARVA MIGRANS

Two forms of larva migrans exist in humans—cutaneous (creeping eruption) and visceral. Both are the result of infection with nonhuman nema-tode larvae that are unable to complete their life cycle in humans.

In cutaneous larva migrans, the infecting organ-ism is commonly the filariform larva of the dog or cat hookworm (Ancylostoma braziliense). Once the larva penetrates the skin of a human, it cannot enter the circulation. It wanders through the sub-cutaneous tissue, creating long, winding tunnels that itch intensely. The infection resolves within several weeks when the larva dies. Diagnosis is primarily based on clinical symptoms.

In visceral larva migrans, a human accidentally ingests the eggs of the dog roundworm (Toxo-cara canis) or cat roundworm (T. cati). The lar-vae hatch in the intestine, penetrate the gut, wander through the abdominal cavity, and pene-trate lungs, eye, liver, or brain. The infection is primarily seen in children 1 to 4 years old. Clin-ical symptoms include fever, pneumonitis, and hepatomegaly. Eosinophilia of 30 to 50% is com-mon. CNS complications may develop. Diagnosis is usually made on the basis of clinical findings and the results of serologic tests.

FILARIAL WORMS

Filarial worms are roundworms of blood and tissue that give birth to larvae referred to as microfilariae. A number of species infect hu-mans. Those considered most pathogenic are *Brugia malayi, Wuchereria bancrofti, Onchocerca volvulus,* and *Loa loa.* In addition, the non-pathogens *Mansonella ozzardi, Mansonella per-stans,* and *Mansonella streptocerca* may be seen. Identification of the various species depends on the morphology of the microfilaria, its periodici-ty, and its location in the host. The morphologic characteristics that are considered include the presence or absence of a sheath (the remnant of the egg from which the larva hatched) and the presence and arrangement of nuclei in the tail. Table 24–9 gives a comparison of the tail mor-phology, periodicity, insect vector, and location for the species of microfilariae commonly found in humans.

Life Cycle. Adults, which may range in size from 2 to 50 cm, live in human lymphatics, mus-cles, or connective tissues. Mature females pro-duce liveborn larvae (microfilariae) that are the infective stage for the insect during the insect's blood meal. Once ingested, microfilariae penetrate the insect's gut wall and develop into infective third-stage (filariform) larvae. These enter the insect proboscis and are introduced into human circulation when the insect feeds. Figure 24–66 shows a generalized life cycle for microfilariae.

Wuchereria bancrofti. The causative agent of bancroftian filariasis and elephantiasis, *W. ban-crofti,* is primarily limited to the tropical and sub-tropical regions. The insect vector is a mosquito, either *Culex* or *Aedes* sp.

Clinical Manifestations. The adult filarial worm lives in the lymphatics and lymph nodes, especial-ly those in the lower extremities. Presence of the adults initiates an immunologic response consist-ing of cellular reactions, edema, and hyperplasia. A strong granulomatous reaction with production of fibrous tissue around the dead worms ensues. The end result of the reaction is that small lym-phatics may be narrowed or closed, with subse-quent development of collateral lymphatics. During this period, the patient may experience generalized symptoms such as fever, headache, and chills as well as localized swelling, redness, and lymphangitis, primarily sites in the male and female genitalia and the extremities. Elephan-tiasis, a debilitating and deforming complication, occurs in less than 10% of infections, usually after many years of continual filarial infection. Chronic obstruction in the lymphatic flow results in lym-phatic varices, fibrosis, and proliferation of der-mal and connective tissue. The enlarged areas eventually develop a hard leathery appearance.

Laboratory Diagnosis. Diagnosis of *W. ban-crofti* should include the examination of a blood specimen obtained at night (10 PM to 2 AM) for the presence of microfilariae. The blood may be examined immediately for live microfilariae or may be pooled on a slide and stained. Filtration of up to 5 mL of blood through a 5 μm Nucleo-pore filter (Nucleopore Corp., Pleasanton, CA) may detect light infections. The microfilariae of *W. bancrofti* are sheathed, and the nuclei do not extend to the tip of the tail (Fig. 24–67).

Brugia malayi. *Brugia malayi,* another noctur-nal microfilarial species, is limited to the Far East, including Korea, China, and the Philippines. Mosquitoes of the genera *Mansonia, Anopheles,*

■■■■ T A B L E 2 4 – 9

COMPARISON OF MICROFILARIAE

Organism	Arthropod vector	Periodicity	Location of Adult/Microfilaria	Tail Morphology	
Wuchereria bancrofti	Mosquito (*Culex, Aedes, Anopheles* sp)	Nocturnal	Lymphatics, blood	Sheathed Nuclei do not extend to tip of tail	
Brugia malayi	Mosquito (*Aedes* sp)	Nocturnal	Lymphatics, blood	Sheathed Terminal nuclei separated	
Loa loa	Fly (*Chrysops* sp)	Diurnal	Subcutaneous tissue, blood	Sheathed Nuclei extend to tip of tail	
Onchocerca volvulus	Fly (*Simulium* sp)	Nonperiodic	Subcutaneous nodule, subcu-taneous tissue	Unsheathed Nuclei do not extend to tip of tail	
Mansonella ozzardi	Midge (*Culicoides* sp)	Nonperiodic	Body cavity, blood, skin	Unsheathed Nuclei do not extend to tip of tail	
Mansonella perstans	Midge (*Culicoides* sp)	Nonperiodic	Mesentery, blood	Unsheathed Nuclei extend to blunt tip of tail	
Mansonella streptocerca	Midge (*Culicoides* sp)	Nonperiodic	Subcutaneous, skin	Unsheathed Nuclei extend to tip of hooked tail	

and *Aedes* have been shown to transmit the organism. Pathology of the disease is the same as that of *W. bancrofti* infections. The distinguishing characteristic of the microfilaria is the presence of a sheath and the arrangement of tail nuclei—the nucei extend to the tip, but a space separates the two terminal nuclei.

Loa loa. Infection with *Loa loa,* the eye worm, is limited to the African equatorial rain forest, where the fly vector *(Chrysops* sp*)* breeds. Adults migrate through the subcutaneous tissue, causing temporary inflammatory reactions called Calabar swellings. These characteristic swellings may cause pain and pruritus that last about a week

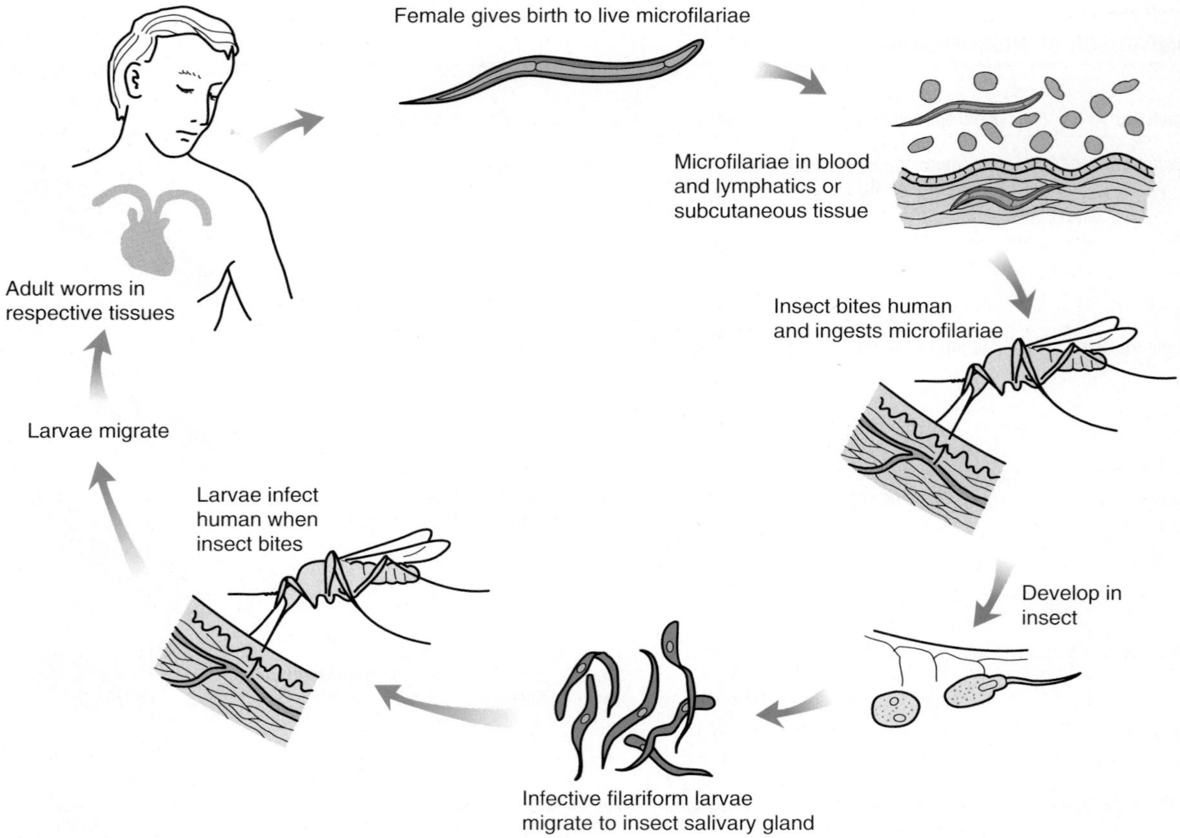

Female gives birth to live microfilariae

Microfilariae in blood and lymphatics or subcutaneous tissue

Adult worms in respective tissues

Insect bites human and ingests microfilariae

Larvae migrate

Develop in insect

Larvae infect human when insect bites

Infective filariform larvae migrate to insect salivary gland

■ F i g u r e   2 4 – 6 6
Generalized life cycle of microfilariae.

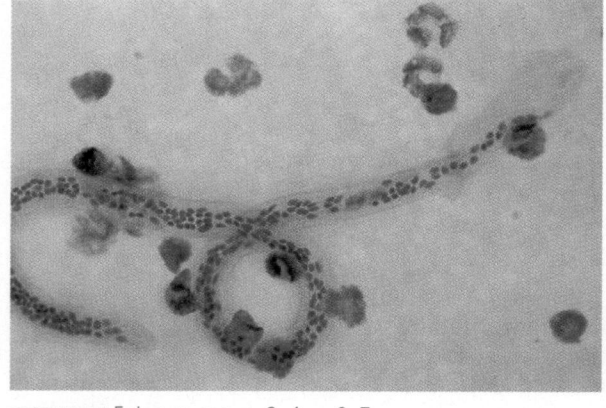

■ F i g u r e   2 4 – 6 7
Wuchereria bancrofti microfilaria. Notice faintly staining sheath extending from both ends of organism.

before disappearing, only to reappear in another part of the body. The adult worm can often be seen as it migrates across the surface of the eye.

Diagnosis may be based on presence of Calabar swellings or of the adult worm in the conjunctiva of the eye. Microfilariae may be seen in a blood specimen if taken during the day, especially around noon, when migration peaks. The microfilaria is sheathed, and nuclei extend to the tip of the tail.

Onchocerca volvulus. Infection with *Onchocerca volvulus* is referred to as onchocercosis or river blindness. The organism may be found in Africa and South and Central America; transmission is by the bite of the black fly *(Simulium* sp). Adults are encapsulated in fibrous tumors in human subcutaneous tissues. Microfilariae may be isolated from the subcutaneous tissue, skin, and the nodule itself but are rarely found in blood or lymphatic fluid.

The nodules in which adults live may measure up to 25 mm in size and can be found on most parts of the body. They are the result of an inflammatory and granulomatous reaction around the adult worms. Figure 24–68 shows a cross-section of tissue containing these organisms. Blindness, the most serious complication, results when microfilariae collect in the cornea and iris, causing keratitis and atrophy of the iris.

Diagnosis involves clinical symptoms such as presence of nodules and microscopic identification of microfilariae. The diagnostic method used is the skin snip, in which a small slice of skin is obtained and placed on a saline mount. Microfilariae with no sheath and with nuclei that do not extend into the tip of the tail are characteristic of this organism.

***Mansonella* Species.** *Mansonella ozzardi, M. streptocerca (Dipetalonema streptocerca),* and *M. perstans (Dipetalonema perstans)* are filarial worms not usually associated with serious infections. They are transmitted by midges belonging to the genus *Culicoides*. The microfilariae of *M. streptocerca* are found in the skin. They are unsheathed and have nuclei that extend to the end of the "shepherd's crook" tail. Microfilariae of *M. ozzardi* and *M. perstans* are found in the blood as unsheathed organisms. *M. ozzardi* microfilariae have tails with nuclei that do not extend to tips, whereas the nuclei in the tail of an *M. perstans* microfilaria extend to the tip.

Dracunculus medinensis

Dracunculus medinensis (guinea worm, fiery serpent of the Israelites) causes serious infections in the Middle East, parts of Africa, and India. It is often found in areas in which "step-down" wells are used. Males and females develop in the human body cavity, and the gravid female migrates through subcutaneous tissues to the legs, feet, and sometimes arms to discharge live-born larvae. Initially a blister-like, inflammatory papule appears in the area of the worm. The papule ulcerates, and when the body comes in contact with water, the female worm exposes her uterus through the ulceration and releases larvae into the water. Patients may experience nausea and vomiting, urticaria, and dyspnea prior to the rupture of the uterus. If the worm is broken during an attempt to remove it, the patient may experience a severe inflammatory reaction and secondary bacterial infection.

Life Cycle. Humans acquire the infection by ingesting a copepod *(Cyclops* sp) that contains an infective larva. The larva is released in the intestine, penetrates the intestinal wall, and migrates to the connective tissue. When mature and gravid, the female migrates through the tissues to the arm or leg to release larvae into the water. The rhabditiform larvae are then ingested by the copepod.

Diagnosis is made from the typical appearance of the lesion. Treatment is metronidazole. The ancient method of removal by rolling the worm a few inches at a time onto a stick is still practiced in some areas of the world.

Bibliography

Acuna-Soto R, Samuelson J, et al: Application of the polymerase chain reaction to the epidemiology of pathogenic and nonpathogenic *Entamoeba histolytica.* Am J Trop Med Hyg 48:58, 1993.

Addiss DG, Mathews HM, et al: Evaluation of a commercially available enzyme-linked immunosorbent assay for *Giardia lamblia* antigen in stool. J Clin Microbiol 29:1137, 1991.

Ahmed M, McAdam KP, et al: Systemic manifestations of invasive amebiasis. Clin Infect Dis 15:974, 1992.

Aikawa M: Human cerebral malaria. Am J Trop Med Hyg 39:3, 1988.

Aikawa M, Miller LH, et al: Erythrocyte entry by malarial parasites. J Cell Biol 77:72, 1978.

Aldeen WE, Whisenant J, et al: Comparison of pooled formalin-preserved fecal specimens with three individual samples for detection of intestinal parasites. J Clin Microbiol 31:144, 1993.

Allason-Jones E, Mindel A, et al: *Entamoeba histolytica* as a commensal intestinal parasite in homosexual men. New Engl J Med 315:353, 1986.

Allason-Jones E, Mindel A, et al: Outcome of untreated infection with *Entamoeba histolytica* in homosexual men with and without HIV antibody. Br Med J 297:654, 1988.

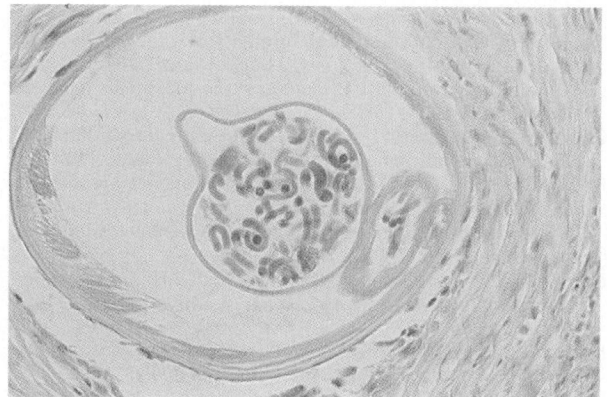

■■ F i g u r e   2 4 – 6 8
Cross-section of tissue infected with *Onchocerca volvulus.*

Anthony RL, Bangs MJ, et al: Onsite diagnosis of *Plasmodium falciparum, P. vivax* and *P. malariae* by using the quantitative buffy coat system. J Parasitol 78:994, 1992.

Arrowood MJ, Sterling CR: Comparison of conventional staining methods and monoclonal antibody–based methods for *Cryptosporidium* oocyst detection. J Clin Microbiol 27:1490, 1989.

Baron EJ, Schenone C, Tanenbaum B: Comparison of three methods for detection of *Cryptosporidium* oocysts in a low-prevalence population. J Clin Microbiol 27:223, 1989.

Baselski VS, Robison MK, et al: Rapid detection of *Pneumocystis carinii* in bronchoalveolar lavage samples by using cellufluor staining. J Clin Microbiol 28:393, 1990.

Beal C, Goldsmith R, et al: The plastic envelope method, a simplified technique for culture diagnosis of trichomoniasis. J Clin Microbiol 30:2265, 1992.

Berger ST, Mondino BJ, et al: Successful medical management of Acanthamoeba keratitis. Am J Ophthalmol 110:395, 1990.

Blumenfield W, Kovacs JA: Use of a monoclonal antibody to detect *Pneumocystis carinii* in induced sputum and bronchoalveolar lavage fluid by immunoperoxidase staining. Arch Pathol Lab Med 112:1233, 1988.

Bottone EJ: Diagnosis of acute pulmonary toxoplasmosis by visualization of invasive and intracellular tachyzoites in giemsa-stained smears of bronchoalveolar lavage fluid. J Clin Microbiol 29:2626, 1991.

Brillman J: Plasmodium vivax malaria from Mexico—a problem in the United States. West J Med 147:469, 1987.

Brown RL: Successful treatment of primary amebic meningoencephalitis. Arch Intern Med 151:1201, 1991.

Carr JM, Emery S, et al: Babesiosis. Diagnostic pitfalls. Am J Clin Pathol 95:774, 1991.

Casemore DP, Roberts C: Guidelines for screening for *Cryptosporidium* in stools: Report of a joint working group. J Clin Pathol 46:2, 1993.

Chan FTH, Guan MX, Mackenzie AMR: Application of indirect immunofluorescence to detection of *Dientamoeba fragilis* trophozoites in fecal specimens. J Clin Microbiol 31:1710, 1993.

Chulay JD, Ockenhouse CF: Host receptors for malaria-infected erythrocytes. Am J Trop Med Hyg 43(suppl 2):6, 1990.

Couroux P, Schieven BC, Hussain Z: *Pneumocystis carinii.* ASM News 59:179, 1993.

Cregan P, Yanamoto A, et al: Comparison of four methods for rapid detection of *Pneumocystis carinii* in respiratory specimens. J Clin Microbiol 28:2432, 1990.

Current WL: The biology of *Cryptosporidium.* In Leech JH, Sande MA, Root RK (eds): Contemporary Issues in Infectious Diseases, Vol 7: Parasitic Infections. New York: Churchill Livingstone, 1988.

DeHovitz JA, Pape JW, et al: Clinical manifestations and therapy of *Isospora belli* in patients with the acquired immunodeficiency syndrome. N Eng J Med 315:87, 1986.

Derouin F, Sarfati C, et al: Laboratory diagnosis of pulmonary toxoplasmosis in patients with acquired immunodeficiency syndrome. J Clin Microbiol 27:1661, 1989.

Doyle PW, Helgason MM, et al: Epidemiology and pathogenicity of *Blastocystis hominis.* J Clin Microbiol 28:116, 1990.

Draper D, Parker R, et al: Detection of *Trichomonas vaginalis* in pregnant women with the InPouch TV culture system. J Clin Microbiol 31:1016, 1993.

Epstein RJ, Wilson LA, et al: Rapid diagnosis of *Acanthamoeba* keratitis from corneal scrapings using indirect fluorescent antibody staining. Arch Ophthalmol 104:1318, 1986.

Esernio-Jenssen D, Scimeca PG, et al: Transplacenta perinatal babesiosis. J Pediatr 110:570, 1987.

Estevez EG, Levine JA: Examination of preserved stool specimens for parasites: Lack of value of the direct wet mount. J Clin Microbiol 22:666, 1985.

Fleck DG: Annotation: Diagnosis of toxoplasmosis. J Clin Pathol 42:191, 1989.

Flores BM, Garcia CA, et al: Differentiation of *Naegleria fowleri* from *Acanthamoeba* species by using monoclonal antibodies and flow cytometry. J Clin Microbiol 28:1999, 1990.

Frenkel JK: Toxoplasmosis. Pediatr Clin North Am 32:917, 1985.

Garcia LS, Shimizu RY, et al: Evaluation of intestinal protozoan morphology in polyvinyl alcohol preservative: Comparison of zinc sulfate– and mercuric chloride–based compounds for use in Schaudinn's fixative. J Clin Microbiol 31:307, 1993.

Garcia LS, Shum AC, Bruckner DA: Evaluation of a new monoclonal antibody combination reagent for direct fluorescence detection of *Giardia* cysts and *Cryptosporidium* oocysts in human fecal specimens. J Clin Microbiol 30:3255, 1992.

Garcia LW, Hemphill RB, et al: Acquired immunodeficiency syndrome with disseminated toxoplasmosis presenting as an acute pulmonary and gastrointestinal illness. Arch Pathol Lab Med 15:459, 1991.

Garfinkel LL, Giladi M, et al: DNA probes specific for *Entamoeba histolytica* possessing pathogenic and nonpathogenic zymodemes. Infect Immun 57:926, 1989.

Gelbart SM, Thomason JL, et al: Growth of *Trichomonas vaginalis* in commercial culture media. J Clin Microbiol 28:962, 1990.

Gellin BG, Soave R: Coccidian infections in AIDS: Toxoplasmosis, cryptosporidiosis, and isosporiasis. Med Clin North Am 76:205, 1992.

Genta RM: Global prevalence of strongyloidiasis: Critical review with epidemiologic insights into the prevention of disseminated disease. Rev Infect Dis 11:755, 1989.

Genta RM: Strongyloidiasis. In Walls KW, Schantz PM, (eds): Immunodiagnosis of Parasitic Disease, Vol 1: Helminthic Diseases. Orlando, FL: Academic Press, 1986.

Genta RM, Weesner R, et al: Strongyloidiasis in US veterans of the Vietnam and other wars. JAMA 258:49, 1987.

Gill VJ, Nelson NA, et al: Optimal use of the cytocentrifuge for recovery and diagnosis of *Pneumocystis carinii* in bronchoalveolar lavage and sputum specimens. J Clin Microbiol 26:1641, 1988.

Gonzalez-Ruiz A, Haque R, et al: A monoclonal antibody for distinction of invasive and noninvasive clinical isolates of *Entamoeba histolytica.* J Clin Microbiol 30:2807, 1992.

Goodgame RW, Genta RM, et al: Intensity of infection in AIDS-associated cryptosporidiosis. J Infect Dis 167:704, 1993.

Goodman PC, Schnap LM: Pulmonary toxoplasmosis in AIDS. Radiology 184:791, 1992.

Gutierrez Y: Diagnostic Pathology of Parasitic Infections with Clinical Correlations. Philadelphia: Lea & Febiger, 1990.

Haque R, Kress K, et al: Diagnosis of pathogenic *Entamoeba histolytica* infection using a stool ELISA based on monoclonal antibodies to the galactose-specifice adhesion. J Infect Dis 167:247, 1993.

Harcourt-Webster JN, Scaravilli F, Darwish AH: *Strongyloides stercoralis* hyperinfection in an HIV positive patient. J Clin Pathol 44:346, 1991.

Ho M, Singh B, et al: Clinical correlates of in vitro *Plasmodium falciparum* cytoadherence. Infect Immun 59:873, 1991.

Holtan N: Amebiasis—the ancient scourge is still with us. Postgrad Med 83:65, 1988.

Institute of Medicine: Malaria: Obstacles and Opportunities. Washington, DC: National Academy Press, 1991.

Iogarashi I, Oo MM, et al: Knob antigen deposition in cerebral malaria. Am J Trop Med Hyg 37:511, 1987.

Irusen EM, Jackson TF, Simjee AE: Asymptomatic intestinal colonization by pathogenic *Entamoeba histolytica* in amebic liver abscess: Prevalence, response to therapy and pathogenic potential. Clin Infect Dis 14:889, 1992.

Janoff EN, Reller LB: *Cryptosporidium* species, a protean protozoan. J Clin Microbiol 25:967, 1987.

Joiner KA, Dubremetz JF: *Toxoplasma gondii:* A protozoan for the nineties. Infect Immun 61:1169, 1993.

Jones JE: Pinworms. Am Fam Physician 38:159, 1988.

Kilvington S, Larkin DFP et al: Laboratory investigation of *Acanthamoeba* keratitis. J Clin Microbiol 28:2722, 1990.

Krause PJ, Teleford SR, et al: Geographical and temporal distribution of babesial infection in Connecticut. J Clin Microbiol 29:1, 1991.

Lanar DE, McLaughlin GL, et al: Comparison of thick films, in vitro culture and DNA hybridization probes for detecting *Plasmodium falciparum* malaria. Am J Trop Med Hyg 40:3, 1989.

Lesho EP: Leishmaniasis: Another threat to Persian Gulf veterans. Postgrad Med 90:213, 1991.

Levy RM, Bredessen DE, Rosenblum ML: Oportunistic central nervous system pathology in patients with AIDS. Ann Neurol 23 (suppl):S7, 1988.

Lindquist TD, Sher NA, Doughman DJ: Clinical signs and medical therapy of early *Acanthamoeba* keratitis. Arch Ophthalmol 106:73, 1988.

List PJ, Dondero RS, et al: Monoclonal-antibody–based enzyme-linked immunosorbent assay for *Trichomonas vaginalis.* J Clin Microbiol 26:1684, 1988.

Long EG, Tsin AT, Robinson BA: Comparison of the FeKal CON-Trate system with the formalin–ethyl acetate technique for detection of intestinal parasites. J Clin Microbiol 22:210, 1985.

Lotter H, Mannweiler E, et al: Sensitive and specific serodiagnosis of invasive amebiasis by using a recombinant surface protein of pathogenic *Entamoeba histolytica.* J Clin Microbiol 30:3163, 1992.

Ma P, Visvesvara GS, et al: *Naegleria* and *Acanthamoeba* infections: review. Rev Infect Dis 12:490, 1990.

Maayan S, Wormer GP, et al: *Strongyloides stercoralis* hyperinfection in a patient with the acquired immunodeficiency syndrome. Am J Med 83:945, 1987.

MacPherson DW, McQueen R: Cryptosporidiosis: Multiattribute evaluation of six diagnostic methods. J Clin Microbiol 31:198, 1993.

MacPherson DW, McQueen WM: Morphological diversity of *Blastocystis hominis* in sodium acetate–acetic acid–formalin–preserved stool samples stained with iron hematoxylin. J Clin Microbiol 32:267, 1994.

Markell EK, Udkow MP: *Blastocystis hominis:* Pathogen or fellow traveler? Am J Trop Med Hyg 35:1023, 1986.

Marti H, Koella JC: Multiple stool examinations for ova and parasites and rate of false-negative results. J Clin Microbiol 31:3044, 1993.

Martinez-Palomo A, Tsutsumi V, et al: Ultrastructure of experimental intestinal invasive amebiasis. Am J Trop Med Hyg 41:273, 1989.

Masur H, Lane HC, et al: *Pneumocystis* pneumonia: From bench to clinic. Ann Intern Med 111:813, 1989.

Mattia AR, Wladron, MA, Sierra LS: Use of the quantitative buffy coat system for detection of parasitemia in patients with babesiosis. J Clin Microbiol 31:2816, 1993.

McDougall RJ, Tandy MW, et al: Incidental finding of a microsporidian parasite from an AIDS patient. J Clin Microbiol 31:436, 1993.

Meredith JT: Toxoplasmosis of the central nervous system Am Fam Physician 35:113, 1987.

Mohr E, Mohr I: Statistical analysis of the incidence of positives in the examination of parasitological specimens. J Clin Microbiol 30:1572, 1992.

Morris AJ, Wilson ML, Reller LB: Application of rejection criteria for stool ovum and parasite examinations. J Clin Microbiol 30:3213, 1992.

Murphy GS, Basri H, et al: Vivax malaria resistant to treatment and prophylaxis with chloroquine. Lancet 341:96, 1993.

Neimeister R, Logan AL, et al: Evaluation of direct wet mount parasitological examination of preserved fecal specimens. J Clin Microbiol 28:1082, 1990.

Neva FA: Biology and immunology of human strongyloidiasis. J Infect Dis 153:397, 1986.

Newman RD, Jaeger KL, et al: Evaluation of an antigen capture enzyme-linked immunosorbent assay for detection of *Cryptosporidium* oocysts. J Clin Microbiol 31:2080, 1993.

Ng VL, Virani NA, et al: Rapid detection of *Pneumocystis carinii* using a direct fluorescent monoclonal antibody stain. J Clin Microbiol 28:2228, 1990.

Ng VL, Yajko DM, et al: Evaluation of an indirect fluorescent-antibody stain for detection of *Pneumocysts carinii* in respiratory specimens. J Clin Microbiol 28:975, 1990.

Norton SA, Frankenburg S, Klaus SN: Cutaneous leishmaniasis acquired during military service in the Middle East. Arch Dermatol 128:83, 1992.

Pawlowski ZS, Schad GA, Stott GJ: Hookworm Infection and Anemia: Approaches to Prevention and Control. Geneva, WHO, 1991.

Peiris JS, Premawansa S, et al: Monoclonal and polyclonal antibodies both block and enhance transmission of human *Plasmodium vivax* malaria. Am J Trop Med Hyg 39:26, 1988.

Perry JL, Matthews JS, Miller GR: Parasite detection efficiencies of five stool concentration systems. J Clin Microbiol 28:1094, 1990.

Peters CS, Sable R, et al: Prevalence of enteric parasites in homosexual patients attending an outpatient clinic. J Clin Microbiol 24:684, 1986.

Petri WA, Jackson TF, et al: Pathogenic and nonpathogenic strains of *Entamoeba histolytica* can be differentiated by monoclonal antibodies to the galactose-specific adherence lectin. Infect Immun 58:1802, 1990.

Petri WA: Invasive amebiasis and the galactose-specific lectin of *Entamoeba histolytica.* ASM News 57:299, 1991.

Pickering LK, Engelkirk PG: *Giardia lamblia.* Pediatr Clin North Am 35:565, 1988.

Pomery C, Filice GA: Pulmonary toxoplasmosis: A review. Clin Infect Dis 14:863, 1992.

Porter JD, Ragazzoni HP, et al: *Giardia* transmission in a swimming pool. Am J Pub Health 78:659, 1988.

Prevention and Control of Intestinal Parasitic Infections: Report of a WHO Expert Committee. World Health Organization Technical Report Series #749. Geneva: WHO, 1987.

Qadri SMH, Al-Okaili GA, Al-Dayel F: Clinical significance of *Blastocystis hominis.* J Clin Microbiol 27:2407, 1989.

Ravdin JI: Amebiasis now. Am J Trop Med Hyg 41(suppl):40, 1989.

Ravdin JI: *Entamoeba histolytica:* Pathogenic mechanisms, human immune response, and vaccine development. Clin Res 38:215, 1990.

Ravdin JI: *Entamoeba histolytica:* From adherence to enteropathy. J Infect Dis 159:420, 1989.

Raviglione MC: Extrapulmonary pneumocystosis: The first 50 cases. Rev Infect Dis 12:1127, 1990.

Recommendations for the prevention of malaria in travelers. MMWR 39: 1–10, 1990.

Reed SL: Amebiasis: An update. Clin Infect Dis 14:385, 1992.

Rickman LS, Long GW: Rapid diagnosis of malaria by acridine orange staining of centrifuged parasites. Lancet 58:69, 1989.

Rieckman KH, Davis DR, Hutton DC: *Plasmodium vivax* resistance to chloroquine? Lancet 2:1183, 1989.

Rosenblatt JE, Sloan LM: Evaluation of an enzyme-linked immunosorbent assay for detection of *Cryptosporidium* spp in stool specimens. J Clin Microbiol 31:1468, 1993.

Rosoff JD, Sanders CA, et al: Stool diagnosis of giardiasis using a commercially available enzyme immunoassay to detect *Giardia*-specific antigen 65 (GSA 65). J Clin Microbiol 27:1997, 1989.

Rusnak J, Hadfield TL, et al: Detection of *Cryptosporidium* oocysts in human fecal specimens by an indirect immunofluorescence assay with monoclonal antibodies. J Clin Microbiol 27:1135, 1989.

Ryan NJ, Sutheralan G, et al: A new technique for detection of microsporidial species in urine, stool, and nasopharyngeal specimens. J Clin Microbiol 31:3264, 1993.

Samuelson J, Acuna-Soto R, et al: DNA hybridization probe for clinical diagnosis of *Entamoeba histolytica*. J Clin Microbiol 27:671, 1989.

Sargeaunt PG, Williams JE: Electrophoretic isoenzyme patterns of the pathogenic and nonpathogenic intestinal amoebae of man. Trans R Soc Trop Med Hyg 73:225, 1979.

Sargeaunt PG, Williams JE: The differentiation of invasive and non-invasive *Entamoeba histolytica* by isoenzyme electrophoresis. Trans R Soc Trop Med Hyg 72:519, 1978.

Schnapp LM, Geaghan SM, et al: *Toxoplasma gondii* pneumonitis in patients infected with the human immunodeficiency virus. Arch Intern Med 152:1073, 1992.

Senay H, MacPherson D: *Blastocystis hominis:* Epidemiology and natural history. J Infect Dis 162:987, 1990.

Shandera WX: From Leningrad to the day-care center—the ubiquitous *Giardia lamblia*. West J Med 153:154, 1990.

Sheehan DJ, Raucher BG, McKitrick JC: Association of *Blastocystis hominis* with signs and symptoms of human disease. J Clin Microbiol 24:548, 1986.

Shetty N, Prabhu T: Evaluation of faecal preservation and staining methods in the diagnosis of acute amoebiasis and giardiasis. J Clin Pathol 41:694, 1988.

Soave R, Johnson WD: *Cryptosporidium* and *Isospora belli* infections. J Infect Dis 157:225, 1988.

Sorvillo FJ, Fujioka K, et al: Swimming-associated cryptosporidiosis. Am J Public Health 82:742, 1992.

Spielman A, Perrone JB, et al: Malaria diagnosis by direct observation of centrifuged samples of blood. Am J Trop Med Hyg 39:337, 1988.

Stanley SL, Jackson TF, et al: Serodiagnosis of invasive amebiasis using a recombinant *Entamoeba histolytica* protein. JAMA 266:1984, 1991.

Stibbs HH, Ongeryth JE: Immunofluorescence detection of *Cryptosporidium* oocysts in fecal smears. J Clin Microbiol 24:517, 1986.

Stibbs HH: Monoclonal antibody-based enzyme immunoassay for *Giardia lamblia* antigen in human stool. J Clin Microbiol 27:2582, 1989.

Strachan WD, Chiodini PL, et al: Immunological differentiation of pathogenic and non-pathogenic isolates of *Entamoeba histolytica*. Lancet 1:561, 1988.

Sullivan PB, Marsh MN, et al: Prevalence and treatment of giardiasis in chronic diarrhoea and malnutrition. Arch Dis Child 65:304, 1990.

Tangermann RH, Gordon S, et al: An outbreak of cryptosporidiosis in a day-care center in Georgia. Am J Epidemiol 133:471, 1991.

Thomason JL, Gelbart SM, et al: Comparison of four methods to detect *Trichomonas vaginalis*. J Clin Microbiol 26:1869, 1988.

Travis WD, Schmidt K, et al: Respiratory cryptosporidiosis in a patient with malignant lymphoma: Report of a case and review of the literature. Arch Pathol Lab Med 114:519, 1990.

Tschirhardt D, Klatt EC: Disseminated toxoplasmosis in the acquired immunodeficiency syndrome. Arch Pathol Lab Med 112:1247, 1988.

Tzipori S: Cryptosporidiosis in animal and humans. Microbiol Rev 47:84, 1983.

Wahlquist SP, Williams RM, et al: Use of pooled formalin-preserved fecal specimens to detect *Giardia lamblia*. J Clin Microbiol 29:1725, 1991.

Walker J, Conner G, et al: Giemsa staining for cysts and trophozoites of *Pneumocystis carinii*. J Clin Pathol 42:432, 1989.

Walsh JA: Problems in recognition and diagnosis of amebiasis: Estimation of the global magnitude of morbidity and mortality. Rev Infect Dis 8:228, 1988.

Wanke C, Tuazon CU: Toxoplasma encephalitis in patients with acquired immunodeficiency syndrome: Diagnosis and response to therapy. Am J Trop Med Hyg 36:509, 1987.

Watson B, Blitzer M, et al: Direct wet mounts versus concentration for routine parasitological examination: Are both necessary? Am J Clin Pathol 89:389, 1988.

Weber R, Bryan RT, Juranek DD: Improved stool concentration procedure for detection of *Cryptosporidium* oocysts in fecal specimens. J Clin Microbiol 30:2869, 1992.

Weber R, Bryan RT, et al: Threshold of detection of *Cryptosporidium* oocysts in human stool specimens: Evidence for low sensitivity of current diagnostic methods. J Clin Microbiol 29:1323, 1991.

Weinke T, Friedrich-Janicke B, et al: Prevalence and clinical importance of *Entamoeba histolytica* in two high-risk groups: Travelers returning from the tropics and male homosexuals. J Infect Dis 161:1029, 1990.

Whiteside ME, Barkin JS, et al: Enteric coccidiosis among patients with the acquired immunodeficiency syndrome. Am J Trop Med Hyg 33:1065, 1984.

Wijdicks EFM, Borleffs JCC, et al: Fatal disseminated hemorrhagic toxoplasmic encephalitis as the initial manifestation of AIDS. Ann Neurol 29:683, 1991.

Wilson M, Ware DA, Walls KW: Evaluation of commercial serodiagnosis kits for toxoplasmosis. J Clin Microbiol 25:2262, 1987.

Wilson RJM: Biochemistry of red cell invasion. Blood Cells 16:237, 1990.

Wongrichanalai C, Pornsilapatip J, et al: Acridine orange fluorescent microscopy and the detection of malaria in populations with low-density parasitemia. Am J Trop Med Hyg 44:17, 1991.

World Health Organization: Drugs Used in Parasitic Diseases. Geneva: WHO, 1990.

Wyler DJ: Malaria chemoprophylaxis for the traveler. N Engl J Med 329:31, 1993.

Xaio L, Herd RP: Quantitation of *Giardia* cysts and *Cryptosporidium* oocysts in fecal samples by direct immunofluorescence assay. J Clin Microbiol 31:2944, 1993.

Yeoh R, Warhurst DC, Falcon MG: *Acanthamoeba* keratitis. Br J Opthalmol 71:500, 1987.

Zierdt CH, Gill VJ, Zierdt WS: Detection of microsporidian spores in clinical samples by indirect fluorescent-antibody assay using whole-cell antisera to *Encephalitozoon cuniculi* and *Encephalitozoon hellem*. J Clin Microbiol 31:3071, 1993.

Zierdt CH: *Blastocystis hominis*—past and future. Clin Microbiol Rev 4:61, 1991.

Clinical Virology

Vee E. Davison • *Gerald L. Alderson*

THE VIRUS

Viruses are the ultimate parasites, functioning only within another cell. Viruses do not grow larger and then divide like other organisms. They attach to and enter host cells, then direct the cells to make viral components. New viruses are assembled from these components much like a car is assembled in a factory.

Viral Structure

The nucleic acid of a virus is called the *genome*. A virus contains only one type of nucleic acid, DNA or RNA, never both. It may be RNA or DNA, single-stranded (ss) or double-stranded (ds), and linear, circular, coiled, or segmented. The genome is surrounded by units of protein called *capsids;* the genome and capsids make up the *virion*. Viruses that contain only the virion are called *naked viruses*. Naked viruses are relatively stable to temperature, pH, and chemicals. The remaining viruses have an envelope that is a typical bilayered membrane; hence they are called *enveloped viruses*. The envelope is acquired as the virus buds throughout a cellular membrane, usually the outer cell membrane. This envelope is primarily of cellular origin with virally coded proteins on the surface.

Often, the viral antigens are glycoproteins that are both functional and antigenic. The glycoproteins may attach to specific receptors on the surfaces of their host cells, or they may be responsible for the fusion of the viral and cellular membranes that precedes the entry of the virus into the cell. The *enveloped* viruses are more fragile than the naked viruses because anything that disrupts their envelopes—temperature, detergents, solvents, etc—inactivates them. Many enveloped viruses contain enough enzymes to start replication without relying on the host cell.

Viruses are helical, icosahedral (a geometric shape with 20 faces), or complex in shape; the shape is primarily determined by the capsids. The envelope masks the shape of the virion, so that most enveloped viruses are *pleomorphic,* or variable in shape. The poxviruses are the largest viruses, 250 by 350 nm, and the smallest human virus is the poliovirus, 25 nm in diameter.

Classification of Viruses

Originally, viruses were classified according to the diseases they caused or where they were found. There were plant, animal, and bacterial viruses and dermatotropic, neurotrophic, and enteric viruses. Now they are classified by the type and structure of their nucleic acids, chemical and physical characteristics, size, type of replication, and host.

Viral Replication

Viral reproduction, called *replication,* is unique to viruses. A virus attaches to the surface of a susceptible cell by means of specialized structures on its surface and specific receptors on the surface of the cell. The virus then enters the cell by endocytosis, fusion of viral membrane and cell membrane, or lysis of the membrane. Once inside the cell, the virus breaks down into its component parts. Then the virus directs the host cell to make viral components. The metabolism of the cell may be shut off completely (polioviruses), or it may continue on a restricted scale (influenza viruses and papovaviruses). The cell churns out viral components until there are enough for viral assembly to start. The components are assembled within the cell and moved to the cell's surface. The virus may be released by lysis of the cell (poliovirus) or by budding through the cell membrane (influenza and parainfluenza viruses).

Whether a virus can productively infect a cell depends on whether the virus can:

- Attach to the cell

- Enter the cell

- Take over the cell's metabolic machinery.

In addition, the right kind of macromolecules to make viral component parts must be present, and the virus must be able to leave the cell.

This chapter discusses the following issues:

- The laboratory diagnosis of viral infections

- The epidemiology and clinical manifestations of various agents of viral infections.

LABORATORY DIAGNOSIS OF VIRAL INFECTIONS

Specimen Collection and Transport

The cardinal rules in specimen collection are as follows:

- **Proper time:** Obtain the specimen early in the disease process.

- **Appropriate means of transport:** Keep the specimen cool and moist.

- **Proper specimen:** Go "where the action is."

Proper Time to Collect Specimen

Studies have shown that in most viral infections, the number of virions shed from the lesion decreases dramatically each day after the lesion manifests. This is true with herpes simplex and varicella/zoster viruses as well as with the respiratory viruses. In fact, it is not uncommon for maximal shedding of the virus to occur during the disease prodrome.

Appropriate Means of Transport

Several viral transport systems are commercially available. Most transport media consist of a buffered isotonic solution with some type of protein such as albumin, gelatin, or serum to protect less stable viruses. Antibiotics are added in some transport systems to inhibit contaminating bacterial flora. It is also important for the transport container to be unbreakable and able to withstand freezing and thawing.

Specimens should be kept "cool" until processing, i.e., refrigerated or on ice. Specimens should not be frozen unless inoculation or processing is significantly delayed, such as for 4 to 7 days; in that case, the specimen should be frozen and held in a −70°C freezer. Freezing in a "household" −20°C freezer is to be avoided. Slow freezing takes place at −20°C, which facilitates the formation of ice crystals and disrupts the host cells, resulting in significant loss of viral viability. In addition, household freezers are "self-defrost" types that go through freeze-thaw cycling, which exposes viruses to constant changes in temperature. To keep viruses viable, it is crucial to store them at a constant temperature.

Although aspirated secretions usually provide maximum recovery, swabs are easier to use for specimen collection. If specimens are taken with swabs, cotton, Dacron, or nylon must be used—never a calcium alginate swab. Calcium alginate is toxic to many viruses. Tissue samples must be kept from drying out. Tissues may be transported in an appropriate viral liquid medium. Saline or trypticase soy broth is also acceptable.

Appropriate Specimen for Maximum Recovery

A common sense approach in selecting specimens for isolation is that if the disease process involves a surface (mucous membrane or skin), a specimen from that surface is most likely to yield useful information. For example, in respiratory infections, secretions from the involved respiratory mucosa are most appropriate. If there is a lesion, a specimen of the involved area is most likely to reveal the etiologic agent. If the intestinal mucosa is involved, a rectal swab or, preferably, a stool specimen is most appropriate. If, however, systemic, congenital, or generalized disease is the clinical picture, then specimens from multiple sites, including blood (buffy coat) and cerebrospinal fluid (CSF), as well as samples of the portals of entry (oral or respiratory tract) or exit (urine or stool) are appropriate. Enteroviruses may cause respiratory infections and may be recovered from stool after the respiratory shedding has ceased. Because enteroviruses have been associated with congenital infections, stool may be an appropriate specimen in congenital infections. Also, enteroviruses are a major cause of "aseptic meningitis" and may be isolated from urine specimens. Table 25–1 lists suggested specimens for viral diagnosis according to the affected body site.

Methods in Diagnostic Virology

There are three major ways for the laboratory to diagnose viral infections:

- Direct detection of the virus in the clinical specimen
- Serologic antibody assays to detect viral antibodies
- Isolation of the virus in culture

Each laboratory must decide which of these methods would fulfill the spectrum of infections encountered. In most circumstances, a combination of these assays is used to determine the viral agent responsible for an infection.

Direct Detection

Methods available for direct detection of either viral particles or viral antigen include the following:

- Immunofluorescence (IF) staining
- Enzyme immunoassay (EIA) or enzyme-linked immunosorbent assay (ELISA)
- Electron microscopy (EM)
- Nucleic acid probe hybridization
- Immunoperoxidase (IP) staining
- Latex agglutination (LA)
- Radioimmunoassay (RIA)

Nucleic acid amplification techniques such as polymerase chain reaction (PCR), an exquisitely sensitive technique, are rapidly evolving and coming into the realm of the clinical diagnostic laboratory. Table 25–2 lists the advantages and disadvantages of direct detection methods.

■■■■■■■T A B L E   2 5 – 1
SUGGESTED SPECIMENS FOR VIRAL DIAGNOSIS BY BODY SYSTEM AFFECTED*

Body System Affected	Suspected Viral Agent	Suggested Specimens	
		For Direct Detection	*For Virus Isolation*
Respiratory	RSV, influenza A, influenza B, parainfluenza, adenovirus, rhinovirus, CMV measles	Nasopharyngeal aspirate, throat washing, bronchoalveolar lavage, lung biopsy	Nasal aspirate, nose or throat swabs, bronchoalveolar lavage, lung biopsy
Gastrointestinal	Rotavirus, adenovirus, enterovirus	Stool	Stool, Rectal swab
Central nervous system	Enteroviruses, mumps, adenovirus, HSV, arboviruses	CSF, brain biopsy	Stool, CSF, NP swabs, urine, brain biopsy heparinized blood (arboviruses, HIV)
Skin	HSV, VZV, enterovirus	Vesicle fluid, ulcer crusts, biopsy	Vesicle fluid, ulcer swabs, NP swabs
Eye	Adenovirus, HSV, VZV, enteroviruses, *Chlamydia*	Conjunctival scrapings	Conjunctival scrapings or corneal swabs
Genital	HSV, *Chlamydia*	Vesicle fluid, scrapings, pus, swabs	Urethral-endocervical swabs, bubo pus, heparinized blood

*CMV, cytomegalovirus; CSF, cereprospinal fluid; HIV, human immunodeficiency virus; HSV, herpes simplex virus; NP, nasopharyngeal; RSV, respiratory syncytial virus; VZV, varicella-zoster virus.

TABLE 25-2

ADVANTAGES AND DISADVANTAGES OF DIRECT DETECTION

Advantages	Disadvantages
Potential rapid diagnosis Detect nonculturable viruses Avoid need for culture	Confined to the specific anti-body used Costly and labor intensive (FA methods) Methods depend on specimen adequacy and quality; specimen quality is not assessed in EIA methods

Used with permission from Costello MJ, Morrow S, et al: Guidelines for specimen collection, transportation, and test selection. Lab Med. 24:19, 1993. Copyright © 1993, by the American Society of Clinical Pathologists.

Serologic Assays

Indications to perform serologic assays for diagnosis of viral infections are limited. With a few exceptions, paired sera (acute and convalescent) demonstrating seroconversion or fourfold rise are required to establish diagnosis of recent infection. Serologic studies are also usually retrospective because of the need for paired sera. In addition, cross-reactions with nonspecific antibodies produced by the host may be detected. Lastly, interpretation is difficult in passive transfer of antibodies (transplacental or transfusion transmission).

Indications for serology are as follows:

■ Diagnosis of infections with nonculturable organisms such as hepatitis viruses

■ Determination of immune status to rubella, measles, varicella-zoster, hepatitis B

■ Monitoring of immunosuppressed and transplant patients

■ Epidemiologic or prevalence studies

There are certain problems inherent with serology. First, serologic assays measure host response rather than detection of organism. Second, there is a wide variation in the antibody-producing capabilities of human hosts. Third, antibody level does not necessarily correlate with "acuteness" or "activity" of infection.

Viral Isolation

In clinical virology, isolation of virus is, in general, still the gold standard against which to measure other methodologies. Traditionally, there are three methods used for isolation of viruses in diagnostic virology: cell culture, animal inoculation, and use of embryonated eggs. Of these three methods, the most commonly used is cell culture. Animal inoculation is extremely costly and is used only as a special resource and in reference or research laboratories. Certain coxsackie A viruses and arboviruses, for example, require suckling mice for isolation. Embryonated eggs are rarely used now; isolation of influenza viruses is enhanced in embryonated eggs but can generally be much more easily accomplished in cell culture. Table 25–3 lists the advantages and disadvantages of traditional viral isolation techniques.

CELL CULTURE FOR VIRAL ISOLATION

The term *cell culture* is technically used to indicate culture of cells in vitro; the cells are not organized into a tissue. The term *tissue culture* or *organ culture* is used to denote growth of tissues or an organ in a way that preserves the architecture or function of the tissue or organ. For practical purposes, however, most clinical virology laboratorians use the two terms, *cell culture* and *tissue culture,* interchangeably.

Cell cultures can be divided into three categories: primary, diploid, and continuous (heteroploid). *Primary* cell cultures are obtained from tissue removed from an animal. The tissue is finely minced and then treated with an enzyme such as trypsin to further disperse individual cells. These cells are then seeded onto a surface to form a monolayer. With primary cell lines, only very minimal cell division occurs; hence the term *primary.* An example of a commonly used primary cell culture is primary monkey kidney (pMK).

Diploid cell lines can divide (passage), but passage is limited to 50 generations. With increasing passage, diploid cells become "senile" and sensitivity decreases. Therefore, indefinite passage in

TABLE 25-3

ADVANTAGES AND DISADVANTAGES OF VIRAL ISOLATION

Advantages	Disadvantages
Sensitive (general "gold standard") Detects broad spectrum of viruses Flexible, adaptable to viral variation Spin amplification (shell vial) possible Susceptibility testing possible* Confirmation and cross-checks Comprehensive quality control Skilled personal available	Time required for isolation and identification: HSV, 1–3 days Enterovirus, 2–5 days Respiratory viruses, 2–5 days Viable organism required Facility required Skilled personnel required

*Still rarely performed; expensive and limited.

diploid cells is not possible. Human neonatal lung (HNL) is an example of a standard diploid cell culture used in diagnostic virology.

Continuous (heteroploid) cell cultures have variable numbers of chromosomes and are capable of indefinite passage; therefore, they are "immortal." HEp2 (derived from a human laryngeal carcinoma), A549 (derived from a human lung carcinoma), and Vero (derived from monkey kidney) are examples of continuous cell lines used in diagnostic virology. Each laboratory must decide which cell lines to use on the basis of the spectrum of viral sensitivity, availability, and cost. Table 25–4 lists cell culture lines most commonly used in clinical virology.

CYTOPATHIC EFFECT ON CELL CULTURE FOR PRESUMPTIVE IDENTIFICATION OF VIRAL AGENTS

Cell cultures provide presumptive identification by a characteristic cytopathic effect (CPE) that viruses produce on certain cells. For example, herpes simplex virus (HSV) grows rapidly on many different types of cell cultures and frequently produces CPE within 24 hours. A predominantly cell-associated virus, HSV produces focal CPE (adjacent cells become infected) and *plaques* (clusters of infected cells). The combination of rapid growth, plaque formation, and growth on

■■■■■ T A B L E   2 5 – 4

CELL CULTURES COMMONLY USED IN THE CLINICAL VIROLOGY LABORATORY

Virus	PMK	HDF	HEp2	RK	A549	CPE
HSV	–	+++	+++	+++	+++	Plaques
CMV	–	+++	–	–	–	Plaques
VZV	–	+++	–	–	+/–	Rough, messy plaques
Enterovirus	+	+	++	–	+	Diffuse cell rounding
Adenovirus	+	++	+++	–	+++	Large cell rounding
RSV	+/–	+	+++	–	++	Syncytial formation
Influenza/ parainfluenza	+++	+/–	–	–	–	Syncytial formation, cell rounding

Abbreviations: A549, human lung carcinoma cell line; CMV, cytomegalovirus; CPE, cytopathic effect; HDF, human diploid fibroblasts; HEp2, human laryngeal carcinoma cell line; PMK, primary monkey kidney; RK, rabbit kidney; RSV, respiratory syncytial virus; VZV, varicella-zoster virus.

Adapted with permission from Costello MJ, Morrow S, et al: Guidelines for specimen collection, transportation, and test selection. Lab Med. 24:19, 1993. Copyright © 1993, by the American Society of Clinical Pathologists.

many different cell types (primary rabbit kidney, human fibroblasts, Vero, HEp2, mink lung, and primary human kidney) is presumptive evidence for the diagnosis of HSV. Herpes simplex virus is one of the few viruses that grows on rabbit kidney, (Fig. 25–1), therefore it is a useful cell line in herpes detection programs.

Cytomegalovirus also forms plaques (Fig. 25–2) but grows much more slowly and only on diploid fibroblasts. Varicella-zoster virus, although not as promiscuous as HSV, grows reasonably well on several types of cells, including diploid fibroblasts, A549, and Vero cells. Plaque formation develops more slowly than with HSV and is somewhat less well defined and more "rough" or "messy." Enteroviruses are less cell associated and characteristically produce rather small round cells diffusely spread on primary monkey kidney, diploid fibroblasts, human embryonal rhabdomyosarcoma (RD) cells, and A549 cells. Adenoviruses also produce cell rounding (Fig. 25–3), usually larger than enteroviruses, on a number of cell types, including diploid fibroblasts, HEp2, A549 and monkey kidney cells. The rounding may be diffuse or more focal (like a cluster of grapes).

The respiratory viruses may or may not produce characteristic CPE on different cell types. Respiratory syncytial virus may produce classic syncytial formation—hence the name—in HEp2 cells or even monkey kidney cells. Also, parainfluenza type 2 (and, to a lesser extent, parainfluenza type 3) viruses may produce syncytial cells. The traditional cell culture for influenza is primary rhesus monkey kidney. LLC-MK2 (a continuous line derived from rhesus monkey kidney) and MDCK (Madin-Darby canine kidney) with trypsin are alternative cell lines. Influenza virus commonly does not exhibit a clear-cut CPE. It is therefore important to add further screening tests to respiratory specimens to detect these "hidden" viruses. Antigenic structures on the surfaces of these viruses hemadsorb red blood cells suspensions; therefore, the hemagglutination and hemadsorption tests are valuable. Fluorescent antibody stains are often used to screen the cell cultures before a final negative result is reported.

CENTRIFUGATION-ENHANCED SHELL VIAL CULTURE

The shell vial culture technique, or spin amplification, is an enhanced viral antigen detection

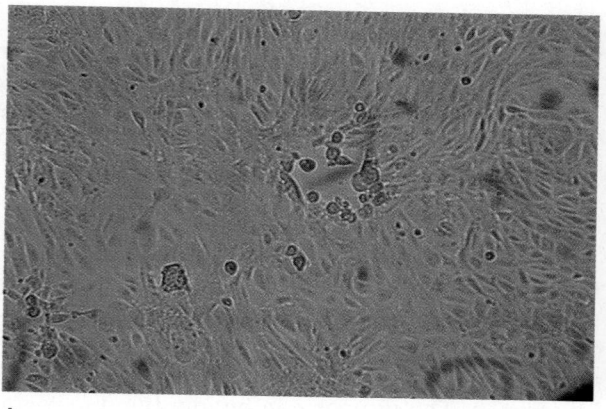

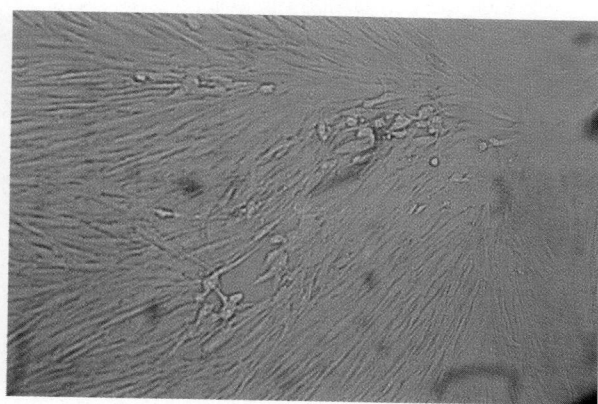

A

B

Figure 25 – 1

A, Herpes simplex virus from the skin, showing CPE in less than 1 day on rabbit kidney cells. *B,* Herpes simplex virus showing CPE in less than 1 day on HEL.

that offers a more rapid identification than the traditional viral culture method. A round coverslip in a shell vial holds the monolayer tissue culture. The shell vial is inoculated with the clinical sample and then spun in a centrifuge to promote viral infection of monolayer cells. It is thought that the prolonged centrifugation (30 to 60 minutes in 100*g*) increases the cell wall sensitivity, thus promoting viral infectivity. The shell vial is incubated for 24 to 48 hours. With a fluorescent antibody (FA) assay the presence of viral antigen is detected. Table 25–5 shows recommended applications of the shell vial culture technique.

AGENTS OF VIRAL INFECTIONS

Agents of Respiratory Infections

Influenza A and B

Influenza viruses are enveloped with ss RNA in eight segments. Their major surface antigens are the glycoproteins hemagglutinin (H) and neuraminidase (N). The hemagglutinin determines infectivity and agglutinates erythrocytes *(hemagglutination)*. The neuraminidase may help disperse

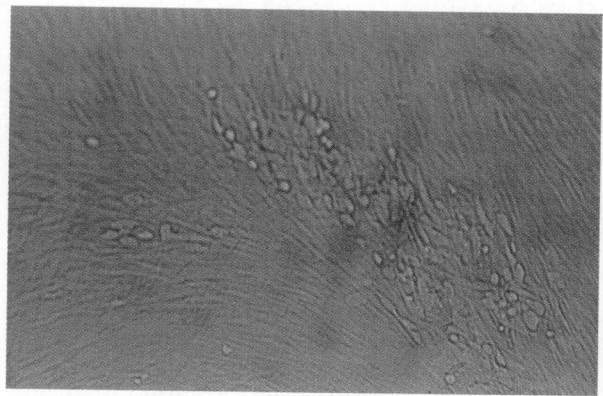

Figure 25 – 2

CMV from cerebrospinal fluid forming CPE on diploid fibroblast cells.

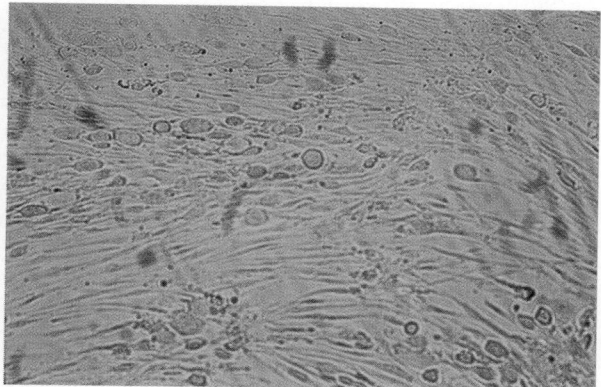

Figure 25 – 3

CPE of adenovirus on HEL.

━━━━T A B L E 2 5 – 5
INDICATIONS FOR SPIN-AMPLIFICATION SHELL VIAL CULTURE

	Indications	
Virus to Isolate	*Sample*	*Identification*
Cytomegalovirus	Urine, respiratory secretions, blood	Stain both vials after 48 hours of incubation
Herpes simplex virus	Vesicles	Stain both vials after 24 hours of incubation
Varicella-zoster	Vesicles	Stain one vial after 3 days and second vial after 5 days of incubation
RSV, parainfluenza 1–3, influenza A & B, adenoviruses	Respiratory secretions	Stain vials after 24–48 hours of incubation

Used with permission from Costello MJ, Morrow S, et al: Guidelines for specimen collection, transportation, and test selection. Lab Med 24:19, 1993. Copyright © 1993, by the American Society of Clinical Pathologists.

the budding viruses and may also activate the hemagglutinin. Influenza A viruses are found in humans, birds, pigs, and horses, but influenza B viruses are found only in humans.

The surface antigens of the influenza viruses change continuously. Periodically, the changes are so great that worldwide pandemics result. The most recent variants are incorporated in the new vaccine each year to provide maximum protection. Over 70% of the persons thus immunized escape the virus; the rest have a milder illness than if they had not been vaccinated.

CLINICAL INFECTION

Influenza viruses are respiratory pathogens that rarely cause viremia. They are spread by aerosols. There is rapid onset of malaise, fever, and myalgia often with a nonproductive cough. Some cases of flu are asymptomatic, but often the patient is incapacitated for days and requires a lengthy convalescence.

The viruses attack the ciliated epithelial cells lining the respiratory tract, causing necrosis and sloughing of these cells. Complications include primary viral pneumonia and secondary bacterial pneumonia, either of which may be life-threatening. If given within 48 hours of exposure, amantadine and rimantadine can halt or reduce the severity of illness caused by influenza type A but not type B. Ribavirin aerosol can kill the influenza virus.

Influenza viruses have worldwide distribution. They are most commonly isolated from November to April in the Northern Hemisphere and from May to October in the Southern Hemisphere. They are endemic in the tropics, where they are more apt to cause symptoms of gastroenteritis.

LABORATORY DIAGNOSIS

Nasopharyngeal swabs, washes, or aspirates taken early in the course of the disease are the best specimens. The viruses are fragile and need to be handled carefully; specimens should *never* be frozen. The viruses can be identified in respiratory secretions by direct immunofluorescent antibody (IFA) and enzyme immunoassay tests. Influenza viruses grow in the amniotic cavity of embryonated chicken eggs, primary monkey kidney cells, and MDCK cells with trypsin added. Hemagglutination inhibition (HI) can be used to identify and type the viruses. Paired sera can be tested by complement fixation (CF), HI, or EIA.

Parainfluenza Viruses

There are four types of parainfluenza viruses. The parainfluenza viruses are enveloped RNA viruses with two surface antigens: the hemagglutinin-neuramindase, or HN, antigen and the fusion, or F, antigen. The HN antigen gives the viruses specificity, and the F antigen is responsible for the fusion of virus to cell and of infected cell to infected cell.

CLINICAL INFECTION

Parainfluenza viruses are found worldwide. Aerosolized ribavirin can be used to treat them. No vaccines are available.

Parainfluenza viruses are the major cause of respiratory disease in young children. Types 1 and 2 cause the most serious illnesses in children between 2 and 4 years of age. Parainfluenza type 3 causes bronchiolitis and pneumonia in infants. The viruses are spread by respiratory secretions, by aerosols, and by contact. Infection of the cells in the respiratory tract leads to cell death and an inflammatory reaction in the upper and lower respiratory tract. Rhinitis, pharyngitis, laryngotracheitis, tracheobronchitis, bronchiolitis, and pneumonia may result.

LABORATORY DIAGNOSIS

Direct examination of nasopharyngeal secretions by IFA and EIA can give rapid results. Parainfluenza viruses are fragile. Specimens for isolation should be taken as early in the illness as possible, kept cold, and inoculated into primary monkey kidney cells or LLC-MK2 cells. The viruses can be identified by hemadsorption, IFA, neutralization, or EIA. Serologic studies are more valuable for epidemiologic studies than for diagnostic purposes.

Respiratory Syncytial Virus

Respiratory syncytial virus (RSV) is closely related to parainfluenza viruses. It also infects the cells lining the respiratory tree, causing acute upper and lower respiratory tract disease in infants and young children. Large amounts of RSV are shed in respiratory secretions, so it is not surprising to find that RSV is the cause of major nosocomial outbreaks in crowded nurseries.

RSV occurs in yearly outbreaks that last 2 to 5 months and usually appear during the winter or early spring in the temperate zones. The virus may be carried in the nares of asymptomatic adults. Testing hospital personnel and infants for RSV, isolating RSV-infected infants, following good handwashing practices, and organizing patients and staff into cohorts can reduce the risk of nosocomial spread.

The virus can be identified in nasopharyngeal swabs and washes directly by IFA or EIA. The virus grows readily in continuous epithelial cell lines such as HEp2 to form large multinucleated cells (see Fig. 25–4). It also grows in primary monkey kidney and human diploid fetal cells. Because the virus is extremely fragile, recovering it from cultures is a major problem. It starts dying as soon as the specimen is taken. Specimens must be kept iced but *not frozen*. From cultures, RSV can be identified by IFA, EIA, and serum neutralization tests.

Adenoviruses

Adenoviruses are naked icosahedral viruses with ds DNA. Of their 10 structural proteins, the most important are the hexons and pentons. There are 42 adenoviruses, and they are classified by their hemagglutination with monkey and rat RBCs, penton fiber length, the content of their DNA, and oncogenicity. Adenoviruses possess a

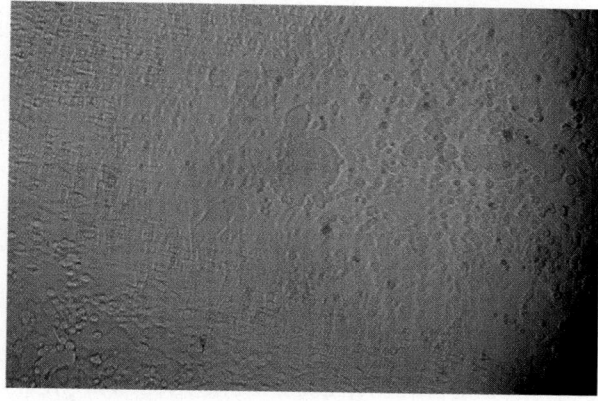

Figure 25–4
CPE of RSV on HEp2.

common group antigen detectable by IFA and complement fixation. The most common serotypes are 1 to 8, 11, 21, 35, 37, and 40.

Adenoviruses are found throughout the body and can remain latent in tissues like in the adenoids or in organs like the kidneys for years. They infect all ages but are more common in school children. Fifty percent of adenovirus infections are asymptomatic.

CLINICAL INFECTION

Adenoviruses types 1 to 7 and 21 cause localized respiratory outbreaks during the fall and winter. Cold symptoms, pharyngitis, tonsillitis, fever, cough, and tender lymph nodes are common. Types 1 to 7 also cause swimming pool–associated pharyngoconjunctivitis that may be accompanied by fever, pharyngitis, cervical lymphadenopathy, headache, diarrhea, and rash.

There are more adenovirus infections than influenza and parainfluenza infections, but the former are less obvious because most are self-resolving or subclinical. Adenovirus infections include conjunctivitis and pharyngitis. The virus is shed in secretions from eyes and respiratory tract but also in stool and urine specimens for days after the disease has disappeared. The viruses can be spread by aerosols, fomites, the oral-fecal route, and personal contact. Adenovirus infections can be lethal in immunosuppressed patients.

LABORATORY DIAGNOSIS

Adenoviruses are quite stable and can be isolated in human embryonic kidney, continuous epithelial cell lines, and Graham 293 transformed cell lines. They have a characteristic cytopathogenic effect, with swollen cells in grape-like clusters. An isolate can be identified as an adenovirus by IFA, EIA, and CF, but serotyping has to be done by serum neutralization or hemagglutination inhibition.

Rhinoviruses

Rhinoviruses are the agents that cause the common cold. There are over than 100 types. Small, naked ss RNA viruses, they are closely related to the enteroviruses. Rhinoviruses are resistant to detergents, lipid solvent, and temperature extremes but are sensitive to pH below 6.

CLINICAL INFECTION

Rhinoviruses infect the nasal epithelial cells and activate inflammatory mediators. There is a profuse watery discharge, nasal congestion, sneezing, headache, sore throat, and cough. In severe cases, bronchitis and asthma may result. Rhinoviruses are most common in the early fall and again in the spring; however, they can be isolated all year. At any one time, multiple types are in circulation.

LABORATORY DIAGNOSIS

Rhinoviruses grow best on human diploid fibroblast cells. They do not have a group antigen, and there are far too many to serotype, so the best method of identifying an isolate as a rhinovirus is to incubate it at pH 3 and with a lipid solvent. Only rhinoviruses are resistant to lipid solvents and sensitive to pH 3.

Unfortunately, there is no cure for the common cold. Treatment with interferon does block rhinovirus infection but has uncomfortable side effects such as nose bleeds.

Coronaviruses

Coronaviruses are enveloped viruses with helical, ss RNA. They have distinctive club-shaped projections on their surfaces. They may be responsible for 15% of upper respiratory infections in adults, but higher seroconversion rates are seen in children. Symptomatically, the disease resembles rhinovirus infections; there is, however, more nasal discharge and malaise with coronavirus infections. At this time, coronavirus identification is not practical for the average clinical laboratory.

Coronaviruses are extremely fragile and difficult to culture, but it is possible to test specimens directly by IFA and EIA. Paired sera can be tested by CF, serum neutralization, and HI if the reagents are available.

Exanthemas

Mumps

The mumps virus is related to the parainfluenza viruses; it is an enveloped virus with ss RNA and both the HN and F surface antigens. Mumps is spread by droplets of infected saliva. Mumps has worldwide distribution. Fortunately, an effective vaccine is available.

CLINICAL INFECTION

Mumps is an acute, usually self-limiting systemic illness marked by unilateral or bilateral swelling of the parotid glands, although other glands such as testes, ovaries, and pancreas can be infected. The primary infection of the ductal epithelial cells in the glands results in cell death and inflammation. The salivary glands recover, but testes and ovaries may be permanently impaired. Orchitis occurs in 5% of all males, adult men having a higher infection rate (18%) than boys. Oophoritis occurs rarely in women (5%), and encephalitis is even less common (0.5 to 2.0%).

LABORATORY DIAGNOSIS

Mumps virus can be isolated from infected saliva and from swabs rubbed over the Stensen duct from 9 days before onset until 8 days after parotitis appears. It can also be recovered from the urine. The virus is relatively fragile and should be handled carefully. Specimens may be examined directly by IFA and EIA. The virus can be isolated in the amniotic cavity of embryonated chicken eggs as well as primary monkey kidney and human embryonic kidney cell cultures.

Isolates can be identified by hemadsorption inhibition, HI, IFA, and EIA. Paired sera can be tested for mumps antibody by FIAX, ELISA, IFA, and HI tests. Paired sera taken as little as 4 to 5 days apart can demonstrate diagnostic or fourfold rise in titer when tested by EIA or HI. Paired sera for CF tests should be taken at least 10 days apart. Cross-reactions between soluble and viral antigens can confuse interpretation of serologic results. Virus isolation is preferable, but physicians rarely have trouble recognizing mumps.

Measles

The measles virus is another virus related to the parainfluenza viruses. It is an enveloped virus with ss RNA. Measles is found worldwide. In the temperate zones, epidemics occur during the winter and spring. It confers lifelong immunity. There is an effective attenuated vaccine that should be given to all children.

CLINICAL INFECTION

Measles was once the most common viral disease of children. It is highly contagious and spreads by aerosol. Measles causes a generalized infection characterized by a maculopapular rash. Initial replication takes place in the mucosal cells of the respiratory tract; next, it replicates in the local lymph nodes, from where it spreads systemically. The virus circulates in the T and B cells and monocytes until eventually lungs, gut, bile duct, bladder, skin, and lymphatic organs are all involved.

Measles has an abrupt onset, with sneezing, runny nose, cough, red eyes, and rapidly rising fever; 2 to 3 days later, a maculopapular rash appears on the head and trunk. The rash remains bright red for about 4 days, and has a brownish stain for another 3 days. Complications such as otitis, pneumonia, and encephalitis may occur. A progressive, highly fatal encephalitis is a rare sequela. In Third World countries, where there is malnutrition and poor hygiene, measles can be fatal.

LABORATORY DIAGNOSIS

Measles is easily diagnosed clinically, so there are few requests for laboratory identification. The virus is fragile and must be handled carefully. Specimens of choice are from the nasopharynx and conjunctiva. Direct examination using IFA or EIA is recommended. The virus grows on primary human kidney cells or primary monkey kidney cells, causing formation of distinctive spindle-shaped cells or multinucleated cells. The virus isolates can be identified by serum neutral-

ization, EIA, or IFA. Serologically, the EIA has replaced the complement fixation test. Demonstration of measles-specific IgM in acute specimens is probably the best diagnostic test.

Rubella

The rubella virus is an enveloped virus with ss RNA. It is sensitive to organic solvents but not to freezing-thawing or ultrasonification. A strictly human virus, rubella is transmitted by infected droplets. It can pass through the placenta to infect infants in utero.

CLINICAL INFECTION

The disease rubella is a mild febrile illness accompanied by an erythematous maculopapular discrete rash with postauricular and suboccipital lymphadenopathy. It starts on the face and spreads to the trunk and limbs. There is no rash on the palms and soles. Like measles, rubella occurs in the winter and spring. As many as 50% of cases are asymptomatic. Transient polyarthralgia and polyarthritis may occur in children and are common in adults.

There would be little concern about rubella if it did not cross the placentas of pregnant women and disseminate to all fetal tissues. The results range from the birth of a normal baby to the birth of a severely impaired infant to fetal death and spontaneous abortion. The impact on the embryo is worst when the infection happens in the earliest stages of pregnancy, because the rubella virus halts or slows the growth of the cells it infects. An effective attenuated vaccine is available that should be administered to all children and particularly to young women before they become sexually active.

LABORATORY DIAGNOSIS

Rubella virus is present in nasopharyngeal specimens or any secretions or tissues of infected infants, who shed the virus in large amounts for long periods. Direct examination of the specimens by IFA or EIA is recommended, because isolation procedures are cumbersome and involve the use of a second or challenge virus. Serologic procedures are effective because the presence of any rubella antibody is presumed protective. The most sensitive assays are the solid-phase assays, passive hemagglutination tests. Latex agglutination and antigen-coated red blood cell (RBC) tests are useful but less sensitive.

Hand, Foot, and Mouth Disease

Hand, foot, and mouth disease (HFMD) is caused by coxsackievirus types A5, 10, and 16 primarily and by enterovirus type 71 occasionally. These are naked viruses with ss RNA. HFMD is generally a disease of young children. It is spread by fomites or the oral-fecal route. There may be a mild prodromal phase, with malaise, headache, and abdominal pains. Suddenly, small, painful sores appear on the tongue, buccal mucosa, and soft palate. Simultaneously, a maculopapular rash appears on hands, feet, and buttocks, followed by bullae on the soles of the feet and palms of the hands. The lesions regress in about a week. If there is a rash, it is transient. The virus can be isolated from mouth swabs and swabs of the bullae. Coxsackievirus A16 grows in primary monkey kidney cells and human diploid fibroblast cells and can be identified by serum neutralization.

B19, a Parvovirus

B19 is a small, naked virus with ss DNA. It is resistant to lipid solvents but sensitive to extremes of pH and heat. It causes fifth disease (so named because it was the fifth infectious rash discovered) and aplastic anemia in cases of hemolytic anemia. It is a mild, febrile illness with lymphopenia and transient rash that lasts only a couple of days. During the period of infection, erythropoiesis stops, creating an aplastic crisis in anemic persons that may require a transfusion. There is a higher risk of fetal death if B19 infects a pregnant woman.

The portal of entry is the respiratory tract. During the viremic phase, erythroid precursors are lost, and then reticulocytes disappear. Lymphocytes, neutrophils, and platelets decrease in number, and no precursors are seen. Within about a week, the patient returns to normal. DNA probes detection of B19 are available from laboratories involved in research.

Human Immunodeficiency Virus Type 1

Human immunodeficiency virus type 1 (HIV-1) causes the acquired immune deficiency syn-

drome (AIDS). HIV-1 is spherical with a three-layered structure (Fig. 25–5). In the center are two identical copies of the ss RNA associated with reverse transcriptase and surrounded by an icosahedral capsid. Finally, there is a viral envelope with viral glycoprotein spikes. The diagnostically important HIV antigens are the structural proteins p24, gp41, gp120, and gp160. The virus is transmitted by blood transfusions and exchange of body fluids. HIV is cell associated, so there is less virus in cell free plasma than in whole blood, and even less virus in saliva, tears, urine, or milk. HIV is not highly contagious, mosquitoes do not carry it, and normal social contact without sex poses no threat.

TRANSMISSION

The major risk for contacting AIDS is anal intercourse, but there is also risk in vaginal intercourse. Transmission by any route is enhanced by the presence of other sexually transmitted diseases. In the early days of the AIDS epidemic, hemophiliacs and other people receiving blood transfusions were infected by contaminated

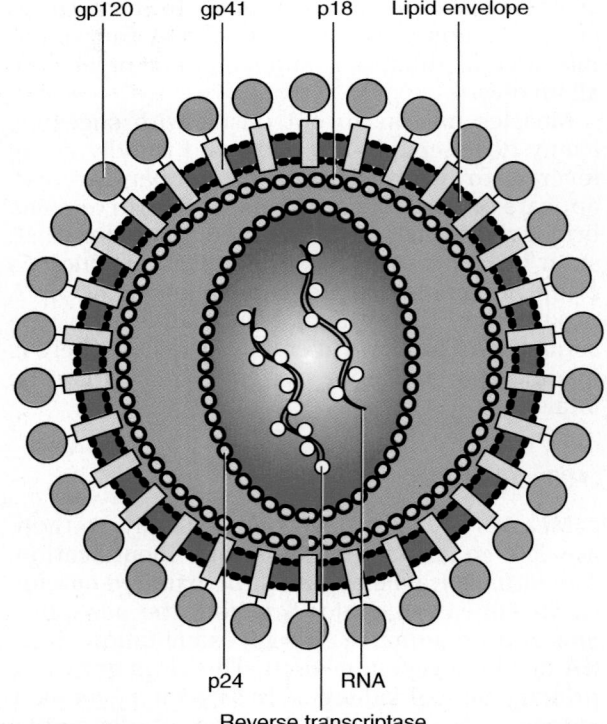

Figure 25–5

Schematic illustration of human immunodeficiency virus (HIV).

blood and blood products. Today, all blood is screened for the HIV virus, and blood products are screened and treated to eliminate any potential virus.

The rate of HIV infection in the homosexual population has declined, and no new cases are seen at this time among hemophiliacs. The virus is still spreading, however, through the intravenous (IV) drug using population. In addition, infection in the women who have sexual contact with infected bisexual men and IV drug users is increasing. HIV-infected mothers can infect their infants perinatally.

EPIDEMIOLOGY

AIDS is a global pandemic. It is spreading in the United States and Europe and in central Africa, where some areas are being depopulated by it. In Africa, the disease is evenly distributed between men and women. In these areas, severe, chronic diarrhea is so common that AIDS is called "slim disease." The prevalence of internal parasites in these areas may play a role in the chronic diarrhea. The major risk groups everywhere are promiscuous homosexual and bisexual men, prostitutes, intravenous drug users, blood recipients, hemophiliacs, the sexual contacts of these groups, and newborn children born to infected mothers.

CLINICAL DIAGNOSIS

Once HIV enters the body, the target cells are the CD4-positive T cells, monocytes, macrophages, regional lymph nodes, and cells of macrophage derivation in the brain. Eventually there is lymphopenia with the greatest losses in the CD4-positive T-cell population. Lymph nodes become enlarged and hyperplastic. Encephalitis and vacuolar myelopathy are common. The virus destroys the cells that are the major defense against viral, fungal, and protozoal infections. Death results from opportunistic infections more often than from the direct effects of the virus. Survival of the patient with AIDS depends on management of the opportunistic infections.

Immunologic markers of AIDS are: the steady decline of the CD4-positive T cells, depression of the T4-T8 ratio to below 0.9 (normal is over 1.5), functional impairment of monocytes and macrophages, decreased natural killer cell activity, and anergy to recall antigens in skin tests. Another marker of AIDS is the presence of any two of the following disorders:

■ Candidiasis of the respiratory tree

■ Cryptococcal meningitis

■ Cryptosporidiosis with persistent diarrhea

■ Cytomegalovirus infections of organs other than liver, spleen, or lymph nodes

■ Persistent herpes simplex virus infections

■ Kaposi's sarcoma or lymphoma of the brain in patients under 60 years

■ Lymphoid interstitial pneumonia and/or pulmonary lymphoid hyperplasia in children under 13 years

■ *Mycobacterium avium, Mycobacterium kansasii,* or *Pneumocystis carinii* pneumonia

■ Progressive multifocal leukoencephalopathy

■ Toxoplasmosis of the brain in patients over 1 month

HIV is fragile. The virus can be isolated from peripheral blood mononuclear cells (PBMCs) of infected patients. Once purified, the PBMCs are stimulated with interleukin-2 and inoculated into cultures of PBMCs from healthy donors that have been grown in the presence of phytohemagglutinin (PHA) or into cell lines that are permissive for HIV. Periodically, the culture fluid can be tested for HIV-specific reverse transcriptase or viral antigens, or the cells can be examined by IFA.

LABORATORY DIAGNOSIS

Serologic tests for HIV are more practical for the average laboratory. The ELISA test used to screen donated blood is one of the most sensitive available. A positive specimen should be retested and, if positive the second time, tested by another method such as immunoblot (Western blot) or IFA. The immunoblot methods test for antibodies to specific viral antigens such as p24, p31, gp41, gp120/gp160 (Fig. 25–6). The presence of HIV antibody is diagnostic, but a negative result simply means that there is no antibody present. It may be 6 weeks after infection before the antibodies appear, and antibodies disappear as immune complexes are formed late in the disease.

There is no successful treatment for AIDS. Research is in progress on vaccines, but none is ready for trial. "Safe sex" using condoms reduces but does not eliminate the risk of infection. The best approach remains comprehensive, community-based education and prevention programs.

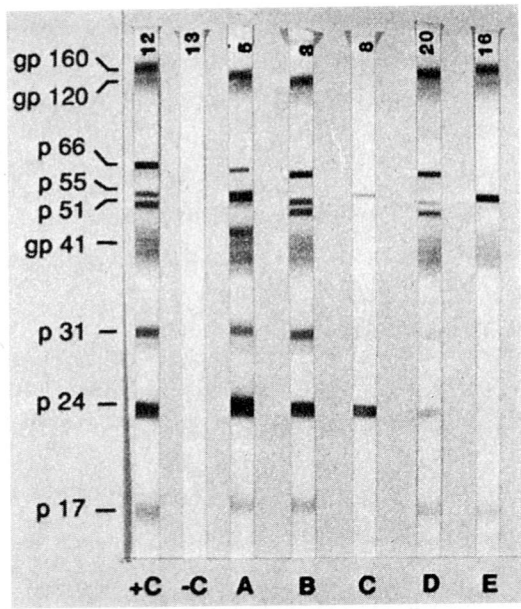

■ Figure 25–6

HIV immunoblot. Reactive protein (p) bands appear as purplish lines across the strip. Proteins with higher molecular weight appear at the top of the strip. Structural and nonstructural proteins are given RNA genome codes. GAG (group-specific antigens), POL (polymerase), and ENV (envelop) are the structural gene codes. ENV codes for gp (glycoprotein precursor) 160, gp120, and gp41–43. POL codes for p65, p51, and p31. GAG codes for p55, p24, and p/1718. Interpretation of results: Negative, indeterminate, or positive, based on the pattern that is present on the strip. Positive: reactivity with a score of + or greater, to any of the following major products of GAG (p24) and ENV (gp120/gp160, gp41). Indeterminate: any appearance of one or more bands, the pattern of which does not satisfy the positive criteria. Negative: the absence of any band on the strip. (Courtesy of Patricia A. Cruse.)

Central Nervous System Viruses

Enteroviruses

Enteroviruses are a large group of viruses, including the following:

- Polioviruses 1 to 3
- Coxsackievirus type A 1 to 23
- Coxsackievirus type B 1 to 6
- Echovirus types 1 to 32
- Enterovirus types 68 to 72

They are small, naked viruses with ss RNA and no envelope. They are relatively resistant to chemical and physical extremes. Enterovirus are responsible for asymptomatic infections such as the following:

- Fever of unknown origin (FUO)
- Aseptic meningitis
- Paralysis
- Sepsis-like illness
- Myopericarditis
- Pleurodynia
- Conjunctivitis
- Enanthemas
- Exanthemas
- Pharyngitis
- Pneumonia

Enteroviruses have also been implicated in early-onset diabetes, cardiomyopathy, and fetal malformations.

CLINICAL INFECTIONS

The portal of entry is the alimentary canal via the mouth. The viruses replicate initially in the lymphoid tissue of the pharynx and gut. When they cause viremia, they can spread to the spinal cord, heart, and skin. Enterovirus infections are commonly accompanied by some nausea and diarrhea. Enteroviruses destroy their host cells. In the intestines, the damage is temporary, because the cells lining the gut are rapidly replaced. Neurons are not replaced, however; hence, when neurons are destroyed, lasting paralysis results.

EPIDEMIOLOGY

Most serotypes of the enteroviruses are distributed worldwide. In the temperate zones, enterovirus epidemics occur in the summer and early fall. Enterovirus infections are more prevalent where there is poverty, overcrowding, and poor hygiene and sanitation. Viruses are spread by aerosol, oral-fecal route, and fomites.

Excellent poliovirus vaccines of either attenuated viruses or killed viruses are available. There are no other vaccines for enteroviruses; good personal and nosocomial hygiene and good sanitation can reduce the incidence of enterovirus infections.

LABORATORY DIAGNOSIS

Enteroviruses can be found in the pharynx immediately before the onset of symptoms and for 1 to 2 weeks afterwards; they can be isolated from the feces for as long as 6 weeks. It is best, however, to obtain specimens early in the infection. Throat cultures, fecal specimens, rectal swabs, CSF, and conjunctival swabs are recommended. Specimens should be selected on the basis of symptoms—more specimens increase the chances of isolation. The viruses are stable and require no special handling.

Polioviruses, type B coxsackieviruses, and the echoviruses grow readily in a number of cell lines: primary monkey kidney, continuous human and primate lines, and human fetal diploid fibroblast lines. (The high-numbered enteroviruses, 68 to 72, require special handling.) The CPE appears quickly and is readily identifiable. There is no group antigen, so the enteroviruses must be identified individually by serum neutralization test. The World Health Organization (WHO) distributes the Lim, Benyesh-Melnick pools of enterovirus antisera that allow identification by neutralization patterns in the pools. The CPE and resistance to detergent, acid, and solvents constitute a presumptive diagnosis of enterovirus infection.

Arboviruses

Arboviruses are a large, diverse group of viruses that share one feature: they are transmitted by vectors or biting arthropods (arthropod-borne viruses) such as mosquitoes, ticks, midges, and sandfleas. The viruses are small, roughly spherical, enveloped viruses with ss RNA. They are fragile and are sensitive to lipid solvents, detergents, and extremes of pH and temperature.

Arboviruses of medical concern in this hemisphere are:

■ Eastern, western, and Venezuelan equine encephalitis (encephalomyelitis) St. Louis, Powassan, California, and La Crosse encephalitides

■ Yellow fever

■ Dengue

The life cycle of arboviruses requires a vertebrate—a bird, rodent, horse, or human—as well as specific types of biting arthropods. The range of the disease depends on the location and numbers of vectors and vertebrate host.

Eastern equine encephalitis is carried by a swamp-dwelling mosquito and normally grows in horses. Outbreaks occur close to rural areas and swamps. St. Louis encephalitis multiplies in spar-

rows and humans; infections occur in urban areas. La Crosse encephalitis is carried by woodland mosquitoes that used to breed in the hollows of trees in the Northeast; the vertebrate hosts are forest rodents. As the suburbs extended into the forests, the mosquitoes found new hosts to bite. La Crosse encephalitis is a disease of children, with more boys than girls being infected.

CLINICAL INFECTION

Arboviruses cause typical viral encephalitis. Neurons and neuroglia of the central nervous system (CNS) are damaged by the viruses. There is fever, partial or complete loss of consciousness that may be accompanied by paresis or spasticity of the arms and legs; convulsions, disorientation, and diffusely abnormal EEGs may result. When the meninges are involved, a stiff neck, headache, and lymphocytosis of the cerebrospinal fluid are observed in addition to the encephalitis; this combination is called *meningoencephalitis*. In the most serious cases, severe convulsions may be fatal. Recovery is slow, and permanent mental and physical disabilities are not unusual.

Vaccines are available for some of the arboviruses. Other methods of prevention include vector control and prevention of insect bites.

LABORATORY DIAGNOSIS

Arboviruses are very fragile and must be handled with care. Identification procedures are difficult and not practical for the average clinical laboratory. For isolation and identification of arboviruses, the World Health Organization for Arbovirus Reference and Research at the Division of Vector-Borne Infectious Diseases of the Centers for Disease Control and Prevention (PO Box 2087, Fort Collins, CO 80521) or the Yale Arbovirus Research Unit (Box 333, New Haven, CT 06510) should be contacted. Some state public health laboratories also perform arbovirus studies. Serologic studies of paired sera can be performed without hazard, but reagents are not easy to find.

Rabies Virus

The rabies virus is a bullet-shaped, enveloped virus with ss RNA. It causes rabies, an infectious disease of the central nervous system that affects warm-blooded animals, including humans. Rabies is usually transmitted by infectious saliva through a bite.

In Africa and Latin America, rabies is found in domestic dogs. In the United States and Western Europe, rabies is found primarily in wild animals such as foxes, raccoons, skunks, and bats. Excellent vaccines are available for dogs, cats, and humans at risk.

CLINICAL INFECTION

After replication in the myocytes, the rabies virus enters the peripheral nerves and begins its trek to the central nervous system. Once the virus is inside the nerve, antibody has little effect. When the virus reaches the central nervous system, death is inevitable. There is risk associated with the treatment, and quick testing of the biting animal is essential.

The virus remains at the bite site and replicates first in the myocytes at the site. The bite should be immediately washed thoroughly with soap and hot water to mechanically reduce the number of viruses present. Then, hyperimmune serum should be instilled into the bite area. Antirabies vaccination should be started immediately to induce immunity as quickly as possible. Interferon can also block virus replication. At this point the virus is still accessible to inactivation by the antibodies.

Rabies virus is easy to detect even in small amounts in the brain by IFA. Any animal suspected of having rabies when it bit a person or domestic animal should be killed, and its head removed and examined as quickly as possible. If the virus is detected, treatment of the bite victim should start immediately. Treatment can be suspended if the head of the biting animal tests negative. A domestic animal or pet that bites a human or pet but is healthy and has had no exposure to rabies or has been vaccinated within the last year should be quarantined for 14 days for observation.

Guidelines for safety should be closely followed by laboratory personnel removing the head and preparing it for transport to the examining laboratory. The head should be wrapped securely in several layers of heavy plastic to make the package waterproof. The sample should be transported on wet ice, *not frozen* or placed in formalin.

LABORATORY DIAGNOSIS

The fastest and most sensitive method of identifying rabies viruses in a specimen is by direct IFA. Impression smears should be made from various area of the brain, primarily the hippocampus, pons, cerebella, and medulla oblongata. In living patients suspected of having rabies, biopsies of skin, especially at the hairline, and impressions of the cornea can be made. The presence of rabies virus in such specimens is diagnostic, but the absence merely means that there is no virus in those specimens, not that the patient does not have rabies.

Rabies viruses can be grown in suckling or young adult mice, murine neuroblastoma, or related cell lines. ELISA tests are currently the most sensitive assays to use for serologic tests.

Dogs and cats should be immunized against rabies annually. Pre-exposure vaccination is available for persons at risk. The current tissue culture vaccine gives better immunity with fewer doses and is far less painful to take than the duck embryo or spinal cord vaccines.

Exotic Viruses

Some viral diseases are relatively rare but are dangerous because of their high mortality rates. Although they are native to other continents, they could easily be introduced into the United States by air travel. The first group of such viruses are arenaviruses. They primarily cause chronic infections with continual shedding of virus in a variety of rodents. Arenaviruses include:

■ Lassa virus (Africa)

■ Junin virus (Argentine hemorrhagic fever)

■ Machupo virus (Bolivian hemorrhagic fever)

When these viruses spill over into the human populations in Africa and South America, they cause major outbreaks with mortality rates as high as 50%. Lassa and Machupo viruses can spread from person to person.

The second group of exotic viruses are two dangerous filoviruses: Marburg virus and Ebola virus. Marburg, apparently a virus of African primates, was first seen in technicians who worked with primate cells and or primate animals in Europe. Since then, a few cases have been seen in Africa. Ebola virus has caused several major outbreaks in Africa, with the mortality running as high as 80%. Both viruses can be spread from person to person; no intermediate host has been found for Ebola virus.

Hantaviruses are the third group of exotic viruses. They are transmitted by rodents, like the arenaviruses. The hantaviruses are found from China, Japan, and Korea across Eurasia and into the Balkans. They cause diseases that range from mild to fatal. The affected persons most seriously ill have hemorrhagic fevers that are accompanied by renal failure and sometimes respiratory collapse.

For a number of years, hantaviruses have been found in rodents in the coastal cities of the United States and in laboratory rodents. Because they did not appear to cause medical problems in people, they were largely ignored. Then, in the summer of 1992, there were deaths in Arizona and New Mexico—mainly among the Navajo. The illness was acute, and death usually came from respiratory collapse. Hantaviruses were isolated from the patients and from the local deer mice. In addition, there was at least one case in East Texas that could not be related to any rodent.

All of these viruses are regarded as dangerous and should be handled only in class IV facilities. (Class IV facilities as classified by the Centers for Disease Control and Prevention (CDC) are used for the most contagious and lethal microorganisms.)

Agents of Gastrointestinal Infections

Rotaviruses

Rotaviruses are naked isometric viruses with a double-layered protein capsid and a genome with 11 segments of ds RNA. Gastroenteritis is a major cause of infant mortality and the failure of young children to thrive. Rotaviruses are the most common cause of gastroenteritis in infants, children, and the elderly. They have worldwide distribution. Outbreaks occur primarily in the winter months in the temperate zones and year round in the tropics.

CLINICAL INFECTION

Rotaviruses are spread by the oral-fecal route. The incubation period is 1 to 2 days. Then there is a sudden onset of vomiting, diarrhea, fever, and occasional abdominal pain and sometimes respiratory symptoms. The vomiting and diarrhea can cause fatal dehydration. The rotavirus replicates in the epithelial cells in the tips of the microvilli of the small intestine. The microvilli are stunted, and adsorption is reduced. The virus is shed in large quantities in stools and can cause nosocomial outbreaks in the absence of good hygiene.

LABORATORY DIAGNOSIS

The rotavirus is present in large numbers in stools but can be isolated only with special procedures. ELISA and latex agglutination tests detect the virus when present in fecal material.

Norwalk and Norwalk-Like Agents

The Norwalk family of agents are small round viruses 27 to 30 nm in diameter. They are most commonly associated with gastroenteritis in older children and adults in developed countries. They have caused outbreaks of acute gastroenteritis in schools, colleges, camps, cruise ships, communities, nursing homes, and family groups. These agents have been found in water, swimming areas, and contaminated food. The incubation period is 24 to 48 hours, and the onset of severe nausea, vomiting, diarrhea, and low-grade fever is abrupt. The attack rate can be as high as 50%. Illness passes as quickly as it starts, usually within 72 hours. Immunity wanes after 4 years. The virus cannot be grown in culture. Diagnosis is usually accomplished by examination of suspect fecal material by electron microscopy. A few laboratories specializing in Norwalk agents can perform serologic procedures.

Enteric Adenoviruses

Adenoviruses type 40 and 41 are called enteric adenoviruses because they cause epidemics of gastroenteritis in young children. Diarrhea is a prominent feature of the illness, but there is less vomiting and fever than with rotaviruses. Enteric adenoviruses have worldwide distribution and are endemic, with an increase in the numbers of cases during the warmer months. These adenoviruses can be identified by not serotyped by IFA and EIA. Kits are available for adenovirus detection.

Coronaviruses

Coronaviruses are described in the discussion of respiratory viruses. They were first recovered from a case of upper respiratory infection, but they appear to be responsible for a small percentage of pediatric diarrhea cases. The illness lasts for about a week, and blood may appear in the stools.

Other Gastroenteritis Viruses

Caliciviruses are small (30 to 35 nm in diameter) and are distinguished by 32 cup-shaped structures on their surfaces. They cause minor outbreaks of pediatric gastroenteritis worldwide in children from 1 to 24 months of age. The outbreaks can be endemic or epidemic and peak in the winter. Vomiting may or may not occur. Illness lasts about 4 days.

Astroviruses are smaller (28 to 30 nm) with a five-point or six-point star-like configuration on their surfaces. These viruses are ubiquitous and cause relatively mild infections in children up to 7 years of age.

Generally, the viruses causing gastroenteritis are extremely small, hard to see, and impossible to grow. They are associated with gastroenteritis because they have been seen in large numbers in stools from gastroenteritis outbreaks by electron microscopy. When convalescent serum from a diarrheal patient is added to a suspension of the patient's stool specimen, the viruses clump together. This procedure is called immune electron microscopy (IEM).

Gastroenteritis viruses are usually fatal only in malnourished or dehydrated children. The best treatment is oral rehydration, with parenteral fluids administered when needed. Prevention requires proper sanitation, clean water supplies, uncontaminated food, and good hygiene.

Papillomaviruses or Wart Viruses

Papillomaviruses are small viruses with circular ds DNA. There are 57 types, and none can be grown in a laboratory with any certainty. There is viral DNA in dividing epithelial cells, but complete, infectious viruses are found only in the most differentiated outer epithelial cells.

Papillomaviruses are widespread. Types 1 to 4 are found in warts on children and adolescents; most disappear spontaneously, giving rise to an incredible array of folk remedies (such as those described in *Tom Sawyer*.) A rare autosomal disease, epidermodysplasia verruciformis (EV), may be caused by human papillomavirus (HPV) types 5 and 8. One third of the patients with EV develop squamous cell carcinoma.

Condylomata acuminata, or genital warts, is one of the most common sexually transmitted diseases. For the most part, genital warts are a nuisance, but evidence is accumulating that they may become neoplastic. HPV types 6 and 11 are most commonly associated with genital warts but have not been found in malignant lesions. Types 16 and 18 are also associated with genital warts, but their DNA has been found in invasive carcinomas of the cervix.

There are geological differences in distribution: HPV 18 is not common in Europe or North America but is found in 25% of invasive lesions of the cervix in Africa and South America. HPV 16 appears to integrate its genome into the host cell chromosome when there are invasive lesions.

Permanent eradication of genital warts is extremely difficult and can involve invasive surgery. Other treatments, such as podophyllin, cryotherapy, electrodiathermy, and laser evaporation, do not guarantee complete eradication of infected cells. If the physician could determine whether or not the warts contained HPV 16 or HPV 18, he or she could better decide what course to follow.

Hepatitis Viruses

There are at least five recognized hepatitis viruses; hepatitis A, hepatitis B, hepatitis C, delta hepatitis, and hepatitis E. These are unrelated viruses that are biologically and morphologically disparate (Fig. 25–7*A* to *D*). Many of the clinical symptoms caused by the different hepatitis viruses are similar, so differentiation on the basis of clinical findings is not reliable. Common features are fatigue, headache, anorexia, nausea, vomiting, abdominal pain (right upper quadrant or diffuse), and, most characteristic, jaundice and dark urine.

Hepatitis A

Hepatitis A virus (HAV) is a small enveloped ss RNA virus in the picornavirus group. In spite of the envelope, it is very stable. It is a human pathogen; the target organ is the liver.

EPIDEMIOLOGY

Hepatitis A virus infects all ages, but more often adults in Northern Europe and the United States. In the temperate zones, the incidence increases in the autumn, then decreases to a minimum by midsummer. In the tropics, the incidence of HAV infections is highest during the rainy season. It is spread by the oral-fecal route, from person to person, and where there is poor sanitation and overcrowding. Outbreaks most

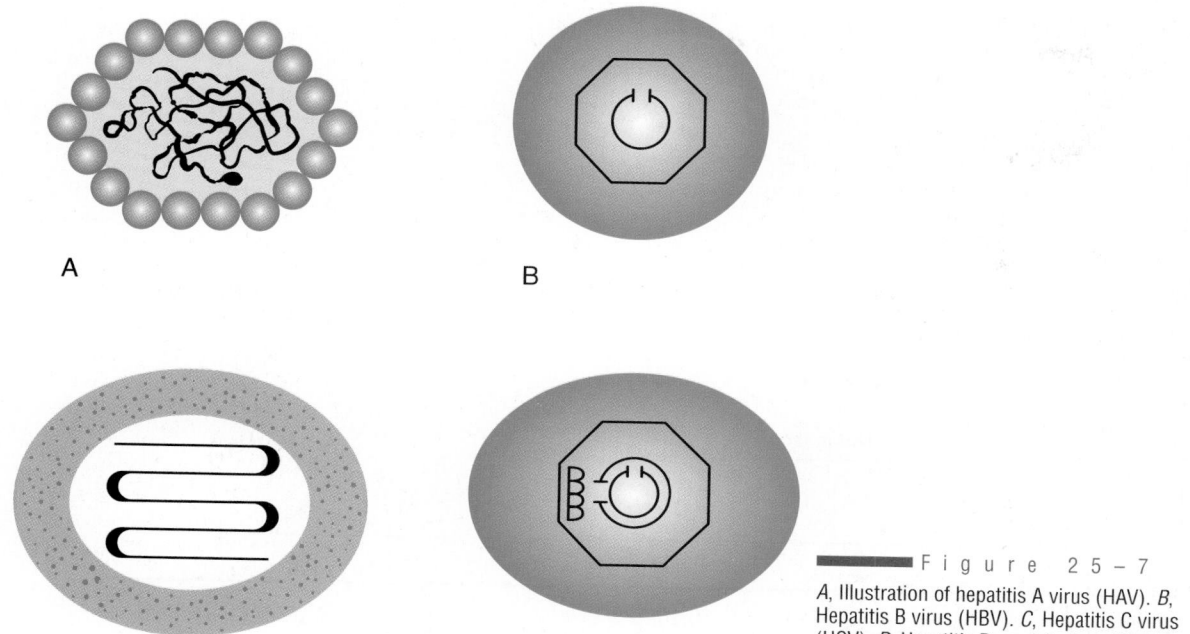

A B C D

Figure 25-7

A, Illustration of hepatitis A virus (HAV). *B,* Hepatitis B virus (HBV). *C,* Hepatitis C virus (HCV). *D,* Hepatitis D, or delta hepatitis virus (HDV).

commonly are caused by fecal contamination of drinking water and food. The virus is shed in large amounts in the feces during the incubation period and early prodromal stage. Sources of infection include: infected food handlers, raw food, food handled wrongly after cooking, and raw or poorly cooked seafood from contaminated water. HAV is endemic, with occasional major outbreaks in tropical or subtropical areas. Travelers in developing countries need to be careful about what they eat and drink. HAV is not transmitted by blood or blood products and rarely parenterally. Control is difficult, because the virus is shed in large amounts before the patient becomes symptomatic. Immune immunoglobulins administered before exposure or during early incubation can prevent or attenuate the infection.

CLINICAL INFECTION

The incubation for HAV is 2 to 6 weeks (average 25 days), and the onset is abrupt, with a variety of nonspecific symptoms such as fever, chills, fatigue, malaise, aches, and pains. A few days later, anorexia, nausea, vomiting, and right quadrant abdominal pain appear, followed by dark urine, clay-colored stools, and the development of jaundice in the sclera and skin. Once the jaundice appears, there is a rapid improvement

in symptoms. Convalescence can last weeks and complete recovery can take months. There is low mortality, no persistence, and no evidence of chronic liver damage by HAV infection.

LABORATORY DIAGNOSIS

Isolation of HAV is not practical, but the virus can be detected in fecal samples. Antibodies appear early and persist for years. The most practical means of diagnosis is the demonstration of IgG and IgM HAV antibodies. For acute infections, the most effective diagnostic method is the identification of HAV-specific IgM. Figure 25–8 shows diagrammatically the serologic evaluation of HAV.

Hepatitis B

Hepatitis B (serum hepatitis, long-incubation hepatitis) is a Hepadnavirus, a DNA virus that is partially double-stranded but has single-stranded areas.

EPIDEMIOLOGY

The main modes of transmission are parenteral, sexual, and perinatal. High-risk groups include IV drug abusers, homosexuals, individuals from endemic areas, household and sexual contacts of

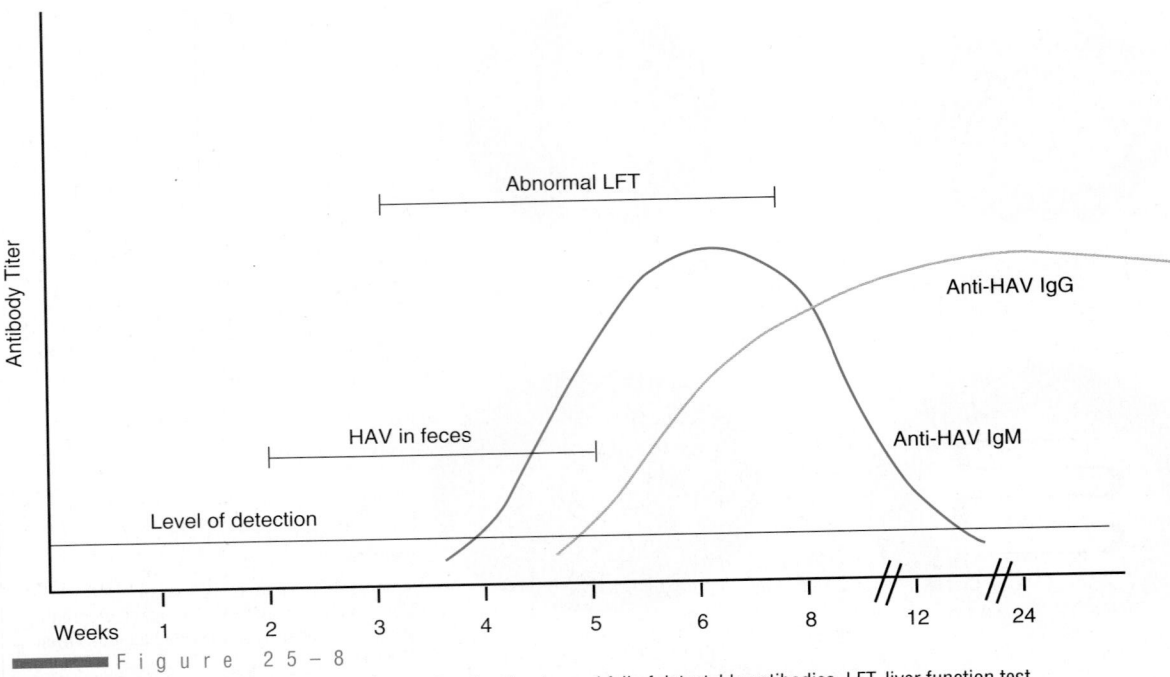

Serologic evaluation of hepatitis A virus infection showing the rise and fall of detectable antibodies. LFT, liver function test.

HBV carriers, health care personnel, and newborns of carrier mothers. Transmission occurs much less commonly by close personal contact.

CLINICAL INFECTIONS

The incubation period for HBV infection varies from 2 to 6 months (average, 45 days), with an insidious onset and symptoms of fever, anorexia, hepatic tenderness, and jaundice. Elevations of serum aminotransferase levels are also observed. The disease represents a much wider spectrum of outcomes. Approximately 50 to 70% of infected persons are asymptomatic; another 20 to 30% demonstrate clinical jaundice but have benign resolution of infection. Therefore, approximately 90% of infections do not cause serious sequelae. Nevertheless, probably 5 to 10% of infections result in a chronic carrier state beyond 6 months. Many of these individuals may progress to chronic disease and have a higher risk of liver disease such as cirrhosis or hepatic carcinoma. In spite of the success of research protocols using alpha-interferon in the treatment of acute and chronic hepatitis B, widespread application and long-term efficacy has not been demonstrated.

LABORATORY DIAGNOSIS

Diagnosis of hepatitis B infections is based on clinical presentation and demonstration of specific serologic markers for hepatitis B virus. A number of hepatitis B antigens and corresponding antibodies have been identified and characterized (Table 25–6):

■ HBsAg: Hepatitis B surface antigen, the covering surface of the virion

■ Anti-HBsAg: Antibody to hepatitis B surface antigen

■ Anti-HBcAg: Total antibody to hepatitis B core antigen

■ IgM anti-HBcAg: IgM antibody to hepatitis B core antigen

■ HBeAg: A part of the core; related to the potential for infectivity

■ Anti-HBeAg: Antibody to hepatitis B e antigen.

The presence of HBsAg in a patient's serum indicates that the patient has an active hepatitis B virus infection, is a chronic carrier, or is in an incubation period. IgM anti-HBcAg appears early

INTERPRETATION OF HEPATITIS B SEROLOGY MARKERS

HBsAg	HBeAg	Anti-HBc	Anti-HBc IgM	Anti-HBs	Anti-HBe	Interpretation
–	NA	–	–	–	NA	No prior infection with HBV or early incubation
–	NA	+	–	+/–	NA	Convalescent or past infection
–	NA	–	–	+	NA	Immunization to HBsAg
+	–	–	+/–	–	–	Acute infection
+	+	+/–	+	–	–	Acute infection, high infectivity
+	–	+/–	+	–	+	Acute infection, low infectivity
+	+	+	–	–	–	Chronic infection, high infectivity
+	–	+	–	–	+	Chronic infection, low infectivity

in the course of the disease and indicates an acute infection. In cases in which HBsAg is not detected and anti-HBs has not appeared, detection of IgM anti-HBcAg confirms the diagnosis of acute HBV infection. The detection of anti-HBs in the serum indicates a convalescent or immune status. When the infection resolves, IgG anti-HBc and anti-HBs become detectable in the patient's serum. The presence of HBsAg after 6 months of acute infection is a strong indication of chronic carrier state. The appearance of HBeAg in this case is indicative of chronic infection with high infectivity. Table 25–6 shows the interpretation of hepatitis B serologic markers. Figure 25–9 shows serologic presentations of acute hepatitis B infection and resolution and of chronic hepatitis B infection.

Hepatitis D (Delta Hepatitis)

Hepatitis D is a defective ss RNA virus that requires hepatitis B for its replication. The hepatitis delta virus utilizes the hepatitis B envelope protein (hepatitis B surface antigen, HBsAg) as its own envelope. Therefore, hepatitis B infection is a prerequisite.

EPIDEMIOLOGY

Delta hepatitis is primarily transmitted by parenteral means, although transmission by mucosal contact has been implicated in epidemics in endemic areas. At-risk groups in the United States are primarily IV drug users, with limited crossover into the homosexual populations in certain parts of the country. Because of overlaps in clinical presentation, a presumed low incidence, and lack of an effective surveillance mechanism, the current epidemiology of delta hepatitis is minimal.

CLINICAL INFECTIONS

Delta hepatitis is commonly severe and can be either acute or chronic in presentation. Delta hepatitis virus infection may occur in two clinical variants: In *coinfection,* there is simultaneous infection by both hepatitis B and hepatitis D viruses. In *superinfection,* a chronic hepatitis B carrier is infected with delta hepatitis (i.e., hepatitis D infection occurs in a patient with chronic hepatitis B infection).

Pathology for hepatitis D virus is similar to that for hepatitis B, but delta hepatitis superinfections have a higher mortality associated with fulminant disease and a higher risk of progression to chronic disease sequelae.

LABORATORY DIAGNOSIS

Diagnosis of delta hepatitis infections requires serologic testing for specific HDV antibody markers. HDV testing is not, however, routinely available in most clinical laboratories.

Table 25–7 shows interpretation of delta hepatitis serologic markers. Figure 25–10 shows serologic presentations of delta hepatitis coinfection and superinfection.

Hepatitis C

Until a few years ago, non-A, non-B hepatitis (NANBH) was diagnosed primarily on the basis of exclusionary findings. With the evolution of highly specific serologic markers, it has been determined that there are at least two viruses responsible for NANBH. The first, hepatitis C, is a parenterally transmitted virus. Hepatitis C is now characterized as a ss RNA virus, probably a flavivirus. The second NANBH agent, now desig-

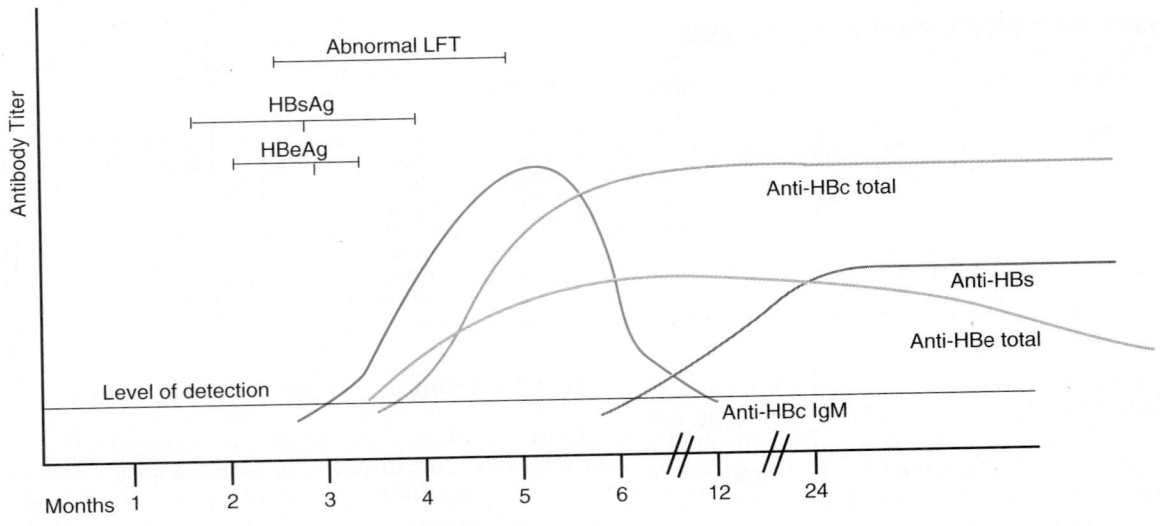

A

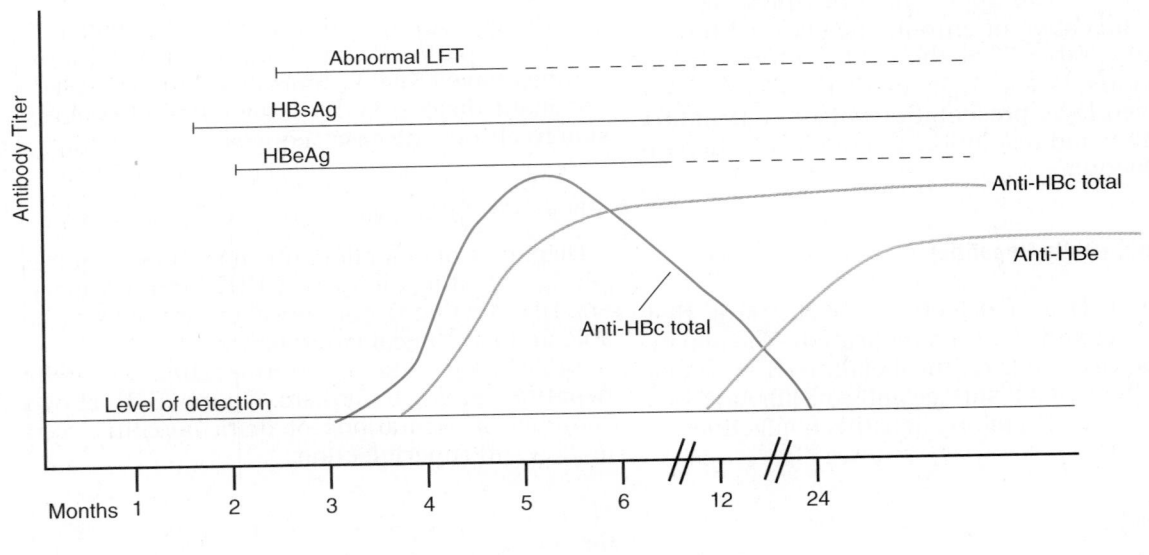

B

■■■■■■ F i g u r e   2 5 – 9

Serologic evaluation of hepatitis B virus infection showing the rise and fall of detectable antibodies. *A,* Serologic presentation in acute hepatitis infection, with resolution. *B,* Serologic presentation of chronic hepatitis infection, with late seroconversion.

nated as hepatitis E virus (HEV), is a waterborne enteric agent believed to be enterically transmitted via the fecal-oral route, similar to HAV; it is discussed in the next section. Currently, it is estimated that in the United States, there are approximately 150 to 170,000 new cases per year.

EPIDEMIOLOGY

Several studies indicate that up to 90% of cases of posttransfusion hepatitis cases are due to hepatitis C virus. Although perinatal and sexual transmissions do occur, and parenteral trans-

INTERPRETATION OF DELTA HEPATITIS INFECTION SEROLOGIC MARKERS

Clinical Variant	Serologic Markers			
	Anti-HBc IgM	HBsAg	Anti-HDV	Anti-HDV IgM
Co-infection	+	+	+	+
Superinfection	–	+	+	+

mission has been identified as a major route for acquiring the infection, hepatitis C antibody has been detected in patients with poorly defined routes of transmission or with no evidence of identifiable risk factor.

CLINICAL INFECTION

The incubation period for hepatitis C infection varies from 6 to 8 weeks. Symptoms may be very subtle and slow to become apparent. Although still undetermined, it is thought that as many as 50% of those infected with hepatitis C may become chronic carriers and that about 20% of patients with chronic infection develop cirrhosis. Also, hepatitis C appears to be a significant risk factor in the development of hepatocellular carcinoma (cancer of the liver).

LABORATORY DIAGNOSIS

ELISA tests based on detection of serum antibodies to proteins c100-3, c33c, and c22-3 are available as standard screening tests. As a supplemental "confirmatory" test, a recombinant immunoblot assay (RIBA) can be performed. The "blot" or "strip" contains separate bands of proteins 5-1-1, c100-3, c33c and c22-3 plus controls to detect antibodies to these proteins by ELISA.
Interpretation of the RIBA is as follows:

- No bands present: Negative.

- 1 band or 5-1-1 and c100-3 present: Indeterminant.

- 2 or more bands (5-1-1 and c100-3 count as one band) present: positive.

In comparison with hepatitis B virus, hepatitis C virus has been reported as not as immunogenic. Hepatitis C infection does not produce persistent, lifelong levels of antibody; rather, persistence of anti-HCV is linked to the presence of replicating HCV. Figure 25–11 shows the diagrammatic representation of HCV infection.

Table 25–8 is a summary table of clinical and epidemiologic differences among hepatitis viruses A, B, delta agent, and non-A, non-B.

Hepatitis E

Hepatitis E (enterically transmitted hepatitis) virus is not as well characterized but appears to be a small (32 to 34 nm), nonenveloped, single-stranded RNA virus.

EPIDEMIOLOGY

HEV has been identified as a cause of epidemics of enterically transmitted hepatitis in developing countries in Asia and Africa. There is also evidence that epidemics have occurred in Central America and Mexico. Although the virus has not been associated with outbreaks in the United States, it has been linked to sporadic cases in travelers returning from endemic areas. HEV is believed to be transmitted via the fecal-oral route through sewage-contaminated drinking water.

CLINICAL INFECTION

Hepatitis E is an acute, self-limiting disease with a clinical presentation similar to that of HAV. Symptoms such as malaise, nausea, and vomiting followed by jaundice occur approximately 2 to 9 weeks after ingestion of the virus. Unlike HAV infections, however, HEV infection has a high mortality rate, particularly in pregnant women, with a reported case-fatality rate of 15 to 20%. The epidemics affect primarily young to middle-aged adults. The overall mortality rate is 1 to 2%.

LABORATORY DIAGNOSIS

One of the first diagnostic tests developed to detect antibodies to HEV was a fluorescent antibody-blocking assay. Although this test showed prevalence of antibodies to HEV during the epidemics in Asia, Africa, and Mexico, the test was not able to differentiate acute from convalescent infections. Most recently, an enzyme-linked immunosorbent assay (ELISA) has been developed that detects IgG and IgM antibodies to HEV. HEV assay testing, however, is not currently performed in diagnostic laboratories in the United States.

Herpesviruses

The herpesviruses are a group of viruses with an extremely wide host range, including many

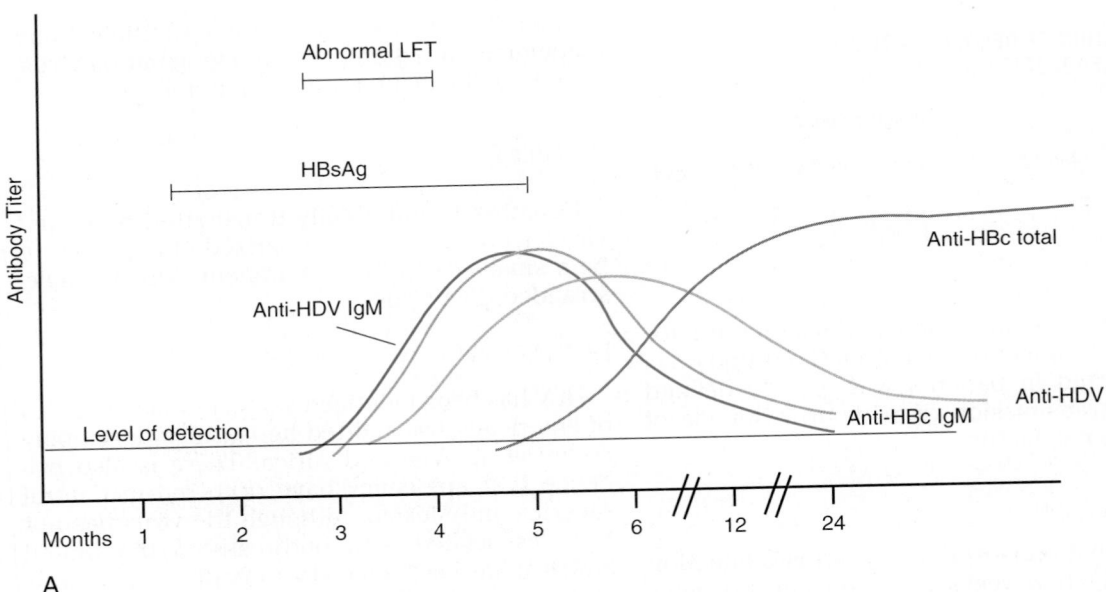

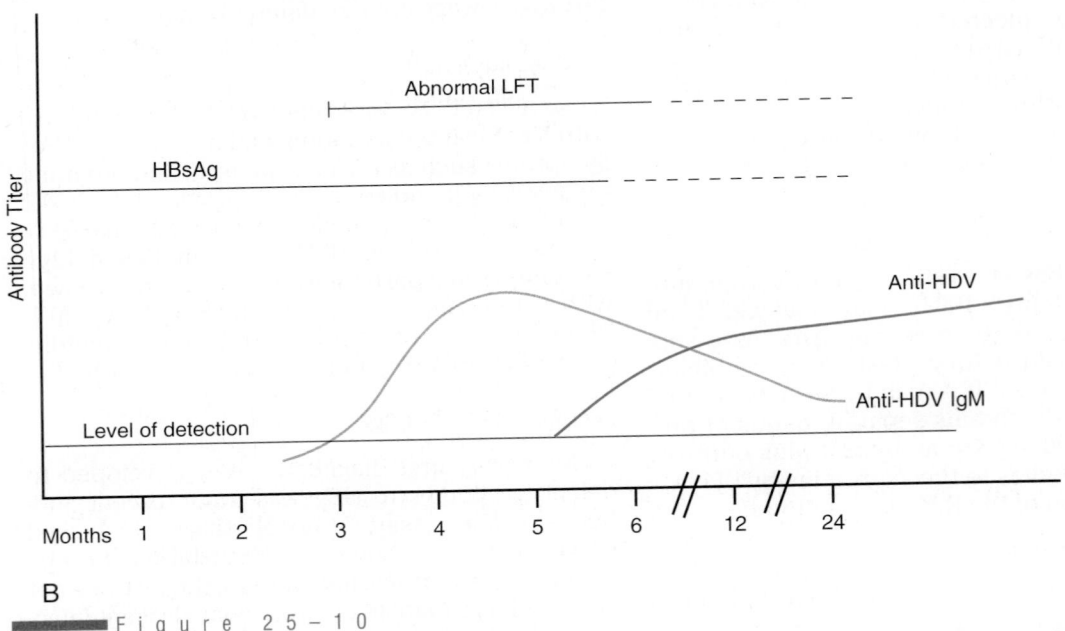

█████ F i g u r e   2 5 – 1 0

Serologic evaluation of hepatitis delta virus infection showing the rise and fall of detectable antibodies. *A*, Serologic presentation in delta hepatitis coinfection. *B*, Serologic presentation of delta hepatitis superinfection.

animal species. All the herpesviruses are morphologically similar; a core consists of linear ds DNA, a capsid surrounding the core, an amorphous tegument surrounding the capsid, and an outer envelope. All the herpesviruses also share the property of latency and lifelong persistence in the host. At least six herpesviruses commonly infect humans:

- Herpes simplex virus type 1 (HSV-1)
- Herpes simplex virus type 2 (HSV-2)
- Varicella-zoster virus (VZV)

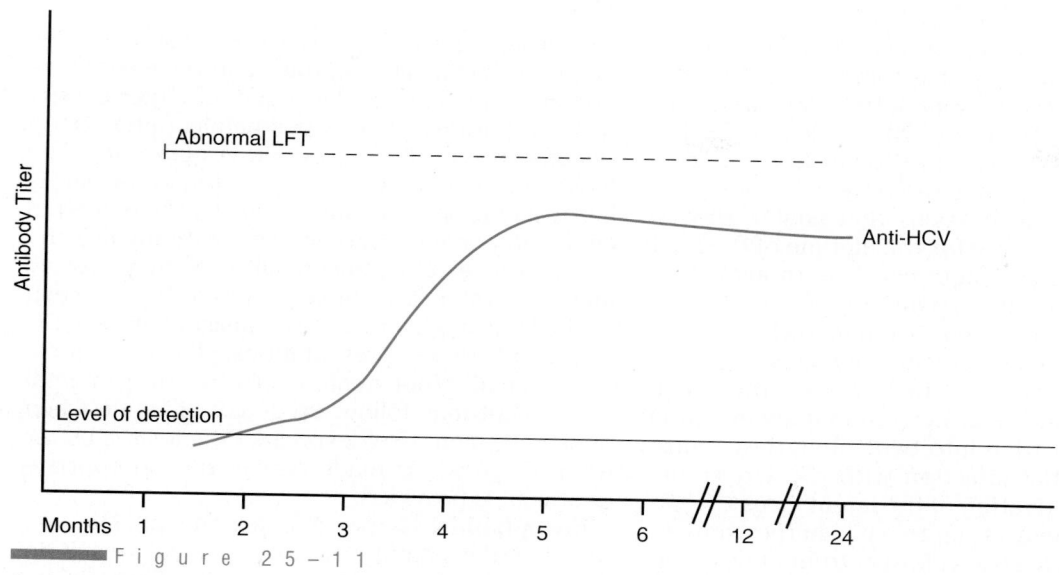

■ F i g u r e   2 5 - 1 1

Serologic evaluation of hepatitis C virus infection showing the rise and fall of detectable antibodies.

- ■ Cytomegalovirus (CMV)

- ■ Epstein-Barr virus (EBV)

- ■ Herpesvirus 6 (HHV-6)

Herpes Simplex Virus

EPIDEMIOLOGY

Most infections with herpes simplex virus are asymptomatic. Host immune competence appears to play a major role. Studies have shown that the majority of the population have antibodies to herpes simplex type 1, which are acquired usually during the first few years in life. Because herpes simplex type 2 is usually acquired by sexual contact, however, antibodies to HSV-2 are usually not present before the age of sexual activity. Although less prevalent than antibodies to HSV-1, antibodies to HSV-2 are common in adults.

CLINICAL INFECTIONS

Herpes simplex infections primarily occur on mucous membranes. Contaminated secretions are an effective means of spread. The incubation period is 2 to 11 days (usually 6 to 7 days). Infected individuals are most infectious during the early days of a primary infection, although

■ T A B L E   2 5 - 8

CLINICAL AND EPIDEMIOLOGIC DIFFERENCES AMONG HEPATITIS A, HEPATITIS B, DELTA AGENT, AND NON-A, NON-B HEPATITIS

Clinical Features	Hepatitis A	Hepatitis B	Delta Hepatitis	Non-A, Non-B Hepatitis
Incubation (days)	15–45	30–120	21–90	40–120
Type of onset	Acute	Insidious	Usually acute	Insidious
Mode of transmission				
Fecal/oral	Usual	Infrequent	Infrequent	Usual for HEV
Parenteral	Increasing	Usual	Usual	Usual for HCV
Other	Food-borne, waterborne	Intimate contact, transmucosal transfer	Intimate contact, less efficient than for HBV	Unknown
Sequelae				
Carrier	No	5–10%	Yes	Yes
Chronic hepatitis	No	Yes	Yes	Yes
Mortality (%)	0.1–0.2	0.5–2.0	30 (chronic form)	Uncertain

the virus may still be transmitted late in the infection or from an asymptomatic person. The virus is shed in secretions 3 to 4 days after primary lesions appear, and the vesicular fluid is rich with viruses. Virus-infected cells are usually found at the edge and in the base of lesions.

The biologic mechanisms that enable HSV to maintain a latent state for the lifetime of the host is an area of intense interest. It is thought that the sensory neurons serving the portal of entry become infected. When, for unknown reasons, the virus becomes reactivated, it travels by axonal transport to the cells innervated by the infected neurons. Disease in herpes simplex infection is classically divided into two categories; primary (first or initial infection with the virus) and recurrent (reactivation of the latent virus).

Although usually asymptomatic, herpes simplex infections can cause a wide spectrum of diseases. Specific clinical manifestations are as follows.

Oral, Labial, and Facial Manifestations. Oral herpes infections are usually but not exclusively caused by herpes simplex virus type 1. The incubation period varies from 2 days to 2 weeks. Primary infections are usually asymptomatic but, when apparent, manifest commonly as intraoral mucosal vesicles (rarely seen) or ulcerations that may be quite widespread and involve the buccal mucosa, posterior pharynx, and gingival and palatal mucosae. In young adults, primary herpes simplex infection may involve the posterior pharynx and present as acute pharyngitis.

Recurrent or reactivation herpes simplex usually occurs on the vermilion border of the lip, the junction of the oral mucosa and skin. A prodrome of burning or pain followed by vesicles, ulcers, and crusted lesions is the typical pattern.

Genitourinary Manifestations. Genital herpes infections are usually but certainly not always caused by herpes simplex virus type 2. As with oral infections, genital primary infections are most commonly asymptomatic. When it manifests, the infection presents in the female as vesicles on the mucosa of the labia, vagina, or both. Cervical and vulvar involvement is not uncommon. In the male, the shaft, glans, and prepuce of the penis are common sites of involvement. The urethra is commonly involved in both sexes.

Recurrent herpes infections involve the same sites as primary infections, but the urethra is less commonly involved. The symptoms are usually less severe in recurrent disease.

Neonatal Herpes Infections. Fortunately, neonatal herpes infections are much less common than genital infections. The risk is greatest when the mother has clinical symptoms and the infection is primary. Infection may be acquired in utero, intrapartum (during birth), or postnatally (after birth). Documented in utero infection is quite rare. The most common mode of transmission by far is intrapartum. Neonatal infections may be limited to cutaneous or mucosal surfaces, may involve the brain with resulting encephalitis, or may involve multiple systems, including internal organs such as the lung and liver, or may consist of any combination of these presentations. If encephalitic involvement is not treated early, death or mental impairment may follow. Prognosis is worse when the involvement is widespread. Prognosis is better when diagnosis is made earlier and appropriate treatment is provided.

Encephalitis. Herpes simplex encephalitis is a rare but devastating disease that can result in death or severe neurologic deficit. There is no clear association with immune defect or active herpes infection elsewhere in the host. No predisposing factors have been identified. Early diagnosis and treatment are the only hope for prevention of the serious sequelae previously described.

Ocular Manifestations. Herpes simplex infection of the conjunctiva may manifest as swelling of the eyelids associated with vesicles. Involvement of the cornea may result in destructive ulceration and even perforation of the cornea.

LABORATORY DIAGNOSIS

Diagnosis of herpes simplex infections is best made by viral isolation or antigen detection. Taken as early as possible, specimens that provide good recovery are aspirate from vesicles or open lesions and other materials from infected sites. Herpes simplex is a rapidly growing virus that often demonstrates cytopathic effect in culture within 24 hours (see Fig. 25–2). Therefore, diagnosis and appropriate therapy can be readily accomplished. An indiscriminate virus, herpes simplex can be isolated by using a number of cell lines, including human embryonic lung, rabbit kidney, HEp2, and A549. Serologic studies are, in general, much less useful.

Cytomegalovirus (CMV)

Cytomegalovirus is a typical herpesvirus but replicates only in human cells and much more slowly than herpes simplex or varicella-zoster.

EPIDEMIOLOGY

Cytomegalovirus is known for widespread distribution. It is spread by close contact with an infected person. Most adults demonstrate antibody against the virus, and those who live in overcrowded living conditions may acquire CMV at an early age. CMV infection may also be acquired perinatally via cervical or vaginal secretions or the blood stream.

CLINICAL INFECTIONS

The vast majority of CMV infections are asymptomatic. Occasionally, in immunocompetent patients, the infection may manifest as a self-limited, infectious mononucleosis–like illness with fever and hepatitis. Congenital infection and infection in immunocompromised patients may be symptomatic and serious.

Cytomegalovirus is the most common viral congenital infection in the United States. Infection in an infant is unlikely to occur if the mother was seropositive at the time of conception. Serious clinical manifestations may develop if the mother acquires the primary infection during pregnancy. Symptomatic congenital infection is characterized by petechiae, hepatosplenomegaly, microcephaly, and chorioretinitis. Other manifestations are low birth weight, CNS involvement, mental retardation, and death.

In immunocompromised hosts, such as transplant recipients and HIV-infected patients, a CMV infection may present as a life-threatening disseminated disease with involvement of almost any organ, including the lungs, liver, intestinal tract, retina, and central nervous system.

LABORATORY DIAGNOSIS

Diagnosis is best confirmed by isolation of the virus from normally sterile body fluids or tissues such as the buffy coat of blood or other internal fluids or tissue. The virus can also be cultured from urine or respiratory secretions, but because shedding of CMV from these sites is common in normal hosts, isolation from such sources must be interpreted with some caution.

Congenital infection is best confirmed by isolation of CMV from the infant within the first 2 weeks of life. Isolation after the first 2 weeks of life does not confirm congenital infection. Urine is the most common source utilized. As with herpes simplex virus, serology is not as helpful in diagnosis as culture.

Cytomegalovirus can be isolated only by using human diploid fibroblasts such as human embryonic lung or human foreskin fibroblasts (see Fig. 25–3). The virus replicates slowly, so it may take up to 3 weeks for a cytopathic effect to appear. The time required for diagnosis can be markedly reduced by use of the shell vial technique. In this technique, the monolayer of cells is on a coverslip in a vial. After the vial (cells on coverslip) is inoculated, it is centrifuged at low speed (2,000 to 2,500 rpm) for 45 minutes. The vial is then incubated for 16 to 24 hours, to allow for replication of any virions present. The coverslip is then removed, rinsed, and fixed. Fluorescent-tagged antibodies against immediate-early and early CMV proteins are added, enabling detection of these proteins, which are produced in CMV infected cells. Using the shell vial technique, the diagnosis may be made within 1 day rather than 3 weeks.

Epstein-Barr Virus

Epstein-Barr virus has the typical herpesvirus morphology. Although in vivo EBV infects epithelial cells of the oral mucosa, salivary gland, and kidney, in vitro propagation is limited to B lymphocytes and is not practical. Infection is probably spread by saliva or other close contact.

CLINICAL INFECTION

The incubation period for EBV varies from 2 weeks to 2 months. As with the other herpes group viruses, infection is very common, and most adults demonstrate antibody against the virus. Infection in young children is almost always asymptomatic. With increasing age, to young adulthood, there is a corresponding increase in the ratio of symptomatic to asymptomatic infections.

The typical picture in young symptomatic adults is infectious mononucleosis. This is a self-limiting disease characterized by sore throat, fever, lymphadenopathy, hepatomegaly, splenomegaly, and general malaise. The signs and symptoms usually resolve within a few weeks, although malaise may be prolonged.

EBV has been associated with Burkitt lymphoma, a malignant disease of the lymphoid tissue seen most commonly in African children. The exact role of the virus has yet to be determined. EBV has also been associated, less clearly, with nasopharyngeal carcinoma. The virus is also increasingly being recognized as an important viral agent in transplant recipients. The most significant clinical effect of EBV infection in these patients is the development of a B-cell lymphoproliferative disorder or lymphoma.

Other complications of EBV infections are splenic hemorrhage and rupture, frank hepatitis, thrombocytopenia purpura with hemolytic anemia, Reye syndrome, encephalitis, and other neurologic syndromes.

LABORATORY DIAGNOSIS

Epstein-Barr virus is not routinely cultured, and the diagnosis is accomplished by serologic studies. Heterophile antibody, also known as Paul-Bunnell (PB) antibody, though nonspecific for any EBV antigen, is associated with acute infection with EBV. Heterophile antibody appears early in infection and may remain for several months. This antibody is the basis of "spot" tests. Other serologic markers are

■ **IgG anti-VCA:** IgG antibodies against the viral capsid antigen.

■ **IgM anti-VCA:** IgM antibodies against the viral capsid antigen.

■ **IgG anti-EA:** antibody to early antigen; it can be separated into D (diffuse) or R (rough) antibody.

■ **anti-EBNA:** antibody to the nuclear antigen.

■ T A B L E   2 5 – 9
INTERPRETATION OF EBV SEROLOGIC MARKERS

PB	Anti-VCA-IgM	Anti-VCA-IgG	Anti-EA-IgG	Anti-EBNA	Interpretation
–	–	–	–	–	No prior exposure to EBV
+	+	+	+/–	–	Acute infectious mononucleosis
+/–	+/–	+	+/–	+	Recent infection
–	–	+	–	+	Remote infection

PB, Paul Bunnell; anti-VCA-IgM, IgM antibodies against the viral capsid antigen; anti-VCA-IgG, IgG antibodies against the viral capsid antigen; anti-EA-IgG, antibody to early antigen; anti-EBNA, antibody to the nuclear antigen.

Table 25–9 shows the interpretation of the EBV serologic markers. Figure 25–12 shows a diagram of serologic evaluation of EBV infection.

Human Herpesvirus 6

Infection with the recently discovered human herpesvirus 6 (HHV-6), as with the other herpes viruses, is widespread. Antibodies against the virus are present in 95% of young adults. Respiratory droplet and close contact are probably the means of transmission.

CLINICAL INFECTION

Most infections are apparently asymptomatic. Human herpes virus 6 has been associated with the childhood disease exanthema subitum (also

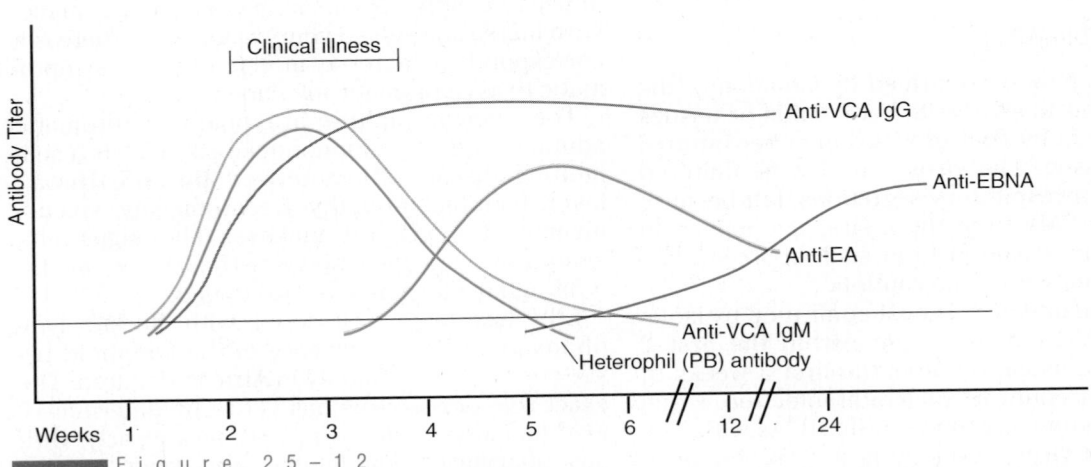

■ F i g u r e   2 5 – 1 2
Serologic evaluation of EBV infection (infectious mononucleosis) showing the rise and fall of detectable antibodies.

called roseola infantum). This disease commonly presents in children 6 months to 3 years of age as an acute, febrile, mild illness. A maculopapular rash appears as the fever resolves. The infection appears to involve the lymphoid tissue. Little is known about site or mechanism of latency.

LABORATORY DIAGNOSIS

The diagnosis of HHV-6 infection is usually made on a clinical basis. Isolation of virus requires lymphocyte culture, which is not practical for routine diagnosis. Serology may not be helpful unless paired sera are available.

Varicella-Zoster Virus

Varicella-zoster virus has the typical structure of the herpesvirus group. It is spread by droplet inhalation or direct contact with infectious lesions. Varicella, or chickenpox, is the primary infection and is highly contagious. Over 90% of adults have antibody to varicella-zoster virus.

CLINICAL INFECTION

In contrast to infections with the other herpes group viruses, which are usually asymptomatic, varicella is usually clinically apparent. The common presentation is in childhood with a mild febrile illness, rash, and vesicular lesions. Usually, the lesions appear first on the head and trunk, then spread to the limbs. The lesions dry, crust over, and heal in 1 to 2 weeks. Painful oral mucosal lesions may occur, particularly in adults.

Zoster, which is the clinical manifestation of reactivation of the varicella-zoster virus, usually occurs in adults. It is thought that the virus remains latent in the dorsal root or cranial nerve ganglia after primary infection with chickenpox. In a small proportion of patients the virus becomes reactivated, travels down the nerve and causes zoster. The most common presentation is rash, followed by vesicular lesions in a unilateral dermatome pattern. These lesions may be associated with prolonged disabling pain that can remain for months, long after the vesicular lesions disappear.

LABORATORY DIAGNOSIS

In the typical case of varicella or zoster, the diagnosis can be made on the basis of the char-

acteristic clinical findings. In atypical cases, such as in immunosuppressed patients, the diagnosis may be more difficult or questionable. In such patients, culture of fresh lesions (vesicles) or use of fluorescent-tagged monoclonal antibodies against VZV confirms the diagnosis. Varicella-zoster virus can be cultured on human embryonic lung or Vero cells. Cytopathic changes may not be evident for 3 to 7 days.

ANTIVIRAL THERAPY

It is paradoxic (if not ironic) that the most simple microorganisms are the most difficult to treat. Viruses are the ultimate parasites, pirating the host cell metabolic processes for their own use. The major challenge in developing antiviral agents is to target an essential viral replicative mechanism without destroying or damaging the host cell. (Perhaps chemotherapy for cancer may be considered an analogous situation, in which the goal is destroying abnormal malignant cells without toxic effects on normal cells.) Functions that are unique to the virus afford the most fruitful targets. These include the steps in the viral replicative cycle discussed here. Many of the currently used antiviral agents are analogs of nucleosides used in viral replication. The virus uses these "counterfeit" nucleoside analogs, which by various mechanisms disrupt the viral replicative cycle. By no means, however, are all antiviral agents nucleoside analogs. Phosphonoformic acid (foscarnet) is an analog of pyrophosphate that acts directly as a DNA polymerase inhibitor.

Not unexpectedly, viruses may develop resistance to antiviral agents, just as bacteria may become resistant to antibacterials. Antiviral susceptibility testing can be done. As more antiviral agents become available, susceptibility testing will become increasingly important. For example, foscarnet is being used in treatment of infections by herpes simplex virus resistant to acyclovir and cytomegalovirus resistant to ganciclovir.

The viral infectious cycle can be thought of as occurring in five steps or phases. These are potential targets for antiviral activity.

1. Absorption. For infection of a cell to occur, there must be absorption or attachment of the virion to the cell. In almost all cases, this depends on the appropriate cell receptors. Most receptors are glycoproteins. The receptors are essential for the cell function and, obviously, are not present for the convenience of the virus. For example, the receptor for poliovirus belongs to the immunglobulin superfamily, rabies virus utilizes the acetylcholine receptor, influenza virus binds to sialic acid, HIV binds to CD4, and Epstein-Barr virus binds to complement receptor C3d.

2. Penetration. Nonenveloped virions may directly penetrate the cell membrane. Once the virus is in the cell, the capsid is removed and the genome unmasked. Enveloped viruses may enter the cell by fusion with the cell membrane. The third, and perhaps the predominant, method is by endocytosis, whereby the virus enters the cell in a cytoplasmic vacuole, which then fuses with cytoplasmic lysosomes (internal fusion).

3. Uncoating. Whichever mechanism of penetration is used, for the viral genome to replicate and be processed by the cellular machinery, it must be released from the capsid into either the cytoplasm (most RNA viruses) or nucleus (most DNA viruses) of the cell.

4. Eclipse/Synthesis. For a short time after infection of the cell, commonly several hours, infectious virions cannot be detected. This is known as the eclipse phase. During and after this time, active synthesis of viral components (utilizing the cell machinery) is occurring.

5. Maturation/Release. The capsid protein subunits aggregate to form capsomers and associate with the genome to form the nucleocapsid. The virion may be released from the cell by budding if enveloped or may cause lysis of the cell.

A central factor in the capability of a virus to multiply is the synthesis of the virus gene product, the proteins. Viral proteins are essential in the replication of the viral genomes and also are the "package" of the viral particle. One should recall that the usual metabolic progression is DNA → (transcription) → RNA → (translation) → protein. The synthesis of the viral proteins is a key event in viral replication. All viruses depend on the cellular machinery in differing degrees for the synthesis of these essential proteins.

Examples of some of the more commonly used antiviral agents follow:

Acyclovir

Mechanism of Action. A guanosine analog that is phosphorylated to acyclovir monophosphate by virus-encoded thymidine kinase. As this is encoded by the viral thymidine kinase, a high degree of selectivity of the drug occurs. The acyclovir monophosphate is further phosphorylated to diphosphate and triphosphate by nonspecific cellular enzymes. Again, the initial activation of acyclovir to acyclovir monophosphate is promoted by herpes thymidine kinase. The cellular thymidine kinase is unable to phosphorylate acyclovir to acyclovir monophosphate. Thus, in uninfected cells, acyclovir is not phosphorylated (activated). Acyclovir binds and is phosphorylated 200 times more efficiently by herpes simplex thymidine kinase than by the nonspecific cellular kinases. Acyclovir triphosphate competes with deoxyguanosine triphosphate, (dGTP), the natural and normal substrate for DNA polymerase. Acyclovir has greater affinity than dGTP for this enzyme. Acyclovir acts as a chain terminator, because additional nucleotides cannot be added (acyclovir lacks the 3' prime hydroxyl group needed to add additional nucleotides). Also, the DNA polymerase ordinarily binds only temporarily to the DNA chain and then dissociates. The DNA polymerase binds irreversibly to acyclovir-associated DNA, and thus the polymerase is inactivated.

Antiviral Activity. Acyclovir is active against herpes simplex types 1 and 2. It is slightly less active but still effective against varicella-zoster virus. Because cytomegalovirus and Epstein-Barr virus do not induce their own thymidine kinase, acyclovir is not generally effective as a therapeutic agent (however, acyclovir is useful as a prophylactic agent in preventing symptomatic infection by cytomegalovirus in transplant patients).

Adverse Effects. Because of the "specificity" provided by the herpes thymidine kinase, as noted previously, acyclovir has minimal toxicity. The therapeutic index for acyclovir in herpes simplex infections has been estimated to be from 300 to 3,000. Side effects are uncommon. There may be slight burning or stinging with the topical ointment. With rapid intravenous administration, reversible crystalline nephropathy can occur.

Ganciclovir

Mechanism of Action. Ganciclovir is an analog of guanosine similar to acyclovir. The active form, as with acyclovir, is ganciclovir triphosphate, which selectively inhibits viral DNA polymerase. Ganciclovir triphosphate is a competitive inhibitor of guanosine triphosphate for DNA polymerase. In herpes simplex–infected cells, ganciclovir is phosphorylated to ganciclovir monophosphate by the same virus-specified thymidine kinase that phosphorylates acyclovir. Cytomegalovirus is not known to code for a viral thymidine kinase, and the mechanism of DNA polymerase inhibition with cytomefalovirus is not yet well worked out.

Antiviral Activity. Ganciclovir is active against cytomegalovirus. Ganciclovir is also active against herpes simlex, but because it requires intravenous administration, acyclovir rather than ganciclovir is usually the antiviral used against herpes simplex.

Adverse Effects. Ganciclovir has a narrow therapeutic to toxic ratio. Bone marrow suppression, usually granulocytopenia, is the most common dose-limiting toxicity. Thrombocytopenia may also occur. These side effects are dose related and are reversible. Other adverse effects involve the central nervous system, with confusion and headache being most common. Also, nausea, diarrhea, and live function abnormalities may be rarely seen.

Ribavirin

Mechanism of Action. Ribavirin is also a guanosine analog. It is converted by cellular enzymes to monophosphate, diphosphate, and triphosphate forms. The major metabolite is ribavirin triphosphate. The exact mechanism of action is unclear. Three mechanisms have been proposed: (1) decrease of intracellular guanosine triphosphate due to competitive inhibition by ribavirin triphosphate; (2) inhibition of "capping" of messenger RNA; and (3) inhibition of the initiation and elongation activity of virus-coded RNA polymerase.

Antiviral Activity. Ribavirin approved for use as an aerosol in selected infections by respiratory syncytial virus. This agent activity against other viruses, including influenza A and B, parainfluenza, and Lassa fever viruses.

Adverse Effects. Side effects of aerosol therapy are apparently minimal and rare. Rash and conjunctivitis may occur. Reversible anemia has been reported with oral or intravenous administration.

Amantadine

Mechanism of Action. Initially, it was proposed that amantadine exerted its effects by inhibiting fusion of the membrane-bound virus with the cell's endosome (the last type of viral penetration discussed in the section on stages of virus infection). It was thought that preventing the lowering of the pH in the endosome prevented activation of the virus hemagglutinin. This process, however, requires very high concentrations of the drug. More recent data indicated that at lower, clinically achievable levels, amantadine interferes with hydrogen ion penetration of M2, a transmembrane viral-encoded protein.

Antiviral Activity. This drug is indicated for influenza A infection. Amantadine is not effective against influenza B.

Adverse Effects. The most common side effects involve the central nervous system, with complaints of nervousness, insomnia, lightheadedness, and difficulty concentrating reported. Also, nausea and vomiting occur. These symptoms often resolve in spite of continued administration of the drug.

Zidovudine

Mechanism of Action. Zidovudine (AZT) is an analog of the pyrimidine thymidine. Cellular enzymes convert zidovudine to monophosphate, diphosphate, and triphosphate forms. Zidovudine monophosphate, probably by competitive inhibition of thymidine kinase, decreases the phosphorylation of thymidine monophosphate to thymidine diphosphate and thymidine triphosphate, the normal substrate for viral reverse transcriptase. Zidovudine triphosphate is the analog of thymidine triphosphate, the normal substrate for viral reverse transcriptase. Zidovudine triphosphate acts as a competitive inhibitor of thymidine triphosphate, the normal substrate or "target" of viral reverse transcriptase.

Antiviral Activity. Zidovudine (AZT) is used in treatment of HIV infection.

Adverse Effects. The major side effect of AZT is on the bone marrow, with hematopoietic depression resulting in anemia and granulocytopenia.

Bibliography

Belshe RB: Textbook of Human Virology, 2nd ed. St. Louis: Mosby Year Book, 1991.

Buchmeier MJ: Marburg and Ebola Viruses: New Agents on the Frontiers of Virology. In Notkins AL, Oldstone MB (ed): Concepts in Viral Pathogenesis. New York: Springer-Verlag, 1984, pp. 338–343.

Costello MJ, Morrow S, et al: Guidelines for specimen collection, transportation, and test selection. Lab Med. 24:19, 1993.

Creager JG, Black JG, Davison VE: Microbiology: Principles and Applications. Englewood Cliffs, NJ: Prentice-Hall, 1990, pp 254–281.

Fields BN, Knipe DM (eds): Flelds Virology, 2nd ed. New York: Raven Press, 1990.

Galasso GJ, Whitley RJ, Merigan TC (eds): A Practical Diagnosis of Viral Infections. New York: Raven Press, 1993.

Haukenes G, Haaheim LR, Pattison JR (eds): A Practical Guide to Clinical Virology. New York: John Wiley & Sons, 1989.

Henig RM: A Dancing Matrix: Voyages Along the Viral Frontier. Alfred A Knopf, New York: 1993.

LeDuc JW, Childs JE et al: Hantaan (Korean hemorrhagic fever) and related rodent zoonoses. In Morse SS (ed): Emerging Viruses, New York: Oxford University Press, 1993.

Lennette DA, Ray CG, Menegus MA: Viruses. In Balows A, Hausler WJ Jr, et al (eds): Manual of Clinical Microbiology, 5th ed. Washington, DC: American Society for Microbiology, 1991, pp 811–1005.

Lennette EH, Halonen P, Murphy FA (eds): Laboratory Diagnosis of Infectious Diseases Principles and Practice, Vol II: Viral, Rickettsial, and Chlamydial Diseases. New York: Springer-Verlag, 1988.

Levins RT, Awerbuch, et al: The emergence of new diseases. Am Sci 82:52, 1994.

Roizman B, Whitley RJ, Lopez C: The Human Herpes Viruses. New York: Raven Press, 1993.

Rose NR (ed): Manual of Clinical Laboratory Immunology, 4th ed. Washington, DC: American Society for Microbiology, 1992.

Schmidt NJ, Emmons RW (eds): Diagnostic Procedures for Viral, Rickettsial and Chlamydial Infections, 6th ed. Washington, DC: American Public Health Association, 1989.

Skidmore SJ, Yarbough PO, et al: Hepatitis E virus: The cause of a waterborne hepatitis outbreak. J Med Virol 37:58, 1992.

Spector S, Lancz G: Clinical Virology Manual, 2nd ed. New York: Elsevier, 1992.

Yanagihara R, Gajdusek, DC: Hantavirus infection: A newly recognized viral enzootic of commensal and wild rodents in the United State. In Notkins AL, Oldstone MD (eds): Concepts in Viral Pathogenesis III. New York: Springer-Verlag, 1989, pp 291–296.

Yarbough PO: Hepatitis E: Diagnosis of Infection. 15:113, 1993.

Zapikan AZ: Viral gastroenteritis. In Notkins AL, Oldstone MB (eds): Concepts in Viral Pathogenesis. New York: Springer-Verlag, 1984, pp 315–324.

Zuckerman AJ, Banatvala JE, Pattison JR: Principles and Practice of Clinical Virology, 2nd ed. New York: John Wiley & Sons, 1990.

PART III

Laboratory Diagnosis of Infectious Diseases: An Organ System Approach to Diagnostic Microbiology

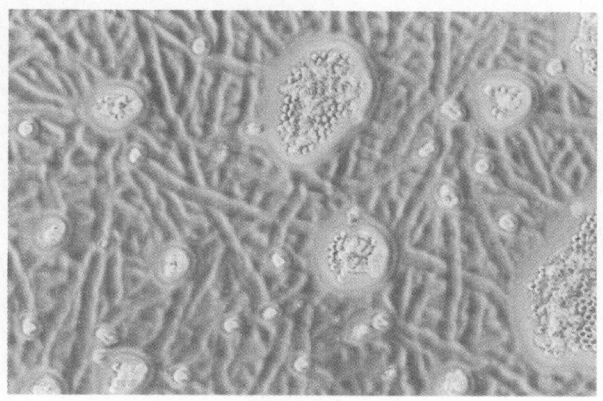

Upper and Lower Respiratory Tract Infections

Patricia M. Simone •
James L. Cook

1. Describe the basic anatomy of the respiratory tract and explain the mechanical defenses of each anatomical site and how alterations of these defenses may lead to infectious diseases of the respiratory tract.

2. Define the importance of normal flora in the respiratory tract and explain how alterations in the normal flora may lead to infectious diseases.

3. Discuss the basic pathogenic mechanisms of infectious diseases of the respiratory tract and associate the virulence factors of the organisms that cause that disease.

4. Given the clinical picture presented and the symptoms, associate the most probable organisms that cause upper and lower respiratory tract infections.

5. Describe the pathogenesis, risk factors, and complications associated with these infections as well as the types of specimens collected for diagnosis.

6. Determine the risk factors in immunocompromised hosts that predispose to infections and provide examples of respiratory tract diseases in different types of immunocompromised hosts.

7. Describe the principles and methods of proper specimen collection and transport of respiratory secretions.

8. Discuss the importance of visual examination and proper culturing of respiratory samples and determine the acceptability for culture of respiratory secretions.

9. Appraise the important aspects of diagnosis of infections of the respiratory tract through case studies.

This chapter describes respiratory tract infections from the perspective of the clinical microbiologist working with the clinician facing the differential diagnosis of these infections. The chapter includes the following:

■ an overview of the anatomy of the upper and lower compartments of the respiratory tract to provide a basis for discussion of defenses against the establishment of infection that may follow mechanical and functional changes in these areas

■ the importance of the normal flora of the respiratory tract in preventing colonization by potential pathogens, which will provide the reader with an understanding of how alterations in the normal flora as well as colonization of normally sterile areas may lead to infection

■ virulence factors of pathogenic microorganisms and how they relate to the mechanisms of establishment and progression of infection

■ clinical syndromes of patients with upper and lower respiratory tract infections to provide the clinical aspects of the microbiology of the respiratory tract

■ pathogenesis, risk factors, and complications of specific disease states (specimen collection and processing of respiratory tract secretions)

■ infections in normal hosts, risk factors that predispose immunocompromised hosts to respiratory tract infections, and how presentations of these infections differ in immunodeficient patients

■ case studies that further illustrate and integrate concepts presented in the chapter

GENERAL CONCEPTS OF INFECTIOUS DISEASES RELATED TO RESPIRATORY TRACT INFECTIONS

There are general concepts that apply to the infectious diseases of each organ system. A brief introductory discussion of some of these concepts provides a better understanding of the different disease entities related to infections of the respiratory tract. These concepts include the following:

■ the role of the normal microbial flora

- the immune status of the host
- seasonal and community trends in communicable respiratory diseases
- the role of empirical antimicrobial therapy

The Role of Normal Flora

In the diagnostic approach to all types of infections, one must consider the types of organisms normally found at the site to be able to determine whether diagnostic microbiology data indicate the presence of a pathogen. This requires a knowledge of the normal flora to be expected at the site, the clinical setting, and the patient's presentation.

The normal microbial flora exists in a symbiotic relationship with the host. These organisms are isolated from the host in the absence of disease; this is referred to as colonization. The normal flora of the respiratory tract plays an important role in protecting the host from infection with pathogenic microorganisms. The normal flora can prevent proliferation and invasion of pathogenic organisms through competition for the same nutrients and the same receptor sites on host cells. In addition, these organisms produce bacteriocins, bacterial products that are toxic to other organisms. The presence of these organisms keeps the immune system primed for a rapid response to invading organisms and stimulates cross-protective immune factors known as natural antibodies. Under normal conditions, a balance is maintained that limits the quantity or dominance of any one organism.

Discriminating Between Normal Flora and Pathogenic Microorganisms

Although there is a standard list of organisms that can be routinely cultured from the upper respiratory tract (Table 26–1), it is interesting that the consensus concerning what is considered "normal flora" can change with time as new associations between organisms and disease states are recognized. For example, in past years, *Moraxella catarrhalis,* formerly *Neisseria catarrhalis,* was considered part of the normal flora of the upper respiratory tract and was considered a relatively nonpathogenic organism that was only rarely associated with serious infec-

T A B L E   2 6 – 1

COMMON NASOPHARYNGEAL AND OROPHARYNGEAL ORGANISMS ISOLATED IN THE NORMAL HOST

Bacteria
  Usually present
    Streptococcus mitis and other α-hemolytic streptococci
    Non-group A β-hemolytic streptococci
    S. pneumoniae
    S. pyogenes
    S. salivarius
    Veillonella spp
    Bacteroides spp
    Fusobacterium spp
    Prevotella spp
    Porphyromonas spp
    Coagulase-negative staphylococci
    Neisseria spp
    Nonhemolytic streptococci
    Diphtheroids
    Micrococcus spp
    Eikenella spp
    Capnocytophaga spp
  Occasionally present
    Haemophilus influenzae
    Haemophilus parainfluenzae
    Peptostreptococcus
    Actinomycetes
    Staphylococcus aureus
    Mycoplasma
Fungus
  Candida spp
Virus
  Herpes simplex

tions. However, since the early 1970s, there has been an increasing awareness that these organisms can be associated with complicated infections of the respiratory tract in children and in adults with chronic lung diseases. This historical anecdote indicates the importance of periodic reevaluation of the pathogenicity of all organisms.

Familiarity with the normal flora of respiratory tract compartments is also important in determining the clinical relevance of an isolate. For example, isolation of α-hemolytic colonies from a pharyngeal culture from a patient with pharyngitis arouses little clinical interest, since such organisms are normal flora in the oropharynx. In contrast, isolation of α-hemolytic colonies from a properly collected sputum specimen or bronchial aspirate in the clinical setting of lobar pneumonia should prompt full identification of the organism and perhaps initiation of empirical therapy for possible pneumococcal infection.

Normal upper respiratory tract flora in an asymptomatic patient may change depending on the clinical setting. Patients who have previously received broad-spectrum antibiotics, have been hospitalized recently, or have chronic illnesses may have a change in the "normal" pharyngeal flora. Gram-negative bacilli are commonly isolated from the pharynx in such patients in the absence of clinical signs of infection. Therefore, it is important to distinguish between a positive culture for an organism that is a potential pathogen and a clinical disease state caused by that pathogen. Table 26–2 lists organisms that are considered primary pathogens and are significant. This table also shows a number of organisms that can be pathogens in the correct clinical setting. In this case, as in many others, proper communication between the clinical microbiologist and the clinician is essential.

Differentiating Colonization From Infection

The isolation of certain organisms from respiratory specimens may represent either colonization or disease, depending on the circumstances. An interpretation must be based on several factors. The method and site of collection of the specimen can influence the risk of contamination with organisms that are part of the normal flora. Characteristics of the specimen such as the presence of white blood cells and the number of organisms in the specimen can help distinguish between colonization and infection. Most important, a compatible clinical syndrome should be present in order to determine whether the pres-

ence of a potential pathogen is clinically relevant. For example, the isolation of a few colonies of Staphylococcus aureus from a sputum specimen with many epithelial cells does not represent S. aureus pneumonia but is more likely to be contamination with an organism that is part of the normal flora of the patient. In contrast, heavy growth of the same organism from a respiratory specimen with many white blood cells from an elderly man with post-influenza pneumonia is highly suggestive of S. aureus pneumonia, especially if the sputum culture is accompanied by a positive culture from a normally sterile site such as blood or pleural fluid.

Other organisms are potentially pathogenic and cause disease only when there is some interruption in the pulmonary defense mechanisms. Streptococcus pneumoniae has been isolated from 5 to 70% of the normal adult population, yet only a very small proportion of these carriers will develop pneumococcal pneumonia. Pneumonia may also result from aspiration of upper respiratory tract secretions in patients with impaired pulmonary defenses, such as those with alcoholism or congestive heart failure.

There are other organisms that are always considered pathogenic when isolated, even in small numbers. Isolation of Mycobacterium tuberculosis in any amount is significant because of the virulence of the organism and the contagiousness of the infected patient.

The Immune Status of the Host

The defenses of the host with a suspected infection must be evaluated when determining whether the identification of a specific microorganism is likely to be significant in causing disease. When considering the likelihood that a microorganism will cause infection in a given host, both the virulence of the organism and the defenses available to the host to counteract the establishment and progression of infection must be considered. At one end of the spectrum of organisms are the normal flora of the respiratory tract that usually exist as commensals for the life of the normal host without causing disease. At the other end are organisms that often establish infection and cause disease in the normal host when present in numbers above a certain threshold. For these purposes, a normal host is considered to be one with mature immunologic defenses but without specific immunity against the

TABLE 26–2

SELECTED NONVIRAL PATHOGENS IN THE RESPIRATORY TRACT

Primary Pathogens	Possible Pathogens
Streptococcus pneumoniae	Acinetobacter spp
Group A β-hemolytic streptococci	Enterics and other gram-negative bacilli
Neisseria meningitidis	Fungi
Neisseria gonorrhoeae	Nocardia spp
Bordetella pertussis	Staphylococcus aureus
Mycobacterium kansasii	
Mycobacterium tuberculosis	Haemophilus influenzae
Legionella pneumophila	β-Hemolytic streptococci, non–group A
Toxin-producing Corynebacterium diphtheriae	Moraxella catarrhalis
Mycoplasma pneumoniae	Anaerobes
Chlamydia trachomatis	Mycobacterium spp
Chlamydia pneumoniae	Actinomycetes
Pneumocystis carinii	

microorganism in question. Examples of microorganisms that cause respiratory tract infections in normal hosts are the common respiratory viruses of childhood. In normal children, these viral infections occur at a high incidence but are relatively benign. The clinical outcomes of these and other respiratory tract infections in normal hosts depend on both the injurious effects of the microorganism and its products at the site of infection and the host immune response to infection.

Previous exposure of a normal host to pathogens such as the respiratory tract viruses is one example in which the host usually develops immunity to reinfection with the same organism. Organism-specific immune responses may prevent or alter the course of subsequent infection. For example, adults previously infected with a given virus serotype as children usually do not manifest the same severity of infection when reexposed to the same pathogen. Evidence of infection may not be present, or clinical signs and symptoms of infection are greatly reduced.

In an immunocompromised host, however, microorganisms that are usually not pathogens in a normal host may cause serious infection. This type of infection is referred to as an "opportunistic" infection to indicate that a combination of a reduced host response and a pathogen of low virulence has resulted in the establishment of infection. Of course, organisms exhibiting high levels of virulence also cause disease in immunocompromised hosts. Since the host response is diminished, infections with virulent pathogens are usually more severe and more rapidly progressive than in the normal host.

Immunocompromise by reason of age is a form of "functional immunodeficiency." Infants and the elderly are more susceptible to certain respiratory tract infections and are more likely to develop complications of these infections. For example, respiratory tract infections with *Haemophilus influenzae* are more commonly complicated by meningitis in infants than in older children or adults. Similarly, in the elderly, complications of common respiratory tract infections occur more frequently than in the younger adult population. One hallmark of seasonal outbreaks of influenza virus infection of the respiratory tract is the increased incidence of death from complicating bacterial pneumonias in the elderly.

It is apparent from these observations that one cannot determine the significance of a respiratory tract microbial isolate without considering the source of the specimen, the age and immunologic status of the host, and the clinical setting of the patient. The isolation of an organism that has great potential to cause disease may represent either simple colonization of the upper respiratory tract or life-threatening disease. In contrast, isolation of a "nonpathogenic" organism that may be part of the normal flora of the upper respiratory tract may be an indication of serious disease if it is found in an unusual location or in a host with decreased defenses against infection. Availability of proper clinical data in these cases facilitates planning by the clinical microbiologist of proper evaluation of the respiratory tract specimen.

In addition to age as a compromising factor, reduced clearance of secretions or obstruction of an area in either the upper or the lower respiratory tract predisposes to infection and can, on occasion, seriously compromise respiratory tract function. Infection behind an obstruction is a common theme in the study of infectious diseases of many organ systems. Obstruction prevents the mechanical clearance that is important in limiting both the numbers of potential pathogens and the associated inflammatory response at a specific site. Decreased clearance of respiratory secretions may result from the following:

- immature anatomical development (e.g., eustachian tube anatomy in young children)
- transient reduction in function of the mucociliary mechanism (e.g., after a viral infection)
- obstruction by a foreign body (e.g., aspirated food or foreign object)
- previous disease that has altered the normal respiratory tract anatomy (e.g., bronchiectasis)
- alterations in the viscosity of mucus (e.g., cystic fibrosis)

Another way in which respiratory tract obstruction can be a factor in infectious disease involves compromise of respiration. In these cases, the obstruction may be a consequence of, rather than the factor that precipitates, infection. The anatomy of the area, rather than the type of pathogen, dictates the urgency with which treatment is initiated. An example of this type of respiratory tract infection is acute epiglottitis caused by infection with *Haemophilus influenzae*. This inflammatory response to bacterial infection can cause life-threatening upper airway obstruction if it is not recognized and treated early in the course of the disease. This is a medical emergency. In contrast, other infections caused by the same pathogen, such as sinusitis or cellulitis, can be treated more deliberately. A search should be undertaken for factors that predispose the host to infection, such as anatomical obstruction or altered clearance of secretions in the respiratory tract.

Seasonal and Community Trends in Infections

Awareness of the patterns of respiratory tract infections at different times of the year and within the community where the patient resides is important for the efficient use of diagnostic microbiology resources. Certain types of respiratory tract infections have peak seasonal incidences and may occur in epidemics in the community. Other infections are observed throughout the year without major seasonal variations. For example, viral respiratory tract infections are more common in the fall and winter months than during other seasons, a fact that can be useful in diagnostic and therapeutic decisions. Therefore, if influenza virus infection is epidemic in the community and a patient presents with symptoms and signs compatible with this viral illness, the likelihood is very high that influenza is the cause of the infection. In such cases, performing extensive bacterial cultures to define the cause of the respiratory tract infection is both wasteful of resources and unnecessarily expensive. In contrast, diseases associated with *Mycoplasma pneumoniae* typically occur throughout the year, without marked seasonal variability. Since there is a reduced incidence of viral infections and secondary bacterial pneumonias during the summer months, *M. pneumoniae* may cause up to 50% of all pneumonias in the summer months. Communication among clinical microbiologists, physicians interested in infectious diseases, and state health department personnel to share information on community trends in the pathogens causing respiratory tract infections helps focus diagnostic and therapeutic efforts. Periodic review of publications such as the *Morbidity and Mortality Weekly Report* from the Centers for Disease Control and Prevention is another means by which clinical microbiologists can review patterns of respiratory tract infections in the community and national setting.

Empiric Antimicrobial Therapy

To properly position the diagnostic microbiology laboratory in the scheme of the patient's care plan, it is important to understand the role of empiric antimicrobial therapy in the care of patients with respiratory tract infections. Although basing antimicrobial therapy on the results of diagnostic microbiologic studies is desirable, certain circumstances dictate that therapy be initiated prior to obtaining these results or without submitting specimens for culture even though an infectious cause is suspected. For example, antimicrobial therapy should be initiated before obtaining microbial identification and susceptibility testing results in patients who are seriously ill with pneumonia. In other cases, cultures should be obtained from the primary site of involvement, if possible (as well as from the blood if bacteremia is suspected), before initiating empirical antimicrobial therapy. In some types of respiratory tract infections, however, it is standard to initiate antimicrobial therapy without obtaining any specimens for culture. This is the procedure in the case of a child who has signs of a middle ear infection (otitis media). It can be predicted that the pathogen will usually be *Streptococcus pneumoniae, Haemophilus influenzae,* or *Moraxella catarrhalis.* Considering the difficulty in obtaining cultures directly from the middle ear and the knowledge of the likely pathogens, treatment without obtaining cultures is reasonable. If treatment fails, it may be necessary to do an invasive procedure to obtain specimens to test for resistant or unexpected pathogens. This same rationale applies to other respiratory tract sites that are difficult to culture directly, such as the sinuses.

When empirical antimicrobial therapy is necessary, it is important to have a working knowledge of the organisms most likely to cause the type of infection observed and of the antibiotics that are most likely to be effective. If the infection is hospital-acquired (so-called nosocomial infection), it is important to know whether the antimicrobial susceptibility patterns of the infectious agent in question as determined in that institution's microbiology laboratory differ from those reported in general. Annual reviews of bacterial antibiotic susceptibility patterns are published by clinical microbiology services to facilitate this type of decision-making process. The empiric use of antibiotics and the adjustment of therapy based on the results of subsequent microbiologic data represent another important interaction between the clinician and the clinical microbiologist.

ANATOMICAL CHARACTERIZATION OF THE RESPIRATORY TRACT

Anatomy of the Respiratory Tract

The function of the respiratory tract is not only to perform respiration (i.e., the exchange of oxygen and carbon dioxide) but also to deliver air from the outside of the body to the alveoli where the gas exchange occurs. When considering the anatomy of the respiratory tract (Fig. 26–1), one must consider the entire course that the air must travel: from the mouth and nose past the sinuses, into the pharynx, past the epiglottis, through the larynx, into the trachea and bronchi, and eventually into the alveoli. In addition to a role in air transport, each of these areas also plays an important role in defending the respiratory tract from infection.

Barriers to Infection

The respiratory tract has many natural barriers to infection that inhaled particles must penetrate in order to cause disease. These mechanisms that normally maintain a sterile environment below the larynx include nasal hair; mucociliary cells that line mucosal surfaces; coughing; normal flora; and phagocytic inflammatory cells.

In the nasopharynx and oropharynx, turbulent air flow causes large particles to impact on mucosal surfaces. Nasal hairs filter air as it passes through the nasal passages. Humidification of the air causes hygroscopic organisms to increase in size, making it more difficult for them to pass to the lower respiratory tract and easier for them to be phagocytosed. The normal flora of the nasopharynx and oropharynx helps protect the host by preventing colonization by pathogenic organisms, and the mucociliary blanket of the sinuses, middle ear, and tracheobronchial tree clears particulate matter and secretes immunoglobulin and other antimicrobial substances. In addition, coughing can aid in the expulsion of particulate matter. If the particles reach the alveoli,

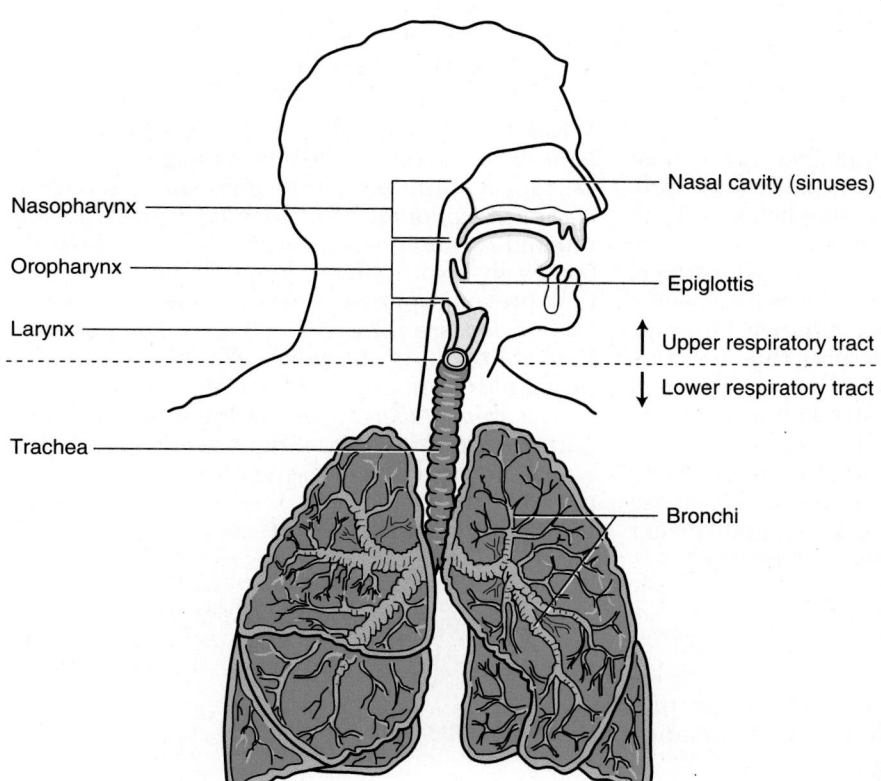

Nasopharynx

Oropharynx

Larynx

Trachea

Nasal cavity (sinuses)

Epiglottis

↑ Upper respiratory tract

↓ Lower respiratory tract

Bronchi

Figure 26–1

Anatomy of the respiratory tract.

macrophages ingest organisms; polymorphonuclear leukocytes and monocytes are called in once the lung becomes inflamed.

Alterations in these barriers may lead to infection. For example, cigarette smoking impairs the ability of the mucociliary blanket to clear particulate matter and interferes with macrophage activity. Structural abnormalities of the bronchial tree such as bronchiectasis or extrinsic compression of the bronchus by a malignancy can alter clearance of mucus that may contain infectious agents.

VIRULENCE FACTORS OF PATHOGENIC ORGANISMS

The disease-producing capability of an organism is the clinical manifestation of its virulence. Microorganisms infect the host by entering the host, interacting with specific target tissues, evading the host's defenses, proliferating, damaging the host, and disseminating or elaborating products that cause systemic disease. Virulence factors involved in disease-producing mechanisms, such as adherence, toxin elaboration, and host evasion, enable the microorganism to complete this process.

Adherence

Attachment of the microorganism to the host tissue is a primary step in the pathogenic process of infection. This is accomplished by adhesins, microbial surface molecules or organelles that bind the organism to a host surface. Specific bacterial adhesins interact with specific cellular receptors. Streptococci possess fimbriae, which are fine, irregular structures that are composed of M protein and lipoteichoic acid and bind to epithelial cells. Group A streptococci release their surface lipoteichoic acid in the presence of sublethal doses of penicillin and lose their ability to bind to epithelial cells. Enterobacteriaceae such as *Escherichia coli* use fimbriae, nonflagellar filamentous structures, to adhere to host cells.

Toxin Elaboration

Microorganisms may elaborate toxins that produce different pathogenic effects, depending on the activity of the toxin and the target cell it interacts with in the host. For example, *Corynebacterium diphtheriae* produces an ADP-ribosylating toxin that interferes with protein synthesis. Locally, the toxin induces necrosis, resulting in a "pseudomembrane" composed of necrotic respiratory epithelial cells, leukocytes, and organisms. Systemically, the toxin preferentially adheres to myocardial, nerve, and kidney tissue, causing myocarditis, neuritis, and renal tubular necrosis. *Pseudomonas* exotoxin A's mechanism of action is similar to that of diphtheria toxin, but it has a different pathogenic effect because of the different target tissue and the involvement of other virulence factors. Another example of the role of toxins is provided by those produced by *Bordetella pertussis*. One is an adenylate cyclase toxin that enters target cells and increases intracellular cAMP levels, causing cell damage or cell death. Another is the pertussis toxin, which interrupts the transduction of signals from cell surface receptors to intracellular systems. During the paroxysmal phase of the illness when patients develop the characteristic whooping cough, the clinical signs and symptoms are attributed to toxin elaborated by the organism.

Evasion of Host Defenses

Evasion of host defenses enables microorganisms to proliferate and cause damage to the host. Certain respiratory pathogens, such as *Streptococcus pneumoniae, Haemophilus influenzae,* and mucoid *Pseudomonas aeruginosa,* evade host defenses by expressing a polysaccharide capsule that prevents phagocytosis by host leukocytes. *Chlamydia* are obligate intracellular parasites that are taken up by host cells, where they are protected from the host immune system. *Mycobacterium tuberculosis,* another intracellular pathogen, survives by inhibiting phagosome-lysosome fusion. Other respiratory pathogens are able to cleave host secretory antibody by producing IgA-specific proteases.

UPPER RESPIRATORY TRACT INFECTIONS

Pharyngitis

CASE STUDY

An 8-year-old girl was brought to an emergency room by her mother because the child was complaining of a sore throat and a low-grade fever. The mother stated that the girl had had a runny nose and cough for the last few days. On examination, the patient had a temperature of 99°F. Her pharynx was red and her tonsils were slightly swollen, but there were no exudates. There were no tender lymph nodes on neck examination.

Diagnosis. Most cases of pharyngitis, like this one, are viral in origin and occur as part of the symptom complex of a common cold. The presence of exudates in the pharynx, fever, and painful adenopathy, or the lack of a cough, would be suggestive of pharyngitis caused by *Streptococcus pyogenes*. This can be diagnosed by throat culture or rapid antigen tests. Other bacterial causes that must be considered are *Corynebacterium diphtheriae* (though rare) and other streptococci. Pharyngitis may also occur with influenza and infectious mononucleosis.

Etiology

Table 26–3 summarizes the clinical syndromes encountered in the upper respiratory tract and the associated etiologic agents. The most commonly encountered etiologic agent of bacterial pharyngitis is group A streptococci. Identified in 30 to 40% of microbial isolates in school-aged children who present with pharyngitis, the isolation of group A streptococci is much lower (less than 10% of microbial isolates) in adults with a similar presentation. Most cases of pharyngitis that are not associated with streptococcal infection are presumed to be of viral etiology. Pharyngitis in these cases usually presents as part of the clinical picture of a common cold or an early case of influenza. In certain cases of presumed viral pharyngitis, a specific pathogen can be isolated. However, there are many cases of pharyngitis without an identifiable pathogen. Whether, as is often concluded, these cases represent viral pharyngitis in which the infectious agents cannot be identified remains to be determined.

Cases of pharyngitis caused by unusual pathogens such as *Neisseria gonorrhoeae, Corynebacterium diphtheriae,* or other bacterial pathogens are suspected in patients with suggestive histories or in cases of pharyngitis refractory to conventional therapy, due to the relative infrequency

■■■■■ T A B L E   2 6 – 3
UPPER RESPIRATORY TRACT INFECTIONS

Clinical Syndrome	Causative Agents	Specimen Collection	Other
Pharyngitis	Group A *Streptococcus* (children), viral (adults)	Swab tonsils and posterior pharynx, and place in transport media or allow to dry; viral cultures not necessary	Culture for *N. gonorrhoeae* or *C. diphtheria* if clinically indicated

Sinusitis	Most common:
	  S. pneumoniae
	  H. influenzae
	Less common:
	  S. pyogenes
	  M. catarrhalis
	  S. aureus
		Direct sinus sampling or nasal culture of sinus ostium	Direct sampling more reliable, indicated for patients who fail empiric therapy or who are severely ill or immunocompromised, or if unusual pathogen is suspected

Otitis media	Most common:
	  S. pneumoniae
	  H. influenzae
	Less common:
	  S. pyogenes
	  M. catarrhalis
	  S. aureus
		Direct culture by tympanocentesis	Direct culture indicated for patients who are severely ill or immunocompromised, or if unusual pathogen is suspected

Epiglottitis	*H. influenzae* type B	Direct swab of epiglottis, blood cultures	Direct swab should be performed only if airway is secure

Pertussis	Most common:
	  B. pertussis
	  B. parapertussis
	Less common:
	  B. bronchiseptica
	  Adenovirus
		Nasopharyngeal swab:
		  a. plate directly onto Bordet-Gengou medium
		  b. direct fluorescent antibody stain
		  c. adenovirus shell vial culture for antigen detection
			—

of these types of infection compared to viral and streptococcal pharyngitis.

Epidemiology

Most pharyngeal infections associated with viral and streptococcal infections occur during the winter and early spring. The increase in person-to-person contact during these seasons favors transmission of the causative pathogens. The pathogens are usually inoculated by contamination of the hands and then gain entrance into the upper respiratory tract of the newly infected patient.

Clinical Manifestations

Clinically, differentiation between viral and streptococcal pharyngitis is difficult. With most cases of viral pharyngitis, other symptoms of a common cold, including rhinorrhea, are present, which is uncommon with streptococcal pharyngitis. Pharyngitis associated with influenza virus infection is accompanied by more numerous and more severe systemic symptoms, including fever, myalgias, and more profound fatigue. In contrast, other forms of viral pharyngitis that are associated with infectious mononucleosis are more commonly accompanied by cervical and generalized lymphadenopathy and enlargement of the spleen.

Streptococcal pharyngitis is associated with marked pharyngeal pain and difficulty in swallowing. Fever is more commonly a major component of the illness with streptococcal than with viral infection. In typical cases, a thick exudate covers the tonsils and posterior pharynx, which is uncommon in viral pharyngitis. It is important to realize, however, that documented streptococcal pharyngitis may also present in a manner that is indistinguishable from that of viral pharyngitis.

Another form of pharyngitis that is uncommon in most clinical practices today is that caused by *Corynebacterium diphtheriae,* the primary clinical presentation of diphtheria. This bacterial pharyngitis is classically associated with the presence of a tightly adherent pharyngeal membrane. Diphtheria is now uncommon in the United States, and it should be suspected only in patient populations in which children have not received DPT (diphtheria/pertussis/tetanus) vaccination. The importance of a good vaccination history in children with pharyngitis is obvious.

Pathogenesis

The reasons for the symptom complex in patients with viral or streptococcal pharyngitis are incompletely understood. Some viruses (e.g., adenoviruses) that infect the pharyngeal mucosa cause cellular destruction (cytopathology) and elicit inflammatory cell responses. The combination of these events is probably responsible for the pharyngeal pain and swelling experienced by patients with pharyngitis. However, in other cases, the viral pathogens (e.g., rhinoviruses) cause symptoms of pharyngitis with little mucosal cell destruction. In some cases, these noncytophatic viruses have been shown to elicit production of inflammatory mediators that can reproduce the symptoms of pharyngitis.

The pathogenesis of streptococcal pharyngitis may be predominantly due to the inflammatory effects of a variety of extracellular products elaborated by the streptococci. These products exhibit a variety of activities, including toxicity to a variety of cells, pyrogenicity, and enhancement of the spread of streptococci through infected tissues.

Complications

Other than the infrequent problem of upper airway obstruction associated with severe soft-tissue swelling, there are few complications that can be attributed directly to either bacterial or viral pharyngitis. The occasional complications of viral infections associated with pharyngitis are usually due to systemic manifestations of the infections or to secondary bacterial infections such as sinusitis, otitis media, or pneumonia. Displacement of either or both of the tonsils or asymmetrical swelling of the soft tissues of the pharynx following pharyngitis should raise suspicion of peritonsillar or pharyngeal abscess. Although group A streptococci have often been associated with these soft-tissue infections, oral anaerobic bacteria should also be considered in the differential diagnosis.

The treatment of streptococcal pharyngitis includes amelioration of the symptoms, limitation of transmission of the infection to contacts (especially in school-aged children), and prevention of the serious complications of acute rheumatic fever and acute glomerulonephritis.

Laboratory Diagnosis

Specimen Collection. The primary goal of obtaining cultures in most cases of acute pharyngitis is to differentiate streptococcal pharyngitis from more common cases of viral pharyngitis. A secondary goal is to be able to detect the uncommon cases of bacterial pharyngitis associated with other pathogens (e.g., diphtheria, gonorrhea) in cases in which the clinical history is suggestive or symptoms are persistent.

In collecting pharyngeal specimens for streptococcal cultures (Fig. 26–2), it is important, using adequate lighting, to vigorously swab the tonsillar areas and the posterior pharynx. It is not uncommon for negative streptococcal cultures to be obtained from specimens collected by personnel who did not swab these areas well due to the discomfort of patients with severe pharyngitis; second cultures by more experienced personnel have been positive. The tongue and other oral structures should be avoided with the swab to minimize contamination with oral flora and dilution of the specimen. If any tonsillar exudate is seen, specific efforts should be made to directly swab the areas where it is present.

After collecting the specimen, swabs may be placed in transport medium.

Direct Microscopic Examination. Direct microscopic examination of pharyngeal secretions is not useful for clinical or laboratory diagnosis. Because respiratory exudates contain a wide variety of organisms, a direct Gram stain will not differentiate suspected pathogens from normal microbial flora and does not produce meaningful results.

Culture. Culture of the pharyngeal area will isolate bacterial pathogens such as β-hemolytic group A streptococci. A nonselective medium such as sheep's blood agar (SBA) is commonly used. The addition of a selective medium with an antimicrobial, such as sulfamethoxazole trimethoprim, that is specific for the isolation of β-hemolytic streptococci may enhance the isolation of this etiologic agent.

Other Methods. Latex agglutination, coagglutination tests, and enzyme immunoassays have been introduced into the clinical practice to detect group A streptococci directly from throat swabs. Although these tests are highly specific, reports suggest that their sensitivity is not satisfactory. Therefore, negative results should be followed with a conventional culture.

Sinusitis

CASE STUDY

A 40-year-old woman went to the doctor's office complaining of fever and nasal drainage. She stated that she had developed cold symptoms approximately 1 week earlier and had been treating herself with over-the-counter medicines with little improvement. In fact, she had gotten worse over the last 48 hours, with increasing headache.

The physical examination was notable for low-grade fever and tenderness over the left maxillary sinus as well as purulent drainage in the left side of the nose. Sinus radiographs showed an air-fluid level in the left maxillary sinus.

Diagnosis. Cultures are not usually helpful in this clinical situation, unless they are obtained by direct sinus puncture. Cultures of nasal drainage are contaminated with normal nasal flora. Because most cases of acute sinusitis are caused by *Streptococcus pneumoniae* or *Haemophilus influenzae,* empirical antibiotic therapy can be used to treat these patients. If the sinusitis is severe, occurs in an immunocompromised host, or is a nosocomial infection, sinus puncture should be performed to attempt to define the etiologic agent by Gram stain and culture.

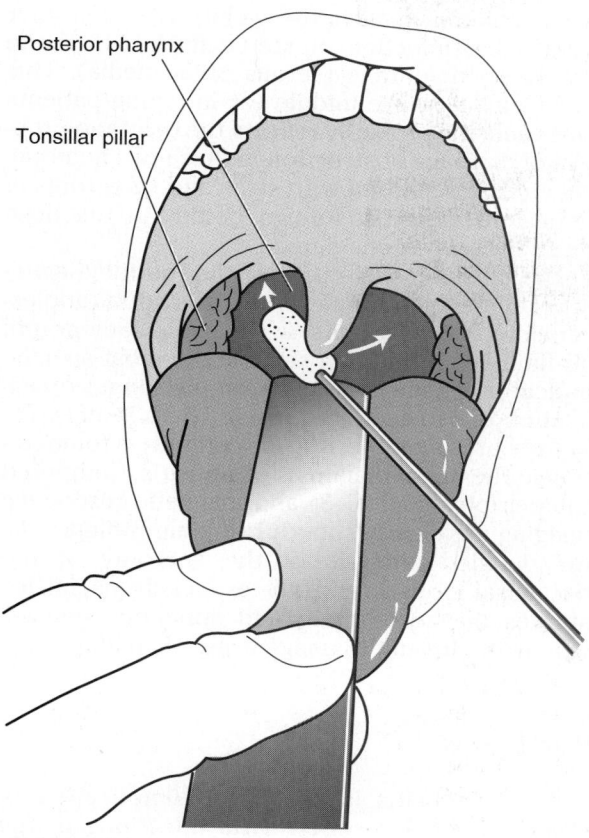
Posterior pharynx
Tonsillar pillar

Figure 26–2
Specimen collection from the throat.

Etiology

Streptococcus pneumoniae and *Haemophilus influenzae* are the pathogens identified in approximately half the cases of sinusitis in which a culture is performed. Diagnosis by culture is done with direct sinus puncture to avoid contamination by the nasopharyngeal flora. In community-acquired infections, *Streptococcus pyogenes, Moraxella catarrhalis,* and *Staphylococcus aureus* account for most of the other pathogens commonly associated with sinusitis. In hospital-acquired infections, *Staphylococcus aureus* and gram-negative bacilli are recovered more frequently. The incidence of *Moraxella catarrhalis* as a pathogen in acute sinusitis is much higher in children than in adults, approaching the incidence of *Haemophilus influenzae.*

The role of viruses in the direct cause of sinusitis is not as clear as with the major bacterial pathogens. Pathogenic viruses that are isolated alone in the setting of acute sinusitis are the most compelling evidence for their role in the disease. Although common respiratory viruses can be recovered from infected sinuses, they are often isolated along with potential bacterial pathogens. This observation raises the possibility that the viral infection preceded a secondary bacterial infection. With the increasing availability of more sensitive and rapid means of viral diagnosis, further evaluation of the association between viruses and acute sinusitis may be possible.

Acute fungal sinusitis is uncommon in normal patients; it occurs predominantly in immunosuppressed hosts treated with cytotoxic or immunosuppressive drugs or in patients with severe underlying illnesses such as uncontrolled diabetes mellitus. In contrast, chronic fungal sinusitis usually occurs in immunocompetent hosts in the clinical setting of other forms of chronic sinusitis (e.g., allergic sinusitis).

Epidemiology

The incidence of acute sinusitis follows that of other upper respiratory tract infections occurring predominantly in the winter and spring months. The clinical diagnosis of acute sinusitis is less common in children than in adults, in part due to the incomplete development of most of the paranasal sinuses until adolescence.

Clinical Manifestations

The presentation of acute sinusitis in older children and adults is often that of a prolonged respiratory tract infection that involves a set of symptoms that are subtly different from those present early in the course of the illness. The most constant features include purulent nasal discharge and pain in the face. When the maxillary sinus is involved, facial pain is worse when leaning forward or when a jarring force is experienced, such as walking down stairs. Patients with maxillary sinusitis may also experience headache and pain that is perceived to come from the upper teeth. Fever is present in only about half the patients with acute sinusitis. In hospitalized patients, sinusitis can result from obstruction of the sinus openings (ostia) by indwelling nasal tubes.

In young children, the only symptoms observed may be persistent rhinorrhea and cough either directly after a viral upper respiratory tract infection or following a brief period of clinical improvement after such an infection. Ear examination is frequently abnormal in these children, with either middle ear infection or sterile fluid behind the tympanic membrane (serous otitis media). This involvement of the middle ear in young patients with sinusitis probably reflects the common problem of drainage obstruction in these two anatomical areas. In children with signs and symptoms of persistent sinusitis, foreign bodies in the nose must always be considered.

The most sensitive test for the routine diagnosis of acute and chronic sinusitis in adult (adolescent or older) patients is the sinus radiograph; air-fluid level seen in a sinus is the most specific indicator of acute sinusitis. Complete sinus opacification is also seen frequently. There is an evolving debate over the role of computed tomography in the initial diagnosis of sinusitis. Computed tomography (Fig. 26–3) and magnetic resonance imaging are clearly superior to plain radiographs for detailed analysis of the anatomy of the paranasal sinuses and are especially useful for studies of the ethmoid and sphenoid sinuses, which are difficult to image using plain films.

Pathogenesis

Acute sinusitis is usually a complication of common colds or other viral infections of the

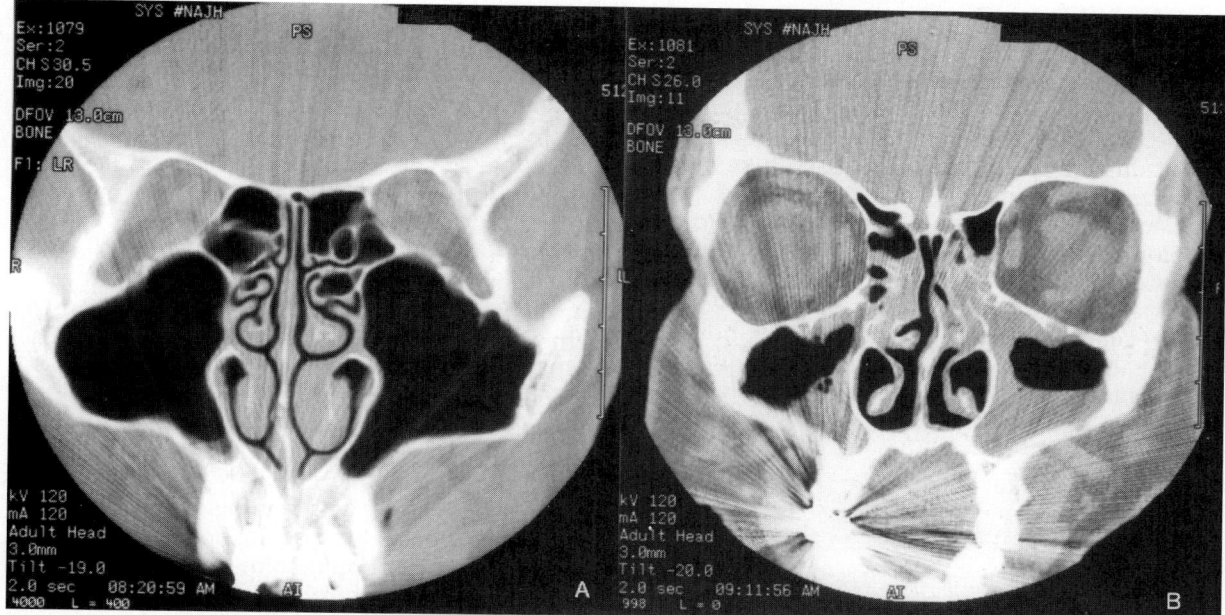

■ Figure 26 – 3

Computed tomography of the paranasal sinuses in a patient with normal maxillary sinuses *(A)* and in a patient with bilateral maxillary sinusitis *(B)*. In *A*, the large (black) cavities on either side of the lobulated midline structures of the nasopharynx are the normal maxillary sinuses in this patient who does not have sinusitis. Of note is the absence of thickening of the mucosal lining of the maxillary sinuses and the absence of opacity within the sinus. In *B*, a tomogram from another patient was taken at a slightly different orientation through the skull. This difference in orientation results in sampling of a smaller cross-section of the maxillary sinuses. The significant difference in the maxillary sinuses between the study in *B* and that in *A* is the presence of a large amount of opacity in both maxillary sinuses. This bilateral partial opacification represents both mucosal thickening and pus within the maxillary sinuses. The maxillary sinus on the right side of the figure also demonstrates an air-fluid level (the relatively straight, horizontal line between the air-filled black space above and the gray, fluid-filled space below) that is characteristic of acute, purulent sinusitis.

upper respiratory tract. Respiratory allergies also predispose individuals to acute sinusitis. When recurrent infectious sinusitis occurs, associated with recurrent pulmonary infection over prolonged periods, it should suggest the possibility of immune dysfunction such as common variable hypogammaglobulinemia.

Sinusitis is a typical example wherein obstruction is a predisposing factor in the development of an infectious disease of the respiratory tract. The sinuses normally undergo a continuous cleansing process through the action of ciliated epithelial cells. These ciliated epithelial cells move the mucous layer lining these areas toward the sinus ostia. This process normally clears the sinuses of bacteria from the adjacent nasopharynx. During acute rhinosinusitis, whether of viral or allergic origin, mucosal swelling may cause partial or complete obstruction of these sinus ostia, interrupting the flow of secretions and predisposing to bacterial overgrowth behind the

obstruction. The function of the normal ciliated epithelial cells (altered by viral infection and bacterial toxins) and the viscosity of the mucous layer (which may be increased by the exudation and inflammation associated with infection or allergic reactions) may also be altered in such a way that the clearance of sinus secretions is less efficient.

Other processes that may limit normal drainage from the sinuses include foreign bodies, tumors, and congenital structural abnormalities of the nasopharynx. Nasotracheal tubes in hospitalized patients in respiratory failure and packing materials used during surgical procedures provide other forms of obstruction to normal sinus drainage that predispose to infection.

Falling oxygen concentration has also been implicated in the pathogenesis of infectious sinusitis. With obstruction of the sinus ostium, the partial pressure of oxygen in the sinus cavity falls, impairing the function of inflammatory cells and facilitating the growth of facultative and true anaerobic bacteria.

After repeated bouts of acute sinusitis, the ciliated epithelium may be replaced by squamous epithelium, compromising the normal movement of the sinus mucous layer. This usually results in colonization of the chronically inflamed sinus with aerobic and anaerobic bacteria that are not usually present in the healthy sinus. The importance of this colonization in chronic sinusitis remains uncertain. It is generally believed that the presence of this colonization does not warrant treatment and that acute flare-ups of chronic sinusitis should be managed essentially the same as acute sinusitis in patients without chronic disease.

Complications

Complications of acute sinusitis can generally be regarded as the result of extension of the infection to adjacent areas or structures. These complications include orbital cellulitis, osteomyelitis, meningitis, brain abscess, and cavernous sinus thrombosis.

Orbital cellulitis or retroorbital abscess can result from direct extension of the infection from adjacent sinuses to the area around and behind the eye. Protrusion of the eye (proptosis) or limitation of ocular movements should suggest the possibility of extension of the infection to the retroorbital space. This is a serious complication that requires aggressive diagnosis and management.

Extension of frontal sinusitis can cause cellulitis in the area of the forehead overlying these sinuses. Osteomyelitis of the frontal bone, referred to as Pott's puffy tumor, may also develop. Further extension of frontal sinusitis may cause an abscess of the frontal lobes of the brain, a serious complication that is difficult to diagnose. Extension of infection from other sinuses (e.g., ethmoid or sphenoid) to the central nervous system in the form of meningitis, brain abscess, or cavernous sinus thrombosis is difficult to diagnose due to the deep location of these sinuses. Suspicion of such severe complications must remain high in patients who appear to have refractory sinusitis in the setting of altered mental status.

In cases in which the adjacent bone is involved (osteomyelitis), active sinusitis can complicate the interpretation of diagnostic radiologic studies, making diagnosis difficult. It is usually necessary to follow serial studies of bone inflammation and integrity during therapy of sinusitis to establish the diagnosis of this complication.

Laboratory Diagnosis

Specimen Collection. Cultures of nasal secretions or of nasal swabs are usually thought to be unreliable indicators of the pathogen causing acute infection within the sinus. Most clinicians caring for adult patients with acute sinusitis believe that the only cultures useful in the etiologic diagnosis of acute sinusitis are those obtained by direct sinus sampling. This is another example in diagnostic microbiology of normal flora possibly interfering with etiologic diagnosis.

In acute sinusitis in children, however, there is a subtle difference of opinion in the pediatric infectious diseases literature about the value of doing nasal cultures instead of direct sinus punctures for the bacteriologic diagnosis. Since studies have shown that *Streptococcus pneumoniae, Haemophilus influenzae,* and *Streptococcus pyogenes* (three of the most common etiologic agents in acute sinusitis in children) are only rarely isolated from nasal cultures from normal children, identification of these agents in cultures of the nose should suggest the etiology of paranasal sinusitis in pediatric patients. It has therefore been suggested that properly collected nasal cultures taken from the area of the sinus ostium are useful in establishing the cause of acute sinusitis. There is general agreement, however, that randomly collected cultures from the nasopharynx of both adult and pediatric patients are of no value in defining the bacterial cause of this disease.

Another issue to be considered is when direct cultures of the sinuses are indicated. In most cases of acute sinusitis in a normal host, studies have shown that the infection is caused by only a few pathogens; sinus puncture is not indicated before initiation of treatment. However, in patients who fail empiric antimicrobial therapy, direct sinus cultures are indicated to look for antibiotic-resistant strains of bacteria or unusual pathogens. Direct culture by sinus puncture is also indicated early in the course of disease in severely ill or immunocompromised patients or in cases in which an unusual pathogen is suspected, such as a hospital-acquired infection.

Direct Microscopic Examination. Gram-stain preparations are useful for only sinus specimens obtained by direct aspiration from the sinus ostia or from sinus puncture. In these cases, finding a predominant bacterial type may be of diagnostic and therapeutic value.

Culture. As mentioned previously, culture of sinus aspirations is useful only when performed on specimens obtained by sinus puncture or by direct aspiration of purulent secretions from sinus ostia. In these cases, samples are inoculated on culture media such as sheep's blood agar, chocolate agar, and MacConkey agar, which are routinely used to isolate respiratory tract pathogens.

Otitis Media

Etiology

As in the case of acute sinusitis, the microbial etiology of the average case of acute otitis media has been clearly defined by cultures of infected middle ear specimens obtained by direct aspiration from the area. Considering the similarities in their pathogenesis, it is not surprising that the same group of pathogens is involved in both acute otitis media and acute sinusitis. *Streptococcus pneumoniae* and *Haemophilus influenzae* account for over 50% of the isolates from cases of acute otitis media. *Streptococcus pyogenes, Moraxella catarrhalis,* and *Staphylococcus aureus* have also been implicated in this disease. In patients who have received multiple courses of broad-spectrum antibiotics for acute otitis media, those who have tympanic membrane perforations, or those with severe underlying disease associated with pharyngeal colonization with gram-negative bacilli, acute otitis media is more likely to be caused by gram-negative pathogens.

The analogy with the etiology of acute sinusitis also extends to the role of viruses in acute otitis media. Since most cases of acute otitis media occur on the heels of a viral infection of the upper respiratory tract, it is not surprising that viruses can be isolated from cultures taken from the middle ears of these patients. The exact role of these viruses in the pathogenesis of acute otitis media, other than inducing inflammation and thereby reducing eustachian tube drainage of the middle ear, remains to be determined. The availability of more sensitive and rapid viral diagnostic studies may allow a clearer definition of the frequency with which viruses can be identi-

fied as primary pathogens in cases of acute otitis media. Such data will be necessary before predominant or cofactor involvement of concomitantly isolated bacteria can be clearly dissected.

Other possible causes of acute otitis media should be considered in unusual cases or circumstances. Theoretically, other pathogens that infect the upper and lower respiratory tract, such as *Mycoplasma pneumoniae* and *Chlamydia trachomatis,* can cause acute otitis media. However, the association between these potential pathogens and the average case of middle ear infection is not clear. In newborns, group B streptococci and gram-negative bacilli can be added to *Streptococcus pneumoniae* and *Haemophilus influenzae* as causes of occasional cases of acute otitis media. Acute otitis media occurs less often than acute sinusitis in the presence of nasal tubes in hospitalized patients with eustachian tube obstruction. Gram-negative bacillary pathogens are isolated more often in this setting than in community-acquired acute otitis media.

Epidemiology

Otitis media is the most common localized infection of the upper respiratory tract in preschool-aged patients. One study showed that almost a third of all visits to pediatricians involved diseases of the middle ear. The increased incidence of these infections in preschoolers has been attributed to the crowding of susceptible hosts in day-care centers. In older children and adults, the syndrome of middle ear infection, although less common, is still a major precipitant of outpatient office visits during the winter and spring seasons, when viral respiratory infections are common.

Clinical Manifestations

Early signs and symptoms of acute otitis media may be nonlocalized, especially in young children. In these cases, fever and irritability may be the only signs of illness. In older children, tugging at the involved ear may be noticed during or at the end of the course of an upper respiratory tract infection. Other symptoms include ear pain, changes in hearing, and, late in the course of the infection, drainage of purulent secretions from the ear canal, associated with perforation of the tympanic membrane. On direct examination of the tympanic membrane, the presence of acute otitis media may be indicated by a red, bulging membrane. However, the changes in the membrane are often more subtle. In these cases, it is important to perform pneumatic otoscopy to determine whether there is evidence of fluid or pus behind the tympanic membrane, as evidenced by its reduced or absent movement in response to changes in air pressure in the canal.

Pathogenesis

The eustachian tube, which is the canal that links the middle ear to the nasopharynx, is shorter and travels a more direct course from the nasopharynx to the middle ear in young children than in older children and adults. This anatomical immaturity appears to predispose young patients to easier contamination of the middle ear with nasopharyngeal bacteria. This factor, coupled with the greater incidence of viral respiratory tract infections in pre-school-aged children, may explain the greater incidence of acute otitis media in this population.

Both the eustachian tube and the middle ear are lined with ciliated epithelium and mucussecreting cells. This creates a situation similar to that seen in the paranasal sinuses, where a mucous layer clearance mechanism can defend against invasion and overgrowth of nasopharyngeal organisms. When this function is altered or impaired, infection may develop. Suppurative otitis media may result from relative or absolute obstruction of this drainage mechanism as a consequence of obstruction of the eustachian tube. Edema of the eustachian tube mucosa due to viral infections is much like the swelling that is induced around the ostia of the sinuses. Similarly, viral infections can cause impairment of the cleansing function of the ciliated epithelium of the eustachian tube–middle ear complex. In a manner similar to that associated with viral infection, the eustachian tube edema associated with allergic nasopharyngitis can cause obstruction and predispose to postobstructive infection of the middle ear. Infection behind the obstruction, irrespective of the nature of the precipitating cause, can cause tissue damage that may affect hearing and can, if therapy is inadequate, lead to progressive infection of adjacent structures.

Complications

Progression of acute otitis media may lead to damage of the tympanic membrane and other associated problems that follow this destructive process. Tympanic membrane damage with subsequent hearing loss can have devastating effects on the speech development and education of young children, but it may be of less consequence in adults. A membrane perforation may heal without long-term effects on hearing or may become chronic, depending on the size of the perforation and the frequency of reinfection. If the perforation destroys crucial areas in the membrane, surgical correction may be necessary to close the perforation and restore membrane function. The necessity of such a procedure depends on multiple factors, the most important of which is the age of the patient. Unfortunately, most such complications occur in children who do not obtain adequate early medical care. Similar considerations apply when considering the approach to chronic adhesive changes in the repeatedly infected, but intact, tympanic membrane and chronic middle ear effusions (serous otitis media).

Acute mastoiditis is an uncommon complication of acute otitis media in the antibiotic era. As a result, surgical treatment (mastoidectomy) to prevent progressive disease and extension to the central nervous system is a rare procedure in modern practice.

All the considerations discussed under acute sinusitis concerning the spread of infection to adjacent areas also apply to acute otitis media that goes untreated. Local extension of infection and spread to the central nervous system are potential problems. These problems are encountered less commonly in the antibiotic era in most clinical settings. However, in situations in which access to health care is limited for any reason, such severe late complications of middle ear infection must be considered.

Laboratory Diagnosis

Since the predominant pathogens for acute otitis media are known, obtaining specimens for culture before initiating therapy is unnecessary in the average case. If the patient is seriously ill or immunocompromised and likely to have rapidly progressive disease or if, for some reason, an unusual or drug-resistant organism is suspected, direct culture by tympanocentesis is indicated. As with acute sinusitis, there is probably no value in culturing the nasopharynx as an indicator of the type of pathogen that may be present in the middle ear.

Epiglottitis

CASE STUDY

A 4-year-old boy was brought to the pediatrician's office with a 6-hour history of fever and trouble swallowing. Inspiratory stridor was noted by the pediatrician. The patient was immediately taken to the emergency room, where the epiglottis was visualized and noted to be red and edematous.

Diagnosis. Visualization of the epiglottis is the most important step in the diagnosis of acute epiglottitis. However, this must be done in a setting where intubation or tracheostomy can be performed immediately because of the risk of airway obstruction during manipulation of the pharynx. Generally, an endotracheal tube is inserted at the time of diagnosis to secure the airway until the inflammation and infection subside with treatment. Blood cultures and cultures of the epiglottis should be obtained, and intravenous antimicrobial therapy should be initiated immediately. *Haemophilus influenza* type b is the most common cause of acute epiglottitis, although isolation of staphylococci, pneumococci, and other streptococci has been described in this setting.

Etiology

Haemophilus influenzae type b is uniquely associated with epiglottitis. This organism is isolated from pharyngeal cultures in most children with this syndrome and from one-fourth of adult cases. Positive blood cultures for *Haemophilus influenzae* are common in both children and adults with acute epiglottitis. Association of this syndrome with other bacteria has been reported but is rare. Viral pathogens have not clearly been implicated in this problem.

Epidemiology

Most cases of epiglottitis occur in pre-school-aged children. Although an uncommon occurrence in the average clinical practice, the fact that it is a potentially life-threatening infection that is almost uniquely associated with a single pathogen, *Haemophilus influenzae* type b, warrants familiarity with the syndrome.

Clinical Manifestations

The epiglottis is a structure positioned at the anterior aspect of the opening of the trachea. It normally protects the airway from aspiration of secretions and food during swallowing. Acute epiglottitis is a rapidly progressive infection of this structure and the adjacent soft tissues in the upper airway. The presence of severe symptoms of pharyngitis and pain on swallowing in the absence of signs of pharyngitis on physical examination should suggest the diagnosis of epiglottitis. The associated edema of the epiglottis can cause the sudden onset of upper airway obstruction and respiratory arrest.

Early manifestations of epiglottitis include fever associated with severe sore throat and difficulty swallowing. Patients often prefer to lean forward and drool rather than to swallow their secretions, due to the extreme pain experienced. The speed with which symptoms may proceed to involve respiratory difficulty and eventually complete airway obstruction is what makes this diagnosis one to remember. There is more likely to be a rapid downhill course in children than in adults with epiglottitis. However, caution should be used in the diagnostic approach to any patient suspected of having this disease. In small children, signs of upper airway obstruction should always prompt consideration of a foreign body lodged in the upper airway in the differential diagnosis of epiglottitis.

Although examining the epiglottis and culturing this area are useful in the differential diagnosis of this infection, the danger of upper airway manipulation for these purposes cannot be underestimated. Since manipulation of the area may precipitate total airway obstruction in patients with epiglottitis, such procedures should be performed only in a setting where the patency of the airway can be maintained if the patient deteriorates rapidly. Specifically, it is not advisable for the clinician to perform such procedures in the average outpatient clinic or clinical laboratory setting. These procedures, especially when done in children, should be performed only in an intensive care or operating room setting.

The syndrome of croup associated with viral respiratory tract infections is the main condition to differentiate from epiglottitis in children. The viral illness that precedes and accompanies croup syndrome is not observed in patients with epiglottitis, and the cough typical of croup syndrome is not common in epiglottitis. In contrast, patients with epiglottitis usually have a relatively short history of illness. Patients with croup typically have predominant inflammation and airway narrowing below the glottis rather than the intense inflammation involving the glottis, as is seen in epiglottitis.

As discussed in the section on the complications of pharyngitis, there are other soft-tissue infections of the upper respiratory tract that are associated with narrowing of the upper airway, such as retropharyngeal abscess. However, these processes usually do not exhibit the same symptom complex and fulminant course seen with *H. influenzae*–induced epiglottitis. In distinction from bacterial pharyngitis (e.g., streptococcal or, rarely, diphtheria pharyngitis), patients with epiglottitis usually do not have tonsillar or pharyngeal exudates.

Pathogenesis

The soft tissues of the epiglottis and the surrounding structures appear to be susceptible to significant accumulation of edema fluid during the inflammatory process incited by *H. influenzae* infection. Combined with the critical location of the epiglottis at the opening of the trachea, this creates the danger of complete airway obstruction as the disease progresses. The reason that *H. influenzae* is so commonly associated with this syndrome, whereas other common respiratory pathogens are not, is unknown. Similarly, it is unclear why inflammation in the area above and including the epiglottis should be prevalent in this condition.

Complications

The main complication is respiratory compromise associated with the sudden onset of airway obstruction in patients with epiglottis. Although bacteremia is common in these patients, development of infections at sites other than the epiglottis is uncommon. Some patients with epiglottitis have simultaneous pneumonia, as evidenced by abnormal chest radiographs. However, the clinical significance of the lower respiratory tract involvement is relatively minor compared to the potential problems with upper airway obstruction early in the course of this infection.

Laboratory Diagnosis

Specimen Collection. Direct swab cultures from the area of the epiglottis are useful in establishing the etiologic diagnosis, but they should be taken only when the airway is secure. Blood cultures should also be performed in all patients in whom epiglottitis is suspected. The incidence of positive blood cultures is high, and a positive culture confirms the diagnosis and justifies antibiotic therapy focused on *H. influenzae* type b infection.

Direct Microscopic Examination. A direct smear from exudates for microscopic examination is a rapid method for early diagnosis. Exudates taken from the epiglottal area may reveal numerous white blood cells and pleomorphic gram-negative bacilli characteristic of *H. influenzae* (Fig. 26–4).

Culture. *H. influenzae* are isolated from blood cultures and from exudates. An enriched medium such as chocolate agar is required to recover this organism. An environment with an enhanced concentration of carbon dioxide (5 to 10%) is also required for recovery. The organism is identified by determining the X and V factor requirements, porphyrin test, and hemolysis on Casman blood agar.

Pertussis

Etiology

Bordetella pertussis and *B. parapertussis* are most frequently associated with the pertussis syndrome. *Bordetella bronchiseptica* has also been associated with a similar clinical syndrome. Low-serotype adenoviruses (types 1, 2, 3, and 5) have been isolated from patients with an illness that may be indistinguishable from pertussis. The recent availability of rapid viral diagnostic tests makes it feasible to distinguish adenovirus infection based on culture data from *B. pertussis* during the acute illness. Formerly, this distinction between pertussis and viral illness required

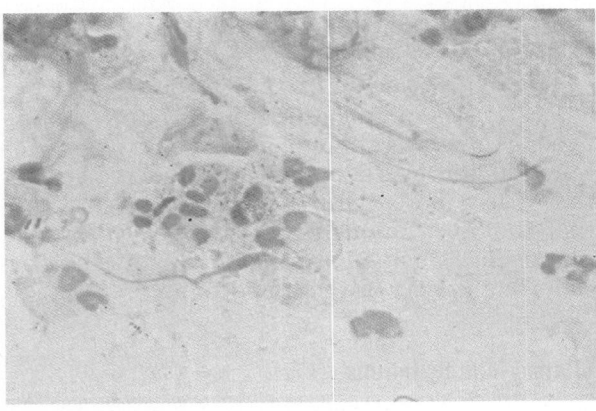

■■■■ F i g u r e 2 6 – 4
Gram-stained smear of sputum/exudate with *Haemophilus.*

serologic studies that were useful only in retrospective studies.

Epidemiology

Pertussis is a highly transmissible respiratory illness in susceptible patient populations that occurs with little seasonal variation. The infection occurs more commonly in infants and young children, and serious complications are seen more often in this age group. In recent years, the incidence of this illness has increased in the teenage population. Adolescents and adults with unusual manifestations of the illness may serve as reservoirs for transmission of the infection to susceptible children.

Clinical Manifestations

In the early phase of the illness, the symptom complex of pertussis is similar to that of a viral upper respiratory tract infection, and the differential diagnosis is difficult. Fever is uncommon throughout the course of the illness unless a secondary bacterial infection has occurred. Following the early phase of the illness, the patient experiences exhausting paroxysms of coughing, often with multiple coughs during one expiratory cycle, which are typically worse at night. The cycle classically ends with an episode of vomiting due to the extreme nature of the cough. The "whooping" sound associated with the forceful inspiration through a narrowed airway after a prolonged episode of coughing is the source of the term "whooping cough," which has been used in common reference to B. pertussis infection. The whooping characteristic of the inspiratory cycle is not uniformly present during pertussis infection, however. Therefore, it should not be used as a diagnostic criterion.

A major public health problem associated with the diagnosis of this infection in a child is limiting its spread within susceptible close contacts. Careful vaccination histories from contacts will allow the formulation of a plan for their protection by active immunization and antibiotic prophylaxis.

Pathogenesis

Although the pathogenesis of pertussis is incompletely understood, numerous studies suggest a significant role for pertussis toxin in the clinical presentation of this infection. Factors produced by these organisms have been implicated in several stages in the disease process, including damage to tracheal epithelial cells, impairment of host immunity, and induction of systemic symptoms of pertussis. It is apparent that pertussis is not a disease that is caused by the effects of a single toxin, as is the case in some other bacterial infections. Rather, it appears that the clinical syndrome may be a manifestation of the sum of several toxins produced by the pathogen and the host response that is elicited.

Complications

The most common complication of pertussis is the pneumonia that occurs in young children. These lower respiratory tract infections are commonly caused by secondary infections with other bacteria, although B. pertussis itself may cause pneumonia. Other secondary bacterial infections such as otitis media also occur.

Many of the complications associated with pertussis are a consequence of the severe and forceful coughing episodes that occur during this illness. Alveolar rupture may induce interstitial and subcutaneous emphysema. Forceful coughing has also been associated with subconjunctival (and other superficial) hemorrhages, epistaxis, rupture of the diaphragm, umbilical and inguinal hernias, and rectal prolapse.

Among the most serious complications of pertussis are those that affect the central nervous system. These problems are most dangerous in infants with pertussis. Seizures during pertussis have been related to fever, cerebral hypoxia, and toxic encephalopathy.

This infection has been implicated as a cause of bronchiectasis later in life. However, the decreasing incidence of pertussis and the increased use of antibiotics for respiratory tract infections in general probably make this infection an uncommon cause of bronchiectasis in the modern era.

Laboratory Diagnosis

Specimen Collection. Recovery of B. pertussis depends to a large extent on proper specimen collection and processing. Pernasal nasopharyngeal swabs using calcium alginate are preferred. Specimens should be plated directly onto selective media such as Bordet-Gengou or Regan-Lowe

(RL). If delay is expected, a transport medium such as RL transport medium should be used.

Direct Microscopic Examination. Complementary data may be obtained by performing direct fluorescent antibody (DFA) staining of secretions from such swabs. Although DFA has limitations regarding false-negative and false-positive results, it can provide a rapid indicator that may be useful in initiating empiric therapy.

Culture. Immunofluorescence studies should not be used in place of culture data. At the same time that nasopharyngeal cultures are being collected for *B. pertussis* culture, specimens should be submitted for adenovirus culture using a rapid diagnostic assay such as the shell vial culture/antigen detection method. Specimens for viral culture are either inoculated directly or placed in a viral carrier medium containing antibiotics to suppress bacterial contamination.

LOWER RESPIRATORY TRACT INFECTIONS

Infections in the lower respiratory tract usually occur when infecting organisms reach the pulmonary parenchyma by bypassing the mechanical and other nonspecific barriers of the upper respiratory tract. Infections may result from inhalation of infectious aerosols, aspiration of oral or gastric contents, or hematogenous spread.

There is a series of host defenses that must be overcome before a potential respiratory tract pathogen can establish infection in the lung parenchyma. The sequence of events is somewhat different for respiratory viruses than for bacterial pathogens. The progression of viral pathogens from the upper to the lower respiratory tract is a process that involves both spread among adjacent cells and distant inoculation of susceptible cells by aspiration of infectious secretions and, to a lesser extent, by bloodborne (hematogenous) transmission of the virus. Lung infections by bacterial pathogens usually occur via direct inoculation of organisms through aspiration from the upper respiratory tract. The ways in which mechanical host defenses are bypassed or suppressed, leading to lower respiratory tract infection, have been considered previously. Table 26–4 summarizes clinical syndromes encountered in the lower respiratory tract and the associated etiologic agents.

Bronchitis and Bronchiolitis

Etiology

Any of the respiratory viruses that cause upper respiratory tract infection can cause cough as a manifestation of acute bronchitis. The severity of the bronchial involvement varies with the different viral respiratory tract pathogens and, to some extent, with the patient population being studied. For example, during seasons when influenza is epidemic in the community, this viral respiratory pathogen is the most common cause of acute bronchitis and bronchiolitis in the general population. Respiratory syncytial virus (RSV) also causes community-wide outbreaks of bronchiolitis in infants. During nonepidemic periods, other respiratory viral pathogens such as rhinovirus, coronavirus, and parainfluenza virus are more likely to be isolated. In populations of young military recruits, adenovirus infections were the primary cause of acute bronchitis prior to the use of adenovirus vaccine. During the summer months, patients with enterovirus infection may exhibit acute bronchitis as part of the clinical syndrome, although this is not a major part of the illness in most cases.

Nonviral respiratory tract pathogens including *Mycoplasma pneumoniae, Chlamydia pneumoniae,* and *Bordetella pertussis* can also induce the syndrome of acute bronchitis. Therefore, the clinical presentation of these infections may be indistinguishable from cases of acute bronchitis caused by viral pathogens.

Epidemiology

Acute bronchitis can be viewed as the lower respiratory tract extension of many of the same viral infections that cause seasonal upper respiratory tract infections. The peak season for acute bronchitis is the winter months, which matches the peak incidence of these viral respiratory tract infections. It is somewhat artificial to separate the clinical syndromes of viral upper respiratory tract infections and acute bronchitis, since in many instances these represent a continuum of the same infection. The difference in the clinical presentations of these two conditions is often simply one of degree.

Clinical Manifestations

Patients with acute bronchitis often begin their illness with a syndrome typical of a nonbacterial

■■■■■ T A B L E 2 6 – 4
LOWER RESPIRATORY TRACT INFECTIONS

Clinical Syndrome	Causative Agents	Specimen Collection	Other

Bronchitis, bronchiolitis	Most common:
	  Respiratory viruses
	Less common:
	  M. pneumoniae
	  C. pneumoniae
	  B. pertussis
		Nasopharyngeal or lower respiratory culture if influenza type A or respiratory syncytial virus infection of the lower respiratory tract is suspected	Diagnostic cultures not indicated in uncomplicated cases

Community-acquired pneumonia	Children:
	  Most common:
	    Respiratory syncytial virus
	    Parainfluenza
	    Adenovirus
	    M. pneumoniae
	  Less common:
	    S. pneumoniae
	    H. influenzae
	    Group B *Streptococcus*
	      (neonates)
	Adults:
	  Most common:
	    S. pneumoniae
	    M. pneumoniae
	  Less common:
	    H. influenzae
	    Gram-negative bacilli
	    S. aureus
	    Legionella spp
		"Deep" expectorated sputum (see text)	Avoid contamination with oropharyngeal flora; specimen collection via fiber-optic bronchoscopy or open lung biopsy may be indicated in some circumstances

Nosocomial pneumonia	Gram-negative bacilli (60%) Gram-positive organisms (16%) Anaerobes *Legionella* spp	"Deep" expectorated sputum	Avoid contamination with oropharyngeal flora; specimen collection via fiber-optic bronchoscopy or open lung biopsy may be indicated in some circumstances
Aspiration pneumonia	Mixed anaerobes and aerobes (50%), anaerobes alone (50%)	Expectorated sputum is of little value; bronchoscopic techniques required for specific diagnosis	Pleural fluid cultures may be useful with anaerobic empyema
Chronic pneumonia	Mycobacteria, fungi	Early morning "deep" expectorated sputum; bronchoscopy or open lung biopsy may be required to identify pathogen	—

Empyema	Community-acquired:
	  S. aureus
	  S. pneumoniae
	  S. pyogenes
	Nosocomial:
	  Gram-negative bacilli
		Pleural fluid should be aspirated directly into a sterile syringe, with excess air removed from syringe immediately	Aliquots of specimen should be distributed to hematology and chemistry labs for other studies

upper respiratory tract infection. The course of the illness may exhibit a fairly rapid progression to lower respiratory tract involvement or, after several days of an upper respiratory tract syndrome, may evolve to symptoms of bronchitis. The typical patient with acute bronchitis has cough and fever as the primary manifestations of the illness. The amount of sputum produced with coughing varies with the individual, with the pathogen causing the illness, with the stage of the illness (the cough usu-

ally becomes more productive later in the course), and with the incidence of secondary bacterial infections that may follow nonbacterial acute bronchitis. Systemic symptoms such as diffuse myalgias and fatigue associated with viremic illnesses may also be seen early in the course of infection.

The distinction between acute and chronic bronchitis is one of both definition and pathogenesis. For purposes of definition, chronic bronchitis is evidenced by the presence of a productive cough on most days for at least 3 months of the year for a minimum of 2 years in succession. Whereas acute bronchitis is usually of infectious etiology, chronic bronchitis is usually caused by long-term cigarette smoking and occasionally other toxic exposures. From the perspective of infectious diseases of the respiratory tract, the main point to remember is that acute exacerbations of chronic bronchitis are usually initiated by the same pathogens that cause acute bronchitis in patients of the same age without underlying chronic respiratory tract disease. The difference is that patients with chronic bronchitis who experience an acute exacerbation of their illness due to intercurrent infection are more likely to have a severe primary illness or a prolonged illness associated with secondary bacterial infection than are otherwise healthy patients with acute bronchitis.

Acute bronchiolitis is a term reserved for an infectious disease of infants that is associated with a typical clinical picture. This syndrome is usually caused by respiratory syncytial virus infection, although other respiratory viruses (e.g., parainfluenza virus) can cause the same clinical picture in infants. In addition to signs of a febrile upper respiratory tract infection, these patients present with signs of lower respiratory tract airway obstruction such as wheezing, respiratory distress, and air trapping. Radiologic studies of the chest do not usually show typical signs of pneumonia. Increased lucency and areas of lung that are underaerated due to airway obstruction (atelectasis) are more common.

Pathogenesis

As mentioned, there is almost always evidence of antecedent or coexistent upper respiratory tract infection in patients with acute bronchitis. The spread of these upper respiratory tract infections to the lower airways, manifested as acute bronchitis, represents infection and damage of respiratory epithelial cells by the same (usually viral) pathogens. The extent of destruction of the respiratory epithelium varies with the pathogen causing the illness. Viruses such as influenza and adenovirus are highly cytopathic and cause significant epithelial cell destruction; other viral infections such as rhinovirus cause epithelial cell dysfunction without inducing much cytopathology.

The inflammatory response, necrotic debris from epithelial cell destruction, and edema of the lower respiratory tract also contribute to the airway abnormalities and symptoms in these infections. There is probably no clinical syndrome where this potential for airway obstruction is more important than in bronchiolitis in infants. The resulting obliteration of the lumen of small airways appears to be the primary pathogenetic mechanism in this infection. Indirect effects of viral infection on airway function are probably also important in the pathogenesis of some forms of acute bronchitis. The results of several studies suggest that alterations in airway β-adrenergic receptors and the production of inflammatory mediators during viral infections of the respiratory tract may be involved in many of the signs and symptoms associated with these infections.

Complications

A complication that is apparent in some patients with acute viral bronchitis is the development of secondary bacterial infections. These secondary infections can present in several ways. Some patients experience persistent or increasing symptoms of bronchitis, with an increase in the volume and purulence of the sputum produced. This usually occurs at a time in the course of the illness when most cases of acute bronchitis would be resolving. A further extension of this scenario is secondary bacterial infection presenting as acute or evolving pneumonia after an episode of acute bronchitis. These secondary bacterial infections of the lower respiratory tract are usually associated with the common pathogens that are found in community-acquired pneumonia (e.g., *Streptococcus pneumoniae, Haemophilus influenzae*), although other bacterial pathogens (e.g., *Moraxella catarrhalis,* gram-negative bacilli) may be secondary pathogens in selected patient populations (e.g., patients with underlying respiratory diseases or who are hospitalized).

The long-term consequences of acute bronchitis continue to be debated. Several epidemiologic studies suggest an association between acute viral infections of the lower respiratory tract and subsequent asthma. Whether the infectious agents that caused acute bronchitis in these patients actually induced a permanent change in their airways that resulted in future airway hyperreactivity, or whether these infections occurred in patients who would have eventually developed asthma irrespective of intercurrent episodes of acute bronchitis, remains an interesting puzzle. There has also been speculation that certain viruses that cause acute bronchitis can cause bronchiectasis. This is an abnormality in which inadequate drainage of respiratory secretions results from airway wall destruction, dilatation, and scarring that are associated with recurrent lower respiratory tract infections. Considering the cytopathic nature of some viral (e.g., adenovirus) infections that can cause acute bronchitis, the suggested association between these pathogens and long-term development of bronchiectasis seems reasonable.

Laboratory Diagnosis

The collection of specimens for culture in cases of acute bronchitis largely follows the procedures outlined in the section on viral pharyngitis. Since acute bronchitis is usually an extension of a syndrome of nonbacterial upper respiratory tract infection, diagnostic cultures are not indicated in uncomplicated cases that follow the expected, self-limited course. If, however, patients develop signs of secondary bacterial bronchitis or pneumonia, culture data may be useful in designing therapy. When such cultures are used to guide the design of a treatment plan, it is important to attempt to obtain lower respiratory tract secretions that are minimally contaminated with oral flora.

In practice, collection of an adequate specimen is usually not difficult in patients with secondary bacterial bronchitis, since there is usually a marked increase in both the volume and the purulence of sputum. The issue of proper collection of lower respiratory tract secretions for the diagnosis of pneumonia is covered is the section on pneumonia.

Other situations in which cultures might be indicated in cases of acute bronchitis or acute bronchiolitis are those in which effective antiviral agents are available against the viral pathogens suspected. For example, in cases of suspected influenza type A or respiratory syncytial virus infection of the lower respiratory tract, the identification of these agents in respiratory tract secretions may be useful in guiding early antiviral therapy. As rapid viral diagnostic techniques become more widely available in diagnostic microbiology laboratories, and as more effective antiviral agents become available to treat these respiratory tract infections, such studies will become a more important part of the daily practice in diagnostic virology laboratories.

Pneumonia

The distinction between acute bronchitis and acute pneumonia may be a subtle one. Both of these conditions are lower respiratory tract infections. The differentiation between acute bronchitis and pneumonia depends on the degree and extent of involvement of the lower respiratory tract with the infectious process. By definition, patients who have bronchitis do not present the physical and chest radiograph findings of pulmonary parenchyma involvement of the infectious and inflammatory process. Such detectable lung tissue involvement defines pneumonia.

Pneumonia can be subdivided into diagnostic categories based on the clinical setting, presentation of the illness, exposure to specific pathogens, and age and type of host infected. The importance of using such a strategy to make a presumptive determination of the infectious etiology of pneumonia is evident when one considers the long list of possible pathogens that can cause this type of infection. Table 26–5 lists the most common etiologic agents of lower respiratory tract infections and the usual affected patient populations. It is important to focus the clinical diagnosis on a subgroup of likely pathogens to allow the institution of reasonable empiric therapy while awaiting a specific etiologic diagnosis

■ T A B L E   2 6 – 5

MOST COMMON PATHOGENS OF LOWER RESPIRATORY INFECTIONS BY AGE

Age	Etiology
Neonates	*Chlamydia trachomatis*
Children	
Infants	Respiratory syncytial virus
	Influenza virus
5–18 months	*S. pneumoniae*
	H. influenzae
3 months–teens	Viruses
	S. aureus
	M. pneumoniae
Young adults (18–45 years)	*M. pneumoniae*
Older adults	*S. pneumoniae*
	Legionella
Institutionalized adults	Gram-negative rods
	S. pneumoniae
	S. aureus

and to make optimal use of the diagnostic microbiology laboratory in planning the diagnostic approach.

Acute Pneumonia

To begin this process of presumptive etiologic diagnosis in pneumonia, one must consider those pneumonias that develop in patients in their normal setting in the community (community-acquired pneumonia) and those that develop in hospitalized patients (nosocomial pneumonia). The highest incidence of both community-acquired and nosocomial pneumonias occurs in very young and very old patients. However, the types of etiologic agents that are most likely to cause pulmonary infections in these two groups are different. In infants and children, respiratory viruses cause the majority of pneumonias; in elderly adults, bacterial pathogens are more likely to be implicated.

COMMUNITY-ACQUIRED PNEUMONIA

CASE STUDY

A 52-year-old, previously healthy woman came to the emergency room complaining of right-sided chest pain with each breath, a cough that produced rust-colored sputum, and fever. She reported that her symptoms had begun abruptly the day before with the onset of shaking chills.

Examination revealed that the patient had a fever of 102°F and coarse breath sounds in the right anterior chest. The chest radiograph (Fig. 26–5) showed a right upper lobe infiltrate; the laboratory analysis included a white blood cell count, which was elevated.

Diagnosis. This presentation is classic for pneumococcal pneumonia. Sputum Gram stain should show gram-positive lancet-shaped diplococci. Sputum culture should also grow *Streptococcus penumonia* if there is not significant overgrowth from oral contaminants. In 25 to 30% of cases of pneumococcal pneumonia, patients have positive blood cultures for this organism (Fig. 26–6).

The differential diagnosis should include other common causes of community-acquired pneumonia such as *Haemophilus influenzae, Moraxella catarrhalis,* or *Legionella. Staphylococcus aureus* must be considered if there is a recent history of influenza. In cases of atypical pneumonia, patients have more constitutional symptoms and less sputum production. The chest radiograph is more likely to show patchy infiltrates rather than lobar consolidation. *Mycoplasma, Chlamydia, Legionella,* and viruses are potential causes of atypical pneumonia.

Etiology. Respiratory syncytial virus is the most commonly identified cause of viral pneumonias in children, especially in infants, in most communities. This pathogen has also been recognized in infections among the elderly, those being cared for in nursing homes, and those who are immunosuppressed. Parainfluenza is the second most commonly recognized viral pathogen to cause childhood pneumonias. Other etiologic agents to consider in pneumonias in this age group are low-serotype adenoviruses and mycoplasma. Viral pneumonias caused by influenza type A and B viruses are prevalent in communities only during periods of epidemic spread.

Although less common than viral pneumonias, bacterial pneumonias also occur in children and must be considered in the differential diagnosis. *Streptococcus pneumoniae* and *Haemophilus influenzae* type B are by far the most common pathogens isolated in childhood bacterial pneumonia. Other bacterial pathogens must also be considered, depending on the clinical situation. For example, group B streptococci are more likely to be associated with pneumonia in neonates than in older children.

In adults, *Streptococcus pneumoniae* remains the most common cause of community-acquired bacterial pneumonia in the general population. In addition, depending on the severity of the illness, the season of the year, and whether the community is experiencing an epidemic respiratory infection such as influenza, other pathogens such as respiratory viruses and *Mycoplasma pneumoniae* are seen in community-acquired pneumonias in adolescents and adults and are responsible for the atypical pneumonia syndrome. It is atypical in that typical signs and symptoms of bacterial pneumonias that are more common in these age groups are absent or less impressive.

The type of patient involved may also help determine which bacterial pathogens are more likely causes of community-acquired pneumonia. For example, in adult patients with chronic lung disease, *Haemophilus influenzae* and *Streptococcus pneumoniae* are frequently colonizers. *H. influenzae* pneumonia is seen more frequently in these patients than in the general population. Patients with alcoholism and patients who have recently been hospitalized or treated with broad-spectrum antibiotics have an increased risk of being colonized with gram-negative bacillary pathogens that may cause pneumonia. Patients with recent influenza infection are at increased risk of developing pneumonia caused by *Staph-*

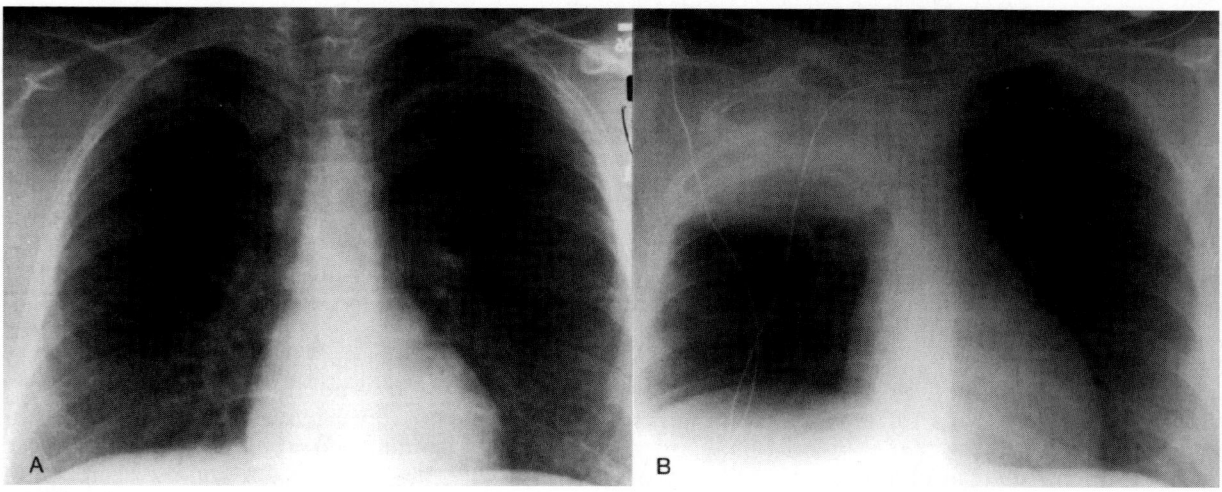

■■■■■Figure 26 – 5

Chest radiographs before *(A)* and after *(B)* development of an acute, community-acquired, pneumococcal pneumonia. The patient is facing toward the reader. In *B,* consolidation of the right upper lobe of the lung is evidenced by the dense, whitish opacification of this lobe, which contrasts with the normal air (black) density of the remainder of the lung.

ylococcus aureus as well as by *Streptococcus pneumoniae.* These types of associations provide general guidelines that are useful for initiating empiric antibiotic therapy and for focusing diagnostic efforts to define the specific pathogen causing pneumonia in a given case.

Another group of pathogens that has emerged in recent years as a cause of both atypical pneumonia and typical bacterial pneumonia is *Legionella* species. These cases can be difficult to diagnose, since unusual efforts are required to identify the organism in the sputum or bronchial aspirate specimen. *Legionella* species do not stain well with Gram stain (Fig. 26–7) and are often

missed in sputum or other types of respiratory secretions. Although patients may initially present with symptoms and signs that are typical of a viral respiratory tract syndrome, there may be a rapid progression of both pulmonary and systemic signs of infection. Chest radiographs may show progressive involvement of multiple areas of the lung, with increasing patchy and lobar consolidation. Other signs and symptoms such as electrolyte abnormalities and gastrointestinal symptoms have also been associated with this infection, but it is usually not possible to distinguish *Legionella* pneumonia from other forms of bacterial pneumonia without the aid of specific

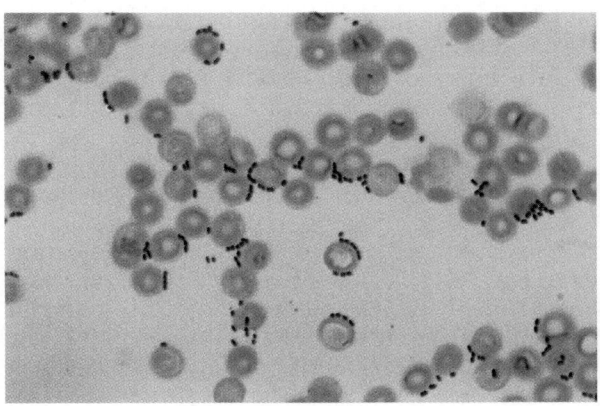

■■■■■Figure 26 – 6

Gram-stained smear of *Streptococcus pneumoniae* isolated from the blood culture of a patient with pneumococcal pneumonia.

■■■■■Figure 26 – 7

Gram-stained smear of *Legionella* species taken from culture.

diagnostic studies. Of primary importance is the consideration of this diagnosis in atypical cases of pneumonia in which the pathogen is not defined early in the course of investigation, so that specific diagnostic studies can be planned to address the possibility of *Legionella* infection.

Epidemiology. Community-acquired pneumonias in children are usually attributable to viral pathogens that cause respiratory tract infections in the community during the winter months. It is useful during such epidemic periods to contact the local state health department virology laboratory to inquire as to the prevalent viral pathogen in the community, since seasonal viral respiratory infections often pass through the community in waves.

Unlike those seen in children, the majority of community-acquired pneumonias in adults are caused by bacterial infections. Although a patient with no other known medical problems can present with an acute bacterial pneumonia, the typical adult patient with community-acquired pneumonia either is elderly or has an underlying disease that predisposes to lower respiratory tract bacterial infection. Examples of such predisposing illnesses include chronic lung diseases, cardiovascular disease, diabetes mellitus, and alcoholism. These patients may either present with a primary pneumonia or develop pneumonia as a secondary infection complicating a primary viral infection. This predisposition to bacterial pneumonias is the reason that the elderly and patients with underlying chronic illnesses are the major targets for influenza and pneumococcal vaccination programs. In the community, the major cause of death in patients with influenza virus infection is secondary bacterial pneumonia. Based on this observation, it has been reasoned that prevention of influenza by vaccination of these high-risk patients will also prevent the serious bacterial pneumonias that may follow.

Clinical Manifestations. The usual onset of nonbacterial pneumonias in children is indistinguishable from an average viral upper respiratory tract infection. However, instead of resolving in the time expected for upper respiratory tract infections, the clinical course proceeds toward increasing severity of illness, with signs of respiratory distress. Few other signs may be seen in young children with viral pneumonia.

In older children and adolescents, the typical signs of systemic viral infection include general fatigue and myalgias associated with early signs of upper respiratory tract infection. The cough usually does not produce sputum early in the course of the illness. Low-grade fever is common. The findings on physical examination are variable, and the chest radiograph may show diffuse interstitial changes, patchy infiltrates, or lobar consolidation (dense localized changes restricted to one lobe of the lung; see Fig. 26–5). Lobar consolidation should always prompt an evaluation for a bacterial pathogen, since this radiographic presentation is atypical for viral pneumonia but common with bacterial infection. Routine laboratory studies are rarely helpful in making the clinical diagnosis of viral pneumonia in either children or adults.

In adolescents and adults, the atypical pneumonia syndrome, represented by *Mycoplasma pneumoniae,* has a symptom complex and clinical presentation that are not dramatically different from those of other nonbacterial pneumonias. Its onset is like that of other nonbacterial upper respiratory tract infections but progresses to produce symptoms and signs of lower tract involvement. Chest radiograph abnormalities are classically more dramatic than would be expected based on the physical examination. The time of the year during which the infection is seen may be helpful in narrowing the list of choices for the etiologic diagnosis. During the winter months, a nonbacterial pneumonia syndrome is often caused by a viral pathogen that is prevalent in the community, for example, influenza. During the summer months, however, when pneumonia caused by most other nonbacterial respiratory pathogens is uncommon, the likelihood that an atypical pneumonia will be caused by *Mycoplasma pneumoniae* approaches 50%.

The clinical presentation of community-acquired pneumonia in adults varies with the age and immunocompetence of the host. Most patients with bacterial pneumonia note a relatively sudden onset of fever associated with chills. In fact, they are often able to report the exact time their illness started. Cough productive of purulent sputum that may be blood-tinged is typical of this type of infection. Chest examination and radiographs typically show a localized area of infiltration in the lung, described as lobar consolidation (see Fig. 26–5). This abnormality in the lung parenchyma is often associated with reduced oxygenation as measured by arterial blood gas monitoring. Routine laboratory studies will usually show a neutrophilic leukocytosis with an increased percentage of immature forms of granulocytes in the differential white blood cell count.

In elderly or immunocompromised patients, the clinical presentation may be less impressive than in patients who are able to mount a normal immune response to pulmonary bacterial infection. These immunologically compromised patients may have few respiratory tract complaints and little or no fever. Nonspecific symptoms such as weakness, loss of appetite, and a minor cough may be the only manifestations of a progressive pneumonia. It is important to maintain a high index of suspicion in these patients and to routinely pursue changes in patterns of behavior with complete evaluations to avoid missing the diagnosis of pneumonia. These patients are also at increased risk for developing pneumonia after influenza infections. Therefore, during influenza epidemics in the community, clinicians should have a heightened awareness of changes in respiratory symptoms that could be compatible with pneumonia in the elderly patient population.

NOSOCOMIAL PNEUMONIA

CASE STUDY

A 60-year-old man with a history of emphysema and chronic bronchitis was admitted to the hospital to have his gallbladder removed. He was given perioperative antibiotic prophylaxis with cefoxitin. Because of his underlying lung disease, he could not be weaned from the ventilator postoperatively. His antibiotics were continued for a few more days. Seven days after surgery, he developed a high fever, increased secretions from his endotracheal tube, and a new infiltrate on his chest radiograph.

Diagnosis. Important considerations in this case are the previous antibiotic therapy, the presence of an endotracheal tube, and the patient's underlying lung disease, which favor colonization with pathogenic (especially gram-negative) bacteria, invasion of the lower respiratory tract with these potential pathogens, and decreased lung defenses against the establishment of infection. In cases such as these, there must be a high suspicion that a gram-negative bacillus (e.g., *Pseudomonas aeruginosa*) is the cause of the pneumonia and that it may be a drug-resistant organism. Gram stain and culture should be performed from the endotracheal secretions, and blood cultures should be done. It is often difficult to distinguish between colonization and infection in patients with endotracheal tubes. In addition to a positive sputum culture, other evidence, such as a new infiltrate on the chest radiograph or positive blood cultures, should be sought. If empyema fluid is present, it should be Gram stained and cultured.

Etiology. Statistics from the Centers for Disease Control and Prevention show that almost 60% of nosocomial (hospital-acquired) lower respiratory tract infections are caused by gram-negative bacilli, including *Klebsiella* sp, *Enterobacter* sp, *Escherichia* sp, *Serratia marcescens,* and *Pseudomonas* sp. A smaller percentage (16%) of cases are associated with gram-positive isolates. Other comparisons of etiologic agents in nosocomial pneumonias have shown that the incidence with which anaerobic bacteria can be isolated varies greatly from study to study, apparently due to differences in sampling techniques and laboratory techniques used to isolate these organisms.

Epidemiology. Development of pneumonia in the hospital setting has different implications for both the diagnosis and the potential outcome of the infection. Colonization of the oropharynx with gram-negative bacillary pathogens is relatively uncommon in otherwise healthy patients but is common in hospitalized patients, especially those with severe underlying illnesses.

Epidemiologic studies have shown that gram-negative bacterial pneumonia is the leading cause of fatal hospital-acquired infections. An aggressive approach to early diagnosis and treatment of these pneumonias is therefore essential. An awareness of risk factors that predispose patients to nosocomial pneumonias sensitizes both the clinician and the clinical microbiologist to this problem so that an etiologic diagnosis can be established as early as possible in the course of the illness. These risk factors can be subdivided into two general categories:

1. increased colonization of the upper respiratory tract with bacterial pathogens
2. compromise of the barriers that normally protect the lower respiratory tract from infection with these pathogens

Increased patient age, more severe illness, previous treatment with antibiotics, and manipulations that increase the gastric pH are all associated with increased oropharyngeal colonization with gram-negative bacilli. Such pathogens may be acquired from the patient's own gastrointestinal tract, but exposure of patients to the microbial flora encountered during hospitalization provides an exogenous source of colonization that can also be involved in subsequent nosocomial pneumonia.

Compromise of normal barriers that prevent invasion of the usually sterile lower respiratory tract with pathogens colonizing the oropharynx predispose hospitalized patients to nosocomial pneumonia. Hospitalized patients with altered levels of consciousness aspirate pharyngeal contents into the lung more often than normal volunteers (70% versus 45%). In addition, intubation of the lower airway, which is commonly used in the intensive care setting for respiratory support of seriously ill patients, greatly increases the risk of developing nosocomial pneumonia by bypassing the normal mechanical defenses provided by the glottis and cough reflex. The longer such intubation is continued, the greater the risk of acquiring an associated nosocomial pneumonia. Nasogastric intubation also increases the risk of aspiration due to interference with glottic function and increased reflux of gastric secretions into the oropharynx.

In addition to an appreciation of the risk factors for colonization with and aspiration of hospital-acquired pathogens, the clinician needs a working knowledge of the antibiotic susceptibility patterns of the common nosocomial pathogens in the institution. Annual reviews of patterns of antibiotic susceptibility of the most common pathogens involved in nosocomial pneumonias should be performed in each institution by the clinical microbiologist in conjunction with the hospital infection control coordinator. The availability of these data provides two important strengths to the institution. Using these data, the clinician can design the best empiric antibiotic treatment regimen while awaiting the results of specific antibiotic susceptibility testing results on the organism isolated from the patient. The infection control coordinator can also use the information to define changing trends in susceptibility patterns of hospital pathogens so that clinicians can be alerted about changes in treatment strategies that may be needed.

Hospitals are also among the institutional settings in which contaminated water supplies have been associated with *Legionella* sp pneumonia. The exact mode of transmission of infections to patients remains a subject of debate. A higher incidence of these infections has been observed among patients with underlying lung disease and among immunocompromised patients (e.g., immunosuppression associated with solid organ transplants or corticosteroid therapy for various underlying conditions). Since the clinical presentation of patients with nosocomial *Legionella* pneumonia is usually not distinct from that of other nosocomial pneumonias, early diagnosis is dependent on a high index of suspicion and specific diagnostic studies.

Aspiration Pneumonia

Etiology. Aspiration pneumonia, which occurs in both children and adults, follows aspiration of oropharyngeal or gastric contents into the lower respiratory tract. Among patients with aspiration pneumonia, 50% have mixed aerobic and anaerobic bacteria isolated, and 50% have only anaerobic bacteria isolated. The primary anaerobes isolated in these cases are usually *Peptostreptococcus* sp, *Fusobacterium* sp, and *Bacteroides* sp; the primary aerobes isolated are usually *Streptococcus* sp, *Eikenella corrodens, Staphylococcus aureus,* Enterobacteriaceae, and *Pseudomonas aeruginosa.*

Patients with nosocomial aspiration pneumonia are more likely to have aerobic pathogens (predominantly gram-negative bacilli) than are patients with community-acquired aspiration pneumonia. Seriously ill hospitalized patients have an increased risk of lower respiratory tract infection with aerobic gram-negative bacilli due to increased colonization with these organisms. Patients who are immunosuppressed by preexisting disease or by chemotherapy and patients with altered lung defenses (e.g., chronic lung disease, cigarette smoking, advanced age) are at increased risk for the development of infection after aspiration of a bacterial inoculum.

In community-acquired aspiration pneumonia, the presence of periodontal disease, associated with an increased burden of oral anaerobic bacteria, increases the risk for the establishment of infection after aspiration.

Epidemiology. Aspiration of oropharyngeal secretions is common in both the general population and hospitalized patients. The likelihood of developing infectious pneumonia after aspiration depends on the frequency of aspiration, the quality and quantity of the material aspirated, and the defenses of the host.

It is useful for diagnosis and treatment planning to attempt to discriminate between chemical aspiration and aspiration of material contaminated with bacteria. Chemical aspiration (Mendelson's syndrome) resulting from aspiration of low-pH gastric contents not contaminated with large numbers of bacteria does not require initial antibiotic treatment. Since the injured lung is at increased risk of secondary bacterial infection, however, one must be vigilant for signs of bacterial superinfection in the setting of "pure" chemical aspiration pneumonia.

Clinical Manifestations. The typical clinical presentation of patients with aspiration pneumonia and its complications may vary from an acute pneumonitis to a chronic, indolent respiratory condition presenting as chronic productive cough. If treatment fails, aspiration pneumonia may progress to a necrotizing pneumonia and lung abscess.

Patients with anaerobic lung abscess usually have a longer history of illness (weeks) with a more subtle onset than patients with common acute bacterial pneumonia whose illness has typically been present for hours to days before they seek medical attention. Patients with lung abscess are also more likely to have putrid sputum and to complain of halitosis due to the involvement of anaerobic bacteria in the lung infection.

On chest radiographs, areas of necrotizing pneumonia and lung abscess resulting from aspiration are more likely to involve parts of the lung that are dependent in the supine position (superior segments of the lower lobes and posterior segments of the upper lobes), since this is the most common position in which aspiration into the lung occurs. Right-sided lung involvement is twice as common as left-sided involvement because the right main-stem bronchus has a more direct course into the lower lobe of the lung than the left main-stem bronchus.

Pathogenesis. Once bacterial pathogens are established in the lung, the outcomes of interactions between the infectious agent and the host's response to infection determine the nature of the pneumonia and depend to some extent on the virulence of the organisms. For example, in pneumonias caused by *Pseudomonas aeruginosa,* there may be extensive tissue destruction due to the production of proteases and cytotoxins by the organism, causing necrotizing pneumonias.

Complications. In patients with extensive viral or bacterial pneumonias, there may be insufficient ventilation to the involved lung to adequately support respiration. In rapidly progressive, multilobe pneumonias, ventilatory support may be needed as a result of overwhelming infection or an injury response of the lung termed "respiratory distress syndrome." In these cases, mechanical support of ventilation and other supportive measures are used to provide time for antimicrobial therapy to control the pneumonia

and to allow recovery of lung function. An example of tissue destruction caused by pneumonia is the development of a lung abscess cavity. In this case, the pulmonary infection, usually an anaerobic infection of the lung associated with aspiration pneumonia, results in destruction of the lung parenchyma in the area of a necrotizing bacterial infection. In most cases, these infections can be treated with antibiotic therapy alone. However, if the abscess cavity is large and refractory to therapy, surgery may be required to eliminate the site of infection.

An example of local extension of a lung infection is empyema, an infection of the pleural space between the lung and the chest wall. Due to the limited ability to deliver antimicrobial agents to this type of space (limited blood supply), bacterial empyema usually requires a drainage procedure, such as insertion of a chest tube, to promote resolution of the infection. Extension of the infection beyond the chest occurs most commonly when bacterial invasion of the blood stream occurs (bacteremia). The same type of spread is possible with viral pneumonias (viremia), although the consequences of viremia are usually much less severe than those of bacteremia. Patients with bacteremia as a complication of pneumonia have an increased mortality compared to patients with pneumonia without bacteremia. In some cases, this may simply be a reflection of the bacterial burden and the difficulty in controlling the infection. In other cases, such as gram-negative bacillary bacteremia, the presence of bacteria and their constituents (e.g., endotoxin) may be associated with severe shock, further complicating the care of the patient and increasing the likelihood of a fatal outcome. In addition to the direct consequences of bacteremia, there are instances in which an infection originally established in the lung may spread via the blood stream to other sites, such as the central nervous system—a so-called metastatic infection.

Laboratory Diagnosis

Specimen Collection. In the lower respiratory tract, there is a constant risk of contaminating the clinical specimen with the normal flora of an adjacent area, primarily the upper respiratory tract. Despite the limited usefulness of this type of specimen from the lower respiratory tract, expectorated sputum specimens can be useful in the diagnosis of pneumonia. In addition, an expectorated sputum, when compared with other procedures, involves no risk to the patient.

Using both the Gram stain and culture, expectorated sputum specimens yield an etiologic diagnosis in almost half the cases of bacterial pneumonia. However, care must be taken to avoid contamination of the specimen with oropharyngeal flora. When possible, patients should be instructed to rinse out their mouths and then to collect a "deep" sputum specimen. The time taken for such communication with patients can be rewarding. It is important that they understand that the purpose of the exercise is to collect lung secretions and not saliva or drainage from the nasopharynx. In addition, patients should remove dentures during the specimen collection.

The expectorated sputum specimen should be examined for its character, which provides a preliminary indication about the type of pneumonia. A patient with typical bacterial pneumonia (e.g., pneumococcal pneumonia) will produce purulent sputum that may be streaked with small amounts of blood or may be rusty in color. In contrast, patients with nonbacterial pneumonia (e.g., mycoplasma pneumonia) may produce no sputum or small amounts of sputum that is relatively clear.

Other Specimen Collection Methods. In some cases of pneumonia, expectorated sputum specimens may be either unavailable or of unsatisfactory quality. Early attempts to sample lower respiratory tract secretions more directly involved transtracheal aspiration and transthoracic needle aspiration directly from the site of pneumonia. Although both these techniques were more successful diagnostic tools than the use of expectorated sputum, there was a significant increase in the morbidity associated with these invasive procedures (e.g., bleeding and pneumothorax). Therefore, the recent trend has been toward the use of the flexible, fiberoptic bronchoscope when direct specimen collection from the lower respiratory tract is required. Proper use of a protected brush extended from the end of the bronchoscope to collect specimens from the area of pneumonia has been valuable for diagnosing both aerobic and anaerobic infections of the lung.

The same general approach used to process expectorated sputum specimens is used to process specimens collected using this method. In addition, quantitative cultures may be performed to provide a further check against the possibility that the pathogen isolated is a contaminant from the upper respiratory tract.

In immunocompromised patients, in whom there are minimal secretions as a result of a weak inflammatory response to infection, bronchoalveolar lavage adds an extra dimension that increases the diagnostic yield. The segment of the lung suspected of being involved in a pneumonic process is lavaged with sterile fluid; this fluid is then collected after aspiration through the bronchoscope and is concentrated before processing. This technique has been highly useful in diagnosing *Pneumocystis carinii* and cytomegalovirus infections in patients with AIDS. There are occasional cases in which an infectious pneumonia is strongly suspected (or must be differentiated from a lung abnormality of another type) and in which the procedures mentioned to this point are unsuccessful in establishing an etiologic diagnosis. In such instances, it may be necessary to proceed to a lung biopsy to obtain tissue for culture and histologic study. Biopsy can be performed through the flexible bronchoscope. However, the size of the specimen taken by this technique is quite small and may be insufficient for diagnosis. Thorascopic lung biopsy has been introduced as a means of sampling pleural-based lung lesions. As a final option, an open lung biopsy performed during a surgical procedure may be necessary to obtain a larger tissue specimen for analysis.

Direct Microscopic Examination. The quality of the specimen is determined by a direct Gram-stained smear. The goal of this evaluation is to attempt to eliminate specimens that are contaminated with oropharyngeal contents. Gram-stained smears of sputum samples show the relative numbers of neutrophils and epithelial cells, which should be quantitated under low-power magnification ($10\times$). As a general guideline, samples with greater than 25 neutrophils and fewer than 10 epithelial cells per field are considered to be relatively free of such contamination (Fig. 26–8); samples containing more than 25 squamous epithelial cells per field (Fig. 26–9) should not be cultured.

Gram stain of samples collected by invasive procedures produce meaningful results. In addition to the presence or absence of microorganisms and inflammatory cells, the microscopic morphology of the organisms present may lead to a presumptive diagnosis.

Culture. Only high-quality specimens should be processed for culture. The protocol should in-

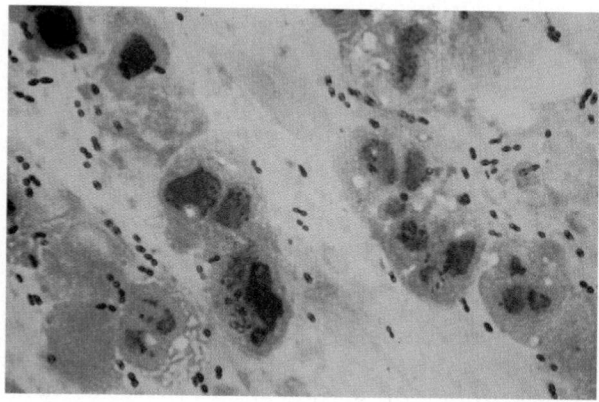

■ F i g u r e   2 6 - 8
Gram-stained smear with white blood cells and gram-positive diplococci—acceptable for culture.

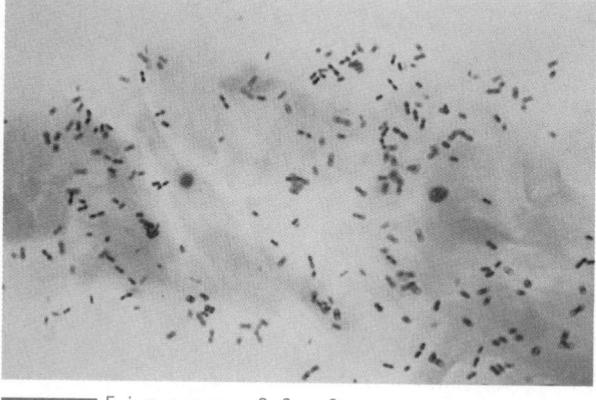

■ F i g u r e   2 6 - 9
Gram-stained smear of sputum with epithelial cells—unacceptable for culture.

clude culture media such as sheep's blood agar, MacConkey, and chocolate agar plates to recover the wide variety of agents that may be suspected. Samples collected by invasive procedures—and therefore unlikely to be contaminated with upper respiratory flora—should be processed for anaerobic culture, particularly if the patient is suspected of having aspiration pneumonia.

Chronic Pneumonia

Bacterial pneumonias usually resolve completely over a period of many weeks. On occasion, however, there is a delayed resolution of pneumonia, with radiographic lung abnormalities persisting far beyond the improvement of clinical symptoms. Some bacteria that typically cause acute pneumonias induce necrotizing processes in the lung that are generally slow to resolve despite a clinical cure. Examples of such infections include anaerobic infections of the lung, gram-negative bacillary pneumonias, and pneumonias caused by *Staphylococcus aureus*. When abnormalities in the lung persist, the clinician is faced with several questions regarding diagnosis and therapy. Is the process the result of a slowly resolving necrotizing pneumonia or a slowly progressing chronic pneumonia? Is there an unsuspected pathogen that is causing progressive infection in the lung? Is there an underlying noninfectious problem in the lung that is causing chest radiograph abnormalities?

CASE STUDY

A 60-year-old man came to an emergency room complaining of cough and fever lasting several weeks. Upon questioning, he reported night sweats and weight loss over the last few months. He admitted to drinking alcohol heavily on weekends and to staying in a shelter for the homeless most of the previous winter.

The patient had a fever and appeared very thin. He coughed frequently during the examination. His chest radiograph revealed a right upper lobe infiltrate.

Diagnosis. Because *Mycobacterium tuberculosis* bacilli stain poorly with Gram stain, the clinician must suspect the diagnosis of tuberculosis as well as other types of pneumonia so that the appropriate diagnostic studies can be performed. Ziehl-Neelsen or Kinyoun stains (Fig. 26–10) or a fluorochrome study may be used to detect acid-fast bacilli in sputum, but the culture is essential in order to diagnose smear-negative, culture-positive patients and to differentiate *M. tuberculosis* from other mycobacteria. Because of the public health implications of *M. tuberculosis* infections, which may be highly contagious, any positive stain or culture for mycobacteria should be reported to the clinician, and proper respiratory isolation should be established until the organism is identified.

Until the acid-fast bacilli studies are completed, other etiologies that should be included in the differential diagnosis are community-acquired bacterial pneumonia and aspiration pneumonia.

Etiology. Among the pathogens that cause clinically apparent chronic pneumonias, mycobacteria are the most common in immunologically intact hosts. The incidence of *M. tuberculosis* infection has stopped the steady decline that

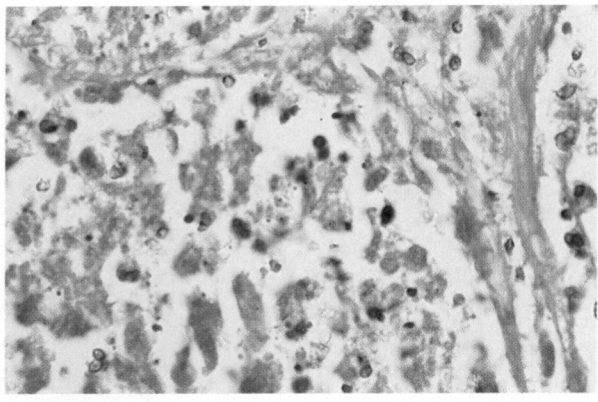

■■■ F i g u r e 2 6 – 1 0
Acid-fast bacillus (AFB) stain of sputum containing mycobacteria.

was observed from the 1950's through the mid-1980's and has increased. This increased incidence is attributed in part to the AIDS epidemic and disproportionately affects the socially disadvantaged, elderly, and immunocompromised subpopulations of our society. Mycobacteria other than tuberculosis (e.g., *M. kansasii, M. avium* complex, *M. fortuitum-chelonei*) may present in a fashion that is indistinguishable from tuberculosis.

Opportunistic fungal pathogens, including *Candida, Aspergillus* (Fig. 26–11) and *Cryptococcus,* can also cause chronic pneumonia. They rarely cause invasive disease in immunologically intact hosts but can cause acute and chronic pneumonias in immunosuppressed patients. In contrast, true fungal pathogens such as *Histoplasma, Coccidioides,* and *Blastomyces,* when acquired in a sufficient inoculum, can cause chronic pneumonias in normal hosts.

Clinical Manifestations. Some infections of the lung are inherently slow in their progress and are chronic in nature. The most common examples of these types of infections are mycobacterial and fungal infections of the lung. Although some of the symptoms seen in the setting of acute pneumonia (e.g., fever, chills, and general weakness) are also observed in patients with chronic pneumonias of mycobacterial or fungal etiology, these and other symptoms may be less dramatic in their onset and less intense in these cases. As a result, there is often a longer delay between the onset of symptoms and the seeking of medical attention. Due to the prolonged period of illness associated with chronic pneumonias, the patient may also exhibit other signs that are not typical-

ly seen with acute infections, such as marked weight loss.

In immunocompromised patients, mycobacterial or fungal infections may be unusually severe, with either local or disseminated infection causing life-threatening complications. For example, *Aspergillus* may cause overwhelming pneumonia in neutropenic leukemia patients, and *Cryptococcus* infections acquired by the respiratory route may cause fatal infections after dissemination to the central nervous system in patients who are immunosuppressed following chemotherapy. Immunosuppressed patients may also have nonspecific symptoms of fever and weakness as manifestations of low-grade chronic pneumonias.

The physical examination of patients with chronic pneumonia usually shows few specific signs other than those associated with general debility. Routine laboratory studies are rarely of help in making a specific diagnosis. In addition to the history, a chest radiograph and, in selected cases, computed tomography of the chest are the most important initial diagnostic studies in establishing a presumptive diagnosis of chronic pneumonia. For example, upper lobe cavitary lung disease is suggestive of tuberculosis or related mycobacterial infection. However, since many infections that cause chronic pneumonia can cause similar radiologic abnormalities, further studies must be planned for a specific diagnosis.

Pathogenesis. Chronic pneumonias caused by mycobacterial or fungal pathogens typically elicit a granulomatous response in the lung that is distinct from the acute inflammatory response

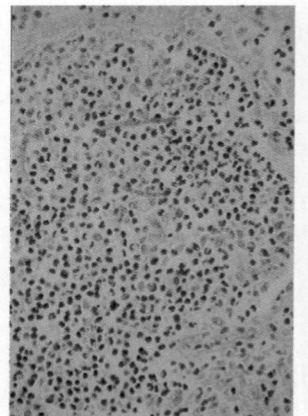
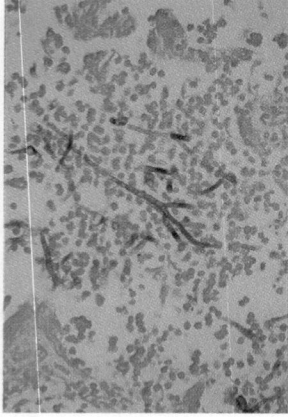

■■■ F i g u r e 2 6 – 1 1
Lung exudate from a patient with hematologic disorder, showing alveoli containing fungal elements. *A,* Hematoxylin & eosin stain; *B,* GMS stain. (Courtesy of Shirlyn B. McKenzie, Ph. D.)

seen with most acute bacterial pneumonias. The interactions between components of the infecting organism and the host immune response that lead to this granulomatous response are still being characterized. Both humoral (e.g., production of a specific antibody) and cellular (e.g., delayed type hypersensitivity) responses are elicited by these pathogens. However, it seems clear that the cellular immune response to infection is primarily involved in both protective and tissue-destructive aspects of these illnesses. One goal of the study of the immunobiology of the infections that can cause chronic pneumonias is to dissect these two general types of cellular immune responses to allow the development of approaches that favor protective cellular immunity while minimizing the potential for lung tissue destruction.

Complications. The complications of chronic infectious pneumonias depend on the extent of the local and systemic spread of the infection and the immunologic status of the host. Patients who develop extensive lung involvement with a chronic infection due to late medical intervention or failed therapy may eventually suffer from respiratory insufficiency. This is usually the result of progressive lung tissue destruction and progression of the fibrotic process that can accompany these chronic pneumonias. When the infection extends beyond the lung (e.g., in immunocompromised patients), other vital organs may be involved, such as central nervous system infection with *Cryptococcus* in an immunosuppressed host. Patients with chronic pneumonias for prolonged periods commonly become cachectic. This appears to be a maladaptive response of components of the host cellular immune response to chronic infection that results from the production of cytokines (e.g., cachectin/tumor necrosis factor α), which interfere with nutrient metabolism.

Laboratory Diagnosis. Among the chronic pneumonias, tuberculosis is usually the easiest cause to diagnose. With the proper clinical picture, the presence of *M. tuberculosis* in an expectorated sputum specimen is diagnostic of infection. Similarly, the discovery of *M. kansasii* in the expectorated sputum of a patient with a clinical picture of chronic cavitary pneumonia is usually sufficient to establish the etiologic diagnosis. In patients with chronic pneumonia caused by mycobacteria other than tuberculosis, repeated sputum specimens in the context of progressive pneumonia and in the absence of another definable pathogen are usually necessary to establish an association between the pathogen and the illness. This is true due to the observation that these pathogens may colonize the respiratory tract in the absence of apparent infection. Therefore, the mere isolation of an organism (e.g., *M. avium*) from a single sputum specimen is insufficient to establish an etiologic diagnosis.

With chronic pneumonias of fungal etiology, it is typically difficult to isolate the pathogen from expectorated sputum. As a result, invasive procedures to obtain either bronchoscopic or open lung biopsies are often required to establish the identity of the pathogen in these cases. Depending on the severity of the clinical illness, empiric therapy may be initiated based on a presumptive diagnosis from the results of histopathologic studies while awaiting culture confirmation from the diagnostic microbiology laboratory.

Direct Microscopic Examination. Smears prepared from concentrated sputum samples for mycobacterial cultures should be stained with acid-fast stains (see Fig. 26–10) such as Kinyoun, Ziehl-Neelsen, or any of the fluorochrome stains. Although the sensitivity of acid-fast smears is approximately 50%—maybe higher with fluorochrome stains—finding mycobacterial organisms on the smear indicates the possible infectiousness of the patient. In addition, rapid diagnosis and initiation of empiric treatment may be established prior to the availability of culture results, which take several weeks.

Direct microscopic examination of respiratory secretions should also include preparations to detect fungal agents. Potassium hydroxide and calcofluor white are commonly used to detect yeast cells and hyphal elements.

Culture. Culture studies in patients with chronic pneumonias require close communication between the clinician and the clinical microbiologist. It is necessary to use special techniques to process respiratory secretions and lung tissue specimens for identification of mycobacterial and fungal pathogens. Furthermore, there are occasions (e.g., in cases of pulmonary tuberculosis and coccidioidomycosis) when isolation of the pathogen creates a potential biohazard in the diagnostic laboratory. Therefore, laboratory personnel must use special precautions when culturing these specimens.

Empyema

Empyema is defined as a collection of purulent fluid in the pleural space between the lung and the chest wall. Although the accumulation of pleural fluid is fairly common in association with acute bacterial pneumonia, most such accumulations are sterile, and only a small percentage qualify as empyemas. The distinction between a sterile pleural effusion and an empyema depends on the presence of a pathogen and an inflammatory response and is often made using chemical parameters of the fluid.

Etiology

In patients with community-acquired pneumonia, *Staphylococcus aureus, Streptococcus pneumoniae,* and *Streptococcus pyogenes* are the most common causes of empyema. Anaerobic bacteria are being isolated with increasing frequency from patients with a history of aspiration pneumonia and lung abscess. Empyema among hospitalized patients is caused primarily by aerobic gram-negative bacilli—the cause of the majority of nosocomial pneumonias. In patients with chronic pneumonia caused by *M. tuberculosis,* empyema may occur as a result of rupture of an underlying cavity into the pleural space.

Clinical Manifestations

The clinical presentation of a patient with empyema is usually an extension of the illness of the underlying lung (e.g., bacterial pneumonia or lung abscess). Patients may have chest pain on the affected side, fever, chills, and night sweats. If the empyema is large, it may be detected during the physical examination. If empyema is not suspected based on the physical examination, the first evidence of it usually comes from chest radiographic studies. Computed tomography of the chest may be necessary to differentiate empyema from underlying lung abscess or lung consolidation.

Empyema may also complicate chest surgery or chest trauma, both of which provide a potential route of infection directly from the exterior to the pleural space. Less commonly, empyema results from direct contamination of the pleural space as a consequence of peritoneal or gastrointestinal disease. Infection may extend to the pleural space from a pyogenic process beneath the diaphragm. Rarely, empyema results from infection of the pleural (and mediastinal) space after esophageal rupture.

Pathogenesis

Once organisms gain access to the pleural space, the resulting inflammatory response to this infection, along with the pleural response to the underlying process that seeded the pleural space, stimulates the exudation of fluid into this area. Typically, accumulation of large numbers of neutrophilic leukocytes occurs, resulting in the purulent appearance of most empyemas. Toward the later (or organizing) stages of an empyemic process, fibroblast infiltration into the area is associated with organization of the contents of the pleural space into a thick capsule that adheres to the lung and chest wall surfaces. If the empyema is not treated before this stage, the process may encase and limit the motion and function of the underlying lung.

Characteristics of the empyema fluid create an environment in which elimination of the offending pathogen is compromised. Opsonins and complement activity, which are necessary for the proper phagocytosis of bacteria by infiltrating granulocytes, are present in reduced concentrations in empyema fluid. In addition to this limitation of the host response, the usefulness of antibiotics in treating empyema is limited by the minimal blood supply to this area, resulting in reduced drug delivery, and by the low pH of empyema fluid, which can reduce the antimicrobial activity of antibiotics.

Complications

The complications associated with empyema depend primarily on the nature of the underlying disease. The main complication of empyema per se is persistence of infection. If the empyema is not drained properly and treated with appropriate antibiotics, the infection causing the empyema may be difficult to eliminate. Persistent, poorly controlled infection may lead to the multiple complications associated with unresolved sepsis. A long-term complication of empyema is encasement of the lung in a thick capsule, which may alter lung function. This complication may necessitate removal of the thickened pleural lining (decortication), which may or may not restore the function of the underlying lung, depending on the duration of the dysfunction.

Laboratory Diagnosis

Pleural fluid should be aspirated directly into an evacuated sterile syringe. If the volume of the pleural fluid is relatively large (greater than 500 mL) as estimated by chest radiograph, this procedure can usually be performed at the patient's bedside. If the volume is smaller or if the space is loculated, pleural fluid aspiration may have to be done in the radiology suite guided by ultrasound. Since the volume of pleural fluid needed for diagnostic studies is small (a few milliliters), it is not necessary to try to remove the majority of the effusion for these purposes. Any excess air in the syringe should be expelled promptly to improve the yield of anaerobic bacteria.

The syringe containing the fluid should be transported directly to the diagnostic microbiology laboratory so that Gram stain and processing for anaerobic and other bacteria can be performed promptly. The character (i.e., purulence, odor, presence of blood) of the fluid should be recorded. Aliquots of the remaining fluid should be distributed to the hematology and chemistry laboratories for studies of cell count and differential count, total protein and lactic dehydrogenase concentration, and pH. These measurements provide parameters that are useful in differentiating an empyema from a transudative effusion.

An aliquot of the pleural fluid should be saved in case special studies are required to detect capsular polysaccharide antigens of pathogenic bacteria such as *Streptococcus pneumoniae* or *Haemophilus influenzae*. If malignancy is in the differential diagnosis, a fluid specimen should also be submitted for cytology. Since it is usually desirable to submit a large specimen for cytology when a malignant pleural effusion is suspected, a larger sample may have to be collected specifically for this purpose. If a diagnosis of tuberculous pleural effusion is suspected, it is usually useful to obtain a pleural biopsy to complement the pleural fluid sample for mycobacterial culture. The combined data from culture of these two specimens can be diagnostic in approximately 85% of cases, whereas culture of pleural fluid alone is positive in less than 25% of confirmed cases of tuberculous empyema.

OPPORTUNISTIC INFECTIONS OF THE RESPIRATORY TRACT

Opportunistic pathogens ordinarily do not cause disease in normal hosts but require an impairment of host defense mechanisms to exhibit pathogenicity. Patients who are susceptible to opportunistic infections are called immunocompromised hosts. Impairment of different components of the immune response can render a host susceptible to infection with different types of microorganisms. To consider the associations between the commonly observed immunodeficiency states, it is convenient to consider three categories of defects: granulocytopenia, cellular immune deficiency, and humoral immune deficiency. It is important, however, to realize that there are many conditions in which the function of more than one component of the immune response is altered.

Granulocytopenic Patients

Severe granulocytopenia is defined as a reduction of the absolute granulocyte count to $500/mm^3$ or less. This state is usually observed in patients with leukemia or those who have been treated with a variety of chemotherapeutic agents. Granulocyte dysfunction associated with conditions such as diabetes mellitus, azotemia, and alcoholism also appears to increase the risk of infection. The linkage between severe infection and secondary granulocyte dysfunction of this type is less clear, however, than the risk observed with severe granulocytopenia.

Granulocytopenic patients are highly susceptible to bacterial infections (e.g., encapsulated bacteria) and to some fungal infections (e.g., invasive *Aspergillus* disease), especially if the granulocytopenia is prolonged. They may present with rapidly progressive sinusitis or otitis media. Diabetic patients with ketoacidosis (and associated granulocyte dysfunction) and leukemic patients with prolonged neutropenia who have been treated with broad-spectrum antibiotics are at risk for both severe bacterial and fungal sinusitis. Nasal and sinus infections with fungal agents such as zygomycetes can progress rapidly into the orbit and even the brain and can result in death.

In addition to causing granulocytopenia, chemotherapeutic agents can damage the mucosal membranes of the mouth and pharynx. This breakdown of normal barriers predisposes the patient to stomatitis and pharyngitis and to damage of the tracheal mucosa and cilia, increasing the risk of both upper respiratory tract infection and pneumonia. In these circumstances, respiratory pathogens colonize the oropharynx and nasopharynx of these patients prior to the development of pneumonia. Since the patients are immunologically compromised, organisms that constitute part of the normal flora that are not pathogenic in normal hosts may also become invasive pathogens. Furthermore, exposure of these patients to prolonged hospitalization and to repeated courses of broad-spectrum antibiotics often results in increased colonization with virulent or drug-resistant microorganisms. Some of these colonizing pathogens, such as *Pseudomonas aeruginosa,* have a high propensity to progress from colonization to invasive disease in immunologically compromised patients. These infections have the potential for rapid progression to sepsis and death. Therefore, empiric therapy is usually initiated at the first sign of infection while awaiting the results of specific diagnostic studies.

Patients who have had bone marrow or organ transplants are also at increased risk for serious pneumonias. During the initial period of severe neutropenia after bone marrow transplantation, the same types of infections seen in other neutropenic states may be observed in these patients. After neutropenia resolves in bone marrow transplant patients, there is an increased risk of cytomegalovirus and *Aspergillus* infections associated with ongoing immunosuppression. The incidence of *Pneumocystis carinii* pneumonia as a complication of bone marrow transplantation has decreased with the introduction of antibiotic prophylaxis with trimethoprim and sulfamethoxazole. In bone marrow transplant patients being treated for chronic graft versus host disease, there is an unusually high rate of late *Streptococcus pneumoniae* infection. Similar infections occur in immunosuppressed recipients of solid organ transplants, with the addition of *Staphylococcus aureus* and *Legionella* to the list of potential pathogens. Pneumonia with either of these organisms or with gram-negative bacilli has a 60% mortality in solid organ transplant recipients. *Nocardia, Strongyloides,* and mycobacterial respiratory infections also occur with significant morbidity and mortality in these patients.

Patients With Defects in Cellular Immunity

CASE STUDY

A 30-year-old homosexual man presented to an emergency room with a 2-week history of shortness of breath. He first noticed having difficulty in climbing stairs at work, but more recently he has felt short of breath even at rest. On questioning, he reported some subjective fever but denied weight loss or a productive cough. The patient denied ever having been tested for human immunodeficiency virus (HIV) and had been monogamous for the last 4 years.

The patient's temperature was slightly elevated, and he appeared mildly short of breath. His lung examination was normal, but his chest radiograph showed hazy infiltrates in both lower lung fields. Arterial blood gas sampling showed hypoxia.

Diagnosis. *Pneumocystis carinii* pneumonia is often the first manifestation of HIV infection. Therefore, the clinician must consider this diagnosis in high-risk patients even if there is no history of previous opportunistic infection. Patients may have severe disease despite a paucity of symptoms and signs. Induced sputum is stained with Giemsa or methenamine silver stain and examined for the presence of pneumocystis cysts. In experienced laboratories, the diagnosis can be made quickly and with excellent sensitivity, eliminating the need for more invasive diagnostic procedures in most cases.

Other potential but less likely causes of pneumonia in patients with HIV infection are community-acquired bacterial pneumonia (e.g., *Streptococcus pneumoniae*), fungal pneumonia (e.g., *Cryptococcus* or *Histoplasma*), or noninfectious causes (e.g., Kaposi sarcoma or lymphoma).

Defects in the network of T-lymphocyte and mononuclear-macrophage interactions can lead to cellular immune dysfunction and increased susceptibility to certain types of infections. These defects may be either congenital or acquired.

Acquired immunologic defects are much more common and are usually caused by immunosuppressive therapy used for cancer chemotherapy or for bone marrow or organ transplantation. A variety of underlying diseases, including lymphomas, other malignancies, and collagen vascular diseases, also appear to impair cellular immune responses independently of chemotherapy.

The types of opportunistic infections observed in patients with defects in cellular immune responses, such as those with acquired immunodeficiency syndrome (AIDS), are typically viral (e.g., herpesvirus, cytomegalovirus), mycobacterial (e.g., tuberculosis, *M. avium*), fungal (e.g., *Cryptococcus, Aspergillus, Histoplasma*), or parasitic (e.g., *Pneumocystis carinii*) in origin. In patients with AIDS due to HIV infection, *Pneumocystis carinii* pneumonia is the most common cause of death.

Tuberculosis tends to occur earlier in the course of HIV infection than do most of the other opportunistic pneumonias seen in these patients. Later in the course of AIDS, tuberculosis may have an atypical presentation, increasing the probability that the diagnosis will be missed and that more contacts will become infected. As AIDS progresses, the likelihood that a mycobacterial infection will be caused by *M. avium* rather than *M. tuberculosis* increases. Encapsulated bacteria, *Cryptococcus neoformans* and other fungi, and cytomegalovirus are other significant respiratory pathogens in the later phase of AIDS.

The oropharynx is susceptible to infection with *Candidia* and herpes simplex virus in patients with AIDS and in patients receiving immunosuppressive chemotherapy. Damage to the oral mucosa from chemotherapeutic agents can cause mucositis followed by secondary bacterial and fungal infections or reactivation of a latent viral infection. The appearance of the mucosa can be misleading, and diagnosis may require culture or biopsy.

Another acquired immunodeficiency state that can be classified loosely with the cellular immune defects is splenic insufficiency. The spleen is an important part of the reticuloendothelial system. In addition to the numerous immunologic responses provided by the cells of the spleen, it appears that a key splenic activity involves clearance of pneumococci *(Streptococcus pneumoniae)* from the blood. Patients with either splenic dysfunction or who lack a spleen (usually from surgical removal after traumatic rupture) have a greatly increased incidence of serious pneumococcal infections.

Patients With Defects in Humoral Immunity

The humoral arm of the immune response to infection consists primarily of the production of specific opsonizing and neutralizing antibodies. The serum complement system also plays a role in humoral immune responses to infection. As with other types of immunodeficiency disease, most defective humoral immune responses are acquired as a result of specific immunosuppressive therapy or underlying disease states (e.g., multiple myeloma, chronic lymphocytic leukemia). Less commonly, congenital defects in both antibody production (hypogammaglobulinemia) and complement synthesis (classical and alternative complement pathway defects) are observed.

Pneumonias in patients with hypogammaglobulinemia are usually caused by encapsulated bacteria (e.g., *Streptococcus pneumoniae, Haemophilus influenzae*) due to the lack of specific opsonizing antibodies to enhance phagocytosis of these organisms by granulocytic cells. Viral, mycobacterial, and fungal infections are usually not a significant problem in patients with humoral immunodeficiency states, unless there is also a defect in some component of the cellular immune response. In addition to problems with other encapsulated bacteria, patients with complement defects may have serious infections with *Neisseria meningitidis*. These pneumonias are often more severe in these patients and may be accompanied by bacteremia in a higher percentage of patients compared to normal hosts. Most episodes of fatal bacteremia in patients with leukemia originate with infections in the lung. Gram-negative bacilli account for most of these infections, with *Klebsiella pneumoniae, Escherichia coli,* and *Pseudomonas aeruginosa* being the most common.

Diagnosis

Diagnosis of opportunistic infections in patients with immunodeficiency diseases requires a high index of suspicion in addition to a working knowledge of the likely pathogens. Pneumonias may be difficult to diagnose due to the broad range of potential pathogens, the atypical presentation of the illness, and the noninfectious conditions that can mimic pneumonia.

In patients with underlying malignancy, open lung biopsy is required in many cases to establish a specific etiologic diagnosis. There is an increasing trend toward using bronchoscopy and bronchoalveolar lavage as the first step in the diagnostic strategy, resulting from the experience with AIDS patients. Bronchoalveolar lavage is diagnostic in most of these patients. Transbronchial biopsy may be required for diagnosis in some cases, but open lung biopsy is rarely necessary. Establishment of a specific etiologic diagnosis can be complicated by the presence of more than one potential pathogen in a specimen or by the coexistence of malignancy and infection.

It has been possible to diagnose cases of *Pneumocystis carinii* pneumonia in some patients with AIDS by direct examination of expectorated sputum for the presence of cysts. This may be due to the presence of large numbers of organisms in

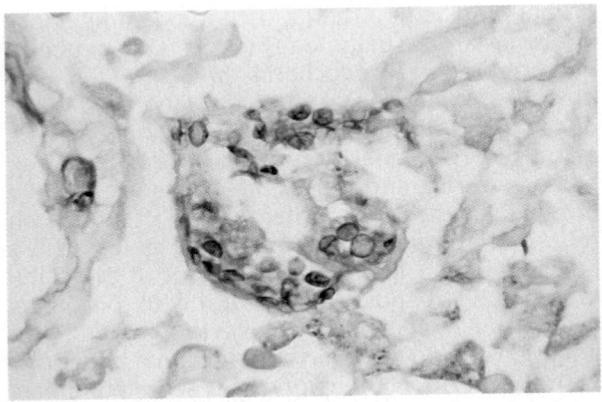

■■■■■■ F i g u r e   2 6 – 1 2
GMS stain of bronchoalveolar lavage (BAL) with *P. carinii* cysts.

these immunocompromised patients. The specimen must be digested, centrifuged, smeared, and fixed and then stained with Giemsa, methenamine silver, or a fluorescent antibody stain. Yield is higher if the sputum is collected in the morning after an overnight fast. Examination of induced sputum is 80% sensitive, whereas bronchoalveolar lavage (Fig. 26–12) increases sensitivity to 85 to 95%, and the addition of transbronchial biopsy increases the diagnostic yield to nearly 100%. Currently, *Pneumocystis carinii* cannot be cultured in the laboratory and must be diagnosed by histologic or immunofluorescent techniques.

The common theme in these and other invasive infections of the lower respiratory tract in immunocompromised patients is that an aggressive approach to diagnosis and empiric therapy is essential. It is important to make an etiologic diagnosis as frequently as possible to allow antimicrobial therapy to be focused on the specific pathogen.

Bibliography

Alkan ML, Beachey EH: Excretion of lipoteichoic acid by group A streptococci: Influence of penicillin on secretion and loss of ability to adhere to human oral epithelial cells. J Clin Invest 61:671, 1978.

Axelsson A, Brorson JE: The correlation between bacteriological findings in the nose and maxillary sinus in acute sinusitis. Laryngoscope 83:2003, 1973.

Bartlett JG, Finegold SM: Anaerobic infections of the lung and pleural space. Am Rev Respir Dis 110:56, 1974.

Beachey EH: Bacterial adherence: Adhesin-receptor interactions mediating the attachment of bacteria to mucosal surfaces. J Infect Dis 143:325, 1981.

Beachey EH, Ofek I: Epithelial cell binding of group A streptococci by lipoteichoic acid on fimbriae denuded of M protein. J Exp Med 143:759, 1976.

Betts RF, Reese RE: Lower respiratory tract infections. In Reese RE, Douglas RG (eds): A Practical Approach to Infectious Diseases, 2nd ed. Boston: Little Brown, 1986, pp 202–257.

Beutler B, Cerami A: Cachectin: More than a tumor necrosis factor. N Engl J Med 316:379, 1987.

Bodey GP, Buckley M, Sathe YS: Quantitive relationships between circulating leukocytes and infection in patients with acute leukemia. Ann Intern Med 64:328, 1966.

Brien JH, Bass JW: Streptococcal pharyngitis: Optimal site for throat culture. J Pediatr 106:781, 1985.

Broaddus C, Dake MD, et al: Bronchoalveolar lavage and transbronchial biopsy for the diagnosis of pulmonary infections in the acquired immunodeficiency syndrome. Ann Intern Med 102:747, 1985.

Broughton WA, Middleton RM III, et al: Bronchoscopic protected specimen brush and bronchoalveolar lavage in the diagnosis of bacterial pneumonia. Infect Dis Clin North Am 5:437, 1991.

Brubaker RR: Mechanisms of bacterial virulence. Annu Rev Microbiol 39:21, 1985.

Buckner CK, Clayton DE, et al: Parainfluenza 3 infection blocks the ability of a beta adrenergic receptor agonist to inhibit antigen-induced contraction of guinea pig isolated airway smooth muscle. J Clin Invest 67:376, 1981.

Caplan ES, Hoyt NJ: Nosocomial sinusitis. JAMA 247:639, 1982.

Carenfelt C, Lundberg C: Purulent and nonpurulent maxillary sinus secretions with respect to pO$_2$, pCO$_2$, and pH. Acta Otolaryngol (Stockh) 84:138, 1977.

Centers for Disease Control: Cases of specified notifiable diseases—United States, weeks ending December 29, 1990, and December 30, 1989 (52nd week) 39:944, 1991.

Chapman SW, Wilson JP: Nocardiosis in transplant recipients. Semin Respir Infect 5:74, 1990.

Cherry JD, Dudley JP: Sinusitis. In Feigin RD, Cherry JD (eds): Textbook of Pediatric Infectious Diseases, 2nd ed. Philadelphia: WB Saunders, 1987, pp 161–168.

Daum RS, Smith AL: Epiglottitis (supraglottitis) In Feigin RD, Cherry JD (eds): Textbook of Pediatric Infectious Diseases, 2nd ed. Philadelphia: WB Saunders, 1987, pp 224–237.

Dobie RA, Tobey DN: Clinical features of diphtheria in the respiratory tract. JAMA 242:2197, 1979.

Douglas RG Jr, Alford BR, Couch RB: Atraumatic nasal biopsy for studies of respiratory virus infection in volunteers. Antimicrob Agents Chemother 8:340, 1968.

Epstein JB, Gangbar SJ: Oral mucosal lesions in patients undergoing treatment for leukemia. J Oral Med 42:132, 1987.

Fishman JA: Diagnostic approach to pneumonia in the immunocompromised host. Semin Respir Infect 1:133, 1986.

Foy HM, Kenny GE, et al: Long-term epidemiology of infections with *Mycoplasma pneumoniae*. J Infect Dis 139:681, 1979.

Friend PA: Pulmonary infection in cystic fibrosis. J Infect 13:55, 1986.

Goldman WE, Klapper DG, Baseman JB: Detection, isolation and analysis of a released *Bordetella pertussis* product toxic to cultured tracheal cells. Infect Immun 36:782, 1982.

Good JA, Taryle DA, et al: The diagnostic value of pleural fluid pH. Chest 78:55, 1980.

Goren MB: Phagocyte lysosomes: Interactions with infectious agents, phagosomes, and experimental perturbations in function. Annu Rev Microbiol 31:507, 1977.

Green GM: In defense of the lung. Am Rev Respir Dis 102:691, 1970.

Green GM, Carolin D: The depressant effect of cigarette smoke on the in vitro antibacterial activity of alveolar macrophages. N Engl J Med 276:421, 1967.

Gross PA: Epidemiology of hospital-acquired pneumonia. Semin Respir Infect 2:2, 1987.

Gwaltney JM Jr: Sinusitis. In Mandell GL, Douglas RG Jr, Bennett JE (eds): Principles and Practice of Infectious Diseases, 3rd ed. New York: Wiley & Sons, 1990, pp 510–514.

Hawkins DB, Miller AH, et al: Acute epiglottitis in adults. Laryngoscope 83:1211, 1973.

Hays GC, Mullard JE: Can nasal bacterial flora be predicted from clinical findings? Pediatrics 49:596, 1972.

Hendley JO, Sande MA, et al: Spread of Streptococcus pneumoniae in families. I. Carriage rate and distribution of types. J Infect Dis 132:55, 1975.

Hooton RW, Haley RW, et al: The joint associations of multiple risk factors with the occurrence of nosocomial infection. Am J Med 70:960, 1981.

Hopewell PC: Tuberculosis and human immunodeficiency virus infection. Semin Respir Infect 4:111, 1989.

Hopewell PC, Luce JM: Pulmonary involvement in the acquired immune deficiency syndrome. Chest 87:104, 1984.

Huxley EJ, Viroslav J, et al: Pharyngeal aspiration in normal adults and patients with depressed consciousness. Am J Med 64:564, 1978.

Ida S, Hook JJ, et al: Enhancement of IgE-mediated histamine released from human basophils by viruses: Role of interferon. J Exp Med 145:892, 1977.

Johanson WG, Pierce AK, Sanford JP: Changing pharyngeal bacterial flora of hospitalized patients. N Engl J Med: 281:1137, 1969.

Johnston RB: Monocytes and macrophages. N Engl J Med 318:747, 1988.

Kasupski GJ, Leers WD: Presumed respiratory syncytial virus pneumonia in three immunocompromised adults. Am J Med Sci 285:28, 1983.

Kendrick PL, Eldering G, Eveland WC: Fluorescent antibody techniques. Methods for identification of Bordetella pertussis. Am J Dis Child 101:149, 1961.

Klein JO: Otitis externa, otitis media, mastoiditis. In Mandell GL, Douglas RG Jr, Bennett JE (eds): Principles and Practice of Infectious Diseases, 3rd ed. New York: Wiley & Sons, 1990, pp 505–510.

Lang WR, Howden CW, et al: Bronchopneumonia with serious sequalae in children with evidence of adenovirus type 21 infection. BMJ 1:73, 1969.

Levy M, Dromer F, et al: Community-acquired pneumonia: Importance of initial non-invasive bacteriologic and radiographic investigations. Chest 92:43, 1988.

Lew P, Zubler R, Vaudauz P: Decreased heat-labile opsonic activity and complement levels associated with evidence of C3 breakdown products in infected pleural effusions. J Clin Invest 63:326, 1979.

Linden BE, Aguilar EA, Allen SJ: Sinusitis in the nasotracheally intubated patient. Arch Otolaryngol Head Neck Surg 114:860, 1988.

Linneman CC Jr, Nasenbeny J: Pertussis in the adult. Annu Rev Med 28:179, 1977.

Lorber B, Swenson RM: Bacteriology of aspiration pneumonia. A prospective study of community and hospital acquired cases. Ann Intern Med 181:329, 1974.

Luce JM, Clement MJ: Pulmonary diagnostic evaluation in patients suspected of having an HIV-related disease. Semin Respir Infect 4:93, 1989.

Mackowiak PA: The normal microbial flora. N Engl J Med 307:83, 1982.

Maki DG: Control of colonization and transmission of pathogenic bacteria in the hospital. Ann Intern Med 89:777, 1978.

Male CJ: Immunoglobulin A₁ protease production by Haemophilus influenzae and Streptococcus pneumoniae. Infect Immun 26:254, 1979.

McDonald CJ, Tierney WM, et al: A controlled trial of erythromycin in adults with nonstreptococcal pharyngitis. J Infect Dis 152:1093, 1985.

McMillian JA, Sandstrom C, et al: Viral and bacterial organisms associated with acute pharyngitis in a school-aged population. J Pediatr 109:747, 1986.

Mermel LA, Maki DG: Bacterial pneumonia in solid organ transplantation. Semin Respir Infect 5:10, 1989.

Meyers BR, Wormser G, Hirschman SZ: Rhinocerebral mucormycosis. Premortem diagnosis and therapy. Arch Intern Med 139:557, 1979.

Murray PR, Washington JA: Microscopic and bacteriologic analysis of expectorated sputum. Mayo Clin Proc 50:339, 1975.

Mustoe T, Strome M: Adult epiglottitis. Am J Otolaryngol 4:393, 1983.

Pappenheimer AM: The diphtheria bacillus and its toxin: A model system. J Hygiene 93:397, 1984.

Pearson RD, Symes P, et al: Inhibition of monocyte oxidative responses by Bordetella pertussis adenylate cyclase toxin. J Immunol 139:2749, 1987.

Pittman M, Furman BL, Wardlaw AC: Bordetella pertussis: Respiratory tract infection in the mouse: Pathophysiological responses. J Infect Dis 142:56, 1980.

Proud D, Reynolds CJ, et al: Nasal provocation with bradykinin induces symptoms of rhinitis and a sore throat. Am Rev Respir Dis 137:613, 1988.

Root RK, Sande MA: New Dimensions in Antimicrobial Therapy. New York: Churchill Livingstone, 1984.

Sakakura Y, Sasaki Y, et al: Mucociliary function during experimentally induced rhinovirus infection in man. Ann Otol Rhinol Laryngol 82:203, 1973.

Scharer L, McClement JH: Isolation of tubercle bacilli from needle biopsy specimens of parietal pleura. Am Rev Respir Dis 97:466, 1968.

Schimpff SC, Young VM, Greene WH: Origin of infection in acute nonlymphocytic leukemia; significance of hospital acquisition of potential pathogens. Ann Intern Med 77:707, 1972.

Schneerson R, Robbins JB: Induction of serum Haemophilus influenzae type b capsular antibodies in adult volunteers fed cross reacting Escherichia coli 075:K100:H5. N Engl J Med 292:1093, 1975.

Schuller DE, Birch JG: The safety of intubation in croup and epiglottitis: An eight-year follow-up. Laryngoscope 85:33, 1975.

Shapiro ED: Milmoe GJ, Wald ER: Bacteriology of the maxillary sinuses in patients with cystic fibrosis. J Infect Dis 146:589, 1982.

Simmons BP, Wong ES: CDC guidelines for the prevention and control of nosocomial infections: Guideline for prevention of nosocomial pneumonia. Am J Infect Control 11:230, 1983.

Sinnott JT, Emmanuel PJ: Mycobacterial infections in the transplant patient. Semin Respir Infect 5:65, 1990.

Sorvillo FJ, Huie SF, et al: An outbreak of respiratory syncytial virus pneumonia in a nursing home for the elderly. J Infect 9:252, 1984.

Stiehm ER, Sztein MB, et al: Deficient antigen expression on human cord blood monocytes. Reversal with lymphokines. Clin Immunol Immunopathol 30:430, 1984.

Stone WJ, Schaffner W: Strongyloides infection in transplant recipients. Semin Respir Infect 5:58, 1990.

Teele DW, Klein JO, et al: Middle ear disease and the practice of pediatrics. JAMA 249:1026, 1983.

Todd JK: Throat cultures in the office laboratory. Bull N Y Acad Med 1:265, 1982.

Uchida T: Diphtheria toxin. In Dorner F, Drews J (eds): Pharmacology of Bacterial Toxins. Tarrytown, NY: 1986, Pergamon, pp 693–708.

Vianna NJ: Nontuberculous bacterial empyema in patients with and without underlying disease. JAMA 215:69, 1971.

Wald ER, Milmoe GJ, et al: Acute maxillary sinusitis in children. N Engl J Med: 304:749, 1981.

Weiss AA, Hewlett EL: Virulence factors of *Bordetella pertussis*. Annu Rev Microbiol 40:661, 1986.

Welliver RC, Wong DT, et al: The development of respiratory syncytial virus-specific IgE and the release of histamine in nasopharyngeal secretions after infection. N Engl J Med 305:841, 1981.

Wells RG, Sty JR, Landers AD: Radiological evaluation of Pott puffy tumor. JAMA 255:1331, 1986.

Winston DJ, Schiffman G, Wang DC: Pneumococcal infections after human bone-marrow transplantation. Ann Intern Med 91:835, 1979.

CHAPTER 27

Skin and Soft-Tissue Infections

Mary Castiglia •
Raymond A. Smego, Jr.

OBJECTIVES

1. Describe the function of the skin as a host defense mechanism.

2. Discuss the role of the microbial skin flora.

3. List the organisms that compose the skin flora.

4. Name the causative agents of primary bacterial infections.

5. Characterize each of the following pyoderma:

   ■ Impetigo
   ■ Folliculitis
   ■ Furuncle and carbuncle
   ■ Cellulitis and erysipelas
   ■ Myonecrosis
   ■ Paronychia
   ■ Erysipeloid
   ■ Erythrasma

6. Describe how each of these infections are diagnosed in the laboratory.

7. Discuss how other agents such as fungal, parasitic, and viral agents cause skin infections.

8. Name the mycobacterial species that cause skin infections and describe how the infection is acquired.

SKIN AND SKIN STRUCTURES

The skin is the body's first line of defense against microbial invasion. As a dynamic physical barrier, it continually undergoes epithelial cell turnover, removing substances as well as potentially pathogenic microorganisms that are present on its surface. In addition, the skin is colonized with a variety of resident microbes that perform a protective function.

This chapter discusses the following:

■ the role of the normal skin flora in the pathogenesis of skin infections

■ the wide variety of skin and soft-tissue infections

■ the etiology of each of these infections

Anatomy of the Skin

The skin consists of three layers: epidermis, dermis, and subcutaneous layer (Fig. 27–1). The epidermis is the outermost layer and is composed of several layers of epithelial cells. The stratum corneum, which is the outermost layer of the epidermis, contains dead cells consisting of a protein called keratin. The second skin layer, the dermis, is a thick layer composed of connective tissue. Sweat gland ducts, hair follicles, and oil gland ducts are found in the dermis and penetrate into the subcutaneous layer. These structures also provide potential passageways through which microbes can enter the skin. Sebum and perspiration are able to provide moisture and nutrients necessary for the growth of certain microbes. However, proliferation of other pathogenic microorganisms can be inhibited by salt and lysozymes contained in perspiration and in the fatty acids found in sebum.

Role of Normal Skin Flora

Normal flora of the skin consists of those microbes able to adapt to the high salt concentration and drying effects of the skin. Important microflora includes gram-positive cocci such as staphylococci and streptococci. Coagulase-negative staphylococci such as *S. epidermidis* are permanent skin residents; coagulase-positive *S. aureus* is a transient colonizer. Other normal flora includes diphtheroids, such as *Propionibacterium acnes* and *Corynebacterium xerosis*, and

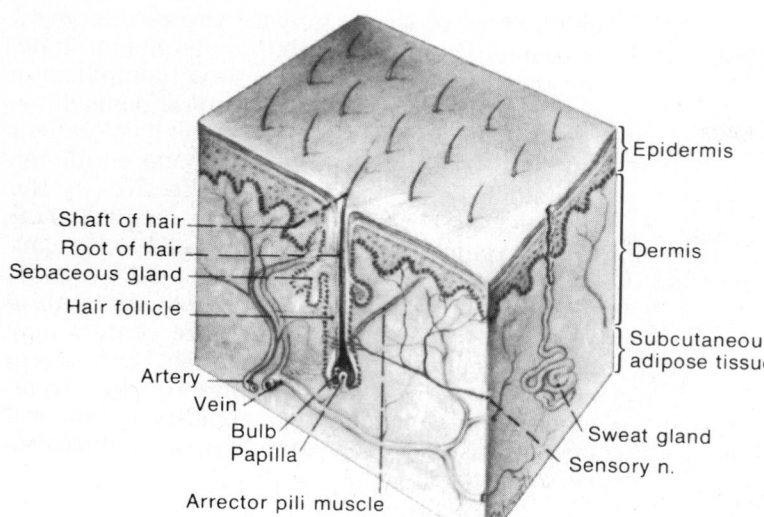

■ F i g u r e 2 7 – 1
Anatomy of the skin. (With permission from Jacob SW, Francone CA: Elements of Anatomy and Physiology, 2nd ed. Philadelphia: WB Saunders, 1989, p 37.)

the yeasts *Candida* and *Pityrosporum*. Although vigorous washing reduces the amount of surface skin flora, it does not eliminate resident flora colonization.

CLINICAL INFECTIONS

There are many infectious diseases of the skin. These can be classified in various ways: according to etiologic organisms (e.g., bacterial, viral, mycobacterial, fungal, parasitic); whether they occur as primary entities, secondary to preexisting skin lesions, or as manifestations of systemic disease; or according to the morphology of the skin lesion produced (Table 27–1). This section examines each of these classifications and describes some of the more clinically important and prevalent infections of skin and soft tissues.

Bacterial Skin Infections

In the management of bacterial infection of the skin, the surface of intact or ulcerated skin is often swabbed for purposes of Gram staining and culture. In most cases, however, this provides little or no clinically useful information because of the lack of correlation between surface colonization and below-the-surface infection. Deep aspirates of involved tissue or specimens taken from closed skin lesions are more interpretable. For example, if pustules or vesi-

cles are present, the roof or crust should be removed with a sterile blade, and any pus or exudate should be examined, Gram stained, and cultured. Obtaining a specimen in a patient with erysipelas or gangrenous or crepitant cellulitis may involve injecting a small amount (about 3 cc) of preservative-free physiologic saline into the advancing margin of the affected skin, aspirating back, and culturing the fluid that is withdrawn. Exuded pus or wound dressings should always be examined for the presence of granules and branching filaments, suggestive of infections with actinomycetes or fungi.

Primary Pyodermas

IMPETIGO

Impetigo is a common pyoderma most often caused by group A streptococci (*S. pyogenes*) (Table 27–2). Less than 10% of cases are caused by *S. aureus*. Group B streptococci may occasionally cause impetigo in a newborn infant, secondary to acquisition of colonizing vaginal flora from the mother. Initially, lesions of impetigo begin as small vesicles that pustulate and rupture, creating a thick, yellow, encrusted appearance. The lesions are superficial and painless but pruritic and are easily spread by scratching. A bullous form of impetigo is caused by phage group II strains of *S. aureus* that produce an exfoliative toxin. The localized and discrete lesions produced are thin-walled and blister-like and

■■■ T A B L E   2 7 – 1

MORPHOLOGIC TYPES OF SKIN LESIONS AND SOME ASSOCIATED INFECTIOUS DISEASES

Macular, Papular, or Maculopapular Rashes

Rubeola (measles)
Rubella (German measles)
Roseola
Other viral exanthems
Scarlet fever
Toxic shock syndrome
Secondary syphilis

Smooth Papules

Molluscum contagiosum
Condyloma latum (secondary syphilis)

Verrucous Papules or Plaques

Condyloma acuminata (venereal warts)
Viral warts
Cutaneous tuberculosis
Blastomycosis
Coccidioidomycosis
Chromomycosis

Wheals

Urticaria
Scabies
Cercarial dermatitis (swimmer's itch)

Pruritic Papules

Scabies
Folliculitis

Erythematous Patches or Nodules

Erythema infectiosum (fifth disease)
Bacterial cellulitis
Necrotizing (gangrenous) cellulitis, fasciitis, myonecrosis
Disseminated mycoses

Serpiginous or Annular Plaques

Erythema multiforme
Erythema chronicum migrans (Lyme borreliosis)
Cutaneous larval migrans (creeping eruption)

Vesicles or Bullae

Herpes simplex
Herpes zoster
Varicella (chickenpox)
Hand-foot-mouth disease
Herpangina
Staphylococcus scalded skin syndrome

Pustules

Folliculitis
Impetigo
Acne
Disseminated gonococcal infection
Furuncles, carbuncles
Kerion
Herpetic whitlow
Ecthyma contagiosum (orf)
Milker's nodule
Hidradenitis suppurativa

Petechiae, Purpura, and Ecchymoses

Rocky Mountain spotted fever
Other rickettsial infections
Meningococcemia
Gonococcemia
Infective endocarditis
Plague
Dengue and other hemorrhagic fever viruses
Enteroviral infections
Leptospirosis

Ulcers or Necrosis

Primary syphilis
Herpes simplex
Chancroid
Lymphogranuloma venereum
Granuloma inguinale
Impetigo
Ecthyma gangrenosum
Sporotrichosis
Atypical mycobacteria
Nocardiosis
Histoplasmosis
Anthrax
Ecthyma contagiosum (orf)
Tularemia
Leishmaniasis

children in order to prevent the development of immune complex–related acute glomerulonephritis, a serious nonsuppurative complication of *S. pyogenes* infections. Parenteral penicillin is the drug of choice, or erythromycin if the patient is allergic to penicillin. Older topical antibiotic-containing ointments were ineffective in the treatment of impetigo. However, a relatively new topical agent, mupirocin, has proved to be clinically useful and as effective as oral systemic therapy. Application of mupirocin to minor abrasions and insect bites in day-care centers may help prevent the spread of impetigo. Healing generally occurs without scarring. Effective treatment of bullous impetigo consists of an oral penicillinase-resistant penicillin or cephalosporin or erythromycin.

FOLLICULITIS

Folliculitis is an inflammation and infection of hair follicles. *S. aureus* is the most common etiologic agent of folliculitis, although *Pseudomonas aeruginosa* has been implicated in cases acquired from contaminated swimming pools or whirlpools. Lesions appear as small, erythematous papules and often evolve to form pustules with a whitish or yellowish central zone. Common sites for folliculitis include points of friction, such as the hips, buttocks, axillae, and scalp. Infection within the ear canal may lead to otitis externa. Scarring rarely develops. Sycosis

■■■ T A B L E   2 7 – 2

MOST COMMON PRIMARY PYODERMAS

Infection	Organism	Comments
Impetigo	*Streptococcus pyogenes*, occasionally; *Staphylococcus aureus*, if bullous	Children affected most; communicable; no fever
Erysipelas	*S. pyogenes*, occasionally; other β streptococci or *S. aureus*	Distinct raised borders; fever common
Cellulitis	*S. pyogenes*, *S. aureus*; *Haemophilus influenzae* in children	Erythema, tenderness, pain, edema, warmth; fever common
Folliculitis	*S. aureus*; gram-negative bacilli or *Candida* if predisposing conditions	Papules around hair follicles; areas exposed to whirlpool bath (*P. aeruginosa*)
Furuncle	*S. aureus*	Fluctuant, painful nodules often in intertriginous areas
Carbuncle	*S. aureus*	Multiple abscesses
Paronychia	*S. aureus*, gram-negative bacilli, *Candida*	Periungual swelling

contain a clear yellow fluid (Fig. 27–2). When the bullae rupture, they dry to form a transparent, varnish-like crust.

Impetigo is common in hot and humid climates and is highly contagious. Prompt treatment of impetigo is especially important in infants and

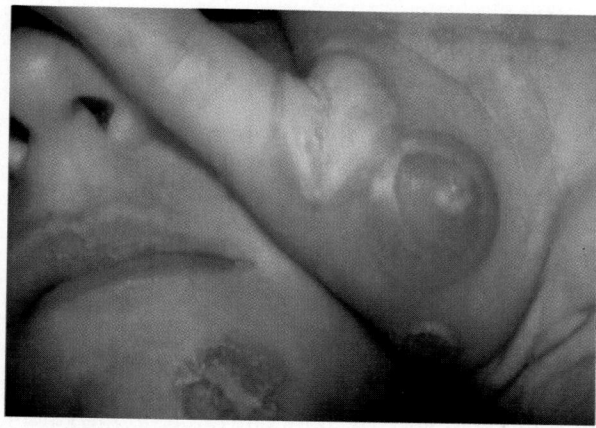

■ F i g u r e   2 7 – 2
Bullous impetigo caused by *Staphylococcus aureus*.

barbae is a form of folliculitis occurring in beard-ed men. A folliculitis caused by *Candida* species or gram-negative bacteria, including *P. aeruginosa,* may occur in immunocompromised hosts.

FURUNCLES AND CARBUNCLES

Lesions of folliculitis may develop into a deep-er inflammatory nodule called a furuncle (Fig. 27–3). A carbuncle is an abscess that extends even more deeply into the subcutaneous fat and may have multiple draining sites. *S. aureus* is the most common causative pathogen. Furuncles are initially red and firm but soon become painful and fluctuant. They generally drain spontaneous-ly. Systemic symptoms such as fever, chills, and malaise may accompany a carbuncle. Carbun-cles commonly occur at the nape of the neck and on the back of the thighs. If a carbuncle or furun-cle is associated with cellulitis or constitutional symptoms, it should be treated with an oral penicillinase-resistant penicillin or cephalospo-rin, erythromycin, or clindamycin. Furuncles can usually be managed by the application of moist heat and antimicrobial therapy. Surgical drainage is generally needed for most carbuncles. In cases of recurrent furunculosis, nasal application of mupirocin ointment and orally administered an-tibiotics such as ciprofloxacin and rifampin may be useful in reducing or eliminating the nasopha-ryngeal carriage of *S. aureus*.

CELLULITIS AND ERYSIPELAS

Cellulitis is a diffuse inflammation and infection of the superficial skin layers. It appears as a local-ized area of mildly painful erythema, warmth, and swelling of the skin with poorly demarcated mar-gins. Depending on its extent and severity, celluli-tis may or may not be accompanied by fever and other clinical or laboratory features of systemic infection (e.g., malaise, rigors, headache, elevat-ed white blood cell count).

In contrast, erysipelas is a deeper form of cel-lulitis that involves not only the superficial epi-dermis but also the underlying dermis and lym-phatic channels. Erysipelas is characterized as a painful, indurated area of cellulitis with a raised, sharply demarcated border and a typical deep crimson hue. Patients generally have fever and leukocytosis. Most uncomplicated cases remain confined to the dermis and lymphatics, but the potential for bacteremia makes erysipelas a po-tentially life-threatening infection. Erysipelas has a strong predilection for involvement of the face and, to a lesser extent, the lower extremities (Fig. 27–4). The majority of cases are caused by group A streptococci, although *S. aureus* can be im-plicated in roughly 10% of cases. Factors predis-posing to erysipelas include diabetes mellitus, venous stasis, alcohol abuse, and lymphatic ob-struction. Diagnostic aspiration of the advancing margins of the erysipelas lesion, as described above, may be useful. With bacteremic disease, the causative organism may be recovered from blood cultures. Intramuscular or oral penicillin or oral erythromycin may be effective in very mild cases. In most cases, however, because of the involvement of lymphatics and the potential for blood stream invasion, intravenous therapy with penicillin G or an antistaphylococcal agent should be given.

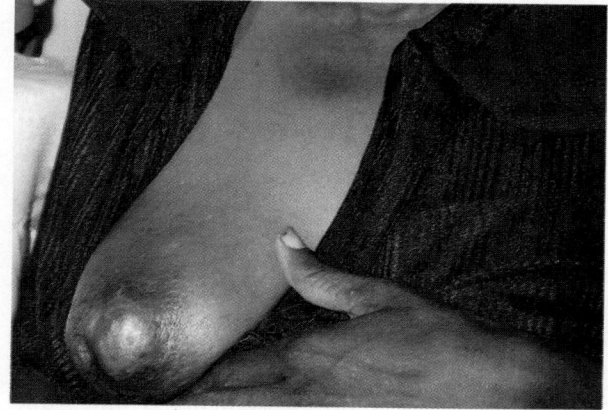

■ F i g u r e   2 7 – 3
Staphylococcus aureus furuncle of the breast.

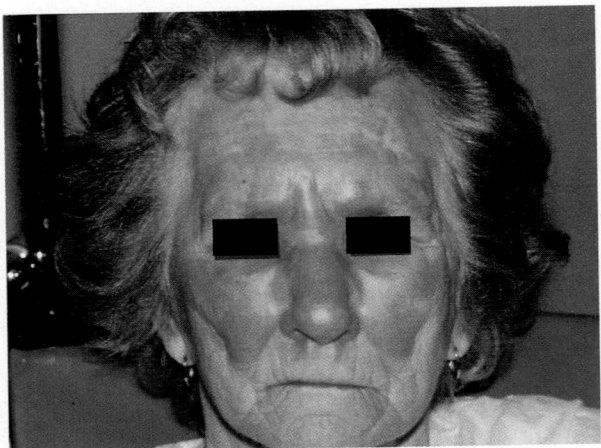

■ F i g u r e   2 7 – 4
Erysipelas due to *S. pyogenes*.

MYONECROSIS

CASE STUDY

An 18-year-old man suffered a severe crush injury to his right forearm in a dune buggy accident on the beach. He was admitted to a local hospital where within 18 hours he developed high fever and systemic toxicity, with rapidly progressive areas of necrosis, crepitus, and hemorrhagic bulla formation on the arm to the mid-shoulder. Broad-spectrum antibiotics were begun, and he was transferred by lifeflight helicopter to the university hospital for hyperbaric oxygen therapy and surgery. Immediately on transfer he was taken to the operating room for extensive soft tissue débridement.

Diagnosis. At surgery, superficial, fascial, and muscle compartments of the lower and upper arm were found to be involved by infection. Gram-staining of bulla fluid and operative specimens revealed fat, non–spore-forming gram-positive rods with few polymorphonuclear leukocytes, and a diagnosis of clostridial myonecrosis was made. A high amputation of the humerus was performed. Three postoperative treatments with hyperbaric oxygen were administered, and high-dose penicillin G was continued for 14 days. Biochemical testing of anaerobic organisms in pure culture revealed *Clostridium perfringens*.

Histotoxic clostridia cause myonecrosis, otherwise known as gas gangrene. *Clostridium septicum* is categorized as a histotoxic clostridia, along with *C. perfringens, C. novyi, C. histolyticum,* and *C. sordellii. C. septicum* is an anaerobic gram-positive rod exhibiting subterminal spores. It grows well in culture with a "swarming" appearance. It is β-hemolytic and can easily be differentiated from other common clostridia. Histotoxic clostridia are found in the soil and are part of the normal flora of the large intestine. Contamination of injured tissue with spore from soil containing these organisms or from bowel flora is the usual means of transmission. In order for myonecrosis to develop, a lowered oxidation-reduction potential must be present. This is a result of a drop in pH due to the reduction of pyruvate to lactate.

Clinical manifestations of myonecrosis include severe pain, edema, cellulitis, production of gas, and foul-smelling purulent discharge—all leading to tissue death. Treatment is extensive surgical débridement and the administration of polyvalent antitoxin and antimicrobials. The use of a hyperbaric chamber has also proved to be effective.

Infections that produce tissue necrosis and soft-tissue gas represent an important subset of cellulitis cases. Serious and frequently life-threatening, these gangrenous and crepitant soft-tissue infections can be classified according to their level of soft-tissue involvement (e.g., superficial epidermal/dermal structures, fascia, or muscle), and according to the causative microbiologic agent or agents. Most infections are polymicrobial, with mixed aerobic-anaerobic, gram-positive and gram-negative bacterial flora being recovered. Anaerobes are the most important gas-producing organisms, with gas being detectable by palpation of the infected area or by soft-tissue radiograph (Fig. 27–5). Other bacteria, however, are capable of gas production, including *Escherichia coli, Proteus, Klebsiella, Aeromonas,* and hemolytic strains of *S. aureus.*

As infection progresses from superficial gangrene to deeper necrotizing fasciitis to still deeper myonecrosis, patients appear more acutely ill and toxic, have more soft-tissue pain, and can exhibit a progressive and fulminant downhill course. Gangrenous and crepitant cellulitides represent true medical-surgical emergencies; surgical exploration is invariably required in order to determine the degree and level of soft-tissue spread, to excise all devitalized tissue, and to obtain useful deep-tissue specimens for accurate microbiologic processing. Aside from cases of infection caused by a single organism (e.g., *Clostridium perfringens* myonecrosis [see Fig. 27–5] or *Streptococcus pyogenes* necrotizing fasciitis), most of these infections can be treated with a combination of agents active against aerobic gram-positive cocci, Enterobacteriaceae, *Pseudomonas,* and anaerobes.

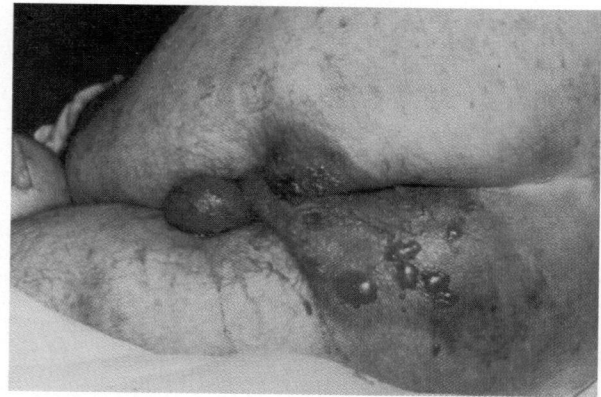

Figure 27-5

Clostridial myonecrosis ("gas gangrene"). (From an Upjohn Monogram, Anaerobic Infections. © 1986 The Upjohn Co., Kalamazoo, MI, p. 48.)

PARONYCHIA

Paronychia is an infection of the cuticle surrounding the nail bed. Cases generally follow minor trauma such as removing a hangnail. The involved part of the finger at the nail margin becomes painful, red, warm, and swollen, and pus may be expressed from around the nail bed. Staphylococci are usually the causative organisms. Paronychia usually responds to warm soaks, which often lead to spontaneous drainage of pus and resolution. Systemic antimicrobials and surgery are typically not required.

ERYSIPELOID

Erysipeloid is a superficial soft-tissue infection caused by *Erysipelothrix rhusiopathiae*. It typically occurs as an occupational hazard in handlers of animals, meat, poultry, and fish. The organism is difficult to Gram stain but may be isolated in culture. Mimicking erysipelas, the erysipeloid lesion is red and painful, with raised borders. The finger and the dorsum of the hand are the most common sites of infection. Penicillin G is the preferred treatment.

ERYTHRASMA

Erythrasma is a chronic, pruritic, reddish-brown macular infection found most commonly in men and in obese patients with diabetes mellitus. The infection is typically located in intertriginous areas such as the groin, toe web, axilla, and inframammary folds. The lesion tends to be finely scaled and wrinkled. *Corynebacterium minutissimum* is the causative organism and is easily observed with a Gram stain of the stratum corneum. The lesions produce a coral red fluores-

cence when examined under a Wood's lamp. Topical clindamycin and oral erythromycin are useful in the management of erythrasma.

Secondary Bacterial Skin Infections

Table 27–3 lists some common secondary bacterial skin infections that complicate preexisting dermatologic lesions. Etiologic organisms and important epidemiologic associations are also shown. Table 27–4 lists several systemic bacterial infections that have important cutaneous manifestations.

Hidradenitis suppurativa is a syndrome associated with a genetic defect of the apocrine sweat glands. Chronic obstruction of these glands, especially in the axillae and groin, predisposes to mixed bacterial superinfection of skin and skin structures, often accompanied by fever and tender lymphadenitis. Multiple recurrences of infection lead to tissue fibrosis, sinus tract formation, and scarring and disfigurement. Individual episodes of infection are treated with local moist heat, broad-spectrum antibiotics, and, frequently, surgical incision and drainage.

TABLE 27–3

INFECTIONS SECONDARY TO PREEXISTING LESIONS

Infection	Major Pathogen
Surgical wound infection	
Clean	*Staphylococcus aureus*, gram-negative bacilli
Contaminated, such as colon	Plus anaerobes, streptococci
Intravenous infusion sites	*S. aureus*, coagulase-negative staphylococci
Trauma	
Soil contamination	*Pseudomonas aeruginosa*, clostridia
Freshwater contamination	*Aeromonas, Plesiomonas*
Saltwater contamination	*Vibrio vulnificus*
Bites	
Human	Oral aerobes and anaerobes, *S. aureus*
Dog, cat	*Pasteurella multocida, S. aureus*, anaerobes
Rat	*Streptobacillus moniliformis, Spirillum minus (minor)*
Decubitus ulcer	Streptococci, *S. aureus*, coliforms, *Pseudomonas*, anaerobes including *Bacteroides fragilis*
Foot ulcer in diabetic patients	*S. aureus*, streptococci, coliforms, *P. aeruginosa*, anaerobes
Hidradenitis suppurativa	*S. aureus*, streptococci, coliforms, *Pseudomonas*, anaerobes
Burns	*S. aureus, Candida, P. aeruginosa*

■ T A B L E 2 7 – 4

CUTANEOUS INVOLVEMENT IN SYSTEMIC BACTERIAL INFECTIONS

Infective endocarditis
Bacteremia and sepsis
Scarlet fever
Toxic shock syndrome
Staphylococcal scalded skin syndrome
Enteric fever

CUTANEOUS MANIFESTATIONS OF SYSTEMIC BACTERIAL INFECTIONS

CASE STUDY

A 55-year-old Hispanic man with non–insulin-dependent diabetes presented to an emergency room with complaints of fever, pain, and swelling of the right lower leg (Fig. 27–6). The patient gave a history of alcohol abuse, and hepatitis as a child. Four weeks prior to admission the patient had eaten raw oysters, and he had experienced diarrhea for the past 2 to 3 weeks. On examination, the patient was acutely ill with tachycardia, hypotension, a slightly enlarged liver, and a right pleural effusion. An ulcerative cellulitis of the right lower leg was present, with overlying exudate. The admitting diagnosis was severe soft tissue infection with sepsis syndrome.

Diagnosis. Cultures of blood, stool, and urine showed no pathogenic growth. A smear made from the soft tissue exudate showed many gram-negative bacilli, and culture yielded *Vibrio vulnificus.* The patient was placed on antimicrobial therapy with ampicillin, gentamicin, and clindamycin; however, the cellulitis continued to worsen and became gangrenous. Surgical exploration revealed necrotizing cellulitis, and an above-the-knee amputation was performed. The patient ultimately had a successful recovery and was discharged from the hospital in stable condition.

Bacteremia with particular organisms can result in various morphologic rashes or lesions. *Vibrio vulnificus,* a halophilic vibrio, has been recognized as a virulent pathogen since 1976, when it was first described. The organism is known to cause severe necrotic cellulitis and primary sepsis. Infections are usually acquired when wounds are exposed to marine water or direct contact with raw shellfish. Ingestion of raw oysters has also been associated with primary septicemia due to *Vibrio vulnificus.*

Other organisms such as *Pseudomonas aeruginosa* may produce vesicles and bullae in some circumstances. Infection with this organism may cause a macular or maculopapular rash, as in

Shanghai fever, a *Pseudomonas* sepsis syndrome seen in the tropics and associated with fever and diarrhea. Also associated with *Pseudomonas* bacteremia is ecthyma gangrenosum; the characteristic lesion appears as a painless ulcer with a central black eschar. Gangrenous cellulitis has also been associated with *Pseudomonas* infection. Both of these latter syndromes are encountered most frequently in neutropenic patients.

Bacteremic infections with *S. aureus,* pneumococci, and *Haemophilus influenzae* can cause a rapidly progressive, frequently fatal syndrome called purpura fulminans, which results in cutaneous bleeding. The initial skin lesions of meningococcemia are erythematous macules, petechiae, and purpura located on the trunk and extremities (Fig. 27–7). They progress to gray, hemorrhagic, necrotic areas. The rash of disseminated *Neisseria gonorrhoeae* consists of a few painful pustules with a thin zone of purpura. There can be macules, papules, or bullae and occasionally purpuric infarcts. Cutaneous infections due to candidemia are characterized by multiple pink maculopapules or subcutaneous nodules on the trunk or extremities. These lesions are typically seen in immunosuppressed hosts such as neutropenic cancer patients and tissue transplant recipients.

Coagulase-positive staphylococci are capable of producing several toxins that affect the skin and result in distinct clinical syndromes. The production of an exfoliative toxin in the staphylococcal scalded skin syndrome (SSSS) results initially in fever, skin tenderness, and a scarlatiniform rash, followed by extensive bullae formation and exfoliation similar to that seen in burn

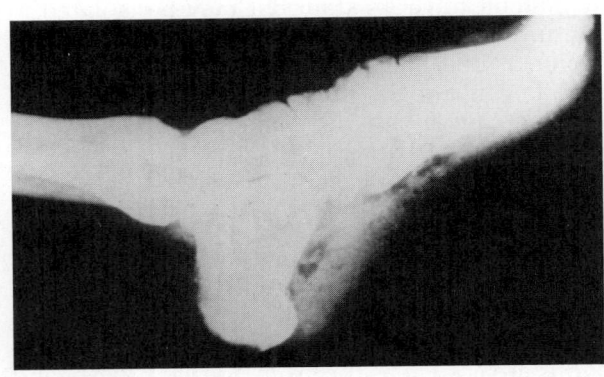

■ F i g u r e 2 7 – 6
Diabetic foot infection with soft-tissue gas formation.

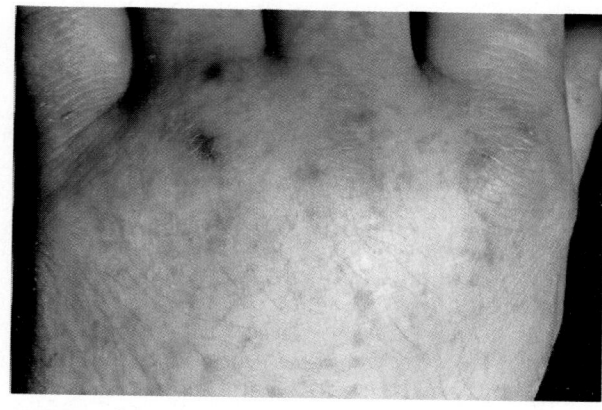

F i g u r e   2 7 – 7
Petechial lesion in meningococcemia.

patients. Large flaccid blisters form and rupture, causing skin to denude and peel off in sheets and leaving areas of bright red underlying skin exposed. Nosocomial epidemics of SSSS have been reported in newborn nurseries. Scarring generally does not occur. SSSS requires intravenous administration of antistaphylococcal agents.

Cutaneous desquamation may be a characteristic of the late stages of toxic shock syndrome, another serious systemic infection caused by noninvasive strains of *S. aureus* that elaborate exotoxin F. A diffuse sunburn-like erythroderma appears early in the course and is accompanied by fever, hypotension, and evidence of multiorgan dysfunction. Desquamation of skin, especially on the palms and soles, occurs during the convalescent stage of the illness. Although originally described in women in epidemiologic association with superabsorbent tampons colonized with *S. aureus,* toxic shock syndrome occurs in many different clinical settings. Examples are surgical wound infections and contaminated nasal packing in patients with nosebleeds. Treatment consists of supportive measures and antistaphylococcal antibiotic therapy.

CASE STUDY

A 48-year-old diabetic man sustained a minor blunt injury to his biceps region. Within 24 hours he developed pain and swelling of the arm, with delirium and high fever, and he was brought to a local emergency room. He appeared acutely ill, and there was edema and tenderness of the upper arm, with several areas of broken skin and black tissue necrosis. No soft tissue gas was detectable by examination or radiography. The patient underwent immediate surgical exploration, and the infection was noted to have tracked extensively along the fascial planes of the arm, from elbow to shoulder. All devitalized tissue was débrided, and antibiotics begun, but the patient expired 12 hours later. Gram stain of the operative deep tissue specimens revealed gram-positive cocci in chains, and all cultures grew β-hemolytic *Streptococcus pyogenes*, confirming the diagnosis of group A streptococcal necrotizing fasciitis.

A streptococcal toxic shock syndrome caused by *Streptococcus pyogenes* has also been well-described, especially in recent years, with the resurgence of invasive and complicated streptococcal infections. Like its staphylococcal counterpart, streptococcal toxic shock syndrome may occur whenever exotoxin-producing strains of group A streptococci cause infection or colonization of the skin or mucous membranes.

Scarlet fever is a characteristic form of group A streptococcal disease that occurs when the infecting strain produces a scarlatiniform toxin. Clinical characteristics may include those symptoms occurring with a streptococcal sore throat (or the disease may be associated with a wound, skin, or puerperal infection) as well as an erythroderma and a "strawberry tongue" with prominent papillae. The bright red, sandpaper-textured rash is often felt better than seen and appears most often on the neck and chest and in skin folds. Typically, the rash does not involve the face, but there is a flushing of the skin with circumoral pallor. During convalescence, desquamation of the skin occurs, especially on the hands and feet.

Infective endocarditis may cause a number of secondary skin manifestations. Immunologically mediated painful skin lesions located on the pads of the fingers and toes and on the thenar eminence are called Osler's nodes. These lesions reflect soft-tissue immune complex deposition. Janeway lesions are flat, painless lesions located on the palms and soles that represent microembolic seeding of the skin. Hemorrhagic, vasculitic skin lesions may also be seen in endocarditis (Fig. 27–8). Staphylococci and streptococci are typically associated with the above-mentioned skin manifestations, although gram-negative bacteria and *Candida* may produce similar lesions.

The rose spots of typhoid fever, caused by *Salmonella typhi*, are erythematous maculopapular lesions measuring 2 to 4 mm that blanch on pressure. They appear characteristically on the upper abdomen in crops of approximately 10 lesions. The lesions are transient and resolve

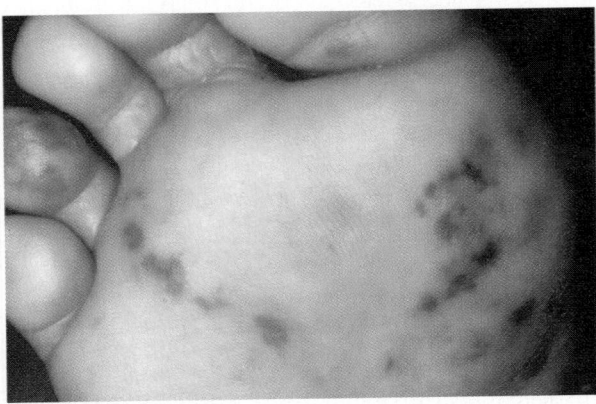

■Figure 27 – 8
Hemorrhagic vasculitic lesion of *S. aureus* endocarditis.

within hours to days. Rose spots are observed in less than 50% of patients with typhoid fever.

Miscellaneous Bacterial Skin Infections

Cutaneous diphtheria is rare in developed countries but may be seen in tropical climates, especially in areas of overcrowding and where immunization of susceptible populations is inadequate or incomplete. Access is gained through wounds or insect bites. The disease appearance of cutaneous diphtheria is variable and may be indistinguishable from impetigo.

Intertrigo is a cutaneous infection occurring in body areas of heat and moisture. As a result of erythema and maceration, skin breakdown may result. This occurs most frequently in the skin folds of obese adults and infants. The most likely organisms include *S. aureus*, *Candida*, and coliforms.

Melioidosis and glanders are diseases caused by *Pseudomonas pseudomallei* and *P. mallei*, respectively. Localized infection with lymphangitis and lymphadenitis may be the result of primary inoculation of organisms into the skin. A generalized papular or pustular rash may be seen in the septicemic form of these diseases. Glanders is rare but is often zoonotic in nature.

Francisella tularensis, a gram-negative bacterium, causes several forms of tularemia, including glandular, oculoglandular, pulmonic, and septicemic or typhoidal forms. The ulceroglandular type, however, is most frequently seen and presents as an indolent ulcer, often on the hand, accompanied by painful swelling of the regional lymph nodes. The disease is zoonotic, and human transmission occurs by direct inoculation of skin through handling of infected rabbits and other wild or, less commonly, domestic animals or by the bite of infected fleas, ticks, deerflies, and mosquitoes. Diagnosis of tularemia is most commonly made by a rise in specific antibodies in the patient's serum, although cross-agglutinations with *Brucella, Proteus,* and heterophile antibodies occur. Examination of ulcer exudates, lymph node aspirates, and other clinical specimens using a fluorescent antibody test may provide a rapid diagnosis. The infectious organism can be identified by culture on special media or by inoculation of laboratory animals with material from lesions, blood, or sputum; great care must be exercised with this approach, however, as this highly infectious agent poses an occupational hazard for laboratory workers. Streptomycin with or without tetracycline is the usual treatment.

Two bacterial diseases, rare in the United States, are included under the general term of rat-bite fever; streptobacillosis is caused by *Streptobacillus moniliformis,* and spirillosis by *Spirillum minus (minor).* They share several clinical and epidemiologic characteristics. An abrupt onset of fever and chills, headache, and muscle pain is followed shortly by a maculopapular or sometimes petechial rash that is most marked in the extremities. One or more joints then become swollen, red, and painful. There is typically a history of a rat bite within 10 days that healed normally. Laboratory confirmation is made by isolation of the causative organism after inoculating material from the primary lesion, lymph node, blood, joint fluid, or pus into culture media or laboratory animals. Serum antibodies may be detected by agglutination tests. Penicillin G is the treatment of choice.

Bartonellosis is a disease geographically restricted to the high-altitude valleys of Peru, Ecuador, and southwest Colombia. This disease, also called verruca peruana, is caused by the aerobic, pleomorphic, poorly staining gram-negative bacterium *Bartonella bacilliformis* and is transmitted to humans by the bite of infected sandflies. Bartonellosis is a febrile systemic infection with associated hepatosplenomegaly, generalized lymphadenopathy, severe anemia, and cutaneous manifestations. The dermal eruption may be miliary, with widely disseminated small, hemangiomatous nodules, or it may be nodular, with fewer but larger deep-seated lesions that are most prominent on the extensor surfaces of the limbs. Individual nodules may develop into tumor-like masses with an ulcerated surface. *Salmonella* infection frequently complicates bartonellosis. The organism may be cultured from the skin and subcutaneous lesions or occasionally from the blood. Diagnosis can also be made by histopathologic demonstration of organisms in tissue specimens using Giemsa staining.

Mycobacterial Skin Infections

Several *Mycobacterium* species are capable of causing skin infections. Primary or secondary infection with *M. tuberculosis* can cause a morphologic array of localized skin lesions, appearing as inflammatory nodules or papules, with lymphangitis and lymphadenitis. Extension from the lymph nodes or bones into the skin is called scrofuloderma. Multiple, diffuse skin lesions may result from hematogenous, or bloodborne, seeding to the skin.

Infection with the marine organism *M. marinum* develops in individuals with a history of cleaning fish tanks or using swimming pools. The microorganism enters through an open wound or via traumatic inoculation of intact skin. The lesions are generally solitary and appear as a tuberculoid granuloma. Alternatively, the organism may produce a lymphocutaneous syndrome characterized by the initial lesion (ulcer or nodule) followed by satellite lesions and nodular lymphangitis.

M. ulcerans infection usually occurs as a single painless ulcer with undermined edges (the so-called Buruli ulcer of the tropics). The disease is associated with swamps and may be a chronic infection in tropical climates. *M. chelonei* can produce subcutaneous nodules as part of either a local or a disseminated infection, the latter being seen most commonly in patients receiving corticosteroids (Fig. 27–9).

The classic skin manifestation of leprosy, caused by *M. leprae,* is a circumscribed, hypopigmented, or, less commonly, hyperpigmented macule. Lesions can occur singly or multiply, and because peripheral nerves are involved and infiltrated with organisms, these lesions are painless. Acid-fast bacilli and granulomas can be seen in variable numbers in tissue biopsy examinations from persons with leprosy. Disfigurement and deformity of the hands, feet, and cooler parts of the body, such as the ears and nose, are frequently seen. This is the result of repeated trauma and

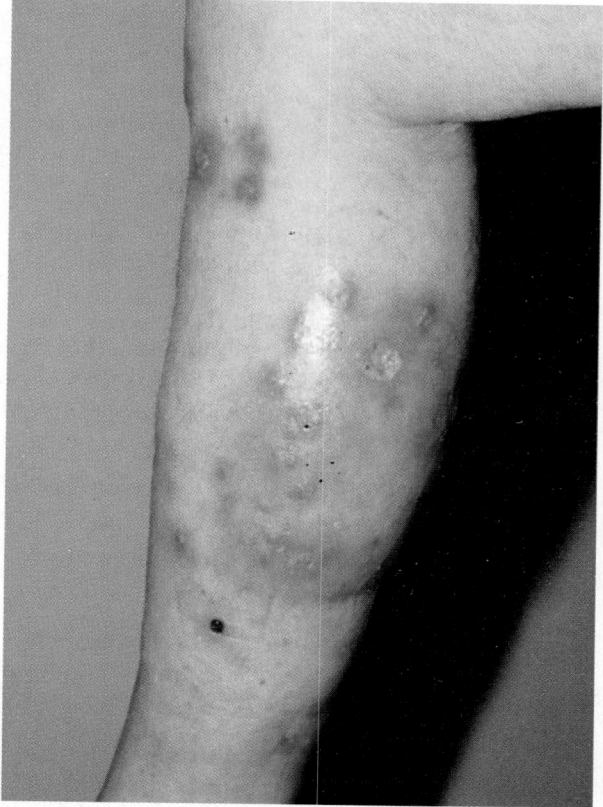

■ Figure 27–9
Subcutaneous nodules and cellulitis due to disseminated *Mycobacterium chelonei.*

secondary bacterial infection of neuropathic and anesthetic body parts. Effective treatment usually involves protracted combination chemotherapy (2 years or more) with agents such as rifampin, dapsone, and clofazimine.

Actinomycetes and the Skin

Actinomycosis is a chronic disease characterized by the formation of abscesses, fibrosis of tissues, and draining sinuses that discharge sulfur granules. It is caused by non–spore-forming anaerobic or microaerophilic bacterial species (especially *Actinomyces israelii*) of the genus *Actinomyces,* order Actinomycetales. Once thought to be fungi because of their branching, *Actinomyces* species and the closely related *Nocardia* species are classified as higher prokaryotic bacteria. *Actinomyces* species are gram-positive, pleomorphic, and diphtheroidal or, more commonly, delicately filamentous. Although there are thoracic, abdominopelvic, and central nervous system forms of the disease, involvement of the face and neck is the most common manifestation and often follows dental sepsis or manipulation, trauma, tonsillitis, otitis, or mastoiditis. Cervicofacial actinomycosis may extend to the underlying mandible or facial bones, leading to osteomyelitis. Penicillin is the preferred treatment.

Nocardia brasiliensis is primarily a cause of skin and soft-tissue infection. Clinical forms of the disease include subcutaneous abscesses, cellulitis, mycetoma, and the lymphocutaneous syndrome. *Nocardia* species can be differentiated from *Actinomyces* by their aerobic growth and partial acid-fastness using a modified Kinyoun stain.

Laboratory Diagnosis of Bacterial Skin and Soft-Tissue Infections

Laboratory diagnosis of bacterial skin infections may consist of direct smear examination and culture of the affected site, which usually reveal the causative agents. Smears prepared from exudative material and subsequently Gram stained may show the presence of inflammatory cells as well as characteristic morphology that may lead to an initial diagnosis, particularly if the infecting agents are gram-positive bacteria. Myonecrosis infections will reveal a mixture of gram-positive and gram-negative bacteria. Results of the direct smear examination are especially important to the clinician, particularly if anaerobic organisms such as *Clostridium* spp are suspected.

Culture is still the most sensitive method of diagnosing cutaneous infections. Agents of primary infections are recovered in routine culture using primary nonselective media such as blood and chocolate agars and selective media such as MacConkey agar. Additional selective media such as PEA (phenylethyl alcohol) or CNA (Columbia colistin-nalidixic acid) must be included if a mixed infection with gram-positive cocci and gram-negative bacilli is suspected. Samples taken from sites suspected of anaerobic organisms must be transported properly, using an anaerobic transport medium or container to maximize recovery. Since most anaerobic infections are polymicrobial, samples must be inoculated on culture media that are selective for gram-positive and gram-negative bacteria. Identifica-

tion of isolates is performed using a variety of methods, both conventional and automated.

Antimicrobial susceptibility testing is performed on isolates to demonstrate variable susceptibility patterns. In other cases, such as in group A streptococcus in impetigo, the infection is treated empirically.

Agents of cutaneous infections that manifest secondary to systemic bacterial invasion, as in the case of meningococcemia, may or may not be isolated from blood cultures or from the site of manifestation. With skin infections that manifest as a result of toxins produced by the invading organism, such as in cases of scalded skin syndrome, toxic shock syndrome, and scarlet fever, the invading organism may not be recovered from the blood or from the desquamated skin. In most of these cases, the invading organisms are colonizing at a distant focus.

Rickettsial Infections

Rickettsiae are gram-negative pleomorphic intracellular pathogens, classified as higher bacteria, that reside within endothelial cells and macrophages. Rickettsial infections are zoonoses; they have various types of animal reservoirs and are transmitted to humans through a number of species-specific insect vectors (e.g., ticks, mites, lice, fleas). Clinical symptoms common to rickettsial infections include high fever, chills, malaise, headache, myalgias, skin rash, and conjunctival injection. Systemic and cutaneous disease manifestations are the pathophysiologic result of diffuse small-vessel vasculitis. Dermatologically, individual rickettsial infections are characterized by the type and distribution of the associated skin rash (e.g., petechial or vesicular, centripetal or centrifugal) and by the presence or absence of a black eschar at the insect bite inoculation site.

In Rocky Mountain spotted fever (RMSF), the most frequently seen rickettsial infection in the United States, a maculopapular rash appears on the extremities on about the third day. It usually includes the palms and soles and spreads rapidly to most of the body. Petechia and purpura represent extravasation of blood out of blood vessels and into the skin and evolve commonly as a result of the cutaneous vasculitis. The clinical syndrome of RMSF may be confused with atypical measles, meningococcemia, other bacterial sepsis, enteroviral infection, and leptospirosis. RMSF is generally diagnosed via serologic testing (e.g., Weil-Felix, complement fixation) or by immunofluorescent staining of skin biopsy specimens. The infection is generally successfully treated with either a tetracycline or chloramphenicol.

Spirochetal Infections

Syphilis is an acute and chronic, venereally transmitted treponemal disease. It is characterized clinically by a primary lesion, a secondary eruption involving skin and mucous membranes, long periods of latency, and late tertiary lesions of the skin, bone, central nervous and cardiovascular systems, and other viscera.

The primary lesion usually appears as a papule at the inoculation site about 3 weeks after the initial exposure. Erosion and ulceration then occur, forming the characteristic indurated, painless chancre (Fig. 27–10) (in contrast to the painful "soft chancre" or chancroid caused by *Haemophilus ducreyi*). Firm, enlarged, nonfluctuant regional lymph nodes, called buboes, commonly follow. Spontaneous resolution of the primary lesion occurs in 4 to 6 weeks, only to be followed by a secondary eruption involving skin, mucous membranes, and internal viscera such as the liver. Mild systemic complaints, including fever and malaise, commonly accompany secondary syphilis. Both primary and secondary lesions are typically teeming with spirochetes and are thus infectious. Secondary manifestations also disappear spontaneously within weeks. Subsequently, the infection remains clinically latent for weeks to years. Within the first few years, latent syphilis may revert to form infectious mucocutaneous lesions.

Laboratory diagnosis is usually based on serologic testing of blood and cerebrospinal fluid, utilizing nontreponemal (e.g., the Venereal Disease Research Laboratory test [VDRL] and Rapid Plasma-Reagin test [RPR]) and treponemal (fluorescent treponemal antibody absorption test [FTA-ABS] and *Treponema pallidum* microhemagglutination test [MHA-TP]) antigens as screening and confirmatory tests, respectively. Lesions of primary and secondary syphilis can also be confirmed by darkfield or phase-contrast microscopic examination of lesion exudates or aspirates from lymph nodes. Penicillin G is the drug of choice for all clinical stages of syphilis.

Yaws (caused by *Treponema pallidum* spp *pertenue*), pinta (*T. carateum*), and bejel (*T. pallidum;* endemic syphilis) are three other important human treponemal diseases. These diseases differ notably from syphilis in certain respects, such as their geographic occurrence (all rural and outside of the United States), mode of transmission (all nonvenereal by direct contact), and tissue involvement (skin and/or mucous membrane and bone). They all exhibit clinical latency, share serologic cross-reactivity, and respond to penicillin.

The tick-borne spirochete of Lyme disease, *Borrelia burgdorferi,* characteristically produces a distinctive serpiginous skin lesion called erythema chronicum migrans (ECM) at the inoculation site. ECM is the most useful clinical diagnostic marker of Lyme disease. Later manifestations include joint, central nervous system, and cardiovascular system involvement. Lyme borreliosis has emerged as the leading vector-borne disease in the United States and is also seen in many other parts of the world. Diagnosis is best made on clinical grounds; serologic testing is currently problematic due to the lack of a standardized test antigen and resultant interlaboratory variation, and due to serologic cross-reactivity. Various antibiotics, including ceftriaxone, penicillin

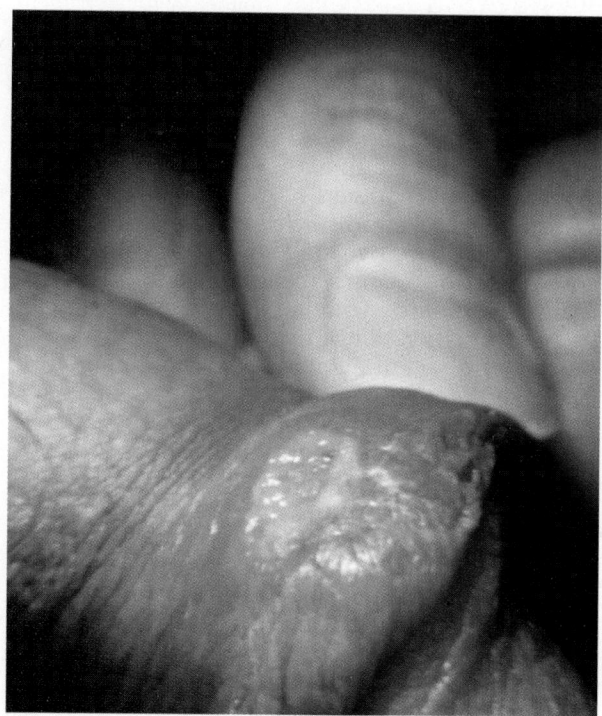

■■■■■ F i g u r e 2 7 – 1 0

Penile syphilitic chancre caused by *Treponema pallidum.*

G, amoxicillin, erythromycin, tetracycline, and cefuroxime axetil, are useful in the treatment of Lyme disease; selection of the most appropriate agent depends on the stage and site of infection.

A nonspecific petechial, macular, or papular skin rash is commonly seen during the primary febrile episode of relapsing fever caused by *Borrelia recurrentis.* The disease can be either epidemic and louse-borne or endemic and tick-borne. Diagnosis of relapsing fever is made by demonstration of borreliae in the peripheral blood of febrile patients using darkfield microscopy or Giemsa- or Wright-stained thick and thin blood smears.

Leptospirosis, a zoonotic disease caused by pathogenic leptospires belonging to the species *Leptospira interrogans,* has protean multisystem disease manifestations that may include a nonspecific maculopapular, or at times hemorrhagic, skin rash. The most common mode of transmission of leptospirosis is through contact of intact or broken skin or mucous membranes with water, moist soil, or vegetation contaminated with the urine of infected animals. Many species of domestic and wild animals, rodents, reptiles, and amphibians may be infected. Diagnosis of leptospirosis is made culturally or serologically; penicillin G and doxycycline have some clinical efficacy for this disease.

Mycoplasmal Infections

Although a primary pathogen of the respiratory tract, *Mycoplasma pneumoniae* can cause maculopapular and vesicular rashes, urticaria, and immunologically mediated erythema nodosum and erythema multiform. The major (Stevens-Johnson syndrome) and minor clinical variants of erythema multiforme involve skin plus mucous membranes or skin only, respectively. Mycoplasma-associated erythema multiforme lesions have been reported both with and without apparent respiratory disease.

Viral Skin Diseases

Warts

Warts can manifest in a variety of skin and mucous membrane lesions. These include common warts (circumscribed, hyperkeratotic, rough-textured, painless papules varying in size from a pinhead to large masses), filiform warts (delicate elongated and pointed lesions that may reach 1 cm in length), laryngeal papillomas (located on the vocal cords and epiglottis of children), flat warts (smooth, slightly elevated, usually multiple lesions varying in size from 1 mm to 1 cm), venereal warts or condyloma acuminata (cauliflower-like fleshy growths seen most often in the moist genital and perianal regions—to be differentiated from condyloma lata of secondary syphilis), flat papillomas of the cervix, and plantar warts (flat, hyperkeratotic lesions of the plantar surface of the feet, which are frequently painful).

Warts are caused by several different types of papillomaviruses. Lesions are the result of an uncontrolled but generally benign growth of skin cells. Some forms of cancer, however, namely skin and cervical cancers, may be associated with oncogenic types of papillomavirus.

Varicella and Zoster

Varicella infection, or chickenpox, is a common childhood illness acquired by respiratory inhalation of the varicella-zoster virus. The skin lesions of primary varicella infection become apparent approximately 2 weeks after initial exposure. The lesions begin as vesicles but quickly rupture, pustulate, and begin to scab in 3 to 4 days. The distribution of the rash is centripetal, starting on the trunk and face and later extending to the extremities. A hallmark of the early stages of chickenpox is the appearance of lesions at all the above stages. When primary varicella infection occurs in adults, it tends to be more severe and systemic, involving lungs, central nervous system, and liver. Treatment with the antiviral agent acyclovir is recommended in adults and immunosuppressed individuals.

Varicella-zoster virus is a herpesvirus that exhibits, like other members of the family Herpesviridae, lifelong latency in the human host. After primary infection, the virus enters peripheral nerves and establishes persistence within the dorsal root ganglia. The virus may be reactivated by imbalances between host and virus that are induced by diverse phenomena such as sunlight, emotional and physiologic stress, intercurrent infection, or immunosuppression associated with age, certain diseases, or drug therapies (e.g., corticosteroids and cytotoxics). With reactivation, the virions move along peripheral sensory nerves of the skin, leading to the appearance of a vesicular eruption in a unilateral dermatomal distribution. The resulting condition is called shingles. The vesicles are similar to those of chickenpox but remain localized along specific sensory nerves. The disease generally heals in a benign fashion, but with facial involvement, it may spread contiguously along the ophthalmic branch of the trigeminal nerve to the eye (zoster ophthalmicus), resulting in severe pain and threatening sight. In immunosuppressed patients, the virus may disseminate widely to the skin and to internal viscera such as lungs, meninges, brain, and liver. In some elderly patients, a chronic pain syndrome called postherpetic neuralgia may follow the acute infection. Acyclovir is the drug of choice for treatment of complicated zoster or to prevent dissemination of localized disease in immunosuppressed hosts.

Herpes Simplex

Herpes simplex is a viral infection characterized by a local primary lesion, latency, and the tendency for localized recurrence. The herpes simplex virus, a DNA virus of the herpesvirus family, causes some of the most common skin, and mucous membrane infections affecting humans. Direct contact, with transmission through infected secretions, is the principal mode of spread of herpes simplex. Transmission can occur both from overtly infected persons and from asymptomatic virus excretors. Herpes simplex type 1 (HSV-1) is transmitted primarily by contact with oral secretions, and HSV-2 is transmitted by contact with sexual secretions. HSV-1 and HSV-2 differ both clinically and epidemiologically; they can be microbiologically differentiated by antigenic analysis using monoclonal antibodies and restriction enzyme technologies.

Primary HSV-1 infection is frequently asymptomatic but may present as severe ulcerative gingivostomatitis and pharyngitis in children under 5 years of age. This initial bout of infection is often accompanied by fever and systemic toxicity. Oral vesicles involving the soft palate, buccal mucosa, tongue, and floor of the mouth quickly ulcerate and may coalesce. Gums are tender and bleed easily, and lesions may spread to the lips and cheek. The breath is fetid, and tender cervical adenopathy is usually present. The differential diagnosis of primary herpetic gingivostomatitis is streptococcal pharyngitis, diphtheritic pharyngitis, herpangina caused by Coxsackie A

virus, aphthous stomatitis, erythema multiforme major (Stevens-Johnson syndrome), Vincent's angina (necrotizing gingivitis), and infectious mononucleosis. Herpetic keratitis can be a serious and sight-threatening ocular complication of HSV-1 infection, with the development of punctate, dendritic epithelial opacities of the cornea.

Herpes simplex type 1 is the most frequent etiologic agent for herpes labialis, a condition commonly known as fever blisters or cold sores. Lesions begin as superficial clear vesicles on an erythematous base and occur on the lips or in the oropharynx. The lesions heal spontaneously, but because of the property of viral latency and endogenous reactivation, they can recur in the same area. Recurrent herpes labialis is generally unaccompanied by systemic complaints.

Cutaneous herpes simplex lesions are indistinguishable from those caused by varicella-zoster virus, although generally they are not as dermatomal in their distribution. In severely immunosuppressed hosts, such as patients with hematologic and lymphoreticular malignancies and tissue transplant recipients, a disseminated and life-threatening form of herpes simplex is associated with diffuse cutaneous lesions and visceral organ involvement. Eczema herpeticum is a generalized vesicular eruption complicating chronic eczema (Fig. 27–11). Up to 75% of cases of erythema multiforme, including Stevens-Johnson syndrome, are preceded by an attack of herpes simplex, either type 1 or type 2. HSV antigen has been identified in skin biopsy specimens from affected lesions.

Primary herpetic lesions of the finger, called herpetic whitlow, can be caused by either HSV-1 or HSV-2. Usually a single digit is involved, with the appearance of one or multiple deep vesicles that may coalesce. Fever and intense local pain are often present. The condition may be misdiagnosed as bacterial paronychia, and unnecessary incision may be performed. Recurrent herpetic whitlow, usually caused by HSV-2, can be a difficult occupational problem among medical, paramedical, and dental personnel. Herpetic whitlow in newborns following fingersucking and serious disseminated infection are two important neonatal herpes syndromes. Infants are infected during passage through the infected maternal genital tract. HSV-2 in adults is transmitted primarily by sexual contact, resulting in primary and recurrent herpes genitalis and perirectal infections.

Herpes simplex can be isolated in tissue culture of material collected from vesicles by unroofing the lesion with a needle or sharp blade, or

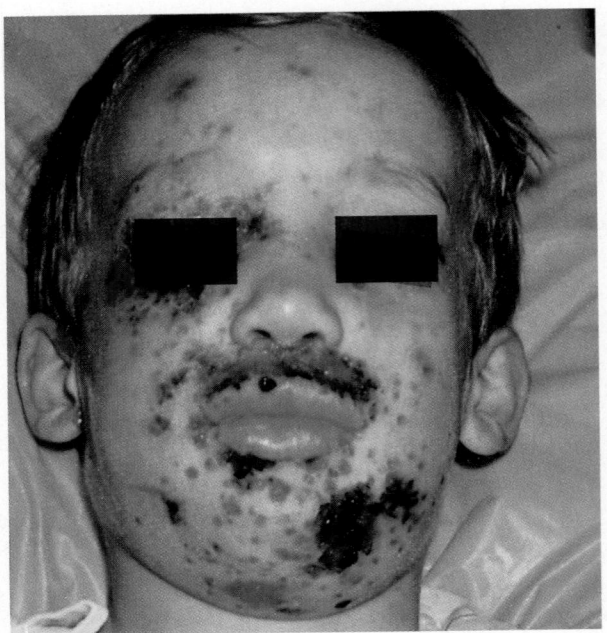

Figure 27–11

Eczema herpeticum due to herpes simplex.

from fresh ulcers. Fluid is collected with a syringe or by swabbing. Specimens can be stained with Giemsa or Wright stains and examined for the presence of multinucleated giant cells (Tzanck prep). Acyclovir is the drug of choice for herpes simplex infections requiring treatment.

Other Herpesviruses

Petechial rashes may be observed in congenital cytomegalovirus (CMV) disease. Rubelliform or maculopapular rashes may occur in CMV-induced mononucleosis in both normal adults and immunosuppressed hosts. Vesicular lesions are distinctly unusual in congenital or acquired CMV infection. Epstein-Barr virus (EBV), the cause of heterophile-positive infectious mononucleosis, may produce a rash that can be macular, petechial, scarlatiniform, urticarial, or erythema multiforme–like in about 5% of patients. Administration of ampicillin results in a pruritic, maculopapular eruption in 90 to 100% of EBV mono patients, although the mechanism is unclear; the rash disappears after cessation of the drug. This phenomenon has been used by some as a clinical marker of the illness.

Molluscum Contagiosum and Orf

Molluscum contagiosum is a common skin disease caused by a poxvirus and is characterized by small, firm, waxy papules, often with umbilicated centers; occasionally giant lesions may be seen (Fig. 27–12). Molluscum contagiosum is transmitted from person to person by direct contact, in many instances by venereal spread. The disease most often appears on the genitalia, face, or perirectal area. It tends to be self-limited and benign, although lesions can be removed by curettage or cryotherapy using liquid nitrogen for cosmetic reasons.

Human orf is another proliferative cutaneous viral disease caused by a poxvirus of the family Poxviridae. The disease is zoonotic, and humans become infected through contact with infected sheep and goats. The skin lesion is usually solitary and located on the hands, arms, or face; it

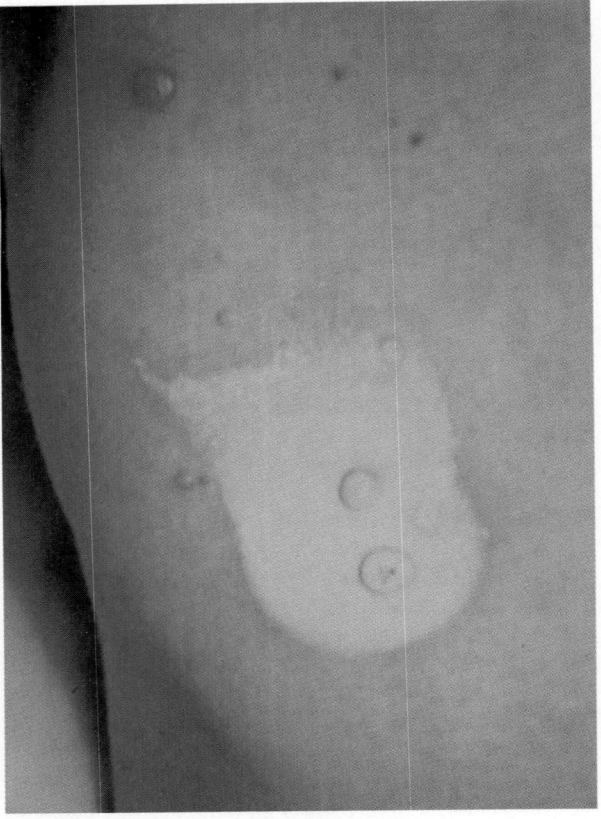

Figure 27–12

Giant molluscum contagiosum.

appears maculopapular or, if secondarily infected with bacteria, may become pustular and be confused with human anthrax. Orf virus, a DNA virus, is closely related to several other parapoxviruses that can be transmitted to humans as occupational diseases (e.g., milker's nodule virus and bovine papular stomatitis virus of cattle). Person-to-person spread is rare, and there is no specific treatment.

Rubeola

Rubeola, or measles, is caused by a paramyxovirus and is spread by direct contact with respiratory secretions of infected persons. Measles is one of the most communicable of all infectious diseases. Although the measles virion is a very labile agent that is sensitive to acid, proteolytic enzymes, strong light, and drying, the virus may remain infective in droplet form in the air for several hours, especially under conditions of low relative humidity.

After an incubation period of 10 to 14 days, the clinical features of coryza, conjunctivitis, and cough develop, followed soon thereafter by a maculopapular rash. The rash extends from the face to the trunk and extremities. Small red patches with central white specks, called Koplik's spots, are characteristically found on the buccal mucosa. Patients with measles are most infectious during the late prodromal phase of the illness, when cough and coryza are at their peak. However, the disease is probably contagious from several days before the onset of the rash to several days after its onset.

Measles can be complicated by secondary bacterial pneumonia, otitis media, or meningoencephalitis. A chronic degenerative neurologic disease called subacute sclerosing panencephalitis (SSPE) has been associated with latent measles infection. This condition occurs more often in immunosuppressed children who had measles in early childhood, usually before 2 years of age. Patients with SSPE have unusually high antibody titers to measles virus in their cerebrospinal fluid and blood, and measles virus has been isolated from brain and lymph nodes of patients with SSPE.

Rubella

Rubella (German measles) is an acute exanthematous viral infection of children and adults caused by an RNA agent of the Togaviridae family. Rubella virus is spread in droplets that are shed from the respiratory secretions of infected persons. The period of contagion extends from about 10 days before the appearance of the rash to 15 days after its onset. Many or most rubella infections are subclinical. For symptomatic persons, the nonspecific maculopapular rash of rubella begins on the face and moves down the body and is often accompanied by cervical and occipital lymphadenopathy and, at times, splenomegaly. The disease is clinically milder than measles, although arthritis and encephalitis can be complications. Maternal infection during the first trimester of pregnancy is associated with a congenital rubella syndrome characterized by serious birth defects in the brain, eyes, ears, and heart. Vaccination is highly effective in preventing rubella in susceptible individuals.

Erythema Infectiosum

Erythema infectiosum, or fifth disease, is one of the common viral exanthematous diseases of childhood. Caused by the human parvovirus B19, epidemic or sporadic disease is characterized by a striking erythema of the face (the so-called slapped cheek appearance), followed in 1 to 4 days by a fine, lace-like rash on the trunk or extremities. The rash may fade quickly only to recur during the ensuing 2 to 3 weeks upon exposure to sunlight or heat. Infection is generally not associated with fever, although mild constitutional symptoms may precede the rash. Scarlet fever and rubella are other childhood diseases to be distinguished from fifth disease. Other disease manifestations of human parvovirus B19 infection include acute polyarthritis, transient aplastic anemia, fetal death in association with intrauterine infection, and severe chronic anemia in immunosuppressed hosts.

Transmission of human parvovirus B19 is through contact with infected respiratory secretions or vertically from infected mother to fetus. The diagnosis of fifth disease is usually made on clinical or epidemiologic grounds; serologic confirmation is made by detection of specific IgM or IgG antibodies against parvovirus B19.

Roseola

Another acute viral infection of childhood is roseola infantum, also called exanthem subitum, caused by human herpesvirus-6. Infection is most common in children aged 2 to 4 years and is associated with a high fever and a centrifugal maculopapular rash, which ordinarily follows lysis of the fever. Symptoms are generally mild, but febrile seizures have been reported. Subclinical infection occurs commonly, and full immunity follows.

Enteroviruses

Coxsackieviruses and echoviruses cause a variety of exanthems, sometimes associated with enanthems of the mucous membranes of the oropharynx. Mucocutaneous features occur more commonly in infants and children than in adults. With the exception of hand-foot-and-mouth disease, these rashes are not sufficiently distinctive to permit reliable etiologic diagnosis on clinical grounds alone. Rashes caused by enteroviruses therefore may be grouped according to the type of exanthem they mimic: morbilliform or rubelliform (maculopapular) (Coxsackie group A, echoviruses), roseoliform (Coxsackie group A, echoviruses), and herpetiform (Coxsackie group A, herpangina and hand-foot-and-mouth disease). At times, enteroviruses may produce a nonblanching petechial or purpuric rash resembling meningococcemia.

Hemorrhagic Fever Viruses

A number of viruses of the Togaviridae (e.g., yellow fever and dengue virus), Arenaviridae (e.g., Junin and Machupo viruses), and Bunyaviridae (e.g., Hantavirus) families, as well as several ungrouped viruses (e.g., Ebola, Marburg, and Lassa fever viruses), may cause a characteristic viral syndrome (fever, headache, myalgias, nausea and vomiting, abdominal pain, prostration) accompanied by severe bleeding manifestations (e.g., gastrointestinal bleeding, hematuria). Bleeding into the skin results in petechiae, purpura, and ecchymoses. Many of these hemorrhagic fever viruses are zoonotic, having a rodent reservoir and/or an arthropod vector, with humans as accidental hosts. Treatment is supportive only, and preventive measures include rodent vector control and isolation for those illnesses in which person-to-person transmission occurs, such as Lassa fever, Congo-Crimean hemorrhagic fever, and Marburg and Ebola viruses.

Fungal Skin Infections

Dermatophytoses

Dermatophytes are fungi that colonize keratinized hair, nails, and skin. *Dermatophytosis* and *tinea* are general terms, essentially synonymous, used to denote superficial fungal diseases of various parts of the body. *Ringworm* is another term often applied to these infections and reflects the tendency of some lesions to expand annularly. The mode of transmission is generally via direct skin-to-skin contact or indirect contact via fomites or environmental surfaces. Various genera and species of fungi known collectively as dermatophytes are the causative agents. The various tinea infections generally have an appellation to describe the site of occurrence. For example, tinea cruris (jock itch) is ringworm of the groin and perianal region; tinea pedis (athlete's foot) is ringworm of the feet; tinea corporis is ringworm of the body (Fig. 27–13); tinea capitis

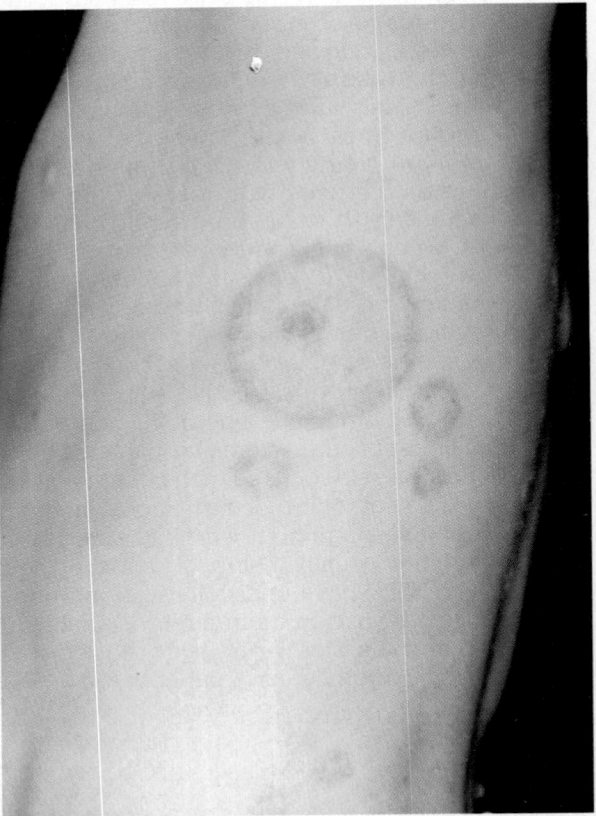

■■■■Figure 27 – 1 3
Tinea corporis.

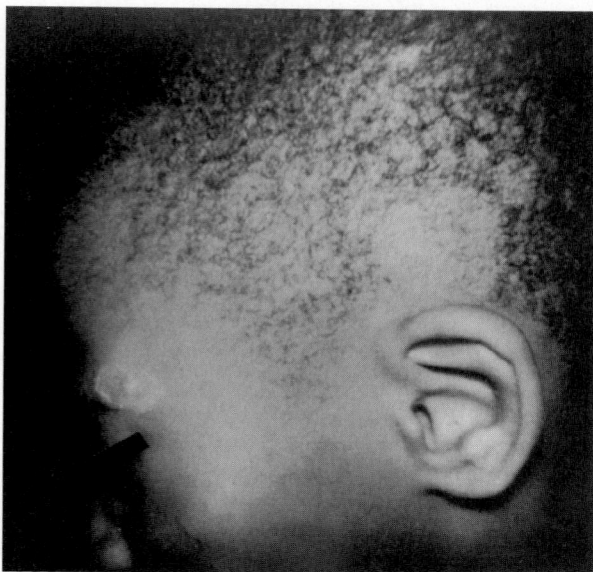

■■■■ F i g u r e   2 7 – 1 4
Tinea capitis.

and nails; *Epidermophyton,* which infects skin and nails; and *Microsporum,* which generally infects only the hair and skin.

Tinea versicolor, or pityriasis versicolor, is an extremely common dermatophytosis that occurs worldwide. The disease is caused by *Malassezia furfur,* a lipophilic yeast and normal skin commensal. The characteristic skin manifestation of tinea versicolor is a diffuse distribution of hypopigmented or, less commonly, hyperpigmented macules principally located on the trunk and proximal portions of the extremities (Fig. 27–15A and *B*). The lesions are usually nonpruritic and often coalesce to form scaly plaques. Spontaneous remission may occur in some patients; for others, topical ketoconazole cream or selenium sulfide lotion is usually curative.

For most of the dermatophytoses, humans are the primary reservoir; occasionally, infections may be acquired from infected domestic animals. In infected tissues, the dermatophytic fungi appear as segmented branching mycelial fragments or as an arrangement of arthrospores. The organisms grow slowly in vitro in 2 to 3 weeks, forming colonies of varying appearance and pigmentation; microscopically they show different arrangements of macro- and microconidia.

is ringworm of the scalp (Fig. 27–14); and tinea unguium, or onychomycosis, is ringworm of the nails. Three genera and over 30 species of dermatophytes cause infection; superficial fungal species that most often cause dermatophytoses include *Trichophyton,* which infects hair, skin,

The diagnosis of dermatophytosis and the identification of its etiologic agents can generally be accomplished by microscopic examination of

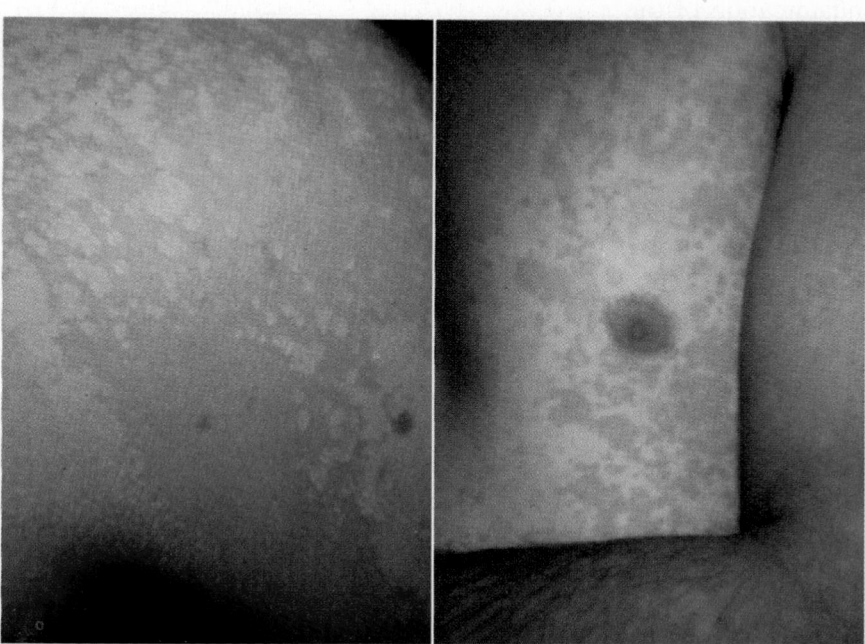

A   B

■■■■ F i g u r e   2 7 – 1 5
Hypopigmented *(A)* and hyperpigmented *(B)* rash of tinea versicolor.

10% potassium hydroxide (KOH) preparations of involved skin, hair, and nails or detritus beneath the nails. Skin or nails can be scraped with a scalpel blade, or nails can be clipped to obtain material for identification. Any deep or suppurative lesions should be aspirated. Precise mycologic etiology is confirmed by culture on Sabouraud agar.

Effective antifungal therapy in managing these superficial infections consists of topical powders, ointments, and creams. Useful preparations contain zinc undecylenate, tolnaftate, miconazole, or clotrimazole. Griseofulvin and ketoconazole, two oral agents, are used to treat more deep-seated nail infections.

Candidiasis

Candidiasis is a superficial mycosis caused by *Candida albicans* and other species. It is generally confined to the superficial layers of the skin or mucous membranes. The term *thrush* is applied to a specific form of oral candidiasis characterized by white, curd-like patches on the tongue or elsewhere on the mucosal surface of the oropharynx. These lesions are typically adherent but can be removed by scraping, leaving a raw, bleeding, and painful surface. Other mucocutaneous candidal syndromes include intertrigo (including diaper rash), an inflammation of the webs of the toes, axillae, umbilicus, groin, and inter- or inframammary and gluteal folds; paronychia, an inflammation of the folds of the skin bordering the nail beds; onychia, an inflammation of the matrix that may lead to loss of the nail; folliculitis, an inflammation of the hair follicles; balanitis, an inflammation of the glans penis; and vulvovaginitis.

Cutaneous lesions of bloodborne disseminated candidiasis can be polymorphic in appearance, but most commonly present as macronodules, petechiae, purpura, and ecthyma gangrenosum. These lesions are most commonly seen in immunosuppressed individuals, such as neutropenic cancer patients. Chronic mucocutaneous candidiasis is a term used to describe a heterogeneous group of *Candida* infections of the skin, mucous membranes, hair, and nails that have a protracted and persistent course despite what is usually adequate therapy. The major manifestations are disfiguring lesions of the face, scalp, and hands. These infections have been associated with defects in T-cell lymphocyte-mediated immunity and with endocrinopathies such as hypoparathyroidism and Addison's disease. Most forms of this disease begin in infancy or early childhood.

A presumptive diagnosis of any of the mucocutaneous *Candida* syndromes is based on the characteristic clinical appearance of the lesions and is supported by demonstration of pseudohyphae and/or yeast cells in Gram-stained or KOH preparations of swabs, scrapings, or biopsies of infected tissue. Culture on blood agar or Sabouraud agar is confirmatory. Gentian violet, topical nystatin, and imidazoles such as clotrimazole and miconazole are effective in many forms of superficial candidiasis. Systemic agents such as ketoconazole, fluconazole, and amphotericin B may be needed for more severe or systemic infections.

Subcutaneous Mycoses

The subcutaneous mycoses include a number of infections of skin and soft tissues caused by diverse fungi. These infections include mycetoma, chromomycosis, sporotrichosis, phaeohyphomycosis, subcutaneous phycomycosis, rhinosporidiosis, rhinoentomophthoromycosis, and Lobo's disease (Table 27–5). Typically, causative

■■■■ T A B L E   2 7 – 5
MAJOR ETIOLOGIC AGENTS OF SUBCUTANEOUS MYCOSES

Disease	Principal Organism(s)
Mycetoma	*Madurella mycetomatis, M. grisea, Scedosporium apiospermum, Leptosphaeria senegalensis, Exophiala jeanselmei, Pyrenochaeta romeroi, Fusarium* spp, *Acremonium* spp, *Pseudoallescheria boydii, Aspergillus nidulans, Neotestadina rosatii, Actinomadura madurae, A. pelletieri, Nocardia brasiliensis, N. asteroides, Nocardia caviae, Streptomyces somaliensis*
Chromomycosis	*Phialophora verrucosa, Fonsecaea pedrosoi, F. compacta, Cladosporium carrionii, Rhinocladiella aquaspersa*
Subcutaneous phycomycosis	*Basidiobolus haptosporus*
Phaeohyphomycosis	Dematiaceous fungi
Sporotrichosis	*Sporothrix schenckii*
Rhinosporidiosis	*Basidiobolus seeberi*
Rhinoentomophthoromycosis	*Conidiobolus*
Lobo's disease	*Loboa loboi*

organisms present in soil and vegetation are traumatically inoculated into the skin and slowly spread to surrounding tissues. As a rule, there are no spontaneous remissions. Mycetoma, chromomycosis, and sporotrichosis are discussed below in some detail.

MYCETOMA

Mycetoma (also called Madura foot or maduromycosis) is a clinical syndrome caused by a variety of aerobic actinomycetes (actinomycetoma) and fungi (eumycetoma) (see Table 27–5). It is characterized by swelling and suppuration of subcutaneous tissues and formation of sinus tracts, with visible granules in the pus draining from these fistulae. The disease process is slowly progressive and destructive, with extension to muscle and bone. Soil and decaying vegetation are reservoirs for the etiologic organisms. The disease develops via subcutaneous implantation of conidial, hyphal, or filamentous elements from a saprophytic source by penetrating trauma (e.g., thorns, splinters). There is no person-to-person transmission. Lesions usually appear on the lower leg or face; sometimes on the hand, shoulder, and back; and rarely at other sites.

Mycetoma is rare in the continental United States but common in tropical and subtropical parts of the world, especially where people go barefoot. Actinomycetoma frequently responds to combination antimicrobial treatment with sulfones and streptomycin or trimethoprim-sulfamethoxazole plus streptomycin or rifampin. Eumycetoma is generally refractory to drug therapy, and radical surgery is usually required.

Mycetoma may be difficult to distinguish from chronic osteomyelitis and botryomycosis, the latter being a clinically and pathologically similar entity caused by a variety of bacteria, including staphylococci and gram-negative bacteria. Specific diagnosis depends on visualizing the granules in fresh preparations or histopathologic sections and isolating the causative actinomycete or fungus in culture.

CHROMOMYCOSIS

Chromomycosis, or chromoblastomycosis, is a chronic spreading mycosis of the skin and subcutaneous tissues, usually of a lower extremity and, less commonly, the hand or back. The geographic occurrence and mode of transmission are similar to those of mycetoma. Progression to contiguous soft tissue is slow, over a period of years, with eventual large verrucous or even cauliflower-like masses and lymphatic stasis. Unlike the case with mycetoma, muscle and bone are generally not involved.

Infectious agents of chromomycosis include *Phialophora verrucosa, Fonsecaea pedrosoi, Cladosporium carrionii,* and *Rhinocladiella aquaspersa.* Microscopic examination of scrapings or biopsies from lesions reveals characteristic brown, thick-walled, rounded cells that divide by fusion in two planes. Confirmation of the diagnosis is made by biopsy and culture of the causative fungus.

SPOROTRICHOSIS

Sporotrichosis is caused by a dimorphic fungus called *Sporothrix schenckii,* which is capable of existing in both yeast (in human tissue) and hyphal (in the environment) forms. Introduction of the fungus through the skin occurs by pricks of thorns or barbs, by the handling of sphagnum moss, or by slivers from wood, lumber, or other material contaminated with the organism. Worldwide in distribution but characteristically sporadic in occurrence, sporotrichosis is most often an occupational disease of gardeners, farmers, and horticulturalists.

The disease begins as a nodule at the inoculation site, which may ulcerate and remain confined (cutaneous form) or grow and develop surrounding satellite nodules and involve draining lymphatic channels as nodular lymphangitis (lymphocutaneous form). Internal visceral spread is rare. Other microorganisms capable of causing a lymphocutaneous syndrome are listed in Table 27–6.

Laboratory confirmation of sporotrichosis is made by culture of pus or exudate, preferably aspirated from an unopened lesion. Organisms are rarely visualized by direct smear of these materials, but Gomori methenamine silver and other fungal stains of biopsied tissue often reveal the causative agent. Oral potassium iodide is an effective treatment in many cases of soft-tissue sporotrichosis; more extensive or refractory infections may be treated with intravenous amphotericin B or oral itraconazole.

Systemic Mycotic Infections

Although the respiratory tract is their portal of entry into the body, almost all the systemic

■■■■■ T A B L E   2 7 – 6

MICROBIOLOGIC CAUSES OF LYMPHOCUTANEOUS SYNDROME

Fungi	**Mycobacteria**
Sporothrix schenckii	*Mycobacterium marinum*
Blastomyces dermatitidis	*M. kansasii*
Histoplasma capsulatum	*M. chelonae*
Coccidioides immitis	
Scopulariopsis blochi	**Spirochetes**
	Treponema pallidum
Actinomycetes	
Nocardia brasiliensis	**Bacteria**
	Staphylococcus aureus
Parasites	*Francisella tularensis*
Leishmania braziliensis	
ssp *braziliensis*	**Viruses**
ssp *panamensis*	Herpes simplex
L. major	

mycoses can produce secondary skin lesions via hematogenous seeding to skin. Lesions most commonly appear as macronodules (e.g., caused by *Candida, Cryptococcus,* or dimorphic fungi), verrucous papules or plaques *(Blastomyces, Coccidioides),* ulcers *(Histoplasma),* or areas of black tissue necrosis (due to *Aspergillus,* Zygomycetes, or dematiaceous and other opportunistic fungi).

Parasitic Skin Infections

A number of parasites have minor dermatologic manifestations. The larvae of certain schistosomes of birds and mammals may penetrate the human skin and cause a dermatitis, sometimes known as swimmer's itch. Infection is acquired from water containing free-living larvae, or cercariae, which have developed in snails. These nonhuman schistosomes do not mature in humans. Such infections may be prevalent among bathers in lakes in many parts of the world, including the Great Lakes region of North America and certain coastal beaches.

The intestinal helminth *Strongyloides stercoralis* may also cause a transient dermatitis when larvae of the parasite penetrate the skin on initial infection. The usual permanent habitat of adult *Strongyloides* worms is the large intestine. With chronic infection, an intensely pruritic dermatitis radiating from the anus may occur. Stationary urticarial wheals lasting 1 to 2 days may appear, as well as a migrating serpiginous rash. Penetration of skin, usually of the feet, by hookworm larvae may cause a "ground-itch" dermatitis.

Onchocerciasis is a chronic, nonfatal filarial disease caused by the tissue nematode *Onchocerca*

volvulus. Confined geographically to parts of West Africa, Mexico, Central and South America, and the Middle East, the disease is spread to humans by the bite of infected blackflies. Infective larvae injected by the bite develop into adult worms and form nodules in subcutaneous tissues (Fig. 27–16), particularly in the head and shoulders, pelvic girdle, and lower extremities. Adult female worms discharge microfilariae, which migrate through the skin, often accompanied by an intensely itchy rash, edema, and atrophy of the skin. The most important manifestation of onchocerciasis, or river blindness, is infiltration of the eye by microfilariae, which causes visual disturbances and blindness. Laboratory diagnosis of the cutaneous disease is made by superficial skin biopsy, with demonstration of microfilariae in fresh preparations by microscopic examination, or by excision of cutaneous nodules and the finding of adult worms. Ivermectin, a microfilaricidal drug given as a single dose and annual retreatment, is useful in reducing morbidity and interrupting transmission of the parasite.

Dracunculiasis, or guinea worm infection, is an infection of the subcutaneous and deeper tissues caused by the tissue nematode *Dracunculus medinensis.* A blister appears—usually on the lower extremity, especially the foot—when the gravid meter-long adult female worm prepares to discharge its larvae. Burning and itching of the overlying skin develop and may be accompanied by fever, nausea and vomiting, diarrhea, headache, generalized urticaria, shortness of breath, and eosinophilia. After the vesicle ruptures, the worm releases larvae whenever the affected part is immersed in water. Local inflammation and adhesions, associated with reactogenic larvae or

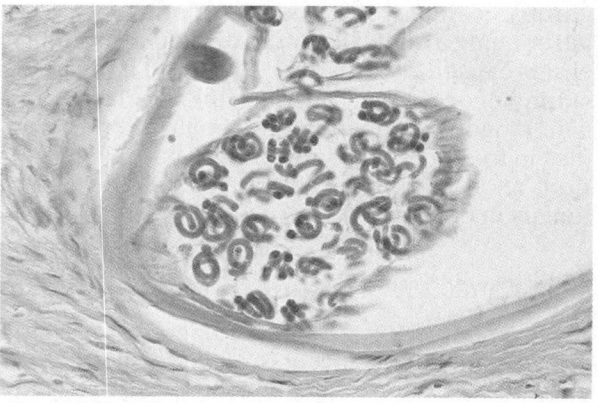

■■■■■ F i g u r e   2 7 – 1 6

Tissue cross-section of a nodule containing *Onchocerca volvulus* microfilaria.

deterioration of the adult worm, and secondary bacterial superinfection may complicate dracunculiasis. Diagnosis is made by microscopic examination of larvae or by recognition of the adult worm. The disease is found in India, Africa, and the Middle East, especially in regions with dry climates. Treatment involves either slow, gradual traction and removal of the emerging worm by winding it around a stick or surgical extraction of the worm before its emergence.

Larvae of the dog and cat hookworms, *Ancylostoma caninum* and *Ancylostoma braziliense,* respectively, can penetrate the skin and produce a self-limited dermatitis ("creeping eruption") characterized by larval migration above the germinative layer with associated serpiginous, elevated tunnels and indurated, itchy papules. Larvae enter the skin and migrate intracutaneously for prolonged periods. The disease is a recreational hazard of children who play in sandboxes frequented by cats and an occupational hazard of workers who crawl or work in areas with damp, sandy soil contaminated by dog or cat feces. Spontaneous cure is the rule, although individual larvae can be killed by freezing with ethyl chloride spray. Oral thiabendazole may be effective for rare systemic infection or as an ointment for topical use.

A number of ectoparasites may cause human infection, including lice (pediculosis), mites (scabies) (Fig. 27–17A and B), fleas, flies, ticks, chiggers, and bedbugs. The dermatologic features of these infestations include severe itching and the formation of papules, vesicles, nodules, linear burrows, and excoriations of the skin and scalp.

Bibliography

Benenson AS (ed): Control of Communicable Diseases in Man. Washington, DC: American Public Health Association, 1990.

Lewis RT: Necrotizing soft tissue infections. Infect Dis Clin North Am 3:693, 1992.

Lipsky BA, Pecoraro RE, Wheel LJ: The diabetic foot: Soft tissue and bone infection. Infect Dis Clin North Am 3:409, 1990.

Norden CW: Osteomyelitis. In Mandell GL, Douglas RG Jr, Bennett JE (eds): Principles and Practice of Infectious Diseases. New York: Churchill Livingstone, 1990, pp 922–930.

Smith JW: Infectious arthritis. In Mandell GL, Douglas RG Jr, Bennett JE (eds): Principles and Practice of Infectious Diseases. New York: Churchill Livingstone, 1990, pp 911–918.

Stevens DL, Tanner MH, et al: Severe group A streptococcal infections associated with a toxic shock–like syndrome and scarlet fever toxin A. N Engl J Med 321:1, 1989.

Swartz MN: Cellulitis and superficial infections. In Mandell GL, Douglas RG Jr, Bennett JE (eds): Principles and Practice of Infectious Diseases. New York: Churchill Livingstone, 1990, pp 796–807.

Swartz MN: Subcutaneous tissue infections and abscesses. In Mandell GL, Douglas RG Jr, Bennett JE (eds): Principles and Practice of Infectious Diseases. New York: Churchill Livingstone, 1990, pp 808–812.

Waldvogel FA, Medoff, G, Swartz MN: Osteomyelitis: A review of clinical features, therapeutic considerations and unusual aspects. N Engl J Med 282:198, 1970.

Waldvogel FA, Vasey H: Osteomyelitis: The past decade. N Engl J Med 303:360, 1980.

Wallace RJ Jr, Tanner D, et al: Clinical trial of clarithromycin for cutaneous (disseminated) infection due to *Mycobacterium chelonae.* Ann Intern Med 119:482, 1993.

Wolfson JS, Sober AJ, Rubin RH: Dermatologic manifestations of infections in immunocompromised patients. Medicine 64:115, 1985.

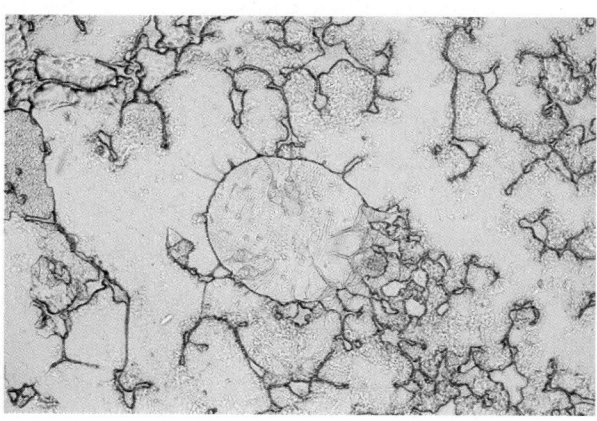
A

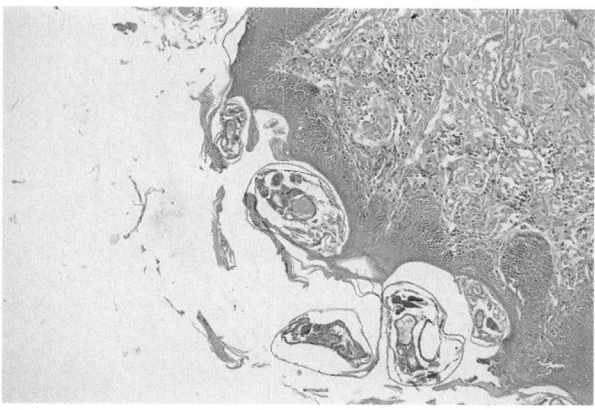
B

Figure 27–17

A, Sarcoptes scabiei adult showing short legs and conical spines. *B,* Tissue cross-section of scabies lesion showing larvae burrowed into the epidermal layer of the skin.

CHAPTER 28

Gastrointestinal Infections and Food Poisoning

Larry J. Goodman •
George Manuselis, Jr. • *Connie R. Mahon*

OBJECTIVES

1. Explain the role of the normal microbial flora of the gastrointestinal tract. Name the organisms that constitute the colon flora.

2. Explain the pathogenic mechanism involved in acute bacterial diarrheas.

3. List the infectious bacterial, viral, and parasitic agents known to cause diarrhea.

4. Give the sources of these infectious agents and describe how they are acquired.

5. Describe the parameters commonly used for presumptive differential diagnosis.

6. For each of the agents described:

■ Give the pathophysiology and clinical manifestations of the infection.
■ Describe the method for diagnosis and general characteristics, including the selective media used for maximum recovery of the agent.

Acute diarrheal illness is among the most common problems presented to the generalist physician (pediatrician, internist, or family doctor). Although most patients have a self-limiting illness typically lasting fewer than 5 days, others have severe chronic symptoms, bacteremia, metastatic infection, or life-threatening dehydration. The early identification and treatment of individuals with these complications are key factors in limiting morbidity and mortality.

This chapter:

■ Reviews the host and organism factors that lead to diarrheal illness

■ Discusses common bacterial, viral, and parasitic pathogens of diarrheal diseases and the mechanisms employed by these pathogens

■ Presents a diagnostic approach to this problem

■ Summarizes various treatment interventions.

GENERAL CONCEPTS IN EVALUATING GASTROINTESTINAL INFECTIONS AND FOOD POISONING

A complete history, clinical evaluation, and careful physical examination are particularly important in the diagnosis of gastrointestinal illnesses. Because nearly all such infections are acquired by ingesting the organism, a history of recent foods ingested as well as of any exposures

to ill persons should be part of the initial evaluation of patients with diarrhea. Patients should also be asked about any recent travel, because travelers are at greater risk for developing diarrheal infections, particularly those visiting countries with inadequate sewage and water treatment facilities. Other important issues are as follows:

■ **Does the patient have a history of previous gastrointestinal symptoms?** A positive history may suggest a more chronic or recurrent illness, such as inflammatory bowel disease.

■ **Does the patient have an underlying illness?** For example, patients with acquired immunodeficiency syndrome (AIDS) may be infected with organisms not routinely considered diarrheal agents in immunocompetent patients.

■ **Is the patient taking any medications?** Some medicines may cause gastrointestinal symptoms or predispose the patient to certain infections.

The differential diagnosis of diarrheal illness is among the broadest of common patient presentations. Viral, bacterial, and parasitic pathogens, as well as food poisonings and noninfectious processes, may cause diarrhea. The cost of evaluating all patients for all of the possible pathogens and other causes is prohibitive. Therefore, physicians must learn an approach to diarrheal illness that is effective in limiting morbidity, mortality, and secondary transmission rates as well as practical from a cost standpoint.

The microbiologist is similarly challenged. Identifying a pathogen from among literally millions of nonpathogens is extremely difficult. Also, as new diagnostic techniques are developed, they must be carefully correlated with clinical cases and evaluated to determine their usefulness in early diagnosis and initiation of therapy.

ANATOMICAL CONSIDERATIONS

Diarrheal pathogens are usually acquired by ingesting the organism as part of a contaminated meal or beverage. Figure 28–1 is a diagram of the gastrointestinal tract with the major host defenses against infection at each level. When organisms reach the stomach, they are exposed to gastric acid. Nearly all organisms are quite sensitive to a

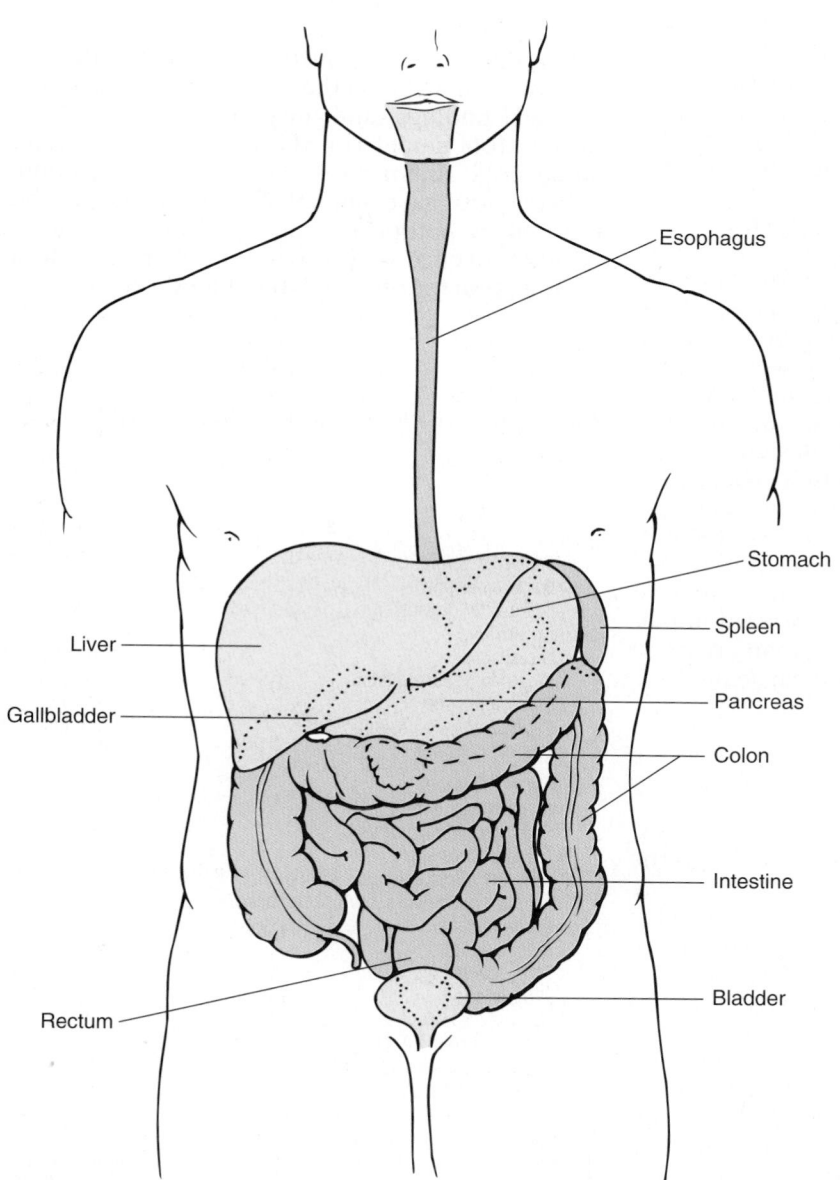

■ F i g u r e 2 8 – 1
Anatomy of gastrointestinal tract.

low pH, resulting in a greatly reduced number of organisms surviving to reach the small bowel. Exceptions include the spore phase of some bacteria and the cyst phase of some parasites.

In the small bowel, the major host defense is motility. The small bowel is constantly in motion. Organisms that must attach to the gut wall to produce symptoms are affected, because the contact time of the organism with the gut surface is limited by the constant peristalsis of the small bowel. Organisms surviving to reach the colon are faced with two other host defenses. First, local antibody (coproantibody), primarily IgA, is secreted and may have an effect against some organisms. Second, organisms reaching the colon must also compete with the huge number of other organisms, such as normal colon flora already present. These other organisms take up attachment sites on the colon wall, compete for nutrients, and, in some cases, produce chemicals (e.g., short-chain fatty acids) toxic to some pathogens.

It is also important to understand that, in most cases, a large number of organisms must be ingested to cause disease. The number of organisms that must be ingested to cause disease in 50% of individuals is termed the *median infectious dose* (ID_{50}). The interplay between the number and virulence of the organism ingested and the host defenses ultimately determines whether an individual becomes ill or not. For example, patients with inadequate stomach acidity (achlorhydria) are more likely than persons with normal stomach acidity to become ill if they ingest an enteric pathogen. Also, organisms present in particularly large numbers in ingested food are more likely to survive host defenses and cause disease.

THE ROLE OF NORMAL FLORA

The stomach contains few organisms. Unless the patient has decreased acid production of the stomach, acidity is buffered by food or reflux of alkaline material from the small bowel. The upper part of the small bowel has small numbers of *Enterococcus* species, lactobacilli, and diphtheroids. *Candida albicans* may be seen in 20 to 40% of persons. In the colon, the flora changes markedly. Here, the number of organisms present approaches 10^{12} per gram of feces and consists primarily of anaerobic, gram-negative, non–spore-forming bacteria. The ratio of anaerobes to aerobes is approximately 1,000:1. *Bacteroides, Fusobacterium, Eubacterium,*

and *Peptococcus* are among the predominant anaerobes present. *Escherichia coli, Klebsiella,* and other members of the Enterobacteriaceae are the most common aerobic gram-negative rods isolated from fecal material. Table 28–1 lists the organisms found as microbial flora in the large intestine.

A PRACTICAL APPROACH TO DIAGNOSIS OF THE PATIENT WITH DIARRHEA

The differential diagnosis of patients with diarrhea may be made on the basis of clinical history, physical findings, and microscopic examination of the stool sample on the day of presentation. Table 28–2 summarizes the commonly encountered clinical gastrointestinal syndromes and the most likely pathogens implicated by incubation period, absence or presence of fever, and stool examination for red and white blood cells.

■■■■ T A B L E   2 8 – 1
MICROBIAL FLORA FOUND IN THE LARGE INTESTINE

Bacterial Species*	Incidence (%)
Strict anaerobes	
Gram-negative	
Bacteroides fragilis	100
Bacteroides spp	100
Fusobacterium spp	100
Gram-positive	
Lactobacilli	20–60
Clostridium perfringens	25–35
Clostridium spp	1–35
Peptostreptococcus spp	Common
Peptococcus spp	Common
Facultative anaerobes	
Gram-positive cocci	
Staphylococcus aureus	30–50
Enterococcus spp	100
β-Hemolytic streptococci, groups B, C, F, and G	0–16
Gram-negative bacilli (Enterobacteriaceae)	
Escherichia coli	100
Klebsiella spp	40–80
Enterobacter spp	5–55
Proteus spp	3–11
Salmonella enteritidis (1400 serotypes)	3–7
Shigella, groups A–D	0–1
Pseudomonas aeruginosa	3–11
Candida albicans	15–30

*Strict anaerobes are present in ratio of 1000:1 with facultative aerobes.
Adapted from Sommers HM The indigenous microbiota of the human host. In Youmans GP, Paterson PY, Sommers HM. (eds): The Biologic and Clinical Basis of Infectious Diseases, 2nd ed. Philadelphia: WB Saunders, 1980, p 92.

■■■■■■ T A B L E   2 8 – 2
CLINICAL SYNDROMES FOR SPECIFIC ENTERIC PATHOGENS*

Syndrome	Incubation Period	Fever	Stool Examination for Leukocytes and Erythrocytes	Pathogens
Nausea and vomiting	5–15 min	No	Negative	Heavy metals, mass psychogenic illness[†]
Nausea, vomiting, and diarrhea	1–18 h	No	Negative	Enterotoxigenic *Escherichia coli*, *Clostridium perfringens*, *Bacillus cereus*, *Staphylococcus aureus*
Nausea, vomiting, diarrhea, myalgias, and headache	12 h–3 d	Yes	Negative	Rotavirus, Norwalk virus, Norwalk-like virus
Diarrhea and abdominal cramps	1–3 d	Yes	Positive[‡]	*Campylobacter jejuni*, *Shigella* spp, *Entamoeba histolytica*, *Salmonella* sp, *Yersinia* sp, *Clostridium difficile*
Gastrointestinal bleed	1–3 d	No	Gross blood	Enterohemorrhagic *E. coli*,[§] *cytomegalovirus*[¶]
Malabsorptive diarrhea with bloating	1–2 wk	No	Negative	*Microsporidium* sp, *Isospora belli*, *Giardia lamblia*

*This table is designed as a guide. Incubation periods and syndromes may overlap.
[†]Mass psychogenic illness is more common in adolescents. Rash may be part of the syndrome.
[‡]Gross blood and pus in stool is a clue for *Campylobacter jejuni*, *Shigella* species, or *Entamoeba histolytica*.
[§]Fever may be seen in approximately one third of cases.
[¶]Syndrome confined to compromised hosts, particularly bone marrow transplant recipients, and occurs 1 and 3 months after transplantation.
Adapted and reprinted with permission from Goodman LJ: Diagnosis, management, and prevention of diarrheal diseases. Curr Opinion Infect Dis 6:88, 1993.

History

Travel and food ingestion history are particularly important. A history of travel to countries with less effective sewage sanitation facilities greatly increases the risk of acquiring an enteric infectious pathogen. Approximately 50% of so-called cases of traveler's diarrhea are caused by enterotoxigenic *E. coli*. This illness has a short incubation period (usually less than 24 hours) and typically lasts 1 to 3 days. Therefore, a traveler who develops diarrhea days to weeks after returning home, or who is complaining of 1 to 2 weeks of diarrhea, most likely has some other pathogen. *Giardia lamblia* and *Entamoeba histolytica* would be among the considerations in this case.

Because most diarrhea pathogens are acquired via ingestion of the organism, a detailed food history (going back at least 3 days) is also useful. Infected food usually tastes and smells normal, so it is important to know which foods are most commonly implicated as vehicles for specific pathogens. Water, unpasteurized milk, poultry, and shellfish are among the most commonly implicated foods. A more complete list of foods and the pathogens they are associated with is given in Table 28–3. A food history is also useful in attempting to ascertain the incubation period of the illness.

The duration of illness prior to presentation is also a useful clue in narrowing the differential diagnosis. Symptoms due to *G. lamblia* and other parasitic infections are often present for 10 days or more before the patient seeks medical attention.

■■■■■■ T A B L E   2 8 – 3
COMMON FOOD VEHICLES FOR SPECIFIC PATHOGENS OR TOXINS

Vehicle	Pathogen or Toxin
Undercooked chicken	*Salmonella* sp, *Campylobacter* sp
Eggs	*Salmonella* sp (especially *S. enteritidis*)
Unpasteurized milk	*Salmonella*, *Campylobacter*, and *Yersinia* sp
Water	*Giardia lamblia*, Norwalk virus, *Campylobacter* sp, *Cryptosporidium* sp, cyclospora
Fried rice	*Bacillus cereus*
Fish	
Shellfish	*Vibrio cholerae*, *V. parahaemolyticus*, *V. vulnificus*, other *Vibrio* sp, neurotoxic shellfish poisoning, paralytic shellfish poisoning, Norwalk virus
Tuna, mackerel, mahi-mahi	Scombroid poisoning
Grouper, amberjack, snapper	Ciguatera
Sushi	*Anisakis* species (anisakiasis)
Beef, gravy	*Salmonella* sp, *Campylobacter* sp, *Clostridium perfringens*

Adapted and reprinted with permission from Goodman LJ: Diagnosis, management, and prevention of diarrheal diseases. Curr Opinion Infect Dis 6:88, 1993.

Patients with diarrhea due to an invasive bacterial pathogen typically present after at least 72 hours of symptoms. Patients with viral or entertoxin-associated disease more commonly present for medical attention within the first 24 to 48 hours of symptoms.

As with other infectious illnesses, it is important to ask patients about any medication they may be taking and their exposure to other persons with a similar illness. Gastrointestinal symptoms are among the most common drug-related side effects. If the patient is taking an antibiotic, *Clostridium difficile* is an important consideration. Finally, the patient should be asked about other illnesses. Diarrhea may be a symptom of some noninfectious illnesses (e.g., inflammatory bowel disease), and in patients with serious underlying illnesses, such as AIDS, may be due to pathogens that would not otherwise be considered.

Physical Examination

The first consideration on physical examination is to assess the patient's state of hydration. The most common cause of death due to diarrhea is dehydration. Signs of dehydration include a drop in blood pressure and an increase in heart rate upon moving from a lying to a sitting position (orthostatic changes), a sunken appearance to the eyes, and a loss of resiliency of the skin, called *skin tenting*. This change in the skin can be detected by gently pinching the skin of the back of the hand. The pinched area in a dehydrated patient remains in a pinched or "tented" position. Patients who are severely dehydrated may also have changes in their mental state and other organ system dysfunction. An elevated temperature is a clue that the pathogen has invasive properties. Examination of the abdomen usually reveals a diffusely tender abdomen. On auscultation with a stethoscope, bowel sounds are present. Localization of pain to one part of the abdomen, severe pain on palpation of the abdomen, and absence of bowel sounds all are indications to evaluate the patient for a complication of diarrhea (e.g., toxic megacolon, intestinal rupture) or a different disease process that may require surgical intervention (e.g., appendicitis).

Laboratory Studies

The most helpful of immediately available laboratory tests in the evaluation of patients with presumed infectious diarrhea is a microscopic evaluation of the stool for red and white blood cells. These cells are common in specimens from patients with invasive infections but are unusual in specimens from patients with most entertoxin-in-mediated illnesses, viral illnesses, and parasitic infections. The duration of illness, travel history, host status, and food ingestion history can then be used to make further decisions concerning cultures.

PATHOGENIC MECHANISMS AND CLINICAL PRESENTATIONS OF ACUTE DIARRHEAS

On the basis of history, physical examination, and laboratory findings, the physician should be able to shorten the list of potential pathogens. This reduction can be accomplished by dividing organisms into groups that reflect the clinical presentations with which they are most commonly associated.

Enterotoxin-Mediated Diarrhea

CASE STUDY

A 32-year-old man from the United States is visiting a small village in Mexico. Four days after arriving, he experiences sudden onset of diarrhea. The diarrhea occurs over 30 times the first day and is accompanied by nausea with several episodes of diarrhea. The stool is watery, without gross blood, pus, or mucus. The patient is afebrile but complains of crampy abdominal pain with stooling and dizziness when standing. His heart rate is rapid (120/minute).

Discussion

This presentation is typical of an enterotoxin-mediated process. Enterotoxic *E. coli* (ETEC) accounts for the largest percentage of cases in travelers. Other organisms associated with enterotoxin-mediated diarrhea are *Vibrio cholerae*, *Staphylococcus aureus*, *Clostridium perfringens*, and *Bacillus cereus*. *Aeromonas* and *Plesiomonas* may produce a similar syndrome, as does viral gastroenteritis, particularly when due to adenovirus or Norwalk virus. The noninvasive parasitic infections, such as *G. lamblia*, *Cryptospor-*

idium, and *Isospora belli,* also produce an afebrile diarrhea with no cells in the stool. Patients with parasitic infections, however, usually look for medical attention after 1 week or more of symptoms, have fewer diarrheal stools per day, and have more prolonged illnesses without treatment.

The patient's rapid heart rate and complaints of dizziness on standing (orthostatic symptoms) are due to loss of intravascular fluid volume. Treatment should initially be aimed at rehydration. The addition of bismuth-subsalicylate with or without an antimicrobial agent may also shorten this man's illness.

Because the toxin may be produced on the food and acts fairly proximally (in the small bowel), the incubation period for enterotoxin-producing organisms is relatively short, typically 6 to 12 hours. Patients present with an illness that began either the day of or the day before presentation. The illness is characterized by watery diarrhea, with a very high frequency of stools (sometimes over 20 per day). Of the infectious diarrhea syndromes, these infections are most associated with nausea and vomiting. Abdominal cramps, particularly during defecation, are common. The bacteria associated with enterotoxin production do not invade the gut wall, and the toxin itself also does not elicit an inflammatory response. Therefore, microscopic examination of the stool does not reveal red or white cells. Similarly, there is no inflammatory response, a fact that explains why affected patients are usually afebrile. Because the organism itself does not invade the tissue, bacteremia and metastatic infection are very rare.

Diarrhea Mediated by Invasion of Bowel Mucosal Surface

CASE STUDY

A 56-year-old woman was hospitalized 9 days ago for gallbladder surgery. Her roommate was hospitalized 2 days ago for fever and diarrhea. Today, the surgical patient is complaining of diarrhea. For the first time since admission, she has a fever. Four other people on the floor also have new onset of diarrhea. Fecal leukocytes are present in all affected patients. Subsequently, in this case, *Shigella* is cultured from all infected patients.

Discussion

The most common cause of an invasive diarrheal syndrome in a hospitalized patient is *C. dif-*

ficile. In this case, however, the patient's roommate has diarrhea, and four others on the same medical floor became sick at the same time. Other invasive syndromes to consider include *Salmonella, Campylobacter,* and *Shigella. Entamoeba histolytica* would also be a consideration, but the incubation period of 2 days is short.

The site of action for invasive organisms is the colon, and the mucosal surface of the bowel is primarily affected. Deeper invasion of the bowel wall and regional lymph nodes are unusual, however. Because organisms must traverse the stomach and small bowel, multiply, and invade, the incubation period is usually longer than for toxin-mediated illness, approximately 1 to 3 days. Organisms in this group elicit an inflammatory response from the bowel wall (usually the colon). Evidence of this is the common finding of white and red cells in the microscopic evaluation of the stool smear. Patients in this group are the most likely to present with a true dysentery syndrome, characterized by gross blood and pus in the stool, as shown in Figure 28–2. There is a systemic inflammatory response, evidenced by fever in some patients. Because the invasion is usually superficial, bacteremia and metastatic infection are not common.

Organisms that produce this syndrome include *Shigella, Campylobacter jejuni,* and *C. difficile.* Enterohemorrhagic *E. coli* often causes gross blood per rectum, although fever and white blood cells in the stool are less common than with the other organisms that produce this clinical manifestation. Of the parasitic infections, *E. histolytica* produces an identical clinical syndrome, except that the incubation period is even longer, ranging from 1 to 3 weeks.

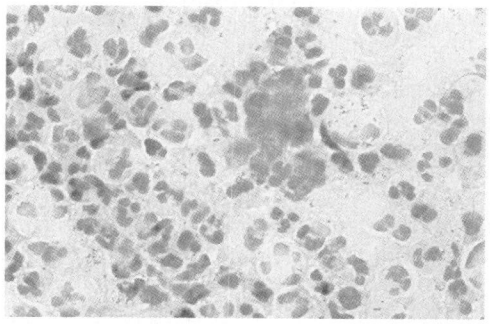

F i g u r e   2 8 – 2

Direct fecal smear gram-stained to show the presence of white blood cells, indicative of an invasive process and not an enterotoxin.

Diarrhea Mediated by Invasion of Full Bowel Thickness with Lymphatic Spread

CASE STUDY

A 37-year-old woman presents with high fever and dry cough. The patient denies any history of vomiting or diarrhea. Other laboratory findings include neutropenia, a relative increase in lymphocytes, a shift to the left, and normocytic, normochromic anemia. Blood culture specimens are taken at the time of hospital admission. During the second week of hospitalization, the patient remains febrile, with temperature reaching 103°F. She also begins to excrete watery, diarrheic stool. Stool cultures are performed.

Salmonella typhi is isolated from both blood and stool cultures. The patient acquired the infection from eating at a local restaurant. Specimens from the cook are cultured for *Salmonella,* and results are positive.

Discussion

Invasive organisms such as *Salmonella typhi* and *Yersinia enterocolitica* act primarily at the colon. Incubation period is approximately 1 to 3 days. Fever is common. Often, the patient experiences constipation rather than diarrhea in the early phase of the disease. Although bacteremia due to a diarrheal pathogen is uncommon, it is most likely to occur with invasive organisms. Because regional adenopathy is commonly seen, these organisms may produce a syndrome of mesenteric adenitis that mimics appendicitis. Microscopic evaluation of the stool usually reveals cells, but gross blood or pus is unusual.

It is important to understand that significant overlap exists among these clinically separated groups, but such a scheme is useful as an initial approach to a common problem.

COMMON BACTERIAL, VIRAL, AND PARASITIC GASTROINTESTINAL INFECTIONS AND THEIR AGENTS

One of the major challenges in the diagnosis of gastrointestinal disease is the recent increase in the number of probable etiologic agents. The majority of diagnosed pathogens are represented by a relative few bacteria, as shown in Table 28–4.

A number of organisms have been added to this list, however, either as suspected or as

■■■■■ T A B L E 2 8 – 4
COMMON BACTERIAL AGENTS ASSOCIATED WITH GASTROINTESTINAL INFECTIONS

Recognized enteropathogens	*Salmonella*
	Shigella
	Enterotoxigenic *Escherichia coli*
	Enterohemorrhagic *E. coli*
	Enteropathogenic *E. coli*
	Enteroinvasive *E. coli*
	Vibrio cholerae
	V. parahaemolyticus
	Campylobacter jejuni
	C. coli
	Yersinia enterocolitica
	Clostridium difficile
Highly associated enteropathogens	*Aeromonas* spp
	Plesoimonas shigelloides
	Edwardsiella tarda
	Enteroaggregative *E. coli*
	Vibrio spp
	Campylobacter spp
Implicated organisms	*Bacteroides fragilis*
	Citrobacter freundii
	Klebsiella pneumoniae
	K. oxytoca
	Providencia alcalifaciens
	Hafnia alvei

Adapted and reprinted by permission of the publisher from Abbott S, Janda JM: Bacterial gastroenteritis. I: Incidence and etiologic agents. Clin Microbiol Newsl 14:17, 1992. Copyright 1992 by Elsevier Science Inc.

strongly associated agents of gastroenteritis. Table 28–4 also shows other groups of bacterial agents that are highly associated with gastrointestinal infections; however, animal models have not been identified to reproduce the infection. A third group of organisms (implicated organisms), as shown in Table 28–4, have been linked to gastrointestinal diseases. Because these organisms normally make up the colon flora, their role in diarrheal disease is currently being questioned. Similarly, animal models have not been identified to reproduce the infection. Table 28–5 shows the characteristics of the organisms that have been implicated in diarrheal diseases. Other etiologic agents include viruses, which make up less than 50% of cases, and parasitic agents, which account for less than 5%.

Bacterial Agents

Campylobacter jejuni

Although several *Campylobacter* species cause disease in humans, *C. jejuni* is by far the most common. In the United States alone, there are an

■ T A B L E 2 8 – 5

CHARACTERISTICS OF ORGANISMS IMPLICATED IN DIARRHEAL DISEASES

Bacterial Species	Symptoms	Association	Virulence Factor
Klebsiella oxytoca	Enterocolitis, which may be bloody	Antibiotics, especially ampicillin	Cytotoxin
Klebsiella pneumoniae	Watery diarrhea	Unknown	Heat-stable enterotoxin
			Heat-labile enterotoxin
Citrobacter freundii	Watery diarrhea	Unknown	Same as above
Hafnia alvei	Watery diarrhea	Unknown	Attachment-effacement
Bacteroides fragilis	Watery diarrhea	Unknown	Heat-labile enterotoxin
	Abdominal cramping		
	Vomiting and bloody stools in children		
Providencia alcalifaciens	Gastroenteritis	Travel in developing countires	Unknown

Adapted and reprinted by permission of the publisher from Abbott S, Janda JM: Bacterial gastroenteritis. I: Incidence and etiologic agents. Clin Microbiol Newsl 14:17, 1992. Copyright 1992 by Elsevier Science Inc.

estimated 2 million cases annually. The organism is a curved, gram-negative rod. This morphology is helpful, because a Gram stain of stool may show these organisms in as many as 50% of cases. The only other curved gram-negative rods associated with diarrheal disease are several of the *Vibrio* species, including *V. cholerae* and *V. parahaemolyticus* (see later).

Campylobacter jejuni is found in many animal species. Outbreaks have been traced to inadequately cooked poultry, untreated water, and unpasteurized milk. Exposure to animals with diarrhea (especially puppies or kittens) has also been implicated as a risk factor.

The organism produces a number of exotoxins, some with cytolytic properties, that contribute to the clinical picture. Symptoms such as fever, abdominal pain, and diarrhea begin approximately 1 to 7 days after ingestion of contaminated food or drink. Feces may contain blood and pus. This diarrheal illness is usually self-limiting, although clinical manifestations may persist for a week or two.

Salmonella Species

There are over 2,000 serotypes of *Salmonella,* although not all cause disease in humans. *Salmonella* are gram-negative rods that are widely distributed in animals. Therefore, exposure to animals, particularly by ingestion of undercooked meat, is an important risk factor for infection. Poultry, beef, unpasteurized milk, and eggs have been identified as vehicles. Hospital outbreaks have been traced to contaminated platelets or contaminated diagnostic materials. Person-to-person spread also occurs.

Salmonellosis manifests as four different syndromes:

- Gastroenteritis (food poisoning)
- Enteric fever
- Bacteremia
- Carrier state

GASTROENTERITIS AND FOOD POISONING

Gastroenteritis and food poisoning result from the ingestion of undercooked meat, poultry, eggs, and dairy products contaminated with serotypes of *Salmonella* found in animal hosts, including turtles, reptiles, and poultry. Other vehicles of transmission are cooking utensils used in preparing contaminated meat.

Symptoms, which begin approximately 6 to 48 hours after ingestion, consist of nausea, vomiting, and diarrhea. Fever, muscle ache, and headache may also follow. The diarrhea may last from a few hours to a few days, depending on the inoculum size and the health status of the individual. The disease is usually self-limited, and the patient recovers without a need for treatment. Antimicrobial therapy is not indicated in this form of the disease, because it is believed to prolong the carrier state.

ENTERIC FEVER

Typhoid fever caused by *S. typhi* is the most severe form of enteric fever. *Salmonella typhi* is host specific, with humans the only identified reservoir. Enteric fever may also be caused by *Salmonella paratyphi A, Salmonella choleraesuis,* and other serotypes.

Enteric fever is usually transmitted through fecally contaminated food or water. The infective dose required to initiate the infection is much lower than for the gastroenteritis form.

In typhoid fever, *Salmonella* invade small bowel and colonic tissue, producing an inflammatory response, but diarrhea is not present. In fact, most patients experience constipation during the early phase of the disease. The organisms are ingested by monocytes, but these cells do not kill *Salmonella* organisms effectively. For this reason, *Salmonella* are termed *intracellular pathogens,* in that they survive inside monocytes. After tissue invasion, they are transported to regional lymph tissue, which undergoes hyperplasia in response. Organisms then reach the blood stream, and the patient experiences fever, malaise, headache, and abdominal tenderness. During the bacteremic spread, the organisms reach the biliary tract, and invade the gallbladder and Peyer patches of the colon. Colonization at these sites leads to the intestinal seeding and the diarrheic phase. The ability of *Salmonella* to survive intracellularly and its invasion of the gallbladder may contribute to this organism's ability to persist in the stool despite antibiotic therapy. *Salmonella* infections of the blood stream or of a metastatic focus (e.g., bone, central nervous system) can be effectively treated with antimicrobial agents. Antibiotic therapy, however, may actually prolong the time of excretion of *Salmonella* in the stool.

BACTEREMIA

Nontyphoidal salmonella, such as *Salmonella typhimurium, S. paratyphi A* and *B,* and *S. choleraesuis,* have been associated with salmonella bacteremia with or without intestinal infection. Persistent fever and intermittent bacteremia are the common manifestations. This form of salmonellosis has been observed in both young children and adults.

CARRIER STATE

Following salmonella infections, individuals may continue to harbor the organisms for a time. Chronic carriage occurs when the organisms infect the gallbladder, which then continuously seeds the gastrointestinal tract. The affected person continuously or intermittently excretes the organisms in the feces and becomes an important source of infection of susceptible persons.

Shigella Species

Shigella organisms, particularly *S. dysenteriae,* may produce a diarrhea syndrome characterized by the presence of gross blood or pus in the stool. Shigellosis is among the most communicable of the bacterial diarrheas. In a study of volunteers, the ingestion of as few as several hundred organisms caused disease. The organism invades the mucosa locally, producing flask-shaped ulcers and an intense inflammatory response. Deeper invasion to regional lymph nodes or the blood stream is uncommon.

In the United States, *Shigella sonnei* is the most common species isolated. This organism is usually associated with milder illness. Other *Shigella* species are *S. boydii, S. dysenteriae,* and *S. flexneri. Shigella flexneri* is the second most common species seen in the United States. *Shigella boydii* is infrequently isolated in this country.

Shigella produce several toxins of clinical importance. A neurotoxin causes paralysis in mice and plays a role in human infection. Infants with shigellosis may transiently lose the reflexive closing of the anal canal with external pressure, as from a cotton swab. The loss of this so-called anal wink reflex in an infant with diarrhea is a soft clue that *Shigella* may be the etiologic agent.

Diarrheogenic *Escherichia coli*

Escherichia coli may cause diarrhea via a number of different mechanisms. Currently, there are five recognized groups of diarrhegenic *E. coli,* as shown in Table 28–6, which lists the types of E. coli and the associated diarrheal diseases. Specific serotypes have been associated with each of these syndromes.

ENTEROPATHOGENIC *E. COLI*

Enteropathogenic *E. coli* (EPEC) was first recognized in the early 1940s as a cause of infantile diarrhea. For many years, its pathogenic role has been controversial, mainly because only certain "H" serotypes have been associated with the infections. Unfortunately, antisera to detect "H" typing are not usually available in the clinical laboratory, and serotyping has been limited to "O" serogroups only. Further studies, however, have reported the presence of a cytotoxin and an EPEC adherence factor (EAF). This factor, which enables EPEC to adhere to epithelial cells, has

■ T A B L E   2 8 - 6

STRAINS OF *ESCHERICHIA COLI* ASSOCIATED WITH DIARRHEAL DISEASE

Strain Group	Designation	Syndrome and Population	Disease Mechanism
Enterotoxigenic	ETEC	Afebrile, watery diarrhea in travelers	Associated with production of enterotoxin, heat labile (LT), heat stable (ST), or both
Enteropathogenic	EPEC	Febrile, invasive diarrhea in children	Adherence-attachment
Enteroinvasive	EIEC	Febrile diarrhea often with dysentery, particularly in developing countries Seen in adults	Invasion
Enterohemorrhagic	EHEC	Afebrile, grossly bloody diarrhea May be complicated by hemolytic-uremic syndrome	Associated with production of a Shiga-like toxin (SLT, SLT_2)
Enteroaggregative	EAggEC	Febrile, grossly bloody diarrhea	Adherence-attachment

Adapted and reprinted by permission of the publisher from Abbott S, Janda JM: Bacterial gastroenteritis. I: Incidence and etiologic agents. Clin Microbiol Newsl 14:17, 1992. Copyright 1992 by Elsevier Science Inc.

been demonstrated both in vivo and in vitro. Investigators believe that this adherence factor plays a significant role in the pathogensis of EPEC.

Newborns in hospital nurseries and daycare centers are most affected. The disease rarely occurs in adults. Primary symptoms are low-grade fever, malaise, vomiting, and diarrhea. Gross evidence of blood is absent from stool, which may contain large amounts of mucus.

ENTEROTOXIGENIC *E. COLI*

Enterotoxigenic *E. coli* (ETEC) excrete enterotoxins, either heat labile (LT) or heat stable (ST), that produce a watery diarrhea syndrome similar to that of cholera. ETEC remain the most common cause of infant diarrhea in developing countries and of traveler's diarrhea in the United States. Clinical manifestations of ETEC infection are similar to those caused by other enterotoxin-producing organisms, such as *S. aureus, C. perfringens, B. cereus,* and *V. cholerae.* Differential diagnosis is usually made by exclusion or by isolation of only lactose-fermenting organisms. DNA probes for ETEC are available but are not widely used in the clinical laboratory.

ENTEROINVASIVE *E. COLI*

Enteroinvasive *E. coli* (EIEC) have invasive properties and produce a syndrome mimicking *Shigella* infection. A Shiga-like toxin found in still other *E. coli* produces a bloody diarrhea syndrome that manifests clinically as a gastrointestinal bleeding disorder. The infection is characterized by fever, severe abdominal cramps, malaise, and watery diarrhea. The stool contains pus and blood.

ENTEROADHERENT/ENTEROAGGREGATIVE *E. COLI*

Enteroadherent *E. coli* or, as most recently termed, enteroaggregative *E. coli* (EAggEC) cause diarrhea by adhering to the mucosal surface, apparently affecting absorption and stimulating peristalsis. These organisms produce symptoms such as watery diarrhea, vomiting, dehydration, and, occasionally, abdominal pain.

ENTEROHEMORRHAGIC *E. COLI*

Enterohemorrhagic *E. coli* serotype 0157:H7 has been associated with hemorrhagic diarrhea, colitis, and hemolytic-uremic syndrome. Other serotypes have also been implicated. The stool typically does not contain any pus, a feature differentiating this infection from shigellosis. Other serotypes have also been implicated. The infection is potentially fatal, especially among young children and among elderly residents of convalescent homes and nursing centers.

Vibrio Species

Vibrio species are gram-negative rods with a curved shape, like *Campylobacter*. Most infections associated with *Vibrio* species occur upon ingestion of contaminated uncooked seafood or upon contamination of wounds with sea water.

Vibrio cholerae is the causative agent of cholera. It produces chitinase, an enzyme that breaks down chitin, part of the shell of various sea animals, providing the organism with an important niche in the marine environment. Outbreaks are seen in coastal areas, where seafood is an important part of the diet. Cases of

V. cholerae infection in the United States have been reported off the Gulf of Mexico and in travelers returning from areas where the disease is endemic. South America and Africa are currently coping with widespread outbreaks. Although direct person-to-person spread is uncommon, contamination of water and food has been the main factor in spreading disease.

Cholera is a form of diarrhea mediated by the elaboration of an enterotoxin. The enterotoxin consists of an A subunit surrounded by five B subunits. The B subunits attach to the mucosa of the small bowel, permitting the A subunit to penetrate to the interior surface of the membrane, where it catalyzes nicotine-adenine dinucleotide to eventually form cyclic adenosine monophosphate. This stimulates the intestinal mucosa to secrete water and electrolytes, characterized as rice-water stool.

The action of the enterotoxin does not elicit an inflammatory response. Therefore, there are no white cells or red cells in the stool, and the patient is usually afebrile. The amount of fluid lost per rectum can be impressive, however, with many liters of replacement fluids needed per day. Death due to dehydration is the major risk, particularly for patients at the extremes of age and those already debilitated by malnourishment or other diseases.

There are more than 60 serovars (serogroups) of *V. cholerae*, but only serovar 01 has been so far implicated in epidemic cholera. An agglutination test for serovar 01 must be performed on isolates identified as *V. cholerae*. In addition, because not all *V. cholerae* serovar 01 produce the enterotoxin that mediates this disease, toxin detection must also be performed. The latex agglutination test and enzyme-linked immunosorbent assay (ELISA) produce comparable results and are currently in use in laboratories located in high incidence areas.

The clinical microbiology laboratory must be notified when *V. cholerae* is suspected in a patient specimen, because special media are required for its isolation.

Other *Vibrio* species also produce disease. *Vibrio parahaemolyticus* causes diarrhea worldwide and is the most commonly identified bacterial pathogen in Japan. Unlike patients with cholera, patients with *V. parahaemolyticus* usually have fever, chills, and cells in their stool, suggesting that this organism has invasive potential.

Other *Vibrio* species of importance are *Vibrio vulnificus* and *Vibrio alginolyticus*. *Vibrio vulnificus* causes bacteremia and an overwhelming sepsis syndrome. Like other *Vibrio* species, *V. vulnificus* is associated with eating shellfish, particularly clams

and oysters. This syndrome is most commonly seen in patients with severe liver disease. The liver is an important filter of organisms that reach the blood stream from the gastrointestinal tract. When this property is diminished, organisms are more likely to reach the systemic circulation, sometimes with disastrous results. *Vibrio alginolyticus* is more commonly associated with wound infections. The exposure history is likely to be a soft-tissue injury that took place with contamination of sea water or a cut or abrasion from a shellfish.

Aeromonas species (*A. hydrophila, A. sobria,* and *A. caviae*) and *Plesiomonas shigelloides* are closely related to *Vibrio*. *Aeromonas* species are usually associated with a fresh-water exposure. Clinical infections described include a watery diarrhea syndrome, cellulitis, and soft-tissue infection after an injury. *Plesiomonas shigelloides,* found in coastal waters, causes gastroenteritis with extraintestinal syndromes similar to those observed with *Aeromonas* infections.

Yersinia enterocolitica

Yersinia enterocolitica is a gram-negative rod that produces a syndrome similar to that of *Salmonella*. This organism may be found in unpasteurized milk and dairy products. It is a common cause of diarrhea in colder climates, such as the Scandinavian countries. In the United States, it is found mostly in the northern states. Infected patients usually have fever, and white and red cells are present in stool. Invasion to the regional lymph nodes may produce a mesenteric adentitis syndrome, which can mimic appendicitis. Severe abdominal pain, fever, and enlarged lymph nodes are common presenting symptoms of *Y. enterocolitica* infections. Acute enteritis, a mild and self-limiting form of the infection, is characterized by fever, abdominal pain, nausea, and diarrhea.

Yersinia has also rarely caused infection via contamination of blood given for transfusion. The ability of this organism to survive refrigeration temperatures may be a contributing factor. Infections have also occurred as a result of consumption of contaminated food, such as market meat and vacuum-packed beef.

Clostridium difficile

Clostridium difficile, the most common cause of infectious diarrhea in hospitalized patients, is a spore-forming, anaerobic organism. It has been isolated from hospital beds, carpeting, and other fomites with which hospitalized patients come into contact. Patients may also acquire this organism in the community. Most antimicrobial agents have a significant effect on the fecal flora. Chemotherapeutic agents such as antitumor drugs have also been associated with pseudomembranous colitis. When other members of the flora are reduced, *C. difficile* may flourish. The organism produces a cytotoxin and an enterotoxin. Pathologically characteristic pseudomembranes are formed, which are nearly diagnostic when seen on colonoscopy.

Viral Agents

CASE STUDY

Seven patrons of a seafood restaurant become ill with nausea, vomiting, and watery diarrhea within 24 hours of eating at the restaurant. An investigation reveals that the only food in common is raw oysters. All patients are afebrile, and symptoms resolve within 1 day. No white or red cells were found in any of the stool specimens.

Discussion

Enteric pathogens associated with raw shellfish include *Vibrio* species and Norwalk virus. This food may also become contaminated with an enterotoxin-producing *Staphylococcus* or hepatitis virus or may harbor the toxin associated with paralytic or neurotoxic shellfish poisoning. Of these, the syndrome is most likely to be due to Norwalk virus.

Approximately 30 to 40% of all cases of gastroenteritis are due to viral pathogens. Included among these are rotavirus, enteric adenovirus, Norwalk virus, calicivirus, and astrovirus. Because viral particles are so small, special techniques are necessary to identify them. Immunoelectronmicroscopy, monoclonal antibody tests, and viral cultures of the stool are used to identify viral causes of diarrhea. Because many microbiology laboratories do not routinely offer these tests, however, many cases go undiagnosed.

Adenoviruses are associated with a number of infectious syndromes, including coryza and tracheitis, pharyngoconjunctival fever, and epidemic keratoconjunctivitis. Adenovirus serotypes 40

and 41 are most associated with diarrhea, particularly in children. These serotypes are rarely implicated in the other infections attributed to adenoviruses.

Rotavirus, a double-stranded RNA virus measuring 70 nm in diameter, is believed to be the major viral cause of gastroenteritis worldwide. Group A rotavirus is the main pathogen and is differentiated from the so-called atypical rotaviruses (groups B and C) by the migration patterns on polyacrylamide gel electrophoresis (PAGE) and by the presence of a group-specific antigen (VP6), which is the inner capsid protein on the group A virus. Infections with rotavirus are more common in the winter months in temperate climates and occur primarily in infants and children. Rotavirus causes diarrhea by interfering with absorption of fluids at the level of the small bowel.

Norwalk virus is probably the most extensively studied of the viruses associated with diarrhea in humans. It is smaller than rotavirus, measuring 20 to 40 nm in diameter. It is a single-stranded RNA virus. Norwalk virus causes disease year round and is the most commonly implicated etiologic agent in outbreaks of acute gastroenteritis. Contaminated drinking water and ingestion of raw or partially cooked shellfish are risk factors for its acquisition.

Caliciviruses and astroviruses are similar in size but differ in the shapes of their outer surfaces. Caliciviruses have cup-shaped cutouts on the outer surface, whereas astroviruses have a surface architecture that forms a five- or six-pointed star. Illness with these viruses is usually characterized by vomiting and a watery diarrhea syndrome. Calicivirus and astrovirus usually produce a milder illness than Norwalk virus.

Parasitic Agents

A number of protozoal organisms also cause diarrhea. Unlike the bacterial and viral causes previously described, these organisms are often associated with more prolonged symptoms. Diarrhea, bloating, and other symptoms may persist for weeks or longer. Protozoal parasites are larger than viruses and bacteria. They are usually detected by direct microscopy of a stained stool smear.

In the United States, *G. lamblia* is the most commonly identified intestinal parasitic pathogen. This organism is most commonly acquired via ingestion of contaminated water or by person-to-person spread in institutions or daycare centers. When water is the vehicle, transmission is usually due either to inadequate chlorination or to a treatment method that lacks a filtration or sedimentation procedure.

Giardia may be detected in the stool as a trophozoite or as a cyst. The trophozoite has a characteristic, face-like appearance on staining. Trophozoites are very sensitive to the acid found in stomach secretions, whereas cysts are relatively resistant and are infectious to humans. Most patients with giardiasis complain of diarrhea, crampy abdominal pain, belching, and flatulence. Blood or pus in the stool is rare. The organism is found primarily in the small bowel, impairing absorption.

Unlike *G. lamblia, E. histolytica* (Fig. 28–3) may cause an invasive syndrome characterized by fever and dysentery. Even when gross blood and pus are not present in stool, nearly all specimens from patients with diarrhea due to *E. histolytica* have red cells seen on microscopy. The appearance of trophozoites ingesting red blood cells is pathognomonic for amebiasis. The ability of this organism to lyse tissue contributed to its being named *histolytica*. This invasive ability contributes to the occasional catastrophic illnesses, including intestinal rupture, liver abscesses, and, rarely, pericardial or pleural disease, caused by *E. histolytica*.

Cryptosporidium parvum has been associated with self-limiting gastrointestinal infections in the immunocompetent population. Sporadic outbreaks of cryptosporidiosis have been reported frequently in daycare centers and nurseries. Watery diarrhea accompanied by nausea, vomiting, and low-grade fever is common. In the immunosuppressed individual, however, such as

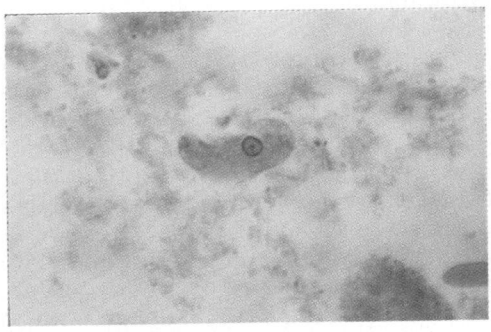

■ F i g u r e   2 8 – 3

E. histolytica trophozoites.

a person with AIDS, the condition these parasites produce is life-threatening. *Isospora belli,* although less commonly encountered, has also been identified as an enteric opportunist in AIDS patients. Both organisms are transmitted via the fecal-oral route.

COMPLICATIONS OF DIARRHEAL INFECTIONS

The vast majority of infectious diarrheal illnesses are self-limiting and uncomplicated. When complications occur, they are often related to severe dehydration. Kidney failure, liver failure, myocardial infarction, and bowel infarction can each be seen in a setting in which the intravascular volume is severely depleted, with less blood going to vital organs.

Toxic megacolon is an acute dilatation of the large bowel that occurs in response to severe inflammation. A postulated pathogenesis for this change is an alteration in the neuromuscular tone of the bowel in response to the inflammatory process. Toxic megacolon may complicate diarrhea due to invasive organisms (*Shigella, Salmonella, Campylobacter, Yersinia, C. difficile,* and *E. histolytica*) and noninfectious processes such as ulcerative colitis. The transverse colon is the most common area dilated on radiography, exceeding 6 cm in diameter. The patient is usually febrile, appears toxic, and has diffuse abdominal pain; bowel sounds are infrequent or absent. Because of the relative atony of colon, diarrhea is absent or infrequent, giving the false impression that the patient is improving. Because of the high risk of bacteremia with colonic flora and the risk of perforation, broad-spectrum antibiotics are recommended. Patients may also require surgical intervention (colectomy).

Reiter syndrome may occur up to 6 weeks after a diarrheal or sexually transmitted infection. Affected patients are usually male and have some combination of arthritis, skin rash (a thick scaly rash involving the palms and soles, as well as a rash around the glans penis), urethritis, and conjunctivitis. This is an inflammatory illness that is treated with antiinflammatory medications, such as aspirin.

The hemolytic-uremic syndrome consists of anemia due to hemolysis or red blood cells and renal failure. This syndrome has been particularly associated with enterohemorrhagic *E. coli*

(O157:H7) infection but may also occur after infection with *Shigella* or enteroviruses (echovirus 22; coxsackievirus A4, B2, and B4).

Occasionally after a severe diarrhea, the villi—absorptive surface of the small bowel—may slough off. Because of the resultant malabsorption, patients have a prolonged diarrhea lasting weeks or months after the pathogen has been eradicated. A biopsy of the small bowel reveals a flattened surface similar to the picture seen in celiac sprue. Unlike that disease, however, this syndrome is reversible. Postinfectious malabsorption syndrome usually responds to a special diet of foods that are absorbed without the need for this specialized surface until the villi can regenerate.

NEWLY RECOGNIZED AGENTS OF ACUTE DIARRHEA

For many patients with acute diarrhea, no pathogen is identified. Because so many organisms are ingested daily, it should not be surprising that new pathogens are regularly identified as possible causes of diarrhea.

Round, acid-fast organisms measuring 8 to 10 μm in diameter have been identified from the stool of AIDS patients with chronic diarrhea. This pathogen has also been associated with diarrhea lasting 3 weeks to 2 months in immunocompetent persons. Several outbreaks have been traced to contaminated water sources. Originally described as Cyanobacteria-like bodies (resembling blue-green algae), the organism is now believed to be a protozoal pathogen in the Cyclospora family.

Enterocytozoon bieneusi (microsporidium) causes a clinical picture similar to that of Cyclospora and *Cryptosporidium. Enterocytozoon bieneusi* is also detected by acid-fast staining but is the smallest of these organisms (1 to 2 μm).

Sexually transmitted diseases (STDs) can also cause diarrhea and other gastrointestinal symptoms. *Neisseria gonorrhoeae, Treponema pallidum* (syphilis), *Chlamydia,* and herpes simplex may cause a proctitis characterized by pain on defecation and loose stools with mucus or pus. This manifestation of a sexually transmitted illness is most commonly seen in persons who are recipients of anal intercourse. Because methods routinely used for culturing stool for bacterial pathogens do not detect the STD organisms, the physician must consider this possibility in the differential diagnosis.

Helicobacter pylori is a small curved gram-negative rod closely related to *Campylobacter* species. Although not a known cause of diarrhea, it is perhaps the most common infecting pathogen of the gastrointestinal tract. It is believed to be a cause of type B gastritis, a common type of inflammation of the stomach. It is also associated with ulcer disease. *Helicobacter pylori* is common, being found in more than 50% of persons over the age of 50 years in developed nations. In developing nations, infection appears to occur earlier. Chronic *H. pylori* infection has even been linked to a higher incidence of gastric cancer. Antibiotic treatment studies suggest that in some patients with duodenal ulcer disease ulcers heal faster and fewer relapses occur if treatment is directed against this organism than in patients given traditional, non–antibiotic-containing ulcer treatment.

AGENTS OF FOOD POISONING

CASE STUDY

A young man has eaten at the restaurant where several other people became ill. Within 1 hour of his meal, he notes the sudden onset of tingling around his mouth, abdominal pain, mild diarrhea, and headache. Later, he experiences a burning in his hands and feet, and generalized weakness. He is seen in the emergency room, where he requires intubation for respiratory arrest. Before intubation, he tells the ER physician that he has eaten no shellfish.

Discussion

Scromboid can cause all of this patient's initial symptoms but is not associated with respiratory arrest. Neurotoxic shellfish poisoning is also a mild illness. Ciguatera, paralytic shellfish poisoning (PSP), and puffer fish intoxication can each produce this syndrome. PSP is unlikely because the patient reported eating no shellfish, and puffer fish is rarely served in the United States. Ciguatera is the most likely diagnosis. Botulism should also be included in the differential diagnosis, although the rate of progression in this case was unusually fast and no focal neurologic findings were present.

A *food-borne outbreak* is defined as the occurrence in two or more persons of gastrointestinal or neurologic symptoms within 72 hours of a common meal. The most commonly identified pathogen has been *Salmonella* species, accounting for over 25% of all reported outbreaks. *Staphylococcus aureus, Clostridium botulinum,* and *C. perfringens* together account for another 25 to 28% of outbreaks in which a pathogen is identified. Of the parasites, *Giardia* and *Cryptosporidium* have each been causes of outbreaks, usually in association with a contaminated water source. Outbreaks of viral gastroenteritis (in particular, Norwalk virus associated with contaminated shellfish) account for nearly 5% of diagnosed outbreaks.

Chemical intoxications are responsible for more than 20% of outbreaks in which a pathogen is identified. Many are associated with a fish exposure. Scromboid is a syndrome consisting of flushing, headache, and diarrhea with crampy abdominal pain. It is due to ingesting a contaminated fish (histamine and enzyme inhibitors are present in the flesh of some fish, including tuna, yellow jack, and mackerel). Symptoms begin within 1 hour of ingestion and usually last several hours only.

Ciguatera is caused by the action of ciguatoxin, produced in dinoflagellates and passed up the food chain. Symptoms of diarrhea, abdominal pain, paresthesias, weakness, and headache begin with 1 to 2 hours of eating contaminated fish (snapper, sea bass, grouper) and may progress to hypotension and respiratory arrest. No antitoxin exists. Patients are observed and treated supportively.

Paralytic shellfish poisoning produces a similar syndrome and has a mortality rate of 10%. It is produced by a different toxin, which is found in contaminated mussels, clams, and scallops. The syndrome is usually seen in the summer months. A much milder syndrome, caused by a different toxin and also associated with shellfish, is termed neurotoxic shellfish poisoning. Puffer fish toxin, (tetrodotoxin) is particularly potent, with a mortality rate of 60% in poisoned patients. Persons who prepare this fish must undergo special training and licensing.

With each of these poisonings, the fish looks, smells, and tastes normal. Prevention is aimed at identifying areas of contaminated fish. Table 28–7 shows common food-borne pathogens and their characteristics.

LABORATORY DIAGNOSIS OF GASTROINTESTINAL PATHOGENS

When a stool specimen is sent to the microbiology laboratory for analysis, the physician must correctly request the appropriate examination and understand the limitations of that test. In the United States, stool specimens are most commonly sent for "bacterial culture." For most laboratories, this means culturing for *C. jejuni, Salmonella,* and *Shigella.* If other organisms are suspected, the laboratory must be notified.

Specimen Collection and Handling

Stool specimens should be transported to the laboratory soon after collection. Refrigeration must be avoided as much as possible. No preservatives can be added to stool samples for bacterial detection. Cary-Blair or other suitable transport medium would be appropriate if samples such as rectal swabs are submitted for culture. Samples for the examination for ova and parasites, however, must be transported in the proper preservatives, such as polyvinyl alcohol (PVA) or formalin, especially if the stool is loose and watery.

Direct Microscopic Examination

Microscopic evaluation of the stool demonstrates white blood cells (see Fig. 28–2) when the etiologic agent is invasive. Such would be observed in *Salmonella, Shigella, Yersinia, Campylobacter* and *Vibrio* sp (other than *V. cholerae*) infections. White blood cells are present because of inflammation, and red blood cells, to bleeding. Colonic biopsy of infected individuals may show tissue invasion.

Characteristic microscopic morphology on Gram-stained smears of certain bacterial pathogens may show gram-negative, curved rods that may align end-to-end and show a "seagull wing" appearance (Fig. 28–4). This morphology may indicate the presence of *Campylobacter* sp or *Vibrio* sp. Hanging-drop preparations or wet mount preparations may demonstrate a "darting" motility, a characteristic of *C. jejuni.* Bloody stools must also be examined immediately for the presence of *E. histolytica* trophozoites.

Culture

The most common method of identifying a bacterial pathogen in the stool is selective culture. This method uses antibiotic(s), chemicals, or environmental changes to inhibit the growth of the predominant fecal flora and selectively

■ T A B L E 2 8 – 7

COMPENDIUM OF COMMON FOOD-BORNE DISEASES

Average Incubation Period	Organism	Average Duration	Implicated Foods	Typical Symptoms	Comments
2–16 hours	Bacillus cereus	1 day	Boiled and fried rice, meats, vegetables	Nausea, vomiting, (emetic) abdominal cramping, watery diarrhea	Produces two toxins; one emetic form that causes nausea and vomiting within hours, and one diarrheic form. Common year round. Isolation of large numbers from implicated foods and patient stool
6–72 hours	Vibrio parahae-molyticus	3 days	Shellfish	Pain, vomiting, fever, watery diarrhea	Blood sometimes in stool. Common spring, summer, fall in the coastal States. Stool culture using TCBS media is recommended
6–72 hours	Vibrio cholerae	3–7 days	Seafood, water	"Rice water" stools, severe diarrhea, no fever	Blood and mucus in stool, mechanism of action in vivo enterotoxin production and tissue invasion. Stool culture using TCBS media is recommended
<8 hours	Staphylococcus aureus	<1 day	Egg salads, meat, poultry, pastries	Abrupt onset of nausea, pain and projectile vomiting, infrequent diarrhea	Mechanism of action is preformed enterotoxin in foods. Common in summer. ELISA or reverse passive latex agglutination enterotoxin test; gel electrophoresis in lieu of phage typing
8–22 hours	Clostridium perfringens	1 day	Beef, poultry, gravy, fish	Abdominal cramping, watery diarrhea; vomiting and fever uncommon	In vivo enterotoxin production; unlike Staphylococcus aureus, viable organisms must be ingested for disease to occur. Common in fall, winter, spring
12–48 hours	Salmonella spp	3 days	Eggs, dairy products, fowl, beef	Fever, abdominal cramping, diarrhea, mild vomiting	WBCs in stool. Common in summer. Culture and serologic identification
16–48 hours	Yersinia enterocolitica	1 day–4 weeks	Milk, pork	Fever, severe abdominal pain, diarrhea	WBCs and RBCs in stool. Common in winter
18–36 hours	Clostridium botulinum	weeks–months	Vegetables, fruits (canned foods), fish, honey (infants)	Nausea, vomiting, diarrhea, paralysis	Mechanism of action is a preformed neurotoxin. Common in summer and fall
24–72 hours	Shigella spp	3 days	Egg and tuna salads, lettuce, milk	Fever, abdominal cramping, diarrhea, occasional vomiting	WBCs, RBCs and mucus in stools, tissue invasion common mechanism of action. Common in summer. Culture and serologic identification
24–72 hours	Enterotoxigenic E. coli (ETEC)	3 days	Fruits, meats, pastries, salads	Abdominal cramping, watery diarrhea, no vomiting or fever	In vivo enterotoxin, major cause of "traveler's" diarrhea, year-round distribution, patient history includes travel to Mexico and other developing countries
24–72 hours	Enterohemor-rhagic E. coli (EHEC)	3 days	Undercooked ground beef, cider	Watery diarrhea progressing to bloody diarrhea, abdominal cramping, no fever or vomiting	Implicated serotype O157:H7, organisms disappear rapidly from stool. Culture of sorbitol-negative E. coli from stool using SMAC plate (sorbitol-negative MAC) recommended

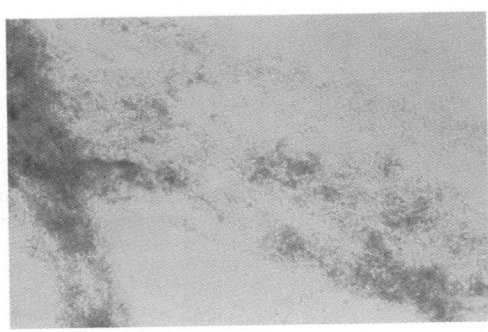

■■■■■ F i g u r e   2 8 – 4

Gram stain of *Campylobacter* colony showing the typical microscopic morphology described as "seagull wings."

permit the pathogen to continue growing. Identification of colonies of the pathogen is then easier, without the growth of the normal flora there to mask them. Table 28–8 lists a variety of selective media commonly used to recover gastrointestinal bacterial pathogens.

Campylobacter

Campylobacter jejuni grows best at 42°C in an atmosphere containing 5 to 10% oxygen. *Campylobacter* is therefore considered a microaerophilic organism. Many laboratories purchase tanks of gas with an appropriate mixture to provide an atmosphere for culturing this organism. In settings where this is not practical (e.g., when evaluating an outbreak in an area without immediate access to a laboratory), an acceptable atmosphere can be created with little expense by using a candle jar.

Campylobacter jejuni has characteristic colonial morphology, described as "running" and "wet-looking" because the colonies seem to run together. Microscopic morphology shows the typical gram-negative curved rods that look like a seagull's wings (see Fig. 28–4). This characteristic morphology of the genus *Campylobacter* differentiates it from *P. aeruginosa,* which also grows at 42°C and is oxidase positive.

Salmonella

Salmonella infections are confirmed by culture (Fig. 28–5). Organisms are most likely to be recovered from blood cultures of patients suspected of typhoid fever if the specimens are obtained during the first week of the infection.

Stool cultures yield the organisms during the third and fourth week of the infection. Routine microbiologic media such as sheep blood agar and MacConkey agar, and highly selective enteric media, such as Hektoen enteric (HE) agar and xylose-lysine-deoxycholate (XLD) agar are used for recovery. Serotyping should be performed whenever possible.

Shigella

Shigellae are fragile organisms that do not survive well outside the host for a long period. These organisms are particularly susceptible to acid pH; therefore, stool samples should be processed as soon as they are received in the laboratory (Fig. 28–6). Diarrheic stools from patients with suspected shigellosis contain pus and blood, a presentation typical of an invasive agent. Bloody stools must be plated as soon as possible on appropriate enteric media.

Escherichia coli

Diarrheogenic *E. coli* do not look different on a growth plate from *E. coli* that do not cause diarrheal disease (Fig. 28–7). Diagnosis of diarrheal disease caused by *E. coli* requires a high index of suspicion from the clinician on the basis of history and physical findings. In addition, because *E. coli* are part of the normal fecal flora, special tests are needed to differentiate these pathogens from the routinely isolated nonpathogenic *E. coli.* For example, many enterohemorrhagic *E. coli* do not ferment sorbitol. Sorbitol-negative *E. coli* can be selected in the laboratory by using a sorbitol plate combined with a color indicator (see Chapter 16). Antisera are also used to screen for specific serotypes. Enteroinvasive *E. coli* produce colonies and biochemical reactions similar to those of *Shigella* species. Tests to differentiate between these pathogens should be performed.

Yersinia

Yersinia sp grow well at 25°C. This characteristic may be used in the laboratory. Plating and incubation at this temperature and the use of selective media (e.g., cefsulodin-iragasan-novobiocin [CIN] agar) permit ready isolation of

■■■■■ T A B L E   2 8 – 8
SELECTIVE MEDIA COMMONLY USED TO RECOVER DIARRHEAL AGENTS

Culture Medium	Purpose	Characteristic Morphology	
		Pathogens	Colon Flora
MacConkey agar	To recover Enterobacteriaceae and other nonfastidious gram-negative bacilli Inhibits gram-positive organisms and some fastidious gram-negative bacilli	Salmonella, Shigella (with few exceptions) Edwardsiella appears clear and colorless	Lactose fermenters, such as Escherichia coli, Klebsiella sp, Enterobacter spp, and certain Citrobacter sp, appear dark pink to red Late or slow lactose fermenters, such as Citrobacter sp, Serratia sp, and Hafnia sp, appear colorless in 24 hours and slightly pink after 24–48 hours Non–lactose-fermenters, such as Citrobacter sp, Proteus sp, Providencia sp, and Morganella sp appear clear and colorless
Hektoen enteric (HE) agar	A highly selective medium to recover primarily Salmonella and Shigella sp Inhibits common colon flora Contains indicators to detect hydrogen sulfide (H$_2$S) production	Salmonella sp appear green to blue-green with black centers because of H$_2$S production Shigella sp appear green without black centers, because they do not produce H$_2$S	Lactose fermenters, such as E. coli, are slightly inhibited and appear orange to salmon-pink Proteus sp are slightly inhibited; small, clear colonies with black centers may appear
Xylose-lysine deoxycholate agar (XLD)	A differential and selective medium to isolate Salmonella sp and Shigella sp from stool Inhibits most colon flora and most gram-positive bacteria Certain Shigella sp (S. dysenteriae and S. flexneri) may be slightly inhibited	Salmonella sp appears red with black centers owing to the production of H$_2$S; Salmonella does not ferment lactose or sucrose but does ferment xylose, which is essential in decarboxylating lysine to revert the acid pH (yellow from sucrose fermentation) to an alkaline pH (red from lysine decarboxylation) Shigella sp do not ferment any of these carbohydrates and appear red or clear	Enterobacteriaceae that may not be completely inhibited, such as Proteus vulgaris, appear yellow (from sucrose) with black centers Citrobacter freundii, which produces H$_2$S, appears yellow with black centers owing to the inability to decarboxylate lysine Other intestinal flora that may grow ferment one or all of the carbohydrates in this medium, resulting in yellow colonies
Campylobacter blood agar (CAMPY-BA)	An enrichment-selective medium primarily to isolate and cultivate Campylobacter sp from stool	Campylobacter jejuni appears pinkish gray, moist, and runny when incubated at 42°C	
Cefsulodin-Irgasan-novobiocin (CIN)	A selective medium to primarily isolate and recover Yersinia enterocolitica Aeromonas and Plesiomonas shigelloides may also be recovered Inhibits most gram-positive cocci, except for enterococci, and most gram-negative bacilli, particularly the Enterobacteriaceae	Y. enterocolitica produce colonies that look like "bull's-eyes:" the center is red and the periphery appears colorless Aeromonas also ferments mannitol present in the medium, like Yersinia; P. shigelloides does not	Except for Pseudomonas aeruginosa, Citrobacter, and Serratia, most colon flora are inhibited
Thiosulfate-citrate–bile salts–sucrose agar (TCBS)	A highly selective medium to recover Vibrio sp, including Vibrio cholerae, from stool and food Inhibits most colon flora because of the high pH (preferred by vibrios) and high bile salts content Aeromonas may be recovered from this medium	TCBS contains sucrose, so sucrose-fermenting Vibrio sp such as V. cholerae and V. alginolyticus produce yellow colonies Non–sucrose fermenters, such as V. parahaemolyticus and V. vulnificus, produce blue-green colonies	Inhibitory to most colon flora, except for occasional Pseudomonas isolates, which may also appear blue-green

Table continued on following page

■■■■■■ T A B L E   2 8 – 8

SELECTIVE MEDIA COMMONLY USED TO RECOVER DIARRHEAL AGENTS *Continued*

		Characteristic Morphology	
Culture Medium	Purpose	*Pathogens*	*Colon Flora*
Cycloserine-cefoxitin-fructose agar (CCFA), anaerobic incubation required	A selective medium to primarily isolates *Clostridium difficile* from stool of patients suspected of antibiotic-associated diarrhea or pseudomembranous colitis Inhibits most colon flora, both gram-positive and gram-negative bacteria	*C. difficile* appears yellow from fructose fermentation	Colon flora are inhibited

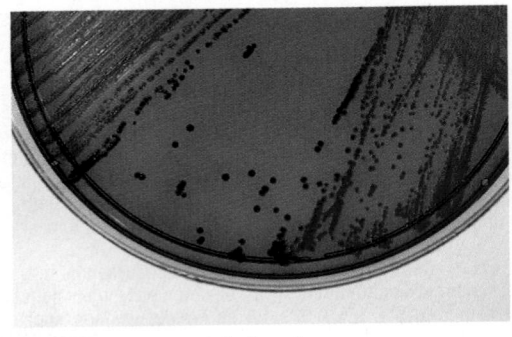

■■■■■■ F i g u r e   2 8 – 5

Salmonella colonies growing on Hektoen enteric (HE) agar showing black centers due to the production of hydrogen sulfide.

■■■■■■ F i g u r e   2 8 – 6

Shigella colonies growing on HE agar showing clear green colonies.

Yersinia. Cold enrichment procedures, such as placing fecal samples on isotonic saline and keeping them at 4°C before the inoculation of selective medium, have increased recovery of the organism.

Vibrio

Vibrio sp requires highly selective medium for maximum recovery. Thiosulfate-citrate–bile salts–sucrose (TCBS) agar inhibits normal colon flora. In addition, TCBS agar differentiates sucrose-fermenting from non–sucrose-fermenting vibrios (Fig. 28–8). When *V. cholerae* is suspected, use of a transport medium such as Cary-Blair or buffered saline in alkaline peptone water is suitable.

Susceptibility to the vibriostatic compound 0:129 is useful in identifying *Vibrio* sp. Salt requirement for growth is usually helpful in differentiating *V. cholerae* from halophilic (non-cholera) vibrios. In endemic areas, antisera to

screen for *V. cholerae* servar 01 should be available. Latex agglutination tests are also commercially available to detect toxin-producing strains.

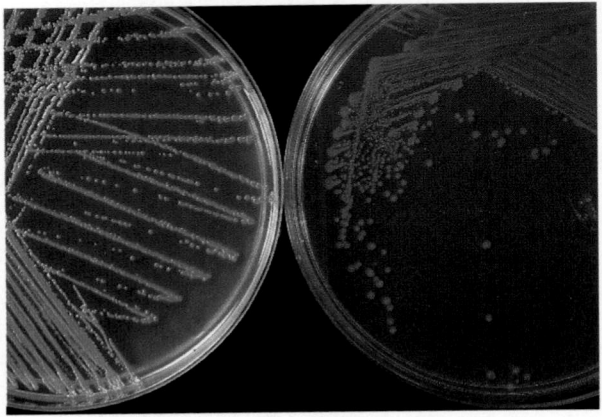

■■■■■■ F i g u r e   2 8 – 7

Left, *E. coli* O157:H7 growing on MacConkey agar. *Right, E. coli* O157:H7 on sorbitol MacConkey agar. *E. coli* O157:H7 does not ferment sorbitol, whereas most other *E. coli* serotypes do ferment sorbitol.

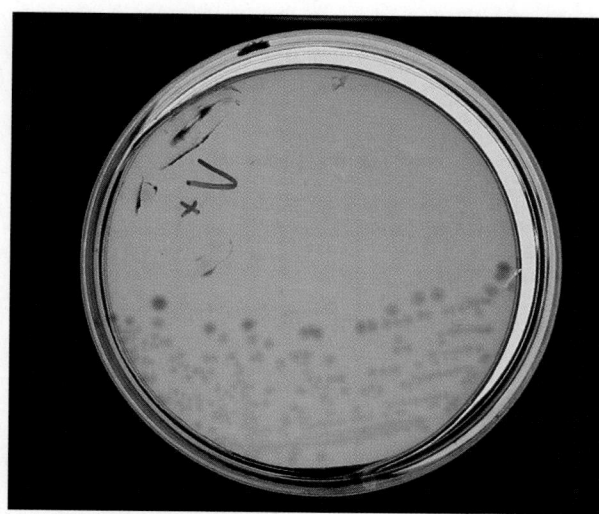

■ F i g u r e 2 8 – 8
Vibrio vulnificus growing on TCBS. Vibrio vulnificus is a non–sucrose-fermenting vibrio.

Clostridium

Clostridium difficile produces yellow, ground-glass colonies on cycloserine-cefoxitin-fructose agar (CCFA). Toxigenic strains may be screened by either enzyme immunoassay (EIA) or latex agglutination assay and confirmed by cell culture.

TREATMENT OF DIARRHEA

The first factor to consider in the assessment of the patient with diarrhea is the state of hydration. Adequate rehydration and maintenance fluids are critical. Patients cannot be adequately rehydrated with water alone, because they have lost a significant amount of electrolyte-rich fluid in the stool. In the 1960s the coupled transport of sodium and glucose across the gut mucosa was first described. It provided the rationale for fluids with the appropriate mix of electrolytes for rehydration. Rehydration fluids can be given orally to all but the most ill patients. When given intravenously, isotonic fluids (e.g., 5% dextrose in 0.9 sodium chloride) are preferred to hypotonic solutions (e.g., 5% dextrose in water).

To date, no effective antibiotic therapy has been developed that shortens diarrheal illness caused by the common viral pathogens. Antibiotics may shorten the clinical illness caused by invasive bacteria or an enterotoxin-mediated process. Clinical studies are under way to further identify which subgroups of patients are most likely to benefit from antibiotic intervention. Antibiotic therapy has also been shown to shorten the clinical illness caused by *G. lamblia, I. belli,* and *E. histolytica.* No effective antimicrobial therapy has yet been identified for *Cryptosporidium, E. bieneusi,* or Cyclospora.

Antidiarrheal medication, such as diphenoxylate with atropine (Lomotil) or loperamide, may also decrease the frequency of stool for some patients. Loperamide appears to be the safer of these agents. Earlier studies with diphenoxylate-atropine suggested that, for patients with an invasive syndrome, the clinical illness might actually worsen with such therapy, theoretically owing to an increase in the contact time of the organism with the mucosa and an impairment of the purging activity of diarrhea to rid the body of the pathogen. Although later studies question the validity of this finding, the use of an antidiarrheal medication may be most effective in the watery diarrhea syndromes associated with an enterotoxin-mediated or viral-induced process.

Pepto-Bismol (bismuth subsalicylate) has also been effective in the treatment of some patients with diarrhea. When taken four to six times a day, Pepto-Bismol shortened clinical illness in patients with traveler's diarrhea, a syndrome mainly due to enterotoxigenic *E. coli.* It has been postulated that the bismuth binds the toxin in the gut. There may also be a beneficial effect of the salicylate component (aspirin) on the mucosa.

Various antibiotics and Pepto-Bismol have each been effective in the prophylaxis of diarrhea when given to travelers who are visiting areas where the risk of acquisition is high. Because of the cost of such therapy and the generally benign and self-limiting nature of most cases of acute diarrhea, such prophylactic therapy is not routinely recommended for all travelers. Travelers are advised to avoid high-risk foods and to drink only bottled beverages. High-risk foods include anything cut up by someone else and served raw (e.g., fruits, salads), dips and other foods left standing out, and raw or partially cooked shellfish. The phrase, "Boil it, peel it, cook it, or forget it," is a good one for travelers to high-risk areas to remember.

In order to prevent secondary infections, all patients should be educated about the most common modes of transmission. Careful hand-washing and not cooking food for others are particularly important preventive measures.

Bibliography

Abbott S, Janda M: Bacterial gastroenteritis. I: Incidence and etiologic agents. Clin Microbiol Newsl 14:17, 1992.

Ashkenazi S, Cleary KR, et al: The association of Shiga toxin and other cytotoxins with the neurologic manifestations of shigellosis. J Infect Dis 161:961, 1990.

Blaser MJ: *Helicobacter pylori:* Its role in disease. Clin Infect Dis 15:386, 1992.

Blaser MJ, Reller LB: Campylobacter enteritis. N Engl J Med 305:1444, 1981.

Blacklow N, Greenberg H: Viral gastroenteritis. N Engl J Med 325:252, 1991.

Craun GF: Waterborne giardiasis in the United States 1965–1984. Lancet 2:513, 1985.

Curran P: Na, Cl, and water transport by rat ileum in vitro. J Gen Physiol 43:1137, 1960.

Drapkin M: Nosocomial infection with *C. difficile.* Infect Dis Clin Prac 1:138, 1992.

DuPont H, Hornick R, et al: The response of man to virulent *Shigella flexneri* 2a. J Infect Dis 119:396, 1969.

DuPont H, Ericsson C, et al: Five versus three days of norfloxacin therapy for traveler's diarrhea: A placebo-controlled study. Antimicrob Agents Chemother 36:87, 1992.

Ericsson C, DuPont H, Mathewson J, et al: Treatment of traveler's diarrhea with sulfamethoxazole and trimethoprim and loperamide. JAMA 263:257, 1990.

Farr B: Diarrhea: A neglected nosocomial hazard? Infect Control Hosp Epidemiol 12:343, 1991.

Glass R, Claeson M, et al: Cholera in Africa: Lessons on transmission and control for Latin America. Lancet 338:791, 1991.

Goodman L, Trenholme G, et al: Empiric antimicrobial therapy of domestically acquired acute diarrhea in urban adults. Arch Intern Med 150:541, 1990.

Goodman LJ: Diagnosis, management, and prevention of diarrheal diseases. Curr Opin Infect Dis 6:88, 1993.

Guerrant RL, Shields DS, et al: Evaluation and diagnosis of acute infectious diarrhea. Am J Med 78(supp 6B):91, 1985.

Harris AA: Hemorrhagic colitis and *Escherichia coli* O157:H7—identifying a messenger while pursuing the message. Mayo Clin Proc 65:884, 1990.

Hodgkin K: Towards Earlier Diagnosis: A Family Doctor's Approach. Baltimore: Williams & Wilkins, 1963.

Jiang Z, Nelson A, et al: Intestinal secretory immune response to infection with *Aeromonas* species and *Plesiomonas shigelloides* among students from the United States in Mexico. J Infect Dis 164:979, 1991.

Koneman E, Allen SD, et al: Color Atlas and Textbook of Diagnostic Microbiology, 3rd ed. Philadelphia: JB Lippincott, 1988.

Levine MM: *Escherichia coli* that cause diarrhea: Enterotoxigenic, enteropathogenic, enteroinvasive, enterohemorrhagic, and enteroadherent. J Infect Dis 155:377, 1987.

Mackowiak PA, Wasserman SS, Levine MM: An analysis of the quantitative relationship between oral temperature and severity of illness in experimental shigellosis. J Infect Dis 166:1181, 1991.

Mishu B, Griffin PM, et al: *Salmonella enteritidis* gastroenteritis transmitted by intact chicken eggs. Ann Intern Med 115:190, 1991.

Ortego YR, Sterling CR, et al: Cyclospora species—a new protozoan pathogen of humans. N Engl J Med 328:1308, 1993.

Peterson WL: *Helicobacter pylori* and peptic ulcer disease. N Engl J Med 324:1043, 1991.

Rabbani G, Islam M, et al: Single-dose treatment of cholera with furazolidone or tetracycline in a double-blind randomized trial. Antimicrob Agents Chemother 33:1447, 1989.

Raj P: Pathogenesis and laboratory diagnosis of *Escherichia coli*–associated Enteritis. Clin Microbiol Newsl 15:89, 1993.

Shadduck JA: Human microsporidiosis and AIDS. Rev Infect Dis 11:203, 1989.

Spika JS, Waterman SH, et al: Chloramphenicol-resistant *Salmonella newport* traced through hamburger to dairy farms. N Engl J Med 316:565, 1987.

Sommers HM: The indigenous microbiota of the human host. In Youmans GP, Paterson PY, Sommers HM (eds): The Biologic and Clinical Basis of Infectious Diseases, 2nd ed. Philadelphia: WB Saunders, 1980.

Steffen R, Rickernbach M, et al: Health problems after travel to developing countries. J Infect Dis 156:84, 1987.

Swerdlow D, Ries A: Cholera in the Americas: Guidelines for the clinician. JAMA 267:1495, 1992.

Taylor PR, Weinstein WM, Bryner JH: Campylobacter fetus infection in human subjects: Association with raw milk. Am J Med 66:779, 1979.

Taylor D, Sanchez J, et al: Treatment of traveler's diarrhea: Ciprofloxacin plus loperamide compared with ciprofloxacin alone. Ann Intern Med 114:731, 1991.

Walker RI, Caldwell MB, et al: Pathophysiology of *Campylobacter* enteritis. Microbiol Rev 50:81, 1986.

Infections of the Central Nervous System

David P. Marmaduke •
Leona W. Ayers

Infections of the CNS are of critical clinical concern, and positive laboratory findings are "critical values" communicated directly to the attending physician or other appropriate members of the patient care team. The symptom complex associated with these infections may be caused by bacteria, fungi, viruses, or parasites. The physician arrives at a presumptive diagnosis based on patient demographics, local epidemiology of CNS infections, physical examination, and radiologic studies. Specific diagnosis by identification of the etiologic agent is the critical task of the laboratory staff.

In the United States, the annual incidence of bacterial meningitis varies from 0.1 to 15 cases per 100,000 population, with a subsequent mortality rate ranging from 0 to 33%. Despite vaccination prevention programs and antibiotic interventions, only a modest overall decrease in new cases of bacterial meningitis has occurred in the United States. Cases of *Haemophilus influenzae* meningitis in children have decreased dramatically where vaccination programs have been fully implemented, but this is countered by the increasing number of opportunistic pathogens involving the CNS in immunocompromised patients, such as those with malignancy or the acquired immunodeficiency syndrome (AIDS). In sub-Saharan Africa ("the meningitis belt"), seasonal epidemics of meningococcal meningitis can reach several hundred cases per 100,000 population. Although mortality rates have decreased as a result of improved medical management of CNS infections, recent studies indicate that some children who have been successfully treated for bacterial or viral meningitis later display learning disabilities, behavioral problems, and developmental delay. Thus, infections of the CNS are a cause of both immediate and long-term health concerns.

This chapter discusses the following:

■ the interplay of host-related risk factors and virulence factors associated with pathogens

■ characteristics of the CSF

■ microbial agents of infections of the CNS

■ laboratory diagnosis of CNS infections

GENERAL CONCEPTS RELATED TO INFECTIONS OF THE CENTRAL NERVOUS SYSTEM

Anatomical Organization

The brain, spinal cord, and cranial nerves constitute the CNS (Fig. 29–1). The brain and spinal cord are protected by the skull, vertebral column, and overlying meninges (coverings). The dura mater is a thick, fibrous, white membrane that is firmly adherent to the overlying skull. Deep to this and covering the brain and spinal cord is the pia mater and pia arachnoid. Between the pia mater and pia arachnoid resides the subarachnoid space occupied by the surface blood vessels and the CSF. The CSF is a unique body fluid produced

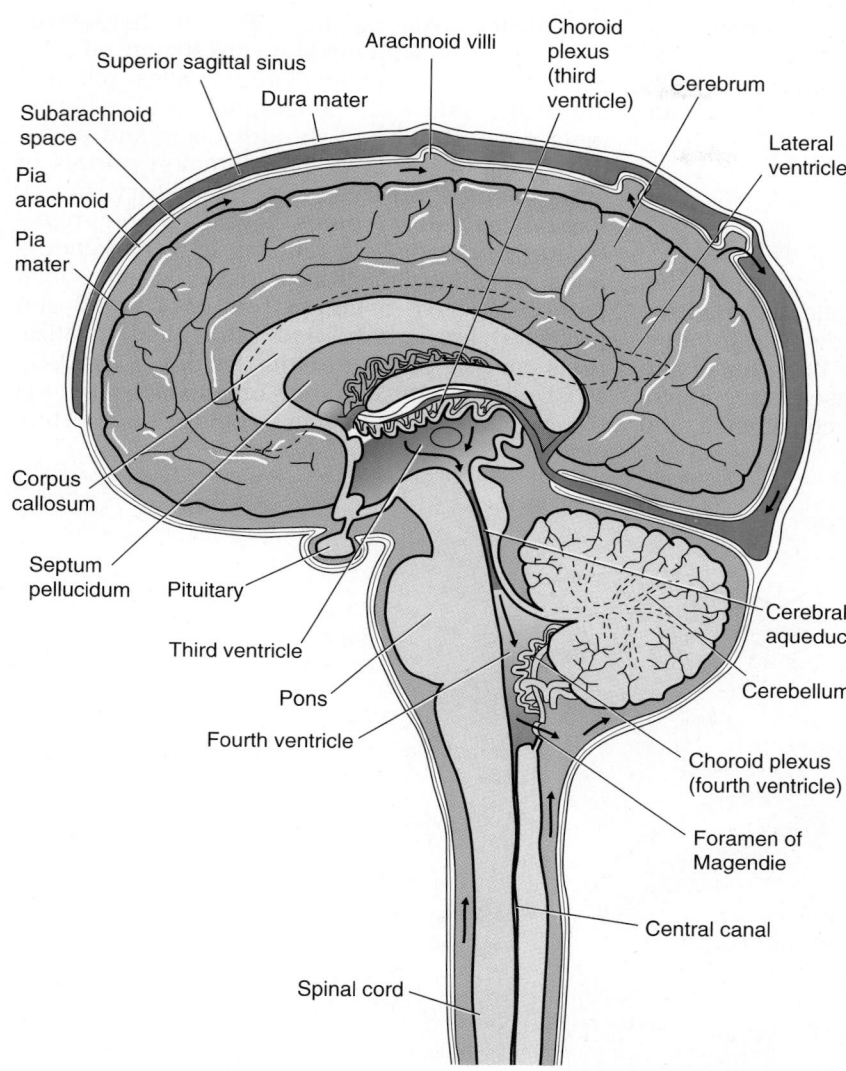

F i g u r e   2 9 – 1
Components of the CNS and flow pattern of CSF.

by both filtration and secretion from specialized capillary tufts of the choroid plexuses within the four ventricles of the brain. CSF enters the subarachnoid space via the cisterna magna and circulates under pressure around the brain and spinal cord to the arachnoid villi, where it is eventually reabsorbed. The CSF provides a protective fluid "cushion" for the brain and spinal cord.

Cerebrospinal Fluid Characteristics

CSF is a clear, colorless, sterile fluid that contains few cells. In normal adults, the CSF volume ranges from 90 to 150 mL, the CSF protein level is 15 to 45 mg/dL, and the CSF glucose level is two-thirds that of plasma (40–80 mg/dL, or CSF glucose:serum glucose ratio of 0.6). In adults, normal CSF contains 0 to 7 leukocytes/mL, with a differential count of 60 to 80% lymphocytes, 10 to 40% monocytes, and 0 to 15% neutrophils. Compared with adults, normal newborns have higher CSF concentrations of protein (15–150 mg/dL) and glucose (30–120 mg/dL). The cell count is somewhat higher as well (0–30 leukocytes/mL), with a greater percentage of monocytes and neutrophils. The paucity of leukocytes and protein (including immunoglobulins) within the CSF provides little initial defense against invading organisms. Infection of the CNS, however, is associated with an inflammatory response or increase in cells (pleocytosis) that varies according to the invading organism.

Host-Pathogen Relationships

Infection results from complex interplay between the host, the organism, and the environment. Host risk factors that predispose to infection include the patient's age, nutritional and immunologic status, and disease states (alcoholism, diabetes mellitus, sickle cell anemia, and malignancy). Among the organism characteristics associated with infection are structural components (e.g., antiphagocytic capsules, pili, and fimbriae); seasonal, geographic, and environmental distribution of the organism; and portal of entry into the host. The common portals of entry include the respiratory, auditory, bloodstream, and neural routes as well as direct penetration from adjacent sinuses or other contiguous sites of infection. Most CNS microbial pathogens have common routes of entry and preferred intracranial and intraspinal localizations causing characteristics lesions (Fig. 29–2). Most community-acquired organisms enter via the respiratory route. Once replication is estab-

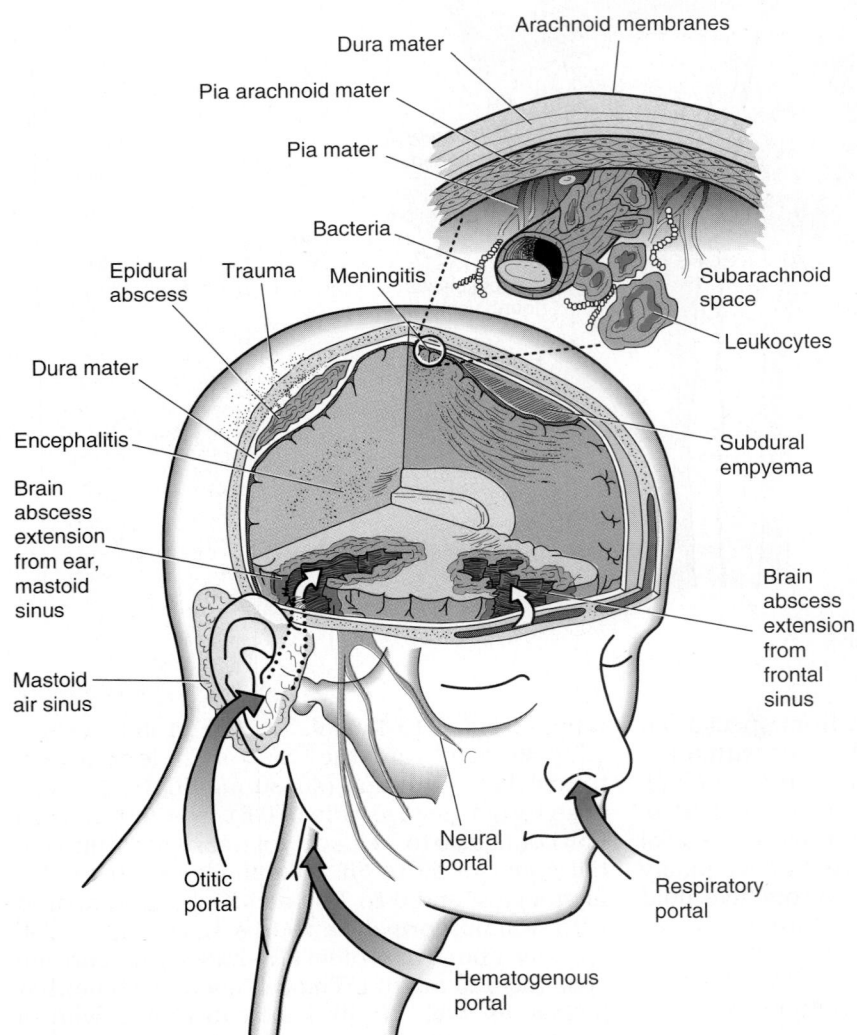

■■■Figure 29–2

Portals of entry resulting in meningitis, meningoencephalitis, and intracranial mass lesions.

lished in the upper respiratory tract, the organism can invade directly through the sinuses or lymphatics or seed the subarachnoid space via the bloodstream. Some viruses spread to the CNS by invading and traveling through cranial nerves (herpesviruses) or peripheral nerves (rabies virus).

Entry of the organism gives rise to inflammation around the blood vessels within the subarachnoid space between the pia mater and pia arachnoid, resulting in meningitis (leptomeningitis). If diffuse inflammation is limited to the brain substance, as is characteristic of many viruses, the disease is termed encephalitis. Inflammation of both the membranes and the brain substance is a meningoencephalitis. If the entry is localized, as may occur following direct extension of an infection from the middle ear, mastoid, or nasal sinuses or with penetrating trauma, then localized areas of inflammation within the dura mater (pachymeningitis), epidural abscess, subdural empyema, or intracerebral abscess may result. Each of these localizations has characteristic associations with select microbial pathogens.

One last mention should be made of CSF samples that show a pleocytosis (usually with predominance of lymphoctyes) but no organisms on direct examination. Such cases are often described as "aseptic" meningitis. As we shall see, many cases of "aseptic" meningitis are not aseptic at all but may be due to viruses, fungi, parasites, or some bacteria.

INFECTIONS OF THE CENTRAL NERVOUS SYSTEM

Bacterial Infections

CASE STUDY

A 3-year-old previously healthy girl complained of headache and vomiting. She was transported to an emergency room, where she was found to have a temperature of 102°F and appeared lethargic. A complete blood count (CBC) with cell differential, urinalysis, blood culture, and serum chemistry profile was obtained. The CBC documented a marked neutrophilia, with a shift to immature forms. A lumbar puncture was performed, and cloudy fluid obtained. The CSF profile showed 1,200 cells/mL, with 95% neutrophils, a glucose level of 40 mg/dL, and a protein level of 150 mg/dL. The urinalysis and serum chemistries were within expected limits. Gram stain of the direct smear confirmed numerous neutrophils, but no organisms were seen; the latex agglutination antigen test for *Haemophilus influenzae* was positive. Antibiotics were begun. A CSF culture grew a few colonies of *H. influenzae* type b after 48 hours of incubation. Blood cultures continued to be negative at 7 days of incubation.

Acute Bacterial Meningitis, Ventriculitis

Haemophilus influenzae type b is the most common cause of meningitis in children 1 to 6 years of age. *H. influenzae* is a gram-negative coccobacillus (Fig. 29–3) associated with acute otitis, pneumonia, and epiglottitis as concomitant or antecedent infections. The increased frequency of *H. influenzae* meningitis in children compared with that in neonates is related to changes in serum antibody levels. Children younger than 6 months of age are protected by transplacentally acquired anti–*H. influenzae* IgG antibody. The level of this antibody progressively decreases following delivery. At approximately 6 months to 1 year of age, children are susceptible to *Haemophilus* infections and remain so until vaccinated or until approximately 5 to 6 years of age, when they are capable of making their own antibodies.

Meningitis results from localization of *H. influenzae* in the meninges. Bacterial proliferation and lysis within the CSF results in escalating neutrophilic inflammation. The inflammation damages adjacent neural tissue along with meningeal vessels resulting in increased CSF pressure and

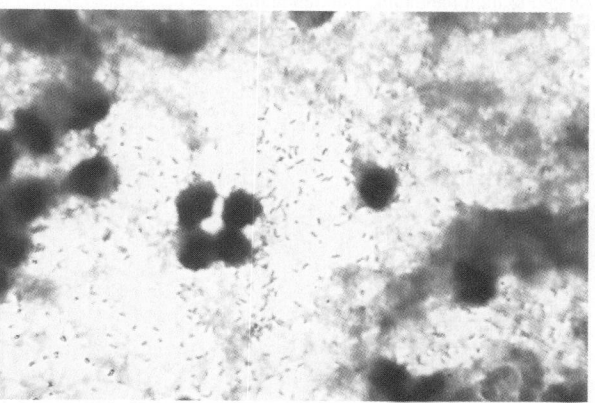

■ F i g u r e 2 9 – 3

Direct smear of CSF from a child, showing abundant gram-negative, pleomorphic coccobacilli characteristic of *H. influenzae*. The background shows degenerating inflammatory cells. Gram stain, High-power view.

protein level, and decreased CSF glucose level. Purulent meningitis is characteristically associated with bacterial infections, although some fungal, viral, and parasitic infections may also cause a neutrophilic pleocytosis.

Adults with acute (neutrophilic) meningitis show characteristic signs and symptoms including fever, headache, nuchal rigidity (meningismus), photophobia, and peripheral blood leukocytosis with neutrophilia. In contrast, the symptoms in infants and children may be exceedingly subtle; lethargy and poor feeding may be the only symptoms. Characteristic test findings in acute bacterial meningitis include increased CSF pressure (up to 250 to 700 mm Hg in adults), decreased glucose level (less than 40 mg/dL), and increased protein level (greater than 100 mg/dL). The most common organisms causing bacterial meningitis include *Haemophilus influenzae, Streptococcus pneumoniae, Neisseria meningitidis, Streptococcus agalactiae* (group B streptococci), and *Listeria monocytogenes* (Table 29–1). *H. influenzae* is the most common cause of bacterial meningitis (45% of reported cases), followed by *S. pneumoniae* (18% of cases), and *N. meningitidis* (14% of cases). *L. monocytogenes* is an uncommon cause of meningitis but is associated with a mortality rate as high as 22%.

Cases of bacterial meningitis show characteristic age-related incidence and routes of entry. In pre-mature infants meningitis is often due to gram-negative bacilli, such as *Escherichia coli,* whereas in newborns meningitis is most commonly due to group B streptococci (GBS) (Fig. 29–4) or less frequently to *L. monocytogenes* (Fig. 29–5). These infections are associated with bloodstream dissemination and neonatal sepsis. If infections are acquired early in pregnancy, the fetus may be naturally aborted. Infections present at birth were acquired in utero. Pregnancy complicated by premature rupture of the membranes is at risk for both maternal and neonatal infections. Neonatal infections occurring 1 to 12 weeks after delivery are usually nosocomial or acquired in the community.

Neisseria meningitidis (Fig. 29–6), the meningococcus, is most frequently identified in children and adolescents or young adults. Meningococcal meningitis is endemic in some developing countries, and epidemic outbreaks occur in the United States. Serogroups A, B, and C account for more than 90% of all cases, and group A is the most common cause of epidemics. The meningococcus enters by way of the respiratory route, colonizes the nasopharynx, and invades the bloodstream. Meningococcemia may be the resulting disease, or meningitis can occur from dissemination through the bloodstream and seeding of the subarachnoid space. Meningococcal disease may pursue a particularly aggressive course with bilateral adrenal gland hemorrhages (Waterhouse-Friderichsen syndrome) and subcutaneous hemorrhage (purpura fulminans) due to disseminated intravascular coagulation. Individuals deficient in terminal components of complement (C5–C9) are at risk for recurrent meningococcal infections and subsequent meningitis. The CSF of patients with meningococcal

■ T A B L E   2 9 – 1

BACTERIA INVOLVING THE CENTRAL NERVOUS SYSTEM

Acute Purulent Meningitis Related to Age
Premature newborns
  Gram-negative bacilli *(Escherichia coli, Klebsiella, Enterobacter, Proteus)*
Infants
  Streptococcus agalactiae (group B)
  Listeria monocytogenes
Children
  Haemophilus influenzae
Adolescents
  Neisseria meningitidis
Adults
  Streptococcus pneumoniae
Elderly
  Gram-negative bacilli

Chronic Meningitis
Mycobacteria
  (*M. tuberculosis* and atypical mycobacteria)
Spirochetes
  Treponema pallidum
  Borrelia burgdorferi
  B. recurrentis
  Leptospira species

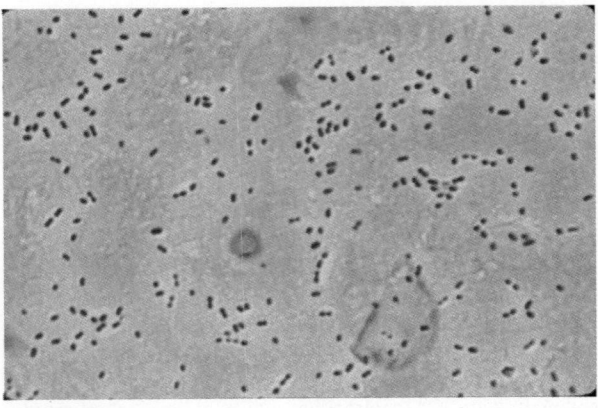

■ F i g u r e   2 9 – 4

Direct smear of CSF from an infant, showing gram-positive cocci in pairs and short chains characteristic of group B streptococci. Gram stain. Noncytocentrifuge preparation. High-power view.

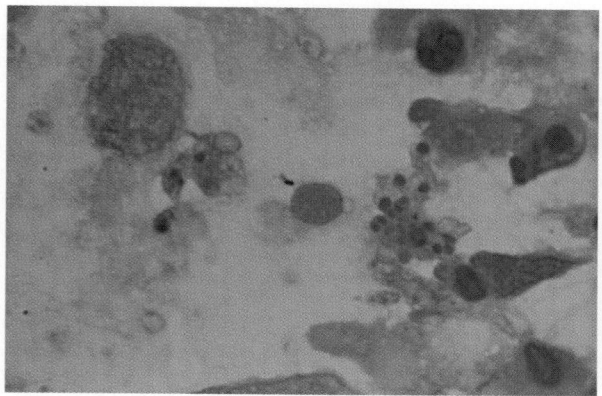

Figure 29-5

Direct smear of CSF from a cancer patient showing a rare gram-positive bacillus consistent with *L. monocytogenes*. The background shows degenerating inflammatory cells. Gram stain. Noncytocentrifuge preparation. High-power view.

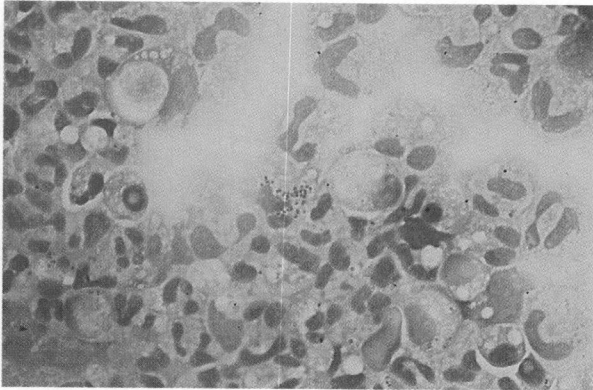

Figure 29-6

Direct smear of CSF from a high-school student showing clusters of gram-negative diplococci consistent with *N. meningitidis* within polymorphonuclear leukocytes. Note the increased cellularity of the smear in this cytocentrifuge preparation. Gram stain. High-power view.

meningitis may initially show scarce neutrophils because of the time delay following initiation of infection in the meninges, proliferation of the organism, and subsequent inflammatory response. All CSF samples must be carefully examined for organisms even in the absence of obvious inflammation. Bacterial antigen tests (discussed later) may be useful in diagnosing such early infections.

Meningitis due to *Streptococcus pneumoniae,* a gram-positive diplococcus (Fig. 29–7), is most commonly seen in adults and may spread to the CNS via the bloodstream or by extension from the middle ear, mastoid, or paranasal sinuses. Predisposing factors associated with pneumococcal meningitis include concurrent pneumo-

nia, alcoholism, liver disease, sickle-cell anemia, and malignancy.

Gram-negative bacilli are occasional causes of meningitis in the elderly, as immunity wanes in the "golden years." The most commonly isolated pathogen is *E. coli,* but infections due to *Klebsiella* (Fig. 29–8), *Enterobacter, Pseudomonas, Salmonella,* and *Proteus* are also seen. Penetrations of the skull, accidental or surgical, are also associated with gram-negative bacillary meningitis as well as with staphylococcal meningitis or abscesses. Cases of meningitis due to anaerobic streptococci, *Bacteroides* (Fig. 29–9), and *Fusobacterium* are uncommon and are usually associated with a concurrent brain abscess or unusual clinical settings.

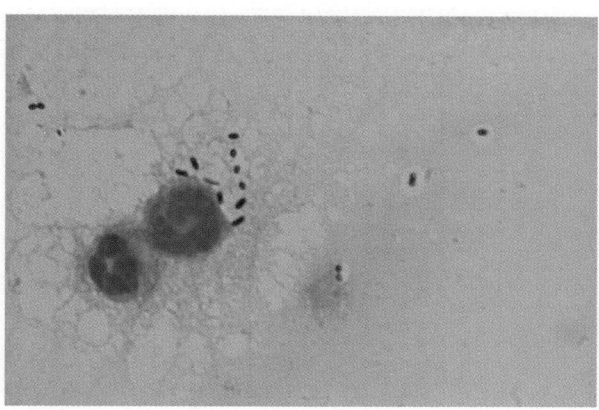

Figure 29-7

Direct smear of acute bacterial meningitis in an adult showing the lancet-shaped gram-positive diplococci characteristic of *S. pneumoniae*. The polysaccharide capsule produces a prominent "halo" around organisms. Gram stain. Noncytocentrifuge preparation. High-power view.

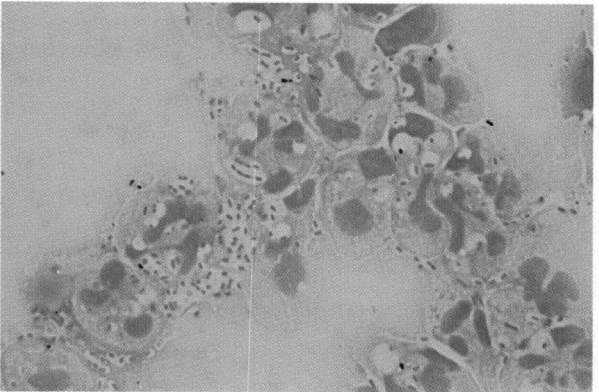

Figure 29-8

Direct smear of posttraumatic acute bacterial meningitis showing numerous intracellular and extracellular gram-negative bacilli with the prominent capsules characteristic of *Klebsiella pneumoniae*. Gram stain. Cytocentrifuge preparation. High-power view.

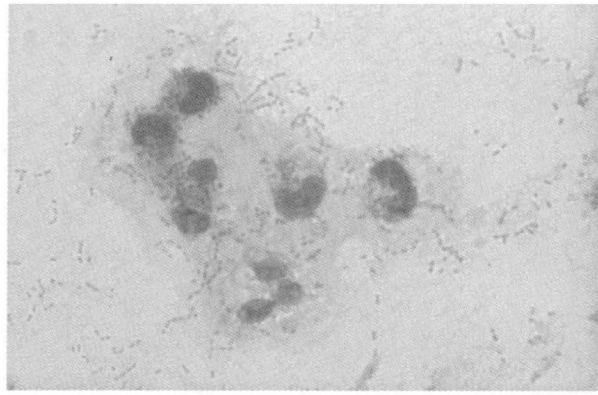

■ F i g u r e   2 9 – 9

Direct smear of CSF from a newborn delivered to a woman with amnionitis secondary to premature rupture of the membranes. Numerous short gram-negative bacilli in chains consistent with a *Bacteroides* species are seen. The organism could easily be confused with a *Streptococcus* species if the Gram stain was improperly decolorized. Gram stain. Noncytocentrifuge preparation. High-power view.

Mycobacterial Infections

The most common mycobacterial infection of the CNS is tuberculous meningitis caused by *Mycobacterium tuberculosis*. Tuberculous meningitis was a scourge in the United States of the early 1900's. The development of antituberculous chemotherapy resulted in a marked decrease in the incidence of this infection; however, the current trend toward increased numbers of immunocompromised patients in this country has resulted in a resurgence of this disease. Tuberculosis has always been a common disease in developing countries.

M. tuberculosis enters the body by the respiratory route on airborne infectious particles that reach alveoli of the lungs. There the organisms multiply within macrophages and then disseminate through the bloodstream. The clinical presentation of tuberculous meningitis is quite variable but usually is subacute. There is frequently involvement of both the meninges and the brain itself, with a resulting thick exudate especially at the base of the brain. Lumbar puncture shows increased CSF pressure, a moderately decreased glucose level, and an increased protein level. In contrast with acute bacterial meningitis, the inflammatory reaction of tuberculous meningitis consists predominantly of lymphocytes and monocytes; however, early infections may show a predominance of neutrophils. Mycobacteria are scarce and difficult to identify in CSF without

concentration of the sample and special staining techniques. Centrifugation for sedimentation must be at high velocity because the fats and waxes of the mycobacterial wall cause the organism to float. Other mycobacteria associated with CNS infections include *M. avium–intracellulare*, *M. kansasii*, and *M. fortuitum*.

Brain Abscesses

In contrast with the superficial meningeal inflammation seen in meningitis, brain abscesses are circumscribed areas of tissue destruction containing organisms and inflammatory cells. The majority of cerebral abscesses result by spread from adjacent sinus or ear infections. Approximately 20% of cerebral abscesses occur by hematogenous spread from distant sites, as may be seen in patients with infective endocarditis. Other conditions associated with brain abscesses include diabetes mellitus, corticosteroid therapy, immunodeficiency (AIDS), and malignancy. Although a frequently cited cause, skull trauma actually accounts for less than 10% of brain abscesses. The most common organism cultured from the latter is *Staphylococcus aureus*. The most common organisms isolated from nontraumatic brain abscesses include aerobic and anaerobic streptococci (Fig. 29–10), *Bacteroides*, gram-negative aerobes, and staphylococci. As a general rule, one-third of brain abscesses are culture-positive for aerobes only; one-third are culture-positive for anaerobes only; and one-third

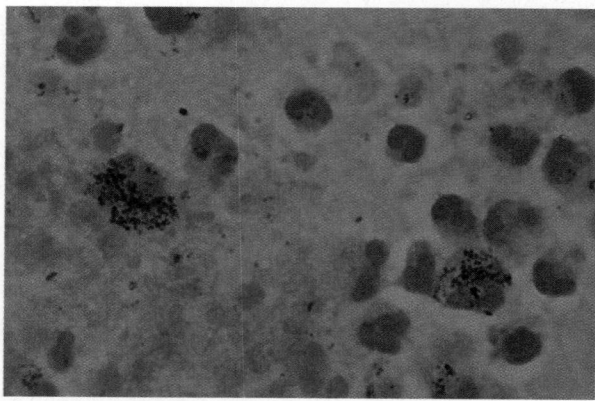

■ F i g u r e   2 9 – 1 0

Direct smear of aspirated brain abscess contents. Clusters of intracellular gram-positive cocci in groups are consistent with microaerophilic streptococci. Gram stain. Noncytocentrifuge preparation. High-power view.

are mixed aerobe-anaerobe infections. *Citrobacter diversus* is a gram-negative bacillus associated with meningitis and brain abscesses in neonates.

A cerebral abscess begins as a focal area of cellulitis with accompanying neutrophilic inflammation. The dead brain tissue is liquefied by the inflammatory response, and the resultant cavity is filled with organisms, necrotic tissue, and inflammatory debris. The expanding brain abscess exerts a mass effect within the closed space of the skull and may shift adjacent vital brain structures sufficiently to cause focal neurologic signs, such as weakness in one extremity or unilateral abnormalities of vision. In this circumstance, the performance of a lumbar tap to obtain CSF is hazardous and can lead to herniation of the brainstem and death of the patient! Occasionally, a brain abscess may rupture into the ventricular system with resultant numerous inflammatory cells seen in the CSF. Brain abscesses may be culture-negative, especially in the face of prior antimicrobial therapy.

Although most brain abscesses contain bacteria, fungal and parasitic organisms may also be causes of brain abscesses. In contrast with yeast infections (see later discussion), CNS infections due to hyphal fungi such as *Aspergillus* species or Zygomycetes, typically produce cerebral abscesses or infarcts rather than meningitis. These organisms are rarely identified by direct examination of CSF samples. Cerebral aspergillosis may be acquired from indwelling catheters, from a focus of infection within the paranasal sinuses, or hematogenously from systemic organ involvement. Polymorphonuclear leukocytes are an important host defense against infections due to *Aspergillus,* and patients with reduced peripheral blood neutrophil counts are at risk for infection. In the brain, *Aspergillus* invades through the cerebral vessels into surrounding neural tissue. The lesion produced is thus a combination of fungal abscess and hemorrhagic infarct. *Aspergillus fumigatus* is the most commonly recovered organism, followed by *A. flavus, A. terreus,* and *A. versicolor.*

The subclass Zygomycetes includes *Mucor, Rhizopus,* and *Absidia.* These organisms are also angioinvasive in patients predisposed to infection because of malignancy or diabetes mellitus. The infection may begin in the paranasal sinuses and erode through the frontal bone to reach the CNS (rhinocerebral mucormycosis). The frequent acidosis in patients with poorly controlled diabetes mellitus predisposes to this infection.

Other fungi occasionally isolated from cerebral abscesses include *Blastomyces dermatitidis, Paracoccidioides brasiliensis, Sporothrix schenckii,* and *Pseudallescheria boydii.*

Abscesses due to *Nocardia* species have been described, especially in patients receiving corticosteroid therapy. Parasites including *Entamoeba histolytica, Toxoplasma gondii,* and hookworm larvae are rare causes of cerebral abscesses.

Spirochetal Infections

Both treponemal and nontreponemal spirochetes may be the cause of CNS diseases. Here we shall discuss only one example of each: Lyme disease and neurosyphilis.

Lyme disease is a recently recognized syndrome caused by the nontreponemal spirochete *Borrelia burgdorferi.* The disease was recognized as a distinct syndrome when a cluster of cases occurred in Lyme, Connecticut, in 1975. The infection is transmitted through the bite of ticks of the genus *Ixodes* (*I. dammini* and *I. pacificus* on the eastern and western coasts of the United States, respectively). As with syphilis, Lyme disease progresses through a series of stages. After a variable incubation period of up to 1 month, approximately 80% of infections demonstrate a peculiar, ring-shaped rash (erythema chronicum migrans) usually at the site of the tick bite. If the infection goes untreated, it progresses in days to weeks to the second stage, characterized by extreme fatigue, less well-defined skin rashes, "aseptic" meningitis, and arthritic or other musculoskeletal symptoms. Cardiac involvement may lead to rhythm disturbances manifested as palpitations or fainting. The third stage is characterized by disturbances in memory or mood, with more pronounced musculoskeletal symptoms.

At all stages of infection, *B. burgdorferi* is present in low numbers. The CSF may be entirely normal or may show only a lymphocytic pleocytosis, but organisms are not seen. The diagnosis of Lyme disease is made on clinical grounds and supported by positive serology (enzyme-linked immunosorbent assay [ELISA] and Western blot test). The infection responds well to oral doxycycline or amoxicillin.

Neurosyphilis is caused by the spirochete *Treponema pallidum.* The natural history of syphilitic infection manifests in three stages. The first, or primary, stage of infection occurs by cutaneous inoculation, with resulting development of a painless ulcer, the chancre. This ulcer is teeming with spirochetes, and the diagnosis may be made at this time by serology or by darkfield microscopy. The chancre is painless and heals spontaneously in 3 to 6 weeks. In patients who have not received antibiotics, the disease progresses to the secondary phase, associated with spirochetemia and characterized by a rash, lymphadenopathy, and oral ulcers. If the infection still goes untreated, patients may become asymptomatic for years during the so-called latent phase. Some patients enter a tertiary, or late, stage with neurologic symptoms (neurosyphilis). The neurologic signs include abnormalities of gait associated with stabbing pain (tabes dorsalis) and absence of pupillary response to light (Argyll Robertson pupil). Other syphilitic patients may suffer progressive paralysis and dementia (general paralysis of the insane). In syphilis the CSF shows a lymphocytic pleocytosis and increased protein level; spirochetes are not seen. The diagnosis of syphilis in the secondary or the tertiary stage is obtained by serology. Serologic diagnosis of syphilis includes both nonspecific and specific tests for *T. pallidum*. Nonspecific tests include the VDRL (venereal disease research laboratory) and RPR (rapid plasma reagin) tests. These tests detect antibody (reagin) that cross-reacts with cardiolipin. The results of these tests are confirmed by specific treponemal assays, such as the TPI (treponema pallidum immobilization) and FTA-ABS (fluorescent treponemal antibody absorption) tests. These tests detect structural components of treponemes.

Fungal Infections

CASE STUDY

A 52-year-old white man arrived at an emergency room in a disoriented and poorly responsive state, with labored breathing. The patient's history included poorly controlled diabetes and chronic obstructive pulmonary disease (COPD) secondary to cigarette smoking. Current medications included steroids for his pulmonary disease. Physical examination showed that the patient was slightly febrile, lethargic, and in respiratory failure. He was believed to have a deteriorating mental status, and a diagnosis of meningitis was considered. A lumbar tap produced a CSF sample that on direct smear showed encapsulated budding yeasts; a cryptococcal antigen test was positive. Culture of CSF identified *Cryptococcus neoformans*. Despite aggressive therapy

with amphotericin B and 5-flucytosine, the patient's condition failed to improve. The patient died on the third day of hospitalization.

Cryptococcus neoformans (Figs. 29–11 and 29–12) is the most common cause of fungal meningitis, especially among immunosuppressed patients. The India ink preparation using nigrosin has traditionally been used to identify these encapsulated yeast in the CSF. However, this test is prone to misinterpretation, as red blood cells and small lymphocytes may look similar to *C. neoformans.* Currently, the India ink preparation is a test of historical interest, but one not much used in modern laboratories. CSF smears (cytopreparation preferred) should be stained with a Gram stain and calcofluor reagent to specifically identify yeast cell walls. A rapid latex agglutination (LA) method is used for the detection of cryptococcal capsular polysaccharide in both CSF and serum. The LA test detects antigen in 90% or more of patients with cryptococcal meningitis. False-positive results may occur as a result of the presence of rheumatoid factor or a cross-reaction with antigen from *Trichosporon beigelii* disseminated infections and some bacterial infections. Cultures for *C. neoformans* are positive in 75 to 85% of cases.

Cryptococcal meningitis may present with nearly normal CSF studies and few organisms (hence the need to directly examine all CSF sam-

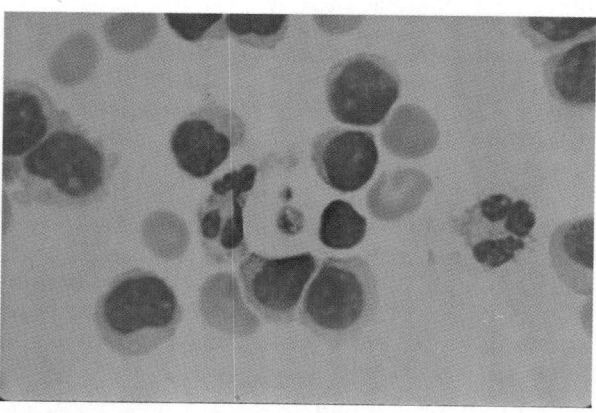

■■■ F i g u r e   2 9 – 1 1

Cytocentrifuge preparation of CSF showing a single yeast with narrow-based budding and prominent surrounding capsule characteristic of *Cryptococcus neoformans*. Cryptococcal meningitis in partially immunocompetent hosts may show only rare organisms mixed with an inflammatory background of lymphocytes, monocytes, and eosinophils. Wright stain. High-power view.

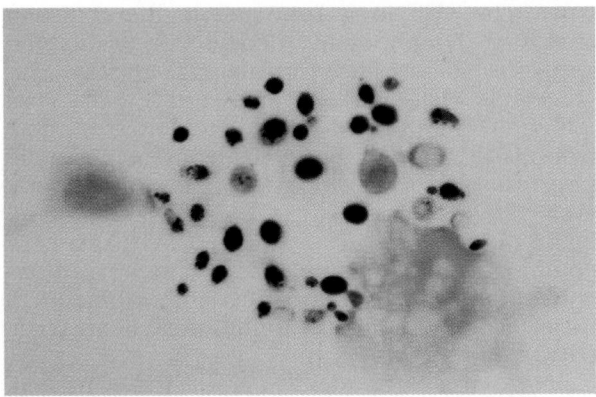

■■■Figure 29–12

In contrast, cryptococcal meningitis in immunosuppressed hosts may show numerous organisms and scarce or absent inflammation. Notice the variation in size, variable gram-staining, and narrow-based budding. The organisms are evenly spaced because of their abundant polysaccharide capsules. Gram stain. Cytocentrifuge preparation. Medium-power view.

ples). During treatment, organisms from chronic cryptococcal meningitis may also become atypical in shape, size, and staining characteristics. At the other extreme, the number of organisms in AIDS patients may be so overwhelming that they present in fluid as clumps. At autopsy, the brain from such a patient may appear covered by a mucinous coating because of the abundant polysaccharide capsule surrounding this organism. Microscopically, masses of yeast penetrate into the brain substance to form flask-shaped cavities.

Despite being widespread in the environment, fungi are uncommon causes of CNS infection (Table 29–2). Chronic meningitis is the common presentation, although fungal CNS infections may present acutely, thus mimicking bacterial meningitis. The predisposing factors to fungal meningitis include age (e.g., *Histoplasma capsulatum* in infants), metabolic disturbances (Zygomycetes), intravenous drug abuse (*Candida* species), and immunodeficiencies (*C. neoformans* and *Candida*). In fungal meningitis, CSF findings vary considerably, but the following are usually present: normal to slightly increased CSF pressure, increased CSF protein level, and decreased CSF glucose level. The CSF usually contains fewer than 500 leukocytes/mL, with a predominance of lymphocytes and monocytes. Infections due to *C. neoformans, C. immitis,* and *Aspergillus*

species are occasionally associated with predominance of neutrophils or eosinophils in the CSF. In such cases, the CSF glucose level may be quite low.

Fungal meningitis due to *H. capsulatum* and *Coccidioides immitis* is frequently secondary to systemic infection. Fungal meningitis with both these organisms most often occurs by the bloodstream route from a pulmonary focus. These organisms show a marked difference in geographic distribution. *H. capsulatum* is endemic to the Ohio and Mississippi River valleys; whereas *C. immitis* is endemic to the southwestern United States. CSF cultures are positive in 50% or fewer of infections due to these organisms. Coccidioidal meningitis may be documented by detection of CSF IgG complement-fixing antibodies.

Candida species are causes of both fungal meningitis and cerebral abscesses. Candidiasis may be acquired as a nosocomial superinfection by patients with indwelling catheters or those receiving antibacterial therapy. Patients with low peripheral blood neutrophil counts secondary to chemotherapy are at risk for candidal infections. *C. albicans, C. tropicalis,* and *C. parapsilosis* are the most commonly identified species.

■■■TABLE 29–2

FUNGAL ORGANISMS INVOLVING THE CENTRAL NERVOUS SYSTEM

Common
Cryptococcus neoformans
Coccidioides immitis

Uncommon
Histoplasma capsulatum
Candida species
Aspergillus species
Blastomyces dermatitidis

Rare
Paracoccidiodes brasiliensis
Pseudallescheria (Allescheria) boydii
Mucorales *(Mucor, Rhizopus, Absidia, Cunninghamella)*
Sporothrix schenckii
Trichosporon beigelii
Penicillium
Fusarium
Alternaria
Curvularia
Acremonium
Fonsecaea
Bipolaris
Drechslera

Viral Infections

CASE STUDY

A 36-year-old HIV-infected homosexual male was admitted to an emergency room with complaints of inability to urinate for 3 days. He additionally reported numbness and weakness in his right leg for 7 months, a 25-lb weight loss, and bowel incontinence. Physical examination revealed an afebrile, dehydrated, emaciated male with bilateral lower extremity weakness and decreased reflexes throughout. Lesions of Kaposi's sarcoma were noted, especially on the lower extremities, and recurrent oral thrush and perianal herpetic vesicles were observed. An MRI scan of the CNS revealed no evidence of spinal cord compression, but because of the high-risk profile, the patient was admitted to the hospital with the presumptive diagnosis of polyradiculopathy secondary to AIDS. Acyclovir was started, pending culture results. A lumbar puncture was performed. The CSF showed an increased protein level (326 mg/dL) and leukocyte count (1,720 cells/mL) with 86% neutrophils. Serology for cytomegalovirus (CMV) was positive, and viral blood cultures were positive for CMV. Following culture isolation of CMV, the patient's medication was switched to ganciclovir. After several weeks of therapy, the patient was sufficiently recovered to be discharged.

Meningitis

Viral infections occur worldwide and are the most common cause of "aseptic" meningitis, defined as clinical or laboratory evidence of meningeal inflammation without demonstrable organisms. The CSF in aseptic meningitis typically shows a lymphocytic pleocytosis associated with normal or nearly normal CSF protein and glucose levels (Figs. 29–13 and 29–14). Early viral infections may show a predominance of neutrophils in the CSF, but the pleocytosis rapidly progresses to a lymphocytosis. The most common viruses producing aseptic meningitis include the enteroviruses and herpesviruses. Less common causes of viral aseptic meningitis include mumps virus, lymphocytic choriomeningitis virus, and the human immunodeficiency virus (HIV) (Table 29–3).

Enteroviruses frequently associated with neurologic illness include coxsackieviruses A and B, echoviruses, and polioviruses. Children and immunocompromised patients are particularly at risk. Enteroviral infections are typically acquired by the fecal-oral route during the summer months. Ingested virus replicates within lymphoid tissue of the gastrointestinal tract and subsequently disseminates by the bloodstream route to the CNS. The enteroviruses associated with aseptic meningitis include coxsackievirus group A (types 1–2, 4–7, 10, 14, 16, 22) and coxsackievirus group B (types 1–6) as well as many of the echoviruses. The majority of enteroviruses can be identified by viral cell culture techniques, serologic testing, or inoculation into suckling mice.

The incidence of poliovirus infection has decreased dramatically in the United States since the implementation of vaccination programs. Paralytic poliomyelitis, however, remains a significant concern in developing countries

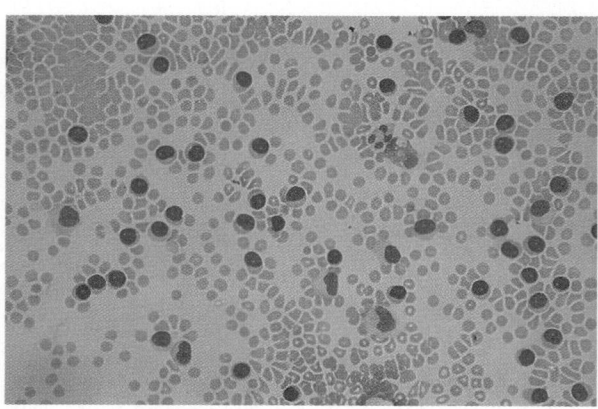

■■■■■ F i g u r e   2 9 – 1 3

Cytocentrifuge preparation of CSF in a case of "aseptic" meningitis. Lymphocytes are present, and in this case the background is bloody. No organisms are seen. Wright stain.

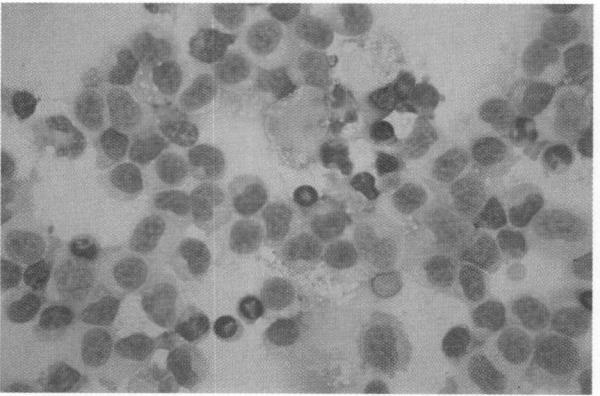

■■■■■ F i g u r e   2 9 – 1 4

Cytocentrifuge preparation of CSF in meningitis due to tularemia. Reactive lymphocytes with "monocytoid" features are the only clue that this is not a viral infection. No organisms are seen. Cultures were positive. Wright stain.

■■■■■■T A B L E   2 9 – 3
VIRAL AGENTS INVOLVING THE CENTRAL NERVOUS SYSTEM

Enteroviruses
    Coxsackieviruses A and B
    Echoviruses
    Polioviruses
Arboviruses (arthropod-borne viruses)
    Alphaviruses
        Eastern equine encephalitis
        Western equine encephalitis
        Venezuelan equine encephalitis
    Flaviviruses (St. Louis encephalitis virus)
    Bunyaviruses (La Crosse virus)
    Colorado tick fever virus
Herpesviruses
    Herpes simplex (HSV-1 and HSV-2)
    Epstein-Barr virus
    Cytomegalovirus
    Varicella-zoster virus
Lymphocytic choriomeningitis virus
Human immunodeficiency virus (HIV)
Mumps virus
Rabies virus

that do not have vaccination programs. The majority of poliovirus infections are asymptomatic or cause an aseptic meningitis. However, a small proportion of infections progress to destruction of anterior spinal cord motor neurons, with accompanying paralysis.

Currently in the United States, herpesviruses account for 10 to 20% of viral meningoencephalitis. The herpesviruses include herpes simplex virus (HSV-1 and HSV-2), Epstein-Barr virus (EBV), cytomegalovirus (CMV), and varicella-zoster virus (VZV). Herpes simplex meningoencephalitis may occur in patients of all ages. Herpes simplex encephalitis occurring in adults may follow primary herpesvirus infection or may result from reactivation of a previous herpesvirus infection. Neonatal herpes simplex meningoencephalitis is typically a consequence of vaginal delivery from a mother with genital herpes and usually reflects disseminated herpetic disease. Less commonly, in utero or postpartum acquisition from a person with active herpetic lesions may occur. Histologic examination of infected brain tissue shows eosinophilic viral inclusions within infected neurons. Death occurs in 75% of untreated cases. Examination of the CSF commonly shows a lymphocytic pleocytosis (typically 100 leukocytes/mL), increased CSF protein level (100 mg/dL), and normal CSF glucose level. Because of the severity of the disease and the possibility of effective antiviral therapy,

brain biopsy with culture may be the diagnostic method of choice. The polymerase chain reaction (PCR) using CSF represents a realistic alternative for diagnosis where available.

The remaining herpesviruses are infrequent causes of CNS illness. Infections due to VZV are spread by aerosols and produce the primary infection varicella (chickenpox). Reactivation of the virus results in herpetic lesions in the skin along the distribution of the infected nerve (dermatome) characteristic of zoster (shingles). Neurologic complications of varicella are estimated to occur in 1 in every 1,000 infections, and the majority of these resolve without serious sequelae. VZV produces a variety of neurologic syndromes, including meningoencephalitis, optic neuritis, and inflammation of the spinal cord (myelitis). VZV has also been associated with a self-limited neurologic illness in children known as Reye syndrome. The symptoms typically occur in the winter months following a bout of chickenpox. Reye syndrome is characterized by increased levels of hepatic transaminases, normal serum bilirubin level, and unremarkable CSF values. Neurologic symptoms include severe vomiting, delirium, and decerebrate posturing. Symptoms and test abnormalities can return to normal within 1 to 2 weeks, but deaths have occurred. There is evidence that administering salicylates (aspirin) during active chickenpox may predispose to this disorder. Aspirin is currently contraindicated for management of fevers in children with viral syndromes.

Neurologic illness is uncommon with both EBV and CMV. Both are endogenous viruses that may be reactivated in or acquired by immunosuppressed patients, particularly HIV-infected individuals. Both EBV and CMV may cause meningoencephalitis or result in an ascending paralysis (Guillain-Barré syndrome).

Lymphocytic choriomeningitis virus, a member of the arenaviruses, is an RNA virus uncommonly a cause of aseptic meningitis in human beings. The virus is transmitted by aerosols or fomites from infected rodents through the respiratory route. Individuals living in rodent-infested dwellings, pet store owners, and laboratory workers who work with rodents are at risk for infection. Aseptic meningitis occurs in only about 15% of patients with lymphocytic choriomeningitis confirmed by culture or serologic techniques. As its name implies, lymphocytic choriomeningitis may show a striking lymphocytic pleocytosis in the CSF (more than 1,000 lymphocytes/mL). Complete recovery is expected.

The mumps virus, a member of the paramyxoviruses, is an RNA virus responsible for epidemic parotitis. Mumps virus is acquired from inhaled respiratory droplets. The virus replicates within the parotid salivary gland, with subsequent viremia and spread to the CNS. Aseptic meningitis is the most common neurologic complication, occurring in up to 30% of cases. The CSF shows a moderate lymphocytic pleocytosis that may persist for weeks. The clinical diagnosis of mumps is confirmed by serologic studies, and neurologic symptoms typically resolve completely.

Human immunodeficiency virus (HIV), a member of the retroviruses, is an RNA virus associated with AIDS. HIV invades and destroys CD4+ (T-helper) lymphocytes, resulting in a loss of cell-mediated immunity. The virus is present in body fluids, secretions, and tissues of infected individuals. Infection can be acquired by those engaging in unprotected sexual (homosexual or heterosexual) contact with infected individuals and by intravenous drug administration using shared contaminated needles. Newborns delivered to infected mothers are also at high risk for infection. Prior to the availability of serologic screening for HIV, some individuals who received blood products or organ transplants from infected individuals acquired infection. Studies have demonstrated that HIV has strong tropism for the CNS. Aseptic meningitis occurs in 10 to 20% of cases and may precede opportunistic infections. Inflammation of the peripheral nerves (inflammatory polyneuropathy) or spinal cord (myelopathy) can occur in 10 to 50% of individuals with AIDS. The majority of individuals with AIDS eventually develop a subacute encephalitis (AIDS dementia complex).

Meningoencephalitis

Arboviruses (arthropod-borne viruses) are RNA viruses that demonstrate strong tropism for the CNS. They are responsible for epidemic encephalitis and, less commonly, meningitis or paralytic disease. There is marked variability in the clinical symptoms produced, ranging from isolated fever to coma and death. All are important pathogens in medical, veterinary, and public health practices in the United States. Included are members of the alphaviruses (eastern equine encephalitis, western equine encephalitis, and Venezuelan equine encephalitis viruses), flaviviruses (St. Louis encephalitis virus), Bunyaviruses (La Crosse virus), and orbivirus (Colorado tick fever virus). The majority of these viruses are transmitted to humans by mosquitoes of the genera *Culex* and *Aedes*. Colorado tick fever virus is transmitted by the bite of ticks from the genus *Dermacentor*. Human beings acquire infection with these viruses while in the geographic range of the aforementioned arthropods. Thus, most infections typically occur from spring to late fall when outdoor activities are increased and climatic and environmental conditions favor mosquito proliferation. Regions of the United States vary in the incidence of these infections.

Eastern equine encephalitis (EEE) is endemic along the entire eastern coast of the United States. Infections are most commonly seen in young children and the elderly. EEE is the most fulminant of the arboviral infections with death occurring in 50 to 75% of cases. Survivors have a high incidence of neurologic sequelae, such as seizures or personality disorders. There is no specific treatment other than supportive care for patients with EEE. Western equine encephalitis (WEE) occurs predominantly within the midwestern and western United States. Venezuelan equine encephalitis (VEE) is endemic in Central America and Florida. WEE and VEE are difficult to distinguish from EEE on clinical grounds, and infants and young children are at greatest risk for infection. Fortunately, the risk of fatal encephalitis is much lower (approximately 10% in WEE, and 0.6% in VEE). Despite its name, St. Louis encephalitis has been reported throughout the United States. The last nationwide epidemic of St. Louis encephalitis occurred in 35 states during the period 1975 to 1976 and resulted in a reported 2,194 cases. Epidemic outbreaks appear to occur at approximately 10-year intervals. In contrast with other arboviruses, St. Louis encephalitis virus appears to be more common in the elderly. The clinical symptoms of St. Louis encephalitis are variable, and the risk of fatal encephalitis is estimated at 22%.

La Crosse virus is the most commonly isolated member of the California serogroup of Bunyaviruses. Encephalitis due to La Crosse virus occurs most commonly in Ohio, Illinois, Wisconsin, and Minnesota; however, infections have been reported across the United States. La Crosse virus is the second most common cause of arboviral encephalitis, following St. Louis encephalitis virus. Children are most commonly afflicted. La Crosse virus and the other California serogroup viruses are relatively benign arboviral infections, with a mortality rate of less than 1%.

Colorado tick fever virus is the only orbivirus that produces CNS illness. Infections are acquired in the mountainous areas of Colorado, Wyoming, Montana, Idaho, Utah, California, and New Mexico, where hiking and camping are popular and the ticks of the genus *Dermacentor* abound. In addition to meningoencephalitis, Colorado tick fever virus shows a peculiar tropism for the bone marrow, and infection is often accompanied by leukopenia and thrombocytopenia. The ability of the virus to invade blood cells aids in evading host defenses. Colorado tick fever virus shows a wide range of clinical symptoms, although most infections allow recovery.

Autopsy examination of the brain from cases of fatal viral encephalitis show similar histologic findings, including perivascular infiltrates of lymphocytes and plasma cells, nodular aggregates of microglia ("glial shrubs"), reactive changes within neurons, and petechiae. Confirmation of the diagnosis is achieved by the isolation of the virus or by the demonstration of neutralizing antibodies (complement fixation or hemagglutination inhibition) in the patient's serum.

Of all the viral infections of the CNS, the encephalitis of rabies is perhaps the most feared. Rabies is caused by a bullet-shaped RNA virus of the genus Lyssavirus. It is a zoonosis with nearly worldwide distribution. Infections occur following bites from infected "rabid" animals, such as skunks, raccoons, bats, and foxes, as well as domesticated animals such as rabid dogs, cats, and cattle. Rabies has rarely been acquired through inhalation of aerosolized virus (usually occurring in caves with a dense bat population) or direct inoculation of live virus in laboratories. Fortunately, human rabies is an uncommon infection, with fewer than five reported cases annually since 1960. Following inoculation, the virus replicates within skeletal muscle and proceeds along the peripheral nerves into the CNS. The incubation period is variable, usually between 1 and 3 months. Initial symptoms include fatigue, gastrointestinal symptoms, and pain at the bite wound. In the United States, the most common neurologic symptoms are those of an acute encephalitis indistinguishable from other viral encephalitides. Less commonly, the classic syndrome of agitation, emotional lability, seizures, and hallucinations occurs (furious rabies). Drinking fluids can initiate painful pharyngeal spasms, resulting in voluntary dehydration

(hydrophobia). Patients also experience "foaming" at the mouth because of increased salivation. Less commonly rabies may manifest as paralysis followed by coma and death (dumb rabies). The mortality of rabies is essentially 100% following the onset of symptoms unless timely administration of rabies immune globulin is received. The diagnosis is confirmed by demonstration of rabies antigen (by immunofluorescence) in neck skin biopsies or by demonstration of rabies antigen or characteristic inclusions (Negri bodies) in neurons of the brains of patients or infected animals.

The determination of a specific etiologic virus associated with a CNS infection is not practical in most cases, particularly in mild or self-limited disease. Specific diagnosis is difficult because (1) a large number of different viruses involve the CNS, (2) virus may come from endogenous reactivation or exogenous infection, (3) many viruses produce a spectrum of neurologic complaints, and (4) the magnitude of any neurologic illness due to viruses may depend on the age and immune status of the patient as well as other undefined factors. The determination of a specific viral etiology can often be made through a careful patient history, selected serologic tests (determination of virus-specific IgM or determination of a fourfold or greater rise in antibody titer between acute and convalescent sera), viral culture or PCR for selected viruses, or tissue biopsy for routine light microscopy, immunofluorescence, or ultrastructural studies. Virus isolated from body sites other than the CNS may be implicated in CNS syndromes. Appropriate specimens for culture include nasopharyngeal swabs, urine, stool, tissue, and occasionally blood. Cultures should generally be obtained within the first 5 days following the onset of a viral syndrome (aseptic meningitis, meningoencephalitis, or paralytic symptoms). Viruses (as well as *Treponema pallidum* and *Toxoplasma gondii*) are also responsible for a variety of well-characterized congenital syndromes. Appropriate diagnostic testing should be used in these clinical situations.

Parasitic Infections

Parasites are an infrequent to rare cause of CNS infection. The parasites most frequently identified as CNS pathogens include *Toxoplasma gondii*, the free-living amebae, and a variety of helminths (Table 29–4). The agent of African

■■■■ T A B L E 2 9 – 4
PARASITES INVOLVING THE CENTRAL NERVOUS SYSTEM

Protozoans
Toxoplasma gondii
Plasmodium falciparum
Naegleria fowleri
Acanthamoeba species

Helminths
    Nematodes
        Strongyloides stercoralis
        Angiostrongylus cantonensis
        Gnathostoma spinigerum
        Toxocara canis and *T. cati*
        Trichinella spiralis
        Loa loa
    Trematodes
        Schistosoma species
    Cestodes
        Cysticercus cellulosae (*Taenia solium* larvae)
        Echinococcus granulosus and *E. multilocularis*
        Paragonimus westermanii and *P. mexicanus*

sleeping sickness, *Trypanosoma brucei gambiense,* also manifests prominent neurologic symptoms but is geographically restricted to limited areas in Africa.

T. gondii is the most common parasitic infection of the CNS. *T. gondii* is a coccidian obligate intracellular protozoan of the family Sarcocystidae. Humans acquire infection by eating raw or undercooked meat containing tissue cysts, or by the ingestion of mature oocysts from the environment. Sporozoites released from the ingested oocyst or tissue cysts invade the human small intestine, spread hematogenously, and invade cells of the viscera and possibly the brain. This process induces focal microscopic areas of cellular necrosis in affected organs. The ability of *T. gondii* to parasitize host macrophages is a major defense mechanism for eluding host immunity. With the development of cell-mediated immunity, the tachyzoites maintain their intracellular position and develop into the less metabolically active bradyzoites. Toxoplasmic cysts seen in tissue sections represent an inactive phase of the infection. Any subsequent loss of or decrease in cell-mediated immunity may result in reactivation of the infection.

Toxoplasmosis is a worldwide zoonosis. In the United States, approximately 20 to 40% of healthy adults are seropositive for *T. gondii.* In European countries, especially France, 90% of the population may be seropositive because of the cultural habits of eating raw or undercooked meat. The majority of primary infections with

T. gondii in healthy adults are asymptomatic or manifest as heterophil-negative infectious mononucleosis. CNS toxoplasmosis in healthy adults may manifest as isolated CNS involvement with the production of necrotic mass lesions or abscesses. In immunocompromised patients, toxoplasmosis may result from primary infection or reactivation of a past infection. Organ transplant recipients may acquire toxoplasmosis from a donated organ. Toxoplasmosis occurs in as many as 40% of AIDS patients and is the most common cause of a focal brain lesion in this patient population. In such an immunocompromised host, toxoplasmosis is inevitably fatal without treatment. The CSF findings (Fig. 29–15) are nonspecific and include a mild lymphocytic pleocytosis and an increased CSF protein level. The diagnosis of toxoplasmosis is usually serologic, although immunocompromised patients may not demonstrate a humoral immune response to the infection. In such cases, and especially if there is a focal brain lesion, diagnosis can be obtained by brain biopsy. In AIDS patients, the brain lesions are radiologically characteristic, and the diagnosis is often confirmed by clinical response to specific therapy.

Primary amebic meningoencephalitis is caused by the free-living amebae *Naegleria* and *Acanthamoeba.* The term primary amebic meningoencephalitis is used to distinguish infections due to *Naegleria* and *Acanthamoeba* from cerebral abscesses due to the lumen-dwelling *E. histolytica,* which also produces abscesses within visceral organs. Infections due to *Naegleria fowleri* and *Acanthamoeba castellanii* show different

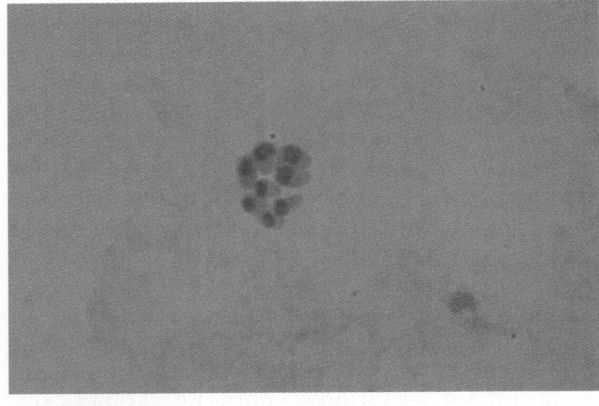

■■■■ F i g u r e 2 9 – 1 5
Touch preparation of brain tissue showing typical "floret" of *T. gondii* trophozoites. The organisms are not typically seen in preparations of CSF. Wright stain. High-power view.

clinical and laboratory features. *N. fowleri* may be found in freshwater lakes and streams worldwide and is the most common cause of acute amebic meningoencephalitis (AAM). The infection is restricted to individuals who swim or bath in infected waters. Trophozoites invade the nasal epithelium and migrate to the CNS via the olfactory nerve. AAM manifests as a purulent meningitis similar to bacterial meningitis, with a neutrophilic pleocytosis (usually more than 500 leukocytes/mL), an increased CSF protein level, and a decreased CSF glucose level. Unlike bacterial meningitis, numerous red blood cells are also present in the CSF. The motile amebae may be visualized in wet preparations, usually first in the cell-counting chamber. Infections due to *A. castellanii* occur most frequently in immunosuppressed individuals. As with *N. fowleri, A. castellanii* may invade directly through the nasal mucosa or may invade the CNS by hematogenous spread from an extracerebral focus. *A. castellanii* causes a subacute meningoencephalitis referred to as granulomatous amebic encephalitis (GAE). The incubation period of GAE is somewhat longer than that of AAM. Focal neurologic signs, rather than meningitis, are commonly seen. The patient with GAE deteriorates over weeks or months. Findings in the CSF are nonspecific. In contrast with infections due to *N. fowleri,* trophozoites of *A. castellanii* are typically not seen in wet preparations of CSF. Serologic tests including complement fixation and indirect fluorescent antibody tests may be helpful in diagnosis. AAM may be successfully treated early in the course of disease. There is no effective treatment for GAE.

Malignant tertian malaria due to *Plasmodium falciparum* is another protozoan infection that may be complicated by CNS disease. Individuals acquire malaria when inoculated with plasmodia by an anopheline mosquito during a blood meal. The inoculated parasites proliferate asexually or differentiate into gametocytes, resulting in lysis of the host's erythrocytes. A high level of parasitemia occurs in falciparum malaria. The cycle of erythrocyte invasion and lysis occurs every 48 hours, corresponding to the recurrent bouts of chills and fever. The numerous parasitized red blood cells are less deformable than normal red blood cells and tend to "sludge" within cerebral capillaries. The occluded vessels produce edema, hemorrhages, and ischemic neural injury. Clinically, cerebral malaria is manifested by delirium, followed eventually by coma and death in untreated cases. A diagnosis of malaria is obtained by examination of peripheral blood smears. Falciparum malaria should be differentiated from malaria due to the other plasmodia because of the possibility of resistance to chloroquine.

A variety of helminths invade the CNS and are the most common pathogens associated with a CSF pleocytosis with predominance of eosinophils (eosinophilic meningitis). Neurocysticercosis, the most common CNS helminth infection, is caused by the larvae *(Cysticercus cellulosae)* of the pig tapeworm *Taenia solium.* Neurocysticercosis occurs when humans become an intermediate host following ingestion of food or water contaminated with eggs of *T. solium.* Within the small intestine, the eggs release their larvae, which migrate through the intestinal wall to the circulation and eventually to the CNS. Initial invasion of the CNS may go unnoticed, and symptoms may not be apparent for several decades. Eventually, focal neurologic signs or seizures develop. The CSF may be normal or may show pleocytosis with predominance of neutrophils or eosinophils and a decreased CSF glucose level. The diagnosis is confirmed by radiographic studies and demonstration of cyst antigen or anticyst antibody within the CSF. A similar disorder occurs in coenurosis due to larvae of the dog tapeworms *T. serialis, T. brauni, T. glomerata,* and *T. multiceps.* Striking eosinophilic meningitis may also be seen with nematode larvae of *Angiostrongylus cantonensis* (the rat lungworm), *Strongyloides stercoralis, Gnathostoma spinigerum,* and *Toxocara canis* or *T. cati.*

CEREBROSPINAL FLUID COLLECTION AND PROCESSING

Specimen Collection: Lumbar Puncture

There are few contraindications to lumbar puncture, with the exception of a space-occupying lesion within the brain. Removal of CSF under this circumstance may produce uneven pressures within the brain, resulting in herniation with catastrophic consequences for the patient. Radiologic studies, including computed tomography (CT), are commonly performed prior to lumbar puncture with removal of CSF. CSF fluid is obtained by inserting a sterile, hollow needle into the spinal subarachnoid space in the lower (lumbar) back (Fig. 29–16). Fluid is collected in three to four sterile

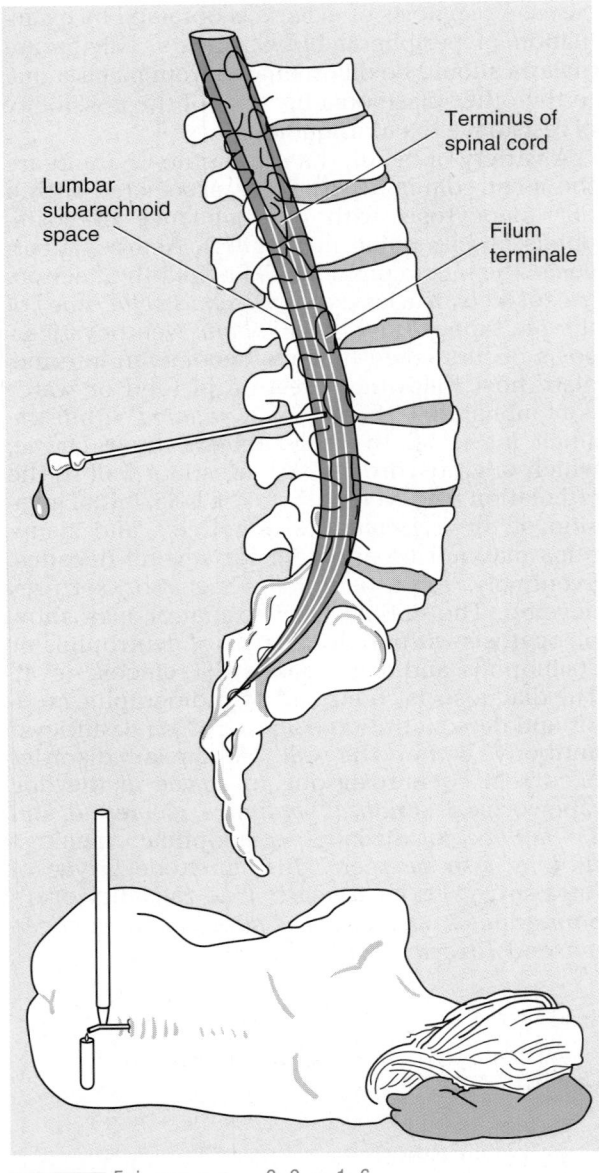

Lumbar subarachnoid space

Terminus of spinal cord

Filum terminale

■ Figure 2 9 – 1 6

Lumbar puncture. The CSF is obtained by inserting a long, sterile, hollow needle into the spinal subarachnoid space in the lower (lumbar) back.

may be used for specialized studies, such as serology. The CSF specimens should be transported to the laboratory without delay. Fastidious organisms such as *N. meningitidis*, *S. pneumoniae*, and *H. influenzae* do not survive prolonged transit times. CSF must never be refrigerated and should be maintained at room temperature until processing.

Nonspecific Indicators of Meningeal Infections

Bacterial meningitis is suspected, based on presenting clinical symptoms and the results of CSF analysis. In acute bacterial meningitis, the CSF is turbid or cloudy, the protein level is increased, the glucose level is markedly decreased, and the cell count is increased (pleocytosis) with predominance of neutrophils. The characteristic physical and chemical findings in the CSF are compared for a variety of infectious etiologies (Table 29–5).

Direct Smear Examination

A Gram-stained sediment of CSF is a rapid and inexpensive method of confirming bacterial as well as some forms of fungal meningitis (positive in 75 to 90% of cases of bacterial meningitis). The sensitivity decreases in patients who have received prior antimicrobial therapy. The diagnostic yield is also dependent on the volume of sample obtained. Whereas 1 to 3 mL of CSF may be sufficient in a case of bacterial meningitis, infections due to fungi or mycobacteria may have few organisms in the CSF and therefore require larger sample volumes (up to 10 to 15 mL) for culture. The diagnostic yield with respect to observing bacteria in the CSF can be increased by using cytocentrifuge preparations rather than conventional centrifugation sediments, smears, or fluid drop smears.

Culture

CSF samples received for culture are routinely plated on 5% sheep's blood agar, enriched chocolate agar, and thioglycolate broth regardless of CSF cell count. Alterations in plating are appropriate for suspected unusual pathogens. It is the practice in our laboratory to examine all CSF specimens submitted, regardless of leukocyte count, as some important pathogens (e.g.,

tubes for analysis. The first tube (which has the highest probability of being contaminated with peripheral blood and skin microbes) is used for chemical studies, including protein and glucose concentrations. The second tube is submitted for direct smear (Gram stain) and culture. The third tube is submitted for red blood cell and leukocyte counts and cellular differential. The fourth tube

■■■■■■■■ T A B L E   2 9 – 5
CHARACTERISTIC FINDINGS IN MENINGITIS

	Bacterial	Fungal	Tuberculous	Syphilitic	Viral	Parasitic
Organisms seen in CSF				None	None	
Cell count (leukocytes/mL)	100–100,000 Neutrophils predominate	Normal–500 Lymphs predominate*	50–500 Lymphs predominate†	100–750 Lymphs predominate	Normal–200 Lymphs predominate†	Normal–200 Lymphs and/or eosinophils predominate
Protein (mg/dL)	100–500	Normal–250	Normal–150	50–250	Frequently normal	Usually increased
Glucose (mg/dL)	<30 usually markedly decreased	Normal to decreased	Usually decreased (<45)	Normal	Normal	Normal to decreased
Additional findings	Bacterial antigen test, Limulus amebocyte lysate, CSF lactate	Calcofluor concentrated specimen; Latex agglutination (i.e., C. neoformans)	Pellicle formation, auramine-rhodamine on concentrated specimen	CSF, positive VDRL	Serology, culture biopsy	Serology, biopsy

*May have normal CSF cell count with *C. neoformans*. Eosinophils may predominate in *C. immitis* infection.
†Neutrophils may predominate in early meningitis.
‡CSF cell count may be more than 1000 leukocytes/mL in lymphocytic choriomeningitis virus infection. Neutrophils may predominate in early meningitis.

C. neoformans) may be present without concurrent CSF pleocytosis.

Rapid Diagnostic Tests

A variety of ancillary tests may be used to aid in the rapid diagnosis of bacterial meningitis. These tests can be separated into antigen-based and non–antigen-based methods.

Antigen-Based Methods. Antigen-based tests include counterimmunoelectrophoresis (CIE), coagglutination (COAG), and latex agglutination (LA). These tests identify soluble bacterial antigens in the CSF or urine of patients with suspected bacterial meningitis and are available for *H. influenzae* type b, *S. pnemoniae*, group B streptococci, and *N. meningitidis* serogroups A, B, C, Y, and W 135. The COAG and LA tests are rapid, easy to perform, and detect 85 to 94% of *H. influenzae* type b infection and up to 100% of infections due to *S. pneumoniae*. Both tests are less sensitive in detecting antigens to *N. meningitidis* in CSF or urine. These tests have been found most useful in pediatric patients with meningitis. Optimal sensitivity and specificity of bacterial antigen tests are obtained when they are employed in children with a CSF pleocytosis of at least 50 leukocytes/mL. The antigen-based tests are also useful in supporting the diagnosis of bacterial meningitis in patients who have received prior antibiotic therapy or who may have no observable organisms in CSF sediment Gram stain. Additionally, rapid antigen tests may be helpful in distinguishing between group B streptococci and *S. pneumoniae*, as these organisms may have a similar appearance on Gram stain. Antigen detection by CIE is not as sensitive as the COAG or LA tests, and this method is now infrequently used. The results of rapid antigen detection should not be substituted for Gram stain and culture in cases of suspected bacterial meningitis.

Non–antigen-based Methods. Non–antigen-based tests include CSF lactate and lactate dehydrogenase isoenzymes, C-reactive protein (CRP), and the *Limulus* amebocyte lysate (LAL) assay. In cases of frankly purulent CSF characteristic of bacterial meningitis (but uncommon with viral meningitis), the increase in anaerobic glycolysis raises the concentration of CSF lactate. Determinations of CSF lactate and lactate dehydrogenase isoenzymes have been used to distinguish between bacterial and viral causes of meningitis. Unfortunately, both of these tests have poor sensitivity and specificity. Determinations of CRP also lack sensitivity and specificity in distinguishing bacterial from viral meningitis. Additionally, the CRP may be increased in a variety of noninfectious CNS disorders. The LAL assay utilizes hemolymph from the horseshoe crab, which clots when exposed to small amounts of lipopolysaccharide (endotoxin) present in the cell wall of gram-negative bacteria. The LAL assay has a reported sensitivity of 93% and a specificity of 99%, compared with CSF culture isolation of gram-negative bacteria. The test has limited clinical utility because (1) the LAL assay detects only gram-negative bacteria and (2) the assay does not distinguish between different gram-negative bacterial pathogens.

A low CSF chloride level has been anecdotally considered characteristic of tuberculous meningitis, but this test has also proved to be unreliable. *M. tuberculosis* is difficult to identify in the CSF of patients with meningitis. In some cases, if the CSF is allowed to stand in a test tube, a delicate pellicle may form on its surface. Direct examination of the pellicle or the use of cytocentrifuge preparations stained with auramine-rhodamine (fluorochrome), rather than more traditional acid-fast stains, may be useful. Culture of *M. tuberculosis* frequently requires several weeks. Thus, newer techniques for rapid confirmation of diagnosis are being developed and include assays for CSF tuberculostearic acid and the use of the polymerase chain reaction.

Bibliography

Agger W, Case KL, et al: Lyme disease: Clinical features, classification, and epidemiology in the upper Midwest. Medicine 70:83, 1991.

Albright RE, Christenson RH, Emlet JL, et al: Issues in cerebrospinal fluid management: CSF venereal disease research laboratory testing. Am J Clin Pathol 95:397, 1991.

Albright RE, Graham CB, et al: Issues in cerebrospinal fluid management: Acid-fast bacillus smear and culture. Am J Clin Pathol 95:418, 1991.

Bailey EM, Domenico P, Cunha BA: Bacterial or viral meningitis? Measuring lactate in CSF can help you know quickly. Postgrad Med 88:217, 1990.

Chun CH, Johnson JD, et al: Brain abscess: A study of 45 consecutive cases. Medicine 65:415, 1986.

Connolly KJ, Hammer SM: The acute aseptic meningitis syndrome. Infect Dis Clin North Am 45:99, 1990.

Coyle PK, Dattwyler R: Spirochetal infection of the central nervous system. Infect Dis Clin North Am 4:731, 1990.

Fishman RA: Cerebrospinal Fluid in Diseases of the Nervous System, 2nd ed. Philadelphia: WB Saunders, 1992.

Gray LD, Fedorko DP: Laboratory diagnosis of bacterial meningitis. Clin Microbiol Rev 5: 130, 1992.

Greenlee JE: Cerebrospinal fluid in central nervous system infections. In Scheld WM, Whitley RJ, Durack DT (eds): Infections of the Central Nervous System. New York: Raven Press, 1991.

Hoban DJ, Witwicki E, Hammond GW: Bacterial antigen detection in cerebrospinal fluid of patients with meningitis. Diagn Microbiol Infect Dis 3:373, 1985.

Hoff GL, Mellon GF, et al: Bats, cats, and rabies in an urban community. South Med J 86:1115, 1993.

Jacob CN, Henein SS, et al: Nontuberculous mycobacterial infection of the central nervous system in patients with AIDS. South Med J 86:638, 1993.

Lambert HP: Infections of the central nervous system. Philadelphia: BC Decker, 1991.

Marmaduke DP, Brandt JT, Theil KS: Rapid diagnosis of cytomegalovirus in cerebrospinal fluid of a patient with AIDS-associated polyradiculopathy. Arch Pathol Lab Med 115:1154, 1191.

McCracken GH: Neonatal septicemia and meningitis. Hosp Pract 11:89, 1976.

Ogawa SK, Smith MA, et al: Tuberculous meningitis in an urban medical center. Medicine 66:317, 1987.

Phillips S, Millan JC: Reassessment of microbiology protocol for cerebrospinal fluid specimens. Lab Med 22:619, 1991.

Saez-Llorens X, McCracken GH: Bacterial meningitis in neonates and children. Infect Dis Clin North Am 4:623, 1990.

Saez-Llorens X, McCracken GH: Mediators of meningitis: Therapeutic implications. Hosp Pract 26:68, 1991.

Steere AC: Current understanding of Lyme disease. Hosp Pract 28:37, 1993.

Snider WD, Simpson DM, et al: Neurological complications of acquired immune deficiency syndrome: Analysis of 50 patients. Ann Neurol 14:403, 1983.

Wenger JD, Hightower AW, et al: Bacterial meningitis in the United States, 1986: Report of a multistate surveillance study. J Infect Dis 162:1316, 1990.

Werner V, Kruger RL: Value of the bacterial antigen test in the absence of CSF fluid leukocytes. Lab Med 22:787, 1991.

Bacteremia

*Ronald R. Trudel • James T. Griffith •
Leonard H. Schleicher*

1. Define bacteremia. Differentiate the types of bacteremia, and describe when these conditions occur.

2. Discuss the epidemiology and pathogenesis of bacteremia.

3. Describe the role of endotoxin in gram-negative septicemia. Explain why gram-negative bacteremia produces more serious consequences than those caused by gram-positive bacteremia.

4. List the organisms most commonly isolated from blood cultures.

5. Describe the proper procedure for blood culture collection.

6. Discuss the methods for the detection of bacteremia, including the following:

 ■ media
 ■ blood culture additives
 ■ methods of removing antimicrobials
 ■ advantages and disadvantages of each procedure described

Bacteremia is the invasion by bacteria of the cardiovascular system (blood stream). The consequences of this event can range from a transient, self-limiting infection to one that is life-threatening. The patients at most serious risk are those who are immunocompromised, either through drug or chemotherapeutic intervention or as the result of preexisting disease and subsequent immunosuppression. Bacteremia is often associated with hospitalization, instrumentation, and other kinds of procedures. Although infection with a wide variety of microorganisms (Table 30–1) can produce bacteremia, this condition is most frequently caused by gram-negative organisms.

This chapter begins with general concepts pertinent to bacteremic infections, including definitions of conditions relating to bacteremia and how each condition manifests. This chapter then does the following:

 ■ discusses the epidemiology and pathogenesis of bacteremia

 ■ describes the predisposing risk factors in different populations

 ■ describes complications of bacteremia that may occur

 ■ presents diagnostic laboratory procedures

 ■ explains treatment modalities

■ T A B L E   3 0 – 1

MICROORGANISMS MOST LIKELY TO BE ISOLATED FROM BACTEREMIC PATIENTS*

Escherichia coli
Staphylococcus aureus
Klebsiella
Enterobacter
Pseudomonas
Acinetobacter
Proteus
Serratia
Citrobacter
Salmonella
Morganella
Xanthomonas (pseudobacteremia often)
Bacteroides (anaerobes)
B. fragilis group
Alcaligenes
Haemophilus influenzae
Fusobacterium nucleatum
Cardiobacterium hominis
Capnocytophaga

*Listed in descending likelihood of isolation.

GENERAL CONCEPTS RELATED TO BACTEREMIC INFECTIONS

Bacteremia versus Septicemia

Bacteremia is the presence of bacteria in the blood stream. Certain infections, such as meningitis, salmonellosis, and endocarditis, have a period of bacteremia as part of the disease process. Somewhere between 10 and 13% of blood cultures are positive for this phenomenon, with two thirds of these representing true clinical positive results. While bacteremia is the mere presence of bacteria in the blood, *septicemia* is bacteremia plus a clinical presentation of physical signs and symptoms of the bacterial invasion and toxin production.

There are several terms associated with isolation of organisms in blood cultures. First described in 1969, *pseudobacteremia* is the term associated with contaminated infusion fluids (e.g., intravenous, hyperalimentation, saline), blood culture bottles, alcohol swabs, syringes, and other materials. Most pseudobacteremias are caused by aerobic gram-negative bacilli. *Occult (unsuspected) bacteremia* predominantly refers to the condition found in children who appear healthy but whose blood culture is positive. This phenomenon is usually observed in children younger than 2 years of age. The most common causes include *Streptococcus pneumoniae, Haemophilus influenzae* type b, and *Haemophilus influenzae* nontypable.

Forms of Bacteremia

The forms of bacteremia include *primary bacteremia,* which is bloodstream invasion by bacteria for which no preceding or simultaneous site of infection with the same microorganism can be identified. *Secondary bacteremia* is isolation of a microorganism from the blood as well as other sites in the same patient prior to or at the same time, such as in pneumonia and urinary tract infections. *Nosocomial bacteremia* occurs on or after the third day of hospitalization; with *polymicrobial bacteremia,* blood cultures yield more than one organism.

Bacteremic Episodes

Bacteremic episodes may be *transient, intermittent,* or *continuous.* The frequency, time, and number of blood cultures to be collected may depend on the type of bacteremic episode the patient is experiencing. *Transient bacteremia* usually occurs after a procedural manipulation of a particular body site that is colonized by indigenous flora. Such sites include the mouth and the gastrointestinal and urogenital tracts. Transient bacteremia may appear for a brief peri-

od following dental, colonoscopic, or cystoscopic procedures. *Intermittent bacteremia* can occur as the result of abscesses present at a particular site or as a clinical manifestation of certain types of infections, such as meningococcemia and gonococcemia. *Continuous bacteremia* occurs when the organisms are coming from an intravascular source and are present in the blood stream consistently. Endocarditis is the most common clinical manifestation associated with continuous bacteremia.

Other Conditions

Warm shock is characterized by fever, increased pulse, hyperventilation, and warm, dry, flushed skin, whereas *cold shock* is associated with decreased blood pressure, increased pulse, and rapid, shallow respirations. *Septic shock* is a syndrome that occurs as a complication of bacteremia. It is characterized by hemodynamic changes, decreased tissue perfusion, and compromised tissue and organ function. Septic shock occurs in about one third of bacteremic cases and is associated with a mortality of 40 to 50%. Mortality in bacteremic cases with no shock ranges from 10 to 15%.

EPIDEMIOLOGY

Brill reported the first case of bacteremia (*Bacillus pyocyaneus,* now *Pseudomonas aeruginosa*) in 1899. Ten years later, fewer than 40 cases had been reported worldwide, with less than 30 additional cases in the 15 years following that. Between 1954 and 1974 the reported incidence of bacteremia increased 20-fold, and bacteremia is now the thirteenth leading cause of death in the United States.

Currently, bacteremia is not an infection reportable to the *Centers for Disease Control and Prevention (CDC&P).* The current incidence of gram-negative bacteremia has been estimated between 70,000 and 330,000 cases per year, with most estimates above 200,000. This represents approximately 1 to 3% of all hospitalized patients. The general mortality rate for bacteremia ranges from 6 to 42%. Mortality rates among patients who are appropriately treated range from 10 to 38%. Patients who are granulocytopenic or inappropriately treated may have a mortality rate that approaches 100%. Moreover, fatalities among patients infected with gram-negative bacilli are higher than those among patients who have gram-positive cocci as the causative agents of their bacteremia.

Risk Factors

The increased incidence of bacteremia over the past 25 years appears to center around the following:

- decreased immune competency of selected patient populations
- increased use of invasive procedures
- age of the patient
- administration of drug therapy

Decreased Immune Competency of Selected Patient Populations

Gram-negative bacteremias seem more increased in frequency in persons who have neoplasias, especially carcinomas, hematologic malignancies, and connective tissue diseases. Persons with other chronic underlying diseases, such as diabetes and cirrhosis, and those who are receiving immunosuppressive therapy are also at increased risk for bacteremia.

Increased Use of Invasive Procedures

Increased use of life support systems, respirators, and invasive diagnostic procedures may play a role in the incidence of gram-negative bacteremia. Indwelling urethral catheters, suprapubic catheters, and intravenous pyelograms predispose patients to catheter infections. These devices penetrate otherwise sterile areas and encourage colonization with bacteria from surrounding tissue. A greater potential of bacteremia also occurs after surgery involving the urinary gastrointestinal, and biliary tracts.

Age

A bimodal distribution of bacteremia has been observed. In the very young, increased bacteremia is observed owing to a defect in humoral immunity, whereas in older populations, bacteremia is due to a general decrease in immune competency.

Elderly patients have a tendency for gastrointestinal and urinary tract infections. This group of patients is expected to have a higher incidence of gram-negative bacteremia, usually in excess of 50%, and a higher recurrence rate due primarily to their immunosuppression and underlying medical conditions. This is especially common in patients who are in intensive care units and who become colonized within 2 to 3 days.

Administrations of Drug Therapy

The administration of broad-spectrum antimicrobials reduces sensitive normal flora and favors colonization and invasion by gram-negative bacteria. Immunosuppressive agents, especially steroids and anticancer chemotherapeutics, place patients at increased risk for bacteremia.

PATHOGENESIS

Sources of Bacteremic Spread

A number of sources are often associated with bacteremic spread, including the following.

Pericarditis and Peritonitis. Pericarditis is often polymicrobial, with *Bacteroides, Clostridium, Streptococcus pneumoniae, Streptococcus pyogenes*, and *Staphylococcus aureus* as the most commonly associated organisms. Bacteremias resulting from peritoneal dialysis are most often linked to *Staphylococcus epidermidis, Staphylococcus aureus,* and gram-negative bacteria.

Pneumonias. The most common organisms in pneumonia that produce a concurrent bacteremia include *Staphylococcus aureus, Streptococcus pneumoniae, Pseudomonas aeruginosa, Haemophilus influenzae,* and *Enterobacter aerogenes.*

Pressure Sores. Almost one half of the cases of bacteremia attributed to pressure sores may be polymicrobial. Some of the most commonly reported offending organisms are *Proteus mirabilis, Staphylococcus aureus, Bacteroides fragilis, Acinetobacter* species, *Bacillus,* and *Corynebacterium.*

Prosthetic Medical Devices. It has been known since the fourteenth century that foreign bodies could potentiate infections. It is currently appreciated that device-related infections constitute a major source of morbidity and mortality in hospitalized patients. The organisms implicated in these infections most often originate from endogenous "skin or gut" flora. Subcutaneous implants are more prone to colonization by gram-positive organisms, whereas devices implanted in the peritoneal cavity are usually colonized by gram-negative and anaerobic organisms.

Production of slime by organisms is one mechanism that has been associated with prosthetic valve endocarditis. The slime may serve as a ligand during the initial surface adhesion, or it may be produced after the organism has established a focal presence by adhering to the surface. The slime also protects the organism from host defenses by inhibiting phagocytosis, chemotaxis, and oxidative metabolism and by suppressing the lymphoproliferative response.

How the prosthetic devices become contaminated in the first place is subject to investigation on a case-by-case basis. Some of the observed instances include breaks in sterile technique, leading to contamination by endogenous skin flora; improperly sterilized materials; or contaminated catheters. Once prosthetic devices are in place, some organisms take advantage of the materials from which the devices are made. For example, *Staphylococcus epidermidis* can hydrolize certain plastics as a food source. Some of the most commonly implicated prosthetic devices include plastic devices, such as catheters (Broviac, Swan-Ganz, Tenckhoff) and shunts (ventricular, chest tubes), and cardiac pacemaker leads.

Pressure transducers have been used since the 1960's as indwelling catheters to monitor cardiovascular system conditions. Reports of bacteremia and fungemia were noted in the literature soon after the initial usage of these devices. Some of the most commonly isolated organisms in these cases include *Serratia marcescens, Klebsiella oxytoca, Pseudomonas, Citrobacter,* and *Enterobacter* species.

Total Hip Replacement. This massive surgical procedure has been associated with acute fulminating infection within 2 or 3 months of surgery. Sepsis between 4 and 26 months of surgery is thought to be due to intraoperative contamination. Late infection occurring 23 to 50 months after surgery is thought to be of hematogenous origin. In many cases removal of the prosthetic hip is necessary to cure these patients. The risk factors for patients who develop bacteremia and colonization of their prosthetic hip include malignancy, steroid therapy, cytotoxic therapy, collagen vascular disease, alcoholism, liver disease, sickle cell disease, hematologic disease, previous surgery, and prior antimicrobial therapy.

Skeletal System. The bones have also been implicated as a source of bacteremia or as a consequence of bacteremic persistence. *Staphylococcus aureus, Pseudomonas aeruginosa,* and various facultative anaerobes have been implicated in these cases. Mixed infections, that is, polymicrobial infections, are uncommon.

In patients with osteomyelitis, however, bacteremia tends to be polymicrobial. *Staphylococcus aureus* is most frequently isolated, with gram-negative bacilli accounting for more than 33% of the polymicrobial infections.

Skin and Soft Tissues. Skin and soft tissue infections as well as wounds are sources of bacteremia. These infections tend to be polymicrobial in 2 to 28% of patients. Soft tissue abscesses, especially *S. aureus* bacteremia β-hemolytic streptococci, *P. aeruginosa,* and *Bacteroides* species, have been reported and tend to be polymicrobial or result in polymicrobial bacteremia. Necrotizing fasciitis has a mortality rate of 20 to 30% when it results in bacteremia.

Clinical Signs and Symptoms

Only about one third of patients experience the classic signs and symptoms of bacteremia, which may include abrupt onset of chills, fever, or hypothermia, and hypotension. Forty percent of patients experience prostration and diaphoresis. Tachypnea is an early sign of bacteremia. Delirium, stupor or agitation, vomiting and nausea, oliguria, or anuria occurs in 50% of patients with gram-negative bacillary bacteremia. Ecthyma gangrenosum, a central necrotic area surrounded by an erythematous base, is strongly correlated with gram-negative bacteremia. The failure of the body to mount an elevated temperature is also associated with increased mortality among neonates and the elderly. Altered clinical laboratory values that may be indicative of bacteremia include the following:

- thrombocytopenia (50–60% of patients with gram-negative bacilli)
- leukocytosis or leukopenia
- acidosis with lactic acidosis
- abnormal liver function tests (especially hyperbilirubinemia)
- coagulopathy (more than 60% of the time)
- disseminated intravascular coagulation (5–10% of patients with gram-negative bacillary bacteremia)

■ elevations in C-reactive protein, haptoglobin, and fibrinogen

Complications

Septic shock, which in most cases is secondary to bacteremia, occurs in about 20 to 40% of bacteremic patients and may appear 2 to 6 hours after the initial manifestations. A hallmark of septic shock is tachycardia with decreased systemic vascular resistance (SVR).

Effects of Endotoxin During Gram-Negative Sepsis

Endotoxin, the lipopolysaccharide component of the outer membrane of gram-negative bacilli, has been implicated as an important mediator of septic shock. It is released from actively dividing as well as dead bacterial cells. Cytotoxins appear in response to endotoxin and amplify the subsequent release of soluble mediators in the activation of different plasma proteins. Increased levels of tumor necrosis factor (TNF), α-interleukin-1 (IL-1), and IL-6 have been correlated with poor outcome in cases of clinical septic shock.

Endotoxin is absorbed onto the surface of platelets, causing them to release 5-hydroxytryptamine, which aggravates and potentiates coagulation. In addition, it activates the fibrinolytic cascade and activates factor XII to initiate clotting.

Polymicrobial Bacteremia

In the 1930's virtually every case of bacteremia involved a single organism. By the middle 1970's, polymicrobial bacteremia was considered the norm rather than the exception.

Multiple organisms are being identified in 20% of cases of bacteremia. Polymicrobial bacteremia is generally associated with a higher mortality than that of monomicrobial bacteremia. The predisposing factors in polymicrobial bacteremia include self-injection, burns, and gastrointestinal tract sources. Especially at risk are immunocompromised patients, particularly patients with alcoholism, granulocytopenia, extensive burns, diabetes mellitus, and chronic renal failure, and patients with vascular insufficiency due to ischemia. *Bacteroides fragilis* has often been associated in polymicrobial infections. Given the resistance of *B. fragilis* to many antimicrobials, its diagnosis carries a high degree of expected mortality.

LABORATORY DIAGNOSIS

Specimen Collection

It is important to remember that even though antiseptic technique is used in the collection of blood, somewhere between 1 and 3% of blood cultures become contaminated with organisms such as coagulase-negative staphylococci, *Corynebacterium* species, α-hemolytic streptococci, and *Propionibacterium acnes,* which are skin colonizers.

Therefore, it is most important to prepare the skin properly prior to venipuncture for blood culture. There are several acceptable protocols. Palpation for the vein can be checked with a gloved finger. Cleansing the skin with 80 to 95% ethanol, followed by an iodine-based compound scrubbed in a concentric fashion about the venipuncture site, is a common practice. Other substances may be substituted for iodine in case of known skin hypersensitivity. For decontamination to be effective, the iodine should be left on the skin for at least 1 minute. After the venipuncture, the disinfecting agent should be removed with an alcohol pad.

General Principles in Determining Volume, Frequency, and Number of Blood Culture Collection

DENSITY OF BACTEREMIA IN ADULTS VERSUS NEONATES

Bacteremia may involve a large number of microorganisms; however, only a relatively small number of bacteria per unit volume of blood (typically fewer than 30 bacteria per mL) are recovered. The detection methods common in most laboratories produce positive results in adult patients who have bacteremias if organisms are present in the range of 10 to 15 bacteria per mL of blood. There are, however, many circumstances in which patients have fewer than this number.

Neonatal patients tend to be more septic as a result of the incomplete development of their defense mechanisms and generally have higher numbers of microorganisms per mL of blood. The following age-volume protocol, therefore, has been recommended.

Age	Amount
Younger than 10 years of age	1 mL of blood for each year of life
10 years old or older	20 mL
10 years old or older (poor veins)	Less than 20 mL

This protocol is based on studies that have demonstrated that as the volume of blood cultured is increased from 2 to 20 mL, the yield of positive culture results increases from 30 to 50%, except in neonatal patients. The optimal ratio of blood to culture medium is about 1:10. The dilution aids in negating the bactericidal effect of normal serum. In cases in which a 1:10 dilution cannot be achieved, most laboratories incorporate 0.25 to 0.50% of sodium polyanetholsulfonate (SPS), which serves as anticomplement and anticoagulant. The pediatric exception to this volume ratio is usually accommodated by collecting 1 to 3 mL into 5-mL culture bottles.

FREQUENCY OF BACTEREMIC EPISODES VERSUS TIME AND FREQUENCY OF COLLECTION

The frequency of bacteremic episodes is another factor to consider in determining the time, frequency, and volume of blood collection for culture. Because bacteremias may be transient, intermittent, or continuous in their production of microorganisms in the peripheral circulation, collection of samples is highly dependent on the type of bacteremia suspected.

Because patients with transient bacteremia rarely present with clinical symptoms, blood cultures are rarely obtained. Organisms are immediately cleared from the peripheral system by the reticuloendothelial system (RES). In patients with continuous bacteremia, on the other hand, the organisms are constantly released into the blood stream and therefore are likely to be isolated whenever the blood culture specimen is taken. Although a single set of blood culture specimens may yield the etiologic agent, a set of three is still highly recommended (a set consists of one bottle for aerobic incubation, another for anaerobic incubation). In cases of intermittent bacteremia, the time interval between collection of samples is critical in the recovery of organisms. In suspected cases of intermittent bacteremia, it is recommended that blood culture specimens be collected before an anticipated temperature rise to ensure maximum recovery. Usually, by the time chills and fever occur the organisms in the blood stream are being cleared by the RES; therefore, fewer organisms may be recovered in blood culture specimens at this time.

RATIONALE FOR MULTIPLE COLLECTION

One study revealed that approximately 80% of bacteremias will be discovered in the first set of blood culture specimens taken, 90% will be detected if two sets of specimens are taken, and up to 99% will be diagnosed if a third set is taken. In most cases of bacterial endocarditis, the causative organism is present in fairly small numbers; therefore, taking three to four sets of culture specimens from three different venipuncture sites within the first 1 to 2 hours of clinical presentation is the best protocol. In subacute bacterial endocarditis, three sets of blood culture specimens taken within the first 24 hours, at 1-hour intervals, are recommended. With bacteremia of unknown origin, taking four to six 10-mL specimens within the first 48 hours ensures the greatest likelihood of recovery of the causative organisms. In brucellosis, blood culture specimens should be obtained during the initial presentation of symptoms and at the anticipated temperature spike.

Blood Culture Methods

Culture Media Used in Conventional Broth Systems

In most laboratories, each set of blood culture specimens includes a bottle or tube designated for aerobic recovery and one for anaerobic recovery of microorganisms. The typical aerobic culture bottle contains a soybean casein digest broth, tryptic or trypticase soy broth, brain-heart infusion, *Brucella* agar, or Columbia broth base. While typical anaerobic broth media may contain the same types of basic media as the aerobic culture systems, 0.5% cysteine may have been added to permit the growth of certain thiol-requiring organisms. Unvented blood culture bottles can generally be used to support anaerobic organisms.

In general, thioglycolate- and thiol-containing broths are not suitable for the recovery of *Pseudomonas* and yeasts from blood.

NEUTRALIZATION OF INHIBITORY PROPERTIES

In specimens from patients receiving clinical amounts of β-lactam antimicrobial agents, penicillinase may be added to the medium to inactivate these agents. Some commercially available automated blood culture systems have blood culture bottles containing antibiotic removal device (ARD), a resin that nonspecifically absorbs any antimicrobial agent present in the patient's blood.

ANTICOAGULANTS AND OTHER ADDITIVES

Sodium polyanetholsulfonate (SPS), one of the commonly used additives, performs the following functions:

■ anticoagulation—it is effective at a 0.03% concentration

■ neutralization of the bactericidal effect of human serum

■ prevention of phagocytosis

■ inactivation of certain antimicrobial agents (streptomycin, kanamycin, gentamicin, polymyxin B)

However, SPS is inhibitory to *Peptostreptococcus anaerobius, Neisseria gonorrhoeae,* and *N. meningitidis* as well as *Gardnerella vaginalis.* If these organisms are suspected, 1.2% gelatin added to the blood culture bottle may help neutralize this inhibitory effect of SPS.

Other anticoagulants and supplements are available for use in blood culture systems. Sodium amylosulfate (SAS) is a structural relative of SPS that is less effective in neutralizing serum bactericidal activity and is inhibitory to *Klebsiella pneumoniae.* Sodium citrate (0.5–1.0%), an anticoagulant, is inhibitory to some gram-positive cocci. Sucrose (10–30%) is sometimes used as an osmotic stabilizer. Sucrose is especially helpful in dealing with bacteria that have undergone cell wall damage, and the resultant hypertonicity counteracts to some extent the normal bactericidal effect of blood. Other supplements, such as the anticoagulants oxylate and ethylenediaminetetraacetic acid (EDTA), are not recommended.

Newer Blood Culture Systems

BIPHASIC BROTH-SLIDE SYSTEM

A broth-slide system (Septi-Chek [Roche Diagnostics, Nutley, NJ]) was designed from the original biphasic blood culture medium Castaneda culture bottle. Septi-Chek consists of a slide paddle containing chocolate, MacConkey, and malt extract agars attached to the top of a standard broth bottle. Once these bottles have been inoculated, they should be tipped daily or at least twice weekly so as to bathe the slide paddle with the broth culture medium. This allows frequent blind subcultures without the use of needles and syringes. Bacterial growth appears either as small, discrete colonies or as a confluent growth on the slide paddle. This system has the advantage of providing more rapid recovery of facultative bacteria and isolated colonies for identification and susceptibility testing. However, there are certain disadvantages, including a slightly higher cost of materials and contamination rate. An additional unvented bottle is still required for adequate isolation of anaerobes.

CONTINUOUS-MONITORING BLOOD CULTURE SYSTEMS

The Bactec System (Becton Dickinson, Towson, MD), first introduced as an automated radiometric growth detection system for blood cultures, incorporated isotopically labeled carbon dioxide (CO_2) in the broth medium. When the organism in the blood culture bottle utilizes $^{14}CO_2$-labeled substrate, $^{14}CO_2$ is released. The instrument monitors CO_2 production by aspirating gas into an ionization chamber using sterile needles. In the ionization chamber, the amount of $^{14}CO_2$ produced is measured as growth index (GI) and is compared with an established threshold level. If the patient's blood culture shows a GI that exceeds the threshold level, the instrument sends off a signal indicating that the culture result is positive. The automated radiometric blood culture system has the advantage of early detection of bacterial growth, especially of slow-growing bacterial species (e.g., *Mycobacterium tuberculosis*). The disadvantages include the high initial cost of the instrument, high contamination rate (due to inadequate needle sterilization between bottles), and the hazards associated with radioisotope disposal. To resolve the problem of radioisotope disposal, newer models of the Bactec system utilize infrared spectroscopy to detect release of CO_2. The contamination rate has been reduced by adjusting the needle sterilization time between sample aspirations.

Recently, several new instruments have been developed to automatically and continuously monitor blood cultures. These instruments include BacT-Alert (Organon Teknika Corp., Durham, NC), the Bactec 9000 (Becton Dickinson, Sparks, MD), and the ESP (Difco Laboratories, Detroit, MI). The advantages of the new instrumentation include the continuous monitoring of the blood culture vials, which results in an improved time-to-detection, and a reduced frequency of bottles with false-positive results. For the Bactec and BacT-Alert systems, an additional advantage is the noninvasive monitoring. After introduction of the blood sample, the bottle is not punctured for monitoring (Table 30–2).

TABLE 30–2

FEATURES OF CONTINUOUS-MONITORING BLOOD CULTURE SYSTEMS

Feature	Bactec 9000	BacT-Alert	ESP
Capacity	120/240	240	128/384
Agitator	30 rpm	60 rpm	280 rpm
Protocol	1–30 days	1–30 days	5 to 7 days
Detection	CO_2, pH, fluorescence	CO_2, pH, colorimetric	CO_2, N_2, pressure
Venting	No	Yes	Yes

Bactec 9000 Series. In response to the changing needs of clinical laboratories, Becton Dickinson introduced the Bactec 9000 series. Two models are currently available, the 9240 and 9120. These are noninvasive, continuous-monitoring blood culture instruments that use fluorescence to detect CO_2. Carbon dioxide is detected using a gas-permeable sensor on the bottom of each vial. When a bottle is placed into the 9240, the instrument takes a base reading of the sensor. This reading is used as a reference for subsequent readings.

Carbon dioxide produced by an organism diffuses into the sensor, generating hydrogen ions. The increase in hydrogen ion concentration increases the fluorescence output of the sensor. Using photodetectors, the 9240 measures the amount of fluorescence, which corresponds to the amount of CO_2 produced by the microorganism. A computer program then interprets these data using several algorithms to determine when to flag a bottle as positive. If a bottle contains a microorganism, the indicator continuously changes. After several increases in absorption by the sensor, the bottle is flagged as positive. The instrument alerts the user of the positive vial by displaying a message on the computer monitor and with an audible alarm. Figure 30–1 illustrates the Bactec 9000 schematic.

Difco ESP. The Difco ESP system differs from the other continuous-monitoring systems in that it detects the consumption and/or production of gases by organisms growing in the culture medium. The consumption and/or production of gases is detected by monitoring changes in headspace pressure. The aerobic bottle is monitored

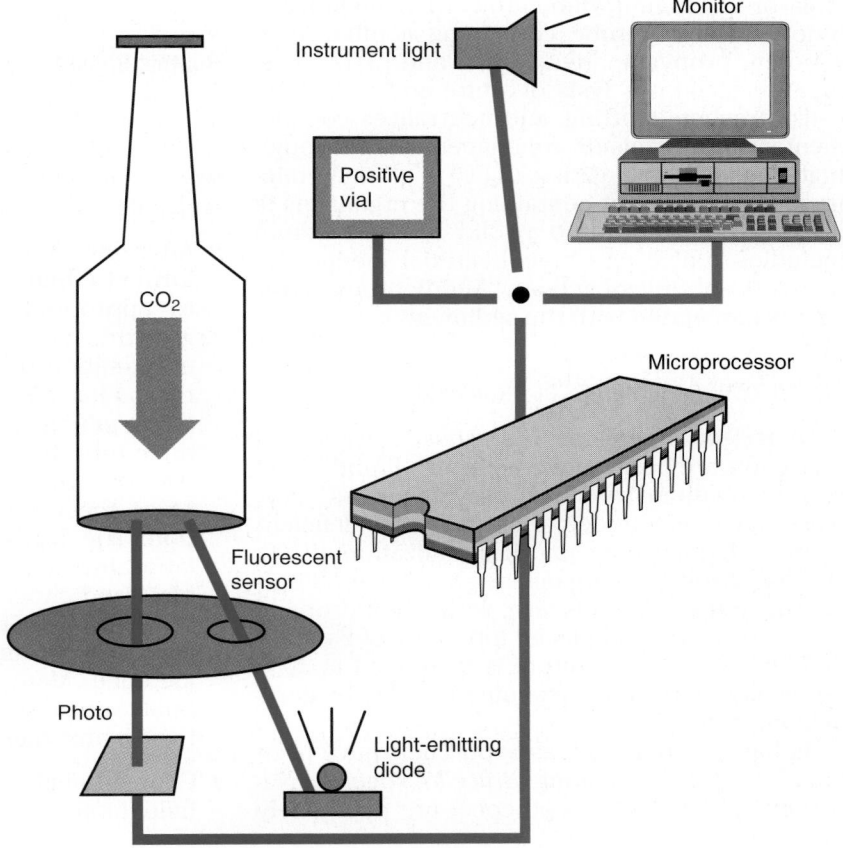

Figure 30–1
Bactec 9000 schematic.

once every 12 minutes, and the anaerobic bottle is monitored once every 24 minutes. An internal computer algorithm monitors the changes and determines when to flag the bottle as positive.

Organon Teknika BacT-Alert System. A fully automated, nonradiometric blood culture system, the Organon BacT-Alert System consists of aerobic and anaerobic bottles with pH-sensitive membranes placed in the bottom of the bottles. Microbial growth causes a pH change resulting from release of CO_2, indicated by a change in color of the growth medium. The instrument measures CO_2 production colorimetrically, without going into the bottles. The major advantage of BacT-Alert over Bactec is fewer false-positive results because of the noninvasive measurement of CO_2.

LYSIS-CENTRIFUGATION METHOD (DuPONT ISOLATOR)

The lysis-centrifugation method has been shown to provide optimal recovery of fungi and mycobacteria from systemic infections. This method provides a concentrated material for direct inoculation of various agar media. The Isolator (DuPont [Wilmington, DE]) includes a blood collection tube containing a mixture of saponin, propylene glycol, SPS, and EDTA. This mixture facilitates lysis of white and red blood cells, prevents clotting, and neutralizes complement. Microorganisms are concentrated through high-speed centrifugation ($3,000 \times g$ for 30 minutes). The sediment containing the organisms is directly inoculated onto a solid culture medium including fungal and mycobacterial media. For recovery of mycobacteria, Middlebrook 7H11 agar is inoculated with the sediment.

RECOVERY OF OTHER TYPES OF ORGANISMS

Francisella organisms, the causative agents of tularemia, are best recovered from a liquid blood culture medium to which L-cysteine and dextrose have been added. Most cell wall–deficient bacteria require the osmotic stabilization of 10% sucrose. *Leptospira* organisms are recovered during the first week of disease, with 1 to 3 drops of freshly drawn blood placed into 5 mL of Fletcher medium. The blood culture is incubated at 30°C for 28 days in the dark, examined weekly by dark-field microscopy.

Biphasic media, such as Septi-Chek, have been found useful in isolating *Brucella* species. Nutritionally deficient streptococci are adequately recovered using standard broth culture bottles because of the vitamin B_6 present in human blood. Pyridoxal-containing blood agar medium or "staph streak," however, is necessary for subculture to recover this group of streptococci. The "staph streak" test is performed with a confluent growth of the organism in a blood agar plate. A single line of *Staphylococcus aureus* is streaked across the middle of the plate. Following 24 hours of incubation, the plate is observed for tiny colonies growing near the staph.

Examination of Conventional Blood Culture Bottles

It is recommended that the sediment not be disturbed during inspection of the blood culture bottle after 6 to 18 hours of incubation. Blood culture bottles are examined with transmitted and reflected light for evidence of turbidity, hemolysis, gas production, or bacterial colonies in or on the blood layer. If visible growth is observed, 0.25 mL of blood should be aspirated with a sterile needle and syringe and then plated to a solid medium.

Routine Blood Culture Work-Up

Not all laboratories are using automated systems for blood cultures. A routine blood culture work-up for nonautomated conventional systems is usually as follows:

■ After 6 to 18 hours, a "blind" subculture is performed ("blind" means subculture all bottles to appropriate media even without visible growth), and then a routine direct smear examination of all bottles is carried out. The reason for this is that many organisms, such as *Haemophilus* and *N. meningitidis,* do not produce turbidity, hemolysis, or gas in the broth medium. Gram stain or acridine orange provides the most useful microscopic examination. The types of medium used vary among laboratories but usually include blood agar plate and chocolate agar, and if gram-negative bacilli are observed on the smear, MacConkey agar and an anaerobic blood agar plate are included. Many laboratories have abandoned blind subcultures because of the risk of infection (by needlesticks, etc.) to personnel.

■ Conventional blood culture bottles must be held for at least 7 days. They may be discarded after 7 days if growth is not observed. Report no aerobic and anaerobic growth after 7 days of incubation.

- Hold blood culture bottles for 2 weeks if bacterial endocarditis, fungemia, brucellosis (hold for 21 to 28 days in CO_2 and subculture weekly), or anaerobic bacteremia is suspected.

- Anaerobic subculture should be performed after 2 days of incubation. Routine subculture beyond 2 days and terminal subculture are of minimal value and are not routinely recommended.

Any presumptive positive finding should be reported immediately by telephone to a physician.

When positive bottles are discovered, rapid identification of the organism and antibiotic susceptibilities are important in the management of a potentially critically ill patient. If organisms can be concentrated through centrifugation and added to small amounts of medium, enzymatic biochemical reactions can be available in 3 to 5 hours. One of these methods includes the use of pellet processing (Procedure 30–1). The alternative is to subculture the organisms and use the growth from an agar plate to inoculate conventional biochemicals. This usually takes an additional day.

Based on the results of Gram stain, a rapid identification procedure should be performed. Table 30–3 describes rapid testing according to cell morphology from pellet or same-day growth.

Sources of Contamination

A contamination rate of 2 to 3% can be expected with *Staphylococcus epidermidis, micrococcus,* diphtheroids, and *Propionibacterium acnes* as common contaminants. However, any organism cultured from two or more blood culture bottles should not be overlooked as a contaminant. Any of the above organisms can also be responsible for bacterial endocarditis.

Some other sources of contamination causing pseudobacteremia include intravenous catheters as well as various skin disinfectant solutions. Benzalkonium chloride has been demonstrated to be occasionally contaminated with *Pseudomonas cepacia* and *Enterobacter* species. Alcohol or iodine as a skin disinfectant has been preferred. However, contamination of certain iodine solutions, such as povidine-iodine, with *Pseudomonas cepacia* has also been reported. Contamination with *Serratia marcescens* and *Moraxella* species has been reported in certain evacuated blood collection tubes.

In the past, contamination was more clear-cut. The media and volume of blood used now are more sensitive and grow more contaminants. It is difficult to determine what is a pathogen and what is contamination. As the volume of blood is increased, the number of positive blood culture results increases. Microbiologists cannot make this determination (true pathogen or contaminant) in the laboratory; physician input and patient history are needed.

PROCEDURE 30–1. PROCEDURE PROCESSING POSITIVE BLOOD CULTURES

Preparation of pellet

- In a biohazard hood, wipe off top of blood bottle and top of serum separator tube with alcohol. Thoroughly mix bottle.
- Vent serum separator tube by carefully puncturing top with a sterile needle. Remove needle and discard in sharps container.
- Remove top of serum separator tube and save. Withdraw 10 to 12 mL from thoroughly mixed blood bottle (10-mL syringe and 18g needle). Remove needle, and place blood in serum separator tube. Replace cap on tube.

- Centrifuge at 2,700 rpm for 10 minutes.
- After centrifugation, use a plastic-backed gauze wipe to remove cap, decant most of the supernatant, and save a portion of it for antigen (Ag) detection if indicated. Organisms will be concentrated on the top of the barrier gel. Resuspend the pellet by adding approximately 1 mL of sterile Tween water.
- Make a smear from the pellet of organisms for Gram stain or acridine orange dye.

■■■■ T A B L E   3 0 – 3
RAPID TESTING ACCORDING TO CELL MORPHOLOGY FROM PELLET OR SAME-DAY GROWTH

Organism	Media Selection	Additional Tests
Gram-negative bacilli	Blood agar–aerobic Blood agar–anaerobic MAC, CHOC–CO_2	Rapid ID* (gram-negative) MIC Oxidase 42° C, TSI, FLO, TECH[†]
Gram-negative bacilli (*Haemophilus*-like)	Blood agar–aerobic Blood agar–anaerobic MAC, CHOC–CO_2	Rapid ID Oxidase β-Lactamase Porphyrin[†] "Satellite" X and V on TSA Supplemented MIC Direct Ag test (HiB)[†]
Gram-positive bacilli (large spore-forming)	Blood agar–aerobic Blood agar–anaerobic CHOC–CO_2, egg yolk agar–anaerobic	Rapid ID Catalase PYG from supernatant MIC
Gram-positive bacilli (non– spore-forming)	Blood agar–aerobic Blood agar–anaerobic CHOC–CO_2	Rapid ID Catalase PYG from supernatant Motility Bile esculin MIC
Gram-positive cocci in groups	Blood agar–aerobic Blood agar–anaerobic CHOC-CO_2	Rapid ID Latex coagulation from blood agar[‡] Tube coagulase MIC
Gram-positive cocci, gram-positive in chains, gram-positive diplococci	Blood agar with A and P disks Blood agar–anaerobic CHOC–CO_2 Bile esculin	Rapid ID Rapid Ag detection kit from supernatant MIC CAMP
Gram-negative diplococci	Blood agar–aerobic Blood agar–anaerobic CHOC–CO_2	Oxidase FA for GC Rapid Ag detection kit from supernatant for meningococci

*Rapid ID includes API (Aryltap Products, Inc., Plainview, NY), Minitek (Becton Dickinson Microbiology Systems, Cockeysville, MD), or any rapid enzymatic tests.
[†]When indicated.
[‡]If from pellet, must wash first in order to remove SPS from blood culture bottle.
CAMP test; CHOC, chocolate agar; FA, fluorescent antibody; FLO, fluorescence; MAC, MacConkey agar; MIC, minimum inhibitory concentration; PYG, peptone, yeast, glucose; TECH agar; TSA, trypticase soy agar; TSI, triple sugar iron agar.

TREATMENT

With such a complex disorder, one would expect that there would be many therapeutic attempts to alleviate this devastating clinical event. In fact, there are several treatment modalities that have been tried; however, some have not shown positive results.

Anticachectin Treatment. Cachectin mediates a substantial portion of endotoxin injury produced in the body. The signs and symptoms of cachectin production include metabolic acidosis, decreased blood pressure, interstitial pneumonitis, acute renal tubular necrosis and mesenteric ischemia, and infarct of the bowel. The chronic production of cachectin leads to anorexia and wasting. When cachectin is produced in large doses, fulminant shock often results. Anticachectin treatment is therefore administered to avoid these clinical complications.

Glucocorticoids. Prospective, randomized studies of severely septic patients involving treatment with glucocorticoids have been conducted. The rationale for antiinflammatory treatment of sepsis is to interfere with the damaging mediators while controlling local and systemic bacterial proliferation with antimicrobial agents. Glucocorticoids (methylprednisolone sodium succinate Solu-Medrol, Upjohn, Kalamazoo, MI) have been shown to almost completely control the septic shock and multiple system organ failure produced by endotoxin with gram-negative bacillus bacteremia.

Combination-Drug Therapy versus Single-Drug Treatment. Various regimens have been tried among classical antimicrobial agents. Improvement generally occurs when broad spectrum antimicrobial agents are administered empirically taking into account the primary site of the infection, the status of the patient and the patient's general immune status. The single-drug regimens seem to work best when they are used for nonneutropenic (more than 500 neutrophils per mm^3) patients.

Treatment for Septic Shock. Initial resuscitation and fluid replacement is commonly followed with chemotherapy. The clinical outcome for septic patients, however, with any of the common treatments may remain unaffected.

Bibliography

Berger, SA: Pseudobacteremia due to contaminated alcohol swabs. J Clin Microbiol 18:874, 1983.

Berger SA, Siegman-Ingra Y, et al: Group Ve-1 septicemia. J Clin Microbiol 17:926, 1983.

Billa JL, Stockman L, et al: Evaluation of lysis-centrifugation system for recovery of yeasts and filamentous fungi from blood. J Clin Microbiol 18:469, 1983.

Cornelis G, Laroche Y, et al: *Yersinia enterocolitica,* a primary model for bacterial invasiveness. Rev Infect Dis 9:64, 1987.

Cros A: *Pseudomonas aeruginosa.* In Mandell GL, Douglas RG, Jr, Bennett JE (eds): Principles and Practice of Infectious Diseases. New York: John Wiley & Sons, 1979, pp 1705–1720.

Dougherty SH: Pathobiology of infection in prosthetic devices. Rev Infect Dis 10:1102, 1988.

Edminston CE, Schmitt DD, Seabrook GR: Coagulase-negative staphylococcal infections in vascular surgery: Epidemiology and pathogenesis. Infect Control Hosp Epidemiol 10:111, 1989.

Hermans PE, Washington JA, II: Polymicrobial bacteremia. Ann Intern Med 73: 387, 1970.

Kreger BE, Cravenn DE, McCabe WR: Gram-negative bacteremia. IV. Re-evaluation of clinical features and treatment of 612 patients. Am J Med 68: 344, 1980.

Peacock JE, Sorrell LA, et al: Nosocomial respiratory tract colonization and infection with aminoglycoside-resistant *Acinetobacter calcoaceticus varanitratus:* Epidemiologic characteristics and clinical significance. Infect Control Hosp Epidemiol 9:302, 1988.

Rackow EC, Astiz ME: Pathophysiology and treatment of septic shock. JAMA 266:548, 1991.

Reuben AG, Musher DM, et al: Polymicrobial bacteremia: Clinical and microbiologic patterns. Rev Infect Dis 11:161, 1989.

Rubin LG, Staiman K, Kamani N: Occult bacteremia with non-typable *Haemophilus influenzae.* J Clin Microbiol 15:1314, 1987.

Stamm WE: Infections related to medical devices. An Intern Med 89(Pt 2): 764, 1978.

Weinstein MP: The clinical significance of blood culture contaminants [editorial]. Clin Microbiol News 7:156, 1985.

Urinary Tract Infections

John G. Thomas

1. Define the different terms associated with urinary tract infections (UTIs).

2. Describe the clinical features and the associated symptoms.

3. Describe and differentiate between the major routes of infection.

4. Explain the prevalence of UTIs in certain age groups and gender populations.

5. Describe predisposing factors to UTIs.

6. Describe the appropriate samples for culture and interpretation of results based on the type of sample submitted.

7. List the organisms commonly associated with UTIs.

8. Discuss the interpretation of urine culture results based on bacterial colony count, pyuria, and symptoms and signs presented by the patient.

9. Name the different methods of laboratory diagnosis:

   ■ conventional methods
   ■ rapid detection and screening tests

OVERVIEW

Urinary tract infection (UTI) is one of the most common infections that afflicts humans. The pathogenesis and course of UTIs are greatly influenced by the anatomy of the organs involved (Fig. 31–1), which include the urethra, bladder, ureters, prostate, and kidneys. It is therefore practical to separate UTIs into upper UTIs and lower UTIs. Upper UTIs involve the renal parenchyma (pyelonephritis) or the ureters (ureteritis). Lower UTIs involve the bladder (cystitis), the urethra (urethritis), and, additionally in males, the prostate (prostatitis).

The heterogeneity of disease presentation, management, and prognosis is reflected in the terminology of UTIs. Table 31–1 lists terms and definitions frequently discussed in connection with UTIs. Each has specific criteria and needs to be used appropriately.

There are two clinical schemas for classifying UTIs. These are single-episode versus recurrent, and complicated versus uncomplicated. A single-episode UTI occurs once, resolves spontaneously or through the use of antibiotics, and does not recur. Patients with chronic or recurrent UTIs have repeated episodes of bacteriuria with or without clinical manifestations. These are arbitrarily divided into relapse and reinfection. The former involves the same organism and implies a focus of infection in the renal or prostatic paren-

chyma; the later implies a different organism and is usually limited to the bladder.

Uncomplicated UTIs occur primarily in sexually active young females without genitourinary (GU) abnormalities and are usually caused by antibiotic-susceptible bacteria. Complicated UTIs occur in individuals who have one or more structural or neurologic GU abnormalities and have indwelling catheters and whose conditions cannot be controlled with therapy. These patients are often hospitalized.

Bacteriuria, which can be symptomatic or asymptomatic, is the presence of bacteria in the urine. Disease occurs when the multiplication of organisms in the urinary tract interferes with the normal function of the involved organ. It is important to remember that infection is defined by clinical parameters and is not defined solely by quantitation and/or identification of microbes.

This chapter discusses the following:

■ variety of infections that occur in the urinary system

■ clinical parameters that define each of the disease manifestations

■ epidemiology and risk factors associated with the development of a UTI

■ laboratory diagnosis of UTIs, including specimen collection, screening methods, and interpretation of colony counts

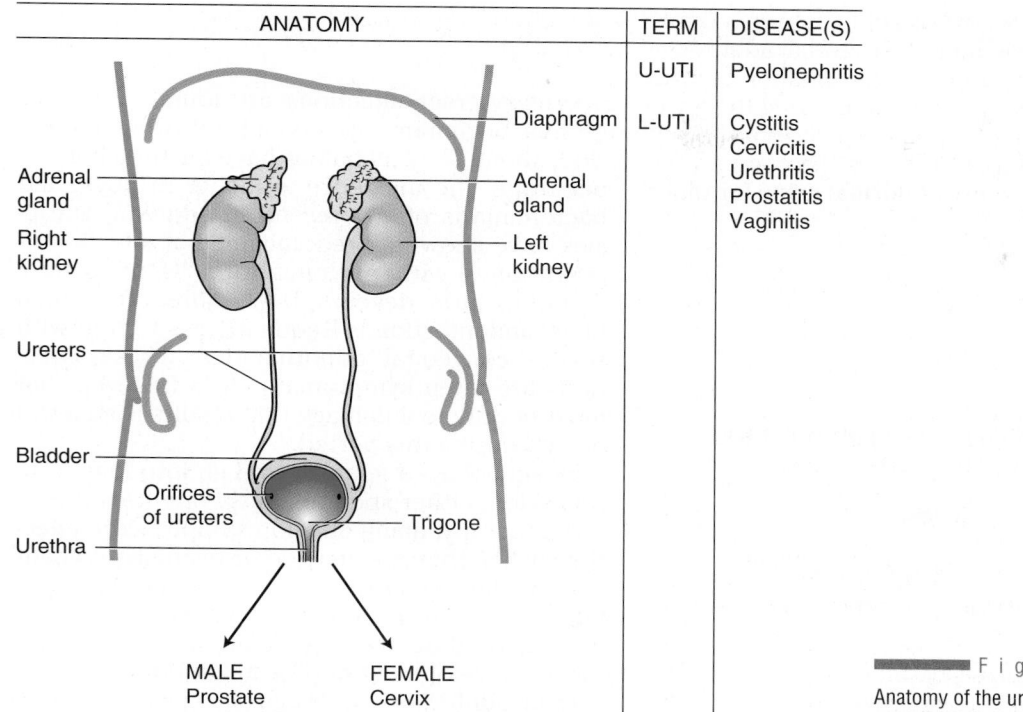

ANATOMY	TERM	DISEASE(S)
	U-UTI	Pyelonephritis
	L-UTI	Cystitis
		Cervicitis
		Urethritis
		Prostatitis
		Vaginitis

■■■■■ F i g u r e   3 1 – 1

Anatomy of the urinary tract with corresponding terms and diseases.

The Urinary System

Except for the urethral mucosa and the renal medulla, which appear to be relatively susceptible to infection, the normal urinary tract is resistant to colonization and subsequent infection by bacteria. The urinary tract efficiently and rapidly eliminates both virulent and avirulent microorganisms.

Although urine is frequently considered a good culture medium, the extremely high urine osmolarity (concentration) and low pH levels are inhibitory for the growth of many uropathogens and almost all normal bacterial flora of the urethra. In addition, very dilute urine fails to support the growth of most bacterial species. In terms of antibacterial activity, urine from men is more inhibitory than urine from women, owing to the presence of prostatic fluids in the urine of men as well as the difference in pH and osmolarity.

However, conditions such as high ammonia concentration, hyperosmolarity, lowered pH, and sluggish blood flow in the renal medulla can contribute to reduced leukocyte chemotaxis and bactericidal activity of white blood cells, resulting in lowered resistance. Urine itself has also been shown to inhibit the migrating, adhering, agitating, and killing function of polymorphonuclear

■■■■■ T A B L E   3 1 – 1

DEFINITIONS OF TERMS AND ABBREVIATIONS COMMONLY USED FOR URINARY TRACT DISEASES

Urinary Tract Infection (UTI)
A spectrum of diseases caused by microbial invasion of the genitourinary (GU) tract that extends from the renal cortex of the kidney to the urethral meatus (see Fig. 31–1)

Upper Urinary Tract Infection (U-UTI)
A GU tract infection that is limited to the renal parenchyma (pyelonephritis) or the ureters (ureteritis). It is often accompanied by lower urinary tract (L-UTI) symptoms in addition to costovertebral (CV) flank pain or tenderness and fever. At times, L-UTI precedes the appearance of fever and U-UTI by 24 to 48 hours

Lower Urinary Tract Infection (L-UTI)
A GU tract infection that is limited to the urethra (urethritis), bladder (cystitis), and, in males, the prostate (prostatitis). These infections generally present in adults with dysuria (pain in urination), increase in frequency, urgency, and occasionally suprapubic tenderness

Acute Urethral Syndrome
Includes dysuria and pyuria. Defined as more than 8 leukocytes per mm^3 of uncentrifuged urine or approximately 2 to 5 leukocytes/hpf in centrifuged urine sediment

Prostatitis
A GU infection in males, involving the prostate; fever often present

Cervicitis
Inflammation of the cervix; may occur as an acute or a chronic presentation. Causative agents include sexually transmitted organisms, such as *N. gonorrhoeae* and *C. trachomatis*. Symptoms include dysuria, urgency, vaginal discharge, and low back pain

Bacteriuria
The presence of detectable bacteria in the urine. Patients may be symptomatic or asymptomatic (geriatrics and pregnancy)

cells. But the presence of acid-labile mucopoly-saccharides, which inhibit bacterial adherence, and the flushing action of the bladder provide additional defensive mechanisms along the lower urinary tract that overcome these immunosuppressive effects. Table 31–2 lists the comparative physiologic parameters of normal urine for different populations.

■■■■ T A B L E   3 1 – 2

COMPARATIVE PARAMETERS FOR URINE IN CONTROL SUBJECTS AND PATIENTS WITH UTIs

Parameter	Ranges		
		Abnormal	
	Normal	CYSTITIS	PYELONEPHRITIS
Chemistries			
Specific Gravity	1.001–1.035		
Vol, average/24 hr			
child (1–14)	500–1,400		
Adult (<60)	600–1,800		
Adult (>60)	250–2,400		
pH range	4.7–8.0 (6.0 average)		
Protein	Negative to trace		Increased
WBC esterase	Negative		
Nitrite	Negative	Positive	Positive
Microscopic			
WBCs			
Male	0–3/hpf	Variable	Elevated to greatly increased
Female/child	0–5/hpf	Variable	Elevated to greatly increased
RBCs	0–2/hpf	Variable	Variable to greatly increased
Epithelial cells			
Squamous	Variable/hpf		
Renal	0–1/hpf		
Transitional	0–2/hpf		
Crystals	Variable	Negative	Negative
Mucus	Variable	Negative	Negative
Casts			
Hyaline	0–2/lpf		
Granular	0–1/lpf		
WBC	Negative	Negative	Positive
Microorganisms			
Bacteria	<1/hpf	Variable	Positive
Yeast	Negative	Variable	Variable
Trichomonas	Negative	Variable	Negative

EPIDEMIOLOGY AND RISK FACTORS

Urinary tract infections are found in all age groups, beginning with neonates. During that period, about 1% of all babies have bacteria in bladder urine; the incidence is higher in boys, and bacteremia is often present. In addition, autopsies have shown a predominance of infection in infant boys with pyelonephritis. Of preschool children, girls develop UTIs more often than boys, and infection is frequently associated with severe congenital abnormalities. These infections are often symptomatic. It is believed that most of the renal damage that results from a UTI occurs during this period.

Of school-aged females who go into long-term remission, either spontaneously or with antibacterial therapy, many develop symptomatic infections when they are married or become pregnant at a rate far higher than that of the general population. Thus, the presence of bacteria in the urine in childhood defines a population at higher risk for the development of UTIs in adulthood.

From adulthood to the age of 65 years, the incidence of UTIs in men is extremely low and is frequently associated with anatomic abnormalities or prostatic disease, with resultant instrumentation. During this time, however, as many as one fifth of women experience a symptomatic UTI. In patients older than 65, the frequency of UTIs increases dramatically for both sexes, and there is a progressive decrease in the female-to-male ratio. The increase in frequency of UTIs in men is related to obstructive uropathology from the prostate and loss of the bactericidal activity of prostate secretions. Bladder prolapse in women and soiling of the perineum from fecal incontinence contribute to the occurrence of infections in women afflicted with dementia, while neuromuscular diseases and increased instrumentation in bladder catheterization are contributing factors in both sexes. The time course of UTIs is diagrammatically shown in Figure 31–2.

The epidemiology of UTIs is influenced by the pathophysiology of the infections as well as the virulence of the isolates and the immune status of the host. The last is complicated by predisposing factors and manipulation, particularly in hospitalized patients. This relationship is diagrammatically represented in Figure 31–3, where a dynamic relationship is established among three factors:

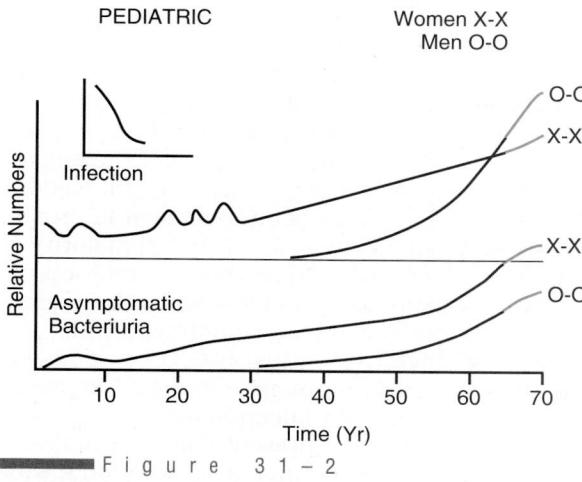

PEDIATRIC   Women X-X
                             Men O-O

Relative Numbers

O-O
X-X
Infection
X-X
O-O
Asymptomatic
Bacteriuria

10   20   30   40   50   60   70

Time (Yr)

■■■■■ F i g u r e   3 1 – 2
Frequency of UTI over time.

- microbes and their inherent mechanisms of pathogenicity

- hosts and their underlying predisposing factors for impaired immunity

- selective external pressures

Table 31–3 lists well-defined microbial virulence factors, and Table 31–4 shows nonspecific and genitourinary specific factors that have an impact on host defense and the ability of the patient's immune system, both humoral and cellular, to resist infection. This dynamic interaction is continually changing and hence cannot be defined by organism or colony count alone; assessment of the total clinical situation is required. The predisposing factors frequently associated include pregnancy; indwelling or intermittent bladder catheterization; urinary tract instrumentation, manipulation, or obstruction; and underlying conditions such as diabetes mellitus. Sexual activity in younger women and incontinence in the geriatric population are additional risk factors. Nonspecific factors that can significantly enhance (either directly or indirectly) the virulence of bacteria are the birth control pill, alcohol, smoking, and antibiotics.

Hospitalized patients develop UTIs more frequently than outpatients. The general ill condition of the hospitalized population and the higher probability of urinary tract instrumentation are major contributors to this difference. While diabetes mellitus has been found to be a predisposing factor in UTIs in some studies, others have found no difference between normal indi-

3. External/selective pressures: predisposing factor

$H_2+O_2 \rightleftharpoons H_2O$

Cohabitation:
no disease
NF

Pathogens:
produce
disease

1. Microorganisms:
   (virulence) bacteria,
   viruses, fungi,
   parasites

2. Host
   resistance to
   infection

■■■■■ F i g u r e   3 1 – 3
Dynamic interactive relationship between microorganisms, host resistance, and external/selective pressures.

viduals and diabetics. The more frequent catheterization that hospitalized diabetic patients undergo may account for the differences. Black women with the sickle-cell trait have been reported to have a higher a prevalence of UTIs during pregnancy than black women who do not have sickle cell trait. This difference may be related to the effects of local tissue hypoxemia that may ensue from the occlusion of renal medullary blood vessels by induced sickled erythrocytes. Chronic potassium deficiency, gout, hypertension, and other conditions causing interstitial renal disease have been associated with UTIs, but without adequate documentation.

■■■■■ T A B L E   3 1 – 3
LIST OF MICROBIAL VIRULENCE FACTORS

Adherence (bacterial adhesions)
Calculi formation (kidney stones)
Toxin production
Lipopolysaccharides
Capsular polysaccharide
Hemolysins

TABLE 31-4

HOST DEFENSES AND EXTERNAL, SELECTED PRESSURES THAT INFLUENCE THE OUTCOME OF INTERACTION BETWEEN BACTERIA AND THE URINARY TRACT

	Nonspecific	Genitourinary Specific
Host defense (resistance)	Intact skin	Diabetes mellitus
	WBC function	Pregnancy
	Age	Sickle cell (black women)
	Neoplasia	Neuromuscular diseases
	Immune competence	Structural abnormality
	Hormonal changes	Gout (?)
	Pregnancy	Hypertension (?)
	Malnutrition	Potassium deficiency (?)
	Malaria, diabetes	
	Psychological state	
	Attitude	
	Immunodeficiency syndrome	
Selective pressures	The birth control pill	Sexual activity
	Smoking	Incontinence
	Alcoholism	Bladder catheterization, indwelling or intermittent
	Anesthesia	
	Drug addiction	
	Immunosuppressive drugs	Instrumentation
	Irradiation, therapeutic	Urinary stone
	Antibiotics	

WBC, white blood cell.

CLINICAL SIGNS AND SYMPTOMS

Neonates and children younger than 2 years of age with UTIs usually present with nonspecific symptoms, including failure to thrive, vomiting, and fever. Children older than 2 years of age are more likely to display localizing symptoms, such as dysuria, frequency, and abdominal or flank pain. Adults with lower UTIs in which infection is limited to the urethra or bladder generally present with dysuria, frequency, urgency, and occasionally superpubic tenderness.

Upper UTIs, particularly in acute pyelonephritis, are often accompanied by lower urinary tract symptoms in addition to flank pain and tenderness and fever. At times, the lower urinary tract symptoms antedate the appearance of fever and upper urinary tract symptoms by 1 or 2 days. Bacteremia, when present, may help confirm a diagnosis of pyelonephritis or prostatitis.

Of critical importance is the recognition that these symptoms are not reliable indicators of infection. Cases of pyelonephritis may be asymptomatic or may present symptoms similar to those in lower UTIs. Pain radiating from the kidneys to one of the lower quadrants of the epigastric region may mimic appendicitis or gall bladder disease; however, fever is often present in prostatitis as well as pyelonephritis. The vast majority of elderly patients with UTIs are asymptomatic, whereas as many as 50% of women with frequency, urgency, and dysuria may not have a UTI.

Acute glomerulonephritis (AGN), a glomerulopathy that results from an immune response to *Streptococcus pyogenes* infection, either respiratory or pyodermal, may present clinical manifestations similar to those of upper UTI. Patients with AGN present with edema around the eyes and produce reddish-brown urine, giving a "Coca-cola" appearance because of hematuria. Red blood cells and red cell casts are usually found in the urine.

Although dysuria is the most common reason for obtaining a urine culture, this clinical presentation is neither sensitive nor specific. Dysuria may be present in infections with herpes simplex virus, *Chlamydia trachomatis,* or *Neisseria gonorrhoeae.* These organisms are not detected by the routine bacteriologic culture of urine. Many noninfectious conditions, including urethral inflammation from physical or chemical agents or from trauma, may also present similar symptoms. Flank pain and fever without lower urinary tract symptoms, and bacteremia without any urinary tract symptoms, are common in patients with indwelling urinary catheters. Figure 31–4 shows an evaluation schema of women with acute dysuria.

ETIOLOGY OF URINARY TRACT INFECTIONS

Pathogenesis of Urinary Tract Infections

Bacteria gain access to the urinary tract by three different routes, the ascending, the hematogenous, and the lymphatic pathways. Women acquire UTIs most frequently via the ascending route. Because of the shorter ureter in women, bacteria are easily introduced into the bladder by sexual intercourse. Once established in the bladder, bacteria ascend the ureters, probably aided in many cases by vesicoureteral reflux or by peristaltic dilated ureters resulting from intraluminal infection, an inflammation of the GU tract musculature.

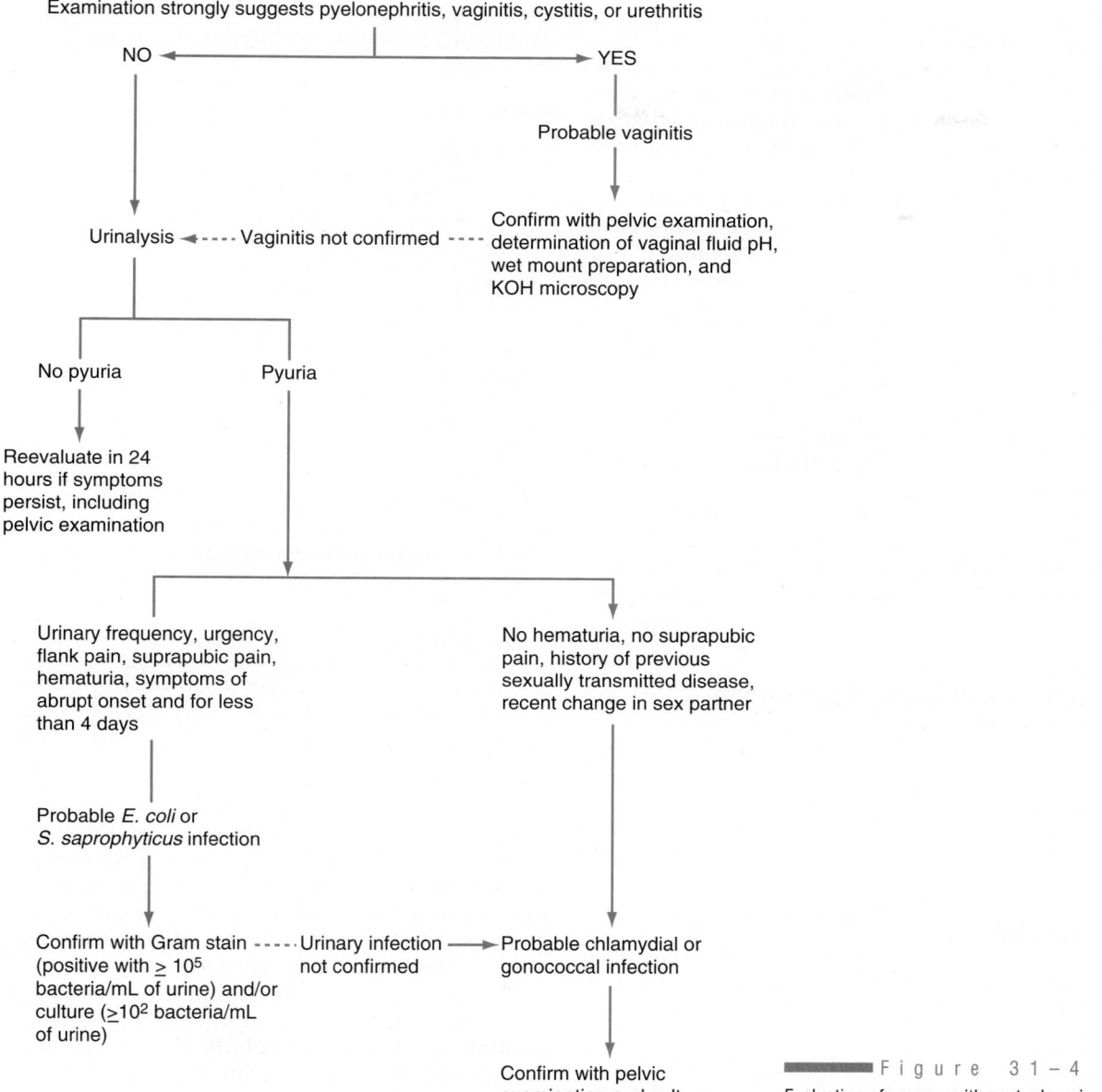

Examination strongly suggests pyelonephritis, vaginitis, cystitis, or urethritis

NO ← → YES

Probable vaginitis

Urinalysis ◄---- Vaginitis not confirmed ---- Confirm with pelvic examination, determination of vaginal fluid pH, wet mount preparation, and KOH microscopy

No pyuria   Pyuria

Reevaluate in 24 hours if symptoms persist, including pelvic examination

Urinary frequency, urgency, flank pain, suprapubic pain, hematuria, symptoms of abrupt onset and for less than 4 days

No hematuria, no suprapubic pain, history of previous sexually transmitted disease, recent change in sex partner

Probable *E. coli* or *S. saprophyticus* infection

Confirm with Gram stain ----Urinary infection → Probable chlamydial or (positive with ≥ 10^5 not confirmed   gonococcal infection
bacteria/mL of urine) and/or culture (≥10^2 bacteria/mL of urine)

Confirm with pelvic examination and cultures

Figure 31–4

Evaluation of women with acute dysuria.

Infection of the renal parenchyma by many species of gram-positive bacteria (particularly in patients with staphylococcal bacteremia or endocarditis, mycobacterial infection, and *Candida* infection) clearly occurs by the hematogenous route. Gram-negative infections rarely occur by the hematogenous route.

Increased pressure on the bladder can cause lymphatic flow to be directed to the kidney. However, evidence for a significant role for renal lymphatics in the pathogenesis of pyelonephritis is unimpressive. Of the three possible routes of infection, the ascending route is of paramount importance, and the hematogenous route offers a less frequent but significant pathway.

Etiologic Agents of Urinary Tract Infections

The normal bacterial flora of the paraurethral area consists mostly of *Staphylococcus epidermidis,*

Lactobacillus, Corynebacterium, Bacteroides, and viridans streptococci, as shown in Table 31–5. These organisms cannot grow well in urine and are not common causes of UTIs. Listed in descending order, by age or patient status, those listed first are most common, whereas those at the end are less frequently isolated.

Table 31–6 lists the more readily recognized agents of UTIs. Enterobacteriaceae and *Pseudomonas* are responsible for more than 85% of the infections. Enterococci and *Staphylococcus saprophyticus* are gram-positive bacteria most commonly responsible for UTIs. Table 31–6 also lists uropathogens that are emerging, and need to be recognized, given the dynamics of urinary tract pathogens, the population at risk, and selected pressures, as outlined earlier. These organisms present unique challenges to clinical microbiologists.

There are other significant organisms that may be encountered in urine cultures. Fungi, particularly *Candida albicans* and *Torulopsis glabrata,* are commonly associated with UTIs. Although

■■■■■■ T A B L E 3 1 – 5
FLORA OF NORMAL VOIDED URINE* DEFINED BY AGE AND PATIENT STATUS

  I. NEWBORN: Sterile
 II. 24 HOURS TO 3 DAYS:
     Staphylococci
     Enterococci
     Diphtheroids
     Mycobacterium smegmatis
III. 3 DAYS TO A FEW WEEKS:
     Lactobacillus acidophilus (Döderlein bacillus)
 IV. PREPUBERTAL:
     Micrococci
     Streptococci (α-hemolytic and nonhemolytic)
     Coliforms
     Diphtheroids
  V. ADULT:
     Lactobacillus acidophilus
     Staphylococcus epidermidis
     Streptococci (α-hemolytic and nonhemolytic)
     Escherichia coli
     Diphtheroids
     Yeasts
     Anaerobic streptococci
     Listeria species
     Clostridium species
 VI. PREGNANCY: Increased numbers of the following:
     Staphylococcus epidermidis
     Lactobacilli
     Yeasts
VII. POSTMENOPAUSAL:
     Similar to prepubertal flora

*Usually sterile or fewer than 1,000 colonies/mL.

■■■■■■ T A B L E 3 1 – 6
RECOGNIZED MICROBIAL AGENTS OF URINARY TRACT INFECTIONS

Most Frequent

Enterococci
Streptococcus agalactiae (group B streptococci)
Enterobacteriaceae
Pseudomonas
Streptococcus pyogenes (group A streptococci)
Staphylococcus aureus
S. saprophyticus
Candida species

Less Frequent

Gardnerella vaginalis
Ureaplasma urealyticum
Mycoplasma hominis
Mobiluncus
Leptospira
Mycobacterium species
Chlamydia trachomatis (males)

Often Associated With Multisystem Diseases

Salmonella (with gastroenteritis)
Schistosoma haematobium
Cryptococcus neoformans
Trichosporon beigelii
Trichomonas vaginalis
Aspergillus
Penicillium
Adenovirus
Herpes simplex virus

anaerobes are the predominant normal bacterial flora in the urethra, they are not commonly responsible for UTIs. These microbes, however, have been determined to be occasionally significant, particularly when isolated from samples obtained by suprapubic aspiration.

Chlamydia trachomatis and *N. gonorrhoeae* can cause the symptoms of urethritis, cystitis, and prostatitis. This is particularly an issue in women, because in them it is clinically difficult to distinguish between urethritis and cervicitis. Recently, *Mycoplasma* and *Ureaplasma* have been associated with UTIs, particularly in neonates from lower socioeconomic groups.

Historically, mycobacteria have been associated with UTIs. This finding may have added significance in patients infected with the human immunodeficiency virus (HIV) or in otherwise immunosuppressed patients.

The isolation of *Gardnerella vaginalis* in urine commonly represents vaginal contamination, but this organism is also an emerging urinary tract pathogen. Although the role of this microbe in

UTIs has not been clearly established, repeated cultures with this organism as the primary isolate should not be ignored.

Certain organisms, particularly *Salmonella,* will be involved with multiple organ systems; these may be part of primary disease or may present secondarily.

Last, recognize that viral agents, particularly adenovirus and herpes simplex virus, have been associated with cystitis and, in selected circumstances, need to be identified when the clinical situation warrants it. Table 31–7 presents organisms associated with UTIs, listing the most frequent uropathogens associated with a particular clinical presentation or disease syndrome.

LABORATORY DIAGNOSIS

Significance of Colony Counts: A Historical Background

Since 1956, interpretation of quantitative urine cultures has been viewed as being among the most straightforward and simplest of laboratory tests to diagnose UTIs. It was "dogma" that a finding of 10^5 colony-forming units per mL (CFU/mL) or more was "positive." In his 1956 classic study, Edward Kass showed a clear separation between the number of bacteria in the urine of asymptomatic or symptomatic women with pyelonephritis and those who were uninfected. Significantly, 95% had colony counts greater than 10^5 CFU/mL when infected.

In retrospect, there were some very definitive parameters in this study:

■ all specimens were collected by catheterization

■ the vast majority of asymptomatic women had counts of fewer than 10^3 CFU/mL

■ prevalence of infection was only 6%

The sensitivity was only 51% for women with clinical pyelonephritis who had colony counts of more than 10^5 CFU/mL.

In the 60's and 70's, additional studies began to erode the "absolutism" of Kass' original work. A number of investigators showed that for women with symptomatic, acute lower UTIs (urethra, bladder, or both), 29 to 45% had colony counts of fewer than 10^5 CFU/mL by suprapubic aspiration. These women had dysuria and pyuria yet unex-

■ T A B L E   3 1 – 7

MOST COMMON UTI ETIOLOGIC AGENTS ASSOCIATED WITH FREQUENT CLINICAL PRESENTATION (DISEASE SYNDROMES)

Clinical Presentation	Common Etiologic Agents
Upper Urinary Tract Infections	
Acute pyelonephritis	Enterobacteriaceae
	Staphylococcus aureus
Subclinical pyelonephritis	Coagulase-negative staphylococci
	Candida spp
	Mycobacterium spp
	Mycoplasma hominis
Lower Urinary Tract Infections	
Acute bacterial cystitis	*E. coli*
	Klebsiella spp
	Other Enterobacteriaceae
	Enterococci
	Coagulase-negative staphylococci
Urethritis	
Acute urethral syndrome	*Chlamydia trachomatis*
	Neisseria gonorrhoeae
	Ureaplasma urealyticum
Other Infections	
Gonococcal urethritis	*Neisseria gonorrhoeae*
Chlamydial urethritis	*Chlamydia trachomatis*
Vaginitis	
Prostatitis	
Symptomatic bacteriuria	
Catheter-associated (hospital-associated) UTI	*E. coli*
	Klebsiella spp
	Proteus mirabilis
	Pseudomonas
	Candida spp
Chronic or recurrent (inpatient/outpatient) UTI	Adherent *E. coli*

pectedly "negative cultures" by traditional "Kass-based" criteria, that is, more than 10^5 CFU/mL.

In 1980 the "acute urethral syndrome," or urethritis, was more clearly defined as one of the three causes of acute dysuria with pyuria. The other causes were vaginitis and cystitis. The causative organisms included classic coliforms and *Staphylococcus saprophyticus* at colony counts of more than 10^3 but less than 10^5 CFU/mL. Simultaneously implicated in sexually active women were the emerging nongonococcal urethritis pathogens *Chlamydia trachomatis* and *Ureaplasma urealyticum.* Thus, in women with UTIs as many as half had urethral syndrome—not cystitis—and were "culture-negative" by traditional "laboratory" methods.

In 1982 Stamm and coworkers restudied the diagnostic criteria for women with acute, symptomatic lower UTIs. In contrast with Kass' classic

■■■■■T A B L E   3 1 – 8

COMPARISON OF COLONY COUNTS FOR ACUTE LOWER URINARY TRACT COLIFORM INFECTION

Investigator	Test Mid-stream Urine (cfu/mL)	Sensitivity	Specificity	Predictive Value	
				Positive	*Negative*
Stamm (1982)	≥10^2 coliforms/mL	0.95	0.85	0.88	0.94
Kass (1956)	≥10^5 coliforms/mL	0.51	0.99	0.98	0.65

work, Stamm's study included coliforms at counts of more than 10^2 CFU/mL. The criteria of 10^2 CFU/mL provided a sensitivity of 0.95 and a specificity of 0.85 (Table 31–8). The presence of pyuria in uncentrifuged urine specimens was a sensitive adjunct. The prevalence of coliform infections was 36% in women evaluated by Stamm and colleagues (1982), compared with Kass' 6%.

Finally, in the late 1980's, because of the emerging significance of pyuria, investigators re-evaluated the accuracy of the classic urinary sediment microscopic examination (UA). All agreed that urinalysis screening for detection of UTI was inherently inaccurate, was not reproducible, and did not correlate with clinically proven UTIs. It was felt that microscopic examination of urine specimens should be reserved for the detection of casts and crystals.

Further, investigators could not correlate UA determinations of white blood cells with the actual leukocyte excretion rate of white cells per mm^3 measured by hemocytometer chamber count. When clinical studies using the latter method of determining pyuria were reviewed, the following conclusions emerged:

■ Ten leukocytes/mm^3 or more occurs in fewer than 1% of asymptomatic, nonbacteriuric patients but in more than 96% of symptomatic men and women with significant bacteriuria.

■ Most symptomatic women with pyuria but without significant bacteriuria either have UTIs with bacterial uropathogens present in colony counts of fewer than 10^5 CFU/mL (>10^2 to <10^5 CFU/mL) or have infections with *Chlamydia trachomatis/Ureaplasma urealyticum*.

■ Women with asymptomatic bacteriuria probably should be divided into those with true asymptomatic infection associated with pyuria and those with transient self-limited bladder colonization and no pyuria.

■ Most patients with catheter-associated bacteriuria also have pyuria, hence, infection.

Simultaneously, an impregnated paper strip that measured urine leukocyte esterase was introduced and found to correlate well with hemocytometer chamber counts. The leukocyte esterase test (LET) was inexpensive and rapid (1 minute) and required no technical skills or equipment, just the classic "dipstick."

What did all this mean and how did it fit into the laboratory faced with a myriad of problems in interpreting uropathogen colony counts? The answer is simple. The universal application of a single breakpoint (≥10^5 CFU/mL) must be recognized as an artifact designed solely to aid the laboratory in the management of a large amount of specimens that are easily contaminated. There are no laboratory parameters that can assist in establishing the diagnosis of significant pyuria. Furthermore, urine cultures are requested not only in connection with the symptoms of acute UTI but also in the absence of specific symptoms, as a test of cure, in evaluating antimicrobial therapy, in detecting asymptomatic bacteriuria in pregnant women, and in the evaluation of bacteremia and/or fever, to name but a few. The fact is that the urine bacterial colony count cannot stand alone as a single criterion for evaluating the presence or absence of UTI.

This dilemma is the result of an inappropriate reliance on an artificial criterion to indicate the presence of a UTI without regard to the presence or absence of symptoms, predisposing factors, patient populations, and/or the type of organism(s) isolated. Until recently, the colony count was regarded as the "gold standard" for determining the presence of real and treatable UTIs; it now appears that this is not the case. In reality, there is no single "gold standard" test because there are several types of urinary tract and vaginal infections that result in dysuria. The outcome of a urine culture must therefore be evaluated together with other laboratory and clinical data; attempts to attach significance to the colony count should be restricted to the original patient population in which the significance was established, that is, asymptomatic individuals with pyelonephritis.

SPECIMEN COLLECTION

Prevention of contamination by normal vaginal, perianal, and interior urethral flora is the most important consideration for collection of a clinically relevant urine specimen. Nonetheless, it is still an important fact that physicians rely on colony counts. Therefore, all necessary precautions should be taken to ensure that the colony count represents the numbers of organisms present at the time the specimen was collected. It is incumbent on the laboratory to define specific criteria for collection and transport and, within their realm of responsibility, to ensure adherence to these protocols.

There is a wide variety of methods for collecting urine samples.

Voided Midstream Specimen Collection. With this most frequently employed method, specimens are collected by the patient. Contamination of the urine occurs with bacteria from the urethra unless the first portion of the voided specimen is discarded. Investigators have found that voided urine collection kits should contain instructions to the patient on proper specimen procurement, and these instructions should be read slowly to the patient rather than merely supplied. Patient education should be part of the specimen processing, since proper collection has much influence on the result of subsequent laboratory procedures. It has also been reported that stick-figure diagrams showing the manner in which a specimen should be collected relay much more meaning to the patient than written instructions, in view of the number of people who cannot read beyond a third-grade level.

Catheterized Specimen Collection. This invasive technique reduces the risk of contamination of urine with the urethral flora; however, because the catheter is passed through the urethra some contamination may occur. When specimens are collected from an indwelling urinary catheter, the catheter collection port should be cleaned with an alcohol pad and punctured directly with a needle and syringe. The specimen should never be collected from the drainage bag. Prior to collecting urine with a single, straight catheter, the urethral opening or vaginal vault is cleansed with a soap solution and rinsed with sterile water. The initial urine flow is discarded because it may contain organisms acquired as the catheter passed through the urethra. Samples obtained from an ileoconduit are collected from the stomal opening after the area has been gently swabbed with an alcohol wipe. The urine on the external appliance is never used for culture, since it is similar to the urine in a drainage bag in patients with indwelling catheters.

The bacteriologic results of separate urine specimens collected by cystoscopy (bilateral urethral catheterization) or by bladder wash-out are used to localize the infection to the upper or lower urinary tract and, in the former case, to the left or right kidney. Specimens obtained by straight catheterization, by bilateral urethral catheterization, by bladder wash-out, or from an ileoconduit may be submitted in a tube or broth unless an assessment of pyuria is necessary. When such a specimen arrives at the laboratory, it must be labeled as to location or timing, and this information must accompany the results for proper interpretation.

Suprapubic Aspiration. This is the definitive method for collecting uncontaminated specimens. Although most consider any organism isolated on these specimens to be clinically significant, this may not be totally correct because transient colonization of the bladder can occur. Suprapubic aspirations are collected primarily from infants and patients in whom the interpretation of the results of voided specimens is difficult. Suprapubic aspiration urine specimens are the only ones suitable for anaerobic culture. With the bladder full, the urine is collected with a needle and syringe following skin antisepsis.

In addition to the manner in which the specimen is collected, other parameters may have an impact on the suitability of the specimen:

Urine Volume. The volume of urine received is rarely a problem for routine bacteriologic culture. The detection of significant pyuria by sediment examination requires at least 10 mL of specimen; however, alternative methods of pyuria detection may be needed when smaller volumes are received. Furthermore, as much as 20 mL of urine may be required to recover mycobacteria or fungi. If these agents are strongly suspected and previous specimens of lesser volume were negative, a request for a single collection of at least 20 mL should be forwarded to the physician. Remember that 24-hour collections for urine specimens in microbiology are totally unsatisfactory.

Number of Specimens and Timing of Collection. The number of specimens and the timing of urine collection depend on the clinical state of the patient and the method used. A single clean voided specimen in a specifically symptomatic patient or a single specimen obtained by catheterization in a patient with specific symptoms is sufficient, if significant pyuria is demonstrable and culture yields a recognized uropathogen. In symptomatic patients antimicrobial therapy is often instituted immediately after urine collection and is not withheld so as to procure subsequent specimens. For optimal purposes of quantitation, it has been recommended that only first morning specimens be processed or, if such specimens are not available, that the urine be allowed to incubate in the bladder for as long as possible before collection to increase the bacterial density (minimum, 4 hours). This procedure appears to be a rather imprecise, and probably unnecessary because of the lack of significance of colony counts in symptomatic patients and the necessity for antimicrobial therapy.

In the absence of symptoms, a single first morning voided urine specimen from a pregnant woman is sufficient to detect asymptomatic bacteriuria. In this case, one or two more first morning specimens may be collected on separate days to demonstrate the significance of the isolates. Because quantitation is necessary for the diagnosis of asymptomatic bacteriuria and because an asymptomatic condition does not require immediate therapeutic intervention, these specimens should be limited to first morning collections. Urine specimens from other asymptomatic patient populations, except those demonstrating significant pyuria, cannot be reliably interpreted.

Significant pyuria in the absence of bacteriuria and symptoms suggest the possibility of renal tuberculosis. For optimal recovery of mycobacteria, three specimens should be collected on 3 to 6 consecutive days.

Surveillance cultures of urine from asymptomatic hospitalized patients with indwelling catheters cannot predict the onset of UTI or the etiologic agent. Knowledge of the specific organisms present on the ward is not necessary to prevent their spread if proper infection control procedures are universally employed during catheter care.

If the initial urine specimen from a symptomatic patient yields significant pyuria and a recognized urinary pathogen, ideally two subsequent "tests-of-cure-specimens" should be processed, one at 48 to 72 hours and another at cessation of therapy. Similarly, a "test-of-cure-specimen" is appropriate subsequent to the treatment of asymptomatic bacteriuria. Given today's concerns about cost and test utilization, this rarely occurs.

ADDITIVES

Additives to urine are designed primarily to preserve the bacterial density present at collection. These additives maintain the original colony count during ambient temperature transport until quantitative cultures can be accomplished, obviating a need for refrigerated transportation. Liquid preservatives have been associated with dilution errors and decreased recovery of organisms at 24 and 48 hours. Lyophilized preservation is less inhibitory and requires a urine volume of at least 3 mL to avoid dilution effect. The effects of preservatives on the detection of pyuria have not been adequately determined. Regardless of the preservative used, the maximum time from collection to processing should not exceed 24 hours. In addition, the effect of preservatives on selected manual or automated methodology—to be discussed later—has not been extensively evaluated.

The importance of the determination of pyuria in many cases, and the lack of evidence that urine preservatives allow an accurate leukocyte count after 2 hours, argue against their use for specimens from symptomatic patients. The dip-slide urine collection container offers a more reliable method of preserving bacterial density in specimens from asymptomatic pregnant women, or from patients for whom quantitative cultures are required, and permits a more rapid detection of the etiologic agent. This method may be particularly suited to doctors' offices. In the current cost-conscious environment, it has been recommended that all urine specimens in physicians' offices be cultured with the dip-slide method and held for 48 hours. If the patient does not respond to empirical therapy, these original specimens can be forwarded to the laboratory for identification and susceptibility testing. For patients in whom the appropriate antibiotic was selected, the specimens can be discarded.

SPECIMEN TRANSPORT

Urine, being an excellent supportive medium for the growth of most uropathogens, must be immediately refrigerated or preserved. Generally, urine should be refrigerated, received, and processed in the laboratory within 2 hours. Longer delays render examination for significant pyuria unreliable, and the extremes of pH and urea concentration and the presence of antimicrobial agents may adversely affect the recovery of uropathogens.

A specimen submitted on a dip-slide transported at room temperature may be received up to 18 hours after collection. Liquid urine submitted for diagnosis of asymptomatic bacteriuria for which an examination for pyuria is not requested may be refrigerated for up to 18 hours prior to processing.

Except for specimens submitted for the sole purpose of isolating mycobacteria or fungi, or the diagnosis of asymptomatic bacteriuria, refrigeration is not an optimal, or to some, an acceptable method of preserving urine specimens. Refrigeration cannot preserve the number of leukocytes beyond 2 hours, and there is no need to stabilize the bacterial density in urine for symptomatic patients for the purpose of quantitative culture.

MICROBIAL DETECTION

Specimen Screening: Rapid, Nonculture Methodologies

The ideal urine-screening system should identify all urine specimens from infected patients at a high negative predictive value and be rapid and cost-effective. Toward this goal, a number of manual and automated methods have been developed. To maximize the benefits of screening methodology, preliminary reports should be generated as soon as possible. Preliminary reports should consist of the degree of pyuria and, depending on the method employed, the presence or absence of bacteria of fungi. These results should reach the physician within 2 hours after collection of a urine specimen.

At this time, too, the report should indicate whether additional identification methods are being utilized and their turn-around time. Ideally, the report should also include guides to appropriate antimicrobial therapy based on "in-house" laboratory antibiograms tabulated over the past 6 months.

However, the need for screening and the extent of identification and susceptibility testing of isolates depend on the method of collection and the purpose for which the urine was submitted. Hence, a uniform and consistent means of communicating that purpose to the laboratory must be developed, so information useful to the physician can be generated in the time frame required.

Additionally, a number of key points need to be remembered when utilizing screening methods:

■ Screening methods capable of detecting bacterial densities of 10^5 CFU/mL or more are appropriate for the detection of asymptomatic bacteriuria in pregnant women.

■ Methodologies that detect both bacteriuria and pyuria may result in false-positive tests for asymptomatic bacteriuria because densities of fewer than 10^5 CFU/mL are detected if significant pyuria unrelated to a UTI is present.

■ False-positive results are more numerous with methods that test for more than one parameter, that is, bacteria and WBC counts.

■ Methods of detecting significant pyuria and bacteriuria with sensitivies of 50 to 100 leukocytes/mm^3 and 10^2 CFU/mL may be appropriate for screening voided urine specimens or specimens from indwelling urinary catheters in symptomatic patients.

■ Screening methods are not appropriate for urine collected by straight catheterization, cystoscopy, suprapubic aspiration, and bladder wash-out or for testing of cure specimens and specimens collected from ileoconduits.

Manual Urine Screening Methods

Table 31–9 lists various manual screening methods that are useful in detecting bacteria and leukocytes in urine.

■ T A B L E 3 1 – 9

VARIOUS MANUAL SCREENING METHODS, PRINCIPLE OF ASSAY, AND THRESHOLD OF DETECTION FOR UTIs

Screens	Principle	Reported Threshold of Detection (cfu/mL)
Manual		
Microscopy Direct, uncentrifuged or centrifuged (cytospin)	Recognition of organism morphotypes and Gram stain	≥1 organism/ OIF = ≥10^5
Chemical		
Enzymatic dipstick		
Nitrate reductase (Griess test)	Gram-negative bacteria reduce nitrates to nitrites	≥10^4
WBC/Leukocyte esterase	Measures presence of WBC enzyme	Equivalent to 5 WBC/hpf
Chemstrip LN	Combination testing of both nitrate and esterase assays	>10^4 to 10^5
Enzyme tube		
Uriscreen	Measures catalase present in both bacteria and somatic cells	>10^4 to 10^5
Colormetric particle filtration		
Filtra Check—UTI	Combination testing of both bacteria and WBCs by membrane filtration and detection utilizing safranin O dye	>10^4

OIF, oil immersion field.

MICROSCOPY

Detection of Bacteria. Gram stain of urine samples should be performed routinely when pyelonephritis is suspected, as it may reveal the etiologic agent, particularly if the urine is submitted as part of a urosepsis work-up. Stains of both uncentrifuged and centrifuged specimens may be prepared. Most recently, the cytospin technology has been found to be remarkably applicable to rapid urine microscopy. The presence of one or more bacterial cells per oil immersion field (OIF) in at least five fields in a smear of uncentrifuged urine correlates with more than 10^5 CFU/mL. If the uncentrifuged preparation is negative, the sedimented preparation for leukocyte examination should be stained. Bacterial cells seen in this preparation indicate a density of fewer than 10^5 organisms/mL and, in the presence of clinical findings of acute pyelonephritis, may suggest urinary obstruction or prenephric abscess. The presence of gram-positive or gram-negative bacteria or fungi assists in the selection of an appropriate antibiotic therapy. Acridine orange stain decreases the detectable threshold from 10^5 CFU/mL to 10^4 CFU/OIF.

Detection of Pyuria. Although more elaborate and precise methods of determining urinary concentation of leukocytes have been evaluated, it is rarely necessary to use these clinically for voided specimens. A sufficient method is microscopic examination of a wet mount of a urinary sediment resulting from centrifugation of 10 mL of a specimen at 200 rpm on a tabletop centrifuge for 5 minutes. At least five fields should be examined, and each leukocyte seen per high-power field (HPF) (40×) represents approximately 5 to 10 cells per cubic millimeter of urine. In this way, 5 to 10 leukocytes per HPF in the sediment is the upper limit of normal, representing 50 to 100 cells/mm^3. If a more precise method is required, the technique described by Brumfit may be used. This involves the examination of a fresh, uncentrifuged specimen in a hemocytometer chamber. More than 8 to 10 leukocytes/mm^3 indicates significant pyuria.

Detection of Fungi and Mycobacteria. The cells of yeasts can usually be readily identified by Gram stain, but the cells of other fungi (because of their varying size and unique forms) may be difficult to discern. If fungi are suspected clinically or from the Gram stain, a smear can be stained utilizing fluorescent microscopy and calcofluor white. This stain preferably binds chitin present in the cell walls of fungi and makes visu-

alization and identification easier. Cotton swabs may not be used to apply the specimen to the microscope slide, however, because the stains bind to the cotton fibers and fluoresce.

Examination of urine for acid-fast bacilli is productive only if restricted to a particular patient population. Because of the presence of nonpathogenic mycobacteria in the smegma, positive smears must be confirmed with culture.

CHEMICAL METHODS

These screening techniques include a variety of procedures, such as the nitrate reductase (Griess) test, WBC leukocyte esterase test, and Chem-strip. As with manual microscopy, these methods may be labor-intensive and insensitive for low-grade significant bacteriuria.

Recently, tube enzyme (catalase) and colorimetric particle filtration (safranin O dye) have been reintroduced into the clinical setting with modified, updated protocols. Both measure a combination of bacteria and extracellular products of white blood cells.

Automated Screening Methods

Table 31–10 summarizes various automated screening methods for bacterial detection in urine samples. Bioluminescence systems detect bacterial ATP. Such systems may be expensive, frequently require batching of specimens (therefore time delays), and have not been adequately evaluated for their efficacy in detecting low-grade bacteriuria and funguria. The popularity of automation has grown, however, as more laboratories search for rapid, same-day results and the means to eliminate negative urine cultures.

A number of photometry methods, including the Vitek system, have been developed to measure growth. If a significant number of organisms are present in the urine specimens rapid growth is detected in the nutrient medium. The clinical evaluations of all these systems are less than optimum because sensitivity for a low-grade bacteriuria had not been assessed. These systems are relatively expensive.

Particle-filtration systems, such as Bac-T-Screen, are used to trap organisms and WBCs on filters and then selectively stain the cells. These systems are very sensitive even for low-grade infections, are somewhat nonspecific, yield many false-positive results, and are relatively expensive.

■■■■■■ T A B L E   3 1 – 1 0

VARIOUS AUTOMATED SCREENING METHODS, PRINCIPLE OF ASSAY, AND THRESHOLD OF DETECTION FOR UTIs

Automated	Principle	Threshold of Detection (CFU/mL)
Bioluminescence UTI screen	Detects bacterial ATP utilizing enzymatic bioluminescent re-action of ATP with luciferin and luciferase	$>10^4$ to 10^5
Photometry Vitek	If significant number of organisms are present in urine specimen, they will grown in me-dium to a detectable concentration utilizing photometry	$>10^4$ to 10^5
Colorimetric particle filtration Bac-T-Screen	Automated combination testing for both bac-teria and WBCs by membrane filtration and detection utilizing safranin O dye	$>10^4$ to 10^5

Rejection Criteria

It is imperative that the laboratory establish, in concert with the various medical services, criteria to obtain optimal specimens. Specimens may be rejected because of inadequate or inappropriate method of collection or transport. These criteria demand strict adherence to guidelines for collection and transport of specimen. If the specimen does not meet these tailored guidelines for each institution, it is incumbent on the laboratory (as rapidly as possible) to inform the service and/or physician of the inadequacy of the specimen and the fact that the specimen will not be processed. Antibiotic therapy may not have been initiated, and a better specimen may be collected. Samples to be rejected include 24-hour urine specimen and Foley catheter tips; these should not be processed.

CULTURE FOR ETIOLOGIC AGENTS OF URINARY TRACT INFECTIONS

Two important factors govern the selection of culture methods for urine specimens in the laboratory. First, low numbers of bacteria may be present in specimens such as pyelonephritis with obstructing, perinephric abscess, prior antimicrobial therapy, and bacterial persistence while the patient is on antimicrobial therapy. In these cases, methods with appropriate sensitivity are necessary to detect these low densities. Second, organisms in deep-seated infections, such as upper UTIs, do not always behave as stock cultures. They are often in a hydrophilic state and do not emerge on direct plating of a specimen on agar. This has been shown in several studies in which the addition of a primary broth culture resulted in the recovery of more than twice as many β-hemolytic streptococci and tripled the number of catheterized kidney urine specimens in which significant organisms were detected.

Generally, routine urine culture should include plating onto one selective and one nonselective medium. Calibrated loops of 0.01 mL should be used—not 0.001 mL (1 μL) loops—because accurate quantitation with a low inoculum is difficult to obtain. Specimens should be routinely incubated 24 to 48 hours at 35°C, depending on the source and the patient diagnosis.

The methodology also needs to be adjusted to the specimen source and the underlying disease process.

Asymptomatic Bacteriuria. In the asymptomatic patient immediate microbial therapy is not necessary. Identification and susceptibility testing of isolates can be achieved by any conventional or automated method. The colony count should accompany any positive culture to indicate the diagnosis of asymptomatic bacteriuria. Since the most frequent organisms identified include *E. coli* and other rapidly growing members of Enterobacteriaceae, 24 hours of incubation at 35°C is sufficient.

Pyelonephritis. Urine specimens submitted from patients suspected of having pyelonephritis generally contain high numbers of bacteria. The microscopic examination of urine for leukocytes and bacteria rapidly provides therapeutically useful information. Because the antibiotic susceptibilities of the responsible organisms are variable and unpredictable, culture should be designed for optimal recovery. Plates should be incubated for 48 hours at 35°C, and methodology may include a drop of specimen inoculated into trypticase soy broth for optimal recovery.

Lower UTIs. Specimens from patients with symptoms of lower UTI should be processed in the same manner as that for suspected pyelonephritis.

If significant pyuria is found in the symptomatic patient, and no recognized urinary pathogen is detected, the laboratory should consider the presence of *Chlamydia trachomatis; Mycoplasma* species, including *Ureaplasma urealyticum;* and/or *Neisseria gonorrhoeae.*

Suprapubic Aspirates. Aspirates may contain bacteria that are likely to be present in low numbers and may include anaerobic species. Such specimens should be routinely inoculated on a blood agar plate, a MacConkey agar plate, and a trypticase soy broth for up to 48 hours. For recovery of *Gardnerella vaginalis,* chocolate agar is acceptable. Remember that anaerobic bacteria are recovered in approximately 1% of cases and therefore need not be sought routinely, but only following consultation.

Catheterized Specimens. Urine specimens obtained by straight catheterization, by bilateral ureteral catheterization, by bladder wash-out, or from ileoconduits require inoculation of both agar plates and liquid medium for maximum recovery.

Test of Cure Specimens. Specimens obtained 48 hours after the initiation of therapy or after the cessation of therapy will be sterile in the absence of indwelling bladder devices if therapy was successful. Growth of an organism other than the one recovered prior to initiation of therapy suggests superinfection if the isolate is a recognized urinary pathogen. Susceptibility tests should be performed routinely on organisms recovered from the test-of-cure specimens if the organism is known to develop resistance during therapy (i.e., *Pseudomonas*).

Prostatic Secretions. The etiologic agent of acute prostatitis is usually recovered from catheterized specimens, which should be cultured in the same manner as that for specimens from symptomatic males. In cases of chronic prostatitis, prostatic secretions are submitted as well as urethral urine and midstream voided urine specimens obtained before and after massage. Quantitative cultures are necessary for proper interpretation.

Certain etiologic agents suspected of being present may also dictate the method of processing for culture. For example, media should include detection of *Neisseria gonorrhoeae* and *Ureaplasma urealyticum* if these organisms are suspected. If fungal cells or hyphae are seen on wet mount or Gram stain, or if a fungal infection is suspected, several options are ideal. A fungal bottle containing Emmons modification of Sabouraud dextrose agar, and a Sabouraud dextrose agar with chloramphenicol, could be inoculated with a few drops of the sediment and incubated at room temperature for up to 2 weeks. If a dimorphic fungi is suspected, brain-heart infusion agar should be held at room temperature for up to 4 weeks.

Historically, yeasts have been detected within 24 hours, but it is also clear, as indicated earlier, that certain forms may need a total of 48 hours of incubation. There has been no clear-cut published study to evaluate which length of incubation is optimal for recovery of yeast forms, but anything less than 24 hours is suboptimal. Further, there has been no clear evaluation determining the clinical significance of colony count of yeast, although custom has dictated that numbers of yeast when isolated be reported. Low colony counts may be just as significant as higher colony counts.

When mycobacteria are suspected, the specimen should be decontaminated and inoculated to one Bactec bottle (Becton Dickinson, Cockeysville, MD) and one Löwenstein-Jensen slant and processed according to the standard acid-fast bacilli standard protocol.

Use of agar dipsticks and biplates should be avoided, if possible, because isolated colonies may not be available, thus delaying final identification and antimicrobial susceptibility testing (AST). These methods should be reserved for clinics or physician's offices where retrieval of specimens following therapy may be less than optimal and only cultures from patients not responding would need detailed identification.

INTERPRETATION OF RESULTS

Routine work-up of isolates and susceptibility testing must be tailored according to the patient at risk and the specimen type submitted. Figure 31–5 represents a flow diagram that takes into account three features considered in all UTIs:

■ colony count of a pure or predominant organism

■ measurement of pyuria

■ presence or absence of symptoms (dysuria, and frequency)

One must recognize, however, that no one scheme can fit all situations. This figure is organized via a cascade scheme utilizing a dichotomous key. This allows the final laboratory selection

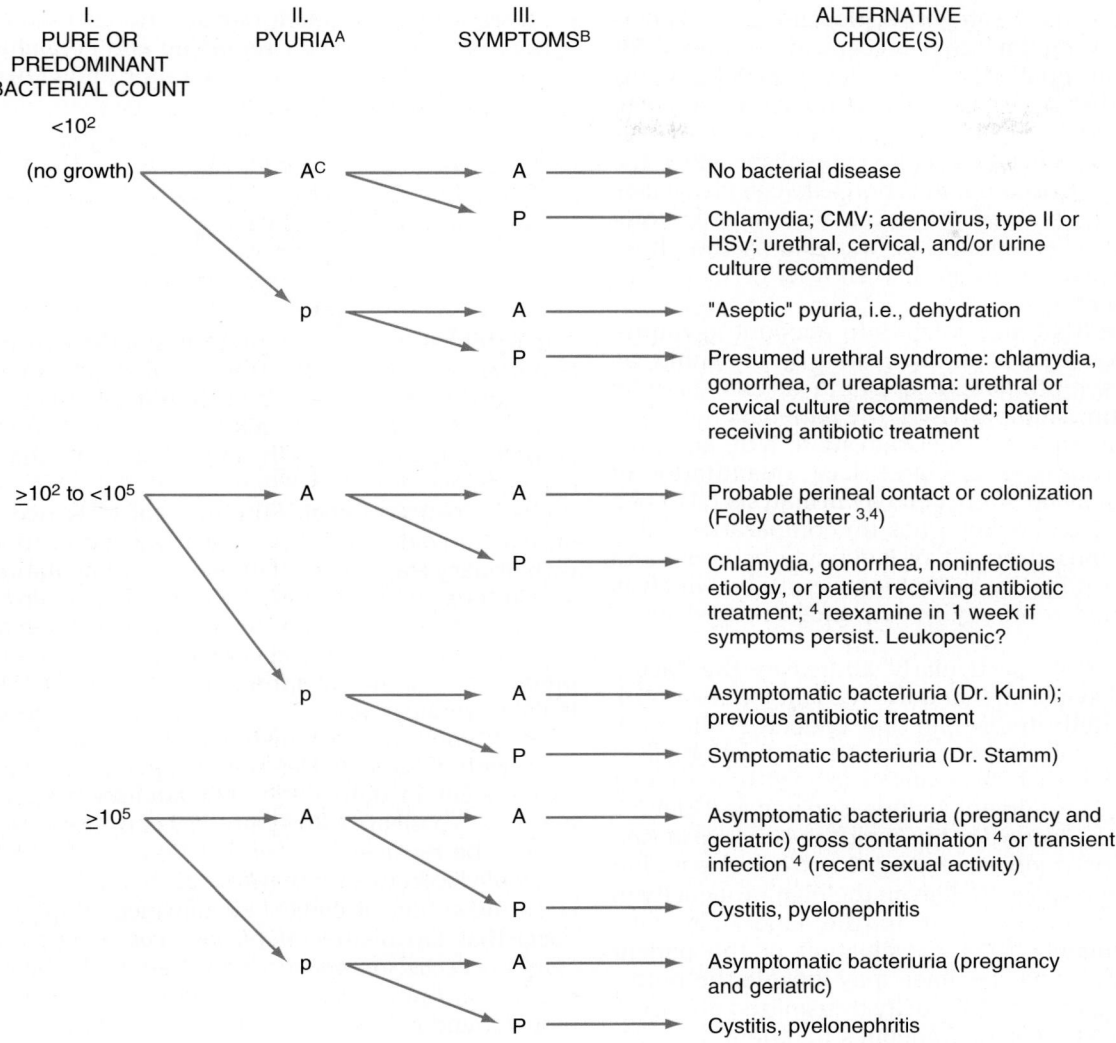

I. PURE OR PREDOMINANT BACTERIAL COUNT	II. PYURIA[A]	III. SYMPTOMS[B]	ALTERNATIVE CHOICE(S)
$<10^2$			
(no growth)	A[C]	A	No bacterial disease
		P	Chlamydia; CMV; adenovirus, type II or HSV; urethral, cervical, and/or urine culture recommended
	p	A	"Aseptic" pyuria, i.e., dehydration
		P	Presumed urethral syndrome: chlamydia, gonorrhea, or ureaplasma: urethral or cervical culture recommended; patient receiving antibiotic treatment
$\geq 10^2$ to $<10^5$	A	A	Probable perineal contact or colonization (Foley catheter [3,4])
		P	Chlamydia, gonorrhea, noninfectious etiology, or patient receiving antibiotic treatment; [4] reexamine in 1 week if symptoms persist. Leukopenic?
	p	A	Asymptomatic bacteriuria (Dr. Kunin); previous antibiotic treatment
		P	Symptomatic bacteriuria (Dr. Stamm)
$\geq 10^5$	A	A	Asymptomatic bacteriuria (pregnancy and geriatric) gross contamination [4] or transient infection [4] (recent sexual activity)
		P	Cystitis, pyelonephritis
	p	A	Asymptomatic bacteriuria (pregnancy and geriatric)
		P	Cystitis, pyelonephritis

[A]Leukocyte esterase (+); equivalent to 5 WBC/hpf

[B]Clinical dysuria and frequency

[C]A, absent; P, present

1. If patient is receiving antibiotic treatment, Gram stain, WBC analysis, and culture may not agree.

2. Quantitation of organisms and white cells by urinalysis on centrifuged urine is of no comparative value.

3. Interpretation for indwelling catheters, not established.

4. Plates will be held for 72 hours for consultation.

■■■■■ F i g u r e 3 1 – 5

Interpretation of urine culture results using algorithm based on bacterial colony count, pyuria, and symptoms.

from 12 clinical categories, depending on knowledge of symptoms and recognizing the need for physician input. Heretofore, most guidelines suggested that laboratory evaluation and culture set-up should be dependent on prior knowledge of symptoms. Since this is often unrealistic, given the general lack of communication between physicians and the laboratory, the schema presented allows the final differentiation to be made by the physician, based on his or her knowledge of patient symptoms.

Figure 31–4 also takes into account asymptomatic bacteriuria, as described by Dr. Kunin, as well as symptomatic bacteriuria, as explained by Dr. Stamm. Additionally, if the patient is receiving antibiotic therapy, the Gram stain, WBC analysis, and culture may not agree. Last, quantitation of organisms and white cells by urinalysis (UA) of a centrifuged specimen has no comparative value for the measurement of leukocyte esterase and bacteria done by microbiologic study, which is performed routinely on a noncentrifuged urine specimen.

This figure particularly addresses the "acute urethral syndrome" and the recognition that cystitis and urethritis are clinically difficult to differentiate, particularly in the female population. It is imperative that clinicians recognize that routine urine cultures do not include isolation and identification of *Chlamydia trachomatis*, *Neisseria gonorrhoeae*, or *Ureaplasma urealyticum*. It is also important to recognize that given the high propensity of negative cultures sent to the laboratory (i.e., approximately 50%) reevaluation of the patient with a negative specimen may include the potential recognition of a sexually transmitted disease.

Table 31–11 lists guidelines for the interpretation of urine cultures and suggests subsequent work-up. This table complements Fig. 31–5. In summary, these guidelines suggest the following:

■ Specimens with multiple uropathogens, that is, three or more, represent likely contamination. The organism should not be identified even if it is present in large numbers, unless this is specifically requested following consultation.

■ One or two significant uropathogens present (i.e., $\geq 10^5$ CFU/mL) should routinely be identified. Susceptibility tests should be performed.

■ One or two uropathogens present in small numbers (i.e., $\geq 10^2$ CFU/mL) should be routinely identified ($\geq 10^2$ to $<10^5$ CFU/mL) if the clinical situation warrants, such as in acute urethral syndrome or previous antibiotic therapy.

■ Plates should be held for an additional 48 to 72 hours following result reporting and consultation with the physician to tailor individual needs for all other criteria.

WHEN TO PERFORM ANTIMICROBIAL SUSCEPTIBILITY TESTING (AST) FOR UTIs

In establishing the interpretation of the culture result and evaluating the numerical significance, the bacterial species isolated should also be considered. The following may serve as helpful guidelines in evaluating the significance of organism species.

Gram-Positive Cocci. Enterococcal UTIs occur primarily in older males, particularly in association with urinary tract manipulation or instrumentation and in those with prostatic hypertrophy. *Staphylococcus* not *aureus* (SNA) occurs in two different populations. *Staphylococcus saprophyticus* is found predominantly in symptomatic sexually active females younger than 40 years of age. The spectrum of symptoms are indistinguishable from those seen with *E. coli* in the same population. Most cases occur in outpatients, but admission specimens from patients with symptoms of pyelonephritis may be received. The verification of SNA such as *Staphylococcus saprophyticus* in this population is an indication of clinical significance. Staphylococci that are neither *Staphylococcus aureus* nor *Staphylococcus saprophyticus* are frequently identified as *Staphylococcus epidermidis*, which are commonly found in hospitalized patients older than 50 years of age. Most often, these individuals have had recent urinary tract surgery, indwelling urinary catheters, or chronic urinary tract disease. *Staphylococcus epidermidis* is associated with disease only about 20% of the time.

On agar, young colonies of *Candida albicans* can resemble colonies of SNA and may be misidentified if Gram-stained smears are not examined. Since *Candida* species are frequently recovered from hospitalized patients with indwelling catheters, this incorrect identification results in a susceptibility report indicating broad antimicrobial resistance.

Rarely, particularly in neonates and in renal transplant patients, *Listeria monocytogenes* is recovered from urine. This organism may be nonhemolytic or β-hemolytic and catalase-positive and may resemble streptococci morphologically. Table 31–12 highlights important features of coagulase-negative staphylococcal UTIs.

■■■■■ T A B L E 3 1 – 1 1

GUIDELINES FOR INTERPRETATION OF URINE CULTURE RESULTS AND SUBSEQUENT WORK-UP

Colony Count (CFU/mL)[1]	Symptoms, Clinical Disease, or Patient Population[2]	Urine Source[3]	No. of Organisms, Types Isolated	Laboratory Work-Up Suggested[4]
$<10^2$		CV/CA	None	None[5]
$\geq10^2$	Pediatric	Suprapubic	≤2 organisms by anaerobic culture	ID & AST
$\geq10^2$	Symptomatic female, urethritis	CV	Pure culture	ID & AST
$\geq10^3$	Symptomatic male, prostatitis	CA	≤2 organisms	ID & AST
$\geq10^3$		CA Bladder wash-out	Pure culture	ID & AST ID & AST
$\geq10^5$	Cystitis/pyelonephritis	CV	Pure culture 2–3 organisms >3 organisms	ID & AST Q & SID Q & M or Q & GS

[1]Inoculation of 0.01 mL of urine is required to detect 10^2 cfu/mL.
[2]See Table 31–1 for description of clinical diseases, symptoms, and patient population.
[3]CV, clean voided; CA, straight catheterized.
[4]Work-up required. Any yeast may be quantitated and reported (regardless of no.); >100,000 needed to identify to species.
[5]See figures and text for suggested comments and educational information helpful to physicians.
ID & AST: Identification and antimicrobial susceptibility testing.
ID & S: Perform identification to genus and species level and do susceptibility testing when appropriate (lactobacilli may be identified on the basis of colony morphology and microscopic morphology: omit susceptibility testing).
Q & SID: Quantitate and perform sight identification. Identification and sensitivity not indicated. Hold plates 72 hours.
Q & M: Quantitate total amount of bacteria and report as "mixed urethral flora."
Q & GS: Quantitate and report Gram stain morphotypes.

Gram-Negative Bacilli. Most uncomplicated UTIs are caused by antibiotic-susceptible strains of *E. coli,* which emanate from the patient's own fecal flora. This may also hold true for most complicated infections, but hospitalized patients with indwelling bladder instruments are more likely to be colonized and subsequently infected with multiple antibiotic-resistant members of the Enterobacteriaceae derived from the hands of hospital personnel and contaminated solutions. As the duration of hospitalization and catheterization increases, *E. coli* is less likely to be isolated than organisms such as *Pseudomonas, Proteus, Klebsiella,* and *Enterobacter.* Complicated infections are often polymicrobic involving renal stones, and can yield urine specimens from which *E. coli* and *Proteus,* or *E. coli* and *K. pneumoniae* are alternatively or concomitantly recovered.

Gram-Positive Bacilli. The isolation of *Bacillus* species can almost without exception be considered contamination. True UTIs with the organism are exceedingly rare. The significance of *Clostridium* species is as difficult to assess in urine as it is in blood. There is insufficient literature concerning clostridial UTI, or the recovery of clostridia in urine in the presence of soft tissue abscess, on which to base a definitive statement of significance. Mycobacteria may infrequently be seen in Gram stains of urine and appear as gram-positive bacilli. *Listeria monocytogenes* may rarely be iso-

lated from the urine of infants with perinatal infection with this organism and in 1% of systemic infections and renal transplant patients.

■■■■■ T A B L E 3 1 – 1 2

URINARY TRACT INFECTIONS CAUSED BY COAGULASE-NEGATIVE STAPHYLOCOCCI

Characteristics of Infections	Organism	
	S. epidermidis	*S. saprophyticus*
Sex and age of affected patients	Men and women equal	Women 95%, 16–35 yr of age
Population at risk	Hospitalized patients with urinary tract complications	Healthy outpatients
Incidence	Common: 20% or more of all UTIs for hospitalized patients older than 50 years of age	Uncommon: 3.5% or fewer of all UTIs in hospitalized patients
Presentation	90% asymptomatic	90% symptomatic; indistinguishable from *E. coli* UTIs
Therapy	Often resistant to multiple	Responds readily to traditional urinary tract antimicrobials except nalidixic acid
Outcome	Bacteriuria often persists after therapy	Relapse rare; occasional reinfection

Diphtheroids, mycobacteria, and *L. monocytogenes* all cause diseases, predominantly in highly selected patient populations and almost always in association with bacteremia. If any of these organisms are recovered from urine or seen on smears, consultation with the physician permits some assessment of significance. Within the selected populations, cultures of blood or specimens from other sites may help determine the significance of the isolate.

Fungi. *Candida* and *Cryptococcus* are usually evident in culture within 2 to 5 days. Unless dimorphic fungi are suspected, fungal urine cultures can be discarded as negative after 2 weeks. Cultures for dimorphic fungi may require 4 to 6 weeks for colonies to appear. While the findings of candiduria may not require antifungal therapy, particularly in catheterized patients, properly collected urine yielding *Candida* species should be considered abnormal. Candiduria may be an indication of bladder or renal parenchymal infection, a urinary tract fungus ball, or disseminated candidiasis. Predisposing factors include diabetes mellitus, antibiotic and corticosteroid therapy, female gender, and disturbance of urine flow. The isolation of any other classic pathogenic fungi, that is, *Cryptococcus neoformans, Blastomyces dermatitidis, Coccidioides immitis,* and *Histoplasma capsulatum,* in the urine is a highly significant finding indicative of possible disseminated infection.

SUSCEPTIBILITY REPORTING

With the growing number of emerging uropathogens and the simultaneous increase in newer antibiotics, it is mandatory that laboratories utilize standardized methodology and report only appropriate antibiotics for UTIs. Table 31–13 lists the suggested grouping of FDA-approved antimicrobial agents for routine testing and reporting by clinical microbiology laboratories for urinary tract isolates. These antimicrobials are listed as group D supplemental for urine only, in the 1993 NCCLS Guideline: Performance Standards for Antimicrobial Disk Susceptibility Tests, 5th ed.; Approved Standard 13(24) NCCLS Document M2-A5.

A number of laboratories have tailored susceptibility testing according to the needs of the clinical setting: outpatient versus inpatient; pediatric versus adult; intensive care versus nonintensive care. Nevertheless, it is imperative to remember that attainable antibiotic blood levels and urine levels are often different; this will have an impact on the interpretation of some semiquantitative susceptibility results or quantitative minimal inhibitory concentration (MIC) of inhibitory antimicrobials.

Antibiograms

One of the primary functions of the clinical microbiology laboratory is to measure antibiotic

---

■■■■ T A B L E   3 1 – 1 3

ANTIMICROBIALS USUALLY TESTED FOR UTI BY CLINICAL MICROBIOLOGY LABORATORIES

Enterobacteriaceae	*Pseudomonas aeruginosa* and Other Non-Enterobacteriaceae	Staphylococci	Enterococci	Streptococci
GROUP D SUPPLEMENTAL FOR URINE ONLY*				
Carbenicillin	Carbenicillin	Lomefloxacin, norfloxacin, or ofloxacin	Ciprofloxacin	Norfloxacin
Cinoxacin	Ceftizoxime		Norfloxacin	Nitrofurantoin
Lomefloxacin, norfloxacin, or ofloxacin	Tetracycline	Norfloxacin	Nitrofurantoin	
	Lomefloxacin, norfloxacin, or ofloxacin	Nitrofurantoin	Tetracycline	
Norfloxacin or ofloxacin	Norfloxacin or ofloxacin	Sulfisoxazole		
Loracarbef	Sulfisoxazole	Trimethoprim		
Nitrofurantoin				
Sulfisoxazole				
Trimethoprim				

*See National Committee for Clinical Laboratory Standards (NCCLS) Guidelines, 1993, vol 13, no. 24, Document M2-A5.

resistance trends. This is most often accomplished utilizing an annual antibiogram that establishes cumulative percentage susceptibilities for selected bacterial-antibiotic combinations. It is important that the laboratory separate urinary tract susceptibility data from susceptibility data for blood and other sources; ideally, this can be further subdivided to unmask low-frequency, high-resistant isolates if the antibiograms reflect the various formularies mentioned earlier.

Bibliography

Clarridge JP, Pezzlo M, Vosti KL: Cumitech 2A. In Wessfield AL (coordinating ed): Laboratory Diagnosis of Urinary Tract Infections. Washington DC: American Society for Microbiology, 1987.

Johnson V, Stamm W: Urinary tract infections in women: Diagnosis and treatment. Ann Intern Med 111:906, 1989.

Kass EH: Asymptomatic infections of the urinary tract. Trans Assoc Am Physicians 69:56, 1956.

Kierkegaard H, Feldt-Rasmussen U, et al: Falsely negative urinary leucocyte counts due to delayed examination. Scand J Clin Lab Invest 40:259, 1980.

Kunin CM: Detection, Prevention, and Management of Urinary Tract Infections. Philadelphia: Lea & Febiger, 1974.

Lennette E, Balows A, Hausler W: Manual of Clinical Microbiology, 5th ed. Washington, DC: American Society of Microbiology, 1991.

Norden C, Kass E: Bacteriuria of pregnancy: A critical appraisal. Annu Rev Med 19:431, 1968.

Pezzlo M: Urine and culture procedure. In Isenberg H et al (eds): Clinical Microbiology Procedures Handbook, Washington, DC: American Society for Microbiology, 1992.

Pezzlo M: Detection of urinary tract infections by rapid methods. Clin Microbiol Rev 1:268, 1988.

Schaeffer AJ: Urinary tract infections in urology. Infect Dis Clin North Am 1(4):875, 1987.

Stamery TA: Pathogenesis and Treatment of Urinary Tract Infections. Baltimore: Williams & Wilkins, 1980.

Stamm W, Counts G, Pumming K: Diagnosis of coliform infections in acutely dysuric women. N Engl J Med 307:463, 1982.

Stamm WE, Wagner KF, Amsel R: Causes of the acute urethral syndrome in women. N Engl J Med 304:409, 1980.

U.S. Preventative Services Task Force: Recommendations on screening for symptomatic bacteriuria by dipstick urinalysis. JAMA 262:1220, 1989.

Winn W: Diagnosis of urinary tract infection: A modern procrustean bed. Am J Clin Pathol 99:117, 1993.

Sexually Transmitted Diseases

William F. Nauschuetz • *Boo H. Kwa* •
Michael A. Pentella

OBJECTIVES

1. Describe the clinical manifestations produced by the following agents:

   ■ *Neisseria gonorrhoeae*
   ■ *Chlamydia trachomatis*
   ■ *Gardnerella vaginalis*
   ■ *Treponema pallidum* subsp *pallidum*
   ■ *Haemophilus ducreyi*
   ■ Herpes simplex virus

2. Discuss the epidemiology and pathogenesis of each of the infections caused by the aforementioned agents.

3. Describe the proper specimen collection and laboratory methods used to diagnose the diseases caused by each of the previously listed organisms.

4. Differentiate the clinical characteristics between gonococcal and nongonococcal urethritis.

5. Interpret both specific and nonspecific serologic test results used to diagnose syphilis.

Sexually transmitted diseases (STDs), also referred to as venereal diseases, are infections acquired and transmitted primarily by sexual contact. Although the initial site of infection is usually the mucous membranes of the genitourinary tract, organisms that cause STDs may affect other body sites. Increased efforts to educate and encourage "safe sex" practices have not had a significant impact on the incidence of STDs in the United States. Sexually transmitted diseases remain a major public health issue as their incidence continues to increase.

The spectrum of diseases transmissible by sexual contact involves a large number of conditions, including acquired immunodeficiency syndrome (AIDS), gonorrhea, genital warts, genital candidiasis, trichomoniasis, genital herpes, syphilis, chancroid, lymphogranuloma venereum, granuloma inguinale, scabies, pediculosis pubis, molluscum contagiosum, hepatitis B, hepatitis A, amebiasis, and giardiasis. This chapter, however, discusses only the most common STDs. These include the exudative mucosal infections gonorrhea, chlamydiosis, and bacterial vaginosis, and ulcerative infections, such as syphilis, genital herpes, and chancroid.

GONORRHEA

CASE STUDY

A 24-year-old Latina female presented with fever, pain, and tenderness around her left elbow joint. The patient had noticed a slight rash around her left hand 10 days earlier. On examination, the patient showed rash on her extremities, swollen joints, and petechiae with vesicular eruptions. Based on the physical findings, the physician obtained an aspirate of the joint fluid and ordered routine and anaerobic cultures along with appropriate cultures for *Neisseria gonorrhoeae.*

The organisms that most commonly cause bacterial arthritis are *Staphylococcus aureus, Haemophilus influenzae, Neisseria gonorrhoeae, Pseudomonas* spp, *Escherichia coli,* and certain anaerobic organisms. With nongonococcal-mediated arthritis, most cases present with a source of the infection, usually skin abscesses, puncture wounds, and intravenous drug use. These presentations were not present in this patient. *Neisseria gonorrhoeae* was isolated from cultures of the joint fluid.

In fewer than 3% of gonococcal infections, the organism can spread through the blood, producing disseminated gonococcal infection characterized by a rash on the extremities and arthritis in one or more joints.

Gonorrhea is caused by the gram-negative coccus *Neisseria gonorrhoeae.* These organisms usually occur in pairs and have characteristic flattened adjacent sides, which is responsible for their kidney-bean appearance on gram-stained smear.

Although the incidence of gonorrhea in the United States has dropped steadily since the middle 1980's, gonorrhea continues to be the most common reportable bacterial infectious disease in the country today. The increasing incidence among adolescents and African American adults is being reported. Worldwide, gonorrhea has also continued to rise during the last 20 years. During recent years, the increased incidence of resistant strains of *N. gonorrhoeae* (penicillinase-producing [PPNG] and tetracycline-resistant [TRNG]) has raised serious concerns among public health officials and medical practitioners. Plasmid-induced penicillin-resistance by β-lactamase production of *N. gonorrhoeae* is detectable by the β-lactamase test. Other *N. gonorrhoeae* strains, however, produce altered penicillin-binding proteins. This chromosomally mediated resistance is detectable only by performing conventional disk diffusion or dilution susceptibility tests.

Clinical Manifestations

Humans are the only known hosts of *N. gonorrhoeae*. This organism attacks the columnar epithelial cells of the genitourinary tract, including the urethra, cervix, endocervix, and Bartholin glands. Columnar epithelial cells in the anal canal, pharynx, and conjunctiva may also be affected. The incubation period is 2 to 7 days after inoculation. The clinical presentations in males and females differ. In males, usually there is urethral inflammation, with pus and pain on urination. In females, the infection is much more likely to be asymptomatic and may remain undetected. The endocervix is the primary site of infection in females, and secondary sites may include Bartholin and Skene glands and the urethra. Vague symptoms including painful intercourse, irregular bleeding, abdominal pain, and light vaginal discharge occur in about 70% of the infected females. Only 10 to 20% of affected females show overt symptoms, such as endocervical mucopurulent discharge. If the disease goes untreated, the organism can ascend the genital tract and cause complications, such as prostatitis, epididymitis, and urethral stricture in males, and salpingitis and pelvic inflammatory disease (PID) in females.

Other clinical presentations in certain patient populations may occur in the rectum, pharynx, and conjunctiva. Rectal gonorrhea is usually asymptomatic and is acquired from receptive anal sex in homosexual males. Females with genital infection can acquire rectal gonorrhea from self-contamination. Pharyngeal gonorrhea is usually self-limiting and asymptomatic. This form of infection occurs in females and in homosexual males via fellatio. Gonorrheal ophthalmia neonatorum is acquired when an infant passes through the birth canal of an infected mother. Ophthalmic gonorrhea may occur in adults via autoinoculation.

Laboratory Diagnosis

Smears from genital specimens from males can be very helpful in the diagnosis of gonorrhea when gram-negative diplococci are found both intracellularly and extracellularly. However, smears of female genital specimens cannot be relied on to diagnose gonorrhea in female patients because certain members of the normal vaginal bacterial flora may resemble gonococci morphologically. Other means of diagnosis (i.e., culture or DNA probes) are required to confirm the diagnosis of suspected gonorrhea in women.

To improve the recovery of *N. gonorrhoeae* in the clinical laboratory, clinical specimens should be inoculated immediately onto appropriate media. Transport or holding media, such as Stuart or Amies, may be used if a delay in medium inoculation is anticipated, but these media should not be held for more than 12 hours or subjected to temperature extremes. *Neisseria gonorrhoeae* is susceptible to drying and requires an environment that contains 3 to 5% carbon dioxide. The Jembec system, a method preferred in many clinics and laboratories, includes a culture medium that supports the growth of *N. gonorrhoeae* and a carbon dioxide–generating tablet. The carbon dioxide–generating tablet is activated by the moisture that evaporates from the medium or by the addition of a drop of water. The culture is transported to the laboratory in a sealed plastic bag or envelope.

For isolation in the laboratory, gonococcal-selective media, such as Thayer-Martin or Martin-Lewis, are commonly used. After inoculation, plates should immediately be placed into an incubator with increased carbon dioxide (5%). Cultures must be examined daily for growth. Small, gray, transluscent, raised colonies become visible within 24 to 48 hours of incubation. Cultures with no visible growth must be held for 72 hours before discarding. In performing identification tests, fresh subculture of the organism must be used. This requirement is necessary because suspected *N. gonorrhoeae* organisms may produce autolytic enzymes that may be responsible for nonviable culture. There are several reliable methods available to identify suspected gonococcal isolates.

Conventional methods, such as carbohydrate degradation (both growth-dependent and pre-formed enzymes) and chromogenic enzyme substrate systems, and fluorescent-antibody and coagglutination (latex) tests are often used in most clinical laboratories. Isolates should be tested for β-lactamase production because of the increasing drug resistance of *N. gonorrhoeae.*

In addition, direct detection of *N. gonorrhoeae* from clinical samples is now available. The enzyme immunoassay (EIA) Gonozyme (Abbott Diagnostics, Abbott Park, IL) is a direct method for detecting *N. gonorrhoeae* from male urethral specimens and endocervical specimens. In men, this test is equal in sensitivity and specificity to a Gram stain. However, this method is not as sensitive and specific for endocervical specimens; therefore, positive EIA test results in women can be regarded only as presumptive and must be confirmed with culture. The direct nucleic acid detection method Pace 2 (Gen-Probe, San Diego, CA) is a nonisotopic DNA probe that hybridizes with gonococcal rRNA. This method, when compared with culture, has a 97% sensitivity.

The recommended antimicrobial therapy for uncomplicated urethral, endocervical, and rectal gonorrhea is a combination of ceftriaxone and doxycycline.

GENITAL CHLAMYDIOSIS

CASE STUDY

A 24-year-old female presented to her gynecologist for a routine pelvic examination. It had been several years since her last examination, and in the interim she had become engaged. Although she was sexually monogamous with her fiancé, she had had several sexual encounters in the past. She also desired to become pregnant. The physician found no abnormalities on the physical examination but decided to culture the cervix for gonorrhea and to perform a direct immunofluorescence test for chlamydia. A serum sample for human immunodeficiency virus (HIV) and syphilis was also obtained. All laboratory test results, except for the chlamydia test, came back negative.

Genital chlamydiosis, caused by *Chlamydia trachomatis,* is the leading bacterial STD in the United States today. *Chlamydia trachomatis* is an obligately intracellular bacterium and requires tissue cells for growth. This microbe causes a wide variety of conditions, including trachoma, lymphogranuloma venereum, and nongonococcal urethritis.

Genital chlamydiosis occurs in approximately 6% of asymptomatic males 18 to 25 years of age. Transmission of *C. trachomatis* is approximately 45% following a single exposure. The classic patient profile of genital chlamydiosis includes African American, unmarried male or female, nonuser of barrier or birth control, history of multiple sex partners, and usually history of other STDs.

The developmental cycle of *Chlamydia* is unique. The elementary body, the infectious but metabolically inert form of the organism, comes into contact with a host cell membrane and enters by pinocytosis. Once within a cytoplasmic vacuole, the elementary body enlarges, becoming an initial body. The initial body divides by binary fission and aggregates into a large inclusion body. The initial bodies then revert to elementary bodies and burst from the infected cell to invade other cells.

Clinical Manifestations

Chlamydia trachomatis causes 30 to 50% of all cases of nongonococcal urethritis (NGU) in males. There is a reported 30% coinfection rate with gonococcal urethritis in some patient populations. The incubation period for *C. trachomatis* urethritis in males is 10 to 20 days, in contrast with that of gonorrhea, which is usually 2 to 7 days. Twenty-five percent of infected patients, however, exhibit no symptoms. When manifestations are present, they are clinically similar to those of gonococcal urethritis (GU). Dysuria, which may be milder than that in GU, may develop, and urethral discharge may generally be less purulent. Nevertheless, the diagnosis should not be made solely on clinical presentation; definitive diagnosis requires laboratory testing. Table 32–1 shows the comparison of clinical characteristics of gonorrhea and NGU.

■■■■■ T A B L E   3 2 – 1

COMPARISON OF CLINICAL CHARACTERISTICS OF GONORRHEA AND NONGONOCOCCAL URETHRITIS

Clinical	Gonococcal	Nongonococcal
Incubation	2–8 days	10–20 days
Onset	Abrupt	Gradual
Dysuria	Prominent	Variable
Discharge	White, purulent	White to clear

In women, the symptoms and signs of *C. trachomatis* genital tract infection resemble those of gonorrhea in that they may be absent or mild enough to ignore. As in gonorrhea, chlamydial infection may cause a mucopurulent cervical discharge associated with erythema and edema, along with suprapubic tenderness. Other likely pathogens that may cause similar symptoms are yeast and *Trichomonas vaginalis*. *Chlamydia* is also implicated in cases of "sterile pyuria"; that is, when urine cultures yield no commonly isolated organisms even when more than 10 white blood cells (WBC) per high-power field (HPF) are seen in centrifuged urine samples.

If genital chlamydiosis is left untreated, it can cause serious complications or sequelae, such as PID, endometritis, infertility, sterility, and ectopic pregnancy. In addition, chlamydial infections in the female genital tract can be transmitted to neonates. Inclusion conjunctivitis of the newborn presents as an acute mucopurulent process occurring 5 to 14 days after delivery. Pneumonia in the newborn can also develop between 4 and 18 weeks. This condition presents as an afebrile illness with respiratory congestion, wheezing, and a cough resembling pertussis. The most effective means of protecting the newborn is detection and treatment of infected pregnant women.

Laboratory Diagnosis

Diagnosis of chlamydial infections in the laboratory relies heavily on the quality of the submitted clinical sample, particularly if direct microscopic examination or culture will be attempted. Samples must contain columnar epithelial cells taken from the infected sites. Culture specimens must be obtained from the anterior urethra or the endocervical canal by vigorous swabbing or scraping. Purulent discharges are considered inadequate. Samples for culture must be transported in the appropriate transport medium for the fragile chlamydiae to be recovered.

Although direct examination of endocervical specimens is technically demanding and requires specimens that contain infected cells, the method offers a rapid diagnosis and allows evaluation of the quality of the collected sample. The laboratory worker may use cytology stains to reveal inclusion bodies in detecting inclusion conjunctivitis or trachoma, or the direct fluorescent antibody (DFA) method to locate the typical apple-green inclusions and elementary bodies.

Direct fluorescent antibody testing and EIA, two antigen detection methods, have proved to be very useful. Compared with culture methods, DFA testing has a sensitivity of 56 to 100% and a specificity of 82 to 100%. As mentioned earlier, DFA testing has an advantage over cultures in that it can evaluate specimen adequacy and quality. Unlike the DFA method, in which the presence or absence of columnar epithelial cells is detected, the EIA detects solubilized outer membrane lipopolysaccharide antigens. The EIA is better designed for multiple batches than is DFA testing; however, when EIA is compared with culture methods, it lacks agreement with culture results, particularly in low-risk groups. The other disadvantage of the EIA is its inability to judge specimen quality.

Culture remains the gold standard for the laboratory diagnosis of chlamydial infections. Under ideal situations, cultures produce a 70 to 80% sensitivity and a 100% specificity. The combination of technical complexity, requirement for special specimen handling, and maintenance of culture media, however, limits performance of chlamydial culture to certain laboratory settings. Because chlamydiae do not survive well in the environment, a commercially available *Chlamydia* transport medium is required to maintain its viability. Chlamydiae also require tissue culture cells, such as McCoy cells and Buffalo Green Monkey kidney cells, HeLA 229 and BHK-21.

A chemiluminescent-labeled DNA probe that binds to chlamydial rRNA is used in many clinical laboratories. Its reported sensitivity and specificity is 88% and 95.8%, respectively. Amplicor (Roche Diagnostics, Nutley, NJ), a newly released polymerase chain reaction (PCR) assay for chlamydiae, promises excellent sensitivity and specificity.

Treatment for chlamydial infections includes administration of doxycycline and tetracycline for uncomplicated urethral and endocervical infections. Erythromycin is recommended for infected pregnant women.

BACTERIAL VAGINOSIS

No single agent causes bacterial vaginosis (formerly called "nonspecific vaginitis"). Rather, the condition results from a disruption of the normal vaginal flora. In 1894, Döderlein determined that the normal flora of the vagina consisted mostly of large, gram-positive bacilli, more specifically, lactobacilli. Almost 30 years later, Schröder was able to correlate vaginal Gram stains either to pathology or to lack of pathology in the vagina. He observed that vaginal Gram stains consisting mostly of large, gram-positive bacilli correlated with female patients with normal vaginas. On the other hand, vaginal Gram stains indicating many gram-negative bacilli and coccobacilli with few or no gram-positive bacilli indicated women undergoing some pathology of the vagina. However, no specific causative agent was determined.

In 1955, Gardner and Dukes revealed that they had found a newly described pathogen, which they called *Haemophilus vaginalis*. This bacterium supposedly fulfilled Koch's postulates as the sole cause of bacterial vaginosis. Immediately after their findings had been reported, they were actively challenged by others who isolated the *H. vaginalis* organism from asymptomatic patients. Subsequent studies showed that while patients with bacterial vaginosis did indeed have *H. vaginalis* in the vagina, so did approximately 50% of the women with apparently normal vaginas. Thus, the claim that this new isolate caused the infection was not completely true.

Gardner's isolate, *H. vaginalis*, was reclassified as *Corynebacterium vaginalis* in the early 1960's and was ultimately reclassified as the new genus *Gardnerella* in 1980 in honor of Dr. Gardner. *Gardnerella vaginalis* is the sole species in the genus.

Bacterial vaginosis is often described as an STD, although the Centers for Disease Control and Prevention do not so classify it at this time. However, it has been shown that male partners of females with bacterial vaginosis can develop urethritis caused by bacterial vaginosis–associated bacteria. If bacterial vaginosis is an STD, its epidemiology differs from those of most other STDs. There is no racial or ethnic bias in women affected; essentially, white women are just as likely to develop bacterial vaginosis as are nonwhite women. Also, bacterial vaginosis is a condition of women in their twenties and thirties, while many other STDs occur primarily in adolescents and young adults.

Clinical Manifestations

The signs and symptoms of bacterial vaginosis include a fish-like odor; a watery, noninflammatory exudate lacking polymorphonuclear neutrophils (PMNs); an increase in pH of the vaginal fluid; and the presence of "clue cells," which are exfoliated vaginal epithelial cells literally covered by gram-negative coccobacilli and curved, gram-negative rods.

Bacterial vaginosis occurs when the hydrogen peroxide–producing lactobacilli, the most common group of bacteria in the healthy vagina, begin to decrease in numbers. As they decrease, the pH of the vagina increases as a result of the loss of the lactic acid produced by the lactobacilli. As the pH increases, other species of bacteria present in the vagina in decreased numbers start to rapidly proliferate. The bacteria commonly associated with this shift include *Prevotella* spp, *G. vaginalis,* and *Mycoplasma hominis*. The anaerobic *Mobiluncus* spp, which can usually be isolated from the rectum, but rarely from the healthy vagina, also appear in many women with bacterial vaginosis. As these anaerobes and *G. vaginalis* increase in concentration, they degrade proteins in the vagina, forming the amines cadaverine, putrescine, and triethylamine, which cause the characteristic unpleasant odor. The odor of the amines is amplified by activation with alkaline compounds, such as potassium hydroxide (KOH). Interestingly enough, male ejaculate is alkaline and thus can activate the odor of the amines; many women with bacterial vaginosis notice the odor almost immediately after intercourse. The combination of the amines and organic acids produced by the anaerobic bacteria, such as acetic and succinic acids, causes an exfoliation of epithelial cells from the vagina, resulting in the noninflammatory exudate. The lack of inflammation separates this vaginosis from vaginitis caused by *Trichomonas vaginalis* or *Candida albicans*. The pH allows the gram-negative and gram-variable bacilli associated with bacterial vaginosis to tightly adhere to the epithelial cells to form the clue cells.

Regardless of its status as an STD, bacterial vaginosis is clearly anything but a minor condition. The sequelae of bacterial vaginosis include recurrence, PID, urinary tract infections, endometritis, preterm labor and membrane rupture, septicemia, and meningitis.

Laboratory Diagnosis

Before discussing laboratory-assisted diagnosis of bacterial vaginosis, it is best to emphasize that bacterial vaginosis is diagnosed best by the primary medical care provider at the patient's bedside. Briefly, if the medical care provider can demonstrate three of the four signs or symptoms of bacterial vaginosis, then the diagnosis is complete. The presence of the exudate can be ascertained with simple examination of the vaginal walls. The exudate is clingy, clear, and watery, and most often PMNs are absent. The increase in pH can be determined by using litmus paper on a vaginal fluid specimen. The presence of amines is demonstrated by adding KOH to a microscope slide containing some vaginal fluid. The health care provider smells the fluid ("whiff test") for the characteristic fish-like odor. Lastly, clue cells can be seen using wet mount preparations of vaginal fluid. In a patient with bacterial vaginosis, approximately 20% of the epithelial cells should be clue cells.

The role of the clinical microbiology laboratory in the diagnosis of bacterial vaginosis is limited. Since *G. vaginalis* can be normal flora in 50 to 60% of adult women, isolating it from a vaginal fluid specimen may provide no clinically significant data. It is essential, however, that clinicians and laboratory workers alike understand the inappropriate nature of trying to diagnose bacterial vaginosis using culture as the sole test. While culture may often be unnecessary, it is nonetheless easy to culture and identify *G. vaginalis*. It grows readily as a nonhemolytic colony on 5% sheep's blood agar (SBA). It does produce β hemolysis on the selective human blood bilayer agar (HBT) and the nonselective vaginalis (V) agar. The purified isolate can be rapidly identified using commercial identification kits, including the RapID NH panel (Innovative Diagnostics, Atlanta, GA). Key characteristics of *G. vaginalis* include catalase production, starch hydrolysis, and the inability to grow on MacConkey agar. The isolate produces zones of inhibition around disks of metronidazole (50 μg) and trimethoprim (5 μg), but not sulfonamide (1 mg).

It has been shown that quantitative Gram stain results, in conjunction with the presence of clinical signs and symptoms, are very sensitive and specific for the diagnosis of bacterial vaginosis. In the method described by Nugent in 1991, Gram-stained vaginal fluid specimens are scored on the basis of the concentrations of different morphotypes (Table 32–2). If the sum of the score is 0 to 3, the fluid is rated as normal. Fluids with scores of 7 to 10 indicate bacterial vaginosis, and scores of 4 to 6 are rated as intermediate.

TABLE 32–2

STANDARDIZED SCORING METHOD FOR GRAM-STAINED SMEARS OF VAGINAL FLUIDS

	Scores for the Number of Each Morphotype Cell per Oil Immersion Field				
	0	*<1*	*1–4*	*5–30*	*>30*
Large gram-positive rods	4	3	2	1	0
Tiny gram-variable or gram-negative rods	0	1	2	3	4
Curved gram-variable or gram-negative rods	0	1	1	2	2

SYPHILIS

CASE STUDY

A black male infant was born at 33 weeks of gestation to a 17-year-old mother. The mother had no prenatal care and had a positive rapid plasma reagin (RPR) test result (1:128) on admission to an obstetric service. The infant was delivered by cesarean section because of fetal distress. On examination, physical findings revealed that the infant was hydropic and showed jaundice and hepatosplenomegaly. Laboratory findings also showed pancytopenia, hypoglycemia, coagulopathy, oliguria, and acidosis. The infant received multiple transfusions of washed, packed red blood cells, platelets, and fresh-frozen plasma. He was treated with penicillin and gentamicin but died 3 days after birth. Examination of the placenta showed spirochetes in the umbilical cord.

The dramatic increase in cases of congenital syphilis has been noticed along with the increase in incidence of syphilis in women. Syphilis, caused by the spirochete *Treponema pallidum* subsp *pallidum,* is currently the second most common ulcerative STD in the United States. Transmission of this infection is approximately 33% after a single exposure.

Figure 32–1 shows the incidence of primary and secondary syphilis from 1983 to 1993, with most new cases occurring in homosexual and bisexual males. The incidence seemed to decrease between 1984 and 1987, probably because of the increased public awareness of the HIV epidemic, and safe sex practices among homosexual men. However, a reversal of this decline began in 1986. In the United States, the incidence of syphilis increased 64% from 1985 to 1989, with the most significant

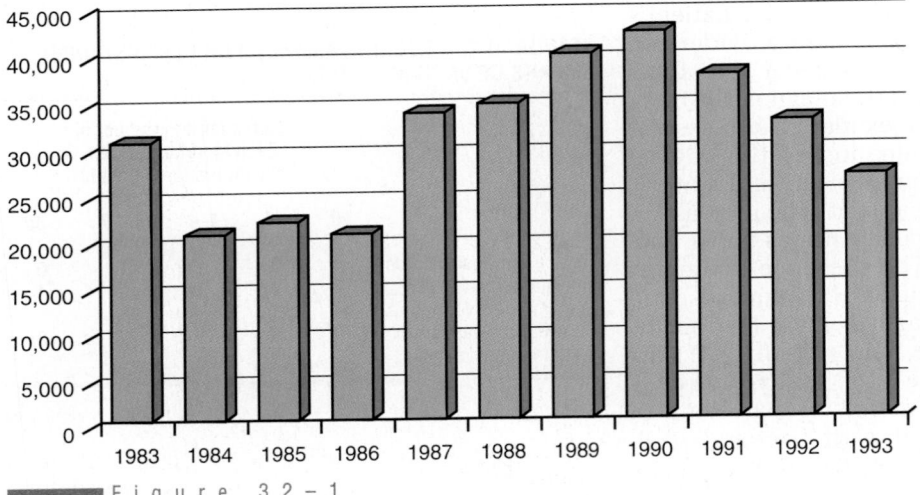

F i g u r e   3 2 – 1

Reported cases of primary and secondary syphilis, 1983 to 1993.
(Data from MMWR, Summary of Notifiable Diseases. Centers
for Disease Control and Prevention, 1983–1993.)

increases appearing among black heterosexuals. The current increase in syphilis in inner city youth is often associated with "drugs for sex."

Clinical Manifestations

There are three stages in the pathogenesis of syphilis: primary, secondary, and tertiary, or late, syphilis. *Treponema pallidum* subsp *pallidum* can enter the host through any site and initiate infection. The incubation period ranges from 1 to 90 days. Primary syphilis begins when a lesion, called the chancre, appears at the site of inoculation. The painless, nonsuppurative lesion teems with treponemes. After several weeks, the chancre heals, but if the condition goes untreated, the organisms can spread throughout the body via the blood stream. Without treatment, approximately 100% of infected patients progress to secondary syphilis.

Secondary syphilis begins approximately 4 to 12 weeks after infection. Because of the hematogenous spread of the treponemes, the patient experiences fever, lymphadenopathy, myalgia, and anorexia. Macular or papular skin lesions involving the trunk, soles of the feet, and palms appear in most patients. Treponemes proliferate in these skin lesions; therefore, fluids from these lesions are contagious. Condylomata lata are infectious mucoid, fleshy wart-like growths that may appear in moist areas, such as

the perianal region, during secondary syphilis. Even without treatment, symptoms disappear after 3 to 12 weeks, and the untreated patient goes into a latent phase of syphilis. During latent syphilis, the patient is asymptomatic. The only evidence of recent infection is the appearance of serologic antibodies.

The incidence of tertiary syphilis has decreased over the years probably as a result of better detection and treatment of early syphilis. If the disease further progresses to tertiary syphilis, however, neurologic and cardiovascular effects and gumma formation may be observed. Gummas present as nodules, tumor-like masses, or large ulcerative lesions. Gumma formation reflects the destructive effects of delayed hypersensitivity to treponemes and involves the skin, bone, or viscera.

Since *T. pallidum* subsp *pallidum* can cross the placenta, congenital infection can occur in newborns. Congenital syphilis occurs when a live or stillborn infant is born to a mother who has a case of untreated or improperly treated syphilis. At birth, the signs of congenital syphilis include low birth weight, hepatosplenomegaly, jaundice, secondary syphilitic lesions, condylomata lata, meningitis, and periostitis. The manifestations of congenital syphilis, however, may not always be present at birth; the newborn may be asymptomatic. In these cases, the clinical signs of congenital syphilis (late congenital syphilis), such as blindness, deafness, and deformed bones or teeth, may not be observed until the child is about 2 years old.

Laboratory Diagnosis

Pathogenic treponemes remain unculturable in vitro. Direct detection by darkfield microscopy may demonstrate the treponemes present in exudate from the chancre. This direct examination, however, requires a fresh specimen, and motility must be observed. Direct detection by fluorescent antibody specific for *T. pallidum* subsp *pallidum* differentiates pathogenic treponemes from nonpathogenic ones. This is the preferred method for examination of anal, rectal, and intestinal lesions.

Infection with *T. pallidum* subsp *pallidum* stimulates the production of two types antibodies: treponemal antibodies specific for the organisms, and nontreponemal (nonspecific) antibodies, referred to as reagin. Rapid plasma reagin (RPR) and VDRL (Venereal Disease Research Laboratory) tests are nontreponemal (nonspecific) antibody tests primarily used for screening purposes; therefore, they lack the specificity needed for a definitive diagnosis. Quantitation (titer) of the nontreponemal serologic test for syphilis is useful in determining the course of the disease and in the assessment of treatment efficacy. Conditions such as febrile infections, pregnancy, and autoimmune disorders may produce biologically false-positive reactions with nontreponemal tests.

For confirmation, specific treponemal antibody tests (SSTS), such as the fluorescent treponemal antibody absorption (FTA-ABS), and microhemagglutination for *T. pallidum* subsp *pallidum* (MHA-TP) are used. The major value of the specific treponemal antibody tests for syphilis is that it can rule out biologic false-positive results, which can occur with the nonspecific tests. The results of treponemal tests cannot be used to monitor therapy or establish reinfection because treponemal antibodies remain positive for life.

The serologic tests for syphilis become reactive approximately 1 week after the chancre appears. The patient who is treated before the treponemal (specific) and nontreponemal (nonspecific) serologic tests for syphilis become reactive will remain nonreactive. Both treponemal and nontreponemal tests are reactive in untreated patients. If the patient receives treatment, the nontreponemal serology becomes nonreactive, but the treponemal serology test remains reactive for life. In tertiary, or late, syphilis, treponemal tests remain reactive, whereas nontreponemal methods are nonreactive.

In congenital syphilis, the neonate's serologic tests, both specific and nonspecific, are reactive. To differentiate maternal antibody from the baby's antibody to syphilis, an IgM-specific FTA-ABS test is appropriate.

CHANCROID

Chancroid, or soft chancre, is caused by a fastidious gram-negative coccobacillus, *Haemophilus ducreyi*. This sexually transmitted infection is endemic in Southeast Asia, Africa, and India; outbreaks have also occurred, however, in developed countries. In the United States, there has been a noted increase in incidence of chancroid, with 5,000 cases per year being reported to the Centers for Disease Control and Prevention. The most recently reported cases are based primarily on clinical diagnosis. The typical patient profile for chancroid includes Latino and African American heterosexuals. The male-to-female ratio is from 3:1 to 53:1. Infection in women usually produces no symptoms.

Clinical Manifestations

An erythematous papule usually develops 3 to 5 days after inoculation. The organisms enter through a trauma site, and the lesions usually develop on the external genitalia in both males and females. The sites of infection in the male include the prepuce, coronal sulcus, glans, and shaft; in the female, the fourchette, labia, clitoris, vaginal wall, and cervix can be involved. Inguinal lymphadenopathy, presenting like buboes, may be present, especially in male patients. The lesion becomes ulcerative after several days, forming an irregular, ragged edge, and a yellow exudate may be found. Differential diagnosis with syphilis has to be made, although the clinical manifestation of chancroid produces a painful, suppurative lesion, unlike the primary lesion of primary syphilis.

Laboratory Diagnosis

Exudates from the ulcer, lymph node aspirates, and material from the base of the ulcer may be submitted for direct smear examination and culture. Microscopically, *H. ducreyi* organisms are pleomorphic, gram-negative coccobacilli that have been described as occurring in chains or "school-of-fish" formations. Gram-stained smear preparation is reported to produce poor sensitivity because of the numerous contaminating bacteria that are present.

Haemophilus ducreyi is a very fastidious organism that requires special growth and environmental supplements. Special media, such as GC agar base (Gibco Laboratories, Grand Island, NY) supplemented with hemoglobin, Isovitalex (BBL Microbiology Systems, Cockeysville, MD), fetal calf serum, and Mueller-Hinton agar with chocolated horse's blood, provide the needed nutrients for recovery. In addition, *H. ducreyi* requires 5 to 10% carbon dioxide environment with a high humidity. Colonies may be found after 48 to 72 hours of incubation. These organisms produce oxidase, reduce nitrate, produce alkaline phosphatase, and do not produce catalase. The requirement for heme is shown by a negative porphyrin test result.

Treatment includes administration of ceftriaxone or erythromycin.

GENITAL HERPES

Genital herpes, caused by herpes simplex virus (HSV), presents as a vesiculoulcerative lesion that frequently appears on mucous membranes. There are two antigenic classifications of HSV, type 1 and type 2. HSV-1 causes 80% of oral infections and 20% of genital infections. The reverse is the case with HSV-2, which causes primarily genital infections. Genital herpes is most predominant during the sexually active years.

Clinical Manifestations

Most primary herpes infections are asymptomatic; however, infection may be transmitted by both asymptomatic individuals who are shedding the virus and those who present clinical evidence of the disease. Transmission is by direct contact with secretions from infected sites. The average incubation period is 3 to 6 days after the sexual encounter. Lesions that appear on the infected surface usually rupture in 24 to 48 hours. Erythematous ulcers with an edematous base form at the site of each vesicle. Ulcers that form on dry surfaces, such as the buttocks or shaft of the penis, resolve by crusting and tend to heal rapidly. But ulcers on moist surfaces, such as the vagina, cervix, and glans penis, become macerated and resolve by slowly healing from the periphery inward. Women have a higher incidence of lesions on moist surfaces, and therefore ulcers of primary infections tend to last longer in women, approximately 4 to 21 days. Lesions in males last approximately 9 days. Approximately 75% of women with primary HSV have a vaginal discharge.

Following recovery, HSV enters its latent period, during which it resides in the dorsal root ganglion until it is stimulated to reactivity by factors such as menses, fever, exposure to sunlight, and emotional stress. Recurrences decrease with time, with the greater frequency occurring during the first year of infection. Virus is shed during recurrence; however, viral shedding is also possible when there are no lesions present. Recurrence is somewhat related to antigenic type. Recurrence with HSV-1 is less likely than with HSV-2.

The greater risk of genital HSV infection is the subsequent exposure of neonates when the mother is clinically symptomatic for primary infection. However, approximately 70% of infected mothers have asymptomatic genital infections during delivery. Fortunately, the neonatal infections occur less commonly than genital infections. Exposure may occur in utero, although this is rare. Neonates acquire the infection most commonly intrapartum, or during birth.

Physicians often culture specimens from HSV-infected women near the end of their pregnancy to determine whether viral shedding is occurring. The symptoms of HSV infection in the newborn begin 24 to 48 hours for in utero infections. Manifestations of neonatal HSV disease acquired during delivery usually appear in the first 4 to 7 days of life. Neonates may present with infection localized to skin, eyes, and mucosa, or the central nervous system. Disseminated neonatal HSV is also possible. If the disease goes untreated, the mortality rate is 70%.

Laboratory Diagnosis

Viral isolation offers the most common method for the diagnosis of herpes simplex infections. Although less sensitive than culture, direct examination of smears taken from infected lesions may be performed by using Giemsa or Papanicolaou stains to observe for intranuclear inclusions or multinucleated giant cells. Specimens for culture must be taken as early in the disease process as possible from mucocutaneous lesions. The likelihood that a herpetic lesion will produce a positive culture diminishes with each day after the appearance of the lesion. Aspirates from fresh vesicles provide the highest recovery, as they contain the greatest number of viruses. Crusted, reepithelialized lesions should be avoided, and when swabbing for samples for culture sufficient force must be used to remove epithelial cells from the base of the lesion. Viral samples should be stored at 4°C and should not be frozen.

Herpes simplex grows rapidly on numerous tissue culture cell lines, and the cytopathic effect in culture can be seen within 24 hours. The most common cell lines used include MRC-5, human embryonic lung, Hep-2, A549, and rabbit kidney.

Treatment of herpes infections with the antiviral agent acyclovir has been shown to shorten the course of current primary and recurrent infections. Sexual contact must be avoided by patients with unhealed lesions.

Bibliography

Balows A, Hausler WJ, Jr, et al: In Balows A, et al (eds): Manual of Clinical MIcrobiology, 5th ed. Washington, DC: American Society for Microbiology, 1991.

Barnes R: Laboratory diagnosis of human *Chlamydia* infections. Clin Microbiol Rev 2:119, 1989.

Benenson AS (ed): Control of Communicable Diseases in Man, 15th ed. Washington, DC: American Public Health Association Publications, 1990.

Centers for Disease Control: Policy guidelines for the detection, management, and control of antibiotic-resistant strains of *Neisseria gonorrhoeae*. MMWR 36(5S):15, 1987.

Centers for Disease Control and Prevention: Summary of notifiable disease, United States 1990. MMWR 39:54, 1990.

Dyckman JD, Storms S, Huber TW: Reactivity of microhemagglutination fluorescent treponemal antibody absorption and Venereal Disease Research Laboratory tests in primary syphilis. J Clin Microbiol 12:629, 1980.

Eschenbach DA: History and review of bacterial vaginosis. Am J Obstet Gynecol 169:441, 1993.

Eschenbach D, Pollock HM, Schachter J: Laboratory diagnosis of female genital tract infections. In Cumitech 17. Washington, DC: American Society for Microbiology, 1983.

Espy MJ, Smith TF: Detection of herpes simplex virus in conventional tube cell cultures and in shell vials with a DNA probe kit and monoclonal antibodies. J Clin Microbiol 26:22, 1988.

Harper M, Johnson R: The predictive value of culture for the diagnosis of gonorrhea and chlamydia infections. Clin Microbiol Newsl 12:54, 1990.

Hillier SL: Diagnostic microbiology of bacterial vaginosis. Am J Obstet Gynecol 169:455, 1993.

Hipp SS, Yangsook H, Murphy D: Assessment of enzyme immunoassay and immunofluorescence tests for detection of *Chlamydia trachomatis*. J Clin Microbiol 25:1938, 1987.

Jungkind DL, Silverman NS, et al: Evaluation of a new polymerase chain reaction test for detection of *Chlamydia trachomatis* in clinical samples. In Abstracts of the American Society for Microbiology Annual Meeting. Washington, DC: American Society for Microbiology, 1992.

Kellog DS, Holmes KK, Hill GA: Laboratory diagnosis of gonorrhea. In Cumitech 4. Washington, DC: American Society for Microbiology, 1976.

LeBar W, Herschman B, et al: Comparison of DNA probe, monoclonal antibody, enzyme immunoassay, and cell culture for the detection of *Chlamydia trachomatis*. J Clin Microbiol 27:826, 1989.

Lombardo JM, Gadol CL: *Chlamydia trachomatis:* A perfect test? Clin Microbiol Newsl 12:100, 1990.

Mardh PA: The vaginal ecosystem. Am J Obstet Gynecol 165:1163, 1991.

Rolfs RT, Nakashima AK: Epidemiology of primary and secondary syphilis in the United States, 1981–1989. JAMA 254:1432, 1990.

Salmon VC, Turner RB, et al: Rapid detection of herpes simplex virus in clinical specimens by centrifugation and immunoperoxidase staining. J Clin Microbiol 23:683, 1986.

Smith SM, Ogbara T, Eng RHK: Involvement of *Gardnerella vaginalis* in urinary tract infections in men. J Clin Microbiol 30:1575, 1992.

Sobel JD: Vaginal infections in adult women. Sex Transm Dis 74:1573, 1990.

Infections in Special Patient Populations

Ronald R. Trudel • James T. Griffith

1. Describe the various conditions that compromise the host's immune status.

2. Discuss the various infections that occur in this patient population.

3. Determine the risk factors associated with each patient condition.

4. Associate the various infectious agents that affect this special patient population with the conditions that predispose them to a particular infection.

Immunocompromised is used to describe patients with serious diseases who are highly predisposed to infections by a variety of opportunistic bacterial, fungal, parasitic, and viral pathogens (Table 33–1). The number of immunocompromised patients has increased steadily during the past 30 years, reflecting major advances in immunosuppressive therapy, instrumentation, and organ transplantation. Unfortunately, the potential gain in years of useful life for many patients through successful management and treatment of diseases is offset by serious effects on the immune system. As a result, infec-

■■■■■ T A B L E   3 3 – 1

OPPORTUNISTIC PATHOGENS MOST FREQUENTLY IDENTIFIED IN THE COMPROMISED HOST

Bacteria	*Staphylococcus aureus*
	Streptococcus spp
	Pseudomonas aeruginosa
	Escherichia coli
	Klebsiella pneumoniae
	Salmonella spp
	Serratia spp
	Haemophilus influenzae
	Legionella pneumophila
	Aeromonas hydrophila
	Nocardia spp
	Mycobacterium tuberculosis
	Marine vibrios (halophilic)
Parasites	*Toxoplasma gondii*
	Strongyloides stercoralis
	Pneumocystis carinii
Yeasts and fungi	*Candida* spp
	Torulopsis spp
	Aspergillus spp
	Zygomycetes
	Cryptococcus neoformans
	Histoplasma capsulatum
Viruses	Herpes simplex
	Varicella-zoster
	Cytomegalovirus
	Epstein-Barr virus
	Hepatitis B virus
	Adenovirus

tion rather than the primary illness becomes the leading cause of death in immunocompromised patients.

The primary predisposing factor in this patient population is the underlying disease state that affects host defense mechanisms. Examples of compromised defense mechanisms are as follows:

■ Leukocyte number or functions, as in aplastic anemia and leukemia

■ Humoral and cell-mediated immune functions, as in B-cell and T-cell abnormalities

■ Reticuloendothelial system function, as in splenectomized patients

Other underlying disease states are complement deficiencies, diabetes mellitus, renal failure, and autoimmune diseases. Table 33–2 summarizes the various conditions that compromise the host's immune status and the pathogens usually encountered in infections associated with them.

Immunosuppression may also predispose patients to severe infections. Immunosuppressive therapy, such as chemotherapy and radiation, particularly in organ transplantation and infection with human immunodeficiency virus (HIV), frequently leads to immunocompromise.

Secondary factors include use of invasive devices, such as indwelling intravenous catheters, which breach the usual skin and mucosal protective barriers and introduce organisms into the blood stream. One or a combination of any of these underlying conditions makes persons less able to cope with infections.

This chapter discusses the conditions that compromise the host's immune status and the various infectious agents that affect this special patient population.

■■■■ T A B L E   3 3 – 2
SUMMARY TABLE SHOWING THE ORGANISMS ASSOCIATED WITH IMMUNOCOMPROMISED PATIENTS

Immune Status	Examples of Conditions	Usually Encountered Pathogen(s)
Decreased leukocyte number or function	Myelocytic leukemia Chronic granulomatous disease Granulocytopenia Acidosis Burns	*Staphylococcus* *Serratia* *Pseudomonas* *Candida* *Aspergillus* *Nocardia* *Legionella*
Decreased humoral immune response and complement deficits	Lymphocytic leukemia Multiple myeloma Nephrosis Antimetabolite therapy Hypogammaglobulinemia	*Pneumococcus* *Haemophilus influenzae* Streptococci *Pseudomonas* *Pneumocystis* Enteroviruses
Decreased cellular immune response	Hodgkin's disease Steroid therapy Uremia Antimetabolite therapy Malnutrition	*Mycobacterium* *Candida* *Coccidioides* *Histoplasma* *Blastomyces* Herpes viruses Adenoviruses *Toxoplasma* *Pneumocystis* *Legionella* *Listeria*
Decreased reticuloendothelial system function	Splenectomy Chronic hemolysis	Pneumococci *Salmonella* *Listeria*

Adapted from Drew L: Infections in the immunocompromised patient. In Ryan K (ed): Sherris Medical Microbiology: An Introduction to Infectious Diseases, 3rd ed. Norwalk, CT: Appleton & Lange, 1994, p 816.

MALIGNANCY

Decrease in Humoral and Cellular Immune Response

Widespread disturbances in the humoral and cellular defense mechanisms occur in patients with malignancy. In addition, cytotoxic drugs administered to cancer patients contribute substantially to the breakdown of mucosal barriers and a decrease in cell-mediated immune (CMI) reactivity in these patients.

Granulocytopenia

Reduction in granulocytes is commonly demonstrated in patients with hematologic malignancies and those receiving chemotherapy. Once the granulocyte count drops to lower than 1000 cells/mm^3, the risk of infections increases steadily with the degree and duration of immunosuppression. In patients with a neutrophil count of less than 1000 cells/mm^3, the mortality rate may be as high as 60%. The fatality rate among patients whose neutrophil count is lower than 100 cells/mm^3 during the first week of infection may be as high as 80%.

Decrease in Leukocyte Function

In addition to the decrease in number of neutrophils, inadequate neutrophil functions, including the inability to migrate to sites of inflammation, impairment of phagocytosis, and reduced killing of ingested organisms, predispose patients suffering from chronic leukemia and Hodgkin's disease to infections. Most bacterial infections are caused by organisms of endogenous origins, that is, gastrointestinal tract, mucosal, or cutaneous flora. Fungi are also likely to invade the granulopenic host.

Infections in Neutropenic Patients

Septicemia is most likely to occur in patients with neutropenia, because neutrophils play an important role in localizing infections. *Escherichia coli, Klebsiella pneumoniae,* and *Pseudomonas aeruginosa* account for most of the gram-negative bacilli infections. *Staphylococcus aureus* is the gram-positive organism most commonly involved in sepsis.

Pneumonia, especially that caused by gram-negative bacilli, is a major problem for a neutropenic patient. Because such a patient cannot mount an adequate inflammatory response, pulmonary infection may spread rapidly and extensively. Infections with these organisms are particularly serious, because they cause extensive necrosis and have a high fatality rate. Owing to the lack of an inflammatory response, neutropenic patients fail to develop the characteristic signs of pneumonia and may remain undiagnosed until after the infection has disseminated.

Infections in Cancer Patients

Factors responsible for the high frequency of infections among cancer patients vary with the underlying malignancy. For example, tumors that outgrow their blood supply may become necrotic and infected. A gastrointestinal tumor may ulcerate, providing a focus for invasion by enteric pathogens. Tumors may also obstruct the drainage of the tracheobronchial tree or the urinary tract, permitting infection to become established distal to the obstruction.

Septicemia accounts for nearly 50% of the fatal infections in patients with genitourinary and gastrointestinal (GI) tumors and for less than 10% in patients with lung cancer. Pneumonia accounts for 75% of fatal infections in patients with cancer of the head, neck, and lungs and for less than 40% in patients with genitourinary and other tumors. Aspiration of oral secretions probably accounts for the pneumonia in these latter patients, whereas pneumonia in those with solid tumors is usually a consequence of primary or metastatic tumor invasion of the lung.

Skin infections are also common in cancer patients, for several reasons. Cancer patients often develop decubitus ulcers because they are constantly bedridden. These patients also develop catheter-associated infections, cellulitis, and local necrosis involving the extremities, some-times associated with venipunctures, intravenous lines, and other procedures.

Patients with hematologic malignancies and those who have undergone surgery for tumors involving the head and spine may suffer from central nervous system infections. In these patients, meningitis accounts for 75% of infections, 30% of which are caused by *Cryptococcus neoformans.* Gram-negative bacilli are responsible for another 40%. Fungal agents such as *Aspergillus* spp and *Mucor* spp are commonly associated with brain abscesses that occur in patients with leukemia. Oropharyngeal candidiasis occurs in about 5% of cancer patients. Superficial gastrointestinal candidiasis is also found in patients with acute leukemia and lymphoma. Although candidiasis can involve any portion of the GI tract, it is found most often in the esophagus and stomach. Other areas sometimes affected are the kidney, liver, spleen, and lungs.

Other types of infections that may occur in cancer patients include those caused by viral and parasitic agents. The most serious viral infections occurring in cancer patients are caused by the herpesvirus group. Characteristically, these viruses infect the young, resulting in lifelong immunity; such immunity may be associated with the latent infections that occur in cancer patients.

Herpes zoster infection represents the reactivation of latent varicella-zoster virus. It is in 3 to 15% of patients with lymphoma, myeloma, and chronic lymphocytic leukemia. Chemotherapy has been shown to cause a two-fold increase in the frequency of zoster infection. Zoster disseminates in 20 to 40% of cancer patients.

Dissemination of varicella, infecting the lungs, liver, pancreas, adrenal, and central nervous system, occurs in nearly 30% of children undergoing chemotherapy. The mortality rate from varicella is about 5%, and death is usually associated with pneumonitis.

Herpes simplex infections of the lip may result in extensive cellulitis due to superinfection with mixed bacterial and fungal organisms (*Candida* and gram-negative bacilli) in cancer patients.

Children with cancer are more susceptible to acquiring cytomegalovirus (CMV) infection than adults. The most common manifestation is pneumonia, which may be unilateral or bilateral and is usually accompanied by a more severe bacterial or fungal infection. Dissemination affects the lungs, kidney, lymph nodes, heart, adrenals, spleen, pancreas, and bone marrow. Death sometimes occurs from myocarditis, renal failure, and adrenal insufficiency.

Pneumocyctis carinii and *Toxoplasma gondii* are parasitic agents that commonly affect patients with malignancy. Infections with either organism may represent reactivation or primary exposure. *Pneumocystis carinii* accounts for as much as 45% of interstitial pneumonia in cancer patients. Similarly, *T. gondii,* an obligate intracellular parasite, causes pneumonia, chorioretinitis, and a mononucleosis-like infection. It is seen most commonly in Hodgkin's disease, but also in patients with lymphoma and leukemia.

Strongyloides stercoralis infections have been reported in patients with chronic lymphocytic leukemia and lymphoma. Serious manifestations due to *Strongyloides* may occur in patients receiving adrenal corticosteroids or antitumor therapy. Ulceration may result when the rhabditiform larvae penetrate the gastrointestinal tract, or larval migration to the lungs may cause a pulmonary infiltrate.

Infections in Patients with Hodgkin's Disease

Hodgkin's disease has historically been associated with impairment of cell-mediated immunity. Patients with Hodgkin's disease are especially susceptible to infections caused by facultative intracellular parasites, such as *Mycobacterium tuberculosis, Listeria monocytogenes*, and *Cryptococcus neoformans*. These intracellular organisms are resistant to bactericidal effects and are able to multiply within phagocytic cells.

As the number of circulating lymphocytes decreases, patients with Hodgkin's disease receiving chemotherapy become more susceptible to infections. Chemotherapy inhibits inflammation, decreases capillary permeability, and decreases cellular exudation. It also interferes with the diapedesis of leukocytes, inhibits antibody production, and impairs reticuloendothelial function.

BURNS AND SURGERY

Each year, more than 2 million Americans are burned. One hundred thousand require hospitalization, and 100,000 more die annually. One third are children less than 1 year old. The risk of infec-tion in a burn patient is proportional to the extent of the burn and reflects the combined effect of the impairment of all aspects of the host defense system. As a result of immunologic impairment, infection in sites *other than* the burn wound itself remains the most common cause of death in burn patients. The balance between host defense capacity and invasiveness of the microorganism in burn patients provides the optimal example of an immunocompromised patient.

In normal patients, bacterial interference (potential pathogens being inhibited by the nonpathogenic resident flora) plays a significant role in controlling cutaneous colonization of the skin. In burn patients, anatomic barriers have been breached. The normal skin flora are destroyed or removed by desquamation. The denatured protein of the burn eschar and the avascularity of the tissue provide an excellent environment for microbial growth.

The immune defense mechanisms, both humoral and cell-mediated defenses, in burn patients are suppressed. Immunoglobulins, especially IgG, are depressed. Fibronectin levels are also reduced. Fibronectin, a dimeric α-glycoprotein found in plasma and the extracellular matrix of most tissues, is necessary for normal reticuloendothelial cell function and is opsonic for *S. aureus*. Decreased fibronectin levels precede sepsis.

The advent of antimicrobial therapy has resulted in a shift of infectious agents recovered in burn patients from primarily gram-positive cocci to gram-negative bacilli, especially *P. aeruginosa*. Risk of fungal infections is increased with wound maceration, acidosis, lack of competitive bacterial pressure, and antibiotic therapy. Methicillin-resistant *S. aureus, Candida, Aspergillus, Mucor,* and herpes simplex are major causes of infections in burn patients.

ANTIMICROBIAL THERAPY

During hospitalization, the normal mucosal flora changes from predominantly gram-positive cocci to gram-negative bacilli. Susceptibility to colonization by these pathogens is usually increased by the extensive use of broad-spectrum antimicrobial agents. These agents alter the patient's endogenous microflora, allowing antimicrobial-resistant organisms to flourish, and may also enhance the susceptibility of patients to fungal infections by altering the normal flora.

ORGAN TRANSPLANTATION

Organ transplant patients may develop specific types of infections at certain intervals after the transplant occurs. During the first month after transplantation, postoperative bacterial infections are most common. One to 6 months after transplantation, opportunistic infections caused by CMV, *M. tuberculosis*, *L. monocytogenes*, *Nocardia*, *Aspergillus*, *P. carinii*, Epstein-Barr virus, varicella-zoster virus, and hepatitis virus are most likely to be reported.

Immunosuppressive therapy reactivates latent CMV infection, which is probably the most significant cause of mortality and morbidity among transplant patients. Sixty to 90% of renal transplant patients develop CMV 1 to 4 months after transplantation.

Infection due to Epstein-Barr virus (EBV) in transplant patients has also increased during recent years. EBV has been recognized as a causative agent of B-cell lymphoma, the pathogenesis of which is related to immunosuppressive therapy. Furthermore, in patients who acquire primary EBV infection after transplantation, EBV-associated lymphoproliferative disease is likely to occur. This has become a major concern, because primary EBV infection develops most commonly in children; hence, children who are transplant recipients are at greatest risk for developing EBV-associated lymphoproliferative disease. Central nervous system infections caused by *Listeria*, *C. neoformans*, *Toxoplasma*, and *Aspergillus* also occur among these patients.

Bone marrow transplant patients are susceptible to similar infections, especially by fungal agents. *Aspergillus* and respiratory viruses are possibly inhaled from the environment, whereas *Pseudomonas* and *Legionella* may be acquired from water sources.

Infections that occur more than 6 months after transplantation are usually associated with community-acquired organisms, such as viral influenza, secondary bacterial pneumonia, food-borne illnesses (acquired during travel to locations with poor sanitary conditions), and mycotic infections from specific geographical areas that may result in dissemination. Precautions about travel to underdeveloped areas and unfamiliar places such as foreign countries are given to transplant patients to help them avoid unnecessary exposure to community-acquired infections.

AGING

Diverse alterations in immune function occur with normal aging. In general, age-induced alterations of the immune system are often more qualitative (i.e., involving lymphocytic function) than quantitative (i.e., involving cell number or immunoglobulin levels).

Infections and malignancy are common among the elderly as a result of immunologic decline. Infections frequently occur in the respiratory or urinary tract, soft tissues, abdominal cavity, or endocardium. Bacteremia of unknown source may also occur. Malignancy is related to decreased tumor surveillance by immune and nonimmune defense mechanisms.

Bibliography

Corey, L: Infections in the immunocompromised patient. In Sherris J (ed): Medical Microbiology, 2nd ed. Elsevier Science, 1990.

Diamond RD: Fungal infections in the compromised host—an overview. Adv Exp Med Biol 202:119, 1986.

Erice, Jordan, MC, et al: Ganciclovir treatment of CMV disease in transplant recipients and other immunocompromised hosts, JAMA, 257:3082, 1987.

Froland S: Bacterial infections in the compromised host. Scand J Infect Dis, Suppl 43:7, 1984.

Hibberd P, Rubin RH: Infections in transplant patients and the role of the microbiology laboratory. Clin Microbiol Newsl 13:161, 1991.

Jacobs P: The immunocompromised host. South Am Med J, 71:371, 1987.

Johanson WG, Pierce AK, et al: Changing pharyngeal bacterial flora of hospitalized patients. N Engl J Med 281:1137, 1969.

Klastersky, J: Infections in immunocompromised patients. I: Pathogenesis, etiology and diagnosis, Clin Ther 8:90, 1985.

Lipschitz DA, Udupa KB, et al: Influence of aging and protein deficiency on neutrophil function. J Gerontol, 41:690, 1986.

Meunier F: Prevention of mycoses in immunocompromised patients. Rev Infect Dis 9:408, 1987.

Munster AM: Immunologic response of trauma and burns—an overview. Am J Med 77:142, 1984.

Neu HC: The patient at risk for infection: A summary. Am J Med 77:1, 1984.

Periti P, Mazzei T, et al: Infections in immunocompromised patients, II: Established therapy and its limitations. Clin Ther 8:100, 1985.

Powers DC, Nagel JE, et al: Immune function in the elderly. Postgrad Med 81:335, 1987.

Pruitt BA, et al: Opportunistic infections in severely burned patients. Am J Med 76:146, 1984.

Rogers TR: Investigation of infection in immunocompromised patients [editorial]. Br J Haematology 61:195, 1985.

Rouse BT, Horohov DW: Immunosuppression in viral infections. Rev Infect Dis 8:850, 1986.

Sheagren JM: Treatment of skin and skin structure infections in the patient at risk. Am J Med 76:180, 1984.

Skinhoj P: Herpesvirus infections in the immunocompromised patient. Scand J Infect Dis Suppl 47:121, 1985.

Sutton RNP, Itzhaki RF, et al: Virus infections in immunocompromised patients: Their importance and their management. J Roy Soc Med 78:100, 1985.

Wang DT, et al: Viral infections in the immunocompromised patients. Med Clin North Am 67:1075, 1983.

Weiner LP, Fleming JO: Viral infections of the nervous system. J Neurosurg 61:207, 1984.

Wolfson JS, Sober AJ, et al: Dermatologic manifestations of infections in immunocompromised patients. Medicine (Balt) 64:115, 1985.

CHAPTER 34

Zoonotic and Rickettsial Infections

William F. Nauschuetz •
Robert G. Whiddon

1. Discuss the pathogenesis and mode of transmission of the following zoonotic infections:

- Anthrax
- Plague
- Erysipeloid
- Leptospirosis
- Tularemia
- Rat-bite fever
- Lyme borreliosis

2. Discuss the three forms of clinical manifestations of anthrax.

3. Describe the morphology and culture characteristics of the following etiologic agents:

- *Brucella* species
- *Yersinia pestis*
- *Bacillus anthracis*
- *Francisella tularensis*
- *Borrelia burgdorferi*
- *Leptospira* species
- *Pasteurella multocida*
- *Erysipelothrix rhusiopathiae*

4. Discuss the appropriate laboratory methods for the diagnosis of these infections and to maximize recovery of the etiologic agents.

5. For the following human rickettsial diseases, give the causative agent and the mode of transmission to humans:

- Epidemic typhus
- Rocky Mountain spotted fever
- Scrub typhus
- Q fever

6. Describe the morphology, growth requirements, and structure of rickettsial organisms and how they differ from bacterial agents.

7. Discuss the pathogenesis of rickettsial infections and the methods of laboratory diagnosis.

A zoonotic infection usually occurs in vertebrate animals, but humans can become infected under certain circumstances. The list of organisms involved in zoonotic infections is constantly increasing. In the United States, people own about 100 million dogs and cats as pets. These animals represent a risk of zoonoses, especially to children. Zoonotic infections comprise an incredibly diverse group of microorganisms, including viruses, bacteria, fungi, and protozoans. Table 34–1 lists a limited sample of the zoonotic infectious agents. The mechanisms of transmission to human hosts can be just as diverse.

This chapter divides zoonotic infections by routes of transmission, such as bites and scratch-es, and by direct contact or inhalation. Also detailed are some classic zoonotic infections, such as the following:

- Anthrax
- Tularemia
- Plague
- Rat-bite fever
- Lyme borreliosis

Other clinically significant zoonotic infections discussed in this chapter that are diagnosed in the clinical laboratory include pasteurellosis, capnocytophagosis, and brucellosis.

TABLE 34-1

SELECTED BACTERIAL, VIRAL, FUNGAL, AND PARASITIC ZOONOTIC AGENTS

Bacteria

Bacillus anthracis
Borrelia spp
Brucella spp
Capnocytophaga canimorsus
Chlamydia psittaci
Campylobacter jejuni
Coxiella burnetii
Ehrlichia spp
Erysipelothrix rhusiopathiae
Francisella tularensis
Leptospira interrogans
Listeria monocytogenes
Mycobacterium leprae
Pasteurella multocida
Pseudomonas mallei
Salmonella enteritidis
Spirillum minus
Streptobacillus moniliformis
Vibrio spp
Yersinia enterocolitica
Yersinia pestis

Viruses

Arboviruses
Herpesvirus simiae
Hepatitis virus
Rhabdovirus
Rubeola virus

Fungi

Trichophyton spp
Microsporum spp

Parasites

Babesia microti
Cryptosporidium spp
Echinococcus granulosus
Strongyloides spp
Toxocara spp
Toxoplasma gondii
Trichinella spiralis
Balantidium coli
Hymenolepis nana
Hymenolepis diminuta
Taenia spp
Trypanosoma spp
Giardia lamblia

This chapter also discusses the rickettsioses. Many of these infections are vector-borne, requiring animals to transmit the disease to humans. Most of the rickettsioses are not true zoonoses, however, since the vectors normally do not show symptoms of infection.

More agents of zoonotic infections will undoubtedly be detected and identified as culture and molecular methods become more sophisticated. Recent technology-assisted discoveries include *Afipia felis* (the agent of cat-scratch disease) and *Rochalimaea henselae* (which causes a tick-borne relapsing fever).

ZOONOTIC INFECTIONS TRANSMITTED BY SCRATCHES AND BITES

Zoonoses transmitted by bites and scratches include plague, pasteurellosis, Lyme borreliosis, and rat-bite fever.

Plague

CASE STUDY

An 8-year-old Native American boy with a 3-day history of malaise and a fever was diagnosed as having acute otitis media. The patient received erythromycin, but the symptoms persisted. Within the next 48 hours, the patient developed nausea, vomiting, and severe abdominal pain. Surgeons performed a laparotomy to rule out acute appendicitis. The appendix was normal. A lymph node biopsy was done, and blood cultures were obtained. The patient received empirical gentamicin and clindamycin. Shortly after surgery, the patient's temperature rose to 106°F, and his white blood cell count increased to 37,000 cells/μL. He developed tachycardia, tachypnea, and hypoxemia. His respiratory distress rapidly progressed, and he developed disseminated intravascular coagulation (DIC), gastrointestinal hemorrhage, and septic shock. He was then given broad-spectrum antibiotic therapy with chloramphenicol and cefotaxime.

Yersinia pestis grew from the patient's lymph node and blood. The state health laboratory confirmed the identification using direct fluorescent antibody (DFA) test. The patient received a full course of chloramphenicol and cefotaxime and recovered fully.

Data from Butler T: The black death, past and present. I. Plague in the 1980s. Trans R Soc Trop Med Hyg 83:458, 1989.

Etiology

Plague, an infamous bacterial disease, is caused by the gram-negative bacillus *Yersinia pseudotuberculosis* subsp *pestis*. This genus is named for Alexander Yersin, a French microbiologist who isolated the plague bacillus during an epidemic in Hong Kong in 1894. *Yersinia pseudotuberculosis* subsp *pestis* is biochemically and genetically similar to *Y. pseudotuberculosis,* although the organism is still commonly referred to as *Y. pestis* to prevent any misunderstanding.

Epidemiology

There have been three pandemics of plague. The first started near Egypt in 542 A.D. and ravaged Europe for 50 years, killing 100 million people. The second pandemic started in the 14th century, when conditions for rat-to-human transfer of *Y. pestis* were excellent. This pandemic, named the Black Death, killed 25 million Europeans, one fourth of the total population at that time. The last pandemic started in the 1890's and is just now subsiding. It was during the last pandemic that the disease was introduced into the United States. The infections started in Burma and spread to many parts of the world by infested rats aboard commerce ships. The majority of cases of plague now occur in

Vietnam. As a result of the last pandemic, however, plague exists on all the continents except Australia.

Rats are the natural host for the vectors that transmit the disease, and humans are accidental hosts. The vectors are fleas *(Xenopsylla cheopis)* that normally infest the brown *(Rattus norvegicus)* and black *(Rattus rattus)* rats. In the United States, most cases are in the Southwest. The organism persists in this region through the sylvatic cycle, being passed among fleas and their rodent hosts. *Yersinia pestis* can survive for months in animal burrows, and uninfected rodents can get the infection from this reservoir. The disease is spread to humans by rodents, when there is contact between contaminated rural areas and areas of human habitation. Humans can also become infected from domestic cats that hunt rodents.

Life Cycle of *Yersinia pestis*

Figures 34–1 and 34–2 illustrate the life cycle of *Yersinia pestis* in the urban and sylvatic modes of transmission.

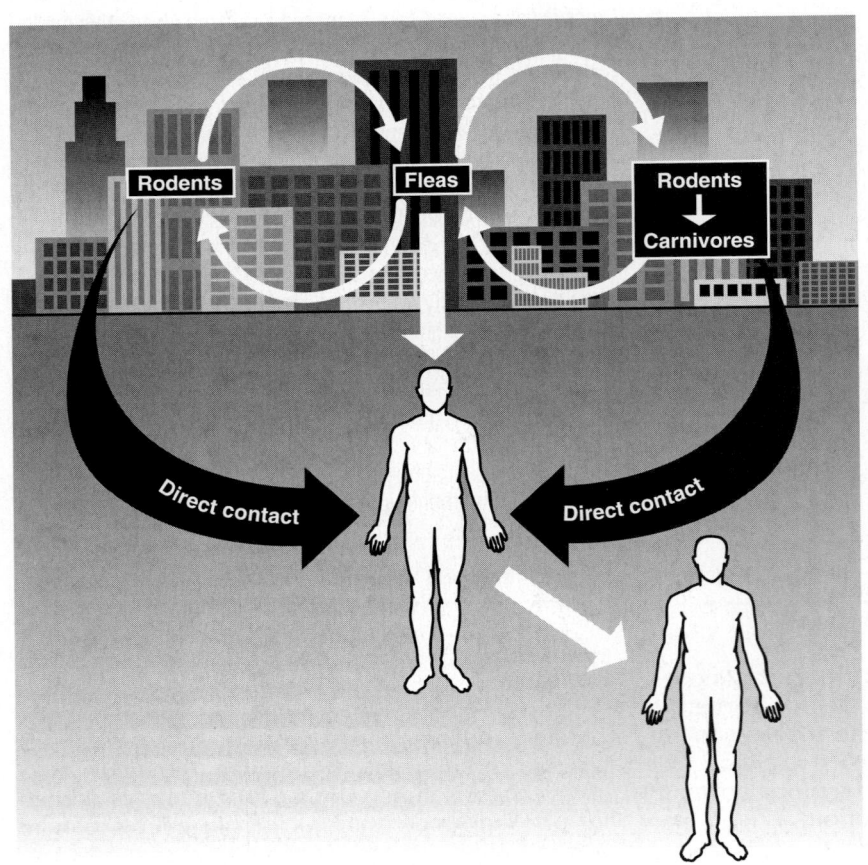

■■■■■■ F i g u r e 3 4 – 1

The urban cycle of plague.

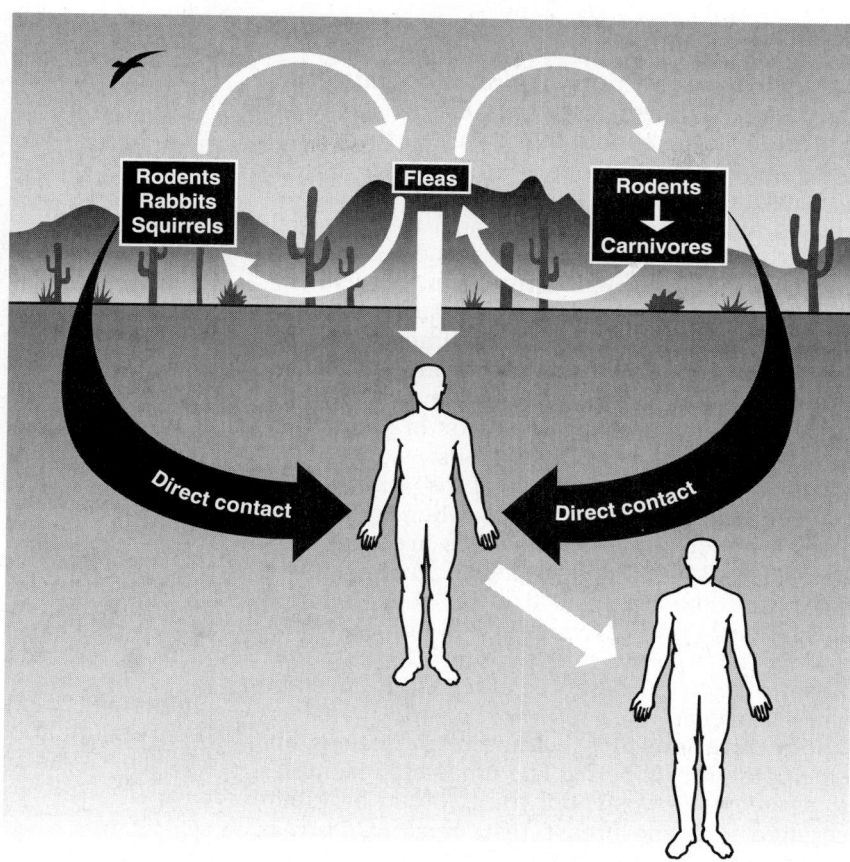

■ Figure 34–2
The sylvatic cycle of plague.

Fleas develop yersiniosis when they take a blood meal from an infected animal host. The yersiniae multiply in the gut of the flea, eventually reaching such a high concentration that they block the flea's gut. This blockage impairs the flea's ability to feed, and it responds by infesting a wider range of hosts and by biting more often. Humans living near rats can be infected by the fleas. Infection is begun when the fleas regurgitate the plague bacilli into the bite wound during feeding.

Clinical Manifestations

Plague has two major manifestations: the bubonic and the pneumonic forms. Bubonic plague is an acute, febrile disease with an incubation time of from 1 to 7 days. The disease is usually characterized by a lesion in a regional lymph node that drains the infected area. The resulting painful buboes usually occur in the groin, axilla, or subauricular area.

Within 3 to 6 days after the onset of infection, the patient develops signs of septic shock. If the disease goes untreated, the organism can be disseminated hematogenously, leading to septicemia. However septicemic plague may occur without the bubonic form. The mortality rate of septicemic plague is nearly 100%, with death occurring within 1 to 3 days.

Septicemic plague may lead to a secondary pneumonia, referred to as pneumonic plague. Pneumonic plague is almost always the result of hematogenously spread bubonic or septicemic plague. There are two forms of pneumonic plague, primary and secondary. Primary pneumonic plague occurs when the patient transmits the organism by infectious droplets, whereas secondary pneumonic plague results from the plague bacillus entering the lungs of the same patient who has either bubonic or septicemic plague. Twenty-two percent of the cases of plague in the United States are secondary pneumonic.

The drug of choice is streptomycin, and alternatives include tetracycline and chloramphenicol.

Laboratory Diagnosis

Specimen Collection. Aspirates of buboes are the best specimens for direct examination and culture. Sputum should be submitted to rule out pneumonic plague. Because of the high risk associated with this organism, clinicians should always alert laboratory personnel to the possibility that a patient may have plague.

Direct Microscopic Examination. *Yersinia pestis* organisms are gram-negative bacilli that can show bipolar, or "safety-pin," staining in tissues on direct examination, as shown in Figure 34–3. The bipolar appearance, however, is neither a sensitive nor a specific indicator of yersiniosis when seen in stained preparations.

Culture. Colonies of *Y. pestis* are slightly opaque, smooth, and round when grown on sheep's blood agar. The colonies are nonhemolytic (Fig. 34–4). The organism grows slowly. Very tiny colonies may form on sheep's blood agar after 24 hours. Even after 48 hours of incubation, the colonies may be only 1.0 to 1.5 mm. *Yersinia pestis* is nonmotile and has an envelope when grown at 37°C.

Identification. Most commercial systems are able to identify *Y. pestis*. The organism resembles *Y. pseudotuberculosis,* but the two can be differentiated with the urease test. *Yersinia pestis* is negative for urease, whereas *Y. pseudotuberculosis* is positive. Isolates that are identified as *Y. pestis* should be confirmed with DFA test or phage sensitivities. Recovery of the organism should always be reported to public health authorities.

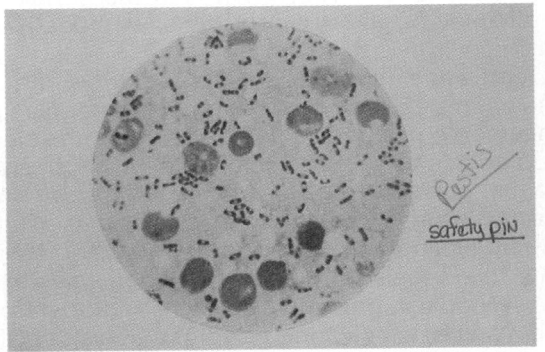

■ F i g u r e   3 4 – 3

Smear from lymph gland of patient with plague. Characteristic bipolar staining of the bacilli is obvious. × 1,000. (With permission from Gillies RR, Dodd TC [eds]: Bacteriology Illustrated, 4th ed. New York: Churchill Livingstone, 1976, p 124.)

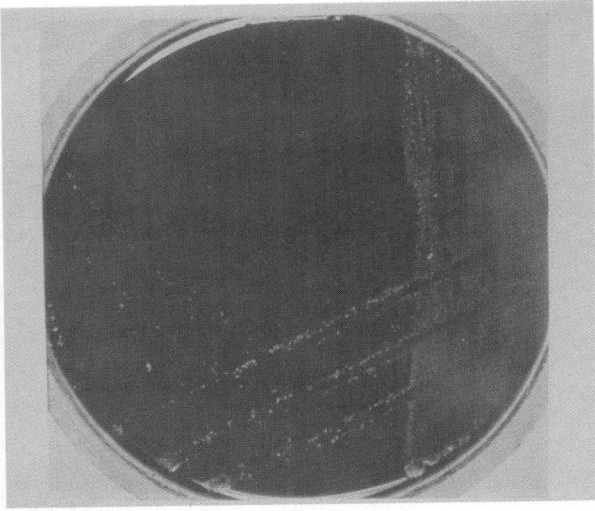

■ F i g u r e   3 4 – 4

Small, translucent, gray colonies of *Y. pestis* after 48 hours of incubation on 5% sheep's blood agar. (With permission from Gillies RR, Dodd TC [eds]: Bacteriology Illustrated, 4th ed. New York: Churchill Livingstone, 1976, p 125.)

Infection is also diagnosed serologically, using a hemagglutination test to detect a four-fold rise in titer between acute and convalescent sera.

Lyme Borreliosis

CASE STUDY

A 28-year-old woman with shaking chills and perspiration was seen at an emergency room. She explained that, before the fever, she had swollen and painful ankles, knees, wrists, and elbows. Her temperature was 100.8°F, and synovitis was noted in the wrists, elbows, knees, and ankles. She had no rash or lymphadenopathy. She admitted to a family history of osteoarthritis and rheumatoid arthritis. She lived in a rural area and had evidence of multiple insect bites, although she could not remember any recent tick bites. The patient was treated with naproxen and doxycycline for polyarthritis with rheumatoid arthritis, systemic lupus erythematosus, and late-stage Lyme disease as causes. During her follow-up examination, she recalled a tick bite about 6 months before the appearance of symptoms but could not recall having had erythema migrans. She received erythromycin, and within a month she was asymptomatic, with negative Lyme serologies. The laboratory results are shown below.

■ Sedimentation rate: 40 mm/h

■ Complete blood count: normal, except platelets of 459,000/μL.

- Rheumatoid arthritis latex: negative

- Lyme IgG titer: 1:320

- Lyme IgM titer: negative

Data from Levin RE: An unusual presentation of Lyme arthritis. J Rheumatol 16:1500, 1989.

Etiology

Lyme borreliosis, caused by *Borrelia burgdorferi,* is an arthropod-borne disease in which humans are accidental hosts. The disease is most commonly transmitted by ticks, including *Ixodes dammini, I. pacificus* (the black-legged tick), *I. ricinus* (the European sheep tick), and *Amblyomma americanum* (the common wood tick, also known as the Lone Star tick), although other insects can also harbor the spirochete. *Borrelia burgdorferi* organisms are gram-negative spirochetes about 0.18 to 0.25 × 4 to 30 μm in size. They exhibit regular coiling and are catalase-negative and microaerophilic.

Epidemiology

The disease that would eventually be named Lyme borreliosis was probably first noted in Sweden in 1908. At that time, erythema chronicum migrans (recently simplified to erythema migrans, or EM) was described as a rash that could expand its margins. Red circles ringed a white, hard center of the rash, resulting in a "bull's-eye" appearance. There were more appearances of erythema chronicum migrans in Europe during the next several decades. Some cases were even associated with tick bites, but no in-depth research was done on the infection. The first report of a similar phenomenon in the United States was in 1970. In 1975, an outbreak of juvenile rheumatoid arthritis occurred in Lyme and Old Lyme, Connecticut. The clustered outbreak of juvenile rheumatoid arthritis appeared suspicious, and thanks to some family members who thought that some infectious agent might be involved, an investigation was begun by health officials. It was the accidental discovery by Dr. Willy Burgdorfer of spirochetes in the blood of ticks recovered from the Old Lyme area that ultimately led to the description of Lyme disease. In 1984, the name *Borrelia burgdorferi* was proposed for these organisms.

Of the 6,876 cases of Lyme borreliosis reported to the Centers for Disease Control and Prevention from 1987 to 1988, 92% were from Wisconsin, Minnesota, New York, New Jersey, Pennsylvania, Connecticut, Massachusetts, and Rhode Island. Only seven states did not report any incidence of Lyme borreliosis.

Life Cycle of *Borrelia burgdorferi*

Ixodid ticks have a 2-year life cycle, requiring blood meals to pass from the larval stage to the nymph stage, and from the nymphal stage to the adult stage. Nymphs and larvae feed primarily on the white-footed mouse, while adult ticks usually infest the white-tailed deer. This horizontal transmission ensures maintenance of the pathogen in the wild. Figure 34–5 shows the presence of the organism in the midgut epithelial cells of the tick. Infected nymphs transmit the organism directly into the tissue of hosts (including humans) by regurgitating during feeding.

Clinical Manifestations

Lyme borreliosis usually has an early stage and a late stage, although patient staging can be difficult. Some patients may not exhibit symptoms during the early phase, whereas other patients may never progress to the late phase. In other patients, the two phases may overlap.

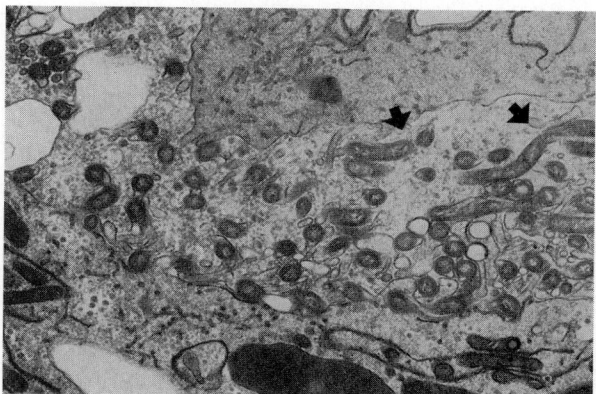

F i g u r e   3 4 – 5

Transmission electron micrograph of *B. burgdorferi* between midgut epithelial cells of *I. dammini.* (With permission from Burgdorfer W, Hayes SF, Corwin D: Pathophysiology of the Lyme disease spirochete, *Borrelia burgdorferi.* Rev Infect Dis 2[Suppl 6]: S1445, 1989.)

Early Stage. In about two thirds of infected patients, the early stage is characterized by a red papule at the site of the bite within the first 30 days of infection. The papules, referred to as erythema migrans or EM, can expand to form erythematous concentric rings with central clearing (Fig. 34–6). Spirochetemia can cause flu-like symptoms, lymphadenopathy, oligoarthritis, carditis, and neurologic manifestations. Secondary lesions (Fig. 34–7) may appear weeks after the initial lesion.

Since the concentration of bacteria in the host remains low, it is possible that many of the effects seen in Lyme borreliosis result from the host immune response, including attraction of macrophages to synovial fluid and production of interleukin-1 by host monocytes.

The Lyme spirochetes are rarely isolated from the blood or other body fluids of infected patients. Apparently the organism prefers solid tissue rather than fluid. The spirochete has a nonspecific adhesin that allows it to attach to the endothelial cells of blood vessels. This ability may enhance the migration of the organism from the blood stream into the basement membrane, resulting in vascular damage that can lead to carditis, arthritis, and central nervous system disease.

Late Stage. Late-stage Lyme disease is characterized by relapsing arthritis, and chronic synovitis can occur months to years after the initial symptoms. The joints most commonly affected are the knees, shoulders, and elbows. Untreated patients may have decreasingly severe attacks with the passage of time, until the symptoms eventually disappear. It is difficult to isolate the Lyme spirochete during the late stage. The arthritis may be mediated by the host immune system, rather than by the organism.

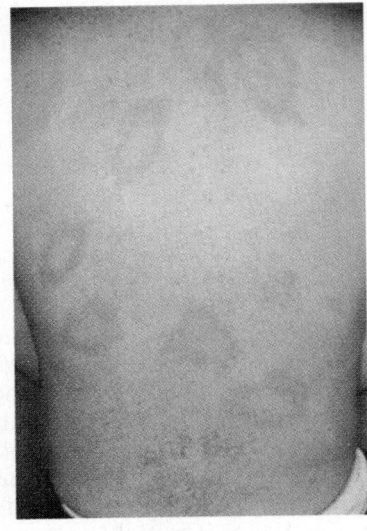

━━━ F i g u r e 3 4 – 7

A patient with multiple EM lesions. (With permission from Berger BW: Dermatologic manifestations of Lyme disease. Rev Infect Dis 2 [Suppl 6]: S1476–S1477, 1989.)

Laboratory Diagnosis

Laboratory diagnosis of Lyme borreliosis is primarily serodiagnostic. Indirect fluorescent antibody (IFA) tests and the enzyme-linked immunosorbent assay (ELISA) are used to detect antibody to the organism. The lack of a rapid or strong immune response, or cross-reactive serum antibodies, can delay serodiagnosis. Antibody testing of cerebrospinal fluid can avoid cross-reactivity problems. Some patients, including those who receive early antimicrobial therapy, may never develop a detectable immune response.

Direct antigen testing is not yet available. Direct examination of peripheral blood smears, adequate for the detection of many borrelioses, is not adequate for detection of Lyme borreliosis. Specialized media are available for culturing *B. burgdorferi* from clinical specimens, but sensitivity is poor. Gene amplification techniques, such as the polymerase chain reaction, can detect the organism in clinical specimens.

Pasteurellosis

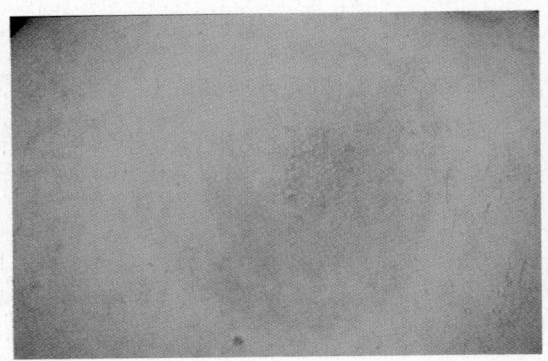

━━━ F i g u r e 3 4 – 6

The annular lesion associated with Lyme borreliosis. (With permission from Berger BW: Dermatologic manifestations of Lyme disease. Rev Infect Dis 2[Suppl 6]: S1476–S1477, 1989.)

CASE STUDY

Parents took their lethargic and irritable 6-month-old daughter to an emergency room. The child had a low-grade fever and a nonerythematous nodule on the right upper arm, but there was no sign of rash or lymphadenopathy. Two shallow abrasions, possibly scratches from the family's pet cat, were near the nodule.

The initial diagnosis was subacute bacterial meningitis. The Gram stain of cerebrospinal fluid showed gram-negative bacilli. *Pasteurella multocida* was isolated from blood cultures.

The patient was treated with ceftriaxone and ampicillin on admission. Therapy was changed to 2 weeks of ampicillin when the bacteriology results were available. The nodule disappeared within 1 week, and the patient recovered without incident. The results of the patient examination and laboratory tests were as follows:

- Temperature: 38.9°C

- Pulse: 200/min

- Respiration: 60/min

- WBC: 22,200/μL
    57% polymorphonuclear neutrophils; 20% band neutrophils
    12% lymphocytes; 6% monocytes

- Hematocrit: 28%

- Cerebrospinal fluid: RBC 3/μL^3; WBC 327/μL
    75% polymorphonuclear neutrophils
    12% lymphocytes
    Glucose: 48 mg/dL
    Protein: 79 mg/dL

Data from Hsu HW, Finberg KW: Infections associated with animal exposure in two infants. Rev Infect Dis 11:108, 1989.

Etiology

Pasteurella multocida is a pleomorphic ovoid to filamentous gram-negative bacillus, about 0.5 to 1.0 μm in size. It can be a primary pathogen or secondary invader. It is pathogenic for a wide range of hosts and occurs in the oral cavities of most dogs and cats. The five serogroups (A, B, D, E, and F) are based on capsular antigens.

Clinical Manifestations

Pasteurellosis occurs worldwide and encompasses a wide range of endemic diseases of fowl and nonhuman mammals. The most common manifestation of human pasteurellosis is cellulitis, primarily from the bites or scratches of dogs and cats. Cats are usually more often involved than dogs. Although most *P. multocida* infections are transmitted directly from the animal bite to the human, animals can infect preexisting abrasions by licking them. The infected site becomes inflamed within 48 hours, but the presence of pus is rare. Typically, the patient with pasteurellosis is afebrile and does not have inflamed lymph nodes.

Other clinical manifestations include upper and lower respiratory tract infections, endophthalmitis, and genitourinary tract infections.

Rare complications of human infection with *P. multocida* can occur in the absence of animal bites or scratches. These include sepsis, meningitis, septic arthritis, peritonitis, and osteomyelitis.

The pathogenesis of *P. multocida* infection in humans is not fully understood, but the antiphagocytic capsule and an outer membrane antiphagocytic protein apparently help the dissemination of the pathogen in avians. These factors are probably responsible for the microbe's virulence in humans as well. Some serogroups of *P. multocida,* including some human isolates, produce an exotoxin.

Laboratory Diagnosis

Specimen Collection. Purulent exudate from infected bites or scratches should be submitted to the laboratory for direct examination and culture. Blood cultures should be submitted for febrile patients. Physicians can submit sputa or bronchial washings for patients with possible respiratory pasteurellosis.

Direct Microscopic Examination. *Pasteurella* is a gram-negative coccobacillus and, in Gram stains of patient specimens, can appear singly, in pairs, or in short chains. The bacteria often show bipolar staining, as shown in Figure 34–8.

Culture. *Pasteurella multocida* produces nonhemolytic colonies (Fig. 34–9), 1 to 3 mm in diameter, on 5% sheep's blood agar within 24 hours. After 48 hours, a narrow green to brown halo can surround the colonies. Colonies of encapsulated strains appear smooth. Gram stains of the isolates reveal small gram-negative coccobacilli to bacilli with bipolar staining. However, *P. multocida* does not grow on MacConkey agar.

Identification. Biochemically, most isolates are oxidase- and catalase-positive, and they acidify the butt and slant of triple sugar iron (TSI) medium. They are positive for glucose and xylose and are usually negative for maltose and lactose.

A selective Mueller Hinton–based agar with amikacin, vancomycin, and amphotericin B is available to detect human pharyngeal carriers of

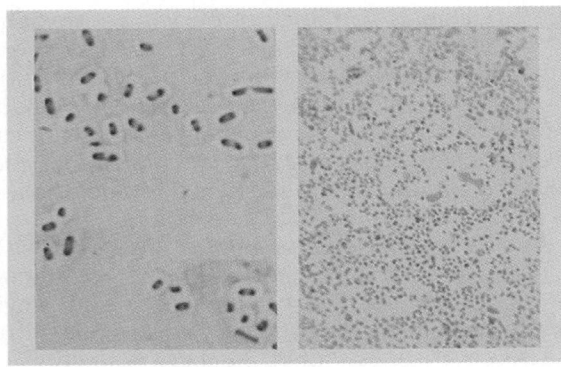

Gram stain of *P. multocida* culture *(left)*. Gram stain of mouse heart blood showing encapsulated *P. multocida* with bipolar staining. (From Bottone EJ, Girolami R, Stamm J [eds]: Schneierson's Atlas of Diagnostic Microbiology, 8th ed. New York: Churchill Livingstone, 1982, p 31.)

P. multocida. This medium may be of benefit for patients who are in professions that are high risk for pasteurellosis.

Erysipeloid

CASE STUDY

A 67-year-old man who fell from a ladder experienced fever and lower back and bilateral leg pain. He went to an emergency room 10 days later and was admitted to the hospital. He was afebrile on admission, but examination revealed a pruritic erythema of the trunk and extremities. The patient explained that the rash had started as a papule that had spread, showing central clearing. The patient also admitted to working in the vicinity of hog pens. Blood cultures were collected, and routine laboratory tests were done. Radiographs showed no bone damage, and the urinalysis was normal.

On hospital day 2, the laboratory reported gram-positive cocci isolated from the blood of the patient. The isolated organism was preliminarily identified as an α-hemolytic streptococcus. The patient received penicillin. On hospital day 4, the laboratory issued an amended report of *Erysipelothrix rhusiopathiae* as the blood culture isolate. The patient then received a 28-day regimen of penicillin. Recovery was complete and unremarkable.

Data from Gorby GL, Peacock JE, Jr: *Erysipelothrix rhusiopathiae* endocarditis: Microbiologic, epidemiologic, and clinical features of an occupational disease. Rev Infect Dis 10:317–325, 1988.

Etiology

Erysipelothrix rhusiopathiae is the causative agent of swine erysipelas, Rosenbach erysipe-

loid, erysipelotrichosis, rose disease, and fish-handler's disease. *Erysipelothrix rhusiopathiae* is a thin, facultatively anaerobic gram-positive bacillus, 0.2 to 0.4 μm by 0.8 to 2.5 μm, which can grow singly, in short chains, or in long filaments.

Epidemiology

Often isolated from contaminated water and soil, *E. rhusiopathiae* is important in veterinary medicine, causing infections in swine, poultry, small mammals, fish, and crustaceans. The organism occurs worldwide. Swine erysipelas is an important economic disease in North America, South America, and Europe.

Clinical Manifestations

The most common clinical manifestation of infection with *E. rhusiopathiae,* erysipeloid, occurs as a result of handling infected animals or animal products. In humans, the site of infection is usually an abrasion or wound of the hands or fingers. The infection is mild, localized, and self-limiting. An edematous lesion forms 1 to 7 days after infection. Erythema and itching may also be present. The lesion usually heals without treatment within 1 month. Patients with structural valvular disease, alcoholism, or other predisposing conditions may develop sepsis and endocarditis.

Laboratory Diagnosis

Specimen Collection. The patient history is extremely valuable when making a diagnosis of

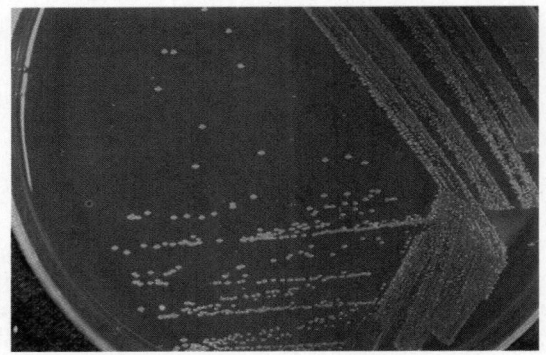

Pasteurella multocida colonies on sheep's blood agar. (From Bottone EJ, Girolami R, Stamm J [eds]: Schneierson's Atlas of Diagnostic Microbiology, 8th ed. New York: Churchill Livingstone, 1982, p 31.)

erysipeloid. Biopsied tissues are better culture specimens than is aspirated lesion fluid.

Direct Microscopic Examination. Gram-stained aspirates or blood reveals smooth forms from patients with acute infections; the rough forms occur in specimens obtained from chronically infected patients.

Culture. Colonies isolated from sheep's blood agar may appear rough or smooth. Smooth colonies are 0.5 to 1.0 mm in diameter, circular, and nonpigmented and may cause a narrow zone of green discoloration of the blood. Gram stains from smooth colonies reveal short, straight gram-positive or gram-variable bacilli. Rough colonies are larger, and Gram stains of these isolates show filamentous gram-positive or gram-variable bacilli.

Identification. *Erysipelothrix rhusiopathiae* organisms are nonmotile, nonsporing, and nonencapsulated. Isolates are catalase- and oxidase-negative and produce hydrogen sulfide in TSI medium. The organism does produce a very characteristic "test-tube brush" growth when stabbed into gelatin deeps.

Capnocytophaga canimorsus Infection (formerly CDC Group DF-2)

CASE STUDY

A 47-year-old woman entered an emergency room with weakness, diarrhea, and a facial rash. The patient had a pulse of 80, temperature of 36.5°C, and blood pressure of 80 mm Hg. She had an eschar on the left hand with no evidence of cellulitis. The facial rash covered the nose and cheeks. There were ecchymoses on her extremities. Her arms and legs were cold to the touch. Lung, cardiac, and central nervous system functions were normal. The patient was admitted to the intensive care unit.

A dog had bitten her on the left hand 5 days prior to admission. She had seen a local physician shortly after the bite, and he treated her for an allergic reaction with corticosteroids. The patient received empirical amoxicillin–clavulanic acid and amikacin, and then later received ceftazidime and amikacin. One aerobic blood culture bottle (using the Bactec NR-660) was positive, with a thin, gram-negative bacillus. The blood was subcultured onto brain-heart infusion agar supplemented with 10% horse's blood, and a small colony appeared after 48 hours of incubation in carbon dioxide. The laboratory identified the isolate as CDC group DF-2.

The patient received heparin and platelets. The ecchymoses and necrosis of the extremities increased in severity. The patient experienced adult respiratory distress syndrome, and died.

Postmortem examination revealed multiple septic microthrombi of the endocardium, lungs, kidneys, and liver.

Data from Hantson P, Gautier PE, et al: Fatal *Capnocytophaga canimorsus* septicemia in previously healthy women. Ann Emerg Med 20:126, 1991.

Etiology

Capnocytophaga canimorsus, previously known as dysgonic fermenter-2 (CDC group DF-2), is a thin, nonsporing, nonmotile, oxidase- and catalase-positive gram-negative bacillus, 1 to 3 μm long. The organism grows poorly on laboratory media. This pathogen causes a wide range of clinical manifestations, ranging from a mild, self-limiting localized infection to fulminant septicemia with involvement of several organs.

Infection occurs as the result of handling dogs. The carrier rates for this organism in dogs seem to be low, but inadequate recovery techniques may have influenced this finding.

Clinical Manifestations

Capnocytophaga canimorsus infections occur as the result of dog bites. Dissemination and septicemia occur more often than the self-limiting lesions. About 90% of infections are found in patients who are splenectomized, have cancer, or abuse drugs. The most severe infections are seen in splenectomized patients, who develop an endotoxin-mediated Shwartzman-like phenomenon with purpura, septic shock, and DIC. Infection of previously healthy individuals is rare, probably because of the susceptibility of the organism to normal serum killing.

Laboratory Diagnosis

Specimen Collection. The organism can be detected in the blood stream of infected patients.

Direct Microscopic Examination. Gram stains show the bacilli to be 1 to 4 μm long, with the longer bacilli usually appearing curved.

Culture. *Capnocytophaga canimorsus* grows poorly on 5% sheep's blood agar and does not grow at all on MacConkey agar. The organism can be successfully subcultured onto chocolate agar and 5% rabbit's blood agar at 37°C and increased carbon dioxide. Colony growth takes 2 to 7 days.

Identification. The organism is oxidase- and catalase-positive and nonmotile. Carbohydrate utilization is useful in identifying the organism, but the best results require use of heavy inocula or serum-supplemented carbohydrates.

Bacillary and Spirillary Rat-Bite Fevers

CASE STUDY

A 28-year-old female graduate student entered the hospital complaining of sudden onset of headache, nausea, vomiting, and myalgia. A rat had bitten her during a recent laboratory experiment. The bite wound, on the left index finger, was slightly indurated and nonsuppurative, and there was no lymphadenopathy. The patient did not have a rash, and she had pain in her right ankle. The patient was febrile (101°F), and her blood pressure was normal.

The laboratory results are shown below.

- Complete blood count:
    Hemoglobin: 13.6 g/dL
    Hematocrit: 44%
    WBCs: 126,000/μL; 74% neutrophils; 9% bands

- Cerebrospinal fluid: clear, colorless
    Protein: 26 mg/dL
    Glucose: 94 mg/dL

The cerebrospinal fluid, urine, and blood were negative for the detection of leptospires. Mice and guinea pigs were inoculated intraperitoneally to rule out spirochete infection.

The laboratory inoculated 2 mL of the patient's blood into bovine serum–supplemented nutrient infusion broths (15% vol/vol) containing sterile glucose and starch. Small, fluffy colonies grew in the broths after 72 hours of incubation at 37°C in 5% carbon dioxide. Gram stains showed polymorphic, gram-negative filamentous bacteria. Laboratory mice injected with purified isolates rapidly developed symptoms of rat-bite fever. Nasopharyngeal cultures of the rat that had bitten the student grew pure cultures of *Streptobacillus moniliformis*.

The patient received primarily empirical tetracycline and then received penicillin after the infective organism was identified.

Data from Holden FA, McKay JC: Rat-bite fever: An occupational hazard. Can Med Assoc J 91:214, 1964.

Etiology

Streptobacillus moniliformis, the causative agent of streptobacillary rat-bite fever, is a facultatively anaerobic, nonmotile, nonsporulating, gram-negative bacillus or fusiform, about 0.2 to 2.0 μm in size. The cells may appear singly or in filaments, often with a beaded-chain appearance.

The cells can lose their cell walls and grow as stable L-forms. This organism also causes a milk- or water-borne disease called Haverhill fever (erythema arthriticum epidemicum). Protein gel electrophoresis can differentiate the Haverhill strains and streptobacillary strains.

Spirillum minus, which causes spirillary rat-bite fever, or sodoku, is a motile gram-negative spirillum, about 0.5 μm by 3.0 μm in size (Fig. 34–10). It is now classified under "Species *incertae sedis*" and may be related to the campylobacteria.

Both of these bacteria are normal inhabitants of the nasopharynx and urine of the rat. Both can cause rat-bite fever, which can present as either a cutaneous disease involving a rash and lymphadenitis or a disease transmitted by infected milk. The disease occurs throughout the world, with a higher incidence in urban areas.

Clinical Manifestations

This disease should be of particular interest to laboratory workers, who can become infected by handling laboratory rodents. As many as 50% of apparently normal laboratory rodents harbor *S. moniliformis* in the oropharynx.

Streptobacillary rat-bite fever has a short incubation period. A healing lesion occurs at the primary site, and the infection is associated with polyarthritis and palmar and plantar petechiae. The mortality rate of untreated patients is about 10%.

Ulceration at the primary site, relapsing fever, and a palmar and plantar maculopapular rash occur in spirillary rat-bite fever. Arthritis usually

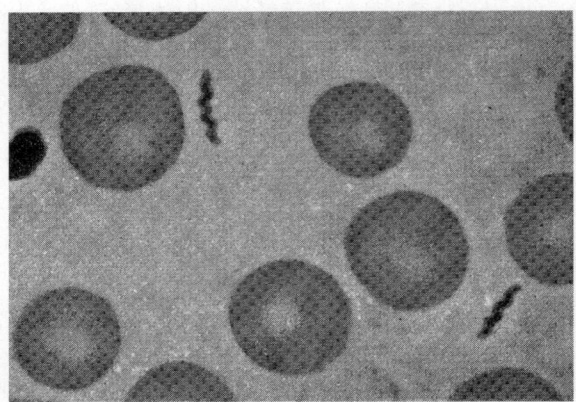

■■■■■ F i g u r e   3 4 – 1 0

Gram stain of *Spirillum* from the blood of an infected mouse. (From Burrows W: Texbook of Microbiology, 19th ed. Philadelphia: WB Saunders, 1968, pp 132, 766.)

does not occur with this infection. Without antibiotic therapy, relapses may occur. The mortality rate is about 7%.

Haverhill fever is a *S. moniliformis* infection transmitted by contaminated milk, milk products, or water. Contaminated milk is often implicated in this presentation of rat-bite fever. All the classic streptobacillary rat-bite fever symptoms may be present, although rash and arthritis may be absent.

Laboratory Diagnosis

Unclotted blood, pus, synovial fluid, and ascites are the specimens of choice for the recovery of *S. moniliformis*. Direct Gram stain preparations reveal pleomorphic gram-negative bacilli, as shown in Figure 34–11. Acridine orange–stained preparations may make it easier to see the organisms through the background material. The growth of the organism is inhibited by SPS-containing media. Good colony growth is possible on Loeffler serum plates, 5% sheep's blood agar, or Columbia colistin–nalidixic acid (CNA) blood agar. Colonies resemble cotton-like masses after 2 to 7 days of incubation. The isolated colonies may change to L-forms, making subculture difficult. Streptobacilli can be identified by using a combination of fatty acid profiles, growth characteristics, and Gram stain morphology.

Laboratory culture of *Spirillum minus* is difficult. The organism grows in vitro in dextrose veal infusion broth and tomato extract–veal infusion broth and can also be grown in vivo using animal inoculations. The organism is best visualized in blood, pus, and lymph node tissue with the use of darkfield microscopy.

ZOONOTIC INFECTIONS TRANSMITTED BY DIRECT CONTACT OR INHALATION

Anthrax

CASE STUDY

A 9-year-old boy admitted to a hospital in Ankara, Turkey, had a swollen wound on the right jaw. The lesion had appeared 3 days earlier. His temperature was 37.5°C. The right side of his face was swollen from the bottom of the right eye to the lower jaw. A large black lesion was in the edematous area.

Interviews revealed that he had recently been in the vicinity of

■Figure 34–11

Gram stain from culture of *S. moniliformis* showing pleomorphic gram-negative bacilli and filaments with characteristic areas of swelling. (From Holroyd KJ, Reiner AP, Dick JD: Streptobacillus moniliformis polyarthritis mimicking rheumatoid arthritis: An urban case of rat bite fever. Am J Med 85:711–714, 1988.)

cows. Laboratory results included a hemoglobin level of 11.4 g/dL and a leukocyte count of 8,400 cells/μL. The white cell count showed 78% polymorphonuclear neutrophils, 2% band neutrophils, 1% eosinophils, 4% monocytes, 15% lymphocytes. Chest radiographs were normal. Gram-positive bacilli, identified as *Bacillus anthracis,* grew from the patient's specimens. The patient received penicillin therapy. On day 10, the patient left the hospital with decreased facial edema.

Data from Gold H: Anthrax: A report of 117 cases. Arch Intern Med 96:387–395, 1955.

Etiology

Anthrax, also known as woolsorter's disease and malignant pustule, is caused by *Bacillus anthracis,* a large (2.5 μm by 10 μm), gram-positive, spore-forming bacillus. This organism occurs naturally in the soil and is a pathogen of herbivores, such as cattle, sheep, and goats. Human infections occur as the result of direct or indirect contact with animals or animal products.

Epidemiology

Bacillus anthracis survives well in soil that is neutral or mildly alkaline. Areas with alternating dry and wet seasons enhance anthrax. Floods tend to concentrate spores, which remain in the grasses after flood waters drain. These spores have been known to last in fields for as long as 20 years. Animals are then infected by grazing in a contaminated area.

Anthrax occurs in nearly every state in the United States, as well as in Central and South America, Africa, and the Middle East. Recurrences of anthrax are prevented by containing the spores and eliminating their spread through the environment. Since the anthrax bacteria in tissues do not sporulate until they are exposed to oxygen, infected animal carcasses are usually incinerated whole or are buried in deep pits and covered with lime.

Vaccines are available for humans and for cattle. A cell-free vaccine, prepared from the protective antigen of *B. anthracis,* is used for people in high-risk occupations. Another vaccine, made with an avirulent, nonencapsulated strain of *B. anthracis,* is available for animals.

Clinical Manifestations

Human anthrax manifests as cutaneous, intestinal, or pulmonary.

Cutaneous Anthrax. The most common form of the infection is the cutaneous form, which mimics many other cutaneous infections. About 95% of cases of anthrax are cutaneous. It is most common in nonindustrialized countries. The spores of the anthrax bacillus from infected animals, or animal products, enter the host through abraded skin and then germinate. After 48 to 72 hours, a small papule develops at the site of entry. The papule darkens and ruptures, creating a painless crater-like ulcer. The ulcer develops into the characteristic necrotic eschar. The infection usually remains localized and self-limiting, and the eschar heals without scarring. In about 20% of the patients with cutaneous anthrax, the immune system is unable to contain the infection, and *B. anthracis* enters the blood stream and disseminates. With treatment, death is rare. In some patients, cutaneous anthrax can manifest as the more life-threatening "malignant edema," in which the initial lesions necrose, and blisters ring the primary site. The resulting edema can impair eating, drinking, and breathing.

Gastrointestinal Anthrax. The gastrointestinal form, the second most common form of anthrax, is found more frequently in nonindustrialized nations. It is caused by the ingestion of meat or meat products contaminated with spores. Once the spores are ingested, they germinate, and the organisms gain entry through preexisting intestinal mucosa lesions. Dissemination of the bacteria then occurs via the lymphatics. Clinically, the patient will be febrile and have bloody stools, and may lose up to 12 L of fluid per 24-hour period. Septicemia and death may result.

Pulmonary (Inhalation) Anthrax. The pulmonary form occurs as a result of inhaling spores, usually from contaminated animal products. This form of anthrax is more common in industrialized countries. Macrophages ingest the spores and then concentrate them in lymph nodes. Eventually, the spores germinate, and sepsis occurs. The result of this form of infection is usually the death of the patient within 24 hours, regardless of the treatment given.

Bacillus anthracis produces three virulence factors, including a poly-D-glutamic acid capsule, edema toxin, and lethal toxin. The edema toxin contains an adenylate cyclase (called edema factor, or EF) and a transport protein (called protective antigen, or PA). The protective antigen carries both edema toxin and the lethal factor. The toxin damages cells and vessels in the infected areas. Necrosis occurs as the result of increased capillary permeability and destruction of the phagocytic cells. Edema can be remarkable and can cause the patient to suffocate by literally swelling the neck shut. Excessive edema of the neck, thorax, and mediastinum signals the beginning of a rapidly fatal course.

The drug of choice for *B. anthracis* infections is penicillin G; secondary choices are tetracycline and erythromycin. Treatment of infected animals includes penicillin or oxytetracycline.

Laboratory Diagnosis

Bacillus anthracis and its spores are dangerous to laboratory workers. Those who work with *B. anthracis* should follow strict infection control measures, including working under a biologic safety hood, using sodium hypochlorite as a disinfectant, wearing protective clothing and gloves, and preventing aerosol formation. Technologists should submit suspected isolates of *B. anthracis* to capable reference laboratories for identification.

Specimen Collection. Specimens for culture include vesicle fluid, blood, and spinal fluid. Recovery of the organism is more likely if the specimen is collected prior to antimicrobial therapy.

Direct Microscopic Examination. Direct examination of blood or edema fluids reveals gram-positive bacilli in short chains. Endospores may be present (Fig. 34–12).

Culture. *Bacillus anthracis* grows as white to gray, nonhemolytic, or slightly hemolytic colonies on sheep's blood agar. The colonies are 2 to 5 mm in diameter. The colonies may have a characteristic "medusa-head" appearance, as shown in Figure 34–13.

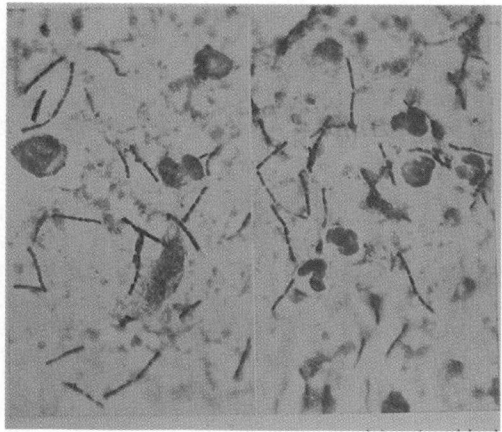

Peripheral blood films collected from a cow dying of anthrax. The preparations are stained with methylene blue. (From Gillies RR, Dodd TC [eds]: Bacteriology Illustrated, 4th ed. New York: Churchill Livingstone, 1976, p 75.)

Identification. The isolate is a nonmotile, encapsulated gram-positive bacillus with square ends. It is susceptible to penicillin and gammaphage. If the bacillus isolate has these characteristics, and if it demonstrates a capsular material on brain-heart infusion agar or nutrient agar with 0.5% sodium bicarbonate in 5% carbon dioxide, it can be presumptively identified as *B. anthracis.* Direct fluorescent antibody testing is used to detect encapsulated organisms in tissue or from culture. Double diffusion assays, in which wells containing antitoxin are placed in media near suspected *B. anthracis* colonies, detect toxigenic strains.

Tularemia

CASE STUDY

A family physician examined a febrile 18-month-old male infant who had diarrhea and nausea. The patient also had pharyngeal pustules. The physician treated the patient with intramuscular penicillin, and a routine throat culture was taken. The laboratory reported "normal flora" on the throat culture.

Despite continued penicillin therapy, the fever persisted. The patient then received Septra (trimethoprim-sulfamethoxazole) but was eventually hospitalized with a temperature of 40°C. The throat lesions were still present, and the patient was lethargic and had a stiff neck. Spinal fluid was collected. Radiographs showed patchy bilateral pneumonia. The patient received ampicillin and chloramphenicol.

The next day the patient had tender, bloody gums, and the lesions had spread to the anterior tonsillar pillars. Physicians also noticed a pseudomembrane forming. A second spinal tap was performed. The patient received isoniazid, rifampin, and streptomycin. Within days, a spinal fluid culture report from the state laboratory indicated recovery of *Francisella tularensis,* and serum collected on day 13 showed a titer of 1:640. Treatment with isoniazid and rifampin was discontinued, and the patient received streptomycin and chloramphenicol for 10 days.

Physicians could not determine the original source of infection. However, the status of the patient improved, and recovery was complete and uneventful. The laboratory results were as follows:

Cerebrospinal Fluid No. 1		Cerebrospinal Fluid No. 2
WBCs	125/µL	3,470/µL
Monocytes	99%	96%
Glucose	62 mg/dL	33 mg/dL
Protein	50 mg/dL	205 mg/dL

Blood No. 1		Blood No. 2
WBCs	28.8×10^9/L	21.7×10^9/L
Bands	9%	28%
Segmented neutrophils	50%	14%

Data from Evans ME, Gregory DW, et al: Tularemia: A 30-year experience with 88 cases. Medicine 64:251, 1985.

Etiology

Tularemia is caused by *Francisella tularensis,* a strictly aerobic gram-negative bacillus about 0.2

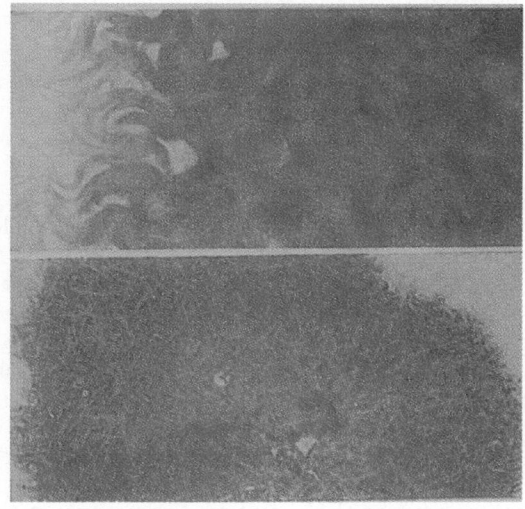

Impression colony of *B. anthracis* stained with methylene blue, demonstrating the "medusa-head" appearance of the colony edge. ×12 and ×75. (From Gillies RR, Dodd TC [eds]: Bacteriology Illustrated, 4th ed. New York: Churchill Livingstone, 1976, p 75.)

μm by 0.2 to 0.7 μm in size. The first isolation of the bacterium was in 1911, during an epizootic outbreak of plague-like ground squirrel disease. The organism was named *Bacterium tularense,* after Tulare County, California, the site of the outbreak. In 1944, researchers defined the role of cottontail and jackrabbits in the transmission of the disease to humans. Since 1945, the proportion of vector-associated tularemia involving deerflies *(Chrysops discalis)* and ticks *(Dermacentor andersoni, D. variabilis,* and *Amblyomma americanum)* has increased, while transmission from vertebrate reservoirs has decreased. In 1959, the genus name of the organism was changed to *Francisella* in honor of Dr. Francis, who first isolated the organism.

Epidemiology

The two biovars *F. tularensis* biovar *tularensis* (type A) and *F. tularensis* biovar *palaearctica* (type B) occur in different parts of the world. In North America, the predominant biovar is type A, the more virulent strain in humans. Humans are usually infected with this strain by rabbits or ticks, although more than 100 species of vertebrate and invertebrate natural reservoirs can transmit the infection. Type B also occurs in North America, but it is more often recovered in Europe and Asia. It is less virulent than type A and is more often transmitted by rodents and mosquitoes. Types A and B differ from each other biochemically and genetically, but not serologically.

Clinical Manifestations

Tularemia is an acute, febrile, granulomatous disease characterized by rapid onset and flu-like symptoms. The most common presentations are ulceroglandular (ulcer and lymphadenopathy), oropharyngeal (pharyngeal ulcers and lymphadenopathy), oculoglandular (conjunctival ulcer and lymphadenopathy), glandular (lymphadenopathy without ulcer), pleuropulmonary (no ulcer, possible lymphadenopathy), and typhoidal (no ulcers or lymphadenopathy). The ulceroglandular, oculoglandular, and glandular types usually occur as the result of direct contact with infected vertebrate or invertebrate reservoirs. Typhoidal and oropharyngeal cases of tularemia usually occur after eating contaminated food.

Pneumonic tularemia results from exposure to aerosols.

The symptoms associated with tularemia include fever, chills, headache, and myalgia. Typhoidal tularemia, especially when complicated with pneumonia, has a high fatality rate.

In the United States, most cases of tularemia are ulceroglandular. These infections are usually caused by direct contact with contaminated game, or by insect bites. The incubation period of tularemia is 3 to 10 days. A papule, which forms at the site of infection, eventually ulcerates. Patients usually have only one lesion, although multiple lesions can occur. The actual site of the lesion is indicative of the transmission; lesions of the hands or arms often indicate infection by direct contact with infected mammals, while lesions on the head or the back indicate transmission by an insect vector.

Francisella tularensis is a facultatively intracellular parasite. Macrophages at the site of infection are able to phagocytose the pathogens, but the bacteria survive intracellularly. Leukocytes containing the bacteria drain to the lymph nodes, resulting in lymphadenopathy. Hematogenous spread then occurs. Extracellular bacteria are, by virtue of their capsule, immune to the antibacterial effects of normal serum. There is focal necrosis and granulocytic changes in parenchymal organs. Many of these granulomatous changes are similar to those seen with tuberculosis.

Humans gain immunity to tularemia in essentially the same ways as they do to other intracellular pathogens, such as *Listeria* and *Mycobacterium.* Lymphokine-activated macrophages and opsonizing antibodies probably are responsible for resolving the infection. However, antibody production does not occur until the second week of infection, and an early IgM response does not occur; IgG, IgA, and IgM all appear at about the same time. Tularemia is unusual in that IgM production to the causative organism lasts for up to 11 years, implying the continued presence of bacterial antigens or whole organisms.

Inappropriate antibiotic therapy promotes relapsing febrile episodes. Broad-spectrum antibiotics often fail to clear the host of *F. tularensis,* whereas gentamicin or streptomycin treatment is bactericidal and allows for complete recovery.

Laboratory Diagnosis

Specimen Collection. While the name of the organism implies bacteremia, blood is not the best specimen for recovery, even though the organism has been isolated using the Bactec blood culture system and the Isolator tubes. Ulcer scrapings, biopsy material from lymph nodes, and sputum may offer better chances of recovering the organism. Figure 34–14*A* shows an infected liver.

Serum should be submitted, as routine detection of tularemia is best done serologically. Only biosafety level 3 reference laboratories should test isolates of *F. tularensis*. Even though *F. tularensis* cannot penetrate unbroken skin, this organism is infamous for its role in causing laboratory-acquired disease.

Direct Microscopic Examination. Microscopically, the organism is a tiny gram-negative coccobacillus that shows bipolar staining, as seen in Figure 34–14*B*.

Culture. The organism grows on IsoVitalex-supplemented chocolate agar, cystine-supplemented blood agar, and buffered charcoal–yeast extract (BCYE) within 24 hours. The opalescent colonies are small and mucoid. Colonies may appear α-hemolytic on blood agar after 24 to 48 hours. Identification of suspected *F. tularensis* is best accomplished at a reference laboratory. *Francisella tularensis* isolated from CYE can give false-positive results with DFA stains and IFA serologies for *Legionella pneumophila*.

Serologic Identification. Conventional tube agglutination, microagglutination, and ELISA can measure the serum antibodies of tularemia patients. Serum titers of more than 1:160 or a fourfold increase in titers indicates tularemia.

Brucellosis

CASE STUDY

A 36-year-old Mexican man went to a hospital emergency room complaining of photophobia, fever, chills, nausea, and vomiting. The patient complained of a month-long headache, a nonproductive cough, and muscle aches. He reported decreased urine but denied dysuria or chest pain. One day before admission, he was diagnosed as having malaria and received a prescription for chloroquine. Three months before that episode, he was diagnosed as having brucellosis. The patient had eaten goat's cheese and drunk goat's milk. He lived in a rural endemic area that had poor sanitation and no safe water supply. On this examination he was alert but toxic. His blood pressure was 80/40, his pulse was 90/min, and he had a temperature of 38.3°C. His neck was not stiff, but he experienced pain in for-

ward movement. There was no lymphadenopathy. He had mild clubbing of the fingers with splinter hemorrhages. The laboratory received blood and cerebrospinal fluid for culture, and serum was tested for febrile agglutinins and *Brucella* antibodies. The laboratory results are shown below:

■ Complete blood count:

WBC: 4.6×10^9/L
  38% segmented neutrophils
   1% band neutrophils
  40% lymphocytes
  11% monocytes

■ Smears negative for presence of parasites

■ Cerebrospinal fluid:

96 WBCs/μL
75 monocytes
21 segmented neutrophils

■ The Gram stain was negative for the presence of bacteria or bacterial antigens

■ Serology tests:

B. abortus antibody titer: 1:160
B. canis titer: Not determined
Febrile agglutinins: positive
Weil-Felix: 1:320 to both Ox2 and Ox19 for patient with brucellosis

Seven days after specimen collection, gram-negative coccobacilli were isolated from the blood and cerebrospinal fluid cultures. The isolates were identified as *B. melitensis*. The patient received tetracycline and rifampin. Recovery was uncomplicated.

This patient is the first to have cross-reacting antibodies between *Brucella* and rickettsiae.

Data from Challoner KR, Riley KB, Larsen RA: *Brucella* meningitis. Am J Emerg Med 8:40, 1990.

Etiology

Brucellosis has many synonyms, including Mediterranean fever, Malta fever, Gibraltar fever, Bang's disease, Neapolitan fever, Cyprus fever, and undulant fever. The genus name comes from Sir David Bruce, who, in 1887, was the first to describe these agents as the cause of undulant fever.

Epidemiology

Four species, which originate from animal reservoirs, are pathogenic to humans:

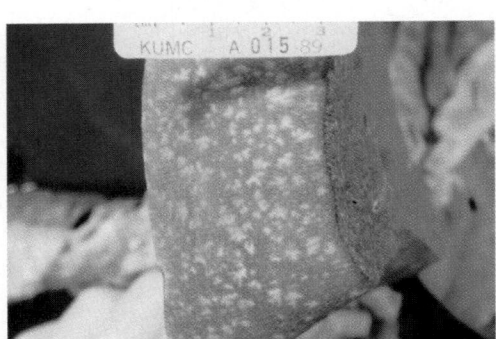

A

B

■ Figure 34 – 14

Francisella tularensis seen in tissue specimens. *A,* Section of an infected liver. The capsular surface shows multiple white, stellate areas of necrosis. *B,* Gram stain from broth culture of *F. tularensis.* (From Hargrave PK, Hulsebus J, Wilson L: Tularemic Shock: A Case Study. Paper presented at the Annual Meeting of the American Society for Medical Technology, 1992.)

■ *Brucella melitensis* (goats)

■ *B. abortus* (cattle)

■ *B. canis* (canines)

■ *B. suis* (swine)

The most pathogenic species for humans, in descending order, are *B. melitensis, B. suis, B. abortus,* and *B. canis.* Other species not known to cause human disease are *B. neotomae* (desert wood rat) and *B. ovis* (sheep). In the animal hosts, the brucellae can induce spontaneous abortion secondary to bacteremia in pregnant females. The urine and milk of infected animals contain the infective organisms.

Brucella melitensis, the most common agent of human brucellosis, occurs in many areas of the world, including Mexico, Central and South America, southeastern Europe, countries bordering the Mediterranean, Africa, southern Russia, India, Iran, and central Saudi Arabia.

Brucella melitensis, B. suis, and *B. abortus* are subdivided into biotypes defined by biochemical reactions and serotypes. This classification defines three biotypes for *B. melitensis,* eight for *B. abortus,* and four for *B. suis.*

Clinical Manifestations

In the United States, the brucellae infect humans primarily through contact with infected animals and animal products. Veterinarians, meat packers, sheepherders, and abattoir workers are at risk for infection.

The organism can enter the body through abraded skin, mucous membranes, or the conjunctiva, and in experiments *B. abortus* has penetrated intact skin. After an incubation period of 1 to 3 weeks, the brucellae are disseminated hematogenously, where circulating monocytes ingest them. The brucellae are intracellular parasites. Monocytes transport the brucellae to lymph nodes. From there, the bacteria are disseminated to the spleen and the liver.

Brucella abortus usually causes granuloma formation, while *B. melitensis* usually causes formation of microabscesses. *Brucella melitensis* is also able to inhibit phagosome-lysosome fusion in phagocytic cells, allowing intracellular bacterial replication. Normal human serum is bactericidal to *B. abortus,* but not to *B. melitensis,* accounting for the relative differences in their pathogenicity. The symptoms of acute brucellosis are chills, fever, sweating, weakness, and fatigue.

The primary virulence factor is apparently the endotoxin, although outer membrane proteins (OMP) may also have a role as virulence factors. There are no detectable exotoxins in the brucellae.

Without antimicrobial treatment, brucellosis can persist from 3 months to 1 year. Relapsing brucellosis is a febrile illness characterized by weight loss, anorexia, and night sweats. Complications include arthritis, vegetative endocarditis, and neurologic disorder. Physically, patients may have lymphadenopathy, hepatomegaly, splenomegaly, orchitis, or epididymitis. Recovery usually occurs within 3 to 6 months with rest and supportive treatment, although tetracycline therapy reduces the convalescent period.

Laboratory Diagnosis

Specimen Collection. Laboratory personnel must be notified when physicians submit specimens that might contain *Brucella,* as specialized media and prolonged incubation periods are required. For laboratory diagnosis of brucellosis, physicians should submit multiple blood cultures and both acute and convalescent sera for serologic testing. Other clinical specimens submitted for examination can include fluids and tissues.

Direct Microscopic Examination. Direct smear prepared from bone marrow, liver biopsies, or exudate from abscesses may reveal minute gram-negative coccobacilli.

Culture. The method of choice for culturing brucellae from blood is the biphasic blood culture bottle. Cultures are incubated in an atmosphere of 5 to 10% carbon dioxide for 30 days. Other blood culture systems are acceptable for the detection of brucellosis; however, blind subcultures should be done about twice a week for 30 days.

Specimens not likely to have excess contamination may be plated onto 5% sheep's blood agar or brucella agar. Specimens with likely contamination may be plated onto selective media, including modified Farrell's Thayer-Martin, BCYE, Kuzdas and Morse, and Farrell.

Brucellae are strictly aerobic gram-negative coccobacilli, although an extended period for counterstaining makes it easier to visualize the organisms on Gram stain. The brucellae can have bipolar staining; are nonsporing, nonmotile, and nonencapsulated; and require carbon dioxide for growth. Cell sizes range from 0.6 to 1.5 μm by 0.5 to 0.7 μm.

Identification. Oxidase- and catalase-positive bacterial isolates that grow on appropriate media within 2 to 3 days, and that are round, 2 to 3 mm in diameter, nonhemolytic, translucent, and opalescent, are possible brucellae (Fig. 34–15). These isolates should be tested with anti-smooth *Brucella* serum.

Preliminary tests needed to differentiate among the species of *Brucella* include carbon dioxide requirements, hydrogen sulfide production, urease, and growth on dye-containing media. Figure 34–16 shows dye inhibition. Additionally, the brucellae reduce nitrates to nitrites, oxidize glucose, but do not oxidize sucrose, lactose, maltose, or mannitol. Table 34–2 shows the differentiating characteristics of brucellae.

Serodiagnosis. Since recovery of *Brucella* from clinical specimens is not highly sensitive, it is recommended to include serologic testing to establish the presence of infection. Titers of more than 100 IU or a four-fold rise in titers indicates recent infection. Cross-reactions with cholera or tularemia (or vaccination against these diseases) are possible.

Leptospirosis

CASE STUDY

A 42-year-old male farmer complained of a flu-like illness, severe headache, photophobia, myalgia, arthralgia, diarrhea, and vomiting. He milked cows regularly and had recently handled an aborted fetus. Cerebrospinal fluid examination revealed 244 cells/μL, with 75% neutrophils. The cerebrospinal fluid protein level was 803 mg/L, and the ratio of cerebrospinal fluid to blood sugar was 2.3:5.7. Both the direct Gram stain and the culture of the cerebrospinal fluid were negative. The patient's white blood cell count was 6.7×10^9/L, with 82% neutrophils.

The patient had abnormal liver function tests. The alanine aminotransferase level was 109 U/L; aspartate aminotransferase was 181 U/L; alkaline phosphatase was 382 U/L, and bilirubin was 19 μmol/L. The patient received ampicillin and cefotaxime.

Serodiagnosis confirmed infection with *Leptospira interrogans* serovar *hardio*. All other family members and his farm dog had no evidence of infection. Serologic tests for *Brucella* in the cow that had aborted were negative, although the cow did have positive serologies for *L. interrogans* serovar *hardio*. Further testing indicated that 35% of the herd had antibodies to that serovar.

No leptospires grew from the cerebrospinal fluid, but paired serum samples confirmed that his infection was due to *L. interrogans* serovar *hardio*.

Data from Shaunak S, Brettle RP, Inglis JM: More on leptospirosis. N Engl J Med 311:261–262, 1984.

Etiology

Leptospires are helical cells, about 6 to 20 μm by 0.1 μm. They are motile and have two subterminal

■ Figure 34–15

Culture of *Brucella melitensis* on sheep's blood agar after 48 hours. (From Gillies RR, Dodd TC [eds]: Bacteriology Illustrated, 4th ed. New York: Churchill Livingstone, 1976, p 125.)

■■■■■F i g u r e   3 4 – 1 6

Dye inhibition tests with streaks of *B. melitensis (top), B. abortus (middle),* and *B. suis (bottom).* Basic fuchsin at concentration of 1:25,000 *(left)* inhibits the growth of *B. suis,* while thionin at a concentration of 1:30,000 *(right)* inhibits the growth of *B. abortus.* Neither dye at those concentrations inhibits the *B. melitensis.* (From Gillies RR, Dodd TC [eds]: Bacteriology Illustrated, 4th ed. New York: Churchill Livingstone, 1976, pp 112, 113, 120.)

flagella. The cells have characteristic hooks on the ends. The genus *Leptospira* contains a large group of serologically diverse organisms. There are two species, the pathogenic *L. interrogans* and the saprophytic *L. biflexa.* There are as many as 250 serovars organized into 23 serogroups.

Epidemiology

Rodents and domestic animals are the primary reservoirs for the organism. Other animals, in- cluding cows, horses, mongoose, and frogs, can harbor the leptospires. Humans may be directly infected from animal urine or indirectly, by con- tact with soil or water that is contaminated with urine from infected animals. Infected humans can shed leptospires in urine for up to 11 months; cows for 3 1/2 months; dogs for 4 years; and infected rodents possibly for the full lifetime. Veterinarians, abattoir workers, fish and poultry processors, and dairy workers are at risk for lep- tospirosis. Agricultural workers and soldiers are also at risk because of contact with soil and mud.

Leptospirosis is endemic in most areas of the world, although the incidence of disease can be underreported because of undiagnosed infec- tions. The infection is more prevalent in areas with warm climates, especially in late autumn and early winter.

The treatment of choice is doxycycline.

Clinical Manifestations

The number of diagnosed leptospiral infec- tions has increased in recent years. Leptospiral infections can range from subclinical to lethal. The organisms enter the host through mucous membranes or abraded skin. The incubation pe- riod ranges from 5 to 14 days.

■■■■T A B L E   3 4 – 2
BIOCHEMICAL REACTIONS USED FOR THE DIFFERENTIATION OF BRUCELLAE

Biotype	CO$_2$ Req	H$_2$S	Urease	Growth in: BF 20 µg	Thionin 20 µg	Thionin 40 µg
B. melitensis						
1	–	–	+/–	+	+	+
2	–	–	+/–	+	+	+
3	–	–	+/–	+	+	+
B. abortus						
1	+/–	+	+	+	–	–
2	+/–	+	+	–	–	–
3	+/–	+	+	+	+	+
4	+/–	+	+/–	–	–	–
5	–	–	+	+	+	–
6	–	+/–	+	+	+	–
7	–	+	+	+	+	–
B. suis						
1	–	+	+	+/–	+	+
2	–	–	+	–	+	–
3	–	–	+	+	+	+
4	–	–	+	+/–	+	+
5	–	–	+	–	+	+
B. canis	–	–	+	–	+	+
B. neotomae	–	+	+	–	–	–
B. ovis	+	–	–	+/–	+	+

BF, basic fuchsin.

Anicteric leptospirosis is usually a biphasic disease. In the first phase, the patient has sudden temperature spikes, severe headaches, nausea, vomiting, and muscle aches. Patients often experience confusion, secondary to dehydration. The majority of patients develop vivid pink eyes. During this period, the leptospires are recoverable from the patient's blood and spinal fluid. This phase of leptospirosis lasts for about 3 weeks.

In the second phase of anicteric leptospirosis, the leptospires disappear from the circulatory system and cerebrospinal fluid of the patient. This change occurs after the appearance of specific IgM antibodies. Symptoms may subside for a few days, but a limited febrile episode may follow. Patients may also develop aseptic meningitis and severe headaches. During this stage of infection, the urine contains leptospires, but the blood of the patient does not. The length of this stage depends on the serotype of the infecting leptospire.

Weil Disease. *Leptospira interrogans* serovar *icterohaemorrhagiae* can cause *icteric leptospirosis,* also known as Weil disease. This form of leptospirosis is more life-threatening than the anicteric form. Weil disease starts in the same way as does anicteric leptospirosis. On about the third day of the illness, however, the patient develops hemolysis, jaundice, and renal failure. These symptoms occur as the leptospires multiply in the liver and the kidney. In Weil disease, mortality ranges from 15 to 40%. In fatal cases, renal failure is the usual cause of death. In nonfatal cases, the leptospires clear from the patient's kidneys, brain, and eyes as antibodies appear.

Laboratory Diagnosis

Specimen Collection. During the first week of the infection, leptospires are present in the patient's blood and spinal fluid. After the first week of the disease, the patient is no longer leptospiremic, so urine is the specimen of choice.

The leptouremia is sporadic, so multiple specimens increase the chance of recovery. Recovery of leptospires from the urine is optimized by diluting the urine 1:10 and 1:100 to dilute the effect of inhibiting substances. One drop of both diluted and undiluted urine should be inoculated into appropriate media.

Direct Microscopic Examination. Microscopic visualization of leptospires requires darkfield microscopy or other staining procedures other than Gram stain or Giemsa-Wright stains.

Culture. Fletcher semisolid media, bovine albumin-Tween 80 (Ba-Tw 80) media enriched with rabbit serum, and Ellinghausen-McCullough-Johnson-Harris (EMJH) media containing fatty acids or albumin can be used to culture leptospires. The addition of 5-fluorouracil, fosfomycin, and nalidixic acid decreases contamination.

THE RICKETTSIAE

The term "rickettsiae" can specifically refer to the genus *Rickettsia* or can refer to a group of organisms included in the order Rickettsiales. The order includes the following genera:

- *Rickettsia*
- *Coxiella*
- *Rochalimaea*
- *Ehrlichia*

Coxiella differs from the other members of the rickettsia in mode of transmission and symptoms caused. *Rochalimaea* is the only member of the group that has been grown on cell-free media.

With few exceptions, rickettsiae are not agents of zoonoses in the same sense as the organisms presented earlier in this chapter. Most of the members of the rickettsial group discussed in this chapter are arthropod-borne, obligately intracellular pathogens.

These bacteria have become extremely well adapted to their arthropod hosts. The primary hosts usually have minimal or no disease from their rickettsial infection. The arthropod host allows rickettsiae to persist in nature in two ways. First, rickettsiae are passed through new generations of arthropods by transovarial transmission. Because of this mechanism, arthropods are not only vectors for rickettsioses but also reservoirs. Second, arthropods directly inoculate new hosts with rickettsiae during feeding. An exception to this pattern occurs with *Rickettsia prowazekii.* In this case, the arthropod vector, the body louse, can die of the rickettsial infection, and humans act as a natural reservoir.

Rickettsia

General Characteristics

Rickettsia are short, nonmotile, gram-negative bacilli about 0.8 to 2.0 μm by 0.3 to 0.5 μm in size. The bacteria of the genus *Rickettsia* have not been grown in cell-free media, but they have been grown in the yolk sacs of embryonated eggs and several monolayer cell lines. The species of *Rickettsia* are divided into groups according to the types of clinical infections they produce (Table 34–3):

- Spotted fever group
- Typhus fever group
- Scrub typhus group

The etiologic agents of these infections differ antigenically and infect different hosts. There are phenotypic and genotypic similarities, however, between the spotted fever rickettsiae and the typhus group. *Rickettsia tsutsugamushi,* on the other hand, differs from the typhus and spotted fever rickettsiae in several key characteristics.

Clinical Infections

SPOTTED FEVER GROUP

Rocky Mountain Spotted Fever. The most severe of the rickettsial infections, Rocky Mountain spotted fever is caused by *R. rickettsii*. It was first described in the western United States during the latter part of the 19th century. It was not until the early 1900's that researchers showed the infectious nature of the disease, when they infected laboratory animals with the blood of infected patients. The nature of the agent was a mystery, because no bacteria were apparent on direct examination or on culture. However, researchers had to discount a viral etiology, since the agent was filterable. The organism was first seen using light microscopy in 1916.

Humans are accidental hosts and acquire the infection by tick bites. The most common tick vectors are *Dermacentor variabilis,* in the southeastern United States, and *D. andersoni,* in the western part of the country. Other species of ticks, however, can be vectors. Ticks transmit the organism into humans via saliva, which is passed into the host during the tick's feeding. Once in the host tissue, the rickettsiae are phagocytosed into endothelial cells, where they live and replicate in the cytoplasm of the host cell. Replication in the nucleus also occurs. The rickettsiae pass directly through the cell membranes of infected cells into adjacent cells without causing damage to the host cells. The rickettsiae are spread throughout the host hematogenously and induce vasculitis in internal organs, including the brain, heart, lungs, and kidneys.

Clinical Manifestations. Clinically, the patient experiences flu-like symptoms for approximately a week, which follows an incubation period of approximately 7 days. The symptoms include fever, headache, myalgia, nausea, vomiting, and rash. The rash, which may be hard to distinguish in individuals of color, begins as erythematous patches on the ankles and wrists during the first week of symptoms. The rash can extend to the palms of the hands and soles of the feet but normally does not affect the face. The maculopapular patches eventually consolidate into larger areas of ecchymoses.

Once disseminated, the organisms cause vasculitis in the blood vessels of the lungs, brain, and heart, leading to pneumonitis, central nervous system manifestations, and myocarditis. The patient experiences symptoms secondary to vasculitis, including decreased blood volume, hypotension, and DIC.

The mechanisms of pathogenesis are not completely understood. *Rickettsia rickettsii* does not produce exotoxins, and the endotoxin is not

■■■■■ T A B L E   3 4 – 3
CHARACTERISTICS OF SPECIES AND BIOTYPES OF RICKETTSIA

	Species	Vector	Reservoir
Spotted fevers			
Rocky Mountain spotted fever	*Rickettsia rickettsii*	Tick	Ticks, dogs, rodents
Boutonneuse fever	*R. conorii*	Tick	Ticks, dogs, rodents
Rickettsialpox	*R. akari*	Mouse mite	House mouse
Typhus group			
Epidemic typhus	*R. prowazekii*	Body louse	Humans
Brill-Zinsser disease	*R. prowazekii*	None	Humans
Endemic (murine) typhus	*R. typhi*	Rat flea	Rats
Scrub typhus	*R. tsutsugamushi*	Mites	Rodents, mites
Trench fever	*Rochalimaea quintana*	Body louse	Humans
Q fever	*Coxiella burnetii*	None	Cattle, goats, sheep, ticks
Ehrlichiosis	*Ehrlichia canis*	Tick	—

potent enough to explain the effects seen in the host. It is possible that host inflammation is responsible for the damage to the blood vessels.

The mortality rates for untreated or incorrectly treated patients can be as high as 20%, although correct antimicrobial therapy with tetracycline or chloramphenicol lowers the rates to 3 to 6%.

Rickettsialpox. Another of spotted fever is rickettsialpox, caused by *R. akari*. The reservoir is the common house mouse, and the vector is the mouse mite *Allodermanyssus sanguineus*. Rickettsialpox occurs in Korea and the Ukraine as well as in the eastern United States, including the cities of New York, Boston, and Philadelphia. The infections occur in crowded urban areas where rodents and their mites exist.

Clinical Manifestations. Rickettsialpox has similarities to Rocky Mountain spotted fever but is a milder infection. The rickettsial organism, *R. akari*, enters the human host following a mite (chigger) bite. The incubation period is about 10 days, after which a papule forms at the site inoculation. The papule progresses to a pustule, then to an indurated eschar. The patient becomes febrile as the rickettsiae are disseminated throughout the body via the blood. The patient also experiences headache, nausea, and chills. Unlike Rocky Mountain spotted fever, the rash of rickettsialpox appears on the face, trunk, and extremities and does not involve the palms of the hands or soles of the feet. Rickettsialpox symptoms resolve without medical attention.

Boutonneuse Fever. Boutonneuse fever, also known as Mediterranean fever, caused by *R. conorii*, occurs in France, Spain, and Italy. *Rickettsia conorii* also causes Kenya tick typhus, South African tick fever, and Indian tick typhus. Like the agent for Rocky Mountain spotted fever, this rickettsia is tick-borne, and its reservoirs include ticks and dogs.

Clinical Manifestations. Boutonneuse fever is also clinically similar to Rocky Mountain spotted fever. The rash involves the palms of the hands and soles of the feet, just as in Rocky Mountain spotted fever. The rash of boutonneuse fever, however, also involves the face. Also in contrast with Rocky Mountain spotted fever, this disease is characterized by the presence of *taches noires* (black spots) at the primary site of infection. Taches noires are lesions caused by the introduction of *R. conorii* into the skin of a nonimmune person. As the organism spreads to the blood vessels in the dermis, damage occurs to the endothelium. Edema, secondary to increased vascular permeability, reduces blood flow to the area and results in local necrosis.

THE TYPHUS GROUP

The typhus group of rickettsia includes the species *R. typhi* (also called *R. mooseri*) (*endemic typhus,* also referred to as *murine typhus*) and *R. prowazekii* (*epidemic louse-borne typhus* and *Brill-Zinsser disease*) (see Table 34–3). Generally, the typhus rickettsiae differ from the other rickettsial groups in that they replicate in the cytoplasm of the host cell. The infection causes host cell lysis, thereby releasing the rickettsiae. Other rickettsiae pass directly through an uninjured cell.

Endemic Typhus. Endemic typhus is caused by *R. typhi*. The arthropod vector is the oriental rat flea *Xenopsylla cheopis,* and the rat (*Rattus exulans*) is the reservoir. Apparently, the cat flea, *Ctenocephalides felis,* can also harbor the organism. Since this flea infests a large number of domestic animals, it may be an important factor in the persistence of infection in urban areas.

The rickettsiae also survive in nature, to a lesser extent, by transovarial transmission. When a flea feeds on an infected host, the rickettsiae enter the flea's midgut, where they replicate in the epithelial cells. They are eventually released into the gut lumen. Humans become infected when fleas defecate on the surface of the skin while feeding. The human host then reacts to the bite by scratching the site, allowing direct inoculation of the infected feces into abrasions. *Rickettsia typhi* can also be transmitted to humans directly from the flea bite itself.

In the 1940's, there were approximately 5,000 cases of endemic typhus annually in the United States. Rigid control measures have reduced that number to fewer than 100 cases annually. The disease essentially occurs only in southern Texas and southern California in this country but continues to be a problem in areas of the world where rats and their fleas are present in urban settings.

Clinical Manifestations. Like the case with Rocky Mountain spotted fever, the clinical course of endemic typhus includes fever, headache, and rash. Unlike Rocky Mountain spotted fever, endemic typhus does not always produce a rash. When the rash is present, however, it usually occurs on the trunk and extremities. Rash on the palms of the hands occurs rarely. Complications are rare, and recovery usually occurs without incident.

Epidemic Louse-Borne Typhus. Epidemic louse-borne typhus is caused by *R. prowazekii*.

The vectors include the human louse *(Pediculus humanis)*, the squirrel flea *(Orchopeas howardii)*, and the squirrel louse *(Neohaematopinus sciuriopteri)*. The reservoirs are primarily humans and flying squirrels located in the eastern United States. The louse often dies of its rickettsemia, unlike vectors of other rickettsiae.

Louse-borne typhus is still found commonly in areas of Africa and Central and South America where unsanitary conditions promote the presence of body lice. As seen during World War II, epidemic typhus can recur even in developed countries when sanitation is disrupted. While there were fewer than 20 cases of epidemic typhus reported in 1991, there were more than 20,000 documented cases of epidemic typhus during the 1980's, with the vast majority originating in Africa.

Clinical Manifestations. Epidemic typhus is similar to the other rickettsioses. Lice are infected with *R. prowazekii* when feeding on infected humans. The organisms invade the cells lining the gut of the louse. They actively divide and eventually lyse the host cells, spilling the organisms into the lumen of the gut. When the louse feeds on another human, it defecates, and the infected feces are scratched into the skin, just as in endemic typhus.

The disease progression is similar to that of Rocky Mountain spotted fever, including involvement of the palms of the hands and soles of the feet with the rash. Unlike the case with Rocky Mountain spotted fever, the face may also be affected by rash. The mortality rates for untreated patients can approach 40%, although mortality rates in treated patients are very low.

Recrudescent typhus, also called Brill-Zinsser disease, is seen in patients who have previously had epidemic typhus. The *R. prowazekii* lies dormant in the lymph tissue of the human host until the infection is reactivated. Recrudescent typhus is a milder infection than epidemic typhus, and death of the patient is rare. These patients with latent infections constitute an important reservoir for the organism.

SCRUB TYPHUS

Scrub typhus is a rickettsial disease that occurs in India, Burma, eastern Russia, Asia, and Australia. The causative agent is the sole member of the scrub typhus group, *R. tsutsugamushi.* The vector is the chigger, *Leptotrombidium deliensis*, and the main reservoir is the rat. The rickettsiae are transmitted transovarially in the chiggers.

Clinical Manifestations. The transmission of the *R. tsutsugamushi* to the human host is followed by an incubation period of approximately 2 weeks. A tache noire, similar to that of boutonneuse fever, forms at the site of inoculation. The normal rickettsial symptoms of fever, headache, and rash are also present. The rash starts on the trunk and spreads to the extremities. Unlike the case with Rocky Mountain spotted fever, the rash does not involve the palms of the hands and soles of the feet, and the face is also not involved.

TRENCH FEVER

In humans trench fever is caused by *Rochalimaea quintana.* This rickettsial organism is not an obligate intracellular pathogen. This, and the fact that it can be cultured in cell-free media, show this organism to be vastly different from the other rickettsial organisms. The vector is the body louse. Since transovarial passage does not occur in the lice, and since the lice infest only humans, it appears that humans are the main reservoir for the organism.

Clinical Manifestations. Trench fever has the classic rickettsial symptoms of fever, myalgia, and rash. The macular rash usually appears on the chest and abdomen. The infection subsides even without antibiotic therapy, although tetracycline allows for rapid resolution.

Q FEVER

Q fever, the disease caused by *Coxiella burnetii*, is named from the word "query" because of the unanswered questions associated with an epidemic in Queensland, Australia, during the 1930's. *Coxiella burnetii*, the only species in the genus, is an obligate intracellular pleomorphic coccobacillus, approximately 0.5 μm long. Q fever joined trench fever and epidemic typhus as major rickettsial epidemics during World War II. *Coxiella burnetii* differs from other members of the rickettsia in mode of transmission, intracellular development, phase variation, and symptoms of disease.

Coxiella is often isolated from cattle, goats, and sheep. These hosts usually have asymptomatic infections, although they can have abortions induced by the organism. The degree of chronic, asymptomatic infection, however, complicates control of the infection. Maintenance of the organism in nature seems to occur by two means. The most common means of transmission of the organism is from infected aerosols. Aerosols from infected body fluids and birth products are especially likely to transmit the disease. With inhalation as the route of entry, the infectious dose in humans can be as small as one bacterium. The disease can also be transmitted by ticks that feed on infected cattle. These can then spread the infection to other animals. Human infection does not seem to occur through this route. Q fever is often an occupational disease, occurring among people who work with livestock or research animals. Q fever is found throughout the world. In the United States, outbreaks occur in states with large numbers of livestock.

Coxiella undergoes a phase variation that is unique among the rickettsial groups. Organisms that are freshly isolated from clinical sources have phase I antigens, which may inhibit phagocytosis by the host. As the organism is passed into the yolk sac, phase II antigens are expressed.

Clinical Manifestations. Q fever shares some vague flu-like symptoms with the other rickettsioses, including fever, headache, myalgia, malaise, nausea, and vomiting. Chest pain is present in as many as one third of patients, and the actual percentage of patients with respiratory symptoms seems to vary widely. The infection can also present as an atypical pneumonia. There is no rash associated with Q fever, however. The average incubation period is 20 days. The disease is normally self-limiting and resolves within 14 days. However, endocarditis is a sequela in a small percentage of Q fever patients. This manifestation occurs 1 to 20 years following the initial infection, and unlike acute Q fever, chronic Q fever endocarditis is often fatal.

EHRLICHIOSIS

CASE STUDY

A 65-year-old white man had experienced fever, headache, myalgia, and anorexia for 5 days. When he went to his personal physician, his temperature was 38.3°C, his blood pressure was 128/60, and he was dehydrated. There was no sign of rash or lymphadenopathy.

Serial blood cultures were negative, as were routine serologies. The patient received intravenous doxycycline for 3 days. He then received oral doxycycline. His fever resolved after 6 hours of therapy. Following are the laboratory test results:

■ WBC: 2,500/µL

■ Platelets: 57,000/µL

■ Hematocrit: 41.1%

■ Hemoglobin: 11.6 g/dL

■ SGOT (aspartate aminotransferase): 167 U/L

■ Creatine phosphokinase: 403 U/L

■ Alkaline phosphatase: 173

■ Total bilirubin: 1.7 mg/dL

Data from Taylor JP, Betz TG, et al: Serological evidence of possible human infection with *Ehrlichia* in Texas. J Infect Dis 158:217–220, 1988.

Ehrlichiosis was first noted in France in the 1930's when dogs infected with brown dog ticks became ill and died. Postmortem examination revealed rickettsial-like inclusions in the monocytes of the dead animals. These newly described rickettsiae were named *Rickettsia canis*. They were obligately intracellular, arthropod-borne gram-negative coccobacilli. They differ from the other members of the rickettsiae in that they multiply in the phagosomes of host leukocytes, and not in the cytoplasm of endothelial cells. The infected leukocytes eventually rupture and release the organisms, to continue the infective cycle.

Since these organisms grew within host cell vacuoles, they were reclassified into the new genus, *Ehrlichia,* in 1945. The ehrlichiae have a developmental cycle similar to that of the chlamydiae. The infective form of the organism is the elementary body, which replicates in the phagosome. These bodies give rise to inclusions with initial bodies inside. As the inclusions mature, they develop morulae (mulberry-like bodies). As the host cell ruptures, the morulae break into many individual elementary bodies that continue the infective cycle.

The role of *E. canis* as a major veterinary pathogen was seen during the Vietnam War, when the organism was found in an epizootic outbreak of tropical canine pancytopenia, which was responsible for the death of approximately 200 military dogs being used in southeast Asia.

Until the mid-1980's, known cases of human ehrlichiosis occurred mostly in Japan, with the mononucleosis-like sennetsu fever, caused by *E. sennetsu.* These infections are rare, and details of the infection, including vectors and reservoirs, are unknown. *Ehrlichia sennetsu,* which occurs in Japan and Malaysia, infects the host's monocytic cells. Patients infected with this organism are febrile and experience lymphadenopathy.

However, in the mid-1980's, a patient from Arkansas with a history of tick bites developed fever, myalgia, and thrombocytopenia. Peripheral blood smears from the patient had rickettsia-like organisms. The preliminary diagnosis indicated a rickettsial infection even though the patient did not have significant titers to rickettsia. Sera from the patient were ultimately tested using antigens from four species of *Ehrlichia*. The sera were strongly positive for *E. canis*. Subsequent prospective and retrospective studies indicated that approximately 10% of all patients diagnosed with rickettsial fevers despite negative serologies had high titers to *E. canis*.

It is possible that the causative organism is not *E. canis,* but rather an antigenically related undescribed species. It infects lymphocytic or myelocytic cells in the host (Fig. 34–17) and causes a disease similar to other rickettsioses, including fever, anorexia, and rash.

The vector for the infection has not yet been determined, although the vector seems to be a tick, possibly *A. americanum.*

Clinical Manifestations

Human ehrlichiosis resembles Rocky Mountain spotted fever in that the patients are febrile, experience myalgia, and are anorexic. Unlike patients with other rickettsial infections, ehrlichiosis patients can be leukopenic, anemic, and thrombocytopenic. However, about 20% of ehrlichiosis patients have rashes. The form of the rash is variable, appearing as maculopapular, vesicular, or petechial. The rash usually appears on the trunk, arms, or legs. The incubation period of the disease ranges from 1 to 21 days.

Laboratory Diagnosis

Most clinical laboratories do not have the facilities or the means to culture specimens for rickettsiae. Molecular diagnostics can be used to detect the presence of the rickettsiae in tissues, but these tests are still available only in research laboratories. However, most laboratories can perform more routine serologic tests to determine whether a patient has a rickettsial infection.

One well-known, simple, but insensitive and nonspecific test is the Weil-Felix agglutination test. In this test, the patient's sera are reacted with antigens of *Proteus vulgaris* OX19, OX2, and OXK.

A simple test, more specific and sensitive than the Weil-Felix, is a commercially available latex agglutination test for the serodiagnosis of Rocky Mountain spotted fever and typhus fever. The agglutination test does not differentiate between IgG and IgM, but IgM shows a stronger reaction. Thus, a strong positive reaction from a single unpaired serum specimen might indicate active infection. The turn-around time for the test is less than 1 hour, and no complex equipment is needed for the procedure.

Also commercially available is a microimmunofluorescence test. Testing sera for IgG only may not help diagnosis, since the patient's IgG can remain relatively high for several years. Running both IgG and IgM tests can provide more specific information for diagnosis. For instance, in serial sera tests an increase in titer of both IgG and IgM is a good indication of active disease. A high IgG titer and a stable or decreasing IgM titer indicate a recent infection.

Rochalimaea quintana can be isolated from blood cultures in the clinical laboratory using the lysis-centrifugation method. The organism forms tiny colonies on media containing whole blood when incubated in carbon dioxide. Growth may take up to 2 weeks.

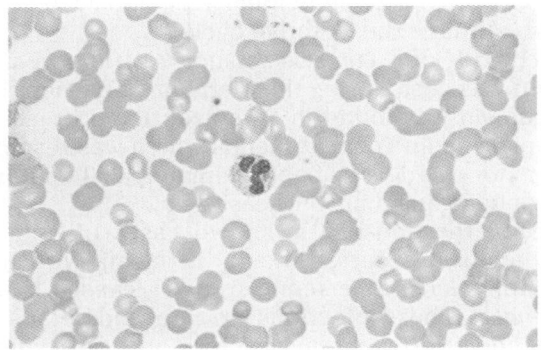

■■■■■ F i g u r e   3 4 – 1 7
Ehrlichia sp in an infected white blood cell.

Bibliography

Barbour AG, Hayes SF: Biology of *Borrelia* species. Microbiol Rev 50:381, 1986.
Brenner DJ, Hollis DG, et al: *Capnocytophaga canimorsus* (formerly CDC Group DF-2), a cause of septicemia following dog-bite, and *C. cynodegmi* sp nov., a cause of localized wound infection following dog bite. J Clin Microbiol 27:231, 1989.

Burgdorfer W, Hayes SF, Corwin D: Pathophysiology of the Lyme disease spirochete, *Borrelia burgdorferi,* in Ixodid ticks. Rev Infect Dis 11(Suppl 6):S1442, 1989.

Butler T: The black death, past and present. I. Plague in the 1980s. Trans R Soc Trop Med Hyg 83:458, 1989.

Challoner KR, Riley KB, Larsen RA: *Brucella* meningitis. Am J Emerg Med 8:40, 1990.

Evans ME, Gregory DW, et al: Tularemia: A 30-year experience with 88 cases. Medicine 64:251, 1985.

Gold H: Anthrax: A report of 117 cases. Arch Intern Med 96:387–395, 1955.

Goldstein EJC: Household pets and human infections. Infect Dis Clin North Am 5:117, 1991.

Gorby GL, Peacock JE, Jr: *Erysipelothrix rhusiopathiae* endocarditis: Microbiologic, epidemiologic, and clinical features of an occupational disease. Rev Infect Dis 10:317–325, 1988.

Hantson P, Gautier PE, et al: Fatal *Capnocytophaga canimorsus* septicemia in previously healthy women. Ann Emerg Med 20:126, 1991.

Harkess JR: Ehrlichiosis. Infect Dis Clin North Am. 5:37, 1991.

Holden FA, McKay JC: Rat-bite fever: An occupational hazard. Can Med Assoc J 91:214, 1964.

Hsu HW, Finberg KW: Infections associated with animal exposure in two infants. Rev Infect Dis 11:108, 1989.

Lang JL: Catching the bug: How scientists found the cause of Lyme disease and why we're not out of the woods yet. Conn Med 53:357, 1989.

Levin RE: An unusual presentation of Lyme arthritis. J Rheumatol 16:1500, 1989.

McCalmont C, Zanolli MD: Rickettsial diseases. Dermatol Clin 7:591, 1989.

McDade JE: Ehrlichiosis: A disease of animals and humans. J Infect Dis 161:609, 1990.

McEnvoy MB, Noah ND, Pilsworth R: Outbreak of fever caused by *Streptobacillus moniliformis.* Lancet 2:361, 1987.

Raffi F, Barrier J, et al: *Pasteurella multocida* bacteremia: Report of 13 cases over 12 years and review of the literature. Scand J Infect Dis 19:385, 1987.

Ramadass P, Jarvis BD, et al: DNA relatedness among strains of *Leptospira biflexa.* Int J Syst Bacteriol 40(3):231, 1990.

Reboli AC, Farrar WE: *Erysipelothrix rhusiopathiae:* An occupational pathogen. Clin Microbiol Rev 2:354, 1989.

Rikihisa Y: The tribe Ehrlichieae and ehrlichial diseases. Clin Microbiol Rev 4:286, 1991.

Sawyer LA, Fishbein DB, McDade JE: Q fever: Current concepts. Rev Infect Dis 9:935, 1987.

Schmid GP: Epidemiology and clinical similarities of human spirochetal diseases. Rev Infect Dis 11(Suppl 6): S1460, 1989.

Shaunak S, Brettle RP, Inglis JM: More on leptospirosis. N Engl J Med 311:261–262, 1984.

Taylor JP, Betz TG, et al: Serological evidence of possible human infection with *Ehrlichia* in Texas. J Infect Dis 158:217–220, 1988.

Tzianobos T, Anderson BE, McDade JE: Detection of *Rickettsia rickettsii* DNA in clinical specimens by using polymerase chain reaction technology. J Clin Microbiol 27:2966, 1989.

Van Eys GJ, Gravekamp C, et al: Detection of leptospires in urine by polymerase chain reaction. J Clin Microbiol 27(10):2258, 1989.

Weber DJ, Walker DH: Rocky Mountain spotted fever. Infect Dis Clin North Am 5:19, 1991.

Ocular Infections

Darlene Miller

OBJECTIVES

OBJECTIVES

1. Provide a brief overview of ocular microbiology.

2. Identify ocular structures and their roles(s) in health and disease.

3. Describe the role of normal ocular flora.

4. Identify common ocular infections.

5. Describe laboratory procedures for the recovery and identification of ocular pathogens.

6. Review ocular therapeutic regimens.

7. Outline unique ocular procedures.

8. Interpret ocular culture results.

The spectrum of ocular infections encompasses relatively mild episodes of conjunctivitis and blepharitis (inflammation of the edges of the eyelids) to the more severe and sight-threatening conditions of keratitis and endophthalmitis (inflammation of the inside of the eye). Adnexal areas and the sclera, lacrimal system, and bony orbit are also subject to microbial invasion. Any organism capable of gaining entrance to ocular structures can cause disease. Bacteria, fungi, viruses, and protozoa all play prominent roles in the pathogenesis of ocular disease.

This chapter will do the following:

■ describe the different ocular structures and their functions

■ discuss the role of normal flora in protecting ocular structures

■ describe the common ocular infections and their causative agents

■ describe the laboratory procedures for recovery and identification of ocular pathogens

■ discuss ocular therapeutic regimens

Because of their location, external ocular structures such as the conjunctivae and cornea are frequently challenged by a variety of microorganisms. Whether an infection or damage ensues depends on the underlying condition of the structure and the character of the invading organism.

The intact epithelia of the external structures provide a protective barrier against invasion by most microorganisms. A few microbes, however, can invade and penetrate the intact epithelium of the conjunctiva or cornea. These include *Neisseria gonorrhoeae*, *N. meningitidis*, *Streptococcus pneumoniae*, *Listeria monocytogenes*, and *Corynebacterium diphtheriae*. For other microbes to enter and establish disease, there must be a break in the protective barrier. Once this barricade is breached (e.g., trauma, insertion or removal of a contact lens) an intrusion by pathogenic and saprophytic organisms can occur. When an infection has started in one layer of the eye, the spread to adjacent layers and tissues can be rapid and can result in devastating and permanent damage to the functional integrity of the eye.

Protection of ocular structures is due in part to a defense system that embraces local and systemic, specific and nonspecific, and humoral and cellular mechanisms that join together to prevent microbial colonization or invasion, and in part to an anatomical arrangement that leaves the inner ocular structures very well sequestered.

Table 35–1 provides a list of organisms recovered from ocular infections. The list is long and varied. Any organism that can gain entrance to the internal structures of the eye is capable of causing infection. The eye does not exist in a vacuum, and many systemic illnesses, such as tuberculosis, diabetes, hypertension, and acquired immunodeficiency syndrome (AIDS), also present ocular manifestations. The organism most likely to be encountered depends on the season, climate, age of the patient, and underlying disease.

ORGANISMS RECOVERED FROM OCULAR DISEASE

Bacteria

Gram-Negative (Aerobic)

Haemophilus influenzae
H. aegyptius
H. parainfluenzae
Neisseria gonorrhoeae
Neisseria meningitidis
Moraxella catarrhalis
Moraxella species
M. lacunata
Pseudomonas aeruginosa
Other pseudomonads
Enterobacteriaceae
Eikenella corrodens
Flavobacterium species
Kingella species
Aeromonas hydrophila
Actinobacillus actinomycetemcomitans
Brucella species
Achromobacter xylosoxidans
Treponema pallidum
Cat-scratch bacillus
Francisella tularensis
Borrelia tularensis
Borrelia burgdorferi (Lyme disease)

Gram-Positive (Aerobic)

Staphylococcus aureus
Streptococcus pneumoniae
Enterococcus faecalis
Enterococcus species
Viridans streptococcal group
β-Hemolytic streptococci (A, B, C, F, G)
Corynebacterium species
Staphylococcus epidermidis
Coagulase-negative Staphylococcal species
Mycobacterium tuberculosis
M. fortuitum
M. chelonae
M. leprae
MOTT, other
Listeria monocytogenes
Bacillus cereus
Bacillus species, other
Micrococcus species

Gram-Negative (Anaerobic)

Capnocytophaga species
Fusobacterium species
Bacteroides species

CHLAMYDIA

Chlamydia trachomatis
C. psittaci
C. pneumoniae

RICKETTSIA

Rickettsia prowazekii
R. tsutsugamushi
R. rickettsii
R. akari
Coxiella burnetii

Gram-Positive (Anaerobic)

Propionibacterium acnes
Actinomyces israelii
Actinomyces species
Peptostreptococcus species
Clostridium species
Propionibacterium propionicus

Parasites

Acanthamoeba species
Microsporidia species
Toxoplasma gondii
Loa loa
Onchocerca volvulus
Wuchereria bancrofti
Oestrus ovis
Taenia solium
Toxocara canis and *cati*
Trypanosoma species
Leishmania species
Ascaris lumbricoides
Schistosoma haematobium
Phthirus pubis
Fly larvae
Trichinella spiralis
Malaria *(Plasmodium)*
Vahlkampfia species

Viruses

Herpes simplex virus I and II
Adenovirus
Enterovirus
Coxsackievirus
Cytomegalovirus
Varicella zoster virus
Epstein-Barr virus
Human papilloma virus
Measles virus

Molluscum contagiosum virus
Vaccinia virus
Mumps virus
Newcastle disease virus
Human immunodeficiency virus
Influenza virus

Fungi

Candida albicans
Candida species (other)
Cryptococcus neoformans
Coccidioides immitis
Acremonium species
Alternaria species
Aspergillus species
Bipolaris species
Blastomyces dermatitidis
Cladosporium species
Curvularia species
Cylindrocarpon species
Drechslera species
Exophiala jeanselmei
Fusarium solani
Fusarium oxysporum
Fusarium species
Helminthosporium species
Histoplasma capsulatum
Lasiodiplodia species
Neurospora species
Paecilomyces species
Penicillium species
Phialophora species
Scedosporium apiospermum
Sporothrix schenckii
Torulopsis glabrata
Volutella species
Zygomycetes
Nocardia species
Streptomyces species

OCULAR STRUCTURES

Figure 35–1 outlines the most important ocular structures. The visual system comprises the eyeball, muscles, fat, nerves, orbital bones, and neural pathways that carry and translate the electrical impulses into vision. This system does not actually "see" but rather acts as a receptor for sensory light stimuli that are translated to neural impulses by the retina and then processed by the occipital lobe of the brain. In this *light*, the eye may be considered a frontal extension of the brain.

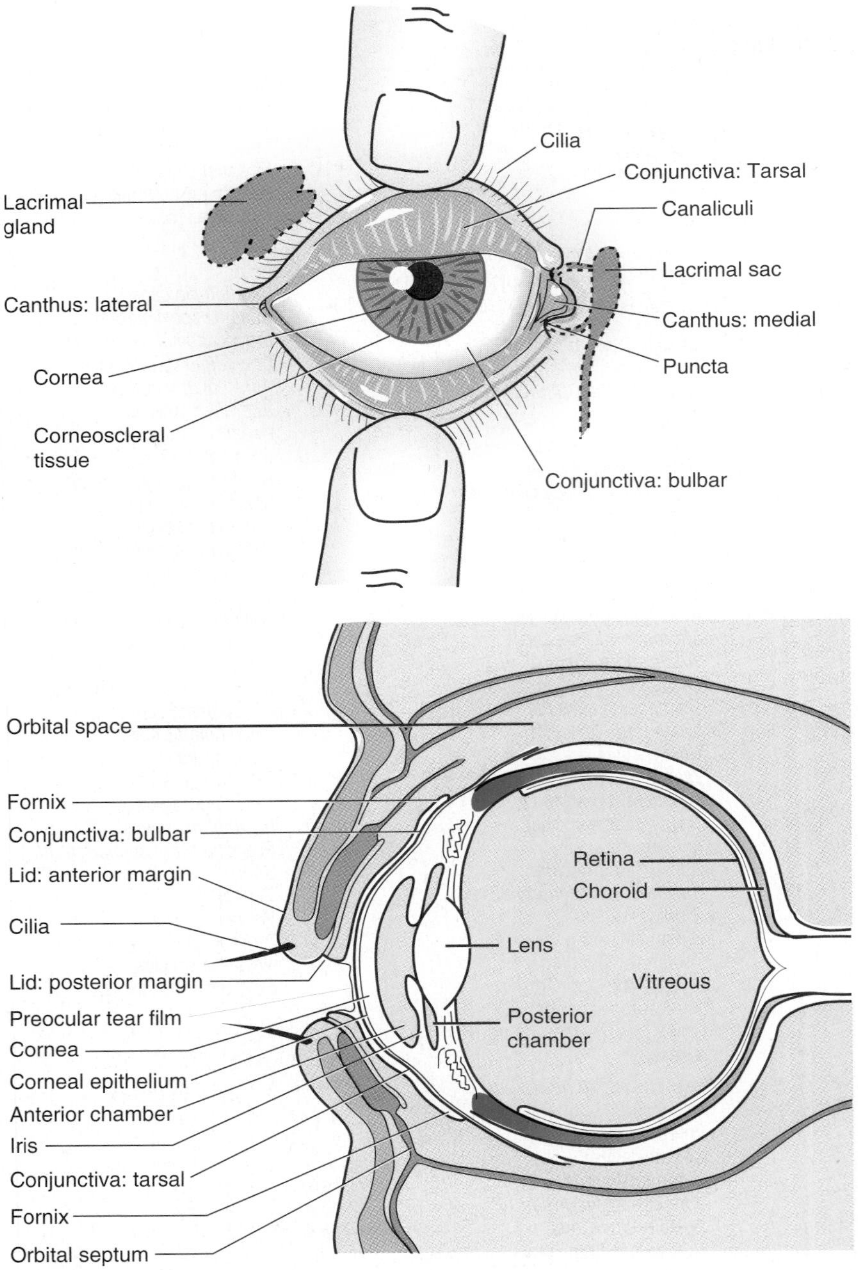
Cilia
Conjunctiva: Tarsal
Canaliculi
Lacrimal gland
Lacrimal sac
Canthus: lateral
Canthus: medial
Cornea
Puncta
Corneoscleral tissue
Conjunctiva: bulbar
Orbital space
Fornix
Conjunctiva: bulbar
Retina
Lid: anterior margin
Choroid
Cilia
Lens
Lid: posterior margin
Vitreous
Preocular tear film
Posterior chamber
Cornea
Corneal epithelium
Anterior chamber
Iris
Conjunctiva: tarsal
Fornix
Orbital septum

Figure 35–1

Common ocular structures. (Modified and reproduced with permission from Jones DB, Leisegang TJ, Robinson NM: Laboratory diagnosis of ocular infections. In Washington JA [ed]: CUMITECH 13. Washington, DC: American Society for Microbiology, 1981, p 2.)

Conjunctiva. The conjunctiva is essentially a mucous membrane, similar to those in the mouth and nose. It lines the upper and lower lids and constitutes the frontline defense in repulsing invading organisms. The tears that keep the conjunctiva moist contain many enzymes and other factors (IgG, IgM, β-lysin) that are antimicrobial in nature.

Lids. The eyelids are thin, elastic layers or folds of tissue that help protect the structures of the orbit. They are the thinnest skin in the body. The blinking actions of the lids and conjunctivae help keep the cornea and sclera lubricated as well as sweep away debris and potential pathogens. The cilia, or lashes, act as a filtering and monitoring system to also alert the brain about potentially harmful agents.

Cornea. The cornea is considered the "window" of the eye. Its size and function are similar to those of a crystal on a wristwatch. The internal structures can be viewed through the cornea. The curved corneal surface is a refractive, collecting and focusing light onto the retina. Many free nerve endings are located in the corneal epithelium. When there is a break in or an injury to the epithelium, the patient usually complains of considerable pain. Infections of and injury to the cornea are considered true ocular emergencies. A layer of tears (the tear film) blankets the cornea to provide optical clarity, lubrication, and nutrition.

Sclera. Known as the "white of the eye," the sclera protects and provides structural support to the internal ocular structures. It consists of tough, interlacing collagen fibers, which give it an opaque appearance.

Orbit. The "closed-box" structure of the bony orbit serves to protect the soft tissues of the eye. These pyramid-shaped bones are connected on several sides with the maxillary, ethmoid, sphenoid, and frontal sinuses. The ocular structures actually occupy only about one fifth of the orbital cavity, with fat and muscle filling the rest of the space. This unit protects the internal structures from blunt trauma.

Lacrimal Apparatus. The lacrimal gland, tear sac, accessory glands, canaliculi, puncta (lacrimal points), and nasolacrimal duct constitute the lacrimal apparatus. The function of this system is to provide tears to lubricate the epithelial lining of the conjunctivae, cornea, and sclera. The tears play a protective role in warding off potential pathogens. When one area of this system is clogged, there is a malfunction that may result in too many or too few tears, exposed epithelium, or gross swelling of the lids. All these conditions can lead to infection.

Anterior Chamber. This is the fluid-filled (aqueous humor) chamber at the front of the eye, divided by the iris into anterior and posterior cavities. Its function is to maintain the intraocular pressure. When there is an alteration in the production of aqueous humor, glaucoma may result. The matrix of the aqueous humor is similar to that of the serum.

Vitreous Chamber. Filled with a gelatinous material (99% water, collagen fibers, and hyaluronic acid), this chamber makes up about two thirds of the volume of the eye. The function of the vitreous humor is to maintain the elliptical shape of the eyeball. There are no blood vessels in the vitreous chamber, and once it is invaded, the vitreous fluid serves as a magnificent culture medium for the growth of microorganisms. Once the vitreous humor has been lost, it is not replaced naturally. For treatment of vitreous loss, saline, gas, or aqueous humor is injected into the vitreous chamber to preserve the spherical shape of the eyeball. If this medium was not replaced, the eye would collapse.

Uveal Tract. The uveal tract, consisting of the iris, the ciliary body, and the choroid, is the middle layer of the ocular system. The ciliary body produces the aqueous humor that fills the chamber at the front of the eye. The iris is an extension of the ciliary body and divides this chamber into the anterior and posterior cavities. It is attached to the lens and controls the amount of light that enters the eye. The choroid is predominantly composed of blood vessels and functions to nourish the retina.

Retina. The retina is the light-sensitive neural tissue of the eye. It is multilayered and functions as a receptor of the light stimuli, transcribing these into electrical impulses and then sending these impulses along the optic nerve to the brain.

ROLE OF NORMAL OCULAR FLORA

Eighty to ninety percent of the flora cultured from noninflamed eyes consists of coagulase-negative staphylococcal species and diphtheroids. However, depending on the age of the patient, season, location, and underlying conditions, *Staphylococcus aureus*, *Streptococcus pneumoniae*, *Haemophilus influenzae*, and other potential pathogens may be recovered from noninfected eyes (Table 35–2). It is important to know the normal ocular flora to better evaluate ocular culture results. The presence of a resident flora on the conjunctivae and lids acts as a protective mechanism, inhibiting invasion and colonization by more harmful organisms. Early studies by Halbert indicated that some ocular flora possessed substances that inhibited the growth of other species. It is important to remember that once the epithelium of the conjunctiva or cornea is compromised, any organism gaining entrance can result in disease.

INFECTIONS OF THE CONJUNCTIVAE (CONJUNCTIVITIS)

Inflammation of the conjunctival tissue with resultant dilation of the blood vessels—red eye—is the commonest of all ocular symptoms. This

■ TABLE 35–2
MICROBES ISOLATED FROM UNINFECTED EYES

Organisms*	Incidence (%)
Coagulase-negative staphylococcal species[†]	34–94
Propionibacterium acnes[†]	40–86
Corynebacterium species[†]	3–83
Staphylococcus aureus	0–30
Haemophilus influenzae	0–25
Micrococcus species	2–22
Streptococcus pneumoniae	0–5
Viridans streptococcal group	0–12
Gram-negative rods (*Proteus* species, *E. coli*, *Klebsiella pneumoniae*, *Enterobacter* species)	0–5
Moraxella species (including *M. catarrhalis*)	0–3
Bacillus species (*B. cereus*, *B. subtilis*)	0–4
Neisseria species (*N. sicca*, *N. flavescens*)	0–7
Fungi (any saprophyte, depends on locale)	0–24
Anaerobic flora other than *Propionibacterium acnes* (*Peptostreptococcus* species, *Bacteroides* species, *Clostridium* species)	1–5
β-Hemolytic streptococcal species, including *Streptococcus pyogenes*	0–3

*Source is usually the conjunctivae and lids; organism isolated is dependent on age of patient, geographic locale, season, previous and current therapy, and underlying condition (e.g., diabetes, epithelial disease). Most common isolates usually reflect that of the surrounding tissue.

[†]These can also cause mild to severe disease in immunocompromised patients.

condition constitutes more than 50% of the complaints that prompt patients to consult an ophthalmologist, resulting in the majority of microbiologic samples submitted. Conjunctivitis may be acute or chronic. The etiologic agents are usually bacterial or viral. A lesser group of patients present with fungal or parasitic infections.

Bacteria

In adults, *Staphylococcus aureus* is the most frequently isolated pathogen in warmer climates, while *Streptococcus pneumoniae* may be the most common isolate in areas with cooler temperatures. *Haemophilus influenzae*, *Staphylococcus aureus*, *Streptococcus pneumoniae*, *Streptococcus* species, and members of the Enterobacteriaceae are the most frequently isolated organisms from infants and children with conjunctivitis. *Neisseria gonorrhoeae* (Fig. 35–2) and *N. meningitidis* initiate a hyperacute conjunctivitis that produces huge amounts of exudate that runs down the face of the patient. Infants acquire *N. gonorrhoeae* as they travel down an infected birth canal. The disease may present within 5 to 7 days after exposure to the pathogen. It is believed that adults acquire gonococcal conjunctivitis through self-inoculation. Meningococcal conjunctivitis may result from contiguous spread from the respiratory tract. Penicillin-resistant ocular isolates reflect the trend in each community. Our present rate (Bascom Palmer Eye Institute) for penicillinase-producing *N. gonorrhoeae* (PPNG) is 22%. Penicillin remains the drug of choice for *N. meningitidis*. The etiologic agents in chronic conjunctivitis are less clear. The microorganisms that have been isolated include coagulase-negative staphylococcal species, *Staphylococcus aureus*, and *Propionibacterium acnes*. Chronic conjunctivitis may be due to an interaction of the organism and the aggressive ocular immune response.

Routine culture and smears from conjunctival scrapings should reveal the etiologic agent in most acute cases. Culture and smears may be of less value in establishing the etiologic agent in chronic conjunctivitis. Therapy is dependent on the isolated organism. Bacterial conjunctivitis can also be caused by instillation of contaminated cosmetics or medications. The organisms encountered in such infections are *S. aureus*, *S. epidermidis*, *Corynebacterium* species, *Pseudomonas aeruginosa*, and *Proteus mirabilis*. Allergic and chemical conjunctivitis can sometimes be confused with microbial infections. Laboratory tests can be of assistance in confirming the diagnosis.

Chlamydia trachomatis causes a myriad of ocular infections, including neonatal conjunctivitis, inclusion conjunctivitis (Fig. 35–3A), lymphogranuloma venereum (LGV), and trachoma. Currently there are fifteen serotypes (A, B, Ba, C–K, L₁–L₃), or serovars,

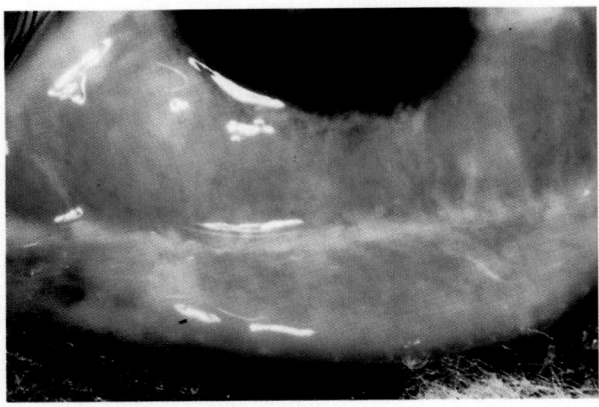

■ Figure 35–2

Gonococcal conjunctivitis. Note the copious discharge in response to invasion by *N. gonorrhoeae*.

of *C. trachomatis*. Certain serovars are associated with certain clinical entities. Trachoma, which is "as old as recorded history," is usually due to serotypes A, B, or C, while immunotypes D to K are usually associated with oculogenital (inclusion conjunctivitis) chlamydiae. Neonatal conjunctivitis occurs when the infant is infected as he or she travels down a contaminated birth canal. Infection becomes apparent within 8 to 10 days. *Lymphogranuloma venereum* (serotypes LGV 1, 2, 3) is strictly an STD, and conjunctival inoculation is accidental. It is more common in tropical climates. We have isolated a case of LGV (serovar 2) in a 13-year-old girl at our institution. *Chlamydia psittaci* infections are rare in ophthalmology, but at least one case has been documented. *Chlamydia pneumoniae* was first isolated from the conjunctiva of a child in Japan, but the disease is more respiratory than ocular in nature.

Direct detection of the chlamydial inclusions or elementary bodies in scrapings or on impression cytology membranes is among the most sensitive methods for confirming ocular chlamydia (see Fig. 35–3*B*). Cultures using McCoy or Hep-2 cells may be used to support the diagnosis, especially in suspected trachoma. Polymerase chain reaction (PCR) and enzyme immunoassay (EIA) procedures for the detection of ocular infection are now available. The preliminary data with such procedures are encouraging. A combination of smear and culture with either PCR or EIA may offer the best system for detecting and confirming ocular chlamydia in populations with high and low rates of sexually transmitted disease (STD) and in areas where trachoma remains endemic.

The prevalence rates for ocular chlamydial infection parallel those of genital disease. In areas with high rates of STD, ocular disease rates also are high. Populations with the highest incidence of disease include neonates and sexually active adolescents and adults. The incidence can range from 20 to 90%, depending on the age group. In areas with low rates of STD, ocular chlamydial infection rates also are low or nonexistent. Tetracyclines, sulfonamides, and erythromycin are the drugs of choice for controlling the spread of these organisms. Both topical and oral preparations are available.

Viruses

Like bacterial conjunctivitis, viral conjunctivitis is quite common. Viral conjunctivitis may indeed be the most common or recognized ocular disease. It too may present as an acute or chronic illness, ranging from a mild, self-limiting condition to a severe, destructive disease resulting in impaired vision. Acute viral infections are attributed mainly to adenoviruses, herpesviruses, or enteroviruses. The adenoviruses are responsible for two distinct ocular viral syndromes. The first, epidemic keratoconjunctivitis (EKC), is quite contagious and is associated with adenovirus types 8 and 19, but recovery of serovars 7, 9, 10, 11, 14 and 16 has also been documented. The second syndrome, pharyngoconjunctival fever (PCF), is

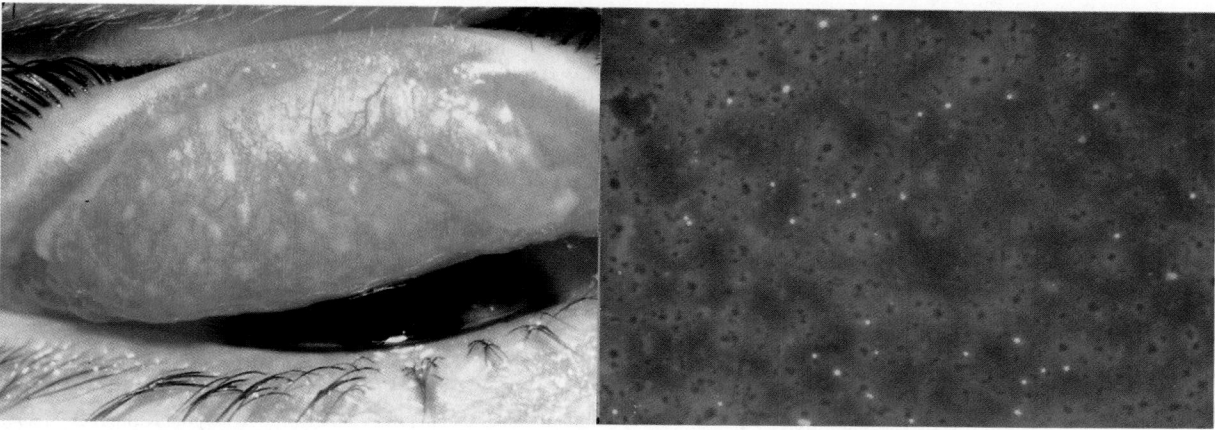

A   B

■■■■■F i g u r e   3 5 – 3

A, White spots on conjunctiva represent pockets of *Chlamydia* organisms in tissue. *B*, Immunofluorescence stain of scrapings from neonatal conjunctivitis, confirming the presence of chlamydial elementary bodies.

caused regularly by adenovirus type 3 and occasionally by serovars 1, 2, 4, 5, 6, 8, and 14. Both syndromes are self-limiting, with no specific treatment; PCF lasts about 10 days, and EKC 3 to 4 weeks at the most. A highly contagious disease, EKC may be spread by physicians, direct contact, or fomites. It is usually responsible for outbreaks in ophthalmic offices and clinics, school locker rooms, or college dormitories.

Acute hemorrhagic conjunctivitis (AHC) (Fig. 35–4), or epidemic hemorrhagic conjunctivitis (EHC), is an acute short-lived infection that is caused by enterovirus 70, coxsackievirus A24, and, on rare occasions, adenovirus type 11. It is often called the Apollo XI conjunctivitis because it was first recognized in Ghana during the time of the Apollo XI moon mission. The onset of symptoms is usually within 8 to 48 hours of exposure, with patients complaining of pain, sensitivity to light, copious tears, and subconjunctival hemorrhages. It is self-limiting, with no outlined treatment program. Recovery occurs within 5 to 7 days. Since this is highly contagious and easily spread from person to person, patients should be isolated or sent home until the condition has resolved.

All of the aforementioned viruses may be grown in tissue culture and serotyped by neutralization tests. Monoclonal antibodies are available for group-specific adenovirus and several members of the enterovirus groups. These antibodies can be used for direct examination or culture confirmation. The more involved neutralization tests should be performed by a reference

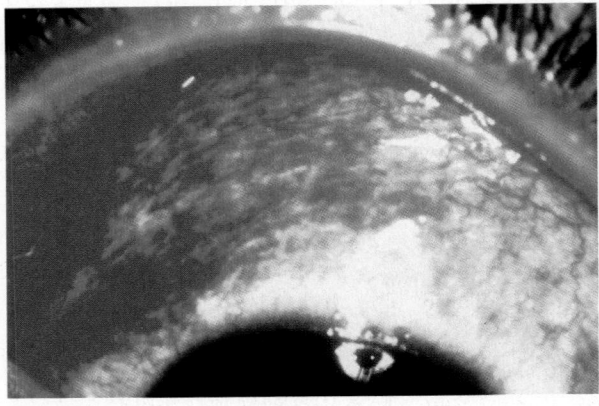

■■■■■■ F i g u r e   3 5 – 4

Acute hemorrhagic conjunctivitis. The etiologic agent is usually enterovirus 70 or coxsackie virus A24. Other members of the enterovirus group may also be recovered. Note the heavy conjunctival hemorrhaging.

laboratory if they are not routinely performed in a particular laboratory.

Herpes simplex blepharoconjunctivitis is responsible for the majority of severe ocular viral infections. This disease usually occurs in young children. It is important to make the distinction between herpes simplex virus (HSV) and adenovirus etiologies, because HSV can be treated with specific ocular antiviral medications, whereas adenovirus infection must run its course. More than 90% of cases of this type of conjunctivitis are due to HSV I, but HSV II has been isolated from both infants and adults. Varicella zoster virus (VZV) may also cause conjunctivitis. This condition is usually a manifestation of the systemic disease chickenpox or a complication of herpes ophthalmicus. Cytomegalovirus can usually cause conjunctivitis when it becomes disseminated or is transmitted through tears from another patient. Epstein-Barr virus conjunctivitis may result as a complication of infectious mononucleosis.

Chronic viral conjunctivitis may result from *Molluscum contagiosum* or vaccinia, a complication of smallpox vaccination.

Rickettsia

All pathogenic human rickettsiae are capable of causing conjunctivitis. The conjunctiva is often the portal of entry for Rocky Mountain spotted fever, scrub typhus, Q fever, endemic murine typhus, and Marseilles fever. This type of conjunctivitis is usually mild but can be a severe complication of the systemic disease. Because isolation of rickettsia is quite difficult, confirmation is by serologic means. The most sensitive and specific tests for confirmation of rickettsial infection are microimmunofluorescence, microagglutination, and complement fixation. Treatment with chloramphenicol or tetracycline can inhibit rickettsial organisms long enough for the body to clear them from the system.

Fungi

Although a variety of fungi can be isolated from the noninflamed eye, fungal conjunctivitis is a relatively rare event. The species that have been recovered from fungal conjunctivitis are *Candida* and *Sporothrix schenckii*. Systemic fungal infections caused by *Coccidioides immitis*, *Histoplasma capsulatum*, and *Cryptococcus neo-*

formans may extend to the conjunctival tissue and cause disease. A high index of suspicion is needed by the physician, and the laboratory must be alerted in order to ensure the proper set-up for recovery of these organisms.

Parasites

Ocular infections due to parasites are, in general, rare; the conjunctiva is among the least affected sites. Conjunctival involvement is usually a secondary complication. Parasites that are involved in ocular tissues include the following.

Loa loa, the "eye worm," is one of the leading causes of blindness in West Africa (river blindness). Clinical symptoms are due to the continued migration of the adult worms into the subcutaneous tissues and blood vessels. Onchocerciasis, which is transmitted by the bite of the blackfly, is also a leading cause of blindness in Africa and is now endemic in South and Central America. Ocular complications come from the discharge of large numbers of microfilariae by the adult female and their subsequent local invasion and tissue damage. Ophthalmyiasis, "fly larvae conjunctivitis" or ocular myiasis, is caused by the deposit of fly larvae (maggots) into the conjunctival sac. This infestation occurs frequently in the tropics but can occur wherever there are people and flies. The maggots may be removed by paralyzing them with 10% cocaine and lifting them out with tweezers.

Detection of ocular parasites depends on observing and removing the actually protozoans, confirming their presence through histologic stains, isolating the organisms from blood or tissues, or confirming their presence by serologic means.

INFECTIONS OF THE LIDS (BLEPHARITIS)

Inflammation of the lid margins and inflammation of the conjunctivae are not mutually exclusive. Conjunctivitis usually presents as a blepharoconjunctivitis. Therefore, any organism that causes conjunctivitis can affect the lids. However, there are unique organisms and conditions for each site. The skin of the lids is among the thinnest on the body. Any organisms capable of initiating skin infections can also cause blepharitis.

Bacteria

Staphylococcus aureus and members of the coagulase-negative staphylococcial family are the most frequently isolated bacteria from the lid margins. Blepharitis involving these organisms is a low-grade inflammation usually associated with functional disease of the seborrheic glands (seborrheic blepharitis). In this mixed infection there are dry (staphylococcal) and greasy (seborrheic) scales attached to the lashes, with various areas of ulcerations that cause the lashes to fall out. Antibiotics are given to cure the staphylococcal disease. Because the scalp, eyebrows, and lids all are involved in seborrhiasis, all must be kept clean with a medicated shampoo. There are four types of glands located in the lids: meibomian gland, the glands of Moll and Zeis, and the accessory lacrimal glands. Acute infection of the glands of Zeis or Moll with staphylococci results in an external hordeolum (stye). This is an abscess with pus formation in the lumen of the affected gland. Hot soaks assist in the continuous draining of the abscess. Topical erythromycin or tetracycline may be applied as supplementary therapy. An internal hordeolum, also due to the staphylococci, is a little larger and affects the meibomian gland. Additional lid glandular disorders include formation of a chalazion, a sterile granulomatous inflammation of the meibomian gland. These usually subside spontaneously. Meibomianitis is inflammation of multiple glands; its etiology is unknown. Only rarely does the laboratory receive a request for lid culture. If such a request is received, it is to confirm the presence of staphylococci and to determine whether therapy is adequate. Other bacteria that can be isolated from blepharitis include *Moraxella* species, especially *M. lacunata,* responsible for "angular blepharitis," infection of the lid angles. Other species of *Moraxella,* however, may be involved. *Bacillus anthracis* causes a caruncle-like lesion on the lid margins, which is eradicated with penicillin. *Actinomyces* species may spread from the skin of the face to the lids. In leprosy the most common ocular finding is total loss of the eyebrow and lashes. An acid-fast stain of biopsy materials reveals typical acid-fast bacilli. No organisms of this mycobacterial species are recovered on Löwenstein-Jensen or other media. Primary infection of the eyelids by *Mycobacterium tuberculosis* results in lid ulceration. Lid involvement is associated with disseminated disease. Differentiation of this organism from *M. leprae* is by growth in culture.

Viruses

Viral blepharitis may be caused by HSV I or II, VZV, poxvirus, papovavirus, or vaccinia. Herpes simplex virus infection usually occurs during early childhood. Vesicles appear on the lid margins and the skin around the eye. The vesicles break open and form crusted secondary lesions, which then may become superinfected by skin organisms. Direct detection with immunofluorescence or immunoperoxidase can be done by scraping the base of a freshly opened vesicle. Vesicular fluids are collected for culture. Ninety-five percent or more of cultures grow HSV within 72 hours or earlier. When the face is involved during episodes of chickenpox (varicella), vesicles may appear on the upper or lower lid margins. Molluscum contagiosum is a wart-like lesion of the lid margins. It is produced by poxvirus. The wart-like lesion is waxy and pearly-white with an umbilicated center. Expression (squeezing) of the white center to allow blood into the lesion is usually adequate management. Vaccinia infection of the lids results from direct inoculation from a smallpox vaccination. Other viruses that produce ocular warts are members of the papillomavirus family.

Fungi

Fungal blepharitis is quite rare and is usually a complication of systemic disease, especially involving *Candida* species and *Blastomyces dermatitidis.*

Parasites

Infestation of the lid margins by parasites is due to complications and spread from adjacent structures. Parasitic lid complications may be observed with cutaneous leishmaniasis, African or American trypanosomiasis, *Loa loa* infections, and dirofilariasis. The crab or pubic louse *(Phthirus pubis)* infests the cilia and lid margins of the eyelids. The patient's main complaint is pruritus, or itching. All members of the patient's family must be treated. Treatment consists of application of a 1% gamma benzene hexachloride (lindane) ointment or shampoo to the affected areas.

INFECTIONS OF THE CORNEA (KERATITIS)

Active invasion of the cornea by microorganisms is considered a true ocular emergency and requires prompt management. *Pseudomonas aeruginosa* as well as some strains of *Staphylococcus aureus* can induce perforation or corneal melt, with resultant loss of the eye within 24 to 48 hours (Fig. 35–5A and B). The cornea contains five layers. Infection begins in the most superficial layer (the epithelium), and, if not checked, it advances through the Bowman zone, the stroma, and the Descemet membrane, to the endothelium (the innermost layer). Many nerve fibers with bare ends are housed in the epithelium, and when exposed, they produce severe pain. Even minor abrasions and insults result in excruciating discomfort. Very few organisms can cross the intact corneal epithelium, but once it has been compromised, any organism can launch an infection. Corneal stromal tissue provides an excellent culture medium for visiting microbes.

Geographic variations in the etiology of bacterial and fungal ulcers are evident. *Pseudomonas aeruginosa* constitutes up to 50% of the cases of keratitis in Florida and the southern United States, while *S. aureus* is the most frequent isolate in the northern part of the country. Laboratory assistance is mandatory in all cases of keratitis or suspected keratitis. Scrapings for smears and cultures are inoculated directly onto slides and culture plates and sent immediately to the laboratory for evaluation. Scrapings for stains can afford the physician an early indication of the offending organism and assist in the selection of appropriate therapy. The cornea may also be invaded via metastatic spread from the conjunctiva or systemic lesions.

Bacteria

Ninety percent of the bacterial corneal ulcers seen at our institute (BPEI) during the last 10 years have included *P. aeruginosa, S. aureus, S. epidermidis, Serratia marcescens, Proteus mirabilis,* and *Streptococcus pneumoniae.* These are listed in descending order of frequency. The order may differ in different regions of the United States, but the list of organisms should be similar. *Pseudomonas aeruginosa* is extremely dangerous

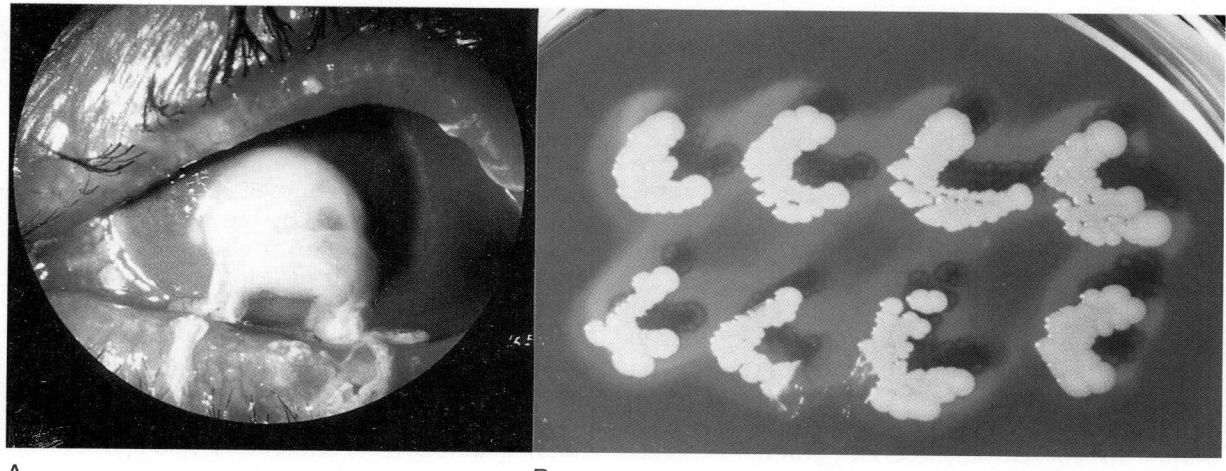

A   B

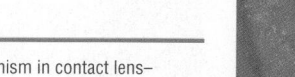

A, Corneal melt due to bacterial invasion. *B,* "C" streaks of *Staphylococcus aureus* from infected cornea. The enzymes produced by some strains of *S. aureus* and *Pseudomonas aeruginosa* can liquefy the cornea within 48 hours.

because it elicits enzymes that liquefy the corneal tissue within 24 to 48 hours. This is the most frequent isolate recovered from contact lens–associated keratitis. These cases include soft, daily, extended-wear, and disposable contact lenses. Table 35–3 indicates other organisms recovered from contact lenses and lens solutions (Fig. 35–6).

TABLE 35 – 3
CONTACT LENS ISOLATES OTHER THAN *PSEUDOMONAS AERUGINOSA**

Serratia marcescens
Achromobacter xylosoxidans
Staphylococcus aureus
Enterobacter cloacae
Klebsiella pneumoniae
Staphylococcus epidermidis
Pseudomonads
Bacillus species
Mycobacterium triviale/other MOTT
Acanthamoeba species
Free-living amebae other than *Acanthamoeba*
Curvularia species
Penicillium species
Aspergillus species
Candida species

* *Pseudomonas aeruginosa* is the prevailing organism in contact lens–associated keratitis.

These organisms have been isolated predominantly from soft (daily-wear, extended-wear, and disposable) lenses, lens cases, and solutions. Hard contact lenses are less involved in this type of keratitis.

The best preventative measure for avoiding this potentially catastrophic ocular disease is to follow the physician's and manufacturer's guidelines for wearing and caring for contact lenses and solutions.

Less frequent corneal isolates include *N. gonorrhoeae, Moraxella* species, *Corynebacterium diphtheriae,* and *Streptococcus* viridans group. Mycobacteria other than tubercle bacilli (MOTT) are being isolated with increasing frequency (Table 35–4; Fig. 35–7). Ulcers with these organisms are chronic and indolent and have been associated with trauma, contact lens wear, and wound contamination with soil and water. *Nocardia* species are also recovered from corneal tissue. The patient histories and disease courses are similar to those of mycobacterial infection. Differential

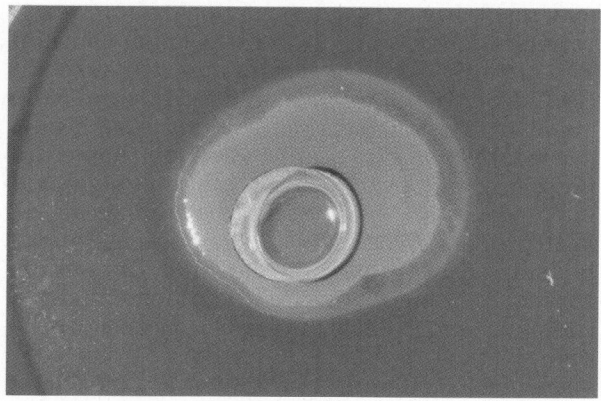

Figure 35 – 6

Growth of *Pseudomonas aeruginosa* from daily-wear (soft) contact lens. Patient had an ulcerative keratitis. The corneal culture also grew *P. aeruginosa.*

■■■■■ T A B L E 3 5 – 4
MOTT KERATITIS ISOLATES

Mycobacteria chelonae
M. fortuitum
M. avium-intracellulare
M. nonchromogenicum
M. triviale

NOTE: *Mycobacterium tuberculosis* and *M. leprae* are ocular pathogens, usually representing an extension of systemic disease.

diagnosis must be made in the laboratory, based on morphology and biochemical and anitmicrobial profiles. All chlamydial infections (inclusion conjunctivitis, LGV, and trachoma) can be accompanied by corneal lesions. The gradual destruction of the cornea by the diseased lids is responsible for the high rate of blindness during chronic infection with these agents.

Any organism gaining entrance into the corneal stroma can produce disease. Laboratory investigation must be prepared to recover both common and uncommon agents and report these findings as swiftly and accurately as possible. Sensitivity profiles are necessary to ensure proper therapeutic management. Selection of antibiotics should be based on the microorganism.

Viruses

Herpes simplex virus is the most common and severe of the ocular antagonists. About 500,000 new cases of ocular disease caused by herpes simplex occur annually. This type of infection may result from direct inoculation or from reactivation of latent virus in the trigeminal ganglion. The majority of the ocular lesions are caused by GHSV I, but HSV II has been recovered from cases involving both infants and adults. The lesions caused by these two are indistinguishable. Impression cytology and direct or indirect immunofluorescence staining confirm the etiologic agent. With proper collection of infected tissue, HSV will be isolated in tissue culture within 72 hours or earlier. Treatment includes débridement, when possible, and topical antivirals. Herpes zoster ophthalmicus, or VZV, is a reactivation of a latent viral infection that first presented as chickenpox in childhood. In the adult, ocular zoster presents as a painful, severe, and sometimes blinding disease. Oral and topical acyclovir is used to reduce the discomfort. Adenovirus infections, measles, rubella, mumps, Epstein-Barr virus infection, and Newcastle disease all have corneal sequelae.

Fungi

Opportunistic or saprophytic organisms are most often recovered from fungal infections of the cornea. This is the rule rather than the exception. There is commonly some history of trauma with soil or plant material. All cases of fungal keratitis must be confirmed by the laboratory.

Recovery and identification should occur within 3 to 7 days in most cases. Composing 90% of the filamentous fungi isolated from corneal ulcers from 1983 through 1991 at our institution (BPEI), in descending order, are *Fusarium solanae, F. oxysporum, Curvularia* sp, *Paecilomyces* sp, and *Aspergillus* sp. Nonfilamentous isolates included *Candida albicans, C. parapsilosis,* and *C. tropicalis.* Samples for culture must be collected from multiple scrapings of the involved actively advancing edges of the ulcer. Sometimes a biopsy is necessary to isolate the fungi. Giemsa and calcofluor white preparations from scrapings can confirm the presence of hyphal elements or budding yeast. Antifungal therapy can be instituted once the fungal etiology has been established. Topical antifungals include amphotericin B, natamycin, nystatin, and imidazoles compounds. Selection is dependent on the organism. Antifungal sensitivity testing is best handled by a reference laboratory.

Parasites

Infection of the cornea by free-living amebae (*Acanthamoeba, Naegleria,* and *Vahlkampfia* species) is a devastating consequence of injury or insult with contaminated water, contact lenses, or soil. *Acanthamoeba* species are the most frequent

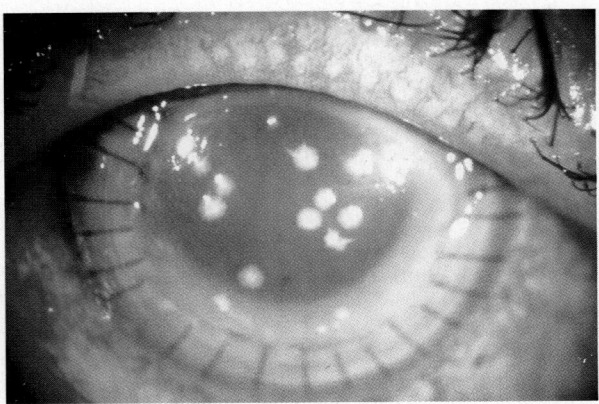

■■■■■ F i g u r e 3 5 – 7
Colonies of *Mycobacterium fortuitum* growing on infected corneal graft tissue.

isolates. Homemade saline solution associated with contact lens wear was the original identified risk factor. This factor seems to play a lesser role in current amebic keratitis. Amebic keratitis is often mistaken for viral or fungal keratitis, with the diagnosis being made only when all other cultures (bacterial, fungal, and viral) are negative. Scrapings are collected and placed in the center of two non-nutrient agar plates. A few drops containing heat-killed or "live" *Escherichia coli* are placed over the tissue. The plates are sealed with tape, and one is placed at room temperature and the second in the incubator (35°C). If only one plate is collected, it is kept at room temperature. Scrapings also may be inoculated into tissue culture, and maintenance media added. The amebae cause a generalized destruction of the tissue culture cells. A third method of cultivation is to grow free-living amebae in broth (axenic) culture. Direct detection of the cysts and trophozoites may be accomplished by utilizing Giemsa or calcofluor white stains.

Depending on the quality of the scrapings and the infectious dose of the organism, trophozoites may be seen within 48 hours. The polygonal, double-walled cysts may be detected on direct smear with either Giemsa or calcofluor white stain. Trophozoites are much less discernible on smear and are best seen in "tracts" on the agar (Fig. 35–8).

Microsporidian keratitis, seen in immunocompromised patients, is a disease that has appeared fairly recently on the horizon. Microsporidiosis is usually a disease of bees or other insects. The microsporidians are intracellular, spore-forming protozoans. In humans, both ocular and nonocular diseases have been reported. Currently, this is an emerging disease of patients

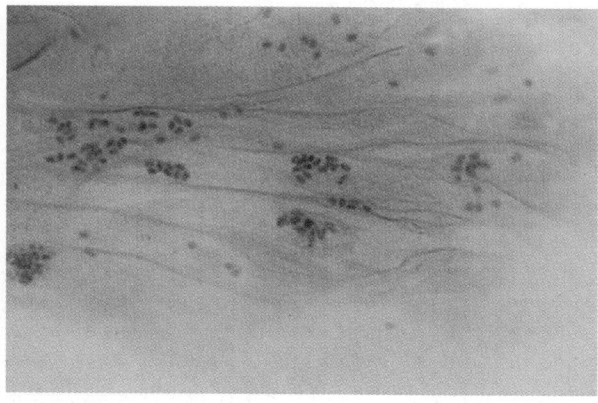

■■■■■ F i g u r e   3 5 – 9
Gram stain revealing the oval cyst of Microsporida species. Organisms can also be detected with Giemsa, acid-fast, and calcofluor white stains.

with AIDS. The protozoan is an obligate intracellular parasite that infects ocular conjunctivae and corneal tissue. Its size may range from 1 to 20 μm, and it has a round or spherical shape (Fig. 35–9). The organism develops in two stages, the schizogenic and the sporulation, or sporogenic, phase. Infection is by the fecal-oral route. Spores may be detected on Gram, Giemsa, acid-fast, or calcofluor white stains. Electron microscopy is usually required for classification and confirmation. There is no uniform treatment.

Detection of ocular parasites is based on (1) observing and removing the actual protozoan, (2) confirming the presence of the organism with histologic or other stains, and (3) isolating the organisms from blood, tissues, or body fluids. Treatment depends on the organism and the availability of an antiinfective.

INFECTIONS OF THE SCLERA AND EPISCLERA (SCLERITIS AND EPISCLERITIS)

The sclera is composed of tough collagen fibers. Few organisms can penetrate this strong protective coat. Infections are protracted, painful, and quite destructive. Scleritis is usually a local manifestation of systemic connective tissue disease (e.g., rheumatoid arthritis, lupus) or the result of contiguous spread from adjacent ocular tissues. The presentation may be acute (pyogenic) or chronic (granulomatous). The episclera is a thin layer of elastic vascular tissue that overlies the sclera. Inflammation of this tissue is

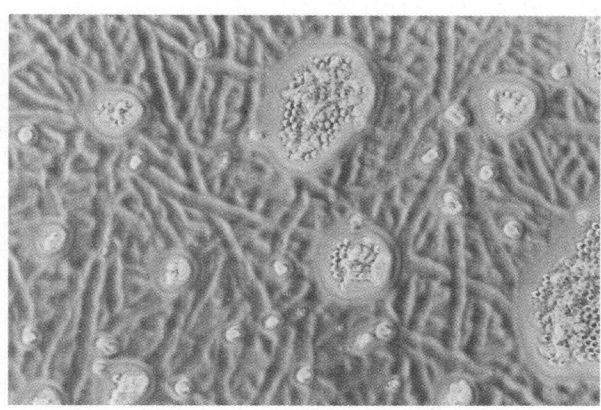

■■■■■ F i g u r e   3 5 – 8
"Tracts" of *Acanthamoeba* trophozoites. The meandering trophozoites are at the end of the tracts. The large clusters of organisms contain both trophozoites and cysts.

termed episcleritis. Episcleritis is actually more common than scleritis, with undetermined etiology in 70% of cases.

Bacteria

Bacterial infections of the sclera are usually caused by *Staphylococcus aureus, Pseudomonas aeruginosa,* and *Serratia marcescens.* Other members of the Enterobacteriaceae, *Streptococcus pneumoniae,* and *Moraxella* species have also been documented as etiologic agents. Members of the *Mycobacterium* genus (*M. tuberculosis, M. leprae,* and MOTT) are involved in chronic granulomatous scleritis. *Treponema pallidum* may also be implicated in chronic disease.

Viruses

Viral infections of the sclera are usually the result of corneal or systemic extension of disease. Not surprising, herpes simplex and varicella zoster are the most frequently recovered viruses. The mumps virus has also been recorded as a cause of viral scleritis.

Fungi

As with bacterial and viral infections, fungal infections are the result of contiguous spread from infected tissue or a localization of a generalized condition. Isolates include *Paecilomyces, Aspergillus, Fusarium* species, *Curvularia* species, *Sporothrix,* and *Blastomyces* species.

Parasites

Progression of infestation of the sclera may be observed with onchocerciasis, ophthalmia nodosa, and intraocular myiasis. In scleritis, laboratory support includes isolation and identification of the microbial agents, with appropriate antiinfective profiles. Parasites must be identified by using histologic preparations or stains or removing the organism from the ocular site.

INFECTIONS OF THE ORBIT (PRESEPTAL AND ORBITAL CELLULITIS)

Infections of the orbit and adnexal structures may have a devastating effect on vision and ocular structural integrity. The close proximity of the orbit and related tissues to the parasinuses, the absence of an effective drainage system from this "closed-box" construction, and the unique structure of the lids all predispose this area to invasion by a variety of microorganisms.

Microorganisms can gain entry into the orbital tissue through trauma or injury to the eyelids or orbit; via surgery, infections of the eyelids and adjacent skin, and upper respiratory tract infections, and by way of dental caries. Parasites may also invade the orbital tissue and cause considerable destruction. Orbital infections are also considered extensions of bacterial or fungal sinal infections.

Bacteria

Because of the close proximity of the orbital cavity to the parasinuses, any organisms that initiate sinusitis will also cause orbital cellulitis. The majority of these infections are due to bacteria, but saprophytic fungi may also be involved. In adults *S. aureus* is the most commonly recovered agent. *Streptococcus pneumoniae, S. pyogenes,* and *Haemophilus influenzae* are frequent isolates. In children, *H. influenzae* remains the causative agent most frequently recovered. Mixed anaerobic organisms are isolated from orbital cellulitis associated with longstanding chronic sinusitis. Usually both aerobic and anaerobic bacteria are isolated from these infections. Chronic orbital cellulitis may be caused by *Mycobacterium* species, *Nocardia* species, and *Actinomyces* species. Bacterial infections associated with orbital implants or prostheses are increasing (BPEI). The organisms recovered include *Mycobacterium chelonae, Nocardia* species, *Staphylococcus epidermidis, Capnocytophaga* species, *Candida parapsilosis,* and *Propionibacterium acnes.*

Fungi

The most frequent mycotic isolates in orbital infections include members of the Zygomycetes subclass and the *Aspergillus* genus. These organisms are often isolated from immunocompromised patients with underlying disease. Other fungi that may be involved in infections of the orbit include *Drechslera, Bipolaris, Curvularia, Sporotrichum, Blastomyces, Penicillium, Histoplasma,* and *Petriellidium* species.

Parasites

Trichinosis, caused by the nematode *Trichinella spiralis,* can invade the extraocular muscles and result in periorbital edema and pain on movement. An ophthalmologist usually makes the diagnosis of trichinosis, because invasion of the ocular muscles is usually the first sign of this disease.

Among the other parasites that may rarely invade or infest the orbits and adnexa are the larvae of *Echinococcus granulosus,* inducing a hydatid cyst that must be surgically exposed and whose contents must be aspirated. Maggots (fly larvae) may occasionally be deposited in the orbit and adnexal tissues.

INFECTIONS OF THE LACRIMAL APPARATUS

The lacrimal glands, accessory glands, puncta, canaliculi, tear sac, and nasolacrimal duct compose what is called the lacrimal apparatus. It has two functions. First, the lacrimal and accessory glands produce the aqueous component of the tear film. Second, the puncta, canaliculi, lacrimal sac, and nasolacrimal duct drain the tears from the conjunctiva cul-de-sac to the nasal cavity. Disorders and infections of the lacrimal apparatus are due to blockage or underproduction or overproduction of tears.

Dacryoadenitis, inflammation of the main lacrimal gland, may be infectious or noninfectious. Organisms are seeded into the gland via the blood stream. Blunt trauma also predisposes the gland to infection. Bacterial isolates include *N. gonorrhoeae, S. aureus,* and *Streptococcus* species. Chronic bacterial infections of the gland involve tuberculosis, syphilis, or leprosy. Mucormycosis and aspergillosis may result from contiguous spread from infections of the orbit. Mumps and infectious mononucleosis are the common viral infections associated with this gland. Subclinical or inapparent infections have occurred with HSV,

VZV, cytomegalovirus, coxsackievirus A, and echovirus. Other viruses implicated in lacrimal gland infections are measles and influenza viruses. Parasitic invasions with *Schistosoma haematobium, Onchocerca volvulus,* and *Cysticercus cellulose* have been reported.

Canaliculitis, a disease exclusively of adults, is a low-grade inflammation that affects the lower canaliculus more than the upper. Purulent, cheesy material may be expressed from the lumen.

Etiologic agents are diverse and include bacteria, fungi, and viruses. Bacterial infections consist of mixed aerobic and anaerobic flora. *Staphylococcus aureus* and streptococci are among the aerobes recovered. Aerobic isolates also include gram-negative rods and *Chlamydia trachomatis.* The predominant flora recovered include the gram-positive anaerobes such as *Actinomyces israelii, Propionibacterium propionicus,* and *P. acnes. Nocardia, Fusobacterium,* and *Capnocytophaga* species may also be recovered. Fungal isolates include *Candida albicans* and *Aspergillus* species. Both herpes simplex and zoster may inflame the canaliculi.

Inflammation of the lacrimal or tear sac—dacryocystitis—is the most common infection of the lacrimal apparatus. Infections are usually associated with obstruction of the nasolacrimal sac. Thus, any organisms that colonize the nasolacrimal sac could be responsible for lacrimal sac infections. The common isolates include *Staphylococcus aureus, Streptococcus pneumoniae, S. pyogenes,* and *Haemophilus influenzae.* Other isolates are *Pseudomonas aeruginosa* and *Proteus mirabilis. Chlamydia trachomatis* may cause a recurrent, chronic inflammation of the tear sac. *Aspergillus, Candida* species, and *Actinomyces* may also be recovered.

Submitted materials for microbiologic evaluation include drainage material, pus, and cheesy exudate. Direct smears for bacteria and fungi should be prepared from this material. Often the etiologic agents (e.g., *Actinomyces*) and fungal hyphae are seen in large numbers. Mixed infections are also evident. Media should be inoculated so that aerobic and anaerobic organisms as well as fungi will be recovered.

INFECTIONS OF THE INTRAOCULAR CHAMBERS (ENDOPHTHALMITIS)

Infectious endophthalmitis, inflammation of intraocular tissues or cavities, is a catastrophic development resulting from complications of surgery, contiguous spread from infected tissues, use of contaminated medications, or penetrating ocular trauma. It is the most serious and sight-threatening of all ocular infections and is an *emergency* in all cases. With endophthalmitis, both anterior and vitreous chambers may be involved. Any organism that gains entry into the inner chambers of the eye can initiate disease. The etiologic agents include bacteria, viruses, fungi, and parasites. Endophthalmitis is also initiated by instillation of contaminated eye drops and implantation of contaminated biomaterials, for example, intraocular lenses (IOLs) and contact lenses. Rapid recovery and identification of the invading organism, complemented by an early, specific, and aggressive therapy, is mandatory in preventing loss of useful vision and preserving internal ocular structures. Specimens include aspirated anterior chamber or vitreous fluids (usually less than 5 mL) and washings from the flushing out of vitreous chambers (usually more than 10 mL). The choice of smears and culture media inoculated must be guided by the suspected organisms as well as by the knowledge of possible organisms.

Bacteria

Gram-positive organisms constitute 70% of all isolated microbes. Coagulase-negative staphylococcal species are recovered from one third of all cases. *Staphylococcus aureus, P. acnes, Streptococcus* viridans group, and *Enterococcus* species round out the top five. *Bacillus cereus* and other *Bacillus* species are most frequently isolated from endophthalmitis resulting from traumatic insults. Presentation is sudden, and the course is fulminant. Release of necrotizing enzymes can result in loss of the eye within 48 hours. A high index of suspicion and aggressive therapy are required to retain useful vision. Staphylococcal and streptococcal species are most often isolated from postsurgical cases. *Haemophilus influenzae* and members of the *Streptococcus* genus are the most commonly isolated organisms after filtering bleb surgery. Gram-negative rods, especially *P. aeruginosa,* are being recovered with increasing frequency, and infection with such microbes can follow surgery, trauma, or instillation of contaminated ocular medications. Other organisms that are isolated with some frequency in cases of endophthalmitis are *Serratia*

marcescens, Proteus mirabilis, Morganella morganii, Citrobacter diversus, and *Mycobacterium chelonae. Nocardia* and *Actinomyces* species have also been isolated.

Viruses

Intraocular infections are usually due to members of the herpesvirus group, HSV, VZV, or cytomegalovirus, and usually manifest as retinitis or chorioretinitis. Because of the small amount of materials available for culture from this site, tissue culture isolation of these viruses has been disappointing. Improved recovery has been documented when the vitreous drops or vitrectomy specimen has been cocultivated with fibroblasts in a 25-mL flask. The vitrectomy sample may also be spun down and the cells collected for cocultivation in routine 12 by 75 mm tubes. The culture must be held for a minimum of 6 weeks, with a range from 2 to 8 weeks. Alternative methods for confirmation include EIA, enzyme-linked immunosorbent assay (ELISA), PCR, and the antigenemia test.

Fungi

Mycotic endophthalmitis is most often due to extension of keratitis. It may also result, however, from hematogenous spread from a remote focus and from implantation of a contaminated intraocular lens. The most frequent isolates are *Candida albicans* and other *Candida* species. Saprophytes that infect the cornea may extend into the intraocular cavities. The filamentous species recovered include *Aspergillus* species, *Fusarium solani, Paecilomyces* species, *Curvularia* species, and *Sporothrix schenckii.* Other reported species recovered in cases of endophthalmitis include *Monosporium apiospermum, Cephalosporium* species, *Volutella* species, *Coccidioides immitis, Histoplasma capsulatum, Cryptococcus neoformans,* and *Blastomyces dermatitidis.*

Parasites

Intraocular parasites usually affect the retina or choroid or are transient invaders from adjacent ocular structures. *Onchocerca, Toxocara* and *Toxoplasma gondii* are the most frequent intraocular pathogens.

INFECTIONS OF THE UVEAL TRACT (UVEITIS)

Uveitis is a general term for inflammatory disorders of one or all three portions (iris, ciliary body, choroid) of the uveal tract. This inflammation results from ocular trauma or insults to the inner structures of the globe from local or systemic inflammatory disease. Connective tissue diseases, such as juvenile rheumatoid arthritis, Reiter syndrome, and systemic lupus, are associated with uveitis. Iritis and iridocyclitis are referred to as anterior uveitis, while choroiditis or chorioretinitis is termed posterior uveitis. The etiology of anterior uveitis is considered nongranulomatous and is nonmicrobial. Posterior uveitis is classified as a granulomatous disease, and its etiology may be bacterial, fungal, viral, or parasitic.

Mycobacterium tuberculosis, T. pallidum, and *M. leprae* all are involved in chronic disease of the uveal tract. Members of the herpes group (HSV, cytomegalovirus, and VZV) may be involved in necrotizing disease. Such disease is usually associated with retinitis. Chorioretinitis may be a better description of the syndrome. *Histoplasma, Aspergillus* species, *Nocardia asteroides,* and *Candida* species all cause granulomatous disease of the uveal tract.

Infections involving the uveal tract usually manifest as chorioretinitis (these will be discussed in the section on infections of the retina which follows). In these cases, the laboratory work-ups are attempts to isolate the offending organisms. Because of the nature of the microbes, a diagnosis may not be readily available. Serologic assessment via ELISA, immunofluorescence, or complement fixation may be of greater value. Genomic amplification (PCR) or DNA probes may also assist in some cases. Viral isolation is attempted by cocultivation of the fluids or tissues with fibroblasts. Precytopathic effect monoclonal antibodies are available to rule out HSV and cytomegalovirus.

INFECTIONS OF THE RETINA (RETINITIS)

The retina is a multilayered neural tissue that transforms images and relays impulses to the brain so the individual can see. Insults to the retina, infectious or noninfectious, are not forgiven—damage is irreversible. Microbial infections are rare but can be devastating. Microbial infections may occur via hematogenous or contiguous spread. The bacterial and fungal infections usually are the same as those found in uveitis. *Pneumocystis carinii* invades the retina, resulting in classic "cottonwood spots" during systemic disease. Viral and parasitic organisms are responsible for syndromes that are peculiar to the retina.

Viruses

Viral syndromes are rare but devastating, resulting in unilateral or bilateral decreased vision. The immunocompromised patient is often at risk for viral retinitis. Increased disease (cytomegalovirus retinitis) is seen in the AIDs population. HSV, VZV, and cytomegalovirus all cause viral retinal disease. In the United States, HSV is the leading cause of blindness. Both the HSV I and II cause a posterior uveitis. Acute necrotizing retinitis is usually seen in the newborn as a result of acquiring the virus during passage through an infected birth canal. Adults may acquire the disease from recurrent episodes of viral keratitis. Varicella zoster retinitis may be a manifestation of reactivation of the virus in herpes zoster ophthalmicus. Cytomegalovirus infections may affect the newborn or the immunocompromised patient, especially transplant and AIDS patients. Cytomegalovirus retinitis, with loss of vision (Fig. 35–10), is often an AIDS-indicating condition. Acyclovir, ganciclovir and foscarnet are antivirals used to treat and modulate these episodes. Both rubeola and rubella introduce retinopathy. About 30% of patients with subacute sclerosing panencephalitis (SSPE), which may appear years after an attack of clinical measles, also have chorioretinitis on examination. The ocular complications of congenital rubella include cataracts and glaucoma.

Parasites

Toxoplasma gondii is a protozoan with a predilection for ocular tissues (Fig. 35–11). Ocular infection is a late manifestation of congenital toxoplasmosis. This infection is now more prevalent in AIDS patients than in the general population. *Cysticercus cellulosae* and *Toxocara* are found in 10 to 13% of intraocular infections. Other protozoal

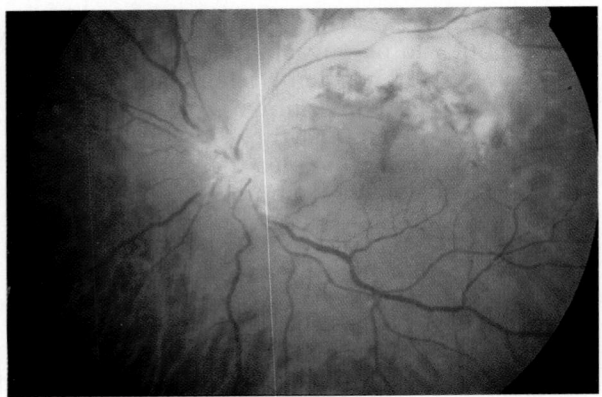

■ Figure 35–10

Acute cytomegalovirus retinitis with optic nerve involvement in a 40-year-old HIV-positive patient. Active viral particles are seen in satellite lesions temporal to the main infection (yellow).

infections that may cause transient retinitis include malaria (retinal hemorrhage); babesiosis, a rare disease similar to malaria but having distinct differences, seen mainly in the northern United States, in the New England area; and infection with *Loa loa,* the eye worm, which can enter the intraocular cavities and cause inflammation in its migration through the ocular tissues. For these infections, the laboratory support is the same as that for uveitis.

SCLERAL BUCKLE INFECTIONS

Scleral buckles, sponges, and bands, biomaterials employed to realign detached retinas, may be partially extruded. Once these materials are exposed, they are subject to bacterial colonization, which can lead to infections of the conjunctivae, sclera, or intraocular cavities. The most frequently isolated bacteria are coagulase-negative staphylococcal species. *Staphylococcus aureus, Proteus mirabilis,* and *Pseudomonas aeruginosa* have also been recovered. Mycobacterial species other than *M. tuberculosis* (MOTT), in particular *M. chelonae* and *M. fortuitum,* have been recovered from these materials at an alarming rate over the last 2 years. Additional isolates include *Candida* species, *Corynebacterium* species, and *Citrobacter diversus.*

Management includes removing the materials along with any remaining necrotic tissue when possible. Whether antibiotics are appropriate is still being debated (Fig. 35–12).

OCULAR MANIFESTATIONS IN PATIENTS WITH HUMAN IMMUNODEFICIENCY VIRUS

Ocular involvement occurs in 50 to 75% of patients infected with human immunodeficiency virus (HIV) during the course of their illness. Patients may not exhibit symptoms early in the disease, but with advancing disease, symptoms appear. The virus has been isolated from tears and conjunctival and cornea tissues; HIV may also be isolated from the contact lenses of infected patients. The route by which HIV reaches the ocular surface tissues and tears remains uncertain. Cytomegalovirus retinitis is the second most common presentation in patients with full-blown AIDS and can affect both eyes even in the absence of systemic disease. This retinitis is listed as an indicator infection in the differential diagnosis of AIDS. Patients may present with a history of sudden loss or impairment of vision. Cotton-wool spots—fluffy, white lesions that usually signal opportunistic infection—are the ocular lesions most often seen in patients with AIDS. Retinal hemorrhages may also be present. Kaposi sarcoma (Fig. 35–13) appears on the ocular tissues of 10% of infected patients. *Pneumocystis carinii* and *T. gondii* retinitis are also ocular diseases associated with AIDS patients. Although toxoplasmosis occurs in the general population, it is currently more prevalent in the AIDS population.

Table 35–5 summarizes the most common isolates from each of these infections; Table 35–6 lists those considered unusual isolates.

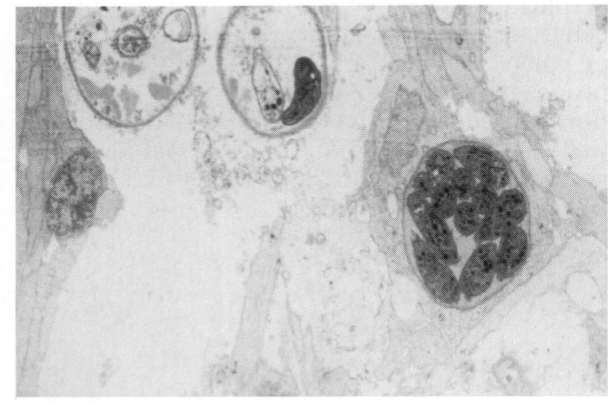

■ Figure 35–11

Toxoplasma gondii trophozoites and cysts in retinal tissue. This protozoan has a predilection for ocular tissue. In AIDS patients, such infections are on the increase.

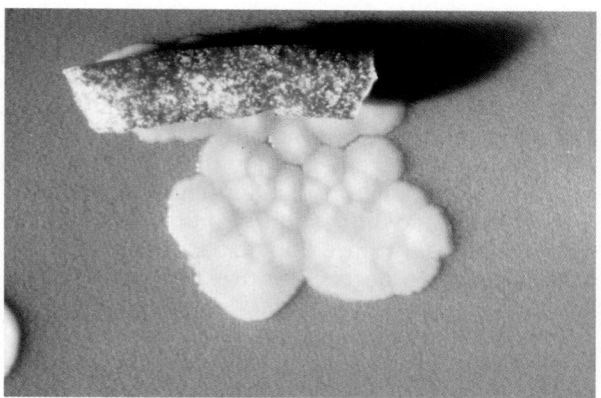

■ F i g u r e   3 5 – 1 2

Extruded scleral buckle on blood agar plate with growth of *Candida albicans.*

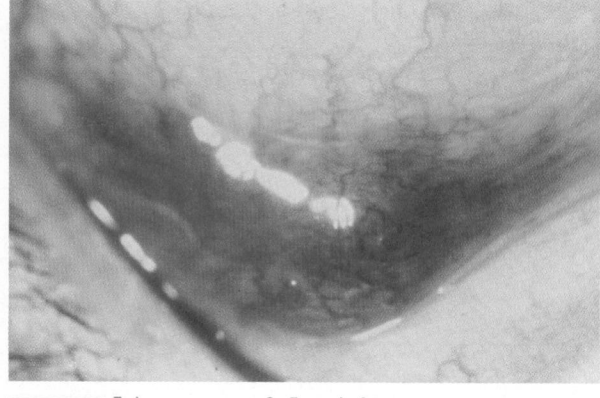

■ F i g u r e   3 5 – 1 3

Kaposi sarcoma on conjunctiva of AIDS patient (raised dark spots).

LABORATORY DIAGNOSIS OF OCULAR INFECTIONS

Specimen Collection

The keys to proper collection of ocular specimens are similar to those for any other microbiologic specimen. Materials or scrapings for cultures should be collected as soon as possible after the onset of infection (24 to 48 hours for bacteria and 3 to 7 days for viruses) and before the instillation of antimicrobials or steroids. The sample must be collected from the actual site of the infection, for example, conjunctiva and lid cultures are inadequate to assess corneal involvement. All ocular fluids, tissues, sponges, and other surgical material must be submitted in a *sterile, leak-proof container,* properly labeled with all pertinent information.

It is important for the ophthalmologist to *communicate* with the hospital or reference laboratory personnel performing the microbiologic evaluation of the specimen to ensure isolation of ocular pathogens. All patient media and the accompanying requisition(s) must be properly labeled with complete patient information. The requisition should list any antibiotics, antivirals, steroids, or antifungals the patient is currently receiving and should indicate whether the patient wears contact lenses.

Unlike routine microbiologic specimens, ocular materials must be inoculated directly onto the appropriate media. Scant recovery or none at all is the norm when culturettes or swabs in transport media are substituted for properly collected specimens. Not only does swab collection in general reduce recovery, but also the type of swab used can further inhibit expected results. Cotton-tipped swabs inhibit the growth of some bacteria and HSV owing to the presence of fatty acids released during sterilization. Dacron or calcium alginate swabs may be used to collect ocular samples when scraping cannot be performed. However, the calcium alginate swab may be lethal to some viral ocular pathogens. Furthermore, "swabbing" often does not collect enough materials from the infected site, especially in chronic conditions. Expressed materials from areas such as canaliculi are preferred to rule out infection (Fig. 35–14A and B). Materials are

■ T A B L E   3 5 – 5

TOP OCULAR ISOLATES AT BPEI (1977–1992) IN DESCENDING ORDER OF FREQUENCY

Conjunctivitis	Keratitis	Endophthalmitis
Staphylococcus aureus	*P. aeruginosa*	*S. epidermidis*
Haemophilus influenzae	*S. aureus*	*S aureus*
Neisseria gonorrhoeae	*Serratia marcescens*	*P. aeruginosa*
Pseudomonas aeruginosa	*Staphylococcus*	Viridans strepto-
Proteus mirabilis	*epidermidis*	coccal group
Chlamydia trachomatis	*P. mirabilis*	*P. mirabilis*
Adenovirus	HSV I	*H. influenzae*
HSV I	HSV II	*Streptococcus*
Enterovirus	VZV	*pneumoniae*
	Fusarium solani	*Enterococcus*
	F. oxysporum	*faecalis*
	Candida parapsilosis	*Candida albicans*
	C. albicans	HSV I
		VZV
		CMV

UNUSUAL ISOLATES

Conjunctivitis	Keratitis	Endophthalmitis
Neisseria meningitidis	*Listeria monocytogenes*	*Actinomyces* species
Eikenella corrodens	*Mycobacterium chelonae,*	*Capnocytophaga* species
Fly larvae (maggots)	*M. fortuitum*	
Cytomegalovirus	Microsporida	*M. chelonae*
Lymphogranuloma	Free-living amebae	*Aspergillus terreus*
venereum (LGV)	(*Acanthamoeba* and *Valkampfia* species)	Enterovirus
	Corynebacterium diphtheriae	

inoculated directly onto the media. Smears are also prepared from the material. Collection by direct aspiration or scraping is performed by the ophthalmologist. The aspirate is transported in a sterile container. Media to recover both aerobic and anaerobic organisms must be included. Vitreous washings (volume greater than 10 mL) may be injected into blood culture bottles or viral transport media or may be sent to the laboratory for concentration.

Biopsy tissue must be minced or ground. Media to recover aerobic and anaerobic bacteria, fungi, and mycobacteria should be inoculated. If free-living amebae are suspected, two nonnutrient (agar) plates should be included.

Direct Smear Examination

Scrapings from the involved ocular site, complemented with the appropriate stains, can af-

ford the physician circumstantial as well as definite information concerning the identity of the invading organisms. Table 35–7 provides a summary of stains and appropriate applications of each procedure. Gram and Giemsa stains (Fig. 35–15) should be collected for all cases of bacterial conjunctivitis, keratitis, and endophthalmitis. The calcofluor white stain should be added to rule out fungal or parasitic disease. The acid-fast stain is used to detect the presence of acid-fast organisms. The acid-fast stain may confirm or rule out *M. leprae* in tissue from chronic infections. These are the stains routinely employed in ocular microbiology. Direct antigen detection in cases of viral or chlamydial disease may be the only confirmation test available (Fig. 35–16*A* and *B*). In addition, impression cytology utilizing 0.45-μm Teflon-coated membrane filters to collect conjunctival and corneal tissue may increase bacterial, fungal, viral, and protozoan detection to as high as 50%. The advantage with this technique is that cells remain intact with characteristic morphology and microbial infestation and invasion (Fig. 35–17*A* to *C*).

Culture

There are numerous media available for isolation and cultivation of microorganisms from ocular sources. Table 35–8 shows a general plating guide that may be followed to recover all possible agents of ocular infections. However, the majority of the bacterial and fungal ocular iso-

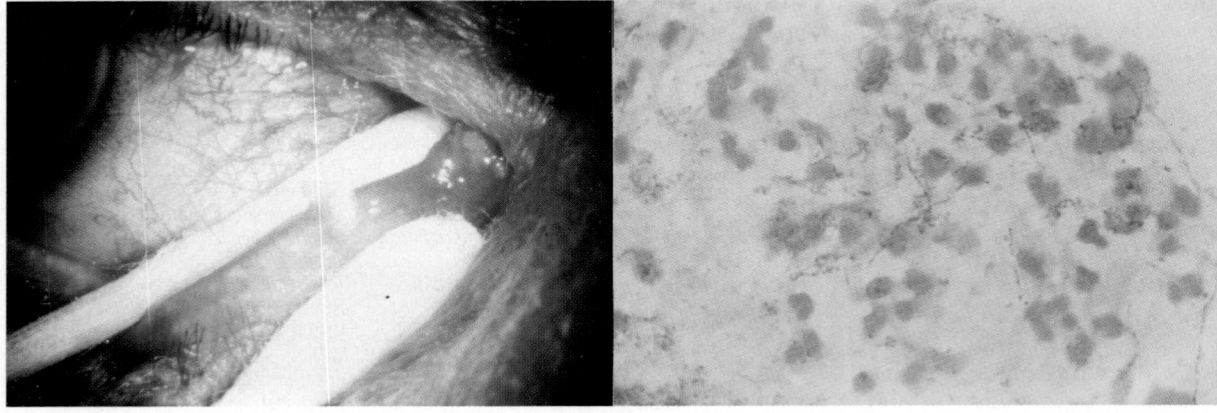

A   B

▬▬▬F i g u r e   3 5 - 1 4

A, Concretions being expressed from canaliculi. *B,* Concretions "smashed" and stained, revealing gram-positive, slender, branching rods (*Actinomyces israelii*).

SMEAR GUIDE

	Gram	Giemsa	IF	IC	CFW
Bacteria	*	*			
Fungi	*	*			*
Viruses		*	*	*	
Chlamydia		*	*	*	
Parasites	*	*		*	*

IF, immunofluorescent stains (monoclonals for chlamydia, HSV, adenovirus, enterovirus, VZV, and cytomegalovirus); IC, impression cytology—collected cells remain intact, and test can locate disease progress; CFW, calcofluor white stain—used to detect fungi or parasites in ocular tissue.

lates may be recovered on chocolate and blood agar when these are incubated under the proper conditions of temperature (35–37°C) and atmosphere (5–10% CO_2, aerobic) or in an anaerobic jar or bag. Table 35–9 shows a minimal plating guide for bacteria and fungi. The addition of thioglycolate broth, Thayer-Martin agar, Sabouraud agar with gentamicin (Fig. 35–18), viral and chlamydial transport media, Löwenstein-Jensen slants, and nonnutrient agar plates allows for the recovery of most pathogens involved in ocular disease. All thioglycolate tubes are held for 10 days or 21 days if *Actinomyces* is suspected.

Culture Interpretation of Ocular Specimens

Interpretation of growth from ocular samples is based on the same sound microbiologic criteria used in general hospital microbiology laboratories. Quantitation of growth is particularly important. Samples from patients receiving anti-infective agents, steroids, or other medications may have reduced flora, and this should be taken into consideration when evaluating the culture.

Special considerations are involved in cornea and intraocular fluid cultures. Corneal scrapings are inoculated in a "C" streak fashion. Each row of "C" streaks represents a separate scraping of a corneal ulcer. The dilution effect is from "left to right." Generally there will be more colonies of bacteria or fungi on the first "C" streaks of each row. Each successive "C" streak progresses from the superficial to the deep layers of the cornea. The greater the number of streaks with growth, the more involved and serious the infection.

Any growth from intraocular fluids appearing on the inoculation sites or filter should be worked up and reported. The physician correlates the organism's pathobiology with the pa-

tient's clinical picture and diagnosis. Remember that those organisms normally considered "contaminants" (Fig. 35–19) are the ones most frequently involved in microbial intraocular infections. Never ignore even a single colony within the inoculation sites or on the filter!

Special Procedures To Recover Ocular Pathogens

***Limulus* Lysate.** The *Limulus* lysate test should be performed for all patients with corneal ulcers receiving medications or who wear contact lenses or for patients with clinical signs suggestive of gram-negative organisms. The *Limulus* lysate is a mixture of the amebocytes of the horseshoe crab. When these cells are mixed with fluids or substances containing endotoxins (cell walls of gram-negative organisms), they form a gel clot, much like that of the coagulase test. Both tests are enzymatic in nature. The assay is set up, incubated, and read in 1 hour. In our laboratory more than 90% of corneal scrapings that grew gram-negative organisms also had positive *Limulus* lysate test results. Ninety percent or more of the corneal ulcers that did not grow gram-negative rods had negative *Limulus* lysate test results. This assay was also evaluated for detection of contaminated contact lenses, solutions, and medications. The preliminary indications are that with modifications, the *Limulus* amebocyte assay (LAL) can assist in detecting gram-negative contamination of contact lens solutions and medications. Chromogenic assays are also available for high-volume laboratories.

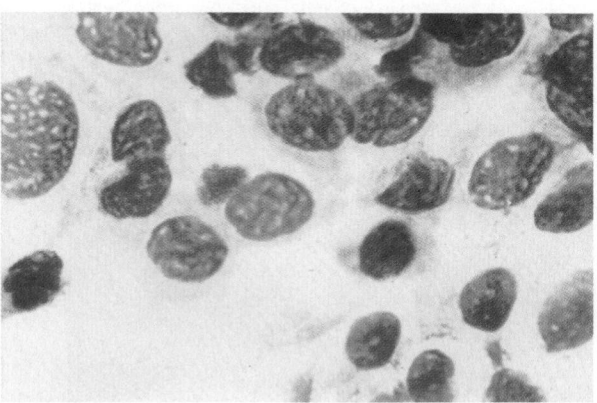

■■■■ F i g u r e   3 5 – 1 5

Giemsa stain of conjunctival epithelial cells with chlamydial inclusions (*center*). The Giemsa stain also provides information on the types and numbers of inflammatory cells and the condition of the epithelial cells.

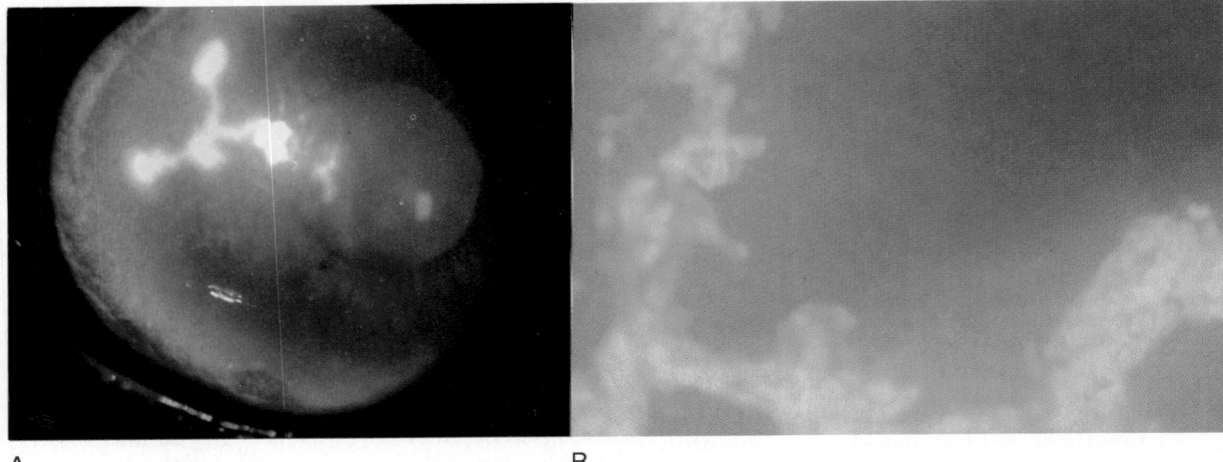

A   B

■■■■■ F i g u r e   3 5 - 1 6

A, Cornea stained with rose bengal to outline HSV-infected dendrite. HSV is the virus most often isolated from corneal dendritic infections. *B,* Dendrite on membrane filter collected by impression cytology. The filter was stained with a monoclonal antibody against HSV I. Almost all the cells of the dendrite "lit up" when stained. The noninfected cells do not stain and appear red.

Agar-Agar Medium. This nonnutrient agar is first inoculated with ocular materials and then overlaid with an aliquot of heat-killed or live *Escherichia coli.* The *E. coli* serve as food for the excysting *Acanthamoeba* organisms. The ame-bic trophozoites are identified by locating them at the end of the "tracts" they generate as they eat through the *E. coli.* Cysts are identified by their refractile, double-walled, polygonal shape and are best seen using calcofluor white and the

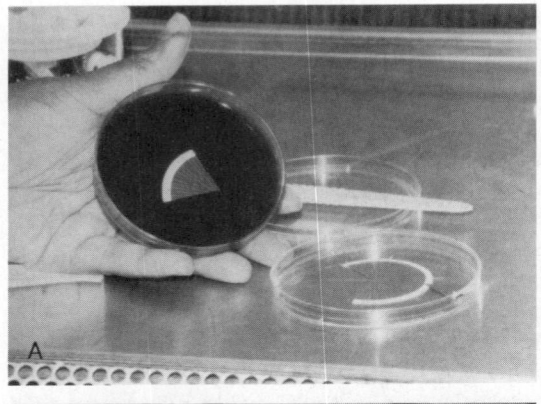

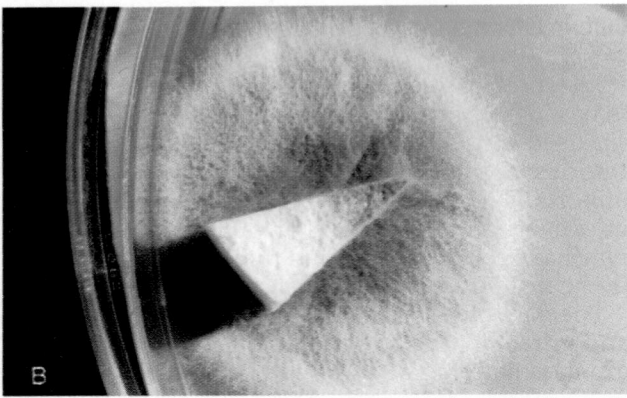

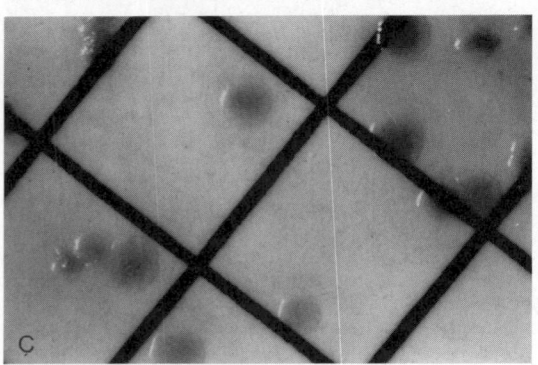

■■■■■ F i g u r e   3 5 - 1 7

A, Once filtered, the 0.45-μm filter is then sectioned and placed on select-ed media. *B, Curvularia* species from intraocular fluids on 0.45-μm filter section. *C, Pseudomonas cepacia* on a filter from a vitrectomy specimen.

■■■ T A B L E   3 5 – 8
GENERAL OCULAR PLATING GUIDELINES

Ocular Clinical Diseases	Media to Inoculate																Smears			
	Chocolate Agar	Blood Agar 5% Sheep	Phenyl Ethyl Alcohol	MacConkey Agar	Blood Agar ANA	Phenyl Ethyl Alcohol ANA	Kanamycin Vancomycin ANA	Brain-Heart Infusion*	Sabouraud With GM*	Thioglycolate	Nonnutrient Agar	Limulus Lysate	Viral Transport	Chlamydial Transport	Löwenstein-Jensen	Blood Culture Bottle	Gram Stain	Giemsa Stain	Calcofluor White Stain	Immunofluorescent Stain
Conjunctivitis																				
Bacterial	♦	♦	♦	♦													♦	♦		♦
Chlamydial	♦													♦				♦		♦
Viral													♦				↑	↑		♦
Blepharitis (same as above)																				
Keratitis																				
Bacterial	♦	♦	♦	♦	♦				♦			♦					♦	♦		
Fungal	♦	♦ */25						♦	♦									♦	♦	
Viral													♦					♦		♦
Acanthamoeba	♦	♦			♦				♦		♦						♦	♦	♦	♦
Chlamydial	♦	♦												♦				♦		♦
Unknown	♦	♦	♦	♦	♦				♦	♦	♦	♦	♦	♦	♦		♦	♦	♦	♦
Lacrimal Apparatus																				
Bacterial	♦	♦	♦	♦	♦	♦	♦	♦		♦							♦	♦		
Fungal		♦ */25						♦	♦									♦	♦	
Viral													♦					♦		♦
Endophthalmitis																				
Bacterial	♦	♦	■	■	■					♦						■	■	■		
Fungal	♦							♦	♦	♦								♦	♦	
Viral	♦									♦			♦					♦		
Unknown	♦	♦			♦					♦						♦		♦	♦	*afb

  * Brain-heart infusion medium or Sabouraud agar, 25/35 C.
  ♦ Media of choice; ↑, optional; ■, if quantity permits.

fluorescent microscope. Recovery of any parasite from corneal scrapings is significant and should be reported. Usually cultures of free-living amebae contain both cysts and trophozoites.

Trypticase Soy Broth (TSB). TSB is used as a wetting agent to collect conjunctiva and lid cultures.

Sabouraud Agar with Gentamicin. The very fungi that traditional mycology media aim to inhibit

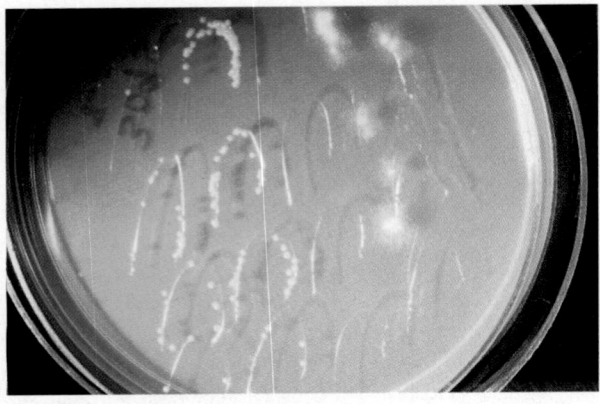

■■■■■ F i g u r e   3 5 – 1 8

Sabouraud plate with mold and yeast. Patient presented with a mixed fungal keratitis. *Right*, The superficial layer of the cornea was infected with a mold (*Fusarium oxysporum*). *Left*, The deeper layers were infected with a yeast (*Candida albicans*).

(saprophytes) are the agents most often recovered in mycotic ocular disease (i.e., *Fusarium, Paecilomyces, Aspergillus, Penicillium* species).

Blood Culture Bottles. While these are used in the traditional manner when blood cultures are requested, they are also used to culture intraocular fluids. Vitreous washings may be placed in the traditional (50- to 70-mL) bottles; fluids from direct vitreous or ac taps should be inoculated into the pediatric (10- to 20-mL) bottles.

Special Culture Techniques

Contaminated Ocular Medications

Ophthalmic drops (e.g., antibiotics, analgesics) are easily contaminated when improperly handled by patients or ophthalmic personnel. Inappropriate handling can contribute to ongoing ocular disorders and to initiation of new disease. Even though most ocular medications are prepared with preservatives to inhibit the growth of microorganisms, once the preservative's threshold has been breached, microbes will survive, multiply, and be dispensed with the next drop. Medications should be collected from patients presenting with conjunctivitis, keratitis, and endophthalmitis and sent to the laboratory for culture.

The medications most often contaminated and associated with concurrent patient infection, in descending order, include the following:

■ steroids

■ β blockers

■ antiglaucoma drugs

■ antibiotics

■ artificial tears

The organisms most frequently recovered from infected patient drops include the following:

■ *Serratia marcescens*

■ *Pseudomonas aeruginosa*

■ Coagulase-negative staphylococcal species

■ *Candida* species

Organisms isolated from ocular medications with concomitant patient disease at our institution (BPEI), in descending order of frequency, include the following:

■ *Serratia marcescens*

■ *Pseudomonas aeruginosa*

■ *Achromobacter xylosoxidans*

■ Coagulase-negative staphylococcal species

■ *Proteus mirabilis*

■ *Morganella morganii*

■ *Staphylococcus aureus*

Contact Lenses and Solutions

Keratitis associated with contact lenses (soft, daily-wear, extended-wear, or disposable) is being documented with increasing frequency. Gram-negative rods (especially and predominantly *P. aeruginosa* and *S. marcescens*) are most often recovered from CTL-associated infections.

■■■■■ T A B L E   3 5 – 9
MINIMAL PLATING GUIDE—BACTERIA AND FUNGI

	CHOC‡	BAP (An)	SAB	THIO	SMEARS
Conjunctivae, lids	+			+	+
Cornea	+	+	+	+	+
Intraocular fluids*	+	+		+	+
Adnexa†	+	+	+	+	+

An, anaerobic; BAP, blood agar plate; CHOC, chocolate agar; THIO, thioglycolate broth; SAB, Sabouraud dextrose agar.

*Intraocular fluids may be injected into pediatric or adult blood culture bottles (bcbs) for recovery of the most frequent endophthalmitis isolates.

†Wounds, orbit, lacrimal apparatus, etc.

‡*Note*: If the ophthalmologist can collect only enough material for one medium, ask him or her to inoculate the chocolate agar plate and rinse the spatula or blade in the thioglycolate broth. Both specimens should be submitted to the microbiology laboratory for processing.

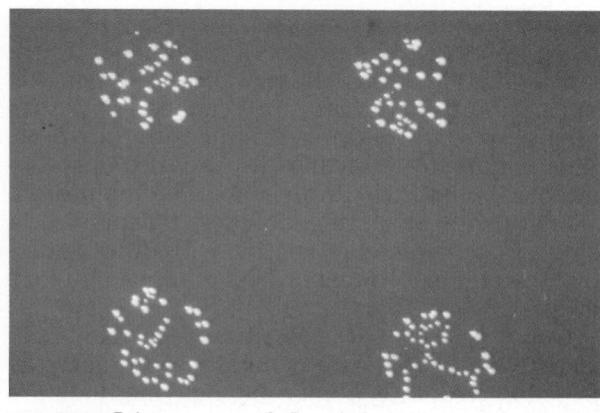

■■■■■■■ F i g u r e 3 5 – 1 9

Staphylococcus epidermidis recovered from vitreous fluids (drops) on blood agar plate. Samples may be inoculated onto a chocolate or blood agar plate and allowed to dry, or may be streaked out as for a routine microbiology specimen.

Acanthamoeba species, MOTT, and gram-positive cocci have also been recovered. Sources include contact lenses as well as contact lens solutions (Fig. 35–20) and cases. Contamination of the soft contact lenses is usually due to failure to follow manufacturer's recommended procedures or poor hygiene. Wearers of hard contact lenses are less prone to infection from microbial contamination.

Cornea Storage Media and Tissue Culture

Replacement of diseased or opacified corneas is a common operation performed by ophthalmic surgeons. The new corneal tissue to be transplanted is obtained from donors within 24 hours of death and is preserved in McCarey and Kaufman (MK) or Dexsol medium. At the time

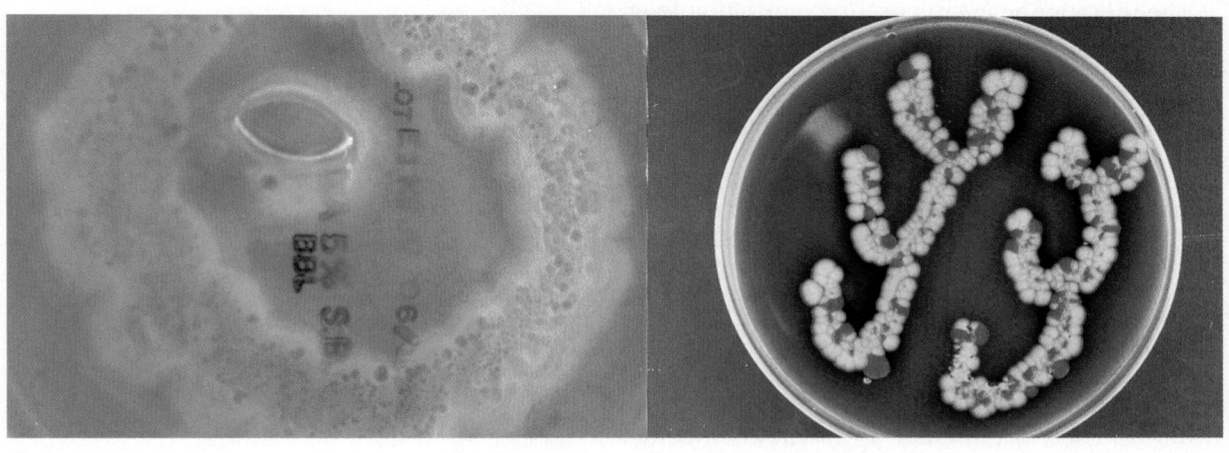

A

B

DILUTION EFFECT ⟶

Left to right

C C C C C
C C C C C
C C C C C
C C C C C

"C" STREAKS

C

NOTE: Each row of "C" streaks represents
a SEPARATE corneal scraping

■■■■■■■ F i g u r e 3 5 – 2 0

A, Contact lens and lens solution on 5% sheep's blood agar surrounded by growth of *Pseudomonas aeruginosa. B,* "C" streaks growing pigmented and nonpigmented *Serratia marcescens. C,* Corneal scrapings.

of surgery the container with the medium and remaining corneal rim is sent to the laboratory for culture. Bacteriologic evaluation of the donor tissue is paramount in reducing the transmission of host-carried disease to the recipient. Viral studies of corneal transplantation tissue have been limited.

Gram-positive organisms (i.e., coagulase-negative staphylococci and *P. acnes*) are the most frequently isolated organisms, but gram-negative organisms, yeasts, or molds may also be recovered from these media. The actual correlation between isolates and resultant disease is less than 1%. All positive cultures should be reported directly to the surgeon, and a final report sent to the EyeBank and the surgeon.

OCULAR THERAPY

The therapeutic agents used in ophthalmology are applied to the diagnosis of ocular disorders, the treatment of ophthalmic disease, and the prevention of postoperative infections. They differ from drugs used in other areas of medicine in delivery routes, antibiotic and antiviral combinations, dosing intervals, and toxicity. The traditional routes of administration of drugs may also be implemented to supplement ocular remedies.

Effective treatment is based on the organisms most likely to be recovered, with an understanding of their pathobiology and the most effective antiinfective agent available.

For ophthalmic drugs to be effective, they must reach ocular tissues in relatively high concentrations. Therefore, ocular formulations include drugs in concentrations 10 to 100 times (i.e., fortified) that of drugs prepared for intramuscular or intravenous delivery. The routes of administration of ophthalmic drugs include the following:

- topical (most common, may include drops or ointments)

- oral

- parenteral

- periocular

- intracameral

- intravitreous

Drugs are dispensed as either drops or ointments, with topical administration the most common route. Alternative or traditional routes are required for the treatment of disease involving the retina, optic nerve, and intraocular cavi-

ties. Ocular medications usually contain preservatives to inhibit the growth of contaminating microorganisms. Nevertheless, when the thresholds of these preservatives are reached, microbes multiply and are dispensed along with the drops. This may result in ocular disease, and contamination of multidose drops should be monitored. Ocular drops or ointments may be dispensed in single doses or in combination with other antibiotics, steroids, or nonsteroidal antiinflammatory agents.

Once the etiologic agent has been identified, sensitivities should be generated as quickly as possible. Traditional microbiologic methods are used to determine susceptibility patterns. Table 35–10 outlines the most common types of antiinfective agents commercially available to ophthalmologists. Pharmacists with training in ocular therapy can prepare ocular dosages for most

■ T A B L E   3 5 – 1 0
OCULAR ANTIINFECTIVES

Antibacterials	Antifungals
Amikacin	Amphotericin B
Ampicillin	Clotrimazole
Bacitracin*	Fluconazole
Carbenicillin	Flucytosine
Cefazolin	Griseofulvin
Cephalothin	Hydroxystilbamidine
Chloramphenicol*	Ketoconazole
Ciprofloxacin*	Miconazole
Clindamycin*	Natamycin*
Colistin*	Potassium iodine
Erythromycin*	
Gentamicin*	**Antiparasitic**
Methicillin	Antimony sodium gluconate
Neomycin*	Diethylcarbamazine
Penicillin G	Iodoquinol
Polymyxin B*	Ivermectin
Sulfacetamide*	Mebendazole
Sulfisoxazole*	Metronidazole
Tetracycline*	Minocycline
Tobramycin*	Niridazole
Trimethoprim-sulfamethoxazole	Paromomycin
Vancomycin	Pentamidine isethionate
	Propamidine isethionate
Antivirals	Physostigmine
Idoxuridine (IDU)*	Pyrimethamine
Trifluridine*	Spiramycin
Vidarabine (ara-A)*	Surinam quassia
Acyclovir*	Thiabendazole
Foscarnet	
Ganciclovir	
Zidovudine (azidothymidine [AZT])	

*Commercially available ocular preparations (drops or ointment).
   Ocular preparations may be single antibiotics or combinations of one or two antibiotics and/or antiinflammatories.
   Only natamycin has been approved by the U.S. Food and Drug Administration for ocular pathogens.

■ T A B L E   3 5 – 1 1
ANTIINFECTIVES FOR OCULAR SUSCEPTIBILITY TESTING

Antibiotic	g+	g–	gndc	haem	pse	other
Amikacin					+	+
Penicillin	+		+			+
Ampicillin		+	+	+		+
Ampicillin/sulbactam	+	+	+	+		+
Oxacillin	+					
Cefaclor				+		
Cefazolin*	+	+	+	+		+
Cephalothin*	+	+	+	+		+
Ceftriaxone	+	+	+	+	+	+
Ciprofloxacin	+	+	+	+	+	+
Clindamycin	+					
Erythromycin	+		+	+		
Gentamicin	+	+			+	+
Tetracycline	+	+	+	+		+
Tobramycin		+			+	+
Trimethoprim-sulfamethoxazole*	+	+	+	+		+
Vancomycin	+					

g+, gram-positive; g–, Enterobacteriaceae; gndc, gram-negative diplococci; haem, *Haemophilus influenzae*; pse, *Pseudomonas aeruginosa*; other, *Moraxella* species, *Kingella* species, *Eikinella* species, etc.

*Most frequently used ophthalmic antibiotics. You must include these on any susceptibility panels. Other antibiotics that may be requested are polymyxin B, chloramphenicol, neomycin, bacitracin, and sulfacetamide. Sulfacetamide is a sulfa drug. There are no disks to evaluate this drug. A general idea of susceptibility to this drug can be gained by running trimethoprim-sulfa and Gantrisin (Roche; sulfisoxazole).

intravenous or intramuscular drugs. Table 35–11 lists the antibiotics that should be included in a general ocular susceptibility profile.

Bibliography

Alfonso E, Miller D: Detection of ocular infections. In Prior RB (ed): Clinical Applications of the *Limulus* Amoebocyte Lysate Test. Boca Raton, FL: CRC Press, 1990, p 121.

Alfonso E, Mandelbaum S, et al: Ulcerative keratitis associated with contact lens wear. Am J Ophthalmol 101:429, 1986.

Allansmith MR: Defense of the ocular surface. Int Ophthalmol Clin 2:93, 1979.

Alvarez H, Tabbara KF: Infections of the eyelids. In Tabbara KF, Hyndiuk RA (eds): Infections of the Eye. Boston: Little, Brown, 1986, p 551.

Antonios S, Tabbara KF: Bacteria conjunctivitis. In Tabbara KF, Hyndiuk RA (eds): Infections of the Eye. Boston: Little, Brown, 1986, pp 413–420.

Asbell P, Stenson S: Ulcerative keratitis: Survey of 30 years' laboratory experience. Arch Ophthalmol 100:77, 1982.

Baron EJ, Finegold SM: Bailey's and Scott's Diagnostic Microbiology. St. Louis: CV Mosby, 1990, Chapter 38.

Bohigan GM: Acquired immune deficiency syndrome: Ocular manifestations. In Handbook of External Disease of the Eye, 3rd ed. Thorofare, NJ: Slack, 1987, pp 84–86.

Brightbill FS (ed): Corneal Surgery: Theory, Technique, and Tissue, 2nd ed. St. Louis: Mosby–Year Book, 1993, Chapters 43–46.

Brinser JH, Burd EM: Principles of diagnostic ocular microbiology. In Tabbara KF, Hyndiuk RA (eds): Infections of the Eye. Boston: Little, Brown, 1986, pp 73–92.

Brod RD, Flynn HW, et al: Endogenous *Candida* endophthalmitis: Management without intravenous amphotericin B. Ophthalmology 97:666, 1992.

Bryan R: Microsporida. In Mandell GL, Douglas RG, Bennett JE (eds): Principles and Practice of Infectious Disease. New York: Churchill Livingstone, 1990, p 47.

Buss DR, Pflugfelder SC et al: Lymphogranuloma venereum conjunctivitis with a marginal corneal perforation.

Chalupa E, Swarbrick HA, et al: Severe corneal infections associated with contact lens wear. Ophthalmology 94:17, 1987.

Chandler JW, Gillette TE: Immunologic defense mechanisms of the ocular surface. Ophthalmology 90:585, 1983.

Davis JL, Koidou-Tsiligianni A, et al: Coagulase-negative staphylococcal endophthalmitis: Increase in antimicrobial resistance. Ophthalmology 95:1404, 1988.

deLuise VP: Viral conjunctivitis. In Tabbara KF, Hyndiuk RA (eds): Infections of the Eye. Boston: Little, Brown, 1986, p 437.

DeVoe AG, Silva-Hunter M: Fungal infections of the eye. In Locatcher-Khorazo DH, Seegal BC (eds): Microbiology of the Eye. St. Louis: CV Mosby, 1972, p 208.

Fedukowicz HB, Stenson S: External Diseases of the Eye: Bacterial, Viral, and Mycotic with Noninfectious and Immunologic Disease. Englewood Cliffs, NJ: Prentice-Hall, Appleton-Century-Crofts, 1987, pp 1–20.

Fischer DH: Viral disease of the retina. In Tabbara KF, Hyndiuk RA (eds): Infections of the Eye. Boston: Little, Brown, 1986, p 487.

Flach AJ: Ophthalmia Neonatorum. In Tabbara KF, Hyndiuk RA (eds): Infections of the Eye. Boston: Little, Brown, 1986, p 461.

Flynn HW, Pulide JS, et al: Endophthalmitis therapy: Changing antibiotic sensitivity patterns and current therapeutic recommendations. Arch Ophthalmol 109:175, 2/1991 corr. 5/1991.

Forster RK: Etiology and diagnosis of bacterial postoperative endophthalmitis: Symposium on postoperative endophthalmitis. Ophthalmology 85:320, 1978.

Freeman LN, Green WR: Periocular Infections. In Mandell GL, Douglas RG, Bennett JE (eds): Principles and Practice of Infectious Diseases, 3rd ed. New York: Churchill Livingstone, 1990, p 995.

Friedman AH (ed): Ocular Infectious Disease: A Guide for the General Practitioner (Seminar Series #7). Chicago: Abbott Laboratories, 1978.

Gardner TW, Soch D (eds): Handbook of Ophthalmology. East Norwalk, CT: Appleton & Lange, 1987, Chapters 1–5 and 9–10.

Gittinger JW: Ophthalmology: A Clinical Introduction. Boston: Little, Brown, 1984, Chapters 2 to 4.

Gonnering RS, Harris GJ: Infections of the orbit. In Tabbara KF, Hyndiuk RA (eds): Infections of the Eye. Boston: Little, Brown, 1986, p 517.

Gutierrez EH: Bacterial infections of the eye. In Locatcher-Khorazo DH, Seegal BC (eds): Microbiology of the Eye. St. Louis: CV Mosby, 1972, pp 63–76.

Halberts SP: Inhibitory properties of the ocular flora. In Locatcher-Khorazo DH, Seegal BC (eds): Microbiology of the Eye. St. Louis: CV Mosby, 1972, p 24.

Holland GN: Prevention of human immunodeficiency virus transmission by ophthalmic examinations and procedures. In Friedlaender MH (ed): Prevention of Eye Disease. New York: Mary Ann Liebert, Inc, 1988, p 69.

Hyndiuk RA, Skorich DN, Burd EM: Bacterial keratitis. In Tabbara KF, Hyndiuk RA (eds): Infections of the Eye. Boston: Little, Brown, 1986, p 303.

Jones DB: Acanthamoeba—the ultimate opportunist? Am J Ophthalmol 102:527, 1986.

Jones DB: Fungal keratitis. In Wilson LA (ed): External Infections of the Eye. New York: Harper & Row, 1979, p 265.

Jones DB, Liesegang TJ, Robinson NM: Laboratory diagnosis of ocular infections. In Washington JA (ed): CUMITECH 13. Washington, DC: American Society for Microbiology, 1981.

Kean BH, Sun T, Ellsworth RM (eds): Color Atlas of Ophthalmic Parasitology. New York: Igaku-Shoin, 1991.

Kervick GN, Flynn HW, et al: Antibiotic therapy for *Bacillus* sp. Ophthalmology 97:666, 1990.

Lewis ML, Culbertson WW, et al: Herpes simplex virus 1: A cause of the acute retinal necrosis syndrome. Ophthalmology 96:875, 1989.

Liesgang TJ, Foster RK: Spectrum of microbial keratitis in South Florida. Am J Ophthalmol 90:38, 1980.

Locatcher-Khorazo DH, Seegal BC: Microbiology of the Eye. St Louis: CV Mosby, 1972, Chapters 2, 4, 8, 14–15.

Ludwig IH, Meisler DM: Acanthamoeba keratitis. In Tabbara KF, Hyndiuk RA (eds): Infections of the Eye. Boston: Little, Brown, 1986, p 665.

McDonnell PJ, Green WR: Conjunctivitis. In Mandell GL, Douglas RG, Bennett JE (eds): Principles and Practice of Infectious Diseases, 3rd ed. New York: Churchill Livingstone, 1990, pp 975–981.

McDonnell PJ, Green WR: Endophthalmitis. In Mandell GL, Douglas RG, Bennett JE (eds): Principles and Practice of Infectious Diseases, 3rd ed. New York: Churchill Livingstone, 1990, p 987.

Ocular therapeutics and management. 2(6):1, 1991.

O'Malley C: Vitreous. In Vaughn D, Asbury T (eds): General Ophthalmology, 11th ed. East Norwalk, CT: Appleton & Lange, 1986, p 151.

Ophthalmic Drug Facts (edited): Facts and Comparison Division, Philadelphia: JB Lippincott, 1990.

Parke DW, II, Brinton GS: Endophthalmitis. In Tabbara KF, Hyndiuk RA (eds): Infections of the Eye. Boston: Little, Brown, 1986, p 563.

Parment O, Ronerstam RA: Soft contact lens keratitis associated with *Serratia marcescens*. Acta Ophthalmol 59:560, 1981.

Pavan-Langston D, Dunkel EC: Handbook of Ocular Drug Therapy and Ocular Side Effects of Systemic Drugs. Boston: Little, Brown, 1991.

Schacter J, Dawson CR: Human Chlamydial Infections. Littleton, IN: PSG Publishing, 1978, pp 1–219.

Shoukrey ND, Tabbara KF: Eye-related parasitic diseases. In Tabbara KF, Hyndiuk RA (eds): Infections of the Eye. Boston: Little, Brown, 1986, p 167.

Sullivan JH: Orbit. In Vaughn D, Asbury T (eds): General Ophthalmology, 11th ed. East Norwalk, CT: Appleton & Lange, 1986, p 223.

Sullivan JH: Lids and lacrimal apparatus. In Vaughn D, Asbury T: General Ophthalmology, 11th ed. East Norwalk, CT: Appleton & Lange, 1986, p 63.

Tabbara KF: Chlamydial conjunctivitis. In Tabbara KF, Hyndiuk RA (eds): Infections of the Eye. Boston: Little, Brown, 1986, p 421.

Tabbara KF, Hyndiuk RA: Infections of the Eye. Boston: Little, Brown, 1986, pp 469–470.

Vaughn D, Asbury T (eds): General Ophthalmology, 11th ed. East Norwalk, CT: Appleton & Lange, 1986, Chapters 4–5, 8–11 and 14.

Vaughn DG, Tabbara KF: Prevention of ocular infections. In Tabbara KF, Hyndiuk RA (eds): Infections of the Eye. Boston: Little, Brown, 1986, pp 13–20.

Veirs ER: The lacrimal system. In Wilson LA (ed): External Diseases of the Eye. New York: Harper & Row, 1979, p 121.

Watson P: Diseases of the sclera and episclera. In Wilson LA (ed): External Diseases of the Eye. New York: Harper & Row, 1979, p 279.

Weinberg RS: Endogenous bacterial and fungal infections of the retina and the choroid. In Tabbara KF, Hyndiuk RA (eds): Infections of the Eye. Boston: Little, Brown, 1986, p 499.

Weinberg RS: Prevention of herpes simplex infections. In Friedlaender MH (ed): Prevention of Eye Disease. New York: Mary Ann Liebert Inc, 1988, p 47.

Whitcher JP: Prevention of bacterial conjunctivitis and keratitis. In Friedlaender MH (ed): Prevention of Eye Disease. New York: Mary Ann Liebert, Inc, 1988, pp 25–40.

Wilson LA: Bacterial corneal ulcers. In Wilson LA (ed): External Diseases of the Eye. New York: Harper & Row, 1979, p 215.

Zambrano W, Flynn HW, et al: Management of options for *Propionibacterium acnes* endophthalmitis. Ophthalmology 96:1100, 1989.

Appendices

Selected Bacteriologic Culture Media, Stains, and Reagents

A wide variety of basal, enrichment, selective, and differential media are available to the clinical microbiology laboratory. Each of these media is intended to aid the laboratory in the isolation, cultivation, and identification of clinically significant organisms from patient specimens. A laboratory's efficient performance of these functions depends not on its maintaining a vast array of media for routine use but, rather, on the laboratory's ability to make wise choices when selecting a routine media menu, choices that should be dictated by the factors discussed in Chapter 7. This appendix provides the reader with information about a select number of bacteriologic media cited in the bacteriology section of this text. Most of the media detailed in this appendix are commercially available in either dehydrated or finished form.

When the medium is prepared "in-house," established formulation and directions must be strictly followed. Accuracy in the calculation of all weights and measures is essential. All glassware should be chemically clean, and only deionized or distilled water should be utilized unless the directions specify otherwise. Media rehydrated from powdered formulations should be monitored for appropriate appearance and pH. Media containing agar should be heated to boiling and allowed to boil for approximately 1 minute before sterilizing.

Heating and short-term boiling ensure that the agar will dissolve completely and guard against charring of the agar during sterilization. Sterilization should be in accordance with the formulation's stated directions. Sterility and performance tests (quality controls) should be performed as discussed in Chapter 4.

A7 Agar

A7 agar is a selective, differential medium useful in the isolation of genital mycoplasmas. While a variety of formulations are available, commonly used formulations contain either penicillin G or penicillin G with amphotericin B to inhibit the normal bacterial flora of the genital tract (selective), while the presence of horse serum and yeast extract makes the medium complex enough to support the growth of mycoplasmal strains. The addition of urea and manganese sulfate enhances this medium's ability to differentiate *Ureaplasma urealyticum* from mycoplasmal strains that do not hydrolyze urea. *Ureaplasma urealyticum* degradation of urea leads to the production of ammonia. The ammonia produced reacts with manganese sulfate to yield a dark-brown product; thus, *U. urealyticum* colonies take on a dark golden-brown to deep, velvet-brown color. The coloration of *U. urealyticum* colonies is in sharp contrast with the clear, light background of the medium. Classic *Mycoplasma, Acholeplasma* species, and *Proteus* L colonies do not react to form the colored end product; thus, colonies remain clear.

Plated completed A7 medium can be stored at 4°C for up to 1 week. (*Note:* While the complete medium can be stored in the refrigerator, the basal medium alone should not be stored.) Any plated medium in which a precipitate is observable under the low-power objective of the light microscope should be discarded. Usable, plated medium should be inoculated, streaked for isolation, and incubated anaerobically at 35°C for 48 hours.

Acetate Agar

Acetate agar is a differential medium used to distinguish *Escherichia coli* from *Shigella* species by monitoring the ability to utilize acetate as the only available carbon source. Organisms capable of utilizing acetate also utilize the medium's ammonium salt as a nitrogen source. The breakdown of the ammonium salt results in a shift of

1049

the pH into the alkaline range. At alkaline pH, the incorporated pH indicator, bromthymol blue, shifts from green to blue.

If the colony to be tested is young, the medium should be inoculated by making a saline suspension from an agar culture and introducing the saline-suspended organisms onto the slant with a needle. Incubate at 37°C for up to 4 days, while monitoring daily for the expected color change.

Alkaline Peptone Water

Alkaline peptone water is an enrichment medium useful in the recovery of *Vibrio* and *Aeromonas* species from stool specimens. The alkaline pH of this medium allows uninhibited replication of these species while temporarily suppressing the replication rate of many commensal intestinal bacteria.

Some formulations recommend adjusting to pH 9.0 specifically for recovery of vibrios. Alkaline peptone water cultures should be incubated at 35°C and subcultured to thiosulfate citrate, bile salts, sucrose (TCBS) agar within 12 to 18 hours.

Bacteroides Bile-Esculin (BBE) Agar

Bacteroides bile-esculin agar is a selective, differential agar used for the isolation and identification of *Bacteroides fragilis*. The incorporation of oxgall (bile salts) separates bile-resistant species from bile-sensitive ones, while the 1% esculin, in conjunction with ferric ammonium citrate, provides information about the isolate's ability to hydrolyze esculin. Products of esculin hydrolysis on reacting with the ferric ammonium citrate form an insoluble iron salt that is deposited within the positive colonies, causing them to turn black.

Plated medium should be inoculated, streaked for isolation, and incubated anaerobically.

Bile-Esculin Agar

Bile-esculin agar is a selective, differential agar used to isolate and identify group D streptococci, including the enterococci. Oxgall (bile salts) is the selective ingredient, while esculin is the differential component. All group D streptococci, including all enterococci, hydrolyze esculin. Products of esculin hydrolysis react with ferric citrate in the medium to produce insoluble iron salts. Deposi-

tion of the iron salts results in a blackening of the tubed medium. Test results must be interpreted in conjunction with Gram stain morphology, as *Listeria monocytogenes* and a small number of other organisms also produce positive reactions.

The medium should be inoculated, incubated aerobically at 35°C, and observed for growth. Darkening of the medium indicates esculin hydrolysis.

Bismuth Sulfite Agar

Bismuth sulfite agar is a selective medium for the isolation of *Salmonella* species. The selective ingredients are bismuth sulfite and brilliant green, which inhibit the growth of gram-positive bacteria, most lactose-fermenting intestinal normal flora, and shigellae. While this is not a differential medium in the strictest sense, the ferrous sulfate in this medium is reactive with hydrogen sulfide to produce ferric sulfide, which is deposited within the bacterial colony as a black, insoluble precipitate. Typically, *Salmonella typhi* colonies are black and surrounded by a metallic sheen, while *S. gallinarum, S. choleraesuis,* and *S. paratyphi* colonies are light green on this medium. When bismuth sulfite agar is used to isolate *Salmonella* species from feces and other clinical specimens, the parallel use of a less inhibitory medium is recommended, as bismuth sulfite agar may inhibit or partially inhibit the growth of some *Salmonella* strains.

This medium cannot be autoclaved and must be used on the day it is prepared. Plated medium should be inoculated with fecal or enrichment broth materials, streaked for isolation, and incubated at 35°C for 48 hours.

Blood Agar, Anaerobic, CDC

CDC anaerobic blood agar is an enrichment medium useful in the isolation and culture of fastidious anaerobes. It contains yeast extract, L-cysteine, hemin, and vitamin K_1.

This medium can be stored for up to 6 weeks if it is sealed in cellophane bags and stored at 4°C. Plates should be inoculated, streaked for isolation, and incubated anaerobically at 35°C for 48 hours.

Blood Agar, Anaerobic, Brucella Base, Wadsworth

Wadsworth Brucella base anaerobic blood agar is a useful enrichment medium for the isolation of moderately fastidious, obligate anaerobes. Sheep's blood provides the enrichment. Vitamin K_1 is also added to this medium.

Plated medium should be sealed in bags and stored at 4°C for up to 2 weeks. Usable plates are inoculated, streaked for isolation, and incubated anaerobically at 35°C for 48 hours.

Blood Agar, Anaerobic, With Kanamycin and Vancomycin (KV Blood Agar)

KV blood agar is a variation of CDC anaerobic blood agar made semiselective by the addition of the antimicrobials kanamycin and vancomycin. It is useful in the primary isolation of obligate anaerobes, particularly *Bacteroides* species, from specimens with a mixed bacterial population.

KV blood agar plates sealed in bags can be stored at 4°C for up to 4 weeks. Usable plates should be inoculated, streaked for isolation, and incubated anaerobically at 35°C for 48 hours. Plates negative at 48 hours may be reincubated, depending on the particular situation.

Blood Agar, Anaerobic, Laked, With Kanamycin, Vancomycin, and Vitamin K (KVKL)

Anaerobic, laked blood agar with kanamycin, vancomycin, and vitamin K (KVKL) is a selective enrichment medium recommended for the isolation of species of *Bacteroides* and *Prevotella* from clinical specimens. While appropriate for the isolation of any *Bacteroides* species, this medium is particularly helpful in the isolation of *Prevotella melaninogenica* as pigment production is enhanced. Laked erythrocytes and vitamin K constitute the enrichment ingredients, while the antibiotics inhibit all cocci and facultative gram-negative bacilli except the pseudomonads.

Plated medium should be inoculated, streaked for isolation, and incubated anaerobically at 35°C for a minimum of 48 hours.

Blood Agar, Rabbit's

Rabbit's blood agar is an enrichment medium particularly useful in the recovery and the demonstration of β hemolysis by *Haemophilus influenzae* and *Gardnerella vaginalis*.

Blood Agar, Sheep's (SBA)

Blood agar is a routine medium used to cultivate a wide variety of moderately fastidious bacterial organisms. An infusion agar or tryptic soy agar base can be enriched by the addition of 5 to 10% defibrinated sheep's, rabbit's, or human's blood. However, sheep's blood has proved the most versatile enrichment additive. Incorporation of the blood not only provides enrichment for growth of the bacterial organisms but also allows the detection and characterization of hemolytic activity.

Usable plates should be inoculated, streaked, and incubated as dictated by the specific application.

Blood Phenylethyl Alcohol Agar, Anaerobic, CDC

Blood phenylethyl alcohol agar (PEA) is a selective enrichment medium useful in the isolation of *Bacteroides, Prevotella,* and other obligate anaerobes from specimens containing a mixture of obligate and facultative anaerobes. Enrichment is provided by yeast extract, hemin, vitamin K, and defibrinated sheep's blood. Selectivity is provided by the incorporation of phenylethyl alcohol, which inhibits facultative anaerobes by suppressing DNA synthesis and cell division.

Plated medium should be sealed in plastic bags for storage. Bagged plates may be stored up to 4 weeks at 4°C. Usable plates should be inoculated, streaked for isolation, and incubated anaerobically at 35°C for at least 48 hours.

Blood Phenylethyl Alcohol Agar, Wadsworth

Wadsworth blood phenylethyl alcohol agar is an alternative to the CDC formulation of blood phenylethyl alcohol agar. Like the CDC formulation, it is a selective medium for the isolation of *Bacteroides* and *Prevotella* species. The inoculation, incubation, and interpretation of growth on this medium are the same as those for the CDC blood phenylethyl alcohol agar discussed previously.

Bordet-Gengou (B-G) Blood Agar

Bordet-Gengou blood agar is a selective enrichment medium for the isolation of *Bordetella pertus-*

sis and *B. parapertussis* from clinical specimens. Formulations for the base medium vary with respect to the use of peptone. The growth on the peptone-free formulation is less luxuriant, and indigenous flora is less likely to overgrow the pathogen; hence, the formulation without peptone should be used for isolation by the "cough plate" method. The base medium formulation with peptone is suitable for isolation if a nasopharyngeal swab specimen is obtained. The complete medium is enriched by the addition of glycerol and sterile, defibrinated sheep's blood. Increased selectivity of the complete medium has been achieved historically by adding penicillin, methicillin, or cephalexin to the medium just be-fore plating.

The antibiotic should be added aseptically just prior to the addition of the blood enrichment. Blood enrichment between 15 and 30% (3–6 mL/20-mL tube) is appropriate; the amount is determined by the preference of the laboratory. The species from which the sterile, defibrinated blood is obtained is not critical, but properly prepared plates should be cherry-red, moist, and bubble-free. Complete plated medium should be used immediately if possible. If for some reason this is not possible, plates can be maintained for about 1 week at 4°C if they are tightly sealed. Usable, complete medium can be inoculated by rolling the nasopharyngeal swab specimen over one third of the plate surface and then streaking for isolation with a platinum loop. The specimen should be inoculated in this fashion onto B-G plates without antibiotic and B-G plates with antibiotic. Inoculated plates should be incubated at 35 to 37°C and then examined at 48 hours. Plates negative at 48 hours should be reincubated. Plates must be held for 5 days before they can be regarded as negative for the organism.

Brain-Heart Infusion Broth

Brain-heart infusion broth is an enriched medium suitable for the cultivation of a number of nonfastidious and moderately fastidious microorganisms. This broth medium is recommended for cultivation of pneumococci for the bile solubility test.

Buffered Charcoal-Yeast Extract Agar With α-Ketoglutarate (BCYE-α)

Buffered charcoal-yeast extract agar with α-ketoglutarate is an enrichment medium useful in the isolation of *Legionella* species from clinical specimens. Yeast extract and L-cysteine enhance the growth of *Legionella* organisms, while activated charcoal absorbs toxic compounds that either accumulate as the result of the organism's metabolism or are present following preparation of this medium.

Usable plated medium may be stored in plastic bags, away from light, at 4°C for up to 4 weeks. Plated medium should be inoculated, streaked for isolation, and incubated at 35°C in a carbon dioxide incubator. Cultures should be checked daily for up to 2 weeks, and incubator humidity monitored to prevent excessive drying of plates. *Legionella* colonies may not be grossly visible until 3 to 5 days after inoculation.

BCYE-α, L-Cysteine–Deficient

This medium is a variation of buffered charcoal-yeast extract agar with α-ketoglutarate (BCYE-α). Its formulation is identical except for the L-cysteine. This medium, deficient in L-cysteine, is helpful in presumptively separating those gram-negative, non–*Legionella* organisms that may grow on BCYE-α from *Legionella* species. Non–*Legionella* gram-negative rods grow in the absence of L-cysteine, whereas *Legionella* species do not. Consequently, growth on both cysteine-containing and cysteine-deficient media suggests that the isolate is not a *Legionella* species, while growth on the cysteine-containing medium suggests only a requirement for L-cysteine, which is characteristic of *Legionella* species.

BCYE-α With Antibiotics

This is a semiselective variation of BCYE-α useful in the isolation of *Legionella* species from body sites that contain a mixed bacterial flora. Three antibiotics are added to the standard BCYE-α formulation to inhibit the growth of other bacteria and fungi. Cefamandole (4 mg/L) inhibits gram-positive organisms, polymyxin B (80,000 units/L) inhibits gram-negative bacilli, especially pseudomonads, and anisomycin (80 mg/L) inhibits fungi.

BCYE-α Modified Wadowsky-Yee (mWY)

This is also a semiselective variation of BCYE-α useful in the isolation of *Legionella* species from body sites that contain a mixed bacterial flora. This variation utilizes glycine and polymyxin B to inhibit gram-negative organisms, vancomycin to inhibit gram-positive cocci, and anisomycin to inhibit fungi.

Campylobacter Blood Agar (Campy-BA)

Campylobacter blood agar is a selective enrichment medium useful in the isolation and cultivation of *Campylobacter* species from stool specimens. Brucella agar serves as the base medium for Campy-BA because it contains sodium bisulfite, which lowers the redox potential, thereby enhancing the recovery of microaerophilic organisms such as *Campylobacter* species. Ten percent sheep's blood enriches the basal medium, and an antibiotic mixture makes the medium selective. While minor variations exist in the composition of this mixture, most formulations have incorporated vancomycin to inhibit gram-positive cocci, trimethoprim to inhibit swarming strains of *Proteus,* polymyxin B to inhibit gram-negative bacilli, and amphotericin B to inhibit filamentous fungi and yeasts. Currently, cefoperazone is being promoted to replace cephalothin. Cefoperazone has antipseudomonal activity (lacked by cephalothin) and is more effective against members of Enterobacteriaceae.

Campylobacter Thioglycolate Broth (Campy-Thio)

Campy-Thio is a selective liquid medium. The base medium is thioglycolate broth with 0.16% agar. The selective component is the antibiotic formulation used in Campy-BA.

Carbohydrate Fermentation Media, Anaerobic

Carbohydrate broths are differential media useful in determining the ability to ferment specific carbohydrates. This formulation, based on a carbohydrate medium base (Difco, Detroit, MI), is useful in the determination of carbohydrate fermentations carried out by anaerobic isolates. Carbohydrate stock solutions are added to the base according to the manufacturer's directions. Prepared media are aseptically dispensed into 15 × 90 mm screw-capped tubes. Tubed medium may be stored at room or refrigerator temperatures, but dissolved oxygen must be removed either before storage or before use. If an anaerobic chamber is available, the tubed medium, loosely capped, should be passed into the anaerobic atmosphere (85% N_2; 10% H_2; 5% CO_2), and the caps securely tightened before removal and storage. If an anaerobic chamber is not available, the tubed medium should be steamed or boiled with the caps loose for 10 minutes, cooled, and inoculated immediately using a capillary pipette to deliver the inoculum near the bottom of the tube without introducing air. One pipette may be used to inoculate multiple tubes.

All cultures should be incubated anaerobically, with the caps loosened, at 35°C for up to 7 days. Cultures should be examined on days 1, 2, and 7 for fermentation with production of acid. At pH 6.0 or lower, the bromthymol blue indicator turns yellow, indicating a positive fermentation reaction. If fermentation does not occur, the bromthymol blue indicator remains blue to blue-green (negative reaction). If the cultured organism can reduce the bromthymol indicator, the medium becomes colorless. If this occurs, a sterile pipette can be used to transfer two to three drops of material from the involved tube(s) to a spot plate. Two to three drops of dilute indicator should then be added to the transferred material, and this observed for color change. *Note:* Rapid test systems are available that include carbohydrate fermentations with a bromcresol purple indicator. Correlation between fermentation and color change is similar, as is the problem of indicator reduction and the method for detecting acid in the presence of reduced indicator.

Carbohydrate Fermentation Media for Gram-Positive Cocci

Carbohydrate broths are differential media useful in determining the ability of a variety of aerobic bacteria to ferment specific carbohydrates. This formulation, based on a heart infusion broth, provides sufficient nutrient to support the growth of gram-positive cocci, including streptococci.

The sterile carbohydrate is added to the sterile basal medium (cooled to 50°C) to give a final concentration of 1%. Tubed carbohydrate fermentation medium should be inoculated and incubated aerobically at 35°C. A positive reaction (acid production) is indicated by a color change from purple to yellow.

Carbohydrate Fermentation Media for Aerobic Gram-Negative Bacilli

Carbohydrate fermentation medium of this formulation does not contain sufficient quantities of complex nutrients to support the growth of fastidious aerobic bacteria. It is, however, sufficient to support the less fastidious Enterobacteriaceae; thus, it is the formulation of choice to determine the fermentative patterns of suspected Enterobacteriaceae isolates. Andrade pH indicator detects acid production. A positive result (acid production) is indicated by a color change from yellow or colorless (alkaline) to pink or red.

Chocolate Agar

Chocolate agar is an enrichment agar especially useful in promoting the growth of *Haemophilus* and other fastidious bacterial species. A variation of sheep's blood agar, this medium may be made by adding sheep's blood while the basal medium is warm enough to release the red cell hemoglobin and nicotinamide-adenine dinucleotide (NAD). Alternatively, sheep's blood may be replaced by 2% hemoglobin and a chemical supplement solution, such as Iso-Vitalex (BBL, Cockeysville, MD). The enrichment used must result in the complete medium containing cell-free hemoglobin and NAD. The temperature at which either enrichment is added results in a chocolate-brown medium.

Plated medium can be stored at 4°C. Usable plates should be inoculated, streaked for isolation, and incubated at 35°C in CO_2.

Citrate Agar, Simmons

Simmons citrate agar is useful in differentiating gram-negative enteric bacilli. Similar in principles of its use to acetate agar, citrate replaces acetate in this medium, and differentiation is based on the isolate's ability or inability to utilize citrate as its sole source of carbon. Several alternative theories have been advanced to explain the biochemical sequences leading to a color change in this medium. One such theory holds that, as with acetate agar, organisms capable of utilizing the citrate also utilize the medium's ammonium salt as a nitrogen source. The breakdown of the ammonium salt results in a shift of the pH into the alkaline range. At alkaline pH, the incorporated pH indicator bromthymol blue shifts from green to blue.

The pH of the medium is approximately 6.9. This medium should be inoculated by making a saline suspension of a young colony from an agar culture, streaking the saline suspended organisms onto the slant with a needle, and stabbing the butt. Inoculated slants should be incubated at 35°C for up to 4 days and monitored daily for the expected color change.

Columbia Agar

Columbia agar is a basal nutrient agar that contains peptones derived from both casein and meat. The basal medium is suitable for cultivation of a number of aerobic and anaerobic bacterial organisms found in clinical materials. Additionally, it provides an efficient base for preparation of a variety of enrichment agars that support the growth of more fastidious aerobes and anaerobes.

Inoculate plated medium, streak for isolation, and incubate at 35°C for up to 7 days as appropriate for the specific organism to be cultured.

Columbia Agar With Antibiotics (Columbia CNA Agar)

Columbia agar with antibiotics is a selective enrichment medium suitable for the isolation of gram-positive cocci from specimens that might also be expected to contain gram-negative bacilli, especially *Proteus* species. Sheep's blood is the usual enrichment ingredient, while colistin and nalidixic acid are incorporated to inhibit gram-negative overgrowth of desired gram-positive isolates.

Plated medium should be inoculated, streaked for isolation, and incubated in an environment appropriate to the isolation of the desired species.

Cooked Meat Medium

Cooked meat medium is useful in the cultivation of anaerobes, especially pathogenic species of *Clostridium*. This medium contains solid meat particles and is excellent for initiating growth from a very small inoculum as well as sustaining culture viability over long periods. It is useful for cultivation of mixed cultures because all organisms are supported, while overgrowth by the more rapid growers is retarded. Dehydrated medium is suspended in tubes according to the manufacturer's directions and allowed to stand until thoroughly moistened (approximately 15 minutes) before sterilizing by autoclaving (121°C for 15 minutes). Tubed medium may be stored at room temperature, but stored tubes should be placed in flowing steam or a boiling water bath for 10 minutes to drive off dissolved gases and then should be cooled rapidly before inoculating. Cooled tubed medium should be inoculated and incubated in a manner appropriate for the species being isolated or subcultured. Proteolytic activity by cultured organisms is usually evidenced by the digestion of the meat particles. Saccharolytic clostridial species typically produce acid with gas. *Note:* A variation of this medium, cooked meat phytone medium, adds calcium carbonate particles and phytone (1.5%) to the original formulation. Cooked meat phytone supports the growth of pathogenic and butyric-butyl clostridial species as well as fusiforms, *Bacteroides,* and other anaerobic non–spore-formers. If subcultures are of strict anaerobes or specimens are suspected of containing strict anaerobes of interest, growth may be expedited by anaerobic incubation. Proteolytic activity by the cultured organism(s) is evidenced by digestion of the meat particles, possibly with darkening of the medium.

Tellurite Blood Agar (With or Without Cystine)

Tellurite blood agar is a selective-differential enrichment agar useful in the isolation of *Corynebacterium diphtheriae.* All formulations include animal blood as a source of enrichment. Some formulations also incorporate cystine to further enhance the growth of fastidious organisms, including *C. diphtheriae.* Potassium tellurite is the selective, differential ingredient responsible for inhibiting the growth of staphylococci and streptococci while allowing the growth of *C. diphtheriae* and diphtheroids, which act on the tellurite, depositing the reduced product within the colonies.

Tubes of base medium may be stored and melted to make the complete medium as culture plates are needed. Freshly plated medium should be inoculated, streaked for isolation, and incubated at 35°C. On this medium, colonies of *C. diphtheriae* are dull gray-black, while diphtheroids are light gray-green with dark centers. Some *Staph-*

ylococcus species, gram-negative bacilli, and yeasts may overcome inhibition and grow on this medium. The *Staphylococcus* colonies are large, glistening, and jet-black, whereas those of the gram-negative bacilli and yeast are dull gray-black but larger than the *C. diphtheriae* colonies.

Cystine Tryptophan Agar, With Sugar (CTA-Sugar)

Cystine tryptophan (tryptic/trypticase) agar is a semisolid base medium that contains no meat or plant extracts and is free from fermentable carbohydrates. It may be made differential by the addition of a carbohydrate (CTA-sugar). CTA-sugars are recommended for the determination of fermentation reactions by fastidious organisms.

Alternatively, carbohydrates are available in the form of differentiation disks that can be aseptically added to the base tubed medium as needed. Sterile, tubed CTA-sugars should be inoculated with a heavy inoculum, stabbing to a depth of approximately 2 mm below the medium surface. With phenol red as the pH indicator, fermentation is indicated by a color change in the medium from red to yellow.

Cycloserine Cefoxitin Fructose Agar (CCFA)

Cycloserine cefoxitin fructose agar is a selective, differential medium useful in the isolation and identification of *Clostridium difficile* from stool specimens of patients suspected of having antibiotic-associated diarrhea with pseudomembranous colitis. The selective antibiotic ingredients, cycloserine and cefoxitin, inhibit the growth of intestinal normal flora by interfering with cell wall synthesis in both gram-positive and gram-negative bacteria. Indigenous bacteria are inhibited, but *C. difficile* is not. While cycloserine and cefoxitin are incorporated for their selective properties, fructose and a pH indicator, neutral red, are included to confirm that the isolates can ferment this sugar. *Note:* A variation of this medium is made with mannitol rather than fructose and utilizes bromthymol blue indicator. A second variation adds egg yolk suspension so lecithinase and lipase activities can be detected.

Decarboxylase Test Medium (Moeller)

Decarboxylase test medium with an incorporated amino acid is a differential medium useful in the identification of fermentative and nonfermentative gram-negative bacteria. The differential ingredient is one of three amino acids, lysine, arginine, or ornithine. Decarboxylation of the amino acids yields alkaline end products detected by a change in the color of an incorporated pH sensitive dye, bromcresol purple. The tubed basal medium (no amino acid) serves as a control for reading the reactions.

Decarboxylase tubes and a control tube should be inoculated from a 24-hour slant culture using a loop. Inoculated tubes should be overlaid with 4 to 5 mm of sterile mineral oil to avoid oxidative deamination of available protein, which would be falsely interpreted as a positive reaction. Inoculated, overlaid tubes should be incubated at 35°C for up to 4 days. Incubating tubes should be checked daily. Early in the incubation, fermentative organisms will ferment glucose, turning the control and all decarboxylase tubes yellow. For these organisms, as the pH drops, the hydrogen ion concentration becomes optimal for decarboxylase activity. In the decarboxylase tubes, the subsequent conversion of the amino acid to amines raises the pH, reversing the yellow to purple, while the control tube remains yellow. Nonfermenters do not produce the initial yellow color change, and utilization of the amino acid is indicated when the amino acid–containing tube becomes a deeper purple than the control.

Deoxycholate Citrate Agar

Deoxycholate citrate agar is a selective, differential medium useful in the isolation of enteric pathogens directly from feces and urine specimens or indirectly from enrichment broths, such as selenite-F. The selective ingredients are sodium citrate and sodium deoxycholate at concentrations that inhibit the majority of nonpathogenic enteric bacilli. The differential ingredient is lactose. Nonfermenting enteric pathogens appear as colorless colonies. Those lactose-fermenters that do overcome the inhibitors appear as pink to red colonies as the result of the pH change that accompanies lactose fermentation.

This medium is not autoclaved. Plates may be stored under refrigeration for several days.

Plated medium should be inoculated, streaked for isolation, and incubated at 35°C in air (not in CO_2) for up to 48 hours. A heavy inoculum is recommended if the specimen is feces, urine, or other direct bodily materials, whereas a light inoculum is recommended if one is subculturing from an initial enrichment broth.

Deoxyribonuclease (DNase) Test Agar, With or Without Indicator Dye

DNase test agar is a differential medium used to detect the production of an active DNase exoenzyme by aerobic bacterial species. The differential ingredient is incorporated DNA. Methods available for detection of DNA degradation include hydrochloric acid precipitation of undegraded DNA and color change of an incorporated metachromatic dye, such as toluidine blue or methyl green.

Sterile plated medium should be inoculated using a 1- to 2-cm streak or a spot inoculum approximately 5 mm in diameter. Inoculated plates should be incubated aerobically at 35°C for 18 to 24 hours. Following incubation, DNase activity is detected in one of the following ways:

1. If a basal medium without a metachromatic dye was inoculated, flood the plate with 1N HCl, and look for a zone of clearing around the bacterial growth. If the incorporated DNA is undegraded, it is precipitated by the 1N HCl, and the medium becomes opaque. If the incorporated DNA is degraded, the nucleotide fragments dissolve in the 1N HCl, and the medium remains clear.

2. If the medium includes toluidine blue, the blue, DNA-bound dye is released from nucleotide fragments, producing a color change to rose. The medium remains clear blue in negative reactions.

3. If the medium includes methyl green, the green, DNA-bound dye is released from nucleotide fragments, resulting in a loss of color. The medium remains green in negative reactions.

E Medium, Diphasic

E medium is one of many media developed for the isolation and culture of mycoplasmal organisms. This medium is enriched with a yeast dialysate, horse serum, penicillin, and thallium acetate.

The basal agar is aliquoted in 65-mL portions. Any basal medium not utilized immediately can be stored under refrigeration and remelted for use as needed. To each 65-mL aliquot add 10 mL of yeast dialysate, 25 mL of sterile horse serum, 2 mL of penicillin (20,000 U), and 1 mL of 3.3% aqueous thallium acetate. Completed agar can be dispensed in 3-mL quantities into 16×125 mm screw-capped tubes, or in 5-mL amounts into 10×35 mm Petri dishes, and allowed to solidify.

The E broth is prepared according to the manufacturer's directions. Broth is dispensed into 65-mL aliquots. To each aliquot add 10 mL of yeast dialysate, 25 mL of sterile horse serum, 2 mL of penicillin (20,000 U), and 1 mL of 3.3% aqueous thallium acetate.

The E agar is overlaid with 3 mL of E broth and stored at room temperature. *Caution:* Tubes must be tightly capped to prevent loss of CO_2 from incorporated horse serum, resulting in elevation of the medium pH.

Egg Yolk Agar, CDC Formulation (EYA)

Egg yolk agar is a differential medium useful in the detection of lecithinase, lipase, and protease activity. Incorporated egg emulsion provides the lecithin, lipids, and proteins to be degraded by these enzymes. On EYA, exoenzyme activity is detected as follows: lecithinase activity produces a zone of opacity immediately around the growth streak; lipase activity results in an iridescent sheen on or around the surface of colonies; and protease activity is seen as a clearing of the medium around and just beyond the streaked growth area. A given organism may produce one or all of these exoenzymes.

Plated medium is inoculated as a single streak across the plate and incubated anaerobically at 35°C for 24 to 72 hours. If the Nagler test is to be performed, one half of the plate surface should be smeared with a few drops of *C. perfringens* type A antitoxin prior to inoculation. The inoculation streak should then extend across both halves (no antitoxin/antitoxin) of the plate. Inoculated plates should be incubated anaerobically at 35°C for 24 to 48 hours. A positive test result is the inhibition of lecithinase activity on the half of the plate with antitoxin. Plated medium not utilized immediately may be stored at 4°C if it is sealed in plastic bags.

Eosin–Methylene Blue (EMB) Agar

Eosin–methylene blue agar is a selective, differential medium useful in the isolation and identification of gram-negative enteric bacteria. Eosin Y and methylene blue dyes, the selective ingredients, are incorporated to inhibit the growth of gram-positive bacteria while allowing the growth of gram-negative ones. The carbohydrates lactose and sucrose are incorporated to allow differentiation of isolates based on lactose fermentation. Fermentation is detected by color changes and precipitation of the incorporated dyes as the pH drops. Sucrose serves as an alternative carbohydrate source for slow lactose-fermenters, allowing their timely elimination from consideration as possible pathogens. *Escherichia coli,* a coliform lactose-fermenter, typically forms blue-black colonies with a metallic greenish sheen. Other coliform fermenters, such as *Enterobacter,* form pink colonies. Nonfermenter colonies are translucent, being either amber-colored or colorless.

Esculin Agar

Esculin agar is a differential medium used to determine the ability of an organism to hydrolyze esculin. The hydrolytic products from esculin react with the ferric salt present in this medium to precipitate iron compounds and produce a gray to black discoloration of the medium.

Slanted medium should be inoculated, incubated aerobically at 35°C, and observed for growth, with darkening of the medium indicative of esculin hydrolysis.

Fletcher Semisolid Medium for Leptospira

Fletcher semisolid medium is an enrichment medium recommended for detection of leptospiral species in blood, spinal fluid, and urine specimens as well as possibly contaminated water and other materials. The enrichment component of this medium is rabbit serum containing some hemoglobin.

The lyophilized rabbit serum with natural hemoglobin is commercially available as Leptospira Enrichment, or sterile, pooled, fresh natural rabbit serum may be added. The medium should be aseptically dispensed into sterile screw-capped tubes (5 mL/tube) and stored at room temperature overnight. The medium *must* be inactivated by placing tubes in a 56°C water bath for 1 hour on the day following preparation. Cooled inactivated medium should be inoculated with one or two drops of the fluid specimen, using a sterile, plugged Pasteur pipette. Small inocula introduced into multiple tubes are recommended to optimize pathogen recovery and minimize any interference by developing antibody titers in the body fluid specimens. Inoculated tubes should be incubated with caps loose at 25 to 30°C for 4 to 5 weeks and examined weekly for growth in the form of turbidity at the top of the medium. A loopful of fluid from any tube showing turbidity should be placed on a clean slide, a coverslip added, and the specimen examined by darkfield microscopy.

Gelatin Medium (Nutrient)

Gelatin medium is a differential medium used to determine a bacterial isolate's ability to produce gelatinase and thereby hydrolyze gelatin. A variety of gelatin-containing media can be utilized for this purpose, including starch-gelatin agar, Kohn modified gelatin for the Kohn gelatin method, and nutrient gelatin.

Sterile tubed medium can be stored at 4°C. The medium is inoculated and incubated, along with an uninoculated control tube, at 35°C for 18 to 24 hours. Following incubation, both the inoculated tube and the control tube are refrigerated for 30 minutes before reading. The control tube should gel, while the consistency of the inoculated tube will depend on the isolate's ability or inability to hydrolyze gelatin. If the inoculated tube gels, the isolate in question is gelatinase negative. If the gelatin in the inoculated tube remains liquid, the isolate in question is gelatinase positive.

Gram-Negative (GN) Broth

Gram-negative (GN) broth is a selective enrichment medium used to enhance the chance of recovering enteric pathogens, such as *Salmonella* and *Shigella* species, from fecal specimens. The selective ingredients are deoxycholate and citrate salts, which retard the growth of gram-positive bacteria while allowing the growth of aerobic gram-negative bacteria. Enrichment is provided by increasing the concentration of

mannitol, which temporarily favors the growth of mannitol-fermenting, gram-negative rods over that of the non–mannitol-fermenters.

Sterile, tubed medium should be inoculated with fecal material and incubated, with caps loosened, at 35°C. Incubated GN broth cultures should be subcultured onto selective, differential plated media after 6 to 8 hours and again after 18 to 24 hours.

Hektoen Enteric (HE) Agar

Hektoen enteric agar is a selective, differential medium used for direct isolation of enteric pathogens from feces and for indirect isolation from selective enrichment broth. The selective ingredients are bile salts at concentrations that not only inhibit the growth of gram-positive bacteria but also retard the growth of many gram-negative organisms that are part of the normal intestinal flora. The differential ingredients include lactose and sucrose to determine fermentation patterns, detected by the pH indicator bromthymol blue, and ferric salts (sodium thiosulfate and ferric ammonium citrate) to detect the production of hydrogen sulfide gas.

This medium should not be autoclaved, and overheating should be avoided. Plated medium should be inoculated, streaked for isolation, and incubated aerobically (not in CO_2) at 35°C for 18 to 24 hours. Most nonpathogens ferment one or both of the sugars, and colonies appear bright orange to salmon-pink owing to the low pH interaction with incorporated dyes. Nonfermenters, such as *Salmonella* and *Shigella* species, typically produce green to blue-green colonies. Hydrogen sulfide gas production is seen as a black precipitate that accumulates within colonies.

Hippurate Broth

Hippurate broth is a differential broth useful in the identification of group B streptococci. The differential ingredient is 1% sodium hippurate, which the group B streptococci hydrolyze to glycine and benzoic acid. In this method, hydrolysis is detected by the addition of ferric chloride, which reacts with the benzoic acid to produce ferric benzoate, which precipitates.

Sterile, tubed medium should be inoculated and incubated, along with an uninoculated control tube, at 35°C for 48 hours. Following incubation, 12% ferric chloride is added to the control tube a drop at a time. A precipitate forms initially but later dissolves as ferric chloride is being added. Count the total number of drops added to the control tube by the time the precipitate redissolves. Add that number of drops to the inoculated tube. A positive reaction yields grossly visible precipitate that persists for 10 minutes or longer after the addition of ferric chloride. A negative reaction produces no precipitate or only a faint precipitate that disappears in less than 10 minutes after the addition of ferric chloride. *Note:* A rapid method available utilizes a 1% aqueous solution of sodium hippurate dispensed in 0.4-mL quantities. Colonies of the isolate are emulsified in the solution until it is very cloudy and are then incubated in a 37°C waterbath for 2 hours. Hydrolysis is detected by adding five drops of Ninhydrin reagent (triketohydrindene hydrate), without shaking the tube, and continuing to incubate for a minimum of 10 minutes but not longer than 30 minutes. Positive reactions are deep-purple in color, while negative reactions show no color change.

Hydrogen Sulfide, Lead Acetate

Lead acetate is one differential method used to detect the production of hydrogen sulfide (H_2S) from sulfur-containing amino acids. The organism is cultured in a nutrient broth or on an agar medium with sufficient protein to ensure the presence of sulfur-containing amino acids. As the organism metabolizes these amino acids, H_2S gas is evolved. The liberated gas is detected by lead acetate–impregnated paper strips that are suspended over the culture during incubation. The produced H_2S reacts with the lead acetate to produce lead sulfide, a black insoluble salt, causing the strip to blacken. Lead acetate strips are available commercially.

Kligler Iron Agar

Kligler iron agar (KIA) can be used to determine whether a gram-negative rod is a glucose-fermenter, a fundamental characteristic in the initial classification of gram-negative rods. The medium also tests for lactose fermentation, gas production during carbohydrate fermentation, and hydrogen sulfide production, all of which are useful in the differentiation of gram-negative

rods belonging to the family Enterobacteriaceae.

KIA contains glucose and lactose (fermentable carbohydrates), phenol red (pH indicator), peptone (carbon/nitrogen source), and iron salt plus sodium thiosulfate (sulfur source and hydrogen sulfide indicator). KIA resembles triple sugar iron agar except that it lacks sucrose. Three carbohydrate fermentation patterns are possible:

1. Acid (yellow) butt and alkaline (red) slant indicate that the organism ferments glucose but not lactose. This organism ferments glucose by the Embden-Meyerhof-Parnas (EMP) pathway to produce organic acids, changing the pH indicator from red to yellow. Once the glucose has been consumed, the organism then breaks down peptones, producing ammonia. This causes a pH rise and the slant reverts to red.

2. Acid (yellow) butt and acid (yellow) slant indicate that the organism ferments both glucose and lactose. The organism ferments glucose, producing acid products. Once the glucose is consumed, it ferments lactose, breaking it down into glucose and galactose. This causes the pH in the slant portion to remain acidic.

3. Alkaline (red) butt and alkaline (red) slant indicate that the organism cannot ferment glucose or lactose and therefore produces no acidic products. The slant may become redder owing to peptone catabolism.

If there are gas bubbles in the butt, splitting of the medium, and/or displacement of the medium from the bottom of the tube, the organism is aerogenic—able to produce carbon dioxide and hydrogen gases during fermentation. Any blackening in the butt indicates that the organism produces hydrogen sulfide gas from thiosulfate. The hydrogen sulfide combines with iron salt to produce ferrous sulfide, a black precipitate.

Inoculation is done by stabbing the butt with an inoculating needle and streaking the slant using pure culture of isolate. The cap should be slightly loose. If the cap is screwed on too tightly, there will not be sufficient air for peptone catabolism. Gram-negative rods able to ferment only glucose may appear as lactose-fermenters. Reactions should be interpreted at 18 to 24 hours. If the medium is read earlier, organisms only able to ferment glucose may appear to be lactose-fermenters. If the medium is read later, lactose-fermenters may consume the lactose and begin to catabolize peptones, with the slant reverting to red. A yellow slant and red butt may indicate failure to stab the butt or inoculation of the medium with gram-positive organisms. Examination of the medium for stab line and/or performance of Gram stain should clarify this situation. The hydrogen sulfide indicator system in KIA is not as sensitive as the lead acetate method or as that found in other media, such as sulfide indole motility agar. A black butt should be read as acid even though yellow color may be obscured. If hydrogen sulfide is reduced, this indicates that an acid condition does exist and can be assumed. Critical to understanding how this medium works is the fact that glucose is present in a much lesser amount than lactose. Organisms use the simplest carbohydrate, glucose, first. Once this is consumed, they attack the more complex carbohydrate, lactose. If they lack the appropriate enzymes, they move on to protein catabolism. There is sufficient lactose in the medium to prevent breakdown of peptone, provided that it is read at the appropriate time.

Loeffler Coagulated Serum Slant

Loeffler coagulated serum slant is used primarily for recovery and identification of *Corynebacterium diphtheriae*. Other issues include detection of proteolysis, pigment formation, and ascospore production as well as cultivation of *Entamoeba histolytica*.

This medium can be used for primary recovery of *Corynebacterium diphtheriae* from nose and throat specimens and for subculture purposes. Because Loeffler medium is so enriched, *C. diphtheriae* grows well within 12 to 16 hours and produces nondistinctive translucent to gray-white colonies. The medium promotes the development of characteristic granules that can be detected microscopically with methylene blue stains. Serum content enables detection of proteolytic activity. Positive organisms produce colonies surrounded by small holes containing liquefied medium. The entire slant may eventually turn to liquid, with production of a foul odor.

Loeffler serum slant should be inoculated as soon as possible after specimen collection, and more selective media containing tellurite should always be used as well. Smears for *C. diphtheriae* should be prepared and examined after 8 to 24 hours of incubation. Although granule formation is typical of *Corynebacterium* species, other organisms can produce a similar microscopic appearance. Therefore, additional testing must

be performed for confirmation of this organism. Proteolysis testing may require 3 or 4 days of aerobic incubation or longer. For ascospore detection, the slants must be dried out. Excess liquid should be poured out of the tube. Slant may be heated until it splits. Screwed caps can be replaced with cotton plugs to maintain a drier environment. Ascospores tend to form on drier areas of the medium.

Löwenstein-Jensen Medium

Löwenstein-Jensen (LJ) medium is used to cultivate *Mycobacterium* spp. Most media contain ingredients that can inhibit the growth of mycobacteria. The potato flour, egg, and glycerol included in LJ medium help detoxify this medium and also supply nutrients required for growth of these organisms. Asparagine is included for maximum production of niacin by certain *Mycobacterium* spp. The malachite green serves as an inhibitor of other bacteria that may be present in specimens.

LJ medium is good for 1 month if tightly capped to prevent moisture loss and stored at 4 to 6°C. LJ medium must be kept out of direct light because malachite green is light-sensitive. Decontaminated or digested or untreated specimens are inoculated onto medium and incubated in a carbon dioxide incubator (5–10% CO_2) for 6 to 10 weeks. It is important to leave caps loose for proper gas exchange.

LJ medium may be prepared as deeps to be used in semiquantitative catalase testing for ascertaining the particular species of *Mycobacterium*. LJ medium with 5% sodium chloride may be prepared to aid in identifying rapid growers. This medium is the same as LJ except for addition of 5 g of sodium chloride per each 100 mL of medium. The additional salt allows for testing the ability of certain mycobacteria to tolerate and grow in presence of high salt concentration. The Gruft modification makes this medium more selective through the addition of penicillin (50 U/mL) and nalidixic acid (35 µg/mL) prior to dispensing into tubes. This formulation also includes 0.05 µg/mL of ribonucleic acid, which increases the rate of mycobacterium isolation over standard LJ formation. In the Petran and Vera modification, cyclohexamide, lincomycin, and nalidixic acid are added to make LJ more selective. Both of these modifications can permit gentler decontamination or digestion procedures.

Lysine-Iron Agar

This medium measures three parameters useful in identifying species of Enterobacteriaceae: lysine decarboxylation, lysine deamination, and hydrogen sulfide production.

Lysine-iron agar (LIA) contains lysine (amino acid), glucose (carbohydrate source), a small amount of protein, bromcresol purple (pH indicator), and sodium thiosulfate/ferric ammonium citrate (sulfur source and hydrogen sulfide indicator).

Three lysine utilization patterns are possible:

1. Alkaline (purple) butt and alkaline (purple) slant indicate that the organism decarboxylates lysine but cannot deaminate it. Initially, the organism ferments glucose, causing production of acid and changing indicator in the butt to yellow. It then decarboxylates lysine to produce cadaverine, an alkaline product. This causes the pH indicator to change back to purple.

2. Acid (yellow) butt and alkaline (purple) slant indicate that the organism fermented the glucose but was unable to deaminate or decarboxylate the lysine.

3. Acid (yellow) butt and Bordeaux red slant indicate that the organism deaminated lysine but could not decarboxylate it. The yellow butt is caused by glucose fermentation. The reason for red slant in cases of lysine deamination has not been clarified. If no indicator is present, one of the products of deamination appears orange. The red slant may be the result of mixing the purple and red colors.

Any blackening in the butt indicates production of hydrogen sulfide from sodium thiosulfate. This gas reacts with ferric salt to produce the black precipitate, ferrous sulfide.

This medium appears purple before use. LIA is inoculated by stabbing the butt twice and streaking the slant with an inoculating needle. Cap should be left slightly loose, since oxygen is required for detection of deamination. Reactions should be read after 18 to 24 hours of incubation at 35°C. The medium may be incubated for up to 48 hours if needed.

LIA is not as sensitive as other media for hydrogen sulfide detection. Typically, hydrogen sulfide–producing *Proteus* sp may appear negative. Also, *Morganella morganii* produces a variable lysine deamination reaction after 24 hours of incubation. This medium can be used only with organisms that can ferment glucose. LIA is not a true replacement for the Moeller decarboxylase tests.

MacConkey Agar

MacConkey agar is a selective, differential primary plating medium. It selects for Enterobacteriaceae and other gram-negative rods in the presence of mixed flora and differentiates them into lactose-fermenters and non–lactose-fermenters.

Bile salts and crystal violet inhibit most gram-positive organisms but permit the growth of gram-negative rods. Lactose serves as the sole carbohydrate source. Gram-negative rods that ferment lactose produce pink or red colonies, which may be surrounded by precipitated bile. Acid production from lactose fermentation causes the neutral red dye absorbed into the colonies to change to red and can also cause the bile salts to become insoluble. Non–lactose-fermenting gram-negative rods produce colorless or transparent colonies.

Plates are streaked for isolation and incubated in ambient air, not a carbon dioxide incubator, for 18 to 24 hours at 35°C. Weak or slow lactose-fermenters may produce colorless colonies at 24 hours or appear slightly pink in 24 to 48 hours. Plates should not be incubated longer than 48 hours, since this can lead to confusing results. Some gram-negative rods may fail to grow on the medium, whereas with prolonged incubation gram-positives such as *Enterococcus* species may produce tiny colonies. Room temperature incubation may enhance recovery of *Yersinia enterocolitica.*

The agar concentration may be increased to prevent swarming of *Proteus* species. A formulation of MacConkey agar without crystal violet has been used to aid in identifying mycobacteria. A closely related medium, MacConkey sorbitol agar, contains the same components as regular MacConkey agar except D-sorbitol is substituted for lactose. This medium has been used for isolating *E. coli* O157:H7. Sorbitol-negative colonies that appear colorless on this medium may indicate possible *E. coli* O157:H7 and should be further tested.

Malonate Broth

Malonate broth is used in the identification of species of Enterobacteriaceae, particularly *Salmonella.*

Malonate broth contains sodium malonate (primary carbon source), small quantities of glucose and yeast extract (nutrients), bromthymol blue (pH indicator), various salts, and a buffering system. Organisms producing a Prussian blue color are able to use malonate as a carbon source. If they can use malonate as a carbon source, they also employ ammonium sulfate as a nitrogen source, thereby producing alkaline products that cause a rise in pH and a change in the color of the medium to blue. Organisms unable to use malonate as a carbon source usually fail to grow, and the medium stays green. Because malonate resembles succinate, it competitively binds succinic dehydrogenase, which catalyzes the succinate to fumarate in the Krebs cycle. The tying up of this enzyme, coupled with the inability to use malonate as a carbon source, prevents growth.

Malonate broth should be inoculated from triple sugar iron agar, KIA, or broth culture of the organism. Inoculum should be light. Cultures should be incubated at 35°C and checked at 18 to 24 hours and at 48 hours for production of a blue color. Some organisms produce only small amounts of alkalinity. Any trace of blue should be considered positive. Comparison with an uninoculated tube may be useful. Production of a yellow color is a negative reaction. This reaction is probably due to fermentation of the small amount of glucose in the medium.

Mannitol Salt Agar

Mannitol salt agar is a selective and differential primary culture medium useful in recovery and identification of staphylococci from specimens containing mixed flora.

High salt concentration (7.5%) inhibits most gram-negative and gram-positive bacteria except *Staphylococcus* species. *Staphylococcus aureus* is able to ferment mannitol, the sole carbohydrate in the medium, to produce acid products. This lowers the pH and changes the color of the pH indicator, phenol red, to yellow. Colonies of *S. aureus* typically appear yellow, surrounded by a yellow zone. Other *Staphylococcus* and *Micrococcus* species usually do not ferment mannitol and therefore produce reddish colonies that may exhibit a red to purple surrounding zone due to peptone breakdown.

Plates are streaked for isolation and incubated at 35°C for 24 to 48 hours but not in a carbon dioxide incubator. *Enterococcus* may be able to grow on mannitol salt agar and produce slight mannitol fermentation. Differentiation is readily accomplished through Gram stain and catalase test. With prolonged incubation, organisms other than staphylococci may begin to grow and produce mannitol fermentation. Some strains of *Staphylococcus aureus* may be slow in fermenting mannitol, so plates should not be discarded until after 48 hours of incubation. All colonies suggestive of *S. aureus* should be further tested for coagulase or with an alternative acceptable procedure. Subculture to less selective agar is preferable prior to performing this testing. Some formulations recommend inclusion of 20 mL of sterile egg yolk. Coagulase-positive staphylococci also produce a lipase that causes formation of an opaque precipitate around the colonies. Non–coagulase-producing staphylococci do not produce this egg yolk lipase and therefore lack these zones.

MES (Ureaplasma Agar)

MES [2-(*N*-morpholino) ethanesulfonic acid] agar is used for the isolation of *Ureaplasma urealyticum*. This medium contains horse serum, which supplies the cholesterol necessary for stabilizing these organisms because they lack cell walls. Yeast dialysate serves as a growth factor and supplies preformed nucleic acid precursors. Urea is a required nutrient for *Ureaplasma*. Phenol red serves as a pH indicator while MES or 2-(*N*-morpholino) ethanesulfonic acid acts as a buffer. The antibiotics, penicillin and lincomycin, inhibit normal flora but permit the growth of *Ureaplasma*. After mixing components, pour the medium into Petri dishes, 5 mL per plate. Store plates in plastic for up to 2 weeks under refrigeration. Inoculated plates should be incubated in a carbon dioxide incubator or a candle jar or under anaerobic conditions. At 48 hours, colonies of *Ureaplasma* show a "fried-egg" appearance. If a solution of 1% urea and 0.8% manganese chloride is poured over these colonies, they will turn dark brown owing to the production of urease.

Methyl Red–Voges-Proskauer (MR-VP) Medium

MRVP broth is used for performing the methyl red and Voges-Proskauer tests. These procedures are useful in distinguishing among numbers of Enterobacteriaceae. For example, *E. coli* is methyl red positive and Voges-Proskauer negative; *Enterobacter aerogenes, Enterobacter cloacae,* and *Klebsiella pneumoniae* show the reverse reactions.

Members of Enterobacteriaceae can be divided into two groups based on the way they metabolize glucose. One group produces large amounts of mixed acids (lactic, formic, succinic, and acetic). When methyl red reagent is added to one of these cultures, a red color is produced owing to the acidic pH. The other group produces predominantly neutral end products, acetoin or acetyl-methylcarbinol, by the butylene glycol pathway. When α-naphthol and 40% potassium hydroxide are added to the broth culture, acetoin, if present, is oxidized to diacetyl in the presence of air and base. α-Naphthol catalyzes a reaction between the diacetyl and guanidine components of peptone to produce a pink-red color.

For the methyl red test, broth culture must be incubated for 48 hours. Avoid using really turbid broth. The Voges-Proskauer test was originally designed to be performed after 5 days of incubation at 30°C. By using 0.5 to 1 mL of broth per tube, the test can be done after 18 to 24 hours of incubation at 35°C. Shaking aerates the broth culture and enhances the reaction.

Middlebrook 7H10 and 7H11 Agars

The purpose of Middlebrook 7H10 and 7H11 agars is to cultivate *Mycobacterium* species. Isoniazid-resistant strains grow better on these media than on egg-based media, such as Löwenstein-Jensen. The Middlebrook agars are also more chemically defined than the Löwenstein-Jensen formulations.

Middlebrook 7H10 and 7H11 are similar except 7H11 contains casein hydrolysate. Both media contain growth factors, such as amino acids and salts, that encourage recovery of mycobacteria. In addition, both formulations include OADC (oleic acid–dextrose-citrate) enrichment, which chemically simulates egg components. Malachite green adds some selectivity.

Antibiotics can be added to this basic formulation to prevent overgrowth with bacteria. Mycobacteria-selective agar contains cycloheximide, lincomycin HCl, and nalidixic acid. Mitchison 7H11 selective agar is more selective owing to the addition of amphotericin B, carbenicillin, polymyxin B, and trimethoprim.

Motility Test Medium

The purpose of the motility test medium is to determine whether an organism is motile or nonmotile. This test is particularly useful in the identification of members of Enterobacteriaceae, in which two genera, *Shigella* and *Klebsiella,* are always nonmotile, and certain *Yersinia* sp show motility at room temperature but not at 35°C. *Listeria monocytogenes* gives a classic umbrella-type motility, and the non–glucose-fermenting gram-negative rods can be differentiated, based in part on their motility.

Nonmotile organisms, which lack flagella, grow only along the stab line, and the surrounding medium remains clear. Motile organisms, which usually possess flagella, move out from the stab line, and the medium appears cloudy. Low agar concentration makes the medium semisolid and permits better detection of motility.

Use an inoculating needle to stab the medium. Be careful to remove the needle along the initial stab line, and do not stab the medium clear to the tube's bottom. Incubate the inoculated medium at 35°C. Because flagellar protein is not formed as well at higher temperatures, some microbiologists prefer incubation at 18 to 20°C. For *Yersinia,* noting motility reaction at room temperature is particularly useful. Triphenyltetrazolium chloride (TTC) may be added to the basic motility medium to enhance detection of motility. A 1% solution of TTC is prepared and filter-sterilized. To 1 L of motility medium add 5 mL of this solution. If TTC is used, bacteria incorporate colorless TTC and reduce it to red formazan pigment. The medium shows reddening where there is growth. Other media such as SIM (sulfide indole motility) and MIO (motility indole ornithine) can be used to detect motility in addition to other reactions.

Mueller-Hinton Agar

Mueller-Hinton agar is a transparent medium, useful in testing susceptibility of organisms to antibiotics. The medium has also been used for testing starch hydrolysis. Because Mueller-Hinton agar contains animal infusion, Casamino acids, and starch, it supports the growth of most organisms. In addition, sheep's blood may be added to the basic formulation in order to perform susceptibility testing on streptococci. The addition of heated or chocolatized sheep's blood to Mueller-Hinton agar makes possible testing for fastidious organisms, such as *Haemophilus* and

Neisseria. Starch is included in the medium for two reasons. It may protect the organisms against toxic substances. Also, it serves as an energy source. Ca^{++} and Mg^{++} concentrations are critical in the testing of *Pseudomonas* isolates with aminoglycoside antibiotics. Usually Mueller-Hinton agar contains sufficient amounts of bivalent cations, but it may be necessary to add these substances to Mueller-Hinton broth.

New York City Medium

The purpose of New York City medium is to isolate *Neisseria gonorrhoeae* and *N. meningitidis* from specimens containing mixed normal flora.

New York City medium contains hemoglobin, yeast dialysate, and horse plasma, making it highly enriched to support the growth of *Neisseria gonorrhoeae* and *N. meningitidis.* Selectivity for these two organisms is accomplished by four antibiotics that inhibit normal flora. Vancomycin prevents the growth of gram-positive bacteria; colistin inhibits gram-negative rods; amphotericin B prevents growth of yeast and molds. Trimethoprim has been included to prevent swarming of *Proteus* species. Cornstarch, another component, absorbs toxic substances that could otherwise inhibit the growth of these *Neisseria* species.

When using New York City medium for recovery of *Neisseria gonorrhoeae* and *N. meningitidis,* the microbiologist should incubate plates under increased carbon dioxide for several days. Also, a nonselective agar such as chocolate agar should be included because 5% of gonococci are inhibited by the antibiotics, particularly vancomycin, found in this medium.

Nitrate Reduction Broth

The purpose of nitrate reduction broth is to determine whether an organism can reduce nitrate to nitrite or to gaseous products, such as nitrogen. The test is very useful in the recognition of members of Enterobacteriaceae, most of which can reduce nitrate to nitrite; non–glucose-fermenting gram-negative rods; *Neisseria;* and *Moraxella catarrhalis.*

The nitrate test is performed in two parts. Sulfanilic acid and α-naphthylamine reagents are added first. If nitrate has been reduced to nitrite, nitrite will react with these reagents to form a red diazonium dye, *p*-sulfobenzene-azo-naphthyl-amine. If there is no color change, zinc dust is added. Zinc reduces the remaining nitrate to nitrite, forming a red color. However, if nitrate was reduced all the way to nitrogen gas, no color change occurs.

When using this medium, incubate broth culture for 34 to 48 hours before testing. Observe the Durham tube for the presence of gas bubbles. Add 1 mL of sulfanilic acid reagent and 1 mL of α-naphthylamine reagent. Interpret the results immediately, since color fades quickly. If it is necessary to add zinc, avoid using large amounts. Too much zinc can result in the formation of hydrogen gas, which can cause reduction and decrease the color reaction. Medium may need to be supplemented with serum and incubated for up to 5 days in testing for *Neisseria.* Because α-naphthylamine is carcinogenic, it is preferable to substitute *N,N*-dimethyl-α-naphthylamine.

Nutrient Agar

Nutrient agar has been used to distinguish between the nonfastidious, less pathogenic *Neisseria* species and pathogenic *Neisseria* species, such as *N. gonorrhoeae* and *N. meningitidis.* The less fastidious neisseriae grow on nutrient agar, whereas the more pathogenic species do not. Nutrient agar has also been used for the maintenance of stock cultures.

Nutrient agar contains minimal nutrients and an especially low concentration of protein. Growth of an isolate on this medium means that it is not very fastidious and does not require special supplements.

Oxidative-Fermentative (OF) Medium (Hugh and Leifson Formulation)

Oxidative-fermentative, or OF, medium is useful in determining an organism's type of carbohydrate utilization—oxidative, fermentative, or inert. This medium is very important in the identification of non–glucose-fermenting gram-negative rods.

Three modifications over traditional media for the detection of fermentation make this medium useful in testing nonfermenting gram-negative rods. A low concentration of peptone prevents formation of alkaline products that may neutralize the small quantities of acid produced through oxidation. The high concentration of carbohydrate increases the potential amount of acid that can be formed. The lower concentration of agar makes the medium semisolid. This permits acids formed on the surface to diffuse throughout the medium. Bromthymol blue serves as the pH indicator for acid detection.

The classic method for using this medium involves the stabbing of two tubes with the organism. The medium in one tube is covered with vaspar (a mixture of petrolatum and paraffin) or melted paraffin. Sterile mineral oil has been used for this purpose, but it is not recommended, since it does not block out oxygen as well. Results are interpreted after incubation at 35°C. Several days of incubation may be required owing to the slower growth of some nonfermenting gram-negative rods. A color change to yellow in both tubes means that the organism is fermentative (can produce acid in the absence of oxygen). Color change to yellow in the uncovered tube only means that the organism is oxidative (requires oxygen to utilize the carbohydrate). If neither tube changes in color or the covered tube shows no change while the uncovered tube turns blue, the organism cannot utilize the carbohydrate oxidatively or fermentatively and is considered inert. A one-tube modification of this test has been described. One tube is stabbed, and it is not covered. Color change to yellow near the top of the medium only indicates oxidative use of glucose. If the entire tube changes to yellow, fermentation is suggested. Because the medium is semisolid, motility may be observed in this medium. *Note:* Sometimes OF medium is used for differentiating staphylococci (fermentative) from micrococci (oxidative). This testing requires a different formulation.

Peptone–Yeast Extract–Glucose (PYG) Broth

Peptone–yeast extract–glucose broth is useful for culturing anaerobes. PYG broth culture of an anaerobic isolate may be used in gas-liquid chromatography procedures that detect metabolic end products.

PYG broth contains several nutrients and supplements that encourage the growth of anaerobes. These enrichments include vitamin K (required for pigment-producing *Prevotella* and *Porphyromonas*), yeast extract, hemin, and glucose. Cysteine helps keep the medium more reduced and anaerobic. Resazurin serves as an anaerobic indicator. Pink color means that oxygen is present.

Phenylalanine Deaminase (PAD) Agar

Phenylalanine deaminase agar is used to detect the organism's ability to deaminate or remove the amino group from phenylalanine. A positive reaction is most useful for distinguishing *Proteus, Providencia,* and *Morganella* from other members of the family Enterobacteriaceae. This test can also be used to distinguish *Moraxella* that are phenylpyruvate-positive from other *Moraxella.*

Phenylalanine deaminase agar includes phenylalanine, the amino acid to be deaminated; yeast extract, a nitrogen and carbon source; various salts; and agar, a solidifying agent. Protein hydrolysates and meat extracts are not included because these substances contain a variable amount of phenylalanine. If an organism produces phenylalanine deaminase, it can convert phenylalanine to the α-keto acid called phenylpyruvic acid. This acid reacts with the added ferric chloride reagent to form a dark-green complex. The immediate appearance of a dark-green slant on addition of ferric chloride reagent is a positive reaction; no color change upon addition of reagent is a negative reaction.

Phenylethyl Alcohol (PEA) Agar

The purpose of phenylethyl alcohol agar is to isolate gram-positive cocci, such as staphylococci and streptococci, from specimens having mixed flora. The anaerobic formulation of this medium selects for gram-negative and gram-positive nonsporulating anaerobes while inhibiting the facultatively anaerobic gram-negative rods and other anaerobes.

PEA agar is similar to sheep's blood agar except that it contains phenylethyl alcohol. This component inhibits facultative gram-negative rods, especially swarming *Proteus,* but permits the growth of gram-positive cocci.

Phenylethyl alcohol is volatile, and plates should be tightly sealed in plastic bags and stored in the refrigerator. Hemolytic reactions are not dependable on this medium because of the action of phenylethyl alcohol on cell membranes. Gram-negative rods may grow on PEA agar, but colonies are smaller than usual and can be readily differentiated from those of gram-positives rods. *Pseudomonas aeruginosa* is not inhibited by this medium. Some gram-positive cocci may require more than 24 hours of incubation to grow well on PEA agar. An anaerobic formulation can be achieved by adding phenylethyl alcohol to CDC anaerobic agar prior to autoclaving or by supplementing the formulation with vitamin K as well as sheep's blood after sterilization of basal medium.

Pseudocel (Cetrimide) Agar

Pseudocel, or cetrimide, agar is used to select for *Pseudomonas aeruginosa* in specimens with mixed flora. Also, because it inhibits other *Pseudomonas* species (except *P. fluorescens*) and closely related organisms, this test can be useful in the differentiation of non–glucose-fermenting gram-negative rods.

This medium contains cetrimide, also called cetyl trimethyl ammonium bromide or hexadecyltrimethylammonium bromide. Produced from bromine, cetrimide is highly inhibitory and has been used as an antiseptic. If the organism can tolerate cetrimide, it will grow on the medium. Magnesium chloride and potassium sulfate stimulate the production of pyocyanin, the green pigment characteristically produced by *Pseudomonas aeruginosa*.

Salmonella-Shigella (SS) Agar

Salmonella-Shigella agar is used to select for *Salmonella* and some strains of *Shigella* from stool specimens. SS agar is also differential in that these organisms produce characteristic colonies on the medium.

SS agar contains bile salts, sodium citrate, and brilliant green, which inhibit the growth of gram-positive and many lactose-fermenting gram-negative rods normally found in feces. Lactose serves as the sole carbohydrate source in the medium; neutral red is the pH indicator. If an organism grows on the medium and ferments lactose, it will produce acid and change the indicator to pink-red. Sodium thiosulfate acts as the source of sulfur for the production of hydrogen sulfide. If hydrogen sulfide is produced, it reacts with the ferric chloride present in the medium, forming a black precipitate in the center of the colony.

A heavy inoculum of stool can be planted on SS agar because the formulation is so inhibitory. However, strains of *Shigella* may not grow on SS agar, and this medium should not be used as the sole primary plating medium when *Shigella* is the potential isolate. *Shigella* colonies appear color-less on SS agar because these organisms do not ferment lactose or produce hydrogen sulfide. *Salmonella* colonies are colorless with a black center because these organisms usually make hydrogen sulfide but do not ferment lactose. Pink to red colonies indicate that the organism ferments lactose; if there is a black center, it also produces hydrogen sulfide. If *Proteus* grows on this medium, swarming is inhibited.

Selenite F Broth

Selenite F broth is an enrichment broth used for the recovery of low numbers of *Salmonella* and some strains of *Shigella* from stool and other specimens containing large amounts of mixed bacteria. The sodium selenite present in this medium inhibits the growth of many gram-negative rods and enterococci but permits recovery of *Salmonella* and some *Shigella* species. Selenite is most effective at a neutral pH. Reduction of selenite during growth of bacteria produces alkaline products, so lactose has also been included in this medium. Lactose-fermenters produce acid, which neutralizes these alkaline products and returns the medium to a neutral pH, at which selenite works most effectively.

When using selenite broth, the microbiologist should subculture the broth to enteric media after it has incubated 8 to 12 hours. Beyond this time frame, overgrowth with normal flora is likely.

Sodium Chloride Broth, 6.5%

Sodium chloride broth is useful in the differentiation of streptococci, particularly those producing α-hemolytic and nonhemolytic colonies. Primarily, it distinguishes *Enterococcus* species (positive) from group D streptococci (negative), both of which produce a positive bile-esculin agar slant. Also, viridans streptococci cannot grow in this medium.

Sodium chloride broth is prepared from heart infusion broth, a general-purpose medium, that already contains 0.5% sodium chloride. By adding 6% sodium chloride to this medium, the salt concentration becomes 6.5%. Sodium chloride broth also contains glucose as a carbohydrate source and bromcresol purple, a pH indicator. If the organism can tolerate this high concentration of salt, it will grow in the medium and produce cloudiness. Fermentation of glucose produces acid and may cause the medium to turn yellow.

Some formulations omit indicator and glucose. To use this medium, inoculate several colonies into broth and incubate culture overnight at 35°C. Any growth in the broth is considered positive even if the indicator does not change color.

To avoid a false-negative result, gently mix broth prior to interpretation. Inoculating the broth too heavily may give a false-positive result. Organisms other than enterococci can produce positive results, that is, group B streptococci and aerococci.

SP-4 Broth/Agar

SP-4 broth and SP-4 agar serve as primary isolation media for *Mycoplasma* species. SP-4 media contain yeast products that serve as growth factors for *Mycoplasma* and supply preformed nucleic acid. Fetal bovine serum supplies the cholesterol necessary for stabilizing these organisms because they lack cell walls. Various antibiotics inhibit normal flora that may be present in the specimen. Penicillin is included to prevent the growth of gram-positive bacteria; amphotericin B inhibits fungi; and polymyxin B inhibits gram-negative rods. Biphasic media provide both microaerophilic and moist conditions, which some *Mycoplasma* species prefer.

SP-4: Arginine, Glucose, and Urea Broths

SP-4 broths are useful in the identification of *Mycoplasma* and *Ureaplasma*. Yeast products included in SP-4 broth supply nutrients required for the growth of *Mycoplasma* and *Ureaplasma*. Fetal bovine serum contains cholesterol, which helps stabilize these organisms, since they lack cell walls. Penicillin prevents the growth of gram-positive cocci but does not affect these organisms because they do not possess cell walls. Phenol red serves as a pH indicator. In SP-4 arginine broth, utilization of arginine results in alkaline products and color change to red. This reaction characterizes *M. hominis*. In SP-4 urea broth, removal of an amino group from urea, characteristic of *Ureaplasma*, causes formation of ammonia and a similar color change to red. SP-4 glucose broth detects glucose fermentation. Acid is produced, which lowers the medium's pH and changes its color to yellow. This reaction typifies *M. pneumoniae* as well as a few other *Mycoplasma* species.

Tetrathionate Broth

Tetrathionate broth is an enrichment medium used for recovery of low numbers of *Salmonella* species from stool specimens. Bile salt in conjunction with thiosulfate and added iodine-iodide solution inhibits the growth of most gram-negative rods and gram-positive organisms except *Salmonella*. Some formulations also include brilliant green or crystal violet, which increases the inhibitory nature of the medium. The medium must be used within 24 hours of preparation. Since the basal medium may be stored in the refrigerator indefinitely, some microbiologists prefer to dispense 10 mL of basal medium per tube. Just before use, 0.2 mL of iodine solution can be added to each tube. Heavy inoculum of stool can be added to the broth. After 12 to 24 hours of incubation at 35°C, the broth should be subcultured to enteric media to prevent overgrowth with normal flora. This medium inhibits most *Shigella* species and should not be used for recovering *Salmonella typhi*.

Thayer-Martin, Modified Agar

Modified Thayer-Martin (MTM) agar is a selective enrichment medium used for recovering *Neisseria gonorrhoeae* and *N. meningitidis* from specimens with mixed flora. Modified Thayer-Martin agar is highly enriched to support the growth of the more fastidious *Neisseria* species. Added growth factors include hemoglobin, vitamins, cocarboxylase, diphosphopyridine nucleotide, and glutamine. Cornstarch is included to absorb any inhibitory substances that might be present. The modified formulation containing more agar may help prevent swarming of *Proteus*. Modified Thayer-Martin agar contains several antibiotics that together inhibit normal flora and prevent the growth of most other organisms. Vancomycin inhibits the growth of gram-positive cocci. Colistin inhibits gram-negative rods, while trimethoprim prevents *Proteus* from swarming. Nystatin prevents the growth of fungi.

Five percent chocolatized, difibrinated sheep's blood may be substituted for the hemoglobin solution. The original Thayer-Martin formulation lacked additional agar and glucose as well as the trimethoprim and is therefore not as effective at inhibiting swarming of *Proteus*. The Martin-Lewis formulation substitutes anisomycin (20 μg/mL) for nystatin as the antifungal agent. In addition, the vancomycin concentration of (4 μg/mL) is higher than in the MTM formulation.

When using MTM plates, the microbiologist should incubate them in a carbon dioxide incubator or a candle jar for several days. Because some strains of *Neisseria gonorrhoeae* may be inhibited by vancomycin, a chocolate plate should also be used.

Thioglycolate Broth, Basal and Enriched

Thioglycolate broth is an all-purpose medium that can be used to isolate a wide range of bacteria. It is often employed as a back-up broth that is inoculated along with culture plates. In this case, it helps detect those organisms that were present in low numbers or anaerobes in the original specimen. When glucose is omitted, thioglycolate broth can be used in fermentation studies of anaerobes.

Thioglycolate, cystine, and sodium sulfite act as reducing agents in this medium, while the low concentration of agar prevents downward diffusion of oxygen. Various supplements can be added to support the growth of more fastidious organisms.

Various supplements can be added to the basic formulation of this broth medium. These include hemin (5 μg/mL), vitamin K (0.1 μg/mL), and sodium bicarbonate (1 mg/mL), which can be autoclaved in the medium. Supplements that must be added after autoclaving include rabbit or horse serum (10% vol/vol) and Fildes enrichment (5% vol/vol). These are given as final concentrations in the medium.

Thioglycolate broth should be stored at room temperature and boiled and cooled prior to use. When used as a back-up broth, the medium is incubated at 35°C for 3 to 7 days and examined for turbidity. Gram stains of broth are compared with growth obtained on primary culture plates. If something different appears to be growing in the broth, subcultures should be done.

Thiosulfate Citrate Bile Salts Sucrose (TCBS) Agar

Thiosulfate citrate bile salts sucrose agar is a selective medium used to isolate *Vibrio* species from stool specimens having mixed flora. TCBS agar is also differential in that *Vibrio* species produce characteristic colonies. *Vibrio* species grow poorly on media designed for isolation of *Salmonella* and *Shigella* but produce colorless colonies on MacConkey agar. TCBS agar includes sodium citrate, sodium thiosulfate, and oxgall (10% solution equivalent to full-strength bile), which together inhibit many gram-positive cocci and gram-negative rods normally present in stool specimens. In addition, the high pH of TCBS agar encourages the growth of *Vibrio* while inhibiting other organisms. Different types of colonies are produced by different species of *Vibrio* owing to the presence of sucrose as a fermentable carbohydrate and bromthymol blue as a pH indicator. For example, *V. cholerae* and *V. alginolyticus* produce yellow colonies because they can ferment sucrose, whereas *V. parahaemolyticus* and *V. vulnificus* usually produce blue-green colonies owing to lack of sucrose fermentation. Organisms that can produce hydrogen sulfide from sodium thiosulfate have black centers because of the reaction of this gas with ferric citrate. *Vibrio* organisms do not produce hydrogen sulfide. Some formulations include a second pH indicator, thymol blue. Oxgall and sodium cholate or bile salts (8 g/L) may be used in place of oxgall alone.

When using TCBS agar, the microbiologist should use a heavy inoculum, since *Vibrio* species die off quickly, and this medium is very inhibitory. Fresh specimen is best because these organisms are sensitive to drying out, sunlight, and acid pH. If there must be a delay in planting, use Cary-Blair semisolid transport medium rather than buffered glycerol transport medium. Plates should be incubated at 35°C for 18 to 24 hours and up to 48 hours. Growth off TCBS agar is not acceptable for performing oxidase testing. Occasionally, strains of *Vibrio cholerae* may produce blue-green colonies on this medium owing to delayed sucrose fermentation. Some *Vibrio* spp do not grow well on this medium. Also, other organisms, such as *Pseudomonas, Plesiomonas,* and *Aeromonas,* can grow on TCBS agar and usually produce blue colonies and must be distinguished from vibrios.

Tinsdale Agar

Tinsdale agar is a selective, differential medium useful in isolating and identifying *Corynebacterium diphtheriae* from specimens containing mixed flora. Tinsdale agar contains a high concentration of potassium tellurite, which inhibits the growth of most normal flora organisms but permits *Corynebacterium* species, especially *Corynebacterium diphtheriae,* to grow. All *Corynebacterium* species growing on the medium produce gray to black colonies owing to the reduction of tellurite to tellurium. In addition, *C. diphtheriae* colonies are surrounded by a brown halo. This brown halo is thought to be produced from tellurite's interacting with the hydrogen sulfide produced by the organism from cystine and thiosulfate. The basal medium can be stored indefinitely; tellurite and serum can be added just prior to use. Once prepared, the medium has a shelf life of 4 days.

When using Tinsdale agar, the microbiologist should streak plates for isolation and stab the medium in several areas. Sometimes browning occurs in these stabbed areas before it can be seen around colonies. Plates should be incubated at 35°C for 24 to 48 hours in ambient air. Increased carbon dioxide can slow down the production of the brown halo. It may require 48 hours for some *C. diphtheriae* strains to produce the characteristic halo. In addition, *C. ulcerans* and *C. pseudodiphtheriticum* may also produce a dark halo on this medium and must be differentiated from *C. diphtheriae*. Other organisms may occasionally grow on Tinsdale agar. *Proteus* produces mucoid colonies and tends to blacken the medium. Rare streptococci and staphylococci can produce dark colonies with a surrounding halo, but could be distinguished by performing a Gram stain.

Triple Sugar Iron (TSI) Agar

Triple sugar iron (TSI) agar can be used to determine whether a gram-negative rod is a glucose-fermenter or non–glucose-fermenter, a fundamental characteristic in the initial classification of gram-negative rods. The medium also tests for sucrose or lactose fermentation, gas production during glucose fermentation, and hydrogen sulfide production, all of which are useful in the differentiation of gram-negative rods belonging to the family Enterobacteriaceae.

TSI agar contains glucose, sucrose, and lactose (fermentable carbohydrates), phenol red (pH indicator), peptone (carbon/nitrogen source), and iron salt plus sodium thiosulfate (sulfur source and hydrogen sulfide indicator). TSI agar resembles Kligler iron agar except that it contains sucrose. Three carbohydrate fermentation patterns are possible:

1. Acid (yellow) butt and alkaline (red) slant indicate that the organism ferments glucose but not lactose. This organism ferments glucose by the Embden-Meyerhof-Parnas pathway to produce organic acids, changing the pH indicator from red to yellow. Once the glucose has been consumed, the organism then breaks down peptones, producing ammonia. This causes a rise in pH and the slant reverts to red.

2. Acid (yellow) butt and acid (yellow) slant indicate that the organism ferments both glucose and sucrose and/or lactose. The organism ferments glucose, producing acid products. Once the glucose is consumed, it ferments sucrose and/or lactose. This causes the pH in the slant portion to remain acidic.

3. Alkaline (red) butt and alkaline (red) slant indicate that the organism cannot ferment glucose or sucrose/lactose and therefore produces no acidic products. The slant may become redder owing to peptone catabolism.

If there are gas bubbles in the butt, splitting of the medium, and/or displacement of medium from the bottom of tube, the organism is aerogenic—able to produce carbon dioxide and hydrogen gases during fermentation. Failure to do this means that the organism is anaerogenic. Any blackening in the butt indicates that the organism produces hydrogen sulfide gas from thiosulfate. The hydrogen sulfide combines with iron salt to produce ferrous sulfide, a black precipitate.

For optimal detection of reactions, it is important that when the medium is dispensed in tubes, the butt be deep and approximately of the same length as the slant. The final medium should be red. Inoculation is done by stabbing the butt with an inoculating needle and streaking the slant using a pure culture of isolate. The cap should be slightly loose. If the cap is screwed on too tightly, there will not be sufficient air for peptone catabolism. Gram-negative rods able to ferment only glucose may appear as sucrose/lactose-fermenters. Reactions should be interpreted at 18 to 24 hours of incubation. If read earlier, organisms only able to ferment glucose may appear to be sucrose/lactose fermenters. If read later, sucrose/lactose-fermenters may consume the more complex carbohydrates and begin to catabolize peptones with the slant reverting to red. A yellow slant and a red butt may indicate the failure to stab butt or inoculation of medium with gram-positive organism. Examination of the medium for stab line and/or performance of Gram stain should clarify this situation. The hydrogen sulfide indicator system in TSI agar is not as sensitive as the lead acetate method or as that found in other media, such as sulfide indole motility agar. A black butt should be read as acid even though the yellow color may be obscured. If hydrogen sulfide is reduced, this indicates that an acid condition does exist and can be assumed. Critical to understanding how this medium works is the fact that glucose is present in a much lesser amount than are sucrose and lactose. Organisms use the simplest carbohydrate, glucose, first. Once this is consumed, they attack the more complex carbohydrate, lactose. If they lack the appropriate enzymes, they move on to protein catabolism. There is sufficient lactose in the medium to prevent breakdown of peptone, provided that the medium is read at the appropriate time.

Trypticase Soy Agar (TSA)

Trypticase soy agar is an all-purpose medium that supports the growth of many organisms. It is frequently used as the basal medium for sheep's blood agar plates. TSA contains protein as a nutrient source and sodium chloride as an osmotic stabilizer. Agar serves as solidifying agent.

If blood agar is to be prepared, cool to 50°C before adding 5% defibrinated sheep's blood agar. In most cases, agar can be added to a broth formulation to produce agar plates. However, the commercial TSA product does not contain glucose, which makes it suitable for a blood agar base. Adding agar to trypticase soy broth does not accomplish the same thing. Trypticase soy broth contains glucose, a fermentable carbohydrate, which can interfere with the expression of β hemolysis on sheep's blood agar plates.

Trypticase Soy Broth (TSB)

Trypticase soy broth is an all-purpose medium that supports the rapid growth of most organisms, including streptococci, without added supplements. Trypticase soy broth contains trypticase and phytone as protein sources, sodium chloride for osmotic stability, glucose as a fermentable carbohydrate, and dipotassium phosphate as a buffer.

Trypticase soy broth contains glucose, which when fermented can lower pH. This can cause acid-sensitive organisms such as *Streptococcus pneumoniae* to die off at 24 hours of incubation.

Tryptophan Broth (1%)

Tryptophan broth is used for performing the indole test, a procedure particularly useful in identifying species of Enterobacteriaceae and in identifying non–glucose-fermenting gram-negative rods. This broth contains trypticase, a peptone rich in tryptophan, and sodium chloride, which serves as an osmotic stabilizer. Some bacteria possess an enzyme system called tryptophanase, which hydrolyzes and deaminates tryptophan, producing indole, pyruvic acid, and ammonia. When Ehrlich or Kovac reagent is added to a tryptophan broth culture, any indole produced by the organism reacts with the aldehyde portion of dimethylaminobenzaldehyde, the primary chemical in these reagents, to form a red color.

Inoculate broth and incubate for 24 hours at 35°C. When performing the indole test, the microbiologist should add 5 drops of Kovac reagent to the medium and look for the appearance of a red color in the reagent layer or at the interface of the reagent and broth. If Ehrlich reagent is used, add 1 mL of xylene or ether to the broth culture, shake, and then add 5 drops of the reagent. Kovac reagent is generally used in testing for members of Enterobacteriaceae, and Ehrlich when testing for non–glucose-fermenting gram-negative rods and anaerobes.

Other media have been used in testing for indole production. These include sulfide indole motility (SIM) agar, indole nitrate broth, and motility indole ornithine (MIO) medium. A spot test employing filter paper saturated with paradimethylaminocinnamaldehyde reagent has also been used for indole determination.

Urea Agar and Broth

Urea media detect an organism's ability to hydrolyze urea. This characteristic is particularly useful in identifying species of Enterobacteriaceae. In these media urea is hydrolyzed to form carbon dioxide, water, and ammonia. The ammonia then reacts with components in the medium to form ammonium carbonate. This compound causes a rise in pH, which changes the pH indicator, phenol red, to pink. Both agar and broth formulations do not contain much protein. This prevents the formation of alkaline products from the breakdown of peptones, which could result in false-positive results. The broth formulation contains monopotassium phosphate and disodium phosphate, which make the medium highly buffered. In addition, the broth formulation lacks glucose and peptone. Only organisms such as *Proteus* species that are strong urease producers and not very fastidious appear positive in this type of medium. The agar formulation is less buffered, so smaller amounts of urease activity can be detected. Also, glucose and peptone are included in these media, which helps support growth.

When using urea agar, streak slant and incubate at 35°C for 18 to 24 hours. If urease is produced, the medium will turn pink. Rapid urease producers such as *Proteus* species turn the entire tube pink and may be detectable in a few hours. Slow urease producers such as *Klebsiella,* only turn the slant pink. If the organism does not produce urease, there will be no color change. Stuart urea broth is incubated at 35°C for 18 to 24 hours. A positive reaction is red color throughout the broth.

Vaginalis Agar (V Agar)

V agar is a nonselective, primary plating enrichment medium useful in the isolation of *Gardnerella vaginalis.* This organism also produces a distinctive colony, which aids in its recognition in mixed culture. V agar is essentially Columbia agar base with added proteose peptone and human blood. The medium contains many protein sources as well as starch, which can be broken down by *Gardnerella.* Human rather than sheep's blood must be used in this medium. *Gardnerella vaginalis* produces diffuse β hemolysis only on media containing human blood.

When using V agar, the microbiologist should incubate inoculated plates in a carbon dioxide incubator or in a candle jar at 35°C. Plates may be observed at 24 hours, but often it requires 48 to 72 hours for *Gardnerella* to grow. The organism produces tiny, dome-shaped colonies surrounded with zones of diffuse hemolysis.

Xylose Lysine and Desoxycholate (XLD) Agar

This agar is a selective, differential primary plating medium used to isolate *Salmonella* and *Shigella* species from stool and other specimens containing mixed flora. *Salmonella* and *Shigella* species produce characteristic colonies on XLD, which aids in their recognition. XLD agar contains sodium desoxycholate, which inhibits gram-positive cocci and some normal flora gram-negative rods. Because XLD agar has a lower concentration of bile salts than other formulations of enteric media, such as SS and HE agars, it is less selective

but permits better recovery of *Shigella.* XLD agar contains three fermentable carbohydrates: sucrose and lactose, which are present in excess concentration, and xylose, which is present in lower amounts. Phenol red serves as a pH indicator. The amino acid lysine is included to detect lysine decarboxylation. Sodium thiosulfate acts as a sulfur source from which organisms can make hydrogen sulfide. The hydrogen sulfide combines with ferric ammonium citrate to produce ferrous sulfide, a black precipitate. Four types of colonies are produced on XLD agar. *Yellow Colonies:* Organisms, such as *E. coli,* that ferment the excess carbohydrates produce a great deal of acid and change the pH indicator to yellow. Because there is excess carbohydrate, they do not decarboxylate the lysine even though they may possess lysine decarboxylase. Also, some bacteria ferment only xylose, but do not decarboxylate lysine, and therefore produce yellow colonies. *Yellow Colonies With Black Centers:* These organisms ferment the excess carbohydrate and also produce hydrogen sulfide. Examples of organisms that produce these organisms are *Citrobacter* and some *Proteus. Colorless or Red Colonies: Shigella* and *Providencia,* which neither ferment xylose, lactose, or sucrose nor produce hydrogen sulfide, have this appearance. *Red Colonies With Black Centers:* This colony type is produced by *Salmonella* and *Edwardsiella.* After fermenting xylose to make acid, these organisms decarboxylate lysine to produce cadaverine, an alkaline product. This causes the pH indicator to turn yellow and then to revert back to red. Blackening is due to hydrogen sulfide production.

When using XLD agar, the microbiologist should incubate plates at 35°C for 24 hours in ambient air. Some authors have recommended incubating plates for up to 48 hours to enhance blackening in *Salmonella* colonies. With any prolonged incubation, the delicate balance of this medium may be altered, and distinguishing normal flora from potential pathogens becomes more difficult. *Shigella dysenteriae* and *S. flexneri* may occasionally be inhibited on XLD agar. Some strains of *Salmonella* may fail to produce hydrogen sulfide and therefore resemble *Shigella* colonies. On this medium, blackening is more likely to occur when alkaline conditions exist.

Bibliography

Balows A, Hausler W, Jr, et al (eds): Manual of Clinical Microbiology, 5th ed. Washington, DC: American Society for Microbiology, 1991.

Baron E, Finegold S: Bailey & Scott's Diagnostic Microbiology, 8th ed. St. Louis: Mosby–Year Book, 1986.

Difco Laboratories: Difco Manual: Dehydrated Culture Media and Reagents for Microbiology, 10th ed. Detroit, MI: Difco Laboratories, 1985.

Finegold S, Martin W: Diagnostic Microbiology, 6th ed. St. Louis: Mosby–Year Book, 1982.

Howard B, Keiser J, et al: Clinical and Pathogenic Microbiology, 2nd ed. St. Louis: Mosby–Year Book, 1994.

Lennette E, Balows A, et al (eds): Manual of Clinical Microbiology, 3rd ed. Washington, DC: American Society for Microbiology, 1980.

Rohde P, Carski T, et al: BBL Manual of Products and Laboratory Procedures, 5th ed. Cockeysville, MD: Becton, Dickinson and Co, 1968.

Shepard M, Lunceford C: Differential agar medium (A7) for identification of *Ureaplasma urealyticum* (human T mycoplasmas) in primary cultures of clinical material. J Clin Microbiol 3:613, 1976.

APPENDIX B

Selected Mycology Media and Stains

Media

A variety of enrichment and selective media are available to the clinical laboratory in the isolation and identification of pathogenic fungi. This appendix is included to provide the reader with information about a select number of mycology media cited in the mycology section of this text. Most of the media detailed in this appendix are commercially available in either dehydrated or finished form. Recommendations as to general usage of the media, either enrichment or selective, or combinations of each type, are outlined in the mycology section of this text.

Birdseed Agar (Modified Staib Agar)

Birdseed agar is a differential enrichment medium designed for the isolation and preliminary identification of *Cryptococcus neoformans*. The ground seeds of *Guizottia abyssinica* provide enrichment, while biphenyl provides a substrate for the detection of phenol oxidase activity. On this medium, *C. neoformans* colonies typically darken to a rich brown as phenol oxidase activity results in the deposition of melanin in yeast cell walls. The colonies of other *Cryptococcus* species and other yeasts remain white.

Blood Agar With Penicillin and Streptomycin

Blood agar with penicillin and streptomycin (pen-strep agar) is a selective enrichment agar useful in the isolation of pathogenic yeast and the more fastidious dimorphic fungi from clinical specimens. Sheep's blood provides the enrichment, while the antibiotic combination inhibits the growth of bacteria.

Casein Medium

Casein medium is a differential medium used to demonstrate proteolytic activity by yeast species. Proteolysis is visualized as a clearing of the medium around the inoculum growth. Sterile plated medium should be inoculated by the cut streak or point inoculation method. Each run should include the set-up of the unknown and three known controls (*Streptomyces* species, *Nocardia asteroides*, *N. brasiliensis*) using one half of the plate for each organism. Inoculated plates should be incubated at room temperature and observed for proteolysis over a 14-day period. *Streptomyces* species typically hydrolyze the incorporated casein within 2 to 5 days. *Nocardia brasiliensis* typically hydrolyzes casein within 7 to 10 days, while *N. asteroides* does not hydrolyze casein.

Corn Meal Agar

Corn meal agar is typically made as any one of three variations on a basal formulation. Each formulation is useful in the cultivation of fungi. Each variation is recommended for the cultivation and/or enhancement of particular fungal characteristics. Corn meal agar without added dextrose is recommended for the cultivation of chlamydospore-bearing *Candida albicans,* with chlamydospore production further enhanced by the addition of 1% Tween-80. On the other hand, corn meal agar with 0.2% added dextrose* is not recommended for production of chlamydospores. Rather, it favors more luxuriant growth and improves pigment production.

To use, inoculate each plate with a known *C. albicans* control and unknown isolate(s) as single streaks cut deep into the agar or as surface streaks to be covered with a flame-warmed cov-

1073

erslip. Incubate at room temperature, and examine with a low-power lens of a microscope daily for up to 1 week. Isolates may become positive for chlamydospores (macroconidia) within 48 hours but cannot be considered negative before the fifth day of culture.

*Alternative Formulation With Dextrose. The aforementioned corn meal agar formulation with 2 g of dextrose is commercially available in dehydrated form. Rehydrate as directed by the manufacturer, and sterilize by autoclaving (121°C for 15 minutes). Inoculate, incubate at room temperature, and observe for growth and pigment production.

Dilute Gelatin Medium (0.4%)

Dilute gelatin medium is a differential medium useful in the differentiation of *Nocardia* species from one another and from *Streptomyces* species on the basis of growth and colonial morphology. In this medium, *N. asteroides* does not grow or grows poorly with a thin, flaky appearance. Conversely, *N. brasiliensis* grows well, forming compact, rounded colonies, while *Streptomyces* species produce poor to good growth with a stringy or flaky morphology.

The medium should be inoculated with a very small fragment of growth from a Sabouraud-dextrose agar slant and incubated at room temperature (or 37°C if the suspected strain grows better at 37°C). Inoculated tubes should be examined daily for growth for up to 21 to 25 days.

Mycosel/Mycobiotic Agar (Cycloheximide-Chloramphenicol Agar)

Mycosel agar is a selective medium useful in the isolation of pathogenic fungi, both dermatophytes and systemic pathogens. The cycloheximide is useful in suppressing the growth of saprophytic fungi, while the chloramphenicol is utilized to inhibit bacterial contaminants.

Potato Dextrose Agar

Potato dextrose agar is a recommended plating medium for cultivation, enumeration, and identification of yeasts and molds from dairy and other food products as well as other specimen types. The potato infusion encourages luxuriant growth and sporulation by fungi. *Note:* Potato dextrose agar to be used in the isolation and enumeration of fungi from milk and food should be acidified to pH 3.5 by aseptically adding 10% sterile tartaric acid to the cooled, sterile medium. Do not attempt to reheat this medium once the tartaric acid has been added, as hydrolysis of the agar will occur and prevent solidification.

Potato Flakes Agar

Potato flakes agar is used to induce sporulation in fungi.

Rice Extract Agar

Rice extract agar without additional dextrose is useful in the cultivation of *C. albicans*, with enhancement of chlamydospore production. Rice extract agar with 2% dextrose has been shown to enhance pigment production by *Trichophyton rubrum*, facilitating its differentiation from *Trichophyton mentagrophytes*.

The medium should be inoculated by cutting through the agar surface. If the medium inoculated is in plated form, the cut streak should be covered with a flame-warmed coverslip to stimulate chlamydospore production. All cultures should be incubated at room temperature for 18 to 72 hours.

Rice Grains Medium

Rice grains medium is useful in the differentiation of *Microsporum audouinii* from other dermatophytes, especially *Microsporum canis*. On this medium, *M. audouinii* grows poorly and discolors the medium. Other dermatophytes and most other fungi grow well and sporulate on this medium with no discoloration of the medium. Sterile rice grains should be spot-inoculated to prevent confusion in differentiating between discoloration and actual growth.

Sabouraud Dextrose Agar or Broth

Sabouraud dextrose agar and Sabouraud dextrose broth are nutrient media suitable for the cultivation of fungi, especially those associated with cutaneous and mucocutaneous infections. The formulations are identical with the exception of the agar. To make a solid plated medium 1.5 to 2% agar is added to the broth formulation.

Yeast Nitrogen Base, Modified: Carbohydrate Assimilation Base

Modified yeast nitrogen base is a synthetic basal medium that provides sufficient sources of nitrogen to support the growth of fungi. Fungal isolates are plated for confluent growth, and carbohydrate disks are dispensed onto the surface to provide the specific carbohydrates for assimilation testing. Incubate at 30°C for 48 to 72 hours and examine for growth around each disk. Good growth around a disk indicates assimilation of that carbohydrate, whereas scant or no growth around a disk indicates no assimilation. A dextrose disk serves as the growth control.

Fungal Mounting Fluids

KOH-Glycerin

Potassium hydroxide (or sodium hydroxide) and glycerin solution is used as a mounting fluid in the preparation of wet mounts to visualize fungi in clinical material.

KOH-Glycerin, Super Quink

This formulation is a variation of the traditional KOH-glycerin mounting solution. It utilizes a permanent blue-black ink to impart color to the system and enhance contrast.

Fungal Stains

Calcofluor White Stain

The use of calcofluor white stain with 10% KOH enhances visualization of fungi in clinical specimens of skin, hair, and nails. Fungal elements take up the fluorescent dye and, depending on the combination of filters utilized, appear brilliant green-yellow or blue-white while the background fluoresces a dim red.

For examination of clinical samples, add one (1) drop of calcofluor white solution with 1 drop of 10% KOH to the specimen on a microscope slide, apply a coverslip, and examine using a fluorescence microscope. The microscope utilized must have either a K532 excitation filter–BG 12 barrier filter or a G-35 excitation filter–LP420 barrier filter combination.

India Ink

The India ink method is useful in demonstrating the presence of a capsule. It is especially recommended for the demonstration of *Cryptococcus neoformans* in clinical specimens. In this method, the capsule displaces the colloidal carbon particles in the ink; thus the capsule appears as a clear halo around the body of the microorganism.

Lactophenol Cotton Blue

Lactophenol cotton blue is a mounting medium useful in examining clinical materials for fungi. It aids in clearing hyphal elements and preserving fungal materials. It enhances visualization by staining all chitin-containing structures a light blue. To use, place a small drop onto the fungal material on a slide, add a coverslip, and examine.

Bibliography

Balows A, Hausler W, Jr, et al (eds): Manual of Clinical Microbiology, 5th ed. Washington, DC: American Society for Microbiology, 1991.

Baron E, Finegold S: Bailey & Scott's Diagnostic Microbiology, 8th ed. St. Louis: Mosby–Year Book, 1986.

Difco Laboratories. Difco Manual: Dehydrated Culture Media and Reagents for Microbiology, 10th ed. Detroit, MI: Difco Laboratories, 1985.

Finegold S, Martin W: Diagnostic Microbiology, 6th ed. St. Louis: Mosby–Year Book, 1982.

Howard B, Keiser J, et al: Clinical and Pathogenic Microbiology, 2nd ed. St. Louis: Mosby–Year Book, 1994.

Lennette E, Balows A, et al (eds): Manual of Clinical Microbiology, 3rd ed. Washington, DC: American Society for Microbiology, 1980.

Rohde P, Carski T, et al: BBL Manual of Products and Laboratory Procedures, 5th ed. Cockeysville, MD: Becton, Dickinson and Co, 1968.

Nomenclature Changes for the Enterobacteriaceae and Nonfermentative Bacilli

CURRENT NOMENCLATURE	PREVIOUS NOMENCLATURE
Enterobacteriaceae	
Citrobacter koseri	*Citrobacter diversus*
Pantoea agglomerans	*Enterobacter agglomerans*
Nonfermentative Bacilli	
Brevundimonas diminuta	*Pseudomonas diminuta*
Brevundimonas vesicularis	*Pseudomonas vesicularis*
Burkholderia cepacia	*Pseudomonas cepacia*
Burkholderia pickettii	*Pseudomonas picketii*
Burkholderia pseudomallei	*Pseudomonas pseudomallei*
Comamonas acidovorans	*Pseudomonas acidovorans*
Comamonas testosteroni	*Pseudomonas testosteroni*
Sphingomonas paucimobilis	*Pseudomonas paucimobilis*
Stenotrophomonas maltophilia	*Xanthomonas maltophilia*

Index

Liquid culture media, colonial morphology on, 317, *317–320*
  for mycobacteria, 647–648
Liquid hybridization, in amplicon detection, 205, *205–206*
Liquid phase, in chromatography, 170
Listeria monocytogenes, 373–376
  characteristics of, 373
  culture of, 375, *376*
  identification of, 375–376, 376t
  in amniotic fluid, microscopy of, 282CP–283CP
  infection(s) caused by, 374
    meningitis as, 920, *921*
    urinary tract, 966–967
  laboratory diagnosis of, 375–376, *375–376*, 376t
  microscopy of, 375, *375*
  physiology of, 373
  versus *Erysipelothrix rhusiopathiae*, 377, *377*, 378t
  versus other gram-positive bacilli, 378t
  virulence factors of, 373–374
Listeriolysin O, of *Listeria monocytogenes*, 373–374
Lithotrophs, nutritional requirements of, 14
Liver abscess, *Entamoeba histolytica* in, 737
Liver flukes, 767, 768t, 769, *769*
Loa loa, 789–790, 789t, 1027
Local materials, in microscopy, 272–273, *274*, 278CP–279CP
Loeffler coagulated serum slant, 1060
Loeffler syndrome, in ascariasis, 782
Log phase, of bacterial growth curve, 16, *16*
Loperamide, in diarrhea, 913
Lo's *Mycoplasma* (*Mycoplasma fermentans*), 625t, 627
Louse, crab and pubic, in eyelid, 1028
Louse-borne diseases, *Borrelia recurrentis* in, 599–600
  rickettsial, 1012t, 1013–1014
Löwenstein-Jensen medium(a), for mycobacterial culture, 645, 646t, 1060
Lumbar puncture, in cerebrospinal fluid collection, 931–932, *932*
Luminometer, in amplicon detection, 206–207
Lung, abscess of, in aspiration, 857
  biopsy of, in pneumonia, 858
  infections of. See *Pneumonia.*
Lung flukes, 768t, 769, *770*
Lutz-Splendore-Almeida disease (paracoccidioidomycosis), 711
Lyme disease, 600–601, 996–998. See also *Borrelia burgdorferi.*
  babesiosis with, 758
  case study of, 996
  central nervous system involvement in, 923
  clinical manifestations of, 997–998, *998*
  cutaneous manifestations of, 881–882
  epidemiology of, 997

etiology of, 997
laboratory diagnosis of, 998
transmission of, 997, *997*
Lymph nodes, *Sporothrix schenckii* infection of, 699
Lymphadenitis, cervical, *Mycobacterium scrofulaceum* in, 667–668
  in tuberculosis, 662
Lymphangitis, in *Wuchereria bancrofti* infection, 788
Lymphocutaneous syndrome, microorganisms causing, 889, 890t
Lymphocytes, in acquired immunity, 143
Lymphocytic choriomeningitis virus, 927
Lymphocytosis-promoting factor, of *Bordetella pertussis*, 441
Lymphogranuloma venereum, *Chlamydia trachomatis* in, 615, *616*, 1025
Lymphokines, in cell-mediated immunity, 143
  in tuberculosis, 661
Lymphoma, Burkitt, Epstein-Barr virus and, 821–822
β-Lysin, antimicrobial activity of, 222
  in CAMP test, 347–348, 348P
Lysine, decarboxylation of, in Enterobacteriaceae identification, 476t, 478t, 483–484, *485*
Lysine iron agar slant test, in Enterobacteriaceae identification, 484, *485*, 486, 486t, 1061
Lysis-centrifugation method (Isolator), for blood culture, 944
  for mycobacteria, 648, 649P
Lysogeny, in transduction, 23, *23*
Lysosomes, antimicrobial activity of, 224
  antimicrobial substances in, 222
  of eukaryotic cells, *6*, 8
Lysozyme, antimicrobial activity of, 142, 222
Lytic pathway, in transduction, 23, *23*

M

M protein, of *Streptococcus pyogenes*, 351–352
MacConkey agar, 1061
  in diarrheal agent recovery, 911t
  in mycobacterial identification, 652t–653t, 655
  nonfermenter growth on, 516, *518*
  organism growth patterns on, 309–310, *309–312*
McFarland standards, for antimicrobial susceptibility testing, 63, *63–64*
Machupo virus, 810
  cutaneous manifestations of, 886
Macroconidia, of dermatophytes. See specific organism.
Macrogametes, of *Cryptosporidium parvum*, 763, *764*
Macrogametocytes, of *Plasmodium*, 752, 755, *756*

Macrolides, mechanisms of action of, 54–55
Macrophages, in phagocytosis, 222–225, *223–225*, 223t, 225t
Macroscopic (gross) examination, of anaerobes, 550, 551t
  of colonies, 311–315, *313–317*
    fungal, 689, *690*
  of specimens, 245–246
    stool, in parasitic infections, 727–728, *728*
Madura foot, 889
Madurella grisea, 702t, 703
Madurella mycetomatis, 702t, 703
Maduromycosis. See *Mycetoma.*
Magnesium, in *Taq* polymerase action, 207–208
MAIS (*Mycobacterium avium-intracellulare-scrofulaceum*) group, 668
Major outer membrane protein (MOMP), of *Chlamydia*, 609, 618
Malabsorption syndrome, after diarrhea, 907
Malachite green, in mycobacterial culture media, 645, 646t
Malaria, cerebral, 757, 931
  clinical manifestations of, 757–758
  epidemiology of. See also *Plasmodium.*
  laboratory diagnosis of, 754–755, 755t, *756–757*, 757–758
  paroxysm of, 757
  pathology of, 757–758
  transmission of, 752
  treatment of, 758
Malassezia furfur, 691–692, *692*, 720, 887, *887*
Malbranchea, versus *Coccidioides immitis*, 711
Malignancy, granulocytopenia in, 985
  immune response defects in, 985
  infections in, 986–987
  leukocyte dysfunction in, 985
  neutropenia in, infections in, 986
Malignant pustule (anthrax), 383–384, 1003–1005, *1005*
Malignant tertian malaria, 931
Malonate broth, 1061–1062
Malonate test, in Enterobacteriaceae identification, 476t, 478t, 482–483, *483*
Malta fever (brucellosis), 1007–1009, *1009*, 1010t
Maltese cross form, of *Babesia microti*, 759, *759*
Maltose, fermentation of, in Enterobacteriaceae identification, 477t, 479t
Mannitol, fermentation of, in Enterobacteriaceae identification, 476t, 478t
Mannitol salt agar, 1062
Mannose, fermentation of, in Enterobacteriaceae identification, 477t, 479t
Mansonella, 789t, 791
Manual, quality control, 107

DNA probes for, 192–193, *193*
entry route for, 922
generation time of, 645
host defenses against, 661
identification of, 652t, 663, *663*
  niacin accumulation test in, 651
  nitrate reduction test in, 651
  nucleic acid hybridization in,
    656–659
immune response to, 144
in sputum, microscopy of,
    284CP–285CP
infections caused by. See *Tuberculosis.*
photoreactivity of, 650
resistance in, 659–660, 660t
transmission of, 661
  in laboratory, 34
Mycobacterium tuberculosis complex,
    species in, 660
Mycobacterium ulcerans, 670
  culture of, 650
  identification of, 652t
  in skin infections, 879
Mycobacterium vaccae, 672
Mycobacterium xenopi, 668–669
  identification of, 652t, 654
Mycolic acids, in bacterial cell wall, 10
  of mycobacteria, chromatographic
    detection of, 656
Mycoplasma, colonial morphology of,
    624, *624*, 625t, 628–630, *629*
  culture media for, 1049, 1056, 1067
  in urinary tract infections, 956
  T (tiny)-strain. See *Ureaplasma ure-*
    alyticum.
  versus *Chlamydia*, 608t
Mycoplasma buccale, in normal flora,
    625t
Mycoplasma faucium, in normal flora,
    625t
Mycoplasma fermentans, 625t, 627
Mycoplasma gallisepticum, morphology
    of, *623*
Mycoplasma genitalium, 625t, 626–628
Mycoplasma hominis, antimicrobial sus-
    ceptibility testing of, 632
  characteristics of, 622–623, 623t–624t
  culture of, 626
  in normal flora, 625t
  infection(s) caused by, 624–626,
    627t–628t
    bacterial vaginosis as, 976
  laboratory diagnosis of, culture in,
    627–628
    identification in, 628–630, *631*, 632t
    methods for, 632t
    results interpretation for, 633, 633t
    serologic, 630
    specimen collection and transport
      for, 627, *629*
Mycoplasma lipophilum, in normal
    flora, 625t
Mycoplasma orale, 628
  in normal flora, 625t
Mycoplasma pneumoniae, antimicrobial
    susceptibility testing of, 632

characteristics of, 622–623, *623–625*,
    623t–624t
colonial morphology of, 623, *624*
DNA probes for, 194
epidemiology of, 623t
in normal flora, 625t
infection(s) caused by, 623–624, *624*
  clinical features of, 627, 628t
  cutaneous manifestations of, 882
  pneumonia as, 852, 854
laboratory diagnosis of, culture of,
    627–628
  DNA probe technology in, 630, 632
  identification in, 628–630, 632t
  methods for, 628t, 632t
  results reporting for, 633, 633t
  serologic, 630, 632t
  specimen collection and transport
    for, 627, *627*, 628t
Mycoplasma pulmonis, morphology of,
    623
Mycoplasma salivarium, 628, *630–631*
  in normal flora, 625t
Mycoplasmataceae, genera of, 622, 623t
Mycosel agar, 1074
Mycoses. See *Fungal infection(s).*
Myonecrosis, 874, *875*
  Clostridium in, 569

Naegleria fowleri, in keratitis, 1030
  in meningoencephalitis, 930–931
Nagler test, direct, for *Clostridium* cul-
    ture, 571
Nails, fungal infections of, 695. See also
    Dermatophytoses.
  specimen of, for mycology, 686
Nalgene filter unit, for specimen prepa-
    ration, 253, 253P
Nalidixic acid, mechanisms of action of,
    56
NAP (p-nitro-acetylamino-β-hydroxy-
    propiophenone) test, in my-
    cobacterial identification, 655
Nasolacrimal duct, infections of, 1033
Nasopharynx, normal flora of, 216, 216t
Natural (nonspecific) immunity,
    142–143
Neapolitan fever (brucellosis),
    1007–1009, *1009*, 1010t
Necator americanus (hookworm), 780t,
    782–784
Necrosis, of skin and soft tissue, 569,
    874, *875*
  diseases associated with, 872t
Necrotic debris, in direct microscopy,
    270
Necrotizing fasciitis, 874
  bacteremia with, 939
Necrotizing pneumonia, 857
Needle-stick injury, disease transmis-
    sion in, 34
Negative predictive value, of labora-
    tory tests, 114

Neisseria, characteristics of, 392, *393*
  classification of, 410, 412t–413t
  colonial morphology of, *315*
  culture media for, 1067–1068
  hosts of, 394t
  nonpathogenic, 392, 409–415
    classification of, 410, 412t–413t
    hosts of, 394t
    identification of, 410, 411t–413t
    infections caused by, 410, 410t
    Kingella denitrificans as, 411t, 412,
      412t–413t
    Neisseria cinerea as, 405t, 410t,
      411–412, 411t–413t, *413*
    Neisseria elongata as, 392,
      411t–413t, 415
    Neisseria flavescens as, 405t,
      410t–413t, 414
    Neisseria lactamica as, 410t–413t,
      413–414, *414*
    Neisseria mucosa as, 410t–413t, 414
    Neisseria polysaccharea as, 410–411,
      411t–413t
    Neisseria sicca as, 410t–413t, 414,
      414
    Neisseria subflava as, 410t–413t,
      414, *414*
    saccharolytic versus asaccha-
      rolytic, 410, 412t–413t
  pathogenic, 392, 394t. See also spe-
    cific organism.
    hosts of, 394t
  rapid identification of, 164–165
Neisseria cinerea, 411–412
  characteristics of, 405t
  colonial morphology of, 411t, 412,
    413
  identification of, 412t–413t
  infections caused by, 410t
Neisseria elongata, 415
  characteristics of, 392
  colonial morphology of, 411t
  identification of, 412t–413t
Neisseria flavescens, 414
  characteristics of, 405t
  colonial morphology of, 411t
  identification of, 412t–413t
  infections caused by, 410t
Neisseria gonorrhoeae, 392–406
  AHU strains of, 395, 398, 403
  antimicrobial resistance in, 404, 406t
  antimicrobial susceptibility of, 74,
    404–405
  cellular structure of, 394, *395*
  characteristics of, 405t
  colonial morphology of, 394, 398, 411t
  culture media for, 309
  culture of, 397, 399t, 973–974
  DNA probes for, 194
  epidemiology of, 394
    plasmid-mediated penicillinase-
      producing strains in, 404
  identification of, 398–404, 412t–413t
    auxotyping in, 403
    carbohydrate utilization in,
      399–401, *403*, 404P, 405t

filtration of, 253, 253P
for microscopic examination, preparation of, 259–262, *259–261,* 259t
freezing of, conditions for, 797
from contaminated equipment, processing of, 255
from implant soak solution, processing of, 254
from intrauterine devices, processing of, 255
from perfusates, processing of, 254
from vascular catheter tips, processing of, 255–256
inoculum counter-streak preparation of, 252–253, *253*
invasive, 243, *244*
labeling of, 243
liquefaction of, 250, 252, *252*
mailing of, 241, *242*
noninvasive, 243, *244*
nonroutine, definition of, 253
   processing of, 253–256, 255P
preservation of, 240
   in parasitic infections, 727
processing of, 245–256
   additives for, 247
   direct examination in, 246–247, *246–247,* 248t
   for *Haemophilus* identification, 422
   for inoculation. See *Inoculation.*
   for latex agglutination testing, 123, 125
   for mycology, 685–686
   gross examination ion, 245–246
   nonroutine, 253–256, 255P
   prioritization of, 245, 245t
   safety in, 34–35, 35t, 242–243
   scheduling of, 245, 245t
   smear preparation in, 246–247, *246–247,* 248t
quality of, for antimicrobial susceptibility testing, 60
rejection of, 243, *244*
   urine, 963
requisitions for, 243
routine, definition of, 253
storage of, 240–241, 241t
transport of, 240–241, *242*
   cerebrospinal fluid as, 932, 933t
   in *Bordetella* infections, 442
   in *Campylobacter* infections, 507
   in Enterobacteriaceae infections, 472
   in *Helicobacter pylori* infections, 507
   in mycobacteriology, 639–641
   in mycology, 685–686
   in *Mycoplasma* infections, 627
   in *Neisseria gonorrhoeae* identification, 396, *397*
   in parasitic infections, 727
   in urinary tract infections, 960–961
   in *Vibrio* infections, 496–497
   in virology, 797
   safety in, 241–242

unacceptable, 243, *244*
unprotected during transport, 240
water sterility, processing of, 254
Spectinomycin, resistance to, in *Neisseria gonorrhoeae,* 404
Spelunker's disease (histoplasmosis), 706t, 707–708
Spherules, morphology of, with Gram stain, 269t
Sphingobacterium, 531
Spicules, of *Blastomyces dermatitidis,* 707
Spin amplification, in shell vial culture, of viruses, 800–801, 802t
Spinal tap, in cerebrospinal fluid collection, 931–932, *932*
Spine, tuberculosis of, 662
Spiral Gradient Endpoint System, in antimicrobial susceptibility testing, 81, *81*
   of anaerobes, 590
Spirillum minus (minor), 878
   infections caused by, 1002–1003, *1002*
Spirochaetaceae, genera of, 596. See also specific organisms.
Spirochaetales, families of, 596
Spirochetemia, *Borrelia recurrentis* in, 599
Spirochetes. See also specific organisms.
   characteristics of, 596
   morphology of, 11, *12*
Spiroplasma, 623t
Splenic insufficiency, opportunistic infections of, 865
Sporangia, morphology of, with Gram stain, 269t
   swollen, of *Bacillus,* 380, 382t
Sporangiophores, 683, *683*
Sporangiospores, 683, *683*
Sporangium, 683, *683*
Spore(s). See also *Endospores.*
   of *Bacillus anthracis,* inhalation of, 1004
   of *Clostridium,* 566, 568, *568*
   of fungi, 682–683, *683*
   of prokaryotic cells, 8
   subterminal, of *Clostridium,* 568, *568*
   terminal, of *Clostridium,* 568, *568*
Spore stain, for *Bacillus,* 380, *381*
Spore-forming gram-positive anaerobes. See *Clostridium.*
Sporocysts, of flukes, 766, *767*
Sporogony, of *Plasmodium,* 753–754
Sporothrix schenckii, 699–700, *700,* 990
   in conjunctivitis, 1026
Sporotrichosis, 699–700, *700,* 990
Sporozoites, of *Plasmodium,* 753
   of *Toxoplasma gondii,* 760, *761*
Spot, in nucleic acid hybridization, 187, *188*
Spot indole test, 164t
   in anaerobe identification, 563
Spotted fevers, 1012–1013, 1012t
Spreading factor (hyaluronidase), production of, by *Streptococcus pyogenes,* 353

Sputum, aspirated, microscopy of, local and contaminating materials in, 278CP–279CP
   polymicrobial population in, 294CP–295CP
   Streptococcus pyogenes in, 280CP–281CP
   Strongyloides stercoralis in, 302CP–303CP
contamination assessment of, 272
examination of, in parasitic infections, 731
   versus bronchoalveolar lavage, 865–866, *866*
expectorated, collection of, in pneumonia, 858
microscopy of, *Actinomyces* in, 284CP–285CP, 286CP–287CP
   Aspergillus in, 298CP–299CP
   Blastomyces dermatitidis, 296CP–297CP
   Candida in, 296CP–297CP
   Corynebacterium pseudodiphtheriticum, 282CP–283CP
   eosinophils in, 296CP–297CP
   Haemophilus influenzae in, 288CP–289CP
   local and contaminating materials in, 278CP–279CP
   Moraxella (Branhamella) catarrhalis in, 288CP–289CP
   Mycobacterium tuberculosis in, 284CP–285CP
   Neisseria meningitidis in, 288CP–289CP
   Nocardia in, 284CP–285CP, 286CP–287CP
   pneumococci in, 280CP–281CP
   polymicrobial population in, 288CP–289CP
   Pseudomonas aeruginosa in, 292CP–293CP
   Rhodococcus equi in, 284CP–285CP
   Stomatococcus mucilaginosus in, 280CP–281CP
   Streptomyces in, 286CP–287CP
microscopy of, *Klebsiella pneumoniae* in, 290CP–291CP
specimen of, collection of, by patient, 238, *238*
   for mycobacteriology, 639–640
   site preparation for, 239t
   for lung fluke infection, 769
   for mycology, 686
Stains, 163, 163t, 262, 262t, 644–645, 730. See also *Gram stain;* specific stain and organism.
   acid-fast. See *Acid-fast stains/staining.*
   diagnostic antibody, 262, 262t
   differential, 262
   for antibodies, in antigen detection, 129–130, *129–130*
   for bacteria, 11–13, *13*
   for fungi, 686–687, *687–688,* 688t, 1075

Y